CONTENTS

PREFACE

"Machine design" is an interdisciplinary subject in mechanical engineering. It draws information from thermal science, mechanics of solids, theory of elasticity, fluid science, material science, engineering mechanics, experimental stress analysis, production engineering, and some other branches of sciences. Machine design, therefore, assimilates more information from these disciplines than does any other engineering subject.

It is indeed gratifying that the first edition of *Machine Design Databook* published by McGraw-Hill in the year 1994 was so well received by the students, teachers, designers, practicing engineers, and working professionals that within a short period after its publication all the copies were sold out. The encouraging reception accorded to the first edition of this Databook has inspired the author to enhance its usefulness further by enlarging it. One of the several reasons for publishing a new second edition of any engineering handbook is to update the technical material and to take advantage of the enhanced receptivity of the readers. The present edition has been enlarged by including several tables, new drawings, and equations and by adding three new chapters–namely–"Elements of Machine Tool Design," "Applied Elasticity," and "Retaining Rings and Circlips". The author hopes that the inclusion of these will be of immense help to students, teachers, designers, practicing engineers, and working professionals in machine design.

The writer of any handbook on an engineering subject has to keep in touch with the developments taking place on the borders of discovery and frontiers of research in the subject, i.e., in all related subjects. This requires extensive study of the burgeoning literature on the subject both in the form of books and in the form of technical papers in professional journals. The author has left no stone unturned in reading all of the latest literature available on the subject to update the Databook and also to enhance its professional value. The present revised edition, therefore, makes a distinct improvement over its predecessor in contents, presentation, and arrangement by adding three new chapters.

Machine Design Databook provides weights and measures both in SI and fps, i.e., U.S. Customary System units, for the benefit of students, teachers, designers, practicing engineers, and working professionals.

The properties of engineering materials, presented in this Databook in Chapter 1, are taken from ANSI, ISO, ASME, Det Norske Veritas Classification, DIN, UNS (SAE), AISI, ASTM, and Indian Standards and are in both SI and U.S. Customary System units. All formulas (empirical, semi-empirical and otherwise), charts, graphs and tables are in both SI and US Customary units.

All elementary deflection, stress, strain, moments and torsion equations under static and dynamic conditions are given in Chapters 2 and 3. Chapter 4, on stress concentration and stress intensity in machine members, includes formulas, graphs, and tables for theoretical and fatigue stress concentration factors and stress intensity factors for bolts, nuts, screws, weldments, plates with and without cutouts (holes) of various shapes, shoulders, and notches in shafts and plates etc. Chapter 5 treats the design of machine elements for strength. The factor of safety, size effect, surface finish, reliability, corrosion, temperature effect, fatigue and stochastic approach are considered in this chapter. A number of design equations and data are given in Chapter 6 for design of cams.

The design of boilers, pressure vessels, boiler tubes and pipes, shells and plates with latest data from ASME Boilers and Pressure Vessels Code and Indian Standards for Unfired Pressure Vessels are presented in Chapters 7, 8, and 9. The design equations, charts, graphs, and tables are presented in a more accessible and useful form for the designers, students, teachers, and practicing engineers. The equations and data in Chapters 10 and 15 provide sufficient information for the design of rotating disks and cylinders and flywheels. Most of the material included in Chapter 11 on metal fits, tolerances, and surface texture uses SI units. Sufficient information is also given in U.S. Customary System units for the benefit of students, designers, and practicing engineers in this chapter.

Chapters 12 and 13, which include stress formulas and exhaustive information on design of welded joints and riveted joints, are written so that students, designers, teachers, and practicing engineers can quickly find the information they search for. Updated information for the design and selection of seals and packings is provided in Chapter 16. Design of shafts is exhaustively dealt with in Chapter 14, as are all theories of failures. This chapter includes the equations with sufficient data to solve shaft problems.

In Chapter 17, design equations for knuckle joint, cotter joint, pins, keys and splines are given in addition to data from standards. Chapter 18, on threaded fasteners and screws for power transmission, covers ISO standards for bolts and nuts in addition to other fps standards. Standard formulas and data for threaded connections, preloaded and tension-loaded bolts joints, and power screws were thoroughly scanned, and updated information is given in this chapter. In Chapter 19 there is extensive inclusion of formulas, standard graphs, and data for the design of couplings, clutches, and brakes. The data and design equations in this chapter are sufficient to design components of couplings, clutches, and brakes. The chapter also includes the latest information on brake- and clutch-facing materials.

Chapter 20, on springs, includes tolerances on helical spring wire diameter, and coil diameter, and free-length and cross-section load, length, and width tolerances of leaf spring for motor vehicle suspension. The spring constant and other design aspects of springs can be studied from this chapter. New materials have been brought into this chapter and rational design of springs can be carried out by using size factor, surface finish factor, statistical data, and other information.

The long felt need for a rational selection of standard materials for flexible machine elements has been, to a large extent, overcome by the presentation of the required number of standard charts, tables, and equations in Chapter 21. Rating formulas for chains and belts have been given in this chapter for the benefit of students, teachers, designers, and practicing engineers.

Chapter 22, on mechanical vibration, is included in this Databook to cover the design of machines so that they work safely within the allowable vibration limits without breakdown during operation.

The topic of sliding contact bearings in Chapter 23 has been thoroughly updated including a large number of charts and tables. The inclusion of this data on the composition and properties of bearing materials and lubricants and guides for the selection of design factors are very useful for students, teachers, designers, and practicing engineers. The dimensions and load rating under static and dynamic conditions of rolling contact bearings such as deep-groove, self-aligning, angular-contact and thrust, ball and roller bearings from different manufacturers catalogues like FAG, SKF, and Indian standards have been listed and compared. Importance is given to the estimation of the equivalent loads in bearings under different operating conditions. The subjects of tribology along with design of bearings were added to help the students, teachers, and practicing engineers to understand and design bearings more effectively. Friction, wear, and lubrication, which constitute the study of tribology, are seldom given any importance in any of the data handbooks of machine design except in *Tribology Handbook* by M. J. Neale and in this Databook. These topics help in the study of the distress condition of bearings and gears to prevent failures of these machine elements in practice.

In Chapter 24, formulas and data for the design of crankshafts, curved beams, connecting and coupling rods, piston and piston rings, friction gearing, pawl and ratchet mechanisms, Geneva mechanisms, universal joints, speed reduction gears, variable speed drives, mechanics of vehicles, and unsymmetrical bending and torsion of noncircular cross-section are included. The information given in this chapter on design are helpful while designing machine components of a complete machine. Design equations and other relevant data are given in Chapters from 1 to 23, and 25 to 28.

Chapter 25, on elements of machine tool design, discusses the fundamental principles involved in machine tool design. It deals with the design of components of machine tools and design practices with practical data and guidelines on various aspects of machine tool and design.

A certain minimum working knowledge of the construction and fabrication of major machines such as engine lathe, turret lathe, capstan lathe, milling machine, grinding machine, shaping machine, slotting machine, planing machine, drilling machine, mechanical and hydraulic presses, broaching machine, and N.C. machine etc. are necessary in order to design even a simple machine tool. A designer cannot learn his work from reading books alone. Reading books and journals in his specialized field must be backed by practical design experience in industries.

Chapter 25 deals with design aspects such as design of spindles, slideways, guides, gear boxes for lathes, drilling machines and other machines, beds for lathes, planing columns for drilling machines, milling frames and selection of bearings etc. In order to design the above parts a designer has to study forces acting on single-point metal cutting tool, Merchant's circle for cutting forces, and specific power requirements. Selection of lubricants and cutting fluids for satisfactory working of a machine is very important. The chip formation, feeds, spindle speeds, tool signature and tool life and other information for satisfactory design of a machine tool are provided here to the designer.

A methodical and judicious selection of material with proper heat-treatment in manufacturing of high quality product at low cost is very important in production of any product. An engineer is considered to be a good designer if he is capable of designing a machine tool capable of economic machining, casting, joining and press working of many engineering materials. The design information in the form of figures, tables, charts etc are collected from ASME, BIS, ASTM, SAE, SKF, FAG, ASTME, ISQ, ANSI, DIN standards and are included in this chapter. This chapter also deals with some mechanical presses in addition to other machines used in production processes.

In the machine tool industry, forming processes are very important production processes. Hence forming machineries accounts for more than one-third of the value/number of the machine tool industry. Both mechanical and hydraulic presses are used in this field of the machine tool industry. Chapter 25 also deals with blanking and piercing press, V-bending and U-bending press, deep drawing and redrawing press, extrusion moulding and impact extrusion press, stamping and coining press, and hot upsetting and drop forging press. Some of the equations for calculation of forces and dimensions are presented for the design of press components etc.

Chapter 26, deals with retaining rings and circlips. Retaining rings are fasteners which are used to secure machine components on shafts, holes, and housings. Retaining rings are stamped from thin carbon spring steel or from other materials with good spring properties. There are more than 40 to 50 varieties of retaining rings on the market. The important series of retaining rings are the horseshoe type rings. They are also called by another name as circlips. The varieties of retaining rings listed in Chapter 26 are internal and external retaining rings, inverted internal and external rings, permanent shoulder ring, crescent-shaped ring, E-ring, heavy duty external retaining ring, reinforced E-ring, miniature high-strength retaining ring, two-part interlocking ring, taper-section internal ring, high-strength radially assembled ring, beveled and double-beveled internal and beveled external rings, bowed E-ring, bowed internal and external retaining rings, reinforced external self-locking ring, triangular self-locking ring, and internal and external self-locking rings. Selection of materials for different types of rings are given. Shear strength of ring materials and tensile strength of grooved materials are given in tabular forms. Calculations of axial and radial loads on shafts and groove wall are also included. Tolerances and corner radii or chamfers are tabulated for the use of the designers. The dimensional units and force units are given in both SI and U.S. Customary System units for the benefits of students, teachers, designers, and practicing engineers

Chapter 27, on applied elasticity, is a very important chapter in this Databook. The theory of elasticity can be used to solve simple problems of regular geometrical-shaped elastic solids subjected to known loads. But elastic solids of irregular shapes and subjected to complicated loads cannot be solved by any mathematical methods. In such cases experimental stress analysis techniques such as photo-elasticity, strain gage techniques, Moire techniques, holographic techniques, etc. can be used successfully to find solutions. The soap-film method may be used to find stresses in prismatic bars subject to torsion and bending. The electrical analogy also can be used to solve stresses developed at the fillets of gear teeth, shoulders of stepped shafts etc. when subjected to torsion and bending.

Chapter 27 deals with stresses and strains in two- and three-dimensional problems, stress functions, volume expansion, equilibrium and compatibility equations, and two- and three-dimensional problems in x,y, and polar coordinates. The chapter also deals with principal stresses and strains, Mohr's circle representation of stress and strain, stress-strain relations, thick cylinders subjected to internal and external pressures, curved bars subject to pure bending, rotating disks and cylinders of uniform and variable thickness, plates with circular, elliptical, square and other shapes of cutouts (holes) subject to uniform unaxial, biaxial and pure shear stress, loads at infinity using first principles and complex variable methods, and solving problems using curvilinear coordinates. Several additional topics covered in Chapter 27 are listed in the detailed table of contents.

The author has obtained design data and equations by consulting authentic sources and by deriving and formulating many of the equations himself. This Databook is written with the latest information.

Students, teachers, designers, and practicing engineers have to use a large number of graphs, tables, equations, and other relevant data to find solutions to machine design problems. Students in mechanical engineering are prone to commit errors if they are not properly guided in designing a machine component or machine. A teacher in machine design finds it very difficult to write all equations (empirical and semi-empirical), charts, figures, tables, and other data on a notice board or to give notes on this subject to cover the syllabus in machine design. He has to set more involved machine design problems in examinations giving data and formulas in the question paper, because it is not possible for students to remember all empirical and semi-empirical formulas and other data in machine design while designing a machine or a machine component. Therefore this Databook is primarily written to meet the reference and study needs of the undergraduate and graduate students in mechanical engineering, researchers, designers, practicing engineers, and others.

Authors who have written textbooks in machine design using fps units (i.e. U.S. Customary System units) can also consult this *Machine Design Databook* with confidence for SI equivalents of empirical and semiempirical formulas and other data which have been drawn from different standards and sources. These equations and data in SI units in this Databook have been tested with numerical examples before incorporating them in the Databook over a period of 20 years during which author has taught machine design to students at undergraduate and graduate level.

Mechanical designers, researchers, and practicing engineers have to design equipment, mechanical devices, and sometime the whole machines in practice. Researchers have to design some component of equipment which they are using in their day-to-day research work. The same is true in case of engineers in industries. In such a situation, the designer, the researcher, or practicing engineer has to refer to source books such as this *Machine Design Databook* from which reliable and updated information can be drawn.

This Databook refers to many standards and data which are in vogue at present in the world. They may change with time and more progress in technology. This has to be taken into consideration in using the standards and data in this Databook.

In a work of this nature there are bound to be errors, and the author would be grateful if these are brought to his notice for future guidance. Any suggestions, for the improvement of this Databook will be gratefully received.

It is sincerely hoped that this Databook will be useful to students, teachers, designers, practicing engineers, and researchers.

K. Lingaiah

ACKNOWLEDGMENTS

The author is grateful to McGraw-Hill for undertaking the complex work of publishing this revised second edition. The author is also grateful to Mr. Robert W. Hauserman who was then Senior Editor at McGraw-Hill for sponsoring the publication of the first edition of the Databook. The author is also grateful to Miss Kimberly A. Golf, Senior Editing Supervisor, Miss Julia A. Dietz, and other supporting staff of McGraw-Hill for having helped and encouraged the author in bringing out this major work.

I am grateful to Ms. Linda R. Ludewig, Mechanical Engineering Editor at McGraw-Hill, who initiated the revision of this *Machine Design Databook*. Mr. Kenneth P. McCombs, Senior Editor, Mechanical and Chemical Engineering, McGraw-Hill deserves a special mention and thanks for his tireless effort. The author also grateful to all supporting staff of McGraw-Hill for having helped the author to complete the work.

The author owes a debt of gratitude to late Professor B.R. Narayana Iyengar, who was a guru, a guide, and a source of inspiration to the author.

The author is grateful to his parents Kenchaiah and Madamma, his wife M.P. Madamma, his children L. Susheela, L. Nagarathna, L. Vasanth Kumar, L. Suma, and L. Vivekananda but for whose patience, constant support and encouragement and help, this publication would have not been possible.

The author owes a debt of gratitude to the Bangalore University and the University Visvesvaraya College of Engineering, Bangalore where the author has taught the students at undergraduate and graduate level in Machine Design as a lecturer, reader and Professor of Mechanical Engineering, Head of the Department, Dean of Engineering and Principal from 1958 up to his retirement. At present he is working on a university level book writing project at Bangalore University sanctioned by University Grants Commission, New Delhi.

With many thanks I acknowledge the permissions granted by the following publishers for reproducing the materials from their publications, which have been credited at appropriate places in the text of *Machine Design Databook* either as footnotes or as references or both:

1. The McGraw-Hill Companies, Inc., New York.
2. International Book Company, Scranton, Pennsylvania.
3. The Macmillan Company, New York.
4. Springer-Verlag, Berlin, Germany.
5. John Wiley and Sons, Inc., New York.
6. Prentice Hall, Inc., Englewood Cliffs, New Jersey.
7. The Ronald Press Company, New York.
8. Allied Publishers Private Ltd, New Delhi.
9. Butterworth-Heinemann, Oxford.
10. Mir Publishers, Moscow.
11. Machine Design, Cleveland, Ohio.
12. DIN-Deutsches Institut fur Normung e.V., Berlin, Germany.
13. Society of Automotive Engineers, Inc., Warrendale, Pennsylvania.
14. CBS Publishers and Distributors, Delhi.
15. The American Society of Mechanical Engineers, New York.

16. MAAG Gear Co., Ltd, Zurich, Switzerland.
17. American Gear Manufacturers Association, Alexandria, Virginia.
18. Bureau of Indian Standards, Manak Bhavan, New Delhi.
19. The Gleason Works, Rochester, New York.
20. International Organization for Standards, Geneva, Switzerland.
21. Det Norske Veritas Classification AS, Hovik, Norway.
22. American Society for Metals, Metals Park, Ohio.
23. Associated Spring, Barnes Group Inc., Bristol, Connecticut.
24. Van Nostrand Reinhold Company, New York.
25. Edward Arnold (Publishers) Ltd, Great Britain.

A large number of extracts from the AGMA, ISO, BIS, BSI, ASME and DIN standards and Gleason Works, etc., have been incorporated in this Databook to make the book more useful to students, teachers, designers, and practicing engineers. Also included are standards from Det Norske Veritas Classification, *MAAG Gear Handbook*, and information from FAG and SKF bearing catalogues.

The author gratefully expresses his gratitude to various other publishers of books and journals and their authors for permission to make use of the materials from their books and technical Journals.

Finally the author wholeheartedly thanks all those persons who have helped directly or indirectly in writing this Databook whose names are not mentioned here.

ABOUT THE AUTHOR

Prof. Dr. K. Lingaiah was former Principal, Professor and Head, Department of Mechanical Engineering (Post-Graduate Studies), University Visvesvaraya College of Engineering, Bangalore University, Bangalore. He has been serving as a member and Chairman of various authorities of many Universities and Institutions in India. His degrees include Bachelor of Engineering (Mechanical) from the University of Mysore, Master of Technology (Machine Design) from the Indian Institute of Technology, Kharagpur (India) and Doctorate Degree in Mechanical Engineering from the University of Saskatchewan, Canada. He was a recipient of Ford Foundation Senior Research Fellow at the University of Toronto, Toronto, Canada. He served as Lecturer, Reader (Assistant Professor), and Professor in Mechanical Engineering, at the University College of Engineering. He has to his credit several papers published in various international and national technical journals. He is a member of SEM (USA), ASME, ISTE (India), and ISTE (India), Fellow of the Institution of Engineers (India),and P.E. He is a member on the Executive Committee of the IFTOMM Gear Committee. He is a consultant to various industries and has delivered seminars and lectures at NASA, University of Toronto, Canada, National University of Singapore, and numerous universities and engineering colleges in India and abroad. He was a Professor Emeritus of Mechanical Engineering, at Adichunchanagiri Institute of Technology from 1994 to 1996, Chikmagalure, Karnataka State, India.

CHAPTER 1

PROPERTIES OF ENGINEERING MATERIALS

SYMBOLS[5,6]

a	area of cross section, m^2 (in^2)*
	original area of cross section of test specimen, mm^2 (in^2)
A_j	area of smallest cross section of test specimen under load F_j, m^2 (in^2)
A_f	minimum area of cross section of test specimen at fracture, m^2 (in^2)
A_0	original area of cross section of test specimen, m^2 (in^2)
A_r	percent reduction in area that occurs in standard test specimen
Bhn	Brinell hardness number
d	diameter of indentation, mm
	diameter of test specimen at necking, m (in)
D	diameter of steel ball, mm
E	modulus of elasticity or Young's modulus, GPa [Mpsi (Mlb/in^2)]
f_ε	strain fringe (fri) value, μm/fri (μin/fri)
f_σ	stress fringe value, kN/m fri (lbf/in fri)
F	load (also with subscripts), kN (lbf)
G	modulus of rigidity or torsional or shear modulus, GPa (Mpsi)
H_B	Brinell hardness number
l_f	final length of test specimen at fracture, mm (in)
l_j	gauge length of test specimen corresponding to load F_j, mm (in)
l_0	original gauge length of test specimen, mm (in)
Q	figure of merit, fri/m (fri/in)
R_B	Rockwell B hardness number
R_C	Rockwell C hardness number
ν	Poisson's ratio
σ	normal stress, MPa (psi)

* The units in parentheses are **US Customary units** [e.g., fps (foot-pounds-second)].

σ_b	transverse bending stress, MPa (psi)
σ_c	compressive stress, MPa (psi)
σ_s	strength, MPa (psi)
σ_t	tensile stress, MPa (psi)
σ_{sf}	endurance limit, MPa (psi)
σ'_{sf}	endurance limit of rotating beam specimen or R R Moore endurance limit, MPa (psi)
σ'_{sfa}	endurance limit for reversed axial loading, MPa (psi)
σ'_{sfb}	endurance limit for reversed bending, MPa (psi)
σ_{sc}	compressive strength, MPa (psi)
σ_{su}	tensile strength, MPa (psi)
σ_u	ultimate stress, MPa (psi)
σ_{uc}	ultimate compressive stress, MPa (psi)
σ_{ut}	ultimate tensile stress, MPt (psi)
$\sigma_{\widehat{su}}$	ultimate strength, MPA (psi)
σ_{suc}	ultimate compressive strength, MPa (psi)
σ_{sut}	ultimate tensile strength, MPa (psi)
σ_y	yield stress, MPa (psi)
σ_{yc}	yield compressive stress, MPa (psi)
σ_{yt}	yield tensile stress, MPa (psi)
σ_{syc}	yield compressive strength, MPa (psi)
σ_{syt}	yield tensile strength, MPa (psi)
τ	torsional (shear) stress, MPa (psi)
τ_s	shear strength, MPa (psi)
τ_u	ultimate shear stress, MPa (psi)
τ_{su}	ultimate shear strength, MPa (psi)
τ_y	yield shear stress, MPa (psi)
τ_{sy}	yield shear strength, MPa (psi)
τ'_{sf}	torsional endurance limit, MPa (psi)

SUFFIXES

a	axial
b	bending
c	compressive
f	endurance
s	strength properties of material
t	tensile
u	ultimate
y	yield

ABBREVIATIONS

AISI	American Iron and Steel Institute
ASA	American Standards Association
AMS	Aerospace Materials Specifications
ASM	American Society for Metals
ASME	American Society of Mechanical Engineers
ASTM	American Society for Testing Materials
BIS	Bureau of Indian Standards
BSS	British Standard Specifications
DIN	Deutsches Institut für Normung
ISO	International Standards Organization

SAE	Society of Automotive Engineers
UNS	Unified Numbering system

Note: σ and τ with subscript s designates strength properties of material used in the design which will be used and observed throughout this *Machine Design Data Handbook*. Other factors in performance or in special aspects are included from time to time in this chapter and, being applicable only in their immediate context, are not given at this stage.

Particular	Formula
For engineering stress-strain diagram for ductile steel, i.e., low carbon steel	Refer to Fig. 1-1
For engineering stress-strain diagram for brittle material such as cast steel or cast iron	Refer to Fig. 1-2
The nominal unit strain or engineering strain	$\varepsilon = \frac{l_f - l_0}{l_0} = \frac{\Delta l}{l_0} = \frac{l_f}{l_0} - 1 = \frac{A_0 - A_f}{A_0}$ (1-1) where l_f = final gauge length of tension test specimen, l_0 = original gauge length of tension test specimen.
The numerical value of strength of a material	$\sigma_s = \frac{F}{A}$ (1-2) where subscript s stands for strength.

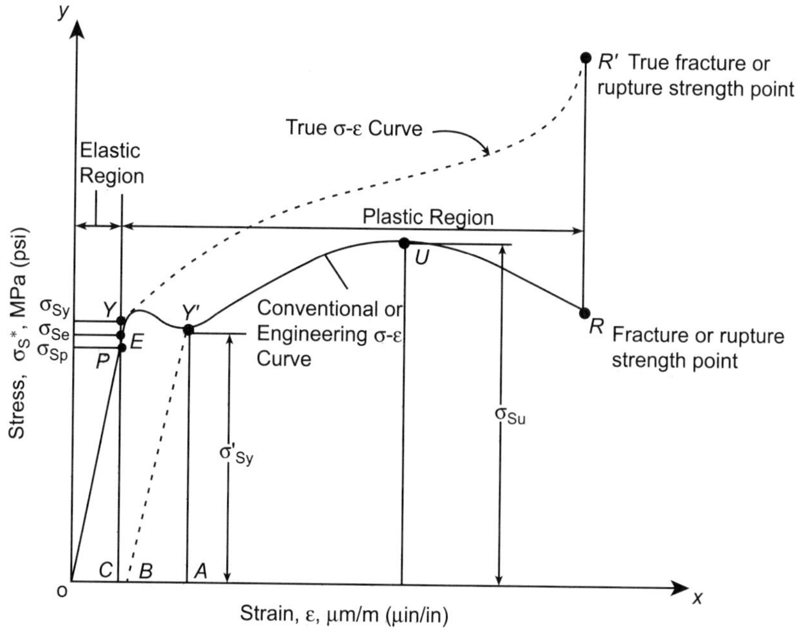

Point P is the proportionality limit. Y is the upper yield limit. E is the elastic limit. Y' is the lower yield point. U is the ultimate tensile strength point. R is the fracture or rupture strength point. R' is the true fracture or rupture strength point.

FIGURE 1-1 Stress-strain diagram for ductile material.
* Subscript s stands for strength.

Particular	Formula
The nominal stress or engineering stress	$$\sigma = \frac{F}{A_0} \quad (1\text{-}3)$$ where F = applied load.
The true stress	$$\sigma_{tru} = \sigma' = \frac{F}{A_f} \quad (1\text{-}4)$$ where A_f = actual area of cross section or instantaneous area of cross-section of specimen under load F at that instant.
Bridgeman's equation for actual stress (σ_{act}) during r radius necking of a tensile test specimen	$$\sigma_{act} = \frac{\sigma_{cal}}{\left(1+\frac{4r}{d}\right)\left[\ln\left(1+\frac{d}{4r}\right)\right]} \quad (1\text{-}5)$$
The true strain	$$\varepsilon_{tru} = \varepsilon' = \frac{\Delta l_1}{l_0} + \frac{\Delta l_2}{l_0+\Delta l_1} + \frac{\Delta l_3}{l_0+\Delta l_1+\Delta l_2} + \cdots \quad (1\text{-}6a)$$ $$= \int_{l_0}^{l_f} \frac{dl_i}{l_i} \quad (1\text{-}6b)$$
Integration of Eq. (1-6) yields the expression for true strain	$$\varepsilon_{tru} = \ln\left(\frac{l_f}{l_0}\right) \quad (1\text{-}7)$$
From Eq. (1-1)	$$\frac{l_f}{l_0} = 1 + \varepsilon \quad (1\text{-}8)$$
The relation between true strain and engineering strain after taking natural logarithm of both sides of Eq. (1-8)	$$\ln\left(\frac{l_f}{l_0}\right) = \ln(1+\varepsilon) \quad \text{or} \quad \varepsilon_{tru} = \ln(1+\varepsilon) \quad (1\text{-}9)$$
Eq. (1-9) can be written as	$$\varepsilon = e^{\varepsilon_{tru}} - 1 \quad (1\text{-}10)$$

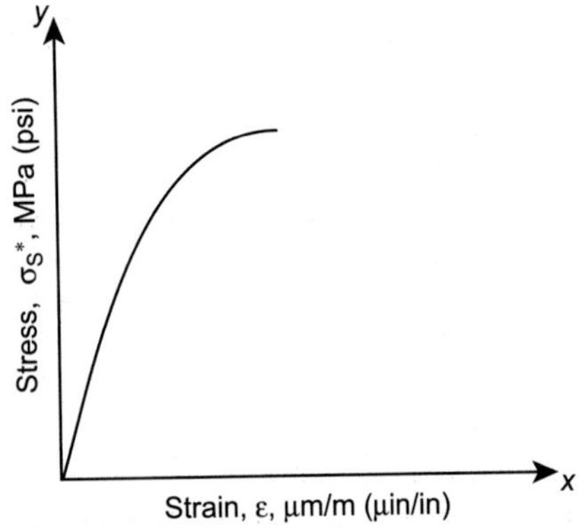

There is no necking at fracture for brittle material such as cast iron or low cast steel.

FIGURE 1-2 Stress-strain curve for a brittle material.

Particular	Formula	
Percent elongation in a standard tension test specimen	$\varepsilon_{100} = \frac{l_f - l_0}{l_0}(100)$	(1-11)
Reduction in area that occurs in standard tension test specimen in case of ductile materials	$A_r = \frac{A_0 - A_f}{A_0}$	(1-12)
Percent reduction in area that occurs in standard tension test specimen in case of ductile materials	$A_{r100} = \frac{A_0 - A_f}{A_0}(100)$	(1-13)
For standard tensile test specimen subject to various loads	Refer to Fig. 1-3.	

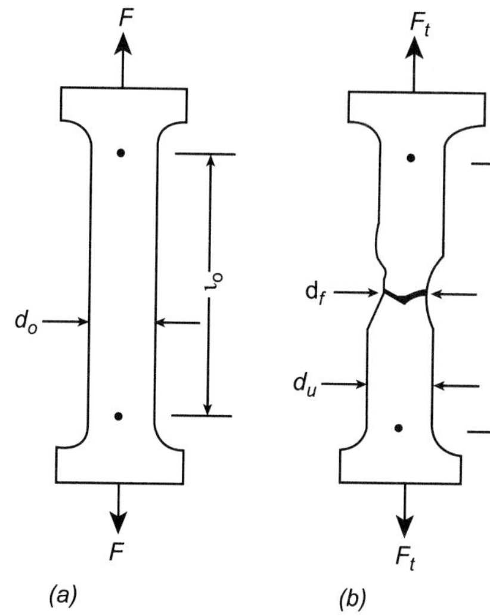

FIGURE 1-3 A standard tensile specimen subject to various loads.

Particular	Formula	
The standard gauge length of tensile test specimen	$l_0 = 6.56\sqrt{a}$	(1-14)
The volume of material of tensile test specimen remains constant during the plastic range which is verified by experiments and is given by	$A_0 l_0 = A_f\, l_f$ or $\frac{l_f}{l_0} = \frac{A_0}{A_f} = \frac{d_0^2}{d_f^2}$	(1-15)
Therefore the true strain from Eqs. (1-7) and (1-15)	$\varepsilon_{tru} = \ln\left(\frac{A_0}{A_f}\right) = \ln\frac{l_f}{l_0} = 2\ln\frac{d_0}{d_f}$ where d_f = minimum diameter in the gauge length l_f of specimen under load at that instant, A_r = minimum area of cross section of specimen under load at that instant.	(1-16)
The true strain at rupture, which is also known as the true fracture strain or ductility	$\varepsilon_{ftru} = \ln\left(\frac{1}{1 - A_r}\right)$ where A_f is the area of cross-section of specimen at fracture.	(1-17)

Particular	Formula
	Refer to Table 1-1A for values of ε_{ftru} of steel and aluminum.
From Eqs. (1-9) and (1-16)	$\frac{A_0}{A_f} = 1 + \varepsilon \quad \text{or} \quad A_f = \frac{A_0}{1 + \varepsilon}$ (1-18)
Substituting Eq. (1-18) in Eq. (1-4) and using Eq. (1-3) the true stress	$\sigma_{tru} = \sigma(1 + \varepsilon) = \sigma e^{\varepsilon_{tru}}$ (1-19)
From experimental results plotting true-stress versus true-strain, it was found that the equation for plastic stress-strain line, which is also called the strain-strengthening equation, the true stress is given by	$\sigma_{tru} = \sigma_0 \varepsilon_{trup}^n$ (1-20) where σ_0 = strength coefficient, n = strain hardening or strain strengthening exponent, ε_{trup} = true plastic strain. Refer to Table 1-1A for σ_0 and n values for steels and other materials.
The load at any point along the stress-strain curve (Fig 1-1)	$F = \sigma_s A_0$ (1-21)
The load-strain relation from Eqs. (1-20) and (1-2)	$F = \sigma_0 A_0 \varepsilon_{tru}^n e^{-\varepsilon_{tru}}$ (1-22)
Differentiating Eq. (1-22) and equating the results to zero yields the true strain equals to the strain hardening exponent which is the instability point	$\varepsilon_u = n$ (1-23)
The stress on the specimen which causes a given amount of cold work W	$\sigma_w = \sigma_0(\varepsilon_w)^n = \frac{F_w}{A_w}$ (1-24) where A_w = actual cross-sectional area of the specimen, F_w = applied load.
The approximate yield strength of the previously cold-worked specimen	$(\sigma_{sy})_w = \frac{F_w}{A'_w}$ (1-25) where $A_w = A'_w$ = the increased cross-sectional area of specimen because of the elastic recovery that occurs when the load is removed.
The approximate yield strength since $A'_w = A_w$	$(\sigma_{sy})_w = \frac{F_w}{A'_w} \approx \sigma_w$ (1-26)
By substituting Eq. (1-26) into Eq. (1-24)	$(\sigma_{sy})_w = \sigma_0(\varepsilon_w)^n$ (1-27)
The tensile strength of a cold worked material	$(\sigma_{su})_w = \frac{F_u}{A'_w}$ (1-28) where $A_w = A_u$, $F_u = A_0(\sigma_{su})_0$, σ_{su} = tensile strength of the original non-cold worked specimen, A_0 = original area of the specimen.
The percent cold work associated with the deformation of the specimen from A_0 to A'_w	$W = \frac{A_0 - A'_w}{A_0}(100) \quad \text{or} \quad w = \frac{A_0 - A'_w}{A_0}$ (1-29) where $w = \frac{W}{100}$

Particular	Formula
For standard tensile specimen at stages of loading A'_w is given by equation	$A'_w = A_0(1 - w)$ (1-30)
Expression for $(\sigma_{su})_w$ after substituting Eq. (1-28)	$(\sigma_{su})_w = \dfrac{(\sigma_{su})_0}{1 - w}$ (1-31)
Eq. (1-31) can also be expressed as	$(\sigma_{su})_w = (\sigma_{su})_0 e^{\varepsilon_{tru}}$ (1-32) Valid for $A_w \le A_u$ or $\varepsilon_w \le \varepsilon_u$.
The modulus of toughness	$T_m = \int_0^{\varepsilon_r} \sigma_s d\varepsilon$ (1-33a) $\approx \dfrac{\sigma_s + \sigma_{su}}{2}\varepsilon_r$ (1-34b) where $\varepsilon_r = \varepsilon_u$ = strain associated with incipient fracture.

HARDNESS

Particular	Formula
The Vicker's hardness number (H_V) or the diamond pyramid hardness number (H_p)	$H_V = \dfrac{2F\sin(\alpha/2)}{d^2} = \dfrac{1.8544F}{d^2}$ (1-35) where F = load applied, kgf, α = face angle of the pyramid, 136°, d = diagonal of the indentation, mm, H_V in kgf/mm^2.
The Knoop hardness number	$H_K = \dfrac{F}{0.07028d^2}$ (1-36) where d = length of long diagonal of the projected area of the indentation, mm, F = load applied, kgf, 0.07028 = a constant which depends on one of angles between the intersections of the four faces of a special rhombic-based pyramid industrial diamond indenter 172.5° and the other angle is 130°, H_K in kgf/mm^2.
The Meyer hardness number, H_M	$H_M = \dfrac{4F}{\pi d^2/4}$ (1-37) where F = applied load, kgf, d = diameter of indentation, mm, H_M in kgf/mm^2.
The Brinell hardness number H_B	$H_B = \dfrac{2F}{\pi D[D - \sqrt{D^2 - d^2}]}$ (1-38) where F in kgf, d and D in mm, H_B in kgf/mm^2.
The Meyer's strain hardening equation for a given diameter of ball	$F = Ad^p$ (1-39) where F = applied load on a spherical indenter, kgf, d = diameter of indentation, mm, p = Meyer strain-hardening exponent.

Particular	Formula
The relation between the diameter of indentation d and the load F according to Datsko[1,2]	$F = 18.8d^{2.53}$ (1-40)
The relation between Meyer strain-hardening exponent p in Eq. (1-39) and the strain-hardening exponent n in the tensile stress-strain Eq. $\sigma = \sigma_0 \varepsilon^n$	$p - 2 = n$ (1-41) where $p = 2.25$ for both annealed pure aluminum and annealed 1020 steel, $p = 2$ for low work hardening materials such as pH stainless steels and all cold rolled metals, $p = 2.53$ experimentally determined value of 70-30 brass.
The ratio of the tensile strength (σ_{su}) of a material to its Brinell hardness number (H_B) as per experimental results conducted by Datsko[1,2]	$K_B = \dfrac{\sigma_{su}}{H_B}$ (1-42)
For the plot of ratio of $(\sigma_{su}/H_B) = K_B$ against the strain-strengthening exponent n* (1)	Refer to Fig. 1-4 for K_B *vs* n for various ratios of (d/D).

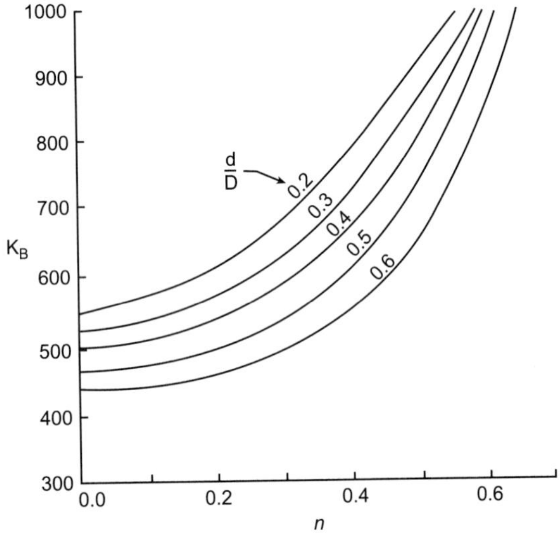

FIGURE 1-4 Ratio of $(\sigma_{su}/H_B) = K_B$ *vs* strain strengthening exponent n.

Particular	Formula
The relationship between the Brinell hardness number H_B and Rockwell C number R_C	$R_C = 88H_B^{0.162} - 192$ (1-43)
The relationship between the Brinell hardness number H_B and Rockwell B number R_B	$R_B = \dfrac{H_B - 47}{0.0074H_B + 0.154}$ (1-44)

* *Courtesy*: Datsko, J., *Materials in Design and Manufacture*, J. Datsko Consultants, Ann Arbor, Michigan, 1978, and *Standard Handbook of Machine Design*, McGraw-Hill Book Company, New York, 1996.

Particular	Formula
The approximate relationship between ultimate tensile strength and Brinell hardness number of carbon and alloy steels which can be applied to steels with a Brinell hardness number between $200H_B$ and $350H_B$ only[1,2]	$\sigma_{sut} = 3.45H_B$ MPa **SI** (1-45a) $= 500H_B$ psi **USCS** (1-45b)
The relationship between the minimum ultimate strength and the Brinell hardness number for steels as per ASTM	$\sigma_{sut} = 3.10H_B$ MPa **SI** (1-46a) $= 450H_B$ psi **USCS** (1-46b)
The relationship between the minimum ultimate strength and the Brinell hardness number for cast iron as per ASTM	$\sigma_{sut} = 1.58H_B - 86.2$ MPa **SI** (1-47a) $= 230H_B - 12500$ psi **USCS** (1-47b)
The relationship between the minimum ultimate strength and the Brinell hardness number as per SAE minimum strength	$\sigma_{sut} = 2.60H_B - 110$ MPa **SI** (1-48a) $= 237.5H_B - 16000$ psi **USCS** (1-48b)
In case of stochastic results the relation between H_B and σ_{sut} for steel based on Eqs. (1-45a) and (1-45b)	$\sigma_{sut} = (3.45, 0.152)H_B$ MPa **SI** (1-49a) $= (500, 22)H_B$ psi **USCS** (1-49b)
In case of stochastic results the relation between H_B and σ_{sut} for cast iron based on Eqs. (1-47a) and (1-47b)	$\sigma_{sut} = 1.58H_B - 62 + (0, 10.3)$ MPa **SI** (1-50a) $= 230H_B - 9000 + (0, 1500)$ psi **USCS** (1-50*b*)
Relationships between hardness number and tensile strength of steel in SI and US Customary units [7]	Refer to Fig. 1.5.
The approximate relationship between ultimate shear stress and ultimate tensile strength for various materials	$\tau_{su} = 0.82\sigma_{sut}$ for wrought steel (1-51a) $\tau_{su} = 0.90\sigma_{sut}$ for malleable iron (1-51b) $\tau_{su} = 1.30\sigma_{sut}$ for cast iron (1-51c) $\tau_{su} = 0.90\sigma_{sut}$ for copper and copper alloy (1-51d) $\tau_{su} = 0.65\sigma_{sut}$ for aluminum and aluminum alloys (1-51e)
The tensile yield strength of stress-relieved (not cold-worked) steels according to Datsko[1,2]	$\sigma_{sy} = (0.072\sigma_{sut} - 205)$ MPa SI (1-52a) $= 1.05\sigma_{sut} - 30$ kpi **USCS** (1-52b)
The equation for tensile yield strength of stress-relieved (not cold-worked) steels in terms of Brinell hardness number H_B according to Datsko (2)	$\sigma_{sy} = (3.62H_B - 205)$ MPa SI (1-53a) $= 525H_B - 30$ kpi **USCS** (1-53b)
The approximate relationship between shear yield strength (τ_{sy}) and yield strength (tensile) σ_{sy}	$\tau_{sy} = 0.55\sigma_{sy}$ for aluminum and aluminum alloys (1-54a) $\tau_{sy} = 0.58\sigma_{sy}$ for wrought steel (1-54b)

Particular	Formula

The approximate relationship between endurance limit (also called *fatigue limit*) for reversed bending polished specimen based on 50 percent survival rate and ultimate strength for nonferrous and ferrous materials

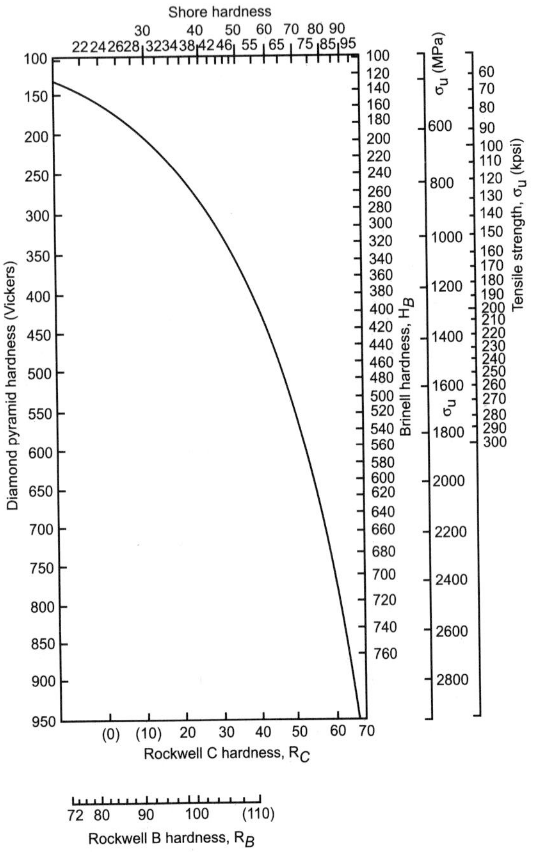

FIGURE 1-5 Conversion of hardness number to ultimate tensile strength of steel σ_{sut}, MPa (kpsi). (*Technical Editor Speaks, courtesy of International Nickel Co., Inc., 1943.*)

For students' use

$\sigma'_{sfb} = 0.50\sigma_{sut}$ for wrought steel having $\sigma_{sut} < 1380$ MPa (200 kpsi) (1-55)

$\sigma'_{sfb} = 690$ MPa for wrought steel having $\sigma_{sut} > 1380$ MPa (1-56a)

$\sigma'_{sfb} = 100$ kpsi for wrought steel having $\sigma_{sut} > 200$ kpsi **USCS** (1-56b)

For practicing engineers' use

$\sigma'_{sfb} = 0.35\sigma_{sut}$ for wrought steel having $\sigma_{sut} < 1380$ MPa (200 kpsi) (1-57)

$\sigma'_{sfb} = 550$ MPa for wrought steel having $\sigma_{sut} > 1380$ MPa **SI** (1-58a)

$\sigma'_{sfb} = 80$ kpsi for wrought steel having $\sigma_{sut} > 200$ kpsi **USCS** (1-58b)

$\sigma'_{sfb} = 0.45\sigma_{sut}$ for cast iron and cast steel when $\sigma_{sut} \le 600$ MPa (88 kpsi) (1-59a)

$\sigma'_{sfb} = 275$ MPa for cast iron and cast steel when $\sigma_{sut} > 600$ MPa **SI** (1-60a)

$\sigma'_{sfb} = 40$ kpsi for cast iron and cast steel when $\sigma_{sut} > 88$ kpsi **USCS** (1-60b)

$\sigma'_{sfb} = 0.45\sigma_{sut}$ for copper-based alloys and nickel-based alloys (1-61)

$\sigma'_{sfb} = 0.36\sigma_{sut}$ for wrought aluminum alloys up to a tensile strength of 275 MPa (40 kpsi) based on 5×10^8 cycle life (1-62)

$\sigma'_{sfb} = 0.16\sigma_{sut}$ for cast aluminum alloys up to tensile strength of 300 MPa (50 kpsi) based on 5×10^8 cycle life (1-63)

$\sigma'_{sfb} = 0.38\sigma_{sut}$ for magnesium casting alloys and magnesium wrought alloys based on 10^6 cyclic life (1-64)

Particular	Formula
The relationship between the endurance limit for reversed axial loading of a polished, unnotched specimen and the reversed bending for steel specimens	$\sigma'_{sfa} = 0.85\sigma'_{sfb} = 0.43\sigma_{sut}$ (1-65)
The relationship between the torsional endurance limit and the reversed bending for reversed torsional tested polished unnotched specimens for various materials	$\tau'_{sf} = 0.58\sigma'_{sfb} = 0.29\sigma_{sut}$ for steel (1-66a) $\tau'_{sf} \approx 0.8\sigma'_{sfb} \approx 0.32\sigma_{sut}$ for cast iron (1-66b) $\tau'_{sf} \approx 0.48\sigma'_{sfb} \approx 0.22\sigma_{sut}$ for copper (1-66c)
For additional information or data on properties of engineering materials	Refer to Tables 1-1 to 1-48

WOOD

Particular	Formula
Specific gravity, G_m, of wood at a given moisture condition, m, is given by	$G_m = \dfrac{W_0}{W_m}$ (1-67) where W_0 = weight of the ovendry wood, N (lbf), W_m = weight of water displaced by the sample at the given moisture condition, N (lbf).
The weight density of wood, D (unit weight) at any given moisture content	$W = \dfrac{\text{weight of ovendry wood and the contained water}}{\text{volume of the piece at the same moisture content}}$ (1-68)
Equation for converting of weight density D_1 from one moisture condition to another moisture condition D_2	$D_2 = D_1 \dfrac{100 + M_2}{100 + M_1 + 0.0135 D_1 (M_2 - M_1)}$ (1-69) where D_1 = known weight density for same moisture condition M_1, kN/m^2 (lbf/ft^2), D_2 = desired weight density at a moisture condition M_2, kN/m^2 (lbf/ft^2). M_1 and M_2 are expressed in percent.
For typical properties of wood of clear material as per ASTM D 143	Refer to Table 1-47.

TABLE 1-1
Hardness conversion (approximate)

Brinell 29.42 kN (3000 kgf) load 10 mm ball		Vickers or Firth hardness number	Rockwell hardness number				Shore scleroscope hardness number	Tensile strength, σ_{sut} approximate	
Diameter (mm)	Hardness number		*A* scale 0.588 kN (60 kgf) load	*B* scale 0.98 kN (100 kgf) load	*C* scale 1.47 kN (150 kgf) load	15-*N* scale 0.147 kN (15 kgf) load		MPa	kpsi
2.25	745	840	84		65	92	91	2570	373
2.30	712	783	83		64	92	87	2455	356
2.35	682	737	82		62	91	84	2350	341
2.40	653	697	81		60	90	81	2275	330
2.45	627	667	81		59	90	79	2227	323
2.50	601	640	80		58	89	77	2192	318
2.55	578	615	79		57	88	75	2124	309
2.60	555	591	78		55	88	73	2020	293
2.65	534	569	78		54	87	71	1924	279
2.70	514	547	77		52	87	70	1834	266
2.75	495	528	76		51	86	68	1750	254
2.80	477	508	76		50	85	66	1675	243
2.85	461	491	75		49	85	65	1620	235
2.90	444	472	74		47	84	63	1532	222
2.95	429	455	73		46	83	61	1482	215
3.00	415	440	73		45	83	59	1434	208
3.05	401	425	72		43	82	58	1380	200
3.10	388	410	71		42	81	56	1338	194
3.15	375	396	71		40	81	54	1296-	188
3.20	363	383	70		39	80	52	1255	182
3.25	352	372	69	110	38	79	51	1214	176
3.30	341	360	69	109	37	79	50	1172	170
3.35	331	350	68	109	36	78	48	1145	166
3.40	321	339	68	108	34	77	47	1103	160
3.45	311	328	67	108	33	77	46	1069	155
3.50	302	319	66	107	32	76	45	1042	151
3 55	293	309	66	106	31	76	43	1010	146
3.60	285	301	65	106	30	75	42	983	142
3.65	277	292	65	105	29	74	41	955	138
3.70	269	284	64	104	28	74	40	928	134
3.75	262	276	64	103	27	73	39	904	131
3.80	255	269	63	102	25	73	38	875	127
3.85	248	261	63	101	24	72	37	855	124
3.90	241	253	62	100	23	71	36	832	120
3.95	235	247	61	99	22	70	35	810	117
4.00	229	241	61	98	21	70	34	790	114
4.05	223	234		97	19			770	111
4.10	217	228		96	18		33	748	108
4.15	212	222		96	16		32	730	106
4.20	207	218		95	15		31	714	103
4.25	201	212		94	14			690	100
4.30	197	207		93	13		30	680	98
4.35	192	202		92	12		29	662	96
4.40	187	196		91	10			645	93

TABLE 1-1
Hardness conversion (approximate) (*Cont.*)

Brinell 29.42 kN (3000 kgf) load 10 mm ball		Vickers or Firth hardness number	Rockwell hardness number				Shore scleroscope hardness number	Tensile strength, σ_{sut} approximate	
Diameter (mm)	Hardness number		*A* scale 0.588 kN (60 kgf) load	*B* scale 0.98 kN (100 kgf) load	*C* scale 1.47 kN (150 kgf) load	15-*N* scale 0.147 kN (15 kgf) load		MPa	kpsi
4.45	183	192		90	9		28	631	91
4.50	179	188		89	8		27	617	89
4.55	174	182		88	7			600	87
4.60	170	178		87	5		26	585	85
4.65	167	175		86	4			576	83
470	163	171		85	3		25	562	81
4.80	156	163		83	1		24	538	78
4.90	149	156		81			23	514	74
5.00	143	150		79			22	493	71
5.10	137	143		76			21	472	68
5.20	131	137		74				451	65
5.30	126	132		72			20	435	63
5.40	121	127		70			19	417	60
5.50	116	122		68			18	400	58
5.60	111	117		65			17	383	55

TABLE 1-1A
Mechanical properties of some metallic materials

Material	Brinell hardness H_B	Process/ Condition	Ultimate strength, σ_{sut}		Yield strength, σ_{sy}		Stress at fracture, σ_f		Reduction in area, A_f	True strain at fracture	Strain hardening exponent	Strength coefficient, σ_0	
			MPa	kpsi	MPa	kpsi	MPa	kpsi	%	ε_f	n	MPa	kpsi
					Steel								
RQC-100[a]	290	HR[b] Plate	931	135	883	128	1331	193	67	1.02	0.06	1172	170
1005-1009	125	CD[c] Sheet	414	60	400	58	841	122	64	1.02	0.05	524	76
1005-1009	90	HR Sheet	345	50	262	38	848	123	80	1.60	0.16	531	77
1015	80	Normalized	414	60	228	33	724	105	68	1.14	0.26		
1020[d]	108	HR Plate	441	64	262	38	710	103	62	0.96	0.19	738	107
1045[e]	225	Q and T[f]	724	105	634	92	1227	178	65	1.04	0.13	1145	166
1045[e]	410	Q and T	1448	210	1365	198	1862	270	51	0.72	0.08	2082	302
5160	430	Q and T	1669	242	1531	222	1931	280	42	0.87	0.06	2124	308
9262	260	Annealed	924	134	455	66	1041	151	14	0.16	0.22	1744	253
9262	410	Q and T	1565	227	1379	200	1855	269	32	0.38	0.06		
950	150	HR Plate	531	77	311	48	1000	145	72	1.24	0.19	903	131
					Aluminum:								
2024-T351		ST, SH[g]	469	68	379	55	558	81	25	0.28	0.03	455	66
2024-T4		ST and RT age[h]	476	69	303	44	636	92	35	0.43	0.20	807	117
7075-T6		ST and AA[i]	579	84	469	68	745	108	33	0.41	0.11	827	120

[a] Tradename, Bethlehem steel Corp. Rolled quenched and tempered carbon steel. Used in structural, heavy applications machinery. [b] Hot-rolled. [c] cold-rolled. [d] low carbon, common machining steels. [e] Bar stock, medium carbon high-strength machining steel. [f] Quenched and tempered. [g] Solution treated, strain hardened. [h] Solution treated and RT age. [i] Solution treated and artificially aged.
Source: SAE j1099, Technical Report on Fatigue properties, 1975.

TABLE 1-1B
Mechanical properties of some typical metallic materials

Material	Form	Condition/Process	Ultimate tensile strength, σ_{sut}		Yield strength: Tensile, σ_{syt}		Yield strength: Compressive, σ_{syc}		Shear (torsional) strength: Ultimate, τ_{su}		Shear (torsional) strength: Yield, τ_{sy}		Fatigue limit, σ_{sf}		Young's modulus, E		Modulus of rigidity, G		Fracture toughness, K_{IC}		Reduction in area	True strain at fracture,
			MPa	kpsi	MPa	kpsi	MPa	kpsi	MPa	kpsi	MPa	kpsi	MPa	kpsi	GPa	Mpsi	GPa	Mpsi	GPa	Mpsi	A, %	ε_f
Steel:																						
1016		CD 0%	455	66	275	40							240	35[d]							70	1.20
		CD 30%	620	90	585	85							315	46[d]							62	0.97
		CD 60%	710	102	605	88							350	51[d]							54	0.78
		CD 80%	790	115	660	96							365	53[d]							26	0.30
1020	25 mm (1 in) bar or plate	HR	448	65	331	48							241	35[d]							59	0.89
1030	25 mm (1 in) WQ bar or plate	(1200°F)	586	85	441	64					241	35	296	43	204	29.6	83	12.0			70	1.20
1040	25 mm (1 in) bar	Annealed	517	75	352	51							269	39							57	0.84
		HR	621	90	414	60							296	43[a]							50	0.69
		CD 20%	805	117	670	97							370	54[d]							44	0.58
		CD 50%	965	140	855	124							410	60[d]							25	0.33
1050	25 mm (1 in) bar	Annealed	634	92	365	53							365	53			79	11.4			40	0.51
		CD 20% + s.r.2h (900°F	876	127	696	101							427	62[d]							31	0.37
4130	25 mm (1 in) bar	WQ + (1200°F)	814	118	703	102	724	105					490	71							64	1.02
4340	25 mm (1 in) bar	OQ + (1000 °F)	1262	183	1172	170	1310	190	876	127	752	109	669	97	207	30.0	81	11.7	110	100	52	0.73
		OQ + (800 °F)	1531	200	1379	200	1517	220	1007	146	855	124	469	68					75	68	47	0.63
18% Ni maraging																						
200	L plate	Aged 482°C	1540	225	1480	215							690	100							55	0.80
250	L plate	Aged 482°C	1760	256	1630	237							690	100							62	0.97
300	L plate	Aged 482°C	1980	288	1920	279							760	110							50	0.69

[a] A description of the materials and typical uses follows the table.
[b] CD = cold drawn (the percentage reduction in area); HR = hot rolled; OQ = oil quenched; WQ = water quenched (temperature following is the tempering temperature); s.r. = stress relieved.
[c] Smooth-specimen rotating-beam results, unless noted A (= axial).
[d] 10^6 cycles.

Source: Extracted from Kenneth S. Edwards, Jr, and Robert B. McKee, *Fundamentals of Mechanical Component Design*, McGraw-Hill, Inc., 1991, which is drawn from the *Structural Alloys Handbook*, published by the Metals and Ceramics Information Center, Battelle Memorial Institute, Columbus, Ohio, 1985.

TABLE 1-2
Poisson's ratio (ν)

Material	ν	Material	ν
Aluminium, cast	0.330	Molybdenum	0.293
Aluminium, drawn	0.348	Monel metal	0.320–0.370
Beryllium copper	0.285	Nickel, soft	0.239
Brass	0.340	Nickel, hard	0.306
Brass, 30 Zn	0.350	Rubber	0.450–0.490
Cast steel	0.265	Silver	0.367
Chromium	0.210	Steel, mild	0.303
Copper	0.343	Steel, high carbon	0.295
Douglas fir	0.330	Steel, tool	0.287
Ductile iron	0.340–0.370	Steel, stainless (18-8)	0.305
Glass	0.245	Tin	0.342
Gray cast iron	0.210–0.270	Titanium	0.357
Iron, soft	0.293	Tungsten	0.280
Iron, cast	0.270	Vanadium	0.365
Inconel x	0.410	Wrought iron	0.278
Lead	0.431	Zinc	0.331
Magnesium	0.291		
Malleable cast iron	0.230		

TABLE 1-3
Mechanical properties of typical cast ferrous materials[a]

Material, class, specification		Ultimate strength: Tension, σ_{sut} MPa	kpsi	Compression, σ_{suc} MPa	kpsi	Shear, τ_{su} MPa	kpsi	Torsional/shear strength, τ_s MPa	kpsi	Yield strength, σ_{sy} MPa	kpsi	Endurance limit in reversed bending, σ_{sfb} MPa	kpsi	Brinell hardness, H_B	Modulation of elasticity: Tension, E GPa	Mpsi	Compression, E GPa	Mpsi	Shear, G GPa	Mpsi	Elongation in 50 mm (2 in), %	Impact strength (Charpy) J	ft-lbf	Typical application
Gray cast iron[b]																								
ASTM class	SAE																							
20	110	152	22	572	83	220	32	179	26			69	10	156	66–97	9.6–14.0			27–39	3.9–5.6		75	55	Soft iron castings
25		179	26	669	97	255	37	220	32			79	11.5	174	79–102	11.5–14.8			32–41	4.6–6.0		75	55	Cylinder blocks and heads, housing
30	111	214	31	752	109	303	44	276	40			97	14	210	90–113	13.0–16.4			36–45	5.2–6.6				Flywheels, brake drums and clutch plates
35	120	252	36.5	855	124	338	49	334	48.5			110	16	212	10–119	14.5–17.2			40–48	5.8–6.9				Heavy-duty brake drums, clutch plates
40	121	293	42.5	965	140	393	57	393	57			128	18.5	235	110–138	16.0–20.0			44–54	6.4–7.8		95	70	Cam shafts, cylinder liners
50		362	52.5	1130	164	448	65	503	73			148	21.5	262	130–157	18.8–22.8			50–55	7.2–8.0		108	80	Special high-strength castings
60		431	62.5	1293	187.5	496	72	610	88.5			169	24.5	302	141–162	20.4–23.5			54–59	7.8–8.5		156	115	Special high-strength castings
Malleable cast iron: Ferrite	Class or grade																							
ASTM A47-52, A338, ANSI G 48-1	32510	345	50	1434	208	324	47			220	32	193	28	156 max	172	25	172	25			10	22	16.5	General purpose at normal and elevated temperature, good machinability, excellent shock resistance.
FED QQ-1-66e	35018	365	53	1517	220	352	51			241	35	214	31	156 max	172	25	172	25			18	22	16.5	
ASTM A197		276	40							207	30			156 max							5			Pipe flanges, valve parts
Perlite and martensite: ASTM A220, ANSI G48-2, MIL-1-11444B	40010	414	60							276	40			149–197							10			General engineering service at normal and elevated temperatures
	45008	448	65							310	45			156–197							8			
	45006	448	65							310	45			156–207							6			
	45010	448	65	1670	242	338	49			310	45	220	32	185	180	26	160	23.2			10	19	14	
	50005	483	70							345	50			179–229							5			
	50007	517	75	1670	242	517	75			345	50	255	37	204	183	26.5	160	23.2			7	19	14	
	60004	552	80							414	60			197–241							4			
	60003	552	80	1670	242	552	80			414	60	270	39	226	186	27	160	23.2			3	19	14	
	70003	586	85							483	70			217–269							3			
	80002	655	95	1670	242	689	100			552	80	276	40	241–285	186	27	160	27			2	19	14	
	90001	724	105							621	90			269–321							1			
Automotive	Grade																							
ASTMA602, SAE J158	M3210[c]	345	50							224	32			156 max							10			Steering gear housing, mounting brackets
	M4504[d]	448	65							310	45			163–217							4			Compressor crankshafts and hubs
	M5003[d]	517	75							345	50			187–241							3			Parts requiring selective hardening, as gears
	M5503[e]	517	75							379	55			187–241							3			For machinability and improved induction hardening
	M7002[e]	621	90							483	70			229–269							2			Connecting rods, universal joint yokes
	M8501[e]	724	105							586	85			269–302							1			Gears with high strength and good wear resistance

TABLE 1-3
Mechanical properties of typical cast ferrous materials[a] (*Cont.*)

Material, class, specification			Ultimate strength: Tension, σ_{sut} MPa	kpsi	Compression, σ_{suc} MPa	kpsi	Shear, τ_{su} MPa	kpsi	Torsional/shear strength, τ_s MPa	kpsi	Yield strength, σ_{sy} MPa	kpsi	Endurance limit in reversed bending, σ_{sfb} MPa	kpsi	Brinell hardness, H_B	Modulation of elasticity: Tension, E GPa	Mpsi	Compression, E GPa	Mpsi	Shear, G GPa	Mpsi	Elongation in 50 mm (2 in), %	Impact strength (Charpy) J	ft-lbf	Typical application
Nodular (ductile) cast iron	Grade	UNS No.																							
ASTM A395-76 ASME SA 395	60-40-18	F32800	414	60							276	40			143–187							18			Valves and fittings for steam and chemical plant equipment
ASTM A476-70(d)	80-60-03	F34100	552	80							414	60			201 min							3			Paper-mill dryer rollers
SAE AMS5316 ASTM A536-72 MIL-I-11466 B(MR)	60-40-18[h]	F32800	461	66.9	359	52.0	472	68.5			329	47.7	241	35	167–178	169	24.5	164	23.8	63–65.5[g]	9.1–9.5[g]	15			Pressure-containing parts such as valve and pump bodies
	65-45-12[h]	F33100	464	67.3	362	52.5	475	68.9			332	48.2			167	168	24.4	163	23.6	64–65[g]	9.3–9.4[g]	15			Machine components subjected to shock and fatigue loads
	80-55-06[h]	F33800	559	81.1	386	56.0	504	73.1			362	52.5	345	50	192	168	24.4	165	23.9	62–64[g]	9.0–9.3[g]	11.2			Crankshafts, gears and rollers
	100-70-03[h]	F34800	758	110	1515	220.0					500	72.5	379	55	257	162	23.5					6-10			High-strength gears and machine components
	120-90-02[h]	F36200	974	141.3	920	133.5	875	126.9			864	125.3	434	63	332	164	23.8	164	23.8	63.5–64[g]	9.2–9.3[g]	1.5			Pinions, gears, rollers and slides
SAE j434C	D4018		414	60							276	40			170 max							18			Steering knuckles
	D4512		448	65							310	45			156–217							12			Disk brake calipers
	D5506		552	80							379	55			187–255							6			Crankshafts
	D7003		689	100							483	70			241–302							3			Gears
Alloy cast irons																									
Medium-silicon gray iron			170–310	24–45	620–1040	90–150									170–250								20–31	15–23	
High chromium gray iron			210–620	30–90	690	100									250–500								27–47	20–35	
High nickel gray iron			170–310	25–45	690–1100	100–160									130–250								80–200	60–150	
Ni-Cr-Si gray iron			140–310	20–45	480–690	70–100									110–210								110–200	80–150	
High-aluminum gray iron			235–620	34–90											180–350										
Medium-silicon ductile iron			415–690	60–100											140–300								7–155	5–115	
High-nickel ductile iron (20Ni)			380–415	55–60	1240–1380	180–200									140–200								16	12	
High-nickel ductile iron (23Ni)			400–450	58–65											130–170								38	28	
Durion			110	16											520	158	23						4	3	
Mechanite			241–380	35–55							193–241	28–35			190	83	12					10			

[a] *Source*: Compiled from *AMS Metals Handbook*, American Society for Metals, Metals Park, Ohio, 1988.
[b] Minimum values of σ_u in MPa (kpsi) are given by class number.
[c] Annealed.
[d] Air-quenched and tempered.
[e] Liquid-quenched and tempered.
[f] Heat-treated and average mechanical properties.
[g] Calculated from tensile modulus and Poisson's ratio in tension.

TABLE 1-4
Typical mechanical properties of gray cast iron

Grade	Tensile strength, σ_{st}		Compressive strength, σ_{sc}		Shear strength, τ_s		Fatigue limit, σ_{sf}		Modulus of elasticity, E: Tension		Modulus of elasticity, E: Compression		Modulus of rigidity, G		Notched tensile strength, σ_{snt}		Elastic strain at failure,	Total elastic strain at fracture,	Brinell hardness	Poisson's ratio,	Density, ρ		Coefficient of the thermal expansion, α, 20 to 200°C		Specific heat capacity at 20 to 200°C, c		Thermal conductivity at 100°C, K	
	MPa	kpsi	MPa	kpsi	MPa	kpsi	MPa	kpsi	GPa	Mpsi	GPa	Mpsi	GPa	Mpsi	MPa	kpsi	%	%	H_B	ν	kg/m³	lb$_m$/ft³	µm/mK	µin/in°F	kJ/kg K	Btu/lb$_m$°F	W/m²K	Btu/ft²h°F
FG 150	150 42[c] 98[d]	21.8 6.0 14.2	600 84 195	87.0 12.2 15.2	173	25.1	68[e] 68[f]	9.9 9.9	100	14.5	100	14.5	40	5.8	120[a] 150[b]	17.4 21.8	0.15	0.6–0.75[g]	130–180	0.26	7050	440.1	11.0	6.1	26.5	0.0640	52.5	9.25
FG 200	200 56[c] 130[d]	29.0 8.1 18.8	720 112 260	104.4 16.2 37.7	230	33.4	90[e] 87[f]	13.1 12.6	114	16.5	114	16.5	46	6.7	160[a] 200[b]	23.2 29.0	0.17	0.48–0.67[g]	160–220	0.26	7100	443.3	11.0	6.1	0.375	0.0896	50.8	8.95
FG 220	220 62[c] 143[d]	32.0 9.0 20.7	768 123 286	111.4 17.8 41.5	253	36.7	99[e] 94[f]	14.4 13.6	120	17.4	120	17.4	48	7.0	176[a] 120[b]	25.5 32.0	0.18	0.39–0.63[g]	180–220	0.26	7150	446.4	11.0	6.1	0.420	0.1003	50.1	8.82
FG 260	260 73[c] 169[d]	37.7 10.6 24.5	864 146 338	125.3 21.2 49.0	299	43.4	117[e] 108[f]	17.0 15.7	128	18.6	128	18.6	51	7.4	208[a] 260[b]	30.2 37.7	0.20	0.57[g]	180–230	0.26	7200	449.5	11.0	6.1	0.460	0.1098	48.8	8.59
FG 300	300 84[c] 195[d]	43.5 12.2 28.3	960 168 390	139.2 24.4 56.6	345	50.0	135[e] 127[f]	19.6 18.4	135	19.6	135	19.6	54	7.8	240[a] 300[b]	34.8 43.5	0.22	0.50[g]	180–230	0.26	7250	452.6	11.0	6.1	0.460	0.1098	47.4	8.35
FG 350	350 98[c] 228[d]	50.8 14.2 33.1	1080 196 455	156.6 28.4 66.0	403	58.5	149[e] 129[f]	21.6 18.7	140	20.3	140	20.3	56	8.1	280[a] 250[b]	40.6 50.8	0.25	0.50[g]	207–241	0.26	7300	455.7	11.0	6.1	0.460	0.1089	45.7	8.05
FG 400	400 112[c] 260[d]	58.0 16.2 37.7	1200 224 520	174.1 32.5 75.4	460	66.7	152[e] 127[f]	2.0 18.4	145	21.0	145	21.0	58	8.4	320[a] 400[b]	46.4 58.0	0.28	0.50[g]	207–270	0.26	7300	455.7	11.0	6.1	0.460	0.1089	44.0	7.75

[a] Circumferential 45° notch-root radius 0.25 mm (0.04 in), notch depth 2.5 mm (0.4 in), root diameter 20 mm (0.8 in), notch depth 3.3 mm (0.132 in), notch diameter 7.6 mm (0.36 in).
[b] Circumferential notch radius 9.5 mm (0.38 in), notch depth 2.5 mm (0.4 in), notch diameter 20 mm (0.8 in).
[c] 0.01% proof stress.
[d] 0.1% proof stress.
[e] Unnotched 8.4 mm (0.336 in) diameter.
[f] V-notched [circumferential 45° V-notch with 0.25 mm (0.04 in) root radius, diameter at notch 8.4 mm (0.336 in), depth of notch 3.4 mm (0.135 in)].
[g] Values depend on the composition of iron.
[h] Poisson's ratio $\nu = 0.26$.

Note: The typical properties given in this table are the properties in a 30 mm (1.2 in) diameter separately cast test bar or in a casting section correctly represented by this size of test bar, where the tensile strength does not correspond to that given. Other properties may differ slightly from those given.
Source: IS (Indian Standards) 210, 1993.

TABLE 1-5
Mechanical properties of spheroidal or nodular graphite cast iron

Grade	Typical casting thickness		Density		Poisson's ratio, ν	Tensile strength, σ_{st} min		0.2% Proof stress, σ_{sy} min		Elongation[a]	Brinell hardness, H_B	Impact values min $(23 \pm 5°C)$		Predominant structural constituent
	mm	in	kg/m³	lbm/ft³		MPa	kpsi	MPa	kpsi	%, min		J	ft-lbf	
						Measured on test pieces from separately cast test samples								
SG 900/2			7150	446.4	0.275	900	130.5	600	87.0	2	280–360			
SG 800/2			7200	449.5	0.275	800	116.0	480	69.6	2	245–335			Pearlite
SG 700/2			7200	449.5	0.275	700	101.5	420	61.0	2	225–305			Pearlite
SG 600/2			7170	447.6	0.275	600	87.0	370	53.7	2	190–270			Ferrite and pearlite
SG 500/7			7100	443.3	0.275	500	72.5	320	46.4	7	160–240			Ferrite and pearlite
SG 450/10			7100	443.3	0.275	450	65.3	310	45.0	10	160–210	9.0[b] (4.3)[c]	6.6 (3.2)	
SG 400/15			7100	443.3	0.275	400	58.0	250	36.3	15	130–180	17.0[b] (15.0)[c]	12.5 (11.0)	Ferrite
SG 400/18			7100	443.3	0.275	400	58.0	250	36.6	18	130–180	14.0[b] (11.0)[c]	10.3 (8.1)	
SG 350/22			7100	443.3	0.275	350	50.8	220	32.0	22	≤150	17.0[b] (14.0)[c]	12.5 (10.3)	Ferrite
						Measured on test pieces from cast-on test samples								
SG 700/2A	30–60	1.2–2.4				700	101.5	400	58.0	2	220–320			Pearlite
	61–200	2.44–8.0				650	94.3	380	55.1	1				
SG 600/3A	30–60	1.2–2.4				600	87.0	360	52.2	2	180–270			Ferrite + pearlite
	61–200	2.44–8.0				550	79.8	340	49.3	1				
SG 500/7A	30–60	1.2–2.4				450	65.3	300	43.5	7	170–240			Ferrite + pearlite
	61–200	2.44–8.0				420	61.0	290	42.0	5				
SG 400/15A	30–60	1.2–2.4				390	56.6	250	36.3	15	130–180			Ferrite
	61–200	2.44–8.0				370	53.7	240	34.8	12				
SG 400/18A	30–60	1.2–2.4				390	56.6	250	36.4	15	130–180	14[b] (11)[c]	10.3 (8.1)	Ferrite
	61–200	2.44–8.0				370	53.7	240	34.8	12		12[b] (9)[c]	8.8 (6.6)	
SG 350/22A	30–60	1.2–2.4				330	47.9	22-	31.9	18	≤150	17[b] (14)[c]	12.5 (10.3)	Ferrite
	61–200	2.44–8.0				320	46.4	210	30.6	15		15[b] (12)[c]	11.1 (8.8)	

Grade	Compression strength, σ_{sc}		Shear strength, σ_{sc}		Fatigue limit, σ_{sc}		Modulus of, Elasticity E		Modulus of rigidity, G				Thermal coefficient of linear expansion, α		Specific heat, c at 20° to 200°C		Thermal conductivity, at 100°C	
									GPa		MPsi			at 20° to 200°C				
	MPa	kpsi	MPa	kpsi	MPa	kpsi	GPa	Mpsi	Ten	Com	Ten	Com	μm/m K	μin/in °F	kJ/kg K	Btu/lbm °F	W/m² K	Btu/ft²h°F
SG 900/2	550	79.8	810	117.5	317	46.0	67.1	9.73	169	169	24.5	24.5	11.0	6.1	0.461	0.1101	33.5	5.90
SG 800/2	362	52.5	720	107.4	304	44.1	68.6	9.95	169	169	24.5	24.5	11.0	6.1	0.461	0.1101	31.40	5.53
SG 700/2	318	46.1	630	91.4	280	40.6	86.6	9.95	169	169	24.5	24.5	11.0	6.1	0.461	0.1101	31.40	5.53
SG 600/2	286	41.5	540	78.3	248	35.0	67.9	9.85	169	169	24.5	24.5	11.0	6.1	0.461	0.1101	32.80	5.72
SG 500/7	272	39.5	45	65.3	224	32.5	65.9	9.56	169	169	24.5	24.5	11.0	6.1	0.461	0.1101	35.50	6.25
SG 450/10	253	36.7	405	58.7	210	30.5	65.9	9.56	174	174	25.2	25.2	11.0	6.1	0.461	0.1101	36.5	6.43
SG 400/15	216	31.3	360	52.2	195	28.3	65.9	9.56	176	176	25.2	25.2	11.0	6.1	0.461	0.1101	36.5	6.43
SG 400/18	216	31.3	360	52.2	195	28.3	65.9	9.86	176	176	25.2	25.2	11.0	6.1	0.461	0.1101	36.5	6.43
SG 350/22	181	31.3	315	45.7	180	26.1	65.9	9.56	169	169	24.5	24.5	11.0	6.1	0.461	0.1101	36.5	6.43

[a] Elongation is measured on an initial gauge length $L = 5d$ where d is the diameter of the gauge length of the test pieces. [b] Mean value from 3 tests on V-notch test pieces at ambient temperature. [c] Individual value.

Source: IS 1865, 1991.

TABLE 1-5A
Chemical composition[a] and mechanical properties[c] of spheroidal graphite austenitic cast iron

Grade	Chemical composition[a], %							Density		Tensile strength, σ_{st} min		0.2% proof stress, σ_{sy} min		Elongation[b]	Brinell hardness,	Impact values[d], min	
	C_{max}	Si	Mn	Ni	Cr	P_{max}	Cu_{max}	kg/m³	lb$_m$/ft³	MPa	kpsi	MPa	kpsi	%, min	H_B	J	ft-lb
ASG Ni 13 Mn 7	3.0	2.0–3.0	6.0–7.0	12.0–14.0	≤0.3	0.080	0.5	7300	455.7	390–460	56.6–66.7	210–260	30.5–37.7	15–25	130–170	15.0–27.5	11.1–20.3
ASG Ni 20 Cr 2	3.0	1.5–3.0	0.5–1.5	18.0–22.0	1.0–2.5	0.080	0.5	7400	462.0	370–470	53.7–68.2	210–250	30.5–36.3	7–20	140–200	13.5–27.5	10.0–20.3
ASG Ni 20 Cr 3	3.0	1.5–3.0	0.5–1.5	18.0–22.0	2.5–3.5	0.080	0.5	7400	462.0	390–490	56.6–71.1	210–260	30.5–37.7	7–15	150–255	12.0	8.9
ASG Ni 20 Si 5 Cr 2	3.0	4.5–5.5	0.5–1.5	18.0–22.0	1.0–2.5	0.080	0.5	7400	462.0	370–430	53.7–62.4	210–260	30.5–37.7	10–18	180–230	14.9	11.0
ASG Ni 22	3.0	1.0–3.0	1.5–2.5	21.0–24.0	<0.5	0.080	0.5	7400	462.0	370–440	53.7–63.8	170–250	24.7–36.3	20–40	130–170	20.0–33.0	14.8–24.3
ASG Ni 23 Mn 4	2.6	1.5–2.5	4.0–4.5	22.0–24.0	<0.2	0.080	0.5	7400	462.0	440–470	63.6–68.2	210–240	30.5–34.8	25–45	150–180	24.0	17.7
ASG Ni 30 Cr 1	2.6	1.5–3.0	0.5–1.5	28.0–32.0	1.0–1.5	0.080	0.5	7400	462.0	370–440	53.7–62.4	210–270	30.5–39.2	13–18	130–190	17.0	8.1
ASG Ni 30 Cr 3	2.6	1.5–3.0	0.5–1.5	28.0–32.0	2.5–3.5	0.080	0.5	7400	462.0	370–470	53.7–68.2	210–260	30.5–37.7	7–18	140–200	8.5	6.3
ASG Ni 30 Si 5 Cr 5	2.6	5.0–6.0	0.5–1.5	28.0–32.0	4.5–5.5	0.080	0.5	7400	462.0	390–490	56.6–70.5	240–310	34.8–45.0	1–4	110–250	3.9–5.9	2.9–4.4
ASG Ni 35	2.4	1.5–3.0	0.5–1.5	34.0–36.0	≤0.2	0.080	0.5	7600	474.5	370–410	53.7–59.5	210–240	30.5–34.8	20–40	130–180	20.5	15.1
ASG Ni 35 Cr 3	2.4	1.5–3.0	0.5–1.5	34.0–36.0	2.0–3.0	0.080	0.5	7600	474.5	370–440	53.7–63.8	210–290	30.5–42.1	7–10	140–190	7.0	5.2

Grade	Modulus of elasticity E		Thermal coefficient of linear expansion, α		Thermal conductivity, K		Properties and applications
	GPa	Mpsi	μm/m K at 20 to 200°C	μin/in °F at 20 to 200°C	W/m² K	Btu/ft²h°F	
ASG Ni 13 Mn 7	140–150	20.3–21.8	18.2	10.1	12.6	2.22	Non-magnetic. Hence used as pressure covers for turbine generator sets, housing for insulators, flanges and switch gears.
ASG Ni 20 Cr 2	112–130	16.2–18.9	18.7	10.4	12.6	2.22	Corrosion and heat resistance. Used in pumps, valves, compressor exhaust gas manifolds, turbo-supercharger housings and bushings.
ASG Ni 20 Cr 3	112–133	16.2–19.3	18.7	10.4	12.6	2.22	Good resistance to corrosion. Used in valves, pump components and components subject to high pressure.
ASG Ni 20 Si 5 Cr 2	112–133	16.2–19.3	18.0	10.0	12.6	2.22	High value of linear expansion and non-magnetic. Used for pumps, valves, compressor and exhaust gas manifold and turbocharge housings.
ASG Ni 22	85–112	12.3–16.2	18.4	10.2	12.6	2.22	High impact properties up to −196°C and non-magnetic. Used in castings for refrigerators, etc.
ASG Ni 23 Mn 4	120–140	17.4–20.3	14.7	8.2	12.6	2.22	Good bearing properties. Used in exhaust manifolds and pumps, valves and turbocharger gas housing.
ASG Ni 30 Cr 1	112–130	16.2–18.9	12.6	7.0	12.6	2.22	Used in boiler pumps, valves, filter parts and exhaust gas manifolds.
ASG Ni 30 Cr 3	92–105	13.3–15.2	12.6	7.0	12.6	2.22	Used in pump components, valves, etc.
ASG Ni 30 Si 5 Cr 5	91	13.2	14.4	8.0	12.6	2.22	Power lower linear coefficient of expansion. Used in machine tool parts, scientific instruments, glass molds, and parts requiring dimensional stability.
ASG Ni 35	112–140	16.2–20.3	5	2.8	12.6	2.22	Possess lower linear thermal expansion. Used in gas turbine housings and glass molds.
ASG Ni 35 Cr 3	112–123	16.2–17.8	5	2.8	12.6	2.22	

[a] Unless otherwise specified, other elements may be present at the discretion of the manufacturer, provided they do not alter the micro-structure substantially, or affect the property adversely.
[b] Elongation is measured on an initial gauge length $L = 5d$ where d is the diameter of the gauge length of the test pieces. [c] Measured on test pieces machined from separately cast test samples.
[d] Mean value from 3 tests on V-notch test pieces at ambient temperature.
Source: IS 2749, 1974.

TABLE 1-5B
Chemical composition[a] and mechanical properties[b] of flake graphite austenitic cast iron

Grade	Chemical composition, %						Density		Tensile strength, σ_{st} min		Elongation	Brinell hardness,	Ultimate compressive strength, σ_{sut}		Modulus of elasticity, E	
	C	Si	Mn	Ni	Cr	Cu	kg/m³	lbm/ft³	MPa	kpsi	%, min	H_B	MPa	kpsi	GPa	Mpsi
AFG Ni 13 Mn 7	≤3.0	1.5–3.0	6.0–7.0	12.0–14.0	0.2	≤0.5	7300	455.7	140–220	20.3–32.0	–	120–150	630–840	91.4–121.8	70–90	10.2–13.1
AFG Ni 15 Cu 6 Cr 2	≤3.0	1.0–2.8	0.5–1.5	13.5–17.5	1.0–2.5	5.5–7.5	7300	455.7	170–210	24.7–30.5	2	140–200	700–840	101.5–121.8	85–105	12.3–15.2
AFG Ni 15 Cu 6 Cr 3	≤3.0	1.0–2.8	0.5–1.5	13.5–17.5	2.5–3.5	5.5–7.5	7300	455.7	190–240	27.6–34.8	1–2	150–250	860–1100	124.7–159.5	98–113	14.2–16.4
AFG Ni 20 Cr 2	≤3.0	1.0–2.8	0.5–1.5	18.0–22.0	1.0–2.5	≤0.5	7300	455.7	170–210	24.7–30.5	2–3	120–215	700–840	101.5–121.8	85–105	12.3–15.2
AFG Ni 20 Cr 3	≤3.0	1.0–2.8	0.5–1.5	18.0–22.0	2.5–3.5	≤0.5	7300	455.7	190–240	27.6–34.8	1–2	160–250	860–1100	124.7–159.5	98–113	14.2–16.4
AFG Ni 20 Si 5 Cr 3	≤2.5	4.5–5.5	0.5–1.5	18.0–22.0	1.5–4.5	≤0.5	7300	455.7	190–280	27.6–40.6	2–3	140–250	860–1100	124.7–159.5	110	16.0
AFG Ni 30 Cr 3	≤2.5	1.0–2.0	0.5–1.5	28.0–32.0	2.5–3.5	≤0.5	7300	455.7	190–240	27.6–34.8	1–3	120–215	700–910	101.5–132.0	98–113	14.2–15.2
AFG Ni 30 Si 5 Cr 5	≤2.5	5.0–6.0	0.5–1.5	29.0–32.0	4.5–5.5	≤0.5	7300	455.7	170–240	27.4–34.8	–	150–210	560	81.2	105	15.2
AFG Ni 35	≤2.4	1.0–2.0	0.5–1.5	34.0–36.0	≤0.2	≤0.5	7300	455.7	120–180	17.4–26.1	1–3	120–140	560–700	81.2–101.5	74	10.7

Grade	Thermal coefficient of linear expansion, α		Specific hear, c		Thermal conductivity, K		Properties and applications
	μm/m K (at 20 to 200°C)	μin/in °F (at 20 to 200°C)	J/kg K	Btu/lbm °F	W/m² K	Btu/ft²h°F	
AFG Ni 13 Mn 7	17.7	9.3	460–500	0.11–0.12	37.7–41.9	6.64–7.38	Non-magnetic. Used in pressure covers for turbine generator sets, housing for switch gears and terminals, and ducts.
AFG Ni 15 Cu 6 Cr 2	18.7	10.4	460–500	0.11–0.12	37.7–41.9	6.64–7.38	Resistance to corrosion, erosion, and heat. Good bearing properties. Used for pumps, valves, piston ring covers for pistons, furnace components, bushings.
AFG Ni 15 Cu 6 Cr 3	18.7	10.4	460–500	0.11–0.12	37.7–41.9	6.64–7.38	Possess high coefficient of thermal expansion, resistance to corrosion and erosion. Used for pumps handling alkalis.
AFG Ni 20 Cr 2	18.7	10.4	460–500	0.11–0.12	37.7–41.9	6.64–7.38	Used in soap, food and plastic industries.
AFG Ni 20 Cr 3	18.7	10.4	460–500	0.11–0.12	37.7–41.9	6.64–7.38	Resistance to erosion, corrosion, heat. Used in high temperature application. Not suitable between 500 and 600°C.
AFG Ni 20 Si 5 Cr 3	18.0	10.0	460–500	0.11–0.12	37.7–41.9	6.64–7.38	Resistance to thermal shock and heat, corrosion at high temperature. Used in pumps, pressure vessels, valves, filters, exhaust gas manifolds, turbine housings.
AFG Ni 30 Cr 3	12.4	6.9	460–500	0.11–0.12	37.7–41.9	6.64–7.38	Resistance to erosion, corrosion, and heat. Possess average thermal expansion. Used in components for industrial furnaces, valves, and pump components.
AFG Ni 30 Si 5 Cr 5	14.6	8.1	460–500	0.11–0.12	37.7–41.9	6.64–7.38	Possess low thermal expansion and resistant to thermal shock. Used for
AFG Ni 35	5.0	2.8	460–500	0.11–0.12	37.7–41.9	6.64–7.38	scientific instruments, glass molds and in such other parts where dimensional stability is required

[a] Unless otherwise specified other elements may be present at the discretion of the manufacturer, provided they do not alter the microstructure substantially, or affect the properties adversely.
[b] Measured on test pieces machined from separately cast test pieces/samples.
Source: IS 2749, 1974.

TABLE 1-6
Carbon steels with specified chemical composition and related mechanical properties

Designation				Tensile strength, σ_{st}		Elongation, % (gauge length 5.56 $\sqrt{a^*}$ round test piece)	Izod impact value, min (if specified)	
New	Old	% C	% Mn	MPa	kpsi		J	ft-lbf
7 C 4	(C 07)	0.12 max	0.50 max	320–400	46.5–58.0	27		
10 C 4	(C 10†)	0.15 max	0.30–0.60	340–420	49.4–70.0	26	55	40.6
14 C 6	(C 14†)	0.10–0.18	0.40–0.70	370–450	53.6–65.0	26	55	40.6
15 C 4	(C 15)	0.20 max	0.30–0.60	370–490	53.6–71.0	25		
15 C 8	(C 15 Mn 75)	0.10–0.20	0.60–0.90	420–500	61.0–72.5	25		
20 C 8	(C 20)	0.15–0.25	0.60–0.90	440–520	63.5–75.4	24		
25 C 4	(C 25)	0.20–0.30	0.30–0.60	440–540	63.5–78.3	23		
25 C 8	(C 25 Mn 75+)	0.20–0.30	0.60–0.90	470–570	68.2–82.7	22		
30 C 8	(C 30+)	0.25–0.35	0.60–0.90	500–600	72.5–87.0	21	55	40.6
35 C 4	(C 35)	0.30–0.40	0.30–0.60	520–620	75.4–90.0	20		
35 C 8	(C 35 Mn 75+)	0.30–0.40	0.60–0.90	550–650	79.8–94.3	20	55	40.6
40 C 8	(C 40+)	0.35–0.45	0.60–0.90	580–680	84.1–98.7	18	41.35	30.5
45 C 8	(C 45+)	0.40–0.50	0.60–0.90	630–710	91.4–103.0	15	41.35	30.5
50 C 4	(C 50+)	0.45–0.55	0.60–0.90	660–780	95.7–113.1	13		
50 C 12	(C 50 Mn 1)	0.45–0.55	1.10–1.40	720 min	104.4 min	11		
55 C 8	(C 55 Mn 75+)	0.50–0.60	0.60–0.90	720 min	104.4 min	13		
60 C 4	(C 60)	0.55–0.65	0.50–0.80	750 min	108.8 min	11		
65 C 6	(C 65)	0.60–0.70	0.50–0.80	750 min	108.8 min	10		

Notes: a^*, area of cross section; † steel for hardening; + steel for hardening and tempering; Mn 75 = average content of Mn is 0.75%.
Source: IS 1570, 1979.

TABLE 1-7
Carbon and carbon - manganese free - cutting steels with specified chemical composition and related mechanical properties

Designation							Tensile strength, σ_{st}		Minimum elongation, % (gauge length $5.65\sqrt{a^{**}}$)	Izod impact value, min (if specified)		Limiting ruling section, mm (in)
New	Old	% C	% Si	% Mn	% S	% P (max)	MPa	kpsi		J	ft-lbf	
10 C 8 S 10	(10 S 11)	0.15 max	0.05–0.30	0.60–0.90	0.08–0.13	0.060	370–490*	53.7–71.0	24*	55	40.6	30 (1.2)
14 C 14 S 14	(14 Mn 1 S 14)	0.10–0.18	0.05–0.30	1.20–1.50	0.10–0.18	0.060	440–540*	63.8–78.3	22*			30 (1.2)
25 C 12 S 14	(25 Mn 1 S 14)	0.20–0.30	0.25 max	1.00–1.50	0.10–0.18	0.060	500–600*	72.5–87.0	20*			
40 C 10 S 18	(40 S 18)	0.35–0.45	0.25 max	0.80–1.20	0.14–0.22	0.060	550–650*	79.8–94.0	17*	41	30.2	60 (2.4)
11 C 10 S 25	(11 S 25)	0.08–0.15	0.10 max	0.80–1.20	0.20–0.30	0.060	370–490*	53.7–71.0	22*			
40 C 15 S 12	(40 Mn 2 S 12)	0.35–0.45	0.25 max	1.30–1.70	0.08–0.15	0.060	600–700*	87.0–101.5	15*	48	35.4	100 (4.0)

Notes: a^{**}, area of cross section; *, steel for case hardening. Minimum values of yield stress may be required in certain specifications, and in such case a minimum yield stress of 55 percent of minimum tensile strength should be satisfactory.
Source: IS 1570, 1979.

TABLE 1-8
Mechanical properties of selected carbon and alloy steels

AISI[a] no.	UNS no.	Treatment	Austenitizing temperature °C	°F	Tensile strength, σ_{st} MPa	kpsi	Yield strength, σ_{sy} MPa	kpsi	Elongation in 50 mm (2 in), %	Reduction in area, %	Brinell hardness, H_B	Izod impact strength J	ft-lbf
1010	G10100	Hot-rolled			320	47	180	26	28	50	95		
		Cold-drawn			370	53	300	44	20	40	105		
1015	G10150	As rolled	–	–	420.6	61.0	313.7	45.5	39.0	61.0	126	110.5	81.5
		Normalized	925	1700	424.0	61.5	324.1	47.0	37.0	69.6	121	115.5	85.2
		Annealed	870	1600	386.1	56.0	284.4	41.3	37.0	69.7	111	115.0	84.8
1020	G10200	As-rolled			448.2	65.0	330.9	48.0	36.0	59.0	143	86.8	64.0
		Normalized	870	1600	441.3	64.0	346.5	50.3	35.8	67.9	131	117.7	86.8
		Annealed	870	1600	394.7	57.3	294.8	42.3	36.5	66.0	111	123.4	91.0
1030	G10300	As-rolled			551.6	80.0	344.7	50.0	32.0	57.0	179	74.6	55.0
		Normalized	925	1700	520.0	75.5	344.7	50.0	32.0	60.8	149	93.6	69.0
		Annealed	845	1550	463.7	67.3	341.3	49.5	31.2	57.9	126	69.4	51.2
1040	G10400	As-rolled			620.5	90.0	413.7	60.0	25.0	50.0	201	48.8	36.0
		Normalized	900	1650	589.5	85.5	374.0	54.3	28.0	54.9	170	65.1	48.0
		Annealed	790	1450	518.8	75.3	353.4	51.3	30.2	57.2	149	44.3	32.7
1050	G10500	As-rolled			723.9	105.0	413.7	60.0	20.0	40.0	229	31.2	23.0
		Normalized	900	1650	748.1	108.5	427.5	62.0	20.0	39.4	217	27.1	20.0
		Annealed	790	1450	636.0	92.3	365.4	53.0	23.7	39.9	187	16.9	12.5
1060	G10600	As-rolled			813.7	118.0	482.6	70.0	17.0	34.0	241	17.6	13.0
		Normalized	900	1650	775.7	112.5	420.6	61.0	18.0	37.2	229	13.2	9.7
		Annealed	790	1450	625.7	90.8	372.3	54.0	22.5	38.2	179	11.3	8.3
1095	G10950	As-rolled			965.3	140.0	572.3	83.0	9.0	18.0	293	4.1	3.0
		Normalized	900	1650	1013.5	147.0	499.9	72.5	9.5	13.5	293	5.4	4.0
		Annealed	790	1450	656.7	95.3	379.2	55.0	13.0	20.6	192	2.7	2.0
1117	G11170	As-rolled			486.8	70.6	305.4	44.3	33.0	63.0	143	81.3	60.0
		Normalized	900	1650	467.1	67.8	303.4	44.0	33.5	63.8	137	85.1	62.8
		Annealed	825	1575	429.5	62.3	279.2	40.5	32.8	58.0	121	93.6	69.0
1144	G11440	As-rolled			703.3	102.0	420.6	61.0	21.0	41.0	212	52.9	39.0
		Normalized	900	1650	667.4	96.8	399.9	58.0	21.0	40.4	197	43.4	32.0
		Annealed	790	1450	584.7	84.8	346.8	50.3	24.8	41.3	167	65.1	48.0
1340	G13400	Normalized	870	1600	836.3	121.3	558.5	81.0	22.0	62.9	248	92.5	68.2
		Annealed	800	1475	703.3	102.0	436.4	63.3	25.5	57.3	207	70.5	52.0

TABLE 1-8
Mechanical properties of selected carbon and alloy steels (*Cont.*)

AISI[a] no.	UNS no.	Treatment	Austenitizing temperature °C	°F	Tensile strength, σ_{st} MPa	kpsi	Yield strength, σ_{sy} MPa	kpsi	Elongation in 50 mm (2 in), %	Reduction in area, %	Brinell hardness, H_B	Izod impact strength J	ft-lbf
3140	G31400	Normalized	870	1600	891.5	129.3	599.8	87.0	19.7	57.3	262	53.6	39.5
		Annealed	815	1500	689.5	100.0	422.6	61.3	24.5	50.8	197	46.4	34.2
4130	G41300	Normalized	870	1600	668.8	97.0	436.4	63.3	25.2	59.5	197	86.4	63.7
		Annealed	865	1585	560.5	81.3	360.6	52.3	28.2	55.6	156	61.7	45.5
4150	G41500	Normalized	870	1600	1154.9	167.5	734.3	106.5	11.7	30.8	321	11.5	8.5
		Annealed	815	1500	729.5	105.8	379.2	55.0	20.2	40.2	197	24.7	18.2
4320	G43200	Normalized	895	1640	792.9	115.0	464.0	67.1	20.8	50.7	235	72.9	53.8
		Annealed	850	1560	579.2	84.0	609.5	61.6	29.0	58.4	163	109.8	81.0
4340	G43400	Normalized	870	1600	1279.0	185.5	861.8	125.0	12.2	36.3	363	15.9	11.7
		Annealed	810	1490	744.6	108.0	472.3	68.5	22.0	49.9	217	51.1	37.7
4620	G46200	Normalized	900	1650	574.3	83.3	366.1	53.1	29.0	66.7	174	132.9	98.0
		Annealed	855	1575	512.3	74.3	372.3	54.0	31.3	60.3	149	93.6	69.0
4820	G48200	Normalized	860	1580	750.0	109.5	484.7	70.3	24.0	59.2	229	109.8	81.0
		Annealed	815	1500	681.2	98.8	464.0	67.3	22.3	58.8	197	92.9	68.5
5150	G51500	Normalized	870	1600	870.8	126.3	529.5	76.8	20.7	58.7	255	31.5	23.2
		Annealed	825	1520	675.7	98.0	357.1	51.8	22.0	43.7	197	25.1	18.5
6150	G61500	Normalized	870	1600	939.8	136.3	615.7	89.3	21.8	61.0	269	35.5	26.2
		Annealed	815	1500	667.4	96.8	412.3	59.8	23.0	48.4	197	27.4	20.2
8630	G86300	Normalized	870	1600	650.2	94.3	429.5	62.3	23.5	53.5	187	94.6	69.8
		Annealed	845	1550	564.0	81.8	372.3	54.0	29.0	58.9	156	95.2	70.2
8740	G87400	Normalized	870	1600	929.4	134.8	606.7	88.0	16.0	47.9	269	17.6	13.0
		Annealed	815	1500	695.0	100.8	415.8	60.3	22.2	46.4	201	40.0	29.5
9255	G92550	Normalized	900	1650	932.9	135.3	579.2	84.0	19.7	43.4	269	13.6	10.0
		Annealed	845	1550	774.3	112.3	486.1	70.5	21.7	41.1	229	8.8	6.5
9310	G93100	Normalized	890	1630	906.7	131.5	570.9	82.8	18.8	58.1	269	119.3	88.0
		Annealed	845	1550	820.5	119.0	439.9	63.8	17.3	42.1	241	78.6	58.0

[a] All grades are fine-grained except for those 1100 series, which are coarse-grained. Heat-treated specimens were oil-quenched unless otherwise indicated.
Values tabulated were averaged and obtained from specimen 12.75 mm (0.505 in) in diameter which were machined from 25 mm (1 in); rounded gauge lengths were 50 mm (2 in).
Source: *ASM Metals Handbook*, American Society for Metals. Metals Park, Ohio. 1988

TABLE 1-9
Mechanical properties of standard steels

Designation		Tensile strength, σ_{st}		Yield stress, σ_{sy}		Elongation in 50 mm (gauge length 5.65 $\sqrt{a^*}$)
New	Old	MPa	kpsi	MPa	kpsi	
Fe 290	(St 30)	290	42.0	170	24.7	27
Fe E 220	–	290	42.0	220	32.0	27
Fe 310	(St 32)	310	45.0	180	26.1	26
Fe E 230	–	310	45.0	230	33.4	26
Fe 330	(St 34)	330	47.9	200	29.0	26
Fe F 250	–	330	47.9	250	36.3	26
Fe 360	(St 37)	360	52.2	220	32.0	25
Fe F 270	–	360	52.2	270	39.2	25
Fe 410	(St 42)	410	59.5	250	36.3	23
Fe E 310	–	410	59.5	310	50.0	23
Fe 490	(St 50)	490	71.1	290	42.0	21
Fe E 370	–	490	71.1	370	53.7	21
Fe 540	(St 55)	540	78.3	320	46.4	20
Fe E 400	–	540	78.3	400	58.0	20
Fe 620	(St 63)	620	90.0	380	55.1	15
Fe E 460	–	620	90.0	460	66.7	15
Fe 690	(St 70)	690	100.0	410	59.5	12
Fe E 520	–	690	100.0	520	75.4	12
Fe 770	(St 78)	770	111.7	460	66.7	10
Fe E 580	–	770	111.7	580	84.1	10
Fe 870	(St 88)	870	126.2	520	75.4	8
Fe E 650	–	870	126.2	650	94.3	8

Note: a^* area of cross-section of test specimen.
Source: IS 1570, 1978.

TABLE 1-10
Chemical composition and mechanical properties of carbon steel castings for surface hardening

Designation	Chemical composition (in ladle analysis) max, %									
	C	Si	Mn	S	P	Cr	Ni	Mo	Cu	Residual elements
Gr 1	0.4-0.5	0.60	1.0	0.05	0.05	0.25	0.40	0.15	0.30	0.80
Gr 2	0.5-0.6	0.60	1.0	0.05	0.05	0.25	0.40	0.15	0.30	0.80

Designation	Tensile strength, σ_{st}		Yield strength, σ_{sy}		Elongation, % min (gauge length 5.65 $\sqrt{a^*}$)	Brinell hardness H_B
	Mpa	kpsi	Mpa	kpsi		
Gr 1	620	90.0	320	46.4	12	460
Gr 2	700	101.5	370	53.7	8	535

Notes: a^* area of cross section of test specimen. All castings shall be free from distortion and harmful defects. They shall be well-dressed, fettled, and machinable. Unless agreed upon by the purchaser and the manufacturer, castings shall be supplied in the annealed, or nomalized and tempered condition.
Source: IS 2707, 1973.

TABLE 1-11
Chemical composition of alloy steel forgings for general industrial use

Designation									
New	**Old**	**% C**	**% Si**	**% Mn**	**% Ni**	**% Cr**	**% Mo**	**% V**	**% Al**
20 C 15	20 Mn 2	0.06–0.24	0.10–0.35	1.30–1.70					
15 Cr 3	15 Cr 65	0.12–0.18	0.10–0.35	0.40–0.60		0.50–0.80			
16 Mn 5 Cr 4	17 Mn 1 Cr 95	0.14–0.19	0.10–0.35	1.00–1.30		0.80–1.10			
20 Mn 5 Cr 5	20 Mn Cr 1	0.18–0.22	0.10–0.35	1.00–1.40		1.00–1.30			
21 Cr 4 Mo 2	21 Cr 1 Mo 28	0.26 max	0.10–0.35	0.50–0.80		0.90–1.20	0.15–0.30		
07 Cr 4 Mo 6	07 Cr 90 Mo 55	0.12 max	0.15–0.60	0.40–0.70	0.30 max	0.70–1.10	0.45–0.65		
10 Cr 9 Mo 10	10 Cr 2 Mo 1	0.15 max	0.50 max	0.40–0.70	0.30 max	2.00–2.50	0.90–1.10		
13 Ni 13 Cr 3	13 Ni 3 Cr 80	0.10–0.15	0.10–0.35	0.40–0.70	3.00–3.50	0.60–1.00			
15 Ni 16 Cr 5	15 Ni 4 Cr 1	0.12–0.18	0.10–0.35	0.40–0.70	3.80–4.30	1.00–1.40			
15 Ni 5 Cr 4 Mo 1	15 Ni Cr 1 Mo 12	0.12–0.18	0.10–0.35	0.60–1.00	1.00–1.50	0.75–1.25	0.80–0.15		
15 Ni 7 Cr 4 Mo 2	15 Ni Cr 1 Mo 15	0.12–0.18	0.10–0.35	0.60–1.00	1.50–2.00	0.75–1.25	0.10–0.20		
16 Ni 8 Cr 6 Mo 2	16 Ni Cr 2 Mo 20	0.12–0.20	0.10–0.35	0.40–0.70	1.80–2.20	1.40–1.70	0.15–0.25		
36 S 17	37 Si 2 Mn 90	0.33–0.40	1.50–2.00	0.80–1.00					
37 C 15	37 Mn 2	0.32–0.42	0.10–0.35	1.30–1.70					
35 Mn 6 Mo 3	35 Mn 2 Mo 28	0.30–0.40	0.10–0.35	1.30–1.80			0.20–0.35		
40 Cr 4	40 Cr 1	0.35–0.45	0.10–0.35	1.60–0.90		0.90–1.70			
40 Cr 4 Mo 3	40 Cr 1 Mo 28	0.35–0.45	0.10–0.35	0.50–0.80		0.90–1.20	0.20–0.35		
35 Ni 5 Cr 2	35 Ni 1 Cr 60	0.30–0.40	0.10–0.35	0.60–0.90	1.00–1.50	0.45–0.75			
40 Ni 6 Cr 4 Mo 3	40 Ni 2 Cr 1 Mo 28	0.35–0.45	0.10–0.35	0.40–0.70	1.25–1.75	0.90–1.30	0.20–0.35		
40 Ni 10 Cr 3 Mo 6	40 Ni 3 Cr 65 Mo 55	0.36–0.44	0.10–0.35	0.40–0.70	2.25–2.75	0.50–0.80	0.40–0.70		
25 Cr 13 Mo 6	25 Cr 3 Mo 55	0.20–0.30	0.10–0.35	0.40–0.70	0.30 max	2.90–3.40	0.45–0.65		
55 Si 7	55 Si 2 Mn 90	0.50–0.60	1.50–2.00	0.80–1.00					
50 Cr 4 V 2	50 Cr 1 V 23	0.45–0.55	0.10–0.35	0.50–0.80		0.90–1.20		0.15–0.30	
20 Ni Cr MO 2		0.18–0.23	0.20–0.35	0.70–0.90	0.40–0.70	0.40–0.60	0.15–0.25		
37 Mn 5 Si 5		0.33–0.41	1.10–1.40	1.10–1.40					

Notes: (1) Sulfur and phosphorus can be ordered as per following limits: (i) S and P – 0.30 max; (ii) S – 0.02–0.035 and P – 0.035 max. (2) When the steel is Al killed, total Al contents shall be between 0.02–0.05 percent.
Source: IS 4367, 1991.

TABLE 1-12
Mechanical properties of alloy steel forgings for general industrial use

Designation	Condition	Tensile strength,[b] σ_{st} MPa	Tensile strength,[b] σ_{st} kpsi	Yield strength,[b] σ_{sy} MPa	Yield strength,[b] σ_{sy} kpsi	Brinell hardness in soft annealed condition, max, H_B	Elongation,[b] %, min (gauge length $5.65\sqrt{a^*}$)[a]	Izod impact[b] value J	Izod impact[b] value ft-lbf	Brinell[b] hardness number H_B	Limiting ruling section mm	Limiting ruling section in
20 C 15	H and T	600–750	87.0–108.8	400	58.0	200	18	50	36.9	178–221	63	2.52
		700–850	101.5–123.3	460	66.7		16	50	36.9	208–252	63	1.20
30 C 15	H and T	600–750	87.0–108.8	440	63.8	220	18	50	36.9	178–221	150	6.00
		700–850	101.5–123.3	540	78.3		18	48	35.4	208–252	100	4.00
		800–950	116.0–137.8	600	87.0		16	48	35.4	235–280	63	2.52
15 Cr 3	R, Q and S.R	600 min	87.0 min			170	13	48	35.4		30	1.20
16 Mn 5 Cr 4	R, Q and S.R	800 min	116.0 min			207	10	35	25.8		30	1.20
20 Mn 5 Cr 5	R, Q and S.R	1000 min	145.0 min			217	8	38	28.0		30	1.20
21 Cr 4 Mo 2	H and T	650–800	94.3–116.0	420	61.0	210	16	60	44.3	190–235	150	6.00
		700–850	101.5–123.3	460	66.7		15	55	40.6	208–252	100	4.00
		800–950	116.0–137.8	580	84.1		14	50	36.9	235–280	40	1.60
7 Cr 4 Mo 6	N and T	380–550	55.1–79.8	225	32.6	170	19	60	44.3		40	1.60
10 Cr 9 Mo 10	N and T	410–590	59.5–85.6	245	35.5	187	18	55	40.6		50	2.00
		520–680	75.4–98.6	310	45.0		18	50	36.9			
13 Ni 13 Cr 3	R, Q and S.R	850 min	123.3 min			229	12	48	35.4		60	2.4
15 Ni 16 Cr 5	R, Q and S.R	1350 min	195.8 min			241	9	35	25.8		30	1.20
15 Ni 5 Cr 4 Mo 1	R, Q and S.R	1000 min	145.0 min			217	9	41	30.2		30	1.20
15 Ni 7 Cr 4 Mo 2	R, Q and S.R	1100 min	159.5 min			217	9	35	25.8		30	1.20
16 Ni 8 Cr 6 Mo 2	R, Q and S.R	1350 min	195.8 min			229	9	35	25.8		30	1.20
20 Ni Cr Mo 2	R, Q and S.R	900 min	130.5 min			213	11	41	30.2			
36 Si 7	H and T	800–950	116.0–137.8			217				235–280	100	4.00
37 Mn 5 Si 5	H and T	780–930	113.1–134.9	590	85.6	217	14					
55 Si 7	H and T	1300–1500	188.6–217.6			245				380–440	100	4.00
35 Mn 6 Mo 3	H and T	700–850	101.5–123.3	540	78.3	220	18	55	40.6	208–252	150	6.00
		900–1050	130.5–152.3	700	101.5		15	50	36.9	268–311	63	2.52
		1000–1150	145.0–166.8	800	116.0		13	45	33.2	295–341	30	1.20
40 Cr 4	H and T	700–850	101.5–123.3	540	78.3	220	18	55	40.6	208–252	100	4.00
		900–1050	130.5–152.3	700	101.5		15	50	36.9	266–311	30	1.20
40 Cr 4 Mo 3	H and T	700–850	101.5–123.3	540	78.3	220	18	55	40.6	208–252	150	6.00
		900–1050	130.5–152.3	700	101.5		15	50	36.9	266–311	63	2.52
		1000–1150	145.0–166.8	800	116.0		13	45	33.2	295–341	30	1.20
35 Ni 5 Cr 2	H and T	700–850	101.5–123.3	540	78.3	220	18	55	40.6	208–252	150	6.00
		900–1050	130.5–152.3	700	101.5		15	50	36.9	266–311	63	2.52

TABLE 1–12
Mechanical properties of alloy steel forgings for general industrial use (*Cont.*)

Designation	Condition	Tensile strength,[b] σ_{st} MPa	Tensile strength,[b] σ_{st} kpsi	Yield strength,[b] σ_{sy} MPa	Yield strength,[b] σ_{sy} kpsi	Brinell hardness in soft annealed condition, max, H_B	Elongation,[b] %, min (gauge length $5.65\sqrt{a^*}$)[a]	Izod impact[b] value J	Izod impact[b] value ft–lbf	Brinell[b] hardness number H_B	Limiting ruling section mm	Limiting ruling section in
25 Cr 13 Mo 6		900–1050	130.5–152.3	700	101.5	230	15	55	40.6	266–311	150	6.00
		1100–1250	159.5–181.3	880	127.6		12	41	30.2	325–370	100	4.00
40 Ni 6 Cr 4 Mo 3	H and T	900–1050	130.5–152.3	700	101.5	230	55	55	40.6	266–311	150	6.00
		1100–1250	159.5–181.3	880	127.6		11	41	30.2	325–370	63	2.52
		1200–1350	174.0–195.8	1000	145.0		10	30	22.1	355–399	30	1.20
40 Ni 10 Cr 3 Mo 6	H and T	1000–1150	145.0–166.8	800	116.0	250	12	48	35.4	295–341	150	6.00
		1200–1350	174.0–195.8	1000	145.0		10	35	25.8	355–399	150	6.00
		1550 min	224.8 min	1300	188.5		8	15	11.1	450 min	100	4.00
50 Cr 4 V 2	H and T	900–1100	130.5–159.5	700	101.5	240	12	45	33.2	266–325	100	4.00
		1000–1200	145.0–174.0	800	116.0		10	45	33.2	295–355	40	1.60

[a] a^*, area of cross section.
[b] Mechanical properties in heat-treated conditions.
Notes: H and T – hardened and tempered; N and T – normalized and tempered; R,Q and S.R – refined quenched and stress-relieved. All properties for guidance only. Other values may be mutually agreed on between the consumers and suppliers.
Source: IS 4367, 1991.

TABLE 1–13
Chemical composition and mechanical properties of alloy steels**

Designation	Percent							Tensile strength, σ_{st}		0.2% proof stress, min, σ_{sy}		Minimum elongation (gauge length $= 5.65\sqrt{a^*}$)[a],	Minimum Izod impact value		Brinell[#] hardness	Limiting ruling section,
	C	Si	Mn	Ni	Cr	Mo	V/Al	MPa	kpsi	MPa	kpsi	%	J	ft-lbf	H_B	mm (in)[+]
20 C 15 (20 Mn 2)[##]	0.16–0.24	0.10–0.35	1.30–1.70					590–740	85.6–107.3	390	56.6	18	48	35.4	170–217	63 (2.5)[+]
								690–840	100.0–121.8	450	65.3	16	48	35.4	201–248	30 (1.2)
27 C 15 (27 Mn 2)	0.22–0.32	0.10–0.35	1.30–1.70					590–740	85.5–107.3	390	56.6	18	48	35.4	170–217	100 (4.0)
								690–840	100.0–121.8	450	65.3	16	48	35.4	201–248	63 (2.5)
37 C 15 (37 Mn 2)	0.32–0.42	0.10–0.35	1.30–1.70					590–740	85.5–107.3	390	56.6	18	48	35.4	170–217	150 (6.0)
								690–840	100.0–121.8	490	71.1	18	48	35.4	201–248	100 (4.0)
								790–940	114.6–136.3	550	79.9	16	48	35.4	229–277	30 (1.2)
								890–1040	129.0–150.8	650	94.3	15	41	30.2	255–311	15 (0.6)
35 Mn 6 Mo 3 (35 Mn 2 Mo 28)	0.30–0.40	0.10–0.35	1.30–1.80			0.20–0.35		690–840	100.0–121.8	490	71.1	14	55	40.6	201–248	150 (6.0)
								790–940	114.6–136.3	550	79.8	12	50	36.8	229–277	100 (4.0)
								890–1040	129.0–150.8	650	94.3	12	50	36.8	255–311	63 (2.5)
								990–1140	143.6–165.3	750	108.8	10	48	35.4	285–341	30 (1.2)
35 Mn 6 Mo 4 (35 Mn 2 Mo 45)	0.30–0.40	0.10–0.35	1.30–1.80			0.35–0.55		790–940	114.6–136.3	550	79.8	16	55	40.6	229–277	150 (6.0)
								890–1040	129.0–150.8	650	94.3	15	55	40.6	255–311	100 (4.0)
								990–1140	143.6–165.3	750	108.9	13	48	35.4	285–341	63 (2.5)
40 Cr 4 (40 Cr 1)	0.35–0.45	0.10–0.35	0.60–0.90		0.90–1.20			690–840	100.0–121.8	490	71.1	14	55	40.6	201–248	100 (4.0)
								790–940	114.6–136.3	550	79.8	12	50	36.8	229–277	63 (2.5)
								890–1040	129.0–150.8	650	94.3	11	50	36.8	255–311	30 (1.2)
40 Cr 4 Mo 2 (40 Cr 1 Mo 28)	0.35–0.45	0.10–0.35	0.50–0.80		0.90–1.20	0.20–0.35		700–850	101.5–123.3	490	71.1	13	55	40.6	201–248	150 (6.0)
								800–950	116.0–137.8	550	79.8	12	50	36.8	229–277	100 (4.0)
								900–1050	130.5–152.3	650	94.3	11	50	36.8	255–311	63 (2.5)
								1000–1150	145.0–166.8	750	108.8	10	48	35.4	285–341	30 (1.2)
15 Cr 13 Mo 6 (15 Cr 3 Mo 55)	0.10–0.20	0.10–0.35	0.40–0.70	0.30 max	2.90–3.40	0.45–0.65		690–840	100.0–121.8	490	71.1	14	55	40.6	201–248	150 (6.0)
								790–940	114.6–136.3	550	79.9	12	50	36.8	229–277	150 (6.0)
25 Cr 13 Mo 6 (25 Cr 3 Mo 55)	0.20–0.30	0.10–0.35	0.40–0.70	0.30 max	2.90–3.40	0.45–0.65		890–1040	129.0–150.8	650	94.3.	11	50	36.8	255–311	150 (6.0)
								990–1140	143.6–165.3	750	108.8	10	48	35.4	285–341	150 (6.0)
								1090–1240	158.1–179.8	830	120.4	9	41	30.2	311–363	100 (4.0)
								1540 min	223.4 min	1240	179.8	8	14	10.3	444 min	63 (2.5)

TABLE 1–13
Chemical composition and mechanical properties of alloy steels** (*Cont.*)

Designation	Percent							Tensile strength, σ_{st}		0.2% proof stress, min, σ_{sy}		Minimum elongation (gauge length $= 5.65\sqrt{a^*}$)[a],	Minimum Izod impact value		Brinell[#] hardness	Limiting ruling section,
	C	Si	Mn	Ni	Cr	Mo	V/Al	MPa	kpsi	MPa	kpsi	%	J	ft-lbf	H_B	mm (in)[+]
40 Cr 13 Mo 10 V 2	0.35 –	0.10–	0.40–	0.30	3.00–	0.90–	V: 0.15–	1340 min	194.4 min	1050	152.2	8	21	15.5	363 min	63 (2.5)
(40 Cr 3 Mo 1 V 20)	0.45	0.35	0.70	max	3.50	1.10	0.25	1540 min	223.4 min	1240	179.8	8	14	10.3	444 min	30 (1.2)
40 Cr 7 Al 10 Mo 2	0.35–	0.10–	0.40–	0.30	1.50–	0.10–	Al: 0.90–	690–840	100.0–121.8	490	71.1	18	55	40.6	201–248	150 (6.0)
(40 Cr 2 Al 1 Mo 18)	0.45	0.45	0.70	max	1.80.	0.25	1.30	790–940	114.6–136.3	550	79.8	16	55	40.6	229–277	100 (4.0)
								890–1040	129.0–150.8	650	94.3	15	48	35.4	255–311	63 (2.5)
40 Ni 14 (40 Ni 31)	0.35–	0.10–	0. 50–	3.20–	0.30			790–940	114.6–136.3	550	79.8	16	55	40.6	229–277	100 (4.0)*
	0.45	0.35	0.80	3.6	max			890–1040	129.0–150.8	650	94.3	15	55	40.6	255–311	63 (2.5)
35 Ni 5 Cr 2	0.30–	0.10–	60–90	1.00–	0.45–			690–840	100.0–121.8	490	71.1	14	55	40.6	201–248	150 (6.0)[+]
(35 Ni 1 Cr 60)	0.40	0.35		1.50	0.75			790–940	114.6–136.3	550	79.8	12	50	36.8	229–277	100 (4.0)
								890–1040	129.0–150.8	650	94.3	10	50	36.8	255–311	63 (2.5)
30 Ni 16 Cr 5	0.26–	0.10–	0.40–	3.90–	1.10–			1540 min	223.4 min	1240	179.9	8	14	10.3	444 min	(air-hardened)
(30 Ni 4 Cr 1)	0.34	0.35	0.70	4.30	1.40											150 (6.0)
																(air-hardened)
40 Ni 6 Cr 4 Mo 2	0.35–	0.10–	0.40–	1.20–	0.90–	0.10–		790–940	114.6–136.3	550	79.8	16	55	40.6	229–277	150 (6.0)
(40 Ni Cr 1 Mo 15)	0.45	0.35	0.70	1.60	1.30	0.20		890–1040	129.0–150.8	650	94.3	15	55	40.6	255–311	100 (4.0)
								990–1140	143.6–165.3	750	108.8	13	48	35.4	285–341	63 (2.5)
								1090–1240	158.1–179.8	830	120.4	13	41	30.3	311–363	30 (1.2)
40 Ni 6 Cr 4 Mo 3	0.35–	0.10–	0.40–	1.25–	0.90–	0.20–		790–940	114.6–136.3	550	79.8	16	55	40.6	229–277	150 (6.0)
(40 Ni 2 Cr 1 Mo 28)	0.45	0.35	0.70	1.75	1.30	0.35		890–1040	129.0–150.8	650	94.3	15	55	40.6	255–311	150 (6.0)
								990–1140	143.6–165.3	750	108.8	11	48	36.8	285–341	100 (4.0)
								1090–1240	158.1–179.8	830	120.4	11	41	30.3	311–363	63 (2.5)
								1190–1340	172.6–194.4	930	134.9	10	30	22.1	341–401	30 (1.2)
								1540 min	223.4 min	1240	179.8	6	11	8.1	444 min	30 (1.2)
31 Ni 10 Cr 3 Mo 6	0.27–	0.10–	0.40–	2.25–	0.50–	0.40–		890–1040	129.0–150.8	650	94.3	15	55	40.6	255–311	150 (6.0)
(31 Ni 3 Cr 65 Mo 55)	0.35	0.35	0.70	2.75	0.80	0.70		990–1140	143.6–165.3	750	108.8	12	48	35.4	285–341	150 (6.0)
								1090–1240	158.1–179.8	830	120.4	11	41	30.3	311–363	100 (4.0)
								1190–1340	172.6–194.4	930	134.9	10	35	25.8	341–401	63 (2.5)
								1540 min	223.4 min	1240	179.8	8	14	10.3	444 min	63 (2.5)

TABLE 1–13
Chemical composition and mechanical properties of alloy steels ** *(Cont.)*

Designation	Percent							Tensile strength, σ_{st}		0.2% proof stress, min, σ_{sy}		Minimum elongation (gauge length $= 5.65\sqrt{a^*})^a$,	Minimum Izod impact value		Brinell[#] hardness	Limiting ruling section,
	C	Si	Mn	Ni	Cr	Mo	V/Al	MPa	kpsi	MPa	kpsi	%	J	ft-lbf	H_B	mm (in)[+]
40 Ni 10 Cr 3 Mo 6 (40 Ni 3 Cr 65 Mo 55)	0.36–(0.44	0.10–0.35	0.40–0.70	2.25–2.75	0.50–0.80	0.40–0.70		990–1140	143.6–165.3	750	108.8	12	48	35.4	285–341	150 (6.0)
								1090–1240	158.1–179.8	830	120.4	11	41	30.3	311–363	150 (6.0)
								1190–1240	172.6–194.4	930	134.9	10	35	25.8	341–401	150 (6.0)
								1540 min	223.4 min	1240	179.8	8	14	10.3	444 min	100 (4.0)

Note: a^*, area of cross section; ** hardened and tempered condition – oil-hardened unless otherwise stated; # hardness given in this table is for guidance only; § steel designations in parentheses are old designations; + numerals in parentheses are in inches.
Source: IS 1750, 1988.

TABLE 1-14
Mechanical properties of case hardening steels in the refined and quenched condition (core properties)

Steel designation	Tensile strength, σ_{st}		Minimum elongation, % (gauge length $= 5.65\sqrt{a^*}$)[a]	Izod impact value, min (if specified)		Limiting ruling section, mm (in)	Brinell hardness number, max, H_B
	MPa	kpsi		J	ft-lbf		
10 C 4 (C 10)	490	71.1	17	54	39.8	15 (0.6)	130
14 C 4 (C 14)	490	71.1	17	54	39.9	>15 (0.6)	143
						<30 (1.2)	
10 C 8 S 11 (10 S 11)	490	71.1	17	54	39.8	30 (1.2)	143
14 C 14 S 14	588	85.4	17	40	29.7	30 (1.2)	154
(14 Mn 1 S 14)							
11 C 15	588	85.4	17	54	39.8	30 (1.2)	154
(11 Mn 2)							
15 Cr 65	588	85.4	13	47	34.7	30 (1.2)	170
17 Mn 1 Cr 95	784	113.8	10	34	25.3	30 (1.2)	207
20 Mn Cr 1	981	142.3	8	37	27.5	30 (1.2)	217
16 Ni 3 Cr 2	686	99.6	15	40	29.7	90 (3.6)	184
(16 Ni 80 Cr 60)							
16 Ni 4 Cr 3	834	121.0	12	40	29.7	30 (1.2)	217
(16 Ni 1 Cr 80)	784	113.8				60 (2.4)	
	735	106.7				90 (3.6)	
13 Ni 13 Cr 3	834	121.0	12	47	34.7	60 (2.4)	229
(13 Ni 3 Cr 80)	784	113.8				100 (4.0)	
15 Ni 4 Cr 1	1324	192.0	9	34	25.3	30 (1.2)	241
	1177	170.7				60 (2.4)	
	1128	163.2				90 (3.6)	
20 Ni 2 Mo 25	834	121.0	12	61	44.8	30 (1.2)	207
	686	99.6				60 (2.4)	
20 Ni 7 Cr 2 Mo 2	882	128.0	11	40	29.7	30 (1.2)	213
(20 Ni 55 Cr 50 Mo	784	113.8				60 (2.4)	
20)	735	106.7				90 (3.6)	
15 Ni 13 Cr 4	981	142.3	9	40	29.7	30 (1.2)	217
(15 Ni Cr 1 Mo 12)	932	135.1				90 (3.6)	
15 Ni 5 Cr 4 Mo 2	1079	156.5	9	34	25.3	30 (1.2)	217
(15 Ni 2 Cr 1 Mo.15)	932	142.3				60 (2.4)	
	932	135.1				90 (3.6)	
16 Ni 8 Cr 6 Mo 2	1324	193.0	9	34	25.3	30 (1.2)	229
(16 Ni Cr 2 Mo 20)	1177	170.7				60 (2.4)	
	1128	163.6				90 (3.6)	

[a] a^* area of cross section.
Source: IS 4432. 1967.

TABLE 1-15
Typical mechanical properties of some carburizing steels[a]

AISI No.	Ultimate tensile strength, σ_{sut} MPa	Ultimate tensile strength, σ_{sut} kpsi	Tensile yield strength, σ_{sy} MPa	Tensile yield strength, σ_{sy} kpsi	Elongation in 50 mm (2 in), %	Reduction of area, %	Hardness: Core Brinell, H_B	Hardness: Case Rockwell, R_C	Hardness: Case Thickness mm	Hardness: Case Thickness in	Izod impact energy J	Izod impact energy ft-lbf	Machin-ability
							Plain carbon						
C1015	503	73	317	46	31	71	149	62	1.22	0.048	123	91	Poor
C1020	517	75	331	48	31	71	156	62	1.17	0.046	126	93	Poor
C1022	572	83	324	47	27	66	163	62	1.17	0.046	110	81	Good
C1117	669	97	407	59	23	53	192	65	1.14	0.045	45	33	Very good
C1118	779	113	531	77	17	45	229	61	1.65	0.065	22	16	Excellent
							Alloy steels						
4320[b]	100	146	648	94	22	56	293	59	1.91	0.075	65	48	
4620[b]	793	115	531	77	22	62	235	59	1.52	0.060	106	78	
8620[b]	897	130	531	77	22	52	262	61	1.78	0.070	89	66	

[a] Average properties for 15 mm (1 in) round section treated, 12.625 mm (0.505 in) round section tested. Water-quenched and tempered at 177°C (350°F), except where indicated.

[b] Core properties for 14.125 mm (0.565 in) round section treated, 12.625 mm (0.505 in) round section tested. Oil-quenched twice, tempered at 232°C (450°F).

Source: *Modern Steels and Their Properties*, Bethlehem Steel Corp., 4th ed., 1958 and 7th ed., 1972.

TABLE 1-16
Minimum mechanical properties of some stainless steels

UNS No.	AISI No.	Tensile strength, σ_{st} MPa	Tensile strength, σ_{st} kpsi	Yield strength[a], σ_{sy} MPa	Yield strength[a], σ_{sy} kpsi	Brinell hardness, H_B	Elongation, %	Reduction in area, %	Weldability	Machinability	Application
							Annealed (room temperatures)				
Austenitic											
S30200	302	515	75	205	30	88	40		Good	Poor	General purpose, springs
S30300	303[b]	585[b]	85[b]	240[b]	35[b]		50[b]	55[b]	Poor	Good	Bolts, rivets, and nuts
S30400	304	515	75	205	30	88	40		Good	Poor	Welded structures
S30500	305	480	70	170	25	88	40		Good		General purpose
S30800	308	515	75	205	30	88	40				
S30900	309	515	75	205	30	95	40				
S31000	310	515	75	205	30	95	40		Good	Poor	Heat-exchange parts
S31008	310 S	515	75	205	30	95	40		Good	Poor	Turbine and furnace
S34800	348	515	75	205	30	88	40				Jet engine parts
S38400	384	415–550	60–80								Fasteners and cold-worked parts
							Annealed high-nitrogen				
Austenitic											
S20200	202	655	95	310	4560		40				
S21600	216	690	100	415	50	100	40				
S30452	304 HN	620	90	345		100	30				
Ferrite											
S40500	405	415	60	170	25	88 max			Excellent		
S43000	430	450	65	205	30	88 max	22[c]		Fair	Fair to good	Screw machine parts, muffler
S44600	446	515	75	275	40	95 max	20		Fair	Fair	Machine parts subjected to high-temperature corrosion
Martensite											
S40300	403	485	70	205	30	88 max	25[c]				Bolts, shafts, and machine parts
S41000	410	450	65	205	30	95 max	22[c]				Bolts, springs, cutlery, and machine parts
S41400	414	795	115	620	90		15	45			
S41800[d]	418[d]	1450[b]	210[b]	1210[b]	175[b]		18[b]	52[b]			
S42000[e]	420[e]	1720	250	1480[b]	215[b]	$52R_C$[b]	8[b]	25[b]			
S43100[d]	431[d]	1370[b]	198[b]	1030[b]	149[b]		16[b]	55[b]			High-strength parts used in aircraft and bolts
S44002	440 A	725[b]	105[b]	415[b]	60[b]	95[b]	20[b]				Cutlery, bearing parts, nozzles and ball bearings
S44003	440 B	740	107[b]	425[b]	62[b]	96[b]	18[b]				
S44004	440 C	760[b]	110[b]	450[b]	65[b]	97[b]	14[b]				
S50200	502[b]	485[b]	70[b]	205[b]	30[b]		30[b]	70[b]			

[a] At 0.2% offset.
[b] Typical values.
[c] 20% elongation for thickness of 1.3 mm (0.050 in) or less.
[d] Tempered at 260°C (500°F).
[e] Tempered at 205°C (400°F).
Source: *ASM Metals Handbook*, American Society for Metals, Metals Park, Ohio, 1988.

TABLE 1-17
Chemical composition and mechanical properties of some stainless, heat resisting and high alloy steels

Designation of steel	Chemical composition, %										Tensile strength, min, σ_{st}		0.2% proof stress, min, σ_{sy}		Hardness number		Elongation in 50 mm (2 in), min, %	Reduction of area, min, %
	C	Si	Mn	Ni	Cr	Mo	Ti	Nb	S max	P max	MPa	kpsi	MPa	kpsi	Brinell H_B	Rockwell R_B		
									Chromium steels									
X 04 Cr 12*	0.08 max	1.0 max	1.0 max		11.5/13.5				0.030	0.040	415 (445)#	60.2 (64.5)	205 (276)#	29.7 (40.0)	183	88	22 (20)#	(45)#
X 12 Cr 12*	0.80/0.15	1.0 max	1.0 max	1.0 max	11.5/13.5				0.030	0.040	450 (483)	65.3 (70.0)	205 (276)	29.7 (40.0)	217	95	20 (20)	(45)
X 07 Cr 17	0.12 max	1.0 max	1.0 max	1.25/2.50	15.0/17.0				0.030	0.040	450 (483)	65.3 (70.0)	205 (276)	29.7 (40.0)	183	88	22 (20)	(45)
X 40 Cr 13	0.35/0.45	1.0 max	1.0 max	1.0 max	12.0/14.0				0.030	0.040	(600 700)	(87.0 101.5)			(225)	–		
X 15 Cr 25 N	0.20 max	1.0 max	1.5 max		23.0/27.0				0.030 and $N = 0.25$ max	0.045	515 (490)	74.7 (71.1)	275 (280)	39.9 (40.6)	217 (212)	–	20 (16)	(45)
									Chromium–nickel steels									
X 02 Cr 19 Ni 10	0.03 max	1.0 max	2.0 max	8.0/12.0	17.5/20.0				0.030	0.045	485 (483)	70.3 (70.0)	170 (172)	24.7 (25.0)	183	88	40 (40)	(50)
X 04 Cr 19 Ni 9	0.08 max	1.0 max	2.0 max	8.0/10.5	17.5/20.0				0.030	0.045	515 (517)	74.7 (75.0)	205 (207)	29.7 (30.0)	183	88	40 (40)	(50)
X 07 Cr 18 Ni 9	0.15 max	1.0 max	2.0 max	8.0/10.0	17.0/19.0				0.030	0.045	515	74.7	205	29.7	183	88	40	
X 04 Cr 18 Ni 10 Nb	0.08 max	1.0 max	2.0 max	9.0/12.0	17.0/19.0			10XC-1.0	0.030	0.045	515 (517)	74.7 (75.0)	205 (207)	29.7 (30.0)	183	88	40 (40)	(50)
X 04 Cr 18 Ni 10 Ti	0.08 max	1.0 max	2.0 max	9.0/12.0	17.0/19.0		5XC-0.80**		0.030	0.045	515 (517)	74.7 (75.0)	205 (207)	29.7 (30.0)	183	88	40 (40)	(50)
X 04 Cr 17 Ni 12 Mo 2	0.08 max	1.0 max	2.0 max	10.0/14.0	16.0/18.0	2.0/3.0			0.030	0.045	515 (517)	74.7 (75.0)	205 (207)	29.7 (30.0)	217	95	40 (40)	(50)
X 02 Cr 17 Ni 12 Mo 2	0.08 max	1.0 max	2.0 max	10.0/14.0	16.0/18.0	2.0/3.0			0.030	0.045	485 (483)	70.3 (70.0)	170 (172)	24.7 (25.0)	217	95	40 (40)	(50)
X 04 Cr 17 Ni Mo 2 Ti 2	0.08 max	1.0 max	2.0 max	10.0/14.0	16.0/18.0	2.0/3.0	5XC-0.80		0.030	0.045	515 (517)	74.7 (75.0)	205 (207)	29.7 (30.0)	217	95	40 (40)	(50)
X 04 Cr 19 Ni 13 Mo 3	0.03 max	1.0 max	2.0 max	11.0/15.0	18.0/20.0	3.0/4.0			0.030	0.045	515 (517)	74.7 {75.0)	205 (207)	29.7 (30.0)	217	95	35 (40)	(50)
X 20 Cr 25 Ni 20	0.25 max	2.5 max	2.0 max	18.0/21.0	24.0/26.0				0.030	0.045	515 (490)	74.7 (71.1)	210 (210)	30.5 (30.5)	217	95	40 (40)	(50)
X 07 Cr 17 Mn 12 Ni 4	0.12 max	1.0 max	10.0/14.0	3.5/5.5	16.0/18.0				0.030	0.045	550	79.8	250	36.3	217	88	45	(50)
X 40 Ni 14 Cr 14 W 3 Si 2	0.35/0.50	2.5 max	1.0 max	12.0/15.0	12.0/15.0			and	0.035 W 2.0/3.0	0.045	(785)	(113.9)	(345)	(50.0)	(269)		(35)	(40)

Notes: Annealed quenched or solution-treated condition; * for free-cutting varieties sulfur and selenium content shall be as agreed between the purchaser and the manufacturer; ** for electrode steel Nb −10C to 1.0 in place of Ti; # the mechanical properties in parentheses are for bars and flats and the properties without parentheses for plates, sheets, and strips.
Source: Compiled from IS 1570 (part 5), 1985.

TABLE 1-18
Mechanical properties of high-strength low-alloy steels

ASTM specification	Type, grade, or condition	UNS designation	Minimum tensile[a] strength,[a] σ_{st} MPa	Minimum tensile[a] strength,[a] σ_{st} kpsi	Minimum yield[a] strength,[a] σ_{sy} MPa	Minimum yield[a] strength,[a] σ_{sy} kpsi	Minimum elongation,[a] % In 200 mm (8 in)	Minimum elongation,[a] % In 50 mm (2 In)	Intended uses
A242	Type 1	K11510	435–480	63–70	290–345	42–50	18	21	Structural members in welded, bolted, or riveted construction
A440		K12810	435–485	63–70	290–345	42–50	18	21	Structural members, primarily in bolted or riveted construction
A441		K12211	415–485	60–70	275–345	42–50	18	21	Welded, bolted, or riveted structures but primarily welded bridges
A572	Grade 42		415	60	290	42	20	24	Welded, bolted, or riveted structures, but used mainly in bolted or riveted bridges and buildings
	Grade 50		450	65	345	50	18	21	
	Grade 60		520	75	415	60	16	18	
	Grade 65		550	80	450	65	15	17	
A606	Hot-rolled		480	70	345	50		22	Structural and miscellaneous purposes where weight saving or added durability is important
	Hot-rolled and annealed or normalized		450	65	310	45		22	
	Cold-rolled		450	65	310	45		22	
A607	Grade 45		410	60	310	45		22–25	Structural and miscellaneous purposes where greater strength or weight saving is important
	Grade 50		450	65	345	50		20–22	
	Grade 60		520	75	415	60		16–18	
	Grade 70		590	85	485	70		14	
A618	Grade I	K02601	483	70	345	50	19	22	General structural purposes including welded, bolted, or riveted bridges and buildings
	Grade II	K12609	483	70	345	50	18	22	
	Grade III	K12700	448	65	345	50	18	20	
A656	Grade 1 and 2		655–793	95–115	552	80	12		Truck frames, brackets, crane booms, railcars. and other applications where weight saving is important
A690		K12249	485	70	345	50	18		Dock walls, sea walls, bulkheads, excavations, and similar structures exposed to sea water
A715	Grade 50		415	60	345	50		22–24	Structural and miscellaneous applications where high strength, weight savings, improved formability, and good weldability are important
	Grade 60		485	70	415	60		20–22	
	Grade 70		550	80	485	70		18–20	
	Grade 80		620	90	550	80		16–18	

[a] May vary with product size and mill form.
Source: *ASM Metal Handbook*, American Society for Metals, Metals Park, Ohio, 1988.

TABLE 1-19
Mechanical properties of some cast alloy, cast stainless, high-strength and iron-based super alloy steels

Materials classification		Tensile strength, σ_{st}		Yield strength, σ_{sy}		Fatigue[c] endurance limit, σ_{sj}		Elongation in 50 mm (2 in)	Modulus of elasticity E		Impact Charpy		Brinell hardness, temperature, °C (°F), H_B	Rupture strength, 100 h at 538°C (1000°F)	
		MPa	kpsi	MPa	kpsi	MPa	kpsi	%	GPa	Mpsi	J	ft-lbf		GPa	Mpsi
Cast Alloy Steels															
ASTM	Grade														
A352-68a	LC1[a]	448	65	241	35	138	20	24			81	60			
A219-6	WC4[a]	483	70	276	40	159	23	20			75	55			
A148-65	80-50[a]	552	80	345	50	172	25	22			65	48			
A148-1	90-60[a]	620	90	414	60	214	31	20			54	40			
A148-65	105-85[b]	724	105	586	85	244	34	17			79	58	217		
A148-65	150-125[b]	1034	150	862	125	303	44	9			41	30	311		
A148-65	120-95[b]	827	120	655	95	255	37	14			61	45	262		
A148-65	175-145[b]	1207	175	1000	145	331	48	6			32	24	352		
Cast Stainless Steels															
ACI[d]	CB-30[d]	655	95	414	60			15			3	2			
	C-50[d]	483–669	70–97	448	65			18	200	29	61	45			
	CE-30[d]	600–669	87–97	448	65			18	172	25	14	10			
	CF-8[d]	517–586	75–85	241–276	35–40			55	193	28	95	70			
	CH-20[d]	552–607	80–88	345	50			38	193	28	20	15			
Ultra-High-Strength Steels															
Medium carbon low alloys 4 140 M, 433OV, D 6AC, 4340		To 2068	To 300	To 1724	To 250			10			23	17			
Mod. 5 Cr-Mo-V tool steels: H-11 (Mod). H-13 (Mod)		To 2144	To 311	To 1703	To 247			6.6–12			20–30	15–22			
Maraging steels (high nickel): 18 Ni (350) Almar 302		1758	255	1689	245			8			31	23			
High-Strength Low-Alloy (HSLA) Steels															
ASTM	SAE												*Composition*		
A607	J410C[g]	414	60	310	45			25					Cb and/or V		
A606 Types 2, 4[h]		448–483	65–70	310–345	45–50			22					(Proprietary) Cu, Cr, Mn, Ni, P, and other additions		
A607	J410CR[g]	448	65	345	50			22					Cb and or V		
715 (sheet)[i] A656 (plate)		414	60	345	50			24					(Proprietary) Cb, Ti, Zr, Si, N, V, and others		
A607	J410C[g]	483	70	379	55			20							
A607	J410C[g]	586	85	483	70			14					Cb and/or V		

TABLE 1-19
Mechanical properties of some cast alloy, cast stainless, high-strength and iron-based super alloy steels (*Cont.*)

Materials classification		Tensile strength, σ_{st}		Yield strength, σ_{sy}		Fatigue[c] endurance limit, σ_{sj}		Elongation in 50 mm (2 in)	Modulus of elasticity E		Impact Charpy		Brinell hardness, temperature, °C (°F), H_B		Rupture strength, 100 h at 538°C (1000°F)	
		MPa	kpsi	MPa	kpsi	MPa	kpsi	%	GPa	Mpsi	J	ft-lbf	Temperature °C	°F	GPa	Mpsi
Iron-Based Superalloys																
Martensitic																
AISI																
601	17-22A	827	120	689	100			30	21	3.08			21	70	338	49
		531	77	372	54			20					538	1000		
604	Chromalloy	682–896	125–138	655–745	95–108			7	22	3.17			21	70	517	75
		758	110	586	85								538	1000		
610	H-11	931–2137	135–310	689–1655	100–240	896	130	3–17	21	3.05	14–43	10–32	21	70	655–793	95–115
		1241	180	965	140			10					538	1000		
616	422	1034–1655	150–240	682–1207	125–175	621–758	90–100	16–19	20	2.90	14–52	10–38	21	70	400	58
		1172	170	869	126			16					538	1000		
Austenitic																
633	AM 350	103–1413	160–205	414–1207	60–175	482–689	70–100	12–38	20.3	2.94	19	14	21	70	710	103
		1130	160	745	108			9					538	1000		
635	Stainless W	1517–1551	220–225	1482–2000	215–290	372–662	54–96	1.5	20.9	3.02	5–144	4–106	21	70	220	32
		517–552	75–80	255–345	37–50			58					538	1000		
650	16-25-G	758–965	110–140	345–689	50–100			20–45	19.5	2.85	20	15	21	70	538	78
		621	90	228	33			58					538	1000		
653	17-24 Cu Mo	593–772	86–112	276–620	40–90			30–45	19.3	2.80	11–35	8–26	21	70	330	48
		448	65	200	29			37					538	1000		
660	A-286	1007–903	146	655	95			25	20	2.88	56–81	41-60	21	70	689	100
			131	607	88			19					538	1000		

[a] Normalized and tempered.
[b] Quenched and tempered.
[c] Polished specimen.
[d] Corrosion resistance.
[e] Heat resistance.
[f] Heat and corrosion resistance.
[g] Semikilled or killed.
[h] Semikilled or killed-improved corrosion resistance.
[i] Inclusion control-improved formability, killed.

Source: *Machine Design*, 1981 Materials Reference Issue, Penton/IPC, Cleveland, Ohio, Vol. 53, No. 6 (March 19, 1981).

TABLE 1-20
Mechanical properties of high tensile cast steel

Grade	Designation	Tensile strength, min, σ_{st}		Yield strength (or 0.5% proof stress), min, σ_{sy}		Reduction in area, min, %	Elongation, min, % (gauge length $5.65\sqrt{a^*}$)[a]	Brinell hardness, min, H_B	Izod impact strength, min	
		MPa	kpsi	MPa	kpsi				J	ft-lbf
1	CS 640	640	92.8	390	56.7	35	15	190	30	22.1
2	CS 700	700	101.5	560	81.2	30	14	207	30	21.1
3	CS 840	840	121.8	700	101.5	28	12	248	29	20.6
4	CS 1030	1030	149.4	850	123.3	20	8	305	20	14.5
5	CS 1230	1230	178.3	1000	145.1	12	5	355		

[a] a^*, area of cross section.
Source: IS 2644, 1979.

TABLE 1-21
Chemical composition of tool steels

Steel designation	% C	% Si	% Mn	% Cr	%Mo	%V	%W	% Ni	% Co
T 140 W 4 Cr 50	1.30–1.50	0.10–0.35	0.25–0.50	0.30–0.70			3.50–4.20		
T 133	1.25–1.40	0.10–0.30	0.20–0.35						
T 118	1.10–1.25	0.10–0.30	0.20–0.35						
T 70	0.65–0.75	0.10–0.30	0.20–0.35						
T 85	0.80–0.90	0.10–0.35	0.50–0.80						
T 75	0.70–0.80	0.10–0.35	0.50–0.80						
T 65	0.60–0.70	0.10–0.35	0.50–0.80						
T 215 Cr 12	2.00–2.30	0.10–0.35	0.25–0.50	11.0–13.0	0.80 max[a]	0.80 max[a]			
T 160 Cr 12	150–1.70	0.10–0.35	0.25–0.50	11.0–13.0	0.80 max[a]	0.80 max[a]			
T 110 W 2 Cr 1	1.00–1.20	0.10–0.35	0.90–1.30	0.90–1.30			1.25–1.75		
T 105 W 2 Cr 60 V 25	0.90–1.20	0.10–0.35	0.25–0.50	0.40–0.80	0.25 max[a]	0.20–0.30	1.25–1.75		
T 90 Mn 2 W 50 Cr 45	0.85–0.95	0.10–0.35	1.25–1.75	0.30–0.60		0.25 max	0.40–0.60		
T 105 Cr 1	0.90–1.20	0.10–0.35	0.20–0.40	1.00–1.60					
T 105 Cr 1 Mn 60	0.90–1.20	0.10–0.35	0.40–0.80	1.00–1.60					
T 55 Cr 70	0.50–0.60	0.10–0.35	0.60–0.80	0.60–0.80					
T 55 Si 2 Mn 90 Mo 33	0.50–0.60	1.50–2.00	0.80–1.00		0.25–0.40	0.12–0.20[a]			
T 50 Cr 2 V 23	0.45–0.55	0.10–0.35	0.50–0.80	0.90–1.20		0.15–0.30			
T 60 Ni 1	0.55–0.65	0.10–0.65	0.50–0.80	0.30 max				1.00–1.50	
T 30 Ni 4 Cr 1	0.26–0.34	0.10–0.35	0.40–0.70	1.10–1.40				3.90–4.30	
T 55 Ni 2 Cr 65 Mo 30	0.50–0.60	0.10–0.35	0.50–0.80	0.50–0.80	0.25–0.35			1.25–1.75	
T 33 W 9 Cr 3 V 38	0.25–0.40	0.10–0.35	0.20–0.40	2.80–3.30		0.25–0.50	8.0–10.0		
T 35 Cr 5 Mo V 1	0.30–0.40	0.80–1.20	0.25–0.50	4.75–5.25	1.20–1.60	1.00–1.20			
T 35 Cr 5 Mo W 1 V 30	0.30–0.40	0.80–1.20	0.25–0.50	4.75–5.25	1.20–1.60	0.20–0.40	1.20–1.60		
T 75 W 18 Co 6 Cr 4 V 1 Mo 75	0.70–0.80	0.10–0.35	0.20–0.40	4.00–4.50	0.50–1.00	1.50–1.50	17.50–19.00		5.00–6.00
T 83 Mo W 6 Cr 4 V 2	0.75–0.90	0.10–0.35	0.20–0.40	3.75–4.50	5.50–6.50	1.75–2.00	5.50–6.50		
T 55 W 14 Cr 3 V 45	0.50–0.60	0.20–0.35	0.20–0.40	2.80–3.30		0.30–0.60	13.00–15.00		
T 16 Ni 85 Cr 60	0.12–0.20	0.10–0.35	0.60–1.00	0.40–0.80				0.60–1.00	
T 10 Cr 5 Mo 75 V 23	0.15 max	0.10–0.35	0.25–0.50	4.75–5.25	0.50–1.00	0.15–0.30			

[a] Optional
Source: IS 1871, 1965.

TABLE 1-22
Mechanical properties of some tool steels

AISI steel designation	Condition[a]	Tensile strength, σ_{st} MPa	Tensile strength, σ_{st} kpsi	Yield strength, σ_{sy} MPa	Yield strength, σ_{sy} kpsi	Elongation, %	Hardness	Hardening temperature °C	Hardening temperature °F	Quenched media	Impact strength Charpy V-notch J	Impact strength Charpy V-notch ft-lbf	Machinability
H-11	Annealed 870°C (1600°F)[b]	690	100	365	53	25	96 R_B	1010	1850	air	14	10	Medium to high
	Tempered 540°C (1000°F)	2034	295	1724	250	9	55 R_C						
L-2	Annealed 775°C (1425°F)	710	103	510	74	25	96 R_B	855	1575	oil	28	21	High
	Tempered 205°C (400°F)	2000	290	1793	260	5	54 R_C						
L-6	Annealed 775°C (1425°F)[a]	665	95	380	55	25	93 R_B	845	1550	oil	12	9	Medium
	Tempered 315°C (600°F)	2000	290	1793	260	4	54 R_C						
P-20	Annealed 775°C (1425°F)	690	100	517	75	17	97 R_B	855	1575	oil	20	15	Medium to high
	Tempered 205°C (400°F)	1860	270	1413	205	10	52 R_C						
S-1	Annealed 800°C (1475°F)	690	100	414	60	24	96 R_B	925	1700	oil	250	184[c]	Medium
	Tempered 205°C (400°F)	2068	300	1896	275	4	57.5 R_C						
S-5	Annealed 790°C (1450°F)	724	105	440	64	25	96 R_B	870	1600	oil	206	152[c]	Medium to high
	Tempered 205°C (400°F)	2344	340	1930	280	5	59 R_C						
S-7	Annealed 830°C (1525°F)	640	93	380	55	25	95 R_R	940	1725	air	244	180	Medium
	Tempered 205°C (400°F)	2170	315	1448	210	7	58 R_C						
A-8	Annealed 845°C (1550°F)[b]	710	103	448	65	25	97 R_B	1010	1850	air	7	5	Medium
	Tempered 565°C (1050°F)	1827	265	1550	225	9	52 R_C						

[a] Single temper, oil-quenched unless otherwise indicated.
[b] Double temper, air-quenched.
[c] Charpy impact unnotched tests made on longitudinal specimens of small cross-sectional bar stock. The heat treatments listed were to develop nominal mechanical properties for hardened and tempered materials for test purposes only and may not be suitable for some applications.
Source: *Machine Design*, 1981 Materials Reference Issue, Penton/IPC, Cleveland, Ohio, Vol. 53, No. 6 (March 19, 1981).

TABLE 1-23
Properties of representative cobalt-bonded cemented carbides

Nominal composition	Grain size	Brinell Hardness H_B	Density		Transverse strength, σ_{sb}		Compressive strength, σ_{sc}		Proportional limit compressive strength, σ_{sp}		Modulus of elasticity, E		Tensile strength, σ_{st}		Impact strength		Thermal conduc-tivity	Coefficient of linear expansion, α µm/m°C at		µin/in°F at	
			Mg/m³	lb/in³	MPa	kpsi	MPa	kpsi	MPa	kpsi	GPa	Mpsi	MPa	kpsi	J	in-lbf	W/m K	200°C	1000°C	400°F	1800°F
94WC-6Co	Fine	92.5–93.1	15.0	0.54	1790	260	5930	860	2550	370	614	89			1.02	9	–	4.3	5.9	2.4	3.3
	Medium	91.7–92.2	15.0	0.54	2000	290	5450	790	1930	280	648	94	1450	210	1.36	12	100	4.3	5.4	2.4	3.0
	Coarse	90.5–91.5	15.0	0.54	2210	320	5170	750	1450	210	641	93	1520	220	1.36	12	121	4.3	5.6	2.4	3.0
90WC-10Co	Fine	90.7–91.3	14.6	0.53	3100	450	5170	750	1590	230	620	90			1.69	15	–	–	–	–	–
	Coarse	87.4–88.2	14.5	0.52	2760	400	4000	580	1170	170	552	80	1340	195	2.03	18	112	5.2	–	2.9	–
84WC-16Co	Fine	89	13.9	0.50	3380	490	4070	590	970	140	524	76			3.05	27	–	–	–	–	–
	Coarse	86.0–87.5	13.9	0.50	2900	420	3860	560	700	100	524	76	1860	270	2.83	25	88	5.8	7.0	3.2	3.8
72WC-8TiC-11.5TaC-8.5Co	Medium	90.7–91.5	12.6	0.45	1720	250	5170	750	1720	250	558	81			0.90	8	50	5.8	7.0	3.2	3.8
64TiC-28WC-2TaC-2Cr$_3$C$_2$-4.0Co	Medium	94.5–95.2	6.6	0.24	690	100	4340	630										–	–		–

Source: *Metals Handbook Desk Edition*, ASM International 1985, Materials Park, OH 44073-0002 (formerly the American Society for Metals, Metals Park, OH 44073, 1985).

TABLE 1-24
Typical uses of tool steel

Steel designation	Type	Typical uses
	Cold-Work Water-Hardening Steels	
T 140 W 4 Cr 50	Fast finishing tool steel	Finishing tools with light feeds, marking tools, etc.
T 133 T 118 T 70	Carbon tool steels	Engraving tools, files, razors, shaping and wood-working tools, heading and press tools, drills, punches, chisels,shear blades, vice jaws, etc.
	Cold-Work Oil and Air-Hardening Steels	
T 215 Cr 12 T 160 Cr 12	High-carbon high-chromium tool steels	Press tools, drawing and cutter dies, shear blade thread rollers. etc.
T 110 W 2 Cr 1 T 105 W 2 Cr 60 V 25 T 90 Mn 2 W 50 Cr 45	Nondeforming tool steels	Engraving tools, press tools, gauge, tape, dies, drills, hard reamers, milling cutters, broaches, cold punches, knives. etc.
T 105 Cr 1 T 105 Cr 1 M 60	Carbon-chromium tool steels	Lathe centers, knurling tools, press tools
T 85 T 75 T 65	Carbon tool steels	Die blocks, garden and agricultural tools, etc.
T 55 Cr 70 T 55 Si 2 Mn 90 Mo 33 T 50 Cr 1 V 23	Shock-resisting tool steels	Pneumatic chisels, rivet shape, shear blades, heavy-duty punches, scarfing tools, and other tools under high shock
T 60 Ni 1 T 30 Ni 4 Cr 1 T 55 Ni 2 Cr 65 Mo 3	Nickel-chrome-molybdenum tool steels	Cold and heavy duty punches, trimming dies, scarfing tools, pneumatic chisels, etc.
	Hot-Work and High-Speed Steel	
T 33, W 9 Cr 3 V 38 T 35 Cr 5 Mo V 1 T 35 Cr 5 Mo W 1 V 30	Hot-work tool steels	Castings dies for light alloys, dies for extrusion, stamping, and forging
T 75 W 18 Co 6 Cr 4 V 1 Mo 75 T 83 Mo W 6 Cr 4 V 2 T 55 W 14 Cr 3 V 45[a]	High-speed tool steels	Drills, reamers, broaches, form cutters, milling cutters, deep-hole drills, slitting saws, high-speed and heavy-cut tools
	Low-Carbon Mold Steel	
T 16 Ni 80 Cr 60 T 10 Cr 5 bee 75 V 23	Carburizing steels	After case hardening for molds for plastic materials

[a] May also be used as hot-work steel.
Source: IS 1871, 1965.

TABLE 1-25
Mechanical properties of carbon and alloy steel bars for the production of machine parts

Steel designation	Ultimate tensile strength, σ_{sut}				Minimum elongation (gauge length $= 5.65\sqrt{a^*}$), %
	MPa##	kpsi	MPa‡	kpsi	
14 C 4 (C 14)**	363	52.6	441	64.0	26
20 C 8 (C 20)	432	62.6	510	74.0	24
30 C 8 (C 30)	490	71.1	588	85.3	21
40 C 8 (C 40)	569	82.5	667	96.7	18
45 C 8 (C 45)	618	89.6	696	101.0	15
55 C 8 (C 55 Mn 75)	706	102.4			13
65 C 6 (C 65)	736	106.7			10
14 C 14 S 14 (14 Mn 1 S 14)	432	62.6	530	76.8	22
11 C 10 S 25 (13 S 25)	363	52.6	481	69.7	23

Notes: a^*, area of cross section; ## minimum; ‡ maximum; ** steel designations in parentheses are old designations
Source: IS 2073, 1970.

TABLE 1-26
Recommended hardening and tempering treatment for carbon and alloy steels

Designation	Hot-working temperature		Normalizing		Hardening		Quenching	Tempering	
	K	°C	K	°C	K	°C	K °C	K	°C
30 C 8 (C 30)	1473–1123	1200–850	1133–1163	860–890	1133–1163	860–890	Water or oil	823–923	550–660
35 C 8 (C 25 Mn 74)	1473–1123	1200–850	1123–1153	850–880	1113–1153	840–880	Water or oil	803–1033	530–760
40 C 8 (C 40)	1473–1123	1200–850	1103–1133	830–860	1103–1133	830–860	Water or oil	823–933	550–660
50 C 8 (C 50)	1473–1123	1200–850	1083–1113	810–840	1083–1113	810–840	Oil	823–933	550–660
55 C 8 (C 55 Ma 75)	1473–1123	1200–850	1083–1113	810–840	1083–1113	810–840	Oil	823–933	550–660
40 C 10 Si 8 (40 S 18)	1473–1123	1200–850	1103–1113	830–860	1103–1133	830–860	Oil	823–933	550–660
40 C 15 Si 2 (40 Mn 2 S 12)	1473–1123	1200–850	1113–1143	840–870	1113–1143	840–870	Oil	823–933	550–660
220 C 15 (20 Mn 2)	1473–1123	1200–850	1133–1173	860–900	1133–1173	860–900	Water or oil	823–933	550–660
27 C 15 (27 Mn 2)	1473–1123	1200–850	1133–1153	840–880	1133–1153	840–880	Water or oil	823–933	550–660
37 C 15 (37 Mn 2)	1473–1123	1200–850	1123–1143	850–870	1123–1143	850–870	Water or oil	823–933	550–660
40 Cr 4 (40 Cr 1)	1473–1123	1200–850	1123–1153	850–880	1123–1153	850–880	Oil	823–933	550–660
35 Mn 6 Mo 3 (35 Mn 2 Mo 28)	1473–1123	1200–850			1113–1133	840–860	Water or oil	823–933	550–660
35 Mn 6 Mo 4 (35 Mn 2 Mo 45)	1473–1123	1200–850			1113–1133	840–860	Oil	823–933	550–660
40 Cr 4 Mo 3 (40 Cr 1 Mo 28)	1473–1123	1200–850	1123–1153	850–880	1123–1153	850–880	Oil	823–933	550–660
40 Ni 14 (40 Ni 3)	1473–1123	1200–850	1103–1133	830–860	1103–1133	830–860	Oil	823–933	550–660
35 Ni Cr 2 Mo (35 Ni Cr Mo 60)	1473–1123	1200–850			1093–1123	820–850	Water or oil	823–933	550–660
40 Ni 6 Cr 4 Mo 2 (40 Ni Cr Mo 15)	1473–1123	1200–850			1103–1123	820–850	Oil	823–933	550–660
40 Ni 6 Cr 4 Mo 3 (40 Ni 2 Cr 1 Mo 28)	1473–1123	1200–850			1103–1123	830–850	Oil	823–933 or 423–473 (depending on hardness required)	550–660 or 150–200
15 Ni Cr 1 Mo 12 (31 Ni 3 Cr 65 Mo 55)	1473–1123	1200–850			1103–1123	830–850	Oil	≤933	≤660
30 Ni 13 Cr 5 (30 Ni 4 Cr 1)	1473–1123	1200–850			1083–1103	810–820	Air or oil	≥523	≥250
15 Cr 13 Mo 6 (15 Cr 3 Mo 55)	1473–1123	1200–850			1163–1183	890–910	Oil	823–973[a]	550–700[a]
25 Cr 13 Mo 6 (25 Cr 3 Mo 55)	1473–1123	1200–850			1163–1183	890–910	Oil	823–973[a]	550–700[a]
40 Cr 13 Mo 10 V 2 (40 Cr 3 Mo 1 V 20)	1473–1123	1200–850			1173–1213	900–940	Oil	843–923	570–650
40 Cr 7 Al 10 Mo 2 (40 Cr 2 Al 1 Mo 18)	1473–1123	1200–850			1123–1173	850–900	Oil	823–973	550–700
55 Cr 70)	1473–1123	1200–850	1073–1123	800–850	1073–1123	800–850	Oil	773–973	500–700
105 Cr 4 (105 Cr 1)	1373–1123	1100–850			1093–1133	820–860	Water or oil	>423	>150 in oil
105 Cr 1 Mn 60	1373–1123	1100–850			1073–1113	800–840	Water or oil	403–453	130–180

[a] Stabilization 823 K (550°C).
Source: IS 1871, 1965.

TABLE 1-27
Mechanical properties of some as-cast austenitic manganese steels

Composition, %					Section		Tensile strength, σ_{st}		Yield strength, σ_{sy} (0.2% offset)		Brinell hardness, H_B	Elongation in 50 mm, %	Reduction in area, %	Impact strength Charpy [b]	
C	Mn	Si	Other	Form	mm	in	MPa	kpsi	MPa	kpsi				J	ft-lbf
								Plain manganese steels							
0.85	11.2	0.57		Round	25	1	440	64	–	–	–	14.5	–	–	–
1.11	12.7	0.54		Round	25	1	450	65	360	52	–	4	–	–	–
1.28	12.5	0.94		Keel block	100	4	330[a]	48[a]	–	–	245	1[a]	–	3.4	2.5
								1 Mo manganese steels							
0.83	11.6	0.38	0.96 Mo	Round	25	1	695	101	345	50	163	30	29	–	–
1.16	13.6	0.60	1.10 Mo	Round	25	1	560	81	400	58	185	13	15	–	–
0.93	13.6	0.67	0.96 Mo	Plate	25	1	510	74	365	53	188	11	16	72	53
0.98	12.6	0.6	0.87 Mo	Plate	50	2	435[a]	63[a]	–	–	–	4[a]	–	–	–
								2 Mo manganese steels							
0.52	14.3	1.47	2.4 Mo	Round	25	1	600	87	370	54	220	15.5	13	–	–
0.75	14.1	0.99	2.0 Mo	Round	25	1	745	108	365	53	183	34.5	27	–	–
1.24	14.1	0.64	3.0 Mo	Round	25	1	600	87	440	64	235	7.5	10	–	–
								3.5 Ni manganese steel							
0.75	13.0	0.95	3.65 Ni	Round	25	1	655	95	295	43	150	36	26	–	–
								6 Mn-1 Mo alloys							
0.90	5.8	0.37	1.46 Mo	Mill liner	100	4	340	49	325	47	181	2	–	9	7
0.89	6.3	0.6	1.20 Mo	Plate	100	4	330[a]	48[a]	–			1[a]	–	–	–

[a] Properties converted from transverse bend tests on 6 × 13 mm ($\frac{1}{4} \times \frac{1}{2}$ in) bars cut from castings and broken by center loading across 25 mm (1 in) span.
[b] Charpy V-notch.
Source: *Metals Handbook Desk Edition*, ASM International, 1985, Materials Park, OH 44073-0002 (formerly the American Society for Metals, Metals Park, OH 44073, 1985).

TABLE 1-28
Mechanical properties, fabrication characteristics,[a] and typical uses of some aluminum alloys[b]

Alloy no.		Ultimate tensile strength, σ_{sut}		Tensile yield strength[d], σ_{syt}		Compressive yield strength,[d] σ_{syc}		Shear strength, τ_s		Endurance limit in reversed bending, σ_{sfb}		Brinell hardness 4.9 kN (500 kgf) load on 10-mm ball, H_B	Modulus of elasticity,[e] E		Elongation in 50 mm (2 in), %	Corrosion resistance	Machi-ability	Welding			Uses
		MPa	kpsi	MPa	kpsi	MPa	kpsi	MPa	kpsi	MPa	kpsi		GPa	Mpsi				Gas	Arc	Resis-tance	
Sand casting alloys																					
201.0	-T 43	414	60	255	37										17	4	1		2		Aircraft structural components
	-T 6	448	65	379	55	386	56	290	42			130			8	4	1		2		
240.0	-F	235	34	200	29	207	30					90			1.0	4	3		4		
295.0	-T 4	221	32	110	16	117	17	179	26	48	7	60	69	10.0	8.5	3	2		2		Crankcases, spring hangers, housing, wheels
	-T 6	250	36	165	24	172	25	217	31	52	7.5	75	69	10.0	5.0	3	2		2		
319.0	-F	186	27	124	18	131	19	152	22	69	10	70	74	10.7	2.0	3	3		2		
	-T 6	250	26	164	24	172	25	200	29	76	11	80	74	10.7	2.0	3	3		2		
C 355.0	-T6	269	39	200	29							85			5.0	3	3		2		Air compressor fitting, crankcase, gear housing
356.0	-T 6	228	33	164	24	172	25	179	26	59	8.5	70	72	10.5	3.5	3	3		2		Cylinder heads, impellers, timing gears, water jackets, meter parts
A 390.0	-F	179	26	179	26							100	82	11.9	<1.0	2	4		2		Automotive engine blocks, pulleys, brake shoes, pumps
	-T 6	278	40	278	40					90	13	140			<1.0	2	4		2		
520.0	-T4	331	48	179	26	186	27	234	34	55	8	75	65	9.5	16	1	1		5		
A 535.0	-F	250	36	124	18							65			9.0	1	1		4		Aircraft fittings and components, levers, brackets
Permanent mold casting																					
355.0	-T 6	290	42	185	27	185	27	235	34	69	10	90			4.0	3	3		2		Timing gears, impellers, compressor and aircraft and missile components requiring high strength
C 355.0	-T61	303	44	234	34	248	36	221	32	97	14	90			3.0	3	3		2		
A 356.0	-T 61	283	41	207	30	221	32	193	28	90	13	90	72	10.5	10.0	2	3		2		Machine-tool parts, aircraft wheels, pump parts, marine hardware, valve bodies
513.0	-F	186	27	110	16	117	17	152	22	69	10	60			7.0	1	1		5		Ornamental hardware and architectural fittings
Wrought alloys																					
1100	-O	90	13	35	5			60	9	35	5	23			35	A	E	A	A	B	Sheet metal work, spun holloware, fin stock
	-H 14	125	18	115	17			75	11	50	7	32			9	A	D	A	A	A	
	-H 18	165	24	150	22			90	13	60	9	44			5	A	D	A	A	A	
2011	-T 3	380	55	295	43			220	32	125	18	95			15	D	A	D	D	D	Screw machine products
	-T 6	395	57	270	39			235	34	125	18	97			17						
2014	-O	185	27	95	14			125	18	90	13	45			18		D	D	D	B	Truck frames, aircraft structures
-T 4.	-T 451	425	62	290	42			260	38	140	20	105			20	C	B	D	B	B	
-T 6.	-T 651	482	70	415	60			290	42	125	18	135			13	C	B	D	B	B	
2017	-O	180	26	70	10			125	18	90	13	45			22						
-T 4.	-T 451	425	62	275	40			260	38	125	18	105			22						
2024	-O	185	27	75	11			125	18	90	13	47			20	–	D	D	D	D	Truck wheels, screw-machine products, aircraft structure
-T 4.	-T 351	470	68	325	47			285	41	140	20	120			20	C	B	C	B	B	
	-T 3	485	70	345	50			280	40	140	20	120			18	D	B	C	B	B	
	-T 86	515	75	490	71			310	45	125	18	135			6	D	B	D	C	B	

TABLE 1-28
Mechanical properties, fabrication characteristics,[a] and typical uses of some aluminum alloys[b] *(Cont.)*

Alloy no.		Ultimate tensile strength, σ_{sut}		Tensile yield strength[d], σ_{syt}		Compressive yield strength,[d] σ_{syc}		Shear strength, τ_s		Endurance limit in reversed bending, σ_{sfb}		Brinell hardness 4.9 kN (500 kgf) load on 10-mm ball, H_B	Modulus of elasticity,[e] E		Elongation in 50 mm (2 in), %	Corrosion resistance	Machi-ability	Welding			Uses
		MPa	kpsi	MPa	kpsi	MPa	kpsi	MPa	kpsi	MPa	kpsi		GPa	Mpsi				Gas	Arc	Resis-tance	
3003	-O	110	16	40	6			75	11	50	7	28			30	A	E	A	A	B	Pressure vessels, storage tanks, heat-exchanger tubes, chemical equipments, cooking utensils
	-H 14	150	22	145	21			95	14	60	9	40			8	A	D	A	A	A	
	-H 18	200	29	185	27			110	16	70	10	55			4	A	D	A	A	A	
3004	-O	180	26	70	10			110	16	95	14	45			20	A	D	B	A	B	Trailer panel sheet, storage tanks, sheet metal works
	-H 34	240	35	200	29			125	18	105	15	63			9	A	C	B	A	A	
	-H 38	285	41	250	36			145	21	110	16	77			5	A	C	B	A	A	
5052	-O	195	28	90	13			125	18	110	16	47			25	A	D	A	A	B	Hydraulic tube, appliances, bus body sheet, sheet metal work, welded structures, boat sheet
	-H 34	260	38	215	31			145	21	125	18	68			10	A	C	A	A	A	
	-H 38	290	42	255	37			165	24	140	20	77			7	A	C	A	A	A	
6061	-O	125	18	55	8			80	12	60	9	30			25	B	D	A	A	B	Heavy-duty structures requiring good corrosion resistance, truck and marine, railroad car, furniture, pipeline applications
	-T 6	310	45	275	40			205	30	95	14	95			12	B	C	A	A	A	
6063	-O	90	13	50	7			70	10	55	8	25				A		A	A	A	Pipe, railing, furniture, architectural extrusions
	-T 6	240	35	215	31			150	22	70	10	73			12	A	C	A	A	A	
7075	-O	230	38	105	15			150	22	115	17	60			17	–	D	D	C	B	Fin stock, cladding alloy
	-T 6	570	83	505	73			330	48	160	23	150			11	C	B	D	C	B	Aircraft and other structures

[a] For ratings of characteristics, 1 is the best and 5 is the poorest of the alloys listed. Ratings A through D are relative ratings in decreasing order of merit.
[b] Average of tensile and hardness values determined by tests on standard 12.5-mm ($\frac{1}{2}$-in) diameter test specimens.
[c] Endurance limits on 500 million cycles of completely reversed stresses using rotating beam-type machine and specimen.
[d] At 0.2% offset.
[e] Average of tension and compression moduli.
Key: Temper designations: F, as cast; O, annealed; Hxx, strain hardened; T1, cooled from an elevated temperature shaping process and naturally aged; T2, cooled from an elevated temperature shaping process, cold-worked and naturally aged; T3, solution heat-treated and cold worked and naturally aged; T4, solution heat-treated and naturally aged; T5, cooled from an elevated temperature shaping process and artificially aged; T6, solution heat-treated and artificially aged; T7, solution heat-treated and stabilized; T8, solution heat-treated, cold-worked and artificially aged; TX 51, stress-relieved by stretching.
Source: *ASM Metals Handbook*, American Society for Metals, Metals Park, Ohio, 1988.

TABLE 1-29
Chemical composition and mechanical properties of cast aluminum alloy[5,6,7]

Designation			Chemical composition, percent													Mechanical properties				
																Tensile strength, σ_{st}			Brinell	
IS new	IS old	BS	Cu	Mg	Si	Fe	Mn	Ni	Zn	Ti	Ti + Nb	Pb	Sn	Al	Condition	MPa	kpsi	Elongation %	hardness, H_B	Test piece
2447	A-1	LM1	6.0–8.0	0.15	2.0–4.0	1.0	0.6	0.5	2.0–4.0	0.2**		0.3	0.20		As cast	124*	18.0			Sand-cast
																154*	22.3			Chill-cast
4520	A-2	LM2	0.7–2.5	0.3	9.0–11.15	1.0	0.5	1.0	1.2		0.2*	0.3	0.20		As cast	124	18.0			Sand-cast
														R		147*	21.3			Chill-cast
4223	A-4	LM4	2.0–4.0	0.15	4.0–6.0	0.8	0.3–0.7	0.3	0.5	0.2*		0.1	0.05		As cast	139*	20.2	2		Sand-cast
														E		154*	22.3	2		Chill-cast
5230	A-5	LM5	0.1	3.0–6.0	0.3	0.6	0.3–0.7	0.1	0.1	0.2**	0.2*	0.05	0.05		As cast	139*	20.2	3		Sand-cast
														M		170*	24.6	5		Chill-cast
4600	A-6	LM6	0.1	0.1	10.0–13.0	0.6	0.5	0.1	0.1	0.2**		0.1	0.05		As cast	162*	23.5	5		Sand-cast
														A		185*	26.9	7		Chill-cast
4250	A-8	LM8	0.1	0.3–0.8	3.5 –6.0	0.6	0.5	0.1	0.1	0.2**	0.2	0.1	0.05		As cast	124	18.0*	2		Sand-cast
														I		162	23.5*	3		Chill-cast
																162	23.5*	2.5		
														N		232	33.6*	5		
																147	21.3*	1		
														D		185	26.9*	2		
																232	33.6*			
														E		278	40.3*	2		
4635	A-9	LM9	0.1	0.2–0.6	10.0–13.0	0.6	0.3–0.7	0.1	0.1	0.2**	0.2	0.1	0.05		Precipitation-	170	24.6	1.5		Sand-cast
														R	treated	231	36.4	2		Chill-cast
5500	A-10	LM10	0.1	9.5–11.0	0.25	0.35	0.1	0.1	0.1	0.2**	0.2	0.05	0.05		Solution-	278*	40.3	8		Sand-cast
															treated	307*	44.8	12		Chill-cast
2280	A-11	LM11	4.0–5.0	0.1	0.25	0.25	0.1	0.1	0.1	0.3*	0.05–0.3	0.05	0.05		Solution-	216	31.3	7		Sand-cast
															treated	263	38.1	13		Chill-cast
															WP	278	40.3*	4		
																309	44.8*	9		
2585	A-12	LM12	9.0–10.5	0.15–0.35	2.0	0.5–1.5	0.6	0.5	0.1			0.1	0.10		Fully heat-				100	
															treated					
	BS 1490		9.0–11.5	0.2–0.4	2.5	1.0	0.6	0.5	0.8	0.2**		0.1	0.1		WP	278	40.3**			
	(LM12)																			
4685	A-13	LM 13	0.5–1.3	0.8–1.5	11.0–13.0	0.8	0.5	2.0–3.0	0.1	0.2**	0.2	0.1	0.10		Fullly heat-	170	24.6		100	Sand-cast
															treated	247	35.9		100	Chill-cast
	A13 (special)														WP	170	24.6			Sand-cast
																278	40.3			Chill-cast
															WP	139	20.2		65	Sand-cast
																201	29.2		65	Chill-cast

TABLE 1-29
Chemical composition and mechanical properties of cast aluminum alloy (*Cont.*)

Designation			Chemical composition, percent													Mechanical properties				
																Tensile strength, σ_{st}			Brinell	
IS new	IS old	BS	Cu	Mg	Si	Fe	Mn	Ni	Zn	Ti	Ti + Nb	Pb	Sn	Al	Condition	MPa	kpsi	Elongation %	hardness, H_B	Test piece
2285	A-14	LM 14	3.5–4.5	1.2–1.7	0.6	0.6	0.6	1.8–2.3	0.1	0.2*		0.05	0.05		Fully heat-treated	216	31.3		100	Sand-cast
																278	40.3		100	Chill-cast
	A-14 (special)														WP	185	26.9		75	Sand-cast
															Solution-treated	232	33.6		75	Chill-cast
4225	A-16	LM16	1.0–1.5	0.4–0.6	4.5–5.5	0.6	0.5	0.25	0.1	0.2**		0.05	0.05			170	24.6	2		Sand-cast
												0.1**			WP	201	29.2	3		Chill-cast
															WP	232	33.6			Sand-cast
															WP	263	38.1			Chill-cast
															WP	232	33.8			Sand-cast
															As cast	278	40.3			Chill-cast
4300	A-18	LM18	0.1	0.1	4.5–6.0	0.6	0.5	0.1	0.1	0.2**	0.2	0.1	0.05			116	16.8	3		Sand-cast
																139	20.2	4		Chill-cast
4223	A-22	LM22	2.8–3.8	0.05	4.0–6.0	0.7	0.3–0.6	0.15	0.15	0.2**	0.2	0.05	0.05		Solution (W)-treated	247	35.9	8		Chill-cast
													0.1**							
4420	A-24	LM24	3.0–4.0	0.1	7.5–9.5	1.3	0.5	0.5	1.0	0.2**		0.3	0.20		As cast	177	25.7	1.5		Chill-cast
									3.0**											

Notes: IS: Sp-1-1967 Specification of Aluminum Alloy Castings and BS 1490 (from LM 1 to LM 24) are same.
* Refer to both Indian Standards and British Standards; ** refer to British Standards, BS 1490 only.
Source IS Sp-1, 1967.

TABLE 1-30
Chemical composition and mechanical properties of wrought aluminum and aluminum alloys for general engineering purposes[6]

	Chemical composition, %										Size		0.2% proof stress, min, σ_{sp}		Tensile strength, min, σ_{st}		
Designation	Al	Cu	Mg	Si	Fe	Mn	Zn	Ti or others	Cr	Condition	Over mm (in)	Up to and including mm (in)	MPa	kpsi	MPa	kpsi	Elongation, % (min)
19000	99 min	0.1	0.2	0.5	0.7	0.1	0.1	–	–	M[a]			20	2.9	65	9.4	18
										O					110#	16.0	25
19500	99.5 min	0.05	–	0.3	0.4	0.05	0.1	–	–	M[a]			18	2.6	65	9.4	23
										O					100#	14.5	25
19600										M[a]			17	2.5	65	9.4	23
										M[a]			90	13	150	21.6	12
										O			175	25.4	240#	34.8	12
24345	Remainder	3.8–5.0	0.2–0.8	0.5–1.2	0.7	0.3–1.2	0.2	0.3*	0.3*	W	10 (0.4)	10 (0.4)	225	32.6	375	54.4	10
											10 (0.4)	75 (3.0)	235	34.1	385	55.8	10
											75 (3.0)	150 (6.0)	235	34.1	385	55.8	8
											150 (6.0)	200 (8.0)	225	32.6	375	54.4	8
										WP	–	10 (0.4)	375	54.4	430	62.4	6
											10 (0.4)	25 (1.0)	400	58.0	460	66.7	6
											25 (1.0)	75 (3.0)	420	60.9	480	69.6	6
											75 (3.0)	150 (6.0)	405	58.7	460	66.7	6
											150 (6.0)	200 (8.0)	380	55.1	430	62.4	6
										M[a]	–	–	90	13.0	150	21.0	12
										O	–	–	175#	25.0	240	34.8	12
											–	10 (0.4)	220	31.9	375	54.4	10
24534	Remainder	3.5–4.7	0.4–1.2	0.2–0.7	0.7	0.4–1.2	0.2	0.3	–	W	10 (0.4)	75 (3.0)	235	34.1	385	55.8	10
											75 (3.0)	150 (6.0)	235	34.1	385	55.8	8
											150 (6.0)	200 (8.0)	225	32.6	375	54.4	8
43000	Remainder	0.1	0.2	4.5– 6.00	0.6	0.5	0.2	–	–	M[a]	–	15 (0.6)	–	–	90	13.0	18
										O	–	15 (0.6)	–	–	130#	18.9	18
46000	Remainder	0.1	0.2	10.0–13.0	0.6	0.5	0.2	–	–	M[a]	–	15 (0.6)	–	–	100	14.5	10
										O	–	15 (0.6)	–	–	150#	21.8	12
52000	Remainder	0.1	1.7–2.6	0.6	0.5	0.5	0.2	0.2	0.25	M[a]	–	150 (6.0)	70	10.2	160	23.2	14
										O	–	50 (2.0)			240#	34.8	18
53000	Remainder	0.1	2.8–4.0	0.6	0.5	0.5	0.2	0.2	0.25	M[a]	–	150 (6.0)	100	14.5	215	31.2	14
											50 (2.0)	150 (6.0)	100	14.5	200	29.0	14
										O	–	150 (6.0)	–	–	260#	37.7	10
54300	Remainder	0.1	4.0–4.9	0.4	0.7	0.5–1.0	0.2	0.2	0.25	M[a]	–	150 (6.0)	130	18.9	275	40.0	11
										O	–	150 (6.0)	125	18.1	350	50.8	13
63400	Remainder	0.1	0.4–0.9	0.3–0.7	0.6	0.3	0.2	0.2	0.1	M[a]	All sizes	–	–	–	110	16.0	13
										O	–	–	–	–	130#	18.8	18
										W	–	150 (6.0)	80	11.6	140	20.3	14
											150 (6.0)	200 (8.0)	80	11.6	125	18.1	13

TABLE 1-30
Chemical composition and mechanical properties of wrought aluminum and aluminum alloys for general engineering purposes *(Cont.)*

	Chemical composition, %										Size		0.2% proof stress, min, σ_{sp}		Tensile strength, min, σ_{st}		
Designation	Al	Cu	Mg	Si	Fe	Mn	Zn	Ti or others	Cr	Condition	Over mm (in)	Up to and including mm (in)	MPa	kpsi	MPa	kpsi	Elongation, % (min)
										P	–	23 (0.12)	140	20.3	170	24.7	7
												12 (0.5)	110	16.0	150	21.8	7
										WP	–	150 (6.0)	150	21.8	185	26.8	7
											150 (6.0)	200 (8.0)	130	18.9	150	21.8	6
64423	Remainder	0.5–1.0	0.5–1.3	0.7–1.3	0.8	1.0				M[a]	–	–	–	–	120	17.4	10
										O	–	–	125#	18.1	215#	31.2	15
										W	–	–	155	22.5	265	38.4	13
										WP	–	–	265	38.4	330	47.9	7
64430	Remainder	0.1	0.4–1.2	0.6–1.3	0.6	0.4–1.0	1.0	0.2	0.25	M[a]	All sizes	–	80	11.6	110	16.0	12
										O	–	–	–	–	150#	21.8	16
										W	–	150 (6.0)	120	18.1	185	26.8	14
										WP	150 (6.0)	200 (8.0)	100	14.5	170	24.7	12
											–	5 (0.2)	255	37.0	295	42.8	7
											5 (0.2)	75 (3.0)	270	39.2	310	45.0	7
											75 (3.0)	150 (6.0)	270	39.2	295	42.8	7
											150 (6.0)	200 (8.0)	240	34.8	280	40.6	6
65032	Remainder	0.15–0.4	0.7–1.2	0.4–0.8	0.7	0.2–0.8	0.2	0.2	0.15–0.35	M[a]	All sizes		50	7.3	110	16	12
										O	–	15 (0.6)	115#	16.7	150#	21.8	16
										W	–	150 (6.0)	115	22.5	185	26.8	14
											150 (6.0)	200 (8.0)	100	14.5	170	24.7	12
										WP	–	150 (6.0)	235	34.0	280	40.6	7
											150 (6.0)	200 (8.0)	200	29.0	245	35.5	6
74530	Remainder	0.2	1.0– 1. 5	0.4–0.8	0.7	0.2–0.7	4–5	0.2	0.2	W	–	6 (0.24)	220	31.4	255	37.0	9
										(Naturally	6 (0.24)	75 (3.0)	230	33.6	275	40.0	9
										aged for 30	75 (3.0)	150 (6.0)	220	31.4	265	38.4	9
										days)							
										WP	–	6 (0.24)	245	35.5	285	41.3	7
											6 (0.24)	75 (3.0)	26-	37.7	310	45.0	7
												150 (6.0)	245	35.5	290	42.1	7
76528										WP	All sizes	–	–	–	290#	42.1	10
											–	6 (0.24)	430	62.4	500	72.5	6
											6 (0.24)	75 (3.0)	455	66.6	530	78.9	6
											15 (0.6)	150 (6.0)	430	62.4	500	72.5	6

[a] Properties in M (as-cast) temper are only typical values and are given for information only.

Key: # Maximum, M – as-cast condition; R – stress-relieved only; P – precipitation-treated; W – solution-treated, WP – solution-treated and precipitation treated; WPS – fully heat treated plus stabilization.

Source: IS 733, 1983.

TABLE 1-31
Typical mechanical properties and uses of some copper alloys[4]

Alloy name	UNS no.	Composition,[a]	Ultimate tensile strength, σ_{sut} MPa	Ultimate tensile strength, σ_{sut} kpsi	Tensile yield[b] strength, σ_{syt} MPa	Tensile yield[b] strength, σ_{syt} kpsi	Elongation in 50 mm (2 in), %	Hardness: Brinell, 4.9 kN (500-kgf load) H_B	Hardness: Rockwell,[a] R	Machinability rating[c]	Typical uses
Cast Alloys											
Leaded red brass	C 83600	85 Cu, 5 Sn, 5 Pb, 5 Zn	255	37	117	17	30	60		84	Valves, flanges, pipe fittings, pump castings, water pump impellers and housings, small gears, ornamental fittings
Leaded yellow brass	C 85400	67 Cu, 1 Sn, 3 Pb, 29 Zn	234	34	83	12	25	50		80	General-purpose yellow casting alloy, furniture hardware, radiator fittings, ship trimmings, clocks, battery clamps, valves, and fittings
Manganese bronze	C 86300	63 Cu, 25 Zn, 3 Fe, 6 Al, 3 Mn	793	115	572	83	15	225[d]		8	Extra-heavy-duty, high-strength alloy, large valve stems, gears, cams, slow heavy-load bearings, screw-down nuts, hydraulic cylinder parts
Silicon bronze	C 87200	89 Cu min, 4 Si	379	55	172	25	30	85		40	Bearings, bells, impellers, pump and valve components, marine fittings, corrosion-resistant castings
Silicon brass	C 87500	82 Cu, 14 Zn, 4 Si	462	67	207	30	21	115 134[d]		50	Bearings, gears, impeller, rocker arms, valve stems, small boat propellers
Tin bronze	C 90500	88 Cu, 10 Sn, 2 Zn	310	45	152	22	25	75		30	Bearings, bushings, piston rings, valve components, steam fittings, gears
Leaded tin bronze	C 92200	88 Cu, 6 Sn, 1.5 Pb, 4.5 Zn	276	40	138	20	30	65		42	Valves, fittings and pressure-containing parts for use up to 288°C (550°F), bolts, nuts, gears, pump piston, expansion joints
Leaded tin nickel bronze	C 92900	84 Cu, 10 Sn, 2.5 Pb, 3.5 Ni	324	47	179	26	20	80		40	Gears, wear plates, cams, guides
High-leaded tin bronze	C 93700	80 Cu, 10 Sn, 10 Pb	241	35	124	18	20	60		80	Bearings for high-speed and heavy-pressure pumps, impellers, pressure-tight castings
Aluminum bronze	C 95500	81 Cu, 4 Ni, 4 Fe, 11 AI	689–827	100–120	303–469	44–68	12–10	192–230[d]		50	Valve guides and seats in aircraft engines, bushings, rolling mill bearings, washers, chemical plant equipment, chains, hooks, marine propellers, gears, worms
Copper-nickel	C 96300	79.3 Cu, 20 Ni, 0.7 Fe	517	75	379	55	10	150		15	Marine fittings, sleeves and seawater corrosion resistance parts
Nickel-silver	C 97800	66 Cu, 5 Sn, 2 Pb, 25 Ni, 2 Zn	379	55	207	30	15	130[d]		60	Valves and valve seats, musical instrument components, sanitary and ornamental hardware
Special alloy	C 99400	90.4 Cu, 2.2 Ni, 2.0 Fe, 1.2 Al, 1.2 Si, 3.0 Zn	455–545	66–79	234–372	34–54	25	125–170[d]		50	Valve stems, marine uses, propeller wheels, mining equipment gears
Wrought Alloys											
Cadmium copper	C 16200	99.0 Cu, 1.0 Cd	241–689	35–100	48–476	7–69	57–1			20	Trolley wire, spring contacts, railbands, high-strength transmission lines, switch gear components, and ware-guide
Beryllium copper	C 17000	99.5 Cu, 1.7 Be, 0.20 Co	483-1310	70-190	221-1172	32-170	45-3		R_B98	20	Bellows, diaphragms, fuse clips, fasteners, lock washers, springs, valves, welding equipment, bourdon tubing
Leaded beryllium copper	C 17300	99.5 Cu, 1.9 Be, 0.4 Pb	469-1479	68-200	172-1255	25-182	48-3		R_B77	50	Bellows, diaphragms, fuse clips, fasteners, lock washers, springs, valves, welding equipment, switch parts, roll pins

TABLE 1-31
Typical mechanical properties and uses of some copper alloys (*Cont.*)

Alloy name	UNS no.	Composition,[a]	Ultimate tensile strength, σ_{sut}		Tensile yield[b] strength, σ_{syt}		Elongation in 50 mm (2 in), %	Hardness		Machinability rating[c]	Typical uses
			MPa	kpsi	MPa	kpsi		Brinell, 4.9 kN (500-kgf load) H_B	Rockwell,[a] R		
Guilding brass (95%)	C 21000	95.0 Cu, 5.0 Zn	234–441	34–64	69–400	10–58	45–4		$64R_B$–$46R_F$	20	Coins, medals, bullet jackets, fuse caps, primers, jewellery base for gold plate
Commercial bronze (90%)	C 22000	90.0 Cu, 10.0 Zn	255–496	37–72	69–427	10–62	50–3		$70R_B$–$53R_F$	20	Etching bronze, grillwork, screen cloth, lipstick cases, marine hardware, screws, rivets
Red brass (85%)	C 23000	85.0 Cu, 15.0 Zn	269–724	39–105	69–434	10–63	55–3		$77R_B$–$55R_F$	30	Conduit, sockets, fasteners, fire extinguishers, condenser and heat-exchanger tubing, radiator cores
Cartridge brass (70%)	C 26000	70.0 Cu, 30.0 Zn	303–896	44–130	76–448	11–65	66–3		$82R_B$–$64R_F$	30	Radiator cores and tanks, flashlight shells, lamp fixtures, fasteners, locks, hinges, ammunition components, rivets
Yellow brass	C 26800	65.0 Cu, 35.0 Zn	317–883	46–128	97–427	14–62	65–3		$80R_B$–$64R_F$	30	Radiator cores and tanks, flashlight shells, lamp fixtures, fasteners, locks, hinges, rivets
Muntz metal	C 28000	60.0 Cu, 41.0 Zn	372– 510	54–74	145–379	21–55	52–10		$85R_F$–$80R_B$	40	Architectural, large nuts and bolts, brazing rods, condenser plates, heat-exchanger and condenser tubing, hot forgings
Medium leaded brass	C 34000	65.0 Cu. 1.0 Pb, 34.0 Zn	324–607	47–88	103–414	15–60	60–7			70	Butts, gears, nuts, rivets, screws, dials, engravings
Free-cutting brass	C 36000	61.5 Cu, 3.0 Pb, 35.5 Zn	338–469	49–68	124–310	18–45	53–18			100	Gears, pinions, automatic high-speed screws, machine parts
Forging brass	C 37700	59.0 Cu, 2.0 Pb, 39.0 Zn	359	52	138	20	45			80	Forgings and pressings of all kinds
Admiralty brass	C 44300 C 44400 C 44500	71.0 Cu, 28.0 Zn, 1.0 Sn	331–379	48–55	124–152	18–22	65–60			30	Ferrules, condenser, evaporator and heat-exchanger tubing, distiller tubing
Naval brass	C 46400 to C 46700	60.0 Cu, 39.25 Zn, 0.75 Sn	379–607	55–88	172–455	25–66	50–17		90-82R_B	30	Aircraft turn buckle barrels, balls, bolts, nuts, marine hardware, propeller, rivets, shafts, valve stems, welding rods, condenser plate
Phosphor bronze (5% A)	C 51000	95.0 Cu. 5.0 Sn, trace P	324– 965	47–140	131–552	19–80	64–2			20	Bellows, bourdon tubing, clutch disks, cotter pins, diaphragms, fasteners, lock washers, chemical hardware, textile machinery
High-silicon bronze -A	C 65500	97.0 Cu, 3.0 Si	386– 1000	56–145	145–483	21–70	63–3			30	Hydraulic pressure liners, anchor screws, bolts, cap screws, machine screws, nuts, rivets, U-bolts, electrical conduits, welding rod
Manganese bronze–A	C 67500	58.5 Cu, 1.4 Fe, 39.0 Zn, 1.0 Sn, 0. 1 Mn	448– 579	65–84	207–414	30–60	33–19			30	Clutch disks, pump rods, shafting, balls, valve stems and bodies
Copper-nickel (30%)	C 71500	70.0 Cu, 30.0 Ni	372–517	54–75	138–483	20–70	45–15			20	Condensers, condenser plates, distiller tubing, evaporator and heat-exchanger tubing, ferrules, salt water piping
Nickel-silver 55-18	C 77000	55.0 Cu, 27.0 Zn, 18.0 Ni	414–1000	60—145	186–621	27–90	40–2			30	Optical goods, springs, and resistance wires

[a] Nominal composition. unless otherwise noted. [b] All yield strengths are calculated by 0.5 percent offset method. [c] Machinability rating expressed as a percentage of the machinability of C 36000, free-cutting brass, based on 100 percent for C 36000. [d] 29.4 kN (3000 kgf) load. [e] R_A, R_B, R_F, Rockwell numbers in *A*, *B*, *F* scales.

Note: Values tabulated are average values of test specimens.

Source: *ASM Metals Handbook*, American Society for Metals, Metals Park., Ohio, 1988.

TABLE 1-32
Nominal compositions and typical room-temperature mechanical properties of some magnesium alloys

Alloy	Composition						Tensile strength, σ_{st}		Yield strength, σ_{sy}						Elongation in 50 mm (2 in), %	Shear strength, τ_s		Brinell[b] hardness, H_B
									Tensile		Compressive		Bearing					
	Al	Mn(a)	Th	Zn	Zr	Others	MPa	kpsi	MPa	kpsi	MPa	kpsi	MPa	kpsi		MPa	kpsi	
							Sand and Permanent Mold Castings											
AZ63A–T6	6.0	0.15		3.0			275	40	130	19	130	19	360	52	5	145	21	73
AZ81A–T4	7.6	0.13		0.7			275	40	83	12	83	12	305	44	15	125	18	55
AZ92A–T6	9.0	0.10		2.0			275	40	150	22	150	22	450	65	3	150	22	84
HK3IA–T6			3.3		0.7		220	32	105	15	105	15	275	40	8	145	21	55
HZ32A–T5			3.3	2.1	0.7		185	27	90	13	90	13	255	37	4	140	20	57
ZE41A–T5				4.2	0.7	1.2 RE	205	30	140	20	140	20	350	51	3.5	160	23	62
ZH62A–T5			1.8	5.7	0.7		240	35	170	25	170	25	340	49	4	165	24	70
ZK61A–T6				6.0	0.7		310	45	195	28	195	28			10	180	26	70
							Die Castings											
AM60A–F	6.0	0.13					205	30	115	17	115	17			6			
AS41A–F[c]	4.3	0.35				1.0 Si	220	32	150	22	150	22			4			
AZ91A and B–F[c]	9.0	0.13	0.7				230	33	150	22	165	24			3	140	20	63
							Extruded Bars and Shapes											
AZ31B and C–F[d]	3.0			1.0			260	38	200	29	97	14	230	33	15	130	19	49
AZ80A–T5	8.5			0.5			380	55	275	40	240	35			7	165	24	82
HM31A–F		1.2	3.0				290	42	230	33	185	27	345	50	10	150	22	
ZK60A–T5				5.5	0.45[a]		365	53	305	44	250	36	405	59	11	180	26	88
							Sheets and Plates											
AZ3IB–H24	3.0			1.0			290	42	220	32	180	26	325	47	15	160	23	73
HK31A–H24			3.0		0.6		255	33	200	39	160	23	285	41	9	140	20	68
HM21A–T8		0.6	2.0				235	34	170	25	130	19	270	39	11	125	18	

[a] Minimum.
[b] 4.9-kN (500-kgf) load, 10-mm ball.
[c] A and B are identical except that 0.30% max residual Cu is allowable in AZ91B.
[d] Properties of B and C are identical, but AZ31C contains 0. 15 min Mn, 0.1 max Cu, and 0.03 max Ni.
Source: *ASM Metals Handbook*, American Society for Metals, Metals Park, Ohio, 1988.

TABLE 1-33
Mechanical properties[a] of some nickel alloys[4]

Name of alloy	Condition	Ultimate tensile strength, σ_{sut} MPa	Ultimate tensile strength, σ_{sut} kpsi	Tensile yield strength, σ_{syt} (0.2% offset) MPa	Tensile yield strength, σ_{syt} (0.2% offset) kpsi	Elongation in 50 mm (2 in), %	Hardness number	Impact strength notched Charpy J	Impact strength notched Charpy ft-lbf	Typical uses
Nickel 200	Bar, cold-drawn	448–758	65–110	276–690	40–100	35–10				Corrosion-resistant parts
	Annealed	379–517	55–75	103–207	15–30	35–40				
Nickel 270	Strip, cold-drawn	655	95	621	90	4	$95R_B$			
	Annealed	345	50	110	16	50	$35B_B$			
Durnickel 301	Bar, cold-drawn, annealed	620–825	90–120	205–415	30–60	55–35	$35R_C$			High strength and hardness, corrosion resistance
	Age-hardened	1275	185	910	132	28				
Monel 400	Bar, annealed, 21°C (70°F)	517–621	75–90	172–345	25–50	60–35				Corrosion-resistant parts
	Wire. annealed	483–655	70–95	205–380	30–50	45–25				
	Spring temper	1000–1240	145–180	862–1172	125–170	5–2				Springs
Monel K-500	Bar, drawn, age-hardened	965–1172	140–170	724–1034	105–150	30–20	$24R_C$	53	39	Corrosion-resistant parts
Inconel 600	Rod, annealed	624	91	210	30.4	49	$75R_B$			Jet engines, missiles, etc. where corrosion resistance and high strength are required
	As rolled	672	98	307	45	46	$86R_B$			
Inconel 825	Bar, annealed, 21°C (70°F)	1276	185	910	132	28				Superalloy, jet engine, turbine, furnace
	871°C (1600°F)	135	19.6	117	17.0	102				
Inconel X-750	Bar, 21°C (70°F)	1120	162	635	92	24				
	760°C (1400°F)	485	70	455	66	9				
Incoloy 800	Bar, annealed	512–690	75–100	207–414	30–60	60–30				
	Hot-finished	552–827	80–120	241–621	35–90	50–25				
	Wire, spring temper	965–1207	140–175	896–1172	130–170	5–2				
Hastelloy W	Bar, solution-treated	725	105	260	38	56.0				
	425°C (800° F)									
	900°C (1650°F)	352	52	220	32	14.5				
Hastelloy G-3	Sheet, 6.4–19 mm (0.25–0.75 in) thick	740	107	365	53	56	$87R_B$			
Hastelloy B	Bar, cast	924	134	462	67	52				
Udimet 700	Bar, 21°C (70°F)	1410	204	965	140	17				Jet engines, missiles, turbines where high-temperature strength and corrosion resistance are important
	870°C (1600°F)	690	100	635	92	27				
Unitemp AF2–IDA	Bar, 21°C (70°F)	1290	187	1050	152	10				
	870°C (1600°F)	830	120	715	104	8				
Rene 95	Bar, forging 21°C (70°F)	1620	235	1310	190	15				
	650°C (1200°F)	1460	212	1220	177	14				
Waspaloy	Bar, 21°C (70°F)	1280	185	795	115	25				
	870°C (1600°F)	525	76	515	75	35				

[a] Values shown represent usual ranges for common sections.
[b] Values tabulated are approximate average ones.
Source: *ASM Metals Handbook*, American Society for Metals, Metals Park, Ohio, 1988.

TABLE 1-34
Mechanical properties of some zinc casting alloys[8]

Designation of alloy				Ultimate tensile strength, σ_{sut}		Tensile yield strength, σ_{syt}		Brinell hardness, H_B	Elongation in 50 mm (2 in), %	Impact strength Charpy		Fatigue endurance limit, σ_{sf}, 10^8 cycles	
Grade	ASTM	SAE	UNS	MPa	kpsi	MPa	kpsi			J	ft-lbf	MPa	kpsi
						Die-Casting Alloys							
Alloy 3	AG 40 A	903	Z 33520	283	41			82	10	58	43	47	6.8
Alloy 5	AC 41 A	925[a]	Z 35531	324	47			91	7	65	48	56	8.2
Alloy 7		903		283	47				14	54	40		
						Zinc Foundry Alloys							
	ZA-12												
	Sand-cast			276–310	40–45	207	30	105–120	1–3				
	Permanent Mold			310–345	45–50	214	31	105–125	1–3				
	Die-cast			393	57	317	46	110–125	2				
	ZA-27												
	Sand-cast			400–440	58–64	365	53	110–120	3–6				
	Sand-cast			310–324	45–47	255	37	90–100	8–11				
	Die-cast			448	65	434	63	110–125	1				

[a] Die-cast.
Note: Values given are average values.
Source: *Machine Design*, 1981 Materials Reference Issue, Penton/IPC, Cleveland, Ohio, Vol. 53, No. 6 (March 19, 1981); *SAE Handbook*, pp. 11–123, 1981.

TABLE 1-35
Mechanical properties of some wrought titanium alloy[4]

Name of alloy	UNS no.	Designation	Ultimate tensile strength, σ_{sut} MPa	Ultimate tensile strength, σ_{sut} kpsi	Tensile yield strength, σ_{syt} MPa	Tensile yield strength, σ_{syt} kpsi	Elongation in 50 mm (2 in), %	Strength impact Charpy J	Strength impact Charpy ft-lbf	Hardness	Machinability	Uses
Commercially pure titanium	R 50520	ASTM Grade 1	240	35	170	25	24	35	26[a]		40[b]	Resistance to temperature effect of structures, easy to fabricate, excellent corrosive resistance, cyrogenic applications
Commercially pure titanium		ASTM Grade 2	340	50	280	40	22					
Commercially pure titanium		ASTM Grade 3 (Ti-65A)	450	65	380	55	20			$30R_C$		
Commercially pure titanium		ASTM Grade 4	550	80	480	70						
Alpha alloy	R 54520	Ti-5Al-2.5Sn	790	115	760	110						Gas turbine engine casting and rings, aerospace structural members, excellent weldability, pressure vessels, excellent corrosive resistance, jet engine blades and wheels, large bulkhead forgings
Alpha alloy	R 54521	Ti-5Al-2.5Sn-ELI	690	100	620	90	19				30[b]	
Alpha alloy	R 54790	Ti-2.25Al-11Sn-5Zr-IMo	1000	145	900	130						
Alpha-beta alloy	R 56400	Ti-6AI-4V[b]	900	130	830	120	10	24.5	18[e]	$34R_C$	22[b]	Most widely used alloy, aircraft gas turbine disks and blades, turbine disks and blades, air frame structural components, gas turbine engines, disks and fan blade, components of compressors
		Ti-6Al-6V-2Sn	1030	150	970	140						
		Ti-10V-2Fe-3Al[a,c]	1170	170	1100	160						
	R 56260	Ti-6A1-2Sn-4Zr-6Mo[e]	1170	170	1100	160						
Beta alloy	R 58010	Ti-13V-11Cr-3Al										Missile applications such as solid rocket motor cases, advanced manned and unmanned airborne systems, springs for airframe applications
Beta alloy		Ti-3Al-8V-6Cr-4Mo-4Zr[b,c]	900	130	830	120	4	13.5	10		40[b]	

[a] At 0.2% offset.
[b] Mechanical and other properties given for annealed conditions.
[c] Mechanical and other properties given for solution-treated and aged condition.
[d] Based on a rating of 100 for B1112 resulfurized steel.
[e] Approximate values of annealed bars at room temperature.
Source: *Metals Handbook*, Desk Edition, ASM International, Materials Park, Ohio 44073-0002, 1985 (formerly the American Society for Metals, Metals Park, Ohio, 1985).

TABLE 1-35A
Mechanical properties of powder metallurgy and wrought titanium and titanium-base alloys

Name of alloy	Processing	Density ρ, %	Ultimate tensile strength, σ_{sut} MPa	σ_{sut} kpsi	Yield strength, σ_{sy} MPa	σ_{sy} kpsi	Fatigue limit notched, σ_{sf} MPa	σ_{sf} kpsi	Fracture toughness, K_{IC} MPA $\sqrt{m}$	K_{IC} kpsi $\sqrt{in}$	Elastic modulus, E GPa	E Mpsi	Elongation, %	Reduction in area, %
Wrought commercial purity titanium Grade II		100	345	50	344	50					103	15	5	35
Sponge commercial purity[a]		95.5	414	60	324	47					103	15	15	14
		94	427	62	338	49								
Powder metallurgy titanium	Forged	100	455	66	365	53							23	30
Wrought Ti-6Al-4V (AMS 4298)		100	896	130	827	120	427	62	55[e]	50[e]	114	16.5	10	20
Powder metallurgy Ti-6Al-4V	Blended elemental alloy, cold	95.5	876	127	786	114	193	28	45[e]	40[e]	117	17	8	14
	Blended elemental alloy, forged,	99 min	937	136	862	125	414	60	61[e]	56[e]	116	16.8	12–18	15-40
	preforms or vacuum hot pressed	99	1103	160	1013	147							4.9	7.6
	Solution treated and aged													
Plasma rotating electrode processed Ti-6Al-4V	Hot isostatically pressed	100	951	138	910	132	414[d]	60[d]	83[f]	76[f]			15	39
Powder metallurgy Ti-6Al-4V[a]		94	827	120	738	107							5	8
	Forged	100	920	133.5	841	122							11.5	25

[a] 0.12% oxygen.
[b] 0.2% oxygen.
[c] Consolidated at 811 MPa (58.8 tpsi), 0.5 s dwell in low-carbon steel fluid dies. Preheat temperature was 940°C (1725°F) held at temperature 0.75 h. Powder was vacuum filled into fluid dies following cold static outgassing for 24 h.
[d] $K_t = 3$.
[e] K_e.
[f] K_{tc}.
Source: *Metals Handbook*, Desk Edition, ASM International, Materials Park, Ohio 44073-0002, 1985 (formerly the American Society for Metals, Metals Park, Ohio, 1985).

TABLE 1-36
Mechanical properties of some lead alloys

UNS No.	Ultimate tensile strength, σ_{sut} MPa	Ultimate tensile strength, σ_{sut} psi	Yield strength, σ_{sy} MPa	Yield strength, σ_{sy} psi	Shear strength, τ_s MPa	Shear strength, τ_s psi	Fatigue strength at 10^7 cycles, σ_{sf} MPa	Fatigue strength at 10^7 cycles, σ_{sf} psi	Hardness number, H_B	Elongation in 50 mm (2 in), %	Creep	Uses
50042	12-13	1740–1885	55	7978	12.5	1810	3.2	464	3.2–4.5	30		Low melting-point chemical process applications, used as solder for the jobs lead alloyed with tin, bismuth cadmium, indium forms alloys with low melting point. Some of these are fusible alloys, used in automotive devices, fire extinguishers, sprinkler heads.
50132	35	5076							13	28	19.5 MPa-1000 h	
	1.8 MPa at 100°C	261 at 212°F								10	7.5 MPa-1000 h at 100°C	
50750	70	10152	66	9570							28 MPa for 100 h	
51120	16-19	2320–2755	6–8	870–1160			4.3	624	4–6		3% per year at 2.07 MPa	
52901	27.6	4002					10.30	1495	8.1	48		
53620	71	10297					30 MPa at 2×10^7	4358 at 2×10^7	20	2		
54321	28	4060	10	1450					8	55		
54520	30	4350							10	10	3.5 MPa for 1000 h	
	8 MPa at 100°C								30% at 100°C		1.1 MPa for 1000 h at 100°C (212°F)	
54820	34	4930			28	4060			12	18	0.790 MPa for 0.01% per day	
54915	37	5367								25		
	6 MPa at 100°C	870			32	4640			12	130	2.1 MPa for 1000 h	Wiping solder for joining lead pipes and cable sheaths; for automobile radiator cores and heating units
55030	32.4	4700	33	4790	36	5200			14	60		For general purpose; most popular of all lead alloys
55111	52.5	7610								30–60	2.9 MPa for 1000 h	
	19 MPa at 100°C	2756			37	5380			16	135–200 at 100°C (212°F)	0.45 MPa for 1000 h at 100°C (212°F)	

Note: Values tabulated are average values obtained from standard test specimens.
Source: *Metals Handbook*, Desk Edition, ASM International, Materials Park, Ohio 44073-0002, 1985 (formerly the American Society for Metals, Metals Park, Ohio, 1985).

TABLE 1-37
Mechanical properties of bronzes

Property	Mode of casting test pieces	Railway, bronze: Class I phosphor[a] bronze[b]	Railway, bronze: Class II gun metal[c]	Railway, bronze: Class III leaded	Railway, bronze: Class IV bronze[d]	Railway, bronze: Class V leaded[e] gun metal	Aluminum bronze: Grade I	Aluminum bronze: Grade II	Aluminum bronze: Grade III	Tin bronze	Silicon bronze
Ultimate strength, min σ_{sut}, MPa (kpsi)	Sand-cast (cast-on)	186 (27.0)[f]	196 (28.4)	137 (19.9)	157 (22.8)	186 (27.0)				216 (31.3)	
	Sand-cast (separately cast)	206 (29.9)[f]	216 (31.3)	157 (22.8)	176 (25.5)	206 (29.9)	647 (93.8)	490 (71.0)	446 (64.7)	226 (32.8)	309 (44.8)
	Chill-cast						647 (93.8)	539 (78.2)	196 (28.4)	245 (35.5)	
Elongation percent, min	Sand-cast (cast-on)	3.0	8.0	2.0	2.0	8.0				8.0	
	Sand-cast (separately cast)	5.0	12.0	4.0	4.0	12.0	15.0	20.0	20.0	12.0	20.0
	Chill-cast						12.0	20.0			

[a] Brinell hardness, H_B for phosphor bronzes: 60 for sand cast (cast-on) test pieces and 65 for sand-cast (separately cast) test pieces.
[b] Used for locomotive side valves, oil-lubricated side rod, pony pivot bushes, steel axle box, oil-lubricated connecting rod.
[c] Used for fusible plugs, relief valves, whistle valve body, stuffing box, nonferrous boxes, oil-lubricated connecting rod, large end bearings.
[d] Used for locomotive grease lubricated non-ferrous axle boxes, side rod and motion bushes.
[e] Used for castings for carriage and wagon bearings shells.
[f] σ_{sut} given in parentheses are the units in US Customary Units (kpsi).

TABLE 1-38
Mechanical properties of rubber and rubber-like materials

Material	Specific gravity	Compressive strength, σ_{sc}		Tensile strength, σ_{st}		Transverse strength, σ_{sb}		Hardness shore durometer	Maximum temperature			Effect of heat
		MPa	kpsi	MPa	kpsi	MPa	kpsi		K	°C	°F	
Duprene	1.27–3.00			1.4–28	0.2–4.0			15–95	422	149	300	Stiffens slightly
Koroseal (hard)	1.30–1.40			14–62	2.0–9.0			80–100	373	100	212	Softens
Koroseal (soft)	1.20–1.30			3.4–17	0.6–2.6			30–80	361	88	190	Softens
Plioform (plastic)	1.06	88	12.8	28–34	4.0–5.0	48	7.0		344–393	71–120	160–250	
Rubber[b] (hard)	1.12–2.00	758	110.0	7–69	1.0–10.0	62	9.0	50[a]/80	328–367	55–71	130–160	Softens
Rubber[c] (soft)	0.97–1.25	14	2.0	3.5	0.6	62	9.0		339–367	65–94	150–200	Softens
Rubber (linings)	0.98–1.35	103	15.1			103	15.1		361	88	190	Softens

[a] Sclerscope.
[b] Coefficient of linear expansion from 0 to 333 K (60°C = 140°F) is 35×10^{-6}.
[c] Coefficient of linear expansion from 0 to 333 K (60°C = 140°F) is 36×10^{-6}.

TABLE 1-39
Properties of some thermoplastics

Name of plastic	Tensile strength, σ_{sut}		Elongation in 50 mm (2 in), %	Modulus of elasticity, E		Izod impact strength		Hardness, Rockwell	Resistance to		Coefficient of friction, μ		Application
	MPa	kpsi		GPa	Mpsi	J	ft-lbf		Heat	Chemical	With plastic	With steel	
ABS (general purpose	41	6	5–20	2.3	0.33	8.8	6.5	103	Available	Fair			Light-duty mechanical and decorative eyeglass frames, automobile-steering wheels, knobs, handles, camera cases, battery cases, phone and flashlight cases, helmets, housing for power tools, pumps
Acrylics	37–72	5.4–10.5	5–50	1.5–3.1	0.22–0.45	0.5–1.6	0.4–1.2	92–100 M	Available	Fair			Light-duty mechanical knobs, pipe fittings, automobile-steering wheels, eyeglass frames, tool handles, camera cases, optical and transparent parts for safety glasses, snowmobile windshields, refrigerator shelves
Acetal	55–69	8–10	40–60	2.8–3.6	0.4–0.52			80–94 M	Good	High			Mechanical gears, cams, pistons, rollers, valves, fan blades, washing-machine agitators, bushings, bearings, chute liners, wear strips, and structural components
Cellulosic (cellulose acetate)	15.2–47.5	2.2–6.9		0.5–2.8	0.065–0.40	1.4–9.9	1.0–7.3	122 R					Decorative knobs, handles, camera cases, pipes, pipe fittings, eyeglass frames, phone and flashlight cases, helmets, pumps and power tool housings, transparent parts for safety glasses, lens signs, refrigerator shelves and snowmobile windshields, extruded and cast film, and sheet for packaging
Epoxy resin (glass-fiber filler)	69–138	10–20	4.0	21.0	3.04	2.7–41	2–30	100–110 M					Filament wound structures, aircraft pressure bottles, oil storage tanks and high-performance tubing, reinforced glass-fiber composites
Fluoroplastic group	3–48	0.5–7.0	100–300			4.1	3	50-80 D	Excellent	Excellent		0.05	Gears, bearings, tracks, bushings, roller-skate wheels, chute liners
Nylon	55–83	8–12	60–200	1.2–2.9	0.18–0.42	1.4–4.5	1.0–3.3	114–120 R	Poor	Good	0.04-0.13		Structural, mechanical components such as gears, fan blades, washing-machine agitator, valve, pump, impeller, pistons, and cams
Phenolic (general purpose)	45–48	6.5–7.0		7.6–9.0	1.1–1.3	0.4–0.5	0.30–0.35	70–95 E					Wall plates, industrial switch gears, handles for appliances, housing for vacuum cleaners, automatic transparent rings, housing for thermostats, small motors, small tools, communication instruments, components for aircraft and computers, used as synthetic rubber for tires

TABLE 1-39
Properties of some thermoplastics (*Cont.*)

Name of plastic	Tensile strength, σ_{sut}		Elongation in 50 mm (2 in), %	Modulus of elasticity, E		Izod impact strength		Hardness, Rockwell	Resistance to		Coefficient of friction, μ		Application
	MPa	kpsi		GPa	Mpsi	J	ft-lbf		Heat	Chemical	With plastic	With steel	
Phenylene oxide	48-123	7.0-17.3	4–60	2.4–6.4	0.35-0.93	2.7-6.8	2-5	115-119 R 106-108 L	Good	Fair			Small housing for power tools, pumps, small appliances, hollow shapes for telephones, flashlight cases, helmets, TV cabinets, cable protective cover, bush-bar sleeves, scrubber-vane mist eliminators
Polycarborate	55-110	8-16	10-125	2.3-5.9	0.34-0.86	2.7-21.7	2-16	62- 91 M	Excellent	Fair	0.52	0.39	Mechanical gears, pistons, rollers, pump impellers, fan blades, rotor housing for pumps, power tools, phone cases, transparent parts for safety glasses, lenses and snowmobile windshields, refrigerator shelves, and flashlight cases
Polyimide	25–345	3.6–50	<1			0.3–23	0.25–17	88–120 M	Excellent	Excellent			Molded polyimides are used in jet-engine vane bushings, high load bearings for business machines and computer printout terminals, gear pump gaskets, hydraulic valve seals, multilayer printed-circuit boards, tubes for oil-well exploration
Polyester	55–159	8–23	1–300	1.9–11.7	0.28–1.7	0.7–2.6	0.5–1.9	65–100 M	Excellent	Poor	0.12 0.22	0.12 0.13	Flashlight and phone cases, housing for pumps, power tools and other appliances, gears, bushings, bearings, tracks, roller-skate wheels and chute liners
Polyethylene	4–38	0.6–5.5	20–1000	0.1–1.2	0.014–0.18	0.7–27.1	0.5–20	10–65 *R*					Decorative knobs, automobile steering wheels, eyeglass frames, tool handles, camera cases, phone and flashlight cases, housings for pump and power tools
Polypropylene	34–100	5–14.5	10–500	0.7–6.2	0.1–0.9	0.7–3.0	0.5–2.2	50–110 *R*					Mechanical cams, pistons, washing machine agitators, fan blades, valves, pump impellers, gears, bushings, chute liners, bearings, tracks, wear strips and other wear-resisting parts
Polysulfone	70	10.2	50–100	25	0.36	1.8	1.3	120 *R*	Excellent	Excellent			Pipe fittings, battery cases, knobs, camera and handle cases, trim moldings, eyeglass frames, tool handles, housings for pumps, power tools, phone cases, transparent parts, safety glasses, lenses, snowmobile windshields, signs, refrigerator shelves, and vandal-resistant glazing

Source: *Machine Design*, 1981 Materials Reference Issue, Penton/IPC, Cleveland, Ohio. Vol. 53, No. 6 (March 19, 1981).

TABLE 1-40
Properties of some thermosets[8]

Name	Tensile strength, σ_{st}		Modulus of elasticity, E		Hardness, Rockwell	Elongation in 50 mm (2 in), %	Impact strength Izod		Resistance		Application
	MPa	kpsi	GPa	Mpsi			J	ft-lbf	Heat	Chemical	
Alkyd	21–66	3–9.5	2–21	0.3–3.0	98E–99 M		0.5–14	0.3–10	Good	Fair	Military switch gear, electrical terminal strips, and relay housings and bases, automotive ignition parts, radio and TV components, switch gear, and small-appliance housings
Allylic	28–69	4–10			103–120 M		0.3–16	0.2–12	Excellent	Excellent	Switch gear and TV components, insulators, circuit boards, and housings, tubing and aircraft parts, copper-clad laminate for high-performance printed-circuit boards
Amino	34–69	5–1 0	9–16	1.3–2.4	110–120 M	0.3–0.9	0.4–24	0.27–18	Excellent	Excellent	Electrical wiring devices and switch housings, toaster and other appliance bases, push buttons, knobs, piano keys and camera parts, dinnerware, utensil handles, food-service trays, housings for electric shavers and mixers, metal blocks, connector plugs, automotive and aircraft ignition parts, coil forms, used as baking enamel coatings, particle-board binders, paper and textile treatment materials
Epoxy	28–138	4–20	2.5–21	0.35–3.04	80–120 M	1–10	0.3–41	0.2–30	Excellent	Excellent	Filament wound structures, aircraft pressure bottles, oil storage tank, used with various reinforcements, glass fibers, asbestos, cotton, synthetic fibers, and metallic foils, imprinted circuits, graphite and carbon-fiber-reinforced laminates used for radomes, pressure vessels, and aircraft components requiring high modulus and light weight, potting and encapsulating electrical and electronic components ranging from miniature coils and switches to large motors and generators
Phenolics	34–62	5–9	7–17	1.0–2.5	70–95 E		0.4–1.5	0.26–1.05	Excellent	Good	Handles for appliances, automotive power-brake systems and industrial terminal strips, industrial switch gear, housing for vacuum cleaners, handles for pots and pans, automotive transmission rings, and electrical components, thermostat housings, housing for small motors and heavy-duty electrical components, small power tools, electrical components for aircraft and computers, pump housings, synthetic rubber for tires and other mechanical rubber goods, dry ingredients for brake linings, clutch facings and other friction products
Silicone	3–45	0.4–6.5			80–90 M	15	0.5–14	0.3–10	Excellent	Excellent	Refrigerator equipment, used as a washing, sealant, laminating parts, injection mold silicon rubber

Source: *Machine Design*, 1981 Materials Reference Issue, Penton/IPC, Cleveland, Ohio. Vol. 53, No. 6 (March 19, 1981).

TABLE 1-41
Optical and mechanical properties of photoelastic material[6]

Material	Elastic limit, σ_{se} MPa	Elastic limit, σ_{se} kpsi	Tensile strength, σ_{st} MPa	Tensile strength, σ_{st} kpsi	Young's modulus, E GPa	Young's modulus, E kpsi	Poisson's ratio, ν	Stress fringe value, f_σ kN/m fri	Stress fringe value, f_σ lbf/in fri	Strain fringe value, f_s µm/fri	Strain fringe value, f_s µin/fri	Figure of merit, $Q = (E/f_\sigma)$ fri/m	Figure of merit, $Q = (E/f_\sigma)$ fri/in	$S = \sigma_e/f\sigma$ fri/m	$S = \sigma_e/f\sigma$ fri/in	Remarks
Glass	60.0	8.68	69.0	10.0	69.0	10,000	0.20	304.724–423.812	1740–2420	4.83	191	226,000–163,000	5747–4132	1970–1415	5–3.5	Low optical sensitivity; rarely used
Cataline (61-893)	38.0–62.0	5.55–9.0	88.2–117.2	12.5–17.0	4.2–4.3	615.0–628.0	0.365	15.236	87	4.83	191	27,600–280,000	7069–7218	2500–4070	63.8–100	Used for 2-dimensional (2-D) and 3-dimensional (3-D) models; susceptible for time edge effect
Methyl methacrylate (unplasticized)			48.2	7.0	2.8	400.0	0.38	154,113	880	91.00	3582	18,200	455			
Polystyrene								54.290	310							Low optical sensitivity
Cellulose nitrate	27.6	4.0	51.7	7.55	2.4	350.0	0.33	42.907	245	25.40	1000	56,000	1428	643	16.3	Free from time edge effect; used for photoplasticity
Castolite					4.3	623.0	0.35	31.873	182	1.00	39	135,000	3423			
Kriston			56.5	8.25	3.7	540.0		14.010	80	4.83	191	264,000	6750			
CR-39 (Columbia resin)	20.6	3.0	48.0–41.4	7.0–6.0	1.7–2.6	250.0–380.0	0.42	14.623–17.338	83.5—99	11.90–12.50	468–492	116,000–150,000	2994–3838	1408–1188	36.0–30.0	Used for 2-D models; free from time edge effect
Epoxy Resin:																
Araldite CN-501	28.3 at 299 K	4.12 at 77°F			3.10 at 298 K	452.0 at 77°F		10.770 at 298 K	61.5 at 77°F	4.57 at 298 K	180 at 77°F	290,000	7350	2628	67	Used for 2-D and 3-D models
Araldite 6020 at 299.7 K (80°F)					3.10	445.0	0.35	10.157	58.0	4.57	180	305,000	7672			Used most commonly for 2-D and 3-D models
Araldite 6020 at 277.4 K (40°F)								10.595	60,5							Used for 2-D and 3-D models
Araldite B			68.9	10.0	3.2 at 294 K	460 at 70°F	0.362	10.332	59.0	4.83	191	310,000	7796			Used most commonly for 2-D and 3-D models
Bakelite ERL 2774 (50 phthalic anhydride)	55.2	8.0			3.30 0.036 at 433 K	478.5 5.2 at 320°F	0.38	10.30 0.435 at 433 K	58.75 2.5 at 320°F	4.3	170	320,000 83,000	8145 2080	5360	136	Used for 2-D and 3-D models
Hysole 4290:																
at 296.9 K (75°F)	48.3	7.0	82.7	12.0	3.45	500.0	0.34	10.508	60.0			328,000	8333	4596	16	Used for 2-D and 3-D models
at 405.2 K (270°F)			1.38	0.20	0.014	2.0	0.50	0.245	1.4			57,000	1428			
Armstrong C-6	22.4 at 296.3 K	3.25 at 74°F			3.3 at 294 K	475 at 70°F		13.222 at 293.6 K	75.5 at 70°F	5.59 at 293.6 K	220 at 70°F	290,000	6291	1694	43	Used for 2-D and 3-D models
Polycarbonate	34.5	5.0			2.55	370.0	0.38	7.355	42.0	4.00	157	347,000	8810	4690	119	Little optical and mechanical creep
Marblette [annealed 72 h at 356 K (181°F)]					3.79 –4.13	550.0 – 600.0	0.41	14.535	83.0	5.33	210	261,000	6626			Good stress-optical relationship; susceptible for time edge effect

TABLE 1-41
Optical and mechanical properties of photoelastic material (*Cont.*)

Material	Elastic limit, σ_{se}		Tensile strength, σ_{st}		Young's modulus, E		Poisson's ratio, ν	Stress fringe value, f_σ		Strain fringe value, f_s		Figure of merit, $Q = (E/f_\sigma)$		$S = \sigma_e/f\sigma$		Remarks
	MPa	kpsi	MPa	kpsi	GPa	kpsi		kN/m fri	lbf/in fri	μm/fri	μin/fri	fri/m	fri/in	fri/m	fri/in	
Marblette (phenofomaldehyde)	18.9	2.70	31.0	4.50	1.65	240.0	0.40	9.982	57.0	7.87	310	165,000	4210	893	47.4	
Catalin 800	6.9	1.0	46.2	6.70	1.72	250.0	0.38	10.087	57.6	5.84	230	170,000	4340	684	17.4	Good stress-optical relationship; susceptible for time edge effect
Natural rubber																
Hard								1.752	10.0							
Soft								0.289	1.7	431.80	17000					
Urethane rubber Hysole 8705 at 26.9 K (75°F)					0.003	0.425	0.467	0.084	0.5	40.60	1598	35,700	850			Used for preparation of models in stress wave propagation and models of dam
Hysole 4485	0.17	2.85 $\times 10^{-2}$			0.003–0.004	0.425–0.625	0.46	0.158	0.9	82.00	3228	19,000–25,000	4722–694	1076	31.6	
Gelatin (15% gelatin, 25% glycerine, 60%water)					75.8 $\times 10^{-6}$	11 $\times 10^{-3}$	0.50	0.025	0.14	483.00	19016	3032	78			Great optical sensitivity; used for model study of body forces

Sources: K. Lingaiah, *Machine Design Data Handbook*, Vol. II (*SI and Customary Metric Units*), Suma Publishers, Bangalore, India, 1986, and K. Lingaiah and B. R. Narayana Iyengar, *Machine Design Data Handbook*, Vol. I (*SI and Customary Metric Units*), Suma Publishers, Bangalore, India, 1986.

TABLE 1-42
Typical mechanical properties of commercial machinable tungsten alloys

Class	Classification of alloy	Density, ρ		Ultimate tensile strength, σ_{sut}		Yield[a] strength, σ_{spl}		Proportional limit, σ_{spl}		Modulus of elasticity, E		Elongation in 50 mm (2 in), %	Hardness	Coefficient of thermal expansion, α		Magnetic properties	Tungsten, % by weight
		Mg/m³	lb/in³	MPa	kpsi	MPa	kpsi	MPa	kpsi	GPa	Mpsi			μm/m°C	μin/in°F		
1	W-Ni-Cu alloy	17.0	0.614	785	114	605	88	205	30	275	40	4	27 R_C	5.4	3.0	Virtually nonmagnetic	Class 1, 89–91
2	W-Ni-Fe alloy	17.0	0.614	895	130	615	89	260	38	275	40	16	27 R_C	5.4	3.0	Slightly magnetic	Class 2, 91–94
3	W-Ni Fe alloy	18.0	0.650	925	134	655	95	350	51	310	45	6	29 R_C	5.3	2.9	Slightly magnetic	Class 3, 94–96
4	W-Ni-Fe alloy	18.5	0.667	795	115	690	100	450	65	345	50	3	32 R_C	5.0	2.6	Slightly magnetic	Class 4, 96–98

[a] 0.2% offset; R_C, Rockwell hardness scale C.
Source: *Metals Handbook* Desk Edition, ASM International, Materials Park, OH 44073-0002 (formerly The American Society for Metals, Metals Park, OH 44073), 1985.

TABLE 1-43
Representative properties for fiber reinforcement

Fiber	Typical fiber diameter		Density, ρ		Tensile strength, σ_{st}		Modulus of elasticity, E		Coefficient of thermal expansion, α		Thermal conductivity, K	
	$\times 10^{-3}$ mm	$\times 10^{-3}$ in	g/cm³	lb/in³	MPa	kpsi	GPa	Mpsi	μm/mK	μin/in°F	W/(m²K/m)	Btu/(ft²h°F)/in
E glass	10.2	0.4	2.48	0.092	3100	450	72.5	10.5	5.0	2.8	1680	7.5
S glass	10.2	0.4	2.43	0.090	4498	650	85	12.3				
970 S glass	10.2	0.4	2.46	0.091	5510	800	100	14.5				
Boron on tungsten	102.0	4	2.56	0.095	2756	400	415	60	5.0	2.8		
Graphite	5–10	0.2–0.4	1.43–1.75	0.053–0.066	1723–3445	250–500	241–689	35–100	2.7	1.5	13440	60
Beryllium	127	5	1.78	0.066	1240	180	310	45	11.5	6.4	19488	87
Silicon carbide on tungsten	102	4	3.40	0.126	2480	360	414	60	4.0	2.2	6496	29
Stainless steel	13	0.5	7.64	0.283	2852–4184	385–600	200	29	54	30	22400–38080	100–170
Asbestos	0.025–0.25	0.001–0.01	2.43	0.090	689–2067	100–300	172	25				
Aluminum	5	0.2	2.62	0.097	1378–2067	200–300	138–413	20–60	6.5	3.7		
Polyamide	5–13	0.2–0.5	1.11	0.041	827	120	2.8	0.4	81–90	45–50	381	1.7
Polyester	20.5	0.8	1.49	0.052	690	100	4	0.6	81–90	45–50	381	1.7

Courtesy: J. E. Ashton, J. C. Halpin, and P. H. Petit, *Primer on Composite Materials—Analysis*, Technomic Publishing Co., Inc., 750 Summer Street, Stanford, Conn. 06901, 1969.

TABLE 1-44
Designation, composition and mechanical properties of ferrous powder metallurgy structural steels

MPIF material designation[a]	MPIF chemical composition limits and ranges[b], %				MPIF density, ρ, g/cm^3	Condition[c]	Tensile strength, σ_{st}		Yield strength, σ_{sy}		Fatigue strength, σ_{sf}		Modulus of elasticity, E		Impact energy		Elongation in 25 mm (1 in), %	Apparent hardness	ASTM designation
	Fe	C	Cu	Ni			MPa	kpsi	MPa	kpsi	MPa	kpsi	GPa	Mpsi	J	ft-lbf			
								Iron and Carbon Steel											
F-0000	97.7–100	0.3 max			<6.0	AS	110	10	75	11	40	6	70	10.5	4.1	3	2	$10R_H$	B 310, Class A
F-0000	97.7–100	0.3 max			6.8–7.2	AS	205	30	150	22	80	11[d]	130	19	20	15	9	$15R_B$	
F-0000	97.7–100	0.3 max			7.2–7.6	AS	275	40	180	26	105	15[d]	160	23	34	25	15	$30R_B$	
								Steel											
F-0005	97.4–99.7	0.3–0.6			6.0–6.4	AS	170	25	140	20	65	10[d]	90	13	4.7	3.5	1.5	$20R_B$	B 310, Class B
F-0005	97.4–99.7	0.3–0.6			6.4–6.8	AS	220	32	160	23	85	12[d]	110	16	6.8	5.0	2.5	$45R_B$	
F-0005	97.4–99.7	0.3–0.6			6.4–6.8	HT	415	60	395	57	155	23[d]	110	16	–	–	0.5	$100R_B$	
F-0008	97.0–99.1	0.6–1.0			6.0–6.4	AS	240	35	205	30	90	13[d]	90	13	4.1	3.0	1.0	$50R_B$	B 310, Class C
F-0008	97.0–99.1	0.6–1.0			6.0–6.4	HT	400	58	–	–	150	22[d]	90	13	–	–	<0.5	$100R_B$	
F-0008	97.0–99.1	0.6–1.0			6.4–6.8	AS	290	42	250	36	100	14[d]	110	16	4.7	3.5	1.5	$65R_B$	
F-0008	97.0–99.1	0.6–1.0			6.4–6.8	HT	510	74	–	–	195	28[d]	110	16	–	–	<0.5	$25R_C$	
F-0008	97.0–99.1	0.6–1.0			6.8–7.2	AS	395	57	275	40	150	22[d]	130	19	9.5	7.0	2.5	$75R_B$	
F-0008	97.0–99.1	0.6–1.0			6.8–7.2	HT	650	94	625	91	245	36[d]	130	19	–	–	<0.5	$30R_C$	
								Iron Copper Steel											
FC-0200	93.8–98.5	0.3 max	1.5–3.9		<6.0	AS	160	23	115	17	60	9[d]	90	13	7.5	5.5	2.5	$80R_H$	
FC-0200	93.8–98.5	0.3 max	1.5–3.9		6.8–7.2	AS	255	37	160	23	95	14[d]	130	19	23	17	7	$30R_B$	
								Copper Steel											
FC-0205	93.5–98.2	0.3–0.6	1.5–3.9		6.4–6.8	AS	345	50	260	38	130	19[d]	110	16	7.5	5.5	1.5	$60R_B$	
FC-0205	93.5–98.2	0.3–0.6	1.5–3.9		6.4–6.8	HT	585	85	560	81	220	31[d]	110	16	–	–	<0.5	$30R_C$	
FC-0205	93.5–98.2	0.3–0.6	1.5–3.9		6.8–7.2	AS	425	62	310	45	160	24[d]	130	19	13	9.5	3.0	$75R_B$	
FC-0205	93.5–98.2	0.3–0.6	1.5–3.9		6.8–7.2	HT	690	100	655	95	260	38[d]	130	19	–	–	<0.5	$35R_C$	
FC-0208	93.1–97.9	0.6–1.0	1.5–3.9		<6.0	AS	225	33	205	30	85	13[d]	70	10.5	3.4	2.5	<0.5	$45R_B$	B 426. Grade 1
PC-0208	93.1–97.9	0.6–1.0	1.5–3.9		<6.0	HT	295	43	–	–	110	16[d]	70	10.5	–	–	<0.5	$95R_B$	
FC-0208	93.1–97.9	0.6–10	1.5–3.9		6.4–6.8	AS	415	60	330	48	155	23[d]	110	16	6.8	5.0	1.0	$70R_B$	
FC-0208	93.1–97.9	0.6–1.0	1.5–3.9		6.4–6.8	HT	550	80	–	–	210	30[d]	110	16	–	–	<0.5	$35R_C$	
FC-0208	93.1–97.0	0.6–1.0	1.5–3.9		6.8–7.2	AS	550	80	395	57	210	30[d]	130	19	11	8.0	1.5	$80R_B$	
FC-0208	93.1–97.9	0.6–1.0	1.5–3.9		6.8–7.2	HT	690	100	655	95	260	38[d]	130	19	–	–	<0.5	$40R_C$	
FC-0505	91.4–95.7	0.3–0.6	4.0–6.0		6.0–6.4	AS	345	50	290	42	130	19[d]	90	13	6.1	4.5	1.0	$60R_B$	
					6.4–6.8	AS	455	66	380	55	170	25[d]	116	16	6.8	5.0	1.5	$75R_B$	
FC-0508	91.0–95.4	0.6–1.0	4.0–6.0		6.0–6.4	AS	425	62	395	57	160	24[d]	90	13	4.7	3.5	1.0	$65R_B$	B 426, Grade 2
					6.0–6.4	HT	480	70	480	70	185	27[d]	90	13	–	–	<0.5	$30R_C$	
FC-0508	91.0–95.4	0.6–1.0	4.0–6.0		6.4–6.8	AS	515	75	480	70	195	29[d]	116	16	6.1	4.5	1.0	$85R_B$	
FC-0808	86.0–93.4	0.6–1.0	6.0–11.0		<6.0	AS	250	36	–	–	–	–	–	–	–	–	<0.5	$55R_B$	B 426, Grade 3

TABLE 1-44
Designation, composition and mechanical properties of ferrous powder metallurgy structural steels (*Cont.*)

MPIF material designation[a]	MPIF chemical composition limits and ranges[b], %				MPIF density, ρ, g/cm^3	Condition[c]	Tensile strength, σ_{st}		Yield strength, σ_{sy}		Fatigue strength, σ_{sf}		Modulus of elasticity, E		Impact energy		Elongation in 25 mm (1 in), %	Apparent hardness	ASTM designation
	Fe	C	Cu	Ni			MPa	kpsi	MPa	kpsi	MPa	kpsi	GPa	Mpsi	J	ft-lbf			
									Iron Copper										
FC-1000	87.2–90.5	0.3 max	9.5–10.5		<60	AS	205	30	–	–	–	–	–	–	–	–	0.5	70R_F	B 222, B439, Grade 3
									Iron Nickel										
FN-0200	92.2–99.0	0.3 max	2.5 max	1.0–3.0	6.4–6.8	AS	195	28	125	18	75	11	115	17	19	14	4	38R_B	B 484, Grade 1, Class A
FN-0200	92.2–99.0	0.3 max	2.5 max	1.0–3.0	7.2–7.6	AS	310	45	205	30	125	18	160	23	36	50	11	51R_B	
									Nickel Steel										
FN-0205	91.9–98.7	0.3–0.6	2.5 max	1.0–3.0	6.4–6.8	AS	255	37	160	23	105	15	115	17	14	10	3.0	50R_B	B 484, Grade 1, Class B
						HT	565	82	450	65	225	33	115	17	8.1	6	0.5	32R_C	
FN-0205	91.9–98.7	0.3–0–6	2.5 max	1.0–3.0	6.8–7.2	SS	345	50	215	31	140	20	145	21	24	18	3.5	70R_B	
						HT	760	110	605	88	305	44	145	21	22	16	1.0	42R_C	
FN-0208	91.9–98.7	0.3–0.6	2.5 max	1.0–3.0	6.4–6.8	AS	330	48	205	30	130	19	115	17	11	8	2.0	62R_B	B 484, Grade 1, Class C
						HT	690	100	650	94	275	40	115	17	8.1	6	0.5	34R_C	
FN-0208	91.6–98.7	0.6–0.9	2.5 max	1.0–3.0	7.2–7.6	AS	545	79	345	50	220	32	160	23	30	22	3.5	87R_B	
						HT	1105	160	1070	155	415	60	160	23	24	18	0.5	47R_C	
									Iron Nickel										
FN-0400	90.2–97.0	0.3 max	2.0 max	3.0–5.5	6.8–7.2.	AS	340	49	205	30	140	20	145	21	47	35	6	60R_B	B 484, Grade 2, Class A
FN-0400					7.2–7.6	AS	400	58	250	36	160	23	160	23	68	50	6.5	67R_B	
									Nickel Steel										
FN-0405	89.9–96.7	0.3–0.6	2.0 max	3.0–5.5	6.4–6.8	AS	310	45	180	26	125	18	115	17	14	10	3.0	63R_B	B 484, Grade 2, Class B
FN-0405	89.9–96.7	0.3–0.6	2.0 max	3.0–5.5	6.4–6.8	HT	770	112	650	94	310	45	115	17	8.1	6	0.5	27R_C	
FN-0405					7.2–7.6	AS	510	74	295	43	205	30	160	23	41	30	6.0	80R_B	
FN-0405					7.2–7.6	HT	1240	80	1060	154	450	65	160	23	19	14	1.5	44R_C	
FN-0408	89.6–96.4	0.6–0.9	2.0 max	3.0–5.5	6.4–6.8	AS	395	57	290	42	160	23	115	17	81	6	1.5	72R_B	B 484, Grade 2, Class C
					7.2–7.6	AS	640	93	470	68	255	37	160	23	22	16	4.5	95R_B	
									Iron Nickel										
FN-0700	87.7–94.0	0.3 max	2.0 max	6.0–8.0	6.8–7.2	AS	490	71	275	40	195	28	145	21	28	21	4	72R_B	B 484, Grade 3, Class A
FN-0700						AS	585	85	330	48	240	34	160	23	35	26	6	83R_B	
									Nickel Steel										
FN-0705	87.4–93.7	0.3–0.6	2.0 max	6.0–8.0	6.4–6.8	AS	370	54	240	35	150	22	115	17	12	9	2.0	69R_B	B 484, Grade 3, Class B
						HT	705	102	550	80	280	41	115	17	11	8	0.5	24R_C	
FN-0705	–				7.2–7.6	AS	620	90	390	57	250	36	160	23	33	24	5.0	90R_B	
						HT	1160	168	895	130	500	65	160	23	27	20	1.5	40R_C	
FN-0708(e)	87.1–93.4	0.6–0.9	2.0 max	6.0–8.0	6.8–7.2	AS	550	80	380	55	220	32	145	21	16	12	2.5	88R_B	B 484, Grade 3, Class C
					7.2–7.6	AS	655	95	455	60	260	38	160	23	22	16	3.0	96R_B	

TABLE 1-44
Designation, composition and mechanical properties of ferrous powder metallurgy structural steels (*Cont.*)

MPIF material designation[a]	MPIF chemical composition limits and ranges[b], %				MPIF density, ρ, g/cm^3	Condition[c]	Tensile strength, σ_{st}		Yield strength, σ_{sy}		Fatigue strength, σ_{sf}		Modulus of elasticity, E		Impact energy		Elongation in 25 mm (1 in), %	Apparent hardness	ASTM designation
	Fe	C	Cu	Ni			MPa	kpsi	MPa	kpsi	MPa	kpsi	GPa	Mpsi	J	ft-lbf			
									Infiltered Steel										
FX-1005 (e)	80.5–91.7	0.3–0.6	8.0–14.9	–	7.2–7.6	AS	570	83	440	64	–	–	135	20	19	14	4.0	$75R_B$	
						HT	830	120	740	107	–	–	135	20	9.5	7.0	1.0	$35R_C$	
FX-1008 (e)	80.1–91.4	0.6–1.0	8.0–14.9	–	7.2–7.6	AS	620	90	515	75	–	–	135	20	16.0	12	2.6	$80R_B$	
						HT	895	130	775	105	–	–	135	20	9.5	7.0	60.5	$40R_C$	
FX-2000 (e)	70.7–85.0	0.3 max	15.0–25.0	–	7.2–7.6	AS	450	65	–	–	–	–	–	–	20	15	1.0	$60R_B$	B 303, Class A
FX-2005 (e)	70.4–84.7	0.3–0.7	15.0–25.0	–	–	AS	515	75	345	50	–	–	125	18	12.9	9.5	1.5	$75R_B$	B 303, Class B
						HT	790	115	655	95	–	–	125	18	8.1	6.0	<0.5	$30R_C$	
FX-2008 (e)	70.0–84.4	0.6–1.0	15.0–25.0	–	7.2–7.6	AS	585	85	515	75	–	–	125	18	14	10	1.0	$80R_B$	B 303, Class C
						HT	860	125	740	107	–	–	125	18	6.8	5.0	<0.5	$92R_C$	

[a] Designation listed are nearest comparable designations, ranges and limits may vary slightly between comparable designations.
[b] Metal Powder Industries Federation (MPIF) standards require that the total amount of all other elements be less then 2.0%, except that the total amount of other elements must be less than 4.0% infiltered steel.
[c] AS, as sintered; SS, sintered and sized; HT, heat treated, typically austenized at 870°C (1600°F), oil quenched and tempered at 200°C (400°F).
[d] Estimated as 38% of tensile strength.
[e] x indicates infiltered steel;
[f] Unnotched Charpy test; R_B = hardness Rockwell B scale; R_C = hardness Rockwell scale C; R_F = hardness Rockwell F scale.
Source: *Metals Handbook* Desk Edition, ASM International, 1985, Materials Park, OH 44073-0002 (formerly The American Society for Metals, Metals Park, OH 44073, 1985).

TABLE 1-45
Nominal composition of some of structural alloys at subzero temperatures (i.e., at cryogenic temperatures)

Alloy designation	UNS No.	Nominal composition, %															
		Si	Cu	Mn	Mg	Cr	V	Ti	S_{max}	Ni	Fe	C	Mo	Al	N	P_{max}	Other
Aluminum Alloys																	
2014		0.8	4.4	0.8	0.5												
2219			6.3	0.3			0.1	0.06									0.18 Zr
5083				0.7	4.4	0.15											
6061		0.6	0.28		1.0	0.20											
7039		0.1	0.05	0.25	2.8	0.20		0.05			0.2						3.0 Zn
7075			1.6		2.5	0.23											5.6 Zn
Copper and Copper Alloys																	
C 17200 Beryllium copper			98.1														1.9 Be
C 26000 Cartridge brass, 70%			70.0														30 Zn
High Nickel Alloys																	
Hastelloy C	No 5500	0.5	0.1	0.5		15.5				Rem	5.5	0.07	16				1.5Co, 4.0W
Inconel 706	No 9706	0.1		0.10		16		1.7		39–44	Rem	0.04		0.35 max			3.0 (Nb+Ta)
Inconel 718	No 7718					18.6		0.9		Rem	18.5	0.04	3.1.	0.4			5.0 Nb
Inconel X 750				0.7		15.0		2.5		Rem	6.8	0.04		0.8			0.85 Nb
Austenitic Stainless Steel																	
304	S 30400 (AISI 304)	0.03		2.0 max		18–20			1.0	8–12		0.08 max				0.045	
310S	S 31008 (AISI 310)	0.03		2.0 max		24–26			1.5	19–22		0.08 max				0.045	
347	S34700 (AISI 347)	0.03		2.0 max		17–19			1.0	9–13		0.08 max				0.045	(10C)(Nb+Tb)
Pyromet 538				9.5		20			0.15	7.0		0.02			0.2		
Nitromic 70				8.0		17			3.5	8.5		0.07			0.1		
Kromare 5S		0.05		9.3		15.5	0.16		0.05	23		0.03	2.2	0.02	0.17	0.005	0.008Zr, 0.016B
A 286				1.4		15	0.3	2.0	0.4	2.6		0.05	1.25	0.20	2.0		0.005B
Titanium and Titanium Alloys																	
								O_{max}	Sn	H_{max}							
Ti-5Al-3.5Sn								0.20	2.0–3.0	0.02	0.5 max	0.15 max		4.0–6.0	0.07 max		$0.30M_{max}$
Ti-5Al-2.5Sn (EL1)								0.12	2–3	0.018		0.2 max	0.08 max	4.7–5.6	0.05 max		
Ti-6Al-4V (EL1)							3.5–4.5	0.13		0.015 max	0.15 max	0.15 max	0.08 max	5.5–6.5	0.05 max		

TABLE 1-46
Typical tensile properties and fracture toughness of structural alloys at sub-zero temperature (i.e. cryogenic temperature)

Alloy and condition	Temperature °C	Temperature °F	Form	Tensile strength, σ_{st} MPa	Tensile strength, σ_{st} kpsi	Yield strength, σ_{sy} MPa	Yield strength, σ_{sy} kpsi	Elongation, %	Reduction in area, %	Room temperature yield strength, σ_{syr} MPa	Room temperature yield strength, σ_{syr} kpsi	Fracture toughness K_{IC} (J) MPa$\sqrt{m}$	Fracture toughness K_{IC} (J) kpsi$\sqrt{m}$	Specimen design	Orientation	Uses and remarks
									Aluminum Alloys							
2014-T651 temper, L.O.	24	75	Plate	310	44.9	290	42.2	16	50	432	62.7	23.2	21.2	Bend	T.L.	Possess high strength at room temperature and sub-zero temperature.
	−196	−320		400	58.3	335	48.9	23	48			28.7	26.1			
	−269	−452		485	70.1	380	55.0	26	42							
2219-T62 temper, L.O.	24	75	Sheet	415	60.2	290	41.9	10		382	55.4			Bend	T.S.	High toughness at room and sub-zero temperatures, used for liquid oxygen and liquid hydrogen tanks for space shuttle.
	−196	−320		460	66.5	345	50.2	11								
	−253	−423		545	78.8	370	53.6	14								
	−269	−452		650	94.5	390	55.2	23								
2219-T87 temper, L.O.	24	75	Plate	465	67.8	380	56.3	11	26	382	55.4	39.9 (28.8)	36.3 (26.2)	Bend, CT	T.S.	Fracture toughness values in parentheses refer to specimen design C.T.
	−196	−320		585	84.5	460	66.6	13	25			46.5 (34.5)	42.4 (31.4)			
	−253	−423		675	97.6	490	71.0	14	21			52.5 (37.2)	48.0 (34.0)			
	−269	−452		675	97.6	510	74.2	15	23							
5083-H113, L.O.	24	75	Plate	335	48.5	235	34.2	15	23	142*	20.6*	27.0*	24.6*	C.T.	T.L.	* Fracture toughness values are for 5083-O, i.e., K_{IC} (J). 5083-O is not heat treatable and used in annealed (O) condition. Used to build liquefied natural gas (LNG) spherical tanks in ships.
	−196	−320		465	67.1	27.5	39.6	31	31			43.4[b]	39.5[b]			
	−253	−423		620	90.0	305	43.9	30	24							
	−269	−452		590	85.8	280	40.5	29	28			48.0*	43.7*			
6061-T6 temper, L.O.	24	75	Sheet	320	46.3	290	42.2	12								Aluminum alloys 6061 in the T6 temper have the same strength and ductility at both room and sub-zero temperatures.
	−196	−320		425	61.8	340	49.6	18								
	−253	−423		495	72.0	365	52.6	26								
6061, T651 temper, L.O.	24	75	Plate	310	44.9	290	42.2	16	50	289	41.9	29.1	26.5	Bend	T.L.	
	−196	−320		400	58.3	335	48.9	23	48			41.6	37.9			
	−269	−452		485	70.1	380	55.0	26	42							
7039, T6 temper, L.O.	24	75	Sheet	455	66.0	410	59.8	11		381	55.3	32.3	29.4	Bend	T.L.	7039 has good combination of strength and fracture toughness at room and at −196 °C
	−196	−320		575	83.2	495	72.0	15				33.5	30.5			
	−253	−423		665	96.2	535	77.4	14								
7075, T 651 temper, L.O.	24	75	Plate	580	84.2	530	77.1	10	14	536	77.7	22.5	20.5	Bend	T.L.	Used for plate of 7075-T6 for the inner tank skirt between the liquid oxygen and liquid hydrogen sections of the external tank of the space shuttle.
	−196	−320		705	102	650	94.4	7	10			27.6	25.1			
	−269	−452		825	120	770	112	6	9							
7075, T 7351 temper, L.O.	24	75	Plate	525	76.2	455	66.2	10	22	403	58.5	35.9	32.7			Improves ductility and notch toughness of 7075 alloy at cryogenic temperature processing it to the T7351 temper.
	−196	−320		675	98.2	570	82.5	11	14			32.1	29.2			
	−269	−452		760	110	605	88.1	11	12							

TABLE 1-46
Typical tensile properties and fracture toughness of structural alloys at sub-zero temperature (i.e. cryogenic temperature) (*Cont.*)

Alloy and condition	Temperature °C	Temperature °F	Form	Tensile strength, σ_{st} MPa	Tensile strength, σ_{st} kpsi	Yield strength, σ_{sy} MPa	Yield strength, σ_{sy} kpsi	Elongation, %	Reduction in area, %	Room temperature yield strength, σ_{syr} MPa	Room temperature yield strength, σ_{syr} kpsi	Fracture toughness K_{IC} (J) MPa$\sqrt{m}$	Fracture toughness K_{IC} (J) kpsi$\sqrt{m}$	Specimen design	Orientation	Uses and remarks
Copper and Copper Alloys (L.O.)																
C26000 03 temper ($\frac{3}{4}$ hard)	24	75	Bar	655	95.2	420	61.0	14	58							Tensile and fatigue properties of copper alloys increase as the testing temperature decreased. Used in stabilizers, components of the windings in superconducting magnets, solenoids and power cables at super temperatures.
	−196	−320		805	117	475	68.5	28	63							
	−253	−423		910	132	505	73.5	32	58							
C17200 TD02 temper (Solution treated, cold worked to $\frac{1}{2}$ hard)	24	75	Sheet	620	90	550	80	15								
	−196	−320		805	117	690	100	37								
	−253	−423		945	137	750	109	45								
High-Nickel Alloys																
Hastelloy C, cold rolled 20%, L.O.	24	75	Sheet	1140	165	1000	145	13								† Fracture toughness values refer to K_{IC} (J). These high nickel alloys exhibit excellent combinations of strength, toughness, and ductility over the entire range of subzero temperatures. Used in energy related equipments such as superconducting motors and generators.
	−196	−320		1520	220	1280	186	32								
	−253	−423		1740	252	1380	200	33								
Inconel 706 (VIM-VAR) STDA	24	75	Forged billets	1260	183	1050	152	24	33	1065	154	133†	121†	C.T.	C.R.	
	−196	−320		1570	228	1200	174	29	33							
	−269	−452		1680	243	1250	181	30	33			157†	143†			
Inconel 718, L.O. [Aged $\frac{3}{4}$h at 982°C (1800°F)]	24	75	Bar	1410	204	1170	170	15	18	1170*	170*	96.4*	87.8*	C.T.		* Refer to STDA alloy. The fatigue strengths of high-nickel alloys are higher at cryogenic temperature than at room temperature.
	−196	−320		1650	239	1340	197	21	20			103	94			
	−269	−452		1810	263	1410	204	21	20			112	102			
GTA weld in Inconel X-750 sheet, X-750 filler metal, aged 20 h at 700°C	24	75	Sheet	1290	187	860	125	22	–	825	120			C.T.		# K_{IJ} (J), fusion zone, gas tungsten arc weld (GTA). ‡ K_{IC} (J), vacuum electron beam weld (VEB). Used in construction of storage tanks for liquefied hydrocarbon gases, structures and machineries in cold regions.
	−196	−320		1540	224	945	137	50								
	−253	−423		1660	241	1020	148	28					134#	122#		
	−269	452	Weldment										176‡	160‡		
Ferritic Nickel Steels																
A645 (5Ni Steel), L.O. quenched, tempered, reversion annealed.	24	75	Plate	715	104	530	76.8	32	72							
	−162	−260		930	135	570	82.9	2.8	68	535	77.5	196	178			
	−196	−320		1130	164	765	111	30	62			87.1	79.3			
	−269	−452										58.4	53.2			
Austenitic Stainless Steels																
304 annealed, L.O.	24	75	Sheet	660	95.5	295	42.5	75	–							Austenitic stainless steels are used extensively for subzero temperature applications to −269°C.
	−196	−320		1650	236	425	55.0	42	–							
	−269	−452		1700	247	570	82.5	30	–							
310 S annealed, T.O.	24	75	Forging	330	48	580	84	40	76	260	37.9				T.L.	† Welding process, SMA, Filler, 310S. ᵇ In weld fusion zone, as welded.
	−196	−320		670	97	1070	155	46	67							

TABLE 1-46
Typical tensile properties and fracture toughness of structural alloys at sub-zero temperature (i.e. cryogenic temperature) *(Cont.)*

Alloy and condition	Temperature °C	Temperature °F	Form	Tensile strength, σ_{st} MPa	Tensile strength, σ_{st} kpsi	Yield strength, σ_{sy} MPa	Yield strength, σ_{sy} kpsi	Elongation, %	Reduction in area, %	Room temperature yield strength, σ_{syr} MPa	Room temperature yield strength, σ_{syr} kpsi	Fracture toughness K_{IC} (J) MPa$\sqrt{m}$	Fracture toughness K_{IC} (J) kpsi$\sqrt{m}$	Specimen design	Orientation	Uses and remarks
310 S annealed, T.O.	−269	−452	Plate,	825	120	1105	160	26	24	–	–	259	236			Strength of these steels is increased by
	−269	−452	Weldment[†,a]									116[b]	106[b]			rolling or cold drawing at −196°C.
347 annealed, L.O.	24	75	Sheet	650	94	255	57	52								Used in liquid hydrogen and liquid
	−196	−320		1365	198	420	61	47								oxygen tank construction in Atlas and
	−253	−423		1610	234	435	63	35								Centaur rockets. Type 304 stainless
304 hard cold rolled, L.O.	24	75	Sheet	1320	191	1190	173	3								steels used in piping, tubing, and valves.
	−196	−320		1900	276	1430	208	29								Used in transfer of oxygen for storage
	−253	−423		2010	292	1560	226	2								tanks. Cast steels are used for Bubble
310, 75% cold reduced, L.O.	24	75	Sheet	1180	171	1100	160	3								Chambers, and for cylindrical magnet
	−196	−320		1720	249	1540	223	10								tubes for superconducting magnets.
	−253	−423		2000	290	1790	259	10								‡ Welding process: GTA.
Pyromet 538 annealed	24	75	Plate	415	60	725	105	51	74	340	49	–	–		T.L.	Filler: Pyromet 538.
Pyromet 538 STQ	−196	−320		1005	146	1455	211	48	61			275[a]	250[a]			Fracture toughness values refer to
																Pyromet 538 STQ.
	−269	452	Plate	1240	180	1650	239	31	24			181	165 at −269°C (−452°F)			[a] Shielded metal arc weld.
			Weldment[c]									82[b]	74[b] at −269°C (−452°F)			[b] In weld fusion zone, as welded.
			Weldment[a]									175[b]	159[b] at −269°C (−452°F)			[c] Gas tungsten arc weld.
Nitronic 60, annealed, L.O.	24	75	Bar	750	109	400	58.1	66	79							
	−196	−320		1500	218	695	101	60	66							
	−253	−452		1410	204	860	125	24	27							
Kromarc 58 annealed	24	75	Plate	495	72	915	133	36	61						T.L.	* Welding process: GTA.
plate, tested as welded*, L.O.	−196	−320		855	124	1325	192	46	41							Filler: Kromrc 58.
Kromarc 58 STQ	−269	−452	Plate	1060	154	1440	209	33	40	370	53.8	214	195			Kromrc is used for structural
			Weldament[c]									155[b]	141[b]			appliances.
A-286 annealed sheet, welded	24	75	Sheet	600	87	860	125	11								Prototype super-conducting
and age hardened, L.O.	−196	−320		745	108	1145	166	16								generators
A-286 STA	−253	−423		870	126	1280	186	15								
	24	75	Bar							610	88.2	125	114		T.S.	A-286 alloy develops good strength,
	−196	−320										123	112			with good ductility and notch
	−269	−452										118	107			toughness in the cryogenic
	24	75	Plate							820	119	161	146		L.T.	temperature range.
	−269	−452										179	163			
	−269	−452	Weldament[c]									247[b]	225[b]			

TABLE 1-46
Typical tensile properties and fracture toughness of structural alloys at sub-zero temperature (i.e. cryogenic temperature) (*Cont.*)

Alloy and condition	Temperature °C	Temperature °F	Form	Tensile strength, σ_{st} MPa	Tensile strength, σ_{st} kpsi	Yield strength, σ_{sy} MPa	Yield strength, σ_{sy} kpsi	Elongation, %	Reduction in area, %	Room temperature yield strength, σ_{syr} MPa	Room temperature yield strength, σ_{syr} kpsi	Fracture toughness K_{IC} (J) MPa$\sqrt{m}$	Fracture toughness K_{IC} (J) kpsi$\sqrt{m}$	Specimen design	Orientation	Uses and remarks
								Titanium and Titanium Alloys								
Ti-5 Al-2.5 Sn, nominal interstitial annealed, L.O.	24	75	Sheet/ Plate	850	123	795	115	16		875	127	71.8	65.4	C.T.	L.T.	Ti-5Al-2.5Sn and Ti-6Al-4V, titanium alloys have high strength to weight ratio at cryogenic temperatures and preferred alloys at temperatures of alloys at temperatures of −196°C to −269°C (−320°F to −452°F). Used in spherical pressure vessels in Atlas and Centaur Rockets, the Apollo and Saturn launch Boosters and Lunar Modules. Should not be used at cryogenic temperatures for storage or transfer of liquid oxygen, since the condensed oxygen will cause ignition during abrasion.
	−196	−320		1370	199	1300	188	14		875	127	53.4	48.6	Bend	L.T.	
	−253	−423		1700	246	1590	231	7		875	127			Bend	L.T.	
Ti-5 Al-2.5 Sn (EL1) annealed, L.O.	24	75	Sheet/ Plate	800	116	740	107	16		705	102			C.T.	L.T.	
	−196	−320		1300	188	1210	175	16		705	102	−111	−101	C.T.	L.T.	
	−253	−423		1570	228	1450	210	10				89.6	81.5	Bend	L.T.	
Ti-6 Al-4 V (EL1) as forged	24	75	Forging	970	141	915	133	14	40	830	120			C.T.	L.T.	
	−196	−320		1570	227	1480	214	11	31			610	55.5			
	−253	−423		1650	239	1570	227	11	24							
	−269	−452										54.1	49.2			

AAM, air arc melted. C.T., compact toughness. GMA, gas metal arc welding process. GTA, gas tungsten arc weld. L.O., longitudinal orientation. S.T., solution treated. STDA, solution treated and double aged. VAR, vacuum arc remelted. VEB, vacuum electron beam weld. VIM, vacuum induction melted. W, weld.
Source: *Metals Handbook* Desk Edition, ASM International, 1985, Materials Park, OH 44073-0002 (formerly The American Society for Metals, Metals Park, OH 44073, 1985).

TABLE 1-47
Typical properties of wood[a] of clear material of section 50 mm × 50 mm (2 in × 2 in), as per ASTM D143

Kind of wood	Specific gravity G_m ovendry volume	Density[b], D		Shrinkage[c]		Static bending: Modulus of rupture		Static bending: Modulus of elasticity, E		Maximum[d] crushing strength, σ_{scr}		Compression[a] proportionality limit, σ_{scp}		Tensile strength[e,f], σ_{st}		Impact bending in a drop of 222 N (50 lbf) hammer		Shear[d] strength, τ_s		Hardness average of R and T
		kN/m³	lbf/ft³	Rad	Tan	MPa	kpsi	GPa	Mpsi	MPa	kpsi	MPa	kpsi	MPa	kpsi	mm	in	MPa	kpsi	
Softwoods																				
Cedar, western red	0.34	3.62	23	2.4	5.0	51.71	7.5	7.65	1.11	31.44	4.56	3.17	0.46	1.52	0.22	430	17	5.93	0.86	350
Cypress	0.48	5.06	32	3.8	6.2	73.09	10.6	9.93	1.44	43.85	6.36	5.38	0.78	1.86	0.27	864	34	6.89	1.00	510
Douglas fir, coast region	0.51	5.37	34	4.8	7.6	85.50	12.4	13.45	1.95	49.92	7.24	5.52	0.80	2.34	0.34	787	31	8.00	1.16	710
Hemlock, eastern	0.43	4.42	28	3.0	6.8	61.37	8.9	8.28	1.20	37.30	5.41	4.48	0.65	–	–	533	21	7.31	1.06	500
Hemlock, western	0.44	4.6	29	4.3	7.0	77.11	11.3	11.31	1.64	49.02	7.71	3.79	0.55	2.34	0.34	660	26	6.62	1.25	540
Larch, western	0.59	6.00	38	4.5	9.1	90.33	13.10	12.90	1.87	52.68	7.64	6.76	0.98	3.00	0.43	889	35	9.38	1.36	830
Pine, red	0.47	4.90	31	3.8	7.2	75.85	11.00	11.24	1.63	41.85	6.07	4.14	0.60	3.17	0.46	660	26	8.76	1.21	560
Pine, ponderosa	0.42	4.42	28	3.9	6.3	64.81	9.4	8.90	1.29	36.68	5.32	4.00	0.58	2.90	0.42	483	19	7.79	1.13	460
Pine, eastern white	0.37	3.62	24	2.1	6.1	59.30	8.6	8.55	1.24	33.10	4.80	3.03	0.44	2.18	0.31	457	18	6.21	0.90	380
Pine, western white	0.42	4.27	27	2.6	5.3	66.88	9.7	10.10	1.46	34.75	5.04	3.24	0.47	–	–	584	23	7.17	1.04	420
Redwood	0.42	4.42	28	2.6	4.4	69.00	10.0	9.24	1.34	42.40	6.15	4.83	0.70	1.66	0.24	483	19	6.48	0.94	480
Spruce, sitka	0.42	4.42	28	4.3	7.5	70.33	10.2	10.83	1.57	38.68	5.61	4.00	0.58	2.55	0.37	635	25	7.93	1.15	510
Spruce, white	0.45	4.42	28	4.7	8.2	67.57	9.8	9.24	1.34	37.72	5.47	3.17	0.46	2.48	0.36	508	20	7.45	1.08	480
Hardwoods																				
Ash, white	0.64	6.64	42	4.9	7.9	106.18	15.4	12.20	1.77	51.10	7.41	8.00	1.16	6.48	0.94	1092	43	13.45	1.95	1320
Beech	0.67	7.11	45	5.1	11.0	102.74	14.9	11.86	1.72	50.33	7.30	6.96	1.01	7.00	1.01	1041	41	13.86	2.01	1300
Birch, yellow	0.66	6.80	43	7.2	9.2	114.46	16.6	14.55	2.11	56.33	8.17	6.69	0.97	6.34	0.92	1397	55	13.00	1.88	1260
Cherry, black	0.53	5.53	35	3.7	7.1	84.80	12.3	10.27	1.49	48.95	7.11	4.76	0.69	3.86	0.56	737	29	11.72	1.70	950
Cottonwood, eastern	0.43	4.42	28	3.9	9.2	58.60	8.5	9.45	1.37	33.85	4.91	2.55	0.37	4.00	0.58	508	20	6.41	0.93	430
Elm, American	0.55	5.53	35	4.2	9.5	81.36	11.8	9.24	1.34	38.06	5.52	4.76	0.69	4.55	0.66	990	39	10.41	1.51	830
Elm, rock	0.66	7.00	44	4.8	8.1	102.05	14.8	10.62	1.54	48.60	7.05	8.48	1.23	–	–	1422	56	13.24	1.92	1320
Sweetgum	0.55	5.69	36	5.4	10.2	86.19	12.5	11.31	1.64	43.58	6.32	4.28	0.62	5.24	0.76	813	32	11.03	1.6	850
Hickory. shagbark	0.77	7.90	50	7.0	10.5	139.28	20.2	14.89	2.16	63.50	9.21	12.14	1.76	–	–	1702	67	16.75	2.43	–
Mahogany (swietenia spp)	0.51	5.37	34	3.5	4.8	79.02	11.46	10.34	1.50	46.88	6.80	7.58	1.10	5.17	0.75	–	–	8.48	1.23	800
Maple, sugar	0.68	6.95	44	4.9	9.5	108.94	15.80	12.62	1.83	53.98	7.83	10.14	1.47	–	–	483	19	16.06	2.33	1450
Oak, red, northern	0.66	6.95	44	4.0	8.2	98.60	14.30	12.55	1.82	46.61	6.76	7.00	1.01	5.52	0.80	1092	43	12.27	1.78	1290
Oak, white	0.71	7.58	48	5.3	9.0	104.80	15.20	12.27	1.78	51.30	7.44	7.38	1.07	5.52	0.80	940	37	13.80	2.00	1360
Tupelo, black	0.55	5.53	35	4.4	7.7	66.20	9.60	8.27	1.20	38.06	5.52	6.41	0.93	3.44	0.50	559	22	9.24	1.34	810
Yellow poplar	0.45	4.58	29	4.2	7.6	69.64	10.10	10.89	1.58	38.20	5.54	3.45	0.50	3.72	0.54	610	24	8.21	1.19	540
Walnut, black	0.56	6.00	38	5.2	7.1	100.67	14.6	11.58	1.68	52.26	7.58	7.00	1.01	4.76	0.69	864	34	9.45	1.37	1010

[a] Seasoned wood at 12% moisture. [b] Seasoned wood at 12% moisture content. [c] Percent from green to ovendry condition based on dimensions when green. [d] Parallel to grain. [e] Perpendicular to grain. [f] Height of drop of 222 N (50 lbf) hammer for failure, mm (in). Rad, radially. Tan, tangentially. [g] Tensile strength parallel to grain may be taken as equal to modulus of rupture in bending.
Source: Extracted from *Wood Handbook* and the U.S. Forest Products Laboratory.

TABLE 1-48
Mechanical and physical properties of typical dense[a] pure refractories

Material	Density		Modulus of rupture, σ_{sr}				Modulus of elasticity, E		Linear coefficient[b] of expansion, α		Thermal conductivity, K at				Specific heat, c		Fusion point		Thermal stress resistance	Electrical resistivity, Ω cm[c]	
			21°C	70°F	982°C	1800°F	21°C	70°F			100°C	982°C	212°F	1800°F						70°C (21°C)	1800°F (982°C)
	g/m^3	lb/in^3	MPa	kpsi	MPa	kpsi	GPa	Mpsi	µm/m°C	µin/in°F	W/(m^2 °C/m)		Btu/(ft^{2}h°F/in)		kJ/kg°C	Btu/lb$_m$ °F	°C	°F			
Alumina (Al_2O_3)	3.97	0.143	689	100	414	60	365	53	9.0	5.0	47040	12320	210	55	1.09	0.26	2032	3690	Good	$>10^{14}$	10^7
Beryllium oxide (BeO)	3.03	0.110	138	20	69	10	310	45	8.8	4.9	324800	29120	1450	130	2.09	0.50	2571	4660	Very good	$>10^{14}$	10^8
Magnesium oxide (MgO)	3.58	0.129	97	14	83	12	214	31	13.5	7.5	53760	10528	240	47	1.05	0.25	2799	5070	Poor	$>10^{14}$	10^7
Thoria (ThO_2)	10.00	0.361	83	12	48	7	145	21	9.0	5.0	13888	4480	62	20	0.25	0.06	3049	5520	Fair	$>10^{14}$	10^5
Zirconia (ZrO_2)	5.60	0.202	138	20	103	15	152	22	9.9	5.5	33600	29120	15	15	0.59	0.14	2538	4600	Fair	10^8	500
Uranium oxide (UO_2)	10.96	0.396	83	12			172	25	10.1	5.6	12992	4480	58	20	0.25	0.06	2799	5070	Fair		
Silicon carbide (SiC)	3.22	0.116	166	24	166	24	469	68	4.0	2.2	87360	32480	390	145	0.84	0.20	2760	5000[d]	Excellent	10	
Boron carbide (BC)	2.52	0.097	345	50	276	40	290	42	4.5	2.5	44800	32480	200	145	1.51	0.36	2449	4440	Good	0.5	
Boron nitride (BN)	2.25	0.0813	48	7	7	1	83	12	4.7	2.6	33600	29120	150	130	1.63	0.39	2760	5000	Good	10^{10}	104
Molybdenum silicide ($MoSi_2$)	6.20	0.224	689	100	276	40	345	50	9.2	5.1	49280	22400	220	100	0.46	0.11	2143	3890	Good	10^{-6}	
Carbon (C)	2.22	0.0802	21	3	28	4	14	2	4.0	2.2	194880	64960	870	290	1.42	0.34	3871	7000	Good	10^{-3}	10^{-2}

[a] Porosity: 0 to 5%.
[b] Between 20°C (65°F) and 982°C (1800°F).
[c] Multiply the values by 0.393 in to obtain electrical resistivity in units of Ω-in.
[d] Stabilized.

Courtesy: Extracted from *Mark's Standard Handbook for Mechanical Engineers*, 8th edition, McGraw-Hill Book Company, New York, 1978, and Norton *Refractories*, 3rd edition, Green and Stewart, ASTM Standards on *Refractory Materials Handbook* (Committee, C-8), 1.

REFERENCES

1. Datsko, J., *Material Properties and Manufacturing Process*, John Wiley and Sons, New York, 1966.
2. Datsko, J. *Material in Design and Manufacturing*, Malloy, Ann Arbor, Michigan, 1977.
3. *ASM Metals Handbook*, American Society for Metals, Metals Park, Ohio, 1988.
4. Machine Design, 1981 Materials Reference Issue, Penton/IPC, Cleveland, Ohio, Vol. 53, No. 6, March 19, 1981.
5. Lingaiah, K., *Machine Design Data Handbook*, Vol. II (*SI and Customary Metric Units*), Suma Publishers, Bangalore, India, 1986.
6. Lingaiah, K., and B. R. Narayana Iyengar, *Machine Design Data Handbook*, Vol. I (*SI and Customary Metric Units*), Suma Publishers, Bangalore, India, 1986.
7. *Technical Editor Speaks*, the International Nickel Company, New York, 1943.
8. Shigley, J. E., *Mechanical Engineering Design*, Metric Edition, McGraw-Hill Book Company, New York, 1986.
9. Deutschman, A. D., W. J. Michels, and C. E. Wilson, *Machine Design—Theory and Practice*, Macmillan Publishing Company, New York, 1975.
10. Juvinall, R. C., *Fundaments of Machine Components Design*, John Wiley and Sons, New York, 1983.
11. Lingaiah, K., and B. R. Narayana Iyengar, *Machine Design Data Handbook*, Engineering College Co-operative Society, Bangalore, India, 1962.
12. Lingaiah, K., *Machine Design Data Handbook*, Vol. II (*SI and Customary Metric Units*), Suma Publishers, Bangalore, India, 1981 and 1984.
13. Lingaiah, K., and B. R. Narayana Iyengar, *Machine Design Data Handbook*, Vol. I (*SI and Customary Metric Units*), Suma Publishers, Bangalore, India, 1983.
14. *SAE Handbook*, 1981.
15. Lessels, J. M., *Strength and Resistance of Metals*, John Wiley and Sons, New York, 1954.
16. Siegel, M. J., V. L. Maleev, and J. B. Hartman, *Mechanical Design of Machines*, 4th edition, International Textbook Company, Scranton, Pennsylvania, 1965.
17. Black, P. H., and O. Eugene Adams, Jr., *Machine Design*, McGraw-Hill Book Company, New York, 1963.
18. Niemann, G., *Maschinenelemente*, Springer-Verlag, Berlin, Erster Band, 1963.
19. Faires, V. M., *Design of Machine Elements*, 4th edition, Macmillan Company, New York, 1965.
20. Nortman, C. A., E. S. Ault, and I. F. Zarobsky, *Fundamentals of Machine Design*, Macmillan Company, New York, 1951.
21. Spotts, M. F., *Design of Machine Elements*, 5th edition, Prentice-Hall of India Private Ltd., New Delhi, 1978.
22. Vallance, A., and V. L. Doughtie, *Design of Machine Members*, McGraw-Hill Book Company, New York, 1951.
23. Decker, K.-H., *Maschinenelemente, Gestalting und Bereching*, Carl Hanser Verlag, Munich, Germany, 1971.
24. Decker, K.-H., and Kabus, B. K., *Maschinenelemente-Aufgaben*, Carl Hanser Verlag, Munich, Germany, 1970.
25. ISO and BIS standards.
26. *Metals Handbook*, Desk Edition, ASM International, Materials Park, Ohio, 1985 (formerly the American Society for Metals, Metals Park, Ohio, 1985).
27. Edwards, Jr., K. S., and R. B. McKee, *Fundamentals of Mechanical Components Design*, McGraw-Hill Book Company, New York, 1991.
28. Shigley, J. E., and C. R. Mischke, *Standard Handbook of Machine Design*, 2nd edition, McGraw-Hill Book Company, New York, 1996.
29. *Structural Alloys Handbook*, Metals and Ceramics Information Center, Battelle Memorial Institute, Columbus, Ohio, 1985.
30. Wood Handbook and U. S. Forest Products Laboratory.
31. SAE J1099, Technical Report of Fatigue Properties.
32. Ashton, J. C., ■. Halpin, and P. H. Petit, *Primer on Composite Materials-Analysis*, Technomic Publishing Co., Inc., 750 Summer Street, Stanford, Conn 06901, 1969.
33. Baumeister, T., E. A. Avallone, and T. Baumeister III, *Mark's Standard Handbook for Mechanical Engineers*, 8th edition, McGraw-Hill Book Company, New York, 1978.
34. Norton, *Refractories*, 3rd edition, Green and Stewart, ASTM Standards on Refractory Materials Handbook (Committee C-8).

BIBLIOGRAPHY

Black, P. H., and O. Eugene Adams, Jr., *Machine Design*, McGraw-Hill Book Company, New York, 1983.
Decker, K.-H., *Maschinenelemente, Gestalting und Bereching*, Carl Hanser Verlag, Munich, Germany, 1971.
Decker, K.-H., and Kabus, B. K., *Maschinenelemente-Aufgaben*, Carl Hanser Verlag, Munich, Germany, 1970.
Deutschman, A. D., W. J. Michels, and C. E. Wilson, *Machine Design—Theory and Practice*, Macmillan Publishing Company, New York, 1975.
Faires, V. M., *Design of Machine Elements*, 4th edition, McGraw-Hill Book Company, New York, 1965.
Honger, O. S. (ed.), (*ASME*) *Handbook for Metals Properties*, McGraw-Hill Book Company, New York, 1954.
ISO standards.
Juvinall, R. C., *Fundaments of Machine Components Design*, John Wiley and Sons, New York, 1983.
Lessels, J. M., *Strength and Resistance of Metals*, John Wiley and Sons, New York, 1954.
Lingaiah, K., and B. R. Narayana Iyengar, *Machine Design Data Handbook*, Engineering College Co-operative Society, Bangalore, India, 1962.
Mark's Standard Handbook for Mechanical Engineers, 8th edition, McGraw-Hill Book Company, New York, 1978.
Niemann, G., *Maschinenelemente*, Springer-Verlag, Berlin, Erster Band, 1963.
Norman, C. A., E. S. Ault, and I. E. Zarobsky, *Fundamentals of Machine Design*, McGraw-Hill Book Company, New York, 1951.
SAE Handbook, 1981.
Shigley, J. E., *Mechanical Engineering Design*, Metric Edition, McGraw-Hill Book Company, New York, 1986.
Siegel, M. J., V. L. Maleev, and J. B. Hartman, *Mechanical Design of Machines*, 4th edition, International Textbook Company, Scranton, Pennsylvania, 1965.
Spotts, M. F., *Design of Machine Elements*, 5th edition, Prentice-Hall of India Private Ltd., New Delhi, 1978.
Vallance, A., and V. L. Doughtie, *Design of Machine Members*, McGraw-Hill Book Company, New York, 1951.

CHAPTER 2

STATIC STRESSES IN MACHINE ELEMENTS

SYMBOLS[3,4,5]

A	area of cross section, m^2 (in^2)
A_w	area of web, m^2 (in^2)
a	constant in Rankine's formula
b	radius of area of contact, m (in)
	bandwidth of contact, m (in)
	width of beam, m (in)
c	distance from neutral surface to extreme fiber, m (in)
D	diameter of shaft, m (in)
C_1	constant in straight-line formula
F	load, kN (lbf)
F_c	compressive force, kN (lbf)
F_t	tensile force, kN (lbf)
F_τ	shear force, kN (lbf)
F_{cr}	crushing load, kN (lbf)
e	deformation, total, m (in)
	eccentricity, as of force equilibrium, m (in)
	unit volume change or volumetric strain
e_t	thermal expansion, m (in)
E	modulus of elasticity, direct (tension or compression), GPa (Mpsi)
E_c	combined or equivalent modulus of elasticity in case of composite bars, GPa (Mpsi)
G	modulus of rigidity, GPa (Mpsi)
M_b	bending moment, N m (lbf ft)
M_t	torque, torsional moment, N m (lbf ft)
i	number of turns
I	moment of inertia, area, m^4 or cm^4 (in^4)
	mass moment of inertia, N s^2 m (lbf s^2 ft)
I_{xx}, I_{yy}	moment of inertia of cross-sectional area around the respective principal axes, m^4 or cm^4 (in^4)
J	moment of inertia, polar, m^4 or cm^4 (in^4)
k	radius of gyration, m (in)
k_0	polar radius of gyration, m (in)
k_t	torsional spring constant, J/rad or N m/rad (lbf in/rad)

l	length, m (in)
l_0	length of rod, m (in)
L	length, m (in)
n	speed, rpm (revolutions per minute)
	coefficient of end condition
n'	speed, rps (revolutions per second)
l, m, n	direction cosines (also with subscripts)
P	power, kW (hp)
	pitch or threads per meter
T	temperature, °C (°F)
ΔT	temperature difference, °C (°F)
r	radius of the rod or bar subjected to torsion, m (in) (Fig. 2-18)
q	shear flow
Q	first moment of the cross-sectional area outside the section at which the shear flow is required
v	velocity, m/s (ft/min or fpm)
V	volume, m^3 (in^3)
	shear force, kN (lbf)
ΔV	volume change, m^3 (in^3)
Z	section modulus, m^3 (in^3)
α	deformation of contact surfaces, m (in)
	coefficient of linear expansion, m/m/K or m/m/°C (in/in/°F)
γ	shearing strain, rad/rad
$\gamma_{xy}, \gamma_{yz}, \gamma_{zx}$	shearing strain components in xyz coordinates, rad/rad
δ	deformation or elongation, m (in)
ε	strain, μm/m (μin/in)
ε_T	thermal strain, μm/m (μin/in)
$\varepsilon_x, \varepsilon_y, \varepsilon_z$	strains in x, y, and z directions, μm/m (μin/in)
θ	angular distortion, rad
	angle, deg
	angular twist, rad (deg)
	angle made by normal to plane nn with the x axis, deg
κ	bulk modulus of elasticity, GPa (Mpsi)
ν	Poisson's ratio
ρ	radius of curvature, m (in)
σ	stress, direct or normal, tensile or compressive (also with subscripts), MPa (psi)
σ_b	bearing pressure, MPa (psi)
	bending stress, MPa (psi)
σ_c	compressive stress (also with subscripts), MPa (psi)
	hydrostatic pressure, MPa (psi)
σ_{sc}	compressive strength, MPa (psi)
σ_{cr}	stress at crushing load, MPa (psi)
σ_e	elastic limit, MPa (psi)
σ_s	strength, MPa (psi)
σ_t	tensile stress, MPa (psi)
σ_{st}	tensile strength, MPa (psi)
$\sigma_x, \sigma_y, \sigma_z$	stress in x, y, and z directions, MPa (psi)
$\sigma_1, \sigma_2, \sigma_3$	principal stresses, MPa (psi)
σ_y	yield stress, MPa (psi)
σ_{sy}	yield strength, MPa (psi)
σ_u	ultimate stress, MPa (psi)
σ_{su}	ultimate strength, MPa (psi)
σ'	principal direct stress, MPa (psi)
σ''	normal stress which will produce the maximum strain, MPa (psi)

σ_θ	normal stress on the plane *nn* at any angle θ to x axis, MPa (psi)
τ	shear stress (also with subscripts), MPa (psi)
τ_s	shear strength, MPa (psi)
τ_{xy}, τ_{yz}, τ_{zx}	shear stresses in xy, yz, and zx planes, respectively, MPa (psi)
τ_θ	shear stress on the plane at any angle θ with x axis, MPa (psi)
ω	angular speed, rad/s

Other factors in performance or in special aspects are included from time to time in this chapter and, being applicable only in their immediate context, are not given at this stage.

(*Note*: σ and τ with initial subscript s designates strength properties of material used in the design which will be used and observed throughout this *Machine Design Data Handbook*.)

Particular	Formula	

SIMPLE STRESS AND STRAIN

The stress in simple tension or compression (Fig. 2-1a, 2-1b)

$$\sigma_t = \frac{F_t}{A}; \quad \sigma_c = \frac{F_c}{A} \tag{2-1}$$

The total elongation of a member of length l (Fig. 2-2a)

$$\delta = \frac{Fl}{AE} \tag{2-2}$$

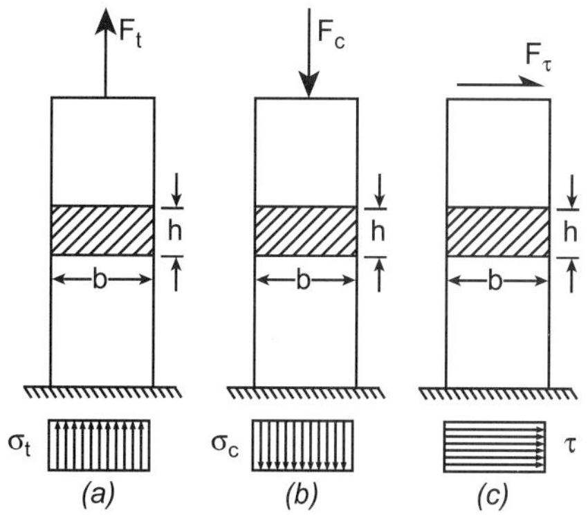

FIGURE 2-1

Strain, deformation per unit length

$$\varepsilon = \frac{\delta}{l} = \frac{\sigma}{E} \tag{2-3}$$

Particular	Formula	

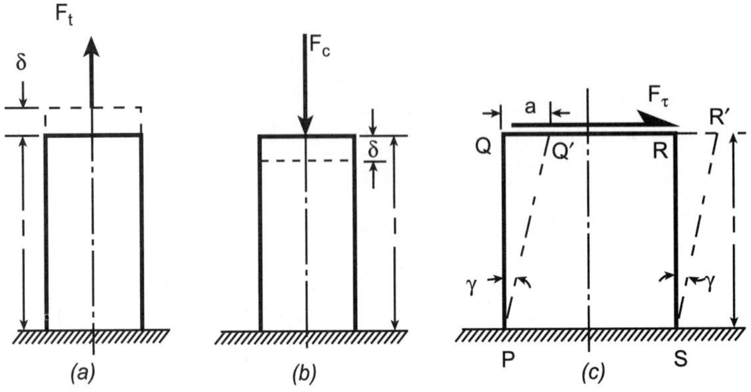

FIGURE 2-2

Particular	Formula	
Young's modulus or modulus of elasticity	$E = \dfrac{\sigma}{\varepsilon}$	(2-4)
The shear stress (Fig. 2-1c)	$\tau = \dfrac{F_\tau}{A}$	(2-5)
Shear deformation due to torsion (Fig. 2-18)	$\theta = \dfrac{\tau L}{G}$	(2-6)
Shear strain (Fig. 2-2c)	$\gamma = \dfrac{\tau}{G} = \dfrac{a}{l}$	(2-7)
The shear modulus or modulus of rigidity from Eq. (2-7)	$G = \dfrac{\tau}{\gamma}$	(2-8)
Poisson's ratio	$\nu = \text{lateral strain/axial strain} = \dfrac{\varepsilon_t}{\varepsilon_a}$	(2-9)
Poisson's ratio may be computed with sufficient accuracy from the relation	$\nu = \dfrac{E}{2G} - 1$	(2-10)
The shear or torsional modulus or modulus of rigidity is also obtained from Eq. (2-10)	$G = \dfrac{E}{2(1+\nu)}$	(2-11)
The bearing stress (Fig. 2-3c)	$\sigma_b = \dfrac{F}{bd_2}$	(2-12)

STRESSES

Unidirectional stress (Fig. 2-4)

Particular	Formula	
The normal stress on the plane at any angle θ with x axis	$\sigma_\theta = \sigma_x \cos^2\theta$	(2-13)

Particular	Formula

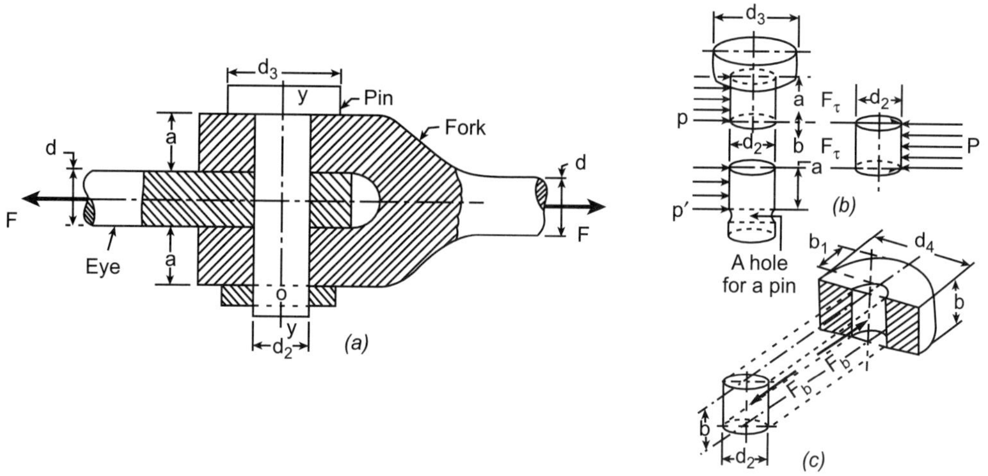

FIGURE 2-3 Knuckle joint for round rods.

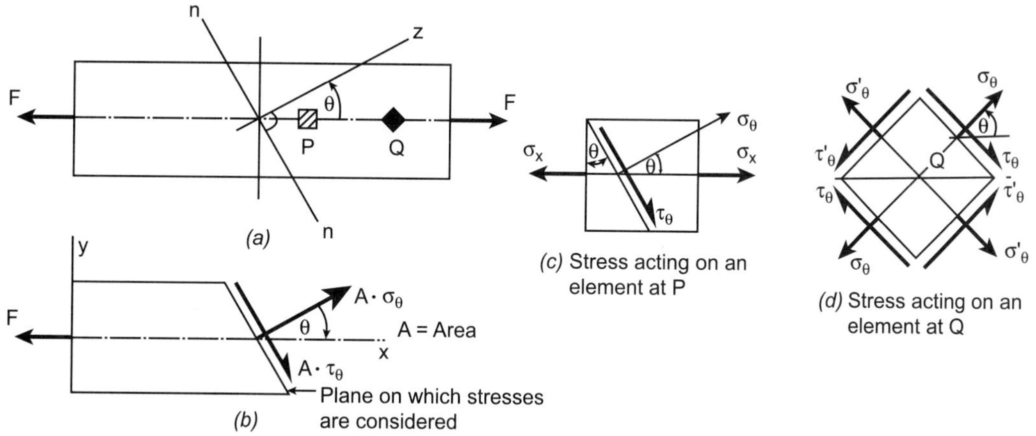

FIGURE 2-4 A bar in uniaxial tension.[3,4]

Particular	Formula	
The shear stress on the plane at any angle θ with x axis	$\tau_\theta = \dfrac{\sigma_x}{2} \sin 2\theta$	(2-14)
Principal stresses	$\sigma_1 = \sigma_x$ and $\sigma_2 = 0$	(2-15)
Angles at which principal stresses act	$\theta_1 = 0^\circ$ and $\theta_2 = 90^\circ$	(2-16)
Maximum shear stress	$\tau_{max} = \dfrac{\sigma_x}{2}$	(2-17)
Angles at which maximum shear stresses act	$\theta_1 = 45^\circ$ and $\theta_2 = 135^\circ$	(2-18)

Particular	Formula	
The normal stress on the plane at an angle $\theta + (\pi/2)$ (Fig. 2-4d)	$\sigma'_\theta = \sigma_x \cos^2\left(\theta + \frac{\pi}{2}\right) = \sigma_x \cos^2\theta$	(2-19)
The shear stress on the plane at an angle $\theta + (\pi/2)$ (Fig. 2-4d)	$\tau'_\theta = \sigma_x \sin\left(\theta + \frac{\pi}{2}\right)\cos\left(\theta + \frac{\pi}{2}\right) = \frac{1}{2}\sigma_x \sin 2\theta$	(2-20)
Therefore from Eqs. (2-13) and (2-19), (2-14), and (2-20)	$\sigma_\theta = \sigma'_\theta$ and $\tau_\theta = -\tau'_\theta$	(2-21)

PURE SHEAR (FIG. 2-5)

The normal stress on the plane at any angle θ	$\sigma_\theta = \tau_{xy} \sin 2\theta$	(2-22)
The shear stress on the plane at any angle θ	$\tau_\theta = \tau_{xy} \cos 2\theta$	(2-23)
The principal stress	$\sigma_1 = \tau_{xy}$ and $\sigma_2 = -\tau_{xy}$	(2-24)
Angles at which principal stresses act	$\theta_1 = 45^\circ$ and $\theta_2 = 135^\circ$	(2-25)
Maximum shear stresses	$\tau_{max} = \tau_{xy} = \sigma$	(2-26)
Angles at which maximum shear stress act	$\theta_1 = 0$ and $\theta_2 = 90^\circ$	(2-27)

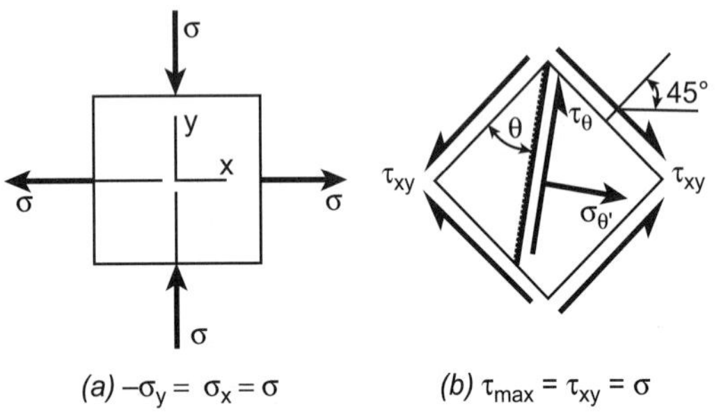

FIGURE 2-5 An element in pure shear.

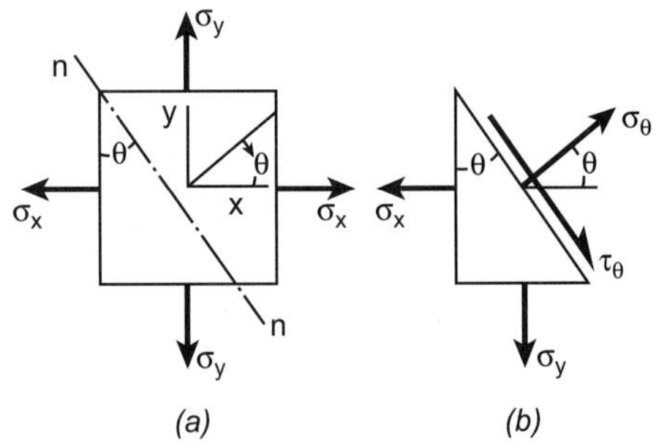

FIGURE 2-6 An element in biaxial tension.

BIAXIAL STRESSES (FIG. 2-6)

The normal stress on the plane at any angle θ	$\sigma_\theta = \frac{\sigma_x + \sigma_y}{2} + \frac{\sigma_x - \sigma_y}{2}\cos 2\theta$	(2-28)
The shear stress on the plane at any angle θ	$\tau_\theta = \frac{\sigma_x - \sigma_y}{2}\sin 2\theta$	(2-29)
The shear stress τ_θ at $\theta = 0$	$\tau_\theta = 0$	(2-30)
The shear stress τ_θ at $\theta = 45^\circ$	$\tau_{max} = (\sigma_x - \sigma_y)/2$	(2-31)

Particular	Formula	
BIAXIAL STRESSES COMBINED WITH SHEAR (FIG. 2-7)		
The normal stress on the plane at any angle θ	$\sigma_\theta = \dfrac{\sigma_x + \sigma_y}{2} + \dfrac{\sigma_x - \sigma_y}{2}\cos 2\theta + \tau_{xy}\sin 2\theta$	(2-32)
The shear stress in the plane at any angle θ	$\tau_\theta = \dfrac{\sigma_x - \sigma_y}{2}\sin 2\theta - \tau_{xy}\cos 2\theta$	(2-33)
The maximum principal stress	$\sigma_1 = \dfrac{\sigma_x + \sigma_y}{2} + \left[\left(\dfrac{\sigma_x - \sigma_y}{2}\right)^2 + \tau_{xy}^2\right]^{1/2}$	(2-34)
The minimum principal stress	$\sigma_2 = \dfrac{\sigma_x + \sigma_y}{2} - \left[\left(\dfrac{\sigma_x - \sigma_y}{2}\right)^2 + \tau_{xy}^2\right]^{1/2}$	(2-35)
Angles at which principal stresses act	$\theta_{1,2} = \frac{1}{2}\arctan\dfrac{2\tau_{xy}}{\sigma_x - \sigma_y}$ where θ_1 and θ_2 are 180° apart	(2-36)
Maximum shear stress	$\tau_{max} = \left[\left(\dfrac{\sigma_x - \sigma_y}{2}\right)^2 + \tau_{xy}^2\right]^{1/2} = \dfrac{\sigma_1 - \sigma_2}{2}$	(2-37)
Angles at which maximum shear stress acts	$\theta = \frac{1}{2}\arctan\dfrac{\sigma_x - \sigma_y}{2\tau_{xy}}$	(2-38)
The equation for the inclination of the principal planes in terms of the principal stress (Fig. 2-8)	$\tan\theta = \dfrac{\sigma_1 - \sigma_x}{\tau_{xy}}$	(2-39)

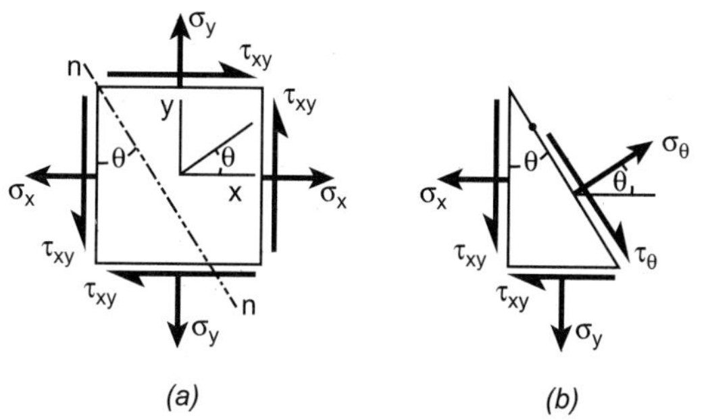

FIGURE 2-7 An element in plane state of stress.

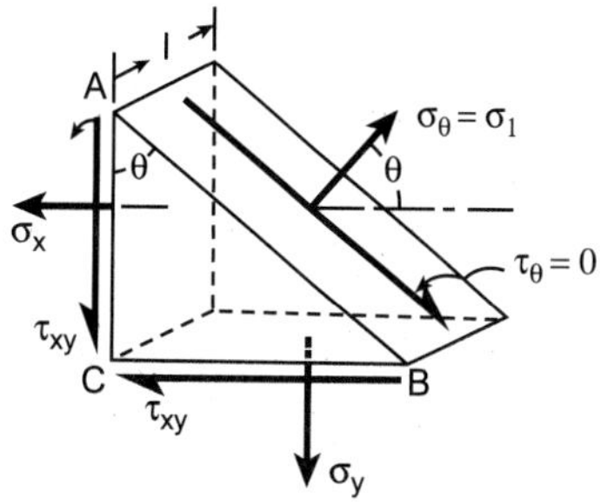

FIGURE 2-8

Particular	Formula

MOHR'S CIRCLE

Biaxial field combined with shear (Fig. 2-9)

Maximum principal stress σ_1	σ_1 is the abscissa of point F
Minimum principal stress σ_2	σ_2 is the abscissa of point G
Maximum shear stress τ_{max}	τ_{max} is the ordinate of point H

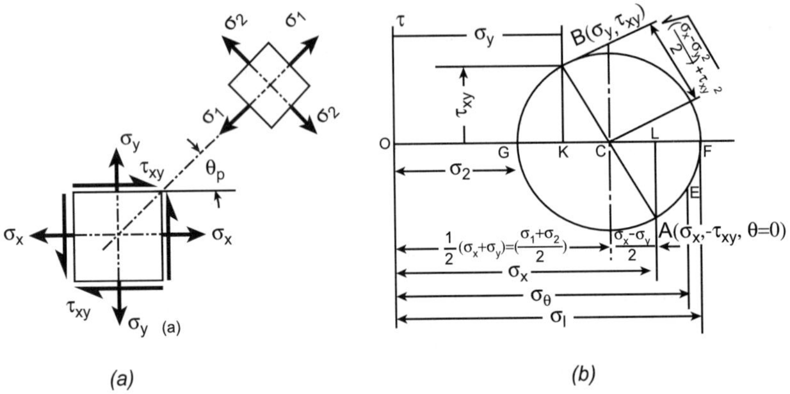

FIGURE 2-9 Mohr's circle for biaxial state of stress.

TRIAXIAL STRESS (Figs. 2-10 and 2-11)

The normal stress on a plane nn, whose direction cosines are l, m, n

$$\sigma_\theta = \sigma_x l^2 + \sigma_y m^2 + \sigma_z n^2 \tag{2-40}$$

The shear stress on a plane normal nn, whose direction cosines are l, m, n

$$\tau_\theta = \sqrt{\sigma_x^2 l^2 + \sigma_y^2 m^2 + \sigma_z^2 n^2} \tag{2-41}$$

The principal stresses

$$\sigma_{1,2,3} = \sigma_x, \sigma_y, \sigma_z \tag{2-42}$$

The cubic equation for general state of stress in three dimensions from the theory of elasticity

$$\sigma^3 - (\sigma_x + \sigma_y + \sigma_z)\sigma^2 + (\sigma_x\sigma_y + \sigma_y\sigma_z + \sigma_z\sigma_x - \tau_{xy}^2 - \tau_{yz}^2 - \tau_{zx}^2)\sigma - (\sigma_x\sigma_y\sigma_z + 2\tau_{xy}\tau_{yz}\tau_{zx} - \sigma_x\tau_{zy}^2 - \sigma_y\tau_{zx}^2 - \sigma_z\tau_{xy}^2) = 0 \tag{2-43}$$

The three roots of this cubic equation give the magnitude of the principal stresses σ_1, σ_2, and σ_3.

The maximum shear stresses on planes parallel to x, y, and z which are designated as

$$(\tau_{max})_1 = \frac{\sigma_2 - \sigma_3}{2}; \quad (\tau_{max})_2 = \frac{\sigma_1 - \sigma_3}{2}; \quad (\tau_{max})_3 = \frac{\sigma_1 - \sigma_2}{2} \tag{2-44}$$

Particular	Formula

MOHR'S CIRCLE

Triaxial field (Figs. 2-10 and 2-11)

Normal stress at point (Fig. 2-11b) on one octahedral plane	$\sigma_t = \frac{1}{3}(\sigma_1 + \sigma_2 + \sigma_3) = \frac{1}{3}(\sigma_x + \sigma_y + \sigma_z)$ or σ_t is the abscissa of point T	(2-45)
Shear stress at point T (Fig. 2-11b) on an octahedral plane	$\tau_t = \frac{1}{3}[(\sigma_x - \sigma_y)^2 + (\sigma_y - \sigma_z)^2 + (\sigma_z - \sigma_x)^2 + 6(\tau_{xy}^2 + \tau_{yz}^2 + \tau_{zx}^2)]^{1/2}$ $= \frac{1}{3}\sqrt{[(\sigma_1 - \sigma_2)^2 + (\sigma_2 - \sigma_3)^2 + (\sigma_3 - \sigma_1)^2]}$ or τ_t is the ordinate of point T	(2-46a)

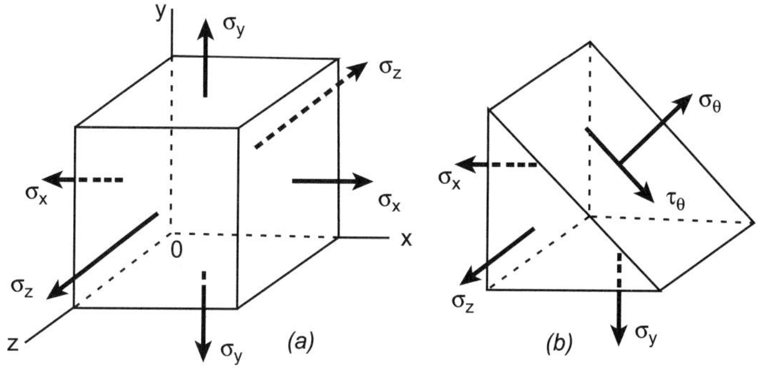

FIGURE 2-10 An element in triaxial state of stress.

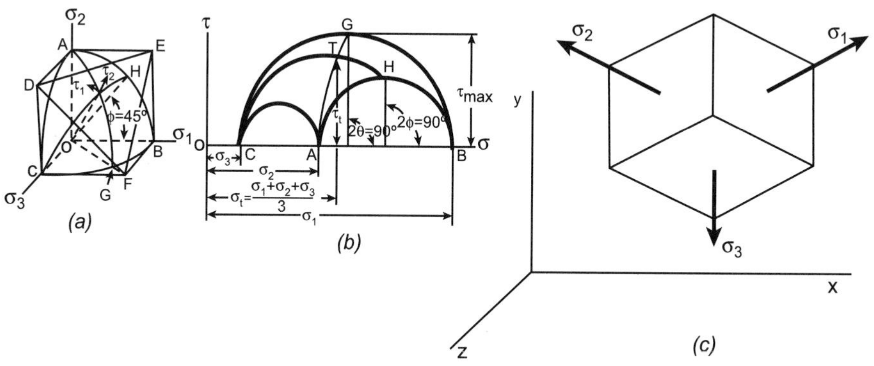

FIGURE 2-11 Mohr's circle for triaxial octahedral stress state.

Particular	Formula	

STRESS-STRAIN RELATIONS

Uniaxial field

Particular	Formula	
Strain in principal direction 1	$\varepsilon_1 = \frac{\sigma_1}{E}; \quad \varepsilon_2 = -\nu\frac{\sigma_1}{E}; \quad \varepsilon_3 = -\nu\frac{\sigma_1}{E}$	(2-47)
The principal stress	$\sigma_1 = E\varepsilon_1$	(2-47a)
The unit volume change in uniaxial stress	$\frac{\Delta V}{V} = \frac{(1-2\nu)\sigma_1}{E} = \varepsilon_1(1-2\nu)$	(2-48)

Biaxial field

Particular	Formula	
Strain in principal direction 1	$\varepsilon_1 = \frac{1}{E}(\sigma_1 - \nu\sigma_2)$	(2-49)
Strain in principal direction 2	$\varepsilon_2 = \frac{1}{E}(\sigma_2 - \nu\sigma_1)$	(2-50)
Strain in principal direction 3	$\varepsilon_3 = -\frac{\nu}{E}(\sigma_1 + \sigma_2)$	(2-51)
The principal stresses in terms of principal strains in a biaxial stress field	$\sigma_1 = \frac{E}{1-\nu^2}(\varepsilon_1 + \nu\varepsilon_2)$	(2-52)
	$\sigma_2 = \frac{E}{1-\nu^2}(\varepsilon_2 + \nu\varepsilon_1)$	(2-53)
	$\sigma_3 = 0$	(2-53a)
The unit volume change in biaxial stress	$\frac{\Delta V}{V} + \frac{(1-2\nu)}{E}(\sigma_1 + \sigma_2)$	(2-54)

Triaxial field

Particular	Formula	
Strain in principal direction 1	$\varepsilon_1 = \frac{1}{E}[\sigma_1 - \nu(\sigma_2 + \sigma_3)]$	(2-55)
Strain in principal direction 2	$\varepsilon_2 = \frac{1}{E}[\sigma_2 - \nu(\sigma_3 + \sigma_1)]$	(2-56)
Strain in principal direction 3	$\varepsilon_3 = \frac{1}{E}[\sigma_3 - \nu(\sigma_1 + \sigma_2)]$	(2-57)
The principal stresses in terms of principal strains in triaxial stress field	$\sigma_1 = \frac{E}{(1-\nu-2\nu^2)}[(1-\nu)\varepsilon_1 + \nu(\varepsilon_2 + \varepsilon_3)]$	(2-58)
	$\sigma_2 = \frac{E}{(1-\nu-2\nu^2)}[(1-\nu)\varepsilon_2 + \nu(\varepsilon_3 + \varepsilon_1)]$	(2-59)
	$\sigma_3 = \frac{E}{(1-\nu-2\nu^2)}[(1-\nu)\varepsilon_3 + \nu(\varepsilon_1 + \varepsilon_2)]$	(2-60)

Particular	Formula	
The unit volume change or volumetric strain in terms of principal stresses for the general case of triaxial stress (Fig. 2-12)	$e = \frac{dV}{V} = \frac{(1-2\nu)}{E}(\sigma_x + \sigma_y + \sigma_z)$	(2-61a)
	$= \frac{(1-2\nu)}{E}(\sigma_1 + \sigma_2 + \sigma_3)$	(2-61b)

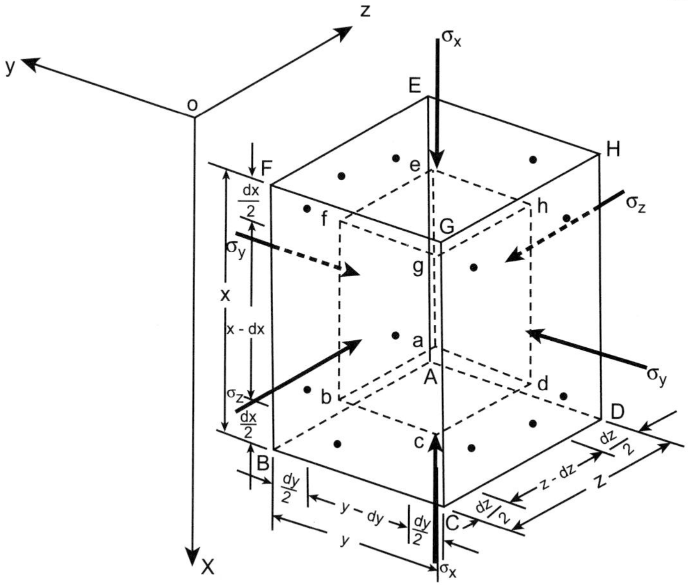

FIGURE 2-12 Uniform hydrostatic pressure.

The volumetric strain due to uniform hydrostatic pressure σ_c acting on an element (Fig. 2-12)	$\frac{\Delta V}{V} = \frac{-3(1-2\nu)\sigma_c}{E} = -\frac{\sigma_c}{\kappa}$	(2-62)
The bulk modulus of elasticity	$\kappa = \frac{E}{3(1-2\nu)}$	(2-63)
The relationship between E, G and K	$E = \frac{9KG}{(3K+G)}$	(2-63a)

STATISTICALLY INDETERMINATE MEMBERS (Fig. 2-13)

The reactions at supports of a constant cross-section bar due to load F acting on it as shown in Fig. 2-13	$R_a = \frac{FL_b}{L_a + L_b} = \frac{FL_b}{L}$	(2-64a)
	$R_B = \frac{FL_a}{L_a + L_b} = \frac{FL_a}{L}$	(2-64b)
The elongation of left portion L_a of the bar	$\delta_a = \frac{R_A L_a}{AE} = \frac{FL_a L_b}{LAE}$	(2-65)

Particular	Formula	
The shortening of right portion L_b of the bar	$\delta_b = -\frac{R_A L_a}{AE} = -\frac{FL_a L_b}{LAE}$	(2-66)

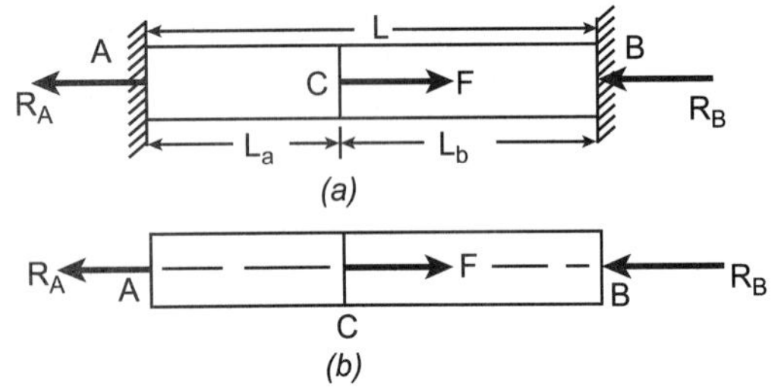

FIGURE 2-13

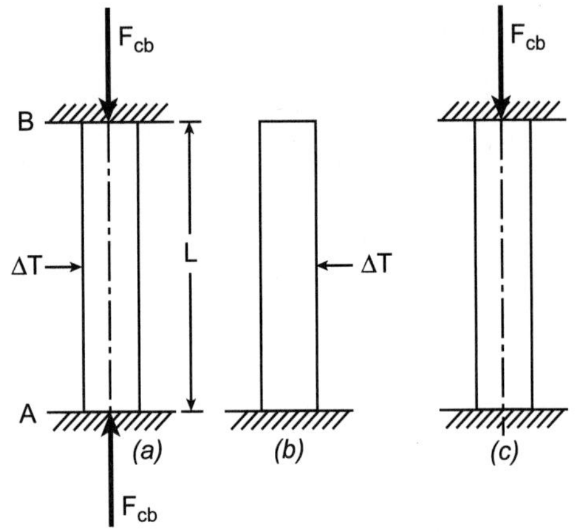

FIGURE 2-14

THERMAL STRESS AND STRAIN

Particular	Formula	
The normal strain due to free expansion of a bar or machine member when it is heated	$\varepsilon_T = \alpha(\Delta T)$	(2-67)
The free linear deformation due to temperature change	$\delta = \alpha L(\Delta T)$	(2-68)
The compressive force F_{cb} developed in the bar fixed at both ends due to increase in temperature (Fig. 2-14)	$F_{cb} = \alpha AE(\Delta T)$	(2-69)
The compressive stress induced in the member due to thermal expansion (Fig. 2-14)	$\sigma_{cT} = \frac{F_{cb}}{A} = -\alpha E(\Delta T)$	(2-70)
The relation between the extension of one member to the compression of another member in case of rigidly joined compound bars of the same length L made of different materials subjected to same temperature (Fig. 2-15)	$\frac{\sigma_s L}{E_s} + \frac{\sigma_c L}{E_c} = (\alpha_c - \alpha_s)L(\Delta T)$	(2-71)
The forces acting on each member due to temperature change in the compound bar	$\sigma_c A_c = \sigma_s A_s$	(2-72)
The relation between compression of the tube to the extension of the threaded member due to tightening of the nut on the threaded member (Fig. 2-16)	$\frac{\sigma_t L}{E_t} + \frac{\sigma_s L}{E_s} =$ [number of turns (i) × (threads/meter) or pitch (P)] $= iP$	(2-73)
The forces acting on tube and threaded member due to tightening of the nut	$\sigma_s A_s = \sigma_t A_t$	(2-74)

Particular	Formula

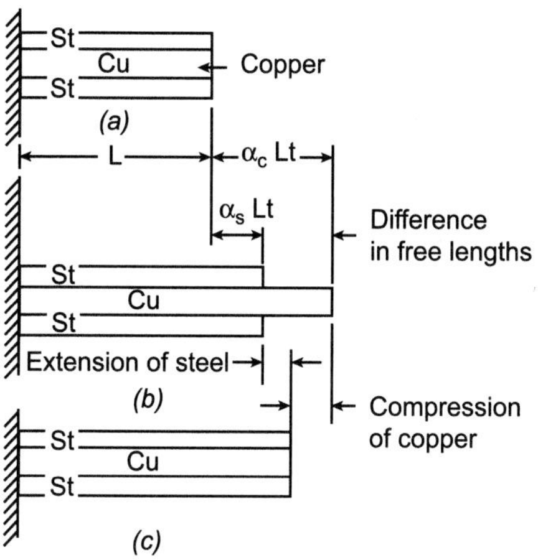

FIGURE 2-15

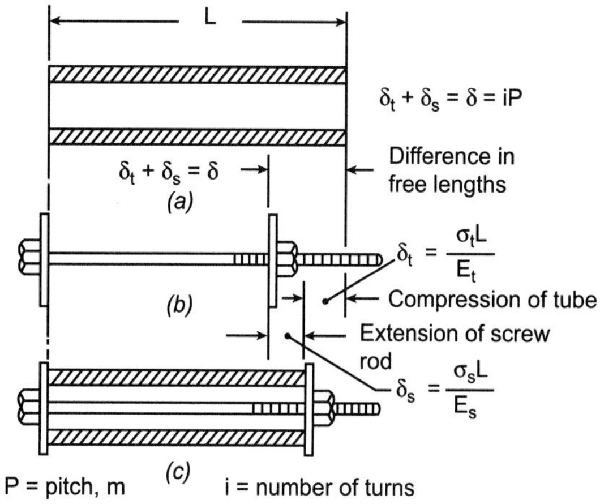

FIGURE 2-16

COMPOUND BARS

The total load in the case of compound bars or columns or wires consisting of i members, each having different length and area of cross section and each made of different material subjected to an external load as shown in Fig. 2-17

$$F = \sum \frac{E_i A_i \delta_i}{L_i} = \delta \sum \frac{E_i A_i}{L_i} \tag{2-75}$$

An expression for common compression of each bar (Fig. 2-17)

$$\delta = \frac{F}{\sum(E_i A_i / L_i)} \tag{2-76}$$

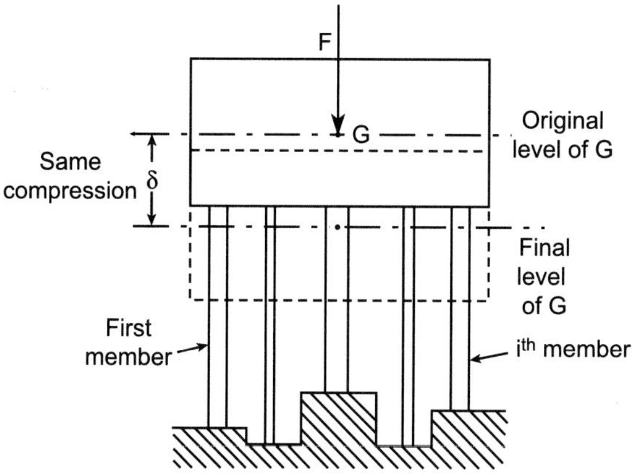

FIGURE 2-17

Particular	Formula	
The load on first bar (Fig. 2-17)	$F_1 = \frac{(E_1A_1/L_1)}{\sum(EA/L)}F$	(2-77)
The load on ith bar (Fig. 2-17)	$F_i = \frac{E_iA_i\delta}{L_i}$	(2-78)

EQUIVALENT OR COMBINED MODULUS OF ELASTICITY OF COMPOUND BARS

The equivalent or combined modulus of elasticity of a compound bar consisting of i members, each having a different length and area of cross section and each being made of different material	$E_c = \frac{E_1A_1 + E_2A_2 + E_3A_3 + \cdots + E_nA_n}{A_1 + A_2 + A_3 + \cdots + A_n}$	(2-79a)
	$= \frac{\sum E_iA_i}{\sum_{i=1,2,\ldots,n} A_i}$	(2-79b)
The stress in the equivalent bar due to external load F	$\sigma = \frac{F}{\sum_{i=1,2,3,\ldots} A_i}$	(2-80)
The strain in the equivalent bar due to external load F	$\varepsilon = \frac{F}{E_c \sum_{i=1,2,3,\ldots} A_i} = \frac{\delta}{L}$	(2-81)
The common extension or compression due to external load F	$\delta = \frac{FL}{E_c \sum_{i=1,2,3,\ldots,n} A_i} = \varepsilon L$	(2-82)

POWER

The relation between power, torque and speed	$P = M_t\omega$	(2-83)
	where M_t in N m (lbf ft), ω in rad/s (rad/min), and P in W (hp)	
	$= \frac{M_tn'}{159}$ **SI**	(2-84a)
	where M_t in kN m, n' in rps, and P in kW	
	$= \frac{M_tn}{9550}$ **SI**	(2-84b)
	where M_t in kN m, n in rpm, and P in kW	
	$= \frac{M_tn}{63030}$ **USCS**	(2-84c)
	where M_t in lbf in, n in rpm, and P in hp	

Particular	Formula
Another expression for power in terms of force F acting at velocity v	$P = \frac{F\nu}{1000}$ **SI** (2-85a) where F in newtons (N), ν in m/s, and P in kW $= \frac{F\nu}{33000}$ **USCS** (2-85b) where F in lbf, ν in fpm (feet per minute), and P in hp (horsepower)

TORSION (FIG. 2-18)

Particular	Formula
The general equation for torsion (Fig. 2-18)	$\frac{M_t}{J} = \frac{G\theta}{L} = \frac{\tau}{\rho}$ (2-86)
Torque	$M_t = \frac{159P}{n'}$ **SI** (2-87a) where M_t in kN m, n' in rps, and P in kW $= \frac{9550P}{n}$ **SI** (2-87b) where M_t in kN m, n in rpm, and P in kW $= \frac{63030P}{n}$ **USCS** (2-87c) where M_t in lbf in, n in rpm, and P in hp
The maximum shear stress at the maximum radius r of the solid shaft (Fig. 2-18) subjected to torque M_t	$\tau_{\max} = \frac{16M_t}{\pi D^3}$ (2-88)
The torsional spring constant	$k_t = \frac{M_t}{\theta} = \frac{GJ}{L}$ (2-89)

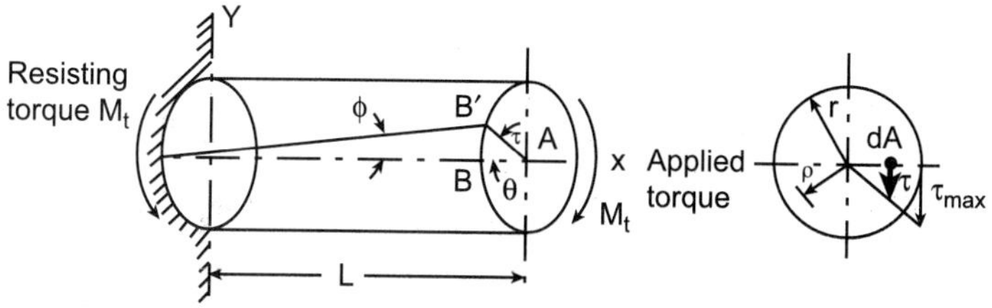

FIGURE 2-18 Cylindrical bar subjected to torque.

Particular	Formula	

BENDING (FIG. 2-19)

The general formula for bending (Fig. 2-19)	$\frac{M_b}{I} = \frac{\sigma_b}{c} = \frac{E}{\rho}$	(2-90)

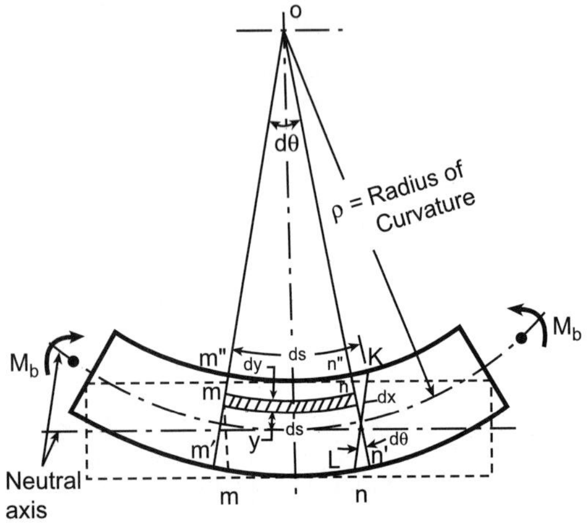

FIGURE 2-19 Bending of beam.

The maximum values of tensile and compressive bending stresses	$\sigma_b = \frac{M_b c}{I}$	(2-91)
The shear stresses developed in bending of a beam (Fig. 2-20)	$\tau = \frac{V}{Ib}\int_{y_0}^{c} y\,dA$	(2-92)
The shear flow	$q = \frac{VQ}{I}$	(2-93)

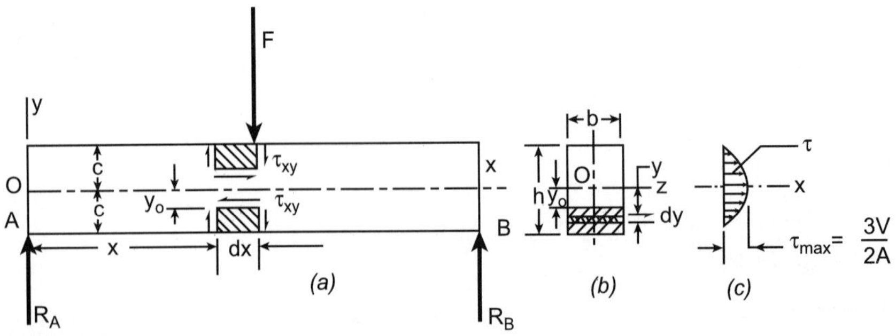

FIGURE 2-20 Beam subjected to shear stress.

Particular	Formula	
The first moment of the cross-sectional area outside the section at which the shear flow is required	$Q = \int_{y_0}^{c} y\,dA$	(2-94)
The maximum shear stress for a rectangular section (Figs. 2-20 and 2-21)	$\tau_{\max} = \dfrac{3V}{2A}$	(2-95)

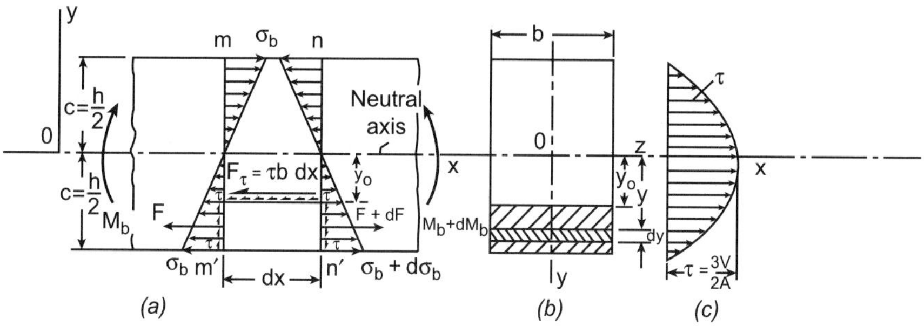

FIGURE 2-21 Element cut out from a beam subjected to shear stress.

For a solid circular section beam, the maximum shear stress	$\tau_{\max} = \dfrac{4V}{3A}$	(2-96)
For a hollow circular section beam, the expression for maximum shear stress	$\tau_{\max} = \dfrac{2V}{A}$	(2-97)
An appropriate expression for $\tau_{\max}$ for structural beams, columns and joists used in structural industries	$\tau_{\max} = \dfrac{V}{A_w}$ where A_w is the area of the web	(2-98)

ECCENTRIC LOADING

The maximum and minimum stresses which are induced at points of outer fibers on either side of a machine member loaded eccentrically (Figs. 2-22 and 2-23)	$\sigma_{\max} = \dfrac{F}{A} + \dfrac{M_b}{Z}$ and $\sigma_{\min} = \dfrac{F}{A} - \dfrac{M_b}{Z}$	(2-99)
The resultant stress at any point of the cross section of an eccentrically loaded member (Fig. 2-24)	$\sigma_z = \pm\dfrac{F}{A} \pm \dfrac{M_{bx}e_y}{I_{xx}} \pm \dfrac{M_{by}e_x}{I_{yy}}$	(2-100)

COLUMN FORMULAS (Fig. 2-25)

Euler's formula (Fig. 2-26) for critical load	$F_{cr} = \dfrac{n\pi^2 EA}{(l/k)^2} = \dfrac{n\pi^2 EI}{l^2}$	(2-101)

Particular	Formula

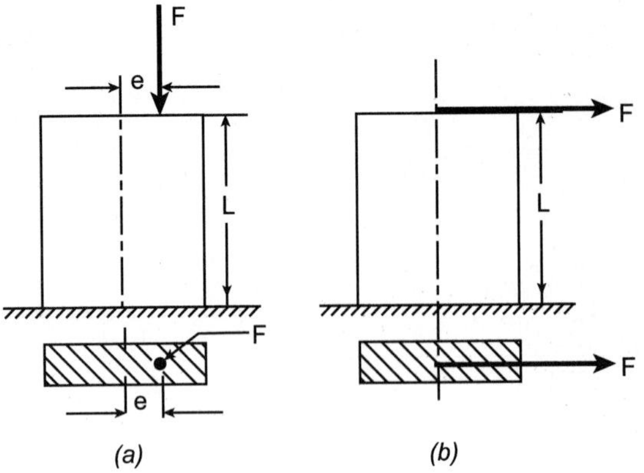

FIGURE 2-22 Eccentric loading.

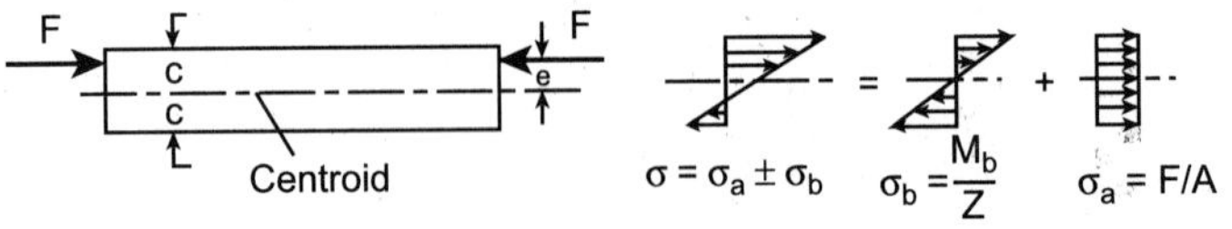

FIGURE 2-23 Eccentrically loaded machine member.

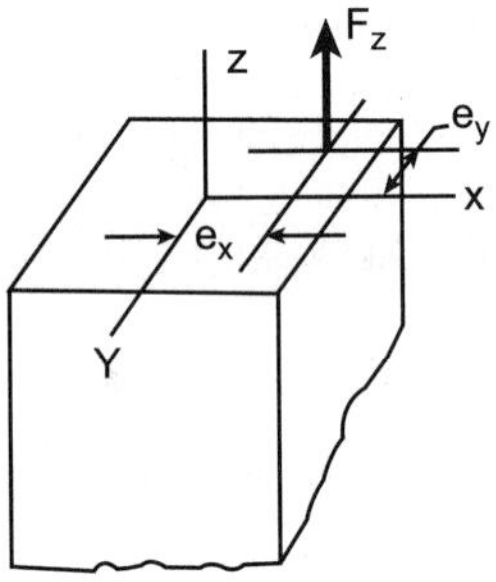

FIGURE 2-24

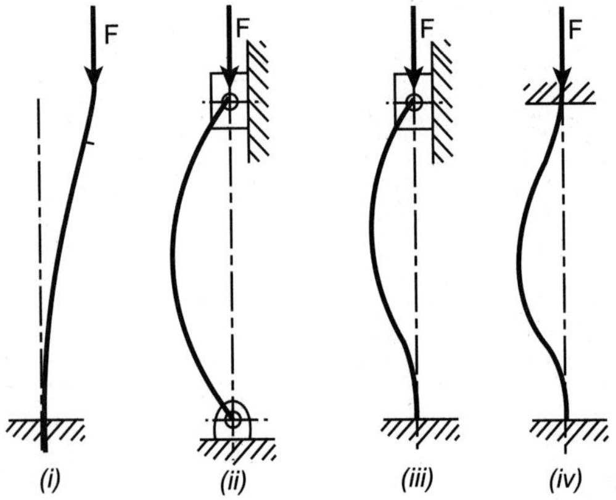

FIGURE 2-25 Column-end conditions. (i) One end is fixed and other is free. (ii) Both ends are rounded and guided or hinged. (iii) One end is fixed and other is rounded and guided or hinged. (iv) Fixed ends.

Particular	Formula	
Johnson's parabolic formula (Fig. 2-26) for critical load	$F_{cr} = A\sigma_y \left[1 - \frac{\sigma_y}{4n\pi^2 E}\left(\frac{l}{k}\right)^2\right]$	(2-102)

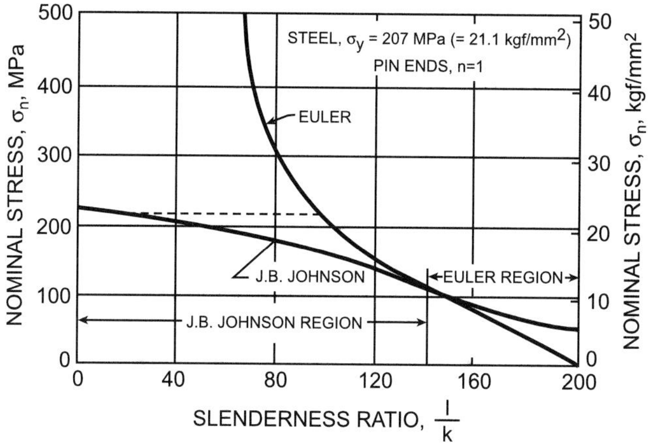

FIGURE 2-26 Variation of critical stress with slenderness ratio.

Straight-line formula for critical load	$F_{cr} = A\left[\sigma_y - \frac{2\sigma_y}{3\pi\sqrt{(\sigma_y/3nE)}}\left(\frac{l}{k}\right)\right]$	(2-103)
Straight-line formula for short column of brittle material for critical load	$F_{cr} = A\left(\sigma - C_1\frac{l}{k}\right)$	(2-104)
Ritter's formula for induced stress	$\sigma_c = \frac{F}{A}\left[1 + \frac{\sigma_e}{n\pi^2 E}\left(\frac{l}{k}\right)^2\right]$	(2-105)
Ritter's formula for eccentrically loaded column (Fig. 2-23) for combined induced stress	$\sigma_c = \frac{F}{A}\left[1 + \frac{\sigma_e}{n\pi^2 E}\left(\frac{l}{k}\right)^2 + \frac{ce}{k^2}\right]$	(2-106)
Rankine's formula for induced stress	$\sigma_c = \frac{F_{cr}}{A}\left[1 + a\left(\frac{l}{k}\right)^2\right]$	(2-107)
The critical unit load from secant formula for a round-ended column	$\frac{F_{cr}}{A} = \frac{\sigma_y}{1 + \frac{ec}{k^2}\sec\frac{l}{k}\sqrt{(F_{cr}/4AE)}}$	(2-108)

Particular	Formula

HERTZ CONTACT STRESS

Contact of spherical surfaces

Sphere on a sphere (Fig. 2-27a)
The radius of circular area of contact

$$a = 0.721 \left[F \frac{\dfrac{1-\nu_1^2}{E_1} + \dfrac{1-\nu_2^2}{E_2}}{\left(\dfrac{1}{d_1} + \dfrac{1}{d_2}\right)} \right]^{1/3} \qquad (2\text{-}109)$$

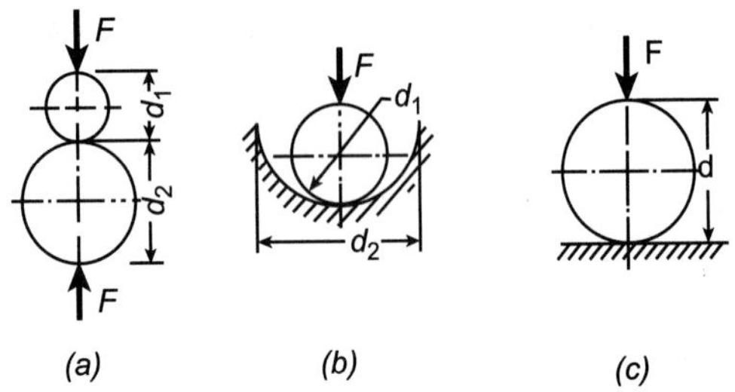

FIGURE 2-27 Hertz contact stress.

The maximum compressive stress

$$\sigma_{c(\max)} = 0.918 \left[F \frac{\left(\dfrac{1}{d_1} + \dfrac{1}{d_2}\right)^2}{\left(\dfrac{1-\nu_1^2}{E_1} + \dfrac{1-\nu_2^2}{E_2}\right)^2} \right]^{1/3} \qquad (2\text{-}110)$$

Combined deformation of both bodies in contact along the axis of load

$$\alpha = 1.04 \left[F^2 \frac{\left(\dfrac{1-\nu_1^2}{E_1} + \dfrac{1-\nu_2^2}{E_2}\right)^2}{\left(\dfrac{d_1 d_2}{d_1 + d_2}\right)} \right]^{1/3} \qquad (2\text{-}111)$$

Spherical surface in contact with a spherical socket (Fig. 2-27b)

The radius of circular area of contact

$$a = 0.721 \left[F \frac{\left(\dfrac{1-\nu_1^2}{E_1} + \dfrac{1-\nu_2^2}{E_2}\right)}{\left(\dfrac{1}{d_1} - \dfrac{1}{d_2}\right)} \right]^{1/3} \qquad (2\text{-}112)$$

The maximum compressive stress

$$\sigma_{c(\max)} = 0.918 \left[F \frac{\left(\dfrac{1}{d_1} - \dfrac{1}{d_2}\right)^2}{\left(\dfrac{1-\nu_1^2}{E_1} + \dfrac{1-\nu_2^2}{E_2}\right)^2} \right]^{1/3} \qquad (2\text{-}113)$$

Particular	Formula
Combined deformation of both bodies in contact along axis of load	$$\alpha = 1.04\left[F^2\frac{\left(\frac{1-\nu_1^2}{E_1}+\frac{1-\nu_2^2}{E_2}\right)^2}{\left(\frac{d_1 d_2}{d_2-d_1}\right)}\right]^{1/3} \quad (2\text{-}114)$$
Distribution of pressure over band of width of contact and stresses in contact zone along the line of symmetry of spheres	Refer to Fig. 2-28a.
Sphere on a flat surface (Fig. 2-27c)	
The radius of circular area of contact	$$a = 0.721\left[Fd_1\left(\frac{1-\nu_1^2}{E_1}+\frac{1-\nu_2^2}{E_2}\right)\right]^{1/3} \quad (2\text{-}115)$$
The maximum compressive stress	$$\sigma_{c(\max)} = 0.918\left[\frac{F}{d_1^2\left(\frac{1-\nu_1^2}{E_1}+\frac{1-\nu_2^2}{E_2}\right)^2}\right]^{1/3} \quad (2\text{-}116)$$ where $d = d_1$ (Fig. 2-27c).

Contact of cylindrical surfaces

Particular	Formula
Cylindrical surface on cylindrical surface, axis parallel (Fig. 2-27a and Fig. 2-28b)	
The width of band of contact	$$2b = 1.6\left[\frac{F}{L}\frac{\left(\frac{1-\nu_1^2}{E_1}+\frac{1-\nu_2^2}{E_2}\right)}{\left(\frac{1}{d_1}+\frac{1}{d_2}\right)}\right]^{1/2} \quad (2\text{-}117)$$
The maximum compressive stress	$$\sigma_{c(\max)} = 0.798\left[\frac{F}{L}\frac{\left(\frac{1}{d_1}+\frac{1}{d_2}\right)}{\left(\frac{1-\nu_1^2}{E_1}+\frac{1-\nu_2^2}{E_2}\right)}\right]^{1/2} \quad (2\text{-}118)$$
Cylindrical surface in contact with a circular groove (Fig. 2-27b)	
The width of band of contact	$$2b = 1.6\left[\frac{F}{L}\frac{\left(\frac{1-\nu_1^2}{E_1}+\frac{1-\nu_2^2}{E_2}\right)}{\left(\frac{1}{d_1}-\frac{1}{d_2}\right)}\right]^{1/2} \quad (2\text{-}119)$$
The maximum compressive stress	$$\sigma_{c(\max)} = 0.798\left[\frac{F}{L}\frac{\left(\frac{1}{d_1}-\frac{1}{d_2}\right)}{\left(\frac{1-\nu_1^2}{E_1}+\frac{1-\nu_2^2}{E_2}\right)}\right]^{1/2} \quad (2\text{-}120)$$
Distribution of pressure over band of width of contact and stresses in contact zone along the line of symmetry of cylinders	Refer to Fig. 2-28b.

Particular	Formula	
Cylindrical surface in contact with a flat surface (Fig. 2-27c):		
The width of band of contact	$2b = 1.6\left[\frac{Fd_1}{L}\left(\frac{1-\nu_1^2}{E_1}+\frac{1-\nu_2^2}{E_2}\right)\right]^{1/2}$	(2-121)
The maximum compressive stress	$\sigma_{c(\max)} = 0.798\left[\frac{F}{Ld_1}\frac{1}{\left(\frac{1-\nu_1^2}{E_1}+\frac{1-\nu_2^2}{E_2}\right)}\right]^{1/2}$ where $d = d_1$ (Fig. 2-27c).	(2-122)
Deformation of cylinder between two plates	$\Delta d_1 = \frac{4F}{L}\left(\frac{1-\nu_1^2}{\pi E}\right)\left(\frac{1}{3}+\log_e\frac{2d_1}{b}\right)$	(2-123)
The maximum shear stress occurs below contact surface for ductile materials		
For sphere	$\tau_{\max} = 0.31\sigma_{c(\max)}$	(2-123a)
For cylinders	$\tau_{\max} = 0.295\sigma_{c(\max)}$	(2-123b)
The depth from contact surface to the point of the maximum shear	$h = 0.786b$	(2-123c)

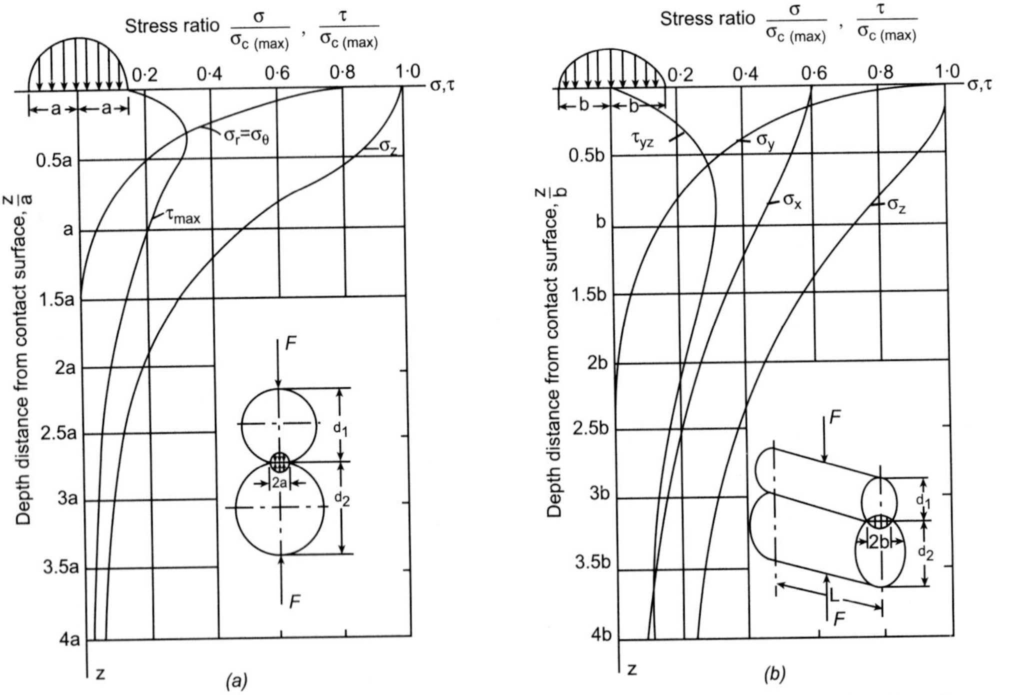

FIGURE 2-28 Distribution of pressure over bandwidth of contact and stresses in contact zone along line of symmetry of spheres and cylinders for $\nu = 0.3$.

Particular	Formula

DESIGN OF MACHINE ELEMENTS AND STRUCTURES MADE OF COMPOSITE

Honeycomb composite

For the components of composite materials which give high strength–weight ratio combined with rigidity — Refer to Fig. 2-29.

For sandwich construction of honeycomb structure — Refer to Fig. 2-30.

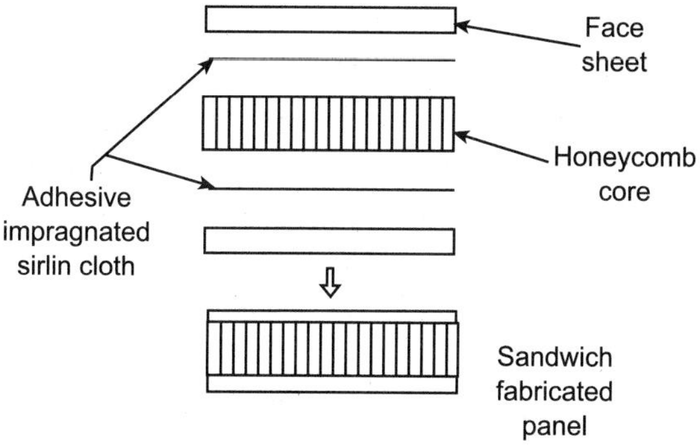

FIGURE 2-29 Sandwich fabricated panel.

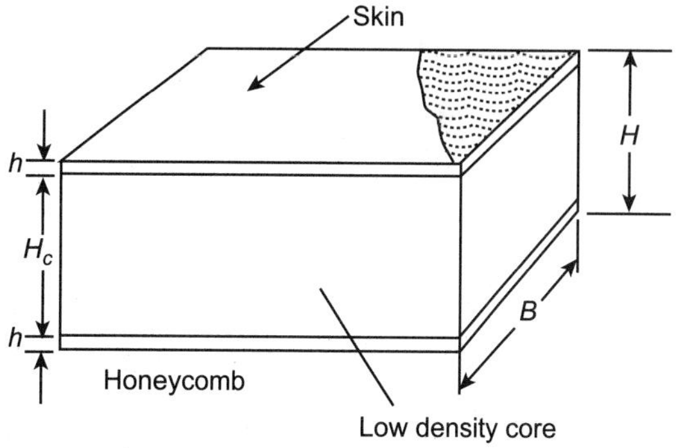

FIGURE 2-30 Honeycomb.

The moment of inertia of sandwich panel, Fig 2-30

$$I = 2\left(\frac{Bh^3}{12}\right) + 2Bh\left(\frac{H_c + h}{2}\right)^2 \quad (2\text{-}124)$$

Simplified Eq. (2-124) after neglecting powers of h

$$I = BhH_c\left(h + \frac{H_c}{2}\right) \quad (2\text{-}125)$$

The flexural rigidity

$$D = EI \quad (2\text{-}126)$$

where E = modulus of elasticity of the facing metal
I is given by Eq. (2-125).

The flexural rigidity of sandwich plate/panel

$$D = \frac{E(H^3 - H_c^3)}{12(1 - \nu^2)} \quad (2\text{-}127)$$

The flexural rigidity of sandwich construction for $(H_c/h) > 5$

$$D = \frac{Eh(H + H_c)^2}{8(1 - \nu^2)} \quad (2\text{-}128)$$

The shear modulus of the core material as per Jones and Hersch

$$G_{core} = \frac{1.5FL_c}{B(H + H_c)^2(11\delta_4 - 8\delta_2)} \quad (2\text{-}129)$$

where δ_4 and δ_2 = deflection at quarter-span and midspan respectively
F = force over a support span L_c

Particular	Formula
The shear modulus G of isotropic material if the modulus of elasticity E is available	$G = \dfrac{E}{2(1-\nu)}$ (2-130)
The modulus of elasticity of the core material (Fig. 2-31)	$E_f = E_m\left(\dfrac{1-V^{2/3}}{1-V^{2/3}+V}\right)$ (2-131) where $V = (H_h/H)^3$, E_f = modulus of elasticity of foam, GPa (psi), E_m = modulus of elasticity of basic solid material, GPa (psi). Subscript f stands for foam/filament, m stands for matrix, and c stands for composite.

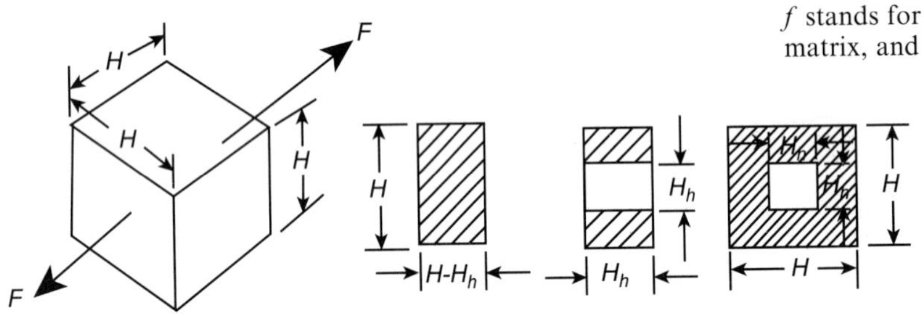

FIGURE 2-31 A unit cube foam subject to a tensile load.

Particular	Formula
The deflection for a beam panel according to Castigliano's theorem	$\delta = \dfrac{\partial U}{\partial F} = \dfrac{\partial}{\partial F}\left(\int \dfrac{M_b^2\,dx}{2EI} + \int \dfrac{V^2\,dx}{2GA}\right)$ (2-132)

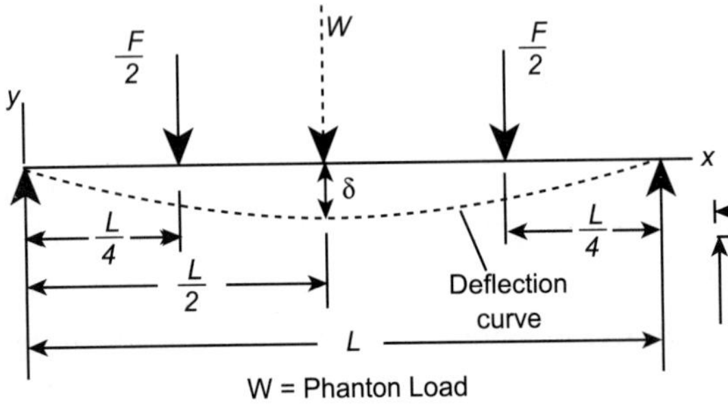

FIGURE 2-32 Phantom load.

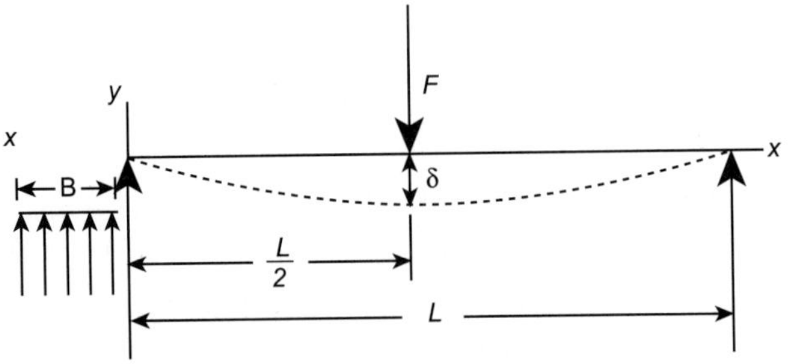

FIGURE 2-33

Particular	Formula
The deflection at midspan (Fig. 2-32)	$\delta_{L/2} = \dfrac{\partial U}{\partial W} = \dfrac{\partial}{\partial W}\left(\int \dfrac{M_b^2\,dx}{2EI} + \int \dfrac{V^2\,dx}{2GA}\right)_{W=0}$ (2-133a) $= \dfrac{5FL^3}{349EI} + \dfrac{FL}{8GA}$ (2-133b) where W is the *phantom load* (Fig. 2-32).

Particular	Formula	
The deflection per unit width for a sandwich panel at midspan (Fig. 2-32) under quarter-point loading	$\delta_{L/2} = \frac{5FL^3}{349DB} + \frac{FL}{8D_cB}$ where $D_c = G_{\text{core}}\left(\frac{H(H+H_c)}{2H_c}\right)$	(2-134)
The deflection per unit width for a sandwich panel at quarter panel (Fig. 2-32) under quarter-point loading	$\delta_{L/4} = \frac{FL^3}{96DB} + \frac{FL}{8D_cB}$	(2-135)
The deflection/unit width for a sandwich panel at center loading (Fig. 2-33)	$\delta_{L/2} = \frac{FL^3}{48DB} + \frac{FL}{4D_cB}$	(2-136)
The maximum normal stress (Fig. 2-32)	$\sigma_{\max} = \frac{M}{Z} = \frac{\left(\frac{F}{2} \times \frac{L}{4}\right)}{\left(\frac{BhH_c(h+H_c/2)}{H/2}\right)} = \frac{FL}{8BhH_c}$	(2-137)
The minimum normal stress	$\sigma_{\min} = \frac{FL}{8BhH}$	(2-138)
The average stress often used in the composite panel design	$\sigma_{\text{av}} = \frac{FL(L+H_c)}{16BhH_cH} \approx \frac{FL}{4Bh(H+H_c)}$	(2-139)
The maximum shear stress in the core	$\tau_{\max} = \frac{V}{[B(H+H_c)]/2} = \frac{2V}{B(H+H_c)}$	(2-140)
The core shear strain	$\gamma_{\text{core}} = \frac{\tau_{\max}}{G_{\text{core}}}$	(2-141)

FILAMENT REINFORCED STRUCTURES (Fig. 2-34)

The strain in the filament is same as the strain in the matrix of composite material if it has to have strain compatibility	$\varepsilon_m = \varepsilon_f$	(2-142)

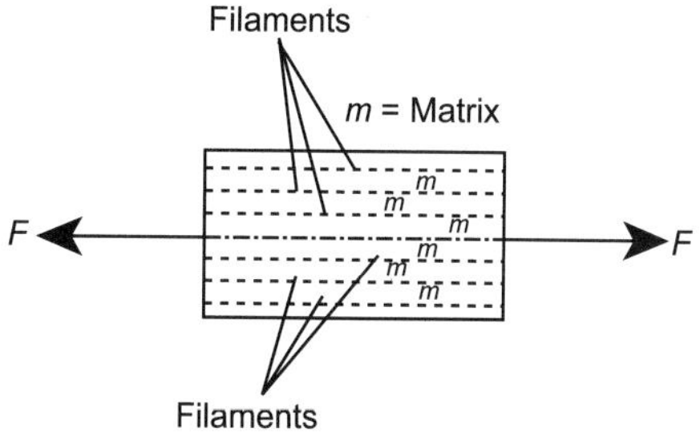

FIGURE 2-34

Particular	Formula
The relation between stress in matrix and stress in filament	$\frac{\sigma_m}{E_m} = \frac{\sigma_f}{E_f}$ (2-143)
For equilibrium	$F = \sigma_m A_m + \sigma_f A_f = \sigma_c A_c$ (2-144)
The stress in the filament	$\sigma_f = \frac{F E_f}{A_f E_f + A_m E_m}$ (2-145)
The stress in the matrix	$\sigma_m = \frac{F E_m}{A_f E_f + A_m E_m}$ (2-146)
The Young's modulus of composite	$E_c = \frac{E_f A_f}{(A_m + A_f)}$ (2-147)
The Young's modulus of chopped-up glass filaments in resin matrix but still oriented longitudinally with respect to load as proposed by Outerwater	$(E_c)_{chpd\text{-}f} = E_f \left[\frac{A_f}{A_c} - \left(\frac{1}{4\nu} \right) \left(\frac{\sigma}{\sigma_{yf}} \right) \left(\frac{D_f^2}{L p_c} \right) \right]$ (2-148) where σ = applied tensile stress, MPa (psi) σ_{yf} = the strength of the fiber, MPa (psi) D_f = diameter of fiber, mm (in) p_c = uniform distance of one fiber from another on circumference, mm (in) L = length of fiber, mm (in) Subscript *chpd-f* stands for chopped-up fiber.
The relation between σ_m and σ_f, which has to satisfy Eq. (2-142) at any location on the curves, Fig. 2-35	$\frac{\sigma_m}{(E_0)_m} = \frac{\sigma_f}{(E_0)_f}$ (2-149) where E_0 = secant modulus, GPa (Mpsi)
From Eq. (2-144), the expression for σ_c	$\sigma_c = \left(\frac{\sigma_f}{A_c} \right) \left(\frac{(E_0)_m A_m}{(E_0)_f} + A_f \right)$ (2-150) $= \frac{(\sigma_u)_f}{A_c} \left(\frac{(\sigma_m)_{max}}{(\sigma_u)_f} A_m + A_f \right)$ (2-151)
For structure with all filament, $A_m = 0$	$\sigma_c = \frac{(\sigma_u)_f A_f}{A_c} = (\sigma_u)_f$ (2-152)
For structure with no filament, $A_f = 0$	$\sigma_c = \frac{(\sigma_u)_f}{A_c} \left(\frac{(\sigma_m)_{max}}{\sigma_u} A_m \right) = (\sigma_m)_{max} = (\sigma_u)_m$ (2-153)

Particular	Formula

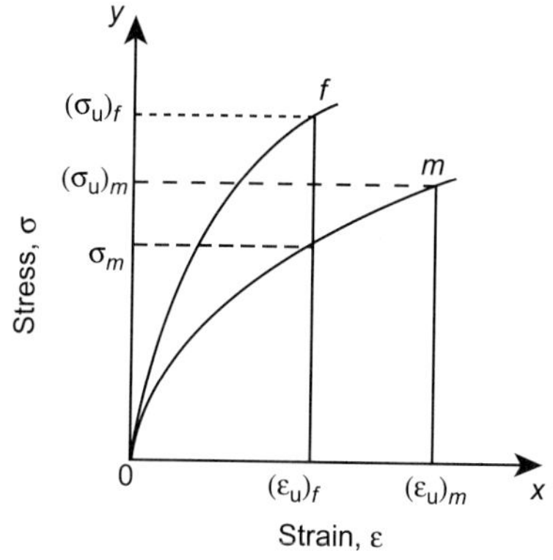

FIGURE 2-35 Stress-strain data for system shown in Fig. 2-34.

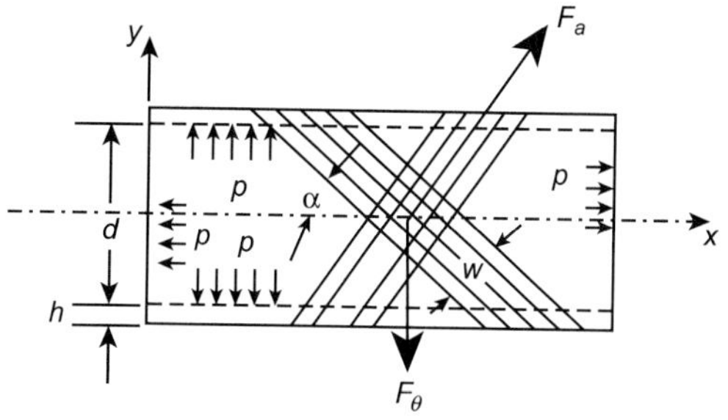

FIGURE 2-36 Filament wound cylindrical pressure vessel.

FILAMENT BINDER COMPOSITE (Fig. 2-36)

Hoop stress for a closed end vessel/cylinder made of filaments winding

$$\sigma_\theta = \frac{pd}{2h} \tag{2-154}$$

Longitudinal/axial stress for a closed end filament wound vessel/cylinder

$$\sigma_a = \frac{pd}{4h} \tag{2-155}$$

The force carried by a helical filament wound on a shell of width w subjected to internal pressure p in the α-direction

$$F_a = \sigma_{so} wh \tag{2-156}$$

where σ_{so} = strength of the filaments

The force in helical filament wound on a shell of width w subjected to internal pressure p in the hoop direction

$$F_\theta = F_\alpha \sin\alpha \tag{2-157}$$

The hoop stress in the vessel wall due to the pressure p

$$\sigma_\theta = \frac{F_\theta}{A} = \sigma_0 \sin^2\alpha \tag{2-158}$$

The stress in the vessel wall in the longitudinal/axial-direction

$$\sigma_a = \sigma_0 \cos^2\alpha \tag{2-159}$$

From Eq. (2-154) to (2-159) the optimum winding angle for closed end cylinders

$$\tan^2\alpha = \frac{\sigma_\theta}{\sigma_a} \quad \text{or} \quad \alpha \approx 55^\circ \tag{2-160}$$

The optimum winding angle for open end cylinders

$$\frac{\sigma_a}{\sigma_\theta} = \frac{\sigma}{\sigma_\theta} = \frac{\cos^2\alpha}{\sin^2\alpha} = \cot^2\alpha \quad \text{or} \quad \alpha = 90^\circ \tag{2-161}$$

Particular	Formula
The stress in the hoop/circumferential direction for the filament wound cylinder/vessel consisting windings in longitudinal, hoop and helical directions to satisfy equilibrium condition	$\sigma_\theta = \dfrac{\sigma'_\theta h_\theta + \sigma_{\theta\alpha} h_\alpha}{h}$ (2-162) where σ'_θ = stress in the circumferential wound layer $\sigma_{\theta\alpha}$ = circumferential component of stress in the helical layer h_t = total thickness $= h_a + h_\theta + h_\alpha$ h_a, h_θ and h_α are the thicknesses in the preceding layers of filament windings
The longitudinal stress for the case of winding under Eq. (2-162)	$\sigma_a = \dfrac{\sigma'_a h'_a + \sigma_{a\alpha} h_\alpha}{h}$ (2-163a) where $\sigma_\theta = \left(\dfrac{\sigma_{so}}{h}\right)(h_\theta + h_\alpha \sin^2 \alpha)$ (2-163b) $\sigma_a = \left(\dfrac{\sigma_{so}}{h}\right)(h_a + h_\alpha \cos^2 \alpha)$ (2-163c) σ_{so} = uniform filament stress $\sigma_{a\alpha}$ = longitudinal component of stress in helical layer
From Eqs. (2-159) and (2-158)	$\sigma_\theta + \sigma_a = \sigma_0(\sin^2 \alpha + \cos^2 \alpha) = \sigma_0$ (2-164)
From Eqs. (2-154) and (2-155)	$h = \dfrac{pd}{4\sigma_a}$ (2-165) $\sigma_\theta = \dfrac{pd}{2h}$ (2-154)
The sum of stresses σ_θ and σ_a	$\sigma_\theta + \sigma_a = 3\sigma_a = \sigma_0$ or $\sigma_a = \dfrac{\sigma_0}{3}$ (2-166)
For the ideal vessel	$h = \dfrac{3pd}{4\sigma_0}$ (2-167) $h_a = \dfrac{h}{3} - h_\alpha \cos^2 \alpha$ (2-168a) $h_\theta = \frac{2}{3} - h_\alpha \sin^2 \alpha$ (2-168b) $h_\alpha = \dfrac{2h_a - h_\theta}{1 - 3\cos^2 \alpha}$ (2-168c)

Particular	Formula
The structural efficiency of the wound vessel/cylinder	$\eta = \frac{W}{V_{en} p_i}$ (2-169) where W = weight of the vessel, kN (lbf) V_{en} = enclosed volume, m^3 (in^3) p_i = internal pressure, MPa (psi)

FILAMENT-OVERLAY COMPOSITE

The stress in the wire which is wound on thin walled shell/cylinder with a wire of the same material (Fig. 2-37)	$\sigma_{wr} = \frac{T}{uw}$ (2-170) where T = tension, kN (lbf) uw = area of the element, m^2 (in^2)
Under equilibrium condition over the length of shell L, the hoop stress	$(\sigma_\theta)_{sh} = -\frac{T}{wh}$ (2-171)

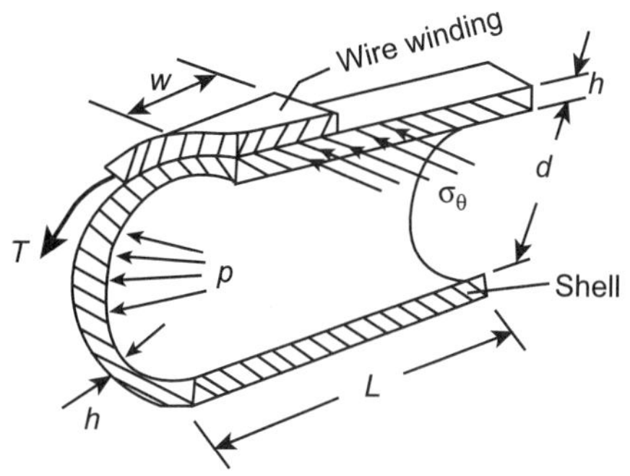

FIGURE 2-37 Shell subjected to an internal pressure.

The tension in the wound wire on the shell under internal pressure	$T_{wr} = \frac{pd}{2(h+u)} + \frac{T}{wu}$ (2-172)
The tension in the shell under the above same condition	$T_{cy} = \frac{pd}{2(h+u)} - \frac{T}{wh}$ (2-173)
The yielding of shell due to internal pressure, i.e., due to plastic flow of material of the shell.	$(\sigma_0)_{shy} = -\frac{T}{wh} = -\sigma_y$ (2-174)
For the above same winding material under the tension equal to compression yield limits, the stress in the wire.	$\sigma_{wr} = \frac{T}{uw} = \sigma_y\left(\frac{h}{u}\right)$ (2-175)
If the vessel material is different from the winding material then stress in the wire and vessel	$\sigma_{cy} = \varepsilon_{sh} E_{sh}$ (2-176a) $\sigma_{wr} = \varepsilon_{wr} E_{wr}$ (2-176b)

Particular	Formula
For uniform distribution of stress in the cylinder/shell and in the wire, strains are proportional to the mean radii	$\frac{\bar{r}_{cy}}{\bar{r}_{wr}} = \frac{\varepsilon_{cy}}{\varepsilon_{wr}} = \frac{\sigma_{cy} E_{wr}}{\sigma_{wr} E_{cy}}$ (2-177)
From Eq. (2-177), the stress in the cylinder and the wire	$\sigma_{sh} = \sigma_{cy} = \sigma_{wr} \frac{E_{cy}\bar{r}_{cy}}{E_{wr}\bar{r}_{wr}}$ (2-178a)
	$\sigma_{wr} = \sigma_{cy} \frac{E_{wr}\bar{r}_{wr}}{E_{cy}\bar{r}_{cy}}$ (2-178b)
	where subscripts cy stands for cylinder, sh for shell and wr for winding. $\bar{r}_{cy}$ and $\bar{r}_{wr}$ are mean radii of cylinder and winding respectively.
The total load on the cylinder and the winding	$\sigma_{cy}(2Lh) + \sigma_{wr}(2Lu) = pdL$ (2-179)
From Eq. (2-179), the stress in the cylinder (σ_{cy}) and the winding (σ_{wr})	$\sigma_{cy} = \frac{pd}{2\left(\frac{E_{wr}\bar{r}_{wr}u}{E_{cy}\bar{r}_{cy}} + h\right)}$ (2-180a)
	$\sigma_{wr} = \frac{pd}{2\left(\frac{E_{cy}\bar{r}_{cy}}{E_{wr}\bar{r}_{wr}} + u\right)}$ (2-180b)
The stress in the cylinder is the sum of results of Eqs. (2-180a) and (2-171)	$\sigma_{Rcy} = \frac{pd}{2\left(\frac{E_{wr}\bar{r}_{wr}u}{E_{cy}\bar{r}_{cy}} + h\right)} - \frac{T}{wh}$ (2-181)
The resultant stress in the winding is the sum of results of Eq. (2-180b) and (2-170)	$\sigma_{Rwr} = \frac{pd}{2\left(\frac{E_{cy}\bar{r}_{cy}h}{E_{wr}\bar{r}_{wr}} + u\right)} + \frac{T}{wu}$ (2-182)
For advanced theory using *Theory of elasticity* and *Plasticity* construction on composite structures and materials	Refer to advanced books and handbooks on composites, structures, handbooks and design data for reinforced plastics and materials.
For representative properties for fiber reinforcement	Refer to Table 2-1

FORMULAS AND DATA FOR VARIOUS CROSS SECTIONS OF MACHINE ELEMENTS

For further data on static stresses, properties and torsion of shafts of various cross-sections: shear, moments, and deflections of beams, strain rosettes, and singularity functions	Refer to Tables 2-2 to 2-12
For summary of stress and strain formulas under various types of loads	Refer to Table 2-13

TABLE 2-1
Representative properties for fiber reinforcement

Fiber	Typical fiber diameter		Density, ρ		Tensile strength, σ_{st}		Modulis of elasticity, E}		Coefficient of thermal expansion, α		Thermal conductivity, K	
	$\times 10^{-3}$mm	$\times 10^{-3}$ in	g/cm^3	lb/in^3	MPa	kpsi	GPa	Mpsi	μm/m K	μin/in°F	W/(m^{2}K/m)	Btu/(ft^{2}h°F)/in)
E glass	10.2	0.4	2.48	0.092	3100	450	72.5	10.5	5.0	2.8	1680	7.5
S glass	10.2	0.4	2.43	0.090	4498	650	85	12.3				
970 S glass	10.2	0.4	2.46	0.091	5510	800	100	14.5				
Boron on tungsten	102.0	4	2.56	0.095	2756	400	415	60	5.0	2.8		
Graphite	5–10	0.2–0.4	1.43–1.75	0.053–0.066	1723–3445	250–500	241–689	35–100	2.7	1.5	13440	60
Beryllium	127	5	1.78	0.066	1240	180	310	45	11.5	6.4	19488	87
Silicon carbide on tungsten	102	4	3.40	0.126	2480	360	414	60	4.0	2.2	6496	29
Stainless steel	13	0.5	7.64	0.283	2852–4184	385–600	200	29	54	30	22400–38080	100–170
Asbestos	0.025–0.25	0.001–0.01	2.43	0.090	689–2067	100–300	172	25				
Aluminum	5	0.2	2.62	0.097	1378–2067	200–300	138–413	20–60	6.5	3.7		
Polyamide	5–13	0.2–0.5	1.11	0.041	827	120	2.8	0.4	81–90	45–50	381	1.7
Polyester	20.5	0.8	1.49	0.052	690	100	4	0.6	81–90	45–50	381	1.7

Courtesy: J. E. Ashton, J. C. Halpin, and P. H. Petit, *Primer on Composite Materials: Analysis*, Technomic Publishing Co., Inc., 750 Summer St., Stanford, Conn. 06901, 1969.

TABLE 2-2
Torsion of shafts of various cross sections

Cross section	Polar section modulus, $Z_0 = J/c$	Polar radius of gyration, k_0	Angular deflection, θ: In terms of torsional moment, M_t	Angular deflection, θ: In terms of maximum stress, τ
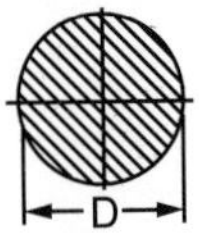	$\frac{\pi D^3}{16}$	$\frac{D}{\sqrt{8}} = 0.354D$	$\frac{32l}{\pi D^4}\frac{M_t}{D}$	$\frac{2l}{D}\frac{\tau}{G}$ τ at circumference
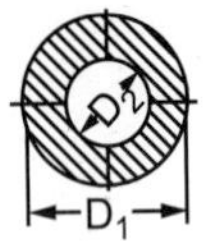	$\frac{\pi(D_1^4 - D_2^4)}{16D_1}$	$\sqrt{\frac{D_1^2 + D_2^2}{8}} = 0.354\sqrt{D_1^2 + D_2^2}$	$\frac{32l}{\pi(D_1^4 - D_2^4)}\frac{M_t}{G}$	$\frac{2l}{D_1}\frac{\tau}{G}$ τ at outer circumference
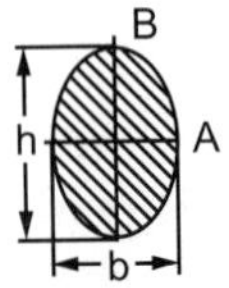	$\frac{\pi b^2 h}{16}$ [a] $h > b$	$\frac{1}{4}\sqrt{b^2 + h^2}$	$\frac{16(b^2 + h^2)l}{\pi b^3 h^3}\frac{M_t}{G}$	$\frac{(b^2 + h^2)l}{bh^2}\frac{\tau}{G}$ τ at A [b]
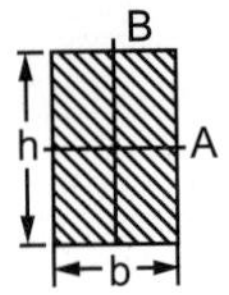	$\frac{2b^2 h}{9}$ [a] $h > b$	$\sqrt{\frac{b^2 + h^2}{12}}$ $= 0.289\sqrt{b^2 + h^2}$	$\frac{m(b^2 + h^2)l}{b^3 h^3}\frac{M_t}{G}$ $\frac{h}{b} = 1$ 2 4 8 $m = 3.56$ 3.50 3.35 3.21 $n = 0.79$ 0.78 0.74 0.71	$\frac{n(b^2 + h^2)l}{bh^2}\frac{\tau}{G}$ τ at A [c]
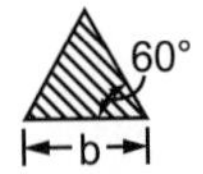	$\frac{b^3}{20}$ [a]	$0.289b$	$\frac{46.2l}{b^4}\frac{M_t}{G}$	$\frac{2.31l}{b}\frac{\tau}{G}$ τ at center of side
	$0.92b^3$ [a]	$0.645b$	$\frac{0.967l}{b^4}\frac{M_t}{G}$	$\frac{0.9l}{b}\frac{\tau}{G}$ τ at center of side

[a] This value is not true value of Z_0 but is the value of Z_0 for a circular section of equal strength and may be used for determining the maximum stress by the formula $\tau = M_t/Z_0$.
[b] At B, shear stress $= 10M_t/\pi bh^2$.
[c] At B, shear stress $= 9M_t/2bh^2$.
Source: V. L. Maleev and J. B. Hartman, *Machine Design*, International Textbook Company, Scranton, Pennsylvania, 1954.

TABLE 2-3
Shear stress in beams, caused by bending

Section	Shear stress at a distance y from neutral axis, τ, MPa (psi)	Maximum shear stress, τ_{max}, MPa (psi)
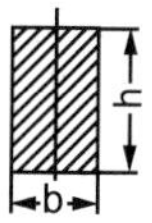	$\frac{3F}{2bh}\left[1-\left(\frac{2y}{h}\right)^2\right]$	$\frac{3F}{2bh}=1.5\frac{F}{A}$ (for $y=0$)
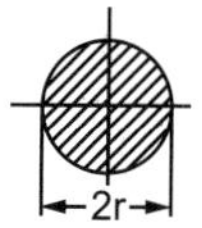	$\frac{4F}{3\pi r^2}\left[1-\left(\frac{y}{r}\right)^2\right]$	$\frac{4F}{3\pi r^2}=1.33\frac{F}{A}$ (for $y=0$)
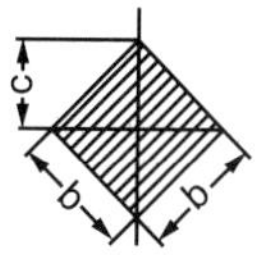	$\frac{F\sqrt{2}}{b^2}\left[1+\frac{y\sqrt{2}}{b}-4\left(\frac{y}{b}\right)^2\right]$	$1.591\frac{F}{A}\left(\text{for } y=\frac{c}{4}\right)$
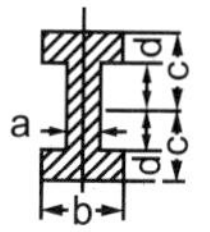		$\frac{3F}{4a}\left[\frac{bc^2-(b-a)d^2}{bc^3-(b-a)d^3}\right]$ (for $y=0$)

TABLE 2-4
The values of constants *a* in Eq. (2-107)

	Yield stress in compression, σ_{yc}		Value of *a* for various end-fixity coefficients			
Material	**MPa**	**kpsi**	**1**	**4**	**2**	***n***
Timber	49	7	$\frac{1}{750}$	$\frac{1}{3000}$	$\frac{1}{1500}$	$\frac{1}{n\times 750}$
Cast iron	549	80	$\frac{1}{1600}$	$\frac{1}{6400}$	$\frac{1}{3200}$	$\frac{1}{n\times 1600}$
Mild steel	324	47	$\frac{1}{7500}$	$\frac{1}{30000}$	$\frac{1}{15000}$	$\frac{1}{n\times 7500}$

TABLE 2-5
End condition coefficient *n* (Fig. 2-25)

Particular	*n*
One end fixed and the other end free	0.25
Both ends rounded and guided or hinged	1
One end fixed, and the other end rounded and guided or hinged	2
Both ends fixed rigidly	4
Both ends flat	1 to 4

TABLE 2-6
End-fixity coefficients for cast iron column to be used in Eq. (2-104)

End conditions	C_1	Maximum, l/k
Round	175	90
Fixed	88	160
One fixed, one round	116	115

TABLE 2-7
Properties of cross sections

Section	Area, A	Moment of inertia, I	Distance to farthest point, c	Section modulus, $Z = I/c$	Radius of gyration, $k = \sqrt{I/A}$
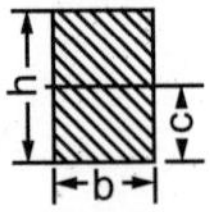	bh	$\frac{bh^3}{12}$	$\frac{h}{2}$	$\frac{bh^2}{6}$	$0.289h$
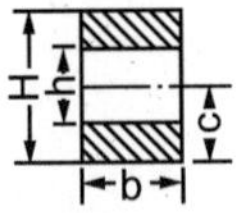	$(H - c)b$	$\frac{b}{12}(H^3 - h^3)$	$\frac{H}{2}$	$\frac{b(H^3 - h^3)}{3H}$	$\sqrt{\frac{H^3 - h^3}{12(H - h)}}$
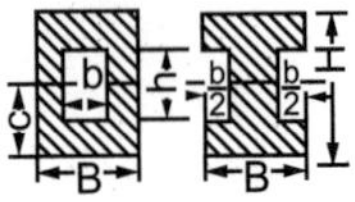	$BH - bh$	$\frac{BH^3 - bh^3}{12}$	$\frac{H}{2}$	$\frac{BH^3 - bh^3}{6H}$	$\sqrt{\frac{BH^3 - bh^3}{12(BH - bh)}}$
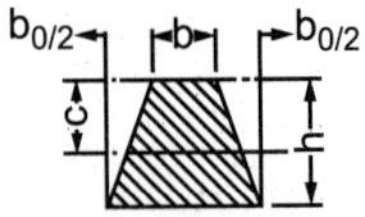	$\left(\frac{2b + b_0}{2}\right)h$	$\frac{(6b^2 + 6bb_0 + b_0^2)h^3}{36(2b + b_0)}$	$\frac{(3b + 2b_0)h}{3(2b + b_0)}$	$\frac{(6b^2 + 6bb_0 + b_0^2)h^2}{12(3b + b_0)}$	$\sqrt{\frac{I}{A}}$
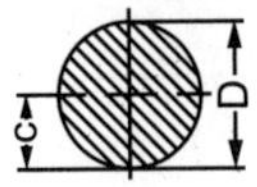	$\frac{\pi D^2}{4}$	$\frac{\pi D^4}{64}$	$\frac{D}{2}$	$\frac{\pi D^3}{32}$	$\frac{D}{4}$
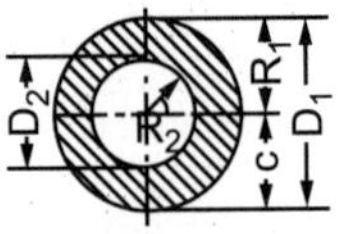	$\frac{\pi}{4}(D_1^2 - D_2^2)$	$\frac{\pi}{64}(D_1^4 - D_2^4) = \frac{\pi}{4}(R_1^4 - R_2^4)$	$\frac{D_1}{2} = R_1$	$\frac{\pi(D_1^4 - D_2^4)}{32D_1}$	$\frac{\sqrt{D_1^2 + D_2^2}}{4} = \frac{\sqrt{R_1^2 + R_2^2}}{2}$
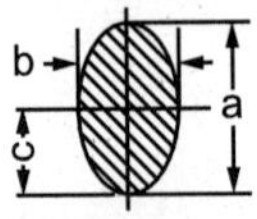	πab	$\frac{\pi ba^3}{64}$	$\frac{a}{2}$	$\frac{\pi ba^2}{32}$	$\frac{a}{4}$
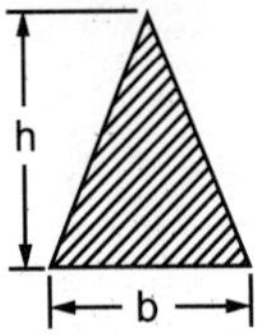	$\frac{bh}{2}$	$\frac{bh^3}{36}$	...	$\frac{bh^2}{24}$	$0.236h$

TABLE 2-8
Shear, moment, and deflection formulas for beams

Loading, support, and reference number	Reactions R_1 and R_2, vertical shear V	Bending moment M_b, and maximum bending moment	Deflection y and maximum deflection
1. Cantilever, end load	$R_2 = +F$	$M_b = -Fx$	$y = -\frac{1}{6}\frac{F}{EI}(x^3 - 3l^2x + 2l^3)$
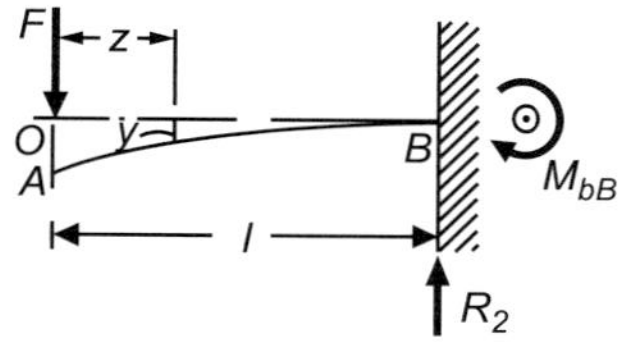	$V = -F$	Max $M_{bB} = Fl$ at B	$y_{max} = -\frac{1}{3}\frac{Fl^3}{EI}$ at A
2. Cantilever, intermediate load	$R_2 = +F$	A to B : $M_b = 0$	A to B: $y = -\frac{1}{6}\frac{F}{EI}(-a^3 + 3a^2l - 3a^2x)$
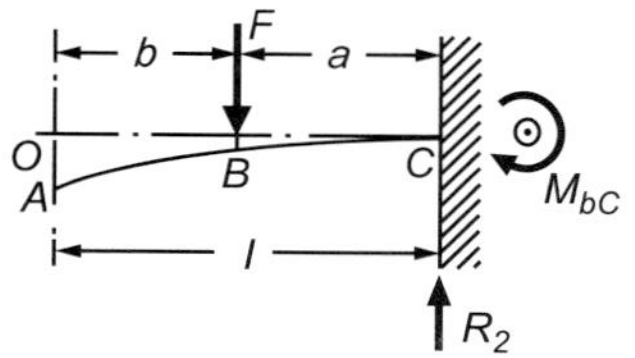	A to B : $V = 0$	B to C : $M_b = -F(x - b)$	B to C: $y = -\frac{1}{6}\frac{F}{EI}[(x-b)^3 - 3a^2(x-b) + 2a^3]$
	B to C : $V = -F$	Max $M_{bC} = -Fa$ at C	$y_{max} = -\frac{1}{6}\frac{F}{EI}(3a^2l - a^3)$
3. Cantilever, uniform load	$R_2 = +W = wl$	$M_b = -\frac{1}{2}\frac{W}{l}x^2$	$y = -\frac{1}{24}\frac{W}{EIl}(x^4 - 4l^3x + 3l^4)$
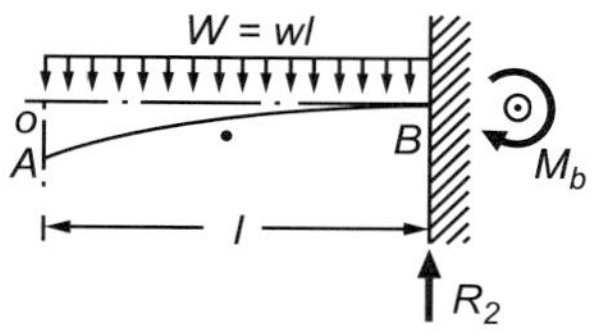	$V = -\frac{W}{l}x$	Max $M_{bB} = -\frac{1}{2}Wl$ at B	$y_{max} = -\frac{1}{8}\frac{Wl^3}{EI}$
4. End supports, center load	$R_1 = +\frac{1}{2}F, R_2 = +\frac{1}{2}F$	A to B : $M_b = +\frac{1}{2}Fx$	A to B: $y = -\frac{1}{48}\frac{F}{EI}(3l^2x - 4x^3)$
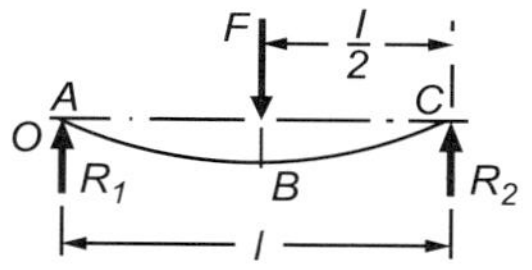	A to B : $V = +\frac{1}{2}F$	B to C : $M_b = +\frac{1}{2}F(l - x)$	$y_{max} = -\frac{1}{48}\frac{Fl^3}{EI}$ at B
	B to C : $V = -\frac{1}{2}F$	Max $M_{bB} = +\frac{1}{4}FL$ at B	

TABLE 2-8
Shear, moment, and deflection formulas for beams (*Cont.*)

Loading, support, and reference number	Reactions R_1 and R_2, vertical shear V	Bending moment M_b, and maximum bending moment	Deflection y and maximum deflection
5. End supports, intermediate load	$R_1 = +F\frac{b}{l},\ R_2 = +F\frac{a}{l}$	A to B: $M_b = +F\frac{b}{l}x$	A to B: $y = -\frac{Fbx}{6EIl}[2l(l-x) - b^2 - (l-x)^2]$
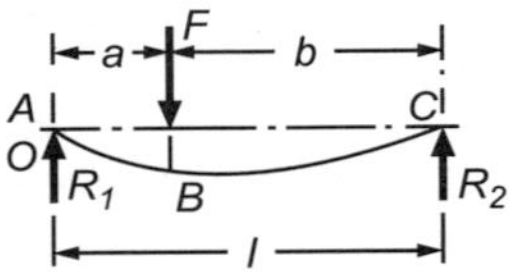	A to B: $V = +F\frac{b}{l}$ B to C: $V = -F\frac{a}{l}$	B to C: $M_b = +F\frac{a}{l}(l-x)$ $\text{Max } M_{bB} = +F\frac{ab}{l}$ at B	B to C: $y = -\frac{Fa(l-x)}{6EIl}[2lb - b^2 - (l-x)^2]$ $y_{\max} = -\frac{Fab}{27EIl}(a+2b)\sqrt{3a(a+2b)}$ at $x = \sqrt{\frac{1}{3}a(a+2b)}$ when $a > b$
6. End supports, uniform load	$R_2 = +\frac{1}{2}W,\ R_2 = +\frac{1}{2}W$	$M_b = \frac{1}{2}W\left(x - \frac{x^2}{l}\right)$	$y = -\frac{1}{24}\frac{Wx}{EIl}(l^3 - 2lx^2 + x^3)$
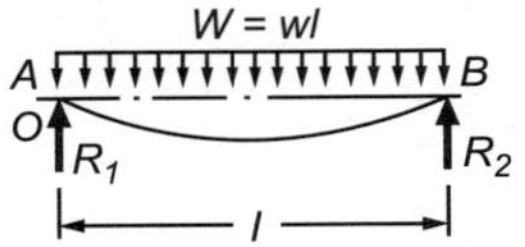	$V = \frac{1}{2}W\left(1 - \frac{2x}{l}\right)$	$\text{Max } M_b = +\frac{1}{8}Wl$ at $x = \frac{1}{2}l$	$\text{Max } y = -\frac{5}{384}\frac{Wl^3}{EI}$ at $x = \frac{1}{2}l$
7. One end fixed, one end supported, center load	$R_1 = \frac{5}{16}F,\ R_2 = \frac{11}{16}F$	A to B: $M_b = \frac{5}{16}Fx$	A to B: $y = \frac{1}{96}\frac{F}{EI}(5x^3 - 3l^2x)$
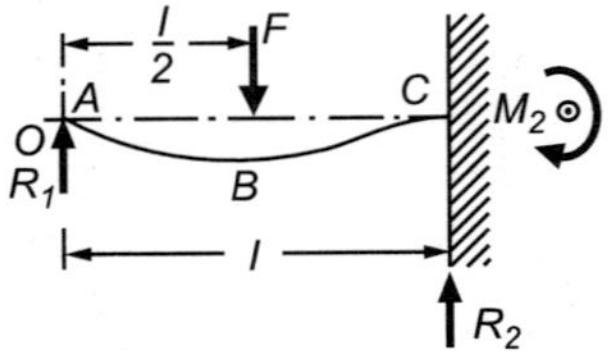	$M_2 = \frac{3}{16}Fl$	B to C: $M_b = F(\frac{1}{2}l - \frac{11}{16}x)$	B to C: $y = \frac{1}{96}\frac{F}{EI}[5x^3 - 16(x - \frac{1}{2}l)^3 - 3l^2x]$
	A to B: $V = +\frac{5}{16}F$	$\text{Max } +M_{bB} = \frac{5}{32}Fl$ at B	$y_{\max} = -0.00932\frac{Fl^3}{EI}$ at $x = 0.4472l$
	B to C: $V = -\frac{11}{16}F$	$\text{Max } -M_{bC} = \frac{3}{16}Fl$ at C	
8. One end fixed, one end supported, uniform load	$R_1 = \frac{3}{8}W,\ R_2 = \frac{5}{8}W$	$M_b = W(\frac{3}{8}x - \frac{1}{2}x^2)$	$y = \frac{1}{48}\frac{W}{EIl}(3lx^3 - 2x^4 - l^3x)$
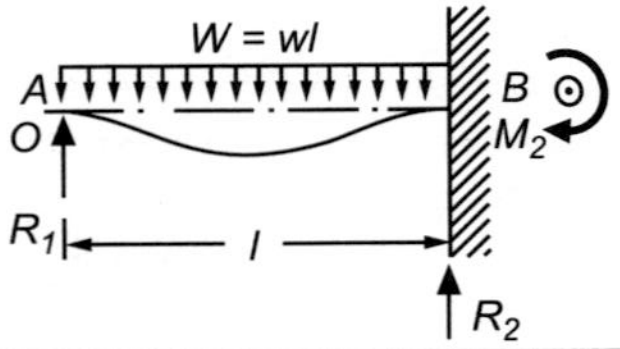	$M_2 = \frac{1}{8}Wl$	$\text{Max } +M_b = \frac{9}{128}Wl$ at $x = \frac{3}{8}l$	$y_{\max} = -0.0054\frac{Wl^3}{El}$ at $x = 0.4215l$
	$V = W\left(\frac{3}{8} - \frac{x}{l}\right)$	$\text{Max } -M_{bB} = -\frac{1}{8}Wl$ at B	

TABLE 2-8
Shear, moment, and deflection formulas for beams (*Cont.*)

Loading, support, and reference number	Reactions R_1 and R_2, vertical shear V	Bending moment M_b, and maximum bending moment	Deflection y and maximum deflection
9. Both ends fixed, center load	$R_1 = \frac{1}{2}F,\ R_2 = \frac{1}{2}F$	A to B: $M_b = \frac{1}{8}F(4x - l)$	A to B: $y = -\frac{1}{48}\frac{F}{EI}(3lx^2 - 4x^3)$
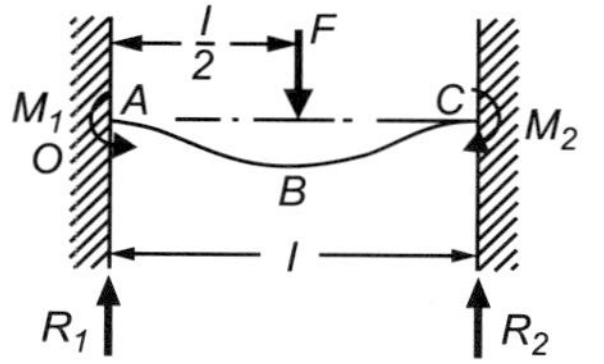	$M_1 = \frac{1}{8}Fl,\ M_2 = \frac{1}{8}Fl$ A to B: $V = +\frac{1}{2}F$ B to C: $V = -\frac{1}{2}F$	B to C: $M_b = \frac{1}{8}F(3l - 4x)$ Max $+M_{bB} = \frac{1}{8}Fl$ at B Max $-M_{bA,C} = -\frac{1}{8}Fl$ at A and C	$y_{max} = -\frac{1}{192}\frac{Fl^3}{EI}$ at B
10. Both ends fixed, intermediate load	$R_1 = \frac{Fb^2}{l^3}(3a + b)$	A to B: $M_b = -F\frac{ab^2}{l^2} + R_1 x$	A to B: $y = \frac{1}{6}\frac{Fb^2x^2}{EIl^3}(3ax + bx - 3al)$
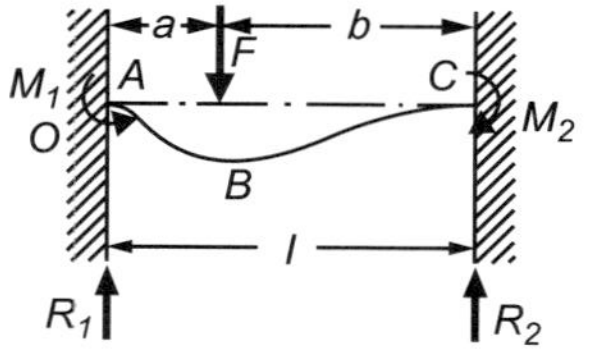	$R_2 = \frac{Fa^2}{l^3}(3b + a)$	B to C: $M_b = -F\frac{ab^2}{l^2} + R_1 x - F(x - a)$	B to C: $y = \frac{1}{6}\frac{Fa^2(l - x)^2}{EIl^3}[(3b + a)(l - x) - 3bl]$
	$M_1 = -F\frac{ab^2}{l_2},\ M_2 = -F\frac{a^2b}{l^2}$	Max $+M_b = -F\frac{ab^2}{l^2} + R_1 a$ at B	$y_{max} = -\frac{2}{3}\frac{F}{EI}\frac{a^3b^2}{(3a + b)^2}$ at $x = \frac{2al}{(3a + b)}$ if $a > b$
	A to B: $V = R_1$ B to C: $V = R_1 - F = -R_2$	Max $-M_b = -M_1$ when $a < b$ Max $-M_b = -M_2$ when $a > b$	$y_{max} = -\frac{2}{3}\frac{F}{EI}\frac{a^2b^3}{(3b + a)^2}$ at $x = l - \frac{2bl}{(3b + a)}$ if $a < b$
11. Both ends fixed, uniform load	$R_1 = \frac{1}{2}W,\ R_2 = \frac{1}{2}W$	$M_b = \frac{1}{2}W\left(x - \frac{x^2}{l} - \frac{1}{6}l\right)$	$y = \frac{1}{24}\frac{Wx^2}{EIl}(2lx - l^2 - x^2)$
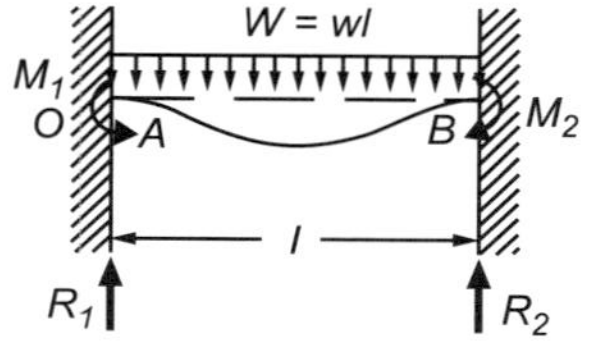	$M_1 = \frac{1}{12}Wl,\ M_2 = \frac{1}{12}Wl$	Max $+M_b = \frac{1}{24}Wl$ at $x = \frac{1}{2}l$	$y_{max} = -\frac{1}{384}\frac{Wl^3}{EI}$ at $x = \frac{1}{2}l$
	$V = \frac{1}{2}W\left(1 - \frac{2x}{l}\right)$	Max $-M_b = -\frac{1}{12}Wl$ at A and B	

Source: J. E. Shigley, *Mechanical Engineering Design*, 3rd. ed., McGraw-Hill Book Company, New York, 1977.

TABLE 2-9
Some equations for use with the Castigliano method

Type of load	General energy equation	Energy equation	General deflection equation
Axial	$U = \int_0^l \frac{F^2}{2AE}\,ds$	$U = \frac{F^2 l}{2AE} = \frac{\sigma^2 Al}{2E}$	$\delta = \int_0^l \frac{F(\partial F/\partial Q)}{AE}\,ds$
Bending	$U = \int_0^l \frac{M_b^2 l}{2EI}\,ds$	$U = \frac{M_b^2 l}{2EI}$	$\delta = \int_0^l \frac{M_b(\partial M_b/\partial Q)}{EI}\,ds$
Combined axial and bending	$U = \int_0^l \frac{F^2}{2AE}\,ds + \int_0^l \frac{M_b^2}{2EI}\,ds$	$U = \frac{F^2 l}{2EA} + \frac{M_b^2 l}{2EI}$	Sum of axial and bending load
Torsion	$U = \int_0^l \frac{M_t^2}{2GJ}\,ds$	$U = \frac{M_t^2 l}{2GJ}$	$\delta = \int_0^l \frac{M_t(\partial M_t/\partial Q)}{GJ}\,ds$
Transverse shear	$U = \int_0^l \frac{V^2\,ds}{2GA}$	$U = \frac{V^2 l}{2GA} = \frac{\tau^2}{2G}Al$	$\delta = \int_0^l \frac{V(\partial V/\partial G)}{GA}\,ds$
Transverse shear (rectangular section)	$U = \int_0^l \frac{3V^2}{5GA}\,ds$	$U = \frac{3V^2 l}{5GA}$	$\delta = \int_0^l \frac{6V(\partial V/\partial Q)}{5GA}\,ds$
Open-coiled helical spring subjected to axial load F	$U = \int_0^l \frac{M_t^2}{2GJ}\,ds + \int_0^l \frac{M_b^2}{2EI}\,ds$	$U = \frac{M_t^2 l}{2GJ} + \frac{M_b^2 l}{2EI}$ $= \frac{LFR^2}{2}\left[\frac{\cos^2\alpha}{GJ} + \frac{\sin^2\alpha}{EI}\right]$	$\delta = 2\pi iFR^3 \sec\alpha\left[\frac{\cos^2\alpha}{GJ} + \frac{\sin^2\alpha}{EI}\right]$ where $R = \frac{D}{2}$ = mean radius of coil, α = helix angle of spring, i = number of coils or turns

TABLE 2-10
Mechanical and physical constants of some materials[1,2]

Material	Modulus of elasticity, E		Modulus of rigidity, G		Poisson's ratio, ν	Density, ρ[a], Mg/m^3	Unit weight, γ[b]			
	GPa	Mpsi	GPa	Mpsi			kfg/m^3	kN/m^3	lbf/in^3	lbf/ft^3
Aluminum	69	10.0	26	3.8	0.334	2.69	2,685	26.3	0.097	167
Aluminum cast	70	10.15	30	4.35			2,650	26.0	0.096	166
Aluminum (all alloys)	72	10.4	27	3.9	0.320	2.80	2,713	27.0	0.10	173
Beryllium copper	124	18.0	48	7.0	0.285	8.22	8,221	80.6	0.297	513
Carbon steel	206	30.0	79	11.5	0.292	7.81	7,806	76.6	0.282	487
Cast iron, gray	100	14.5	41	6.0	0.211	7.20	7,197	70.6	0.260	450
Malleable cast iron	170	24.6	90	13.0			7,200			
Inconel	214	31.0	76	11.0	0.290	8.42	8,418	83.3	0.307	530
Magnesium alloy	45	6.5	16	2.4	0.350	1.80	1,799	17.6	0.065	117
Molybdenum	331	48.0	117	17.0	0.307	10.19	10,186	100.0	0.368	636
Monet metal	179	26.0	65	9.5	0.320	8.83	8,830	86.6	0.319	551
Nickel-silver	127	18.5	48	7.0	0.332	8.75	8,747	85.80	0.316	546
Nickel alloy	207	30	79	11.5	0.30	8.3			0.300	518
Nickel steel	207	30.0	79	11.5	0.291	7.75	7,751	76.0	0.280	484
Phosphor bronze	111	16.0	41	6.0	0.349	8.17	8,166	80.1	0.295	510
Steel (18-8), stainless	190	27.5	73	10.6	0.305	7.75	7,750	76.0	0.280	484
Titanium (pure)	103	15.0				4.47	4,470	43.9	0.16	279
Titanium alloy	114	16.5	43	6.2	0.33	6.6				
Brass	106	15.5	40	5.8	0.324	8.55	8,553	83.9	0.309	534
Bronze	96	14.0	38	5.5	0.349	8.30	8,304	81.4		
Bronze cast	80	11.6	35	5.0			8,200			
Copper	121	17.5	46	6.6	0.326	8.90	8,913	87,4	0.322	556
Tungsten	345	50.0	138	20.0		18.82	18,822	184.6		
Douglas fir	11	1.6	4	0.6	0.330	4.43	443	4.3	0.016	28
Glass	46	6.7	19	2.7	0.245	2.60	2,602	25.5	0.094	162
Lead	36	5.3	13	1.9	0.431	11.38	11,377	111.6	0.411	710
Concrete (compression)	14–28	2.0–4.0				2.35	2,353	23.1		147
Wrought iron	190	27.5	70	10.2			7,700			
Zinc alloy	83	12	31	4.5	0.33	6.6			0.24	415

[a] ρ = mass density.
[b] γ = weight density; w is also the symbol used for unit weight of materials.
Sources: K. Lingaiah and B. R. Narayana Iyengar, *Machine Design Data Handbook*, Vol. I (*SI and Customary Metric Units*), Suma Publishers, Bangalore, India, and K. Lingaiah, *Machine Design Data Handbook*, Vol. II (*SI and Customary Metric Units*), Suma Publishers, Bangalore, India. 1986.

TABLE 2-11
Relations between strain rosette readings and principal stresses

	Rosette type			
Required solutions	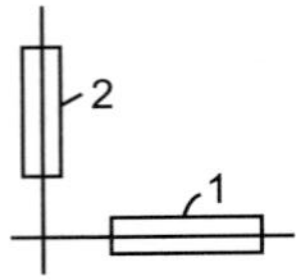 Two-gage	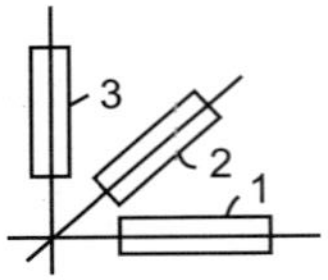 Rectangular	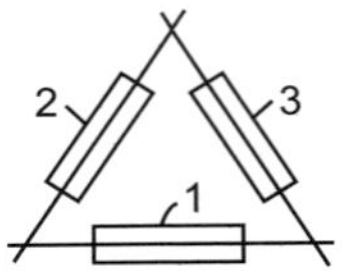 Delta	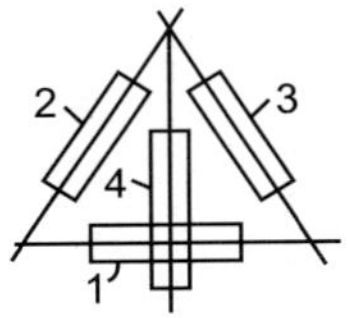 T-Delta
Maximum normal stress, $\sigma_{\max}$	$\dfrac{E}{1-\mu^2}(\varepsilon_1 + \mu\varepsilon_2)$	$\dfrac{E}{2}\left\{\dfrac{\varepsilon_1+\varepsilon_3}{1-\mu} + \dfrac{1}{1+\mu} \times \sqrt{(\varepsilon_1-\varepsilon_3)^2 + [2\varepsilon_2 - (\varepsilon_1+\varepsilon_3)]^2}\right\}$	$E\left[\dfrac{\varepsilon_1+\varepsilon_2+\varepsilon_3}{3(1-\mu)} + \dfrac{1}{1+\mu} \times \sqrt{\left(\varepsilon_1 - \dfrac{\varepsilon_1+\varepsilon_2+\varepsilon_3}{3}\right)^2 + \left(\dfrac{\varepsilon_2-\varepsilon_3}{\sqrt{3}}\right)^2}\right]$	$\dfrac{E}{2}\left[\dfrac{\varepsilon_1+\varepsilon_4}{1-\mu} + \dfrac{1}{1+\mu} \times \sqrt{(\varepsilon_1-\varepsilon_4)^2 + \dfrac{4}{3}(\varepsilon_2-\varepsilon_3)^2}\right]$
Minimum normal stress, $\sigma_{\min}$	$\dfrac{E}{1-\mu^2}(\varepsilon_2 + \mu\varepsilon_1)$	$\dfrac{E}{2}\left\{\dfrac{\varepsilon_1+\varepsilon_3}{1-\mu} - \dfrac{1}{1+\mu} \times \sqrt{(\varepsilon_1-\varepsilon_3)^2 + [2\varepsilon_2 - (\varepsilon_1+\varepsilon_3)]^2}\right\}$	$E\left[\dfrac{\varepsilon_1+\varepsilon_2+\varepsilon_3}{3(1-\mu)} - \dfrac{1}{1+\mu} \times \sqrt{\left(\varepsilon_1 - \dfrac{\varepsilon_1+\varepsilon_2+\varepsilon_3}{3}\right)^2 + \left(\dfrac{\varepsilon_2-\varepsilon_3}{\sqrt{3}}\right)^2}\right]$	$\dfrac{E}{2}\left[\dfrac{\varepsilon_1+\varepsilon_4}{1-\mu} - \dfrac{1}{1+\mu} \times \sqrt{(\varepsilon_1-\varepsilon_4)^2 + \dfrac{4}{3}(\varepsilon_2-\varepsilon_3)^2}\right]$
Maximum shearing stress, $\tau_{\max}$	$\dfrac{E}{2(1+\mu)}(\varepsilon_1 - \varepsilon_2)$	$\left\{\dfrac{E}{2(1+\mu)} \times \sqrt{(\varepsilon_1-\varepsilon_3)^2 + [2\varepsilon_2 - (\varepsilon_1+\varepsilon_3)]^2}\right\}$	$\left[\dfrac{E}{1+\mu} \times \sqrt{\left(\varepsilon_1 - \dfrac{\varepsilon_1+\varepsilon_2+\varepsilon_3}{3}\right)^2 + \left(\dfrac{\varepsilon_2-\varepsilon_3}{\sqrt{3}}\right)^2}\right]$	$\left[\dfrac{E}{2(1+\mu)} \times \sqrt{(\varepsilon_1-\varepsilon_4)^2 + \dfrac{4}{3}(\varepsilon_2-\varepsilon_3)^2}\right]$
Angle from gauge 1 axis to maximum normal stress angle, φ_p	0	$\dfrac{1}{2}\tan^{-1}\left[\dfrac{2\varepsilon_2 - (\varepsilon_1+\varepsilon_3)}{\varepsilon_1-\varepsilon_3}\right]$	$\dfrac{1}{2}\tan^{-1}\left[\dfrac{\dfrac{1}{\sqrt{3}}(\varepsilon_2-\varepsilon_3)}{\varepsilon_1 - \dfrac{\varepsilon_1+\varepsilon_2+\varepsilon_3}{3}}\right]$	$\dfrac{1}{2}\tan^{-1}\dfrac{2(\varepsilon_2-\varepsilon_3)}{\sqrt{3}(\varepsilon_1-\varepsilon_4)}$

* Poisson's ratio. The author has used μ as symbol for Poisson's ratio.
Source: Perry, C. C., and H. R. Lissner, *The Strain Gage Primer*, 2nd ed., McGraw-Hill Publishing Company, New York, p. 147, 1962.

Table 2-12
Singularity functions

Function	Graph of $f_n(x)$	Meaning
Concentrated moment	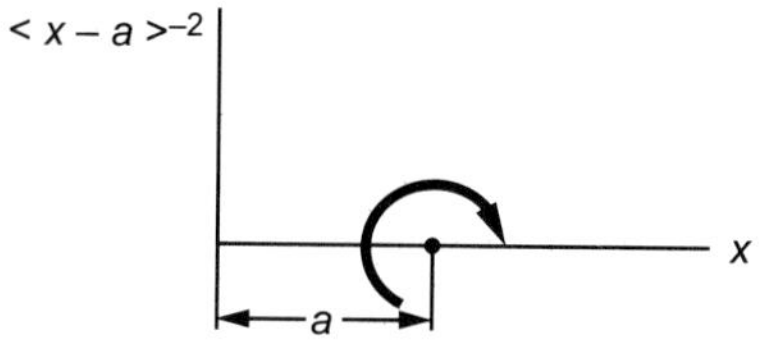	$\langle x-a\rangle^{-2} = \begin{cases} 1 & x = a \\ 0 & x \neq a \end{cases}$ $\int_{-\infty}^{x} \langle x-a\rangle^{-2}\,dx = \langle x-a\rangle^{-1}$
Concentrated force	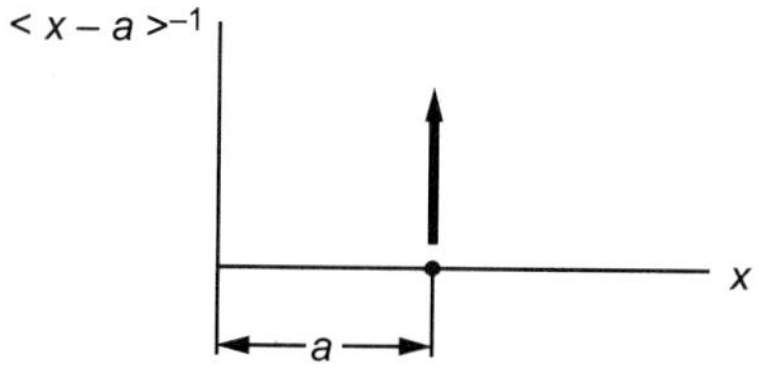	$\langle x-a\rangle^{-1} = \begin{cases} 1 & x = a \\ 0 & x \neq a \end{cases}$ $\int_{-\infty}^{x} \langle x-a\rangle^{-1}\,dx = \langle x-a\rangle^{0}$
Unit step	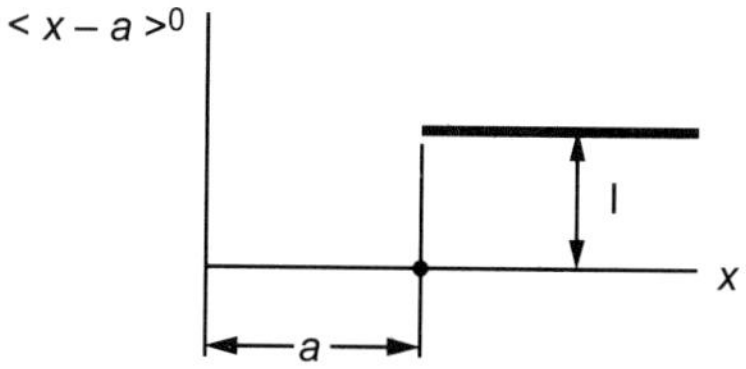	$\langle x-a\rangle^{0} = \begin{cases} 0 & x < a \\ 1 & x \geq a \end{cases}$ $\int_{-\infty}^{x} \langle x-a\rangle^{0}\,dx = \langle x-a\rangle^{1}$
Rump	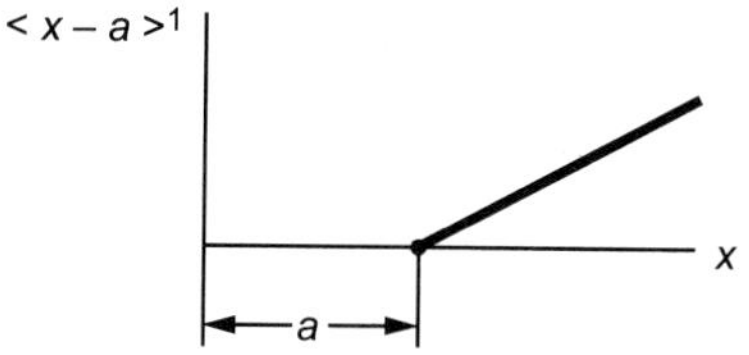	$\langle x-a\rangle^{1} = \begin{cases} 0 & x < a \\ x-a & x \geq a \end{cases}$ $\int_{-\infty}^{x} \langle x-a\rangle^{1}\,dx = \frac{\langle x-a\rangle^{2}}{2}$
Parabolic	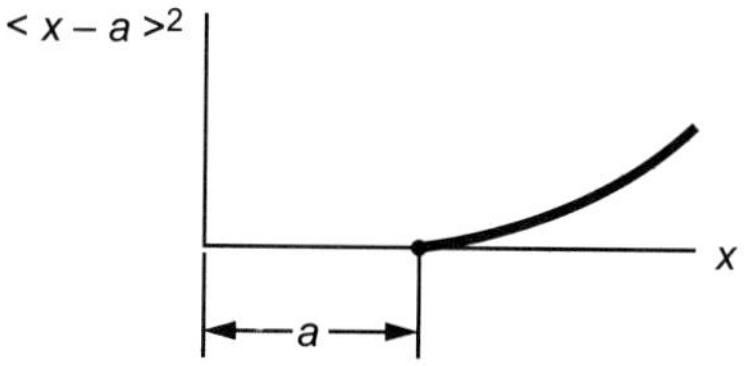	$\langle x-a\rangle^{2} = \begin{cases} 0 & x < a \\ (x-a)^2 & x \geq a \end{cases}$ $\int_{-\infty}^{x} \langle x-a\rangle^{2}\,dx = \frac{\langle x-a\rangle^{3}}{3}$

TABLE 2-13
Summary of strain and stress equations due to different types of loads

Symbols: a = major semi-axis of ellipse of area of contact, mm (in), and also radius of band of contact in case of spheres, mm (in); b = minor semi-axis of ellipse of area of contact, mm (in), and also half-bandwidth of rectangle contact between cylinders with parallel axis, mm (in); d_1, d_2 = diameters of small and large spheres respectively, mm (in); E_1, E_2 = moduli of elasticities of bodies in contact respectively, GPa (psi); F = load, kN (lbf); $F' = (F/\ell)$ = load per unit length, kN/m (lbf/in); k_1, k_2 = material constants for small and large solid elastic bodies in contact; L = length of cylinder, m (in); p_{max} = maximum pressure on surfaces of contact, MPa (psi); $\sigma_{c\,max}$ = maximum contact compressive stress, MPa (psi); σ = normal stress, also with subscripts, MPa (psi); τ = shear stress, also with subscripts MPa (psi); ν_1, ν_2 = Poisson's ratio of materials of small and large elastic bodies in contact respectively; δ = approach distance along the line of action of the load between two points on the elastic bodies in contact, mm (in); σ_θ = hoop or circumferential or tangential stress, MPa (psi); σ_a = axial or longitudinal stress, MPa (psi); h = thickness of cylinder/vessel/shell, mm (in). Meaning of other symbols used in this Table are given under Symbols introduced at the beginning of this Chapter.

$$d_o = \frac{d_1 d_2}{d_1 + d_2};\ d'_o = \frac{d_1 d_2}{d_1 - d_2};\ k_1 = \frac{1-\nu_1^2}{E_1};\ k_2 = \frac{1-\nu_2^2}{E_2};\ \beta = \frac{b}{a};\ e = \frac{\sqrt{a^2-b^2}}{a}$$

Figure showing loads	Type of loads	Figure showing stress	Applied stresses σ_x	σ_y	σ_z	τ	Strain equations/Area /Approach distance	Maximum stress produced σ_{max}	τ_{max}	Principal stresses
1 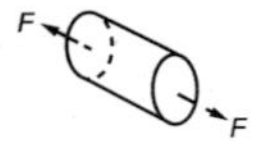	Axial load		$\frac{F}{A}$	0	0	0	$\varepsilon_x = \varepsilon_1 = \frac{\sigma}{E} = \frac{\sigma_x}{E}$ $\varepsilon_y = \varepsilon_2 = -\nu\varepsilon_1 = -\nu\varepsilon_x,$ $\varepsilon_z = \varepsilon_3 = -\nu\varepsilon_1 = -\nu\varepsilon_z,$	$\sigma_x = \sigma_{max}$	$\frac{\sigma_x}{2} = \frac{\sigma_{max}}{2}$	$\sigma_1 = E\varepsilon_1,\ \sigma_2 = 0,\ \sigma_3 = 0$
2 	Bending load	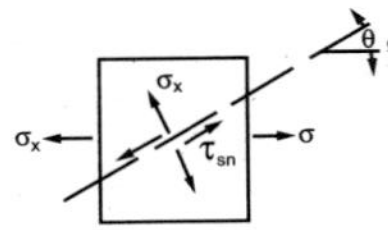	$\sigma_b = \frac{M_b c}{I}$	0	0	0	$\varepsilon_x = \varepsilon_1 = \frac{\sigma_{bx}}{E};$ $\varepsilon_y = \varepsilon_2 = -\nu\varepsilon_1,$ $\varepsilon_z = \varepsilon_3 = -\nu\varepsilon_1$	σ_{bx}	$\frac{\sigma_{bx}}{2}$	
3 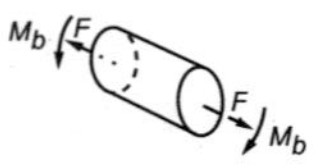	Bending and axial load		$\sigma_R = \left(\frac{F}{A} + \frac{M_b}{I}\right)$	0	0	0	$\varepsilon_x = \varepsilon_1 = \frac{\sigma}{E};$ $\varepsilon_y = \varepsilon_2 = -\upsilon\varepsilon_1,$ $\varepsilon_z = \varepsilon_3 = -\upsilon\varepsilon_1$	σ_{bx}	$\frac{\sigma_{bx}}{2}$	
4 	Torsion	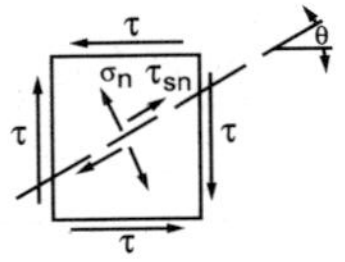	0	0	0	$\frac{M_t r}{J} = \frac{M_t}{Z_p}$	$\gamma = \frac{\tau}{G} = \frac{r\theta}{L}$	τ	τ	$\sigma_{1t} = -\sigma_{1c} = \tau$ at 45° to the shaft axis

TABLE 2-13
Summary of strain and stress equations due to different types of loads *(Cont.)*

	Figure showing loads	Type of loads	Figure showing stress	Applied stresses σ_x	Applied stresses σ_y	Applied stresses σ_z	Applied stresses τ	Strain equations/Area /Approach distance	Maximum stress produced σ_{max}	Maximum stress produced τ_{max}	Principal stresses
5	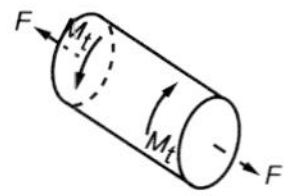	Torsion and axial load	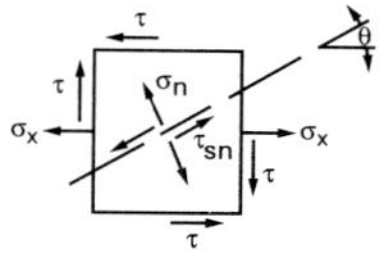	$\frac{F}{A}$	0	0	$\frac{M_t r}{J} = \frac{M_t}{Z_p}$	$\varepsilon_x = \frac{\sigma_x}{E}$ and $\gamma = \frac{\tau}{G} = \frac{r\theta}{L}$	$\sigma_{max} = \left[\frac{\sigma_x}{2} + \frac{1}{2}\sqrt{\sigma_x^2 + 4\tau^2}\right]$	$\tau_{max} = \frac{1}{2}\sqrt{\sigma_x^2 + 4\tau^2}$	$\sigma_{1,2} = \frac{1}{2}\left[\sigma_{at} + \sqrt{\sigma_{at}^2 + 4\tau^2}\right]$
6	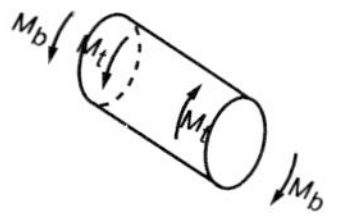	Torsion and bending load		$\frac{M_b c}{I}$	0	0	$\frac{M_t r}{J} = \frac{M_t}{Z_p}$	$\varepsilon_x = \frac{\sigma_x}{E}$ and $\gamma = \frac{\tau}{G} = \frac{r\theta}{L}$	$\sigma_{max} = \left[\frac{\sigma_x}{2} + \frac{1}{2}\sqrt{\sigma_x^2 + 4\tau^2}\right]$	$\tau_{max} = \frac{1}{2}\sqrt{\sigma_x^2 + 4\tau^2}$	$\sigma_1 = \frac{1}{2}\left[\sigma_{ab} + \sqrt{\sigma_{ab}^2 + 4\tau^2}\right] = \frac{16}{\pi D^3}\left[M_b \pm \sqrt{M_b^2 + M_t^2}\right]$
7	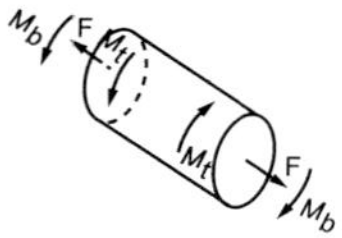	Axial, bending and torsion load		$\frac{M_b c}{I} + \frac{F}{A}$	0	0	$\frac{M_t r}{J} = \frac{M_t}{Z_p}$	$\gamma = \frac{\tau}{G} = \frac{r\theta}{L}$	$\sigma_{max} = \left[\frac{\sigma_x}{2} + \tau_{max}\right]$	$\tau_{max} = \frac{1}{2}\sqrt{\sigma_x^2 + 4\tau^2}$	$\sigma_{1,2} = \left[\frac{1}{2}[\sigma_{at} + \sigma_{ab}] \pm \frac{1}{2}\sqrt{(\sigma_{at} - \sigma_{ab})^2 + 4\tau^2}\right]$
8	P= pressure, d= diameter 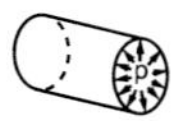	Thin-walled cylinder under internal pressure with closed ends		$\sigma_\theta = \frac{pd}{2h}$	$\sigma_a = \frac{pd}{4h}$	0	0	General biaxial $\varepsilon_a = \frac{1}{E}(\sigma_a - \nu\sigma_\theta)$ $\varepsilon_\theta = \frac{1}{E}(\sigma_\theta - \nu\sigma_a)$	σ_x	$\frac{\sigma_x}{2}$	General biaxial: $\sigma_1 = E\frac{(\varepsilon_1 - \nu\varepsilon_2)}{1 - \nu^2}$ $\sigma_2 = E\frac{(\varepsilon_2 - \nu\varepsilon_1)}{1 - \nu^2}$ $\sigma_3 = 0$
9	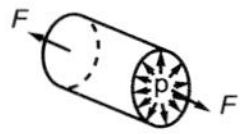	Thin-walled cylinder under internal pressure and axial tensile load with closed ends	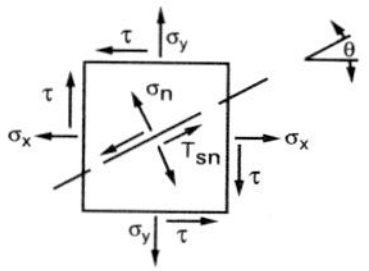	$\sigma_\theta = \frac{pd}{2h}$	$\sigma_a = \frac{pd}{4h} + \frac{F}{A}$	0	0	$\varepsilon_a = \frac{1}{E}(\sigma_a - \nu\sigma_\theta)$ $\varepsilon_\theta = \frac{1}{E}(\sigma_\theta - \nu\sigma_a)$	σ_x or σ_y whichever is larger	$\frac{\sigma_{max}}{2}$	$\sigma_1 = E\frac{(\varepsilon_1 - \nu\varepsilon_2)}{1 - \nu^2}$ $\sigma_2 = E\frac{(\varepsilon_2 - \nu\varepsilon_1)}{1 - \nu^2}$ $\sigma_3 = 0$
10		Thin walled cylinder under internal pressure and compressive load with closed ends.		$\sigma_\theta = \frac{pd}{2h}$	$\sigma_a = \frac{pd}{4h} - \frac{F}{A}$	0	0	$\varepsilon_a = \frac{1}{E}(\sigma_a - \nu\sigma_\theta)$ $\varepsilon_\theta = \frac{1}{E}(\sigma_\theta - \nu\sigma_a)$	σ_x	$\frac{\sigma_x}{2}$, if $\sigma_y > 0$ $\frac{\sigma_x - \sigma_y}{2}$, if $\sigma_y < 0$	$\sigma_1 = E\frac{(\varepsilon_1 - \nu\varepsilon_2)}{1 - \nu^2}$ $\sigma_2 = E\frac{(\varepsilon_2 - \nu\varepsilon_1)}{1 - \nu^2}$ $\sigma_3 = 0$

TABLE 2-13
Summary of strain and stress equations due to different types of loads (*Cont.*)

Figure showing loads	Type of loads	Figure showing stress	Applied stresses σ_x	σ_y	σ_z	τ	Strain equations/Area /Approach distance	Maximum stress produced σ_{max}	τ_{max}	Principal stresses
11 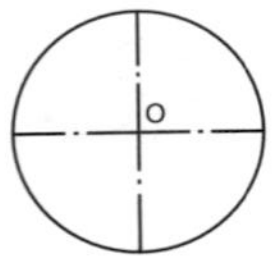	Closed walled spherical shell under internal pressure	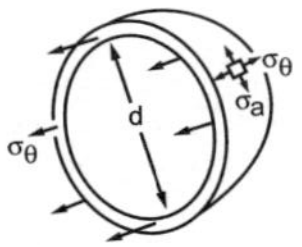	$\sigma_\theta = \frac{pd}{4h}$	$\sigma_a = \frac{pd}{4h}$	0	0	Biaxial hoop stress: Volume strain $\varepsilon_\nu = \frac{3pd}{4hE}(1-\nu)$	σ		
12 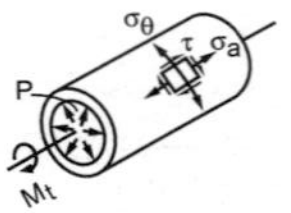	A thin-walled cylinder under internal pressure and torsion with closed ends	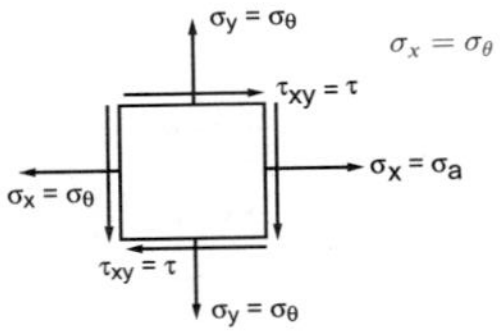	$\sigma_x = \sigma_\theta$	$\sigma_y = \sigma_a$	0	$\frac{M_t r}{J}$	$\varepsilon_\theta = \frac{1}{E}(\sigma_\theta - \nu\sigma_a)$ $\varepsilon_a = \frac{1}{E}(\sigma_a - \nu\sigma_\theta)$ $\gamma = \frac{\tau}{G} = \frac{r\theta}{L}$	σ_θ	$\tau_{max} = \frac{\sigma_\theta - \theta_a}{2}$	$\sigma_{12} = \frac{1}{2}\Big[(\sigma_\theta + \sigma_a)$ $\pm.\sqrt{(\sigma_\theta - \sigma_a)^2 + 4\tau^2}\Big]$
13 h: THICKNESS 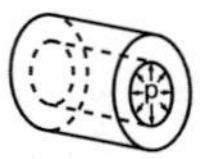	Thick-walled cylinder under internal and external pressure with closed ends	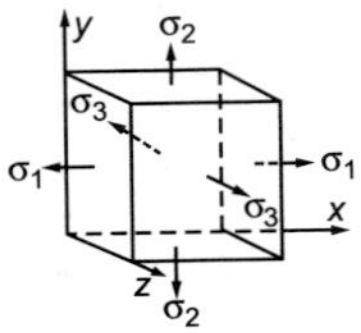	$\sigma_\theta = a + \frac{b}{r^2}$ where $a = \frac{p_1 d_1^2 - p_0 d_0^2}{d_0^2 - d_1^2}$ $b = \frac{(p_1 - p_0) d_1^2 d_0^2}{4(d_0^2 - d_1^2)}$	$\sigma_y = a - \frac{b}{r^2}$	-p		General triaxial $\varepsilon_\theta = \frac{1}{E}[\sigma_\theta - \nu(\sigma_r + \sigma_a)]$ $\varepsilon_a = \frac{1}{E}[\sigma_a - \nu(\sigma_\theta + \sigma_r)]$	σ_θ	$\frac{\sigma_\theta - \sigma_r}{2}$	$\sigma_1 = \frac{E[(1-\nu)\varepsilon_1 + \nu(\varepsilon_2 + \varepsilon_3)]}{1 - \nu - 2\nu^2}$ $\sigma_2 = \frac{E[(1-\nu)\varepsilon_2 + \nu(\varepsilon_3 + \varepsilon_1)]}{1 - \nu - 2\nu^2}$ $\sigma_3 = \frac{E[(1-\nu)\varepsilon_3 + \nu(\varepsilon_2 + \varepsilon_1)]}{1 - \nu - 2\nu^2}$
14 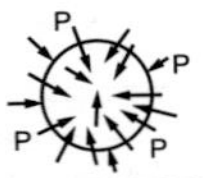 SOLID MEMBER	Hydraulic pressure		$-p$	$-p$	$-p$	0		σ_θ	0	

TABLE 2-13
Summary of strain and stress equations due to different types of loads *(Cont.)*

	Figure showing loads	Type of loads	Figure showing stress	Applied stresses σ_x	σ_y	σ_z	τ	Strain equations/Area /Approach distance	Maximum stress produced σ_{max}	τ_{max}	Principal stresses
15	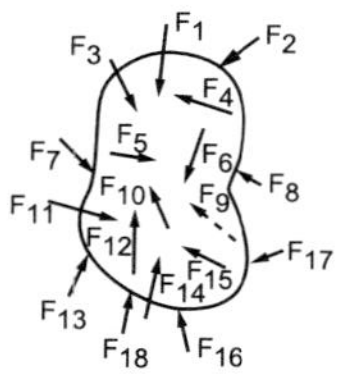	+ + General case of loading (Triaxial stress including shear stress)	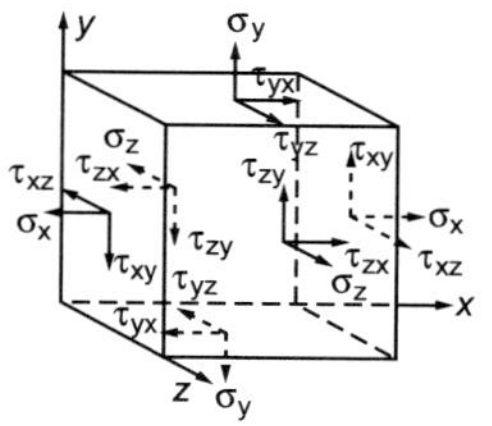		The three principal stresses σ_1, σ_2 and σ_3 are given by the three roots of the cubic equation in σ $\sigma^3 - (\sigma_x + \sigma_y + \sigma_x)\sigma^2 - (\sigma_x\sigma_y + \sigma_y\sigma_z + \sigma_z\sigma_x - \tau_{xy}^2 - \tau_{yz}^2 - \tau_{zx}^2)\sigma - (\sigma_x\sigma_y\sigma_z + 2\tau_{xy}\tau_{yz}\tau_{zx} - \sigma_x\tau_{yz}^2 - \sigma_y\tau_{zx}^2 - \sigma_z\tau_{xy}^2) = 0$ The direction of each principal stress is defined by the cosines of the angles it makes with $0x$, $0y$, and $0z$. The maximum shear stress occurs in each two planes inclined at 45° to the principal stress. The maximum shear stresses are: $(\tau_{max})_1 = \frac{\sigma_2 - \sigma_3}{2}$, $(\tau_{max})_2 = \frac{\sigma_3 - \sigma_1}{2}$, $(\tau_{max})_3 = \frac{\sigma_2 - \sigma_1}{2}$ For details of general cases of stress, strains and direction cosines refer to *Handbook* and *Theory of Elasticity*.						
16	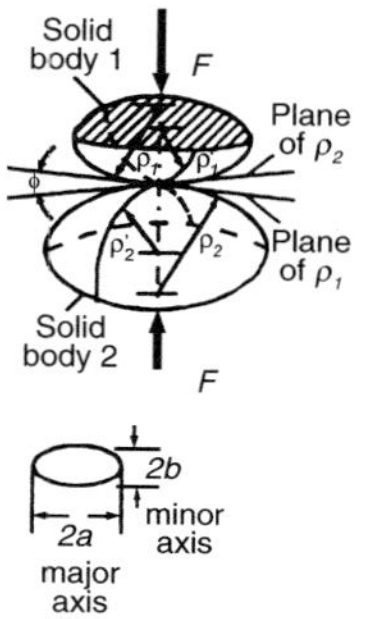	General case of contact of two elastic bodies under compressive load	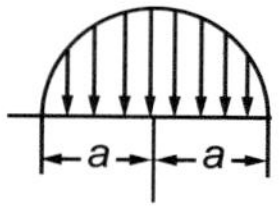 The auxiliary angle χ defines the two coefficients p and q and is given by $\chi = \cos^{-1}(\eta/\xi)$. p, q and λ are obtained from table given below for various values of χ.	$\sigma_{av} = \frac{F}{\pi ab}$				$A = \pi ab$ $a = p\left[\frac{3}{4}\frac{F(k_1+k_2)}{\xi}\right]^{1/3}$; $b = q\left[\frac{3}{4}\frac{F(k_1+k_2)}{\xi}\right]^{1/3}$ $\xi = (A+B) = \frac{1}{2}\left(\frac{1}{\rho_1} + \frac{1}{\rho_1'} + \frac{1}{\rho_2} + \frac{1}{\rho_2'}\right)$ $\eta = (B-A) = \frac{1}{2}\left[\left(\frac{1}{\rho_1} - \frac{1}{\rho_1'}\right)^2 + \left(\frac{1}{\rho_2} - \frac{1}{\rho_2'}\right)^2 + 2\left(\frac{1}{\rho_1} - \frac{1}{\rho_1'}\right) + \left(\frac{1}{\rho_2} - \frac{1}{\rho_2'}\right)\cos 2\phi\right]^{1/2}$ $\delta = \lambda\left[\frac{9\xi F^2}{128}(k_1+k_2)^2\right]^{1/3}$	$\sigma_{c\,max} = \frac{3}{2}\frac{F}{\pi ab}$		$[(\tau_{max})_{z=0.63a} = 0.34\sigma_{c\,max}]$ $\sigma_1 = (\sigma_x) = \left[-2\nu\sigma_{c\,max} - \left\{(1-2\nu)\sigma_{c\,max}\frac{b}{a+b}\right\}\right]$ $\sigma_2 = (\sigma_y) = \left[-2\nu\sigma_{c\,max} - \left\{(1-2\nu)\sigma_{c\,max}\frac{a}{a+b}\right\}\right]$ $\sigma_3 = (\sigma_z) = -\sigma_{c\,max}$. $(\tau)_{x=\pm a, y=0} = \left[(1-2\nu)\sigma_{c\,max} \times \frac{\beta}{e^2}\left(\frac{1}{e}\tanh^{-1} e - 1\right)\right]$ $(\tau)_{x=0, y=\pm b} = \left[(1-2\nu)\sigma_{c\,max} \times \frac{\beta}{e^2}\left(1 - \frac{\beta}{e}\tan^{-1}\frac{\varepsilon}{\beta}\right)\right]$

χ, deg	10	20	30	35	40	45	50	55	60	65	70	75	80	85	90
p	6.612	3.778	2.731	2.397	2.136	1.926	1.754	1.611	1.486	1.378	1.286	1.202	1.128	1.061	1.000
q	0.319	0.408	0.493	0.530	0.567	0.604	0.641	0.678	0.717	0.750	0.802	0.845	0.893	0.944	1.000
λ	0.851	1.220	1.453	1.550	1.637	1.709	1.772	1.828	1.875	1.912	1.944	1.967	1.985	1.996	2.000

TABLE 2-13
Summary of strain and stress equations due to different types of loads (*Cont.*)

Figure showing loads	Type of loads	Figure showing stress	Applied stresses σ_x	Applied stresses σ_y	Applied stresses σ_z	Applied stresses τ	Strain equations/Area /Approach distance	Maximum stress produced σ_{max}	Maximum stress produced τ_{max}	Principal stresses
17 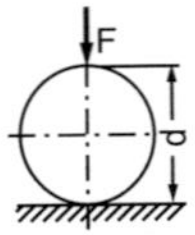	Contact of a solid sphere on a solid plane surface under compressive load				$\sigma_{av} = \frac{F}{\pi a^2}$ $\sigma_{c\,max} = -\frac{3}{2}\frac{F}{\pi a^2}$		$A = \pi a^2$ $a = 0.721[Fd(k_1 + k_2)]^{1/3}$ $\delta = 1.78\sqrt[3]{\frac{2F^2}{d}(k_1 + k_2)^2}$	$\sigma_{c(max)} = \left\{0.918 \times \left[\frac{F}{d^2(k_1 + k_2)^2}\right]^{1/3}\right\}$		
18 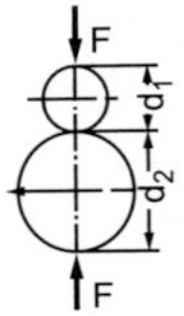	Contact of a solid sphere on solid sphere under compressive load				$\sigma_{av} = \frac{F}{\pi a^2}$ $\sigma_{e\,max} = -\frac{3}{2}\frac{F}{\pi a^2}$		$A = \pi a^2$ $a = \left[\frac{3}{8}Fd_0(k_1 + k_2)\right]^{1/3}$ $\delta = \left[\frac{9F^2(k_1 + k_2)^2}{8d_0}\right]^{1/3}$	$\sigma_{c(max)} = \left[\frac{24F}{a^3 d_0^2(k_1 + k_2)^2}\right]^{1/3}$		$-\left(\sigma_{x\,max}\right)_{z=0} = [(\sigma_{y\,max})_{z=0} = -\frac{1+2\nu}{2}p_{max} = -\frac{1+2\nu}{2}\sigma_{c\,max}]$ $\tau_{13\,max} = \tau_{xz\,max} = \frac{p_{max}}{2}\left[\left(\frac{1-2\nu}{2}\right) + \frac{2}{9}(1+\nu)\sqrt{2(1+\nu)}\right]$

The three principal stresses at the point of contact are: $\sigma_r(=\sigma_x) = \sigma_\theta(=\sigma_y) = -\frac{p_{max}}{2}\left[(1+2\nu) - 2(1+\nu)\left(\frac{z}{\sqrt{a^2+z^2}}\right) + \left(\frac{z}{\sqrt{a^2+z^2}}\right)^3\right]$

$$\sigma_z(=\sigma_c) = -p_{max}\left[1 - \frac{z^3}{(a^2+z^2)^{3/2}}\right]$$

The maximum Hertz contact stress is $(\sigma_{z\,max})_{z=0} = -p_{max} = \sigma_{c\,max} = -\frac{3F}{2\pi a^2}$

The maximum shear stress is $\tau_{13} = \tau_{xz} = \frac{p_{max}}{2}\left[\left(\frac{1-2\nu}{2}\right) + (1+\nu)\left(\frac{z}{\sqrt{a^2+z^2}}\right) - \frac{3}{2}\left(\frac{z}{\sqrt{a^2+z^2}}\right)^3\right]$

The distance from the surface of contact on the line of action of the load at which the maximum shear stress accurse $z = a\sqrt{\frac{2+2\nu}{7-2\nu}}$

The maximum sub-surface shear stress at $z = 0.63a$ is $(\tau_{max})_{z=0.63a} = 0.34\sigma_{c\,max}$
For principal stresses and variation of stresses along the line of action of load refer to Figure 2-28

Figure showing loads	Type of loads	Figure showing stress	Applied stresses σ_x	Applied stresses σ_y	Applied stresses σ_z	Applied stresses τ	Strain equations/Area /Approach distance	Maximum stress produced σ_{max}	Maximum stress produced τ_{max}	Principal stresses
19 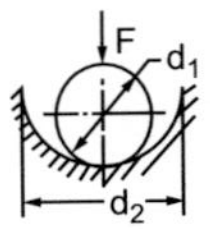	Contact of a solid sphere with a spherical socket subject to compressive load.				$\sigma_{av} = \frac{F}{\pi a^2}$ $\sigma_{c\,max} = \frac{3}{2}\frac{F}{\pi a^2}$		$a = 0.721[Fd_0'(k_1 + k_2)]^{1/3}$ $\delta = 1.04\left[F^2\frac{(k_1 + k_2)}{d_0}\right]^{1/3}$	$\sigma_{c(max)}\left[\frac{24}{\pi^3}\frac{F}{d_0'^2}\frac{1}{(k_1 + k_2)^2}\right]^{1/3}$		

TABLE 2-13
Summary of strain and stress equations due to different types of loads (*Cont.*)

	Figure showing loads	Type of loads	Figure showing stress	Applied stresses σ_x	σ_y	σ_z	τ	Strain equations/Area /Approach distance	Maximum stress produced σ_{max}	τ_{max}	Principal stresses
20	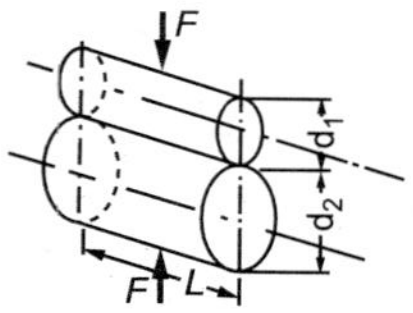	Contact of a solid cylinder on a solid cylinder under compressive load with axes parallel.	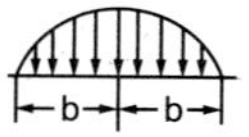			$\sigma_{av} = \sigma_z = \dfrac{F}{2Lb}$		$A = 2Lb$ $b = 1.6\left[\dfrac{F}{L} d_0(k_1 + k_2)\right]^{1/2}$ $\delta = \dfrac{2F}{\pi L}\left[k_1\left(\ln\dfrac{d_1}{b} + 0.41\right) + k_2\left(\ln\dfrac{d_2}{b} + 0.41\right)\right]$	$\sigma_{c(max)} = 0.798\left[\dfrac{F}{Ld_\theta}\dfrac{1}{k_1 + k_2}\right]^{1/2}$		
21	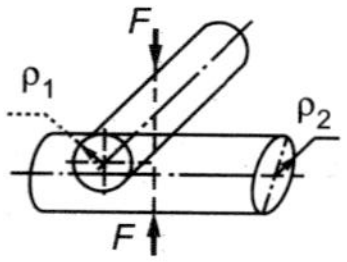	Contact of solid cylinder on a solid cylinder under compressive load with axes perpendicular				$\sigma_{av} = \sigma_z = \dfrac{F}{\pi ab}$		$A = \pi ab$ $\delta = 1.41C_{rv}\sqrt[3]{2F^2 \dfrac{2d_2 + d_1}{d_1 d_2}(k_1 + k_2)^2}$ *Refer to Table C_{rv} for various ratios of $i_p = \left(\dfrac{1}{d_2}\right)\Big/\left(\dfrac{1}{d_1}\right)$	$\sigma_{c\,max} = \sigma_{s\theta} = -\dfrac{1.5F}{\pi ab}$		$(\tau_{max})_{z=0.63a} = .034\sigma_{c\,max}$
22	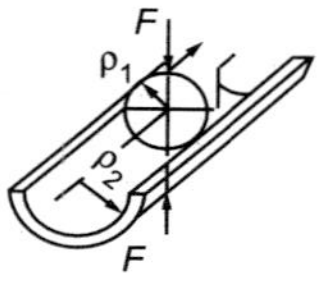	Contact of a solid sphere with cylindrical groove/socket under compressive load.	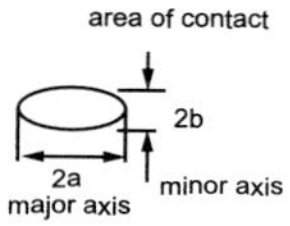			$\sigma_{av} = \sigma_z = \dfrac{F}{\pi ab}$		$A = \pi ab$ $2b = 1.6\left[\dfrac{F}{L}\dfrac{k_1 + k_2}{d_0''}\right]^{1/2}$ $\delta = 1.41C_{rv}\sqrt[3]{2F^2 \dfrac{2d_2 - d_1}{d_1 d_2}(k_1 + k_2)^2}$ *Refer to Table C_{rv} for various ratios of $i_p = \left(\dfrac{1}{d_1} - \dfrac{1}{d_2}\right)\Big/\left(\dfrac{1}{d_1}\right)$	$\sigma_{c(max)} = 0.798\left[\dfrac{F}{L}\dfrac{d_0'}{k_2 + k_2}\right]^{1/2}$		
23	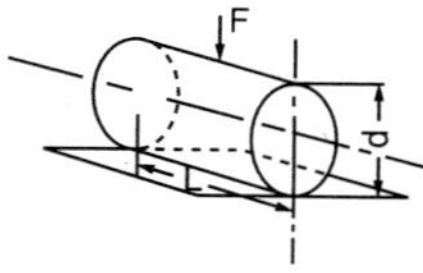	Contact of a solid cylinder with a flat surface subject to compressive load				$\sigma_{av} = \sigma_a = \dfrac{F}{2Lb}$		$A = 2Lb$ $b = 1.6\left[\dfrac{Fd}{L}(k_1 + k_2)\right]^{1/2}$ $\delta = 4\dfrac{F}{L}k_1\left(\ln\dfrac{2d}{b} + 0.41\right)$	$\sigma_{c(max)} = 0.798\left[\dfrac{F}{Ld}\dfrac{1}{k_1 + k_2}\right]^{1/2}$		

TABLE 2-13
Summary of strain and stress equations due to different types of loads (*Cont.*)

Figure showing loads	Type of loads	Figure showing stress	Applied stresses σ_x	σ_y	σ_z	τ	Strain equations/Area /Approach distance	Maximum stress produced σ_{max}	τ_{max}	Principal stresses
24 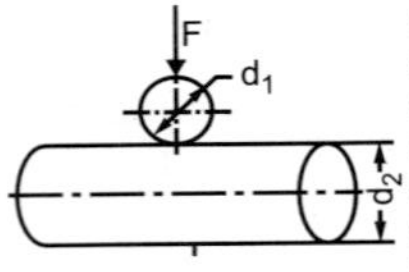	Contact of a solid sphere on a solid cylinder under compressive load	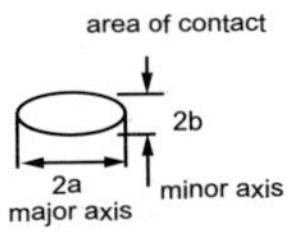			$\sigma_{av} = \sigma_z = \frac{F}{\pi ab}$		$A = \pi ab$ $a = \left[\frac{3}{8} F d_0 (k_1 + k_2)\right]^{1/3}$ $\delta = 1.4 C_{rv} \sqrt[3]{2F^2 \frac{(2d_2 + d_1)}{d_1 d_2} (k_1 + k_2)^2}$ **Refer to Table C_{rv} for various ratios of $i_\rho = 2\left(\frac{1}{d_1}\right) \Big/ \left(\frac{1}{d_1} + \frac{1}{d_2}\right)$	$\sigma_{c\,max} = \left[\frac{2AF}{\pi^3 d_0^4} \frac{1}{(k_1 + k_2)^2}\right]^{1/3}$		
25 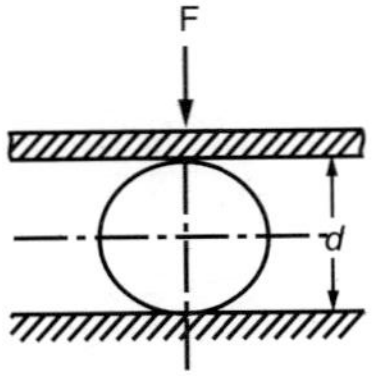	Cylinder between two flat plates under compressive load				$\sigma_{av} = \sigma_z = \frac{F}{2Lb}$ Total deformation due to compression of cylinder is δ_{cy} Approach distance of two points along the line of action of load in two plates is δ, if $E_1 = E_2 = E$ and $\nu_1 = \nu_2 = \nu$		$A = 2Lb$ $b = 1.6\left[\frac{Fd}{L}(k_1 + k_2)\right]^{1/2}$ $\delta_{cy} = 4\frac{F}{L} k_1 \left(0.41 + \ln \frac{2d}{b}\right)$ $\delta = 4\frac{F}{L}\left(\frac{1 - \nu^2}{\pi E}\right) \ln \frac{\pi EL}{F(1 - \nu^2)}$	$\sigma_{c\,max} = 0.798\sqrt{\frac{F}{Ld(k_1 + k_2)}}$		

****TABLE: C_{rv} values for various ratios of i_ρ**

i_ρ	1.00	0.404	0.250	0.160	0.085	0.067	0.044	0.032	0.020	0.015	0.003
C_{rv}	1.00	0.957	0.905	0.845	0.751	0.716	0.655	0.607	0.546	0.510	0.358

** *Source:* Roark, R.J., and W. C. Young, *Formulas for Stress and Strain*, McGraw-Hill Publishing Company, New York, 1975.
+ Hertz. H., *On the Contact of Elastic Solids*, J. Math. (Crelle's J.) vol. 92, pp 156-171, 1981
Hertz. H., *On Gesammelte werke*, Vol 1., p 155, Leipzig, 1895.

REFERENCES

1. Maleev, V. L. and J. B. Hartman, *Machine Design*, International Textbook Company, Scranton, Pennsylvania, 1954.
2. Shigley, J. E., *Mechanical Engineering Design*, 3rd edition, McGraw-Hill Book Company, New York, 1977.
3. Lingaiah, K., and B. R. Narayana Iyengar, *Machine Design Data Handbook*, Vol. 1 (*SI and Customary Metric Units*), Suma Publishers, Bangalore, India, 1986.
4. Lingaiah, K., *Machine Design Data Handbook*, Vol II (*SI and Customary Metric Units*), Suma Publishers, Bangalore, India, 1986.
5. Lingaiah, K., *Machine Design Data Handbook*, McGraw-Hill Publishing Company, New York, 1994.
6. Ashton, J. E, J. C. Halpin and P. H. Petit, *Primer on Composite Materials-Analysis*, Technomic Publishing Co., Inc., 750 Summer St., Stanford, Conn. 06901, 1969.
7. Roark, R. J., and W. C. Young, *Formulas for Stress and Strain*, McGraw-Hill Publishing Company, New York, 1975.
8. Hertz, H., *On the Contact of Elastic Solids*, J. Math. (Crelle's J.) Vol. 92, pp. 156–171, 1981.
9. Hertz, H., *On Gesammelte werke*, Vol I., p. 155, Leipzig, 1895.
10. Timoshenko, S., and J. N. Goodier, *Theory of Elasticity*, McGraw-Hill Book Company, New York, 1951.

BIBLIOGRAPHY

1. Black, P. H., and O. Eugene Adams, Jr., *Machine Design*, McGraw-Hill Book Company, New York, 1965.
2. Lingaiah, K, and B. R. Narayana Iyengar, *Machine Design Data Handbook* (fps units), Engineering College Co-Operative Society, Bangalore, India, 1962.
3. Norman, C. A., E. S. Ault, and I. F. Zarobsky, *Fundamentals of Machine Design*, The Macmillan Company, New York, 1951.
4. Vallance, A. E., and V. L. Doughtie, *Design of Machine Members*, McGraw-Hill Book Company, New York, 1951.
5. Timosheko, S., and J. N. Goodier, *Theory of Elasticity*, McGraw-Hill Book Company, New York, 1951.
6. Timoshenko, S., and J. M. Gere, *Mechanics of Materials*, Van Nostrand Reinhold Company, New York, 1972.
7. George Lubin, Editor, *Handbook of Composites*, Van Nostrand Reinhold Company, New York, 1982.
8. John Murphy, *Reinforced Plastic Handbook*, 2nd edition, Elsevier, Advanced Technology, 1998.
9. Hamcox, N. L., and R. M. Mayer, *Design Data for Reinforced Plastics*, Chapman and Hall, 1994.

CHAPTER

3

DYNAMIC STRESSES IN MACHINE ELEMENTS[2]

SYMBOLS[2,3]

A	area of cross-section, m^2
a, b	coefficients
b	width of bar or beam, m
c	distance from neutral axis to extreme fibre, m
	velocity of propagation of plane wave along a thin bar, m/s
c_L	velocity of propagation of plane longitudinal waves in an infinite plate, m/s
c_T	velocity of propagation of plane transverse waves in an infinite plate, m/s
d	diameter of bar, m
E	modulus of elasticity, GPa
F	force or load, kN
	force acting on piston due to steam or gas pressure corrected for inertia effects of the piston and other reciprocating parts, kN
F_σ	centrifugal force per unit volume, kN/m^3
F_c	the component of F acting along the axis of connecting rod, kN
F_d	dynamic load, kN
F_g	gas load, kN
F_i	inertia force, kN
F_{ic}	inertia force due to connecting rod, kN
F_{ir}	inertia force due to reciprocating parts of piston, kN
F_s	static load, kN
g	acceleration due to gravity, 9.8066 m/s^2
h	depth of bar or beam, m
	height of fall of weight, m
J	polar moment of inertia, m^4 (cm^4)
k	radius of gyration, m
k_p	radius of gyration, polar, m
K	kinetic energy, N m
l	length, m
m	mass, kg
	moving mass, kg
$M = m/A$	ratio of moving mass to area of cross-section of bar
M_b	bending moment, N m

M_t	torque, m N
n	speed, rpm
n'	speed, rps
$n' = l/r$	ratio of length of connecting rod to radius of crank
p	pressure
P	power, kW
r	radius of crank, m
	radius of curvature of the path of motion of mass, m
	the moment arm of the load, m
t	time, s
u	displacement in x-direction
	modulus of resilience, N m/m^3
u, v, w	displacement components in x, y, and z-directions respectively, m
U	resilience, N m
	internal elastic energy, N m
U_i	work done in case of suddenly applied load, N m
U_{max}	maximum internal elastic energy, N m
U_p	potential energy, N m
v	velocity, m/s
V	velocity of particle in the stressed zone of the bar, m/s
	volume, m^3
V_0	initial velocity at the time of impact, m/s
w	specific weight of material, kN/m^3
W	total weight, kN
Z	section modulus, m^3 (cm^3)
α	angle between the crank and the centre line of connecting rod, deg
γ	unit shear strain, rad/rad
	weight density, kN/m^3
δ	deflection/deformation, m (mm)
δ_i	deformation/deflection under impact action, m (mm)
δ_s	static deformation/deflection under the action of weight, m (mm)
ε	unit strain also with subscripts, μm/m
$\varepsilon_x, \varepsilon_y, \varepsilon_z$	strains in x, y, and z-directions, μm/m
$\gamma_{xy}, \gamma_{yz}, \gamma_{zx}$	shearing-strains in rectangular coordinates, rad/rad
θ	angle between the crank and the centre line of the cylinder measured from the head-end dead-centre position, deg
	static angular deflection, deg
	angle of twist, deg
θ_i	angular deflection under impact load, deg
λ, μ	Lamé's constants
ν	Poisson's ratio
ρ	mass density, kg/m^3
σ	normal stress (also with subscripts), MPa
σ_i	impact stress (also with subscripts), MPa
σ_0	initial stress at the time of impact and velocity V_0, MPa
$\sigma_x, \sigma_y, \sigma_z$	normal stress components parallel to x, y, and z-axis
τ	shearing stress, MPa
τ_l	time of load application, s
τ_n	period of natural frequency, s
$\tau_{xy}, \tau_{yz}, \tau_{zx}$	shearing stress components in rectangular coordinates, MPa
ω	angular velocity, rad/s

Note: σ_s and τ_s with first subscript s designate strength properties of material used in the design which will be used and followed throughout the book. Other factors in performance or in special aspects which are included from time to time in this book and being applicable only in their immediate context are not given at this stage.

Particular	Formula

INERTIA FORCE

Power	$P = \dfrac{Fv}{1000}$ **SI** (3-1a)
	where F is in newtons (N), v in m/s, and P in kW.
	$= \dfrac{Fv}{33000}$ **US Customary System units** (3-1b)
	where F is in lbf, v in ft/min, and P in hp.
Velocity	$v = \dfrac{2\pi rn}{12}$ **US Customary System units** (3-2a)
	where r in in, v in ft/min, and n in rpm.
	$v = \dfrac{2\pi rn}{60}$ **SI** (3-2a)
	where r in m, v in m/s, and n in rpm.
Centrifugal force per unit volume	$F_{cv} = \dfrac{wv^2}{rg}$ (3-3)

ENERGY METHOD

The internal elastic energy or work done when a machine member is subjected to a gradually applied load, Fig. 3.1.	$U_\rho = \frac{1}{2}F\delta$ (3-4)
The work done in case of suddenly applied load on an elastic machine member (Fig. 3-2)	$U_d = F_d\delta$ (3-5)

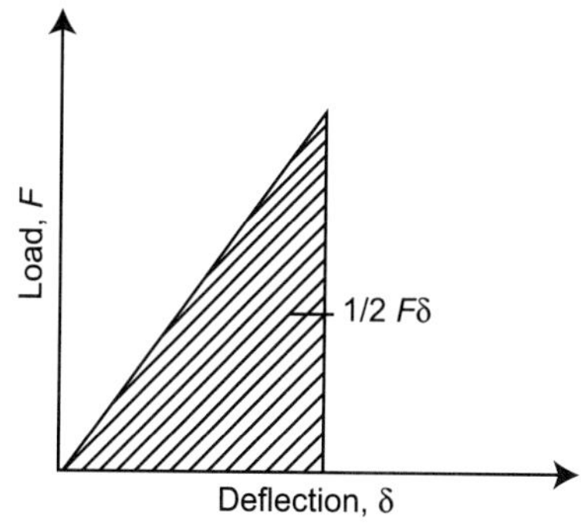

FIGURE 3-1 Plot of force against deflection in case of elastic machine member subject to gradually applied load.

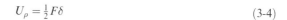

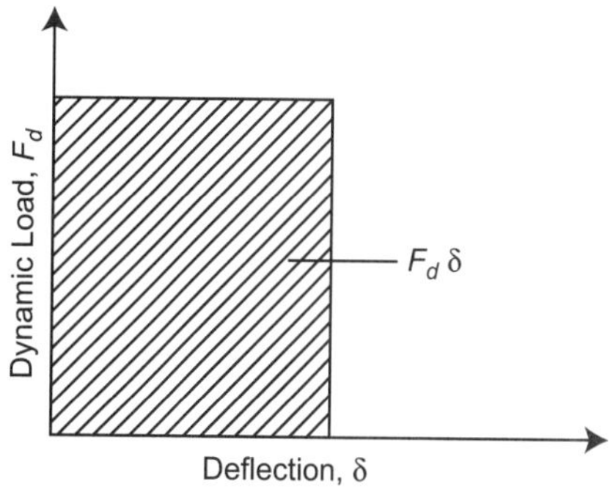

FIGURE 3-2 Plot of force against deflection in case of suddenly applied load on a machine member.

The relation between suddenly applied load and gradually applied load on an elastic machine member to produce the same magnitude of deflection.	$U_p = U_d$ (3-6a)
	$F_d = \frac{1}{2}F$ (3-6b)

Particular	Formula
The static deformation or deflection	$\delta_{st} = \dfrac{W}{k}$ (3-7) where k = spring constant of the elastic machine member, kN/m (lbf/in).

IMPACT STRESSES

Impact from direct load

Kinetic energy	$K = \dfrac{Wv^2}{2g}$ (3-8)
Impact energy of a body falling from a height h	$K = Wh$ (3-9)
The height of fall of a body that would develop the velocity v.	$h = \dfrac{v^2}{2g}$ (3-10)
The maximum stresses produced due to fall of weight W through the height h from rest without taking into account the weight of shaft and collar (Fig. 3-3)	$\sigma_i = \sigma_{max} = \dfrac{W}{A}\left[1 + \sqrt{1 + \dfrac{2hEA}{WL}}\right]$ (3-11a) $= \sigma_{st}\left[1 + \sqrt{1 + \dfrac{2hEA}{WL}}\right]$ (3-11b) $= \sigma_{st}\left[1 + \sqrt{1 + \dfrac{2h}{\delta_{st}}}\right]$ (3-11c)

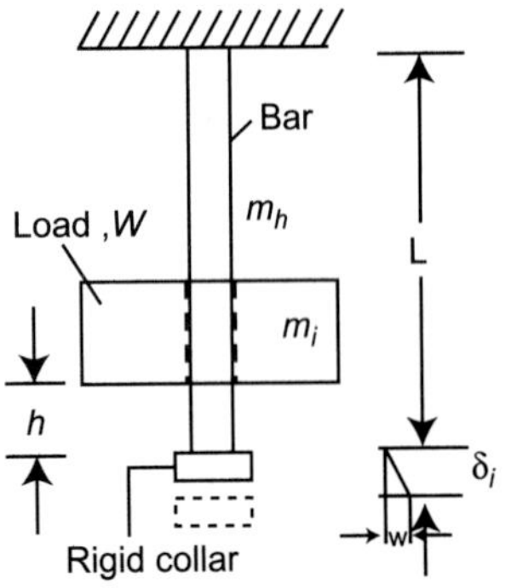

FIGURE 3-3 Striking impact of an elastic machine member by a body of weight W falling through a height h.

The maximum deflection or deformation of shaft due to fall of weight W through the height h from rest neglecting the weight of shaft and collar	$\delta_{max} = \delta_i = \delta_{st}\left[1 + \sqrt{1 + \dfrac{2hAE}{WL}}\right]$ (3-12a) $= \delta_{st}\left[1 + \sqrt{1 + \dfrac{2h}{\delta_{st}}}\right]$ (3-12b)
The stress produced due to suddenly applied load	$(\sigma_{max})_{sud} = 2(\sigma_{max})_{stat}$ (3-13)
The maximum deflection or deformation produced by suddenly applied load	$(\delta_{max})_{sud} = 2\delta_{st}$ (3-14) where subscript $stat$ = st = static and sud = suddenly

Particular	Formula
The kinetic energy taking into account the weight of shaft or bar and collar	$K = \frac{WV_c^2}{2g}\left[1 + \frac{W_b}{3W}\right]$ (3-15) where V_c = velocity of collar and weight W after the load striking the collar, m/s. where W_b = weight of shaft or bar
The relation between σ, δ, F and W	$\frac{F}{W} = \frac{\sigma_{max}}{\sigma_{st}} = \frac{\delta_{max}}{\delta_{st}} = \left[1 + \sqrt{1 + \frac{2hAE}{WL}}\right]$ (3-16a) $= \left[1 + \sqrt{1 + \frac{2h}{\delta_{st}}}\right]$ (3-16b)
The maximum stress due to fall of weight W through the height h from rest taking into account the weight of shaft/bar and collar	$\sigma_i = \sigma_{max} = \frac{W}{A}\left[1 + \sqrt{1 + \frac{2EAh}{WL}\left\{\frac{1}{1 + (W_b/3W)}\right\}}\right]$ (3-17a) $= \frac{W}{A}\left[1 + \sqrt{1 + \frac{2EAh\alpha}{WL}}\right]$ (3-17b) $= \sigma_{st}\left[1 + \sqrt{1 + \frac{2h\alpha}{\delta_{st}}}\right]$ (3-17c) where $\alpha = \frac{1}{1 + (\zeta/3)}$ and $\zeta = \frac{W_b}{W}$
The maximum deflection due to fall of weight W through the height h from rest taking into consideration the weight of shaft/bar and collar	$\delta_{max} = \frac{WL}{AE}\left[1 + \sqrt{1 + \frac{2hEA}{WL}\left\{\frac{1}{1 + (W_b/3W)}\right\}}\right]$ (3-18a) $= \frac{WL}{AE}\left[1 + \sqrt{1 + \frac{2hAE\alpha}{WL}}\right]$ (3-18b) $= \delta_{st}\left[1 + \sqrt{1 + \frac{2h\alpha}{\delta_{st}}}\right]$ (3-18c)
Internal elastic energy of weight W whose velocity v is horizontal	$U = \frac{Wv^2}{2g}$ (3-20)
Internal elastic energy of weight W whose velocity has random direction	$U = \frac{Wv^2}{2g} + W\delta \sin\beta$ (3-21) where β = angle of velocity, v, to the horizontal plane, deg.

Particular	Formula

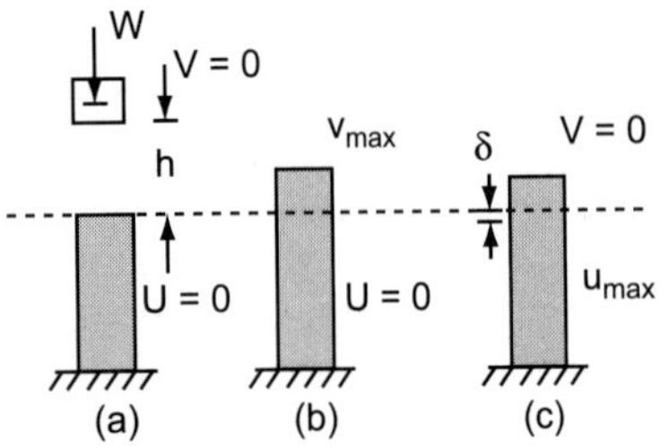

FIGURE 3-4 Impact by a falling body

Energy	Fig. 3-4a	Fig. 3-4b	Fig. 3-4c	Equation
U_p	$W(h+\delta)$	$W\delta$	0	(3-22a)
K	0	$\dfrac{Wv^2}{2g}$	0	(3-22b)
U	0	0	$W(h+\delta)$	(3-22c)

The equation for energy balance for an impact by a falling body (Fig. 3-4)

$$(U_p + K + U)_a = (U_p + K + U)_b = (U_p + K + U)_c \tag{3-23}$$

Another form of equation for deformation or deflection in terms of velocity v at impact

$$\delta_{max} = \delta_{st}\left(1 + \sqrt{1 + \frac{v^2}{g\delta_{st}}}\right) \tag{3-24}$$

Equivalent static force that would produce the same maximum values of deformation or deflection due to impact δ

$$F_{eq} = W\left(1 + \sqrt{1 + \frac{2h}{\delta_{st}}}\right) = W\left(1 + \sqrt{1 + \frac{v^2}{g\delta_{st}}}\right) \tag{3-25}$$

BENDING STRESS IN BEAMS DUE TO IMPACT

Impact stress due to bending

$$(\sigma_b)_{max} = \sigma_{bi} = \frac{Wlc}{I}\left[1 + \sqrt{1 + \frac{6hEI}{Wl^3}}\right] \tag{3-26a}$$

$$= \frac{Wlc}{I}\left[1 + \sqrt{1 + \frac{2h}{\delta_{st}}}\right] \tag{3-26b}$$

$$= (\sigma_b)_{st}\left(1 + \sqrt{1 + \frac{2h}{\delta_{st}}}\right) \tag{3-26c}$$

where $(\sigma_b)_{st} = \dfrac{Wlc}{I} = \dfrac{M_b c}{I} = \dfrac{M_b}{Z_b}$.

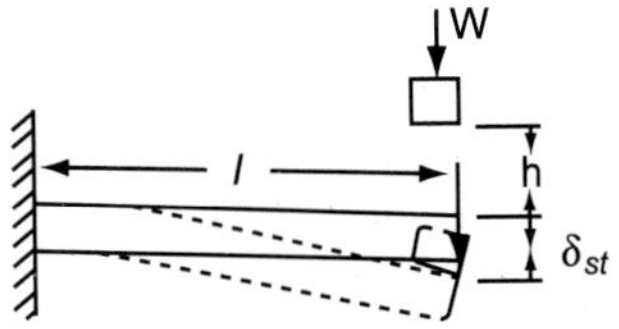

FIGURE 3-5 Impact by a falling body on a cantilever beam

Deflection of the end of cantilever beam under impact (Fig. 3-5)

$$\delta_{max} = \delta_{st}\left(1 + \sqrt{1 + \frac{2h}{\delta_{st}}}\right) \tag{3-27}$$

The maximum bending stress for a cantilever beam taking into account the total weight of beam

$$(\sigma_b)_{max} = (\sigma_b)_{st}\left[1 + \frac{2h\alpha}{\delta_{st}}\right] \tag{3-28}$$

where $\zeta = \dfrac{m_b}{m} = \dfrac{W_b}{W}$ and $\alpha = \dfrac{1}{1 + (33\zeta/140)}$

Particular	Formula

The maximum deflection at the end of a cantilever beam due to fall of weight W through the height h from rest taking into consideration thc wcight of beam

$$\delta_{\max} = \delta_{st}\left[1 + \sqrt{1 + \frac{2h\alpha}{\delta_{st}}}\right] \tag{3-28a}$$

The maximum bending stress for a simply supported beam due to fall of a load/weight W from a height h at the midspan of the beam taking into account the total weight of the beam (Fig. 3-6)

$$(\sigma_b)_{\max} = (\sigma_b)_{st}\left[1 + \sqrt{1 + \frac{2h}{\delta_{st}}\left\{\frac{1}{1 + (17\zeta/35)}\right\}}\right] \tag{3-29a}$$

$$= (\sigma_b)_{st}\left[1 + \sqrt{1 + \frac{2h\alpha}{\delta_{st}}}\right] \tag{3-29b}$$

where $\alpha = \dfrac{1}{1 + (17\zeta/35)}$ and $\zeta = \dfrac{W_b}{W}$

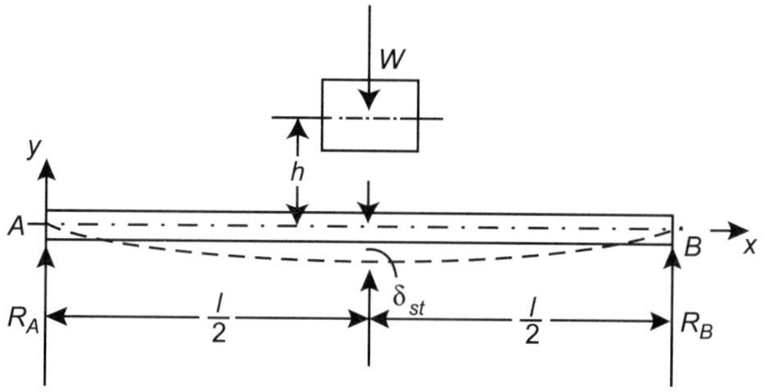

FIGURE 3-6 Simply supported beam

The maximum deflection for a simply supported beam due to fall of a weight W from a height h at the mid-span of the beam taking into account the weight of beam. (Fig. 3-6)

$$\delta_{\max} = \delta_{st}\left[1 + \sqrt{1 + \frac{2h\alpha}{\delta_{st}}}\right] \tag{3-30}$$

TORSION OF BEAM/BAR DUE TO IMPACT (Fig. 3-7)

The equation for maximum shear stress in the bar due to impact load at a radius r of a falling weight W from a height h neglecting the weight of bar

$$\tau_{\max} = \tau_{st}\left[1 + \sqrt{1 + \frac{2h}{r\theta_{st}}}\right] \tag{3-31}$$

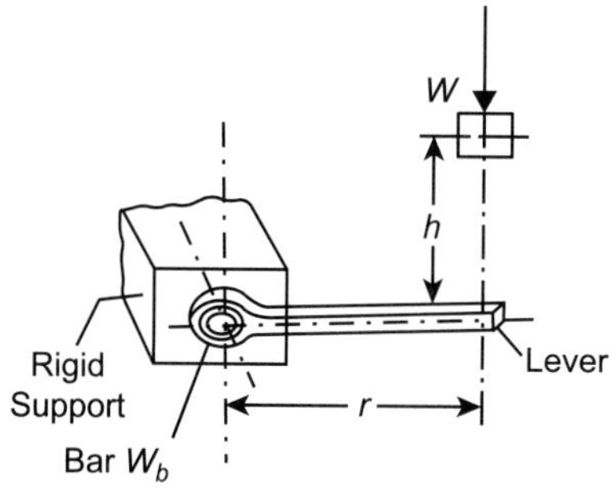

FIGURE 3-7 Twist of a beam/bar

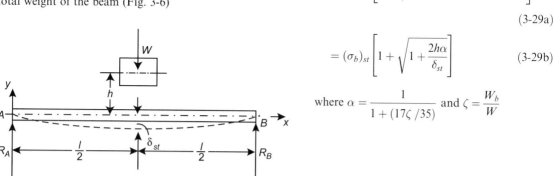

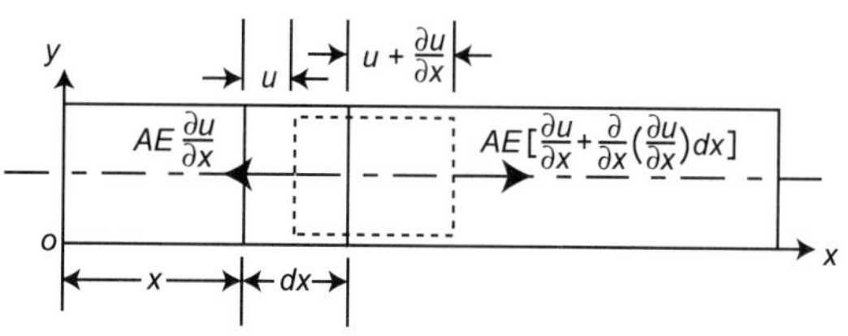

FIGURE 3-8 Displacements due to forces acting on an element of an elastic media.

Particular	Formula
The equation for angular deflection or angular twist of bar due to impact load W at radius r and falling through a height h neglecting the weight of bar	$\theta_{max} = \theta_{st}\left[1 + \sqrt{1 + \frac{2h}{r\theta_{st}}}\right]$ (3-32)

LONGITUDINAL STRESS-WAVE IN ELASTIC MEDIA (Fig. 3-8)

One-dimensional stress-wave equation in elastic media (Fig. 3-8)	$\frac{\partial^2 u}{\partial t^2} = c^2 \frac{\partial^2 u}{\partial x^2}$ (3-33a)
	where $c = \sqrt{\frac{Eg}{\gamma}} = \sqrt{\frac{E}{\rho}}$ (3-33b)
	= velocity of propagation of stress waves, m/s.
For velocity of propagation of longitudinal stress-wave in elastic media	Refer to Table 3-1.
The solution of stress-wave Eq. (3-33a)	$x = \left(A \sin\frac{p}{c}x + B\cos\frac{p}{c}x\right)(c \sin pt + D\cos pt)$ (3-34)
	where A, B, C and D are arbitrary constants which can be found from initial or boundary condition of the problem.
The value of circular frequency p	$p = \frac{n\pi c}{l} = \frac{n\pi}{l}\sqrt{\frac{Eg}{\gamma}} = \frac{n\pi}{l}\sqrt{\frac{E}{\rho}}$ (3-35a)
	where n is an integer $= 1, 2, 3, \ldots$
The frequency	$f = \frac{p}{2\pi} = \frac{n}{2l}\sqrt{\frac{E}{\rho}} = \frac{c}{\lambda}$ (3-35b)
	where λ = wave length $= 2l/n$, c = speed of sound or stress wave velocity, m/s.

LONGITUDINAL IMPACT ON A LONG BAR

The velocity of particle in the compression zone	$V = \sigma\sqrt{\frac{g}{E\gamma}} = \frac{\sigma}{\sqrt{E\rho}}$ (3-36)
The uniform initial compressive stress on the free end of a bar (Fig. 3-9)	$\sigma_0 = V_0\sqrt{\frac{E\gamma}{g}} = V_0\sqrt{E\rho}$ (3-37)
	where V_0 = initial velocity of the moving weight/mass at the time of impact, m/s.
The variation of stress at the end of bar at any time t	$\sigma = \sigma_0 \exp\left(-\frac{\sqrt{E\rho}}{M}t\right) \quad 0 < t < \frac{2l}{c}$ (3-38)

Particular	Formula	
The equations of motion in terms of three displacement components assuming that there are no body forces.	$(\lambda + G)\dfrac{\partial \varepsilon}{\partial x} + G\nabla^2 u = \rho \dfrac{\partial^2 u}{\partial t^2}$	(3-39a)
	$(\lambda + G)\dfrac{\partial \varepsilon}{\partial y} + G\nabla^2 v = \rho \dfrac{\partial^2 v}{\partial t^2}$	(3-39b)
	$(\lambda + G)\dfrac{\partial \varepsilon}{\partial z} + G\nabla^2 w = \rho \dfrac{\partial^2 w}{\partial t^2}$	(3-39c)
	where $\varepsilon = \varepsilon_x + \varepsilon_y + \varepsilon_z$	
	$\nabla = \dfrac{\partial^2}{\partial x^2} + \dfrac{\partial^2}{\partial y^2} + \dfrac{\partial^2}{\partial z^2}$ = the Laplacian operator	
	$\lambda = \dfrac{\nu E}{(1+\nu)(1-2\nu)}$ and	
	$\mu = G = \dfrac{E}{2(1+\nu)}$ are Lamé's constants	

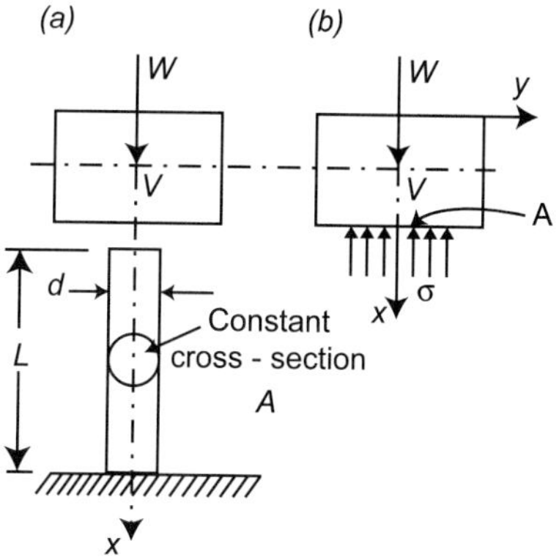

FIGURE 3-9 Prismatic bar subject to suddenly applied uniform compressive stress

Dilatational and distortional waves in isotropic elastic media

From the classical theory of elasticity equations for irrotational or dilatational waves	$\dfrac{\partial^2 u}{\partial t^2} = \dfrac{\lambda + 2G}{\rho}\nabla^2 u$	(3-40a)
	$\dfrac{\partial^2 v}{\partial t^2} = \dfrac{\lambda + 2G}{\rho}\nabla^2 v$	(3-40b)
	$\dfrac{\partial^2 w}{\partial t^2} = \dfrac{\lambda + 2G}{\rho}\nabla^2 w$	(3-40c)
Equations for distortional waves	$\dfrac{\partial^2 u}{\partial t^2} = \dfrac{G}{\rho}\nabla^2 u$	(3-41a)
	$\dfrac{\partial^2 v}{\partial t^2} = \dfrac{G}{\rho}\nabla^2 v$	(3-41b)
	$\dfrac{\partial^2 w}{\partial t^2} = \dfrac{G}{\rho}\nabla^2 w$	(3-41c)
Equations (3-40) to (3-41) are one-dimensional stress wave equations of the form	$\dfrac{\partial^2 \theta}{\partial t^2} = a^2\nabla^2 \theta$	(3-42)
The velocity of stress wave propagation for the case of no rotation	$a = c_1 = \sqrt{\dfrac{\lambda + 2G}{\rho}} = \sqrt{\dfrac{E(1-\nu)}{(1+\nu)(1-2\nu)\rho}}$	(3-43)

Particular	Formula	
The velocity of stress wave propagation for the case of zero volume change	$a = c_2 = \sqrt{\frac{G}{\rho}} = \sqrt{\frac{E}{2(1-\nu)\rho}}$	(3-44)
The ratio of c_1 to c_2	$\frac{c_1}{c_2} = \sqrt{\frac{2(1-\nu)}{(1-2\nu)}} = \sqrt{3}$ for Poisson's ratio of $\nu = 0.25$	(3-45)
The velocity of stress wave propagation for a transverse stress wave, i.e. distortional wave in an infinite plate	$c_T = \sqrt{\frac{G}{\rho}} = \sqrt{\frac{E}{2(1+\nu)\rho}}$	(3-46)
The velocity of stress wave propagation for plane longitudinal stress wave in case of an infinite plate	$c_L = \sqrt{\frac{4G(\lambda+G)}{\rho(\lambda+2G)}} = \sqrt{\frac{E}{\rho(1-\nu^2)}}$	(3-47)

TORSIONAL IMPACT ON A BAR

Particular	Formula	
Equation of motion for torsional impact on a bar (Fig. 3-10)	$\frac{\partial^2\theta}{\partial t^2} = c_t^2 \frac{\partial^2\theta}{\partial x^2}$	(3-48)
Torsional wave propagation in a bar subjected to torsion.	$c_t = \sqrt{\frac{Gg}{\gamma}} = \sqrt{\frac{G}{\rho}}$	(3-49)
For velocity of propagation of torsional stress-wave in an elastic bar	Refer to Table 3-1.	

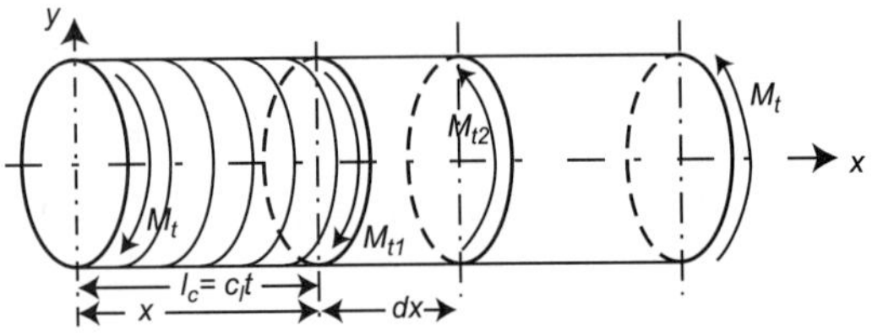

FIGURE 3-10 Torsional impact on a uniform bar showing torque on two faces of an element

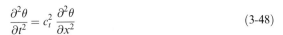

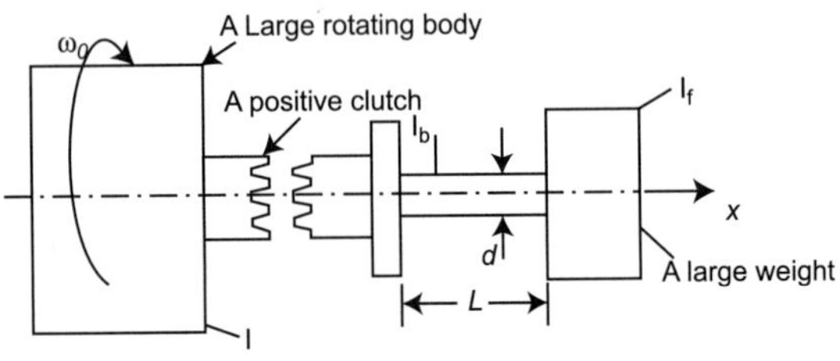

FIGURE 3-11 Torsional striking impact

Particular	Formula	
The angular velocity of the end of a bar subject to torsion relative to the unstressed region	$\omega = \frac{\theta}{t} = \frac{\left(\frac{2\tau_t}{d\sqrt{\rho G}}\right)t}{t} = \frac{2\tau_t}{d\sqrt{\rho G}}$	(3-50)
The shear stress from Eq. (3-50)	$\tau = \frac{\omega d}{2}\sqrt{\rho G}$	(3-51a)
The initial shear stress, if the rotating body strikes the end of the bar with an angular velocity ω_0	$\tau_0 = \frac{\omega_0 d}{2}\sqrt{\rho G}$	(3-51b)

Particular	Formula
The maximum shear stress for the case of a shaft fixed or attached to a very large mass/weight at one end and suddenly applied rotational load at the other end by means of some mechanical device such as a jaw clutch (Fig. 3-11)	$\tau_{\max} = \tau_0 \left[\sqrt{\frac{1}{\zeta}\left(\frac{1}{1+\frac{\zeta}{3}}\right)} \right] = \tau_0 \left[\sqrt{\frac{\alpha}{\zeta}} \right]$ (3-52)
	where $\zeta = \frac{I_b}{I}$.
	I_b = mass moment of inertia of bar $= m_b \left(\frac{d^2}{8}\right)$
	I = mass moment of inertia of striking rotating weight
	I_b and I correspond to W_b and W of the weight of the bar and the rotating mass or weight respectively.

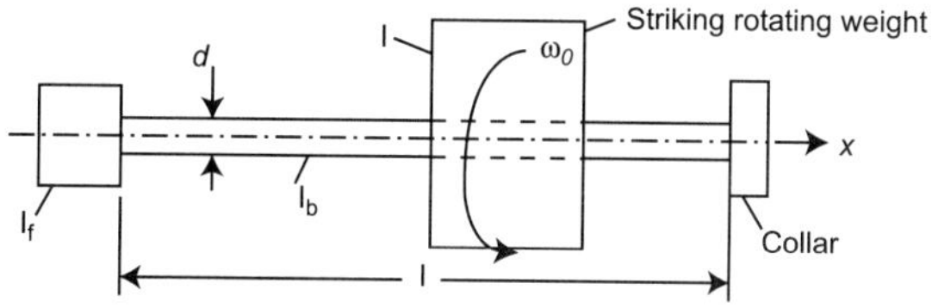

FIGURE 3-12 A striking rotating weight with mass-moment of inertia I rotating at ω_0 engages with one end of shaft and the other end of shaft fixed to a mass-moment of inertia I_f

Particular	Formula
The more accurate equation for the $\tau_{\max}$ which is based on stress wave propagation	$\tau_{\max} = \tau_0 \left[1 + \sqrt{\frac{1}{\zeta} + \frac{2}{3}} \right]$ (3-53)
The initial/maximum ($\tau_i = \tau_{\max}$) shear stress for the case of a system shown in Fig. 3-12	$\tau_i = \tau_{\max} = \tau_0 \sqrt{\frac{\lambda(1+\zeta)}{\zeta(1+\zeta+\lambda)}}$ (3-54)
	where $\zeta = \frac{I_b}{I}$, $\lambda = \frac{I}{I_f}$ and $I_b = \rho J l$.
A similar equation to Eq. (3-54) for maximum stress for longitudinal impact	$\sigma_i = \sigma_{\max} = \sigma_0 \sqrt{\frac{\lambda(1+\zeta)}{\zeta(1+\zeta+\lambda)}}$ (3-55)
	where $\zeta = \frac{W_b}{W} = \frac{m_b}{m}$ and $\lambda = \frac{m}{m_f}$
Accurate maximum stress for longitudinal impact stress based on stress wave propagation as suggested by Prof. Burr	$\sigma_i = \sigma_{\max} = \sigma_0 \left[1.1 + \sqrt{\frac{\lambda(1+\zeta)}{\zeta(1+\zeta+\lambda)}} \right]$ (3-56)
Accurate maximum stress for torsional impact shear stress based on stress-wave propagation as suggested by Prof. Burr	$\tau_i = \tau_{\max} = \tau_0 \left[1.1 + \sqrt{\frac{\lambda(1+\zeta)}{\zeta(1+\zeta+\lambda)}} \right]$ (3-57)

Particular	Formula
INERTIA IN COLLISION OF ELASTIC BODIES	
When a body having weight W strikes another body that has a weight W', impact energy Wh is reduced to nWh, according to law of collision of two perfectly inelastic bodies, the formula for the value of n	$n = \dfrac{1 + am}{(1 + bm)^2}$ (3-58) where $m = \dfrac{W'}{W}$; a and b are taken from Table 3-3
RESILIENCE	
The expression for resilience in compression or tension	$U = \dfrac{\sigma^2}{2}\dfrac{V}{E} = \dfrac{1}{2}\dfrac{\sigma^2 AL}{E}$ (3-59)
The modulus of resilience	$u = \dfrac{\sigma^2}{2E}$ (3-60)
The area under the stress-strain curve up to yielding point represents the modulus of resilience (Fig. 1.1)	$u = \frac{1}{2}\sigma\varepsilon$ (3-61)
The resilience in bending	$U_b = \left(\dfrac{k}{c}\right)^2 \dfrac{\sigma_b^2 AL}{6E}$ (3-62)
The modulus of resilience in bending	$u_b = \left(\dfrac{k}{c}\right)^2 \dfrac{\sigma_b^2}{6E}$ (3-63) where $(k/c)^2 = \frac{1}{3}$ for rectangular cross-section $= \frac{1}{4}$ for circular section c = distance from extreme fibre to neutral axis
Resilience in direct shear	$U_\tau = \dfrac{\tau_e^2 V}{2G}$ (3-64)
The modulus of resilience in direct shear	$u_\tau = \dfrac{\tau_e^2}{2G}$ (3-65)
Resilience in torsion	$U_\tau = \dfrac{\tau_e^2 AL}{2G}\left(\dfrac{k_0}{c}\right)^2$ (3-66) where $k_0 = \sqrt{D_1^2 - D_0^2/8}$ and $c = \frac{1}{2}D_0$ for hollow shaft.
The modulus of resilience in torsion	$u_\tau = \dfrac{\tau_e^2}{2G}\left(\dfrac{k_0}{c}\right)^2$ (3-67)
The equation for strain energy due to shear in bending	$U_{\tau b} = \displaystyle\int_0^l \dfrac{k_\tau F_\tau^2}{2GA}\,dx$ (3-68)
The modulus of resilience due to shear in bending	$u_{\tau b} = \dfrac{k_\tau \tau_e^2}{2G}$ (3-69)

Particular	Formula
The equation for shear or distortional strain energy per unit volume associated with distortion, without change in volume	$U_\tau = \frac{1}{6G}[\sigma_1^2 + \sigma_2^2 + \sigma_3^2 - (\sigma_1\sigma_2 + \sigma_2\sigma_3 + \sigma_3\sigma_1)]$ (3-70a)
	$= \frac{1}{12G}[(\sigma_1 - \sigma_2)^2 + (\sigma_2 - \sigma_3)^2 + (\sigma_3 - \sigma_1)^2]$ (3-70b)
The equation for dilatational or volumetric strain energy per unit volume without distortion, only a change in volume	$U_v = \frac{(1-2\nu)}{6E}[(\sigma_1 + \sigma_2 + \sigma_3)^2]$ (3-71)
For maximum resilience per unit volume (i.e., for modulus of resilience), resilience in tension for various engineering materials and coefficients a and b; velocity of propagation c and c_t.	Refer to Tables 3-1 to 3-4.

TABLE 3-1
Longitudinal velocity of longitudinal wave c and torsional wave c_t propagation in elastic media

	Density			Modulus of elasticity, E		Modulus of rigidity, G		$c = \sqrt{\frac{E}{\rho}} = \sqrt{\frac{Eg}{\gamma}}$ #		$c_t = \sqrt{\frac{G}{\rho}} = \sqrt{\frac{Gg}{\gamma}}$ #	
	ρ		γ								
Material	g/cm³	lb$_m$/in³	kN/m³	GPa	Mpsi	GPa	Mpsi	m/s	ft/s	m/s	ft/s
Aluminum alloy	2.71	0.098	26.6	71.0	10.3	26.2	3.8	5116	16785	3110	10466
Brass	8.55	0.309	83.9	106.2	15.4	40.1	5.82	3523	11560	2165	7106
Carbon steel	7.81	0.282	76.6	206.8	30.0	79.3	11.5	5145	16887	3200	10485
Cast iron, gray	7.20	0.260	70.6	100.0	14.5	41.4	6.0	3727	12223	2407	7865
Copper	8.91	0.320	87.4	118.6	17.7	44.7	6.49	3648	12176	2240	7373
Glass	2.60	0.094	25.5	46.2	6.7	18.6	2.7	4214	13823	2675	8775
Lead	11.38	0.411	111.6	36.5	5.3	13.1	1.9	1796	5879	1073	3520
Inconel	8.42	0.307	83.3	213.7	31.0	75.8	11.0	5016	16452	2987	9800
Stainless steel	7.75	0.280	76.0	190.3	27.6	73.1	10.6	4955	15972	3071	10074
Tungsten	18.82	0.680	184.6	344.7	50.0	137.9	20.0	4279	14039	2707	8880

Note: ρ = Mass density, g/cm³ (lb$_m$/in³), γ = weight density (specific weight), kN/m³ (lbf/in³), $g = 9.8066\ \text{m/s}^2$ in SI units, $g = 980\ \text{in/s}^2 = 32.2\ \text{ft/s}^2$ in fps units.

TABLE 3-2
Maximum resilience per unit volume (2, 1)

Type of loading	Modulus of resilience, J (in lbf)
Tension or compression	$\frac{\sigma_e^2}{2E}$
Shear, simple transverse	$\frac{\tau_e^2}{2G}$
Bending in beams	
With simply supported ends:	
Concentrated center load and rectangular cross-section	$\frac{\sigma_e^2}{18E}$
Concentrated center load and circular cross-section	$\frac{\sigma_e^2}{24E}$
Concentrated center load and I-beam section	$\frac{3\sigma_e^2}{32E}$
Uniform load and rectangular section	$\frac{4\sigma_e^2}{45E}$
Uniform-strength beam, concentrated load, and rectangular section	$\frac{\sigma_e^2}{6E}$
Fixed at both ends:	
Concentrated load and rectangular cross-section	$\frac{\sigma_e^2}{18E}$
Uniform load and rectangular cross-section	$\frac{\sigma_e^2}{30E}$
Cantilever beam:	
End load and rectangular cross-section	$\frac{\sigma_e^2}{18E}$
Uniform load and rectangular cross-section	$\frac{\sigma_e^2}{30E}$
Torsion	
Solid round bar	$\frac{\tau_e^2}{4G}$
Hollow round bar with D_0 greater than D_i	$\left[1+\left(\frac{D_i}{D_o}\right)^2\right]\frac{\tau^2}{4G}$
Springs	
Laminated with flat leaves of uniform strength	$\frac{\sigma_e^2}{6E}$
Flat spiral with rectangular section	$\frac{\sigma_e^2}{24E}$
Helical with round section and axial load	$\frac{\tau_e^2}{4G}$
Helical with round section and axial twist	$\frac{\sigma_e^2}{8E}$
Helical with rectangular section and axial twist	$\frac{\sigma_e^2}{6E}$

Sources: K. Lingaiah and B. R. Narayana Iyengar, *Machine Design Data Handbook*, Vol I (*SI and Customary Metric Units*), Suma Publishers, Bangalore, India, 1986; K. Lingaiah, *Machine Design Data Handbook*, Vol II (*SI and Customary Metric Units*), Suma Publishers, Bangalore, India, 1986; V. L. Maleev and J. B. Hartman, *Machine Design*, International Textbook Company, Scranton, Pennsylvania, 1954.

TABLE 3-3
Coefficients in Eq. (3-58) (1)

Type of impact	a	b
Longitudinal impact on bar	$\frac{1}{3}$	$\frac{1}{2}$
Center impact on single beam	$\frac{17}{35}$	$\frac{5}{8}$
Center impact on beam with fixed ends	$\frac{13}{35}$	$\frac{1}{2}$
End impact on cantilever beam	$\frac{4}{17}$	$\frac{3}{8}$

TABLE 3-4
Resilience in tension

Material	Elastic limit, σ		Modulus of elasticity, E		Modulus of resilience, u		Impact strength (Izod no.)
	MPa	kpsi	GPa	Mpsi	J	in lbf	
Cast iron:							
Class 20 (ordinary)	42.8[a]	62	68.9	10	0.22	1.9	
Class 25	68.9[a]	10.0	89.2	13	0.43	3.8	7.9
Nickel, Grade II	117.2[a]	17.0	24.5	18	0.90	8.0	
Malleable	137.9	20.0	172.6	25	0.90	8.0	2.7
Aluminum alloy, SAF 33	48.3	7.0	66.7	9.7	0.28	2.5	
Brass, SAE 40 or SAE 41	68.9	10.0	82.4	12	0.45	4.0	
Bronze, SAE 43	193.0	28.0	110.8	16	2.77	24.5	
Monel metal:							
Hot-rolled	206.9	30.0	176.5	25.5	1.96	17.6	120
Cold-rolled, normalized	482.6	70.0	176.5	25.5	10.79	96	100
Steel:							
SAE 1010	206.9	30.0		30.3	1.69	15	
SAE 1030	248.2	36.5	206.9	30	2.45	22	20
SAE 1050, annealed	330.9	48.5	204.8	29.7	4.27	38	
SAE 1095, annealed	413.7	60.0	204.8	29.7	6.77	60	
SAE 1095, tempered	517.1	75.0	204.8	29.7	16.08	94	
SAE 2320, annealed	310.3	45.0	204.8	29.7	3.82	34	52
SAE 2320, tempered	689.5	100.0	204.8	29.7	18.83	167	40
SAE 3250, annealed	551.6	80.0	213.7	31	21.58	193	
SAE 3250, tempered	1379.0	200.0	213.7	31.0	72.57	645	30
SAE 6150, annealed	427.6	62.0	213.7	31	6.96	62	
SAE 6150, tempered	1102.3	160.0	213.7	31	52.47	466	
Rubber	2.1	0.3	1034×10^{-9}	150×10^{-6}	33.89	300	

[a] Cast iron has no well-defined elastic limit, but the values may be safely used anyway for all practical purposes.
Source: Reproduced courtesy of V. L. Maleev and J. B. Hartman, *Machine Design*, International Textbook Co., Scranton, Pennsylvania, 1954.

REFERENCES

1. Maleev, V. L., and J. B. Hartman, *Machine Design*, International Textbook Co., Scranton, Pennsylvania, 1954.
2. Lingaiah, K., and B. R. Narayana Iyengar, *Machine Design Data Handbook*, Vol I (*SI and Customary Metric Units*), Suma Publishers, Bangalore, India, 1986.
3. Lingaiah, K., *Machine Design Data Handbook*, Vol II (*SI and Customary Metric Units*), Suma Publishers, Bangalore, India, 1986.
4. Lingaiah, K., *Machine Design Data Handbook*, McGraw-Hill Book Company, New York, 1994.
5. Burr, Arthur H., and John B. Cheatham, *Mechanical Analysis and Design*, 2nd edition, Prentice Hall, Englewood Cliffs, NJ, 1995.
6. Spotts, Merhyle F.,"Impact Stress in Elastic Bodies Calculated by the Energy Method", *Engineering Data for Product Design*, edited by Douglas C. Greenwood, McGraw-Hill Book Company, New York, 1961.
7. Burr, A. H.,"Longitudinal and Torsional Impact in Uniform Bar with a Rigid Body at One End", *J. Appl. Mech.*, Vol. 17, No. 2 (June 1950), pp. 209–217; *Trans. ASME*, Vol. 72 (1950).
8. Timoshenko, S., and J. N. Goodier, *Theory of Elasticity*, 3rd edition, McGraw-Hill Book Company, New York, pp. 485–513, 1970.
9. Kolsky, H., *Stress Waves in Elastic Solids*, Dover Publications, New York, 1963.
10. Durellli, A. J., and W. F. Riley, *Introduction to Photomechanics*, Prentice Hall Inc, Englewood Cliffs, NJ, 1965.
11. Arnold, "Impact Stresses in a Simply Supported Beam", *Proc. I.M.E.*, Vol. 137, p. 217.
12. Dohrenwend and Mehaffy, "Dynamic Loading", *Machine Design*, Vol. 15 (1943), p. 99.

BIBLIOGRAPHY

Lingaiah, K., and B. R. Narayana Iyengar, *Machine Design Data Handbook*, Engineering College Co-operative Society, Bangalore, India, 1962.

Norman, C. A., E. S. Ault, and E. F. Zarobsky, *Fundamentals of Machine Design*, The Macmillan Company, New York, 1951.

CHAPTER

4

STRESS CONCENTRATION AND STRESS INTENSITY IN MACHINE MEMBERS

SYMBOLS[6,7,8]

a	diameter of circular hole (cut-out), m (in)*
	semimajor axis of elliptical hole (cut-out), m (in)
	half of the length of the slot, m (in)
$2a$	length of straight crack, m (in) (Figs. 4-26A and 4-28)
A	area of cross section, m^2 $(in)^2$
b	semiminor axis of elliptical hole (cut-out), m (in)
	maximum breadth of section of curved bar, m (in)
	width of notch at the edge, m (in)
$b = (w - a)$	effective width of plate across the hole, m (in) or net width of plate, m (in)
$2b$	total width of plate with a crack, m (in) (Fig. 4-28)
B	constant in curved bar equation
	outside diameter of reinforced ring in an asymmetrically reinforced circular hole
c	distance from centroidal axis to extreme fiber of beam or inside edge of curved bar, m (in)
C	spring index
d	effective width of plate, m (in)
	width of U-piece at dangerous section, m (in)
	diameter of shaft at reduced section, m (in)
	reduced width of shoulder plate, m (in)
d_i	diameter of hole (cut-out), m (in)
d_o	outside diameter of reinforcement, m (in)
D	total diameter of shaft, m (in)
	total width of plate, m (in)
F	load, kN (lbf)
	force, kN (lbf)
	diametrically opposite concentrated loads on ring or hollow roller, kN (lbf)
h	thickness of plate, m (in)
	thickness of ring or roller, m (in)
	lever arm from critical section of tooth, m (in)

$2h$	length of plate with a crack, m (in) (Fig. 4-28)
h_1	depth of groove, m (in) (Figs 4-11, 4-12 and 4-13)
	depth of shoulder, m (in)
H	thickness of reinforcement including plate thickness, m (in)
I	moment of inertia, area, m^4 (in^4)
J	moment of inertia, polar, m^4 (in^4)
K_I	opening mode or mode I of stress intensity factor, $MPa\sqrt{m}$ ($kpsi\sqrt{in}$)
K_{II}	mode II of stress intensity factor, $MPa\sqrt{m}$ ($kpsi\sqrt{in}$)
K_{IC}	opening mode or mode I of critical stress intensity factor or fracture toughness factor, $MPa\sqrt{m}$ ($kpsi\sqrt{in}$)
$K_\sigma = \dfrac{\sigma_{max}}{\sigma_{nom}}$	theoretical stress-concentration factor for normal stress
K'_σ	combined stress-concentration factor (K'_σ is a theoretical factor)
K_τ	theoretical stress-concentration factor for shear stress (torsion)
$K_{f\sigma}$	fatigue stress-concentration factor for normal stress or fatigue notch factor for axial or bending (normal) (Fig. 14-13) or fatigue strength reduction factor

$$K_{f\sigma} = \frac{\sigma_f}{\sigma_{nf}} = \frac{\text{fatigue limit of unnotched specimen (axial or bending)}}{\text{fatigue limit of notched specimen (axial or bending)}}$$

$K_{\sigma n}$	stress-concentration factor (normal) based on net area (nominal) of cross section (i.e., net nominal stress)
$K_{\sigma g}$	stress-concentration factor (normal) based on gross area of cross section (i.e., gross stress)
$K_{f\tau}$	fatigue stress-concentration factor for shear stress (torsion) or fatigue notch factor (torsion)

$$K_{f\tau} = \frac{\tau_f}{\tau_{nf}} = \frac{\text{fatigue limit of unnotched specimen (torsion)}}{\text{fatigue limit of notched specimen (torsion)}}$$

$K_{\sigma u}$	stress-concentration factor for U-grooved plate
$K_{\sigma v}$	stress-concentration factor for a V-grooved plate
$K_{\sigma e}$	effective stress-concentration factor under a static load, equivalent stress-concentration factor
m	module, mm
MF	magnification factor
M_b	bending moment, N m (lbf ft)
M_t	torsional moment, N m (lbf ft)
n	safety factor
$q = \dfrac{K_{f\sigma} - 1}{k_\sigma - 1}$	index of sensitivity or notch sensitivity factor
r	radius of curvature of groove or notch of curved bar, m (in)
	minimum radius of curvature of an ellipse, m (in)
	polar coordinate
r	distance of a point in a plate from the crack tip (Fig. 4-26C), m (in)
r_j	minimum fillet radius of gear tooth, m (in)
r_t	cutter tip radius, m (in)
R_o	outside radius of ring or hollow roller, m (in)
R_i	inside radius of ring or hollow roller, m (in)
s	thickness of the tooth at critical section, m (in)
V_H	volume of hole, m^3 (in^3)
V_R	volume of reinforcement, m^3 (in^3)

w	depth of U-piece-arm, m (in)
W	total width of flat plate, m (in)
σ_x, σ_y, τ_{xy}	stress components in x, y coordinates, MPa (kpsi)
σ_{max}	maximum normal stress, MPa (kpsi)
σ_{nom}	nominal normal stress computed from F/A or $M_b c/I$ or from an elementary formula which does not take into account the stress concentration, MPa (kpsi)
σ_0	average stress at the root of gear tooth, MPa (kpsi)
τ_{max}	maximum shear stress, MPa (kpsi)
τ_{nom}	nominal shear stress computed from $M_t r/J$, MPa (kpsi)
α	angle of a shallow U-groove, deg
β	angle of V-groove, deg
θ	polar coordinate, deg
	angle made by r the distance of a point from tip of crack with x axis (Fig. 4-26C)
	Other factors in performance or in special aspects are included from time to time in this chapter, and being applicable only in their immediate context, are not given at this stage

Particular	Stress concentration factor theoretical/empirical or otherwise: Extreme value	Stress concentration factor theoretical/empirical or otherwise: Formula
(a) Keyway (Fig. 4-1 and Tables 4-1 and 4-2:		
Profile keyway	$K_{f\tau} = 1.68$	
Sled-runner keyway	$K_{f\tau} = 1.44$	

TABLE 4-1
Shear stress-concentration factor for a keyway in a shaft subjected to torsion (by Leven)

r/d	0.0052	0.0104	0.0208	0.0417	0.0833
K_τ	3.92	3.16	2.62	2.30	2.06

TABLE 4-2
Stress-concentration factor $K_{f\tau}$ for keyways

	Annealed		Hardened	
Type of keyway	Bending	Torsion	Bending	Torsion
Profile	1.6	1.3	2.0	1.6
Sled runner	1.3	1.3	1.6	1.6

(b) Curved bar (Fig. 4.1a):
For curved bar

$$K_\sigma = 1.00 + B\left(\frac{I}{bc^2}\right)\left(\frac{1}{r-c} + \frac{1}{r}\right) \tag{4-1}$$

where

$B = 1.05$ for circular or elliptical cross-section
$= 0.5$ for other cross-section

Particular	Stress concentration factor theoretical/empirical or otherwise	
	Extreme value	Formula

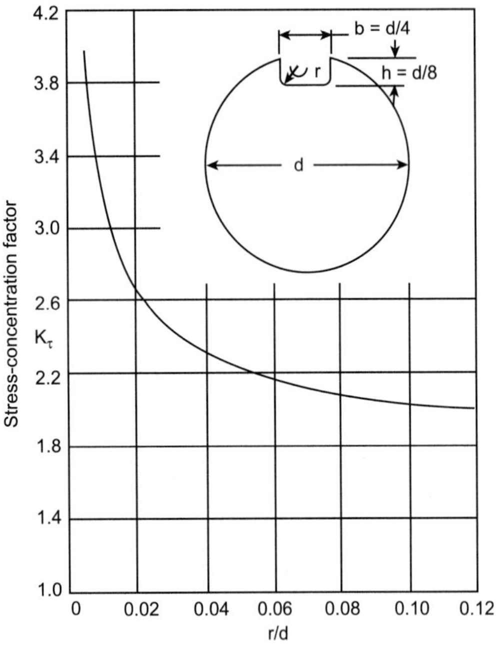

FIGURE 4-1 Stress-concentration factor for the straight portion of a keyway in a shaft in torsion.

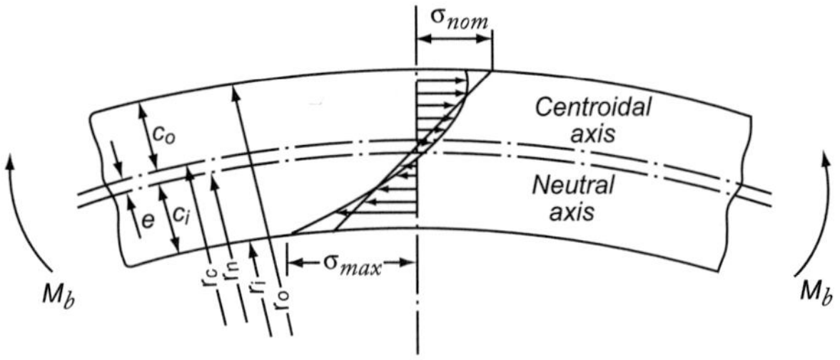

FIGURE 4-1a Stress distribution in curved bar under bending.

(c) Spur gear tooth (Figs. 4-2 and 4-3, Table 4-3):
At root fillet of an involute tooth profile of 14.5° pressure angle

$$K_\tau = 2 \text{ to } 2.5 \quad K_\sigma = 0.22 + \frac{1}{\left(\frac{r_f}{s}\right)^{0.2}\left(\frac{h}{s}\right)^{0.4}} \tag{4-2}$$

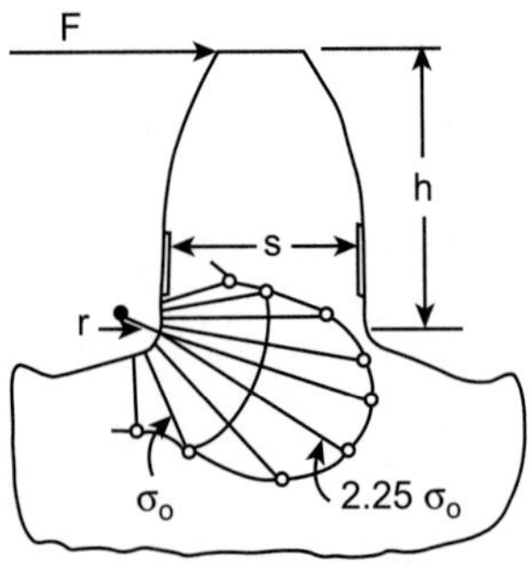

FIGURE 4-2 Stress-concentration factor at root of gear tooth.

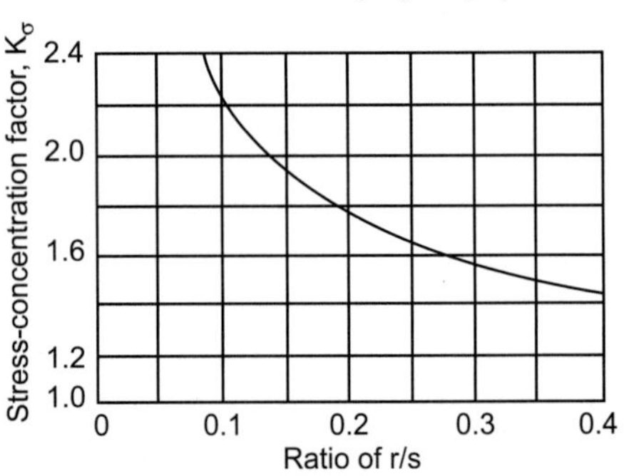

FIGURE 4-3 Effect of fillet radius on stress-concentration at root of gear tooth.

Particular	Stress concentration factor theoretical/empirical or otherwise: Extreme value	Formula	

TABLE 4-3
Stress-concentration factors at root fillet of gear

m, mm	$K_{\sigma t}$	$K_{\sigma c}$
6.24	1.47	1.61
5	1.47	1.61
4.25	1.42	1.57
3.5	1.35	1.50
3	1.345	1.50

At root fillet of a full depth involute tooth profile of 20° pressure angle

$$K_\sigma = 0.18 + \frac{1}{\left(\frac{r_f}{s}\right)^{0.15}\left(\frac{h}{s}\right)^{0.45}} \tag{4-3}$$

At root fillet of an involute tooth profile of 25° pressure angle

$$K_\sigma = 0.14 + \left(\frac{s}{r_f}\right)^{0.11}\left(\frac{s}{h}\right)^{0.50}$$

(d) Circular cut-outs (holes) in plates (Fig. 4-4c):
For infinite plate in:

(i) Uniaxial tension (Fig. 4-4c) — $K_\sigma = 3$

$$K_\sigma = \frac{1}{2}\left(2 + \frac{a^2}{r^2} + \frac{3a^4}{r^4}\right)_{r=a} = 3 \tag{4-4}$$

(ii) Biaxial tension — $K_\sigma = 2$

$$K_\sigma = \left(1 + \frac{a^2}{r^2}\right)_{r=a} = 2 \tag{4-5}$$

(iii) Pure shear — $K_\tau = 4$

$$K_\tau = \left(1 + \frac{3a^4}{r^4}\right)_{r=a} = 4 \tag{4-6}$$

For stress concentration factor for a semi-infinite plate with a circular hole near the edge under tension. — $K_\sigma = 3$

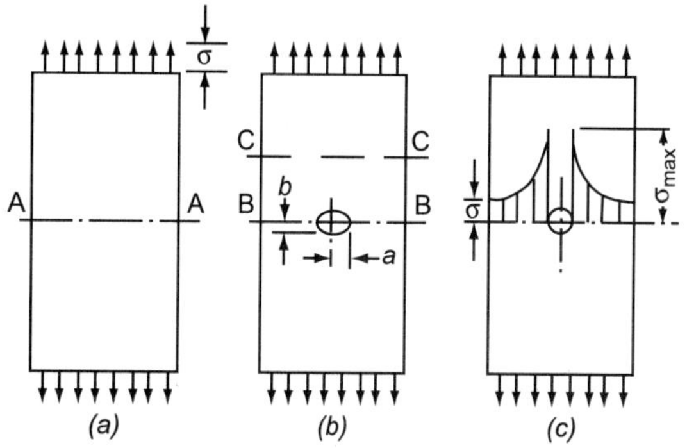

FIGURE 4-4 Stress distribution around holes (cut-outs) in plate in tension.

Particular	Stress concentration factor theoretical/empirical or otherwise: Extreme value	Formula	
For finite plate in:			
(i) Uniaxial tension (Fig. 4-5)	$K_\sigma = 3$	$K_\sigma = 2 + \left(\frac{b}{w}\right)^3$	(4-7)
(ii) Bending (Fig. 4-6)	$K_\sigma = 2$	$K_\sigma = 2\,\frac{b}{w}$	(4-8)
(e) Filled circular hole:			
For filled circular holes in plate subjected to tension	$K_\sigma = 2.5$		
(f) Reinforced circular holes:			
(i) For stress-concentration factor for a symmetrically reinforced circular hole in a flat plate under uniform uniaxial tension (ii) For stress-concentration factor for an asymmetrically reinforced circular hole in a flat plate under uniform uniaxial tension	Refer to Fig. 4-7a, b, and c.		

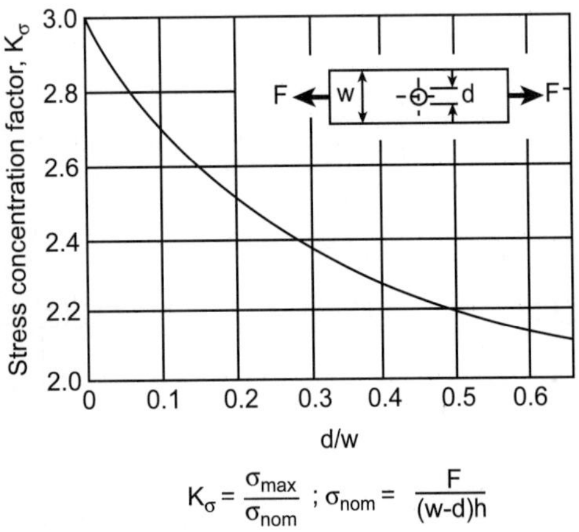

FIGURE 4-5 Reproduced with permission. Stress-concentration factor for a plate of finite width with a circular hole (cut-out) in tension. ("Design Factors for Stress Concentration," *Machine Design*, Vol. 23, Nos. 2 to 7, 1951.)

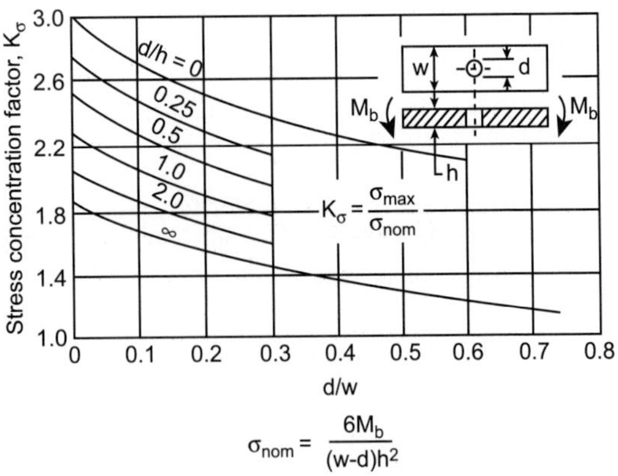

FIGURE 4-6 Reproduced with permission. Stress-concentration factor for a plate of finite width with transverse circular hole (cut-out) subjected to bending. ("Design Factors for Stress Concentration," *Machine Design*, Vol. 23, Nos. 2 to 7, 1951.)

Particular	Stress concentration factor theoretical/empirical or otherwise: Extreme value	Formula
(g) Noncircular holes (cut-outs) in a plate:		
(i) An infinite plate with an elliptical hole (cut-out) in uniaxial tension (load at right angles to major axis) (Fig. 4-4b)		$K_\sigma = 1 + 2\dfrac{a}{b}$ (4-9a)
		$K_\sigma = 1 + 2\sqrt{\dfrac{a}{r}}$ (4-9b)
(ii) An infinite plate with elliptical hole (cut-out) in uniaxial tension (load parallel to major axis)		$K_\sigma = 1 + \dfrac{2b}{a}$ (4-9c)
(iii) An infinite plate with elliptical hole (cut-out) in pure shear		$K_\tau = 2\left(1 + \dfrac{a}{b}\right)$ (4-10)
(iv) An infinite plate with an elliptical cut-out in biaxial tension		$K_\sigma = \dfrac{2a}{b}$ (4-11)
(v) Transverse bending of a plate containing an elliptical cut-out (or hole)		$K_\sigma = \dfrac{(1+v)\left(3 - v + \dfrac{2a}{b}\right)}{(3+v)}$ (4-12)
(vi) Slotted plate loaded in tension or bending		$K_\sigma = 1.064 + 0.788\dfrac{a}{b}$ for $v = 0.3$ (4-13a)
		$K_\sigma = 2 + \left(\dfrac{b}{w}\right)^3$ for $\dfrac{a}{r} = 1$ (4-13b)
For reduction of endurance strength of steel	Refer to Fig. 4-8.	
(h) U-shaped member subjected to bending (Fig. 4-9):		
(1) At 0° with horizontal axis		$K_{\sigma A} = 1 + \dfrac{d}{4r}$ (4-14)
(2) At 70° with horizontal axis		$K_{\sigma B} = 1 + \dfrac{w}{5r}$ (4-15)
(i) Helical spring:		
Stress concentration or Wahl's correction factor for helical spring		$K_\tau = \dfrac{4c-1}{4c-4} + \dfrac{0.615}{c}$ (4-16)

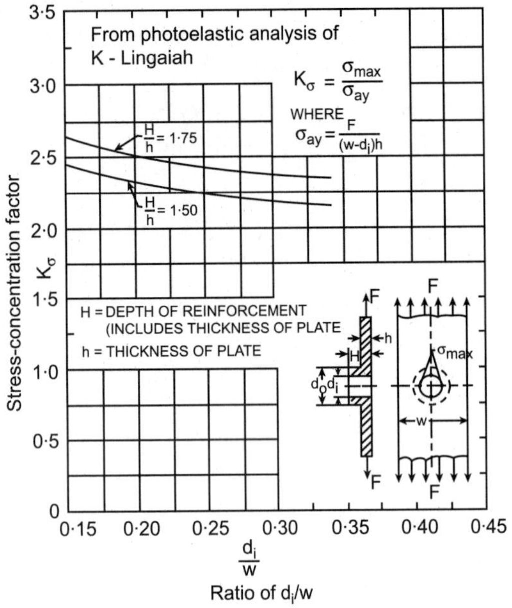

FIGURE 4-7(a) Stress-concentration factor for an asymmetrically reinforced circular hole (cut-out) in a flat plate subjected to tension. (PhD work of the author.)

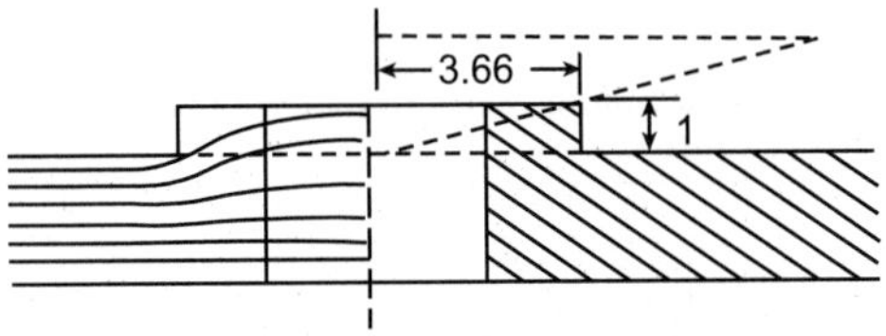

FIGURE 4-7(c)

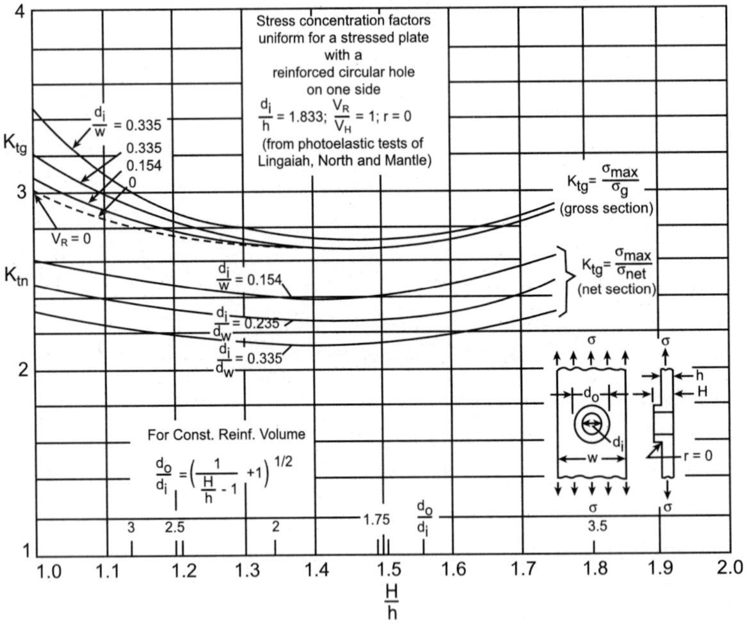

FIGURE 4-7(b) Stress-concentration factor for an asymmetrically reinforced circular hole in a flat plate subjected to uniform unidirectional tensile stress. (PhD work of the author, and R. E. Peterson, *Stress Concentration Factors*, John Wiley and Sons, Inc., 1974.)

Particular	Stress concentration factor theoretical/empirical or otherwise: Extreme value	Formula

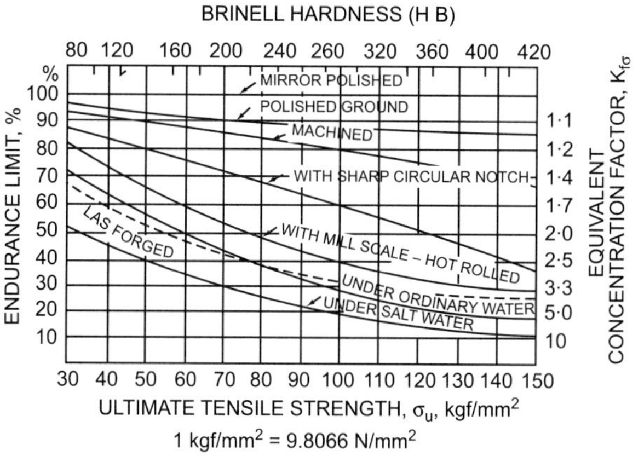

FIGURE 4-8 Reduction of endurance strength of steel, σ_f.

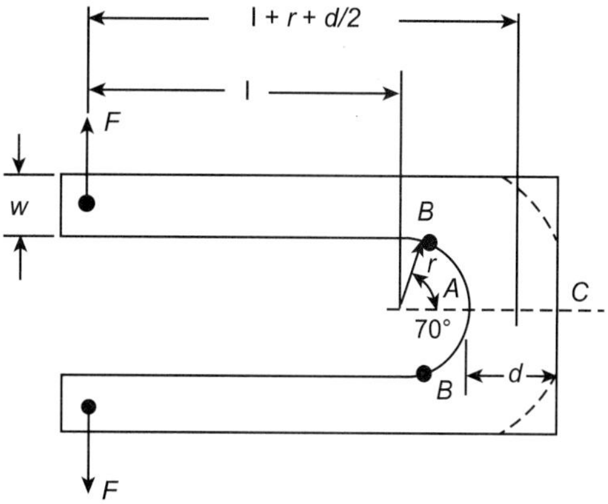

FIGURE 4-9 U-shaped member. (R. E. Peterson, *Stress Concentration Factors*, John Wiley and Sons, 1974.)

(j) Ring or hollow roller:

For the ring loaded internally

$$K_\sigma = \frac{\sigma_{\max}[2h(R_o - R_i)]}{F\left[1 + 3\dfrac{(R_o + R_i)\left(1 - \dfrac{2}{\pi}\right)}{R_o - R_i}\right]} \tag{4-17}$$

For the ring loaded externally

$$K_\sigma = \frac{\sigma_{\max}[\pi h(R_o - R_i)]}{3F(R_o + R_i)} \tag{4-18}$$

(k) Shafts with transverse holes (Fig. 4-10):

(i) Shaft with a circular hole subjected to transverse bending for $a/d \to 0$ — $K_\sigma = 3.0$

(ii) Shaft with a circular hole subjected to torsion for $a/d \to 0$ — $K_\tau = 2.0$

(l) Shafts with grooves:

Shaft with U and V circumferential groove in:

(i) Tension or bending (Figs. 4-11 to 4-16 and 4-18)

$$K_\sigma = 1 + 2\sqrt{\frac{h_1}{r}} \tag{4-19}$$

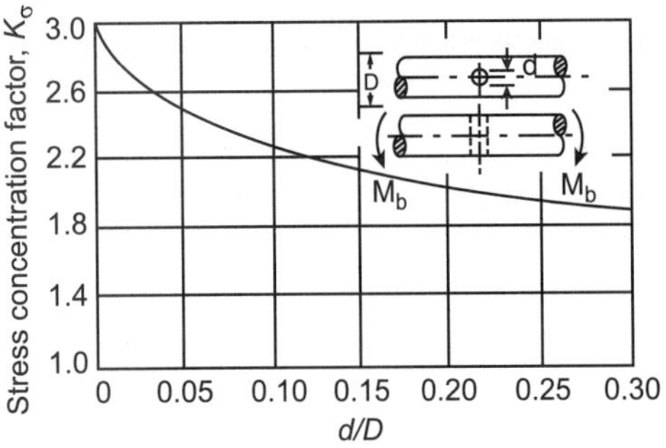

$$\frac{I}{c} = \frac{\pi d^3}{32}$$

$$\sigma_{nom} = \frac{M_b}{\left(\frac{\pi D^3}{32}\right) - \left(\frac{dD^2}{6}\right)} \quad \text{(approx.)}$$

FIGURE 4-10 Reproduced with permission. Stress-concentration factor K_σ for a shaft with transverse circular hole subjected to bending. (R. E. Peterson, "Stress Concentration Factors," *Machine Design*, Vol. 23, Nos. 2 to 7, 1951.)

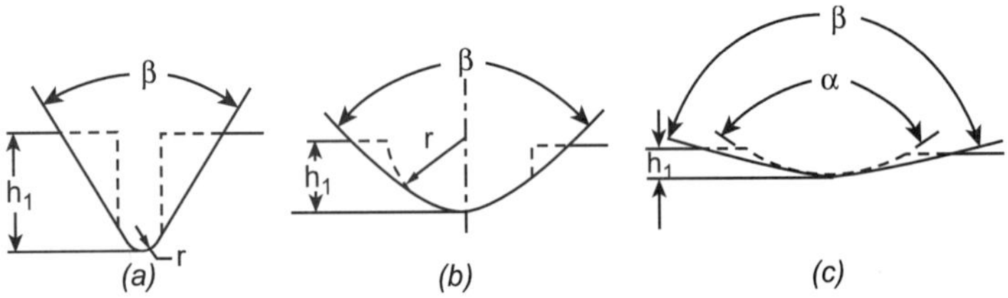

FIGURE 4-11 Types of V-grooves.

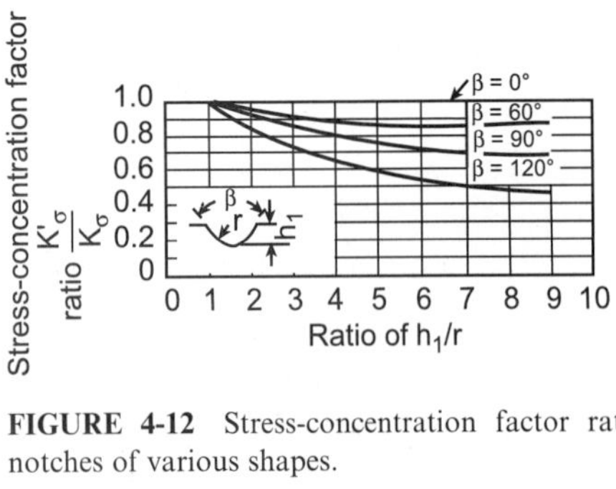

FIGURE 4-12 Stress-concentration factor ratio due to notches of various shapes.

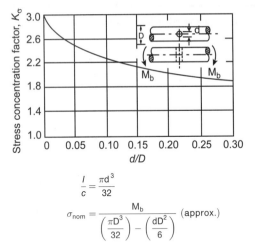

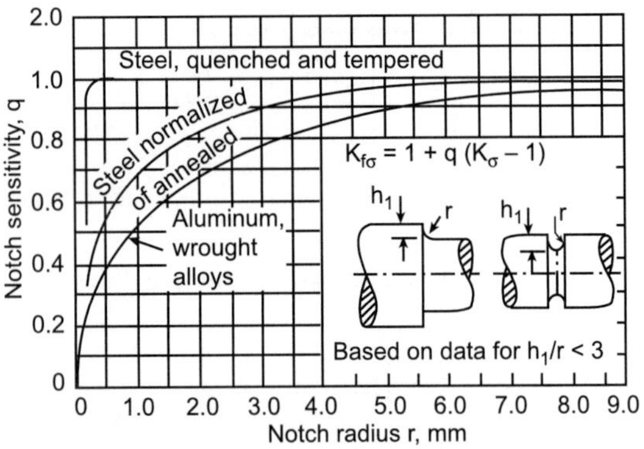

FIGURE 4-13 Average notch sensitivity.

Particular	Stress concentration factor theoretical/empirical or otherwise	
	Extreme value	Formula

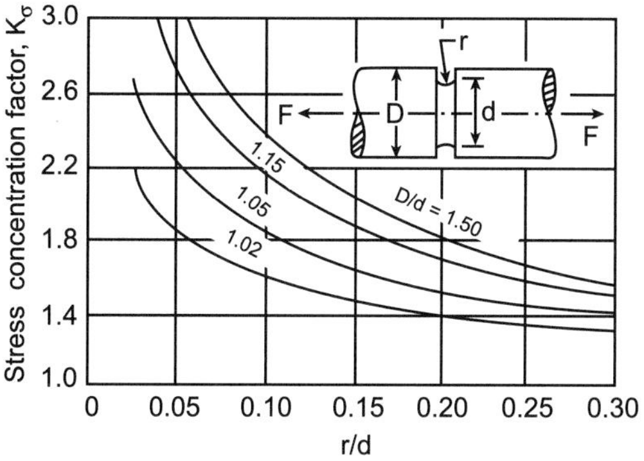

$$A = \frac{\pi d^2}{4};\ \sigma_{nom} = \frac{F}{A}$$

FIGURE 4-14 Reproduced with permission. Stress-concentration factor K_σ for a grooved shaft in tension. (R. E. Peterson, "Design Factors for Stress Concentration," *Machine Design*, Vol. 23, Nos. 2 to 7, 1951.)

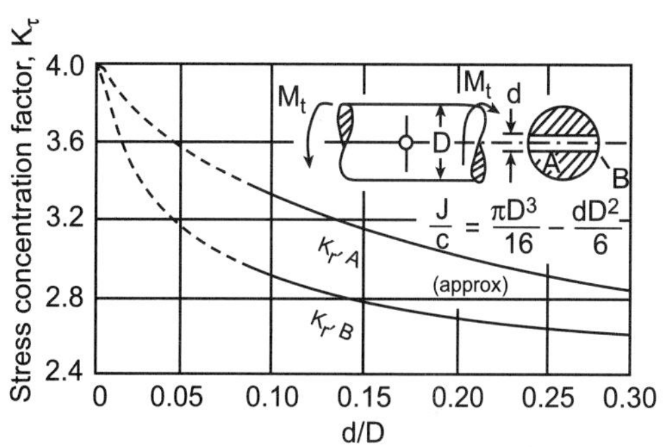

FIGURE 4-15 Reproduced with permission. Round shaft in torsion with transverse hole. (R. E. Peterson, "Design Factors for Stress Concentration," *Machine Design*, Vol. 23, Nos. 2 to 7, 1951.)

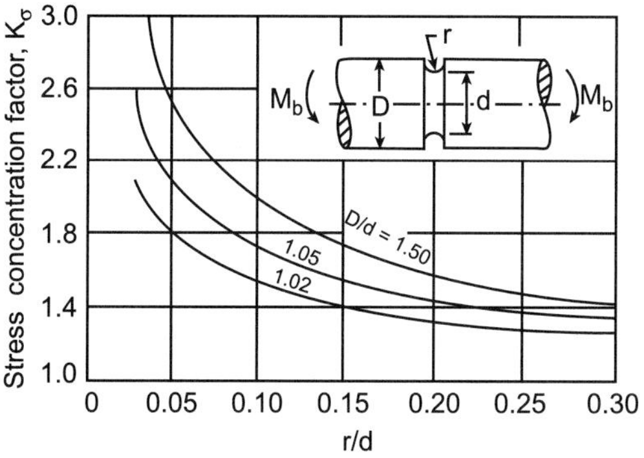

$$\frac{I}{c} = \frac{\pi d^3}{32};\ \sigma_{nom} = \frac{M_b c}{I};\ I = \frac{\pi d^4}{64}$$

$$c = d/2$$

FIGURE 4-16 Reproduced with permission. Stress-concentration factor K_σ for grooved shaft in bending. (R. E. Peterson, "Design Factors for Stress Concentration," *Machine Design*, Vol. 23, Nos. 2 to 7, 1951.)

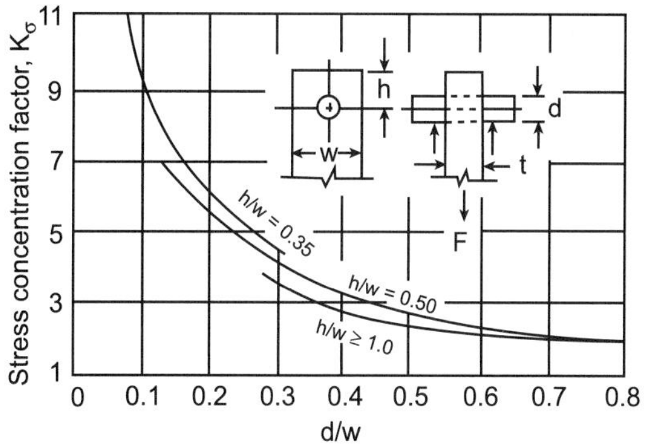

FIGURE 4-17 Plate loaded in tension by a pin through a hole, $\sigma_0 = F/A$, where $A = (w - d)t$. When clearance exists, increase K_t 35 to 50 percent. (M. M. Frocht and H. N. Hill, "Stress Concentration Factors around a Central Circular Hole in a Plate Loaded through a Pin in Hole," J. Appl. Mechanics, vol. 7, no. 1, March 1940, p. A-5.)

Particular	Stress concentration factor theoretical/empirical or otherwise: Extreme value	Formula	
(ii) Torsion (Fig. 4-18)		$K_\tau = 1 + \sqrt{\dfrac{h_1}{r}}$	(4-20)
(iii) Shaft with a small elliptical groove in torsion		$K_\tau = 1 + \sqrt{\dfrac{h_1}{r}}$	(4-21a)
		or	
		$K_\tau = 1 + \dfrac{h_1}{b}$	(4-21b)
		$K_\tau = 1 + \dfrac{b}{r}$	(4-21c)
(m) Shouldered shaft in torsion (Fig. 4-19):		$K_\tau - 1 + \left(S\dfrac{d}{r}\right)^{0.65}$	(4-22a)
		where S is some function of $\dfrac{D}{d}$	
		$K_\tau = 1 + \dfrac{d}{12r}\left[1 - \dfrac{\left(1 + 2\dfrac{r}{d}\right)}{\dfrac{D}{d}\left(1 + 6\dfrac{r}{d}\right)}\right]$	(4-22b)
For stress-concentration factor and combined factor for stepped-shaft in tension and bending		Refer to Figs. 4-14, 4-16, and 4-18 to 4-21.	

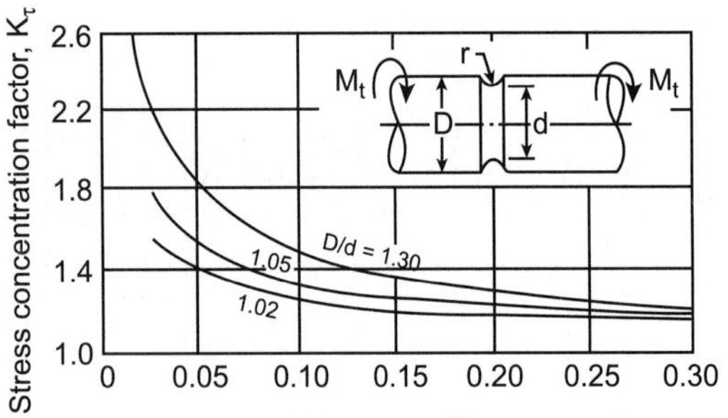

$\dfrac{J}{c} = \dfrac{\pi d^3}{16}$; $\tau_{nom} = \dfrac{M_t c}{J}$; $J = \dfrac{\pi d^4}{32}$

and $c = d/2$

FIGURE 4-18 Reproduced with permission. Stress-concentration factor K_σ for grooved shaft in tension. (R. E. Peterson, "Design Factors for Stress Concentration," *Machine Design*, Vol. 23, Nos. 2 to 7, 1951.)

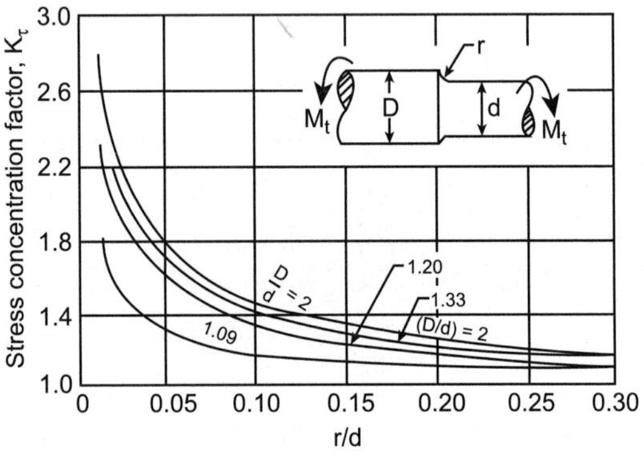

$\dfrac{J}{c} = \dfrac{\pi d^3}{16}$; $\tau_{nom} = \dfrac{M_t c}{J}$

$c = d/2$; $J = \dfrac{\pi d^4}{32}$

FIGURE 4-19 Reproduced with permission. Stress-concentration factor K_τ for stepped shaft in torsion. (R. E. Peterson, "Design Factors for Stress Concentration," *Machine Design*, Vol. 23, Nos. 2–7, 1951.)

Particular	Stress concentration factor theoretical/empirical or otherwise: Extreme value	Formula	
(n) Bar containing grooves:			
(i) Bar with U, semicircular or shallow grooves symmetrically placed in tension (Figs. 4-11, 4-12, 4-22)		$K_\sigma = 1 + \left[\dfrac{1}{\left(1.55\dfrac{D}{d} - 1.3\right)}\dfrac{h_1}{r}\right]^n$	(4-23a)
		or	
		$K_\sigma = 1 + \left[\dfrac{\left(\dfrac{D}{d} - 1\right)}{2\left(1.55\dfrac{D}{d} - 1.3\right)}\dfrac{d}{r}\right]^n$	(4-23b)
		where $n = \dfrac{\left(\dfrac{D}{d} - 1\right) + 0.3\sqrt{\dfrac{h_1}{r}}}{\left(\dfrac{D}{d} - 1\right) + \sqrt{\dfrac{h_1}{r}}}$	
(ii) Bar with deep V-groove in tension for $\dfrac{r}{h_1} < 1$ (Fig. 4-11a)		$K_{\sigma v} = 1 + (K_{\sigma v} - 1)\left\{1 - \left(\dfrac{\beta}{180}\right)^{1+2.4\sqrt{r/h_1}}\right\}$	(4-24)

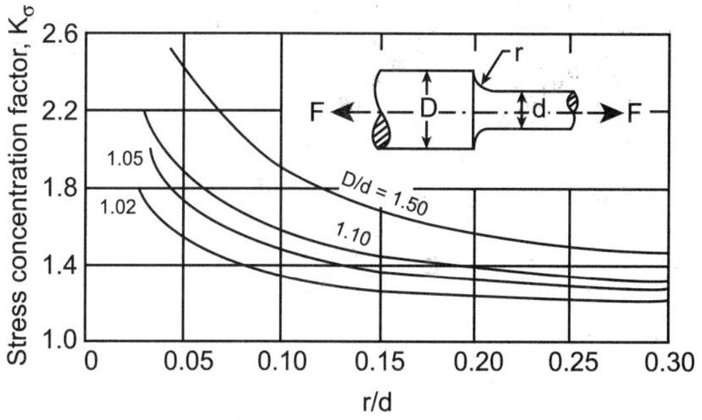

$$A = \frac{\pi d^2}{4}\ ;\ \sigma_{nom} = \frac{F}{A}$$

FIGURE 4-20 Reproduced with permission. Stress-concentration factor K_σ for stepped shaft in tension. (R. E. Peterson, "Design Factors for Stress Concentration," *Machine Design*, Vol. 23, Nos. 2 to 7, 1951.)

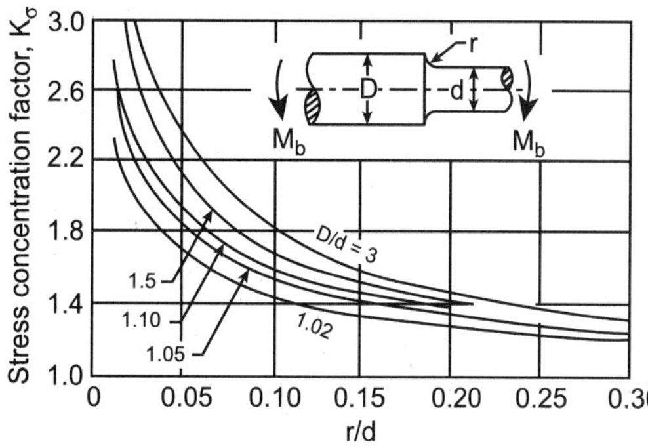

$$\frac{I}{c} = \frac{\pi d^3}{32}$$

$$\sigma_{nom} = \frac{M_b c}{I}\ ;\ I = \frac{\pi d^4}{64}$$

FIGURE 4-21 Reproduced with permission. Stress-concentration factor K_σ for stepped shaft in bending. (R. E. Peterson, "Design Factors for Stress Concentration," *Machine Design*, Vol. 23, Nos. 2–7, 1951.)

Particular	Stress concentration factor theoretical/empirical or otherwise: Extreme value	Formula	

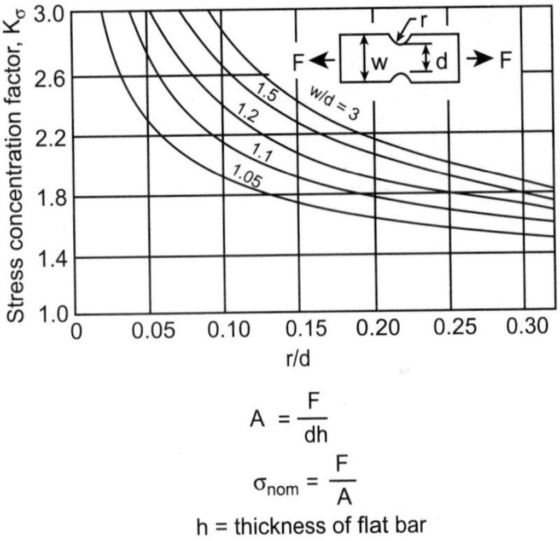

FIGURE 4-22 Reproduced with permission. Stress-concentration factor K_σ for notched flat bar in tension. (R. E. Peterson, "Design Factors for Stress Concentration," *Machine Design*, Vol. 23, Nos. 2 to 7, 1951.)

Particular	Extreme value	Formula	
(iii) Bar with shallow V-groove in tension for $\frac{r}{h_1} > 1$		$K_{\sigma v} = 1 + (K_{\sigma v} - 1)\left\{1 - \left(\frac{\beta - \alpha}{180 - \alpha}\right)^{1+2.4\sqrt{r/h_1}}\right\}$	(4-25)
(iv) Elliptical groove at the edge of plate in tension		$K_\sigma = 1 + \frac{2h_1}{b}$	(4-26a)
		$K_\sigma = 1 + 2\sqrt{\frac{h_1}{r}}$	(4-26b)
(v) Bar with symmetrical U, semicircular shallow grooves in bending (Fig. 4-23).		$K_\sigma = 1 + \left[\frac{1}{4.27\frac{D}{d} - 4}\frac{h_1}{r}\right]^{0.85}$	(4-27a)
		or	
		$K_\sigma = 1 + \left[\frac{\left(\frac{D}{d} - 1\right)}{2\left(4.27\frac{D}{d} - 4\right)}\frac{d}{r}\right]^{0.85}$	(4-27b)

Particular	Stress concentration factor theoretical/empirical or otherwise: Extreme value	Formula

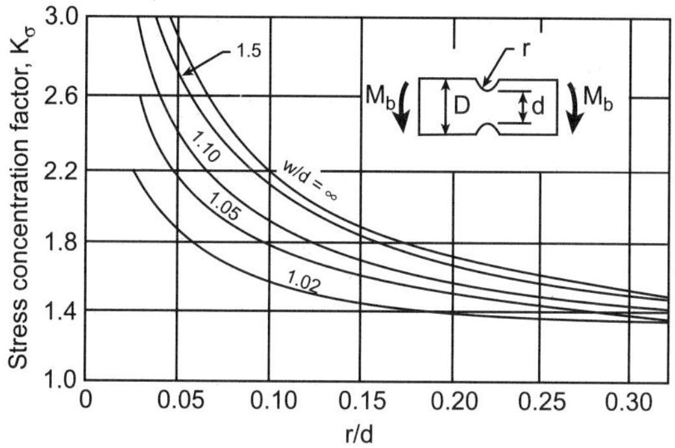

$$\frac{I}{c} = \frac{hd^2}{6}; \quad \sigma_{nom} = \frac{M_b c}{I}$$

FIGURE 4-23 Reproduced with permission. Stress-concentration factor K_σ for notched flat bar in bending. (R. E. Peterson, "Design Factors for Stress Concentration," *Machine Design*, Vol. 23, Nos. 2 to 7, 1951.)

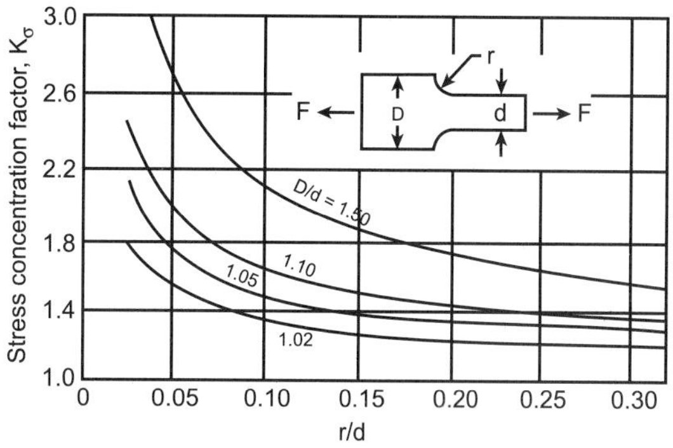

$$A = dh\,; \quad \sigma_{nom} = \frac{F}{A}$$

h = thickness of bar

FIGURE 4-24 Reproduced with permission. Stress-concentration factor K_σ for filleted flat bar in tension. (R. E. Peterson, "Design Factors for Stress Concentration," *Machine Design*, Vol. 23, Nos. 2–7, 1951.)

For stress-concentration factors for small grooves in a shaft subjected to torsion.

Refer to Table 4-4.

(o) Bar containing shoulders

(i) Bar with shoulders in tension (Fig. 4-24)

$$K_\sigma = 1 + \left[\frac{1}{2.8\dfrac{D}{d} - 2}\,\frac{h_1}{r}\right]^{0.85} \tag{4-28a}$$

or

$$K_\sigma = 1 + \left[\frac{\left(\dfrac{D}{d} - 1\right)}{2\left(2.8\dfrac{D}{d} - 2\right)}\,\frac{d}{r}\right]^{0.85} \tag{4-28b}$$

TABLE 4-4
Stress-concentration factors for relatively small grooves in a shaft subject to torsion, K_τ

	h_1/r				
Included angle of V, deg	**0.5**	**1**	**3**	**5**	**2**
0	1.85	2.01	2.66	3.23	4.54
60	1.84	2.00	2.54	3.06	3.99
90	1.81	1.95	2.40	2.40	3.12
120	1.66	1.75	1.95	2.00	2.13

Particular	Stress concentration factor theoretical/empirical or otherwise: Extreme value	Formula	
(ii) Bar with shoulders in bending (Fig. 4-25)	$K_\sigma = 1 + \left[\frac{1}{5.37\frac{D}{d} - 4.8}\frac{h_1}{r}\right]^{0.85}$		(4-29a)
	or		
	$K_\sigma = 1 + \left[\frac{\left(\frac{D}{d} - 1\right)}{2\left(5.37\frac{D}{d} - 4.8\right)}\frac{d}{r}\right]^{0.85}$		(4-29b)
(p) Press-fitted or shrink-fitted members (Table 4-5):			
(i) Plain member	$K_\sigma = 1.95$		
(ii) Grooved member	$K_\sigma = 1.34$		
(iii) Plain member	$K_{f\sigma} = 2.00$		
(iv) Grooved member	$K_{f\sigma} = 1.70$		
(q) Bolts and nuts (Tables 4-6 and 4-7)			
Bolt and nut of standard proportions	$K_\sigma = 3.85$		
Bolt and nut having lip	$K_\sigma = 3.00$		

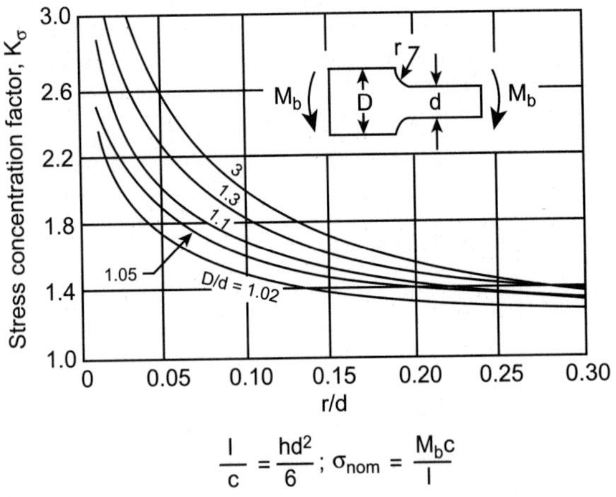

FIGURE 4-25 Reproduced with permission. Stress-concentration factor K_σ for stepped bar in bending. (R. E. Peterson, "Design Factors for Stress Concentration," *Machine Design*, Vol. 23, Nos. 2 to 7, 1951.)

TABLE 4-5
Stress-concentration factors in shrink-fitted members

Particular	K_σ	$K_{f\sigma}$
Plain	1.95	2.00
Grooved	1.34	1.70

TABLE 4-6
Stress-concentration factors for screw threads

	Analysis			
Types of thread	**Seely and Smith**	**Black (8)**	**Peterson (1)**	**Suggested value**
Square	2.0			
Sharp V	3.0			
Whitworth	2.0		3.35	5 to 6
US standard				
Medium				
Carbon steel	2.5			
National coarse thread				
Heat-treated		2.84		
Nickel steel		3.85		

TABLE 4-7
Stress-concentration factors $K_{f\sigma}$ for screw threads

	Annealed		Hardened	
Type of thread	**Rolled**	**Cut**	**Rolled**	**Cut**
Sellers, American National, square thread	2.2	2.8	3.0	3.8
Whitworth rounded roots	1.4	1.8	2.6	3.3

TABLE 4-8
Stress-concentration factors for welds

Location	K_σ
End or parallel fillet weld	2.7
Reinforced butt	1.2
Tee of transverse fillet weld	1.5
T-butt weld with sharp corners	2.0

TABLE 4-9
Index of sensitivity for repeated stress

	Average index of sensitivity q		
Material	**Annealed or soft**	**Heat-treated and drawn at 921 K (648°C)**	**Heat-treated and drawn at 755 K (482°C)**
Armco iron, 0.02% C	0.15–0.20		
Carbon steel			
0.10% C	0.05–0.10		
0.20% C (also cast steel)	0.10		
0.30% C	0.18	0.35	0.45
0.50% C	0.26	0.40	0.50
0.85% C		0.45	0.57
Spring steel, 0.56% C, 2.3 Si, rolled		0.38	
SAE 3140, 0.73 C; 0.6 Cr; 1.3 Ni.	0.25	0.45	
Cr–Ni steel 0.8 Cr; 3.5 Ni		0.25	0.70
Stainless steel, 0.3 C; 8.3 Cr, 19.7 Ni	0.16		
Cast iron	0–0.05		
Copper, electrolitic	0.07		
Duraluminum	0.05–0.13		

Particular	Stress concentration factor theoretical/empirical or otherwise: Extreme value	Formula	
(r) Crane hook:			
For crane hook under tensile load	$K_\sigma = 1.56$		
(s) Rotating disk:			
For rotating disk with a hole for $\frac{R_i}{R_o} \to 0$	$K_\sigma = 2$		
For thin disk (ring)	$K_\sigma = 1$		
(t) Eye bar:			
For eye bar subjected to tensile load	$K_\sigma = 2.8$		
Stress concentration factors for welds	Refer to Table 4-8.		
(u) Notch sensitivity factors (Table 4-9):			
(i) Notch sensitivity factor for normal stress	$q_\sigma = \frac{K_{f\sigma} - 1}{K_\sigma - 1}$		(4-30a)
	$q_\sigma = \frac{K_{f\sigma} - 1}{K'_\sigma - 1}$		(4-30b)
For index of sensitivity for repeated stresses.	Refer to Table 4-9.		
(ii) Fatigue stress concentration factor for normal stress	$K_{f\sigma} = 1 + q_\sigma(K_\sigma - 1)$		(4-31a)
	$K_{f\sigma} = 1 + q_\sigma(K'_\sigma - 1)$		(4-31b)
(iii) Notch sensitivity factor for shear stress	$q_\tau = \frac{K_{f\tau} - 1}{K_\tau - 1}$		(4-32)
(iv) Fatigue stress-concentration factor for shear stress	$K_{f\tau} = 1 + q_\tau(K_\tau - 1)$		(4-33)

STRESS CONCENTRATION IN FLANGED PIPE SUBJECTED TO AXIAL EXTERNAL FORCE

The stress in the pipe due to external load F (Fig. 4-25A)

$$\sigma = \sigma_f + \frac{F}{A} \tag{4-33a}$$

where σ_f = depends on the distance x from the flange of the pipe, MPa (psi)

σ_{fm} = maximum stress at $x = 0$, MPa (psi)

A = area of the cross section of pipe, m^2 (in^2)

F = external load, kN (lbf)

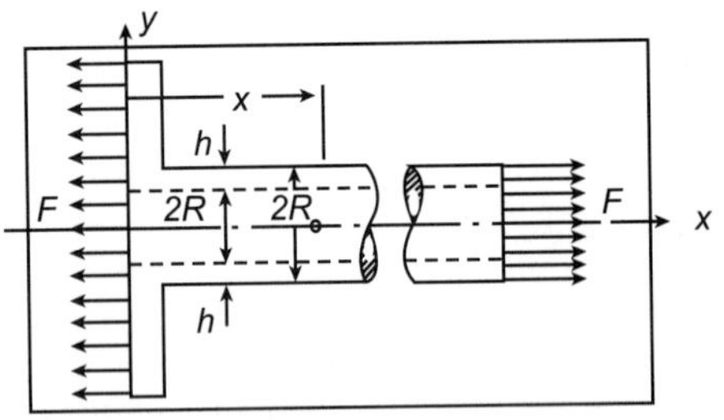

FIGURE 4-25A Pipe and flange under the axial force F

Particular	Formula
The value of constant β	$\beta = 10\sqrt[4]{\dfrac{3(1-\nu^2)}{R^2h^2}}$ (4-33b)
	where $2R = 2R_i + h$ = mean diameter of pipe, m (in)
	$2R_o$ = outer diameter of pipe, m (in)
	$2R_i$ = inner diameter of pipe, m (in)
	h = thickness of pipe, m (in)
	ν = Poisson's ratio of material
For plot of the stress ratio $\dfrac{\sigma_f}{\sigma_{fm}}$ versus βx	Refer to Fig. 4-25B.

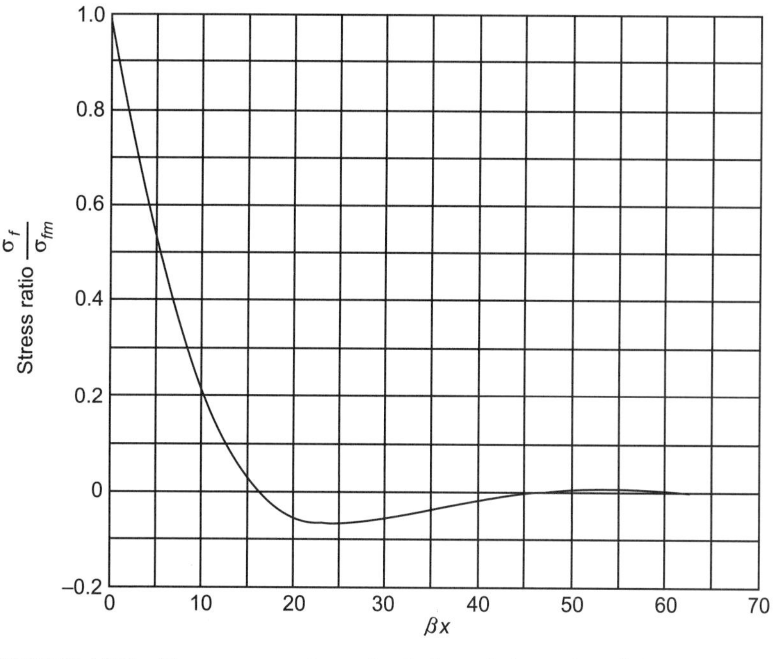

FIGURE 4-25B Stress concentration region in flanged pipe under axial external force F.
Courtesy: Douglas C. Greenwood, *Engineering Data for Product Design*, McGraw-Hill Publishing Company, New York, 1961.

REDUCTION OR MITIGATION OF STRESS CONCENTRATIONS

In designing a machine part, one has to take into consideration the stress concentration occurring in such parts and eliminate or reduce stress concentration. Fig. 4-25C shows various methods used to reduce stress concentration. Stream line flowing analogy in a channel can be applied to force flow lines of a flat plate without any type of flow subject to uniform uniaxial tensile stress σ as shown in Fig. 4-25C i(a). The stream line flow of water or any fluid is smooth and straight as shown in Fig. 4-25C i(*a*). If there is any obstruction such as a heavy iron ball or a pipe or stone boulder in the path of flow of water, the flow of water or fluid will not be smooth and straight as shown in Fig. 4-25C i(*e*). Similarly the force flow lines will not be straight as in case of plate with a circular or elliptical or any shape of holes in a plate as shown in Figs. 4-25C i(*b*), i(*c*), i(*d*) and i(*f*). By providing some geometric changes, abrupt change of force-flow lines are smoothened. Fatigue strength of parts with stress raiser can be increased by cold working operation such as shot peening or pressing by balls which creates a nature of stress in thin layers of the part just opposite to the one induced in it. Press fit stress concentration can be reduced by making the gripping portion conical in case of hardening steel parts. Nitriding and plating the parts eliminate the corrosion effect, which combined with stress concentration reduces the fatigue strength of the machine part.

(i) Plates:

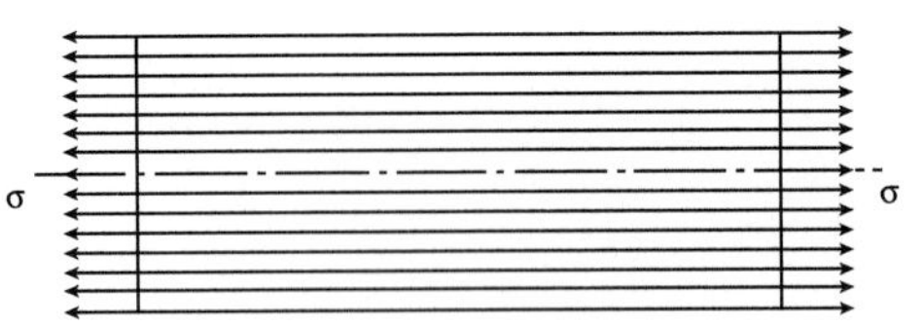

Force flow lines in a flat plate without cut-out under uniform uniaxial tension

(a)

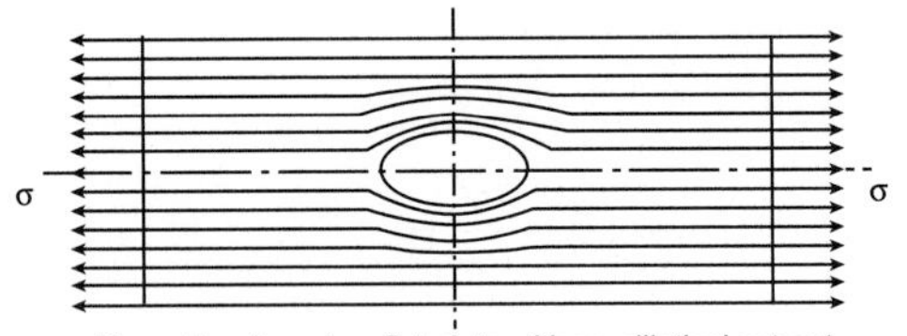

Force flow lines in a flat plate with an elliptical cut-out major axis parallel to line of action of load

(b)

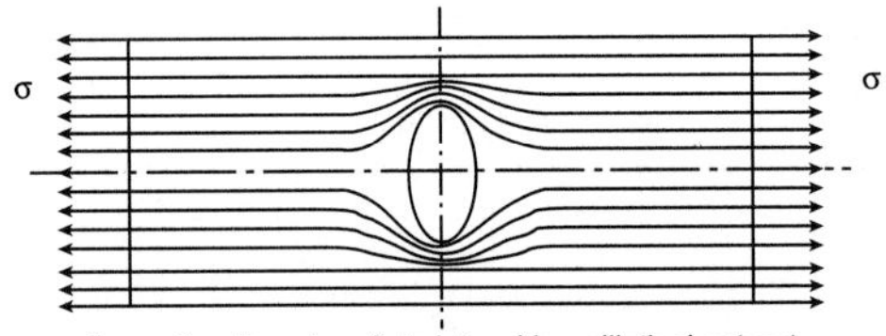

Force flow lines in a flat plate with a elliptical cut-out major axis perpendicular line of action of load.

(c)

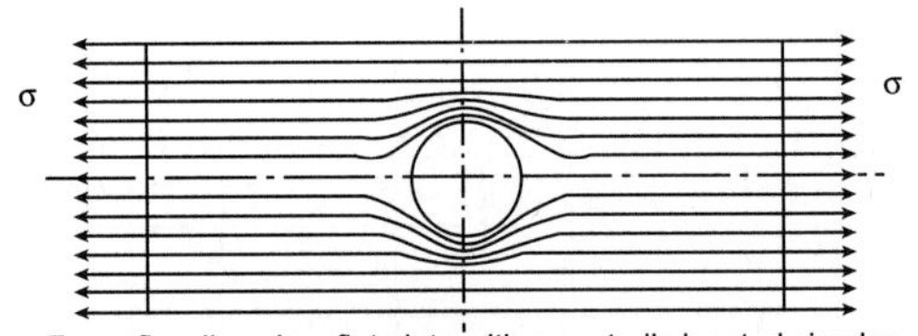

Force flow lines in a flat plate with a centrally located circular cut-out subject to uniform uniaxial stress.

(d)

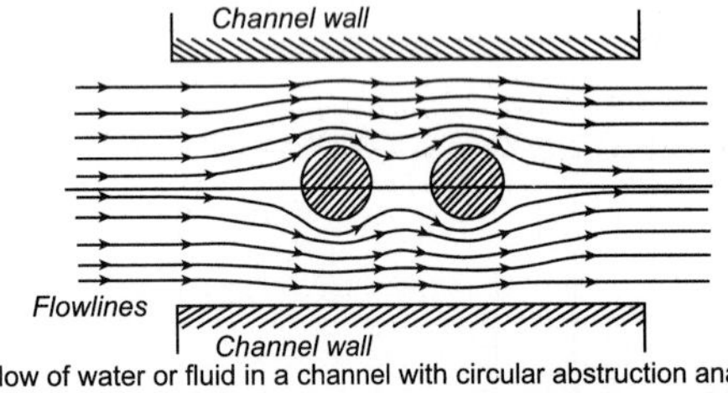

Flow of water or fluid in a channel with circular abstruction analogues to force flow lines with circular or elliptical or other shapes of cut-outs.

(e)

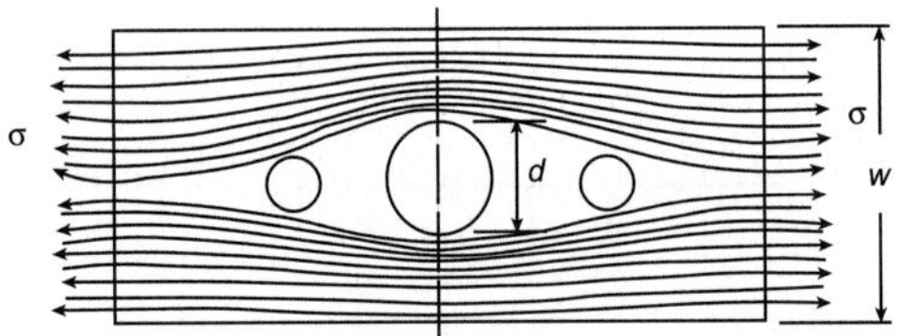

Force flow lines with cut-out in a flat plate under uniform uniaxial force. Two drilled small circular holes on both sides of a big circular cut-out make force flow lines smooth and reduced stress concentration

(f)

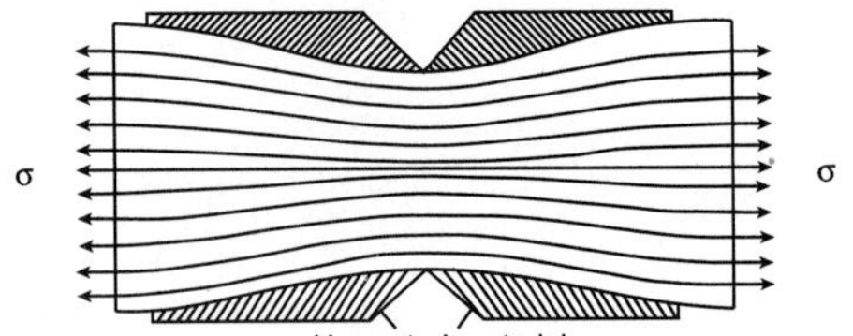

Removal of unwanted material reduces stress concentration in plate

(g)

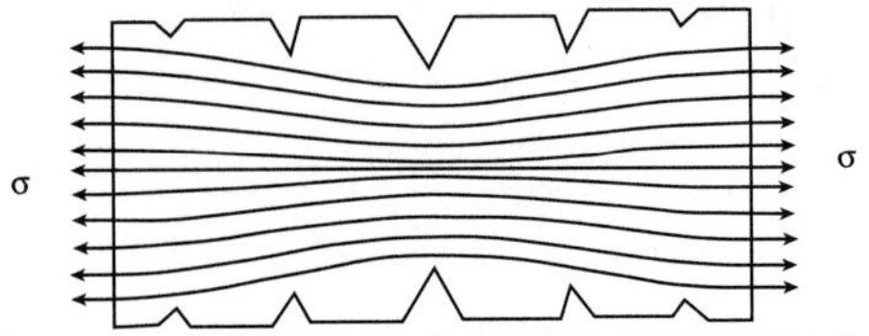

Stress concentration is reduced by adding notches in case of plates and adding grooves in case of circular bar as shown above

(h)

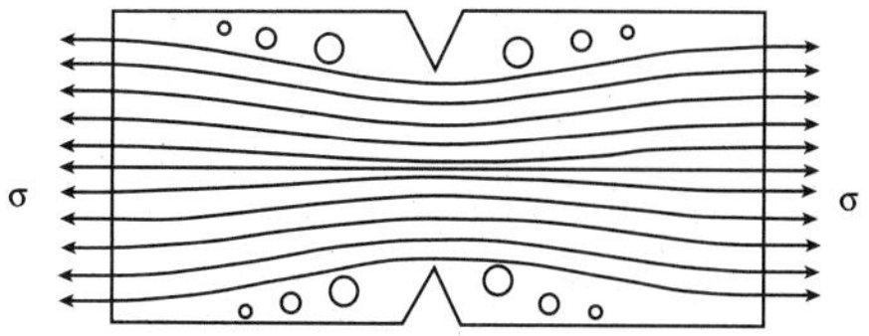

By drilling holes on both sides of notch or V-groove stress concentration can be reduced

(i)

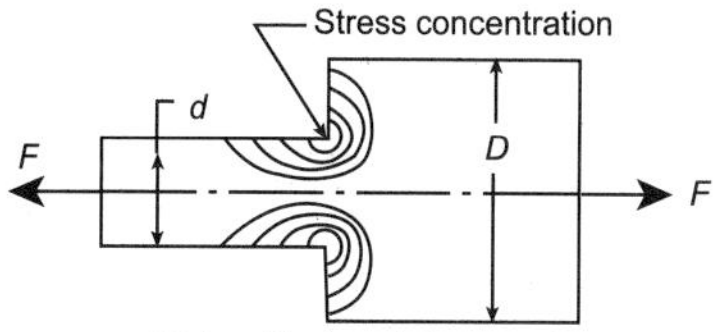

Plate with abrupt change of cross-sectional area

(j)

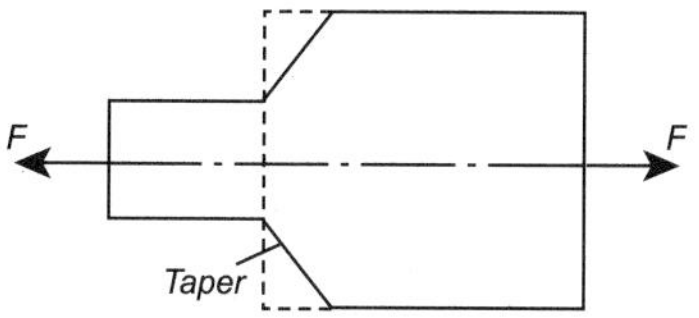

Taper reduces stress concentration

(k)

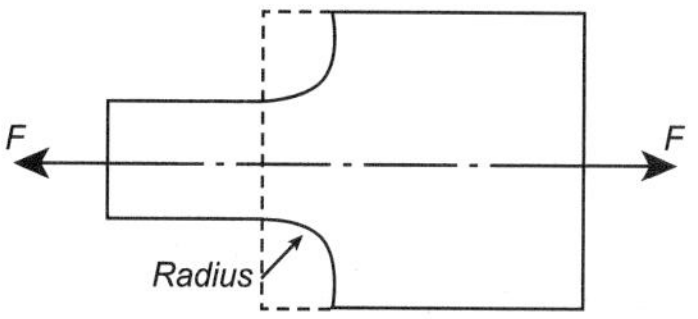

Reduction in stress concentration at corners due to rounding corner by large or optimum radius

(l)

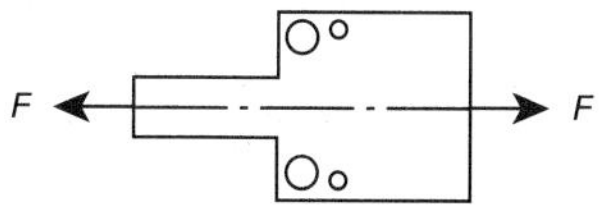

By drilling holes K_σ reduced

(m)

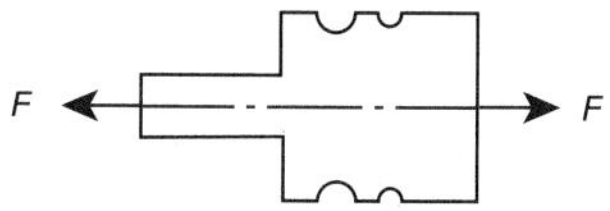

By providing grooves as shown K_σ is reduced

(n)

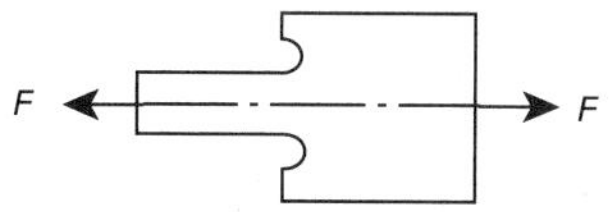

By providing undercut as shown in plate or shaft K_σ is reduced

(o)

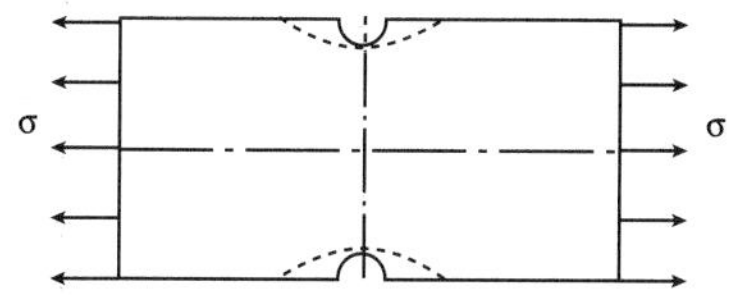

Provide large radius as shown in dotted lines to reduce K_σ in notched plate or shaft

(p)

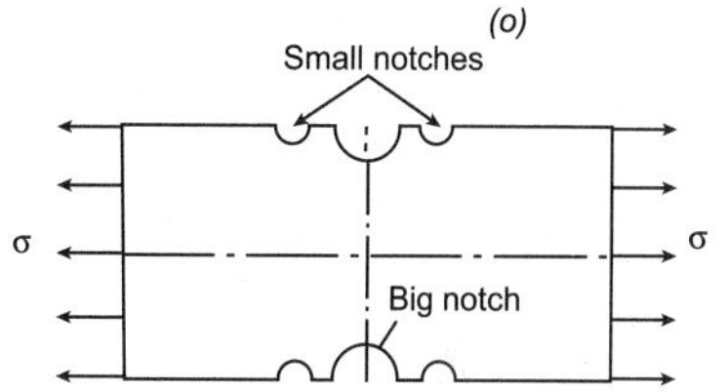

Small notches are provided on both sides of a big notch to relieve stress concentration

(q)

(ii) Stepped shafts subject to tensile force:

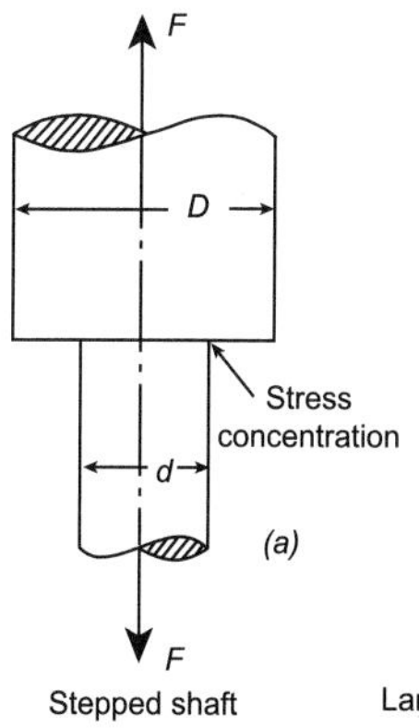

Stepped shaft

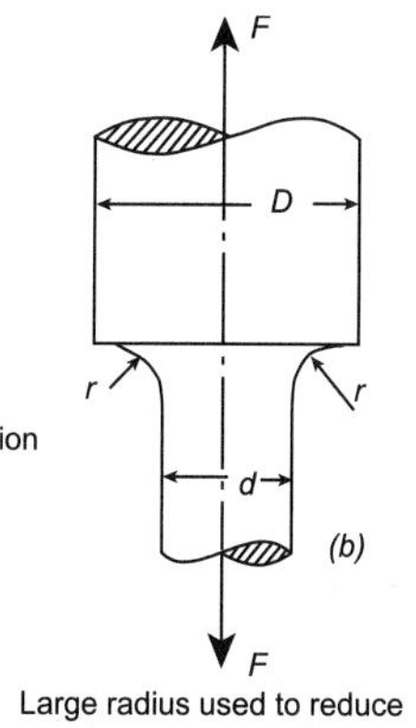

Large radius used to reduce stress concentration

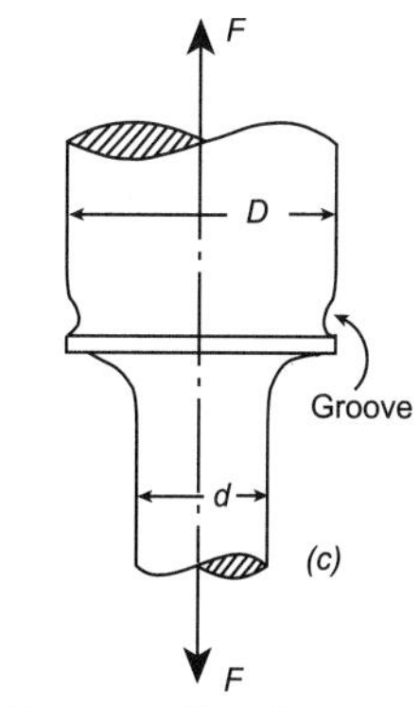

Groove used to reduce stress concentration

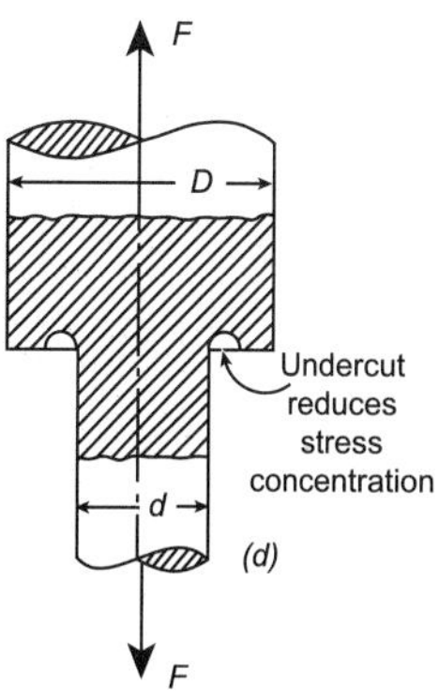

(iii) Shafts with narrow collar, cylindrical holes and grooves subject to tensile force:

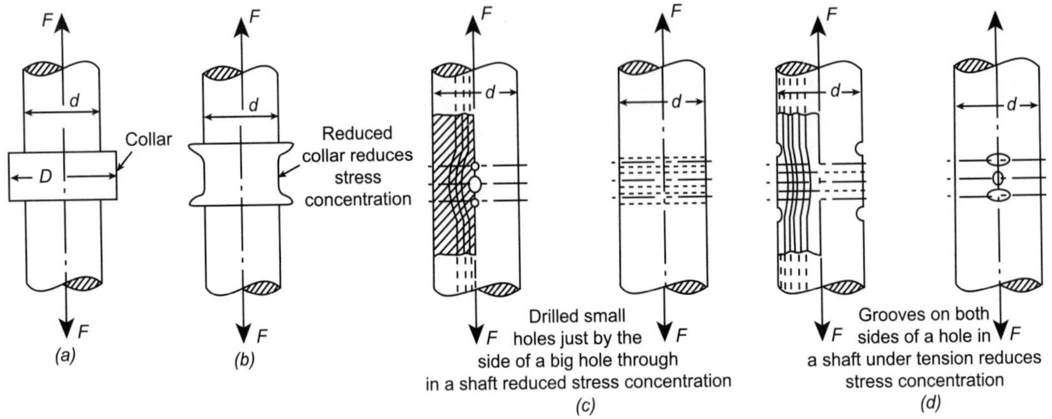

(iv) Shafts subject to bending and torque:

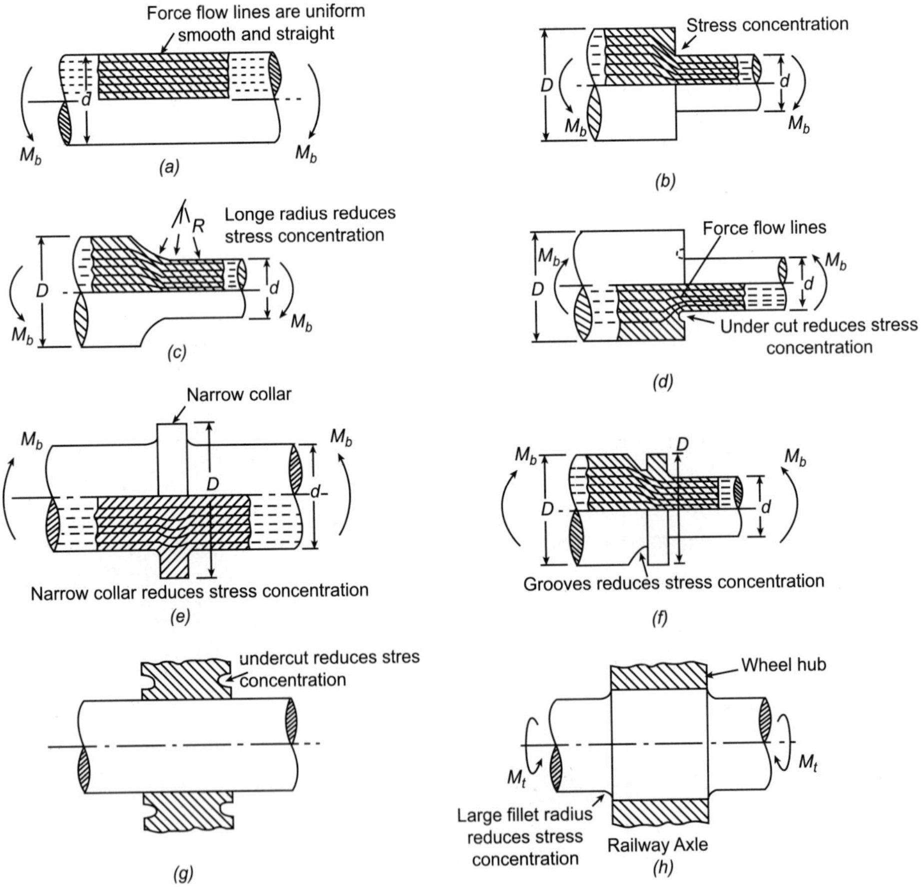

(v) Screws and nuts under torque:

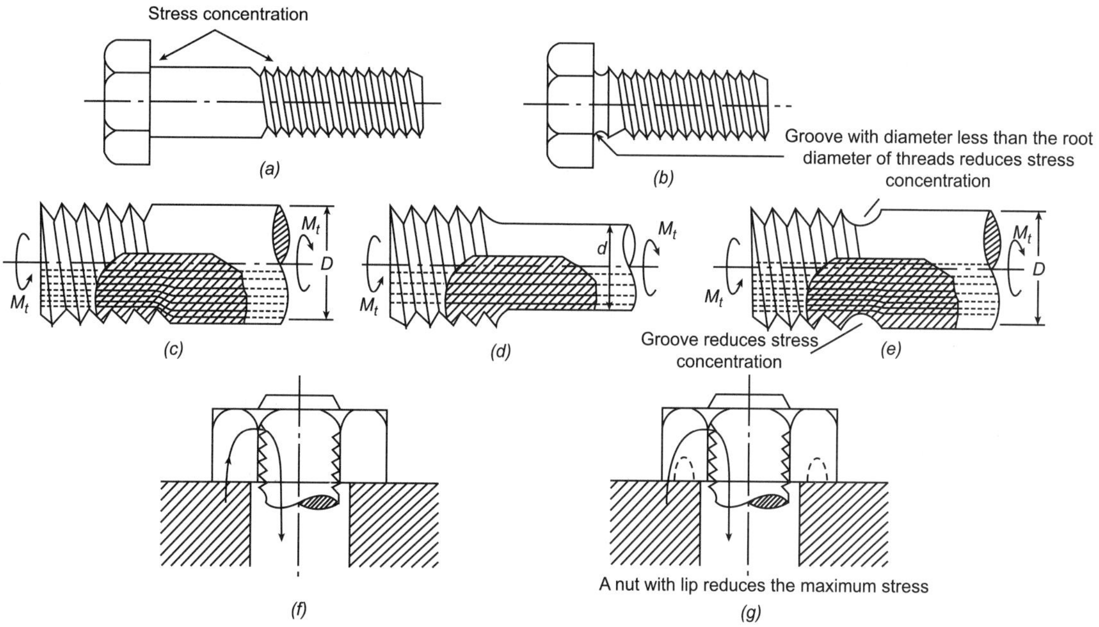

(vi) Keyways in shafts subject to torque:

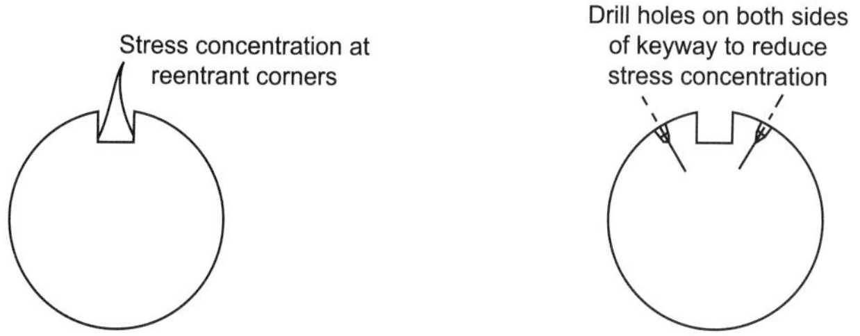

(vii) Gears:

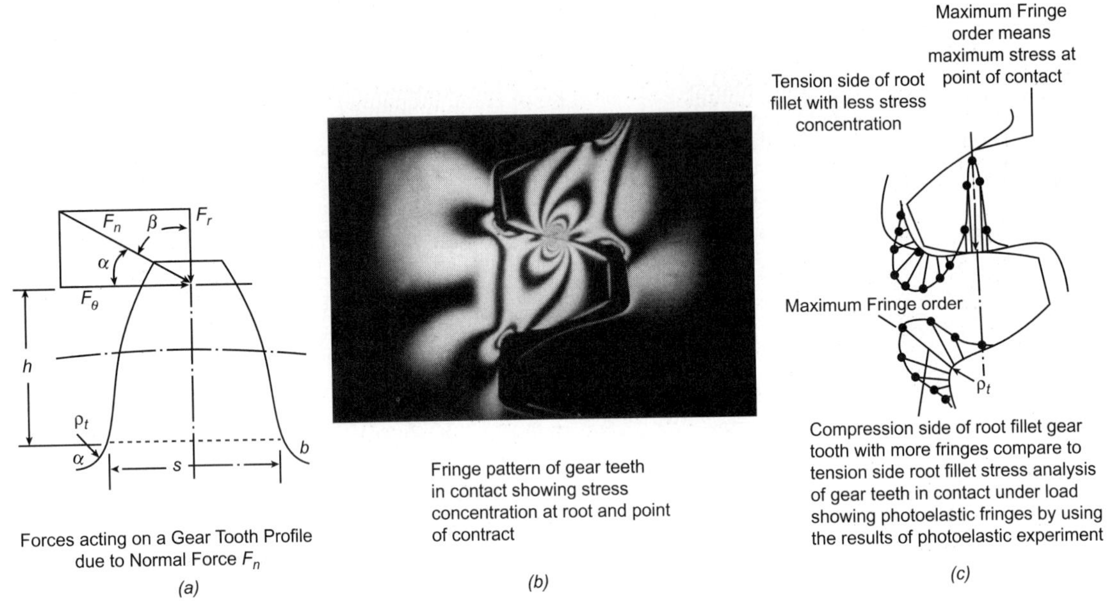

Stress concentration at fillet and at point of contact are shown in photoelastic fringe pattern and also in Fig. *c*. The stress concentration at fillet can be reduced by providing suitable large fillet radius.

Source: From the photoelastic work of K. Lingaiah, *Fringe Pattern of Gear-Teeth Showing Stress Concentration at Root and Contact Point*, Department of Mechanical Engineering, University Visvesvaraya College of Engineering, Bangalore University, Bangalore, 1973.

(viii) Flate plate with and without asymmetrically reinforced circular cutout subjected to uniform uniaxial stress:

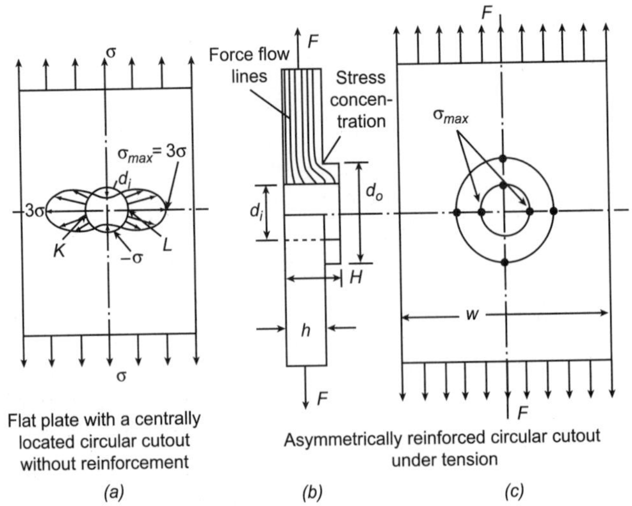

FIGURE 4-25C Mitigation of stress concentration in machine members.

Particular	Formula

STRESS INTENSITY FACTOR OR FRACTURE TOUGHNESS FACTOR

The energy criterion approach in the fracture mechanics analysis:

The energy release rate in case of a crack of length $2a$ in an infinite plate subject to tensile stress at infinity σ (Fig. 4-26A).

$$G = \frac{\pi\sigma^2 a}{E} \quad \text{(4-34a)}$$

The energy release rate is defined as the rate of change in potential energy with crack area for a linear elastic material.

where G = energy release rate
E = modulus of elasticity, GPa (psi)
a = half crack length, mm (in)

The critical energy release rate.

$$G_c = \frac{\pi\sigma_f^2 a_c}{E} \quad \text{(4-34b)}$$

where σ_f = failure stress, MPa (psi)
G_c = material resistance to fracture or critical fracture toughness

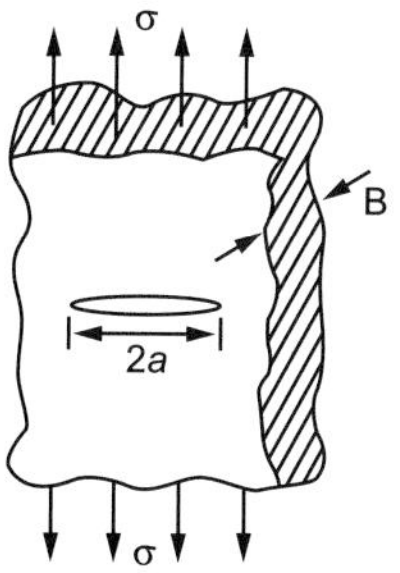

FIGURE 4-26A A flat infinite plate with a through thickness crack subject to tensile stress at infinity.

The stress intensity factor for a centrally located straight crack in an infinite plate subjected to uniform uniaxial tensile stress σ perpendicular to the plane of the crack.

$$K_I = \sqrt{\pi}\sigma\sqrt{a} \quad \text{(4-34c)}$$

The definition and unit of critical stress intensity factor K_{Ic}.

K_{Ic} is the critical stress intensity factor for static loading and plane-strain conditions of maximum constraints and is also referred to as the fracture toughness factor of the material at the onset of rapid fracture and has dimension of (stress $\sqrt{\text{length}}$), i.e. MPa $\sqrt{\text{m}}$ (kpsi $\sqrt{\text{in}}$).

The relation between K_I and G.

$$G = \frac{K_I}{E} \quad \text{(4-34d)}$$

Particular	Formula

Three modes of loading to analyse stress fields in cracks:

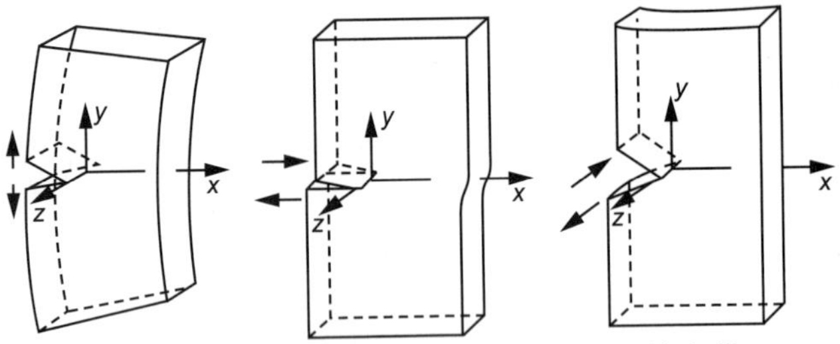

FIGURE 4-26B Three modes of loading for deformation of crack tip.

First mode of loading and stress components at crack tip, Fig. 4-26B (*a*):

The localized stress components at the vicinity of the "opening mode" or "mode I" crack tip in a flat plate subjected to uniform applied stress σ at infinity from the theory of fracture mechanics (Fig. 4-26C).

$$\sigma_x = \frac{K_I}{\sqrt{2\pi r}}\cos\frac{\theta}{2}\left[1 - \sin\frac{\theta}{2}\sin\frac{3\theta}{2}\right] \quad (4\text{-}35a)$$

$$\sigma_y = \frac{K_I}{\sqrt{2\pi r}}\cos\frac{\theta}{2}\left[1 + \sin\frac{\theta}{2}\sin\frac{3\theta}{2}\right] \quad (4\text{-}35b)$$

$$\sigma_z = \nu(\sigma_x + \sigma_y) \quad \text{for plane strain} \quad (4\text{-}35c)$$

$$= 0 \quad \text{for plane stress} \quad (4\text{-}35d)$$

$$\tau_{xy} = \frac{K_I}{\sqrt{2\pi r}}\cos\frac{\theta}{2}\sin\frac{\theta}{2}\cos\frac{3\theta}{2} \quad (4\text{-}35e)$$

The crack tip displacement fields for "first mode" (Mode I) in case of linear elastic, isotropic materials.

$$u_x = \frac{K_I}{2G}\sqrt{\frac{r}{2\pi}}\cos\frac{\theta}{2}\left[\kappa - 1 + 2\sin^2\frac{\theta}{2}\right] \quad (4\text{-}35f)$$

$$u_y = \frac{K_I}{2G}\sqrt{\frac{r}{2\pi}}\sin\frac{\theta}{2}\left[\kappa + 1 - 2\cos^2\frac{\theta}{2}\right] \quad (4\text{-}35g)$$

where G = modulus of shear, GPa (psi)

$\kappa = 3 - 4\nu$ for plane strain

$\kappa = (3 - \nu)/(1 + \nu)$ for plane stress

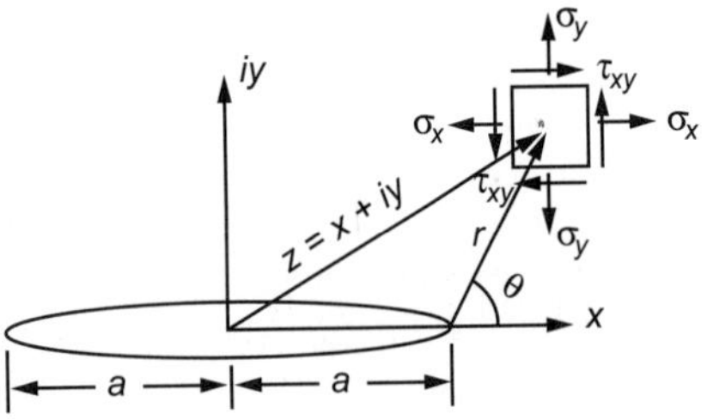

FIGURE 4-26C State of stress in the vicinity of a crack tip.

Particular	Formula

Second mode of loading and stress components in the vicinity of crack tip, Fig. 4-26B (*b*):

The localized stress components at the vicinity of the "second mode" or "sliding mode" crack tip in a flat plate subjected to in-plane shear, Fig. 4-26B (*b*).

$$\sigma_x = -\frac{K_{II}}{\sqrt{2\pi r}}\sin\frac{\theta}{2}\left[2+\cos\left(\frac{\theta}{2}\right)\cos\left(\frac{3\theta}{2}\right)\right] \quad \text{(4-35h)}$$

$$\sigma_y = \frac{K_{II}}{\sqrt{2\pi r}}\sin\frac{\theta}{2}\cos\frac{\theta}{2}\cos\left(\frac{3\theta}{2}\right) \quad \text{(4-35i)}$$

$$\tau_{xy} = \frac{K_{II}}{\sqrt{2\pi r}}\cos\frac{\theta}{2}\left[1-\sin\frac{\theta}{2}\sin\frac{3\theta}{2}\right] \quad \text{(4-35j)}$$

$$\sigma_z = 0 \quad \text{for plane stress} \quad \text{(4-35k)}$$

$$\sigma_z = \nu(\sigma_x - \sigma_y) \quad \text{for plane strain} \quad \text{(4-35l)}$$

$$\tau_{xz} = \tau_{yz} = 0 \quad \text{(4-35m)}$$

The crack tip displacement fields for the "second mode" (Mode II) in case of linear elastic, isotropic materials.

$$u_x = \frac{K_{II}}{2G}\sqrt{\frac{r}{2\pi}}\sin\frac{\theta}{2}\left[\kappa+1+2\cos^2\frac{\theta}{2}\right] \quad \text{(4-35n)}$$

$$u_y = -\frac{K_{II}}{2G}\sqrt{\frac{r}{2\pi}}\cos\frac{\theta}{2}\left[\kappa-1-2\sin^2\frac{\theta}{2}\right] \quad \text{(4-35o)}$$

Third mode of loading and stress components in the vicinity of crack tip, Fig. 4-26B (*c*):

The localized stress components at the vicinity of the "third mode" or "tearing mode III" crack tip in a flat plate subjected to out-of-plane shear, Fig. 4-26B (*c*), in case of linear elastic, isotropic materials.

$$\tau_{xz} = \frac{K_{III}}{\sqrt{2\pi r}}\sin\frac{\theta}{2} \quad \text{(4-35p)}$$

$$\tau_{yz} = \frac{K_{III}}{\sqrt{2\pi r}}\cos\frac{\theta}{2} \quad \text{(4-35q)}$$

The crack tip displacement field for the "third mode" (Mode III) in case of linear elastic, isotropic materials.

$$u_z = \frac{K_{III}}{G}\sqrt{\frac{r}{2\pi}}\sin\frac{\theta}{2} \quad \text{(4-35r)}$$

$$w = \nu = 0$$

Stress intensity factor:

The stress intensity factor for a center cracked tension plate (CCT), according to Fedderson (Fig. 4-27*a*).

$$K_I = \sigma\sqrt{\pi a}\left[\sec\left(\frac{\pi a}{2b}\right)\right] \quad \text{(4-35s)}$$

The stress intensity factor for a double edge cracked plate according to Keer and Freedman (Fig. 4-27*b*).

$$K_i = \sigma\sqrt{\pi a}\left[1.12 - 0.61\left(\frac{a}{b}\right) + 0.13\left(\frac{a}{b}\right)^3\right] \times \left[1-\frac{a}{b}\right]^{-1/2} \quad \text{(4-35t)}$$

Particular	Formula

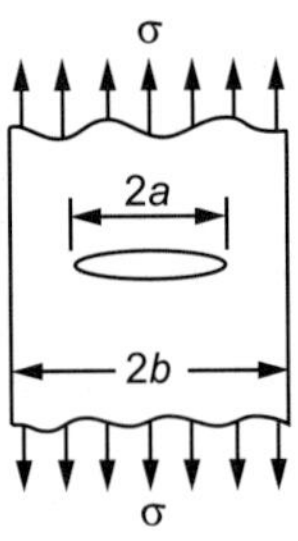

FIGURE 4-27*a*

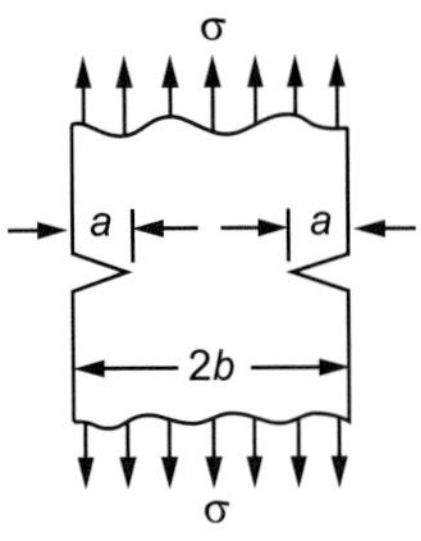

FIGURE 4-27*b*

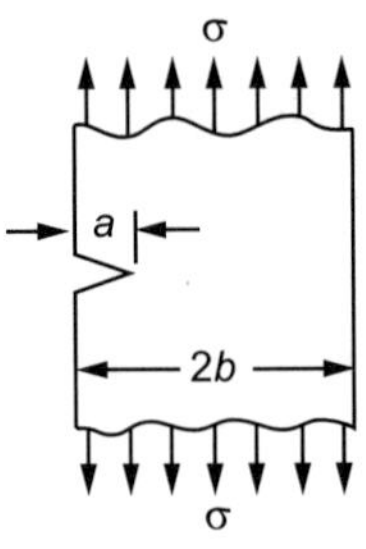

FIGURE 4-27*c*

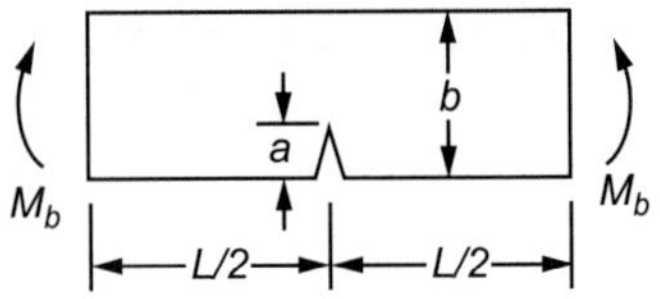

FIGURE 4-27*d*

The stress intensity factor for the plate with a single edge crack, according to Gross, Srawley and Brown (Fig. 4-27*c*).

$$K_I = \sigma\sqrt{\pi a}\left[1.12 - 0.23\left(\frac{a}{b}\right) + 10.6\left(\frac{a}{b}\right)^2 - 21.7\left(\frac{a}{b}\right)^3 + 30.4\left(\frac{a}{b}\right)^4\right] \quad \text{(4-35u)}$$

The stress intensity factor for single edged cracked plate/specimen subjected to bending (M_b) (Fig. 4-27*d*).

$$K_I = \sigma\sqrt{\pi a}\left[1.112 - 1.40\left(\frac{a}{b}\right) + 7.33\left(\frac{a}{b}\right)^2 - 13.08\left(\frac{a}{b}\right)^3 + 14.0\left(\frac{a}{b}\right)^4\right] \quad \text{(4-35v)}$$

Stress intensity factor for the case of angled crack (Fig. 4-27A):

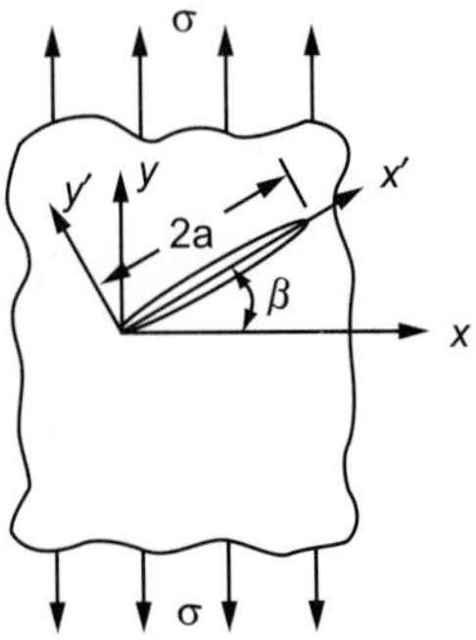

FIGURE 4-27A Through crack in an infinite plate for the general case where the crack plane is inclined at $90° - \beta$ angle from the applied normal stress σ acting at infinity.

The stress intensity factors for Modes I and II.

$$K_I = K_{I(0)} \cos^2\beta \quad \text{(4-36a)}$$

$$K_{II} = K_{I(0)} \cos\beta \sin\beta \quad \text{(4-36b)}$$

where $K_{I(0)}$ is the Mode I stress intensity factor when $\beta = 0$

Particular	Formula

Equations for stress and displacement components in terms of polar coordinates:

The localized stress components at the vicinity of Mode I crack tips in terms of polar coordinates.

$$\sigma_r = \frac{K_I}{4\sqrt{2\pi r}}\left[5\cos\frac{\theta}{2} - \cos\frac{3\theta}{2}\right] \quad (4\text{-}36c)$$

$$\sigma_\theta = \frac{K_I}{4\sqrt{2\pi r}}\left[3\cos\frac{\theta}{2} + \cos\frac{3\theta}{2}\right] \quad (4\text{-}36d)$$

$$\tau_{r\theta} = \frac{K_I}{4\sqrt{2\pi r}}\left[\sin\frac{\theta}{2} + \sin\frac{3\theta}{2}\right] \quad (4\text{-}36e)$$

$$\sigma_z = \nu_1(\sigma_r + \sigma_\theta) \quad (4\text{-}36f)$$

where $\nu_1 = 0$ for plane stress and ν_1 is Poisson's ratio, ν, for plane strain. These singular fields only apply as $r \to 0$.

The crack tip displacement fields for "first mode" (Mode I) in case of linear elastic, isotropic materials.

$$u_r = \frac{K_I}{2E}\sqrt{\frac{r}{2\pi}}(1+\nu)\left[(2\kappa - 1)\cos\frac{\theta}{2} - \cos\frac{3\theta}{2}\right] \quad (4\text{-}36g)$$

$$u_\theta = \frac{K_I}{2E}\sqrt{\frac{r}{2\pi}}(1+\nu)\left[-(2\kappa - 1)\sin\frac{\theta}{2} + \sin\frac{3\theta}{2}\right] \quad (4\text{-}36h)$$

$$u_z = -\left(\frac{\nu_2 z}{E}\right)(\sigma_r + \sigma_0) \quad (4\text{-}36i)$$

where

$\kappa = \left(\frac{3-\nu}{1+\nu}\right)$, $\nu_1 = 0$, and $\nu_2 = \nu$ for plane stress

$\kappa = (3 - 4\nu)$, $\nu_1 = \nu$, and $\nu_2 = 0$ for plain strain

K_I is given by Eq. (4-36a).

The localized stress components at the vicinity of Mode II crack tip in terms of polar coordinates.

$$\sigma_r = \frac{K_{II}}{\sqrt{2\pi r}}\sin\frac{\theta}{2}\left(1 - 3\sin^2\frac{\theta}{2}\right) \quad (4\text{-}36j)$$

$$\sigma_\theta = \frac{3K_{II}}{\sqrt{2\pi r}}\sin\frac{\theta}{2}\cos^2\frac{\theta}{2} \quad (4\text{-}36k)$$

The crack tip displacement fields for Mode II.

$$u_r = \frac{K_{II}}{2E}\sqrt{\frac{r}{2\pi}}(1+\nu)\left[-(2\kappa - 1)\sin\frac{\theta}{2} + 3\sin\frac{3\theta}{2}\right] \quad (4\text{-}36l)$$

$$u_\theta = \frac{K_{II}}{2E}\sqrt{\frac{r}{2\pi}}(1+\nu)\left[-(2\kappa - 1)\cos\frac{\theta}{2} + 3\cos\frac{3\theta}{2}\right] \quad (4\text{-}36m)$$

Particular	Formula	
The localized stress components and crack tip displacement fields for Mode III in terms of polar coordinates.	$\sigma_r = \dfrac{K_{III}}{\sqrt{2\pi r}} \sin\dfrac{\theta}{2}$	(4-36n)
	$\sigma_\theta = \dfrac{K_{III}}{\sqrt{2\pi r}} \cos\dfrac{\theta}{2}$	(4-36p)
	$u_z = \dfrac{2K_{III}}{G}\sqrt{\dfrac{r}{2\pi}} \sin\dfrac{\theta}{2}$	(4-36q)
The critical applied tensile stress necessary for crack extension according to Griffith theory for brittle metals.	$\sigma_c \propto \sqrt{\dfrac{EU}{a}}$ where σ_c = critical applied stress E = Young's modulus U = surface energy per unit area a = crack length.	(4-36r)
The modified Griffith's equation for a small amount of plastic deformation according to Orowan which can be applied to ductile materials at low temperature, high strain rate and localized geometric constraint.	$\sigma = \sqrt{\dfrac{E(U+p)}{a}}$ where p = plastic deformation energy per unit area for metallic solid, $p \gg U$.	(4-36s)
The elastic energy release rate for Mode I.	$G_I = \left(\dfrac{1-\nu^2}{E}\right) K_I^2$ for plane strain	(4-36t)
	$= K_I^2/E$ for plane stress	(4-36u)
The elastic energy release rate for Mode II.	$G_{II} = \dfrac{(1-\nu^2)}{E} K_{II}^2$	(4-36v)
The elastic energy release rate for Mode III.	$G_{III} = \dfrac{(1+\nu)}{E} K_{III}^2$	(4-36w)
The stress-intensity factor for a centrally located straight crack in an infinite plate subjected to uniform shear stress τ.	$K_I - iK_{II} = -i\sqrt{\pi}\tau\sqrt{a}$	(4-37a)
The stress-intensity magnification factor for a centrally located straight crack of length $2a$ in a flat plate whose length $2h$ and width $2b$ are very large compared with the crack length subjected to uniform uniaxial tensile stress σ.	$MF = \dfrac{K_I}{\sqrt{\pi}\sigma\sqrt{a}}$	(4-37b)
For stress-intensity magnification factors of plates with straight crack located at various positions in the plate and cylinders subjected to various types of rate of loadings and for various values of a/b, a/d, a/h, $a/(r_o - r_i)$, and other ratios.	Refer to Figs. from 4-28, 4-29 to 4-34.	
The factor of safety.	$n = \dfrac{K_{Ic}}{K}$	(4-38)

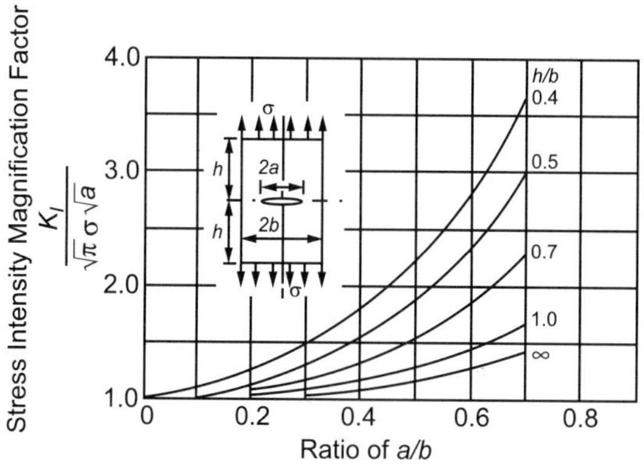

FIGURE 4-28 Stress intensity magnification factor $K_I/\sqrt{\pi}\,\sigma\sqrt{a}$ for various ratios a/b of a flat plate with a centrally located straight crack under the action of uniform uniaxial tensile stress σ.

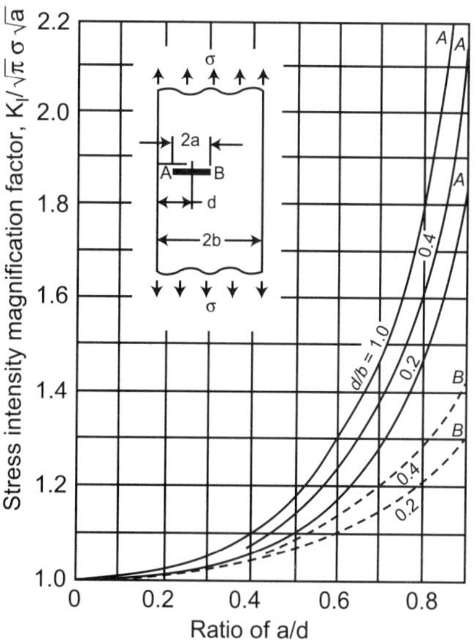

FIGURE 4-29 Stress intensity magnification factor $K_I/\sqrt{\pi}\sigma\sqrt{a}$ for an off-center straight crack in a flat plate subjected to uniform unidirectional tensile stress σ; solid curves are for the crack tip at A; dashed curves for tip at B.

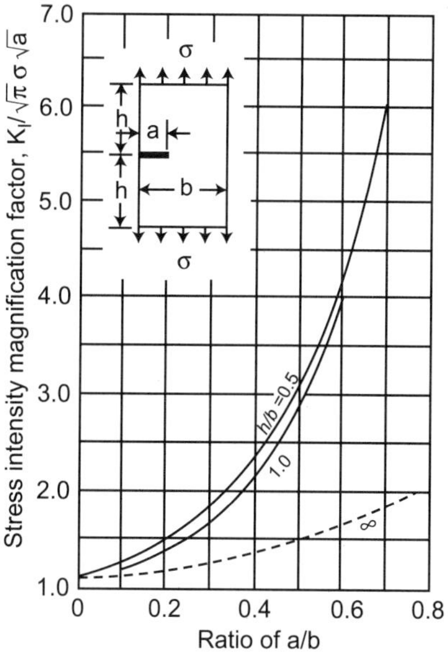

FIGURE 4-30 Stress intensity magnification factor $K_I/\sqrt{\pi}\sigma\sqrt{a}$ for an edge straight crack in a flat plate subjected to uniform uniaxial tensile stress σ for solid curves there are no constraints to bending; the dashed curve was obtained with bending constraints added.

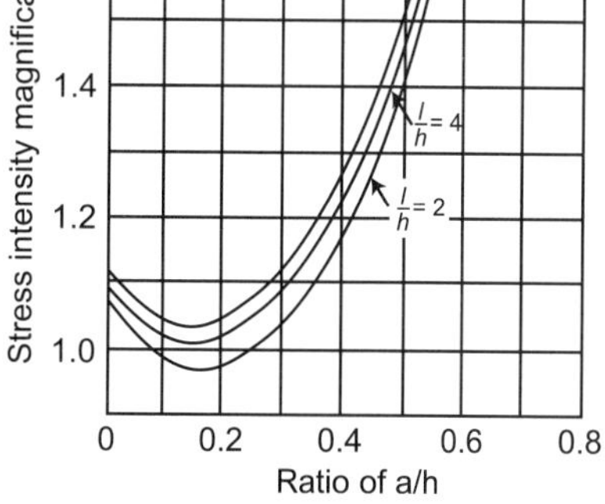

FIGURE 4-31 Stress intensity magnification factor $K_I/\sqrt{\pi}\sigma\sqrt{a}$ for a rectangular cross-sectional beam subjected to bending M_b.

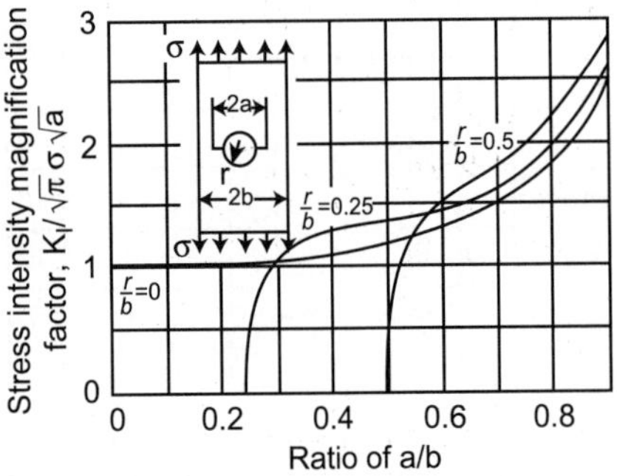

FIGURE 4-32 Stress intensity magnification factor $K_I/\sqrt{\pi}\sigma\sqrt{a}$ for a flat plate with a centrally located circular hole with two straight cracks under uniform uniaxial tensile stress σ.

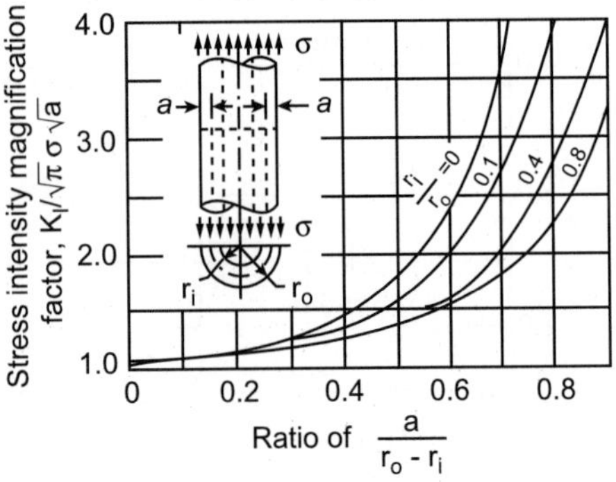

FIGURE 4-33 Stress intensity magnification factor $K_I/\sqrt{\pi}\sigma\sqrt{a}$ for axially tensile loaded cylinder with a radial crack of a depth extending completely around the circumference of the cylinder.

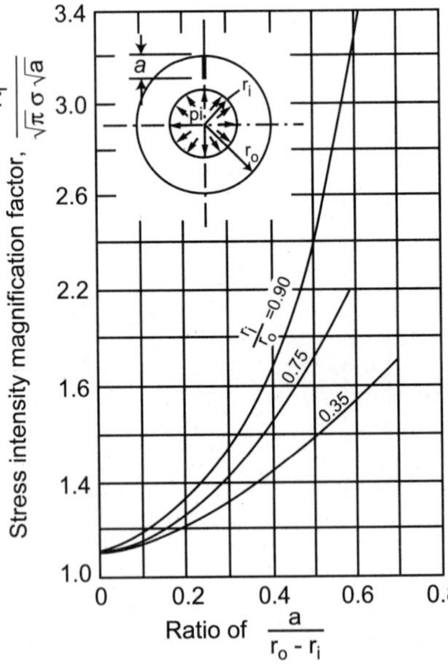

FIGURE 4-34 Stress intensity magnification factor $K_I/\sqrt{\pi}\sigma\sqrt{a}$ for a cylinder subjected to internal pressure p_i having a radial crack in the longitudinal direction of depth a. Use equation of tangential stress of thick cylinder subjected to internal pressure to calculate the stress σ_θ at $r = r_o$.

Particular	Formula
Critical crack length	$a_c = \dfrac{1}{\pi}\left(\dfrac{K_{Ic}}{\sigma_{sy}/2}\right)^2$
For values of critical stress-intensity factor (K_{Ic}) for some engineering materials.	Refer to Table 4-10.

TABLE 4-10
Plane-strain fracture toughness or stress intensity factor K_{Ic} for some engineering materials

Material		K_{Ic}		Yield strength, σ_{xy}		Critical crack length, a_c	
Previous designation	UNS designation	MPa $\sqrt{m}$	kpsi $\sqrt{in}$	MPa	kpsi	mm	in
Aluminum							
2014-T651		24.2	22	455	66	3.6	0.14
2024-T851	A92024-T851	26	24	455	66	4.3	0.17
7075-T651	A97075-T651	24	22	495	72	3.0	0.12
7178		13	30	490	71	5.8	0.23
Titanium							
Ti-6Al-4V	R56401	115	105	910	132	20.5	0.81
Ti-6Al-4V*	R56401*	55	50	1035	150	3.6	0.14
Steel							
4340	G43400	99	90	860	125	16.8	0.66
4340*	G43400*	60	55	1515	220	2	0.08
H-11	–	38.5	35	1790	260	<0.6	<0.02
H-11	–	27.8	27	2070	300	0.23	0.009
52100	G52986	14	13	2070	300	<0.06	<0.002

* Heat treated to a higher strength.

REFERENCES

1. Lingaiah, K., Solution of an Asymmetrically Reinforced Circular Cut-out in a Flat Plate Subjected to Uniform Unidirectional Stress, Ph.D. Thesis, Department of Mechanical Engineering, University of Saskatchewan, Saskatoon, Sask., Canada, 1965.
2. Lingaiah, K., W. P. T. North, and J. B. Mantle, "Photoelastic Analysis of an Asymmetrically Reinforced Circular Cut-out in a Flat Plate Subjected to Uniform Unidirectional Stress," *Proc. SESA*, Vol. 23, No. 2 (1966), p. 617.
3. Peterson, R. E., "Design Factors for Stress Concentration," *Machine Design*, Vol. 23, No. 27, Pentagon Publishing, Cleveland, Ohio, 1951.
4. Lingaiah, K., "Effect of Contact Stress on Fatigue Strength of Gears," M.Tech. Thesis, Indian Institute of Technology, Kharagpur, India, 1958.
5. Lingaiah, K., "Photoelastic Stress Analysis of Gear Teeth Under Load," Department of Mechanical Engineering, University Visveswaraya College of Engineering, Bangalore University, Bangalore, 1980.
6. Lingaiah, K., and B. R. Narayana Iyengar, *Machine Design Data Handbook*, Vol. I (*SI and Customary Metric Units*), Suma Publishers, Bangalore, India, 1986.
7. Lingaiah, K., *Machine Design Data Handbook*, Vol. II (*SI and Customary Metric Units*), Suma Publishers, Bangalore, India, 1986.

8. Lingaiah, K., *Machine Design Data Handbook* (*SI and Customary US Units*), McGraw-Hill Publishing Company, New York, 1994.
9. Aidad, T., and Y. Terauchi, "On the Bending Stress in a Spur Gear," 3 Reports, *Bull. JSME*, Vol. 5 (1962), p. 161.
10. Dolan, T. J., and E. L. Broghamer, "A Photoelastic Study of Stresses in Gear Tooth Fillets," *Univ. Illinois Exptl. Station. Bull.*, 335 (1942).
11. Hetenyi, M., The Application of Hardening Resins in Three-Dimensional Photoelastic Studies, *J. Appl. Phys.*, Vol. 10 (1939), p. 295.
12. Shigley, J. E., and L. D. Mitchell, *Mechanical Engineering Design*, McGraw-Hill Publishing Company, New York, 1983.
13. Greenhood, D. C., *Engineering Data for Production Design*, McGraw-Hill Publishing Company, New York, 1961.
14. Carlson, R. L., and G. A. Kardomateas, *An Introduction to Fatigue in Metals and Composites.*
15. Anderson, T. L., *Fracture Mechanics—Fundamentals and Application*, 2nd edition, CRC Press, New York, 1995.
16. Fedderson, C., "Discussion", in *Plane Strain Crank Toughness Testing of High Strength Metallic Materials*, ASTM STP410, American Society for Testing Materials, Philadelphia (1967), p. 77.
17. Keer, L. M., and J. M. Freedman, "Tensile Strip with Edge Cracks," *Int. J. Engineering Science*, Vol. 11 (1973), pp. 1965–1075.
18. Gross, B., and J. E. Srawley, "Stress Intensity Factors for Bend and Compact Specimens," *Engineering Fracture Mechanics*, Vol. 4 (1972), pp. 587–589.
19. Gross, B., J. E. Srawley, and W. E. Brown Jr., *Stress Intensity Factors for a Single Edge Notch Tension Specimen by a Boundary Collocation of a Stress Function*, NASA Technical Note D-2395, 1964.
20. Damage Tolerant Design Handbook, *MICIC-HB-01, Air Force Materials Laboratory*, Wright-Patterson Air Force Base, Ohio, December 1972, and supplements.

CHAPTER

5

DESIGN OF MACHINE ELEMENTS FOR STRENGTH

SYMBOLS[5,6]

A	area of cross-section, m^2 (in^2)
b	a shape factor ($b > 0$)
B	a constant
e_{sz}	size coefficient
$e_{s\tau}$	surface coefficient in case of tension and bending
$e'_{s\tau}$	surface coefficient in case of torsion
E	Young's modulus, GPa (Mpsi)
F	normal load (also with suffixes and primes), kN (lbf)
F'_m	static equivalent of cyclic load, kN (lbf)
G	modulus of rigidity, GPa (Mpsi)
h	thickness, m (in)
K_{sz}	size factor
$K_{s\tau}$	surface factor
K_σ	theoretical normal stress-concentration factor
K_τ	theoretical shear stress-concentration factor
$K_{f\sigma}$	fatigue normal stress-concentration factor
$K_{f\tau}$	fatigue shear stress-concentration factor
M_b	bending moment (also with suffixes and primes), N m (lbf in)
M'_{bm}	static equivalent of cyclic bending moment, N m (lbf in)
M_t	twisting moment (also with suffixes and primes), N m (lbf in)
M'_{tm}	static equivalent of cyclic twisting moment, N m (lbf in)
n	safety factor
	a constant
n_a	actual safety factor (also with suffixes)
n_d	design safety factor (also with suffixes)
q	index of sensitivity
q_f	index of notch sensitivity for alternating stresses
r	notch radius, mm (in)
t	time, h
x_0	the guaranteed value of x ($x_0 \geq 0$)
y_{max}	maximum deflection
Z_b	flexural section modulus, m^3 or cm^3 (in^3)
Z_t	polar section modulus, m^3 or cm^3 (in^3)
θ	characteristic or scale value ($\theta \geq x_0$)

σ	normal stress (also with suffixes and primes), MPa (psi)
σ_0	initial stress, MPa (psi)
σ_{su}	ultimate strength, MPa (psi)
σ_e	elastic limit for standard specimen for 12.5 mm ($\frac{1}{2}$ in), MPa (psi)
σ_d	design stress (also with suffixes), MPa (psi)
σ_x	normal stress in x direction, MPa (psi)
σ_y	yield stress, MPa (psi)
	normal stress in y direction, MPa (psi)
σ_{nom}	nominal normal stress, MPa (psi)
σ_{max}	maximum normal stress, MPa (psi)
σ'_e	elastic limit for any thickness h between 12.5 mm ($\frac{1}{2}$ in) and 75 mm (3 in), MPa (psi)
σ''_e	elastic limit for 75 mm (3 in) specimen, MPa (psi)
σ_{fb}	endurance limit in bending, MPa (psi)
τ	shear stress (also with suffixes and primes), MPa (psi)
τ_e	elastic limit in shear, MPa (psi)
τ_{sy}	yield strength in shear, MPa (psi)
τ_{xy}	shear stress in xy plane, MPa (psi)
τ_{nom}	nominal shear stress, MPa (psi)
τ_f	endurance limit in torsion, MPa (psi)
ε	engineering or average strain, μm/m (μin/in)
ε'	true strain, μm/m (μin/in)
ε_t	total creep, after a time t, μm/m (μin/in)
ε_0	initial creep, μm/m (μin/in)
$\dot{\varepsilon}$	creep rate (μ/m)/h [(μin/in)/h]
v_0	a constant

Suffixes	**for**
s	static strength (σ_u or σ_y)
u	ultimate strength
y	yield strength
e	elastic limit
a	amplitude
b	bending
m	mean
t	tension
max	maximum
min	minimum
f	endurance limit (also used for reversed cycle)
o	endurance limit repeated cycle

Primes	**for**
′ (single)	static equivalent
″ (double)	combined stress

Particular	Formula

STATIC LOADS

Influence of size

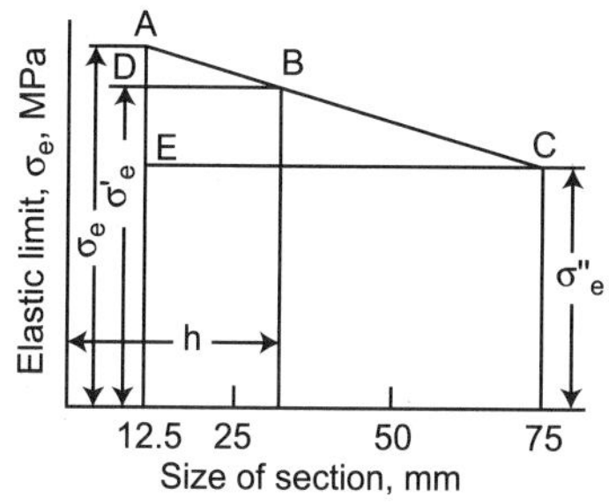

FIGURE 5-1 Change of elastic limit with size of section.

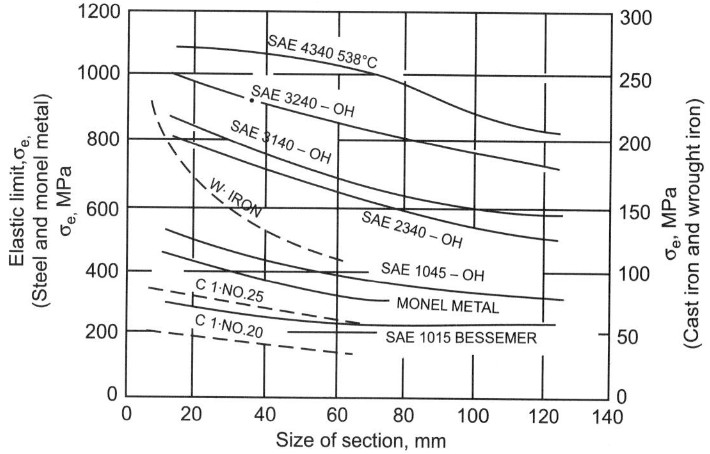

FIGURE 5-2 Influence of size on elastic limits.

The size coefficient (Fig. 5-1, Fig. 5-2, and Table 5-1)

$$e_{sz} = 1 - 0.016\left(1 - \frac{\sigma_e''}{\sigma_e}\right)(h - 12.5) \quad (5\text{-}1)$$

where σ_e = elastic limit for 12.5 mm (0.5 in)
σ_e'' = elastic limit for 75 mm (3.0 in)

TABLE 5-1
Strength ratios of various materials for use in Eqs. (5-1) and (5-2)

	Values of σ_e''/σ_e				
Material	**Natural state**	**Annealed**	**Drawn at 650°C**	**Drawn at 535°C**	**Drawn at 425°C**
Aluminum, strong, wrought	0.93	—	—	—	—
Tobin bronze	0.90	—	—	—	—
Monel metal, forged	0.80	—	—	—	—
Ductile iron	0.80	0.98	—	—	—
Low-carbon steel, $C < 0.20\%$	0.84	—	—	—	—
Medium-carbon steel, 0.30 to 0.50% C	—	0.85	0.72	0.59	0.53
Nickel steel, SAE 2340	—	0.86	0.80	0.74	—
Cr–Ni steel, SAE 3140	—	0.86	0.75	0.70	0.65
Cast iron, Class no. 20	0.55	—	—	—	—
Cast iron, Class no. 25	0.73	—	—	—	
Cast iron, Class no. 35	0.60	—	—	—	
Wrought iron	0.55	—	—	—	

Particular	Formula	
The size factor	$K_{sz} = \dfrac{250}{300 - 4h + \dfrac{\sigma''_e}{\sigma_e}(4h - 50)}$	(5-2)
The relation between size coefficient and size factor	$e_{sz} = \dfrac{1}{k_{sz}}$	(5-3)
The elastic limit for any thickness h between 12.5 mm and 75 mm can be determined from the relation (Fig. 5-1)	$\sigma''_e = \sigma_e - \dfrac{(\sigma_e - \sigma''_e)(h - 12.5)}{(75 - 12.5)}$	(5-4)

INDEX OF SENSITIVITY

The index of sensitivity	$q = \dfrac{K_{\sigma a} - 1}{K_\sigma - 1}$	(5-5)
The actual or real stress-concentration factor	$K_{\sigma a} = 1 + q(K_\sigma - 1)$	(5-6)

SURFACE CONDITION (Fig. 5-3)

The surface factor for the case of tension and bending	$K_{s\tau} = \dfrac{1}{e_{s\tau}}$	(5-7)
The surface coefficient in case of torsion	$e'_{s\tau} = 0.425 + 0.575 e_{s\tau}$	(5-8)

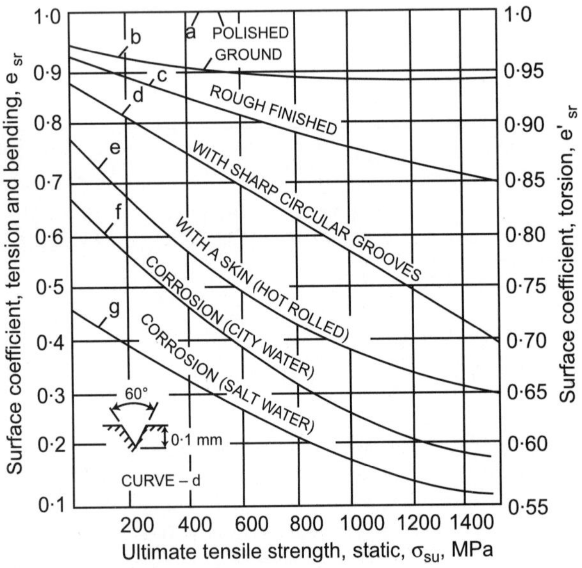

FIGURE 5-3 Reciprocals of stress-concentration factors caused by surface conditions.

Particular	Formula

SAFETY FACTOR

The general equation for design safety factor (Table 5-2)

$$n = k_1 k_2 k_3 k_4 \ldots k_m n_a \tag{5-9}$$

where $k_1 = K_{sz}$ = size factor
$k_2 = K_{sT}$ = surface factor
$k_3 = K_l$ = load factor (Table 14-3)
k_4 = material factor
⋮
n_a = actual safety factor (Table 5-2).

TABLE 5-2
Actual safety factor[a]

Circumstance	Actual factor of safety or reliability factor, n_a
Strength properties of material well known, load accurately predictable, parts produced with close dimensional control and brought to close tolerance specifications, and low-weight criteria	1
Load accurately predictable and low-weight and low-cost criteria	1.1–1.5
Load accurately predictable and low-cost criteria (low-weight–no criteria)	1.5–2
Overloads expected, materials ordinary but reliability important	2–3
Strength properties not well defined, loading uncertain, human life at stake if failure occurred, high maintenance and shutdown cost	≥3

[a] These values are recommended for use in design, in the absence of specific reliability data.

The design safety factor based on ultimate strength

$$n_{ud} = K_{sz} K_{\sigma a} n_{ua} \tag{5-10}$$

The relationships between allowable stress and specified minimum yield strength using the AISC Code are given here:

Tension

$$0.45\sigma_{sy} \le \sigma_a \le 0.60\sigma_{sy} \tag{5-11}$$

Shear

$$\tau_a = 0.40\sigma_{sy} \tag{5-12}$$

Bearing

$$\sigma_a = 0.90\sigma_{sy} \tag{5-13}$$

Bending

$$0.60\sigma_{sy} \le \sigma_a \le 0.75\sigma_{sy} \tag{5-14}$$

The expression for forces or loads used to find stresses in machine members or structures as per AISC Code.

$$F = \sum W_d + \sum W_l + \sum KF_l + F_w + \sum F_{me} \tag{5-15}$$

where $\sum W_d$ = sum of dead loads
$\sum W_l$ = sum of all stationary or static live loads
F_l = impact or dynamic live load
F_w = wind load on the structure
$\sum F_{me}$ = load which accounts for earthquakes, hurricanes, etc.
K = service factor obtained from Table 5-3.

The value of design normal stress

$$\sigma_d \le \sigma_a \tag{5-16}$$

Particular	Formula

TABLE 5-3
AISC service factor K for use in Eq. (5-15)

Particular	K
For support of elevators	2
For cab-operated traveling-crane support girders and their connections	1.25
For pendant-operated traveling-crane support girders and their connections	1.10
For support of light machinery, shaft- or motor-driven	$\geq$1.20
For supports of reciprocating machinery or power-driven units	$\geq$1.50
For hangers supporting floors and balconies	1.33

The value of design shear stress $\quad \tau_d \leq \tau_a \quad$ (5-16a)

The design safety factor $\quad n_d = \dfrac{\text{strength}}{\text{stress}} = n_s n_L \quad$ (5-17)

where n_s = safety factor to take into account the uncertainty of strength
n_L = safety factor to take into account the uncertainty of load.

The equation for design safety factor $\quad n_d = \dfrac{\text{strength in force units}}{\text{applied force or load}} \quad$ (5-18)

The realized safety factor $\quad n_r = \dfrac{\sigma_s}{\sigma} \quad \text{or} \quad n_r = \dfrac{\tau_s}{\tau} \quad$ (5-19)

The design safety factor based on elastic limit $\quad n_{ed} = K_{sz} K_{\sigma a} n_{ea} \quad$ (5-20)

The design safety factor based on yield strength $\quad n_{yd} = K_{sz} K_{\sigma a} n_{ya} \quad$ (5-21)

The design safety factor based on endurance limit on bending $\quad n_{fd} = K_{sz} K_{ST} K_{ld} n_{fa} \quad$ (5-22)

where K_{ld} = load factor

Design stress based on elastic limit $\quad \sigma_{ed} = \dfrac{\sigma_e}{n_{ed}} \quad$ (5-23)

Design stress based on ultimate strength $\quad \sigma_{ud} = \dfrac{\sigma_{su}}{n_{ud}} \quad$ (5-24)

Design stress based on yield strength $\quad \sigma_{yd} = \dfrac{\sigma_{sy}}{n_{yd}} \quad$ (5-25)

Design stress based on yield strength in shear $\quad \tau_{yd} = \dfrac{\tau_{sy}}{n_{yd}} \quad$ (5-26)

Static design stress $\quad \sigma_{sd} = \dfrac{\sigma_{su}}{n_{ud}} \quad \text{or} \quad \dfrac{\sigma_{sy}}{n_{yd}}$ as the case may be $\quad$ (5-27)

Particular	Formula
Design stress based on endurance limit	$\sigma_{fd} = \dfrac{\sigma_{sf}}{n_{fd}}$ (5-28)
The corrected fatigue strength or design fatigue strength	$\sigma_{sf} = K_{sr}K_{sz}K_{ld}K_R K_T K_{me}\sigma'_{sf}$ (5-28a)
The corrected endurance limit or design endurance limit	$\sigma_{se} = k_{sr}k_{sz}k_{ld}k_R k_T k_{me}\sigma'_{se}$ (5-28b) where σ'_{se} = endurance limit of test specimen σ'_{sf} = fatigue strength of test specimen K_{sr} = surface factor K_{sz} = size factor K_{ld} = load factor K_R = reliability factor K_T = temperature factor K_{me} = miscellaneous-effect factor also known as fatigue strength reduction factor $\approx 1/K_{\sigma f}$ (5-28c)
The size facto k_{sz} for bending or torsion of round bars made of ductile materials according to Juvinall	$K_{sz} = \begin{cases} 1 & d < 10\,\text{mm (0.4 in)} \\ 0.9 & 10\,\text{mm (0.4 in)} < d < 50\,\text{mm (2 in)} \\ 0.8 & 50\,\text{mm (2 in)} < d < 100\,\text{mm (4 in)} \\ 0.7 & 100\,\text{mm (4 in)} < d < 150\,\text{mm (5 in)} \end{cases}$ (5-28d)
The size factor for axial force	$K_{sz} = 0.7$ to 0.9 (5-28e)
The size factor as suggested by the ASME national standard on "Design of Transmission Shafting"	$K_{sz} = \begin{cases} d^{-0.19} & 2 < d < 10\,\text{in} \\ 1.85d^{-0.19} & 50 < d < 250\,\text{mm} \end{cases}$ (5-28f)
The surface factor	$K_{sr} = \begin{cases} 1.00 & \text{for longitudinal hand polish} \\ 0.90 & \text{for hand burnish} \\ 0.87 & \text{for smooth mill cut} \\ 0.79 & \text{for rough mill cut} \end{cases}$ (5-28g) Also refer to Fig. 5-3 for surface coefficient $e_{sr} = \dfrac{1}{K_{sr}}$ or $K_{sr} = \dfrac{1}{e_{sr}}$
For a rectangular cross-section in bending	$d = 0.81\sqrt{A}$ (5-28h) where A = area of the cross section
The effective diameter of round-section corresponding to a nonrotating solid or hollow round-section	$d_e = 0.370D$ (5-28i) where D = diameter
The effective diameter of a rectangular section of dimensions $h \times b$ which has $A_{0.95\text{cr}} = 0.05bh$	$d_e = 0.808(hb)^{1/2}$ (5-28j)

Particular	Formula
The equivalent diameter rotating-beam specimen for any cross-section according to Shigley and Mitchell	$d_{eq} = \sqrt{\dfrac{A_{95}}{0.0766}}$ (5-28j) where A_{95} is the portion of the cross sectional area of the nonround part that is stressed between 95% and 100% of the maximum stress.
The load factor according to Shigley	$k_{Id} = \begin{cases} 0.923 & \text{axial loading } \sigma_{sut} \le 1520\text{ MPa (220 kpsi)} \\ 1 & \text{axial loading } \sigma_{sut} \ge 1520\text{ MPa (220 kpsi)} \\ 1 & \text{bending} \\ 0.577 & \text{torsion and shear} \end{cases}$ (5-28k)
The fatigue stress concentration factor which is used here as the fatigue strength reduction factor at endurance limit 10^6 cycles	$K_{\sigma f} = 1 + q(k_{\sigma t} - 1)$ (5-28l) where $K_{\sigma f}$, $K_{\sigma t}$ and q have the same meaning as given in Chapter 4.
The fatigue strength reduction factor for lives less than $N = 10^6$ cycles is $K'_{\sigma f}$ and is given by	$K'_{\sigma f} = aN^b$ (5-28m) where $a = \left(\dfrac{1}{K_{\sigma f}}\right)$ and $b = -\dfrac{1}{3}\log\dfrac{1}{K_{\sigma f}}$ (5-28n) $K'_{\sigma f} = 1$ at 10^3 cycles.
For reliability factor K_R	Refer to Table 5-3A.

TABLE 5-3A
Reliability correction factor based on a standard deviation equal to 8% or the mean fatigue limit.

Reliability, %	K_R
50	1.000
90	0.897
99	0.814
99.9	0.743
99.999	0.659

Particular	Formula
The temperature factor as suggested by Shigley and Mitchell	$K_T = \begin{cases} 1 & \text{for } T \le 450^\circ\text{C } (840^\circ\text{F}) \\ 1 - 0.0058\,(T - 450) & \text{for } 450^\circ\text{C} < T < 550^\circ\text{C} \\ 1 - 0.0032\,(T - 840) & \text{for } 840^\circ\text{F} < T < 1020^\circ\text{F} \end{cases}$ (5-28p) These equations are applicable to steel. These cannot be used for Al, Mg, and Cu alloys.
For typical fracture surfaces for laboratory test specimens subjected to range of different loading conditions	Refer to Fig. 5-3A.

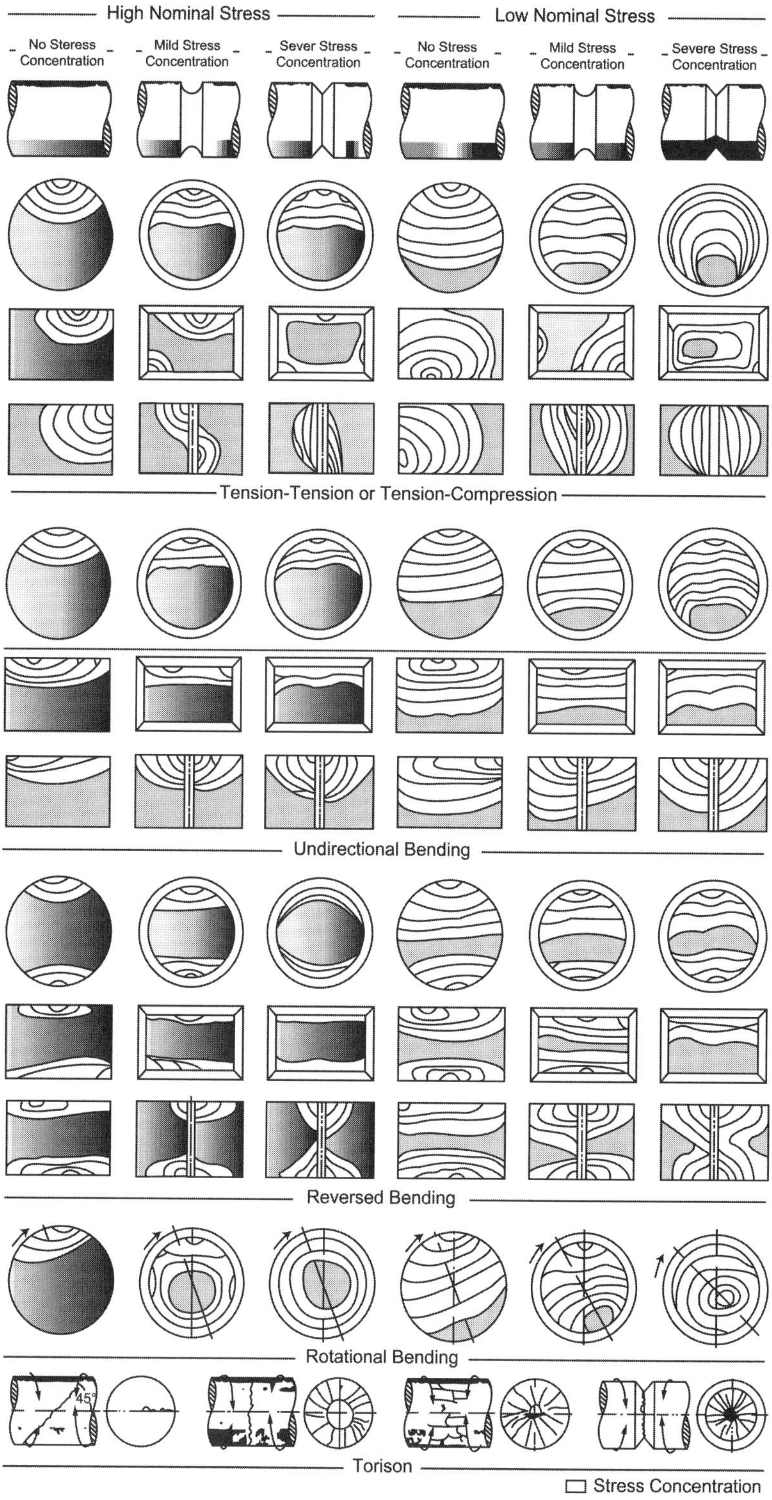

FIGURE 5.3A Typical fracture surfaces for laboratory test specimens subjected to a range of different loading conditions. *Courtesy*: Reproduced from *Metals Handbook*, Vol. 10, 8th edition, p. 102, American Society for Metals, Metals Park, Ohio, 1975.

Particular	Formula	
THEORIES OF FAILURE		
The maximum normal stress theory or Rankine's theory	$\sigma_e = \frac{1}{2}\left[(\sigma_x + \sigma_y) + \sqrt{(\sigma_x - \sigma_y)^2 + 4\tau_{xy}^2}\right]$	(5-29)
The maximum shear stress theory or Guest's theory	$\sigma_e = \sqrt{(\sigma_x - \sigma_y)^2 + 4\tau_{xy}^2}$	(5-30)
The shear-energy theory or constant energy-of-distortion or Hencky–von Mises theory	$\sigma_e = \sqrt{(\sigma_x - \sigma_y)^2 + 3\tau_{xy}^2}$	(5-31)
The maximum strain theory or Saint Venant's theory	$\sigma_e = \frac{1}{2}\left[(1-\nu)(\sigma_x + \sigma_y) + (1+\nu)\sqrt{(\sigma_x - \sigma_y)^2 + 4\tau_{xy}^2}\right]$	(5-32)
The bearing stress which causes failure for no friction at the surface of contact	$\sigma_b = 1.81\sigma_e$	(5-33)
The bearing stress which causes failure for the friction at the surface of contact	$\sigma_b = 2\sigma_e$	(5-34)
CYCLIC LOADS (Figs. 5-4 and 5-5)		
The fatigue stress-concentration factor for normal stress	$K_{f\sigma} = q_f(K_\sigma - 1) + 1$	(5-35)
The fatigue stress-concentration factor for shear stress	$K_{f\tau} = q_f(K_\tau - 1) + 1$	(5-36)
The empirical formula for notch sensitivity for alternating stress of steel	$q_f = 1 - \exp\left[-\frac{r\sigma_u^2}{0.904 \times 10^6}\right]$	(5-37)
Notch sensitivity curves for steel and aluminum alloys	Refer to Fig. 5-6.	
The empirical formula for notch sensitivity for alternating stress for high-strength aluminum alloys having $\sigma_u = 415$ to 550 MPa (60 to 80 kpsi)	$q_f = 1 - \exp\left(\frac{-r}{0.01}\right)$	(5-38)
Endurance strength for finite life	$\sigma_f' = \sigma_f\left(\frac{10^6}{N}\right)^{0.09}$ where N = required life in cycles.	(5-39)
The empirical relation between ultimate strength and endurance limits for various materials	Refer to Tables 5-4 and 5-5.	

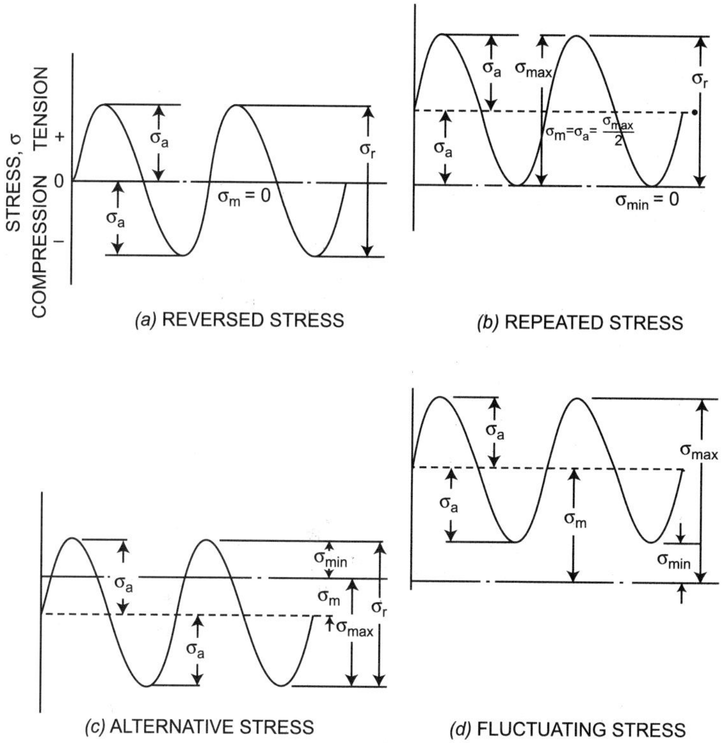

FIGURE 5-4 Types of fatigue stress variations.

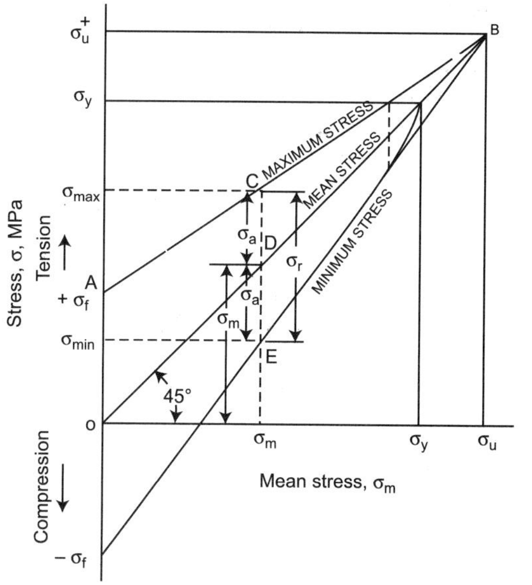

FIGURE 5-5 Modified Goodman diagram.

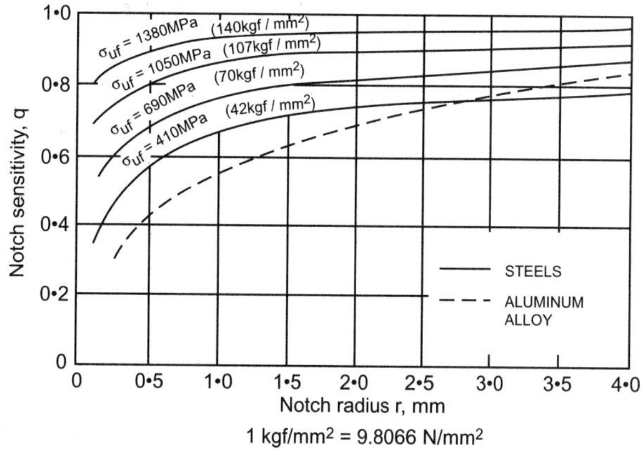

FIGURE 5-6 Notch-sensitivity curves for steel and aluminum alloys.

TABLE 5-4
Empirical relationship between ultimate strength and endurance limits for various materials (approximate)

Material	Tension, compression, and bending (reversed or repeated cycle)[a]	Torsion (reversed or repeated cycle)[a]
Gray cast iron	$\sigma_{ft} = 0.6\sigma_{fb}$ to $0.7\sigma_{fb}$ $\sigma_b = 1.2\sigma_{fb}$ to $1.5\sigma_{fb}$	$\tau = 0.75\sigma_{fb}$ to $0.9\sigma_{fb}$ $\tau = 1.2\tau_f$ to $1.3\tau_f$
Carbon steels	$\sigma_{ot} = 1.6\sigma_{fb}$ $\sigma_{ob} = 1.5\sigma_{fb}$	$\tau_o = 1.8\tau_f$ to $2\tau_f$
Steels (general)	$\sigma_{ft} = 0.7\sigma_{fb}$ to $0.8\sigma_{fb}$ $\sigma_{ft} = 0.36\sigma_u$; $\sigma_{ot} = 0.5\sigma_u$ $\sigma_{fb} = 0.46\sigma_u$; $\sigma_{ob} = 0.6\sigma_u$	$\tau_f = 0.55\sigma_{fb}$ to $0.58\sigma_{fb}$ $\tau_f = 0.22\sigma_u$ $\tau_o = 0.3\sigma_u$
Alloy steels	$\sigma_{ft} = 0.95\sigma_{fb}$ $\sigma_{ot} = 1.5\sigma_{ft}$ to $1.6\sigma_{ft}$ $\sigma_{ob} = 1.6\sigma_{fb}$	$\tau_o = 1.8\tau_f$ to $2\tau_f$
Aluminum alloys	$\sigma_{ot} = 0.7\sigma_{fb}$ $\sigma_{ob} = 1.8\sigma_{fb}$	$\tau_f = 0.55\tau_{fb}$ to $0.58\tau_{fb}$ $\tau_o = 1.4\tau_f$ to $2\tau_f$
Copper alloys		$\tau_f = 0.58\sigma_{fb}$ $\tau_o = 1.4\tau_f$ to $2\tau_f$
Endurance strength for finite life		$\sigma'_f = \sigma_f \left(\frac{10^6}{N}\right)^{0.09}$

[a] f—ensurance limit (also for reversed cycle); o—endurance for repeated cycle; t—tension; b—bending; u—ultimate; N—number of cycles

TABLE 5-5
The empirical relation for endurance limit

Material	Endurance limit, σ_f		
	Bending	Axial	Torsion
For steel and other ferrous materials [for $\sigma_u < 1374$ MPa (199.5 kpsi)]	$1/2–5/8\sigma_u$	$7/20–5/8\sigma_u$	$7/80–5/32\sigma_u$
For nonferrous materials	$1/4–1/3\sigma_u$	$7/40–1/3\sigma_u$	$7/160–1/12\sigma_u$

STRESS-STRESS AND STRESS-LOAD RELATIONS

Axial load

The maximum stress
$$\sigma_{max} = \frac{F_{max}}{A} \tag{5-40}$$

The minimum stress
$$\sigma_{min} = \frac{F_{min}}{A} \tag{5-41}$$

The load amplitude
$$F_a = \frac{F_{max} - F_{min}}{2} \tag{5-42}$$

The mean load
$$F_m = \frac{F_{max} + F_{min}}{2} \tag{5-43}$$

Particular	Formula	
The stress amplitude (Figs. 5-4 and 5-5)	$\sigma_a = \frac{F_a}{A}$	(5-44)
The mean stress	$\sigma_m = \frac{F_m}{A}$	(5-45)
The ratio of amplitude stress to mean stress	$\frac{\sigma_a}{\sigma_m} = \frac{F_a}{F_m}$	(5-46)
The static equivalent of cyclic load $F_m \pm F_a$	$F'_m = F_m + \frac{\sigma_{sd}}{\sigma_{fd}} F_a$	(5-47)
The static equivalent of mean stress $\sigma_m \pm \sigma_a$	$\sigma'_m = \frac{F'_m}{A}$	(5-48)
The Gerber parabolic relation (Fig. 5-7)	$\frac{\sigma_a}{\sigma_{fd}} + \left(\frac{\sigma_m}{\sigma_{ud}}\right)^2 = 1$	(5-49)

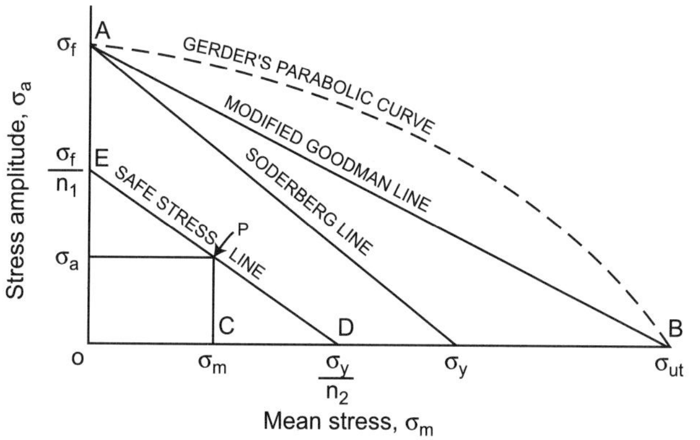

FIGURE 5-7 Graphical representation of steady and variable stresses.

The Goodman relation (Figs. 5-5, 5-7, and 5-9)	$\frac{\sigma_a}{\sigma_{fd}} + \frac{\sigma_m}{\sigma_{ud}} = 1$	(5-50)
The Soderberg relation (Figs. 5-7 and 5-8)	$\frac{\sigma_a}{\sigma_{fd}} + \frac{\sigma_m}{\sigma_{yd}} = 1$	(5-51)

Bending loads

The maximum stress	$\sigma_{max} = \frac{M_{b(max)}}{Z_b}$	(5-52)
The minimum stress	$\sigma_{min} = \frac{M_{b(min)}}{Z_b}$	(5-53)
The bending moment amplitude	$M_{ba} = \frac{M_{b(max)} - M_{b(min)}}{2}$	(5-54)

Particular	Formula

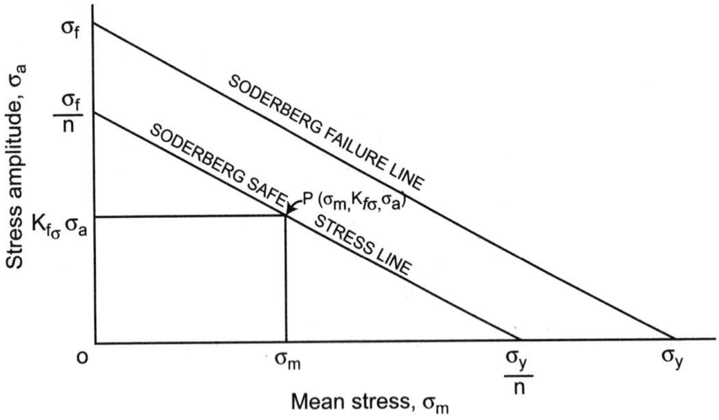

FIGURE 5-8 Representation of safe limit of mean stress and stress amplitude by Soderberg criterion.

The mean bending moment $$M_{bm} = \frac{M_{b(\max)} + M_{b(\min)}}{2} \quad (5\text{-}55)$$

The bending stress amplitude $$\sigma_{ba} = \frac{M_{ba}}{Z_b} \quad (5\text{-}56)$$

The mean bending stress $$\sigma_{bm} = \frac{M_{bm}}{Z_b} \quad (5\text{-}57)$$

The ratio of stress amplitude to mean stress $$\frac{\sigma_{ba}}{\sigma_{bm}} = \frac{M_{ba}}{M_{bm}} \quad (5\text{-}58)$$

The static equivalent of cyclic bending moment $M_{bm} \pm M_{ba}$ $$M'_{bm} = M_{bm} + \frac{\sigma_{sd}}{\sigma_{fd}} M_{ba} \quad (5\text{-}59)$$

The static equivalent of cyclic stress $$\sigma'_{bm} = \frac{M'_{bm}}{Z_b} \quad (5\text{-}60)$$

The Gerber parabolic relation (Fig. 5-7) $$\frac{\sigma_{ba}}{\sigma_{fd}} + \frac{\sigma_{bm}^2}{\sigma_{ud}^2} = 1 \quad (5\text{-}61)$$

The Goodman straight-line relation (Figs. 5-5, 5-7, and 5-9) $$\frac{\sigma_{ba}}{\sigma_{fd}} + \frac{\sigma_{bm}}{\sigma_{ud}} = 1 \quad (5\text{-}62)$$

The Soderberg straight-line relation (Figs. 5-7 and 5-8) $$\frac{\sigma_{ba}}{\sigma_{fd}} + \frac{\sigma_{bm}}{\sigma_{yd}} = 1 \quad (5\text{-}63)$$

Torsional moments

The maximum shear stress $$\tau_{\max} = \frac{M_{t(\max)}}{Z_t} \quad (5\text{-}64)$$

Particular	Formula	

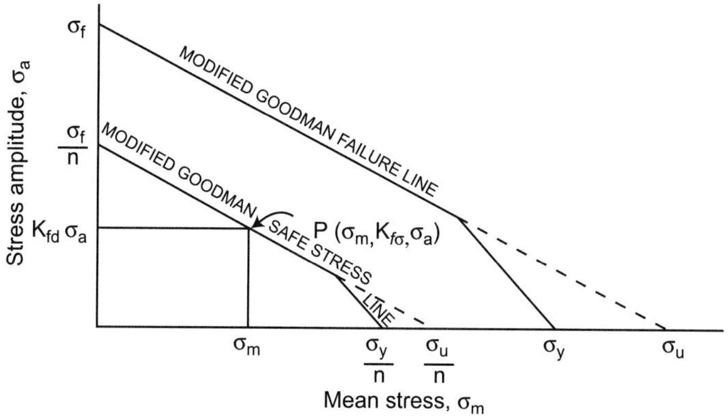

FIGURE 5-9 Representation of safe limit of mean stress and stress amplitude by Goodman criterion.

The minimum shear stress	$\tau_{\min} = \dfrac{M_{t(\min)}}{Z_t}$	(5-65)
The load amplitude	$M_{ta} = \dfrac{M_{t(\max)} - M_{t(\min)}}{2}$	(5-66)
The mean load	$M_{tm} = \dfrac{M_{t(\max)} + M_{t(\min)}}{2}$	(5-67)
The shear stress amplitude	$\tau_a = \dfrac{M_{ta}}{Z_t}$	(5-68)
The mean shear stress	$\tau_m = \dfrac{M_{tm}}{Z_t}$	(5-69)
The ratio of stress amplitude to mean stress	$\dfrac{\tau_a}{\tau_m} = \dfrac{M_{ta}}{M_{tm}}$	(5-70)
The static equivalent of cyclic twisting moment $M_{tm} \pm M_{ta}$	$M'_{tm} = M_{tm} + \dfrac{\tau_{sd}}{\tau_{fd}} M_{td}$	(5-71)
The static equivalent of cyclic stress	$\tau'_m = \dfrac{M'_{tm}}{Z_t}$	(5-72)
The Gerber parabolic relation (Fig. 5-7)	$\dfrac{\tau_a}{\tau_{fd}} + \dfrac{\tau_m^2}{\tau_{ud}^2} = 1$	(5-73)
The Goodman straight-line relation (Figs. 5-5, 5-7, and 5-9)	$\dfrac{\tau_a}{\tau_{fd}} + \dfrac{\tau_m}{\tau_{ud}} = 1$	(5-74)
The Soderberg straight-line relation (Figs. 5-7 and 5-8)	$\dfrac{\tau_a}{\tau_{fd}} + \dfrac{\tau_m}{\tau_{yd}} = 1$	(5-75)

Particular	Formula	

THE COMBINED STRESSES

Method 1

Particular	Formula	
The static equivalent of $\sigma_m \pm \sigma_a$	$\sigma'_m = \sigma_m + \dfrac{\sigma_{sd}}{\sigma_{fd}}\,\sigma_a$	(5-76)
The static equivalent of $\tau_m \pm \tau_a$	$\tau'_m = \tau_m + \dfrac{\tau_{sd}}{\tau_{fd}}\,\tau_a$	(5-77)
The maximum normal stress theory or Rankine's theory	$\sigma_e = \frac{1}{2}\left[\sigma'_m + \sqrt{\sigma'^2_m + 4\tau'^2_m}\right]$	(5-78)
The maximum shear theory or Coulomb's or Tresca criteria or Guest's theory	$\tau_e = \sqrt{\sigma'^2_m + 4\tau'^2_m}$	(5-79)
The distortion energy theory or Hencky–von Mises theory	$\sigma_e = \sqrt{\sigma'^2_m + 3\tau'^2_m}$	(5-80)
The maximum strain theory or Saint Venant's theory	$\sigma_e = \frac{1}{2}\left[(1-\nu)\sigma'_m + (1+\nu)\sqrt{\sigma'^2_m + 4\tau'^2_m}\right]$	(5-81)

Method 2

Particular	Formula	
The combined maximum normal stress	$\sigma''_{max} = \frac{1}{2}\left[\sigma_{max} + \sqrt{\sigma^2_{max} + 4\tau^2_{max}}\right]$	(5-82)
The combined minimum normal stress	$\sigma''_{min} = \frac{1}{2}\left[\sigma_{min} + \sqrt{\sigma^2_{min} + 4\tau^2_{min}}\right]$	(5-83)
The combined maximum shear stress	$\tau''_{max} = \frac{1}{2}\sqrt{\sigma^2_{max} + 4\tau^2_{max}}$	(5-84)
The combined minimum shear stress	$\tau''_{min} = \frac{1}{2}\sqrt{\sigma^2_{min} + 4\tau^2_{min}}$	(5-85)
The combined maximum normal stress according to strain theory	$\sigma''_{max} = \frac{1}{2}\left[(1-\nu)\sigma_{max} + (1+\nu)\sqrt{\sigma^2_{max} + 4\tau_{max}}\right]$	(5-86)
The combined minimum normal stress according to strain theory	$\sigma''_{min} = \frac{1}{2}\left[(1-\nu)\sigma_{min} + (1+\nu)\sqrt{\sigma^2_{min} + 4\tau^2_{min}}\right]$	(5-87)
The combined maximum octahedral shear stress	$\tau''_{max} = \frac{1}{2}\left[\sqrt{\sigma^2_{max} + 3\tau^2_{max}}\right]$	(5-88a)
The combined minimum octahedral shear stress	$\tau''_{min} = \frac{1}{2}\left[\sqrt{\sigma^2_{min} + 3\tau^2_{min}}\right]$	(5-88b)
The combined mean stress	$\sigma''_m = \dfrac{\sigma''_{max} + \sigma''_{min}}{2}$	(5-88c)

Particular	Formula	
The combined stress amplitude	$\sigma''_a = \dfrac{\sigma''_{\max} - \sigma''_{\min}}{2}$	(5-88d)
The Gerber parabolic relation (Fig. 5-7)	$\dfrac{\sigma''_a}{\sigma_{fd}} + \left(\dfrac{\sigma''_m}{\sigma_{ud}}\right)^2 = 1$	(5-88e)
The Goodman straight-line relation (Figs. 5-5, 5-7, and 5-9)	$\dfrac{\sigma''_a}{\sigma_{fd}} + \dfrac{\sigma''_m}{\sigma_{ud}} = 1$	(5-88f)
The Soderberg straight-line relation (Figs. 5-7 and 5-8)	$\dfrac{\sigma''_a}{\sigma_{fd}} + \dfrac{\sigma''_m}{\sigma_{yd}} = 1$	(5-88g)

COMBINED STRESSES IN TERMS OF LOADS

Method 1

Maximum shear stress theory

$$\frac{\sigma_e}{n_{ed}} = \sqrt{\left(\frac{M'_{bm}}{Z_b} + \frac{F'_m}{A}\right)^2 + 4\left(\frac{M'_{tm}}{Z_t}\right)^2} \tag{5-89a}$$

The shear energy theory

$$\frac{\sigma_e}{n_{ed}} = \sqrt{\left(\frac{M'_{bm}}{Z_b} + \frac{F'_m}{A}\right)^2 + 3\left(\frac{M'_{tm}}{Z_t}\right)^2} \tag{5-89b}$$

where

$$Z_b = \frac{\pi d^3}{32} \quad \text{and} \quad Z_t = \frac{\pi d^3}{16}$$

$$A = \frac{\pi d^2}{4}$$

for solid shafts

Method 2

Maximum shear stress theory

$$\left[\sqrt{\left(\frac{M_{b(\max)}}{Z_b} + \frac{F_{\max}}{A}\right)^2 + 4\left(\frac{M_{t(\max)}}{Z_t}\right)^2}\,\right]\left[\frac{1}{\tau_{fd}} + \frac{1}{\tau_d}\right] + \left[\sqrt{\left(\frac{M_{b(\min)}}{Z_b} + \frac{F_{\min}}{A}\right)^2 + 4\left(\frac{M_{t(\min)}}{Z_t}\right)^2}\,\right] \times \left[-\frac{1}{\tau_{fd}} + \frac{1}{\tau_d}\right] = 2 \tag{5-90a}$$

The shear energy theory

$$\left[\sqrt{\left(\frac{M_{b(\max)}}{Z_b} + \frac{F_{\max}}{A}\right)^2 + 3\left(\frac{M_{t(\max)}}{Z_t}\right)^2}\,\right]\left[\frac{1}{\tau_{fd}} + \frac{1}{\tau_d}\right] + \left[\sqrt{\left(\frac{M_{b(\min)}}{Z_b} + \frac{F_{\min}}{A}\right)^2 + 3\left(\frac{M_{t(\min)}}{Z_t}\right)^2}\,\right] \times \left[-\frac{1}{\tau_{fd}} + \frac{1}{\tau_d}\right] = 2 \tag{5-90b}$$

Particular	Formula

CREEP

Creep in tension

When the curve for total creep ε_t is approximated as a straight line its equation is

$$\varepsilon_t = \varepsilon_0 + \varepsilon t \tag{5-91a}$$

The creep rate $\dot{\varepsilon}$ can be approximated by the equation

$$\dot{\varepsilon} = B\sigma^n \tag{5-91b}$$

Refer to Table 5-6 for creep constants B and n.

Creep rate $\dot{\varepsilon}$, when extrapolated into the region of lower stresses, can be determined with greater accuracy by the hyperbolic sine term

$$\dot{\varepsilon} = \nu_0 \sinh\left(\frac{\sigma}{\sigma_1}\right) \tag{5-91c}$$

True strain

$$\varepsilon' = \ln(1+\varepsilon) \tag{5-91d}$$

Creep life of aluminum

$$\varepsilon_{cr} = \frac{1}{\dot{\varepsilon}^n} \tag{5-92}$$

Time for the stress to decrease from an initial value of σ_0 to a value of σ

$$t = \frac{1}{EB(n-1)\sigma_0^{n-1}}\left[\left(\frac{\sigma_0}{\sigma}\right)^{n-1} - 1\right] \tag{5-93}$$

Creep in bending

The maximum stress at the extreme fibers in case of bending of beam is given by the relation

$$\sigma = \left(\frac{C_1}{BD}\right)^{1/n} M_b \tag{5-94}$$

The maximum deflection of a cantilever beam loaded at free end by a load F

$$y_{max} = \frac{tF^n l^{n+2}}{D(n+2)} \tag{5-95}$$

$$\text{where } D = \frac{1}{B}\,\frac{(2b)^n\left(\dfrac{h}{2}\right)^{2n+1}}{\left(2+\dfrac{1}{n}\right)^n}$$

Creep constants B and n are taken from Table 5-6.

TABLE 5-6
Creep constants for various steels for use in Eqs. (5-91b) to (5-95)

Steel	Temperature °C	B	n
0.39% C	400	14×10^{-36}	8.6
0.30% C	400	44×10^{-30}	6.9
0.45% C	475	—	6.5
2% Ni, 0.8% Cr, 0.4% Mo	450	10×10^{-19}	3.2
2% Ni, 0.3% C, 1.4% Mn	450	21×10^{-22}	4.7
12% Cr, 3% W, 0.4% Mn	550	24×10^{-14}	1.9
Ni-Cr-Mo	500	12×10^{-16}	2.7
Ni-Cr-Mo	500	16×10^{-12}	1.3
12% Cr	455	12×10^{-22}	4.4

Particular	Formula	

RELIABILITY

Particular	Formula	
The probability function or frequency function	$p = f(x)$	(5-96)
The cumulative probability function	$F(x_j) = \sum_{x_i \leq x_j} f(x_i)$	(5-97)
	where $f(x)$ is the probability density	
The sample mean or arithmetic mean of a sample	$\bar{x} = \dfrac{x_1 + x_2 + x_3 + x_4 + \cdots + x_n}{n}$	(5-98a)
	$= \dfrac{1}{n} \sum_{i=1}^{n} x_i$	(5-98b)
	where x_i is the ith value of the quantity n is the total number of measurements or elements	
The population mean of a population consisting of n elements	$\mu = \dfrac{x_1 + x_2 + x_3 + x_4 + \cdots + x_n}{n}$	(5-99a)
	$= \dfrac{1}{n} \sum_{i=1}^{n} x_i$	(5-99b)
The sample variance	$s_x^2 = \dfrac{(x_1 - \bar{x})^2 + (x_2 - \bar{x})^2 + \cdots + (x_n - \bar{x})^2}{n-1}$	(5-100a)
	$= \dfrac{1}{n-1} \sum_{i=1}^{n} (x_i - \bar{x})^2$	(5-100b)
A suitable equation for variance for use in a calculator	$s_x^2 = \dfrac{\sum x^2}{n} - \bar{x}^2$	(5-101)
The sample standard deviation (the symbol used for true standard deviation is $\hat{\sigma}$)	$s_x = \left[\dfrac{1}{n-1} \sum_{i=1}^{n} (x_i - \bar{x})^2 \right]^{1/2}$	(5-102)
A suitable equation for standard deviation for use in a calculator	$s_x = \left\{ \dfrac{\sum x^2 - \dfrac{(\sum x)^2}{n}}{n-1} \right\}^{1/2}$	(5-103)
The coefficient of variation	$cv = (s_x/\bar{x})100$	(5-104)
The normal, or Gaussian, distribution (Fig. 5-10)	$f(x) = \dfrac{1}{\hat{\sigma}\sqrt{2\pi}} e^{-(x-\mu)^2/2\hat{\sigma}^2} \quad -\infty < x < \infty$	(5-105)
The normal distribution as defined by parameters, the mean μ and standard deviation $\hat{\sigma}$ according to the relation for the relative frequency $f(t)$, which is the ordinate at t	$f(t) = \dfrac{1}{\sqrt{(2\pi)}} e^{-(t^2/2)}$	(5-106)
	where $t = (x - \mu)/\hat{\sigma}$. Refer to Table 5-7 for ordinate $f(t)$ [i.e. $y = f(t)$] for various values of t.	

Particular	Formula

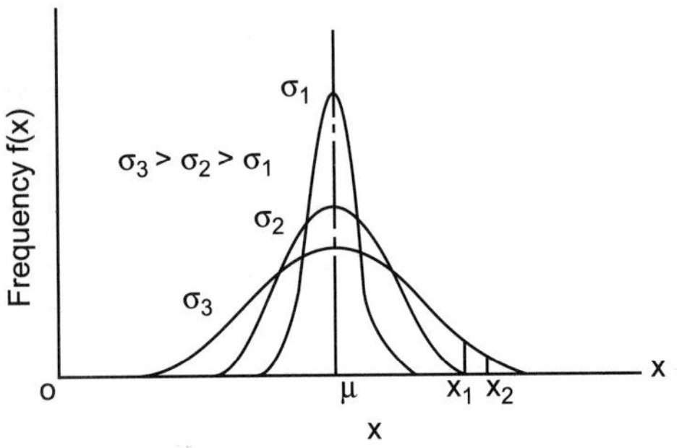

FIGURE 5-10 The shapes of normal distribution curves for various σ and constant μ.

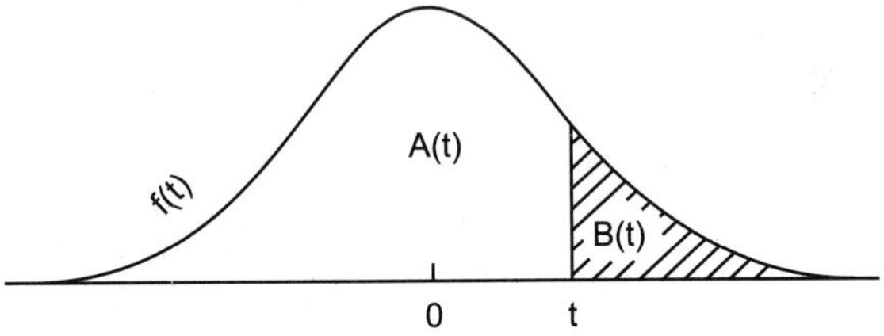

FIGURE 5-11 The Gaussian (normal) distribution curve.

The area under normal distribution curve to the right of t (Fig. 5-11)

Refer to Table 5-8 for area under the standard normal distribution curve.

$$B(t) = 1 - A(t) \tag{5-107}$$

where $A(t)$ is the area to the left of t.

The area under the entire normal distribution curve is $A(t) + B(t)$ and is equal to unity. The term $B(t)$ can be found from Table 5-8 or by integrating the area under the curve.

Error function or probability integral

$$\text{erf}(x) = \frac{2}{\sqrt{\pi}} \int_0^x e^{-t^2}\, dt \tag{5-108}$$

Refer to Table 5-9 for erf(x) for various values of x.

The resultant mean of adding the means of two populations (Fig. 5-12)

$$\mu = \mu_s + \mu_\sigma \tag{5-109}$$

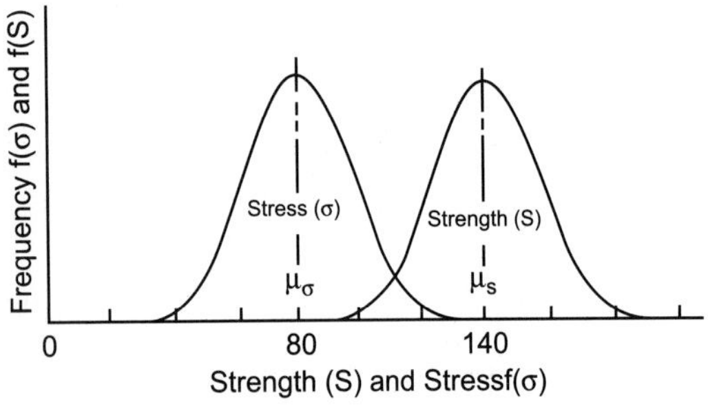

FIGURE 5-12 Distribution curves for two means of populations.

The resultant mean of subtracting the means of two populations

$$\mu = \mu_s - \mu_\sigma \tag{5-110}$$

The resultant standard deviation for both subtraction and addition of two standard deviations $\hat{\sigma}_s$ and $\hat{\sigma}_\sigma$

$$\hat{\sigma} = \sqrt{\hat{\sigma}_s^2 + \hat{\sigma}_\sigma^2} \tag{5-111}$$

TABLE 5-7
Standard normal curve ordinates

$$y = \frac{1}{\sqrt{2\pi}} e^{-t^2/2}$$

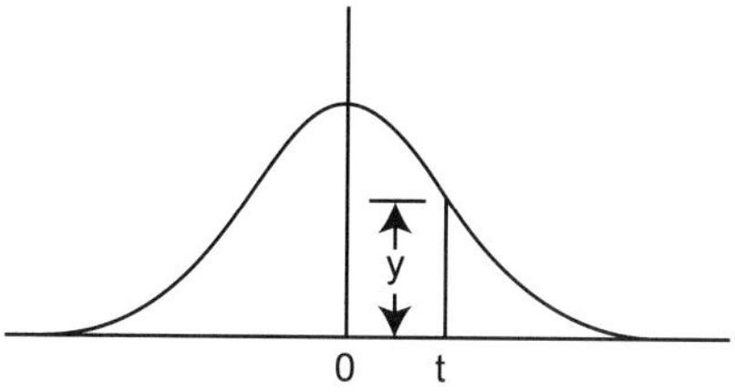

t	0	1	2	3	4	5	6	7	8	9
0.0	.3989	.3989	.3989	.3988	.3986	.3984	.3982	.3980	.3977	.3973
0.1	.3970	.3965	.3961	.3956	.3951	.3945	.3939	.3932	.3925	.3918
0.2	.3910	.3902	.2894	.3885	.3876	.3867	.3857	.3847	.3836	.3815
0.3	.3814	.3802	.3790	.3778	.3765	.3752	.3739	.3725	.3712	.3697
0.4	.3683	.3668	.3653	.3637	.3621	.3605	.3589	.3572	.3555	.3538
0.5	.3521	.3503	.3485	.3467	.3448	.3429	.3410	.3391	.3372	.3352
0.6	.3332	.3312	.3292	.3271	.3251	.3230	.3209	.3187	.3166	.3144
0.7	.3123	.3101	.3079	.3056	.3034	.3011	.2989	.2966	.2943	.2920
0.8	.2897	.2874	.2850	.2827	.2803	.2780	.2756	.2932	.2709	.2685
0.9	.2661	.2637	.2613	.2589	.2565	.2541	.2516	.2492	.2468	.2444
1.0	.2420	.2396	.2371	.2347	.2323	.2299	.2275	.2251	.2227	.2203
1.1	.2179	.2155	.2131	.2107	.2083	.2059	.2036	.2012	.1989	.1965
1.2	.1942	.1919	.1895	.1872	.1849	.1826	.1804	.1781	.1758	.1736
1.3	.1714	.1691	.1669	.1647	.1626	.1604	.1528	.1561	.1539	.1518
1.4	.1497	.1476	.1456	.1435	.1415	.1394	.1374	.1354	.1334	.1315
1.5	.1295	.1276	.1257	.1238	.1219	.1200	.1182	.1163	.1145	.1127
1.6	.1109	.1092	.1074	.1057	.1040	.1023	.1006	.0989	.0973	.0957
1.7	.0940	.0925	.0909	.0893	.0878	.0863	.0848	.0833	.0818	.0804
1.8	.0790	.0775	.0761	.0748	.0734	.0721	.0707	.0694	.0681	.0669
1.9	.0656	.0644	.0632	.0620	.0608	.0596	.0584	.0573	.0562	.0551
2.0	.0540	.0529	.0519	.0508	.0498	.0488	.0487	.0468	.0459	.0449
2.1	.0440	.0431	.0422	.0413	.0404	.0396	.0387	.0379	.0371	.0363
2.2	.0355	.0347	.0339	.0332	.0325	.0317	.0310	.0303	.0297	.0290
2.3	.0283	.0277	.0270	.0264	.0258	.0252	.0246	.0241	.0235	.0229
2.4	.0224	.0219	.0213	.0208	.0203	.0198	.0194	.0189	.0184	.0180
2.5	.0175	.0171	.0167	.0163	.0158	.0154	.0151	.0147	.0143	.0139
2.6	.0136	.0132	.0129	.0126	.0122	.0119	.0116	.0113	.0110	.0107
2.7	.0104	.0101	.0099	.0096	.0093	.0091	.0088	.0086	.0084	.0081
2.8	.0079	.0077	.0075	.0073	.0071	.0069	.0067	.0065	.0063	.0061
2.9	.0060	.0058	.0056	.0055	.0053	.0051	.0050	.0048	.0047	.0046
3.0	.0044	.0043	.0042	.0040	.0039	.0038	.0037	.0036	.0035	.0034
3.1	.0033	.0032	.0031	.0030	.0029	.0028	.0027	.0026	.0025	.0025
3.2	.0024	.0023	.0022	.0022	.0021	.0020	.0020	.0019	.0018	.0018
3.3	.0017	.0017	.0016	.0016	.0015	.0015	.0014	.0014	.0013	.0013
3.4	.0012	.0012	.0012	.0011	.0011	.0010	.0010	.0010	.0009	.0009
3.5	.0009	.0008	.0008	.0008	.0008	.0007	.0007	.0007	.0007	.0006
3.6	.0006	.0006	.0006	.0005	.0005	.0005	.0005	.0005	.0005	.0004
3.7	.0004	.0004	.0004	.0004	.0004	.0004	.0003	.0003	.0003	.0003
3.8	.0003	.0003	.0003	.0003	.0003	.0002	.0002	.0002	.0002	.0002
3.9	.0002	.0002	.0002	.0002	.0002	.0002	.0002	.0002	.0001	.0001

TABLE 5-8
Areas under the standard normal distribution curve

$$A(t) = \int_0^t \frac{1}{\sqrt{2\pi}} e^{-t^2/2}\, dt$$

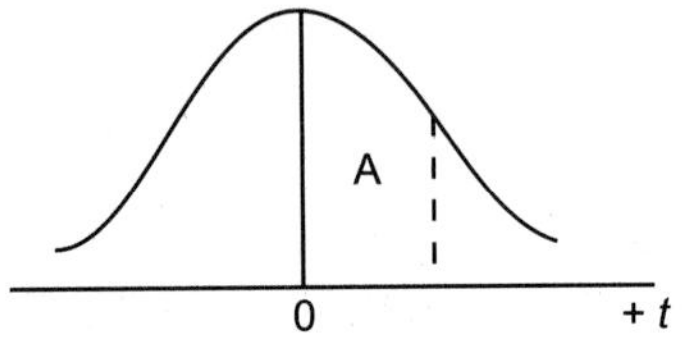

t	0	1	2	3	4	5	6	7	8	9
0.0	.0000	.0040	.0080	.0120	.0160	.0199	.0239	.0279	.0319	.0359
0.1	.0398	.0438	.0478	.0517	.0557	.0596	.0636	.0675	.0714	.0754
0.2	.0793	.0832	.0871	.0910	.0948	.0987	.1026	.1064	.1103	.1141
0.3	.1179	.1217	.1255	.1293	.1331	.1368	.1406	.1443	.1480	.1517
0.4	.1554	.1591	.1628	.1664	.1700	.1736	.1772	.1808	.1844	.1879
0.5	.1915	.1950	.1985	.2019	.2054	.2088	.2123	.2157	.2190	.2224
0.6	.2258	.2291	.2324	.2357	.2389	.2422	.2454	.2486	.2518	.2549
0.7	.2580	.2612	.2642	.2673	.2704	.2734	.2764	.2794	.2823	.2852
0.8	.2881	.2910	.2939	.2967	.2996	.3023	.3051	.3078	.3106	.3133
0.9	.3159	.3186	.3212	.3238	.3264	.3289	.3315	.3340	.3365	.3389
1.0	.3413	.3438	.3461	.3485	.3508	.3531	.3554	.3577	.3599	.3621
1.1	.3643	.3665	.3686	.3708	.3729	.3749	.3770	.3790	.3810	.3830
1.2	.3849	.3869	.3888	.3907	.3925	.3944	.3962	.3980	.3997	.4015
1.3	.4032	.4049	.4066	.4082	.4099	.4115	.4131	.4147	.4162	.4177
1.4	.4192	.4207	.4222	.4236	.4251	.4265	.4279	.4292	.4306	.4319
1.5	.4332	.4345	.4357	.4370	.4382	.4394	.4406	.4418	.4429	.4441
1.6	.4452	.4463	.4474	.4484	.4495	.4506	.4515	.4525	.4535	.4545
1.7	.4554	.4564	.4573	.4582	.4591	.4599	.4608	.4616	.4625	.4633
1.8	.4641	.4649	.4656	.4664	.4671	.4678	.4686	.4693	.4699	.4706
1.9	.4713	.4719	.4726	.4732	.4738	.4744	.4750	.4756	.4761	.4767
2.0	.4772	.4778	.4783	.4788	.4793	.4798	.4803	.4808	.4812	.4817
2.1	.4821	.4826	.4830	.4834	.4838	.4842	.4846	.4850	.4854	.4857
2.2	.4861	.4864	.4868	.4871	.4875	.4878	.4881	.4884	.4887	.4890
2.3	.4893	.4896	.4898	.4901	.4904	.4906	.4909	.4911	.4913	.4916
2.4	.4918	.4920	.4922	.4925	.4927	.4929	.4931	.4932	.4934	.4936
2.5	.4938	.4940	.4941	.4943	.4945	.4946	.4948	.4949	.4951	.4952
2.6	.4953	.4955	.4956	.4957	.4959	.4960	.4961	.4962	.4963	.4964
2.7	.4965	.4966	.4967	.4968	.4969	.4970	.4971	.4972	.4973	.4974
2.8	.4974	.4975	.4976	.4977	.4977	.4978	.4979	.4979	.4980	.4981
2.9	.4981	.4982	.4982	.4983	.4984	.4984	.4985	.4985	.4986	.4986
3.0	.4987	.4987	.4987	.4988	.4988	.4989	.4989	.4989	.4990	.4990
3.1	.4990	.4991	.4991	.4991	.4992	.4992	.4992	.4992	.4993	.4993
3.2	.4993	.4993	.4994	.4994	.4994	.4994	.4994	.4995	.4995	.4995
3.3	.4995	.4995	.4995	.4996	.4996	.4996	.4996	.4996	.4996	.4997
3.4	.4997	.4997	.4997	.4997	.4997	.4997	.4997	.4997	.4997	.4998
3.5	.4998	.4998	.4998	.4998	.4998	.4998	.4998	.4998	.4998	.4998
3.6	.4998	.4998	.4999	.4999	.4999	.4999	.4999	.4999	.4999	.4999
3.7	.4999	.4999	.4999	.4999	.4999	.4999	.4999	.4999	.4999	.4999
3.8	.4999	.4999	.4999	.4999	.4999	.4999	.4999	.4999	.4999	.4999
3.9	.5000	.5000	.5000	.5000	.5000	.5000	.5000	.5000	.5000	.5000

TABLE 5-9
Error function or probability integral

$$\mathrm{erf}(x) = \frac{2}{\sqrt{\pi}} \int_0^x e^{-t^2}\, dt$$

x	0	1	2	3	4	5	6	7	8	9
0.0		.01128	.02256	.03384	.04511	.05637	.06762	.07886	.09008	.10128
0.1	.11246	.12362	.13476	.14587	.15695	.16800	.17901	.18999	.20094	.21184
0.2	.22270	.23352	.24430	.25502	.26570	.27633	.28690	.29742	.30788	.31828
0.3	.32863	.33891	.34913	.35928	.36936	.37938	.38933	.39921	.40901	.41874
0.4	.42839	.43797	.44747	.45689	.46623	.47548	.48466	.49375	.50275	.51167
0.5	.52050	.52924	.53790	.54646	.55494	.56332	.57162	.57982	.58792	.59594
0.6	.60386	.61168	.61941	.62705	.63459	.64203	.64938	.65663	.66378	.67084
0.7	.67780	.68467	.69143	.69810	.70468	.71116	.71754	.72382	.73001	.73610
0.8	.74210	.74800	.75381	.75952	.76514	.77067	.77610	.78144	.78669	.79184
0.9	.79691	.80188	.80677	.81156	.81627	.82089	.82542	.82987	.83243	.83851
1.0	.84270	.84681	.85084	.85478	.85865	.86244	.86614	.86977	.87333	.87680
1.1	.88021	.88353	.88679	.88997	.89308	.89612	.89910	.90200	.90484	.90761
1.2	.91031	.91296	.91553	.91805	.92051	.92290	.92524	.92751	.92973	.93190
1.3	.93401	.93606	.93807	.94002	.94191	.94376	.94556	.94731	.94902	.95067
1.4	.95229	.95385	.95538	.95686	.95830	.95970	.96105	.96237	.96365	.96490
1.5	.96611	.96728	.96841	.96952	.97059	.97162	.97263	.97360	.97455	.97546
1.6	.97635	.97721	.97804	.97884	.97962	.98038	.98110	.98181	.98249	.98315
1.7	.98379	.98441	.98500	.98558	.98613	.98667	.98719	.98769	.98817	.98864
1.8	.98909	.98952	.98994	.99035	.99074	.99111	.99147	.99182	.99216	.99248
1.9	.99279	.99309	.99338	.99366	.99392	.99418	.99443	.99466	.99489	.99511
2.0	.99532	.99552	.99572	.99591	.99609	.99626	.99642	.99658	.99673	.99688
2.1	.99702	.99715	.99728	.99741	.99753	.99764	.99775	.99785	.99795	.99805
2.2	.99814	.99822	.99831	.99839	.99846	.99854	.99861	.99867	.99874	.99880
2.3	.99886	.99891	.99897	.99902	.99906	.99911	.99915	.99920	.99924	.99928
2.4	.99931	.99935	.99938	.99941	.99944	.99947	.99950	.99952	.99955	.99957
2.5	.99959	.99961	.99963	.99965	.99967	.99969	.99971	.99972	.99974	.99975
2.6	.99976	.99978	.99979	.99980	.99981	.99982	.99983	.99984	.99985	.99986
2.7	.99987	.99987	.99988	.99989	.99989	.99990	.99991	.99991	.99992	.99992
2.8	.99992	.99993	.99993	.99994	.99994	.99994	.99995	.99995	.99995	.99996
2.9	.99996	.99996	.99996	.99997	.99997	.99997	.99997	.99997	.99997	.99998
3.0	.99998									

The standard variable t_R (deviation multiplication factor) in order to determine the probability of failure or the reliability

$$t_R = \frac{\mu_s - \mu_\sigma}{\sqrt{\hat{\sigma}_s^2 + \hat{\sigma}_\sigma^2}} = \frac{\mu_s - \mu_\sigma}{\hat{\sigma}} \tag{5-112}$$

where subscripts s and σ refer to strength and stress, respectively.

The reliability associated with t_R

$$R = 0.5 + A(t_R) \tag{5-113}$$

where $A(t_R)$ is the area under a standard normal distribution curve.

Refer to Table 5-10 for typical values of R as a function of standardized variable t_R.

Particular	Formula

TABLE 5-10
Reliability R as a function of t_R

Survival rate (R) %	t_R
50	0
90.00	1.288
95.00	1.645
98.00	2.050
99.00	2.330
99.90	3.080
99.99	3.700

A safety factor of 1 is taken into account in determining the reliability from Eq. (5-113).

The fatigue strength reduction factor based on reliability

$$C_R = 1 - 0.08(t_R) \quad (5\text{-}114)$$

where t_R is also called the *deviation multiplication factor* (DMF), taken from Table 5-10.

If a factor of safety n' is to be specified together with reliability, then Eq. (5-112) is rewritten to give a new expression for t_R

$$t_R = \frac{\mu_s - n'\mu_\sigma}{\sqrt{\hat{\sigma}_s^2 + \hat{\sigma}_\sigma^2}} = \frac{\mu_s - n'\mu_\sigma}{\hat{\sigma}} \quad (5\text{-}115)$$

The expression for safety factor n' from Eq. (5-115)

$$n' = \frac{1}{\mu_\sigma}\left[\mu_s - t_R\sqrt{\hat{\sigma}_s^2 + \hat{\sigma}_\sigma^2}\right] \quad (5\text{-}116a)$$

$$= \frac{1}{\mu_\sigma}(\mu_s - t_R\hat{\sigma}) \quad (5\text{-}116b)$$

The best-fitting straight line which fits a set of scattered data points as per linear regression

$$y = mx + b \quad (5\text{-}117)$$

where m is the slope and b is the intercept on the y axis

The equations for regression

$$m = \frac{\sum xy - \frac{\sum x \sum y}{n}}{\sum x^2 - \frac{(\sum x)^2}{n}} \quad (5\text{-}118a)$$

$$b = \frac{\sum y - m\sum x}{n} \quad (5\text{-}118b)$$

The correlation coefficient

$$r = \frac{ms_x}{s_y} \quad (5\text{-}119)$$

where r lies between -1 and $+1$.

If r is negative, it indicates that the regression line has a negative slope.

If $r = 1$, there is a perfect correlation, and if $r = 0$, there is no correlation.

The equation for frequency or density function according to Weibull

$$f(x) = \frac{b}{\theta - x_0}\left(\frac{x - x_0}{\theta - x_0}\right)^{b-1}\left\{\exp\left[-\left(\frac{x - x_0}{\theta - x_0}\right)^{b}\right]\right\} \quad (5\text{-}120)$$

The cumulative distribution function

$$F(x) = \int_{x_0}^{x} f(x)\,dx = 1 - \exp\left[-\left(\frac{x - x_0}{\theta - x_0}\right)^{b}\right] \quad (5\text{-}121)$$

Equation (5-121) after simplification

$$F(x) = 1 - \exp\left[-\left(\frac{x}{\theta}\right)^{b}\right] \quad (5\text{-}122)$$

REFERENCES

1. Maleev, V. L., and J. B. Hartman, *Machine Design*, International Textbook Company, Scranton, Pennsylvania, 1954.
2. Shigley, J. E., and L. D. Mitchell, *Mechanical Engineering Design*, McGraw-Hill Book Company, New York, 1983.
3. Faires, V. M., *Design of Machine Elements*, The Macmillan Company, New York, 1965.
4. Lingaiah, K., and B. R. Narayana Iyengar, *Machine Design Data Handbook*, Engineering Co-operative Society, Bangalore, India, Bangalore, India, 1962.
5. Lingaiah, K., and B. R. Narayana Iyengar, *Machine Design Data Handbook*, Vol. I (*SI and Customary Units*), Suma Publishers, Bangalore, India, 1986.
6. Lingaiah, K., *Machine Design Data Handbook*, Vol. II (*SI and Customary Metric Units*), Suma Publishers, Bangalore, India, 1986.
7. Juvinall, R. C., *Fundamentals of Machine Component Design*, John Wiley and Sons, New York, 1983.
8. Deutschman, A. D., W. J. Michels, and C. E. Wilson, *Machine Design—Theory and Practice*, Macmillan Publishing Company, New York, 1975.
9. Edwards, Jr., K. S., and R. B. McKee, *Fundamentals of Mechanical Component Design*, McGraw-Hill Publishing Company, New York, 1991.
10. Norton, R. L., *Machine Design—An Integrated Approach*, Prentice Hall International, Inc., Upper Saddle River, New Jersey, 1996.
11. Lingaiah, K. *Machine Design Data Handbook*, McGraw-Hill Publishing Company, New York, 1994.
12. *Metals Handbook*, American Society for Metals, Vol. 10, 8th edition, p. 102, Metals Park, Ohio, 1975.

CHAPTER 6

CAMS

SYMBOLS[3,4]

a	radius of circular area of contact, m (in)
A	acceleration of the follower, m/s^2 (in/s^2)
	follower overhang, m (in)
A_c	arc of pitch circle, m (in)
b	half the band of width of contact, m (in)
B	follower bearing length, m (in)
$a_o = \rho_o + \rho_i$	distance between centers of rotation, m (in)
d	diameter of shaft, m (in)
d_h	hub diameter, m (in)
D_o	minimum diameter of the pitch surface of cam, m (in)
E_1, E_2	moduli of elasticity of the materials which are in contact, GPa (Mpsi)
f	cam factor, dimensionless
$f(\theta)$	the desired motion of follower, as a function of cam angle
F	applied load, kN (lbf)
F	total external load on follower (includes weight, spring force, inertia, friction, etc.), kN (lbf)
F_n	force normal to cam profile (Fig. 6-6), kN (lbf)
F_t	side thrust, kN (lbf)
h	depth to the point of maximum shear, m (in)
K_i, K_o	constants for input and output cams, respectively
L	length of cylinder in contact, m (in)
	total distance through which the follower is to rise, m (in)
n	cam speed, rpm
N_1, N_2	forces normal to follower stem, kN (lbf)
r	radius of follower, m (in)
R_c	radius of the circular arc, m (in)
R_o	minimum radius of the pitch surface of the cam, m (in)
R_p	pitch circle radius, m (in)
R_r	radius of the roller, m (in)
R, S	functions of θ_i and θ_o, in basic spiral contour cams
S	displacement of the follower corresponding to any cam angle θ, m (in)
S_1	initial compression spring force with weight w, at zero position, kN (lbf)
v	velocity of the follower, m/s (in/s)
w	equivalent weight at follower ends, kN (lbf)

x, y	cartesian coordinates of any point on the cam surface
y	actual lift at follower end, m (in)
y_c	rise of cam, m (in)
ρ	radius of curvature of the pitch curve, m (in)
ρ_1, ρ_2	radii of curvature of the contact surfaces, m (in)
α	pressure angle, deg
α_m	maximum pressure angle, deg
β	angle through which cam is to rotate to effect the rise L, rad
θ	cam angle corresponding to the follower displacement S, rad
θ_o	angle rotated by the output-driven member, deg
θ_i	angle rotated by the input driver, deg
ω	angular velocity of cam, rad/s
μ	coefficient of friction between follower stem and its guide bearing
ν_1, ν_2	Poisson's ratios for the materials of contact surfaces
$\sigma_{c,\max}$	maximum compressive stress, MPa (kpsi)
τ	shear stress, MPa (kpsi)

Particular	Formula	
Cam factor	$f = \dfrac{A_c}{L}$	(6-1)
The length of arc of the pitch circle	$A_c = R_p\beta$	(6-2)
The pitch circle radius	$R_p = \dfrac{fL}{\beta}$	(6-3)

RADIUS OF CURVATURE OF DISK CAM WITH ROLLER FOLLOWER

Particular	Formula	
The displacement of the center of the follower from the center of cam (Fig. 6-1)	$R = R_o + f(\theta)$	(6-4)
For pointed cam, the radius of curvature of the pitch curve to roller follower	$\rho = R_r$	(6-5)
For roller follower, the radius of curvature of the pitch curve must always be greater than the roller radius to prevent points or undercuts	$\rho > R_r$	(6-6)
The radius of curvature for concave pitch curve	$\rho = -\dfrac{\{R^2 + [f'(\theta)]^2\}^{3/2}}{R^2 + 2[f'(\theta)]^2 - R[f''(\theta)]}$	(6-7)
	where	
	$R = R_o + f(\theta); \quad \dfrac{dR}{d\theta} = f'(\theta); \quad \dfrac{d^2R}{d\theta^2} = f''(\theta)$	(6-7a)
The minimum radius of curvature	$\rho_{\min} = \dfrac{R_o^2}{R_o - f''(\theta)_o}$	(6-8)
	where $f''(\theta)_o$ is the acceleration at $\theta = 0$	

Particular	Formula	
The minimum radius of curvature of the cam curve ρ_c	$\rho_{c,\min} = \rho_{\min} \pm R_r$	(6-9)

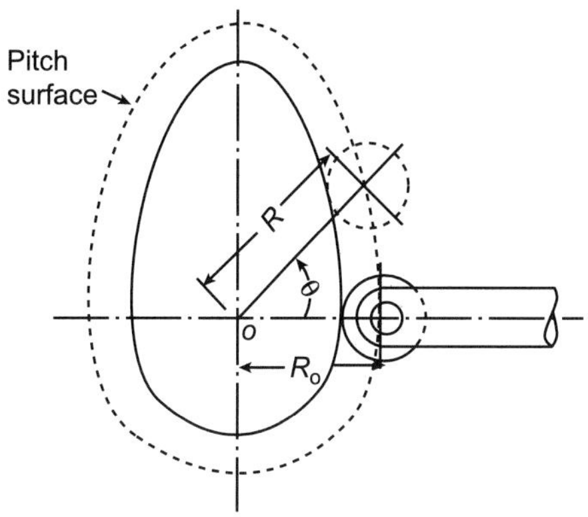

FIGURE 6-1

The radius of curvature for convex pitch curve (Fig. 6-2)	$\rho = \dfrac{\{R^2 + [f'(\theta)]^2\}^{3/2}}{R^2 + 2[f'(\theta)]^2 - R[f''(\theta)]}$	(6-10)
The minimum radius of a mushroom cam for harmonic motion	$R_o = \left(\dfrac{16200}{\beta^2} - 1\right)L$	(6-11)
The minimum radius of a mushroom cam for uniformly accelerated and retarded motion	$R_o = \left(\dfrac{13131}{\beta^2} - \dfrac{1}{2}\right)L$	(6-12)
For cast-iron cam, the hub diameter	$d_h = 1.75d + 13.75\,\text{mm}\ (1.75d + 0.55\,\text{in})$	(6-13)

Plate cam design – radius of curvature:

For cycloidal motion	Refer to Fig. 6-10.
For harmonic motion	Refer to Fig. 6-11.
For eight-power polynomial motion	Refer to Fig. 6-12.

RADIUS OF CURVATURE OF DISK CAM WITH FLAT-FACED FOLLOWER

The displacement of the follower from the origin (Fig. 6-2)	$R = a + f(\theta)$	(6-14)
The parametric equations of the cam contour (Fig. 6-2)	$x = [a + f(\theta)]\cos\theta - f'(\theta)\sin\theta$	(6-15a)
	$y = [a + f(\theta)]\sin\theta + f'(\theta)\cos\theta$	(6-15b)

Particular	Formula	
The cam contour given by equations will be free of cusps if	$a + f(\theta) + f''(\theta) > 0$	(6-16)
Half of the minimum length of the flat-faced follower or the minimum length of contact of the follower	$b = f'(\theta)$	(6-17)

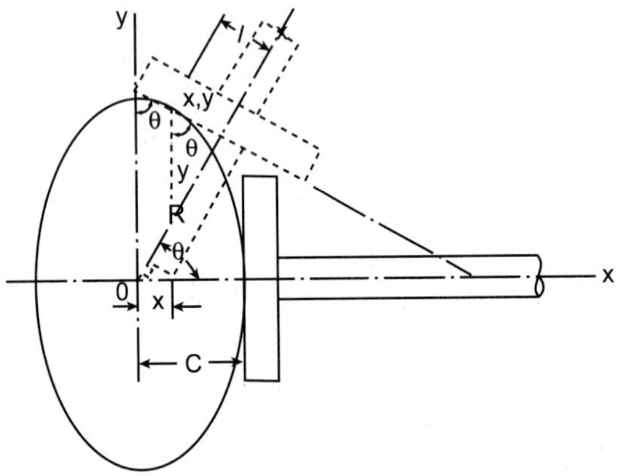

FIGURE 6-2 (Courtesy of H. H. Mabie and F. W. Ocvivk, *Dynamics of Machinery*, John Wiley and Sons, 1957.)

PRESSURE ANGLE (Figs. 6-3 and 6-4)

The pressure angle for roller follower	$\alpha = \tan^{-1} \dfrac{1}{R} \dfrac{dR}{d\theta}$	(6-18)
The pressure angle for a plate cam or any cylindrical cam giving uniform velocity to the follower	$\tan\alpha = \dfrac{360L}{2\pi\beta R_o} = \dfrac{360L}{\pi\beta D_o}$	(6-19)
The pressure angle for a plate cam giving uniformly accelerated and retarded motion to the follower	$\tan\alpha = \dfrac{360 \times 2L}{\pi\beta(D_o + L)}$ when $L > D_o$	(6-20a)
	$= \dfrac{180 \times 2}{\pi\beta}\sqrt{\dfrac{L}{D_o}}$ when $L > D_o$	(6-20b)
A precise pressure angle equation for a plate cam giving harmonic motion to the follower or a tangential cam	$\tan\alpha = \dfrac{90L}{\beta\sqrt{R_o^2 + R_o L}}$	(6-21)
For measuring maximum pressure angle of a parabolic cam with radially moving roller follower	Refer to Fig. 6-3 for nomogram of parabolic cam with radially moving follower	

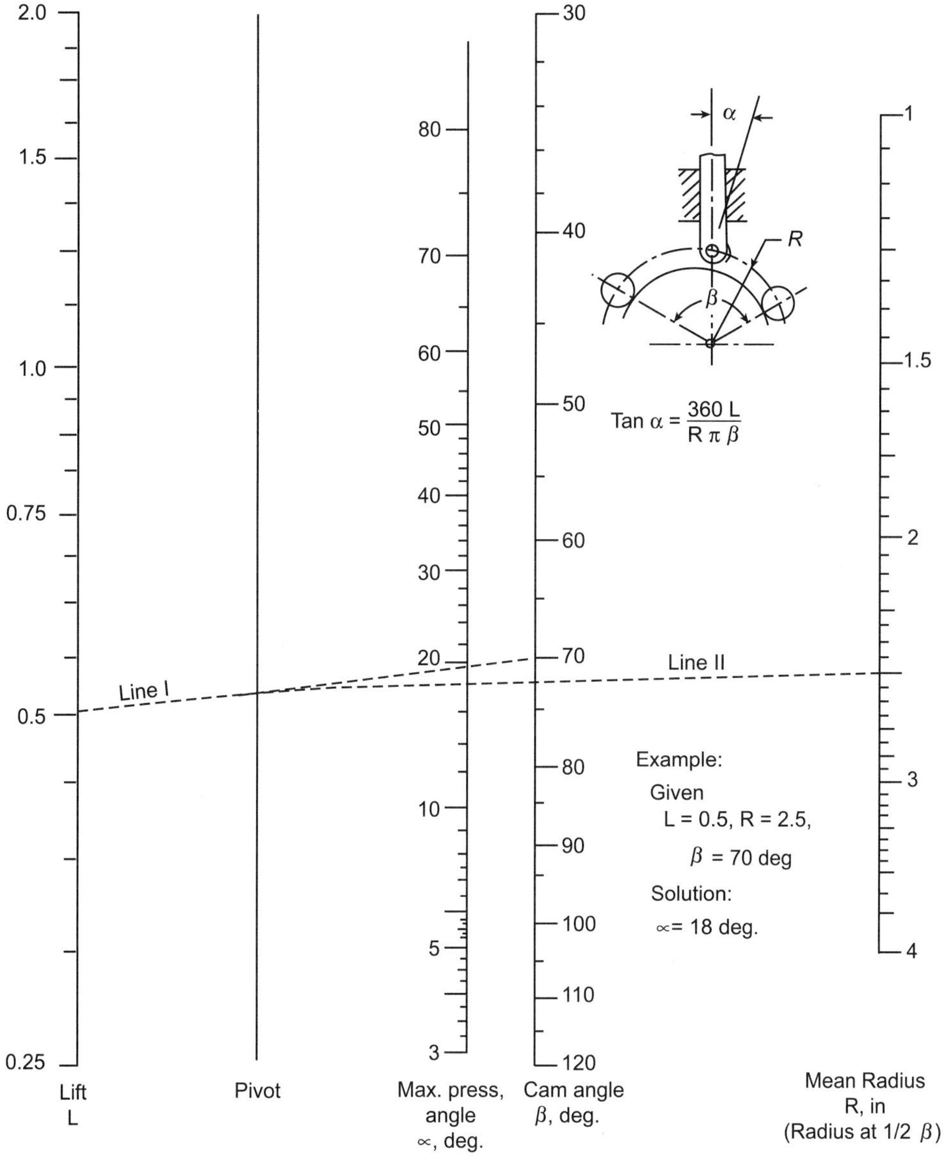

FIGURE 6-3 Nomogram for parabolic cam with radially moving follower.
Source: Rudolph Gruenberg, "Nomogram for Parabolic Cam with Radially Moving Follower," in Douglas C. Greenwood, Editor, *Engineering Data for Product Design*, McGraw-Hill Book Company, New York, 1961.

Particular	Formula

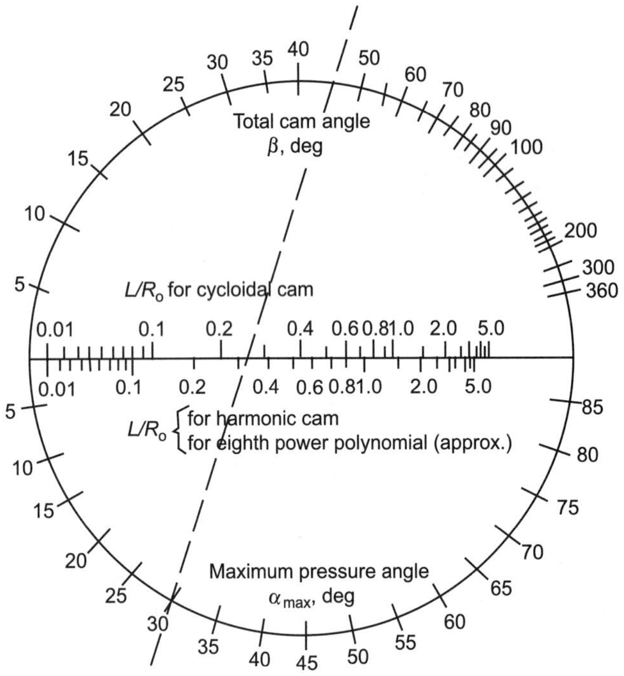

FIGURE 6-4 Nomogram to determine maximum pressure angle. (Courtesy of E. C. Varnum, Barber-Coleman Co.) Reproduced with permission from *Machine Design*, Cleveland, Ohio.

RADIAL CAM-TRANSLATING ROLLER-FOLLOWER-FORCE ANALYSIS (Fig. 6-5)

The forces normal to follower stem (Fig. 6-5)

$$F_R = \frac{l_r}{l_g} F_n \sin\alpha \quad (6\text{-}22)$$

$$F_L = \frac{l_r + l_g}{l_g} F_n \sin\alpha \quad (6\text{-}23)$$

The total external load

$$F = F_n\left[\cos\alpha - \mu\left(\frac{2l_r + l_g}{l_g}\right)\sin\alpha\right] \quad (6\text{-}24)$$

The force normal to the cam profile

$$F_n = \frac{F}{\cos\alpha - \mu\left(\dfrac{2l_r + l_g}{l_g}\right)\sin\alpha} \quad (6\text{-}25)$$

The maximum pressure angle for locking the follower in its guide

$$\alpha_m = \tan^{-1}\frac{l_g}{\mu(2l_r + l_g)} \quad (6\text{-}26)$$

Particular	Formula

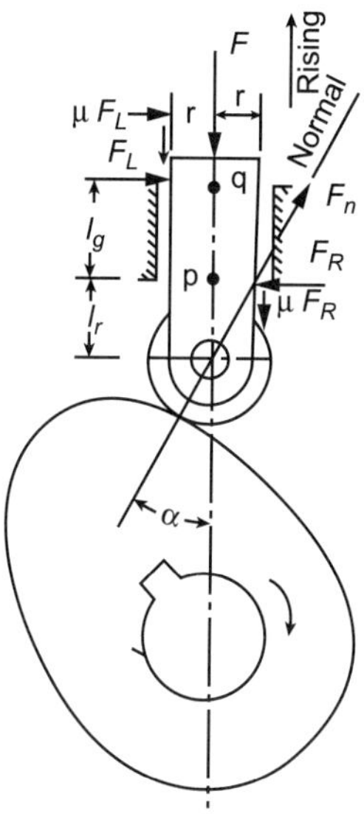

FIGURE 6-5 Radial cam-translating roller-follower force analysis.

SIDE THRUST (Fig. 6-5)

The side thrust produced on the follower bearing $\quad F_i = F \tan \alpha \quad$ (6-27)

BASIC SPIRAL CONTOUR CAM

The radius to point of contact at angle θ_o (Fig. 6-6)

$$\rho_o = \frac{a_o}{1 + \dfrac{d\theta_o}{d\theta_i}} \tag{6-28}$$

The radius to point of contact at angle θ_i (Fig. 6-6)

$$\rho_i = \frac{a_o \dfrac{d\theta_o}{d\theta_i}}{1 + \dfrac{d\theta_o}{d\theta_i}} \tag{6-29}$$

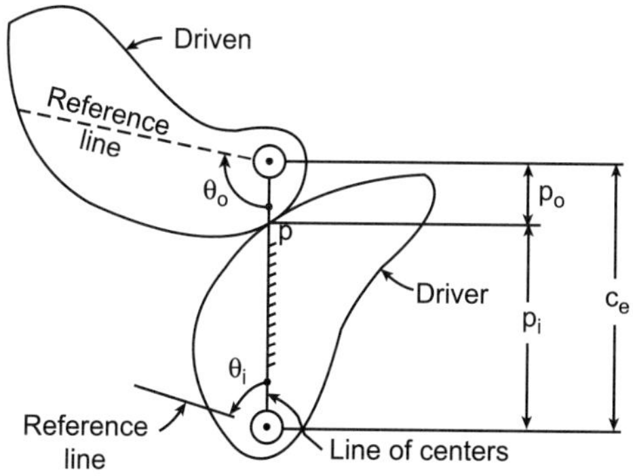

FIGURE 6-6 Basic spiral contour cam.

Particular	Formula

BASIC SPIRAL CONTOUR CAM CONSTANTS

The radius to point of contact at angle θ_o

$$\rho_o = \frac{a_o}{1 + \dfrac{K_o}{K_i}\left(\dfrac{dS}{dR}\right)} \tag{6-30}$$

The radius to point of contact at angle θ_i

$$\rho_i = \frac{a_o \dfrac{K_o}{K_i}\left(\dfrac{dS}{dR}\right)}{1 + \dfrac{K_o}{K_i}\left(\dfrac{dS}{dR}\right)} \tag{6-31}$$

where $R = \dfrac{\theta_i}{K_i}$, $S = \dfrac{\theta_o}{K_o}$, $\dfrac{d\theta_i}{dR} = k_i$, and $\dfrac{d\theta_o}{dS} = k_o$.

For characteristic curves of cycloidal, harmonic, and eight-power polynomial motions

Refer to Figs. 6-7 to 6-12

HERTZ CONTACT STRESSES

Contact of sphere on sphere

The radius of circular area of contact

$$a = \sqrt[3]{\frac{3F\left[\left(\dfrac{1-\nu_1^2}{E_1}\right) + \left(\dfrac{1-\nu_2^2}{E_2}\right)\right]}{4\left(\dfrac{1}{\rho_1} + \dfrac{1}{\rho_2}\right)}} \tag{6-32}$$

The maximum compressive stress

$$\sigma_{c,\max} = \frac{3F}{2\pi a^2} \tag{6-33}$$

Contact of cylindrical surface on cylindrical surface

Width of band of contact

$$2b = \sqrt{\frac{16F\left[\left(\dfrac{1-\nu_1^2}{E_1}\right) + \left(\dfrac{1-\nu_1^2}{E_2}\right)\right]}{\pi L\left(\dfrac{1}{\rho_1} + \dfrac{1}{\rho_2}\right)}} \tag{6-34}$$

The maximum compressive stress

$$\sigma_{c,\max} = \frac{2F}{\pi b L} \tag{6-35}$$

The maximum compressive stress for $\nu_1 = \nu_2 = 0.3$

$$\sigma_{c,\max} = \sqrt{\frac{0.35F\left(\dfrac{1}{\rho_1} + \dfrac{1}{\rho_2}\right)}{L\left(\dfrac{1}{E_1} + \dfrac{1}{E_2}\right)}} \tag{6-36}$$

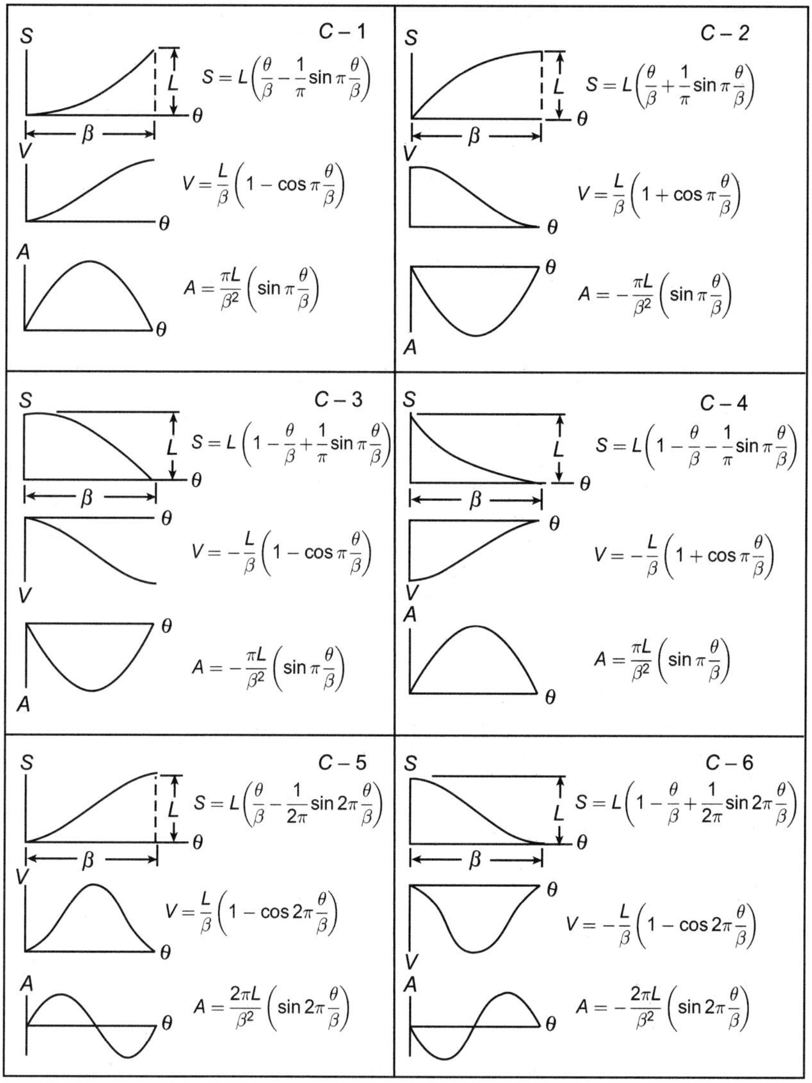

FIGURE 6-7 Cycloidal motion characteristics. S = displacement, inches; V = velocity, inches per degree; A = acceleration, inches per degree^2. (From "Plate Cam Design—with Emphasis on Dynamic Effects," by M. Kloomok and R. V. Muffley, *Product Eng.*, February 1955.) Reproduced with permission from *Machine Design*, Cleveland, Ohio.

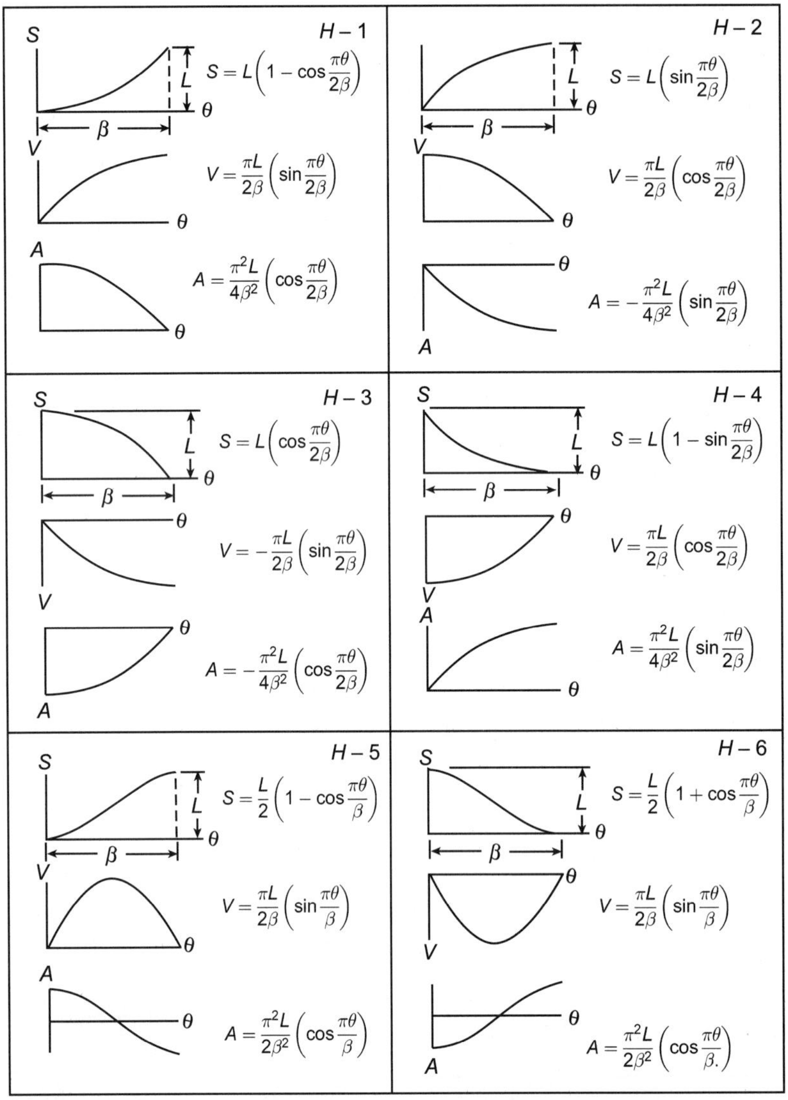

FIGURE 6-8 Harmonic motion characteristics. S = displacement, inches; V = velocity, inches per degree; A = acceleration, inches per degree^2. (From "Plate Cam Design—with Emphasis on Dynamic Effects," by M. Kloomok and R. V. Muffley, *Product Eng.*, February 1955.) Reproduced with permission from *Machine Design*, Cleveland, Ohio.

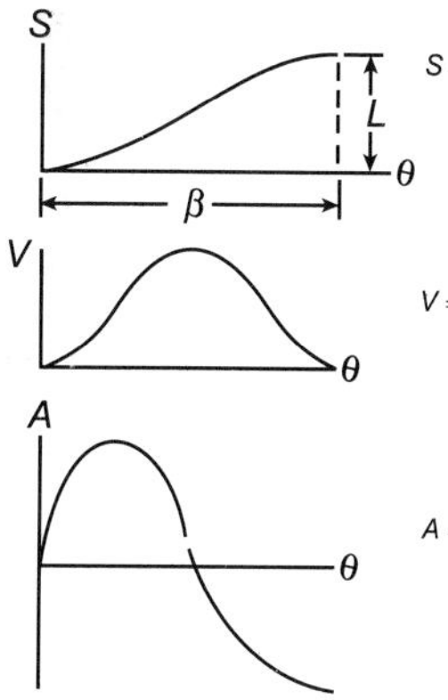

$$S = L\left[6.09755\left(\frac{\theta}{\beta}\right)^3 - 20.78040\left(\frac{\theta}{\beta}\right)^5 + 26.73155\left(\frac{\theta}{\beta}\right)^6 - 13.60965\left(\frac{\theta}{\beta}\right)^7 + 2.56095\left(\frac{\theta}{\beta}\right)^8\right]$$

$$V = \frac{L}{\beta}\left[18.29265\left(\frac{\theta}{\beta}\right)^2 - 103.90200\left(\frac{\theta}{\beta}\right)^4 + 160.38930\left(\frac{\theta}{\beta}\right)^5 - 95.26755\left(\frac{\theta}{\beta}\right)^6 + 20.48760\left(\frac{\theta}{\beta}\right)^7\right]$$

$$A = \frac{L}{\beta^2}\left[36.58530\left(\frac{\theta}{\beta}\right) - 415.60800\left(\frac{\theta}{\beta}\right)^3 + 801.94650\left(\frac{\theta}{\beta}\right)^4 - 571.60530\left(\frac{\theta}{\beta}\right)^5 + 143.41320\left(\frac{\theta}{\beta}\right)^6\right]$$

P – 1

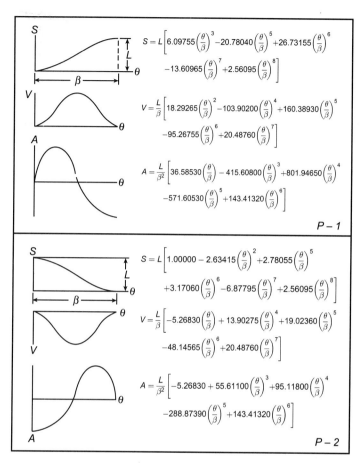

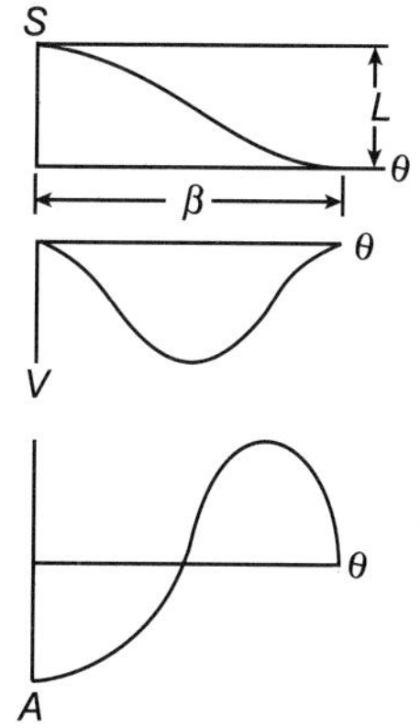

$$S = L\left[1.00000 - 2.63415\left(\frac{\theta}{\beta}\right)^2 + 2.78055\left(\frac{\theta}{\beta}\right)^5 + 3.17060\left(\frac{\theta}{\beta}\right)^6 - 6.87795\left(\frac{\theta}{\beta}\right)^7 + 2.56095\left(\frac{\theta}{\beta}\right)^8\right]$$

$$V = \frac{L}{\beta}\left[-5.26830\left(\frac{\theta}{\beta}\right) + 13.90275\left(\frac{\theta}{\beta}\right)^4 + 19.02360\left(\frac{\theta}{\beta}\right)^5 - 48.14565\left(\frac{\theta}{\beta}\right)^6 + 20.48760\left(\frac{\theta}{\beta}\right)^7\right]$$

$$A = \frac{L}{\beta^2}\left[-5.26830 + 55.61100\left(\frac{\theta}{\beta}\right)^3 + 95.11800\left(\frac{\theta}{\beta}\right)^4 - 288.87390\left(\frac{\theta}{\beta}\right)^5 + 143.41320\left(\frac{\theta}{\beta}\right)^6\right]$$

P – 2

FIGURE 6-9 Eighth-power polynomial motion characteristics. S = displacement, inches; V = velocity, inches per degree; A = acceleration, inches per degree^2. (From "Plate Cam Design—with Emphasis on Dynamic Effects," by M. Kloomok and R. V. Muffley, *Product Eng.*, February 1955.) Reproduced with permission from *Machine Design*, Cleveland, Ohio.

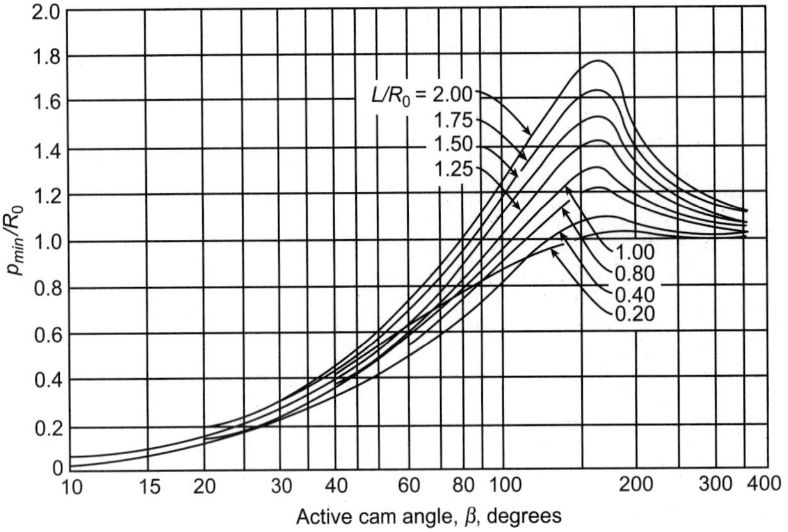

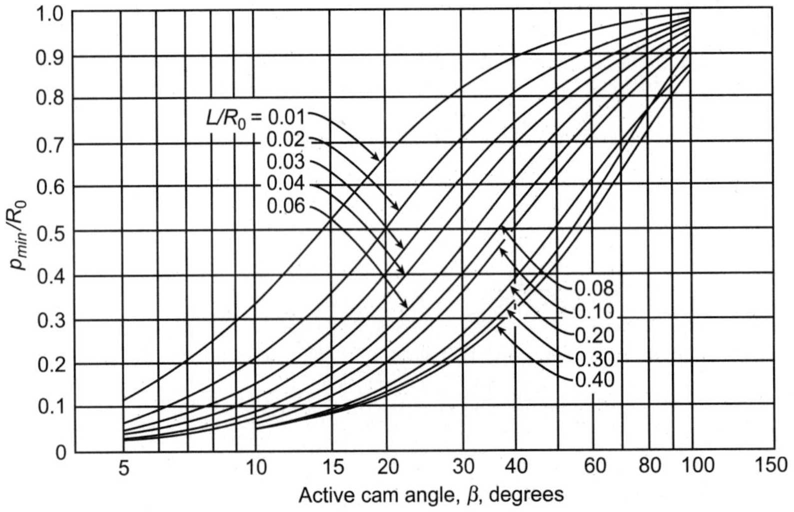

FIGURE 6-10 Cycloidal motion. (From "Plate Cam Design—Radius of Curvature," by M. Kloomok and R. V. Muffley, *Product Eng.*, September 1955.) Reproduced with permission from *Machine Design*, Cleveland, Ohio.

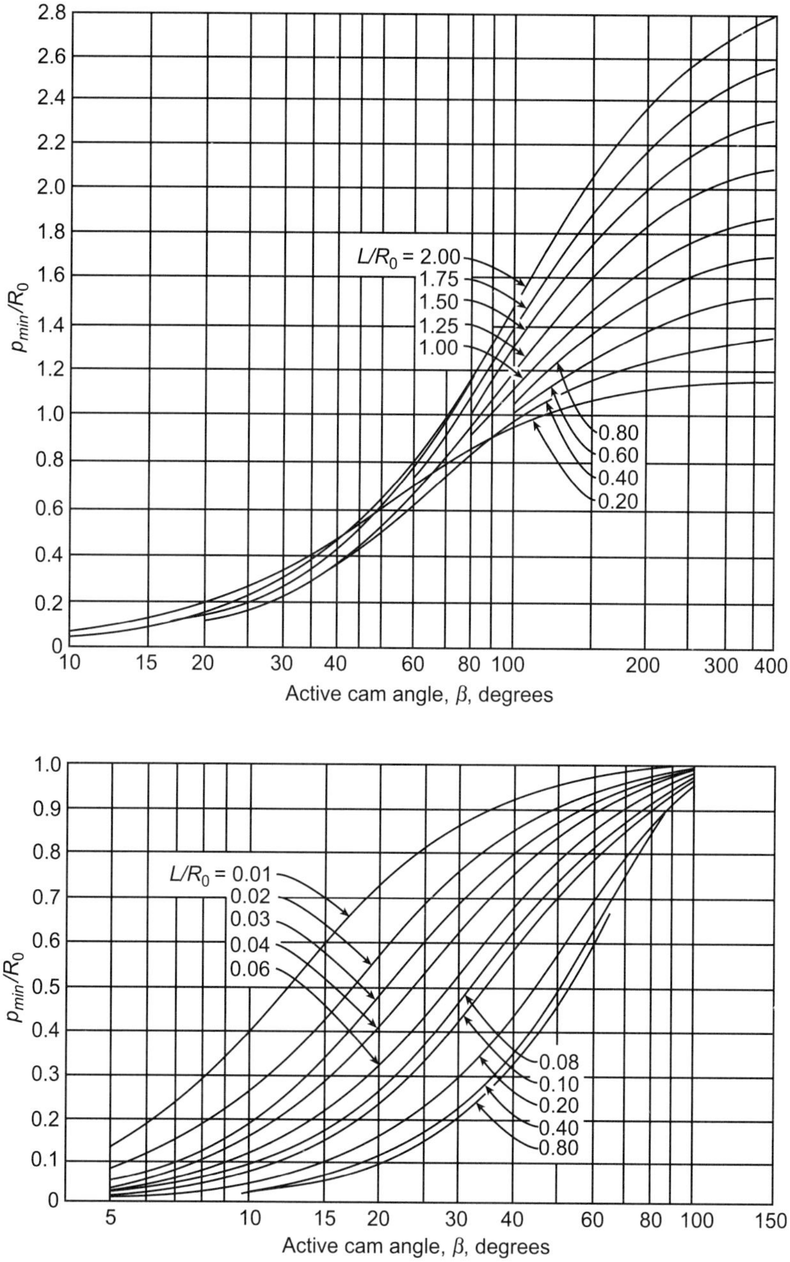

FIGURE 6-11 Harmonic motion. (From "Plate Cam Design—Radius of Curvature," by M. Kloomok and R. V. Muffley, *Product Eng.*, September 1955.) Reproduced with permission from *Machine Design*, Cleveland, Ohio.

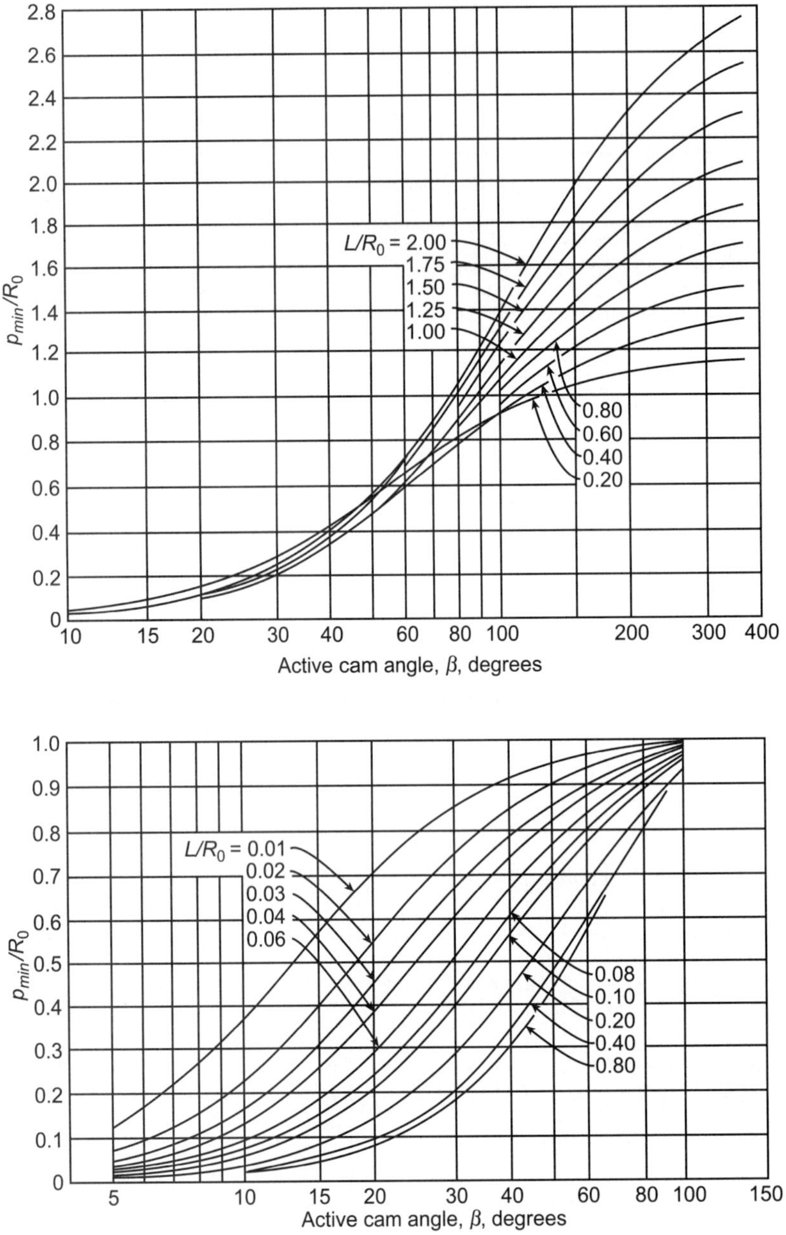

FIGURE 6-12 Eighth-power polynomial motion. (From "Plate Cam Design—Radius of Curvature," by M. Kloomok and R. V. Muffley, *Product Eng.*, September 1955.) Reproduced with permission from *Machine Design*, Cleveland, Ohio.

Particular	Formula

TABLE 6-1
Cam factors for basic curves

Pressure angle α, deg	Types of motion			
	Uniform	Modified uniform	Simple harmonic	Parabolic and cycloidal
10	5.67	5.84	8.91	11.34
15	3.73	3.99	5.85	7.46
20	2.75	3.10	4.32	5.50
25	2.14	2.58	3.36	4.28
30	1.73	2.27	2.72	3.46
35	1.43	2.06	2.24	2.86
40	1.19	1.92	1.87	2.38
45	1.00	1.82	1.57	2.00

The maximum shear stress

$$\tau_{\max} = 0.295\sigma_{c,\max} \tag{6-37}$$

The depth to the point of maximum shear

$$h = 0.786b \tag{6-38}$$

For further data on characteristic equations of basic curves, different motion characteristics, cam factors, materials for cams and followers, and displacement ratios

Refer to Tables 6-1 and Figures 6-7, 6-8 and 6-9. For materials of cams refer to Chapter 1 on "Properties of Engineering Materials."

REFERENCES

1. Rothbart, H. A., *Cams*, John Wiley and Sons, New York, 1956.
2. Marks, L. S., *Mechanical Engineers' Handbook*, McGraw-Hill Book Company, New York, 1951.
3. Lingaiah, K., and B. R. Narayana Iyengar, *Machine Design Data Handbook*, Engineering College Co-operative Society, Bangalore, India, 1962.
4. Lingaiah, K., *Machine Design Data Handbook*, Vol. II (*SI and Customary Metric Units*), Suma Publishers, Bangalore, India, 1986.
5. Rothbart, H. A., *Mechanical Design and Systems Handbook*, McGraw-Hill Book Company, New York, 1964.
6. Shigley, J. E., *Theory of Machines*, McGraw-Hill Book Company, New York, 1961.
7. Mabie, H. H., and F. W. Ocvirk, *Mechanisms and Dynamics of Machinery*, John Wiley and Sons, New York, 1957.
8. Kent, R. T., *Mechanical Engineers' Handbook—Design and Production*, Vol. II. John Wiley and Sons, New York, 1961.
9. Klcomok, M., and R. V. Muffley, "Plate Cam Design—with Emphasis on Dynamic Effects," *Product Eng.*, February 1955.
10. Klcomok, M., and R. V. Muffley, "Plate Cam Design—Radius of Curvature," *Product Eng.*, February 1955.
11. Varnum, E. C., "Circular Nomogram—Theory and Practice Construction Technique," Barber-Coleman Co., *Product Eng.*
12. Gruenberg, R., "Nomogram for Parabolic Cam with Radially Moving Follower,", in Douglas C. Greenwood, Editor, *Engineering Data for Product Design*, McGraw-Hill Book Company, New York, 1996.

CHAPTER

7

PIPES, TUBES, AND CYLINDERS

SYMBOLS[5,6,9]

d	diameter of cylinder, m (in)
d_c	diameter of contact surface in compound cylinder, m (in)
d_i	inside diameter of cylinder or pipe or tube, m (in)
d_o	outside diameter of cylinder or pipe or tube, m (in)
e	factor for expanded tube ends
E	modulus of elasticity, GPa (Mpsi)
h or t	thickness of cylinder or pipe or tube, m (in)
I	moment of inertia, area, m^4 or cm^4 (in^4)
K	constant
L	maximum distance between supports or stiffening rings, m (in)
p	maximum allowable working pressure, MPa (psi)
p_c	unit pressure between the compound cylinders, MPa (psi)
p_{cr}	collapsing pressure, MPa (psi)
p_i	internal pressure, MPa (psi)
p_o	external pressure, MPa (psi)
r_i	inside radius of tube or pipe, m (in)
σ	permissible working stress, from Table 7-1, MPa (psi)
σ_c	crushing stress, MPa (psi)
σ_r	radial stress (also with primes), MPa (psi)
$\sigma_{r(\max)}$	maximum radial stress, MPa (psi)
σ_{sa}	maximum allowable stress value at design condition, MPa (psi)
σ_{su}	ultimate strength, MPa (psi)
σ_θ	tangential stress (also with primes), MPa (psi)
$\sigma_{\theta(\max)}$	maximum tangential stress, MPa (psi)
$\tau_{\max}$	maximum shear stress, MPa (psi)
ν	Poisson's ratio
η	efficiency, from Table 7-4

Note: The initial subscript s, along with σ, which stands for strength, is used throughout this book.

Particular	Formula

LONG THIN TUBES WITH INTERNAL PRESSURE

The permissible steam pressure in steel and iron pipes (Table 7-1) according to *ASME Power Boiler Code*

$$p = \frac{2\sigma_{sa}}{d_o}(h - 1.625 \times 10^{-3}) - 0.9 \qquad \textbf{SI} \quad (7\text{-}1a)$$

where h, d_o in m, and p and σ in MPa.

$$p = \frac{2\sigma_{sa}}{d_o}(h - 0.065) - 125 \qquad \textbf{USCS} \quad (7\text{-}1b)$$

where h, d_o in in, and p and σ in psi.

For tubes from 6.35 mm (0.25 in) to 127 mm (5 in) nominal diameter

$$p = \frac{2\sigma_{sa}}{d_o}(h - 2.54 \times 10^{-3}) \qquad \textbf{SI} \quad (7\text{-}2a)$$

where h, d_o in m, and p and σ in MPa.

$$p = \frac{2\sigma_{sa}}{d_o}(h - 0.1) \qquad \textbf{USCS} \quad (7\text{-}2b)$$

where h, d_o in in, and p and σ in psi.

For over 127 mm (5 in) diameter

The minimum required thickness of ferrous tube up to and including 125 mm (5 in) outside diameter subjected to internal pressure as per *ASME Power Boiler Code*

$$h = \frac{pd_o}{2\sigma_{sa} + p} + 0.005d_o + e \qquad (7\text{-}3)$$

where σ_{sa} is the maximum allowable stress value at design condition and e is the thickness factor for expanded tube ends.

Refer to Table 7-1 for σ_{sa}.
Refer to table 7-2 for e.

The maximum allowable working pressure (MAWP) from Eq. (7-3) as per *ASME Power Boiler Code*

$$p = \sigma_{sa}\left[\frac{2h - 0.01d_o - 2e}{d_o - (h - 0.005d_o - e)}\right] \quad \text{or} \qquad (7\text{-}4)$$

$$= \sigma_{sa}\left[\frac{2h - 0.01d_o - 2e}{1.005d_0 - h + e}\right]$$

For maximum allowable working pressure

Refer to Table 9-1.

The minimum required thickness of ferrous pipe under internal pressure as per *ASME Power Boiler Code*

$$h = \frac{pd_o}{2\sigma_{sa}\eta + 2yp} + C = \frac{pr_i}{\sigma_{sa}\eta - (1-y)p} + C \qquad (7\text{-}5)$$

where

η = efficiency (refer to Table 7-4 for η)
y = temperature coefficient (refer to Table 7-3 for y)
C = minimum allowance for the threading and structural stability, mm (in) (refer to Table 7-5 for h values and Table 7-6 for C values).

TABLE 7-1
Maximum allowable stress values in tension of metals for tubes and pipes, σ_{sa}

Specification number	Grade, alloy designation and temper	UNS number	Nominal composition and size, mm (in)	Product form	Specified minimum yield strength, σ_{sy}		Specified minimum tensile strength, σ_{st}		Maximum allowable stress, σ_{sa}: 38 (100)		93 (200)		150 (300)		205 (400)		260 (500)	
					MPa	kpsi	MPa	kpsi	MPa	kpsi	MPa	kpsi	MPa	kpsi	MPa	kpsi	MPa	kpsi
1	2	3	4	5	6	7	8	9	10	11	12	13	14	15	16	17	18	19
(A) *Carbon and Low Alloy Steels*																		
Carbon Steel:																		
SA-106[c]	A		C-si	Smls[†].Pp*	207	30	331	48										
SA-210[c]	C		C-Mn-Si	Smls.Tb**	276	40	483	70										
SA-557[b,f]	C		C-Mn	Smls.Tb	276	40	483	70										
Low Alloy Steel:																		
SA-209[g]	T1			Smls.Tb	207	30	379	55										
SA-213	T12		1Cr-$\frac{1}{2}$Mo	Smls.Tb	207	30	414	60										
SA-369	Fp11		1$\frac{1}{4}$Cr-$\frac{1}{2}$Mo-Si	Smls.Pp	207	30	414	60										
SA-250	T1		C-$\frac{1}{2}$Mo*	Wld•Pp & Tb	207	30	379	55										
(B) *High Alloy Steels*																		
SA-268	TP410	S4 1 000	13Cr	Smls.Tb	207	30	414	60	103	15.0	99	14.3	95	13.8	92	13.3	89	12.9
SA-268	TP405	S40500	12Cr-1Al	Wld.Tb[f]	207	30	414	60	88	12.8	84	12.2	81	11.8	78	11.3	75	10.9
SA-268	TpxM-8	S43035	18Cr-Ti	Wld.Tb[d,f]	207	30	414	60	88	12.8	84	12.)	81	11.8	78	11.3	75	10.9
SA-268	TpxM-8	S43035	18Cr-Ti	Smls.Tb[d,f]	207	30	414	60	103	15.0	98	14.3	95	13.8	92	13.3	89	12.9
SA-268	18Cr-2Mo	S44400	18Cr-2Mo	Wld.Tb[d,f]	276	40	414	60	88	12.8	84	12.2	81	11.8	78	11.3	75	10.9
SA-268	18Cr-2Mo	S44400	18Cr-2Mo	Smls.Tb[d,f]	276	40	414	60	103	15.0	99	14.3	95	13.8	92	13.3	88	12.8
SA-249. SA-312	TP304L	S30403	18Cr-8Ni	Wld.Tb[f] & Pp	172	25	483	70	92	13.3	78	11.4	70	10.2	64	9.3	60	8.7
SA-213, SA-312	TP304H, TP304	S30409, S30400	18Cr-8Ni	Smls.Tb[g,h] & Pp	207	30	517	75	130	18.8	123	17.8	115	16.6	112	16.2	110	15.9
SA-213, SA-312	TP304N	S30451	18Cr-8Ni-N	Smls.Tb & Pp[g,h]	241	35	552	80	138	20.0	138	131	20.0	19.0	126	18.3	123	17.8
SA-249, SA-312	TP304N	S30451	18Cr-8Ni-N	Wld.Th & Pp[f,g,h]	241	35	552	80	117	17.0	117	17.0	111	16.1	108	15.6	104	15.1
SA-213, SA-312	TP316L	S31603	16Cr-12Ni-2Mo	Smls.Tb & Pp	172	25	483	70	108	15.7	92	13.3	82	11.9	75	10.8	69	10.0
SA-312, SA-688	TP316L	S31603	16Cr-12Ni-2Mo	Wld.P & Tb[f]	172	25	483	70	92	13.3	78	11.3	70	10.1	63	9.2	59	8.5
SA-452	TP316H	S31609	16Cr-12Ni-2Mo	Cast. Pp[g]	207	30	517	75	130	18.8	130	18.8	127	18.4	125	18.1	124	18.0
SA-312	XM-15	S31800	18Cr-18Ni-2Si	Wld.Tb[f,g]	207	30	517	75	110	15.9	104	15.1	97	14.1	95	13.7	93	13.5
SA-213	XM-15	S38100	18Cr-18Ni-2Si	Smls.Tb[g]	207	30	517	75	130	18.8	122	17.7	115	16.6	111	16.1	110	15.9
SA-213	TP316N	S31651	16Cr-12Ni-2Mo-N	Smls.Tb[g,h]	241	35	552	80	138	20.0	138	20.0	132	19.2	130	18.8	128	18.6
SA-312	TP316N	S3i651	16Cr-12Ni-2Mo-N	Wld.Pp[f,g,h]	241	35	552	80	117	17.0	117	17.0	112	16.3	110	16.0	109	15.8
SA-312, SA-688	XM-29	S24000	18Cr-3Ni-12Mn	Wld.Pp[f] & Tb	379	55	689	100	146	21.2	143	20.8	132	19.2	119	17.3	110	16.0
SA-213, SA-312	TP321	S32100	18Cr-10Ni-Ti	Smls.Tb[g,h] & Pp	207	30	517	75	130	18.8	127	18.4	119	17.3	118	17.1	118	17.1
SA-249, SA-312	TP321H	S32109	18Cr-10Ni-Ti	Wld.Tb & Pp[f]	207	30	517	75	110	16.0	93	13.5	83	12.1.	76	11.0	70	10.2
SA-430	FP347H	S34700	18Cr-10Ni-Cb	Smls.Pp[g]	207	30	483	70	130	18.8	123	17.9	113	16.4	107	15.5	103	14.9
SA-213	TP348	S34800	18Cr-10Ni-Cb	Smls.Tb[g,h]	207	30	517	75	130	18.8	123	17.9	113	16.4	107	15.3	103	14,9
SA-249, SA-312	TP348H, TP347H	S34809, S34709	18Cr-10Ni-Cb	Wld.Tb & Pp[f,g]	207	30	517	75	110	16.0	105	15.2	97	14.0	91	13.2	88	12.7
SA-213, SA-312	S30815	S30815	21Cr-11Ni-N	Smls.Tb & Pp[f]	310	45	600	87	150	21.8	149	21.6	141	20.4	135	19.6	127	18.4
SA-789, SA-790		S32550	25.5Cr-5.5Ni-3.5Mo-Cu	Smls.Tb & Pp[d]	552	80	758	110	190	27.5	189	27.4	177	25.7	170	24.7	170	24.7
SA-789, SA-790		S32550	25.5Cr-5.5Ni-3.5Mo-Cu	Wld.Tb & Pp[d]	552	80	758	110	161	23.4	161	23.3	151	21.9	145	21.0	145	21.0
SA-789, SA-790, SA-669	S31500	S31500	18Cr-5Ni-3Mo	Smls.Tb[d,f]	441	64	634	92	159	23.0	153	22.2	147	21.3	146	21.2	146	21.2
SA-789, SA-790	S31500	S31500	18Cr-5N-3Mo	Wld.Tb[d,f]	441	64	634	92	135	19.6	130	18.9	125	18.1	124	18.0	124	18.0

TABLE 7-1
Maximum allowable stress values in tension of metals for tubes and pipes, σ_{sa} (*Cont.*)

for metal temperature, °C (°F), not exceeding																				
315 (600)		370 (700)		427 (800)		482 (900)		538 (1000)		593 (1100)		650 (1200)		704 (1300)		760 (1400)		815 (1500)		
MPa	kpsi	MPa	kpsi	MPa	kpsi	MPa	kpsi	MPa	kpsi	MPa	kpsi	MPa	kpsi	MPa	kpsi	MPa	kpsi	MPa	kpsi	Specification number
20	21	22	23	24	25	26	27	28	29	30	31	32	33	34	35	36	37	38	39	40
(A) *Carbon and Low Alloy Steels*																				
Carbon Steel:																				
83	12.0	81	11.7	64	9.3	45	6.5	17	2.5											SA- 106[c]
121	17.5	115	16.6	83	12.0	35	5.0	10	1.5											SA-210[c]
103	15.0	97	14.1	70	10.2	38	5.5	15	2.1											SA-557[b,f]
Low Alloy Steels																				
95	13.8	95	13.8	93	13.5	86	12.7	33	4.8											SA-209[g]
103	15.0	103	15.0	99	14.4	76	11.0	38	5.5	27	4.0	8	1.2							SA-213
		98	14.2	93	13.5	86	12.5	48	6.2	18	2.6	7	1.0							SA-369
88	12.8	88	12.8	84	12.2	76	11.0	98	4.1											SA-250
86	12.4	83	12.1	77	11.1	67	9.7	44	6.4	20	2.9	7	1.0							SA-268
(B) *High Alloy Steels*																				
73	10.6	71	10.3	65	9.4	57	8.2	23	3.4											SA-268
73	10.6	71	10.3	65	9.4															SA-268
86	12.4																			SA-268
72	10.5																			SA-268
86	12.4																			SA-268
57	8.3	55	8.0	53	7.7															SA-249, SA-312
110	15.9	110	15.9	105	15.2	101	14.7	95	13.8	68	9.8	42	6.1	25	3.7	16	2.3	10	1.4	SA-213, SA-312
120	17.4	118	17.1	115	16.6	110	15.9	103	15.0	67	9.7	41	6.0							SA-213, SA-312
102	14.8	101	14.6	98	14.2	93	13.5	86	12.7	57	8.3	35	5.1							SA-249, SA-312
51	7.4	62	9.0	59	8.6															SA-213, SA-312
55	8.0	52	7.6	50	7.3															SA-312, SA-688
117	17.0	112	16.3	110	15.9	107	15.5	106	15.3	85	12.4	51	7.4	28	4.1	16	2.3	9	1.3	SA-452
93	13.5	93	13.5	89	12.9	85	12.4													SA-312
110	15.9	110	15.9	104	15.1	101	14.6	95	13.7											SA-213
128	18.6	128	18.6	127	18.4	125	18.1	120	17.4	85	12.4	51	7.4							SA-213
109	15.8	109	15.8	108	15.6	106	15.4	102	14.8	72	10.5	43	6.3							SA-312
106	15.4	101	14.7	97	14.1															SA-312, SA-688
112	16.4	109	15.8	107	15.5	106	15.3	95	13.8	48	6.9	24	3.6	12	1.7	5	0.8	2	0.3	SA-213, SA-312
67	9.7	64	9.3	63	9.2	62	9.0	61	8.9	52	7.5	32	4.6	19	2.7	11	1.6	7	1.0	SA-249, SA-312
101	14.7	101	14.7	101	14.7	101	14.7	99	14.4	90	13.0	55	7.9	30	4.4	17	2.5	9	1.3	SA-430
101	14.7	101	14.7	101	14.7	101	14.7	97	14.0	63	9.1	30	4.4	15	2.2	8	1.2	5	0.8	SA-213
86	12.5	86	12.5	86	12.5	86	12.5	84	12.3	76	11.1	46	6.7	25	3.7	15	2.1	8	1.1	SA-249, SA-312
122	17.7	119	17.3	116	16.8	112	16.3	103	14.9	62	9.0	36	5.2	21	3.1	13	1.9	9	1.3	SA-312, SA-213
																				SA-789, SA-790
146	21.2	146	21.2																	SA-789, SA-790
124	18.0	124	18.0																	SA-789, SA-790, SA-669
																				SA-789, SA-790

TABLE 7-1
Maximum allowable stress values in tension of metals for tubes and pipes, σ_{sa} (*Cont.*)

Specification number	Grade, alloy designation and temper	UNS number	Nominal composition and size, mm (in)	Product form	Specified minimum yield strength, σ_{sy} MPa	Specified minimum yield strength, σ_{sy} kpsi	Specified minimum tensile strength, σ_{st} MPa	Specified minimum tensile strength, σ_{st} kpsi	Maximum allowable stress 38 (100) MPa	38 (100) kpsi	93 (200) MPa	93 (200) kpsi	150 (300) MPa	150 (300) kpsi	205 (400) MPa	205 (400) kpsi	260 (500) MPa	260 (500) kpsi
1	2	3	4	5	6	7	8	9	10	11	12	13	14	15	16	17	18	19
(C) *Non-ferrous Metals*																		
Aluminum and Aluminum Alloys:																		
SB-210	1060-1114[d]		0.250–12.500 (0.010–0.500)	Drawn	69	10	83	12	21	3.0	21	3.0	18	2.6	8	1.2		
SB-210	6061-T6[e]		0.625–12.50 (0.025–0.50)	Smls.Tb	241	35	290	42	72	10.5	72	10.5	58	8.4	31	4.5		
SB-241	3003-H118[e]		Under 25 (under 1)	Smls.Pp	165	24	186	27	47	6.8	46	6.7	37	5.4	17	2.5		
SB-241	5083-H111[d,p]			Smls. extruded Tb	131	19	228	33	57	8.3	57	8.3	38	5.5	21	3.0		
SB-234	1060-H14[e]		Up to 125 (up to 5.00)	Condenser and heat exchanger Tb	69	10	83	12	21	3.0	21	3.0	18	2.6	8	1.2		
SB-234	3003-H25[e]		0.250–12.50 (0.010–0.5000)	Condenser and heat exchanger Tb	131	19	145	21	38	5.5	38	5.5	30	4.3	17	2.4		
SB-234	6061-T6[e]		0.625–6.225 (0.025–0.249)	Condenser and heat exchanger Tb	241	35	290	42	72	10.5	72	10.5	58	8.4	31	4.5		
Copper and Copper Alloys:																		
SB-111	102, 120, 122, 142[i]		Ann	Smls. Copper condenser, Tb.	62	9	207	30	41	6.0	33	4.8	32	4.7	21	3.0		
			LD[‡]		207	30	248	36	62	9.0	62	9.0	60	8.7	57	8.2		
			HD**		276	40	310	45	78	11.3	78	11.3	78	11.3	30	4.3		
SB-1 1 1	192 Ann			Smls. Copper, iron alloy condenser Tb	83	12	262	38	52	7.5	46	6.7	42	6.1				
SB-315	655. Ann[g]			Smls. Cu-Si Alloy Pp and Th	103	15	345	50	69	10.0	69	10.0	69	10.0	35	5.0		
SB-466	C71500 Ann[p]			Smls. Cu-Ni 70/30 Pp & Tb.	124	18	345	50	83	12.0	78	11.3	75	10.8	71	10.3	68	9.9
SB-467	C71500 Ann[pp]		(Up to 112.5 incl) (up to $4\frac{1}{2}$ incl)	Wld. Cu-Ni-70/30 Pp	138	20	345	50	87	12.6	61	8.9	61	8.8	61	8.8	61	8.8
SB-543	C700-Ann[p]			Wld.Cu-Ni-90/10Tb	103	15	270	40	59	8.5	56	8.1	52	7.6	50	7.2	43	6.3
	LCW***				241	35	310	45	59	8.5	56	8.1	52	7.6	50	7.2	43	6.3
Nickel and High Nickel Alloys:																		
SB-161	201 Ann	N02201	Ni Low C	Pp & Tb	69	10	345	50	46	6.7	44	6.4	43	6.3	43	6.2	43	6.2
SB-163	800H Ann[k]	N08810	Ni-Fe-Cr	Pp & Tb	172	25	448	65	112	16.2	112	16.2	112	16.2	112	16.2	110	16.0
SB-163	825 Ann[k]	N08825	Ni-Fe-Cr-Mo-Cu	Pp & Tb	241	35	586	85	146	21.2	146	21.2	146	21.2	146	21.2	146	21.2
SB-144	625 Ann[p]	N06625	Ni-Cr-Mo-Cb		414	60	827	120	207	30.0	207	30.0	207	30.0	194	28.2	186	27.0
SB-468	20 cb.Wld. Ann[k,p]	N08020	Cr-Ni-Fe-Mo-Cu-Cb	Pp & Tb	241	35	552	80	117	17.0	117	17.0	115	16.8	110	15.9	107	15.5
SB-619	C-276 Sol. Ann[p]	N10276	Ni-Mo-Cr (All sizes)	Pp & Tb	283	41	689	100	146	21.2	146	21.2	146	21.2	143	20.7	140	20.3
SB-619	G. Sol. Ann[k,p]	N06007	Ni-Cr-Fe-Mo-Cu (All sizes)	Pp & Tb	242	35	620	90	132	19.1	132	19.1	132	19.1	128	18.6	126	18.3

TABLE 7-1
Maximum allowable stress values in tension of metals for tubes and pipes, σ_{sa} (*Cont.*)

for metal temperature, °C (°F), not exceeding																				
315 (600)		370 (700)		427 (800)		482 (900)		538 (1000)		593 (1100)		650 (1200)		704 (1300)		760 (1400)		815 (1500)		
MPa	kpsi	MPa	kpsi	MPa	kpsi	MPa	kpsi	MPa	kpsi	MPa	kpsi	MPa	kpsi	MPa	kpsi	MPa	kpsi	MPa	kpsi	Specification number
20	21	22	23	24	25	26	27	28	29	30	31	32	33	34	35	36	37	38	39	40
(C) *Non-ferrous Metals*																				
Aluminum and Aluminum Alloys:																				SB-210
																				SB-210
																				SB-241
																				SB-241
																				SB-234
																				SB-234
																				SB-234
Copper and Copper Alloys:																				
																				SB-111
67	9.6																			SB-111
61	8.8																			SB-315
30	4.3																			SB-466
30	4.3																			SB-467
																				SB-543
Nickel and High Nickel Alloy																				
43	6.2	43	6.2	41	5.9	31	4.5	21	3.0	14	2.0	8	1.2							SB-161
110	16.0	108	15.7	105	15.3	102	14.8	99	14.4	80	11.6	51	7.4	32	4.7	21	3.0	13	1.9	SB-163
146	21.2	145	21.0	143	20.8	141	20.5	36	19.7											SB-163
182	26.4	179	26.0	179	26.0	179	26.0	179	26.0	179	26.0	91	13.2							SB-444
104	15.1	101	14.7	99	14.3															SB-468
138	20.0	135	19.6	134	19.4	130	18.9	128	18.5	88	12.7	57	8.3							SB-619
123	17.9	123	17.8	120	17.4	117	17.0	111	16.1											SB-619

Source: The American Society of Mechanical Engineers, *Boiler and Pressure Vessel Code*, Section VIII, Division 1, July 1986.
* Pp = pipe; ** Tb = tube; *** LCW = light cold worked; Smls = seamless; • Wld = welded; ‡ LD = light drawn; •• HD = hard drawn; Ann = annealed; Soln Ann = solution annealed.
Notes: The superscript letters a, b, c, etc., refer to notes under each category of (A) Carbon and Low Alloy Steels, (B) High Alloy Steels, and (C) Non-ferrous Metals in Tables 8-9, 8-10, and 8-11 in Chapter 8.

TABLE 7-2
Thickness factor for expanded tube ends *e* for use in Eqs. (7-3) and (7-4)

Particular	Value of *e*
Over a length at least equal to the length of the seat plus 25 mm (1 in) for tubes expanded into tube seats, except	0.04
For tubes expanded into tube seats provided the thickness of the tube ends over a length of the seat plus 25 mm (1 in) is not less than the following: 2.375 mm (0.095 in) for tubes $\leq$31.25 mm (1.25 in) OD 2.625 mm (0.105 in) for tubes >31.25 mm (1.25 in) OD and $\leq$50 mm (2 in) OD, including 3.000 mm (0.120 in) for tubes >50 mm (2 in) and $\leq$75 mm (3 in) OD, including 3.375 mm (0.135 in) for tubes >75 mm (3 in) OD and $\leq$100 mm (4 in) OD, including 3.75 mm (0.150 in) for tubes >100 mm (4 in) and $\leq$125 mm (5 in) OD, including	0
For tubes strength-welded to headers and drums	0

Source: *ASME Boiler and Pressure Vessel Code*, Section 1, 1983.

TABLE 7-3
Temperature coefficient *y*

	Temperature, °C (°F)[a]					
Material	**$\leq$482 (900)[a]**	**510 (950)**	**540 (1000)**	**565 (1050)**	**595 (1100)**	**$\geq$620 (1150)**
Ferrite steels	0.4	0.5	0.7	0.7	0.7	0.7
Austenitic steels	0.4	0.4	0.4	0.4	0.5	0.7
For nonferrous materials	0.4	0.4	0.4	0.4	0.4	0.4

[a] Temperatures in parentheses are in Fahrenheit (°F). Values of *y* between temperatures not listed may be determined by interpolation.
Source: *ASME Boiler and Pressure Vessel Code*, Section 1, 1983.

TABLE 7-4
Efficiency of joints, η

Particular	Efficiency, η
Longitudinal welded joints or of ligaments between openings, whichever is lower	
Seamless cylinders	1.00
For welded joints provided all weld reinforcement on the longitudinal joints is removed substantially flush with the surface of the plate	1.00
For welded joints with the reinforcement on the longitudinal joints left in place	0.90
Riveted joints	Refer to Table 13-4 (Chap. 13)
Ligaments between openings	Refer to Eqs. under Ligament (Chap. 8)
Welded joint efficiency factor	Refer to Table 8-3 (Chap. 8)

Source: *ASME Boiler and Pressure Vessel Code*, Section 1, 1983.

Particular	Formula

TABLE 7-5
The depth of thread h (formula $h = 0.8/i$)

Number of threads per mm (in), i	Depth of thread, h mm (in)
0.32 (8)	2.5 (0.100)
0.46 (11.5)	1.715 (0.0686)

Source: ASME Boiler and Pressure Vessel Code, Section 1, 1983.

The maximum allowable working pressure from Eq. (7-5) as per *ASME Power Boiler Code*

$$p = \frac{2\sigma_{sa}\eta(h-C)}{d_o - 2y(h-C)} \text{ or } p = \frac{\sigma_{sa}\eta(h-C)}{r_i + (1-y)(h-C)} \quad (7\text{-}6)$$

The minimum required thickness of nonferrous seamless tubes and pipes for outside diameters 12.5 mm (0.5 in) to 150 mm (6 in) inclusive and for wall thickness not less than 1.225 mm (0.049 in) as per *ASME Power Boiler Code*

$$h = \frac{pd_o}{2\sigma_{sa}} + C \quad (7\text{-}7)$$

Refer to Table 7-6 for values of C.

The maximum allowable working pressure as per *ASME Power Boiler Code*

$$p = \frac{2\sigma_{sa}}{d_o}(h-C) \quad (7\text{-}8)$$

The minimum required thickness of tubes made of steel or wrought iron subjected to internal pressure which are used in watertube and firetube boilers as per *ASME Power Boiler Code*

$$h = 0.0251 d_o \quad (7\text{-}9)$$

The minimum required thickness of tubes made of nonferrous materials such as copper, red brass, admiralty and copper-nickel alloys used in watertube and firetube boilers with a design pressure over 207 kPa (30 psi) but not greater than 414 kPa (60 psi)

$$h = \frac{d_o}{30} + 0.75 \quad \textbf{SI} \quad (7\text{-}10a)$$

$$h = \frac{d_o}{30} + 0.03 \quad \textbf{USCS} \quad (7\text{-}10b)$$

The minimum required thickness of tubes made of nonferrous materials such as copper, red brass, admiralty and copper-nickel alloys used in steam boilers of less than 103 kPa (15 psi) and water boilers of less than 207 kPa (30 psi)

$$h = \frac{d_o}{45} + 0.75 \quad \textbf{SI} \quad (7\text{-}11a)$$

$$h = \frac{d_o}{45} + 0.03 \quad \textbf{USCS} \quad (7\text{-}11b)$$

The minimum required thickness of tubes when made of nonferrous materials but assembled with fittings, which are based on materials used, and based on whether the pressure is over 207 kPa (30 psi), but not in excess of 1013 kPa (160 psi) or whether the pressure does not exceed 207 kPa (30 psi)

$$h = \frac{d_o}{\text{factor}} + 0.75 \text{ except for copper} = 0.027 \quad \textbf{SI} \quad (7\text{-}12a)$$

$$h = \frac{d_o}{\text{factor}} + 0.03 \quad \textbf{USCS} \quad (7\text{-}12b)$$

The formula for permissible pressure in wrought-iron and steel tubes for watertube boilers according to *ASME Power Boiler Code*

$$p = 125\left(\frac{h - 1\times 10^{-3}}{d_o}\right) - 0.32 \quad \textbf{SI} \quad (7\text{-}13a)$$

where h, d_o in m, and p in MPa.

$$p = 18000\left(\frac{h - 0.039}{d_o}\right) - 250 \quad \textbf{USCS} \quad (7\text{-}13b)$$

where h, d_o in in, and p in psi.

Particular	Formula
	$p = 96.5\left(\frac{h - 1 \times 10^{-3}}{d_o}\right)$ **SI** (7-14a)
	where h, d_o in m, and p in MPa.
	$p = 14000\left(\frac{h - 0.039}{d_o}\right)$ **USCS** (7-14b)
	where h, d_o in in, and p in psi.
	$p = 73\left(\frac{h - 1 \times 10^{-3}}{d_o}\right)$ **SI** (7-15a)
	where h, d_o in m, and p in MPa.
	$p = 10600\left(\frac{h - 0.039}{d_o}\right)$ **USCS** (7-15b)
	where h, d_o in in, and p in psi.
	Formula (7-13) applies to seamless tubes at all pressures, to welded steel tubes at pressure below 6 MPa (875 psi), and to lap-welded wrought-iron tubes at pressures below 2.5 MPa (358 psi).
	Formula (7-14) applies to welded steel tubes at pressures of 6 MPa (875 psi) and above.
	Formula (7-15) applies to lap-welded wrought-iron tubes at pressures of 2.5 MPa (358 psi) and above.

ENGINES AND PRESSURE CYLINDERS

Particular	Formula
The wall thickness of engines and pressure cylinders	$h = \frac{pd_i}{2\sigma_{sta}} + 7.5 \times 10^{-3}$ **SI** (7-16a)
	where p, σ_{st} in MPa, and d_i and h in m.
	$h = \frac{pd_i}{2\sigma_{sta}} + 0.3$ **USCS** (7-16b)
	where p, σ_t in psi, and d_i and h in in.
	$\sigma_{sta} = 9$ MPa (1250 psi) for ordinary grades of cast iron.

OPENINGS IN CYLINDRICAL DRUMS

Particular	Formula
The largest permissible diameter of opening according to D. S. Jacobus	$d' = 0.81\sqrt[3]{d_o h(1.0 - K)}$ **SI** (7-17a)
	where d_o and h in m
	$d' = 2.75\sqrt[3]{d_o h(1.0 - K)}$ **USCS** (7.17b)
	where d_o and h in in.

Particular	Formula
	$K = \left(\frac{pd_o}{2h}\right)\left(\frac{5}{\sigma_{su}}\right)$ **USCS** (7-17b)
	The maximum diameter of the unreinforced hole should be limited to 0.203 m (8 in) and should not exceed $0.6d_o$.

THIN TUBES WITH EXTERNAL PRESSURE

Particular	Formula
Professor Carman's formulas for the collapsing pressure for seamless steel tubes	$p_{cr} = 346120\left(\frac{h}{d_o}\right)^3$ **SI** (7-18a) where h, d_o in m, and p_{cr} in MPa. $p_{cr} = 50200000\left(\frac{h}{d_o}\right)^3$ **USCS** (7-18b) where h, d_o in in, and p_{cr} in psi when $\frac{h}{d_o} < 0.025$. $p_{cr} = 658.5\left(\frac{h}{d_o}\right) - 1.50$ **SI** (7-19a) where h, d_o in m, and p_{cr} in MPa $p_{cr} = 95520\left(\frac{h}{d_o}\right) - 2090$ **USCS** (7-19b) where h, d_o in in, and p_{cr} in psi when $\frac{h}{d_o} > 0.03$
Professor Carman's formula for the collapsing pressure for lap-welded steel tubes	$p_{cr} = 574\left(\frac{h}{d_o}\right) - 0.72$ **SI** (7-20a) where h, d_o in m, and p_{cr} in MPa $p_{cr} = 83290\left(\frac{h}{d_o}\right) - 1025$ **USCS** (7-20b) where h, d_o in in, and p_{cr} in psi when $\frac{h}{d_o} > 0.03$
Professor Carman's formula for the collapsing pressure for lap-welded brass tubes	$p_{cr} = 173385\left(\frac{h}{d_o}\right)^3$ **SI** (7-21a) where h, d_o in m, and p_{cr} in MPa $p_{cr} = 25150000\left(\frac{h}{d_o}\right)^3$ **USCS** (7-21b) where h, d_o in in, and p_{cr} in psi when $\frac{h}{d_o} < 0.025$ $p_{cr} = 644\left(\frac{h}{d_o}\right) - 1.75$ **SI** (7-22a) where h, d_o in m, and p_{cr} in MPa

Particular	Formula
	$$p_{cr} = 93365\left(\frac{h}{d_o}\right) - 2474 \qquad \textbf{USCS} \quad (7\text{-}22b)$$ where h, d_o in in, and p_{cr} in psi when $\frac{h}{d_o} > 0.03$
SHORT TUBES WITH EXTERNAL PRESSURES	
Sir William Fairbairn's formula for collapsing pressure for length less than six diameters	$$p_{cr} = 66580\left(\frac{h^{2.19}}{Ld_o}\right) \qquad \textbf{SI} \quad (7\text{-}23a)$$ where h, L, d_o in m, and p_{cr} in MPa $$p_{cr} = 9657600\left(\frac{h^{2.19}}{Ld_o}\right) \qquad \textbf{USCS} \quad (7\text{-}23b)$$ where h, L, d_o in in, and p_{cr} in psi
Thickness of tubes, and pipes when used as tubes under external pressure as per Indian Standards	Refer to Fig. 7-1 to determine the standard thickness of tubes and pipes; see also Table 7-7.

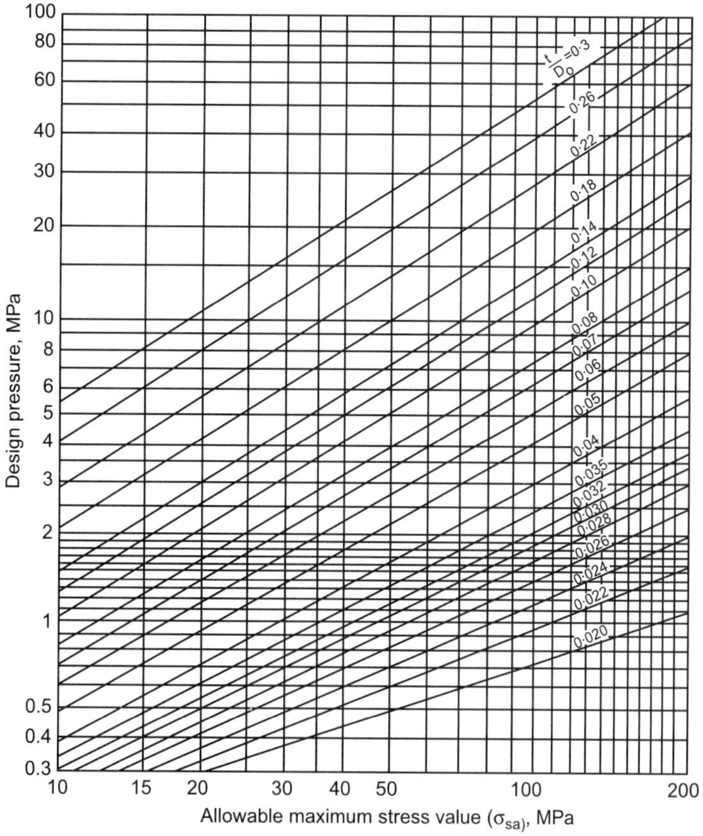

FIGURE 7-1 Thickness of tubes and pipes under external pressure.

TABLE 7-6
Values of C for use in Eqs. (7-5) to (7-8)

Type of pipe	Value of C,[b] mm (in)
Threaded steel, wrought iron, or nonferrous pipe[a]	
19 mm (0.75 in), nominal and smaller	1.625 (0.065)
25 mm (1 in), nominal and larger	Depth of thread h[c]
Plain-end[d] steel, wrought iron, or nonferrous pipe	
87.5 mm (3.5 in), nominal and smaller	1.625 (0.065)
100 mm (4 in), nominal and larger	0

[a] Steel, wrought iron, or nonferrous pipe lighter than schedule 40 of the American National Standard for wrought iron and steel pipe, ANSI B36.10-1970, shall not be threaded.
[b] The values of C stipulated above are such that the actual stress due to internal pressure in the wall of the pipe is no greater than the value of S (i.e. σ_{sa}) given in Table PG 23.1 of ASME Power Boiler Code as applicable in the formulas.
[c] The depth of thread h in inches may be determined from the formula $h = 0.8/i$, where i is the number of threads per inch or from Table 7-5.
[d] Plain-end pipe includes pipe joined by flared compression coupling, lap (Van Stone) joints, and by welding, i.e., by any method which does not reduce the wall thickness of pipe at the joint.
Source: *ASME Boiler and Pressure Vessel Code*, Section 1, 1983.

Particular	Formula

LAMÉ'S EQUATIONS FOR THICK CYLINDERS

General equations

The tangential stress in the cylinder wall at radius r when subjected to internal and external pressures

$$\sigma_\theta = \frac{p_i d_i^2 - p_o d_o^2}{d_o^2 - d_i^2} + \frac{d_i^2 d_o^2 (p_i - p_o)}{4r^2(d_o^2 - d_i^2)} \tag{7-24a}$$

$$= a + \frac{b}{r^2} \tag{7-24b}$$

The radial stress in the cylinder at radius r when subjected to internal and external pressures

$$\sigma_r = \frac{p_i d_i^2 - p_o d_o^2}{d_o^2 - d_i^2} - \frac{d_i^2 d_o^2 (p_i - p_o)}{4r^2(d_o^2 - d_i^2)} \tag{7-25a}$$

$$= a - \frac{b}{r^2} \tag{7-25b}$$

where

$$a = \frac{p_i d_i^2 - p_o d_o^2}{d_o^2 - d_i^2} \tag{7-25c}$$

$$b = \frac{d_i^2 d_o^2 (p_i - p_o)}{4(d_o^2 - d_i^2)} \tag{7-25d}$$

Particular	Formula
Cylinder under internal pressure only	
The tangential stress in the cylinder wall at radius r	$\sigma_\theta = \dfrac{p_i d_i^2}{d_o^2 - d_i^2}\left(1 + \dfrac{d_o^2}{4r^2}\right)$ (7-26)
The radial stress in the cylinder wall at radius r	$\sigma_r = \dfrac{p_i d_i^2}{d_o^2 - d_i^2}\left(1 - \dfrac{d_o^2}{4r^2}\right)$ (7-27)
The maximum tangential stress at the inner surface of the cylinder at $r = d_i/2$	$\sigma_{\theta(\max)} = \dfrac{p_i(d_i^2 + d_o^2)}{d_o^2 - d_i^2}$ (7-28)
The maximum radial stress	$\sigma_{r(\max)} = -p_i$ (7-29)
The maximum shear stress at the inner surface of the cylinder under internal pressure	$\tau_{\max} = \dfrac{p_i d_o^2}{d_o^2 - d_i^2}$ (7-30)
Cylinder under external pressure only	
The tangential stress in the cylinder wall at radius r	$\sigma_\theta = -\dfrac{p_o d_o^2}{d_o^2 - d_i^2}\left(1 + \dfrac{d_i^2}{4r^2}\right)$ (7-31)
The radial stress in the cylinder wall at radius r	$\sigma_\tau = -\dfrac{p_o d_o^2}{d_o^2 - d_i^2}\left(1 - \dfrac{d_i^2}{4r^2}\right)$ (7-32)
DEFORMATION OF A THICK CYLINDER	
The radial displacement of a point at radius r in the wall of the cylinder subjected to internal and external pressures	$u = \left(\dfrac{1-\nu}{E}\right)\dfrac{p_i d_i^2 - p_o d_o^2}{d_o^2 - d_i^2}\, r + \left(\dfrac{1+\nu}{E}\right)\dfrac{d_i^2 d_o^2 (p_i - p_o)}{4r(d_o^2 - d_i^2)}$ (7-33)
Cylinder under internal pressure only	
The radial displacement at $r = d_i/2$ of the inner surface of the cylinder	$u_i = \dfrac{p_i d_i}{2E}\left(\dfrac{d_i^2 + d_o^2}{d_0^2 - d_i^2} + \nu\right)$ (7-34)
The radial displacement at $r = d_o/2$ of the outer surface of the cylinder	$v_o = \dfrac{p_i d_i^2 d_o}{E(d_o^2 - d_i^2)}$ (7-35)
Cylinder under external pressure only	
The radial displacement at $r = d_i/2$ of the inner surface of the cylinder	$u_i = -\dfrac{p_o d_i d_o^2}{E(d_o^2 - d_i^2)}$ (7-36)

Particular	Formula
The radial displacement at $r = d_o/2$ of the outer surface of the cylinder	$$u_o = -\frac{p_o d_o}{2}\frac{1}{E}\left(\frac{d_i^2 + d_o^2}{d_o^2 - d_i^2} - \nu\right) \quad (7\text{-}37)$$

COMPOUND CYLINDERS

Birnie's equation for tangential stress at any radius r for a cylinder open at ends subjected to internal pressure	$$\sigma_\theta = (1-\nu)\frac{p_i d_i^2}{d_o^2 - d_i^2} + (1+\nu)\frac{d_i^2 d_o^2 p_i}{4r^2(d_o^2 - d_i^2)} \quad (7\text{-}38)$$
The tangential stress at the inner surface of the inner cylinder in the case of a compound cylinder (Figs. 11-1 and 11-2)	$$\sigma_{\theta-i} = -\frac{2p_c d_c^2}{d_c^2 - d_i^2} \quad (7\text{-}39)$$
The tangential stress at the outer surface of the inner cylinder	$$\sigma_{\theta-ic} = -p_c\left(\frac{d_c^2 + d_i^2}{d_c^2 - d_i^2} - \nu\right) \quad (7\text{-}40)$$
The tangential stress at the inner surface of the outer cylinder	$$\sigma_{\theta-oc} = p_c\left(\frac{d_o^2 + d_c^2}{d_o^2 - d_c^2} + \nu\right) \quad (7\text{-}41)$$
The tangential stress at the outer surface of the outer cylinder	$$\sigma_{\theta-o} = \frac{2p_c d_c^2}{d_o^2 - d_c^2} \quad (7\text{-}42)$$

THERMAL STRESSES IN LONG HOLLOW CYLINDERS

The general expressions for the radial σ_r, tangential σ_θ, and longitudinal σ_z stresses in the cylinder wall at radius r when the temperature distribution is symmetrical with respect to the axis and constant along its length, respectively	$$\sigma_r = \frac{\alpha E}{(1-\nu)r^2}\left[\frac{4r^2 - d_i^2}{d_o^2 - d_i^2}\int_{r_i}^{r_o} Tr\,dr - \int_{r_i}^{r} Tr\,dr\right] \quad (7\text{-}43)$$ $$\sigma_\theta = \frac{\alpha E}{(1-\nu)r^2}\left[\frac{4r^2 + d_i^2}{d_o^2 - d_i^2}\int_{r_i}^{r_o} Tr\,dr + \int_{r_i}^{r} Tr\,dr - Tr^2\right] \quad (7\text{-}44)$$ $$\sigma_z = \frac{\alpha E}{1-\nu}\left[\frac{8}{d_o^2 - d_i^2}\int_{r_i}^{r_o} Tr\,dr - T\right] \quad (7\text{-}45)$$ where $d_o = 2r_o$ and $d_i = 2r_i$
The expressions for radial (σ_r), tangential (σ_θ), longitudinal (σ_z) stresses in the cylinder at r when the cylinder is subjected to steady-state temperature distribution, i.e., logarithmic temperature distribution throughout the wall thickness of the cylinder by using equation $T = T_i[\ln R_o / \ln R]$	$$\sigma_r = \frac{\alpha E T_i}{2(1-\nu)\ln(R)} \times\left[-\ln(R_o) - \frac{1}{R^2 - 1}(1 - R_o^2)\ln(R)\right] \quad (7\text{-}46)$$ $$\sigma_\theta = \frac{\alpha E T_i}{2(1-\nu)\ln(R)} \times\left[1 - \ln(R_o) - \frac{1}{R^2 - 1}(1 + R_o^2)\ln(R)\right] \quad (7\text{-}47)$$

Particular	Formula
	$$\sigma_z = \frac{\alpha E T_i}{2(1-\nu)\ln(R)}\left[1 - 2\ln(R_o) - \frac{2}{R^2-1}\ln(R)\right] \quad (7\text{-}48)$$ where $R = \frac{d_o}{d_i} = \frac{r_o}{r_i}$; $R_o = \frac{r_o}{r} = \frac{d_o}{2r}$; $R_i = \frac{r_i}{r} = \frac{d_i}{2r}$ T_i = temperature at inner surface of cylinder, °C (°F)
The expressions for maximum values of tangential (hoop) and longitudinal stresses at inner and outer surfaces of the cylinder under logarithmic temperature distribution. respectively	$$\sigma_{\theta i} = \sigma_{zi} = \frac{\alpha E T_i}{2(1-\nu)\ln R}\left[1 - \frac{2R^2}{R^2-1}\ln R\right] \quad (7\text{-}49)$$ $$\sigma_{\theta o} = \sigma_{zo} = \frac{\alpha E T_i}{2(1-\nu)\ln R}\left[1 - \frac{2}{R^2-1}\ln R\right] \quad (7\text{-}50)$$
The simplified expressions for maximum values of tangential and longitudinal stresses at inner and outer surfaces of the cylinder under logarithmic temperature distribution when the thickness of cylinder is small in comparison with the inner radius of the cylinder, respectively	$$\sigma_{\theta i} = \sigma_{zi} = -\frac{\alpha E T_i}{2(1-\nu)}\left(1 + \frac{n}{3}\right) \quad (7\text{-}51)$$ $$\sigma_{\theta o} = \sigma_{zo} = \frac{\alpha E T_i}{2(1-\nu)}\left(1 - \frac{n}{3}\right) \quad (7\text{-}52)$$ where $d_o/d_i = 1 + n$ and $\ln(d_o/d_i) = \ln(1+n)$
The simplified expressions for maximum tangential and longitudinal stresses for thin cylinders under the logarithmic temperature distribution, respectively	$$\sigma_{\theta i} = \sigma_{zi} = -\frac{\alpha E T_i}{2(1-\nu)} \quad (7\text{-}53)$$ $$\sigma_{\theta o} = \sigma_{zo} = \frac{\alpha E T_i}{2(1-\nu)} \quad (7\text{-}54)$$
The expressions for radial (σ_r), tangential (hoop) (σ_θ), and longitudinal (σ_z) stresses in a cylinder at radius r subject to linear thermal temperature distribution throughout the wall thickness of the cylinder by using equation $T = T_i(r_o - r)/(r_o - r_i)$ when the thickness of the cylinder wall is small in comparison with the outside radius	$$\sigma_r = \frac{\alpha E T_i}{(1-\nu)r^2}\left[\frac{(r^2 - r_i^2)(r_o + 2r_i)}{6(r_o + r_i)} + \frac{2(r^3 - r_i^3) - 3r_o(r^2 - r_i^2)}{6(r_o - r_i)}\right] \quad (7\text{-}55)$$ $$\sigma_\theta = \frac{\alpha E T_i}{(1-\nu)r^2}\left[\frac{(r^2 + r_i^2)(r_o + 2r_i)}{6(r_o + r_i)} - \frac{2(r^3 - r_i^3) - 3r_o(r^2 - r_i^2)}{6(r_o - r_i)} - \frac{(r_o - r)r^2}{r_o - r_i}\right] \quad (7\text{-}56)$$ $$\sigma_z = \frac{\alpha E T_i}{1-\nu}\left[\frac{r_o + 2r_i}{2(r_o + r_i)} - \frac{r_o - r}{r_o - r_i}\right] \quad (7\text{-}57)$$
The expressions for maximum tangential (hoop), (σ_θ) and longitudinal (σ_z) stresses at inner and outer surfaces of the cylinder under the linear thermal gradient as per equation $T = T_i(r_o - r)/(r_o - r_i)$	$$\sigma_{\theta i} = \sigma_{zi} = -\frac{\alpha E T_i}{1-\nu}\left[\frac{2r_o + r_i}{3(r_o + r_i)}\right] \quad (7\text{-}58)$$ $$\sigma_{\theta o} = \sigma_{zo} = \frac{\alpha E T_i}{1-\nu}\left[\frac{r_o + 2r_i}{3(r_o + r_i)}\right] \quad (7\text{-}59)$$

Particular	Formula
The expressions for maximum tangential and longitudinal stresses at inner and outer wall surfaces of thin cylinder (i.e., $r_o \approx r_i$) under the linear thermal gradient as per equation $T = T_i(r_o - r)/(r_o - r_i)$	$\sigma_{\theta i} = \sigma_{zi} = -\dfrac{\alpha E T_i}{2(1-\nu)}$ (7-60) $\sigma_{\theta o} = \sigma_{zo} = \dfrac{\alpha E T_i}{2(1-\nu)}$ (7-61) Eqs. (7-60) and (7-61) for the linear thermal gradient are the same as Eqs. (7-53) and (7-54) for a logarithmic thermal gradient.
The wall thickness of a cylinder made of brittle materials	$h = \dfrac{d_i}{2}\left\{\left(\dfrac{\sigma_\theta + p_i}{\sigma_\theta - p_i}\right)^{1/2} - 1\right\}$ (7-62)
The wall thickness of a cylinder made of ductile materials	$h = \dfrac{d_i}{2}\left\{\left(\dfrac{\sigma_\theta}{\sigma_\theta - 2p_i}\right)^{1/2} - 1\right\}$ (7-63) where σ_θ = permissible working stress in tension, MPa (psi).

CLAVARINO'S EQUATION FOR CLOSED CYLINDERS

Particular	Formula
(Based on the maximum strain energy)	
The general equation for equivalent tangential stress at any radius r	$\sigma'_\theta = (1-2\nu)a + \dfrac{(1+\nu)b}{r^2}$ (7-64)
The general equation for equivalent radial stress at any radius r	$\sigma'_r = (1-2\nu)a - \dfrac{(1+\nu)b}{r^2}$ (7-65) where a and b have the same meaning as in Eqs. (7-25c) and (7-25d)
The wall thickness for cylinders with closed ends	$h = \dfrac{d_i}{2}\left[\left\{\dfrac{\sigma'_\theta + (1-2\nu)p_i}{\sigma'_\theta - (1+\nu)p_i}\right\}^{1/2} - 1\right]$ (7-66) where σ'_θ = permissible working stress in tension, MPa (psi).

BIRNIE'S EQUATIONS FOR OPEN CYLINDERS

Particular	Formula
The equation for equivalent tangential stress at any radius r	$\sigma'_\theta = (1-\nu)a + (1+\nu)\dfrac{b}{r^2}$ (7-67)
The equation for equivalent radial stress at any radius r	$\sigma'_r = (1-\nu)a - (1+\nu)\dfrac{b}{r^2}$ (7-68) where a and b have the same meaning as in Eqs. (7-25c) and (7-25d)

Particular	Formula	
The wall thickness of cylinders with open ends	$$h = \frac{d_i}{2}\left[\left\{\frac{\sigma'_\theta + (1-\nu)p_i}{\sigma'_\theta - (1-\nu)p_i}\right\}^{1/2} - 1\right]$$	(7-69)

BARLOW'S EQUATION

The tangential stress in the wall thickness of cylinder	$$\sigma_\theta = \frac{p_i d_o}{2h}$$ For σ_θ refer to Table 7-1.	(7-70)

TABLE 7-7
Standard thickness of tubes

Diameter, mm (in)	Minimum thickness, mm (in)
25 (1) and over but less than 62.5 (2.5)	2.37 (0.095)
62.5 (2.5) and over but less than 87.5 (3.25)	2.625 (0.105)
87.5 (3.25) and over but less than 100 (4)	3.000 (0.120)
100 (4) and over but less than 125 (5)	3.375 (0.135)
125 (5) and over but less than 150 (6)	3.750 (0.150)
150 (6) and over	$h = 0.0251d_o$

Source: *ASME Boiler and Pressure Vessel Code*, Section 1, 1983.

TABLE 7-8
Comparison of various thick cylinder formulas

Symbols:
$d_o = 2r_o$ = outside diameter of thick cylinder, in; $d_i = 2r_i$ = inside diameter of thick cylinder, in; $h = (d_o - d_i)/2$ = cylinder wall thickness, in; p_i = internal pressure, psi; ν = Poisson's statio (for steel $\nu = 0.3$); σ_θ = tangential stress, psi; σ_r = radial stress, psi; $(\sigma'_\theta)_{\substack{p_o=0\\ r=r_i}}$ = tangential stress, psi

$$R = \frac{d_o}{d_i} = \frac{r_o}{r_i};\ a = \frac{p_i d_i^2 - p_o d_o^2}{d_o^2 - d_i^2};\ b = \frac{d_i^2 d_o^2(p_i - p_o)}{4(d_o^2 - d_i^2)};\ (a')_{p_o=0} = \frac{p_i d_i^2}{d_o^2 - d_i^2};\ (b')_{p_o=0} = \frac{p_i d_i^2 d_o^2}{4(d_o^2 - d_i^2)}$$

Author	Particular	Formula	Remark
1. Birnie	The equation for an equivalent tangential stress at any radius r of a thick cylinder under internal pressure p_i and external pressure p_o	$\sigma'_\theta = (1-\nu)a + (1+\nu)\frac{b}{r^2}$	Eqn. (7-67)* Used for ductile materials
	The equation for an equivalent tangential stress at inner radius r_i of a thick cylinder subject to internal pressure p_i only when $\nu = 0.3$ for steel	$(\sigma'_\theta)_{\substack{p_o=0\\ r=r_i}} = (1-\nu)a' + (1+\nu)\frac{b'}{r_i}$ $= \frac{(1-\nu)p_i + (1+\nu)\left(\frac{d_o}{d_i}\right)^2 p_i}{\left(\frac{d_o}{d_i}\right)^2 - 1}$ $= \frac{p_i[(1-\nu) + (1+\nu)R^2]}{R^2 - 1}$ $= \frac{p_i[0.7 + 1.3R^2]}{R^2 - 1}$	Open ends thick cylinder
2. Clavarino	The general equation for an equivalent tangential stress at any radius r of a thick cylinder under internal pressure p_i and external pressure p_o	$\sigma'_\theta = (1-2\nu)a + (1+\nu)\frac{b}{r^2}$	Eqn. (7-64)* Used for ductile materials
	The equation for an equivalent tangential stress at inner radius r_i of a thick cylinder subject to internal pressure p_i only when $\nu = 0.3$ for steel	$(\sigma'_\theta)_{\substack{p_o=0\\ r=r_i}} = p_i\left[\frac{(1-2\nu)}{\left(\frac{d_o}{d_i}\right)^2 - 1} + \frac{(1+\nu)\left(\frac{d_o}{d_i}\right)^2}{\left(\frac{d_o}{d_i}\right)^2 - 1}\right]$ $= p_i\left[\frac{(1-2\nu)}{R^2-1} + \frac{(1+\nu)R^2}{R^2-1}\right]$ $= p_i\left[\frac{0.4}{R^2-1} + \frac{1.3R^2}{R_2-1}\right]$ $= p_i\left[\frac{0.4 + 1.3R^2}{R^2-1}\right]$	Closed ends cylinder
3. Barlow	The tangential stress in the wall thickness of cylinder under internal pressure p_i	$\sigma_\theta = \frac{p_i d_o}{2h} = p_i \frac{d_0}{d_o - d_i}$ $= p_i \frac{\left(\frac{d_0}{d_i}\right)}{\left(\frac{d_0}{d_i}\right) - 1} = \frac{p_i R}{R - 1}$	Eqn. (7-70)* Open ends cylinder
4. Lamé	The tangential stress in the thick cylinder wall at any radius r subject to internal pressure p_i and external pressure p_o	$\sigma_\theta = a + \frac{b}{r^2}$	Eqn. (7-24a)* Used for brittle materials
	The tangential stress in the thick cylinder wall at inside radius r_i of cylinder subject to internal pressure p_i only when $\nu = 0.3$ for steel	$(\sigma'_\theta)_{\substack{p_o=0\\ r=r_i}} = \frac{p_i d_i^2}{d_o^2 - d_i^2}\left(1 + \frac{d_o^2}{d_i^2}\right)$ $= \frac{p_i}{(d_o/d_i)^2 - 1}\left[1 + \left(\frac{d_o}{d_i}\right)^2\right] = p_i\left[\frac{1+R^2}{R^2-1}\right]$	Closed ends cylinder

* Refer to equations in Lingaiah, K., *Machine Design Data Handbook*, McGraw-Hill Book Company, New York, 1994

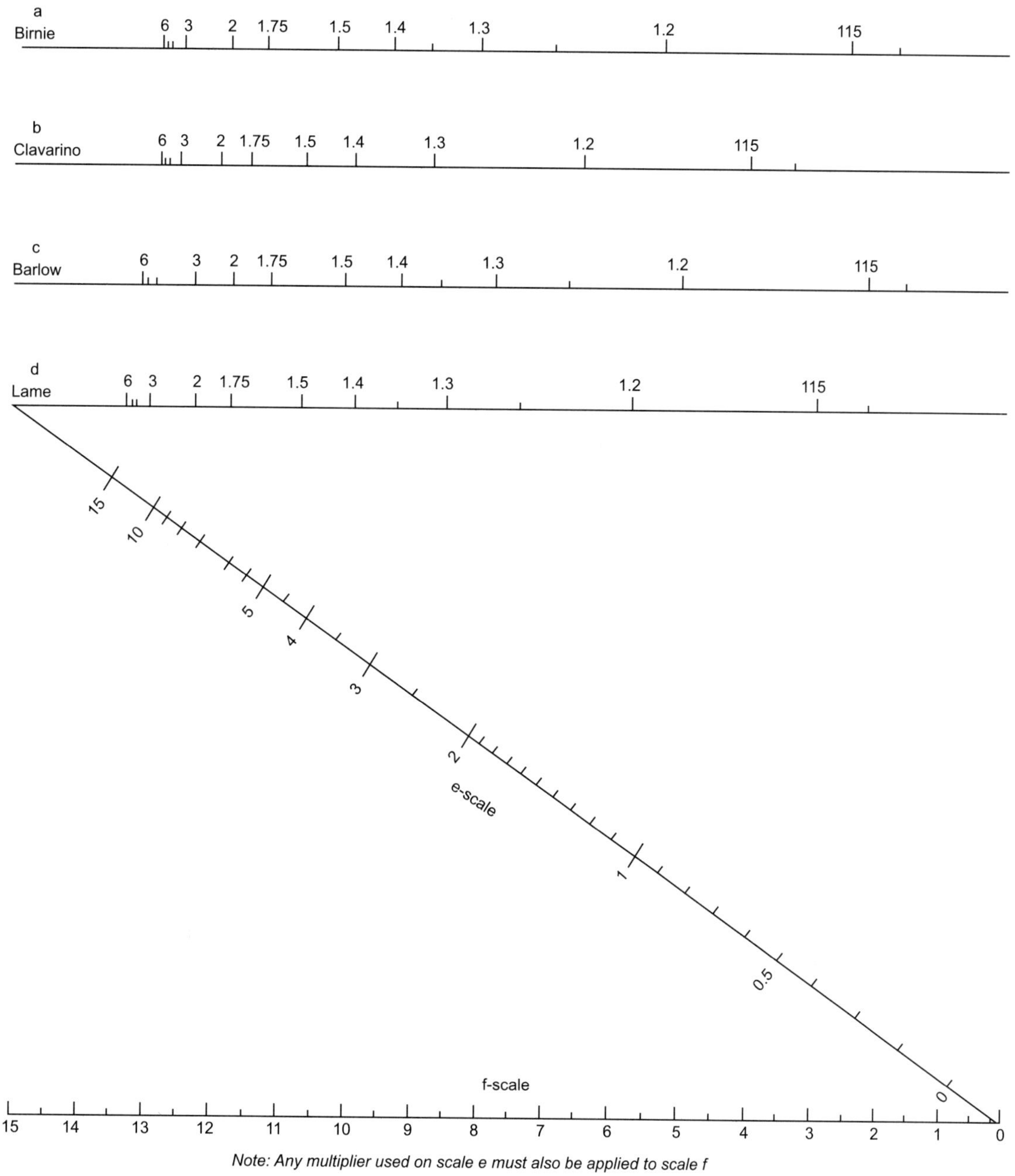

FIGURE 7-2 Nomogram to find the stress in thick cylinder subject to internal pressure using four formulas given in Table 7-8.

Particular	Formula

PROBLEM A closed end cylinder made of ductile material has inner diameter of 10 in (250 mm) and outside diameter of cylinder is 25 in (625 mm). The pressure inside the cylinder is 5000 psi. Use Clavarino's equation from Table 7-8

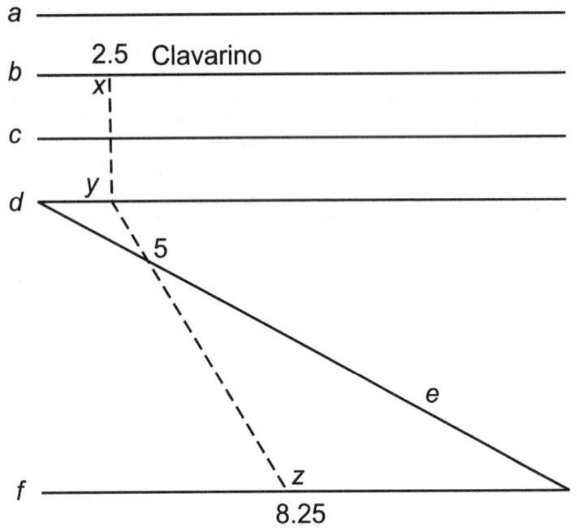

$$R = \frac{d_o}{d_i} = \frac{25}{10} = 2.5$$

Mark on scale b at 2.5

Draw a perpendicular from x and this perpendicular meets scale d at y

Join y and 5 (5000 psi) on scale e. Produce y–5 to meet scale f at z. y–5–z meets scale f at 8.25

Stress = 8.25 = 8250 psi

Stress in SI units = $8250 \times 6.894 \times 10^3 = 56.88$ MPa

Check by using Clavarino's equation from Table 7-8

$$\sigma = p_1 \left[\frac{0.4 + 1.3R^2}{R^2 - 1}\right] = 5000\left[\frac{0.4 + 1.3(2.5)^2}{(2.5)^2 - 1}\right]$$

$$= 5000\left[\frac{0.4 + 8.125}{6.25 - 1}\right] = \frac{4.2625}{5.25} \times 10^4$$

$$= 8120 \text{ psi } (56 \text{ MPa})$$

The stress obtained from nomogram 8250 psi (56.88 MPa) is very close to stress value found from Clavarino's equation

REFERENCES

1. "Rules for Construction of Power Boilers," Section I, *ASME Boiler and Pressure Vessel Code*, American Society of Mechanical Engineers, New York, 1983.
2. "Rules for Construction of Pressure Vessels," Section VIII, Division 1, *ASME Boiler and Pressure Vessel Code*, American Society of Mechanical Engineers, New York, July 1, 1986.
3. "Rules for Construction of Pressure Vessels," Section VIII, Division 2—Alternative Rules, *ASME Boiler and Pressure Vessel Code*, American Society of Mechanical Engineers, New York, July 1, 1986.
4. Nicholas, R. W., *Pressure Vessel Codes and Standards*, Elsevier Applied Science Publications, Crown House, Linton Road, Barking, Essex, England.
5. Lingaiah, K., and B. R. Narayana Iyengar, *Machine Design Data Handbook*, Engineering College Co-operative Society, Bangalore, India, 1962.
6. Lingaiah, K., *Machine Design Data Handbook*, Vol. II (*SI and Customary Metric Units*), Suma Publishers, Bangalore, India, 1986.
7. Courtesy: Durham, H. M., *Stress Chart for Thick Cylinders.*
8. Greenwood, D. C., Editor, *Engineering Data for Product Design*, McGraw-Hill Book Company, New York, 1961.
9. Lingaiah, K., *Machine Design Data Handbook (SI and U.S. Customary Systems Units)*, McGraw-Hill Book Company, New York, 1994.

CHAPTER 8

DESIGN OF PRESSURE VESSELS, PLATES, AND SHELLS

SYMBOLS[13,14,15]

a	length of the long side of a rectangular plate, m (in)
	pitch or distance between stays, m (in)
	major axis of elliptical plate, m (in)
	long span of noncircular heads or covers measured at perpendicular distance to short span, m (in) (see Fig. 8-10)
A	factor determined from Fig. 8-3
A	total cross-sectional area of reinforcement required in the plane under consideration, m^2 (in^2) (see Fig. 8-17) (includes consideration of nozzle area through shell for $\sigma_{sna}/\sigma_{sva} < 1.0$)
A	outside diameter of flange or, where slotted holes extend to the outside of the flange, the diameter to the bottom of the slots, m (in)
A_1	area in excess thickness in the vessel wall available for reinforcement, m^2 (in^2) (see Fig. 8-17) (includes consideration of nozzle area through shell if $\sigma_{sna}/\sigma_{sva} < 1.0$)
A_2	area in excess thickness in the nozzle wall available for reinforcement, m^2 (in^2) (see Fig. 8-17)
A_3	area available for reinforcement when the nozzle extends inside the vessel wall, m^2 (in^2) (see Fig. 8-17)
A_{41}, A_{42}, A_{43}	cross-sectional area of various welds available for reinforcement (see Fig. 8-17), m^2 (in^2)
A_5	cross-sectional area of material added as reinforcement (see Fig. 8-17), m^2 (in^2)
A_b	cross-sectional area of the bolts using the root diameter of the thread or least diameter of unthreaded portion, if less, Eq. (8-111), m (in)
A_m	total required cross-sectional area of bolts taken as the greater of A_{m1} and A_{m2}, m^2 (in^2)
$A_{m1} = W_{m1}/\sigma_{sb}$	total cross-sectional area of bolts at root of thread or section of least diameter under stress, required for the operating condition, m^2 (in^2)
$A_{m2} = W_{m2}/\sigma_{sa}$	total cross-sectional area of bolts at root of thread or section of least diameter under stress, required for gasket seating, m^2 (in^2)

b	length of short side or breadth of a rectangular plate, m (in)
	short span of noncircular head, m (in) (see Fig. 8-10 and Eq. 8-86a)
b	effective gasket or joint-contact-surface seating width, m (in)
b_o	basic gasket seating width, m (in) (see Table 8-21 and Fig. 8-13)
B	factor determined from the application material–temperature chart for maximum temperature, psi
B	inside diameter of flange, m (in)
c	corrosion allowance, m (in)
c	basic dimension used for the minimum sizes of welds, mm (in), equal to t_n or t_x, whichever is less
c_1	empirical coefficient taking into account the stress in the knuckle [Eq. (8-68)]
c_2	empirical coefficient depending on the method of attachment to shell [Eqs. (8-82) and (8-85)]
c_4, c_5	empirical coefficients depending on the mode of support [(Eqs. (8-92) to (8-94)]
C	bolt-circle diameter, mm (in)
d	finished diameter of circular opening or finished dimension (chord length at midsurface of thickness excluding excess thickness available for reinforcement) of nonradial opening in the plane under consideration in its corroded condition, m (in) (see Fig. 8-17)
d	diameter or short span, m (in)
	diameter of the largest circle which may be inscribed between the supporting points of the plate (Fig. 8-11), m (in)
	diameter as shown in Fig. 8-9, m (in)
d	factor, m^3 (in^3)
$d = \frac{U}{V} h_o g_o^2$	for integral-type flanges
$d = \frac{U}{V_L} h_o g_o^2$	for loose-type flanges
d'	diameter through the center of gravity of the section of an externally located stiffening ring, m (in);
	inner diameter of the shell in the case of an internally located stiffening ring, m (in) [Eq. (8-55)]
d_e	outside diameter of conical section or end (Fig. 8-8(A)d), m (in)
d_i, D_i	inside diameter of shell, m (in)
d_o, D_o	outside diameter of shell, m (in)
d_k	inside diameter of conical section or end at the position under consideration (Fig. 8-8(A)d), m (in)
D	inside shell diameter before corrosion allowance is added, m (in)
D_p	outside diameter of reinforcing element, m (in) (actual size of reinforcing element may exceed the limits of available reinforcement)
e	factor, m^{-1} (in^{-1})
$e = \frac{F}{h_o}$	for integral-type flanges
$e = \frac{F_L}{h_o}$	for loose-type flanges
E	modulus of elasticity at the operating temperature, GPa (Mpsi)
E_{am}	modulus of elasticity at the ambient temperature, GPa (Mpsi)

f	hub stress correction factor for integral flanges from Fig. 8-25 (When greater than one, this is the ratio of the stress in the small end of the hub to the stress in the large end. For values below limit of figure, use $f = 1$.)
f_r	strength reduction factor, not greater than 1.0
f_{r1}	$\sigma_{sna}/\sigma_{sva}$
f_{r2}	(lesser of σ_{sna} or $\sigma_{spa})/\sigma_{sva}$
f_{r3}	$\sigma_{spa}/\sigma_{sva}$
F	total load supported, kN (lbf)
	total bolt load, kN (lbf)
F	correction factor which compensates for the variation in pressure stresses on different planes with respect to the axis of a vessel (a value of 1.00 shall be used for all configurations, except for integrally reinforced openings in cylindrical shells and cones)
F	factor for integral-type flanges (from Fig. 8-21)
F_L	factor for loose-type flanges (from Fig. 8-23)
g_a	thickness of hub at small end, m (in)
g_1	thickness of hub at back of flange, m (in)
G	diameter, m (in), at location of gasket load reaction; except as noted in Fig. 8-13, G is defined as follows (see Table 8-22): When $b_o \leq 6.3$ mm (1/4 in), G = mean diameter of gasket contact face, m (in).
	When $b_o > 6.3$ mm (1/4 in), G = outside diameter of gasket contact face less $2b$, m (in).
h	distance nozzle projects beyond the inner or outer surface of the vessel wall, before corrosion allowance is added, m (in)
	(Extension of the nozzle beyond the inside or outside surface of the vessel wall is not limited; however, for reinforcement calculations the dimension shall not exceed the smaller of $2.5t$ or $2.5t_n$ without a reinforcing element and the smaller of $2.5t$ or $2.5t_n + t_e$ with a reinforcing element or integral compensation.)
h	hub length, m (in)
h, t	minimum required thickness of cylindrical or spherical shell or tube or pipe, m (in)
	thickness of plate, m (in)
	thickness of dished head or flat head, m (in)
h_a	actual thickness of shell at the time of test including corrosion allowance, m (in)
h_c	thickness for corrosion allowance, m (in)
h_D	radial distance from the bolt circle, to the circle on which H_D acts, m (in)
$h_G = (C - G)/2$	radial distance from gasket load reaction to the bolt circle, m (in)
$h_o = \sqrt{Bg_o}$	factor, m (in)
h_T	radial distance from the bolt circle to the circle on which H_T acts as prescribed, m (in)
$H = \pi G^2 P/4$	total hydrostatic end force, kN (lbf)
$H_D = \pi B^2 P/4$	hydrostatic end force on area inside of flange, kN (lbf)
$H_G = W - H$	gasket load (difference between flange design bolt load and total hydrostatic end force), kN (lbf)
$H_P = 2b \times \pi GmP$	total joint-contact-surface compression load, kN (lbf)

$H_T = H - H_D$	difference between total hydrostatic end force and the hydrostatic end force on area inside of flange, kN (lbf)
I_s	required moment of inertia of the stiffening ring cross-section around an axis extending through the center of gravity and parallel to the axis of the shell, m^4 or cm^4 (in^4)
I'_s	required moment of inertia of the combined ring-shell cross-section about its neutral axis parallel to the axis of the shell, m^4 (in^4)
I	available moment of inertia of the stiffening ring cross-section about its neutral axis parallel to the axis of the shell, m^4 (in^4)
I'	available moment of inertia of combined ring shell cross-section about its neutral axis parallel to the axis of the shell, m^4 or cm^4 (in^4)
k_1, k_2, k_3, k_4, k_5	coefficients
k_6	factor for noncircular heads depending on the ratio of short span to long span b/a (Fig. 8-10)
$K = A/B$	ratio of outside diameter of flange to inside diameter of flange (Fig. 8-20)
K	ratio of the elastic modulus E of the material at the design material temperature to the room temperature elastic modulus, E_{am}, [Eqs. (8-26) to (8-31), (8-55)]
K_1	spherical radius factor (Table 8-18)
l	length of flange of flanged head, m (in)
L	effective length, m (in)
	distance from knuckle or junction within which meridional stresses determine the required thickness, m (in)
	perimeter of noncircular bolted heads measured along the centers of the bolt holes, m (in)
	distance between centers of any two adjacent openings, m (in)
	length between the centers of two adjacent stiffening rings, m (in) (Fig. 8-1)
$L = \dfrac{te+1}{T} + \dfrac{t^3}{d}$	factor
m	gasket factor, obtained from Table 8-20
$m = 1/\nu$	reciprocal of Poisson's ratio
M_b	longitudinal bending moment, N m (lbf in)

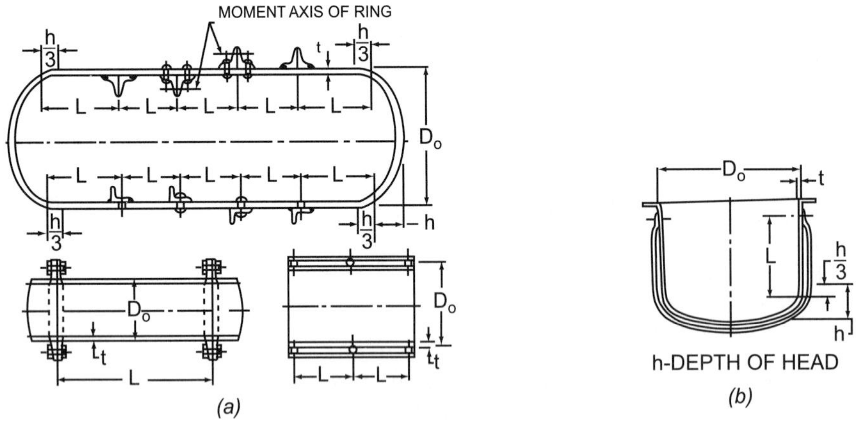

FIGURE 8-1 Cylindrical pressure vessels under external pressure.

M_t	torque about the vessel axis, N m (lbf in)
$M_D = H_D h_D$	component of moment due to H_D, m N (in-lbf)
$M_G = H_G h_G$	component of moment due to H_G, m N (in-lbf)
M_o	total moment acting on the flange, for the operating conditions or gasket seating as may apply, m N (in-lbf)
$M_T = H_T h$	component of moment due to H_T, m N (in-lbf)
N	width, m (in), used to determine the basic gasket seating with b_o, based on the possible contact width of the gasket (see Table 8-21)
p_i	internal design pressure, MPa (psi)
p	maximum allowable working pressure or design pressure, MPa (psi)
p_o	load per unit area, MPa (psi)
	external design pressure, MPa (psi)
P	total pressure on an area bounded by the outside diameter of gasket, kN (lbf)
	design pressure (or maximum allowable working pressure for existing vessels), MPa (psi)
P_a	calculated value of allowable external working pressure for assumed value of t or h, MPa (psi)
r	radius of circle over which the load is distributed, m (in)
r_i	inner radius of a circular plate, m (in)
	inside radius of transition knuckle which shall be taken as $0.01d_k$ in the case of conical sections without knuckle transition, m (in)
R	inner radius of curvature of dished head, m (in)
R_i	inner radius of shell or pipe, m (in)
r_o, R_o	outer radius of a circular plate, m (in)
	outer radius of shell, m (in)
$R = [(C - B)/2] - g_1$	radial distance from bolt circle to point of intersection of hub and back of flange, m (in) (for integral and hub flanges)
R	inside radius of the shell course under consideration, before corrosion allowance is added, m (in)
R_n	inside radius of the nozzle under consideration, before corrosion allowance is added, m (in)
t or h	minimum required thickness of spherical or cylindrical shell, or pipe or tube, m (in)
t	flange thickness, m (in)
t	nominal thickness of the vessel wall, less corrosion allowance, m (in)
t_c	weld dimensions
t_e	thickness or height of reinforcing element, m (in)
t_n	nominal thickness of shell or nozzle wall to which flange or lap is attached, irrespective of product form less corrosion allowance, m (in)
t_r	required thickness of a seamless shell based on the circumferential stress, or of a formed head, computed by the rules of this chapter for the designated pressure, m (in)
t_{rn}	required thickness of a seamless nozzle wall, m (in)
t_s	nominal thickness of cylindrical shell or tube exclusive of corrosion allowance, m (in)
t_w	weld dimensions
t_x	two times the thickness g_o, when the design is calculated as an integral flange, m (in), or two times the thickness, m (in), of shell nozzle wall required for internal pressure, when the design is calculated as a loose flange, but not less than 6.3 mm

(1/4 in)	
T	factor involving K (from Fig. 8-20)
U	factor involving K (from Fig. 8-20)
V	factor for integral-type flanges (from Fig. 8-22)
V_L	factor for loose-type flanges (from Fig. 8-24)
w	width, m (in), used to determine the basic gasket seating width b_o, based on the contact width between the flange facing and the gasket (see Table 8-21)
W	weight, kN (lbf)
W	total load to be carried by attachment welds, kN (lbf)
W	flange design bolt load, for operating conditions or gasket seating, as may apply, kN (lbf)
W_{m1}	minimum required bolt load for the operating conditions, kN (lbf) (For flange pairs used to contain a tubesheet for a floating head for a U-tube type of heat exchanger, or for any other similar design, W_{m1} shall be the larger of the values as individually calculated for each flange, and that value shall be used for both flanges.)
W_{m2}	minimum required bolt load for gasket seating, kN (lbf)
y	gasket or joint-contact-surface unit seating load, MPa (psi)
y	deflection of the plate, m (in)
y_{max}	maximum deflection of the plate, m (in)
Y	factor involving K (from Fig. 8-20)
Z	factor involving K (from Fig. 8-20)
	a factor for non-circular heads [Eq. (8-86b)]
α, α_1, α_2	angles of conical section to the vessel axis, deg (Fig. 8-8(A)d)
ψ	difference between angle of slope of two adjoining conical sections, deg (Fig. 8-8(A)d)
σ	normal or direct stress, MPa (psi)
σ_{sy}	0.2 percent proof stress, MPa (psi)
σ_{sa}	maximum allowable stress value, MPa (psi)
σ_e	equivalent stress (based on shear strain energy), MPa (psi)
σ_{sam}	allowable stress at ambient temperature, MPa (psi)
σ_{sd}	design stress value, MPa (psi)
σ_{sa}	allowable stress value as given in Tables 8-9 to 8-12, MPa (psi)
σ_{sna}	allowable stress in nozzle, MPa (psi)
σ_{sva}	allowable stress in vessel, MPa (psi)
σ_{spa}	allowable stress in reinforcing element (plate), MPa (psi)
σ_{sbat}	allowable bolt stress at atmospheric temperature, MPa (psi)
σ_{sbd}	allowable bolt stress at design temperature, MPa (psi)
σ_{sfd}	allowable design stress for material of flange at design temperature (operating condition) or atmospheric temperature (gasket seating), as may apply, MPa (psi)
σ_{snd}	allowable design stress for material of nozzle neck, vessel or pipe wall, at design temperature (operating condition) or atmospheric temperature (gasket seating), as may apply, MPa (psi)
σ_H	calculated longitudinal stress in hub, MPa (psi)
σ_R	calculated radial stress in flange, MPa (psi)
σ_θ	calculated tangential stress in flange, MPa (psi)
σ_0	hoop stress, MPa (psi)
σ_r	radial stress, MPa (psi)
σ_s	strength, MPa (psi)
σ_{su}	ultimate strength, MPa (psi)
σ_z or σ_l	longitudinal stress, MPa (psi)

σ_{zt}	tensile longitudinal stress, MPa (psi)
σ_{zc}	compressive longitudinal stress, MPa (psi)
τ	shear stress (also with subscripts), MPa (psi)
ν	Poisson's ratio
η	joint factor (Table 8-3) or efficiency
$\eta = 1$	(see definitions for t_r and t_{rn})
$\eta_1 = 1$	when an opening is in the solid plate or joint efficiency obtained from Table 8-3 when any part of the opening passes through any other welded joint

Note: σ and τ with initial subscript s designates strength properties of material used in the design which will be used and observed throughout this *Machine Design Data Handbook*.

Other factors in performance or in special aspect are included from time to time in this chapter and, being applicable only in their immediate context, are not given at this stage.

Particular	Formula	
PLATES[13,14,15]	Refer to Table 8-1	
For maximum stresses and deflections in flat plates		
Plates loaded uniformly		
The thickness of a plate with a diameter d supported at the circumference and subjected to a pressure p distributed uniformly over the total area	$h = k_1 d \left[\frac{p}{\sigma_{sd}}\right]^{1/2}$ Refer to Table 8-2 for values of k_1.	(8-1)
The maximum deflection	$y = k_2 d^4 \frac{p}{Eh^3}$ Refer to Table 8-2 for values of k_2.	(8-2)
Plates loaded centrally		
The thickness of a flat cast-iron plate supported freely at the circumference with diameter d and subjected to a load F distributed uniformly over an area $(\pi d_o^2/4)$	$h = 1.2\left[\left(1 - \frac{0.67d_o}{d}\right)\frac{F}{\sigma_{sd}}\right]^{1/2}$	(8-3)
The deflection	$y = \frac{0.12d^2F}{Eh^3}$	(8-4)
Grashof's formula for the thickness of a plate rigidly fixed around the circumference with the above given type of loading	$h = 0.65\left(\frac{F}{\sigma_{sd}} \ln \frac{d}{d_o}\right)^{1/2}$	(8-5)
	$y = \frac{0.055d^2F}{Eh^3}$	(8-6)

TABLE 8-1
Maximum stresses and deflections in flat plates

Form of plate	Type of loading	Type of support	Eq.	Total load, F	Maximum stress, σ_{max}	Location of σ_{max}	Maximum deflection, y_{max}
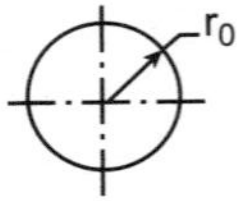	Distributed over the entire surface	Edge supported	8-129	$\pi r_o^2 p$	$\sigma_r = \sigma_\theta = \dfrac{-3F(3m+1)}{8\pi m h^2}$	Center	$\dfrac{3F(m-1)(5m+1)r_o^2}{16\pi E m^2 h^3}$
		Edge fixed	8-130	$\pi r_o^2 p$	$\sigma_r = \dfrac{3F}{4\pi h^2}$	Edge	$\dfrac{3F(m^2-1)r_o^2}{16\pi E m^2 h^3}$
		Edge supported	8-131	$\pi r^2 p$	$\sigma_r = \sigma_\theta = \dfrac{-3F}{2\pi m h^2}\left[(m+1)\log_e \dfrac{r_o}{r} - (m-1)\dfrac{r^2}{4r_o^2} + m\right]$	Center	$\dfrac{3F(m^2-1)}{16\pi E m^2 h^3}\left[\dfrac{(12m+4)r_o^2}{m+1} - 4r^2 \log_e \dfrac{r_o}{r} - \dfrac{(7m+3)}{m+1} r^2\right]$
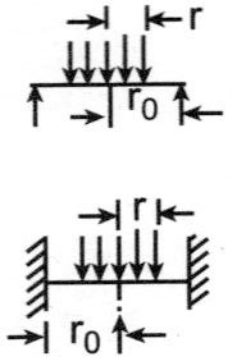	Distributed over a concentric circular area of radius r				$\sigma_r = \dfrac{3F}{2\pi h^2}\left(1 - \dfrac{r^2}{2r_o^2}\right)$	Edge	$\dfrac{3F(m^2-1)}{16\pi E m^2 h^3}\left[4r_o^2 - 4r^2 \log_e \dfrac{r_o}{r} - 3r^2\right]$ when r is very small (concentrated load)
		Edge fixed	8-132	$\pi r^2 p$	$\sigma_r = \sigma_\theta = \dfrac{-3F}{2\pi m h^2}\left[(m+1)\log_e \dfrac{r}{r_o} + (m+1)\dfrac{r^2}{4r_o^2}\right]$	Center	$\dfrac{3F(m^2-1)r_o^2}{4\pi E m^2 h^3}$
		Edge supported	8-133	$2\pi r p$	$\sigma_r = \sigma_\theta = \dfrac{-3F}{2\pi m h^2}\left[\dfrac{m-1}{2} + (m+1)\log_e \dfrac{r_o}{r} - (m-1)\dfrac{r^2}{r_o^2}\right]$	All points inside the circle of radius r	$\dfrac{3F(m^2-1)}{2\pi E m^2 h^3}\left[\dfrac{(3m+1)(r_o^2-r^2)}{2(m+1)} - r^2 \log_e \dfrac{r_o}{r}\right]$
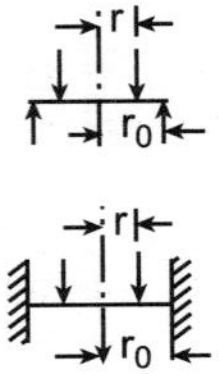	Distributed on circumference of a concentric circle of radius r	Edge fixed	8-134	$2\pi r p$	$\sigma_r = \sigma_\theta = \dfrac{-3F}{4m\pi h^2}\left[(m+1) \times \left(2\log_e \dfrac{r_o}{r} + \dfrac{r^2}{r_o^2} - 1\right)\right]$ $\sigma_r = \dfrac{3F}{2\pi h^2}\left(1 - \dfrac{r^2}{r_o^2}\right)$	Center when $r < 0.31r_o$ Edge when $r > 0.31r_o$	$\dfrac{3F(m^2-1)}{2\pi E m^2 h^3}\left[\tfrac{1}{2}(r_o^2 - r^2) - r^2 \log_e \dfrac{r_o}{r}\right]$

TABLE 8-1
Maximum stresses and deflections in flat plates (*Cont.*)

Form of plate	Type of loading	Type of support	Eq.	Total load, F	Maximum stress, σ_{max}	Location of σ_{max}	Maximum deflection, y_{max}
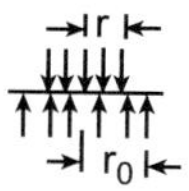	Distributed over a concentric circular area of radius r	Uniform pressure over entire lower surface	8-135	$\pi r^2 p$	$\sigma_r = \sigma_\theta = \frac{-3F}{2\pi m h^2}\left[(m+1)\log_e \frac{r_o}{r} + \frac{m-1}{4}\left(1 - \frac{r^2}{r_o^2}\right)\right]$	Center	$\frac{3F(m^2-1)}{16\pi E m^2 h^3}\left[4r^2 \log_e \frac{r_o}{r} + 2r^2\left(\frac{3m+1}{m+1}\right) - r_o^2\left(\frac{7m+3}{m+1}\right) + \frac{(r_o^2 - r^2)r^4}{r^2 r_o^2} + \frac{r^4}{r_o^2}\right]$ where r is very small (concentrated load) $3F(m-1)(7m+3)r_o^2/16\pi E m^2 h^3$
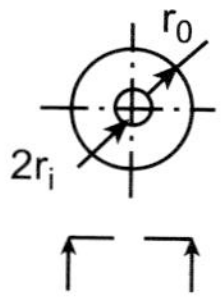	Distributed over the entire surface	Outer edge supported	8-136	$F = \pi(r_o^2 - r_i^2)p$	$\sigma_\theta = \frac{-3P}{4mh^2(r_o^2 - r_i^2)}\left[r_o^4(3m+1) + r_i^4(m-1) - 4mr_o^2 r_i^2 - 4(m+1)r_o^2 r_i^2 \log_e \frac{r_o}{r}\right]$	Inner edge	$\frac{3F(m^2-1)}{2Em^2h^3}\left[\frac{r_o^4(5m+1)}{8(m+1)} + \frac{r_i^4(m+3)}{8(m+1)} - \frac{r_o^2 r_i^2(3m+1)}{2(m+1)} + \frac{r_o^2 r_i^2(3m+1)}{2(m-1)}\log_e \frac{r_o}{r_i} - \frac{2r_o^2 r_i^4(m+1)}{(r_o^2 - r_i^2)(m-1)}\left(\log_e \frac{r_o}{r}\right)^2\right]$
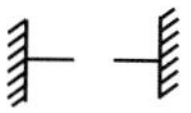	Distributed over the entire surface	Outer edge fixed and supported	8-137	$F = \pi(r_o^2 - r_i^2)p$	$\sigma_\theta = \frac{-3p(m^2-1)}{4mh^2} \times \left[\frac{r_o^2 - r_i^4 - \frac{1}{2}r_o^2 r_i^2 \log_e \frac{r_o}{r_i}}{r_o^2(m-1) + r_i^2(m+1)}\right]$	Inner edge	. . .
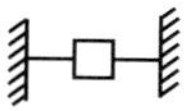	Distributed over the entire surface	Outer edge fixed and supported, inner edge fixed	8-138	$F = \pi(r_o^2 - r_i^2)p$	$\sigma_r = \frac{3p}{4h^2}\left[(r_o^2 + r^2) - \frac{4r_o^2 r^2}{r_o^2 - r^2}\left(\log_e \frac{r_o}{r}\right)^2\right]$	Inner edge	$\frac{3p(m^2-1)}{16Em^2h^3}\left[r_o^4 + 3r_i^4 - 4r_o^2 r_i^2 - 4r_o^2 r_i^2 \log_e \frac{r_o}{r_i} + \frac{16r_o^2 r_i^4}{r_o^2 - r_i^2}\left(\log_e \frac{r_o}{r_i}\right)^2\right]$

TABLE 8-1
Maximum stresses and deflections in flat plates (*Cont.*)

Form of plate	Type of loading	Type of support	Eq.	Total load, F	Maximum stress, σ_{max}	Location of σ_{max}	Maximum deflection, y_{max}
	Distributed over the entire surface	Inner edge fixed and supported	8-139	$F = \pi(r_o^2 - r_i^2)p$	$\sigma_r = \frac{3p}{4h^2}\left[4r_o^4(m+1)\log_e\frac{r_o}{r} - \frac{r_o^4(m+3)+r_o^4(m-1)+4r_o^2r_i^2}{r_o^2(m+1)+r_i^2(m-1)}\right]$	Inner edge	...
	Uniform over entire surface	All edges supported	8-140	$F = abp$	$\sigma_b = \frac{-0.75b^2p}{h^2\left(1+1.61\frac{b^3}{a^3}\right)}$	Center	$\frac{0.1422b^4p}{Eh^3\left(1+2.21\frac{b^3}{a^3}\right)}$
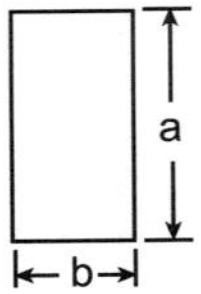	Uniform over entire surface	All edges fixed	8-141	$F = abp$	$\sigma_b = \frac{0.5b^2p}{h^2\left(1+0.623\frac{b^6}{a^6}\right)}$	Center of long edge	$\frac{0.0284b^4p}{Eh^3\left(1+1.056\frac{b^5}{a^5}\right)}$
	Uniform over entire surface	Short edges fixed, long edges supported	8-142	$F = abp$	$\sigma_b = \frac{0.75b^2p}{h^2\left(1+0.8\frac{b^4}{a^4}\right)}$	Center of short edge	...
	Uniform over entire surface	Short edges supported, long edges fixed	8-143	$F = abp$	$\sigma_b = \frac{-b^2p}{2h^2\left(1+0.2\frac{a^4}{b^4}\right)}$	Center of long edge	...

Note: Positive sign for σ indicates tension at upper surface and equal compression at lower surface; negative sign indicates reverse condition.

TABLE 8-2
Coefficients in formulas for cover plates[13,14,15]

		Circular plate		Rectangular plate		Elliptical plate
Material of cover plate	**Methods of holding edges**	k_1	k_2	k_3	k_4	k_5
Cast iron	Supported, free	0.54	0.038	0.75	1.73	1.5
	Fixed	0.44	0.010	0.62	1.4; 1.6[a]	1.2
Mild steel	Supported, free	0.42	. . .	0.60	1.38	1.2
	Fixed	0.35	. . .	0.49	1.12; 1.28	0.9

[a] With gasket.

Particular	Formula

The deflection

Rectangular plates

UNIFORM LOAD

The thickness of a rectangular plate according to Grashof and Bach

$$h = abk_3 \left[\frac{p}{\sigma_{sd}(a^2 + b^2)} \right]^{1/2} \tag{8-7}$$

where k_3 = coefficient, taken from Table 8-2

CONCENTRATED LOAD

The thickness of a rectangular plate on which a concentrated load F acts at the intersection of diagonals

$$h = k_4 \left[\frac{abF}{\sigma_{sd}(a^2 + b^2)} \right]^{1/2} \tag{8-8}$$

where k_4 = coefficient, taken from Table 8-2

Elliptical plate

The thickness of uniformly loaded elliptical plate

$$h = abk_5 \left[\frac{p}{\sigma_{sd}(a^2 + b^2)} \right]^{1/2} \tag{8-9}$$

where k_5 = coefficient, taken from Table 8-2

SHELLS (UNFIRED PRESSURE VESSEL)

Shell under internal pressure—cylindrical shell

CIRCUMFERENCE JOINT

The minimum thickness of shell exclusive of corrosion allowance as per Bureau of Indian Standards[11]

$$h = \frac{pd_i}{2\sigma_{sa}\eta - p} = \frac{pd_o}{2\sigma_{sa}\eta + p} \tag{8-10}$$

Refer to Tables 8-3 and 8-8 for values of η and σ_{sa}, respectively.

TABLE 8-3
Joint efficiency factor (η)[13,14,15]

Requirement	Class 1	Class 2	Class 3		
Weld joint efficiency factor (η)	1.00	0.85	0.70	0.60	0.50
Shell or end plate thickness	No limitation on thickness	Maximum thickness 38 mm after adding corrosion allowance	Maximum thickness 16 mm before corrosion allowance is added	Maximum thickness 16 mm before corrosion allowance is added	Maximum thickness 16 mm before corrosion allowance is added
Type of joints	Double-welded butt joints with full penetration excluding butt joints with metal backing strips which remain in place	Double-welded butt joints with full penetration excluding butt joints with metal backing strips which remain in place	Double-welded butt joints with full penetration excluding butt joints with metal backing strips which remain in place	Single-welded butt joints with backing strip not over 16 mm thickness or over 600 mm outside diameter	Single full fillet lap joints for circumferential seams only
	Single-welded butt joints with backing strip $\eta = 0.9$	Single-welded butt joints with backing strip $\eta = 0.80$	Single-welded butt joints with backing strip $\eta = 0.65$	Single-welded butt joints without backing strip $\eta = 0.55$	

Source: K. Lingaiah and B. R. Narayana Iyengar, *Machine Design Data Handbook*, Engineering College Cooperative Society, Bangalore, India, 1962; K. Lingaiah and B. R. Narayana Iyengar, *Machine Design Data Handbook*, Vol. I (*SI and Customary Metric Units*), Suma Publishers, Bangalore, India, 1983; K. Lingaiah, *Machine Design Data Handbook*, Vol. II (*SI and Customary Metric Units*), Suma Publishers, Bangalore, India, 1986; and IS: 2825-1969.

Particular	Formula
Note: A minimum thickness of 1.5 mm is to be provided as corrosive allowance unless a protective lining is employed. The design pressure or maximum allowable working pressure	$p = \dfrac{2\sigma_{sa}\eta h}{d_i + h} = \dfrac{2\sigma_{sa}\eta h}{d_o - h}$ (8-11)
The minimum thickness of shell exclusive of corrosion allowance as per *ASME Boiler and Pressure Vessel Code**	$t = \dfrac{pR_i}{2\sigma_{sa}\eta + 0.4p}$ (8-12) when the thickness of shell does not exceed one-half the inside radius (R_i)
The maximum allowable working pressure as per *ASME Boiler and Pressure Vessel Code** [from Eq. (8-12)][1,2]	$p = \dfrac{2\sigma_{sa}\eta t}{R_i - 0.4t}$ (8-13) when the pressure p does not exceed $1.25\sigma_{sa}\eta$. σ_{sa} is taken from Tables 8-9, 8-11, and 8-12.

* Rules for construction of pressure vessel, section VIII, Division 1, *ASME Boiler and Pressure Vessel Code*, July 1, 1986.

Particular	Formula
LONGITUDINAL POINT	
The minimum thickness of shell exclusive of corrosive allowance as per *ASME Boiler and Pressure Vessel Code*. * [1-10]	$t = \frac{pR_i}{\sigma_{sa}\eta - 0.6p} = \frac{pR_o}{\sigma_{sa}\eta + 0.4p}$ (8-14) when the thickness of shell does not exceed one-half the inside radius R_i
The maximum allowable working pressure as per *ASME Boiler and Pressure Vessel Code* [from Eq. 8-14)]	$p = \frac{\sigma_{sa}\eta t}{R_i + 0.6t} = \frac{\sigma_{sa}\eta t}{R_o - 0.4t}$ (8-15) when the pressure p does not exceed $0.385\sigma_{sa}\eta$
The design stress for the case of welded cylindrical shell assuming a Poisson ratio of 0.3	$\sigma_d = 0.87 \frac{p_i r_o}{h}$ (8-16)
The allowable stress for plastic material taking into consideration the combined effect of longitudinal and tangential stress (*Note*: The design stress for plastic material is 13.0 percent less compared with the maximum value of the main stress.)	$\sigma_a = \frac{p_i d_o}{2.3h}$ (8-17)
The thickness of shell from Eq. (8-17) without taking into account the joint efficiency and corrosion allowance	$h = \frac{pd_o}{2.3\sigma_{sa}}$ (8-18)
The design thickness of shell taking into consideration the joint efficiency η and allowance for corrosion, negative tolerance, and erosion of the shell (h_c)	$h_d = \frac{pd_o}{2.3\sigma_{sa}\eta} + h_c$ (8-19)
The design formula for the thickness of shell according to Azbel and Cheremisineff [10]	$h_d = \frac{pd_i}{2.3\eta\sigma_{sa} - p} + h_c$ (8-20)
The factor of safety as per pressure vessel code, which is based on yield stress of material used for shell	$n = \frac{\sigma_{sy}}{\sigma_a}$ (8-21) The factor of safety n should not be less than 4, which is based on yield strength σ_{sy} of material.

Shell under internal pressure—spherical shell

The minimum thickness of shell exclusive of corrosion allowance as per Bureau of Indian Standards	$h = \frac{pd_i}{4\sigma_{sa}\eta - p} = \frac{pd_o}{4\sigma_{sa}\eta + p}$ (8-22)
The design pressure as per Bureau of Indian Standards	$p = \frac{4\sigma_{sa}\eta h}{d_i + h} = \frac{4\sigma_{sa}\eta h}{d_o - h}$ (8-23)

* Rules for construction of pressure vessel, section VIII, Division 1, *ASME Boiler and Pressure Vessel Code*, July 1, 1986.

Particular	Formula
The minimum thickness of shell exclusive of corrosion allowance as per *ASME Boiler and Pressure Vessel Code*	$t = \dfrac{pR_i}{2\sigma_{sa}\eta - 0.2p}$ (8-24) when thickness of the shell of a wholly spherical vessel does not exceed $0.356R_i$
The design pressure (or maximum allowable working pressure) as per *ASME Boiler and Pressure Vessel Code*	$p = \dfrac{2\sigma_{sa}\eta t}{R_i + 0.2t}$ (8-25) when the maximum allowable working pressure p does not exceed $0.655\sigma_{sa}\eta$

Shells under external pressure—cylindrical shell (Fig. 8-1)

(a) The minimum thickness of cylindrical shell exclusive of corrosion allowance as per Bureau of Indian Standards

$$h = d_o\left[\frac{1.15p}{\sigma} + 1.1570 \times 10^{-4}\left(\frac{K\sigma L}{d_o}\right)^{2/3}\right] \quad \textbf{SI} \quad (8\text{-}26a)$$

where h, d_o, and L in m; σ and p in MPa and $h = t =$ thickness of shell.

$$h = d_o\left[\frac{1.15p}{\sigma} + 4.19 \times 10^{-6}\left(\frac{K\sigma L}{d_o}\right)^{2/3}\right] \quad \textbf{USCS} \quad (8\text{-}26b)$$

where h, d_o, and L in in; σ and p in psi

The design pressure as per Bureau of Indian Standards

$$p = \frac{\sigma}{1.15}\left[\frac{h}{d_o} - 1.157 \times 10^{-4}\left(\frac{K\sigma L}{d_o}\right)^{2/3}\right] \quad \textbf{SI} \quad (8\text{-}27a)$$

where p and σ in MPa; h, d_o, and L in m

$$p = \frac{\sigma}{1.15}\left[\frac{h}{d_o} - 4.19 \times 10^{-6}\left(\frac{K\sigma L}{d_o}\right)^{2/3}\right] \quad \textbf{USCS} \quad (8\text{-}27b)$$

where p and σ in psi; h, d_o, and L in in

$$\text{for } \frac{L}{d_o} < \frac{5.7(10p/\sigma)^{5/2}}{pK} \text{ or } < \frac{372.65 \times 10^3 (h/d_o)^{3/2}}{K\sigma} \quad \textbf{SI} \quad (8\text{-}27c)$$

where σ and p in MPa; d_o, h, and L in m

$$\text{for } \frac{L}{d_o} < \frac{5.7(10p/\sigma)^{5/2}}{pK} \text{ or } < \frac{5.41 \times 10^7 (h/d_o)^{3/2}}{K\sigma} \quad \textbf{USCS} \quad (8\text{-}27d)$$

where σ and p in psi; L, d_o and h in in

Particular	Formula
(b) The minimum thickness of cylindrical shell exclusive of corrosion allowance according to Bureau of Indian Standards[11]	$h = 2.234 \times 10^{-4} d_o (pK)^{1/3}$ but not less than $(3.5/2)(pd_o/\sigma)$ **SI** (8-28a) where d_o and h in m and p in MPa
	$h = 4.25 \times 10^{-3} d_o (pK)^{1/3}$ but not less than $(3.5/2)(pd_o/\sigma)$ **USCS** (8-28b) where d_o and h in in and p in psi
	or
The design pressure as per Bureau of Indian Standards from Eq. (8-28)	$p = \dfrac{8.97 \times 10^{10}}{K}\left(\dfrac{h}{d_o}\right)^3$ but not greater than $\dfrac{2h\sigma}{3.5d_o}$ **SI** (8-29a) where p in MPa and h and d_o in m
	$p = \dfrac{13 \times 10^6}{K}\left(\dfrac{h}{d_o}\right)^3$ but not greater than $\dfrac{2}{3.5}\dfrac{h\sigma}{d_o}$ **USCS** (8-29b) where p in psi and h and d_o in in
	for $\dfrac{L}{d_o} > \dfrac{97.78}{(pK)^{1/6}}$ or $> \dfrac{14.6}{(100h/d_o)^{1/2}}$ **SI**
	for $\dfrac{L}{d_o} > \dfrac{22.4}{(pK)^{1/6}}$ or $> \dfrac{1.46}{(h/d_o)^{1/2}}$ **USCS**
	or $5.7\dfrac{(10p/\sigma)^{5/2}}{pK} > \dfrac{97.78}{(pK)^{1/6}}$ **SI**
	$0.58\dfrac{(10p/\sigma)^{5/2}}{pK} > \dfrac{22.4}{(pK)^{1/6}}$ **USCS**
	or $372.65 \times 10^3 \dfrac{(h/d_o)^{3/2}}{K\sigma} > \dfrac{1.46}{(h/d_o)^{1/2}}$ **SI**
	$54.1 \times 10^6 \dfrac{(h/d_o)^{3/2}}{K\sigma} > \dfrac{1.46}{(h/d_o)^{1/2}}$ **USCS**
(c) In other cases, the minimum thickness of the shell exclusive of corrosion allowance as per Bureau of Indian Standards	$h = 3.576 \times 10^{-5} d_o \left(p\dfrac{L}{d_o}K\right)^{2/5}$ **SI** (8-30a) where h, d_o, and L in m; p in MPa

Particular	Formula
	$$h = 1.227 \times 10^{-3} d_o \left(p \frac{L}{d_o} K\right)^{2/5}$$ **USCS** (8-30b) where h, L, and d_o in in; p in psi or
The design pressure as per Bureau of Indian Standards	$$p = \frac{3.162 \times 10^{12} (h/d_o)^{5/2}}{LK/d_o}$$ **SI** (8-31a) where h, L, and d_o in m; p in MPa $$h = \frac{189.58 \times 10^{6} (h/d_o)^{5/2}}{LK/d_o}$$ **USCS** (8-31b) where h, d_o, and L in in; p in psi
Reference Chart for *ASME Boiler and Pressure Vessel Code,* Section VIII, Division 1[12]	Refer to Fig. 8-2.
(d) Maximum allowable stress values	
(1) The maximum allowable stress values in tension for ferrous and nonferrous materials σ_{sa}	Refer to Tables 7-1, 8-8 and 8-13 for σ_{sa}.
The maximum allowable stress values (σ_{sa}) for bolt, tube, and pipe materials	Refer to Tables 7-1, 8-8, 8-12 and 8-17.

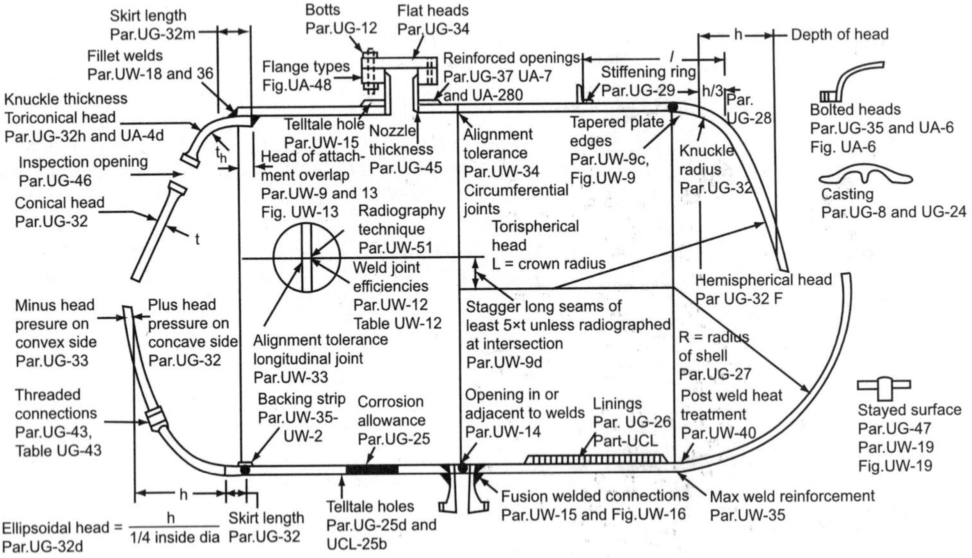

FIGURE 8-2 Reference chart for *ASME Boiler and Pressure Vessel Code*, Section VIII, Division 1. (*By permission*, Robert Chuse, *Pressure Vessels—The ASME Code Simplified*, 5th edition, McGraw-Hill, 1977.)[12]

Particular	Formula
(2) The maximum allowable longitudinal compressive stress (σ_{ac}) to be used in the design of cylindrical shells or tubes, either seamless or butt-welded subjected to loadings that produce longitudinal compression in shell or tube. shall be as given in either Eq. (a) or (b).	$\sigma_{ac} < \sigma_{sa}$ from Tables 7-1, 8-9 to 8-13 (a) $\sigma_{ac} < B$ (b) where B = a factor determined from the applicable material/temperature chart for maximum design temperature, psi, Figs. 8-4, 8-5. [*Note*: **US Customary units** (i.e., fps system of units) were used in drawing Figs. 8-3 to 8-5 of *ASME Pressure Vessel and Boiler Code*, which is now used to find the thickness of walls of cylindrical and spherical shells and tubes, unless it is otherwise mentioned to use both **SI** and **US Customary units**. Figures 8-3 to 8-5 are in US Customary units. The values from these figures and others can be used in the appropriate equation to find the values or results in SI units, if these values and equations are converted into SI units beforehand.]
(3) The procedure for determining the value of the factor B	Select the thickness t (= h) and outside diameter D_o or outside radius R_o of a cylindrical shell or tube in the corroded condition. Then calculate the value of A from Eq. (8-32)
The value of factor A	$A = \dfrac{0.125}{R_o/t}$ (8-32) Using this value of A enter the applicable material/temperature chart for the material (Figs. 8-4 and 8-5) under consideration to find B. In case the value of A falls to the right of the end of the material/temperature line (Figs. 8-4 and 8-5), assume an intersection with the horizontal projection of the upper end of the material/temperature line. From the intersection move horizontally to the right and find the value of B. This is the maximum allowable compressive stress for the value of t and R_o assumed. If the value of A falls to the left of the applicable material/temperature line, the value of B, psi, shall be calculated from Eq. (8-33).
The expression for value of factor B	$B = \dfrac{AE}{2}$ (8-33) where E = modulus of elasticity of material at design temperature, psi Compare the value of B determined from Eq. (8-33) or from the procedure outlined above with the computed longitudinal compressive stress in the cylindrical shell or tube using the selected values of t and R_o. If the value of B is smaller than the computed, compressive stress, a greater value of t must be

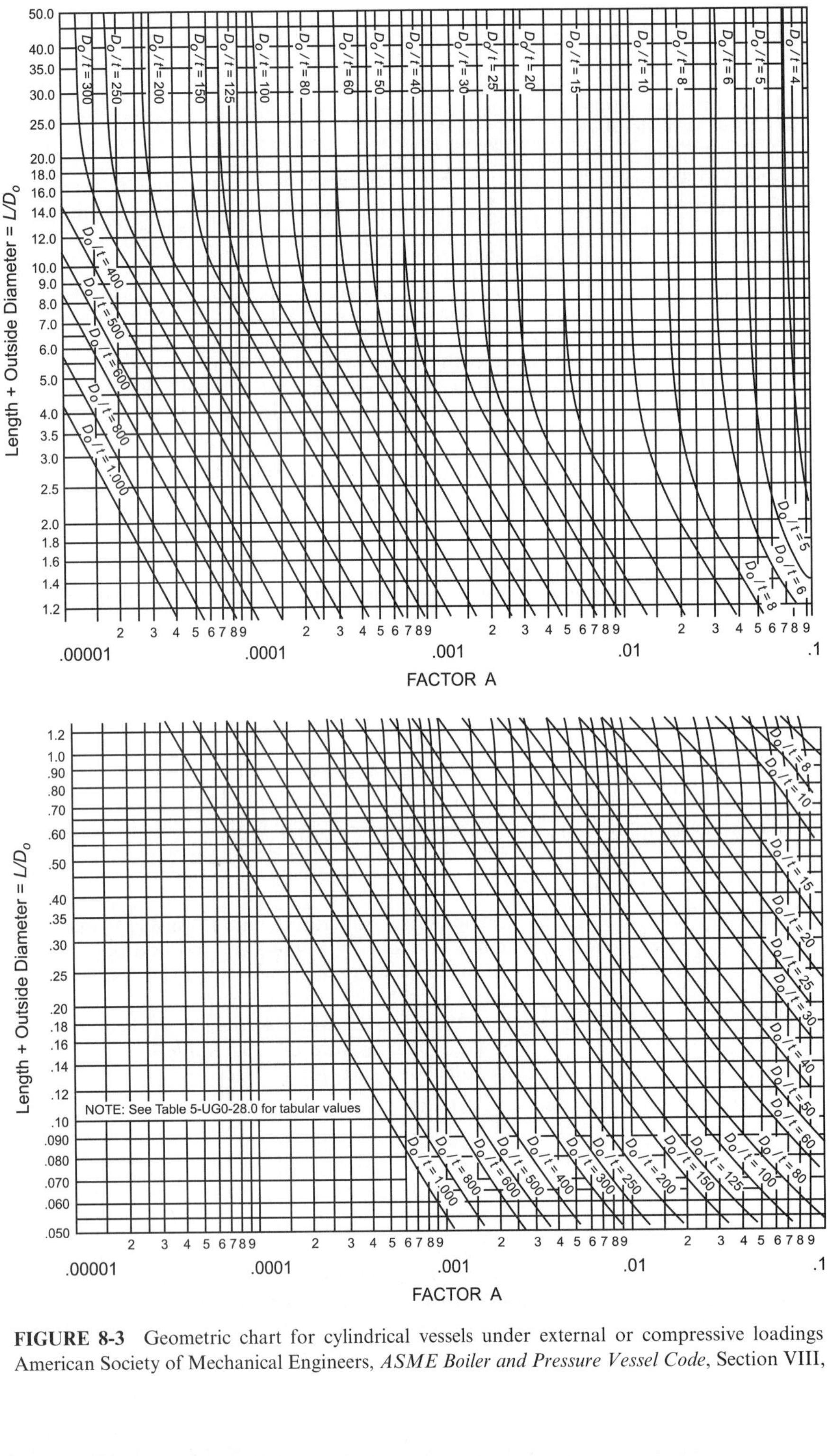

FIGURE 8-3 Geometric chart for cylindrical vessels under external or compressive loadings (for all materials). (*Source*: American Society of Mechanical Engineers, *ASME Boiler and Pressure Vessel Code*, Section VIII, Division 1, July 1, 1986.)[1,2,3]

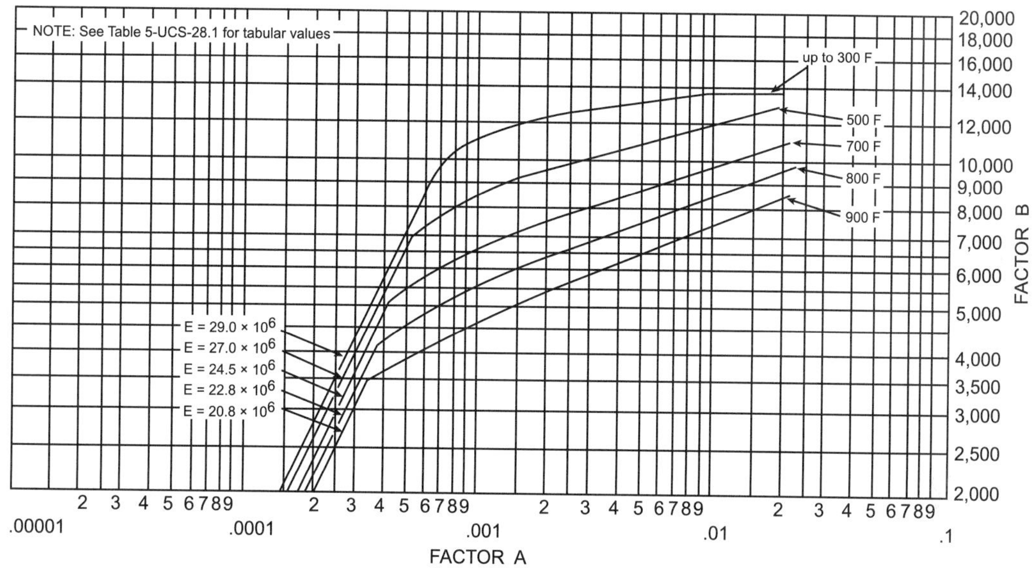

FIGURE 8-4 Chart for determining shell thickness of cylindrical and spherical vessels under external pressure when constructed of carbon or low-alloy steels (specified minimum yield strength 24,000 psi to, but not including, 30,000 psi); (1 kpsi = 6.894757 MPa).[1,2,3]

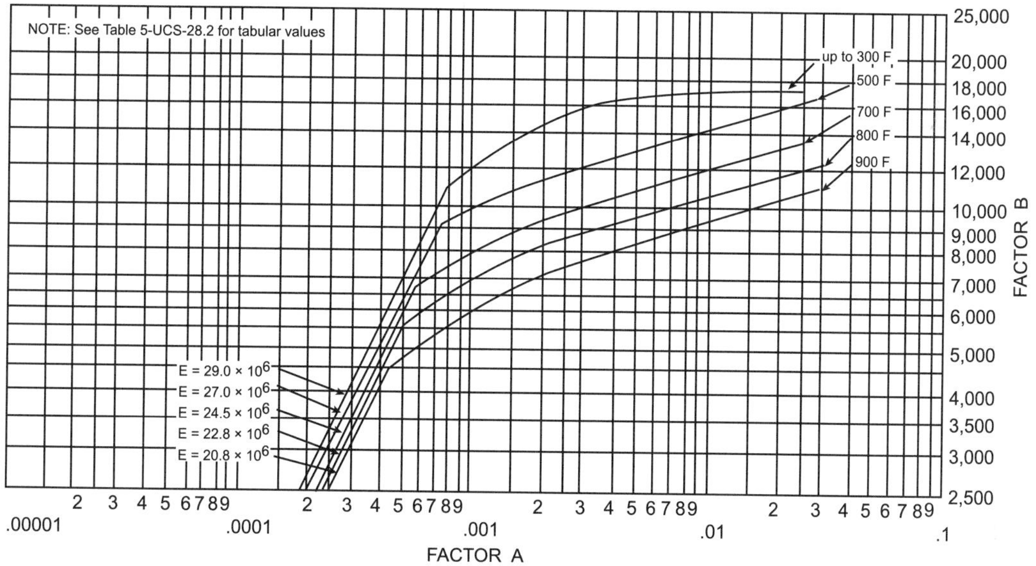

FIGURE 8-5 Chart for determining shell thickness of cylindrical and spherical vessels under external pressure when constructed of carbon or low-alloy steels (specified minimum yield strength 30,000 psi and over except for materials within this range where other specific charts are referenced) and type 405 and type 410 stainless steels (1 kpsi = 6.894757 MPa). (*Source*: American Society of Mechanical Engineers, *ASME Boiler and Pressure Vessel Code*, Section VIII, Division 1, July 1, 1986.)[1,2,3]

Particular	Formula
	selected and the procedure outlined above is repeated until a value of B is obtained, which is greater than the compressive stress computed for the loading on the cylindrical shell or tube.
(e) Cylindrical shells and tubes. The required thickness of cylindrical shell or tube exclusive of corrosion allowance under external pressure either seamless or with longitudinal butt-welded joint as per *ASME Boiler and Pressure Vessel Code* can be determined by the following procedure:	
(1) Cylinders having (D_o/t) values ≥ 10. Assume the thickness t of shell or tube. Determine D_o/t and L/D_o. Use Fig. 8-3 to find A. Find the value of A from Fig. 8-3 by following the procedure explained in paragraph (d) (3)	In cases where the value of A falls to the right of the end of the material/temperature line, assume an intersection with the horizontal projection of the upper end of the material/temperature line. Using this value of A enter the applicable material/temperature chart for material (Figs. 8-4 and 8-5) under consideration and find the value of B. This value of B is the maximum allowable compressive stress for the value of t and R_o assumed, Pa (psi).
The equation for maximum allowable external pressure (P_a) by using this value of B	$P_a = \dfrac{4B}{3(D_o/t)}$ (8-34)
The equation for maximum allowable external pressure P_a for values of A falling to the left of the applicable material/temperature line.	$P_a = \dfrac{2AE}{3(D_o/t)}$ (8-35) where P_a obtained from Eq. (8-35) is equal to or greater than P. P is the external design pressure, psi. This external allowable pressure is 15 psi (103.4 kPa) or less. The maximum external pressure is 15 psi (103.4 kPa) or 25% more than the maximum possible external pressure, whichever is smaller.
(2) Cylinders having (D_o/t) values <10. Using the procedure as outlined in section (d)(3), obtain the value of B. For values of (D_o/t) less than 4, the value of A can be calculated using Eq. (8-36)	$A = \dfrac{1.1}{(D_o/t)^2}$ (8-36) For values of A greater than 0.10, use a value of 0.10
The formula to calculate the value of P_{a1}	$P_{a1} = \left[\dfrac{2.167}{D_o/t} - 0.0833\right] B$ (8-37)
The formula to calculate the value of P_{a2}	$P_{a2} = \dfrac{2\sigma_s}{D_o/t}\left[1 - \dfrac{1}{D_o/t}\right]$ (8-38) where σ_s is the lesser of two times the maximum allowable stress value at design metal temperature, from the applicable Tables 8-9 to 8-12 or 0.9 times the yield strength of the

Particular	Formula
	material at design temperature. The yield strength values are twice the B value obtained from the applicable material/temperature chart.
	The smaller of the values of P_{a1} or P_{a2} shall be used for the maximum allowable external pressure P_a. Thus P_a obtained is equal to or greater than the design pressure P.

Shell under external pressure—spherical shell

Particular	Formula
The thickness of a spherical shell as per Bureau of Indian Standards	$h = \dfrac{pd_o}{0.80\sigma_{sa}}$ (8-39)
The design pressure as per Indian Standards	$p = \dfrac{0.80\sigma_{sa}h}{d_o}$ (8-40)
The minimum required thickness of a spherical shell exclusive of corrosion allowance under external pressure, either seamless or of built-up construction with butt joints, shall be determined by the following procedure as per *ASME Boiler and Pressure Vessel Code*.	
Select a value for t. Determine D_o/L and D_o/t. Find the value of A by using Fig. 8-3.	
The value of the factor A is also calculated from Eq. (8-41). Using this value of A, find the value of B from the applicable material/temperature chart as done in case of the cylindrical shell	$A = \dfrac{0.125}{R_o/t}$ (8-41) where R_o is the outside radius of spherical shell in the corroded condition, in
The maximum allowable external pressure P_a for values of A falling to the right of the applicable material/temperature line	$P_a = \dfrac{B}{R_o/t}$ (8-42) where P_a is the calculated value of allowable external working pressure for the assumed value of t, Pa (psi), and P is the external design pressure, Pa (psi)
The maximum allowable external pressure P_a for values of A falling to the left of the applicable material/temperature line	$P_a = \dfrac{0.0625E}{(R_o/t)^2}$ (8-43) The smaller value of P_a from Eq. (8-42) or (8-43) shall be used for the maximum allowable external pressure P_a. P_a obtained is equal to or greater than the design pressure P.

Particular	Formula
For finding the thickness of a shell in the design of a longitudinal lap joint in a cylindrical or any lap joint in a spherical shell under external pressure	The thickness of the shell shall be determined by the rules already narrated for the longitudinal butt joint of the cylindrical and spherical shell, except that $2P$ shall be used instead of P in the calculations for the required thickness.
Cylindrical shell under combined loading as per Indian Standards	
The longitudinal stress	$\sigma_z = \dfrac{\dfrac{\pi}{4} p d_i^2 + W \pm 4 \dfrac{M_b}{d_i}}{\pi h(d_i + h)}$ (8-44)
The hoop stress	$\sigma_\theta = \dfrac{p(d_i + h)}{2h}$ (8-45)
The shear stress	$\tau = \dfrac{2M_t}{\pi h d_i (d_i + h)}$ (8-46)
The Huber-Hencky equation for equivalent stress based on the shear strain energy criterion	$\sigma_e = \sqrt{\sigma_\theta^2 - \sigma_\theta \sigma_z + \sigma_z^2 + 3\tau^2}$ (8-47)
The basic design stress based on distortion energy theory	$\sigma_d = [\sigma_\theta^2 + \sigma_z^2 + \sigma_r^2 - 2(\sigma_\theta \sigma_z + \sigma_z \sigma_r + \sigma_r \sigma_\theta)]^{1/2}$ (8-48)
Requirements are:	
(a) At design conditions	$\sigma_e \le \sigma_{sa}$ (8-49)
	$\sigma_{zt} \le \sigma_{sa}$ (8-50)
	$\sigma_{ze} \le 0.125E(h/d_o)$ (8-51)
	Refer to Table 8-14 for values of E.
(b) At test conditions	$\sigma_e \le 1.3\sigma_{sam}$ (8-52)
	$\sigma_{zt} \le 1.3\sigma_{sam}$ (8-53)
	$\sigma_{zc} \le 0.125E_{sam}(h_a/d_o)$ (8-54)
Stiffening rings for cylindrical shells under external pressure	
The moment of inertia of the stiffening rings as per Indian Standards	$I_s = 0.714 \times 10^{-6} p L_s d'^3 K$ **SI** (8-55a) where I_s in m^4, p in Pa, L_s and d' in m
	$I_s = 4.29 \times 10^{-3} p L_s d'^3 K$ **USCS** (8-55b) where I_s in in^4, p in psi, L_s and d' in in

Particular	Formula
The required moment of inertia of a circumferential stiffening ring shall be not less than that determined by one of the formulas given in Eqs. (8-56) and (8-57) as per *ASME Boiler and Pressure Vessel Code*	$I_s = \dfrac{D_o^2 L_s[t + (A_s/L_s)A]}{14}$ **USCS** (8-56) $I'_s = \dfrac{D_o^2 L_s[t + (A_s/L_s)A]}{10.9}$ **USCS** (8-57)
Select a member to be used for stiffening a ring after knowing D_o, L_s, and t of a shell designed already. Then calculate factor B using Eq. (8-58)	
The expression for factor B	$B = \dfrac{3}{4}\left(\dfrac{PD_o}{t + A_s/L_s}\right)$ (8-58)
For calculating factor A	Use the applicable material/temperature chart to find A
For values of B falling below the left end of the material/temperature chart line for the design temperature the value of A can be determined from Eq. (8-59)	$A = \dfrac{2B}{E}$ (8-59)

FORMED HEADS UNDER PRESSURE ON CONCAVE SIDE

For domed ends of hemispherical, semiellipsoidal, or dished shape	Refer to Figs. 8-6 for domed end.
The required thickness at the thinnest point after forming of ellipsoidal, torispherical, hemispherical, conical, and toriconical heads under pressure on the concave side of the shell shall be computed by the appropriate formulas	
The thickness of the ends and/or heads under pressure on concave side (plus heads) as per Indian Standards	$h = \dfrac{pd_oC}{2\eta\sigma_{sa}}$ (8-60) where C is a shape factor taken from Fig. 8-7
The allowable pressure as per Indian Standards	$p = \dfrac{2\eta h\sigma_{sa}}{d_oC}$ (8-61)

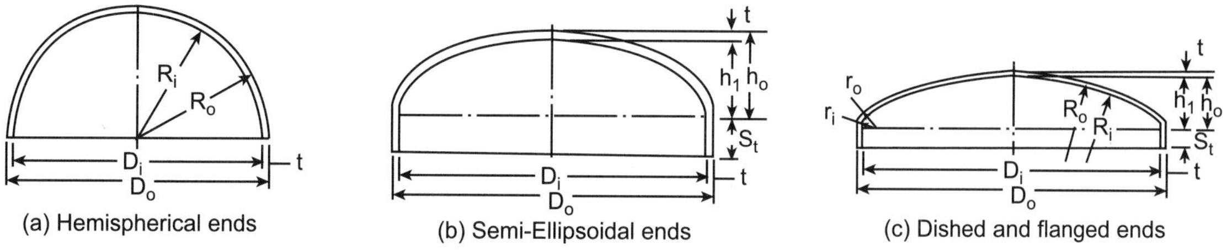

FIGURE 8-6 Domed ends.

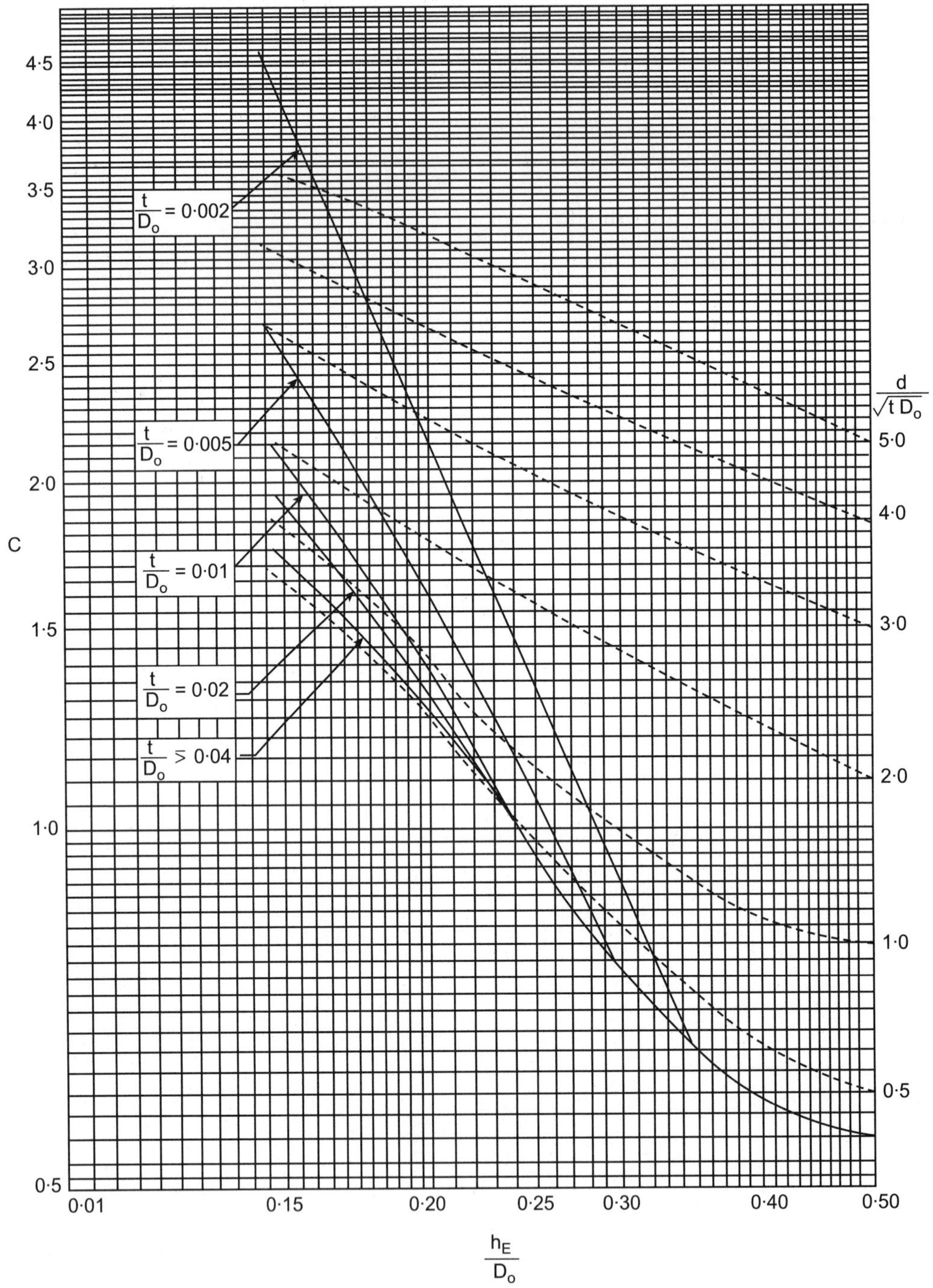

FIGURE 8-7 Shape factor *C* for domed ends.[11]

Particular	Formula	
Ellipsoidal heads		
The required thickness of a dished head of semi-ellipsoidal form. in which half the minor axis (inside depth of the head minus the skirt) equals one-fourth of the inside diameter of the head skirt, shall be determined by Eq. (8-62) as per *ASME Boiler and Pressure Vessel Code*	$t = \dfrac{PD}{2\sigma_{sa}\eta - 0.2P}$	(8-62)
The maximum allowable working pressure or design pressure as per *ASME Boiler and Pressure Vessel Code*	$p = \dfrac{2\sigma_{sa}\eta t}{D + 0.2t}$	(8-63)
Torispherical heads		
The required thickness of a torispherical head for the case in which the knuckle radius is 6 percent of the inside crown radius and the inside crown radius equals the outside diameter of the skirt, shall be determined by Eq. (8-64) as per *ASME Boiler and Pressure Vessel Code*	$t = \dfrac{0.885PL}{\sigma_{sa}\eta - 0.1P}$	(8-64)
The maximum allowable working pressure as per *ASME Boiler and Pressure Vessel Code*	$P = \dfrac{\sigma_{sa}\eta t}{0.885L + 0.1t}$	(8-65)
Hemispherical heads		
The required thickness of a hemispherical head when its thickness does not exceed $0.36L$ or P does not exceed $0.665\sigma_{sa}\eta$, shall be determined by Eq. (8-66) as per *ASME Boiler and Pressure Vessel Code*	$t = \dfrac{PL}{2\sigma_{sa}\eta - 0.2P}$	(8-66)
The design pressure	$P = \dfrac{2\sigma_{sa}\eta t}{L + 0.2t}$	(8-67)
Conical ends subject to internal pressure (Fig. 8-8d) as per Indian Standards		
KNUCKLE OR CONICAL SECTION The thickness of cylinder and conical section (frustrum) within the distance L from the junction $(L = 0.5\sqrt{d_e h}/\cos\alpha)$	$h = \dfrac{pd_e c_1}{2\sigma_{sa}\eta}$ Refer to Table 8-4 for values of c_1.	(8-68)
The thickness of those parts of conical sections not less than a distance L away from the junction with the cylinder or other conical section	$h = \dfrac{pd_k}{2\sigma_{sa}\eta - p}\left(\dfrac{1}{\cos\alpha}\right)$	(8-69)

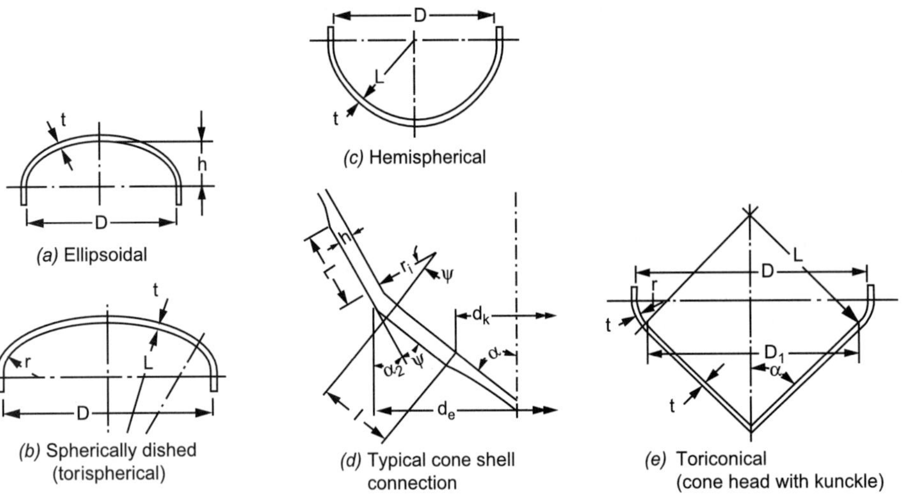

FIGURE 8-8(A) (a) Ellipsoidal; (b) spherically dished (torispherical); (c) hemispherical; (d) typical conical shell connection; and (e) toriconical (cone head with knuckle).

Particular	Formula
SHALLOW CONICAL SECTIONS The thickness of conical sections having an angle of inclination to the vessel axis more than 70°	$h = 0.5(d_e - r_i)\,\dfrac{\alpha}{90}\,\sqrt{p/\sigma_{sa}}$ (8-70) The lower of values given by Eqs. (8-69) and (8-70) shall be used.

Conical heads (without transition knuckle) as per *ASME Boiler and Pressure Vessel Code*

The required thickness of conical heads or conical shell sections that have a half-apex angle α not greater than 30° shall be determined by Eq. (8-71)	$t = \dfrac{PD}{2\cos\alpha(\sigma_{sa}\eta - 0.6P)}$ (8-71) where D = inside diameter η = minimum joint efficiency, percent

TABLE 8-4
Values of c_1 for use in Eq. (18-68) (as function of Ψ and r_i/d_e)

r_i/d_e / Ψ	0.01	0.02	0.03	0.04	0.06	0.08	0.10	0.15	0.20	0.30	0.40	0.50
10°	0.70	0.65	0.60	0.60	0.55	0.55	0.55	0.55	0.55	0.55	0.55	0.55
20°	1.00	0.90	0.85	0.80	0.70	0.65	0.60	0.55	0.55	0.55	0.55	0.55
30°	1.35	1.20	1.10	1.00	0.90	0.85	0.80	0.70	0.55	0.55	0.55	0.55
45°	2.05	1.85	1.65	1.50	1.30	1.20	1.10	0.95	0.90	0.70	0.55	0.55
60°	3.20	2.85	2.55	2.35	2.00	1.75	1.60	1.40	1.25	1.00	0.70	0.55
75°	6.80	5.85	5.35	4.75	3.85	3.50	3.15	2.70	2.40	1.55	1.00	0.55

Source: IS 2825, 1969.

Particular	Formula
The design pressure	$$P = \frac{2\sigma_{sa}\eta t \cos\alpha}{D + 1.2t\cos\alpha} \quad (8\text{-}72)$$

Toriconical heads

The required thickness of the conical portion of a toriconical head, in which the knuckle radius is neither less than 6 percent of the outside diameter of the head skirt nor less than three times the knuckle thickness and pressure shall be determined by Eqs. (8-73) and (8-74)	$$t_c = \frac{PD_i}{2\cos\alpha(\sigma_{sa}\eta - 0.6P)]} \quad (8\text{-}73)$$ $$P = \frac{2\sigma_{sa}\eta t \cos\alpha}{D_i + 1.2t\cos\alpha} \quad (8\text{-}74)$$ where D_i = inside cone diameter at point of tangency to knuckle
The required thickness of the knuckle and pressure shall be determined by Eqs. (8-75) and (8-76)	$$t_k = \frac{PLM}{2\sigma_{sa}\eta - 0.2P} \quad (8\text{-}75)$$ or refer to Eqs. (8-66) and (8-67) where M = factor depending on head proportion, L/r; $L = D_i/2\cos\alpha$. Toriconical heads may be used when the angle $\alpha \le 30^\circ$
The design pressure	$$P = \frac{2\sigma_{sa}\eta t_k}{ML + 0.2t_k} \quad (8\text{-}76)$$

FORMED HEADS UNDER PRESSURE ON CONVEX SIDE

The thickness of heads and ends under pressure on convex side (minus heads) as per Indian Standards

(a) Spherically dished ends and heads	The thickness of the spherically dished heads and ends shall be the greater of the following thicknesses: (1) The thickness of an equivalent sphere, having a radius r_o or R_o equal to the outside crown radius of the end as determined from Eq. (8-39) (2) The thickness of the end under an internal pressure equal to 1.2 times the external pressure
(b) Ellipsoidal heads	The thickness of ends of a semiellipsoidal shape shall be the greater of the following: (1) The thickness of an equivalent sphere, having a radius r_o or R_o calculated from the values of r_o/d_o or R_o/D_o in Table 8-5, determined as per Eq. (8-39)

Particular	Formula
	(2) The thickness of the end under an internal pressure equal to 1.2 times the external pressure
(c) Conical heads under external pressure: For a conical end or conical section under external pressure, whether the end is of seamless or butt-welded construction as per Indian Standards.	Use Eqs. (8-68), (8-69), and (8-70). Equations (8-68) to (8-70) are applicable, except that the shell thickness shall be no less than as prescribed below: (1) The thickness of a conical end or conical section under external pressure, when the angle of inclination of the conical section to the vessel axis is not more than 70°, shall be made equal to the required thickness of cylindrical shell, in which the diameter is $(d_e/\cos\alpha)$ and the effective length is equal to the slant height of the cone or conical section, or slant height between the effective stiffening rings, whichever is less. (2) The thickness of conical ends having an angle of inclination to the vessel axis of more than 70° shall be determined as a flat cover.

The thickness of formed heads under pressure on convex side (minus heads) as per *ASME Boiler and Pressure Vessel Code*

The required thickness at the thinnest point after forming of ellipsoidal or torispherical heads under pressure on the convex side (minus heads) shall be the greater of the thicknesses given here

(1) The thickness as computed by the procedure given for the heads with the pressure on the concave side of the previous section using a design pressure 1.67 times the design pressure of the convex side, assuming the joint efficiency $\eta = 1.00$ for all cases, *or*

(2) The thickness as determined by the appropriate procedure given in *Ellipsoidal heads* or *Torospherical heads* as per *ASME Boiler and Pressure Vessel Code*

HEMISPHERICAL HEADS

The required thickness of a hemispherical head having pressure on the convex side shall be determined by Eqs. (8-41) to (8-43)

$$A = \frac{0.125}{R_o/t} \tag{8-41}$$

$$P_a = \frac{B}{R_o/t} \tag{8-42}$$

TABLE 8-5
Values of spherical radius factor $K_o = R_o/D_o$ for ellipsoidal head with pressure on convex side as a function of h_o/D_o for use in Eq. (8-41)

h_o/D_o	0.167	0.178	0.192	0.208	0.227	0.25	0.276	0.313	0.357	0.417	0.5
$K_o = R_o/D_o$	1.36	1.27	1.182	1.08	0.99	0.90	0.81	0.73	0.65	0.57	0.5

Particular	Formula

$$P_a = \frac{0.0625\eta}{(R_o/t)^2} \qquad (8\text{-}43)$$

where R_o = outside radius in corroded condition

The procedure outlined in this section for finding the thickness of a spherical shell can be used to find the thickness of a hemispherical head by using Eqs. (8-41) to (8-43).

ELLIPSOIDAL HEADS

The minimum required thickness of ellipsoidal head having pressure on the convex side either seamless or of built-up construction with butt joints shall not be less than that determined by the procedure given here

Assume a value of t and calculate the value of factor A using the following equation:

$$A = \frac{0.125}{R_o/t} \qquad (8\text{-}41)$$

The factor A is given by Eq. (8-41)

Using the value of A calculated from Eq. (8-41) follow the procedure as that given for spherical shells to find the thickness of ellipsoidal heads.

where R_o = the equivalent outside spherical radius taken as K_oD_o in the corroded condition
K_o = factor depending on the ellipsoidal head proportion R_o/D_o (Table 8-5)

TORISPHERICAL HEADS

The required minimum thickness of a torispherical head having pressure on the convex side, either seamless or of built-up construction with butt joint

The required thickness shall not be less than that determined by the same design procedure as is used for ellipsoidal heads given in the *Ellipsoidal heads* section, using appropriate values for R_o. For torispherical head, the outside radius of the crown portion of the head (R_o) in corroded condition is taken for design purposes.

CONICAL HEADS AND SECTIONS

The required minimum thickness of a conical head or section under pressure on the convex side, either seamless or the built-up construction with butt joints

This thickness can be determined following the procedure outlined under cylindrical shells in the *Torispherical heads* section to find factors A and B by assuming a value for t_e.

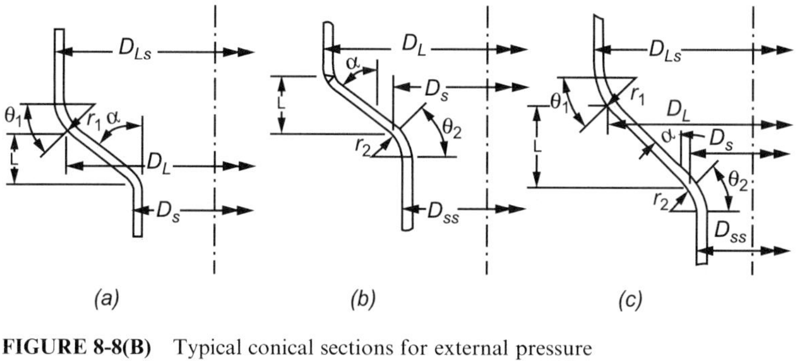

FIGURE 8-8(B) Typical conical sections for external pressure

Particular	Formula
The symbols involved in design calculations are	t_e = effective thickness of conical section L_e = equivalent length of conical section $= (L/2)(1 + D_s/L)$ L = axial length of cone or conical section (Fig. 8-8(B)) D_s = outside diameter at small end of conical section under consideration
(1) When $\alpha \leq 60°$ and for cones having D_L/t_e, values ≥ 10: Assume a value of t_e and determine ratios L_e/D_L and D_L/t_e. The equation for calculating the maximum allowable external pressure P_a for the case of values of factor A falling to the right of the end of the material/temperature line.	$P_a = \dfrac{4B}{3(D_L/t_e)}$ (8-76a)
Equation for calculating the maximum allowable external pressure P_a for the case of values of factor A falling to the left of the applicable material/temperature line	$P_a = \dfrac{2AE}{3(D_L/t_e)}$ (8-77) where D_L = outside diameter at large end of conical section under consideration (Fig. 8-8(B)) α = one-half the apex angle in conical heads and section, deg. (Compare the value of P_a with design pressure P. If $P_a < P$, then select a new value for t_e and repeat the design procedure.)
(2) When $\alpha \leq 60°$ and for cones having D_L/t_e, values <10: For values of D_L/t_e less than 4, the value of factor A can be calculated by using Eq. (8-78).	$A = \dfrac{1.1}{(D_L/t_e)^2}$ (8-78)
For values of factor A greater than 0.10, use a value of 0.10	
The equation for calculating P_{a1} using the value of factor B obtained from material/temperature chart	$P_{a1} = \left[\dfrac{2.167}{D_L/t_e} - 0.0833\right]B$ (8-79)
The equation for calculating P_{a2}	$P_{a2} = \dfrac{2\sigma_s}{D_L/t_e}\left[1 - \dfrac{1}{D_L/t_e}\right]$ (8-80) where σ_s is less than two times the maximum allowable stress value at design metal temperature, from the applicable table or 0.9 times the yield strength of the material at design temperature. The yield strength is twice the value of B determined from the applicable material/temperature chart. The smaller of the values of P_{a1} or P_{a2} shall be used for the maximum allowable external pressure P_a. P_a is equal to or greater than P. (Design pressure P is obtained from appropriate table for material.)
(3) When $\alpha > 60°$	The thickness of the cone shall be the same as the required thickness for a flat head under external pressure, the diameter of which equals the largest diameter of the cone.

Particular	Formula

Toriconical head having the pressure on the convex side

Particular	Formula
The required thickness of a toriconical head having pressure on the convex side, either seamless or of built-up construction with butt joints within the head	The thickness shall not be less than that determined using the procedure followed in the case of a cone having D_L/t_e values ≤ 10 for conical heads and sections with exception that L_e shall be determined using Eqs. (8-81):
The length L_e (Fig. 8-8B, panel a)	$L_e = r_1 \sin\theta_1 + \frac{L}{2}\left(\frac{D_L + D_s}{D_{Ls}}\right)$ (8-81a)
The length L_e (Fig. 8-8B, panel b)	$L_e = r_2 \frac{D_{ss}}{D_L} \sin\theta_2 + \frac{L}{2}\left(\frac{D_s + D_L}{D_L}\right)$ (8-81b)
The length L_e (Fig. 8-8B, panel c)	$L_e = r_1 \sin\theta_1 + r_2 \frac{D_{ss}}{D_{Ls}} \sin\theta_2 + \frac{L}{2}\left(\frac{D_L + D_s}{D_{Ls}}\right)$ (8-81c)
To find the thickness when lap joints are used in formed head construction or for longitudinal joints in a conical header under external pressure	The rules in this section, except the design pressure $2P$, shall be used instead of P in the calculations for the design of required thickness.

UNSTAYED FLAT HEADS AND COVERS (Fig. 8-9, Table 8-6)

Particular	Formula
The thickness h of that unstayed circular heads, covers, and blind flanges as per Indian Standards	$h = c_2 d\sqrt{(p/\sigma_{sa})}$ (8-82) Refer to Table 8-6 for values of c_2.
The minimum required thickness t of unstayed circular heads, covers and blind flanges as per *ASME Boiler and Pressure Vessel Code*	$t = d\sqrt{(CP/\sigma_{sa}\eta)}$ (8-83) Refer to Table 8-6 for values of C.
The minimum required thickness t of flat unstayed circular heads, covers, and blind flanges which are attached by bolts, causing an edge moment as per *ASME Boiler and Pressure Vessel Code*	$t = d\left[\frac{CP}{\sigma_{sa}\eta} + 1.9\frac{Wh_G}{\sigma_{sa}\eta d^3}\right]^{1/2}$ (8-84) where C is taken from Table 8-6 W = flange design bolt load, lbf $W = W_{m1}$ = the minimum bolt load for operating condition, lbf $t = 0.785 D_G^2 P + (2b \times 3.14 D_G mp)$ (8-84a) where $W = W_{m2}$ = the minimum required bolt load for gasket seating, lbf $t = 3.14 b D_G y$ (8-84b)

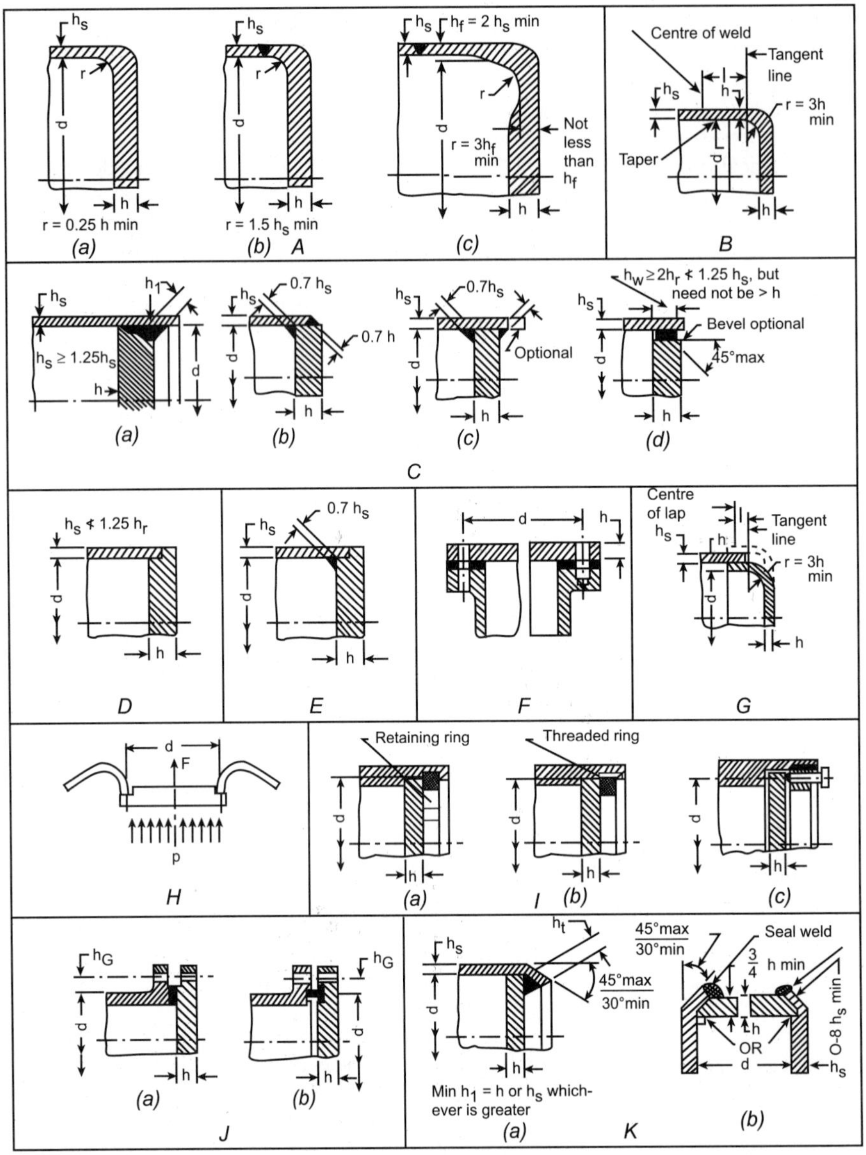

FIGURE 8-9 Types of unstayed flat heads. (K. Lingaiah and B. R. Narayana Iyengar, *Machine Design Data Handbook*, Engineering College Cooperative Society, Bangalore, India, 1962; K. Lingaiah and B. R. Narayana Iyengar, *Machine Design Data Handbook*, Vol. I (*SI and Customary Metric Units*), Suma Publishers, Bangalore, India, 1983; K. Lingaiah, *Machine Design Data Handbook*, Vol. II (*SI and Customary Metric Units*), Suma Publishers, Bangalore, India, 1986.)

TABLE 8-6
Coefficients c_2 and C for determining head thickness for typical unstayed flat heads (Fig. 8-9)

Type of head		Coefficient, c_2 and C		
IS (Fig. 8-9)[a]	ASME	c_2, IS (Fig. 8-9)	C, ASME	Remarks
$A(a)$	(b-2)	0.50	0.33 m but $\nless 0.20$	Forged circular and noncircular heads integral with or butt-welded to the vessel
$A(b)$		0.50		
$A(c)$	(b-1)	0.45	0.17	
B	(a)		0.17	Flanged circular and noncircular heads forged integral with or butt-welded to the vessel with an inside corner radius not less than three times the required head thickness
		0.35	0.10	For circular heads when the flange length: 1. $l \geq (1.1 - 0.8h_s^2/h^2)\sqrt{d_i h}$; $r \geq 2h$, $d = d_i - r$ and taper is 1 : 4 (Fig. 8-9) 2. $l = [1.1 - 0.8((t_s/t_h)^2]\sqrt{t_d}d$ and taper is 1 : 3 (1)
		0.45		When $r \geq 2h$, $d = d_i - r$ and taper is 1 : 4 (Fig. 8-9)
		0.50		When $d = d_i$ and $0.25h \leq r < 2h$.
			0.1	For circular heads, when the flange length $l : l < [1.1 - 0.8(t_s/t_h)^2\sqrt{t_h}d]$ but the shell thickness: $t_s \nless 1.12t_h\sqrt{1.1 - 1/\sqrt{t_h}d}$; taper is at least 1 : 3 (2)
$C(a)$ to $C(d)$	(e), (f) and (g)	$0.7\sqrt{h_r/h_s}$ but $\nless 0.55$	0.33 m but $\nless 0.20$	Circular plates welded to inside of the pressure vessel
			0.33	Noncircular plates, welded to the inside of a vessel and otherwise meeting the requirements for the respective types of welded vessels
D	(h)	0.7	0.33	For circular plates welded to the end of the shell when t_s is at least $1.25t_r$
E	(i)	$0.7\sqrt{h_r/h_s}$ but $\nless 0.55$	0.33 m but $\nless 0.20$	For circular plates, if an inside fillet weld with minimum throat thickness of $0.7t_s$ is used
F	(p)	0.42	0.25	For circular and noncircular covers bolted with a full-face gasket, to shells, flanges or side plates
G	(c)	0.45	0.13	Circular heads lap welded or brazed to the shell with corner radius not less than the $3h$ or $3t$ and l not less than required by formula (2)
		0.55 in other cases	0.20	Circular and noncircular lap welded or brazed construction as above, but with no special requirement with regard to l
			0.30	Circular flanged plates screwed over the end of the vessel. with inside corner radius not less than $3t$ or $3h$ in which the design of threaded joint is based on a safety factor of 4
H		$\sqrt{0.31 + 95(F/Pd^2)}$		Autoclave manhole covers $d \ngtr 610$ mm (24 in)
$I(a)$	(m)	0.55	0.30	Circular plates inserted in the end of a pressure vessel and held in place by a positive mechanical locking arrangement, and when all possible means of failure are resisted with a safety factor of at least 4; seal welding may be used, if desired
$I(b)$	(n)	0.55	0.30	
$I(c)$	(o)	0.55	0.30	

TABLE 8-6
Coefficients c_2 and C for determining head thickness for typical unstayed flat heads (Fig. 8-9) (*Cont.*)

Type of head		Coefficient, c_2 and C		
IS (Fig. 8-9)[a]	ASME	c_2, IS (Fig. 8-9)	C, ASME	Remarks
				Circular and noncircular head and covers bolted to the vessel as shown in Fig. 8-9
$J(a)$	(j)	$\sqrt{0.31 + 190(Fbh_G/Pd^3)}$	0.3	
$J(b)$	(k)		0.3	
$K(a)$	(r)	0.7	0.33	Circular plates having a dimension d not exceeding 450 mm (18 in) inserted into the vessel as shown in Fig. 8-9 and the end of the vessel shall be crimped over at least 30°, but not more than 45°; the crimping may be done cold only when this operation will not injure the metal in case of (r); in case of (s) the crimping shall be done when the entire circumference of the cylinder is uniformly heated to the proper forging temperature for the material used
$K(b)$	(s)	0.7	0.33	

[a] Symbols in this column refer to Fig. 8-9.
[b] Where F (or W) is load on bolt.
Sources: K. Lingaiah and B. R. Narayana Iyengar, *Machine Design Data Handbook,* Engineering College Cooperative Society, Bangalore, India, 1962; K. Lingaiah and B. R. Narayana Iyengar, *Machine Design Data Handbook*, Vol. I (*SI and Customary Metric Units*), Suma Publishers, Bangalore, India, 1983; and K. Lingaiah, *Machine Design Data Handbook*, Vol. II (*SI and Customary Metric Units*), Suma Publishers, Bangalore, India, 1986.

Particular	Formula
The minimum thickness of noncircular heads and covers as per Indian Standards	$h = c_2 k_6 b\sqrt{(p_i/\sigma_{sa})}$ (8-85) Refer to Fig. 8-10 for values of k_6 and Table 8-6 for values of c_2.
The minimum heads, covers, or blind flanges of square, rectangular, oblong, segmental, or otherwise noncircular shape as per *ASME Boiler and Pressure Vessel Code*	$t = d\sqrt{(ZCP)/\sigma_{sa}\eta}$ (8-88a) where Z = a factor for noncircular heads depending on the ratio of long span to short span a/b $Z = 3.4 - \dfrac{2.4d}{D}$ (8-86b) Refer to Fig. 8-10 for values of Z ($Z = k_6$). Z need not be greater than 2.5. d = diameter or short span as indicated in Fig. 8-9. t, d in in and p and σ_{sa} in psi

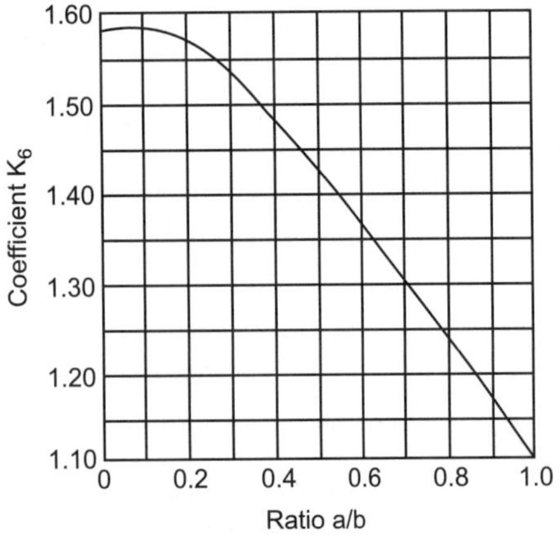

FIGURE 8-10 Coefficient k_6 for noncircular flat heads.

Particular	Formula
The minimum thickness of flat unstayed non-circular heads, covers, or blind flanges attached by bolts causing a bolt load edge moment as per *ASME Boiler and Pressure Vessel Code* (*Note*: A stayed flat plate and types of stays are shown in Figs. 8-11 and 8-12.)	$t = d\left[\frac{ZCP}{\sigma_{sa}\eta} + \frac{6Wh_G}{\sigma_{sa}\eta Ld^2}\right]^{1/2}$ (8-87) where h_G = gasket moment arm, equal to the radial distance from the center line of the bolts to the line of the gasket reaction as shown in Fig. 8-13

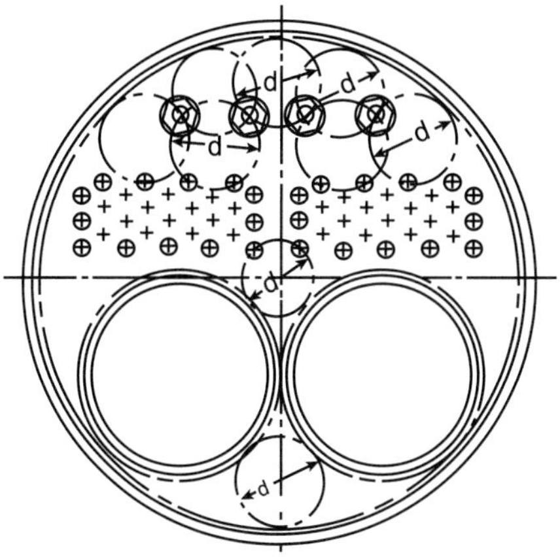

FIGURE 8-11 Stayed flat plate.

Particular	Formula
The net cover plate thickness under the groove or between the groove and outer edge (t_g) of the cover plate	$t_g \nless d\sqrt{1.9Wh_G/\sigma_{sa}d^3}$ (8-88) for circular heads and covers $t_g \nless d\sqrt{6Wh_G/\sigma_{sa}Ld^2}$ (8-89) for noncircular heads and covers
The thickness of spherically dished ends and heads secured to the shell through a flange connection by means of bolts as per Indian Standards	$h = \frac{3pd_i}{2\sigma_{sa}\eta}$ for $R \ngtr 1.3d_i$ and $\frac{100h}{R} \ngtr 10$ (8-90)
The thickness of a dished head that is riveted or welded to a cylindrical shell according to *ASME Boiler and Pressure Vessel Code*	$h = \frac{8.33PR}{2\sigma_{su}}$ (8-91)

STAYED FLAT AND BRACED PLATES OR SURFACES (Figs. 8-11 and 8-12)

Particular	Formula
The thickness of stayed and braced plate as per Indian Standards	$h = c_4 d\sqrt{(p/\sigma_{sa})}$ (8-92) Refer to Table 8-7 for values of c_4 and Tables 8-8, 8-9, and 8-11 for allowable stress values σ_{sa}.

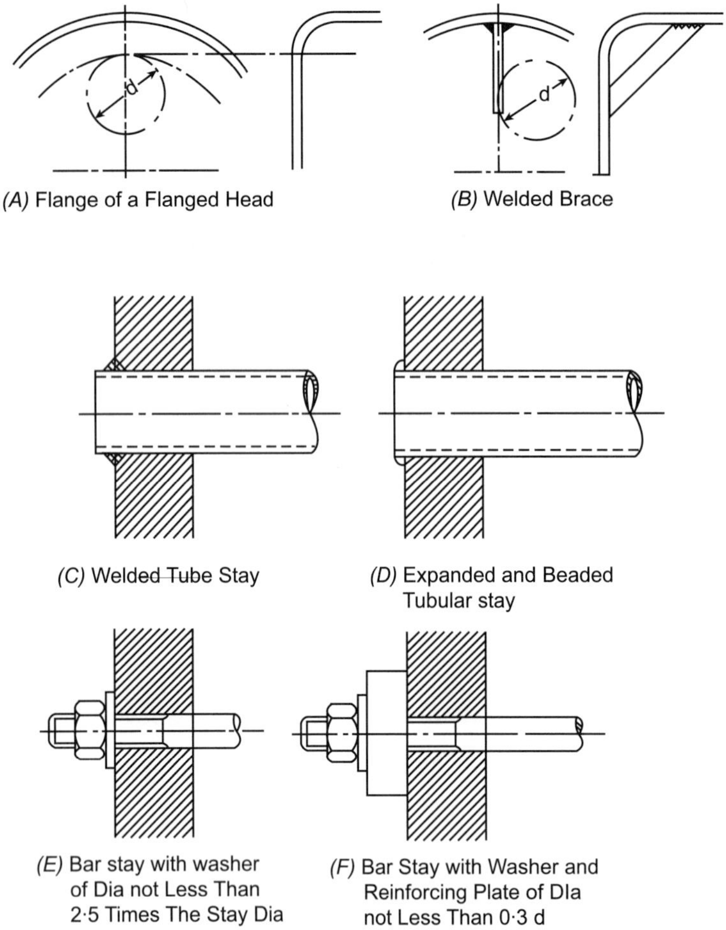

FIGURE 8-12 Types of stays.

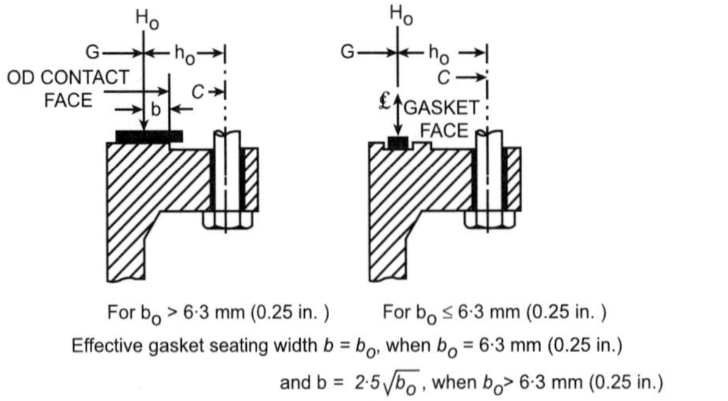

NOTE — The gasket factors listed only apply to flanged joints in which the gasket is contained entirely within the inner edges of the bolt holes.

FIGURE 8-13 Location of bolt load reaction.

TABLE 8-7
Coefficients c_4 for determining head thickness for stayed and braced plates

Type of stay	c_4	Remarks
A	0.45	Flange of a flanged head
B	0.45	Welded brace
C	0.55	Welded tube stay
D	0.55	Expanded and beaded tubular stay
E	0.45	Bar stay with washer of diameter not less than 2.5 times the stay diameter
F	0.40	Bar stay with washer and reinforcing plate of diameter not less than $0.3d$

Particular	Formula

The minimum thickness for braced and stayed flat plates with braces or stay bolts of uniform diameter symmetrically spaced as per *ASME Boiler and Pressure Vessel Code*

$$t = p_t\sqrt{(P/\sigma_{sa}c_5)} \quad (8\text{-}93)$$

where

t = minimum thickness of plate, exclusive of corrosion allowance, in

σ_{sa} = maximum allowable stress, MPa (psi), taken from Tables 8-9 to 8-13 for shell plates and Table 8-23 for bolts

The maximum allowable working pressure for braced and stayed flat plates as per *ASME Boiler and Pressure Vessel Code*

$$P = \frac{t^2\sigma_{sa}c_5}{p_t^2} \quad (8\text{-}94)$$

where c_5 = a factor depending on the plate thickness and type of stay taken from Table 8-15

Stayed flat plates with uniformly distributed load

Grashof's formula for maximum stress

$$\sigma = 0.2275\,\frac{p_t^2 p}{h^2} \quad (8\text{-}95)$$

The deflection

$$y = 0.0284\,\frac{p_t^4 p}{Eh^3} \quad (8\text{-}96)$$

OPENINGS AND REINFORCEMENT

For flanged-in and unreinforced openings in cylindrical or conical shell or spherical shell or heads and ends

Refer to Figs. 8-14 and 8-15. Holes cut in domed ends shall be circular, elliptical, or oblong. The radius r of flanged-in openings (Fig. 8-14) shall not be less than 25 mm. Flanged-in and other openings shall be arranged so that the distance from the edge of the end is not less than that shown in Fig. 8-14. In all cases the projected width of the ligament between any two adjacent openings shall be at least equal to the diameter of the smaller openings as shown in Fig. 8-15.

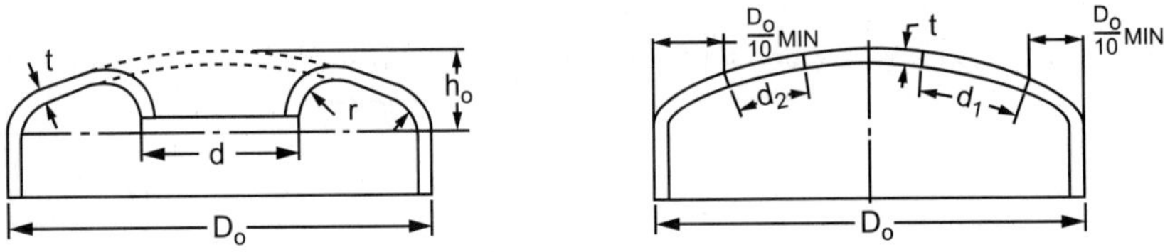

FIGURE 8-14 Flanged-in unreinforced opening.

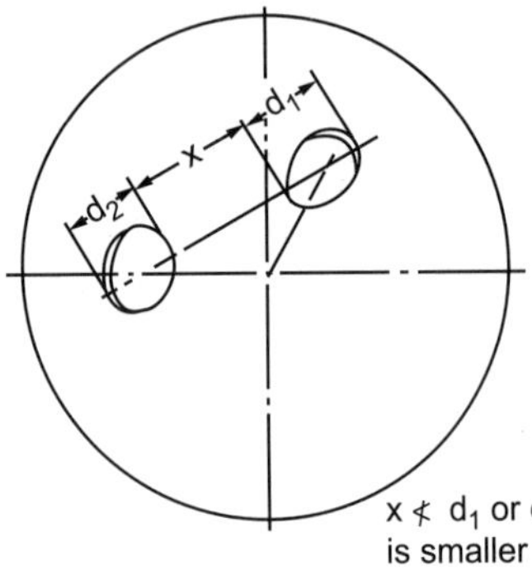

FIGURE 8-15 Unreinforced opening.

Particular	Formula
The distance between openings spaced apart, L_o, in a cylindrical or conical or spherical shell as per Indian Standards	$$L_o \nless L = \frac{d(h_a/h_r)}{(h_a/h_r - 0.95)} \quad (8\text{-}97)$$ Refer to Tables 8-16 and 8-17 for flange bolting.
For size of openings in cylinders or conical shells or spherical shells subject to a maximum of 200 mm (8 in) which do not require reinforcement	Refer to Fig. 8-16. where h = distance between centers of any two adjacent openings, mm (in) d = diameter or largest opening = mean value of the major and minor axes in case of elliptical or obround openings, mm (in) h_a = actual thickness of vessel before corrosion allowance is provided, mm (in) h_r = required thickness of vessel putting $\eta = 1.0$ before corrosion allowance is added, mm (in)
The total cross-sectional area of reinforcement, A_r, as per Indian Standards	$$A_r \nless A = dh_r \quad (8\text{-}98)$$ where d = nominal internal diameter of the branch plus twice the corrosion allowance, mm (in) h_r = thickness of an unpierced shell or end calculated from Eq. (8-10)

Particular	Formula
The expression for K factor	$K = pD_o/1.82\sigma_a h_a$ (8-99) Refer to Fig. 8-16*a* and *b*, where K has a value of unity or greater, the maximum size of an unreinforced opening shall be 50 mm (2 in)

Design for internal pressure

Particular	Formula
The total cross-sectional area of reinforcement A required in any given plane through the opening for a shell or formed head under internal pressure shall not be less than as per *ASME Boiler and Pressure Vessel code*	
The total cross-sectional area of reinforcement in flat heads that have an opening with a diameter that does not exceed one-half of the head diameter or shortest span, shall not be less than that given by Eq. (8-100) as per *ASME Boiler and Pressure Vessel Code*	$A = 0.5td$ (8-100) where t = minimum required thickness of flat head or cover, exclusive of corrosion allowance, m (in)
For nomenclature and formulas for reinforced openings as per *ASME Boiler and Pressure Vessel Code*	Refer to Fig. 8-17.
For values of spherical radius factor K_1	Refer to Table 8-18.
The length of tapped hole l_s to engage a stud	$l_s = 0.75d_s \dfrac{\text{maximum allowable stress value of stud material at design temperature}}{\text{maximum allowable stress value of tapped material at design temperature}}$ (8-101) and also $l_s \nless d_s$, where d_s = nominal diameter of the stud, except that the thread engagement need not exceed $1.5d_s$

TABLE 8-8
Allowable stresses σ_{sa} for various ferrous and nonferrous materials

Materials with grade or designation and product	Mechanical properties: Tensile strength, σ_{st}, min R_{20}, MPa (kpsi)	Mechanical properties: Yield stress, σ_{sy} min, E_{20} MPa (kpsi)		Allowable stress, σ_{sa} at design temperature, K (°C): 323 K (50°C), MPa (kpsi)	373 K (100°C), MPa (kpsi)	423 K (150°C), MPa (kpsi)	473 K (200°C) MPa (kpsi)	≤523 K (250°C), MPa (kpsi)	≤573 K (300°C), MPa (kpsi)	≤623 K (350°C), MPa (kpsi)	≤648 K (375°C), MPa (kpsi)	≤673 K (400°C), MPa (kpsi)	≤698 K (425°C), MPa (kpsi)	≤723 K (450°C), MPa (kpsi)	≤748 K (475°C), MPa (kpsi)	≤773 K (500°C), MPa (kpsi)	≤798 K (525°C), MPa (kpsi)	≤823 K (550°C), MPa (kpsi)	≤848 K (575°C) MPa (kpsi)
Carbon and Low-Alloy Steel in Tension																			
Plates																			
1	363 (52.6)	0.55 R_{20} [a]	26	121 (17.5)	121 (17.5)	113 (16.4)	102 (14.8)	93 (13.5)	85 (12.3)	77 (11.2)	74 (10.7)	71 (10.3)	58 (8.4)	42 (6.1)	35 (5.1)				
2A	412 (59.8)	0.50 R_{20}	25	137 (19.8)	126 (18.3)	117 (17.0)	106 (15.4)	96 (13.9)	88 (12.8)	79 (11.5)	76 (11.0)	73 (10.6)	58 (8.4)	42 (6.1)	35 (5.1)				
2B	510 (74.0)	0.50 R_{20}	20	170 (24.7)	156 (22.6)	144 (20.8)	130 (18.8)	119 (17.3)	109 (15.8)	98 (14.2)	93 (13.5)	81 (11.8)	58 (8.4)	42 (6.1)	35 (5.1)				
20 Mo 6	471 (68.3)	275 (39.9)	20	157 (22.8)	157 (22.8)	157 (22.8)	150 (21.8)	140 (20.3)	129 (18.7)	121 (17.6)	117 (17.0)	113 (16.4)	110 (16.0)	106 (15.4)	76 (11.0)	55 (8.0)	36 (5.2)		
20 C 15	510 (74.0)	294 (42.6)	20	170 (24.7)	170 (24.7)	167 (24.2)	150 (21.9)	137 (19.9)	126 (18.3)	114 (16.5)	108 (15.7)	81 (11.8)	58 (8.4)	42 (6.1)	35 (5.1)				
15 Cr 4 Mo 6	490 (71.1)	294 (42.6)	20	163 (23.6)	163 (23.6)	163 (23.6)	163 (23.6)	157 (22.8)	149 (21.6)	141 (20.5)	135 (19.6)	128 (18.6)	128 (18.6)	124 (18.0)	114 (16.5)	84 (12.2)	57 (8.3)	34 (5.0)	
15 C 8	412 (59.8)	226 (32.8)	25	137 (19.9)	137 (19.9)	127 (18.4)	116 (16.8)	105 (15.2)	96 (13.9)	87 (17.6)	82 (11.9)	79 (11.5)	58 (8.4)	42 (6.1)	35 (5.1)				
Forgings																			
20 Mo 6	471 (68.3)	275 (39.9)	20	157 (22.8)	157 (22.8)	157 (22.8)	150 (21.8)	140 (20.3)	130 (18.9)	121 (17.6)	107 (15.5)	113 (16.4)	110 (16.0)	106 (15.4)	76 (11.0)	55 (8.0)	56 (8.1)		
15 Cr 4 Mo 6	490 (71.1)	294 (42.6)	20	163 (23.6)	163 (23.6)	163 (23.6)	163 (23.6)	157 (22.8)	149 (21.6)	141 (20.5)	135 (19.6)	131 (19.0)	128 (18.6)	124 (18.0)	114 (16.5)	84 (12.2)	57 (8.3)	34 (5.0)	
10 Cr 9 Mo 10	490 (71.1)	314 (45.5)	20	163 (23.6)	163 (23.6)	163 (23.6)	163 (23.6)	176 (25.5)	170 (24.6)	161 (23.4)	158 (22.9)	155 (22.5)	150 (21.8)	146 (21.2)	125 (18.1)	94 (13.6)	69 (10.0)	48 (7.0)	31 (4.5)
Tubes and pipes																			
1% Cr ½% Mo	432 (62.7)	235 (34.1)	950R_{20}	143 (20.7)	143 (20.7)	139 (20.2)	133 (19.3)	127 (18.4)	119 (17.3)	113 (16.4)	109 (15.8)	105 (15.2)	102 (14.8)	98 (14.2)	95 (13.8)	84 (12.2)	57 (8.3)	34 (5.0)	
20 Mo 6	451 (65.4)	245 (35.5)	950R_{20}	150 (21.8)	150 (21.8)	143 (20.7)	133 (19.3)	127 (18.4)	115 (16.7)	108 (15.7)	104 (15.1)	101 (14.7)	98 (14.2)	94 (13.6)	76 (11.0)	55 (8.0)	36 (5.2)		
Fe 170	310 (45.0)	173 (25.1)		103 (15.0)	103 (15.0)	97 (14.0)	88 (12.8)	80 (11.6)	74 (10.7)	66 (9.6)	63 (9.1)	61 (8.8)	58 (8.4)	42 (6.1)	35 (5.1)				
Fe 240	414 (60.0)	241 (35.0)		137 (19.9)	137 (19.9)	136 (19.7)	124 (18.0)	113 (16.4)	103 (15.0)	93 (13.5)	88 (12.8)	81 (11.8)	58 (8.4)	42 (6.1)	35 (5.1)				
Fe 290	414 (60.0)	290 (42.1)		137 (19.9)	137 (19.9)	137 (19.9)	137 (19.9)	135 (19.6)	124 (18.0)	113 (16.4)	106 (15.4)	81 (11.8)	58 (8.4)	42 (6.1)	35 (5.1)				
Castings																			
Grade 1	539 (78.2)	343 (49.8)	17	134 (19.4)	134 (19.4)	134 (19.4)	132 (19.1)	120 (17.4)	110 (16.0)	99 (14.4)	94 (13.6)	61 (8.8)	43 (6.2)	31 (4.5)	26 (3.8)				
Grade 2	461 (66.9)	245 (35.5)	17	115 (16.7)	114 (16.5)	108 (15.7)	100 (14.5)	94 (13.6)	86 (12.5)	80 (11.6)	79 (11.5)	76 (11.0)	74 (10.7)	71 (10.3)	57 (8.8)	41 (5.9)	27 (3.9)		
Grade 3	510 (74.0)	304 (44.1)	15	127 (18.4)	127 (18.4)	127 (18.4)	125 (18.1)	117 (17.0)	108 (15.7)	100 (14.5)	98 (14.2)	94 (13.6)	91 (13.2)	82 (11.9)	57 (8.3)	41 (5.9)	27 (3.9)		
Grade 4	481 (69.8)	275 (39.9)	17	120 (17.4)	120 (17.4)	120 (17.4)	117 (17.0)	110 (17.3)	104 (15.1)	99 (14.4)	95 (13.8)	91 (13.2)	89 (12.9)	86 (12.5)	83 (12.0)	64 (9.3)	43 (6.2)	25 (3,6)	
Grade 5	510 (74.0)	304 (44.1)	17	127 (18.4)	127 (18.4)	127 (18.4)	127 (18.4)	128 (18.6)	127 (18.4)	117 (17.0)	115 (16.7)	116 (16.9)	109 115.8)	106 (15.4)	94 (13.6)	71 (10.3)	51 (7.5)	36 (5.2)	33 (4.8)
Grade 6	618 (89.6)	422 (61.2)	15	154 (22.3)	154 (22.3)	154 (22.3)	154 (22.3)	169 (24.5)	160 (23.2)	152 (22.0)	146 (21.2)	141 (20.5)	137 (19.9)	132 (19.1)	66 (9.6)	48 (7.0)	34 (4.9)	25 (3.6)	16 (2.3)
Sections, plates, and bars																			
Grade 1	363 (52.6)	0.55 R_{20}	26	121 (17.5)	111 (16.1)	102 (14.8)	93 (13.5)	84 (12.2)	76 (11.0)	70 (10.2)	67 (9.7)	64 (9.3)	58 (8.4)	42 (6.1)	35 (5.0)				
Grade 2	412 (59.8)	0.55 R_{20}	25	137 (19.9)	126 (18.3)	117 (17.0)	106 (15.4)	96 (13.9)	88 (12.8)	79 (11.5)	76 (11.0)	73 (10.6)	58 (8.4)	42 (6.1)	35 (5.0)				
Grade 3	432 (62.7)	0.55 R_{20}	23	143 (20.7)	131 (19.0)	122 (17.7)	111 (16.1)	100 (14.5)	91 (13.2)	83 (12.0)	79 (11.5)	76 (11.0)	58 (8.4)	42 (6.1)	35 (5.0)				
Grade 4	461 (66.9)	0.55 R_{20}	22	153 (22.2)	141 (20.5)	130 (18.9)	119 (17.3)	115 (16.7)	105 (15.2)	94 (13.6)	89 (12.9)	81 (11.8)	58 (8.4)	42 (6.1)	35 (5.0)				
Grade 5	491 (71.2)	0.55 R_{20}	21	163 (23.6)	150 (21.8)	138 (20.0)	126 (18.3)	119 (17.3)	109 (15.8)	98 (14.2)	93 (13.5)	81 (11.8)	58 (8.4)	42 (6.1)	35 (5.0)				
Grade A-N	432 (62.7)	235 (34.0)	23	143 (20.7)	143 (20.7)	133 (19.2)	121 (17.5)	96 (13.9)	88 (12.8)	79 (11.5)									
Grade B-N	490 (71.1)	280 (40.6)	20	163 (13.6)	163 (23.6)	158 (23.0)	143 (20.7)	115 (16.7)	105 (15.2)	94 (13.6)									
High-Alloy Steels in Tension																			
Plates, bars, forgings, seamless tubes																			
X04 Cr 19 Ni 9	540 (78)	235 (34)	28	157 (22.8)		139 (20.2)		122 (17.7)		104 (15.0)	97 (14)	92 (13.3)	85 (12.3)	79 (11.5)					
X04 Cr 19 Ni 9 Ti 20	540 (78)	235 (34)	28	157 (22.8)		140 (20.3)		124 (18.0)		106 (15.4)	104 (15)	104 (15)	104 (15.1)	101 (14.6)					
X04 Cr 19 Ni 9 No 40	540 (78)	235 (34)	28	157 (22.8)		140 (20.3)		124 (18.0)		106 (15.4)	104 (15)	104 (15)	10.4 (15.1)	104 (15.1)					
X05 Cr 18 Ni 11 Mo 3	540 (78)	235 (34)	28	157 (22.8)		142 (20.6)		127 (18.4)		113 (16.4)	110 (16)	110 (16)	110 (16.0)	109 (15.8)					
X05 Cr 19 Ni 9 Mo 3	540 (78)	235 (34)	28	157 (22.8)		142 (20.6)		127 (18.4)		113 (16.4)	110 (16)	110 (16)	110 (16.0)	109 (15.8)					
Ti 20																			
Castings																			
Grades 7, 8	461 (66.9)	205 (30)	21	137 (19.9)		127 (18.4)		117 (18.4)		106 (15.4)	104 (15.1)	104 (15.1)	104 (15.1)	104 (15.1)					

TABLE 8-8
Allowable stresses σ_{sa} for various ferrous and nonferrous materials (*Cont.*)

Materials with grade or designation and product	Mechanical properties: Tensile strength, σ_{st}, min R_{20}, MPa (kpsi)	Mechanical properties: Yield stress, σ_{sy} min, E_{20} MPa (kpsi)		Allowable stress, σ_{sa} at design temperature, K (°C): 323 K (50°C), MPa (kpsi)	373 K (100°C), MPa (kpsi)	423 K (150°C), MPa (kpsi)	473 K (200°C) MPa (kpsi)	≤523 K (250°C), MPa (kpsi)	≤573 K (300°C), MPa (kpsi)	≤623 K (350°C), MPa (kpsi)	≤648 K (375°C), MPa (kpsi)	≤673 K (400°C), MPa (kpsi)	≤698 K (425°C), MPa (kpsi)	≤723 K (450°C), MPa (kpsi)	≤748 K (475°C), MPa (kpsi)	≤773 K (500°C), MPa (kpsi)	≤798 K (525°C), MPa (kpsi)	≤823 K (550°C), MPa (kpsi)	≤848 K (575°C) MPa (kpsi)
				Aluminum and Aluminum Alloys in Tensions															
Plates																			
PIB—M	64 (9.3)		30	12 (1.7)	13 (1,9)	11 (1.6)	10 (1.5)	9 (1.3)	8 (1.2)	7 (1.1)									
NP4—M	186 (27.0)		12	43 (6.2)	42 (6.1)	42 (6.1)	41 (5.9)	37 (5.4)	32 (4.6)	24 (3.5)									
Sheet, strip																			
S1B—½H	98 (14.2)		8	21 (3.0)	20 (2.9)	19 (2.8)	18 (2.6)	16 (2.3)	14 (2.0)	11 (1.6)									
NS4—¼H	196 (28.4)		8	54 (7.8)	53 (7.7)	52 (7.5)	49 (7.1)	44 (6.4)	37 (5.4)	24 (3.5)									
Bars, rods, and sections																			
NE5—M	216 (31.3)		88⁺	54 (7.8)															
NE6—M	265 (38.4)		18⁺	66 (9.6)															
NE8—0	265 (38.4)		16	69 (10.0)															
HE30—W	186 (27.0)	108 (15.7)	18	51 (7.4)	49 (7.1)	47 (6.8)	46 (6.7)	44 (6.4)	39 (6.7)	28 (4.1)									
HE30—WP	294 (42.6)	245 (35.5)	10	71 (10.3)	70 (10.2)	67 (9.7)	65 (9.4)	54 (7.8)	43 (6.2)	30 (4.4)									
Drawn tubes																			
HT30—W	216 (31.3)	108 (15.7)	16	51 (7.4)	50 (7.3)	48 (7.0)	46 (6.7)	44 (6.38)	39 (5.7)	28 (4.1)									
HT30—WP	309 (44.8)	245 (35.5)	7	72 (10.4)	70 (10.2)	67 (9.7)	65 (9.4)	55 (8.0)	43 (6.2)	30 (4.4)									
				Copper and Copper Alloys															
Plate sheet and strips																			
Cu Zn 30	275 (40.0)		45	69 (10.0)	69 (10.0)	69 (10.0)	69 (10.0)	68 (9.9)	56 (8.1)	38 (5.5)									
Cu Zn 40	275 (40.0)		30	86 (12.5)	85 (12.3)	81 (11.7)	77 (11.2)	71 (10.3)	53 (7.7)	19 (2.8)									
Bars and rods																			
	392 (56.9)		22	69 (10.0)	69 (10.0)	69 (10.0)	69 (10.0)	68 (9.9)	56 (8.1)	38 (5.5)									
Tubes																			
Alloy 1	284 (41.2)			69 (10.0)	69 (10.0)	69 (10.0)	69 (10.0)	68 (9.9)	56 (8.1)	38 (5.5)									
Alloy 2	284 (41.2)			86 (12.5)	85 (12.3)	81 (11.7)	77 (11.2)	71 (10.3)	53 (7.7)	19 (2.8)									

[a] These values have been used on a quality factor of 0.75.

[b] $0.55R_{20} = 0.55 \times 363 = 199.7$ MPa (29.0 kpsi).

Notes: ⁺ The elongation values are based on 50.8-mm test piece; a^* area of cross-section; † tube normalized and tempered.

Sources: K. Lingaiah and B. R. Narayana Iyengar, *Machine Design Data Handbook*, Engineering College Cooperative Society, Bangalore, India, 1962; K. Lingaiah and B. R. Narayana Iyengar, *Machine Design Data Handbook*, Vol. I (*SI and Customary Metric Units*), Suma Publishers. Bangalore. India. 1983; and K. Lingaiah, *Machine Design Data Handbook*, Vol. II (*SI and Customary Metric Units*), Suma Publishers, Bangalore, India, 1986

TABLE 8-9
Maximum allowable stress values, σ_{sa}, in tension for carbon and low-alloy steel

Specification no.	Grade	Nominal composition	Specified minimum yield strength, σ_{sy} MPa	Specified minimum yield strength, σ_{sy} kpsi	Specified minimum tensile strength σ_{st} MPa	Specified minimum tensile strength σ_{st} kpsi	−19 to 345 (−30 to 650) MPa	−19 to 345 (−30 to 650) kpsi	370 (700) MPa	370 (700) kpsi	400 (750) MPa	400 (750) kpsi	427 (800) MPa	427 (800) kpsi	455 (850) MPa	455 (850) kpsi	482 (900) MPa	482 (900) kpsi	510 (950) MPa	510 (950) kpsi	538 (1000) MPa	538 (1000) kpsi	566 (1050) MPa	566 (1050) kpsi	593 (1100) MPa	593 (1100) kpsi	620 (1150) MPa	620 (1150) kpsi	650 (1200) MPa	650 (1200) kpsi	Specification no.
													Carbon Steel																		
Plates and sheets																															
SA 285[b,c,d]	A	C	165	24	310	45	78	11.3	76	11.0	71	10.3	62	9.0	54	7.8	45	6.5	(Fig. 8-5)												SA 285[b,c,d]
SA 299[c]		C-Mn-Si	276/290	40/42	517	75	130	18.8	121	17.7	108	15.7	87	12.6	66	9.6	45	6.5	31	4.5	17.0	2.5									SA 299[c]
SA 414[b,c]	F	C-Mn	290	42	483	70	121	17.5	114	16.6	101	14.7	83	12.0	63	9.2	45	6.5													SA 414[b,c]
SA 515[c]	60	C-Si	220	32	413	60	103	15.0	99	14.4	90	13.0	75	10.8	60	8.7	45	6.5	31	4.5	17.0	2.5									SA 515[c]
	70	C-Si	262	38	483	70	121	17.5	114	16.6	102	14.8	83	12.0	64	9.3	45	6.5	31	4.5	17.0	2.5									
SA 516[c]	65	C-Mn-Si	241	35	448	65	115	16.3	107	15.5	96	13.9	79	11.4	62	9.0	45	6.5	31	4.5	17.0	2.5	(Fig. 8-5)								SA 516[c]
	70	C-Mn-Si	262	38	483	70	121	17.5	114	16.6	102	14.8	83	12.0	64	9.3	45	6.5	31	4.5	17.0	2.5	(Fig. 8-5)								
SA 537[c]	Cl-1 up to 62.5 mm (2.5 in) incl.	C-Mn-Si	345	50	483	70																									SA 537[c]
SA 620	1 and 2	C-,Mn	138	20	276	40	69	10.0																							SA 620
SA 812	80	C-Mn-Si-Cb-V	552	80	689	100	147	21.3	(Fig.	8-5)																					SA 812
Carbon steel forgings, castings, and bars																															
SA 36[b,c] bars and shapes		C-Mn-Si	248	36	400	58	100	14.5	56	13.9	87	12.6	72	10.5	57	8.5	45	6.5													SA 36[b,c] bars and shapes
SA 216[c] cast	WCA	C-Si	207	30	413	60	103	15.0	99	14.4	90	13.0	75	10.8	60	8.7	45	6.5	31	4.5	17.0	2.5									SA 216[c] cast
	WCC	C-Mn-Si	276	40	483	70	121	17.5	115	16.6	102	14.8	83	12.0	64	9.3	45	6.5	31	4.5	17.0	2.5									
SA 350[c] forge	LFI	C-Mn-Si	207	30	413	60	103	15.0	99	14.4	90	13.0	75	10.8	54	7.8	35	5.0	21	3.0	10.0	1.5									SA 350[c] forge
	LF2	C-Mn-Si	248	36	483	70	121	17.5	115	16.6	102	14.8	83	12.0	54	7.8	35	5.0	21	3.0	10.0	1.5									
SA 675[b,c] bar	45	C	155	22.5	310	45	78	11.3	76	11.0	71	10.3	62	9.0	54	7.8	45	6.5													SA 675[b,c] bar
	60	C	190	27.5	379	55	95	13.8	92	13.3	83	12.1	70	10.2	60	8.4	45	6.5													
	70	C	207	30.0	483	70	121	17.5	115	16.6	102	14.8	83	12.0	64	9.3	45	6.5	31	4.5	17.0	2.5									
													Low-Alloy Steel																		
Plate																															
SA 202	A	0.5 Cr-1.25 Mn-Si	310	45	517	75	130	18.8	122	17.7	108	15.7	83	12.0	54	7.8	35	5.0	21	3.0	10.0	1.5									SA 202
	B	0.5 Cr-1.25 Mn-Si	324	47	586	85	147	21.3	136	19.8	122	17.7	83	12.0	54	7.8	35	5.0	21	3.0	10.0	1.5									
SA 203	F	3.5 Ni, ≤ 50 mm (2 in.)	379	55	552	80	138	20.0																							SA 203 SA 204[gb]
SA 204[g]	A	C-0.5 Mo	255	37	448	65	112	16.3	112	16.3	112	16.3	109	15.8	106	15.3	94	13.7	56	8.2	33.0	4.8									SA 204[g]
	C	C-0.5 Mo	296	43	517	75	130	18.8	130	18.8	130	18.8	130	18.8	126	18.3	94	13.7	56	8.2	33.0	4.8									SA 225[h]
SA 225[h]	C	Mn-0.5 Ni-V	483	70	724	105	181	26.3	181	26.3																					SA 225[h]
SA 302	C	Mn-0.5 Mo-0.5 Ni	345	50	552	80	138	20.0	138	20.0	135	19.6	130	18.8	123	17.9	94	13.7	56	8.2	33.0	4.8									SA 302

TABLE 8-9
Maximum allowable stress values, σ_{sa}, in tension for carbon and low-alloy steel (*Cont.*)

Specification no.	Grade	Nominal composition	Specified minimum yield strength, σ_{sy} MPa	σ_{sy} kpsi	Specified minimum tensile strength σ_{st} MPa	σ_{st} kpsi	Maximum allowable stress, σ_{sa} for metal temperature, °C (°F), not exceeding: −19 to 345 (−30 to 650) MPa	−19 to 345 (−30 to 650) kpsi	370 (700) MPa	370 (700) kpsi	400 (750) MPa	400 (750) kpsi	427 (800) MPa	427 (800) kpsi	455 (850) MPa	455 (850) kpsi	482 (900) MPa	482 (900) kpsi	510 (950) MPa	510 (950) kpsi	538 (1000) MPa	538 (1000) kpsi	566 (1050) MPa	566 (1050) kpsi	593 (1100) MPa	593 (1100) kpsi	620 (1150) MPa	620 (1150) kpsi	650 (1200) MPa	650 (1200) kpsi	Specification no.
SA 387	11 Cl.2	1.25 Cr-0.5 Mn-Si	310	45	517	75	130	18.8	130	18.8	130	18.8	130	18.8	126	18.3	110	15.9	76	11.0	48.0	6.9	32	4.6	19	2.8	15	2.1	8	1.2	SA 387
	5 Cl.1	5 Cr-0.5 Mo	207	30	413	60			95	13.7	91	13.2	88	12.8	83	12.1	75	10.9	55	8.0	40.0	5.8	29	4.2	20	2.9	14	2.0	9	1.3	
Forgings, castings, and bars																															
SA 182 forge	F12	1 Cr-0.5 Mo	276	40	483	70	121	17.5	121	17.5	121	17.5	121	17.5	118	17.1	110	15.9	76	11.0	45.0	6.6	30	4.3	18	2.6	10	1.4	7	1.0	SA 182 forge
	F11b	1.25 Cr-0.5 Mo-Si	276	40	483	70	121	17.5	121	17.5	121	17.5	121	17.5	118	17.1	110	15.9	76	11.0	48.0	6.9	32	4.6	19	2.8	15	2.1	8	1.2	
SA 217 cast	WC1	C-0.5 Mo	241	35	448	65	112	16.3	112	16.3	112	16.2	109	15.8	105	15.3	90	13.7	56	8.2	33.0	4.8									SA 217 cast
	WC4	1 Ni-0.5 Cr-0.5 MO	276	40	483	70	121	17.5	121	17.5	121	17.5	121	17.5	118	17.1	103	15.0	63	9.2	40.0	5.9									
	C12	9 Cr-1 Mo	413	60	620	90			141	20.5	136	19.8	132	19.1	125	18.2	114	10.5	76	11.0	51.0	7.4	35	5.0	23	3.3	15	2.2	10	1.5	
SA 336 forge	F11	1.25 Cr-0.5 Mo-Si	276	40	483	70	121	17.5	121	17.5	121	17.5	121	17.5	118	17.1	110	15.9	76	11.0	48.0	6.9	32	4.6	19	2.8					SA 336 forge
SA 487 cast	4N	0.5 Ni-0.5 Cr-0.25 Mo V	413	60	620	90																									SA 487 cast
SA 541 forge	3	0.5 Ni-0.5 Mo V	345	50	552	80	138	20.0	138	20.0	138	20.0	132	19.1																	SA 541 forge
SA 739 bar	B11	1.25 Cr-0.5 Mo	310	45	483	70	121	17.5	121	17.5	121	17.5	121	17.5	118	17.1	110	15.9	76	11.0	48.0	6.9	32	4.6	19	2.8	15	2.1	8	1.2	SA 739 bar
	B22	2.25 Cr- 1 Mo	310	45	517	75			121	17.5	119	17.2	116	16.9	113	16.4	109	15.8	76	11.0	52.0	7.6	40	5.8	30	4.4	17	2.5	9	1.3	

Notes:
[a] These stress values are one-fourth the specified minimum strength multiplied by a quality factor of 0.92, except for SA 283, grade D and SA-36.
[b] For service temperature above 455°C (850°), it is recommended that killed steels containing not less than 10% residual silicon be used.
[c] Upon prolonged exposure to temperature above 426°C (800°F) the carbide phase of carbon steel may be converted to graphite.
[d] The material shall not be used in thickness above 50 mm (2 in).
[e] The material shall not be used in thickness above 62 mm (2.5 in).
[f] Only killed steel shall be used above 455°C (850°F).
[g] Upon prolonged exposure to temperature above 468°C (875°F), the carbide phase of carbon molybdenum steel may be converted to graphite.
[h] The maximum nominal plate thickness shall not exceed 14.75 mm (0.58 in).
[i] These stress values apply to normalized and drawn materials only.
[j] For other conditions and specifications, the reader is referred to the general notes given for Table UCS-23 of *ASME Boiler and Pressure Vessel Code*, Section VIII, Division 1, July 1, 1986.
Source: The American Society of Mechanical Engineers, *ASME Boiler and Pressure Vessel Code*, Section VIII, Division 1, July 1, 1986.

TABLE 8-10
Maximum allowable stress values, σ_{sa} in tension for nonferrous metals

Specification no.	Alloy designation	Temper condition	Nominal composition	UNS no.	Size or thickness mm (in)	Specified minimum tensile strength, σ_{st} MPa	Specified minimum tensile strength, σ_{st} kpsi	Specified minimum yield strength, σ_{sy} MPa	Specified minimum yield strength, σ_{sy} kpsi	Maximum allowable stress, σ_{sa}, for metal temperature °C (°F), not exceeding: 38 (100) MPa	38 (100) kpsi	65 (150) MPa	65 (150) kpsi	93 (200) MPa	93 (200) kpsi
												Aluminum and Aluminum Alloys			
Sheet and plate															
SB 209	1100[d]	-H 12			1.275–50.0 (0.051–2.000)	96	14	76	11	24	3.5	24	3.5	24	3.5
		-H 14			0.225–25.0 (0.009–1.000)	110	16	96	14	26	4.0	26	4.0	26	4.0
SB 209	3003[d]	-H 14			0.15–25.00 (0.006–1.000)	138	20	117	17	35	5.0	35	5.0	35	5.0
		-H 112			6.25–12.475 (0.250–0.499)	117	17	69	10	30	4.3	30	4.3	30	4.3
SB 209	3004[d]	-H 32			1.275–50.00 (0.051–2.000)	193	28	145	21	48	7.0	48	7.0	48	7.0
SB 209	5052[d]	-H 34			1.275–25.00 (0.051–1.000)	234	34	179	26	58	8.5	58	8.5	58	8.5
SB 209	5454[d]	-O			1.275–75.00 (0.051–3.000)	214	31	83	12	54	7.8	54	7.8	54	7.8
		-H 32			1.275–50.00 (0.051–2.000)	248	36	179	26	62	9.0	62	9.0	62	9.0
SB 209	6061[e,f]	T 4			1.275–6.225 (0.051–0.249)	207	30	110	16	52	7.5	52	7.5	52	7.5
		T 6 Wld			1.275–6.225 (0.051–0.249)	165	24			41	6.0	41	6.0	41	6.0
Rods, bars, and shapes															
SB 221	2024[e]	-T 4			3.125–12.475 (0.125–0.499)	427	62	290	42	107	15.5	107	15.5	107	15.5
					162.54–200.00 (6.501–8.000)	400	58	262	38	100	14.5	100	14.5	100	14.5
SB 221	5086[e]	-H 112			≤125.00 (5.000)	241	35	96	14	61	8.8	61	8.8		
SB 221	3003[d]	-H 112			All	96	14	35	5	23	3.4	23	3.4	23	3.4
SB 221	5456[d]	-O			≤125.00 (5.00)	282	41	131	19	71	10.3	71	10.3		
		-H 111			≤125.00 (5.00)	290	42	179	26	72	10.5	72	10.5		
SB 308	6061[e]	-T 6				262	38	241	35	65	9.5	63	9.2	62	9.0
		-T 6 Wld				165	24			41	6.0	40	5.9	39	5.7
Die and hand forgings															
SB 247	2014 Die[e]	-T 4			≤100.0 (4.000)	379	55.0	207	30	95	13.8	95	13.8	92	13.3
		-T 6			≤50.0 (2.00)	441	64.0	379	55	110	16.0	110	16.0	109	15.9
					50.0–100.00 (2.001–4.000)	434	63.0	372	54		15.8		15.8		15.8
SB 247	6061 Die[e]	-T 6			≤100.0 (4.00)	262	38.0	241	35	65	9.5	65	9.5	65	9.5
SB 247	6061 Hand[e]	-T 6			≤100.0 (4.000)	255	37.0	228	33	64	9.3	64	9.3	64	9.3
					100.025–200.0 (4.001–8.000)	241	35.0	220	32	61	8.8	61	8.8	61	8.8
Castings															
SB 26	SG 70 A(356)[e]	-T 6				207	30.0	138	20	52	7.5	52	7.5	52	7.5
		-T 71				172	25.0	124	18	43	6.3	43	6.3	43	6.3
SB 108	204.0	-T 4			≤50.0 (2.000)	331	48.0	200	29	65	9.5	52	7.5		
													Copper and Copper Alloys		
Sheet and plates															
SB 96	655	Annealed	Cu-Si alloy		≤50 mm (2 in)	345	50.0	124	18	83	12.0	83	12.0	82	11.9
SB 169	610	Annealed	Al-bronze		≤50 mm (2 in)	345	50.0	138	20	86	12.5	86	12.5		
	614	Annealed	Al-bronze		≤12. 5 mm ($\frac{1}{2}$ in)	496	72.0	220	32	124	18.0	124	18.0	124	18.0
SB 171	C 36500, C 36600	Annealed	Lead-Muntz metal		≤50 mm (2 in)	345	50.0	138	20	86	12.5	86	12.5	86	12.5
	C 36700, C 36800	Annealed			>50 mm (2 in) −87.5 mm (3.5 in)	310	45.0	103	15	69	10.0	69	10.0	69	10.0
SB 171	443, 444, 445	Annealed	Admiralty		≤100 mm (4 in)	310	45.0	103	15	69	10.0	69	10.0	69	10.0
SB 171	C 46400, C 46500	Annealed	Naval brass		>75(3)–125(5)	345	50.0	138	20	86	12.5	86	12.5	86	12.5
	C 46600, C 46700	Annealed				345	50.0	124	18	83	12.0	83	12.0	83	12.0
SB 171	715	Annealed	Cu-Ni 70130		≤62.5 (2.5)	345	50.0	138	20	86	12.5	78	11.3	72	10.5
		Annealed			≤62.5 (2.5) −125(5) incl	310	45.0	124	18	78	11.3	70	10.1	65	9.4
SB402	706	Annealed	Cu-Ni 90/10		≤62.5 (2.5)	276	40.0	103	15	70	10.1	67	9.7	66	9.5
Die forgings (hot pressed)															
SB 283[h]	C 37700[h]	As forged	Forging brass		≤ 37.5 (1.5)	345	50.0	124	18	83	12.0	78	11.3	75	10.9
					>37.5 (1.5)	317	46.0	103	15	69	10.0	66	9.5	63	9.1
	C 64200	As forged	Forgings, Al-Si bronze		≤ 37.5 (1.5)	482	70.0	172	25	115	16.7	100	14.5	97	14.0
					>37.5 (1.5)	469	68.0	159	23	105	15.3	93	13.5	90	13.0
Rods and bars															
SB 98[g]	655, 661[g]	Soft[h]	Cu-Si			358	52	103	15	69	10.0	69	10.0	69	10.0
		Half hard[i]				482	70	262	38	121	17.5	121	17.5	121	17.5
SB 98	651[j]	Soft	Cu-Si			276	40	83	12	55	8.0	55	8.0	55	8.0
		Half hard				379	55	138	20	92	13.3	92	13.3	92	13.3

TABLE 8-10
Maximum allowable stress values, σ_{sa} in tension for nonferrous metals (*Cont.*)

Maximum allowable stress, σ_{sa}, for metal temperature, °C (°F), not exceeding

120 (250)		150 (300)		176 (350)		205 (400)		232 (450)		260 (500)		288 (550)		315 (600)		343 (650)		370 (700)		
MPa	kpsi	MPa	kpsi	MPa	kpsi	MPa	kpsi	MPa	kpsi	MPa	kpsi	MPa	kpsi	MPa	kpsi	MPa	kpsi	MPa	kpsi	Spec. No.
																				Sheet and plate
22	3.2	19	2.8	14	2.0	8	1.2													SB 209
25	3.7	19	2.8	14	2.0	8	1.2													
34	4.9	30	4.3	21	3.0	16	2.4	(Fig. 8-9)												SB 209
28	4.0	25	3.6	21	3.0	16	2.4	(Fig. 8-8)												
48	7.0	40	5.8	26	3.8	16	2.4													SB 209
58	8.5	43	6.2	32	4.1	16	2.4													SB 209
51	7.4	38	5.5	28	4.1	21	3.0													SB 209
52	7.5	38	5.5	28	4.1	21	3.0													
51	7.4	48	6.9	43	6.3	31	4.5													SB 209
41	5.9	38	5.5	32	4.6	24	3.5													
																				Rods, bars, and shapes
95	13.7	72	10.4	49	6.5	31	4.5													SB 221
88	12.8	69	9.7	42	6.1	29	4.2													
21	3.0	16	2.4	12	1.8	10	1.4													SB 221
																				SB 221
																				SB 221
59	8.5	50	7.2	39	5.6	28	4.0													SB 308
37	5.4	35	5.0	29	4.2	22	3.2													
																				Die and hand forgings
86	12.5	79	11.5	47	6.8	27	3.9													SB 247
102	14.8	79	11.5	47	6.8	27	3.9													
102	14.8	79	11.5	47	6.8	27	3.9													
63	9.1	54	7.9	43	6.3	31	4.5													SB 247
61	8.8	53	7.7	43	6.3	31	4.5													SB 247
58	8.4	51	7.4	42	6.1	31	4.5													
																				Castings
43	6.3																			SB 26
42	6.1	37	5.4	28	4.1	16	2.4													
																				SB 108
																				Sheet and plates
81	11.7	69	10.0	38	5.0															SB 96[g]
124	18.0	124	18.0	124	18.0	121	17.5	117	17.0	114	16.5									SB 169
86	12.5	85	12.3	75	10.8	36	5.3													SB 171
69	10.0	69	10.0	69	10.0	36	5.3													
69	10.0	69	10.0	68	9.8	24	3.5	14	2.0											SB 171
																				SB 171
86	12.5	86	12.5	43	6.3	17	2.5													
83	12.0	83	12.0	43	6.3	17	2.5													
72	10.4	72	10.4	72	10.4	72	10.4	72	10.4	72	10.4	72	10.4	72	10.4	72	10.4	72	10.4	SB 171
64	9.3	64	9.3	64	9.3	64	9.3	64	9.3	64	9.3	64	9.3	64	9.3	64	9.3	64	9.3	
64	9.3	62	9.0	60	8.7	59	8.5	57	8.2	55	8.0	48	7.0	41	6.0					SB 204
																				Die forgings (hot pressed)
93	13.5	93	13.5	90	13.0	76	11.0	52	7.5	36	5.2									SB 283[h]
96	12.5	86	12.5	83	12.0	76	11.0	52	7.5	36	5.2									
																				Rods and bars
69	10.0	69	10.0	35	5.0															SB 98[g]
121	17.5	121	17.5	69	10.0															
55	8.0	48	7.0	35	5.0															SB 98
88	12.8	69	10.0	55	8.0															

TABLE 8-10
Maximum allowable stress values, σ_{sa} in tension for nonferrous metals (*Cont.*)

Specification no.	Alloy designation	Temper condition	Nominal composition	UNS no.	Size or thickness mm (in)	Specified minimum tensile strength, σ_{st}		Specified minimum yield strength, σ_{sy}		Maximum allowable stress, σ_{sa}, for metal temperature °C (°F), not exceeding					
										38 (100)		65 (150)		93 (200)	
						MPa	kpsi	MPa	kpsi	MPa	kpsi	MPa	kpsi	MPa	kpsi
Castings															
SB 61	922	As Cast				234	34	110	16	59	8.5	59	8.5	59	8.5
SB 148	954	As Cast				517	75	207	30	13	18.8	130	18.8	129	18.7
SB 271	952	As Cast				448	65	172	25	108	15.7	108	15.7	103	14.9
SB 584	976	As Cast				276	40	117	17	52	7.5	52	7.2	48	7.0
												Titanium and Titanium Alloys			
Sheet, strip, plate, bar, billet, and casting															
SB 265	Grade 1 (F1)		Sheets, strips, plate			241	35	172	25	61	8.8	59	8.1	50	7.3
SB 381	2 (F2)	Annealed	Forging			345	50	276	40	86	12.5	82	12.0	75	10.9
SB 348	3 (F3)	Annealed	(F stands for forging)			448	65	379	55	112	16.3	107	15.6	99	14.3
	12 (F12)		Bar, billet			482	70	345	50	121	17.5	121	17.5	113	16.4
SB 367[h]	Grade C-2		Casting[b]			345	50	276	40	86	12.5	81	11.7	74	10.7
														Zirconium	
Flat-rolled products and bars															
SB 551	Grade R 60702		Hot-rolled products			358	52	207	30	90	13.0			76	11.0
SB 550	R 60705		Bars			552	80	379	55	138	20.0			114	16.6
												Nickel and Nickel Alloys			
Plate, sheet, and strip										38°C (100°F)		93°C (200°F)		150°C (300°F)	
SB 127[j]	400	Annealed[j]	Ni-Cu	N04400		482	70	193	28	128	18.6	113	16.4	106	15.4
		Hot-rolled				517	75	276	40	129	18.7	129	18.7	129	18.7
SB 168	600	Annealed	Ni-Cr-Fe	N06600		552	80	241	35	138	20.0	138	20.0	138	20.0
SB 168[j]	600[j]	Hot-rolled[j]	Ni-Cr-Fe	N06600		586	85	241	35	146	21.2	146	21.2	146	21.2
SB 333[k]	B2	Sol. ann. [k]	Ni-Mo	N10665	All	758	110	352	51	190	27.5	190	27.5	190	27.5
SB 424[k]	825	Annealed	Ni-Fe-Cr-Mo-Cu	N08825		586	85	241	35	148	21.5	148	21.5	141	20.4
SB 435[k]	X	Annealed[k]	Ni-Cr-Mo-Fe	N06002	0.063 (1/16) 0.188 (3/16)	689	100	276	40	161	23.3	144	20.9	132	19.2
SB 435	X	Annealed	Ni-Cr-Mo-Fe	N06002	0.063 (1/16) 0.188 (3/16)[k]	689	100	276	40	161	23.3	161	23.3	116	23.3
SB 435[k]	X	Annealed	Ni-Cr-Mo-Fe	N06002	>0.188 (3/16)	655	95	241	35	161	23.3	144	20.9	132	19.2
SB 443	625	Annealed	Ni-Cr-Mo-Cb	N06625	>100 (4)	758	110	379	55	190	27.5	190	27.5	190	27.5
SB 463	20Cb	Annealed	Cr-Ni-Fe-Mo-Cu-Cb	N08020		552	80	241	35	138	20.0	138	20.0	136	19.8
SB 575[k]	C22	Sol. ann. [k]	Ni-Mo-Cr	N06022		689	100	310	45	172	25.0	172	25.0	171	24.8
SB 582	G	Sol. ann.	Ni-Cr-Fe-Mo-Cu	N06007	19.3 (3/4)	620	90	241	35	155	22.5	144	22.9	134	19.5
SB 582[k]	G	Sol. ann. [k]	Ni-Cr-Fe-Mo-Cu	N06007	>19.3 (3/4) [k]	586	85	207	30	138	20.0	138	20.0	138	20.0
SB 709	28	Annealed	Ni-Fe-Cr-Mo-Cu Low C	N08028		503	73	213	31	125	18.2	125	18.2	117	17.0
Bars, rods, shapes, and forgings															
SB 164	400	Annealed	Ni-Cu	N04400	All sizes	482	70	172	25	114	16.6	101	14.6	94	13.6
SB 166[k]	600[k]	Annealed[k]	Ni-Cr-Fe	N06600		552	80	241	35	138	20.0	138	20.0	138	20.0
SB 166	600	Hot fin	Ni-Cr-Fe	N06600		586	85	241	35	146	21.2	146	21.2	146	21.2
SB 425[k]	825	Annealed[k]	Ni-Fe-Cr-Mo-Cu	N08825		586	85	241	35	146	21.2	146	21.2	141	20.4
SB 462[k]	20Cb	Annealed[k]	Cr-Ni-Fe-Mo-Cu-Cb	N08020		552	80	141	35	138	20.0	138	20.0	136	19.8
SB 511[l]	330		Ni-Fe-Cr-Si	N08330[l]		482	70	207	30	121	17.5	121	17.5	112	16.3
SB 564	625	Annealed	Ni-Cr-Mo-Cb	N06625		758	110	345	50	190	27.5	190	27.5	190	27.5
SB 574[k]	C-4	Sol. ann.	Ni-Mo-Cr	N06455[k]	All sizes	689	100	276	40	172	25.0	172	25.0	172	25.0
Castings															
SA 494[h]	B	Annealed[h]	Ni-Mo	N-12 WV		524	76	276	40	131	19.0	123	17.8	123	17.8
SA 494	C	Annealed	Ni-Mo-Cr	CW-12MW		496	72	276	40	124	18.0	118	17.1	112	16.2

[a] The stress values in this table may be interpolated to determine values for intermediate temperatures.
[b] Stress values in restricted shear shall be 0.8 times the values in this table.
[c] Stress values in bearing shall be 1.60 times the values in the table.
[d] For weld construction, stress values for this material shall be used.
[e] The stress values given for this material are not applicable when either welding or thermal cutting is employed.
[f] Allowable stress values shown are 90 percent those for the corresponding core material.
[g] Copper-silicon alloys are not always suitable when exposed to certain media and high temperature, particularly steam above 100°C (212°F).
[h] No welding is permitted.
[i] If welded, the allowable stress values for annealed condition shall be used.
[j] For plates only.

TABLE 8-10
Maximum allowable stress values, σ_{sa} in tension for nonferrous metals (*Cont.*)

Maximum allowable stress, σ_{sa}, for metal temperature, °C (°F), not exceeding																				
120 (250)		150 (300)		176 (350)		205 (400)		232 (450)		260 (500)		288 (550)		315 (600)		343 (650)		370 (700)		
MPa	kpsi	MPa	kpsi	MPa	kpsi	MPa	kpsi	MPa	kpsi	MPa	kpsi	MPa	kpsi	MPa	kpsi	MPa	kpsi	MPa	kpsi	Spec. No.
																				Castings
59	8.5	59	8.5	59	8.5	57	8.3	53	7.7	50	7.2	34	5.0							SB 61
129	18.7	129	18.7	125	18.1	120	17.4	110	16.0	96	13.9	76	11.0	59	8.5					SB 148
100	14.5	98	14.2	98	14.2	98	14.2	98	14.2	98	14.2	81	11.7	51	7.4					SB 271
48	6.9	46	6.7																	SB 584
																				Sheet, strip, plate, bar, billet, and casting
45	6.5	40	5.8	36	5.2	33	4.8	31	4.5	28	4.1	25	3.6	21	3.1					SB 265
68	9.9	62	9.0	58	8.4	53	7.7	50	7.2	46	6.6	43	6.2	39	5.7					SB 381
90	13.0	81	11.7	72	10.4	64	9.3	57	8.3	52	7.5	46	6.7	41	6.0					SB 348
105	15.2	98	14.2	92	13.3	86	12.5	82	11.9	79	11.4	77	11.1	75	10.8					Flat-rolled products and bars
68	9.8	61	8.9	55	8.0	50	7.2													SB 367[h]
																				Plate, sheet, and strip
		64	9.3			48	7.0			42	6.1			41	6.0			33	4.8	SB 551
		98	14.2			86	12.5			78	11.3			72	10.4			68	4.9	SB 550
205°C (400°F)		**260°C (500°F)**		**315°C (600°F)**		**370°C (700°F)**		**426°C (800°F)**		**482°C (900°F)**		**539°C (1100°F)**		**593°C (1100°F)**		**648°C (1200°F)**		**704°C (1300°F)**		**760°C (1400°F)**
102	14.8	101	14.7	101	14.7	101	14.7	98	14.2	55	8.0									SB 127[j]
129	18.7	129	18.7	129	18.7	124	1 8.0	98	14.2	28	4.0					14	2.0			
138	20.0	138	20.0	138	20.0	135	19.6	132	19.1	110	16.0	48	7.0	21	3.0	38	5.5			SB 168
146	21.2	146	21.2	146	21.2	145	21.1	141	20.4	135	19.6	100	14.5	50	7.2					SB 168[j]
190	27.5	190	27.5	189	27.2	187	27.1	137	19.8											SB 333[k]
132	19.2	126	18.3	123	17.8	119	17.3	118	17.1	116	16.8	115	16.6							SB 424[k]
123	17.8	114	16.5	108	15.6	103	15.6	101	14.7	100	14.5	99	14.3	98	14.2	78	11.3	53	7.7	33 4.8
158	22.9	154	22.3	146	21.1	140	20.3	136	19.7	135	19.6	131	19.3	121	17.5	78	11.3	53	7.7	33 4.8
123	17.8	114	16.5	108	15.6	103	15.0	101	14.7	100	14.5	99	14.3	98	14.2	78	11.3	53	7.7	33 4.8
185	26.8	180	26.1	175	25.4	172	25.0	170	24.6	165	24.0	163	23.7	166	23.4	91	13.2			SB 443
129	18.7	125	18.2	121	17.5	119	17.3	116	16.8											SB 463
165	23.9	160	23.2	157	22.7	154	22.4	153	22.2											SB 575[k]
125	18.2	120	17.4	116	16.8	113	16.4	111	16.1	110	16.0	109	15.8							SB 582
138	20.0	138	20.0	134	19.4	131	19.0	128	18.6	127	18.4	126	18.3							SB 582[k]
109	15.8	100	14.5	92	13.3															SB 709
																				Bars, rods, shapes, and forgings
91	13.2	98	13.1	98	13.1	98	13.1	88	12.7	55	8.0					14	2.0			SB 164
138	20.0	138	20.0	138	20.0	138	20.0	138	20.0	110	16.0	48	7.0	21	3.0	38	5.5			SB 166[k]
146	21.2	146	21.2	146	21.2	146	21.1	141	20.4	134	19.5	100	14.5	50	7.2					SB 166
132	19.2	126	18.3	123	17.8	119	17.3	118	17.1	116	16.8	114	16.6							SB 425[k]
129	18.7	125	18.2	122	17.7	119	17.3	116	16.8							32	4.7	21	3.1	SD 462[k]
105	15.3	101	14.6	94	13.7	92	13.4	89	12.9	85	12.3	82	11.9	54	7.8	91	13.2			12 1.8
185	26.8	180	26.1	175	25.4	172	25.0	170	24.6	165	24.0	163	23.7	166	23.4					SB 564
172	25.0	170	24.7	168	24.4	165	24.0	158	23.0											SB 574[k]
																				Castings
123	17.8	123	17.9	123	17.8	122	17.7	119	17.3	114	16.6	108	15.7							SA 494[h]
112	16.2	112	16.2	112	16.2	111	16.1	105	15.2	99	14.4	95	13.8							SA 494

[k] Nickel alloys have low yield strength. The stress values of these alloys used are slightly on the high side. These higher stress values exceed 2/3 but do not exceed 90 percent of the yield strength at temperature. These stress values are not recommended for the flanges of gasket joints where a slight amount of distortion can cause leakage. Sol. ann. = Solution annealed.

[l] At temperature above 538°C (1000°F), these stress values may be used only if the material is annealed at a minimum temperature of 1038°C (1900°F) and has a carbon content of 0.04% or higher.

[m] These stress values multiplied by a joint efficiency factor of 0.85.

[n] A joint efficiency factor of 0.85 has been applied in arriving at the maximum allowable stress values in tension for this material.

[o] Alloy NO6225 in the annealed condition is subject to severe loss of impact strength at room temperature after exposure in the range of 538° to 760°C (1000° to 1400°F).

[p] For other conditions and specifications, it is suggested to refer to the General Notes given for Table UNF-23.1 of *ASME Boiler and Pressure Vessel Code*, Section VIII, Division 1, July 1, 1986.

Source: The American Society of Mechanical Engineers, *ASME Boiler and Pressure Vessel Code*, Section VIII, Division 1, July 1, 1986.

TABLE 8-11
Maximum allowable stress values (σ_{sa}) in tension for high-alloy steel

					Specified minimum yield strength, σ_{sy}		Specified minimum tensile strength, σ_{st}		Maximum allowable stress, σ_{sa}, for metal temperature, °C (°F), not exceeding							
									−30 to 38 (−20 to 100)		93 (200)		150 (300)		205 (400)	
Spec. no.	Grade	UNS no.	Nominal composition	Product form	MPa	kpsi	MPa	kpsi	MPa	kpsi	MPa	kpsi	MPa	kpsi	MPa	kpsi
SA-240, SA-479	405	S 40500	12 Cr-1 Al[d]	Plate. bar	172	25	414	60	103	15.0	99	14.3	95	13.8	92	13.3
SA-240	410 S	S 41008	13 Cr	Plate	207	30	414	60	103	15.0	99	14.3	95	13.8	92	13.3
SA-240	TP 409	S 40900	11 Cr-Ti	Plate	207	30	379	55	95	13.8	90	13.1	97	12.7	84	12.2
SA-240	18 Cr-2 Mo	S 44400	18 Cr-2 Mo[d]	Plate	276	40	414	60	103	15.0	99	14.3	95	13.8	92	13.3
SA-240	430	S 43000	17 Cr[d]	Plate	207	30	448	65	112	16.3	107	15.5	103	15.0	99	14.4
SA-479	410	S 41000	13 Cr	Bar, forge	276	40	483	70	111	16.2	106	15.4	103	14.9	99	14.4
SA-182	F6 ACI.1	S 41000	13 Cr	Bar, forge	276	40	483	70	111	16.2	106	15.4	103	14.9	99	14.4
SA-217	CA 15	J 91150	13 Cr[d]	Cast	448	65	620	90	155	22.5	148	21.5	143	20.7	138	20.0
SA-479	430, XM8	S 43000, S 43035	17 Cr[d,e]; 18 Cr-Ti[d,e]	Bar[e,g]	276	40	483	70	121	17.5	114	16.6	111	16.1	107	15.5
SA-412	201	S 20100	17 Cr-4 Ni-6 Mn	Plate	310	45	655	95	158	23.0	143	20.8	132	19.1		
SA-182	F 304 L	S 30403	18 Cr-8 Ni	Forge[g]	172	25	448	65	108	15.6	106	15.4	98	14.2	94	13.6
SA-240, SA-479	304 L	S 30403	18 Cr-8 Ni	Plate[g], bar[e,g]	172	25	483	70	108	15.7	108	15.7	105	15.3	101	14.7
SA-351	CF 3	J 92500	18 Cr-S Ni	Cast[g]	207	30	483	70	121	17.5	114	16.6	105	15.3	104	5.1
SA-351	CF 8	J 92600	18 Cr-8 Ni	Cast[g,h]	207	30	483	70	121	17.5	114	16.6	104	15.1	103	15.0
SA-351	CF 8 M	J 92900	18 Cr-9 Ni-2 Mo	Cast[g,h]	207	30	483	70	121	17.5	121	17.5	118	17.1	116	16.8
SA-336	CI-F 304 H	S 30409	18 Cr-8 Ni	Forge[g]	207	30	483	70	121	17.5	114	16.6	107	15.5	104	15.1
SA-240, SA-479	302	S 30200	18 Cr-8 Ni	Plate, bar[e,g]	207	30	517	75	130	18.8	123	17.8	114	16.6	112	16.2
SA-182	F 304	S 30400	18 Cr-8 Ni	Forge[e,g]	207	30	517	75	130	18.8	123	17.8	114	16.6	112	16.2
SA-479	304 H	S 30400	18 Cr-8 Ni	Bar[g,e]	207	30	517	75	130	18.8	123	17.8	114	16.6	112	6.2
SA-240	304	S 30400	18 Cr-8 Ni	Plate	207	30	517	75	130	18.8	123	17.8	114	16.6	112	16.2
SA-351	CF 3A	J 92500	18 Cr-8 Ni	Cast[g]	241	35	534	77.5	134	19.4	125	18.2			116	16.9
SA-240	304 N	S 30451	18 Cr-8 Ni-N	Plate[g,h]	241	35	552	80	138	20.0	138	20.0	131	19.0	126	18.3
SA-336	F 304 N	S 30451	18 Cr-8 Ni-N	Forge	241	35	552	80	138	20.0	138	20.0	131	19.0	126	18.3
SA-240	316 L	S 31603	16 Cr-12 Ni-2 Mo	Plate[g]	172	25	483	70	108	15.7	108	15.7	108	15.7	107	15.5
SA-182	F 316 L	S 31603	16 Cr-12 Ni-2 Mo	Forge[g]	172	25	448	65	108	15.7	108	15.7	108	15.7	107	15.5
SA-479	316 L	S 31603	16 Cr- 1 2 Ni-2 Mo	Bar[g,f]	172	25	483	70	108	15.7	108	15.7	108	15.7	107	15.5
SA-351	CF 8 M	J 92900	16 Cr- 1 2 Ni-2 Mo	Cast	207	30	483	70	121	17.5	121	17.5	118	17.1	116	16.8
SA-182	F 316	S 31600	16 Cr-12 Ni-2 Mo	Forge[g,h,j]	207	30	483	70	121	17.5	121	17.5	118	17.1	116	16.8
SA-336	CI-F 316 H	S 31609	16 Cr-12 Ni-2 Mo	Forge	207	30	483	70	121	17.5	111	16.2	100	14.6	92	13.4
SA-240	316 Ti	S 31635	16 Cr-12 Ni-2 Mo	Plate[g,h,i]	207	30	517	75	130	18.8	130	18.8	127	18.4	125	18.1
SA- 1 82	F 316 H	S 31609	16 Cr-12 Ni-2 Mo	Forge[g]	207	30	517	75	130	18.8	130	18.8	127	18.4	125	18.1
SA-479	316	S 31600	16 Cr-12 Ni-2 Mo	Bar[e,g,h]	207	30	517	75	130	18.8	130	18.8	127	18.4	125	18.1
SA-240	317 L	S 31703	18 Cr-13 Ni-3 Mo	Plate[g]	207	30	517	75	130	18.8	112	16.2	98	14.2	92	13.4
SA-240	XM-15	S 38100	18 Cr-18 Ni-2 Si	Plate[g]	207	30	517	75	130	18.8	122	17.7	114	16.6	111	16.1
SA-240	316 M	S 31651	16 Cr-12 Ni-2 Mo-N	Plate[g,h]	241	35	552	80	138	20.0	138	20.0	132	19.2	130	18.8
SA-479, SA-240	XM-29	S 24000	18 Cr-3 Ni-12 Mn	Plate, bar[f,g]	379	55	689	100	172	25.0	169	24.5	156	22.6	149	21.6
SA-182, SA-336	F 321 H	S 32100	18 Cr-10 Ni-Ti	Forge[g,i]	207	30	483	70	121	17.5	118	17.1	111	16.1	110	16.0
SA-240, SA-479	321	S 32100	18 Cr-10 Ni-Ti	Plate[g,h], bar[g,h,e]	207	30	517	75	130	18.8	127	18.4	119	17.3	118	17.1
SA-182, SA-336	F 347	S 34700	18 Cr-10 Ni-Cb	Forge[g,h,i]	207	30	483	70	121	17.5	115	16.7	105	15.3	99	14.4
SA-351	CFBC	J 92710	18 Cr-10 Ni-Cb	Cast[g,h]	207	30	483	70	121	17.5	114	16.6	105	15.3	96	13.9
SA-240, SA-182	347,348	S 34700	18 Cr-10 Ni-Cb	Plateg[g,h], forge[g,h]	207	30	517	75	130	18.8	123	17.9	113	16.4	107	15.5
SA-479	F 347, F 348	S 34800	18 Cr-10 Ni-Cb	Bar[g,h,e]	207	30	517	75	130	18.8	123	17.9	113	16.4	107	15.5
SA-351	CG 8 M		19 Cr-11 Ni-Mo	Cast[g]	241	35	517	75	121	17.5	121	17.5	118	17.1	116	16.8
SA-182, SA-240	F 44	S 31254	20 Cr-18 Ni-6 Mo	Forge, plate	303	44	648	94	162	23.5	162	23.5	147	21.4	137	19.9
SA-182, SA-240.	F 45	S 30815	21 Cr- 1 1 Ni-N	Forge, plate, bar	310	45	600	87	150	21.8	149	21.6	141	20.4	135	19.6
SA-479		S 30815	21 Cr-11 Ni-N	Forge. plate, bar	310	45	600	87	150	21.8	149	21.6	141	20.4	135	19.6
SA-240, SA-479		S 32550	25.5 Cr-5.5 Ni-3.5 Mo	Plate, bar	552	80	758	110	190	27.5	189	27.4	177	25.7	170	24.7
SA-351	CH 8	J 93400	25 Cr-12 Ni	Cast[g,h]	193	28	448	65	112	16.3	103	14.9	98	14.2	95	13.8
SA-351	CH 20	J 93402	25 Cr-12 Ni	Cast[h]	207	30	483	70	121	17.5	111	16.1	105	15.3	102	14.8
SA-240	309 S, 309Cb	S 30908, S 30940	23 Cr-12 Ni	Plate[g,h,j]	207	30	517	75	130	18.8	118	17.2	113	16.4	110	15.9
SA-240, SA-182	310 Cb	S 31040,	25 Cr-20 Ni	Plate[g,k,h,j], forge[g,k,h]	207	30	517	75	130	18.8	118	17.2	113	16.4	110	15.9
SA-479	CI-F310 310 S	S 31000 S 310 S	25 Cr-20 Ni	Bar[g,k,h,e]	207	30	517	75	130	18.8	118	17.2	113	16.4	109	15.8
SA-240	TP 329	S 32900	26 Cr-4 Ni-Mo	Plate[d]	483	70	620	90	155	22.5	151	21.9	141	20.5	136	19.8
SA-182, SA-336	FXM-27 Cb	S 44625	27 Cr-Mo	Forge[d]	241	35	414	60	103	15.0	103	15.0	101	14.6	98	14.2
SA-240, SA-479	XM-27	S 44627	27 Cr-Mo	Plate[d], bar, shape[d,e]	276	40	448	65	112	16.2	112	16.2	110	15.9	110	15.9
SA-240	XM-33	S 44626	27 Cr-Mo-Ti	Plate[d]	310	45	469	68	117	17.0	117	17.0	116	16.8	114	16.6
SA-240, SA-479	844800		29 Cr-4 Mo-2 Ni	Plate[d], bar[d,e]	414	60	552	50	138	20.0	134	19.4	126	18.3	125	18.1
SA-564	630 H 1100	S 17400	17Cr-4 Ni-4 Cu	Bar[d,l]	793	115	965	140	241	35.0	241	35.0	241	35.0	235	34.1
SA-182, SA-336, SA-41Z	FMX-11, NM-11	S-21904	20 Cr-6 Ni-9 Mn	Forge, plate	345	50	620	90	155	22.5	154	22.4	148	21.4	136	19.7

TABLE 8-11
Maximum allowable stress values (σ_{sa}) in tension for high-alloy steel (*Cont.*)

Maximum allowable stress, σ_{sa}, for metal temperature, °C (°F), not exceeding																						
260 (500)		315 (600)		370 (700)		427 (800)		482 (900)		538 (1000)		593 (1100)		650 (1200)		704 (1300)		760 (1400)		815 (1500)		
MPa	kpsi	MPa	kpsi	MPa	kpsi	MPa	kpsi	MPa	kpsi	MPa	kpsi	MPa	kpsi	MPa	kpsi	MPa	kpsi	MPa	kpsi	MPa	kpsi	Spec. no.
89	12.9	85	12.4	83	12.1	76	11.1	69	9.7	27	4.0	(Fig. 8-5)										SA-240, SA-479
89	12.9	85	12.4	83	12.1	76	11.1	69	9.7	44	6.4	20	2.9	7	1.0	(Fig. 8-5)						SA-240
81	11.8	79	11.4	76	11.1	70	10.2	(Fig. 8-5)														SA-240
88	12.8	85	12.4																			SA-240
96	13.9	93	13.5	90	13.1	82	12.0	72	10.5	45	6.5	22	3.2	12	1.8							SA-240
96	13.9	92	13.4	90	13.1	82	12.0	72	10.4	44	6.4	(Fig. 8-5)										SA-479
96	13.9	92	13.4	90	13.1	82	12.0	72	10.4	44	6.4	(Fig. 8-5)										SA-182
133	19.3	129	18.7	125	18.1	115	16.7	76	11.0	34	5.0	15	2.2	7.0	1.0	(Fig. 8-5)						SA-217
103	15.0	100	14.5	97	14.1	89	12.9	76	11.0	45	6.5											SA-479
																						SA-412
92	13.4	92	13.3	90	13.1	89	12.9															SA-182
99	14.4	96	14.0	93	13.5	90	13.0															SA-240, SA-479
102	14.8	102	14.8	102	14.8	100	14.6															SA-351
102	14.8	102	14.8	102	14.8	100	14.6	92	13.4	83	12.0	52	7.5	33	4.8	23	3.3	16	2.3	12	1.7	SA-351
116	16.8	116	16.8	112	16.3	109	15.8	107	15.5	103	14.9	61	8.9	37	5.4	23	3.4	16	2.3	11	1.6	SA-351
102	14.8	102	14.8	102	14.8	100	14.6	98	14.2	92	13.4	68	9.8	42	6.1	25	3.7	16	2.3	10	1.4	SA-336
110	15.9	110	15.9	110	15.9																	SA-240, SA-479
110	15.9	110	15.9	110	15.9	105	15.2	101	14.7	95	13.8	68	9.8	42	6.1	25	3.7	16	2.3	10	1.4	SA-182
110	15.9	110	15.9	110	15.9	105	15.2	101	14.7	95	13.8	68	9.8	42	6.1	25	3.7	16	2.3	10	1.4	SA-479
110	15.9	110	15.9	110	15.9	105	15.2	101	14.7	95	13.8	68	9.8	42	6.1	25	3.7	16	2.3	10	1.4	SA-240
114	16.5	112	16.3	112	16.3	100	14.6															SA-351
123	17.8	120	17.4	118	17.1	114	16.6	110	15.9	103	15.0	67	9.7	41	6.0							SA-240
123	17.8	120	17.4	118	17.1	114	16.6	110	15.9	103	15.0	67	9.7	41	6.0							SA-336
99	14.4	93	13.5	89	12.9	85	12.4	83	12.1													SA-240
99	14.4	93	13.5	89	12.9	85	12.4	83	12.1													SA-182
99	14.4	93	13.5	89	12.9	85	12.4	93	12.1													SA-479
116	16.8	116	16.8	112	16.3	109	15.8	107	15.5	103	14.9	65	9.4	41	6.0	27	4.0	16	2.4	10	1.5	SA-351
116	16.8	116	16.8	112	16.3	110	15.9	107	15.6	103	15.0	85	12.4	51	7.4	28	4.1	17	2.5	8	1.2	SA-182
86	12.5	81	11.8	78	11.3	76	11.0	74	10.8	73	10.6	71	10.3	51	7.4	28	4.1	16	2.3	9	1.3	SA-336
124	18.0	117	17.0	112	16.3	110	15.9	103	15.5	105	15.3	85	12.4	51	7.4	28	4.1	16	2.3	9	1.3	SA-240
124	18.0	117	17.0	112	16.3	110	15.9	103	15.5	105	15.3	85	12.4	51	7.4	28	4.1	16	2.3	9	1.3	SA-182
124	18.0	117	17.0	112	16.3	110	15.9	103	15.5	105	15.3	85	12.4	51	7.4	28	4.1	16	2.3	9	1.3	SA-479
86	12.5	81	11.8	78	11.3	76	11.0															SA-240
110	15.9	110	15.9	110	15.9	104	15.1	101	14.6	94	13.7											SA-240
128	18.6	128	18.6	128	18.6	127	18.4	125	18.1	120	17.4	85	12.4	51	7.4							SA-240
148	21.4	144	20.9	138	20.0	131	19.0															SA-479, SA-240
110	16.0	110	16.0	109	15.8	107	15.5	105	15.3	96	14.0	62	9.0	37	5.4	22	3.2	13	1.9	8	1.1	SA-182, SA-336
118	17.1	113	16.4	109	15.8	107	15.5	105	15.3	95	13.8	48	6.9	25	3.6	12	1.7	5	0.8	2	0.3	SA-240, SA-479
96	13.9	94	13.7	94	13.7	94	13.7	94	13.7	91	13.2	63	9.1	30	4.4	15	2.2	8	1.2	5	0.8	SA-182, SA-336
94	13.7	94	13.7	94	13.7	94	13.7	94	13.7	91	13.2	72	10.5	34	5.0	19	2.7	11	1.6	7	1.0	SA-351
103	14.9	101	14.7	101	14.7	101	14.7	101	14.7	96	14.0	63	9.1	30	4.4	15	2.2	8	1.2	5	0.8	SA-240, SA-182
103	14.9	101	14.7	101	14.7	101	14.7	101	14.7	96	14.0	63	9.1	30	4.4	15	2.2	8	1.2	5	0.8	SA-479
116	16.8																					SA-351
128	18.5	123	17.9	121	17.5																	SA-182, SA-240
127	18.4	122	17.7	86	12.5	116	16.8	112	16.3	103	14.9	62	9.0	36	5.2	21	3.1	13	1.9	9	1.3	SA-182, SA-240
127	18.4	122	17.7	86	12.5	116	16.8	112	16.3	103	14.9	62	9.0	36	5.2	21	3.1	13	1.9	9	1.3	SA-479
170	24.7																					SA-240, SA-479
93	13.5	92	13.3	90	13.0	90	13.0	86	12.5	72	10.5	45	6.5	26	3.8	16	2.3	9	1.3	5	0.8	SA-351
97	14.1	92	13.4	88	12.7	94	12.2	81	11.7	70	10.2	45	6.5	26	3.8	16	2.3	9	1.3	5	0.8	SA-351
107	15.5	105	15.3	104	15.1	103	14.9	96	13.9	72	10.5	45	6.5	26	3.8	16	2.3	9	1.3	5	0.8	SA-240
107	15.5	105	15.3	104	15.1	103	14.9	96	13.9	76	11	59	8.5	41	6.0	24	3.5	11	1.6	5	0.8	SA-240, SA-182
107	15.5	105	15.3	104	15.1	103	14.9	95	13.8	76	11											SA-479
136	19.8																					SA-240
98	14.2	98	14.2																			SA-182
																						SA-336
110	15.9	110	15.9																			SA-440, SA-479
113	16.4	111	16.1																			SA-240
125	18.1	125	18.1																			SA-440, SA-479
230	33.3	226	32.8																			SA-564
123	17.9	117	17.0																			SA-182, SA-336

TABLE 8-11
Maximum allowable stress values (σ_{sa}) in tension for high-alloy steel (*Cont.*)

Spec. no.	Grade	UNS no.	Nominal composition	Product form	Specified minimum yield strength, σ_{sy} MPa	kpsi	Specified minimum tensile strength, σ_{st} MPa	kpsi	Maximum allowable stress, σ_{sa}: −30 to 38 (−20 to 100) MPa	kpsi	93 (200) MPa	kpsi	150 (300) MPa	kpsi	205 (400) MPa	kpsi
SA-351	CG 6 MM	J 93790	22 Cr-13 Ni-5 Mn	Cast	241	35	517	75	130	18.8	116	16.9	103	14.9	94	13.6
SA-240, SA-412, SA-479, SA-182	XM-19	S 20910	22 Cr-13 NI-5 Mn	Plate, bar, forge[f]	379	55	689	100	172	25.0	172	24.9	163	23.6	156	22.7

[a] The stress value in this table may be interpolated to determine values for intermediate temperatures.
[b] Stress values in restricted shear shall be 0.8 times the values in this table.
[c] Stress values in bearing shall be 1.60 times the values in this table.
[d] This steel may be expected to develop embrittlement after service at moderately elevated temperature.
[e] Use of external pressure charts for material in the form of barstock is permitted for stiffening rings only.
[f] These stress values are the basic values multiplied by a joint efficiency factor of 0.85.

TABLE 8-11
Maximum allowable stress values (σ_{sa}) in tension for high-alloy steel (*Cont.*)

For metal temperature, °C (°F), not exceeding																						
260 (500)		315 (600)		370 (700)		427 (800)		482 (900)		538 (1000)		593 (1100)		650 (1200)		704 (1300)		760 (1400)		815 (1500)		
MPa	kpsi	MPa	kpsi	MPa	kpsi	MPa	kpsi	MPa	kpsi	MPa	kpsi	MPa	kpsi	MPa	kpsi	MPa	kpsi	MPa	kpsi	MPa	kpsi	Spec. no.
90	13.0	87	12.6	85	12.3	83	12.0	81	11.8	79	11.4											SA-351
154	22.3	151	21.9	149	21.6	146	21.2	142	20.6	137	19.9	131	19.0	57	8.3							SA-240, SA-412, SA-479, SA-182

[g] Alloy steels have low yield strength. The stress values of these alloy steels used are slightly on the high side. These higher stress values exceed 2/3 but do not exceed 90 percent of the yield strength at temperature. These stress values are not recommended for the flanges of gasket joints where a slight amount of distortion can cause leakage.
[h] At temperature above 540°C (1000°F), these stress values apply only when carbon is 0.04% or higher on heat analysis.
[i] These stress values shall be applicable to forging over 125 mm (5 in) in thickness.
[j] For temperature above 540°C (1000°F), these stress values may be used only if the material is heat-treated by heating it to a minimum temperature of 1040°C (1900°F) and quenching in water or rapidly cooling by other means.
[k] These stress values at 565°C (1050°F) and above shall be used only when the grain size is ASTM 6 or coarser.
[l] These stress values are established from a consideration of strength only and shall be satisfactory for average service.
Source: The American Society of Mechanical Engineers, ASME Boiler and Pressure Vessel Code, Section VIII, Division 1, July 1, 1986.

TABLE 8-12

Maximum allowable stress values, σ_{sa}, in tension for ferrite steels with properties enhanced by heat treatment

Spec. no.	Grade and size	Specified minimum yield strength, σ_{sy}		Specified minimum tensile strength, σ_{st}		Maximum allowable stress values, σ_{sa}, for metal temperatures, °C (°F) not exceeding																			
						≤66 (150)		93 (200)		120 (250)		150 (300)		205 (400)		260 (500)		315 (600)		345 (650)		370 (700)		400 (750)		427 (800)	
		MPa	kpsi	MPa	kpsi	MPa	kpsi	MPa	kpsi	MPa	kpsi	MPa	kpsi	MPa	kpsi	MPa	kpsi	MPa	kpsi	MPa	kpsi	MPa	kpsi	MPa	kpsi	MPa	kpsi
												Plates															
SA-353[a,b]		517	75	690	100	172	25.0	161	23.4	157	22.7																
SA-517	*A, B, D, J* 31 25 mm (1.25 in)	690	100	792	115	198	28.8	198	28.8	198	28.8	198	28.8	198	28.8	198	28.8	198	28.8	197	28.7						
SA-517	62.5 mm (2.5 in) *E* ≤ 62.5 (2½ in) ≯50 mm (6 in) ≯100 mm (4 in)	690	100	792	115	198	28.8	198	28.8	198	28.8	198	28.8	198	28.8	198	28.8	198	28.8	197	28.7						
SA-533	*A, B, C, D, Cl* 2	482	70	620	90	155	22.5	155	22.5	155	22.5	155	22.5	155	22.5	155	22.5	155	22.5	155	22.5	155	22.5	152	22.2	146	21.2
	B, D, Cl 3 ≯62.5 mm (2½ in)	572	83	690	100	172	25.0	172	25.0	172	25.0	172	25.0	172	25.0	172	25.0	172	25.0	172	25.0	172	25.0	169	24.5		
SA-553	I, II[a,b,d]	586	85	690	100	172	25.0	161	23.4	156	22.7																
SA-645[a]		448	65	655	95	163	23.7	163	23.7	161	23.3																
SA-724	*B*	517	75	655	95	164	23.8	164	23.8	164	23.8	164	23.8	163	23.5	163	23.5	163	23.5	161	23.4	159	23.1				
	A, C	482	70	620	90	155	22.5	155	22.5	155	22.5	155	22.5	154	22.3	154	22.3	154	22.3	153	22.2	151	21.9				
												Castings															
SA-487	*Cl.* 4 Q[e]	586	85	724	105	181	26.3	181	26.3	181	26.3	181	26.3	181	26.3	181	26.3	181	26.3	181	26.3	181	26.3				
SA-487	*Cl.* 4 QA[e]	655	95	792	115	198	28.8	199	28.8	198	28.8	198	28.8	198	28.8	198	28.8	198	28.8	198	28.8	198	28.8				
SA-487	*Cl. CA* 6 NM[e]	551	80	758	110	190	27.5	190	27.5	187	27.2	185	26.9	181	26.3	178	25.8	174	25.3	172	24.9	168	24.4	165	23.9	161	23.3
												Pipes and tubes															
SA-333	8[a,b]	517	75	690	100	172	25.0	161	23.4	156	22.7																
SA-334	8[a,c,f]	517	75	690	100	146	21.3	137	19.9	133	19.3																
												Forgings															
SA-508	*Cl* 4	586	85	724	105	181	26.2	181	26.2			181	26.2	179	26.0	178	25.8	175	25.4	173	25.1						
SA-522	I[e,g]	517	75	690	100	163	23.7	153	22.2	148	21.5																
SA-592	*A* ≤ 37.6 mm (1½ in)	690	100	792	115	198	28.8	198	28.8	198	28.8	198	28.8	198	28.8	198	28.8	198	28.8	197	28.7						
	E, F ≤ 62.5 mm (2½ *in*)																										
SA-592	*E, F* > 62.5[h] mm (2½ in)	620	90	724	105	181	26.3	181	26.3	181	26.3	181	26.3	181	26.3	181	26.3	181	26.3	180	26.2						

[a] Minimum thickness after forming any section subject to pressure shall be 4.6875 mm (3/16 in).
[b] Not welded or welded if the tensile strength of the Section IX reduced section tension test is not less than 600 MPa (100 kpsi).
[c] Welded with the tensile strength of the Section IX reduced tension test less than 690 MPa (100 kpsi) but not less than 655 MPa (95 kpsi).
[d] Grade II of SA-533 shall not he used for minimum allowable temperature below −170°C (275°F).
[e] To these stress values a quality factor as specified in UG-24 shall be applied for castings.
[f] These stress values are the basic values multiplied by a joint efficiency factor of 0.85.
[g] The maximum section thickness shall not exceed 75 mm (3 in) for double normalized and tempered forgings, or 125 mm (5 in) for quenched and tempered forgings.
[h] The maximum thickness of non-heat-treated forgings shall not exceed 93.75 mm (3¾ in). The maximum thickness as heat treated may be 100 mm (4 in).
Source: The American Society of Mechanical Engineers, *ASME Boiler and Pressure Vessel Code*, Section VIII, Division 1, July 1, 1986.

TABLE 8-13
Maximum allowable stress values, σ_{sa}, in tension for cast iron

		Specified minimum tensile strength, σ_{st}		Maximum allowable stress, σ_{sa}, for metal temperature, °C (°F) not exceeding			
				Subzero to 232 (450)		345 (650)	
Spec. no.	Class	MPa	kpsi	MPa	kpsi	MPa	kpsi
SA-667	—	138	20	13.8	2.0		
SA-278	20	138	20	13.8	2.0		
SA-278	25	172	25	17.2	2.5		
SA-278	30	207	30	20.7	3.0		
SA-278	35	241	35	24.1	3.5		
SA-278	40	276	40	27.6	4.0	27.6	4.0
SA-278	45	310	45	31.0	4.5	31.0	4.5
SA-278	50	345	50	34.5	5.0	34.5	5.0
SA-47	(Grade 3-2510)	345	50	34.5	5.0	34.5	5.0
SA-278	55	379	55	37.9	5.5	37.9	5.5
SA-278	60	414	60	41.4	6.0	41.4	6.0
SA-476	—	552	80	55.2	8.0		—
SA-748	20	138	20	13.8	2.0		—
SA-748	25	172	25	17.2	2.5		—
SA-748	30	207	30	20.7	3.0		
SA-748	35	241	35	24.1	3.5		

Source: *ASME Boiler and Pressure Vessel Code*, Section VIII, Division 1, July 1, 1986.

TABLE 8-14
Modulus of elasticity for various materials

Material	Design temperature																															
	73 K (−200°C)		173 K (−100°C)		273 K (0°C)		293 K (20°C)		323 K (50°C)		348 K (75°C)		373 K (100°C)		398 K (125°C)		423 K (150°C)		473 K (200°C)		573 K (300°C)		673 K (400°C)		773 K (500°C)		973 K (600°C)		973 K (700°C)		1023 K (750°C)	
	GPa	Mpsi	GPa	Mpsi	GPa	Mpsi	GPa	Mpsi	GPa	Mpsi	GPa	Mpsi	GPa	Mpsi	GPa	Mpsi	GPa	Mpsi	GPa	Mpsi	GPa	Mpsi	GPa	Mpsi	GPa	Mpsi	GPa	Mpsi	GPa	Mpsi	GPa	Mpsi
Ferrous Materials																																
Low-carbon steel C ≤ 0.03%					192	27.8	192	27.8					191	27.7							186	27.0	179	26	169	24.5						
High-carbon steel C > 0.3%					206	29.9	206	29.9					203	29.4							195	28.3	186	27.0	170	24.7						
Carbon molybdenum and chrome molybdenum steel up to 3% Cr					206	29.9	206	29.9					203	29.4							197	28.6	190	27.6	181	26.3	17	2.5				
Aluminum and Aluminum Alloys																																
1B, N3, N4	77	11.2	73	10.6	70	10.2			69	10.0	69	10.0	68	9.9	67	9.7	66	9.6	65	9.4												
H9	73	10.6	70	10.2	67	9.7			65	9.4	65	9.4	64	9.3	64	9.3	63	9.1	59	8.6												
H15	81	11.7	78	11.3	74	10.7			73	10.6	73	10.6	72	10.4	71	10.3	70	10.2	67	9.7												
A6	87	12.6	84	12.2	80	11.6			79	11.5	79	11.5	78	11.3	77	11.2	76	11.0	75	10.9												
Nickel and Nickel Alloy																																
Nickel							207	30.0													200	29.0	184	26.7	162	23.5	137	19.9	115	16.7	107	15.5
Nickel-copper alloy—Ni 70%, Cu 30%							184	26.3													176	25.5	173	25.0	166	24.0	159	23.0	152	22.0	147	21.3
Nickel-chromium ferrous alloy-Ni 75%, Cr 14%,Fe 10%							214	31.0													203	29.4	197	28.6	172	25.0	157	22.8	128	18.6	118	17.0
Copper and Its Alloys																																
Copper—Cu 99.98% Commercial brass—Cu 66%, Zn 34%							110	16.0	109	15.8			108	15.7			106	15.4	104	15.0	99	14.4										
Leaded tin bronze—Cu 88%, Sn 6%,							96	13.9	95	13.8			94	13.6			93	13.5	89	12.9	87	12.6										
Pb-1.5%, Zn-4.5%							89	12.9	88	12.8			87	12.6			85	12.3	82	11.9	85	12.3										
Phosphor bronze—Cu 85.5%. Sn 12.5%, Zn 10%							103	14.9	101	14.6			100	14.5			96	13.9	93	13.5	83	12.0										
Muntz—Cu 59%, Zn 39%							105	15.2	100	14.5			96	13.9			89	12.9	81	11.7												
Cupronickel—Cu 80%. Ni 20%							130	18.8	128	18.6			127	18.4			124	18.0	122	17.7	116	16.8										

TABLE 8-15
Values of coefficient c_5

	Coefficient c_5	Types of stays
1	112	Stays screwed through plates $\leq$1.1 cm thick, with the ends riveted over
2	120	Stays screwed through plates $>$1.1 cm thick, with the ends riveted over
3	135	Stays screwed through plates and provided with single nuts outside the plate or with inside and outside nuts, but no washers
4	150	With heads $\nless$1.3 times the stay diameter, screwed through the plates, or made with a taper fit and having heads formed before installing and not riveted over; these heads have a true bearing on the plate
5	175	Stays with inside and outside nuts and outside washers, when the washer diameter is $\geq$0.4a, and the thickness n

TABLE 8-16
Design stresses for bolted flanged beads, σ_d

Maximum temperature		Minimum of specified range of tensile strength of flange material at room temperature											
		3170		3520		3870		4220		4920		Alloy bolt steel	
K	°C	MPa	kpsi	MPa	kpsi	MPa	kpsi	MPa	kpsi	MPa	kpsi	MPa	kpsi
643	370	74.0	10.5	81.9	12.0	90.7	13.2	97.6	14.2	115.2	16.5	97.6	14.2
672	399	63.3	8.2	73.1	10.6	77.0	11.2	87.3	12.7	102.0	14.8	87.3	12.6
696	423	55.8	8.0	62.8	9.0	68.6	10.0	75.5	11.0	87.3	12.0	75.5	11.0
727	454	47.3	6.9	52.5	7.6	57.4	8.3	62.8	9.0	73.5	10.6	62.8	9.0
755	482	38.3	5.5	41.4	6.0	45.5	6.5	50.5	7.3	58.8	8.5	50.5	7.3
783	510	27.9	4.0	31.1	4.5	31.2	4.5	38.3	5.5	44.2	6.4	38.3	5.5

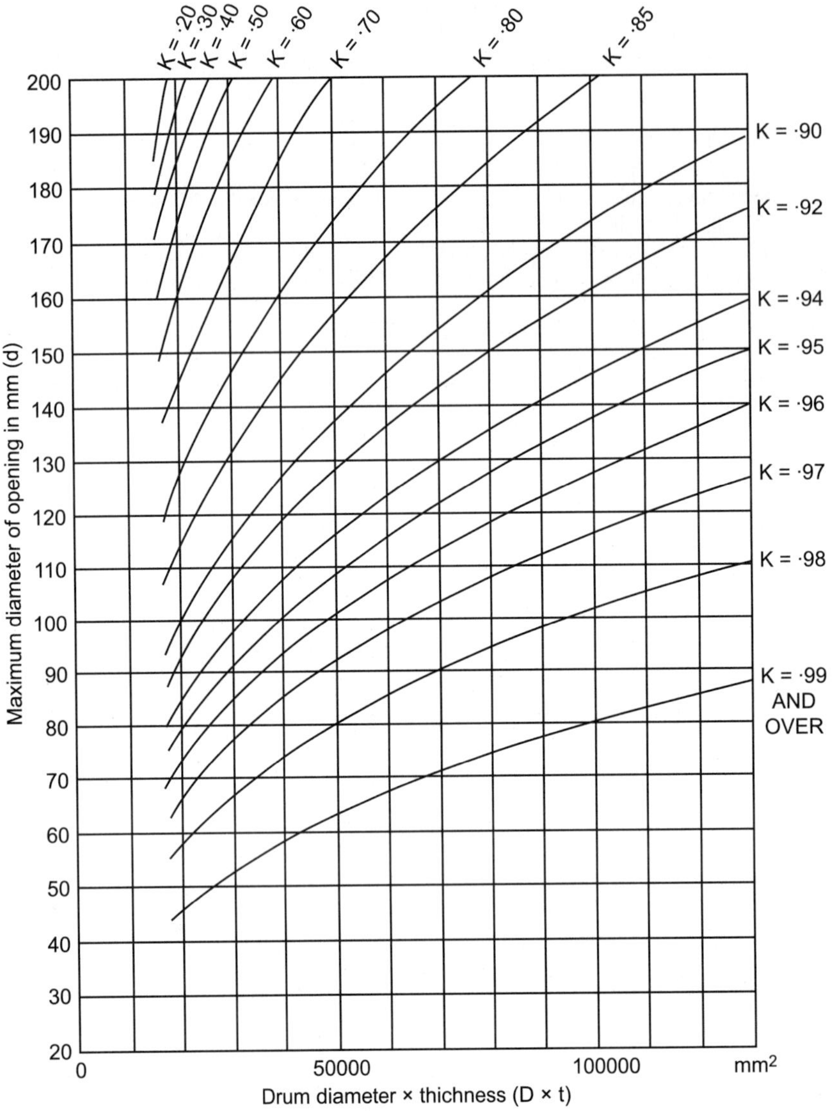

FIGURE 8-16(a) Maximum diameter of nonreinforced openings. (*Source*: IS 2825, 1969.)

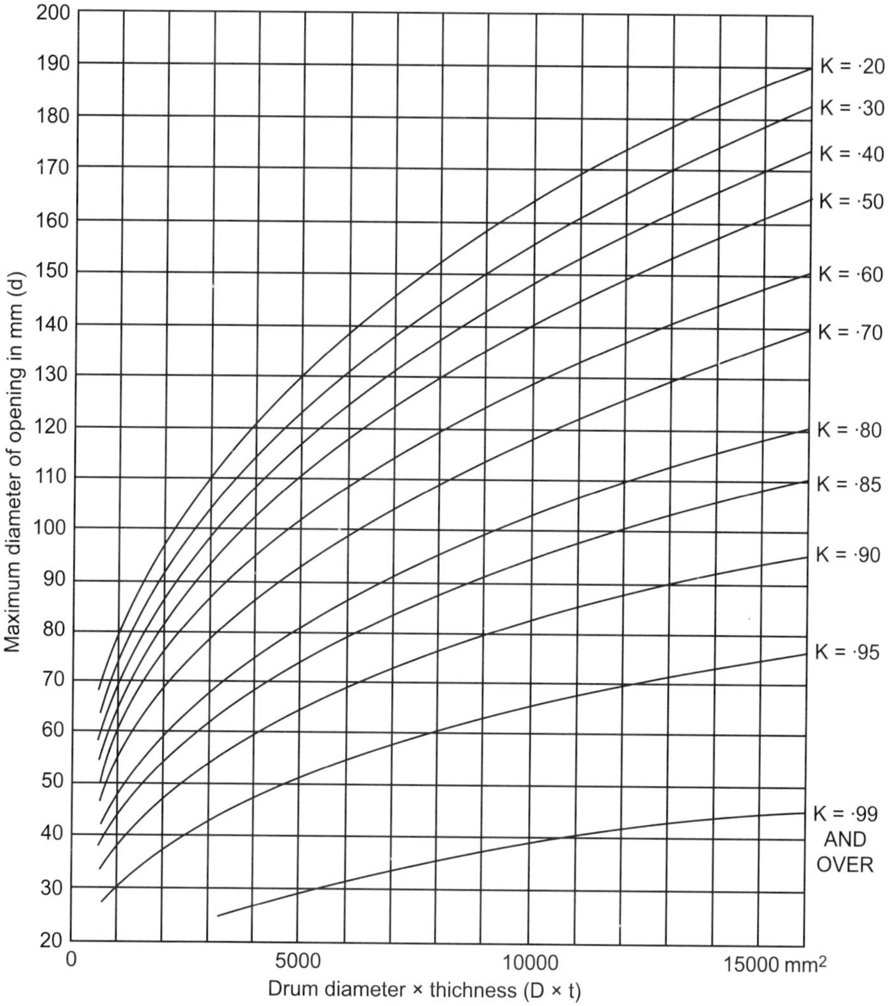

FIGURE 8-16(b) Maximum diameter of nonreinforced openings. (*Source*: IS 2825, 1969.)

TABLE 8-17
Allowable stresses (σ_{sa}) for flange bolting material

		Specified tensile strength, σ_{st}	Allowable stress, σ_{sa}, for design metal temperature not exceeding (°C)						
			50°C	100°C	200°C	250°C	300°C	350°C	400°C
Material	Diameter, mm (in)**	MPa (kpsi)	MPa (kpsi)	MPa (kpsi)	MPa (kpsi)	MPa (kpsi)	MPa (kpsi)	MPa (kpsi)	MPa (kpsi)
Hot-rolled carbon steel	≤150 (6)**	431–510 (62.5–74.0)*	57 (8.3)*	55 (8.0)*	53 (7.7)*	48 (6.9)*			
1% Cr Mo steel	≤63.5 (2.5)	843 min (122.3)	193 (28.0)	181 (26.3)	168 (24.3)	159 (23.0)	154(22.4)	148 (21.5)	140 (20.0)
	>63.5 (2.5) to 102 (4)	775 min (112.4)	174 (25.2)	163 (23.6)	152 (22.0)	145 (21.0)	141(20.5)	134 (19.4)	127 (18.4)
5% Cr Mo steel	≤63.5 (2.5)	896 min (130.0)	138 (20.0)	138 (20.0)	138 (20.0)	138 (20.0)	138(20.0)	138 (20.0)	138 (20.0)
	>63.5 (2.5) to 102 (4)	647 min (93.8)							
1% Cr V steel	≤63.5 (2.5)	843 min (122.3)	193 (28.0)	187 (27.1)	181 (26.3)	176 (25.5)	170 (24.7)	165 (23.9)	157 (22.8)
	>63.5 (2.5) to 102 (4)	804 min (116.6)	174 (25.2)	169 (24.5)	163 (23.6)	159 (23.1)	152 (22.0)	150 (21.8)	143 (20.7)
13% Cr Ni steel	≤ 102(4)	696 min (101.0 min)	176 (25.5)	161 (23.4)	141 (20.5)	134 (19.4)	126 (18.3)	119 (11.3)	104 (15.1)
18/8 Cr Ni steel	All (1) (2)	539 min (78.2 min)	129 (18.7)	109 (15.7)	85 (12.3)	78 (11.3)	76 (11.0)	73 (10.6)	72 (10.4)
18/8 Cr Ni Ti stabilized steel	All (1) (2)	In softened condition or ≤863 min (125.2) if cold-drawn	129 (18.7)	113 (16.4)	100 (14.5)	93 (13.5)	90 (13.0)	86 (12.5)	84 (12.2)
18/9 Cr Ni Nb stabilized steel	All (1) (2)		129 (18.7)	113 (16.4)	100 (14.5)	93 (13.5)	90 (13.0)	86 (12.5)	84 (12.2)
17/10/2$\frac{1}{2}$Cr Ni Mo steel	All (1) (2)		129 (18.7)	110 (16.0)	94 (13.6)	87 (12.6)	83 (12.0)	79 (11.5)	78 (11.3)
18/Cr 2 Ni steel	≤102 (4)	843 min (122.3 min)	212 (30.8)	195 (28.3)	169 (24.5)	160 (23.2)	152 (22.0)	144 (20.9)	127 (18.4)

1. Austenitic steel bolts for use in pressure joints shall not be less than 10 mm in diameter.
2. For bolts of up to 38 mm diameter use torque spanners.
3. High strength is obtainable in bolting materials by heat treatment of the ferritic and martensitic steels and by cold working of austenitic steels.

* Values in parentheses are in US Customary units (i.e., fps system of units).

** Sizes in parentheses are in inches and outside parentheses are in millimeters.

Source: IS 2825, 1969.

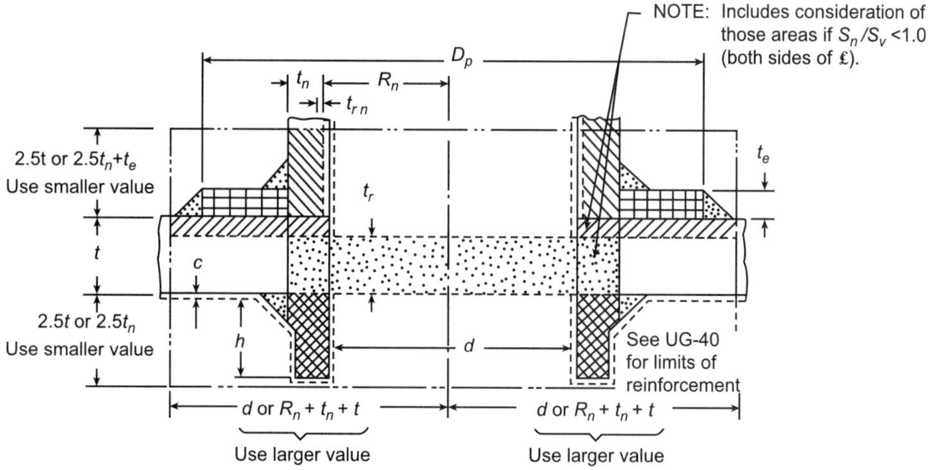

Without Reinforcing Element

$= A$	$= d\,t_r F + 2t_n t_r F(1 - f_{r1})$	Area required
$= A_1$	$= d\,(E_1 t - Ft_r) - 2t_n (E_1 t - Ft_r)(1 - f_{r1})$ $= 2\,(t + t_n)(E_1 t - Ft_r) - 2t_n (E_1 t - Ft_r)(1 - f_{r1})$	Area available in shell; use larger value
$= A_2$	$= 5\,(t_n - t_{rn}) f_{r1} t$ $= 5\,(t_n - t_{rn}) f_{r1} t_n$	Area available in nozzle projecting outward; use smaller value
$= A_3$	$= 2(t_n - c) f_{r1} h$	Area available in inward nozzle
$= A_{41}$	= outward nozzle weld = $(\text{leg})^2 f_{r1}$	Area available in outward weld
$= A_{43}$	= inward nozzle weld = $(\text{leg})^2 f_{r1}$	Area available in inward weld

If $A_1 + A_2 + A_3 + A_{41} + A_{43} > A$ — Opening is adequately reinforced

If $A_1 + A_2 + A_3 + A_{41} + A_{43} < A$ — Opening is not adequately reinforced so reinforcing elements must be added and / or thicknesses must be increased

With Reinforcing Element Added

A	= same as A, above	Area required
A_1	= same as A_1, above	Area available
A_2	$= 5\,(t_n - t_{rn}) f_{r1} t$ $= 2\,(t_n - t_{rn})(2.5t_n + t_e) f_{r1}$	Area available in nozzle projecting outward; use smaller area
A_3	= same as A_3, above	Area available in inward nozzle
$= A_{41}$	= outward nozzle weld = $(\text{leg})^2 f_{r2}$	Area available in outward nozzle
$= A_{42}$	= outward element weld = $(\text{leg})^2 f_{r3}$	Area available in outer weld
$= A_{43}$	= outward nozzle weld = $(\text{leg})^2 f_{r1}$	Area available in inward weld
$= A_5$	$= (D_p - d - 2t_n)\, t_e f_{r3}$	Area available in element

If $A_1 + A_2 + A_3 + A_{41} + A_{42} + A_{43} + A_5 > A$ — Opening is adequately reinforced

FIGURE 8-17 Nomenclature and formulas for reinforced openings. (This figure illustrates a common-nozzles configuration and is not intended to prohibit other configurations permitted by the code.) (American Society of Mechanical Engineers, *ASME Boiler and Pressure Vessel Code*, Section VIII, Division 1, July 1, 1986.)

TABLE 8-18
Values of spherical radius factor K_1 equivalent to spherical radius = $K_1 D$, $D/2h$ = axis ratio

$D/2h$	3.0	2.8	2.6	2.4	2.2	2	1.8	1.6	1.4	1.2	1.0
K_1	1.36	1.27	1.18	1.08	0.99	0.90	0.81	0.73	0.65	0.57	0.50

Particular	Formula

LIGAMENTS

The efficiency η of the ligament between the tube holes, when the pitch of the tube holes on every row is equal

$$\eta = \frac{p-d}{p} \quad (8\text{-}102)$$

where p = longitudinal pitch of tube holes, m (in)
d = diameter of tube holes, m (in)

The efficiency η of the ligament between the tube holes, when the pitch of tube holes on any one row is unequal (Fig. 8-18)

$$\eta = \frac{p_1 - nd}{p_1} \quad (8\text{-}103)$$

where p_1 = unit length of ligament, m (in)
n = number of tube holes in length, p_1

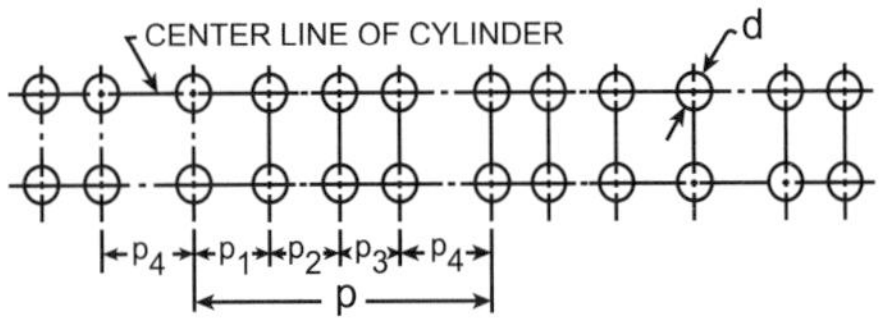

FIGURE 8-18 Irregular drilling.

The efficiency η of the ligament, when bending stress due to weight is negligible and the tube holes are arranged along a diagonal line with respect to the longitudinal axis or to a regular sawtooth pattern as shown in Fig. 8-19*a* to *d*

$$\eta = \frac{2}{A + B + \sqrt{(A-B)^2 + 4C_2}} \quad (8\text{-}104)$$

where $A = \dfrac{\cos^2\alpha + 1}{2[1 - (d\cos\alpha)/2a]}$

$$B = \frac{1}{2}\left(1 - \frac{d\cos\alpha}{a}\right)(\sin^2\alpha + 1)$$

$$C = \frac{\sin\alpha\cos\alpha}{2\left(1 - \dfrac{d\cos\alpha}{a}\right)}$$

$$\cos\alpha = \frac{1}{\sqrt{1 + (b^2/a^2)}}; \quad \sin\alpha = \frac{1}{\sqrt{1 + a^2/b^2}}$$

The smallest value of efficiency η of all the ligaments (longitudinal, circumferential, and diagonal) in the case of regular staggered spacing of tube holes

$$\eta = \frac{p_c}{p_L} = \frac{P_L - d}{P_L} \quad \text{or} \quad \frac{d}{a} \quad (8\text{-}105)$$

The symbols are as shown in Fig. 8-19d.

For minimum number of pipe threads for connections as per *ASME Boiler and Pressure Vessel Code*

Refer to Table 8-19.

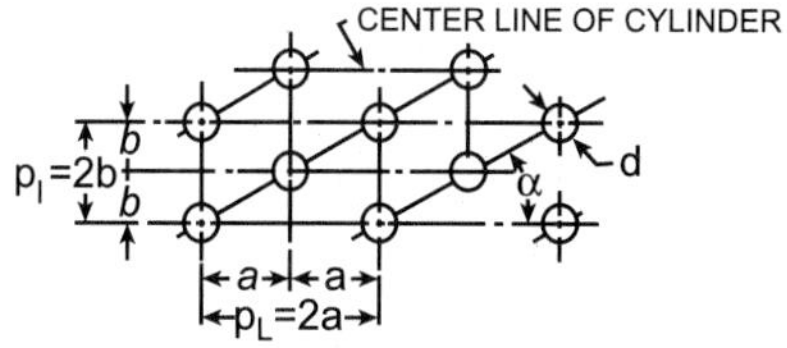

(a) A regular staggering of holes

FIGURE 8-19(a) A regular staggering of holes.

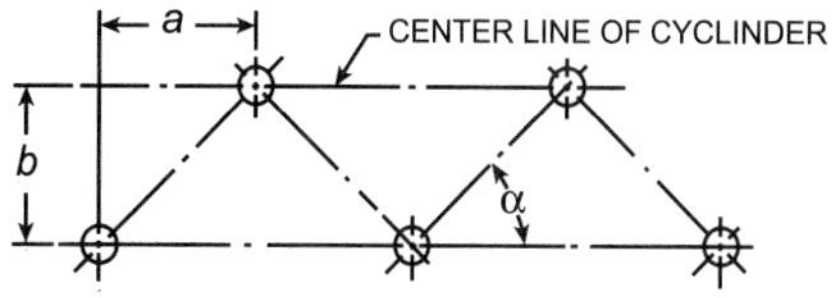

(c) Regular sawtooth pattern of holes

FIGURE 8-19(c) Regular sawtooth pattern of holes.

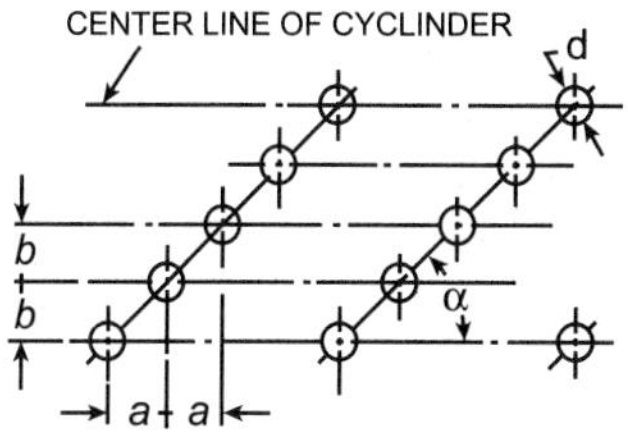

(b) Spacing of holes on a diagonal line

FIGURE 8-19(b) Spacing of holes on a diagonal line.

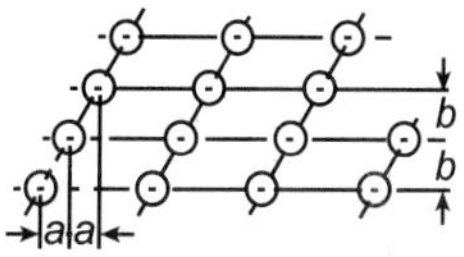

(d)

FIGURE 8-19(d)

Particular	Formula

BOLTED FLANGE CONNECTIONS

Bolt loads

The required bolt load under operating conditions sufficient to contain the hydrostatic end force and simultaneously to maintain adequate compression on the gasket to ensure seating

$$W_{m1} = H + H_P = \frac{\pi}{4} G^2 P + 2b\pi GmP \qquad (8\text{-}106)$$

For additional gasket criteria

Refer to Tables 8-20 and 8-21.

TABLE 8-19
Minimum number of threads for connections

Size of pipe connection, mm (in)	12.5 and 18.75 ($\frac{1}{2}$ and $\frac{3}{4}$)	25.0, 31.25, and 37.5 (1, $1\frac{1}{4}$, and $1\frac{1}{2}$)	50.0 (2)	62.5 and 75 ($2\frac{1}{2}$ and 3)	100–150 (4–6)	200 (8)	250 (10)	300 (12)
Threads engaged	6	7	8	8	10	12	13	14
Minimum plate thickness required, mm (in)	10.75 (0.43)	15.25 (0.62)	17.50 (0.70)	25.0 (1.0)	31.25 (1.25)	37.50 (1.5)	40.5 (1.62)	43.75 (1.75)

TABLE 8-20
Gasket materials and contact facings[a]

Dimension N mm (in) (min)	Gasket material		Gasket factor, m	Minimum design seating stress, y MPa (kpsi)	Sketches and and notes	Refer to Table 8-21: Use facing sketch	Refer to Table 8-21: Use column
10	Rubber without fabric or a high percentage of asbestos fiber:				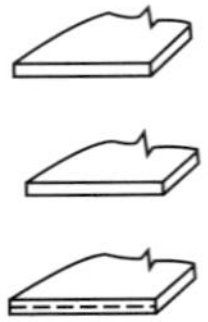	1 (a, b, c, d), 4, 5,	II
	<70 IRHD* (75A Shore Durometer)		0.50	0			
	70 IRHD (75A)* or higher		1.00	1.37 (0.2)			
	Asbestos with a suitable binder for the operating conditions	3.2 mm (0. 125 in)	2.00	11.0 (1.6)			
		1.6 mm (0.062 In)	2.75	25.5 (3.7)			
		0.8 mm (0.031 in) thickness	3.50	44.8 (6.5)			
	Rubber and elastomers with cotton fabric insertion		1.25	2.75 (0.40)			
	Rubber and elastomers with asbestos fabric insertion, with or without wire reinforcement	3-ply	2.25	15.2 (2.2)	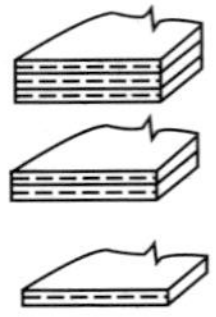	1 (a, b, c, d), 4, 5	II
		2-ply	2.50	20.0 (2.9)			
		1-ply	2.75	25.5 (3.7)			
	Vegetable fiber		1.75	7.55 (1.1)			
10	Spiral-wound metal, asbestos-filled	Carbon steel, stainless steel or monel metal	2.50	68.9 (10.0)	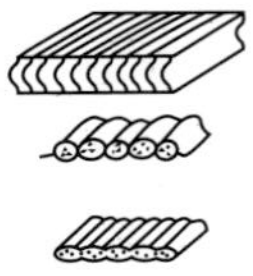		
			3.00	68.9 (10.0)			
	Corrugated metal, asbestos inserted or Corrugated metal, jacketed asbestos filled	Soft aluminum	2.50	20.0 (2.9)		1 (a, b)	II
		Soft copper or brass	2.75	25.5 (3.7)			
		Iron or soft steel	3.00	31.0 (4.5)			
		Monel metal or 4–6% chrome steel	3.25	38.0 (5.5)			
		Stainless steels	3.50	44.8 (6.5)			
	Corrugated metal	Soft aluminum	2.75	25.5 (3.7)		1 (a, b, c, d)	II
		Soft copper or brass	3.00	31.0 (4.5)			
		Iron or soft steel	3.25	38.0 (5.5)			
		Monel metal or 4–6% chrome steel	3.50	44.0 (6.5)			
		Stainless steel	3.75	52.4 (7.6)			
	Flat-metal-jacketed, asbestos-filled	Soft aluminum	3.25	38.0 (5.5)	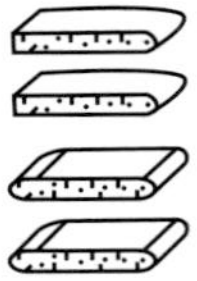	1a, 1b, 1c*, 1d* 2*	II
		Soft copper or brass	3.50	44.0 (6.5)			
		Iron or soft steel	3.75	52.4 (7.6)			
		Monel metal or 4–6% chrome steel	3.50	55.1 (8.0)			
		Stainless steels	3.75	62.1 (9.0)			

TABLE 8-20
Gasket materials and contact facings[a] (*Cont.*)

Dimension N mm (in) (min)	Gasket material		Gasket factor, m	Minimum design seating stress, y MPa (kpsi)	Sketches and and notes	Refer to Table 8-21	
						Use facing sketch	Use column
10	Grooved metal	Soft aluminum	3.25	3.80 (5.5)		1 (*a*, *b*, *c*, *d*), 2, 3	II
		Soft copper or brass	3.50	44.8 (6.5)			
		Iron or soft steel	3.75	52.4 (7.6)			
		Monel metal or 4–6% chrome steel	3.75	62.1 (9.0)			
		Stainless steels	4.25	69.6 (10.1)			
6	Solid flat metal	Soft aluminum	4.00	60.7 (8.8)	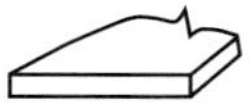	1 (*a*, *b*, *c*, *d*), 2, 3, 4, 5	I
		Soft copper or brass	4.75	89.6 (13.0)			
		Iron or soft steel	5.50	124.2 (18.0)			
		Monel metal or 4–6% chrome steel	6.00	150.3 (21.8)			
		Stainless steels	6.50	179.3 (26.0)			
	Ring joint	Iron or soft steel	5.50	124.2 (18.0)	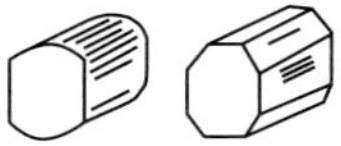	6	
		Monel metal or 4–6% chrome steel	6.00	150.3 (21.8)			
		Stainless steels	6.50	179.3 (26.0)			
	Rubber O-rings:				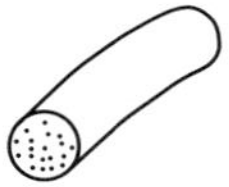	7 only	
	<75 IRHD (75A Shore Dur)		3[c]	0.69 (0.10)			
	75 (75A) to 85 IRHD (85A)		6[c]	1.42 (0.2)			
	Rubber square section rings:				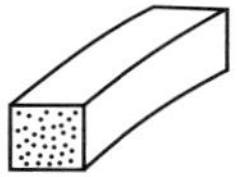	8 only	II
	<75 IRHD (75A Shore Dur)		4[c]	0.98 (0.14)			
	75 (75A) to 85 IRHD (85A)		9[c]	2.75 (0.40)			
	Rubber T-section rings:					9 only	
	Below 75 IRHD (75A Shore Dur)		4[c]	0.98 (0.14)			
	Between 75 (75A) and 85 IRHD (85A)		9[c]	2.75 (0.40)			

[a] Gasket factors (m) for operating conditions and minimum design seating stress (y).
[b] or * The surface of a gasket having a lap should not be against the nubbin.
[c] These values have been calculated.
Note: This table gives a list of many commonly used gasket materials and contact facings with suggested design values of m and y that have generally proved satisfactory in actual service when using effective gasket seating with b given in Table 8-21 and Fig. 8-13. The design values and other details given in this table are suggested only and are not mandatory.
Source: IS 2825, 1969.

TABLE 8-21
Effective gasket width

		Basic gasket seating width, b_σ	
Facing sketch (exaggerated)		**Column I**	**Column II**
1*a*	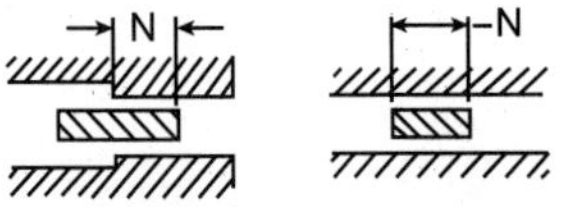	$\frac{N}{2}$	$\frac{N}{2}$
1*b*[a]	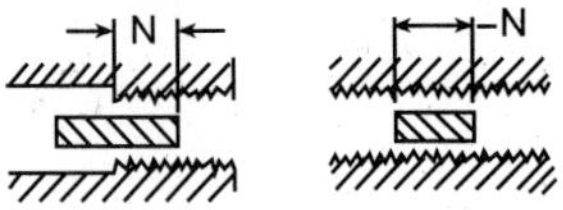		
1*c*	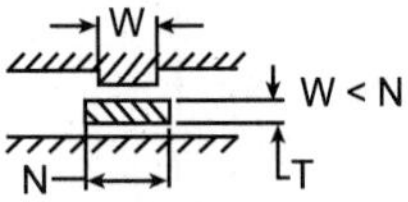	$\frac{w+25T}{3}; \left(\frac{w+N}{4} \text{ max}\right)$	$\frac{w+25T}{3}; \left(\frac{w+N}{2} \text{ max}\right)$
1*d*[a]	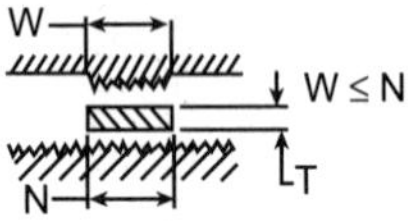		
2	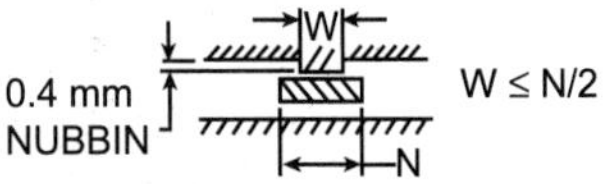	$\frac{w+N}{4}$	$\frac{w+3N}{8}$
3	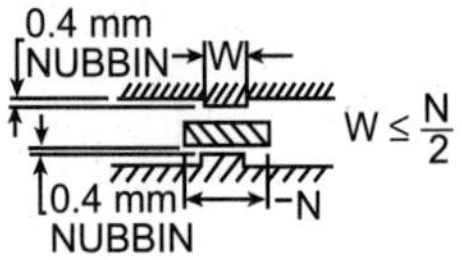	$\frac{w}{2}; \left(\frac{N}{4} \text{ min}\right)$	$\frac{w+N}{4}; \left(\frac{3N}{8} \text{ min}\right)$
4[a]	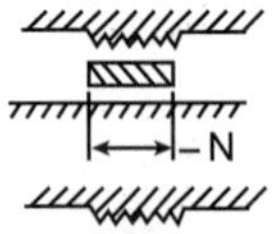	$\frac{3N}{8}$	$\frac{7N}{16}$
5*	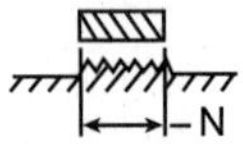	$\frac{N}{4}$	$\frac{3N}{8}$
6	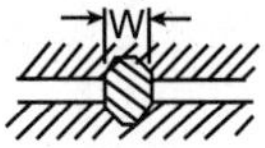	$\frac{w}{8}$	—
7	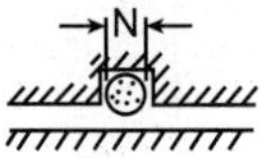	—	$\frac{N}{2}$
8	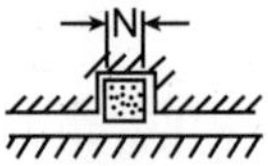	—	$\frac{N}{2}$
9	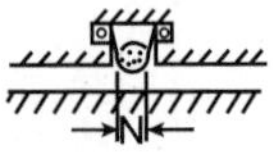	—	$\frac{N}{2}$

[a] Where serrations do not exceed 0.4 mm depth and 0.8 mm width spacing, sketches 1*b* and 1*d* shall be used.

Particular	Formula	
The required initial bolt load to seat the gasket joint-contact surface properly at atmospheric temperature condition without internal pressure	$W_{m2} = \pi bGy$ Refer to Table 8-20 for y.	(8-107)
Total required cross-sectional area of bolts at the root of thread	$A_m > A_{m1}$ or A_{m2}	(8-108)
Total cross-sectional area of bolt at root of thread or section of least diameter under stress required for the operating condition	$A_{m1} = \dfrac{W_{m1}}{\sigma_{sbd}}$ Refer to Tables 8-17 and 8-23 for σ_{sbd}.	(8-109)
Total cross-sectional area of bolt at root of thread or section of least diameter under stress required for gasket seating	$A_{m2} = \dfrac{W_{m2}}{\sigma_{sbat}}$	(8-110)

TABLE 8-22
Moment arms for flange loads under operating conditions

Type of flange	h_D	h_T	h_G
Integral-type flanges	$R + 0.5g_1$	$\dfrac{R + g_1 + h_G}{2}$	$\dfrac{C - G}{2}$
Loose-type except lap joint flanges and optional-type flanges	$\dfrac{C - B}{2}$	$\dfrac{h_D + h_G}{2}$	$\dfrac{C - G}{2}$
Lap joint flanges	$\dfrac{C - B}{2}$	$\dfrac{C - G}{2}$	$\dfrac{C - G}{2}$

TABLE 8-23
Maximum allowable stresses in stays and stay bolts, σ_{sa}

	Stress			
	For lengths between support not exceeding 120 × diameter		For lengths between support exceeding 120 × diameter	
Type of stay	MPa	kpsi	MPa	kpsi
(a) Unwelded or flexible stays less than 20 × diameter long, screwed through plates with ends riveted over	51	7.5		
(b) Hollow steel stays less than 20 × diameter long, screwed through plates with ends riveted over	55	8.0		
(c) Unwelded stays and unwelded portions of welded stays, except as specified in (a) and (b)	66	9.5	58	8.5
(d) Steel through stays exceeding 38 mm diameter	71	10.4	62	9.0
(e) Welded portions of stays	41	6.0	51	7.5

Source: *ASME Boiler and Pressure Vessel Code.*

Particular	Formula
The actual cross-sectional area of bolts using the root diameter of thread or least diameter of unthreaded portion (if less), to prevent damage to the gasket during bolting up	$A_b = \dfrac{2\pi yGN}{\sigma_{sbat}} \nless A_m$ (8-111)

Flange design bolt load W

The bolt load in the design of flange for operating condition	$W = W_{m1}$ (8-112)
The bolt load in the design of flange for gasket seating	$W = \left(\dfrac{A_m + A_b}{2}\right)\sigma_{sbat}$ (8-113)
The relation between bolt load per bolt (W_b), diameter of bolt D and torque M_t	$W_b = 0.17DM_t$ for lubricated bolts **USCS** (8-114a) where W_b in lbf, D in in, M_t in lbf in $W_b = 263.5DM_t$ **SI** (8-114b) where W_b in N, D in m, M_t in N m $W_b = 0.2DM_t$ for unlubricated bolts **USCS** (8-114c) where W_b in lbf, D in in, M_t in lbf in $W_b = 310DM_t$ **SI** (8-114d) where W_b in N, D in m, M_t in N m

Flange moments

The total moment acting on the flange M_o for operating condition	$M_o = M_D + M_t + M_G$ (8-115a) $= H_D h_D + H_T h_T + H_G h_G$ (8-115b) This is based on the flange design load of Eq. (8-112) with moment arms as given in Table 8-22.
The total flange moment M_o for gasket seating, which is based on the flange design bolt load of Eq. (8-113)	$M_o = Wh_G = \left(\dfrac{A_m + A_b}{2}\right)\left(\dfrac{C - G}{2}\right)\sigma_{sbat}$ (8-116)

Flange stresses

The stress in the flange shall be determined for both the gasket seating condition and the operating condition.	The larger of these two controls with the following formulas:

Particular	Formula

INTEGRAL-TYPE FLANGES AND LOOSE-TYPE FLANGES WITH A HUB

There are three types of stress:

Longitudinal hub stress	$\sigma_H = \dfrac{fM_o}{Lg_1^2 B}$	(8-117)
Radial flange stress	$\sigma_R = \dfrac{(1.33te + 1)M_o}{Lt^2 B}$	(8-118)
Tangential stress	$\sigma_\theta = \left(\dfrac{YM_o}{t^2 B}\right) - Z\sigma_R$	(8-119)
For flange factors values	Refer to Figs. 8-20 to 8-25.	

LOOSE-TYPE FLANGES WITHOUT HUB AND LOOSE-TYPE FLANGES WITH HUB WHICH THE DESIGNER CHOOSES TO CALCULATE

(a) Stresses without considering the hub

(1) Tangential stress	$\sigma_\theta = \dfrac{YM_o}{t^2 B}$	(8-120)
(2) The radial and longitudinal stress	$\sigma_H = \sigma_R = 0$	(8-121)

(b) Allowable flange design stresses:

The flange stresses calculated by Eqs. (8-117) to (8-121) shall not exceed the values of stresses given by Eqs. (8-122) to (8-126).

(1) The longitudinal hub stress	$\sigma_H \nless \sigma_{sfd}$ for cast iron	(8-122a)
	$\sigma_H \ngtr 1.5\sigma_{sfd}$ for other materials	(8-122b)
(i) The longitudinal hub stress for optional-type flanges designed as integral and also integral type where the neck material constitutes the hub of the flange	(a) $\sigma_H \ngtr 1.5\sigma_{sfd}$ or $1.5\sigma_{snd}$ The smaller of σ_{sfd} and σ_{snd} is to be selected.	(8-123a)
(ii) The longitudinal hub stress for integral-type flanges with hub welded to the neck, pipe, or vessel wall	(b) $\sigma_H \ngtr 1.5\sigma_{sfd}$ or $2.5\sigma_{snd}$ The smaller of σ_{sfd} and σ_{snd} is to be selected.	(8-123b)
(2) The radial stress	$\sigma_R \ngtr \sigma_{sfd}$	(8-124)
(3) The tangential stress	$\sigma_\theta \ngtr \sigma_{sfd}$	(8-125)

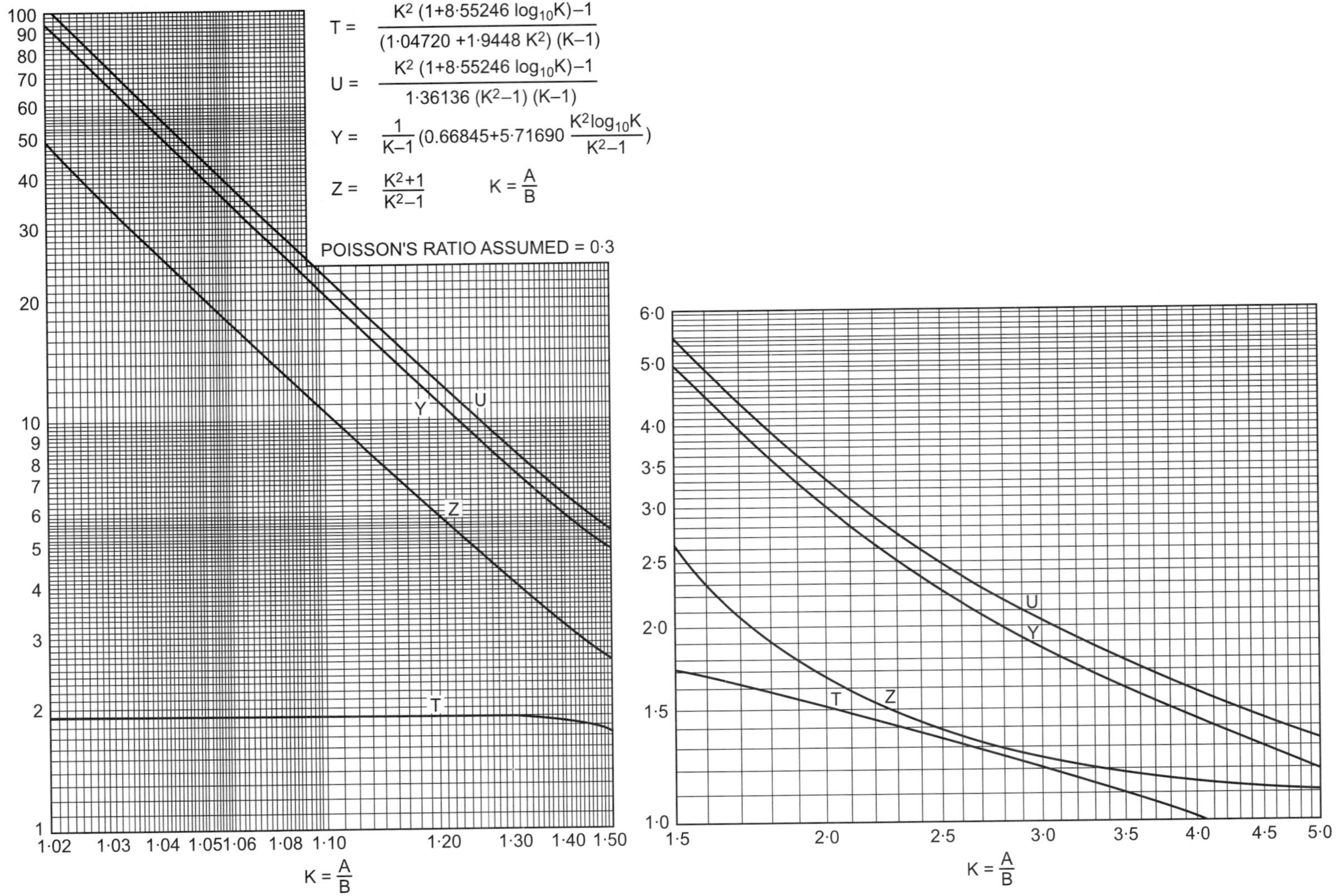

FIGURE 8-20 Values of T, U, Y, and Z for $K = (A/B) > 1.5$. (*Source*: IS 2825, 1969.)

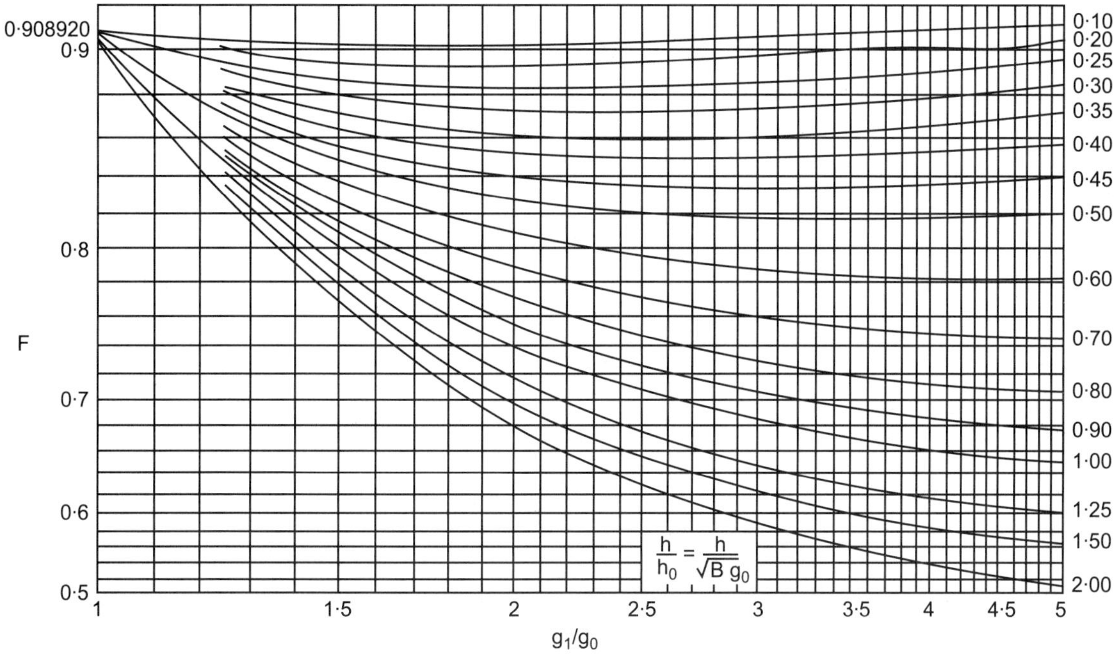

FIGURE 8-21 Values of F (integral flange factors). (*Source*: IS 2825, 1969.)

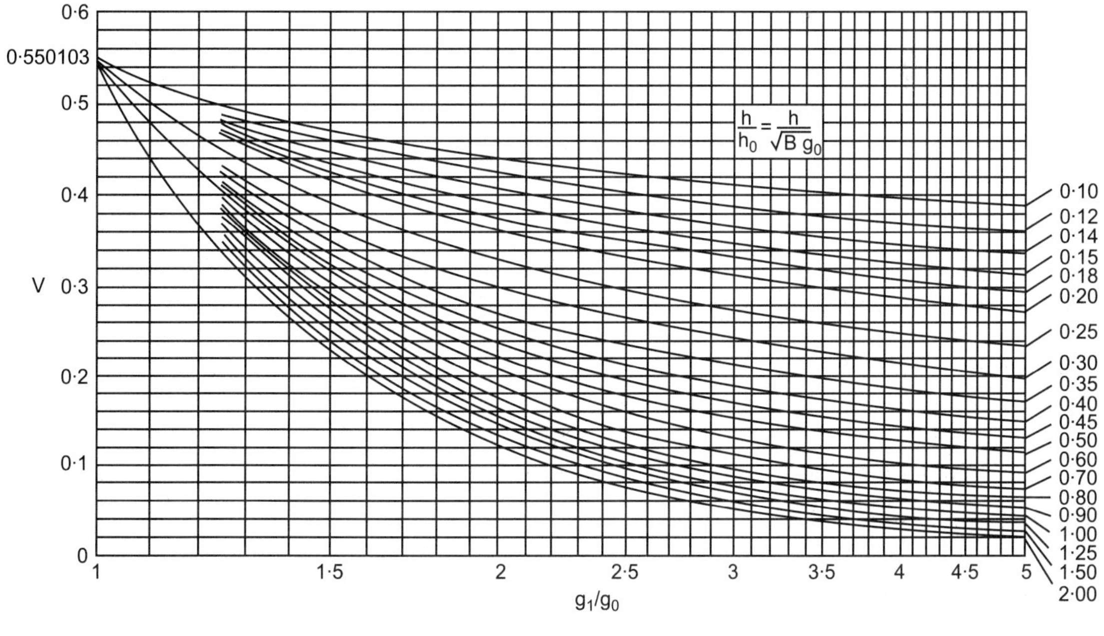

FIGURE 8-22 Values of V (integral flange factors). (*Source*: IS 2825, 1969.)

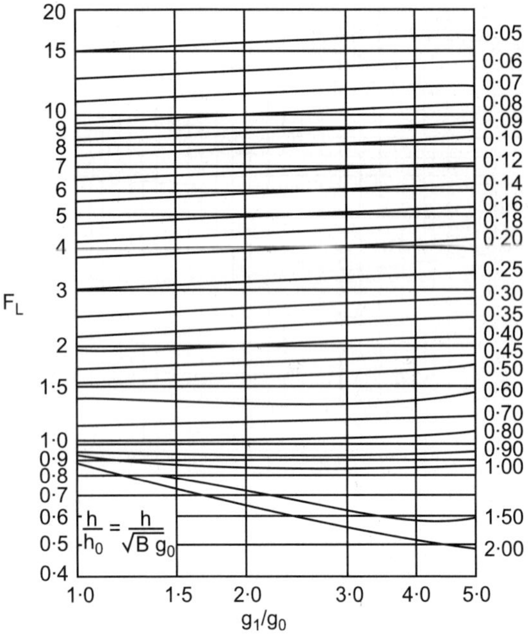

FIGURE 8-23 Values of F_L (loose hub flange factors). (*Source*: IS 2825, 1969.)

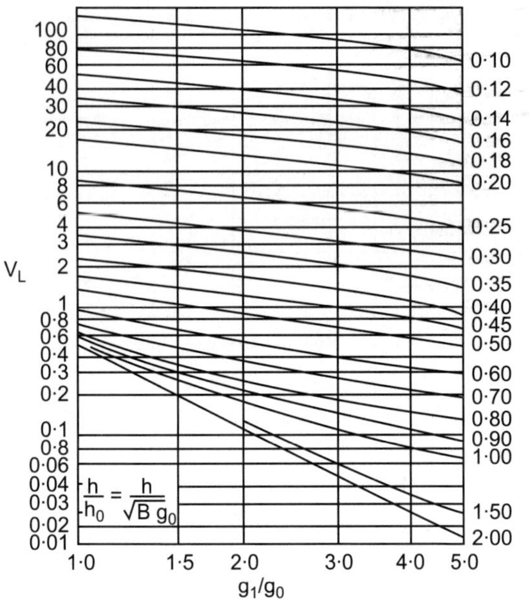

FIGURE 8-24 Values of V_L (loose hub flange factors). (*Source*: IS 2825, 1969.)

Particular	Formula	
(4) The average of σ_H and σ_R, and σ_H and σ_θ	$(\sigma_H + \sigma_R)/2 \not> \sigma_{sfd}$	(8-126a)
	$(\sigma_H + \sigma_\theta)/2 \not> \sigma_{sfd}$	(8-126b)

Flanges under external pressure

The design of flanges for external pressure only shall be based on the formulas given for internal pressure except that for operating conditions.	$M_o = H_D(h_D - h_G) + H_T(h_T - h_G)$ for operating conditions	(8-127a)
	$M_o = Wh_G$ for gasket seating	(8-127b)
	where $W = \sigma_{sbat}(A_{m2} + A_b)/2$	(8-128)

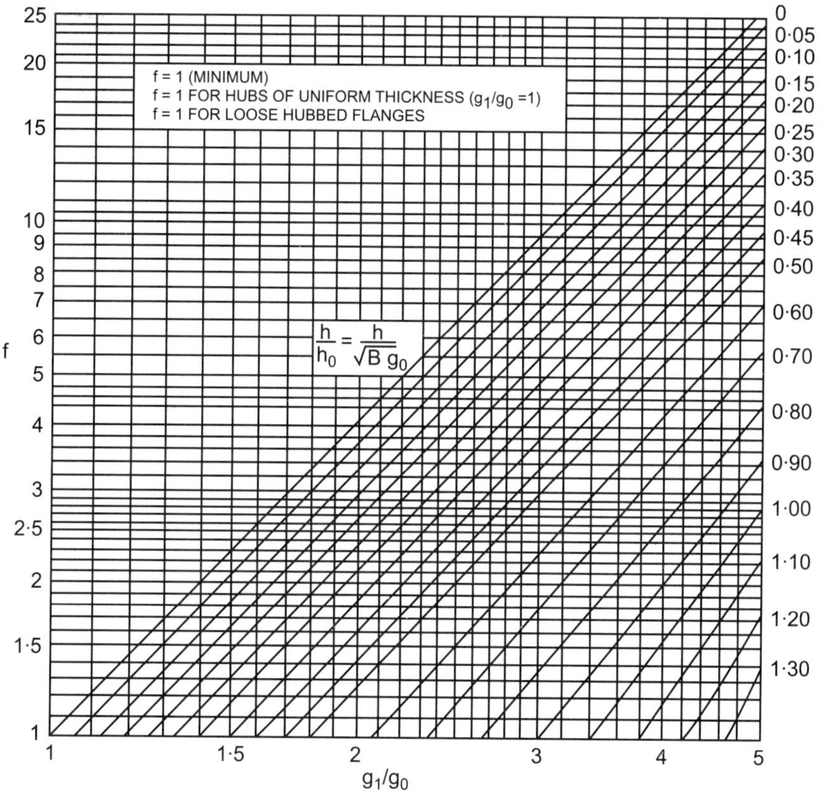

FIGURE 8-25 Values of *f* (hub stress correction factor). (*Source*: IS 2825, 1969.)

REFERENCES

1. "Rules for Construction of Pressure Vessels," Section VIII, Division 1, *ASME Boiler and Pressure Vessel Code,* The American Society of Mechanical Engineers (ASME), New York, 1986 ed., July 1, 1986.
2. "Rules for Construction of Pressure Vessels," Section VIII, Division 2, Alternative Rules, *ASME Boiler and Pressure Vessel Code,* ASME, New York, 1986 ed., July 1, 1986.
3. "Rules for Construction of Power Boiler," Section 1, *ASME Boiler and Pressure Vessel Code,* ASME, New York, 1983 ed., July 1, 1971.
4. "Recommended Rules for Care of Power Boilers," Section VII, *ASME Boiler and Pressure Vessel Code,* ASME, New York, 1983.
5. "Rules for in Service Inspection of Nuclear Power Plant Components," Section XI, *ASME Boiler and Pressure Vessel Code,* 1971.
6. "Heating Boilers," Section IV, *ASME Boiler and Pressure Vessel Code,* ASME, New York, 1983.
7. "Recommended Rules for Care and Operation of Heating Boilers," Section VI, *ASME Boiler and Pressure Vessel Code,* ASME, New York, 1983.
8. "Part A: Ferrous Materials," Section II, *ASME Boiler and Pressure Vessel Code,* ASME, New York, 1983.
9. "Part B: Non-ferrous Materials," Section II, *ASME Boiler and Pressure Vessel Code,* ASME, New York, 1983.
10. Azbel, D. S., and N. P. Cheremisinoff, *Chemical and Process Equipment Design—Vessel Design and Selection,* Ann Arbor Science Publishers, Ann Arbor, Michigan, 1982.

11. Bureau of Indian Standards, ZS 2825-1969 (under revision).
12. Chuse, R., *Pressure Vessels—The ASME Code Simplified,* 5th edition, McGraw-Hill Book Company, New York, 1977.
13. Lingaiah, K., and B. R. Narayana Iyengar, *Machine Design Data Handbook,* Engineering College Cooperative Society, Bangalore, India, 1962.
14. Lingaiah, K., and B. R. Narayana lyengar, *Machine Design Data Handbook*, Vol. I (*SI and Customarv Metric Units*), Suma Publishers, Bangalore, India, 1983.
15. Lingaiah, K., *Machine Design Data Handbook,* Vol. II (*SI and Customary Metric Units*), Suma Publishers, Bangalore, India, 1986.

CHAPTER 9

DESIGN OF POWER BOILERS

SYMBOLS[6,7]

C	smoke area consisting of the total internal transverse area of the tube, m^2 (ft^2)
d	diameter of cylinder or shell, in (in)
	diameter or short span, measured as shown in Fig. 8-9 (Chap. 8)
d_o	maximum allowable diameter of opening, m (in) outside diameter of cylinder or shell or tube or pipe, m (in)
D_o	outside diameter of furnace or flue, m (in)
$D.S.$	disengaging surface or area of water surface through which steam bubbles must be discharged, the water being considered at the middle-gauge cock, m^2 (ft^2)
E	modulus of elasticity, GPa (Mpsi)
G	area of the grate as finally adopted, m^2 (ft^2)
h or t	thickness of tube or shell wall, m (in)
H	total heating surface in contact with the fire, m^2 (ft^2)
l	length of the flue sections, m (in)
n	factor of safety to be taken as 5 for usual cases
L	radius to which the head is formed, measured on the concave side of the head, m (in)
P	rated power of boiler
p or P	maximum allowable working pressure, Pa or MPa (psi)
R_i	inside radius of cylindrical shell, m (in)
S	volume of steam included between the shell and a horizontal line through the position of the central gauge as finally determined, m^2 (ft^2)
t (or h)	thickness of tube or pipe or cylinder or shell or plate, m (in)
SHS	total area of superheating surface based on the actual area in contact with the fire, m^2 (ft^2)
W	net water volume in the boiler below the line of the central gauge cock, m^2 (ft^2)
WHS	total area of water heating surface based on the actual area in contact with the fire, m^2 (ft^2)
σ_{sa}	maximum allowable stress value, MPa (kpsi) from Tables 7-1 (Chapter 7), 8-9 to 8-11, and 8-17 (Chapter 8)
η	efficiency of joint

Other factors in performance or in special aspects are included from time to time in this chapter and, being applicable only in their immediate context, are not given at this stage.

Note: σ and τ with initial subscript s designates strength properties of material used in the design, which will be used and observed throughout this *Machine Design Data Handbook*.

Particular	Formula

BOILER TUBES AND PIPES

For calculation of the minimum required thickness (t) and maximum allowable working pressure (p *or* P) of ferrous and nonferrous tubes and pipes from 12.5 mm ($\frac{1}{2}$ in) to 150 mm (6 in) outside diameter used in power boilers as per *ASME Boiler and Pressure Vessel Code*[2,3]	Refer to Eqs. (7-1) to (7-15) (Chap. 7).
For efficiency of joints (η), temperature coefficient (y), minimum allowance for threading, and structural stability (C) as per *ASME Boiler and Pressure Vessel Code*	Refer to Tables from 7-2 to 7-6 (Chap. 7).
For maximum allowable stress value (σ_{sa}) for the materials of tubes and pipes as per *ASME Boiler and Pressure Vessel Code*[3]	Refer to Table 7-1.
The maximum allowable working pressure for steel tubes or flues of fire tube boilers for different diameters and gauges of tubes as per *ASME Power Boiler Code*[2]	$p = \frac{96.5}{d_o}(h - 1.625 \times 10^{-3})$ **SI** (9-1a) where p in MPa, h and d_o in m $p = \frac{14000}{d_o}(h - 0.065)$ **USCS** (9-1b) where p in psi, h and d_o in in
For maximum allowable working pressure and thickness of steel tubes	Refer to Tables 7-7, 9-1, 9-2 and 9-4 and Fig. 7-1.
The maximum allowable working pressure for copper tubes for firetube boilers subjected to internal or external pressure as per *ASME Power Boiler Code*[2]	$p = \frac{83}{d_o}(h - 1 \times 10^{-3}) - 1.7$ **SI** (9-2a) where p in MPa, d_o and h in m $p = \frac{12000}{d_o}(h - 0.039) - 250$ **USCS** (9-2b) where p in psi, d_o and h in in
For maximum allowable working pressure and thickness of copper tubes	Refer to Tables 9-3 and 9-5.

TABLE 9-1
Maximum allowable working pressures for seamless steel and electric resistance welded steel tubes or nipples for watertube boilers [from Eq. (7-4)]

Wall thickness			Tube outside diameter, mm (in)																									
			12.5	(0.5)	19.0	(1.75)	25.0	(1.0)	31.25	(1.25)	37.5	(1.5)	43.75	(1.75)	50.0	(2.0)	62.5	(2.5)	75.0	(3.0)	87.5	(3.5)	100.1	(4.0)	112.5	(4.5)	125.0	(5.0)
mm	in	Nearest Bwg no.	MPa	psi	MPa	psi	MPa	psi	MPa	psi	MPa	psi	MPa	psi	MPa	psi	MPa	psi	MPa	psi	MPa	psi	MPa	psi	MPa	psi	MPa	psi
1.375	0.055	17–	3.38	590	2.41	350																						
1.625	0.065	16	7.52	1090	4.62	670	3.24	470	2.42	350																		
1.875	0.075	15+	11.03	1600	6.90	1000	5.00	720	3.80	550	3.0	430																
2.125	0.085	14+			9.24	1340	6.62	960	5.10	740	4.06	590	3.38	490	2.83	410												
2.375	0.095	13							12.13	1760	5.24	760	4.34	630	3.65	530	2.76	400										
2.625	0.105	12–							13.65	1980	11.03	1600	9.24	1340	7.93	1150	3.45	500	2.34	390								
3.000	0.120	11									12.90	1870	10.82	1570	9.24	1340	7.17	1040	5.80	840	2.90	420						
3.375	0.135	10+											12.34	1790	10.62	1540	8.20	1190	6.62	960	5.52	800	4.68	680				
3.750	0.150	9+											13.92	2020	12.00	1740	9.24	1340	7.52	1090	6.27	910	5.38	780	4.62	670	4.70	590
4.125	0.165	8													13.38	1940	10.34	1500	8.34	1210	7.03	1020	6.00	870	5.24	760	4.62	670
4.500	0.180	7															11.45	1660	9.24	1340	7.72	1120	6.62	960	5.80	840	5.10	740
5.000	0.200	6–															12.90	1870	10.48	1520	8.76	1270	7.52	1090	6.55	950	7.80	840
5.500	0.220	5																	11.65	1690	9.80	1420	8.34	1210	7.31	1060	6.48	940
6.000	0.240	4+																	12.90	1870	10.68	1550	9.24	1340	8.07	1170	7.17	1040
6.500	0.260	3+																			11.86	1720	10.14	1470	8.90	1290	7.86	1140
7.000	0.280	2–																			12.90	1870	11.03	1600	9.65	1400	8.55	1240
7.500	0.300																				13.92	2020	12.00	1740	10.48	1520	9.24	1340
8.000	0.320																						12.90	1870	12.24	1630	10.00	1450
8.500	0.340																						13.78	2000	12.06	1750	10.68	1550
9.000	0.360																								12.90	1870	11.45	1660
9.500	0.380																								13.72	1990	12.13	1760
10.000	0.400																										12.90	1870
10.500	0.420																										13.65	1980

* Bwg = Birmingham wire gauge
Source: *ASME Power Boiler Code*, Section I, 1983.

TABLE 9-2
Maximum allowable working pressures for steel tubes or flues for firetube boilers [from Eq. (9-1)]

Wall thickness			Size outside diameter mm (in)																					
			25.00 (1)		37.50 (1.50)		50.00 (2)		62.50 (2.50)		75.00 (3)		87.50 (3.50)		200 (4)		112.50 (4.50)		125.00 (5)		137.50 (5.50)		150.0 (6)	
mm	In	Nearest Bwg no.	MPa	psi	MPa	psi	MPa	psi	MPa	psi	MPa	psi	MPa	psi	MPa	psi	MPa	psi	MPa	psi	MPa	psi	MPa	psi
2.375	0.095	13	2.90	420	1.93	280	1.45	210	1.17	170														
2.625	0.105	12	3.86	560	2.62	380	1.93	280	1.59	230	1.31	190	1.10	160										
3.000	0.120	11	5.31	770	3.58	520	2.69	390	2.14	310	1.80	260	1.52	220	1.38	200	1.24	180						
3.375	0.135	10+	6.76	980	4.55	660	3.38	490	2.76	400	2.28	330	1.93	280	1.72	250	1.52	220	1.38	200				
3.375	0.150	9+			5.52	800	4.14	600	3.30	480	2.76	400	2.34	340	2.06	300	1.86	270	1.65	240	1.52	220		
4.125	0.165	8			6.48	940	4.83	700	3.86	560	3.24	470	2.76	400	2.41	350	2.21	320	1.93	280	1.80	260	1.65	240
4.500	0.180	7					5.58	810	4.48	650	3.72	540	3.17	460	2.83	410	2.48	360	1.28	330	2.07	300	1.86	270
5.000	0.200	6−					6.55	950	5.24	760	4.34	630	3.72	540	3.31	480	2.90	420	2.62	380	2.41	350	2.21	320
5.500	0.220	5					7.52	1090	6.00	870	5.03	730	4.27	620	3.79	550	3.38	490	3.03	440	2.76	400	2.55	370
6.000	0.240	4+					8.40	1230	6.83	990	5.65	820	4.83	700	4.28	260	4.80	550	3.38	490	3.10	450	2.83	410

Source: *ASME Power Boiler Code*, Section I, 1983.

Particular	Formula
The external working pressure, for plain lap-welded or seamless tubes up to and including 150 mm (6 in) external diameter, and if the thickness is greater than the standard one	$$p = \frac{1}{n}\left[\frac{596h}{d_o} - 9.6\right] \qquad \textbf{SI} \quad (9\text{-}3a)$$ where p in Pa, h and d in m $$p = \frac{1}{n}\left[\frac{86670h}{d_o} - 1386\right] \qquad \textbf{USCS} \quad (9\text{-}3b)$$ where p in psi, h and d in in
For proportion of standard boiler tubes	Refer to Table 9-6.

TABLE 9-3
Maximum allowable working pressure for copper tubes for firetube boilers[a] [from Eq. (9-2)]

Outside diameter of tube		Gauge, Bwg																	
		12		11		10		9		8		7		6		5		4	
mm	in	MPa	psi	MPa	psi	MPa	psi	MPa	psi	MPa	psi	MPa	psi	MPa	psi	MPa	psi	MPa	psi
50.00	2	1.17	170	1.65	240	1.72	250	1.72	250	1.72	250	1.72	250	1.72	250	1.72	250	1.72	250
81.25	3.25					0.76	110	1.03	150	1.52	220	1.72	250	1.72	250	1.72	250	1.72	250
100.00	4									0.90	130	1.10	160	1.72	250	1.72	250	1.72	250
125.00	5													1.03	150	1.31	190	1.59	230

[a] For use at pressure not to exceed 1.7 MPa (250 psi) or temperature not to exceed 208°C (406°F).
Source: *ASME Power Boiler Code*, Section I, 1983.

TABLE 9-4
Maximum boiler pressures for use of ANSI B16.5 standard steel pipe flanges and flanged valves and fittings

Primary service pressure rating		Maximum allowable boiler pressure			
		Steam service at saturation temperature		Boiler feed and blow-off line service	
Mpa	psi	MPa	psi	MPa	psi
1.14	164.7	1.41	204.7	1.20	174.7
2.17	314.7	4.44	644.7	3.65	529.7
2.86	414.7	5.75	834.7	4.68	679.7
4.23	614.7	8.10	1174.7	6.79	984.7
6.30	914.7	11.40	1654.7	10.10	1464.7
10.44	1514.7	17.23	2514.7	16.13	2339.7
17.33	2514.7	22.10	3206.0	22.20	3220.7

Notes: Adjusted pressure ratings for steam service at saturated temperature corresponding to the pressure, derived from Table 2 to 8 ANSI B 16.5–1968. Pressures shown include the factor for boiler feed and blow-off line service required by ASME corrected for saturation temperature corresponding to this pressure.
Source: *ASME Power Boiler Code*, Section I, 1983.

TABLE 9-5
Maximum external working pressures for use with lap-welded and seamless boiler tubes[a]

Nominal diameter, external diameter, mm (in)	Standard thickness, mm	Maximum allowable pressure		Nominal diameter, external diameter, mm (in)	Standard thickness, mm	Maximum allowable pressure	
		MPa	psi			MPa	psi
51 (2)	2.4	2.84	427	89 (3.5)	3.1	2.16	308
58 (2.25)	2.4	2.55	380	96 (3.75)	3.1	1.96	282
64 (2.5)	2.8	2.65	392	102 (4)	3.4	2.06	303
70 (2.75)	2.8	2.45	356	115 (4.5)	3.4	1.67	238
76 (3)	2.8	2.26	327	127 (5)	3.8	1.67	235
83 (3.25)	3.1	2.26	327	153 (6)	4.2	1.37	199

[a] External diameter 50 to 150 mm (2 to 6 in).

TABLE 9-6
Proportions of standard boiler tubes

Nominal diameter, actual external diameter mm (in)	Actual internal diameter, mm	Thickness, mm	External circumference, mm	Internal circumference, mm	External transverse area, mm^2	Internal transverse area, mm^2	Length of tube m^{-2} of internal heating surface, m	Weight per meter	
								N	lbf
45 (1.76)	38	2.4	140	125	1600	1200	7.58	24.5	1.679
51 (2)	46	2.4	160	144	2000	1700	6.58	28.2	1.932
58 (2.25)	50	2.4	181	165	2000	2100	5.78	32.0	2.186
64 (2.5)	56	2.8	200	183	3200	2600	5.24	40.7	2.783
70 (2.75)	64	2.8	220	200	3800	3200	4.74	44.9	3.074
76 (3)	71	2.8	240	221	4500	3900	4.38	49.1	3.365
83 (3.25)	76	3.0	260	241	5400	4500	3.98	58.5	4.011
89 (3.5)	81	3.0	280	260	6200	5400	3.71	63.0	4.331
96 (3.75)	89	3.0	300	280	7000	6200	3.45	68.0	4.652
102 (4)	94	3.3	320	290	8000	6900	3.25	80.8	5.532
115 (4.5)	107	3.3	360	340	10000	9000	2.86	91.2	6.248
127 (5)	120	3.8	400	370	12800	11100	2.58	112.3	7.669
153 (6)	142	4.2	480	450	18300	16300	2.15	150.0	10.282

Particular	Formula
The external pressure, for plain lap-welded, or seamless tubes or flues over 50 mm (2 in) and not exceeding 150 mm (6 in) external diameter	Refer to Table 9-5.
The minimum required thickness of component when it is of riveted construction or does require staying as per *ASME Power Boiler Code*[2]	$h = \dfrac{pR_i}{0.8\sigma_{sa}\eta - 0.6p}$ (9-4)
The maximum allowable working pressure as per *ASME Power Boiler Code*	$p = \dfrac{0.8\sigma_{sa}\eta}{R_i + 0.6h}$ (9-5)

DISHED HEADS

Particular	Formula
The thickness of a blank unstayed dished head with the pressure on the concave side, when it is a segment of a sphere as per *ASME Power Boiler Code*	$h = \dfrac{5pL}{4.8\sigma_{sa}\eta}$ (9-6) where $L =$ radius to which the head is dished, measured on the concave side of the head, m (in) $\eta =$ efficiency of weakest joint used in forming the head. (Refer to Table 8-3 for η.)
The minimum distance between the centers of any two openings, rivet holes excepted, shall be determined by Eq. (9-7)	$L = \dfrac{A + B}{2(1 - K)}$ (9-7) where $L =$ distance between the centers of the two openings measured on the surface of the head, m (in) $A, B =$ diameters of two openings, m (in) $K =$ same as defined in Eqs. (9-8a) and (9-8b)
The expression for K	$K = \dfrac{pd_o}{1.6\sigma_{sa}h}$ (9-8a) $K = \dfrac{pd_o}{1.82\sigma_{sa}h}$ (9-8b) Equation (9-8a) shall be used with shells and headers designed by using Eqs. (9-4) and (9-5). Equation (9-8b) shall be used with shells and headers designed by using Eqs. (9-9) and (9-10):
The minimum required thickness of ferrous drums and headers based on strength of weakest course as per *ASME Power Boiler Code*	$h = \dfrac{pd_o}{2\sigma_{sa}\eta + 2yp} + C$ or $\dfrac{pR_i}{\sigma_{sa}\eta - (1 - y)p} + C$ (9-9)
The maximum allowable working pressure as per *ASME Power Boiler Code*	$p = \dfrac{2\sigma_{sa}\eta(h - C)}{d_o - 2y(h - C)}$ or $\dfrac{\sigma_{sa}\eta(h - C)}{R_i + (1 - y)(h - C)}$ (9-10)

Particular	Formula
	For values y, C, and σ_{sa} refer to Tables 7-1, 7-3, and 7-6.
The thickness of a blank unstayed full-hemispherical head with the pressure on the concave side	$$h = \frac{pL}{1.6\sigma_{sa}\eta} \quad (9\text{-}11a)$$ $$h = \frac{pL}{(2\sigma_{sa}\eta - 0.2p)} \quad (9\text{-}11b)$$ Equation (9-11b) may be used for heads exceeding 12.5 mm (0.5 in) in thickness that are to be used with shells or headers designed under Eqs. (9-9) and (9-10) and that are integrally formed on seamless drums or are attached by fusion welding and do not require staying.
The formula for the minimum thickness of head when the required thickness of the head given by Eqs. (9-9) and (9-10) exceeds 35 percent of the inside radius	$$h = L(y^{1/3} - 1) \quad (9\text{-}12)$$ where $$y = \frac{2(\sigma_{sa}\eta + p)}{2\sigma_{sa}\eta - p} \quad (9\text{-}12a)$$

UNSTAYED FLAT HEADS AND COVERS

Particular	Formula
The minimum required thickness of flat unstayed circular heads, covers and blind flanges as per *ASME Power Boiler Code*	$$h = d\sqrt{Cp/\sigma_{sa}} \quad (9\text{-}13)$$ where C = a factor depending on the method of attachment of head on the shell, pipe or header (refer to Table 8-6 for C) d = diameter or short span, measured as shown in Fig. 8-9
The minimum required thickness of flat unstayed circular heads, covers or blind flange which is attached by bolts causing edge moment Fig. 8-9(j) as per *ASME Power Boiler Code*	$$h = d[Cp/\sigma_{sa} + 1.78Wh_G/\sigma_{sa}d^3]^{1/2} \quad (9\text{-}14)$$ where W = total bolt load, kN (lbf) h_G = gasket moment arm, Fig. 8-13 and Table 8-22.
For details of bolt load H_G, bolt moments, gasket materials, and effect of gasket width on it	Refer to Tables 8-20 and 8-22 and Fig. 8-13
The minimum required thickness of unstayed heads, covers, or blind flanges of square, rectangular, elliptical, oblong segmental, or otherwise noncircular as per *ASME Power Boiler Code*	$$t \text{ or } h = d\sqrt{ZCp/\sigma_{sa}\eta} \quad (9\text{-}15)$$

Particular	Formula
	where $Z = 3.4 - 2.4d/a$ (9-15a) a = long span of noncircular heads or covers measured perpendicular to short span, m (in) Z need not be greater than 2.5 Equation (9-15) does not apply to noncircular heads, covers, or blind flanges attached by bolts causing bolt edge moment
The minimum required thickness of unstayed non-circular heads, covers, or blind flanges which are attached by bolts causing edge moment Fig. 8-9 as *per ASME Power Boiler Code*	$h = d[ZCp/\sigma_{sa} + 6Wh_G/\sigma_{sa}Ld^2]^{1/2}$ (9-16)
The required thickness of stayed flat plates (Figs. 8-10 and 8-11) as per *ASME Power Boiler Code*	$h = p_t\sqrt{[p/\sigma_{sa}c_5]}$ (9-17) where p_t = maximum pitch, m (in), measured between straight lines passing through the centers of the stay bolts in the different rows (Refer to Table 9-7 for pitches of stay bolts.) c_5 = a factor depending on the plate thickness and type of stay (Refer to Table 8-15 for values of c_5.) For σ_{sa} refer to Tables 8-8, 8-23, and 8-11
The maximum allowable working pressure for stayed flat plates as per *ASME Power Boiler Code*	$p = \dfrac{h^2\sigma_{sa}c_5}{p_i^2}$ (9-18)
For all allowable stresses in stay and stay bolts	Refer to Chapter 8
Also for detail design of different types of heads, covers, openings and reinforcements, ligaments, and bolted flanged connection	

COMBUSTION CHAMBER AND FURNACES

Combustion chamber tube sheet

Particular	Formula
The maximum allowable working pressure on tube sheet of a combustion chamber where the crown sheet is suspended from the shell of the boiler as per *ASME Power Boiler Code*	$P = 27000\dfrac{h(D - d_i)}{wD}$ USCS (9-19a) where h = thickness of tube, in w = distance from the tube sheet to opposite combustion chamber sheet, in

TABLE 9-7
Maximum allowable pitch for screwed staybolts, ends riveted over

Pressure		Thickness of plate, mm (in)													
		7.8125	(0.3125)	9.375	(0.375)	10.9375	(0.4375)	12.500	(0.50)	14.0625	(0.5625)	15.6250	(0.625)	17.1875	(0.6875)
MPa	psi	Maximum pitch of staybolts, mm (in)													
0.67	100	131.25	(5.25)	159.375	(6.375)	184.375	(7.375)								
0.76	110	125.000	(5.000)	150.000	(6.000)	175.000	(7.000)	209.375	(8.375)						
0.83	120	118.750	(4.75)	143.750	(5.75)	168.750	(6.75)	200.000	(8.000)						
0.86	125	118.750	(4.75)	140.625	(5.625)	165.625	(6.625)	193.750	(7.75)						
0.90	130	115.625	(4.625)	137.500	(5.50)	162.500	(6.50)	190.625	(7.625)						
0.96	140	112.500	(4.50)	134.375	(5.375)	156.250	(6.25)	184.375	(7.375)	209.375	(8.375)				
1.03	150	106.250	(4.25)	128.125	(5.125)	150.000	(6.000)	178.125	(7.125)	200.00	(8.000)				
1.10	160	103.125	(4.125)	125.000	(5.000)	146.875	(5.875)	171.875	(6.875)	193.750	(7.75)				
1.17	170	100.000	(4.000)	121.875	(4.875)	140.625	(5.625)	168.150	(6.75)	187.500	(7.500)	209.375	(8.375)		
1.24	180			118.750	(4.75)	137.500	(5.50)	162.500	(6.50)	184.375	(7.375)	203.125	(8.125)		
1.31	190			115.625	(4.625)	134.375	(5.375)	159.375	(6.375)	178.125	(7.125)	196.875	(7.875)		
1.38	200			112.500	(4.50)	131.25	(5.25)	153.125	(6.125)	175.000	(7.000)	193.750	(7.750)	212.500	(8.50)
1.55	225			106.25	(4.25)	121.875	(4.875)	146.875	(5.875)	162.500	(6.50)	181.250	(7.25)	200.000	(8.00)
1.72	250			100.000	(4.000)	115.625	(4.625)	137.50	(5.50)	156.250	(6.25)	171.875	(6.875)	175.625	(7.625)
2.07	300					106.250	(4.25)	125.000	(5.000)	140.625	(5.625)	156.250	(6.25)	175.00	(7.000)

Source: *ASME Power Boiler Code*, Section I, 1983.

Particular	Formula
	D = least horizontal distance between tube centers on a horizontal row, in d_i = inside diameter of tube, in P = maximum allowable working pressure, psi
	$P = 186 \dfrac{h(D - d_i)}{wD}$ **SI** (9-19b) where p in MPa; h, D, d_i, and w in m
The vertical distance between the center lines of tubes in adjacent rows where tubes are staggered	$D_{va} = (2d_i D + d_i^2)^{1/2}$ (9-20) where d_i and D have the same meaning as given under Eq. (9-19)
For minimum thickness of shell plates, dome plates, and tube plates and tube sheet for firetube boiler	Refer to Table 9-8
For mechanical properties of steel plates of boiler	Refer to Table 9-9

TABLE 9-8
Minimum thickness of shell plates, dome plates, and tube sheet for firetube boiler

Diameter of				Minithickness			
Shell and dome plates		Tube sheet		Shell and dome plates		Tube sheet	
m	in	m	in	mm	in	mm	in
≥0.9	≥36	1.05	42	6.25	0.25	9.375	0.375
>0.9–1.35	>36–54	>1.05–1.35	>42–54	7.81	0.3125	10.94	0.4375
>1.35—1.8	>54–72	>1.35–1.8	>54–72	9.375	0.375	12.5	0.500
>1.8	>72	>1.8	>72	12.5	0.50	14.06	0.5625

Source: *ASME Power Boiler Code*, Section I, 1983.

TABLE 9-9
Mechanical properties of steel plates for boilers

	Tensile strength			
Grade	MPa	kpsi	Yield stress, percent min of tensile strength	Elongation percent gauge length, $5.65\sqrt{a^*}$ [a]
1	333.4–411.9	48.4–59.7	55	26
2A	362.8–480.5	52.6–69.7	50	25
2B	509.9–608.0	74.0–88.2	50	20

[a] a^* area of cross section.
Source: IS 2002-1, 1962.

Particular	Formula
Plain circular furnaces	
FURNACES 300 mm (12 in) TO 450 mm (18 in) OUTSIDE DIAMETER, INCLUSIVE	
Maximum allowable working pressure for furnaces not more than $4\frac{1}{2}$ diameters in length or height where the length does not exceed 120 times the thickness of the plate	$p = \frac{0.36(18.75T - 1.03L)}{D}$ **SI** (9-21a) where p in MPa; T, D, and L in m $p = \frac{51.5(18.75T - 1.03L)}{D}$ **USCS** (9-21b) where p in psi D = outside diameter of furnace, in L = total length of furnace between centers of head rivet seams, in T = thickness of furnace walls, sixteenth of an inch
The maximum allowable working pressure for furnaces not more than $4\frac{1}{2}$ diameter in length of height where the length exceeds 120 times the thickness of the plate	$p = \frac{29.3T^2}{LD}$ **SI** (9-22a) where p in MPa; T, L, and D in m $p = \frac{4250T^2}{LD}$ **USCS** (9-22b) where p in psi; T, L, and D in in
Circular flues	
The maximum allowable external pressure for riveted flues over 150 mm (6 in) and not exceeding 450 mm (18 in) external diameter, constructed of iron or steel plate not less than 6 mm (0.25 in) thick and put together in sections not less than 600 mm (24 in) in length	$p = \frac{56h}{d}$ **SI** (9-23a) where p in Pa; h and d in m $p = \frac{8100h}{d}$ **USCS** (9-23b) where p in psi; h and d in in d = external diameter of flue, in
The formula for maximum allowable external pressure for riveted, seamless, or lap-welded flues over 450 mm (18 in) and not exceeding 700 mm (28 in) external diameter, riveted together in sections not less than 600 mm (24 in) nor more than $3\frac{1}{2}$ times the flue diameter in length, and subjected to external pressure only	$p = \frac{6.7h - 0.4l}{d}$ **SI** (9-24a) where p in Pa; h, l, and d in m $p = \frac{966h - 53l}{d}$ **USCS** (9-24b)

Particular	Formula
	where p in psi and d in in
	h = thickness of wall in 1.5 mm (0.06 in) $l > 600$ mm (24 in) and $< 3\frac{1}{2}d$
The maximum allowable working pressure for seamless or welded flues over 125 mm (5 in) in diameter and including 450 mm (18 in)	
(a) Where the thickness of the wall is not greater than 0.023 times the diameter as per *ASME Power Boiler Code*	$p = \dfrac{68948h^3}{D^3}$ **SI** (9-25a)
	where p in MPa; h and D in m
	p = maximum allowable working pressure D = outside diameter of flue h = thickness of wall of flue
	$p = \dfrac{10^7 h^3}{D^3}$ **USCS** (9-25b)
	where p in psi; h and D in in
(b) Where the thickness of the wall is greater than 0.023 times the diameter.	$p = \dfrac{119h}{D} - 1.9$ **SI** (9-26a)
	where p in MPa; h and D in m
	$p = \dfrac{17300h}{D} - 275$ **USCS** (9-26b)
	where p in psi; h and D in in
Equations (9-24) and (9-25) may applied to riveted flues of the size specified provided the section are not over 0.91 m (3 ft) in length and the efficiency (η) of the joint	$\eta \nless \dfrac{pD}{138h}$ **SI** (9-27a)
	where p in MPa; D and h in m
	$\eta \nless \dfrac{pD}{20000h}$ **USCS** (9-27b)
	where p in psi; D and h in in

Particular	Formula

THE MAXIMUM ALLOWABLE PRESSURE FOR SPECIAL FURNACES HAVING WALLS REINFORCED BY RIBS, RINGS, AND CORRUGATIONS

(a) Furnaces reinforced by Adamson rings

$$p = \frac{6.7h - 0.4l}{d} \quad \textbf{SI} \quad (9\text{-}28a)$$

where p in Pa; h and d in m

$$p = \frac{1080h - 59l}{d} \quad \textbf{USCS} \quad (9\text{-}28b)$$

where p in psi

h = thickness of wall, 1.5 mm (0.06 in) not to be less than 8 mm ($\frac{5}{16}$ in)
l = length of flue section, not to be less than 450 mm (18 in)

(b) Another expression for the maximum allowable working pressure when plain horizontal flues are made in sections not less than 450 mm (18 in) in length and not less than 8 mm ($\frac{5}{16}$ in) in thickness (Adamson-type rings)

$$p = \frac{0.4(300h - 1.03L)}{D} \quad \textbf{SI} \quad (9\text{-}29a)$$

where p in MPa; h, L, and D in m

$$p = \frac{57.6(300h - 1.03L)}{D} \quad \textbf{USCS} \quad (9\text{-}29b)$$

where p in psi; h, L, and D in in

(c) Corrugated rings

$$p = 68.5\,\frac{h}{d} \quad \textbf{SI} \quad (9\text{-}30a)$$

where p in Pa; h and d in m

$$p = 10000\,\frac{h}{d} \quad \textbf{USCS} \quad (9\text{-}30b)$$

where p in psi; h and d in in

h = thickness of tube wall, mm (in), not to be less than 11 mm (0.44 in)

(d) Plain circular flues riveted together in sections

$$p = \frac{6.7d - 0.4l}{d} \quad \textbf{SI} \quad (9\text{-}31a)$$

where p in Pa; d and l in m

$$p = \frac{966h - 53l}{D} \quad \textbf{USCS} \quad (9\text{-}31b)$$

where p in psi; l and d in in

Particular	Formula

Ring-reinforced type

The required wall thickness of a ring-reinforced furnace of flue shall not be less than that determined by the procedure given here

Assume a value for h (or t) and L. Determine the ratios L/D_o and D_o/t.

Following the procedure explained in Chap. 8, determine B by using Fig. 9-1. Compute the allowable working pressure P_a by the help of Eq. (9-32)

The allowable working pressure (P_a)

$$P_a = \frac{B}{(D_o/t)} \tag{9-32}$$

where D_o = outside diameter of furnace or flue, in

Compare P_a with P. If P_a is less than P select greater value of t (or h) or smaller value of L so that P_a is equal to or greater than P, psi

The required moment of inertia (I_s) of circumferential stiffening ring

$$I_s = \frac{LD_o^2\left(t + \dfrac{A_s}{L}\right)A}{14} \tag{9-33}$$

where

I_s = required moment of inertia of stiffening ring about its neutral axis parallel to the axis of the furnace, in^4
A_s = area of cross section of the stiffening ring, in^2
A = factor obtained from Fig. 9-1

The required moment of inertia of a stiffening ring shall be determined by the procedure given here

Assume the values of D_o, L, and t (or h) of furnace. Select a rectangular member to be used for stiffening ring and find its area A_s and its moment of inertia I. Then find the value of B from Eq. (9-34)

The expression for B

$$B = \frac{PD_o}{t + A_s/L} \tag{9-34}$$

where P, D_o, t, A_s, and L are as defined under Eq. (9-33)

The value of factor A

After computing B from Eq. (9-34), determine the value of factor A by the help of Fig. 9-1 and B. If the required I_s is greater than the moment of inertia I, for the section selected above, select a new section with a larger moment of inertia and determine a new value of I_s. If the required I_s is smaller than the moment of inertia I selected as above, then that section should be satisfactory.

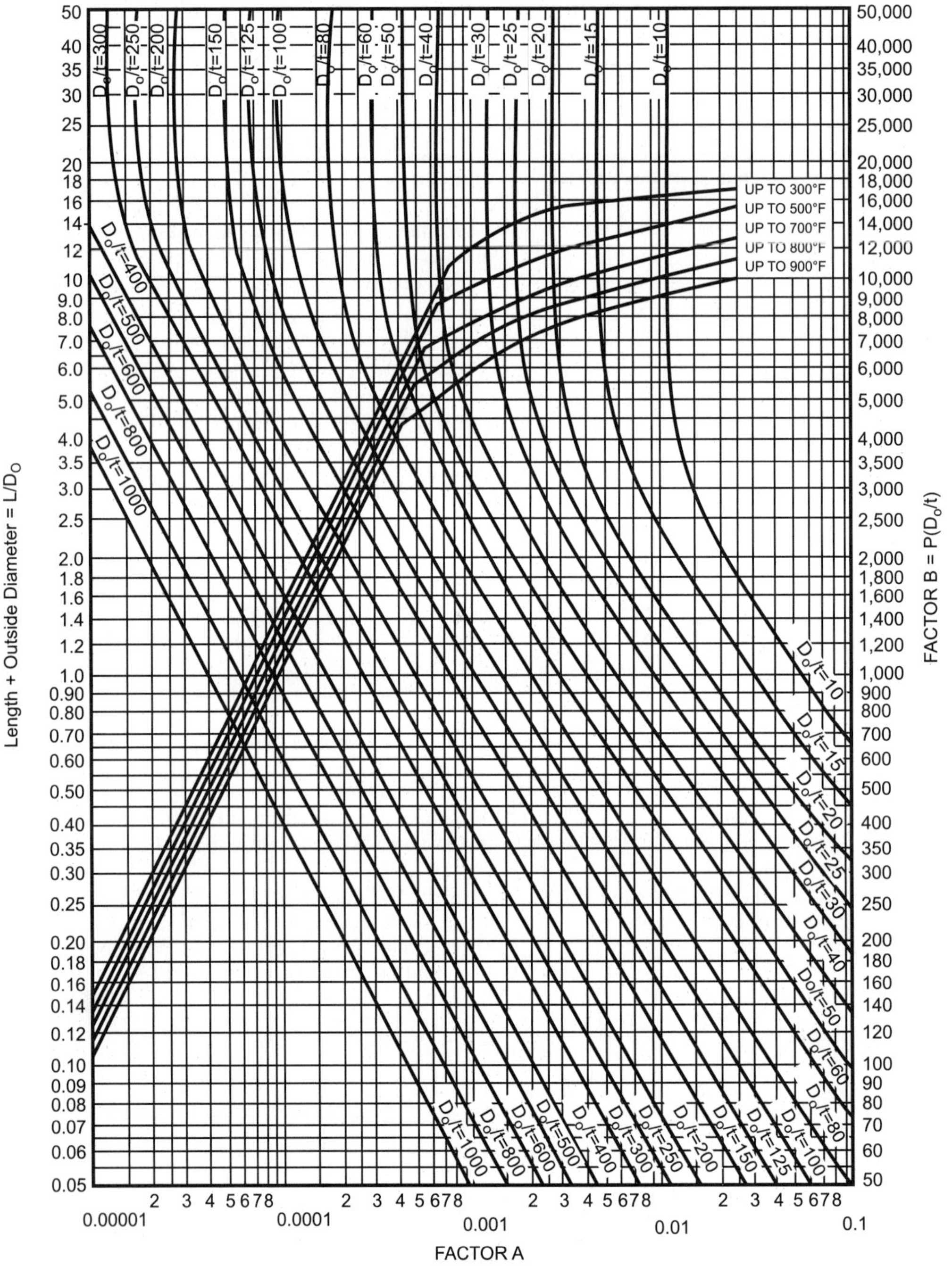

FIGURE 9-1 Chart for determining wall thicknesses of ring reinforced furnaces when constructed of carbon steel (specified yield strength, 210 to 262 MPa (30 to 38 kpsi) (1 kpsi = 6.894757 MPa). (*Source*: "Rules for Construction of Power Boilers," ASME Boiler and Pressure Vessel Code, Section I, *1983* and "Rules for Construction of Pressure Vessels," Section VIII, *Division 1*, ASME Boiler and Pressure Vessel Code, July 1, 1986.)[1,2]

Particular	Formula

Corrugated furnaces

The maximum allowable working pressure (P) on corrugated furnace having plain portion at the ends not exceeding 225 mm (9 in) in length

$$P = \frac{tC_6}{D} \quad (9\text{-}35)$$

where

t = thickness, in, not less than 7.8 mm ($\frac{5}{16}$ in) for Leeds, Morison, Fox and Brown, and not less than 11 mm ($\frac{7}{16}$ in) for Purves and other furnaces corrugated by sections not over 450 mm (18 in) long.

D = mean diameter, in

Values of C_6 are taken from Table 9-10

TABLE 9-10
Values of C_6 for use in Eq. (9–35)

		C_6
1.	For Leeds furnaces, when corrugations are not more than 200 mm (8 in) from center and not less than 56.25 mm (2.25 in) deep	17,300
2.	For Morison furnaces, when corrugations are not more than 200 mm (8 in) from center to center and the radius of the outer corrugation is not more than one-half of the suspension curve	15,600
3.	For Fox furnaces, when corrugations are not more than 200 mm (8 in) from center to center and not less than 37.5 mm (1.5 in) deep	14,000
4.	For Purves furnaces, when rib projections are not more than 225 mm (9 in) from center to center and not less than 34.375 mm (1.375 in) deep	14,000
5.	For Brown furnaces, when corrugations are not more than 225 mm (9 in) from center to center and not less than 40.625 mm (1.625 in) deep	14,000
6.	For furnaces corrugated by sections not more than 450 mm (18 in) from center to center and not less than 37.5 mm (1.5 in) deep, measured from the least inside greatest outside diameter of the corrugations and having the ends fitted into the other and substantially riveted together, provided the plain parts at the ends do not exceed 300 mm (12 in) in length	

Source: *ASME Power Boiler Code*, Section I, 1983.

Stayed surfaces

The maximum allowable working pressure (P) for a stayed wrapper sheet of a locomotive-type boiler

$$P = \frac{11000t\eta}{R - \sum s \sin\alpha} \quad \textbf{USCS} \quad (9\text{-}36a)$$

where

t = thickness of wrapper sheet, in

R = radius of wrapper sheet, in

η = minimum efficiency of wrapper sheet through joints or stay holes

Particular	Formula
	$\sum s \sin\alpha$ = summated value of transverse spacing ($s \sin\alpha$) for all crown stays considered in one transverse plane and on one side of the vertical axis of the boiler s = transverse spacing of crown stays in the crown sheet, in α = angle any crown stay makes with the vertical axis of boiler
	$$P = \frac{76t\eta}{R - \sum s \sin\alpha} \quad \textbf{SI} \quad (9\text{-}36b)$$ where P in MPa; s, t, and R in m
The longitudinal pitch between stay bolts or between the nearest row of stay bolts and the row of rivets at the joints between the furnace sheet and the tube sheet or the furnace sheet and the mud ring	$$L = \left(\frac{56320t^2}{PR}\right)^2 \quad \textbf{USCS} \quad (9\text{-}37a)$$ where t = thickness of furnace sheet, in R = outside radius of furnace, in P = maximum allowable working pressure, psi
	$$L = \left(\frac{2.535 \times 10^9 t^2}{PR}\right)^2 \quad \textbf{SI} \quad (9\text{-}37b)$$ where P in Pa; t, L, and R in m
Cross-sectional area of diagonal stay (A)	$$A = \frac{aL}{l} \quad (9\text{-}38)$$ where a = sectional area of direct stay, m (in) L = length of diagonal stay, m (in) l = length of line drawn at right angles to boiler head or a projection of L on a horizontal surface parallel to boiler drum, m (in)
The total cross-sectional area of stay tubes which support the tube plates in multitubular boilers	$$A_t = \frac{(A - a)P}{\sigma_{sa}} \quad (9\text{-}39)$$ where A = area of that portion of tuber plate containing the tubes, m (in) a = aggregate area of holes in the tube plate, m^2 (in^2) P = maximum allowable working pressure, Pa (psi) σ_{sa} = maximum allowable stress value in the tubes, MPa (psi) $\not>$48 MPa (7 kpsi) σ_{sa} is also taken from Table 8-23 The pitch of stay tubes shall conform to Eqs. (9-17) and (9-18) and using the values of C_7 as given in Table 9-11

Particular	Formula
The pitch from the stay bolt next to the corner to the point of tangency to the corner curve for stays at the upper corners of fire boxes shall be as given in Eq. (9-40)	$p = \dfrac{90[C_7(T^2/P)]^{1/2}}{\text{angularity of tangent lines } (\beta)}$ **USCS** (9-40a) where T = thickness of plate in sixteenths of an inch P = maximum allowable working pressure, psi C_7 = factor for the thickness of plate and type of stay used $p_t = 7592 \dfrac{\sqrt{C_7(T^2/p)}}{\text{angularity of tangent lines } (\beta)}$ **SI** (9-40b) where p_t and T in m, and p in Pa
For various values of C_7	Refer to Table 9-11

TABLE 9-11
Values of C_7 for determining pitch of stay tubes

Pitch of stay tubes in the bounding rows	When tubes have nuts not outside of plates	When tubes are fitted with nuts outside of plates
Where there are two plain tubes between two stay tubes	120	130
Where there is one plain tube between two stay tubes	140	150
Where every tube in the bounding rows is a stay tube and each alternate tube has a nut	—	170

Source: *ASME Power Boiler Code*, Section I, 1983.

FINAL RATIOS[1]

Design of a *horizontal return tubular boiler*	$\dfrac{H}{G}$ ranges from 35 to 45 in firetube boilers; 37 is a good working value (9-41a) $\dfrac{S}{W}$ lines between $\frac{1}{2}$ and $\frac{1}{3}$ for most types of cylindrical boilers (9-41b) $\dfrac{C}{G}$ varies from $\frac{1}{6}$ to $\frac{1}{8}$ (9-41c) $\dfrac{S}{P} = 16.7 \times 10^{-3}\ (0.6)^* \text{ to } 19.5 \times 10^{-3}\ (0.7)^*$ (9-41d) $\dfrac{H}{P} = 0.92 \text{ to } 1.12\,\text{m}^2\ (10 \text{ to } 12\,\text{ft}^2)$ for externally fired boiler per hp $= 0.74\,\text{m}^2\ (8\,\text{ft}^2)$ for Scotch boiler per hp (9-41e)

* The units in parentheses are in US Customary System units.

Particular	Formula	
	$\frac{P}{G} = 53\ (5.22)^*$	(9-41f)
	$\frac{DS}{N} = 64 \times 10^{-3}$ to 73×10^{-3}	(9-41g)
Design of a vertical *straight shell multitubular boiler*	$\frac{H}{G} = 60$	(9-42a)
	$\frac{WHS}{G} = 45$	(9-42b)
	$\frac{SHS}{WHS} = \frac{1}{3}$	(9-42c)
	$\frac{S}{W} = \frac{1}{3}$	(9-42d)
	$\frac{S}{P} = 22.3 \times 10^{-3}\ (0.80)^*$	(9-42e)

A = Total area of steam segment

D = Diameter of shell or drum

h = Height of the segment to be occupied by steam

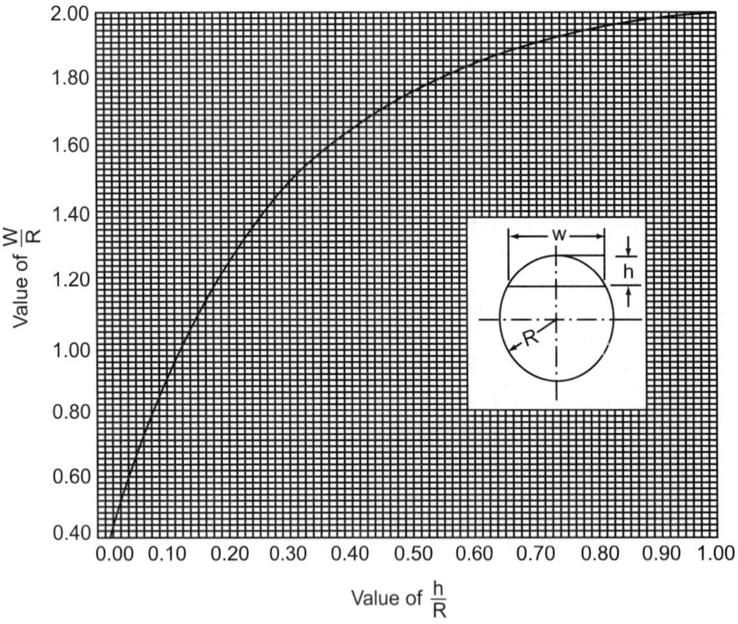

FIGURE 9-2 Disengaging surface in horizontal cylindrical shell. (*Source*: Reproduced from G. B. Haven and G. W. Swett, *The Design of Steam Boilers and Pressure Vessels*, John Wiley and Sons, Inc., 1923.)[1]

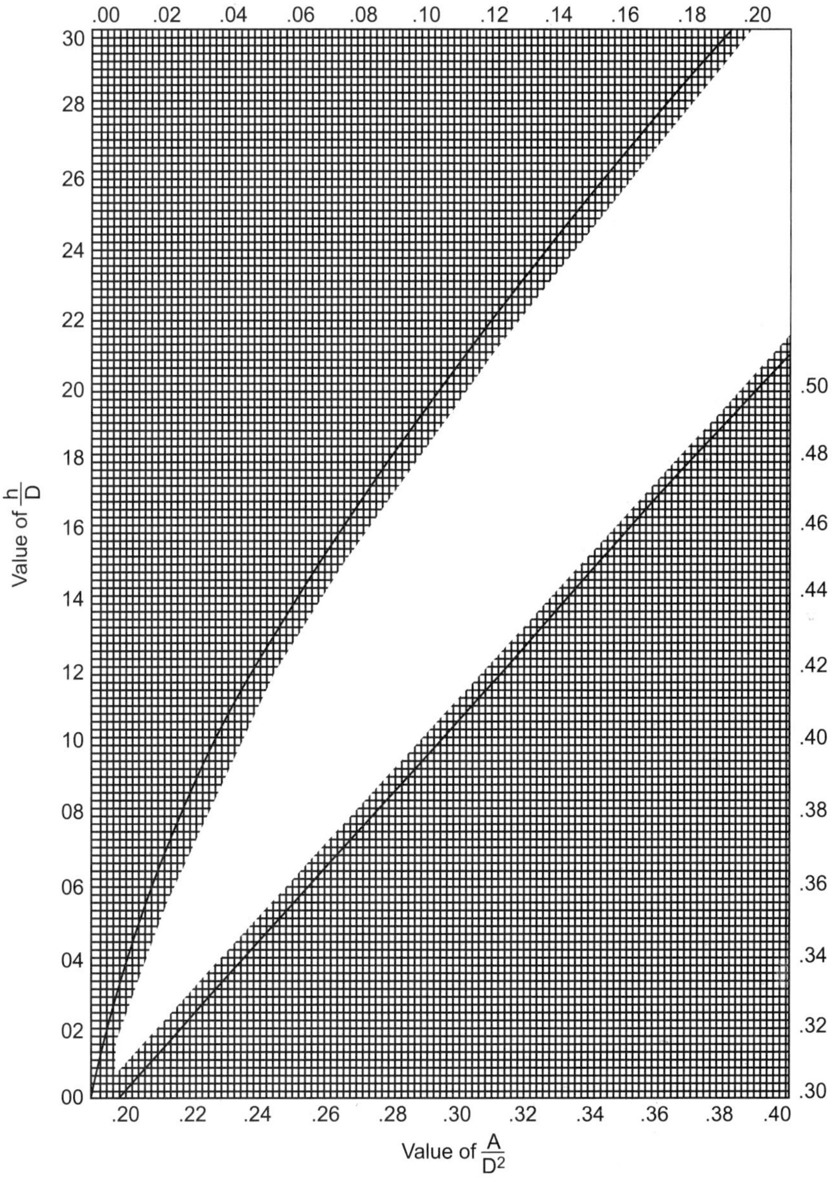

FIGURE 9-3 Areas of circular segments. (Reproduced from G. B. Haven and G. W. Swett, *The Design of Steam Boilers and Pressure Vessels*, John Wiley and Sons, Inc., 1923.)[1]

Particular	Formula	
	$\frac{C}{G} = \frac{1}{5.5}$	(9-42f)
	$\frac{H}{P} = 1.12\ (12)^*$	(9-42g)
	$\frac{G}{P} = 18.3 \times 10^{-3}\ (20)^*$ or $\frac{P}{G} = 51\ (5)$	(9-42h)
	$\frac{P}{DS} = 51\ (5)^*$	(9-42i)
Watertube boiler design	$\frac{H}{G} = 50$	(9-43a)
	$\frac{S}{p} = 11.2 \times 10^{-3}\ (0.424)^*$	(9-43b)
	$\frac{H}{P} = 0.92\ (10)^*$	(9-43c)
	$\frac{P}{G} = 51\ (4.37)^*$	(9-43d)
	$\frac{DS}{P} = 27.5 \times 10^{-3}\ (0.308)^*$	(9-43e)
For mechanical properties of carbon and carbon manganese steel plates, sections and angles for marine boilers pressure vessels and welded machinery and mechanical properties of steel plates for boilers	Refer to Table 9-12	
For properties of boilers	Refer to Table 9-13	
For evaporation of water, average rate of combustion of fuels, and minimum rate of steam produced	Refer to Tables 9-14 to 9-16	

TABLE 9-12
Mechanical properties of carbon and carbon manganese steel plates, sections, and angles steel for marine boilers, pressure vessels, and welded machinery

Grade	Tensile strength		Elongation percentage min on gauge length		Bond test diameter of former
	MPa	kpsi	$5.65\sqrt{a^*}$ [a]	200 mm	
1	362.8–441.3	52.6–64.0	26	25	$2t$
2	411.9–490.3	59.7–71.1	25	23	$2t$
3	431.5–529.6	62.6–76.8	23	21	$3t$
4	460.9–559.0	66.8–81.0	22	20	$3t$
5	490.3–588.4	71.1–85.3	21	19	$3\frac{1}{2}t$

[a] Area of cross section.
Source: IS 3503, 1966.

TABLE 9-13
Properties of boilers
Horizontal return tubular boilers

Diameter of shell, mm	Diameter of tubes, mm			Length of tubes, m				
910	64	76		2.44	3.66			
1070	64	76		3.05	3.66	4.28		
1220	64	76		3.66	4.28	4.58	4.88	
1370	76	89		3.97	4.28	4.58	4.88	5.5
1520	76	89		4.28	4.88	5.50		
1680	76	89		4.88	5.18	5.50	6.10	
1830	76	89		4.88	5.50	6.10		
1980	76	89	102	4.88	5.18	5.50	6.10	
2130	76	89	102	4.88	5.50	6.10		
2290	76	89	102	4.88	5.18	5.50	6.10	
2440	76	89	102	4.88	5.18	5.50	6.10	

Dry-back scotch boilers

Diameter of shell, m	Diameter of tubes, mm	Length of tubes, m	Inside diameter of furnace, mm	Length of grate, m
		Short Types		
1.19	76	2.90	920	1.22
1.99	89	3.50	920	1.53
2.14	89	3.80	970	1.93
2.29	89	3.80	1150	2.03
2.44	89	3.97	1270	2.21
2.90	89	3.80	970	1.93
3.20	89	3.80	1150	2.03
3.50	89	3.89	1270	2.21
		Long Types		
2.06	102	4.88	970	1.93
2.21	102	4.88	1040	2.24
2.36	102	4.88	1140	2.44
2.84	102	4.88	970	1.93
3.05	102	4.88	1040	2.24
3.28	102	4.88	1140	2.44

Locomotive-type boilers without dome				Vertical firetube boiler for power plant use	
		Dimensions of grate			
Diameter of waist, mm	Length of 7.5 mm tubes, m	Width, mm	Length, m	Diameter of tubes, mm	Length of tubes, m
925	2.14	0.76	1.22	50	3.96
1070	2.44	0.92	1.27	62	4.27
1220	3.20	1.07	1.38		4.57
1370	3.36	1.22	1.53		4.88
1520	3.97	1.38	1.53		
1680	4.58	1.53	1.68		
1830	4.58	1.68	1.83		

Particular	Formula
For permissible strain rates of steam plant equipments	Refer to Table 9-17
For water level requirements of boilers	Refer to Table 9-18
For minimum allowable thickness of plates for boilers	Refer to Table 9-19
For disengaging surface per horsepower	Refer to Table 9-20
For heating boiler efficiency	Refer to Table 9-21

TABLE 9-14
Evaporation kg (lb) of water per kg (lb) of fuel reduced to standard condition [from and at 373 K (100°C)]

	Approximate		Evaporation
Type of fuel	**kJ/kg**	**Btu/lb**	**per kg (lb) of fuel, kg (lb)**
Anthracite	29,038.3–27842.2	12,500–1 2000	9.5–9
Coke	30,228.7	13,000	9.5
Semibituminous	33,703.7	14,500	10
Bituminous	29,098.3	12,500	9
Lignite	22,106.3	9,500	6
Fuel oil	41,868.0	19,000	14.5

TABLE 9-15
Average rates of combustion [kg/m^2 (lb/ft^2) of grate surface per hour] draft 12.55 mm ($\frac{1}{2}$ in) water column

Fuel used	Stationary grate
Anthracite	44–68.5 (9.14)
Semibituminous	98 (20)
Bituminous	68.5 (14)
Lignite	58.5 (12)

TABLE 9-16
Minimum kilograms (pounds) of steam per h per ft^2 of surface

	Firetube boilers		Watertube boilers	
Particulars	**kg**	**lb**	**kg**	**lb**
Boiler heating surface				
Hand-fired	11.0	5	13.2	6
Stoker-fired	15.4	7	17.6	8
Oil-, gas-, or powder-fired	17.6	8	22.1	10
Water wall heating surface				
Hand-fired	17.6	8	17.6	8
Stoker-fired	22.1	10	26.5	12
Oil-, gas-, or powder-fired	30.9	14	35.3	16

Source: *ASME Power Boiler Code*. Section I, 1983.

TABLE 9-17
Permissible strain rates for steam plant equipment

Machine part	Strain rate per hour
Turbine disk (pressed on shaft)	10^{-9}
Bolted flanges, turbine cylinders	10^{-8}
Steam piping, welded joints, and boiler tubes	10^{-7}
Superheated tubes	10^{-6}

TABLE 9-18
Water level requirements[a]

Horizontal return tubular boilers		Vertical firetube boilers	
Boiler diameters, mm	Distance between gauge cocks, mm	Boiler diameters mm	Distance between gauge cocks, mm
910, 1070, 1220	75	910–1220	100
1370, 1520	100	1250–1680	125
1680, 1830, 1980, 2130	125	1700–2410	150
		2460–3100	175
Dry-back Scotch boilers		**Locomotive-type boilers**	
Low water level 89 mm above surface of tubes for all diameters: distance between gauge cocks may be reduced to a minimum of 75 mm		Low water level must be 75–125 mm above the water surface of the crown sheet; distance between gauge cocks is usually 75 mm for all diameters	

[a] Low water level 890 mm above surface of tubes.

TABLE 9-19
Minimum allowable thickness of plates for boilers (all dimensions in mm)

	Power boilers		Heating boilers	
Minimum thickness	Shell and dome plate diameter	Tube sheet diameter	Shell or other plate diameter	Tube sheet or head diameter
6.5	≤910		≤1065	
8.0	>910–1370		>1065–1530	≤1065
9.5	>1370–1830	≤1065	>1530–1980	>1065–1530
11.0		>1065–1370	>1980	>1530–1980
12.5	>1830	>1370–1830		>1980
14.5		>1830		

TABLE 9-20
Disengaging surface per horsepower mean water level

	Disengaging surface	
Type of boiler	m^2/kW	m^2/hp
Horizontal return tubular	0.087–0.10	0.065–0.0745
Dry-back Scotch	0.075–0.087	0.056–0.0650
Vertical straight shell	0.020–0.025	0.0149–0.0186
Vertical (Manning)	0.011–0.013	0.0084–0.0093
Locomotive type	0.100–0.125	0.0745–0.093
Sectional water tube	0.037–0.0500	0.0279–0.0372

TABLE 9-21
Heating boiler efficiency

Firing method	Efficiency, %
Hand-Fired Coal	
Lignite	49
Subbituminous	44–63
Bituminous	50–65
Low-volatile bituminous	44–61
Anthracite	60–75
Coke	75–76
Stoker Conversion	
Bituminous	55–69
Anthracite	63
Burner Conversion	
Natural gas	69–76
Oil	51; 65; 70
Designed for Burner	
Stoker	60–75
$\leq$45 kg	65
>45 kg	70
Gas	70–80
Oil	70–80
Cast-iron boilers	68
Steel boilers	70
Package units	75

REFERENCES

1. Haven, G. B., and G. W. Swett, *The Design of Steam Boilers and Pressure Vessels*, John Wiley and Sons, Inc., New York, 1923.
2. "Rules for Construction of Power Boilers," *ASME Boiler and Pressure Vessel Code*, Section I, 1983.
3. "Rules for Construction of Pressure Vessels, " *ASME Boiler and Pressure Vessel Code*, Section VIII, Division I, July 1, 1986.
4. *Code of Unfired Pressure Vessels, Bureau of Indian Standards*, IS 2825, 1969, New Delhi, India.
5. Nichols, R. W., *Pressure Vessel Codes and Standards*, Elsevier Applied Science Publishing Ltd., Barking, Essex, England, 1987.
6. Lingaiah, K., and B. R. Narayana Iyengar, *Machine Design Data Handbook*, Engineering College Cooperative Society, Bangalore, India, 1962.
7. Lingaiah, K., *Machine Design Data Handbook*, Vol. II (*SI and Customary Metric Units*), Suma Publishers, Bangalore, India, 1986.
8. Lingaiah, K., *Machine Design Data Handbook*, (*SI and U.S. Customary Units*), McGraw-Hill Publishing Company, New York, 1994.

CHAPTER 10

ROTATING DISKS AND CYLINDERS[1]

SYMBOLS[1]

g	acceleration due to gravity, m/s^2 (ft/s^2)
r	any radius, m (in)
r_i	inside radius, m (in)
r_o	outside radius, m (in)
h	thickness of disk at radius r from the center of rotation, m (in)
h_2	thickness of disk at radius r_2 from the center of rotation, m (in)
σ	uniform tensile stress in case of a disk of uniform strength, MPa (psi)
σ_θ	tangential stress, MPa (psi)
σ_r	radial stress, MPa (psi)
σ_z	axial stress or longitudinal stress, MPa (psi)
ρ	density of material of the disk, kg/m^3 (lb_m/in^3)
ω	angular speed of disk, rad/s
ν	Poisson's ratio

Particular	Formula	

DISK OF UNIFORM STRENGTH ROTATING AT ω rad/s (Fig. 10-1)

The thickness of a disk of uniform strength at radius r from center of rotation

$$h = h_2 \exp\left[\frac{\rho\omega^2}{2\sigma}(r_2^2 - r^2)\right] \quad (10\text{-}1)$$

SOLID DISK ROTATING AT ω rad/s

The general expression for the radial stress of a rotating disk of uniform thickness

$$\sigma_r = \frac{3+\nu}{8}\rho\omega^2(r_o^2 - r^2) \quad (10\text{-}2)$$

Particular	Formula

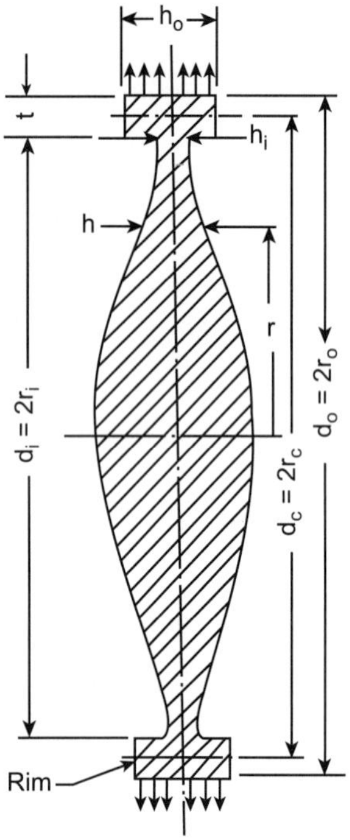

FIGURE 10-1 High-speed rotating disk of uniform strength.

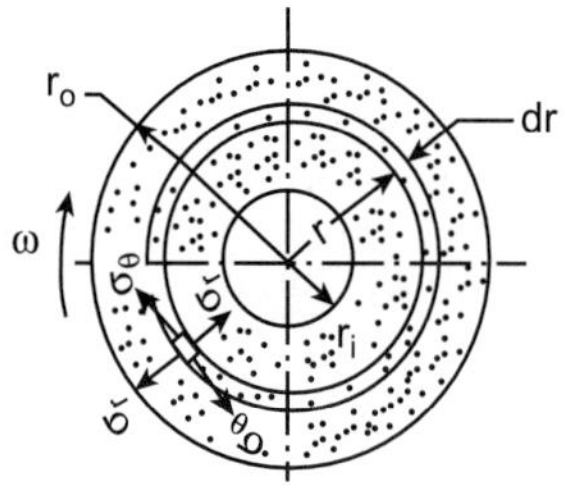

FIGURE 10-2 Rotating disk of uniform thickness.

The general expression for the tangential stress of a rotating disk of uniform thickness

$$\sigma_\theta = \frac{3+\nu}{8}\rho\omega^2\left(r_o^2 - \frac{1+3\nu}{3+\nu}r^2\right) \tag{10-3}$$

The maximum values of stresses are at the center, where $r = 0$, and are equal to each other

$$\sigma_{r(\max)} = \sigma_{\theta(\max)} = \frac{3+\nu}{8}\rho\omega^2 r_o^2 \tag{10-4}$$

HOLLOW DISK ROTATING AT ω rad/s (Fig. 10-2)

The general expression for the radial stress of a rotating disk of uniform thickness

$$\sigma_r = \frac{3+\nu}{8}\rho\omega^2\left(r_i^2 + r_o^2 - \frac{r_o^2 r_i^2}{r^2} - r^2\right) \tag{10-5}$$

The general expression for the tangential stress of a rotating disk of uniform thickness

$$\sigma_\theta = \frac{3+\nu}{8}\rho\omega^2\left(r_i^2 + r_o^2 + \frac{r_o^2 r_i^2}{r^2} - \frac{1+3\nu}{3+\nu}r^2\right) \tag{10-6}$$

The maximum radial stress occurs at $r^2 = r_o r_i$

$$\sigma_{r(\max)} = \frac{3+\nu}{8}\rho\omega^2(r_o - r_i)^2 \tag{10-7}$$

Particular	Formula
The maximum tangential stress occurs at inner boundary where $r = r_i$	$\sigma_{\theta(\max)} = \frac{3+\nu}{4}\rho\omega^2\left(r_o^2 + \frac{1-\nu}{3+\nu}r_i^2\right)$ (10-8)

SOLID CYLINDER ROTATING AT ω rad/s

The tangential stress	$\sigma_\theta = \frac{\rho\omega^2}{8(1-\nu)}[(3-2\nu)r_o^2 - (1+2\nu)r^2]$ (10-9)
The radial stress	$\sigma_r = \frac{\rho\omega^2}{8}\left(\frac{3-2\nu}{1-\nu}\right)(r_o^2 - r^2)$ (10-10)
The maximum stress occurs at the center	$\sigma_{r(\max)} = \sigma_{\theta(\max)} = \frac{\rho\omega^2}{8}\left(\frac{3-2\nu}{1-\nu}\right)r_o^2$ (10-10a)
The axial strain in the z direction (ends free)	$\varepsilon_z = \frac{-\nu}{2}\frac{\rho\omega^2 r_o^2}{E}$ (10-11)
The axial stress under plane strain condition (ends free)	$\sigma_z = \frac{\rho\omega^2}{4}\left(\frac{\nu}{1-\nu}\right)(r_o^2 - 2r^2)$ (10-12a)
The axial stress under plane strain condition (ends constrained)	$\sigma_z = \frac{\rho\omega^2\nu}{4(1-\nu)}\left[\frac{1}{2}(3-2\nu)r_o^2 - 2r^2\right]$ (10-12b)

HOLLOW CYLINDER ROTATING AT ω rad/s

The tangential stress at any radius r	$\sigma_\theta = \frac{\rho\omega^2}{8}\left(\frac{3-2\nu}{1-\nu}\right)\left[r_i^2 + r_o^2 + \frac{r_i^2 r_o^2}{r^2} - \left(\frac{1+2\nu}{3-2\nu}\right)r^2\right]$ (10-13)
The radial stress at any radius r	$\sigma_r = \frac{\rho\omega^2}{8}\left(\frac{3-2\nu}{1-\nu}\right)\left[r_i^2 + r_o^2 - \frac{r_i^2 r_o^2}{r^2} - r^2\right]$ (10-14)
The axial stress (ends free) at any radius r	$\sigma_z = \frac{\rho\omega^2}{4}\left(\frac{\nu}{1-\nu}\right)[r_i^2 + r_o^2 - 2r^2]$ (10-15)
The axial stress under plane strain conditions (ends constrained) at any radius r	$\sigma_z = \frac{\nu\rho\omega^2}{4}\left(\frac{3-2\nu}{1-\nu}\right)\left[r_i^2 + r_o^2 - \frac{2r^2}{3-2\nu}\right]$ (10-16)
The maximum stress occurs at the inner surface where $r = r_i$	$\sigma_{\theta(\max)} = \frac{\rho\omega^2}{4}\left(\frac{3-2\nu}{1-\nu}\right)\left[r_o^2 + \left(\frac{1-2\nu}{3-2\nu}\right)r_i^2\right]$ (10-17)

Particular	Formula
The axial strain in the z direction (ends free)	$\varepsilon_z = -\frac{\nu\rho\omega^2}{2E}(r_i^2 + r_o^2)$ (10-18)
The displacement u at any radius r of a thin hollow rotating disk	$u = \left[\frac{\rho\omega^2 r}{E}\frac{(3+\nu)(1-\nu)}{8} \times \left(r_o^2 + r_i^2 + \frac{1+\nu}{1-\nu}\frac{r_o^2 r_i^2}{r^2} - \frac{1+\nu}{3+\nu}r^2\right)\right]$ (10-19)

SOLID THIN UNIFORM DISK ROTATING AT ω rad/s UNDER EXTERNAL PRESSURE p_o (Fig. 10-3)

Particular	Formula
The radial stress at any radius r	$\sigma_r = -p_o + \rho\omega^2\left(\frac{3+\nu}{8}\right)(r_o^2 - r^2)$ (10-20)
The tangential stress at any radius r	$\sigma_\theta = -p_o + \rho\omega^2\left(\frac{3+\nu}{8}\right)\left(r_o^2 - \frac{1+3\nu}{3+\nu}r^2\right)$ (10-21)
The maximum radial stress at $r = 0$	$\sigma_{r(\max)} = -p_o + \rho\omega^2\left(\frac{3+\nu}{8}\right)r_o^2$ (10-22)
The maximum radial stress at $r = r_o$	$\sigma_r = -p_o$ (10-23)
The maximum tangential stress at $r = 0$	$\sigma_{\theta(\max)} = \sigma_{r(\max)}$ (10-24)
The displacement u at any radius r	$u = \frac{r}{E}(1-\nu)\left\{-p_o + \frac{\rho\omega^2}{8}[(3+\nu)r_o^2 - (1+\nu)r^2]\right\}$ (10-25)

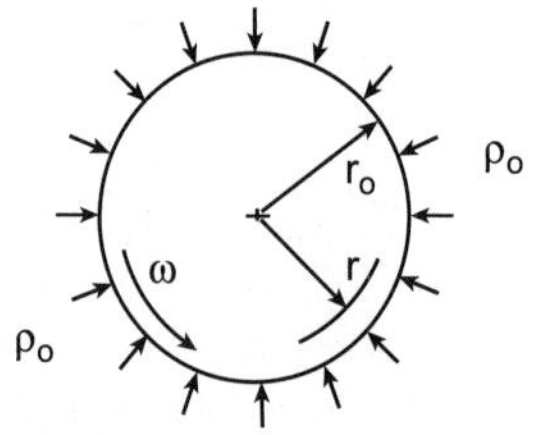

FIGURE 10-3 Rotating disk of uniform thickness under external pressure.

Particular	Formula

HOLLOW CYLINDER OF UNIFORM THICKNESS ROTATING AT ω rad/s. SUBJECT TO INTERNAL (p_i) AND EXTERNAL (p_o) PRESSURES (Fig. 10-4)

The general expression for the radial stress of a hollow cylinder of uniform thickness rotating at ω rad/s under internal (p_i) and external (p_o) pressure at any radius r

$$\sigma_r = A - \frac{B}{r^2} + \frac{\rho\omega^2}{8}\left(\frac{3-2\nu}{1-\nu}\right) \times \left[r_i^2 + r_o^2 - \frac{r_i^2 r_o^2}{r^2} - r^2\right] \quad (10\text{-}26)$$

The general expression for the tangential or hoop stress of a hollow cylinder of uniform thickness rotating at ω rad/s under internal (p_i) and external (p_o) pressure at any radius r.

$$\sigma_\theta = A + \frac{B}{r^2} + \frac{\rho\omega^2}{8}\left(\frac{3-2\nu}{1-\nu}\right) \times \left[r_i^2 + r_o^2 + \frac{r_i^2 r_o^2}{r^2} - \left(\frac{1+2\nu}{3-2\nu}\right)r^2\right] \quad (10\text{-}27)$$

where $A = \dfrac{p_i r_i^2 - p_o r_o^2}{r_o^2 - r_i^2}$; $\quad B = \dfrac{r_i^2 r_o^2 (p_i - p_o)}{r_o^2 - r_i^2}$

The tangential or hoop stress in a hollow cylinder rotating at ω rad/s under p_o and p_i at $r = r_i$ (Fig. 10-4)

$$\sigma_{(\theta\ \max)r=r_i} = \frac{p_i(r_i^2 + r_o^2) - 2p_o r_o^2}{r_o^2 - r_i^2} + \frac{\rho\omega^2}{8}\left(\frac{3-2\nu}{1-\nu}\right)\left[2r_o^2 + \left(\frac{2-4\nu}{3-2\nu}\right)r_i^2\right] \quad (10\text{-}28a)$$

$$= \frac{p_i(r_i^2 + r_o^2) - 2p_o r_o^2}{r_o^2 - r_i^2} + \frac{\rho\omega^2}{4}\left(\frac{3-2\nu}{1-\nu}\right)\left[r_o^2 + \left(\frac{1-2\nu}{3-2\nu}\right)r_i^2\right] \quad (10\text{-}28b)$$

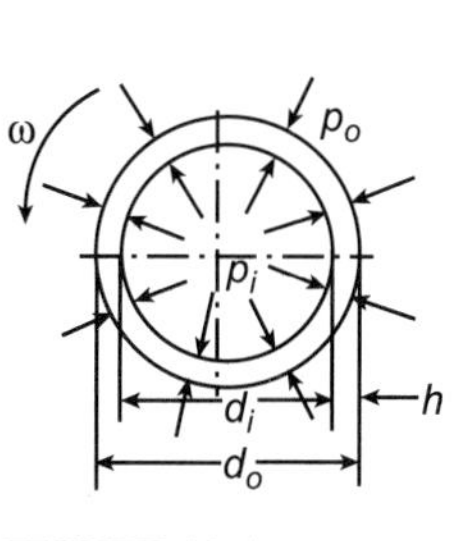

FIGURE 10-4

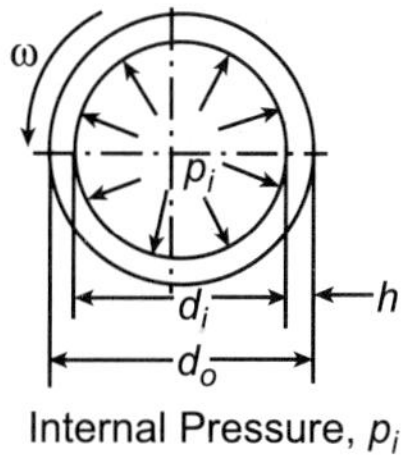

FIGURE 10-5

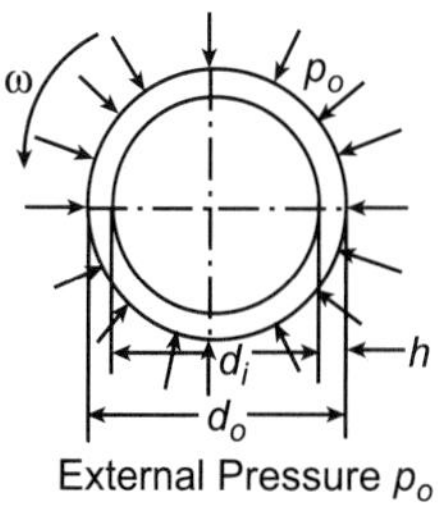

FIGURE 10-6

Particular	Formula
The tangential or hoop stress in a hollow cylinder rotating at ω rad/s under p_o and p_i at $r = r_o$ (Fig. 10-4)	$\sigma_{(\theta\ \max)r=r_o} = \dfrac{2p_i r_i^2 - p_o(r_o^2 + r_i^2)}{r_o^2 - r_i^2} + \dfrac{\rho\omega^2}{4}\left(\dfrac{3-2\nu}{1-\nu}\right)\left[r_i^2 + \dfrac{1-2\nu}{3-2\nu} r_o^2\right]$ (10-29)
The tangential stress in a cylinder rotating at ω rad/s at any radius r when subjected to internal pressure (p_i) only (Fig. 10-5)	$(\sigma_\theta)_{p_o=0} = \dfrac{p_i r_i^2(r_o^2 + r^2)}{r^2(r_o^2 - r_i^2)} + \dfrac{\rho\omega^2}{4}\left(\dfrac{3-2\nu}{1-\nu}\right) \times \left[r_i^2 + r_o^2 + \dfrac{r_i^2 r_o^2}{r^2} - \left(\dfrac{1+2\nu}{3-2\nu}\right) r^2\right]$ (10-30)
The tangential stress in a cylinder rotating at ω rad/s at any radius r when subject to external pressure (p_o) only (Fig. 10-6)	$(\sigma_\theta)_{p_i=0} = \dfrac{-p_o r_o^2(r^2 + r_i^2)}{r^2(r_o^2 - r_i^2)} + \dfrac{\rho\omega^2}{4}\left(\dfrac{3-2\nu}{1-\nu}\right) \times \left[r_i^2 + r_o^2 + \dfrac{r_i^2 r_o^2}{r^2} - \left(\dfrac{1+2\nu}{3-2\nu}\right) r^2\right]$ (10-31)

ROTATING THICK DISK AND CYLINDER WITH UNIFORM THICKNESS SUBJECT TO THERMAL STRESSES

Particular	Formula
The hoop or tangential stress in thick disk or cylinder at any radius r rotating at ω rad/s subject to pressure p_o and p_i	$\sigma_\theta = A + \dfrac{B}{r^2} - (3+\nu)\dfrac{\rho\omega^2}{8}\left[r_o^2 - \left(\dfrac{1+3\nu}{3+\nu}\right) r^2\right] - E\alpha T + \dfrac{E\alpha}{r^2}\int Tr\,dr$ (10-32)
The radial stress in thick disk or cylinder at any radius r rotating at ω rad/s subject to pressure p_o and p_i	$\sigma_r = A - \dfrac{B}{r^2} - \dfrac{\rho\omega^2}{8}(3+\nu)(r_o^2 - r^2) - \dfrac{E\alpha}{r^2}\int Tr\,dr$ (10-33) where A and B are Lamé's constants and can be found from boundary or initial conditions α = linear coefficient of thermal expansion, mm/°C (in/°F) T = temperature, °C or K (°F) ρ = density of rotating cylinder or disk material, kg/m^3 (lb$_m$/in^3) E = modulus of material of disk or cylinder, GPa (Mpsi)

Particular	Formula

ROTATING LONG HOLLOW CYLINDER WITH UNIFORM THICKNESS ROTATING AT ω rad/s SUBJECT TO THERMAL STRESS

The general expression for the radial stress in the cylinder wall at any radius r when the temperature distribution is symmetrical with respect to the axis and constant along its length.

$$\sigma_r = \frac{\rho\omega^2}{8}\left(\frac{3-2\nu}{1-\nu}\right)\left[r_i^2 + r_o^2 - \frac{r_i^2 r_o^2}{r^2} - r^2\right] + \frac{\alpha E}{(1-\nu)r^2}\left[\frac{4r^2 - d_i^2}{d_o^2 - d_i^2}\int_{r_i}^{r_o} Tr\,dr - \int_{r_i}^{r_o} Tr\,dr\right] \quad (10\text{-}34)$$

The general expression for the tangential stress in the cylinder wall at any radius r when the temperature distribution is symmetrical with respect to the axis and constant along its length.

$$\sigma_\theta = \frac{\rho\omega^2}{8}\left(\frac{3-2\nu}{1-\nu}\right)\left[r_i^2 + r_o^2 + \frac{r_i^2 r_o^2}{r^2} - \left(\frac{1+2\nu}{3-2\nu}\right)r^2\right] + \frac{\alpha E}{(1-\nu)r^2}\left[\frac{4r^2 + d_i^2}{d_o^2 - d_i^2}\int_{r_i}^{r_o} Tr\,dr + \int_{r_i}^{r_o} Tr\,dr - Tr^2\right] \quad (10\text{-}35)$$

The general expression for the axial stress in the cylinder wall at any radius r when the temperature distribution is symmetrical with respect to the axis and constant along its length.

$$\sigma_\theta = \frac{\rho\omega^2}{4}\left(\frac{\nu}{1-\nu}\right)[r_i^2 + r_o^2 - 2r^2] + \frac{\alpha E}{1-\nu}\left[\frac{8}{d_o^2 - d_i^2}\int_{r_i}^{r_o} Tr\,dr - T\right] \quad (10\text{-}36)$$

where $d_o = 2r_o$ and $d_i = 2r_i$

DEFLECTION OF A ROTATING DISK OF UNIFORM THICKNESS IN RADIAL DIRECTION WITH A CENTRAL CIRCULAR CUTOUT

The tangential stress within elastic limit, σ_θ, in a rotating disk of uniform thickness (Fig. 10-7)

$$\sigma_\theta = \frac{\delta E}{h} \quad (10\text{-}37)$$

The expression for the inner deflection δ_i, of rotating thin uniform thickness disk with centrally located circular cut-out as per Stodala[a] (Fig. 10-7)

$$\delta_i = 3.077 \times 10^{-6}\left(\frac{n}{1000}\right)^2 (7.5K^2 + 5) \quad (10\text{-}38)$$

[a] *Source*: Stodala "*Turbo-blower and compressor*"; Kearton, W. J. and Porter, L. M., Design Engineer, Pratt and Whitney Aircraft; McGraw-Hill Publishing Company, New York, U.S.A. Douglas C. Greenwood, Editor, *Engineering Data for Product Design*, McGraw-Hill Publishing Company, New York, 1961.

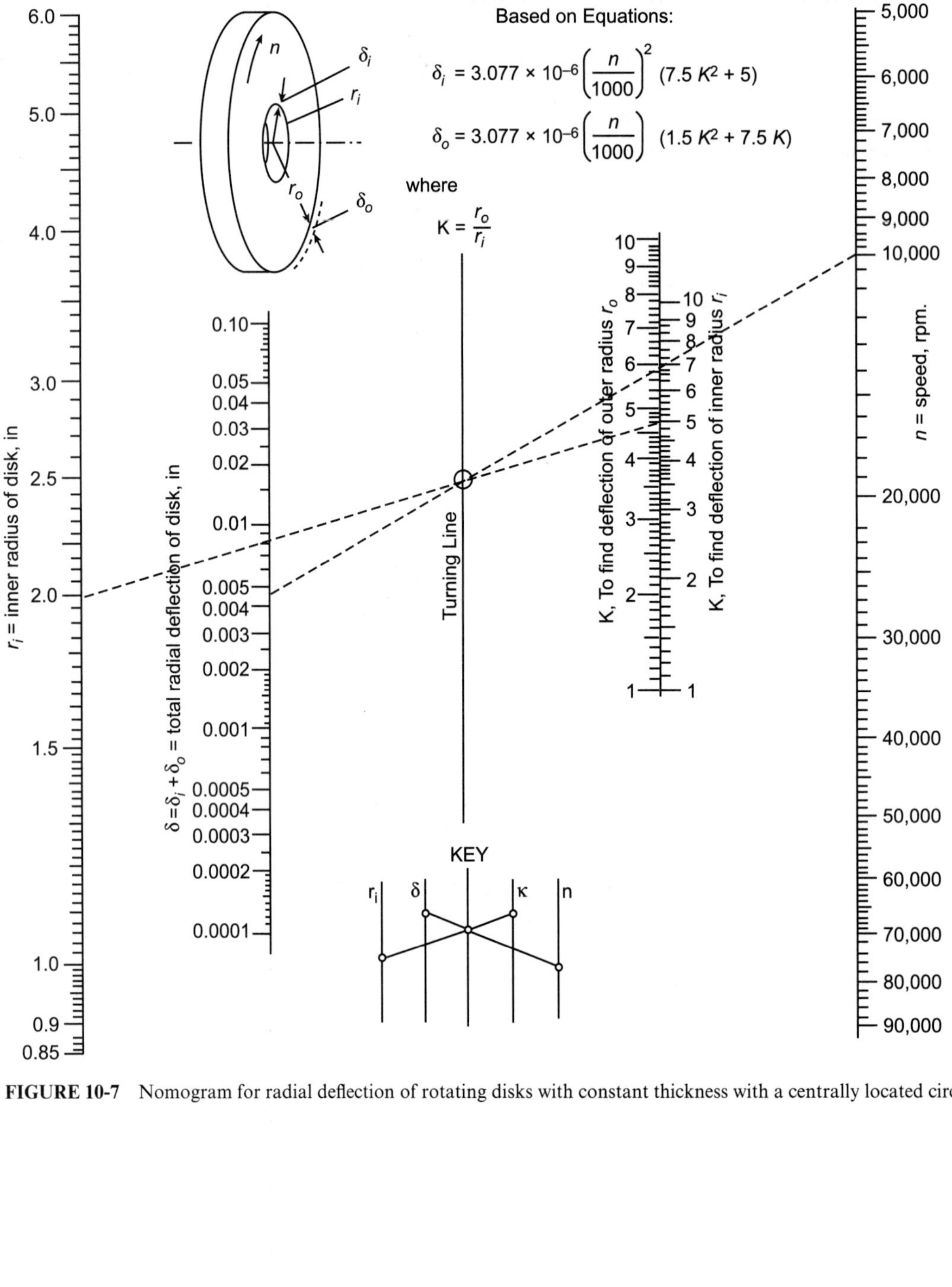

FIGURE 10-7 Nomogram for radial deflection of rotating disks with constant thickness with a centrally located circular hole.

Particular	Formula
The expression for the outer deflection δ_o of rotating thin uniform thickness disk with centrally located circular cut-out as per Stodala[a] (Fig. 10-7)	$$\delta_o = 3.077 \times 10^{-6}\left(\frac{n}{1000}\right)^2(1.5K^2 + 7.5K) \quad (10\text{-}39)$$ where $K = r_o/r_i$; σ_θ = tangential stress, psi; $\delta = \delta_i + \delta_o$ = total deflection of disk, in; r_i = inner radius of disk, in; r_o = outer radius of disk, in; n = speed, rpm

The Nomogram can be used for steel, magnesium and aluminum since the modulus of elasticity $E = 29 \times 10^6$ psi (200 MPa) for steel and Poisson's ratio $\nu = 1/3$. The error involved in using this equation with E and ν of steel for aluminum is about 0.5% and for magnesium is 2.5%.

REFERENCES

1. Lingaiah, K., and B. R. Narayana Iyengar, *Machine Design Data Handbook*, Volume I (*SI and Customary Metric Units*), Suma Publishers, Bangalore, 1986.
2. Lingaiah, K., *Machine Design Data Handbook*, McGraw-Hill Publishing Company, New York, 1994.
3. Douglas C. Greenwood, *Engineering Data for Product Design*, McGraw-Hill Publishing Company, New York, 1961.

[a] *Source*: Stodala "*Turbo-blower and compressor*"; Kearton, W. J. and Proter, L. M., Design Engineer, Pratt and Whitney Aircraft; McGraw-Hill Publishing Company, New York, U.S.A. Douglas C. Greenwood, Editor, *Engineering Data for Product Design*, McGraw-Hill Publishing Company, New York, 1961.

CHAPTER 11

METAL FITS, TOLERANCES, AND SURFACE TEXTURE

SYMBOLS[1,2,3]

A	area of cross section, m^2 (in^2)
d	diameter of shaft, m (in)
	diameter of cylinder, m (in)
E	modulus of elasticity, GPa (Mpsi)
E_c	modulus of elasticity of cast iron, GPa (Mpsi)
E_s	modulus of elasticity of steel, GPa (Mpsi)
F	force, kN [lbf or tonf (pound force or tonne force)]
l	length, m (in)
	length of hub, m (in)
	effective length of anchor, m (in)
L	original length of slot, m (in)
M_t	torque or twisting moment, N m (lbf in)
p	pressure, MPa (psi)
p_c	contact pressure MPa (psi)
t	temperature, °C (°F)
α	coefficient of linear expansion, (m/m)/°C [(in/in)/°F]
δ	total change in diameter (interference), m (in)
Δd	change in diameter, m (in)
ν	Poisson's ratio
σ	stress, MPa (psi)
μ	coefficient of friction
n	factor of safety

SUFFIXES

a	axial
b	bearing surface
c	contact surface, compressive
d	design
f	final
h	hub
i	internal, inner
o	original, external, outer
r	radial, rim

s	shaft
θ	tangential or hoop
1	initial
2	final

Particular	Formula

PRESS AND SHRINK FITS

Change in cylinder diameter due to contact pressure

The change in diameter

$$\Delta d = d\varepsilon_\theta \tag{11-1}$$

The change in diameter of the inner member when subjected to contact pressure p_c (Fig. 11-1)

$$\Delta d_i = -\frac{p_c d_c}{E}\left(\frac{d_c^2 + d_i^2}{d_c^2 - d_i^2} - \nu\right) \tag{11-2}$$

The change in diameter of the outer member when subjected to contact pressure p_c (Fig. 11-1)

$$\Delta d_o = \frac{p_c d_c}{E}\left(\frac{d_o^2 + d_c^2}{d_o^2 - d_c^2} + \nu\right) \tag{11-3}$$

The original difference in diameters of the two cylinders when the material of the members is the same

$$\delta = \Delta d_o + \Delta d_i = \frac{p_c d_c}{E}\left(\frac{d_o^2 + d_c^2}{d_o^2 - d_c^2} + \nu\right) + \frac{p_c d_c}{E}\left(\frac{d_c^2 + d_i^2}{d_c^2 - d_i^2} - \nu\right) \tag{11-4}$$

The total change in the diameters of hub and hollow shaft due to contact pressure at their contact surface when the material of the members is the same

$$\delta = \Delta d_s + \Delta d_h = d_s - d_h = \frac{p_c d_s}{E_s}\left(\frac{d_s^2 + d_i^2}{d_s^2 - d_i^2} - \nu_s\right) + \frac{p_c d_h}{E_h}\left(\frac{d_o^2 + d_h^2}{d_o^2 - d_s^2} + \nu_h\right) \text{ exactly} \tag{11-5a}$$

$$\delta = p_c d_c\left(\frac{d_c^2 + d_i^2}{E_s(d_c^2 - d_i^2)} + \frac{d_0^2 + d_c^2}{E_h(d_o^2 - d_c^2)} - \frac{\nu_s}{E_s} + \frac{\nu_h}{E_h}\right) \text{ (approx.)} \tag{11-5b}$$

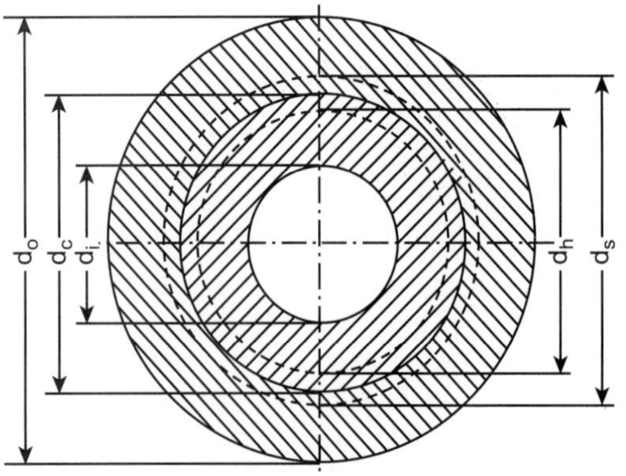

FIGURE 11-1

Particular	Formula	
The shrinkage stress in the band	$\sigma_\theta = \dfrac{E\delta}{d_c}$	(11-6)
The contact pressure between cylinders at the surface of contact when the material of both the cylinders is same (Fig. 11-2)	$p_c = \dfrac{\delta E(d_c^2 - d_i^2)(d_o^2 - d_c^2)}{2d_c^3(d_o^2 - d_i^2)}$	(11-7)
The tangential stress at any radius r of outer cylinder (Fig. 11-2a)	$\sigma_{\theta-o} = \dfrac{p_c d_c^2}{d_o^2 - d_c^2}\left(1 + \dfrac{d_o^2}{4r^2}\right)$	(11-8)
The tangential stress at any radius r of inner cylinder (Fig. 11-2a)	$\sigma_{\theta-i} = -\dfrac{p_c d_c^2}{d_o^2 - d_c^2}\left(1 + \dfrac{d_i^2}{4r^2}\right)$	(11-9)
The radial stress at any radius r of outer cylinder (Fig. 11-2a)	$\sigma_{r-o} = -\dfrac{p_c d_c^2}{d_o^2 - d_c^2}\left(\dfrac{d_o^2}{4r^2} - 1\right)$	(11-10)
The radial stress at any radius r of inner cylinder (Fig. 11-2a)	$\sigma_{r-i} = \dfrac{p_c d_c^2}{d_c^2 - d_i^2}\left(1 - \dfrac{d_i^2}{4r^2}\right)$	(11-11)
The tangential stress at outside diameter of outer cylinder (Fig. 11-2)	$\sigma_{\theta-oo} = \dfrac{2p_c d_c^2}{d_o^2 - d_c^2}$	(11-12)
The tangential stress at inside diameter of outer cylinder (Fig. 11-2)	$\sigma_{\theta-oi} = p_c\left(\dfrac{d_o^2 + d_c^2}{d_o^2 - d_c^2}\right)$	(11-13)
The tangential stress at outside diameter of inner cylinder (Fig. 11-2)	$\sigma_{\theta-io} = -\dfrac{p_c(d_c^2 + d_i^2)}{d_c^2 - d_i^2}$	(11-14)
The tangential stress at inside diameter of inner cylinder (Fig. 11-2)	$\sigma_{\theta-ii} = -\dfrac{2p_c d_c^2}{d_c^2 - d_i^2}$	(11-15)
The radial stress at outside diameter of outer cylinder (Fig. 11-2)	$\sigma_{r-oo} = 0$	(11-16)

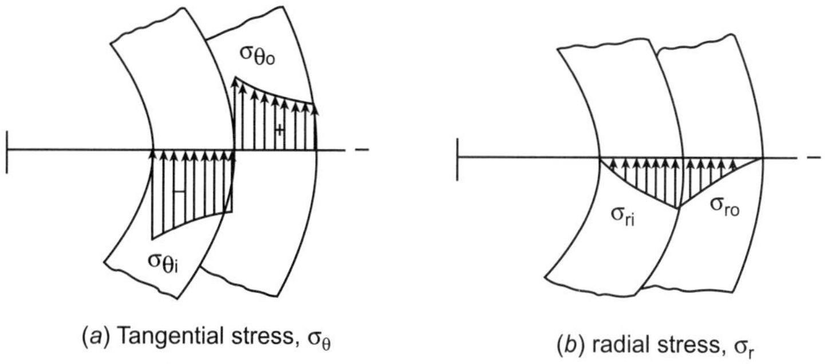

FIGURE 11-2 Distribution of stresses in shrink-fitted assembly.

Particular	Formula
The radial stress at inside diameter of outer cylinder (Fig. 11-2)	$\sigma_{r-oi} = -p_c$ (11-17)
The radial stress at outside diameter of inner cylinder (Fig. 11-2)	$\sigma_{r-io} = -p_c$ (11-18)
The radial stress at inside diameter of inner cylinder (Fig. 11-2)	$\sigma_{r-ii} = 0$ (11-19)
The semiempirical formula for tangential stress for cast-iron hub on steel shaft	$\sigma_\theta = \dfrac{E_o\delta}{d_c + 0.14d_o}$ (11-20)
Timoshenko equation for contact pressure in case of steel shaft on cast-iron hub	$p_c = \dfrac{E_c\delta}{d_c}\left(\dfrac{1-(d_c/d_o)^2}{1.53+0.47(d_c/d_o)^2}\right)$ for $\dfrac{E_s}{E_c} = 3$ (11-21a)
The allowable stress for brittle materials	$\sigma_{all} = \dfrac{\sigma_{su}}{n} = \dfrac{E_c\delta[1+(d_c/d_o)^2]}{d_c[1.53+0.47(d_c/d_o)^2]}$ (11-21b)

INTERFERENCE FITS

Press

Particular	Formula
The axial force necessary to press shaft into hub under an interface pressure p_c	$F_a = \pi d_c l \mu p_c$ (11-22a) where $\mu = 0.085$ to 0.125 for unlubricated surface $= 0.05$ with special lubricants
The approximate value of axial force to press steel shaft into cast-iron hub with an interference	$F = 4137 \times 10^4 \dfrac{(d_o + 0.3d_c)l\delta}{d_o + 6.33d_c}$ **SI** (11-23a) where d_o, d_c, l and δ in m, and F in N $F = 6000 \dfrac{(d_o + 0.3d_c)l\delta}{d_o + 6.33d_c}$ **USCS** (11-23b) where d_o, d_c, l and δ in in, and F in tonf
The approximate value of axial force to press steel shaft in steel hub	$F = 28.41 \times 10^4 \dfrac{(d_o^2 - d_c^2)l\delta}{d_o^2}$ **SI** (11-24a) where d_o, d_c, l and δ in m, and F in N $F = 4120 \dfrac{(d_o^2 - d_c^2)l\delta}{d_o^2}$ **USCS** (11-24b) where d_o, d_c, l and δ in in, and F in tonf

Particular	Formula
The transmitted torque by a press fit or shrink fit without slipping between the hub and shaft	$M_t = \dfrac{\pi d_c^2 l \mu p_c}{2}$ (11-25) where $\mu = 0.10$ for press fit $= 0.125$ for shrink fits
The temperature t_2 in °C to which the shaft or shrink link must be heated before assembly	$t_2 \geq \left(\dfrac{2\delta}{\alpha d_c} + t_1\right)$ (11-26) where t_1 = temperature of hub or larger part to which shaft or shrink link to be shrunk on, °C

Shrink links or anchors (Fig. 11-3)

The average compression in the part of rim affected according to C. D. Albert	$\sigma_c = \dfrac{F}{\sqrt{A_b A_r}}$ (11-27)

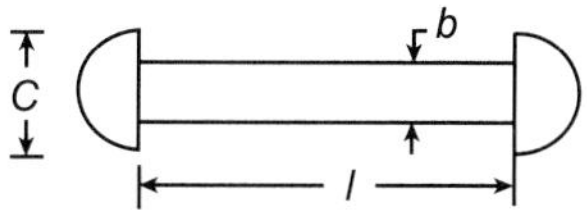

FIGURE 11-3 Shrink link.

The tensile stress in link	$\sigma_t = \dfrac{L_f - L_o}{L_o} E$ (11-28)
The total load on link	$F = \dfrac{(L_f - L_o)EA}{L_o}$ (11-29)
The compressive stress in rim	$\sigma_c = \dfrac{L_f - L_o}{L_o} \dfrac{EA}{\sqrt{A_b A_r}}$ (11-30)
The original length of link	$L_o = \dfrac{L}{1 + \left(1 + \dfrac{AE}{E_r\sqrt{A_b A_r}}\right)\dfrac{\sigma_r}{E}}$ (11-31)
The necessary linear interference δ for shrink anchors	$\delta = \dfrac{\sigma_d l}{E}$ (11-32)
The force exerted by an anchor	$F = ab\sigma_d$ (11-33) $\dfrac{b}{a} = 2$ to 3 σ_d = design stress based on a reliability factor of 1.25

Particular	Formula
For letter symbols for tolerances, basic size deviation and tolerance, clearance fit, transition fit, interference fit	Refer to Figs. 11-4 to 11-8
For press-fit between steel hub and shaft, cast-iron hub and shaft and tensile stress in cast-iron hub in press-fit allowance	Refer to Figs. 11-9 to 11-11

TOLERANCES AND ALLOWANCES

The tolerance size is defined by its value followed by a symbol composed of a letter (in some cases by two letters) and a numerical value as	$45\,g7$
A fit is indicated by the basic size common to both components followed by symbols corresponding to each component, the hole being quoted first, as	$\frac{45H8}{g7}$ or $45H8-g7$ or $45\,\frac{H8}{g7}$
For grades 5 to 16 tolerances have been determined in terms of standard tolerance unit i in micrometers (Refer to Table 11-l).	$i = 0.45D^{1/3} + 0.001D$ (11-34) where D is expressed in mm
Values of standard tolerances corresponding to grades 01, 0, and 1 are (values in μm for D in mm)	IT 01 $0.3 + 0.008\,D$ IT 0 $0.5 + 0.012\,D$ IT 1 $0.8 + 0.020\,D$ (11-35)

TABLE 11-1
Relative magnitudes of standard tolerances for grades 5 to 16 in terms of standard tolerance unit "i" [Eq. (11-34)]

Grade	IT 5	IT 6	IT 7	IT 8	IT 9	IT 10	IT 11	IT 12	IT 13	IT 14	IT 15	IT 16
Values	7 i	10 i	16 i	25 i	40 i	64 i	100 i	160 i	250 i	400 i	640 i	1000 i

Source: IS 919, 1963.

TABLE 11-1A
Coefficient of friction, μ (for use between conical metallic surfaces)

Contacting surface	Nature of surfaces	Coefficient of friction, μ
Any metal in contact with another metal	Lubricated with oil	0.15
Any metal in contact with another metal	Greased	0.15
Cast iron on steel	Shrink-fitted	0.33
Steel on steel	Shrink-fitted	0.13
Steel on steel	Dry	0.22
Cast iron on steel	Dry	0.16

Source: Courtesy J. Bach, "Kegelreibungsverbindungen," *Zeitschrift Verein Deutscher Ingenieure*, Vol. 79, 1935.

TABLE 11-2
Formulas for fundamental shaft deviations (for sizes $\leq$500 mm)

Upper deviations (*es*)		Lower deviation (*ei*)	
Shaft designation	**In μm (for *D* in mm)**	**Shaft designation**	**In μm (for *D* in mm)**
	$= -(265 + 1.3D)$	$j5$–$j8$	No formula
a	for $D \leq 120$	$k4$–$k7$	$= +0.6\sqrt[3]{D}$
	$= -3.5D$	k for grades ≤ 3 and ≥ 8	$= 0$
	for $D < 120$		
	$\simeq -(140 + 0.85D)$	m	$= +(\text{IT } 7 - \text{IT } 6)$
	for $D \leq 160$	n	$= +5D^{0.34}$
	$\simeq -1.8D$	p	$= \text{IT } 7 + 0$ to 5
	for $D > 160$		
	$= -52D^{0.2}$	r	= geometric mean of values
c	for $D \leq 40$		*ei* for p and s
			$= +\text{IT } 8 + 1$ to 4
	$= -(95 + 0.8D)$		for $D \leq 50$
	for $D > 40$	s	$= +\text{IT } 7 + 0.4D$
d	$= -16D^{0.44}$		for $D > 50$
e	$= -11D^{0.41}$	t	$= \text{IT } 7 + 0.63D$
f	$= -5.5D^{0.41}$	u	$= +\text{IT } 7 + D$
g	$= -2.5D^{0.34}$	v	$= +\text{IT } 7 + 1.25D$
		x	$= +\text{IT } 7 + 1.6D$
		y	$= +\text{IT } 7 + 2D$
h	$= 0$	z	$= +\text{IT } 7 + 2.5D$
		za	$= +\text{IT } 8 + 3.15D$
		zb	$= +\text{IT } 9 + 4D$
		zc	$= +\text{IT } 10 + 5D$

For js: The two deviations are equal to $\pm\frac{\text{IT}}{2}$

Source: IS 919, 1963.

TABLE 11-3
Rules for rounding off values obtained by the use of formulas

Values in μm	Above / Up to	5 / 45	45 / 60	60 / 100	100 / 200	200 / 300	300 / 560	560 / 600	600 / 800	800 / 1000	1000 / 2000	2000
Rounded in multiples of	For standard tolerances for Grades II and finer	1	1	1	5	10	10					
	For deviations *es*, from *a* to *g*	1	2	5	5	10	10	20	20	20	50	
	For deviations *ei*, from *k* to *zc*	1	1	1		2	5	5	10	20	50	1000

Source: IS 919, 1963.

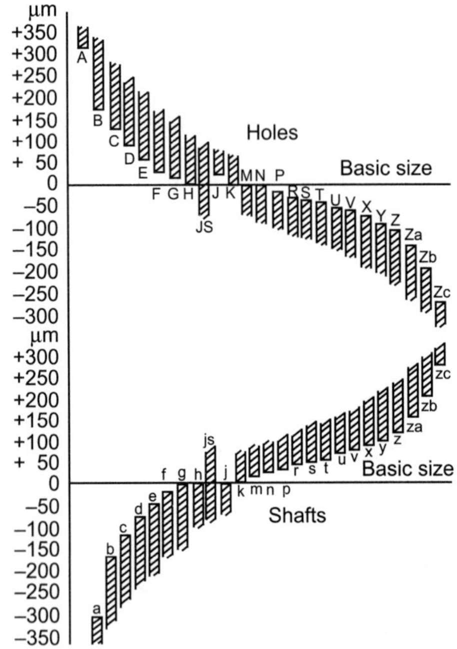

FIGURE 11-4 Letter symbols for tolerances.

TABLE 11-4
Fundamental tolerances of grades 01, 0, and 1 to 16

Diameter steps in mm	Values of tolerances in μm (1 μm = 0.001 mm)																	
	Tolerance grades																	
	01	0	1	2	3	4	5	6	7	8	9	10	11	12	13	14[a]	15[a]	16[a]
≤3	0.3	0.5	0.8	1.2	2	3	4	6	10	14	25	40	60	100	140	250	400	600
>3 ≤6	0.4	0.6	1	1.5	2.5	4	5	8	12	18	30	48	75	120	180	300	480	750
>6 ≤10	0.4	0.6	1	1.5	2.5	4	6	9	15	22	36	58	90	150	220	360	580	900
>10 ≤18	0.5	0.8	1.2	2	3	5	8	11	18	27	43	70	110	180	270	430	700	1100
>18 ≤30	0.6	1	1.5	2.5	4	6	9	13	21	33	52	84	130	210	330	520	840	1300
>30 ≤50	0.6	1	1.5	2.5	4	7	11	16	25	39	62	100	160	250	390	620	1000	1600
>50 ≤80	0.8	1.2	2	3	5	8	13	19	30	46	74	120	190	300	460	740	1200	1900
>80 ≤120	1	1.5	2.5	4	6	10	15	22	35	54	87	140	220	350	540	870	1400	2200
>120 ≤180	1.2	2	3.5	5	8	12	18	25	40	63	100	160	250	400	630	1000	1600	2500
>180 ≤250	2	3	4.5	7	10	14	20	29	46	72	115	185	290	460	720	1150	1850	2900
>250 ≤315	2.5	4	6	8	12	16	23	32	52	81	130	210	320	520	810	1300	2100	3200
>315 ≤400	3	5	7	9	13	18	25	36	57	89	140	230	360	570	890	1400	2300	3600
>400 ≤500	4	6	8	10	15	20	27	40	63	97	155	250	400	630	970	1550	2500	4000

[a] Up to 1 mm grades 14 to 16 are not provided.
Source: IS 919, 1963.

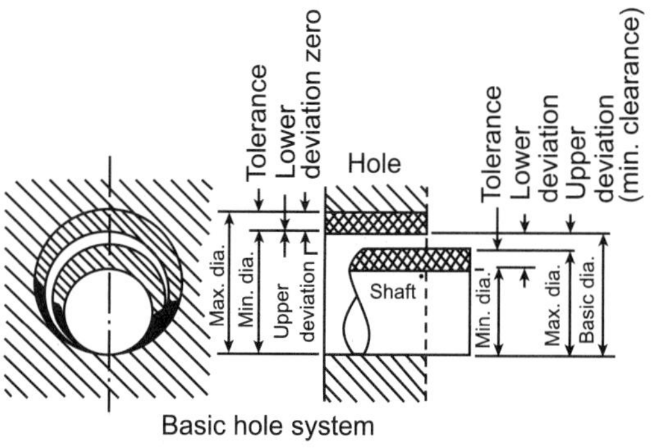

FIGURE 11-5 Basic size deviation and tolerances.

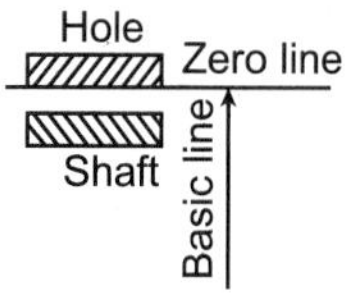

FIGURE 11-6 Clearance fit.

TABLE 11-5
Clearance fits (Fig. 11-6) (hole basis)

Quality of fit	Combination of shaft and hole	Remarks and uses
Large clearance	*H* 11 *a* 9, *H* 11 *b* 9 } coarse	Not widely used
	H 11 *a* 11 normal	
	H 9 *a* 9, *H* 8 *b* 8 } fine	
Slack running	*H* 11 *c* 9 coarse	Not widely used
	H 11 *c* 11, *H* 9 *c* 9 } normal	
	H 8 *c* 8, *H* 7 *c* 8 } fine	
Loose running	*H* 11 *d* 11, *H* 9 *d* 9 } coarse	Suitable for plummer block bearings and loose pulleys
	H 8 *d* 9 normal	
	H 8 *d* 8, *H* 7 *d* 8 } fine	
Easy running	*H* 8 *e* 9, *H* 9 *e* 9 } coarse	Recommended for general clearance fits, used for properly lubricated bearings requiring appreciable clearance; finer grades for high speeds, heavily loaded bearings such as turbogenerator and large electric motor bearings
	H 8 *e* 8, *H* 7 *e* 8 } normal	
	H 7 *e* 7, *H* 6 *e* 7 } fine	
Normal running	*H* 8 *f* 8 coarse *H* 7 *f* 7 normal *H* 6 *f* 6 fine	Widely used as a normal grease lubricated or oil-lubricated bearing having low temperature differences, gearbox shaft bearings, bearings of small electric motor and pumps, etc.
Close running or sliding	*H* 8 *g* 7 coarse *H* 7 *g* 6 normal	Expensive to manufacture, small clearance. Used in bearings for accurate link work, and for piston and slide valves; also used for spigot or location fits
	H 6 *g* 6, *H* 6 *g* 5 } fine	
Precision sliding	*H* 11 *h* 11 *H* 8 *h* 7 *H* 8 *h* 8 *H* 7 *h* 6 *H* 6 *h* 5	Widely used for nonrunning parts; also used for fine spigot and location fit

TABLE 11-6
Values of standard tolerances for sizes >500 to 3150 mm

IT 6	IT 7	IT 8	IT 9	IT 10	IT 11	IT 12	IT 13	IT 14	IT 15	IT 16
10 I*[a]	16 I	25 I	40 I	64 I	100 I	160 I	250 I	400 I	640 I	1000 I

[a] * Standard Tolerance Unit I (in μm) $-0.004D + 2.1$ for D in mm. *Source*: IS: 2101-1962.

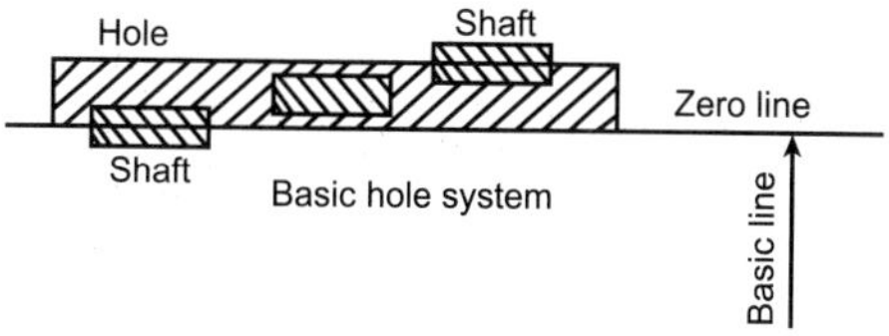

FIGURE 11-7 Transition fit.

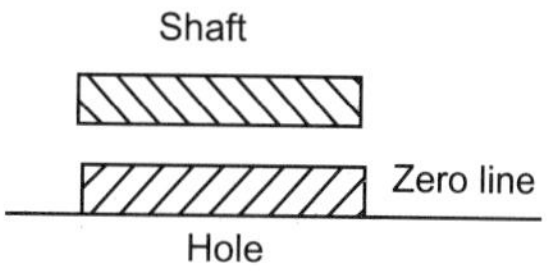

FIGURE 11-8 Interference fit.

TABLE 11-7
Transition and interference fits (hole basis)

Quality of fit	Combination of shaft and hole	Remarks and uses
		Transition fit (Fig. 11-7)
Push	*H* 8 *j* 7 coarse *H* 7 *j* 6 normal *H* 6 *j* 5 fine	Slight clearance—recommended for fits where slight interference is permissible, coupling spigots and recesses, gear rings clamped to steel hubs
True transition	*H* 8 *k* 7 coarse *H* 7 *k* 6 normal *H* 6 *k* 5 fine	Fit averaging virtually no clearance-recommended for location fits where a slight interference can be tolerated, with the object of eliminating vibration; used in clutch member keyed to shaft, gudgeon pin in piston bosses, hand wheel, and index disk on shaft
Interference transition	*H* 8 *m* 7 coarse *H* 7 *m* 6 normal *H* 6 *m* 5 fine	Fit averages a slight interference suitable for general tight-keying fits where accurate location and freedom from play are necessary; used for the cam holder, fitting bolt in reciprocating slide
True interference	*H* 8 *n* 7, *H* 7 *n* 6 } coarse *H* 6 *n* 5 fine	Suitable for tight assembly of mating surfaces
		Interference fit (Fig. 11-8)
Light press fit	*H* 7 *p* 6 normal *H* 6 *p* 5 fine	Light press fit for nonferrous parts which can be dismantled when required; standard press fit for steel, cast iron, or brass-to-steel assemblies, bush on to a gear, split journal bearing
Medium drive fit	*H* 7 *r* 6 normal *H* 6 *r* 5 fine	Medium drive fit with easy dismantling for ferrous parts and light drive fit with easy dismantling for nonferrous parts assembly; pump impeller on shaft, small-end bush in connecting rod, pressed in bearing bush, sleeves, seating, etc.
Heavy drive fit	*H* 8 *s* 7, *H* 7 *s* 6 } normal *H* 6 *s* 5 fine	Used for permanent or semipermanent assemblies of steel and cast-iron members with considerable gripping force; for light alloys this gives a press fit; used in collars pressed on to shafts, valve seatings, cylinder liner in block, etc.
Force fit	*H* 8 *t* 7, *H* 7 *t* 6 } normal *H* 6 *t* 5 fine	Suitable for the permanent assembly of steel and cast-iron parts; used in valve seat insert in cylinder head, etc.
Heavy force fit or shrink fit	*H* 8 *u* 7, *H* 7 *u* 6 } normal *H* 6 *u* 5 fine	High interference fit; the method of assembly will be by power press

TABLE 11-8
Preferred basic and design sizes
Linear dimensions (in mm)

Shaft basis		Hole basis							
A	**B**	**Priority 1**			**Priority 2**			**Priority 3**	
1.6	5.0	1.0	22.0	110.0	1.2	34.0	170.0	145.0	440.0
2.5	8.0	1.6	25.0	125.0	2.0	38.0	190.0	155.0	460.0
4.0	12.0	2.5	28.0	140.0	3.2	42.0	210.0	165.0	490.0
6.0	14.0	4.0	32.0	160.0	4.5	48.8	230.0	175.0	
10.0	18.0	5.0	36.0	180.0	5.5	53.0	240.0	185.0	
16.0	20.0	6.0	40.0	200.0	7.0	58.0	260.0	195.0	
25.0	22.0	8.0	45.0	220.0	9.0	65.0	270.0	290.0	
40.0	32.0	10.0	50.0	250.0	11.0	75.0	300.0	310.0	
63.0	50.0	12.0	56.0	280.0	13.0	85.0	340.0	330.0	
100.0	80.0	14.0	63.0	320.0	15.0	95.0	380.0	350.0	
		16.0	71.0	360.0	17.0	105.0	420.0	370.0	
		18.0	80.0	400.0	19.0	115.0	430.0	390.0	
		20.0	90.0	450.0	21.0	120.0	470.0	410.0	
			100.0	500.0	23.0	130.0	480.0		
					26.0	135.0			
					30.0	150.0			

Angular dimensions (in deg)

Priority	Preferred angles															
1	1		3		6		10		16		30		45	60	90	120
2		2		4		5		8		12		20				

TABLE 11-9
Formulas for shaft and hole deviations (for sizes >500 to 3150 mm)

Shafts			Formulas for deviations in μm (for D in mm)	Holes		
d	*es*	—	$16\ D^{0.44}$	+	*EI*	*D*
e	*es*	—	$11\ D^{0.41}$	+	*EI*	*E*
f	*es*	—	$5.5\ D^{0.41}$	+	*EI*	*F*
(*g*)	*es*	—	$2.5\ D^{0.34}$	+	*EI*	(*G*)
h	*es*	—	0		*EI*	*H*
js	*ei*	—	$0.5\ \mathrm{IT}_n$	+	*ES*	*JS*
k	*ei*		0		*ES*	*K*
m	*ei*	+	$0.024\ D + 12.6$	—	*ES*	*M*
n	*ei*	+	$0.04\ D + 21$	—	*ES*	*N*
p	*ei*	+	$0.072\ D + 37.8$	—	*ES*	*P*[a]
r	*ei*	+	geometric mean between *p* and *s* or *P* and *S*	—	*ES*	*R*[a]
s	*ei*	+	$\mathrm{IT}\ 7 + 0.4D$	—	*ES*	*S*[a]
t	*ei*	+	$\mathrm{IT}\ 7 + 0.63D$	—	*ES*	*T*[a]
u	*ei*	+	$\mathrm{IT}\ 7 + D$	—	*ES*	*U*

[a] It is assumed that associated shafts and holes are of the same grade contrary to what has been allowed for the dimensions up to 500 mm (see IS 919, 1959).
Source: IS 2101, 1962.

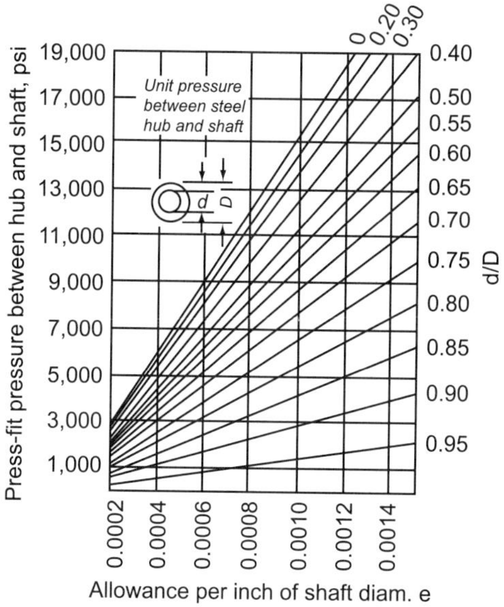

FIGURE 11-9 Press-fit pressures between steel hub and shaft (1 psi = 6894.757 Pa; 1 in = 25.4 mm). (Baumeister, T., *Marks' Standard Handbook for Mechanical Engineers*, 8th ed., McGraw-Hill, 1978.)

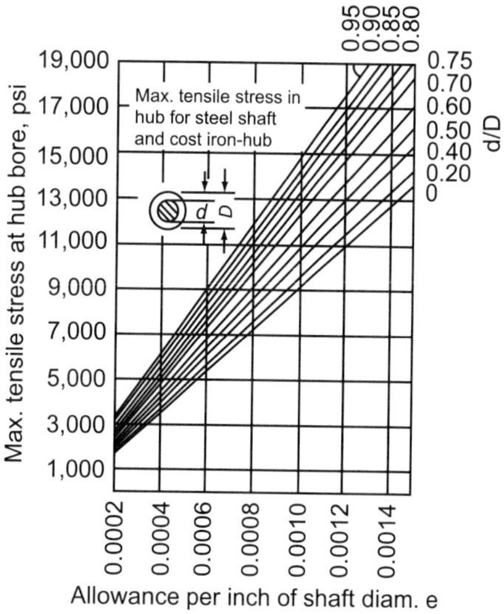

FIGURE 11-10 Variation in tensile stress in cast-iron hub in press-fit allowance (1 psi = 6894.757 Pa; 1 in = 25.4 mm). (Baumeister, T., *Marks' Standard Handbook for Mechanical Engineers*, 8th ed., McGraw-Hill, 1978.)

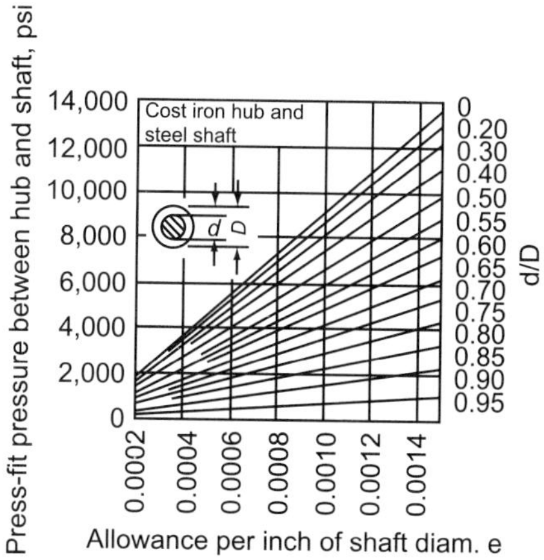

FIGURE 11-11 Press-fit pressure between cast-iron hub and shaft (1 psi = 6894.757 Pa; 1 in = 25.4 mm). (Baumeister, T., *Marks' Standard Handbook for Mechanical Engineers*, 8th ed., McGraw-Hill, 1978.)

TABLE 11-10
Tolerances[a] for shafts for sizes up to 500 mm

System of basic shaft	Limits	— 3	3 6	6 10	10 18	18 24 / 24 30	30 40	40 50	50 65	65 80	80 100	100 120	120 140	140 160	160 180	180 200	200 225	225 250	250 280	280 315	315 355	355 400	400 450	450 500
		Diameter steps, mm																						
a9	*es*[b]	−270	−270	−280	−290	−300	−310	−320	−340	−360	−380	−410	−460	−520	−580	−660	−740	−820	−920	−1050	−1200	−1350	−1500	−1650
	ei[c]	−295	−300	−316	−333	−352	−372	−382	−414	−434	−467	−497	−560	−620	−680	−775	−855	−935	−1050	−1180	−1340	−1490	−1655	−1805
b9	*es*	−140	−140	−150	−150	−160	−170	−180	−190	−200	−220	−240	−260	−280	−310	−340	−380	−420	−480	−540	−600	−680	−760	−840
	ei	−165	−170	−186	−193	−212	−232	−242	−264	−274	−307	−327	−360	−380	−410	−455	−495	−535	−610	−670	−740	−820	−915	−995
c8	*es*	−60	−70	−80	−95	−110	−120	−130	−140	−150	−170	−180	−200	−210	−230	−240	−260	−280	−300	−330	−360	−400	−440	−480
	ei	−74	−88	−102	−122	−143	−159	−169	−186	−196	−224	−234	−263	−273	−293	−312	−332	−352	−381	−411	−449	−489	−537	−577
c9	*es*	−60	−70	−80	−95	−110	−120	−130	−140	−150	−170	−180	−200	−210	−230	−240	−260	−280	−300	−330	−360	−400	−440	−480
	ei	−85	−100	−116	−138	−162	−182	−192	−214	−224	−257	−267	−300	−310	−330	−355	−375	−395	−430	−460	−500	−540	−595	−635
c11	*es*	−60	−70	−80	−95	−110	−120	−130	−140	−150	−170	−180	−200	−210	−230	−240	−260	−280	−300	−330	−360	−400	−440	−480
	ei	−120	−145	−170	−205	−240	−280	−290	−330	−340	−390	−400	−450	−460	−480	−530	−550	−570	−620	−650	−720	−760	−840	−880
d8	*es*	−20	−30	−40	−50	−65		−80		−100		−120		−145			−170			−190		−210		−230
	ei	−34	−48	−62	−77	−98		−119		−146		−174		−208			−242			−271		−299		−327
d9	*es*	−20	−30	−40	−50	−65		−80		−100		−120		−145			−170			−190		−210		−230
	ei	−45	−60	−76	−93	−117		−142		−174		−207		−245			−285			−320		−350		−385
d10	*es*	−20	−30	−40	−50	−65		−80		−100		−120		−145			−170			−190		−210		−230
	ei	−60	−78	−98	−120	−149		−180		−220		−260		−305			−355			−400		−440		−480
e6	*es*	−14	−20	−25	−32	−40		−50		−60		−72		−85			−100			−110		−125		−135
	ei	−20	−28	−34	−43	−53		−66		−79		−94		−110			−129			−142		−161		−175
e7	*es*	−14	−20	−25	−32	−40		−50		−60		−72		−85			−100			−110		−125		−135
	ei	−24	−32	−40	−50	−61		−75		−90		−107		−125			−146			−162		−182		−198
e8	*es*	−14	−20	−25	−32	−40		−50		−60		−72		−85			−100			−110		−125		−135
	ei	−28	−38	−47	−59	−73		−89		−106		−126		−148			−172			−191		−214		−232
e9	*es*	−14	−20	−25	−32	−40		−50		−60		−72		−85			−100			−110		−125		−135
	ei	−39	−50	−61	−75	−92		−112		−134		−159		−185			−215			−240		−265		−290
f6	*es*	−06	−10	−13	−16	−20		−25		−30		−36		−43			−50			−56		−62		−68
	ei	−12	−18	−22	−27	−33		−41		−49		−58		−68			−79			−88		−98		−108
f7	*es*	−06	−10	−13	−16	−20		−25		−30		−36		−43			−50			−56		−62		−68
	ei	−16	−22	−28	−34	−41		−50		−60		−71		−83			−96			−108		−119		−131
f8	*es*	−06	−10	−13	−16	−20		−25		−30		−36		−43			−50			−56		−62		−68
	ei	−20	−28	−35	−43	−53		−64		−76		−90		−106			−122			−137		−151		−165
g4	*es*	−02	−04	−05	−06	−07		−09		−10		−12		−14			−15			−17		−18		−20
	ei	−05	−08	−09	−11	−13		−16		−18		−22		−26			−29			−33		−36		−40
g5	*es*	−02	−04	−05	−06	−07		−09		−10		−12		−14			−15			−17		−18		−20
	ei	−06	−09	−11	−14	−16		−20		−23		−27		−32			−35			−40		−43		−47

TABLE 11-10
Tolerances[a] for shafts for sizes up to 500 mm (*Cont.*)

System of basic shaft	Limits	— 3	3 6	6 10	10 18	18 24 / 24 30	30 40 / 40 50	50 65 / 65 80	80 100 / 100 120	120 140 / 140 160 / 160 180	180 200 / 200 225 / 225 250	250 280 / 280 315	315 355 / 355 400	400 450 / 450 500
g6	*es*	−02	−04	−05	−06	−07	−09	−10	−12	−14	−15	−17	−18	−20
	ei	−08	−12	−14	−17	−20	−25	−29	−34	−39	−44	−49	−54	−60
h5	*es*	−00	−00	−00	−00	−00	−00	−00	−00	−00	−00	−00	−00	−00
	ei	−04	−05	−06	−08	−09	−11	−13	−15	−18	−20	−23	−25	−27
h6	*es*	−00	−00	−00	−00	−00	−00	−00	−00	−00	−00	−00	−00	−00
	ei	−06	−08	−09	−11	−13	−16	−19	−22	−25	−29	−32	−36	−40
h7	*es*	−00	−00	−00	−00	−00	−00	−00	−00	−00	−00	−00	−00	−00
	ei	−10	−12	−15	−18	−21	−25	−30	−35	−40	−46	−52	−57	−63
h8	*es*	−00	−00	−00	−00	−00	−00	−00	−00	−00	−00	−00	−00	−00
	ei	−14	−18	−22	−27	−33	−39	−46	−54	−63	−72	−81	−89	−97
h9	*es*	−00	−00	−00	−00	−00	−00	−00	−00	−00	−00	−00	−00	−00
	ei	−25	−30	−36	−43	−52	−62	−74	−87	−100	−115	−130	−140	−155
h10	*es*	−00	−00	−00	−00	−00	−00	−00	−00	−00	−00	−00	−00	−00
	ei	−40	−48	−58	−70	−84	−100	−120	−140	−160	−185	−210	−230	−250
h11	*es*[b]	−00	−00	−00	−00	−00	−00	−00	−00	−00	−00	−00	−00	−00
	ei[c]	−60	−75	−90	−110	−130	−160	−190	−220	−250	−290	−320	−360	−400
j5	*es*	+02	+03	+04	+05	+05	+06	+06	+06	+07	+07	+07	+07	+07
	ei	−02	−02	−02	−03	−04	−05	−07	−09	−11	−13	−16	−18	−20
j6	*es*	+04	+06	+07	+08	+09	+11	+12	+13	+14	+16	+16	+18	+20
	ei	−02	−02	−02	−03	−04	−05	−07	−09	−11	−13	−16	−18	−20
j7	*es*	+06	+08	+10	+12	+13	+15	+18	+20	+22	+25	+26	+29	+31
	ei	+04	−04	−05	−06	−08	−10	−12	−15	−18	−21	−26	−28	−32
k6	*es*	+06	+09	+10	+12	+15	+18	+21	+25	+28	+33	+36	+40	+45
	ei	+00	+01	+01	+01	+02	+02	+02	+03	+03	+04	+04	+04	+05
k7	*es*	+10	+13	+16	+19	+23	+27	+32	+38	+43	+50	+56	+61	+68
	ei	+01	+01	+01	+01	+02	+02	+02	+03	+03	+04	+04	+04	+05
m6	*es*	+08	+12	+15	+18	+21	+25	+30	+35	+40	+46	+52	+57	+63
	ei	+02	+04	+06	+07	+08	+09	+11	+13	+15	+17	+20	+21	+23
m7	*es*	+02	+16	+21	+25	+29	+34	+41	+48	+55	+63	+72	+78	+86
	ei	—	+04	+06	+07	+08	+09	+11	+13	+15	+17	+20	+21	+23
n4	*es*	+07	+12	+14	+17	+21	+24	+28	+33	+39	+45	+50	+55	+60
	ei	+04	+08	+10	+12	+15	+17	+20	+23	+27	+31	+34	+37	+40
n5	*es*	+08	+13	+16	+20	+24	+28	+33	+38	+45	+51	+57	+62	+67
	ei	+04	+08	+10	+12	+15	+17	+20	+23	+27	+31	+34	+37	+40

TABLE 11-10
Tolerances[a] for shafts for sizes up to 500 mm (*Cont.*)

System of basic shaft	Limits	Diameter steps, mm — 3	3 6	6 10	10 18	18 24	24 30	30 40	40 50	50 65	65 80	80 100	100 120	120 140	140 160	160 180	180 200	200 225	225 250	250 280	280 315	315 355	355 400	400 450	450 500
p5	*es*	+10	+17	+21	+26	+31		+37		+45		+52		+61			+70			+79		+87		+95	
	ei	+06	+12	+15	+18	+22		+26		+32		+37		+43			+50			+56		+62		+68	
p6	*es*	+12	+20	+24	+29	+35		+42		+51		+59		+68			+79			+88		+98		+108	
	ei	+06	+12	+15	+18	+22		+26		+32		+37		+43			+50			+56		+62		+68	
r6	*es*	+16	+23	+28	+34	+41		+50		+62		+76		+90			+109			+130		+150		+172	
	ei	+10	+15	+19	+23	+28		+34		+43		+54		+65			+80			+98		+114		+132	
t7	*es*	—	—	—	—	+62		+79		+105		+139		+174			+226			−292		+351		+423	
	ei					+41		+54		+75		+104		+134			+180			−240		+294		+360	
u5	*es*	+22	+28	+34	+41	+50	+57	+71	+81	+100	+115	+139	+159	+188	+208	+228	+256	+278	+304	+338	+373	+415	+460	+517	+567
	ei	+18	+23	+28	+33	+41	+48	+60	+70	+87	+102	+124	+144	+170	+190	+210	+236	+258	+284	+315	+350	+390	+435	+490	+540
u8	*es*	+32	+41	+50	+60	+74	+81	+99	+109	+133	+148	+178	+198	+233	+253	+273	+308	+330	+356	+396	+431	+479	+524	+587	+637
	ei	+18	+23	+28	+33	+41	+48	+60	+70	+87	+102	+124	+144	+170	+190	+210	+236	+258	+284	+315	+350	+390	+435	+490	+540
v5	*es*	—	—	—	— +47	+56	+64	+79	+92	+115	+133	+161	+187	+220	+246	+270	+304	+330	+360	+408	+448	+500	+555	+622	+687
	ei	—	—	—	— +39	+47	+55	+68	+81	+102	+120	+146	+172	+202	+228	+252	+284	+310	+340	+385	+425	+475	+530	+595	+660
x8	*es*	+34	+46	+56	+67 +72	+87	+97	+119	+136	+168	+192	+232	+264	+311	+343	+373	+422	+457	+497	+556	+606	+679	+749	+837	+917
	ei	+20	+28	+34	+40 +45	+54	+64	+80	+97	+122	+146	+176	+210	+248	+280	+310	+350	+385	+425	+475	+525	+590	+660	+740	+820
y6	*es*	—	—	—	—	+76	+88	+110	+130	+163	+193	+236	+276	+325	+365	+405	+454	+499	+549	+612	+682	+766	+856	+960	+1040
	ei	—	—	—	—	+63	+75	+94	+114	+144	+174	+214	+254	+300	+340	+380	+425	+470	+520	+580	+650	+730	+820	+920	+1000
z7	*es*	+36	+47	+57	+68 +78	+94	+109	+137	+161	+202	+240	+293	+345	+405	+455	+505	+566	+621	+686	+762	+842	+957	+1057	+1163	+1313
	ei	+26	+35	+42	+50 +60	+73	+88	+112	+136	+172	+210	+258	+310	+365	+415	+465	+520	+575	+640	+710	+790	+900	+1000	+1100	+1250
za6	*es*	+38	+50	+61	—	—	—	—	—	—	—	—	—	—	—	—	—	—	—	—	—	—	—	—	—
	ei	+32	+42	+52	—	—	—	—	—	—	—	—	—	—	—	—	—	—	—	—	—	—	—	—	—
zb7	*es*	+50	+62	+82	—	—	—	—	—	—	—	—	—	—	—	—	—	—	—	—	—	—	—	—	—
	ei	+40	+50	+67	—	—	—	—	—	—	—	—	—	—	—	—	—	—	—	—	—	—	—	—	—
zc8	*es*	+74	+98	+119	—	—	—	—	—	—	—	—	—	—	—	—	—	—	—	—	—	—	—	—	—
	ei	+60	+80	+97	—	—	—	—	—	—	—	—	—	—	—	—	—	—	—	—	—	—	—	—	—

[a] Tolerances in micrometers (1 μm = 10^{-3} mm).
[b] *es* = upper deviation.
[c] *ei* = lower deviation.

TABLE 11-11
Tolerances[a] for holes for sizes up to 500 mm

System of basic hole	Limits	Diameter steps, mm — 3	3 6	6 10	10 14	14 18	18 24	24 30	30 40	40 50	50 65	65 80	80 100	100 120	120 140	140 160	160 180	180 200	200 225	225 250	250 280	280 315	315 355	355 400	400 450	450 500
A9	*ES*[b]	+295	+300	+316	+333		+352		+372	+382	+414	+434	+467	+497	+560	+620	+680	+775	+855	+925	+1050	+1180	+1340	+1490	+1655	+1805
	EI[c]	+270	+270	+280	+290		+300		+310	+320	+340	+360	+380	+410	+460	+520	+580	+660	+740	+820	+920	+1050	+1200	+1350	+1500	+1650
B9	*ES*	+165	+170	+186	+193		+212		+232	+242	+264	+274	+307	+327	+360	+380	+410	+455	+495	+535	+610	+670	+740	+820	+915	+995
	EI	+140	+140	+150	+150		+160		+170	+180	+190	+200	+220	+240	+260	+280	+310	+340	+380	+420	+480	+540	+600	+680	+760	+840
B11	*ES*	+200	+215	+240	+260		+290		+330	+340	+380	+390	+440	+460	+510	+530	+560	+630	+670	+710	+800	+860	+960	+1040	+1160	+1240
	EI	+140	+140	+150	+150		+160		+170	+180	+190	+200	+220	+240	+260	+280	+310	+340	+380	+420	+480	+540	+600	+680	+760	+840
C8	*ES*	−74	+88	+102	+122		+143		+159	+169	+186	+196	+224	+234	+263	+273	+293	+312	+332	+352	+381	+411	+449	+489	+537	+577
	EI	+60	+70	+80	+95		+110		+120	+130	+140	+150	+170	+180	+200	+210	+230	+240	+260	+280	+300	+330	+360	+400	+440	+480
C11	*ES*	+120	+145	+170	+205		+240		+280	+290	+330	+340	+390	+400	+450	+460	+480	+530	+550	+570	+620	+650	+720	+760	+840	+880
	EI	+60	+70	+80	+95		+110		+120	+130	+140	+150	+170	+180	+200	+210	+230	+240	+260	+280	+300	+330	+360	+400	+440	+480
D8	*ES*	+34	+48	+62	+77		+98		+119		+146		+174		+208			+242			+271		+299		+327	
	EI	+20	+30	+40	+50		+65		+80		+100		+120		+145			+170			+190		+210		+230	
D9	*ES*	+45	+60	+76	+93		+117		+142		+174		+207		+245			+285			+320		+350		+385	
	EI	+20	+30	+40	+50		+65		+80		+100		+120		+145			+170			+190		+210		+230	
E5	*ES*	+18	+25	+31	+40		+49		+61		+73		+87		+103			+120			+133		+150		+162	
	EI	+14	+20	+25	+32		+40		+50		+60		+72		+85			+100			+110		+125		+135	
F6	*ES*	+12	+18	+22	+27		+33		+41		+49		+58		+68			+79			+88		+98		+108	
	EI	+6	+10	+13	+16		+20		+25		+30		+36		+43			+50			+56		+62		+68	
F8	*ES*	+20	+28	+35	+43		+53		+64		+76		+90		+106			+122			+137		+151		+165	
	EI	+6	+10	+13	+16		+20		+25		+30		+36		+43			+50			+56		+62		+68	
G7	*ES*	+12	+16	+20	+24		+28		+34		+40		+47		+54			+61			+69		+75		+83	
	EI	+2	+4	+5	+6		+7		+9		+10		+12		+14			+15			+17		+18		+20	
H5	*ES*	+4	+5	+6	+8		+9		+11		+13		+15		+18			+20			+23		+25		+27	
	EI	0	0	0	0		0		0		0		0		0			0			0		0		0	
H6	*ES*	+6	+8	+9	+11		+13		+16		+19		+22		+25			+29			+32		+36		+40	
	EI	0	0	0	0		0		0		0		0		0			0			0		0		0	
H7	*ES*	+10	+12	+15	+18		+21		+25		+30		+35		+40			+46			+52		+57		+63	
	EI	0	0	0	0		0		0		0		0		0			0			0		0		0	
H8	*ES*	+14	+18	+22	+27		+33		+39		+46		+54		+63			+72			+81		+89		+97	
	EI	0	0	0	0		0		0		0		0		0			0			0		0		0	
H9	*ES*	+25	+30	+36	+43		+52		+62		+74		+87		+100			+115			+130		+140		+155	
	EI	0	0	0	0		0		0		0		0		0			0			0		0		0	
H10	*ES*	+40	+48	+58	+70		+84		+100		+120		+140		+160			+185			+210		+230		+250	
	EI	0	0	0	0		0		0		0		0		0			0			0		0		0	
H11	*ES*	+60	+75	+90	+110		+130		+160		+190		+220		+250			+290			+320		+360		+400	
	EI	0	0	0	0		0		0		0		0		0			0			0		0		0	
J7	*ES*	+4	+6	+8	+10		+12		+14		+18		+22		+26			+30			+36		+39		+43	
	EI	−6	−6	−7	−8		−9		−11		−12		−13		−14			−16			−16		−18		−20	

TABLE 11-11
Tolerances[a] for holes for sizes up to 500 mm (*Cont.*)

System of basic hole	Limits	Diameter steps, mm — 3	3 6	6 10	10 14	14 18	18 24	24 30	30 40	40 50	50 65	65 80	80 100	100 120	120 140	140 160	160 180	180 200	200 225	225 250	250 280	280 315	315 355	355 400	400 450	450 500
K6	*ES*	0	+2	+2		+2		+2		+3		+4		+4		+4			+5			+5		+7		+8
	EI	−6	−6	−7		−9		−11		−13		−15		−18		−21			−24			−27		−29		−32
K7	*ES*	0	+3	+5		+6		+6		+7		+9		+10		+12			+13			+16		+17		+18
	EI	−10	−9	−10		−12		−15		−18		−21		−25		−28			−33			−36		−40		−45
M7	*ES*	−2	0	0		0		0		0		0		0		0			0			0		0		0
	EI	−12	−12	−15		−18		−21		−25		−30		−35		−40			−46			−52		−57		−63
N7	*ES*	−4	−4	−4		−5		−7		−8		−9		−10		−12			−14			−14		−16		−17
	EI	−14	−16	−19		−23		−28		−33		−39		−45		−52			−60			−66		−73		−80
P7	*ES*	−6	−8	−9		−11		−14		−17		−21		−24		−28			−33			−36		−41		−45
	EI	−16	−20	−24		−29		−35		−42		−51		−59		−68			−79			−88		−98		−108
S6	*ES*	−14	−16	−20		−25		−31		−38	−47	−53	−64	−72	−85	−93	−101	−113	−121	−131	−149	−161	−179	−197	−219	−239
	EI	−20	−24	−29		−36		−44		−54	−66	−72	−86	−94	−110	−118	−126	−142	−150	−160	−181	−193	−215	−233	−259	−279
S7	*ES*	−14	−15	−17		−21		−27		−34	−42	−48	−58	−66	−77	−85	−93	−105	−113	−123	−138	−150	−169	−187	−209	−229
	EI	−24	−27	−32		−39		−48		−59	−72	−78	−93	−101	−117	−125	−133	−151	−159	−169	−190	−202	−226	−244	−272	−292
T6	*ES*	—	—	—		—	—	−37	−43	−49	−60	−69	−84	−97	−115	−127	−139	−157	−171	−187	−209	−231	−257	−283	−317	−347
	EI	—	—	—		—	—	−50	−59	−65	−79	−88	−106	−119	−140	−152	−164	−186	−200	−216	−241	−263	−293	−319	−357	−387
U7	*ES*	−18	−19	−22		−26	−33	−40	−51	−61	−76	−91	−111	−131	−155	−175	−195	−219	−241	−267	−295	−330	−369	−414	−467	−517
	EI	−28	−31	−37		−44	−54	−61	−76	−86	−106	−121	−146	−166	−195	−215	−235	−265	−287	−313	−347	−382	−426	−471	−530	−580
V6	*ES*	—	—	—	—	−36	−43	−51	−63	−76	−96	−114	−139	−165	−195	−221	−245	−275	−301	−331	−376	−416	−464	−519	−582	−647
	EI	—	—	—	—	−47	−56	−64	−79	−92	−115	−133	−161	−187	−220	−246	−270	−304	−330	−360	−408	−448	−500	−555	−622	−687
X7	*ES*	−20	−24	−28	−33	−38	−46	−56	−71	−88	−111	−135	−165	−197	−233	−265	−295	−333	−368	−408	−455	−505	−569	−639	−717	−797
	EI	−30	−36	−43	−51	−56	−67	−77	−96	−113	−141	−165	−200	−232	−273	−305	−335	−379	−414	−454	−507	−557	−626	−696	−780	−860
Y7	*ES*	—	—	—	—	—	−55	−67	−85	−105	−133	−163	−201	−241	−285	−325	−365	−408	−453	−503	−560	−630	−709	−799	−897	−977
	EI	—	—	—	—	—	−76	−88	−110	−130	−163	−193	−236	−276	−325	−365	−405	−454	−499	−549	−612	−682	−766	−856	−960	−1040
Z8	*ES*	−26	−35	−42	−50	−60	−73	−88	−112	−136	−172	−210	−258	−310	−365	−415	−465	−520	−575	−640	−710	−790	−900	−1000	−1100	−1250
	EI	−40	−53	−64	−77	−87	−106	−121	−151	−175	−218	−256	−312	−364	−428	−478	−528	−592	−647	−712	−791	−871	−989	−1089	−1197	−1347
ZA7	*ES*	−32	−38	−46	—	—	—	—	—	—	—	—	—	—	—	—	—	—	—	—	—	—	—	—	—	—
	EI	−42	−50	−61	—	—	—	—	—	—	—	—	—	—	—	—	—	—	—	—	—	—	—	—	—	—
ZB8	*ES*	−40	−50	−67	—	—	—	—	—	—	—	—	—	—	—	—	—	—	—	—	—	—	—	—	—	—
	EI	−54	−68	−89	—	—	—	—	—	—	—	—	—	—	—	—	—	—	—	—	—	—	—	—	—	—
ZC9	*ES*	−60	−80	−97	—	—	—	—	—	—	—	—	—	—	—	—	—	—	—	—	—	—	—	—	—	—
	EI	−85	−110	−133	—	—	—	—	—	—	—	—	—	—	—	—	—	—	—	—	—	—	—	—	—	—

[a] Tolerances in μm; 1 μm = 10^{-3} mm
[b] *ES* = upper deviation.
[c] *EI* = lower deviation.

TABLE 11-12
Tolerances[a] for shafts for sizes 500 to 3150 mm

System of basic shaft	Limits	Diameter steps, mm 500–560	560–630	630–710	710–800	800–900	900–1000	1000–1120	1120–1250	1250–1400	1400–1600	1600–1800	1800–2000	2000–2250	2250–2500	2500–2800	2800–3150
*d*10	*es*[b]	−260		−290		−320		−350		−390		−430		−480		−520	
	ei[c]	−540		−610		−680		−770		−890		−1030		−1180		−1380	
*e*8	*es*	−145		−160		−170		−195		−220		−240		−260		−290	
	ei	−255		−285		−310		−360		−415		−470		−540		−620	
*f*9	*es*	−76		−80		−86		−98		−110		−120		−130		−145	
	ei	−251		−280		−316		−358		−420		−490		−570		−685	
*g*6	*es*	−22		−24		−26		−28		−30		−32		−34		−38	
	ei	−66		−74		−82		−94		−108		−124		−140		−173	
*g*7	*es*	−22		−24		−26		−28		−30		−32		−34		−38	
	ei	−92		−103		−115		−133		−155		−182		−209		−248	
*h*6	*es*	0		0		0		0		0		0		0		0	
	ei	−44		−50		−56		−66		−78		−92		−110		−135	
*h*7	*es*	0		0		0		0		0		0		0		0	
	ei	−70		−80		−90		−105		−125		−150		−175		−210	
*h*8	*es*	0		0		0		0		0		0		0		0	
	ei	−110		−125		−140		−165		−195		−230		−280		−330	
*h*9	*es*	0		0		0		0		0		0		0		0	
	ei	−175		−200		−230		−260		−310		−370		−440		−540	
*h*10	*es*	0		0		0		0		0		0		0		0	
	ei	−280		−320		−360		−420		−500		−600		−700		−860	
*h*11	*es*	0		0		0		0		0		0		0		0	
	ei	−440		−500		−560		−660		−780		−920		−1100		−1350	
*js*9	*es* *ei*	±87.5		±100		±115		±130		±155		±185		±220		±270	
*k*6	*es*	+44		+50		+56		+66		+78		+92		+110		+135	
	ei	0		0		0		0		0		0		0		0	
*m*6	*es*	+70		+80		+90		+106		+126		+150		+178		+211	
	ei	+26		+30		+34		+40		+48		+58		+68		+76	
*n*6	*es*	+88		+100		+112		+132		+156		+184		+220		+270	
	ei	+44		+50		+56		+66		+78		+92		+110		+135	
*p*6	*es*	+122		+139		+156		+186		+218		+262		+305		+375	
	ei	+78		+88		+100		+120		+140		+170		+195		+240	
*r*7	*es*	+220	+225	+255	+265	+300	+310	+355	+365	+425	+455	+520	+550	+615	+635	+760	+790
	ei	+150	+155	+175	+185	+210	+220	+250	+260	+300	+330	+370	+400	+440	+460	+550	+580
*s*7	*es*	+350	+380	+420	+460	+520	+560	+625	+685	+765	+845	+970	+1070	+1175	+1275	+1460	+1610
	ei	+280	+310	+340	+380	+430	+470	+520	+580	+640	+720	+820	+920	+1000	+1100	+1250	+1400
*t*7	*es*	+470	+520	+580	+640	+710	+770	+885	+945	+1085	+1175	+1350	+1500	+1675	+1825	+2110	+2310
	ei	+400	+450	+500	+560	+620	+680	+780	+840	+960	+1050	+1200	+1350	+1500	+1650	+1900	+2100
*u*7	*es*	+570	+730	+820	+920	+1031	+1140	+1255	+1405	+1575	+1725	+2000	+2150	+2475	+2675	+3110	+3410
	ei	+600	+660	+740	+840	+940	+1050	+1150	+1300	+1450	+1600	+1850	+2000	+2300	+2500	+2900	+3200

[a] Tolerances in μm (1 μm = 10^{-3} mm).
[b] *es* = upper deviation.
[c] *ei* = lower deviation.
Source: IS 2101, 1962.

TABLE 11-13
Tolerances[a] for holes for sizes 500 to 3150 mm

System of basic hole	Limits	Diameter steps, mm															
		500–560	560–630	630–710	710–800	800–900	900–1000	1000–1120	1120–1250	1250–1400	1400–1600	1600–1800	1800–2000	2000–2240	2240–2500	2500–2800	2800–3150
*D*10	*ES*[a]	+540		+610		+680		+770		+890		+1030		+1180		+1380	
	ES[b]	+260		+290		+320		+350		+390		+430		+480		+520	
*E*8	*ES*	+255		+285		+310		+360		+415		+470		+540		+620	
	EI	+145		+160		+170		+195		+220		+240		+260		+290	
*F*9	*ES*	+251		+280		+316		+358		+420		+490		+570		+685	
	EI	+76		+80		+86		+98		+110		+120		+130		+145	
*G*6	*ES*	+66		+74		+82		+94		+108		+124		+144		+173	
	EI	+22		+24		+26		+28		+30		+32		+34		+38	
*G*7	*ES*	+92		+103		+115		+133		+155		+182		+209		+248	
	EI	+22		+24		+26		+28		+30		+32		+34		+38	
*H*6	*ES*	+40		+50		+56		+66		+78		+92		+110		+135	
	EI	0		0		0		0		0		0		0		0	
*H*7	*ES*	+70		+80		+90		+105		+125		+150		+175		+210	
	EI	0		0		0		0		0		0		0		0	
*H*8	*ES*	+110		+125		+140		+165		+195		+230		+280		+330	
	EI	0		0		0		0		0		0		0		0	
*H*9	*ES*	+175		+200		+230		+260		+310		+370		+440		+540	
	EI	0		0		0		0		0		0		0		0	
*H*10	*ES*	+280		+320		+360		+420		+500		+600		+700		+860	
	EI	0		0		0		0		0		0		0		0	
*H*11	*ES*	+440		+500		+560		+660		+780		+920		+1100		+1350	
	EI	0		0		0		0		0		0		0		0	
*JS*9	*ES* *EI*	±87.5		±100		±115		±130		±155		±185		±220		±270	
*K*6	*ES*	0		0		0		0		0		0		0		0	
	EI	−44		−50		−56		−66		−78		−92		−110		−135	
*M*6	*ES*	−26		−30		−34		−40		−48		−58		−68		−76	
	EI	−70		−80		−90		−106		−126		−150		−178		−211	
*N*6	*ES*	−44		−50		−56		−66		−78		−92		−110		−135	
	EI	−88		−100		−112		−132		−156		−184		−220		−270	
*P*6	*ES*	−78		−88		−100		−120		−140		−170		−195		−240	
	EI	−122		−138		−156		−186		−218		−262		−305		−375	
*R*7	*ES*	−150	−155	−175	−185	−210	−200	−250	−260	−300	−330	−370	−400	−440	−460	−550	−580
	EI	−220	−225	−255	−265	−300	−310	−355	−365	−425	−455	−520	−550	−615	−635	−760	−790
*S*7	*ES*	−280	−310	−340	−380	−430	−470	−520	−580	−640	−720	−820	−920	−1000	−1100	−1250	−1400
	EI	−350	−380	−420	−460	−520	−560	−625	−685	−765	−845	−970	−1070	−1175	−1275	−1460	−1610
*T*7	*ES*	−400	−450	−500	−560	−620	−680	−780	−840	−960	−1050	−1200	−1350	−1500	−1650	−1900	−2100
	EI	−470	−520	−580	−640	−710	−770	−885	−945	−1085	−1175	−1350	−1500	−1675	−1825	−2110	−2310
*U*7	*ES*	−600	−660	−740	−840	−940	−1050	−1150	−1300	−1450	−1600	−1850	−2000	−2300	−2500	−2900	−3200
	EI	−670	−730	−820	−920	−1030	−1140	−1255	−1405	−1575	−1725	−2000	−2150	−2475	−2675	−3110	−3410

[a] Tolerances in ×μm (1 μm = 10^{-3} mm).
[b] *ES* = upper deviation.
[c] *EI* = lower deviation.
Source: IS 2101, 1962.

TABLE 11-14
Mean fit and variation about the mean fit for holes for sizes up to 400 mm

Quality of fit	Combination of shaft and hole	Diameter steps, mm —	3	6	10	18	24	30	40	50	65	80	100	120	140	160	180	200	225	250	280	315	355
		3	6	10	18	24	30	40	50	65	80	100	120	140	160	180	200	225	250	280	315	355	400
Clearance Fit (Fig. 11-6)																							
Precision sliding	*H7 g6* Normal	+11 ±8	+14 ±10	+17 ±12	+20.5 ±14.5	+24 ±17		+29.5 ±20.5		+34.5 ±24.5		+40.5 ±28.5		+46.5 ±32.5			+52.5 ±37.5			+59 ±42		+64.5 ±46.5	
Normal running	*H7 f7* Normal	+16 ±9	+22 ±12	+28 ±15	+34 ±18	+41 ±21		+50 ±25		+60 ±30		+71 ±35		+83 ±40			+96 ±46			+108 ±52		+119 ±57	
Easy running	*H8 e8* Normal	+28 ±14	+38 ±18	+47 ±22	+59 ±27	+73 ±33		+89 ±39		+106 ±46		+126 ±54		+148 ±63			+172 ±72			+191 ±81		+214 ±89	
Loose running	*H8 d9* Normal	+39.5 ±19.5	+54 ±24	+69 ±29	+85 ±35	+107.5 ±42.5		+130.5 ±50.5		+160 ±60		+190.5 ±70.5		+226.5 ±81.5			+263.5 ±93.5			+295.5 ±105.5		+324.5 ±114.5	
Slack running	*H9 c9* Normal	+85 ±25	+100 ±30	+116 ±36	+138 ±43	+162 ±52		+182 ±62	+192 ±62	+214 ±74	+224 ±74	+257 ±74	+267 ±87	+300 ±100	+310 ±100	+330 ±100	+355 ±115	+375 ±115	+395 ±115	+420 ±130	+460 ±130	+500 ±140	+540 ±140
Location and Assembly Fit																							
Position fits	*H8 a9*	+289.5 ±19.5	+294 ±24	+309 ±29	+325 ±35	+342.5 ±42.5		+360.5 ±50.5	+370.5 ±50.5	+400 ±60	+420 ±60	+450.5 ±70.5	+480.5 ±70.5	+541.5 ±81.5	+601.5 ±81.5	+661.5 ±81.5	+753.5 ±93.5	+833.5 ±93.5	+913.5 ±93.5	+1025.5 ±105.5	+1155.5 ±105.5	+1314.5 ±114.5	+1454.5 ±114.5
Position fits	*H8 b9*	+159.5 ±19.5	+164 ±24	+179 ±29	+185 ±35	+202.5 ±42.5		+220.5 ±50.5	+230.5 ±50.5	+250 ±60	+260 ±60	+290.5 ±70.5	+310.5 ±70.5	+341.5 ±81.5	+361.5 ±81.5	+391.5 ±81.5	+433.5 ±93.5	+473.5 ±93.5	+513.5 ±93.5	+585.5 ±105.5	+645.5 ±105.5	+714.5 ±114.5	+794.5 ±114.5
Precision location	*H6 h6*	+7 ±7	+8 ±8	+9 ±9	+11 ±11	+13 ±13		+16 ±16		+19 ±19		+22 ±22		+25 ±25			+29 ±29			+32 ±32		+36 ±36	
Normal location	*H8 h8*	+14 ±14	+18 ±18	+22 ±22	+27 ±27	+33 ±33		+39 ±39		+46 ±46		+54 ±54		+63 ±63			+72 ±72			+81 ±81		+89 ±89	
Loose location	*H9 h9*	+25 ±25	+30 ±30	+36 ±36	+43 ±43	+52 ±52		+62 ±62		+74 ±74		+87 ±87		+100 ±100			+115 ±115			+130 ±130		+140 ±140	
Slack assembly	*H11 h11*	+60 ±60	+75 ±75	+90 ±90	+110 ±110	+130 ±130		+160 ±160		+190 ±190		+220 ±220		+250 ±250			+290 ±290			+320 ±320		+360 ±360	
Transition Fits (Fig. 11-7)																							
Push	*H7 k6* Normal	+2 ±8	+3 ±10	+5 ±12	+6.5 ±14.5	+8 ±17		+9.5 ±20.5		+12.5 ±24.5		+15.5 ±28.5		+18.5 ±32.5			+21..5 ±37.5			+26 ±42		+28.5 ±46.5	
True transition	*H7 h6* Normal	—	—	+2 ±12	+2.5 ±14.5	+2 ±17		+2.5 ±20.5		+3.5 ±24.5		+3.5 ±28.5		+4.5 ±32.5			+4.5 ±37.5			+6 ±42		+6.5 ±46.5	
Interference transition	*H7 m6* Normal	−1 ±8	−2 ±10	−3 ±12	−3.5 ±14.5	−4 ±17		−4.5 ±20.5		−5.5 ±24.5		−6.5 ±28.5		−7.5 ±32.5			−8.5 ±37.5			−10 ±42		−10.5 ±46.5	
Interference Fits (Fig. 11-8)																							
Light press fit	*H7 p6* Normal	−8 ±8	−10 ±10	−12 ±12	−14.5 ±14.5	−18 ±17		−21.5 ±20.5		−26.5 ±24.5		−30.5 ±28.5		−35.5 ±32.5			−41.5 ±37.5			−46 ±42		−51.5 ±46.5	
Medium drive fit	*H7 r6* Normal	−11 ±8	−13 ±10	−16 ±12	−19.5 ±14.5	−24 ±17		−29.5 ±20.5		−35.5 ±24.5	−37.5 ±24.5	−44.5 ±28.5	−47.5 ±28.5	−55.5 ±32.5	−57.5 ±32.5	−60.5 ±32.5	−66.5 ±37.5	−71.5 ±37.5	−75.5 ±37.5	−84 ±42	−88 ±42	−97.5 ±46.5	−103.5 ±46.5
MHeavy drive fit	*H7 s6* Normal	−12 ±8	−17 ±10	−20 ±12	−26.5 ±14.5	−31 ±17		−38.5 ±20.5		−47.5 ±24.5	−53.5 ±24.5	−64.5 ±28.5	−72.5 ±28.5	−84.5 ±32.5	−92.5 ±32.5	−100.5 ±32.5	−113.5 ±37.5	−121.5 ±37.5	−131.5 ±37.5	−148 ±42	−160 ±42	−179.5 ±46.5	−197.5 ±46.5
Heavy force Fir or shrink fit	*H7 u6* Normal	−17 ±8	−21 ±10	−25 ±12	−29.5 ±14.5	−37 ±17	−44 ±17	−55.5 ±20.5	−65.5 ±20.5	−81.5 ±24.5	−96.5 ±24.5	−117.5 ±28.5	−137.5 ±28.5	−162.5 ±32.5	−182.5 ±32.5	−202.5 ±32.5	−227.5 ±37.5	−249.5 ±37.5	−275.5 ±37.5	−305 ±42	−340 ±42	−379.5 ±46.5	−424.5 ±46.5

Tolerance in microns; 1 micron = 10^{-3} mm = μm = 10^{-6} m

TABLE 11-15
International tolerance grades

Basic sizes		Grades											
		IT6		IT7		IT8		IT9		IT10		IT11	
mm	in	mm	in	mm	in	mm	in	mm	in	mm	in	mm	in
0–3	0–0.12	0.006	0.0002	0.010	0.0004	0.014	0.0006	0.025	0.0010	0.040	0.0016	0.060	0.0024
3–6	0.12–0.24	0.008	0.0003	0.012	0.0005	0.018	0.0007	0.030	0.0012	0.048	0.0019	0.075	0.0030
6–10	0.24–0.40	0.009	0.0004	0.015	0.0006	0.022	0.0009	0.036	0.0014	0.058	0.0023	0.090	0.0035
10–18	0.40–0.72	0.011	0.0004	0.018	0.0007	0.027	0.0011	0.043	0.0017	0.070	0.0028	0.110	0.0043
18–30	0.72–1.20	0.013	0.0005	0.021	0.0008	0.033	0.0013	0.052	0.0020	0.084	0.0033	0.130	0.0051
30–50	1.20–2.00	0.016	0.0006	0.025	0.0010	0.039	0.0015	0.062	0.0024	0.100	0.0039	0.160	0.0063
50–80	2.00–3.20	0.019	0.0007	0.030	0.0012	0.046	0.0018	0.074	0.0029	0.120	0.0047	0.190	0.0075
80–120	3.20–4.80	0.022	0.0009	0.035	0.0014	0.054	0.0021	0.087	0.0034	0.140	0.0055	0.220	0.0087
120–180	4.80–7.20	0.025	0.0010	0.040	0.0016	0.063	0.0025	0.100	0.0039	0.160	0.0063	0.250	0.0098
180–250	7.20–10.00	0.029	0.0011	0.040	0.0018	0.072	0.0028	0.115	0.0045	0.185	0.0073	0.290	0.0114
250–315	10.00–12.60	0.032	0.0013	0.052	0.0020	0.081	0.0032	0.130	0.0051	0.210	0.0083	0.320	0.0126
315–400	12.60–16.00	0.036	0.0014	0.057	0.0022	0.089	0.0035	0.140	0.0055	0.230	0.0091	0.360	0.0142

Source: Preferred metric limits and fits—BSI 4500.

TABLE 11-16
Fundamental tolerance[a] (µm and µin) for shafts for sizes up to 400 mm (16 in)

System of basic shaft	Limits		Diameter steps																							
		mm	0	3	6	10	14	18	24	30	40	50	65	80	100	120	140	160	180	200	225	250	280	315	355	400	450
		in	0	0.12	0.24	0.40	0.56	0.72	0.96	1.20	1.60	2.00	2.60	3.20	4.00	4.80	5.60	6.40	7.20	8.00	9.00	10.00	11.20	12.60	14.20	16.00	18.00
		mm	3	6	10	14	18	24	30	40	50	65	80	100	120	140	160	180	200	225	2.50	280	315	355	400	4.50	500
		in	0.12	0.24	0.40	0.56	0.72	0.96	1.20	1.60	2.0	2.60	3.20	4.00	4.80	5.60	6.40	7.20	8.00	9.00	10.00	11.20	12.60	14.20	16.00	18.00	20.00
a	*es*[b]	µm	−270	−270	−280	−290	−290	−300	−300	−310	−320	−340	−360	−380	−410	−460	−520	−580	−660	−740	−820	−920	−1,050	−1,200	−1,350	− 1,500	−1,650
		µin	−10,600	10,600	11,000	−11,400	−11,400	−11,800	−11,800	− 12200	−12,600	−13,400	−14,200	−14,900	−16,100	−18,100	−20,500	−22,800	−26,000	−29,100	−32,300	−36,200	−41,300	−47,200	−53,200	−59,000	−64,900
j	*ei*[c]	µm	−2	−2	−2	−2	−2	−3	−4	5	5	−7	−7	−9	−9	−11	−11	−11	−13	−13	−13	−16	−16	−18	−18	−18	−20
		µin	−80	−80	−80	−80	−80	−100	−160	−200	−200	−280	−280	−360	−360	−450	−450	−450	−510	−500	−500	−600	−600	−700	−700	−700	−800
c	*es*	µm	−60	−70	−80	−95	−95	−110	−110	−120	−130	−140	−150	−170	−180	−200	−210	−230	−240	−260	−280	−300	−330	−360	−400	−440	−480
		µin	−2,400	−2,800	−3,100	−3,700	−3,700	−4,300	−4,300	−4,700	−5,100	−5,500	−5,900	−6,700	−7,100	−7,900	−8,300	−9,100	−9,400	−10,200	−11,000	− 11,800	−13,000	−14,200	−15,700	−17,300	−18,900
k	*ei*	µm	0	+1	+1	+1	+1	+2	+2	+2	+2	+2	+2	+3	+3	+3	+3	+3	+4	+4	−+4	+4	4−4	+4	+4	+4	+5
		µin	0	+40	+40	+40	+40	+100	+100	+100	+100	+100	+100	+100	+100	+100	+100	+100	+160	+160	+160	+160	+160	+160	+160	+160	+200
d	*es*	µm	−20	−30	−40	−50	−50	−65	−65	−80	−80	−100	−100	−120	−120	−145	−145	−145	−170	−170	−170	−190	−190	−210	−210	−230	−230
		µin	−800	−1,200	−1,600	−2,000	−2,500	−2,600	−2,600	−3,100	−3,100	−3,900	−3,900	−4,700	−4,700	−5,700	−5,700	−5,700	−6,700	−6,700	−6,700	−7,500	−7,500	−8.300	−8,300	−9,100	−9,100
n	*ei*	µm	+4	+8	+10	+12	+12	+15	+15	+17	+17	+20	+20	+23	+23	+27	+27	+27	+31	+31	+31	+34	+34	+37	+37	+40	+40
		µin	+200	+300	+400	+500	+500	+600	+600	+700	+700	+800	+800	+900	+900	+1,100	+1,100	+1,100	+1,200	+1,200	+1,200	+1,300	+1,300	+1,500	+1,500	+1,600	+1,600
f	*es*	µm	−5	−10	−13	−16	−16	−20	−20	−20	−25	−25	−30	−30	−36	−36	−43	−43	−43	−50	−50	−56	−56	−62	−62	−68	−68
		µin	−200	−400	−500	−600	−600	−800	−800	−1,000	−1,000	−1,200	−1,200	−1,400	−1,400	−1,700	−1,700	−1,700	−2,000	−2,000	−2,000	−2,200	−2200	−2,400	−2,400	−2,680	−2,680
p	*ei*	µm	+6	+12	+15	+18	+18	+22	+22	+26	+26	+32	+32	+37	+37	+43	+43	+43	+.50	+50	+50	+.56	+.56	+62	+62	+68	+68
		µin	+200	+500	+600	+700	+700	+900	+900	+1,000	+1,000	+1,300	+1,300	+ 1,500	+1,500	+1,700	+1,700	+1,700	+2,000	+2,000	+2,000	+2,200	+2,200	+2,400	+2,400	+2,680	+2,680
g	*es*	µm	−2	−4	−5	−6	−6	−7	−7	−9	−9	−10	−10	−12	−12	−14	−14	−14	−15	−15	−15	−17	−17	−18	−18	−20	−20
		µin	−100	−200	−200	−200	−200	−300	−300	−400	−400	−400	−400	−500	−500	−600	−600	−600	−600	−600	−600	−700	−700	−700	−700	−800	−800
s	*ei*	µm	+14	+19	+23	+28	+28	+35	+35	+43	+43	+53	+59	+71	+79	+92	+100	+108	+122	+130	+140	+158	+170	+190	+208	+232	+252
		µin	+600	+700	+900	+1,100	+1,100	+1,400	+1,400	+1,700	+1,700	+2,100	+2,300	+2,800	+3,100	+3,600	+3,900	+4,300	+4,800	+5,100	+5,500	+6,200	+6,700	+7,500	+8,200	+9,100	+9,100
h	*es*	µm	0	0	0	0	0	0	0	0	0	0	0	0	0	0	0	0	0	0	0	0	0	0	0	0	0
		µin	0	0	0	0	0	0	0	0	0	0	0	0	0	0	0	0	0	0	0	0	0	0	0	0	0
u	*ei*	µm	+18	+23	+28	+33	+33	+41	+48	+60	+70	+87	+102	+124	+144	+170	+190	+210	+236	+258	+284	+315	+350	+390	+435	+490	+540
		µin	+700	+900	+1,100	+1.300	+1,300	+1,600	+1,900	+2,400	+2,800	+3,400	+4,000	+4,900	+5,700	+6,700	+7,500	+8,300	+9,300	+10,200	+11,200	+12,400	+13,000	+15,400	+17,100	+19,300	+21,300

[a] Tolerance in µm (1 µm = 10^{-6} m: 1 µin = 10^{-6} in).
[b] *es* = upper deviations.
[c] *ei* = lower deviations.
Source: Preferred limits and fits—BSI 4500; IS 2101, 1962.

TABLE 11-17
Relation between machine processes and geometry tolerances

Machining processes	Order of tolerance							
	Roundness[a] (circularity) of cylinders	Expressed as mm/mm length of surface or cylinder						
		Flatness of surfaces	Parallelism of cylinders on diameter	Straightness of cylinders, gaps and tongues	Angularity			
					Flat surface		Cylinders, gaps, tongues	
					Parallelism squareness	Any[b] other angle	Parallelism squareness	Any[b] other angle
Drill							10^{-3}	10^{-3}
Mill, slot, plane	—	5×10^{-5}	—	10^{-4}	10^{-4}	3×10^{-4}	10^{-4}	3×10^{-4}
Turn, bore	IT 4	5×10^{-5}	10^{-4}	10^{-4}	10^{-4}	3×10^{-4}	10^{-4}	3×10^{-4}
Fine turn, fine bore	IT 2	3×10^{-5}	4×10^{-5}	4×10^{-5}	5×10^{-5}	3×10^{-4}	5×10^{-5}	3×10^{-4}
Cylindrical grind	IT 3	—	5×10^{-5}	5×10^{-5}	—	—	5×10^{-5}	3×10^{-4}
Fine cylindrical grind	IT 1	—	2×10^{-5}	2×10^{-5}	—	—	2×10^{-5}	10^{-4}
Surface grind	—	3×10^{-5}	—	—	5×10^{-5}	3×10^{-4}	5×10^{-5}	3×10^{-4}
Fine surface grind	—	10^{-5}	—	—	2×10^{-5}	10^{-4}	2×10^{-5}	10^{-4}

[a] A roundness tolerance of 0.016 corresponds to a permissible diametrical variation of 0.032 (ovality).
[b] The values quoted are for good class of machine tools. Thrice or twice the above values, i.e., tolerance may have to be allowed for worn machine tools.

TABLE 11-18
Formulas for recommended allowances and tolerances (all dimensions in mm)

Class of fit	Method of assembly	Allowance	Selected average interference of metal	Hole tolerance	Shaft tolerance	Uses
Loose	Strictly interchangeable	$0.0075\ D^{2/3}$		$0.02\ D^{1/3}$	$0.02\ D^{1}$	Suitable for running fit; considerable freedom permissible; used in agricultural, mining, and general-purpose machinery
Free	Strictly interchangeable	$0.004\ D^{2/3}$		$0.01\ D^{1/3}$	$0.01\ D^{1/3}$	Suitable for running fit; suitable for shafts of motors, generators, engines, and some automotive parts
Medium	Strictly interchangeable	$0.0025\ D^{2/3}$		$0.007\ D^{1/3}$	$0.007\ D^{1/3}$	Accurate automotive parts and machine tools; suitable for running fit
Snug	Strictly interchangeable	0.0000		$0.005\ D^{1/3}$	$0.0035\ D^{1/3}$	Closest fit; zero allowance; suitable where no perceptible shake is permissible under load
Wringing	Selective assembly		0.0000	$0.005\ D^{1/3}$	$0.0035\ D^{1/3}$	A metal-to-metal contact fit
Tight	Selective assembly		$0.00025\ D$	$0.005\ D^{1/3}$	$0.005\ D^{1/3}$	Slightly negative allowance; suitable for semipermanent assembly and shrink fits
Medium force	Selective assembly		$0.0005\ D$	$0.005\ D^{1/3}$	$0.005\ D^{1/3}$	Suitable for press fits on locomotive wheels, car wheels, generator and motor armature, and crank discs
Heavy force or shrink	Selective assembly		$0.001\ D$	$0.005\ D^{1/3}$	$0.005\ D^{1/3}$	Used for steel external members that have a high yield stress

TABLE 11-19
Surface finish[a] values (CLA)

	High quality		Normal quality		Coarse quality	
Machining process	**Tolerance grade**	**Finish (μm)**	**Tolerance grade**	**Finish (μm)**	**Tolerance grade**	**Finish (μm)**
Drill	11	1.6–3.2	12			
Mill, slot, plane	9	0.4–0.8	11	0.8–1.6	12	1.6–3.2
Turn, bore	8	0.4–0.8	9	0.8–1.6	11	1.6–3.2
Ream	7	0.4–0.8	8	0.8–1.6		
Commercial grind	7	0.4–0.8	8	0.8–1.6	9	1.6–3.2
Fine turn, bore	6	0.2–0.4	7	0.4–0.8		
Hone	6	0.1–0.2	7	0.2–0.4		
Broach	6	0.1–0.2	7	0.2–0.4		
Fine grind	5	0.1–0.2	6	0.2–0.4		
Lap	3	0.05–0 1	4	0.1–0.2		

[a] The Roughness Number represents the average departure of the surface from perfection over a prescribed "sampling length" normally 0.8 mm, and is expressed in micrometers (μm). The measurements are normally made along a line at right angles to the general directions of tool marks or scratches on the surface.

1 μ = 0.001 mm

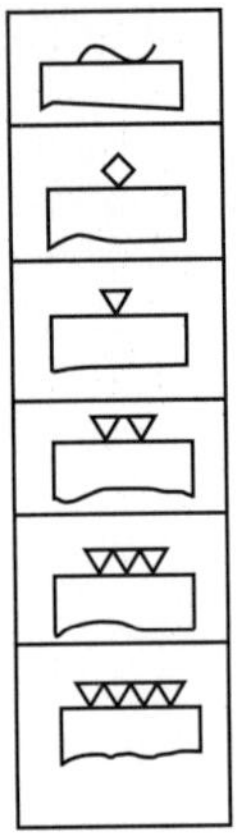

Old machining symbols	Description	Surface roughness
	Unmachined surface. cleaned up by sand blasting, brushing, etc.	5–80 μ
	Surface to be rough machined if found necessary (to prevent fouling)	
	Surface obtained by rough machining under turning, planing, milling etc. Quality coarser than 9	8–25 μ
	Finish-machined surface obtained by turning, milling etc. Quality 12–7	1.6–8 μ
	Fine finish-machined surface obtained by boring, reaming, grinding etc. Quality 9–6	0.25–1.6 μ
	Super finish-machined surface obtained by honing, lapping, super finish grinding. Quality 7–4	0–0.25 μ

FIGURE 11-12 Machining symbols.

TABLE 11-20
Lay symbols

Lay symbol	Interpretation	Example showing direction of tool marks
=	Lay parallel to the line representing the surface to which the symbol is applied	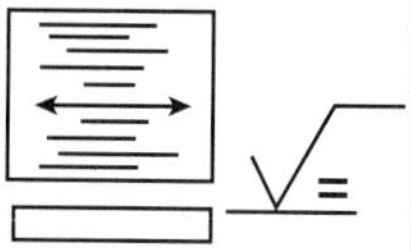
⊥	Lay perpendicular to the line representing the surface to which the symbol is applied	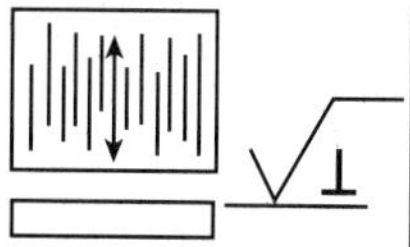
X	Lay angular in both directions to line representing the surface to which symbol is applied	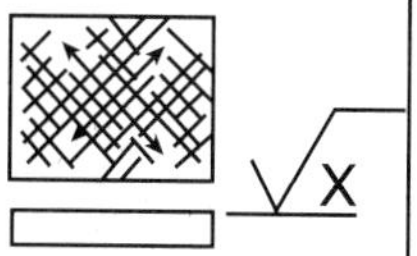
M	Lay multidirectional	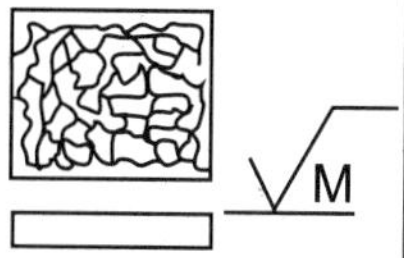
C	Lay approximately circular relative to the center of the surface to which the symbol is, applied	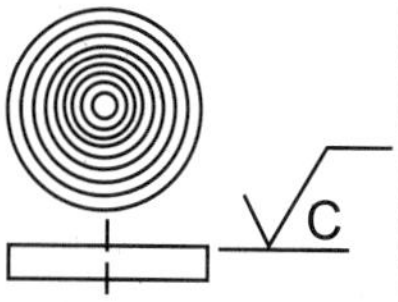
R	Lay approximately radial relative to the center of the surface to which the symbol is applied	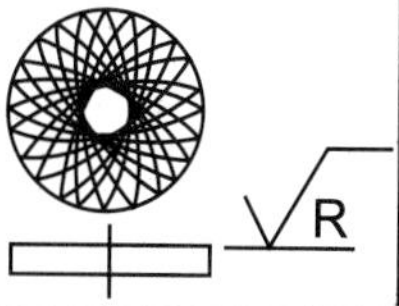
P	Pitted, protuberant, porous, or particulate nondirectional lay	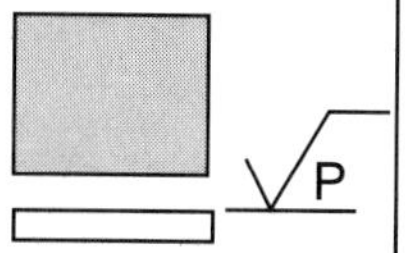

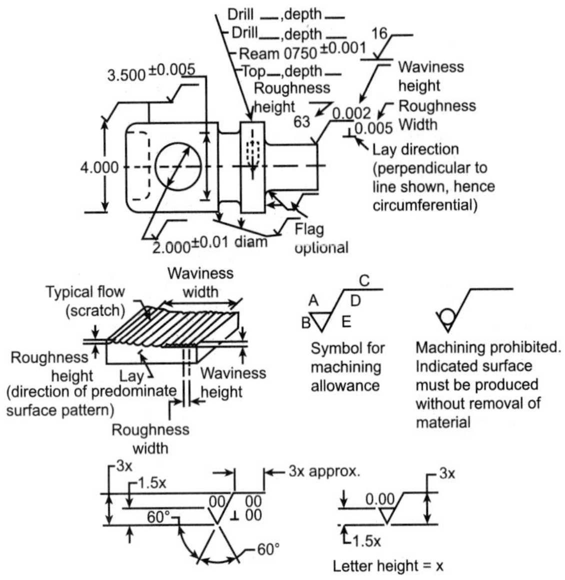

FIGURE 11-13 Application and use of surface-texture symbols. (Baumeister, T., *Marks' Standard Handbook for Mechanical Engineers*, 8th ed., McGraw-Hill, 1978.)

TABLE 11-21
Preferred series roughness average values (R_a) (in μm and μin)

μm	μin	μm	μin	μm	μin	μm	μin	μm	μin
0.012	0.5	0.125	5	0.50	20	2.00	80	8.0	320
0.025	1	0.15	6	0.63	25	2.50	100	10.0	400
0.050	2	0.20	8	0.80	32	3.20	125	12.5	500
0.075	3	0.25	10	1.00	40	4.0	160	15.0	600
0.10	4	0.32	13	1.25	50	5.0	200	20.0	800
		0.40	16	1.60	63	6.3	250	25.0	1000

Source: Reproduced from Baumeister, T., *Marks' Standard Handbook for Mechanical Engineers*, 8th ed., with permission from McGraw-Hill Book Company, New York, 1978.

TABLE 11-22
Preferred series maximum waviness height values

mm	in	mm	in	mm	in
0.0005	0.00002	0.008	0.0003	0.12	0.005
0.0008	0.00003	0.012	0.0005	0.20	0.008
0.0012	0.00005	0.020	0.0008	0.25	0.010
0.0020	0.00008	0.025	0.001	0.38	0.015
0.0025	0.0001	0.05	0.002	0.50	0.020
0.005	0.0002	0.08	0.003	0.80	0.030

Source: Reproduced from Baumeister, T., *Marks' Standard Handbook for Mechanical Engineers*, 8th ed., with permission from McGraw-Hill Book Company, New York, 1979.

TABLE 11-23
Surface roughness ranges of production processes

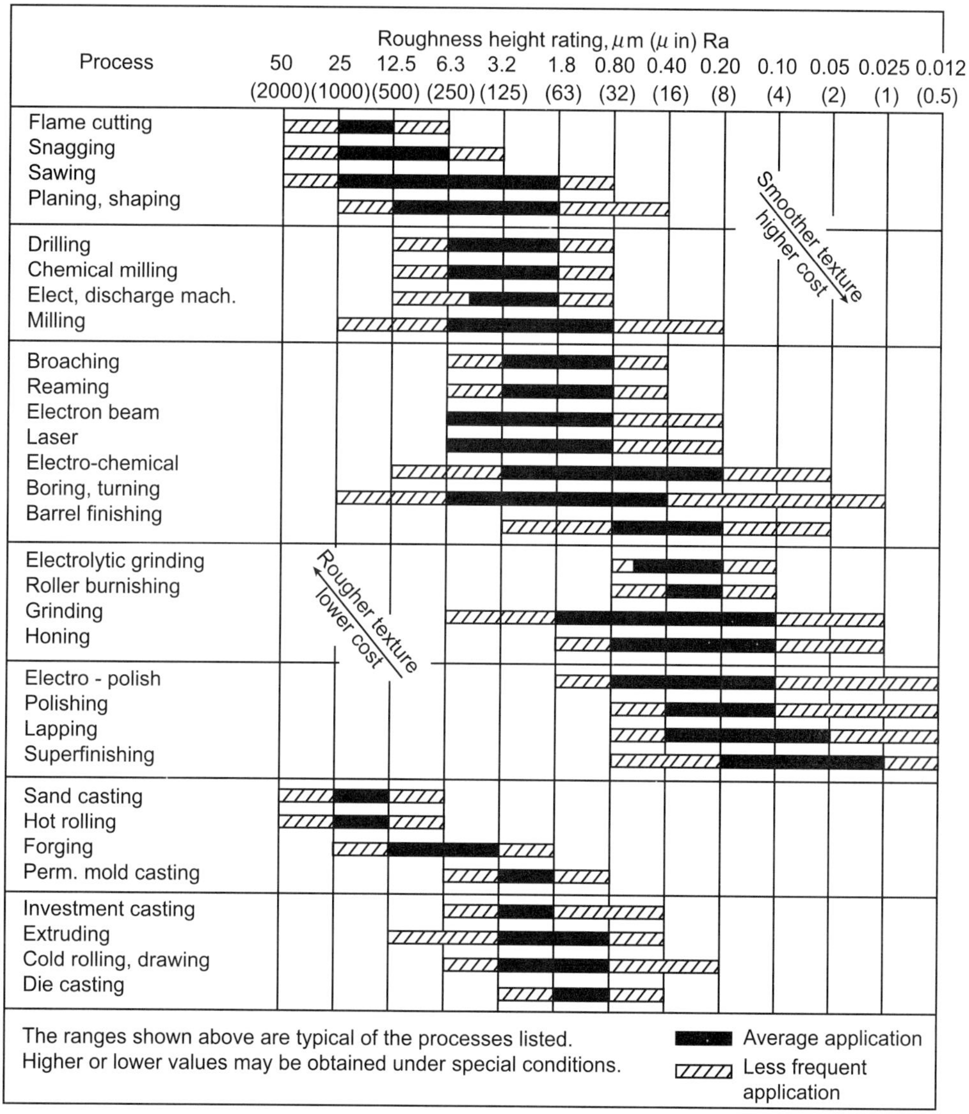
Process
Roughness height rating, μm (μ in) Ra
50 (2000)
25 (1000)
12.5 (500)
6.3 (250)
3.2 (125)
1.8 (63)
0.80 (32)
0.40 (16)
0.20 (8)
0.10 (4)
0.05 (2)
0.025 (1)
0.012 (0.5)
Flame cutting
Snagging
Sawing
Planing, shaping
Drilling
Chemical milling
Elect, discharge mach.
Milling
Broaching
Reaming
Electron beam
Laser
Electro-chemical
Boring, turning
Barrel finishing
Electrolytic grinding
Roller burnishing
Grinding
Honing
Electro - polish
Polishing
Lapping
Superfinishing
Sand casting
Hot rolling
Forging
Perm. mold casting
Investment casting
Extruding
Cold rolling, drawing
Die casting
Smoother texture higher cost
Rougher texture lower cost
The ranges shown above are typical of the processes listed.
Higher or lower values may be obtained under special conditions.
Average application
Less frequent application

Source: Reproduction from Baumeister, T., *Marks' Standard Handbook for Mechanical Engineers*, 8th ed., with permission from McGraw-Hill Book Company, New York, 1979.

TABLE 11-24
Application of surface texture values to surface symbols

(63)	1.6 √‾‾	Roughness average rating is placed at the left of the long leg; the specification of only one rating shall indicate the maximum value and any lesser value shall be acceptable	(63)	3.5 1.6 √‾‾	Machining is required to produce the surface; the basic amount of stock provided for machining is specified at the left of the short leg of the symbol
(63) (32)	1.6 0.8 √‾‾	The specification of maximum value and minimum value roughness average ratings indicates permissible range of value rating	(63)	1.6 √‾‾ (with circle)	Removal of material by machining is prohibited
			(32)	0.8 √‾⊥‾	Lay designation is indicated by the lay symbol placed at the right of the long leg
(32)	0.8 √‾‾ with 0.05 above extension	Maximum waviness height rating is placed above the horizontal extension; any lesser rating shall be acceptable	(32)	0.8 √‾2.5‾ (0.100)	Roughness sampling length or cutoff rating is placed below the horizontal extension; when no value is shown, 0.80 mm is assumed
(32)	0.8 √‾‾ with 0.05 – 100 above extension	Maximum waviness spacing rating is placed above the horizontal extension and to the right of the waviness height rating; any lesser rating shall be acceptable	(32)	0.8 √‾⊥ 0.5‾	Where required, maximum roughness spacing shall be placed at the right of the lay symbol; any lesser rating shall be acceptable

Source: Reproduction from Baumeister, T., *Marks' Standard Handbook for Mechanical Engineers*, 8th ed., with permission from McGraw-Hill Book Company, New York, 1979.

TABLE 11-25
Typical surface texture design requirements

(250 μin)	6.3 √	Clearance surfaces Rough machine parts
(125 μin)	3.2 √	Mating surfaces (static) Chased and cut threads Clutch-disk faces Surfaces for soft gaskets
(63 μin)	1.60 √	Piston-pin bores Brake drums Cylinder block, top Gear locating faces Gear shafts and bores Ratchet and pawl teeth Milled threads Rolling surfaces Gearbox faces Piston crowns Turbine-blade dovetails
(32 μin)	0.80 √	Broached holes Bronze journal bearings Gear teeth Slideways and gibs Press-fit parts Piston-rod bushings Antifraction-bearing seats Sealing surfaces for hydraulic tube fittings
(16 μin)	0.40 √	Motor shafts Gear teeth (heavy loads) Spline shafts O-ring grooves (static) Antifraction-bearing bores and faces Camshaft lobes Compressor-blade airfoils Journals for elastomer lip seals
(13 μin)	0.32 √	Engine cylinder bores Piston outside diameters Crankshaft bearings
(8 μin)	0.20 √	Jet-engine stator blades Valve-tappet cam faces Hydraulic-cylinder bores Lapped antifriction bearings
(4 μin)	0.10 √	Ball-bearing races Piston pins Hydraulic piston rods Carbon-seal mating surfaces
(2 μin)	0.050 √	Shop-gauge faces Comparator anvils
(1 μin)	0.025 √	Bearing balls Gauges and mirrors Micrometer anvils

TABLE 11-26
Range of surface roughness[a]

Manufacturing process	With difficulty	Normally	Roughing
Manual			
Hack saw cut		6.3–50	
Chipping		3.2–50	
Filing	0.8–1.6	1.6–12.5	
Emery polish	0.1–0.4	0.4–1.6	1.6–3.2
Casting			
Sand casting		6.3–12.5	12.5–25
Permanent mold	0.8–1.6	1.6–6.3	
Die casting		0.8–3.2	
Forming			
Forging	1.6–3.2	3.2–25	
Extrusion	0.4–0.8	0.8–6.3	
Rolling	0.4–0.8	0.8–3.2	
Machining			
Drilling	3.2–6.3	6.3–25	
Planing and shaping		1.6–12.5	
Face milling	0.8–1.6	1.6–12.5	12.5–50
Turning	0.2–1.6	1.6–6.3	6.3–50
Boring	0.2–1.6	1.6–6.3	6.3–50
Reaming	0.4–0.8	0.8–6.3	6.3–12.50
Cylindrical grinding	0.025–0.4	0.4–3.2	3.2–6.3
Centerless grinding	0.05–0.4	0.4–3.2	
Surface grinding	0.025–0.4	0.4–3.2	3.2–6.3
Broaching	0.2–0.8	0.8–3.2	3.2–6.3
Superfinishing	0.025–0.1	0.1–0.4	
Honing	0.025–0.1	0.1–0.4	
Lapping	0.006–0.05	0.05–0.4	
Gear manufacture			
Milling with form cutter	1.6–3.2	3.2–12.5	12.5–50
Milling, spiral bevel	1.56–3.2	3.2–12.5	12.5–25
Hobbing	0.8–3.2	3.2–12.5	12.5–50
Shaping	0.4–1.6	1.6–12.5	12.5–250
Shaving	0.4–0.8	0.8–3.2	
Grinding	0.1–0.4	0.4–0.8	
Lapping	0.05–0.2	0.2–0.8	
Surface process			
Shot blast	1.6–3.2	3.2–50	
Abrasive belt		0.1–6.3	
Fiber wheel brushing	0.1–0.2	0.2–0.8	0.8–1
Cloth buffing	0.012–0.05	0.05–0.1	

[a] Surface roughness in μm ($1 \mu m = 10^{-3} mm = 10^{-6} m$).

Characteristics to be toleranced	Symbol
Straightness Flatness	—
Circularity	○
Accuracy of any profile Accuracy of any surface	⌒
Parallelism	//
Perpendicularity	⊥
Angularity	∠
Position	⌖
Concentricity or coaxiality	◎
Symmetry	⌯

IS : 696–1960

FIGURE 11-14 Symbols for tolerances of form and position.

Rivet	Symbol
Shop snap headed rivets	+
Shop *Csk* (near side) rivets	
Shop *Csk* (far side) rivets	
Shop *Csk* (both sides) rivets	
Site snap headed rivets	⊕
Site *Csk* (near side) rivets	
Site *Csk* (far side) rivets	
Site *Csk* (both sides) rivets	
Open hole	

IS : 696–1960

FIGURE 11-15 Rivet symbols

REFERENCES

1. Lingaiah, K., and B. R. Narayana Iyengar, *Machine Design Data Handbook*, Engineering College Co-operative Society, Bangalore, India, 1962.
2. Lingaiah, K., and B. R. Narayana Iyengar, *Machine Design Data Handbook*, Vol. I, Suma Publishers, Bangalore, India, 1986.
3. Lingaiah, K., *Machine Design Data Handbook*, Vol. II (*SI and Customary Metric Units*), Suma Publishers, Bangalore, India, 1986.
4. Black, P. H., and O. Eugene Adams, Jr., *Machine Design*, McGraw-Hill Publishing Company, New York.
5. Baumeister, T., *Marks' Standard Handbook for Mechanical Engineers*, 8th ed., McGraw-Hill Publishing Company, New York, 1978.
6. Maleev, V. L., and J. B. Hartman, *Machine Design*, International Textbook Company, Scranton, Pennsylvania, 1954.
7. Shigley, J. E., *Machine Design*, McGraw-Hill Publishing Company, New York, 1956.
8. Vallance, A., and V. L. Doughtie, *Design of Machine Members*, McGraw-Hill Publishing Company, New York, 1951.
9. British Standard Institution.
10. Bureau of Indian Standards.

CHAPTER 12

DESIGN OF WELDED JOINTS

SYMBOLS[2,3,4]

A	area of flange material held by welds in shear, m^2 (in^2)
$A' = l_\omega$	length of weld when weld is treated as a line, m (in)
b	width of connection, m (in)
c	distance to outer fiber (also with suffixes), m (in)
c_x	distance of x axis to face, m (in)
c_y	distance of y axis to face, m (in)
c_1	distance of weld edge parallel to x-axis from the center of weld, to left, m (in)
c_2	distance of weld edge from parallel to x-axis from the center of weld, to right, m (in)
c_3	distance from farthest weld corner, Q, to the center of gravity of weld, m (in) (Fig. 12-8)
d	depth of connection, m (in)
e_x	eccentricity of P_z and P_y about the center of weld, m (in)
e_y	eccentricity of P_x about the center of weld, m (in)
h	thickness of plate (also with suffixes), m (in)
i	number of welds
I_x, I_y, I_z	moment of inertia of weld about x, y, and z axes respectively, m^4, cm^4 (in^4)
J	moment of inertia, polar, m^4, cm^4 (in^4)
J_ω	polar moment of inertia of weld, when weld is treated as a line, m^3, cm^3 (in^3)
$K_{f\sigma}$	fatigue stress-concentration factor (Table 12-7)
l	effective length of weld, m (in)
l_t	total length of weld, m (in)
M_b	bending moment, N m (lbf in)
M_t	twisting moment, N m (lbf in)
n_a	actual factor of safety or reliability factor
N_a	fatigue life (for which σ_{sfa} is known) for fatigue strength σ_{sfa}, cycle
N_b	fatigue life (required) for fatigue strength σ_{sfb}, cycle
P	load on the joint, kN (lbf)
P_x	component of P in x direction, kN (lbf)
P_y	component of P in y direction, kN (lbf)

P_z	component of P in z direction, kN (lbf)
r	distance to outer fiber, m (in)
R	ratio of calculated leg size for continuous weld to the actual leg size to be used for intermittent weld
t	throat dimension of weld, m (in)
V	shear load, kN (lbf)
w	size of weld leg, m (in)
Z	section modulus, m^3 (in^3)
Z_ω	section modulus of weld, when weld is treated as line (also with suffixes, m^2 (in^2)
σ	normal stress in the weld (in standard design formula), MPa (psi)
σ'	force per unit length of weld (in standard design formula) when weld treated as a line, kN/m (lbf/in)
σ_{sfa}	fatigue strength (known) for fatigue life N_a, MPa (psi)
σ_{sfb}	fatigue strength (allowable) for fatigue life N_b, MPa (psi)
σ_d	design stress, MPa (psi)
σ_e	elastic limit, MPa (psi)
τ	shear stress in the weld (in standard design formula), MPa (psi)
τ'	shear force per unit length of weld (in standard design formula) when weld is treated as a line, kN/m (lbf/in)
θ	angle, deg
η	efficiency of joint

Particular	Formula	

FILLET WELD

The throat thickness t, for case with equal legs, of weld (Fig. 12-1)	$t = w \sin 45° = 0.707w$	(12-1a)
The allowable load on the weld	$P = 0.707\, i\tau\, wl$	(12-1b)

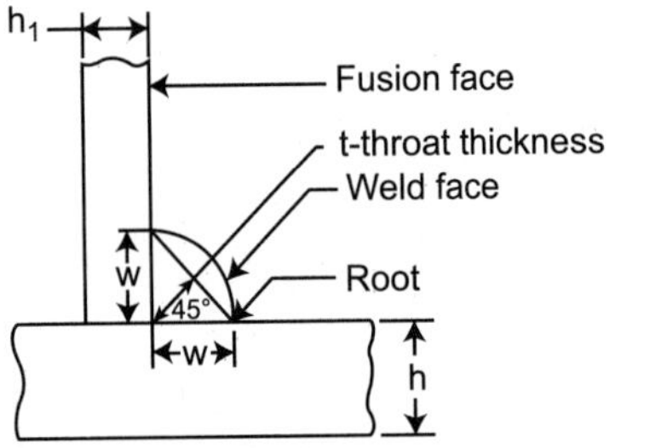

FIGURE 12-1 Fillet weld.

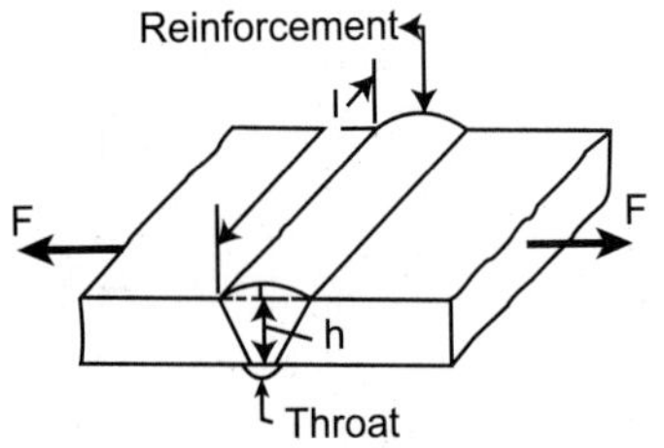

FIGURE 12-2 A typical butt-weld joint.

BUTT WELD

The average normal stress in a butt weld subjected to tensile or compression loading (Fig. 12-2)	$\sigma = \dfrac{F}{hl}$ where h is the throat dimension. The dimensions of throat (t) are the same as the thickness of plate (h).	(12-2)

Particular	Formula
	The throat dimension (h) does not include the reinforcement.
The average shear stress in butt weld	$$\tau = \frac{F}{hl} \quad (12\text{-}3)$$
The allowable load on the weld	$$F_a = \eta\sigma_a hl \quad (12\text{-}4)$$

TRANSVERSE FILLET WELD

Particular	Formula
The average normal tensile stress	$$\sigma = \frac{F}{wl\cos 45^\circ} = \frac{F}{0.707wl} \quad (12\text{-}5)$$
The average normal tensile stress for the case of transverse fillet weld shown in Fig. 12-3.	$$\sigma = \frac{F}{0.707hl} \quad (12\text{-}6)$$ The stress concentration occurs at A and B on the horizontal leg and at B on the vertical leg in the weld according to photoelastic tests conducted by Norris.[1]
A double fillet lap weld joint.	Refer to Fig. 12-4.

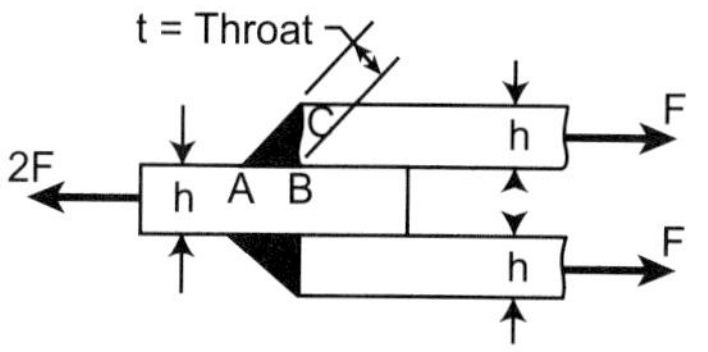

FIGURE 12-3 A transverse fillet weld.

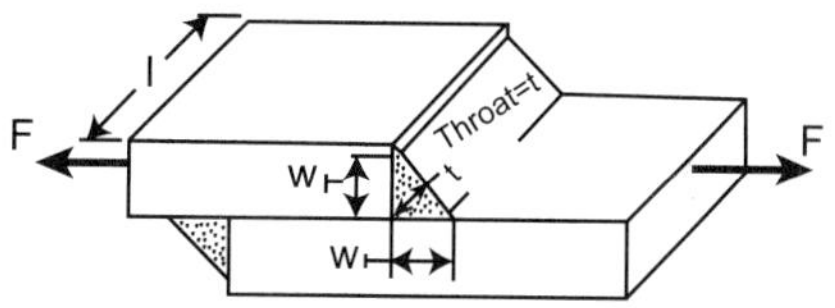

FIGURE 12-4 A double-fillet lap-weld joint.

PARALLEL FILLET WELD (Fig. 12-5)

Particular	Formula
The average shear stress in the weld	$$\tau = \frac{P}{0.707wl} \quad (12\text{-}7a)$$ where w = dimension of leg of weld. w can be replaced by h (thickness of plate) when w and h are of same dimension. Either symbol F or P can be used for force or load depending on symbols used in figures in this chapter.
The shear stress in a reinforced fillet weld	$$\tau = \frac{P}{0.85wl} \quad (12\text{-}7b)$$ where throat t is taken as $0.85w$

Particular	Formula

LENGTH OF WELD

The effective length of weld (Fig. 12-5)

$$l = l_t - \frac{i}{4} \tag{12-8}$$

where i = total number of free ends

The total length of weld (Fig. 12-5)

$$l_t = \frac{P}{0.707\,w\sigma_a} \quad \text{where } l_t = 2(l_1 + l_2) \tag{12-9}$$

The relation between the length l_1 and l_2 (Fig. 12-5)

$$\frac{l_1}{L - \bar{x}} = \frac{l_2}{\bar{x}} = \frac{l_1 + l_2}{L} = \frac{l_t}{2L} \tag{12-10}$$

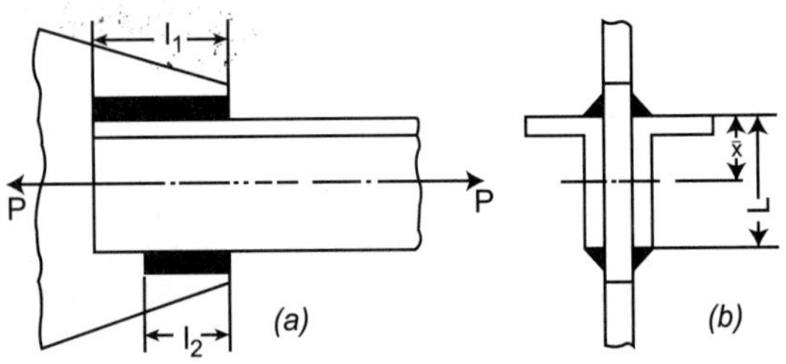

FIGURE 12-5 Parallel fillet weld.

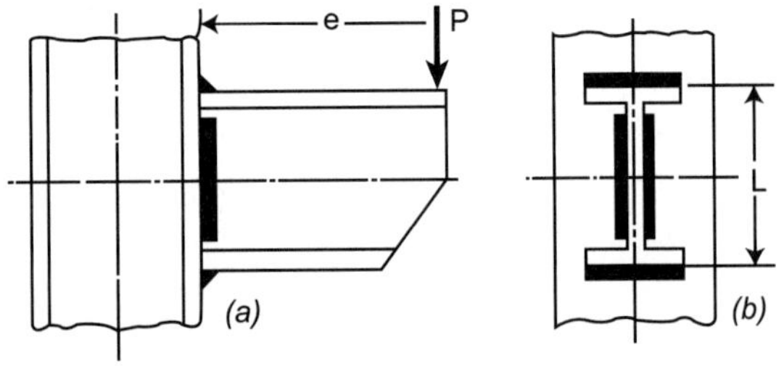

FIGURE 12-6

ECCENTRICITY IN A FILLET WELD

The bending stress due to fillet weld placed on only one side of the plate (Fig. 12-6)

$$\sigma_b = \frac{4M_b}{(0.707w)^2 l} = \frac{4Pw}{4(0.707w)^2 l} = \frac{2P}{wl} \tag{12-11}$$

The stress due to tensile load

$$\sigma_t = \frac{P}{1.414wl} \tag{12-12}$$

The combined normal stress at the root of the weld

$$\sigma_n = \sigma_t + \sigma_b = \frac{P}{1.414wl} + \frac{2P}{wl} \tag{12-13}$$

The shear stress

$$\tau = \frac{P}{0.707wl} \tag{12-14}$$

The maximum normal stress

$$\sigma_{max} = \tfrac{1}{2}\left(\sigma_n + \sqrt{\sigma_n^2 + 4\tau^2}\right) \tag{12-15}$$

The maximum shear stress

$$\tau_{max} = \tfrac{1}{2}\sqrt{\sigma_n^2 + 4\tau^2} \tag{12-16}$$

Particular	Formula

ECCENTRIC LOADS

Moment acting at right angles to the plane of welded joint (Fig. 12-6)

Direct load per unit length of weld

$$P_d = \frac{P}{l} \qquad (12\text{-}17)$$

Load due to bending per unit length of weld

$$P_n = \frac{Pey}{I} \qquad (12\text{-}18)$$

The resultant load or force

$$P_R = \sqrt{P_d^2 + P_n^2} \qquad (12\text{-}19)$$

Moment acting in the plane of the weld (Fig. 12-7)

Load due to twisting moment per unit length of weld

$$P_n = \frac{Per}{J} \qquad (12\text{-}20)$$

The resultant load (Fig. 12-7)

$$P_R = \sqrt{P_d^2 + P_n^2 + 2P_dP_n\cos\theta} \qquad (12\text{-}21)$$

where $\cos\theta = \dfrac{l_2}{\sqrt{l_1^2 + l_2^2}}$

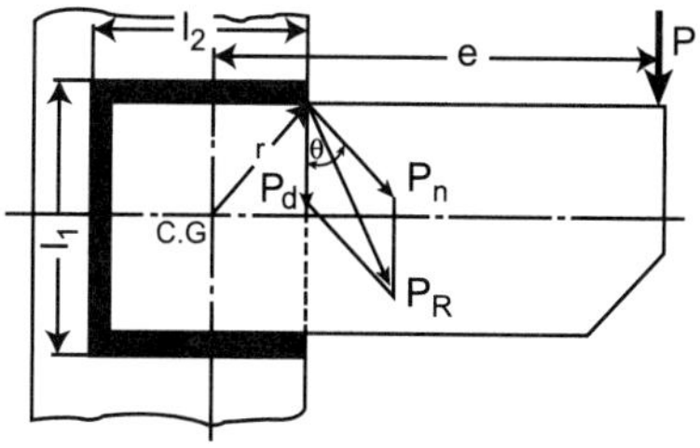

FIGURE 12-7

STRESSES

Bending

The bending stress

$$\sigma_b = \frac{M_b}{wZ_w} \qquad (12\text{-}22a)$$

or

$$\sigma_b' = \frac{M_b}{Z_w} \text{ (treating weld as a line)} \qquad (12\text{-}22b)$$

Particular	Formula

Torsion

The shear stress due to torsion

$$\tau = \frac{M_t r}{wJ_w} \tag{12-23a}$$

or

$$\tau' = \frac{M_t r}{J_w} \text{ (treating weld as a line)} \tag{12-23b}$$

Combined bending and torsion

The resultant or maximum induced normal force per unit throat of weld

$$\sigma'_{\max} = \frac{1}{2}\left[\frac{M_b}{Z_w} + \sqrt{\left(\frac{M_b}{Z_w}\right)^2 + 4\left(\frac{M_t r}{J_w}\right)^2}\right] \tag{12-24}$$

The resultant induced torsional force per unit throat of weld

$$\tau'_{\max} = \frac{1}{2}\sqrt{\left(\frac{M_b}{Z_w}\right)^2 + 4\left(\frac{M_t r}{J_w}\right)^2} \tag{12-25}$$

The required leg size of the weld when weld is treated as a line

$$w = \frac{\text{actual force}}{\text{permissible force}} = \frac{\sigma'_{\max} \text{ or } \tau'_{\max}}{\sigma'_a \text{ or } \tau'_a} \tag{12-26}$$

The resultant normal stress induced in the weld

$$\sigma_{\max} = \frac{1}{2}\left[\frac{M_b}{wZ_w} + \sqrt{\left(\frac{M_b}{wZ_w}\right)^2 + 4\left(\frac{M_t r}{wJ_w}\right)^2}\right] \tag{12-27}$$

The resultant shear stress induced in the weld

$$\tau_{\max} = \frac{1}{2}\sqrt{\left(\frac{M_b}{wZ_w}\right)^2 + 4\left(\frac{M_t r}{wJ_w}\right)^2} \tag{12-28}$$

The required leg size of weld when the weld area is considered

$$w = \frac{\text{actual maximum stress induced in the weld}}{\text{permissible stress}}$$

$$= \frac{\sigma_{\max} \text{ or } \tau_{\max}}{\sigma_a \text{ or } \tau_a}$$

FATIGUE STRENGTH

The fatigue strength related to fatigue life can be expressed by the empirical formula

$$\sigma_{sfa} = \sigma_{sfb}\left(\frac{N_b}{N_a}\right)^k \tag{12-29}$$

or

$$N_a = N_b\left(\frac{\sigma_{sfb}}{\sigma_{sfa}}\right)^{1/k} \tag{12-30}$$

where

$k = 0.13$ for butt welds
$= 0.18$ for plates in bending, axial tension, or compression

Particular	Formula

DESIGN STRESS OF WELDS

The design stress

$$\sigma_d = \frac{\sigma_a}{n_a} \tag{12-31}$$

where

n_a = actual safety factor or reliability factor = 3 to 4

The design stress for completely reversed load

$$\sigma_{fd} = \frac{\sigma_f}{n_a K_{f\sigma}} \tag{12-32}$$

THE STRENGTH ANALYSIS OF A TYPICAL WELD JOINT SUBJECTED TO ECCENTRIC LOADING (Fig. 12-8)[2,3,4]

Throughout the analysis of a weld joint, the weld is treated as a line

Area of cross section of weld

$$A = (2b + d)w \tag{12-33}$$

The distance of weld edge parallel to x axis from the center of weld, to left

$$c_1 = \frac{b^2}{2b + d} \tag{12-34}$$

The distance of weld edge parallel to x axis from the center of weld, to right

$$c_2 = \frac{b(b + d)}{2b + d} \tag{12-35}$$

The distance from farthest weld corner, Q, to the center of gravity of weld

$$c_3 = \sqrt{c_2^2 + \left(\frac{d}{2}\right)^2} \tag{12-36}$$

The moment of inertia of weld about x axis

$$I_x = \frac{wd^2}{12}(d + 6b) \tag{12-37}$$

The moment of inertia of weld about y axis

$$I_y = \frac{wb^3(2d + b)}{3(d + 2b)} \tag{12-38}$$

The moment of inertia of weld about z axis

$$I_z = I_x + I_y \tag{12-39}$$

The section modulus of weld, about x axis

$$Z_{wx} = \frac{I_x}{(d/2)} = \frac{wd}{6}(d + 6b)$$

The section modulus of weld, about y axis

$$Z_{wy} = \frac{I_y}{c_2} = \frac{wb^2(2d + b)}{3(b + d)} \tag{12-40}$$

The section modulus of weld, about z axis

$$Z_{wz} = \frac{I_z}{c_3} \tag{12-41}$$

Particular	Formula

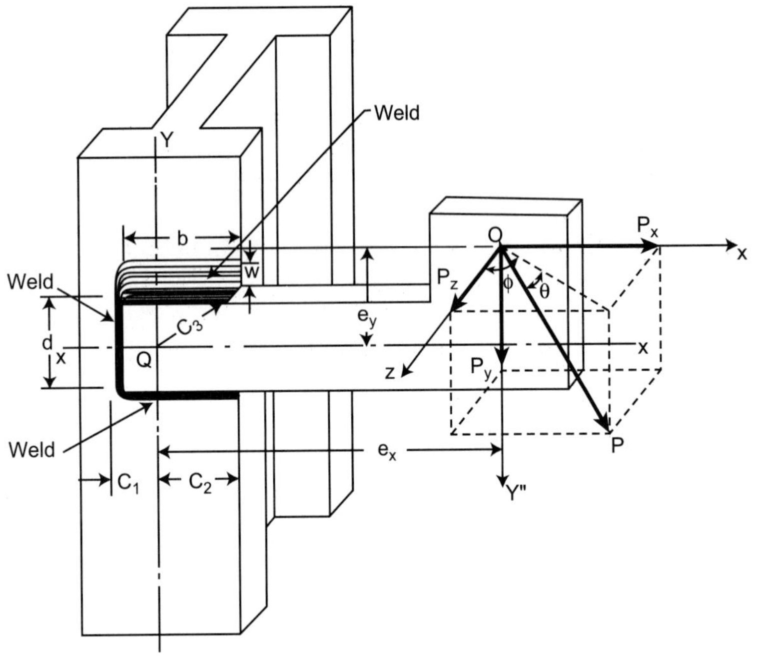

FIGURE 12-8 A typical weld joint subjected to Eccentric Loading. K. Lingaiah and B. R. Narayana Iyengar, *Machine Data Handbook (fps Units)*, Engineering College Cooperative Society, Bangalore, India, 1962; K. Lingaiah and B. R. Narayana Iyengar, *Machine Design Data Handbook*, Vol. I (*SI and Customary Metric Units*), Suma Publishers, Bangalore, India, 1986; and K. Lingaiah, *Machine Design Handbook*, Vol. II (*SI and Customary Metric Units*), Suma Publishers, Bangalore, India, 1986.

P_z component

Throughout the analysis of this problem the weld is considered as a line

The force per unit length of weld due to direct force P_z	$\sigma'_{zd} = \dfrac{P_z}{A'}$	(12-42)
The force per unit length of weld an account of bending at the farthest weld corner, Q, due to eccentricity e_x of load P_z	$\sigma'_{zb1} = \dfrac{P_z e_x}{Z_{wy}}$	(12-43)
The force per unit length of weld an account of bending at the farthest weld corner, Q, due to eccentricity e_y of load P_z	$\sigma'_{zb2} = \dfrac{P_z e_y}{Z_{wx}}$	(12-44)
The total force per unit length of weld due to bending	$\sigma'_{zb} = \sigma'_{zb1} + \sigma'_{zb2}$	(12-45)
The combined force per unit length of weld due to load P_z	$\sigma'_z = \sigma'_{zd} + \sigma'_{zb}$	(12-46)

Particular	Formula	
P_x component		
The force per unit length of weld due to direct shear force P_x which acts in the horizontal direction (Fig. 12-8)	$\tau'_{xd} = \frac{P_x}{A'}$	(12-47)
The twisting moment	$M_{tx} = P_x e_y$	(12-48)
The shear force per unit length due to twisting moment M_{tx}	$\tau'_{tx} = \frac{M_{tx} c_3}{J_{wz}}$	(12-49)
The vertical component of τ'_{tx}	$\tau'_{txv} = \frac{M_{tx} c_3}{J_{wz}} \cos \psi$	(12-50)
The horizontal component of τ'_{tx}	$\tau'_{txh} = \frac{M_{tx} c_3}{J_{wz}} \sin \psi$ where c_3 = distance from the center of gravity of the weld to the point being analyzed (i.e., Q) $\cos \psi = \frac{c_2}{c_3}$ (Fig. 12-8)	(12-51)
The resultant shear force per unit length of weld in the horizontal direction due to P_x only	$\tau'_{txrh} = \tau'_{xd} = \tau'_{txh}$	(12-52)
P_y component		
The direct shear force per unit length of weld parallel to y direction due to force P_y (Fig. 12-8)	$\tau'_{yd} = \frac{P_y}{A'}$	(12-53)
The twisting moment	$M_{ty} = P_y e_x$	(12-54)
The shear force per unit length of weld due to twisting moment M_{ty}	$\tau'_{ty} = \frac{M_{ty} c_3}{J_{wz}}$	(12-55)
The vertical component of τ'_{ty}	$\tau'_{tyv} = \tau'_{ty} \cos \psi$	(12-56)
The horizontal component of τ'_{ty}	$\tau'_{tyh} = \tau'_{ty} \sin \psi$	(12-57)
The resultant shear force per unit length of weld in the vertical direction due to P_y only	$\tau'_{tyrv} = \tau'_{yd} + \tau'_{tyv}$	(12-58)

Particular	Formula

COMBINED FORCE DUE TO P_x, P_y, AND P_z AT POINT Q (Fig. 12-8)

From Eqs. (12-46), (12-50), (12-52), (12-57), and (12-58)

The total shear force per unit length of weld in the x direction (Fig. 12-8) from Eqs. (12-52) and (12-57)

$$\tau'_x = \tau'_{tzrh} + \tau'_{tyh} \tag{12-59}$$

The total shear force per unit length of weld in the y direction (Fig. 12-8) from Eqs. (12-50) and (12-58)

$$\tau'_y = \tau'_{txv} + \tau'_{tyrv} \tag{12-60}$$

The resultant shear force per unit length of weld at point Q due to P_x and P_y forces (Fig. 12-8) from Eqs. (12-59) and (12-60)

$$\tau' = \sqrt{\tau'^2_x + \tau'^2_y} \tag{12-61}$$

The resultant actual force per unit length of weld (treating weld as a line) due to components P_x, P_y, and P_z at point Q from Eqs. (12-46) and (12-61)

$$\sigma'_{\text{actual}} = \sqrt{\sigma'^2_z + \tau'^2} \tag{12-62}$$

The leg size of the weld

$$w' = \frac{\sigma'_{\text{actual}}}{\sigma'_{\text{allowable}}} \tag{12-63}$$

For the AWS standard location of elements of welding symbol, weld symbols and direction for making weld

Refer to Figs. 12-9 to 12-11.

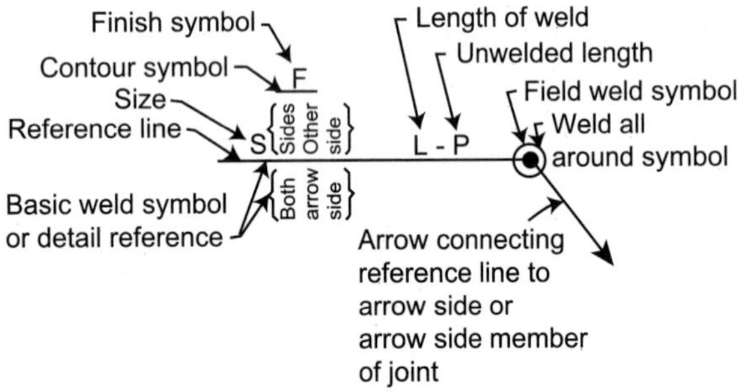

FIGURE 12-9 The AWS Standard location of elements of a welding symbol.

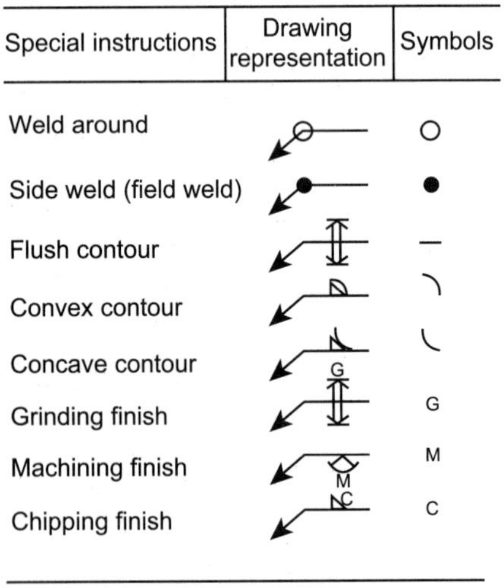

FIGURE 12-10 Weld symbols

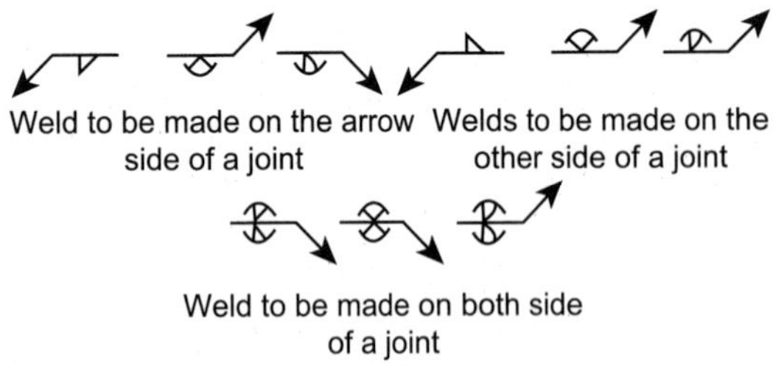

FIGURE 12-11

Particular	Formula

GENERAL

For further data on welded joint design	Refer to Tables 12-1 to 12-16.

REFERENCES

1. Norris, C. H., *Photoelastic Investigation of Stress Distribution in Transverse Fillet Welds*, *Welding Journal*, Vol. 24, p. 557, 1945.
2. Lingaiah, K., and B. R. Narayana Iyengar, *Machine Design Data Handbook*, Engineering College Cooperative Society, Bangalore, India, 1962.
3. Lingaiah, K., and B. R. Narayana Iyengar, *Machine Design Data Handbook*, Vol. I (*SI and Customary Metric Units*), Suma Publishers, Bangalore, India, 1986.
4. Lingaiah, K., *Machine Design Data Handbook*, Vol. II (*SI and Customary Metric Units*), Suma Publishers, Bangalore, India, 1986.
5. *Welding Handbook*, 3rd ed., American Welding Society, 1950.
6. Bureau of Indian Standards.
7. Lingaiah, K., *Machine Design Data Handbook*, McGraw-Hill Publishing Company, New York, 1994.

BIBLIOGRAPHY

Design of Weldments, The James F. Lincoln Arc Welding Foundation, Cleveland, Ohio, 1968.

Design of Welded Structures, The James F. Lincoln Arc Welding Foundation, Cleveland, Ohio, 1966.

Maleev, V. L. and J. B. Hartman, *Machine Design*, International Textbook Company, Scranton, Pennsylvania, 1954.

Procedure Handbook of Arc Welding Design and Practice, The James F. Lincoln Arc Welding Foundation, Cleveland, Ohio, 1950.

Salakian, A. G., and G. E. Claussen, *Stress Distribution in Fillet Welds: A Review of the Literature*, *Welding Journal*, Vol. 16, pp. 1–24, May 1937.

Shigley, J. E., *Machine Design*, McGraw-Hill Publishing Company, New York, 1956.

Spotts, M. F., *Design of Machine Elements*, 5th ed., Prentice-Hall of India Private Ltd., New Delhi, 1978.

Vallance, A., and V. L. Doughtie, *Design of Machine Members*, McGraw-Hill Publishing Company, New York, 1951.

TABLE 12-1
Weld-stress formulas

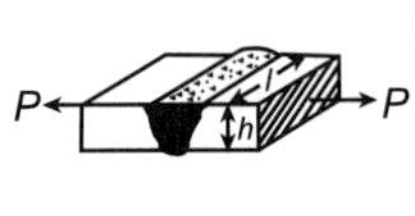 $\sigma = \frac{P}{hl}$	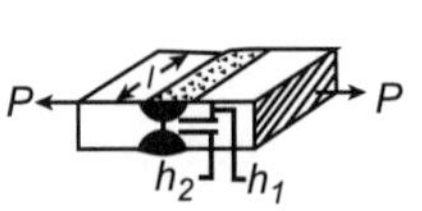 $\sigma = \frac{P}{(h_1+h_2)l}$	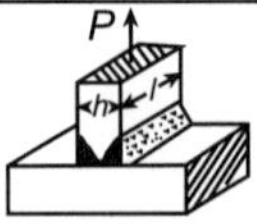 $\sigma = \frac{P}{hl}$	 $\sigma_b = \frac{6M_b}{lh^2}$	 $\sigma_b = \frac{6PL}{lh^2}$ $\tau = \frac{P}{lh}$
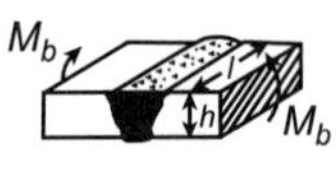 $\sigma_b = \frac{M_b}{lh}$	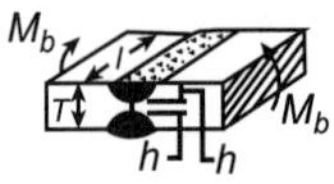 $\sigma_b = \frac{3T\,M_b}{lh(3T^2-6Th+4h^2)}$	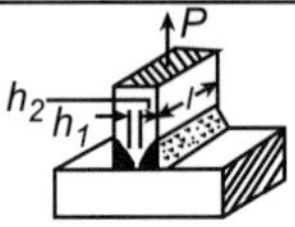 $\sigma = \frac{P}{(h_1+h_2)l}$	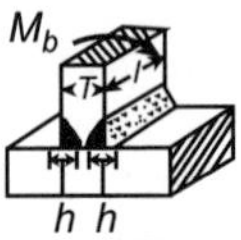 $\sigma_b = \frac{3TM_b}{lh(3T^2-6Th+4h^2)}$	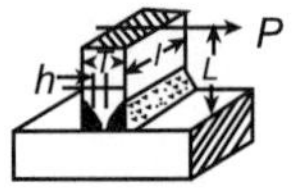 $\sigma_b = \frac{3TPL}{lh(3T^2-6Th+4h^2)}$ $\tau = \frac{P}{2lh}$
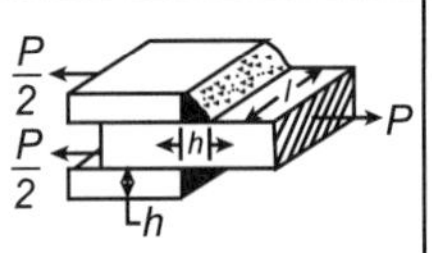 $\sigma = \frac{0.707\,P}{hl}$	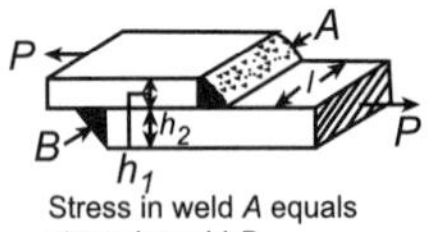 Stress in weld A equals stress in weld B $\sigma = \frac{1.414\,P}{(h_1+h_2)l}$	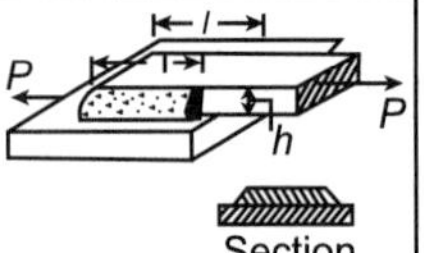 $\sigma = \frac{0.707\,P}{hl}$	 Both plates same thickness $\sigma = \frac{0.707\,P}{hl}$	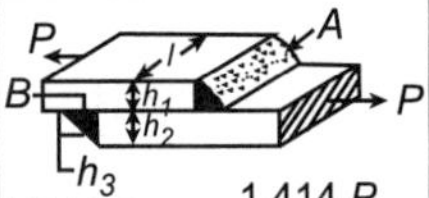 Weld A $\sigma = \frac{1.414\,P}{(h_1+h_2)l}$ Weld B $\sigma = \frac{1.414\,Ph_2}{h_3 l\,(h_1+h_2)}$
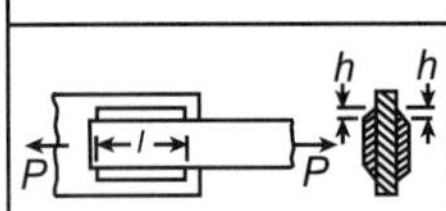 $\sigma = \frac{0.354\,P}{hl}$	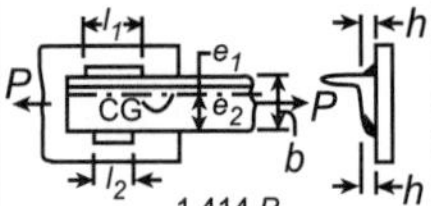 $\sigma = \frac{1.414\,P}{h\,(l_1+l_2)}$ or $l_1 = \frac{1.414\,Pe_2}{\sigma\,hb}$, $l_2 = \frac{1.414\,Pe_1}{\sigma\,hb}$	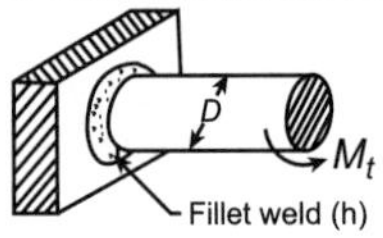 $\tau = \frac{2.83\,M_t}{hD^2\pi}$	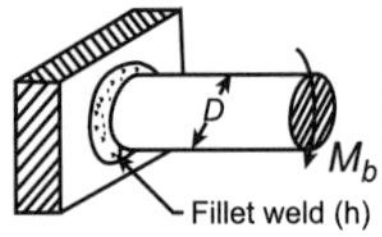 $\tau = \frac{5.66\,M_b}{hD^2\pi}$	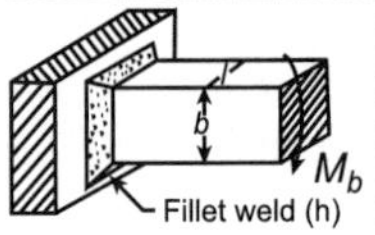 $\sigma_b = \frac{4.24M_b}{h[b^2+3l\,(b+h)]}$
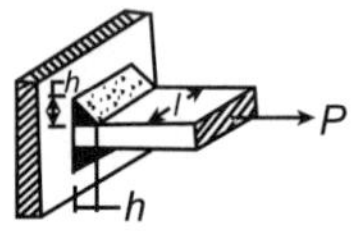 $\sigma = \frac{0.707\,P}{hl}$	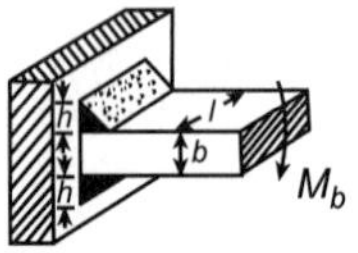 $\sigma_b = \frac{1.414\,M_b}{hl\,(b+h)}$	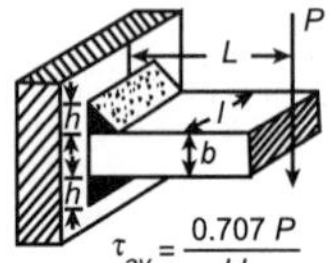 $\tau_{av} = \frac{0.707\,P}{hl}$ $\sigma_{max} = \frac{P}{hl\,(b+h)}\sqrt{2L^2 + \frac{(b+h)^2}{2}}$	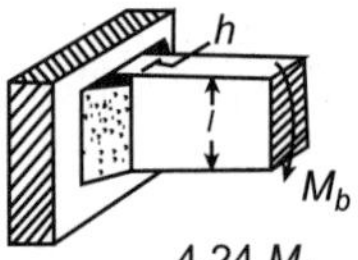 $\sigma_b = \frac{4.24\,M_b}{hl^2}$	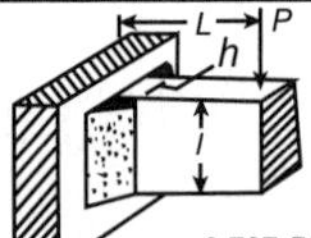 $\tau_{av} = \frac{0.707\,P}{hl}$ $\sigma_{max} = \frac{4.24\,PL}{hl^2}$
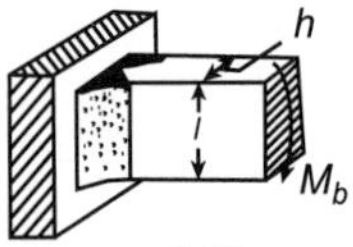 $\sigma_b = \frac{6\,M_b}{hl^2}$	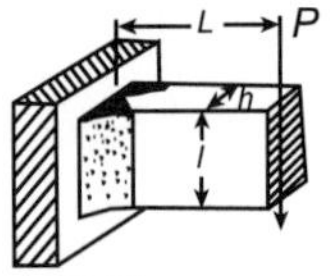 $\sigma_b = -\frac{6\,PL}{hl^2}$ $\tau = \frac{P}{hl}$	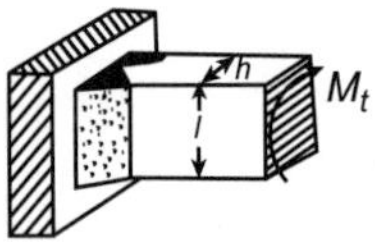 $\tau = \frac{M_t\,(3l+1.8\,h)}{h^2l^2}$	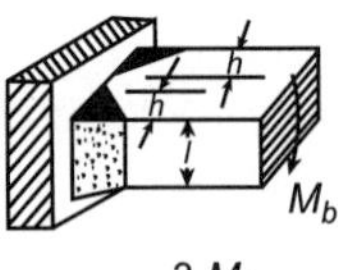 $\sigma_b = \frac{3\,M_b}{hl^2}$	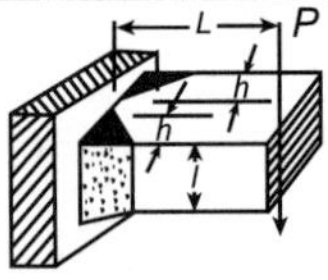 $\sigma_b = \frac{3\,PL}{hl^2}$ $\tau = \frac{P}{2\,hl}$
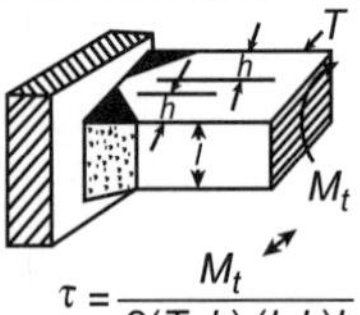 $\tau = \frac{M_t}{2(T-h)\,(l-h)h}$	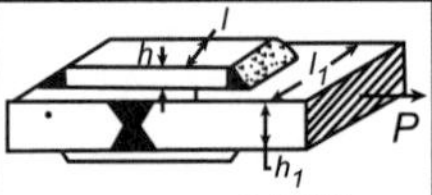 Fillet weld $\sigma = \frac{1.414\,P}{2hl+h_1l_1}$ Butt weld $\sigma = \frac{P}{2hl+h_1l_1}$	σ Normal stress, MPa(psi); τ Shear stress, MPa(psi); M_b Bending moment, N m(lbf in); M_t twisting moment, N m(lbf in);		P External load, kN(lbf); L Linear distance, m(in); h Size of weld, m(in); l Length of weld, m(in).

Source: *Welding Handbook*, 3rd edition, American Welding Society, 1950.

TABLE 12-2
Design formulas used to obtain stress in weld

Type of loading		**Standard design formula, MPa (psi)**	**Treating the weld as a line, kN/m (lbf/in)**
		Primary Welds (transmit entire load)	
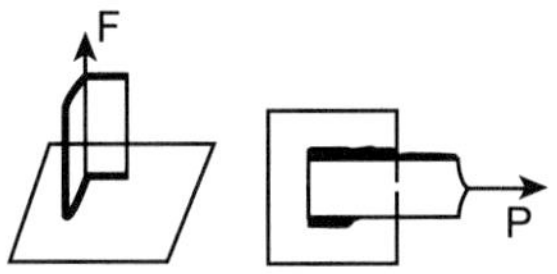	Tension or compression	$\sigma = \frac{P}{A}$	$\sigma' = \frac{P}{l_w}$
	Vertical shear	$\tau = \frac{V}{A}$	$\tau' = \frac{V}{l_w}$
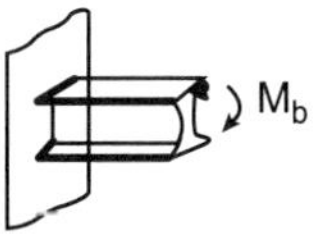	Bending	$\sigma_b = \frac{M_b}{Z}$	$\sigma' = \frac{M_b}{Z_w}$
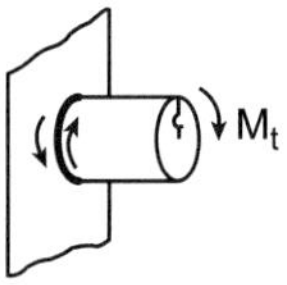	Twisting	$\tau = \frac{M_b c}{J}$	$\tau' = \frac{Mc}{J_w}$
		Secondary Welds (hold section together; low stress)	
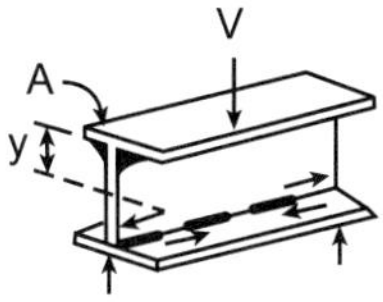	Horizontal shear	$\tau = \frac{VAy}{Ih}$	$\tau' = \frac{VAy}{I}$
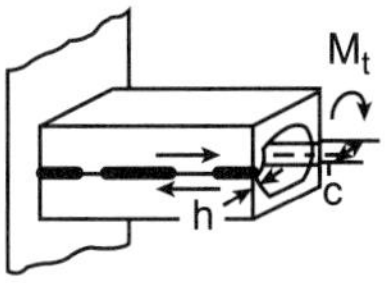	Torsional horizontal shear	$\tau = \frac{M_t c}{J}$	$\tau' = \frac{M_t ch}{J}$

TABLE 12-3
Properties of weld—treating weld as line

Outline of welded joint b = width, d = depth	Bending (about horizontal axis x–x)	Twisting
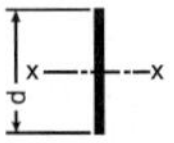	$Z_w = \dfrac{d^2}{6}$	$J_w = \dfrac{d^3}{12}$
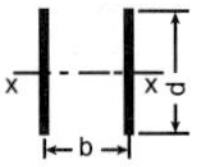	$Z_w = \dfrac{d^2}{3}$	$J_w = \dfrac{d(3b^2 + d^2)}{6}$
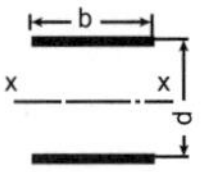	$Z_w = bd$	$J_w = \dfrac{b^3 + 3bd^2}{6}$
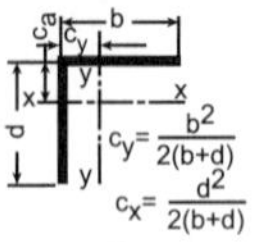	$Z_w = \dfrac{4bd + d^2}{6} = \dfrac{d^2(4bd + d)}{6(2b + d)}$ top bottom	$J_w = \dfrac{(b + d)^4 - 6b^2d^2}{12(b + d)}$
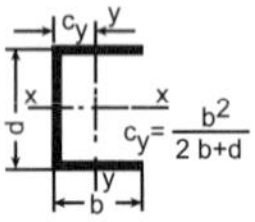	$Z_w = bd + \dfrac{d^2}{6}$	$J_w = \dfrac{(2b + d)^3}{12} - \dfrac{b^2(b + d)^2}{2b + d}$
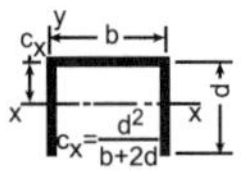	$Z_w = \dfrac{2bd + d^2}{3} = \dfrac{d^2(2b + d)}{3(b + d)}$ top bottom	$J_w = \dfrac{(b + 2d)^3}{12} - \dfrac{d^2(b + d)^2}{b + 2d}$
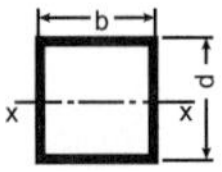	$Z_w = bd + \dfrac{d^2}{3}$	$J_w = \dfrac{(b + d)^3}{6}$
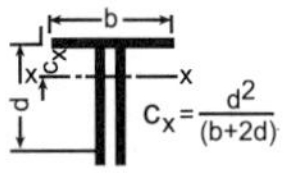	$Z_w = \dfrac{2bd + d^2}{3} = \dfrac{d^2(2b + d)}{2(b + d)}$ top bottom	$J_w = \dfrac{(b + 2d)^3}{12} - \dfrac{d^2(b + d)^2}{b + 2d}$
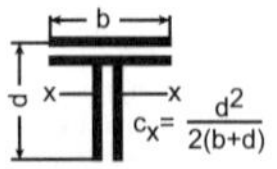	$Z_w = \dfrac{4bd + d^3}{3} = \dfrac{4bd^2 + d^3}{6b + 3d}$ top bottom	$J_w = \dfrac{d^3(4b + d)}{6(b + d)} + \dfrac{b^3}{6}$
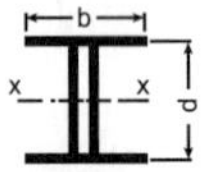	$Z_w = bd + \dfrac{d^2}{3}$	$J_w = \dfrac{b^3 + 3bd^2 + d^3}{6}$
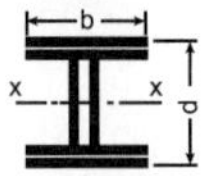	$Z_w = 2bd + \dfrac{d^2}{3}$	$J_w = \dfrac{2b^3 + 6bd^2 + d^3}{6}$
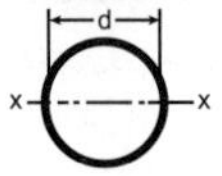	$Z_w = \dfrac{\pi d^2}{4}$	$J_w = \dfrac{\pi d^3}{4}$
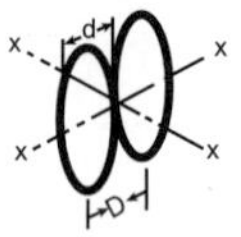	$Z_w = \dfrac{\pi d^2}{2} + \pi D^2$	—
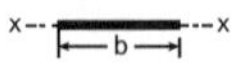	—	$J_w = \dfrac{b^3}{12}$

Note: Multiply the values J_w by the size of the weld w to obtain polar moment of inertia J_o of the weld.

TABLE 12-4
Types of welds and symbols

Form of weld	Sectional representation	Appropriate symbol	Form of weld	Sectional representation	Appropriate symbol
Fillet			Plug or slot	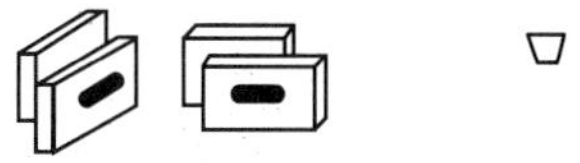	
Square butt			Backing strip		
Single-V butt			Spot	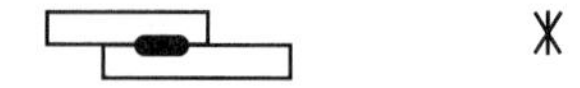	
Double-V butt			Seam		
Single-U butt			Mashed seam		
Double-U butt			Stitch		
Single-bevel butt			Mashed stitch		
Double-bevel butt			Projection		
Single-J butt			Flash	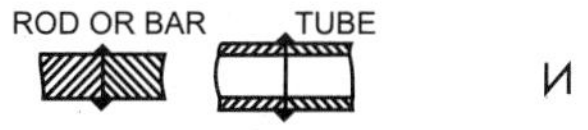	
Double-J butt			Butt resistance or Pressure (upset)	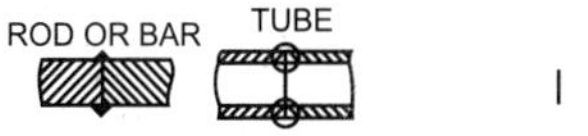	
Stud					
Bead (edge or seal)					
Sealing run					

IS: 696-1960(b) Bureau of Indian Standards.

TABLE 12-5A
Properties of common welding rods

Rods	Melting point		Tensile strength		Elongation in 50 mm (2 in), %
	°F	°C	MPa	kpsi	
Copper-coated mild steel	2750	1510	358.5	52	23
High-tensile low-alloy steel	2750	1510	427.5	62	20
Cast iron	2200	1204	275.5	40	—
Stainless steel	2550	1399	551.5	80	30
Bronze	1600–1625	870–885	379.0	55	—
Ever dur	1870	1019	344.5	50	20
Aluminum	1190	643	110.5	16	25
White metal	715	379	358.5	52	8
Low-temperature brazing rod	1170–1185	632–640	Varies with parent metal		

TABLE 12-5
Allowable loads on mild-steel fillet welds

Size of weld, mm	Allowable static load per linear cm of weld							
	Bare welding rod				Shielding arc			
	Normal weld		Parallel weld		Normal weld		Parallel weld	
	N	lbf	N	lbf	N	lbf	N	lbf
2 × 3	1667.1	375	1323.9	298	2059.4	462	1667.1	375
5 × 5	2745.8	617	2186.9	491	3432.3	772	2745.8	617
6 × 6	3285.2	738.5	2628.2	590	4118.8	926	3285.2	738.5
8 × 8	4373.7	983	3501.0	787	5491.7	1235	4373.7	983
10 × 10	5491.7	1235	4079.5	983	6864.6	1543	5491.7	1235
12 × 12	6570.4	1477	5263.3	1182	8237.5	1852	6570.4	1477
14 × 14	7659.0	1722	6129.1	1378	9581.0	2154	7659.0	1722
15 × 15	8237.5	1852	6570.4	1477	10296.9	2315	8237.5	1852
18 × 18	9855.6	2216	7884.5	1772	12326.9	2772	9855.6	2216
20 × 20	10944.2	2460	8757.3	1968	13680.2	3075	10944.2	2460

Note: For intermediate sizes interpolate the values.
Source: *Welding Handbook*, American Welding Society, 1950.

TABLE 12-6
Design stresses for welds made with mild-steel electrodes

Type of load		Bare electrodes σ_u = 274.6–380.5 MPa (40–55 kpsi)		Covered electrodes σ_u = 416.8–519.7 MPa (60–75 kpsi)	
		Static loads	Dynamic loads	Static loads	Dynamic loads
Butt Welds					
Tension	MPa	89.70	34.30	110.30	55.10
	kpsi	13.0	5.0	16.0	8.0
Compression	MPa	103.40	34.30	124.10	55.10
	kpsi	15.0	5.0	19.5	8.0
Shear	MPa	55.10	20.60	68.90	83.40
	kpsi	8.0	3.0	10.0	12.0
Fillet Welds					
Shear	MPa	78.0	20.60	96.50	34.30
	kpsi	11.5	3.0	14.0	5.0

Source: *Welding Handbook*, American Welding Society, 1950.

TABLE 12-7
Fatigue stress-concentration factors $K_{f\sigma}$

Type of weld	Stress-concentration factors, $K_{f\sigma}$
Reinforced butt weld	1.2
Toe of transverse fillet weld or normal fillet weld	1.5
End of parallel weld or longitudinal weld	2.7
T-butt joint with sharp corners	2.0

TABLE 12-8
Strength of shielded-arc flush steel welds

	Limit stress			Recommended design stress		
		Deposited metal				
Type of stress	**Base metal elastic limit, σ_e**	**Elastic limit, σ_e**	**Endurance limit, σ_f**	**Static load**	**Load varies from O to F**	**Load varies from $+F$ to $-F$**
Tension						
MPa	220.60	275.80	151.70	110.30	100.00	55.20
kpsi	32	40	22	16	14.5	8.0
Compression						
MPa	241.20	303.40	—	124.20	110.30	55.23
kpsi	35.0	44.0	—	10.0	16.0	8.0
Bending						
MPa	241.20	303.40	179.30	124.20	110.30	62.10
kpsi	35	44	26	18	16	9.0
Shear						
MPa	137.90	165.40	—	75.80	68.90	34.50
kpsi	20	24	—	11	10	5
Shear and tension						
MPa	—	—	—	75.80	68.90	34.50
kpsi	—	—	—	11	10	5

For bare electrode welds, the allowable stress must be multiplied by 0.8 and for gas welds, they should be multiplied by 0.8 to 0.85.

TABLE 12-9
Length and spacing of intermittent welds

***R*, % of continuous weld**	**Length of intermittent welds and distance between centers, mm**		
75		75–100[a]	
66			100–150
60		75–125	
57			100–175
50	50–100	75–150	100–200
44			100–225
43		75–175	
40	50–125		100–250
37		75–200	
33	50–160	75–225	100–300
30		75–250	
25	50–200	75–300	
20	50–250		
16	50–300		

[a] 75–100 means a weld 75 mm long with a distance of 100 mm between the centers of two consecutive welds.

$$R \text{ in } \% = \frac{\text{calculated leg size (continuous)}}{\text{actual leg size used (intermittent)}}$$

TABLE 12-10
Fatigue data on butt weld joints (average strength values)

Material and joint		Base metal σ_u	Base metal σ_y	Endurance strength, σ_f: $K = -1$[a]		$K = 0$[a]		$K = 0.5$[a]	
				No. of cycles 2×10^6					
Carbon steel	MPa	423	235						
	kpsi	61.3	34.0						
With bead, or welded	MPa			100.0	152.0	155.9	227.5	253.0	368.7
	kpsi			14.5	22.0	22.5	33	37	53.5
With bead, tempered	MPa			98.0	148.0	160.8	214.7	264.7	379.5
923 K (650°C)	kpsi			14	21.5	23	31	38	55
Bead machined off	MPa			121.6	198.0	198.0	335.3	304.0	
	kpsi			17.5	28.5	28.5	48.5	44	
Bead machined off, tempered 923 K (650°C)	MPa			114.7	193.1	132.3	340.2	292.2	
	kpsi			16.5	28	19	49.3	42.4	
Alloy steel	MPa	745.6	672.0						
	kpsi	108	97.5						
As welded	MPa			400.1	539.3				
	kpsi			58	78				
Stress-relieved	MPa			456.0	593.2				
	kpsi			66	86				

[a] $K = +1$ steady; $K = -1$ complete reversal; $K = 0$ repeated; $K = \frac{1}{2}$ fluctuating; $K = \dfrac{\text{min stress}}{\text{max stress}}$.

Source: *Design of Weldments*, The James F. Lincoln Arc Welding Foundation, Cleveland, Ohio, 1968.

TABLE 12-11
Stresses as per the AISC Code for weld metal

Load type	Weld type	Allowable stress, σ_a
Tension	Butt	$0.60\ \sigma_y$
Compression	Butt	$0.60\ \sigma_y$
Shear	Butt or fillet	$0.40\ \sigma_y$
Bending	Butt	$0.90\ \sigma_y$
Bending	Butt	$0.60\ \sigma_y$–$0.66\ \sigma_y$

TABLE 12-12
Properties of weld metal

AWS electrode number[a]	Elongation %	Tensile strength MPa	Tensile strength kpsi	Yield strength MPa	Yield strength kpsi
E 60xx	17–25	427	62	345	50
E 70xx	22	483	70	393	57
E 80xx	19	550	80	462	67
E 90xx	14–17	620	90	530	77
E 100xx	13–16	690	100	600	87
E 120xx	14	828	120	738	107

[a] The American Welding Society (AWS) Specification Code numbering system for electrodes.

TABLE 12-13
Selection of fillet weld sizes by rule-of-thumb (all dimensions in mm)

Plate thickness, h mm	Designing for strength, full-strength weld ($w = 3/4h$)	Designing for rigidity: 50% of full-strength weld ($w = 3/8h$)	Designing for rigidity: 33% of full-strength weld ($w = 1/4h$)
6	4.5	4.5	4.5
8	6	4.5	4.5
9.5	8	4.5	4.5
11	9.5	4.5	4.5
12.5	9.5	4.5	4.5
14	11	6	6
15.5	12.5	6	6
19	14	8	6
22	15.5	9.5	8
25	19	9.5	8
28.5	22	11	8
31.5	25	12.5	8
35	25	12.5	9.5
37.5	28.5	14	9.5
41	31.5	15.5	11
44	35	19	11
50	37.5	19	12.5
54	41	22	14
57	44	22	14
60	44	25	15.5
62.5	47.5	25	15.5
66.5	50	25	19
70	50	25	19
75	56	28.5	19

Source: *Welding Handbook*, 3rd edition, American Welding Society, 1950.

TABLE 12-14
Equivalent length of fillet weld to replace rivets

Rivet diameter, mm	Rivet shear value at 100 MPa ($10.2\ kgf/mm^2$)		Length of fillet welds[a] "Fusion Code" (structural) shielded arc welding, mm				
	MPa	kgf/mm^2	6-mm fillet	8-mm fillet	9.5-mm fillet	12.5-mm fillet	15.5-mm fillet
12.5	20.0	2.07	37.5	31.5	28.5	22	19
15.5	31.5	3.23	56	44.0	37.5	31.5	25
19	45.5	4.66	75	61.5	54	41	35
22	61.0	6.34	105	85.5	73	54	44
25	81.2	8.28	133	108.0	92	70	56

[a] 6 mm is added to calculated length of bead for starting and stopping the arc.

TABLE 12-15
Stress concentration factor, K_σ

Weld type and metal	Stress concentration factor, K_σ	
	Low-carbon steel	Low-alloy steel
Weld metal		
Butt welds with full penetration	1.2	1.4
End fillet welds	2	2.5
Parallel fillet welds	3.5	4.5
Base metal		
Toe of machined butt weld	1.2	1.4
Toe of unmachined butt weld	1.5	1.9
Toe of machined end fillet weld with leg ratio 1:1.5	2	2.5
Toe of unmachined end fillet weld with leg ratio 1:1.5	2.7	3.3
Parallel fillet weld	3.5	4.5
Stiffening ribs and partitions welded with end fillet welds having smooth transitions at the toes	1.5	1.9
Butt and T-welded corner plates	2.7	3.3
Butt and T-welded corner plates, but with smooth transitions in the shape of the plates and with machined welds	1.5	1.9
Lap-welded corner plates	2.7	3.3

TABLE 12-16
Allowable stresses for welds under static loads

Weld type and process	Allowable stresses		
	Tension, σ_{ta}	Compression, σ_{ca}	Shear, τ_a
Automatic and hand welding with shielded arc and butt welding	σ_t [a]	σ_t	$0.65\sigma_t$
Hand welding with ordinary quality electrodes	$0.9\sigma_t$	σ_t	$0.6\sigma_t$
Resistance spot welding	$0.9\sigma_t$	σ_t	$0.5\sigma_t$

[a] σ_t is the allowable stress in tension of the base metal of the weld.

CHAPTER 13

RIVETED JOINTS

SYMBOLS[2,3,4]

A	area of cross-section, m^2 (in^2)
	the cross-sectional area of rivet shank, m^2 (in^2)
b	breadth of cover plates (also with suffixes), m (in)
c	distance from the centroid of the rivet group to the critical rivet, m (in)
d	diameter of rivet, m (in)
D_i	internal diameter of pressure vessel, m (mm)
e or l	eccentricity of loading, m (in)
F	force on plate or rivets (also with suffixes), kN (lbf)
h	thickness of plate or shell, m (in)
h_c, h_1, h_2	thickness of cover plate (butt strap), m (in)
i	number of rivets in a pitch fine (also with suffixes 1 and 2, respectively, for single shear and double shear rivets)
I	moment of inertia, area, m^4, cm^4 (in^4)
J	moment of inertia, polar, m^4, cm^4 (in^4)
$K = \dfrac{F}{F'}$	coefficient (Table 13-11)
m	margin, m (in)
M_b	bending moment, N m (lbf in)
p	pitch on the gauge line or longitudinal pitch, m (in)
p_c	pitch along the caulking edge, m (in)
p_d	diagonal pitch, m (in)
p_t	transverse pitch, m (in)
P_f	intensity of fluid pressure, MPa (psi)
Z	section modulus of the angle section, m^3, cm^3 (in^3)
σ_θ	hoop stress in pressure vessel or normal stress in plate, MPa (psi)
σ_a	allowable normal stress, MPa (psi)
σ_c	crushing stress in rivets, MPa (psi)
τ	shear stress in rivet, MPa (psi)
τ_a	allowable shear stress, MPa (psi)
η	efficiency of the riveted joint
θ	angle between a line drawn from the centroid of the rivet group to the critical rivet and the horizontal (Fig. 13-5)

Particular	Formula

PRESSURE VESSELS

Thickness of main plates

The thickness of plate of the pressure vessel with longitudinal joint

$$h = \frac{P_f D_i}{2\eta\sigma_\theta} \quad (13\text{-}1)$$

For thickness of boiler plates and suggested types of joints

Refer to Tables 13-1 and 13-2.

The thickness of plate of the pressure vessel with circumferential joint

$$h = \frac{P_f D_i}{4\eta\sigma_\theta} \quad (13\text{-}2)$$

For allowable stress and efficiency of joints

Refer to Tables 13-3, 13-4, 13-5, and 13-6.

PITCHES

Lap joints

The diagonal pitch (staggered) (Fig. 13-1) for p, p_t, and p_d

$$p_d = \frac{2p + d}{3} \quad (13\text{-}3)$$

Refer to Tables 13-7 and 13-8 for rivets for general purposes and boiler rivets.

The distance between rows or transverse pitch or back pitch (staggered)

$$p_t = \sqrt{\left(\frac{2p+d}{3}\right)^2 - \left(\frac{p}{2}\right)^2} \quad (13\text{-}4)$$

The rivet diameter

$$d = 0.19\sqrt{h} \text{ to } 0.2\sqrt{h} \qquad \textbf{SI} \quad (13\text{-}5a)$$

where h and d in m

$$d = 1.2\sqrt{h} \text{ to } 1.4\sqrt{h} \qquad \textbf{USCS} \quad (13\text{-}5b)$$

where h and d in in

$$d = 6\sqrt{h} \text{ to } 6.3\sqrt{h} \qquad \textbf{CM} \quad (13\text{-}5c)$$

where h and d on mm

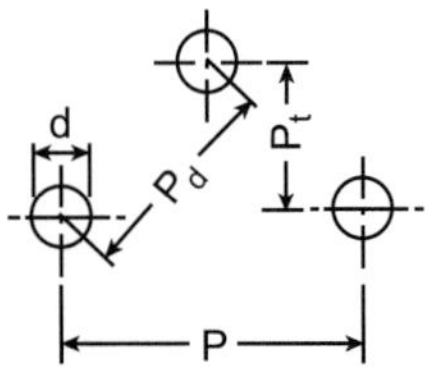

FIGURE 13-1 Pitch relation

TABLE 13-1
Suggested types of joint

Diameter of shell, mm (in)	Thickness of shell, mm (in)	Type of joint
600–1800 (24–72)	6–12 (0.25–0.5)	Double-riveted
900–2150 (36–84)	7.5–25 (0.31–1.0)	Triple-riveted
1500–2750 (60–108)	9.0–44 (0.375–1.75)	Quadruple-riveted

TABLE 13-2
Minimum thickness of boiler plates

Shell plates		Tube sheets of firetube boilers	
Diameter of shell, mm (in)	**Minimum thickness after flanging, mm (in)**	**Diameter of tube sheet, mm (in)**	**Minimum thickness, mm (in)**
≤900 (36)	6.0 (0.25)	≤1050 (42)	9.5 (0.375)
900–1350 (36–54)	8.0 (0.3125)	1050–1350 (42–54)	11.5 (0.4375)
1350–1800 (54–72)	9.5 (0.375)	1350–1800 (54–72)	12.5 (0.50)
≥1800 (72)	12.5 (0.5)	≤1800 (72)	14.0 (0.5625)

TABLE 13-3
Efficiency of riveted joints (η)

	% Efficiency, η	
Type of joint	**Normal range**	**Maximum**
Lap joints		
Single-riveted	50–60	63
Double-riveted	60–72	77
Triple-riveted	72–80	86.6
Butt joints (with two cover plates)		
Single-riveted	55–60	63
Double-riveted	76–84	87
Triple-riveted	80–88	95
Quadruple-riveted	86–94	98

TABLE 13-4
Allowable stresses in structural riveting (σ_b)

			Rivets acting in single shear		Rivets acting in double shear	
Load-carrying member	**Type of stress**	**Rivet-driving method**	**MPa**	**kpsi**	**MPa**	**kpsi**
Rolled steel SAE 1020	Tension		124	18.0	124	18.0
	Shear	Power	93	13.5	93	13.5
Rivets, SAE 1010	Shear	Hand	68	10.0	68	10.0
	Crushing	Power	165	24.0	206	30.0
	Crushing	Hand	110	16.0	137	20.0

TABLE 13-5
Allowable stress for aluminum rivets, σ_a

		Allowable stress[a], σ_a			
		Shear		Bearing	
Rivet alloy	Procedure of drawing	MPa	kpsi	MPa	kpsi
2S (pure aluminum)	Cold, as received	20	3.0	48	7.0
17S	Cold, immediately after quenching	68	10.0	179	26.0
17S	Hot, 500–510°C	62	9.0	179	26.0
615–T6	Cold, as received	55	8.0	103	15.0
53S	Hot, 515–527°C	41	6.0	103	15.0

[a] Actual safety factor or reliability factor is 1.5.

TABLE 13-6
Values of working stress[a] at elevated temperatures

Maximum temperatures		Minimum of the specified range of tensile strength of the material, MPa (kpsi)									
		(45)	311	(50)	344	(55)	380	(60)	413	(75)	517
°F	°C	MPa	kpsi	MPa	kpsi	MPa	kpsi	MPa	kpsi	MPa	kpsi
0–700	0–371	61	9.0	68	10.0	76	11.00	82	12.00	103	15.00
750	399	56	8.22	62	9.11	68	10.00	77	11.20	89	13.00
800	427	45	6.55	53	7.33	54	8.00	61	9.00	70	10.20
850	455	37	5.44	41	6.05	46	6.75	51	7.40	57	8.30
900	482	29	4.33	33	4.83	37	5.50	38	5.60	41	6.00
950	511	22	3.20	26	3.60	27	4.00	27	4.00	27	4.00

[a] Design stresses of pressure vessels are based on a safety factor of 5.

TABLE 13-7
Pitch of butt joints

Type of joint	Diameter of rivets, d, mm	Pitch, p
Double-riveted— use for $h \leq 12.5$ mm (0.5 in)	Any	$5.5d$ (approx.)
Triple-riveted—	1.75–23.80	$8d$–$8.5d$
use for $h \leq 25$ mm	27.00	$7.5d$
(1 in)	30.15–36.50	$6.5d$–$7d$
Quadruple-riveted—	17.50–23.80	$16d$–$17d$
use for $h \leq 31.75$ mm	27.00	$15d$ (approx.)
(1.25 in)	30.15	$14d$ (approx.)
	33.30–36.50	$13d$–$14d$

TABLE 13-8
Transverse pitch (p_t) as per ASME Boiler Code

Value of p/d	1	2	3	4	5	6	7
Value of p_t	$2d$	$2d$	$2d$	$2d$	$2d$	$2.2d$	$2.3d$

Particular	Formula

Butt joint

The transverse pitch

$$p_t = 2d \text{ to } 2.5d \tag{13-6a}$$

$$p_t \geq \sqrt{0.5pd + 0.25d^2} \tag{13-6b}$$

For rivets, rivet holes, and strap thick

Refer to Tables 13-9, 13-10, and Fig. 13-2.

TABLE 13-9
Rivet hole diameters

Diameter of rivet, mm	Rivet hole diameters, mm (min)
12	13
14	15
16	17
18	19
20	21
22	23
24	25
27	28.5
30	31.5
33	34.5
36	37.5
39	41.0
42	44
48	50

TABLE 13-10
Rivet hole diameters and strap thickness

Plate thickness, h, mm	Minimum strap thickness, h_c mm	Hole diameter, d, mm	Plate thickness, h, mm	Minimum strap thickness, h_c mm	Hole diameter, d, mm
6.25			14.25	11.10	
7.20	6.25	17.50			27.0
8.00			15.90	12.50	
8.75		20.50	19.00		30.15
9.50					
10.30	8.00		22.25	15.90	33.30
11.10			25.00	12.50	
12.00	9.50	24.00	28.50	19.00	36.50
12.50			31.75	22.25	
13.50	11.10		83.10	25.00	39.70

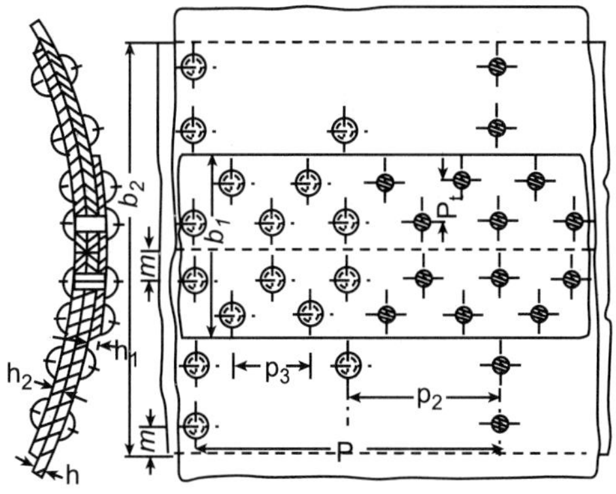

FIGURE 13-2 Quadruple-riveted double-strap butt joint.

Particular	Formula
Minimum transverse pitch as per *ASME Boiler Code*	$p_t = 1.75d \quad \text{if } \frac{p}{d} \leq 4$ (13-7a)
	$p_t = 1.75d + 0.001(p - d) \quad \text{if } \frac{p}{d} > 4$ **SI** (13-8a)
	where p_t, p, and d in m
	$p_t = 1.75d + 0.1(p - d) \quad \text{if } \frac{p}{d} > 4$ **USCS** (13-8b)
	where p_t, d, and p in in
For transverse pitches	Refer to Table 13-8.
Haven and Swett formula for permissible pitches along the caulking edge of the outside cover plate	$p_c - d = 14\sqrt[4]{\frac{h_c^3}{P_f}}$ **CM** (13-9a)
	where p_c, d, h_c in cm, and P_f in kgf/cm^2
	$p_c - d = 21.38\sqrt[4]{\frac{h_c^3}{P_f}}$ **USCS** (13-9b)
	where p_c, d, h_c in in, and P_f in psi
	$p_c - d = 77.8\sqrt[4]{\frac{h_c^3}{P_f}}$ **SI** (13-9c)
	where p_c, d, h_c in m, and P_f in N/m^2
Diagonal pitch, p_d, is calculated from the relation	$2(p_d - d) \geq (p - d)$ (13-10)

MARGIN

Margin for longitudinal seams of all pressure vessels and girth seams of power boiler having unsupported heads	$m = 1.5d$ to $1.75d$ (13-11a)
Margin for girth seams of power boilers having supported heads and all unfired pressure vessels	$m \geq 1.25d$ (13-11b)

COVER PLATES

The thickness of cover plate	$h_c = 0.6h + 0.0025 \quad \text{if } h \leq 0.038\,\text{m}$ **SI** (13-12a)
	where h_c and h in m
	$h_c = 0.6h + 0.1 \quad \text{if } h \leq 1.5\,\text{in}$ **USCS** (13-12b)
	where h_c and h in in
	$h_c = 0.67h \quad \text{if } h > 0.038\,\text{m}$ **SI** (13-12c)
	where h_c and h in m
	$h_c = 0.67h \quad \text{if } h > 1.5\,\text{in}$ **USCS** (13-12d)
	where h_c and h in in

TABLE 13-11
Rivet groups under eccentric loading value of coefficient *K*

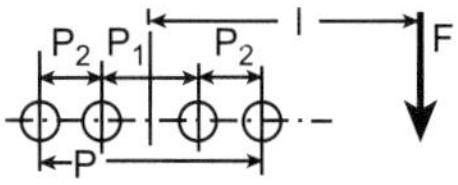

Case 1

$$K = \frac{1}{\dfrac{lp}{p_1^2 + p^2} + \dfrac{1}{4}}$$

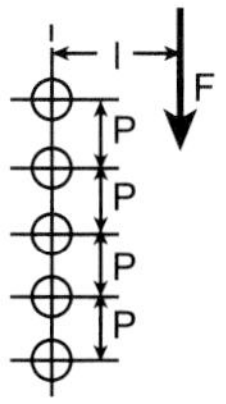

Case 2

$$K = \frac{n}{\sqrt{\left[\dfrac{6l}{(n+1)p_t}\right]^2 + 1}}$$

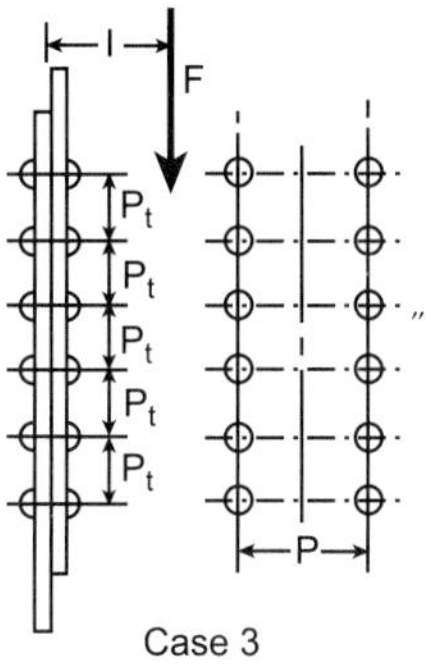

Case 3

$$K = \frac{n}{\dfrac{Alcn}{2I} + \sqrt{\left(\dfrac{Alcn}{2I}\right)^2 + 1}}$$

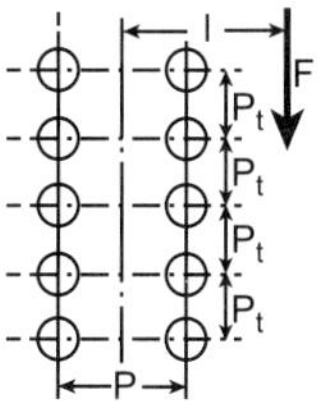

Case 4

$$K = \frac{n}{\sqrt{\left[\dfrac{l(n-1)p_t}{p^2 + \frac{1}{3}(n^2-1)p_t^2}\right]^2 + \left[\dfrac{lp}{p^2 + \frac{1}{3}(n^2-1)p_t^2} + \dfrac{1}{2}\right]^2}}$$

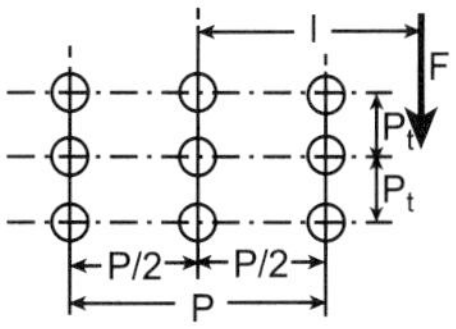

Case 5

$$K = \frac{n}{\sqrt{\left[\dfrac{l(n-1)p_t}{p^2 + \frac{1}{2}(n-1)p_t^2}\right]^2 + \left[\dfrac{lp}{p^2 + \frac{1}{3}(n^2-1)p_t^2} + \dfrac{1}{3}\right]^2}}$$

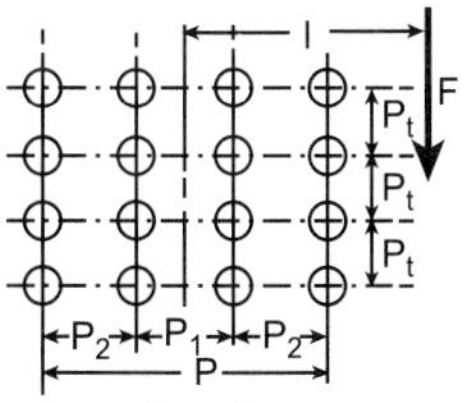

Case 6

$$K = \frac{n}{\sqrt{\left(\dfrac{l(n-1)p_t}{p_1^2 + p^2 + \frac{2}{3}(n^2-1)p_t^2}\right)^2 + \left(\dfrac{lp}{p_1^2 + p^2 + \frac{2}{3}(n^2-1)p_t^2} + \dfrac{1}{4}\right)^2}}$$

Key:
n = total number of rivets in a column
F = permissible load, acting with lever arm, l, kN (lbf)
F' = permissible load on one rivet, kN (lbf)
$K = F/F'$, coefficient
Source: K. Lingaiah and B. R. Narayana Iyengar, *Machine Design Data Handbook* (*fps Units*), Engineering College Cooperative Society, Bangalore, India, 1962; K. Lingaiah and B. R. Narayana Iyengar, *Machine Design Data Handbook*, Vol. I (*SI and Customary Metric Units*), Suma Publishers, Bangalore, India, 1983; and K. Lingaiah, *Machine Design Data Handbook*, Vol. II (*SI and Customary Metric Units*), Suma Publishers, Bangalore, India, 1986.

Particular	Formula	
Thickness of the cover plate according to Indian Boiler Code		
Thickness of single-butt cover plate	$h_1 = 1.125h$	(13-13)
Thickness of single-butt cover plate omitting alternate rivet in the over rows	$h_2 = 1.25h \dfrac{p-d}{p-2d}$	(13-14)
Thickness of double-butt cover plates of equal width	$h_c = h_1 = h_2 = 0.625h$	(13-15)
Thickness of double-butt cover plates of equal width omitting alternate rivet in the outer rows	$h_c = h_1 = h_2 = 0.625h \dfrac{p-d}{p-2d}$	(13-16)
Thickness of the double-butt cover plates of unequal width	$h_1 = 0.625h$ for narrow strap	(13-17a)
	$h_2 = 0.750h$ for wide strap	(13-17b)
For thickness of cover plates	Refer to Table 13-10.	
The width of upper cover plate (narrow strap)	$b_1 = 4m + 2p_{t1}$	(13-18)
The width of lower cover plate (wide strap)	$b_2 = b_1 + 2p_{t2} + 4m$	(13-19)
STRENGTH ANALYSIS OF TYPICAL RIVETED JOINT (Fig. 13-2)		
The tensile strength of the solid plate	$F_\theta = ph\sigma_\theta$	(13-20)
The tensile strength of the perforated strip along the outer gauge line	$F_\theta = (p-d)h\sigma_\theta$	(13-21)
The general expression for the resistance to shear of all the rivets in one pitch length	$F_\tau = (2i_2 + i_1) \dfrac{\pi d^2}{4} \tau$	(13-22)
The general expression for the resistance to crushing of the rivets	$F_c = (i_2 h + i_1 h_2) d\sigma_c$	(13-23)
The resistance against failure of the plate through the second row and simultaneous shearing of the rivets in the first row	$F_{\tau 1} = (p-2d)h\sigma_\theta + \dfrac{\pi d^2}{4} \tau$	(13-24)
The resistance against failure of the plate through the second row and simultaneous crushing of the rivets in the first row	$F_{c1} + (p-2d)h\sigma_\theta + dh\sigma_c$	(13-25)
The resistance against shearing of the rivets in the outer row and simultaneous crushing of the rivets in the two inner rows	$F_{\tau c} = \dfrac{\pi}{4} d^2\tau + idh\sigma_c$	(13-26)

Particular	Formula

EFFICIENCY OF THE RIVETED JOINT

The efficiency of plate

$$\eta = \frac{p-d}{p} \tag{13-27}$$

The efficiency of rivet in general case

$$\eta = \frac{\pi d^2 \tau (i_1 + 2i_2)}{4ph\sigma_\theta} = \frac{\left(i_2 + i_1 \frac{h_2}{h}\right)\sigma_c}{\left(i_2 + i_1 \frac{h_2}{h}\right)\sigma_c + \sigma_\theta} \tag{13-28}$$

For efficiency of joints — Refer to Table 13-3.

The diameter of the rivet in general case

$$d = \frac{4hi_2 + i_1 h_2 \sigma_c}{\pi(i_1 + 2i_2)\tau} \tag{13-29}$$

Note: for lap joint $i_2 = 0$
for butt joint $i_1 = 0$

The pitch in general case

$$p = \frac{(2i_2 + i_1)\pi d^2 \tau}{4h\sigma_\theta} + d \tag{13-30}$$

For pitch of joint — Refer to Table 13-7.

THE LENGTH OF THE SHANK OF RIVET (Fig. 13-3)

$$L = h + h_1 + h_2 + (1.5 \text{ to } 1.7)D \tag{13-31a}$$

$$L = h + h_c + (1.5 \text{ to } 1.7)D \tag{13-31b}$$

for butt joint with single cover plate

$$L = 2h + (1.5 \text{ to } 1.7)D \tag{13-31c}$$

for lap joint

where D = diameter of rivet

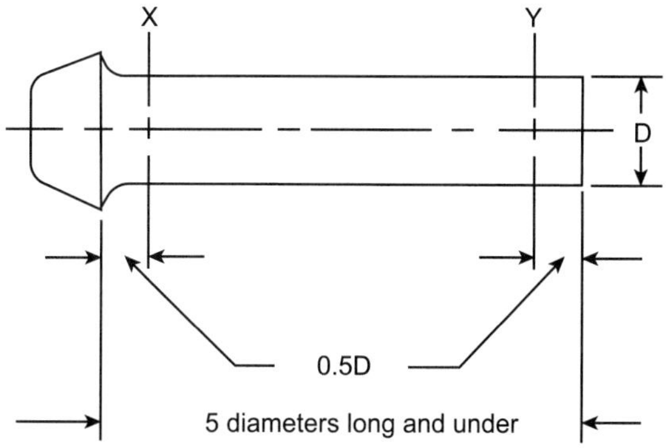

FIGURE 13-3

Particular	Formula

STRUCTURAL JOINT

Riveting of an angle to a gusset plate (Fig. 13-4)

The resultant normal stress

$$\sigma = \frac{F}{A} + \frac{Fe}{Z} \tag{13-32}$$

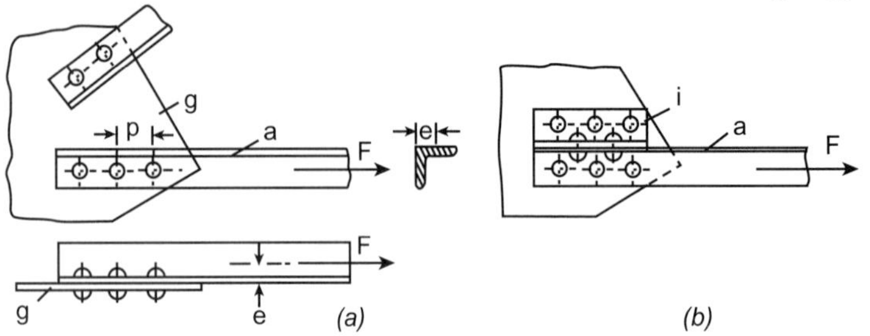

FIGURE 13-4 Riveting of an angle to a gusset plate.

RIVETED BRACKET (Fig. 13-5)

The resultant load on the farthest rivet whose distance is c from the center of gravity of a group of rivets (Fig. 13-5)

$$F_R = \left[\left(\frac{F}{nn'}\right)^2 + \left(\frac{M_b c}{\sum x^2 + \sum y^2}\right)^2 + 2\left(\frac{F}{nn'}\right)\left(\frac{M_b c}{\sum x^2 + \sum y^2}\right)\cos\theta\right]^{1/2} \tag{13-33}$$

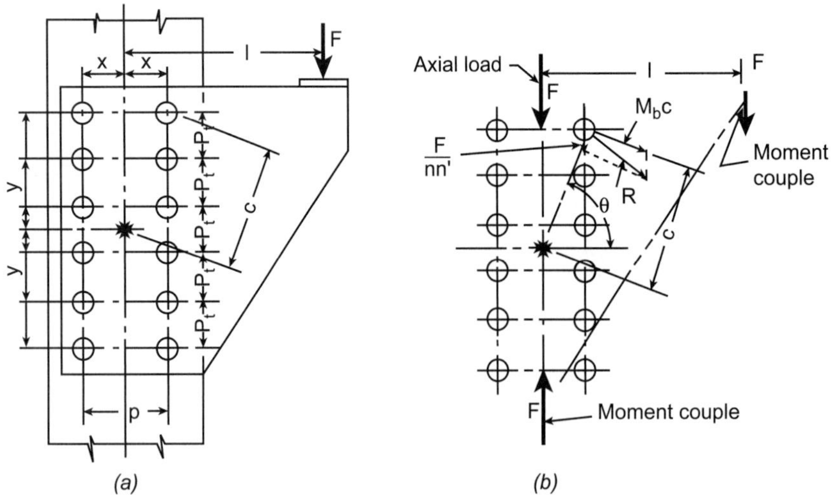

FIGURE 13-5 Riveted bracket. (Bureau of Indian Standards.)

Particular	Formula
	where n = number of rivets in one column n' = number of rivets in one row x, y have the meaning as shown in Fig. 13-5
For rivet groups under eccentric loading value of coefficient K	Refer to Table 13-11.
For preferred length and diameter of rivets	Refer to Figs. 13-6 to 13-8 and Tables 13-12 to 13-13.
For collected formulas of riveted joints	Refer to Table 13-14.

REFERENCES

1. Maleev, V. L., and J. B. Hartmen, *Machine Design,* International Textbook Company, Scranton, Pennsylvania, 1954.
2. Lingaiah, K., and B. R. Narayana Iyengar, *Machine Design Data Handbook* (*fps Units*), Engineering College Cooperative Society, Bangalore, India, 1962.
3. Lingaiah, K., and B. R. Narayana Iyengar, *Machine Design Data Handbook,* Vol. I (*SI and Customary Metric Units*), Suma Publishers, Bangalore, India, 1983.
4. Lingaiah, K., *Machine Design Data Handbook,* Vol. II (*SI and Customary Metric Units*), Suma Publishers, Bangalore, India, 1986.
5. Bureau of Indian Standards.
6. Lingaiah, K., *Machine Design Data Handbook*, McGraw-Hill Publishing Company, New York, 1994.

BIBLIOGRAPHY

Faires, V. M., *Design of Machine Elements,* The Macmillan Company, New York, 1965.

Norman, C. A., E. S. Ault, and I. F. Zarobsky, *Fundamentals of Machine Design,* The Macmillan Company, New York, 1951.

Vallance, A., and V. L. Doughtie, *Design of Machine Members,* McGraw-Hill Publishing Company, New York, 1951.

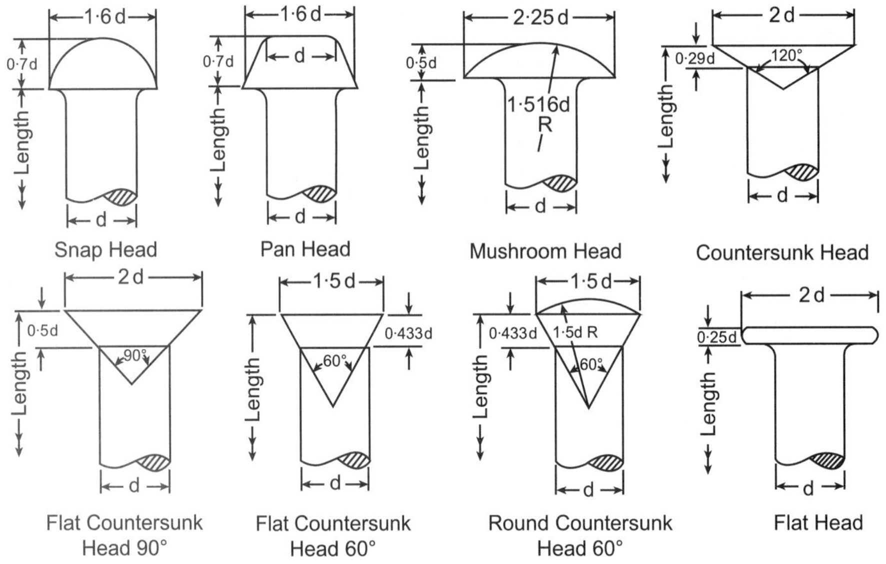

FIGURE 13-6 Rivets for general purposes (less than 12 mm diameter). For preferred length and diameter combination, refer to Table 13-12.

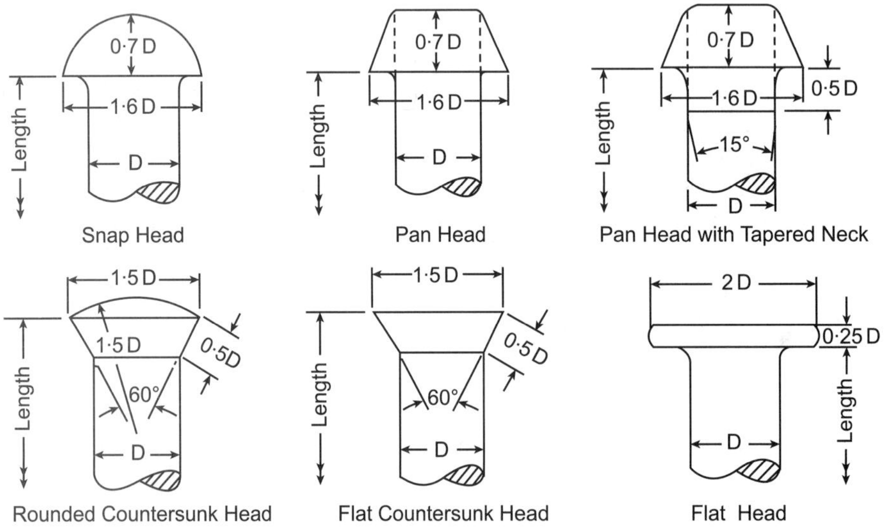

FIGURE 13-7 Rivets for general purposes (12 to 48 mm diameter). For preferred length and diameter combination, refer to Table 13-13.

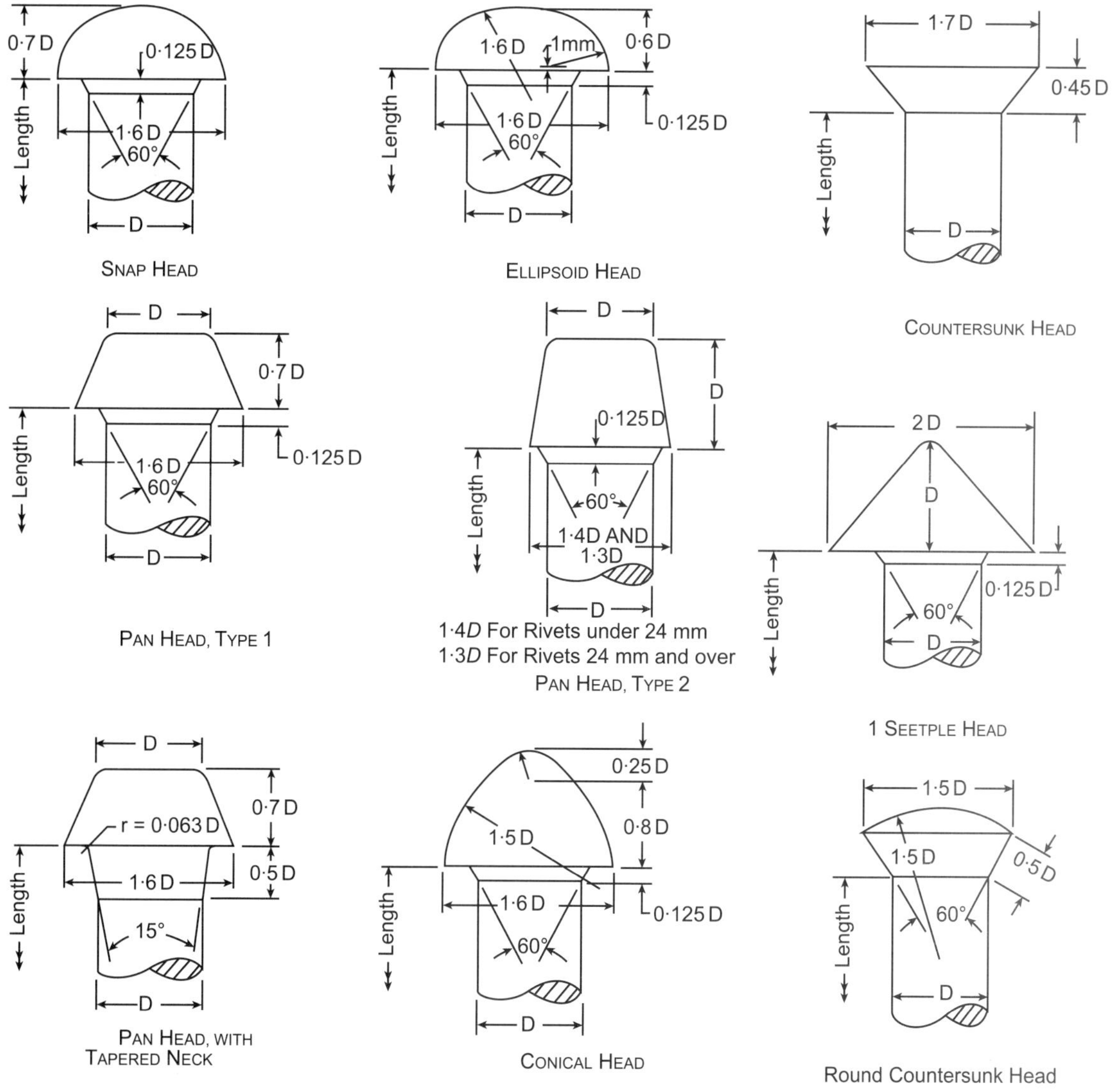

FIGURE 13-8 Boiler rivets (12 to 48 mm diameter). For preferred length and diameter combination, refer to Table 13-13.

TABLE 13-12
Preferred length (×) and diameter combinations for rivets (Fig. 13-6)

	Diameter, mm								
Length, mm	**1.6**	**2**	**2.5**	**3**	**4**	**5**	**6**	**8**	**10**
5	×	—	—	—	—	—	—	—	—
6	×	×	×	×	—	—	—	—	—
7	×	×	×	×	—	—	—	—	—
8	×	×	×	×	×	—	—	—	—
9	×	×	×	×	×	—	—	—	—
10	×	×	×	×	×	×	—	—	—
12	—	×	×	×	×	×	×	—	—
14	—	—	×	×	×	×	×	×	—
16	—	—	×	×	×	×	×	×	—
18	—	—	—	×	×	×	×	×	×
20	—	—	—	×	×	×	×	×	×
22	—	—	—	×	×	×	×	×	×
24	—	—	—	×	×	×	×	×	×
26	—	—	—	×	×	×	×	×	×
28	—	—	—	×	×	×	×	×	×
30	—	—	—	×	×	×	×	×	×
35	—	—	—	×	×	×	×	×	×
40	—	—	—	—	×	×	×	×	×
45	—	—	—	—	×	×	×	×	×
50	—	—	—	—	—	—	×	×	×
55	—	—	—	—	—	—	×	×	×
60	—	—	—	—	—	—	—	×	×
65	—	—	—	—	—	—	—	×	×
70	—	—	—	—	—	—	—	×	×

Source: Bureau of Indian Standards, IS: 2155, 1962.

TABLE 13-13
Preferred lengths (×) and diameter combinations of rivets (Fig. 13-7)

	Diameter, mm													
Length, mm	**12**	**14**	**16**	**18**	**20**	**22**	**24**	**27**	**30**	**33**	**36**	**39**	**42**	**48**
28	×	—	—	—	—	—	—	—	—	—	—	—	—	—
31.5	×	×	—	—	—	—	—	—	—	—	—	—	—	—
35.5	×	×	×	—	—	—	—	—	—	—	—	—	—	—
40	×	×	×	×	—	—	—	—	—	—	—	—	—	—
45	×	×	×	×	×	—	—	—	—	—	—	—	—	—
50	×	×	×	×	×	×	—	—	—	—	—	—	—	—
56	×	×	×	×	×	×	×	—	—	—	—	—	—	—
63	×	×	×	×	×	×	×	×	—	—	—	—	—	—
71	×	×	×	×	×	×	×	×	×	—	—	—	—	—
80	×	×	×	×	×	×	×	×	×	—	—	—	—	—
85	—	×	×	×	×	×	×	×	×	×	—	—	—	—
90	—	×	×	×	×	×	×	×	×	×	—	—	—	—
95	—	×	×	×	×	×	×	×	×	×	×	—	—	—
100	—	—	×	×	×	×	×	×	×	×	×	—	—	—
106	—	—	×	×	×	×	×	×	×	×	×	×	—	—
112	—	—	×	×	×	×	×	×	×	×	×	×	—	—
118	—	—	—	×	×	×	×	×	×	×	×	×	×	—
125	—	—	—	—	×	×	×	×	×	×	×	×	×	×
132	—	—	—	—	—	×	×	×	×	×	×	×	×	×
140	—	—	—	—	—	×	×	×	×	×	×	×	×	×
150	—	—	—	—	—	—	×	×	×	×	×	×	×	×
160	—	—	—	—	—	—	×	×	×	×	×	×	×	×
180	—	—	—	—	—	—	—	×	×	×	×	×	×	×
200	—	—	—	—	—	—	—	—	×	×	×	×	×	×
224	—	—	—	—	—	—	—	—	—	×	×	×	×	×
250	—	—	—	—	—	—	—	—	—	—	—	—	×	×

Source: Bureau of Indian Standards, IS: 1929, 1961.

TABLE 13-14
Formulas for riveted joints[2,3,4]

Type of joint	Figure	Efficiency of plate, η_p	Efficiency of rivets, η_r	Combined efficiency, η_c	Longitudinal pitch, p, mm	Transverse pitch, p_t, mm	Margin, m, mm	Thickness of cover plate, mm: Inner h_2 (wider)	Thickness of cover plate, mm: Outer, h_1 (narrower)
				LAP JOINT					
One rivet per pitch, Type a	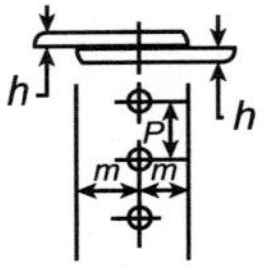 (a)	$\frac{p-d}{p}$	$\left(\frac{\pi d^2}{4}\right)\frac{\tau}{ph\sigma_\theta}$		$1.13h + 40$		$1.5d$		
Two rivets per pitch Type b	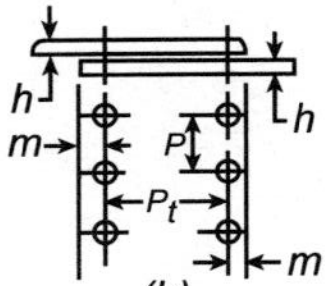 (b)	$\frac{p-d}{p}$	$2\left(\frac{\pi d^2}{4}\right)\frac{\tau}{ph\sigma_\theta}$		$2.62h + 40$	$2d$	$1.5d$		
Type c	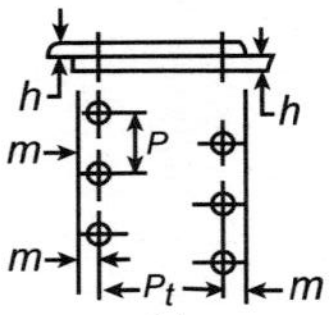 (c)	$\frac{p-d}{p}$	$2\left(\frac{\pi d^2}{4}\right)\frac{\tau}{ph\sigma_\theta}$		$2.62h + 40$	$0.33p + 0.67d$	$1.5d$		
Three rivets per pitch Type d	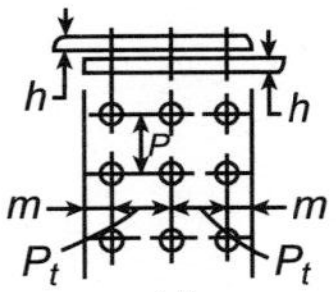 (d)	$\frac{p-d}{p}$	$3\left(\frac{\pi d^2}{4}\right)\frac{\tau}{ph\sigma_\theta}$		$3.47h + 40$	$2d$	$1.5d$		

TABLE 13-14
Formulas for riveted joints[2,3,4] (*Cont.*)

Type of joint	Figure	Efficiency of plate, η_p	Efficiency of rivets, η_r	Combined efficiency, η_c	Longitudinal pitch, p, mm	Transverse pitch, p_t, mm	Margin, m, mm	Thickness of cover plate, mm: Inner h_2 (wider)	Thickness of cover plate, mm: Outer, h_1 (narrower)
Type *e*	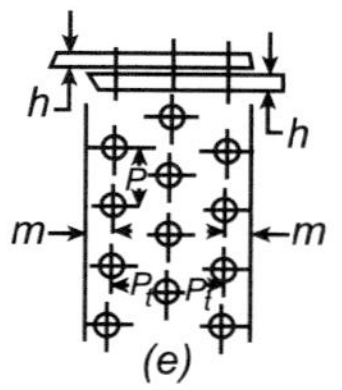	$\frac{p-d}{p}$	$3\left(\frac{\pi d^2}{4}\right)\frac{\tau}{ph\sigma_\theta}$		$3.47h+40$	$0.33p+0.67d$	$1.5d$		
Four rivets per pitch Type *f*	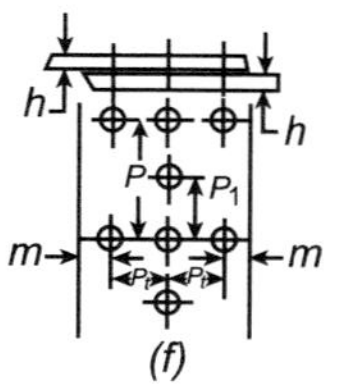	$\frac{p-d}{p}$	$4\left(\frac{\pi d^2}{4}\right)\frac{\tau}{ph\sigma_\theta}$	$\left\{\frac{p-2d}{p} + \left[\left(\frac{\pi d^2}{4}\right)\frac{\tau}{ph\sigma_\theta}\right]\right\}$	$4.14h+40$	$0.33p+0.67d$ or $2d$ (whichever is greater)	$1.5d$		
Type *g*	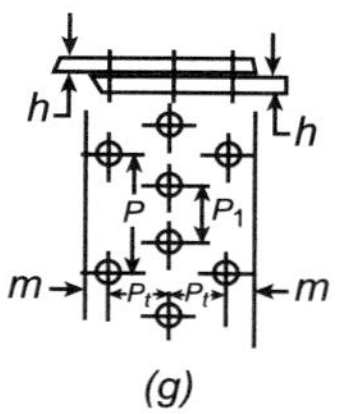	$\frac{p-d}{p}$	$4\left(\frac{\pi d^2}{4}\right)\frac{\tau}{ph\sigma_\theta}$	$\left\{\frac{p-2d}{p} + \left[\left(\frac{\pi d^2}{4}\right)\frac{\tau}{ph\sigma_\theta}\right]\right\}$	$4.14h+40$	$0.2p+1.15d$	$1.5d$		

TABLE 13-14
Formulas for riveted joints[2,3,4] *(Cont.)*

Type of joint	Figure	Efficiency of plate, η_p	Efficiency of rivets, η_r	Combined efficiency, η_c	Longitudinal pitch, p, mm	Transverse pitch, p_t, mm	Margin, m, mm	Thickness of cover plate, mm: Inner h_2 (wider)	Thickness of cover plate, mm: Outer, h_1 (narrower)
				BUTT JOINT					
Single butt strap One rivet per pitch Type *a*	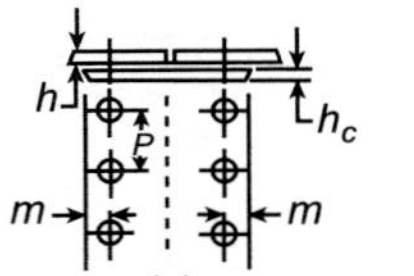 (a)	$\dfrac{p-d}{p}$	$\left(\dfrac{\pi d^2}{4}\right)\dfrac{\tau}{ph\sigma_\theta}$		$1.53h+40$		$1.5d$		$1.125h$
Two rivets per pitch Type *b*	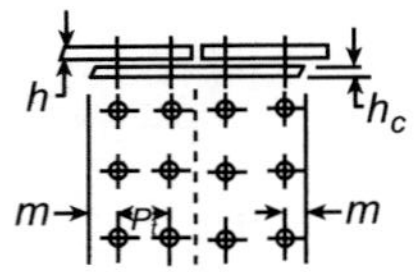 (b)	$\dfrac{p-d}{p}$	$2\left(\dfrac{\pi d^2}{4}\right)\dfrac{\tau}{ph\sigma_\theta}$		$3.06h+40$	$2d$	$1.5d$		$1.125h$
Type *c*	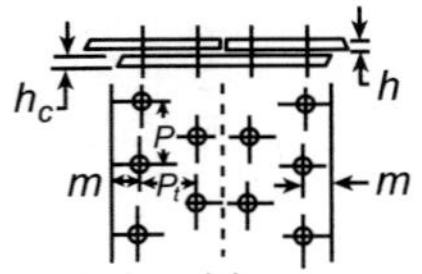 (c)	$\dfrac{p-d}{p}$	$2\left(\dfrac{\pi d^2}{4}\right)\dfrac{\tau}{ph\sigma_\theta}$		$3.06h+40$	$0.33p+0.67d$	$1.5d$		$1.125d$
Three rivets per pitch Type *d*	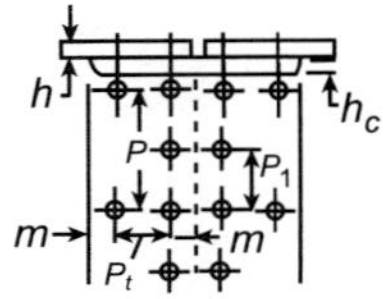 (d)	$\dfrac{p-d}{p}$	$3\left(\dfrac{\pi d^2}{4}\right)\dfrac{\tau}{ph\sigma_\theta}$	$\left\{\dfrac{p-2d}{p} + \left(\dfrac{\pi d^2}{4}\right)\dfrac{\tau}{ph\sigma_\theta}\right\}$	$4.05h+40$	$0.33p+0.67d$ or $2d$ (whichever is greater)	$1.5d$		$1.125h\,\dfrac{p-d}{p-2d}$

TABLE 13-14
Formulas for riveted joints[2,3,4] *(Cont.)*

Type of joint	Figure	Efficiency of plate, η_p	Efficiency of rivets, η_r	Combined efficiency, η_c	Longitudinal pitch, p, mm	Transverse pitch, p_t, mm	Margin, m, mm	Thickness of cover plate, mm: Inner h_2 (wider)	Thickness of cover plate, mm: Outer, h_1 (narrower)
Two rivets per pitch Type *e*	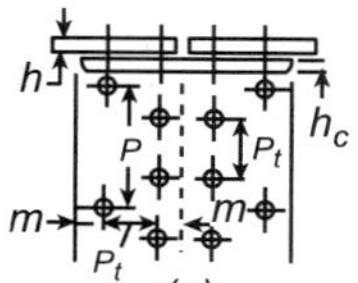 (e)	$\frac{p-d}{p}$	$3\left(\frac{\pi d^2}{4}\right)\frac{\tau}{ph\sigma_\theta}$	$\left\{\frac{p-2d}{p} + \left(\frac{\pi d^2}{4}\right)\frac{\tau}{ph\sigma_\theta}\right\}$	$4.05h + 40$	$0.2p + 1.15d$	$1.5d$		$1.125h\,\frac{p-d}{p-2d}$
Double-butt strap (equal widths) One rivet per pitch Type *f*	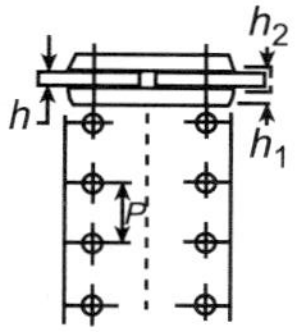 (f)	$\frac{p-d}{p}$	$1.875\left(\frac{\pi d^2}{4}\right) \times \frac{\tau}{ph\sigma_\theta}$		$1.75h + 40$		$1.5d$	$0.625h$	$0.625h$
Two rivets per pitch Type *g*	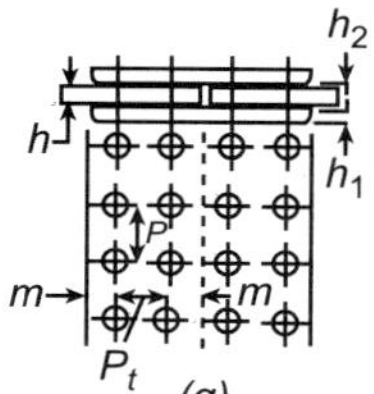 (g)	$\frac{p-d}{p}$	$3.75\left(\frac{\pi d^2}{4}\right) \times \frac{\tau}{ph\sigma_\theta}$		$3.5h + 40$	$2d$	$1.5d$	$0.625h$	$0.625h$
Type *h*	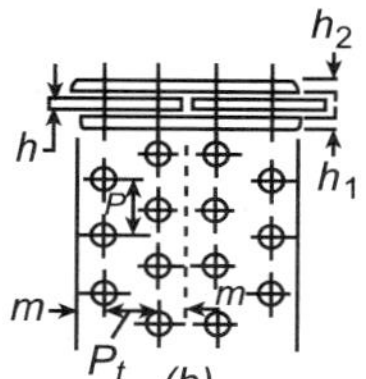 (h)	$\frac{p-d}{p}$	$3.75\left(\frac{\pi d^2}{4}\right) \times \frac{\tau}{ph\sigma_\theta}$		$3.5h + 40$	$0.33p + 0.67d$	$1.5d$	$0.625h$	$0.625h$

TABLE 13-14
Formulas for riveted joints[2,3,4] *(Cont.)*

Type of joint	Figure	Efficiency of plate, η_p	Efficiency of rivets, η_r	Combined efficiency, η_c	Longitudinal pitch, p, mm	Transverse pitch, p_t, mm	Margin, m, mm	Thickness of cover plate, mm: Inner h_2 (wider)	Thickness of cover plate, mm: Outer, h_1 (narrower)
Three rivets per pitch Type i	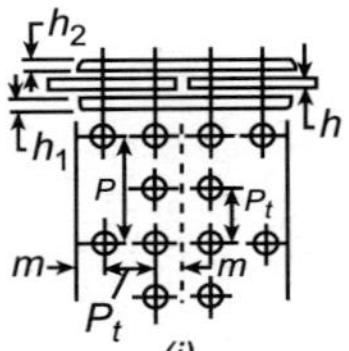 (i)	$\frac{p-d}{p}$	$5.625\left(\frac{\pi d^2}{4}\right) \times \frac{\tau}{ph\sigma_\theta}$	$\left\{\frac{p-2d}{p} + \left[1.875 \times \left(\frac{\pi d^2}{4}\right)\frac{\tau}{ph\sigma_\theta}\right]\right\}$	$4.63h + 40$	$0.33p + 0.67d$ or $2d$ (whichever is greater)	$1.5d$	$\left[0.615h \times \frac{p-d}{p-2d}\right]$	$\left[0.625h \times \frac{p-d}{p-2d}\right]$
Type j	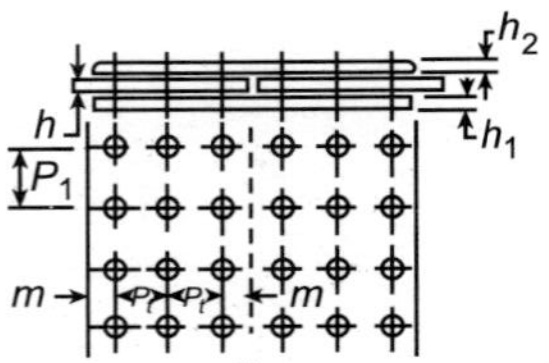 (j)	$\frac{p-d}{p}$	$5.625\left(\frac{\pi d^2}{4}\right) \times \frac{\tau}{ph\sigma_\theta}$		$4.63h + 40$	$2d$	$1.5d$	$0.625h$	$0.625h$
Type k	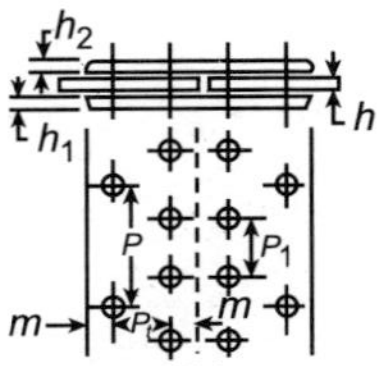 (k)	$\frac{p-d}{p}$	$5.625\left(\frac{\pi d^2}{4}\right) \times \frac{\tau}{ph\sigma_\theta}$		$4.63h + 40$	$0.2p + 1.15d$	$1.5d$	$\left[0.625h \times \frac{p-d}{p-2d}\right]$	$\left[0.625h \times \frac{p-d}{p-2d}\right]$
Type l	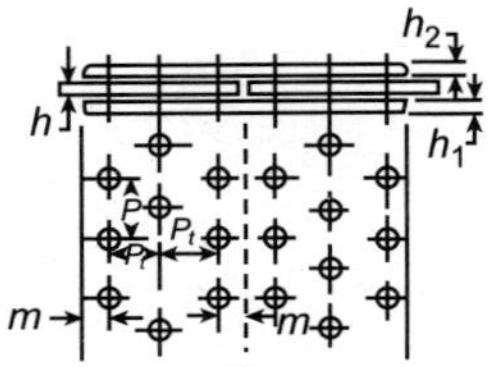 (l)	$\frac{p-d}{p}$	$5.625\left(\frac{\pi d^2}{4}\right) \times \frac{\tau}{ph\sigma_\theta}$		$4.63h + 40$	$0.33p + 0.67d$	$1.5d$	$0.625h$	$0.625h$

TABLE 13-14
Formulas for riveted joints[2,3,4] *(Cont.)*

Type of joint	Figure	Efficiency of plate, η_p	Efficiency of rivets, η_r	Combined efficiency, η_c	Longitudinal pitch, p, mm	Transverse pitch, p_t, mm	Margin, m, mm	Thickness of cover plate, mm: Inner h_2 (wider)	Thickness of cover plate, mm: Outer, h_1 (narrower)
Four rivets per pitch Type *m*	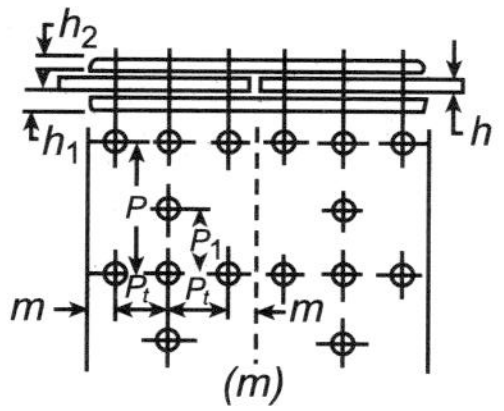	$\frac{p-d}{p}$	$7.5\left(\frac{\pi d^2}{4}\right) \times \frac{\tau}{ph\sigma_\theta}$	$\frac{p-2d}{d} + \left[1.875 \times \left(\frac{\pi d^2}{4}\right)\frac{\tau}{ph\sigma_\theta}\right]$	$5.52h + 40$	$0.33p + 0.67d$ or $2d$ (whichever is greater)	$1.5d$	$0.625h$	$0.625h$
Type *n*	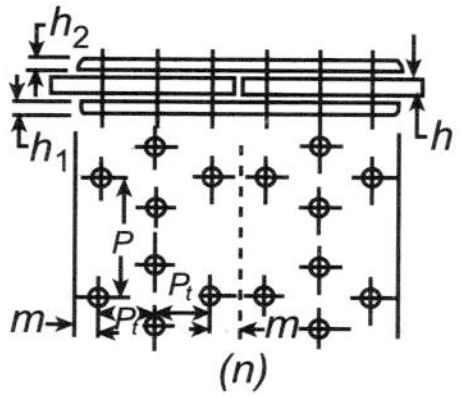	$\frac{p-d}{p}$	$7.5\left(\frac{\pi d^2}{4}\right) \times \frac{\tau}{ph\sigma_\theta}$	$\frac{p-2d}{d} + \left[1.875 \times \left(\frac{\pi d^2}{4}\right)\frac{\tau}{ph\sigma_\theta}\right]$	$5.52h + 40$	$0.2p + 1.15d$	$1.5d$	$0.625h$	$0.625h$
Double butt (unequal widths) Two rivets per pitch Type *o*	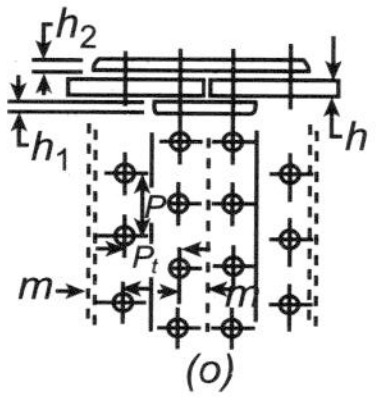	$\frac{p-d}{p}$	$2.875\left(\frac{\pi d^2}{4}\right) \times \frac{\tau}{ph\sigma_\theta}$		$3.5h + 40$	$0.33p + 0.67d$	$1.5d$	$0.75h$	$0.625h$
Type *p*	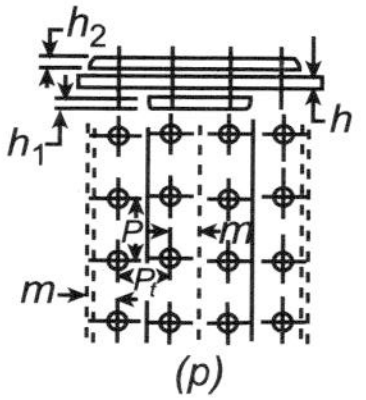	$\frac{p-d}{p}$	$2.875\left(\frac{\pi d^2}{4}\right) \times \frac{\tau}{ph\sigma_\theta}$		$3.5h + 40$	$2d$	$1.5d$	$0.75h$	$0.625h$

TABLE 13-14
Formulas for riveted joints[2,3,4] *(Cont.)*

Type of joint	Figure	Efficiency of plate, η_p	Efficiency of rivets, η_r	Combined efficiency, η_c	Longitudinal pitch, p, mm	Transverse pitch, p_t, mm	Margin, m, mm	Thickness of cover plate, mm: Inner h_2 (wider)	Thickness of cover plate, mm: Outer, h_1 (narrower)
Three rivets per pitch Type q	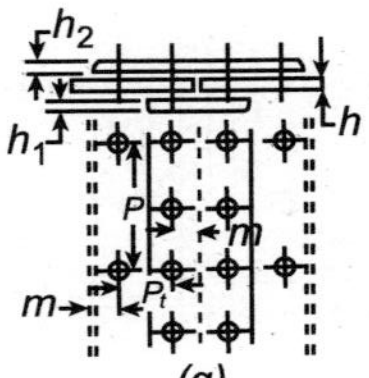	$\frac{p-d}{p}$	$4.75\left(\frac{\pi d^2}{4}\right) \times \frac{\tau}{ph\sigma_\theta}$	$\left[\frac{p-2d}{d} + \left(\frac{\pi d^2}{4}\right)\frac{\tau}{ph\sigma_\theta}\right]$	$4.63h + 40$	$0.33p + 0.67d$ or $2d$ (whichever is greater)	$1.5d$	$0.75h$	$0.625h$
Type r	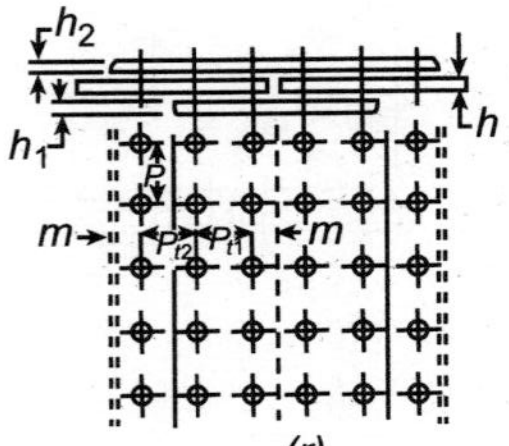	$\frac{p-d}{p}$	$4.75\left(\frac{\pi d^2}{4}\right) \times \frac{\tau}{ph\sigma_\theta}$		$4.63h + 40$	$2d$	$1.5d$	$0.75h$	$0.625h$
Type s	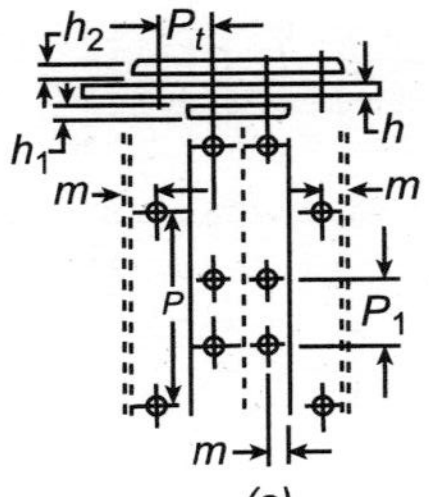	$\frac{p-d}{p}$	$4.75\left(\frac{\pi d^2}{4}\right) \times \frac{\tau}{ph\sigma_\theta}$	$\left[\frac{p-2d}{d} + \left(\frac{\pi d^2}{4}\right)\frac{\tau}{ph\sigma_\theta}\right]$	$4.63h + 40$	$0.2p + 1.15d$	$1.5d$	$0.75h$	$0.625h$

TABLE 13-14
Formulas for riveted joints (*Cont.*)

Type of joint	Figure	Efficiency of plate, η_p	Efficiency of rivets, η_r	Combined efficiency, η_c	Longitudinal pitch, p, mm	Transverse pitch, p_t, mm	Margin, m, mm	Thickness of cover plate, mm: Inner h_2 (wider)	Thickness of cover plate, mm: Outer, h_1 (narrower)
Type t	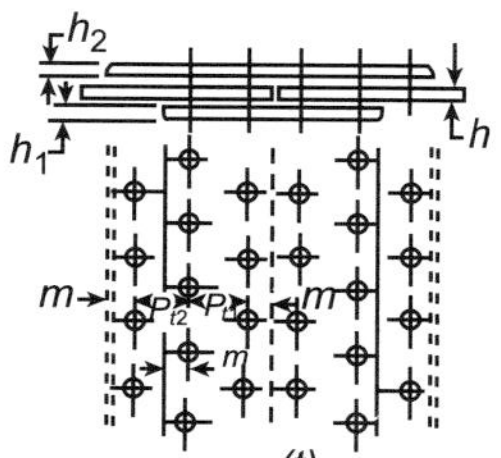 (t)	$\frac{p-d}{p}$	$4.75\left(\frac{\pi d^2}{4}\right) \times \frac{\tau}{ph\sigma_\theta}$		$4.63h + 40$	$0.33p + 0.67d$	$1.5d$	$0.75h$	$0.625h$

Particular	Formula	
Common Formula: The thickness of the main plate of a longitudinal joint	$h = \frac{P_f D_i}{2\eta\sigma_\theta}$	
Unwin's formula for diameter of rivet	$d = 0.19\sqrt{h}$ to $0.2\sqrt{h}$ where d and h in m	**SI**
	$= 1.2\sqrt{h}$ to $1.4\sqrt{h}$ where d and h in m	**USCS**

Key: d = diameter of rivet, m (in); h = thickness of main plate, m (in); σ_θ = hoop stress, MPa (psi); D_i = inside diameter of pressure vessel, m (in); P_f = internal fluid pressure, MPa (psi); η = efficiency of the riveted joint.

Source: K. Lingaiah and B. R. Narayana Iyengar, *Machine Design Data Handbook*, Engineering College Cooperative Society, Bangalore, India, 1983; and K. Lingaiah, *Machine Design Data Handbook*, Vol. II (*SI and Customary Metric Units*), Suma Publishers, Bangalore, India, 1986.

CHAPTER 14

DESIGN OF SHAFTS

SYMBOLS[1,2,3]

b	width of keyway, m (in)
c	machine cost, \$/m (\$/in) (US dollars)
D	diameter of shaft (also with subscripts), m (in)
D_i	inside diameter of hollow shaft, m (in)
D_o	outside diameter of hollow shaft, m (in)
E	modulus of elasticity, GPa (Mpsi)
F	axial load (tensile or compressive), kN (lbf)
F'_m	the static equivalent of cyclic load, $(= F_m \pm F_a)$, kN (lbf)
G	modulus of rigidity, GPa (Mpsi)
h	depth of keyway, m (in)
k	radius of gyration, m (in) material cost (also with subscripts), \$/kg
$K = \dfrac{D_i}{D_o}$	ratio of inner to outer diameter of hollow shaft
K_b	numerical combined shock and fatigue factor to be applied to computed bending moment
K_t	numerical combined shock and fatigue factor to be applied to computed twisting moment
l	length, m (in)
M_b	bending moment, N m (lbf in)
M_t	twisting moment, N m (lbf in)
M'_{bm}	static equivalent of cyclic bending moment $M_{bm} \pm M_{ba}$, N m (lbf in)
M'_{tm}	static equivalent of cyclic twisting moment $M_{tm} \pm M_{ta}$, N m (lbf in)
P	power, kW (hp)
n	speed, rpm; safety factor
n'	speed, rps
ρ	specific weight of material, kN/m^3 (lbf/in)
σ	stress (tensile or compressive) also with subscripts, MPa (psi)
τ	shear stress (also with subscripts), MPa (psi)
α	ratio of maximum intensity of stress to the average value from compressive stress only
θ	angular deflection, deg

SUFFIXES

a	amplitude
b	bending
d	design
e	elastic limit
h	hollow
m	mean
sc	static strength (σ_{su} or σ_{sy}), solid
t	twisting
u	ultimate
y	yield strength
max	maximum
min	minimum
f	endurance

Other factors in performance or in special aspect are included from time to time in this chapter and, being applicable in their immediate context, are not given at this stage.

Note: σ and τ with the initial subscript s designates strength properties of material used in the design which will be used and observed throughout this handbook. In some books on machine design and in this *Machine Design Data Handbook* the ratios of design stresses σ_{sd}/σ_{fd} and τ_{sd}/τ_{fd}; and design stresses σ_{yd}, $\tau_{yd'}$, σ_{fd}, and τ_{fd} have been used instead of σ_{sy}/σ_{sf}, τ_{sy}/τ_{sf}; and yield strengths σ_{sy}, τ_{sy} and fatigue strengths, σ_{sf}, τ_{sf} in the design equations for shafts [Eqs. (14-1) to (14-65)]. This has to be taken into consideration in the design of shafts while using Eqs. (14-1) to (14-65).

Particular	Formula

SOLID SHAFTS

(1) Stationary shafts with static loads

The diameter of shaft subjected to simple torsion

$$D = \left(\frac{16M_t}{\pi\tau_{yd}}\right)^{1/3} \quad (14\text{-}1)$$

The diameter of shaft subjected to simple bending

$$D = \left(\frac{32M_b}{\pi\sigma_{yd}}\right)^{1/3} \quad (14\text{-}2)$$

The diameter of shaft subjected to combined torsion and bending:

(a) According to maximum normal stress theory

$$D = \left[\frac{16}{\pi\sigma_{yd}}\{M_b + (M_b^2 + M_t^2)^{1/2}\}\right]^{1/3} \quad (14\text{-}3)$$

(b) According to maximum shear stress theory

$$D = \left\{\frac{16}{\pi\tau_{yd}}(M_b^2 + M_t^2)^{1/2}\right\}^{1/3} \quad (14\text{-}4)$$

(c) According to maximum shear energy theory

$$D = \left\{\frac{16}{\pi\tau_{yd}}\left(M_b^2 + \frac{3}{4}M_t^2\right)^{1/2}\right\}^{1/3} \quad (14\text{-}5)$$

Particular	Formula
The diameter of shaft subjected to axial load, bending, and torsion:[1–3]	
(a) According to maximum normal theory	$D = \left[\frac{16}{\pi\sigma_{yd}}\left\{\left(M_b + \frac{\alpha FD}{8}\right) + \left\{\left(M_b + \frac{\alpha FD}{8}\right)^2 + M_t^2\right\}^{1/2}\right\}\right]^{1/3}$ (14-6)
(b) According to maximum shear stress theory	$D = \left[\frac{16}{\pi\tau_{yd}}\left\{\left(M_b + \frac{\alpha FD}{8}\right)^2 + M_t^2\right\}^{1/2}\right]^{1/3}$ (14-7)
(c) According to maximum shear energy theory	$D = \left[\frac{16}{\pi\tau_{yd}}\left\{\left(M_b + \frac{\alpha FD}{8}\right)^2 + \frac{3}{4}M_t^2\right\}^{1/2}\right]^{1/3}$ (14-8)
(2) Rotating shafts with dynamic loads, taking dynamic effect indirectly into consideration[1–3]	
For empirical shafting formulas	Refer to Table 14-1.
The diameter of shaft subjected to simple torsion	$D = \left\{\frac{16}{\pi\tau_{yd}}(K_t M_t)\right\}^{1/3}$ (14-9)
The diameter of shaft subjected to simple bending	$D = \left\{\frac{32}{\pi\sigma_{yd}}(K_b M_b)\right\}^{1/3}$ (14-10)
The diameter of shaft subjected to combined bending and torsion	
(a) According to maximum normal stress theory	$D = \left\{\frac{16}{\pi\sigma_{yd}}[K_b M_b + \{(K_b M_b)^2 + (K_t M_t)^2\}^{1/2}]\right\}^{1/3}$ (14-11)
(b) According to maximum shear stress theory	$D = \left[\frac{16}{\pi\tau_{yd}}\{(K_b M_b)^2 + (K_t M_t)^2\}^{1/2}\right]^{1/3}$ (14-12)
(c) According to maximum shear energy theory	$D = \left[\frac{16}{\pi\tau_{yd}}\{(K_b M_b)^2 + \frac{3}{4}(K_t M_t)^2\}^{1/2}\right]^{1/3}$ (14-13)
The diameter of shaft subjected to axial load, bending, and torsion	
(a) According to maximum normal stress theory	$D = \left\{\frac{16}{\pi\sigma_{yd}}\left(K_b M_b + \frac{\alpha FD}{8}\right) + \left[\left(K_b M_b + \frac{\alpha FD}{8}\right)^2 + (K_t M_t)^2\right]^{1/2}\right\}^{1/3}$ (14-14)

Particular	Formula
(b) According to maximum shear stress theory	$D = \left[\frac{16}{\pi\tau_{yd}}\left\{\left(K_b M_b + \frac{\alpha F D}{8}\right)^2 + (K_t M_t)^2\right\}^{1/2}\right]^{1/3}$ (14-15)
(c) According to maximum shear energy theory	$D = \left[\frac{16}{\pi\tau_{yd}}\left\{\left(K_b M_b + \frac{\alpha F D}{8}\right)^2 + \frac{3}{4}(K_t M_t)^2\right\}^{1/2}\right]^{1/3}$ (14-16)
The diameter of shaft based on torsional rigidity	$D = \left\{\frac{584 M_t L}{G\theta}\right\}^{1/4}$ (14-17) where K_b and K_t are taken from Table 14-2
(3) Rotating shafts and fluctuating loads, taking fatigue effect directly into consideration[1-3]	
The diameter of shaft subjected to fluctuating torsion	$D = \left\{\frac{16}{\pi}\left(\frac{M_{tm}}{\tau_{yd}} + \frac{M_{ta}}{\tau_{fd}}\right)\right\}^{1/3}$ (14-18)
The diameter of shaft subjected to fluctuating bending	$D = \left\{\frac{32}{\pi}\left(\frac{M_{bm}}{\sigma_{yd}} + \frac{M_{ba}}{\sigma_{fd}}\right)\right\}^{1/3}$ (14-19)
The diameter of shaft subjected to combined fluctuating torsion and bending:	
(a) According to maximum normal stress theory	$D = \left[\frac{16}{\pi\sigma_{yd}}\{M'_{bm} + (M'^2_{bm} + M'^2_{tm})^{1/2}\}\right]^{1/3}$ (14-20)
(b) According to maximum shear stress theory	$D = \left\{\frac{16}{\pi\tau_{yd}}(M'^2_{bm} + M'^2_{tm})^{1/2}\right\}^{1/3}$ (14-21)
(c) According to maximum shear energy theory	$D = \left\{\frac{16}{\pi\tau_{yd}}\left(M'^2_{bm} + \frac{3}{4}M'^2_{tm}\right)^{1/2}\right\}^{1/3}$ (14-22) where $M'_{bm} = M_{bm} + \frac{\sigma_{sd}}{\sigma_{fd}} M_{ba}$ (14-22a) $M'_{tm} = M_{tm} + \frac{\tau_{sd}}{\tau_{fd}} M_{ta}$ (14-22b)

Particular	Formula
The diameter of shaft subjected to combined fluctuating axial load, bending, and torsion	
(a) According to maximum normal stress theory	$$D = \left\{ \frac{16}{\pi \sigma_{yd}} \left[\left(M'_{bm} + \frac{\alpha F'_m D}{8} \right) + \left\{ \left(M'_{bm} + \frac{\alpha F'_m D}{8} \right)^2 + M'^2_{tm} \right\}^{1/2} \right] \right\}^{1/3} \quad (14\text{-}23)$$
(b) According to maximum shear stress theory	$$D = \left[\frac{16}{\pi \tau_{yd}} \left\{ \left(M'_{bm} + \frac{\alpha F'_m D}{8} \right)^2 + M'^2_{tm} \right\}^{1/2} \right]^{1/3} \quad (14\text{-}24)$$
(c) According to maximum shear energy theory	$$D = \left[\frac{16}{\pi \tau_{yd}} \left\{ \left(M'_{bm} + \frac{\alpha F'_m D}{8} \right)^2 + \frac{3}{4} M'^2_{tm} \right\}^{1/2} \right]^{1/3} \quad (14\text{-}25)$$ where M'_{bm} and M'_{tm} have the same meaning as in Eqs. (14-22a) and (14-22b) and $F'_m = F_m + \frac{\sigma_{sd}}{\sigma_{fd}} F_a$ (14-25a)

HOLLOW SHAFTS

Particular	Formula
(1) Stationary shafts with static loads	
The outside diameter of shaft subjected to simple torsion	$$D_o = \left(\frac{16 M_t}{\pi \tau_{yd}(1 - K^4)} \right)^{1/3} \quad (14\text{-}26)$$
The outside diameter of shaft subjected to simple bending	$$D_o = \left(\frac{32 M_b}{\pi \sigma_{yd}(1 - K^4)} \right)^{1/3} \quad (14\text{-}27)$$
The diameter of shaft subjected to combined torsion and bending	
(a) According to maximum normal stress theory	$$D_o = \left[\frac{16}{\pi \sigma_{yd}(1 - K^4)} \{ M_b + (M_b^2 + M_t^2)^{1/2} \} \right]^{1/3} \quad (14\text{-}28)$$
(b) According to maximum shear stress theory	$$D_o = \left\{ \frac{16}{\pi \tau_{yd}(1 - K^4)} (M_b^2 + M_t^2)^{1/2} \right\}^{1/3} \quad (14\text{-}29)$$
(c) According to maximum shear energy theory	$$D_o = \left\{ \frac{16}{\pi \tau_{yd}(1 - K^4)} \left(M_b^2 + \frac{3}{4} M_t^2 \right)^{1/2} \right\}^{1/3} \quad (14\text{-}30)$$

Particular	Formula
The outside diameter of shaft subjected to axial load, bending, and torsion	
(a) According to maximum normal stress theory	$D_o = \left\{ \dfrac{16}{\pi\sigma_{yd}(1-K^4)} \left(\left[M_b + \dfrac{\alpha F D_o}{8}(1+K^2) \right] + \left[\left(M_b + \dfrac{\alpha F D_o(1+K^2)}{8} \right)^2 + M_t^2 \right]^{1/2} \right) \right\}^{1/3}$ (14-31)
(b) According to maximum shear stress theory	$D_o = \left\{ \dfrac{16}{\pi\tau_{yd}(1-K^4)} \left[\left(M_b + \dfrac{\alpha F D_o}{8}(1+K^2) \right) + M_t^2 \right]^{1/2} \right\}^{1/3}$ (14-32)
(c) According to maximum shear energy theory	$D_o = \left\{ \dfrac{16}{\pi\tau_{yd}(1-K^4)} \left[\left(M_b^2 + \dfrac{\alpha F D_o}{8}(1+K^2) \right) + \dfrac{3}{4} M_t^2 \right]^{1/2} \right\}^{1/3}$ (14-33)
(2) Rotating shafts with dynamic loads, taking dynamic effect indirectly into consideration[1–3]	
The outside diameter of shaft subjected to simple torsion	$D_o = \left(\dfrac{16}{\pi\tau_{yd}(1-K^4)} K_t M_t \right)^{1/3}$ (14-34)
The outside diameter of shaft subjected to simple bending	$D_o = \left(\dfrac{32}{\pi\sigma_{yd}(1-K^4)} K_b M_b \right)^{1/3}$ (14-35)
The outside diameter of shaft subjected to combined bending and torsion	
(a) According to maximum normal stress theory	$D_o = \left\{ \dfrac{16}{\pi\sigma_{yd}(1-K^4)} [K_b M_b + \{(K_b M_b)^2 + (K_t M_t)^2\}^{1/2}] \right\}^{1/3}$ (14-36)
(b) According to maximum shear stress theory	$D_o = \left[\dfrac{16}{\pi\tau_{yd}(1-K^4)} \{(K_b M_b)^2 + (K_t M_t)^2\}^{1/2} \right]^{1/3}$ (14-37)

Particular	Formula	
COMPARISON BETWEEN DIAMETERS OF SOLID AND HOLLOW SHAFTS OF SAME LENGTH		
For equal strength in bending, torsion, and/or combined bending and torsion, the diameter		
(a) When materials of both shafts are same	$D = D_o(1 - K^4)^{1/3}$	(14-51)
(b) When materials of shafts are different	$D = D_o \dfrac{\sigma_{eh}}{\sigma_{es}}(1 - K^4)^{1/3}$	(14-52)
For torsional rigidity		
(a) When torsional rigidities are equal	$D = D_o(1 - K^4)^{1/4}$	(14-53)
(b) When torsional rigidities are different	$D = D_o \left\{ \dfrac{G_h}{G_s}(1 - K^4) \right\}^{1/4}$	(14-54)
For equal weight		
(a) When material of both shafts is same	$D = D_o(1 - K^2)^{1/2}$	(14-55)
(b) When materials of both shafts are different	$D = D_o \left\{ (1 - K^2) \dfrac{w_h}{w_s} \right\}^{1/2}$	(14-56)
For equal cost		
(a) For same material and machining cost for both shafts	$D = D_o(1 - K^2)^{1/2}$	(14-57)
(b) For no machining cost for both shafts but with different material cost	$D = D_o \left\{ (1 - K^2) \dfrac{w_h k_h}{w_s k_s} \right\}^{1/2}$	(14-58)
(c) When machining costs are different and material cost negligible	$D = \left\{ \dfrac{c_h}{c_s} \right\}^{1/2}$	(14-59)
(d) When machining and material costs are different	$D = \left\{ \dfrac{\pi D_o^2(1 - K^2) w_h k_h + c_h}{\pi w_s k_s + \dfrac{c_s}{D^2}} \right\}^{1/2}$	(14-60)

Note: If the axial load does not produce column action, the constant α need not be used to multiply the term $[FD_o(1 + K^2)/8]$ throughout this chapter

Particular	Formula

STIFFNESS

Instead of computing the transverse deflection, the maximum distance between the bearings (in meters) may be computed by the empirical formula to limit the transverse deflection to 0.8 mm/m of length

$$L = \frac{1500}{n + 1500} cD^{2/3} \tag{14-61}$$

where c is a constant from Table 14-3

RIGIDITY

Moor's formula for the increase of the angle of twist θ due to the keyway and applies only to the keyseated length of shaft

$$K_1 = 1 + \frac{0.4b + 0.7h}{D} \tag{14-62}$$

EFFECT OF KEYWAYS

The lowering of the strength of shaft by keyways may be taken into account by introducing a factor similar to a stress-concentration factor (or Moor's formula for lowering the strength of shaft)

$$K = 1 + \frac{0.2b + 1.1h}{D} \tag{14-63}$$

THE BUCKLING FACTOR

For short columns or when $l/k \leq 115$

$$\alpha = \frac{1}{1 - 0.0044(l/k)} \tag{14-64}$$

For long columns or when $l/k \geq 115$ (Euler's formula)

$$\alpha = \frac{\sigma_{sy}}{\pi^2 nE}\left(\frac{l}{k}\right) \tag{14-65}$$

where

$n = 1$ for hinged ends

$= 2.25$ for fixed ends

$= 1.6$ for both ends pinned or guided and partly restrained

($\alpha = 1$ for tensile load)

SHAFTS SUBJECTED TO VARIOUS STRESSES

(1) Shaft subjected to steady torque and reversed bending moment taking into consideration stress concentration:

Particular	Formula
Diameter of solid shaft:	
(a) According to maximum shear stress failure theory using Soderberg [4] criterion for fatigue strength $\frac{\sigma_a}{\sigma_{sf}} + \frac{\sigma_m}{\sigma_{sy}} = \frac{1}{n}$	$D = \left\{ \frac{32n}{\pi\sigma_{sy}} \left[\left(K_{f\sigma} \frac{\sigma_{sy}}{\sigma_{sf}} M_{ba} \right)^2 + (K_{f\tau} M_{tm})^2 \right]^{1/2} \right\}^{1/3}$ (14-66) where $K_{f\sigma}$ = fatigue stress-concentration factor due to bending, tension, or compression; K_{fr} = fatigue stress-concentration factor due to torsion; $K_{f\sigma} = K_{f\tau} = 1$ for ductile material under steady state of stress
(b) According to maximum shear stress theory of failure using modified Goodman criterion for fatigue strength $\frac{\sigma_a}{\sigma_{sf}} + \frac{\sigma_m}{\sigma_{sut}} = \frac{1}{n}$	$D = \left\{ \frac{32n}{\pi\sigma_{sut}} \left[\left(K_{f\sigma} \frac{\sigma_{sut}}{\sigma_{sf}} M_{ba} \right)^2 + (K_{f\tau} M_{tm})^2 \right]^{1/2} \right\}^{1/3}$ (14-67)
Diameter of hollow shaft:	
(c) According to distortion-energy theory of failure using modified Goodman criterion for fatigue strength	$D_o = \left[\frac{16n}{\pi\sigma_{sut}(1 - K^4)} \left(2K_{f\sigma} \frac{\sigma_{sut}}{\sigma_{sf}} M_{ba} + \sqrt{3} K_{f\tau} M_{tm} \right) \right]^{1/3}$ (14-68)
(d) According to distortion-energy theory of failure combined with Gerber parabolic relation $\left(\frac{n\sigma_a}{\sigma_{sf}} \right) + \left(\frac{n\sigma_m}{\sigma_{sut}} \right)^2 = 1$	$D_o = \frac{16n}{\pi\sigma_{sut}(1 - K^4)} \left\{ K_{f\sigma} \frac{\sigma_{sut}}{\sigma_{sf}} M_{ba} + \left[\left(K_{f\sigma} \frac{\sigma_{sut}}{\sigma_{sf}} M_{ba} \right)^2 + 3(K_{f\tau} M_{tm})^2 \right]^{1/2} \right\}^{1/3}$ (14-69)
(e) According to distortion-energy theory of failure using ASME elliptic locus for fatigue strength $\left(\frac{n\sigma_a}{\sigma_{sf}} \right)^2 + \left(\frac{n\sigma_m}{\sigma_{sy}} \right)^2 = 1$	$D_o = \left\{ \frac{16n}{\pi\sigma_{sy}(1 - K^4)} \left[\left(2K_{f\sigma} \frac{\sigma_{sy}}{\sigma_{sf}} M_{ba} \right)^2 + 3(K_{f\tau} M_{tm})^2 \right]^{1/2} \right\}^{1/3}$ (14-70)
Bagci failure locus equation in quartic (fourth-degree) form	i.e., $\frac{n\sigma_a}{\sigma_{sf}} + \left(\frac{n\sigma_m}{\sigma_{sy}} \right)^4 = 1$
and yielding criterion (Langer) equation combined with any theories of failure can be used to predict the fatigue strength of shaft	i.e., $\frac{\sigma_a + \sigma_m}{\sigma_{sy}} = \frac{1}{n}$

Particular	Formula
(2) Shaft subjected to fluctuating loads, i.e., reversed bending and reversed torque, taking into consideration stress concentration	
(a) The diameter of solid shaft according to maximum shear stress theory of failure using Soderberg criterion for fatigue strength	$D = \left[\dfrac{32n}{\pi\sigma_{sy}}(M_{be}^2 + M_{te}^2)^{1/2}\right]^{1/3}$ (14-71) where M_{be} = static equivalent of cyclic bending moment $= K_{f\sigma}M_{bm} + K_{f\sigma}\dfrac{\sigma_{sy}}{\sigma_{sf}}M_{ba}$ M_{te} = static equivalent of cyclic torsional moment $= K_{f\tau}M_{tm} + K_{f\tau}\dfrac{\sigma_{sy}}{\sigma_{sf}}M_{ta}$
(b) The diameter of hollow shaft according to distortion-energy theory of failure combined with Soderberg criterion for fatigue strength	$D_o = \left[\dfrac{16n}{\pi\sigma_{sy}(1-K^4)}(4M_{be}^2 + 3M_{te}^2)^{1/2}\right]^{1/3}$ (14-72) where M_{be} and M_{te} have the same meaning as given under Eq. (14-71)
(3) Shaft subjected to constant bending and torsional moments and reversed torsional and bending moments at the same frequency taking into consideration stress concentration	
(a) The diameter of solid shaft according to maximum distortion energy theory of failure using modified Goodman criterion for fatigue strength	$D = \left(\dfrac{16n}{\pi\sigma_{sut}}\{[4(K_{f\sigma}M_{bm})^2 + 3(K_{f\tau}M_{tm})^2]^{1/2}\} + \dfrac{\sigma_{sut}}{\sigma_{sf}}\{[4(K_{f\sigma}M_{ba})^2 + 3(K_{f\tau}M_{ta})^2]^{1/2}\}\right)^{1/3}$ (14-73) where $K_{f\sigma} = K_{f\tau} = 1$ for constant torsional and bending moments
(b) The diameter of solid shaft according to maximum shear stress theory of failure combined with modified Goodman criterion for fatigue strength	$D = \left\{\dfrac{32n}{\pi\sigma_{sut}}\left[\left(M_{bm} + K_{f\sigma}\dfrac{\sigma_{sut}}{\sigma_{sf}}M_{ba}\right)^2 + \left(M_{tm} + K_{f\tau}\dfrac{\sigma_{sut}}{\sigma_{sf}}M_{ta}\right)^2\right]^{1/2}\right\}^{1/3}$ (14-74)
(c) The diameter of hollow shaft according to maximum shear stress theory of failure using Soderberg criterion for fatigue strength	$D_o = \left\{\dfrac{32n}{\pi\sigma_{sy}(1-K^4)}\left[\left(K_{f\sigma}M_{bm} + K_{f\sigma}\dfrac{\sigma_{sy}}{\sigma_{sf}}M_{ba}\right)^2 + \left(K_{f\tau}M_{tm} + K_{f\tau}\dfrac{\sigma_{sy}}{\sigma_{sf}}M_{ta}\right)^2\right]^{1/2}\right\}^{1/3}$ (14-75) where $K_{f\sigma} = K_{f\tau} = 1$ for constant bending and torsional moments

Particular	Formula
(4) Cyclic axial load combined with reversed bending and torsional moments taking into consideration stress concentration as per *ASME Code for Design of Transmission Shafting*	
(a) The diameter of solid shaft according to maximum shear stress theory of failure and Soderberg relation for fatigue strength	$$D = \left\{\frac{32n}{\pi\sigma_{sy}}\left[\left(M_{be} + \frac{F_{ae}D}{8}\right)^2 + M_{te}^2\right]^{1/2}\right\}^{1/3} \quad (14\text{-}76)$$ where M_{be} and M_{te} have the same meaning as given under Eq. (14-71) F_{ae} = static equivalent axial load $= K_{f\sigma}F_{am} + K_{f\sigma}\frac{\sigma_{sy}}{\sigma_{sf}}F_{aa}$
(b) The diameter of hollow shaft according to distortion-energy theory of failure combined with modified Goodman relation for fatigue strength	$$D_o = \left[\frac{32n}{\pi\sigma_{sut}(1-K^4)}\left\{\left[M'_{be} + \frac{F'_{ae}D_o(1+K^2)}{8}\right]^2 + \frac{3}{4}M'^2_{te}\right\}^{1/2}\right]^{1/3} \quad (14\text{-}77)$$ where $M'_{be} = K_{f\sigma}M_{bm} + K_{f\sigma}\frac{\sigma_{sut}}{\sigma_{sf}}M_{ba}$ $M'_{te} = K_{f\tau}M_{tm} + K_{f\tau}\frac{\sigma_{sut}}{\sigma_{sf}}M_{ta}$ $F'_{ae} = K_{f\sigma}F_{am} + K_{f\sigma}\frac{\sigma_{sut}}{\sigma_{sf}}F_{aa}$ When $K = 0$, this equation reduces to an equation for a solid shaft. The value of α is given by Eq. (14-65)
(5) The diameter of solid shaft subjected to axial, bending, and torsional alternating loads according to distortion-energy theory of failure combined with Soderberg relation for fatigue as per *ASME Code for Design of Transmission Shafting*[5]	$$D = \left(\frac{32n}{\pi\sigma_{sf}}\left[\left(M_{ba} + \frac{F_aD}{2}\right)^2 + \frac{3M_{ta}^2}{4}\right]^{1/2} + \left\{\frac{32n}{\pi\sigma_{sut}}\left[\left(M_{bm} + \frac{F_mD}{2}\right)^2 + \frac{3M_{tm}^2}{4}\right]^{1/2}\right\}\right)^{1/3} \quad (14\text{-}78)$$ Not explicit in D, use iterative methods to solve

Although ASME has withdrawn the *ASME Code for Design of Transmission Shafting*, some of the ASME equations given here have historic interest and hence are retained in this book.

Particular	Formula
(6) The diameter of shaft made of brittle material, which is subjected to reversed bending and torsional moments taking into consideration stress concentration as per maximum normal stress theory of failure combined with modified Goodman relation for fatigue strength	$D = \left\{ \frac{16n}{\pi\sigma_{sut}} [M'_{be} + (M'^2_{be} + M'^2_{te})^{1/2}] \right\}^{1/3}$ (14-79) for solid shaft $D_o = \left\{ \frac{16n}{\pi\sigma_{sut}(1-K^4)} [M'_{be} + (M'^2_{be} + M'^2_{te})^{1/2}] \right\}^{1/3}$ (14-80) for hollow shaft, where M'_{be} and M'_{te} have the same meaning as given under Eq. (14-77)
(7) Shaft subjected to combined axial, bending, and torsional reversed loads taking into consideration stress concentration and shock	
(a) The diameter of hollow shaft according to distortion-energy theory of failure using Soderberg relation	$D_o = \left(\frac{32n}{\pi\sigma_{sy}(1-K^4)} \left\{ K_{sb} \left[M_{be} + \frac{F_{ae} D_o (1+K^2)}{8} \right]^2 + \frac{3}{4} K_{st} M^2_{te} \right\}^{1/2} \right)^{1/3}$ (14-81)
The symbols used in Eqs. (14-80) to (14-85) and Figs. 14-1 and 14-2 are different than that of the ANSI/ASME standard B106. IM-1985 in order to remain consistent with the symbols used in this Handbook.	where F_{ae}, M_{be} and M_{te} have the same meaning as given under Eqs. (14-71) and (14-76) Refer to Table 14-4 for K_{sb} and K_{st}

New ASME Code for design of transmission shafting:

The diameter of shaft subjected to fully reversed bending i.e., zero mean bending component and torsional fluctuating loads, i.e. alternating loads taking into consideration stress concentration according to distortion energy theory of failure combined with modified Goodman relation for fatigue as per new ANSI/ASME code for transmission shafting.	$D = \left[\frac{32n}{\pi\sigma_{sf}} \left\{ (K_{f\sigma} M_{ba})^2 + \frac{3}{4} (K_{f\tau} M_{ta})^2 \right\}^{1/2} + \frac{32n}{\pi\sigma_{sut}} \left\{ (K_{f\sigma} M_{bm})^2 + \frac{3}{4} (K_{f\tau m} M_{tm})^2 \right\}^{1/2} \right]^{1/3}$ (14-82) when the axial load $= F_a$ is zero
The factor of safety, n	$\frac{1}{n} = \frac{32}{\pi D^3} \left[\frac{1}{\sigma_{sf}} \left\{ (K_{f\sigma} M_{ba})^2 + \frac{3}{4} (K_{f\tau} M_{ta})^2 \right\}^{1/2} + \frac{1}{\sigma_{sut}} \left\{ (K_{f\sigma} M_{ba})^2 + \frac{3}{4} (K_{f\tau m} M_{tm})^2 \right\}^{1/2} \right]$ (14-83)

Particular	Formula
The diameter of shaft made of brittle material subjected to reversed bending and torsional moments taking into consideration stress concentration as per maximum normal stress theory of failure combined with modified Goodman relation for fatigue strength	$$D = \left[\frac{16n}{\pi\sigma_{sut}}\{M'_{be} + (M'^2_{be} + M'^2_{te})^{1/2}\}\right]^{1/3}$$ for solid shaft (14-84)
	$$D_o = \left[\frac{16n}{\pi\sigma_{sut}(1-K^4)}\{M'_{be} + (M'^2_{be} + M'^2_{te})^{1/2}\}\right]^{1/3}$$ for hollow shaft (14-85)
	where, $M'_{be} = K_{f\sigma}M_{bm} + K_{f\sigma}\frac{\sigma_{sul}}{\sigma_{sf}}M_{ba}$
	$M'_{te} = K_{f\tau}M_{tm} + \frac{\sigma_{sul}}{\sigma_{sf}}M_{ta}$
For combined fatigue test data for reversed bending combined torsion and combined with reversed torsion on steel specimens.	Refer to Fig. 14-1.

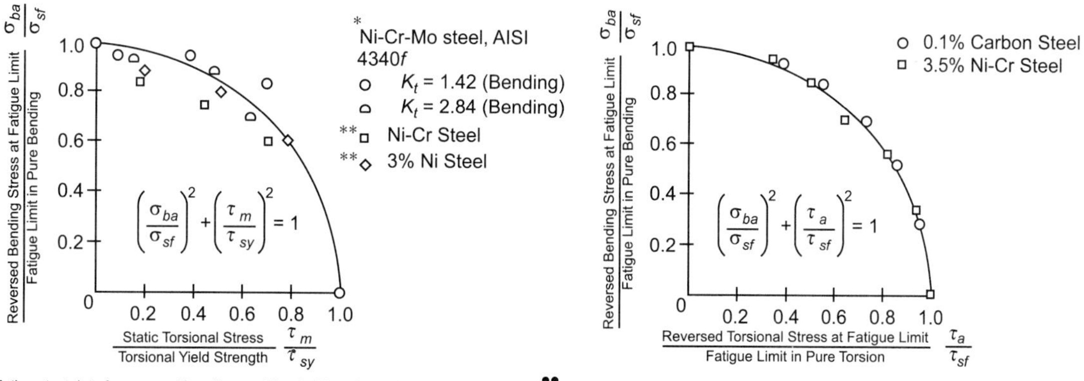

•(a) Fatigue test data for reversed bending combined with static torsion

••(b) Fatigue test data for reversed bending combined with reversed torsion

FIGURE 14-1 Results of fatigue Tests of steel specimens subjected to Reversed Bending and Torsion.
Source: *Design of Transmission Shafting*, American Society for Mechanical Engineers, New York, ANSI/ASME standard B106-IM, 1985.

* Kececioglu, D. B., and V. R. Lalli, *Reliability Approach to Rotating Component Design*, Technical Note TND-7846, NASA, 1975.

** Davies, V. C., H. T. Gough, and H. V. Pollard, Discussion to the Strength of Metals under Combined Alternating stresses, *Proc of the Inst. Mech. Eng.*, **131**(3), pp. 66–69, 1935.

• Loewenthal, S. H., *Proposed Design Procedure for Transmission Shafting under Fatigue Loading*, Technical Note TM-7802, NASA, 1978.

•• Gough, H. J., and H. V. Pollard, The Strength of Metals under Combined Alternating Stresses, *Proc of the Inst. Mech. Eng.*, **131**(3), pp. 3–103, 1935.

Particular	Formula

GENERAL

See Tables 14-1 to 14-6 and Fig. 14-2 for further details on shafting design;[3] refer to Table 14-4 for shock load factors K_{sb} and K_{st}

For further design details on shafting	Refer to Tables 14-5 to 14-7.

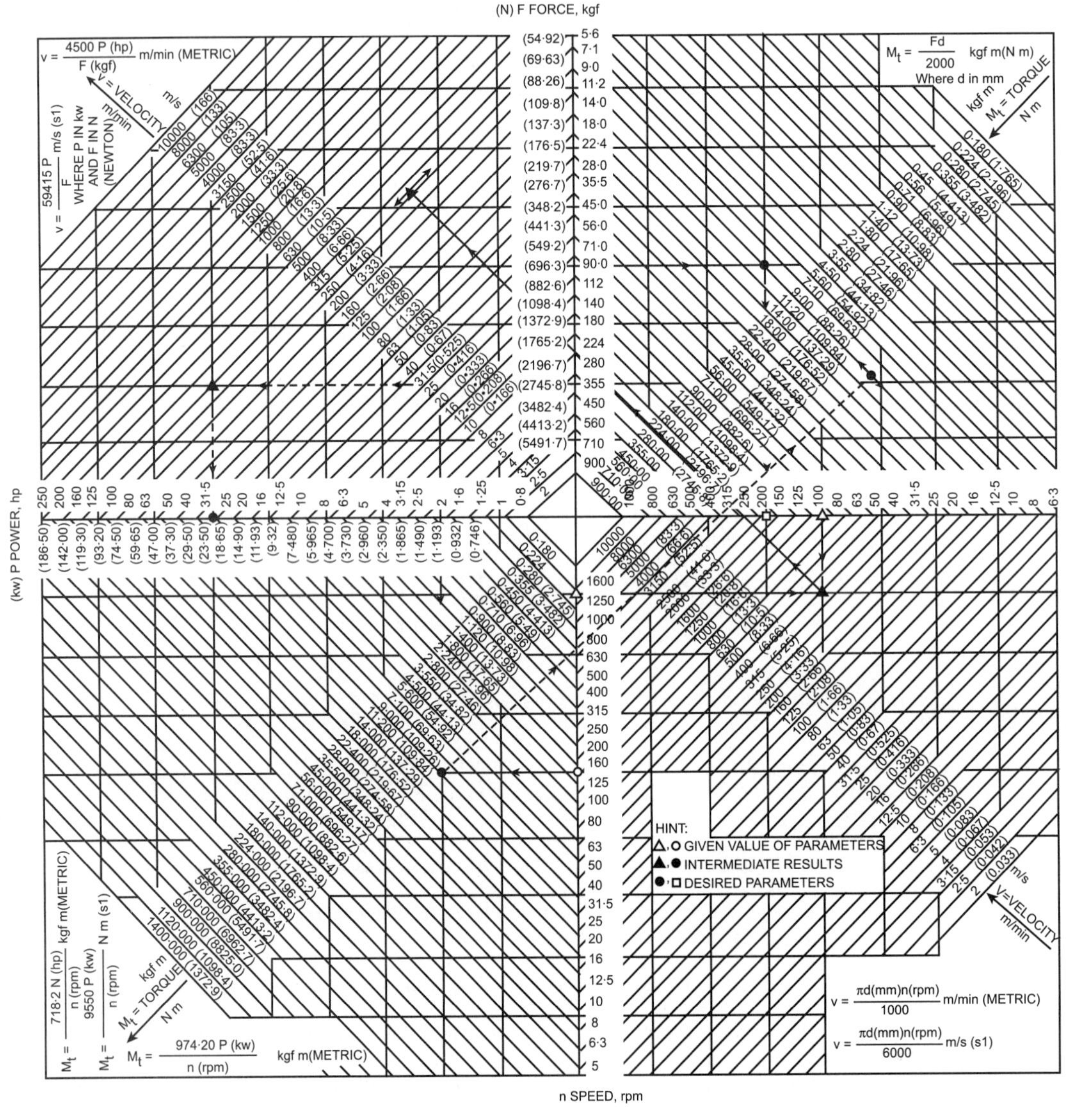

FIGURE 14-2 Nomogram for determining diameter (d), speed (n), force (F), torque (M_t), and power (P) in Customary Metric units and System International units. (K. Lingaiah, *Machine Design Data Handbook*, Vol. II, Suma Publishers, Bangalore, India, 1986.)

REFERENCES

1. Lingaiah, K., and B. R. Narayana Iyengar, *Machine Design Data Handbook,* Engineering College Cooperative, Bangalore, India, 1962.
2. Lingaiah, K., and B. R. Narayana Iyengar, *Machine Design Data Handbook,* Vol. I (*SI Units and Customary Metric Units*), Suma Publishers, Bangalore, India, 1986.
3. Lingaiah, K., *Machine Design Data Handbook,* Vol. II (*SI Units and Customary Metric Units*), Suma Publishers, Bangalore, India, 1986.
4. Soderberg, C. R., "Working Stresses," *J. Appl. Mechanics,* Vol. 57, p. A-106, 1935.
5. *ASME Code for Design of Transmission Shafting,* Standard ANS/ASME B106.1M, 1985.
6. Shigley, J. E., *Machine Design,* McGraw-Hill Publishing Company, New York, 1956.
7. Kececioglu, D. B., and V. R. Lalli, Reliability Approach to Rotating Component Design, Technical Note TND-7846, NASA, 1975
8. Davies, V. C., H. T. Gough, and H. V. Pollard, Discussion to the Strength of Metals under Combined Alternating stresses, *Proc of the Inst. Mech. Eng.*, **131**(3), pp. 66–69, 1935.
9. Loewenthal, S. H., Proposed Design Procedure for Transmission Shafting under Fatigue Loading, Technical Note TM-7802, NASA, 1978.
10. Gough, H. J., and H. V. Pollard, The Strength of Metals under Combined Alternating stresses, *Proc of the Inst. Mech. Eng.*, **131**(3), pp. 3–103, 1935.

BIBLIOGRAPHY

Berchard, H. A., "A Comprehensive Method for Designing Shafts to Insure Fatigue Life," *Machine Design,* April 25, 1963.

Black, P. H., and O. Eugene Adams, Jr., *Machine Design,* McGraw-Hill Publishing Company, New York, 1983.

British Standards Institution.

Deutschman, A. D., W. J. Michels, and C. E. Wilson, *Machine Design—Theory and Practice,* Macmillan Publishing Company, New York, 1975.

Maleev, V. L., and J. B. Hartman, *Machine Design,* International Textbook Company, Scranton, Pennsylvania, 1954.

Marks' Standard Handbook for Mechanical Engineers, 8th ed., McGraw-Hill Publishing Company, New York, 1978.

Vallance, A., and V. L. Doughtie, *Design of Machine Members,* McGraw-Hill Publishing Company, New York, 1951.

TABLE 14-1
Empirical shafting formulas

	Load factors considered		Power capacity, P	
Kind of service	**Torsion, K_t**	**Bending, K_b**	**kW**	**hp**
Transmission shafts in torsion only	1.0	1.0	$54{,}831D^3n'$	$1.225 \times 10^{-6}D^3n$
Line shafting with limited bending	1.0	1.5	$34{,}532D^3n'$	$7.715 \times 10^{-7}D^3n$
Head or main shafts with heavy bending loads	1.0	2.5	$20{,}715D^3n'$	$4.628 \times 10^{-7}D^3n$

TABLE 14-2
Shock and endurance factors

Nature of loading	K_b	K_t
Stationary shafts		
Gradually applied load	1.0	1.0
Suddenly applied load	1.5–2.0	1.5–2.0
Rotating shafts		
Steady or gradually applied loads	1.5	1.0
Suddenly applied loads, minor shocks only	1.5–2.0	1.0–1.5
Suddenly applied loads, heavy shocks	2.0–3.0	1.5–3.0

TABLE 14-3
Values of constant c

		Allowable stress	
Type of shaft loading	**Coefficient c in Eq. (14-61)**	**MPa**	**kpsi**
Shaft heavily loaded, subjected to shock, or reversed under full load	0.82	17	2.5
Line shafts and countershafts, loaded in bending but not reversed	1.1	27	4.0
Line shafts or bar with pulleys close to the bearings	1.56	44	6.4

TABLE 14-4
Shock load factors[a] for use in Eq. (14-81)

Nature of load	K_{sb}, K_{st}
Gradually applied load	1.00
Loads applied with minor shocks	1.0–1.5
Loads applied with heavy shocks	1.5–2.0

[a] Data from Berchard, H. A., "A Comprehensive Method for Designing Shafts to Insure Fatigue Life," *Machine Design*, April 25, 1963.

TABLE 14-5
Spacing[a] for fine shaft bearings

Diameter of shaft, mm	Transmission shaft stressed in torsion only, mm		Line shaft carrying pulleys or gears and subjected to usual bending loads, mm	
	1–250 rpm	251–400 rpm	1–250 rpm	251–400 rpm
36.5	274.5	244.0	213.5	198.0
49.0	305.0	274.5	229.0	213.5
62.0	335.5	305.0	244.0	228.5
74.5	366.0	335.5	259.0	244.0
87.5	396.0	366.0	274.5	259.0
100.0	427.0	396.0	289.5	274.5
112.5	457.0	427.0	305.0	289.5

[a] Center-to-center distance in millimeters.

TABLE 14-6
Sizes of shafts

Diameters, mm (in)					
4 (0.16)	12 (0.48)	40 (1.6)	75 (3.0)	110 (4.4)	180 (7.2)
5 (0.20)	15 (0.60)	45 (1.8)	80 (3.2)	120 (4.8)	190 (7.6)
6 (0.24)	17 (0.68)	50 (2.0)	85 (3.4)	130 (5.2)	200 (8.0)
7 (0.28)	20 (0.80)	55 (2.2)	90 (3.6)	140 (5.6)	220 (8.8)
8 (0.32)	25 (1.0)	60 (2.4)	95 (3.8)	150 (6.0)	240 (9.6)
9 (0.36)	30 (1.2)	65 (2.6)	100 (4.0)	160 (6.4)	260 (10.4)
10 (0.4)	35 (1.4)	70 (2.8)	105 (4.2)	170 (6.8)	280 (11.2)

TABLE 14-7
Load factors for various machines, k_l [a]

Driver	Driven machinery	Factor, k_l
Steam turbine	Electric generator, steady load; turbine blower	1.00
	Electric generator, uneven load; centrifugal pump	1.25
	Induced-draft fan; line shaft; gear drive	1.50
	Rolling mill, gear drive	2.00
Electric motor	Turbine blower; metalworking machinery	1.25
	Centrifugal pump; wood working machinery	1.50
	Line shaft; ship propeller; double acting pump	1.75
	Triplex single-acting pump; elevator; crane	1.75
	Compressor, air or ammonia	1.75
	Rolling mill; rubber mill	2.50
Steam engine	Values for electric-motor drive multiplied by 1.2–1.5	
Gas and oil engines	Values for electric-motor drive multiplied by 1.3–1.6 the factor depending on the coefficient of steadiness of the flywheel	

[a] To be used also in Eqs. (5–9) and (19–79).

CHAPTER 15

FLYWHEELS

SYMBOLS[1,2]

a	major axis of ellipse, m (in)
	negative acceleration or deceleration, m/s^2 (ft/s^2)
A	cross-sectional area of the rim, m^2 (in^2)
b	minor axis of ellipse, m (in)
	width of rim, m (in)
C_f	coefficient of fluctuation of rotation
d	diameter of shaft, m (in)
d_h	hub diameter, m (in)
D	flywheel diameter, m (in)
D_o	outside diameter of rim, m (in)
E	excess energy, J (ft lbf)
F_c	centrifugal force, kN (lbf)
F'_c	centrifugal force per unit width of rim, kN (lbf)
g	acceleration due to gravity, 9.8066 m/s^2 (32.2 ft/s^2)
h	depth of rim, m (in)
i	number of arms
k_o	polar radius of gyration of the rim, m (in)
I	mass moment of inertia, N s^2 m (lbf s^2 ft)
J	polar second moment of inertia, m^4 (in^4)
k_t	torsional stiffness of shaft, N m/rad (lbf in/rad)
M_{tm}	mean torque, N m (lbf ft)
M_t	transmitted torque, N m (lbf ft)
m	coefficient of steadiness
n	mean speed, rpm
n_1	maximum speed, rpm
n_2	minimum speed, rpm
r	mean radius of the flywheel, m (in)
t	time, s
T_1	tension in belt on tight side, kN (lbf)
T_2	tension in belt on slack side, kN (lbf)
v	mean rim velocity, m/s (ft/min)
v_1	maximum rim velocity, m/s (ft/min)
v_2	minimum rim velocity, m/s (ft/min)
W	rim weight, kN (lbf)
ρ	specific weight of material or weight density, N/m^3 (lbf/in^3)
Z	sectional modulus of the arm cross section at the hub, m^3 (in^3)

σ	stress (also with subscripts), MPa (psi)
θ_1, θ_2	maximum and minimum angular displacement of flywheel from constant speed deviation, rad (deg)
ω	average angular speed, rad/s
ω_1, ω_2	maximum and minimum angular speed, respectively, rad/s

Particular	Formula
The equation of motion of ith rotor of I_i inertia in a multirotor system connected by $(i-1)$ number of shafts of various inertias subjected to external torque	$I_i\theta_i = M_{ti} - M_{t(i-1)}$ (15-1)
The equation of motion of a flywheel, which is mounted on a shaft between two supports and rotates with an angular velocity and subjected to an input external torque M_{ti}	$I\theta = M_{ti} - M_{to} = k_t(\theta_2 - \theta_1)$ (15-2) where M_{to} = output torque, N m (lbf ft) θ = angular displacement of flywheel, rad (deg)

KINETIC ENERGY

Kinetic energy (Fig. 15-1)

$$K = \frac{1}{2}mv^2 = \frac{Wv^2}{2g} = \frac{1}{2}I\omega^2 \tag{15-3}$$

For variation of torque with crank angle for two-cylinder engine

Refer to Fig. 15-1.

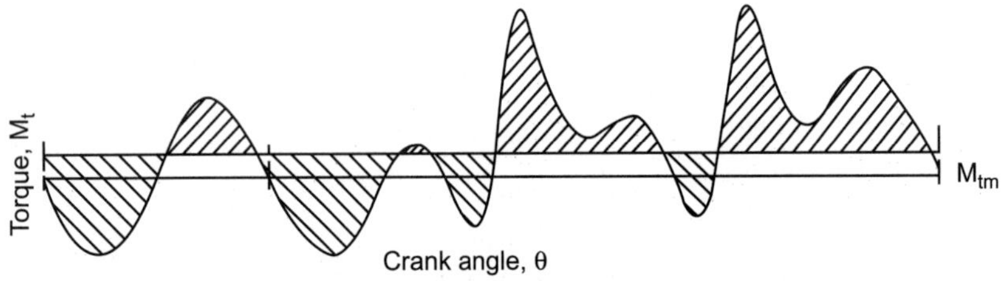

FIGURE 15-1 Torque-crank shaft angle curve for a two-cylinder engine.

The kinetic energy of flywheel at an angular displacement θ_1 and at angular velocity ω_1 during one cycle

$$K_1 = \frac{1}{2}I\omega_1^2 = \frac{Wv_1^2}{2g} \tag{15-4}$$

The kinetic energy of flywheel at an angular displacement θ_2 and at angular velocity ω_2

$$K_2 = \frac{1}{2}I\omega_2^2 = \frac{Wv_2^2}{2g} \tag{15-5}$$

The change in kinetic energy or energy fluctuation due to change in angular velocity ω_1 to ω_2 in one cycle

$$E = K_2 - K_1 = \frac{1}{2}I(\omega_2^2 - \omega_1^2) = \frac{W(v_2^2 - v_1^2)}{2g}$$
$$= \tfrac{1}{2}I(\omega_2 - \omega_1)(\omega_2 + \omega_1)$$
$$= I(\omega_2 - \omega_1)\omega = W(v_2 - v_1)\frac{v}{g} \tag{15-6}$$

Particular	Formula	
The coefficient of fluctuation of speed or rotation	$C_f = \frac{\omega_2 - \omega_1}{\omega} = \frac{v_2 - v_1}{v} = \frac{n_2 - n_1}{n}$	(15-7)
The change in kinetic energy or excess energy	$E = K_2 - K_1 = I\omega^2 C_f = \frac{Wv^2 C_f}{g}$	(15-8)

FLYWHEEL EFFECT OR POLAR MOMENT OF INERTIA

$$Wk^2 = \frac{182.40gE}{n_1^2 - n_2^2} \qquad (15\text{-}9)$$

The mean angular velocity

$$\omega = \frac{\omega_2 + \omega_1}{2} \qquad (15\text{-}10)$$

The coefficient of steadiness

$$m = \frac{1}{C_f} \qquad (15\text{-}11)$$

Refer to table 15-1 for C_f.

STRESSES IN RIM (Figs. 15-2 and 15-3)

Particular	Formula		
The component of the centrifugal force normal to any diameter of the flywheel	$F_c = \frac{2\rho bhr^2\omega^2}{g}$		(15-12)
The tangential force due to hoop stress in the flywheel rim (Fig. 15-3)	$F_\theta = \frac{\rho bhr^2\omega^2}{g}$		(15-13)
The tensile stress created in each cross section of the rim by the centrifugal force	$\sigma = 0.01095 \frac{\rho}{g} r^2 n^2$	SI	(15-14)
The centrifugal force per unit width of rim (Fig. 15-3)	$F_c' = 0.01095 \frac{\rho r^2 n^2 h}{g}$	SI	(15-15)

TABLE 15-1
Coefficient of fluctuation of rotation, C_f

Driven machine	Type of drive	C_f
AC generators, single or parallel	Direct-coupled	0.01
AC generators, single or parallel	Belt	0.0167
DC generators, single or parallel	Direct-coupled	0.0143
DC generators, single or parallel	Belt	0.029
Spinning machinery	Belt	0.02–0.015
Compressure, pumps	Gears	0.02
Paper, textiles, and flour mills	Belt	0.025–0.02
Woodworking and metalworking machinery	Belt	0.0333
Shears and pumps	Flexible coupling	0.05–0.04
Concrete mixers, excavators, and compressors	Belt	0.143–0.1
Crushers, hammers, and punch presses	Belt	0.2

Particular	Formula
The bending stress	$\sigma_b = 0.2146\,\dfrac{\rho r^3 n^2}{ghi^2}$ **SI** (15-16)
The combined tensile stress	$\sigma_R = 0.75\sigma + 0.25\sigma_b$ (15-17)

STRESSES IN ARMS (Fig. 15-2)

The stresses in the arm	$\sigma_1 = \dfrac{M_t(D - d_h)}{iZD}$ (15-18)

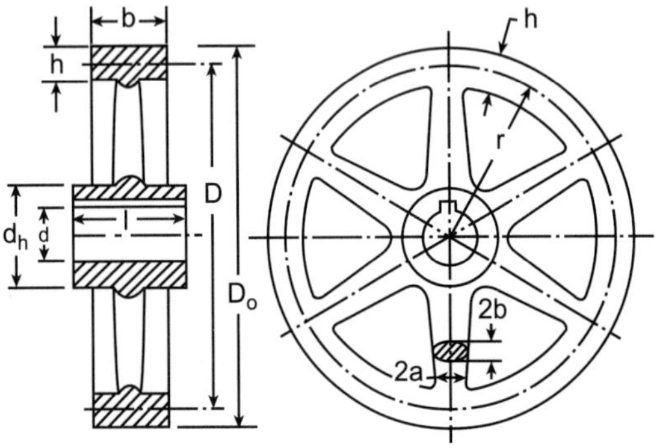

FIGURE 15-2 Flywheel.

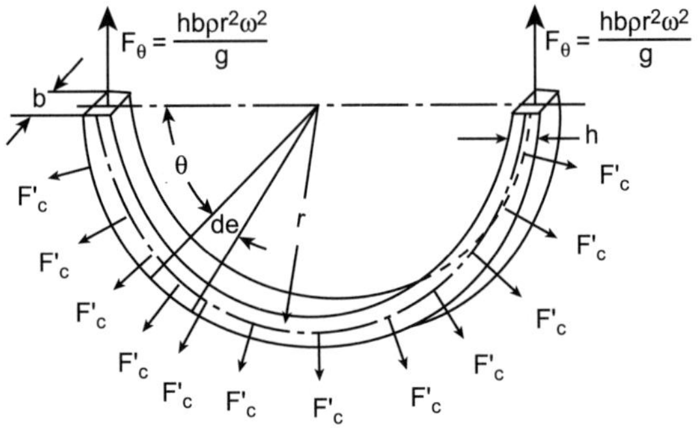

FIGURE 15-3 Centrifugal force acting on the rim of a flywheel.

When the flywheel is used as a belt pulley, the stresses at the hub	$\sigma_2 = \dfrac{(T_1 - T_2)(D - d_h)}{2iZ}$ (15-19)
In case of thin-rim flywheel, the stress	$\sigma_2' = \dfrac{(T_1 - T_2)(D - d_h)}{iZ}$ (15-20)
Stress due to centrifugal force	$\sigma_3 = 0.01095\,\dfrac{\rho r^2 n^2}{g}$ **SI** (15-21)
The maximum tensile stress in an arm is at hub	$\sigma_{max} = \sigma_1 + \sigma_2 + \sigma_3$ (15-22)
The force necessary to stop the flywheel	$F = \dfrac{Wa}{g}$ (15-23)

RIM DIMENSIONS (Fig. 15-2)

The relation between k_o in cm and the outside diameter D of the rim in m	$k_o^2 = 0.125[D_o^2 + (D_o - 2h)^2]$ (15-24)
Cross-sectional area of the rim	$A = \dfrac{W}{2\pi k\rho}$ (15-25)

Particular	Formula	
The relation between depth and width of rim	$\frac{b}{h} = 0.65 \text{ to } 2$	(15-26)
The outside diameter of rim	$D_o = 2k_o + h$ (approx.)	(15-27)
The hub diameter in m	$d_h = 1.75d + 6.35 \times 10^{-3} = 2d$	(15-28)
The hub length	$l = 2d \text{ to } 2.5d$	(15-29)

ARMS (Fig. 15-2)

The major axis in case of elliptical section can be computed from the relation	$a = \sqrt[3]{\frac{64Z}{\pi}}$	(15-30)
	where $z = \frac{\pi ba^2}{32}$ and $a = 2b$	(15-31)

REFERENCES

1. Lingaiah, K., and B. R. Narayana Iyengar, *Machine Design Data Handbook*, Vol. I (*SI and Customary Metric Units*), Suma Publishers, Bangalore, India, 1986.
2. Lingaiah, K., *Machine Design Data Handbook* (*SI and U.S. Customary Units*), McGraw-Hill Publishing Company, New York, 1994.

CHAPTER 16

PACKINGS AND SEALS

SYMBOLS[1,2]

A	area of seal in contact with the sliding member, m^2 (in^2)
A_g	gasket area over which the bolt loads are distributed, m^2 (in^2)
A_1, A_2	area of cross section of unthreaded and threaded portions of bolt, m^2 (in^2)
b	width of U-collar, m (in)
	gland width or depth of groove, m (in)
c	radial clearance between rod and the bushing,
	radial deflection of the ring, m (in)
d	nominal diameter of the bolt, m (in)
	diameter of sliding member, m (in)
d_1	outside diameter of packing material, m (mm)
	outside diameter of seal ring (Fig. 16-3), m (in)
d_2	minor diameter of bolt, m (in)
d_a	actual diameter of wire, m (in)
d_i	inside diameter of packing material, m (in)
D_m	estimated mean diameter of conical spring, m (in)
D_{am}	actual mean diameter of conical spring, m (in)
E	modulus of elasticity, GPa (psi)
F_b	bolt load, kN (lbf)
F_μ	frictional force, kN (lbf)
$F_{\mu o}$	frictional force of the stuffing box when there is no fluid pressure, kN (lbf)
g	acceleration due to gravity, 9.8066 m/s^2 (9806.6 mm/s^2) (32.2 ft/s^2)
h	radial ring wall thickness, m (in)
h_i	uncompressed gasket thickness, m (in)
h_μ	loss of head, m/m (in/in)
i	number of bolts
l	depth of U-collar (Fig. 16-2a), m (in)
l_1, l_2	length of joint, m (in)
(dl)	incremental length in the direction of velocity [Eq. (16-15)], m (in)
	bolt elongation [Eq. (16-24)], m (in)
M_t	twisting moment, N m (lbf in)
M_{ti}	initial bolt torque, N m (lbf in)

p	fluid pressure, MPa (psi)
p_f	flange pressure on the gasket, MPa (psi)
P_s	minimum per cent compression to seal
(dp)	pressure differential in the direction of velocity [Eq. (16-15)], MPa (psi)
Q	discharge, m^3/s (cm^3/s, mm^3/s) (in^3/s)
r	equivalent radius, m (in)
v	velocity, m/s (ft/min)
w	nominal packing cross section, m (in)
y	deflection of spring, m (in)
η	absolute viscosity of fluid, Pa s (cP)
σ_d	design stress, MPa (psi)
μ	coefficient of friction

Particular	Formula

ELASTIC PACKING[1–3]

Frictional force exerted by a soft packing on the reciprocating rod

$$F_\mu = kpd \tag{16-1}$$

where $k = 0.005$ and $p = 0.343$ MPa **SI**
$k = 0.2$ and $p = 50$ psi **USCS**

FRICTION

Friction resistance

$$F_\mu = F_o + \mu A p \tag{16-2}$$

where $\mu = 0.01$ for rubber and soft lubricated leather
$\mu = 0.15$ for hard leather

Torsional resistance in a rotary motion friction

$$M_t = \frac{F_\mu d}{2} = \frac{kd^2 p}{2} \tag{16-3}$$

where $k = 0.005$ **SI**
$k = 0.2$ **USCS**

METALLIC GASKETS (Fig. 16-1)

The empirical relations[3]

$$c = 0.2d + 5\,\text{mm if } d \leq 100\,\text{mm} \tag{16-4}$$

$$c = 0.08\sqrt{d} \quad \text{if } d > 0.1\,\text{mm} \quad \textbf{SI} \tag{16-5a}$$

$$c = 0.5\sqrt{d} \quad \text{if } d > 4 \quad \textbf{USCS} \tag{16-5b}$$

$$h = \frac{d}{8} + 12.54\,\text{mm or } 0.5\,\text{in} \tag{16-6}$$

$$a = d + 2c \tag{16-7}$$

$$\alpha = 10^\circ \text{ to } 15^\circ \tag{16-8}$$

$$d_2 = 0.2(d + 0.102)/\sqrt{i} \quad \textbf{SI} \tag{16-9a}$$

$$d_2 = 0.2(d + 4)/\sqrt{i} \quad \textbf{USCS} \tag{16-9b}$$

Particular	Formula

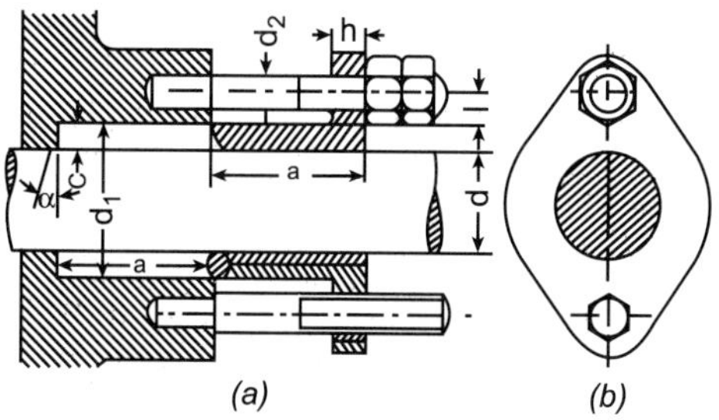

FIGURE 16-1 Stuffing box with bolted gland. (V. L. Maleev and J. B. Hartman, *Machine Design*, International Textbook Company, Scranton, Pennsylvania, 1954.)

Diameter of bolt is also found by equating the working strength of the bolts to the pressure p exerted by the fluid on the gland and the frictional force F_μ

$$d_2 = \sqrt{\frac{(d_1^2 - d^2)p}{i\sigma_d} + \frac{4F_\mu}{\pi i \sigma_d}} \qquad (16\text{-}10)$$

where

d_2 = minor diameter of bolt, m (in)
σ_d = 68.7 to 83.3 MPa (10 to 12 kpsi)

SELF-SEALING PACKING (Fig. 16-2)

Houghton, Welch, and Jenkin's formula for an approximate thickness of a U-shaped collar for great pressure[3]

$$h = 6.36 \times 10^{-3} d^{0.2} \qquad \textbf{SI} \quad (16\text{-}11a)$$

where h and d in m

$$h = 1.6d^{0.2} \qquad \textbf{SI} \quad (16\text{-}11b)$$

where h and d in mm

$$h = 0.12d^{0.2} \qquad \textbf{USCS} \quad (16\text{-}11c)$$

where d and d in in

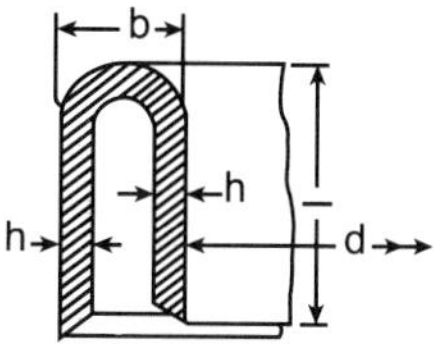

FIGURE 16-2 U-collar.

Width

$$b = 4h \qquad (16\text{-}12a)$$

Depth

$$l = 1.2b \text{ to } 1.8b \qquad (16\text{-}12b)$$

Particular	Formula

PACKINGLESS SEALS

Leakage of the fluid past a rod can be computed with fair accuracy by the formula

$$Q = \frac{\pi c^3}{12}(p_1 - p_2)\frac{d}{l\eta} \quad \textbf{SI} \quad (16\text{-}13a)$$

$$Q = 1.79(100c)^3 \frac{(p_1 - p_2)d}{l\eta} \quad \textbf{USCS} \quad (16\text{-}13b)$$

Refer to Table 16-1 for values of η.

TABLE 16-1
Absolute viscosities η

Fluid	Temperature K	Temperature °C	Absolute viscosity, η MPa s	Absolute viscosity, η cP	Temperature K	Temperature °C	Absolute viscosity, η MPa s	Absolute viscosity, η cP
Steam	293	20	0.0097	0.0097	539	266	0.018	0.018
Air	293	20	0.018	0.018	366	93	0.022	0.022
Water	273	0	1.79	1.79	311	38	0.69	0.69
Water	293	20	1.0	1.0	333	60	0.40	0.40
Gasoline	293	20	0.6	0.6	355	82	0.30	0.30
Kerosene	293	20	2.7	2.7	355	82	1.30	1.30
Fuel oil, 30° Baumé	293	20	5.0	5.0	355	82	1.60	1.60
Fuel oil, 24° Baumé	293	20	40	40	355	82	4	4
Spindle oil	293	20	20–35	20–35	355	82	3–4	3–4
Machine oil	293	20	200–500	200–500	372	99	1.5–16	5.5–16
Castor oil	293	20	1000	1000	316	43	200	200

STRAIGHT-CUT SEALINGS (Fig. 16-3a)

The equation for loss of liquid head

$$h_\mu = 64\eta v / 2g\rho d_1^2 \quad (16\text{-}14)$$

Leakage velocity

$$v = \frac{(dp)r^2}{8(dl)\eta} \quad (16\text{-}15)$$

Quantity of leakage

$$Q = vA \quad (16\text{-}16)$$

Stress in a seal ring

$$\sigma = \frac{0.4815cE}{h\left(\frac{d_1}{h} - 1\right)^2} \quad (16\text{-}17)$$

For allowable temperatures for materials and surface treatment

Refer to Table 16-2.

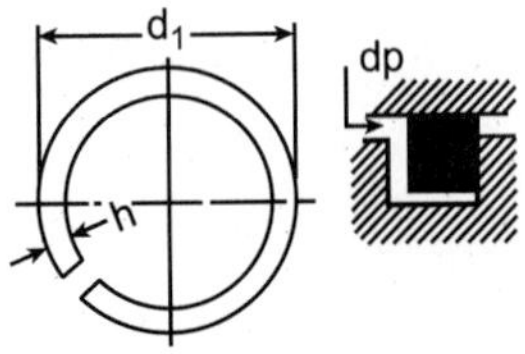

FIGURE 16-3(a) Straight-cut seal.

Particular	Formula

V-RING PACKING

Single-spring installations

The estimated mean diameter of conical spring	$D_m = d_i + \frac{3w}{2}$ (16-18)
The wire size (Table 16-3)	$d = \left(\frac{\pi D_m^2}{139300}\right)^{1/3}$ **SI** (16-19a) where d and D_m in m $d = \left(\frac{\pi D_m^2}{3535}\right)^{1/3}$ **USCS** (16-19b) where d and D_m in in $d = \left(\frac{\pi D_m^2}{193.3}\right)^{1/3}$ **Customary Metric** (16-19c) where d and D_m in mm
The actual mean diameter of conical spring	$D_{am} = d_1 - \frac{1}{2}(w + d_a)$ (16-20)
The deflection of spring	$y = \frac{0.0123 D_{am}^2}{d_a}$ (16-21)

Multiple-spring installations

Two standard cylindrical spring sizes are generally used, depending on packing size.

BOLTS AND STRESSES IN FLANGE JOINTS

The bolt load in gasket joint	$F_b = \frac{11 m_{ti}}{d}$ (16-22)
The flange pressure developed due to tightening of bolts that hold the gasket joint mechanical assembly together	$p_f = \frac{iF_b}{A_g C_u} = \frac{2iM_t}{A_g C_u d_b}$ (16-23) where C_u = torque friction coefficient
The load on the bolt when it is tightened	$F_b = \frac{E(dl)}{(l_1/A_1) + (l_2/A_2)}$ (16-24)

STRESSES IN GROOVED JOINTS

The uncompressed gasket thickness that will provide the minimum sealing compression when the flanges are tightened into face-to-face contact	$h_i = \frac{100b}{100 - P_s}$ (16-25)

Particular	Formula

BOLT LOADS IN GASKET JOINT ACCORDING TO *ASME BOILER AND PRESSURE VESSEL CODE* (Fig. 16-3b)[4]

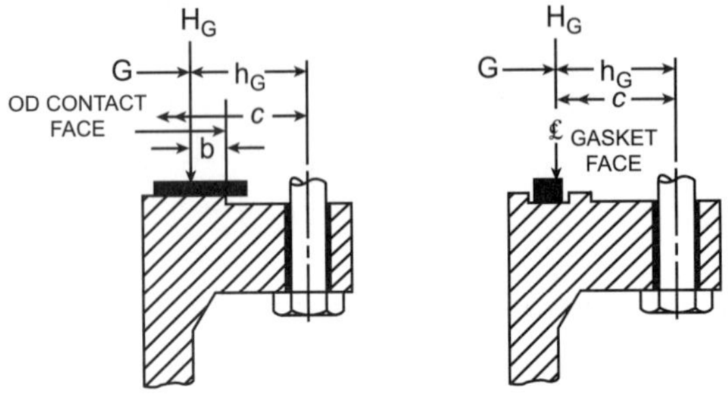

For $b_o > 6{\cdot}3$ mm (0.25in .)   For $b_0 \leq 6{\cdot}3$ mm (0.25in .)

Effective gasket seating width $b = b_{o1}$, when $b_o = 6{\cdot}3$ mm (0.25 in .)
and $b = 2{\cdot}5\sqrt{b_{o1}}$, when $b_o > 6{\cdot}3$ mm (0.25 in .)

NOTE– The gasket factors listed only apply to flanged joints in which the gasket is contained entirely within the inner edges of the bolt holes.

FIGURE 16-3(b) Location of gasket load reaction.

The required bolt load under operating condition sufficient to contain the hydrostatic end force and simultaneously to maintain adequate compression on the gasket to ensure seating

$$W_{m1} = H + H_P = (\pi/4G^2P) + 2b\pi GmP \tag{16-26}$$

The required initial bolt load to seat the gasket joint-contact surface properly at atmospheric temperature condition without internal pressure

$$W_{m2} = \pi bGy \tag{16-27}$$

Refer to Tables 8-20 and 8-21 for gasket factor m and minimum design seating stress, y, b, and b_o

Total required cross-sectional area of bolts at the root of thread

$$A_m > A_{m1} \text{ or } A_{m2} \tag{16-28}$$

Total cross-sectional area of bolt at root of thread or section of least diameter under stress required for the operating condition

$$A_{m1} = \frac{W_{m1}}{\sigma_{sbd}} \tag{16-29}$$

Refer to Table 8-17 for σ_{sbd}

Total cross-sectional area of bolt at root of thread or section of least diameter under stress required for gasket seating

$$A_{m2} = \frac{W_{m2}}{\sigma_{sbat}} \tag{16-30}$$

The actual cross-sectional area of bolts using the root diameter of thread or least diameter of unthreaded portion (if less), to prevent damage to the gasket during bolting-up

$$A_b = \frac{2\pi yGN}{\sigma_{sbat}} \not< A_m \tag{16-31}$$

Particular	Formula

FLANGE DESIGN BOLT LOAD W

The bolt load in the design of flange for operating condition	$W = W_{m1}$ (16-32)
The bolt load in the design of flange for gasket seating	$W = \left(\frac{A_m + A_b}{2}\right)\sigma_{sbat}$ (16-33)
The relation between bolt load per bolt (W_b), diameter of bolt (D) and torque (M_t)	$W_b = 0.17DM_t$ for lubricated bolts **USCS** (16-34) where W_b in lbf, D in in, M_t in lbf in
(*Note*: The meanings of symbols given in Eqs. (16-26) to (16-37) are defined in Chap. 8.)	$W_b = 263.5DM_t$ **SI** (16-35) where W_b in N, D in m, M_t in N m $W_b = 0.2DM_t$ for unlubricated bolts **USCS** (16-36) where W_b in lbf, D in in, M_t in lbf in $W_b = 310DM_t$ **SI** (16-37) where W_b in N, D in m, M_t in N m
For location of gasket load reaction due to tightening of flange bolts	Refer to Fig. 16-3b
The total load on bolts in the gasket joint according to Whalen[5]	$F_b = \sigma_g A_g$ (16-38) where A_g = contact area of gasket, m^2 (in^2) σ_g = gasket seating stress, MPa (psi), taken from Table 16-35
The load on bolts, which is based on hydrostatic end force	$F_b = nP_tA_m$ (16-39) where P_t = test pressure or internal pressure if no test pressure is available, MPa (psi) A_m = hydrostatic area (based on mean diameter of gasket) on which internal pressure acts, m^2 (in^2) n = factor of safety taken from Table 16-36
For more information on design data, selection of packing and seals, properties of sealants and packing materials, dimensions and tolerances of seals, and chamfers on shaft, operating temperatures of various types of seals, data for metallic o-rings, q-rings and o-ring gaskets, static and dynamic seals, lip seals, and safety factors, etc.	Refer to Tables 16-4 to 16-36

Particular	Formula

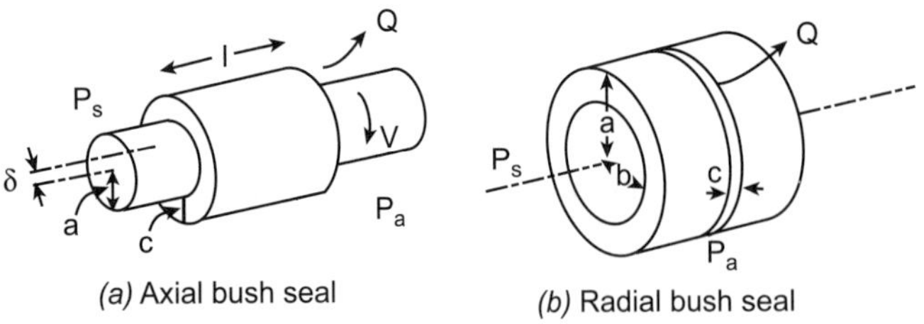

FIGURE 16-3(c) Plain bush seals. (Panels a and b courtesy of J. M. Neale, *Tribology Handbook*, Butterworths, London, 1973.)

Leakage through bush seals (Fig. 16-3c):

The oil flow (Q) through plain axial bush seal due to leakage under laminar flow condition, Fig. 16-3c, panel *a*

$$Q = \frac{2\pi a(P_s - P_a)}{l} q \tag{16-40}$$

where Q in m^3/s (in^3/s)

η = absolute viscosity, Pa s (cP)

The symbols used in Eqs. (16-40) to (16-45) have the meaning as defined in Fig. 6-13c, panels *a* and *b*.

The volumetric flow rate per unit pressure per unit periphery (q) under laminar flow condition for axial bush seal, Fig. 16-3c, panel *a*

$$q = \frac{c^3}{12\eta}(1 + 1.5\varepsilon^2)^{\text{a}} \tag{16-14}$$

for incompressible fluid

where $\left(\varepsilon = \frac{\delta}{c}\right)$

$$q = \frac{c^3}{24\eta}\frac{P_s + P_a}{P_a} \tag{16-42}$$

for compressible fluid[b]

The oil flow (Q) through plain radial bush seal due to leakage under laminar flow condition, Fig. 16-3c, panel *b*

$$Q = \frac{2\pi a(P_s - P_a)}{a - b} q \tag{16-43}$$

The volumetric flow rate per unit pressure per unit periphery (q) under laminar flow condition for radial bush seal, Fig. 16-3c, panel *b*

$$q = \frac{c^3}{12\eta}\frac{a - b}{a \log_e \frac{a}{b}} \tag{16-44}$$

for incompressible fluid

$$q = \frac{c^3}{24\eta}\frac{a - b}{a}\frac{P_s + P_a}{P_a} \tag{16-45}$$

for compressible fluid

---

[a] If shaft rotates, onset of Taylor vortices limits validity of formula to $(V_c/\upsilon)\sqrt{c/a} < 41.3$ (where υ = kinematic viscosity).
[b] For Mach number <1.0, i.e., fluid velocity < local velocity of sound.

Particular	Formula
The radial pressure distribution for laminar flow condition between smooth parallel surfaces in case face seal	$p - p_1 = \dfrac{3\rho\omega^2}{20g}(r^2 - R_1^2) - \dfrac{6v}{\pi h^3}\ln\dfrac{r}{R}$ (16-46) where p = pressure at radial position r, MPa (lbf/in^2) p_1 = pressure at seal inside radius, MPa (psi) p_2 = internal hydraulic pressure MPa (lbf/in^2) r = radial position, m (in) v = kinematic viscosity N s/m^2 (lbf s/in^2) ρ = fluid density, lb/in^3 (kg/mm^3) ω = rotational speed, rad/s R_1 = inside radius of rotating member, m (in) R_2 = outside radius of rotating member, m (in) h = thickness of fluid between members, m (in)
The amount of leakage of fluid through face seal	$Q = \dfrac{\pi h^3}{6v\ln(R_2/R_1)}\left[\dfrac{3\rho\omega^2}{20g}(R_2^2 - R_1^2) - p_2 - p_1\right]$ (16-47) where Q = volumetric leakage rate of fluid, m^3/s (in^3/s)
The theoretical equation for zero leakage of fluid through face seal	$p_2 - p_1 = \dfrac{3}{20}\rho\omega^2(R_2^2 - R_1^2)$ (16-48)
The power loss or consumed due to leakage of fluid through face seal	$P = \dfrac{\pi v w^2}{13200h}(R_2^4 - r_1^4)$ (16-49) where P = power loss, hp
The shape factor (S_{pf}) for a circular or annular gasket which is the ratio of the area of one load face to the area free to bulge[6]	$S_{pf} = \dfrac{D_o - D_i}{4h}$ (16-50) where D_o = outside diameter of gasket, m (in) D_i = inside diameter of gasket, m (in)
For further design and selection of various types of seals, packings and gaskets, etc.	Refer to Figs. 16-4 to 16-14.
For nomenclature of gasketed joint	Refer to Fig. 16-15.
For packing assembly for a mechanical piston rod	Refer to Fig. 16-16.
For shape factor for various gasket materials[6]	Refer to Fig. 16-17.
For power absorption and starting torque for unbalanced and balanced seals	Refer to Fig. 16-18.

FIGURE 16-4 Single radial lip seal.

FIGURE 16-5 Exclusion seal.

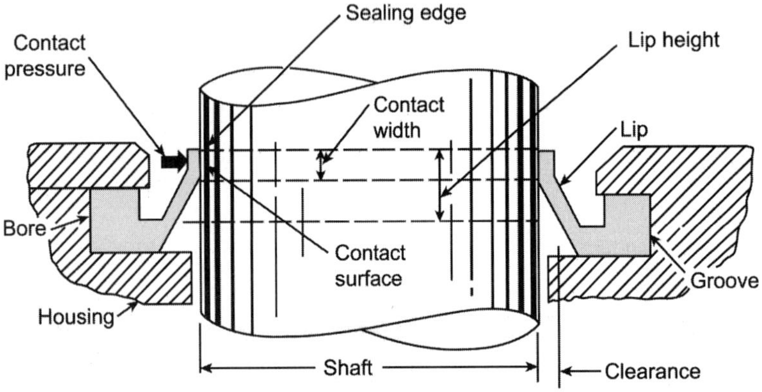

FIGURE 16-6 Radial exclusion seal. (Produced from "Packings and Seals" Issue, *Machine Design*, Jan. 20, 1977.)

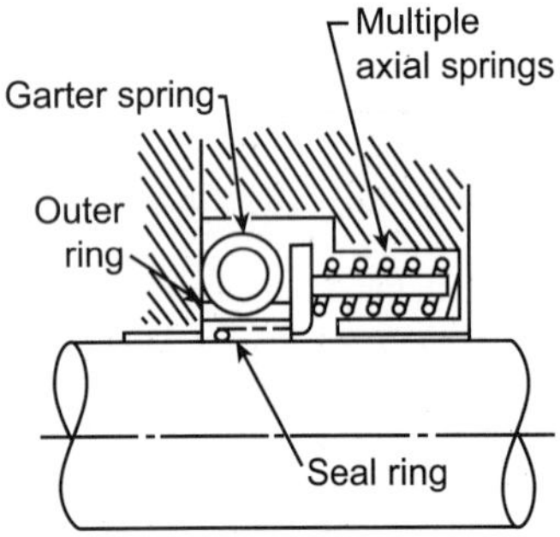

FIGURE 16-7 Two-piece rod seal. (Produced from "Packings and Seals" Issue, *Machine Design*, Jan. 20, 1977.)

FIGURE 16-8 Clearance seal idealized labyrinth.

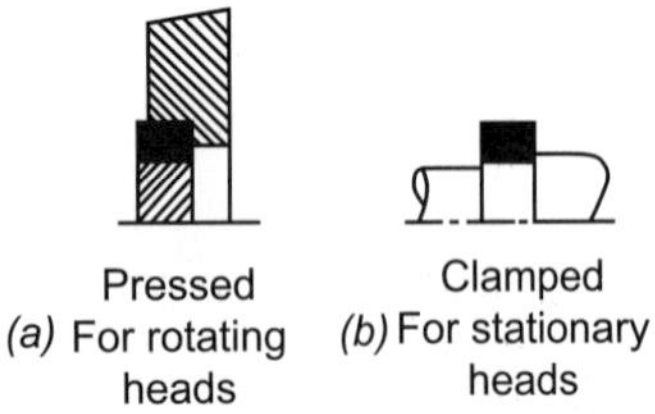

FIGURE 16-9 Face seal.

FIGURE 16-10 Compression packing.

TABLE 16-2
Allowable temperatures for materials and surface treatments

Material or surface treatment	Temperature °F	Temperature °C	Material or surface treatment	Temperature °F	Temperature °C
Material					
Low-alloy gray irons	650	343	Carbon (high-temperature)	950	510
Malleable iron	720	382	K-30 (filled teflons)	450–500	232–260
Ductile iron	720	382	S-Monel	950	510
Ni-Resist	800	427	Polymide	750	399
Ductile Ni-Resist	1000	538	Surface treatment		
410 Stainless Steel	900	482	Chromium plate	500	260
17-4 PH Stainless Steel	900	482	Tin plate	720	382
Bronze	500	260	Silver plate	600	315.5
Stellite no. 31	1200	649	Cadmium nickel plate	1000	538
Inconel X	1200	649	Flame plate LW1	1000	538
Tool steel, Rc 62–65	900	482	Flame plate LC-1A	1600	871
			Flame plate LA-2	1600	871

TABLE 16-3
Standard wire sizes for V-packing expanders

Wire gauge[a]	Wire diameter, mm	Wire gauge	Wire diameter, mm	Wire gauge	Wire diameter, mm
19	1.04	13	2.31	$\frac{5}{32}$	3.82
18	1.20	12	2.67	8	4.11
17	1.37	11	2.05	7	4.49
16	1.57	$\frac{1}{8}$	3.17	$\frac{3}{16}$	4.77
15	1.83	10	3.31	6	4.89
14	2.03	9	3.60	5	5.25

[a] American Wire Gauge (AWG).

TABLE 16-4
Dimensions (in mm) for chamfer on the shaft for mounting the seals

d_1 $h11$	d_3	d_1 $h11$	d_3	d_1 $h11$	d_3	d_1 $h11$	d_3	d_1 $h11$	d_3	d_1 $h11$	d_3
6	4.8	24	21.5	52	48.3	85	80.4	160	153	340	329
7	5.7	25	22.5	55	51.3	90	85.3	170	163	360	349
8	6.6	26	23.4	56	52.3	95	90.1	180	173	380	369
9	7.5	28	25.3	58	54.2	100	95.0	190	183	400	389
10	8.4	30	27.3	60	56.1	105	99.9	200	193	420	409
11	9.3	32	29.2	62	58.1	110	104.7	210	203	440	429
12	10.2	35	32.0	63	59.1	115	109.6	220	213	460	449
14	12.1	36	33.0	65	61.0	120	114.5	230	223	480	469
15	13.1	38	34.9	68	63.9	125	119.4	240	233	500	489
16	14.0	40	36.8	70	65.8	130	124.3	250	243		
17	14.9	42	38.7	72	67.7	135	129.2	260	252		
18	15.1	45	41.6	75	70.7	140	133.0	280	269		
20	17.7	48	44.5	78	73.6	146	138.0	300	289		
22	19.6	50	46.4	80	75.5	150	143.0	320	309		

TABLE 16-5
Selection of guide for packing materials

Condition	Leather (natural and synthetic)	Homogeneous	Fabricated
Oil	Good	Good	Good
Air	Good	Good	Good
Water	Good	Good	Good
Steam	Not recommended	Good	Good
Solvents	Not recommended	Good	Good
Acids	Not recommended	Good	Good
Alkalis	Not recommended	Good	Fair
Temperature range	−55°C +82°C[a]	−55°C +200°C[a]	−40°C +260°C[a]
Types of metal	Ferrous and nonferrous	Chrome-plated steel and nonferrous alloys with hard, smooth surfaces	Chrome-plated steel and nonferrous alloys with hard, smooth surfaces
Metal finish, rms (max.)	63	16	32
Clearances	Medium	Very close	Close
Extrusions or cold flow	Good	Poor	Fair
Friction coefficient	Low	Medium and high	Medium
Resistance to abrasion	Good	Fair	Fair
Maximum pressure, MPa (kpsi)	861.7 (125)	343.4 (50)	549.4 (80)
Concentricity	Medium	Very close	Close
Side loads	Fair	Poor	Fair
High shock loads	Good	Poor to fair	Fair

[a] Depending on specification or combination of materials.

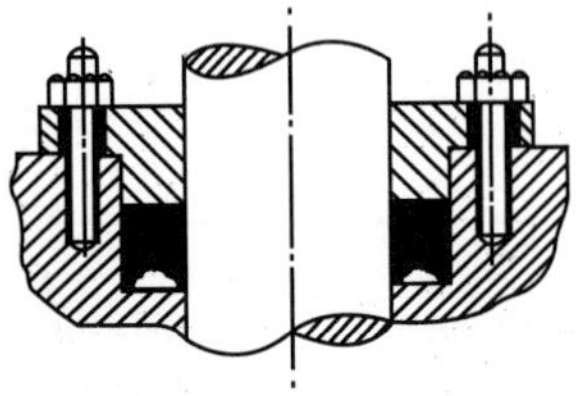

FIGURE 16-11 Molded packing. Typical U-ring packing.

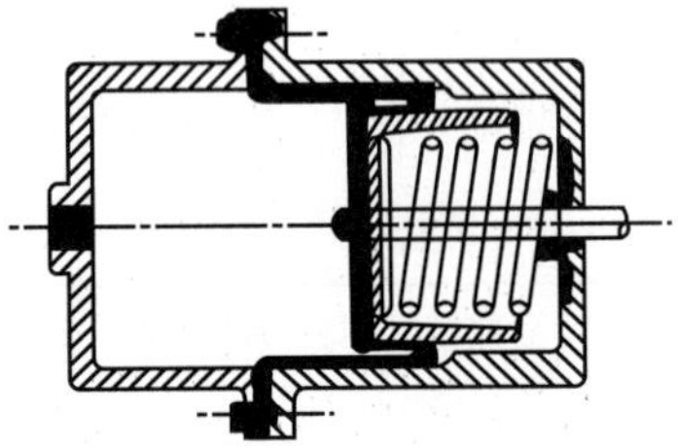

FIGURE 16-12 Diaphragm seals-rolling diaphragm.

TABLE 16-6
Types of seals and their uses

Type	Uses
Radial lip seals	For retaining lubricants in equipments having rotating, reciprocating oscillating shafts, to exclude foreign matter
Single lip (Fig. 16-4)	For containing highly viscous materials at low speeds
Single lip—spring-loaded	For containing lubricants of lower viscosity at higher speeds in clean atmosphere
Double lip with one lip spring-loaded	For excluding contaminants such as dust and dirt
Dual lip with both lips spring-loaded	For containing lubricant on one side and for excluding fluid on the other
Split seal	For splash system of lubrication
External seal	For fixed shaft and rotating bore
Hydrodynamic seal	For directing oil flow back into the area to be sealed
Exclusion seals (Figs. 16-5 and 16-6)	
Wipers, scrapers, axial seals, bellows, and boots	To prevent entry of foreign materials into moving parts of machinery—to avoid contamination of lubricants
Clearance seals (Fig. 16-8)	
Labyrinths, bushing, and ring seals	Dynamic seals-to prevent leakage from a high-pressure station at one end of bushing to a region of low-pressure station at the other end of bushing
Ring seals—split ring seals	To seal reciprocating components
Expanding split ring	Used in compressors, pumps, and internal-combustion engines
Contracting split ring	Linear actuators where high-pressure, high-temperature radiation and fatigue are expected
Straight-cut seal ring (Fig. 16-3a)	Piston seal for low-grade actuators
Step seal ring	Devices where free-passage leakage is not permissible
Circumferential seal	For rotary applications with low leakage and high performance
Face seals (Fig. 16-9)	
Stationary, rotating, and metal bellows type	Running seal between two flat precision finished surfaces, for high-speed applications, stuffing boxes, and temperature applications
Compression packing (Fig. 16-10)	For the throat of a stuffing box and its gland, dynamic seal
Molded packing (Fig. 16-11)	For automatic-hydraulic or mechanical packings
Lip type	
Single and multiple spring-loaded packings	For sealing reciprocating parts
Squeeze type	Fitted in rectangular grooves machined in hydraulic or pneumatic mechanisms and used as a piston seal in hydraulic actuating cylinder, valve seat, or valve stem packing
Felt radial type	Used at high speeds from 10 to 20 m/s
Diaphragm seals (Fig. 16-12)	To prevent interchange of a fluid or contaminant between two separated areas, dynamic sealing and force transmitter
Nonmetallic gaskets (Fig. 16-13)	Static sealing
Metallic gaskets (Table 16-7)	
Corrugated, metal-jacketed, plain or machined (flat metal) round, heavy, or light cross-section (solid metal)	Static sealing, for high pressures and severe conditions, cast iron flanges, ammonia fittings, hydraulic cylinders, gas mains, heat exchangers, boiler openings, vacuum and cryogenic lines, and valve bonnets
Sealants	
Hardening (rigid or flexible), non-hardening and tapes	To exclude dust, dirt, moisture, and chemicals or contain a liquid or gas-surface coatings to protect against mechanical or chemical attack, to exclude noise, to improve appearance and to perform a joining function, thermal insulating, vibration damping

TABLE 16-7
Properties and uses of nonmetallic gasket materials

Classification	Special characteristics	General uses
Rubber asbestos	Tough and durable, relatively incompressible, good steam and hot water resistance	Heavy duty bolted and threaded joints as in water and steam pipe fittings; temperatures up to 260°C
Cork and rubber	Provides fluid barrier and resilience with compressibility; does not extrude from joint; die cuts well; high coefficient of friction	General-purpose gasketing; enables design of metal-to-metal joints; high friction keeps gasket positioned even where closing pressure is not perpendicular to flange faces
Cork composition	General purpose material compressible; high friction, low cost; excellent oil and solvent resistance; poor resistance to alkalis and corrosive acids	Mating rough or irregular parts; oil sealing at low cost in normal range of temperatures and pressures
Rubber, plastics	Highly adjustable according to compounding, hardness, modulus, fabric reinforcement, etc.; generally impervious, not compressible	Installations involving stretching over projections or where flow of gasket into threads or recesses is desired; for lowest compression set and maximum resistance to fluids such as alkalis, hot water, and certain acids
Paper		
Untreated	Low cost, noncorrosive	Spacers, dust barriers, splash seals where breathing and wicking not objectionable
Treated	General-purpose material; good oil, gasoline and water resistance	Machined or reasonably uniform flanges where adequate bolt pressures can be applied
Combination constructions	Innumerable modifications available, depending on materials used and methods of combining	Usually employed for extreme conditions and special purposes

TABLE 16-8
Minimum metallic gasket seating stress

Type	Material	Thickness, mm	Minimum seating stress[a]	
			MPa	kpsi
 (a) Flat metal	Aluminum	3	109.8	16.0
		1.5 and 0.75	137.3	20.0
	Copper	3	248.1	36.0
		1.5 and 0.75	309.9	45.0
	Soft steel (iron)	3	379.0	55.0
		1.5 and 0.75	474.1	69.0
	Monel	3	448.2	65.0
		1.5 and 0.75	559.9	81.0
	Stainless steel	3	577.3	84.0
		1.5 and 0.75	646.2	94.0
 (b) Flat metal, serrated or grooved	Aluminum	3[b]	172.1	25.0
		1.5[b]	206.9	30.0
		0.75[b]	241.2	35.0
	Copper	3[b]	241.2	35.0
		1.5[b]	275.6	40.0
		0.75[b]	309.9	45.0
	Soft steel (iron)	3[b]	379.0	55.0
		1.5[b]	413.8	60.0
		0.75[b]	448.2	65.0
	Monel	3[b]	448.2	65.0
		1.5[b]	482.5	70.0
		0.75[b]	557.6	80.0
	Stainless steel	3[b]	517.3	75.0
		1.5[b]	557.6	80.0
		0.75[b]	655.1	95.0
 (c) Corrugated	Aluminum	3	10.3	1.5
	Copper	3	13.7	2.0
	Soft steel (iron)	3	27.4	4.0
	Monel	3	30.9	4.5
	Stainless steel	3	41.2	6.0
 (d) Corrugated coat	Aluminum	3	13.7	2.0
	Copper	3	17.2	2.5
	Soft steel (iron)	3	20.6	3.0
	Monel	3	24.0	3.5
	Soft steel	3	27.4	4.0
 (e) Corrugated jackected, soft filler	Lead	3	3.4	0.5
	Aluminum	3	6.9	1.0
	Copper	3	17.1	2.5
	Soft steel (iron)	3	24.0	3.5
	Monel	3	30.9	4.5
	Stainless steel	3	41.2	6.0
	Inconel	3	51.5	7.5
	Hastelloy c	3	68.6	10.0

[a] Seating stress values shown do not apply to ASME Code. Also they are based on optimum surface finish and clean flange surface, i.e., no grease, oil or gasket compound.
[b] Figures indicated are pitch, and the values of stress apply for all thicknesses.

TABLE 16-9
Compression packing for various service conditions

Fluid medium	Service condition: Reciprocating shafts	Rotating shafts	Piston or cylinders	Valve stems
Acids and caustics	Asbestos, metallic, plastic (pliable), semimetallic, TFE fluorocarbon resins and yarns	Asbestos, plastic (pliable), semimetallic, TFE fluorocarbon resins and yarns	TFE fluorocarbon resins	Asbestos, plastic (pliable), semimetallic TFE fluorocarbon resins and yarns
Air, gas	Asbestos, metallic, semimetallic	Asbestos, semimetallic	Leather, metallic	Asbestos, semimetallic
Ammonia, low-pressure steam	Duck and rubber, metallic, semimetallic	Asbestos, semimetallic	Duck and rubber	Asbestos, duck and rubber, semimetallic
Cold and hot gasoline and oils	Asbestos, plastic (pliable), semimetallic	Asbestos, plastic (pliable), semimetallic		Asbestos, plastic (pliable), semimetallic
High-pressure steam	Asbestos, metallic, plastic (pliable), semimetallic	Asbestos, metallic, plastic (pliable), semimetallic	Metallic	Asbestos, metallic, plastic (pliable), semimetallic
Cold and hot water	Duck and rubber, leather, plastic (pliable), semimetallic	Asbestos, plastic (pliable), semimetallic	Duck and rubber	Asbestos, duck and rubber, plastic (pliable), semimetallic

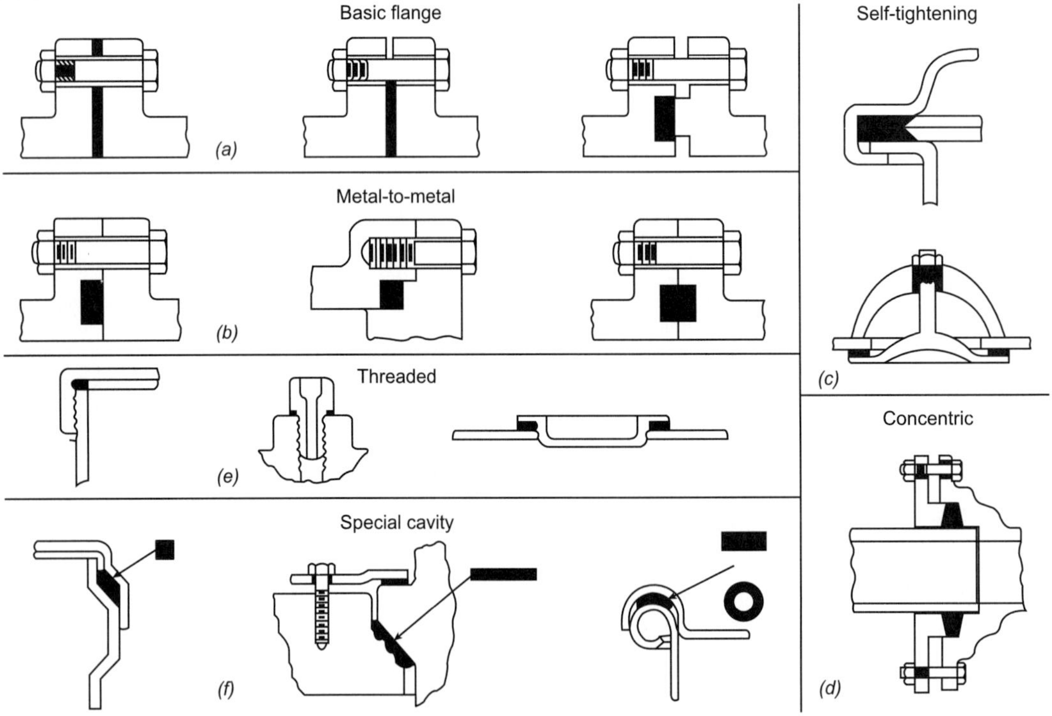

FIGURE 16-13 Common types of gasketed joints.

TABLE 16-10
Dimensions of oil seals

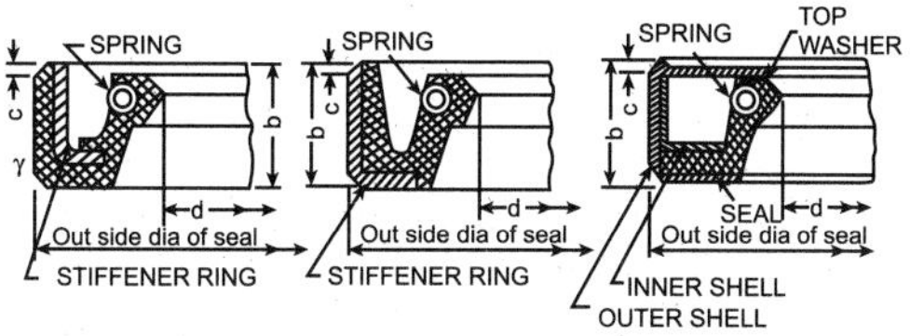

Type A Type B Type C

Shaft diameter d_1, mm	Nominal[a] bore diameter of housing, mm	$b \pm 0.2$, mm Types *A* and *B*	$b \pm 0.2$, mm Type *C*	c^b min, mm
6	16	7		0.3
	22			
7	16	7		0.3
	22			
8	16	7		0.3
	22			
	24			
9	22	7		0.3
	24			
	26			
10	19	7		0.3
	22			
	24			
	26			
11	22	7		0.3
	26			
12	22	7		0.3
	24			
	28			
	30			
14	24	7		0.3
	28			
	30			
	35			
15	24	7		0.3
	26			
	30			
	32			
	35			
16	28	7		0.3
	30			
	32			
	35			
17	28	7		0.3
	30			
	32			
	35			
	40			
18	30	7		0.3
	32			
	35			
	40			
20	30	7		0.3
	32			
	35			
	40			
	47			
22	32	7	—	0.3
	35		9	
	40			
	47			
24	35	7	—	0.3
	37		9	
	40			
	47			
25	35	7	—	0.4
	40		9	
	42			
	47			
	52			
26	37	7	—	0.4
	42		9	
	47			
28	40	7	—	0.4
	47		9	
	52			
30	40	7	—	0.4
	42		9	
	47			
	52			
	62			
32	45	7	—	0.4
	47			
	52		9	

TABLE 16-10
Dimensions for oil seals (*Cont.*)

Shaft diameter d_1, mm	Nominal[a] bore diameter of housing, mm	$b \pm 0.2$, mm Types *A* and *B*	$b \pm 0.2$, mm Type *C*	c^b min, mm
35	47	7	—	0.4
	50		9	
	52			
	62			
36	47	7	—	0.4
	50		9	
	52			
	62			
38	52	7	—	0.4
	55		9	
	62			
40	52	8	—	0.4
	55		9	
	62			
	72			
42	55	8	—	0.4
	62		10	
	72			
45	60	8	—	0.4
	62		10	
	65			
	72			
48	62	8	—	0.4
	72		10	
50	65	8	—	0.4
	68		10	
	72			
	80			
52	68	8	—	0.4
	72		10	
55	70	8	—	0.4
	72		10	
	80			
	85			
56	70	8	—	0.4
	72		10	
	80			
	85			
58	72	8	—	0.4
	80		10	
60	80	8	—	0.4
	85		10	
	90			
62	85	10	12	0.5
	90			
63	85	10	12	0.5
	90			
65	85	10	12	0.5
	90			
	100			
68	90	10	12	0.5
	100			
70	90	10	12	0.5
	100			
72	95	10	12	0.5
	100			
75	95	10	12	0.5
	100			
78	100	10	12	0.5
	100			
80	100	10	12	0.5
	110			
85	110	12	15	0.8
	120			
90	110	12	15	0.8
	120			
95	120	12	15	0.8
	125			
100	120	12	15	0.8
	125			
	130			
105	130	12	15	0.8
	140			
110	130	12	15	0.8
	140			
115	140	12	15	0.8
	150			
120	150	12	15	0.8
	160			
125	150	12	15	0.8
	160			
130	160	12	15	0.8
	170			
135	170	11	15	0.8
140	170	15	15	1
145	175	15	15	1

TABLE 16-10
Dimensions for oil seals (*Cont.*)

Shaft diameter d_1, mm	Nominal[a] bore diameter of housing, mm	$b \pm 0.2$, mm Types *A* and *B*	Type *C*	c^b min, mm	Shaft diameter d_1, mm	Nominal[a] bore diameter of housing, mm	$b \pm 0.2$, mm Types *A* and *B*	Type *C*	c^b min, mm
150	180	15	15	1	280	320	20	20	1
160	190	15	15	1	300	340	20	20	1
170	200	15	15	1	320	360	20	20	1
180	210	15	15	1	340	380	20	20	1
190	220	15	15	1	360	400	20	20	1
200	230	15	15	1	380	420	20	20	1
210	240	15	15	1	400	440	20	20	1
220	250	15	15	1	420	460	20	20	1
230	260	15	15	1	440	480	20	20	1
240	270	15	15	1	460	500	20	20	1
250	280	15	15	1	480	520	20	20	1
260	300	20	20	1	500	540	20	20	1

[a] For limits of housing, see Tables 16-11 and 16-12.
[b] The edges may be chamfered or rounded according to the manufacturer's discretion.
Source: Bureau of Indian Standards: 5129, 1969.

TABLE 16-11
Press-fit allowances and tolerances[a] for type *A* seals

Nominal bore diameter of housing, mm	Housing bore, mm		Outside diameter of seal, mm		Possible press-fit variation, mm	
	High limit	Low limit	High limit	Low limit	Maximum interference	Minimum interference
≤25	+0.03	−0.03	+0.20	+0.10	0.23	0.07
25–55	+0.03	−0.03	+0.25	+0.15	0.28	0.12
55–125	+0.03	−0.03	+0.30	+0.20	0.33	0.17
125–200	+0.04	−0.04	+0.38	0.22	0.42	0.18
≥200	+0.05	−0.05	+0.48	0.32	0.53	0.27

[a] All tolerances are relative to nominal bore diameter of housing.
Source: IS 5129, 1969.

TABLE 16-12
Press-fit allowances and tolerances' for types *B* and *C* seals

Nominal bore diameter of housing, mm	Housing bore, mm		Outside diameter of seal, mm		Possible press-fit variation, mm	
	High limit	Low limit	High limit	Low limit	Maximum interference	Minimum interference
≤ 50	Nominal	−0.03	+0.12	+0.04	0.15	0.04
50–90	Nominal	−0.03	+0.14	+0.06	0.17	0.06
90–115	+0.03	−0.03	+0.18	+0.08	0.21	0.05
115–170	+0.03	−0.03	+0.20	+0.10	0.23	0.07
170–215	+0.04	−0.04	+0.23	+0.13	0.27	0.09
215–230	+0.04	−0.04	+0.25	+0.15	0.29	0.11
≥ 230	+0.04	−0.04	+0'30	+0.20	0.34	0.16

[a] All tolerances are relative to nominal bore diameter of housing.
Source: IS 5129, 1969.

TABLE 16-13
Depth of the housing bore (all dimensions in mm)

b	t (0.85b) Min	t_2 (b to 0.3) Min
7	5.95	7.3
8	6.80	8.3
9	7.65	9.3
10	8.50	10.3
12	10.30	12.3
15	12.75	15.3
20	17.00	20.3

Source: Indian Standards 5129, 1969.

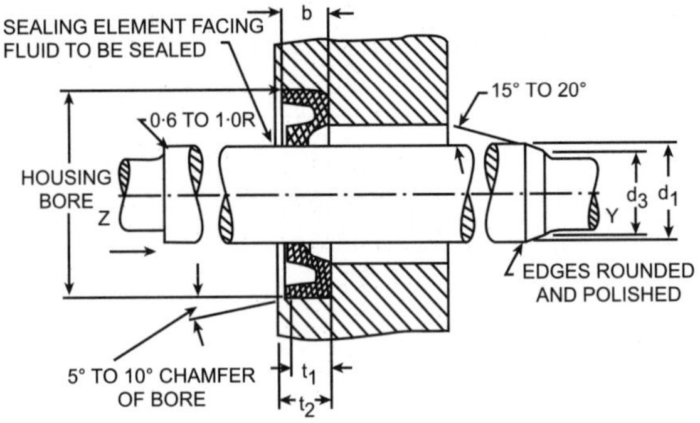

Z-AND Y-DIRECTION MOUNTING OF THE SHAFT

TABLE 16-14
Types of hollow, metallic O-rings[8,9]

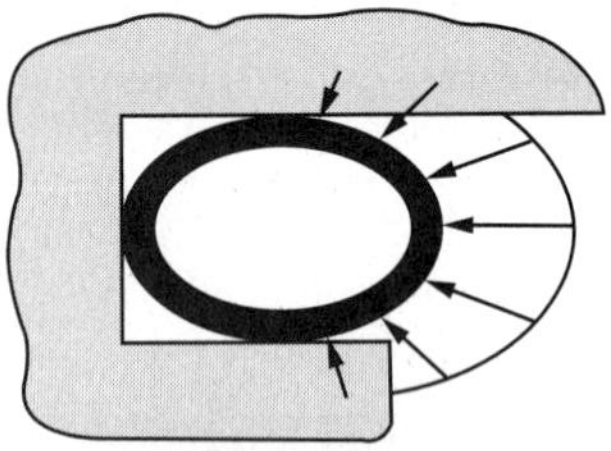

Plain, sealed metalic O-ring:
For fully confined or semi-confined ring joints-sealing vacuum, pressure, corrosive liquids and gases

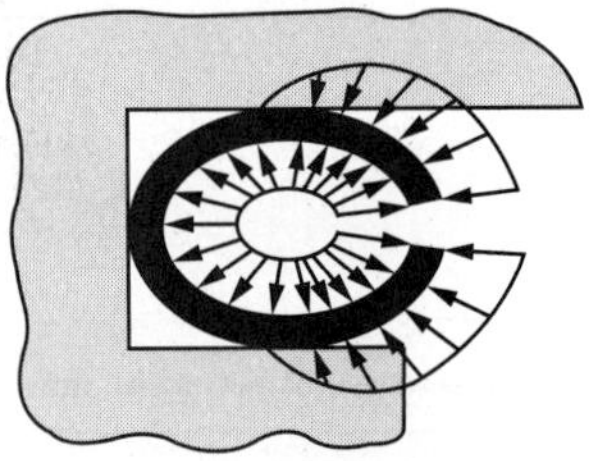

Self-energizing metallic O-ring:
For semiconfined designs increase in internal pressure causes ring to be crammed into groove, increases sealing effectiveness

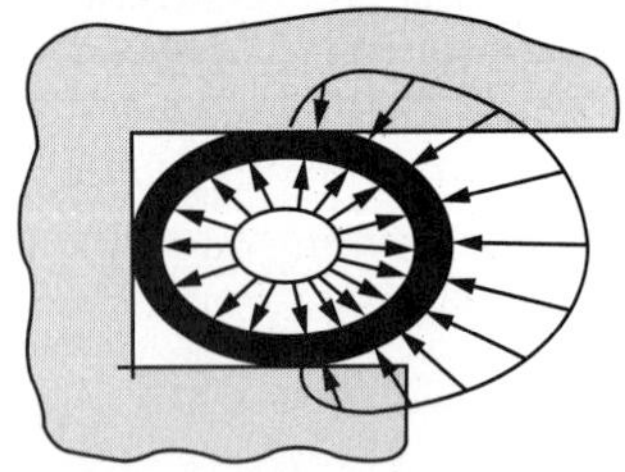

Pressure-filled metallic O-ring:
For fully confined or semiconfined designs. Ring is filled with an inert gas at 412 MPa (42 kgf/mm^2) useful in higher temperature range from 533 K (260°C) to 733 K (427°C)

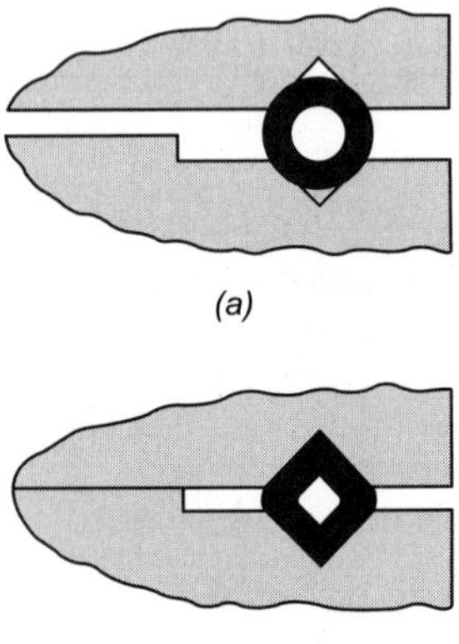

FIGURE 16-14 Fully confined hollow-metal O-ring: (a) before bolting and (b) after bolting down.

Source: Wes J. Ratelle, "Seal Selection, Beyond Standard Practice,", *Machine Design*, Jan. 20, 1977 and, "Packings and Seals" Issue, *Machine Design*, Jan. 1977.

TABLE 16-15
Recommended groove dimensions for metallic Q-ring sealing inside pressure

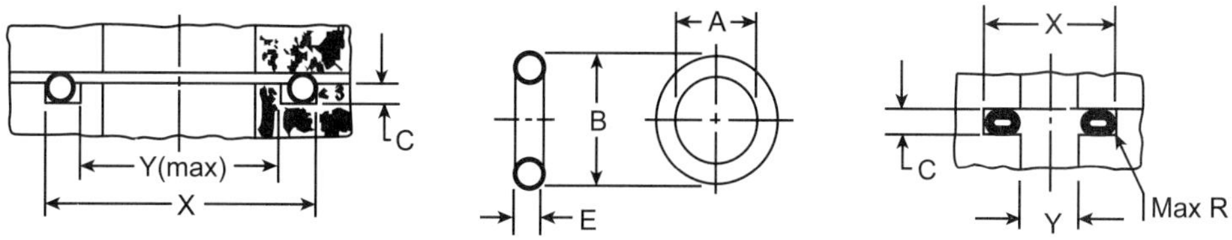

	Nominal O-ring OD		Actual O-ring dimensions		Open-groove dimensions				
Nominal tubing OD mm	Min *B*, mm	Incremental increase, *I*, mm	Tubing OD, *B*, mm	O-Ring OD, mm	Depth, *C*, mm	Groove OD, *X*, mm	Minimum groove ID, *Y*, mm	Maximum ID of closed groove, *Y*, mm	Maximum, radius of groove corner, *R*, mm
0.8	6.30	0.8 up to 25	0.74–0.96	$B + 0.075$ $- 0.000$	0.510–0.500	$B + 0.0105$ to 0.01525	$B - 2.160$	$B - 3.000$	0.125
1.6	11.0	1.6 thereafter	0.14–0.16	$B + 0.075$ $- 0.0006$	1.066–1.145	$B + 0.0105$ to 0.01525	$B - 3.600$	$B - 4.825$	0.255
2.4	19	1.6	0.20–0.24	$B + 0.100$ $- 0.000$	1.0650–1.750	$B + 0.0127$ to 0.0225	$B - 5.665$	$B - 6.860$	0.510
3.2	44	1.6	0.30–0.31	$B + 0.125$ $- 0.000$	2.290–2.415	$B + 0.0178$ to 0.0305	$B - 7.495$	$B - 8.635$	0.760
4.0	75	1.6	0.37–0.40	$B + 0.150$ $- 0.000$	2.920–3.050	$B + 0.0203$ to 0.0355	$B - 9.245$	$B - 10.410$	0.760
4.8	100	1.6	0.44–0.48	$B + 0.175$ $- 0.000$	3.685–3.810	$B + 0.0228$ to 0.0380	$B - 11.170$	$B - 12.190$	0.760
6.3	125	1.6	0.59–0.81	$B + 0.200$ $- 0.000$	4.955–5.080	$B + 0.0280$ to 0.0480	$B - 14.730$	$B - 16.000$	0.760
9.5	250	No limit	0.90–0.95	$B + 0.300$ $- 0.000$	7.495–7.620	$B + 0.0355$ to 0.0735	$B - 22.600$	$B - 23.110$	0.760
12.7	250	No limit	1.20–1.25	$B + 0.400$ $- 0.000$	9.910–10.160	$B + 0.0510$ to 0.0965	$B - 30.480$	$B - 30.480$	0.760

Source: Wes J. Ratelle, "Seal Selection. Beyond Standard Practice," *Machine Design*, Jan. 20, 1977, and "Packings and Seals" Issue, *Machine Design*, Jan. 20, 1977.

TABLE 16-16
Rectangular groove dimensions for O-ring gaskets

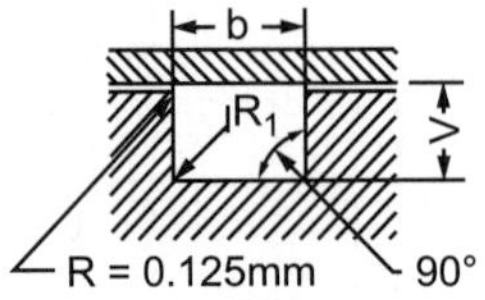

O-ring nominal cross section, mm	Actual O-ring cross section, mm	Maximum groove depth, V, mm	Groove width, b, mm	Minimum diametral squeeze mm	Bottom radius, R_1, mm
		For Flange Gaskets (Axial)			
1.6	0.100 ± 0.0075	0.070–0.0050	0.160 ± 0.0050	0.025	0.0125
1.6	0.125 ± 0.0075	0.090–0.0050	0.185 ± 0.0075	0.030	0.0200
1.6	0.150 ± 0.0075	0.110–0.0050	0.210 ± 0.0075	0.035	0.0300
1.6	0.175 ± 0.0075	0.130–0.0100	0.240 ± 0.0075	0.040	0.0380
1.6	0.175 ± 0.0075	0.125–0.0100	0.240 ± 0.0075	0.045	0.0380
2.4	0.260 ± 0.0075	0.205–0.0105	0.270 ± 0.0125	0.050	0.0500
3.2	0.350 ± 0.0100	0.280–0.0200	0.470 ± 0.0125	0.060	0.0750
4.8	0.530 ± 0.0125	0.445–0.0250	0.725 ± 0.0125	0.075	0.1250
6.4	0.700 ± 0.0150	0.585–0.0250	0.960 ± 0.0125	0.100	0.1500
		For Nonflange Gaskets (Radial)			
1.6	0.100 ± 0.0075	0.075–0.0025	0.140 ± 0.0050	0.0175	0.0125
1.6	0.125 ± 0.0075	0.095–0.0025	0.160 ± 0.0075	0.025	0.0200
1.6	0.150 ± 0.0075	0.115–0.0025	0.190 ± 0.0075	0.030	0.0300
1.6	0.175 ± 0.0075	0.135–0.0025	0.230 ± 0.0075	0.035	0.0380
1.6	0.175 ± 0.0075	0.130–0.0050	0.230 ± 0.0075	0.038	0.0380
2.4	0.260 ± 0.0075	0.210–0.0075	0.315 ± 0.0125	0.043	0.0500
3.2	0.350 ± 0.0100	0.290–0.0100	0.430 ± 0.0125	0.050	0.0750
4.8	0.530 ± 0.0125	0.455–0.0125	0.600 ± 0.0125	0.065	0.1250
6.4	0.700 ± 0.0150	0.595–0.0150	0.800 ± 0.0125	0.090	0.1500

TABLE 16-17
Triangular groove dimensions for O-ring flange gaskets

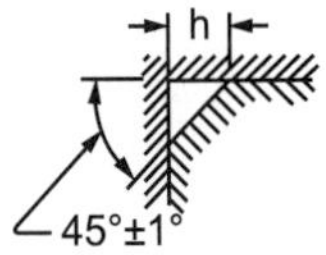

O-ring nominal cross section, mm	Actual O-ring cross section, mm	Width, h, mm
1.6	0.175 ± 0.0075	0.240 + 0.0075 − 0.000
2.4	0.260 ± 0.0075	0.345 + 0.0125 − 0.000
3.2	0.350 ± 0.0100	0.470 + 0.0175 − 0.000
4.8	0.530 ± 0.0125	0.710 + 0.0255 − 0.000
6.4	0.700 ± 0.0150	0.950 + 0.0375 − 0.000

TABLE 16-18
Packing sizes recommended for various shaft diameters

Shaft diameter, mm	Packing size, mm
12.70–15.85	7.95
17.45–38.10	9.50
39.70–50.80	11.10
52.40–63.50	12.70
65.10–76.20	14.30
77.80–101.60	15.85

TABLE 16-19
Temperature limits for gasket materials

Material	Maximum sustained temperature K	Maximum sustained temperature °C
Asbestos fiber and rubber	673	400
Cellulose-fiber and rubber	423	150
Cork and rubber	393	120
Synthetic rubber	393	120
Cork composition	393	120

TABLE 16-21
Selection of shaft piston seals

Type name	Distributor	U-Ring	Cup	O-Ring
External-fitted to piston, sealing in bore				
Internal-fitted in housing, sealing on piston or rod				
Simple housing design	Good	Good	Poor	Very good
Low wear rate	Very good	Good	Good	Poor
High stability	Good	Fair	Very good	Poor
Low friction	Fair	Fair	Fair	Good
Resistance to extrusion	Good	Good	Good	Fair
Availability in small sizes	Fair	Good	Poor	Very good
Availability in large sizes	Good	Fair	Good	Good
Bidirectional sealing		Single acting only		Effective but usually used in pairs

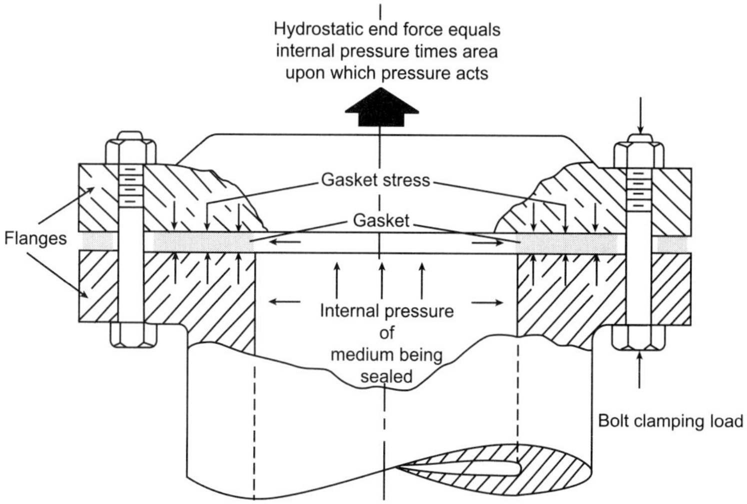

FIGURE 16-15 Nomenclature of gasketed joint. (J. E. Shigley and C. R. Mischke, *Standard Handbook of Machine Design*, McGraw-Hill, 1986.)

TABLE 16-20
Properties of sealants

Sealant base test method	Tensile strength, ASTM D412		Elongation %, ASTM D412	Adhesion in tension, ASA 1161-1960		Shear strength, ASTM D1002		Moisture resistance, % ASTM D570	Abrasion resistance	Operating tempera-ture °C	Shore A hardness ASTM 676	Shrinkage
	MPa	psi		MPa	psi	MPa	psi					
Polysulfide	0.39–0.86	56.5–125	150–500	0.34–0.69	50–100	0.55–1.20	80–175	0.25–1.5	Fair to good	–50 to 120	15–60	0–3.0
Polyurethane[a]	6.86–20.50	1000–3000		0.15–0.44	15–65	1.72–2.40	240–350	1–3.0	Excellent	–55 to 205	45–90	0–10.0
Silicone	1.96–5.39	285–780	50–750	0.34–0.59	50–85.5	1.03–1.37	150–200	0–1–0.25	Fair to good	–75 to 370	25–80	0–10.0
Ncoprene	6.86–10.29	1000–1500	250–350	0.44–0.69	65–100			0.5–1.5	Excellent	–40 to 150	30–80	0–10.0
Hypalon	3.43–4.11	500–600	75–125			0.27–0.69	40–100	1.0–5.0	Excellent	–40 to 150	30–80	10–20
Viton	8.33	1210	325			0.85–1.20	125–175	0–3.0	Fair to good	–55 to 230	40–60	
Epoxy	27.46–89.73	4000–13000	3.0–6.0			10.29–24.0	1500–3500	0.04–0.10	Good to excellent	–35 to 150	40–100	0–3.0
											Shore "D"	0–3.0
Epoxy–modified	8.33–24.02	1200–3500	10–20			10.29–19.1	1500–2750	0.27–0.50	Good to excellent	–35 to 150	40–60	0–3.0
											Shore "D"	
Acrylic	0.34–2.94	50–425	100–270					1.0–5.0	Fair to good	–25 to 150	40–100	5.0–15.0
Polyester	27.46–48.05	4000–7000	3–15			10.79–23.54	1500–3400	0.25–0.75	Good	–55 to 150	10–70	2.0–10.0
											Shore "D"	
Polyurethane-bitumen modified	0.29–0.53	42–75	250–400	0.17–0.27	24.5–40			0.75–1.50	Good	–35 to 95	10–45	0–3.0
Butyls—mastic type			5–150					0.5–5.0	Good	–20 to 95	5–70	15.0–15.0
Butyls—curing type	17.16–20.59	2500–3000	650–800			1.03–1.873	150–270	0.25–1.5	Good	–60 to 150	15–75	
Polybutene									Good	–30 to 120	20–40	0–3.0
Olcoresin	0.49–4.90	70–710	5–20						Good	–25 to 95	5–70	15–45

[a] Compounds built specifically for plotting and molding, where high strength and abrasion resistance are required.

TABLE 16-22
Recommended maximum temperature for materials (supplement to Table 16-2)

Material	Temperature °F	°C
Coil spring material		
Phosphor-bronze ASTM B159	200	93
Silicon bronze ASTM B99	200	93
Ni-span C902	200	93
Music wire ASTM A228	250	121
Hard-drawn spring wire ASTM A227	250	121
Oil-tempered wire ASTM A229	300	149
Valve spring wire ASTM A230	300	149
Beryllium-copper ASTM B197	400	204
Chrome-vanadium alloy steel AISI 6150	425	218
Silicon-manganese alloy steel AISI 9260	450	232
Chrome-silicon alloy steel AISI 9254	475	246
Martensite AISI 410	500	260
Martensite AISI 420	500	260
Austenitic AISI 301	600	315
Austenitic AISI 302	600	315
17-7 PH Stainless Steel	590	311
Inconel6OO	700	371
Nickel-chrome alloy steel A286	950	510
Inconel 718	1200	649
Inconel X-750	1300	704
L-605	1400	760
S-816	1400	760
Rene 41	1400	760
Flat spring material		
Ni-span C902	200	93
Phosphor-bronze ASTM B103	200	93
High-carbon AISI 1050	200	93
High-carbon AISI 1065	200	93
High-carbon AISI 1075	250	121
High-carbon AISI 1095	250	121
Beryllium-copper ASTM B194	400	204
Austenitic AISI 301	600	315
Austenitic AISI 302	600	315
17-7 PH Stainless Steel	700	371
Inconel 600	700	371
Beryleo-nickel	700	371
Titanium 6-6-2	750	399
Sandvik 11 R51	800	427
Duranickel 301	800	427
Permanickel	800	427
Elgiloy	900	482
Havar	900	482
Inconel 718	1200	649
Inconel X-750	1300	704
Rene 41	1400	760

TABLE 16-22
Recommended maximum temperature for material (supplement to Table 16-2)

Material	Temperature °F	°C
Formed metal bellows materials		
Brass CDA 240	300	149
Phosphor-bronze CDA 510	300	149
Beryllium-copper CDA 172	350	177
Monel 404	450	232
Unstabilized 300 series stainless steel	500	260
Inconel 600	750	399
Inconel X-750	800	427
Welded metal bellows materials		
Ni-span C	500	260
AM-350 Stainless Steel	800	427
410 Stainless Steel	800	427
Commercially pure titanium	800	427
Stabilized 300 series Stainless Steel	1220	659
Inconel X-750	1500	815
Inconel 625	1500	815
Hastelloy-C	1800	982
Rene 41	1800	982

TABLE 16-23
***pv* values for seal face material (life of 8000 h)**

		pv Value			
		Unbalanced		Balanced	
Product	Combination face material	MPa, m/s	kpsi fpm	MPa, m/s	kpsi fpm
Water	Stainless steel }	0.9	25.5	Seldom used	Seldom used
Oil	Carbon[a]	1.8	51.0	Seldom used	Seldom used
Water	Lead bronze }	1.8	51.0	Seldom used	Seldom used
Oil	Carbon[a]	3.5	100	Seldom used	Seldom used
Water	Stellite carbon[a]	3.5	100	10	285
Oil		9		70	2000
Water	Tungsten carbide }	9	255	25	710
Oil	Carbon[b]	9	255	150	4280
Water	Solid ceramic	15	430	Seldom used	Seldom used
Water	Sprayed ceramic	15	430	90	2570
Oil		20	560	150	4280

[a] Metal-impregnated carbon.
[b] Retain impregnated carbon.
Source: Courtesy M. J. Neale, *Tribology Handbook*, Butterworths, London, 1973.

TABLE 16-24
Spring arrangements for various sizes of shaft and speeds

		Spring arrangement			
		Stationary		Rotary	
Shaft diameter, mm	Speed, rpm	Single	Multiple	Single	Multiple
$\le$100	$\le$3000	Yes	Yes	Yes	Yes
>100	$\le$3000	No	Yes	No	Yes
$\le$75	$\le$4500	Yes	Yes	Yes	Yes
$\le$100	>4500	Yes	Yes	No	No
>100	>4500	No	Yes	No	No

Source: Courtesy of M. J. Neale, *Tribology Handbook*, Butterworths, London, 1973.

TABLE 16-25
Types of static and dynamic seals

	Dynamic seals			
	Clearance seals		Contact seals	
Static seals	**Reciprocating**	**Rotary**	**Reciprocating**	**Rotary**
Fibrous gasket	Labyrinth[a] (Fig. 16-8)	Labyrinth (Fig. 16-8)	U-ring (Fig. 16-11)	Lip seal (Fig. 16-4)
Metallic gasket	Fixed bushing	Viscoseal	O-ring (Table 16-15)	Face seal (Fig. 16-9*a*)
Elastomeric gasket	Floating bushing	Fixed bushing	Lobed O-ring	Packed gland (Fig. 10-10)
Plastic gasket		Floating bushing	Rectangular ring	O-ring[b] (Fig. 16-14)
Sealant, setting		Centrifugal seal	Packed gland	Felt ring
Sealant, nonsetting			Piston ring	
O-ring			Bellows	
Inflatable gasket			Diaphragm (Fig. 16-12)	
Pipe coupling				
Bellows				

[a] Usually for steam or gas.
[b] Only for very slow speeds.
Source: Courtesy M. J. Neale, *Tribology Handbook*, Butterworths, London, 1973.

TABLE 16-26
Operating conditions of lip seals

Particular	**Shaft diameter and housing**	**Remarks**
Maximum pressure of fluid	≤75 mm diameter	60 kPa (8.7 psi)
	>75 mm diameter	30 kPa (4.35 psi)
Maximum speed	≤35 mm diameter	8000 rpm
	75 mm diameter	4000 rpm
	>75 mm diameter	15 m/s
Surface finish	Housing	Fine-turned
	Shaft	Grind and polish to better than 0.5 μm
Eccentricity	Housing	0.25 mm total indicator reading
	Shaft	Depends on speed, 0.25 mm
Temperature		Varies from −20°C to 200°C (−68°F to 266°F)

Source: M. J. Neale, *Tribology Handbook*, Butterworths, London, 1973

TABLE 16-27
Types of seals for reciprocating shafts

Type of packing	**Remarks**
Cups and hats	Semiautomatic, leather and rubber/fabric used
U-packing	Used for piston rod application up to 10 MPa (1.5 kpsi) (rubber) or 20 MPa (3.0 kpsi) (rubber/fabric)
Nylon-supported	Used up to 25 MPa (3.6 kpsi)
Composite	Used with rubber sealing lips, rubber/fabric supporting portions and nylon wearing portions—used for pressure varying from 15 to 20 MPa (2.2 to 3.0 kpsi)

Source: M. J. Neale, *Tribology Handbook*, Butterworths, London, 1973.

TABLE 16-28
Materials for lip seals (rubber)

Type of rubber	Trade names	Resistance to		Temperature	
		Mineral oil	Chemical fluids	°F	°C
Acrylate	Thiacril Cyanacryl	Excellent	Fair	−68 to +266	−20 to +130
Fluoropolymer	Viton Fluorel	Excellent	Excellent	−77 to +392	−25 to +200
Polysiloxane	Silastic Silastomer	Fair	Poor	−158 to +392	−70 to +200
Nitrile	Hycar Polysar	Excellent	Fair	−104 to +212	−40 to +100

Source: Courtesy M. J. Neale, *Tribology Handbook*, Butterworths, London, 1973.

TABLE 16-29
Seal materials for reciprocating shafts

Material	Remarks
Rubber (nitrile)	Highest scaling efficiency; low cost; easily formed to shape; limited to a pressure of 10 MPa (1.5 kpsi); excellent wear resistance
Rubber-impregnated fabric	Great toughness; resistance to extrusion and cutting; wear resistance inferior to rubber
Leather	Good wear and extrusion resistance; poor resistance to permanent set; limited shaping capability
Nylon	Resist extrusion; provide a good bearing surface

Source: Courtesy M. J. Neale, *Tribology* Handbook, Butterworths, London, 1973.

TABLE 16-30
Extrusion clearance for reciprocating shafts—dimensions in mm (in)

Material	≤10 MPa (1.5 kpsi)		10–20 MPa (1.5–3.0 kpsi)		>20 MPa (3.0 kpsi)	
	Normal	Short life	Normal	Short life	Normal	Short life
Rubber	0.25 (0.01)	0.50 (0.02)	—	—	—	—
Rubber/fabric leather	0.40 (0.015)	0.60 (0.025)	0.25 (0.01)	0.50 (0.02)	0.10 (0.005)	0.25 0.01)
Polyurethane	0.40 (0.015)	0.60 (0.025)	0.25 (0.01)	0.50 (0.02)	0.10 (0.005)	0.25 (0.01)
Nylon support	—	—	0.25 (0.01)	1.00 (0.04)	0.10 (0.005)	0.25 (0.01)

Source: Courtesy M. J. Neale, *Tribology*, *Handbook*, Butterworths, London, 1973.

TABLE 16-31
Operation conditions of packed glands (Fig. 16-1)

Type of gland	Pressure		Temperature		Velocity,	Remarks
	MPa	psi	°F	°C	m/s (fpm)	
Graphited asbestos—rotary type	0.105	15	200	93	17.75 (4000)	No latern or jacket ring cooling required
Graphited asbestos with latern ring cooling arrangement—rotary type	0.280	40	240	115	17.75 (4000)	Cooling liquid used below 34.5 kPa sealing pressure
Graphited asbestos with latern ring and jacket cooling arrangement—rotary type	0.700	100	320	160	17.75 (4000)	Latern ring cooling liquid and water to jacket cooler used below sealing pressure of 34.5 kPa
Graphited asbestos with PTFE antiextrusion ring hand surface replaceable sleeve, jacket cooling arrangement—rotary type	0.525	75	290	143	306 (6100)	Cooling as per type 3; special packing and accurate assembly is required
Graphited asbestos and PTFE yarn with PTFE antiextrusion ring, jacket cooling arrangement—rotary type	7.000	1000	545	285	5.5 (1080)	Water to jacket coolant used
Reciprocating, steam-graphited asbestos	1.750	250	500	260	0.75	Steam
Reciprocating, water-greased cotton packing	2.100	300	500	260	0.75	Water
Reciprocating, oil-graphited hemp yam	2.100	300	200	93	0.75 (150)	Oil

Source: Courtesy M. J. Neale, *Tribology Handbook*, Butterworths, London, 1973.

TABLE 16-32
Axial stress in packed glands

Type of packing	Minimum axial stress required for seal packing	
	MPa	psi
Teflon-impregnated braided asbestos	1.40	200
Plastic	1.12	160
Braided vegetable fiber, lubricated	1.75	255
Plaited asbestos, lubricated	2.8	405
Braided metallic	3.5	505

Source: Courtesy M. J. Neale, *Tribology Handbook*, Butterworths, London, 1973.

TABLE 16-33
Selection of number of sealing rings

Pressure		Number of sets of sealing rings
MPa	psi	
$\leq$1.0	150	3
1.0–2.0	(150–250)	4
2.0–5.0	250–500	5
3.5–17.0	500–1000	6
7.0–15.0	1000–2000	8
>15.0	above 2000	9–12

Source: Courtesy M. J. Neale, *Tribology Handbook*, Butterworths, London, 1973.

TABLE 16-34
Selection of packing materials

Material	Hardness of rod, H_B	Axial clearance, mm	Application
Lead bronze	250 min	0.08–0.12 (0.003–0.005 in)	Optimum material with good lubricated bearing property High thermal conductivity; used where chemical condition exists and suited for pressure up to 300 MPa (50 kpsi)
Flake graphite gray cast iron	400 min	0.08–0.12	Cheaper suitable up to a pressure of 7 MPa (1.0 kpsi)
White metal (Babbitt)		0.08–0.12	Used where lead-bronze and flake graphite gray cast iron are not suitable because of chemical condition; used up to a maximum pressure of 35 MPa (5.0 kpsi) and maximum temperature 120°C (250°F)
Filled PTFE	400 min	0.4–0.5	Suitable for unlubricated; very good chemical resistance; suited above 2.5 MPa (400 psi)
Reinforced *pf* resin		0.25–0.5	Used with sour hydrocarbon gases and where lubricant may be thinned by solvents in gas stream
Carbon-graphite	400 min	0.030–0.06	Used with carbon-graphite piston rings; must be kept oil free; used up to 350°C (660°F)
Graphite/metal sinter	250 min	0.08–0.12	Alternative to filled PTFE and carbon-graphite

Source: Courtesy M. J. Neale, *Tribology Handbook*, Butterworths, London, 1973.

TABLE 16-35
Minimum recommended seating stresses for various gasket materials (Supplement to Table 16-8)

	Material, mm (in)	Gasket type	Minimum seating stress range (psi[a]) MPa
Nonmetallic	Asbestos fiber sheet	Flat	
	3.125 ($\frac{1}{8}$ in) thick		(1400–1600) 9.7–11.0
	1.563 ($\frac{1}{16}$ in) thick		(3500–3700) 24.1–25.5
	0.78 ($\frac{1}{32}$ in) thick		(6000–6500) 41.4–44.8
	Asbestos fiber sheet 0.78 ($\frac{1}{32}$ in) thick	Flat with rubber beads	(1000–1500 lb/in) on beads 175–263 kN/m
	Asbestos fiber sheet 0.78 ($\frac{1}{32}$ in) thick	Flat with metal grommet	(3000–4000 lb/in) on grommet 525.4–700.5 kN/m
	Asbestos fiber sheet 0.78 ($\frac{1}{32}$ in) thick	Flat with metal grommet and metal wire	(2000–3000 lb/in) on wire 350.2–525.4 kN/m
	Cellulose fiber sheet	Flat	(750–1100) 5.2–7.6
	Cork composition	Flat	(400–500) 2.8–3.5
	Cork-rubber	Flat	(200–300) 1.4–2.1
	Fluorocarbon (TFE)	Flat	
	3.125 ($\frac{1}{8}$ in) thick		(1500–1700) 10.3–11.7
	1.563 ($\frac{1}{16}$ in) thick		(3500–3800) 24.1–26.2
	0.78 ($\frac{1}{32}$ in) thick		(6200–6500) 42.8–44.8
	Nonasbestos fiber sheets (glass, carbon, aramid, and ceramics)	Flat	(1500–3000) depending on composition 10.3–20.7
	Rubber	Flat	(100–200) 0.7–1.4
	Rubber with fabric or metal reinforcement	Flat with reinforcement	(300–500) 2.1–3.5
Metallic	Aluminum	Flat	(10,000–20,000) 68.9–137.9
	Copper	Flat	(15,000–45,000) 103.4–310.3 depending on hardness
	Carbon steel	Flat	(30,000–70,000) 207–483 depending on alloy and hardness
	Stainless steel	Flat 241–655	(35,000–95,000) 241–655 depending on alloy and hardness
	Aluminum (soft)	Corrugated	(1000–3700) 6.9–25.5
	Copper (soft)	Corrugated	(2500–4500) 17.2–31.0
	Carbon steel (soft)	Corrugated	(3500–5500) 24.1–37.9
	Stainless steel	Corrugated	(6000–8000) 41.4–55.2
	Aluminum	Profile	(25,000) 172.4
	Copper	Profile	(35,000) 241.3
	Carbon steel	Profile	(55,000) 379.2
	Stainless steel	Profile	(75,000) 517.1
Jacketed metal-asbestos	Aluminum	Plain	(2500) 17.2
	Copper	Plain	(4000) 27.6
	Carbon steel	Plain	(6000) 41.4
	Stainless steel	Plain	(10,000) 68.9
	Aluminum	Corrugated	(2000) 13.8
	Copper	Corrugated	(2500) 17.2
	Carbon steel	Corrugated	(3000) 20.7
	Stainless steel	Corrugated	(4000) 27.6
	Stainless steel	Spiral-wound	(3000–30,000) 20.7–206.8

[a] Stresses in pounds per square inch except where otherwise noted.
Source: J. E. Shigley and C. R. Mischke, *Standard Handbook of Machine Design*, McGraw-Hill Book Company, New York, 1986.

TABLE 16-36
Safety factors for gasketed joints, *n*, for use in Eq. (16-39)

Safety factor, *n*	When to apply
1.2 to 1.4	For minimum-weight applications where all installation factors (bolt lubrication, tension, parallel seating, etc.) are carefully controlled; ambient to 250°F (121°C) temperature applications; where adequate proof pressure is applied
1.5 to 2.5	For most normal designs where weight is not a major factor, vibration is moderate and temperatures do not exceed 750°F (399°C); use high end of range where bolts are not lubricated
2.6 to 4.0	For cases of extreme fluctuations in pressure, temperature, or vibration; where no test pressure is applied; or where uniform bolt tension is difficult to ensure

Source: J. E. Shigley and C. R. Mischke, *Standard Handbook of Machine Design*, McGraw-Hill Book Company, New York, 1986.

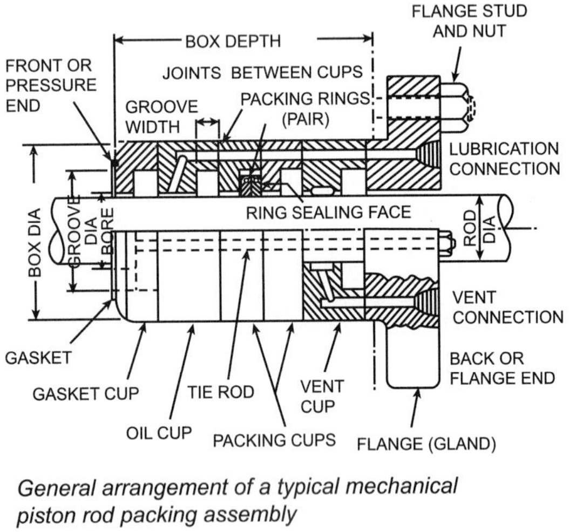

FIGURE 16-16 Packing assembly for a mechanical piston rod. (M. J. Neale, *Tribology Handbook*, Butterworths, London, 1973.)

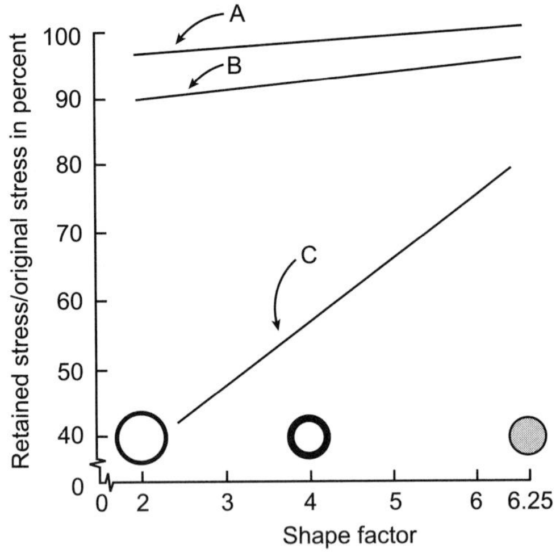

FIGURE 16-17 Ratio of retained stress to origins versus shape factor for, various materials: *A*—asbestos sheet; *B*—cellulose; *C*—cork-rubber. (J. E. Shigley and Mischke, *Standard Handbook of Machine Design*, McGraw-Hill, 1986.)

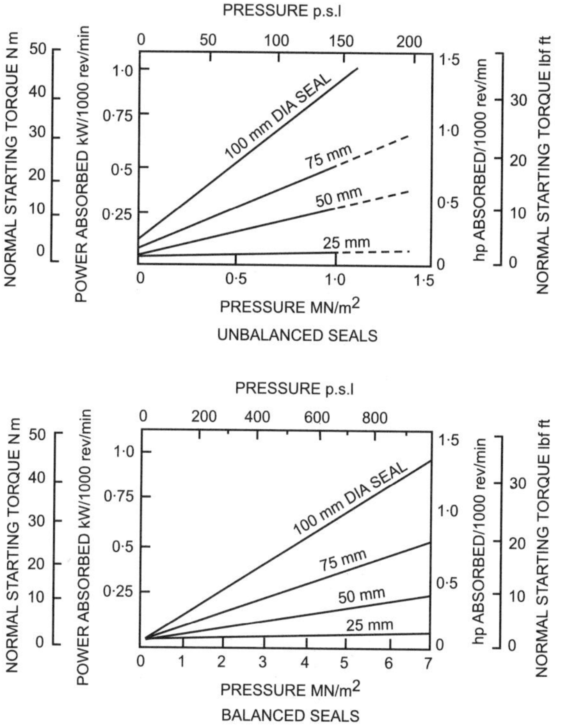

Power absorption and starting torque for aqueous solutions, light oils and medium hydrocarbons, use above values. For light hydrocarbons use $\frac{2}{3}$ of above values. For heavy hydrocarbons use $1\frac{1}{3}$ of above Values. Allow ± 25% on all values

FIGURE 16-18 Power absorption and starting torque for balanced and unbalanced seals. (M. J. Neale, *Tribology Handbook*, Butterworths, London, 1973.)

REFERENCES

1. Lingaiah, K., and B. R. Narayana Iyengar, *Machine Design Data Handbook*, Vol. I (*SI and Customary Units*), Suma Publishers, Bangalore, India, 1986.
2. Lingaiah, K., *Machine Design Data Handbook*, Vol. II (*SI and Customary Metric Units*), Suma Publishers, Bangalore, India, 1986.
3. Maleev, V. L., and J. B. Hartman, *Machine Design*, International Textbook Company, Scranton, Pennsylvania, 1954.
4. The American Society of Mechanical Engineers, *ASME Boilers and Pressure Vessel Code*, Section VIII, Division I, 1986.
5. Whalen, J. J., "How to Select the Right Gasket Material," *Product Engineering*, Oct. 1860.
6. Shigley, J. E., and C. R. Mischke, *Standard Handbook of Machine Design*, McGraw-Hill Book Company, 1986.
7. Neale, M. J., *Tribology Handbook*, Butterworths, London, 1975.
8. Ratelle, W. J., "Seal Selection, Beyond Standard Practice," *Machine Design*, Jan. 20, 1977.
9. "Packings and Seals" Issue, *Machine Design*, Jan. 1977.
10. Faires, V. M., *Design of Machine Elements*, Macmillan Book Company, 1955.
11. Bureau of Indian Standards.
12. Rothbart, H. A., *Mechanical Design and Systems Handbook*, McGraw-Hill Book Company, New York, 1985.
13. Lingaiah, K., *Machine Design Data Handbook*, McGraw-Hill Book Company, New York, 1994.

CHAPTER

17

KEYS, PINS, COTTERS, AND JOINTS

SYMBOLS[4,5,6]

a	addendum for a flat root involute spline profile, m (in)
A	area, m^2 (in^2)
b	breadth of key, m (in)
	effective length of knuckle pin, m (in)
	dedendum for a flat root involute spline profile, m (in)
d	diameter, m (in)
d_1	major diameter of internal spline, m (in)
d_2	minor diameter of internal spline, m (in)
d_3	major diameter of external spline, m (in)
d_4	minor diameter of external spline, m (in)
d_c	core diameter of threaded portion of the taper rod, m (in)
d_{pl}	large diameter of taper pin, m (in)
d_m (or d_{pm})	mean diameter of taper pin, m (in)
d_{nom}	nominal diameter of thread portion, m (in)
D	diameter of shaft, m (in)
	pitch diameter, m (in)
F	force, kN (lbf)
	force on the cotter joint, kN (lbf)
	pressure between hub and key, kN (lbf)
F', F''	force applied in the center of plane of a feather keyed shaft which do not change the existing equilibrium but give a couple, kN (lbf)
F'_2, F''_2	two opposite forces applied on the center plane of a double feather keyed shaft which give two couples, but tending to rotate the hub clockwise, kN (lbf)
F_t	tangential force, kN (lbf)
F_μ	frictional force, kN (lbf)
h	thickness of key, m (in)
	minimum height of contact in one tooth, m (in)
l	length of key (also with suffixes), m (in)
	length of couple (also with suffixes), m (in)
	length of sleeve, m (in)
L	length of spline, m (in)
l_o, s_o	space width and tooth thickness of spline, m (in)
m	module, mm, m (in)
M_b	bending moment, N m (lbf in)
M_t	twisting moment, N m (lbf in)

p	pressure, MPa (psi)
	tangential pressure per unit length, MPa (psi)
p_1	maximum pressure where the shaft enters the hub, MPa (psi)
p_2	pressure at the end of key, MPa (psi)
p_d (or P)	diametral pitch
Q	external load, kN (lbf)
R	resistance on the key and on the shaft to be overcome when the hub is shifted lengthwise, kN (lbf)
t	thickness of cotter, m (in)
xm	profile displacement, m (in)
z	number of teeth, number of splines
σ	stress tensile or compressive (also with suffixes), MPa (psi)
σ_{b1}	nominal bearing stress at dangerous point, MPa (psi)
τ	shear stress, MPa (psi)
α	angle of cotter slope, deg
θ	angle of friction, deg
μ	coefficient of friction (also with suffixes)

SUFFIXES

b	bearing
c	compressive
d	design
m	mean
p	pin
s	small end
t	tensile, tangential

Particular	Formula

ROUND OR PIN KEYS

The large diameter of the pin key

$$d = 3.035\sqrt{D} \text{ to } 3.45\sqrt{D} \quad \textbf{SI} \quad (17\text{-}1a)$$

where d and D are in mm

$$d = 0.6\sqrt{D} \text{ to } 0.7\sqrt{D} \quad \textbf{USCS} \quad (17\text{-}1b)$$

where d and D are in in

$$d = 0.096\sqrt{D} \text{ to } 0.11\sqrt{D} \quad \textbf{SI} \quad (17\text{-}1c)$$

where d and D are in m

STRENGTH OF KEYS

Rectangular fitted key (Fig. 17-1, Table 17-1)

Pressure between key and keyseat

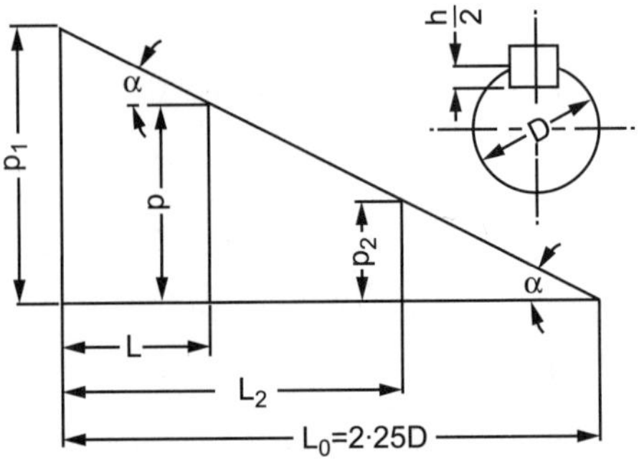

FIGURE 17-1

TABLE 17-1
Dimensions (in mm) of parallel keys and keyways

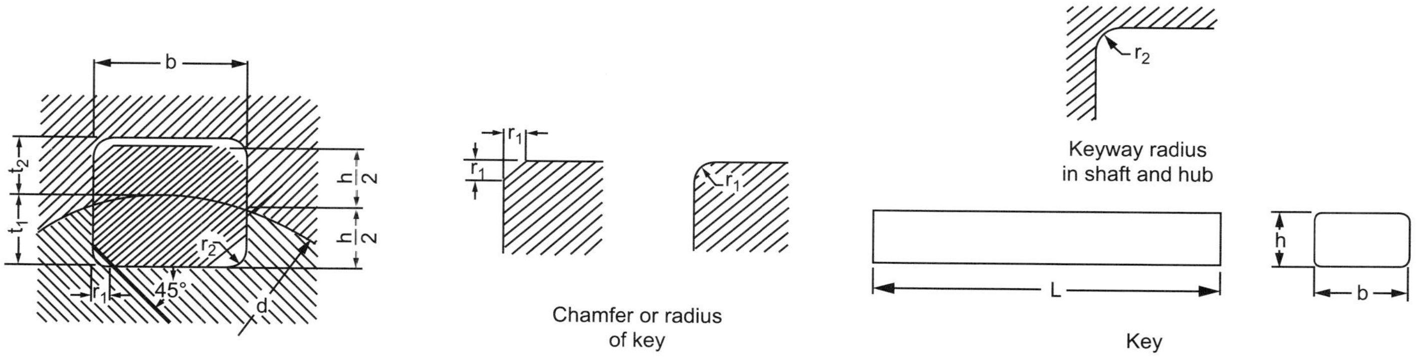

For shaft diameters	Above	6	8	10	12	17	22	30	38	44	50	58	65	75	85	95	110	130	150	170	200	230	260	290	330	380	440
	Up to	8	10	12	17	22	30	38	44	50	58	65	75	85	110	110	130	150	170	200	230	260	290	330	380	440	500
Key cross section	Width b	2	3	4	5	6	8	10	12	14	16	18	20	22	25	28	32	36	40	45	50	56	63	70	80	90	100
	Height h	2	3	4	5	6	7	8	8	9	10	11	12	14	14	16	18	20	22	25	28	32	32	36	40	45	50
Keyway depth (nominal)	In shaft t_1	1.2	1.8	2.5	3.0	3.5	4.0	5	5	5.5	6	7	7.5	8.5	9.0	10	11	12	13	15	17	19	20	22	25	28	31
	In hub t_2	1	1.4	1.8	2.3	2.8	3.3	3.3	3.8	3.8	4.3	4.4	4.9	5.4	5.9	6.4	7.4	8.4	9.4	10.4	11.4	12.4	13.4	14.4	15.4	17.4	19.5
Tolerance on keyway depth	t_1		+0.05 −0.00					+0.1 −0.0												+0.15 −0.00							
	t_2		+0.05 −0.00					+0.1 −0.0												+0.15 −0.00							
Chamfer or radius of key	r max		0.25			0.35			0.55					0.80						1.30			2.00				2.95
	r min		0.16			0.25			0.40					0.60						1.00			1.60				2.50
Keyway radius	r_2 max		0.16			0.25			0.40					0.60						1.00			1.60				2.50
Length of key	L min	6	6	8	10	14	18	22	28	36	45	50	56	63	71	80	90	100	110	125	140	160	180	200	220	250	280
	L max	20	36	45	50	71	90	110	140	160	180	200	220	250	280	320	360	400	400	400	400	400	400	400	400	400	400

Source: IS 2048, 1962.

Particular	Formula

Crushing strength

The tangential pressure per unit length of the key at any intermediate distance L from the hub edge (Fig. 17-1, Table 17-2)	$p = p_1 - L \tan\alpha$ where $\tan\alpha = \dfrac{p_1 - p_2}{L_2} = \dfrac{p_1}{L_0}$	(17-2)
The torque transmitted by the key (Fig. 17-1)	$M_t = \frac{1}{2} p_1 D L_2 - D L_2^2 \tan\alpha$	(17-3)
The general expression for torque transmitted according to practical experience	$M_t = \frac{1}{4}\sigma_{b1} h D L_2 - \frac{1}{18}\sigma_{b1} b L_2^2$ where $p_2 = 0$, when $L_2 = L_o = 2.25D$; $\tan\alpha = \dfrac{p_1}{L_o} = \dfrac{\sigma_{b1} h}{4.5D}$	(17-4)
For dimensions of tangential keys given here.	Refer to Table 17-2.	

Shearing strength

The torque transmitted by the key (Fig. 17-1)	$M_t = \frac{1}{2}\tau_1 b D L_2 - \frac{1}{9}\tau_1 b L_2^2$ where $\tan\alpha = \dfrac{p_1}{L_o} = \dfrac{\tau_1 b}{2.25D}$	(17-5)
The shear stress at the dangerous point (Fig. 17-1)	$\tau_1 = \dfrac{M_t}{L_2 b(0.5D - 0.11L_2)}$	(17-6)

TAPER KEY (Fig. 17-2, Table 17-3)

The relation between the circumferential force F_t and the pressure F between the shaft and the hub	$F_t = \mu_1 F$	(17-7)
The pressure or compressive stress between the shaft and the hub	$F = blp$	(17-8)
The torque	$M_t = \frac{1}{2}\mu_1 blpD$ where μ_1 = coefficient of friction between the shaft and the hub $= 0.25$	(17-9)

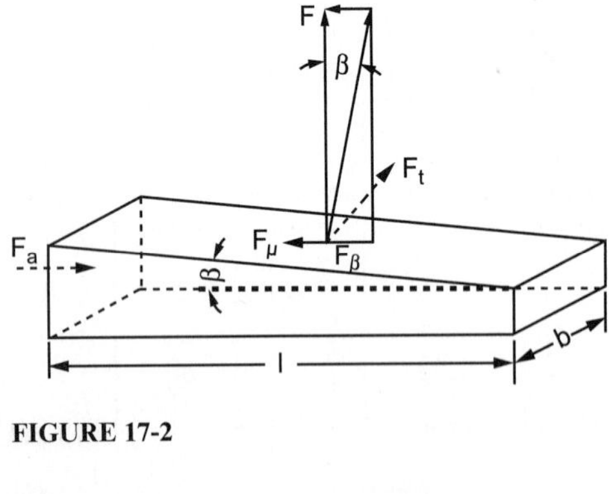

FIGURE 17-2

TABLE 17-2
Dimensions (in mm) of tangential keys and keyways

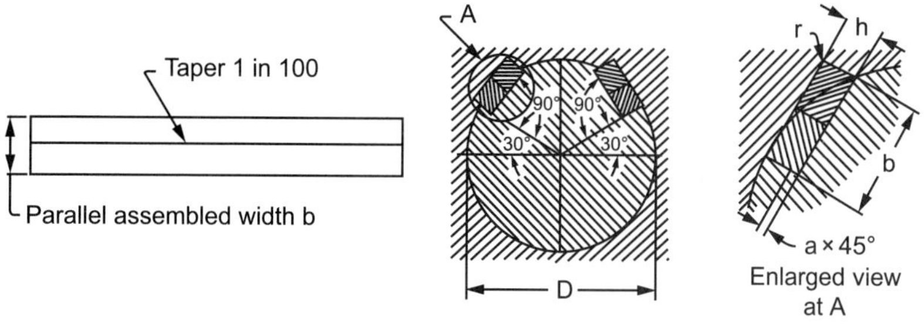

Shaft diameter, *D*	Keyway Height, *h*	Keyway Width, *b*	Keyway Radius, *r*	Key chamfer, *a*	Shaft diameter, *D*	Keyway Height, *h*	Keyway Width, *b*	Keyway Radius, *r*	Key chamfer, *a*
100	10	30	2	3	460	46	138	4	5
110	11	30	2	3	480	48	144	5	6
120	12	36	2	3	500	50	150	5	6
130	13	39	2	3	520	52	156	5	6
140	14	42	2	3	540	54	162	5	6
150	15	45	2	3	560	56	168	5	6
160	16	48	2	3	580	58	174	5	6
170	17	51	2	3	600	60	180	6	7
180	18	54	2	3	620	62	186	6	7
190	19	57	2	3	640	64	192	6	7
200	20	60	2	3	660	66	198	6	7
210	21	63	2	3	680	68	204	6	7
220	22	66	2	4	700	70	210	6	7
230	23	69	3	4	720	72	216	6	7
240	24	72	3	4	740	74	222	6	7
250	25	75	3	4	760	76	228	6	7
260	26	78	3	4	780	78	234	6	7
270	27	81	3	4	800	80	240	6	7
280	28	84	3	4	820	82	246	6	7
290	29	87	3	4	840	84	252	6	7
300	30	90	3	4	860	86	258	6	7
320	32	95	3	4	880	88	264	8	9
340	34	102	3	4	900	90	270	8	9
360	36	108	3	4	920	92	276	8	9
380	38	114	4	5	940	94	282	8	9
400	40	129	4	5	960	96	288	8	9
420	42	126	4	5	980	98	294	8	9
440	44	132	4	5	1000	100	300	8	9

Notes: (1) The dimensions of the keys are based on the formula: width 0.3 shaft diameter, and thickness = 0.1 shaft diameter; (2) if it is not possible to fix the keys at 120°, they may be fixed at 180°; (3) it is recommended that for an intermediate diameter of shaft, the key section shall be the same as that for the next larger size of the shaft in this table.
Source: IS 2291, 1963.

TABLE 17-3
Dimensions (in mm) of taper keys and keyways

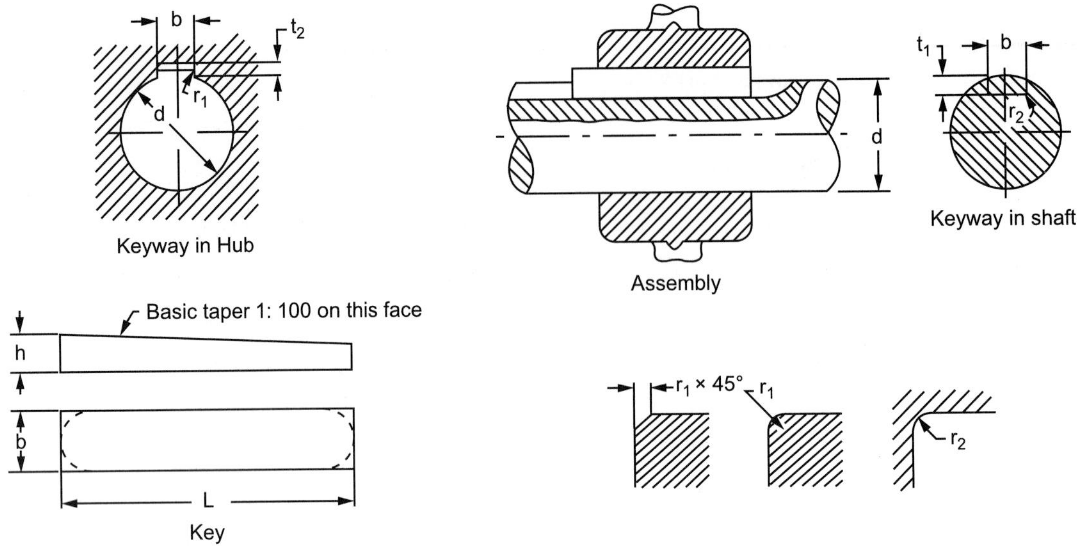

Shaft		Key			Keyway in shaft and hub					
Above	Up to and including	Width, b ($h9$)	Height, h	Chamfer or radius r_1, min	Keyway width, b ($D10$)	Depth in shaft, t_1	Tolerance on t_1	Depth in hub, t_2	Tolerance on t_2	Radius, r_2, max
6	8	2	2	0.16	2	1.2	+0.05	0.5		0.16
8	10	3	3	—	3	1.8	—	0.9		
10	12	4	4		4	2.5		1.2		
12	17	5	5		5	3.0		1.7	+0.1	0.25
17	22	6	6	0.25	6	3.5		2.1		
22	30	8	7	—	8	4.0	+0.10	2.5		—
30	38	10	8		10	5.0		2.5		
38	44	12	8		12	5.0		2.5	—	
44	50	14	9	0.40	14	5.5		2.9		0.40
50	58	16	10	—	16	6.0	—	3.4		
58	65	18	11		18	7.0		3.3		—
65	75	20	12		20	7.5		3.8		
75	85	22	14	0.60	22	8.5		4.8		
85	95	25	14	—	25	9.0		4.3	+0.2	0.60
95	110	28	18		28	10.0		5.3		—
110	130	32	10		32	11.0		6.2		
130	150	36	25		36	12.0		7.2		
150	170	40	22	1.00	40	13.0	+0.15	8.2		1.00
170	200	45	25	—	45	15.0		9.2	—	—
200	230	50	28		50	17.0		10.1		
230	260	56	32		56	19.0		12.1		
260	290	63	32	1.60	63	20.0		11.1		1.60
290	330	70	36	—	70	22.0		13.1	+0.3	—
330	380	80	40		80	25.0		14.1		
380	440	90	45	2.50	90	28.0		16,1		2.50
440	500	100	50		100	31.0		18.1		

Source: IS 2292, 1963.

Particular	Formula
The necessary length of the key	$l = \dfrac{2M_t}{\mu_1 bpD}$ (17-10)
The axial force necessary to drive the key home (Fig. 17-2)	$F_a = F_\mu + F_\beta = 2\mu_2 F + F \tan\beta$ (17-11) where $\mu_2 = 0.10$, $\tan\beta = 0.0104$ if the taper is 1 in 100
The axial force is also given by the equation	$F_a = 0.21pbl$ (17-12)

FRICTION OF FEATHER KEYS (Fig. 17-3)

Particular	Formula
The circumferential force (Fig. 17-3)	$F_t = \dfrac{M_t}{a}$ (17-13)
The resistance to be overcome when a hub connected to a shaft by a feather, Fig. 17-3a and subjected to torque M_t, is moved along the shaft	$R = \mu F_t + \mu_2 F'$ (17-14) $= (\mu + \mu_2)F_t$ (17-15) and $F' = F'' = F_t$ = force assumed to be acting at the shaft axis without changing the equilibrium Fig. 17-3a
The equation for resistance R, if μ and μ_2 are equal	$R = 2\mu F_t$ (17-16)
The equation for torque if two feather keys are used, Fig. 17-3b	$M_t = 2F_2 a$ (17-17)
The force F_2 applied at key when two feather keys are used, Fig. 17-3b	$F_2 = \dfrac{M_t}{2a} + \dfrac{F_t}{2}$ (17-18)
The resistance to be overcome when the hub connected to the shaft by two feather keys Fig. 17-3b and subjected to torque M_t is moved along the shaft	$R_2 = 2\mu F_2 = \dfrac{R}{2}$ (17-19)
For Gib-headed and Woodruff keys and keyways	Refer to Tables 17-4 and 17-5.

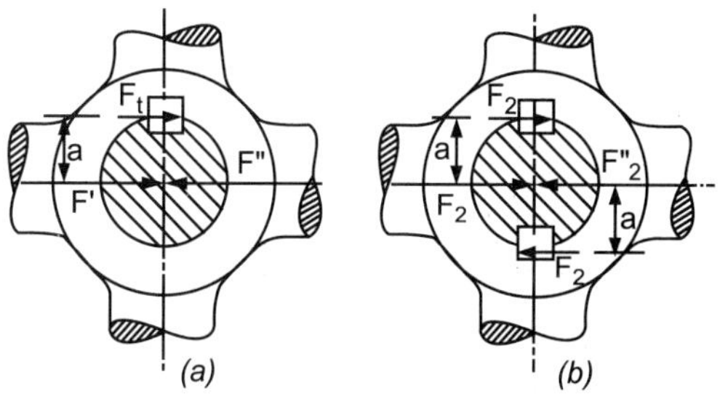

FIGURE 17-3 Feather key.

TABLE 17-4
Gib-head keys and keyways (all dimensions in mm)

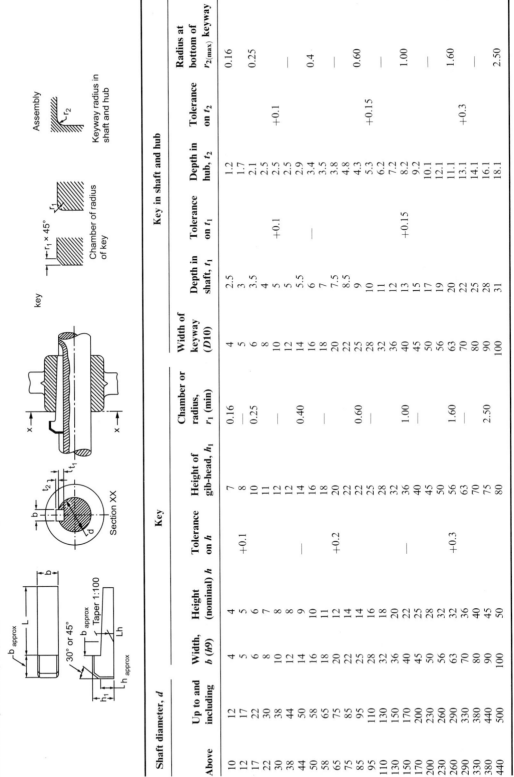

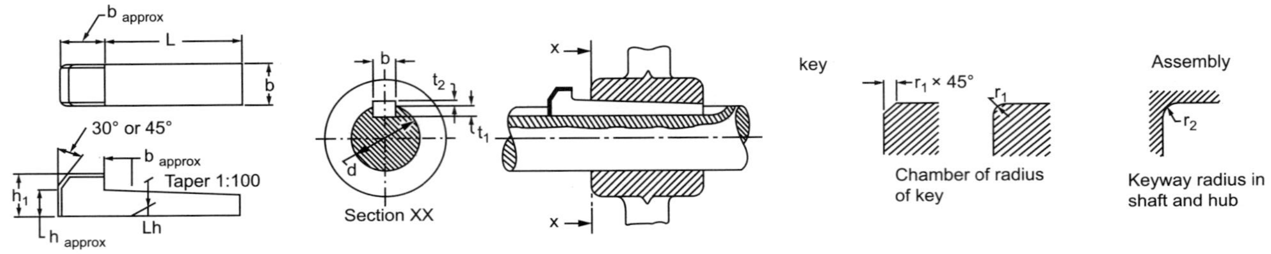

Shaft diameter, d		Key					Key in shaft and hub					
Above	Up to and including	Width, b ($h9$)	Height (nominal) h	Tolerance on h	Height of gib-head, h_1	Chamber or radius, r_1 (min)	Width of keyway ($D10$)	Depth in shaft, t_1	Tolerance on t_1	Depth in hub, t_2	Tolerance on t_2	Radius at bottom of $r_{2(max)}$ keyway
10	12	4	4		7	0.16	4	2.5		1.2		0.16
12	17	5	5	+0.1	8	—	5	3		1.7		
17	22	6	6		10	0.25	6	3.5		2.1		0.25
22	30	8	7		11		8	4		2.5		
30	38	10	8		12	—	10	5	+0.1	2.5	+0.1	
38	44	12	8		12		12	5		2.5		—
44	50	14	9	—	14	0.40	14	5.5		2.9		
50	58	16	10		16		16	6	—	3.4		0.4
58	65	18	11		18	—	18	7		3.5		
65	75	20	12	+0.2	20		20	7.5		3.8		—
75	85	22	14		22		22	8.5		4.8		
85	95	25	14		22	0.60	25	9		4.3		0.60
95	110	28	16		25	—	28	10		5.3	+0.15	
110	130	32	18		28		32	11		6.2		—
130	150	36	20		32		36	12		7.2		
150	170	40	22	—	36	1.00	40	13	+0.15	8.2		1.00
170	200	45	25		40	—	45	15		9.2		
200	230	50	28		45		50	17		10.1		—
230	260	56	32		50		56	19		12.1		
260	290	63	32	+0.3	56	1.60	63	20		11.1		1.60
290	330	70	36		63	—	70	22		13.1	+0.3	
330	380	80	40		70		80	25		14.1		—
380	440	90	45		75	2.50	90	28		16.1		
440	500	100	50		80		100	31		18.1		2.50

Source: IS 2293, 1963.

TABLE 17-5
Woodruff keys and keyways (all dimensions in mm)

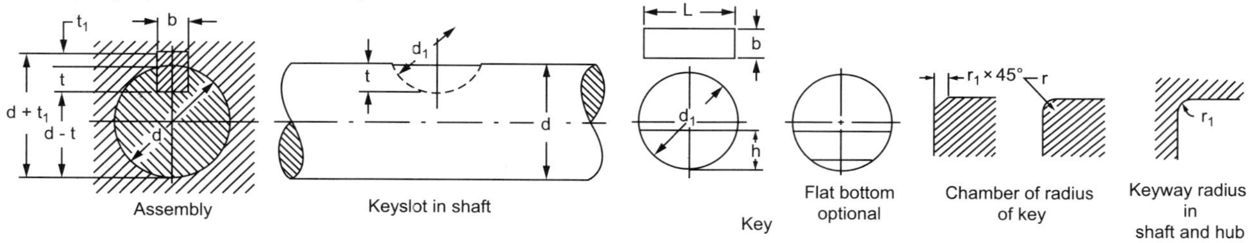

Key section		Range of shaft dia, d				Key					Keyslot in shaft			Keyslot in hub			Radius, r_1	
		Group I		Group II		Diameter of tolerance		Chamfer or			Depth, t			Depth t_1				
b ($h9$)	h ($h12$)	Over	Up to and including	Over	Up to and including	d_1	Tolerance on d_1	radius, r	Tolerance on r	Length L	Series A	Series B	Tolerance	Series A	Series B	Tolerance	Nominal	Tolerance
1	1.4	3	4	6	8	4.0		0.2		3.82	1.0	1.0		0.6	0.6		0.2	
1.5	2.6	4	6	8	10	7.0		0.2		6.76	2.0	2.0		0.8	0.8		0.2	
2	2.6	6	8	10	12	7.0		0.2		6.76	1.8	1.8		1.0	1.0		0.2	
2	3.7	6	8	10	12	10.0		0.2		9.66	2.9	2.9		1.0	1.0		0.2	
2.5	3.7	8	10	12	17	10.0		0.2		9.66	2.9	2.9		1.0	1.0		0.2	
3	3.7	8	10	12	17	10.0		0.2		9.66	2.5	2.8		1.4	1.1		0.2	
3	5	8	10	12	17	13.0		0.2	+0.1	12.65	3.8	4.1		1.4	1.1		0.2	−0. 1
3	6.5	—	—	16	17	16.0		0.2		15.72	5.3	5.6	+0.1	1.4	1.1		0.2	
4	5	10	12	17	22	13.0	−0.1	0.2		12.65	3.5	4.1		1.7	1.1		0.2	
4	6.5	10	12	17	22	16.0		0.2		15.72	5.0	5.6		1.7	1.1		0.2	
4	7.5	—	—	17	22	19.0		0.2		18.57	6.0	6.6		1.8	1.1		0.2	
5	6.5	12	17	22	30	16.0		0.2		15.72	4.5	5.4		2.2	1.3	+0.1	0.2	
5	7.5	12	17	22	30	19.0		0.2		18.57	5.5	6.4		2.2	1.3		0.2	
5	9	—	—	22	30	22.0		0.2		21.63	7.0	7.9	+0.2	2.2	1.3		0.2	
6	7.5	17	22	30	38	19.0		0.4		18.57	5.1	6.0		2.6	1.7		0.4	
6	9	17	22	30	38	22.0		0.4		21.63	6.6	7.5	+0.1	2.6	1.7		0.4	
6	(10)	17	22	30	38	25.0	−0.2	0.4		24.49	7.6	8.5		2.6	1.7		0.4	
6	11	—	—	30	38	28.0	−0.1	0.4		27.35	8.6	9.5		2.6	1.7		0.4	
8	9	22	30	38	—	22.0		0.4	+ 0.2	21.63	6.2	7.5		3.0	1.7		0.4	−0.2
8	11	22	30	38	—	28.0		0.4		27.35	8.2	9.5	+0.2	3.0	1.7		0.4	
8	13	—	—	38	—	32.0		0.4		31.43	10.2	11.5		3.0	1.7		0.4	
10	11	30	38	38	—	28.0	−0.2	0.4		27.35	7.8	9.1		3.4	2.1		0.4	
10	13	30	38	38	—	32.0		0.4		31.43	9.8	11.1		3.4	2.1		0.4	
10	16	—	—	38	—	45.0		0.4		43.08	12.8	14.1		3.4	2.1		0.4	

Notes:
(1) The dimensions $d - t$ and $d + t_1$ may be specified on workshop drawings; (2) the key size 6×10 is nonpreferred; (3) the key size 2.5×3.7 shall be used in automobile industries only.
Source: IS 2294, 1963.

Particular	Formula	

SPLINES

Parallel-sided or straight-sided spline

The torque which an integral multispline shaft can transmit (Tables 17-6 to 17-12)

$$M_t = \frac{1}{2} phli(D - h) \quad (17\text{-}20)$$

TABLE 17-6
Proportions of SAE standard parallel side splines

Types of spline fittings	Symbols	Proportions	Fit	Bearing pressure, p MPa	Bearing pressure, p kpsi
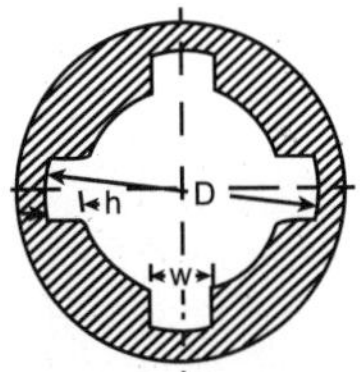	w	$w = 0.241D$			
	h	$4A$, $h = 0.075D$	A	20.6	3.00
	h	$4B$, $h = 0.125D$	B	13.7	2.00
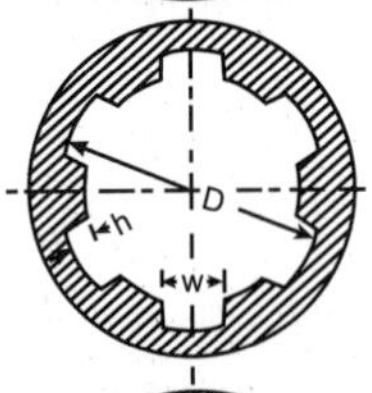	w	$w = 0.250D$			
	h	$6A$, $h = 0.050D$	A	20.6	3.00
	h	$6B$, $h = 0.075D$	B	13.7	2.00
	h	$6C$, $h = 0.100D$	C	6.9	1.00
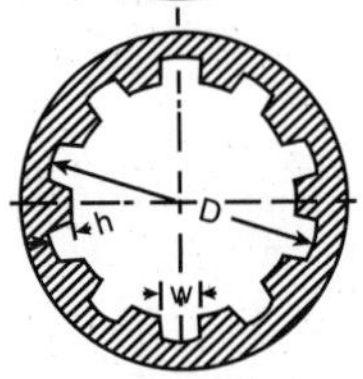	w	$w = 0.156D$			
	h	$10A$, $h = 0.045D$	A	20.6	3.00
	h	$10B$, $h = 0.070D$	B	13.7	2.00
	h	$10C$, $h = 0.095D$	C	6.9	1.00
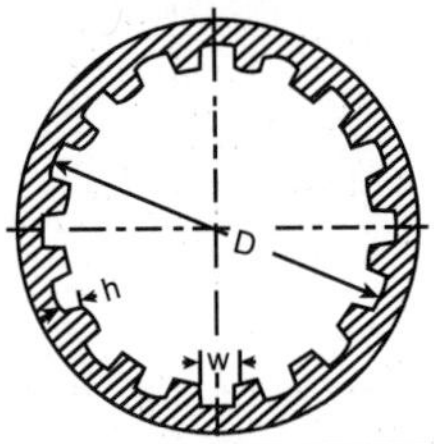	w	$w = 0.098D$			
	h	$16A$, $h = 0.045D$	A	20.6	3.00
	h	$16B$, $h = 0.070D$	B	13.7	2.00
	h	$16C$, $h = 0.095D$	C	6.9	1.00

TABLE 17-7
Proportions of involute spline profile (American Standard)

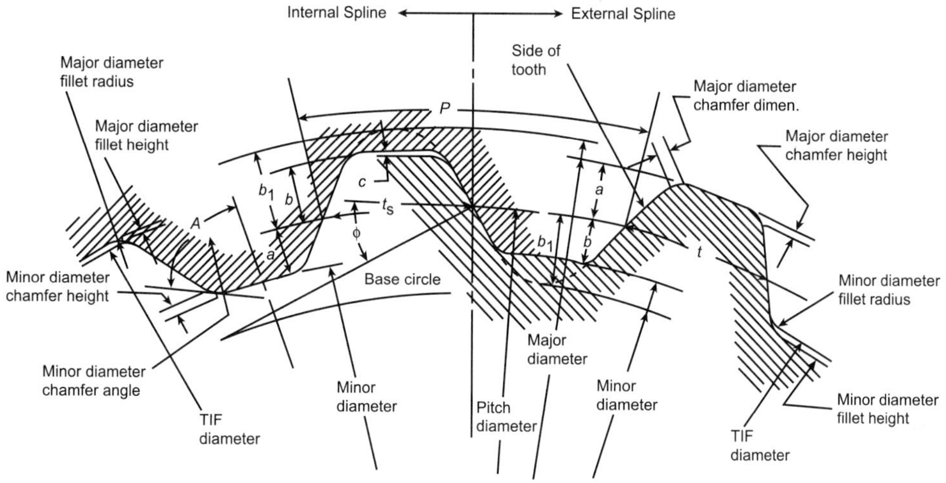

		Proportions	
Spline characteristics	**Symbols**	$P = \frac{1}{2}$ **through** $\frac{12}{24}$	$P = \frac{6}{32}$ **through** $\frac{48}{96}$
Pitch diameter	D	$D = zm = \dfrac{z}{P}$	$D = zm = z/P$
Circular pitch	p	$p = (\pi/P)$	$p = (\pi/P)$
Tooth thickness	t	$t = \dfrac{\pi m}{2} = \dfrac{\pi}{2P}$	$t = (\pi m/2) = (\pi/2P)$
Diametral pitch	P	$P = (\pi/p)$	$P = (\pi/p)$
Addendum	a	$a = 0.5m = \dfrac{0.500}{P}$	$a = 0.5m = 0.500/P$
Dedendum (internal)	b_1	$b_1 = 0.90m = \dfrac{0.900}{P}$	$b_1 = 0.9m = 0.900/P$
Dedendum	b	$b = 0.5m = \dfrac{0.500}{P}$	$b = 0.5m = 0.500/P$
Dedendum (external)	b_1	$b_1 = 0.9m = 0.900/P$	$b_1 = 1.0m = 1.000/P$
Major diameter (internal)	D_{oi}	$D_{oi} = (z + 1.8)m = (z + 1.8)/P$	$D_{oi} = (z + 1.8)m = (z + 1.8)/P$
Minor diameter (external)	D_{me}	$D_{me} = (z - 1.8)m = (z - 1.8)/P$	$D_{me} = (z - 2.0)m = (z - 2.0)/P$

Source: Courtesy H. L. Horton, ed., *Machinery's Handbook*, 15th ed., The Industrial Press, New York, 1957.

TABLE 17-8
Straight sided splines (all dimensions in mm)

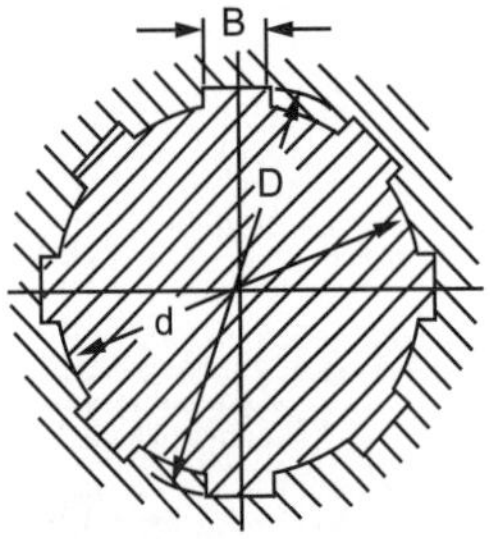

Splined shaft and hub profile

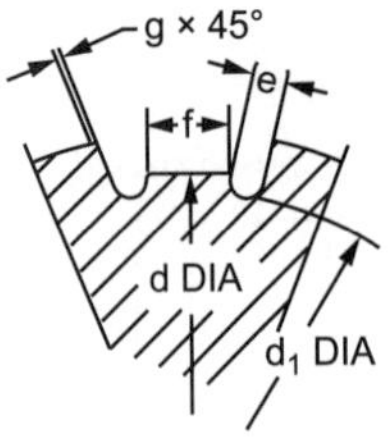

Splined shaft

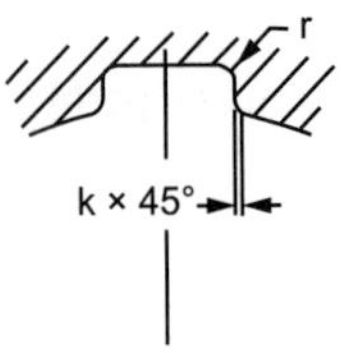

Splined hub

Nominal size $i \times d \times D$	No. of splines, i	Minor diameter, d	Major diameter, D	Width, B	d_1,[a] min	e,[a] max	f[a]	g, max	k, mix	r, max	Centering on
				Light-Duty Series							
6 × 23 × 26	6	23	26	6	22.1	1.25	3.54	0.3	0.3	0.2	Inside diameter[a]
6 × 26 × 30	6	26	30	6	24.6	1.84	3.85	0.3	0.3	0.2	
6 × 28 × 32	6	28	32	7	26.7	1.77	4.03	0.3	0.3	0.2	
8 × 32 × 36	8	32	36	6	30.4	1.89	2.71	0.4	0.4	0.3	
8 × 36 × 40	8	36	40	7	34.5	1.78	3.46	0.4	0.4	0.3	Inside diameter or flanks[b]
8 × 42 × 46	8	42	46	8	40.4	1.68	5.03	0.4	0.4	0.3	
8 × 46 × 50	8	46	50	9	44.6	1.61	5.75	0.4	0.4	0.3	
8 × 52 × 58	8	52	58	10	49.7	2.72	4.89	0.5	0.5	0.5	
8 × 56 × 62	8	56	62	10	53.6	2.76	6.38	0.5	0.5	0.5	
8 × 62 × 68	8	62	68	12	59.8	2.48	7.31	0.5	0.5	0.5	
10 × 72 × 78	10	72	78	12	69.6	2.54	5.45	0.5	0.5	0.5	
10 × 82 × 88	10	82	88	12	79.3	2.67	8.62	0.5	0.5	0.5	
10 × 92 × 98	10	92	98	14	89.4	2.36	10.08	0.5	0.5	0.5	
10 × 102 × 108	10	102	108	16	99.9	2.23	11.49	0.5	0.5	0.5	
10 × 112 × 120	10	112	120	18	108.8	3.23	10.72	0.5	0.5	0.5	
				Medium-Duty Series							
6 × 11 × 14	6	11	14	3	9.9	1.55		0.3	0.3	0.2	Inside diameter[a]
6 × 13 × 16	6	13	16	3.5	12.0	1.50	0.32	0.3	0.3	0.2	
6 × 16 × 20	6	16	20	4	14.5	2.10	0.16	0.3	0.3	0.2	
6 × 18 × 22	6	18	22	5	16.7	1.95	0.45	0.3	0.3	0.2	
6 × 21 × 25	6	21	25	5	19.5	1.98	1.95	0.3	0.3	0.2	
6 × 23 × 28	6	23	28	6	21.3	2.30	1.34	0.3	0.3	0.2	
6 × 26 × 32	6	26	32	6	23.4	2.94	1.65	0.4	0.4	0.3	
6 × 28 × 34	6	28	34	7	25.9	2.94	1.70	0.4	0.4	0.3	
8 × 32 × 38	8	32	38	6	29.4	3.30	0.15	0.4	0.4	0.3	Inside diameter or flanks[b]
8 × 36 × 42	8	36	42	7	33.5	3.01	1.02	0.4	0.4	0.3	
8 × 42 × 48	8	42	48	8	39.5	2.91	2.54	0.4	0.4	0.3	
8 × 46 × 54	8	46	54	9	42.7	4.10	0.86	0.5	0.5	0.3	
8 × 52 × 60	8	52	60	10	48.7	4.00	2.44	0.5	0.5	0.5	
8 × 56 × 65	8	56	65	10	52.2	4.74	2.50	0.5	0.5	0.5	
8 × 62 × 72	8	62	72	12	57.8	5.00	2.40	0.5	0.5	0.5	
10 × 72 × 82	10	72	82	12	67.4	5.43	2.70	0.5	0.5	0.5	
10 × 82 × 92	10	82	92	12	77.1	5.40	3.00	0.5	0.5	0.5	
10 × 92 × 102	10	92	102	14	87.3	5.20	4.50	0.5	0.5	0.5	
10 × 102 × 112	10	102	112	16	97.7	4.90	6.30	0.5	0.5	0.5	
10 × 112 × 125	10	112	125	18	106.3	6.40	4.40	0.5	0.5	0.5	

[a] These values are based on the generating process.

[b] Inside centering is not always possible with generating processes.

Source: IS 2327, 1963.

TABLE 17-9
Tolerances for straight-sided splines (all dimensions in mm)

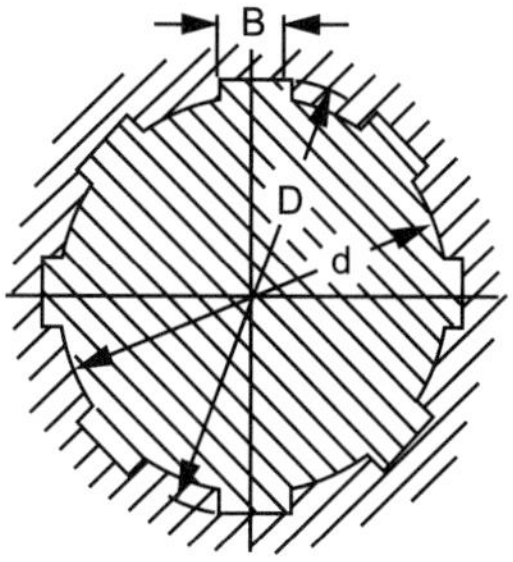

Assembly of splined hub and shaft			Width of hub B (Soft or hardened)		Tolerance on: Minor diameter of hub, d (Soft or hardened)	Tolerance on: Major diameter of hub, D (Soft or hardened)
Splined hub	For centering on inner diameter or flanks	Shaft sliding or fixed	D9	F10	H7	H11
	For centering on inner diameter	Shaft sliding inside hub	*h*8	*e*8	*f*7	a11
Splined shaft		Shaft fixed in hub	*p*6	*h*6	*j*6	a11
		Shaft sliding inside hub	*h*8	*e*8	—	a11
	For centering on flanks	Shaft fixed in hub	*u*6	*k*6	—	a11

Particular	Formula	

Involute-sided spline

AMERICAN STANDARD (Table 17-7) The addendum a and dedendum b for a flat root, Table 17-7

$$a = b = m = \frac{1}{P} \tag{17-21}$$

The area resisting shear, Table 17-7

$$A_\tau = \frac{\pi DL}{2} \tag{17-22}$$

The minimum height of contact on one tooth

$$h = 0.8m = \frac{0.8}{P} = \frac{0.8D}{z} \tag{17-23}$$

The corresponding area of contact of all z teeth

$$A = \left(\frac{0.8D}{z}\right) zL = 0.8DL \tag{17-24}$$

The torque capacity of teeth in shear

$$M_t = \left(\frac{\pi DL}{z}\right)\frac{D}{2}\tau_d = 0.7854D^2L\tau_d \tag{17-25}$$

The torque capacity of the spline in bearing with $\sigma_b = 2\sigma_{dc}$

$$M_{tb} = 0.8D^2L\sigma_{dc} \tag{17-26}$$

TABLE 17-10
Straight-sided splines for machine tools (all dimensions in mm)

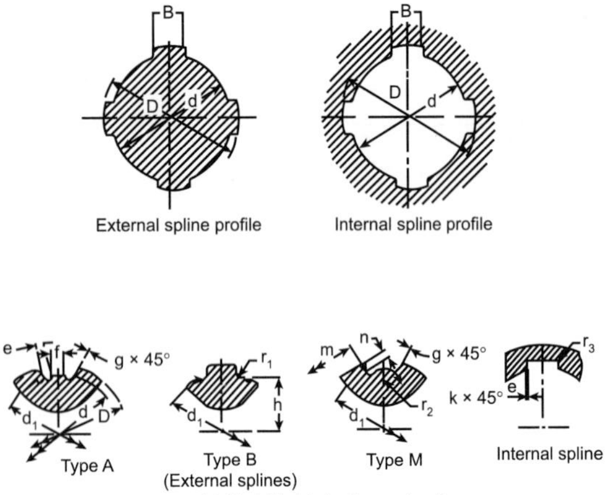

(a) Straight sided splines - 4 splines

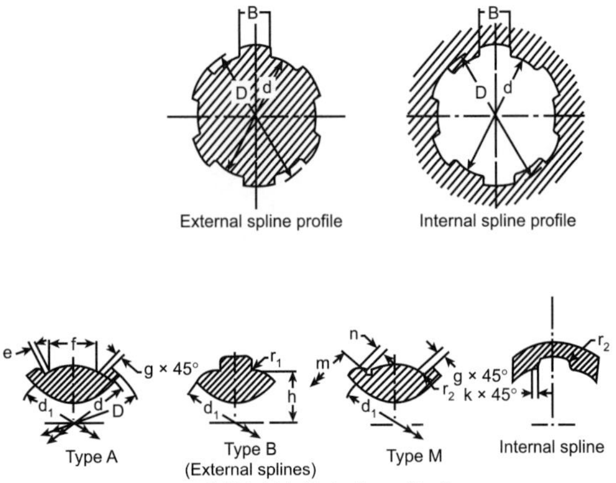

(b) Straight sided splines - 6 splines

4 Splines

Nominal size, $i^a \times d \times D$	Minor diameter, d	Major diameter, D	Width, B
4 × 11 × 15	11	15	3
4 × 13 × 17	13	17	4
4 × 16 × 20	16	20	6
4 × 18 × 22	18	22	6
4 × 21 × 25	21	25	8
4 × 24 × 28	24	28	8
4 × 28 × 32	28	32	10
4 × 32 × 38	32	38	10
4 × 36 × 42	36	42	12
4 × 42 × 48	42	48	12
4 × 46 × 52	46	52	14
4 × 52 × 60	52	60	14
4 × 58 × 65	58	65	16
4 × 62 × 70	62	70	16
4 × 68 × 78	68	78	16

[a] i = number of splines
Source: IS 2610, 1964.

6 Splines

Nominal size, $i^a \times d \times D$	Minor diameter, d	Major diameter, D	Width, B
6 × 21 × 25	21	25	5
6 × 23 × 28	23	28	6
6 × 26 × 32	26	32	6
6 × 28 × 34	28	34	7
6 × 32 × 38	32	38	8
6 × 36 × 42	36	42	8
6 × 42 × 48	42	48	10
6 × 46 × 52	46	52	12
6 × 52 × 60	52	60	14
6 × 58 × 65	58	65	14
6 × 62 × 70	62	70	16
6 × 68 × 78	68	78	16
6 × 72 × 82	72	82	16
6 × 78 × 90	78	90	16
6 × 82 × 95	82	95	16
6 × 88 × 100	88	100	16
6 × 92 × 105	92	105	20
6 × 98 × 110	98	110	20
6 × 105 × 120	105	120	20
6 × 115 × 130	115	130	20
6 × 130 × 145	130	145	24

TABLE 17-11
Undercuts, chamfers, and radii for straight-sided splines[a] (all dimensions in mm)

		External splines								Internal splines		
		Type A			Type B		Type M					Projected
Designation, $i \times d \times D$	B	d_1, min	g, max	f, min	h	r_1, max	m	n	r_2	k, max	r_3, max	tip width of hub
4 × 11 × 15	3	9.6	0.2	1.50	5.0	0.10	2.82	1.70	0.3	0.2	0.15	0.5
4 × 13 × 17	4	11.8	0.2	2.37	5.5	0.10	3.76	1.70	0.3	0.2	0.15	0.5
4 × 16 × 20	6	15.0	0.3	2.87	6.7	0.15	5.64	1.70	0.3	0.3	0.25	0.7
4 × 18 × 22	6	16.9	0.3	4.35	7.7	0.15	5.64	1.70	0.3	0.3	0.25	0.7
4 × 21 × 25	8	20.1	0.3	5.00	8.9	0.15	7.52	1.70	0.6	0.3	0.25	0.7
4 × 24 × 28	8	23.0	0.3	7.30	10.4	0.15	7.52	1.70	0.6	0.3	0.25	0.7
4 × 28 × 32	10	26.8	0.5	7.39	12.1	0.25	9.40	1.63	0.6	0.5	0.40	1.0
4 × 32 × 38	10	30.3	0.5	9.56	14.2	0.25	9.40	2.55	0.6	0.5	0.40	1.0
4 × 36 × 42	12	34.5	0.5	11.03	15.9	0.25	11.28	2.55	0.6	0.5	0.40	1.0
4 × 42 × 48	12	40.2	0.5	15.41	19.0	0.25	11.28	2.55	1.0	0.5	0.40	1.0
4 × 46 × 52	14	44.4	0.5	16.79	20.7	0.25	13.16	2.55	1.0	0.5	0.40	1.3
4 × 52 × 60	14	49.5	0.5	21.63	23.7	0.25	13.16	3.40	1.0	0.5	0.40	1.3
4 × 56 × 65	16	56.2	0.5	23.26	26.4	0.25	15.04	2.98	1.0	0.5	0.40	1.6
4 × 62 × 70	16	59.5	0.5	23.61	28.3	0.25	15.04	3.40	1.0	0.5	0.40	1.6
4 × 68 × 78	16	64.4	0.5	27.57	31.2	0.25	15.04	4.25	1.0	0.5	0.40	1.6

[a] Four splines; see Fig. 17-4a.
Source: IS 2610, 1964

TABLE 17-12
Undercuts, chamfers, and radii for straight-sided splines[a] (all dimensions in mm)

		External splines								Internal splines		
		Type A			Type B		Type M					Projected
Designation, $i \times d \times D$	B	d_1, min	g, max	f, min	h	r_1, max	m	n	r_2	k, max	r_3, max	tip width of hub
6 × 21 × 25	5	19.50	0.3	1.95	9.7	0.15	4.70	1.70	0.6	0.3	0.2	0.7
6 × 23 × 28	6	21.30	0.3	1.34	11.0	0.15	5.64	2.13	0.6	0.3	0.2	0.7
6 × 26 × 32	6	23.40	0.4	1.65	11.8	0.15	5.64	2.55	0.6	0.4	0.3	1.0
6 × 28 × 34	7	25.90	0.4	1.70	12.9	0.25	6.58	2.55	0.6	0.4	0.3	1.0
6 × 32 × 38	8	29.90	0.5	2.83	14.8	0.25	7.52	2.55	0.6	0.5	0.4	1.0
6 × 36 × 42	8	33.70	0.5	4.95	16.5	0.25	7.52	2.55	0.6	0.5	0.4	1.0
6 × 42 × 48	10	39.94	0.5	6.02	19.3	0.25	9.40	2.55	1.0	0.5	0.4	1.0
6 × 46 × 52	12	44.16	0.5	5.81	21.1	0.25	11.28	2.55	1.0	0.5	0.4	1.3
6 × 52 × 60	14	49.50	0.5	5.89	23.9	0.25	13.16	3.40	1.0	0.5	0.4	1.3
6 × 58 × 65	14	55.74	0.5	8.29	26.7	0.25	13.16	3.98	1.0	0.5	0.4	1.6
6 × 62 × 70	16	59.50	0.5	8.03	28.6	0.25	15.04	3.40	1.0	0.5	0.4	1.6
6 × 68 × 78	16	64.40	0.5	9.73	31.4	0.25	15.04	4.25	1.0	0.5	0.4	1.6
6 × 72 × 82	16	68.30	0.5	12.67	33.4	0.25	15.04	4.25	1.6	0.5	0.4	2.0
6 × 78 × 90	16	73.00	0.5	13.07	36.2	0.25	15.04	5.10	1.6	0.5	0.4	2.0
6 × 82 × 95	16	79.60	0.5	13.96	38.0	0.25	15.04	5.53	1.6	0.5	0.4	2.0
6 × 88 × 100	16	82.90	0.5	17.84	41.3	0.25	15.04	5.10	1.6	0.5	0.4	2.0
6 × 92 × 105	20	87.10	0.6	18.96	43.1	0.30	18.80	5.53	1.6	0.6	0.5	2.0
6 × 98 × 110	20	93.40	0.6	19.22	46.4	0.30	18.80	5.10	2.0	0.6	0.5	2.0
6 × 105 × 120	20	98.80	0.6	19.25	49.2	0.30	18.80	6.38	2.0	0.6	0.5	2.4
6 × 115 × 130	20	108.4	0.6	24.75	54.2	0.30	18.80	6.38	2.5	0.6	0.5	2.4
6 × 130 × 145	24	123.9	0.6	29.20	61.8	0.30	22.56	6.38	2.5	0.6	0.5	2.4

[a] Six splines see Fig. 17-4*b*.

Particular	Formula
The theoretical torque capacity of straight-sided spline with sliding according to SAE	$M_t = 6.895 \times 10^6 i\left(\frac{D+d}{4}\right)hL$ **SI** (17-26a) where i = number of splines D, d = diameter as shown in Table 17-7, m d = inside diameter of spline, m D = pitch diameter of spline, m L = length of spline contact, m h = minimum height of contact in one tooth of spline, m M_t in N m
	$M_t = 1000i\left(\frac{D+d}{4}\right)hL$ **USCS** (17-26b) where M_t in lb in; d, D, L, and h in in
Equating the strength of the spline teeth in shear to the shear strength of shaft, the length of spline for a hollow shaft	$L = \frac{D_{me}^3(1 - D_i^4/D_{me}^4)}{4D^2}$ (17-26c) where D_i = internal diameter of a hollow shaft, m (in) D_{me} = minor diameter (external), m (in)
The length of spline for a solid shaft	$L = \frac{D_{me}^3}{4D^2}$ (17-26d)
The effective length of spline for a hollow shaft used in practice according to the SAE	$L_e = \frac{D_{me}^3(1 - D_i^4/D_{me}^4)}{D^2}$ (17-26e) For solid shaft $D_i = 0$.
For diametrical pitches used in involute splines (SAE and ANSI)	Refer to Table 17-13.

TABLE 17-13
Diametral pitches[a] used in involute splines (SAE and ANSI)

2.5/5	3/6	4/8	5/10	6/12	8/16	10/20	12/24	16/32	20/40	24/48	32/64	40/80	48/96

[a] Diametral pitches are designated as fractions; the numerator of these fractions is the diametral pitch, P.

INDIAN STANDARD (Figs. 17-4 and 17-5, Tables 17-14 and 17-15)

Particular	Formula
The value of profile displacement (Fig. 17-4)	$xm = \frac{1}{2}(d_1 - mz - 1.1m)$ (17-27) The value xm varies from $-0.05m$ to $+0.45m$

Particular	Formula	
The number of teeth	$z = \frac{1}{m}(d_1 - 2xm - 1.1m)$	(17-28)
The minor diameter of the internal spline (Fig. 17-4a)	$d_2 = mz + 2xm - 0.9m = d_1 - 2m$	(17-29)
The major diameter of the external spline (Fig. 17-4a)	$d_3 = mz + 2xm + 0.9m = d_1 - 0.2m$	(17-30)
The minor diameter of the external spline (Fig. 17-4a)	$d_4 - mx + 2xm - 1.1m = d_1 - 2.2m$	(17-31)

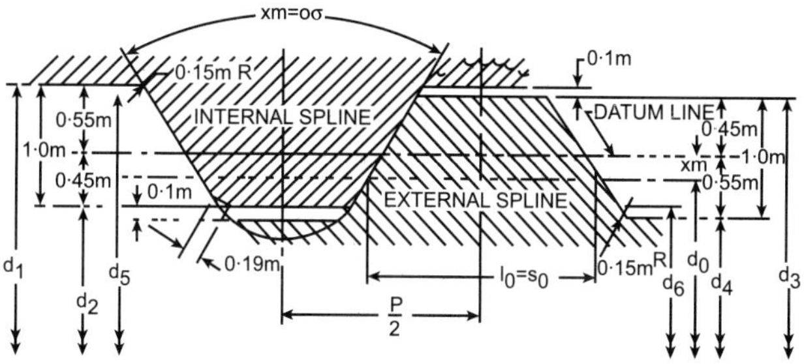

FIGURE 17-4(a) Reference profile of an involute-sided spline. (*Source*: IS 3665, 1966.)

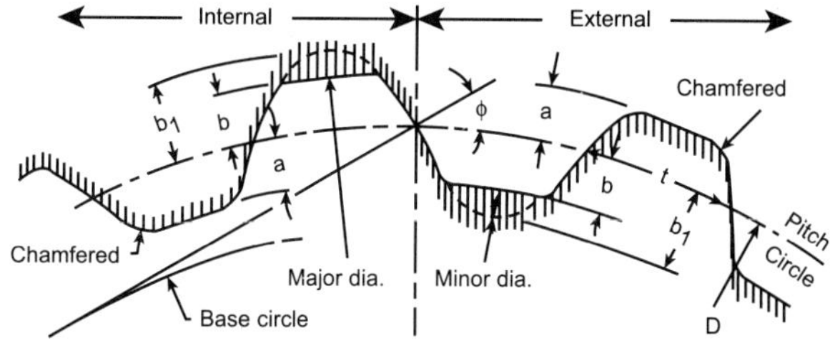

FIGURE 17-4(b) Nomenclature of the involute spline profile.

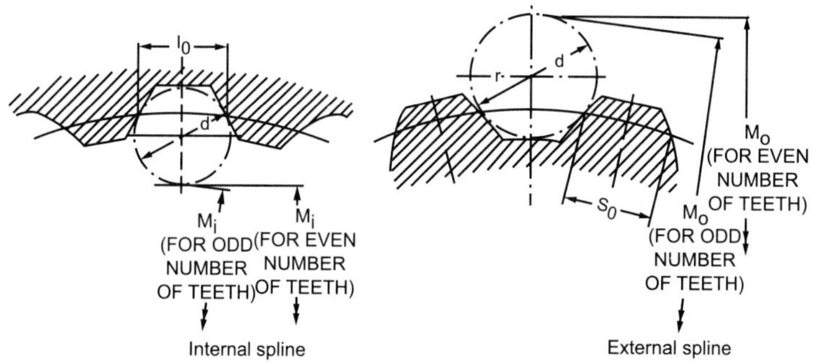

FIGURE 17-5 Measurement between pins and measurement over pins of an involute-sided spline. (*Source*: IS 3665, 1966.)

TABLE 17-14
Dimensions (in mm) for involute splines of module 2

										Internal spline			External spline				Tooth thickness over z' teeth
Nominal size $d_1 \times d_2$	z	d_o	d_b	d_3	d_4	d_5, min	d_6, max	xm	$l_o = s_o$	Pin diameter, d	Measurement between pins, Mi	Deviation factor, fi	Pin diameter d	Measurement over pins, Ma	Deviation factor, fa	z'	Tooth thickness deviation factor, 0.866
15 × 11	6	12	10.392	14.6	10.6	14.68	10.92	+0.4	3.603	3.5	7.629	2.42	5.5	22.212	1.11	2	9.121
17 × 13	7	14	12.124	16.6	12.6	16.68	12.92	+0.4	3.603	3.5	9.324	2.19	5.0	22.695	1.13	2	9.214
18 × 14	7	14	12.124	17.6	13.6	17.68	13.92	+0.9	4.181	3.5	10.379	1.61	6.0	25.588	1.06	2	9.714
20 × 16	8	16	13.856	19.6	15.6	19.68	15.92	+0.9	4.181	3.5	12.736	1.66	6.0	28.206	1.11	2	9.807
22 × 18	9	18	15.588	21.6	17.6	21.68	17.92	+0.9	4.181	3.5	14.460	1.64	5.5	28.790	1.13	—	—
25 × 21	11	22	19.053	24.6	20.6	24.68	20.92	+0.4	3.603	3.5	17.478	1.96	4.5	29.898	1.28	—	—
28 × 24	12	24	20.785	27.6	23.6	27.68	23.92	+0.9	4.181	3.5	20.738	1.68	5.0	34.161	1.23	3	15.621
30 × 26	14	28	24.299	29.6	25.6	29.69	25.91	−0.1	3.326	3.5	22.484	2.41	4.0	34.144	1.46	3	14.807
32 × 28	14	28	24.249	31.6	27.6	31.69	27.91	+0.9	4.681	3.5	24.738	1.69	4.5	37.016	1.30	3	15.807
35 × 31	16	32	27.713	34.6	30.6	34.69	30.91	+0.4	3.603	3.5	27.711	1.88	4.0	39.000	1.42	3	15.493
38 × 33	17	34	29.445	36.6	32.6	36.69	32.91	+0.4	3.603	3.5	29.571	1.86	4.0	40.857	1.42	4	21.028
37 × 34	18	36	31.177	37.6	33.6	37.69	33.91	−0.1	3.026	3.5	30.566	2.15	4.0	42.181	1.50	3	15.179
40 × 36	18	36	31.177	39.6	35.6	39.69	35.91	+0.9	4.181	3.5	32.739	1.70	4.0	45.137	1.15	4	21.621
42 × 38	20	40	34.641	41.6	37.6	41.69	37.91	−0.1	3.026	3.5	34.589	2.08	4.0	46.195	1.52	4	20.807
45 × 41	21	42	36.373	44.6	40.6	44.69	40.91	+0.4	3.603	3.5	37.604	1.84	4.0	48.938	1.46	4	21.400
47 × 43	22	44	38.105	46.6	42.6	46.69	42.91	+0.4	3.603	3.5	39.720	1.84	4.0	51.074	1.47	4	21.493
48 × 44	22	44	38.105	47.6	43.6	47.69	43.91	+0.9	4.181	3.5	40.740	1.70	4.0	51.912	1.43	5	27.435
50 × 46	24	48	41.569	49.6	45.6	49.69	45.91	−0.1	3.026	3.5	42.621	2.00	4.0	54.218	1.54	4	21.179
(52 × 48)	24	48	41.569	51.6	47.6	51.69	47.91	+0.9	4.181	3.5	44.740	1.71	4.0	55.939	1.44	5	27.621
55 × 51	26	52	45.033	54.6	50.6	54.70	50.90	+0.4	3.603	3.5	47.724	1.82	4.0	59.109	1.50	5	27.307
(58 × 54)	28	56	48.497	57.6	53.6	57.70	53.90	−0.1	3.026	3.5	50.624	1.95	4.0	62.235	1.56	5	26.993
60 × 56	28	56	48.497	59.6	55.6	59.70	55.90	+0.9	4.181	3.5	52.740	1.71	4.0	63.984	1.47	6	33.435
(62 × 58)	30	60	51.962	61.6	57.6	61.70	57.90	−0.1	3.026	3.5	54.650	1.93	4.0	66.242	1.57	5	27.179
65 × 61	31	62	53.694	64.6	60.6	64.70	60.90	+0.4	3.600	3.5	57.648	1.80	4.0	69.058	1.53	6	33.214
(68 × 64)	32	64	55.426	67.6	63.6	67.70	63.90	+0.9	4.181	3.5	60.740	1.71	4.0	72.021	1.49	6	33.807
70 × 66	34	68	58.890	69.6	65.6	69.70	65.90	−0.1	3.026	3.5	62.663	1.90	4.0	74.253	1.59	6	32.993
(72 × 68)	34	68	58.890	71.6	67.6	71.70	67.90	+0.9	4.181	3.5	64.740	1.71	4.0	76.036	1.50	7	39.435
75 × 71	36	72	62.354	74.6	70.6	74.70	70.90	+0.4	3.603	3.5	67.729	1.79	4.0	79.166	1.55	7	39.121
(78 × 74)	38	76	65.818	77.6	73.6	77.70	73.90	−0.1	3.026	3.5	70.672	1.88	4.0	82.263	1.60	7	38.807
80 × 76	38	76	65.818	79.6	75.6	79.70	75.90	+0.9	4.181	3.5	72.740	1.72	4.0	84.063	1.52	7	39.807
(82 × 78)	40	80	69.283	81.6	77.6	81.70	77.90	−0.1	3.026	3.5	74.676	1.87	4.0	86.267	1.61	7	38.993

Note: Values within parentheses are nonpreferred.

TABLE 17-15
Dimensions (in mm) for involute spline of module 2.5

										Internal spline			External spline				
Nominal size $d_1 \times d_2$	z	d_o	d_b	d_3	d_4	d_5, min	d_6, max	xm	$l_o = s_o$	Pin diameter, d	Measurement between pins, Mi	Deviation factor, fi	Pin diameter d	Measurement over pins, Ma	Deviation factor, fa	z'	Tooth thickness over z' teeth / Tooth thickness deviation factor, 0.866
20 × 15	6	15.0	12.990	19.5	14.5	19.58	14.92	+1.125	5.226	4.6	10.552	1.71	9.0	33.258	1.03	2	12.026
22 × 17	7	17.5	15.155	21.5	16.5	21.58	16.92	+0.875	4.937	4.5	12.105	1.85	7.0	30.558	1.08	2	11.892
25 × 20	8	20.0	17.321	24.5	19.5	24.58	19.92	+1.125	5.226	4.5	15.552	1.72	7.0	34.113	1.13	2	12.252
28 × 23	10	25.0	21.651	27.5	22.5	27.58	22.92	+0.125	4.071	4.55	19.116	2.30	5.0	33.006	1.37	2	11.491
30 × 25	10	25.0	21.651	29.5	24.5	29.58	24.92	+1.125	5.226	4.5	20.552	1.72	6.5	38.151	1.19	3	19.293
32 × 27	11	27.5	23.816	31.5	26.5	31.59	26.91	+0.875	4.937	4.5	22.265	1.81	6.0	38.835	1.23	3	19.160
35 × 30	12	30.0	25.981	34.5	29.5	34.59	29.91	+1.125	5.226	4.5	25.552	1.72	6.0	42.093	1.25	3	19.526
37 × 32	13	32.5	28.146	36.5	31.5	36.59	31.91	+0.875	4.937	4.5	27.308	1.80	5.5	42.764	1.30	3	19.392
38 × 33	14	35.0	30.311	37.5	32.5	37.59	32.91	+0.125	4.071	4.5	28.316	2.26	5.0	43.096	1 43	3	18.759
40 × 35	14	35.0	30.311	39.5	34.5	39.59	34.91	+1.125	5.226	4.5	30.552	1.72	6.0	47.204	1.28	3	19.759
42 × 37	15	37.5	32.476	41.5	36.5	41.59	36.91	+0.875	4.937	4.5	32.340	1.79	5.5	47.881	1.33	3	19.625
45 × 40	16	40.0	34.641	44.5	39.5	44.59	39.91	+1.125	5.226	4.5	35.552	1.73	5.5	51.035	1.33	4	26.793
47 × 42	17	42.5	36.806	46.5	41.5	46.59	41.91	+0.875	4.937	4.5	37.365	1.78	5.5	52.974	1.36	4	26.660
48 × 43	18	45.0	38.971	47.5	42.5	47.59	42.91	+0.125	4.071	4.5	38.387	2.07	5.0	53.156	1.47	4	26.026
50 × 45	18	43.0	38.971	49.5	44.5	49.59	44.91	+1.125	5.226	4.5	40.552	1.73	5.5	56.100	1.36	4	27.026
(52 × 47)	19	47.5	41.136	51.5	46.5	51.59	46.91	+0.875	4.937	4.5	42.384	1.78	5.5	58.052	1.38	4	29.892
55 × 50	20	50.0	43.301	54.5	49.5	54.59	49.91	+1.125	5.226	4.5	45.552	1.73	5.5	61.157	1.38	4	27.259
(58 × 53)	22	55.0	47.631	57.5	52.5	57.60	52.90	+0.125	4.071	4.5	48.424	1.99	5.0	63.198	1.51	4	26.491
60 × 55	22	55.0	47.631	59.5	54.5	59.60	54.90	+1.125	5.226	4.5	50.552	1.73	5.5	66.206	1.40	5	34.193
(62 × 57)	23	57.5	49.796	61.5	56.5	61.60	56.10	+0.875	4.937	4.5	52.413	1.77	5.0	66.846	1.45	5	34.160
65 × 60	24	60.0	51.962	64.5	59.5	64.60	56.90	+1.125	5.226	4.5	55.552	1.73	5.0	69.924	1.44	5	34.526
(68 × 63)	26	65.0	56.292	67.5	62.5	67.60	59.90	+0.125	4.071	4.5	58.448	1.94	5.0	73.229	1.53	5	33.759
70 × 65	26	65.0	56.292	69.5	64.5	69.60	64.90	+1.125	5.226	4.5	60.552	1.73	5.0	74.954	1.46	5	34.759
(72 × 67)	27	67.5	58.457	71.5	66.5	71.60	66.90	+0.875	4.937	4.5	62.434	1.77	5.0	76.920	1.48	5	34.625
75 × 70	28	70.0	60.622	74.5	69.5	74.60	69.90	+1.125	5.226	4.5	65.552	1.73	5.0	79.981	1.47	6	41.793
(78 × 73)	30	75.0	64.952	77.5	72.5	77.60	72.90	+0.125	4.071	4.5	62.464	1.90	5.0	83.253	1.55	6	41.026
80 × 75	30	75.0	64.952	79.5	74.5	79.60	74.90	+1.125	5.226	4.5	70.552	1.73	5.0	85.004	1.48	6	42.026
(82 × 77)	31	77.5	67.117	81.5	77.5	81.60	77.90	+0.875	4.937	4.5	72.449	1.76	5.0	86.978	1.50	6	41.892
85 × 80	32	80.0	69.282	84.5	79.5	84.60	79.90	+1.125	5.226	4.5	75.552	1.73	5.0	90.026	1.49	6	42.259
(88 × 83)	34	85.0	73.612	87.5	82.5	87.60	82.90	+0.125	4.071	4.5	78.476	1.88	5.0	93.273	1.57	6	41.491
90 × 85	34	85.0	73.612	89.5	84.5	89.60	84.90	+1.125	5.226	4.5	80.552	1.73	5.0	95.045	1.50	7	49.293
(92 × 87)	35	87.5	75.777	91.5	86.5	91.60	86.90	+0.875	4.937	4.5	82.461	1.76	5.0	97.024	1.52	7	49.160
95 × 90	36	90.0	77.942	94.5	89.5	94.60	89.90	+1.125	5.226	4.5	85.552	1.73	5.0	100.063	1.51	7	49.526
(98 × 93)	38	95.0	82.272	97.5	92.5	97.60	92.90	+0.125	4.071	4.5	88.485	1.86	5.0	103.288	1.58	7	48.759
100 × 95	38	95.0	82.272	99.5	94.5	99.60	94.60	+1.125	5.226	4.5	90.552	1.73	5.0	105.079	1.52	7	49.759
105 × 100	40	100.0	86.603	104.5	99.5	104.60	99.90	+1.125	5.226	4.5	95.552	1.73	5.0	110.094	1.53	8	56.793

Note: Values within brackets are nonpreferred.

Particular	Formula	
The value of tooth thickness and space width of spline	$l_o = s_o = m\,\dfrac{\pi}{2} + 2xm\tan\alpha$	(17-32)

PINS

Taper pins

The diameter at small end (Figs. 17-6 and 17-7, Tables 17-16 and 17-17)	$d_{ps} = d_{pl} - 0.0208l$	(17-33)
The mean diameter of pin	$d_m = 0.20D$ to $0.25D$	(17-34)

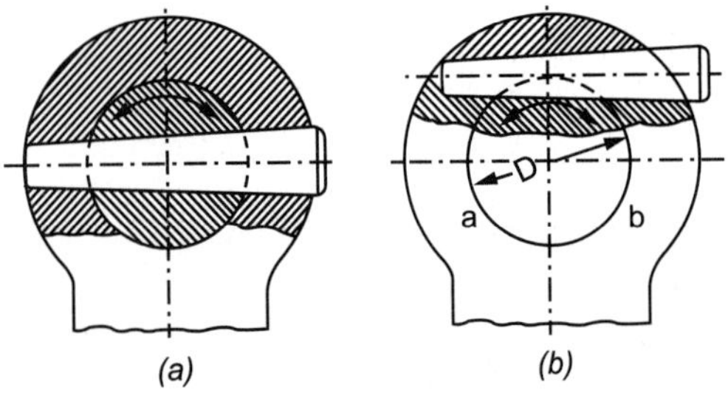

FIGURE 17-6 Tapered pin.

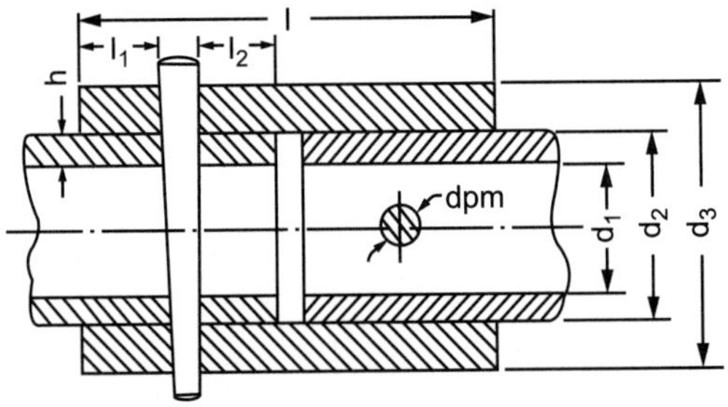

FIGURE 17-7 Sleeve and tapered pin joint for hollow shafts.

Sleeve and taper pin joint (Fig. 17-7)

AXIAL LOAD

The axial stress induced in the hollow shaft (Fig. 17-7) due to tensile force F	$\sigma = \dfrac{F}{\frac{\pi}{4}(d_2^2 - d_1^2) - 2(d_2 - d_1)d_m}$	(17-35)
The bearing stress in the pin due to bearing against the shaft an account of force F	$\sigma_c = \dfrac{F}{2(d_2 - d_1)d_m}$	17-36)
The bearing stress in the pin due to bearing against the sleeve	$\sigma_c = \dfrac{F}{2(d_3 - d_2)d_m}$	(17-35)
The shear stress in pin	$\tau = \dfrac{2F}{\pi d_m^2}$	(17-38)
The shearing stress due to double shear at the end of hollow shaft	$\tau = \dfrac{F}{2(d_2 - d_1)l_2}$	(17-39)
The shear stress due to double shear at the sleeve end	$\tau = \dfrac{F}{2(d_3 - d_2)l_1}$	(17-40)

TABLE 17-16
Dimensions (in mm) for cylindrical pins

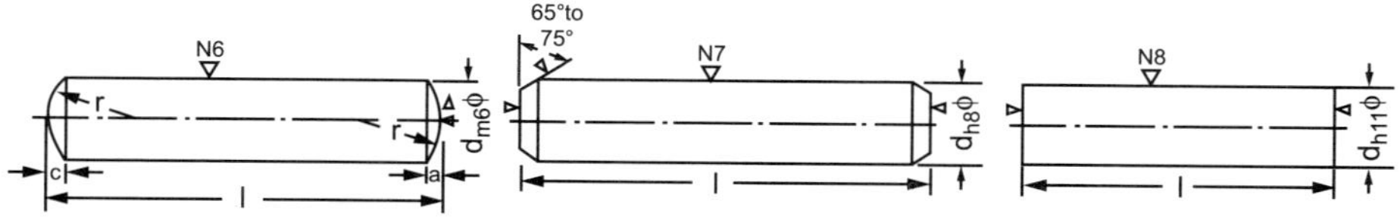

		Nominal diameter, d_{nom}, mm															
		1.5	2	2.5	3	4	5	6	8	10	12	16	20	25	32	40	50
d_{m6}	Max	1.61	2.01	2.51	3.01	4.01	5.01	6 01	8.02	10.02	12.02	16.02	20.02	25.02	32.02	40.02	50.02
	Min	1.60	2.00	2.50	3.00	4.00	5.00	6.00	8.01	10.01	12.01	16.01	20.01	25.01	32.01	40.01	50.01
d_{h6}	Max	1.60	2.00	2.50	3.00	4.00	5.00	6.00	8.00	10.00	12.00	16.00	20.00	25.00	32.00	40.00	50.00
	Min	1.59	1.99	2.49	2.99	3.98	4.98	5.98	7.98	9.98	11.97	15.97	19.97	24.97	31.96	39.96	49.96
d_{h11}	Max	1.60	2.00	1.50	3.00	4.00	5.00	6.00	8.00	10.00	12.00	16.00	20.00	25.00	32.00	40.00	50.00
	Min	1.54	1.94	2.44	2.94	3.92	4.92	5.92	7.91	9.91	11.89	15.89	19.87	24.87	31.84	39.84	49.84
	a_{max}	0.20	0.25	0.30	0.40	0.50	0.63	0.80	1.00	1.20	1.60	2.00	2.50	3.00	4.00	5.00	6.30
	r_{nom}	1.50	2.00	2.50	3.00	4.00	5.00	6.00	8.00	10.00	12.00	16.00	20.00	25.00	32.00	40.00	50.00
	c	0.30	0.35	0.40	0.50	0.63	0.80		1.60	2.00	2.50	3.00	3.50	4.00	5.00	6.30	8.00

Source: IS 2393, 1980.

TABLE 17-17
Dimensions (in mm) for solid and split taper pins

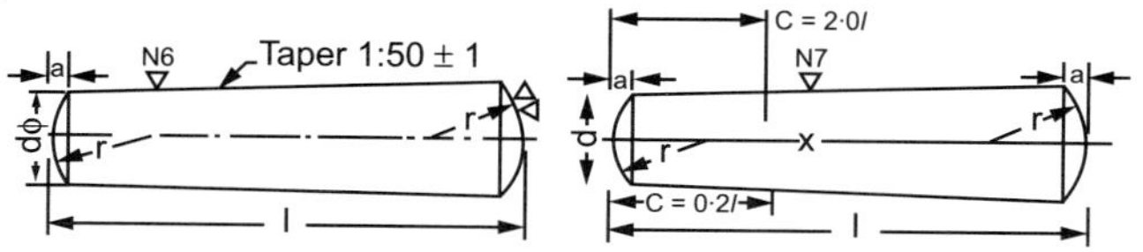

(r = d appro.)
split taper pin

Type A

(r = d appro. c = split length = 0.2l)
split taper pin

Type B

	d_{nom}	1.5	2	2.5	3	4	5	6	8	10	12	16	20	25	32	40	50
d_{h10}	Max	1.50	2.00	2.50	3.00	4.00	5.00	6.00	8.00	10.00	12.00	16.00	20.00	25.00	32.00	40.00	50.00
	Min	1.46	1.96	2.46	2.94	3.95	4.95	5.95	7.94	9.94	11.93	15.63	19.92	24.92	31.90	39.90	49.90
a		0.20	0.25	0.30	0.35	0.40	0.63	0.80	1.00	1.20	1.60	2.00	2.50	3.00	4.00	5.00	6.30

Source: IS 549, 1974.

Particular	Formula	
The axial stress in the sleeve	$\sigma = \dfrac{F}{\dfrac{\pi}{4}(d_3^2 - d_2^2) - 2(d_3 - d_2)d_m}$	(17-41)
TORQUE		
The shear due to twisting moment applied	$\tau = \dfrac{M_t}{\dfrac{\pi}{4} d_m^2 d_2}$	(17-42)
For the design of hollow shaft subjected to torsion	Refer to Chapter 14.	

Taper joint and nut

Particular	Formula	
The tensile stress in the threaded portion of the rod (Fig. 17-8) without taking into consideration stress concentration	$\sigma_t = \dfrac{F}{\dfrac{\pi}{4} d_c^2}$	(17-43)

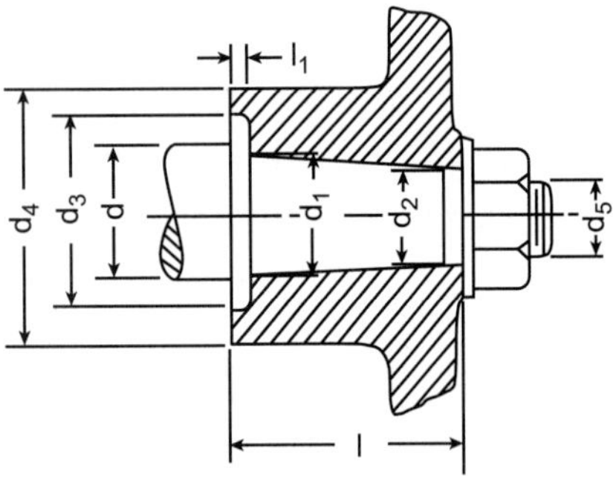

FIGURE 17-8 Tapered joint and nut.

Particular	Formula	
The bearing resistance offered by the collar	$\sigma_c = \dfrac{F}{\dfrac{\pi}{4}(d_3^2 - d_2^2)}$	(17-44)
The diameter of the taper d_2	$d_2 > d_{nom}$	(17-45)
Provide a taper of 1 in 50 for the length $(l - l_1)$		

Knuckle joint

Particular	Formula	
The tensile stress in the rod (Fig. 17-9)	$\sigma_t = \dfrac{4F}{\pi d^2}$	(17-46)
The tensile stress in the net area of the eye	$\sigma_t = \dfrac{F}{(d_4 - d_2)b}$	(17-47)
Stress in the eye due to tear of	$\sigma_{tn} = \dfrac{F}{b(d_4 - d_2)}$	(17-48)

Particular	Formula

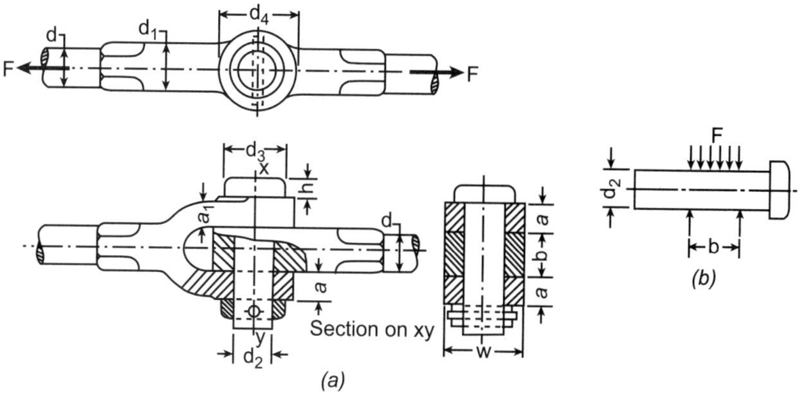

FIGURE 17-9 Knuckle joint for round rods.

Particular	Formula	
Tensile stress in the net area of the fork ends	$\sigma_i = \dfrac{F}{2a(d_4 - d_2)}$	(17-49)
Stress in the fork ends due to tear of	$\sigma_{tr} = \dfrac{F}{2a(d_4 - d_2)}$	(17-50)
Compressive stress in the eye due to bearing pressure of the pin	$\sigma_e = \dfrac{F}{d_2 b}$	(17-51)
Compressive stress in the fork due to the bearing pressure of the pin	$\sigma_c = \dfrac{F}{2d_2 a}$	(17-52)
Shear stress in the knuckle pin	$\tau = \dfrac{2F}{\pi d_2^2}$	(17-53)
The maximum bending moment, Fig. 17-9 (panel *b*)	$M_b = \dfrac{Fb}{8}$	(17-54)
The maximum bending stress in the pin, based on the assumption that the pin is supported and loaded as shown in Fig. 17-9b and that the maximum bending moment M_b occurs at the center of the pin	$\sigma_b = \dfrac{4Fb}{\pi d_2^3}$	(17-55)
The maximum bending moment on the pin based on the assumption that the pin supported and loaded as shown in Fig. 17-10b, which occurs at the center of the pin	$M_b = \dfrac{F}{2}\left(\dfrac{b}{4} + \dfrac{a}{3}\right)$ (approx.)	(17-56)
The maximum bending stress in the pin based on the assumption that the pin is supported and loaded shown in Fig. 17-10b	$\sigma_b = \dfrac{4(3b + 4a)F}{3\pi d_2^3}$	(17-57)

Particular	Formula	
COTTER		
The initial force set up by the wedge action	$F = 1.25Q$	(17-58)
The force at the point of contact between cotter and the member perpendicular to the force F	$H = F\tan(\alpha + \theta)$	(17-59)
The thickness of cotter	$t = 0.4D$	(17-60)
The width of the cotter	$b = 4t = 1.6D$	(17-61)
Cotter joint		
The axial stress in the rods (Fig. 17-10)	$\sigma = \dfrac{4F}{\pi d^2}$	(17-62)
Axial stress across the slot of the rod	$\sigma = \dfrac{4F}{\pi d_1^2 - 4d_1 t}$	(17-63)
Tensile stress across the slot of the socket	$\sigma = \dfrac{4F}{\pi(d_3^2 - d_1^2) - 4t(d_3 - d_1)}$	(17-64)
The strength of the cotter in shear	$F = 2bt\tau$	(17-65)
Shear stress, due to the double shear, at the rod end	$\tau = \dfrac{F}{2ad_1}$	(17-66)
Shear stress induced at the socket end	$\tau = \dfrac{F}{2c(d_4 - d_1)}$	(17-67)
The bearing stress in collar	$\sigma_c = \dfrac{4F}{\pi(d_2^2 - d_1^2)}$	(17-68)
Crushing strength of the cotter or rod	$F = d_1 t\sigma_c$	(17-69)

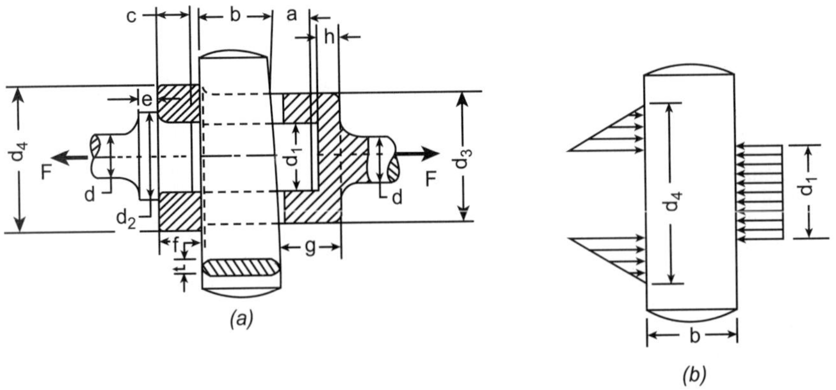

FIGURE 17-10 Cotter joint for round rods.

Particular	Formula	
Crushing stress induced in the socket or cotter	$\sigma_c = \dfrac{F}{(d_4 - d_1)t}$	(17-70)
The equation for the crushing resistance of the collar	$F = \dfrac{\pi(d_2^2 - d_1^2)}{4}\sigma_c$	(17-71)
Shear stress induced in the collar	$\tau = \dfrac{F}{\pi d_1 e}$	(17-72)
Shear stress induced in the socket	$\tau = \dfrac{F}{\pi d_1 h}$	(17-73)
The maximum bending stress induced in the cotter assuming that the bearing load on the collar in the rod end is uniformly distributed while the socket end is uniformly varying over the length as shown in Fig. 17-10*b*	$\sigma_b = \dfrac{F(d_1 + 2d_4)}{4tb^2}$	(17-74)

Gib and cotter joint (Fig. 17-11)

The width *b* of both the Gib and Cotter is the same as far as a cotter is used by itself for the same purpose (Fig. 17-11). The design procedure is the same as done in cotter joint Fig. 17-10.

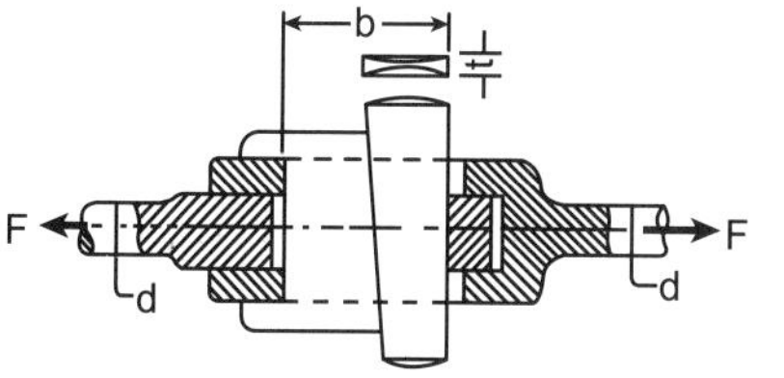

FIGURE 17-11 Gib and cotter joint for round rods.

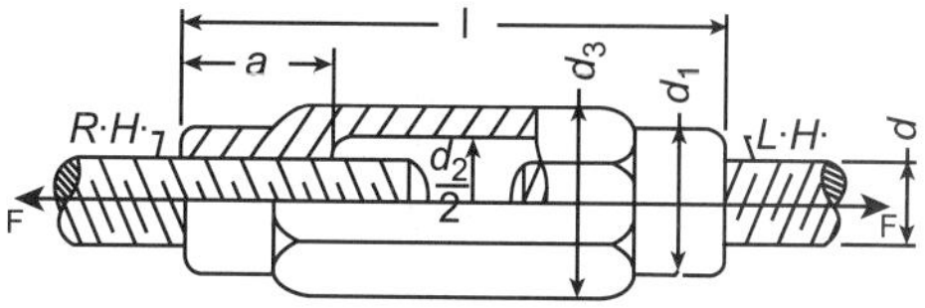

FIGURE 17-12 Coupler or turn buckle.

Threaded joint

COUPLER OR TURN BUCKLE

Particular	Formula	
Strength of the rods based on core diameter d_c, (Fig. 17-12)	$F = \dfrac{\pi}{4} d_c^2 \sigma_t$	(17-75)
The resistance of screwed portion of the coupler at each end against shearing	$F_\mu = \pi a d \tau$	(17-76)
From practical considerations the length *a* is given by	$a = d$ to 1.25d for steel nuts	(17-77a)
	$a = 1.5d$ to $2d$ for cast iron	(17-77b)
The strength of the outside diameter of the coupler at the nut portion	$F = \dfrac{\pi}{4}(d_1^2 - d^2)\sigma_t$	(17-78)

Particular	Formula	
The outside diameter of the turn buckle or coupler at the middle is given by the equation	$F = \frac{\pi}{4}(d_3^2 - d_2^2)\sigma_t$	(17-79)
The total length of the coupler	$l = 6d$	(17-80)

REFERENCES

1. Maleev, V. L., and J. B. Hartman, *Machine Design*, International Textbook Company, Scranton, Pennsylvania, 1954.
2. Shigley, J. E., and L. D. Mitchell, *Mechanical Engineering Design*, McGraw-Hill Book Company, New York, 1983.
3. Faires, V. M., *Design of Machine Elements*, The Macmillan Company, New York, 1965.
4. Lingaiah, K., and B. R. Narayana Iyengar, *Machine Design Data Handbook*, Engineering College Cooperative Society, Bangalore, India, 1962.
5. Lingaiah, K., and B. R. Narayana Iyengar, *Machine Design Data Handbook*, Vol. I (*SI and Customary Metric Units*), Suma Publishers, Bangalore, India, 1986.
6. Lingaiah, K., *Machine Design Data Handbook*, Vol. II (*SI and Customary Metric Units*), Suma Publishers, Bangalore, India, 1986.
7. Juvinall, R. C., *Fundamentals of Machine Component Design*, John Wiley and Sons, New York, 1983.
8. Deutschman, A. D., W. J. Michels, and C. E. Wilson, *Machine Design—Theory and Practice*, Macmillan Publishing Company, New York, 1975.
9. Bureau of Indian Standards.
10. *SAE Handbook*, 1981.

CHAPTER

18

THREADED FASTENERS AND SCREWS FOR POWER TRANSMISSION

SYMBOLS[5,6,7]

A_b	area of cross section of bolt, m^2 (in^2)
A_{br}	area of base of preloaded bracket, m^2 (in^2)
A_c	core area of thread, m^2 (in^2)
A_g	loaded area of gasket, m^2 (in^2)
A_r	stress area, m^2 (in^2)
A_τ	shear area, m^2 (in^2)
d	nominal diameter of screw m (in)
	major diameter of external thread (bolt), m (in)
d_2	pitch diameter of external thread (bolt), m (in)
d_1	minor diameter of external thread (bolt), m (in)
d_c	mean diameter of thrust collar, m (in)
$d_m = d_2$	mean diameter of square threaded power screw, m (in)
D	diameter of shaft, m (in)
	major diameter of internal thread (nut), m (in)
D_1	minor diameter of internal thread (nut), m (in)
D_2	pitch diameter of internal thread (nut), m (in)
D_b	diameter of bolt circle, m (in)
D_i	inside diameter of a pressure vessel or cylinder, m (in)
	mean diameter of inside screw of differential or compound screw, m (in)
D_o	mean diameter of outside screw of differential or compound screw, m (in)
e	eccentricity, m (in)
E_b, E_g	moduli of elasticity of bolt and gasket, respectively, GPa (Mpsi)
F	permissible load on bolt, kN (lbf)
	tightening load on the nut, kN (lbf)
F_a	applied or external load, kN (lbf)
F_f	final load on the bolt, kN (lbf)
F_i	initial load due to tightening, kN (lbf)
	preload in each bolt, kN (lbf)
F_t	tangential force, kN (lbf)
h	thickness of a pressure vessel, m (in)
	thickness of a cylinder, m (in)

h_2	thickness of the flange of the cylindrical pressure vessel, m (in)
h_o	depth of tapped hole (Fig. 18-1), m (in)

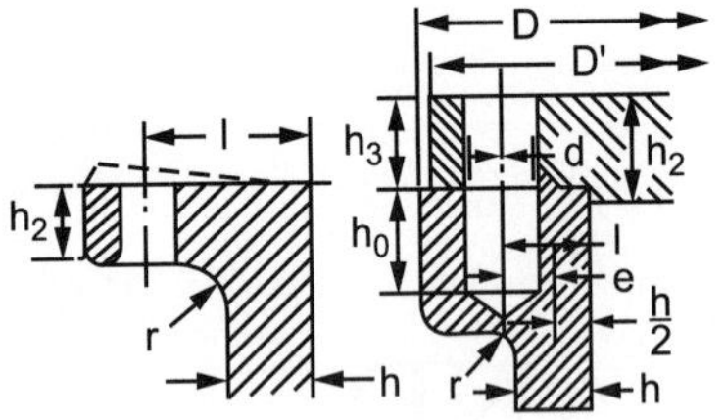

FIGURE 18-1 Flanged bolted joint.

i	number of threads in a nut
	number of bolts
I	moment of inertia of bracket base, area (Fig. 18-6), m^4 or cm^4 (in^4)
K	constant (Eq. (18-4a))
K_σ	stress concentration factor
l	lever arms (with suffixes), m (in)
	distance from the inside edge of the cylinder to the center line of bolt, m (in)
	lead, m (in)
l_c	required length of engagement of screw or nut (also with suffixes), m (in)
l_g	gasket thickness, m (in)
L	length of bolt nut to head (Fig. 18-2), m (in)
M_b	bending moment, N m (lbf in)
M_t	twisting moment, N m (lbf in)
n	factor of safety
p	pressure, MPa (psi)
p_c	circular pitch of bolts or studs on the bolt circle of a cylinder cover, m (in)
P	pitch of thread, m (in)
t	thread thickness at major diameter, m (in)
t_1	thread thickness at minor diameter, m (in)
W	axial load, kN (lbf)
α	helix angle, deg
α_o, α_i	respective helix angles of outside and inside screws of differential or compound screws, deg
ϕ	friction angle, deg
θ	half apex angle, deg
μ	coefficient of friction between nut and screw
μ_c	coefficient of collar friction
μ_i, μ_o	respective coefficient of friction in case of differential or compound screw
η	efficiency
σ	stress (normal), MPa (psi)
σ_a	allowable stress, MPa (psi)
σ_b	bending stress, MPa (psi)
σ_b'	bending stress due to eccentric load [Eq. (18-61)]
	allowable bearing pressure between threads of nut and screw, MPa (psi)

σ_c	compressive stress, MPa (psi)
σ_d	design stress, MPa (psi)
σ_w	working stress, MPa (psi)
τ	applicable shear stress, MPa (psi)
τ_a	allowable shear stress, MPa (psi)
τ_w	permissible working shear stress, MPa (psi)

SUFFIXES

v	vertical
h	horizontal

Particular	Formula

SCREWS

The empirical formula for the proper size of a set screw

$$d = \frac{D}{8} + 8\,\text{mm where } D \text{ in mm} \qquad (18\text{-}1)$$

The maximum safe holding force of a set screw

$$F = 54{,}254d^{2.31} \qquad \textbf{SI} \quad (18\text{-}2a)$$

where F in kN and d in m

$$F = 2500d^{2.31} \qquad \textbf{USCS} \quad (18\text{-}2b)$$

where F in lbf and d in in

Applied torque

$$M_t = 0.2F_a \text{ nominal diameter of bolt)} \qquad (18\text{-}3)$$

Gasket joint (Fig. 18-2)

Final load on the bolt

$$F_f = KF_a + F_i \qquad (18\text{-}4)$$

$$\text{where } K = \left[\frac{\dfrac{E_b A_b}{L}}{\dfrac{E_b A_b}{L} + \dfrac{E_g A_g}{l_g}}\right] \qquad (18\text{-}4a)$$

Refer also to Table 18-1 for values of K

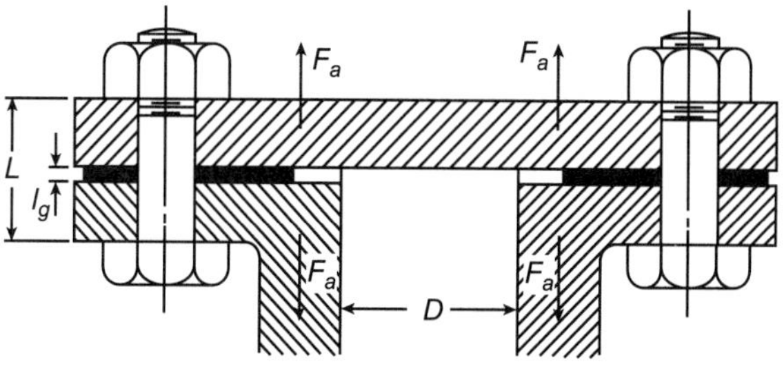

FIGURE 18-2 Gasket joint.

Particular	Formula

TABLE 18-1
Values of K for use in Eq. (18-4)

Type of joint	K
Soft, elastic gasket with studs	1.00
Soft gasket with through bolts	0.90
Copper asbestos gasket	0.60
Soft copper corrugated gasket	0.40
Lead gasket with studs	0.10
Narrow copper ring	0.01
Metal-to-metal joint	0.00

According to Bart, the tightening load for a screw of a steamtight, metal-to-metal joint

$$F = 2804.69d \quad \textbf{SI} \quad (18\text{-}5a)$$

where F in kN and d in m

$$F = 1600d \quad \textbf{USCS} \quad (18\text{-}5b)$$

where F in lbf and d in in

Tightening load for screw of a gasket joint

$$F = 1402.34d \quad \textbf{SI} \quad (18\text{-}6a)$$

where F in kN and d in m

$$F = 8000d \quad \textbf{USCS} \quad (18\text{-}6b)$$

where F in lbf and d in in

Cordullo's equation for the tightening load on the nuts

$$F = \sigma_w(0.55d^2 - 6.45 \times 10^{-3}d) \quad \textbf{SI} \quad (18\text{-}7a)$$

where F in kN, σ_w in MPa, and d in m

$$F = \sigma_w(0.55d^2 - 0.036d) \quad \textbf{USCS} \quad (18\text{-}7b)$$

where F in lbf, σ_w in psi, and d in in

Bolted joints (Fig. 18-2)

The flange thickness of the cylinder or pressure vessel

$$h_2 = 1.25d \text{ to } 1.5d \nless 1.1h \text{ to } 1.25h \quad (18\text{-}8)$$

The bolt diameter

$$d = 0.67h \text{ to } 0.8h \quad (18\text{-}9)$$

Circular pitch of the bolts or studs on the cylinder cover to ensure water and steamtight joint

$p_c = 7d$ for pressure from 0 to 0.33 MPa (0 to 48 psi) as per American Navy Standards (18-10)

$p_c = 3.5d$ for pressure from 1.2 MPa (175 psi) to 1.37 MPa (200 psi) (18-11)

$p_c = 3d$ for tight joint (18-12)

Particular	Formula
The average stress for screw for sizes from 12.5 to 75 mm	$\sigma_{av} = \frac{490.33}{d}$ **SI** (18-13a) where σ_{av} in MPa and d in m $\sigma_{av} = \frac{2,800,000}{d}$ **USCS** (18-13b) where σ_{av} in psi and d in in
Unwin's formula for allowable stresses in bolts of ordinary steel to make a fluidtight joint	$\sigma_d = 17,537.4d^2 + 11$ for rough joint **SI** (18-14a) where σ_d in MPa and d in m $\sigma_d = 6030d^2 + 1600$ **USCS** (18-14b) where σ_d in psi and d in in $\sigma_s = 33,828.9d^2 + 17.3$ for faced joint **SI** (18-14c) where σ_d in MPa and d in m $\sigma_d = 3070d^2 + 2500$ **USCS** (18-14d) where σ_d in psi and d in in

TENSION BOLTED JOINT UNDER EXTERNAL LOAD

Spring constant of clamped materials and bolt (Fig. 18-3A)

Particular	Formula
The spring constant or stiffness of the threaded and unthreaded portion of a bolt is equivalent to the stiffness of two springs in series.	$\frac{1}{k} = \frac{1}{k_1} + \frac{1}{k_2}$ (18-15a)
The basic equations for deflection (δ), and spring constant (k) of a tension bar/bolt subject to tension load.	$\delta = \frac{Fl}{AE}$ (18-15b) $k = \frac{F}{\delta} = \frac{AE}{l}$ (18-15c)
The effective spring constant/total spring rate in case of long bolt consisting of the threaded and unthreaded portion having different area of cross-sections, the clamped two or more materials of two or more different elasticities which act as spring with different stiffness sections in series.	$\frac{1}{k_{eff}} = \frac{1}{k_1} + \frac{1}{k_2} + \frac{1}{k_3} + \cdots + \frac{1}{k_n}$ (18-15d)
Spring constant of the clamped material	$k_m = \frac{A_m E_m}{l} = \frac{\pi D_{eff}^2}{4} \frac{E_m}{l}$ (18-15e)
Spring constant of the threaded fastener	$\frac{1}{k_b} = \frac{l_t}{A_t E_b} + \frac{l - l_t}{A_b E_b} = \frac{l_t}{A_t E_b} + \frac{l_{unt}}{A_b E_b}$ (18-15f)

Particular	Formula

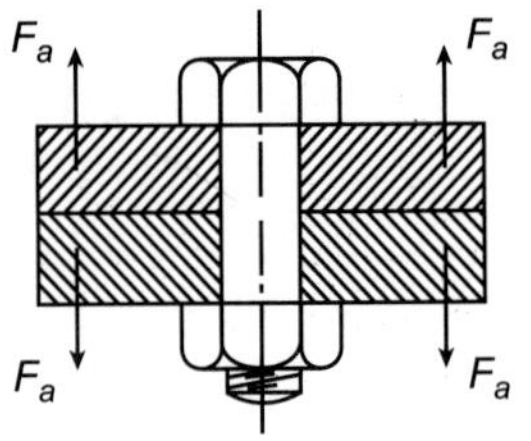

FIGURE 18-3A

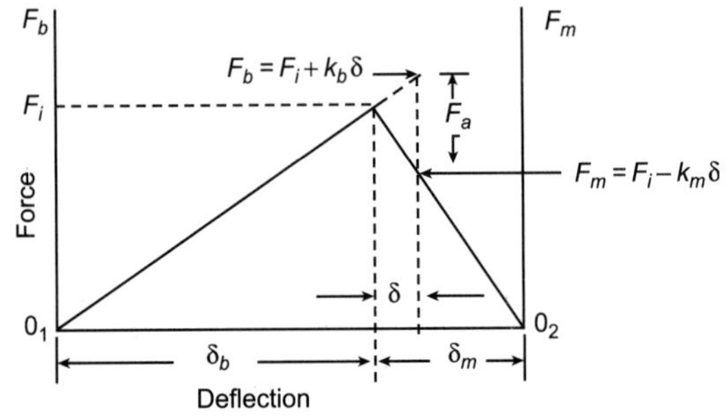

FIGURE 18-3B Bolt joint diagram due to external load acting on the joint.

Approximate effective area of clamped material

$$A_m = \frac{\pi}{4}(D_{eff}^2 - d^2)$$

where

D_{eff} = effective diameter, m
d = round bolt of diameter equal to shank, m
l_t = threaded length of bolt, m
l_{unt} = unthreaded portion of bolt length, m

PRELOADED BOLT (Fig. 18-3B)

The external load

$$F_a = F_{ab} + F_{am} \tag{18-15g}$$

The bolted joint in Fig 18-3A subjected to external load F_a is such that the common deflection is given by

$$\delta = \frac{F_b}{k_b} = \frac{F_m}{k_m} \tag{18-15h}$$

The load shared by bolt

$$F_{ab} = \frac{k_b}{k_b + k_m} F_a \tag{18-15i}$$

The resultant/total bolt load

$$F_b = F_{ab} + F_i = F_i + k_b\delta = \frac{k_b}{k_b + k_m} F_a + F_i \tag{18-15j}$$

$$= CF_a + F_i \tag{18-15k}$$

The resultant load on the clamped material

$$F_m = F_{ab} - F_i = F_i - k_m\delta = \frac{k_m}{k_b + k_m} F_a - F_i \tag{18-15l}$$

$$= (1 - C)F_a - F_i$$

where C is called the joint constant or stiffness parameter

F_m = portion of load F_a taken by member/material, kN
F_t = preload, kN

Particular	Formula
The joint constant or stiffness parameter	$C = \frac{k_b}{k_b + k_m}$ or $C = \frac{1}{1 + \frac{k_m}{k_b}}$ (18-15m)
The preload to prevent joint separation occurs when $F_m = 0$	$F_i = (1 - C)F_{ao}$ where F_{ao} = external load that cause separation of joint
The external load required to separate joint	$F_{ao} = \frac{F_i}{1 - C}$ (18-15n)
The tensile stress in the bolt	$\sigma_b = \frac{F_b}{A_t} = \frac{CF_a}{A_t} + \frac{F_i}{A_t}$ (18-15o) where A_t = tensile stress area, m^2 or mm^2
Preload under static and fatigue loading as per the recommendation of R, B and W,[a] and Bowman	$F_t = \begin{cases} 0.75F_p & \text{for reused bolt connections} \\ 0.90F_p & \text{for permanent bolt connections} \end{cases}$ (18-15p) where F_p is proof load, N
The proof stress load that has to be used in Eq. (18-15p)	$F_p = A_t\sigma_{sp}$ (18-15q) where σ_{sp} = proof strength, taken from tables 18-5c and 18-5d $\sigma_{sp} \approx 0.85\sigma_{sy}$
The load factor	$n = \frac{F_{ao}}{F_a}$ or $F_{ao} = nF_a$ (18-15r)
The load factor guarding against joint separation	$n = \frac{F_i}{F_a(1 - C)}$ (8-15s)

GASKET JOINTS

For design of gasket bolted joint	Refer to Chapter 16 under *Bolt loads in gasket joints.*

PRELOADED BOLTS UNDER DYNAMIC LOADING

The mean forces felt by the bolt	$F_{mn} = \frac{F_b + F_i}{2}$ (18-15t)
The alternating forces felt by the bolt	$F_{al} = \frac{F_b - F_i}{2}$ (18-15u)

[a] Russel, Bardsall and Ward Corp., *Helpful Hints for Fastener Design and Application*, Mentor, Ohio 1976, p. 42.

Particular	Formula	
The stress due to the preload F_i	$\sigma_i = K_{fm} \dfrac{F_i}{A_i}$	(18-15v)
The fatigue safety factor by using modified Goodman criterion	$n_f = \dfrac{\sigma_{se}(\sigma_{sut} - \sigma_i)}{\sigma_{se}(\sigma_m - \sigma_i) + \sigma_{sut}\sigma_a}$	(18-15w)
The alternating component of bolt stress	$\sigma_a = \dfrac{F_b - F_i}{2A_t} = \dfrac{k_b}{k_b + k_m} \dfrac{F_a}{2A_t} = \dfrac{CF_a}{2A_t}$	(18-16a)
The mean stress	$\sigma_m = \sigma_a + \dfrac{F_i}{A_t} = \dfrac{CP}{2A_t} + \dfrac{F_i}{A_t}$	(18-16b)
The factor of safety according to the Goodman criterion	$n = \dfrac{\sigma_{sa}}{\sigma_a}$	(18-16c)
	$\sigma_{sa} = \sigma_{sm} - \dfrac{F_i}{A_t}$	(18-16d)
	$\sigma_{sm} = \sigma_{sut}\left(1 - \dfrac{\sigma_{sa}}{\sigma_{se}}\right)$	(18-16e)
Solving of Eqs. (18-16c) and (18-16d) simultaneously	$\sigma_{sa} = \dfrac{\sigma_{sut} - \dfrac{F_i}{A_t}}{1 + \dfrac{\sigma_{sut}}{\sigma_{se}}}$	(18-16f)
The factor of safety on the basis of yield strength	$n = \dfrac{\sigma_{sy}}{\sigma_{max}} = \dfrac{\sigma_{sy}}{\sigma_m + \sigma_a}$	(18-16g)
For specification of SAE, ASTM and ISO standard steel bolts	Refer to Tables 18-5c and 18-5d	
The depth of tapped hole (Fig. 18-2)	$h_o = 1.25d$ in steel castings	(18-16h)
	$h_o = 1.50d$ to $1.75d$ in cast iron	(18-17)
	$h_o = 1.75d$ to $2d$ in aluminum	(18-18)
The distance l from the inside edge of the cylinder to the center line of bolts (Fig. 18-2)	$l = 1.25d$ to $1.5d$	(18-19)
The diameter of bolt circle	$D_b = D_1 + 2d$	(18-20)
The safe load on each bolt	$F' = A_r\sigma_d$	(18-21)
The number of bolts	$i = \dfrac{\pi D_b^2 p}{4F'}$	(18-22)
Another expression for the number of bolts	$i = \dfrac{\pi D_b}{p_c}$	(18-23)

Particular	Formula

TABLE 18-2
Approximate bolt tension and torque values

Bolt size, mm	Minimum bolt tension		Equivalent torque	
	kN	lbf	kN m	lbf ft
12.7	51.2	11,500	1.353	1,000
15.9	76.9	17,300	2.442	1,800
19.6	113.9	25,600	4.835	3,570
22.2	139.7	31,400	6.374	4,700
25.4	189.1	42,500	9.620	7,090
21.6	225.4	50,600	13.013	9,600
31.8	286.9	64,500	18.289	13,500

TABLE 18-3
Load and working stress for metric coarse threads

Major diameter, d, mm	Stress area, A_r, mm^2	Design stress, σ_w		Permissible load	
		MPa	psi	kN	lbf
16	0.016	18.9	2,740	2.97	667
20	0.025	22.8	3,300	5.59	1,260
24	0.035	27.2	3,950	9.59	2,160
30	0.056	32.2	4,670	18.04	4,060
36	0.082	37.1	5,380	30.89	6,940
42	0.112	43.1	6,250	48.35	10,870
48	0.147	48.3	7,000	71.10	16,000
56	0.203	55.2	8,000	111.80	25,130

Stress in tensile bolt

Seaton and Routhwaite formula for working stress for bolt made of steel containing 0.08 to 0.25% carbon and with diameter of 20 mm and over

$$\sigma_w = C(A_r)^{0.418} \tag{18-24}$$

Refer to Table 18-2 for bolt tension and torque values and Table 18-3 for σ_w.

Applied load

$$F_a = C(A_r)^{1.418} \tag{18-25}$$

where

$C = 7.8 \times 10^8$ (5000) for carbon steel bolts of $\sigma_u = 414$ MPa (60 kpsi)
$= 23.3 \times 10^8$ (15,000) for alloy–steel bolts
$= 0.33 \times 10^8$ (1000) for bronze bolts

The values of C inside parentheses are for **US Customary System** units, and values without parentheses are for **SI** units.

Rotsher's pressure-cone method for stiffness calculation of Fastener[a]

The elongation of frustum of a cone (Fig. 18-3C)

$$\delta = \frac{F_a}{\pi E d \tan\alpha} \ln \frac{(2t\tan\alpha + D - d)(D + d)}{(2t\tan\alpha + D + d)(D - d)} \tag{19-25a}$$

The spring stiffness of the frustum

$$k = \frac{F_a}{\delta} = \frac{\pi E d \tan\alpha}{\ln \dfrac{(2t\tan\alpha + D - d)(D + d)}{(2t\tan\alpha + D + d)(D - d)}} \tag{18-25b}$$

[a] Courtesy: Shigley J. E. and C. R. Mischke, *Mechanical Engineering Design*, 5th Edn., McGraw-Hill Publishing Company, New York, 1989.

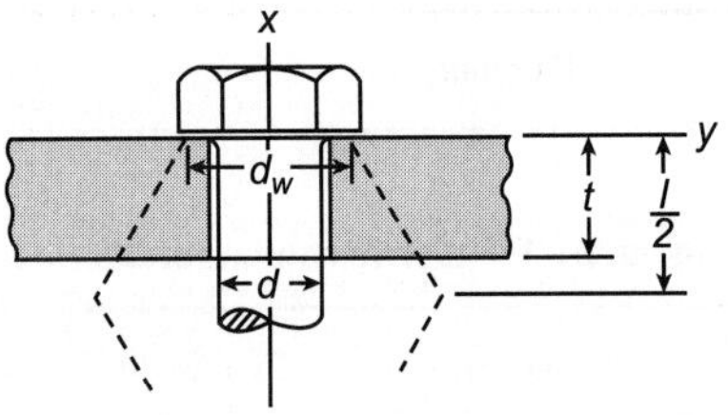

FIGURE 18-3C Compression of a member assumed to be confined to the frustum of a hollow cone.

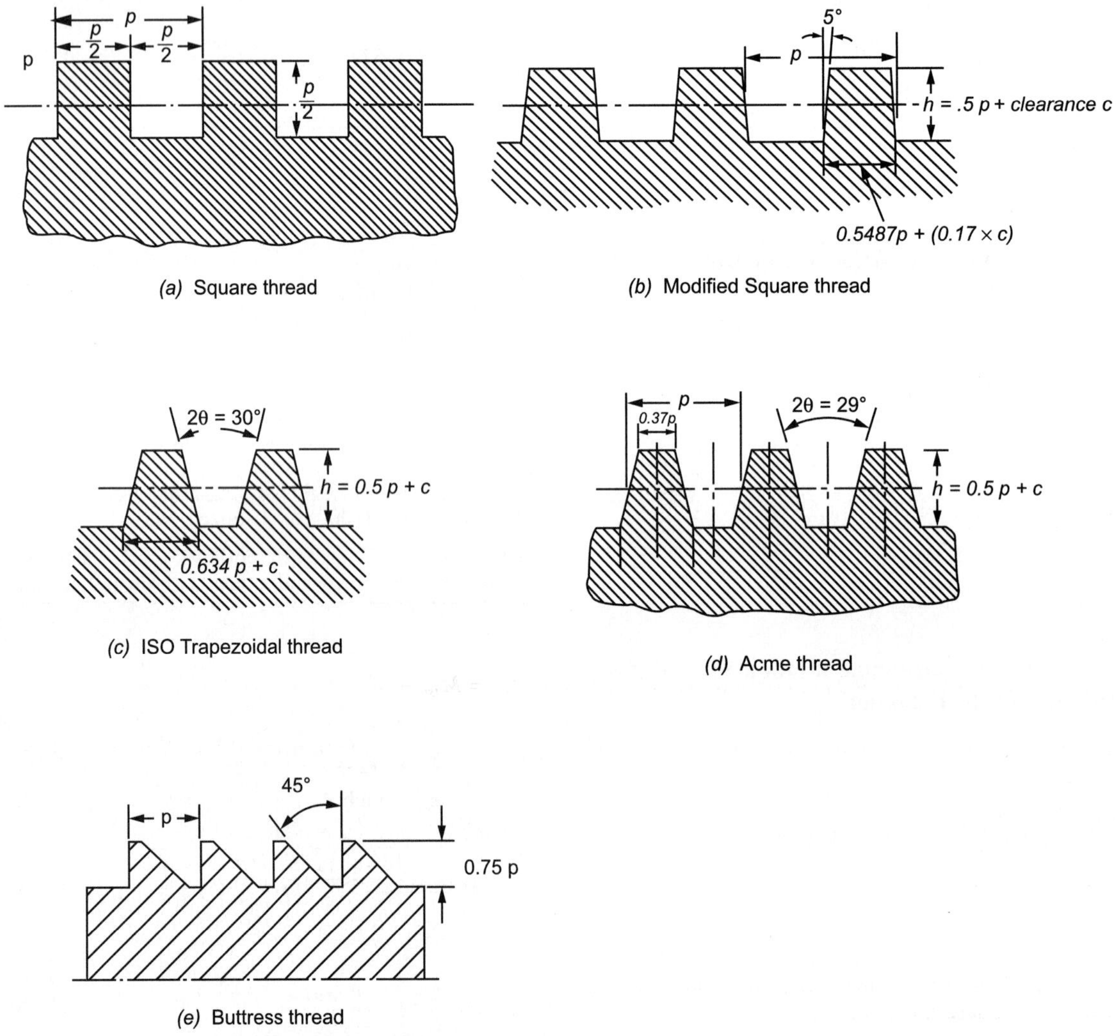

FIGURE 18-3D Forms of threads for power screw.

Particular	Formula

The spring stiffness of the frustum when cone angle of frustum $\alpha = 30^\circ$

$$k = \frac{0.577\pi dE}{\ln \dfrac{(1.15t + D - d)(D + d)}{(1.15t + D + d)D - d)}} \qquad \text{(18-35c)}$$

For the members of the joint having same modulus of elasticity E with symmetrical frusta back to back which constitute as two springs in series and using the grip as $l = 2t$ and d_w as the diameter of the washer face, the effective spring constant for the system.

$$k_e = \frac{\pi E d \tan\alpha}{2 \ln \dfrac{(l \tan\alpha + d_w - d)(d_w + d)}{(l \tan\alpha + d_w + d)(d_w - d)}} \qquad \text{(18-25d)}$$

The effective spring constant for the case of back to back cone frusta with a washer face $d_w = 1.5d$ and $\alpha = 30^\circ$ from Eq. (18-25d).

$$k_e = \frac{0.577\pi dE}{2 \ln \left(5 \dfrac{0.577l + 0.5d}{0.577l + 2.5d} \right)} \qquad \text{(18-25e)}$$

Power screw

The helix angle of a V-thread (Fig. 18-3E)

$$\alpha = \tan^{-1} \frac{l}{\pi d_2} \qquad \text{(18-26)}$$

The tangential force for a square thread at mean radius of screw

$$F_t = W \frac{\tan\alpha + \mu}{1 - \mu \tan\alpha} \qquad \text{(18-27)}$$

Refer to Table 18-4 for μ.

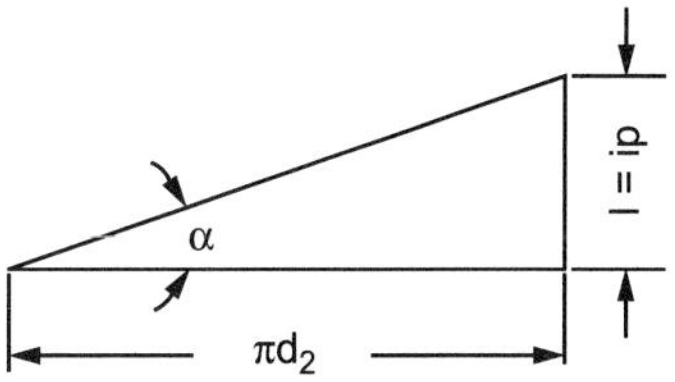

FIGURE 18-3E Helix angle of a single-start thread.

TABLE 18-4
Coefficient of friction for power screws

Lubricant	Coefficient of friction, μ
Machine oil and graphite	0.07
Lard oil	0.11
Heavy machine oil	0.14

Torque required to raise the load by a power screw

$$M_{tu} = M_{tsu} + M_{te}$$
$$= \frac{Wd_2}{2} \left(\frac{\mu \pi d_2 + l \cos\alpha}{\pi d_2 \cos\alpha - \mu l} \right) + \mu_c W \frac{d_c}{2} \qquad \text{(18-28)}$$

where d_2 = pitch diameter of thread

The tangential force for V-thread or angular thread at mean radius (Fig. 18-4)

$$F_t = W \frac{\tan\alpha + \dfrac{\mu}{\cos\theta}}{1 - \mu \dfrac{\tan\alpha}{\cos\theta}} \qquad \text{(18-28a)}$$

The total frictional torque including collar friction torque for square thread

$$M_t = W \left[\frac{d_2}{2} \left(\frac{\tan\alpha + \mu}{1 - \mu \tan\alpha} \right) + \mu_c \frac{d_c}{2} \right] \qquad \text{(18-29)}$$

Refer to Table 18-4 for μ and Table 18-5a for μ_c.

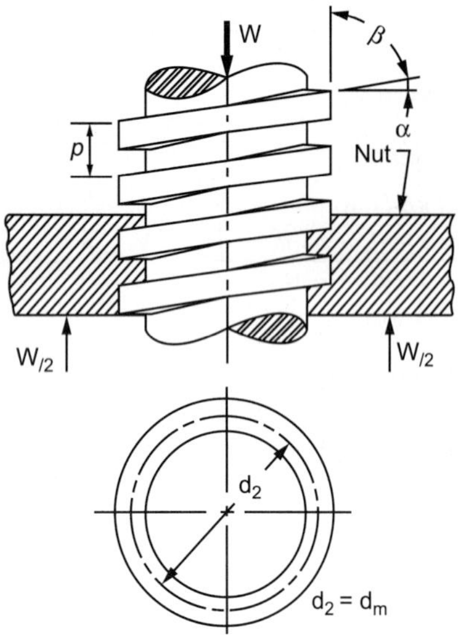

FIGURE 18-3F Power screw.

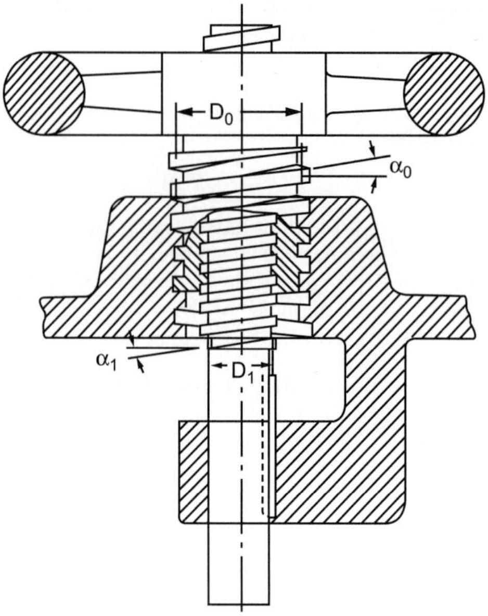

FIGURE 18-3G Differential screw.

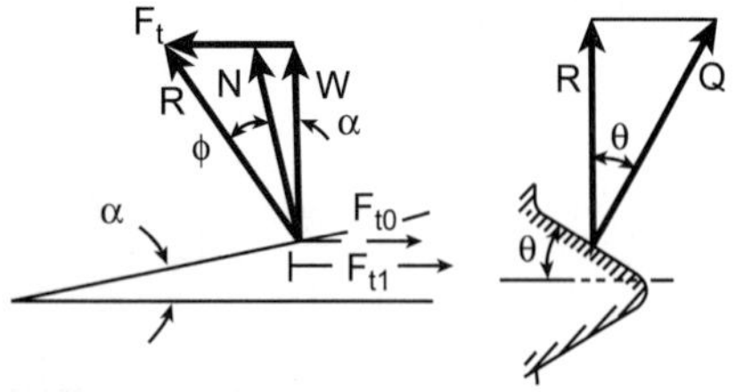

FIGURE 18-4 Forces acting on a triangular thread.

TABLE 18-5a
Coefficient of friction on thrust collar, μ_c

Material	Coefficient of running friction	Coefficient of starting friction
Soft steel on cast iron	0.121	0.170
Hardened steel on cast iron	0.092	0.147
Soft steel on bronze	0.084	0.101
Hardened steel on bronze	0.063	0.081

TABLE 18-5b
Torque factor K_μ for use in Eq. (18-30c)

Bolt condition	K_μ
Nonplated, black finish	0.30
Zinc-plated	0.20
Lubricated	0.18
Cadmium-plated	0.16
With Bowman anti-seize	0.12
With Bowman-grip nuts	0.09

Particular	Formula	
The total frictional torque for V-thread, including collar friction torque	$M_t = W\left[\frac{d_2}{2}\left(\frac{\tan\alpha + \frac{\mu}{\cos\theta}}{1-\mu\frac{\tan\alpha}{\cos\theta}}\right) + \mu_c\frac{d_c}{2}\right]$	(18-30a)
The mean diameter of collar*	$d_c = (d + 1.5d)/2$	(18-30b)
Substituting the value of d_c in Eq. (18-30a) and after simplifying	$M_t = K_\mu F_i d$	(18-30c)
	where K_μ is the torque factor	
	$W = F_i =$ preload, N (lbf)	
The torque factor	$K_\mu = \frac{d_2}{2d}\left(\frac{\tan\alpha + \frac{\mu}{\cos\theta}}{1-\mu\frac{\tan\alpha}{\cos\theta}}\right) + \mu_c \times 0.625$	(18-30d)
	where $d_2 = d_m$	
	Refer to Table 18-5b for K_μ.	
The efficiency of square thread neglecting collar friction	$\eta = \frac{\tan\alpha}{\tan(\alpha+\phi)} = \frac{Wl}{2\pi M_t}$	(18-31)
The efficiency formula for an angular-type thread with half apex angle θ and an allowance for nut or end friction on a radius r_c	$\eta = \frac{d_2\tan\alpha}{\frac{\tan\alpha + \mu/\cos\theta}{1-\mu\tan\alpha/\cos\theta}d_2 + \mu_c d_c}$	(18-32)
The efficiency formula for square thread	$\eta = \frac{d_2\tan\alpha}{\frac{\tan\alpha+\mu}{1-\mu\tan\alpha}d_2 + \mu_c d_c}$	(18-33)
	$\eta = \frac{l}{\pi[d_2\tan(\alpha+\phi) + \mu_c d_c]}$	(18-34)

LOADING

Lowering the load

The tangential force at mean or pitch radius $r_2 = r_m$	$F_t = W\tan(\phi+\alpha)$	(18-35)
The frictional torque at mean or pitch radius $r_2 = r_m$	$M_t = \frac{Wd_2}{2}\tan(\phi-\alpha)$	(18-36)
The condition for overhauling for square threads	$\tan\alpha \geq \frac{\mu d_2 + \mu_c d_c}{d_2 - \mu\mu_c d_c}$	(18-37)

* Since the flat faces of hexagonal nut is same as the diameter of washer face which is 1.5 times the nominal diameter d.

Particular	Formula

Differential screws (Fig. 18-3G)

The loading efficiency of a differential screw, not including the collar friction

$$\eta = \frac{D_o \tan\alpha_o - D_i \tan\alpha_i}{D_o \dfrac{\tan\alpha_o + \mu_o}{1 - \mu_o \tan\alpha_o} - D_i \dfrac{\tan\alpha_i + \mu_i}{1 - \mu_i \tan\alpha_i}} \qquad (18\text{-}38)$$

Compound screws

The loading efficiency of a compound screw, not including collar friction

$$\eta = \frac{D_o \tan\alpha_o + D_i \tan\alpha_i}{D_o \dfrac{\tan\alpha_o + \mu_o}{1 - \mu_o \tan\alpha_o} + D_i \dfrac{\tan\alpha_i + \mu_i}{1 - \mu_i \tan\alpha_i}} \qquad (18\text{-}39)$$

The number of threads necessary in the nut

$$i = \frac{4W}{\sigma'_b \pi (d^2 - d_i^2)} \qquad (18\text{-}40)$$

The length of nut

$$l_n = iP = \frac{4WP}{\sigma'_b \pi (d^2 - d_1^2)} \qquad (18\text{-}41)$$

TABLE 18-5c
Metric mechanical-property classes for steel bolts, screws, and studs[a]

Property class	Size range inclusive	Minimum proof strength, σ_{sp} MPa	Minimum tensile strength, σ_{st} MPa	Minimum yield strength, σ_{sy} MPa	Material	Head marking
4.6	M5–M36	225	400	240	Low or medium carbon	4.6
4.8	M1.6–M16	310	420	340	Low or medium carbon	4.8
5.8	M5–M24	380	520	420	Low or medium carbon	5.8
8.8	M16–M36	600	830	660	Medium carbon, Q and T	8.8
9.8	M1.6–M16	650	900	720	Medium carbon, Q and T	9.8
10.9	M5–M36	830	1040	940	Low-carbon martensite, Q and T	10.9
12.9	M1.6–M36	970	1220	1100	Alloy, Q and T	12.9

[a] The thread length for bolts and cap screws is

$$L_T = \begin{cases} 2d + 6 & L \le 125 \\ 2d + 12 & 125 < L \le 200 \\ 2d + 25 & L > 200 \end{cases}$$

where L is the bolt length. The thread length for structural bolts is slightly shorter than given above.

TABLE 18-5d
Grade identification marks and mechanical properties of bolts and screws

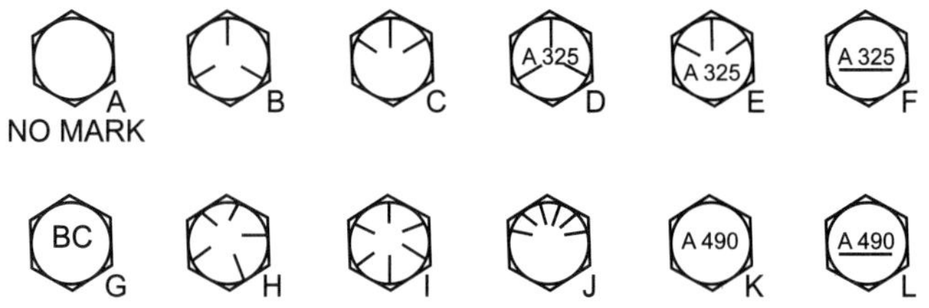

Identifier	Grade	Size range (in)	Min. strength (10^3 psi) Proof	Tensile	Yield	Material and treatment
A	SAE Grade 1	$\frac{1}{4}$ to $1\frac{1}{2}$	33	60	36	Low or medium carbon
	ASTM A307	$\frac{1}{4}$ to $1\frac{1}{2}$	33	60	36	Low carbon
	SAE Grade 2	$\frac{1}{4}$ to $\frac{3}{4}$	55	74	57	Low or medium carbon
		$\frac{7}{8}$ to $1\frac{1}{2}$	33	60	36	
	SAE Grade 4	$\frac{1}{4}$ to $1\frac{1}{2}$	65	115	100	Medium carbon, cold drawn
B	SAE Grade 5 and ASTM A449	$\frac{1}{4}$ to 1	85	120	92	Medium carbon, Q and T
	SAE Grade 5, ASTM A449	$1\frac{1}{8}$ to $1\frac{1}{2}$	74	105	81	
	ASTM A449	$1\frac{3}{4}$ to 3	55	90	58	
C	SAE Grade 5.2	$\frac{1}{4}$ to 1	85	120	92	Low-carbon martensite, Q and T
D	ASTM A325, Type 1	$\frac{1}{2}$ to 1	85	120	92	Medium carbon, Q and T
		$1\frac{1}{8}$ to $1\frac{1}{2}$	74	105	81	
E	ASTM A325, Type 2	$\frac{1}{2}$ to 1	85	120	92	Low carbon martensite, Q and T
		$1\frac{1}{8}$ to $1\frac{1}{2}$	74	105	81	
F	ASTM A325, Type 3	$\frac{1}{2}$ to 1	85	120	92	Weathering steel, Q and T
		$1\frac{1}{8}$ to $1\frac{1}{2}$	74	105	81	
G	ASTM A354, Grade BC	$\frac{1}{4}$ to $2\frac{1}{2}$	105	125	109	Alloy-steel, Q and T
		$2\frac{3}{4}$ to 4	95	115	99	
H	SAE Grade 7	$\frac{1}{4}$ to $1\frac{1}{2}$	105	133	115	Medium carbon alloy, Q and T
I	SAE Grade 8	$\frac{1}{4}$ to $1\frac{1}{2}$	120	150	130	Medium carbon alloy, Q and T
	ASTM A354, Grade BD	$\frac{1}{4}$ to $1\frac{1}{2}$	120	150	130	Alloy-steel, Q and T
J	SAE Grade 8.2	$\frac{1}{4}$ to 1	120	150	130	Low-carbon martensite, Q and T
K	ASTM A490, Type 1	$\frac{1}{2}$ to $1\frac{1}{2}$	120	150	130	Alloy-steel, Q and T
L	ASTM A490, Type 3	$\frac{1}{2}$ to $1\frac{1}{2}$	120	150	130	Weathering steel, Q and T

Particular	Formula
The required length of engagement for adequate shear strength (assuming that the load is distributed over the threads in contact)	$l_c = \frac{nPF}{A_\tau \tau}$ (18-42)
Neglecting the radial clearance between threads, or allowance at the major and minor diameters and considering the threads as a series of collars the equation for thread engagement	$l_{e(\text{screw})} = \frac{nPF}{\pi d_1 t_1 \tau_{(\text{screw})}}$ (18-43)
The normal length of thread engagement as per Indian standard	$l_{e(\text{nut})} = \frac{nPF}{\pi d t \tau_{(\text{nut})}}$ (18-44)
	$l_{eN(\text{min})} = 8.92Pd^{0.2}$ SI (18-45a) where l_{eN}, P, and d in m
	$l_{eN(\text{min})} = 2.24Pd^{0.2}$ SI (18-45b) where l_{eN}, P, and d in mm
	$l_{eN(\text{max})} = 26.67Pd^{0.2}$ SI (18-46a) where l_{eN}, P, and d in m
	$l_{eN(\text{max})} = 6.7Pd^{0.2}$ SI (18-46b) where l_{eN}, P, and d in mm

Note:

If l_{eN} has to be between the limits, the length of the thread is said to be normal (N)

If l_{eN} has to be below the minimum level, length of thread is said to be short (S)

If l_{eN} has to be above the maximum level, length of thread is said to be long (L)

Eccentric loading

Particular	Formula
The load on bolt 1, Fig. 18-5 (panel a)	$F_1 = \frac{Fll_1}{l_1^2 + l_2^2 + l_3^2 + l_4^2} = F\frac{l(a - b\cos\alpha)}{4a^2 + 2b^2}$ (18-47)
The general expression for the load carried by ith bolt, F_i	$F_i = F\frac{2l(a - b\cos\alpha)}{(2a^2 + b^2)i}$ (18-48)
The maximum load on the bolt, Fig. 18-5(b)	$F_{\max} = \frac{2Fl(a + b)}{(2a^2 + b^2)i}$ (18-49)
The maximum load on the bolt, Fig. 18-5(c)	$F_{\max} = \frac{2Fl\left[a + b\cos\left(\frac{180^\circ}{i}\right)\right]}{(2a^2 + b^2)i}$ (18-50)

Particular	Formula

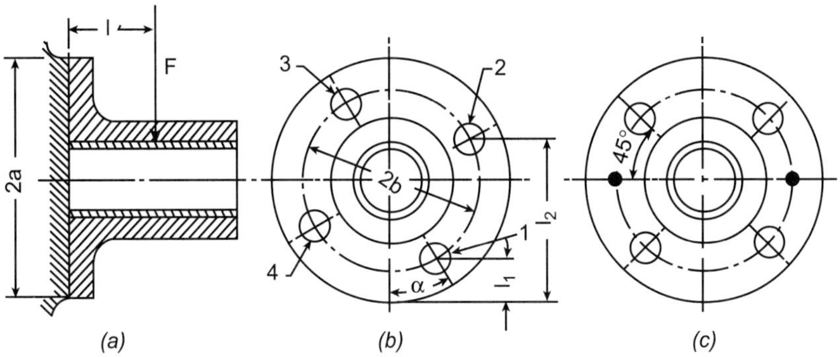

FIGURE 18-5 Fastening of a flanged bearing.

Fastening of a bracket

Bracket with no preload

Tensile load taken by the bolts, Fig. 18-6(a)

$$F_1 = \frac{Fll_1}{2(l_1^2 + l_2^2 + l_3^2)} \quad (18\text{-}51)$$

$$F_2 = \frac{Fll_2}{2(l_1^2 + l_2^2 + l_3^2)} \quad (18\text{-}52)$$

$$F_3 = \frac{Fll_3}{2(l_1^2 + l_2^2 + l_3^2)} \quad (18\text{-}53)$$

Shear stresses

(i) If shear load is taken completely by the lug, shear load on lug is given by

$$F_1 = F \quad (18\text{-}54)$$

(ii) If shear load is taken completely by the bolt shear load on each bolt is given by

$$F_b = \frac{F}{i} \quad (18\text{-}55)$$

(iii) If shear load is shared equally between the bolt and the lug

$$F_1' = \frac{F}{2} \quad (18\text{-}56)$$

$$F_b' = \frac{F}{2i} \quad (18\text{-}57)$$

Shear load due to the eccentricity e, Fig. 18-6(b), in each bolt is given by

$$F_{ei}' = \frac{Fex_i}{\sum x_i^2} \quad (18\text{-}58)$$

where x_i = distance between the center of bolts and the center of the particular bolt

Resultant shear load

$$F_r = F_b \text{ (or } F_b') + \frac{Fex_i}{\sum x_i^2} \quad (18\text{-}59)$$

Particular	Formula

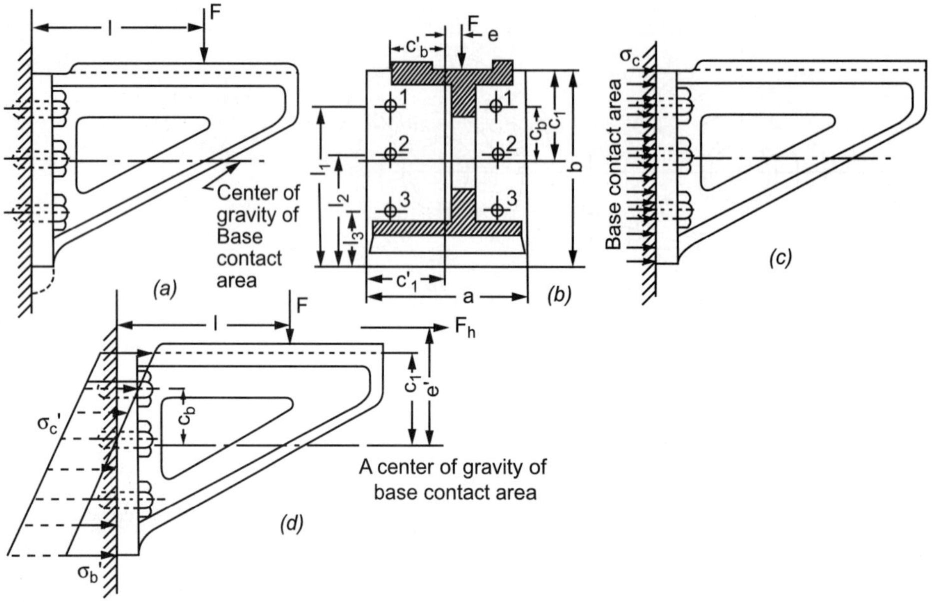

FIGURE 18-6 Preloaded bracket.

Preloaded bracket

Compression stress in contact area between the bracket base and the wall, Fig. 18-6(c)	$\sigma_c = \frac{iF_i}{A_c}$	(18-60)
Bending stress due to eccentric load, Fig. 18-6(d)	$\sigma'_b = \frac{M_b c_1}{I_c} = \frac{Flc_1}{I_c}$	(18-61)
Resultant compressive stress in the contact area	$\sigma'_c = \frac{iF_i}{A_c} - \frac{M_b c_1}{I_c} = \frac{iF_i}{A_c} - \frac{Flc_1}{I_c}$	(18-62)
Tensile stress in any individual bolt is given by	$\sigma'_b = \frac{F_i}{A_b} + \frac{M_b c_b}{I_c}$	(18-63)
Condition to avoid separation of the base and wall	$F_i > \frac{M_b c_1 A_c}{iI_c}$	(18-64)
With a 25% margin on the preload to account for overloads, condition to avoid separation of the base and wall	$F_i = \frac{1.25 M_b c_1 A_c}{iI_c}$	(18-65)
Bolt load taking into consideration 25% margin on the preload to account for overloads	$F_b = \frac{1.25 M_b c_1 A_c}{iI_c} + \frac{M_b c_b}{I_c}$	(18-66)

Particular	Formula
With an additional horizontal load F_h, the preload F_i is given by	$F_i = \frac{1.25 M_b c_1 A_c}{i I_c} \pm \frac{F_h}{i}$ (18-67) where (+) is used when F_h is away from the wall and (−) when F_h is toward the wall
With the addition of a horizontal load F_h, the bolt load is given by	$F_b = \frac{1.25 M_b c_1 A_c}{i I_c} \pm \frac{F_h}{i} + \frac{M_b c_b}{I_c} \pm \frac{F_h A_b}{A_c}$ (18-68)
Moment on the bracket	$M_b = Fl \pm F_h e'$ (18-69)

Shear loads

Particular	Formula
Shear load due to the eccentricity e in each of the bolts with no horizontal load	$F_{\tau_i} = \frac{M_1 x_i}{\sum x_i^2}$ (18-70) where $M_1 = Fe - \left(\frac{M_b c_1}{16 I_c} \sqrt{a^2 + b^2} - \frac{0.25 \mu M_b \sum x_i' c_1 A_b^2}{I_c A_c} \right)$ (88-70a) where x_i' = distance of the center of a particular bolt to the center of the base of the bracket
Shear load due to eccentricity e in each of the bolts with a horizontal load, F_h	$F_{\tau_i} = \frac{M_1 x_i}{\sum x_i^2}$ (18-71) where $M_1 = Fe \left[\frac{\mu}{4} \left(\frac{0.25 M_b c_1}{I_c} \pm \frac{F_h}{A_c} \sqrt{a^2 + b^2} \right) - \frac{\mu A_b}{A_c} \left(\frac{0.25 M_b c_1}{I_c} \pm \frac{F_h}{A_c} \right) \left(\sum x_i' \right) \right]$
Vertical applied load due to the friction component of the preload	$F_v = \mu \left(\frac{1.25 M_b c_1 A_c \pm F_h}{i I_c} \right)$ (18-72)
Condition for the nonexistence of the support for the shearload	$F < \mu \left(\frac{1.25 M_b c_1 A_c}{i I_c} \pm F_h \right)$ (18-73)

Particular	Formula
GENERAL	
See Tables 18-6 to 18-22 and Figs. 18-7 to 18-16 for further particulars on threaded fasteners and screws for power transmission.	
For British Standard ISO metric precision hexagon bolts, screws and nuts, and machine screws and machine screw nuts.	Refer to Tables 18-23 and 18-24.
For hexagon bolts, finished hexagon bolts, regular square nuts, hexagon and hexagon jam nuts, finished hexagon slotted nuts, regular hexagon and hexagon jam nuts, carriage bolts, countersunk, buttonhead and step bolts, machine screw heads, pan, truss and 100° flat heads, slotted head cap screws, square head setscrews, slotted headless setscrews, etc.	Refer to Tables from 18-25 to 18-42. All dimensions in inches.
For bolts, screws and nuts metric series—American National Standards hexagon cap screws, formed hex screws, heavy hex screws, recommended diameter–length combinations for screws, hexagon bolts, heavy hex bolts, heavy hex structural bolts, hexagon nuts, slotted hex nuts, etc.	Refer to Tables from 18-43 to 18-52.

TABLE 18-6
Allowable bearing pressure for screws, σ'_b

	Material		Safe bearing pressure, σ'_b		
Type	Screw	Nut	MPa	psi	Rubbing velocity, m/s [fpm = (ft/min)]
Hand press	Steel	Bronze	17.2–24.0	2500–3500	Low speed, well lubricated
Jack screw	Steel	Cast iron	12.3–17.2	1800–2500	Low speed, not over 0.04 (8)
Jack screw	Steel	Bronze	10.8–17.2	1600–2500	Low speed, not over 0.05 (10)
Hoisting screw	Steel	Cast iron	4.4–6.9	600–1000	Medium speed, 0.1 to 0.2 (20–40)
Hoisting screw	Steel	Bronze	5.4–9.8	800–1400	Medium speed, 0.1 to 0.2 (20–40)
Lead screw	Steel	Bronze	1.0–1.5	150–240	High speed, 0.25 and over (50)

Particular	Formula
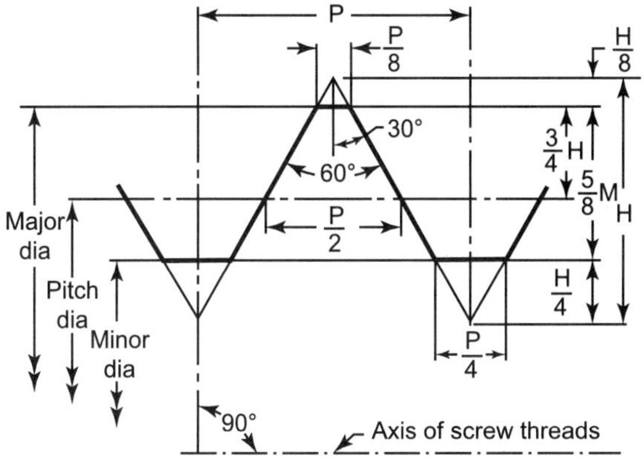 **FIGURE 18-7** Basic profile ISO metric screw threads.	$H = 0.86603\,P;$ $D_1 = d_2 - \frac{H}{2} = d - 2H_1 = d - 1.082\,P;$ $D_2 = d_2 = d - \frac{3}{4}H = d - 0.64952\,P;$ $d_1 = d_2 - \frac{H}{3} = d - 1.22687\,P;$ $H_1 = \frac{D - D_1}{2} = \frac{5}{8}H = 0.54127\,P;$ $h_3 = \frac{d - d_1}{2} = \frac{17}{24}H = 0.61343\,P;$ $r = \frac{H}{6} = 0.1443\,P; \quad r_c = 0.10825\,P;$ stress area $= A_c = \frac{\pi}{4}\left(\frac{d_1 + d_2}{2}\right)^2$
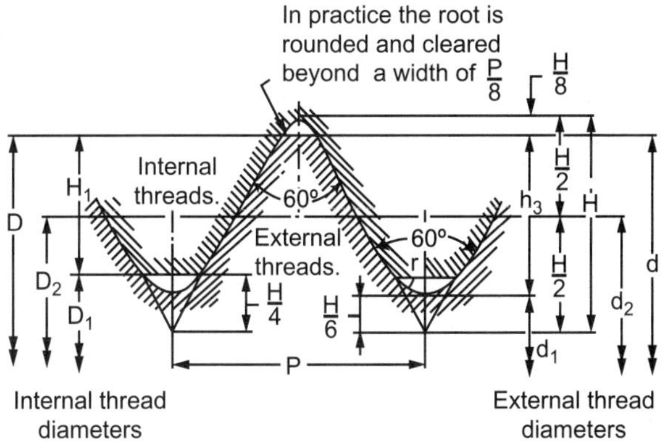 **FIGURE 18-8** ISO metric screw thread design profiles of external and internal threads.	*Designation*: A pitch diameter combination of thread size 8 mm and pitch 1 mm shall be designated as M8 × 1. M8 shall designate pitch diameter combination of thread size 8 mm and pitch 1.25 mm.

TABLE 18-7
Basic dimensions for design profiles of ISO metric screw threads

Basic diameter, mm	Pitch, P, mm	Major diameter, d, mm	Pitch diameter, d_2, mm	Minor diameter, mm: External threads, d_1	Minor diameter, mm: Internal threads, D_1	Lead angle at basic pitch diameter: deg	Lead angle at basic pitch diameter: min	Tensile stress area, A_c, mm^2
1	0.25	1.0	0.837620	0.693283	0.729367	5	27	0.46
	0.20	1.0	0.870096	0.754626	0.783494	4	11	0.53
2	0.40	2.0	1.740192	1.509252	1.566987	4	11	2.07
	0.25	2.0	1.837620	1.693283	1.729367	2	29	2.45
2.5	0.45	2.5	2.207716	1.947909	2.012861	3	43	3.39
	0.35	2.5	2.272668	2.070596	2.121114	2	20	3.70
3.0	0.50	3.0	2.675240	2.386565	2.458734	3	24	5.03
	0.35	3.0	2.772668	2.570596	2.621114	2	18	5.61
4.0	0.70	4.0	3.545337	3.141191	3.242228	3	36	8.78
	0.50	4.0	3.675240	3.386565	3.458734	2	29	9.79
5.0	0.80	5.0	4.480385	4.018505	4.133975	3	15	14.2
	0.50	5.0	4.675240	4.386565	4.458734	2	57	16.1
6.0	1.00	6.0	5.350481	4.773131	4.917468	3	24	20.1
	0.75	6.0	5.512861	5.079848	5.188101	2	29	22.0
7.0	1.00	7.0	6.350481	5.773131	5.917408	2	52	28.9
	0.75	7.0	6.512861	6.079848	6.188101	2	6	31.3
8.0	1.25	8.0	7.188101	6.466413	6.646835	3	10	36.6
	1.00	8.0	7.350481	6.773131	6.917468	2	29	39.2
10	1.50	10.0	9.025721	8.159696	8.376202	3	2	58.0
	1.25	10.0	9.188101	8.466413	8.646835	2	29	61.2
	1.00	10.0	9.350481	8.773131	8.917468	1	57	64.5
12	1.75	12	10.863342	9.852979	10.105569	2	56	84.3
	1.50	12	11.025721	10.159686	10.376202	2	29	88.1
	1.25	12	11.188101	10.466413	10.646835	2	2	92.1
	1.00	12	11.350481	10.773131	10.917468	1	36	96.1
14	2.00	14	12.700962	11.546261	11.834936	2	52	115
	1.50	14	13.025721	12.159696	12.376202	2	6	125
	1.25	14	13.188101	12.466413	12.646835	1	44	129
16	2.00	16	14.700962	13.546261	13.834936	2	29	157
	1.50	16	15.025721	14.159696	14.376202	1	49	167
18	2.50	15	16.376202	14.932827	15.293671	2	47	192
	2.00	18	16.700962	15.546261	15.834936	2	11	204
	1.50	18	17.025721	15.159696	16.376202	1	36	216
20	2.50	20	18.376202	16.932827	17.293671	2	29	245
	2.00	20	18.700962	17.516261	17.834936	1	57	258
	1.50	20	19.025721	18.159696	18.376202	1	26	272
22	2.50	22	20.376202	18.932827	19.293671	2	14	303
	2.00	22	20.700962	19.546261	19.834936	1	46	318
	1.50	22	21.025721	20.159696	20.376202	1	18	333
24	3.00	24	22.051443	20.319392	20.752405	2	49	353
	2.00	24	22.700962	21.556261	21.834936	1	39	384
	1.50	24	23.025721	22.159696	22.376202	1	11	401
25	3.00	25	23.051443	21.319392	21.752405	2	36	385

TABLE 18-7
Basic dimensions for design profiles of ISO metric screw threads (*Cont.*)

Basic diameter, mm	Pitch, P, mm	Major diameter, d, mm	Pitch diameter, d_2, mm	Minor diameter, mm: External threads, d_1	Minor diameter, mm: Internal threads, D_1	Lead angle at basic pitch diameter: deg	Lead angle at basic pitch diameter: min	Tensile stress area, A_c, mm^2
30	3.50	30	27.726683	25.705957	26.211139	2	18	561
	3.00	30	28.051443	26.319392	26.752405	1	57	581
	2.00	30	28.700962	27.546261	27.834936	1	16	621
	1.50	30	29.025721	28.159696	28.376202	0	57	642
35	1.50	35	34.055721	33.159696	33.376202	0	48	860
42	4.5	42	39.072114	36.479088	37.128607	2	6	1120
	4.0	42	39.401924	37.092523	37.669873	1	51	1150
	3.0	42	40.051443	38.319392	38.752405	1	22	1210
	2.0	42	40.700962	39.546261	39.834936	0	52	1260
	1.5	42	41.025771	40.159696	40.376202	0	40	1290
45	4.5	45	42.077164	39.479088	40.128607	1	57	1300
	4.0	45	42.401924	40.092523	40.669873	1	43	1340
	3.0	45	43.051443	41.319392	41.752405	1	16	1400
	2.0	45	43.700962	42.546261	42.834936	0	50	1460
	1.5	45	44.025771	43.159696	43.376202	0	37	1490
52	5.0	52	48.752405	45.865653	46.587341	1	52	1760
	4.0	52	49.401924	47.092523	47.669873	1	29	1830
	3.0	52	50.051443	48.319392	48.752405	1	6	1900
	2.0	52	50.700962	49.546261	49.834936	0	43	1970
	1.5	52	51.025721	50.159696	50.376202	0	32	2010
60	5.5	60	56.427645	53.252219	54.046075	1	47	2360
	4.0	60	57.401924	55.092523	55.669873	1	16	2490
	3.0	60	58.051443	56.319392	56.752405	0	57	2570
	2.0	60	58.700962	57.546261	57.834936	0	37	2650
	1.5	60	59.025721	58.159696	58.376202	0	28	2700
72	6	72	68.102886	64.638784	66.504809	1	36	3460
	4	72	69.401924	67.092523	67.669873	1	3	3660
	3	72	70.051443	68.319392	68.752405	0	47	3760
	2	72	70.700962	69.546261	69.834936	0	31	3860
80	6	80	76.102886	72.638724	73.504809	1	26	4340
	4	80	77.401924	75.092523	75.669873	0	57	4570
	3	80	78.051443	76.319392	76.752405	0	42	4680
	2	80	78.700962	77.546261	77.834936	0	28	4790
90	6	90	86.102886	82.638784	83.504809	1	16	5590
	4	90	87.401924	85.092523	85.669873	0	50	5840
	3	90	88.051449	86.319292	86.752405	0	37	5970
	2	90	88.700962	87.546261	87.834936	0	25	6100
100	6	100	96.102886	92.638784	93.504809	1	8	7000
	4	100	97.401924	95.092523	95.669873	0	45	7280
	3	100	98.051443	96.319392	96.752405	0	33	7420
	2	100	98.700962	97.546261	97.834936	0	22	7560
110	6	110	106.102886	102.638784	103.504809	1	2	8560
	4	110	107.401924	105.092523	105.669873	0	41	8870
	3	110	108.051443	106.319392	106.752405	0	30	9020

TABLE 18-7
Basic dimensions for design profiles of ISO metric screw threads (*Cont.*)

Basic diameter, mm	Pitch, P, mm	Major diameter, d, mm	Pitch diameter, d_2, mm	Minor diameter, mm: External threads, d_1	Minor diameter, mm: Internal threads, D_1	Lead angle at basic pitch diameter: deg	Lead angle at basic pitch diameter: min	Tensile stress area, A_c, mm^2
120	6	120	116.102886	112.638784	113.504819	0	57	10300
	4	120	117.401924	115.092523	115.669873	0	37	10600
	3	120	118.051443	116.319392	116.752405	0	28	10800
150	6	150	146.102886	142.538784	143.504809	0	45	16400
	4	150	147.401924	145.092523	145.669873	0	30	16800
	3	150	148.051443	146.319392	146.752405	0	22	17000
160	6	160	156.102886	152.638784	153.504809	0	42	18700
	4	160	157.401924	155.092523	155.669873	0	28	19200
	3	160	158.051443	156.319392	156.752405	0	21	19400
180	6	180	176.102886	172.638784	173.504809	0	37	23900
	4	180	177.401924	175.092523	175.669873	0	25	24400
	3	180	178.051443	176.319392	176.752405	0	18	24700
200	6	200	196.102886	192.638784	193.504809	0	33	29700
	4	200	197.401924	195.092523	195.669873	0	22	30200
	3	200	198.051453	196.319392	196.752405	0	17	30500
250	6	250	246.102886	242.638784	243.504809	0	27	46900
	4	250	247.401924	245.092523	245.669873	0	18	47600
	3	250	248.051443	246.319392	246.752405	0	13	48000
300	6	300	296.102886	295.638784	293.504809	0	22	68100
	4	300	297.401924	292.092523	295.669873	0	15	68900

Source: IS: 4218-1967 (Part III).

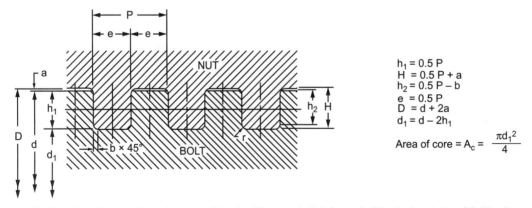

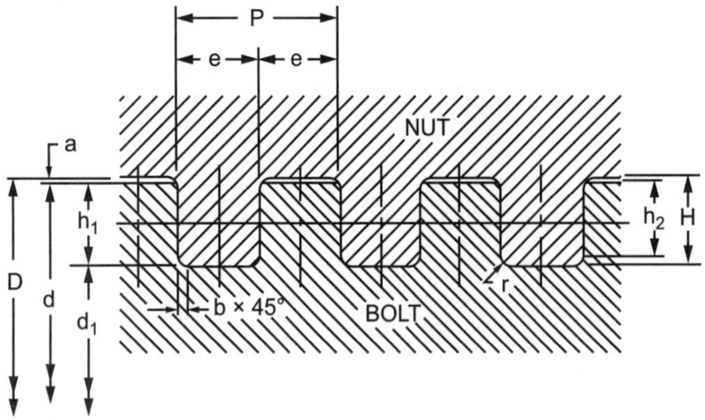

$h_1 = 0.5\,P$
$H = 0.5\,P + a$
$h_2 = 0.5\,P - b$
$e = 0.5\,P$
$D = d + 2a$
$d_1 = d - 2h_1$

Area of core $= A_c = \dfrac{\pi d_1^2}{4}$

Designation : A square thread nominal diameter 30mm and pitch 6mm shall be designated as SQ 30 × 6

FIGURE 18-9 Basic profile of square threads.

TABLE 18-8
Basis dimensions (in mm) for square threads

Nominal diameter	Major diameter		Minor diameter, d_1	Pitch, P	e	r	h_2	b	h_1	a	H	Area of core, A_c, mm^2
	Bolt, d	Nut, D										
10	10	10.5	8									50.3
14	14	14.5	12	2	1	0.12	0.75	0.25	1	0.25	1.25	113
20	20	20.5	18									201
26	26	26.5	23									415
30	30	30.5	27									573
36	36	36.5	33	3	1.5	0.12	1.25	0.25	1.5	0.25	1.75	855
40	40	40.5	37									1075
44	45	44.5	41									1320
50	50	50.5	47									1735
55	55	55.5	52	3	1.5	0.12	1.25	0.25	1.5	0.25	1.75	2124
60	60	60.5	57									2552
75	65	65.5	61									2922
80	70	70.5	66									3421
85	75	75.5	71									3959
90	80	80.5	76									4536
95	85	85.5	84	4	2	0.12	1.75	0.25	2	0.25	2.25	5153
90	90	90.5	86									5809
95	95	95.5	91									5504
100	100	100.5	96									7248
110	110	110.5	106									8825
120	120	120.5	114									10207
130	130	130.5	124									12076
140	140	140.5	134	6	3	0.25	2.5	0.5	3	0.25	3.25	14103
150	150	150.5	144									16286
160	160	160.5	154									18627
170	170	170.5	164									21124
180	180	180.5	172									23235
190	190	190.5	182									26016
200	200	200.5	192	8	4	0.25	3.5	0.5	4	0.25	4.25	28953
220	220	220.5	212									35299
240	240	240.5	232									42273
				Normal Series								
22	22	22.5	17									227
24	24	24.5	19	5	2.5	0.25	2	0.5	2.5	0.25	2.75	284
26	26	26.5	21									346
28	28	28.5	23									415
30	30	30.5	24	6	3	0.25	2.5	0.5	3	0.25	3.25	452
36	36	36.5	30									707
40	40	40.5	33	7	3.5	0.25	3	0.5	3.5	0.25	3.75	855
44	44	44.5	37									1075
50	50	50.5	42	8	4	0.25	3.5	0.5	4	0.25	4.25	1385
52	52	52.5	44									1521
55	55	55.5	46	9	4.5	0.25	4	0.5	4.5	0.25	4.75	1662
60	60	60.5	51									2043
65	65	65.5	55									2376
70	70	70.5	60	10	5	0.25	4.5	0.5	5	0.25	5.25	2827

TABLE 18-8
Basis dimensions (in mm) for square threads (*Cont.*)

Nominal diameter	Major diameter, Bolt, d	Major diameter, Nut, D	Minor diameter, d_1	Pitch, P	e	r	h_2	b	h_1	a	H	Area of core, A_c, mm^2
75	75	75.5	65									3318
80	80	80.5	70									3848
85	85	85.5	73									4185
90	90	90.5	78									4778
95	95	95.5	83	12	6	0.25	5.5	0.5	6	0.25	6.25	5411
100	100	100.5	88									6082
110	110	110.5	98									7543
120	120	121	106									8825
130	130	131	116	14	7	0.5	6	1	7	0.5	7.5	10568
140	140	141	126									12469
150	150	151	134									14103
160	160	161	144	16	8	0.5	7	1	8	0.5	8.5	16286
170	170	171	154									18627
180	180	181	162									20612
190	190	191	172	18	9	0.5	8	1	9	0.5	9.5	23235
200	200	201	182									26016
300	300	301	274	26	13	0.5	12	1	13	0.5	13.5	58965
				Coarse Series								
22	22	22.5	14									164
24	24	24.5	16	8	4	0.25	3.5	0.5	4	0.25	4.25	201
26	26	26.5	18									254
28	28	28.5	20									314
30	30	30.5	20									314
36	36	36.5	26	10	5	0.25	4.5	0.5	5	0.25	5.25	531
40	40	40.5	28									616
50	50	50.5	38	12	6	0.25	5.5	0.5	6	0.25	6.25	1134
60	60	61	46	14	7	0.5	6	1	7	0.5	7.5	1662
70	70	71	54									2290
75	75	76	59	16	8	0.5	7	1	8	0.5	8.5	2734
80	80	81	64									3217
90	90	91	72	18	9	0.5	8	1	9	0.5	9.5	4072
120	120	121	98	22	11	0.5	10	1	11	0.5	11.5	8332
150	150	151	126	24	12	0.5	11	1	12	0.5	12.5	12469
180	180	181	152	28	14	0.5	13	1	14	0.5	14.5	18146
200	200	201	168	32	16	0.5	15	1	16	0.5	16.5	22167
300	300	301	256	44	24	0.5	21	1	22	0.5	22.5	51472
400	400	401	352	48	24	0.5	23	1	24	0.5	24.5	97314

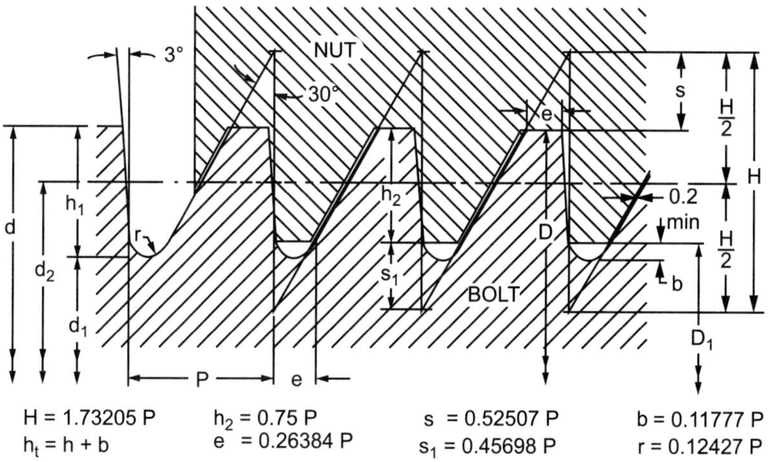

Pitch, mm	Depth of thread, mm	Depth of engagement, mm	*e*, mm	*b*, mm	*r*, mm
2	1.736	1.5	0.528	0.236	0.249
3	2.603	2.25	0.792	0.353	0.373
4	3.471	3	1.055	0.471	0.497
5	4.339	3.75	1.319	0.589	0.621
6	5.207	4.5	1.583	0.707	0.746
7	6.074	5.25	1.847	0.824	0.870
8	6.942	6	2.111	0.942	0.994
9	7.810	6.75	2.375	1.060	1.118
10	8.678	7.5	2.638	1.178	1.243
12	10.413	9	3.166	1.413	1.491
14	12.149	10.5	3.694	1.649	1.740
16	13.884	12	4.221	1.884	1.988
18	15.620	13.5	4.749	2.120	2.237
20	17.355	15	5.277	2.355	2.485
22	19.091	16.5	5.804	2.591	2.734
24	20.826	18	6.332	2.826	2.982
26	22.562	19.5	6.860	3.062	3.231
28	24.298	21	7.388	3.298	3.480
32	27.769	24	8.443	3.769	3.977
36	31.240	27	9.498	4.240	4.474
40	34.711	30	10.554	4.711	4.971
44	38.182	33	11.609	5.182	5.468
48	41.653	36	12.664	5.653	5.965

Designation: A sawtooth thread of nominal diameter 48 mm and pitch 3 mm shall be designated as ST 48 $\times$ 3.

FIGURE 18-10 Basic profile of sawtooth threads. (*Source*: IS 4696, 1968.)

TABLE 18-9
Basic dimensions (in mm) for sawtooth threads

	Bolt					Nut	
Nominal diameter	**Major diameter, d**	**Minor diameter, d_1**	**Area of core, mm^2**	**Pitch diameter, d_2**	**Pitch, P**	**Major diameter, D**	**Minor diameter, D_1**
			Fine Series				
10	10	6.528	33.5	8.636	2	10	7
12	12	8.528	57.1	10.636	2	12	9
14	14	10.538	87.1	12.636	2	14	11
16	16	12.528	123	14.636	2	16	13
20	20	16.528	215	18.636	2	20	17
22	22	16.794	222	19.954	3	22	17.5
30	30	24.794	483	27.954	3	30	25.5
36	36	30.794	745	32.954	3	36	31.5
40	40	34.794	951	37.954	3	40	35.5
50	50	44.794	1576	42.954	3	50	45.5
55	55	49.794	1947	57.954	3	55	50.5
60	60	54.794	2358	57.954	3	60	55.5
65	65	58.058	2647	62.272	4	65	59
70	70	63.058	3123	67.272	4	70	64
75	75	68.058	3638	72.272	4	75	69
80	80	73.058	4192	77.272	4	80	74
85	85	78.058	4785	82.272	4	85	79
90	90	83.058	5418	87.272	4	90	84
95	95	88.058	6090	92.272	4	95	89
100	100	93.058	6801	97.272	4	100	94
120	120	109.586	9432	115.909	6	120	111
150	150	139.586	15303	145.909	6	150	141
180	180	166.116	21673	174.545	8	180	168
200	200	186.116	27206	194.545	8	200	188
			Normal Series				
22	22	13.322	139	18.590	5	22	14.5
24	24	15.322	184	20.590	5	24	16.5
26	26	17.322	236	22.590	5	26	18.5
30	30	19.586	301	25.909	6	30	21
36	36	25.586	514	31.909	6	36	27
40	40	27.852	709	35.227	7	42	31.5
44	44	31.852	797	39.227	7	44	33.5
50	50	36.116	1024	44.545	8	50	38
55	55	39.380	1218	48.863	9	55	41.5
60	60	44.380	1547	53.863	9	60	46.5
70	70	52.644	2177	63.181	10	70	55
80	80	62.644	3082	73.181	10	80	65
90	90	69.174	3758	81.817	12	90	72
100	100	79.174	4923	91.817	12	100	82
110	110	89.174	6246	101.817	12	110	92
130	130	102.702	8775	120.459	14	130	109
150	150	122.232	11734	139.089	16	150	126
180	180	148.760	17381	167.726	18	180	153
200	200	168.760	22368	187.726	18	200	173

TABLE 18-9
Basic dimensions (in mm) for sawtooth threads (*Cont.*)

	Bolt					Nut	
Nominal diameter	Major diameter, d	Minor diameter, d_1	Area of core, mm^2	Pitch diameter, d_2	Pitch, P	Major diameter, D	Minor diameter, D_1
			Coarse Series				
22	22	8.116	51.4	16.545	8	22	10
24	24	10.116	80.7	18.545	8	24	12
26	26	12.116	115	20.545	8	26	14
30	30	12.644	126	23.181	10	30	15
40	40	19.174	289	31.817	12	40	22
50	50	29.174	668	41.817	12	50	32
60	60	35.702	1001	50.453	14	60	39
70	70	42.232	1401	59.089	16	70	46
80	80	52.232	2143	69.089	16	80	56
90	90	58.760	2712	77.726	18	90	63
100	100	65.290	3348	86.362	20	100	70
150	150	108.348	9220	138.634	24	150	114
200	200	144.462	16391	178.179	32	200	152

$H = 1.866\,P$
$h_1 = 0.5\,P$
$h_2 = 0.084\,P$
$a = 0.05\,P$
$b = 0.683\,P$
$R = 0.256\,P$
$R_1 = 0.221\,P$

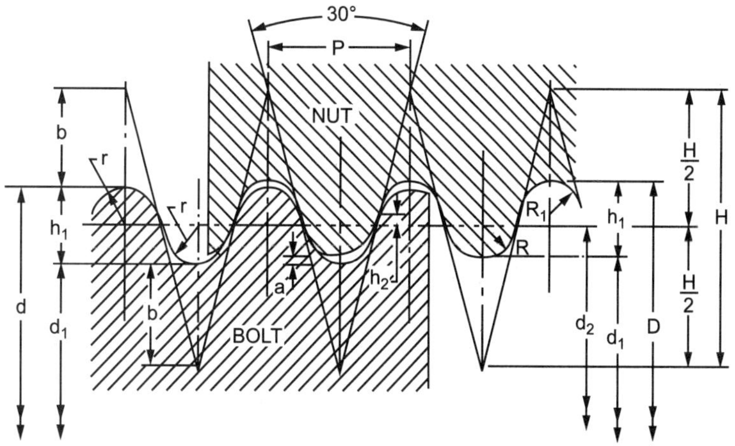

Nominal diameter, d, mm	Pitch, P, mm	Depth of thread, h_1, mm	Depth of engagement, h_2, mm	Radii, mm: Bolt, r	Radii, mm: Nut R	Radii, mm: Nut R_1
8–12	2.550	1.270	0.212	0.606	0.650	0.561
14–38	3.175	1.588	0.265	0.757	0.813	0.702
40–100	4.233	2.117	0.353	1.010	1.084	0.936
105–200	6.350	3.175	0.530	1.515	1.625	1.404

Designation: A knuckle thread of nominal diameter 10 mm and pitch of 2.54 mm shall be designated as $K10 \times 2.54$.

FIGURE 18-11 Basic profile of knuckle threads. (*Source*: IS 4695: 1968.)

TABLE 18-10
Basic dimensions (in mm) for knuckle threads

	Bolt				Nut	
Nominal diameter	Major diameter, d	Minor diameter, d_1	Area of core, mm^2	Pitch diameter, d_2	Major diameter, D	Minor diameter, D_1
8	8	5.460	23.4	6.730	8.254	5.714
9	9	6.460	32.8	7.730	9.254	6.714
10	10	7.460	43.7	8.730	10.254	7.714
12	12	9.460	70.3	10.730	12.254	9.714
14	14	10.825	92.0	12.412	14.318	11.142
16	16	12.825	129.2	14.412	16.318	16.142
20	20	16.825	222.3	18.412	20.318	17.142
24	24	20.825	340.6	22.412	24.318	21.142
30	30	26.825	565.2	28.412	30.318	27.142
36	36	32.825	846.3	34.412	36.318	33.142
40	40	35.767	1005	37.883	40.423	36.190
44	44	39.767	1242	41.883	44.423	40.190
50	50	45.767	1645	47.883	50.423	46.190
55	55	50.767	2024	52.883	55.423	51.190
60	60	55.767	2443	57.883	60.423	56.190
65	65	60.767	2900	62.883	65.423	61.190
70	70	65.767	3397	67.883	70.423	66.190
75	75	70.767	3933	72.883	75.423	71.190
80	80	75.767	4509	77.883	80.423	76.190
85	85	80.767	5123	82.883	85.423	81.190
90	90	85.767	5777	87.883	90.423	86.190
95	95	90.767	6471	92.883	95.423	91.190
100	100	95.767	7203	97.883	100.423	96.190
110	110	103.650	8438	106.825	110.635	104.285
120	120	113.650	10145	116.885	120.635	114.985
130	130	123.650	12008	126.825	130.635	124.285
140	140	133.650	14029	136.825	140.635	134.285
150	150	143.650	16207	146.825	150.635	144.285
160	160	153.650	18542	156.825	160.635	154.285
170	170	163.650	21034	166.825	170.635	164.285
180	180	173.650	23683	176.825	180.635	174.285
190	190	183.650	26489	186.825	190.635	184.285
200	200	193.650	29453	196.825	200.635	194.285

Source: IS 4695, 1968.

TABLE 18-11
Pitch-diameter combinations for ISO metric threads

Pitch, P, mm	Maximum diameter, mm
0.5	22
0.75	33
1.00	80
1.50	150
2.00	200
3.00	300

TABLE 18-12
Tolerance grades 3, 4, 5 for precision; 6 for medium; and 7, 8, and 9 for coarse qualities for bolts and nuts

Minor diameter of nut threads		4	5	6	7	8	
Major diameter of bolt threads		4		6		8	
Pitch diameter of nut threads		4	5	6	7	8	
Pitch diameter of bolt threads	3	4	5	6	7	8	9

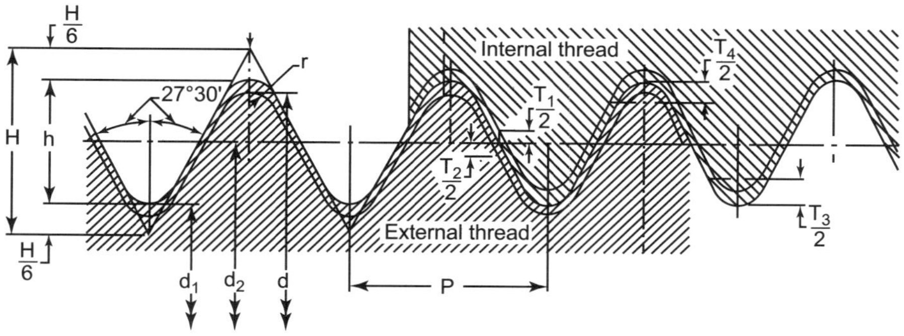

Pipe threads for fastening
$H = 0.9605P$; $h = 0.6404P$; $r = 0.1374P$

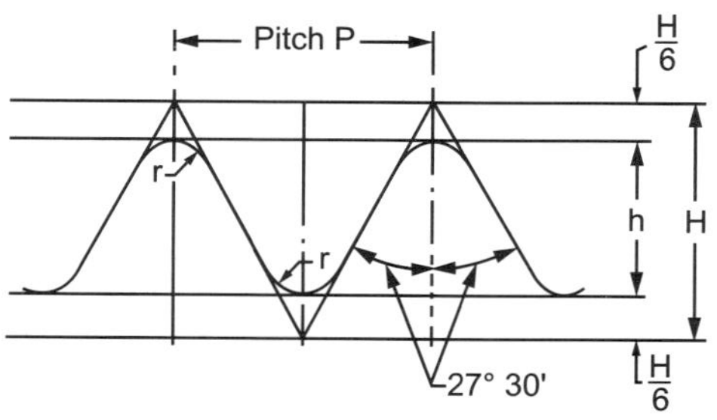

Parallel pipe thread

All dimensions in mm

Designation	Pitch, P	Basic diameter, internal and external threads			Designation	Pitch, P	Basic diameter, internal and external threads		
		Major, d_2	Pitch, P	Minor, d_1			Major, d_2	Pitch, P	Minor, d_1
FP $\frac{1}{8}$	0.907	9.728	9.147	8.566	FP 2	2.309	59.614	58.135	56.656
FP $\frac{1}{4}$	1.337	13.157	15.301	11.445	FP $2\frac{1}{4}$	2.309	62.710	64.231	62.752
FP $\frac{3}{8}$	1.337	16.662	19.806	14.940	FP $2\frac{1}{4}$	2.309	75.184	73.705	72.226
FP $\frac{1}{2}$	1.814	20.955	25.793	18.631	FP 3	2.309	87.884	86.407	84.926
FP $\frac{3}{4}$	1.814	26.441	25.279	24.117	FP $3\frac{1}{2}$	2.309	100.330	98.851	97.372
FP 1	2.309	33.249	31.770	30.291	FP 4	2.309	113.030	111.551	110.072
FP $1\frac{1}{4}$	2.309	41.910	40.431	38.952	FP 5	2.309	138.430	136.951	135.472
FP $1\frac{1}{2}$	2.309	47.803	46.324	44.845	FP 6	2.309	193.830	162.351	160.872

Designation: An external pipe thread for fastening purposes of size 2 with class B tolerance shall be designated as Ext-FP 2B, and an internal pipe thread of size 2 shall be designated as Int-FP 2.

FIGURE 18-12 Pipe threads for fastening purposes. (*Source*: IS 2643, 1964.)

TABLE 18-13
Tolerance for crest and pitch diameters of bolts and nuts[a]

Diameter	Bolt/nut	Unit of tolerance	Value of tolerance unit	Tolerance grades 3	4	5	6	7	8	9
Crest diameter	Bolt	Td (6)	$180P^{2/3} - \frac{3.15}{\sqrt{P}}$	—	0.63 Td (6)	—	Td (6)	—	1.6 Td (6)	—
	Nut	Td_1 (6)	$433P - 190P^{1.22}$ for P from 0.2 to 0.8 mm $230P^{0.7}$ for P from 1 mm and above	—	—	—	—	—	—	—
Pitch diameter	Bolt	Td_2 (6)	$90P^{0.4}\ d^{0.1}$	0.5 Td_2 (6)	0.63 Td_2 (6)	0.8 Td_2 (6)	Td_2 (6)	1.25 Td_2 (6)	1.6 Td_2 (6)	2 Td_2 (6)
	Nut	Td_2 (6)	$90P^{0.4}\ d^{0.1}$	—	0.85 Td_2 (6)	1.06 Td_2 (6)	1.32 Td_2 (6)	1.7 Td_2 (6)	2.12 Td_2 (6)	—

[a] T_d in μm; P in μm; Td_2 in μm; d in mm.
Source: IS 4218 (Part IV), 1967.

TABLE 18-14
Preferred tolerance classes for nuts

Tolerance quality	Small allowance position *G* S	N	L	No allowance position *H* S	N	L
Fine				4H	5H	6H
Medium	5G	6G	7G	5H	6H	7H
Coarse		7G	8G		7H	8H

Source: IS 4218 (Part IV), 1967.

TABLE 18-15
Preferred tolerance classes for bolts

Tolerance quality	Large allowance position *e* S	N	L	Small allowance position *g* S	N	L	No allowance position *h* S	N	L
Fine							3*h* 4*h*	4*h*	5*h* 4*h*
Medium		6*e*	7*e* 6*e*	7*g* 6*g*	6*g*	7*g* 6*g*	5*h* 6*h*	6*h*	7*h* 6*h*
Coarse					8*g*	9*g* 8*g*			

Source: IS 4218 (Part IV), 1967.

TABLE 18-16
Coarse-threaded series—UNC and NC (dimensions in inches)

Sizes[a]	Basic major (nominal) diameter, D	Threads per inch	Basic pitch diameter	Basic minor diameter external thread	Root area[b] in in^2, A	Minor diameter internal thread classes 1B, 2B, and 3B for engagement $\frac{2}{3}D$ to $\frac{3}{2}D$	
						Minimum	Maximum
1	0.0730	64	0.0629	0.0538	0.0023	0.0585	0.0623
2	0.0860	56	0.0744	0.0641	0.0032	0.0699	0.0737
3	0.0990	48	0.0855	0.0734	0.0042	0.0805	0.0845
4	0.1120	40	0.0958	0.0813	0.0052	0.0894	0.0939
5	0.1250	40	0.1088	0.0943	0.0070	0.1021	0.1062
6	0.1380	32	0.1177	0.0997	0.0078	0.1091	0.1140
8	0.1640	32	0.1437	0.1257	0.0124	0.1346	0.1389
10	0.1900	24	0.1629	0.1389	0.0152	0.1502	0.1555
12	0.2160	24	0.1889	0.1649	0.0214	0.1758	0.1807
$\frac{1}{4}$ UN	0.2500	20	0.2175	0.1887	0.0280	0.2013	0.2067
$\frac{5}{16}$ UN	0.3125	18	0.2764	0.2443	0.0469	0.2577	0.2630
$\frac{3}{8}$ UN	0.3750	16	0.3344	0.2983	0.0699	0.3128	0.3182
$\frac{7}{16}$ UN	0.4375	14	0.3911	0.3499	0.0962	0.3659	0.3717
$\frac{1}{2}$	0.5000	13	0.4500	0.4056	0.1292	0.4226	0.4284
$\frac{1}{2}$ UN	0.5000	12	0.4459	0.3978	0.1243	0.4160	0.4223
$\frac{9}{16}$ UN	0.5625	12	0.5084	0.4603	0.1664	0.4783	0.4843
$\frac{5}{8}$ UN	0.6250	11	0.5660	0.5135	0.2071	0.5329	0.5391
$\frac{3}{4}$ UN	0.7500	10	0.6850	0.6273	0.3091	0.6481	0.6545
$\frac{7}{8}$ UN	0.8750	9	0.8028	0.7387	0.4286	0.7614	0.7681
1 UN	1.0000	8	0.9188	0.8466	0.5629	0.8722	0.8797
$1\frac{1}{8}$ UN	1.1250	7	1.0322	0.9497	0.7178	0.9789	0.9875
$1\frac{1}{4}$ UN	1.2500	7	1.1572	1.0747	0.9071	1.1039	1.1125
$1\frac{3}{8}$ UN	1.3750	6	1.2667	1.1705	1.0760	1.2046	1.2146
$1\frac{1}{2}$ UN	1.5000	6	1.3917	1.2955	1.3182	1.3296	1.3396
$1\frac{3}{4}$ UN	1.7500	5	1.6201	1.5046	1.7780	1.5455	1.5575
2 UN	2.0000	$4\frac{1}{2}$	1.8557	1.7274	2.3436	1.7728	1.7861
$2\frac{1}{4}$ UN	2.2500	$4\frac{1}{2}$	2.1057	1.9774	3.0610	2.0228	2.0361
$2\frac{1}{2}$ UN	2.5000	4	2.3376	2.1933	3.7782	2.2444	2.2594
$2\frac{3}{4}$ UN	2.7500	4	2.5876	2.4433	4.6886	2.4944	2.5094
3 UN	3.0000	4	2.8376	2.6933	5.6972	2.7444	2.7594
$3\frac{1}{4}$ UN	3.2500	4	3.0876	2.9433	6.8039	2.9944	3.0094
$3\frac{1}{2}$ UN	3.5000	4	3.3376	3.1933	8.0088	3.2444	3.2594
$3\frac{3}{4}$ UN	3.7500	4	3.5876	3.4433	9.3119	3.4944	3.5094
4 UN	4.0000	4	3.8376	3.6933	10.7132	3.7444	3.7594

[a] Unified diameter-pitch relationships are marked UN.

[b] The actual root area of a screw will be somewhat less than A, but, since the tensile strength of a screw of ductile material is greater than that of a plain specimen of the same material and of a diameter equal to the root diameter of the screw, the tensile strength of a screw may be assumed to correspond to A as given.

For complete manufacturing information and tolerances, see ASA Standard B1.1, 1949.

TABLE 18-17
Fine-thread series UNF and NF (dimensions in inches)

Sizes[a]	Basic major (nominal) diameter, D	Threads per inch	Basic pitch diameter	Basic minor diameter external thread	Root area[b] in in^2, A	Minor diameter internal thread classes 1B, 2B, and 3B for engagement $\frac{2}{3}D$ to $\frac{3}{2}D$	
						Minimum	Maximum
0	0.0600	80	0.0519	0.0447	0.0016	0.0479	0.0514
1	0.0730	72	0.0640	0.0560	0.0025	0.0602	0.0635
2	0.0860	64	0.0759	0.0668	0.0035	0.0720	0.0753
3	0.0990	56	0.0874	0.0771	0.0047	0.0831	0.0865
4	0.1120	48	0.0985	0.0864	0.0059	0.0931	0.0968
5	0.1250	44	0.1102	0.0971	0.0074	0.1042	0.1079
6	0.1380	40	0.1218	0.1073	0.0090	0.1147	0.1186
8	0.1640	36	0.1460	0.1299	0.0133	0.1358	0.1416
10	0.1900	32	0.1697	0.1517	0.0181	0.1601	0.1641
12	0.2160	28	0.1928	0.1722	0.0233	0.1815	0.1857
$\frac{1}{4}$ UN	0.2500	28	0.2268	0.2062	0.0334	0.2150	0.2190
$\frac{5}{16}$ UN	0.3125	24	0.2854	0.2614	0.0541	0.2714	0.2754
$\frac{3}{8}$ UN	0.3750	24	0.3479	0.3239	0.0824	0.3332	0.3372
$\frac{7}{16}$ UN	0.4375	20	0.4050	0.3762	0.1112	0.3875	0.3916
$\frac{1}{2}$ UN	0.5000	20	0.4675	0.4387	0.1512	0.4497	0.4537
$\frac{9}{16}$ UN	0.5625	18	0.5264	0.4943	0.1919	0.5065	0.5106
$\frac{5}{8}$ UN	0.6250	18	0.5889	0.5568	0.2435	0.5690	0.5730
$\frac{3}{4}$ UN	0.7500	16	0.7094	0.6733	0.3560	0.6865	0.6908
$\frac{7}{8}$ UN	0.8750	14	0.8286	0.7874	0.4869	0.8023	0.8068
1 UN	1.000	12	0.9459	0.8978	0.6331	0.9148	0.9198
$1\frac{1}{8}$ UN	1.1250	12	1.0709	1.0228	0.8216	1.0398	1.0448
$1\frac{1}{4}$ UN	1.2500	12	1.1959	1.1478	1.0347	1.1648	1.1698
$1\frac{3}{8}$ UN	1.3750	12	1.3209	1.2728	1.2724	1.2893	1.2948
$1\frac{1}{2}$ UN	1.5000	12	1.4459	1.3978	1.5346	1.4148	1.4198

[a] Unified diameter-pitch relationships are marked UN.

[b] The actual root area of a screw will be somewhat less than A, but, since the tensile strength of a screw of *ductile* material is greater than that of a plain specimen of the same material and of a diameter equal to the root diameter of the screw, the tensile strength of a screw may be assumed to correspond to A as given.

For complete manufacturing information and tolerances, see ASA Standard B1.1, 1949.

TABLE 18-18
Extra-fine thread series—NEF

Sizes[a]	Basic major (nominal) diameter, D, in	Threads per inch	Basic pitch diameter, in	Basic minor diameter external thread, in	Root area[b] in in^2, A	Minor diameter internal thread classes 1B, 2B, and 3B for engagement $\frac{2}{3}D$ to $\frac{3}{2}D$, in	
						Minimum	Maximum
12	0.2160	32	0.1957	0.1777	0.0248	0.1855	0.1895
$\frac{1}{4}$	0.2500	32	0.2297	0.2117	0.0352	0.2189	0.2229
$\frac{5}{16}$	0.3125	32	0.2922	0.2742	0.0591	0.2807	0.2847
$\frac{3}{8}$	0.3750	32	0.3547	0.3367	0.0890	0.3429	0.3469
$\frac{7}{16}$ UN	0.4375	28	0.4143	0.3937	0.1217	0.4011	0.4051
$\frac{1}{2}$ UN	0.5000	28	0.4768	0.4562	0.1635	0.4636	0.4676
$\frac{9}{16}$	0.5625	24	0.5354	0.5114	0.2054	0.5204	0.5244
$\frac{5}{8}$	0.6250	24	0.5979	0.5739	0.2587	0.5829	0.5869
$\frac{11}{16}$	0.6875	24	0.6604	0.6364	0.3181	0.6454	0.6494
$\frac{3}{4}$ UN	0.7500	20	0.7175	0.6887	0.3725	0.6997	0.7037
$\frac{13}{16}$ UN	0.8125	20	0.7800	0.7512	0.4432	0.7622	0.7662
$\frac{7}{8}$ UN	0.8750	20	0.8425	0.8137	0.5200	0.8247	0.8287
$\frac{15}{16}$ UN	0.9375	20	0.9050	0.8762	0.6030	0.8872	0.8912
1 UN	1.0000	20	0.9675	0.9387	0.6921	0.9497	0.9537
$1\frac{1}{16}$	1.0625	18	1.0264	0.9943	0.7765	1.0064	1.0105
$1\frac{1}{8}$	1.1250	18	1.0889	1.0568	0.8772	1.0689	1.0730
$1\frac{3}{16}$	1.1875	18	1.1514	1.1193	0.9840	1.1314	1.1355
$1\frac{1}{4}$	1.2500	18	1.2139	1.1818	1.0969	1.1939	1.1980
$1\frac{5}{16}$	1.3125	18	1.2764	1.2443	1.2160	1.2564	1.2605
$1\frac{3}{8}$	1.3750	18	1.3389	1.3068	1.3413	1.3189	1.3230
$1\frac{7}{16}$	1.4375	18	1.4014	1.3693	1.4726	1.3814	1.3855
$1\frac{1}{2}$	1.5000	18	1.4639	1.4318	1.6101	1.4439	1.4480
$1\frac{9}{16}$	1.5625	18	1.5264	1.4943	1.7538	1.5064	1.5105
$1\frac{5}{8}$	1.6250	18	1.5889	1.5568	1.9035	1.5689	1.5730
$1\frac{11}{16}$	1.6875	18	1.6514	1.6193	2.0594	1.6314	1.6355
$1\frac{3}{4}$ UN	1.7500	16	1.7094	1.6733	2.1991	1.6865	1.6908
2 UN	2.0000	16	1.9594	1.9233	2.9053	1.9365	1.9408

[a] Unified diameter-pitch relationships are marked UN.
[b] The actual root area of a screw will be somewhat less than A, but, since the tensile strength of a screw of ductile material is greater than that of a plain specimen of the same material and of a diameter equal to the root diameter of the screw, the tensile strength of a screw may be assumed to correspond to A as given.
For complete manufacturing information and tolerances, see ASA Standard B1.1, 1949.

TABLE 18-19
8-pitch thread series—8N (dimensions in inches)

Size[a] also basic major (normal) diameter, D	Basic minor diameter external thread	Minor diameter internal thread classes 1B, 2B, and 3B for engagement $\frac{2}{3}D$ to $\frac{3}{2}D$		Size[a] also basic major (nominal) diameter, D	Basic minor diameter external thread	Minor diameter internal thread classes 1B, 2B, and 3B for engagement $\frac{2}{3}D$ to $\frac{3}{2}D$	
		Minimum	Maximum			Minimum	Maximum
1 UN	0.8466	0.8722	0.8797	3	2.8466	2.8722	8.8797
$1\frac{1}{8}$	0.9716	0.9972	1.0047	$3\frac{1}{4}$	3.0966	3.1222	3.1297
$1\frac{1}{4}$	1.0966	1.1222	1.1297	$3\frac{1}{2}$	3.3466	3.3722	3.3797
$1\frac{3}{8}$	1.2216	1.2472	1.2547	$3\frac{3}{4}$	3.5966	3.6222	3.6297
$1\frac{1}{2}$	1.3466	1.3722	1.3797	4	3.8466	3.8722	3.8797
$1\frac{5}{8}$	1.4716	1.4972	1.5047	$4\frac{1}{4}$	4.0966	4.1222	4.1297
$1\frac{3}{4}$	1.5966	1.6222	1.6297	$4\frac{1}{2}$	4.3466	4.3722	4.3797
$1\frac{7}{8}$	1.7216	1.7472	1.7547	$4\frac{3}{4}$	4.5966	4.6222	4.6297
2	1.8466	1.8722	1.8797	5	4.8466	4.8722	4.8797
$2\frac{1}{8}$	1.9716	1.9972	2.0047	$5\frac{1}{4}$	5.0966	5.1222	5.1297
$2\frac{1}{4}$	2.0966	2.1222	2.1297	$5\frac{1}{2}$	5.3466	5.3722	5.3797
$2\frac{1}{2}$	2.3466	2.3722	2.3797	$5\frac{3}{4}$	5.5966	5.6222	5.6297
$2\frac{3}{4}$	2.5966	2.6222	2.6297	6	5.8466	5.8722	5.8797

[a] Unified diameter-pitch relationships are marked UN.
For complete manufacturing information and tolerances, see ASA Standard B1.1, 1949.

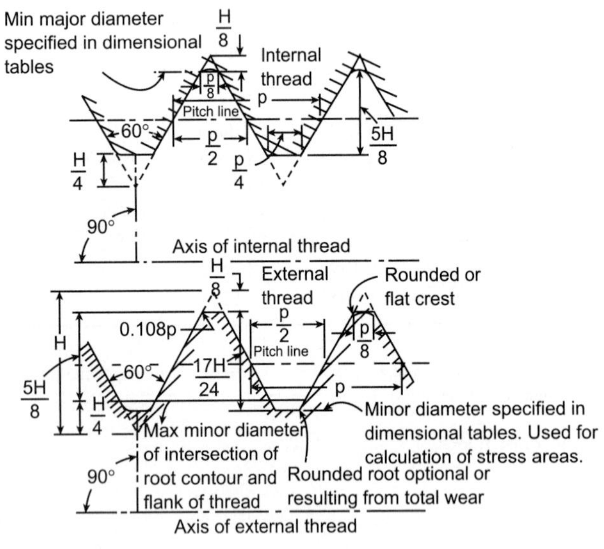

FIGURE 18-13 60° unified and American Standard screw-thread forms.

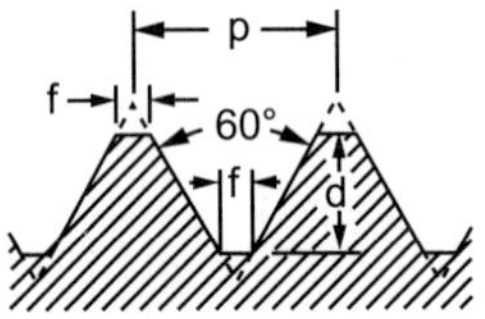

FIGURE 18-14 American Standard screw thread.

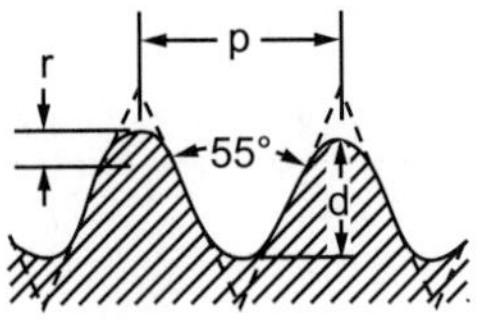

FIGURE 18-15 Whitworth screw thread.

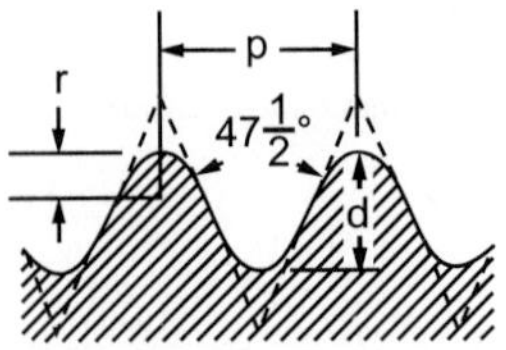

FIGURE 18-16 British Association screw thread.

TABLE 18-20
12-pitch thread series—12N (dimensions in inches)

Size[a] also basic major (normal) diameter, D	Basic minor diameter external thread	Minor diameter internal thread classes 1B, 2B, and 3B for engagement $\frac{2}{3}D$ to $\frac{3}{2}D$		Size[a] also basic major (nominal) diameter, D	Basic minor diameter external thread	Minor diameter internal thread classes 1B, 2B, and 3B for engagement $\frac{2}{3}D$ to $\frac{3}{2}D$	
		Minimum	Maximum			Minimum	Maximum
$\frac{1}{2}$	0.3978	0.4160	0.4223	2 UN	1.8978	1.9148	1.9198
$\frac{9}{16}$	0.4603	0.4783	0.4843	$2\frac{1}{8}$	2.0228	2.0398	2.0448
$\frac{5}{8}$	0.0228	0.5405	0.5463	$2\frac{1}{4}$ UN	2.1478	2.1648	2.1698
$\frac{11}{16}$	0.5853	0.6029	0.6085	$2\frac{3}{8}$	2.2728	2.2898	2.2948
$\frac{3}{4}$	0.6478	0.6653	0.6707	$2\frac{1}{2}$ UN	2.3978	2.4148	2.4198
$\frac{13}{16}$	0.7103	0.7276	0.7329	$2\frac{5}{8}$	2.5228	2.5398	2.5M8
$\frac{7}{8}$	0.7728	0.7900	0.7952	$2\frac{3}{4}$ UN	2.6478	2.6648	2.6698
$\frac{15}{16}$ UN	0.8353	0.8524	0.8575	$2\frac{7}{8}$	2.7728	2.7898	2.7948
1	0.8978	0.9148	0.9198	3 UN	2.8978	2.9148	2.9198
$1\frac{1}{16}$ UN	0.9603	0.9773	0.9823	$3\frac{1}{8}$	3.0228	3.0398	3.0448
$1\frac{1}{8}$	1.0228	1.0398	1.0448	$3\frac{1}{4}$ UN	3.1478	3.1648	3.1698
$1\frac{3}{16}$ UN	1.0853	1.1023	1.1073	$3\frac{3}{8}$	3.2728	3.2898	3.2948
$1\frac{1}{4}$	1.1478	1.1648	1.1698	$3\frac{1}{2}$ UN	3.3978	3.4148	3.4198
$1\frac{5}{16}$ UN	1.2103	1.2273	1.2323	$3\frac{5}{8}$	3.5228	3.5398	3.5448
$1\frac{3}{8}$	1.2728	1.2898	1.2948	$3\frac{3}{4}$ UN	3.6478	3.6648	3.6698
$1\frac{7}{16}$ UN	1.3353	1.3523	1.3573	$3\frac{7}{8}$	3.7728	3.7898	3.7948
$1\frac{1}{2}$	1.3978	1.4148	1.4198	4 UN	3.8978	3.9148	3.9198
$1\frac{5}{8}$	1.5228	1.5398	1.5448	$4\frac{1}{4}$ UN	4.1478	4.1648	4.1698
$1\frac{3}{4}$ UN	1.6478	1.6648	1.6698	$4\frac{1}{2}$ UN	4.3978	4.4148	4.4198
$1\frac{7}{8}$	1.7728	1.7898	1.7948	$4\frac{3}{4}$ UN	4.6478	4.6648	4.6698
				5 UN	4.8978	4.9148	4.9198
				$5\frac{1}{4}$ UN	5.1478	5.1648	5.1698
				$5\frac{1}{2}$ UN	5.3978	5.4148	5.4198
				$5\frac{3}{4}$ UN	5.6478	5.6648	5.6698
				6 UN	5.8978	5.9148	5.9198

[a] Unified diameter-pitch relationships are marked UN.
For complete manufacturing information and tolerances, see ASA Standard B1.1, 1949.

TABLE 18-21
16-pitch thread series—16N (dimensions in inches)

Size[a] also basic major (nominal) diameter, D	Basic minor diameter external thread	Minor diameter internal thread classes 1B, 2B, and 3B for engagement $\frac{2}{3}D$ to $\frac{3}{2}D$		Size[a] also basic major (nominal) diameter, D	Basic minor diameter external thread	Minor diameter internal thread classes 1B, 2B, and 3B for engagement $\frac{2}{3}D$ to $\frac{3}{2}D$	
		Minimum	Maximum			Minimum	Maximum
$\frac{3}{4}$	0.6733	0.6865	0.6908	$2\frac{1}{4}$ UN	2.1733	2.1865	2.1908
$\frac{13}{16}$ UN	0.7358	0.7490	0.7553	$2\frac{5}{16}$	2.2358	2.2490	2.2533
$\frac{7}{8}$ UN	0.7983	0.8115	0.8158	$2\frac{3}{8}$	2.2983	2.3115	2.3158
$\frac{15}{16}$ UN	0.8608	0.8740	0.8783	$2\frac{7}{16}$	2.3608	2.3740	2.3783
1 UN	0.9233	0.9365	0.9408	$2\frac{1}{2}$ UN	2.4233	2.4365	2.4408
$1\frac{1}{16}$ UN	0.9853	0.9990	1.0033	$2\frac{5}{8}$	2.5483	2.5615	2.5658
$1\frac{1}{8}$ UN	1.0483	1.0615	1.0658	$2\frac{3}{4}$ UN	2.6733	2.6865	2.6908
$1\frac{3}{16}$ UN	1.1108	1.1240	1.1283	$2\frac{7}{8}$	2.7983	2.8115	2.8158
$1\frac{1}{4}$ UN	1.1733	1.1865	1.1908	3 UN	2.9233	2.9365	2.9408
$1\frac{5}{16}$ UN	1.2358	1.2490	1.2533	$3\frac{1}{8}$	3.0483	3.0615	3.0658
$1\frac{3}{8}$ UN	1.2983	1.3115	1.3158	$3\frac{1}{4}$ UN	3.1733	3.1865	3.1908
$1\frac{7}{16}$ UN	1.3608	1.3740	1.3783	$3\frac{3}{8}$	3.2983	3.3115	3.3158
$1\frac{1}{2}$ UN	1.4233	1.4365	1.4408	$3\frac{1}{2}$ UN	3.4233	3.4365	3.4408
$1\frac{9}{16}$	1.4858	1.4990	1.5033	$3\frac{5}{8}$	3.5483	3.5615	3.5658
$1\frac{5}{8}$	1.5483	1.5615	1.5658	$3\frac{3}{4}$ UN	3.6733	3.6865	3.6908
$1\frac{11}{16}$	1.6108	1.6240	1.6283	$3\frac{7}{8}$	3.7983	3.8115	3.8158
$1\frac{3}{4}$ UN	1.6733	1.6865	1.6908	4 UN	3.9233	3.9365	3.9408
$1\frac{13}{16}$	1.7358	1.7490	1.7533	$4\frac{1}{4}$ UN	4.1733	4.1865	4.1908
$1\frac{7}{8}$	1.7983	1.8115	1.8158	$4\frac{1}{2}$ UN	4.4233	4.4365	4.4408
$1\frac{15}{16}$	1.8608	1.8740	1.8783	$4\frac{3}{4}$ UN	4.6733	4.6865	4.6908
2 UN	1.9233	1.9365	1.9408	5 UN	4.9233	4.9365	4.9408
$1\frac{1}{16}$	1.9858	1.9990	2.0033	$5\frac{1}{4}$ UN	5.1733	5.1865	5.1908
$2\frac{1}{8}$	2.0483	2.0615	2.0658	$5\frac{1}{2}$ UN	5.4233	5.4365	5.4408
$2\frac{3}{16}$	2.1108	2.1240	2.1283	$5\frac{3}{4}$ UN	5.6733	5.6865	5.6908
				6 UN	5.9233	5.9365	5.9408

[a] Unified diameter-pitch relationships are marked UN.
For complete manufacturing information and tolerances, see ASA Standard B1.1, 1949.

TABLE 18-22
Proportions of power threads (dimensions in inches)

	Square threads		Acme threads		
Size in	Threads per inch	Minor diameter	Threads per inch	Regular minor diameter	Stub minor diameter
$\frac{1}{4}$	10	0.163	16	0.188	0.213
$\frac{5}{16}$	9	0.2153	14	0.241	0.270
$\frac{3}{8}$	8	0.266	12	0.292	0.325
$\frac{7}{16}$	7	0.3125	12	0.354	0.388
$\frac{1}{2}$	$6\frac{1}{2}$	0.366	10	0.400	0.440
$\frac{5}{8}$	$5\frac{1}{2}$	0.466	8	0.500	0.550
$\frac{3}{4}$	5	0.575	6	0.583	0.650
$\frac{7}{8}$	$4\frac{1}{2}$	0.681	6	0.708	0.775
1	4	0.783	5	0.800	0.880
$1\frac{1}{8}$	$3\frac{1}{2}$	0.8750	5	0.925	1.005
$1\frac{1}{4}$	$3\frac{1}{2}$	1.000	5	1.050	1.130
$1\frac{3}{8}$	3	1.0834	4	1.125	1.225
$1\frac{1}{2}$	3	1.284	4	1.250	1.350
$1\frac{3}{4}$	$2\frac{1}{2}$	1.400	4	1.500	1.600
2	$2\frac{1}{4}$	1.612	4	1.750	1.850
$2\frac{1}{4}$	$2\frac{1}{4}$	1.862	3	1.917	2.050
$2\frac{1}{2}$	2	2.063	3	2.167	2.300
$2\frac{3}{4}$	2	2.313	3	2.417	2.550
3	$1\frac{3}{4}$	2.500	2	2.500	2.700
$3\frac{1}{2}$	$1\frac{5}{8}$	2.962	2	3.000	3.200
4	$1\frac{1}{2}$	3.168	2	3.500	3.700
$4\frac{1}{2}$			2	4.000	4.200
5			2	4.000	4.700

TABLE 18-23
British Standard ISO Metric Precision Hexagon Bolts, Screws and Nuts (BS 3692: 1967)

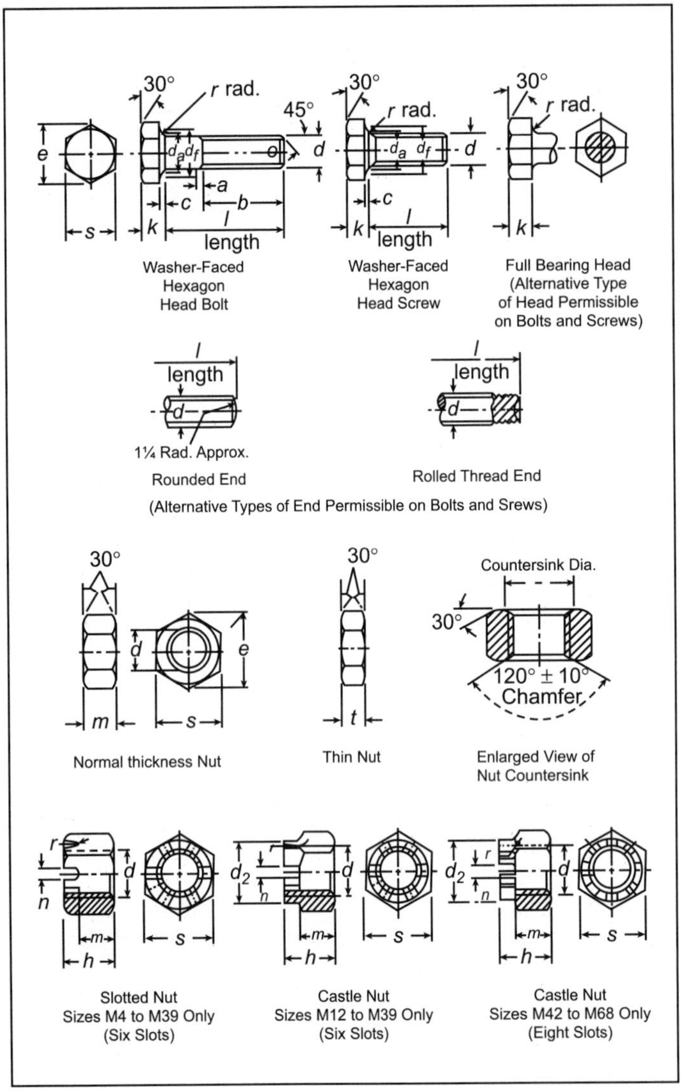

For general dimensions see Tables 2, 3, 4 and 5.
Source: Courtesy British Standards Institution, 2 Park Street, London W1A 2BS, 1986.

TABLE 18-24
British standard machine screws and machine screw nuts—metric series

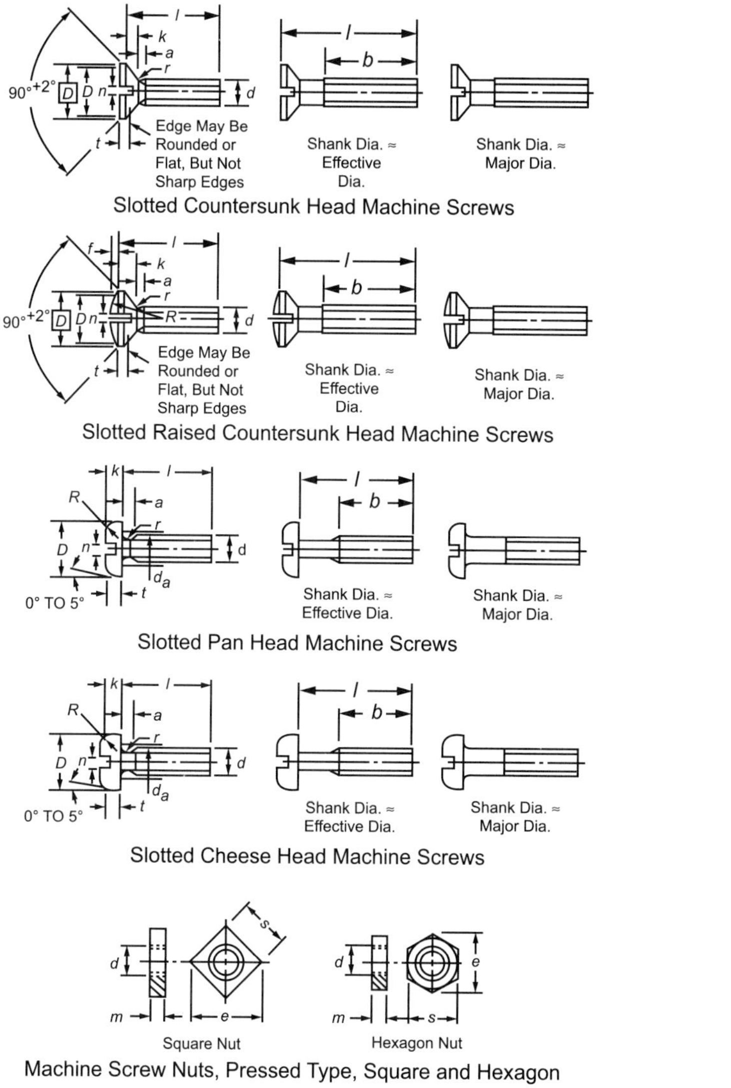

For dimensions, see Tables 1 through 5.
Source: Courtesy British Standards Institution, 2 Park Street, London W1A 2BS, 1986.

TABLE 18.25
Hexagon Bolts[21]

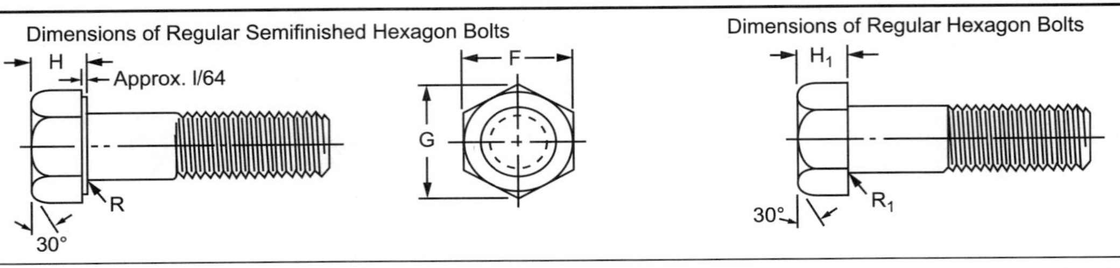

Nominal size or basic major diam of thread		Body diam, Max	Width across flats F			Width across corners G		Semifinished height H			Radius of fillet R		Regular height H_1			Radius of fillet R_1
			Max (basic)		Min	Max	Min	Nom	Max	Min	Max	Min	Max	Nom	Min	Max
3/4	0.2500	0.280	7/16	0.4375	0.425	0.505	0.484	3/32	0.163	0.150	0.031	0.010	11/64	0.188	0.150	0.031
3/16	0.3125	0.342	1/2	0.5000	0.484	0.577	0.552	13/64	0.211	0.195	0.031	0.010	7/32	0.235	0.195	0.031
3/8	0.3750	0.405	9/16	0.5625	0.544	0.650	0.620	13/64	0.243	0.226	0.031	0.010	1/4	0.268	0.226	0.031
7/16	0.4375	0.468	3/8	0.6250	0.603	0.722	0.687	3/32	0.291	0.272	0.031	0.016	19/64	0.316	0.272	0.031
1/2	0.5000	0.530	3/4	0.7500	0.725	0.866	0.826	3/16	0.323	0.302	0.031	0.016	11/32	0.364	0.302	0.031
5/8	0.6250	0.675	15/16	0.9375	0.906	1.083	1.033	23/64	0.403	0.378	0.031	0.016	27/64	0.444	0.378	0.062
3/4	0.7500	0.800	1 3/8	1.1250	1.088	1.299	1.240	13/32	0.483	0.455	0.047	0.031	1/2	0.524	0.455	0.062
7/8	0.8750	0.938	1 3/16	1.3125	1.269	1.516	1.447	35/64	0.563	0.531	0.047	0.031	37/64	0.604	0.531	0.062
1	1.0000	1.063	1 1/2	1.5000	1.450	1.732	1.653	30/64	0.627	0.591	0.047	0.031	42/64	0.700	0.591	0.062
1 1/4	1.1250	1.188	1 11/16	1.6875	1.631	1.949	1.859	11/16	0.718	0.658	0.062	0.047	3/4	0.780	0.658	0.125
1 1/4	1.2500	1.313	1 7/8	1.8750	2.812	1.165	2.066	23/32	0.813	0.749	0.062	0.047	27/32	0.876	0.749	0.125
1 3/8	1.3750	1.469	2 1/16	2.0625	1.994	2.382	2.273	23/32	0.878	0.810	0.062	0.047	20/32	0.940	0.810	0.125
1 1/2	1.5000	1.594	2 1/4	2.2500	2.175	2.598	2.480	13/16	0.974	0.902	0.062	0.047	1	1.036	0.902	0.125
1 3/8	1.6250	1.719	2 7/16	2.4375	2.356	2.815	2.686	1	0.038	0.962	0.062	0.047	1 1/16	1.100	0.962	0.125
1 3/4	1.7500	1.844	2 3/8	2.6250	2.538	3.031	2.893	1 3/32	1.134	1.054	0.062	0.047	1 5/32	1.196	1.054	0.125
1 7/8	1.8750	1.969	2 13/16	2.8125	2.719	3.248	3.100	1 3/32	1.198	1.114	0.062	0.047	1 7/32	1.260	1.114	0.125
2	2.0000	2.094	3	3.0000	2.900	3.464	3.306	1 7/32	1.263	1.175	0.062	0.047	1 11/32	1.388	1.175	0.125
2 1/4	2.2500	2.375	3 3/8	3.3750	3.262	3.897	3.719	1 3/8	1.423	1.327	0.062	0.047	1 1/2	1.548	1.327	0.188
2 1/2	2.5000	2.625	3 3/4	3.7500	3.625	4.330	4.133	1 17/32	1.583	1.479	0.062	0.047	1 21/32	1.708	1.479	0.188
2 3/4	2.7500	2.875	4 1/8	4.1250	3.988	4.763	4.546	1 11/16	1.744	1.632	0.062	0.047	1 13/16	1.869	1.632	0.188
3	3.0000	3.125	4 1/2	4.5000	4.350	5.196	4.959	1 7/8	1.935	1.815	0.062	0.047	2	2.060	1.815	0.188
3 1/4	3.2500	3.438	4 7/8	4.8750	4.712	5.629	5.372	2	2.064	1.936	0.062	0.047	2 3/16	2.251	1.936	0.188
3 1/2	3.5000	3.688	5 1/4	5.2500	5.075	6.062	5.786	2 1/8	2.193	2.057	0.062	0.047	2 5/16	2.380	2.057	0.188
3 3/4	3.7500	3.938	5 3/8	5.5250	5.437	6.495	6.198	2 5/16	2.385	2.241	0.062	0.047	2 1/2	2.572	2.241	0.188
4	4.0000	4.188	6	6.0000	5.800	6.928	6.612	2 1/2	2.576	2.424	0.062	0.047	2 11/16	2.764	2.424	0.188

Courtesy: Viegas, J. J., "Standards for Mechanical Elements", Horald A. Rothbart, Editor, *Mechanical Design and Systems Handbook*, McGraw-Hill Publishing company, New York, 1964.

TABLE 18-26
Regular unfinished square bolts[21]

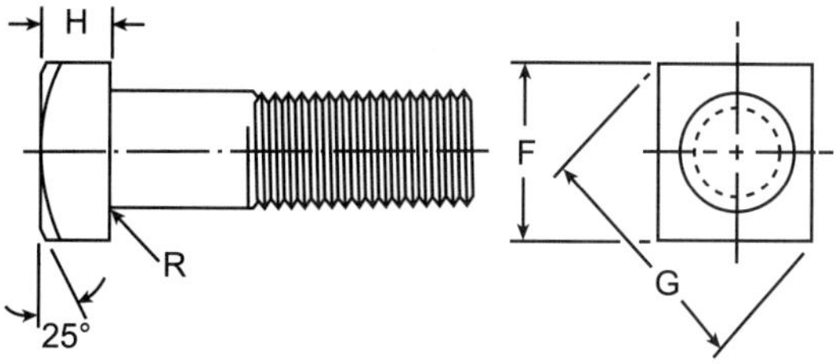

Nominal size or basic major diam of thread		Body diam, max	Width across flats F			Width across corners G		Height H			Radius of fillet R, Max
			Max (basic)		Min	Max	Min	Nom	Max	Min	
$\frac{1}{4}$	0.2500	0.280	$\frac{3}{8}$	0.3750	0.362	0.530	0.498	$\frac{11}{64}$	0.188	0.156	0.031
$\frac{5}{16}$	0.3125	0.342	$\frac{1}{2}$	0.5000	0.484	0.707	0.665	$\frac{13}{64}$	0.220	0.186	0.031
$\frac{3}{8}$	0.3750	0.405	$\frac{9}{16}$	0.5625	0.544	0.795	0.747	$\frac{1}{4}$	0.268	0.232	0.031
$\frac{7}{16}$	0.4375	0.468	$\frac{5}{8}$	0.6250	0.603	0.884	0.828	$\frac{19}{64}$	0.316	0.278	0.031
$\frac{1}{2}$	0.5000	0.530	$\frac{3}{4}$	0.7500	0.725	1.061	0.995	$\frac{21}{64}$	0.348	0.308	0.031
$\frac{5}{8}$	0.6250	0.675	$\frac{15}{16}$	0.9375	0.906	1.326	1.244	$\frac{27}{64}$	0.444	0.400	0.062
$\frac{3}{4}$	0.7500	0.800	$1\frac{1}{8}$	1.1250	1.088	1.591	1.494	$\frac{1}{2}$	0.524	0.476	0.062
$\frac{7}{8}$	0.8750	0.938	$1\frac{15}{16}$	1.3125	1.269	1.856	1.742	$\frac{19}{32}$	0.620	0.568	0.062
1	1.0000	1.063	$1\frac{1}{2}$	1.5000	1.450	2.121	1.991	$\frac{21}{32}$	0.684	0.628	0.062
$1\frac{1}{8}$	1.1250	1.188	$1\frac{11}{16}$	1.6875	1.631	2.386	2.239	$\frac{3}{4}$	0.780	0.720	0.125
$1\frac{1}{4}$	1.2500	1.313	$1\frac{7}{8}$	1.8750	1.812	2.652	2.489	$\frac{27}{32}$	0.876	0.812	0.125
$1\frac{3}{8}$	1.3750	1.469	$2\frac{1}{16}$	2.0625	1.994	2.917	2.738	$\frac{29}{32}$	0.940	0.872	0.125
$1\frac{1}{2}$	1.5000	1.594	$2\frac{1}{4}$	2.2500	2.175	3.182	2.986	1	1.036	0.964	0.125
$1\frac{5}{8}$	1.6250	1.719	$2\frac{7}{16}$	2.4375	2.356	3.447	3.235	$1\frac{3}{32}$	1.132	1.056	0.125

Note: Bolt is not finished on any surface.
Minimum thread length shall be twice the diameter plus $\frac{1}{4}$ in for length up to and including 6 in and twice the diameter plus $\frac{1}{2}$ in for lengths over 6 in. Thread shall be coarse thread series class 2A.

TABLE 18-27
Finished hexagon bolts[21]

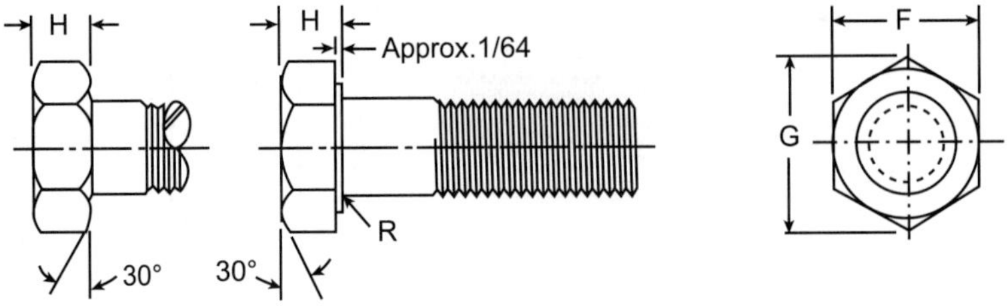

Nominal size or basic major diam of thread		Body diam, min (max equal to nominal size)	Width across flats *F*			Width across corners *G*		Height *H*			Radius of fillet *R*	
			Max (basic)		Min	Max	Min	Nom	Max	Min	Max	Min
$\frac{1}{4}$	0.2500	0.2450	$\frac{7}{16}$	0.4375	0.428	0.505	0.488	$\frac{5}{32}$	0.163	0.150	0.023	0.009
$\frac{5}{16}$	0.3125	0.3065	$\frac{1}{2}$	0.5000	0.489	0.577	0.557	$\frac{13}{64}$	0.211	0.195	0.023	0.009
$\frac{3}{8}$	0.3750	0.3690	$\frac{9}{16}$	0.5625	0.551	0.650	0.628	$\frac{15}{64}$	0.243	0.226	0.023	0.009
$\frac{7}{16}$	0.4375	0.4305	$\frac{5}{8}$	0.6250	0.612	0.722	0.698	$\frac{9}{32}$	0.291	0.272	0.023	0.009
$\frac{1}{2}$	0.5000	0.4930	$\frac{3}{4}$	0.7500	0.736	0.866	0.840	$\frac{5}{16}$	0.323	0.302	0.023	0.009
$\frac{9}{16}$	0.5625	0.5545	$\frac{13}{16}$	0.8125	0.798	0.938	0.910	$\frac{23}{64}$	0.371	0.348	0.041	0.021
$\frac{5}{8}$	0.6250	0.6170	$\frac{15}{16}$	0.9375	0.922	1.083	1.051	$\frac{25}{64}$	0.403	0.378	0.041	0.021
$\frac{3}{4}$	0.7500	0.7410	$1\frac{1}{8}$	1.1250	1.100	1.299	1.254	$\frac{15}{32}$	0.483	0.455	0.041	0.021
$\frac{7}{8}$	0.8750	0.8660	$1\frac{5}{16}$	1.3125	1.285	1.516	1.465	$\frac{35}{64}$	0.563	0.531	0.062	0.047
1	1.0000	0.9900	$1\frac{1}{2}$	1.5000	1.469	1.732	1.675	$\frac{39}{64}$	0.627	0.591	0.062	0.047
$1\frac{1}{8}$	1.1250	1.1140	$1\frac{11}{16}$	1.6875	1.631	1.949	1.859	$\frac{11}{16}$	0.718	0.658	0.125	0.110
$1\frac{1}{4}$	1.2500	1.2390	$1\frac{7}{8}$	1.8750	1.812	1.165	1.066	$\frac{25}{32}$	0.813	0.749	0.125	0.110
$1\frac{3}{8}$	1.3750	1.3630	$2\frac{1}{16}$	2.0625	1.994	2.382	2.273	$\frac{27}{32}$	0.878	0.810	0.125	0.110
$1\frac{1}{2}$	1.5000	1.4880	$2\frac{1}{4}$	2.2500	2.175	2.598	2.480	$\frac{15}{16}$	0.974	0.902	0.125	0.110
$1\frac{5}{8}$	1.6250	1.6130	$2\frac{1}{16}$	2.4275	2.356	2.815	2.686	1	1.038	0.962	0.125	0.110
$1\frac{3}{4}$	1.7500	1.7380	$2\frac{5}{8}$	2.6250	2.538	2.031	2.893	$1\frac{3}{32}$	1.134	1.054	0.125	0.110
$1\frac{7}{8}$	1.8750	1.8630	$2\frac{13}{16}$	2.8125	2.719	2.248	3.100	$1\frac{5}{32}$	1.198	1.114	0.125	0.110
2	2.0000	1.9880	3	3.0000	2.900	3.464	3.306	$1\frac{7}{32}$	1.263	1.175	0.125	0.110
$2\frac{1}{4}$	2.2500	2.2380	$3\frac{3}{8}$	3.3750	3.262	3.897	3.719	$1\frac{3}{8}$	1.423	1.327	0.188	0.173
$2\frac{1}{2}$	2.5000	2.4880	$3\frac{3}{4}$	3.7500	3.625	4.330	4.133	$1\frac{17}{32}$	1.583	1.479	0.188	0.173
$2\frac{3}{4}$	2.7500	2.7380	$4\frac{1}{8}$	4.1250	3.988	4.763	4.546	$1\frac{11}{16}$	1.744	1.632	0.188	0.173
3	3.0000	2.9880	$4\frac{1}{2}$	4.5000	4.350	5.196	4.959	$1\frac{7}{8}$	1.935	1.815	0.188	0.173

Note: Bold type indicates unified thread.

TABLE 18-28
Regular square nuts[21]

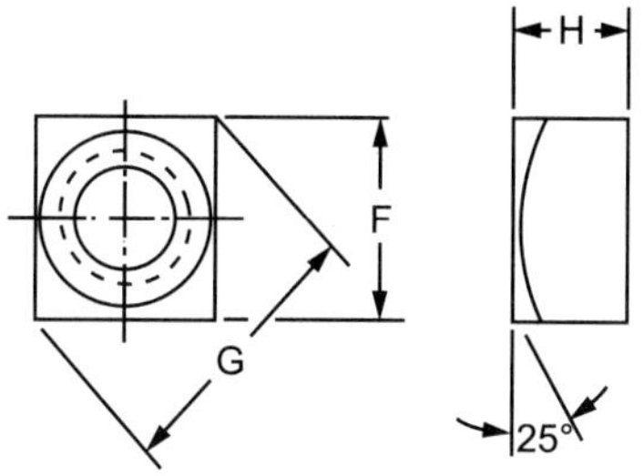

Nominal size or basic major diam of thread		Width across flats F			Width across corners G		Thickness H		
			Max (basic)	Min	Max	Min	Nom	Max	Min
$\frac{1}{4}$	0.2500	$\frac{7}{16}$	0.4375	0.425	0.619	0.584	$\frac{7}{32}$	0.235	0.203
$\frac{5}{16}$	0.3125	$\frac{9}{16}$	0.5625	0.547	0.795	0.751	$\frac{17}{64}$	0.283	0.249
$\frac{3}{8}$	0.3750	$\frac{3}{8}$	0.6250	0.606	0.884	0.832	$\frac{31}{64}$	0.346	0.310
$\frac{7}{16}$	0.4375	$\frac{3}{4}$	0.7500	0.728	1.061	1.000	$\frac{3}{8}$	0.394	0.356
$\frac{1}{2}$	0.5000	$\frac{13}{16}$	0.8125	0.788	1.149	1.082	$\frac{7}{16}$	0.458	0.418
$\frac{5}{8}$	0.6350	1	1.0000	0.969	1.414	1.330	$\frac{35}{64}$	0.569	0.525
$\frac{3}{4}$	0.7500	$1\frac{1}{8}$	1.1250	1.088	1.591	1.494	$\frac{21}{32}$	0.680	0.632
$\frac{7}{8}$	0.8750	$1\frac{5}{16}$	1.3125	1.269	1.856	1.742	$\frac{49}{64}$	0.792	0.740
1	1.0000	$1\frac{1}{2}$	1.5000	1.450	2.121	1.991	$\frac{7}{8}$	0.903	0.847
$1\frac{1}{8}$	1.1250	$1\frac{11}{16}$	1.6875	1.631	2.386	2.239	1	1.030	0.970
$1\frac{1}{4}$	1.2500	$1\frac{7}{8}$	1.8750	1.812	2.652	2.489	$1\frac{3}{32}$	1.126	1.062
$1\frac{3}{8}$	1.3750	$2\frac{1}{16}$	2.0625	1.994	2.917	2.738	$1\frac{13}{64}$	1.237	1.169
$1\frac{1}{2}$	1.5000	$2\frac{1}{4}$	2.2500	2.175	3.182	2.986	$1\frac{5}{16}$	1.348	1.276
$1\frac{5}{8}$	1.6250	$2\frac{7}{16}$	2.4375	2.356	3.447	3.235	$1\frac{27}{64}$	1.460	1.384

TABLE 18-29
Hexagon and hexagon jam nuts[21]

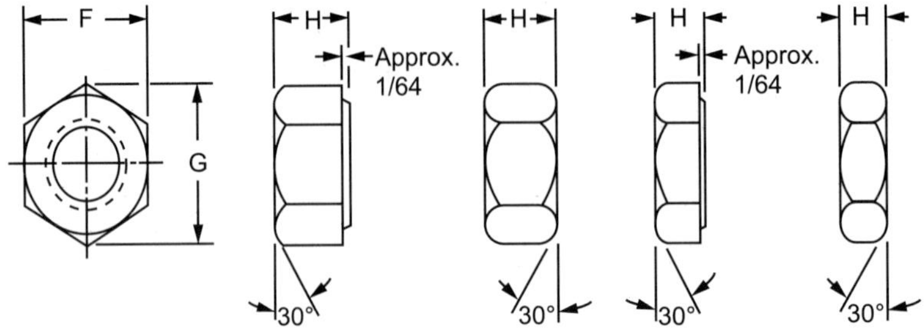

Nominal size or basic major diam of thread		Width across flats *F*			Width across corners *G*		Finished and regular semi-finished hexagon nuts thickness *H*			Finished and regular semi-finished jam nuts thickness *H*		
		Max (basic)		Min	Max	Min	Nom	Max	Min	Nom	Max	Min
$\frac{1}{4}$	0.2500	$\frac{7}{16}$	0.4375	0.428	0.505	0.488	$\frac{7}{32}$	0.226	0.212	$\frac{5}{32}$	0.163	0.150
$\frac{5}{16}$	0.3125	$\frac{1}{2}$	0.5000	0.489	0.577	0.557	$\frac{17}{64}$	0.273	0.258	$\frac{3}{16}$	0.195	0.180
$\frac{3}{8}$	0.3750	$\frac{9}{16}$	0.5625	0.551	0.650	0.628	$\frac{21}{64}$	0.337	0.320	$\frac{7}{32}$	0.227	0.210
$\frac{7}{16}$	0.4375	$\frac{11}{16}$	0.6875	0.675	0.794	0.768	$\frac{3}{8}$	0.385	0.365	$\frac{1}{4}$	0.260	0.240
$\frac{1}{2}$	0.5000	$\frac{3}{4}$	0.7500	0.736	0.866	0.840	$\frac{7}{16}$	0.448	0.427	$\frac{5}{16}$	0.323	0.302
$\frac{9}{16}$	0.5625	$\frac{7}{8}$	0.8750	0.861	1.010	0.982	$\frac{31}{64}$	0.496	0.473	$\frac{6}{16}$	0.324	0.301
$\frac{5}{8}$	0.6250	$\frac{15}{16}$	0.9375	0.922	1.083	1.051	$\frac{35}{64}$	0.559	0.535	$\frac{3}{8}$	0.387	0.363
$\frac{3}{4}$	0.7500	$1\frac{1}{8}$	1.1250	1.088	1.299	1.240	$\frac{41}{64}$	0.665	0.617	$\frac{27}{64}$	0.446	0.398
$\frac{7}{8}$	0.8750	$1\frac{5}{16}$	1.3125	1.269	1.516	1.447	$\frac{3}{4}$	0.776	0.724	$\frac{31}{64}$	0.510	0.458
1	1.0000	$1\frac{1}{2}$	1.5000	1.459	1.732	1.653	$\frac{55}{64}$	0.887	0.831	$\frac{35}{64}$	0.575	0.519
$1\frac{1}{8}$	1.1250	$1\frac{11}{16}$	1.6875	1.631	1.949	1.859	$\frac{31}{32}$	0.999	0.939	$\frac{39}{64}$	0.639	0.579
$1\frac{1}{4}$	1.2500	$1\frac{7}{8}$	1.8750	1.812	2.165	2.066	$1\frac{1}{16}$	0.094	1.030	$\frac{23}{32}$	0.751	0.687
$1\frac{3}{8}$	1.3750	$2\frac{1}{16}$	2.0625	1.994	2.382	2.273	$1\frac{11}{64}$	1.206	1.138	$\frac{25}{32}$	0.815	0.747
$1\frac{1}{2}$	1.5000	$2\frac{1}{4}$	2.2500	2.175	2.598	2.480	$1\frac{9}{32}$	1.317	1.245	$\frac{27}{32}$	0.880	0.808
$1\frac{5}{8}$	1.6250	$2\frac{7}{16}$	2.4375	2.356	2.815	2.686	$1\frac{25}{64}$	1.429	1.353	$\frac{29}{32}$	0.944	0.868
$1\frac{3}{4}$	1.7500	$2\frac{5}{8}$	2.6250	2.538	3.031	2.893	$1\frac{1}{2}$	1.540	1.460	$\frac{31}{32}$	1.009	0.929
$1\frac{7}{8}$	1.8750	$2\frac{13}{16}$	2.8125	2.719	3.248	3.100	$1\frac{39}{64}$	1.651	1.567	$1\frac{1}{32}$	1.073	0.989
2	2.0000	3	3.0000	2.900	3.464	3.306	$1\frac{23}{32}$	1.763	1.675	$1\frac{3}{32}$	1.138	1.050
$2\frac{1}{4}$	2.2500	$3\frac{3}{8}$	3.3750	3.262	3.897	3.719	$1\frac{59}{64}$	1.970	1.874	$1\frac{13}{64}$	1.251	1.155
$2\frac{1}{2}$	2.5000	$3\frac{3}{4}$	3.7500	3.625	4.330	4.133	$2\frac{9}{64}$	2.193	2.089	$1\frac{29}{64}$	1.505	1.401
$2\frac{3}{4}$	2.7500	$4\frac{1}{8}$	4.1250	3.988	4.763	4.546	$2\frac{23}{64}$	2.415	2.303	$1\frac{37}{64}$	1.634	1.522
3	3.0000	$4\frac{1}{2}$	4.5000	4.350	5.196	4.959	$2\frac{37}{64}$	2.638	2.518	$1\frac{45}{64}$	1.763	1.643

Note: Bold type indicates unified threads.

TABLE 18-30
Finished hexagon slotted nuts[21]

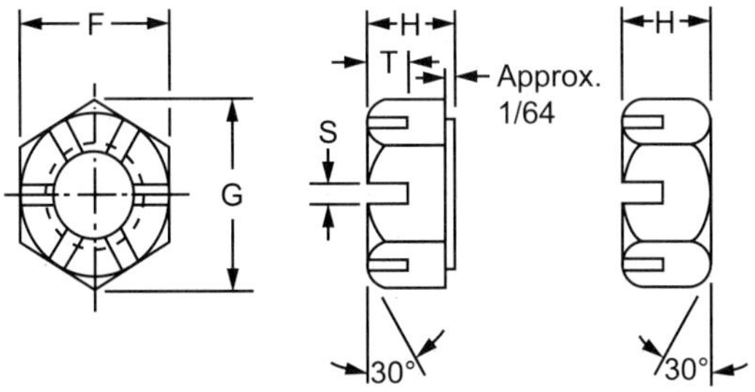

Nominal size or basic major diam of thread		Width across flats F			Width across corners G		Thickness H			Slot	
		Max (basic)		Min	Max	Min	Nom	Max	Min	Width, S	Depth, T
$\frac{1}{4}$	0.2500	$\frac{7}{16}$	0.4375	0.428	0.505	0.488	$\frac{7}{32}$	0.226	0.212	0.078	0.094
$\frac{5}{16}$	0.3125	$\frac{1}{2}$	0.5000	0.489	0.577	0.557	$\frac{17}{64}$	0.273	0.258	0.094	0.094
$\frac{3}{8}$	0.3750	$\frac{9}{16}$	0.5625	0.551	0.650	0.628	$\frac{21}{64}$	0.337	0.320	0.125	0.125
$\frac{7}{16}$	0.4375	$\frac{11}{16}$	0.6875	0.675	0.794	0.768	$\frac{3}{8}$	0.385	0.365	0.125	0.156
$\frac{1}{2}$	0.5000	$\frac{3}{4}$	0.7500	0.736	0.866	0.840	$\frac{7}{16}$	0.448	0.427	0.156	0.156
$\frac{9}{16}$	0.5625	$\frac{7}{8}$	0.8750	0.861	1.010	0.982	$\frac{31}{64}$	0.496	0.473	0.156	0.188
$\frac{5}{8}$	0.6250	$\frac{15}{16}$	0.9375	0.922	1.083	1.051	$\frac{35}{64}$	0.559	0.535	0.156	0.219
$\frac{3}{4}$	0.7500	$1\frac{1}{8}$	1.1250	1.088	1.299	1.240	$\frac{41}{64}$	0.665	0.617	0.188	0.250
$\frac{7}{8}$	0.8750	$1\frac{5}{16}$	1.3125	1.269	1.516	1.447	$\frac{3}{4}$	0.776	0.724	0.188	0.250
1	1.0000	$1\frac{1}{2}$	1.5000	1.450	1.732	1.653	$\frac{55}{64}$	0.887	0.831	0.250	0.281
$1\frac{1}{8}$	1.1250	$1\frac{11}{16}$	1.6875	1.631	1.949	1.859	$\frac{31}{32}$	0.999	0.939	0.250	0.344
$1\frac{1}{4}$	1.2500	$1\frac{7}{8}$	1.8750	1.812	2.165	2.066	$1\frac{1}{16}$	1.094	1.030	0.312	0.375
$1\frac{3}{8}$	1.3750	$2\frac{1}{16}$	2.0625	1.994	2.382	2.273	$1\frac{11}{64}$	1.206	1.138	0.312	0.375
$1\frac{1}{2}$	1.5000	$2\frac{1}{4}$	2.2500	2.175	2.598	2.480	$1\frac{9}{32}$	1.317	1.245	0.375	0.438
$1\frac{5}{8}$	1.6250	$2\frac{7}{16}$	2.4375	2.356	2.815	2.686	$1\frac{25}{64}$	1.429	1.353	0.375	0.438
$1\frac{3}{4}$	1.7500	$2\frac{5}{8}$	2.6250	2.538	2.031	2.893	$1\frac{1}{2}$	1.540	1.460	0.438	0.500
$1\frac{7}{8}$	1.8750	$2\frac{13}{16}$	2.8125	2.719	3.248	3.100	$1\frac{39}{64}$	1.651	1.567	0.438	0.562
2	2.0000	3	3.0000	2.900	3.464	3.306	$1\frac{23}{32}$	1.763	1.675	0.438	0.562
$2\frac{1}{4}$	2.2500	$3\frac{3}{8}$	3.3750	3.262	3.897	3.719	$1\frac{59}{64}$	1.970	1.874	0.438	0.562
$2\frac{1}{2}$	2.5000	$3\frac{3}{4}$	3.7500	3.625	4.330	4.133	$2\frac{9}{64}$	2.193	2.089	0.562	0.688
$2\frac{3}{4}$	2.7500	$4\frac{1}{8}$	4.1250	3.988	4.763	4.546	$2\frac{23}{64}$	2.415	2.303	0.562	0.688
3	3.0000	$4\frac{1}{2}$	4.5000	4.350	5.196	4.959	$2\frac{37}{64}$	2.638	2.518	0.625	0.750

Note: Bold type indicates unified threads.

TABLE 18-31
Regular hexagon and hexagon jam nuts

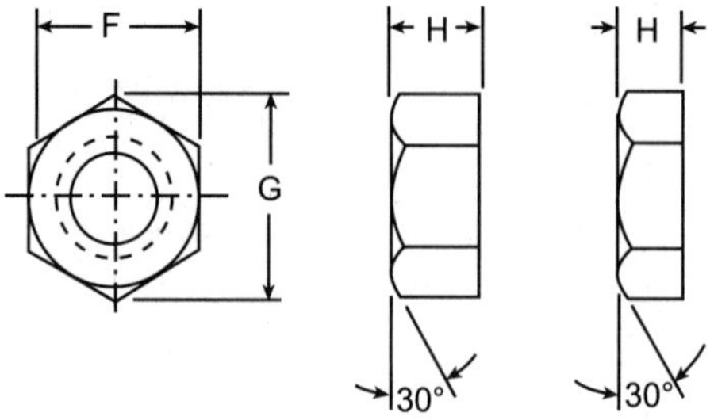

Nominal size or basic major diam of thread		Width across flats *F*			Width across corners *G*		Thickness regular nuts *H*			Thickness regular jam nuts *H*		
		Max (basic)		Min	Max	Min	Nom	Max	Min	Nom	Max	Min
$\frac{1}{4}$	0.2500	$\frac{7}{16}$	0.4375	0.425	0.505	0.484	$\frac{7}{32}$	0.235	0.203	$\frac{5}{32}$	0.172	0.140
$\frac{5}{16}$	0.3125	$\frac{9}{16}$	0.5625	0.547	0.650	0.624	$\frac{17}{64}$	0.283	0.249	$\frac{3}{16}$	0.204	0.170
$\frac{3}{8}$	0.3750	$\frac{5}{8}$	0.6250	0.606	0.722	0.691	$\frac{21}{64}$	0.346	0.310	$\frac{7}{32}$	0.237	0.201
$\frac{7}{16}$	0.4375	$\frac{3}{4}$	0.7500	0.728	0.866	0.830	$\frac{3}{8}$	0.394	0.356	$\frac{1}{4}$	0.260	0.231
$\frac{1}{2}$	0.5000	$\frac{13}{16}$	0.8125	0.788	0.938	0.898	$\frac{7}{16}$	0.458	0.418	$\frac{5}{16}$	0.332	0.292
$\frac{9}{16}$	0.5625	$\frac{7}{8}$	0.8750	0.847	1.010	0.966	$\frac{1}{2}$	0.521	0.479	$\frac{11}{32}$	0.365	0.323
$\frac{5}{8}$	0.6250	1	1.0000	0.969	1.155	1.104	$\frac{35}{64}$	0.569	0.525	$\frac{3}{8}$	0.397	0.353
$\frac{3}{4}$	0.7500	$1\frac{1}{8}$	1.1250	1.088	1.299	1.240	$\frac{21}{32}$	0.680	0.632	$\frac{7}{16}$	0.462	0.414
$\frac{7}{8}$	0.8750	$1\frac{5}{8}$	1.3125	1.269	1.516	1.447	$\frac{49}{64}$	0.792	0.740	$\frac{1}{2}$	0.526	0.474
1	1.0000	$1\frac{1}{2}$	1.5000	1.450	1.732	1.653	$\frac{7}{8}$	0.903	0.847	$\frac{9}{16}$	0.590	0.534
$1\frac{1}{8}$	1.1250	$1\frac{11}{16}$	1.6875	1.631	1.949	1.859	1	1.030	0.970	$\frac{5}{8}$	0.655	0.595
$1\frac{1}{4}$	1.2500	$1\frac{7}{8}$	1.8750	1.812	2.165	2.066	$1\frac{0}{32}$	1.126	1.062	$\frac{3}{4}$	0.782	0.718
$1\frac{5}{8}$	1.3750	$2\frac{1}{16}$	2.0625	1.994	2.382	2.273	$1\frac{13}{64}$	1.237	1.169	$\frac{13}{16}$	0.846	0.778
$1\frac{1}{2}$	1.5000	$2\frac{1}{4}$	2.2500	2.175	2.598	2.480	$1\frac{3}{16}$	1.348	1.276	$\frac{7}{8}$	0.911	0.839

TABLE 18-32
Carriage bolts

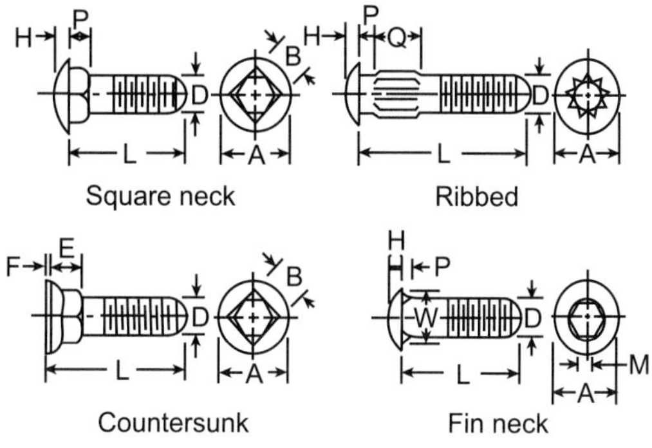

Nominal diam of bolt D	Countersunk: Diam of head max, A	Countersunk: Feed thickness F	Countersunk: Depth of square and countersink max, E	Countersunk: Width of square, max, B	Fin neck: Diam of head, max, A	Fin neck: Height of head, max, H	Fin neck: Depth of fins, max, P	Fin neck: Distance across fins, max, W	Fin neck: Thickness of fins, max, M
No. 10	0.520	0.016	0.250	0.199	0.469	0.114	0.088	0.395	0.098
$\frac{1}{4}$	0.645	0.016	0.312	0.260	0.594	0.145	0.104	0.458	0.114
$\frac{3}{16}$	0.770	0.031	0.375	0.324	0.719	0.176	0.135	0.551	0.145
$\frac{3}{8}$	0.895	0.031	0.437	0.388	0.844	0.208	0.151	0.645	0.161
$\frac{7}{16}$	1.020	0.031	0.500	0.452	0.969	0.239	0.182	0.739	0.192
$\frac{1}{2}$	1.145	0.031	0.562	0.515	1.094	0.270	0.198	0.833	0.208
$\frac{5}{8}$	1.400	0.031	0.687	0.642					
$\frac{3}{4}$	1.650	0.047	0.812	0.768					

Nominal diam of bolt D	Square neck: Diam of head, max, A	Square neck: Height of head, max, H	Square neck: Depth of square, max, P	Square neck: Width of square, max, B	Ribbed neck: Ribs below head P, $L \leq \frac{7}{8}$	Ribbed neck: Ribs below head P, $L \geq 1$	Ribbed neck: Length of ribs Q, $L \leq \frac{7}{8}$	Ribbed neck: Length of ribs Q, $L = 1, L = 1\frac{1}{8}$	Ribbed neck: Length of ribs Q, $L \geq 1\frac{1}{4}$	Number of ribs
No. 10	0.469	0.114	0.125	0.199	0.031	0.063	0.188	0.313	0.500	9
$\frac{1}{4}$	0.594	0.145	0.156	0.260	0.031	0.063	0.188	0.313	0.500	10
$\frac{5}{16}$	0.719	0.176	0.187	0.324	0.031	0.063	0.188	0.313	0.500	12
$\frac{3}{8}$	0.844	0.208	0.219	0.388	0.031	0.063	0.188	0.313	0.500	12
$\frac{7}{16}$	0.969	0.239	0.250	0.452	0.031	0.063	0.188	0.313	0.500	14
$\frac{1}{2}$	1.094	0.270	0.281	0.515	0.031	0.063	0.188	0.313	0.500	16
$\frac{5}{8}$	1.344	0.344	0.344	0.642	0.094	0.094	0.188	0.313	0.500	19
$\frac{5}{8}$	1.594	0.406	0.406	0.768	0.094	0.094	0.188	0.313	0.500	22
$\frac{7}{8}$	1.844	0.469	0.469	0.895						
1	2.094	0.531	0.531	1.022						

TABLE 18-33
Countersunk, Buttonhead, and Step bolts

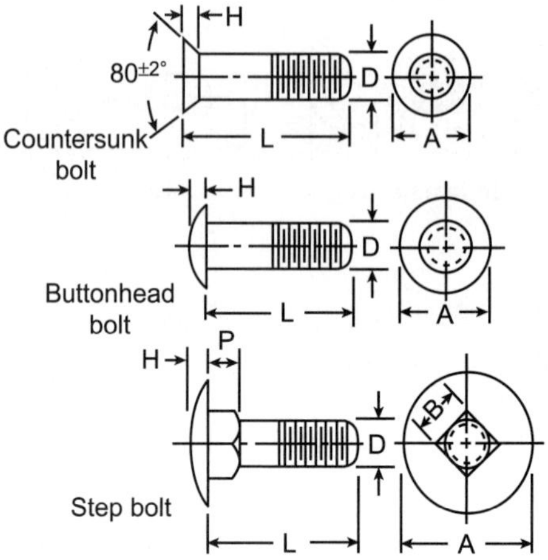

		Countersunk bolt		Buttonhead		Step bolt			
Nominal diam of bolt *D*		Head diam, max, *A*	Head depth *H*	Head diam, max, *A*	Head height, max, *H*	Head diam, max, *A*	Head height, max, *H*	Depth of square, max, *P*	Width of square, max, *B*
No. 10	24	—	—	0.469	0.114	0.656	0.114	0.125	0.199
$\frac{1}{4}$	20	0.493	0.140	0.594	0.145	0.844	0.145	0.156	0.260
$\frac{5}{16}$	18	0.618	0.176	0.719	0.176	1.031	0.176	0.187	0.324
$\frac{3}{8}$	16	0.740	0.210	0.844	0.208	1.219	0.208	0.219	0.388
$\frac{7}{16}$	14	0.803	0.210	0.969	0.239	1.406	0.239	0.250	0.452
$\frac{1}{2}$	13	0.935	0.250	1.094	0.270	1.594	0.270	0.281	0.515
$\frac{5}{8}$	11	1.169	0.313	1.344	0.344				
$\frac{3}{4}$	10	1.402	0.375	1.594	0.406				
$\frac{7}{8}$	9	1.637	0.438	1.844	0.469				
1	8	1.869	0.500	2.094	0.531				
$1\frac{1}{8}$	7	2.104	0.563						
$1\frac{1}{4}$	7	2.337	0.625						
$1\frac{3}{8}$	6	2.571	0.688						
$1\frac{1}{2}$	6	2.804	0.750						

TABLE 18-34
Machine-screw heads

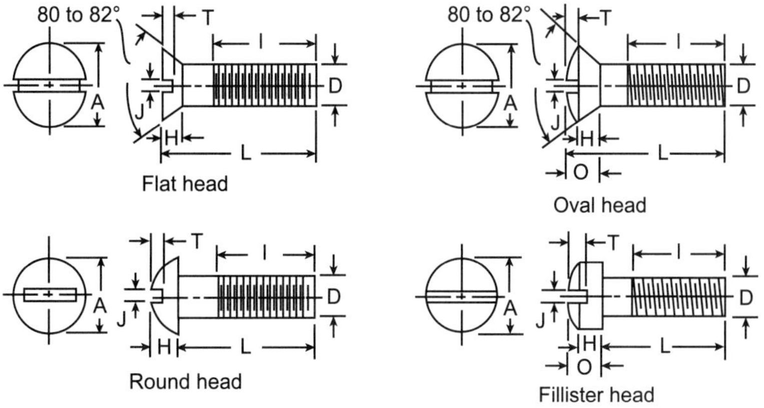

		Flat head					Round head				
Nominal size	Max diam *D*	Head diam, max, *A*	Height of head, max, *H*	Width of slot, min, *J*	Depth of slot, min, *T*	Total height of head, max, *O*	Head diam, max, *A*	Height of head, max, *H*	Width of slot, min, *J*	Depth of slot, min, *T*	Total height of head, max, *O*
No. 0	0.060	0.119	0.035	0.016	0.010		0.113	0.053	0.016	0.029	
No. 1	0.073	0.146	0.043	0.019	0.012		0.138	0.061	0.019	0.033	
No. 2	0.086	0.172	0.051	0.023	0.015		0.162	0.069	0.023	0.037	
No. 3	0.099	0.199	0.059	0.027	0.017		0.137	0.078	0.027	0.040	
No. 4	0.112	0.225	0.067	0.031	0.020		0.211	0.086	0.031	0.044	
No. 5	0.125	0.252	0.075	0.035	0.022		0.236	0.095	0.035	0.047	
No. 6	0.138	0.279	0.083	0.039	0.024		0.260	0.103	0.039	0.051	
No. 8	0.164	0.332	0.100	0.045	0.029		0.309	0.120	0.045	0.058	
No. 10	0.190	0.385	0.116	0.050	0.034		0.359	0.137	0.050	0.065	
No. 12	0.216	0.438	0.132	0.056	0.039		0.408	0.153	0.056	0.072	
$\frac{1}{4}$	0.250	0.507	0.153	0.064	0.046		0.472	0.175	0.064	0.082	
$\frac{5}{16}$	0.3125	0.635	0.191	0.072	0.058		0.590	0.216	0.072	0.099	
$\frac{3}{8}$	0.375	0.762	0.230	0.081	0.070		0.708	0.256	0.081	0.117	
$\frac{7}{16}$	0.4375	0.812	0.223	0.081	0.066		0.750	0.328	0.081	0.148	
$\frac{1}{2}$	0.500	0.875	0.223	0.091	0.065		0.813	0.355	0.091	0.159	
$\frac{9}{16}$	0.5625	1.000	0.260	0.102	0.077		0.938	0.410	0.102	0.183	
$\frac{5}{8}$	0.625	1.125	0.298	0.116	0.088		1.000	0.438	0.116	0.195	
$\frac{3}{4}$	0.750	1.375	0.372	0.131	0.111		1.250	0.547	0.131	0.242	

TABLE 18-34
Machine-screw heads (*Cont.*)

		Oval head					Fillister head				
Nominal Size	Max diam *D*	Head diam, max, *A*	Height of head, max, *H*	Width of slot, min, *J*	Depth of slot, min, *T*	Total height of head, max, *O*	Head diam, max, *A*	Height of head, max, *H*	Width of slot, min, *J*	Depth of slot, min, *T*	Total height of head, max, *O*
No. 0	0.060	0.119	0.035	0.016	0.025	0.056	0.098	0.045	0.016	0.015	0.059
No. 1	0.073	0.146	0.043	0.019	0.031	0.068	0.118	0.053	0.019	0.020	0.071
No. 2	0.086	0.172	0.051	0.023	0.037	0.080	0.140	0.062	0.023	0.025	0.083
No. 3	0.099	0.199	0.059	0.027	0.043	0.092	0.161	0.070	0.027	0.030	0.095
No. 4	0.112	0.225	0.067	0.031	0.049	0.104	0.183	0.079	0.031	0.035	0.107
No. 5	0.125	0.252	0.075	0.035	0.055	0.116	0.205	0.088	0.035	0.040	0.120
No. 6	0.138	0.279	0.083	0.039	0.060	0.128	0.226	0.096	0.039	0.045	0.132
No. 8	0.164	0.332	0.100	0.045	0.072	0.152	0.270	0.113	0.045	0.054	0.156
No. 10	0.190	0.385	0.116	0.050	0.084	0.176	0.313	0.130	0.050	0.064	0.180
No. 12	0.216	0.438	0.132	0.056	0.096	0.200	0.357	0.148	0.056	0.074	0.205
$\frac{1}{4}$	0.250	0.507	0.153	0.064	0.112	0.232	0.414	0.170	0.064	0.087	0.237
$\frac{5}{16}$	0.3125	0.635	0.191	0.072	0.141	0.290	0.518	0.211	0.072	0.110	0.295
$\frac{3}{8}$	0.375	0.762	0.230	0.081	0.170	0.347	0.622	0.253	0.081	0.133	0.355
$\frac{7}{16}$	0.4375	0.812	0.223	0.081	0.174	0.345	0.625	0.265	0.081	0.135	0.368
$\frac{1}{2}$	0.500	0.875	0.223	0.091	0.176	0.354	0.750	0.297	0.091	0.151	0.412
$\frac{9}{16}$	0.5625	1.000	0.260	0.102	0.207	0.410	0.812	0.336	0.102	0.172	0.466
$\frac{5}{8}$	0.625	1.125	0.298	0.116	0.235	0.467	0.875	0.375	0.116	0.193	0.521
$\frac{3}{4}$	0.750	1.375	0.372	0.131	0.293	0.578	1.000	0.441	0.131	0.226	0.612

Note: Edges of head on flat- and oval-head machine screws may be rounded.
Radius of fillet at base of flat- and oval-head machine screws shall not exceed twice the pitch of the screw thread.
Radius of fillet at base of round- and fillister-head machine screws shall not exceed one-half the pitch of the screw thread.
All four types of screws in this table may be furnished with cross-recessed heads.
Fillister-head machine screws in sizes No. 2 to $\frac{3}{8}$ in, inclusive, may be furnished with a drilled hole through the head along a diameter at right angles to the slot but not breaking through the slot.

TABLE 18-35
Machine screw heads—pan, hexagon, truss, and 100° Flat heads

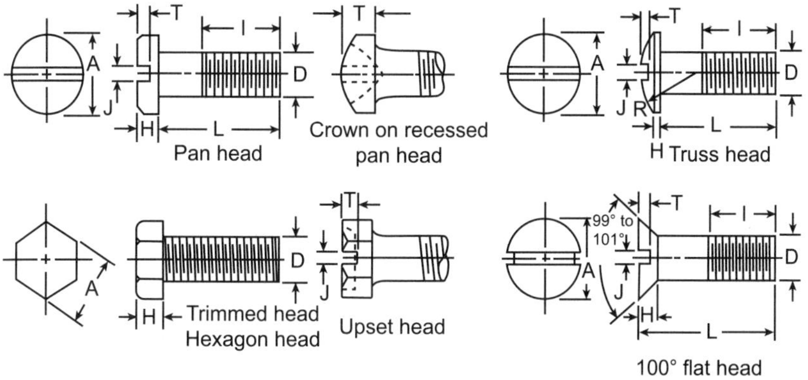

Nominal size	Max diam *D*	Pan head: Head diam, max, *A*	Pan head: Height of slotted head, max, *H*	Pan head: Width of slot, min, *J*	Pan head: Depth of slot, min, *T*	Pan head: Radius *R*	Pan head: Height of recessed head max, *O*	Hexagon head: Head diam, max, *A*	Hexagon head: Height of head, max, *H*	Hexagon head: Width of slot, min, *J*	Hexagon head: Depth of slot, min, *T*
No. 2	0.086	0.167	0.053	0.023	0.023	0.035	0.062	0.125	0.050		
No. 3	0.099	0.193	0.060	0.027	0.027	0.037	0.071	0.187	0.055		
No. 4	0.112	0.219	0.068	0.031	0.030	0.042	0.080	0.187	0.060	0.031	0.025
No. 5	0.125	0.245	0.075	0.035	0.032	0.044	0.089	0.187	0.070	0.035	0.030
No. 6	0.138	0.270	0.082	0.039	0.038	0.046	0.097	0.250	0.080	0.039	0.033
No. 8	0.164	0.322	0.096	0.045	0.043	0.052	0.115	0.250	0.110	0.045	0.052
No. 10	0.190	0.373	0.110	0.050	0.050	0.061	0.133	0.312	0.120	0.050	0.057
No. 12	0.216	0.425	0.125	0.056	0.060	0.078	0.151	0.312	0.155	0.056	0.077
$\frac{1}{4}$	0.250	0.492	0.144	0.064	0.070	0.087	0.175	0.375	0.190	0.064	0.083
$\frac{5}{16}$	0.3125	0.615	0.178	0.072	0.092	0.099	0.218	0.500	0.230	0.072	0.100
$\frac{3}{8}$	0.375	0.740	0.212	0.081	0.113	0.143	0.261	0.562	0.295	0.081	0.131

TABLE 18-35
Machine screw heads—pan, hexagon, truss, and 100° Flat heads (*Cont.*)

		Truss head						100° flat head			
Nominal size	Max diam *D*	Head diam, max, *A*	Height of slotted head, max, *H*	Width of slot, min, *J*	Depth of slot, min, *T*	Radius *R*	Height of recessed head max, *O*	Head diam, max, *A*	Height of head, max, *H*	Width of slot, min, *J*	Depth of slot, min, *T*
No. 2	0.086	0.194	0.053	0.023	0.022	0.129					
No. 3	0.099	0.226	0.061	0.027	0.026	0.151					
No. 4	0.112	0.257	0.069	0.031	0.030	0.169		0.225	0.048	0.031	0.017
No. 5	0.125	0.289	0.078	0.035	0.034	0.191					
No. 6	0.138	0.321	0.086	0.039	0.037	0.211		0.279	0.060	0.039	0.022
No. 8	0.164	0.384	0.102	0.045	0.045	0.254		0.332	0.072	0.045	0.027
No. 10	0.190	0.448	0.118	0.050	0.053	0.283		0.385	0.083	0.050	0.031
No. 12	0.216	0.511	0.134	0.056	0.061	0.336					
$\frac{1}{4}$	0.250	0.573	0.150	0.064	0.070	0.375		0.507	0.110	0.064	0.042
$\frac{5}{16}$	0.3125	0.698	0.183	0.072	0.085	0.457		0.635	0.138	0.072	0.053
$\frac{3}{8}$	0.375	0.823	0.215	0.081	0.100	0.538		0.762	0.165	0.081	0.064
$\frac{7}{16}$	0.4375	0.948	0.248	0.081	0.116	0.619					
$\frac{1}{2}$	0.500	1.073	0.280	0.091	0.131	0.701					
$\frac{9}{16}$	0.5625	1.198	0.312	0.102	0.146	0.783					
$\frac{5}{8}$	0.625	1.323	0.345	0.116	0.162	0.863					
$\frac{3}{4}$	0.750	1.573	0.410	0.131	0.182	1.024					

Note: Radius of fillet at base of truss- and pan-head machine screws shall not exceed one-half the pitch of the screw thread.
Truss-, pan-, and 100° flat-head machine screws may be furnished with cross-recessed heads.
Hexagon-head machine screws are usually not slotted; the slot is optional. Also optional is an upset-head type for hexagon-head machine screws of sizes 4, 5, 8, 12, and $\frac{1}{4}$ in.

TABLE 18-36
Machine-screw heads—binding head

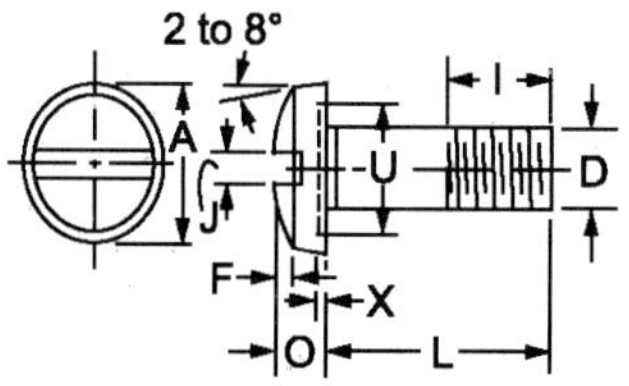

Binding head

Nominal size	Max diam *D*	Head diam, max, *A*	Total height of head, max, *O*	Width of slot, min, *J*	Depth of slot, min, *T*	Height of oval, max, *F*	Diam of undercut,[a] min, *U*	Depth of undercut, min, *X*
No. 2	0.086	0.181	0.046	0.023	0.024	0.018	0.124	0.005
No. 3	0.099	0.208	0.054	0.027	0.029	0.022	0.143	0.006
No. 4	0.112	0.235	0.063	0.031	0.034	0.025	0.161	0.007
No. 5	0.125	0.263	0.071	0.035	0.039	0.029	0.180	0.009
No. 6	0.138	0.290	0.080	0.039	0.044	0.032	0.199	0.010
No. 8	0.164	0.344	0.097	0.045	0.054	0.039	0.236	0.012
No. 10	0.190	0.399	0.114	0.050	0.064	0.045	0.274	0.015
No. 12	0.216	0.454	0.130	0.056	0.074	0.052	0.311	0.018
$\frac{1}{4}$	0.250	0.513	0.153	0.064	0.088	0.061	0.360	0.021
$\frac{5}{16}$	0.3125	0.641	0.193	0.072	0.112	0.077	0.450	0.027
$\frac{3}{8}$	0.375	0.769	0.234	0.081	0.136	0.094	0.540	0.034

[a] Use of undercut is optional.
Note: Binding-head machine screws may be furnished with cross-recessed heads.

TABLE 18-37
Slotted-head cap screws[28]

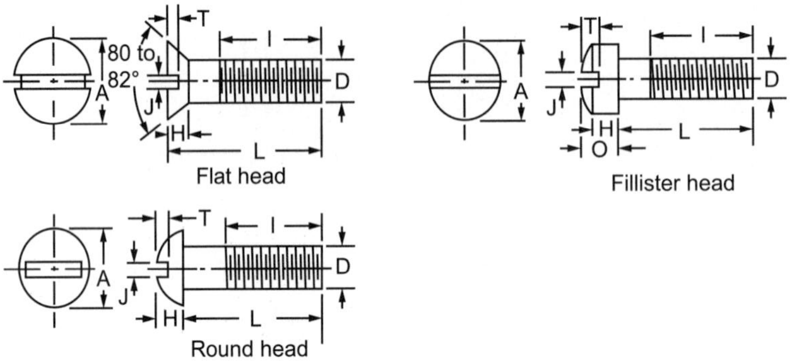

Nominal size (body diam, max) *D*	Width of slot min, *J*	Fillister head: Head diam, max, *A*	Fillister head: Height of head, max, *H*	Fillister head: Total height of head, max, *O*	Fillister head: Depth of slot, min, *T*	Flat head: Head diam, max, *A*	Flat head: Height of head average, *H*	Flat head: Depth of slot, min, *T*	Round head: Head diam, max, *A*	Round head: Height of head, max, *H*	Round head: Depth of slot, min, *T*
1/4	0.064	0.375	0.172	0.216	0.077	0.500	0.140	0.046	0.437	0.191	0.097
3/16	0.072	0.437	0.203	0.253	0.090	0.625	0.176	0.057	0.562	0.246	0.126
3/8	0.081	0.562	0.250	0.314	0.113	0.750	0.210	0.069	0.625	0.273	0.135
7/16	0.081	0.625	0.297	0.368	0.133	0.8125	0.210	0.069	0.750	0.328	0.167
1/2	0.091	0.750	0.328	0.412	0.148	0.875	0.210	0.069	0.812	0.355	0.179
9/16	0.102	0.812	0.375	0.466	0.169	1.000	0.245	0.080	0.937	0.410	0.208
5/8	0.116	0.875	0.422	0.521	0.190	1.125	0.281	1.092	1.000	0.438	0.220
3/4	0.131	1.000	0.500	0.612	0.233	1.375	0.352	0.115	0.125	0.547	0.227
7/8	0.147	1.125	0.594	0.720	0.264	1.625	0.423	0.139			
1	0.166	1.312	0.656	0.802	0.292	1.875	0.494	0.162			
1 1/8	0.178	—	—	—	—	2.062	0.529	0.173			
1 1/4	0.193	—	—	—	—	2.312	0.600	0.197			
1 3/8	0.208	—	—	—	—	2.562	0.665	0.220			
1 1/2	0.240	—	—	—	—	2.812	0.742	0.244			

TABLE 18-38
Socket-head cap screws

					Hexagon[a]	Fluted socket			
Nominal size	Body diam, max, *D*	Head diam, max, *A*	Head height max, *H*	Head side height, max, *S*	Socket width across flats, min, *J*	Number of flutes	Socket diam minor, min, *K*	Socket diam major, min, *M*	Width of socket land, max, *N*
No. 0	0.060	0.096							
No. 1	0.073	0.118							
No. 2	0.0860	0.140	0.086	0.0803	$\frac{1}{16}$	6	0.063	0.073	0.016
No. 3	0.0990	0.161	0.099	0.0923	$\frac{5}{64}$	6	0.080	0.097	0.022
No. 4	0.1120	0.183	0.112	0.1043	$\frac{5}{64}$	6	0.080	0.097	0.022
No. 5	0.1250	0.205	0.125	0.1163	$\frac{3}{32}$	6	0.096	0.113	0.025
No. 6	0.1380	0.226	0.138	0.1284	$\frac{3}{32}$	6	0.096	0.113	0.025
No. 8	0.1640	0.270	0.164	0.1522	$\frac{1}{8}$	6	0.126	0.147	0.032
No. 10	0.1900	$\frac{5}{16}$	0.190	0.1765	$\frac{5}{32}$	6	0.161	0.186	0.039
No. 12	0.2160	$\frac{11}{32}$	0.216	0.2005	$\frac{5}{32}$	6	0.161	0.186	0.039
$\frac{1}{4}$	0.2500	$\frac{3}{8}$	$\frac{1}{4}$	0.2317	$\frac{3}{16}$	6	0.188	0.219	0.050
$\frac{5}{16}$	0.3125	$\frac{7}{16}$	$\frac{5}{16}$	0.2894	$\frac{7}{32}$	6	0.219	0.254	0.060
$\frac{3}{8}$	0.3750	$\frac{9}{16}$	$\frac{3}{8}$	0.3469	$\frac{5}{16}$	6	0.316	0.377	0.092
$\frac{7}{16}$	0.4375	$\frac{5}{8}$	$\frac{7}{16}$	0.4046	$\frac{5}{16}$	6	0.316	0.377	0.092
$\frac{1}{2}$	0.5000	$\frac{1}{4}$	$\frac{1}{2}$	0.4620	$\frac{3}{8}$	6	0.383	0.460	0.112
$\frac{9}{16}$	0.5625	$\frac{13}{16}$	$\frac{9}{16}$	0.5196	$\frac{3}{8}$	6	0.383	0.460	0.112
$\frac{5}{8}$	0.6250	$\frac{7}{8}$	$\frac{5}{8}$	0.5771	$\frac{1}{2}$	6	0.506	0.601	0.138
$\frac{3}{4}$	0.7500	1	$\frac{3}{4}$	0.6920	$\frac{9}{16}$	6	0.531	0.627	0.149
$\frac{7}{8}$	0.8750	$1\frac{1}{8}$	$\frac{7}{8}$	0.8069	$\frac{9}{16}$	6	0.600	0.705	0.168
1	1.0000	$1\frac{5}{16}$	1	0.9220	$\frac{5}{8}$	6	0.681	0.797	0.189
$1\frac{1}{8}$	1.1250	$1\frac{1}{2}$	$1\frac{1}{8}$	1.0372	$\frac{3}{4}$	6	0.824	0.966	0.231
$1\frac{1}{4}$	1.2500	$1\frac{3}{4}$	$1\frac{1}{4}$	1.1516	$\frac{3}{4}$	6	0.824	0.966	0.231
$1\frac{3}{8}$	1.3750	$1\frac{7}{8}$	$1\frac{3}{8}$	1.2675	$\frac{3}{4}$	6	0.824	0.966	0.231
$1\frac{1}{2}$	1.5000	2	$1\frac{1}{2}$	1.3821	1	6	1.003	1.271	0.298

[a] Maximum socket depth *T* should not exceed three-fourths of minimum head height *H*.
Note: Head chamfer angle *E* is 28 to 32°, the edge between flat and chamfer being slightly rounded.
Screw point chamfer angle 35 to 40°, the chamfer extending to the bottom of the thread. Edge between flat and chamfer is slightly rounded.

TABLE 18-39
Square-head set screws[28]

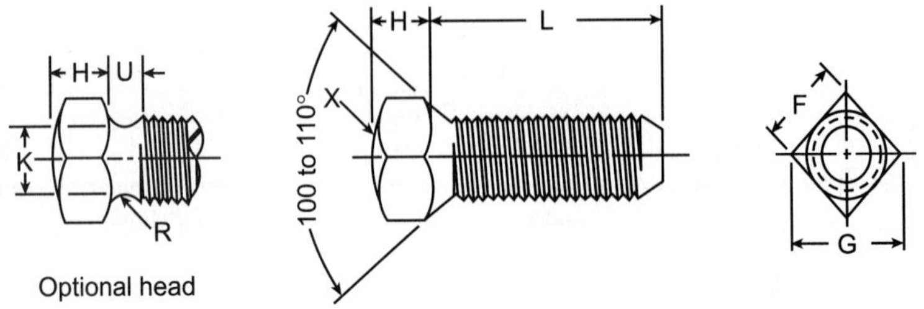

Nominal size		Width across flats *F* Max	Width across flats *F* Min	Width across corners *G* Min	Height of head *H* Nom	Height of head *H* Max	Height of head *H* Min	Diam of neck relief *K* Max	Diam of neck relief *K* Min	Radius of head *X* Nom	Radius of neck relief *R* Max	Width of neck relief *U* Max
No. 10	0.190	0.1875	0.180	0.247	$\frac{9}{64}$	0.148	0.134	0.145	0.140	$\frac{15}{32}$	0.027	0.083
No. 12	0.216	0.216	0.208	0.292	$\frac{5}{32}$	0.163	0.147	0.162	0.156	$\frac{35}{64}$	0.029	0.091
$\frac{1}{4}$	0.250	0.250	0.241	0.331	$\frac{3}{16}$	0.196	0.178	0.185	0.170	$\frac{5}{8}$	0.032	0.100
$\frac{5}{16}$	0.3125	0.3125	0.302	0.415	$\frac{15}{64}$	0.245	0.224	0.240	0.225	$\frac{25}{32}$	0.036	0.111
$\frac{3}{8}$	0.3750	0.375	0.362	0.497	$\frac{7}{32}$	0.293	0.270	0.294	0.279	$\frac{15}{16}$	0.041	0.125
$\frac{7}{16}$	0.4375	0.4375	0.423	0.581	$\frac{31}{64}$	0.341	0.315	0.345	0.330	$1\frac{3}{32}$	0.046	0.143
$\frac{1}{2}$	0.500	0.500	0.484	0.665	$\frac{3}{8}$	0.398	0.361	0.400	0.385	$1\frac{1}{4}$	0.050	0.154
$\frac{9}{16}$	0.5625	0.5625	0.545	0.748	$\frac{27}{64}$	0.437	0.407	0.454	0.439	$1\frac{13}{32}$	0.054	0.167
$\frac{5}{8}$	0.6250	0.625	0.606	0.833	$\frac{15}{32}$	0.485	0.452	0.507	0.492	$1\frac{9}{16}$	0.059	0.182
$\frac{3}{4}$	0.750	0.750	0.729	1.001	$\frac{9}{16}$	0.582	0.544	0.620	0.605	$1\frac{7}{8}$	0.065	0.200
$\frac{7}{8}$	0.875	0.875	0.852	1.170	$\frac{21}{32}$	0.678	0.635	0.731	0.716	$2\frac{3}{16}$	0.072	0.222
1	1.000	1.000	0.974	1.337	$\frac{3}{4}$	0.774	0.726	0.838	0.823	$2\frac{1}{2}$	0.081	0.250
$1\frac{1}{8}$	1.125	1.125	1.096	1.505	$\frac{27}{32}$	0.870	0.817	0.939	0.914	$2\frac{13}{16}$	0.092	0.283
$1\frac{1}{4}$	1.250	1.250	1.219	1.674	$\frac{15}{16}$	0.966	0.908	1.064	1.039	$3\frac{1}{8}$	0.092	0.283
$1\frac{3}{8}$	1.376	1.375	1.342	1.843	$1\frac{1}{32}$	1.063	1.000	1.159	1.134	$3\frac{7}{16}$	0.109	0.333
$1\frac{1}{2}$	1.500	1.500	1.464	2.010	$1\frac{1}{8}$	1.159	1.091	1.284	1.259	$3\frac{3}{4}$	0.109	0.333

TABLE 18-40
Square-head setscrew points[28]

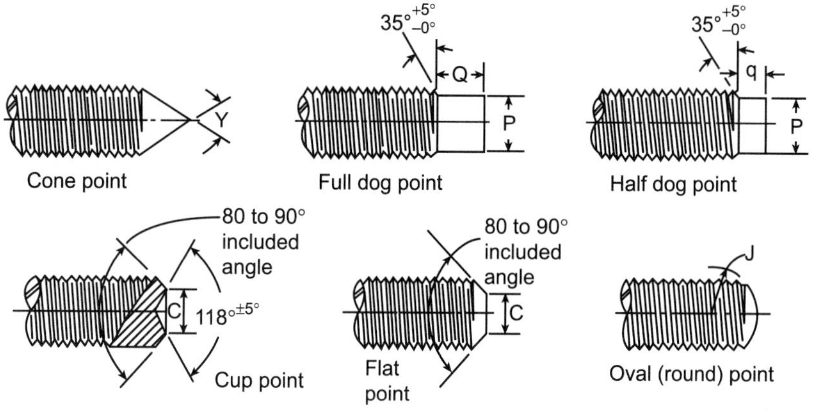

Nominal size	Diam of cap and flat points C: Nom	Max	Min	Oval (round) point radius J, nom	Full-dog, half-dog, and pivot point[a]: Diam P Max	Diam P Min	Full dog and pivot Q	Half dog q
No. 10	$\frac{3}{32}$	0.102	0.088	0.141	0.127	0.120	0.090	0.045
No. 12	$\frac{7}{64}$	0.115	0.101	0.156	0.144	0.137	0.110	0.055
$\frac{1}{4}$	$\frac{1}{8}$	0.132	0.118	0.188	0.156	0.149	0.125	0.063
$\frac{5}{16}$	$\frac{11}{64}$	0.172	0.156	0.234	0.203	0.195	0.156	0.078
$\frac{3}{8}$	$\frac{13}{64}$	0.212	0.194	0.281	0.250	0.241	0.188	0.094
$\frac{7}{16}$	$\frac{16}{64}$	0.252	0.232	0.328	0.297	0.287	0.219	0.109
$\frac{1}{2}$	$\frac{9}{32}$	0.291	0.270	0.375	0.344	0.334	0.250	0.125
$\frac{9}{16}$	$\frac{5}{16}$	0.332	0.309	0.422	0.391	0.379	0.281	0.140
$\frac{5}{8}$	$\frac{22}{64}$	0.371	0.347	0.469	0.469	0.456	0.313	0.156
$\frac{3}{4}$	$\frac{7}{16}$	0.450	0.425	0.563	0.563	0.549	0.375	0.188
$\frac{7}{8}$	$\frac{33}{64}$	0.530	0.502	0.656	0.656	0.642	0.438	0.219
1	$\frac{19}{32}$	0.609	0.579	0.750	0.750	0.734	0.500	0.250
$1\frac{1}{8}$	$\frac{43}{64}$	0.689	0.655	0.844	0.844	0.826	0.562	0.281
$1\frac{1}{4}$	$\frac{3}{4}$	0.767	0.733	0.938	0.938	0.920	0.625	0.312
$1\frac{3}{8}$	$\frac{53}{64}$	0.848	0.808	1.031	1.031	1.011	0.688	0.344
$1\frac{1}{2}$	$\frac{29}{32}$	0.926	0.886	1.125	1.125	1.105	0.750	0.375

[a] Pivot points are similar to full-dog point except that the point is rounded by a radius equal to J.
Where usable length of thread is less than the nominal diameter, half-dog point shall be used.
When length equals nominal diameter or less, $Y = 118° \pm 2°$; when length exceeds nominal diameter, $Y = 90° \pm 2°$
Note: All dimensions are given in inches.

TABLE 18-41
Slotted headless setscrews[28]

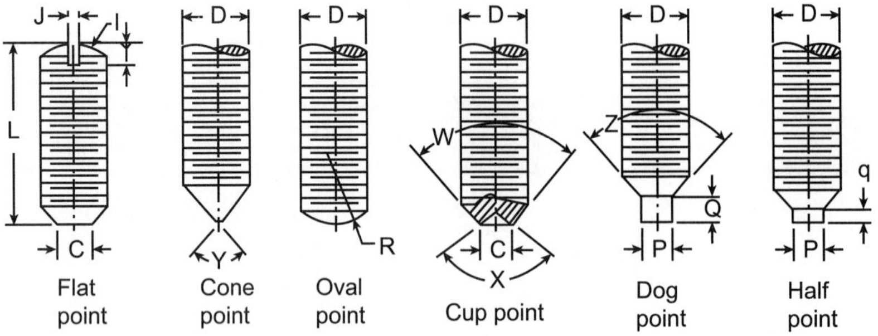

Nominal size	D	Radius of headless crown I	Width of slot J	Depth of slot T	Oval-points radius R	Diam of cup and flat points C		Diam of dog point P		Length of dog point[a]	
						Max	Min	Max	Min	Fill Q	Half q
5	0.125	0.125	0.023	0.031	0.094	0.067	0.057	0.083	0.078	0.060	0.030
6	0.138	0.138	0.025	0.035	0.109	0.074	0.064	0.092	0.087	0.070	0.035
8	0.164	0.164	0.029	0.041	0.125	0.087	0.076	0.109	0.103	0.080	0.040
10	0.190	0.190	0.032	0.048	0.141	0.102	0.088	0.127	0.120	0.090	0.045
12	0.216	0.216	0.036	0.054	0.156	0.115	0.101	0.144	0.137	0.110	0.055
$\frac{1}{4}$	0.250	0.250	0.045	0.063	0.188	0.132	0.118	0.156	0.149	0.125	0.063
$\frac{5}{16}$	0.3125	0.313	0.051	0.078	0.234	0.172	0.156	0.203	0.195	0.156	0.078
$\frac{3}{8}$	0.375	0.375	0.064	0.094	0.281	0.212	0.194	0.250	0.241	0.188	0.094
$\frac{7}{16}$	0.4375	0.438	0.072	0.109	0.328	0.252	0.232	0.297	0.287	0.219	0.109
$\frac{1}{2}$	0.500	0.500	0.081	0.125	0.375	0.291	0.270	0.344	0.344	0.250	0.135
$\frac{9}{16}$	0.5625	0.563	0.091	0.141	0.422	0.332	0.309	0.391	0.379	0.281	0.140
$\frac{5}{8}$	0.625	0.625	0.102	0.156	0.469	0.371	0.347	0.469	0.456	0.313	0.156
$\frac{3}{4}$	0.750	0.750	0.129	0.188	0.563	0.450	0.425	0.563	0.549	0.375	0.188

[a] Where usable length thread is less than the nominal diameter, half-dog point shall be used.
When L (length of screw) equals nominal diameter or less, $Y = 118° \pm 2°$; when L exceeds nominal diameter, $Y = 90° \pm 2°$.
Point angles $\alpha = 80$ to $90°$; $X = 118° \pm 5°$; $Z = 100$ to $110°$.
Allowable eccentricity of dog point axis with respect to axis of screw shall not exceed 3% of nominal screw diameter with maximum 0.005 in.
Note: All dimensions given in inches.

TABLE 18-42
Fluted and hexagon socket-headless setscrews[24]

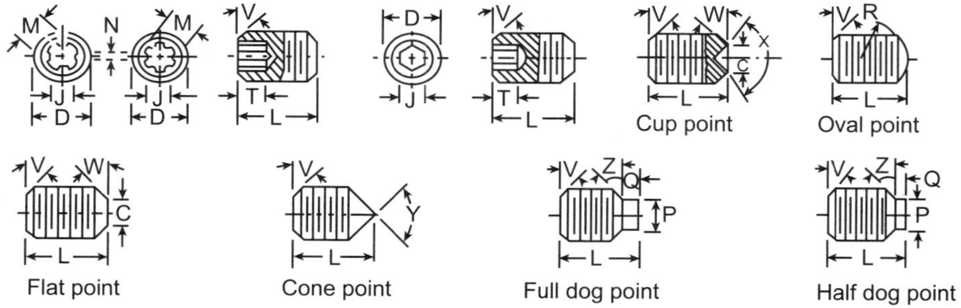

Screw size nominal diam *D*	Cup- and flat-point diameter *C*		Oval-point radius *R*	Cone-point angle *Y*: 118 ± 2° for these lengths and under	Cone-point angle *Y*: 90 ± 2° for these lengths and over	Dog point: Diam *P*		Dog point		Key engagement[†] min	Hexagon type: Socket width across flats *J*		Fluted type[a]: Socket diam, minor, *J*		Fluted type[a]: Socket diam, major, *M*		Fluted type[a]: Socket land width *N*	
	Max	Min				Max	Min	Full *Q*	Half *q*		Max	Min	Max	Min	Max	Min	Max	Min
No. 0	0.033	0.027	3/64	1/16	5/64	0.040	0.037	0.030	0.015	0.022	0.0285	0.028	0.026	0.0255	0.035	0.034	0.012	0.0115
No. 1	0.040	0.033	0.055	5/64	3/32	0.049	0.045	0.037	0.019	0.028	0.0355	0.035	0.026	0.0255	0.035	0.034	0.012	0.0115
No. 2	0.047	0.039	1/16	3/32	7/64	0.057	0.053	0.043	0.022	0.028	0.0355	0.035	0.038	0.0375	0.050	0.049	0.017	0.016
No. 3	0.054	0.045	5/64	7/64	1/8	0.066	0.062	0.050	0.025	0.040	0.051	0.050	0.038	0.0375	0.050	0.049	0.017	0.016
No. 4	0.061	0.051	0.084	1/8	5/32	0.075	0.070	0.056	0.028	0.040	0.051	0.050	0.051	0.050	0.062	0.061	0.014	0.013
No. 5	0.067	0.057	3/32	1/8	3/16	0.083	0.078	0.06	0.03	0.050	0.0635	1/16	0.053	0.052	0.071	0.070	0.022	0.021
No. 6	0.074	0.064	7/64	1/8	3/16	0.092	0.087	0.07	0.035	0.050	0.0635	1/16	0.056	0.055	0.079	0.078	0.023	0.022
No. 8	0.087	0.076	1/8	3/16	1/4	0.109	0.103	0.08	0.04	0.062	0.0791	5/64	0.082	0.080	0.098	0.097	0.022	0.021
No. 10	0.102	0.088	9/64	3/16	1/4	0.127	0.120	0.09	0.045	0.075	0.0947	3/32	0.098	0.096	0.115	0.113	0.025	0.023
No. 12	0.115	0.101	5/32	3/16	1/4	0.144	0.137	0.11	0.055	0.075	0.0947	3/32	0.098	0.096	0.115	0.113	0.025	0.023

TABLE 18-42
Fluted and hexagon socket-headless setscrews[24] *(Cont.)*

Screw size nominal diam *D*	Fluted and hexagon socket types: Cup- and flat-point diameter *C*		Oval-point radius *R*	Cone-point angle *Y*		Dog point				Key engagement[b] min	Hexagon type: Socket width across flats *J*		Fluted type[a]: Socket diam, minor, *J*		Socket diam, major, *M*		Socket land width *N*	
				118 ± 2° for these lengths and under	90 ± 2° for these lengths and over	Diam *P*												
	Max	Min				Max	Min	Full *Q*	Half *q*		Max	Min	Max	Min	Max	Min	Max	Min
1/4	0.132	0.118	3/16	1/4	5/16	5/32	0.149	1/8	1/16	0.100	0.1270	1/8	0.128	0.126	0.149	0.147	0.032	0.030
5/16	0.172	0.156	15/64	5/16	3/8	13/64	0.195	3/32	5/64	0.125	0.1582	5/32	0.163	0.161	0.188	0.186	0.039	0.037
3/8	0.212	0.194	9/32	3/8	7/16	1/4	0.241	3/16	3/32	0.150	0.1895	3/16	0.190	0.188	0.221	0.219	0.050	0.048
7/16	0.252	0.232	21/64	7/16	1/2	19/64	0.287	7/32	7/64	0.175	0.2207	7/32	0.221	0.219	0.256	0.254	0.060	0.058
1/2	0.291	0.270	3/8	1/2	9/16	11/32	0.334	1/4	1/8	0.200	0.2520	1/4	0.254	0.252	0.298	0.296	0.068	0.066
9/16	0.332	0.309	27/64	9/16	5/8	25/64	0.379	9/32	5/64	0.200	0.2520	1/4	0.254	0.252	0.298	0.296	0.068	0.066
5/8	0.371	0.347	15/32	5/8	3/4	15/32	0.456	5/16	3/32	0.250	0.3155	5/16	0.319	0.316	0.380	0.377	0.092	0.089
3/4	0.450	0.425	9/16	3/4	7/8	9/16	0.549	3/8	3/16	0.300	0.3780	3/8	0.386	0.383	0.463	0.460	0.112	0.109
7/8	0.530	0.502	21/32	7/8	1	21/32	0.642	7/16	7/32	0.400	0.5030	1/2	0.509	0.506	0.604	0.601	0.138	0.134
1	0.609	0.579	3/4	1	1 1/8	3/4	0.734	1/2	1/4	0.450	0.5655	9/16	0.535	0.531	0.631	0.627	0.149	0.145
1 1/8	0.689	0.655	27/32	1 1/8	1 1/4	27/32	0.826	9/16	9/16	0.450	0.5655	9/16	0.604	0.600	0.709	0.705	0.168	0.164
1 1/4	0.767	0.733	15/16	1 1/4	1 1/2	15/16	0.920	5/8	5/16	0.500	0.6290	3/8	0.685	0.681	0.801	0.797	0.189	0.185
1 3/8	0.848	0.808	1 1/32	1 3/8	1 5/8	1 1/32	1.011	11/16	11/32	0.500	0.6290	3/8	0.744	0.740	0.869	0.865	0.207	0.203
1 1/2	0.926	0.886	1 1/8	1 1/2	1 3/4	1 1/8	1.105	3/4	3/8	0.600	0.7540	3/4	0.828	0.824	0.970	0.966	0.231	0.227
1 3/4	1.086	1.039	1 5/16	1 3/4	2	1 5/16	1.289	7/8	7/16	0.800	0.0040	1	1.007	1.003	1.275	1.271	0.298	0.294
2	1.244	1.193	1 1/2	2	2 1/4	1 1/2	1.474	1	1/2	0.800	1.0040	1	1.007	1.003	1.275	1.271	0.298	0.294

[a] The number of flutes for setscrews Nos. 0, 1, 2, 3, 5, and 6 is four. The number of flutes for Nos. 4, 8, and larger is six.

[b] These dimensions apply to cup- and flat-point screws one diameter in length or longer. For screws shorter than one diameter in length, and for other types of points, socket to be as deep as practicable.

Note: All dimensions are given in inches.

Source: Courtesy: John J. Viegas, *Standards for Mechanical Elements*, Harold A. Rothbart, Editor-in-Chief; *Mechanical Design and Systems Handbook*, McGraw-Hill Publishing Company, New York, 1964.

TABLE 18-43
American National Standard metric hex cap screws (ANSI B18.2.3.1M-1979, R1989)

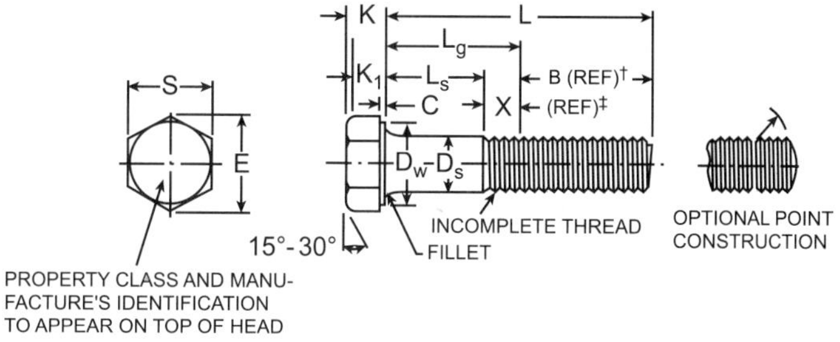

Nominal screw diam, *D* and thread pitch	Body diam, D_s		Width across flats, *S*		Width across corners, *E*		Head height, *K*		Wrenching height, K_1	Washer face thickness, *C*	
	Max	Min	Max	Min	Max	Min	Max	Min	Min	Max	Min
M5 × 0.8	5.00	4.82	8.00	7.78	9.24	8.79	3.65	3.35	2.4	0.5	0.2
M6 × 1	6.00	5.82	10.00	9.78	11.55	11.05	4.15	3.85	2.8	0.5	0.2
M8 × 1.25	8.00	7.78	13.00	12.73	15.01	14.38	5.50	5.10	3.7	0.6	0.3
M10 × 1.5*	10.00	9.78	15.00	14.73	17.32	16.64	6.63	6.17	4.5	0.6	0.3
M10 × 1.5	10.00	9.78	16.00	15.73	18.48	17.77	6.63	6.17	4.5	0.6	0.3
M12 × 1.75	12.00	11.73	18.00	17.73	20.78	20.03	7.76	7.24	5.2	0.6	0.3
M14 × 2	14.00	13.73	21.00	20.67	24.25	23.35	9.09	8.51	6.2	0.6	0.3
M16 × 2	16.00	15.73	24.00	23.67	27.71	26.75	10.32	8.68	7.0	0.8	0.4
M20 × 2.5	20.00	19.67	30.00	29.16	34.64	32.95	12.88	12.12	8.8	0.8	0.4
M24 × 3	24.00	23.67	36.00	35.00	41.57	39.55	15.44	14.56	10.5	0.8	0.4
M30 × 3.5	30.00	29.67	46.00	45.00	53.12	50.85	15.48	17.92	13.1	0.8	0.4
M36 × 4	36.00	35.61	55.00	53.80	63.51	60.79	23.38	21.62	15.8	0.8	0.4
M42 × 4.5	42.00	41.38	65.00	62.90	75.06	71.71	26.97	25.03	18.2	1.0	0.5
M48 × 5	48.00	47.38	75.00	72.60	86.60	83.76	31.07	28.93	21.0	1.0	0.5
M56 × 5.5	56.00	55.26	85.00	82.20	98.15	93.71	36.20	33.80	24.5	1.0	0.5
M64 × 6	64.00	63.26	95.00	91.80	109.70	104.65	41.32	38.68	28.0	1.0	0.5
M72 × 6	72.00	71.26	105.00	101.40	121.24	115.60	46.45	43.55	31.5	1.2	0.6
M80 × 6	80.00	79.26	115.00	111.00	132.72	126.54	51.58	48.42	35.0	1.2	0.6
M90 × 6	90.00	89.13	130.00	125.50	150.11	143.07	57.74	54.26	39.2	1.2	0.6
M100 × 6	100.00	99.13	145.00	140.00	167.43	159.60	63.90	60.10	43.4	1.2	0.6

All dimensions are in millimeters.

* This size with width across flats of 15 mm is not standard. Unless specifically ordered hex cap screws with 16 mm width across flats will be furnished.

† Transition thread length, *X*, includes the length of incomplete threads and tolerances gaging length and body length. It is intended for calculation purposes.

‡ Basic thread lengths, *B*, are the same as given in Table 18-47.

For additional manufacturing and acceptance specifications, reference should be made to Standard.

Courtesy: American National Standards Institution, New York, USA. (ANSI B18.2.1M-1979, R1989)

TABLE 18-44
American National Standard metric formed hex screws (ANSI B18.2.3.2M-1979, R1989)

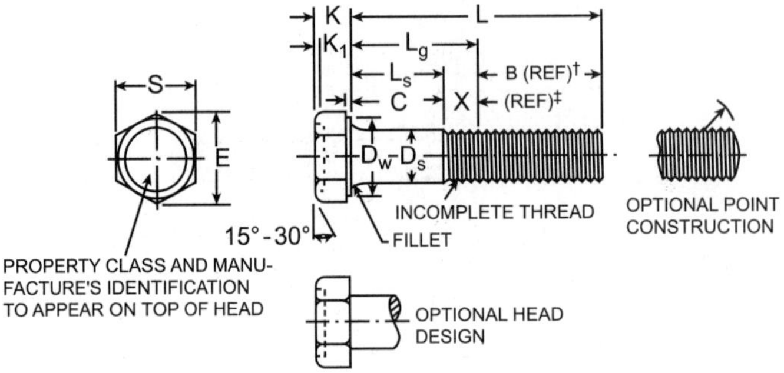

Nominal screw diam, *D*, and thread pitch	Body diam, D_s		Width across flats, *S*		Width across corners, *E*		Head height, *K*		Wrenching height, K_1	Washer face thickness, *C*		Washer face diam, D_w
	Max	Min	Max	Min	Max	Min	Max	Min	Min	Max	Min	Min
M5 × 0.8	5.00	4.82	8.00	7.64	9.24	8.56	3.65	3.35	2.4	0.5	0.2	6.9
M6 × 1	5.00	5.82	10.00	9.64	11.55	10.80	4.15	3.85	2.0	0.5	0.2	8.9
M8 × 1.25	8.00	7.78	13.00	12.57	15.01	14.08	5.50	5.10	3.7	0.6	0.3	11.6
M10 × 1.5*	10.00	9.78	15.00	14.57	17.32	16.32	6.63	6.17	4.5	0.6	0.3	13.6
M10 × 1.5	10.00	9.78	16.00	15.57	18.48	17.43	6.63	6.17	4.5	0.6	0.3	14.6
M12 × 1.75	12.00	11.73	18.00	17.57	20.78	19.68	7.76	7.24	5.2	0.6	0.3	16.6
M14 × 2	14.00	13.73	21.00	20.16	24.25	22.58	9.09	8.51	6.2	0.6	0.3	19.6
M16 × 2	16.00	15.73	24.00	23.16	27.71	25.94	10.32	9.68	7.0	0.8	0.4	22.5
M20 × 2.5	20.00	19.67	30.00	29.16	34.64	32.66	12.88	12.12	8.8	0.8	0.4	27.7
M24 × 3	24.00	23.67	36.00	35.00	41.57	39.20	15.44	14.56	10.5	0.8	0.4	32.2

All dimensions are in millimeters.
* This size with width across flats of 15 mm is not standard. Unless specifically ordered M10 formed hex screws with 16 mm width across flats will be furnished.
† Transition thread length, *X*, includes the length of incomplete threads and tolerances on the grip gaging length and body length. It is intended for calculation purposes.
‡ Basic thread lengths, *B* are the same as given in Table 18-47.
For additional manufacturing and acceptance specifications, reference should be made to the Standard.
Courtesy: American National Standards Institution, New York, USA. (ANSI B18.2.3.2M-1979, R1989)

TABLE 18-45
American National Standard metric heavy hex screws (ANSI B18.2.3.3M-1979, R1989)

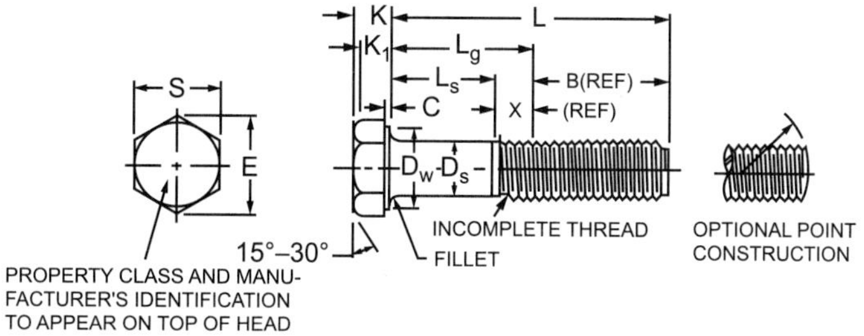

Nominal screw diam, *D*, and thread pitch	Body diam, D_s		Width across flats, *S*		Width across corners, *E*		Head height, *K*		Wrenching height, K_1	Washer face thickness, *C*		Washer face diam, D_w
	Max	Min	Max	Min	Max	Min	Max	Min	Min	Max	Min	Min
M12 × 1.75	12.00	11.73	21.00	20.67	24.25	23.35	7.76	7.24	5.2	0.6	0.3	19.0
M14 × 2	14.00	13.73	24.00	23.67	27.71	26.75	9.09	8.51	6.2	0.6	0.3	22.0
M16 × 2	16.00	15.73	27.00	26.67	31.18	30.14	10.32	9.68	7.0	0.8	0.4	25.0
M20 × 2.5	20.00	19.67	34.00	33.00	39.26	37.29	12.88	12.12	8.8	0.8	0.4	31.0
M24 × 3	24.00	23.67	41.00	40.00	47.34	45.20	15.44	14.56	10.5	0.8	0.4	38.0
M30 × 3.5	30.00	29.67	50.00	49.00	57.74	55.37	19.48	17.92	13.1	0.8	0.4	46.0
M36 × 4	36.00	35.61	60.00	58.80	69.28	66.44	23.38	21.72	15.8	0.8	0.4	55.0

All dimensions are in millimeters
Basic thread lengths, *B*, are the same as given in Table 18-47.
Transition thread length, *X*, includes the length of incomplete threads and tolerances on the grip gaging length and body length. It is intended for calculation purposes.
For additional manufacturing and acceptance specifications, reference should be made to the Standard.

TABLE 18-46
American National Standard metric hex flange screws (ANSI/ASME B18.2.3.4M-1984)

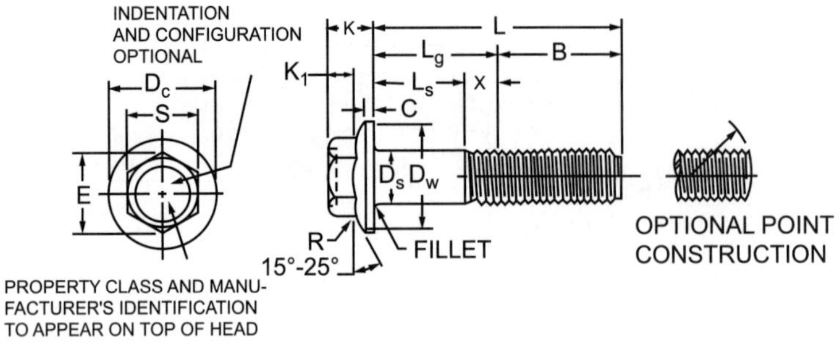

Nominal screw diam, *D*, and thread pitch	Body diam, D_s		Width across flats, *S*		Width across corners, *E*		Flange diam, D_c	Bearing circle diam, D_w	Flange edge thickness *C*	Head height, *K*	Wrenching height, K_1	Fillet radius, *R*
	Max	Min	Max	Min	Max	Min	Max	Min	Min	Max	Min	Max
M5 × 0.8	5.00	4.82	7.00	6.64	8.08	7.44	11.4	9.4	1.0	5.6	2.30	0.3
M6 × 1	6.00	5.82	8.00	7.64	9.24	8.56	13.6	11.6	1.1	6.8	2.90	0.4
M8 × 1.25	8.00	7.78	10.00	9.64	11.55	10.80	17.0	14.9	1.2	8.5	3.80	0.5
M10 × 1.5	10.00	9.78	13.00	12.57	15.01	14.08	20.8	18.7	1.5	9.7	4.30	0.6
M12 × 1.75	12.00	11.73	15.00	14.57	17.32	16.32	24.7	22.0	1.8	11.9	5.40	0.7
M14 × 2	14.00	13.73	18.00	17.57	20.78	19.68	28.6	25.9	2.1	12.9	5.60	0.8
M16 × 2	16.00	15.73	21.00	20.48	24.25	22.94	32.8	30.1	2.4	15.1	6.70	1.0

All dimensions are in millimeters.
Basic thread lengths, *B*, are the same as given in Table 18-47.
Transition thread length, *X*, includes the length of incomplete threads and tolerances on grip gaging length and body length.
This dimension is intended for calculation purposes only.
For additional manufacturing and acceptance specifications, reference should be made to the Standard.

TABLE 18-47
Received diameter-length combinations for metric hex cap screws, formed hex screws, heavy hex screws, hex flange screws and heavy hex flange screws

	Diameter—Pitch										
Nominal Length[a]	M5 × 0.8	M6 × 1	M8 × 1.25	M10 × 1.5	M12 × 1.75	M14 × 2	M16 × 2	M20 × 2.5	M24 × 3	M30 × 3.5	M36 × 4
8	×	—	—	—	—	—	—	—	—	—	—
10	×	×	—	—	—	—	—	—	—	—	—
12	×	×	×	—	—	—	—	—	—	—	—
14	×	×	×	×[b]	—	—	—	—	—	—	—
16	×	×	×	×[b]	×[b]	×[b]	—	—	—	—	—
20	×	×	×	×	×	×	—	—	—	—	—
25	×	×	×	×	×	×	×	—	—	—	—
30	×	×	×	×	×	×	×	×	—	—	—
35	×	×	×	×	×	×	×	×	×	—	—
40	×	×	×	×	×	×	×	×	×	×	—
45	×	×	×	×	×	×	×	×	×	×	—
50	×	×	×	×	×	×	×	×	×	×	×
(55)	—	×	×	×	×	×	×	×	×	×	×
60	—	×	×	×	×	×	×	×	×	×	×
(65)	—	—	×	×	×	×	×	×	×	×	×
70	—	—	×	×	×	×	×	×	×	×	×
(75)	—	—	×	×	×	×	×	×	×	×	×
80	—	—	×	×	×	×	×	×	×	×	×
(85)	—	—	—	×	×	×	×	×	×	×	×
90	—	—	—	×	×	×	×	×	×	×	×
100	—	—	—	×	×	×	×	×	×	×	×
110	—	—	—	—	×	×	×	×	×	×	×
120	—	—	—	—	×	×	×	×	×	×	×
130	—	—	—	—	—	×	×	×	×	×	×
140	—	—	—	—	—	×	×	×	×	×	×
150	—	—	—	—	—	—	×	×	×	×	×
160	—	—	—	—	—	—	×	×	×	×	×
(170)	—	—	—	—	—	—	—	×	×	×	×
180	—	—	—	—	—	—	—	×	×	×	×
(190)	—	—	—	—	—	—	—	×	×	×	×
200	—	—	—	—	—	—	—	×	×	×	×
220	—	—	—	—	—	—	—	—	×	×	×
240	—	—	—	—	—	—	—	—	×	×	×
260	—	—	—	—	—	—	—	—	—	×	×
280	—	—	—	—	—	—	—	—	—	×	×
300	—	—	—	—	—	—	—	—	—	×	×

All dimensions are in millimeters.

[a] Lengths in parentheses are not recommended. Recommended lengths of formed Hex Screws, Hex Flange Screws and Heavy Hex Flange Screws do not extend above 150 mm. Recommended lengths of Heavy Hex Screws do not extend below 20 mm. Standard sizes for government use. Recommended diameter-length combinations are indicated by the symbol ×. Screws with lengths above cross lines are threaded full length.

[b] Does not apply to Hex Flange Screws and Heavy Hex Flange Screws.

For available diameters of each type of screw, see respective dimensional table.

TABLE 18-48
American National Standard Metric Hex Bolts (ANSI B18.2.3.5M-1979, R1989)

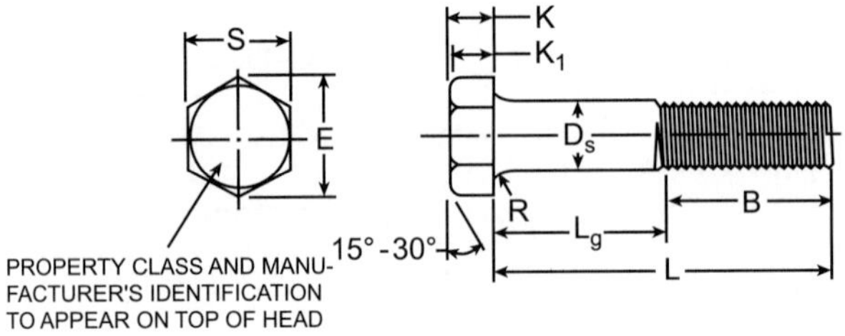

Nominal bolt diam, *D*, and thread pitch	Body diam, D_s		Width across flats, *S*		Widths across corners, *E*		Head height, *K*		Wrenching height, K_1	For bolt lengths (mm) <125	>125 and <200	>200
	Max	Min	Max	Min	Max	Min	Max	Min	Min	Basic thread length,[b] *B*		
M5 × 0.8	5.48	4.52	8.00	7.64	9.24	8.63	3.58	3.35	2.4	16	22	35
M6 × 1	6.19	5.52	10.00	9.64	11.55	10.89	4.18	3.55	2.8	18	24	37
M8 × 1.25	8.58	7.42	13.00	12.57	15.01	14.20	5.68	5.10	3.7	22	28	41
[a]M10 × 1.5	10.58	9.42	15.00	14.57	17.32	16.46	6.85	6.17	4.5	26	32	45
M10 × 1.5	10.58	9.42	16.00	15.57	18.48	17.59	6.85	6.17	4.5	26	32	45
M12 × 1.75	12.70	11.30	18.00	17.57	20.78	19.85	7.95	7.24	5.2	30	36	49
M14 × 2	14.70	13.30	21.00	20.16	24.25	22.78	9.25	8.51	6.2	34	49	53
M16 × 2	16.70	15.30	24.00	23.16	27.71	26.17	10.75	9.68	7.0	38	44	57
M20 × 2.5	20.84	19.16	30.00	29.16	34.64	32.95	13.40	12.12	8.8	46	52	65
M24 × 3	24.84	23.16	36.00	35.00	41.57	39.55	15.90	14.56	10.5	54	60	73
M30 × 3.5	30.84	29.16	46.00	45.00	53.12	50.55	19.75	17.92	13.1	66	72	85
M36 × 4	37.00	35.00	55.00	53.80	63.51	60.79	23.55	21.72	15.8	78	84	97
M42 × 4.5	43.00	41.00	65.00	62.90	75.06	71.71	27.05	25.03	18.2	90	96	109
M48 × 5	49.00	47.00	75.00	72.60	86.60	82.76	31.07	28.93	21.0	102	108	121
M56 × 5.5	57.20	54.80	85.00	82.20	98.15	93.71	36.20	33.80	24.5	—	124	137
M64 × 6	65.52	63.80	95.00	91.80	109.70	104.65	41.32	38.68	28.0	—	140	153
M72 × 6	73.84	70.80	105.00	101.40	121.24	115.60	46.45	43.55	31.5	—	156	169
M80 × 6	82.16	78.80	115.00	111.00	132.79	126.54	51.58	48.42	35.0	—	172	185
M90 × 6	92.48	88.60	130.00	125.50	150.11	143.07	57.74	54.26	39.2	—	192	205
M100 × 6	102.80	98.60	145.00	140.00	167.43	159.60	63.90	60.10	43.4	—	212	225

All dimensions are in millimeters.
[a] This size with width across flats of 15 mm is not standard. Unless specifically ordered, M10 set bolts with 16 mmm width across flats will be furnished.
[b] Basic thread length, *B*, is a reference dimension.
For additional manufacturing and acceptance specifications, reference should be made to the Standard.

TABLE 18-49
American National Standard Heavy Hex Bolts (ANSI B18.2.3.6M-1979, R1989)

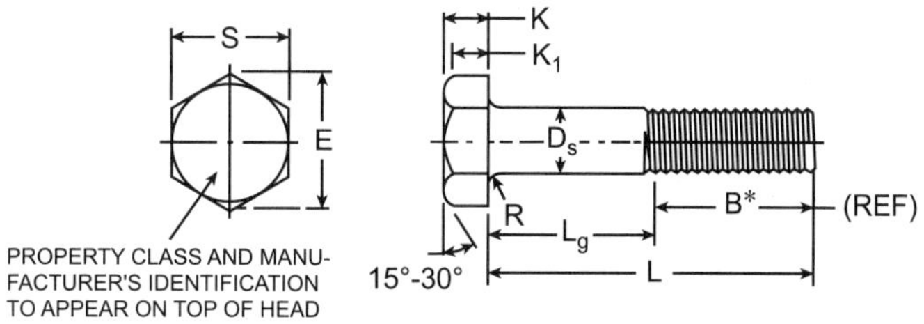

Nominal diam, D and thread pitch	Body diam, D_s		Width across flats, S		Width across corners, E		Head height, K		Wrenching height, K_1
	Max	Min	Max	Min	Max	Min	Max	Min	Min
M12 × 1.75	12.70	11.30	21.00	29.16	24.25	22.78	7.95	7.24	5.2
M14 × 2	14.70	13.30	24.00	23.16	27.71	26.17	9.25	8.51	6.2
M16 × 2	16.70	15.30	27.00	26.16	31.18	29.56	10.75	9.68	7.2
M20 × 2.5	20.84	19.16	34.00	33.00	39.26	37.29	13.40	12.12	8.2
M24 × 3	24.84	23.16	41.00	40.00	47.34	45.20	15.90	14.56	10.5
M30 × 3.5	30.84	29.16	50.00	49.00	57.74	55.37	19.75	17.92	13.1
M36 × 4	37.00	35.00	60.00	58.80	69.28	66.44	23.55	21.72	15.1

All dimensions are in millimeters.
[a] Basic thread lengths, B, are the same as given in Table 18-47.
For additional manufacturing and acceptance specifications, reference should be made to the Standard.

TABLE 18-50
Recommended clearance holes for metric hex screws and bolts[a]

Nominal diam, D and thread pitch	Clearance hole diam, basic, D_h			Nominal diam, D and thread pitch	Clearance hole diam, basic, D_h		
	Close	Normal, preferred	Loose		Close	Normal, preferred	Loose
M5 × 0.8	5.3	5.5	5.8	M30 × 3.5	31.0	33.0	35.0
M6 × 1	6.4	6.6	7.0	M36 × 4	37.0	39.0	42.0
M8 × 1.25	8.4	9.0	10.0	M42 × 4.5	43.0	45.0	48.0
M10 × 1.5	10.5	11.0	12.0	M48 × 5	50.0	52.0	56.0
M12 × 1.75	13.0	13.5	14.5	M56 × 5.5	58.0	62.0	66.0
M14 × 2	15.0	15.5	16.5	M64 × 6	66.0	70.0	74.0
M16 × 2	17.0	17.5	18.5	M72 × 6	74.0	78.0	82.0
M20 × 2.5	21.0	22.0	24.0	M80 × 6	82.0	86.0	91.0
M22 × 2.5[b]	23.0	24.0	26.0	M90 × 6	93.0	96.0	101.0
M24 × 3	25.0	26.0	28.0	M100 × 6	104.0	107.0	112.0
M27 × 3[b]	28.0	30.0	32.0	—	—	—	—

All dimensions are in millimeters.
[a] Does not apply to hex lag screws, hex socket headless screws, or round head square neck bolts.
[b] Applies only to heavy hex structural bolts.
Normal Clearance: This is preferred for general purpose applications and should be specified unless special design considerations dictate the need for either a close or loose clearance hole.
Close Clearance: This should be specified only where conditions such as critical alignment of assembled parts, wall thickness or other limitations necessitate use of a minimum hole. When close clearance holes are specified, special provision (e.g. countersinking) must be made at the screw used bolt entry side to permit proper seating of the screw or bolt head.
Loose Clearance: This should be specified only for applications where maximum adjustment capability between components being assembled is necessary.
Recommended Tolerances: The clearance hole diameters given in this table are minimum. Recommended tolerances are: for screw or bolt diameter M5, +0.2 mm; for M6 through M8, +0.3 mm; for M20 through M24, +0.4 mm; for M48 through M72, +0.5 mm; and for M80 through M100, +0.6 mm.

TABLE 18-51
American National Standard metric hex screws and bolts—reduced body diameters

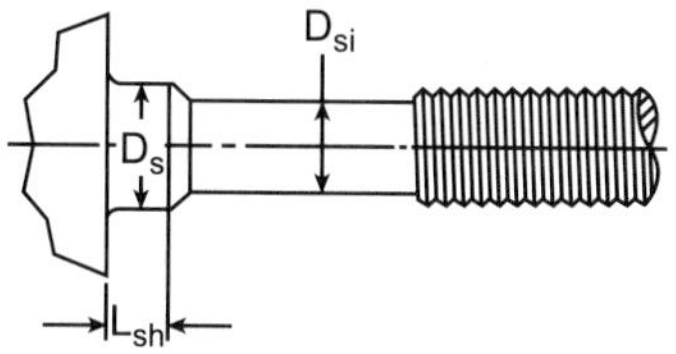

Nominal diam, *D*, and thread pitch	Shoulder diam,[b] D_s		Body diam, D_{si}		Shoulder length,[b] L_{sh}		Nominal diam, *D*, and thread pitch	Shoulder diam,[b] D_s		Body diam, D_{si}		Shoulder length,[b] L_{sh}	
	Max	Min	Max	Min	Max	Min		Max	Min	Max	Min	Max	Min
Metric Formed Hex Screws (ANSI B18.2.3.2M-1979, R1989)													
M5 × 0.8	5.00	4.82	4.46	4.36	3.5	2.5	M14 × 2	14.00	13.73	12.77	12.50	8.0	7.0
M6 × 1	6.00	5.82	5.39	5.21	4.0	3.0	M16 × 2	16.00	15.73	14.77	14.50	9.0	8.0
M8 × 1.25	8.00	7.78	7.26	5.04	5.0	4.0	M20 × 2.5	20.00	19.67	18.49	18.16	11.0	10.0
M10 × 1.5	10.00	9.78	9.08	8.86	6.0	5.0	M24 × 3	24.00	23.67	22.13	21.80	13.0	12.0
M12 × 1.75	12.00	11.73	10.95	10.68	7.0	6.0	—	—	—	—	—	—	—
Metric Hex Flange Screws (ANSI B18.2.3.4M-1984)													
M5 × 0.8	5.00	4.82	4.46	4.36	3.5	2.5	M12 × 1.75	12.00	11.73	10.95	10.68	7.0	6.0
M6 × 1	6.00	5.82	5.39	5.21	4.0	3.0	M14 × 2	14.00	13.73	12.77	12.50	8.0	7.0
M8 × 1.25	8.00	7.78	7.26	7.04	5.0	4.0	M16 × 2	16.00	15.73	14.77	14.50	9.0	8.0
M10 × 1.5	10.00	9.78	9.08	8.86	6.0	5.0	—	—	—	—	—	—	—
Metric Hex Bolts (ANSI B18.2.3.5M-1979, R1989)													
M5 × 0.8	5.48	4.52	4.46	4.36	3.5	2.5	M14 × 2	14.70	13.30	12.77	12.50	8.0	7.0
M6 × 1	6.48	5.52	5.39	5.21	4.0	3.0	M16 × 2	16.70	15.30	14.77	14.50	9.0	8.0
M8 × 1.25	8.58	7.42	7.26	7.04	5.0	4.0	M20 × 2.5	20.84	19.16	18.49	18.16	11.0	10.0
M10 × 1.5	10.58	9.42	9.08	8.86	6.0	5.0	M24 × 3	24.84	23.16	22.13	21.80	13.0	12.0
M12 × 1.75	12.70	11.30	10.95	10.68	7.0	6.0	—	—	—	—	—	—	—
Metric Heavy Hex Bolts (ANSI B18.2.3.6M-1979, R1989)													
M12 × 1.75	12.70	11.30	10.95	10.68	7.0	6.0	M20 × 2.5	20.84	19.16	18.49	18.16	11.0	10.0
M14 × 2	14.70	13.30	12.77	12.50	8.0	7.0	M24 × 3	24.84	23.16	22.13	21.80	13.0	12.0
M16 × 2	16.70	15.30	14.77	14.50	9.0	8.0	—	—	—	—	—	—	—
Metric Heavy Hex Flange Screws (ANSI B18.2.3.9M- 1984)													
M10 × 1.5	10.00	9.78	9.08	8.86	6.0	5.0	M16 × 2	16.00	15.73	14.77	14.50	9.0	8.0
M12 × 1.75	12.00	11.73	10.95	10.68	7.0	6.0	M20 × 2.5	20.00	19.67	18.49	18.16	11.0	10.0
M14 × 2	14.00	13.73	12.77	12.50	8.0	7.0	—	—	—	—	—	—	—

All dimensions are in millimeters.

[a] Shoulder is mandatory for formed hex screws, hex flange screws, and heavy hex flange screws. Shoulder is optional for hex bolts and heavy hex bolts.

TABLE 18-52
American National Standard metric heavy hex structural bolts (ANSI B18.2.3.7M-1979, R1989)

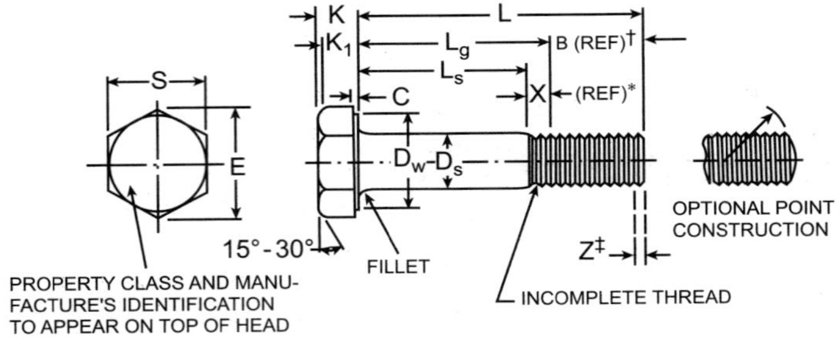

Nominal bolt diam, *D*, thread pitch	Body diam, D_S		Width across flats, *S*		Width across corners, *E*		Head height, *K*		Wrenching height, K_1	Washer face diam, D_w	Washer face thickness, *C*		Thread length, B^a: Bolt lengths, ≤100	Thread length, B^a: Bolt lengths, >100	Transition thread length, X^b
	Max	Min	Max	Min	Max	Min	Max	Min	Min	Min	Max	Min	Basic	Basic	Max
M16 × 2	16.70	15.30	27.00	26.16	31.18	29.56	10.75	9.26	6.5	24.9	0.8	0.4	31	38	6.0
M20 × 2.5	20.84	19.16	34.00	33.00	39.26	37.29	13.40	11.60	8.1	31.4	0.8	0.4	36	43	7.5
M22 × 2.5	22.84	21.16	36.00	35.00	41.57	39.55	14.90	13.10	9.2	33.3	0.8	0.4	38	45	7.5
M24 × 3	24.84	23.16	41.00	40.00	47.34	45.20	15.90	14.10	9.9	38.0	0.8	0.4	41	48	9.0
M27 × 3	27.84	26.16	46.00	45.00	53.12	50.85	17.90	16.10	11.3	42.8	0.8	0.4	44	51	9.0
M30 × 3.5	30.84	29.16	50.00	49.00	57.74	55.37	19.75	17.65	12.4	46.5	0.8	0.4	49	56	10.5
M36 × 4	37.00	35.00	60.00	58.80	69.28	66.44	23.55	21.45	15.0	55.9	0.8	0.4	56	63	12.0

TABLE 18-53
Recommended diameter-length combinations for metric heavy hex structural bolts

Nominal length, L	Nominal diameter and thread pitch						
	M16 × 2	M20 × 2.5	M22 × 2.5	M24 × 3	M27 × 3	M30 × 3.5	M36 × 4
45	×	—	—	—	—	—	—
50	×	×	—	—	—	—	—
55	×	×	×	—	—	—	—
60	×	×	×	×	—	—	—
65	×	×	×	×	×	—	—
70	×	×	×	×	×	×	—
75	×	×	×	×	×	×	—
80	×	×	×	×	×	×	×
85	×	×	×	×	×	×	×
90	×	×	×	×	×	×	×
95	×	×	×	×	×	×	×
100	×	×	×	×	×	×	×
110	×	×	×	×	×	×	×
120	×	×	×	×	×	×	×
130	×	×	×	×	×	×	×
140	×	×	×	×	×	×	×
150	×	×	×	×	×	×	×
160	×	×	×	×	×	×	×
170	×	×	×	×	×	×	×
180	×	×	×	×	×	×	×
190	×	×	×	×	×	×	×
200	×	×	×	×	×	×	×
210	×	×	×	×	×	×	×
220	×	×	×	×	×	×	×
230	×	×	×	×	×	×	×
240	×	×	×	×	×	×	×
250	×	×	×	×	×	×	×
260	×	×	×	×	×	×	×
270	×	×	×	×	×	×	×
280	×	×	×	×	×	×	×
290	×	×	×	×	×	×	×
300	×	×	×	×	×	×	×

All dimensions are in millimeters.
Recommended diameter-length combinations are indicated by the symbol ×.
Bolts with lengths above the heavy cross lines are threaded full length.

TABLE 18-54
American National Standard metric hex nuts, Styles 1 and 2 (ANSI B18.2.4.1M and B18.2.4.2M-1979, R1989)

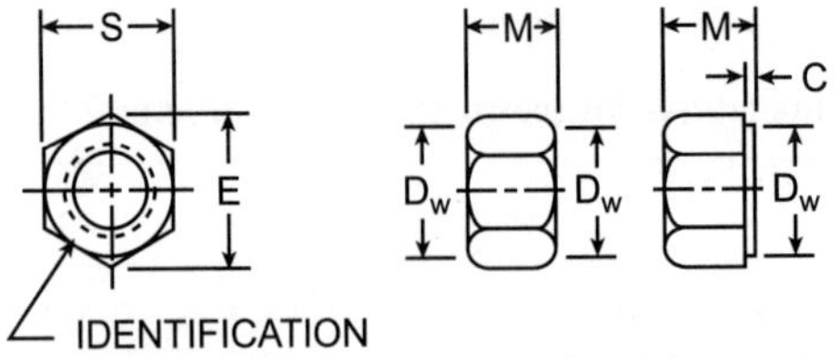

Nominal nut diam, and thread pitch	Width across flats, *S*		Width across corners, *E*		Thickness, *M*		Bearing face diam, D_w	Washer face thickness *C*	
	Max	Min	Max	Min	Max	Min	Min	Max	Min
				Metric Hex Nuts—Style 1					
M1.6 × 0.35	3.20	3.02	3.70	3.41	1.30	1.05	2.3	—	—
M2 × 0.4	4.00	3.82	4.62	4.32	1.60	1.35	3.1	—	—
M2.5 × 0.45	5.00	4.82	5.77	5.45	2.00	1.75	4.1	—	—
M3 × 0.5	5.50	5.32	6.35	6.01	2.40	2.15	4.6	—	—
M3.5 × 0.6	6.00	5.82	6.93	6.58	2.80	2.55	5.1	—	—
M4 × 0.7	7.00	6.78	8.08	7.66	3.20	2.90	6.0	—	—
M5 × 0.8	8.00	7.78	9.24	8.79	4.70	4.40	7.0	—	—
M6 × 1	10.00	9.78	11.55	11.05	5.20	4.90	8.9	—	—
M8 × 1.25	13.00	12.73	15.01	14.38	6.80	6.44	11.6	—	—
[a]M10 × 1.5	15.00	14.73	17.32	16.64	9.1	8.7	13.6	0.6	0.3
M10 × 1.5	16.00	15.73	18.48	17.77	8.40	8.04	14.6	—	—
M12 × 1.75	18.00	17.73	20.78	20.03	10.80	10.37	16.6	—	—
M14 × 2	21.00	20.67	24.25	23.36	12.80	12.10	19.4	—	—
M16 × 2	24.00	23.67	27.71	26.75	14.80	14.10	22.4	—	—
M20 × 2.5	30.00	29.16	34.64	32.95	18.00	16.90	27.9	0.8	0.4
M24 × 3	36.00	35.00	41.57	39.55	21.50	20.20	32.5	0.8	0.4
M30 × 3.5	46.00	45.00	53.12	50.85	25.60	24.30	42.5	0.8	0.4
M36 × 4	55.00	53.80	63.51	60.79	31.00	29.40	50.8	0.8	0.4
				Metric Hex Nuts—Style 2					
M3 × 0.5	5.50	5.32	6.35	6.01	2.90	2.65	4.6	—	—
M3.5 × 0.6	6.00	5.82	6.93	6.58	3.30	3.00	5.1	—	—
M4 × 0.7	7.00	6.78	8.08	7.66	3.80	3.50	5.9	—	—
M5 × 0.8	8.00	7.78	9.24	8.79	5.10	4.80	6.9	—	—
M6 × 1	10.00	9.78	11.55	11.05	5.70	5.40	8.9	—	—
M8 × 1.25	13.00	12.73	15.01	14.38	7.50	7.14	11.6	—	—
[a]M10 × 1.5	15.00	14.73	17.32	16.64	10.0	9.6	13.6	0.6	0.3
M10 × 1.5	16.00	15.73	18.48	17.77	9.30	8.94	14.6	—	—
M12 × 1.75	18.00	17.73	20.78	20.03	12.00	11.57	16.6	—	—
M14 × 2	21.00	20.67	24.25	23.35	14.10	13.40	19.6	—	—
M16 × 2	24.00	23.67	27.71	26.75	16.40	15.70	22.5	—	—
M20 × 2.5	30.00	29.16	34.64	32.95	20.30	19.00	27.7	0.8	0.4
M24 × 3	36.00	35.00	41.57	39.55	23.90	22.60	33.3	0.8	0.4
M30 × 3.5	46.00	45.00	53.12	50.85	28.60	27.30	42.7	0.8	0.4
M36 × 4	55.00	53.80	63.51	60.79	34.70	33.10	51.1	0.8	0.4

All dimensions are in millimeters.
[a] This size with width across flats of 15 mm is not standard. Unless specifically ordered, metric hex nuts with 16 mm width across flats will be furnished.

TABLE 18-55
American National Standard metric slotted hex nuts (ANSI B18.2.4.3M-1979, R1989)

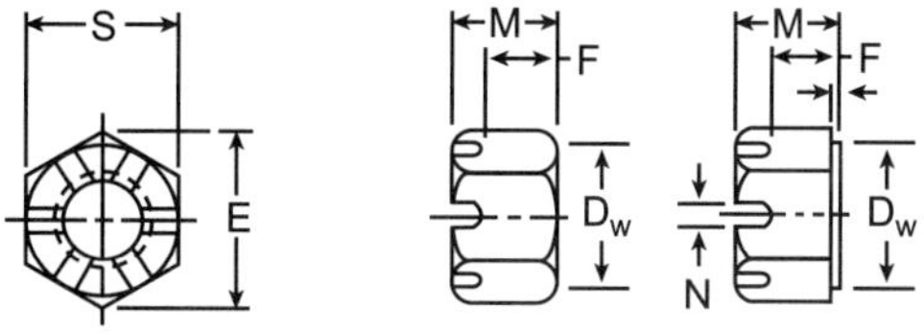

Nominal nut diam, D, and thread pitch	Width across flats, S		Width across corners, E		Thickness, M		Bearing face diam, D_w	Unslotted thickness, F		Width of slot, N		Washer face thickness, C	
	Max	Min	Max	Min	Max	Min	Min	Max	Min	Max	Min	Max	Min
M5 × 0.8	8.00	7.78	9.24	8.79	5.10	4.80	6.9	3.2	2.9	2.0	1.4	—	—
M6 × 1	10.00	9.78	11.55	11.05	5.70	5.40	8.9	3.5	3.2	2.4	1.8	—	—
M8 × 1.25	13.00	12.73	15.01	14.38	7.50	7.14	11.6	4.4	4.1	2.9	2.3	—	—
[a]M10 × 1.5	15.00	14.73	17.32	16.64	10.0	9.6	13.6	5.7	5.4	3.4	2.8	0.6	0.3
M10 × 1.5	16.00	15.73	18.48	17.77	9.30	8.94	14.6	5.2	4.9	3.4	2.8	—	—
M12 × 1.75	18.00	17.73	20.78	20.03	12.00	11.57	16.6	7.3	6.9	4.0	3.2	—	—
M14 × 2	21.00	20.67	24.25	23.35	14.10	13.40	19.6	8.6	8.0	4.3	3.5	—	—
M16 × 2	24.00	23.67	27.71	26.75	16.40	15.70	22.5	9.9	9.3	5.3	4.5	—	—
M20 × 2.5	30.00	29.16	34.64	32.95	20.30	19.00	27.7	13.3	12.2	5.7	4.5	0.8	0.4
M24 × 3	36.00	35.00	41.57	39.55	23.90	22.60	33.2	15.4	14.3	6.7	5.5	0.8	0.4
M30 × 3.5	46.00	45.00	53.12	50.85	28.60	27.30	42.7	18.1	16.8	8.5	7.0	0.8	0.4
M36 × 4	55.00	53.80	63.51	60.79	34.70	33.10	51.1	23.7	22.4	8.5	7.0	0.8	0.4

All dimensions are in millimeters.
[a] This size with width across flats of 15 mm is not standard. Unless specifically ordered, M10 slotted hex nuts with 16 mm width across flats will be furnished.

REFERENCES

1. Norman, C. A., E. S. Ault, and E. F. Zarobsky, *Fundamentals of Machine Design*, The Macmillan Company, New York, 1951.
2. Maleev, V. L., and J. B. Hartman, *Machine Design*, International Textbook Company, Scranton, Pennsylvania, 1954.
3. Black, P. H., and O. E. Adams, Jr., *Machine Design*, McGraw-Hill Publishing Company, New York, 1955.
4. Baumeister, T., ed., *Marks' Standard Handbook for Mechanical Engineers*, 8th ed., McGraw-Hill Publishing Company, New York, 1978.
5. Lingaiah, K., and B. R. Narayana Iyengar, *Machine Design Data Handbook*, Engineering College Cooperative Society, Bangalore, India, 1962.
6. Lingaiah, K., and B. R. Narayana Iyengar, *Machine Design Data Handbook*, Vol. I (*SI and Customary Metric Units*), Suma Publishers, Bangalore, India, 1986.
7. Lingaiah, K., *Machine Design Data Handbook*, Vol. II (*SI and Customary Metric Units*), Suma Publishers, Bangalore, India, 1986.
8. Bureau of Indian Standards.
9. ISO Standards.
10. John, J. Viegas, *Standards for Mechanical Elements*, Harold A. Rothbart, Editor-in-Chief, *Mechanical Design and Systems Handbook*, McGraw-Hill Publishing Company, New York, 1964.
11. Russel, Bardsall and Ward Corp., *Helpful Hints for Fastener Design and Application*, Mentor, Ohio, 1976, p. 42.
12. Shigley J., E., and C. R. Mischke, *Mechanical Engineering Design*, 5th ed., McGraw-Hill Publishing Company, New York, 1989.
13. Burr, J. H., and J. B. Cheatham, *Mechanical Analysis and Design*, 2nd ed., Prentice Hall, Englewood Cliffs, New Jersey, 1995.
14. Edwards, K. S., Jr., and R. B. Mckee, *Fundamentals of Mechanical Component Design*, McGraw-Hill Publishing Company, New York, 1991.
15. Ito, Y., J. Toyoda, and S. Nagata, *Interface Pressure Distribution in a Bolt-Flange Assembly*, ASME Paper No. 77-WA/DE-11, 1977.
16. Little, R. E., Bolted Joints: How Much Give? *Machine Design*, Nov. 9, 1967.
17. Osgood, C. C., *Saving Weight on Bolted Joints*, *Machine Design*, Oct. 25, 1979.
18. Bowman Distribution-Barnes Group, *Fastener Facts*, Cleveland, Ohio, 1985, p. 90.
19. American National Standards, ANSI B18.2.3.5M-1979, R1989.
20. British Standards Institution, 2 Park Street, London, 1986.
21. Machinery Handbook, 20th ed., 1999, Industrial Press, U.S.A.
22. Shigley J. E., and C. R. Mischke, *Standard Handbook of Machine Design*, McGraw-Hill Publishing Company New York, 1996.
23. Lingaiah, K., *Machine Design Data Handbook*, McGraw-Hill Publishing Company, New York, USA, 1994.

CHAPTER 19

COUPLINGS, CLUTCHES, AND BRAKES

SYMBOLS[8,9,10]

a	distance between center lines of shafts in Oldham's coupling, m (in)
A	area, m^2 (in^2)
	external area, m^2 (in^2)
A_r	radiating surface required, m^2 (in^2)
A_c	contact area of friction surface, m^2 (in^2)
b	width of key, m (in)
	width of shoe, m (in)
	width of inclined face in grooved rim clutch, m (in)
	width of spring in centrifugal clutch, m (in)
	width of wheel, m (in)
	width of operating lever (Fig. 19-16), m (in)
c	heat transfer coefficient, kJ/m^2 K h (kcal/m^2/°C/h)
c_1	specific heat of material, kJ/kg K (kcal/kg/°C)
c_2	radiating factor for brakes, kJ/m^2 K s (kcal/m^2/min/°C)
d	diameter of shaft, m (in)
	diameter of pin, roller pin, m (in)
d_1	diameter of bolt, m (in)
	diameter of pin at neck in the flexible coupling, m (in)
d_2	diameter of hole for bolt, m (in)
d'	outside diameter of bush, m (in)
D	diameter of wheel, m (in)
	diameter of sheave, m (in)
	outside diameter of flange coupling, m (in)
D_1	inside diameter of disk of friction material in disk clutches and brakes, m (in)
D_2	outside diameter of disk of friction material in disk clutches and brakes, m (in)
D_i	inside diameter of hollow rigid type of coupling, m (in)
D_o	outside diameter of hollow rigid type of coupling, m (in)
D_m	mean diameter, m (in)
e_1, e_2, e_3	dimensions shown in Fig. 19-16, m (in)
E	energy (also with suffixes), N m (lbf in)
	Young's modulus of elasticity, GPa (Mpsi)

F	operating force on block brakes, kN (lbf); force at each pin in the flexible bush coupling, kN (lbf)
	total pressure, kN (lbf)
	force (also with suffixes), kN (lbf)
	actuating force, kN (lbf)
F_1	tension on tight side of band, kN (lbf)
	the force acting on disks of one operating lever of the clutch (Fig. 19-16), kN (lbf)
F_2	tension on slack side of band, kN (lbf)
F_a'	total axial force on i number of clutch disks, kN (lbf)
F_b	tension load in each bolt, kN (lbf)
F_c	centrifugal force, kN (lbf)
F_n	total normal force, kN (lbf)
F_x, F_y	components of actuating force F acting at a distance c from the hinge pin (Figs. 19-25 and 19-26), kN (lbf)
F_θ	tangential force at rim of brake wheel, kN (lbf)
	tangential friction force, kN (lbf)
g	acceleration due to gravity, 9.8066 m/s^2 (9806.6 mm/s^2) (32.2 ft/s^2)
h	thickness of key, m (in)
	thickness of central disk in Oldham's coupling, m (in)
	thickness of operating lever (Fig. 19-16), m (in)
	depth of spring in centrifugal clutch, m (mm)
H	rate of heat to be radiated, J (kcal)
H_g	heat generated, J (kcal)
H_d	the rate of dissipation, J (kcal)
i	number of pins,
	number of bolts,
	number of rollers,
	pairs of friction surfaces
	number of shoes in centrifugal clutch
	number of times the fluid circulates through the torus in one second
i_1	number of driving disks
i_2	number of driven disks
i'	number of operating lever of clutch
I	moment of inertia, area, m^4, cm^4 (in^4)
k_l	load factor or the ratio of the actual brake operating time to the total cycle of operation
k_s	speed factor
l	length (also with suffixes), m (in)
	length of spring in centrifugal clutch measured along arc, m (in)
	length of bush, m (in)
L	dimension of operating lever as shown in Fig. 19-16
M_t	torque to be transmitted, N m (lbf in)
M_{ta}	allowable torque, N m (lbf in)
n	speed, rpm
n_1, n_2	speed of the live load before and after the brake is applied, respectively, rpm
n	number of clutching or braking cycles per hour
P	power, kW (hp)
N	normal force (Figs. 19-25 and 19-26), kN (lbf)
μN	frictional force (Figs. 19-25 and 19-26), kN (lbf)
p	unit pressure, MPa (psi)

p	unit pressure acting upon an element of area of the frictional material located at an angle θ from the hinge pin (Figs. 19-25 and 19-26), MPa (psi)
	maximum pressure between the fabric and the inside of the rim, MPa (psi)
p_a	allowable pressure, MPa (psi)
	maximum pressure located at an angle θ_a from the hinge pin (Figs. 19-25 and 19-26), MPa (psi)
p_b	bearing pressure, MPa (psi)
P	total force acting from the side of the bush on operating lever (Fig. 19-16), kN (lbf)
P'	the force acting from the side of the bush on one operating lever, kN (lbf)
r	radius, m (in)
	distance from the center of gravity of the shoe from the axis of rotation, m (in)
r_m	mean radius, m (in)
r_{mi}	mean radius of inner passage of hydraulic coupling, m (in)
r_{mo}	mean radius of outer passage in hydraulic coupling, m (in)
R	reaction (also with suffixes), kN (lbf)
R_c	radius of curvature of the ramp at the point of contact (Fig. 19-21), m (in)
R_d	radius of the contact surface on the driven member (Fig. 19-21), m (in)
R_r	radius of the roller (Fig. 19-21), m (in)
R_x, R_y	hinge pin reactions (Figs. 19-25 and 19-26), kN (lbf)
t	time of single clutching or braking operation (Eq. 19-198), s
T_a	ambient or initial temperature, °C (°F)
T_{av}	average equilibrium temperature, °C (°F)
ΔT	rise in temperature of the brake drum, °C (°F)
t_c	cooling time, s (min)
v	velocity, m/s
v_1, v_2	speed of the live load before and after the brake is applied, respectively, m/s
w	axial width in cone brake, m (in)
	width of band, m (in)
W	work done, N m (lbf in)
	weight of the fluid flowing in the torus, kN (lbf)
	weight lowered, kN (lbf)
	weight of parts in Eq. (19-136), kN (lbf)
	weight of shoe, kN (lbf)
y	deflection, m (in)
σ	stress (also with suffixes), MPa (psi)
σ_b	allowable or design stress in bolts, MPa (psi)
σ'_b	design bearing stress for keys, MPa (psi)
$\sigma_{c(\max)}$	maximum compressive stress in Hertz's formula, MPa (psi)
σ_{db}	design bending stress, MPa (psi)
τ	shear stress, MPa (psi)
τ_b	allowable or design stress in bolts, MPa (psi)
τ_{d1}	design shear stress in sleeve, MPa (psi)
τ_{d2}	design shear stress in key, MPa (psi)
τ_f	design shear stress in flange at the outside hub diameter, MPa (psi)
τ_s	design shear stress in shaft, MPa (psi)
α	one-half the cone angle, deg
	pressure angle, deg

ϕ	friction angle, deg
θ	one-half angle of the contact surface of block, deg
μ	coefficient of friction
η	factor which takes care of the reduced strength of shaft due to keyway
ω_1	running speed of centrifugal clutch, rad/s
ω_2	speed at which the engagement between the shoe of centrifugal clutch and pulley commences, rad/s

SUFFIXES

a	axial
d	dissipated, design
g	generated
1, i	inner
2, o	outer
n	normal
x	x direction
y	y direction
θ	tangential
μ	friction

Other factors in performance or in special aspects are included from time to time in this chapter and, being applicable only in their immediate context, are not included at this stage.

Particular	Formula

19.1 COUPLINGS

COMMON FLANGE COUPLING (Fig. 19-1)

The commonly used formula for approximate number of bolts

$$i = 20d + 3 \quad \text{SI} \quad (19\text{-}1a)$$

where d in m

$$i = 0.5d + 3 \quad \text{USCS} \quad (19\text{-}1b)$$

where d in in

The torque transmitted by the shaft

$$M_t = \frac{\pi d^3}{16}\eta\tau_s \quad (19\text{-}2)$$

The torque transmitted by the coupling

$$M_t = \frac{1000P}{\omega} \quad \text{SI} \quad (19\text{-}3a)$$

where M_t in N m; P in kW; ω in rad/s

$$M_t = \frac{63{,}000P}{n} \quad \text{USCS} \quad (19\text{-}3b)$$

where M_t in lbf in; P in hp, n in rpm

$$M_t = \frac{9550P}{n} \quad \text{SI} \quad (19\text{-}3c)$$

Particular	Formula

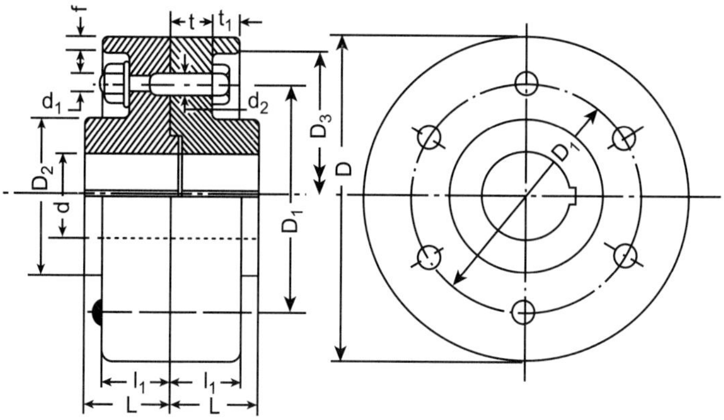

FIGURE 19-1 Flange coupling.

Particular	Formula
	where M_t in N m; P in kW; n in rpm
	$M_t = \dfrac{159P}{n'}$ **SI** (19-3d)
	where M_t in N m; P in kW; n' in rps
The torque transmitted through bolts	$M_t = i\left(\dfrac{\pi d_1^2}{4}\right)\tau_b \dfrac{D_1}{2}$ (19-4)
The torque capacity which is based on bearing of bolts	$M_t = i(d_1 l_1)\sigma_b \dfrac{D_1}{2}$ (19-5)
The torque capacity which is based on shear of flange at the outside hub diameter	$M_t = t(\pi D_2)\tau_f \dfrac{D_2}{2}$ (19-6)
The friction-torque capacity of the flanged coupling which is based on the concept of the friction force acting at the mean radius of the surface	$M_t = i\mu F_b r_m$ (19-7)
	where $r_m = \dfrac{D+d}{2}$ = mean radius
	F_b = tension load in each bolt, kN (kgf)
The preliminary bolt diameter may be determined by the empirical formula	$d_1 = \dfrac{0.5d}{\sqrt{i}}$ (19-8)
The bolt diameter from Eqs. (19-2) and (19-4)	$d_1 = \sqrt{\dfrac{d^2 \tau_s \eta}{2i\tau_b D_1}}$ (19-9)
The bolt diameter from Eqs. (19-3) and (19-4)	$d_1 = \sqrt{\dfrac{8000P}{\pi i\omega\tau_b D_1}}$ **SI** (19-10a)

Particular	Formula
	where d_1, D_1 in m; P in kW; τ_b in Pa; ω in rad/s
	$d_1 = \sqrt{\dfrac{1273P}{\pi i n' D_1 \tau_b}}$ **SI** (19-10b)
	where d_1, D_1 in m; P in kW; τ_b in Pa; n' in rps
	$d_1 = \sqrt{\dfrac{76{,}400P}{\pi i n \tau_b D_1}}$ **SI** (19-10c)
	where d_1, D_1 in m; P in kW; τ_b in Pa; n in rpm
	$d_1 = \sqrt{\dfrac{50{,}400P}{\pi i n D_1 \tau_b}}$ **USCS** (19-10d)
	where d_1, D_1 in in; P in hp; τ_b in psi; ω in rpm where i = effective number of bolts doing work should be taken as all bolts if they are fitted in reamed holes and only half the total number of bolts if they are not fitted into reamed holes
The diameter of shaft from Eqs. (19-2) and (19-3)	$d = \sqrt[3]{\dfrac{16{,}000P}{\pi \eta \omega \tau_s}}$ **SI** (19-11a)
	where P in kW; d in m
	$d = \sqrt[3]{\dfrac{100{,}800P}{\pi \eta n \tau_s}}$ **USCS** (19-11b)
	where P in hp; d in in
	$d = \sqrt[3]{\dfrac{152{,}800P}{\pi \eta n \tau_s}}$ **SI** (19-11c)
	where P in kW; d in m
	$d = \sqrt[3]{\dfrac{2546P}{\pi \eta n' \tau_s}}$ **SI** (19-11d)
	where P in kW; d in m; n' in rps
The average value of diameter of the bolt circle	$D_1 = 2d + 0.05$ **SI** (19-12a)
	where D_1 in m
	$D_1 = 2d + 2$ **USCS** (19-12b)
The hub diameter	$D_2 = 1.5d + 0.025$ **SI** (19-13a)
	where D_2 in m
	$D_2 = 1.5d + 1$ **USCS** (19-13b)
The outside diameter of flange	$D = 2.5d + 0.075$ **SI** (19-14a)
	where D in m
	$D = 2.5d + 3$ **USCS** (19-14b)

Particular	Formula		
The hub length	$l = 1.25d + 0.01875$ where l in m and d in m	**SI**	(19-14c)
	$l = 1.25d + 0.75$ where l and d in in	**USCS**	(19-14d)

MARINE TYPE OF FLANGE COUPLING

Solid rigid type [Fig. 19-2(a), Table 19-1]

The number of bolts	$i = 33d + 5$ where d in in	**SI**	(19-15a)
	$i = 0.85d + 5$ where d in in	**USCS**	(19-15b)
The diameter of bolt	$d_1 = \sqrt{\dfrac{\eta d^3 \tau_s}{2iD_1 \tau_b}}$ based on torque capacity of the shaft		(19-16a)
	$d_1 = \sqrt{\dfrac{tD_2^2 \tau_f}{4iD_1 \tau_b}}$ based on torque capacity of flange		(19-16b)

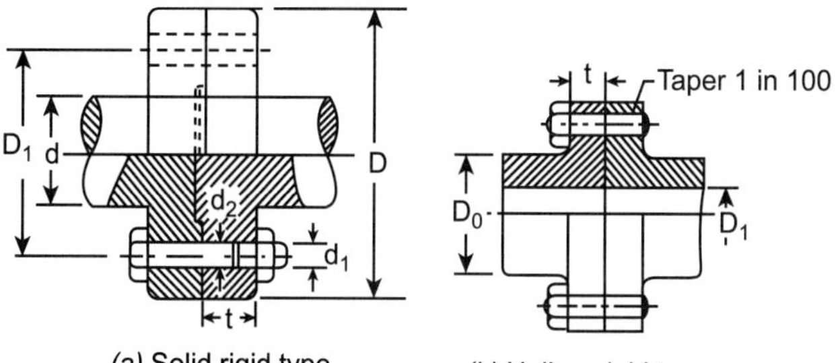

(a) Solid rigid type (b) Hollow rigid type

FIGURE 19-2 Rigid marine coupling.

The thickness of flange	$t = 0.25$ to $0.28d$	(19-17)
The diameter of the bolt circle	$D_1 = 1.4d$ to $1.6d$	(19-18)
The outside diameter of flange	$D = D_1 + 2d$ to $3d$	(19-19)
Taper of bolt	1 in 100	

TABLE 19-1
Forged end type rigid couplings (all dimensions in mm)

Number coupling		Shaft diameter		Flange outside diameter, D	Flange width, t	Locating diameter, D_2	Recess depth, c_1	Spigot depth, c_2	Pitch circle diameter, D_1	Bolt size, d_1	Bolt hole diameter, d_2 $H8$	Number of bolts
Recessed flange	Spigot flange	Max	Min									
R_1	S_1	53	—	100	17	50	6	4	70	M10	11	4
$R2$	$S2$	45	36	120	22	60	6	4	85	M12	13	4
$R3$	$S3$	55	46	140	22	75	7	5	100	M14	15	4
$R4$	$S4$	70	55	175	27	95	7	5	125	M16	17	6
$R5$	$S5$	80	71	195	32	95	7	5	140	M18	19	6
$R6$	$S6$	90	81	225	32	125	7	5	160	20	21	6
$R7$	$S7$	110	91	265	36	150	9	7	190	24	25	6
$R8$	$S8$	130	111	300	46	150	9	7	215	30	32	6
$R9$	$S9$	150	131	335	50	195	9	7	240	33	34	8
$R10$	$S10$	170	151	375	55	195	10	8	265	36	38	6
$R11$	$S11$	190	171	400	55	240	10	8	290	36	38	8
$R12$	$S12$	210	191	445	65	240	10	8	315	42	44	8
$R13$	$S13$	230	211	475	70	280	10	8	340	45	46	8
$R14$	$S14$	250	231	500	70	280	10	8	370	45	46	10
$R15$	$S15$	270	251	560	80	330	10	8	400	52	55	10
$R16$	$S16$	300	271	600	85	330	10	8	410	56	60	10
$R17$	$S17$	330	301	650	90	400	10	8	480	60	65	10
$R18$	$S18$	360	331	730	100	400	10	8	520	68	72	10
$R19$	$S19$	390	361	775	105	480	11	9	570	72	76	10
$R20$	$S20$	430	391	875	110	480	11	9	620	76	80	12
$R21$	$S21$	470	431	900	115	560	11	9	670	80	85	12
$R22$	$S22$	520	471	925	120	560	12	10	730	90	95	12
$R23$	$S23$	571	521	1000	125	640	12	10	790	110	105	12
$R24$	$S24$	620	571	1090	130	720	12	10	850	110	115	12

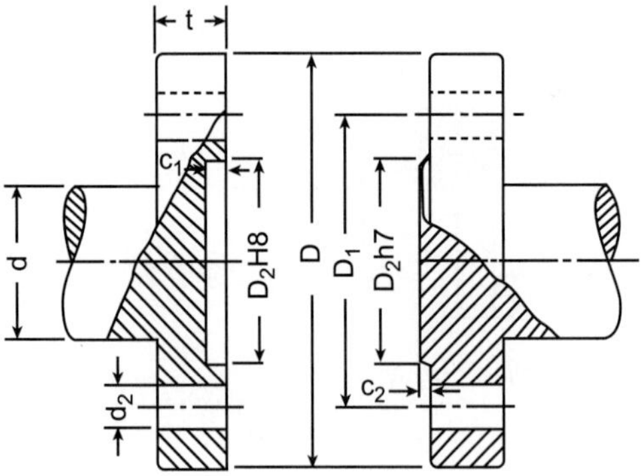

Particular	Formula

Hollow rigid type [Fig. 19-2(b)]

The minimum number of bolts	$i = 50D_o$ **SI** (19-20a) where D_o in m $i = 1.25D_o$ **USCS** (19-20b) where D_o in in
The mean diameter of bolt	$$d_1 = \sqrt{\frac{(1-K^4)D_o^3\tau_s}{2iD_1\tau_b}} \quad (19\text{-}21)$$ where $K = \dfrac{D_i}{D_o}$
The thickness of flange	$$t = \frac{(1-K^4)D_o^3\tau_s}{8D_2^2\tau_f} \quad (19\text{-}22)$$
The empirical formula for thickness of flange	$t = 0.25$ to $0.28D_o$ (19-23)
The diameter of bolt circles	$D_1 = 1.4D_o$ (19-24)
For design calculations of other dimensions of marine hollow rigid type of flange coupling	The method of analyzing the stresses and arriving at the dimensions of the various parts of a marine hollow flange coupling is similar to that given for the marine solid rigid type and common flange coupling.
For dimensions of fitted half couplings for power transmission	Refer to Table 19-2.

PULLEY FLANGE COUPLING (Fig. 19-3)

The number of bolts	$i = 20d + 3$ **SI** (19-25a) where d in m $i = 0.5d + 3$ **USCS** (19-25b) where d in in
The preliminary bolt diameter	$$d_t = \frac{0.5d}{\sqrt{i}} \quad (19\text{-}26)$$

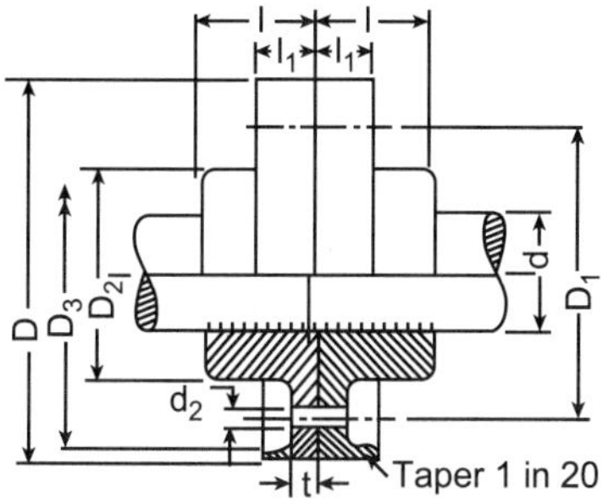

FIGURE 19-3 Pulley flange coupling.

TABLE 19-2
Fitted half couplings (all dimensions in mm)

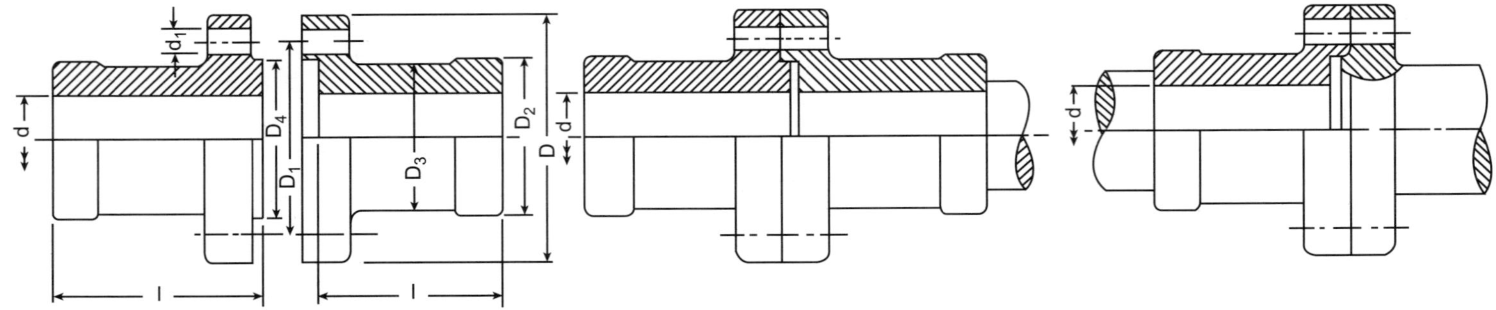

(a) (b) (c)

Nominal diameter, d $H7$	Locating diameter, D_4 $H8/h7$	Pitch circle diameter D_1	Bolt: No. of bolts	Bolt: Diameter of hole, d_2H7	Bolt: Bolt size, d_1	Hub diameter, D_2	Shoulder diameter, D_2	Flange diameter, D	Length of shaft end, l: Long	Length of shaft end, l: Short
30	75	100	4	13	M12	70	80	125	80	58
40, 45, 50, 56	95	125	6	17	M16	90	100	160	110	82
63, 71	125	160	6	21	M20	120	180	200	140	105
80, 90	150	190	6	25	M24	145	155	240	170	130
100, 110, 125	195	240	6	25	M24	190	200	300	210	165
140	240	290	8	32	M30	230	240	360	250	200
160, 180	280	340	8	38	M36	270	285	420	300	240
200, 220	330	400	10	44	M32	320	335	500	350	280
250	400	480	10	50	M48	380	400	600	410	330
280, 320	480	570	10	60	M56	460	480	710	470	380
360	580	670	12	68	M64	540	570	850	550	450
400, 450, 500	720	850	12	95	M90	690	720	1050	650	540

Particular	Formula
The width of flange l_1 (Fig. 19-3)	$l_1 = \frac{1}{2}d + 0.025$ **SI** (19-27a) where l_1 and d in m $l_1 = \frac{1}{2}d + 1.0$ **USCS** (19-27b) where d in in
The hub length l	$l = 1.4d + 0.0175$ **SI** (19-28a) where l and d in m $l = 1.4d + 0.7$ **USCS** (19-28b) where l and d in in
The thickness of the flange	$t = 0.25d + 0.007$ **SI** (19-29a) where t and d in m $t = 0.25d + 0.25$ **USCS** (19-29b) where t and d in in
The hub diameter	$D_2 = 1.8d + 0.01$ **SI** (19-30a) where D_2 and d in m $D_2 = 1.8d + 0.4$ **USCS** (19-30b) where D_2 and d in in
The average value of the diameter of the bolt circle	$D_1 = 2d + 0.025$ **SI** (19-31a) where D_1 and d in m $D_1 = 2d + 1.0$ **USCS** (19-31b) where D_1 and d in in
The outside diameter of flange	$D = 2.5d + 0.075$ **SI** (19-32a) where D and d in m $D = 2d + 3.0$ **USCS** (19-32b) where D and d in in

PIN OR BUSH TYPE FLEXIBLE COUPLING (Fig. 19-4, Table 19-3)

Torque to be transmitted

$$M_t = iF\,\frac{D_1}{2} \tag{19-33a}$$

$$M_t = ip_b l d'\left(\frac{D_1}{2}\right) \tag{19-33b}$$

where

p_b = bearing pressure, MPa (psi)
F = force at each pin, kN (lbf) $= p_b l d'$
d' = outside diameter of the bush, m (in)

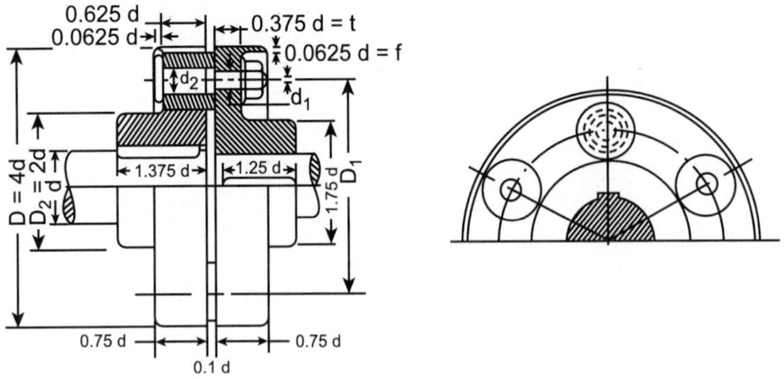

Shear stress in pin

$$\tau_p = \frac{F}{0.785 d_p^2} \tag{19-34}$$

where

τ_p = allowable shearing stress, MPa (psi)
$d_p = d_1$ = diameter of pin at the neck, m (in)

Bending stress in pin

$$\sigma_b = \frac{F\left(\frac{l}{2} + b\right)}{\frac{\pi}{32} d_p^3} \tag{19-35}$$

OLDHAM COUPLING (Fig. 19-5)

The total pressure on each side of the coupling

$$F = \tfrac{1}{4} pDh \tag{19-36}$$

where h = axial dimension of the contact area, m (in)

The torque transmitted on each side of the coupling

$$M_t = 2Fl = \frac{pD^2 h}{6} \tag{19-37}$$

where

$l = \frac{1}{3}D$ = the distance to the pressure area centroid from the center line, m (in)
p = allowable pressure $\ngtr$ 8.3 MPa (1.2 kpsi)

Power transmitted

$$P = \frac{pD^2 hn}{57{,}277} \quad \textbf{SI} \tag{19-38a}$$

where P in kW

$$P = \frac{pD^2 hn}{378{,}180} \quad \textbf{USCS} \tag{19-38b}$$

where P in hp; D, h in in; p in psi

The diameter of the disk

$$D = 3d + a \tag{19-39}$$

The diameter of the boss

$$D_2 = 2d \tag{19-40}$$

TABLE 19-3
Cast-iron flexible couplings (all dimensions in mm)

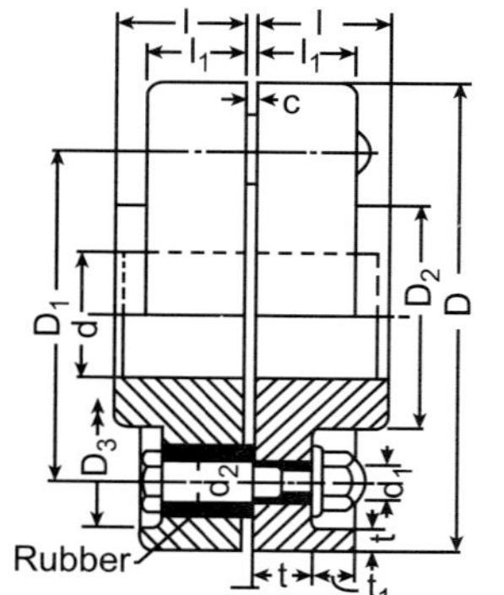

Bush type flexible coupling

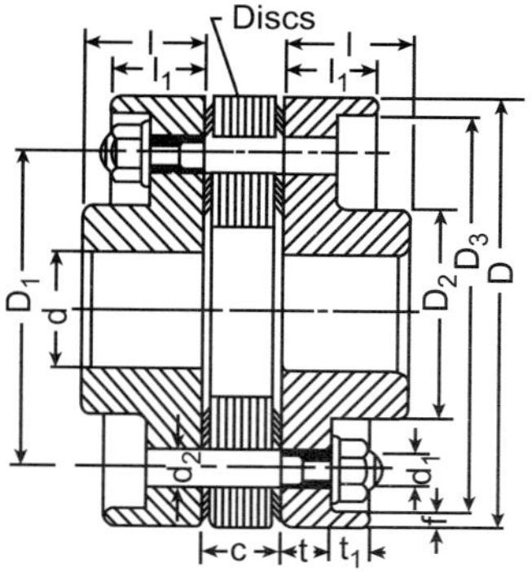

Disc type flexible coupling

Type of flexible couplings	Coupling number	Bore, d Min	Bore, d Max	Outside diameter, D, min	Hub diameter, D_2 min	Hub length, l, min	Flange width, l_1	Thickness of disk, C	Diameter of bolt, d_1	Number of bolt holes	Pitch circle diameter of bolts, D_1	Bolt recess, t_1	Bush diameter	Nominal gap between coupling holes, c	Maximum rating per 100 rpm, kW
Bush type flexible coupling	B_1	12	16	80		28	18	—	8	3	53	10	20	2	0.4
	B_2	16	22	100		30	20	—	10	3	63	12	22	2	0.6
	B_3	22	30	112		32	22	—	10	3	73	12	22	2	0.8
	B_4	30	45	132		40	30	—	12	4	90	15	25	4	2.5
	B_5	45	56	170	80	45	35	—	12	4	120	15	25	4	4.0
	B_6	56	75	200	100	56	40	—	12	4	150	15	30	4	6.0
	B_7	75	85	250	140	63	45	—	16	6	190	22	40	5	16.0
	B_8	85	110	315	180	80	50	—	16	6	250	22	40	5	25.0
	B_9	110	130	400	212	90	56	—	18	8	315	28	45	6	52.0
	B_{10}	130	150	500	280	100	60	—	18	8	400	28	45	6	74.0
Disc type flexible coupling	D_1	12	16	80		28	18	15, 16	8	6	55	10	—	—	0.4
	D_2	16	22	100		30	20	16, 18	10	6	63	12	—	—	0.6
	D_3	22	30	110		32	22	18, 25	10	6	73	12	—	—	0.8
	D_4	30	45	132		40	30	25, 30	12	8	90	15	—	—	2.5
	D_5	45	56	165	80	45	35	30, 35	12	8	120	15	—	—	4.0
	D_6	56	75	200	100	56	40	35, 40	12	8	150	15	—	—	6.0
	D_7	75	85	250	140	63	45	40, 45	16	12	190	22	—	—	16.0
	D_8	85	110	315	180	80	50	45, 50	16	12	250	22	—	—	25.0
	D_9	110	130	400	212	90	56	50, 55	18	16	315	28	—	—	52.0
	D_{10}	130	150	500	280	100	60	55	18	16	400	28	—	—	74.0

Particular	Formula	

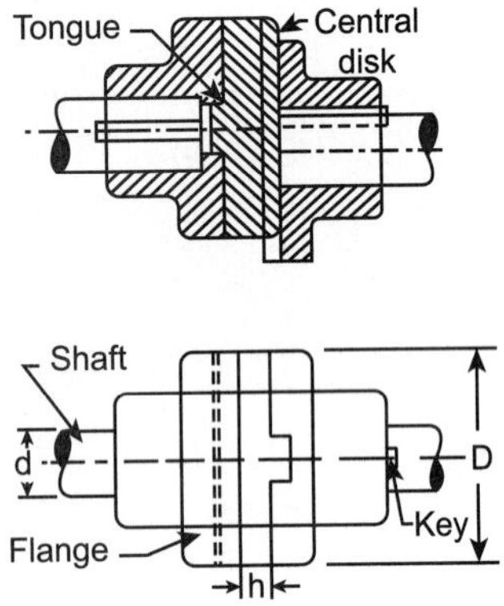

FIGURE 19-5 Oldham's coupling.

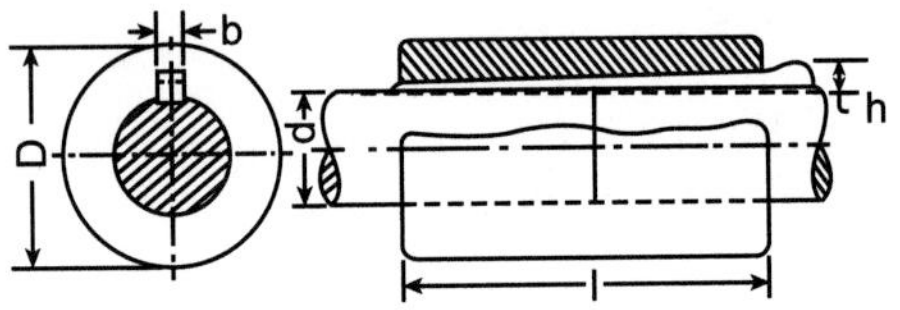

FIGURE 19-6 Muff or sleeve coupling.

Particular	Formula	
Length of the boss	$l = 1.75d$	(19-41)
Breadth of groove	$w = \dfrac{D}{6}$	(19-42)
The thickness of the groove	$h_1 = \dfrac{w}{2}$	(19-43a)
The thickness of central disk	$h = \dfrac{w}{2}$	(19-43b)
The thickness of flange	$t = \frac{3}{4}d$	(19-44)

MUFF OR SLEEVE COUPLING (Fig. 19-6)

Particular	Formula	
The outside diameter of sleeve	$D = 2d + 0.013$ where D, d in m	**SI** (19-45a)
	$D = 2d + 0.52$ where D, d in in	**USCS** (19-45b)
The outside diameter of sleeve is also obtained from equation	$D = \sqrt[3]{\dfrac{16M_t}{\pi\tau_{d1}(1 - K^4)}}$ where $K = \dfrac{d}{D}$	(19-46)
The length of the sleeve (Fig. 19-6)	$l = 3.5d$	(19-47)
Length of the key (Fig. 19-6)	$l = 3.5d$	(19-48)
The diameter of shaft	$d = \sqrt[3]{\dfrac{16M_t}{\eta\pi\tau_d}}$ where M_t is torque obtained from Eq. (19-2)	(19-49)

Particular	Formula	
The width of the keyway	$$b = \frac{2M_t}{\tau d_2 l d}$$	(19-50)
The thickness of the key	$$h = \frac{2M_t}{\sigma'_b l d}$$	(19-51)

FAIRBAIRN'S LAP-BOX COUPLING (Fig. 19-7)

Particular	Formula	
The outside diameter of sleeve	Use Eqs. (19-45) or (19-46)	
The length of the lap	$l = 0.9d + 0.003$ where l, d in m	**SI** (19-52a)
	$l = 0.9d + 0.12$ where l, d in in	**USCS** (19-52b)
The length of the sleeve	$L = 2.25d + 0.02$ where L, d in m	**SI** (19-53a)
	$L = 2.25d + 0.8$ where L, d in in	**USCS** (19-53b)

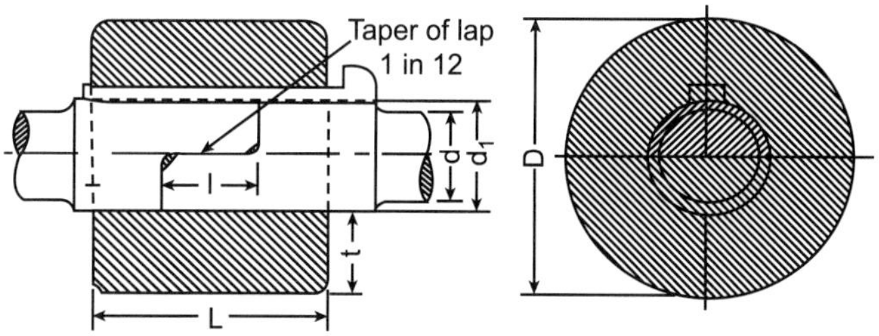

FIGURE 19-7 Fairbairn's lap-box coupling.

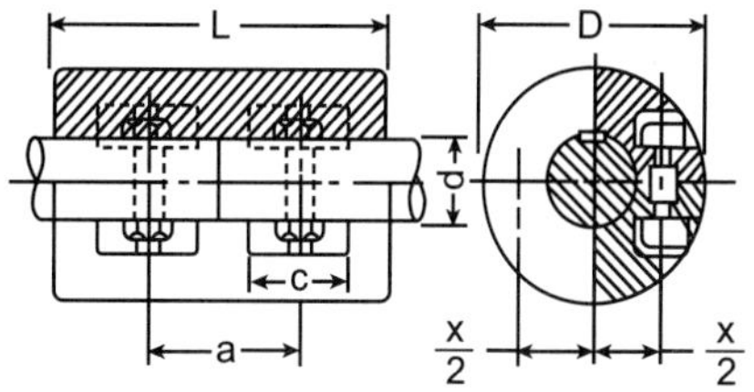

FIGURE 19-8 Split muff coupling.

SPLIT MUFF COUPLING (Fig. 19-8)

Particular	Formula	
The outside diameter of the sleeve	$D = 2d + 0.013$ where D, d in m	**SI** (19-54a)
	$D = 2d + 0.52$ where D, d in in	**USCS** (19-54b)
The length of the sleeve (Fig. 19-8)	$l = 3.5d$ or $2.5d + 0.05$ where l, d in m	**SI** (19-55a)

Particular	Formula
	$l = 3.5d$ or $2.5d + 2.0$ **USCS** (19-55b) where l, d in in
The torque to be transmitted by the coupling	$M_t = \dfrac{\pi d_c^2 \sigma_t \mu i d}{16}$ (19-56) where d_c = core diameter of the clamping bolts, m (in) i = number of bolts

SLIP COUPLING (Fig. 19-9)

Particular	Formula
The axial force exerted by the springs	$F_a = \dfrac{\pi}{4}(D_2^2 - D_1^2)p$ (19-57)
With two pairs of friction surfaces, the tangential force	$F_\theta = 2\mu F_a$ (19-58)
The radius of applications of F_θ with sufficient accuracy	$r_m = \dfrac{D_m}{2} = \dfrac{D_2 + D_1}{4}$ (19-59)
The torque	$M_t = 0.000385(D_2^2 - D_1^2)(D_2 + D_1)\mu p$ **SI** (19-60a) $M_t = 0.3927(D_2^2 - D_1^2)(D_2 + D_1)\mu p$ **USCS** (19-60b) where the values of μ and p may be taken from Table 19-4
The relation between D_1 and D_2	$\dfrac{D_2}{D_1} = 1.6$ (19-61) where D_1 and D_2 are the inner and outer diameters of disk of friction lining

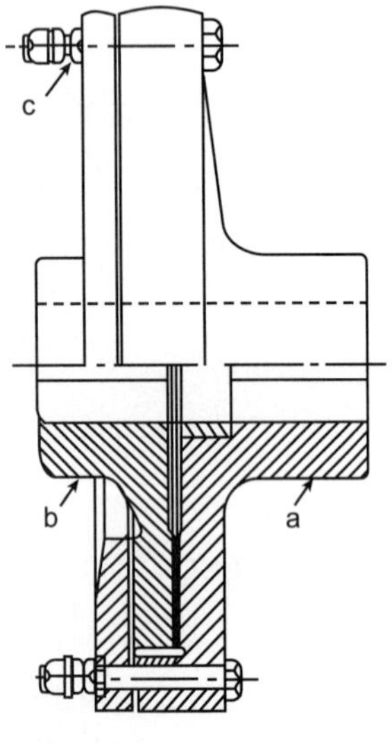

FIGURE 19-9 Slip coupling.

Particular	Formula	

SELLERS' CONE COUPLING (Fig. 19-10)

The length of the box	$L = 3.65d$ to $4d$	(19-62)
The outside diameter of the conical sleeve	$D_1 = 1.875d$ to $2d + 0.0125$ **SI** where D, d in m	(19-63a)
	$D_1 = 1.875d$ to $2d + 0.5$ **USCS** where D, d in in	(19-63b)
Outside diameter of the box	$D_2 = 3d$	(19-64)
The length of the conical sleeve	$l = 1.5d$	(19-65)

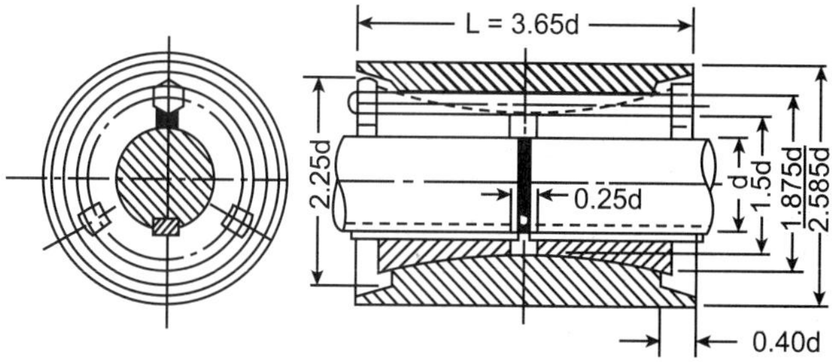

FIGURE 19-10 Sellers, cone coupling.

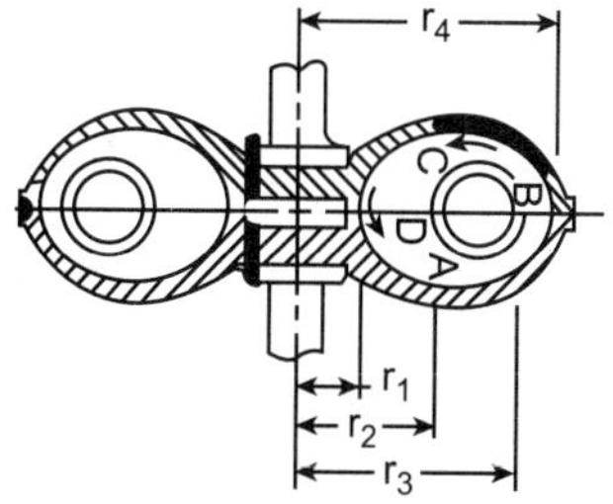

FIGURE 19-11 Hydraulic coupling.

HYDRAULIC COUPLINGS (Fig. 19-11)

Torque transmitted	$M_t = Ksn^2 W(r_{mo}^2 - r_{mi}^2)$	(19-66)
	where K = coefficient $= \dfrac{1.42}{10^7}$ (approx.)	
Percent slip between primary and secondary speeds	$s = \dfrac{n_p - n_s}{n_p} \times 100$	(19-67)
	where n_p and n_s are the primary and secondary speeds of impeller, respectively, rpm	
The mean radius of inner passage (Fig. 19-11)	$r_{mi} = \dfrac{2}{3}\left(\dfrac{r_2^3 - r_1^3}{r_2^2 - r_1^2}\right)$	(19-68a)
The mean radius of outer passage (Fig. 19-11)	$r_{mo} = \dfrac{2}{3}\left(\dfrac{r_4^3 - r_3^3}{r_4^2 - r_3^2}\right)$	(19-68b)
The number of times the fluid circulates through the torus in one second is given by	$i = \dfrac{13{,}000 M_t}{nW(r_{mo}^2 - r_{mi}^2)}$	(19-69)

TABLE 19-4
Friction materials for clutches

Contact surfaces		Friction coefficient,[a] μ		Maximum temperature		Maximum pressure, p		Relative cost	Comment
Wearing	**Opposing**[b]	**Wet**	**Dry**	**K**	**°C**	**MPa**	**kgf/mm²**		
Cast bronze	Cast iron or steel	0.05		422	149	0.5521–0.8277	0.0563–0.0844	Low	Subject to seizing
Cast iron	Cast iron	0.05	0.15–0.2	589	316	1.0346–1.7240	0.1055–0.1755	Very low	Good at low speeds
Cast iron	Steel	0.06		533	260	0.8277–1.3788	0.0844–0.1406	Very low	Fair at low speeds
Hard steel	Hard steel	0.05		533	260	0.6894	0.0703	Moderate	Subject to galling
Hard steel	Hard steel, chromium plated	0.03		533	260	1.3788	0.1406	High	Durable combination
Hard-drawn phosphor bronze	Hard steel, chromium plated	0.03		533	260	1.0346	0.1055	High	Good wearing qualities
Powder metal[c]	Cast iron or steel	0.05–0.1	0.1–0.4	811	538	1.0346	0.1055	High	Good wearing qualities
Powder metal[c]	Hard steel, chromium plated	0.05–0.1	0.1–0.3	811	538	2.0682	0.2109	Very high	High energy absorption
Wood	Cast iron or steel	0.16	0.2–0.35	422	149	0.4138–0.6208	0.0422–0.0633	Lowest	Unsuitable at high speed
Leather	Cast iron or steel	0.12–0.15	0.3–0.5	363.3	90.3	0.0686–0.2746	0.0070–0.0284	Very low	Subject to glazing
Cork	Cast iron or steel	0.15–0.25	0.3–0.5	363.3	90.3	0.0549–0.0981	0.0056–0.01	Very low	Cork-insert type preferred
Felt	Cast iron or steel	0.18	0.22	411	138	0.0343–0.0686	0.0035–0.0070	Low	Resinent engagement
Vulcanized fiber or paper	Cast iron or steel		0.3–0.5	363.3	90.3	0.0686–0.2746	0.3070–0280	Very low	Low speeds, light duty
Woven asbestos[c]	Cast iron or steel	0.1–0.2	0.3–0.6	444–533	171–260	0.03432–0.6894	0.0350–0.0703	Low	Prolonged slip service ratings given
Woven asbestos	Cast iron or steel	0.1–0.2		533	260	0.6894–1.3788	0.0703–0.1406	Low	This rating for short infrequent engagements
Woven asbestos	Hard steel, chromium plated	0.1				8.2738	0.8437	Moderate	Used in Napier Sabre engine
Molded asbestos[c]	Cast iron or steel	0.08–0.12	0.2–0.5	533	260	0.3452–1.0346	0.0352–0.1055	Very low	Wide field of applications
Impregnated asbestos	Cast iron or steel	0.12	0.32	533–659	260–386	1.0346	0.1055	Moderate	For demanding applications
	Steel	0.05–0.1	0.25	632–811	359–538	2.0682	0.2109	High	For critical requirements
Carbon graphite	Cast iron	0.1–0.15	0.25	422	149	0.6894	0.0703	Low	For light special service
Molded phenolic plastic, macerated cloth base									

[a] Conservative values should be used to allow for possible glazing of clutch surfaces in service and for adverse operating conditions.
[b] Steel, where specified, should have a carbon content of approximately 0.70%. Surfaces should be ground true and smooth.
[c] For a specific material within this group, the coefficient usually is maintained within plus or minus 5%.
Note: 1 kpsi = 6.894757 MPa or 1 Pa = 145×10^{-6} psi or 1 MPa = 145 psi.

Particular	Formula
Power transmitted by torque converter	$M_t - Kn^2 D^5$ (19-70) where K = coefficient—varies with the design n = speed of driven shaft, rpm D = outside diameter of vanes, m (in)

19.2 CLUTCHES

POSITIVE CLUTCHES (Fig. 19-12)

Jaw clutch coupling

SI	US Customary
$a = 2.2d + 0.025$ m	$a = 2.2d + 1.0$ in
$c = 1.2d + 0.03$ m	$c = 1.2d + 1.2$ in
$f = 1.4d + 0.0055$ m	$f = 1.4d + 0.3$ in
$g = d + 0.005$ m	$g = d + 0.2$ in
$h = 0.3d + 0.0125$ m	$h = 0.3d + 0.5$ in
$i = 0.4d + 0.005$ m	$i = 0.4d + 0.2$ in
$j = 0.2d + 0.0375$ m	$j = 0.2d + 0.15$ in
$k = 1.2d + 0.02$ m	$k = 1.2d + 0.8$ in
$l = 1.7d + 0.0584$ m	$l = 1.7d + 2.3$ in

(19-71)

The area in shear

$$A = \frac{0.5(a-b)h}{\sin\alpha} \tag{19-72}$$

The shear stress assuming that only one-half the total number of jaws i is in actual contact

$$\tau = \frac{4F_\theta \sin\alpha}{i(a-b)h\cos\alpha} \tag{19-73}$$

$$\tau = \frac{2.8F_\theta}{i(a-b)h} \quad \text{for } \tan\alpha = 0.7 \tag{19-74}$$

where α = angle made by the shearing plane with the direction of pressure

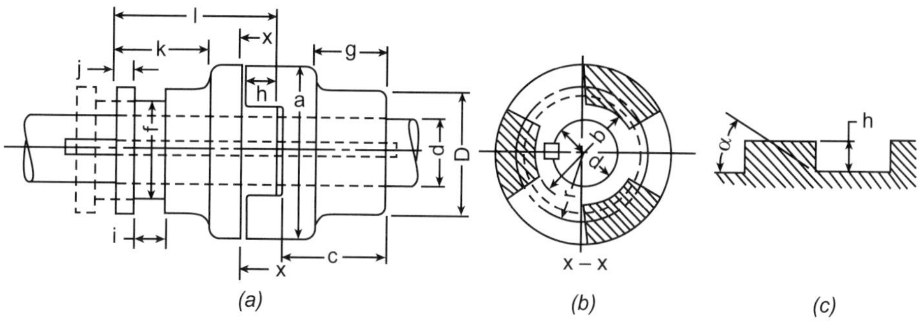

FIGURE 19-12 Square-jaw clutch.

Particular	Formula

FRICTION CLUTCHES

Cone clutch (Fig. 19-13)

The axial force in terms of the clutch dimensions	$F_a = \pi D_m pb \sin\alpha$ (19-75) where $D_m = \frac{1}{2}(D_1 + D_2)$ (approx.) $\alpha =$ one-half the cone angle, deg = ranges from 15° to 25° for industrial clutches faced with wood = 12.5° for clutches faced with asbestos or leather or cork insert
Axial force in terms of normal force (Fig. 19-13)	$F_a = F_n \sin\alpha$ (19-76)
The tangential force due to friction	$F_\theta = \dfrac{\mu F_a}{\sin\alpha}$ (19-77)
Torque transmitted through friction	$M_t = \dfrac{\mu F_a D_m}{2\sin\alpha}$ (19-78)
Power transmitted	$P = \dfrac{\mu F_a D_m n}{19{,}100 \sin\alpha k_l}$ **SI** (19-79a)
	$P = \dfrac{\mu F_a D_m n}{126{,}000 \sin\alpha k_l}$ **USCS** (19-79b)
	$P = \dfrac{\pi\mu p D_m^2 bn}{19{,}100 k_l}$ **SI** (19-79c)

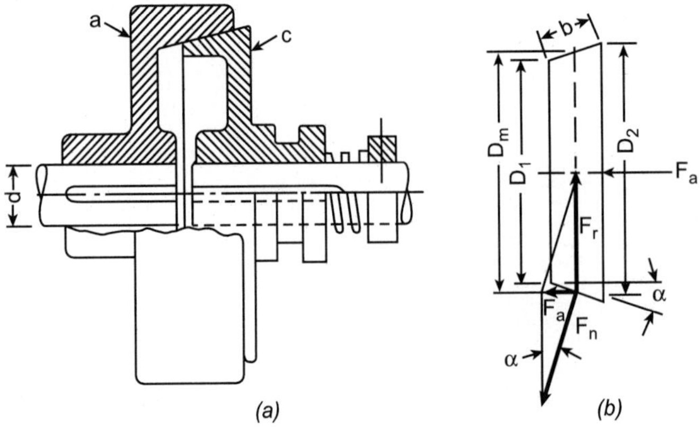

FIGURE 19-13 Cone clutch.

Particular	Formula	
	$P = \dfrac{\pi \mu p D_m^2 bn}{126{,}000 k_l}$ **USCS**	(19-79d)
	where k_l = load factor from Table 14-7	
	Refer to Table 19-4 for p.	
The force necessary to engage the clutch when one member is rotating	$F'_a = F_n(\sin\alpha + \mu\cos\alpha)$	(19-80)
The ratio (D_m/b)	$q = \dfrac{D_m}{b} = 4.5$ to 8	(19-81)
The value of D_m in commercial clutches	$D_m = 18.2\sqrt[3]{\dfrac{Pk_l q}{\mu pn}}$ **SI**	(19-82a)
	$D_m = 34.2\sqrt[3]{\dfrac{Pk_l q}{\mu pn}}$ **USCS**	(19-82b)
	$D_m = 5d$ to $10d$	(19-82c)

DISK CLUTCHES (Fig. 19-14)

Particular	Formula	
The axial force	$F_a = \frac{1}{2}\pi p D_1(D_2 - D_1)$	(19-83)
	Refer to Table 19-4 for p.	
The torque transmitted	$M_t = \frac{1}{2}\mu F_a D_m$	(19-84)
	where	
	$D_m = \dfrac{2}{3}\dfrac{(D_2^3 - D_1^3)}{(D_2^2 - D_1^2)}$	(19-85a)
	for uniform pressure distribution and	
	$D_m = \frac{1}{2}(D_2 + D_1)$	(19-85b)
	for uniform wear	

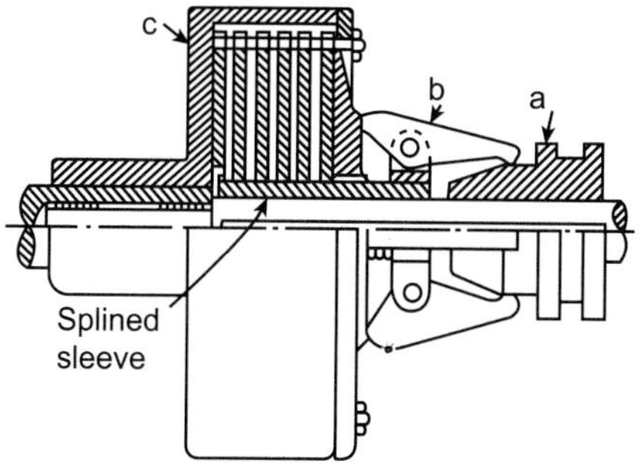

FIGURE 19-14 Multidisk clutch.

Particular	Formula	
Power transmitted	$P = \dfrac{i\mu F_a n}{28{,}650 k_l}\left(\dfrac{D_2^3 - D_1^3}{D_2^2 - d_1^2}\right)$ **SI**	(19-86a)
	$P = \dfrac{i\mu F_a n}{189{,}000 k_l}\left(\dfrac{D_2^3 - D_1^3}{D_2^2 - d_1^2}\right)$ **USCS**	(19-86b)
	for uniform pressure	

Particular	Formula
	where $F_a = \pi p \dfrac{D_2^2 - D_1^2}{4}$
	$P = \dfrac{\pi i \mu p n D_1 (D_2^2 - D_1^2)}{76{,}400 k_l}$ **SI** (19-87a)
	$P = \dfrac{\pi i \mu p n D_1 (D_2^2 - D_1^2)}{504{,}000 k_l}$ **SI** (19-87a)
	for uniform wear
The clutch capacity at speed n_1	$P_1 = \dfrac{P n_1}{n k_s}$ (19-88)
	where P = design power at speed, n k_s = speed factor obtained from Eq. (19-89)
The speed factor	$k_s = 0.1 + 0.001n$ (19-89)
	where n = speed at which the capacity of clutch to be determined, rpm

DIMENSIONS OF DISKS (Fig. 19-15)

Particular	Formula
The maximum diameter of disk	$D_2 = 2.5 \text{ to } 3.6 D_1$ (19-90)
The minimum diameter of disk	$D_1 = 4d$ (19-91)
The thickness of disk	$h = 1 \text{ to } 3\,\text{mm}$ (19-92)
The number of friction surfaces	$i = i_1 + i_2 - 1$ (19-93)
The number of driving disks	$i_1 = \dfrac{i}{2}$ (19-94)
The number of driven disks	$i_2 = \dfrac{i}{2} + 1$ (19-95)

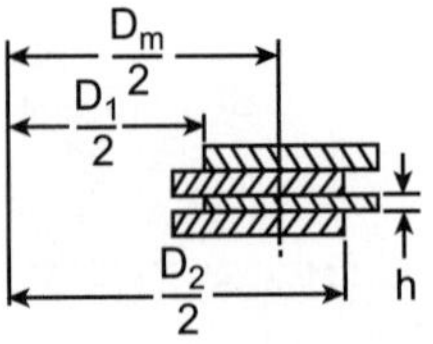

FIGURE 19-15 Dimensions of disks.

Particular	Formula

DESIGN OF A TYPICAL CLUTCH OPERATING LEVER (Fig. 19-16)

The total axial force on i number of clutch disk or plates

$$F'_a = i\pi p' D_1(D_1 - D_2) \tag{19-96}$$

where p' = actual pressure between disks

$$F'_a = \frac{4M_{ta}}{i\pi\mu(D_1 - D)D_m^2}, \quad \text{MPa (psi)}$$

M_{ta} = allowable torque, N m (lbf in)

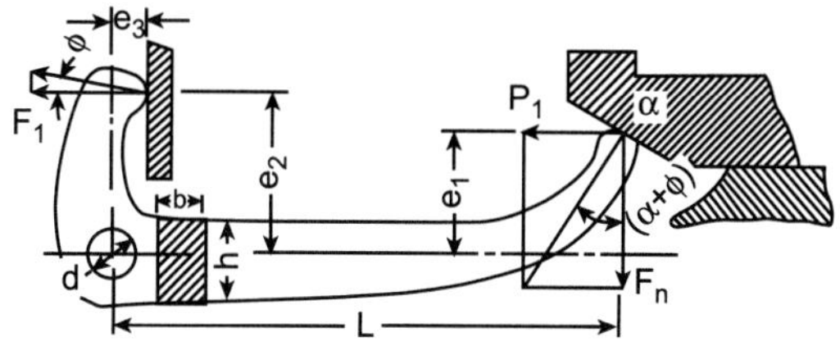

FIGURE 19-16 A typical clutch operating lever.

The force acting on disks of one operating lever of the clutch (Fig. 19-16)

$$F_1 = \frac{F'_a}{i'} \tag{19-97}$$

where i' = number of operating levers

The total force acting from the side of the bushing (Fig. 19-16)

$$P = i' p_1 \tag{19-98}$$

The force acting from the side of the bushing on one operating lever (Fig. 19-16)

$$P_1 = F_1 \frac{\left(L\cot(\alpha+\phi) - e_1 - \mu\dfrac{d}{2}\right)}{e_2 + \mu\left(e_3 + \dfrac{d}{2}\right)} \tag{19-99}$$

The thickness of the !ever very close to the pin (Fig. 19-16)

$$h = \left[\frac{6F'_a e_3}{\left(\dfrac{b}{h}\right) i' \sigma_{db}}\right]^{1/3} \tag{19-100}$$

where σ_{db} = design bending stress for the material of the levers, MPa (psi)

Ratio of $b/h = 0.75$ to 1

The diameter of the pin (Fig. 19-16)

$$d = \sqrt{\frac{2F_r}{\pi\tau_d}} \tag{19-101}$$

where

F_r = resultant force due to F_1 and $P_1\cot(\alpha+\phi)$ on the pin, kN (lbf)

τ_d = design shear stress of the material of the pin, MPa (psi)

Particular	Formula

EXPANDING-RING CLUTCHES (Fig. 19-17)

Torque transmitted [Fig. 19-17(a)]

$$M_t = 2\mu pwr^2\theta \quad (19\text{-}102)$$

where

θ = one half the total arc of contact, rad
w = width of ring, m (in)

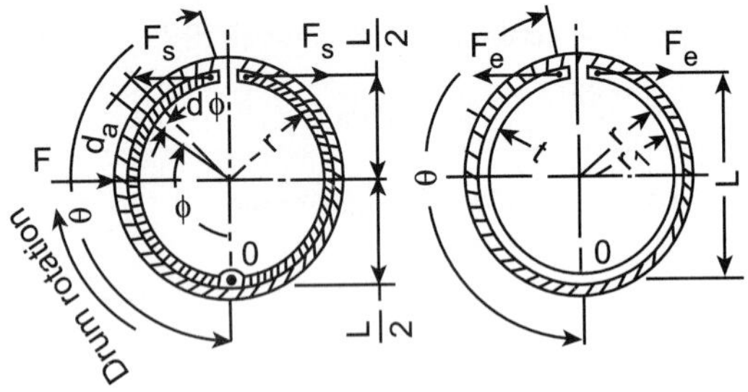

FIGURE 19-17 Expanding-ring clutch.

The moment of the normal force for each half of the band [Fig. 19-17(a)]

$$M_o = pwrL \quad (19\text{-}103)$$

when $\theta \approx \pi$ rad

$$F_s = pwr \quad (19\text{-}104)$$

The force applied to the ends of the split ring to expand the ring [Fig. 19-17(a)]

$$F_e = \frac{Ewt^3}{6L}\left(\frac{1}{d_1} - \frac{1}{d}\right) \quad (19\text{-}105)$$

If the ring is made in one piece (Fig. 19-7(b)] an additional force required to expand the inner ring before contact is made with inner surface of the shell

where

d_1 = original diameter of ring, m (in)
d = inner diameter of drum, m (in)
w = width of ring, m (in)
t = thickness of ring, m (in)

$$F = F_s + F_e \quad (19\text{-}106)$$

The total force required to expand the ring and to produce the necessary pressure between the contact surfaces

$$F = pwr + \frac{Ewt^3}{6L}\left(\frac{1}{d_1} - \frac{1}{d}\right) \quad (19\text{-}107)$$

RIM CLUTCHES (Fig. 19-18)

When the grooved rim clutch being engaged, the equation of equilibrium of forces along the vertical axis

$$F_n = F_n'(\sin\alpha + \mu\cos\alpha) \quad (19\text{-}108)$$

After the block is pressed on firmly the equation of equilibrium of forces along the vertical axis

$$F_n = F_n'\sin\alpha \quad (19\text{-}109)$$

Torque transmitted

$$M_t = \tfrac{1}{2} i_1 i_2 F_\theta D = i_1 i_2 \mu\beta D^2 bp \quad (19\text{-}110)$$

where

i_1 = number of grooves in the rim
i_2 = number of shoes
b = inclined face, m (in)
2β = angle of contact, rad

Particular	Formula

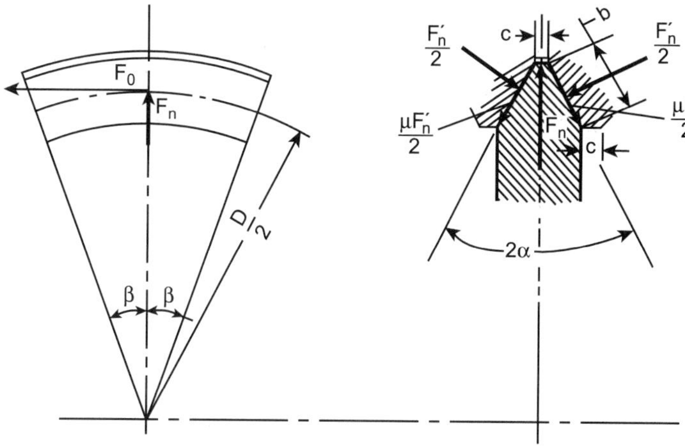

FIGURE 19-18 Grooved rim clutch.

Particular	Formula		
	D = pitch diameter, m (in) 2α = V-groove angle, deg		
The width of the inclined face	$b = 0.01D + 0.006\,\text{m}$	SI	(19-111a)
	$b = 0.01D + 0.25\,\text{in}$	USCS	(19-111b)
Frictional force	$F_\theta = \mu F'_n$		(19-112a)
	where $F'_n = 2\beta Dbp$		(19-112b)
Torque transmitted in case of a flat rim clutch when $i_1 = 1$ and the number of sides b is only one-half that of a grooved rim	$M_t = \frac{i}{2}\mu\beta D^2 bp$		(19-113)

CENTRIFUGAL CLUTCH (Fig. 19-19)

Design of shoe

Particular	Formula	
Centrifugal force for speed ω_1 (rad/s) at which engagement between shoe and pulley commences	$F_{c1} = \frac{w}{g}\omega_1^2 r$	(19-114)
Centrifugal force for running speed ω_2 (rad/s)	$F_{c2} = \frac{w}{g}\omega_2^2 r$	(19-115)
The outward radial force on inside rim of the pulley at speed ω_2	$F_c = F_{c2} - F_{c1}$	(19-116a)
	$F_c = \frac{w}{g}(\omega_2^2 - \omega_1^2)r$	(19-116b)
The centrifugal force for $\omega_1 = 0.75\omega_2$	$F'_c = \frac{7w}{16g}\omega_2^2 r$	(19-117)

Particular	Formula

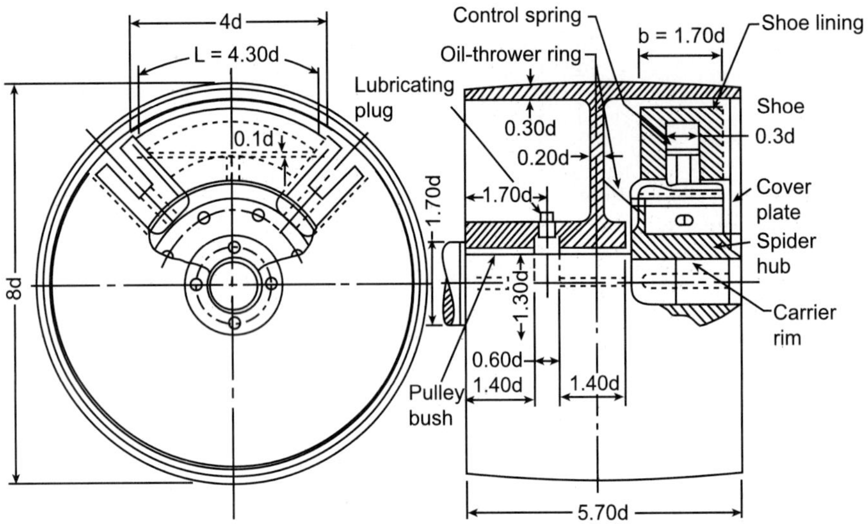

FIGURE 19-19 Centrifugal clutch.

Torque required for the maximum power to be transmitted

$$M_t = 4\mu r' F_c = 4\mu \frac{w}{g}(\omega_2^2 - \omega_1^2) r r' \quad (19\text{-}118)$$

where r' = inner radius of the rim

The equation to calculate the length of the shoe (Fig. 19-19)

$$l = \frac{F_c}{bp} = \frac{w}{gbp}(\omega_2^2 - \omega_1^2) r \quad (19\text{-}119)$$

Spring

The central deflection of flat spring (Fig. 19-19) which is treated as a beam freely supported at the points where it bears against the shoe and loaded centrally by the adjusting screw

$$y = \frac{1 W l^3}{48EI} \quad (19\text{-}120)$$

The maximum load exerted on the spring at speed ω_1

$$W = F_{c1} = \frac{w}{g}\omega_1^2 r \quad (19\text{-}121)$$

The cross section of spring can be calculated by the equation

$$I = \frac{bh^3}{12} = \frac{Wl^3}{48Ey} \quad (19\text{-}122)$$

For other proportionate dimensions of centrifugal clutch

Refer to Fig. 19-19.

Particular	Formula

OVERRUNNING CLUTCHES

Roller clutch (Fig. 19-20)

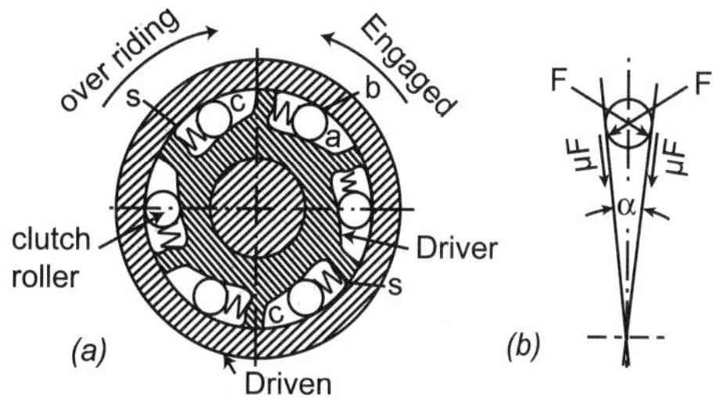

FIGURE 19-20 Roller clutch.

The condition for the operation of the clutch

$$\alpha < 2\phi \quad (19\text{-}123)$$

where ϕ = angle of friction, μ varies from 0.03 to 0.005

For ϕ = angle 1°43′, the angle $\alpha < 3°26'$

The force crushing the roller

$$F = \frac{F_\theta}{\tan\alpha} \quad (19\text{-}124)$$

where F_θ = tangential force necessary to transmit the torque at pitch diameter D

The torque transmitted

$$M_c = \tfrac{1}{2}F_\theta D \quad (19\text{-}125)$$

The allowable load on roller

$$F_a \le i\sigma_c k' ld$$

where

k' = coefficient of the flattening of the roller

$$= \frac{4.64\sigma_c}{E} \quad (19\text{-}126)$$

for σ_c = allowable crushing stress = 1035.0 MPa (150 kpsi)

The roller diameter

$$d = 0.1D \text{ to } 0.15D$$

The number of roller

$$i = \frac{\pi(D+d)}{2d} \quad (19\text{-}126a)$$

LOGARITHMIC SPIRAL ROLLER CLUTCH (Fig. 19-21)

The radius of curvature of the ramp at the point of contact (Fig. 19-21)

$$R_c = 2(R_d - R_r)\frac{\sin 2\phi}{\sin 2\psi} \quad (19\text{-}127)$$

The radius vector of point C (Fig. 19-21)

$$R_v = \frac{\sin 2\phi}{\cos(2\phi+\psi)}R_r \quad (19\text{-}128)$$

The radius of the contact surface on the driven member in terms of the roller radius and functions angles ψ and ϕ (Fig. 19-21)

$$R_d = R_c\left(1 + \frac{\cos\phi}{\cos(2\phi+\psi)}\right) \quad (19\text{-}129)$$

The tangential force

$$F_\theta = F\sin\phi \quad (19\text{-}130)$$

Particular	Formula

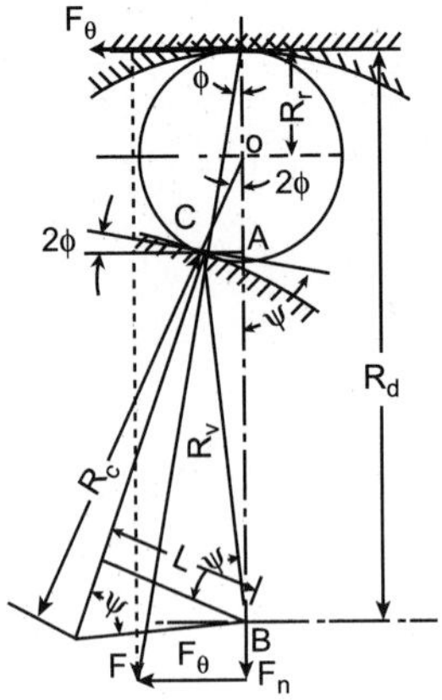

FIGURE 19-21 Logarithmic spiral roller-clutch.

The normal force

$$F_n = \frac{F_\theta}{\tan\phi} = F\cos\phi \tag{19-130a}$$

The torque transmitted

$$M_t = \frac{iF_nR_d}{\cot\phi} \tag{19-130b}$$

where

2ϕ = angle of wedge, deg (usually ϕ varies from 3° to 12°)

i = number of rollers in the clutch

The maximum compressive stress at the surface area of contact between the roller and the cage made of different materials

$$\sigma_{c(\max)} = 0.798\left[\frac{F}{2l}\frac{\left(\frac{1}{R_r}+\frac{1}{R_c}\right)}{\left(\frac{1-v_r^2}{E_r}+\frac{1-v_c^2}{E_c}\right)}\right]^{1/2} \tag{19-131}$$

The maximum compressive stress at the surface area of contact between the roller and the cage for $v_c = v_r = 0.3$

$$\sigma_{c(\max)} = \left[\frac{0.35F\left(\frac{1}{R_r}+\frac{1}{R_c}\right)}{l\left(\frac{1}{E_r}+\frac{1}{E_c}\right)}\right]^{1/2} \tag{19-132}$$

The maximum compressive stress at the surface area of contact between the roller and the cage made of same material ($E_c = E_r = E$) and $v_c = v_r = 0.3$

$$\sigma_{c(\max)} = 0.418\left[\frac{FE\left(\frac{1}{R_r}+\frac{1}{R_c}\right)}{l}\right]^{1/2} \tag{19-133a}$$

$$\sigma_{c(\max)} = 0.418\sqrt{\frac{FE}{lR_r}} \quad \text{if } R_c \gg R_r \tag{19-133b}$$

$$\sigma_{c(\max)} = 0.418\sqrt{\frac{2FE}{ld}} \tag{19-133c}$$

where

$d = 2R_r$ = diameter of roller, m (mm)

l = length of the roller, m (mm)

Particular	Formula
The design torque transmitted by the clutch	$$M_{td} = \frac{ildR_d\sigma_{c(\max)} \tan\phi}{0.35E} \quad (19\text{-}134)$$ where 2ϕ varies from 3 to 6 deg.
For further design data for clutches	Refer to Tables 19-5, 19-6, 19-7.

TABLE 19-5
Preferred dimensions and deviations for clutch facings (all dimensions in mm)

Outside diameter	Deviation	Inside diameter	Deviation	Thickness	Deviation
120, 125, 130	0	80, 85, 90	+0.5	3, 3.5, 4	±0. 1
135, 140, 145	−0.5	95, 100, 105	0		
150, 155, 150		110			
170, 180, 190					
200, 210, 220	0	120, 130, 140	+0.8		
230, 240, 250	−0.8	150	0		
260, 270, 280		175, 203	+1.0		
290, 300			0		
	0				
325, 350	−1.0				

19.3 BRAKES

ENERGY EQUATIONS

Case of a hoisting drum lowering a load:

The decrease of kinetic energy for a change of speed of the live load from v_1 to v_2

$$E_k = \frac{F(v_1^2 - v_2^2)}{2g} \quad (19\text{-}135a)$$

where v_1, v_2 = speed of the live load before and after the brake is applied respectively, m/s

F = load, kN (lbf)

The change of potential energy absorbed by the brake during the time t

$$E_p = \frac{F}{2}(v_1 + v_2)t \quad (19\text{-}135b)$$

The change of kinetic energy of all rotating parts such as the hoist drum and various gears and sheaves which must be absorbed by the brake

$$E_r = \sum \frac{Wk_o^2(\omega_1^2 - \omega_2^2)}{2g} \quad (19\text{-}136)$$

where

k_o = radius of gyration of rotating parts, m (mm)

ω_1, ω_2 = angular velocity of the rotating parts, rad/s

TABLE 19-6
Service factors for clutches

Type of service	Service factor not including starting factor
Driving machine	
Electric motor steady load	1.0
Fluctuating load	1.5
Gas engine, single cylinder	1.5
Gas engine, multiple cylinder	1.0
Diesel engine, high-speed	1.5
Large, low-speed	2.0
Driven machine	
Generator, steady load	1.0
Fluctuating load	1.0
Blower	1.0
Compressor depending on number of cylinders	2.0–2.5
Pumps, centrifugal	1.0
Pumps, single-acting	2.0
Pumps, double-acting	1.5
Line shaft	1.5
Wood working machinery	1.75
Hoists, elevators, cranes, shovels	2.0
Hammer mills, ball mills, crushers	2.0
Brick machinery	3.0
Rock crushers	3.0

TABLE 19-7
Shear strength for clutch facings

		Shear strength	
Type	Facing material	MPa	kgf/mm^2
A	Solid woven or plied fabric with or without metallic reinforcement	7.4	0.75
B	Molded and semimolded compound	4.9	0.50

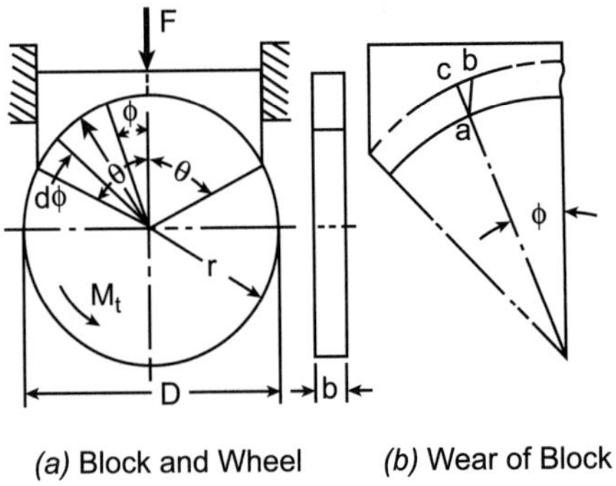

FIGURE 19-22 Single-block brake.

Particular	Formula	
The work to be done by the tangential force F_θ at the brake sheave surface in t seconds	$W_k = \dfrac{F_\theta \pi D(n_1 + n_2)t}{2 \times 60}$	(19-137)
The tangential force at the brake sheave surface	$F_\theta = \dfrac{38.2(E_k + E_p + E_r)}{D(n_1 + n_2)}$	(19-138)
Torque transmitted when the blocks are pressed against flat or conical surface	$M_t = \mu F_n \dfrac{D_m}{2}$	(19-139)
	where F_n = total normal force, kN (lbf)	
The operating force on block in radial direction (Fig. 19-22)	$F = \dfrac{F_\theta}{\mu}\left(\dfrac{2\theta + \sin 2\theta}{4 \sin \theta}\right)$	(19-140)
Torque applied at the braking surface, when the blocks are pressed radially against the outer or inner surface of a cylindrical drum (Fig. 19-22)	$M_t = \mu F \dfrac{D}{2}\left(\dfrac{4 \sin \theta}{2\theta + \sin 2\theta}\right)$	(19-141)

Particular	Formula

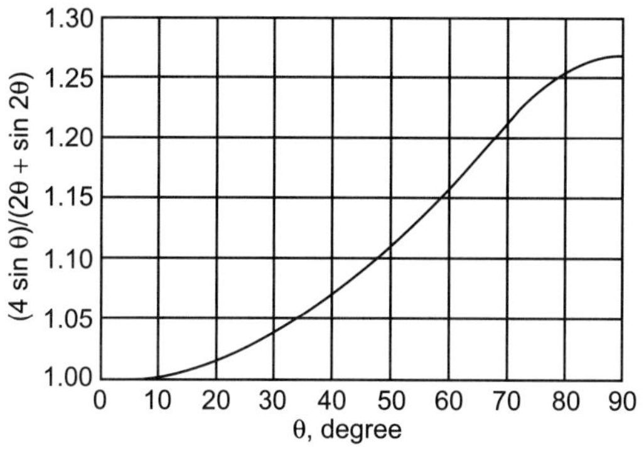

FIGURE 19-23 $(4\sin\theta)/(2\theta+\sin 2\theta)$ plotted against the semiblock angle θ.

The tangential frictional force on the block (Fig. 19-22)	$F_\theta = \mu F\left(\dfrac{4\sin\theta}{2\theta+\sin 2\theta}\right)$	(19-142)
	Refer to Fig. 19-23 for values of $\dfrac{4\sin\theta}{2\theta+\sin 2\theta}$.	
Torque applied when θ is less than 60°	$M_t = \mu F\,\dfrac{D}{2}$ (approx.)	(19-143)
	where $F = \mu p_a(br\theta)$	

BRAKE FORMULAS

Block brake formulas

For block brake formulas	Refer to Table 19-8 for formulas from Eqs. (19-144) to (19-148)	

Band brake formulas

For band brake formulas	Refer to Table 19-8 for formulas from Eqs. (19-149) to (19-157)	
The magnitude of pressure between the band and the brake sheave	$p = \dfrac{F_1+F_2}{Dw}$	(19-158)
The practical rule for the band thickness	$h = 0.005D$	(19-159)
Width of band	$w = \dfrac{F_1}{h\sigma_d}$	(19-160)

TABLE 19-8
Formulas for block, simple, and differential band brakes

Type of brake and rotation			Force at the end of brake handle, kN (kgf)	
Block brake	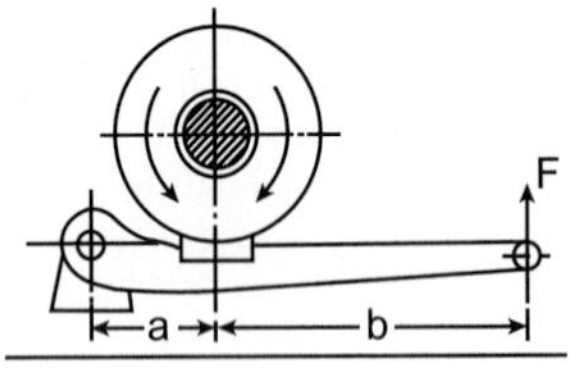 (a)	Rotation in either direction	$F = F_\theta \dfrac{a}{\mu(a+b)}$	(19-144)
Block brake	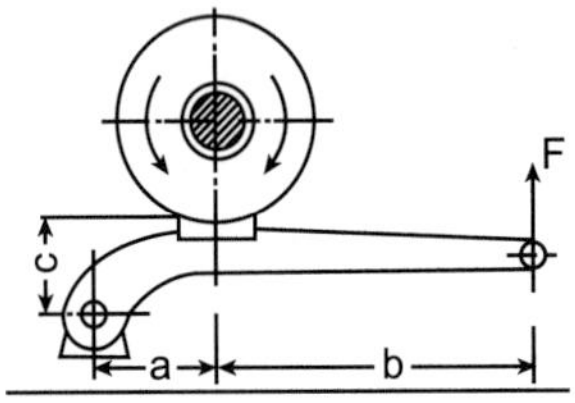 (b)	Clockwise	$F = \dfrac{F_\theta a}{a+b}\left(\dfrac{1}{\mu} - \dfrac{c}{a}\right)$	(19-145)
		Counterclockwise	$F = \dfrac{F_\theta a}{a+b}\left(\dfrac{1}{\mu} + \dfrac{c}{a}\right)$	(19-146)
Block brake	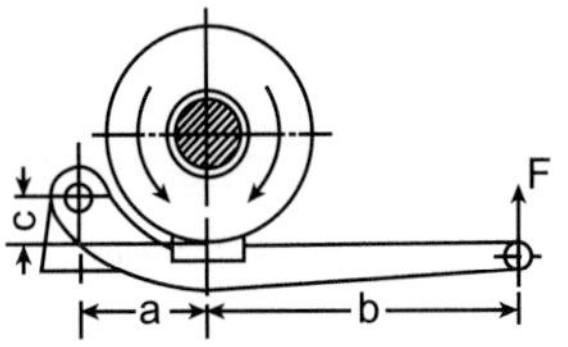 (c)	Clockwise	$F = \dfrac{F_\theta a}{a+b}\left(\dfrac{1}{\mu} + \dfrac{c}{a}\right)$	(19-147)
		Counterclockwise	$F = \dfrac{F_\theta a}{a+b}\left(\dfrac{1}{\mu} - \dfrac{c}{a}\right)^*$	(19-148)
Simple band brake	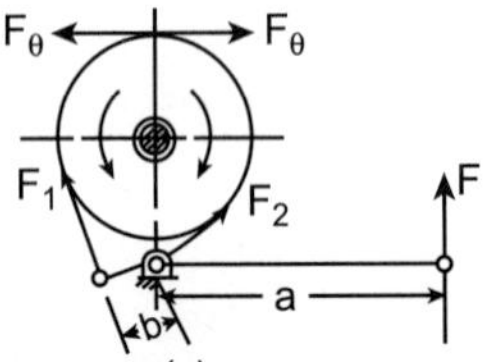 (a)	Clockwise	$F = \dfrac{F_\theta b}{a}\left(\dfrac{e^{\mu\theta}}{e^{\mu\theta} - 1}\right)$	(19-149)
		Counterclockwise	$F = \dfrac{F_\theta b}{a}\left(\dfrac{1}{e^{\mu\theta} - 1}\right)$	(19-150)
Simple band brake	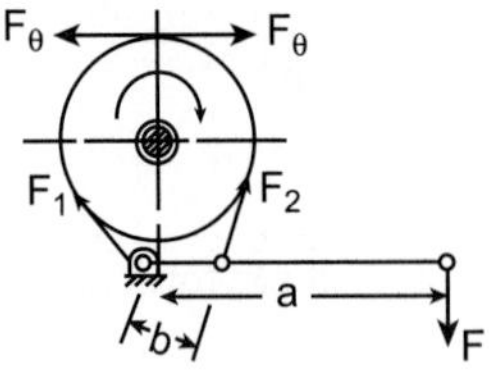 (b)	Clockwise	$F = \dfrac{F_\theta b}{a}\left(\dfrac{1}{e^{\mu\theta} - 1}\right)$	(19-151)
		Counterclockwise	$F = \dfrac{F_\theta b}{a}\left(\dfrac{e^{\mu\theta}}{e^{\mu\theta} - 1}\right)$	(19-152)

* For counterclockwise direction (c/a) must be less than $(1/\mu)$ or brake will be self-locking.

TABLE 19-8
Formulas for block, simple, and differential band brakes (*Cont.*)

Type of brake and rotation		Force at the end of brake handle, kN (kgf)	
Differential band brake	Clockwise	$F = \dfrac{F_\theta}{a}\left(\dfrac{b_2 e^{\mu\theta} + b_1}{e^{\mu\theta} - 1}\right)$	(19-153)
	Counterclockwise	$F = \dfrac{F_\theta}{a}\left(\dfrac{b_1 e^{\mu\theta} + b_2}{e^{\mu\theta} - 1}\right)$	(19-154)
	If $b_2 = b_1$ F is the same for rotation in either direction	$F = \dfrac{F_\theta b}{a}\left(\dfrac{b_1 e^{\mu\theta} + 1}{e^{\mu\theta} + 1}\right)^*$	(19-155)
Differential band brake	Clockwise	$F = \dfrac{F_\theta}{a}\left(\dfrac{b_2 e^{\mu\theta} - b_1}{e^{\mu\theta} - 1}\right)$	(19-156)
	Counterclockwise	$F = \dfrac{F_\theta}{a}\left(\dfrac{b_2 - b_1 e^{\mu\theta}}{e^{\mu\theta} - 1}\right)^{**}$	(19-157)

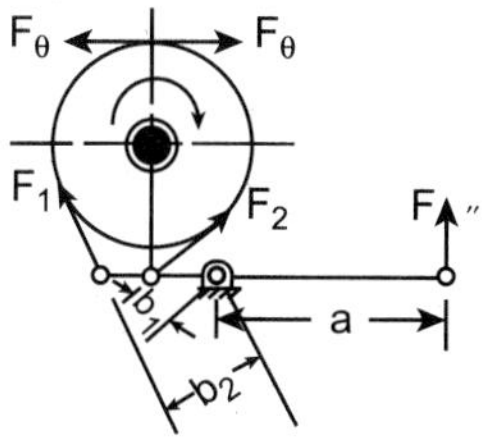

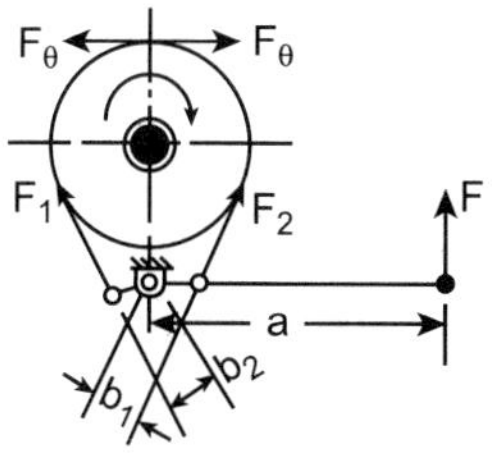

* For the above two cases, if $b_2 = b_1 = b$.
** In this case if $b_2 \le b_1 e^{\mu\theta}$, F will be negative or zero and the brake works automatically or the brake is "self-locking."

Particular	Formula	
Suitable drum diameter according to Hagenbook	$\left(\dfrac{M_t}{69}\right)^{1/3} < 10D < \left(\dfrac{M_t}{54}\right)^{1/3}$ **SI**	(19-161)
	where M_t in N m and D in m	
	$\left(\dfrac{M_t}{5}\right)^{1/3} < D < \left(\dfrac{M_t}{4}\right)^{1/3}$ **USCS**	(19-162)
	where M_t in lbf and D in in	
Suitable drum diameter in terms of frictional horsepower	$(79.3\mu P)^{1/3} < 100D < (105.8\mu P)^{1/3}$ **SI**	(19-163a)
	where P in kW and D in m	
	$(60\mu P)^{1/3} < D < (80\mu P)^{1/3}$ **USCS**	(19-163b)
	where P in hp and D in in	
	μP is taken as the maximum horsepower to be dissipated in any 15-min period	

Particular	Formula	

CONE BRAKES (Fig. 19-24)

Particular	Formula	
The normal force	$F_n = \dfrac{F_a}{\sin\alpha}$	(19-164)
The radial force	$F_r = \dfrac{F_a}{\tan\alpha}$	(19-165)
The tangential force or braking force	$F_\theta = \mu F_n = \dfrac{\mu F_a}{\sin\alpha}$	(19-166)
The braking torque	$M_t = \dfrac{\mu F_a D}{2\sin\alpha}$ where D = mean diameter, m (mm)	(19-167)

CONSIDERING THE LEVER (Fig. 19-24)

Particular	Formula	
The axial force	$F_a = \dfrac{aF}{h}$	(19-168)
The relation between the operating force F and the braking force F_θ	$F = \dfrac{hF_\theta \sin\alpha}{\mu a}$	(19-169)
The area of the contact surface using the designation given in Fig. 19-24	$A = \dfrac{\pi D w}{\cos\alpha}$ where w = axial width, m (mm) α = half the cone angle, deg	(19-170)
The average pressure between the contact surfaces	$F_{av} = \dfrac{F_n}{A} = \dfrac{F_a}{\pi D w \tan\alpha}$ Take α = from 10° to 18° w = 0.12D to 0.22D	(19-171)

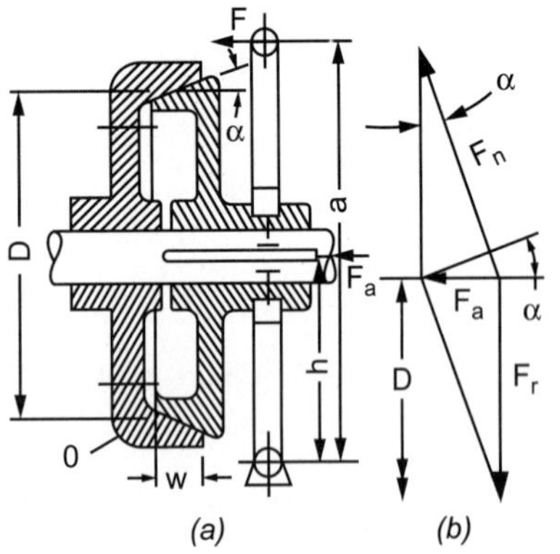

FIGURE 19-24 Cone brake.

Particular	Formula

DISK BRAKES

The torque transmitted for i pairs of friction surfaces

$$M_t = \frac{\pi i \mu p_1 D_1 (D_2^2 - D_1^2)}{8} \tag{19-172}$$

The axial force transmitted

$$F_a = \tfrac{1}{2}\pi p_1 D_1 (D_2 - D_1) \tag{19-173}$$

where p_1 = intensity of pressure at the inner radius, MPa (psi)

For design values of brake facings — Refer to Table 19-9.

TABLE 19-9
Design values for brake facings

Facing material	Design coefficient of friction μ	Permissible unit pressure 1 m/s MPa	Permissible unit pressure 1 m/s kgf/mm^2	Permissible unit pressure 10 m/s MPa	Permissible unit pressure 10 m/s kgf/mm^2
Cast iron on cast iron					
Dry	0.20				
Oily	0.07				
Wood on cast iron	0.25–0.30	0.5521–0.6824	0.0563–0.0703	0.1383–0.1726	0.0141–0.0176
Leather on cast iron					
Dry	0.40–0.50	0.0549–0.1039	0.0056–0.0106		
Oily	0.15				
Asbestos fabric on metal					
Dry	0.35–0.40	0.6209–0.6894	0.0633–0.0703	0.1726–0.2069	0.0176–0.0211
Oily	0.25				
Molded asbestos on metal	0.30–0.35	1.0395–1.2062	0.106–0.123	0.2069–0.2756	0.0211–0.0281

Note: 1 kpsi = 6.894757 MPa or 1 MPa = 145 psi.

INTERNAL EXPANDING-RIM BRAKE

Forces on Shoe (Fig. 19-25)

FOR CLOCKWISE ROTATION

The maximum pressure

$$p_a = p\,\frac{\sin\theta_a}{\sin\theta} \tag{19-174}$$

The moment $M_{t\mu}$ of the frictional forces

$$M_{t\mu} = \frac{\mu p_a b r}{\sin\theta_a}\int_{\theta_1}^{\theta_2} \sin\theta (r - a\cos\theta)\, d\theta \tag{19-174a}$$

The moment of the normal forces

$$M_{tn} = \frac{p_a b r a}{\sin\theta_a}\int_{\theta_1}^{\theta_2} \sin^2\theta\, d\theta \tag{19-175}$$

Particular	Formula

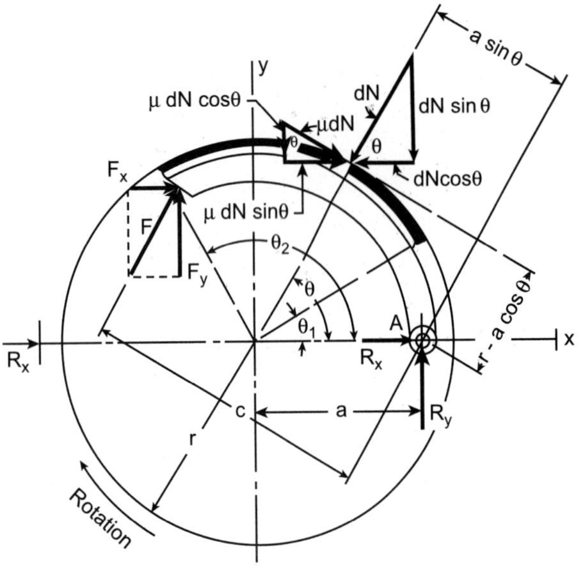

FIGURE 19-25 Forces on the shoe. (J. E. Shigley, *Mechanical Engineering Design*, 1962, courtesy of McGraw-Hill.)

Particular	Formula	
The actuating force	$F = \dfrac{M_{tn} - M_{t\mu}}{c}$	(19-176)
The torque M_t applied to the drum by the brake shoe	$M_t = \dfrac{\mu p_a b r^2 (\cos\theta_1 - \cos\theta_2)}{\sin\theta_a}$	(19-177)
The hinge-pin horizontal reaction	$R_x = \dfrac{p_a b r}{\sin\theta_a}\left(\int_{\theta_1}^{\theta_2} \sin\theta\cos\theta\, d\theta - \mu\int_{\theta_1}^{\theta_2} \sin^2\theta\, d\theta\right) - F_x$	(19-178)
The hinge-pin vertical reaction	$R_y = \dfrac{p_a b r}{\sin\theta_a}\left(\int_{\theta_1}^{\theta_2} \sin^2\theta\, d\theta + \mu\int_{\theta_1}^{\theta_2} \sin\theta\cos\theta\right) - F_y$	(19-179)

Particular	Formula
FOR COUNTERCLOCKWISE ROTATION (Fig. 19-25)	$F = \dfrac{M_{tn} + M_{t\mu}}{c}$ (19-180)
	$R_x = \dfrac{p_a br}{\sin\theta_a}\left(\int_{\theta_1}^{\theta_2} \sin\theta\cos\theta\,d\theta + \mu\int_{\theta_1}^{\theta_2}\sin^2\theta\,d\theta\right) - F_x$ (19-181)
	$R_y = \dfrac{p_a br}{\sin\theta_a}\left(\int_{\theta_1}^{\theta_2} \sin^2\theta\,d\theta - \mu\int_{\theta_1}^{\theta_2}\sin\theta\cos\theta\,d\theta\right) - F_y$ (19-182)

EXTERNAL CONTRACTING-RIM BRAKE

Forces on shoe (Fig. 19-26)

Particular	Formula
FOR CLOCKWISE ROTATION The moment $M_{t\mu}$ of the friction forces Fig. 19-26	$M_{t\mu} = \dfrac{\mu p_a br}{\sin\theta_a}\int_{\theta_1}^{\theta_2} \sin\theta(r - a\cos\theta)\,d\theta$ (19-183)
The moment of the normal force	$M_{tn} = \dfrac{p_a br}{\sin\theta_a}\int_{\theta_1}^{\theta_2} \sin^2\theta\,d\theta$ (19-184)

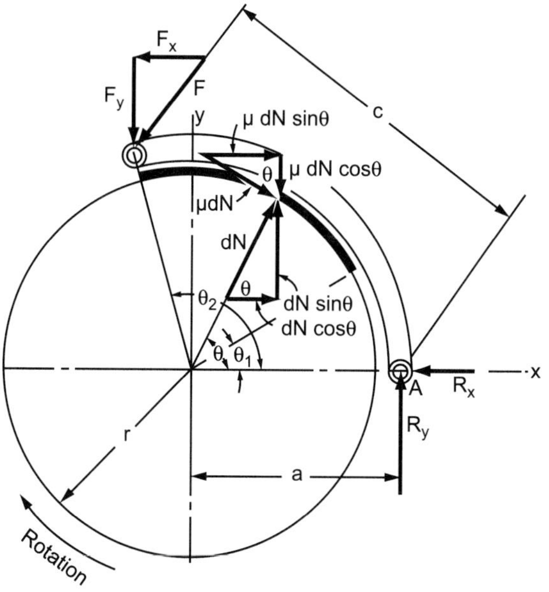

FIGURE 19-26 Forces and notation for external-contracting shoe. (J. E. Shigley, *Mechanical Engineering Design*, 1962, courtesy of McGraw-Hill.)

Particular	Formula
The actuating force	$$F = \frac{M_{tn} + M_{t\mu}}{c} \quad (19\text{-}185)$$
The horizontal reaction at the hinge-pin	$$R_x = \frac{p_a br}{\sin\theta_a}\left(\int_{\theta_1}^{\theta_2} \sin\theta\cos\theta\, d\theta + \mu\int_{\theta_1}^{\theta_2} \sin^2\theta\, d\theta\right) - F_x \quad (19\text{-}186)$$
The vertical reaction at the hinge-pin	$$R_y = \frac{p_a br}{\sin\theta_a}\left(\mu\int_{\theta_1}^{\theta_2} \sin\theta\cos\theta\, d\theta - \int_{\theta_1}^{\theta_2} \sin^2\theta\, d\theta\right) + F_y \quad (19\text{-}187)$$
FOR COUNTERCLOCKWISE ROTATION	$$F = \frac{M_{tn} - M_{t\mu}}{c} \quad (19\text{-}188)$$
	$$R_x = \frac{p_a br}{\sin\theta_a}\left(\int_{\theta_1}^{\theta_2} \sin\theta\cos\theta\, d\theta - \mu\int_{\theta_1}^{\theta_2} \sin^2\theta\, d\theta\right) - F_x \quad (19\text{-}189)$$
	$$R_y = \frac{p_a br}{\sin\theta_a}\left(-\mu\int_{\theta_1}^{\theta_2} \sin\theta\cos\theta\, d\theta - \int_{\theta_1}^{\theta_2} \sin^2\theta\, d\theta\right) + F_y \quad (19\text{-}190)$$

HEATING OF BRAKES

Particular	Formula
Heat generated from work of friction	$H_g = \mu p A_c v$ J (joules) **SI** (19-191a)
	$H_g = \frac{\mu p A_c v}{778}$ **USCS** (19-191b)
Heat to be radiated for a brake lowering the load	$H = Wh$ J (joules) **SI** (19-192a)
	$H = \frac{Wh}{778}$ **USCS** (19-192b) where h = total height or distance, m (ft)
The heat generated is also given by the equation	$H_g = 754 k_l P$ **SI** (19-193a) where P in kW and H_g in J/s
	$H_g = 42.4 k_l P$ **USCS** (19-193b) where P in hp

Particular	Formula
The rise in temperature in °C of the brake drum or clutch plates	$\Delta T = \dfrac{H}{mC}$ (19-194) where m = mass of brake drum or clutch plates, kg C = specific heat capacity = 500 J/kg °C for cast iron or steel = 0.13 Btu/lb_m °F for cast iron = 0.116 Btu/lb_m °F for steel
The rate of heat dissipation	$H_d = C_2 \Delta T A_r$ **SI** (19-195a) where H_d in J. $H_d = 0.25 C_2 \Delta T A_r$ **Metric** (19-195b) where C_2 = radiating factor from Table 19-13 H_d in kcal.
The required area of radiating surface	$A_r = \dfrac{754 k_l N}{C_2 \Delta T}$ **SI** (19-196a) where A_r in m^2 $A_r = \dfrac{0.18 k_l N}{C_2 \Delta T}$ **SI** (19-196b) where A_r in mm^2
Approximate time required for the brake to cool	$t_c = \dfrac{W_r C_2 \ln \Delta T}{K A_r}$ (19-197) where K = a constant varying from 0.4 to 0.8
Gagne's formula for heat generated during a single operation	$H_g = \dfrac{AC}{n_c}(T_{av} - T_a)\left[\dfrac{n_c t}{3600} + 1.5\left(1 - \dfrac{N_c t}{3600}\right)\right]$ (19-198) where $(T_{av} - T_a)$ = temperature difference between the brake surface and the atmosphere, °C Refer to Table 19-15 for values of C.
For additional design data for brakes	Refer to Tables 19-11 to 19-17.

TABLE 19-10
Working pressure for brake blocks

Rubbing velocity, m/s	Pressure					
	Wood blocks		Asbestos fabric		Asbestos blocks	
	MPa	kgf/mm^2	MPa	kgf/mm^2	MPa	kgf/mm^2
1	0.5521	0.0563	0.6894	0.0703	1.1032	0.1125
2	0.4482	0.0457	0.5521	0.0563	1.0346	0.1055
3	0.3452	0.0352	0.4138	0.0422	0.8963	0.0914
4	0.2412	0.0246	0.2756	0.0281	0.6894	0.0703
5	0.1726	0.0176	0.2069	0.0211	0.4825	0.0492
10	0.1726	0.0176	0.2069	0.0211	0.2756	0.0281

Note: 1 kpsi = 6.894754 MPa or 1 MPa = 145 psi.

TABLE 19-11
Comparison of hoist brakes

Brake characteristics	Block brakes		Band brakes		Axial brakes	
	Double block	V-grooved sheave	Simple	Both directions of rotation	Cone	Multidisk
Force ratio $\frac{F}{F_\theta}$	$\frac{b}{\mu a}$	$\frac{b \sin \alpha}{\mu a}$	$\frac{b}{a(e^{\mu\theta}-1)}$	$\frac{b(e^{\mu\theta}+1)}{a(e^{\mu\theta}-1)}$	$\frac{b \sin \alpha}{\mu a}$	$\frac{b}{n\mu a}$
Average numerical value	0.667	0.282	0.0323	0.165	0.161	0.097
Relative value	20.6	8.7	1	5.1	5.0	3.00
Travel at lever end	$\frac{h_1 a}{b}$ [a]	$\frac{h_1 a}{b \sin \alpha}$	$\frac{h_1 a\theta}{2\pi b}$	$\frac{h_1 a\theta}{4\pi b}$	$\frac{h_1 a}{b \sin \alpha}$ [b]	$\frac{ih'_1 a}{b}$
Average travel, mm (in)	8.0 (0.313)	18.8 (0.74)	74.5 (2.943)	37.36 (1.471)	32.8 (1.292)	5.56 (0.219)
Maximum capacity P, kW (hp)	1512.7 (2000)	18.9 (25)	227.0 (300)	75.6 (100)	37.8 (50)	90.8 (120)

[a] h_1 = the normal distance between the sheave and the stationary braking surface to prevent dragging.
[b] $h = b$ in Fig. 19-21.

TABLE 19-12
Service factors for typical machines

Type of driven machine	Electric motor steam or water turbine	Service factors for prime movers					
		High-speed steam or gas engine	Petrol engine		Oil engine		
			$\geq$4 Cyl[a]	$\leq$4 Cyl	$\geq$6 Cyl	$\leq$4 Cyl
Alternators and generators (excluding welding generators), induced-draft fans, printing machinery, rotary pumps, compressors, and exhausters, conveyors	1.5	2.0	2.5	3.0	3.5	5.0
Woodworking machinery, machine tools (cutting) excluding planing machines, calenders, mixers, and elevators	2.0	2.5	3.0	3.5	3.0	5.5
Forced-draft fans, high-speed reciprocating compressors, high speed crushers and pulverizers, machine tools (forming)	2.5	3.0	3.5	4.0	4.5	6.0
Rotary screens, rod mills, tube, cable and wire machinery, vacuum pumps	3.0	3.5	4.0	4.5	5.0	6.5
Low-speed reciprocating compressors, haulage gears, metal planing machines, brick and tile machinery, rubber machinery, tube mills, generators(welding)	3.5	4.0	4.5	5.0	5.5	7.0

TABLE 19-13
Radiating factors for brakes

Temperature difference, ΔT	Radiating factor, C_2		$C_2\Delta T$	
	W/m^2 K	cal/m^2 s °C	W/m^2	cal/m^2 s
55.5	12.26	2.93	681.36	162.73
111.5	15.33	3.66	1703.41	406.83
166.5	16.97	4.05	2827.66	675.34
222.6	18.40	4.39	4088.19	976.40

TABLE 19-14
***pv* values as recommended by Hutte for brakes**

Service	*pv*	
	SI	Metric
Intermittent operations with long rest periods and poor heat radiation, as with wood blocks	26.97	2.75
Continuous service with short rest periods and with poor radiation	13.73	1.40
Continuous operation with good radiation as with an oil bath	40.70	4.15

TABLE 19-15
Values of beat transfer coefficient *C* for rough block surfaces

Velocity, *v*, m/s	Heat-transfer coefficient, *C*	
	W/m^2 K	kcal/m^2 h °C
0.0	8.5	7.31
6.1	14.1	12.13
12.2	18.8	16.20
18.3	22.5	19.30
24.4	25.6	22.00
30.5	29.0	24.90

TABLE 19-16
Values of $e^{\mu\theta}$

		Leather belt on			
		Wood		Cast iron	
Proportion of contact to whole circumference	Steel band on cast iron $\mu = 0.18$	Slightly greasy $\mu = 0.47$	Very greasy $\mu = 0.12$	Slightly greasy $\mu = 0.28$	Damp $\mu = 0.38$
0.1	1.12	1.34	1.08	1.19	1.27
0.2	1.25	1.81	1.16	1.42	1.61
0.3	1.40	2.43	1.25	1.69	2.05
0.4	1.57	3.26	1.35	2.02	2.60
0.425	1.62	3.51	1.38	2.11	2.76
0.45	1.66	3.78	1.40	2.21	2.93
0.475	1.71	4.07	1.43	2.31	3.11
0.500	1.76	4.38	1.46	2.41	3.30
0.525	1.81	4.71	1.49	2.52	3.50
0.6	1.97	5.88	1.57	2.81	4.19
0.7	2.21	7.90	1.66	3.43	5.32
0.8	2.47	10.60	1.83	4.09	6.75
0.9	2.77	14.30	1.97	4.87	8.57
1.0	3.10	19.20	2.12	5.81	10.90

TABLE 19-17
Coefficient of friction and permissible variations on dimensions for automotive brakes lining

			Tolerance on width for sizes, mm		Tolerance on thickness for sizes, mm	
Type and class of brake lining	Range of coefficient of friction, μ	Permissible variation in μ, %	≤ 5 mm thickness	>5 mm thickness	≤ 5 mm thickness	>5 mm thickness
Type I—rigid molded sets or flexible molded rolls or sets						
Class A—medium friction	0.28–0.40	+30, −20	+0	+0	+0	+0
Class B—high friction	0.36–0.45	+30, −20	−0.2	−0.3	−0.8	−0.8
Type II—rigid woven sets or flexible woven rolls or sets						
Class A—medium friction	0.33–0.43	+20, −30				
Class B—high friction	0.43–0.53	+20, −30				

REFERENCES

1. Shigley, J. E., *Machine Design,* McGraw-Hill Book Company, New York, 1962.
2. Maleev, V. L. and J. B. Hartman, *Machine Design,* International Textbook Company, Scranton, Pennsylvania, 1954.
3. Black, P. H., and O. E. Adams, Jr., *Machine Design,* McGraw-Hill Book Company, New York, 1968.
4. Norman, C. A., E. S. Ault, and I. F. Zarobsky, *Fundamentals off Machine Design,* The Macmillan Company, New York, 1951.

5. Spotts, M. F., *Machine Design Analysis,* Prentice-Hall, Englewood Cliffs, New Jersey, 1964.
6. Spotts, M. F., *Design of Machine Elements,* Prentice-Hall of India Ltd., New Delhi, 1969.
7. Vallance, A., and V. L. Doughtie, *Design of Machine Members,* McGraw-Hill Book Company, New York, 1951.
8. Lingaiah, K., and B. R. Narayana Iyengar, *Machine Design Data Handbook,* Engineering College Cooperative Society, Bangalore, India, 1962.
9. Lingaiah, K., and B. R. Narayana Iyengar, *Machine Design Data Handbook,* Vol. I (*SI and Customary Metric Units*), Suma Publishers, Bangalore, India, 1986.
10. Lingaiah, K., *Machine Design Data Handbook*, Vol. II (*SI and Customary Metric Units*), Suma Publishers, Bangalore, India, 1986.
11. Lingaiah, K., *Machine Design Data Handbook*, McGraw-Hill Publishing Company, New York, 1994.
12. Bureau of Indian Standards, New Delhi, India.

CHAPTER 20

SPRINGS

SYMBOLS

A	area of loading, m^2 (in^2)
b	width of rectangular spring, m (in)
	width of laminated spring, m (in)
b'	width of each strip in a laminated spring, m (in)
c	spring index
c_1, c_2	constants taken from Table 20-1 and to be used in Eqs. (20-1) to (20-36)
C_1, C_2	constants to be used in Eqs. (20-20) and (20-21) and taken from Fig. 20-3
d	diameter of spring wire, m (in)
	diameter of torsion bar, m (in)
d_1, d_2	diameter of outer and inner wires of concentric spring, m (in)
D	mean or pitch diameter of spring, m (mm) overall diameter of the absorber, m (in)
D_1	mean or pitch diameter of outer concentric spring, m (in)
	smallest mean diameter of conical spring, m (in)
D_2	mean or pitch diameter of inner concentric spring, m (in)
	largest mean diameter of conical spring, m (in)
e_{sz}	size coefficient
e'_{sr}	surface influence coefficient
E	modulus of elasticity, GPa (psi)
F	frequency, cycles per minute, Hz
F	load, kN (lbf)
	steady-state load [Eq. (20-84)]
F_{max}	maximum force that can be imposed on the housing, kN (lbf)
k_o	force to compress the spring one meter (in)
	N/m (lbf/in) [spring rate, N/m (lbf/in)]
F_{cr}	critical load, kN (lbf)
g	acceleration due to gravity, 9.8066 m/s^2
	9806.06 mm/s^2 (32.2 ft/s^2; 386.4 in/s^2)
G	modulus of rigidity, GPa (psi)
h	height (thickness) of laminated spring, m (in) axial height of a rectangular spring wire, m (in)
i	total number of strips or leaves in a leaf spring number of coils in a helical spring
i'	total number of full-length blunt-ended leaves in a leaf spring

I	moment of inertia, area, m^4, cm^4 (in^4)
k, k_1, k_2	stress factor (Wahl factor)
k_4	correction factor
K_l	factor depends on the ratio l_o/D as shown in Fig. 20-8
	reduced stress correction factor or Wahl stress factor or fatigue stress correction factor
k_r	shear stress correction factor
l	length, m (in)
l_f or l_o	free length of helical spring, m (in)
L	$i\pi D$ length of the coil part of torsion spring, m (in)
	effective length of bushing, m (in)
	overall length of the absorber (Fig. 20-15), m (in)
M	constant depends on d_o/d_i as indicated in Fig. 20-3
M_t	twisting moment, N m (lbf in)
n	factor of safety
n_a	actual factor of safety or reliability factor
U	resilience, N m (lbf in)
	energy to be absorbed by a rubber spring, N m (lbf in)
V	volume of spring, m^3, mm^3 (in^3)
γ	specific weight of the spring material, N/m^3 (lbf/in^3)
W	weight of spring, kN (lbf)
	weight of effective number of coils i involved in the operation of the spring [Eq. (20-77)], kN (lbf)
y	deflection, m (in)
y_{cr}	critical deflection, m (in)
Z	section modulus, m^3, cm^3 (in^3)
Z_o	polar section modulus, m^3, cm^3(in^3)
σ	stress, normal, MPa (psi)
τ	shear stress, MPa (psi)
α, α'	constant from Table 20-3
β, β'	constants from Table 20-3
θ	angular deflection, rad
ν	Poisson's ratio

SUFFIXES

1	outside
2	inside
a	amplitude
m	mean
max	maximum
min	minimum
f	endurance lirnit (also used for reversed cycle)
o	endurance limit for repeated cycle

Particular	Formula	
LEAF SPRINGS (Table 20-1)[1,2,3]		
The general equation for the maximum stress in springs	$\sigma = \dfrac{c_1 Fl}{bh^2}$	(20-1)

Particular	Formula	
The general equation for the maximum deflection springs	$y = \dfrac{c_2 F l^3}{E b h^3}$	(20-2)
The thickness of spring	$h = \dfrac{c_2 \sigma l^2}{c_1 y E}$	(20-3)
	Refer to Table 20-1 for values of c_1 and c_2.	
For sizes and tolerances for leaf springs for motor vehicle suspension [4]	Refer to Tables 20-2 to 20-5.	

TABLE 20-1
Deflection formula for beams of rectangular cross section and constants in beam Eqs. (20-1) to (20-3)

Particular	Maximum deflection, y_{max}	c_1, for the stress	c_2, for the deflection	Unit resilience, Nm/m^3 (kgf mm/mm^3)
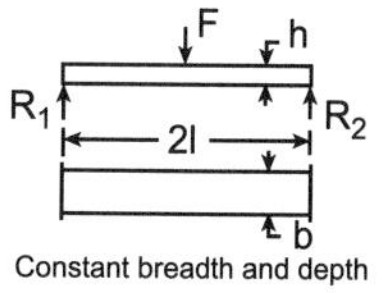 Constant breadth and depth	$y_{max} = \dfrac{2F}{bE}\left(\dfrac{1}{h}\right)^3$	3	2	$\dfrac{\sigma^2}{18E}$
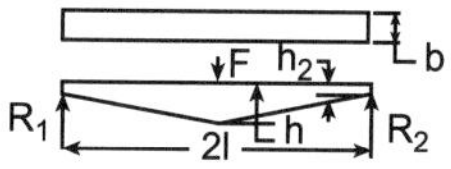 Constant breadth, varying depth	$y_{max} = \dfrac{4F}{bE}\left(\dfrac{1}{h}\right)^3$	3	4	$\dfrac{\sigma^2}{6E}$
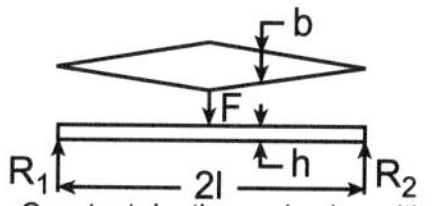 Constant depth, varying breadth	$y_{max} = \dfrac{3F}{bE}\left(\dfrac{1}{h}\right)^3$	3	3	$\dfrac{\sigma^2}{6E}$
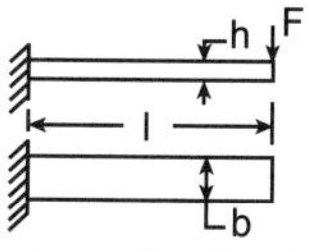 Constant depth and breadth	$y_{max} = \dfrac{4F}{bE}\left(\dfrac{1}{h}\right)^3$	6	4	$\dfrac{\sigma^2}{18E}$
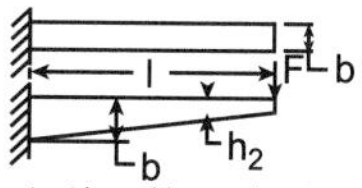 Constant breadth, varying depth	$y_{max} = \dfrac{8F}{bE}\left(\dfrac{1}{h}\right)^3$	6	8	$\dfrac{\sigma^2}{6E}$
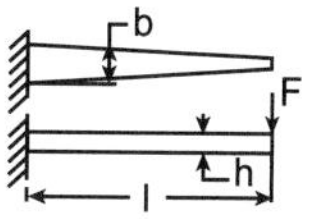 Constant depth, varying breadth	$y_{max} = \dfrac{6F}{bE}\left(\dfrac{1}{h}\right)^3$	6	6	$\dfrac{\sigma^2}{6E}$

Source: K. Lingaiah and B. R. Narayana Iyengar, *Machine Design Data Handbook*, Engineering College Cooperative Society, Bangalore, India, 1962; K. Lingaiah and B. R. Narayana Iyengar, *Machine Design Data Handbook*, Vol. I, Suma Publishers, Bangalore, India, 1986; and K. Lingaiah, *Machine Design Data Handbook*, Vol. II, Suma Publishers, Bangalore, India, 1986.

Particular	Formula

LAMINATED SPRING (Fig. 20-1)[5]

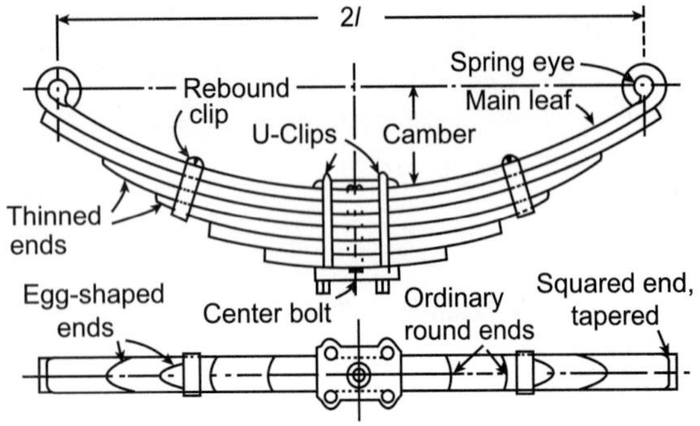

FIGURE 20-1 Laminated springs for automobiles.

Particular	Formula
The load on the spring	$F = \dfrac{\sigma i b' h^2}{c_1 l}$ (20-4) where $ib' = b$
The maximum deflection	$y = \dfrac{c_2 F l^3}{E i b' h^3}$ (20-5)
The maximum deflection in case of laminated semi-elliptical spring for heavy loads	$y = \dfrac{c_2 F l^3 k_4}{E i b' h^3}$ (20-6)
The correction factor to be used in Eq. (20-6)	$k_4 = \dfrac{1 - 4r + 2r^2(1.5 - \ln r)}{(1 - r)^3}$ (20-7) where $r = \dfrac{i'}{l}$
For standard sections of steel plates for laminated springs	Refer to Tables 20-2 to 20-6.
The correction factor k_4 can also be obtained from	$k_4 = \begin{cases} 0.73r^{0.1} & \text{for } 2 < r < 20 \\ 1 & \text{for } r > 20 \end{cases}$ (20-8)
Size coefficient	$e_{sz} = 0.8 + \dfrac{0.0025}{h}$ **SI** (20-9a) where h in mm $e_{sz} = 0.8 + \dfrac{0.1}{h}$ **USCS** (20-9b) where h in in $e_{sz} = 0.8 + \dfrac{2.5}{h}$ **SI** (20-9c) where h in mm

Particular	Formula

TABLE 20-2
Cross section tolerances for leaf springs for motor vehicle suspension—metric bar sizes—SAE J1123a

	Width tolerance, mm	Tolerance, mm in thickness (+)[a] and in flatness (−)[b]			Maximum difference in thickness[c]		
		For thickness			For thickness, mm		
Width, mm	Minus 0.00	5.00–9.50	10.00–21.20	22.40–37.50	5.00–9.50	10.00–21.20	22.40–37.50
40.0	+0.75	0.13	0.15	—	0.05	0.05	—
45.0	+0.75	0.13	0.15	—	0.05	0.05	—
50.0	+0.75	0.13	0.15	—	0.05	0.05	—
56.0	+0.75	0.13	0.15	—	0.05	0.05	—
63.0	+0.75	0.13	0.15	—	0.05	0.05	—
75.0	+1.15	0.15	0.20	0.30	0.08	0.10	0.15
90.0	+1.15	0.15	0.20	0.30	0.08	0.10	0.15
100.0	+1.15	0.15	0.20	0.30	0.08	0.10	0.15
125.0	+1.65	0.18	0.25	0.40	0.10	0.13	0.20
150.0	+2.30	—	0.30	0.50	—	0.15	0.25

[a] Thickness measurements shall be taken at the edge of the bar where the flat surfaces intersect the rounded edge.
[b] This tolerance represents the maximum amount by which the thickness of the center of the bar may be less than the thickness at the edges. Thickness of the center may never exceed the thickness at the edges.
[c] Maximum difference in thickness between the two edges of each bar.
Source: Reproduced from *SAE Handbook*, Vol. I, 1981, courtesy SAE.

Size factor

$$k_{sz} = \frac{1}{e_{sz}} = 4.66h^{0.35} \quad \textbf{SI} \quad (20\text{-}10a)$$

where k in m

$$k_{sz} = 1.27h^{0.35} \quad \textbf{USCS} \quad (20\text{-}10b)$$

where h in in

$$k_{sz} = 0.415h^{0.35} \quad \textbf{SI} \quad (20\text{-}10c)$$

where h in mm

LAMINATED SPRINGS WITH INITIAL CURVATURE

The load shared by graduated leaves of the spring

$$F_g = \frac{2(1-r)}{2+r}F \quad (20\text{-}11)$$

The load shared by full-length leaves of the spring

$$F_f = \frac{3r}{2+r}F \quad (20\text{-}12)$$

The maximum stress in the graduated leaves

$$\sigma_g = \frac{\alpha Fl}{lb'h^2} \quad (20\text{-}13)$$

The maximum stress in the full-length leaves

$$\sigma_f = \frac{1.5\alpha Fl}{ib'h^2} \quad (20\text{-}14)$$

Particular	Formula	
The maximum deflection of the leaves (in both graduated and full-length leaves)	$y = \dfrac{\beta F l^3}{E i b' h^3}$	(20-15)
The camber to be provided for equalization of stress in both graduated and full-length leaves	$c = \dfrac{\beta' F l^3}{i b' E h^3}$	(20-16)
The load on the clip bolt to be applied to provide camber	$F_b = \dfrac{\alpha' F l}{i b' h^2}$	(20-17)
The maximum equalized stress	$\sigma = \dfrac{\alpha' F l}{i b' h^2}$	(20-18)
	The values of constants α, α', β and β' can be obtained from Table 20-7.	

TABLE 20-3
Cross section tolerances for leaf spring for motor vehicle suspension—SAE J510c

Nominal width				Tolerance in width		For thickness		Tolerance in thickness[a]		Tolerance in flat surfaces[b]		Maximum difference in thickness[c]	
Over		to and including		in	mm								
in	mm	in	mm	−0.00	−0.0	in	mm	±in	±mm	in	mm	in	mm
0.00	0.0	2.50	63.5	+0.030	−0.076	≤0.375	9.52	0.005	0.13	−0.005	−0.13	0.002	0.05
						>0.375–0.875	9.52–22.22	0.006	0.15	−0.006	−0.15	0.002	0.05
2.50	63.5	4.00	101.6	+0.045	+1.14	≤0.375	9.52	0.006	0.15	−0.006	−0.15	0.003	0.08
						>0.375–0.875	9.52–22.22	0.008	0.20	−0.008	−0.20	0.004	0.10
						>0.875–1.500	22.22–38.10	0.012	0.30	−0.012	−0.30	0.006	0.15
4.00	101.6	5.00	127.0	+0.065	+1.65	≤0.375	9.52	0.007	0.18	−0.007	−0.18	0.004	0.10
						>0.375–0.875	9.52–22.22	0.010	0.25	−0.010	−0.25	0.005	0.13
						>0.875–1.500	22.22–38.10	0.016	0.41	−0.016	−0.41	0.008	0.20
5.00	127.0	6.00	152.4	+0.090	+2.90	>0.375–0.875	9.52–22.22	0.012	0.30	−0.012	−0.30	0.006	0.15
						>0.875–1.500	22.22–38.10	0.020	0.51	−0.020	−0.51	0.010	0.25

[a] Thickness measurements shall be taken at the edge of the bar where the flat surfaces intersect the rounded edge.
[b] This tolerance represents the maximum amount by which the thickness of the center of the bar may be less than the thickness at the edges. Thickness at the center may never exceed the thickness at the edges.
[c] Maximum difference in thickness between the two edges of each bar.
Source: Reproduced from SAE Handbook, Vol. I, 1981.

DISK SPRINGS (BELLEVILLE SPRINGS)

The relation between the load F and the axial deflection y of each disk is given by the equation (Fig. 20-2)

$$F = \frac{4Ey}{(1-\nu^2)Md_o^2}\left[(h-y)\left(h-\frac{y}{2}\right)t + t^3\right] \qquad (20\text{-}19)$$

where

t = thickness, m (mm)
h = height, m (mm)
M is a constant from Fig. 20-3

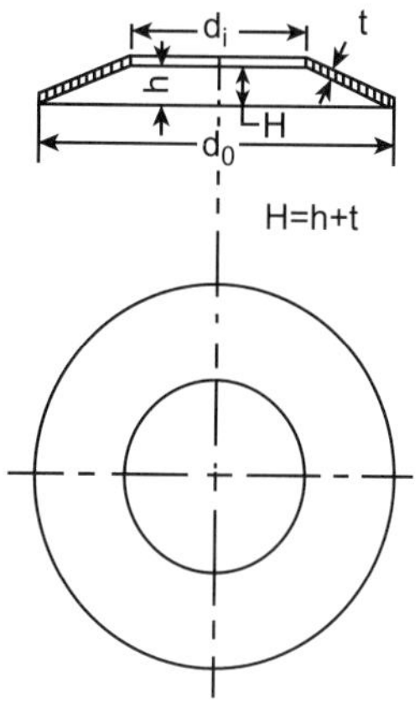

FIGURE 20-2 Disk spring.

TABLE 20-4
Width and thickness of leaf springs for motor vehicle suspension—SAE J1123a

Width, mm		Thickness mm					
40.0	75.0	5.00	7.10	10.00	14.00	20.00	28.00
45.0	90.0	5.30	7.50	10.60	15.00	21.20	30.00
50.0	100.0	5.60	8.00	11.20	16.00	22.40	31.50
56.0	125.0	6.00	8.50	11.80	17.00	23.60	33.50
63.0	150.0	6.30	9.00	12.50	18.00	25.00	35.50
		6.70	9.50	13.20	19.00	25.50	37.50

Source: Reproduced from *SAE Handbook*, Vol. I, 1981.

The maximum stress induced at the inner edge

$$\sigma_i = \frac{4Ey}{(1-v^2)d_o^2}\left[C_1\left(h-\frac{y}{2}\right)+C_2 t\right] \tag{20-20}$$

The maximum stress induced at the outer edge

$$\sigma_o = \frac{4Ey}{(1-v^2)d_o^2}\left[C_1\left(h-\frac{y}{2}\right)C_2 t\right] \tag{20-21}$$

where C_1 and C_2 are constants taken from Fig. 20-3

For spring design stresses

Refer to Table 20-8

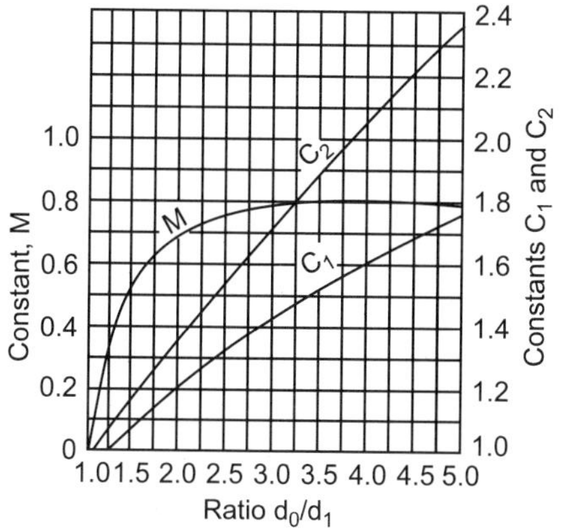

FIGURE 20-3 Constants C_1, C_2, and M for a disk spring. (V. L. Maleev and J. B. Hartman, *Machine Design*, International Textbook Company, Scranton, Pennsylvania, 1954.)

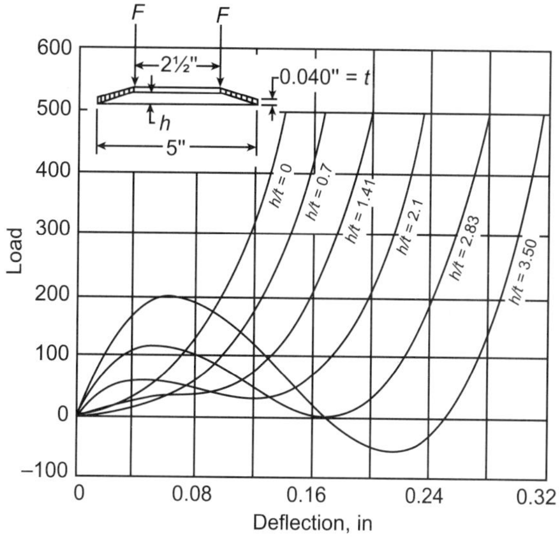

FIGURE 20-3a Load-deflection curves for Belleville springs. *Courtesy*: Shigley, J. E. and L. D. Mitchell, *Mechanical Engineering Design*, McGraw-Hill Publishing Company, New York, 1983.

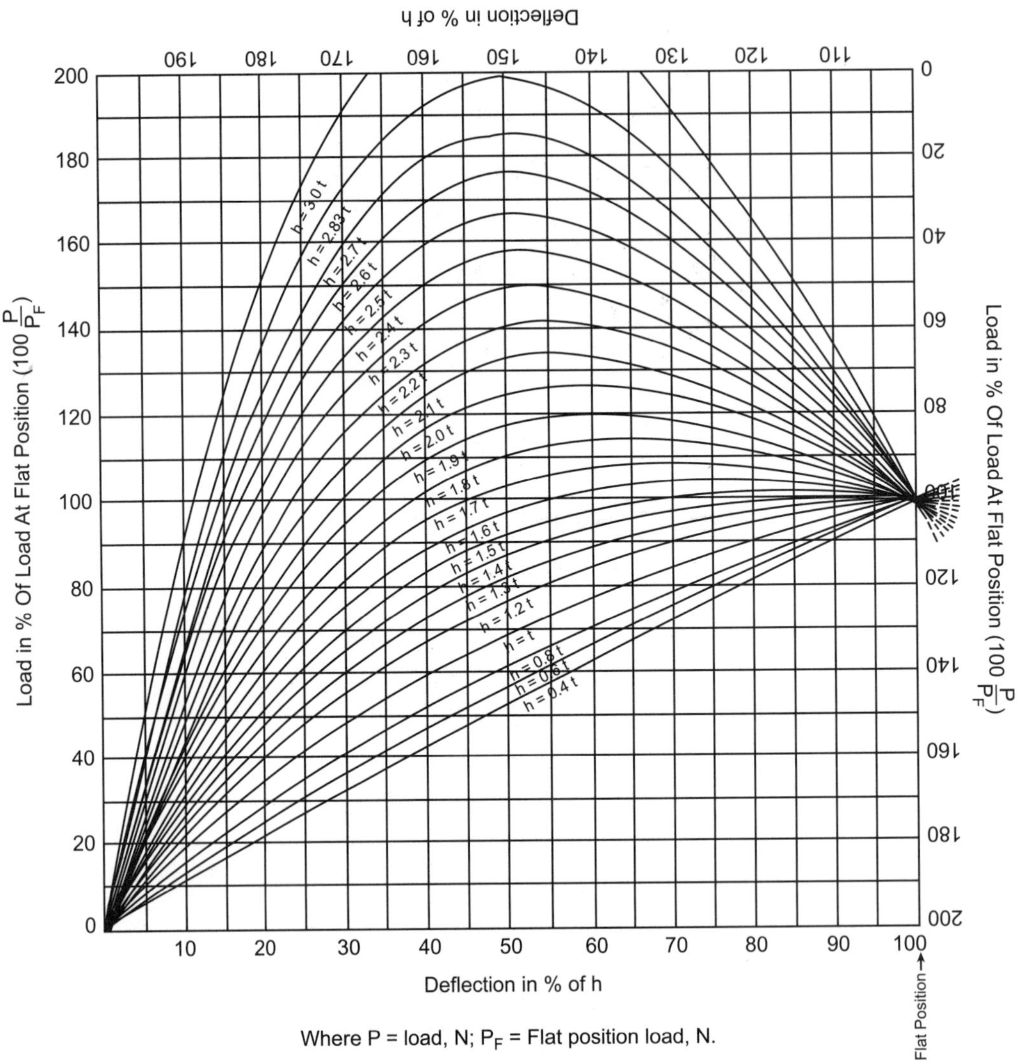

FIGURE 20-3b Load deflection characteristics for Belleville washers *Courtesy*: Jorres, R. L., *Springs*; Chapter 24 in Shigley, J. E. and C. R. Mischke, eds, *Standards Handbook of Machine Design*, McGraw-Hill Publishing Company, New York, 1986.

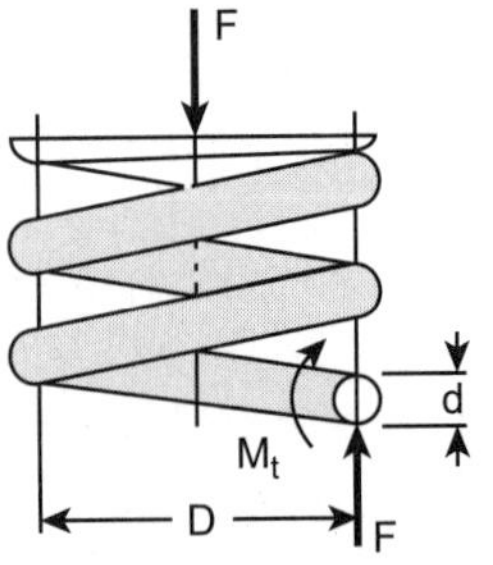

FIGURE 20-4 Helical spring under axial load.

Particular	Formula

HELICAL SPRINGS (Fig. 20-4)[5]

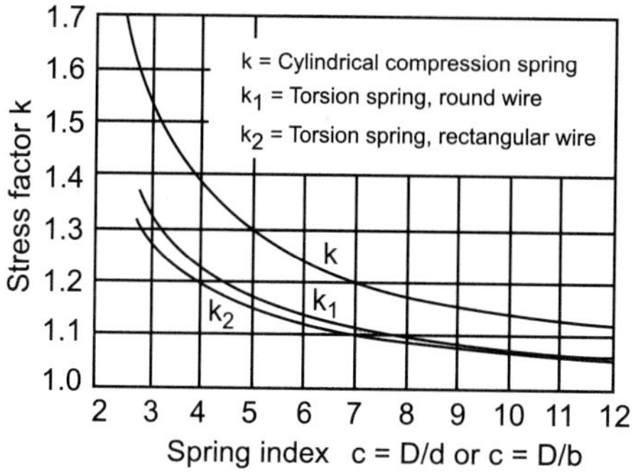

FIGURE 20-5 Stress factors for a helical spring. (V. L. Maleev and J. B. Hartman, *Machine Design*, International Textbook Company, Scranton, Pennsylvania, 1954.)

The more accurate formula for shear stress

$$\tau = \frac{8FDk}{\pi d^3} \quad (20\text{-}22)$$

Refer to Fig. 20-5 for values of k.

The stress factor or Wahl factor[6] k to be used in Eq. (20-2)

$$k = \frac{4c-1}{4c-4} + \frac{0.615}{c} \quad (20\text{-}23)$$

The spring index

$$c = \frac{D}{d} \quad (20\text{-}24)$$

The value of stress factor k may be approximated very closely (between $c = 2$ and 12) by the relation

$$k = \frac{2}{c^{0.25}} = 2\left(\frac{d}{D}\right)^{0.25} \quad (20\text{-}25)$$

Size factor

$$k_{sz} = \frac{1}{e_{sz}} = \frac{d^{0.25}}{0.335} \quad \textbf{SI} \quad (20\text{-}26a)$$

where d in mm

$$k_{sz} = \frac{d^{0.25}}{0.85} \quad \textbf{USCS} \quad (20\text{-}26b)$$

where d in in

$$K_{sz} = \frac{d^{0.25}}{1.89} \quad \textbf{SI} \quad (20\text{-}26c)$$

where d in mm

The shear stress taking into consideration k from Eq. (20-25)

$$\tau = \frac{16FD^{0.75}}{\pi d^{2.75}} = \frac{5.1Fc^{0.75}}{d^2} \quad (20\text{-}27)$$

The angular deflection

$$\theta = \frac{16FDl}{\pi d^4 G} \quad (20\text{-}28)$$

The length of the spring wire

$$l = i\pi D \text{ (approx.)} \quad (20\text{-}28a)$$

The axial deflection of the whole spring

$$y = \frac{8FD^3 i}{d^4 G} = \frac{\pi i \tau D^2}{kdG} = \frac{\pi i \tau D^{2.25}}{2d^{1.25}G} \quad (20\text{-}29)$$

Particular	Formula

SPRING SCALE (Fig. 20-6)

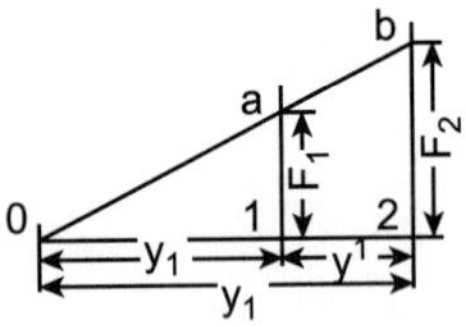

FIGURE 20-6 Relation between loads and deflection in a helical spring.

Force to compress the spring 1 m (mm) (stiffness of spring)

$$F_o = \frac{F}{y} = \frac{d^4 G}{8iD^3} = \frac{GD}{8ic^4} = \frac{Gd}{8ic^3} \quad (20\text{-}30)$$

The total deflection (Fig. 20-6)

$$y_2 = \frac{F_2}{F_o} = \frac{y' F_2}{F_2 - F_1} \quad (20\text{-}31)$$

RESILIENCE

The resilience U of a spring is equal to energy absorbed

$$U = \frac{Fy}{2} = \frac{4F^2 D^3 i}{d^4 G} = \frac{\pi^2 d^2 i D \tau^2}{16k^2 G} \quad (20\text{-}32a)$$

$$U = \frac{\pi^2 d^{1.5} D^{1.5} i \tau^2}{64G} = \frac{1.55 d^2 c^{1.5} i \tau^2}{G} \quad (20\text{-}32b)$$

$$U = \frac{V\tau^2}{4k^2 G} = \frac{V\tau^2}{16G}\left(\frac{D}{d}\right)^{0.5} = \frac{V\tau c^{0.5}}{16G}$$

$$= \frac{y^2 d^4 G}{16iD^3} \quad (20\text{-}32c)$$

Volume of the spring

$$V = \pi Di\left(\tfrac{1}{4}\pi d^2\right) \quad (20\text{-}33)$$

RECTANGULAR SECTION SPRINGS (Fig. 20-7a)[5]

Shear stress

$$\tau' = \frac{kFD(1.5h + 0.9b)}{b^2 h^2} = \frac{FD^{0.75}(3h + 1.8b)}{b^{1.75} h^2} \quad (20\text{-}34a)$$

$$\tau' = \frac{kFD(1.5 + 0.9m)}{m^2 h^3} = \frac{FD^{0.75}(3 + 1.8m)}{m^{1.75} h^2} \quad (20\text{-}34b)$$

where $m = \frac{b}{h}$

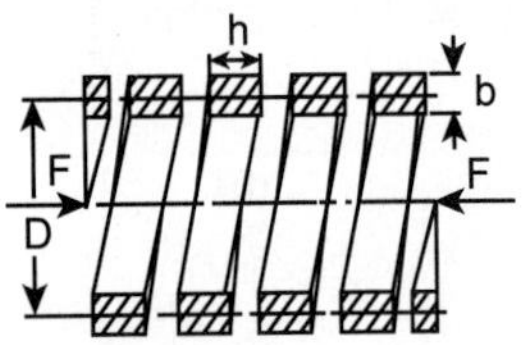

FIGURE 20-7(a) Spring with rectangular section.

The uncorrected shear stress for a rectangular section spring

$$\tau' = \frac{FD}{k_l bh^2} \quad (20\text{-}35)$$

where k_1 = factor depending on $b/h = m$, which is given in Table 20-9

The stress factor

$$k = \frac{4c - 1}{4c - 4} + \frac{0.615}{c} \quad (20\text{-}36)$$

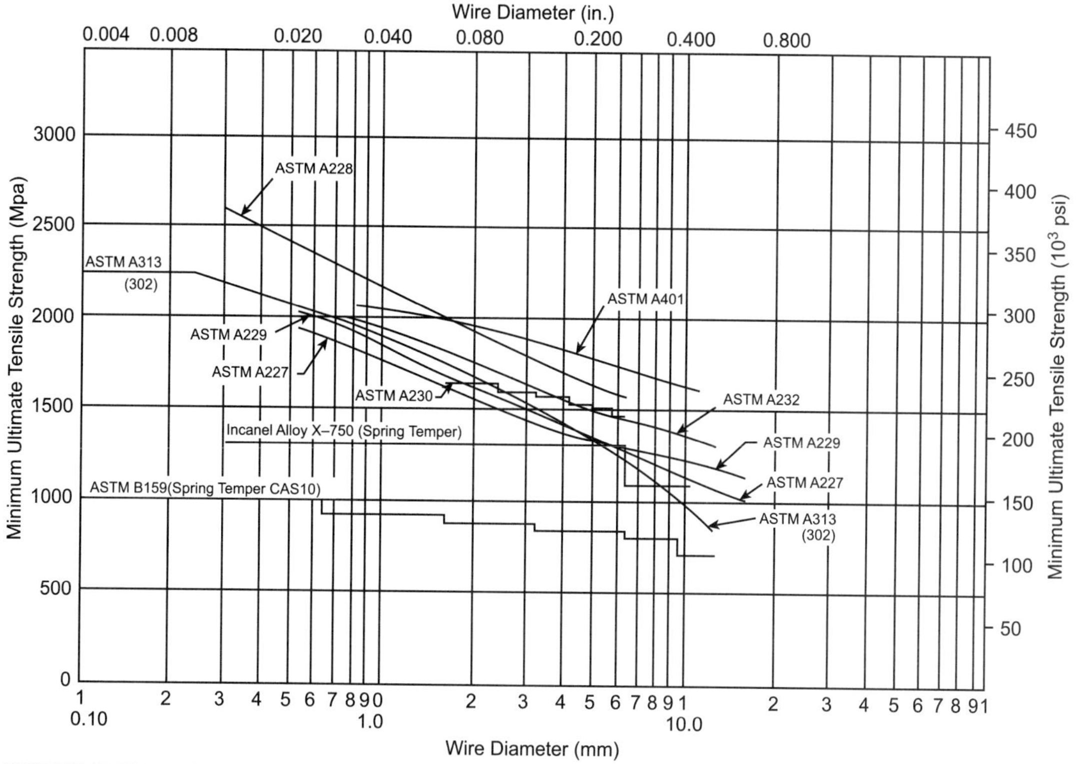

FIGURE 20-7(b) Minimum tensile strengths of spring wire. (Associated Spring, Barnes Group Inc., Bristol, Connecticut)

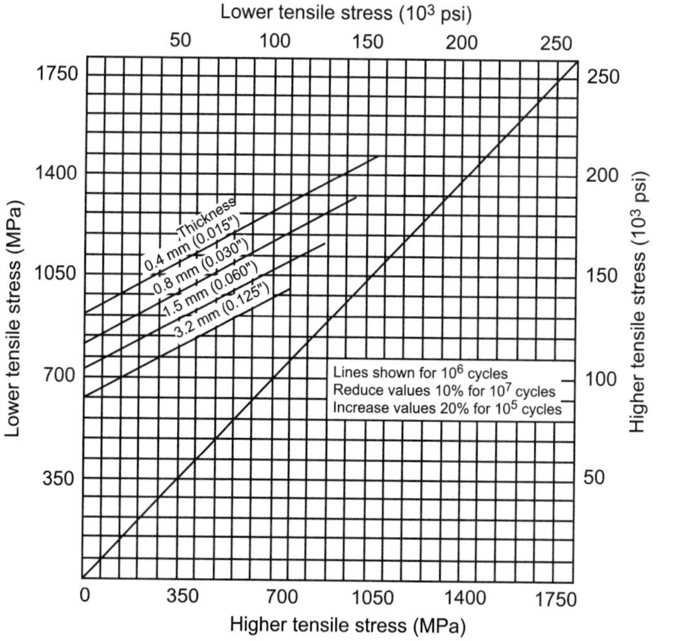

FIGURE 20-7(c) Modified Goodman diagram for Belleville washers; for carbon and alloy steels at 47 to 49 R_c with set removed, but not shotpeened (Associated Spring, Barnes Group Inc., Bristol, Connecticut.)

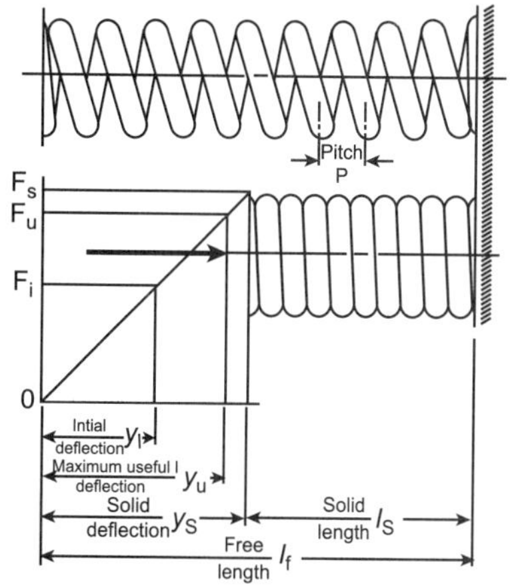

FIGURE 20-7(d) Deflection for helical compression spring at various loads. (Richard M. Phelan, *Fundamentals of Mechanical Engineering Design*, New Delhi, 1975, courtesy of Tata-McGraw-Hill.)

TABLE 20-5
Width of leaf springs for motor vehicle suspension—SAE J510C

in	mm	in	mm	in	mm
1.75	44.4	2.50	63.5	4.00	101.6
2.00	50.8	3.00	76.2	5.00	127.0
2.25	57.2	3.50	88.9	6.00	152.4

Source: Reproduced from *SAE Handbook*, Vol. I, 1981.

TABLE 20-6
Standard sections of steel plates for laminated springs (railway rolling stocks)

Width, mm	50	63	75	90	100	115	120	125	140	150
Thickness, mm	10, 13	6, 8, 10, 11, 13	6, 8, 10, 11, 13, 16	6, 8, 10, 11, 16, 19	8, 10, 11, 13, 16, 19	10, 11, 13, 16, 19	16, 19	10, 13, 16	11, 13	11, 13, 16

TABLE 20-7
Constant in Eqs. (20-13) to (20-18)

Constant	Cantilever	Simply supported
α	$\frac{12}{2+r}$	$\frac{3}{2+r}$
β	$\frac{12}{2+r}$	$\frac{3}{4(2+r)}$
α'	6	1.5
β'	2	$\frac{1}{8}$

Particular	Formula
The value of stress k may be approximated very closely by the relation	$k = \frac{2}{c^{0.25}}$ (20-37)
The spring index	$c = \begin{cases} \frac{D}{b} & \text{if } b < h \\ \frac{D}{h} & \text{if } h < b \end{cases}$ (20-38) where b = breadth of spring wire, m (mm) h = thickness of spring wire, m (mm)
The deflection	$y = \frac{2.83 i F D^3 (b^2 + h^2)}{b^3 h^3 G}$ (20-39)

Particular	Formula
The deflection for an uncorrected spring of rectangular cross section	$y = \dfrac{iFD^3}{k_2 bh^3 G}$ for $h < b$ and $m > 8$ (20-40)
Force required to compress the spring by one meter (millimeter) (i.e., the spring rate)	$F_o = \dfrac{m^3 h^4 G}{2.83 iD^3(1+m^2)}$ (20-41)
The spring rate for an uncorrected rectangular section spring	$F_o = \dfrac{4Gb^3 h}{k_1 \pi D^3 i}$ (20-42) Refer to Table 20-9 for k_1.

SQUARE SECTION SPRING

Particular	Formula
The shear stress, for $m = 1$	$\tau' = \dfrac{2.4kFD}{h^3} = \dfrac{4.8FD^{0.75}}{h^{2.75}}$ (20-43)
The deflection	$y = \dfrac{5.66 iFD^3}{h^4 G}$ (20-44)
The approximate equivalent rectangular dimension of a rectangular cross section wire spring to restrict the solid length, which is equivalent to spring of round-wire cross section	$h = \dfrac{2d}{1+(b/h)}$ (20-44a) where d = diameter of round wire
The larger dimension of a keystone shape of rectangular wire after coiling	$h_1 = h\dfrac{C+0.5}{C}$ (20-44b) where h = wider end of keystone section h = original, smaller dimension of rectangular section
The estimated solid height or length of a uniformly tapered, but not telescoping, spring with squared and ground ends made from round wire	$l_s = i(d^2 - u^2)^{1/2} + 2d$ (20-44c) where u = outside diameter of large end minus outside diameter of small end divided by $2i$
The increase in coil diameter due to compression of a helical spring	$D_{o\,\text{at solid}} = \left(\dfrac{D^2 + p^2 - d^2}{\pi^2 + d}\right)^{1/2}$ (20-44d)
The size coefficient for sections above 12.5 mm in section for round wires	$e_{sz} = 0.86 + \dfrac{0.0018}{d}$ **SI** (20-45a) for steel, where d in m $e_{sz} = 0.986 + \dfrac{0.0001}{d}$ **SI** (20-45b) for monel metal, where d in m

Particular	Formula
	$e_{sz} = 0.86 + \frac{0.07}{d}$ **USCS** (20-45c) for steel, where d in in
	$e_{sz} = 0.986 + \frac{0.0043}{d}$ **USCS** (20-45d) for monel metal, where d in in
	$e_{sz} = 0.86 + \frac{1.8}{d}$ **SI** (20-45e) for steel, where d in mm
	$e_{sz} = 0.986 + \frac{0.1}{d}$ **SI** (20-45f) for monel metal, where d in mm
The general expression for size factor	$k_{sz} = 4.66h^{0.35}$ where h in m **SI** (20-46a)
	$k_{sz} = 1.27h^{0.35}$ where h in in **USCS** (20-46b)
	$k_{sz} = 0.415h^{0.35}$ where h in mm **SI** (20-46c)
Wire diameter	$d = \sqrt[3]{\frac{8kFD}{\pi\sigma_d e_{sz}}}$ (20-47)

SELECTION OF MATERIALS AND STRESSES FOR SPRINGS

For materials for springs[7]	Refer to Tables 20-8 and 20-10 and Figs. 20-7b and 20-7c.
The torsional yield strength	$0.35\sigma_{sut} \leq \tau_{sy} \leq 0.52\sigma_{sut}$ for steels (20-47a)
The maximum allowable torsional stress for static applications according to Joerres[8,9,11]	$\tau_{sy} = \tau_a = \begin{cases} 0.45\sigma_{sut} & \text{cold-drawn carbon steel} \\ 0.50\sigma_{sut} & \text{hardened and tempered carbon and low-alloy steel} \\ 0.35\sigma_{sut} & \text{austenitic stainless steel and nonferrous alloys} \end{cases}$ (20-47b) where τ_{sy} = torsional yield strength, MPa (psi)
The maximum allowable torsional stress according to Shigley and Mischke[9]	$\tau_{sy} = \tau_a = 0.56\sigma_{sut}$ (20-47c)
The shear endurance limit according to Zimmerli[10]	$\tau_{sf} = 310$ MPa (45 kpsi) (20-47d) for unpeened springs $\tau_{sf} = 465$ MPa (67.5 kpsi) (20-47e) for peened springs
The torsional modulus of rupture	$\tau_{su} = 0.67\sigma_{sut}$ (20-47f)

TABLE 20-8
Spring design stress, σ_d, MPa (kpsi)

	Severe service		Average service		Light	
Wire diameter, mm	**MPa**	**kpsi**	**MPa**	**kpsi**	**MPa**	**kpsi**
$\leq$2.15	413.8	60	517.3	75	641.4	93
2.15–4.70	379.0	55	476.6	69	585.4	85
4.70–8.10	331.0	48	413.8	60	510.0	74
8.10–13.45	289.3	42	358.4	52	448.2	65
13.45–24.65	248.1	36	310.4	45	385.9	56
24.65–38.10	220.6	32	275.6	40	344.7	50

TABLE 20-9
Factors for helical springs with wires of rectangular cross section

Ratio $b/h = m$	1	1.2	1.5	2.0	2.5	3	5	10	∞
Factor k	0.416	0.438	0.462	0.492	0.516	0.534	0.582	0.624	0.666
Factor k_2	0.180	0.212	0.250	0.292	0.317	0.335	0.371	0.398	0.424

Particular	Formula
The weight of the active coil of a helical spring	$$W = \frac{\pi^2 d^2 D i \gamma}{4} \quad (20\text{-}47g)$$ where γ = weight of coil of helical spring per unit volume
For free-length tolerances, coil diameter tolerances, and load tolerances of helical compression springs	Refer to Tables 20-11 to 20-13.

DESIGN OF HELICAL COMPRESSION SPRINGS

Design stress

The size factor

$$k_{sz} = \frac{d^{0.35}}{0.355} \quad \text{where } d \text{ in m} \qquad \textbf{SI} \quad (20\text{-}48a)$$

$$k_{sz} = \frac{d^{0.25}}{0.84} \quad \text{where } d \text{ in in} \qquad \textbf{USCS} \quad (20\text{-}48b)$$

$$k_{sz} = \frac{d^{0.25}}{1.89} \quad \text{where } d \text{ in mm} \qquad \textbf{SI} \quad (20\text{-}48c)$$

The design stress

$$\sigma_{ds} = \frac{\sigma_e}{n_a k_{sz}} = \frac{0.335\sigma_e}{n_a d^{0.25}} \qquad \textbf{SI} \quad (20\text{-}49a)$$

where σ_e in MPa and d in m

$$\sigma_{ds} = \frac{\sigma_e}{n_a k_{sz}} = \frac{0.84\sigma_e}{n_a d^{0.25}} \qquad \textbf{USCS} \quad (20\text{-}49b)$$

where σ_e in psi and d in in

TABLE 20-10
Chemical composition and mechanical properties of spring materials

Material	Analysis		Tensile properties: Ultimate strength		Tensile properties: Elastic limit		Modulus of elasticity, E		Rockwell hardness	Torsional properties of wire: Ultimate strength		Torsional properties of wire: Elastic limit		Modulus in torsion, G		Chief uses
	Element	%	Mpa	kpsi	GPa	kpsi	GPa	Mpsi		MPa	kpsi	GPa	kpsi	GPa	Mpsi	
Flat Cold-rolled Spring Steel																
Watch spring steel	C	1.10–1.19	2274–2412	330–350	2.14–2.28	310–330	220	32	C55–55	Not used		Not used		Not used		Main springs for watches and similar uses
	Mn	0.15–0.25														
Clock spring steel																Clock and motor springs,
AS 100	C	0.90–1.05	1240–2343	180–340	1.03–2.14	150–310	207	30	C40–52	Not used		Not used		Not used		miscellaneous flat springs
SAE 1095	Mn	0.20–0.50														for high stress
Flat spring steel																
AS 101	C	0.65–0.80	1103–2206	160–320	0.86–1.93	125–280	207	30	Annealed, B70–85	Not used		Not used		Not used		Miscellaneous flat springs
SAE 1074	Mn	0.50–0.90							tempered C38–50							
Carbon Steel Wires																
High–carbon wire	C	0.85–0.95	1382–1725	200–250	1.10–1.45	160–210	207	30	C44–48	1103	160–200	0.76	110–150	79	11.5	High-grade helical springs
AS 8	Mn	0.25–0.60								1377		1.03				or wire forms
Oil-tempered wire																
(ASTM A229–41)	C	0.60–0.70	1068–2059	155–300	0.83–1.73	120–250				794	115–200	0.55	80–130			General spring use
AS10	Mn	0.60–0.90					200	29	C42–46	1377		0.90		79	11. 5	
Music wire (ASTM																
A228–47)	C	0.70–1.00	1725–3790	250–500	1.03–2.41	1 50–350				1034	150–300	0.62	90–180	79	11.5	Miscellaneous small
AS 5	Mn	0.30–0.60					207	30		2069		1.24		82	12.0	springs of various types—
															depending on size	high quality
Hard-drawn spring wire																
(ASTM A227–47)	C	0.60–0.70	150–300	0.69–1.38	100–200					828	120–220	0.51	75–130			Same uses as music wire
		1034–2068														
AS 20	Mn	0.90–1.20					200	29		1515		0.90		79	11.5	but lower-quality wire
Hot-rolled Special Steel																
Hot-rolled bars																
SAE 1095,	C	0.90–1.05	1206–1377	175–200	0.73–0.97	105–140				760	110–140	0.51	75			Hot-rolled heavy coil or
ASTM A14–42	Mn	0.25–0.50					196	28.5	C40–46	965		0.76	110	72	10.5	flat springs
Alloy and Stainless Spring Materials																
Chrome-vanadium	C	0.45–0.55	1377	200–250	1.24	180–230				965		0.69	100–130			
alloy steel	Mn	0.50–0.80					207	30	C42–48		140–175			79	11.5	Cold–rolled or drawn:
(SAE 6150)	Cr	0.80–1.10	1725		1.58					1206		0.90				special applications
AS 32	V	0.15–0.18														
Silico-manganese	C	0.55–0.65														Used as a lower–cost
alloy steel	Mn	0.60–0.90														material in place of
(SAE 9260)	Si	1.80–2.20	About the same as chrome vanadium							About the same as chrome vanadium						chrome vanadium
Type 18–8 stainless	C	17–20	1103	160–330	0.41	60–260										
(Type 302,	Ni	7–10					193	28	C35–45	828	120–240	0.31	45–140			
SAE 30915)	C	0.08–0.15	2275		1.79									69	10	Best corrosion resistance,
	Mn	2 max								1653		0.97				fair temperature
	Si	0.30–0.75														resistance

Cutlery-type																
stainless	Cr	12–14	1171	170–250	0.90	130–200				828	120–180	0.55	80–120	76		Resists corrosion when
(Type 420)	C	0.25–0.40	1725		1.38		193	28	C42–47	1240		0.83			11	polished; good
																temperature resistance
Nonferrous Spring Materials																
Spring brass																For electrical
AS 55	Cu	64–74	691	100–130	0.27					308	45–90	0.21	30–60			conductivity at low
AS 155	Zn	balance	897		0.41		107	15	B90	622		0.41		38	5.5	stresses; for corrosion
																resistance
Nickel silver	Cu	56	897	130–150	0.55	80–110				588	85–100	0.41	60–70			Used for its color;
	Zn	25	1034		0.76		110	16	B95–100	691		0.48		38	5.5	corrosion resistance
	Ni	18														
Phosphor bronze	Cu		91–93	691	100–150	0.41	60–110									
AS 60	Sn	7–9								554		0.35				Used for corrosion
AS 160	or						103	15	B90–100		80–105		50–85	43	6.25	resistance and electrical
	Cu	94–96	102		0.76					725		0.59				conductivity
	Sn	4–6														
Nonferrous Spring Materials																
Silicon bronze (made	Si	2–3														
under various trade	Sn or	Small														Used as substitute for
names)	Mn	amounts		Properties similar to those of phosphor bronze							Properties similar to those of phosphor bronze					phosphor bronze
AS 46	Cu	balance														
AS 146																
Monel	Ni	64	691	100–140	0.55	80–120				519	75–110	0.31	45–70			Resists corrosion;
AS 40	Cu	26	964		0.83		179	26	C23–28	760		0.48		65	9.5	moderate stresses to
AS 140	Mn	2.5														204.5°C
	Fe	2.25														
Inconel	Ni	80	965	140–175	0.76	110–135				651	95–120	0.38	55–80			Resists corrosion; high
AS 40	Cr	14	1206		0.93		213	31	C30–40	828		0.55		76	11	stresses to 343°C
AS140	Fe	Balance														
K-Monel	Ni	66	1103	160–180	0.79	115–145				725	105–125	0.45	65–85			Resists corrosion; high
AS 40	Cr	29	1241		1.00		179	26	C33–40	862		0.58		65	9.5	stresses to 232°C
AS 140	Al	2.75														
	Fe	0.90														
Z-nickel	Ni	98	1241		0.90					828		0.41				
	Cu			180–230		130–170	207	30	C36–46		120–150		60–90	76	11	Resists corrosion; high
	Mn	Small	1583		1.17					1034		0.68				stresses to 288°C
	Fe	amounts														
	Si															
Beryllium-copper	Cu	98	1103	160–200	0.69	100–150	110	16–18.5		691	100–130	0.45	65–95	41	6–7	Corrosion resistance like
AS 45	Be	2	1377		1.03		127	Subject	to C35–42	897		0.66		48		copper; high physical
AS 145								heat						Subject to		properties for electrical
								treatment						heat		work; low hysteresis
														treatment		

Note: The property values given in this table do not specify the minimum properties.
Source: Handbook of Mechanical Spring Design, courtesy Associated Spring, Barnes Group Inc., Bristol, Connecticut.

TABLE 20-11
Free-length tolerances of squared and ground helical compression springs[a]

Number of active coils per mm (in)	Tolerances: ±mm/mm (in/in) of free length — Spring index (D/d) 4	6	8	10	12	14	16
0.02 (0.5)	0.010	0.011	0.012	0.013	0.015	0.016	0.016
0.04 (1)	0.011	0.013	0.015	0.016	0.017	0.018	0.019
0.08 (2)	0.013	0.015	0.017	0.019	0.020	0.022	0.023
0.2 (4)	0.016	0.018	0.021	0.023	0.024	0.026	0.027
0.3 (8)	0.019	0.022	0.024	0.026	0.028	0.030	0.032
0.5 (12)	0.021	0.024	0.027	0.030	0.032	0.034	0.036
0.6 (16)	0.022	0.026	0.029	0.032	0.034	0.036	0.038
0.8 (20)	0.023	0.027	0.031	0.034	0.036	0.038	0.040

[a] For springs less than 12.7 mm (0.500 in) long, use the tolerances for 12.7 mm (0.500 in). For closed ends not ground, multiply above values by 1.7.
Source: Associated Spring, Barnes Group Inc., Bristol, Connecticut.

TABLE 20-12
Coil diameter tolerances of helical compression and extension springs

Wire diameter, mm (in)	Tolerances: ± mm (in) — Spring index (D/d) 4	6	8	10	12	14	16
0.38	0.05	0.05	0.08	0.10	0.13	0.15	0.18
(0.015)	(0.002)	(0.002)	(0.003)	(0.004)	(0.005)	(0.006)	(0.007)
0.58	0.05	0.08	0.10	0.15	0.18	0.20	0.25
(0.023)	(0.002)	(0.003)	(0.004)	(0.006)	(0.007)	(0.008)	(0.010)
0.89	0.05	0.10	0.15	0.18	0.23	0.28	0.33
(0.035)	(0.002)	(0.004)	(0.006)	(0.007)	(0.009)	(0.011)	(0.013)
1.30	0.08	0.13	0.18	0.25	0.30	0.38	0.43
(0.051)	(0.003)	(0.005)	(0.007)	(0.010)	(0.012)	(0.015)	(0.017)
1.93	0.10	0.18	0.25	0.33	0.41	0.48	0.53
(0.076)	(0.004)	(0.007)	(0.010)	(0.013)	(0.016)	(0.019)	(0.021)
2.90	0.15	0.23	0.33	0.46	0.53	0.64	0.74
(0.114)	(0.006)	(0.009)	(0.013)	(0.018)	(0.021)	(0.025)	(0.029)
4.34	0.20	0.30	0.43	0.58	0.71	0.84	0.97
(0.171)	(0.008)	(0.012)	(0.017)	(0.023)	(0.028)	(0.033)	(0.038)
6.35	0.28	0.38	0.53	0.71	0.90	1.07	1.24
(0.250)	(0.011)	(0.015)	(0.021)	(0.028)	(0.035)	(0.042)	(0.049)
9.53	0.41	0.51	0.66	0.94	1.17	1.37	1.63
(0.375)	(0.016)	(0.020)	(0.026)	(0.037)	(0.046)	(0.054)	(0.064)
12.70	0.53	0.76	1.02	1.57	2.03	2.54	3.18
(0.500)	(0.021)	(0.030)	(0.040)	(0.062)	(0.080)	(0.100)	(0.125)

Source: Associated Spring, Barnes Group Inc., Bristol, Connecticut.

TABLE 20-13
Load tolerances of helical compression springs

	Tolerance: ±% of load, start with tolerance from Table 20-11 multiplied by L_F														
	Deflection from free length to load, mm (in)														
Length tolerance ± mm (in)	1.27 (0.050)	2.54 (0.100)	3.81 (0.150)	5.08 (0.200)	6.35 (0.250)	7.62 (0.300)	10.2 (0.400)	12.7 (0.500)	19.1 (0.750)	25.4 (1.00)	38.1 (1.50)	50.8 (2.00)	76.2 (3.00)	102 (4.00)	152 (6.00)
0.13 (0.005)	12	7	6	5	—	—	—	—	—	—	—	—	—	—	—
0.25 (0.010)	—	12	8.5	7	6.5	5.5	5	—	—	—	—	—	—	—	—
0.51 (0.020)	—	22	15.5	12	10	8.5	7	6	5	—	—	—	—	—	—
0.76 (0.030)	—	—	22	17	14	12	9.5	8	6	5	—	—	—	—	—
1.0 (0.040)	—	—	—	22	18	15.5	12	10	7.5	6	5	—	—	—	—
1.3 (0.050)	—	—	—	—	22	19	14.5	12	9	7	5.5	—	—	—	—
1.5 (0.060)	—	—	—	—	25	22	17	14	10	8	6	5	—	—	—
1.8 (0.070)	—	—	—	—	—	25	19.5	16	11	9	6.5	5.5	—	—	—
2.0 (0.080)	—	—	—	—	—	—	22	18	12.5	10	7.5	6	5	—	—
2.3 (0.090)	—	—	—	—	—	—	25	20	14	11	8	6	5	—	—
2.5 (0.100)	—	—	—	—	—	—	—	22	15.5	12	8.5	7	5.5	—	—
5.1 (0.200)	—	—	—	—	—	—	—	—	—	22	15.5	12	8.5	7	5.5
7.6 (0.300)	—	—	—	—	—	—	—	—	—	—	22	17	12	9.5	7
10.2 (0.400)	—	—	—	—	—	—	—	—	—	—	—	21	15	12	8.5
12.7 (0.500)	—	—	—	—	—	—	—	—	—	—	—	25	18.5	14.5	10.5

First load test at not less than 15% of available deflection; final load test at not more than 85% of available deflection.
Source: Associated Spring, Barnes Group Inc., Bristol, Connecticut.

TABLE 20-14
Equations for springs with different types of ends[2,3]

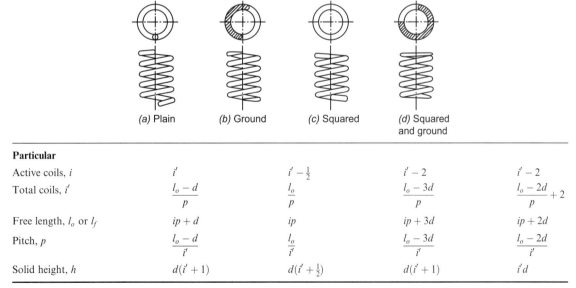

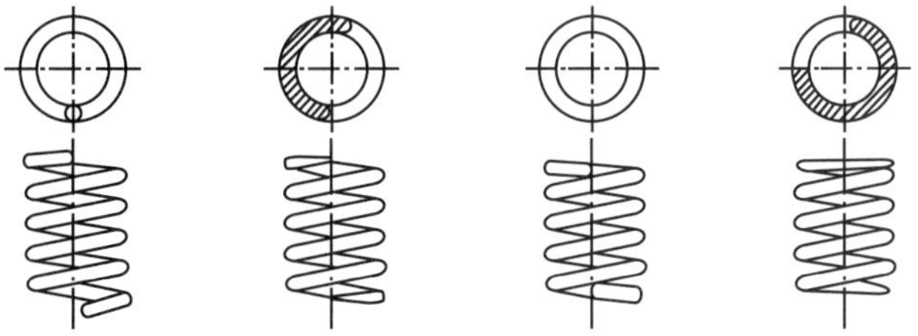

(*a*) Plain (*b*) Ground (*c*) Squared (*d*) Squared and ground

Particular	(a) Plain	(b) Ground	(c) Squared	(d) Squared and ground
Active coils, i	i'	$i' - \frac{1}{2}$	$i' - 2$	$i' - 2$
Total coils, i'	$\frac{l_o - d}{p}$	$\frac{l_o}{p}$	$\frac{l_o - 3d}{p}$	$\frac{l_o - 2d}{p} + 2$
Free length, l_o or l_f	$ip + d$	ip	$ip + 3d$	$ip + 2d$
Pitch, p	$\frac{l_o - d}{i'}$	$\frac{l_o}{i'}$	$\frac{l_o - 3d}{i'}$	$\frac{l_o - 2d}{i'}$
Solid height, h	$d(i' + 1)$	$d(i' + \frac{1}{2})$	$d(i' + 1)$	$i'd$

Source: K. Lingaiah and B. R. Narayana Iyengar, *Machine Design Data Handbook*, Vol. I, Suma Publishers, Bangalore, India, 1986, and K. Lingaiah, *Machine Design Data Handbook*, Vol. 11, Suma Publishers, Bangalore, India, 1986.

Particular	Formula

TABLE 20-15
Curvature factor k_c

c	3	4	6	7	8	9	10
k_c	1.35	1.25	1.15	1.13	1.11	1.1	1.09

$$\sigma_{ds} = \frac{\sigma_e}{n_a k_{sz}} = \frac{1.89\sigma_e}{n_a d^{0.25}} \quad \textbf{Metric} \quad (20\text{-}49c)$$

where σ_e in kgf/mm^2 and d in mm

where n_a = actual factor of safety or reliability factor

The actual factor of safety or reliability factor

$$n_a = \frac{F(\text{compressed})}{F(\text{working})} \quad (20\text{-}50a)$$

$$n_a = \frac{\text{free length} - \text{fully compressed length}}{\text{free length} - \text{working length}}$$

$$= \frac{y+a}{y} \quad (20\text{-}50b)$$

where y is deflection under working load, m (mm), a is the clearance which is to be added when determining the free length of the spring and is made equal to 25% of the working deflection

Generally n_a is chosen at 1.25.

The wire diameter for static loading

$$d = 1.445\left(\frac{6n_aF}{\sigma_e}\right)^{0.4} D^{0.3}$$

$$= 2.945\left(\frac{n_aF}{\sigma_e}\right)^{0.4} D^{0.3} \quad \textbf{SI} \quad (20\text{-}51a)$$

where F in N, σ_e in MPa, D in m, and d in m

$$d = 0.724\left(\frac{6n_aF}{\sigma_e}\right)^{0.4} D^{0.3}$$

$$= 1.48\left(\frac{n_aF}{\sigma_e}\right)^{0.4} D^{0.3} \quad \textbf{Metric} \quad (20\text{-}51b)$$

where F in kgf, σ_e in kgf/mm^2, D in mm, and d in mm

$$d = \left(\frac{6n_aF}{\sigma_e}\right)^{0.4} D^{0.3}$$

$$= 2.05\left(\frac{n_aF}{\sigma_e}\right)^{0.4} D^{0.3} \quad \textbf{USCS} \quad (20\text{-}51c)$$

where F in lbf, σ_e in psi, D in in, and d in in

The wire diameter where there is no space limitation ($D = cd$)

$$d = 4.64\left(\frac{n_aF}{\sigma_e}\right)^{0.57} c^{0.43} \quad \textbf{SI} \quad (20\text{-}51d)$$

where d in m, F in N, σ_e in Pa

$$d = \left(\frac{6n_aF}{\sigma_e}\right)^{0.57} c^{0.43} \quad \textbf{USCS} \quad (20\text{-}51e)$$

where d in in, F in lbf, σ_e in psi

Particular	Formula
	$d = 1.77\left(\frac{n_a F}{\sigma_e}\right)^{0.57} c^{0.43}$ **Metric** (20-51f) where d in mm, F in kgf, σ_e in kgf/mm^2

Final dimensions (Fig. 20-7d)

The number of active coils	$i = \frac{yd^4 G}{8FD^3} = \frac{ydG}{8Fc^3} = \frac{kydG}{\pi\tau D^2}$ (20-52)
The minimum free length of the spring	$l_f \geq (i + n)d + y + a$ (20-53) where a = clearance, m (mm) n = 2 if ends are bent before grinding = 1 if ends are either ground or bent = 0 if ends are neither ground nor bent
Outside diameter of cod of helical spring	$D_o = D + d$ (20-53a)
Solid length (or height) of helical spring	$l_s = i_t d$ (20-53b)
Pitch of spring	$p = \frac{y_s}{i} + d$ (20-53c)
Free length of helical spring l_f or l_o	$l_f - l_s + y_s$ (20-53d)
Maximum working length of helical spring	$l_{max} = l_f - y_{max}$ (20-53e)
Minimum working length of helical spring	$l_{min} = l_f - y_{min}$ (20-53f) where i_t = total number of coild in the spring
Springs with different types of ends[1,2,3]	Refer to Table 20-14.

STABILITY OF HELICAL SPRINGS

The critical axial load that can cause buckling	$F_{cr} = F_o K_l l_f$ (20-54) where K_l is factor taken from Fig. 20-8

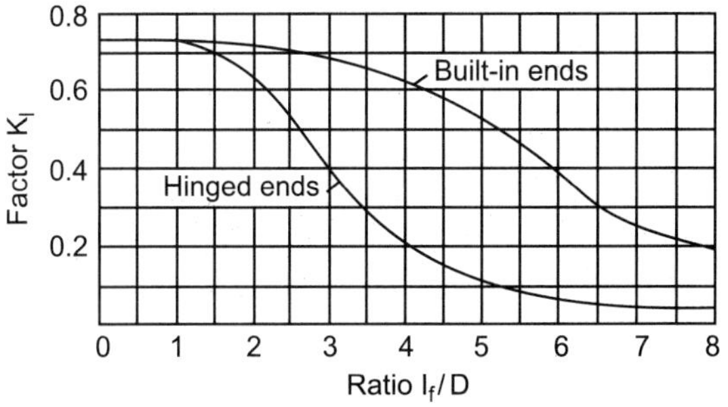

FIGURE 20-8 Buckling factor for helical compression springs. (V. L. Maleev and J. B. Hartman, *Machine Design*, International Textbook Company, Scranton, Pennsylvania, 1954.)

Particular	Formula	
The equivalent stiffness of springs	$$(EI)_{\text{spring}} = \frac{Ed^4 l}{32iD(2+v)}$$	(20-55)
The critical load on the spring	$$F_{cr} = \frac{\pi^2 Ed^4}{32(2+v)iD(l_f - y_{cr})}$$	(20-56)
The critical deflection is explicitly given by	$$\left(\frac{y_{cr}}{l_f}\right)^2 - \frac{y_{cr}}{l_f} + \frac{\pi^2}{2}\frac{1+v}{2+v}\left(\frac{D}{l_f}\right)^2 = 0$$ where $l = (l_f - y_{cr})$	(20-57)

REPEATED LOADING (Fig. 20-9)

The variable shear stress amplitude	$$\tau_a = k_w \frac{8D}{\pi d^3}\frac{F_{\max} - F_{\min}}{2}$$ where $k_w = k_\tau k_c$ Refer to Table 20-15 for k_c.	(20-58)

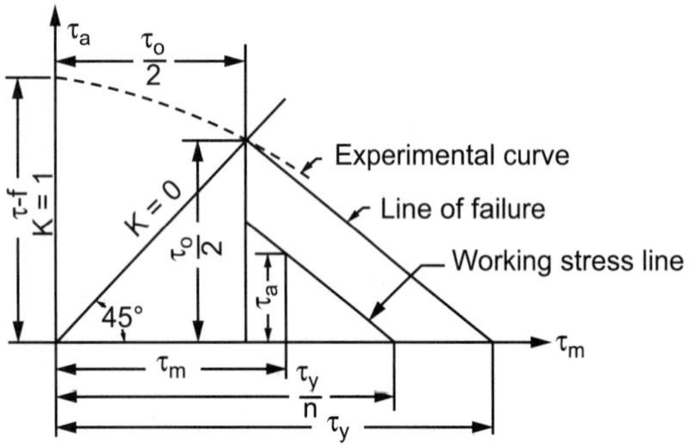

FIGURE 20-9 Cyclic stresses in spring. (K. Lingaiah and B. R. Narayana Iyengar, *Machine Design Data Handbook*, Engineering College Cooperative Society, Bangalore, India, 1962; K. Lingaiah and B. R. Narayana Iyengar, *Machine Design Data Handbook*, Vol. I, Suma Publishers, 1986; K. Lingaiah, *Machine Design Data Handbook*, Vol. II, Suma Publishers, Bangalore, India, 1986.)

The mean shear stress	$$\tau_m = k_\tau \frac{8D}{\pi d^3}\frac{F_{\max} + F_{\min}}{2}$$ where $k_\tau = 1 + 0.5/c$	(20-59)

Design equations for repeated loadings[1,2,3]

Method 1 The Gerber parabolic relation	$$\frac{\tau_a}{\tau_{od}} + \left(\frac{\tau_m}{\tau_{ud}}\right)^2 = 1$$	(20-60)

Particular	Formula
The Goodman straight-line relation	$\frac{\tau_a}{\tau_{od}} + \frac{\tau_m}{\tau_{ud}} = 1$ (20-61)
The Soderberg straight-line relation	$\frac{\tau_a}{\tau_{od}} + \frac{\tau_m}{\tau_{yd}} = 1$ (20-62)
Method 2	
The static equivalent of cyclic load $F_m \pm F_a$	$F'_m = F_m + \frac{\sigma_{sd}}{\sigma_o} F_a$ (20-63a) or $F'_m = F_m + \frac{\sigma_{sd}}{\sigma_{fd}} F_a$ (20-63b)
The relation between σ_e and σ_f for brittle material	$\sigma_e = 2\sigma_f$ (20-64)
The static equivalent of cyclic load for brittle material	$F'_m = F_m + 2F_a$ (20-65)
The relation between F'_m, $F_{\max}$ and $F_{\min}$	$F'_m = \frac{1}{2}(3F_{\max} - F_{\min})$ (20-66)
The diameter of wire for static equivalent load	$d = 1.45\left(\frac{3n_a(3F_{\max} - F_{\min})}{\sigma_e}\right)^{0.4} D^{0.3}$ **SI** (20-67a) where F in N, σ_e in MPa, D in m, and d in m $d = \left(\frac{3n_a(3F_{\max} - F_{\min})}{\sigma_e}\right)^{0.4} D^{0.3}$ **USCS** (20-67b) where F in lbf, σ_e in psi, D in in, and d in in $d = 0.724\left(\frac{3n_a(3F_{\max} - F_{\min})}{\sigma_e}\right)^{0.4} D^{0.3}$ **Metric** (20-67c) where F in kgf, σ_e in kgf/mm^2, D in mm, and d in mm
The wire diameter when there is no space limitation ($D = cd$)	$d = 1.67\left(\frac{3n_a(3F_{\max} - F_{\min})}{\sigma_e}\right)^{0.57} c^{0.43}$ **SI** (20-68a) where F in N, σ_e in MPa, and d in m $d = \left(\frac{3n_a(3F_{\max} - F_{\min})}{\sigma_e}\right)^{0.57} c^{0.43}$ **USCS** (20-68b) where F in lbf, σ_e in psi, and d in in $d = 0.64\left(\frac{3n_a(3F_{\max} - F_{\min})}{\sigma_e}\right)^{0.57} c^{0.43}$ **Metric** (20-68c) where F in kgf, σ_e in kgf/mm^2, and d in mm

Particular	Formula

CONCENTRIC SPRINGS (Fig. 20-10)

The relation between the respective loads shared by each spring, when both the springs are of the same length

$$\frac{F_1}{F_2} = \left(\frac{D_3}{D_1}\right)^3 \left(\frac{d_1}{d_2}\right)^4 \frac{i_2}{i_1}\frac{G_1}{G_2} \quad (20\text{-}69)$$

The relation between the respective loads shared by each spring, when both are stressed to the same value

$$\frac{F_1}{F_2} = \frac{D_2}{D_1}\left(\frac{d_1}{d_2}\right)^3 \frac{k_1}{k_2} \quad (20\text{-}70)$$

The approximate relation between the sizes of two concentric springs wound from round wire of the same material

$$\frac{F_1}{F_2} = \left(\frac{D_2}{D_1}\right)^{0.75}\left(\frac{d_1}{d_2}\right)^{2.5} \quad (20\text{-}71)$$

where suffixes 1 and 2 refer, respectively, to springs 1 and 2 (Fig. 20-10)

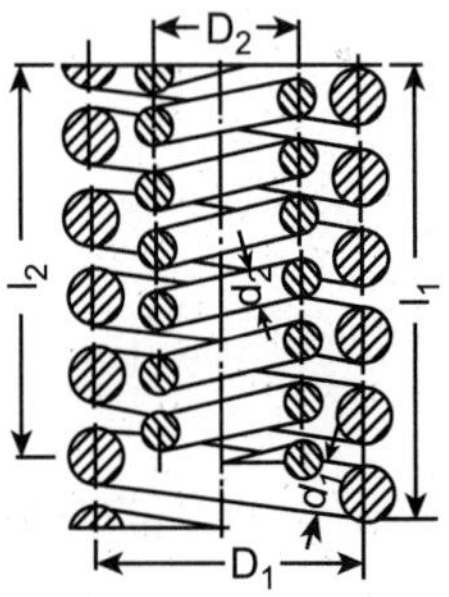

FIGURE 20-10 Concentric spring.

Total load on concentric springs

The total maximum load on the spring

$$F = F_1 + F_2 \quad (20\text{-}72)$$

The load on the inner spring

$$F_2 = mF_1 \quad (20\text{-}73)$$

The load on the outer spring

$$F_1 = \frac{F}{1+m} \quad (20\text{-}74)$$

where $m \leq 1$ and F maximum spring load, kN (lbf)

VIBRATION OF HELICAL SPRINGS

The natural frequency of a spring when one end of the spring is at rest

$$f_n = \frac{1}{2\pi}\sqrt{\frac{2k_0 g}{W}} = 0.705\sqrt{\frac{k_0}{W}} \quad \textbf{SI} \quad (20\text{-}75)$$

where

f_n = natural frequency, Hz
W = weight of vibrating system, N
k_0 = scale of spring, N/m
$g = 9.8066\ \text{m/s}^2$

Particular	Formula
	$f_n = 22.3\left(\frac{k_0}{W}\right)^{1/2}$ **SI** (20-75a) where k_0 in N/mm, W in N, f_n in Hz, $g = 9086.6\,\text{mm/s}^2$
	$f_n = 4.42\left(\frac{k_0}{W}\right)^{1/2}$ **USCS** (20-75b) where k_0 in lbf/in, W in lbf, f_n in Hz, $g = 32.2\,\text{ft/s}^2$
	$f_n = 1.28\left(\frac{k_0}{W}\right)^{1/2}$ **USCS** (20-75c) where k_0 in lbf/in, W in lbf, f_n in Hz, $g = 386.4\,\text{in/s}^2$
The natural frequency of a spring when both ends are fixed	$f_n = \frac{1}{\pi}\sqrt{\frac{2k_0 g}{W}} = 1.41\sqrt{\frac{k_0}{W}}$ **SI** (20-76) where k_0 in N/m, W in N, f_n in Hz, $g = 9.0866\,\text{mm/s}^2$
	$f_n = 44.6\left(\frac{k_0}{W}\right)^{1/2}$ **SI** (20-76a) where k_0 in N/mm, W in N, f_n in Hz, $g = 9086.6\,\text{mm/s}^2$
	$f_n = 2.56\left(\frac{k_0}{W}\right)^{1/2}$ **USCS** (20-76b) where k_0 in lb/ft, W in lbf, f_n in Hz, $g = 32.2\,\text{ft/s}^2$
	$f_n = 8.84\left(\frac{k_0}{W}\right)^{1/2}$ **USCS** (20-76c) where k_0 in lbf/in, W in lbf, f_n in Hz, $g = 386.4\,\text{in/s}^2$
The natural frequency for a helical compression spring one end against a flat plate and free at the other end according to Wolford and Smith[7]	$f_n = 0.25\left(\frac{k_0 g}{W}\right)^{1/2}$ (20-76d)
Another form of equation for natural frequency of compression helical spring with both ends fixed without damping effect	$f_n = \frac{1.12(10^3)d}{D^2 i}\left(\frac{Gg}{\gamma}\right)^{1/2}$ **SI** (20-76e) where G = shear modulus, MPa $g = 9.8006\,\text{m/s}^2$ d and D in mm, f_n in Hz, γ in g/cm^3
	$f_n = \frac{3.5(10^5)d}{D^2 i}$ for steel **SI** (20-76f)

Particular	Formula
	$f_n = \dfrac{0.11d}{D^2 i}\left(\dfrac{Gg}{\gamma}\right)^{1/2}$ **USCS** (20-76g)
	where
	G = modulus of rigidity, psi $g = 386.4\,\text{in/s}^2$ d and D in in, f_n in Hz, γ in lbf/in^2
	$f_n = \dfrac{14(10^3)d}{D^2 i}$ for steel **USCS** (20-76h)

STRESS WAVE PROPAGATION IN CYLINDRICAL SPRINGS UNDER IMPACT LOAD

The velocity of torsional stress wave in helical compression springs	$V_\tau = 10.1\left(\dfrac{Gg}{\gamma}\right)^{1/2}$ **SI** (20-76i)
	where V_τ in m/s, G in MPa, $g = 9.8066\,\text{m/s}^2$, γ in g/cm^3
	$V_\tau = \left(\dfrac{Gg}{\gamma}\right)^{1/2}$ **USCS** (20-76j)
	where V_τ in in/s, G in psi, $g = 386.4\,\text{in/s}^2$, γ in lbf/in^3
The velocity of surge wave (V_s)	(It varies from 50 to 500 m/s.)
The impact velocity (V_{imp})	$V_{imp} = 10.1\sigma\left(\dfrac{g}{2\gamma G}\right)^{1/2}$ **SI** (20-76k)
	$V_{imp} = \dfrac{\sigma}{35.5}$ m/s for steel **SI**
	$V_{imp} = \sigma\left(\dfrac{g}{2\gamma G}\right)^{1/2}$ **USCS** (20-76l)
	$V_{imp} = \dfrac{\sigma}{131}$ in/s for steel **USCS**
The frequency of vibration of valve spring per minute	$f_n = 84.627\sqrt{\dfrac{k_0}{W}}$ **SI** (20-77a)
	where k_0 in N/m, W in N
	$f_n = 2676.12\sqrt{\dfrac{k_0}{W}}$ **Metric** (20-77b)
	where k_0 in kgf/mm, W in kgf
	$f_n = 530\sqrt{\dfrac{k_0}{W}}$ **USCS** (20-77c)
	where k_0 in lbf/in, W in lbf

Particular	Formula	

HELICAL EXTENSION SPRINGS (Fig. 20-11 to 20-13)

For typical ends of extension helical springs	Refer to Fig. 20-11.	
The maximum stress in bending at point A (Fig. 20-12)	$\sigma_A = \dfrac{16K_1DF}{\mu d^3} + \dfrac{4F}{\mu d^2}$	(20-78a)

Type	Configurations	Recommended length min.–max.
Twist loop or hook		0.5–1.7 I.D.
Cross center loop or hook	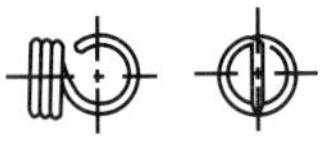	I.D.
Side loop or hook		0.9–1.0 I.D.
Extended hook	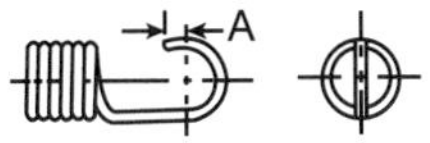	1.1 I.D. and up, as required by design
Special ends	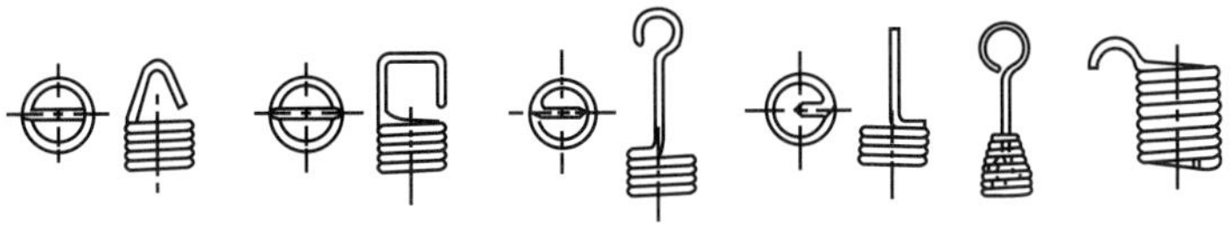	As required by design

FIGURE 20-11 Common-end configuration for helical extension springs. Recommended length is distance from last body coil to inside of end. ID is inside diameter of adjacent coil in spring body. (Associated Spring, Barnes Group, Inc.)

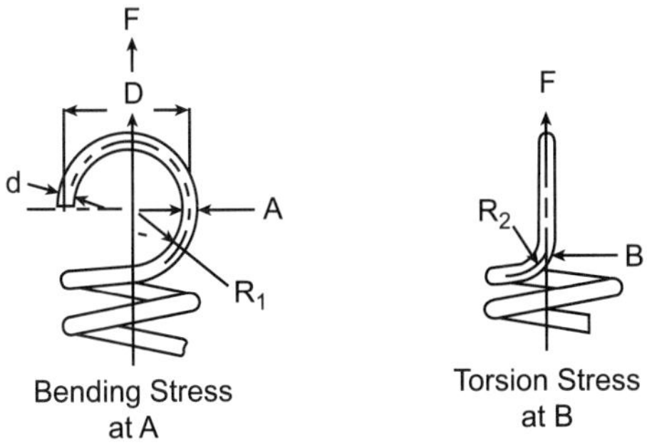

FIGURE 20-12 Location of maximum bending and torsional stresses in twist loops. (Associated Spring, Barnes Group, Inc.)

The constant K_1 in Eq. (20-78a)	$K_1 = \dfrac{4C^2 - C_1 - 1}{4C_1(C_1 - 1)}$	(20-78b)
The constant C_1 in Eq. (20-78b)	$C_1 = \dfrac{2R_1}{d}$	(20-78c)

Particular	Formula
	For R_1, refer to Fig. 20-12.
The maximum stress in torsion at point B (Fig. 20-12)	$$\sigma_B = \frac{8DF}{\mu d^3}\frac{4C_2 - 1}{4C_2 - 4} \quad (20\text{-}78d)$$
The constant C_2 in Eq. (20-78d)	$$C_2 = \frac{2R_2}{d} \quad (20\text{-}78e)$$ For R_2, refer to Fig. 10-12. In practice C_2 may be taken greater than 4.
For extension helical spring dimensions	Refer to Fig. 20-13.

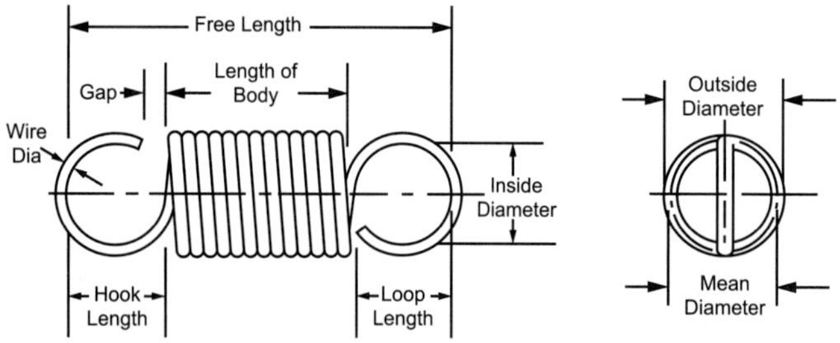

FIGURE 20-13 Typical extension-spring dimensions. (Associated Spring, Barnes Group, Inc.)

Particular	Formula
For design equations of extension helical springs	The design equations of compression springs may be used.
The spring rate	$$k_0 = \frac{F - F_i}{y} = \frac{Gd^4}{8D^3 i} \quad (20\text{-}78f)$$ where F_i = initial tension
The stress	$$\sigma = \frac{k8FD}{\mu d^3} \quad (20\text{-}78g)$$ where k = stress factor for helical springs. Refer to Fig. 20-5 for k.

CONICAL SPRINGS [Fig. 20-14(a)]

Particular	Formula
The axial deflection y for i coils of round stock may be computed by the relation [Fig. 20-14(a)]	$$y = \frac{2iF(D_2^3 + D_2^2 D_1 + D_2 D_1^2 + D_1^3)}{d^4 G} \quad (20\text{-}79)$$ $$y = \frac{\pi i \tau (D_2^3 + D_2^2 D_1 + D_2 D_1^2 + D_1^3)}{4dD_2 kG} \quad (20\text{-}80)$$
The axial deflection of a conical spring made of rectangular stock with radial thickness b and an axial dimension h [Fig. 20-14(c)]	$$y = \frac{0.71iF(b^2 + h^2)(D_2^3 + D_2^2 D_1 + D_2 D_1^2 + D_1^3)}{b^3 h^3 G} \quad (20\text{-}81)$$

Particular	Formula

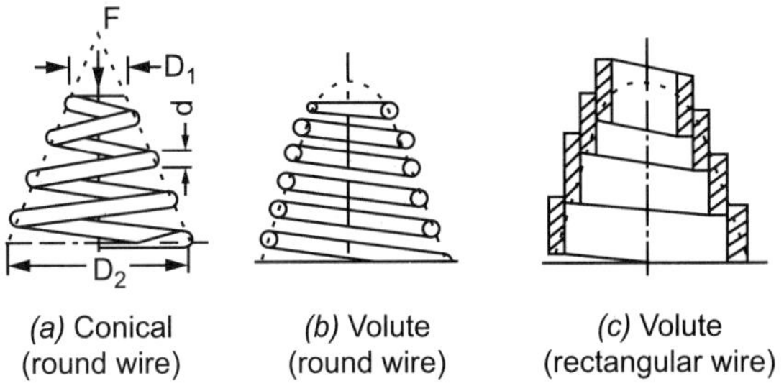

FIGURE 20-14 Conical and volute springs.

NONMETALLIC SPRINGS

Rectangular rubber spring (Fig. 20-15)

Approximate overall dimension of the shock absorber can be obtained by (Fig. 20-15)

$$\frac{L}{D^2} = \frac{\pi E}{2F^2}\left(\frac{U}{(F_{max}/F)^2 - 1}\right) \quad (20\text{-}82)$$

Spring constant K of an absorber

$$K = \frac{\pi D^2 E}{L} \quad (20\text{-}83)$$

Dimensions of sleeve and core are found by empirical relations

$$L_1 = 0.75L \quad (20\text{-}84)$$

$$D_1 = 0.70D \quad (20\text{-}85)$$

$$D_2 = 1.12D_1 \quad (20\text{-}86)$$

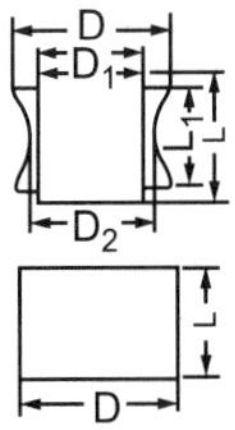

FIGURE 20-15 Rectangular rubber spring.

TORSION SPRINGS (Fig. 20-16)[7]

The maximum stress in torsion spring

$$\sigma = \frac{M_t}{Z} + \frac{F}{A} \quad (20\text{-}87)$$

The stress in torsion spring taking into consideration the correction factor k'

$$\sigma = \frac{k' M_t}{Z} + \frac{2M_t}{DA} \quad (20\text{-}88)$$

The deflection

$$y = \frac{M_t LD}{2EI} \quad (20\text{-}89)$$

The stress in round wire spring

$$\sigma = \frac{8M_t(4k'D + d)}{\pi d^3 D} \quad (20\text{-}89a)$$

where $k' = k_1$ can be taken from curve k_1 in Fig. 20.5

The torsional moment M_t is numerically equal to bending moment M_b.

Particular	Formula

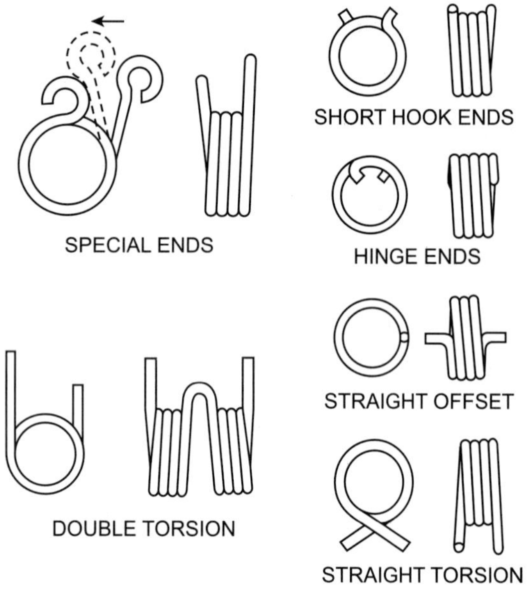

FIGURE 20-16 Common helical torsion-spring end configurations. (Associated Spring, Barnes Group, Inc.)

Particular	Formula	
The stress is also given by Eq. (20-90) without taking into consideration the direct stress (F/A)	$\sigma = k\dfrac{M_b c}{I}$ where $M_b = Fr$	(20-90)
The expressions for k for use in Eq. (20-90)	$k = k_o = \dfrac{4C^2 + C - 1}{4C(C+1)}$ for outer fiber	(20-91a)
	$k = k_i = \dfrac{4C^2 - C - 1}{4C(C-1)}$ for inner fiber	(20-91b)
Equation (20-90) for stress becomes	$\sigma = k_i \dfrac{32Fr}{\pi d^3}$	(20-92)
The angular deflection in radians	$\theta = \dfrac{64M_b Di}{Ed^4}$	(20-93)
The spring rate of torsion spring	$k_0 = \dfrac{M_b}{\theta} = \dfrac{d^4 E}{64Di}$	(20-94)
The spring rate can also be expressed by Eq. (20-95), which gives good results	$k'_0 = \dfrac{d^4 E}{10.8Di}$	(20-95)

Particular	Formula
The allowable tensile stress for torsion springs	$\sigma_{sy} = \sigma_a = \begin{cases} 0.78\sigma_{sut} & \text{cold-drawn carbon steel} \\ 0.87\sigma_{sut} & \text{hardened and tempered carbon and low-alloy steels} \\ 0.61\sigma_{sut} & \text{stainless steel and nonferrous alloys} \end{cases}$
The endurance limit for torsion springs	$\sigma_{sf} = 538\,\text{MPa}\ (78\,\text{kpsi})$

Torsion spring of rectangular cross section

The stress in rectangular wire spring	$\sigma = \dfrac{6k'M_t}{b^2h} + \dfrac{2M_t}{Dbh}$ (20-96) where $k' = k_2$ can be taken from curve k_2 in Fig. 20-5
	$c = \dfrac{D}{h}$ (20-97)
Axial dimension b after keystoning	$b_1 = b\dfrac{C - 0.5}{C}$ (20-98)
Another expression for stress for rectangular cross-sectional wire torsion spring without taking into consideration the direct stress ($\sigma = F/A$)	$\sigma = \dfrac{6k_iM_b}{bh^2}$ (20-99) where $k_i = \dfrac{4C}{4C - 3}$
The spring rate	$k_0 = \dfrac{M_b}{\theta} = \dfrac{Ebh^3}{66Di}$ (20-100)

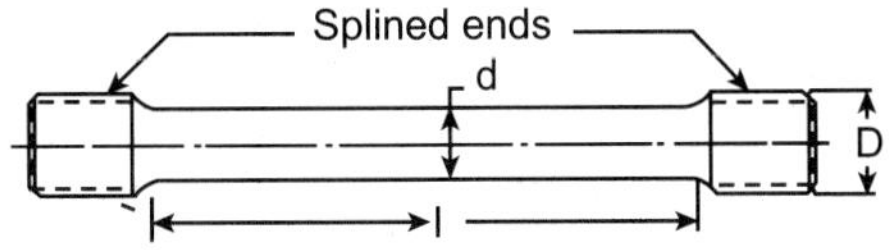

FIGURE 20-17 Torsion bar spring

Torsion bar springs

Refer to Tables 20-16 and 20-17 and Fig. 20-17.

For allowable working stresses for rubber compression springs	Refer to Table 20-18.

TABLE 20-16
Design formulas for bar springs

Cross section of bar	Angular deflection, θ, rad	Maximum shear stress, τ
Solid circular bar	$\frac{584M_t l}{d^4 G}$	$\frac{16M_t}{\pi d^3}$
Hollow circular bar	$\frac{584M_t l}{(d_1^4 - d_2^4 G}$	$\frac{16M_t d_1}{\pi(d_1^4 - d_2^4)}$
Square bar	$\frac{407M_t l}{b^4 G}$	$\frac{4.81M_t}{b^3}$
Rectangular bar	$\frac{57.3M_t l}{k_1' bh^3 G}$ [a]	$\frac{M_t}{k_2' 2bh^2}$ [a]

[a] Values of k_1' and k_2' can be obtained from Table 20-9.

TABLE 20-17
Factors for computing rectangular bars in torsion

b/h	k'	k_1'	k_2'
1.0	0.675	0.140	0.208
1.2	0.759	0.166	0.219
1.5	0.848	0.196	0.231
2.0	0.930	0.229	0.246
2.5	0.968	0.249	0.258
3.0	0.985	0.263	0.267
4.0	0.997	0.281	0.282
5.0	0.999	0.291	0.291
10.0	1.000	0.312	0.231
∞	1.000	0.333	0.333

TABLE 20-18
Suggested allowable working stresses for rubber compression springs

		Limits of allowable stress			
		Occasional loading		Cont. or freq. loading[b]	
Durometer hardness	Area[a] ratio	MPa	psi	MPa	psi
30	5	2.76	400	0.97	140
30	3	2.48	360	0.93	135
30	2	2.24	325	0.86	125
30	1	1.79	260	0.73	105
30	0.5	1.45	210	0.62	90
50	4	4.82	700	1.86	270
50	2	3.73	540	1.58	230
50	1	2.69	390	1.24	180
50	0.5	2.07	300	1.03	150
80	2	6.13	890	2.69	390
80	1	4.14	600	2.07	300
80	0.5	2.90	420	1.65	240

[a] Ratio of load-carrying area available for bulging or lateral expansion

REFERENCES

1. Lingaiah, K. and B. R. Narayana Iyengar, *Machine Design Data Handbook,* Engineering College Co-operative Society, Bangalore, India, 1962.
2. Lingaiah, K., and B. R. Narayana Iyengar, *Machine Design Data Handbook,* Vol. I (*SI and Customary Metric Units*), Suma Publishers, Bangalore, India, 1986.
3. Lingaiah, K., *Machine Design Data Handbook*, Vol. II (*SI and Customary Metric Units*), Suma Publishers, Bangalore, India, 1986.
4. *SAE Handbook*, *Springs*, Vol. I, 1981.
5. Maleev, V. L., and J. B. Hartman, *Machine Design*, International Textbook Company, Scranton, Pennsylvania, 1954.
6. Wahl, A. M., *Mechanical Springs*, McGraw-Hill Book Company, New York, 1963.
7. Associated Spring, Barnes Group Inc., Bristol, CT, USA.
8. Jorres, R. E., *Springs*; Chap. 24 in J. E. Shigley and C. R. Mischke, eds., *Standard Handbook of Machine Design*, McGraw-Hill Book Company, New York, 1986.
9. Shigley, J. E., and C. R. Mischke, *Mechanical Engineering Design*, 5th ed. McGraw-Hill Company, New York, 1989.
10. Zimmerli, F. P., *Human Failures in Springs Applications*, *The Mainspring*, No. 17, Associated Spring Corporation, Bristol, Connecticut, Aug.-Sept. 1957.
11. Shigley, J. E., and C. R. Mischke, *Standard Handbook of Machine Design*, McGraw-Hill Book Company, New York, 1986.
12. Phelan, R. M., *Fundamentals of Mechanical Design*, Tata-McGraw-Hill Publishing Company Ltd, New Delhi, 1975.
13. Lingaiah, K., *Machine Design Data Handbook of Machine Design*, 2nd edition, McGraw-Hill Publishing Company, New York, 1996).
14. Shigley, J. E., and C. R. Mischke, *Standard Handbook of Machine Design*, 2nd edition, McGraw-Hill Publishing Company, New York, 1996.

BIBLIOGRAPHY

Baumeister, T., ed., *Marks' Standard Handbook for Mechanical Engineers*, McGraw-Hill Book Company, New York, 1978.

Black, P. H., and O. Eugene Adams, Jr., *Machine Design*, McGraw-Hill Book Company, New York, 1968.

Bureau of Indian Standards.

Chironis, N. P., *Spring Design and Application*, McGraw-Hill Book Company, 1961.

Norman, C. A., E. S. Ault, and I. F. Zarobsky, *Fundamentals of Machine Design*, The Macmillan Company, New York, 1951.

Shigley, J. E., *Machine Design*, McGraw-Hill Book Company, 1962.

CHAPTER 21

FLEXIBLE MACHINE ELEMENTS

SYMBOLS[11,12,13]

a	width of pulley face, m (in)
	pivot arm length in Rockwood drive, m (in)
a_1	width of belt, m (in)
$A = 0.4(\pi d^2/4)$	useful area of cross-section of the wire rope, m^2 (in^2)
b	thickness of arm, m (in)
	dimension in Rockwood drive (Fig. 21-5), m (in)
c	dimension in Rockwood drive (Fig. 21-5), m (in)
C	center distance between sprockets (also with suffixes), m (in)
	center distance between pulleys, m (in)
	capacity of conveyor, m^3 (ft^3)
	constant depends on the rope diameter, sheave diameter, chain, the bearing, and coefficient of friction [Eqs. (21-59) to (21-62) and (21-86) to (21-103)] (also with suffixes)
C_1	tooth width in precision roller and bush chains, m (in)
d	size of chain, m (in)
	diameter of shaft, m (in)
	diameter of idler bearing, m (in)
	diameter of smaller pulley, m (in)
	diameter of rope, m (in)
	pitch diameter of sprocket, m (in)
d_1	diameter of small sprocket, m (in)
	hub diameter of pulley, m (in)
d_2	diameter of large sprocket, m (in)
d_a	tip diameter of sprocket, m (in)
d_{a1}	tip diameter of small sprocket, m (in)
d_{a2}	tip diameter of large sprocket, m (in)
$d_c = f_p F_b$	equivalent pitch diameter, m (in)
d_f	root diameter of sprocket, m (in)
d_p	pitch diameter of the V-belt small pulley, m (in)
d_r	diameter of roller pin, m (in)
D	pitch diameter of sheave, m (in)
	diameter of large pulley, m (in)
	wire rope drum diameter, m (in) (Fig. 21-4)
D_r	diameter of reel barrel, m (in) Eq. (21-76)
D_d	diameter of the drum in mm as measured over the outermost layer filling the reel drum

D_o	diameter of the sheave pin, m (in)
e	unit elongation of belt
E'	corrected elasticity modulus of steel ropes (78.5 GPa = 11.4 Mpsi), GPa (psi)
F	force, load, kN (lbf)
	tension in belt, kN (lbf)
	minimum tooth side radius, m (in)
F_a	correction factor for instructional belt service from Table 21-27
F_c	correction factor for belt length from Table 21-26
F_{ct}	centrifugal tension, kN (lbf)
F_d	correction factor for arc of contact of belt from Table 21-25
F_θ	tangential force in the belt, required chain pull, kN (lbf)
F_s	tension due to sagging of chain, kN (lbf)
F_1	tension in belt on tight side, kN (lbf)
F_2	tension in belt on slack side, kN (lbf)
F_c	centrifugal force, kN (lbf)
	values of coefficient for manila rope, Table 21-32
FR_1	the minimum value of tooth flank radius in roller and bush chains, m (in)
FR_2	the maximum value of tooth flank radius in roller and bush chains, m (in)
g	acceleration due to gravity, 9.8066 m/s^2 (32.2 ft/s^2)
G	tooth side relief in bush and roller chain, m (in)
h	the thickness of wall of rope drum, m (in)
	crown height, m (in)
h_1	depth of groove in rope drum, m (in)
$H = (D_d - D_r)/2$	depth of rope layer in reel drum, m (in)
i	number of arms in the pulley,
	number of V-belts,
	number of strands in a chain,
	transmission ratio
$k = (e^{\mu\theta} - 1)/e^{\mu\theta}$	variable in Eqs. (21-2d), (21-4a), (21-6), and (21-123), which depends on $(z_1 - z_2)/C_p$
k_d	duty factor
k_l	load factor
$K_{\min}$	center distance constant from Table 21-57
k_s	service factor
k_{sg}	coefficient for sag from Table 21-55
l	width of chain or length of roller, m (in)
	minimum length of boss of pulley, m (in)
	minimum length of bore of pulley, m (in)
	length of conveyor belt, m (in)
	length of cast-iron wire rope drum, m (in)
	outside length of coil link chain, m (in)
K_1	tooth correction factor for use in Eq. (21-116a)
K_2	multistrand factor for use in Eq. (21-116a)
L	length of flat belt, m (in)
	pitch length of V-belt, m (in)
	rope capacity of wire rope reel, m (in)
L_p	length of chain in pitches
M_t	torque, N m (lbf in)
n	number of times a rope passes over a sheave, number of turns on the drum for one rope member
	speed, rpm
	factor of safety

n_1	speed of smaller pulley, rpm or rps
	speed of smaller sprocket, rpm or rps
n_2	speed of larger pulley, rpm or rps
	speed of larger sprocket, rpm or rps
$n' = nk_d$	stress factor
P	power, kW (hp)
P_T	power required by tripper, kW (hp)
p	pitch of chain, m (in)
	pitch of the grooves on the wire rope drum, m (in)
p_1	distance between the grooves of two-rope pulley, m (in)
P	effort, load, kN (lbf)
P_b	bending load, kN (lbf)
P_s	service load, kN (lbf)
P_t	tangential force due to power transmission, kN (lbf)
P_u	ultimate load, kN (lbf)
	breaking load, kN (lbf)
P_w	working load, kN (lbf)
Q	load, kN (lbf)
r	radius near rim (with subscripts), m (in)
	radius, m (in)
s	the amount of shift of the line of action of the load from the center line on the raising load side of sheave, m (in)
s	the average shift of the center line in the load on the effort side of the sheave, m (in)
S	the distance through which the load is raised, m (in)
SA_1	the minimum value of roller or bush seating angle, deg
SA_2	the maximum value of roller or bush seating angle, deg
SR_1	the minimum value of roller or bush seating radius, m (in)
SR_2	the maximum value of roller or bush seating radius, m (in)
t	nominal belt thickness, m (in)
	thickness of rim, m (in)
T	tension in ropes, chains, kN (lbf)
TD_{min}	minimum limit of the tooth top diameter, m (in)
TD_{max}	maximum limit of the tooth top diameter, m (in)
v	velocity of belt chain, m/s (ft/min)
w	specific weight of belt, kN/m^3 (lbf/in^3)
W	width between reel drum flanges, m (in)
W_B	weight of belt, kN/m (lbf/in)
w_c	weight of chain, kN/m (lbf/in)
W_I	weight of revolving idler, kN/m (lbf/in) belt
W_L	load, kN/m (lbf/in)
z_1	number of teeth on the small sprocket
z_2	number of teeth on the large sprocket
σ	stress, MPa (psi)
σ_1	unit tension in belt on tight side, MPa (psi)
σ_2	unit tension in belt on slack side, MPa (psi)
σ_c	centrifugal force coefficient for leather belt, MPa (psi)
σ_{br}	breaking stress for hemp rope, MPa (psi)
τ	shear stress, MPa (psi)
θ	arc of contact, rad
α	angle between tangent to the sprocket pitch circle and the center line, deg
μ	coefficient of friction between belt and pulley
	coefficient of journal friction
μ_c	coefficient of chain friction

η	efficiency
ω_1	angular speed of small sprocket, rad/s
ω_2	angular speed of large sprocket, rad/s

SUFFIXES

b	bending
br	breaking
t	torque
c	compressive
d	design
min	minimum
max	maximum

Other factors in performance or in special aspects of design of flexible machine elements are included from time to time in this chapter and being applicable only in their immediate context, are not given at this stage.

Particular	Formula

BELTS

Flat belts

The ratio of tight side to slack side of belt at low velocities

$$\frac{F_1}{F_2} = e^{\mu\theta} \quad (21\text{-}1)$$

The power transmitted by belt

$$P = \frac{F_\theta v}{1000c_s} \quad \textbf{SI} \quad (21\text{-}2a)$$

where $F_\theta = F_1 - F_2$, P in kW, and v in m/s; F_θ in N

$$P = \frac{F_\theta v}{33{,}000c_s} \quad \textbf{USCS} \quad (21\text{-}2b)$$

where F_θ in lbf; P in hp; v in ft/min

$$P = \frac{F_\theta \omega r}{1000c_s} \quad \textbf{SI} \quad (21\text{-}2c)$$

where F_θ in N, P in kW, r in m, and ω in rad/s

Refer to Table 21-1 for c_s.

Power transmitted per m^2 (in^2) of belt at low velocities

$$P = \frac{\sigma_1 kv}{1000} \quad \textbf{SI} \quad (21\text{-}2d)$$

where $k = (e^{\mu\theta} - 1)/e^{\mu\theta}$, and also from Table 21-2
σ_1 in N/m^2, v in m/s, and P in kW

$$P = \frac{\sigma_1 kv}{33{,}000} \quad \textbf{USCS} \quad (21\text{-}2e)$$

where σ_1 in psi, v in ft/min, and P in hp

TABLE 21-1
Service correction factors, c_s

Atmospheric condition	Clean, scheduled maintenance on large drives	1.2
	Normal	1.0
	Oily, wet, or dusty	0.7
Angle of center line	Horizontal to 60° from horizontal	1.0
	60°–75° from horizontal	0.9
	75°–90° from horizontal	0.8
Pulley material	Fiber on motor and small pulleys	1.2
	Cast iron or steel	1.0
Service	Temporary or infrequent	1.2
	Normal	1.0
	Intermittent or continuous	0.8
Peak loads	Light, steady load, such as steam engines, steam turbines, diesel engines, and multicylinder gasoline engines	1.0
	Jerky loads, reciprocating machines such as normal-starting-torque squirrel-cage motors, shunt-wound, DC motors, and single-cylinder engines	0.8
	Shock and reversing loads, full-voltage start such as squirrel-cage and synchronous motors	0.6

TABLE 21-2
Values of $(e^{\mu\theta} - 1)/e^{\mu\theta} = k$ for various coefficients of frictions and arcs of contact

	Arc of contact between the belt and pulley (θ, deg)										
Value of μ	**90**	**100**	**110**	**120**	**130**	**140**	**150**	**160**	**170**	**180**	**200**
0.28	0.356	0.387	0.416	0.444	0.470	0.496	0.520	0.542	0.564	0.585	0.502
0.30	0.376	0.408	0.438	0.467	0.494	0.520	0.544	0.567	0.590	0.610	0.553
0.33	0.404	0.438	0.469	0.499	0.527	0.554	0.579	0.602	0.624	0.645	0.684
0.35	0.423	0.457	0.489	0.520	0.548	0.575	0.600	0.624	0.646	0.667	0.705
0.38	0.449	0.485	0.518	0.549	0.578	0.605	0.630	0.654	0.676	0.697	0.735
0.40	0.467	0.502	0.536	0.567	0.597	0.624	0.649	0.673	0.695	0.715	0.753
0.43	0.491	0.528	0.562	0.593	0.623	0.650	0.676	0.699	0.721	0.741	0.777
0.45	0.507	0.544	0.579	0.610	0.640	0.667	0.692	0.715	0.737	0.757	0.792
0.48	0.529	0.567	0.602	0.634	0.663	0.690	0.715	0.738	0.759	0.779	0.813
0.50	0.544	0.582	0.617	0.649	0.678	0.705	0.730	0.752	0.773	0.792	0.825
0.53	0.565	0.603	0.638	0.670	0.700	0.726	0.750	0.772	0.793	0.811	0.843

TABLE 21-3
Values of coefficients σ_c for leather belts for use in Eqs. (21-3) and (21-4)

Belt velocity, m/s (ft/min)	7.5 (1500)	10.0 (1950)	12.70 (2500)	15.0 (2950)	17.5 (3500)	20.0 (3950)	22.5 (4450)	25.0 (4950)
Coefficient, σ_c, kgf/cm^2	0.57	1.05	1.63	2.35	3.10	4.07	5.14	6.36
MPa	0.0559	0.1030	0.1598	0.2305	0.3040	0.3991	0.5041	0.5237
psi	8.0	15.0	23.2	33.5	45.0	58.0	73.0	76.0

Particular	Formula
The ratio of tight to slack side of belt at high velocities	$\dfrac{\sigma_1 - \sigma_c}{\sigma_2 - \sigma_c} = e^{\mu\theta}$ (21-3a)
	where $\sigma_c = \dfrac{wv^2}{g}$ (21-3b)
Power transmitted per m^2 (in^2) of belt at high velocities	$P = \dfrac{(\sigma_1 - \sigma_2)kv}{1000}$ **SI** (21-4a)
	where σ_1 and σ_c in N/m^2; v in m/s; P in kW
	$P = \dfrac{(\sigma_1 - \sigma_c)kv}{33{,}000}$ **USCS** (21-4b)
	where σ_1 and σ_c in psi; v in ft/min; P in hp
	Refer to Table 21-3 for values of σ_c.
Equation (21-3a) in terms of tension on tight side (F_1) and slack side of belt (F_2), and centrifugal force (F_c)	$\dfrac{F_1 - F_c}{F_2 - F_c} = e^{\mu\theta}$ (21-4c)
	where $F_1 = \sigma_1 A$; $F_2 = \sigma_2 A$; $F_c = \sigma_c A$; $A = a_1 t$ = area of cross section of belt, m^2 (in^2)
The relation between the initial tension in the belt (F_0) and the tension in the belt on the tight side ($F_{1,max}$) to obtain maximum tension in the belt	$F_{1,max} = 2F_0$ (21-4d)
The power transmitted at maximum tension in belt, i.e., when $F_1 = 2F_0$, from Eq. (21-1)	$P = \dfrac{F_{1,max}v}{33{,}000} = \dfrac{2F_0 v}{33{,}000}$ **USCS** (21-4e)
	$P = \dfrac{F_{1,max}v}{1000} = \dfrac{2F_0 v}{1000}$ **SI** (21-4f)
The power transmitted in actual practice taking into consideration pulley correction factor (K_p), velocity correction factor (K_v), and service factor (C_s) at maximum tension in belt.	$P = \dfrac{2K_p K_v F_a v}{33{,}000 C_s}$ **USCS** (21-4g)
	$P = \dfrac{2K_p K_v F_a v}{1000 C_s}$ **SI** (21-4h)
	where F_a = allowable tension in belt, N (lbf); v = velocity of belt, m/s (ft/min)
Stresses in belt (Fig. 21-1c)	
Tensile stress due to tension on tight side of belt $F_1(S_1)$	$\sigma_1 = \dfrac{F_2}{a_1 t}$ (21-4i)
Tensile stress due to tension on slack side of belt $F_2(S_2)$	$\sigma_2 = \dfrac{F_2}{a_1 t}$ (21-4j)
Tensile stress due to tangential force (effective stress)	$\sigma_\theta = \dfrac{F_\theta}{a_1 t}$ (21-4k)

Particular	Formula
The tensile stress due to belt tension on account of centrifugal force	$\sigma_c = \dfrac{F_c}{a_1 t} = \dfrac{\gamma v^2}{9810}$* (21-4l) where γ = specific weight of belt material N/dm^3 (lbf/in^3)
The bending stress	$\sigma_b = \dfrac{F_b}{d}$ (21-4m)
The maximum belt stress	$\sigma_{max} = \sigma_1 + \sigma_c + \sigma_b + \sigma_{tw} \le \sigma_a$ (21-4n)
Stress due to twist in belt	$\sigma_{tw} = E\left(\dfrac{a_1}{a}\right)^2$ for crossed belt (21-4p) $= 0$ for open belt $= \left(\dfrac{Ea_1 D}{2a^2}\right)$ for half-crossed belt where a = distance from centre of bigger pulley diameter to the point of twist of half-crossed belt and crossed belt $>2D$ σ_a = allowable stress in belt, MPa (psi)
For distribution of various stresses in belt	Refer to Fig. 21-1C. Refer to Table 21-4B for most commonly used belt materials in practice. The values of K_p and C_s are Table from Tables 21-4C and 21-4D, and K_v from Fig. 21-1B, and also Table 21-4E for minimum pulley sizes. F_a = allowable tension in belt, N (lbf) v = velocity of belt, m/s (ft/min)
Coefficient of friction (μ)	$\mu = 0.54 - \dfrac{0.7}{2.4 + v}$ **SI** (21-5) μ may also be obtained from Tables 21-4A and 21-5. v = velocity of belt, m/s. $\mu = 0.54 - \dfrac{140}{500 + v}$ **USCS** (21-5a) where v = velocity of belt, ft/min

* For leather belts and belts of similar material σ_c is of importance only if $v > 15\%$.

TABLE 21-4A
Coefficients of frictions of leather belts on iron pulleys depending on velocity of belt

Velocity of belt, v, m/s	Coefficient of friction, μ	Velocity of belt, v, m/s	Coefficient of friction, μ	Velocity of belt, v, m/s	Coefficient of friction, μ
	0.360	4.0	0.432	15.0	0.500
0.25	0.285	4.5	0.440	17.5	0.505
0.50	0.307	5.0	0.446	20.0	0.509
1.00	0.340	6.0	0.458	22.5	0.512
1.50	0.365	7.0	0.456	25.0	0.514
2.00	0.384	8.0	0.473	27.5	0.517
2.50	0.400	9.0	0.479	30.0	0.519
3.00	0.413	10.0	0.494	32.5	0.520
3.50	0.423	12.5	0.493		

TABLE 21-4B
Properties of some flat and round materials

Material	Specification	Size, in	Minimum pulley diameter, in	Allowable tension per unit width at 600 ft/min, lb/in	Weight, lb/in^3	Coefficient of friction
Leather	1 ply	$t = \frac{11}{64}$	3	30	0.035–0.045	0.4
		$t = \frac{13}{64}$	$3\frac{1}{2}$	33	0.035–0.045	0.4
	2 ply	$t = \frac{18}{64}$	$4\frac{1}{2}$	41	0.035–0.045	0.4
		$t = \frac{20}{64}$	6[a]	50	0.035–0.045	0.4
		$t = \frac{23}{64}$	9[a]	60	0.035–0.045	0.4
Polyamide[b]	F-0[c]	$t = 0.03$	0.60	10	0.035	0.5
	F-1[c]	$t = 0.05$	1.0	35	0.035	0.5
	F-2[c]	$t = 0.07$	2.4	60	0.051	0.5
	A-2[c]	$t = 0.11$	2.4	60	0.037	0.8
	A-3[c]	$t = 0.13$	4.3	100	0.042	0.8
	A-4[c]	$t = 0.20$	9.5	175	0.039	0.8
	A-5[c]	$t = 0.25$	13.5	275	0.039	0.8
Urethane[d]	$w = 0.50$	$t = 0.062$	See	5.2[e]	0.038–0.045	0.7
	$w = 0.75$	$t = 0.078$	Table	9.8[e]	0.038–0.045	0.7
	$w = 0.125$	$t = 0.090$	17-4E	18.9[e]	0.038–0.045	0.7
	Round	$d = \frac{1}{4}$	See	8.3[e]	0.038–0.045	0.7
		$d = \frac{3}{8}$	Table	18.6[e]	0.038–0.045	0.7
		$d = \frac{1}{2}$	17-4E	33.0[e]	0.038–0.045	0.7
		$d = \frac{1}{4}$		74.3[e]	0.038–0.045	0.7

[a] Add 2 in to pulley size for belts 8 in wide or more.
[b] *Source*: Habasit Engineering Manual, Habasit Belting, Inc., Chamblee (Atlanta), Ga.
[c] Friction cover of acrylonitrile-butadiene rubber on both sides.
[d] *Source*: Eagle Belting Co., Des Plaines, Ill.
[e] At 6% elongation; 12% is maximum allowable value.
Notes: d = diameter, t = thickness, w = width. The values given in this table for the allowable tension are based on a belt speed of 600 ft/min. Take $K_v = 1.0$ for polyamide and urethane belts.
Source: Eagle Belting Co., Des Plaines, Illinois; table reproduced from J. E. Shigley and C. R. Mischke, *Mechanical Engineering Design*, McGraw-Hill Book Company, New York, 1989.

TABLE 21-4C
Pulley correction factor K_P for flat belts[a]

Material	Small-pulley diameter, in					
	1.6–4	4.5–8	9–12.5	14, 16	18–31.5	>31.5
Leather	0.5	0.6	0.7	0.8	0.9	1.0
polyamide, F-0	0.95	1.0	1.0	1.0	1.0	1.0
F-1	0.70	0.92	0.95	1.0	1.0	1.0
F-2	0.73	0.86	0.96	1.0	1.0	1.0
A-2	0.73	0.86	0.96	1.0	1.0	1.0
A-3	—	0.70	0.87	0.94	0.96	1.0
A-4		—	0.71	0.80	0.85	0.92
A-5			—	0.72	0.77	0.91

[a] Average values of K_P for the given ranges were approximated from curves in the Habasit Engineering Manual, Habasit Belting, Inc., Chamblee (Atlanta), Ga.
Source: Eagle Belting Co., Des Plaines, Illinois; table reproduced from J. E. Shigley and C. R. Mischke, *Mechanical Engineering Design*, McGraw-Hill Book Company, New York, 1989.

TABLE 21-4D
Service factors C_s for V-belt and flat belt drives

Driven machinery	Power source	
	Normal torque characteristic	High or nonuniform torque
Uniform	1.0–1.2	1.1–1.3
Light shock	1.1–1.3	1.2–1.4
Medium shock	1.2–1.4	1.4-1.6
Heavy shock	1.3–1.5	1.5–1.8

Source: Eagle Belting Co., Des Plaines, Illinois; table reproduced from J. E. Shigley and C. R. Mischke, *Mechanical Engineering Design*, McGraw-Hill Book Company, New York, 1989.

TABLE 21-4E
Minimum pulley sizes for flat and round urethane belts (pulley diameters in inches)

Belt style	Belt size, in	Ratio of pulley speed to belt length, rev/(ft-min)		
		Up to 250	250 to 499	500 to 1000
Flat	0.50×0.062	0.38	0.44	0.50
	0.75×0.078	0.50	0.63	0.75
	1.25×0.090	0.50	0.63	0.75
Round	$\frac{1}{4}$	1.50	1.75	2.00
	$\frac{3}{8}$	2.25	2.62	3.00
	$\frac{1}{2}$	3.00	3.50	4.00
	$\frac{3}{4}$	5.00	6.00	7.00

Source: Eagle Belting Co., Des Plaines, Illinois; table reproduced from J. E. Shigley and C. R. Mischke, *Mechanical Engineering Design*, McGraw-Hill Book Company, New York, 1989.

TABLE 21-5
Coefficient of friction for belts depending on materials of pulley and belt

	Pulley material						
	Cast iron/steel						
Belt material	**Dry**	**Wet**	**Greasy**	**Wood**	**Compressed paper**	**Leather face**	**Rubber face**
Leather, oak-tanned	0.25	0.20	0.15	0.30	0.33	0.38	0.40
Leather, chrome-tanned	0.35	0.32	0.22	0.40	0.45	0.48	0.50
Canvas, stitched	0.20	0.15	0.12	0.23	0.25	0.27	0.30
Cotton, woven	0.22	0.15	0.12	0.25	0.28	0.27	0.30
Camel hair, woven	0.35	0.25	0.20	0.40	0.45	0.45	0.45
Rubber	0.30	0.18	—	0.32	0.35	0.40	0.42
Balata	0.32	0.20	—	0.35	0.38	0.40	0.42

TABLE 21-6A
Thickness and width of leather belts

	Average thickness, mm				Width, mm	
Grade	**Single**	**Double**	**Triple**	**Quadruple**	**Range**	**Increment**
Light	3	6	—	—	12–24	3
					24–102	6
					102–198	12
Medium	4	8	12.5	17.5	200–800	25
					800–1400	50
Heavy	5	10	15	20	800–1400	50
					1500–2100	100

TABLE 21-6B
Relative strength of belt joints

Type of joint	**Relative strength of joint to an equal section of solid leather, efficiency, %**
Cemented, endless Cemented at factory	90–100
Cemented in shop	80–90
Laced, wire	
By machine	75–85
By hand	70–80
Rawhide, small holes	60–70
Rawhide, large holes	50–60
Hinged	
Wire hooks	40
Metal hooks	35–40

Particular	Formula
The cross section of the belt is given	$$a_1 t = \frac{1000P}{v\left(\sigma_d - \dfrac{wv^2}{g}\right)k} \qquad \textbf{SI} \quad (21\text{-}6a)$$ where P in kW, v in m/s, $g = 9.8066\,\text{m/s}^2$, w in N/m^3, and σ_d in MPa
	$$a_1 t = \frac{33{,}000P}{v\left(\sigma_d - \dfrac{wv^2}{g}10^4\right)k} \qquad \textbf{USCS} \quad (21\text{-}6b)$$ where P in hp, v in ft/min, $g = 386.4\,\text{in/s}^2 = 32.2\,\text{ft/s}^2$, w in lbf/in^3, and σ_d in psi
For cross section and properties of belts	Refer to Tables 21-6A to 21-14.

TABLE 21-7
Standard widths of transmission belting for different plies

	Standard width, mm																			
Ply	**25**	**32**	**40**	**44**	**50**	**63**	**76**	**90**	**100**	**112**	**125**	**140**	**152**	**180**	**200**	**224**	**250**	**305**	**355**	**400**
3	p^a	q^b	p	q	p	p	p	q	q	—	—	—	—	—	—	—	—	—	—	—
4	q	q	p	q	p	p	p	p	p	p	p	q	p	—	q	—	—	—	—	—
5	—	—	—	—			p	q	p	p	p	—	p	r^c	q	r	r	—	—	—
6	—	—	—	—	—	—	—	—	q	p	p	—	p	p	p	—	r	—	—	—
8	—	—	—	—	—	—	—	—	—	—	—	—	—	—	r	—	r	r	r	r

[a] p = these sizes are available in Hi-speed and Fort.
[b] q = these sizes are available in Hi-speed only.
[c] r = these sizes are available in Fort only.

TABLE 21-8
Widths of friction surface—rubber transmission belting

Nominal belt width $\times 10^{-3}$ m	Tolerance $\times 10^{-3}$ m
25, 32, 40, 50, 63	±2.0
71, 80, 90, 100, 112, 125	±3.0
140, 160, 180, 200, 224, 250	±4.0
280, 315, 355, 400, 450, 500	±5.0

Source: IS 1370, 1965.

TABLE 21-9
Thickness of friction surface—rubber transmission belting

Ply construction	Nominal thickness hard-type fabric $\times 10^{-3}$ m	Tolerance $\times 10^{-3}$ m
3	3.9	±0.5
4	5.1	±0.7
5	6.4	±0.8
6	7.7	±0.9
7	9.1	±1.0
8	10.4	±1.1

Source: IS 1370, 1964.

TABLE 21-10
Properties of leather belting for various purposes

			Purpose					
			Power transmission			Round belting for small machine		
Properties		General	Single belts	Double belts	Splices single and double	Heavy (5)	Regular (6)	Heavy (7)
Tensile strength, min	MPa	20.6	24.5	24.5	20.6			
	kpsi	3.0	3.5	3.5	3.0			
Breaking strength, min	N					441	667	755
	lbf					100	150	170
Temporary elongation, %, max		6						
Permanent elongation, %, max		2						
Stitch tear resistance thickness, min	N/m	83,356						
	lbf/in	475						
Grain strength		Shall not crack	—					

TABLE 21-11
Tensile strength of fabric in finished rubber transmission belting

			Tensile strength, N/m (kgf/mm) of width			
	Weight of fabric per square meter		Warp		Weft	
Type of fabric	N/m^2	kgf/m^2	N/m	kgf/mm	N/m	kgf/mm
Soft	8.0	0.815	61,291.3	6.25	29,419.8	3.00
Hard	8.8	0.900	61,291.3	6.25	35,303.8	3.60
Soft	9.1	0.930	69,626.9	7.10	32,361.8	3.30
Hard	3.6	0.975	73,549.7	7.50	44,129.7	4.50

Source: IS 1370, 1965.

TABLE 21-12
Properties of ply woven fire-resistant conveyor belting for use in coal mines

Belt designation		1A		1AA		1B		1C		2A		2B		2C		3A		3B		3C	
Direction		A[a]	B[b]	A	B	A	B	A	B	A	B	A	B	A	B	A	B	A	B	A	B
Tensile strength in kgf/mm width for number of plies	3	—	—	—	—	—	—	—	—	—	—	—	—	39.3	21.4	—	—	62.5	21.4	89.3	28.6
	4	23.0	11.2	26.4	12.1	32.1	14.8	38.6	18.6	39.3	21.3	44.7	24.1	51.1	27.9	57.2	21.4	81.3	27.9	116.1	37.2
	5	28.0	13.7	32.1	14.8	39.3	18.0	47.1	22.7	48.0	26.1	54.3	29.5	62.2	34.4	87.7	26.1	99.1	34.0	141.1	45.0
	6	32.7	15.9	37.5	17.4	45.7	21.1	55.0	26.4	—	—	—	—	—	—	—	—	—	—	—	—
Tensile strength in N-m × 10^{-3} width for number of plies	3	—	—	—	—	—	—	—	—	—	—	—	—	385.4	209.9	—	—	612.9	209.9	875.7	280.5
	4	225.6	109.8	258.9	118.7	314.8	145.1	378.5	182.4	385.4	209.9	438.4	236.3	501.1	273.6	560.9	209.9	797.3	273.6	1138.5	364.8
	5	274.0	134.4	314.8	145.1	385.4	176.5	461.9	222.6	470.7	255.0	532.5	289.3	610.0	333.4	860.0	256.0	971.8	333.4	1383.7	441.3
	6	320.7	155.9	397.7	170.6	448.2	206.9	539.4	258.9	—	—	—	—	—	—	—	—	—	—	—	—
Tear strength in kgf for the number of plies	3	—		—		—		—		—		—		90.8		—		—		—	
	4	20.4		25.0		29.5		36.3		90.8		104.3		117.9		—		—		—	
	5	27.2		31.8		36.3		45.4		113.4		131.4		149.7		—		—		—	
	6	34.0		38.6		43.1		54.4		—		—		—		—		—		—	
Tear strength in N for the number of. plies	3	—		—		—		—		—		—		890.4		—		—		—	
	4	200.1		245.2		289.3		356.0		890.4		1022.8		1156.2		—		—		—	
	5	266.7		311.8		356.0		445.2		1112.1		1288.6		1468.0		—		—		—	
	6	333.4		378.5		422.7		533.5		—		—		—		—		—		—	
Percentage elongation at break		15	8	15	8	15	8	15	8	17	18	17	18	17	18	—		—		—	

[a] A = warp. [b] B = weft.

TABLE 21-13
Allowable tension in width of belt

Belt material	Ply or number of thickness of belt																							
	kN/m	kgf/mm	kN/m	kgf/mm	kN/m	kgf/mm	kN/m	kgf/mm	kN/m	kgf/mm	kN/m	kgf/mm	kN/m	kgf/mm	kN/m	kgf/mm	kN/m	kgf/mm	kN/m	kgf/mm	kN/m	kgf/mm	kN/m	kgf/mm
Leather																								
Light	—	—	16.7	1.7	—	—	—	—	—	—	—	—	—	—	—	—	—	—	—	—	—	—	—	—
Medium	14.7	1.5	24.5	2.5	—	—	—	—	—	—	—	—	—	—	—	—	—	—	—	—	—	—	—	—
Heavy	17.7	1.8	28.4	2.9	35.3	3.6	—	—	—	—	—	—	—	—	—	—	—	—	—	—	—	—	—	—
Canvas-stitched	—	—	—	—	—	—	6.9	0.7	8.8	0.9	10.8	1.1	—	—	11.8	1.2	—	—	13.7	1.4	—	—	15.7	1.6
Balata	—	—	—	—	4.9	0.5	6.9	0.7	8.8	0.9	10.8	1.1	11.8	1.2	13.7	1.4	25.5	2.6	28.4	2.9	30.4	3.1	33.3	3.4
Rubber	—	—	—	—	7.8	0.8	10.8	1.1	12.7	1.3	15.7	1.6	18.6	1.9	22.6	2.3	25.5	2.6	28.4	2.9	30.4	3.1	33.3	3.4

Particular	Formula

BELT LENGTHS AND CONTACT ANGLES FOR OPEN AND CROSSED BELTS (Fig. 21-1A)

Length of belt for open drive (Fig. 21-1(A)a)

$$L = \sqrt{4C^2 - (D-d)^2} = \tfrac{1}{2}(D\theta_L + d\theta_s) \qquad (21\text{-}7)$$

Length of belt for crossed drive (Fig. 21-1(A)b)

$$L = \sqrt{4C^2 - (D+d)^2} + \frac{\theta}{2}(D+d) \qquad (21\text{-}8)$$

Length of belt for quarter turn drive

$$L = \frac{\pi}{2}(D+d) + \sqrt{C^2+D^2} + \sqrt{C^2+d^2} \qquad (21\text{-}9)$$

For two-pulley open drive the center distance between the two pulleys when the length of the belt is known

$$C = \frac{L}{4} - 0.393(D+d) + \left[\left\{\frac{L}{4} - 0.393(D+d)\right\}^2 - \frac{(D-d)^2}{8}\right]^{1/2} \qquad (21\text{-}10)$$

where

$$\theta_l = \pi + 2\sin^{-1}\left(\frac{D-d}{2C}\right) \qquad (21\text{-}10a)$$

$$\sigma_s = \pi - \sin^{-1}\left(\frac{D-d}{2C}\right) \qquad (21\text{-}10b)$$

$$\theta = \pi + 2\sin^{-1}\left(\frac{D+d}{2C}\right) \qquad (21\text{-}10c)$$

The unit elongation of belt is given by the equation

$$e = \frac{\sqrt{\sigma}}{69{,}000} \qquad \textbf{SI} \quad (21\text{-}11a)$$

where σ in MPa

$$e = \frac{\sqrt{\sigma}}{21{,}000} \qquad \textbf{USCS} \quad (21\text{-}11b)$$

where σ in psi

$$e = \frac{\sqrt{\sigma}}{22} \qquad \textbf{Metric} \quad (21\text{-}11c)$$

where σ in kgf/mm^2

The relation between initial belt tension and final belt tension

$$2\sqrt{F_0} = \sqrt{F_1} + \sqrt{F_2} \qquad (21\text{-}12)$$

where F_0 = initial belt tension, kN (lbf)

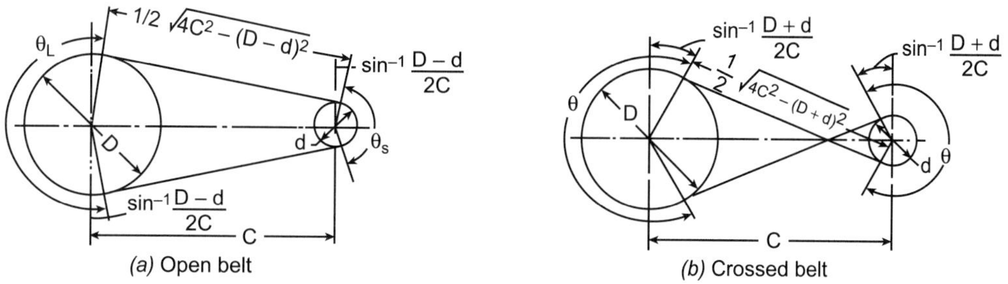

FIGURE 21-1(A) Open and crossed belts.

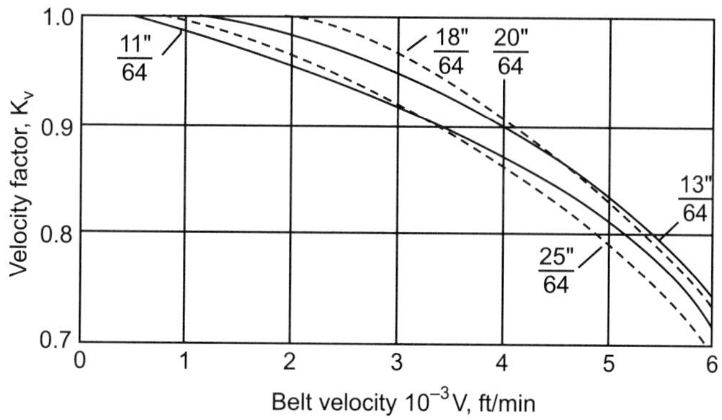

FIGURE 21-1(B) Velocity correction factor for K_v for use in Eq. (21-4g) for leather belts.

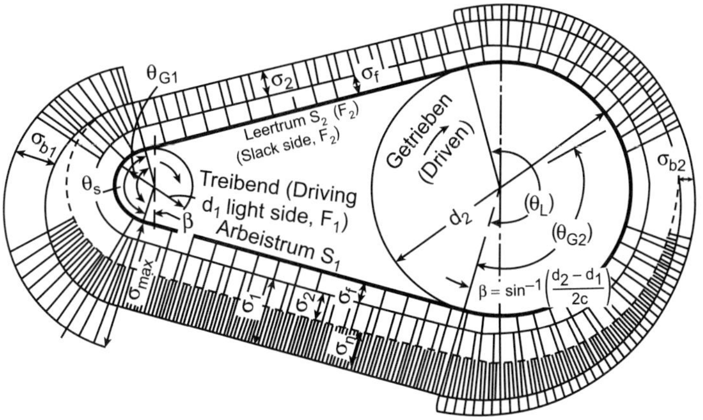

Belt stresses in open drive: $\sigma_f = \sigma_c$ centrifugal stress; σ_2 slack side stress; σ_1 tight side stress $= \sigma_2 + \sigma_n$; σ_n effective stress $= \sigma_u$; σ_{b1}, σ_{b2} bending stresses on pulleys 1 and 2 respectively; σ_G creep angle (angle over which creep takes place between belt and pulley). Lectrum S_2 = slack side F_2; treibend = driving; Arbeitstrum S_1 = tight side F_1; getrieben = driven

FIGURE 21-1(C) Stress distribution in belt. (G. Niemann, *Maschinenelemente*, Springer International Edition, Allied Publishers Private Ltd., New Delhi, 1978.)

Particular	Formula
PULLEYS (Fig. 21-2 and Fig. 21-3)	
C. G. Barth's formula for the width of the pulley face	$a = 1.19a_1 + 10\,\text{mm}$ for single belt **SI** (21-13a)
	$a = 1.1a_1 + 5\,\text{mm}$ for double belt **SI** (21-13b)
	Refer to Table 21-15 for width of pulley.
	$a = 1.1875a_1 + \frac{3}{8}\,\text{in}$ **USCS** (21-13c) where a and a_1 in in for a single belt
	$a = 1.09375a_1 + \frac{3}{16}\,\text{in}$ **USCS** (21-13d) where a and a_1 in in for double belt
C. G. Barth's empirical formula for the crown height for wide belts	$h = 0.00426\sqrt[3]{a^2}$ **SI** (21-14a) where a in m
	$h = 0.013125\sqrt[3]{a^2}$ **USCS** (21-14b) where a in in
For rubber belts on well-aligned shafts, the crown height	$h = \dfrac{a}{200}$ **Customary Metric Units** (21-14c)
	$h = \dfrac{a}{2}$ **SI** (21-14d)
For poorly aligned shafts, the crown height	$h = \dfrac{a}{120}$ **Customary Metric Units** (21-14e)
	$h = \dfrac{a}{0.12}$ **SI** (21-14f) Refer to Tables 21-16, 21-17A, and 21-17B for crown height.
The rim thickness at edge for light-duty pulley	$t = 0.25\sqrt{D} + 1.5\,\text{mm}$ (21-15a)
The rim thickness at edge for heavy-duty pulley for a triple belt	$t = 0.375\sqrt{D} + 3.2\,\text{mm}$ (21-15b)
The hub diameter of the pulley (Fig. 21-2)	$d_1 = 1.5d + 25\,\text{mm}$ (21-16)
Arms	
The bending moment on each arm	$M_b = \dfrac{F_\theta D}{i}$ (21-17)
The section modulus of the arm at the hub	$Z = \dfrac{F_\theta D}{i\sigma_d}$ (21-18)

Particular	Formula

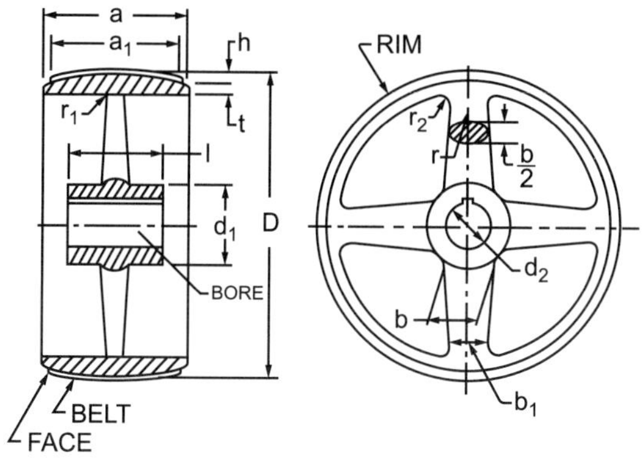

FIGURE 21-2 Cast-iron pulley.

INDIAN STANDARD SPECIFICATION

Cast-iron pulley

Particular	Formula	
Minimum length of bore (Fig. 21-2)	$l = \frac{2}{3}a$ It should not exceed a	(21-19)
Half of the difference in diameters d_1 and d_2 (Fig. 21-2)	$\frac{d_1 - d_2}{2} = 0.412\sqrt[3]{aD} + 6\,\text{mm}$ for a single belt	(21-20)
	$\frac{d_1 - d_2}{2} = 0.529\sqrt[3]{aD} + 6\,\text{mm}$ for a double belt	(21-21)
The radius r_1 near rim (Fig. 21-2)	$r_1 = b/2$	(21-22)
The radius r_2 near rim (Fig. 21-2)	$r_2 = b/2$	(21-23)

TABLE 21-14
Properties of solid woven fire-resistance conveyor belting for use in coal mines

Belt designation	Direction	Tensile strength/width kN/m	Tensile strength/width kgf/mm	Percentage elongation at break	Tear strength kN	Tear strength kgf
4*A*	Warp	385.4	39.3	18	1.3	136.1
	Weft	209.9	21.4	19		
4*B*	Warp	525.6	53.6	18	1.3	136.1
	weft	262.8	26.8	19		
4*C*	Warp	665.9	67.9	18	1.3	136.1
	Weft	262.8	26.8	19		

TABLE 21-15
Width of flat cast-iron and mild steel pulleys

Width, mm	Tolerance, mm
20, 25, 32 40, 50, 63, 71	±2
80, 90, 100, 112, 125, 140	±1.5
160, 180, 200, 224, 250, 280, 315	±2
355, 400, 450, 500, 560, 630	±3

TABLE 21-16
Crown of cast iron and mild steel flat pulleys of diameters up to 355 mm

Nominal diameter, *D*, mm	Crown, *h*, mm
40–112	0.3
125, 140	0.4
160, 180	0.5
200, 224	0.6
250, 280	0.8
315, 355	1.0

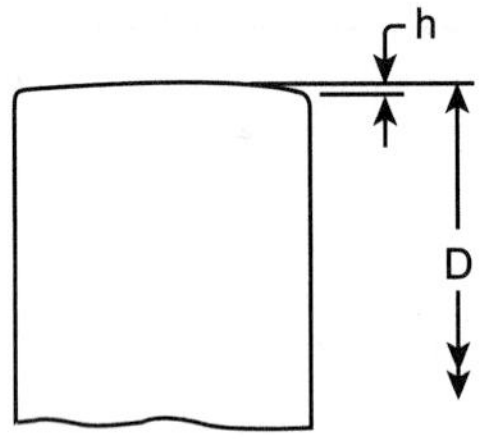

TABLE 21-17A
Crown of cast iron and mild steel flat pulleys of diameters 400 to 2000 mm[a]

Nominal diameter, *D*, mm	Crown *h* of pulleys of width						
	≤125	140, 160	180, 200	224, 250	280, 315	355	≥400
400	1	1.2	1.2	1.2	1.2	1.2	1.2
450	1	1.2	1.2	1.2	1.2	1.2	1.2
500	1	1.5	1.5	1.5	1.5	1.5	1.5
560	1	1.5	1.5	1.5	1.5	1.5	1.5
630	1	1.5	2	2	2	2	2
710	1	1.5	2	2	2	2	2
800	1	1.5	2	2.5	2.5	2.5	2.5
900	1	1.5	2	2.5	2.5	2.5	2.5
1000	1	1.5	2	2.5	3	3	3
1120	1.2	1.5	2	2.5	3	3	3.5
1250	1.2	1.5	2	2.5	3	3.5	4
1400	1.5	2	2.5	3	3.5	4	4
1600	1.5	2	2.5	3	3.5	4	5
1800	2.0	2.5	3	3.5	4	4.5	5
2000	2.0	2.5	3	3.5	4	4.5	6

[a] All dimensions in mm.
Source: IS 1691, 1968.

TABLE 21-17B
Crown height and ISO pulley diameters for flat belts

ISO pulley diameter, in	Crown height, in	ISO pulley diameter, in	Crown height, in	
			$w \leq 10$ in	$w > 10$ in
1.6, 2, 2.5	0.012	12.5, 14	0.03	0.03
2.8, 3.15	0.012	12.5, 14	0.04	0.04
3.55, 4, 4.5	0.012	22.4, 25, 28	0.05	0.05
5, 5.6	0.016	31.5, 35.5	0.05	0.06
6.3, 7.1	0.020	40	0.05	0.06
8, 9	0.024	45, 50, 56	0.06	0.08
10, 11.2	0.030	63, 71, 80	0.07	0.10

Crown should be rounded, not angled; maximum roughness is R_o = AA 63 μin.

Particular	Formula
Arms	Use webs for pulleys up to 200 mm diameter
The number of arms	$i = 4$ (21-24a)
	for pulleys above 200 mm diameter and up to 400 mm diameter
	$i = 6$ (21-24b)
	for pulleys above 450 mm diameter
Cross section of arms	Use elliptical section
Thickness of arm near boss (Fig. 21-2)	$b = 0.294\sqrt[3]{\dfrac{aD}{4i}}$ **SI** (21-25a)
	$b = 1.6\sqrt[3]{\dfrac{aD}{i}}$ for single belt **USCS** (21-25b)
	$b = 0.294\sqrt[3]{\dfrac{aD}{2i}}$ **SI** (21-26a)
	$b = 1.25\sqrt[3]{\dfrac{aD}{i}}$ for double belt **USCS** (21-26b)
The diameter of pulleys and arms in pulleys	Refer to Tables 21-18 to 21-21.
The thickness of arm near rim	b_1—give a taper of 4 mm per 100 mm
The radius of the cross-section of arms	$r = \frac{3}{4}b$ (21-27)

TABLE 21-18
Minimum pulley diameters for given belt speeds and plies[a]

	Maximum belt speeds, m/s				
No. of plies	**10**	**15**	**20**	**25**	**30**
2	50	63	80	90	112
3	90	100	112	140	180
4	140	160	180	200	250
5	200	224	250	315	355
6	250	315	355	400	450
7	355	400	450	500	560
8	450	500	560	630	710
9	560	630	710	800	900
10	630	710	800	900	1000

[a] All dimensions in mm.
Source: IS 1370, 1965.

TABLE 21-19
Diameters of flat pulley and tolerances

Nominal diameter, mm	Tolerance, mm	Nominal diameter, mm	Tolerance, mm
40	±0.5	280, 315, 355	±3.0
45, 50	±0.6	400, 450, 500	±4.0
56, 63	±0.8	560, 630, 710	±5.0
71, 80	±1.0	800, 900, 1000	±6.3
90, 100, 112	±1.2	1120, 1250, 1400	±8.0
125, 140	±1.9	1600, 1800, 2000	±10.22
160, 180, 200	±2.0	—	—
224, 250	±2.5	—	—

Source: IS 1691, 1968.

TABLE 21-20
Minimum pulley diameters for conveyor belting

Running	No. of plies	Fabric 28			Fabric 32			Fabric 36			Fabric 42			Fabric 48		
		A	*B*	*C*	*A*	*B*	*C*	*A*	*B*	*C*	*A*	*B*	*C*	*A*	*B*	*C*
>75–100% rated max working tension	2	205	155	155	255	205	155	305	255	205	305	255	205	—	—	—
	3	305	255	205	360	305	205	460	36	305	460	360	305	530	460	330
	4	410	305	255	460	360	305	610	460	360	610	510	410	710	610	510
	5	510	410	360	610	460	360	690	610	460	765	610	510	890	760	635
	6	610	460	410	690	510	460	915	690	610	915	765	610	1065	915	760
	7	690	610	460	765	690	510	1070	765	690	1070	915	690	1245	1065	890
	8	765	690	500	915	765	610	1220	915	690	1220	1020	765	1420	1220	1015
	9	915	690	610	1070	915	610	1375	1070	765	1375	1070	915	1600	1370	1145
	10	1070	765	690	1220	915	690	1525	1220	915	1525	1220	1070	1780	1525	1245
>50–75% rated max working tension	2	205	155	155	205	155	155	255	205	155	305	255	205	—	—	—
	3	305	205	205	305	255	205	410	305	255	460	360	305	430	355	305
	4	360	305	255	410	305	255	510	410	360	610	460	410	560	485	405
	5	460	360	305	510	410	360	690	510	410	765	610	460	710	610	510
	6	510	460	360	610	510	410	765	610	510	915	690	610	865	735	610
	7	610	510	410	690	610	460	915	690	610	1070	915	690	990	865	710
	8	765	610	510	915	690	610	1070	915	690	1220	915	765	1145	965	815
	9	915	690	610	915	690	610	1220	915	765	1375	1070	915	1270	1090	915
	10	915	765	610	1070	915	690	1375	1070	915	1525	1220	915	1420	1220	1015
≤50% rated max working tension	2	155	155	155	205	155	155	255	205	155	255	205	155	—	—	—
	3	255	205	155	305	205	205	360	305	255	410	305	255	380	330	280
	4	305	255	205	360	305	255	460	410	360	510	410	360	510	430	355
	5	410	360	255	460	360	305	610	460	410	690	510	410	635	530	455
	6	510	410	360	510	460	360	690	510	510	765	610	510	735	635	535
	7	610	460	410	610	510	410	765	690	610	915	690	610	865	735	635
	8	690	510	460	765	610	510	915	705	690	1070	765	690	990	865	710
	9	765	610	510	915	690	610	1070	915	765	1220	915	765	1220	965	815
	10	915	690	510	915	765	610	1220	915	915	1220	1070	915	1245	1065	890

Source: IS 1891 (Part 1), 1968.

TABLE 21-21
Number of arms in mild steel pulley

	Details of spokes	
Diameter, mm	**No.**	**Of diameter**
250–500	6	19
560–710	8	19
800–1000	10	22
1120	12	22
1250	14	22
1400	16	22
1600	18	22
1800	18	22
2000	22	22

Source: IS 1691, 1968.

Particular	Formula

Mild Steel Pulley

Minimum length of boss (Fig. 21-3)

$$l = a/2 \tag{21-28}$$

$\nless$ 100 mm for 19-mm-diameter spokes
$\nless$ 138 mm for 22-mm-diameter spokes

The thickness of rim

$t = 5$ mm for diameters from 400 to 2000 mm

$$t = \frac{D}{200} + 0.003\,\text{m} \quad \textbf{SI} \tag{21-29a}$$

$$t = \frac{D}{200} + 0.12\,\text{in} \quad \text{for single belt} \quad \textbf{USCS} \tag{21-29b}$$

$$t = \frac{D}{200} + 0.006\,\text{m} \quad \textbf{SI} \tag{21-29c}$$

$$t = \frac{D}{200} + 0.24\,\text{in} \quad \text{for double belt} \quad \textbf{USCS} \tag{21-29d}$$

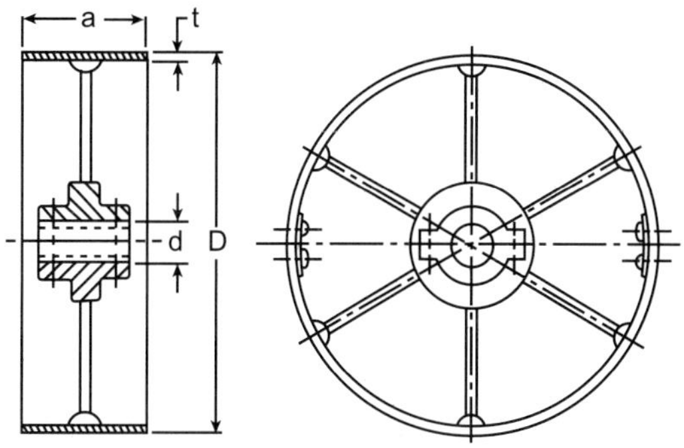

FIGURE 21-3 Mild steel pulley.

The crown height — Refer to Tables 21-16, 21-17A, and 21-17B.

Arms for mild steel pulleys — Refer to Table 21-21.

Particular	Formula

V-BELT

The formula to obtain the maximum power in kilowatt which the V-belts of sections A, B, C, D, and E can transmit (Table 21-22 and 21-23)

Belt cross section	Power rating per strand, kW	Max. value of d_c, mm	SI
A	$P^* = v\left(\frac{0.45}{v^{0.09}} - \frac{19.62}{d_e} - \frac{0.765v^2}{10^4}\right)$	125	(21-30)
B	$P^* = v\left(\frac{0.79}{v^{0.09}} - \frac{51.33}{d_e} - \frac{1.31v^2}{10^4}\right)$	175	(21-31)
C	$P^* = v\left(\frac{1.47}{v^{0.09}} - \frac{143.27}{d_e} - \frac{2.34v^2}{10^4}\right)$	300	(21-32)
D	$P^* = v\left(\frac{3.16}{v^{0.09}} - \frac{507.50}{d_e} - \frac{4.77v^2}{10^4}\right)$	425	(21-33)
E	$P^* = v\left(\frac{4.57}{v^{0.09}} - \frac{952}{d_e} - \frac{7.05v^2}{10^4}\right)$	700	(21-34)

where P^* = maximum power in kW at 180° arc of contact for a belt of average length

The equivalent pitch diameter

$$d_e = d_p F_b \tag{21-35}$$

Refer to Table 21-24 for F_b, the small-diameter factor.

TABLE 21-22
Classification of V-belts

Cross-sectional symbol	Nominal top width, W, mm	Nominal thickness, T, mm
A	13	8
B	17	11
C	22	14
D	30	12
E	33	23

Source: IS 2494, 1964.

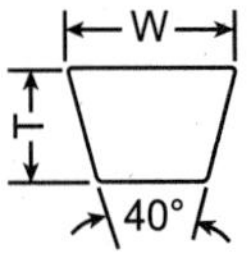

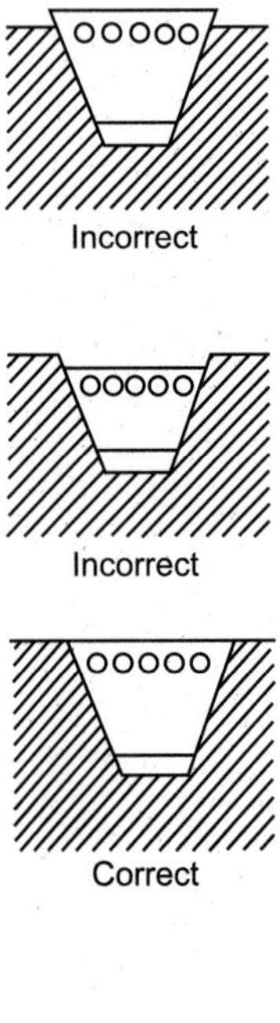

Particular	Formula
The formulas to obtain the maximum horsepower of V-belts of A, B, C, D, and E sections	Refer to Eqs. (21-35a) to (21-35e).

Belt section	Horsepower rating per strand (equations in USCS)	
A	$P = V\left(\frac{1.95}{V^{0.09}} - \frac{3.80}{kd} - 0.0136V^2\right)$	(21-35a)
B	$P = V\left(\frac{3.43}{V^{0.09}} - \frac{9.83}{kd} - 0.0234V^2\right)$	(21-35b)
C	$P = V\left(\frac{6.37}{V^{0.09}} - \frac{27.0}{kd} - 0.0416V^2\right)$	(21-35c)
D	$P = V\left(\frac{13.6}{V^{0.09}} - \frac{93.9}{kd} - 0.0848V^2\right)$	(21-35d)
E	$P = V\left(\frac{19.9}{V^{0.09}} - \frac{17.8}{kd} - 0.122V^2\right)$	(21-35e)

where

V = belt speed, thousands of ft/min
k = small-diameter factor for speed ratio of drive from Fig. 21-4b
d = pitch diameter of small sheave, in

TABLE 21-23
Ratings for V-belts in kilowatts

Belt speed, m/s	Equivalent pitch diameter, d_e, mm																																
	Cross section, *A*						Cross section, *B*						Cross section, *C*							Cross section, *D*								Cross section, *E*					
	80	90	100	110	120	125	130	140	150	160	170	180	180	200	220	240	260	280	300	300	320	340	360	380	400	420	430	450	500	550	600	650	700
0.5	0.13	0.13	0.14	0.14	0.15	0.15	0.22	0.22	0.22	0.22	0.29	0.29	0.37	0.44	0.44	0.51	0.51	0.51	0.51	0.81	0.88	0.96	0.96	1.03	1.03	1.10	1.10	1.40	1.47	1.54	1.62	1.69	1.77
1	0.22	0.24	0.25	0.27	0.28	0.29	0.37	0.44	0.54	0.44	0.51	0.51	0.64	0.73	0.81	0.88	0.88	0.96	0.96	1.47	1.54	1.69	1.77	1.84	1.91	1.91	1.99	2.43	2.65	2.87	2.94	3.09	3.24
2	0.37	0.40	0.43	0.46	0.49	0.51	0.66	0.74	0.81	0.81	0.88	0.88	1.15	1.32	1.47	1.85	1.69	1.77	1.84	2.50	2.79	2.94	3.09	3.24	3.38	3.53	3.53	4.34	4.78	5.15	5.44	5.66	5.88
3	0.51	0.58	0.64	0.68	0.72	0.74	0.96	1.03	1.10	1.17	1.25	1.39	1.62	1.84	2.06	2.21	2.35	2.50	2.57	3.46	3.75	4.04	4.34	4.56	4.71	4.92	5.00	6.03	6.69	7.21	7.65	8.02	8.38
4	0.58	0.74	0.81	0.88	0.93	0.96	1.18	1.32	1.40	1.47	1.50	1.62	2.06	2.35	2.57	2.76	3.02	3.16	3.31	4.34	4.78	5.15	5.44	5.74	6.03	6.25	6.40	7.58	8.46	9.19	9.78	10.22	12.74
5	0.74	0.85	0.95	1.04	1.13	1.18	1.47	1.54	1.69	1.84	1.91	1.99	2.35	2.79	3.09	3.38	3.60	3.75	3.97	5.07	5.66	6.10	6.55	6.91	7.21	7.58	7.65	9.05	10.15	11.03	11.77	12.36	13.02
6	0.81	0.94	1.05	1.15	1.25	1.32	1.62	1.84	1.99	2.06	2.21	2.28	2.72	3.16	3.60	3.90	4.19	4.41	4.63	5.81	6.55	7.06	7.51	7.94	8.38	8.75	8.90	10.44	11.69	12.80	13.68	14.49	15.07
7	0.88	1.08	1.25	1.39	1.50	1.54	1.91	2.13	2.21	2.35	2.50	2.65	3.02	3.60	4.05	4.41	4.71	5.00	5.22	6.47	7.28	7.94	8.46	8.97	9.41	9.86	10.08	11.77	13.32	14.49	15.51	16.40	17.14
8	0.90	1.17	1.35	1.50	1.62	1.69	2.06	2.28	2.43	2.65	2.79	2.94	3.31	3.97	4.49	4.85	5.22	5.59	5.87	7.13	7.94	8.68	9.34	10.00	10.52	10.96	11.25	13.02	14.78	16.11	17.28	18.31	19.05
9	1.03	1.32	1.54	1.69	1.77	1.84	2.21	2.43	2.65	2.87	3.02	3.16	3.60	4.27	4.95	4.37	5.81	6.10	6.40	7.65	8.68	9.49	10.22	10.96	11.47	12.06	12.28	14.12	16.11	17.72	18.98	20.15	21.11
10	1.10	1.40	1.62	1.77	1.91	1.99	2.35	2.65	2.87	3.09	3.31	3.46	3.82	4.56	5.22	5.81	6.25	6.62	6.99	8.23	9.34	10.22	11.03	11.84	12.43	13.02	13.31	15.22	17.43	19.12	20.59	21.85	22.95
11	1.18	1.47	1.69	1.91	2.06	2.13	2.50	2.79	3.09	3.31	3.53	3.95	4.04	4.92	5.59	6.18	6.69	7.13	7.58	8.68	9.86	10.88	11.84	12.58	13.39	14.05	14.34	16.25	18.61	20.59	22.20	23.61	24.71
12	1.25	1.54	1.84	2.06	2.21	2.28	2.65	3.02	3.31	3.53	3.75	3.67	4.19	5.15	5.96	6.62	7.13	7.72	8.00	9.12	10.37	11.47	12.50	13.39	14.19	14.93	15.30	17.21	19.78	21.92	23.68	25.15	26.40
13	1.32	1.62	1.91	2.13	2.35	2.43	2.79	3.16	3.46	3.75	3.97	4.19	4.34	5.44	6.25	6.91	7.58	8.16	8.46	9.49	10.86	12.06	13.16	14.05	15.00	15.74	16.11	18.02	20.81	23.17	25.08	26.69	28.02
14	1.32	1.69	1.99	2.28	2.50	2.50	2.87	3.16	3.60	3.90	4.19	4.56	4.49	5.59	6.55	7.28	7.94	8.53	8.97	9.79	11.25	12.88	13.75	14.56	15.74	16.55	16.99	18.84	21.85	24.35	26.40	28.10	29.57
15	1.32	1.77	2.06	2.35	2.57	2.65	2.94	3.38	3.75	4.05	4.34	4.63	4.63	5.81	6.77	7.58	8.31	8.90	9.41	10.00	11.62	13.09	14.34	15.44	16.40	17.28	17.72	19.56	22.80	25.45	27.65	29.49	31.04
16	1.40	1.84	2.13	2.43	2.65	2.79	3.02	3.46	3.90	4.19	4.49	4.78	4.71	5.96	7.06	7.87	8.61	9.27	9.78	10.30	11.91	13.46	14.78	15.96	16.99	18.02	18.46	20.15	23.61	26.40	28.83	30.82	32.44
17	1.40	1.84	2.21	2.50	2.79	2.87	3.09	3.66	3.97	4.34	4.71	4.92	4.78	6.10	7.21	8.09	8.90	9.56	10.15	10.44	12.21	13.83	15.22	16.55	17.58	18.61	19.05	20.74	24.42	27.36	29.86	31.99	33.76
18	1.40	1.84	2.28	2.57	2.87	2.94	3.16	3.60	4.04	4.19	4.78	5.07	4.78	6.25	7.35	8.38	9.19	9.93	10.51	10.51	12.43	14.12	15.59	16.99	18.09	19.20	19.71	21.18	25.08	28.24	30.89	33.10	35.01
19	1.40	1.84	2.28	2.65	2.94	3.02	3.16	3.68	4.19	4.46	4.92	5.22	4.78	6.33	7.58	8.53	9.41	10.15	10.81	10.51	12.50	14.27	15.89	17.36	18.53	19.86	20.23	21.55	25.60	28.97	31.77	34.13	36.19
20	1.32	1.91	2.35	2.72	3.02	3.00	3.16	3.75	4.19	4.63	5.00	5.44	4.78	6.33	7.65	8.68	9.63	10.44	11.11	10.51	12.58	14.49	16.11	17.65	18.98	20.15	20.14	21.77	26.11	29.71	32.58	35.08	37.22
21	1.32	1.91	2.35	2.72	3.02	3.16	3.16	3.75	4.27	4.71	5.07	5.44	4.71	6.33	7.72	8.83	9.86	10.66	11.33	10.37	13.31	14.56	16.40	17.87	19.27	20.52	21.11	21.99	26.48	30.23	33.17	35.97	38.17
22	1.25	1.91	2.35	2.72	3.09	3.24	3.16	3.75	4.27	4.78	5.15	5.52	4.56	6.33	7.80	8.90	10.00	10.81	11.62	10.22	12.58	14.56	16.47	18.02	19.49	20.81	21.48	22.06	26.77	30.74	33.19	36.70	38.98
23	1.25	1.84	2.35	2.79	3.09	3.24	3.09	3.75	4.27	4.78	5.22	5.55	4.34	6.25	7.80	8.97	10.08	11.03	11.77	9.93	12.43	14.56	16.40	18.17	19.71	21.11	21.70	21.99	27.07	31.04	34.42	37.29	39.72
24	1.18	1.84	2.35	2.79	3.16	3.31	3.02	3.68	4.27	4.78	5.22	5.66	4.19	6.18	7.72	9.05	10.15	11.11	11.91	9.63	12.21	14.34	16.40	18.24	19.78	21.26	21.99	21.92	27.07	31.33	34.79	37.88	40.38
25	1.10	1.77	2.28	2.79	3.16	3.31	2.94	3.60	4.19	4.78	5.22	5.66	4.15	6.03	7.72	9.05	10.15	11.18	11.99	9.19	11.91	14.27	16.25	18.17	19.78	21.33	22.06	21.71	26.99	31.48	35.08	38.25	40.89
26	1.03	1.69	2.28	2.72	3.16	3.31	2.79	3.53	4.19	4.71	5.22	5.66	3.82	5.88	7.58	8.97	10.15	11.18	12.06	8.68	11.47	13.97	16.11	17.87	19.78	21.33	21.99	21.25	26.92	31.48	35.30	38.54	41.46
27	0.88	1.62	2.20	2.72	3.09	3.31	2.65	3.45	4.12	4.63	5.15	5.66	3.60	5.66	7.43	8.90	10.15	11.18	12.13	8.16	11.03	13.53	15.81	17.80	19.56	21.26	21.99	20.74	26.62	31.41	35.38	38.69	41.56
28	0.81	1.54	2.13	2.65	3.09	3.24	2.50	3.31	3.97	4.56	5.15	5.59	3.24	5.44	7.35	8.75	10.08	11.18	12.13	7.51	10.51	13.09	15.44	17.43	19.34	21.11	21.84	20.15	25.89	31.18	35.30	38.83	41.78
29	0.66	1.47	2.06	2.57	3.02	3.24	2.43	3.16	3.82	4.49	5.00	5.52	3.19	5.15	7.06	8.60	9.93	11.03	12.06	6.77	9.86	12.58	15.00	17.14	19.12	20.89	21.62	19.49	25.74	30.89	35.08	39.76	41.78
30	0.51	1.32	1.99	2.50	2.94	3.16	2.13	2.94	3.68	4.34	4.92	5.44	2.43	4.18	6.77	8.38	9.71	10.80	11.99	5.96	9.12	11.91	14.42	16.70	18.68	20.52	21.33	18.61	25.08	30.45	34.79	38.54	41.70

Source: IS 2494, 1964.

TABLE 21-24
Small-diameter factor, F_b

Speed ratio range	Small diameter factor
1.000–1.019	1.00
1.020–1.032	1.01
1.033–1.055	1.02
1.056–1.081	1.03
1.082–1.109	1.04
1.110–1.142	1.05
1.143–1.178	1.06
1.179–1.222	1.07
1.223–1.274	1.08
1.275–1.340	1.09
1.341–1.429	1.10
1.430–1.562	1.11
1.563–1.814	1.12
1.815–2.948	1.13
≥1.949	1.14

Source: IS 2494, 1964.

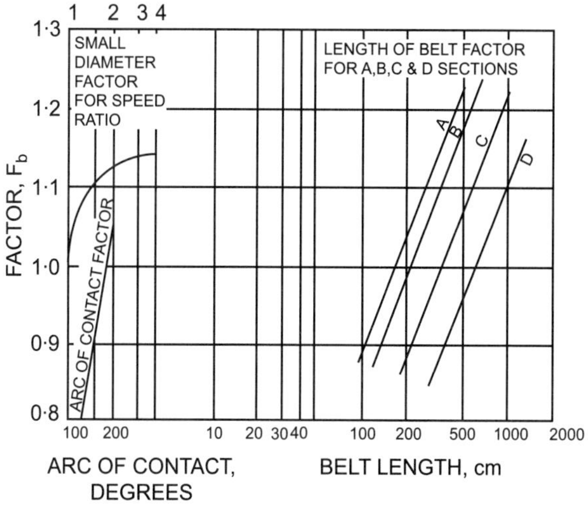

FIGURE 21-4(a) Factors for power rating of V-belt for use with Eqs. (21-30) to (21-35).

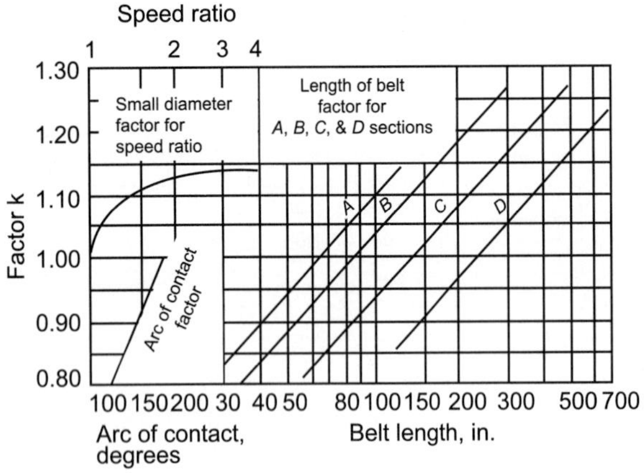

FIGURE 21-4(b) Factors for horsepower ratings of V-belts for use with Eqs. (21-35a) to (21-35e).

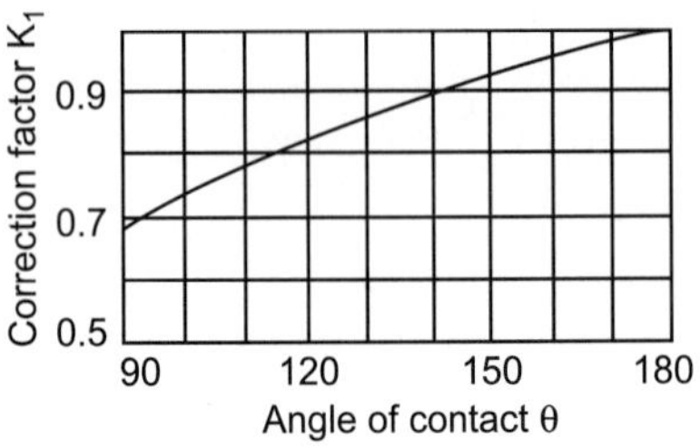

FIGURE 21-4(c) Correction factor K_1 for angle of contact.

TABLE 21-25
Correction factors for arc of contact, F_d

Arc of contact on smaller pulley, deg	Correction factor (proportion of 180° rating)		Arc of contact on smaller pulley, deg	Correction factor (proportion of 180° rating)	
	VV	V-flat		VV	V-flat
180	1.00	0.75	133	0.87	0.86
177	0.99	0.76	130	0.86	0.86
174	0.99	0.76	127	0.85	0.85
171	0.98	0.77	123	0.83	0.83
169	0.97	0.78	120	0.82	0.82
166	0.97	0.79	117	0.81	0.81
163	0.96	0.79	113	0.80	0.80
160	0.95	0.80	110	0.78	0.78
157	0.94	0.81	106	0.77	0.77
154	0.93	0.81	103	0.75	0.75
151	0.93	0.82	99	0.73	0.73
148	0.92	0.83	95	0.72	0.72
145	0.91	0.83	91	0.70	0.70
142	0.90	0.84	87	0.68	0.68
139	0.89	0.85	83	0.65	0.65
136	0.88	0.85			

Source: IS 2494, 1964.

TABLE 21-26
Correction factors for belt length, F_c

Nominal inside length, mm	Belt cross-section					Nominal inside length, mm	Belt cross-section				
	A	*B*	*C*	*D*	*E*		*A*	*B*	*C*	*D*	*E*
610	0.80	—	—	—	—	2159	1.05	0.99	0.90	—	—
660	0.81	—	—	—	—	2286	1.06	1.00	0.91	—	—
711	0.82	—	—	—	—	2438	1.08	—	0.92	—	—
787	0.84	—	—	—	—	2464	—	1.02	—	—	—
813	0.85	—	—	—	—	2540	—	1.03	—	—	—
889	0.87	0.81	—	—	—	2667	1.10	1.04	0.94	—	—
914	0.87	—	—	—	—	2845	1.11	1.05	0.95	—	—
965	0.88	0.83	—	—	—	3048	1.13	1.07	0.97	0.86	—
991	0.88	—	—	—	—	3150	—	—	0.97	—	—
1016	0.89	0.84	—	—	—	3251	1.14	1.08	0.98	0.87	—
1067	0.90	0.85	—	—	—	3404	—	—	0.99	—	—
1092	0.90	—	—	—	—	3658	—	1.11	1.00	0.90	—
1168	0.92	0.87	—	—	—	4013	—	1.13	1.02	0.92	—
1219	0.93	0.88	—	—	—	4115	—	1.14	1.03	0.92	—
1295	0.94	0.89	0.80	—	—	4394	—	1.15	1.04	0.93	—
1372	—	0.90	—	—	—	4572	—	1.16	1.05	0.94	—
1397	0.96	0.90	—	—	—	4953	—	1.18	1.07	0.96	—
1422	0.96	0.90	—	—	—	5334	—	1.19	1.08	0.96	0.94
1473	0.97	—	—	—	—	6045	—	—	1.11	1.00	0.96
1524	0.98	0.92	0.82	—	—	6807	—	—	1.14	1.03	0.99
1600	0.99	—	—	—	—	7569	—	—	1.16	1.05	1.01
1626	0.99	—	—	—	—	8331	—	—	1.19	1.07	1.03
1651	1.00	0.94	—	—	—	9093	—	—	1.21	1.09	1.05
1727	1.00	0.95	0.85	—	—	9855	—	—	1.23	1.11	1.07
1778	1.01	9.95	—	—	—	10617	—	—	1.24	1.12	1.09
1905	1.02	0.97	0.87	—	—	12141	—	—	—	1.16	1.12
1981	1.03	0.98	—	—	—	13665	—	—	—	1.18	1.14
2032	1.04	—	—	—	—	15189	—	—	—	1.20	1.17
2057	1.04	0.98	0.89	—	—	16713	—	—	—	1.23	1.19

Source: IS 2494, 1964.

TABLE 21-27
Correction factors for industrial service, F_a

		Type of driving unit					
		AC motors; normal torque, squirrel cage, synchronous and split phase DC motors; shunt-wound, multiple cylinder internal combustion engines >600 rpm			AC motors; high torque, high slip repulsion induction, single phase, series-wound and slip-ring DC motors; series-wound and compound wound; single-cylinder internal-combustion engines; multicylinder internal-combustion engines <600 rpm, line shafts, clutches, brakes, direct on-line starting		
Severity of service	**Type of driven machines**	**≤10 h**	**>10 to 16 h**	**>16 h and continuous service**	**≤10 h**	**>10 to 16 h**	**>16 h and continuous service**
Light-duty	Agitators for liquids, blowers, and exhausters, centrifugal pumps and compressors, fans up to 7.5 kW (10 hp) and light-duty conveyors	1.0	1.1	1.2	1.1	1.2	1.3
Medium-duty	Belt conveyors for sand, grain, etc; dough mixers; fans over 7.5 kW (10 hp); generators; line shafts; laundry machinery; machine tools; punches, presses and shears; printing machinery; positive-displacement rotary pumps; revolving and vibrating screens	1.1	1.2	1.3	1.2	1.3	1.4
Heavy-duty	Brick machinery, bucket elevators, exciters, piston compressors, conveyors (drag-pan-screw), hammer mills, paper mill beaters, piston pumps, positive displacement blowers, pulverizers, saw mill and woodworking machinery, and textile machinery	1.2	1.3	1.4	1.4	1.5	1.6
Extra-heavy-duty	Crushers (gyratory-jaw-roll), mills (ball-rod-tube), hoists, and rubber (calenders-extruders-mills) machinery	1.3	1.4	1.5	1.5	1.6	1.8

Note: This table gives only a few examples of particular machines. If an idler pulley is used, the following values must be added to the service factors:

Idler pulley on the slack side { inside: 0; outside: 0.1 }   Idler pulley on the tight side { inside: 0.1; outside: 0.2 }

Source: IS 2494, 1964.

TABLE 21-28
Nominal inside length, nominal pitch lengths and permissible length variations for V-belts

Nominal inside length, mm	Nominal pitch length, mm Cross-section					Pitch length variation	
	A	*B*	*C*	*D*	*E*	PLL[a]	MVL[b]
610	645						
660	696					+11.4	
711	747					−6.4	
787	823						
813	848					+12.5	
889	925	932				−7.5	
914	950						2.5
965	1001	1008					
991	1076						
1016	1051	1059				+14.0	
1067	1102	1110				−8.9	
1092	1128						
1168	1204	1212					
1219	1255	1262					
1295	1331	1339	1351			+16.0	
1372		1415				−9.0	
1397	1433	1440					
1422	1451	1466					
1473	1509						
1524	1560	1567	1580				
							5.0
1600	1636						
1626	1661					+17.8	
1651	1687	1694				−12.5	
1727	1763	1770	1783				
1778	1814	1821					
1905	1941	1948	1991				
1981	2017	2024					
2032	2068						
2057	2093	2101	2113			+30	
2159	2195	2202	2215			−16	7.5
2286	2322	2329	2342				
2438	2474		2494				
2464		2507					
2540		2583				+34	
2667	2703	2710	2723			−18	
2845	2880	2888	2901				
3048	3084	3091	3104	3127			10
3150			3205				
3251	3287	3294	3307	3330		+38	
3404			3459			−21	
3658	3693	3701	3713	3736			
4013		4056	4069	4092			
4115		4158	4171	4194		+43	
4394		4437	4450	4473		−24	
4572		4615	4628	4651			12.5

TABLE 21-28
Nominal inside length, nominal pitch lengths and permissible length variations for V-belts (*Cont.*)

Nominal inside length, mm	Nominal pitch length, mm, Cross section					Pitch length variation	
	A	*B*	*C*	*D*	*E*	PLL[a]	MVL[b]
4953		4996	5009	5032		+49	
5334		5377	5390	5413	5426	−28	
6045			6101	6124	6137		
6807			6863	5886	6899	+56	
7569			7625	7648	7661	−32	
8331			8387	8410	8423	+65	15
9093			9149	9172	9185		
9855				9934	9947	−37	
10617				10696	10709	+76	
12141				12220	12233	−43	
							17.5
13665				13744	13757	+89	
15189				15268	15281	−50	
16713				16792	16805	+105	
							−59

[a] Pitch length limit.
[b] Maximum variation in length within a matched set.
Source: IS 2494, 1964.

TABLE 21-29
Dimensions for standard V-grooved pulleys

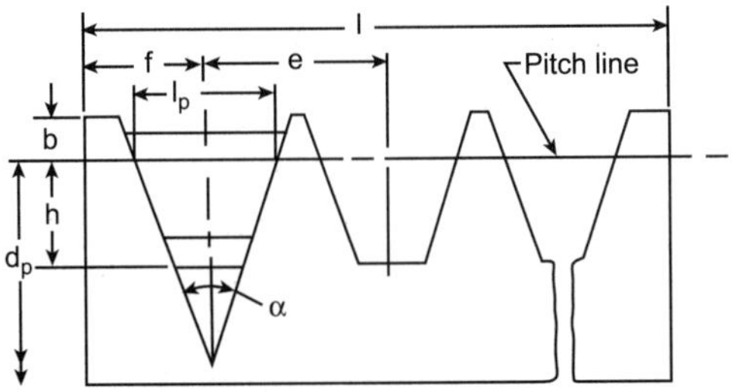

Groove section	Pitch width, l_p, min	Minimum height of groove above pitch line, b_{min}, mm	Minimum depth of groove below pitch line, *h*, min, mm	Center to center distance of grooves, *e*, mm	Edge of pulley to first groove center, *f*, mm
A	11	3.3	8.7	15 ± 0.3	10^{+2}_{-1}
B	14	4.2	10.1	19 ± 0.4	12.5^{+2}_{-1}
C	19	5.7	14.3	25.5 ± 0.5	17^{+2}_{+1}
D	27	8.1	19.9	37 ± 0.6	24^{+3}_{-1}
E	32	9.6	23.4	44.5 ± 0.7	29^{+4}_{-1}

Source: IS: 3142-1965.

TABLE 21-30A
Recommended standard pulley pitch diameters

Series of pitch diameters			Degree of preference[a] for pitch diameters, according to groove section					Series of pitch diameters			Degree of preference[a] for pitch diameters, according to groove section				
Nominal value,	Pitch diameter limits							Nominal value	Pitch diameter limits						
mm	Min, mm	Max, mm	*A*	*B*	*C*	*D*	*E*	mm	Min, mm	Max, mm	*A*	*B*	*C*	*D*	*E*
75	75	76.3	3					375	375	381.0		2	2	2	
80	80	81.3	3					400	400	406.4	1	1	1	1	
85	85	86.4	3					425	425	431.8			2		
90	90	91.4	1					450	450	457.2	2	2	2	1	
95	95	96.5	2					475	475	482.6				2	
100	100	101.6	1					500	500	508.8	1	1	1	1	1
106	106	107.7	2					530	530	538.5		3	3	3	2
112	112	113.8	1					560	560	569.0	2	2	2	2	1
118	118	119.9	2					600	600	609.6		2	2	2	2
125	125	127.0	1	2				630	630	640.0	1	1	1	1	1
132	132	134.1	2	2				670	670	680.7					2
140	140	142.2	1	1				710	710	721.4	2	2	2	2	1
150	150	152.4	2	2				750	750	762.0		2	2	2	3
160	160	162.6	1	1				800	800	812.8	3	1	1	1	1
170	170	172.7	3	2				900	900	914.4		2	2	2	2
180	180	182.9	1	1				1000	1000	1016.0		1	1	1	1
190	190	193.0	3	3				1060	1060	1077.0				2	
200	200	203.2	1	1	1			1120	1120	1137.9		3	2	2	2
212	212	215.4			2			1250	1250	1270.0			1	1	1
224	224	227.6	2	2	1			1400	1400	1422.4			2	2	2
236	236	239.8			2			1500	1500	1524.0				2	2
250	250	354.0	1	1	1			1600	1600	1625.6			1	1	1
265	265	269.2			2			1800	1800	1828.4				2	2
280	280	284.5	2	2	1			1900	1900	1930.4					2
300	300	304.8	2	2	2			2000	2000	2032.0				1	1
315	315	320.0	1	1	1			2240	2240	2275.8					2
355	355	360.7	2	2	2	1		2500	2500	2540.0					1

[a] *Key*: 1—first preference; 2—second preference; 3—not recommended
Source: IS 3142, 1965.

TABLE 21-30B
Standard V-belt sections

Belt section	Width, *a*, in	Thickness, *b*, in	Minimum sheave diameter, in	hp range, one or more belts
A	$\frac{1}{2}$	$\frac{11}{32}$	3.0	$\frac{1}{4}$–10
B	$\frac{21}{32}$	$\frac{7}{16}$	5.4	1–25
C	$\frac{7}{8}$	$\frac{17}{32}$	9.0	15–100
D	$1\frac{1}{4}$	$\frac{3}{4}$	13.0	50–250
E	$1\frac{1}{2}$	1	21.6	≥100

TABLE 21-30C
Inside circumferences of standard V-belts

Section	Circumference, in
A	26, 31, 33, 35, 38, 42, 46, 48, 51, 53, 55, 57, 60, 62, 64, 66, 68, 71, 75, 78, 80, 85, 90, 96, 105, 112, 120, 128
B	35, 38, 42, 46, 48, 51, 53, 55, 57, 60, 62, 64, 65, 66, 68, 71, 75, 78, 79, 81, 83, 85, 90, 93, 97, 100, 103, 105, 112, 120, 128, 131, 136, 144, 158, 173, 180, 195, 210, 240, 270, 300
C	51, 60, 68, 75, 81, 85, 90, 96, 105, 112, 120, 128, 136, 144, 158, 162, 173, 180, 195, 210, 240, 270, 300, 330, 360, 390, 420
D	120, 128, 144, 158, 162, 173, 180, 195, 210, 240, 270, 300, 330, 360, 390, 420, 480, 540, 600, 660
E	180, 195, 210, 240, 270, 300, 330, 360, 390, 420, 480, 540, 600, 660

TABLE 21-30D
Length conversion dimensions[a]

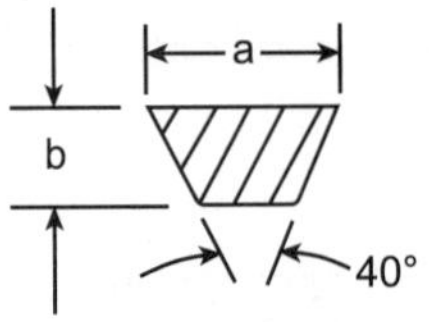

Belt section	*A*	*B*	*C*	*D*	*E*
Quantity to be added	1.3	1.8	2.9	3.3	4.5

[a] Add the values given above to the inside circumference to obtain the pitch length in inches.

TABLE 21-30E
Horsepower rating of standard V-belts

Belt section	Sheave pitch diameter, in	Belt speed, ft/min				
		1000	2000	3000	4000	5000
A	2.6	0.47	0.62	0.53	0.15	
	3.0	0.66	1.01	1.12	0.93	0.38
	3.4	0.81	1.31	1.57	1.53	1.12
	3.8	0.93	1.55	1.92	2.00	1.71
	4.2	1.03	1.74	2.20	2.38	2.19
	4.6	1.11	1.89	2.44	2.69	2.58
	≥5.0	1.17	2.03	2.64	2.96	2.89
B	4.2	1.07	1.58	1.68	1.26	0.22
	4.6	1.27	1.99	2.29	2.08	1.24
	5.0	1.44	2.33	2.80	2.76	2.10
	5.4	1.59	2.62	3.24	3.34	2.82
	5.8	1.72	2.87	3.61	3.85	3.45
	6.2	1.82	3.09	3.94	4.28	4.00
	6.6	1.92	3.29	4.23	4.67	4.48
	≥7.0	2.01	3.46	4.49	5.01	4.90
C	6.0	1.84	2.66	2.72	1.87	
	7.0	2.48	3.94	4.64	4.44	3.12
	8.0	2.96	4.90	6.09	6.36	5.52
	9.0	3.34	5.65	7.21	7.86	7.39
	10.0	3.64	6.25	8.11	9.06	8.89
	11.0	3.88	6.74	8.84	10.0	10.1
	≥12.0	4.09	7.15	9.46	10.9	11.1
D	10.0	4.14	6.13	6.55	5.09	1.35
	11.0	5.00	7.83	9.11	8.50	5.62
	12.0	5.71	9.26	11.2	11.4	9.18
	13.0	6.31	10.5	13.0	13.8	12.2
	14.0	6.82	11.5	14.6	15.8	14.8
	15.0	7.27	12.4	15.9	17.6	17.0
	16.0	7.66	13.2	17.1	19.2	19.0
	≥17.0	8.01	13.9	18.1	20.6	20.7
E	16.0	8.68	14.0	17.5	18.1	15.3
	18.0	9.92	16.7	21.2	23.0	21.5
	20.0	10.9	18.7	24.2	26.9	26.4
	22.0	11.7	20.3	26.6	30.2	30.5
	24.0	12.4	21.6	28.6	32.9	33.8
	26.0	13.0	22.8	30.3	35.1	36.7
	≥28.0	13.4	23.7	31.8	37.1	39.1

TABLE 21-30F
Belt-length correction factor, K_2 [a]

	Nominal belt length, in				
Length factor	***A* belts**	***B* belts**	***C* belts**	***D* belts**	***E* belts**
0.85	≤35	≤46	≤75	≤ 128	
0.90	38–46	48–60	81–96	144–162	≤195
0.95	48–55	62–75	105–120	173–210	210–240
1.00	60–75	78–97	128–158	240	270–300
1.05	78–90	105–120	162–195	270–330	330–390
1.10	96–112	128–144	210–240	360–420	420–480
1.15	120 and up	158–180	270–300	480	540–600
1.20		195 and up	330 and up	540 and up	660

[a] Multiply the rated horsepower per belt by this factor to obtain the corrected horsepower.

Particular	Formula
Number of belts	$i = \dfrac{PF_a}{P^* F_c F_d}$ (21-36) where P = drive power in kW Obtain F_d, F_c, and F_a from Tables 21-25, 21-26, and 21-27, respectively.
The diameter of larger pulley	$D = \dfrac{dn_1}{n_2}\eta$ (21-37)
Nominal pitch length of belt	$L = 2C + \dfrac{\pi}{2}(D+d) + \dfrac{(D-d)^2}{4C}$ (21-38)
For nominal inside length, nominal pitch lengths and permissible length variations for standard sizes of V-belts	Refer to Table 21-28.
Dimensions for standard V-grooved pulley	Refer to Table 21-29.
For small-diameter factor, for speed ratio and length of belt factor	Refer to Figs. 21-4a and 21-4b.
Recommend standard pitch diameters of pulleys	Refer to Table 21-30A.
For further data for design of V-belts in US Customary system units for use with Eqs (21-35a) to (21-35e)	Refer to Tables 21-30B and 21-30F, and Figs. 21-4b and 21-4c.
Center distance for a given belt length and diameters of pulleys	$C = \dfrac{L}{4} - \dfrac{\pi(D+d)}{8} + \sqrt{\left(\dfrac{L}{4} - \dfrac{\pi(D+d)}{8}\right)^2 - \dfrac{(D-d)^2}{8}}$ (21-39)
Maximum center distance	$C_{max} = 2(D+d)$ (21-40)
Minimum center distance	$C_{min} = 0.55(D+d) + t$ (21-41)

Particular	Formula

MINIMUM ALLOWANCES FOR ADJUSTMENT OF CENTERS FOR TWO TRANSMISSION PULLEYS

Lower limiting value	$C_L = C_{\text{nominal}} - 1.5\%L$	(21-42)
Higher limiting value	$C_H = C_{\text{nominal}} + 3\%L$	(21-43)

INITIAL TENSION

In order to give the initial tension, the belts may be stretched to	$\Delta L = 0.5 \text{ to } 1\%L$	(21-44)
Arc of contact angle	$\theta = 2\cos^{-1}\dfrac{D-d}{2C}$	(21-45)
	$\theta = 180^\circ - 60^\circ\left(\dfrac{D-d}{C}\right)$	(21-46)
For V-belt and pulley dimensions as per SAE J 636C standard	Refer to Table 21-31A and Fig. 21-5A, Tables 21-31B and 21-31C.	

SYNCHRONOUS BELT DRIVE ANALYSIS

The transmission ratio of synchronous belt drive

$$i = \frac{n_1}{n_2} = \frac{z_2}{z_1} = \frac{d'_2}{d'_1} \qquad \text{(21-46a)}$$

where

z_1, z_2 = number of teeth in smaller and larger pulley, respectively

d'_1, d'_2 = pitch diameter of smaller and larger pulley, respectively, m (in).

Datum length of synchronous belt

$$l = 2C\sin\frac{\theta}{2} + \frac{p}{2}\left(z_1 + z_2 + \frac{\theta}{90^\circ}(z_2 - z_1)\right) \qquad \text{(21-46b)}$$

$$l \approx 2C\,\frac{p}{2}(z_1 + z_2) + \left(\frac{p}{2\pi}\right)^2 \frac{(z_2 - z_1)^2}{l} \qquad \text{approximate (21-46c)}$$

$$l \approx pz_b \qquad \text{(21-46d)}$$

where

θ = angle of contact of belt, deg

p = pitch, m (in)

z_b = number of teeth in belt

The minimum number of meshing teeth: $z_b = 6$ to 8 teeth

Particular	Formula
For S1 synchronous belts and pulley dimensions and tolerances	Refer to Figs. 21-5B, 21-5C, 21-5D, 21-5E, 21-5F and Tables 21-31D(a) to 21-31D(i).
For the standard pitch according to ISO 5296 Standard	Refer to Table 21-31D(j).

TABLE 21-31D(j)
Standard pitch value

	Extra light *XL*	Light *L*	Heavy *H*	Extra heavy *XH*	Double extra heavy *XXH*
Belt pitch, in	$\frac{1}{4}$	$\frac{3}{8}$	$\frac{1}{2}$	$\frac{7}{8}$	$1\frac{1}{4}$
Nominal power kW	0.15	1.0	10	40	107

TABLE 21.31B
Standard belt center distance tolerances

Belt length		Tolerance on center distance	
mm	in	mm	in
≤1270	≤50	±3.0	±0.12
>1270 to 1524, incl	>50 to 60, incl	±4.1	±0.16
>1524 to 2032, incl	>60 to 80, incl	±4.8	±0.19
>2032 to 2540, incl	>80 to 100, incl	±5.6	±0.22

TABLE 21.31C
Maximum center distance for belts in a set

SAE size			
SI units	fps units	mm	in
6A	0.250	0.8	0.03
8A	0.315	0.8	0.03
10A	0.380	1.0	0.04
11A	0.440	1.0	0.04
13A	0.500	1.0	0.04
15A	11/16 (0.600)	1.5	0.06
17A	3/4 (0.660)	1.5	0.06
20A	7/8 (0.790)	1.5	0.06
23A	1 (0.910)	1.5	0.06

Source: *V-belts and Pulleys*, SAE J 636 C. Reprinted with permission from SAE Handbook, Part I, 1977, Society of Automotive Engineers, Inc.

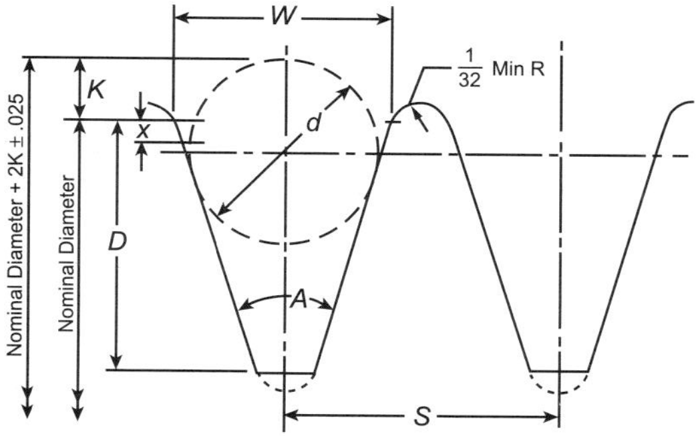

FIGURE 21-5A V-belt pulley dimensions.

Notes:

1. The sides of the groove are to be 125 μin (3.2 μm) A. A. maximum.
2. Radial run-out not to exceed 0.015 in (0.38 mm) full indicator movement (FIM). Axial run-out is not to exceed 0.015 in (0.38 mm) FIM. Run-out in the two directions is measured separately with a ball mounted under spring pressure to follow the groove as the pulley is rotated. Diameter, load, and overhang conditions may require or permit variations in the above specified run-out limits.
3. Bottom corner radii optional but, if used, it shall be below the depth, D.
4. In pulleys for use with belts in multiple on common centers, the diameters over the ball gages are not to vary from groove to groove in the same pulley more than 0.002 in/in (0.05 mm/25 mm) of diameter, with top limit of 0.012 in (0.30 mm) for diameters 6 in (152 mm) and above.
5. Centerline of groove is to be $90 \pm 2°$ with pulley axis.
6. The X dimension is radial. $2X$ is to be subtracted from the effective diameter to obtain "pitch diameter" for speed ratio calculation.

SAE size		Recommended minimum effective diameter		Groove angle deg. ±0.5 A,	Effective groove width W		Groove depth minimum D		Ball or rod diameter d		$2K$[d]		$2X$[b]		Groove spacing[a] ±0.38 S	
SI units	fps units	mm	in	deg	mm	in	mm	in	±0.013 mm	±0.0005 in	mm	in	mm	in	mm	in
6A	0.250	57	2.25	36	6.3	0.248	7	0.276	5.558	0.2188	4.16	0.164	1.0	0.04	8.00	0.315
8A	0.315	57	2.25	36	8.0	0.315	9	0.345	7.142	0.2812	5.63	0.222	1.3	0.05	10.49	0.413
10A	0.380	61	2.40	36	9.7	0.380	11	0.433	7.938	0.3125	3.77	0.154	1.5	0.06	13.71	0.541
11A	0.440	70	2.75	36	11.2	0.441	13	0.512	9.525	0.3750	5.88	0.231	1.8	0.07	15.01	0.591
13A	0.500	76	3.00	36	12.7	0.500	14	0.551	11.113	0.4375	7.99	0.314	2.0	0.08	16.79	0.661
15A	11/16	76	3.00	34	—	0.597	—	0.551	—	0.500	6.42	0.258	—	0.00	—	0.778
	(0.600)	>102	>4.00	36	15.2	—	14	—	12.70	—	7.02	0.280	0	—	19.76	—
		>152	>6.00	38	—	—	—	—	—	—	7.56	0.302	—	—	—	—
17A	3/4	76	3.00	34	—	0.660	—	0.630	—	0.5625	8.21	0.328	—	0.02	—	0.84
	(0.660)	>102	>4.00	36	16.8	—	15	—	14.288	—	8.82	0.352	6.5	—	21.36	—
		>152	>6.00	38	—	—	—	—	—	—	9.38	0.374	—	—	—	—
20A	7/8	89	3.50	34	—	0.785	—	0.709	—	0.6875	11.77	0.472	—	0.04	—	0.966
	(0.79)	>114	>4.50	36	20.0	—	18	—	17.463	—	12.42	0.496	1.0	—	24.54	—
		>152	>6.00	38	—	—	—	—	—	—	13.02	0.520	—	—	—	—
23A	1	102	4.00	34	—	0.910	—	0.827	—	0.8125	15.67	6.616	—	0.06	—	1.091
	(0.900)	>152	>6.00	36	23.1	—	21	—	20.638	—	16.33	0.642	1.5	—	27.71	—
		>203	>8.00	38	—	—	—	—	—	—	16.94	0.666	—	—	—	—

[a] Pulley effective diameters below those recommended should be used with caution, because power transmission and belt life may be reduced.
[b] $2X$ is to be subtracted from the effective diameter to obtain "pitch diameter" for speed ratio calculation.
[c] These values are intended for adjacent grooves of the same effective width (W). Choice of pulley manufacture or belt design parameter may justify variance from these values. The S dimension shall be the same on all multiple groove pulleys in a drive using matched belts.
[d] $2K$ dimensions are calculated in millimeters.

Particular	Formula
For determining the center distance of synchronous belt pulleys.[a]	Refer to Fig. 21-5F.[a]
The distance from belt pitch line to the pulley—tip circle radius (Fig. 21-5C)	$a = \frac{d'}{2} - \frac{d_o}{2}$ (21-46e)
The permissible initial tensioning force range F_A	$F_u \leq F_A \leq 1.5F_w$ (21-46f) where F_u = the transmissible peripheral force, kN (lbf) F_w = the effective shaft tensioning force, kN (lbf)
The belt side-force ratio	$\frac{F_1}{F_2} \cong 5$ (21-46g) where F_1 = tension belt on tight side of synchronous belt, kN (lbf) F_2 = tension belt on slack side of synchronous belt, kN (lbf)

[a] Courtesy: J. E. Shigley and C. R. Mischke, *Standard Handbook of Machine Design*, 2nd edition, McGraw-Hill Publishing Company, New York, 1996.

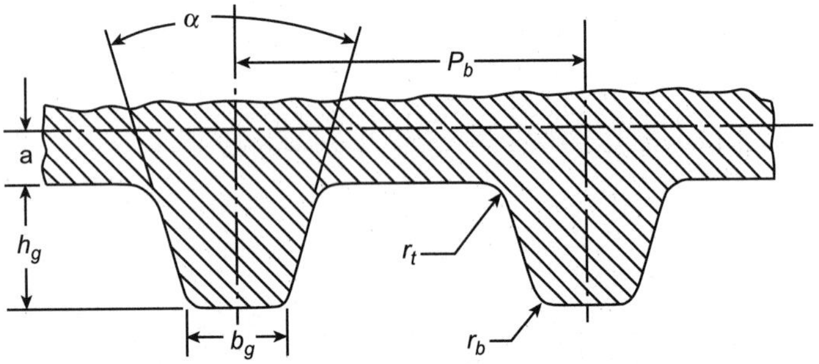

FIGURE 21-5B Pulley generating tool rack form

TABLE 21.31D (a)
Pulley generating tool rack form dimensions (mm)

Pulley section	Diameter range (No. of grooves)	P_b Pitch ±0.003	α ±0.25 deg	h_g +0.05 −0.00	b_g +0.05 −0.00	r_b ±0.03	r_t ±0.03	2a
ST	≥10	9.525	40	2.13	3.10	0.86	0.53	0.762
SU	14–19	12.700	40	2.59	4.24	1.47	1.04	1.372
SU	> 19	12.700	40	2.59	4.24	1.47	1.42	1.372
STA	≥ 19	9.525	40	2.13	3.10	0.86	0.71	1.372

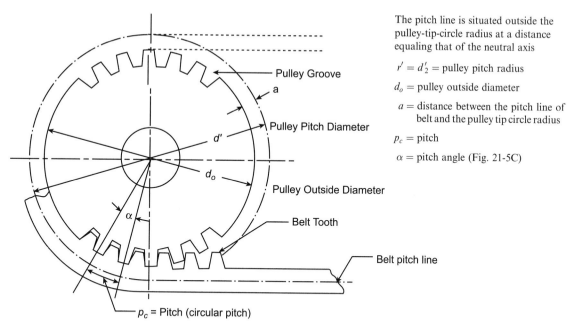

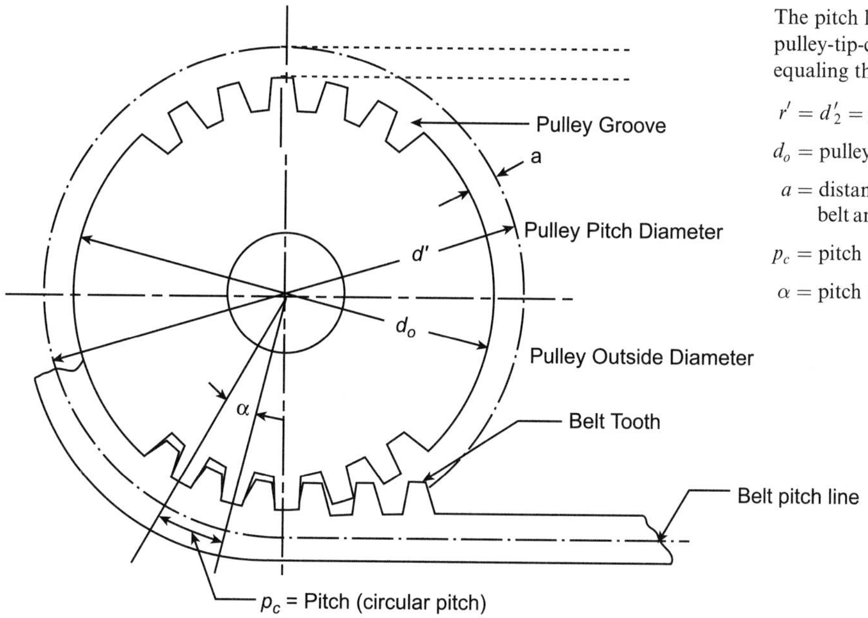

FIGURE 21-5C Pulley dimensions

The pitch line is situated outside the pulley-tip-circle radius at a distance equaling that of the neutral axis

$r' = d'_2$ = pulley pitch radius

d_o = pulley outside diameter

a = distance between the pitch line of belt and the pulley tip circle radius

p_c = pitch

α = pitch angle (Fig. 21-5C)

TABLE 21.31D (b)
Pulley tolerance (mm)

	Pitch to pitch tolerance	
Outside diameter range	**Adjacent grooves**	**Accumulative over 90°**
≤50, incl	±0.03	±0.09
>50 to 100, incl	±0.03	±0.11
>100 to 175, incl	±0.03	±0.13
>175 to 300, incl	±0.03	±0.15

Outside diameter	*Tolerance*
Up to 50 mm, incl	+0.05 to 0.00 mm
For each additional 25 mm or portion thereof	+0.025 to 0.00 mm
Outside diameter runout	
Up to 75 mm, incl outside diameter	0.08 mm (max)
For each additional 25 mm or portion thereof	0.01 mm (max)
Axial runout[a] *(side wobble)*	
Up to 250 mm, incl outside diameter	0.02 mm per 25 mm of diameter
For each additional 25 mm outside diameter over 220 mm ad 0.01 mm	add 0.01 mm
Diametrical taper	
0.01 mm per 10 mm of face width	
Groove helix	
0.01 mm per 10 mm of face width	

[a] Full indicator movement

TABLE 21.31D (c)
Nominal belt dimensions (mm) (Fig. 21-5C)

Belt section	Pitch	h_b	2β deg	h_t	b_t	r_{bb}	r_{bt}
ST	9.525	3.6	40	1.9	0.5	0.5	
SU	12.700	4.1	40	2.3	4.4	1.0	1.0
STA	9.525	4.1	40	1.9	3.2	0.5	0.5

TABLE 21.31D (d)
Belt width tolerances (mm)

	Belt length range	
Belt width	≤840, incl	>840 to 1680, incl
≤40, incl	+0.6 −0.6	+0.6 −0.6
>40 to 50, incl	+0.8 −0.8	+1.0 −1.0

TABLE 21.31D (e)
Measuring pulley dimensions, (mm)

Belt section	No. of grooves	Pitch circumference	Outside diam, ±0.013	Outside diam, runout FIM,[a] max	Axial runout (side wobble) FIM,[a] max	Min clearance[b]
ST	16	152.40	47.748	0.013	0.025	0.33
SU	20	254.00	79.479	0.013	0.025	0.38
STA	20	190.50	59.266	0.013	0.025	0.33

[a] Full indicator movement.
[b] See Fig. 21.5.

TABLE 21.31D (f)
Total measuring force (N)

Belt section	Belt width (mm)																
	8	10	12	14	16	18	19	20	22	25	28	30	33	35	40	45	50
ST	55	75	100	125	145	165	175	185	210	240	275	295	330	355	410	470	530
SU	—	—	245	300	370	420	445	475	530	610	700	750	840	900	1050	1200	1350
STA	—	—	245	300	370	420	445	475	530	610	700	750	840	900	1050	1200	1350

TABLE 21.31D (g)
Minimum recommended pulley diameters and flange dimensions (mm)

Pulley section	Pitch diam	Min. grooves	Min. pitch diam	Min. outside diam	Min. flange thickness	Min. flange height
ST	9.525	10	30.32	29.56	1.3	1.6
SU	12.700	14	56.60	55.23	1.3	2.0
STA	9.525	19	57.61	56.23	1.3	2.4

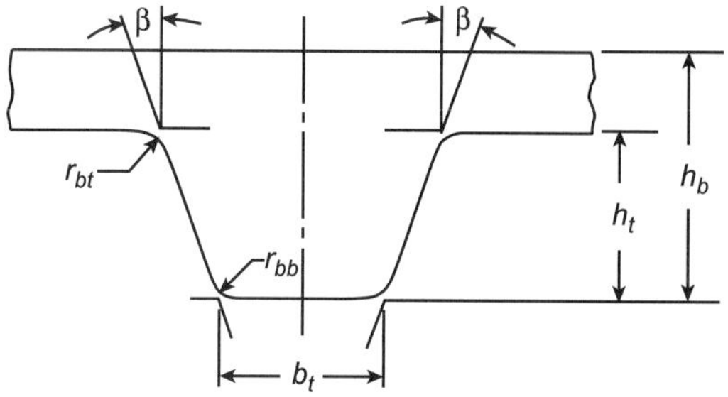

FIGURE 21-5D Belt section

TABLE 21.31D (h)
Belt length tolerances (mm)

Belt length range	Tolerance on belt pitch length
≤400, incl	±0.46
>400 to 520, incl	±0.51
>520 to 770, incl	±0.61
>770 to 1020, incl	±0.66
>1020 to 1270, incl	±0.76
>1270 to 1525, incl	±0.81
>1525 to 1780, incl	±0.86
>1780 to 2040, incl	±0.91
>2040 to 2300, incl	±0.97
>2300 to 2560, incl	±1.02
>2560 to 3050, incl	±1.12

TABLE 21.31D (i)
Pulley groove tolerances (mm) (Fig. 21-5D)

Pulley section	Top curvature band width	Max. top radius tolerance	Flank band width	Bottom curvature band width	Depth band width	Upper reference depth
ST	0.04	±0.1 −0.0	0.05	0.05	0.05	0.5
SU	0.04	±0.1 −0.0	0.05	0.05	0.05	0.8
STA	0.04	±0.1 −0.0	0.05	0.05	0.05	0.5

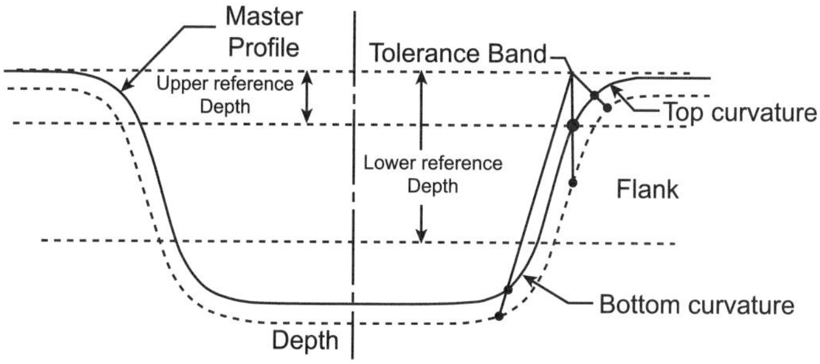

FIGURE 21-5E Pulley groove profile.

Source: *Synchronous Belts and Pulleys*, SAE J 1313 Oct. 80. Reprinted with permission from SAE Handbook, Part I, Society of Automotive Engineers, Inc., 1997.

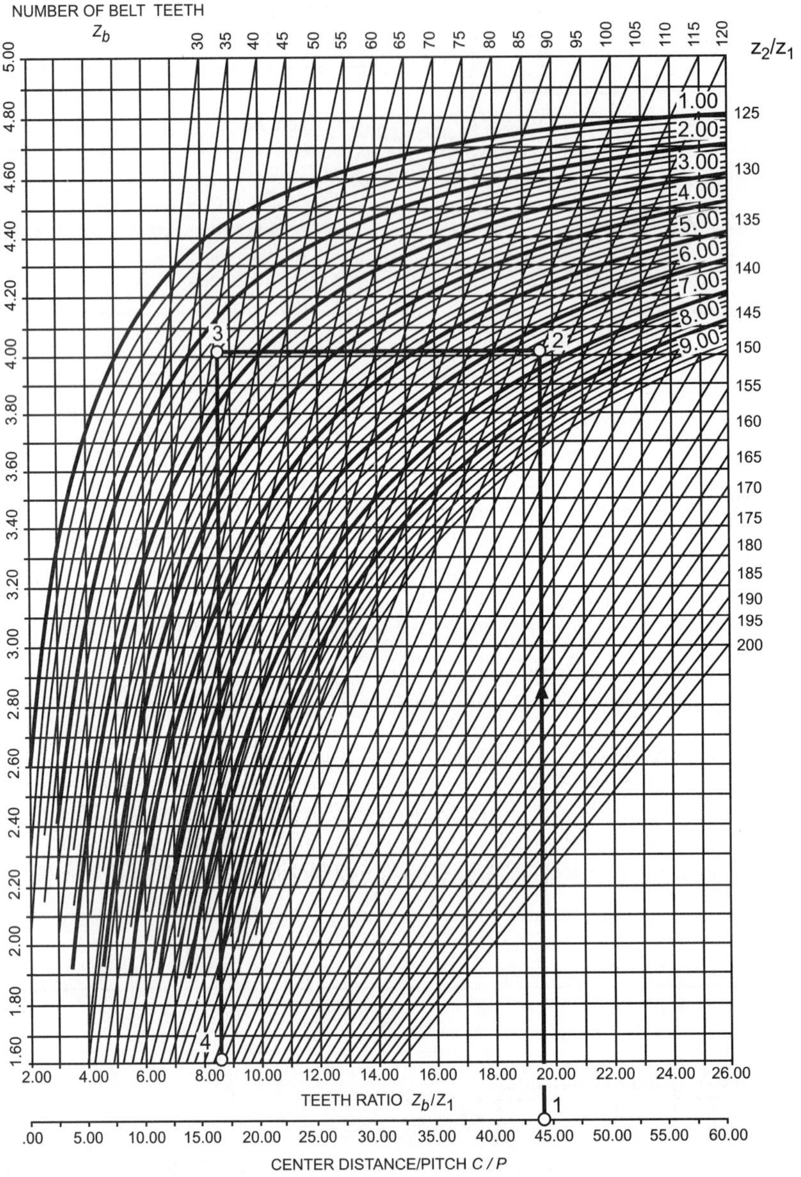

FIGURE 21-5F Determination of center distance of synchronous belts.

Source: J. E. Shigley and C. R. Mischke, *Standard Handbook of Machine Design*, 2nd edition, McGraw-Hill Book Company, New York, 1996.

Particular	Formula
The power transmitted by synchronous belt	$P = \frac{P_s}{C_s}$ (21-46h) where P_s = standard capacity of the selected belt, kW (hp) C_s = service correction factor

CONVEYOR (Tables 21-12, 21-14, 21-20, and 21-31)

The average capacity, C, of conveyor in m^3 (in)^3 per hour at 0.5 m/s (100 fpm) speed

For flat belts	when a_1 in m	$C = 70a_1^2$	SI	(21-47a)
	when a_1 in in	$C = 2756a_1^2$	USCS	(21-47b)
	when a_1 in mm	$C = 0.7 \times 10^5 a_1^2$	SI	(21-47c)
For belts on idlers	when a_1 in m	$C = 88a_1^2$	SI	(21-48a)
	when a_1 in in	$C = 3465a^2$	USCS	(21-48b)
	when a_1 in mm	$C = 0.88 \times 10^5 a_1^2$	SI	(21-48c)
For belts on three- to five-step idlers	when a_1 in m	$C = 132a_1^2$ to $154a_1^2$	SI	(21-49a)
	when a_1 in in	$C = 5158a_1^2$ to $6063a_1^2$	USCS	(21-49b)
	when a_1 in mm	$C = 1.32 \times 10^5 a_1^2$ to $1.54 \times 10^5 a_1^2$	SI	(21-49c)

TABLE 21-31
Maximum inclination of belt conveyors

Material conveyed	Maximum inclination, deg	Material conveyed	Maximum inclination, deg
Briquets and egg-shaped material	12	Glass batch	20
Wet-mixed concrete	15	Run-of-mine coal	22
Sized coal	8	Run-of-bank gravel	22
Washed and screened gravel	18	Crushed ore	25
Loose cement	20	Crushed stone	20
Crushed and screened coke	20	Tempered foundry sand	25
Sand	20	Wood chips	28

Particular	Formula
The power required by a horizontal belt conveyor	$$P = \left[(W_I + 2W_B + W_L)\mu \frac{d}{D}\right] \frac{vL}{1000} + P_T$$ **SI** (21-50a) where W in N/m, v in m/s, L in m, and P in kW $$P = \left[(W_I + 2W_B + W_L)\mu \frac{d}{D}\right] \frac{vL}{102} + P_T$$ **Metric** (21-50b) where W in kgf/m, v in m/s, L in m, and P in kW $$P = \left[(W_I + 2W_B + W_L)\mu \frac{d}{D}\right] \frac{vL}{33{,}000} + P_T$$ **USCS** (21-50c) where W in lbf/in, v in ft/min, L in in, and P in hp where μ = coefficient of friction of idler bearing = 0.15 for roller bearings = 0.35 for grease lubricated idlers

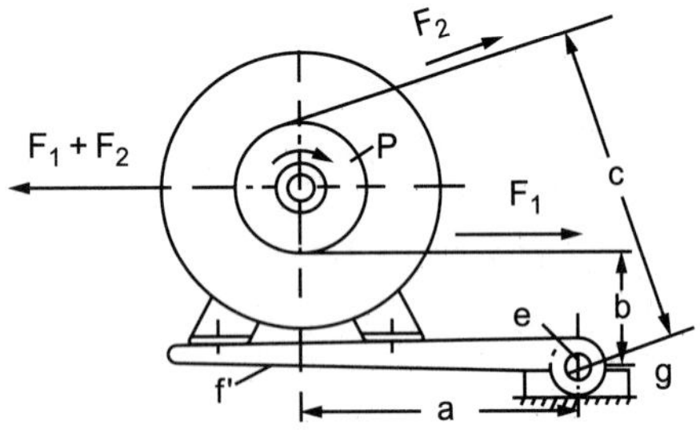

FIGURE 21-5 Rockwood pivoted motor base.

SHORT CENTER DRIVE

Rockwood drive (Fig. 21-5)

Particular	Formula
The value of F_1	$$F_1 = \frac{aW + cF_n}{c+b} \quad (21\text{-}51)$$
The value of F_2	$$F_2 = \frac{aW - bF_n}{c+b} \quad (21\text{-}52)$$
The pivot-arm length for motor of weight W	$$a = \frac{F_n\left(b\frac{F_1}{F_2} + c\right)}{W\left(\frac{F_1}{F_2} - 1\right)} \quad (21\text{-}53)$$ where F_n = required net pull, kN (lbf) W = weight of the motor, kN (lbf)

Particular	Formula

ROPES

Manila rope (Tables 21-32 and 21-34)

Particular	Formula
The ultimate load	$P_u = 48053d^2$ **SI** (21-54a) where d in m and P_u in kN $P_u = 7000d^2$ **USCS** (21-54b) where d is diameter of rope in in and P_u in lbf
The maximum tension on the tight side	$F_1 = 137.5 \times 10^4 d^2 = F + \frac{F}{2} + F_c$ **SI** (21-55a) where d in m and F_1 in N $F_1 = 200d^2 = F + \frac{F}{2} + F_c$ **USCS** (21-55b) where d in in and F_1 in lbf $F_1 = 0.14d^2$ **Customary Metric** (21-55c) where d in mm and F_1 in kgf
Power transmitted	$P = v(0.6 - 6.7 \times 10^{-4} F_c)$ **SI** (21-56a) where F_c in N, P in kW, and v in m/s $P = \frac{2v}{10^5}(200 - F_c)$ **USCS** (21-56b) where F_c in lbf and P in hp Refer to Table 21-32 for F_c = values of coefficients for manila rope

Hemp ropes

Particular	Formula
The load on the hemp rope	$F = \frac{\pi d^2}{4}\sigma_{br}$ (21-57) where σ_{br} = breaking stress, MPa (psi) = 9.81 MPa (1.42 kpsi) for white rope = 8.82 MPa (1.28 kpsi) for tarred rope

TABLE 21-32
Value of coefficient F_c for manila rope

Velocity, mps	7.50	10.00	12.50	15.00	17.50	20.00	22.50	25.00	27.50	30.00	32.50	35.00
Coefficient, F_c	2.96	5.40	8.44	12.60	16.10	21.00	26.55	32.89	39.69	41.17	55.34	64.40

Particular	Formula
The load on the hemp rope in terms of nominal diameter of rope	$F = 7.7 \times 10^6 d^2$ for white rope **SI** (21-58a) where d in m and F in N $F = 1120 d^2$ **USCS** (21-58b) where d in in and F in lbf $F = 7 \times 10^6 d^2$ for tarred rope **SI** (21-58c) where d in m and F in N $F = 1020 d^2$ **USCS** (21-58d) where d in in and F in lbf

HOISTING TACKLE

The effort on the rope in case of single-sheave pulley (Fig. 21-6)

$$P = \left(\frac{D + \mu d + 2s}{D - \mu d - 2s'}\right) Q = CQ \quad (21\text{-}59)$$

Refer to Table 21-33 for C.

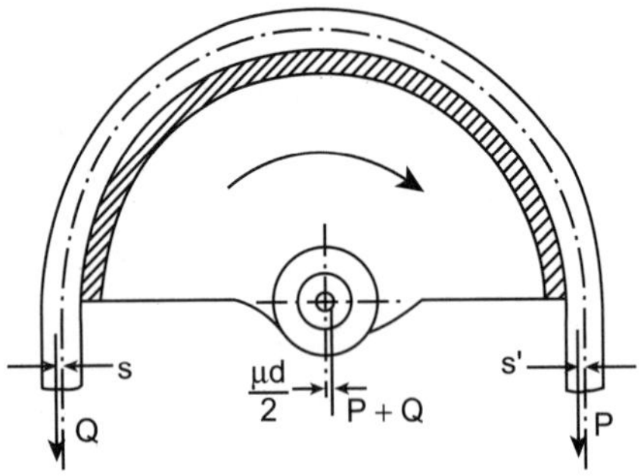

FIGURE 21-6 Rope passing over sheave.

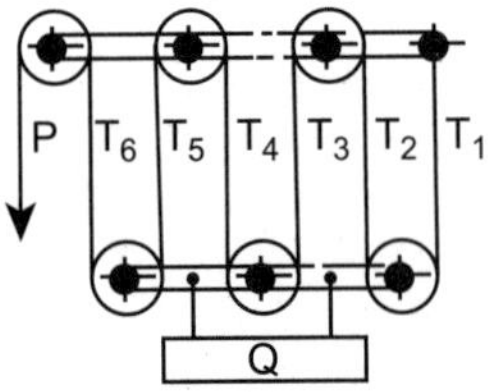

FIGURE 21-7 Load on a hoist.

The effort on the rope in a hoist for raising the load (Fig. 21-7)

$$P = \frac{C^n (C - 1)}{C^n - 1} Q \quad (21\text{-}60)$$

The pull required on the rope in a hoist for lowering the load

$$P' = \frac{C - 1}{C(C^n - 1)} Q \quad (21\text{-}61)$$

TABLE 21-33
Value of C

Manila rope	1.15
Wire rope	1.07
Dry chain	1.10
Greased chain	1.04

Particular	Formula
Efficiency of hoist	$$\eta = \frac{C^n - 1}{nC^n(C-1)} \qquad (21\text{-}62)$$ where n = number of times a rope passes over a sheave

Continuous system Fig. (21-8)

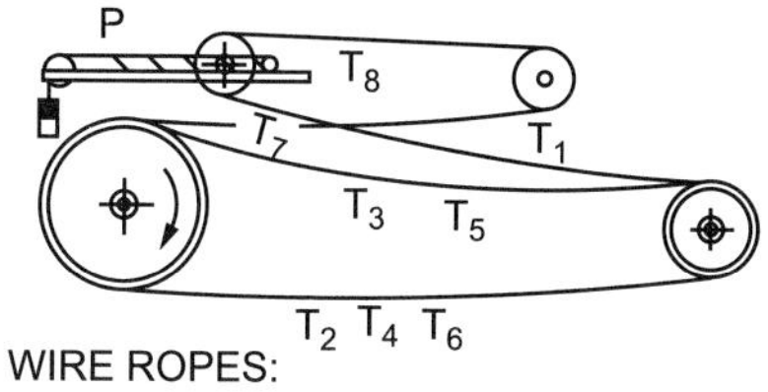

FIGURE 21-8 Continuous system.

In the continuous system one continuous rope passes around the driving and driven sheaves several times, in addition to making one loop about tension pulley located on a traveling carriage.

Particular	Formula
The relation between ultimate load, bending and service load in wire rope	$$\frac{P_u}{n} \geq P_b + P_s \qquad (21\text{-}63a)$$
The bending load	$$P_b = kA\frac{d_w}{D} \qquad (21\text{-}63b)$$ where $k = 82728.5\,\text{MPa}$ (12 Mpsi)
Another formula connecting ultimate strength of rope, tensile load on rope (P), dimensions of the rope, wire, and sheave diameter	$$P_u = \frac{P}{\dfrac{1}{n'} - \left(\dfrac{d}{D}\right)\left(\dfrac{d_w}{d}\right)\dfrac{E'}{\sigma_u}} \qquad (21\text{-}63c)$$ where D = minimum diameter of sheave or pulley, m (in); n' = stress factor = nk_d; n = safety factor; k_d = duty factor obtainable from Table 21-35
Area of useful cross-section of the rope	$$A = \frac{P}{\dfrac{\sigma_u}{n'} - \left(\dfrac{d}{D}\right)\left(\dfrac{d_w}{d}\right)E'} \qquad (21\text{-}63d)$$
The approximate ultimate strength of plow-steel ropes	$P_u = 524{,}000d^2$ for 6×7 and 6×19 ropes **SI** (21-64a) where P_u in kN and d in m
	$P_u = 76d^2$ **USCS** (21-64b) where P_u in lbf and d in in
	$P_u = 517{,}800d^2$ for 6×37 ropes **SI** (21-64c) where P_u in kN and d in m

TABLE 21-34
Manila rope

Size designation $(C)^a$ mm	Number of yards per strand	Linear density kilotex	Pitch		Breaking load					
					Grade 1		Grade 2		Grade 3	
			$2.6C^a$ π mm	$3.2C^a$ π mm	kN	kgf	kN	kgf	kN	kgf
25	3	53	20.7–25.5		5.4	546	4.7	483	4.1	419
32	4	66	26.5–32.6		6.9	711	6.2	635	5.5	559
35	5	89	29.0–36.7		8.9	902	7.8	800	6.9	699
38	6	107	31.5–38.7		10.5	1,067	9.3	953	8.2	838
41	7	120	34.0–41.8		12.3	1,257	11.0	1,118	9.6	978
44	8	138	36.4–44.8		14.2	1,448	12.6	1,283	11.0	1,118
51	11	191	42.2–52.0		19.9	2,032	17.7	1,803	15.4	1,575
57	13	226	47.2–58.1		23.9	2,439	21.2	2,159	18.4	1,880
64	17	294	53.0–65.2		31.6	3,226	28.1	2,870	24.7	2,515
70	20	346	58.0–71.3		37.6	3,836	33.4	3,404	29.1	2,972
76	24	413	62.9–77.4		44.8	4,572	39.9	4,064	34.9	3,556
83	28	489	68.7–M.6		52.1	5,309	46.3	4,725	40.6	4,140
89	33	569	73.7–90.7		59.5	6,071	53.1	5,410	46.3	4,725
95	37	635	78.7–96.8		68.0	6,935	60.5	5,172	52.8	5,383
102	43	742	84.5–104.0		76.5	7,798	68.0	6,935	59.5	6,071
108	48	831	89.4–110.1		85.2	8,687	75.7	7,722	66.3	6,757
114	54	933	94.4–116.2		95.4	9,729	84.7	8,636	74.2	7,570
121	60	1,090	100.2–123.3		105.1	10,719	93.4	9,525	81.7	8,332
127	67	1,159	105.2–129.4		116.1	11,837	103.1	10,516	95.2	9,703
140	81	1,329	116.0–142.7		139.0	14,174	123.6	12,599	108.1	11,024
152	96	1,661	125.9–154.9		163.9	16,714	145.5	14,834	127.0	12,955
165	113	1,954	136.6–168.2		190.8	19,457	169.4	17,273	148.0	15,088
178	131	2,265	147.4–181.4		219.7	22,404	195.3	19,915	170.9	17,425
203	171	2,958	168.1–206.9		282.5	28,805	251.1	25,604	219.7	22,404
229	216	3,736	189.6–233.8		353.2	36,019	313.9	32,005	274.5	27,992
254	267	4,620	210.3–258.8		432.9	44,147	384.6	39,219	336.3	34,292
279	323	5,583	231.0–284.3		520.1	53,038	462.3	47,145	404.5	41,252
305	384	6,640	252.5–360.5		616.8	62,893	548.0	55,883	479.3	48,872
330	451	7,800	273.2–336.3		719.9	73,409	639.7	65,230	559.5	57,051
356	523	9,044	294.8–362.8		829.5	84,586	737.3	75,188	645.2	65,789
381	600	10,376	315.5–388.3		953.1	97,185	846.9	86,364	740.8	75,543
406	683	11,811	336.2–413.8		1081.6	10,292	961.8	98,049	841.5	85,805
432	771	13,335	357.7–440.3		1216.1	24,009	1081.1	110,241	946.1	96,474
457	864	14,943	378.4–465.7		1362.1	38,894	1210.6	123,450	1059.2	108,006

[a] C stands for nominal circumference of the rope.

TABLE 21-35
Duty factor and life of mechanism of electric wire rope hoists

Mechanism class	Duty factor		Average life	
	Strength	Wear	Running h/day	Total life *h*, over
1	1.0	0.4	0.5	2500
2	1.2	0.5	0.5	9000
3	1.4	0.6	3.0	20000
4	1.6	0.7	over 6	40000

Source: IS 3938, 1967.

Particular	Formula
	$P_u = 75d^2$ USCS (21-64d) where P_u in lbf and d in in
The nominal bearing pressure	$p = \dfrac{2P_t}{D_r D_i} \leq C\sigma_u$ (21-65) where $C = 0.0015$ Refer to Table 21-33 for C.

DRUMS

Wire rope drum

The number of turn on the drum for one rope member (Fig, 21-9)	$n = \dfrac{iS}{\pi D} + 2$ (21-66)
The length of the drum	$l = \left(\dfrac{2iS}{\pi D} + 7\right)p$ for one rope (21-67a) $l = \left(\dfrac{2iS}{\pi D} + 12\right)p + p_1$ for two ropes (21-67b) where S = height to which the load is raised, m (in)

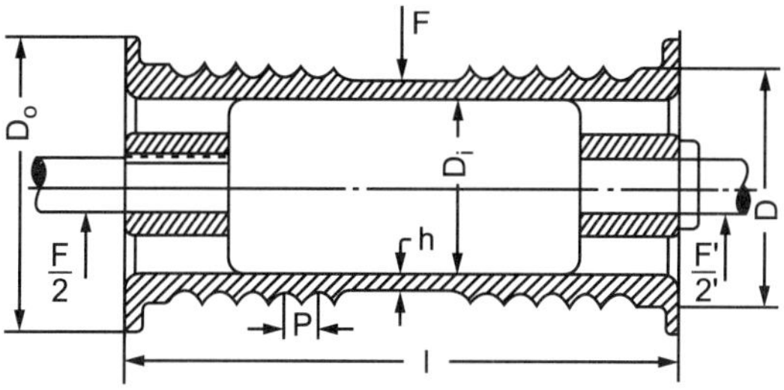

FIGURE 21-9 Wire rope drum

The minimum diameter of groove of sheaves and drums (d)	$d_{gs} = d + 0.8\,\text{mm}$ to $d_0 + 3.2\,\text{mm}$
The thickness of wall of drum made of cast iron	$h = 0.02D + 0.6$ to $1.0\,\text{cm}$ (21-68)
The outside diameter of the drum (Fig. 21-9)	$D_o = (D + 6d)$ (21-69)
The depth of groove in drum or sheave	$h_1 < 1 - 1.5d$ (21-70)
The outside diameter of sheave (d_{os})	$d_{os} = d_s + 2h_1$ where d_s = minimum diameter of sheave, m

Stresses developed in drum

The maximum bending stress	$\sigma_b = \dfrac{8FlD}{\pi(D^4 - D_i^4)}$ (21-71)
The maximum torque on the drum	$M_t = F\left(\dfrac{D+d}{2}\right)$ (21-72) where d = diameter of rope

Particular	Formula
The maximum shear stress	$\tau = \frac{16M_t D}{\pi(D^4 - D_i^4)}$ (21-73)
The crushing stress	$\sigma_c = \frac{F}{ph}$ (21-74) where p = pitch of the grooves on the drum
The combined stress according to normal stress theory	$\sigma = \sqrt{\sigma_b^2 + \sigma_c^2 + 4\tau^2} \le \sigma_d$ (21-75) where σ_d = design stress

HOLDING CAPACITY OF WIRE ROPE REELS

The rope capacity (L) in meters in any size length may be calculated by the formula	$L = \frac{\pi(H + D_r)WH}{1000d}$ (21-76)

WIRE ROPE CONSTRUCTION

For wire rope strand construction, diameter, weight, breaking load for different purposes	Refer to Tables 21-36 to 21-39 and Figs. 21-10 to 21-16.
For wire rope data, factor of safety, values of C, and application	Refer to Tables 21-40 to 21-45.

CHAINS

Hoisting chains

The working load for the ordinary steel common coil chain	$P_w = 84{,}800d^2$ **SI** (21-77a) where d in m and P_u in kN $P_w = 12{,}300d^2$ **USCS** (21-77b) where d in in and P_u in lbf $P_w = 8.65d^2$ **Customary Metric** (21-77c) where d in mm and P_u in kgf
The working load for stud chain	$P_w = 60{,}310d^2$ **SI** (21-78a) where d in m and P_u in kN $P_w = 8750d^2$ **USCS** (21-78b) where d in in and P_u in lbf $P_w = 6.15d^2$ **Customary Metric** (21-79) where d in mm and P_u in kgf

TABLE 21-36
Steel wire ropes (from Indian standards)

Strand construction	Diameter of rope mm	Approx. weight N/m	Approx. weight kgf/m	Nominal breaking strength of rope — Tensile strength of wire 1568–1716 MPa (160–175 kgf/mm²) kN	1568–1716 MPa tf	1716–1863 MPa (175–190 kgf/mm²) kN	1716–1863 MPa tf
General Engineering Purposes							
Group 6 × 19	8	2.4	0.24	33.3	3.4	36.3	3.7
6 × 12/6/1	10	4.3	0.44	64.7	6.6	70.6	7.2
6 × 1216 + 6F/ 1	12	5.3	0.54	84.3	8.6	92.2	9.4
6 × 9/9/1	14	7.5	0.76	106.9	10.9	116.7	11.9
	16	9.2	0.94	131.4	13.4	144.2	14.7
	18	12.3	1.25	189.3	19.3	206.9	21.1
6 × 10/5 + 5F/1	20	14.4	1.47	221.6	22.6	241.2	24.6
(Fig. 21-10)	22	18.0	1.84	254.0	25.9	278.5	28.4
	24	20.9	2.13	294.2	30.0	323.6	33.0
	25	23.6	2.41	333.4	34.0	368.7	37.6
	29	29.9	3.05	423.6	43.2	462.9	47.2
	32	36.8	3.75	522.7	53.3	570.7	58.2
	35	44.6	4.55	623.5	64.6	692.3	70.6
	38	53.3	5.43	752.2	76.7	826.7	84.3
	41	62.5	6.37	886.5	90.4	971.8	99.1
	44	72.4	7.38	1026.8	104.7	1125.8	114.8
	48	83.2	8.48	1175.8	119.9	1295.5	132.1
	51	94.5	9.64	1345.6	137.2	1474.9	150.4
	54	106.8	10.89	1514.1	154.4	1664.2	269.7
Group 6 × 37	10	4.4	0.45	60.8	6.2	66.7	6.8
6 × 14/7 and	12	5.9	0.60	79.4	8.1	87.3	8.9
7/7/1;	14	7.3	0.74	101.0	10.3	110.8	11.3
6 × 14/7+	16	9.0	0.92	124.5	12.7	136.3	13.9
7F/7/1;	18	12.3	1.32	179.5	18.3	196.1	20.0
6 × 1618+	20	15.5	1.58	209.9	21.4	230.5	23.5
8F6/1	22	17.7	1.81	241.2	24.6	263.8	26.9
	24	20.6	2.10	278.5	28.4	304.0	31.0
	25	32.2	2.37	318.7	32.5	349.1	35.5
6 × 15/15/6/ 1;	29	29.3	2.99	398.2	40.6	438.4	44.7
	32	36.2	3.69	493.3	50.3	543.3	55.4
6 × 18/12/6/1;	35	43.9	4.48	598..2	61.0	658.0	67.1
6 × 16/8 and	38	52.2	5.32	712.0	72.6	782.6	79.8
8/1/1	41	61.3	6.25	836.5	85.3	916.9	93.5
(Fig. 21-11)	44	71.0	7.24	971.8	99.1	1065.9	108.7
	48	81.6	8.32	1116.0	113.8	1225.8	125.0
	51	92.8	9.46	1266.0	129.1	1394.5	142.2
	54	104.7	10.68	1434.1	146.3	1574.0	160.5
	57	117.5	11.98	1604.4	163.6	1763.2	179.8
	64	145.0	14.79	1982.2	202.2	2172.2	221.5
	70	175.3	17.98	2401.6	244.9	2630.1	288.0

Ropes in the group have six strands in one of the following constructions: $6\times12/6l$; $6\times12/6+6F/l$; $6\times9/9/l$; and $6\times10/5+5F/l$.

Ropes in this group have six strands in one of the following constructions: $6\times14/7$ and 7/7/1; $6\times14/7+7F/7/1$; $6\times16/8+8F/6/1$; $6\times15/15/6/1$; $6\times18/12/6/1$; $6\times16/8$ and 8/8/1

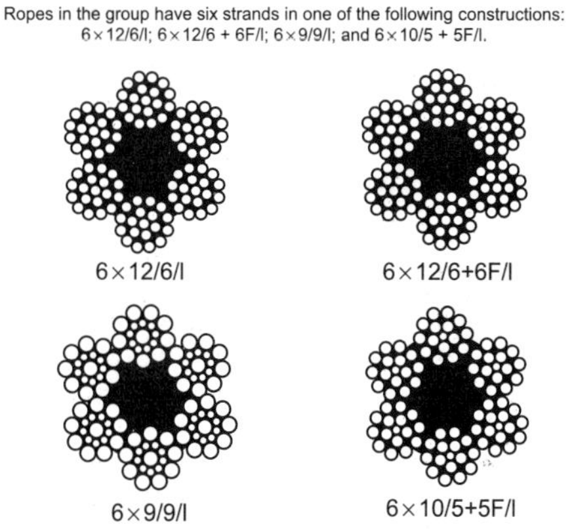

FIGURE 21-10 Round strand group 6 × 19.

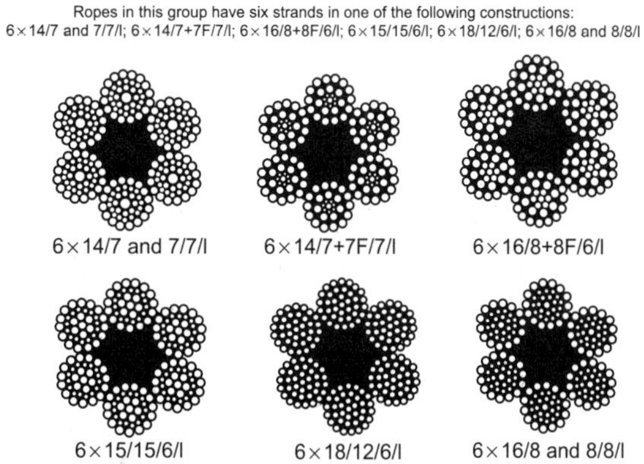

FIGURE 21-11 Round strand group 6 × 37.

TABLE 21-36
Steel wire ropes (from Indian standards) (*Cont.*)

				Nominal breaking strength of rope			
				Tensile strength of wire			
				1568–1716 MPa ($160–175\ kgf/mm^2$)		1716–1863 MPa ($175–190\ kgf/mm^2$)	
	Diameter of rope	Approx. weight					
Strand construction	**mm**	**N/m**	**kgf/in**	**kN**	**tf**	**kN**	**tf**
6 × 24 Fiber Core (Fig. 21-12)	8	2.1	0.21	29.4	3.0	32.4	3.3
	10	3.1	0.32	53.9	5.5	59.8	6.1
	12	5.3	0.54	74.5	7.6	81.4	8.3
	14	6.6	0.67	92.2	9.4	102.0	10.4
	16	7.8	0.80	112.8	11.5	123.6	12.6
	18	11.7	1.19	164.8	16.8	181.4	18.5
	20	13.8	1.41	196.1	20.0	214.8	21.9
	22	16.1	1.64	228.5	23.3	249.1	25.4
	24	18.2	1.86	258.9	26.4	278.5	28.4
	25	20.4	2.08	289.3	29.5	313.8	32.0
	29	26.3	2.68	368.7	37.6	403.1	41.1
	32	31.8	3.24	448.2	45.7	493.3	50.3
	35	38.8	3.96	548.2	55.9	603.1	61.5
	38	46.7	4.76	662.9	67.6	722.7	73.7
	41	54.0	5.51	762.0	77.7	836.5	85.3
	44	63.1	7.43	891.4	90.9	976.7	99.6
	48	73.0	7.44	1025.0	104.6	1125.8	114.8
	51	82.0	8.36	1166.0	118.9	1274.9	130.0
	54	93.4	9.52	1315.1	134.1	1443.5	147.3
Group 11 F 6 × 9/12/Δ; 6 × 10/12/Δ; 6 × 12/12/Δ; (Fig. 21-14)	14	8.3	0.85	112.8	11.5	121.6	12.4
	16	10.2	1.04	143.2	14.6	155.9	15.9
	18	13.7	1.40	208.9	21.3	224.6	22.9
	20	16.3	1.66	246.1	25.1	263.8	26.9
	22	20.1	2.05	284.4	29.0	308.9	31.5
	24	23.2	2.37	323.6	33.0	349.1	35.6
	25	26.4	2.69	363.8	37.1	393.2	40.1
	29	33.2	3.39	462.9	47.2	498.2	50.8
	32	41.2	4.20	572.7	58.4	622.7	53.5
	35	49.6	5.05	692.5	69.6	737.5	75.2
	38	59.0	6.02	816.9	83.3	886.5	90.4
	41	69.1	7.05	966.9	98.6	1036.6	105.7
	44	81.0	8.26	1116.0	113.8	1216.0	124.0
	48	92.7	9.45	1275.2	130.1	1374.9	140.2
	51	105.0	10.71	1454.3	148.3	1574.0	160.5

6×15/9/Fibre

FIGURE 21-12 Round strand group 6 × 24 fiber core.

Ropes in this group have six strands in one of the following constructions:
6×9/12/Δ, 6×10/12/Δ, 6×12/12/Δ.
*6×9/12/Δ
*6×10/12/Δ
*6×12/12/Δ

FIGURE 21-14 Compound flattened strand, group II F.

TABLE 21-36
Steel wire ropes (from Indian standards) (*Cont.*)

				Nominal breaking strength of rope			
				Tensile strength of wire			
				1568–1716 MPa ($160–175\ kgf/mm^2$)		1716–1863 MPa ($175–190\ kgf/mm^2$)	
Strand construction	Diameter of rope mm	Approx. weight N/m	Approx. weight kgf/in	kN	tf	kN	tf
17 × 7, 18 × 7 (Fig. 21-15)	8	2.5	0.25			35.3	3.6
	10	4.1	0.42			68.6	7.0
	12	5.6	0.57			87.3	8.9
	14	7.8	0.80			113.8	11.6
	16	9.6	0.98			142.2	14.5
	18	12.9	1.32			201.0	20.5
	20	15.2	1.55			237.3	24.2
	22	18.9	1.93			268.7	27.4
	24	21.9	2.23			313.8	32.0
	25	24.8	2.53			359.9	36.6
	29	31.4	3.20			443.3	45.2
	32	38.8	3.96			548.2	55.9
	35	46.8	4.77			672.7	68.6
	39	55.9	5.70			802.2	81.8
34 × 7 (Fig. 21-13)	15	10.2	1.04			134.4	13.7
	18	13.4	1.37			193.2	19.7
	20	16.0	1.63			225.6	23.0
	22	19.8	2.02			263.8	26.9
	24	22.8	2.32			299.1	30.5
	25	26.0	2.65			344.2	35.1
	29	32.9	3.35			433.5	44.2
	32	40.6	4.14			538.4	54.9
	35	49.0	5.00			647.2	66.0
	38	58.3	5.95			771.8	78.7
	44	79.5	8.21			1025.8	104.6
	51	103.9	10.59			1334.7	136.1

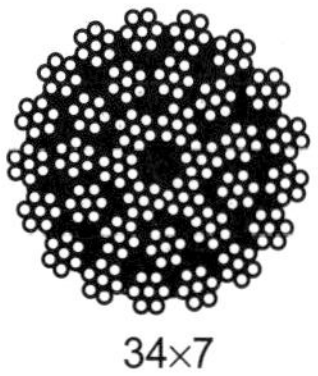

FIGURE 21-13 Multistrand nonrotating ropes 34 × 7.

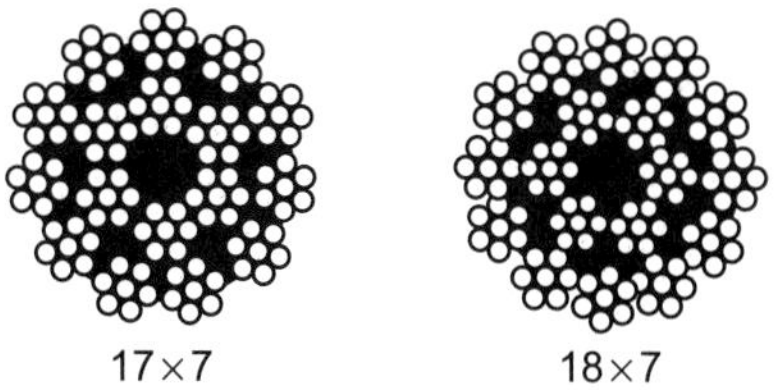

FIGURE 21-15 Multistrand nonrotating ropes 17 × 7 and 18 × 7.

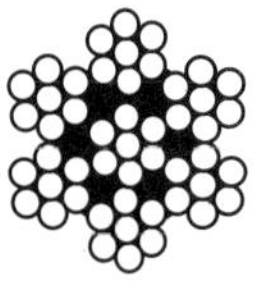

FIGURE 21-16(a) Metal core.

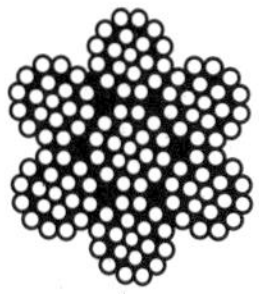

FIGURE 21-16(b) Metal core.

TABLE 21-36
Steel wire ropes (from Indian standards) (*Cont.*)

				Nominal breaking strength of rope			
				Tensile strength of wire			
				1079–1226 MPa ($110–125\ kgf/mm^2$)		1226–1372 MPa ($125–140\ kgf/mm^2$)	
Strand construction	**Diameter of rope mm**	**Approx. weight N/m**	**Approx. weight kgf/m**	**kN**	**tf**	**kN**	**tf**
			Lifts and Hoists				
Group 6 × 19	6	1.5	0.15	14.7	1.5	16.7	1.7
6 × 19 (12/6/1);	8	2.5	0.25	22.6	2.3	26.5	2.7
6 × 19 filler	10	3.9	0.40	39.2	4.0	44.1	4.5
wire,	12	5.4	0.55	53.9	5.5	58.8	6.0
6 × 19 (9/9/1)	14	7.4	0.75	75.5	7.7	86.3	8.8
Seale	16	9.3	0.95	94.1	9.6	107.9	11.4
	18	12.2	1.25	124.5	12.7	139.5	14.2
	20	14.2	1.45	147.1	15.0	166.7	17.0
	21	18.1	1.85	184.4	18.8	207.9	21.2
	25	22.1	2.25	225.6	23.3	255.0	26.0
Group 8 × 19	8	2.0	0.20	21.3	2.2	24.5	2.5
8 × 19 filler	10	3.4	0.35	37.6	3.8	42.2	4.3
wire;	12	4.9	0.50	49.0	5.0	53.9	5.5
8 × 19 (9/9/1)	14	6.9	0.70	68.6	7.0	79.4	8.1
Seale	16	8.3	0.85	88.3	9.0	98.1	10.0
	18	10.9	1.10	112.8	11.1	132.4	13.5
	20	13.2	1.35	137.3	14.0	152.0	15.5
	22	16.7	1.70	181.4	18.5	205.9	21.0
	25	19.6	2.00	01.0	20.6	235.4	24.0
6 × 25	10	4.4	0.45	42.2	4.3	49.0	5.0
flattened	12	5.9	0.60	56.9	5.8	64.7	6.6
strand	14	8.3	0.85	79.4	8.1	90.2	9.2
	16	10.3	1.05	102.9	10.5	117.7	12.0
	18	13.7	1.40	137.3	14.0	151.0	15.4
	20	16.2	1.65	161.8	16.5	184.4	18.8
	22	19.6	2.00	203.0	20.7	230.5	23.5
	25	24.5	2.50	243.2	24.8	272.6	27.8

TABLE 21-36
Steel wire ropes (from Indian Standard) (*Cont.*)

				Nominal breaking strength of rope							
				Tensile strength of wire							
				1225.8–1373.0 MPa (125–140 kgf/mm²)		1373.0–1520.0 MPa (140–155 kgf/mm²)		1520.0–1667.0 MPa (155–170 kgf/mm²)		1667.0–1814.2 MPa (170–185 kgf/mm²)	
Strand construction	Diameter of rope, mm	Approx. weight N/m	Approx. weight kgf/m	kN	tf	kN	tf	kN	tf	kN	tf
				Winding purposes in mines							
6 × 7	19	12.8	1.31	166.7	17.0	183.5	18.9	199.1	20.3	213.8	21.8
	20	15.0	1.53	192.2	19.6	211.8	21.6	230.4	23.5	250.1	25.5
	22	17.7	1.80	224.6	22.9	250.1	25.5	268.7	27.4	289.3	29.5
	24	29.3	2.07	254.0	25.9	283.4	28.9	309.1	31.5	333.4	34.0
	25	23.0	2.35	283.3	29.5	325.6	33.2	349.1	35.6	378.6	38.6
	26	24.6	2.51	310.0	31.6	341.3	34.8	399.1	39.9	402.1	41.0
	27	26.3	2.68	332.4	33.9	366.7	37.4	391.9	40.7	430.5	40.9
	28	29.2	2.98	368.7	37.6	410.0	41.8	443.3	45.2	478.6	43.8
	31	35.9	3.66	453.1	46.2	512.8	52.3	548.2	55.9	598.2	71.2
	35	43.4	4.43	553.1	56.4	618.9	63.1	662.9	67.6	717.8	73.2

				Nominal breaking strength of rope							
				Tensile strength of wire							
				1226–1373 MPa (125–140 kgf/mm²)		1373–1520 MPa (140–155 kgf/mm²)		1520–1667 MPa (155–170 kgf/mm²)		1667–1814 MPa (170–185 kgf/mm²)	
Strand construction	Diameter of rope, mm	Approx. weight N/m	Approx. weight kgf/m	kN	tf	kN	tf	kN	tf	kN	tf
6 × 19	19	13.2	1.35	154.9	15.8	171.6	17.5	189.3	19.3	206.9	21.2
	20	14.6	1.49	179.5	18.3	199.1	20.3	221.6	22.6	243.2	24.8
	21	16.4	1.67	193.2	19.7	213.2	21.8	237.3	24.2	260.8	26.6
	22	18.0	1.84	206.2	21.1	229.5	23.4	254.0	25.9	278.5	28.4
	23	19.5	1.99	222.6	22.7	246.1	25.1	273.6	27.9	301.1	30.7
	24	20.9	2.13	237.3	24.2	263.8	26.9	294.2	30.0	323.6	33.0
	25	23.6	2.41	268.7	27.4	300.1	30.6	334.4	34.1	368.7	37.7
	26	26.6	2.71	291.3	29.7	326.5	33.3	365.8	37.3	399.1	40.7
	27	28.3	2.89	318.7	32.5	352.1	35.9	394.2	40.2	436.4	44.5
	28	31.3	3.19	348.1	35.5	383.4	39.1	423.6	43.2	462.9	47.2
	29	33.8	3.45	372.6	38.0	413.8	42.2	456.0	46.5	502.1	51.2
	30	35.6	3.63	400.1	40.8	443.3	45.2	483.5	49.3	536.4	54.7
	31	38.2	3.90	428.5	43.7	473.7	48.3	522.7	53.3	572.7	58.4
	32	39.7	4.05	447.1	45.6	498.2	50.8	545.2	55.6	603.1	61,5
	33	41.6	4.24	471.7	48.1	522.7	53.3	572.7	58.4	632.5	64.5
	34	43.1	4.39	493.3	50.3	548.2	55.9	608.1	61.5	663.9	67.7
	35	44.6	4.55	518.8	52.9	572.7	58.4	632.5	64.5	692.3	73.6
	36	47.3	4.82	548.2	55.9	608.0	62.0	672.7	68.6	732.5	74.7
	37	50.2	5.12	580.5	59.2	641.3	65.4	707.1	72.1	773.7	78.9
	38	53.3	5.43	611.2	62.4	678.6	69.2	752.2	76.7	826.7	84.3
	39	55.9	5.70	629.6	64.2	696.3	71.0	772.8	78.8	849.3	86.6
	40	59.2	6.04	650.2	66.3	714.9	72.8	792.4	80.8	868.9	88.6

TABLE 21-36
Steel wire ropes (from Indian Standard) (*Cont.*)

Strand construction	Diameter of rope, mm	Approx. weight		Nominal breaking strength of rope							
				Tensile strength of wire							
				1226–1373 MPa (125–140 kgf/mm^2)		1373–1520 MPa (140–155 kgf/mm^2)		1520–1667 MPa (155–170 kgf/mm^2)		1667–1814 MPa (170–185 kgf/mm^2)	
		N/m	kgf/m	kN	tf	kN	tf	kN	tf	kN	tf
	41	62.5	6.37	726.7	74.1	803.2	81.9	886.5	90.4	771.8	99.2
	42	65.4	6.67	781.6	79.7	863.9	88.1	955.2	97.4	1048.3	106.9
	44	72.4	7.38	836.5	85.3	926.7	94.5	1025.8	104.6	1125.8	114.8
	46	78.1	7.96	893.3	91.1	995.4	101.5	1101.3	112.3	1210.1	123.4
	48	83.2	8.48	950.3	96.9	1057.2	107.8	1175.8	119.9	1225.5	192.1
	51	94.5	9.64	1100.3	112.2	1217.0	124.1	1345.6	157.2	1475.0	150.4
	54	106.8	10.89	1230.7	125.5	1365.1	139.2	1514.1	155.4	1664.2	169.7
6 × 37	19	12.9	1.32	145.1	14.8	162.8	16.6	179.5	18.3	196.1	20.0
	21	15.5	1.58	170.6	17.4	190.2	19.4	209.8	21.4	230.5	23.5
	22	17.7	1.81	195.8	19.9	218.7	22.3	241.2	24.6	263.8	26.9
	24	20.6	2.10	222.5	23.4	254.0	25.9	278.6	28.4	304.0	31.0
	25	23.2	2.37	260.0	26.4	289.3	29.5	318.4	32.5	349.1	35.6
	29	29.3	2.99	318.7	32.5	359.0	36.6	398.1	40.6	438.4	44.7
	22	36.2	3.69	343.7	35.0	393.2	48.1	493.3	50.3	543.3	55.4
	25	43.9	4.48	478.6	48.8	548.2	55.9	598.2	61.0	658.0	67.1
	31	52.2	5.32	572.7	58.4	642.3	65.5	712.0	72.6	782.6	79.8
	41	61.3	6.25	665.1	67.8	757.1	70.2	836.5	85.3	916.9	93.5
	44	71.0	7.24	676.9	69.0	857.1	87.4	871.8	99.1	1066.0	108.7
	48	81.6	8.32	896.3	91.4	1006.2	102.6	1116.0	113.8	1225.8	125.0
	51	92.8	9.45	1006.2	102.6	1135.6	115.8	1226.0	129.8	1394.0	142.2
	54	104.8	10.68	1156.2	117.9	1295.5	132.1	1434.7	146.3	1574.0	160.5
	57	117.6	11.98	1285.6	131.1	1444.5	147.3	1604.4	163.6	1763.2	179.8
	64	145.0	14.79	1624.0	165.6	1793.6	182.9	1912.9	202.2	2172.2	221.5
	70	175.3	17.88	1932.9	197.3	2172.2	221.5	2401.6	244.9	2630.0	268.2
6 × 7	19	15.0	1.53	181.4	18.5	199.1	20.3	216.7	22.1	235.4	24.0
Triangular core	21	17.6	1.79	205.9	21.0	228.5	23.3	249.1	25.4	272.6	27.8
	22	20.1	2.05	244.2	24.9	268.7	27.4	294.2	30.0	313.7	32.6
Group IF	24	23.24	2.37	278.5	28.4	306.9	31.3	333.4	34.0	363.8	37.1
6 × 7/Δ	25	26.28	2.68	313.8	32.0	347.1	35.4	378.5	38.6	413.8	42.2
	28	33.24	3.39	403.1	41.1	443.3	45.2	483.5	49.3	528.6	53.1
	31	41.19	4.20	498.2	50.8	553.1	56.4	608.0	62.9	658.0	67.1
	36	49.62	5.06	598.2	61.0	662.9	67.6	727.6	74.2	792.4	80.8
Group IIF	19	15.00	1.53	179.5	18.3	194.2	19.8	208.9	21.3	224.6	22.9
6 × 8/Δ;	21	17.55	1.79	209.9	21.4	228.5	23.3	246.1	25.1	263.8	26.9
6 × 8/12	22	20.10	2.05	234.4	23.9	258.9	26.4	284.4	29.0	308.9	31.5
Or less/Δ;	24	23.24	2.37	273.6	27.9	299.1	30.5	323.6	33.0	349.1	35.6
6 × 9/12	25	26.28	2.68	304.0	31.0	333.4	34.0	363.8	37.1	393.2	40.1
Or less/Δ;	29	33.24	3.39	393.2	40.1	428.5	43.7	492.9	47.2	498.2	50.8
6 × 10/12	32	41.18	4.20	473.6	48.3	522.7	53.3	572.7	58.4	622.7	63.5

TABLE 21-36
Steel wire ropes (from Indian Standard) (*Cont.*)

				Nominal breaking strength of rope							
				Tensile strength of wire							
				1226–1373 MPa (125–140 kgf/mm^2)		1373–1520 MPa (140–155 kgf/mm^2)		1520–1667 MPa (155–170 kgf/mm^2)		1667–1814 MPa (170–185 kgf/mm^2)	
Strand construction	**Diameter of rope, mm**	**Approx. weight N/m**	**kgf/m**	**kN**	**tf**	**kN**	**tf**	**kN**	**tf**	**kN**	**tf**
Or less/Δ;	35	49.62	5.06	572.7	58.4	627.6	64.0	682.5	69.6	737.5	75.2
6 × 12/12	38	59.03	6.02	677.6	69.1	766.9	78.2	816.9	83.3	886.5	90.4
Or less/Δ	41	69.14	7.05	825.2	84.3	896.3	91.4	966.9	98.6	1036.6	105.7
	44	81.00	8.26	916.2	93.6	1016.0	103.6	1116.0	113.8	1216.0	124.0
	48	92.67	9.45	1075.8	109.7	1175.8	119.9	1275.8	138.1	1375.0	140.2
	51	105.03	10.71	1216.0	124.0	1334.7	136.1	1454.3	140.3	1574.0	160.5
Group IIIF	19	15.00	1.53	156.9	16.0	174.6	17.8	193.2	19.7	210.8	21.5
6 × 15/12/A	21	17.55	1.79	184.4	18.8	205.0	20.9	226.5	23.1	247.1	25.2
6 × 18/12/A	22	20.10	2.05	208.9	21.3	234.4	23.9	258.9	26.4	284.4	29.0
	24	23.05	2.35	234.4	23.9	263.8	26.9	294.2	30.0	323.6	33.0
	25	26.28	2.68	273.6	27.9	304.0	31.0	333.4	34.0	363.8	37.1
	29	33.24	3.39	354.0	36.1	388.3	39.6	423.6	43.2	458.0	46.7
	32	41.19	4.20	443.3	45.2	488.4	49.8	533.5	54.4	577.6	58.9
	35	49.62	5.06	517.8	52.8	577.6	58.9	537.4	65.0	656.3	71.0
	38	59.04	6.02	627.6	64.0	682.5	69.6	757.1	77.2	821.8	83.8
	41	69.14	7.05	747.3	76.2	816.9	83.3	886.5	90.4	986.1	97.5
	44	81.00	8.26	857.1	87.4	946.3	96.5	1036.6	105.7	1125.8	114.8
	48	92.67	9.45	986.5	100.6	1085.6	110.7	1185.6	120.9	1285.6	131.1
	51	105.03	10.71	1125.7	114.8	1235.4	126.0	1348.5	137.2	1452.8	148.3
	54	118.17	12.05	1255.2	128.0	1385.7	141.3	1514.1	154.4	1643.7	167.6
	57	133.57	13.62	1448.5	147.3	1584.2	161.6	1724.0	175.8	1863.3	190.0
	64	164.26	16.75	1793.6	189.9	1954.8	199.1	2112.6	215.4	2278.2	231.7
	70	198.58	20.25	2152.5	219.5	2341.8	238.8	2550.7	260.1	2740.0	279.4

				Minimum break load			
		Approx. weight		For tensile designation			
Strand construction	**Diameter rope, mm**	**N/100 m**	**kgf/100 m**	**1569.3 MPa**	**160 kgk/mm^2**	**1765.2 MPa**	**180 kgf/mm^2**
		Haulage purposes in mines					
6 × 7 (6 × 1)	8	217.7	22.2	33.3	3400	37.6	3830
Round	9	275.6	28.1	42.3	4300	47.5	4840
	10	340.3	38.7	52.2	3320	58.6	5980
	11	411.9	42.0	63.1	6430	71.0	7240
	12	490.3	50.0	75.1	7660	84.4	8610

TABLE 21-36
Steel wire ropes (from Indian Standard) (*Cont.*)

				Minimum breaking load of rope			
				For tensile designation			
		Approx. weight		1569 MPa (160 kgf/mm^2)		1765 MPa (180 kgf/mm^2)	
Strand construction	Diameter of rope, mm	N/100 m	kgf/100 m	kN	kgf	kN	kgf
6 × 7 (6 × 1) Round	13	574.7	58.6	88.1	8980	99	10100
	14	666.9	68.0	102	10400	115	11700
	16	870.8	88.8	133	13600	150	15300
	18	1098.3	112.0	169	17200	190	19400
	19	1225.8	125.0	188	19200	212	21600
	20	1363.1	139.0	209	21300	234	23900
	21	1500.4	153.0	229	23400	259	26400
	22	1647.5	168.0	252	25700	283	28900
	24	1961.3	200.0	300	30600	337	34400
	25	2128.0	217.0	326	33200	367	37400
	26	2304.6	235.0	352	35900	396	40400
	27	2481.1	253.0	380	38700	428	43600
	28	2667.4	272.0	409	41700	460	46900
	29	2863.5	292.0	438	44700	493	50300
	31	3275.9	334.0	501	51100	564	57500
	35	4167.8	425.0	638	65100	719	73300
6 × 19 (9/9/1) Round	13	599.2	61.1	87.8	8950	99	10100
	14	695.3	70.9	102	10400	115	11700
	16	908.1	92.6	133	13600	150	15300
	18	1147.4	117	169	17200	189	19300
	19	1284.7	131	187	19100	211	21500
	20	1422.0	145	208	21200	233	23800
	21	1569.1	160	229	23400	258	26300
	22	1716.2	175	251	25600	282	28800
	24	2039.8	208	299	30.500	336	34300
	25	2216.3	226	325	33100	365	37200
	26	2422.2	247	351	35800	395	40300
	28	2785.5	284	407	41500	459	46700
	29	2981.2	304	436	44500	491	50100
	32	3628.4	370	532	54200	598	61000
	35	4344.3	443	636	64900	716	73000
	36	4599.3	469	673	68600	757	77200
	38	5413.2	552	750	76500	843	86000

TABLE 21-36
Steel wire ropes (from Indian Standard) (*Cont.*)

Strand construction	Diameter of rope, mm	Approx. weight N/100 m	Approx. weight kgf/100 m	Minimum breaking load of rope, for tensile designation 1569 MPa (160 kgf/mm^2) kN	1569 MPa (160 kgf/mm^2) kgf	1765 MPa (180 kgf/mm^2) kN	1765 MPa (180 kgf/mm^2) kgf
6 × 8 (7/Δ) Triangular	13	675.7	68.9	95.9	9780	106	10800
	15	783.5	79.9	111	11300	124	12600
	16	1019.9	104	145	14800	161	16400
	18	1294.5	132	183	18700	204	20800
	19	1441.6	147	205	20900	228	23200
	20	1598.5	163	227	23100	252	25700
	21	1765.2	180	250	25500	278	28300
	22	1931.9	197	275	28007	305	31100
	24	2304.5	235	327	33306	363	37000
	25	2500.1	255	354	36100	393	40100
	26	2696.8	275	383	39100	426	43400
	28	3128.3	319	445	45400	493	50300
	29	3363.7	343	478	48700	530	54000
	31	3844.2	392	545	55600	605	61700
	35	4893.5	499	695	70900	771	78600
6 × 22 (9/12/Δ) Triangular	13	685.5	69.9	93.1	9490	104	10610
	14	794.3	81	108	11000	120	12200
	16	1039.5	106	141	14400	157	16000
	18	1314.1	134	178	18200	198	20200
	19	1461.2	149	199	20300	222	22600
	20	1618,1	165	221	22500	245	25000
	21	1784.8	182	243	24800	270	27500
	22	1961.3	200	267	27200	296	30200
	24	2334.0	238	317	32300	353	36000
	25	2530.1	258	343	35000	384	39100
	26	2736.0	279	372	37900	414	42200
	28	3137.3	324	431	44000	481	49000
	29	3412.7	348	463	47200	515	52500
	32	4148.2	423	564	57500	629	64000
	35	4962.1	506	675	68800	750	76500
	38	5854.5	597	795	81100	885	90200

TABLE 21-36
Steel wire ropes (from Indian Standard) (*Cont.*)

Strand construction	Diameter of wire, mm	Approx. weight N/m	Approx. weight kgf/m	Maximum breaking load of rope kN	Maximum breaking load of rope kgf
		Small Wire Ropes (Fiber Core)			
6 × 7 (6/1)	2	0.147	0.015	2.6	260
	3	0.324	0.033	5.9	600
	4	0.559	0.057	10.4	1060
	5	0.873	0.099	16.3	1660
	6	1.255	0.128	23.5	2400
	7	1.696	0.172	32.0	3260
6 × 12 (12/fiber)	3	0.235	0.024	3.7	380
	4	0.412	0.042	6.5	670
	5	0.637	0.065	10.3	1050
	6	0.922	0.094	14.9	1520
	7	1.255	0.128	20.3	2070
6 × 19 (12/6/1)	3	0.314	0.035	4.9	500
	4	0.539	0.052	8.7	890
	5	0.843	0.086	13.5	1880
	6	1.206	0.124	19.6	2000
	7	1.648	0.168	26.6	2710
6 × 24 (15/9/fiber)	4	0.530	0.054	8.6	880
	5	0.834	0.085	13.3	1360
	6	1.206	0.122	19.3	1970
	7	1.618	0.165	29.3	2680

Strand construction	Diameter of wire Max, mm	Diameter of wire Min, mm	Approx. weight, max N/m	Approx. weight, max kgf/m	Minimum breaking load kN	Minimum breaking load kgf
		Preferred Galvanized Steel Wire Ropes for Aircraft Controls				
7 × 7	1.8	1.6	0.108	0.011	2.2	220
7 × 7	2.7	2.4	0.235	0.024	4.1	420
7 × 19	3.5	3.2	0.422	0.043	8.9	910
7 × 19	4.4	4.0	0.657	0.067	12.5	1270
7 × 19	5.2	4.8	0.804	0.082	18.6	1900
7 × 19	6.0	5.6	1.236	0.126	24.9	2540
7 × 10	6.8	6.4	1.608	0.164	31.1	3170

Note: kgf = kilogram – force; tf = ton – force.

TABLE 21-37
Round strand galvanized steel wire ropes for shipping purposes

Diameter of wire, mm	Tensile strength of wire, 1373–1569 MPa (140–160 kgf/mm²)															
	Approx. weight		Breaking strength of rope, min		Approx. weight		Breaking strength of rope, min		Approx. weight		Breaking strength of rope, min		Approx. weight		Breaking strength of rope, min	
	N/m	kgf/m	kN	kgf	N/m	kgf/m	kN	kgf	N/m	kgf/m	kN	kgf	N/m	kgf/m	kN	kgf
	6 × 7		Fiber core		16 × 12		Fiber core		6 × 13		Fiber core		7 × 7		Fiber core	
8	2.2	0.22	31.0	3150	1.5	0.15	18.1	1850								
9	2.8	0.28	38.8	3950	2.1	0.21	26.0	2650								
10	3.3	0.34	47.1	4800	2.5	0.25	30.4	3100								
11	4.0	0.41	56.9	5800	2.8	0.29	35.3	3650					3.7	0.38	52.0	5300
12	5.1	0.52	72.6	7400	3.7	0.38	46.1	4700	5.1	0.52	72.6	7400	4.4	0.45	62.8	6400
14	5.8	0.69	96.1	9800	4.7	0.48	58.4	5950	7.8	0.80	109.8	11200	5.7	0.58	80.4	8200
16	8.7	0.89	123.6	12600	6.4	0.65	79.4	8100	9.4	0.96	132.4	13500	7.6	0.77	106.9	10900
18	10.9	1.11	154.0	15700	7.6	0.78	95.6	9750	12.4	1.26	174.6	17800	9.7	0.99	137.3	14000
20	13.9	1.42	198.1	20200	9.8	1.00	122.1	12450	14.9	1.52	209.9	21400	12.1	1.23	171.6	17500
22	16.6	1.69	236.3	24100	11.4	1.16	141.7	14450	18.5	1.89	261.8	26700	15.5	1.58	219.7	22400
24	19.5	1.99	278.5	28400	13.9	1.42	173.6	17700	21.0	2.14	295.2	30100	18.5	1.89	262.8	26800
26	22.8	2.32	323.6	33000	16.8	1.70	208.9	21300	25.5	2.60	361.9	36900	21.9	2.23	308.9	33500
28	27.1	2.76	385.4	39300	18.7	1.91	234.4	23900	30.5	3.11	429.5	43900	25.4	2.59	359.9	36700
32	34.7	3.54	494.3	50400	24.3	2.48	304.0	31000	39.4	4.02	555.1	56600	30.2	3.08	427.6	43600
36	44.5	4.54	634.5	64700	30.6	3.12	382.5	39000	40.1	4.09	703.1	71700	38.7	3.95	548.2	56000
40	54.3	5.54	773.9	78900	39.0	3.98	489.4	49900	61.6	6.28	867.9	88500	49.7	5.07	704.1	71800
	6 × 19		Fiber core		6 × 24		Fiber core		6 × 37		Fiber core		7 × 19		Wire core	
8	1.9	0.20	28.0	2850	2.2	0.22	28.4	2900	2.3	0.23	31.9	3250				
9	2.8	0.29	40.2	4100	2.6	0.27	34.8	3550	2.9	0.30	41.2	4200				
10	3.4	0.35	47.3	4800	3.1	0.32	42.2	4300	3.3	0.34	47.1	4800				
11	3.9	0.40	53.9	5500	3.7	0.38	50.0	5100	4.1	0.42	58.4	5950				
12	5.1	0.52	71.1	7250	4.4	0.45	58.8	6000	5.0	0.51	70.6	7200	7.3	0.74	101.5	10350
14	6.5	0.66	90.2	9200	5.9	0.60	78.5	8000	7.0	0.71	98.1	10000	9.9	1.01	138.3	14100
16	8.8	0.90	122.6	12500	8.4	0.86	112.8	11500	9.3	0.95	31.4	13400	11.9	1.21	165.7	16900
18	10.6	1.08	147.1	15000	10.4	1.06	140.2	14300	12.0	1.22	168.7	17200	15.2	1.55	211.8	21600
20	13.5	1.38	188.3	19200	12.7	1.29	169.7	17300	14.9	1.52	256.9	26200	17.7	1.80	245.2	25000
22	15.7	1.60	218.7	22300	15.0	1.53	201.0	20500	18.2	1.86	256.9	26200	20.8	2.12	289.3	29500
24	19.2	1.96	267.7	27300	17.7	1.80	238.3	24100	20.0	2.04	282.4	29900	26.0	2.65	361.9	36900
26	23.1	2.36	321.6	32800	22.0	2.24	294.2	30000	23.8	2.43	336.4	34300	29.1	2.97	406.0	41400
28	26.0	2.65	360.9	36800	25.1	2.56	336.9	34300	28.0	2.85	394.2	40200	37.9	3.86	526.6	53700
32	33.7	3.44	468.8	47800	32.0	3.26	428.6	43700	37.2	3.79	524.7	53500	47.6	4.85	662.9	67600
36	42.4	4.32	588.4	60000	41.8	4.26	599.0	57000	47.8	4.87	674.7	68800	60.8	6.20	847.3	86400
40	54.1	5.52	664.9	67800	50.5	5.15	676.7	69000	56.6	5.77	798.3	81400	73.1	7.45	1027.7	104800
44	65.1	6.64	905.2	92300	60.1	6.13	806.1	82200	69.5	7.09	981.6	100100	86.3	8.80	1203.3	122700
48	78.0	7.95	1084.6	110600	73.3	7.47	982.6	100200	83.9	8.55	1183.7	120700	101.0	10.30	1407.3	143500
52	90.0	9.18	1251.3	127600	84.7	8.64	1136.6	115900	95.2	9.71	1344.5	137100	116.6	11.89	1625.0	165700
60	104.0	10.61	1446.5	147500	97.0	9.90	1301.3	132700	111.7	11.39	1577.9	160900	133.8	13.59	1857.4	189400

Source: IS 2581, 1968.

TABLE 21-38
Dimensions and breaking strength of flat balancing wire ropes

Constructions	Nominal size $b \times s$,[a] mm: Double-stitched	Nominal size $b \times s$,[a] mm: Single-stitched	Diameter of the wire, mm	Cross section of the strand, mm^2	Approximate weight, Double-stitched: N/m	Approximate weight, Double-stitched: kgf/m	Approximate weight, Single-stitched: N/m	Approximate weight, Single-stitched: kgf/m	Minimum breaking strength of rope: kN	Minimum breaking strength of rope: kgf
6 × 4 × 7	70 × 17	70 × 15	1.60	338	34.3	3.5	33.3	3.4	463.9	47300
	74 × 18	74 × 16	1.70	381	39.2	4.0	37.3	3.8	522.7	53300
	78 × 19	78 × 17	1.80	427	44.1	4.5	42.2	4.3	585.5	59700
	82 × 20	82 × 18	1.90	477	49.0	5.0	47.1	4.8	654.1	66700
	87 × 21	87 × 19	2.00	528	53.9	5.5	52.0	5.3	724.7	73900
	91 × 22	91 × 20	2.20	581	59.8	6.1	56.9	5.8	797.2	81300
	95 × 23	95 × 21	2.20	638	65.7	6.7	62.8	6.4	875.7	89300
8 × 4 × 7	110 × 20	110 × 18	1.90	636	65.7	6.7	62.8	6.4	872.8	89000
	113 × 20	113 × 18	1.95	670	68.7	7.0	65.7	6.7	919.9	93800
	116 × 21	116 × 19	2.00	703	72.6	7.4	68.7	7.0	956.0	98400
	119 × 21	119 × 19	2.05	739	76.5	7.8	72.6	7.4	1014.0	103400
	122 × 22	122 × 20	2.10	775	79.4	8.1	76.5	7.8	1064.0	108500
	125 × 22	125 × 20	2.15	812	83.4	8.5	79.4	8.1	1116.0	113800
	128 × 23	128 × 21	2.20	851	87.3	8.9	83.4	8.5	1168.0	119100
6 × 4 × 12	112 × 26	112 × 23	1.90	818	84.3	8.6	80.4	8.2	1122.9	114500
	115 × 26	115 × 23	1.95	861	88.3	9.0	84.3	8.6	1181.7	120500
	118 × 27	118 × 24	2.00	904	98.2	9.5	88.3	9.0	1240.5	126500
	121 × 27	121 × 24	2.05	950	98.1	10.0	93.2	9.5	1304.3	133000
	124 × 28	124 × 25	2.10	996	103.0	10.5	98.1	10.0	1367.0	139400
	127 × 28	127 × 25	2.15	1045	107.9	11.0	103.0	10.5	1439.6	146300
	130 × 29	130 × 26	2.20	2094	112.8	11.5	106.9	10.9	1483.7	151300
8 × 4 × 12	146 × 26	146 × 23	1.90	1091	112.8	11.5	106.9	10.9	1497.5	152700
	149 × 26	149 × 23	1.95	1148	118.7	12.1	112.8	11.5	1575.9	160700
	154 × 27	154 × 24	2.00	1206	124.5	12.7	118.7	12.1	1655.4	168800
	157 × 27	157 × 24	2.05	1267	130.4	13.3	124.5	12.7	1738.7	177300
	160 × 28	160 × 25	2.10	1329	137.3	14.0	130.4	13.3	1824.0	186000
	165 × 28	165 × 25	2.15	1394	143.2	14.6	136.3	13.9	1913.3	195100
	168 × 29	168 × 26	2.20	1459	150.0	14.3	143.2	14.6	2002.5	204200
8 × 4 × 14	160 × 27	160 × 24	1.90	1272	131.4	13.4	124.6	12.7	1745.5	178000
	164 × 28	164 × 25	1.95	1340	138.3	14.1	131.4	13.4	1842.2	187800
	168 × 28	168 × 25	2.00	1407	145.1	14.8	138.3	14.1	1930.9	196900
	172 × 29	172 × 26	2.05	1478	152.0	15.5	145.1	14.8	2029.0	206900
	176 × 29	176 × 26	2.10	1550	159.8	16.3	152.0	15.5	2188.0	217000
	180 × 30	180 × 27	2.15	1626	167.7	17.1	159.9	16.3	2232.0	227600
	184 × 30	184 × 27	2.20	1702	175.5	17.9	166.7	17.0	2335.9	238200
8 × 4 × 91	186 × 31	186 × 28	1.90	1727	177.5	18.1	169.7	17.3	2377.3	251700
	190 × 32	190 × 29	1.95	1818	187.3	19.1	178.5	18.2	2495.8	254500
	194 × 33	194 × 30	2.00	1909	191.1	20.1	187.3	19.1	2620.3	267200

[a] b = width of rope, s = thickness of rope.
Source: IS 5203, 1969.

TABLE 21-39
Dimensions and breaking strength of flat hoisting wire ropes

Construction	Nominal size, $b \times s$, mm	Nominal wire diameter, mm	Cross section of strand, mm^2	Weight		Minimum breaking strength of rope[a]	
				N/m	kgf/m	kN	kgf
6 × 4 × 7	52 × 10	1.20	190	18.6	1.9	298.1	30400
	56 × 11	1.30	223	21.6	2.2	349.1	35600
	60 × 12	1.40	259	25.5	2.6	406.0	41400
	65 × 14	1.50	297	29.4	3.0	465.8	47500
	70 × 15	1.60	338	33.3	3.4	529.6	54000
	74 × 16	1.70	381	37.3	3.8	597.2	60900
	78 × 16	1.80	427	42.2	4.3	669.8	68300
	82 × 18	1.90	477	47.1	4.8	748.2	76300
	87 × 19	2.00	528	52.0	5.3	827.7	84400
	91 × 20	2.10	581	56.9	5.8	911.0	92900
	95 × 21	2.20	638	62.8	6.4	1000.3	102000
8 × 4 × 7	70 × 10	1.20	253	24.5	2.5	396.2	40400
	75 × 11	1.30	298	29.4	3.0	466.8	47600
	80 × 12	1.40	345	34.3	3.5	541.3	55200
	86 × 14	1.50	396	39.2	4.0	620.8	63300
	92 × 15	1.60	450	44.1	4.5	706.1	72000
	98 × 16	1.70	508	50.0	5.1	796.3	81200
	104 × 17	1.80	569	55.9	5.7	892.4	91000
	110 × 18	1.90	636	62.8	6.4	997.3	101700
	116 × 19	2.00	703	68.6	7.0	1102.3	112400
	122 × 20	2.10	775	76.5	7.8	1216.0	124400
	128 × 21	2.20	851	83.4	8.5	1333.7	136600

[a] Rope having wires of tensile strength of 1569 MPa (160 kgf/mm^2).
Source: IS 5202, 1269.

TABLE 21-40
Tensile grade

Grade of wire	Tensile strength range MPa	Tensile strength range kgf/mm^2
120	1176.8–1471.0	120–150
140	1372.9–1078.7	140–170
160	1569.1–1863.3	160–190
180	1765.2–2059.4	180–210
200	1961.3–2353.6	200–240

TABLE 21-41
Values of C for wire ropes

Rope diameter, mm	C	Rope diameter, mm	C
9.50	1.090	15.90	1.064
11.11	1.083	19.00	1.054
12.70	1.076	22.20	1.046
14.30	1.070	25.40	1.040

TABLE 21-42A
Approximate wire rope and sheave data

Rope construction	Ultimate strength, F_u MN	Ultimate strength, F_u lbf × 10^3	Weight kN/m	Weight lbf/ft	Wire, diameter d_w, mm (in)	Area A, mm^2 (in^2)	Recommended sheave diameter, mm (in) Average	Recommended sheave diameter, mm (in) Minimum
6 × 19	$500.8d^2$	$72d^2$	$36.3d^2$	$1.60d^2$	$0.063d$	$0.38d^2$	$45d$	$30d$
6 × 37	$473.1d^2$	$68d^2$	$35.3d^2$	$1.55d^2$	$0.045d$	$0.38d^2$	$27d$	$18d$
8 × 19	$431.3d^2$	$62d^2$	$34.3d^2$	$1.50d^2$	$0.050d$	$0.35d^2$	$31d$	$21d$
6 × 7	$473.0d^2$	$68d^2$	$32.4d^2$	$1.45d^2$	$0.106d$	$0.38d^2$	$72d$	$42d$

SI units: d = diameter of rope, m.
US Customary units: d = diameter of rope, in.

TABLE 21-42B
Wire rope data

Rope	Weight per foot, lb	Minimum sheave diameter, in	Standard sizes, *d*, in	Material	Size of outer wires	Modulus of elasticity,[a] Mpsi	Strength,[b] kpsi
6 × 7 haulage	$1.50d^2$	$42d$	$\frac{1}{4}$–$1\frac{1}{2}$	Monitor steel	$d/9$	14	100
				Plow steel	$d/9$	14	88
				Mild plow steel	$d/9$	14	76
6 × 19 standard hoisting	$1.60d^2$	$26d$–$34d$	$\frac{1}{4}$–$2\frac{3}{4}$	Monitor steel	$d/13$–$d/16$	12	106
				Plow steel	$d/13$–$d/16$	12	93
				Mild plow steel	$d/13$–$d/16$	12	80
6 × 37 special flexible	$1.55d^2$	$18d$	$\frac{1}{4}$–$3\frac{1}{2}$	Monitor steel	$d/22$	11	100
				Plow steel	$d/22$	11	88
8 × 19 extra flexible	$1.45d^2$	$21d$–$26d$	$\frac{1}{4}$–$1\frac{1}{2}$	Monitor steel	$d/15$–$d/19$	10	92
				Plow steel	$d/15$–$d/19$	10	80
7 × 7 aircraft	$1.70d^2$	—	$\frac{1}{16}$–$\frac{3}{8}$	Corrosion-resistant steel	—	—	124
				Carbon steel	—	—	124
7 × 9 aircraft	$1.75d^2$	—	$\frac{1}{8}$–$1\frac{3}{8}$	Corrosion-resistant steel	—	—	135
				Carbon steel	—	—	143
19-wire aircraft	$2.15d^2$	—	$\frac{1}{32}$–$\frac{5}{16}$	Corrosion-resistant steel	—	—	165
				Carbon steel	—	—	165

[a] The modulus of elasticity is only approximate: it is affected by the loads on the rope and, in general, increases with the life of the rope.
[b] The strength is based on the nominal area of the rope. The figures given are only approximate and are based on 1-in rope sizes and $\frac{1}{4}$-in aircraft-cable sizes.
Source: Compiled from American Steel and Wire Company Handbook.

TABLE 21-43
Common wire rope application

		Sheave diameter, cm	
Type of service	Rope construction	Recommended	Minimum
Haulage rope	6 × 7	$72d$	$42d$
Mine haulage			
Factory-yard haulage			
Inclined planes			
Tramways			
Power transmission			
Guy wires			
Standard hoisting rope	6 × 19	$45d$	$30d$
(Most commonly used rope)		60–$100d$	
Mine hoists			
Quarries			
Ore docks			
Cargo hoists			
Car pullers			
Cranes			
Derricks		20–$30d$	
Tramways			
Well drilling			
Elevators			
Extraflexible hoistings rope	8 × 19	$31d$	$21d$
Special flexible hoisting rope	6 × 37	$27d$	$18d$
Steel-mill ladles			
Cranes			
High speed elevators			

TABLE 21-44
Recommended safety factors for wire ropes

Rope application	Safety factor		
	100 or other figure laid down by the statutory authority		
	Class 1	Classes 2, 3	Class 4
From Indian Standards			
Mining ropes	3.5	4.0	4.5
Wire ropes used on the cranes and other hoisting equipment			
Fixed guys			
Unreeved rope bridles of jib cranes or ancillary appliances, such as lifting beams			
Ropes which are straight between terminal fittings			
Hoisting, luffing and reeved bridle systems of inherently flexible crances (e.g., mobile crawler tower, guy derrick, stiffleg derrick) where jibs are supported by ropes or where equivalent shock absorbing devices are incorporated in jib supports	4.0	4.5	5.5
Cranes and hoists in general hoist blocks	4.5	5.0	6.0
From Other Sources			
Mine Shafts			
Depths to 152 m	8		
305–610 m	7		
610–915 m	6		
>915 m	5		
Haulage ropes	6		
Small electric and air hoists	7		
Hot ladle cranes	8		
Slings	8		

Source: IS 3973, 1967.

TABLE 21-45
Ratio of drum and sheave diameter to rope diameter

Purpose	Construction	Minimum, ratio[a]		
		100		
Mining Installation	All	Class 1	Classes 2, 3	Class 4
Cranes and allied hoisting equipment	6 × 15	15	17	22
	8 × 19 filler wire			
	8 × 19	17	18	24
	8 × 19 Warrington			
	8 × 19 Seale			
	34 × 7 nonrotating			
	6 × 24	18	19	25
	6 × 19 filler wire	18	20	23
	6 × 19	19	23	27
	6 × 19 Warrington			
	17 × 7 nonrotating			
	18 × 7 nonrotating			
	6 × 19 Seale	24	28	35

[a] The ratio of the sheave diameters specified are valid for rope speeds up to 50 m/min. For speeds above 50 m/min, the drum or sheave diameter should be increased pro rata by 8% for each additional 50 m/min of rope speed where practicable.

Particular	Formula
The working load for the ordinary steel BB crane chain	$P_w = 93{,}750d^2$ **SI** (21-79a) where d in m and P_w in kN
	$P_w = 13{,}600d^2$ **USCS** (21-79b) where d in in and P_w in lbf
	$P_w = 9.56d^2$ **Customary Metric** (21-79c) where d in mm and P_w in kgf
The sheave diameter	$D = 20d$ to $30d$ (21-80)

Round steel short link and round steel link chain

LENGTH AND WIDTH (Figs. 21-17 and 21-18):

Particular	Formula
The outside dimensions of the links shall fall between the following limits:	
Outside link length limits (Fig. 21-17)	$l \not> 5d_n$ for uncalibrated chain (21-81a)
	$l \not< 6d_n$ for calibrated chain (21-81b)
Maximum outside link width (Fig. 21-18)	$W_{max} \not> 3.5d_n$ away from weld (21-82a)
	$W_{max} \not> 1.05$ (adjacent width) at weld for noncalibrated chains (21-82b)
Minimum inside link width	$W_{max} = 3.25d_n$ for calibrated chain (21-83)
	$W_t \not< 1.25d_n$ except at the weld for noncalibrated chain (25-84)
Pitch (i.e., inside length)	$p = 3d_n$ for calibrated chain (21-85)
Dimensions and lifting capacities and properties of noncalibrated and calibrated chains	Refer to Tables 21-46 to 21-51.

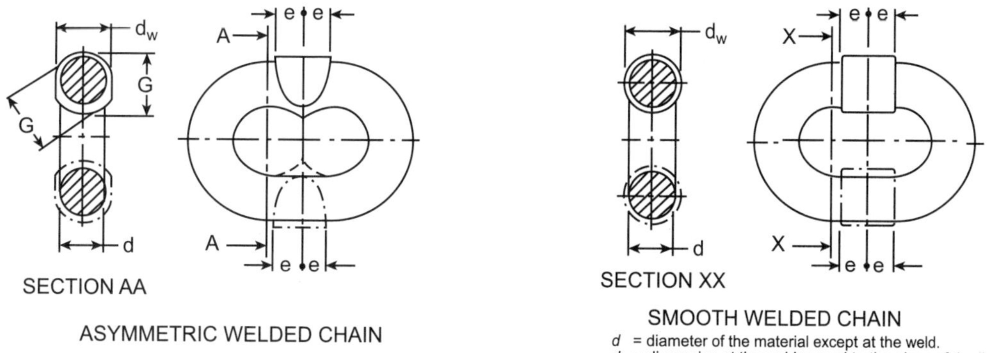

d = diameter of the material except at the weld.
d_w = dimension at the weld normal to the plane of the link.
G = dimension in other planes.
e = length effected by welding on either side of the link.

FIGURE 21-17 Diameter of material and welded chain.

TABLE 21-46
Dimensions and lifting capacities of grade 30 noncalibrated chain (Figs. 21-17 and 21-18)

Nominal size, d_n, mm	Diameter tolerance $d_n \not> 16$, $+0.02s$ $-0.06s$; $d_n \geq 16$, $\pm 0.05s$	Maximum additional weld dimensions $(d_w - d)$ max	$(G - d)$ max	Outside link length limits $5d_n$	$4.75d_n$	Maximum outside link width, W: Away from weld, W_{max} $3.5d_n$	Max extra at weld, $0.5W_{max}$	Minimum inside link width, $1.25d_n$	Guaranteed minimum breaking load stress 30h bar, kN	Minimum energy absorption factor (energy absorption 0.054 kJ m^{-1} mm^{-2}), kJ/m	Minimum safe working load (stress 7.5h bar), kN	Lifting capacity (stress 7.6h bar), tonnes
6.3	+0.12, −0.36	1.2	2.1	32	30	22	1.1	7.9	18.9	3.4	4.8	0.50
7.1	+0.14, −0.42	1.4	2.4	36	34	25	1.25	8.9	23.6	4.3	5.9	0.63
8.0	+0.16, −0.48	1.6	2.8	40	38	28	1.4	10	30.2	5.5	7.5	0.80
9.0	+0.18, −0.54	1.8	3.1	45	43	31	1.6	1.1	38.1	6.9	9.5	1.00
10.0	+0.20, −0.60	2.0	3.5	50	48	35	1.8	12	47.1	8.5	11.8	1.25
11.2	+0.22, −0.66	2.2	3.9	56	53	39	2.0	14	59.2	10.7	14.8	1.6
12.5	+0.25, −0.75	2.5	4.4	62	59	44	2.2	16	73.8	13.4	18.5	2.0
14.0	+0.28, −0.84	2.8	4.9	70	66	49	2.5	18	93.0	16.7	23.2	2.5
16.0	+0.32, −0.96	3.2	5.6	80	76	56	2.8	20	120.0	22.0	30.0	3.2
18.0	+0.90	3.6	6.3	90	86	63	3.1	22	153.0	39.0	38.2	4.0
20.0	+1.0	4.0	7.0	100	95	70	3.5	25	189.0	32.0	49.0	5.0
22.0	+1.1	4.4	7.7	110	105	77	3.9	28	228.0	42.0	57.0	6.3
25.0	±1.2	5.0	8.7	125	120	87	4.4	31	296.0	53.0	74.0	8.0
28.0	±1.4	5.6	9.8	140	130	98	4.9	35	372.0	67.0	93.0	10.0
32.0	±1.6	6.4	1.1	160	150	110	5.5	40	483.0	87.0	121.0	12.5
36.0	±1.9	7.2	1.2	180	170	120	6.0	45	610.0	112.0	152.0	16.0
40.0	±2.0	8.0	1.4	200	190	140	7.0	50	757.0	136.0	189.0	20.0
45.0	±2.2	9.0	1.6	225	215	160	8.0	56	953.0	173.0	228.0	22.5

Source: IS 2429 (Part I), 1970.

TABLE 21-47
Dimensions and lifting capacities of grade 30 calibrated chain (Figs. 21-17 and 21-18)

Nominal size, d_n, mm	Diameter tolerance $d_n \not> 16$, $+0.02s$ $-0.06s$; $d_n \geq 16$, $\pm 0.05s$	Maximum additional weld dimensions $(d_w - d)$ max	Maximum additional weld dimensions $(G - d)$ max	Preferred pitch (inside length), $3d_n$	Pitch tolerance (one link), $0.00396d_n$	Preferred outside width, $w = 3.25d_n$	Outside width tolerance away from weld zone $+0.075d_n$ 0	At weld zone $+0.15d_n$ 0	Guaranteed minimum breaking load (stress $30h$ bar), kN	Minimum energy absorption factor (energy absorption 0.054 kJ m^{-1} mm^{-2}), kJ/m	Maximum safe working load (stress $7.5h$ bar), kN	Lifting capacity (stress $7.5h$ bar), tonnes
6.3	+0.12, −0.36	0.48	2.1	19	0.26	20	0.45	0.90	18.9	3.4	4.8	0.50
7.1	+0.14, −0.42	0.56	2.4	21	0.30	23	0.52	1.0	23.6	4.3	5.9	0.63
8.0	+0.16, −0.48	0.64	2.8	24	0.33	26	0.59	1.1	30.2	5.5	7.5	0.80
9.0	+0.18, −0.54	0.72	3.1	27	0.36	29	0.67	1.3	38.1	6.9	9.5	1.00
10.0	+0.20, −0.60	0.80	3.5	30	0.40	32	0.75	1.5	47.1	8.5	11.8	1.60
11.2	+0.22, −0.66	0.88	3.9	34	0.44	36	0.84	1.7	59.2	10.7	14.8	1.60
12.5	+0.25, −0.75	1.0	4.4	37	0.49	41	0.93	1.9	73.8	13.4	18.5	2.0
14.0	+0.28, −0.80	1.1	4.9	42	0.55	46	1.05	2.1	93.0	16.7	23.2	2.5
16.0	+0.32, −0.96	1.2	5.6	48	0.63	52	1.20	2.4	120	22.0	30.0	3.2
18.0	±0.90	1.4	6.3	54	0.71	58	1.35	2.7	153	26.0	38.2	4.0
20.0	±1.0	1.6	7.0	60	0.79	65	1.50	3.0	189	39.0	49.0	5.0
22.0	±1.1	1.8	7.7	66	0.87	73	1.70	3.4	228	42.0	57.0	6.3
25.0	±1.2	2.0	8.7	75	0.99	82	1.90	3.8	296	53.0	74.0	8.0
28.0	±1.4	2.2	9.8	84	1.1	91	2.10	4.2	372	67.0	93.0	10.0
32.0	±1.6	2.5	1.1	96	1.2	100	2.40	4.8	483	87.0	121.0	12.5
36.0	±1.9	2.8	12	108	1.4	110	2.70	5.4	610	112.0	152.0	16.0
40.0	±2.0	3.2	14	120	1.6	130	3.00	6.0	757	136.0	189.0	20.0
45.0	±2.2	3.6	16	155	1.8	150	3.40	6.8	953	173.0	228.0	22.5

Source: IS 2429 (Part II), 1970.

TABLE 21-48
Dimensions and lifting capacities of grade 40 noncalibrated chain (Figs. 21-17 and 21-18)

Nominal size, d_n, mm	Diameter tolerance $d_n \not> 16$, $+0.02s$ $-0.06s$; $d_n \geq 16$, $\pm 0.05s$	Maximum additional weld dimensions $(d_w - d)$ max	$(G - d)$ max	Outside link length limits $5d_n$	$4.75d_n$	Maximum outside link width, W: Away from weld, W_{max} $3.5d_n$	Max extra at weld, $0.05W_{max}$	Minimum inside link width, $1.25d_n$	Guaranteed minimum breaking load stress 30h bar, kN	Minimum energy absorption factor (energy absorption 0.072 kJ m^{-1} mm^{-2}), kJ/m	Minimum safe working load (stress 10h bar), kN	Lifting capacity (stress 10h bar), tonnes
6.3	0.12, −0.36	1	2.1	30	28	21	1.0	7.5	24.9	4.50	6.2	0.63
7.1	0.14, −0.42	1.4	2.4	35	33	24	1.24	8.8	31.6	4.70	7.9	0.80
8.0	0.16, −0.48	1.6	2.8	40	38	28	1.4	10	40.2	7.25	10.0	1.00
9	0.18, −0.54	1.8	3.1	45	43	31	1.6	1.1	50.9	9.18	12.7	1.25
10	0.20, −0.60	2.0	3.5	50	48	35	1.8	12	62.8	11.30	15.7	1.6
11	0.22, −0.66	2.2	3.9	55	52	39	2.0	14	79.0	14.20	19.7	2.0
12.5	0.25, −0.75	2.5	4.4	62	59	44	2.2	16	98.4	17.7	24.5	2.5
14.0	0.28, −0.84	2.8	4.9	70	66	49	2.5	18	124.0	22.2	30.8	3.2
16.0	0.32, −0.96	3.2	5.6	80	76	56	2.8	20	161.0	29.0	40.3	4.0
18.0	+0.90	3.6	6.3	90	86	63	3.1	22	204.0	37.7	50.5	5.0
20.0	+1.0	4.0	7.0	100	95	70	3.5	25	252.0	45.3	63.0	6.3
22.0	+1.1	4.4	7.7	110	105	77	3.9	28	304.0	55.0	76.0	8.0
25.0	±1.2	5.0	8.7	125	120	87	4.4	31	394.0	70.7	98.5	10.0
28.0	±1.4	5.6	9.8	140	130	98	4.9	35	492.0	89.0	123.0	12.5
32.0	±1.6	6.4	11	160	150	110	5.5	40	644.0	116.0	161.0	16.0
36.0	±1.9	7.2	12	180	170	120	6.0	45	814.0	147.0	204.0	20
40.0	±2.0	8.0	14	200	190	140	7.0	50	1010.0	181.0	252.0	25
45.0	±2.2	9.0	16	225	215	160	8.0	56	1270.0	230.0	318.0	32

Source: IS 3109 (Part I), 1970.

TABLE 21-49
Dimensions and lifting capacities of grade 40 calibrated chain (Figs. 21-17 and 21-18)

Nominal size, d_n, mm	Diameter tolerance $d_n \not> 16$, $+0.02s$ $-0.06s$; $d_n \geq 16$, $\pm 0.05s$	Maximum additional weld dimensions $(d_w - d)$ max	$(G - d)$ max	Preferred pitch (inside length), $3d_n$	Pitch tolerance (one link), $0.0396d_n$	Preferred outside width, $w = 3.25d_n$	Tolerance on outside width: Away from weld zone $+0.075d_n$ 0	At weld zone $+0.167d_n$ 0	Guaranteed minimum breaking load (stress 40h bar), kN	Minimum energy absorption factor (energy absorption 0.072 kJ m^{-1} mm^{-2}), kJ/m	Maximum safe working load (stress 10h bar), kN	Lifting capacity (stress 10h bar), tonnes
6.3	0.12, −0.36	0.48	2.1	19	0.26	20	0.45	1.1	24.9	4.50	6.20	0.63
7.1	0.14, −0.42	0.56	2.4	21	0.30	23	0.52	1.2	31.6	5.70	7.80	0.70
8.0	0.16, −0.48	0.64	2.8	24	0.33	26	0.59	1.4	42.2	7.25	10.00	1.0
9.0	0.18, −0.54	0.72	3.1	27	0.36	29	0.57	1.9	50.9	9.18	12.7	1.25
10.0	0.20, −0.60	0.80	3.5	30	0.40	32	0.75	1.7	62.8	11.3	15.7	1.60
11.2	0.22, −0.66	0.88	3.9	34	0.44	36	0.88	1.9	79.0	14.2	19.7	2.00
12.5	0.25, −0.75	1.0	4.4	37	0.49	41	0.93	2.1	98.4	17.7	24.5	2.5
14.0	0.28, −0.80	1.1	4.9	42	0.55	46	1.05	2.4	124	22.2	30.8	3.2
16.0	0.32, −0.96	1.2	5.6	48	0.63	52	1.20	2.7	161	29.0	40.3	4.0
18.0	±0.90	1.4	6.3	54	0.71	58	1.35	3.0	204	37.7	50.5	5.0
20.0	±1.0	1.6	7.0	60	0.79	65	1.50	3.4	252	45.3	63.0	6.3
22.0	±1.1	1.8	7.7	66	0.87	73	1.70	3.8	304	55.0	76.0	8.0
25.0	±1.2	2.0	8.7	75	0.99	82	1.90	4.3	394	70.7	98.5	10.0
28.0	±1.4	2.2	9.8	84	1.1	91	2.10	4.8	492	890	123	12.5
32.0	+1.6	2.5	11	96	1.2	100	2.40	5.0	644	116	161	16.0
36.0	+1.9	2.8	12	108	1.4	110	2.70	6.1	814	147	204	20.0
40.0	+2.0	3.2	14	120	1.6	130	3.00	6.8	1010	181	252	25.0
45.0	+2.2	3.6	16	135	1.8	150	3.40	7.6	1270	230	318	32.0

Source: IS 3102 (Part II), 1970.

TABLE 21-50
Requirements of arc welded grade 30 chain for lifting purposes

Size (nominal diameter), mm	Proof load based on a stress of 98.1 MPa (10 kgf/mm^2)		Minimum breaking load based on a stress of 294.2 MPa (30 kgf/mm^2)		Minimum energy absorption factor for 1-m gauge length based on an energy absorption of 58.8 MN-m/m^2 (6 kgf-m/mm^2)		Maximum safe working load for nominal working condition based on a stress of 49 MPa (5 kgf/mm^2)	
	kN	kgf	kN	kgf	N m	kgf-m	kN	kgf
6	8.6	570	16.7	1700	3.3	340	2.8	285
8	9.8	1000	29.5	3010	5.9	602	4.9	500
9	12.5	1270	37.5	3820	7.5	764	6.2	635
10	15.4	1570	46.2	4710	9.2	942	7.7	785
12	22.2	2260	66.5	6780	13.3	1356	11.1	1130
14	30.2	3080	90.6	9140	18.1	1848	15.1	1540
16	39.4	4020	118.3	12060	23.7	2412	19.7	2010
18	49.9	5090	149.8	15270	30.0	3054	25.0	2545
20	61.6	6280	184.9	18850	37.0	3770	30.8	3140
22	74.5	7600	223.7	22810	44.7	4562	37.3	3800
24	88.8	9050	266.2	27140	53.2	5428	44.4	4525
27	102.5	10450	336.9	34350	67.4	6870	56.14	5725
30	138.7	14140	415.9	42410	83.2	8482	69.3	7070
33	167.7	17100	503.3	51320	100.7	10264	83.9	8550
36	192.7	20360	598.8	61070	119.8	12214	99.8	10180
39	234.4	23900	702.9	71680	140.6	14336	117.2	11950

TABLE 21-51
Requirements for electrically welded steel chain grade 30 chain for lifting purposes

Size (nominal diameter), mm	Proof load based on a stress of 157 MPa (16 kgf/mm^2)		Minimum breaking load based on a stress of 392.3 MPa (40 kgf/mm^2)		Minimum energy absorption factor for 1-m gauge length based on an energy absorption of 78.5 MN-m/m^2 (8 kgf-m/mm^2)		Maximum safe working load for nominal working condition based on a stress of 49 MPa (5 kgf/mm^2)	
	kN	kgf	kN	kgf	N m	kgf-m	kN	kgf
5	6.1	628	15.4	1571	3.1	314	3.1	314
6	8.9	904	22.2	2262	4.4	452	4.4	452
7	12.1	1232	30.2	3079	6.0	616	6.0	616
8	15.8	1608	39.4	4021	7.9	804	7.9	804
9	20.0	2036	49.9	5089	10.0	1018	10.0	1018
9.5	22.2	2268	55.6	5671	11.1	1134	11.1	1134
10	24.7	2514	61.6	6283	12.3	1257	12.3	1257
11	29.8	3042	74.6	7603	14.9	1521	14.9	1521
12	38.5	3928	96.3	9818	19.3	1964	19.3	1964
14	48.3	4926	120.8	12315	24.2	2463	24.2	2463
16	63.1	6434	157.7	16085	31.6	3217	31.6	3217
18	79.9	8144	199.6	20358	39.9	4072	39.9	4072
20	98.6	10054	246.5	25133	49.2	5027	49.3	5027
22	119.3	12164	298.2	30411	59.6	6082	59.6	6082
24	142.0	14476	354.9	36191	71.0	7238	71.0	7238
26	166.6	16990	416.5	42474	83.3	8495	83.3	8495
28	193.2	19704	483.1	49260	96.6	9852	96.6	9852
30	221.8	22620	554.6	56549	110.9	11310	110.9	11310
33	268.4	27370	671.0	68424	134.2	13685	134.2	13685
36	319.4	32572	798.6	81430	159.7	16286	153.7	16286
39	374.9	38228	937.2	95567	187.4	19114	187.4	19114
42	434.8	44334	1086.9	110836	217.4	22167	217.4	22167

Particular	Formula

Chain passing over a sheave (Fig. 21-19)

The effort on the chain in case of single-sheave pulley (Fig. 21-19)

$$P = \left(\frac{D + \mu D_o + \mu_c d}{D - \mu D_o - \mu_c d}\right) Q = CQ \quad (21\text{-}86)$$

where $C = 1.04$ for lubricated chains
$C = 1.10$ for chains running dry

The efficiency of the chain sheave

$$\eta = \frac{1}{C} \quad (21\text{-}87)$$

where $\eta = 0.96$ for lubricated chains
$\eta = 0.91$ for chain running dry

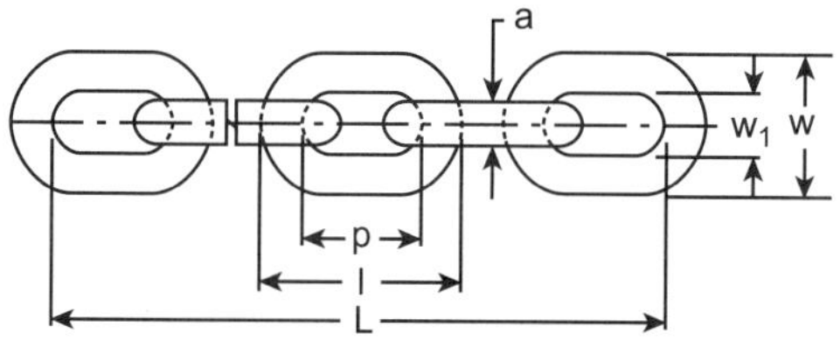

FIGURE 21-18 Pitch length and width of link.

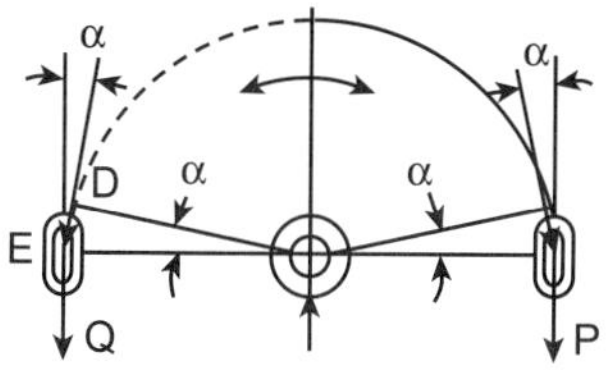

FIGURE 21-19 Chain passing over sheave.

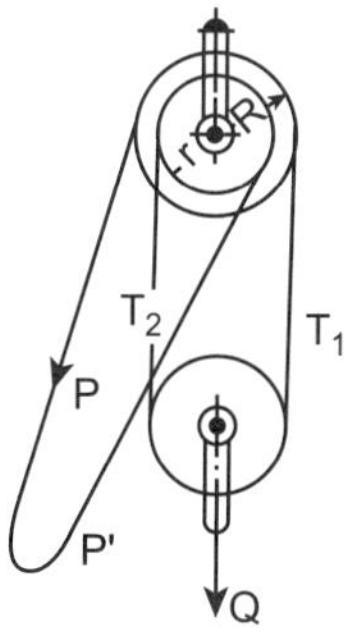

FIGURE 21-20 Differential chain block.

Differential chain block (Fig. 21-20)

RAISING THE LOAD Q

The effort required for raising the load without friction

$$P_o = \frac{Q}{2}(1 - n) \quad (21\text{-}88)$$

where $n = \dfrac{d}{D} = \dfrac{r}{R}$

The relation between the tension in the running-off and running-on chains

$$T_1 = C_1 T_2 \quad (21\text{-}89)$$

where C_1 depends on the size of the chain and diameter of the lower sheave

The tension in the running-off chains

$$T_1 = \frac{C_t}{1 + C_1} Q \quad (21\text{-}90)$$

The tension in the running-on chain

$$T_2 = \frac{Q}{1 + C_1} \quad (21\text{-}91)$$

Particular	Formula
The relation between effort (P), load (Q), T_1 and T_2	$PR + T_2 r = C_2 T_1 R$ (21-92) where C_2 depends on the size of the chain and diameter of upper sheave
The effort required for raising the load with friction	$P = \left(\dfrac{C_2 C_2 - n}{1 + C_1}\right) Q$ (21-93) when C_1 and C_2 are different Or $P = \left(\dfrac{C^2 - n}{1 + C}\right) Q$ (21-94) where C is the average value of C_1 and C_2
The efficiency for the differential chain hoist	$\eta = \left(\dfrac{1 - n}{2}\right)\left(\dfrac{1 + C}{C^2 - n}\right)$ (21-95)
Lowering the load	
The equations for the tension in the running-on running-off and pull (P') required on the chain so as to prevent running down of the load	$T_1 = \dfrac{Q}{1 + C}$ (21-96) $T_2 = \dfrac{CQ}{1 + C}$ (21-97) $T_1 R = CP'R + CT_2 r$ (21-98)
The pull required on the chain so as to prevent running down of the load	$P' = \dfrac{Q}{C}\left(\dfrac{1 - nC^2}{1 + C}\right)$ (21-99)
The efficiency for the reversed motion	$\eta' = \dfrac{2}{C}\left(\dfrac{1 - nC^2}{(1 - n)(1 + C)}\right)$ (21-100) where C varies from 1.054 to 1.09 or obtained from Table 21-33
For mechanical properties of the coil link chain and the strength of hoisting chains in terms of bar from which they are made	Refer to Tables 21-52 and 21-53.

TABLE 21-52
Mechanical properties of the coil link chain

Properties	Requirement	
	Grade 30	Grade 40
Mean stress at guaranteed minimum breaking load, F_w min, h bar	30	40
Mean stress at proof load, F_e, h bar	15	20
Ratio of proof load of guaranteed minimum breaking load	50%	50%
Guaranteed minimum elongation at fracture, A min	14.4%	14.4%
Guaranteed minimum energy absorption factor, $F_w \times A$	0.054 kJ m^{-1} mm^{-2}	0.054 kJ m^{-1} mm^{-2}
Maximum safe working load mean stress, h bar	7.5	10

TABLE 21-53
The strength of hoisting chains in terms of the bars from which they are made

Particular	% of bar
Standard close link	138
Coil chain	120
BB crane chain	145
Stud chain	165

Particular	Formula

Conditions for self-locking of differential chain block

The condition for self-locking

$$P' = \frac{Q}{C}\left(\frac{1 - nC^2}{1 + C}\right) \leq 0 \qquad (21\text{-}101)$$

For self-locking differential chain block

$$n > \frac{1}{C^2} \qquad (21\text{-}102)$$

The initial value of the ratio $\frac{r}{R}$

$$n = \frac{1}{C^2} \qquad (21\text{-}103)$$

Power chains

Roller chains

The transmission ratio

$$i - \frac{\omega_1}{\omega_2} = \frac{n_1}{n_2} = \frac{d_2}{d_1} = \frac{z_2}{z_1} \qquad (21\text{-}104)$$

The average speed of chain

$$v = \frac{pz_1 n_1}{60}\ \text{m/s} \quad \text{or} \quad v = \frac{pz_1 n_1}{12}\ \text{ft/min} \qquad (21\text{-}105)$$

where z_1 = number of teeth on the small sprocket and p in m (in)

Particular	Formula
The empirical formula for pitch	$p \le 0.25\left(\frac{900}{n_1}\right)^{2/3}$ **SI** (21-106a) where p in m $p \le \left(\frac{900}{n_1}\right)^{2/3}$ **USCS** (21-106b) where p in in $p \le 250\left(\frac{900}{n_1}\right)^{2/3}$ **Customary Metric** (21-106c) where p in mm, n_1 = speed of the small sprocket, rpm
Bartlett formula relating speed (n_1) and pitch (p) based on allowable amount of impact between a roller and a sprocket	$n_1 = \frac{1170}{p}\sqrt{\frac{A}{w_f p}}$ **SI** (21-107a) where n_1 in rpm, p in m, w_f in N/m, and A in m^2 $n_1 = \frac{11{,}800}{p}\sqrt{\frac{A}{w_f p}}$ **Customary Metric** (21-107b) where n_1 in rpm, p in mm, w_f in kgf/m, and A in mm^2 $n_1 = \frac{1920}{p}\sqrt{\frac{A}{w_f p}}$ **USCS** (21-107c) where n_1 in rpm, p in in, w_f in lbf/ft, and A in in^2 $A = ld_r$ = projected area of the roller d_r = diameter of rollers l = width of chain or length of roller
Maximum allowable chain velocity based on Eq. (21-107)	$v_{max} \le 19.48 z_1 \sqrt{\frac{A}{w_f p}}$ **SI** (21-108a) where v_{max} in m/s, A in m^2, p in m, and w_f in N/m $v_{max} \le 160 z_1 \sqrt{\frac{A}{w_f p}}$ **USCS** (21-108b)

Particular	Formula
	where v_{max} in ft/min, A in in^2, p in in, and w_f in lbf/ft
	$v_{max} \le 0.196 z_1 \sqrt{\dfrac{A}{w_f p}}$ **Customary Metric** (21-108c)
	where v_{max} in m/s, A in mm^2, p in mm, and w_f in kgf/m
Maximum speed based on the energy of impact per tooth per minute	$n \le \dfrac{1437}{p} \sqrt[3]{\dfrac{A}{w_f}}$ **SI** (21-109a)
	where A in m^2, p in m, and w_f in N/m
	$n \le \dfrac{2000}{p} \sqrt[3]{\dfrac{A}{w_f}}$ **USCS** (21-109b)
	where A in in^2, p in in, and w_f in lbf/ft
	$n \le \dfrac{6712}{p} \sqrt[3]{\dfrac{A}{w_f}}$ **Customary Metric** (21-109c)
	where A in mm^2, p in mm, and w_f in kgf/m
Maximum chain velocity based on Eq. (21-109), m/s	$v_{max} \le 24 z_1 \sqrt[3]{\dfrac{A}{w_f}}$ **SI** (21-110a)
	where v_{max} in m/s, A in m^2, and w_f in N/m
	$v_{max} \le 166 z_1 \sqrt[3]{\dfrac{A}{w_f}}$ **USCS** (21-110b)
	where v_{max} in ft/min, A in in^2, and w_f in lbf/ft
	$v_{max} \le 0.11 z_1 \sqrt[3]{\dfrac{A}{w_f}}$ **Customary Metric** (21-110c)
	where v_{max} in m/s, A in mm^2, and w_f in kgf/m
Maximum sprocket speed based on the effect of centrifugal force	$n \le \dfrac{36350}{p} \sqrt{\dfrac{A}{z_1 w_f}}$ **SI** (21-111a)
	where p in m, A in m^2, and w_f in N/m
	$n \le \dfrac{9516}{p} \sqrt{\dfrac{A}{z_1 w_f}}$ **USCS** (21-111b)
	where p in in, A in in^2, and w_f in lbf/ft

Particular	Formula
Maximum velocity based on Eq. (21-111)	$v_{max} \leq 600\sqrt{\frac{Az_1}{w_f}}$ **SI** (21-112a)
	where v_{max} in m/s, A in m^2, and w_f in N/m
	$v_{max} \leq 793\sqrt{\frac{Az_1}{w_f}}$ **USCS** (21-112b)
	where v_{max} in ft/min, A in in^2, and w_f in lbf/ft
	$v_{max} \leq 0.2\sqrt{\frac{Az_1}{w_f}}$ **Customary Metric** (21-112c)
	where v_{max} in m/s, A in mm^2, and w_f in kgf/m

Chain pull

For preliminary computation, the allowable pull	$F_a = \frac{F_u}{n_o}$ (21-113)
	where
	F_u = ultimate strength from Tables 21-35B and 21-42
	n_o = working factor,
	n_o = 5 for sprocket having over 40 teeth and a speed of 0.5 m/s
	n_o = 18 for sprocket having 10 or 11 teeth and a speed of 6 m/s
AGMA formula for allowable pull based on velocity factor $C_v = 3/(3+v)$ and bearing pressure of 29.4 MPa (4333 psi) for the pin	$F_a = \frac{90 \times 10^6 l d_r}{3+v} - \frac{v^2 w_f}{9.8}$ **SI** (21-114a)
	where l and d_r in m, v in m/s, and w_f in N/m
	$F_a = \left(\frac{l d_r}{600+v} - \frac{v^2 w_f}{3(10^{11})}\right)2{,}600{,}000$ **USCS** (21-114b)
	where l and d_r in in, v in ft/min, and w_f in lbf/ft
	where
	l = length of roller pins, m (in)
	$v = \frac{z_1 p n_1}{60}$ m/s
	d_r = roller pin diameter, m (in)
For dimensions of American Standard Roller Chains—single-strand	Refer to Tables 21-54A.

TABLE 21-54A
Dimensions of American Standard roller chains—single-strand

ANSI chain number	Pitch, in (mm)	Width, in (mm)	Minimum tensile strength, lb (N)	Average weight, lb/ft (N/m)	Roller diameter, in (mm)	Multiple-strand spacing, in (mm)
25	0.250	0.125	780	0.09	0.130	0.252
	(6.35)	(3.18)	(3470)	(1.31)	(3.30)	(6.40)
35	0.375	0.188	1760	0.21	0.200	0.399
	(9.52)	(4.76)	(7830)	(3.06)	(5.08)	(10.13)
41	0.500	0.25	1500	0.25	0.306	—
	(12.70)	(6.35)	(6670)	(3.65)	(7.77)	—
40	0.500	0.312	3130	0.42	0.312	0.566
	(12.70)	(7.94)	(13920)	(6.13)	(7.92)	(14.38)
50	0.625	0.375	4880	0.69	0.400	0.713
	(15.88)	(9.52)	(21700)	(10.1)	(10.16)	(18.11)
60	0.750	0.500	7030	1.00	0.469	0.897
	(19.05)	(12.7)	(31300)	(14.6)	(11.91)	(22.78)
80	1.000	0.625	12500	1.71	0.625	1.153
	(25.40)	(15.88)	(55600)	(25.0)	(15.87)	(29.29)
100	1.250	0.750	19500	2.58	0.750	1.409
	(31.75)	(19.05)	(86700)	(37.7)	(19.05)	(35.76)
120	1.500	1.000	28000	3.87	0.875	1.789
	(38.10)	(25.40)	(124500)	(56.5)	(22.22)	(45.44)
140	1.750	1.000	38000	4.95	1.000	1.924
	(44.45)	(25.40)	(169000)	(72.2)	(25.40)	(48.87)
160	2.000	1.250	50000	6.61	1.125	2.305
	(50.80)	(31.75)	(222000)	(96.5)	(28.57)	(58.55)
180	2.250	1.406	63000	9.06	1.406	2.592
	(57.15)	(35.71)	(280000)	(132.2)	(35.71)	(65.84)
200	2.500	1.500	78000	10.96	1.562	2.817
	(63.50)	(38.10)	(347000)	(159.9)	(39.67)	(71.55)
240	3.00	1.875	112000	16.4	1.875	3.458
	(76.70)	(47.63)	(498000)	(239.0)	(47.62)	(87.83)

Source: Compiled from ANSI B29.1-1975.

TABLE 21-54B
Rated horsepower capacity of single-strand single-pitch roller chain for a 17-tooth sprocket

Sprocket speed, rpm	ANSI chain number					
	25	35	40	41	50	60
50	0.05	0.16	0.37	0.20	0.72	1.24
100	0.09	0.29	0.69	0.38	1.34	2.31
150	0.13[a]	0.41[a]	0.99[a]	0.55[a]	1.92[a]	3.32
200	0.16[a]	0.54[a]	1.29	0.71	2.50	4.30
300	0.23	0.78	1.85	1.02	3.61	6.20
400	0.30[a]	1.01[a]	2.40	1.32	4.67	8.03
500	0.37	1.24[a]	2.93	1.61	5.71	9.81
600	0.44[a]	1.46[a]	3.45[a]	1.90[a]	6.72[a]	11.6
700	0.50	1.68	3.97	2.18	7.73	13.3
800	0.56[a]	1.89[a]	4.48[a]	2.46[a]	8.71[a]	15.0
900	0.62	2.10	4.98	2.74	9.69	16.7
1000	0.68[a]	2.31[a]	5.48	3.01	10.7	18.3
1200	0.81	2.73	6.45	3.29	12.6	21.6
1400	0.93[a]	3.13[a]	7.41	2.61	14.4	18.1
1600	1.05[a]	3.53[a]	8.36	2.14	12.8	14.8
1800	1.16	3.93	8.96	1.79	10.7	12.4
2000	1.27[a]	4.32[a]	7.72[a]	1.52[a]	9.23[a]	10.6
2500	1.56	5.28	5.51[a]	1.10[a]	6.58[a]	7.57
3000	1.84	5.64	4.17	0.83	4.98	5.76
Type A	Type B				Type C	

[a] Estimated from ANSI tables by linear interpolation.
Note: Type A—manual or drip lubrication, type B—bath or disk lubrication; type C—oil-stream lubrication.
Source: Compiled from ANSI B29.1-1975 information only section, and from B29.9-1958.

TABLE 21-54C
Rated horsepower capacity of single-strand single-pitch roller chain for a 17-tooth sprocket

Sprocket speed, rpm		ANSI chain number 80	100	120	140	160	180	200	240
50	Type B	2.88	5.52	9.33	14.4	20.9	28.9	38.4	61.8
100		5.38	10.3	17.4	26.9	39.1	54.0	71.6	115
150		7.75	14.8	25.1	38.8	56.3	77.7	103	166
200		10.0	19.2	32.5	50.3	72.9	101	134	215
300		14.5	27.7	46.8	72.4	105	145	193	310
400		18.7	35.9	60.6	93.8	136	188	249	359
500		22.9	43.9	74.1	115	166	204	222	0
600		27.0	51.7	87.3	127	141	155	169	
700		31.0	59.4	89.0	101	112	123	0	
800		35.0	63.0	72.8	82.4	91.7	101		
900	Type A	39.9	52.8	61.0	69.1	76.8	84.4		
1000		37.7	45.0	52.1	59.0	65.6	72.1		
1200		28.7	34.3	39.6	44.9	49.0	0		
1400		22.7	27.2	31.5	35.6	0			
1600		18.6	22.3	25.8	0				
1800		15.6	18.7	21.6					
2000		13.3	15.9	0					
2500		9.56	0.40						
3000		7.25	0						
Type C						Type C′			

Note: Type A—manual or drip lubrication; type B—bath or disk lubrication; type C–oil-stream lubrication; type C′—type C, but this is a galling region; submit design to manufacturer for evaluation.
Source: Compiled from ANSI B29.1-1975 information only section, and from B29.9-1958.

TABLE 21-54D
Tooth correction factors, K_1

Number of teeth on driving sprocket	Tooth correction factor, K_1	Number of teeth on driving sprocket	Tooth correction factor, K_1
11	0.53	22	1.29
12	0.62	23	1.35
13	0.70	24	1.41
14	0.78	25	1.46
15	0.85	30	1.73
16	0.92	35	1.95
17	1.00	40	2.15
18	1.05	45	2.37
19	1.11	50	2.51
20	1.18	55	2.66
21	1.26	60	2.80

TABLE 21-54E
Multistrand factors K_2

Number of strands	K_2
1	1.0
2	1.7
3	2.5
4	3.3

TABLE 21-54F
Service factor for roller chains, k_s

Operating characteristics	Intermittent few hours per day, few hours per year	Normal 8 to 10 hours per day 300 days per year	Continuous 24 hours per day
Easy starting, smooth, steady load	0.06–1.00	0.90–1.50	0.90–2.00
Light medium shock or vibrating load	0.90–1.40	1.20–1.90	1.50–2.40
Medium to heavy shock or vibrating load	1.20–1.80	1.50–2.30	1.80–2.80

Particular	Formula

Power

For the rated horsepower capacity of single-strand-single-pitch roller chains for 17-tooth sprocket and values of K_1 and K_2

Refer to Tables 21-54B to 21-54E.

Power required

$$P = \frac{F_\theta v}{1000 k_l k_s} \quad \textbf{SI} \quad (21\text{-}115a)$$

where F_θ in N and P in kW

$$P = \frac{F_\theta v}{33{,}000 k_l k_s} \quad \textbf{USCS} \quad (21\text{-}115b)$$

where F_θ in lbf and P in hp

$$P = \frac{F_\theta v}{102 k_l k_s} \quad \textbf{Customary Metric} \quad (21\text{-}115c)$$

where

F_θ = required chain pull in kgf and P in kW
k_l = load factor from 1.1 to 1.5 and also obtained from Chap. 14
k_s = service factor
= 1 for 10 h service per day
= 1.2 for 24 h operation and also obtained from Table 21-54F

Particular	Formula
The rated horsepower of roller chain per strand	$P = p^2\left[\dfrac{v}{0.75} - \dfrac{v^{1.41}}{3.7}\left(1 + 5o\sin^2\dfrac{90}{z_1}\right)\right]$ (21-116)
The corrected horsepower (P_c)	$P_c = K_1 K_2 P_r$ (21-116a) where P_r = rated horsepower and K_1 and K_2 from Tables 21-54D and 21-54E

CHECK FOR ACTUAL SAFETY FACTOR

The actual safety factor checked by the formula	$n_a = \dfrac{F_u}{F_\theta + F_{cs} + F_s}$ (21-117) where $F_{cs} = \dfrac{wv^2}{g}$, $F_s = k_{sg}wC$ (21-117a) $F_\theta = \dfrac{33{,}000P}{v}$ (21-117b) where F_θ in lbf, P in hp, and v in ft/min $F_\theta = \dfrac{1000P}{v}$ where F_θ in N, P in kW, and v in m/s w = weight per meter of chain, N (lbf) v = velocity of chain, m/s (ft/min) C = center distance, m (in) k_{sz} = coefficient for sag from Table 21-55
The number of strand in a chain, if $F_\theta > F_a$	$i = \dfrac{F_\theta}{F_a}$ (21-118)

Center distance of chain length

The proper center distance between sprockets in pitches	$C_p = 20p$ to $30p$ or $C_p = 40 \pm 10$ pitches (21-119) where $pC_p = C$
The minimum center distance	$C_{min} = K_{min}C$ (21-120) where $C = \dfrac{d_{a1} + d_2}{2}$

TABLE 21-55
Coefficient for *sag*, k_{sg}

	Position of chain drive			
k_{sg}	**Horizontal**	**<40°**	**>40°**	**Vertical**
	6	4	2	1

Particular	Formula

TABLE 21-56
Values of k to he used in Eq. (21-123)

$(z_1 - z_2)/C_p$	0.1	1.0	2.0	3.0	4.0	5.0	6.0
k	0.02533	0.02538	0.02555	0.02584	0.02631	0.02704	0.02828

TABLE 21-57
Minimum center distance constant, K_{min}

Transmission ratio, i	Minimum center distance constant, K_{min}
3	$1 + (30\text{–}50/c')$
3–4	1.2
4–5	1.3
5–6	1.4
6–7	1.5

$$d_a = \frac{p}{\tan\left(\frac{180}{z}\right)} + 0.6p$$

Refer to Table 21-56 for values of k [used in Eq. (21-123)] and Table 21-57 for K_{min}.

The maximum center distance

$$C_{max} = 80p \tag{21-121}$$

where p = pitches of chain, mm

The chain length in pitches

$$L_p = 2C_p \cos\alpha + \frac{z_1 + z_2}{2} + \alpha\frac{z_1 - z_2}{180} \quad \text{(exact)} \tag{21-122}$$

$$L_p = 2C_p + \frac{z_1 + z_2}{2} + \frac{k(z_1 - z_2)^2}{C_p} \tag{21-123}$$

The chain length, m or in

$$L = 2C\cos\alpha + \frac{z_1 p(180 + 2\alpha)}{360} + \frac{z_2 p(180 - 2\alpha)}{360} \tag{21-124}$$

where

z_1 = number of teeth on a small sprocket
z_2 = number of teeth on a large sprocket
α = angle between tangent to the sprocket pitch circle and the center line

$$\alpha = \sin^{-1}\left(\frac{d_2 - d_1}{2C}\right)$$

k = a variable which depends on $\frac{z_1 - z_2}{C_p}$ and obtained from Table 21-56

Particular	Formula	
The chain length	$L = pL_p$	(21-125)
The pitch diameter of a sprocket	$d = \dfrac{p}{\sin\left(\dfrac{180}{z}\right)}$	(21-126)

Roller chain sprocket

Minimum number of teeth	$z_{min} = \dfrac{4d_r}{p} + 5$ for pitches of 25 mm	(21-127a)
	$z_{min} = \dfrac{4d_r}{p} + 4$ for pitches 32 to 58 mm	(21-127b)

Silent chain sprocket

Minimum number of teeth	$z_{min} = \dfrac{4d_r}{p} + 6$ for pitches to 51 mm	(21-128)
The root diameter of sprocket	$d_f = d - d_r$ where d_r = diameter of roller pin, m (in)	(21-129)
The width of sprocket tooth (Fig. 21-22)	$C_1 = l - 0.05p$ where l = chain width or roller length	(21-130)
Maximum hub diameter	$D_h = d\cos\dfrac{180}{z} - (H + 0.1270)$ where H = height of link plate, m or in $= 0.3p$	(21-131)
Power per cm of width in hp	$P = \dfrac{pv}{6.80}\left(1 - \dfrac{v}{2.16(z_1 - 8)}\right)$ where $v = \dfrac{pz_1n_1}{60}$ = chain speed, m/s; p in m	(21-132)
The relationship between depth of sag, and tension due to weight of chain in the catenary (approx.)	$h = 0.433\sqrt{S^2 - L^2}$	(21-133a)
	$F = w\left(\dfrac{S^2}{8h} + \dfrac{h}{2}\right)$ where h = depth of sag, m (in) L = distance between points of support, m (in) S = catenary length of chain, m (in) F = tension or chain pull, kN (lbf) w = weight of chain, kN/m (lbf/in)	(21-133b)

Particular	Formula

Tension chain linkages

Allowable load

$$F_a = 13.1 \times 10^6 p^2 \quad \textbf{SI} \quad (21\text{-}134a)$$

where p in m and F_u in N

$$F_a = 1900p^2 \quad \textbf{USCS} \quad (21\text{-}134b)$$

where p = pitch of chain, in, and F_u in lbf

Allowable load for lightweight chain

$$F_a = 7 \times 10^6 p^2 \quad \textbf{SI} \quad (21\text{-}134c)$$

where p in m

$$F_a = 1020p^2 \quad \textbf{USCS} \quad (21\text{-}134d)$$

where p in in, F in lbf

Indian Standards

PRECISION ROLLER CHAIN (Figs 21-21 to 21-25, Tables 21-58, 21-59, 21-60)

Pitch circle diameter (Fig. 21-21)

$$PCD = \frac{P}{\sin \frac{180}{z}} \quad (21\text{-}135)$$

Bottom diameter

$$BD = PCD - D_r \quad (21\text{-}136)$$

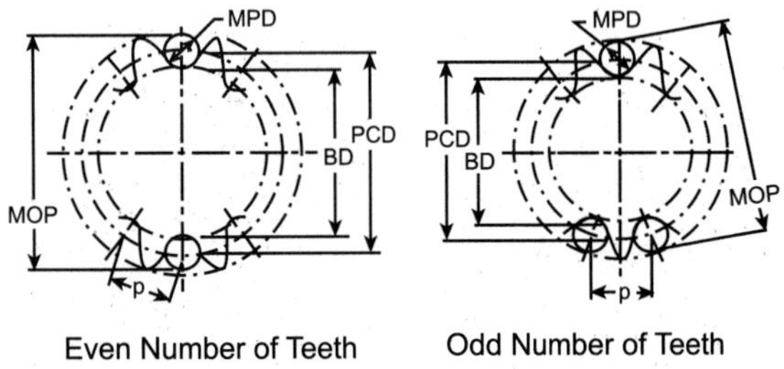

FIGURE 21-21 Notation for wheel rim of chain.

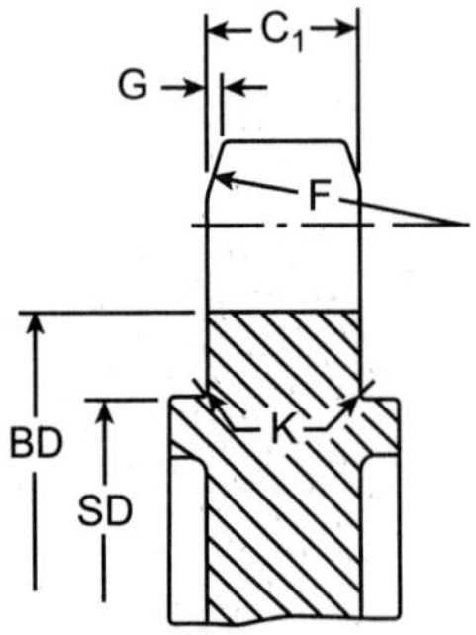

FIGURE 21-22 Notation for wheel rim profile of roller chain.

Wheel tooth gap form

The minimum value of roller seating radius, mm (Fig. 21-24)

$$SR_1 = 0.505D_r \quad (21\text{-}137)$$

The maximum value of roller seating radius, mm (Fig. 21-25)

$$SR_2 = 0.505D_r + 0.069 \sqrt[3]{D_r} \quad (21\text{-}138)$$

where D_r = roller diameter, mm

TABLE 21-58
Extended pitch transmission roller chain dimensions, measuring loads and breaking loads

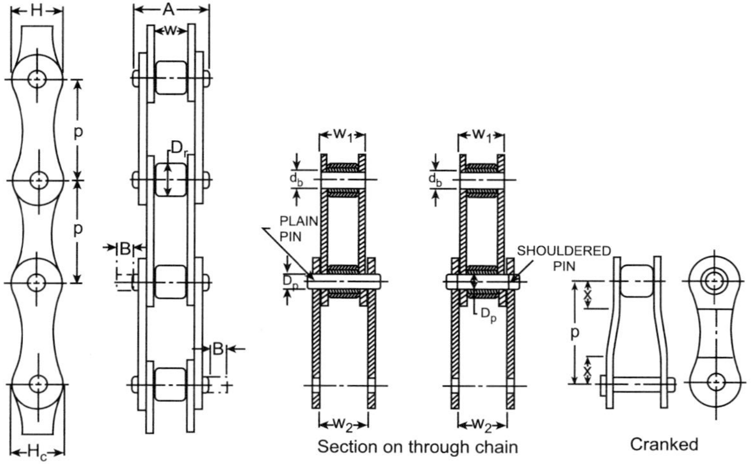

Chain no.	Pitch, P, mm	Roller diameter max, D_r, mm	Width between inner plates, W, min, mm	Bearing in diameter max, D_p, mm	Bush bore, min, d_b, mm	Chain path depth, max, h_c, mm	Plate depth, H, min, mm	Crank linked dimension max, X, mm	Width over inner link, min, W, mm	Width between outer plates max, mm	Width over bearing pin, min, A, mm	Addition width for joint fastener, max, B, mm	Measuring load N	Measuring load kgf	Breaking load, min kN	Breaking load, min kgf
208*A*	25.40	7.92	7.95	3.96	4.01	12.33	12.07	6.9	11.18	11.31	17.8	3.9	127.5	13	13.8	1410
208*B*	25.40	8.51	7.75	4.45	4.50	12.07	11.81	6.9	11.30	11.43	17.0	3.9	127.5	13	17.9	1820
210*A*	31.70	10.16	9.53	5.08	5.13	15.35	15.09	8.4	13.84	13.97	21.8	4.1	196.1	20	21.8	2220
210*B*	31.75	10.16	9.65	5.08	5.13	14.99	14.73	8.4	13.28	13.41	19.6	4.1	196.1	20	22.3	2270
212*A*	38.10	11.91	12.70	5.94	5.99	18.34	18.08	9.9	17.75	17.88	26.9	4.6	284.4	29	31.2	3180
212*B*	38.10	12.07	11.68	5.72	5.77	16.39	16.13	9.9	13.62	15.76	22.7	4.6	284.4	29	28.9	2950
216*A*	50.80	15.88	15.88	7.92	7.97	24.39	24.13	13.0	22.61	22.74	33.5	5.4	500.2	51	55.6	5670
216*B*	50.80	15.88	17.02	8.28	8.33	21.34	21.08	13.0	25.45	25.58	36.1	5.4	500.1	51	42.3	4310
220*A*	63.50	19.05	19.05	9.53	9.58	30.48	30.18	16.0	27.46	27.59	41.1	6.1	774.7	79	86.8	8850
220*B*	63.50	19.05	19.56	10.19	10.24	26.68	26.42	16.0	29.01	29.14	43.2	6.1	774.7	79	64.5	6580
224*A*	76.20	22.23	25.40	11.10	11.15	36.55	36.20	19.1	35.46	35.59	50.8	6.6	1108.2	113	124.5	12700
224*B*	76.20	25.40	25.40	14.63	14.68	33.73	33.40	19.1	37.92	38.05	53.4	6.6	1108.2	113	97.9	9980
228*B*	88.90	27.94	30.99	15.90	15.95	36.46	37.08	21.3	46.58	46.71	65.1	7.4	1510.2	164	129.1	13160
232*B*	101.60	29.21	30.99	17.81	17.86	42.72	42.29	24.4	45.57	45.70	64.7	7.9	2000.6	204	169.1	17240

Notes: (1) The chain path depth H_c is the minimum depth of channel through which the assembled chain will pass; (2) the overall width of chain with joint fastener is $-A + B$ for riveted pin end and fastener on one side; $A + 1.6B$ for headed pin end and fastener on one side; and $A + 2B$ for fastener on both sides. The actual dimensions will depend on the type of fastener used, but they should not exceed the dimensions in this column.

Source: IS 3542, 1966.

Particular	Formula

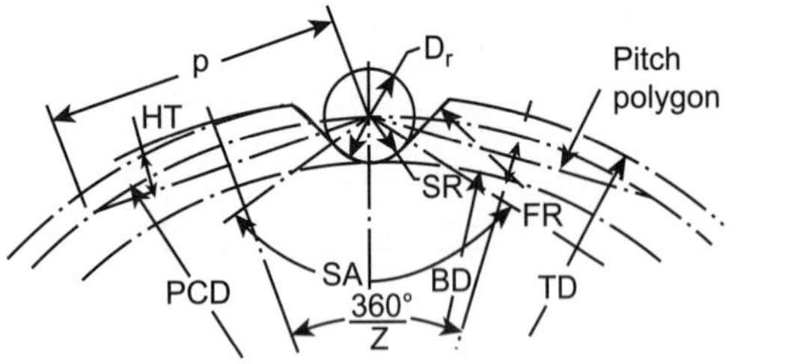

FIGURE 21-23 Notation for tooth gap form of roller chain.

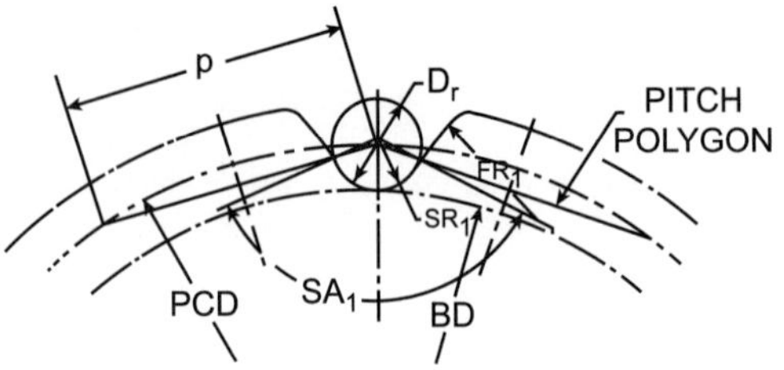

FIGURE 21-24 Notation for minimum tooth gap form of roller chain.

The minimum value of roller seating angle, deg (Fig. 21-24)

$$SA_1 = 140^\circ - \frac{90^\circ}{z} \quad (21\text{-}139)$$

The maximum value of roller seating angle, deg (Fig. 21-25)

$$SA_2 = 120^\circ - \frac{90^\circ}{z} \quad (21\text{-}140)$$

The minimum value of tooth flank radius, mm (Fig. 21-24)

$$FR_1 = 0.12D_r(z+2) \quad (21\text{-}141)$$

The maximum value of tooth flank radius, mm (Fig. 21-25)

$$FR_2 = 0.008D_r(z^2+180) \quad (21\text{-}142)$$

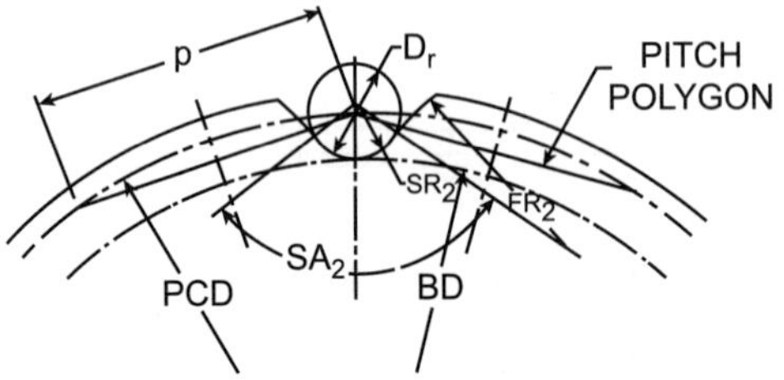

FIGURE 21-25 Notation for maximum tooth gap form of roller chain.

Tooth heights and top diameters (Fig. 21-23)

The maximum limit of the tooth height above the pitch polygon

$$HT_{\max} = p\left(0.3125 + \frac{0.8}{z}\right) - 0.5D_r \quad (21\text{-}143)$$

The minimum limit of the tooth height above the pitch polygon

$$HT_{\min} = p\left(0.25 + \frac{0.6}{z}\right) - 0.5D_r \quad (21\text{-}144)$$

The maximum limit of the tooth top diameter, mm

$$TD_{\max} = PCD + 0.625p - D_r \quad (21\text{-}145)$$

The minimum limit of the tooth top diameter, mm

$$TD_{\min} = PCD + p\left(0.5 - \frac{0.4}{z}\right) - D_r \quad (21\text{-}146)$$

TABLE 21-59
Pitch circle diameters[a] for extended pitch transmission roller chain wheels

No. of teeth, z	Pitch circle diameter	No. of teeth, z	Pitch circle diameter	No. of teeth, z	Pitch circle diameter	No. of teeth, z	Pitch circle diameter	No. of teeth, z	Pitch circle diameter	No. of teeth, z	Pitch circle diameter
5	1.7013	17	5.4422	29	9.2491	41	13.0635	53	16.8803	65	20.6982
$5\frac{1}{2}$	1.8496	$17\frac{1}{2}$	5.6005	$29\frac{1}{2}$	9.4080	$41\frac{1}{2}$	13.2225	$53\frac{1}{2}$	17.0393	$65\frac{1}{2}$	20.8575
6	2.0000	18	5.7588	30	9.5668	42	13.3815	54	17.1984	66	21.0164
$6\frac{1}{2}$	2.1519	$18\frac{1}{2}$	5.9171	$30\frac{1}{2}$	9.7256	$42\frac{1}{2}$	13.5405	$54\frac{1}{2}$	17.3575	$66\frac{1}{2}$	21.1757
7	2.3048	19	6.0755	31	9.8845	43	13.6995	55	17.5166	$66\frac{1}{2}$	21.3346
$7\frac{1}{2}$	2.4586	$19\frac{1}{2}$	6.2340	$31\frac{1}{2}$	10.0434	$43\frac{1}{2}$	13.8585	$55\frac{1}{2}$	17.6756	$67\frac{1}{2}$	21.4939
8	2.6131	20	6.3925	32	10.2023	44	14.0176	56	17.8347	68	21.6528
$8\frac{1}{2}$	2.7682	$20\frac{1}{2}$	6.5509	$32\frac{1}{2}$	10.3612	$44\frac{1}{2}$	14.1765	$56\frac{1}{2}$	17.9938	$68\frac{1}{2}$	21.8121
9	2.9238	21	6.7095	33	10.5201	45	14.3356	57	18.1529	69	21.9710
$9\frac{1}{2}$	3.0798	$21\frac{1}{2}$	6.8681	$33\frac{1}{2}$	10.6790	$45\frac{1}{2}$	14.4946	$57\frac{1}{2}$	18.3119	$69\frac{1}{2}$	22.1303
10	3.2361	22	7.0266	34	10.8380	46	14.6537	58	18.4710	70	22.2892
$10\frac{1}{2}$	3.3927	$22\frac{1}{2}$	6.1853	$34\frac{1}{2}$	10.9969	$46\frac{1}{2}$	14.8127	$58\frac{1}{2}$	18.6301	$70\frac{1}{2}2$	22.4485
11	3.5494	23	7.3439	35	11.1558	47	14.9717	59	18.7892	71	22.6074
$11\frac{1}{2}$	3.7065	$23\frac{1}{2}$	7.5026	$35\frac{1}{2}$	11.3148	$47\frac{1}{2}$	15.1308	$59\frac{1}{2}$	18.9482	$71\frac{1}{2}$	22.7667
12	3.8637	24	7.6613	36	11.4737	48	15.2898	60	19.1073	72	22.9256
$12\frac{1}{2}$	4.0211	$24\frac{1}{2}$	7.8200	$36\frac{1}{2}$	11.6327	$48\frac{1}{2}$	15.4488	$60\frac{1}{2}$	19.2665	$72\frac{1}{2}$	23.0849
13	4.1786	25	7.9787	37	11.7916	49	15.6079	61	19.4255	73	23.2438
$13\frac{1}{2}$	4.3362	$25\frac{1}{2}$	8.1375	$37\frac{1}{2}$	11.9506	$49\frac{1}{2}$	15.7669	$61\frac{1}{2}$	19.5847	$73\frac{1}{2}$	23.4031
14	4.4940	26	8.2962	38	12.1095	50	15.9260	62	19.7437	74	23.5620
$14\frac{1}{2}$	4.6518	$26\frac{1}{2}$	8.4550	$38\frac{1}{2}$	12.2685	$50\frac{1}{2}$	16.0850	$62\frac{1}{2}$	19.6029	$74\frac{1}{2}$	23.7213
15	4.8097	27	8.6138	39	12.4275	51	16.2441	63	20.0619	75	23.8802
$15\frac{1}{2}$	4.9677	$27\frac{1}{2}$	8.7726	$39\frac{1}{2}$	12.5865	$51\frac{1}{2}$	16.4031	$63\frac{1}{2}$	20.2210		
16	5.1258	28	8.9314	40	12.7455	52	16.5622	64	20.3800		
$16\frac{1}{2}$	5.2840	$28\frac{1}{2}$	9.0902	$40\frac{1}{2}$	12.9045	$52\frac{1}{2}$	16.7212	$64\frac{1}{2}$	20.5393		

[a] The values given are for a unit pitch (e.g., 1 mm).
Source: IS 3542, 1966.

TABLE 21-60
Maximum speed (rpm), recommended of sprockets for roller chains

No. of teeth	Pitch												
	6.35	9.50	12.70	15.80	19.05	25.40	31.75	38.10	44.45	50.80	57.15	63.50	76.20
11	4310	2260	1690	1220	920	580	415	325	235	200	165	145	110
12	4960	2590	1940	1400	1050	670	475	375	270	230	190	165	125
13	5540	2900	2180	1570	1110	750	535	415	305	260	215	186	140
14	6070	3170	2380	1720	1290	820	585	455	335	280	255	205	155
15	6500	3420	2560	1850	1390	880	630	490	360	305	255	220	165
16	6940	3630	2720	1969	1480	935	670	520	380	325	270	235	175
17	7290	3810	2860	2060	1550	985	700	550	400	340	285	245	185
18	7590	3970	2980	2150	1610	1020	730	750	415	355	295	255	195
19	7840	4100	3080	2220	1670	1060	755	590	430	365	305	265	200
20	8050	4210	3160	2280	1720	1090	755	605	440	375	315	270	205
21	8230	4300	3230	2330	1750	1110	790	620	450	385	320	280	210
22	8380	4380	3290	2370	1780	1130	805	630	460	390	325	280	215
23	8480	4480	3330	2400	1800	1150	875	640	405	395	330	285	215
24	8560	4410	3360	2420	1820	1160	825	645	470	400	300	290	220
25	8610	4510	3380	2440	1830	1100	830	650	475	400	335	290	220
30	8780	4490	3370	2430	1830	1160	825	645	470	400	335	290	220
35	8200	4290	3220	2320	1740	1110	790	615	450	380	320	275	210
40	7580	3970	2970	2140	1610	1020	730	570	415	355	295	255	195
45	6820	3570	2670	1930	1450	920	655	515	375	320	265	230	175
50	5950	3110	2330	1680	1270	805	575	450	325	275	230	200	150
55	5010	2620	1970	1420	1070	675	410	375	275	235	195	170	125
60	4020	2100	1580	1140	860	545	390	305	220	185	155	135	100

Particular	Formula

Wheel rim profile (Fig. 21-22)

Tooth width	$C_1 = 0.95W$ with a tolerance of $h/4$	(21-147)
The minimum tooth side radius	$F = 0.5p$	(21-148)
The tooth side relief	$G = 0.05p$ to $0.075p$	(21-149)
Absolute maximum shroud diameter	$D = p \cot \frac{180^\circ}{z} - 1.05H - 1.00 - 2 \times K_{oct}$, mm	(21-150)
For leaf chain dimension, breaking load, anchor clevises and chain sheaves	Refer to Tables 21-61, 21-62, and 21-63.	

Leaf chains

PRECISION BUSH CHAINS (Figs. 21-26 to 21-29, Tables 21-64 to 21-68)

The pitch circle diameter (Fig. 21-21, Table 21-62)	$PCD = \dfrac{p}{\sin \dfrac{180}{z}}$	(21-151)
Bottom diameter	$BD = PCD - D_b$	(21-152)
The minimum value of bush seating radius, mm (Fig. 21-28)	$SR_1 = 0.505D_b$	(21-153)
The maximum value of bush seating radius, mm (Fig. 21-29)	$SR_2 = 0.505D_b + 0.0693\sqrt{D_b}$	(21-154)
The minimum value of bush seating angle, deg (Fig. 21-28)	$SA_1 = 140^\circ - \frac{90^\circ}{z}$	(21-155)
The maximum value of bush seating angle, deg (Fig. 21-29)	$SA_2 = 120^\circ - \frac{90^\circ}{z}$	(21-156)
The minimum value of tooth flank radius, mm (Fig. 21-28)	$FR_1 = 0.12D_b(z + 2)$	(21-157)
The maximum value of tooth flank radius, mm (Fig. 21-29)	$FR_2 = 0.008D_b(z^2 + 180)$	(21-158)

TOOTH TOP DIAMETERS AND TOOTH HEIGHT (Fig. 21-27)

The maximum limit of the tooth top diameter	$TD_{\max} = PCD + 1.25p - D_b$	(21-159)
The minimum limit of the tooth top diameter	$TD_{\min} = PCD + p\left(1 - \frac{1.6}{z}\right) - D_b$	(21-160)
The maximum limit of the tooth height above the pitch polygon	$HT_{\max} = 0.625p - 1.5D_b + \frac{0.8p}{z}$	(21-161)

TABLE 21-61
Leaf chain dimensions, measuring loads, and breaking loads

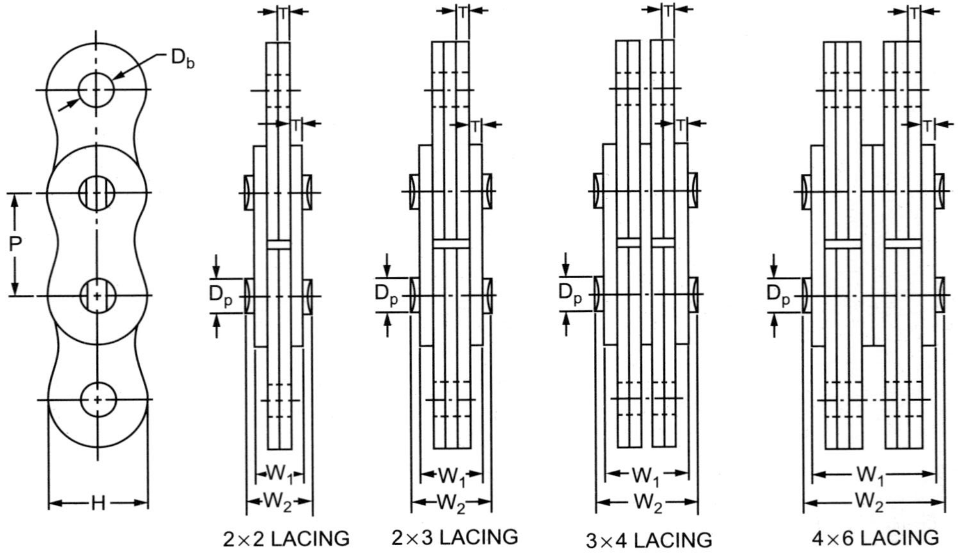

Chain number	Pitch mm	Lacing	Chain width, W_1 mm	Width over bearing pins, W_2 max, mm	Pin body diameter, max, D_p max mm	Articulating plates bore, diameter, min, D_p max mm	Plate depth, min, H mm	Plate thickness, max, T mm	Measuring load N	Measuring load kgf	Breaking load, min kN	Breaking load, min kgf
0822	12.70	2 × 2	6.45	8.69	4.45	4.48	11.81	1.57	190.0	19.10	18.7	1910
0823	12.70	2 × 3	8.08	10.31	4.45	4.48	11.81	1.57	190.0	19.10	18.7	1910
0834	12.70	3 × 4	11.30	13.54	4.45	4.48	11.81	1.57	280.0	28.60	26.3	2860
0846	12.70	4 × 6	16.13	18.36	4.45	4.48	11.81	1.57	370.0	38.10	37.4	3810
1022	15.88	2 × 2	7.26	9.80	5.08	5.10	14.73	1.78	250.0	25.40	24.9	2540
1023	15.88	2 × 3	9.09	11.63	5.08	5.10	14.73	1.78	250.0	25.40	24.9	2540
1034	15.88	3 × 4	12.73	15.27	5.08	5.10	14.73	1.78	390.0	39.90	39.1	3990
1046	15.88	4 × 6	18.16	20.70	5.08	5.10	14.73	1.78	500.0	50.80	49.8	5080
1222	19.05	2 × 2	12.50	15.90	6.78	6.80	16.13	3.07	450.0	45.40	44.5	4540
1223	19.05	2 × 3	15.62	19.02	6.78	6.80	16.13	3.07	450.0	45.40	44.5	4540
1234	19.05	3 × 4	21.87	25.27	6.78	6.80	16.13	3.07	670.0	68.00	66.7	6800
1246	19.05	4 × 6	31.24	34.65	6.78	6.80	16.13	3.07	890.0	90.70	82.0	9070
1623	25.40	2 × 3	21.34	25.48	8.28	8.30	21.08	4.22	630.0	63.50	62.3	6350
1634	25.40	3 × 4	29.87	34.01	8.28	8.30	21.08	4.22	1020.0	104.30	102.3	10430
1646	25.40	4 × 6	42.67	46.81	8.28	8.30	21.08	4.22	1250.0	127.00	124.5	12700
2023	31.75	2 × 3	23.24	28.35	10.19	10.22	26.42	4.60	980.0	99.80	97.9	9980
2034	31.76	3 × 4	32.54	37.64	10.19	10.22	26.42	4.60	1510.0	154.20	151.2	15420
2046	31.75	4 × 6	46.68	51.59	10.19	10.22	26.42	4.60	1960.0	199.60	195.7	19960
2423	38.10	2 × 3	30.73	38.05	14.63	14.66	33.40	6.10	1600.0	163.30	160.1	16330
2434	38.10	3 × 4	43.03	50.34	14.63	14.66	33.40	6.10	2400.0	244.90	240.2	24490
2446	38.10	4 × 6	61.47	68.78	14.63	14.66	33.40	6.10	3200.0	326.60	320.3	32660
2823	44.45	2 × 3	35.94	43.89	15.90	15.92	37.08	7.14	2400.0	217.70	213.5	21770
2834	44.45	3 × 4	50.32	58.27	15.90	15.92	37.08	7.14	3200.0	326.60	320.3	32660
2846	44.45	4 × 6	71.88	79.83	15.90	15.92	37.08	7.14	4300.0	435.50	427.1	43550
3223	50.80	2 × 3	40.51	49.43	17.81	17.84	42.29	8.05	2800.0	281.20	275.8	28120
3234	50.80	3 × 4	56.72	65.63	17.81	17.84	42.29	8.05	4100.0	421.80	413.6	42180
3246	50.80	4 × 6	81.03	89.94	17.81	17.84	42.29	8.05	5500.0	562.50	551.9	56280

Source: IS: 1072-1967.

TABLE 21-62
Dimensions of anchor clevises for leaf chains (all dimensions in mm)

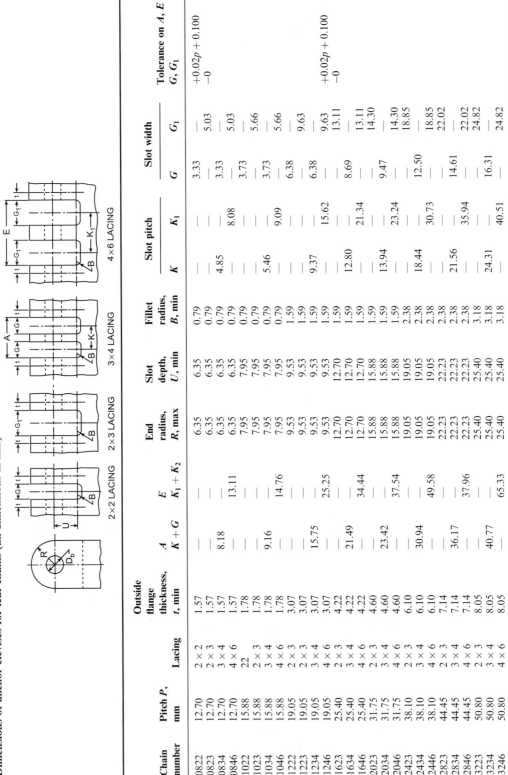

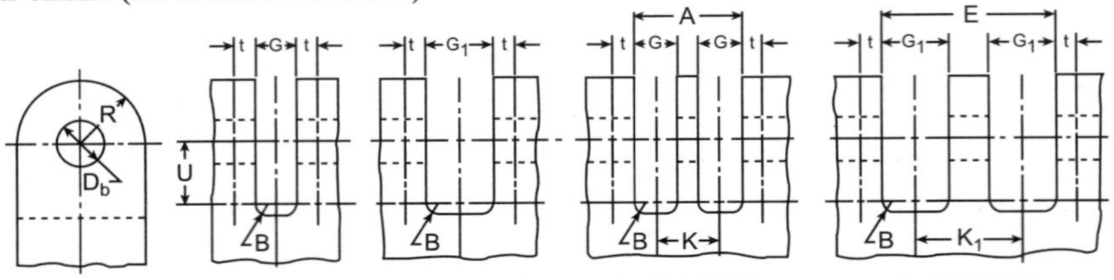

Chain number	Pitch P, mm	Lacing	Outside flange thickness, t, min	A $K+G$	E K_1+K_2	End radius, R, max	Slot depth, U, min	Fillet radius, B, min	Slot pitch K	Slot pitch K_1	Slot width G	Slot width G_1	Tolerance on A, E G, G_1
0822	12.70	2 × 2	1.57	—	—	6.35	6.35	0.79	—	—	3.33	—	$+0.02p+0.100$
0823	12.70	2 × 3	1.57	—	—	6.35	6.35	0.79	—	—	—	5.03	-0
0834	12.70	3 × 4	1.57	8.18	—	6.35	6.35	0.79	4.85	—	3.33	—	
0846	12.70	4 × 6	1.57	—	13.11	6.35	6.35	0.79	—	8.08	—	5.03	
1022	15.88	22	1.78	—	—	7.95	7.95	0.79	—	—	3.73	—	
1023	15.88	2 × 3	1.78	—	—	7.95	7.95	0.79	—	—	—	5.66	
1034	15.88	3 × 4	1.78	9.16	—	7.95	7.95	0.79	5.46	—	3.73	—	
1046	15.88	4 × 6	1.78	—	14.76	7.95	7.95	0.79	—	9.09	—	5.66	
1222	19.05	2 × 3	3.07	—	—	9.53	9.53	1.59	—	—	6.38	—	
1223	19.05	2 × 3	3.07	—	—	9.53	9.53	1.59	—	—	—	9.63	
1234	19.05	3 × 4	3.07	15.75	—	9.53	9.53	1.59	9.37	—	6.38	—	
1246	19.05	4 × 6	3.07	—	25.25	9.53	9.53	1.59	—	15.62	—	9.63	$+0.02p+0.100$
1623	25.40	2 × 3	4.22	—	—	12.70	12.70	1.59	—	—	—	13.11	-0
1634	25.40	3 × 4	4.22	21.49	—	12.70	12.70	1.59	12.80	—	8.69	—	
1646	25.40	4 × 6	4.22	—	34.44	12.70	12.70	1.59	—	21.34	—	13.11	
2023	31.75	2 × 3	4.60	—	—	15.88	15.88	1.59	—	—	—	14.30	
2034	31.75	3 × 4	4.60	23.42	—	15.88	15.88	1.59	13.94	—	9.47	—	
2046	31.75	4 × 6	4.60	—	37.54	15.88	15.88	1.59	—	23.24	—	14.30	
2423	38.10	2 × 3	6.10	—	—	19.05	19.05	2.38	—	—	—	18.85	
2434	38.10	3 × 4	6.10	30.94	—	19.05	19.05	2.38	18.44	—	12.50	—	
2446	38.10	4 × 6	6.10	—	49.58	19.05	19.05	2.38	—	30.73	—	18.85	
2823	44.45	2 × 3	7.14	—	—	22.23	22.23	2.38	—	—	—	22.02	
2834	44.45	3 × 4	7.14	36.17	—	22.23	22.23	2.38	21.56	—	14.61	—	
2846	44.45	4 × 6	7.14	—	37.96	22.23	22.23	2.38	—	35.94	—	22.02	
3223	50.80	2 × 3	8.05	—	—	25.40	25.40	3.18	—	—	—	24.82	
3234	50.80	3 × 4	8.05	40.77	—	25.40	25.40	3.18	24.31	—	16.31	—	
3246	50.80	4 × 6	8.05	—	65.33	25.40	25.40	3.18	—	40.51	—	24.82	

Source: IS 1072, 1967.

TABLE 21-63
Dimensions (in mm) for leaf chain sheaves

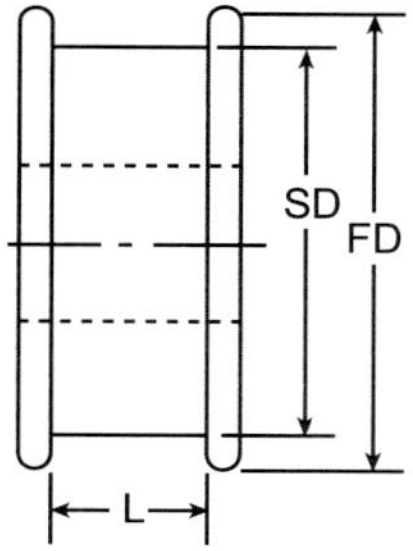

Chain number	Distance between flanges, *L*, min	Sheave diameter, *SD*, min	Flange diameter, *FD*, min	Chain number	Distance between flanges, *L*, min	Sheave diameter, *SD*, min	Flange diameter, *FD*, min
0822	9.12	63.50	88.90	1646	49.15	127.00	152.40
0823	10.80	63.50	88.90	2023	29.77	158.75	184.15
0834	14.20	63.50	88.90	2034	39.52	158.75	184.15
0846	19.28	63.50	88.90	2046	51.18	158.75	184.15
1022	10.29	79.38	104.78	2423	39.95	190.50	215.90
1028	12.22	79.38	104.78	2434	52.86	190.50	215.90
1034	16.03	79.38	104.78	2446	72.21	190.50	215.90
1046	21.74	79.38	104.78	2823	46.08	222.25	247.65
1222	16.69	95.25	120.65	2834	61.19	222.25	247.65
1223	19.96	95.25	120.65	2846	83.82	222.25	247.65
1234	26.54	95.25	120.65	3223	51.89	254.00	279.40
1246	36.37	95.25	120.65	3234	68.92	254.00	279.40
1623	26.75	127.00	152.40	3246	94.44	254.00	279.40
1634	35.71	127.00	152.40				

Source: IS 1072, 1967.

TABLE 21-64
Short pitch transmission precision bush chain dimensions, measuring loads, and breaking loads (all dimensions in mm)

	Chain No.			
	04C		**06C**	
Pitch, p	6.35		9.525	
Bush diameter, D_b, max	3.30		5.08	
Width between inner plates, min, W	3.18		4.77	
Bearing pin body diameter, max, D_p	2.29		3.59	
Bush bore, d_b, min	2.34		3.63	
Chain path depth, H_d, min	6.27		9.30	
Inner plate depth, H_i, max	6.02		9.05	
Outer or immediate plate depth, H_o, max	5.21		7.80	
Cranked link dimensions				
X, min	2.64		3.96	
Y, min	3.06		4.60	
Z, min	0.08		0.08	
Transverse pitch, Y_p	6.40		10.13	
Width over inner link, W_1, max	4.80		7.47	
Width between outer plates, W_2, min	4.93		7.60	
Width over bearing pins				
A, max	9.10		13.20	
A_2, max	15.5		23.4	
A_3, max	21.8		33.5	
Additional width for joint fasteners, B, max	2.5		3.3	
Measuring load				
Simplex	0.05 kN	5 kgf	0.07 kN	7 kgf
Duplex	0.10 kN	10 kgf	0.14 kN	14 kgf
Triplex	0.15 kN	15 kgf	0.20 kN	21 kgf
Breaking load, min				
Simplex	3.4 kN	350 kgf	7.8 kN	790 kgf
Duplex	6.9 kN	700 kgf	15.5 kN	1580 kgf
Triplex	10. 3 kN	1050 kgf	23.2 kN	2370 kgf

Notes: (1) Dimension C represents clearance between the cranked link plates and the straight plates available during articulation; (2) the chain path depth H_c is the minimum depth of channel through which the assembled chain passes; (3) width over bearing pins for chains wider than triplex $= A_1 + T_p$ (No. of strands in chain—1); (4) cranked links are not recommended for use on chains which are intended for onerous applications.
Source: IS 3563, 1966.

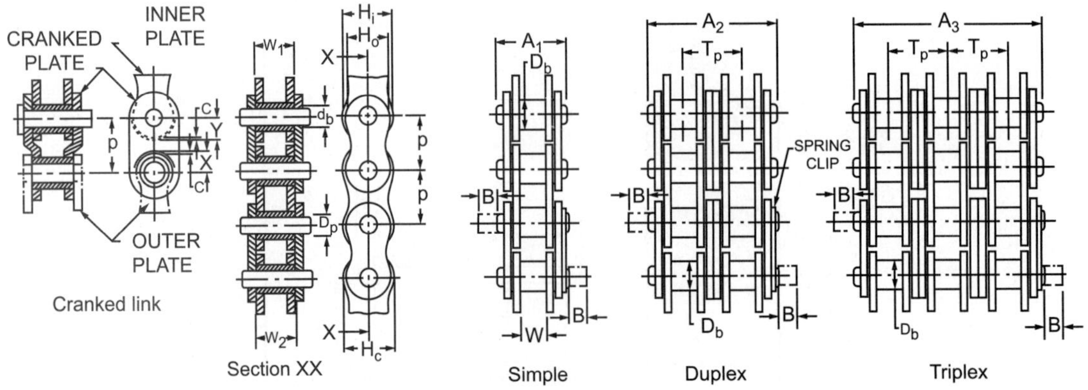

TABLE 21-65
Pitch circle diameters[a] for short pitch transmission precision bush chain wheels

No. of teeth	Pitch circle diameter	No. of teeth	Pitch circle diameter	No. of teeth	Pitch circle diameter	No. of teeth	Pitch circle diameter	No. of teeth	Pitch circle diameter	No. of teeth	Pitch circle diameter
9	2.9238	33	10.5201	57	18.1529	81	26.7896	105	33.4275	129	41.0660
10	3.2361	34	10.8380	58	18.4710	82	26.1078	106	33.7458	130	41.3843
11	3.5494	35	11.1558	59	18.7892	83	26.4260	107	34.0648	131	41.7026
12	3.8637	36	11.3747	60	19.1073	84	26.7443	108	34.3823	132	42.0209
13	4.1786	37	11.7916	61	19.9255	85	27.0625	109	34.7006	133	42.3291
14	4.4940	38	12.1096	62	19.7437	86	27.3807	110	35.0188	134	42.6574
15	4.8097	39	12.4275	63	20.0619	87	27.6990	111	35.3371	135	42.9757
16	5.1258	40	12.7455	64	20.3800	88	28.0172	112	35.6554	136	43.2940
17	5.4422	41	13.0635	65	20.6982	89	28.3355	113	35.9737	137	43.6123
18	5.7588	42	13.3815	66	21.0164	90	28.6537	114	36.2919	138	43.9306
19	6.0755	43	13.6995	67	21.3246	91	28.9719	115	36.6102	139	44.2488
20	6.3925	44	14.0176	68	21.6528	92	29.2902	116	36.9285	140	44.5671
21	6.7095	45	14.3356	69	21.9710	93	29.6084	117	37.2467	141	44.8854
22	7.0266	46	14.6537	70	22.2892	94	29.9267	118	37.5650	142	45.2037
23	7.3439	47	14.9717	71	22.6074	95	30.2449	119	37.8833	143	45.5220
24	7.6613	48	15.2868	72	22.9256	96	30.5632	120	38.2016	144	45.8403
25	7.9787	49	15.6079	73	23.2438	97	30.8815	121	38.5198	145	46.1585
26	8.2962	50	15.9260	74	23.5620	98	31.9097	122	38.8381	146	46.4768
27	8.6138	51	16.2441	75	24.8802	99	31.5180	123	39.1564	147	46.7951
28	8.9314	52	16.5622	76	24.1985	100	31.8362	124	39.4776	148	47.1134
29	9.2491	53	16.8803	77	24.5167	101	32.1545	125	39.7929	149	47.4317
30	9.5668	54	17.1984	78	24.8349	102	32.4727	126	40.1112	150	47.7500
31	9.8845	55	17.5166	79	25.1513	103	32.7910	127	40.4295		
32	10.2023	56	17.8347	80	25.4713	104	33.1093	128	43.7478		

[a] The values given are for a unit pitch length (e.g., 1 mm).
Source: IS 3560, 1966.

TABLE 21-66
Recommended design data for silent chains

Chain pitch, mm	Speed of small sprocket	No. of teeth		Min center distance, mm
		Driver	Driven	
9.3	2000–4000	17–25	21–120	152.4
12.7	1500–2000	17–25	21–130	228.6
15.8	1200–1500	19–25	21–150	304.8
19.0	1000–1200	19–25	23–150	381.0
22.2	900–1000	19–25	23–150	457.2
25.4	800–900	19–25	23–150	533.4
31.7	650–800	21–25	25–150	685.8
38.1	500–650	25–27	27–150	914.4
50.8	300–500	25–27	27–150	1219.2
76.2	≤ 300	25–27	27–150	1676.4

TABLE 21-67A
Maximum speed of small sprocket for inverted tooth chains

Pitch, mm	Max width, mm	Number of teeth	Speed, rpm							
9.50	101.6	17	4000	3500	2500	2000	1200			
12.70	177.8	19	5000	3500	2500	2500	1500	1200	1000	700
15.88	203.2	21	6000	3000	3000	2500	1800	1200	1000	700
19.05	254.0	23	6000	4000	3000	2500	1800	1800	1200	800
25.40	355.6	25	6000	4000	3500	2500	1800	1800	1200	900
31.75	508.0	27	6000	4000	3500	2500	2000	1800	1200	900
38.10	609.6	29	6000	4000	3500	2500	2000	1800	1200	900
50.80	762.0	31	6000	4000	3500	2500	2000	1800	1200	900
		33	6000	4000	3500	2500	2000	1800	1200	900
		35	6000	4000	3500	2500	2000	1800	1200	900
		37	5000	3500	3000	2500	1800	1200	1000	800
		45	4000	3000	2000	2000	1500	1000	900	700
		40	5000	3500	2500	2500	1500	1200	900	800
		50	3500	2500	2000	1800	1200	1000	800	600

TABLE 21-67B
Maximum velocity for various types of chains, rpm

Type of chain		Number of sprocket teeth						
		Bush roller chain						Silent chains
			15	19	23	27	30	17.35
	12	2300	2400	2530	2550	2600	12.7	3300
Chain	15	1900	2000	2100	2150	2200	15.87	2650
pitch, *p*,	20	1350	1450	1500	1550	1550	19.05	2200
mm	25	1150	1200	1250	1300	1300	25.40	1650
	30	1000	1050	1100	1100	1100	31.75	1300

TABLE 21-68
Safety factor

	Speed of smaller sprocket, rpm								
Chains	**50**	**260**	**400**	**600**	**800**	**1000**	**1200**	**1600**	**2000**
Bush roller chains									
$p = 12$, 15 mm	7.0	7.8	8.55	9.35	10.2	11.0	11.7	13.2	1.48
$p = 20$, 25 mm	7.0	8.2	9.35	10.3	10.7	12.9	14.0	16.3	
$p = 30$, 35 mm	7.0	8.55	10.2	13.2	14.8	16.3	19.5		
Silent chains									
$p = 12.7$, 15.87 mm	20	22.2	24.4	28.7	29.0	31.0	33.4	37.8	42.0
$p = 19.05$, 25.4 mm	20	23.4	26.7	30.0	33.4	36.8	40.0	46.5	53.5

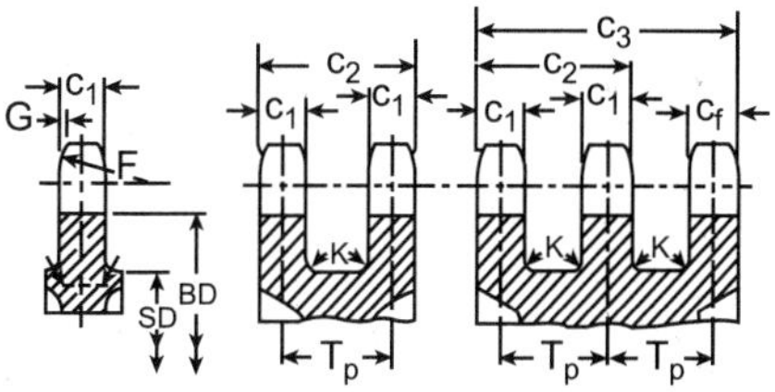

FIGURE 21-26 Notation for wheel rim profiles of bush chain.

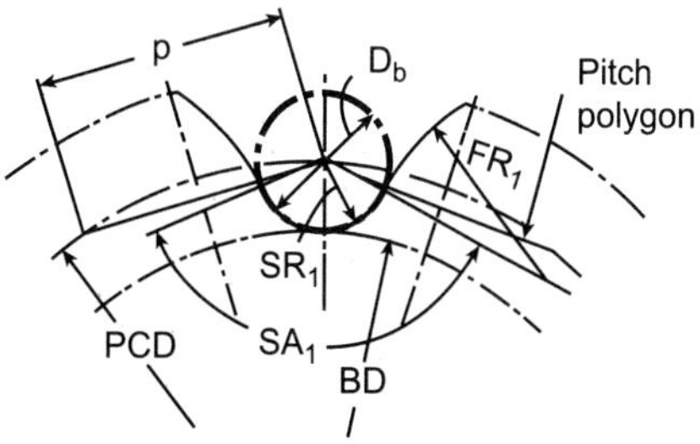

FIGURE 21-28 Notation for minimum tooth gap form for bush chain.

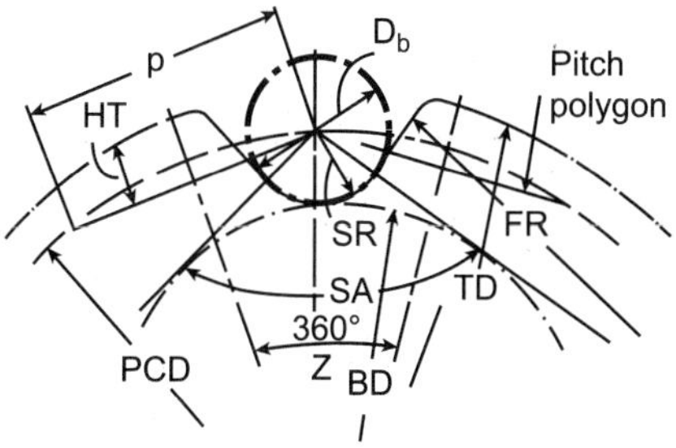

FIGURE 21-27 Notation for tooth gap form of bush chain.

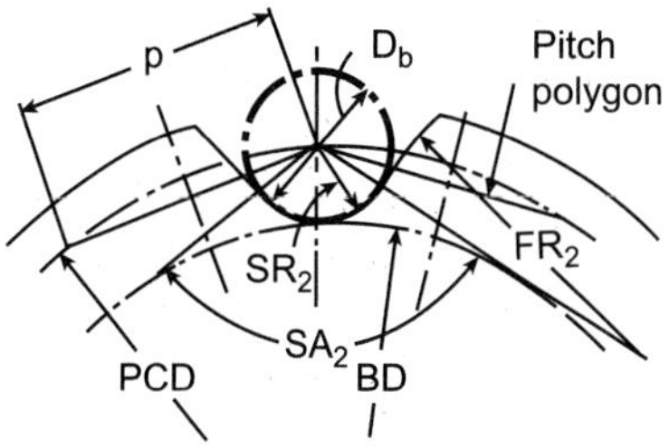

FIGURE 21-29 Notation for maximum tooth gap form for bush chain.

Particular	Formula	
The minimum limit of the tooth height above the pitch polygon	$HT_{\min} = 0.5(p - D_b)$	(21-162)
WHEEL RIM PROFILE (Fig. 21-26) The value of tooth width for simple chain wheels (Fig. 21-26)	$C_1 = 0.93w$	(21-163)
The value of tooth width for duplex and triplex chain wheels	$C_1 = 0.91w$	(21-164)
The value of tooth width for quadruplex chain wheels and above	$C_1 = 0.88w$	(21-165)
	The value of tolerance shall be $h/4$.	
The value of width over tooth	C_2 (or C_3) = number of strands $- 1T_p + C_1$	(21-166)
	with a tolerance value of $h/4$ where T_p = transmission pitch of strands	
The minimum tooth side radius	$F = p$	(21-167)
The tooth side relief	$G = 0.1p$ to $0.15p$	(21-168)
Absolute maximum shroud diameter	$SD = p \cot \dfrac{180^\circ}{z} - 1.05H_i - 1.00 - 2K_{ort}$, mm	(21-169)
For bush chains dimensions, breaking load, pitch circle diameters, etc.	Refer to Tables 21-64 to 21-68.	

REFERENCES

1. Maleev, V. L., and J. B. Hartman, *Machine Design*, International Textbook Company, Scranton, Pennsylvania, 1954.
2. Black, P. H., and O. E. Adams, Jr., *Machine Design*, McGraw-Hill Book Company, New York, 1968.
3. Norman, C. A., E. S. Ault, and I. F. Zarobsky, *Fundamentals of Machine Design*, The Macmillan Company, New York, 1951.
4. Shigley, J. E., *Machine Design*, McGraw-Hill Book Company, New York, 1962.
5. Shigley, J. E., and C. R. Mischke, *Mechanical Engineering Design*, McGraw-Hill Book Company, New York, 1989.
6. Shigley, J. E., and C. R. Mischke, *Standard Handbook of Machine Design*, McGraw-Hill Book Company, New York, 1986.
7. Baumeister, T., ed., *Marks' Standard Handbook for Mechanical Engineers*, McGraw-Hill Book Company, New York, 1978.
8. Niemann, G., *Maschinenelemente*, Springer-Verlag, Berlin; Zweiter Band, Munich, 1965.
9. Niemann, G., *Machine Elements—Design and Calculations in Mechanical Engineering*, Vol. II, Allied Publishers Private Ltd., New Delhi, 1978.
10. Decker, K. H., *Maschinenelemente, Gestaltung and Berechnung*, Carl Hanser Verlag, Munich, 1971.

11. Lingaiah, K. and B. R. Narayana Iyengar, *Machine Design Data Handbook*, Engineering College Co-operative Society, Bangalore, India, 1962.
12. Lingaiah, K. and B. R. Narayana Iyengar, *Machine Design Data Handbook*, Vol. I (*SI and Customary Metric Units*), Suma Publishers, Bangalore, India, 1973.
13. Lingaiah, K., *Machine Design Data Handbook*, Vol. II (*SI and Customary Metric Units*), Suma Publishers, Bangalore, India, 1986.
14. Bureau of Indian Standards.
15. Albert, C. D., Machine Design Drawing Room Problems, John Wiley and Sons, New York, 1949.
16. *V-Belts and Pulleys*, SAE J 636C, SAE Handbook, Part I, Society of Automotive Engineers, Inc., 1997.
17. *SI Synchronous Belts and Pulleys*, SAE J 1278 Oct.80, SAE Handbook, Part I, Society of Automotive Engineers, Inc., 1997.
18. *Synchronous Belts and Pulleys*, SAE J 1313 Oct.80, SAE Handbook, Part I, Society of Automotive Engineers, Inc., 1997.
19. Wolfram Funk, 'Belt Drives,' J. E. Shigley and C. R. Mischke, *Standard Handbook of Machine Design*, 2nd edition, McGraw-Hill Publishing Company, New York, 1996.

CHAPTER 22

MECHANICAL VIBRATIONS

SYMBOLS

a	coefficients with subscripts
	flexibility
	acceleration, m/s^2 (ft/s^2)
A	area of cross section, m^2 (in^2)
	constant
B	constant
C	coefficient of viscous damping, N s/m or N/ν (lbf s/in or lbf/ν)
	constant
C_c	critical viscous damping, N s/m (lbf s/in)
C_t	coefficient of torsional viscous damping, N m s/rad (lbf in s/rad)
C_1, C_2	coefficients
	constants
d	diameter of shaft, m (in)
D	flexural rigidity $[= Eh^3/12(1 - \nu^2)]$
e	displacement of the center of mass of the disk from the shaft axis, m (in)
E	modulus of elasticity, GPa (Mpsi)
f	frequency, Hz
F	exciting force, kN (lbf)
F_o	maximum exciting force, kN (lbf)
F_T	transmitted force, kN (lbf)
g	acceleration due to gravity, 9.8066 m/s^2 (32.2 ft/s^2 or 386.6 in/s^2)
G	modulus of rigidity, GPa (Mpsi)
h	thickness of plate, m (in)
i	integer (= 0, 1, 2, 3, ...)
I	mass moment of inertia of rotating disk or rotor, N s^2 m (lbf s^2 in)
J	polar second moment of inertia, m^4 or cm^4 (in^4)
k	spring stiffness or constant, kN/m (lbf/in)
k_e	equivalent spring constant, kN/m (lbf/in)
k_t	torsional or spring stiffness of shaft, J/rad or N m/rad (lbf in/rad)
K	kinetic energy, J (lbf/in)

l	length of shaft, m (in)
m	mass, kg (lb)
m_e	equivalent mass, kg (lb)
M	total mass, kg (lb)
M_t	torque, N m (lbf ft)
p	circular frequency, rad/s
q	damped circular frequency $(= \sqrt{1-\zeta^2})$
r	radius, m (in)
$R = 1 - T_R$	percent reduction in transmissibility
$R_2 = D_2/2$	radius of the coil, m (in)
t	time (period), s
T	temperature, K or °C (°F)
T_R	transmissibility
U	vibrational energy, J or N m (lbf in) potential energy, J (lbf in)
v	velocity, m/s (ft/min)
w	weight per unit volume, kN/m^3 (lbf/in^3)
W	total weight, kN (lbf)
x	displacement or amplitude from equilibrium position at any instant t, m (in)
x_1, x_2	successive amplitudes, m (in)
x_o	maximum displacement, m (in)
$\dot{x}$	linear velocity, m/s (ft/min)
$\ddot{x}$	linear acceleration, m/s^2 (ft/s^2)
X_{st}	static deflection of the system, m (in)
y	deflection of the disk center from its rotational axis, m or mm (in)
γ	weight density, kN/m^3 (lbf/in^3)
$\zeta = \dfrac{C}{C_c}$	damping factor
δ	logarithmic decrement, deflection, m (in)
δ_{st}	static deflection, m (in)
θ	phase angle, deg
λ	wavelength, m (in)
ν	Poisson's ratio
ρ	mass density, kg/m^3 (lb/in^3)
σ	normal stress, MPa (psi)
τ	shear stress, MPa (psi) period, s
ϕ	angular deflections, rad (deg)
$\dot{\phi}$	angular velocity, rad/s
$\ddot{\phi}$	angular acceleration, rad/s^2
ω	forced circular frequency, rad/s

Particular	Formula	
SIMPLE HARMONIC MOTION (Fig. 22-1)		
The displacement of point P on diameter RS (Fig. 22-1)	$x = x_o \sin pt$	(22-1)
The wavelength	$\lambda = 2\pi$	(22-2)

Particular	Formula

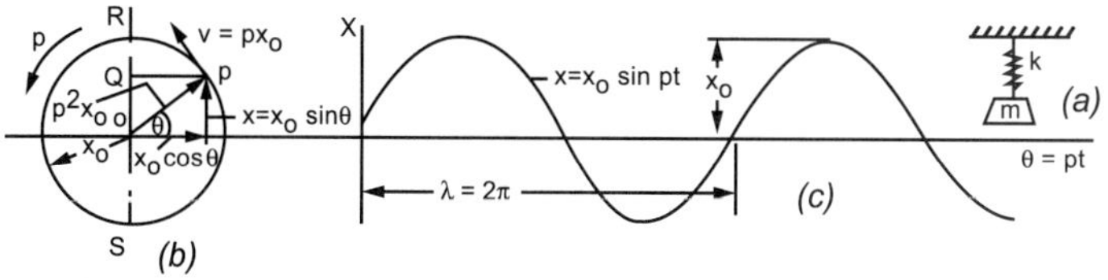

FIGURE 22-1 Simple harmonic motion.

The periodic time
$$\tau = \frac{2\pi}{p} \tag{22-3}$$

The frequency
$$f = \frac{1}{\tau} = \frac{p}{2\pi} \tag{22-4}$$

The maximum velocity of point Q
$$v_{\max} = px_o \tag{22-5}$$

The maximum acceleration of point Q
$$a_{\max} = \dot{v}_{\max} = p^2 x_o \tag{22-6}$$

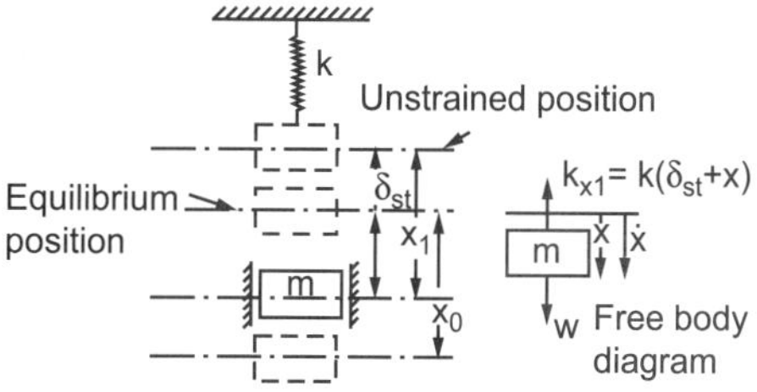

FIGURE 22-2 Spring-mass system.

Single-degree-of-freedom system without damping and without external force (Fig. 22-2)

Linear system

The equation of motion
$$m\ddot{x} + kx = 0 \tag{22-7}$$

The general solution for displacement
$$x = A \sin pt + B \cos pt \tag{22-8}$$

$$x = C \sin(pt - \phi) \tag{22-9}$$

where ϕ = phase angle of displacement

The equation for displacement of mass for the initial condition $x = x_o$ and $\dot{x} = 0$ at $t = 0$
$$x = x_o \cos pt \tag{22-10}$$

The natural circular frequency
$$p_n = \sqrt{\frac{k}{m}} = \sqrt{\frac{g}{\delta_{st}}} \tag{22-11}$$

The natural frequency of the vibration
$$f_n = \frac{p_n}{2\pi} = \frac{1}{2\pi}\sqrt{\frac{k}{m}} \tag{22-12}$$

The natural frequency in terms of static deflection δ_{st}
$$f_n = \frac{1}{2\pi}\sqrt{\frac{g}{\delta_{st}}} \tag{22-13}$$

$$f_n = \frac{3.132}{2\pi}\left(\frac{1}{\delta_{st}}\right)^{1/2} \approx 0.5\left(\frac{1}{\delta_{st}}\right)^{1/2} \tag{22-13a}$$

where δ_{st} in m and f_n in Hz

Particular	Formula

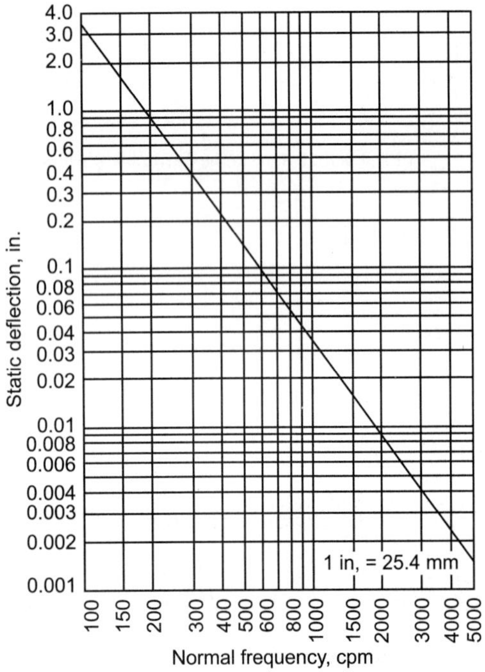

FIGURE 22-3 Static deflection (δ_{st}) vs. natural frequency. (Courtesy of P. H. Black and O. E. Adams, Jr., *Machine Design*, McGraw-Hill, New York, 1955.)

$$f_n = \frac{99}{2\pi}\left(\frac{1}{\delta_{st}}\right)^{1/2} \approx 15.76\left(\frac{1}{\delta_{st}}\right)^{1/2} \qquad \textbf{SI} \qquad (22\text{-}13b)$$

where δ_{st} in mm and f_n in Hz

$$f_n = \frac{5.67}{2\pi}\left(\frac{1}{\delta_{st}}\right)^{1/2} \approx 0.9\left(\frac{1}{\delta_{st}}\right)^{1/2} \qquad \textbf{USCS} \qquad (22\text{-}13c)$$

where δ_{st} in ft and f_n in Hz

$$f_n = \frac{19.67}{2\pi}\left(\frac{1}{\delta_{st}}\right)^{1/2} = \frac{3.127}{\sqrt{\delta_{st}}} \qquad \textbf{USCS} \qquad (22\text{-}13d)$$

where δ_{st} in in and f_n in Hz

$$f_n = \frac{187.6}{\sqrt{\delta_{st}}} \qquad \textbf{USCS} \qquad (22\text{-}13e)$$

where δ_{st} in in and f_n in cpm (cycles per minute)

The plot of natural frequency vs. static deflection — Refer to Fig. 22-3.

Simple pendulum

The equation of motion for simple pendulum (Fig. 22-4)

$$\ddot{\theta} = \frac{g}{l}\sin\theta = \ddot{\theta} + \frac{g}{l}\theta = 0 \qquad (22\text{-}14)$$

The angular displacement for $\theta = \theta_o$ and $\dot{\theta} = 0$ at $t = 0$

$$\theta = \theta_o \sin\sqrt{\frac{g}{l}}t \qquad (22\text{-}15)$$

The circular frequency for simple pendulum for small oscillation

$$p = \sqrt{\frac{g}{l}} \qquad (22\text{-}15a)$$

ENERGY

The total energy in the universe is constant according to conservation of energy

$$K + U = \text{constant} \qquad (22\text{-}16)$$

Kinetic energy

$$K = \tfrac{1}{2}mv^2 = \tfrac{1}{2}m\dot{x}^2 \qquad (22\text{-}17)$$

Particular	Formula	
Potential energy	$U = \frac{1}{2}kx^2$	(22-18)
Maximum kinetic energy is equal to maximum potential energy according to conservation of energy	$K_{max} = U_{max}$	(22-19)

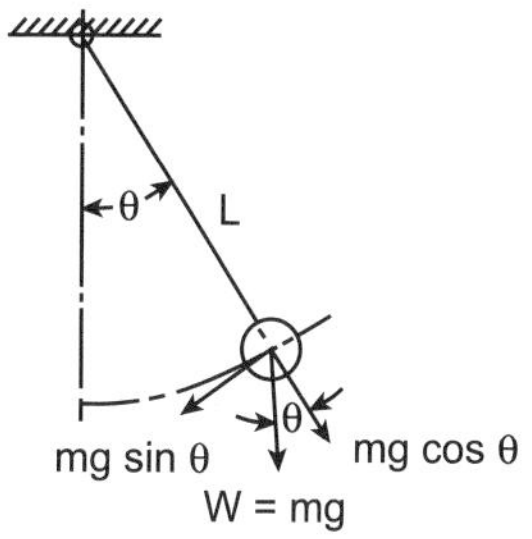

FIGURE 22-4 Simple pendulum.

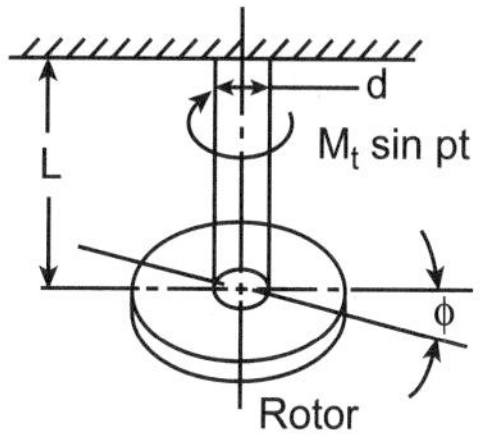

FIGURE 22-5 Single rotor system subject to torque.

Torsional system (Fig. 22-5)

Particular	Formula	
The equation of motion of torsional system (Fig. 22-5) with torsional damping under external torque $M_t \sin pt$	$I\ddot{\phi} + C_t\dot{\phi} + k_t x = M_t \sin pt$ where C_t = coefficient of torsional viscous damping, N m s/rad	(22-20)
The equation of motion of torsional system without considering the damping and external force on the rotor	$I\ddot{\phi} + k_t\phi = 0$	(22-21)
The equation for angular displacement	$\phi = A\sin pt + B\cos pt$	(22-22a)
	$\phi = C\sin(pt - \theta)$ where θ = phase of displacement	(22-22b)
The angular displacement for $\phi = \phi_o$ and $\dot{\phi} = 0$ at $t = 0$	$\phi = \phi_o \cos(\sqrt{k_t/I})t$	(22-23)
The natural circular frequency	$p_n = \sqrt{k_t/I}$	(22-24)
The natural circular frequency taking into account the shaft mass	$p_n = \left[k_t\Big/\left(I + \frac{I_s}{3}\right)\right]^{1/2}$	(22-25)
The natural frequency	$f_n = \frac{p_n}{2\pi} = \frac{1}{2\pi}\sqrt{k_t/I}$	(22-26)
The expression for torsional stiffness	$k_t = \frac{JG}{l} = \frac{\pi d^4}{32}\frac{G}{l}$ where $J = \pi d^4/32$ = moment of inertia, polar, m^4 or cm^4	(22-27)

Particular	Formula

Single-degree-freedom system with damping and without external force (Fig. 22-6)

The equation of motion

$$m\ddot{x} + c\dot{x} = kx = 0 \tag{22-28}$$

The general solution for displacement

$$x = C_1 e^{s_1 t} + C_2 e^{s_2 t} \tag{22-29}$$

$$x = C_1 e^{(-\zeta - \sqrt{\zeta^2 - 1})p_n t} + C_2 e^{(-\zeta + \sqrt{\zeta^2 - 1})p_n t} \tag{22-30}$$

$$x = Ae^{-\zeta p_n t} \sin(qt + \phi) \tag{22-31}$$

where C_1, C_2, and A are arbitrary constants of integration. (They can be found from initial conditions.)

$$s_{1,2} = -\frac{C}{2m} \pm \left[\left(\frac{C}{2m}\right)^2 - \frac{k}{m}\right]^{1/2} \tag{22-32}$$

$$s_{1,2} = \left(-\zeta \pm \sqrt{\zeta^2 - 1}\right) p_n \tag{22-33}$$

where $\zeta = \dfrac{C}{C_c}$ = damping ratio,

$$C_c = 2mp_n = 2\sqrt{km}$$

q = frequency of damped oscillation

$$= \left(\frac{2\pi}{\tau_d}\right) = \left(\sqrt{1 - \zeta^2}\right) p_n = \left(\frac{k}{m} - \frac{c^2}{4m^2}\right)^{1/2} \tag{22-33a}$$

ϕ = phase angle or phase displacement with respect to the exciting force

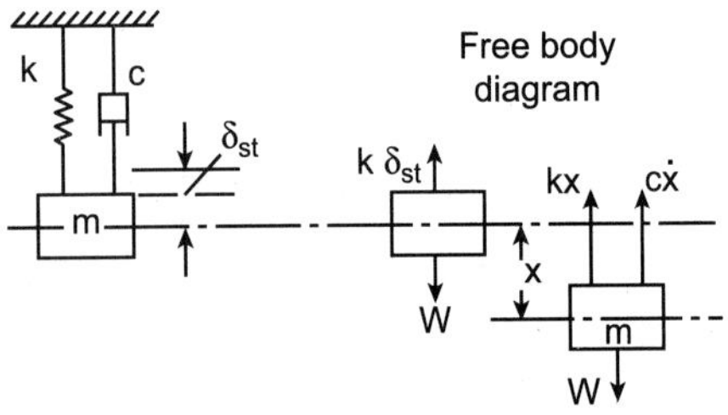

FIGURE 22-6 Single-degree-of-freedom spring-mass-dashpot system.

For the damped oscillation of the single-degree-freedom system with time for damping factor $\zeta < 1$

Refer to Figs. 22-7 and 22-8.

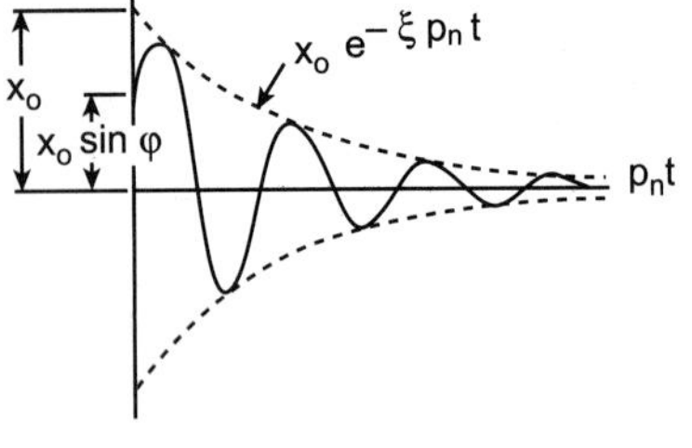

FIGURE 22-7 Damped motion $\zeta < 1.0$.

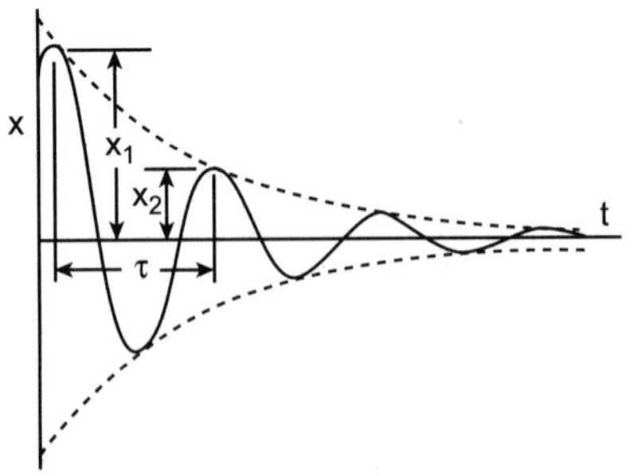

FIGURE 22-8 Logarithmic decrement. (Reproduced from *Marks' Standard Handbook for Mechanical Engineers*, 8th edition, McGraw-Hill, New York, 1978.)

Particular	Formula

LOGARITHMIC DECREMENT (Fig. 22-8)

The equation for logarithmic decrement	$\delta = \ln \frac{x_o}{x_1} = \ln \frac{x_1}{x_2} = \frac{\Delta U}{U} = \frac{2\pi\zeta}{\sqrt{1-\zeta^2}} \approx 2\pi\zeta$	(22-34)

EQUIVALENT SPRING CONSTANTS (Fig. 22-9)

The spring constant or stiffness	$k = \frac{F}{x}$	(22-35)
The flexibility	$a = \frac{x}{F}$	(22-36)
The equivalent spring constant for springs in series (Fig. 22-9a)	$k_e = \frac{1}{\frac{1}{k_1} + \frac{1}{k_2}}$	(22-37)
The equivalent spring constant for springs in parallel (Fig. 22-9b)	$k_e = k_1 + k_2$	(22-38)
For spring constants of different types of springs, beams, and plates	Refer to Table 22-1	

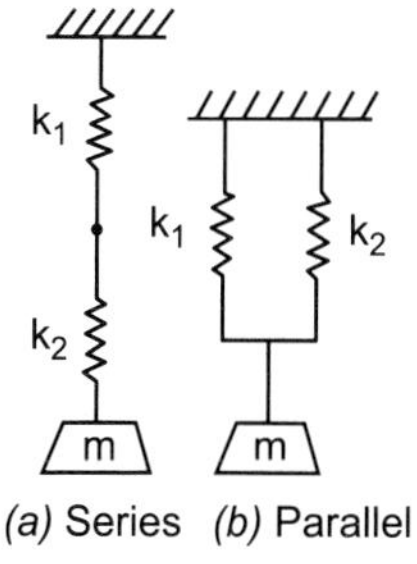

FIGURE 22-9 Springs in series and parallel.

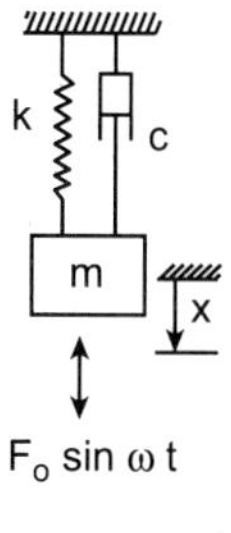

FIGURE 22-10 Spring-mass-dashpot system subjected to external force.

Single-degree-of-freedom system with damping and external force (Fig. 22-10)

The equation of motion	$m\ddot{x} + c\dot{x} + kx = F_o \sin \omega t$	(22-39)
	$\ddot{x} + 2\zeta p_n \dot{x} + p_n^2 x = \frac{F_o}{m} \sin \omega t$	(22-40)

TABLE 22-1
Spring constants or spring stiffness of various springs, beams, and plates

Particular	Formula for spring constant, k	Figure	Equation
Linear Spring Stiffness or Constants [Load per mm (in) Deflection]			
Helical spring subjected to tension with i number of turns	$k = \dfrac{Gd^4}{64iR^3}$	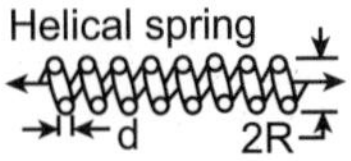	(22-41)
Bar under tension	$k = \dfrac{EA}{l}$	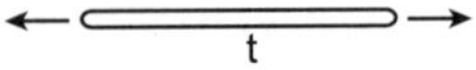	(22-42)
Cantilever beam subjected to transverse load at the free end	$k = \dfrac{3EI}{l^3}$	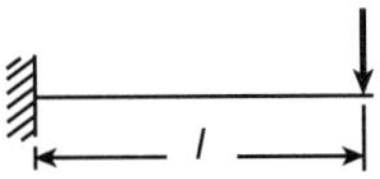	(22-43)
Cantilever beam subjected to bending at the free end	$k = \dfrac{2EI}{l^2}$	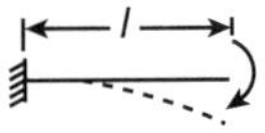	(22-44)
Simply supported beam with concentrated load at the center	$k = \dfrac{48EI}{l^3}$	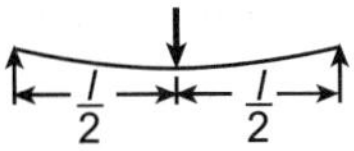	(22-45)
Simply supported beam subjected to a concentrated load not at the center	$k = \dfrac{3EIl}{a^2b^2}$	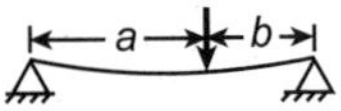	(22-46)
Beam fixed at both ends subjected to a concentrated load at the center	$k = \dfrac{192EI}{l^3}$	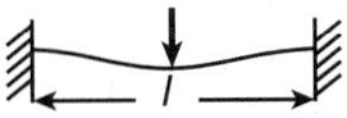	(22-47)
Beam fixed at one end and simply supported at another end subjected to concentrated load at the center	$k = \dfrac{768EI}{7l^3}$	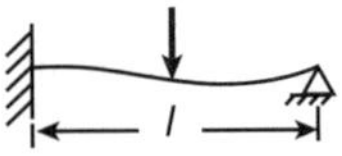	(22-48)
Circular plate clamped along the circumferential edge subjected to concentrated load at the center whose flexural rigidity is $D = Eh^3/12(1 - \nu^2)$, thickness h and Poisson ratio ν	$k = \dfrac{16\pi D}{R^2}$	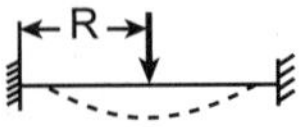	(22-49)
Circular plate simply supported along the circumferential edge with concentrated load at the center	$k = \dfrac{16\pi D}{R^2}\left(\dfrac{1+\nu}{3+\nu}\right)$ where ν = Poisson's ratio	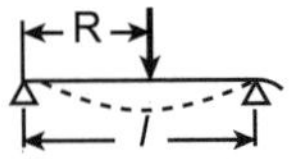	(22-50)
String fixed at both ends subjected to tension T	$k = \dfrac{4T}{l}$ String tension T	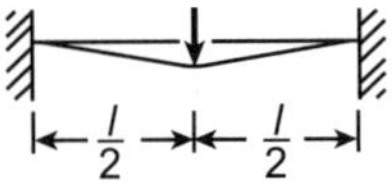	(22-51)
Torsional or Rotational Spring Stiffness or Constants (Load per Radian Rotation)			
Spiral spring whose total length is l and moment of inertia of cross section I	$k_t = \dfrac{EI}{l}$		(22-52)
Helical spring with i turns subjected to twist whose wire diameter is d, the coil diameter is D			

$$k_t = \frac{Ed^4}{64iD}$$

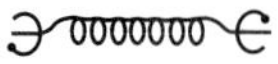

(22-53)

TABLE 22-1
Spring constants or spring stiffness of various springs, beams, and plates (*Cont.*)

Particular	Formula for spring constant, k	Figure	Equation
Bending of helical spring of i number of turns	$k_t = \frac{Ed^4}{32iD} \frac{1}{1+(E/2G)}$	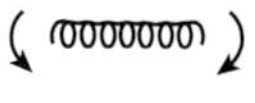	(22-54)
Twisting of bar of length l	$k_t = \frac{JG}{l}$	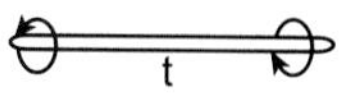	(22-55)
Twisting of a hollow circular shaft with length l, whose outside diameter is D_o, and inside diameter is D_i	$k_t = \frac{GI_p}{l} = \frac{\pi G}{32} \frac{D_o^4 - D_i^4}{l}$		(22-56)
Twisting of cantilever beam	$k_t = \frac{GJ}{l}$	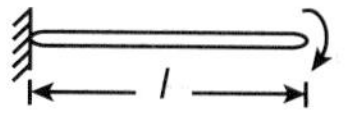	(22-57)
Simply supported beam subjected to couple at the center	$k_t = \frac{12EI}{l}$		

Particular	Formula	
The complete solution for the displacement	$x = Ae^{-\zeta p_n t} \sin(qt + \phi_1) + X_o \sin(\omega t - \phi)$	(22-60a)
	$x = Ae^{-\zeta p_n t} \sin(qt + \phi_1) + \frac{(F_o/k)\sin(\omega t - \phi)}{[\{1-(\omega/p_n)^2\}^2 + (2\zeta\omega/p_n)^2]^{1/2}}$	(22-60b)
The steady-state solution for amplitude of vibration	$X = \frac{F_o}{\sqrt{(k - m\omega^2)^2 + (c\omega)^2}} = \frac{F_o/k}{[\{1-(\omega/p_n)^2\}^2 + (2\zeta\omega/p_n)^2]^{1/2}}$	(22-60c)
The phase angle	$\phi = \tan^{-1}\left[\frac{2\zeta(\omega/p_n)}{1-(\omega/p_n)^2}\right]$	(22-61)
The magnification factor	$\frac{X_o}{X_{st}} = \frac{1}{[\{1-(\omega/p_n)^2\}^2 + (2\zeta\omega/p_n)^2]^{1/2}}$	(22-62)
The plot of magnification factor (X_o/X_{st}) vs. frequency ratio (ω/p_n) and phase angle ϕ vs. (ω/p_n)	Refer to Figs. 22-11 and 22-12.	

Particular	Formula

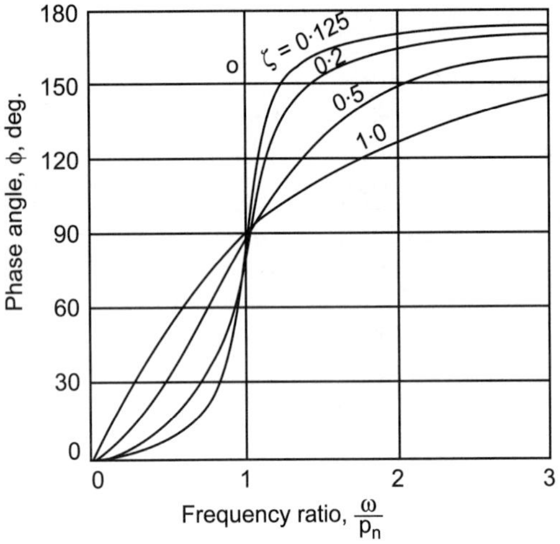

FIGURE 22-11 Phase angle ϕ vs. frequency ratio (ω/p_n).

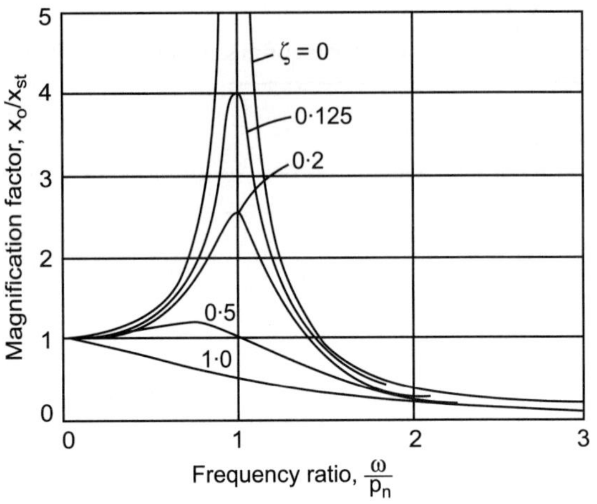

FIGURE 22-12 Magnification factor (X_o/X_{st}) vs. frequency ratio (ω/p_n).

The amplitude at resonance (i.e. for $\omega/p_n = 1$)

$$X_{res} = \frac{F_o}{cp_n} = \frac{F_o}{2\zeta k} = \frac{X_{st}}{2\zeta} \tag{22-63}$$

UNBALANCE DUE TO ROTATING MASS (Fig. 22-13)

The equation of motion

$$M\ddot{x} + c\dot{x} + kn = (me\omega^2)\sin\omega t \tag{22-64}$$

The steady-state solution for displacement

$$X = \frac{me\omega^2}{\sqrt{(k - M\omega^2)^2 + (c\omega)^2}} \tag{22-65a}$$

$$X = \frac{(m/M)\,e(\omega/p_n)^2}{[\{1 - (\omega/p_n)^2\}^2 + (2\zeta\omega/p_n)^2]^{1/2}} \tag{22-65b}$$

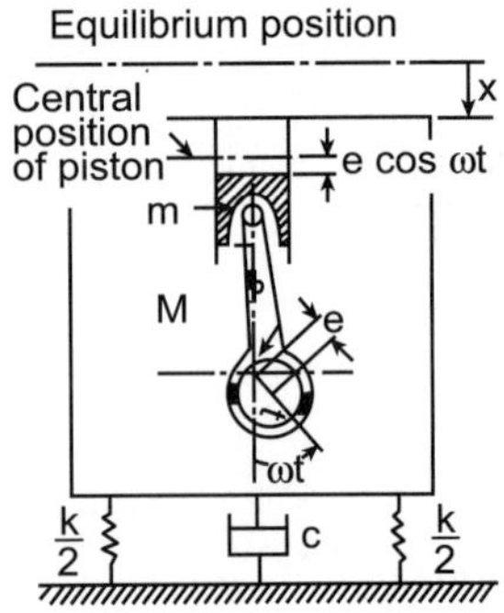

FIGURE 22-13 External force due to rotating unbalanced mass. (Produced with some modification from N. O. Myklestad, *Fundamentals of Vibration Analysis*, McGraw-Hill, New York, 1956.)

Particular	Formula
The complete solution for the displacement	$x = Ae^{-\zeta p_n t}\sin(qt+\phi_1) + \dfrac{e(m/M)(\omega/p_n)^2}{[\{1-(\omega/p_n)^2\}^2 + (2\zeta\omega/p_n)^2]^{1/2}}\sin(\omega t - \phi)$ (22-66)
Nondimensional form of expression for Eq. (22-65b)	$\dfrac{M}{m}\dfrac{X}{e} = \dfrac{(\omega/p_n)^2}{[\{1-(\omega/p_n)^2\}^2 + (2\zeta\omega/p_n)^2]^{1/2}}$ (22-67)
The phase angle	$\phi = \tan^{-1}\left[\dfrac{2\zeta(\omega/p_n)^2}{1-(\omega/p_n)^2}\right]$ (22-68)
For a schematic representation of Eqs. (22-67) and (22-68) or harmonically disturbing force due to rotating unbalance	Refer to Figs. 22-14 and 22-15

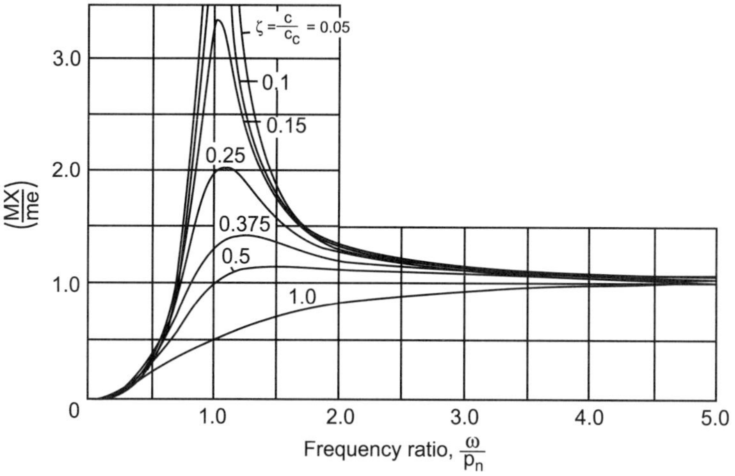

FIGURE 22-14 MX/me vs. frequency ratio (ω/p_n).

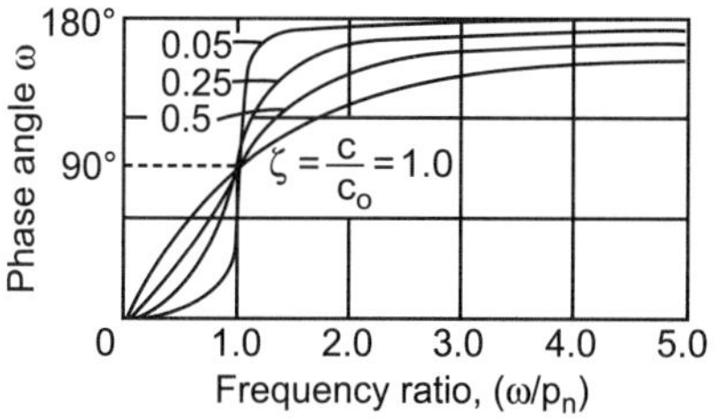

FIGURE 22-15 Phase angle ϕ vs. frequency ratio (ω/p_n).

Particular	Formula

WHIPPING OF ROTATING SHAFT (Fig. 22-16)

The equation of motion of shaft due to unbalanced mass

$$m\ddot{x}_c + c\dot{x}_c + kx_c = me\omega^2 \cos \omega t \quad (22\text{-}69a)$$

$$m\ddot{y}_c + c\ddot{y}_c + ky_c = me\omega^2 \sin \omega t \quad (22\text{-}69b)$$

where x_c and y_c are coordinates of position of center of shaft with respect to x and y coordinates

The solution

$$x_c = \frac{me\omega^2 \cos(\omega t - \phi)}{\sqrt{(k - m\omega^2)^2 + (c\omega)^2}} \quad (22\text{-}70a)$$

$$y_c = \frac{me\omega^2 \sin(\omega t - \phi)}{\sqrt{(k - m\omega^2)^2 + (c\omega)^2}} \quad (22\text{-}70b)$$

The displacement of the center of the disk from the line joining the centers of bearings

$$r = \sqrt{x_c^2 + y_c^2} = \frac{me\omega^2}{\sqrt{(k - m\omega^2)^2 + (c\omega)^2}} \quad (22\text{-}71a)$$

$$r = \frac{e(\omega/p_n)^2}{[\{1 - (\omega/p_n)^2\}^2 + (2\zeta\omega/p_n)^2]^{1/2}} \quad (22\text{-}71b)$$

The phase angle

$$\phi = \tan^{-1}\left(\frac{c\omega}{k - m\omega^2}\right) = \tan^{-1}\left[\frac{2\zeta(\omega/p_n)}{1 - (\omega/p_n)^2}\right] \quad (22\text{-}72)$$

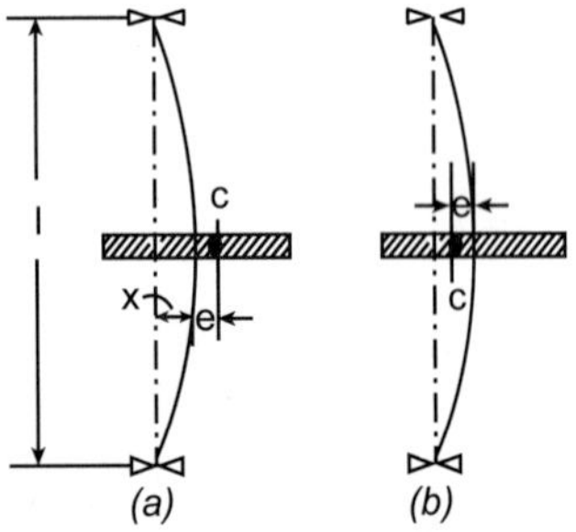

FIGURE 22-16 Whipping of shaft. (Reproduced from *Marks' Standard Handbook for Mechanical Engineers*, 8th edition, McGraw-Hill Book Company, New York, 1978.)

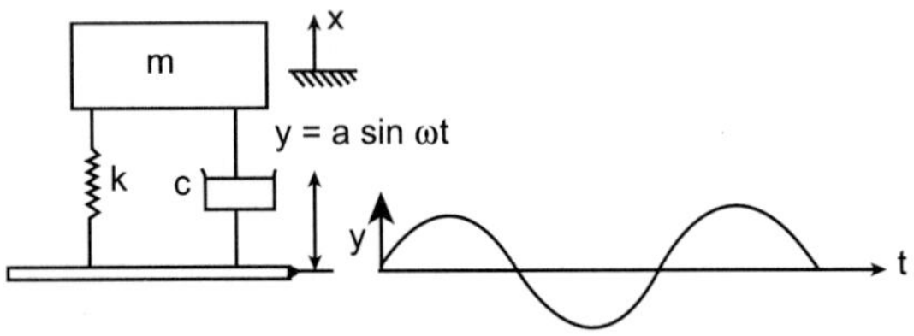

FIGURE 22-17 Excitation of a system by motion of support.

EXCITATION OF A SYSTEM BY MOTION OF SUPPORT (Fig. 22-17)

The equation of motion

$$m\ddot{x} + c\dot{x} + kx = ky + c\dot{y} \quad (22\text{-}73)$$

The absolute value of the amplitude ratio of x and y

$$\left|\frac{X}{Y}\right| = \left[\frac{1 + (2\zeta\omega/p_n)^2}{[1 - (\omega/p_n)^2]^2 + (2\zeta\omega/p_n)^2}\right]^{1/2} \quad (22\text{-}74)$$

Particular	Formula	
The phase angle	$\phi = \tan^{-1}\left[\dfrac{2\zeta(\omega/p_n)^3}{\{1-(\omega/p_n)^2\}^2 + (2\zeta\omega/p_n)^2}\right]$	(22-75)
The plot of Eq. (22-55) for motion due to support	Refer to Fig. 22-20 for $\lvert X/Y \rvert$ vs. ω/p_n. The curves are similar.	

INSTRUMENT FOR VIBRATION MEASURING (Fig. 22-18)

Particular	Formula	
The equation of motion	$m\ddot{z} + c\dot{z} + kz = -m\ddot{y} = mY\omega^2 \sin \omega t$	(22-76)
The steady-state solution for relative displacement Z	$Z = \dfrac{Y(\omega/p_n)^2}{[\{1-(\omega/p_n)^2\}^2 + (2\zeta\omega/p_n)^2]^{1/2}}$	(22-77)
The phase angle	$\phi = \tan^{-1}\left[\dfrac{2\zeta(\omega/p_n)}{1-(\omega/p_n)^2}\right]$	(22-78)
The plot of absolute value of $\lvert Z/Y \rvert$ vs. frequency ratio (ω/p_n) and the phase angle ϕ vs. frequency ratio (ω/p_n)	Refer to Figs. 22-14 and 22-15. The curves for $\lvert Z/Y \rvert$ vs. ω/p_n and ϕ vs. ω/p_n are identical.	

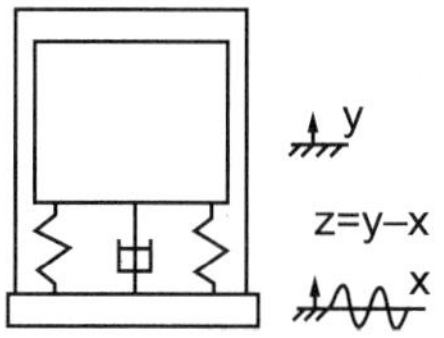

FIGURE 22-18 Instrument for vibration measuring. (Reproduced from *Marks' Standard Handbook for Mechanical Engineers*, 8th edition, McGraw-Hill, New York, 1978.)

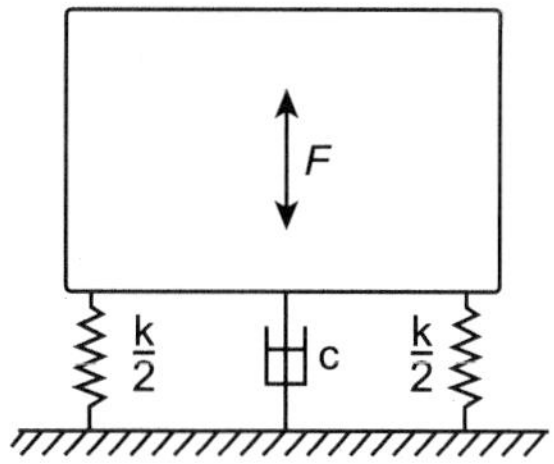

FIGURE 22-19 External force transmitted to foundation through damper and springs. (Reproduced from *Marks' Standard Handbook for Mechanical Engineers*, 8th edition, McGraw-Hill, New York, 1978.)

ISOLATION OF VIBRATION (Fig. 22-19)

Particular	Formula	
The force transmitted through the springs and damper	$F_T = \sqrt{(kX)^2 + (c\omega X)^2}$	(22-79)
	$F_T = \dfrac{F_o[1 + (2\zeta\omega/p_n)^2]^{1/2}}{[\{1-(\omega/p_n)^2\}^2 + (2\zeta\omega/p_n)^2]^{1/2}}$	(22-80)

Particular	Formula
Transmissibility	$$T_R = \frac{F_T}{F_o} = \frac{\sqrt{1+(2\zeta\omega/p_n)^2}}{[\{1-(\omega/p_n)^2\}^2+(2\zeta\omega/p_n)^2]^{1/2}} \quad (22\text{-}81)$$ Refer to Fig. 22-20 for T_R and $\lvert X/Y \rvert$.
Comparison of Eqs. (22-81) and (22-85) indicates that the plot of F_T/F_o is identical to $\lvert X/Y \rvert$. Transmissibility when damping is negligible	$$T_R = \frac{1}{(\omega/p_n)^2 - 1} \quad (22\text{-}82)$$
The transmissibility in terms of static deflection δ_{st}	$$T_R = \frac{1}{\dfrac{(2\pi f_n)^2\delta_{st}}{g} - 1} \quad (22\text{-}83)$$
The frequency from Eq. (22-83)	$$f_n = \frac{3.132}{2\pi}\left[\frac{1}{\delta_{st}}\left(\frac{1}{T_R}+1\right)\right]^{1/2} = 0.5\left[\frac{1}{\delta_{st}}\left(\frac{2-R}{1-R}\right)\right]^{1/2} \quad \textbf{SI} \quad (22\text{-}84a)$$ where f_n in Hz and δ_{st} in m
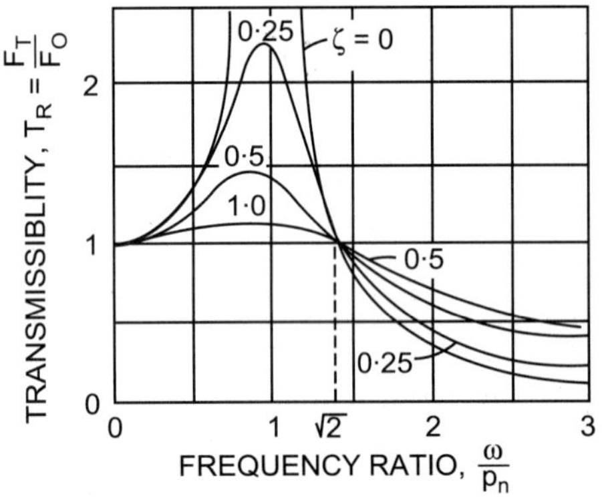 **FIGURE 22-20** Transmissibility (T_R) vs. frequency ratio (ω/p_n).	The percent reduction in the transmissibility is defined as $R = 1 - T_R$ $$f_n = \frac{99}{2\pi}\left[\frac{1}{\delta_{st}}\left(\frac{2-R}{1-R}\right)\right]^{1/2} = 15.76\left[\frac{1}{\delta_{st}}\left(\frac{2-R}{1-R}\right)\right]^{1/2} \quad \textbf{SI} \quad (22\text{-}84b)$$ where f_n in Hz and δ_{st} in mm $$f_n = \frac{19.67}{2\pi}\left[\frac{1}{\delta_{st}}\left(\frac{2-R}{1-R}\right)\right]^{1/2} \quad \textbf{USCS} \quad (22\text{-}84c)$$ where δ_{st} in in and f_n in Hz $$f_n = 187.6\left[\frac{1}{\delta_{st}}\left(\frac{2-R}{1-R}\right)\right]^{1/2} \quad \textbf{USCS} \quad (22\text{-}84d)$$ where f_n in rpm and δ_{st} in in
For the plot of static deflection δ_{st} vs. R	Refer to Fig. 22-21.

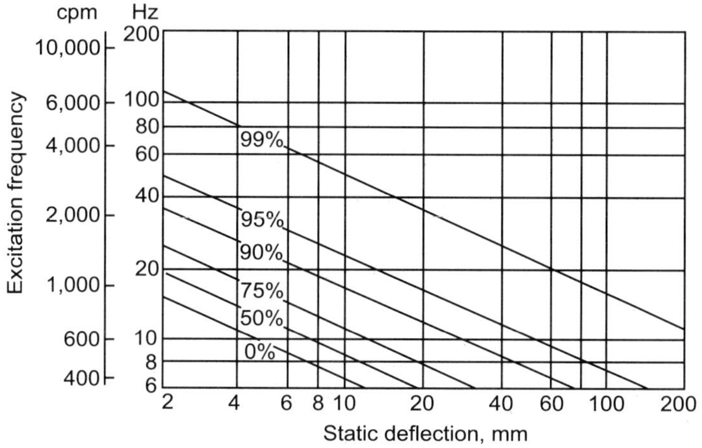

FIGURE 22-21 Static deflection (δ_{st}) vs. disturbing frequency for various percent reduction in transmissibility (T_R) for $\zeta = 0$. (Courtesy of F. S. Tes, I. E. Morse, and R. T. Hinkle, *Mechanical Vibration—Theory and Applications*, CBS Publishers and Distributors, New Delhi, India, 1983.)

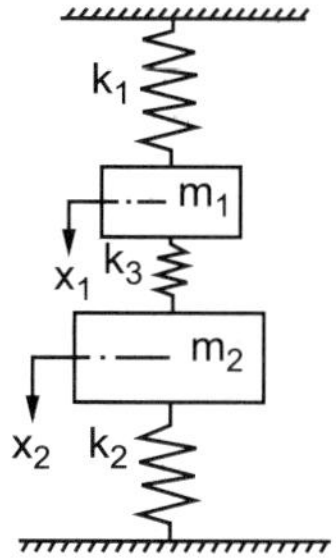

FIGURE 22-22 Undamped two-degree-of freedom system.

Particular	Formula

UNDAMPED TWO-DEGREE-OF-FREEDOM SYSTEM (Fig. 22-22) WITHOUT EXTERNAL FORCE

Equations of motion

$$m_1\ddot{x}_1 + (k_1 + k_3)x_1 - k_3x_2 = 0 \quad (22\text{-}85a)$$

$$m_2\ddot{x}_2 + (k_2 + k_3)x_2 - k_3x_1 = 0 \quad (22\text{-}85b)$$

The frequency of equation which gives two values for p^2

$$p^4 - p^2\left(\frac{k_1 + k_3}{m_1} + \frac{k_2 + k_3}{m_2}\right) + \frac{k_1k_2 + k_2k_3 + k_1k_3}{m_1m_2} = 0 \quad (22\text{-}86)$$

The amplitude ratio

$$\frac{a_1}{a_2} = \frac{-k_3}{m_1p^2 - k_1 - k_3} = \frac{m_2p^2 - k_2 - k_3}{-k_3} \quad (22\text{-}87)$$

DYNAMIC VIBRATION ABSORBER (Fig. 22-23)

Equations of motion

$$M\ddot{x}_1 + (K + k)x_1 - kx_2 = F_o \sin\omega t \quad (22\text{-}88a)$$

$$m\ddot{x}_2 + k(x_2 - x_1) = 0 \quad (22\text{-}88b)$$

The solution of the forced vibration of the absorber will be of the form

$$x_1 = a_1 \sin pt \quad (22\text{-}89a)$$

$$x_2 = a_2 \sin pt \quad (22\text{-}89b)$$

Particular	Formula
The ratio of amplitudes a_1 and a_2 to the static deflection of the main system x_{st}	$$\frac{a_1}{x_{st}} = \frac{1-\frac{\omega^2}{p_a^2}}{\left(1-\frac{\omega^2}{p_a^2}\right)\left(1+\frac{k}{K}-\frac{\omega^2}{p_m^2}\right)-\frac{k}{K}} \quad \text{(22-90a)}$$ $$\frac{a_2}{x_{st}} = \frac{1}{\left(1-\frac{\omega^2}{p_a^2}\right)\left(1+\frac{k}{K}-\frac{\omega^2}{p_m^2}\right)-\frac{k}{K}} \quad \text{(22-90b)}$$ where $x_{st} = F_o/K =$ static deflection of main system; $p_a^2 = K/m =$ natural circular frequency of absorber; $p_m^2 = k/M =$ natural circular frequency of main system; $R_m = \frac{m}{M} =$ mass ratio $= \frac{\text{absorber mass}}{\text{main mass}}$
If the main system is in resonance, then considering $p_a = p_m$ or $\frac{k}{m} = \frac{K}{M}$ or $\frac{k}{K} = \frac{m}{M} = R_m$. Eqs. (7-90a) and (7-90b) become	$$\frac{x_1}{x_{st}} = \frac{1-(\omega/p_a)^2}{[1-(\omega/p_a)^2][1+R_m-(\omega/p_a)^2]-R_m}\sin\omega t \quad \text{(22-91a)}$$ $$\frac{x_2}{x_{st}} = \frac{1}{[1-(\omega/p_a)^2][1+R_m-(\omega/p_a)^2]-R_m}\sin\omega t \quad \text{(22-91b)}$$
The natural frequencies	$$\left(\frac{\omega}{p_a}\right)^2 = \left(1+\frac{R_m}{2}\right) \pm \left(R_m+\frac{R_m^2}{4}\right)^{1/2} \quad \text{(22-92)}$$
The mass equivalent for the absorber	$$\frac{m_{eq}}{m} = \frac{1}{1-(\omega/p_a)^2} \quad \text{(22-93)}$$ where $m_{eq} =$ equivalent mass solidly attached to the main mass M

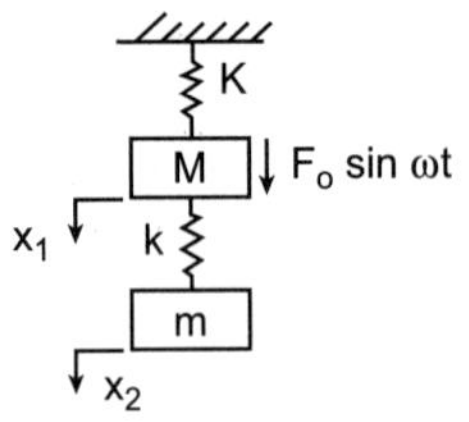

FIGURE 22-23 Dynamic vibration absorber.

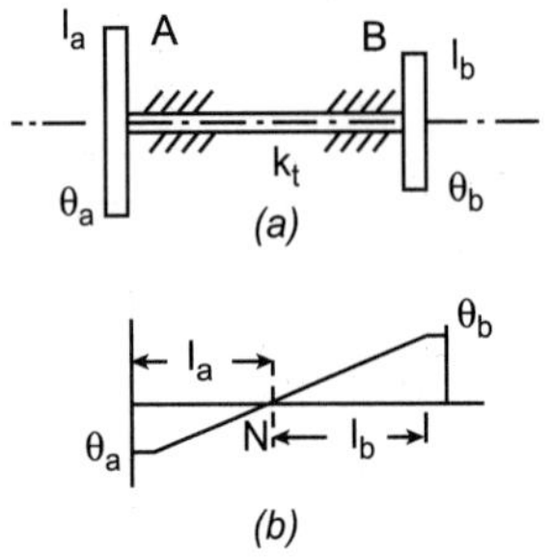

FIGURE 22-24 Two-rotor system.

TORSIONAL VIBRATING SYSTEMS

Two-rotor system (Fig. 22-24)

Particular	Formula
The torque on rotor A	$M_{ta} = I_a p^2 \theta_a$ (22-94)
The total torque on two rotors	$M_{ti} = M_{ta} + M_{tb} = I_a p^2 \theta_a + I_b p^2 \theta_b = 0$ (22-95) where $i =$ imaginary
The angular displacement or angle of twist of rotor B	$\theta_b = \theta_a - \frac{M_{ta}}{k_t} = \theta_a\left(1 - \frac{I_a p^2}{k_t}\right)$ (22-96)

Particular	Formula	
The frequency equation	$p^2\theta_a\left(I_a + I_b - \frac{I_a I_b p^2}{k_t}\right) = 0$	(22-97a)
The natural circular frequency	$p_n = \left(\frac{(I_a + I_b)k_t}{I_a I_b}\right)^{1/2}$	(22-97b)
The natural frequency	$f_n = \frac{1}{2\pi}\left(\frac{(I_a + I_b)k_t}{I_a I_b}\right)^{1/2}$	(22-98)
The amplitude ratio	$\frac{\theta_a}{\theta_b} = -\frac{I_b}{I_a} = -\frac{l_a}{l_b}$	(22-99)
The relation between I_a, I_b, l_a, and l_b	$I_a l_a = I_b l_b$	(22-100)
The distance of node point from left end of rotor A	$l_a = \frac{I_b l}{I_a + I_b}$	(22-101)

Two rotors connected by shaft of varying diameters

The length of torsionally equivalent shaft of diameter d whose varying diameters are d_1, d_2, and d_3	$l_e = d^4\left(\frac{l_1}{d_1^4} + \frac{l_2}{d_2^4} + \frac{l_3}{d_3^4}\right)$	(22-102)

Three-rotor torsional system (Fig. 22-25)

The algebraic sum of the inertia torques of rotors A, B, and C	$M_{ti} = M_{ta} + M_{tb} + M_{tc} = I_a p^2\theta_a + I_b p^2\theta_b + I_c p^2\theta_c$ where θ_a, θ_b, and θ_c are angular displacement or angular twist at rotors, A, B, and C, respectively	(22-103)

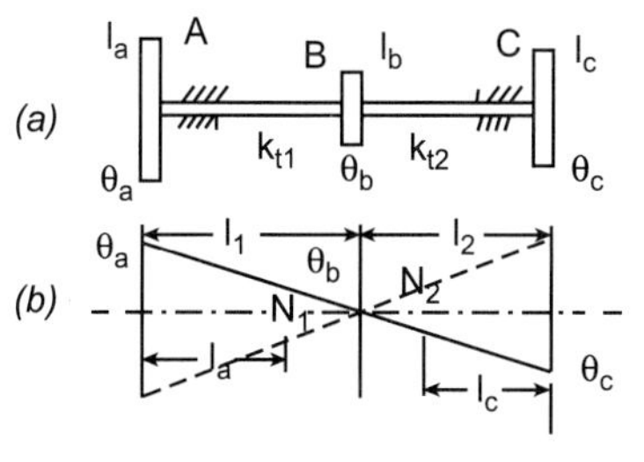

FIGURE 22-25 Three-rotor system.

Particular	Formula
The frequency equation	$$p^2\theta_a\left[(I_a+I_b+I_c)-p^2\left(\frac{I_aI_b}{k_{t1}}+\frac{I_aI_c}{k_{t1}}+\frac{I_aI_c}{k_{t2}}+\frac{I_bI_c}{k_{t2}}\right)+p^4\left(\frac{I_aI_bI_c}{k_{t1}k_{t2}}\right)\right]=0 \quad (22\text{-}104a)$$
	$$p^2=\frac{1}{2}\left(\frac{k_{t1}}{I_a}+\frac{k_{t2}}{I_c}+\frac{k_{t1}+k_{t2}}{I_b}\right)\pm\frac{1}{2}\left[\left(\frac{k_{t1}}{I_a}+\frac{k_{t2}}{I_c}+\frac{k_{t1}+k_{t2}}{I_b}\right)^2-4\frac{k_{t1}k_{t2}}{I_aI_bI_c}(I_a+I_b+I_c)\right]^{1/2} \quad (22\text{-}104b)$$ where k_{t1} and k_{t2} are torsional stiffness of shafts of lengths l_1 and l_2
The amplitude ratio	$$\frac{\theta_b}{\theta_a}=1-\frac{I_ap^2}{k_{t1}} \quad (22\text{-}105a)$$
	$$\frac{\theta_c}{\theta_a}=1-p^2\left(\frac{I_a}{k_{t1}}+\frac{I_c}{k_{t2}}+\frac{I_b}{k_{t2}}\right)+\frac{p^4I_aI}{k_{t1}k_{t2}} \quad (22\text{-}105b)$$
The relation between I_a, I_c, l_a, and l_c	$$I_al_a=I_cl_c \quad (22\text{-}106)$$
The relation between I_a, I_b, l_a, and l_c	$$\frac{1}{I_al_a}=\frac{1}{I_b}\left(\frac{1}{l_1-l_a}+\frac{1}{l_2-l_c}\right) \quad (22\text{-}107)$$
Frequency can also be found from Eqs. (22-108) and (22-109)	$$f_c=\left(\frac{1}{2\pi}\right)\sqrt{\frac{k_{tc}}{I_c}} \quad (22\text{-}108)$$ where $k_{tc}=\frac{GJ_2}{l_c}$
	$$f_b=\left(\frac{1}{2\pi}\right)\sqrt{\frac{k'_{tb}}{I_b}} \quad (22\text{-}109)$$ where $k'_{tb}=\frac{GJ_1}{l_1-l_a}+\frac{GJ_2}{l_2-l_c}$
For collection of mechanical vibration formulas to calculate natural frequencies	Refer to Table 22-2.
For analogy between different wave phenomena	Refer to Table 22-3.
For analogy between mechanical and electrical systems	Refer to Table 22-4.

TABLE 22-2
A collection of formulas

	Particular	Formula	Equation
		Natural Frequencies of Simple Systems	
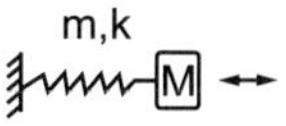	End mass M, spring mass m, spring stiffness k	$p_n = \sqrt{\dfrac{k}{M + m/3}}$	(22-110)
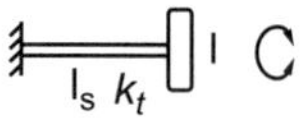	End inertia I, shaft inertia I_s, shaft stiffness k_t	$p_n = \sqrt{\dfrac{k_t}{I + I_s/3}}$	(22-111)
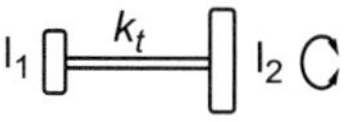	Two disks on a shaft	$p_n = \sqrt{\dfrac{k_t(I_1 + I_2)}{I_1 I_2}}$	(22-112)
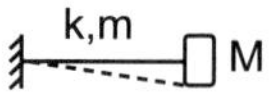	Cantilever; end mass M, beam mass m, stiffness by formula (22-93)	$p_n = \sqrt{\dfrac{k}{M + 0.23m}}$	(22-113)
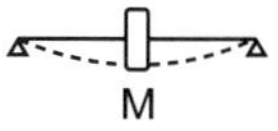	Simply supported beam central mass M; beam mass m; stiffness by formula (22-95)	$p_n = \sqrt{\dfrac{k}{M + 0.5m}}$	(22-114)
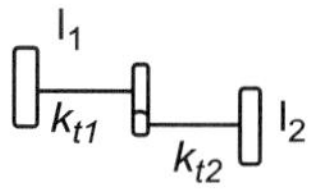	Massless gears, speed of I_2 n times as as speed of I_1	$p_n = \sqrt{\dfrac{1}{\dfrac{1}{k_{t1}} + \dfrac{1}{n^2 k_{t2}}} \left(\dfrac{I_1 + n^2 I_2}{I_1 I_2 n^2} \right)}$	(22-115)
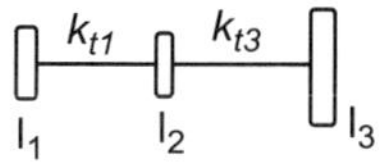		$p_n^2 = \dfrac{1}{2}\left(\dfrac{k_{t1}}{I_1} + \dfrac{k_{t3}}{I_3} + \dfrac{k_{t1} + k_{t3}}{I_2}\right) \pm \dfrac{1}{2}\sqrt{\left(\dfrac{k_{t1}}{I_1} + \dfrac{k_{t3}}{I_3} + \dfrac{k_{t1} + k_{t3}}{I_2}\right)^2 - 4\dfrac{k_{t1}k_{t3}}{I_1 I_2 I_3}(I_1 + I_2 + I_3)}$	(22-116)
		Uniform Beams (Longitudinal and Torsional Vibration)	
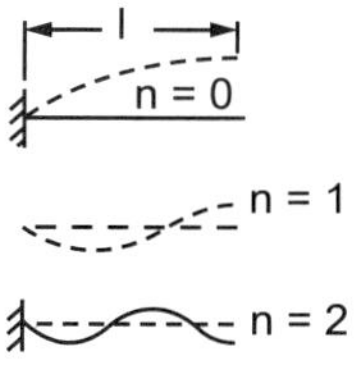	Longitudinal vibration of cantilever: A = cross section, E = modulus of elasticity	$p_n = \left(n + \dfrac{1}{2}\right)\pi\sqrt{\dfrac{AE}{\mu_1 l^2}}$	(22-117)
	μ_1 = mass per unit length, $n = 0, 1, 2, 3$ = number of nodes	For steel and l in inches, this becomes $f = \dfrac{p_n}{2\pi} = (1 + 2n)\dfrac{1295}{l}$ Hz	(22-118)
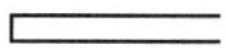	Organ pipe open at one end, closed at the other	For air at atm. pressure, l in m $f = \dfrac{p_n}{2\pi} = (1 + 2n)\dfrac{84}{l}$ Hz $n = 0, 1, 2, 3, \ldots$	(22-119)
	Water column in rigid pipe closed at one end (l in m)	$f = \dfrac{p_n}{2\pi} = (1 + 2n)\dfrac{360}{l}$ Hz $n = 0, 1, 2, 3, \ldots$	(22-120)
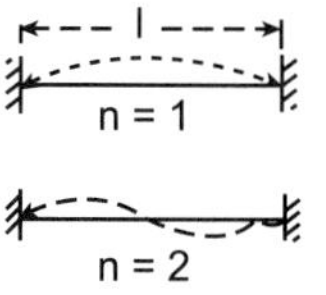	Longitudinal vibration of beam clamped or free at both ends; n = number of half waves along length	$p_n = n\pi\sqrt{\dfrac{AE}{\mu_1 l^2}}$ $n = 1, 2, 3, \ldots$	(22-121)

TABLE 22-2
A collection of formulas (*Cont.*)

	Particular	Formula	Equation
	For steel, l in m	$f = \frac{p_n}{2\pi} = \frac{2590}{l}$ Hz	(22-122a)
	For steel, l in inches	$f = \frac{p_n}{2\pi} = \frac{102{,}000}{l}$ Hz	(22-122b)
	Organ pipe closed (or open) at both ends (air at 60°F, 15.5°C)	$f = \frac{p_n}{2\pi} = \frac{n168}{l}$ Hz $n = 1, 2, 3, \ldots$	(22-123)
	Water column in rigid pipe closed (or open) at both ends	$f = \frac{n721}{l}$ Hz $n = 1, 2, 3, \ldots$	(22-124)
	For water columns in nonrigid pipes	$\frac{f_{\text{nonrigid}}}{f_{\text{rigid}}} = \frac{1}{\sqrt{1 + \frac{206D}{tE_{\text{pipe}}}}}$ E_{pipe} = elastic modulus of pipe, MPa D, t = pipe diameter and wall thickness, same units	(22-125a)
	For water columns in nonrigid pipes ...	$\frac{f_{\text{nonrigid}}}{f_{\text{rigid}}} = \frac{1}{\sqrt{1 + \frac{300{,}000D}{tE_{\text{pipe}}}}}$ E_{pipe} = elastic modulus of pipe, psi D, t = pipe diameter and wall thickness, same units	(22-125b)
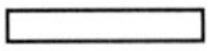	Torsional vibration of beams ...	Same as (22-117) and (22-118); replace tensional stiffness AE by torsional stiffness GI_p; replace μ_1 by the moment of inertia per unit length $i_1 = I_{\text{bar}}/l$	

Uniform Beams (Transverse or Bending Vibrations)

The same general formula holds for all the following cases,

$$p_n = a_n \sqrt{\frac{EI}{\mu_1 l^4}} \qquad \text{(22-126)}$$

where EI is the bending stiffness of the section, l is the length of the beam, μ_1 is the mass per unit length $= W/gl$, and a_n is a numerical constant, different for each case and listed below.

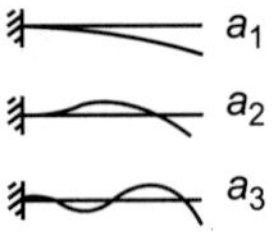

Cantilever or "clamped-free" beam ...

$a_1 = 3.52$
$a_2 = 22.0$
$a_3 = 61.7$
$a_4 = 121.0$
$a_5 = 200.0$

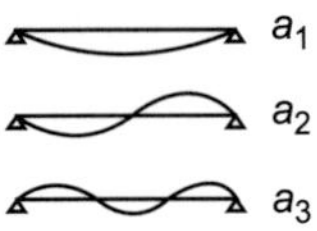

Simply supported or "hinged-hinged" beam

$a_1 = \pi^2 = 9.87$
$a_2 = 4\pi^2 = 39.5$
$a_3 = 9\pi^2 = 88.9$
$a_4 = 16\pi^2 = 158$
$a_5 = 25\pi^2 = 247$

TABLE 22-2
A collection of formulas (*Cont.*)

	Particular	Formula	Equation
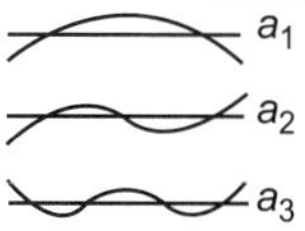	"Free-free" beam or floating ship . . .	$a_1 = 22.0$ $a_2 = 61.7$ $a_3 = 121.0$ $a_4 = 200.0$ $a_5 = 298.2$	
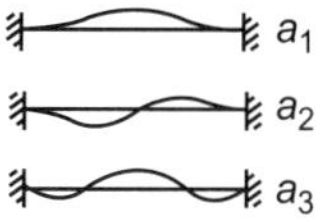	"Clamped-clamped" beam has same frequencies as "free-free"	$a_1 = 22.0$ $a_2 = 61.7$ $a_3 = 121.0$ $a_4 = 200.0$ $a_5 = 298.2$	
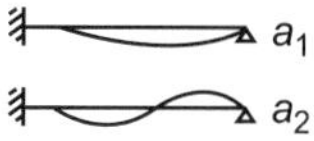	"Clamped-hinged" beam may be considered as half a "clamped-clamped" beam for even *a*-numbers	$a_1 = 15.4$ $a_2 = 50.0$ $a_3 = 104$ $a_4 = 178$ $a_5 = 272$	
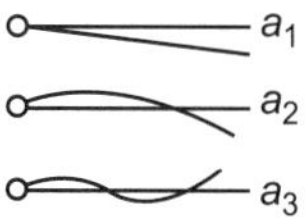	"Hinged-free" beam or wing of autogyro may be considered as half a "free-free" beam for even *a*-numbers	$a_1 = 0$ $a_2 = 15.4$ $a_3 = 50.0$ $a_4 = 104$ $a_5 = 178$	

Rings, Membranes, and Plates

Extensional vibration of a ring, radius r, weight density γ

$$p_n = \frac{1}{r}\sqrt{\frac{Eg}{\gamma}} \tag{22-127}$$

Bending vibrations of ring, radius r, mass per unit length, μ_1, in its own plane with n full "sine waves" of disturbance along circumference

$$p_n = \frac{n(n^2-1)}{\sqrt{1+n^2}}\sqrt{\frac{EI}{\mu_1 r^4}} \tag{22-128}$$

Circular membrane of tension T, mass per unit area μ_1, radius r

$$p_n = a_{cd}\sqrt{\frac{T}{\mu_1 r^2}} \tag{22-129}$$

The constant a_{cd} is shown below, the subscript c denotes the number of nodal circles, and the subscript d the number of nodal diameters:

	c		
d	**1**	**2**	**3**
0	2.40	5.52	8.65
1	3.83	7.02	10.17
2	5.11	8.42	11.62
3	6.38	9.76	13.02

TABLE 22-2
A collection of formulas (*Cont.*)

Membrane of any shape of area A roughly of equal dimensions in all directions, fundamental mode:

$$p_n = \text{const}\sqrt{\frac{T}{\mu_1 A}} \tag{22-130}$$

Circle	const $= 2.40\pi = 4.26$
Square	const $= 4.44$
Quarter circle	const $= 4.55$
2×1 rectangle	const $= 4.97$

Circular plate of radius r, mass per unit area μ_1; the "plate constant D" defined in Eq (22-49)

$$p_n = a\sqrt{\frac{D}{\mu_1 r^4}} \tag{22-131}$$

For free edges, 2 perpendicular nodal diameters	$a = 5.25$
For free edges, one nodal circle, no diameters	$a = 9.07$
Clamped edges, fundamental mode	$a = 10.21$
Free edges, clamped at center, umbrella mode	$a = 3.75$

Rectangular plate, all edges simply supported, dimensions l_1 and l_2:

$$p_n = \pi^2\left(\frac{m^2}{l_1^2} + \frac{n^2}{l_1^2}\right)\sqrt{\frac{D}{\mu_1}} \qquad m = 1, 2, 3, \ldots;\ n = 1, 2, 3, \ldots \tag{22-132}$$

Square plate, all edges clamped, length of side l, fundamental mode:

$$p_n = \frac{36}{l^2}\sqrt{\frac{D}{\mu_1}} \tag{22-133}$$

Source: Formulas (Eqs.) (7-110) to (7-133) extracted from J. P. Den Hartog, *Mechanical Vibrations*, McGraw-Hill Book Company, New York, 1962.

TABLE 22-3
Analogy between different wave phenomena

Quantity	Phenomenon: String	Transverse wave	Longitudinal wave	Acoustic wave	Torsional wave in bar	Electric cable
Particle velocity	$\dot{x}$	$\dot{x}$	$\dot{x}$	$\dot{x}$	$\dot{\theta}$	e voltage
Mass per unit length	$\rho . A$	$\rho . A$	$\rho . A$	$\rho_a . A$	$\rho . J$	C capacitance/cm
Inverse spring constant per unit length	$1/T$	$1/G . A$	$1/E . A$	$\frac{1}{p_n . k . A}$	$\frac{1}{J . G}$	L self-inductance/cm
Elastic force on a mass-element	$T_? = T . \frac{\partial x}{\partial y}$	$\tau A = G . A . \frac{\partial x}{\partial y}$	$\sigma A = E . A . \frac{\partial x}{\partial y}$	$pA = p_n . k . A . \frac{\partial x}{\partial y}$	$M_t = J . G . \frac{\partial \theta}{\partial y}$	i current
Velocity of propagation c	$\sqrt{\frac{T}{pA}}$	$\sqrt{\frac{G}{p}}$	$\sqrt{\frac{E}{p}}$	$\sqrt{\frac{p_n . k}{p_n}}$	$\sqrt{\frac{G}{p}}$	$\sqrt{\frac{1}{LC}}$
Ratio of force to velocity	$\dot{x} = T_? . \sqrt{\frac{A}{pT}}$	$\dot{x} = \frac{\tau A}{\sqrt{pG}}$	$\dot{x} = \frac{\sigma A}{\sqrt{pE}}$	$\dot{x} = \frac{pA}{\sqrt{p_n . p_n k}}$	$\dot{\theta} = \frac{M_t}{\sqrt{pG}}$	$c = \frac{i}{\sqrt{L/C}}$
Intensity I	$\frac{(\dot{x}_o)^2}{2} . p . C$	$\frac{(\dot{x}_o)^2}{2} . p . C$	$\frac{(\dot{x}_o)^2}{2} . p . C$	$\frac{(\dot{x}_o)^2}{2} . p . C$	energy per sec total $\frac{(\dot{\theta}_o)^2}{2} . J . p . c$	energy per sec $\frac{c^2}{2} . C . c$
Wave impedance	$p . c = \sqrt{\frac{pT}{A}}$	$p . c = \sqrt{p . G}$	$p . c = \sqrt{p . G}$	$p_n . c = \sqrt{p_n . p_n k}$	$p . c = \sqrt{p . G}$	inverse wave impedance $\frac{1}{Z_{\text{wave}}} = \sqrt{\frac{C}{L}}$

Source: *Courtesy* G. W. van Santen, *Introduction to Study of Mechanical Vibration*, 3rd edition, Philips Technical Library, 1961.
Key: c = capacitance; e = voltage; i = current, A; I = intensity, W/m^2; J = polar moment of inertia, m^4 or cm^4; $k = c_p/c_v$ = ratio of specific heats; L = inductance, H; n = any integer = 1, 2, 3, 4, ...; p = pressure of gas, sound pressure, MPa; p_n = average pressure of gas, MPa; R = resistance, Ω; T = tension; $T_?$ = component of tension T which returns the string to the position of equilibrium, kN; ρ = specific mass of the material of string, density of air, kg/m^3; ρ_n = average density of gas, kg/m^3; σ = normal stress, MPa; τ = shear stress, MPa; λ = wavelength, m.
The meaning of other symbols in Table 7-3 are given under symbols at the beginning of this chapter.

TABLE 22-4
Analogy between mechanical and electrical systems

Mechanical system		Electrical system: Force—current	Electrical system: Force—voltage
D'Alembert's principle Force applied		Kirchhoff's current law Switch closed	Kirchhoff's voltage law Switch closed
Rectilinear system	*Torsional system*	*Electrical network*	*Electrical network*
Mass 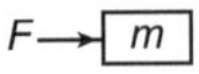		Capacitance 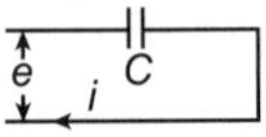	Inductance 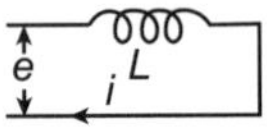
$F = m\dot{\nu} = m\ddot{x}$, Kinetic energy $= \frac{1}{2}m\nu^2$		$i = C\dot{c}$, $q = C\ddot{c}$, Energy $= \frac{1}{2}C_4^2$	$e = L\frac{di}{dt} = L\ddot{q}$ Energy $= \frac{1}{2}Li^2$
Viscous Damping 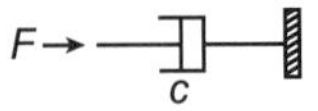		Resistance 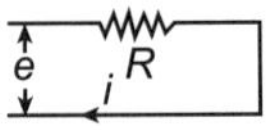	Resistance 
$F = c\dot{x}$, Power $= F\dot{x} = c\nu^2$		$i = \frac{c}{R}$; $q = \frac{1}{R}\dot{e}$ Power $= ci = \frac{c^2}{R}$	$e = Ri = R\dot{q}$, Power $= ci = Ri^2 = R\dot{q}^2$
Spring 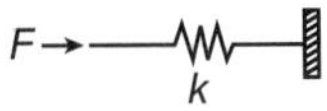		Inductance 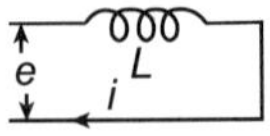	Capacitance 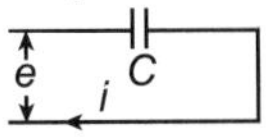
$F = kx = k\int \dot{x}\,dt$ Potential energy $= \frac{1}{2}\frac{F_0^2}{k}$		$i = \frac{1}{L}\int e\,dt$; $q = \frac{e}{L}$ Energy $= \frac{1}{2}Li^2$	$e = \frac{1}{C}q = \frac{1}{C}\int i\,dt$ Energy $= \frac{1}{2}Ce^2$
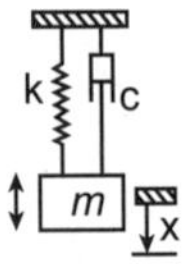 $F(t) = kx = \int \dot{x}\,dt$	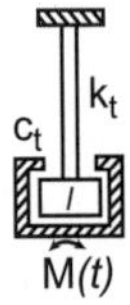	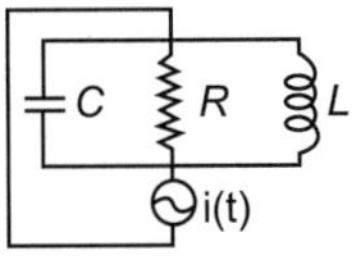	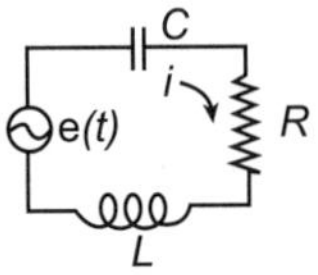
(a) *Spring-mass-dashpot elements*	*Shaft-rotor-dashpot elements*	(b) *Parallel connected electrical elements*	(c) *Series connected electrical elements*
Differential equation of motion		Differential equation for current	Differential equation for voltage
$m\dot{\nu} + c\nu + c\nu + k\int \nu\,dt = F(t)$ $m\ddot{x} + c\dot{x} + kx = F(t)$	$I\ddot{\phi} - c_t\dot{\phi} + k_t\phi = M_t(t)$	$C\dot{e} = \frac{r}{R} + \frac{1}{L}\int e\,dt = i(t)$ $C\ddot{e} + \frac{1}{R}\dot{e} + \frac{d}{L} = \frac{e_i(t)}{ex}$	$L\frac{dI}{dt} + Ri + \frac{1}{C}\int i\,dt = e(t)$ $L\ddot{q} + R\dot{q} + \frac{q}{C} = e(t)$

REFERENCES

1. Den Hartog, J. P., *Mechanical Vibrations,* McGraw-Hill Book Company, New York, 1962.
2. Thomson, W. T., *Theory of Vibration with Applications,* Prentice-Hall, Englewood Cliffs, New Jersey, 1981.
3. Baumeister, T., ed., *Marks' Standard Handbook for Mechanical Engineers,* 8th ed., McGraw-Hill Book Company, New York, 1978.
4. Black, P. H., and O. E. Adams, Jr., *Machine Design,* McGraw-Hill Book Company, New York, 1955.
5. Lingaiah, K., and B. R. Narayana Iyengar, *Machine Design Data Handbook,* Engineering College Co-Operative Society, Bangalore, India, 1962.
6. Myklestad, N. O., *Fundamentals of Vibration Analysis*, McGraw-Hill Book Company, New York, 1956.
7. Tse, F. S., I. E. Morse, and R. T. Hinkle, *Mechanical Vibration—Theory and Applications,* CBS Publishers and Distributors, New Delhi, India, 1983.

CHAPTER

23

DESIGN OF BEARINGS AND TRIBOLOGY

23.1 SLIDING CONTACT BEARINGS[1,2,11]

SYMBOLS

a	distance between bolt centers [Eqs. (23-70) to (23-72)], m (in)
$a = \dfrac{h_2}{B}$	dimensionless quantity
$A = Ld$	projected area of the journal bearing (Fig. 23-6), m^2 (in^2) effective area of the bearing, m^2 (in^2) projected area at full pool pressure in case of hydrostatic journal bearing (Fig. 23-47), m^2 (in^2)
A'	projected area of the region having a linear pressure gradient in case of hydrostatic journal bearing (Fig. 23-47), m^2 (in^2)
B	width of slider bearing in the direction of motion, m (in) length of journal bearing in the direction of motion, m (in)
$c = D - d$	diametral clearance, m (in)
C	combined coefficient of radiation and convection, W/m^2 K (kcal/mm^2 s°C)
C_1, C_2	constants in Eq. (23-23)
$C_F = \dfrac{F\mu}{F_{\mu\infty}}$	friction leakage factor in Eq. (23-54)
$C_{P\mathcal{F}1}$, $C_{P\mathcal{F}2}$, $C_{P\mathcal{F}3}$, $C_{P\mathcal{F}4}$	constants in Eqs. (23-77b), (23-78b), (23-79b), and (23-80b)
C_{PFm}, C_{PFs}	friction resistance factor for moving and stationary member, respectively, in pivoted shoe slider bearing in Eqs. (23-96b) and (23-97b)
C_{PW}	load factor in Eq. (23-95b)
C_Q	flow correction factor from (Fig. 23-42) and Eq. (23-65)
C_{S1} to C_{S7}	constants in Eqs. (23-86b), (23-87b), (23-88b), (23-89b), (23-90b), (23-91b), and (23-92b)
$C_W = \dfrac{W}{W_\infty}$	load leakage factor in Eqs. (23-52)
$C_\mu = \dfrac{\mu_\infty}{\mu}$	coefficient of friction factor in Eq. (23-53)
$C_{P\mu}$	coefficient of friction factor in Eqs. (23-98) and Table 23-17

d	diameter of journal, m (in)
d_i, d_2	inside and outside diameters of thrust, pivot, and collar bearings, m (in)
d_c	diameter of capillary in case of hydrostatic journal bearing, m (in)
D	diameter of bearing, m (in)
$e = c - h_{\min}$	eccentricity, m (in)
E	Young's modulus, GN/m^2 or GPa (Mpsi)
E_t^o	Engler, deg
F	force (also with subscripts), kN (lbf)
$F_{P\mathcal{F}W}$	load factor in Eqs. (23-83) and (23-84)
F_μ	friction force, kN (lbf)
F'_μ	$\frac{F_\mu}{dL}$ friction force per unit area of bearing, MPa (Psi)
$F_{\mu m}$	friction force on the moving member of bearing (i.e., slider), kN (lbf)
$F_{\mu mp}$	friction force on the moving member of pivoted slider bearing (i.e., slider), kN (lbf)
$F_{\mu s}$	friction force on the stationary member of bearing (i.e., shoe), kN (lbf)
$F_{\mu sp}$	friction force on the stationary member of pivoted slider bearing (i.e., shoe), kN (lbf)
$F_{\mu\infty}$	friction force acting on the moving surface of the same bearing with the same oil-film shape but without end leakage, kN (lbf)
G	flow factor given by Eq. (23-82)
h	oil film thickness, m (in)
h_1, h_2	thickness of oil film at entrance and exit, respectively, of a slider bearing (Fig. 23-48 and Fig. 23-52), m (in)
h_c	thickness of bearing cap, m (in)
$h_{\min} = h_o$	minimum thickness of oil film, m (in)
$h_{\max}$	maximum thickness of oil film, m (in)
H_d	heat dissipating capacity of bearing, kJ/s (kcal/s)
H_g	heat generated in bearing. kJ/s (kcal/s)
i	number of collars
k	characteristic number of the given crude oil ($\simeq$1.4 to 2.8), constant (also with subscripts)
	heat dissipating coefficient
$k = (h)_{P(\max) \atop P(\min)}$	thickness of the oil film where the pressure has its maximum or minimum values, m (in)
K	constant for a given grade of oil (varies from 1.000 to 1.004)
K_1, K_2, K_3, K_4	constants in Eqs. (23-73b), (23-74b), (23-75b), and (23-76b) respectively
K_5, K_6	constants in Eqs. (23-143b) and (23-144b), respectively
K_{LP1}, K_{LP2}, K_{LP3}	constants in Eqs. (23-116b), (23-118b), and (23-119b) for parallel surface thrust bearing
K_{lt}	constant in Eq. (23-121b) for a tilting-pad bearing
K_{Pt}	constant in Eq. (23-120b) for a tilting-pad bearing
$K_{\mu t}$	coefficient of friction factor in Eq. (23-126b) for a tilting-pad bearing
l_1	length of bearing pressure pad in case of hydrostatic journal bearing (Fig. 23-47), m (in)
l_c	length of capillary, m (in)
L	axial length of the journal (or of the bearing) normal to the direction of motion, m (in)
$m = \frac{h_1}{h_2}$	ratio of the film thicknesses at the entrance to exit in the slider bearing
M_t	torque, N m (lbf in)
n	speed, rpm
n'	speed, rps

P power (also with subscripts), kW (hp)
P intensity of pressure, MPa (psi)
$P = \dfrac{W}{Ld}$ load per projected area of the bearing, MPa (psi)
P_u unit load supported by a parallel surface thrust bearing, MPa (psi)
P_1 lower pool pressure in hydrostatic journal bearing (Fig. 23-47), MPa (psi)
P_2, P_4 left and right pool pressure in hydrostatic journal bearing (Fig. 23-47), MPa (psi)
P_3 upper pool pressure in hydrostatic journal bearing (Fig. 23-47), MPa (psi)
$P'_1 = P'_2 = P'_3 = P'_4 = P'$ the pressure in first, second, third and fourth quadrant of the pool, respectively, when the journal is concentric ($e = o$) in hydrostatic journal bearing, MPa (psi)
P_i inlet pressure, MPa (psi)
P_o constant manifold pressure, MPa (psi),
pressure in the oil film in journal bearing at the point when $\theta = 0$, MPa (psi)
$q = \dfrac{h_1}{h_2} - 1$ constant used in Eqs. (23-95b) and (23-97b) for a slider bearing
Q flow of lubricant through the bearings, m^3/s
r radius of journal, m (in)
r_1, r_2 inside and outside radii of thrust bearing, m (in)
R number of Redwood seconds in Eqs. (23-15) and (23-16)
$S = \dfrac{\eta n'}{P} \dfrac{1}{\psi^2}$ Sommerfeld number or bearing characteristic number
$S' = \dfrac{60\eta n'}{P} \dfrac{1}{\psi^2}$ bearing characteristic number (Fig. 23-40)
$S'' = \dfrac{\eta_1 n}{P}$ bearing modulus (Tables 23-2 and 23-7)
t running temperature of the bearing, K (°C),
number of seconds, Saybolt, in Eqs. (23-7) and (23-8)
$\Delta T = (t_b - t_a)$ difference in temperature between bearing housing and surrounding air, K (°C)
u average velocity, m/s (ft/min)
velocity in the oil film at height y (Fig. 23-1), m/s (ft/min)
U maximum velocity (Fig. 23-1), m/s (ft/min)
v velocity, m/s (ft/min)
v_m mean velocity, m/s (ft/min)
surface speed of journal, m/s (ft/min)
V rubbing velocity, m/s (ft/min)
W load on the bearing, kN (lbf)
load acting on the journal bearing with end leakage, kN (lbf)
W_∞ load acting on the journal bearing without end leakage, kN (lbf)
X_0 factors used with Eqs. (23-162), (23-165)
$\bar{x}$ the distance of the pivoted point from the lower end of the shoe (Fig. 23-48), i.e., the distance of the pressure center from the origin of the coordinate, m (mm)
y distance from the stationary surface (Fig. 23-1), m (in)
y_0 factors used with Eqs. (23-162) and (23-165)
$\kappa = -qa$ a constant in equation of pivoted-shoe slider bearing [Eqs. (23-86b) and (23-86c)]

β	angular length of bearing or circumferential length of bearing, deg
γ_t	specific weight (weight density) at temperature t, °C, kN/m^3 (lbf/in^3)
$\delta = 1 - \varepsilon$	the minimum film thickness variable
$\varepsilon = \dfrac{2e}{c}$	attitude or eccentricity ratio or relative eccentricity
$= 1 - \dfrac{h_{\min}}{d\psi}$	
η	absolute viscosity (dynamic viscosity), Pa s
η'	absolute viscosity (dynamic viscosity), kgf s/m^2
η_1	absolute viscosity (dynamic viscosity), cP
η_2	absolute viscosity (dynamic viscosity), kgf s/cm^2
η_p	dynamic viscosity of oil above atmospheric pressure P, N s/m^2 or Pa s (cP, kgf s/m^2)
η_o	dynamic viscosity of oil at atmospheric pressure, i.e., when $P = 0$, N s/m^2 (cP, kgf s/m^2)
θ	the angle measured from the position of minimum of oil film to any point of interest in the direction of rotation or the angle from the line of centers to any point of interest in the direction of rotation around the journal, deg
μ	coefficient of friction (also with subscripts)
μ_o	viscosity, reyn
$\nu = \dfrac{\eta}{\gamma g}$	kinematic viscosity, m^2/s (cSt)
ρ	density of oil or specific gravity of oil used, kg/m^3 (g/mm^3)
σ	stress (normal), MPa (psi)
τ	shear stress in lubricant, MPa (psi)
ϕ	attitude angle or angle of eccentricity, deg
$\psi = \dfrac{c}{d}$	diametral clearance ratio or relative clearance
ω	angular speed, rad/s

Other factors in performance or in special aspects are included from time to time in this chapter and being applicable only in their immediate context, are not included at this stage.

Particular	Formula

SHEAR STRESS[1,2]

The shearing stress in the lubricant (Fig. 23-1)

$$\tau = \frac{F}{A} = \eta \frac{U}{h} = \eta \frac{u}{y} = \eta \frac{du}{dy} \qquad \text{(23-1)}$$

VISCOSITY

The absolute viscosity (dynamic viscosity) in **SI** units

$$\eta = 10^{-3}\eta_1 \qquad \textbf{SI} \quad \text{(23-2a)}$$

where η in Pa s or (N s/m^2) and η_1 in cP

$$\eta = 9.8066\eta' \qquad \text{(23-2b)}$$

$$\eta = 9.8066 \times 10^4 \eta_2 \qquad \text{(23-2c)}$$

where η in Pa s, η' in kgf s/m^2, and η_2 in kgf s/cm^2

$$\eta = \frac{10^4}{1.45}\mu_o \qquad \text{(23-2d)}$$

where η is Pa s and μ_o in reyn

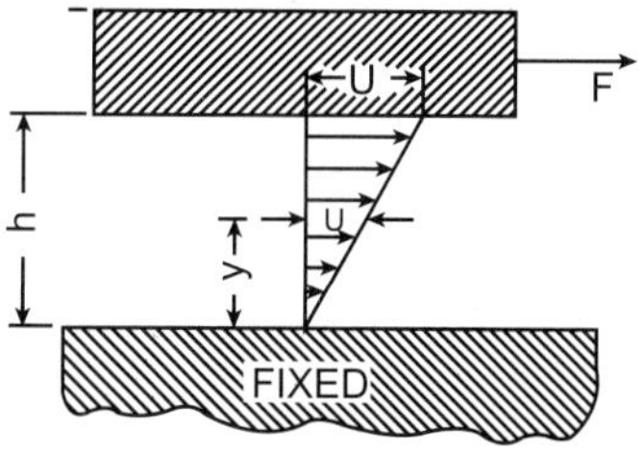

FIGURE 23-1 Shearing stress in lubricant.

The absolute viscosity (dynamic viscosity) in **Customary Metric units**

$$\eta' = 0.102\eta \qquad \textbf{Customary Metric} \quad \text{(23-3a)}$$

where η' in $\frac{\text{kgf s}}{\text{m}^2}$ and η in Pa s

$$\eta' = 1.02 \times 10^{-4}\eta_1 \qquad \text{(23-3b)}$$

where η' in $\frac{\text{kgf s}}{\text{m}^2}$ and η_1 in cP

$$\eta' = \frac{10^3}{1.422}\mu_o \qquad \text{(23-3c)}$$

where η' in $\frac{\text{kgf s}}{\text{m}^2}$ and μ_o in reyn

For absolute viscosity (dynamic viscosity) in centipoise and **SI units**

Refer to Figs 23-2a and 23-2b

The absolute viscosity (dynamic viscosity) in centipoise

$$\eta_1 = 10^3\eta \qquad \textbf{Customary Metric} \quad \text{(23-4a)}$$

where η_1 in cP and η in Pa s

$$\eta_1 = \frac{10^8}{1.02}\eta_2 \qquad \text{(23-4b)}$$

where η_1 in cP and η_2 in $\frac{\text{kgf s}}{\text{cm}^2}$

$$\eta_1 = \frac{10^4}{1.02}\eta' \qquad \text{(23-4c)}$$

where η_1 in cP and η' in $\frac{\text{kgf s}}{\text{m}^2}$

Particular	Formula
	$\eta_1 = \frac{10^7}{1.45}\mu_o$ (23-4d) where η_1 in cP and μ_o in reyn
The viscosity in reyn (lbf s/in^2)	$\mu_o = 1.45 \times 10^{-4}\eta$ **USCS** (23-5a) where μ_o in reyn and η in Pa s
	$\mu_o = 1.45 \times 10^{-7}\eta_1$ (23-5b) where μ_o in reyn and η_1 in cP
	$\mu_o = 14.22\eta_2$ (23-5c) where μ_o in reyn and η_2 in kgf s/cm^2

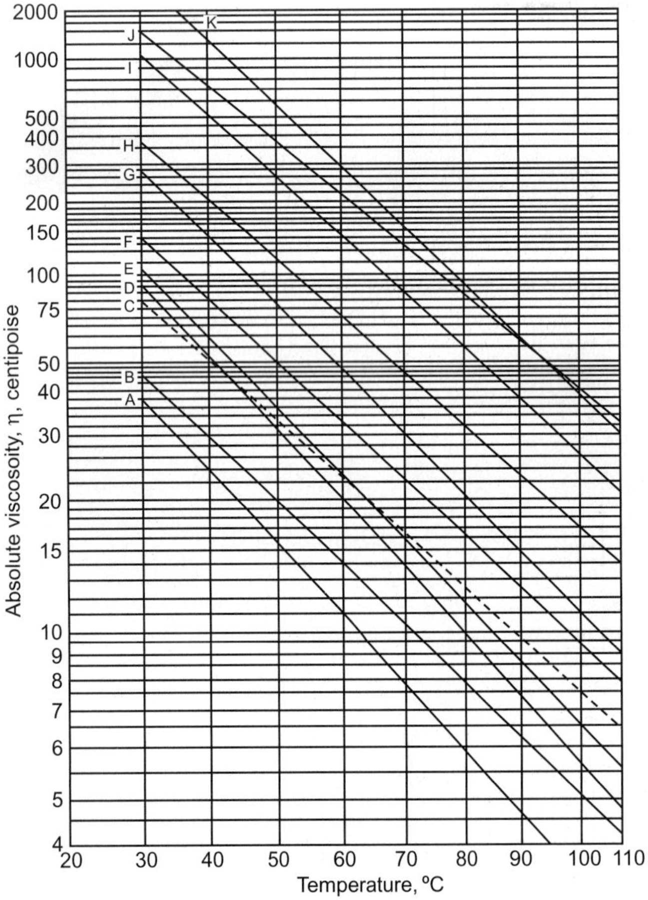

FIGURE 23-2a Absolute viscosity versus temperature.

Particular	Formula	
	$\mu_o = 1.422 \times 10^{-3}\eta'$ where μ_o in reyn and η' in $\dfrac{\text{kgf s}}{\text{m}^2}$	(23-5d)
Kinematic viscosity	$\nu = \dfrac{\eta}{\text{density}} = \dfrac{\eta_2 g}{\gamma}$ **Customary Metric**	(23-6a)

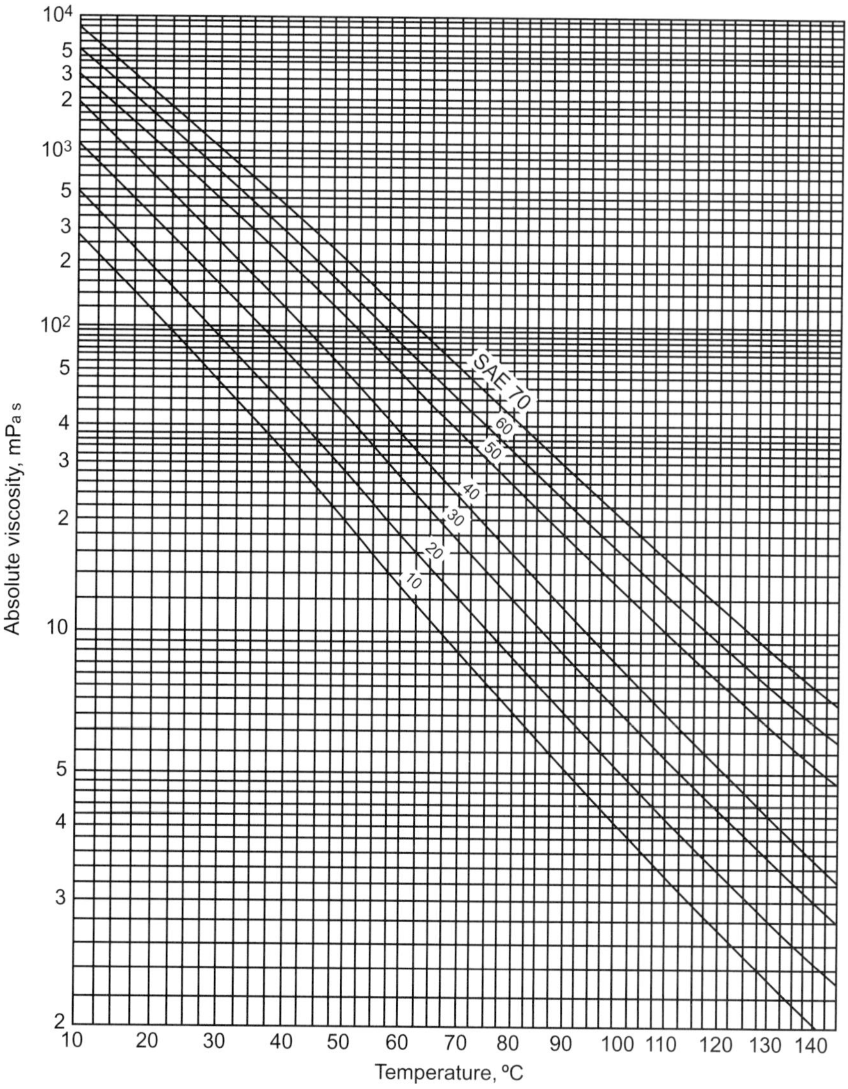

FIGURE 23-2b Absolute viscosity versus temperature.

Particular	Formula
	where v in cm^2/s and η_2 in $\dfrac{kgf\,s}{cm^2}$, $g = 980.66\,cm/s^2$ and γ in $\dfrac{kgf}{cm^3}$
Kinematic viscosity	$v = \dfrac{\eta g}{\gamma} 10^{-4}$ **SI** (23-6b) where η in $\dfrac{N\,s}{m^2}$ or (Pa s), γ in N/m^3, and v in m^2/s
Saybolt to centipoises (Fig. 23-3)[3] or mPa s	$\eta = \gamma_t\left(0.22t - \dfrac{180}{t}\right)$ **SI/Customary Metric** (23-7) where η in cP and γ_t in gf/cm^3 or N/m^3, t in s. Refer to Table 23-1 for γ_t.
Saybolt to reyn	$\mu_o = 0.145\gamma_t\left[0.22t - \left(\dfrac{180}{t}\right)\right]$ **USCS** (23-7a)
Kinematic viscosity in centistokes from Saybolt universal seconds (Figs. 23-3 and 23-4)[3]	$v_k = \left(0.22t - \dfrac{180}{t}\right)$ (23-8a) where v_k in cSt and t in s
Kinematic viscosity	$v = 10^{-6} v_k$ **SI** (23-8b) where v in m^2/s and v_k in cSt

TABLE 23-1
Specific gravity of oils at 15.5°C (60°F)

No.	Oil characteristics	$\gamma_{15.5}$
A	Turbine oil, ring-oiled bearing	0.8877
B	Turbine oil, ring-oiled bearing, SAE 10	0.8894
C	All-year automobile oil, SAE 20	0.9036
D	Ring-oiled bearing oil, high-speed machinery	0.9346
E	Automobile oil, SAE 20	0.9254
F	Automobile oil, SAE 30	0.9263
G	Automobile oil, SAE 40, medium-speed machinery	0.9275
H	Airplane oil 100, SAE 60	0.8927
I	Transmission oil, SAE 110, spur and bevel gears	0.9328
J	Gear oil, slow-speed worm gears	0.9153
K	Transmission oil. SAE 60, slow-speed gears	0.9365

Particular	Formula
	$$v = \left(0.22t - \frac{180}{t}\right)10^{-6} \quad \text{(23-8c)}$$ where t in Saybolt seconds and v in m^2/s
Specific weight at 15.5°C	$$\gamma_{15.5} = \frac{141.5}{131.5 + {}^\circ\text{API}} \quad \textbf{Customary Metric} \quad \text{(23-9a)}$$ where $\gamma_{15.5}$ in gf/ml (gram force/milliliter)
	$$\gamma_{15.5} = \left(\frac{141.5}{131.5 + {}^\circ\text{API}}\right)9807 \quad \textbf{SI} \quad \text{(23-9b)}$$ where $\gamma_{15.5}$ in N/m^3

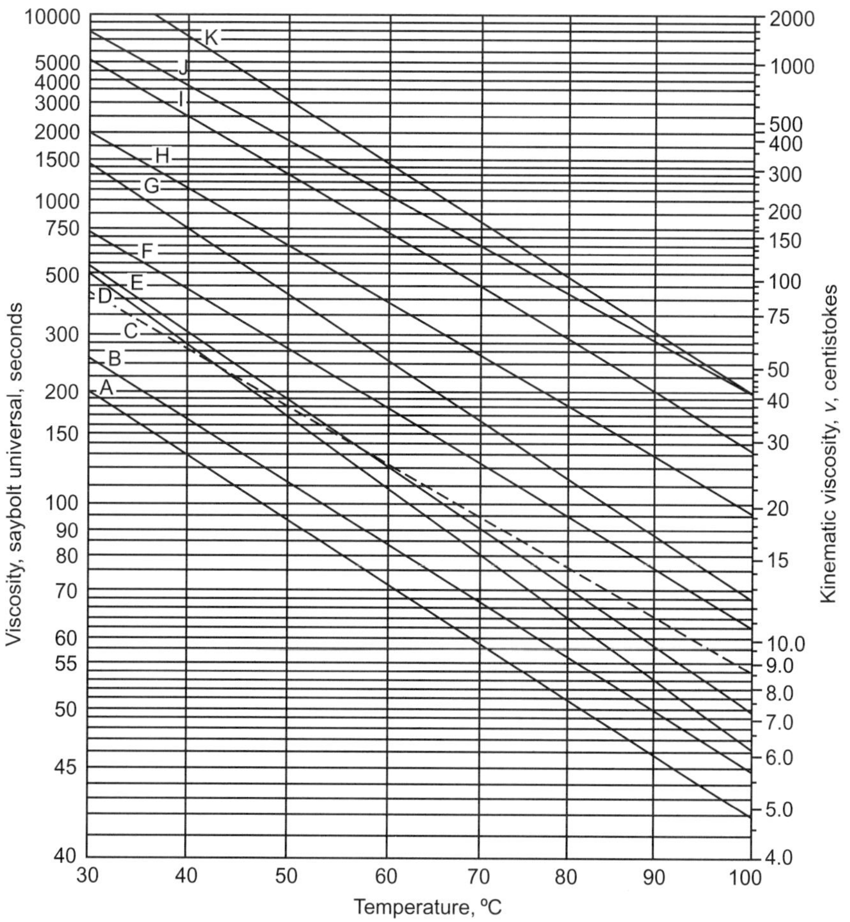

FIGURE 23-3 Viscosity Saybolt universal seconds and kinematic viscosity versus temperature.

Particular	Formula
	API = American Petroleum Institute gravity constant
Specific weight at any temperature	$\gamma_t = \gamma_{15.5} - 0.000637(t - 15.5)$ (23-10)
	Refer to Table 23-1 for $\gamma_{15.5}$
	$\gamma_t = \gamma_{60} - 0.000365(t - 60)$ **USCS** (23-10a)

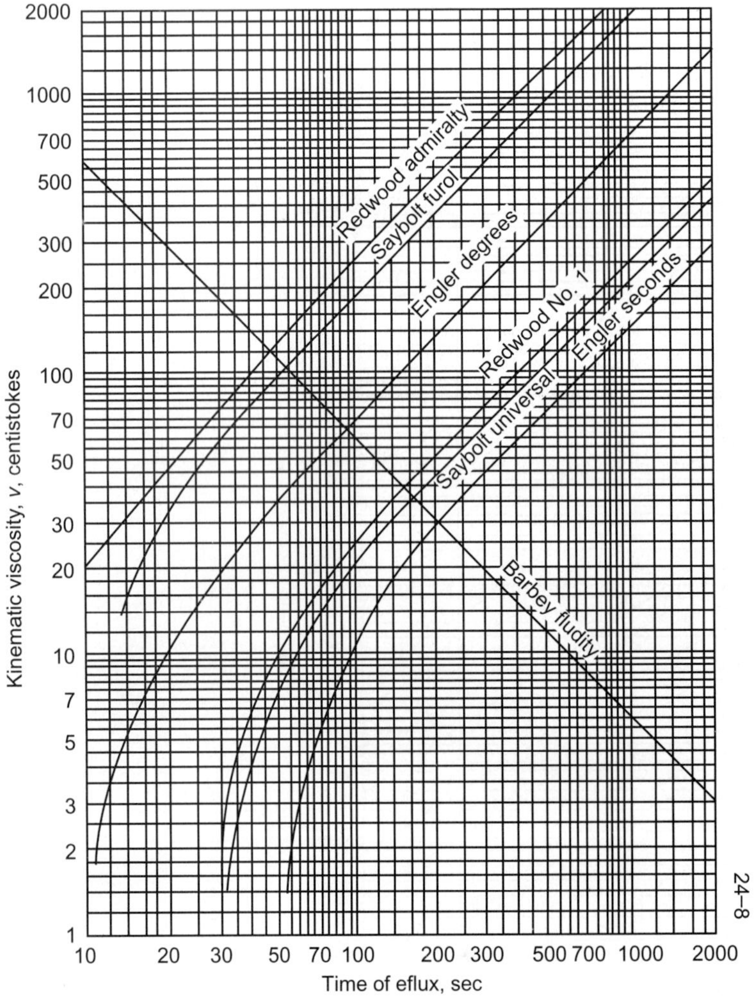

FIGURE 23-4 Viscosity conversion chart.

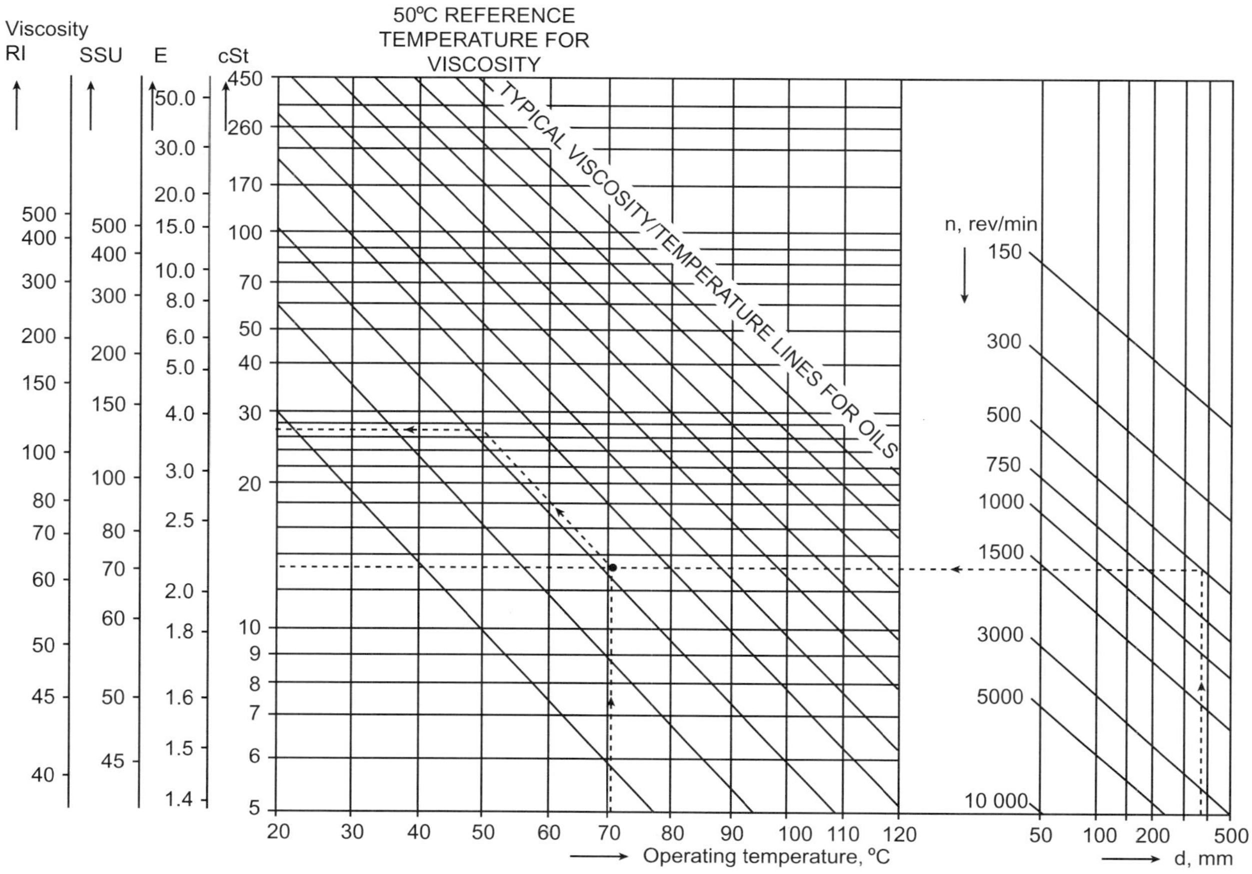

In the figure d = bearing bore diameter mm

n = rotational speed rev/min

An example is given below and shown on the graph by means of the lines of dashes.

Example : A bearing having a bore diameter d = 340mm and operating at a speed n = 500 rev/min requires an oil having a viscosity of 13·2 centistokes at the operating temperature. If this operating temperature is assumed to be 70°C an oil having a viscosity of 26 centistokes at 50°C should be selected.

FIGURE 23-4a Viscosity conversion chart and a guide to suitable oil viscosities for rolling contact bearings (Courtesy: SKF Rolling Bearings).

Particular	Formula
The dynamic viscosity	$\eta = \rho(0.22t - 180/t)10^{-6}$ where t in °F where η in Pa s, $\rho = \gamma/g$, and ρ in kg/m^3 The density (ρ) of oil and its specific gravity (γ) relative to water have the same numerical value.
The absolute viscosity (dynamic viscosity) in terms of Engler degree, E_t°	$\eta' = 10^{-6}\gamma_t\left(0.737E_t^\circ - \frac{0.635}{E_t^\circ}\right)$ **Customary Metric** (23-11) where η' in kgf s/m^2
The relation between arbitrary viscosity in Engler degree (V in E_t°) and the absolute viscosity (dynamic viscosity) in kgf s/m^2	$V = k\eta'$ (23-12) where $k \simeq 14.9 \times 10^3 E_t^\circ/(\text{kgf s/m}^2)$ = proportionality factor
The change in viscosity η' depending on temperature is expressed by formula	$\eta' = \frac{i}{(0.1t^\circ)^3}$ **Customary Metric** (23-13) where i = characteristic number of the given grade of oil $i \simeq 1.4$ to 2.8 η' in kgf s/m^2
The relation between viscosity and pressure	$\eta_p = n_o K^P$ **Customary Metric** (23-14) where P = pressure, kgf/cm^2 K = constant for the given grade of oil $\simeq$ varies from 1.001 to 1.004 for pressure P up to 400 kgf/cm^2 (39 MPa). (Changes in oil viscosity due to change in pressure can be neglected.)
Kinematic viscosity in centistokes from Redwood No	$v = 0.260R - \frac{179}{R}$ when $34 < R < 100$ **Customary Metric** (23-15a) where v in cSt and R in number of Redwood seconds $v = 0.247R - \frac{50}{R}$ when $R > 100$ (23-15b)
Kinematic viscosity in centistokes from Redwood Admiralty	$v = 2.7R - \frac{2000}{R}$ **Customary Metric** (23-16) where R = the number of Redwood seconds

HAGEN-POISEUILLE LAW

Particular	Formula
The rate of laminar flow of lubricant in tubes	$Q = \frac{\pi d^4}{128\eta}\frac{dp}{dz}$ (23-17)

Particular	Formula

VERTICAL SHAFT ROTATING IN A GUIDE BEARING (Fig. 23-5)

The surface velocity of shaft	$U = \pi d n'$	(23-18)
The length of bearing in the direction of motion	$B = \pi d \dfrac{\beta^\circ}{360}$	(23-19)
The torque (Fig. 23-5)	$M_t = \mu(Ld)P\dfrac{d}{2} = \dfrac{\pi^2 d^2 L\eta n'}{\psi}$	(23-20)
	Refer to Fig. 23-6 for projected area (Ld).	
Petroffs equation for coefficient of friction (Fig. 23-5)	$\mu = 2\pi^2\left(\dfrac{\eta n'}{P}\right)\left(\dfrac{1}{\psi}\right)$	(23-21)
Design practice for journal bearing[3]	Refer to Table 23-2.	
The coefficient of friction can also be obtained from expression	$\mu = K_a\left(\dfrac{\eta n'}{P}\right)\left(\dfrac{1}{\psi}\right)10^{-10} + \Delta\mu$	(23-22)
	where	
	$K_a = 5.53\beta = 1980$ for $\beta = 360^\circ$ **Customary Metric**	(23-22a)
	where η in cP, n' in rps, and P in kgf/cm^2	
	$K_a = 1.31\beta = 473$ for $\beta = 360^\circ$ **USCS**	(23-22b)
	where η in cP, n in rpm, and P in psi	
	$K_a = 9.23 \times 10^4\beta = 0.33$ for $\beta = 360^\circ$ **Customary Metric**	(23-22c)
	where η in cP, n in rpm, and P in kgf/mm^2	

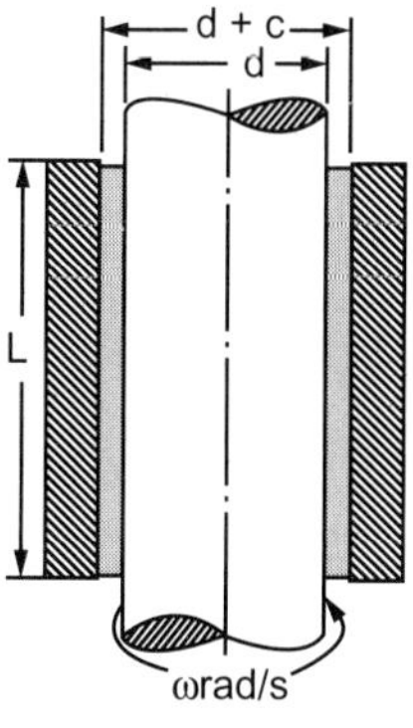

FIGURE 23-5 Vertical shaft rotating in a cylindrical bearing.

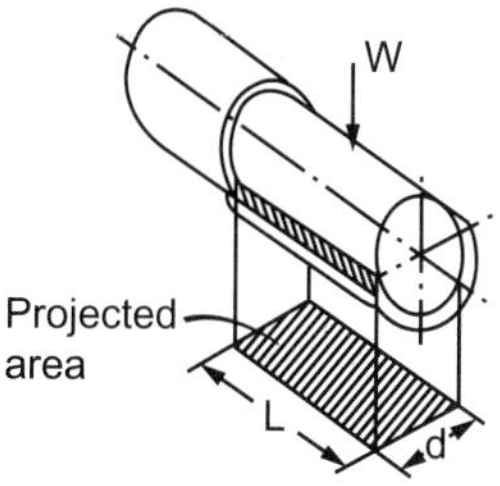

FIGURE 23-6 Projected area of a bearing.

Particular	Formula
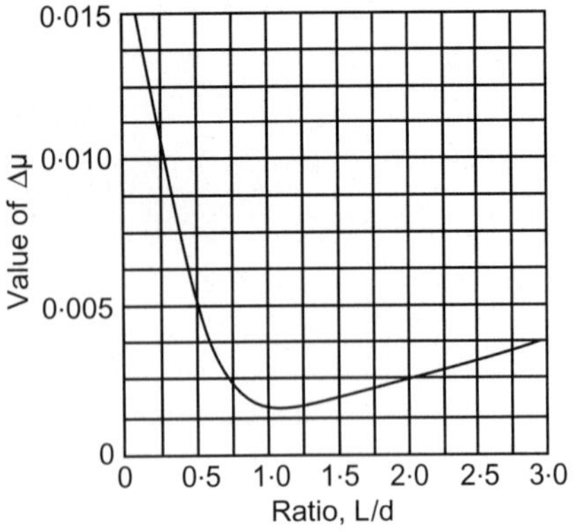 **FIGURE 23-7** Correction factor for use in Eq. (23-22).	$K_a = 0.0553\beta = 19.8$ for $\beta = 360°$ **Customary Metric** (23-22d) where η in cP, n' in rps, and P in kgf/mm^2 $K_a = 5.4 \times 10^8\beta = 1.95 \times 10^{11}$ for $\beta = 360°$ **SI** (23-22e) where η in Pa s, n' in rps, and P in N/m^2 $\Delta\mu$ = factor to correct for end leakage = 0.002 for L/d ranging from 0.75 to 2.8 Refer also to Fig. 23-7 for $\Delta\mu$.
Louis Illmer equation for coefficient of friction in case of imperfect lubrication	$\mu = 0.00012 C_1 C_2 \sqrt[4]{\frac{P}{v_m}}$ **SI** (23-23a) where P in N/m^2 and v_m in m/s $\mu = 0.0066 C_1 C_2 \sqrt[4]{\frac{P}{v_m}}$ **Customary Metric** (23-23b) where P in kgf/mm^2 and v_m in m/s $\mu = 0.004 C_1 C_2 \sqrt[4]{\frac{P}{v_m}}$ **USCS** (23-23c) where P in psi and v_m in ft/min Refer to Tables 23-3 and 23-4 for C_1 and C_2, respectively.
For behaviour of journal at stand still, at start and running in its bearing	Refer to Fig. 23-8.

TABLE 23-2
Journal bearing design practices

		Maximum pressure, P			Diameter clearance ratio		Viscosity, η_1	Viscosity, η	Bearing modulus (minimum) $S'' = \frac{\eta_1 n}{P}$	Bearing modulus (minimum) $S'' = \frac{\eta n'}{P}$
Machinery	**Bearing**	**kgf/mm^2**	**kpsi**	**MPa**	$\psi = \frac{c}{d}$	**Ratio** $\frac{L}{d}$	**cP**	**Pa s $\times 10^{-3}$**	**USCSU**	**SI Units, $\times 10^{-9}$**
Automobile and aircraft engines	Main	0.56–1.19	0.8–1.7	5.50–11.70	—	0.1–1.8	7	7	15	36.3
	Crankpin	1.06–2.47	1.5–3.5	10.40–24.40		0.7–1.4	to	to	10	24.2
	Wrist pin	1.62–3.62	2.3–5.0	15.00–34.80		1.5–2.2	8	8	8	19.3
Gas and oil engines (four-stroke)	Main	0.49–0.85	0.7–1.2	4.85–8.35	0.001	0.6–2.0	20	20	20	48.4
	Crankpin	0.90–1.27	1.4–1.8	8.80–12.40	<0.001	0.6–1.5	to	to	10	24.2
	Wrist pin	1.27–1.55	1.8–2.2	12.40–15.20	<0.001	1.5–2.0	65	65	5	12.1
Gas and oil engines (two-stroke)	Main	0.35–0.56	0.5–0.8	3.42–5.50	0.001	0.6–2.0	20	20	25	60.4
	Crankpin	0.70–1.06	1.0–1.5	6.85–10.40	<0.001	0.6–1.5	to	to	12	29.0
	Wrist pin	0.85–1.07	1.2–1.8	8.35–12.50	<0.001	1.5–2.0	65	65	10	24.2
Marine steam engines	Main	0.35	0.5	3.42	<0.001	0.7–1.5	30	30	20	48.4
	Crankpin	0.42	0.6	4.14	<0.001	0.7–1.2	40	40	15	36.3
	Wrist pin	1.06	1.5	10.40	<0.001	1.2–1.7	30	30	10	24.2
Stationary, slow-speed steam engines	Main	0.28	0.4	2.75	<0.001	1.0–2.0	60	60	20	48.4
	Crankpin	1.06	1.5	10.40	<0.001	0.9–1.3	80	80	6	14.5
	Wrist pin	1.27	1.8	12.50	<0.001	1.2–1.5	60	60	5	12.1
Stationary, high-speed steam engines	Main	0.17	0.25	1.66	<0.001	1.5–3.0	15	15	25	60.4
	Crankpin	0.42	0.6	4.14	<0.001	0.9–1.5	30	30	6	14.5
	Wrist pin	1.27	1.8	12.50	<0.001	1.3–1.7	25	25	5	12.1
Steam locomotives	Driving axle	0.39	0.55	3.72	0.001	1.6–1.8	100	100	30	72.5
	Crankpin	1.40	2.0	13.70	<0.001	0.7–1.1	40	40	5	12.1
	Wrist pin	2.82	4.0	27.60	<0.001	0.8–1.3	30	30	5	12.1
Reciprocating pumps and compressors	Main	0.17	0.25	1.66	<0.001	1.0–2.2	30	30	30	72.5
	Crankpin	0.42	0.6	4.14	<0.001	0.9–1.7	to	to	20	48.4
	Wrist pin	0.70	1.0	6.85	<0.001	1.5–2.0	80	80	10	24.2
Railway cars	Axle	0.35	0.45	3.42	0.001	1.8–2.0	100	100	50	120.9
Steam turbines	Main	0.07–0.19	0.1–0.275	0.69–1.87	0.001	1.0–2.0	2–16	2–16	100	241.8
Generators, motors, centrifugal pumps	Rotor	0.07-0.14	0.1-0.2	0.69-1.37	0.0013	1.0–2.0	25	25	200	483.5
Gyroscope	Rotor	0.60	0.85	5.90	0.0013	—	30	30	55	133.0
Transmission shafting	Light, fixed	0.08	0.025	0.17	0.001	2.0–3.0	25	25	100	241.8
	Self-aligning	0.106	0.15	1.04	0.001	2.5–4.0	to	to	30	72.5
	Heavy	0.106	0.15	1.04	0.001	2.0–3.0	60	60	30	72.5
Cotton mill	Spindle	0.0007	0.001	0.0069	0.005	—	2	2	10000	24177.5
Machine tools	Main	0.21	0.3	2.06	0.001	1.0–1.4	40	40	40	96.7
Punching and shearing machine	Main	2.82	4.0	27.80	0.001	1.0–2.0	100	100	—	—
	Crankpin	5.62	8.0	55.60	0.001	1.0–2.0	100	100	—	—
Rolling mills	Main	2.11	3.0	20.60	0.0015	1.1–1.5	50	50	10	24.2

Key: $\eta(\eta_1)$ = absolute viscosity, Pa s (cP); n = speed, rpm; n' = speed, rps; P = pressure, N/m^2 or MPa (psi); MPa = megapascal = 10^6 N/m^2; Pa = Pascal = 1 N/m^2; 1 psi = 6894.757 Pa; 1 kpsi = 6.89475 MPa; USCSU = US Customary System units.

TABLE 23-3
Values of factor C_1 in Eq. (23-23)

Lubrication	Workmanship	Attendance	Operating condition	Constant C_1
Oil bath or flooded	High grade	First class	Clean and protected	1
Oil, free drop (constant feed)	Good	Fairly good	Favorable (ordinary condition)	2
Oil cup or grease (intermittent feed)	Fair	Poor	Exposed to dirt, grit or other unfavorable conditions	4

TABLE 23-4
Values of factor C_2 in Eq. (23–23)

Type of bearing	Constant C_2
Rotating journals, such as rigid bearing and crankpins	1
Oscillating journals, such as rigid wrist pin and Pintle blocks	1
Rotating bearings lacking ample rigidity, such as eccentric and the like	2
Rotating flat surfaces lubricated from the center to the circumference, such as annular step or pivot bearings	2
Sliding flat surfaces wiping over the guide ends, such as reciprocating crossheads; use 2 for relatively long guides and 3 for short guides	2–3
Sliding or wiping surfaces lubricated from the periphery or outer wiping edge, such as marine thrust bearings and worm gears	3–4
Long power-screw nuts and similar wiping parts over which it is difficult to effect a uniform distribution of lubricant or load	4–6

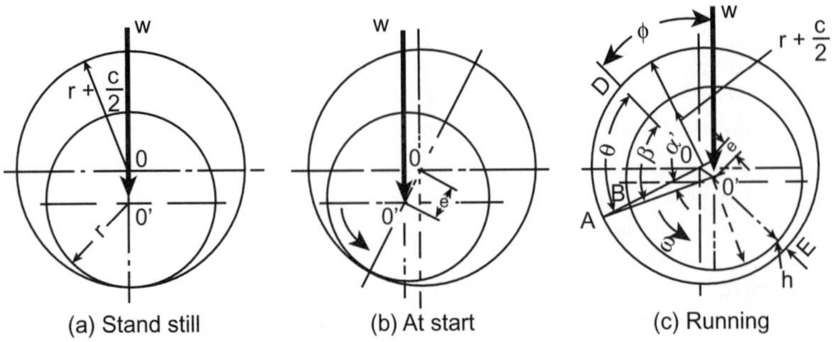

FIGURE 23-8 Behaviour of a journal in its bearing.

Particular	Formula

BEARING PRESSURE (Fig. 23-9)

General Electric Company's formula for bearing pressure in the design of motor and generator bearing	$P_a = 6.2 \times 10^5 \sqrt[3]{v_m}$ **SI** (23-24a) where P_a in N/m^2 and v_m in m/s
	$P_a = 15.5 \sqrt[3]{v_m}$ **USCS** (23-24b) where P_a in psi and v_m in ft/min
	$P_a = 0.0635 \sqrt[3]{v_m}$ **Customary Metric** (23-24c) where P_a in kgf/mm^2 and v_m in m/s
Victor Tatarinoff's equation for safe operating load	$W = \dfrac{\eta_1 n d^3 (L/d)^2}{127(10^6) h\psi \left(1 + \dfrac{L}{d}\right)}$ **USCS** (23-25a) where η_1 in cP, n in rpm, L, d, h, and c in in, W in lbf
	$W = \dfrac{\eta n' d^3 (L/d)^2}{0.295 h\psi \left(1 + \dfrac{L}{d}\right)}$ **SI** (22-25b) where η in Pa s, n' in rps, L, d, h, and c in m, W in N
Victor Tatarinoff's equation for permissible unit pressure	$P = \dfrac{\eta_1 n'}{3175(10^4)\psi^2} \left(\dfrac{L}{L+d}\right)$ **USCS** (23-26a) where P in psi, η_1 in cP, n in rpm, L and d in in
	$P = 13.5 \dfrac{\eta n'}{\psi^2} \left(\dfrac{L}{L+d}\right)$ **SI** (23-26b) where P in Pa, η in Pa s, n' in rps, and L and d in m

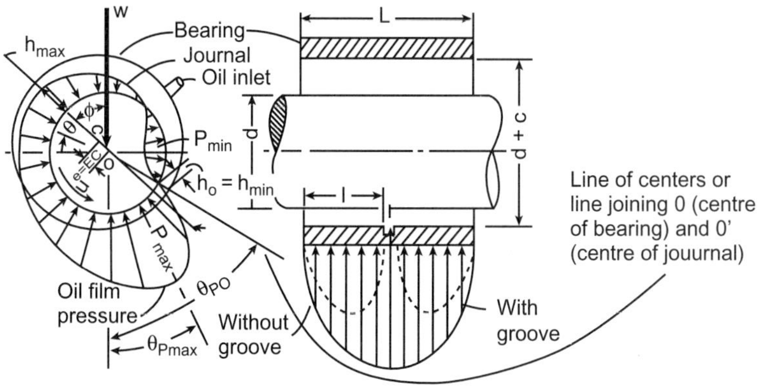

FIGURE 23-9 Oil film pressure distribution in the full journal bearing.

Particular	Formula
H. F. Moore's equation for critical pressure	$P_c = 7.23 \times 10^5 \sqrt{v}$ **SI** (23-27a) where P_c in N/m^2 and v in m/s
	$P_c = 0.0737\sqrt{v}$ **Customary Metric** (23-27b) where P_c in kgf/mm^2 and v in m/s
	$P_c = 7.5\sqrt{v}$ **USCS** (23-27c) where P_c in psi and v in ft/min
The critical unit pressure for any given velocity should not exceed according to Louis Illmer	$P_c = 4.6 \times 10^6 \sqrt[3]{\dfrac{v_m}{t - 288.5}}$ **SI** (23-28a) where P_c in N/m^2, v_m in m/s, and t in K
	$P_c = 0.47\sqrt[3]{\dfrac{v_m}{t - 15.5}}$ **Customary Metric** (23-28b) where P_c in kgf/mm^2, v_m in m/s, and t in °C
	$P_c = 140\sqrt[3]{\dfrac{v_m}{t - 15.5}}$ **USCS** (23-28c) where P_c in psi, v_m in ft/min, t in °F
Stribeck's equation for the critical pressure when the speed does not exceed 2.5 m/s (500 ft/min)	$P_c = 9.7 \times 10^5 \sqrt{v}$ **SI** (23-28d) where P_c in N/m^2 and v in m/s
	$P_c = 10\sqrt{v}$ **USCS** (23-28e) where P_c in psi and v in ft/min
	$P_c = 0.0986\sqrt{v}$ **Customary Metric** (23-28f) where P_c in kgf/mm^2 and v in m/s
Stribeck's equation for the critical pressure when the speed exceeds 2.5 m/s (500 ft/min)	$P_c = 2.9 \times 10^6 \sqrt{v}$ **SI** (23-28g) where P_c in N/m^2 and v in m/s
	$P_c = 30\sqrt{v}$ **USCS** (23-28h) where P_c in psi and v in ft/min
	$P_c = 0.296\sqrt{v}$ **Customary Metric** (23-28i) where P_c in kgf/mm^2 and v in m/s Refer to Table 23-5 for allowable pressures for reciprocating motion.
For permissible Pv values	Refer to Table 23-6.
For values S'' for various combinations of journal bearing materials, abrasion pressure for bearings, allowable bearing pressures for semi-fluid lubricants and diametral clearances in bearing dimensions.	Refer to Tables 23-7 to 23-10.

TABLE 23-5
Allowable bearing pressure, reciprocating motion

Type of bearing	Type of machinery	Pressure, P psi	Pressure, P MPa
Crosshead	Steam engine, stationary	35–60	0.24–0.412
	Steam engine, marine	55–100	0.378–0.688
	Steam engine, locomotive	70–90	0.48–0.62
	Gas and oil engines, stationary	40–70	0.275–0.48
	Compressors and pumps	50–90	0.342–0.62
Trunk pin	Gas and oil engines, stationary	20–25	0.136–0.172
	Automotive and aircraft engines	25–40	0.172–0.275

TABLE 23-6
Permissible Pv values

Class of bearing or journal	Values psi ft/s	Values N/m s
Mill shafting, with self-aligning cast-iron bearings, grease, or imperfect oil-lubrication, maximum value	12,000	4.2×10^5
Mill shafting, self-aligning ring-oiled babbitt bearings, maximum	24,000	8.45×10^5
Self-aligning ring-oiled bearings, continuous load in one direction	35,000–40,000	12.3×10^5 to 14×10^5
Crankshaft journals with bronze bearings	22,000	7.7×10^5
Crankshaft bearings with babbitted bearings, maximum	59,000	20.8×10^5
For excellent radiating condition	133,000	46.5×10^5

Key: US Customary unit: P = pressure, psi, v = velocity, ft/s; SI unit: P = pressure, N/m^2, v = velocity, m/s

TABLE 23-7
Values S'' for various combinations of journal bearing materials

Shaft	Bearing	Bearing-modulus $S'' = \frac{\eta_1 n}{P}$ Metric	Bearing-modulus $S'' = \frac{\eta n'}{P}$ SI $\times 10^{-9}$
Hardened and ground steel	Babbitt	28,500	48.5
Machined, soft steel	Babbitt	36,000	61.2
Hardened and ground steel	Plastic bronze	42,700	72.6
Machined, soft steel	Plastic bronze	35,800	60.9
Hardened and ground steel	Rigid bronze	56,900	96.7
Machined, soft steel	Rigid bronze	71,100	120.8

TABLE 23-8
Abrasion pressures for bearings

Materials in contact	Pressure psi	Pressure MPa	Remarks
Hardened tool steel on lumen or phosphor bronze	10,000	68.8	Values applies to rigid, polished and accurately fitted rubbing surface
0.50 C machine steel on lumen or phosphor bronze	8,000	55.0	When not worn to a fit or well lubricated reduce to 4.22 kgf/mm^2 (41.4 MPa)
Hardened tool steel on hardened tool steel	7,000	48.0	
0.50 C machine steel or wrought iron on genuine hard babbitt	6,000	41.5	
Cast iron on cast iron (close grained or chilled)	4,500	31.0	
Case-hardened machine steel on case-hardened machine steel	4,000	27.5	
0.30 C machine steel on cast iron (close-grained)	3,500	24.0	
0.40 C machine steel on soft common babbitt	3,000	20.6	
Soft machine steel on machine steel (not case-hardened)	2,000	13.8	
Machine steel on lignum vitae (water-lubricated)	1,500	10.2	

TABLE 23-9
Allowable bearing pressures for semifluid lubrication

Bearing material	Journal material	Allowable pressure, P_a psi	Allowable pressure, P_a MPa
Lumen of phosphor bronze	Hardened tool steel	2500	17.30
Hardened steel	Hardened alloy steel	2000	14.40
Hard babbitt	SAE 1050 steel	1500	10.30
Bronze	Hardened alloy steel	1300	8.90
Cast iron	Cast iron	1100	7.58
Bronze	Alloy steel	850	5.90
Babbit, soft	SAE 1040 steel	750	5.20
Bronze	Mild steel, smooth finish	540	3.70
Bronze	Mild steel, ordinary finish	400	2.75
Bronze	Cast iron	400	2.75
Cast iron	Mild steel	350	2.40
Lignum vitae, water lubricated	Mild steel	350	2.40

TABLE 23-10
Diametral clearance in bearings dimension in micrometers (1 µm = 10^{-6} m)

Particular about bearing and journal	Diametral clearances, c in µm				
	$d = 12$	$d = 25$	$d = 50$	$d = 100$	$d = 140$
Precision spindle, hardened and ground steel, lapped into bronze bearing–$v_m < 25$ m/s; $P < 500$ psi (3.43 N/m^2); 0.2–0.4 µm rms	7–19	19–38	38–63	63–88	88–125
Precision spindle, hardened and ground steel, lapped into bronze bearing–$v_m > 25$ m/s; $P > 500$ psi (3.43 N/m^2); 0.2–0.4 µm rms	13–25	25–50	50–75	75–113	113–163
Electric motors and generators, ground journals in broached or reamed bronze or babbitt bearings; 0.4–0.8 µm rms	13–38	25–50	38–85	50–100	75–150
General machinery, intermittent or continuous motion, turned or cold-rolled journal in reamed and bored bronze or babbitt bearings; 0.8–1.5 µm rms	50–100	63–113	75–125	100–175	125–200
Rough machinery, turned or cold-rolled steel journals in poured babbitt bearings; 1.5–3.8 µm rms	77–150	125–225	200–300	275–400	350–500
Automotive crankshaft					
Babbitt-lined bearing			38	63	
Cadmium silver copper			50	75	
Copper lead			36	88	

Particular	Formula

IDEALIZED JOURNAL BEARING (Figs. 23-8 and 23-9)

The diametral clearance ratio or relative clearance

$$\psi = \frac{c}{d} \tag{23-29}$$

Attitude or eccentricity ratio or eccentricity coefficient

$$\varepsilon = \frac{2e}{c} = 1 - \frac{2h_{\min}}{d\psi} \tag{23-30}$$

Refer to Fig. 23-10 for ε.

Oil film thickness at any position θ

$$h = \frac{c}{2}(1 + \varepsilon \cos\theta) \tag{23-31}$$

For position of minimum oil thickness and max oil film pressure

Refer to Figs. 23-11 and 23-11A

Minimum oil film thickness

$$h_{\min} = h_o = \frac{c}{2}(1 - \varepsilon) \tag{23-32}$$

The minimum oil film thickness variable

$$\delta = \frac{2h_{\min}}{c} = (1 - \varepsilon) \tag{23-33}$$

Refer to Figs. 23-12 to 23-14 and 23-15 for δ.

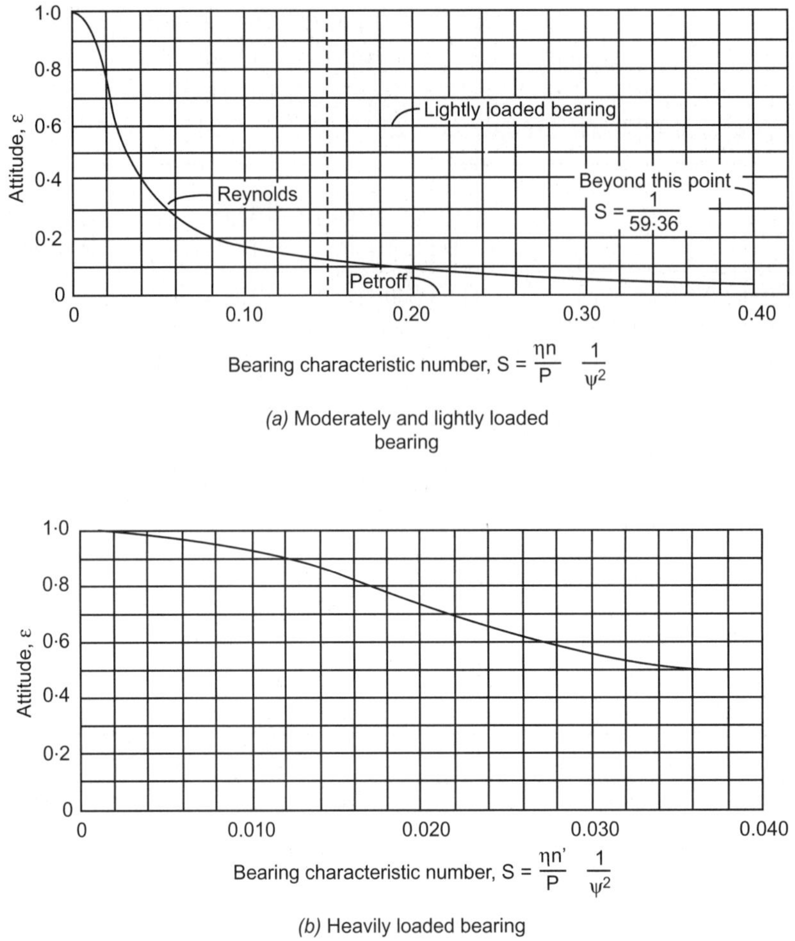

FIGURE 23-10 Variation of attitude ε of full journal bearing with characteristic number S. [Radzimosvksy[4]]

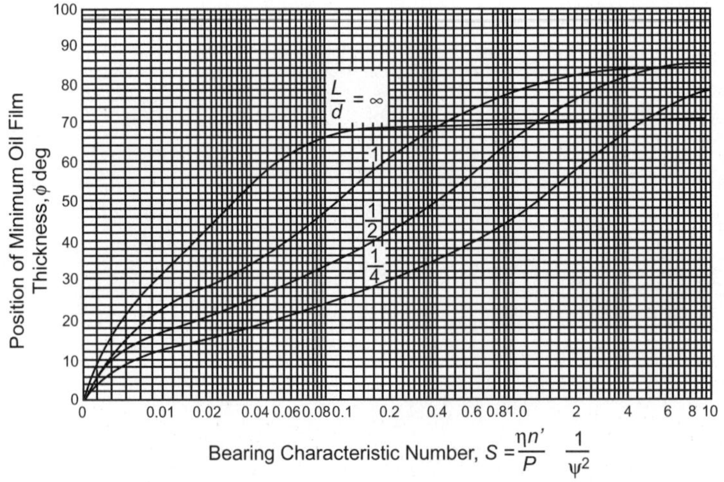

FIGURE 23-11 Position of minimum oil film thickness vs. bearing characteristic number S for full journal bearing. (Refer to Fig. 23-9 for definition of ϕ.) [Boyd and Raimondi[5]]

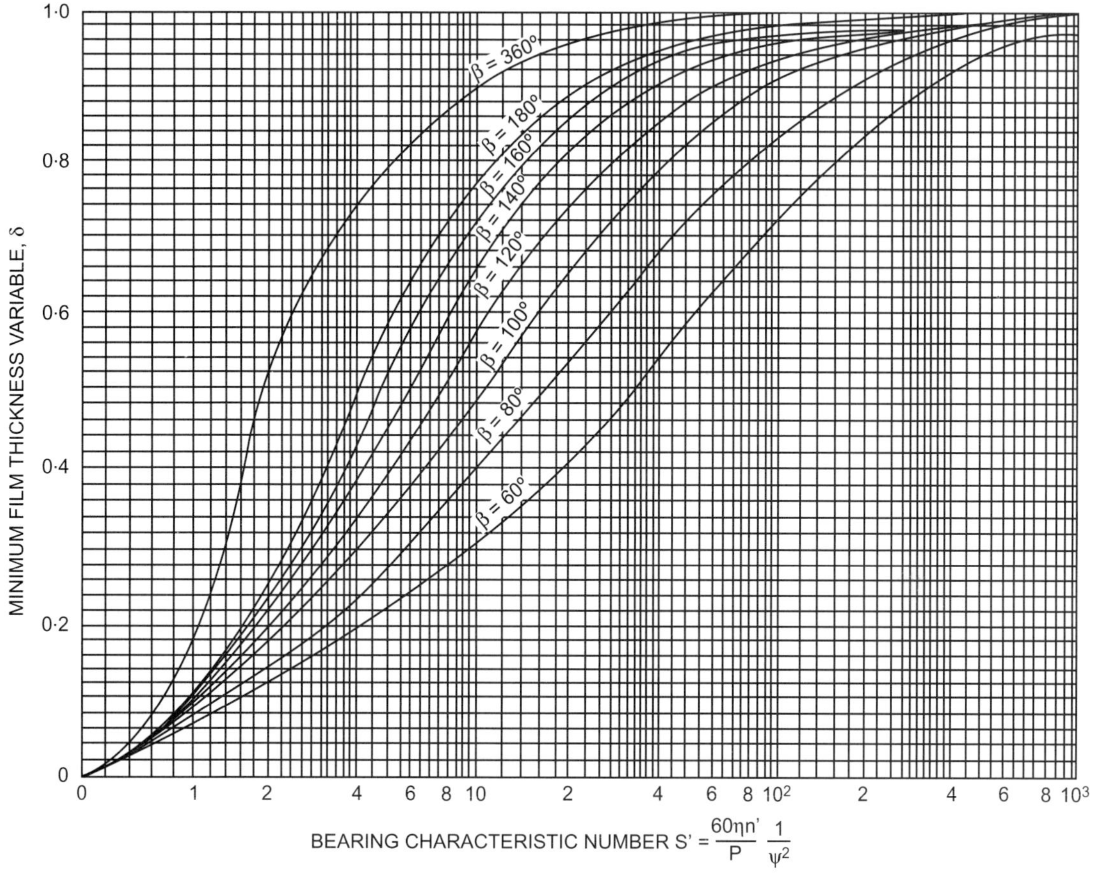

FIGURE 23-12 Minimum oil film thickness variable δ based on no side flow. [Boyd and Raimondi[5]]

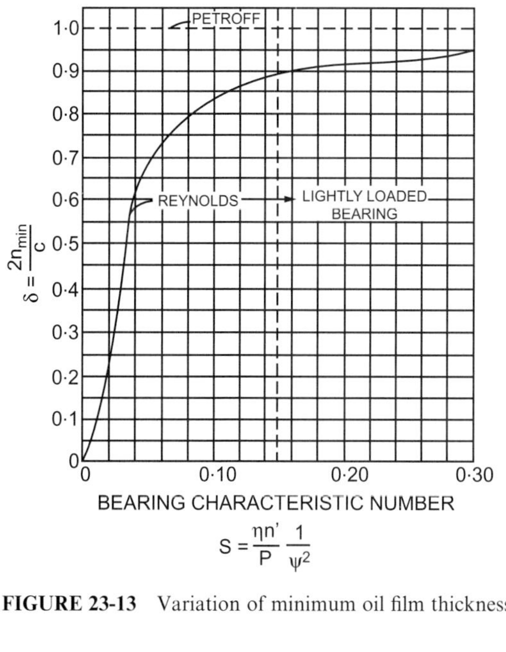

FIGURE 23-13 Variation of minimum oil film thickness variable δ of full journal bearing with S.

Particular	Formula
The safe oil film thickness for a bearing in good condition and $v_m \geq 1$ m/s (200 ft/min)	$h_{min} = h_o = 2.37 \times 10^{-5} v_m^{0.4} A^{0.2}$ **SI** (23-34a) where h_{min} in m, A in m^2, and v_m in m/s $h_{min} = 0.0015 v_m^{0.4} A^{0.2}$ **Customary Metric** (23-34b) where h_{min} in mm, A in mm^2, and v_m in m/s $h_{min} = 0.000026 v_m^{0.4} A^{0.2}$ **USCS** (23-34c) where h_{min} in in, A in in^2, and v_m in in
The thickness of oil film where the pressure is maximum or minimum	$(h)_{\substack{P(max)\\P(min)}} = k = \frac{2c(1-\varepsilon^2)}{2+\varepsilon^2}$ (23-35)
The resultant pressure distribution around a journal bearing excluding P_o the oil film pressure at the point where $\theta = 0$ or $\theta = 2\pi$	$P_r = (P - P_o) = \frac{12\eta U}{\psi^2 d}\left(\frac{\varepsilon(2+\varepsilon\cos\theta)\sin\theta}{(2+\varepsilon^2)(1+\varepsilon\cos\theta)^2}\right)$ (23-36)
The pressure at any point θ (Figs. 23-8 and 23-9)	$P = P_r + P_o$ (23-37)
The load carrying capacity of the bearing [Fig. 23-8 (panel c)]	$W = \frac{\eta UL}{\psi^2}\left(\frac{2\pi\varepsilon}{(2+\varepsilon^2)\sqrt{2-\varepsilon^2}}\right)$ (23-38)
The bearing characteristic number or Sommerfeld number	$S = \frac{\eta n'}{P}\frac{1}{\psi^2}$ (23-39)
For Sommerfeld number S	Refer to Tables 23-10 to 23-12 for Sommerfeld numbers S for full and partial bearings.

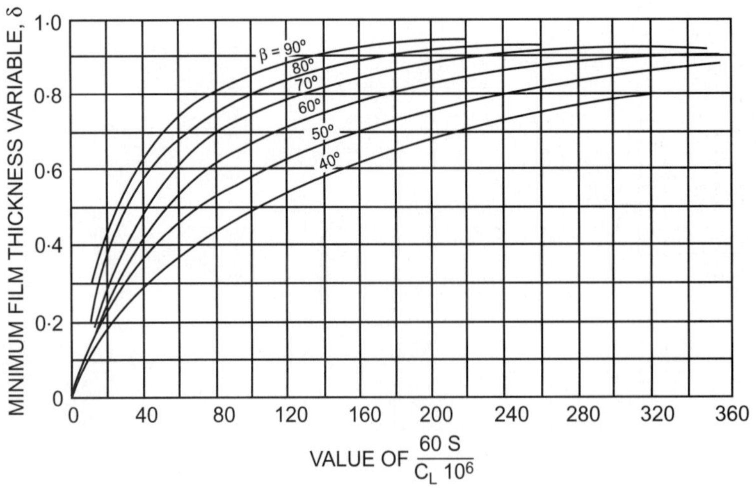

FIGURE 23-14 Variation of minimum oil film thickness variable δ with S/C_L.

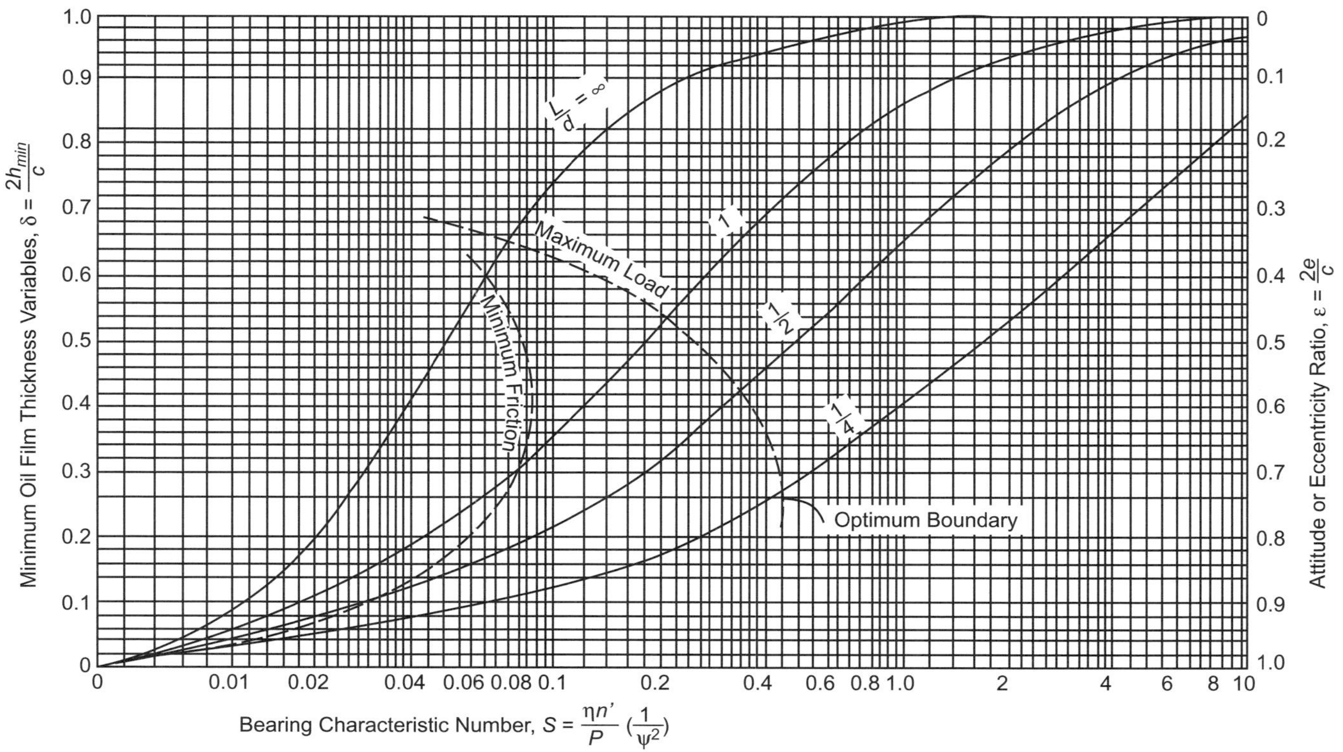

FIGURE 23-15 Variation of minimum oil film thickness variable δ and attitude ε of full journal bearing with bearing characteristic number S. [Boyd and Raimondi[5]]

Particular	Formula
The constant of the bearing or bearing modulus	$S'' = \dfrac{\eta n}{P}$ (23-40) where η in Pa s (cP) Refer to Table 23-7 for bearing modulus.
The calculation of minimum oil film thickness from Figs 23-14 and 23-16	*Hint*: S is determined from Eq. (23-39) and C_L from Fig. 23-16 for a given (L/d) ratio. Calculate $60S/(C_L 10^6)$. Knowing $60S/(C_L 10^6)$, you can then obtain the minimum film thickness variable δ from Fig. 23-14. From δ and Eq. (23-33), you can then determine the minimum oil film thickness.
The bearing characteristic number or Sommerfeld number as a function of attitude	$S = \dfrac{(2+\varepsilon^2)\sqrt{1-\varepsilon^2}}{12\pi^2\varepsilon}$ (23-41) Refer to Fig. 23-10 for ε for various values of S.
The angular positions of points where the maximum or minimum pressure in the oil film occur [Fig. 23-8c and Fig. 23-9]	$\theta = \cos^{-1}\left(-\dfrac{3\varepsilon}{\varepsilon^2+2}\right)$ (23-42)
For positions of maximum oil film pressure and oil film termination vs. bearing characteristic number S	Refer to Fig. 23-15A.

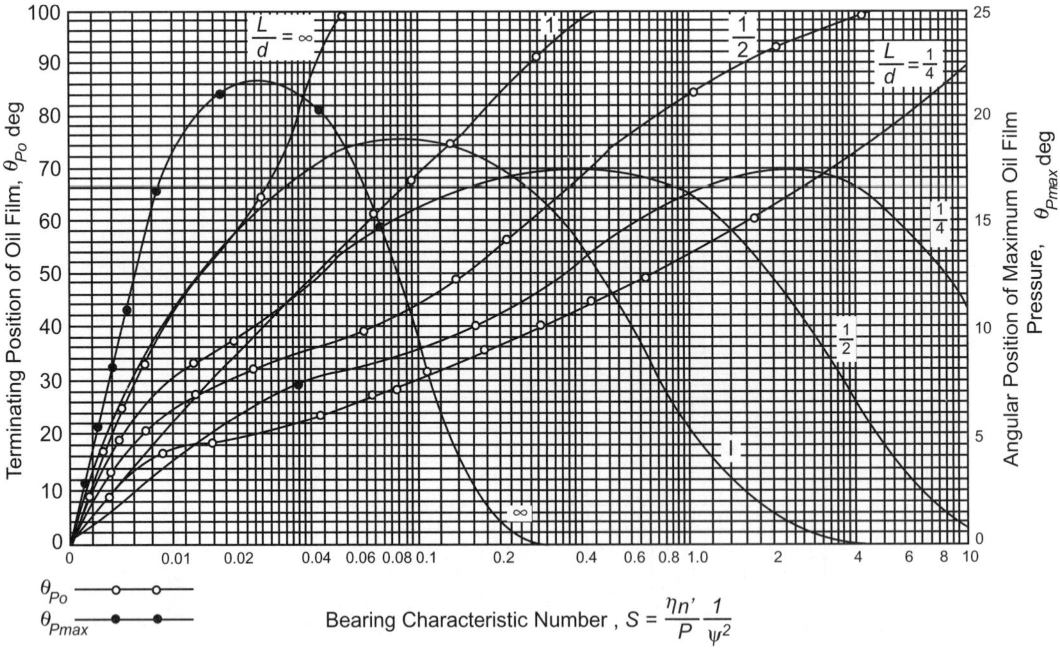

FIGURE 23-15A Position of maximum oil film pressure and oil film termination versus bearing characteristic number S. [Boyd and Raimondi[24]] (Refer to Fig. 23-9 for definition of $\theta_{P_{max}}$, and θ_{P_0}.)

TABLE 23-11
Dimensionless performance parameters for full journal bearings with side flow

Values of δ				
L/d ratio	0.25	0.5	1.0	∞
For maximum load	0.27	0.43	0.53	0.66
For minimum friction	0.03	0.12	0.3	0.6

$\frac{L}{d}$	ε	δ	S	ϕ	$\frac{\mu}{\psi}$	$\frac{4Q}{\psi d^2 n' L}$	$\frac{Q_s}{Q}$	$\frac{\gamma c_{sp} T_0}{P}$	$\frac{P}{P_{\max}}$
0.25	0	1.0	∞	(89.5)	∞	π	0	∞	—
	0.1	0.9	16.2	82.31	322.0	3.45	0.180	1287.0	0.515
	0.2	0.8	7.57	75.18	153.0	3.76	0.330	611.0	0.489
	0.4	0.6	2.83	60.86	61.1	4.37	0.567	245.0	0.415
	0.6	0.4	1.07	46.72	26.7	4.99	0.746	107.6	0.334
	0.8	0.2	0.261	31.04	8.80	5.60	0.884	35.4	0.240
	0.9	0.1	0.0736	21.85	3.50	5.91	0.945	14.1	0.180
	0.97	0.03	0.0101	12.22	0.922	6.12	0.984	3.73	0.108
	1.0	0	0	0	0	—	1.0	0	0
0.5	0	1.0	∞	(88.5)	∞	π	0	∞	—
	0.1	0.9	4.31	81.62	85.6	3.43	0.173	343.0	0.523
	0.2	0.8	2.03	74.94	40.9	3.72	0.318	164.0	0.506
	0.4	0.6	0.779	61.45	17.0	4.29	0.552	68.6	0.441
	0.6	0.4	0.319	48.14	8.10	4.85	0.730	33.0	0.365
	0.8	0.2	0.0923	33.31	3.26	5.41	0.874	13.4	0.267
	0.9	0.1	0.0313	23.66	1.60	5.69	0.939	6.66	0.206
	0.97	0.03	0.00609	13.75	0.610	5.88	0.980	2.56	0.126
	1.0	0	0	0	0	—	1.0	0	0
1	0	1.0	∞	(85)	∞	π	0	∞	—
	0.1	0.9	1.33	79.5	26.4	3.37	0.150	106	0.540
	0.2	0.8	0.631	74.02	12.8	3.59	0.280	52.1	0.529
	0.4	0.6	0.264	63.10	5.79	3.99	0.497	24.3	0.484
	0.6	0.4	0.121	50.58	3.22	4.33	0.680	14.2	0.415
	0.8	0.2	0.0446	36.24	1.70	4.62	0.842	8.0	0.313
	0.9	0.1	0.0188	26.45	1.05	4.74	0.919	5.16	0.247
	0.97	0.03	0.00474	15.47	0.514	4.82	0.973	2.61	0.152
	1.0	0	0	0	0	—	1.0	0	0
∞	0	1.0	∞	(70.92)	∞	π	0	∞	—
	0.1	0.9	0.240	69.10	4.80	3.03	0	19.9	0.826
	0.2	0.8	0.123	67.26	2.57	2.83	0	11.4	0.814
	0.4	0.6	0.0626	61.94	1.52	2.26	0	8.47	0.764
	0.6	0.4	0.0389	54.31	1.20	1.56	0	9.73	0.667
	0.8	0.2	0.021	42.22	0.961	0.760	0	15.9	0.495
	0.9	0.1	0.0115	31.62	0.756	0.411	0	23.1	0.358
	0.97	0.03	—	—	—	—	0	—	—
	1.0	0	0	0	0	0	0	∞	0

Key: Q_s = flow of lubricant with side flow, cm^3/s; γ = weight per unit volume of lubricant whose specific gravity is 0.90 = 8.83 kN/m^3 (0.0325 lbf/in^3); c_{sp} = specific heat of the lubricant, kJ/NK (Btu/lbf °F) = 0.19 kJ/NK (0.42 Btu/lbf °F); T_0 = difference in temperature, °C.
Source: A. A. Raimondi and J. Boyd, "A Solution for the Finite Journal Bearings and Its Applications to Analysis and Design" ASME, J. Lubrication Technol., Vol. 104, pp. 135–148, April 1982.

TABLE 23-12
Dimensionless performance parameters for 180° bearing centrally loaded with side flow[a]

	Values of δ			
L/d ratio	0.25	0.5	1.0	∞
For maximum load	0.28	0.42	0.52	0.64
For minimum friction	0.03	0.23	0.44	0.60

$\frac{L}{d}$	ε	δ	S	ϕ	$\frac{\mu}{\psi}$	$\frac{4Q}{\psi d^2 n' L}$	$\frac{Q_s}{Q}$	$\frac{\gamma c_{sp} T_0}{P}$	$\frac{P}{P_{max}}$
0.25	0	1.0	∞	90.0	∞	π	0	∞	—
	0.1	0.9	16.3	81.40	163.0	3.44	0.176	653.0	0.513
	0.2	0.8	7.60	73.70	79.4	3.71	0.320	320.0	0.489
	0.4	0.6	2.84	58.99	35.1	4.11	0.534	146.0	0.417
	0.6	0.4	1.08	44.96	17.6	4.25	0.698	79.8	0.336
	0.8	0.2	0.263	30.43	6.88	4.07	0.837	36.5	0.241
	0.9	0.1	0.0736	21.43	2.99	3.72	0.905	18.4	0.180
	0.97	0.03	0.0104	12.28	0.877	3.29	0.961	6.46	0.110
	1.0	0	0	0	0	—	1.0	0	0
0.50	0	1.0	∞	90.0	∞	π	0	∞	—
	0.1	0.9	4.38	79.97	44.0	3.41	0.167	177.0	0.518
	0.2	0.8	2.06	72.14	21.6	3.64	0.302	87.8	0.499
	0.4	0.6	0.794	58.01	9.96	3.93	0.506	42.7	0.438
	0.6	0.4	0.321	45.01	5.41	3.93	0.665	25.9	0.365
	0.8	0.2	0.0921	31.29	2.54	3.56	0.806	15.0	0.273
	0.9	0.1	0.0314	22.80	1.38	3.17	0.886	9.80	0.208
	0.97	0.03	0.00635	13.63	0.581	2.62	0.951	5.30	0.132
	1.0	0	0	0	0	—	1.0	0	0
1	0	1.0	∞	90.0	—	π	0	∞	—
	0.1	0.9	1.40	78.50	14.1	3.34	0.139	57.0	0.525
	0.2	0.8	0.670	68.93	7.15	3.46	0.252	29.7	0.513
	0.4	0.6	0.278	58.86	3.61	3.49	0.425	16.5	0.466
	0.6	0.4	0.128	44.67	2.28	3.25	0.572	12.4	0.403
	0.8	0.2	0.0463	32.33	1.39	2.63	0.721	10.4	0.313
	0.9	0.1	0.0193	24.14	0.921	2.14	0.818	9.13	0.244
	0.97	0.03	0.00483	14.57	0.483	1.60	0.915	6.96	0.157
	1.0	0	0	0	0	—	1.0	0	0
∞	0	1.0	∞	90.0	∞	π	∞	∞	—
	0.1	0.9	0.347	72.90	3.55	3.04	0	14.7	0.778
	0.2	0.8	0.179	61.32	2.01	2.80	0	8.99	0.759
	0.4	0.6	0.898	49.99	1.29	2.20	0	7.34	0.700
	0.6	0.4	0.0523	43.15	1.06	1.52	0	8.71	0.607
	0.8	0.2	0.0253	33.35	0.859	0.767	0	14.1	0.459
	0.9	0.1	0.0128	25.57	0.681	0.380	0	22.5	0.337
	0.97	0.03	0.00384	15.43	0.416	0.119	0	44.0	0.190
	1.0	0	0	0	0	0	0	∞	0

[a] See *Key* and *Source* under Table 23-11.

TABLE 23-13
Dimensionless performance parameters for 120° for centrally loaded bearing with side flow[a]

	Values of δ			
L/d ratio	0.25	0.5	1.0	∞
For maximum load	0.26	0.38	0.46	0.53
For minimum friction	0.06	0.28	0.4	0.5

$\frac{L}{d}$	ε	δ	S	ϕ	$\frac{\mu}{\psi}$	$\frac{4Q}{\psi d^2 n' L}$	$\frac{Q_s}{Q}$	$\frac{\gamma c_{sp} T_0}{P}$	$\frac{P}{P_{max}}$
0.25	0	1.0	∞	90.0	∞	π	0	∞	—
	0.10	0.9044	18.4	76.97	124.0	3.34	0.143	502.0	0.456
	0.20	0.8011	8.45	65.97	60.4	3.44	0.260	254.0	0.438
	0.40	0.6	3.04	51.23	26.6	3.42	0.442	125.0	0.389
	0.6	0.4	1.12	40.42	13.5	3.20	0.599	75.8	0.321
	0.8	0.2	0.268	28.38	5.65	2.67	0.753	42.7	0.237
	0.9	0.1	0.0743	20.55	2.63	2.21	0.846	25.9	0.178
	0.97	0.03	0.0105	12.11	0.852	1.69	0.931	11.6	0.112
	1.0	0	0	0	0	—	1.0		0
0.50	0	1.0	∞	90.0	∞	π	0	—	—
	0.1	0.9034	5.42	74.99	36.6	3.29	0.124	149.0	0.431
	0.2	0.8003	2.51	63.38	18.1	3.32	0.225	77.2	0.424
	0.4	0.6	0.914	48.07	8.20	3.15	0.386	40.5	0.389
	0.6	0.4	0.354	38.50	4.43	2.80	0.530	27.0	0.336
	0.8	0.2	0.0973	28.02	2.17	2.18	0.684	19.0	0.261
	0.9	0.1	0.0324	21.02	1.24	1.70	0.787	15.1	0.203
	0.97	0.03	0.00631	13.00	0.550	1.19	0.899	10.6	0.136
	1.0	0	0	0	0	—	1.0	0	0
1	0	1.0	∞	90.0	∞	π	0	∞	—
	0.1	0.9024	2.14	72.43	14.5	3.20	0.0876	59.5	0.427
	0.2	0.8	1.01	58.25	7.44	3.11	0.157	32.6	0.420
	0.4	0.6	0.385	43.98	3.60	2.75	0.272	19.0	0.396
	0.6	0.4	0.162	35.65	2.16	2.24	0.384	15.0	0.356
	0.8	0.2	0.0531	27.42	1.27	1.57	0.535	13.9	0.290
	0.9	0.1	0.0208	21.29	0.855	1.11	0.657	14.4	0.233
	0.97	0.03	0.00498	13.49	0.461	0.694	0.812	14.0	0.162
	1.0	0	0	0	0	—	1.0	0	0
∞	0	1.0	∞	90.0	∞	π	0	∞	—
	0.1	0.9007	0.877	66.69	6.02	3.02	0	25.1	0.610
	0.2	0.8	0.431	52.60	3.26	2.75	0	14.9	0.599
	0.4	0.6	0.181	39.02	1.78	2.13	0	10.5	0.566
	0.6	0.4	0.0845	32.67	1.21	1.47	0	10.3	0.509
	0.8	0.2	0.0328	26.80	0.853	0.759	0	14.1	0.405
	0.9	0.1	0.0147	21.51	0.653	0.388	0	21.2	0.311
	0.97	0.03	0.00406	13.86	0.399	0.118	0	42.3	0.199
	1.0	0	0	0	0	0	0	∞	0

[a] See *Key* and *Source* under Table 23-11.

TABLE 23-14
Dimensionless performance parameters for 60° centrally loaded bearing with side flow[a]

Values of δ				
L/d ratio	0.25	0.5	1.0	∞
For maximum load	0.15	0.20	0.23	0.25
For minimum friction	0.10	0.16	0.22	0.23

$\frac{L}{d}$	ε	δ	S	ϕ	$\frac{\mu}{\psi}$	$\frac{4Q}{\psi d^2 n' L}$	$\frac{Q_s}{Q}$	$\frac{\gamma c_{sp} T_0}{P}$	$\frac{P}{P_{max}}$
0.25	0	1.0	∞	90.0	∞	π	0	∞	—
	0.1	0.9251	35.8	71.55	121.0	3.16	0.0666	499.0	0.251
	0.2	0.8242	16.0	58.51	58.7	3.04	0.131	260.0	0.249
	0.4	0.6074	5.20	41.01	24.5	2.57	0.236	136.0	0.242
	0.6	0.4	1.65	30.14	11.2	1.98	0.346	86.1	0.228
	0.8	0.2	0.333	21.70	4.27	1.30	0.496	54.9	0.195
	0.9	0.1	0.0844	16.87	2.01	0.894	0.620	41.0	0.159
	0.97	0.03	0.0110	10.81	0.713	0.507	0.786	29.1	0.107
	1.0	0	0	0	0	—	1.0	0	0
0.5	0	1.0	∞	90.0	∞	π	0	∞	—
	0.1	0.9223	14.2	69.00	48.6	3.11	0.0488	201.0	0.239
	0.2	0.8152	6.47	52.60	24.2	2.91	0.0883	109.0	0.239
	0.4	0.6039	2.14	37.00	10.3	2.38	0.160	59.4	0.233
	0.6	0.4	0.695	26.98	4.93	1.74	0.236	40.3	0.225
	0.8	0.2	0.149	19.57	2.02	1.05	0.350	29.4	0.201
	0.9	0.1	0.0422	15.91	1.08	0.664	0.464	26.5	0.172
	0.97	0.03	0.00704	10.85	0.490	0.329	0.650	27.8	0.122
	1.0	0	0	0	0	—	1.0	0	0
1	0	1.0	∞	90.0	∞	π	0	∞	—
	0.1	0.9212	8.52	67.92	29.1	3.07	0.0267	121.0	0.252
	0.2	0.8133	3.92	50.96	14.8	2.82	0.0481	67.4	0.251
	0.4	0.6010	1.34	33.99	6.61	2.22	0.0849	39.1	0.247
	0.6	0.4	0.450	24.56	3.29	1.56	0.127	28.2	0.239
	0.8	0.2	0.101	18.33	1.42	0.883	0.200	22.5	0.220
	0.9	0.1	0.0309	15.33	0.822	0.519	0.287	23.2	0.192
	0.97	0.03	0.00584	10.88	0.422	0.226	0.465	30.5	0.139
	1.0	0	0	0	0	—	1.0	0	0
∞	0	1.0	∞	90.0	∞	π	0	∞	—
	0.1	0.9191	5.75	65.91	19.7	3.01	0	82.3	0.337
	0.2	0.8109	2.66	48.91	10.1	2.73	0	46.5	0.336
	0.4	0.6002	0.931	31.96	4.67	2.07	0	28.4	0.329
	0.6	0.4	0.322	23.21	2.40	1.40	0	21.4	0.317
	0.8	0.2	0.0755	17.39	1.10	0.722	0	19.2	0.287
	0.9	0.1	0.0241	14.94	0.667	0.372	0	22.5	0.243
	0.97	0.03	0.00495	10.58	0.372	0.115	0	40.7	0.163
	1.0	0	0	0	0	0	0	∞	0

[a] See *Key* and *Source* under Table 23-11.

Particular	Formula

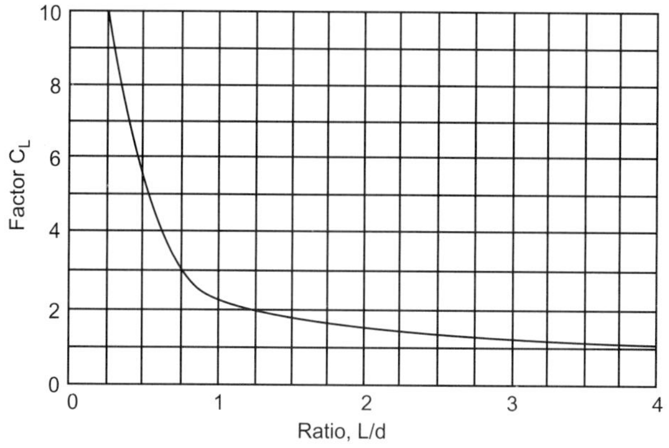

FIGURE 23-16 Variation of factor C_L with L/d ratio.

The total frictional resistance on an idealized journal bearing surface

$$F_\mu = \frac{4\pi\eta UL}{\psi}\left(\frac{1+2\varepsilon^2}{(2+\varepsilon^2)\sqrt{1-\varepsilon^2}}\right) \tag{23-43}$$

or

$$F_\mu = \frac{4\pi^2\eta n' Ld(1+2\varepsilon^2)}{\psi(2+\varepsilon^2)\sqrt{1-\varepsilon^2}} \tag{23-44}$$

The total frictional resistance on an idealized lightly loaded journal bearing

$$F_\mu = \frac{2\pi^2\eta n' Ld}{\psi} \tag{23-45}$$

For the relation between dimensionless quantity $\frac{\eta n'}{F'_\mu}\left(\frac{1}{\psi}\right)$ and Sommerfeld numbers S

Refer to Fig. 23-17.

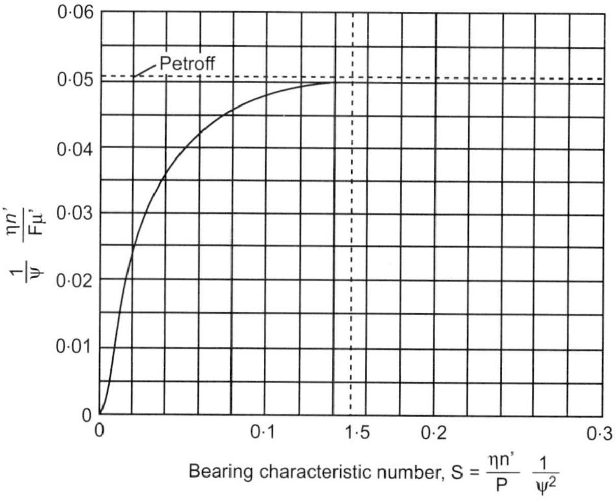

FIGURE 23-17 Variation of dimensionless quantity $\frac{1}{\psi}\frac{\eta n'}{F'_\mu}$ with Sommerfeld number S for an idealized full journal bearing.

Particular	Formula	
The relation between coefficient of friction and bearing characteristic number	$\mu = 2\pi^2 \frac{\eta n'}{P}\left(\frac{1}{\psi}\right)$	(23-46)
The relation between the coefficient of friction and attitude ε	$\mu = \psi\left(\frac{1+2\varepsilon^2}{3\varepsilon}\right)$	(23-47)
For average coefficient of friction at very high pressures	Refer to Table 23-15 for coefficient of friction	
The friction coefficient variable	$\lambda_\mu = \frac{\mu}{\psi} = \frac{1+2\varepsilon^2}{3\varepsilon}$	(23-48)
	Refer to Figs. 23-20 to 23-24.	

TABLE 23-15
Average coefficient of friction at very high pressure

	Angular displacement, deg	
Material	**10°**	**50°**
Stearic acid	0.022	0.029
Tungsten disulphide	0.032	0.037
Molybdenum disulphide	0.032	0.033
Graphite	0.036	0.058
Silver sulphate	0.055	0.054
Turbine oil plus 1% MoS_2	0.060	0.068
Lead iodide	0.061	0.071
Palm oil	0.063	0.075
Castor oil	0.064	0.081
Grease (zinc-oxide base)	0.071	0.080
Lard oil	0.072	0.084
Grease (calcium base)	0.073	0.082
Residual	0.076	0.083
Sperm oil	0.077	0.085
Turbine oil plus 1% graphite	0.081	0.105
Turbine oil plus 1% stearic acid	0.087	0.096
Turbine oil	0.088	0.108
Capric acid	0.089	0.109
Turbine oil plus 1% mica	0.091	0.105
Oleic acid	0.093	0.119
Machine oil	0.099	0.115
Soapstone (powdered)	0.169	0.306
Mica (powdered)	0.257	0.305
Boron (not a lubricant)	0.482	0.710

Particular	Formula

INFLUENCE OF MISALIGNMENT OF SHAFT IN BEARING

Minimum oil film thickness corresponding to the materials factor (k_m), the surface roughnesses (R_p) and amount of misalignment of the journal and bearing (M_a)

$$h_{\min} = k_m(R_{pj} + R_{pb}) + \frac{M_a L}{2} \quad \text{(23-48a)}$$

where k_m = material factor from Table 23-15a
L = length of bearing
R_{pb} = surface roughness of bearing from Table 23-15b
$M_a = x/L$ amount of misalignment

Refer to Table 23-15c and Fig. 23-18 for M_a

Dimensionless oil feed rate

$$Q' = \frac{2Q}{L\pi dn'c} \quad \text{(23-48b)}$$

where Q in m^3/s, L, d, and c in m, n' in rps

TABLE 23-15a
Material factor, k_m

Bearing lining material	k_m
Phosphor bronze	1
Leaded bronze	0.8
Tin aluminium	0.8
White metal (Babbitt)	0.5
Thermoplastic (bearing grade)	0.6
Thermosetting plastic	0.7

Courtesy: Neale, M. J., *Tribology Handbook*, Newnes-Butterworths, 1973

TABLE 23-15b
Surface finish, predominant peak height, R_p

				R_p	
Surface type	Micro-inch *cla*	μm *RMS*	Class	μm	μin
Turned or rough ground	100	2.8	6	12	480
Ground or fine bored	20	0.6	8	3	120
Fine ground	7	0.19	10	0.8	32
Lapped or polished	1.5	0.04	12	0.2	9

Courtesy: Neale, M. J., *Tribology Handbook*, Newnes-Butterworths, 1973

TABLE 23-15c
Values of misalignment factor, M_a at two ratios of ($h_{\min}/c$)

	($h_{\min}/c$)	
$M_a \times (L/c)$	**0.1**	**0.01**
0	100	100
0.05	65	33
0.25	25	7
0.50	12	3
0.75	8	1

Courtesy: Neale, M. J., *Tribology Handbook*, Newnes-Butterworths, 1973

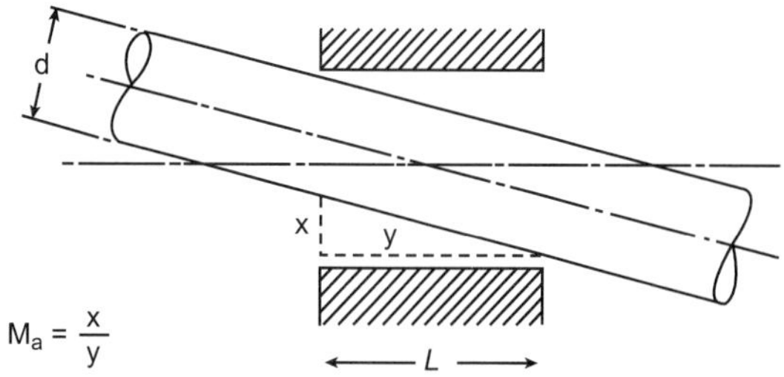

FIGURE 23-18 Misaligned journal inside the bearing under load

Particular	Formula
Bearing load capacity number	$W' = \dfrac{W}{\eta_e n' L d}\psi^2$ (23-48c) where $W'(d/L)^2$ = dimensionless load number, W in N, n' in rps, η_e in N s/m^2, L, d, and c in m
The required grease supply rate per hour for grease lubricated bearing	$Q_g = k_g c\pi dL$ (23-48d) where k_g = a factor for grease lubrication at various rotational speeds. Taken from Table 23-15d.
The coefficient of friction	$\mu = \lambda_\mu \psi$ (23-48e) where $\lambda_\mu = \mu/\psi$
The diameter of journal bearing for speeds below 2.5 m/s	$d = 3.2 \times 10^{-3}\sqrt[5]{\dfrac{W^2}{i^2 n'}}$ (23-48f) where W in N, d in m; $l/d = i$ and n' in rps
The diameter of journal bearing for speeds exceeding 2.5 m/s	$d = 2 \times 10^{-3}\sqrt[7]{\dfrac{W^2}{i^3 n'}}$ (23-48g)

POWER LOSS

Particular	Formula
The power loss in the bearing due to viscous friction	$P = \dfrac{F_\mu U}{33{,}000}$ **USCS** (23-49a) where P in hp, F_μ in lbf, and U in ft/min $P = \dfrac{F_\mu U}{102}$ **Customary Metric** (23-49b) where P in kW, F_μ in kgf, $U = \pi dn'$ = velocity in m/s, d in m, and n' in rps $P = \dfrac{F_\mu U}{1000}$ **SI** (23-49c) where P in kW, F_μ in N, and U in m/s

TABLE 24-15d
Values of factor k_g for grease lubrication at various rotational speeds

Journal speed, n in rpm		k_g
up to	100	0.1
	250	0.2
	500	0.4
	1000	1.0

Courtesy: Neale, M. J., *Tribology Handbook*, Newnes Butterworth

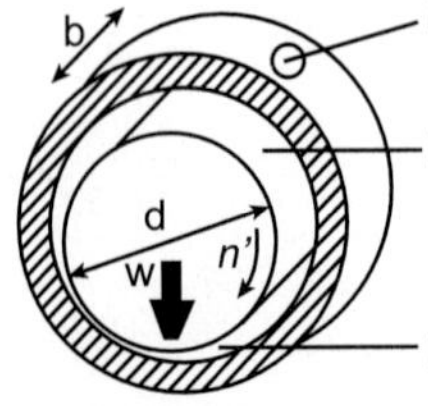

FIGURE 23-19 Journal inside the bearing under Load (W) at speed (n').

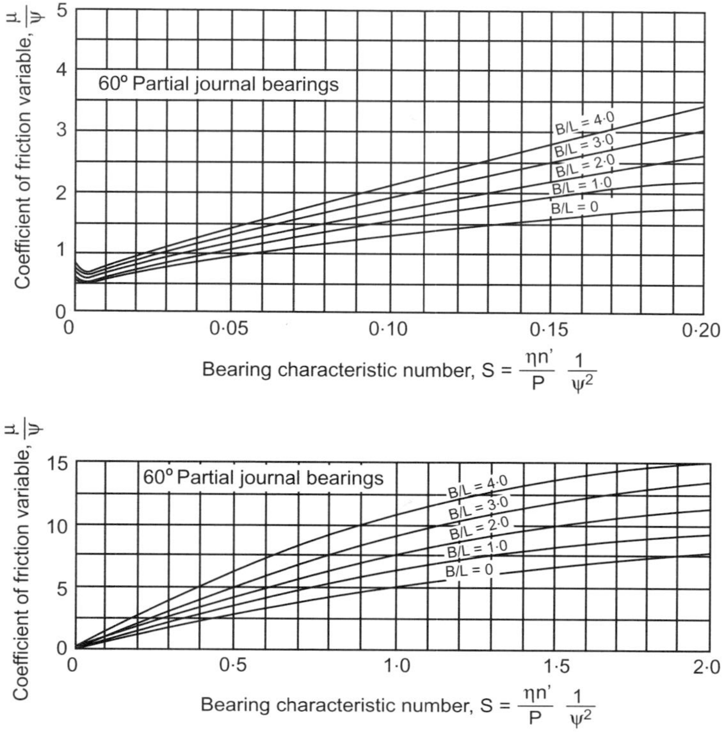

FIGURE 23-20 Variation of coefficient of friction variable μ/ψ with S for 60° partial journal bearing.

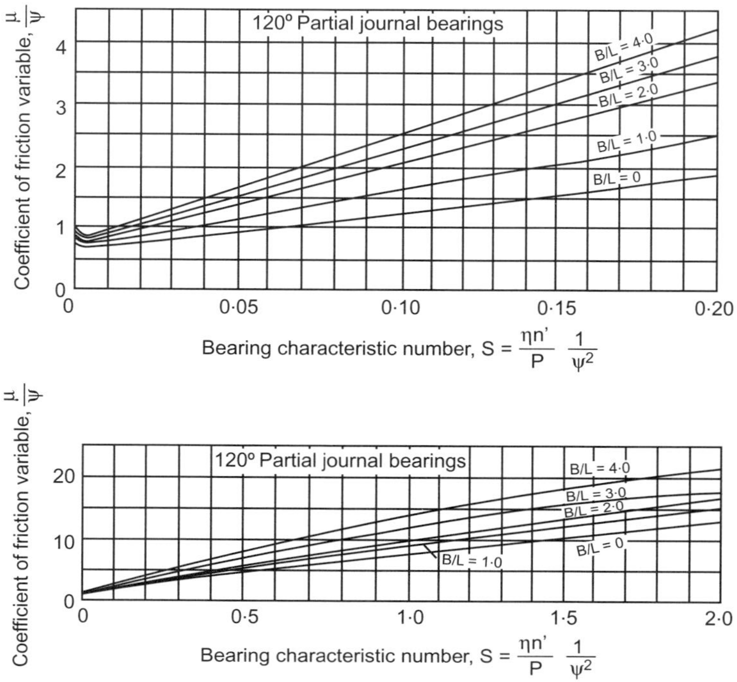

FIGURE 23-21 Variation of coefficient of friction variable μ/ψ with S for 120° partial journal bearing.

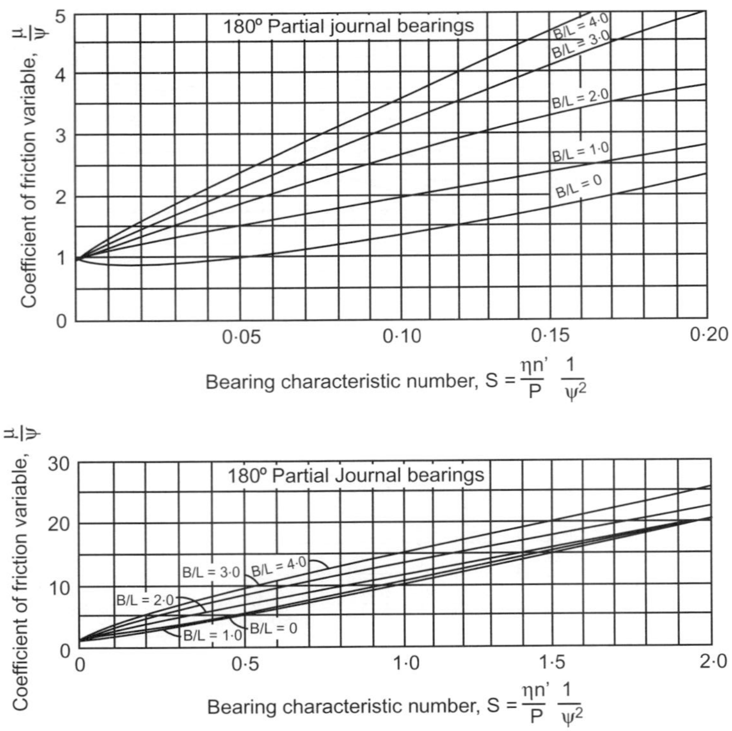

FIGURE 23-22 Variation of coefficient of friction variable μ/ψ with S for 180° partial journal bearing.

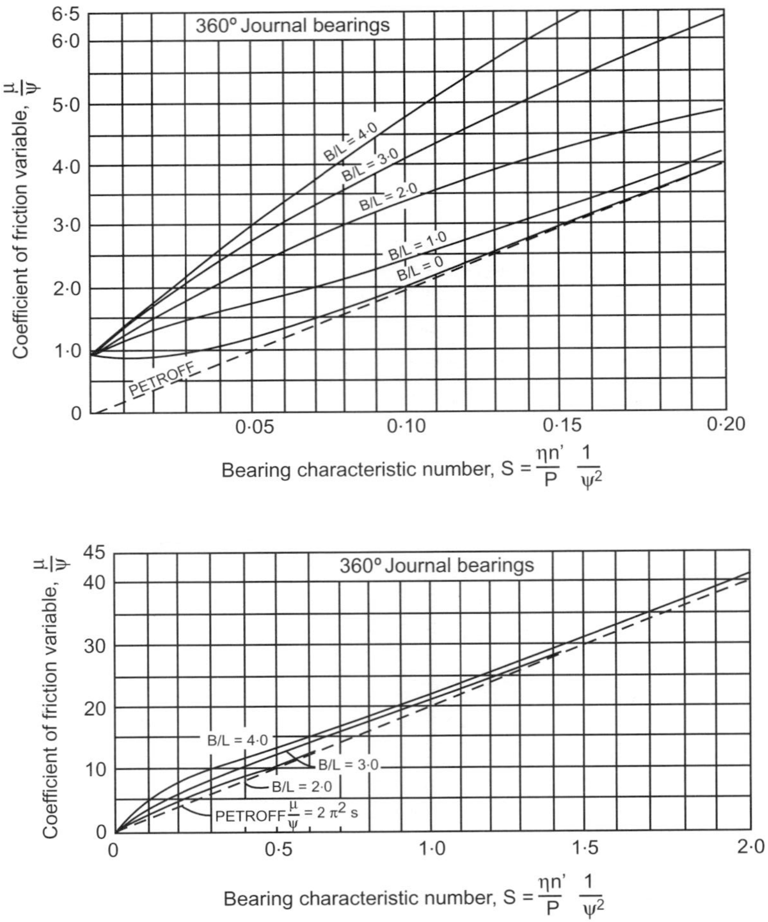

FIGURE 23-23 Variation of coefficient of friction variable μ/ψ with S for 360° partial journal bearing.

Particular	Formula

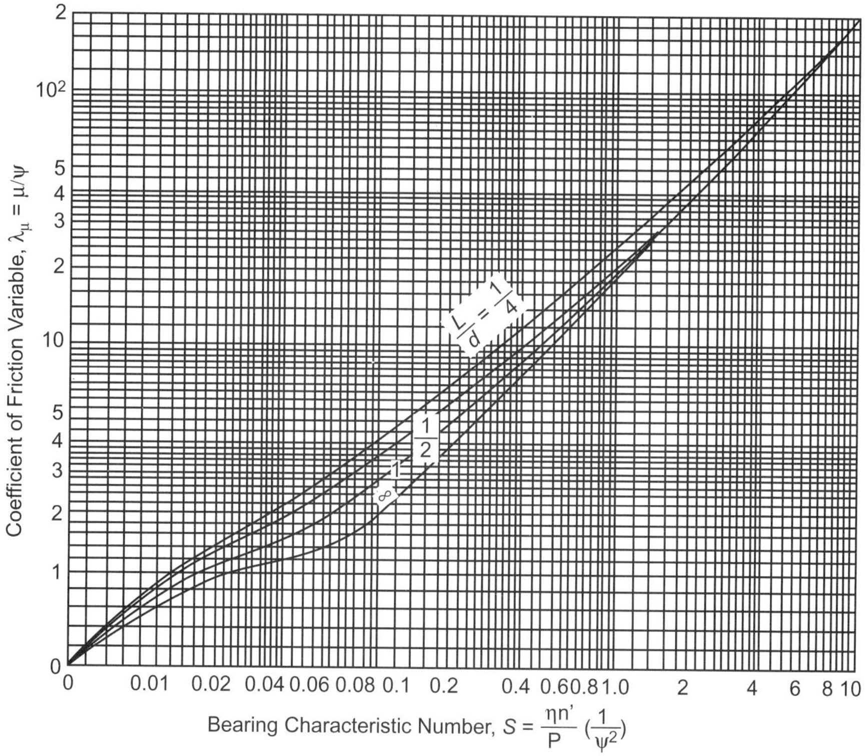

FIGURE 23-24 Variation of the coefficient of friction variable $\lambda_\mu = \mu/\psi$ with S for 360° journal bearing. [Boyd and Raimondi[5]]

PARTIAL JOURNAL BEARING (Fig. 23-25)

The resultant pressure distribution around the partial journal bearing excluding, P_o oil film pressure at the point where $\theta = 0$

$$P_r = P - P_o \tag{23-50}$$

where

$$P - P_o = \frac{12\eta U}{\psi^2 d}\left[\frac{(1-\varepsilon^2) - (2+\varepsilon^2)(k/c)}{(1-\varepsilon^2)^{2/5}} \times \arctan\left(\sqrt{\frac{1-\varepsilon}{1+\varepsilon}}\tan\frac{\theta}{2}\right) + \frac{(k/2c)\varepsilon\sin\theta}{2(1-\varepsilon^2)(1+\varepsilon\cos\theta)^2} + \frac{\varepsilon\sin\theta\{(3k/2c) - 2(1-\varepsilon^2)\}}{2(1-\varepsilon^2)^2(1+\varepsilon\cos\theta)}\right]$$

where $k = h$ is the thickness of oil film at maximum pressure value

Particular	Formula

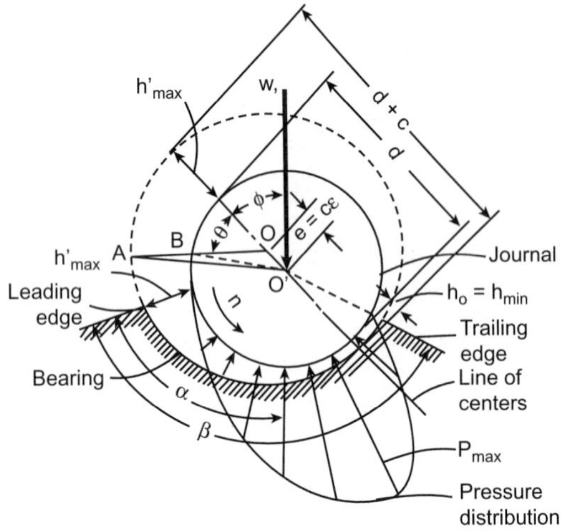

FIGURE 23-25 Partial journal bearing.

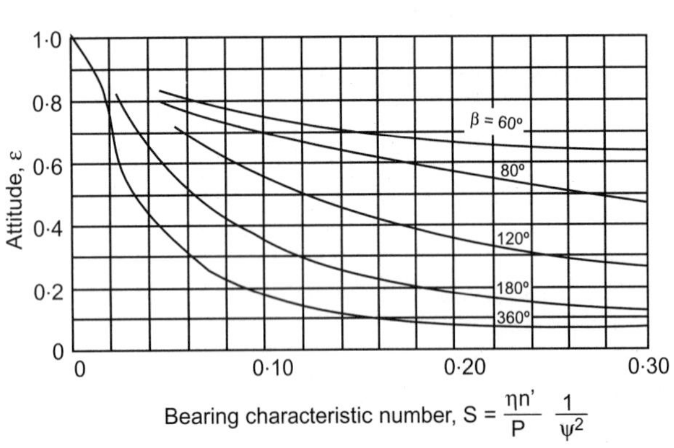

FIGURE 23-26 Variation of attitude ε with S for an idealized offset partial bearing having the maximum load capacity corresponding to a given attitude.

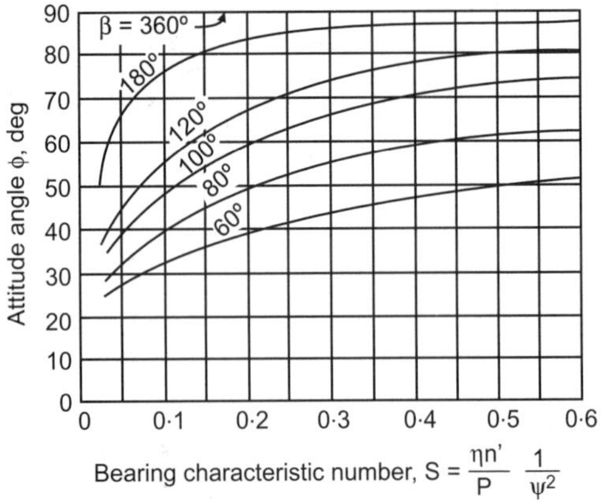

FIGURE 23-27 Variation of attitude angle ϕ with S for an idealized offset partial bearing.

Pressure at any point in a partial journal bearing

$$P = P_o + P_r \tag{23-51}$$

To determine the attitude ε and attitude angle ϕ for various values of S and for an idealized offset partial bearing having the maximum load capacity corresponding to a given attitude

Refer to Figs. 23-26 and 23-27 respectively.

INFLUENCE OF END LEAKAGE

Leakage factors C_W, C_F, and C_μ

Refer to Fig. 23-28 for C_W, C_F, and C_μ for various values of B/L ratios

Particular	Formula	
Load leakage factor according to Kingsbury[6]	$C_W = \dfrac{W}{W_\infty}$	(23-52)
	Refer to Fig. 23-28 for C_W.	
Load leakage factor C_W as a function of B/L ratio for a slider bearing having $q = (h_1/h_2) - 1 = 1$ or $h_1 = 2h_2$	Refer to Table 23-16.	
Load leakage factor for 120°, centrally loaded partial bearing according to Needs[7]	Refer to Fig. 23-29 for C_W for various attitudes ε.	
Load correction factor for side flow according to Boyd and Raimondi[24]	Refer to Fig. 23-30 for C_W for various minimum oil film thickness variables δ.	
Coefficient of friction leakage factor according to Kingsbury[6]	$C_\mu = \dfrac{\mu_\infty}{\mu}$	(23-53)
	Refer to Fig. 23-28 for C_μ	
Friction leakage factor according to Kingsbury[6]	$C_F = \dfrac{F_\mu}{F_{\mu_\infty}}$	(23-54)
	Refer to Fig. 23-28 for C_F.	
Friction leakage factor for 120°, centrally loaded partial bearing according to Needs[7]	Refer to Fig. 23-31 for C_F for various attitudes ε.	

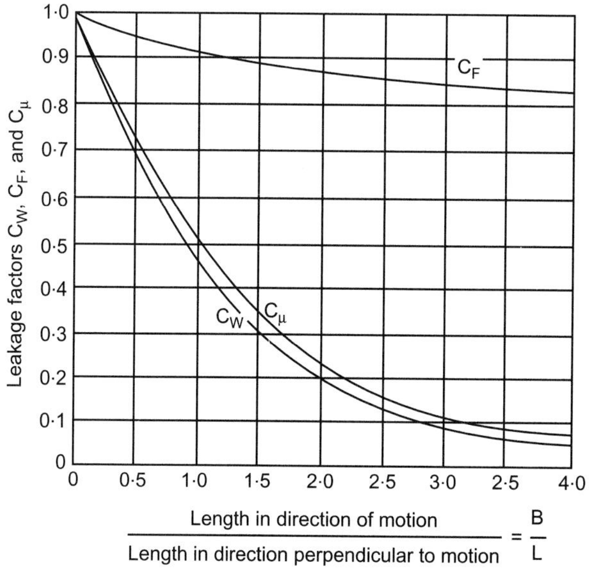

FIGURE 23-28 Kingsbury's leakage factors as function of B/L ratios under minimum friction. [Kingsbury[6]]

TABLE 23-16
Load leakage factor C_W as a function of B/L ratio for a slider bearing having the quality q equal to unity

B/L	C_W	B/L	C_W
0.00	1.00	1.00	0.44
0.175	0.92	1.50	0.278
0.25	0.835	2.00	0.185
0.50	0.68	3.00	0.090
0.75	0.55	4.00	0.060

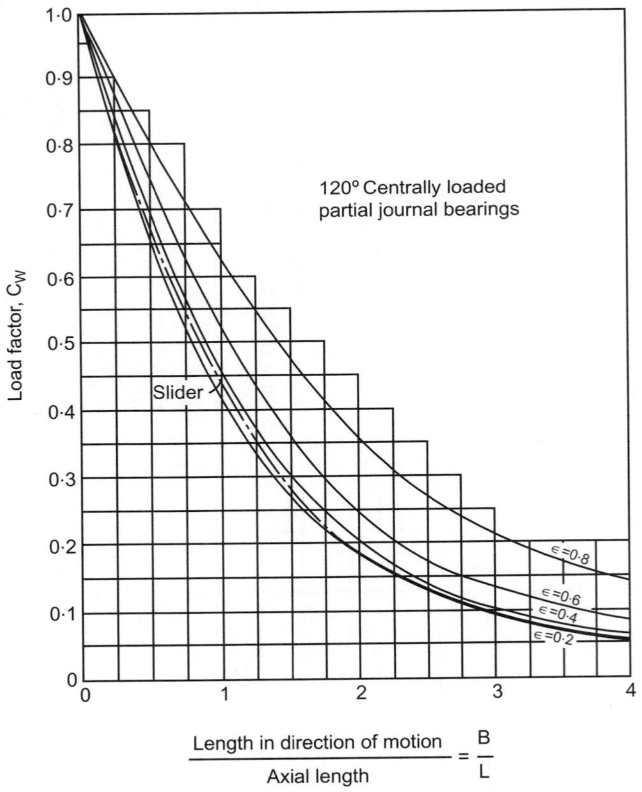

FIGURE 23-29 Leakage factors for load for 120° centrally partial journal bearings for various attitudes. [Needs[7]]

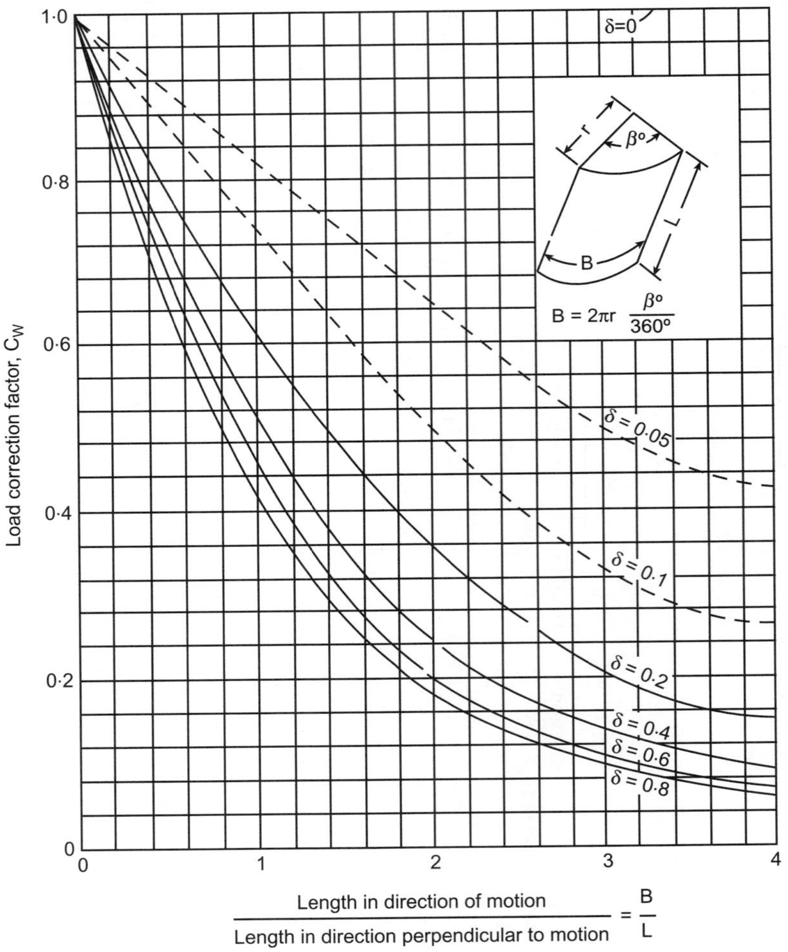

FIGURE 23-30 Load correction factor (C_W) for side flow. [Boyd and Raimondi[5]]

Particular	Formula

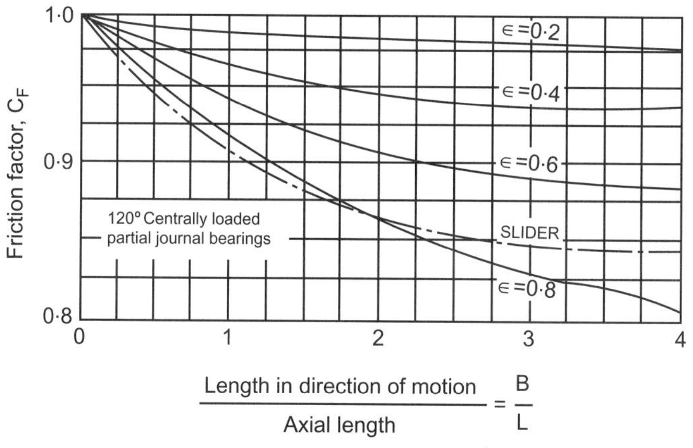

FIGURE 23-31 Leakage factors for friction force for 120° centrally loaded partial journal bearings for various attitudes. [Needs[7]]

Friction correction factor for side flow according to Boyd and Raimondi[5]	Refer to Fig. 23-32 for C_F for various minimum oil film thickness variables δ and B/L ratios.
The ratios of the maximum pressure in the oil film P_{max}, and the unit load, P, with B/L ratios for various values of attitude, ε, for 120° central partial journal bearing according to Needs[7]	Refer to Fig. 23-33 for P_{max}/P and Fig. 23-34 for P/P_{max} for various values of B/L ratios and attitudes ε.
The variation of attitude, ε, with bearing characteristic number, S, for various values of B/L ratios for 60°, 120°, 180° partial and full journal bearings	Refer to Figs. 23-35 to 23-38 for ε for various values of S and B/L ratios.
The variation of coefficient of friction variable, $\lambda = \mu/\psi$, with bearing characteristic number, S, for various values of B/L ratios for 60°, 120°, 180°, partial and full journal bearing	Refer to Figs. 23-20 to 23-24 and 23-39 for $\lambda_\mu = \dfrac{\mu}{\psi}$ for various values of S and B/L ratios.
The friction curves illustrating boundary conditions	Refer to Fig. 23-39.

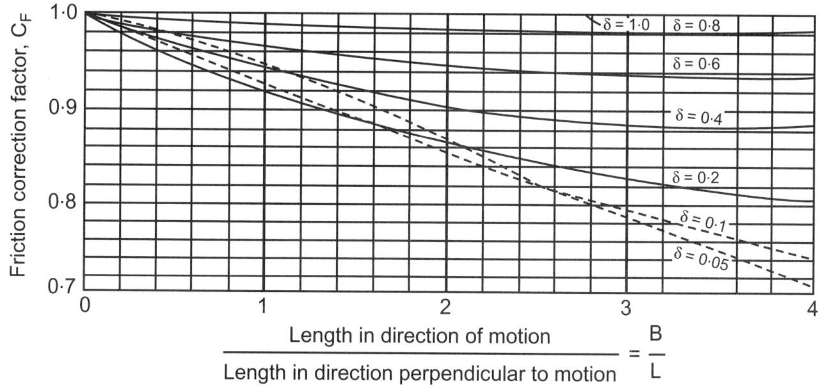

FIGURE 23-32 Friction correction factor for side flow. [Boyd and Raimondi[5]]

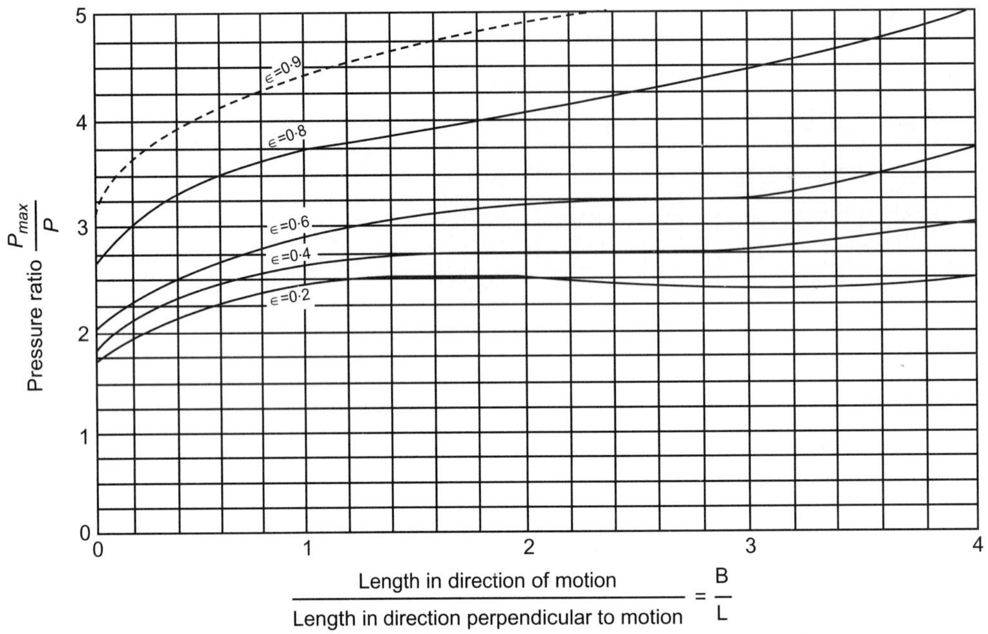

FIGURE 23-33 The ratio of the maximum pressure (P_{max}) and the unit load $P(= P_u)$ with B/L ratios for various attitudes for a 120° central partial bearing. [Needs[7]]

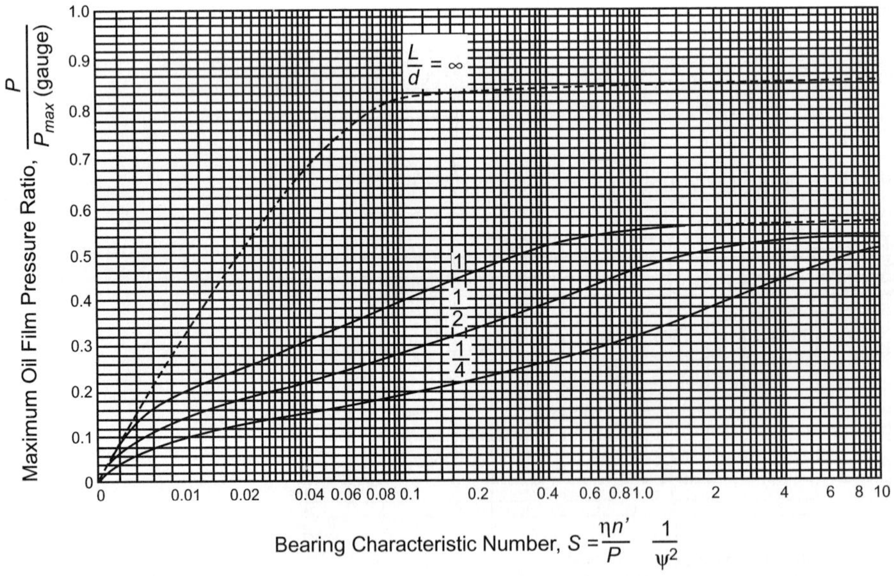

FIGURE 23-34 Chart for maximum oil film pressure ratio $P/P_{max(gauge)}$ with bearing characteristic number S for full journal bearing. [Boyd and Raimondi[5]]

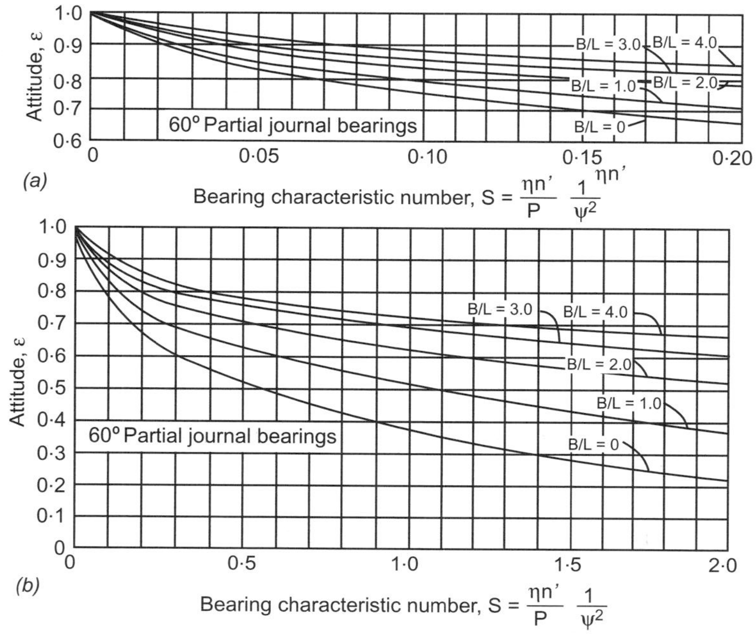

FIGURE 23-35 Variation of attitude ε with S for 60° partial journal bearing.

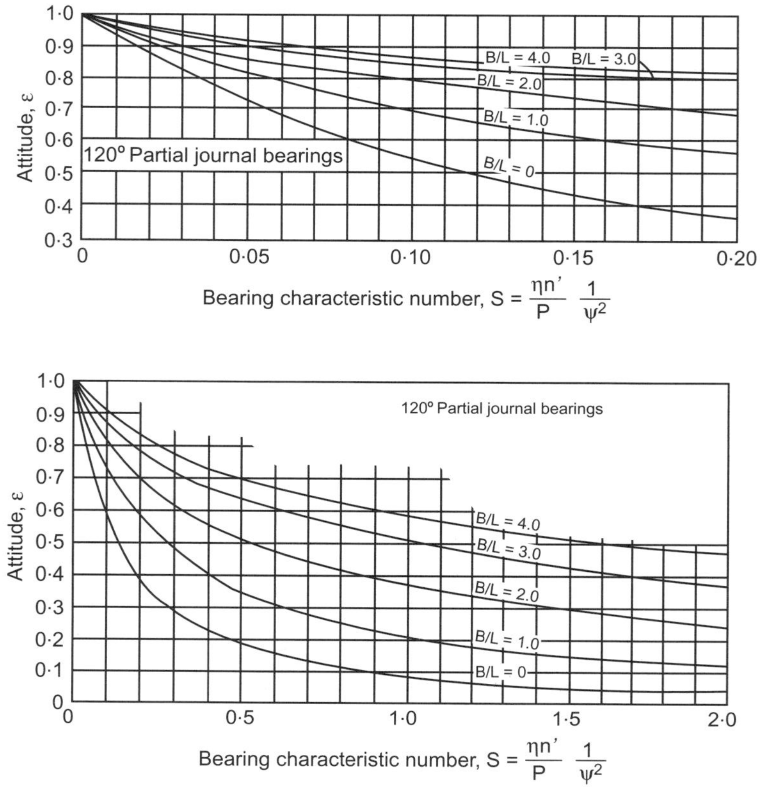

FIGURE 23-36 Variation of attitude ε with S for 120° partial journal bearing.

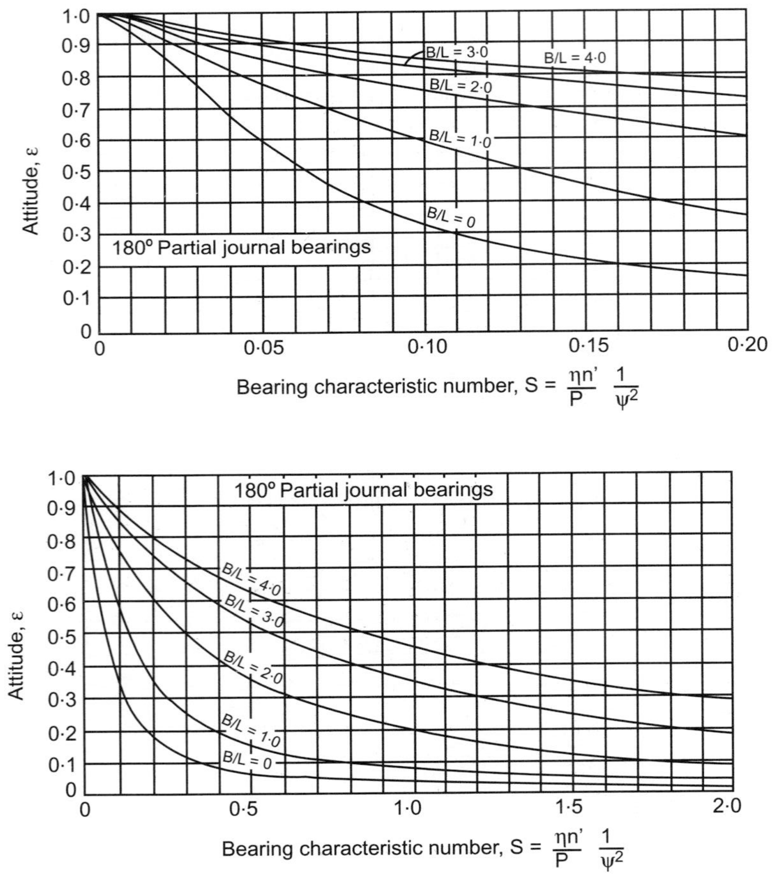

FIGURE 23-37 Variation of attitude ε with S for 180° partial journal bearing.

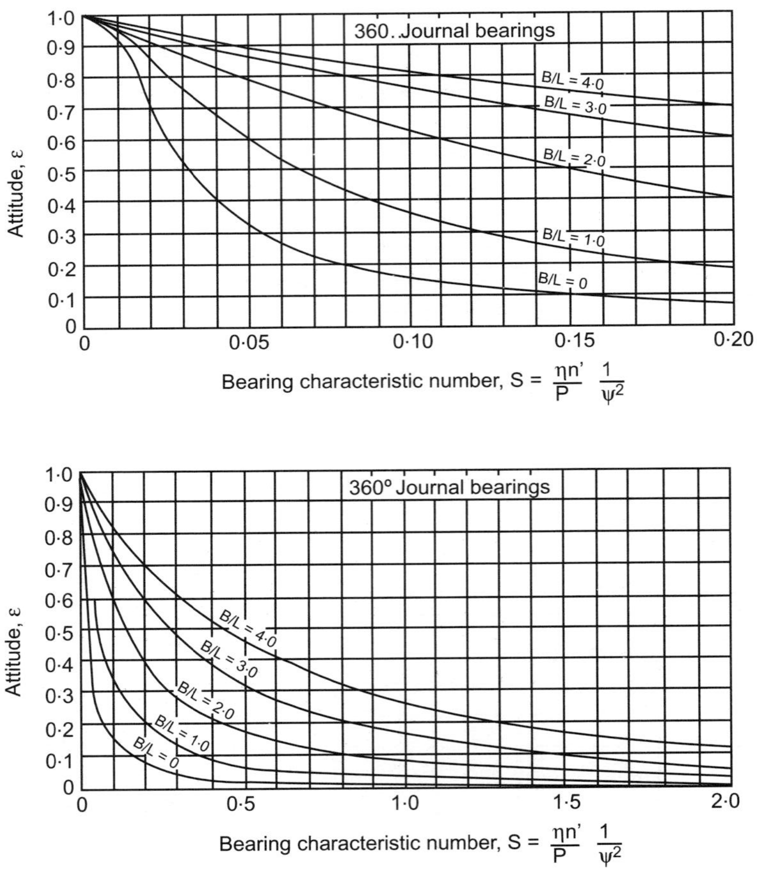

FIGURE 23-38 Variation of attitude ε with S for 360° partial journal bearing.

Particular	Formula

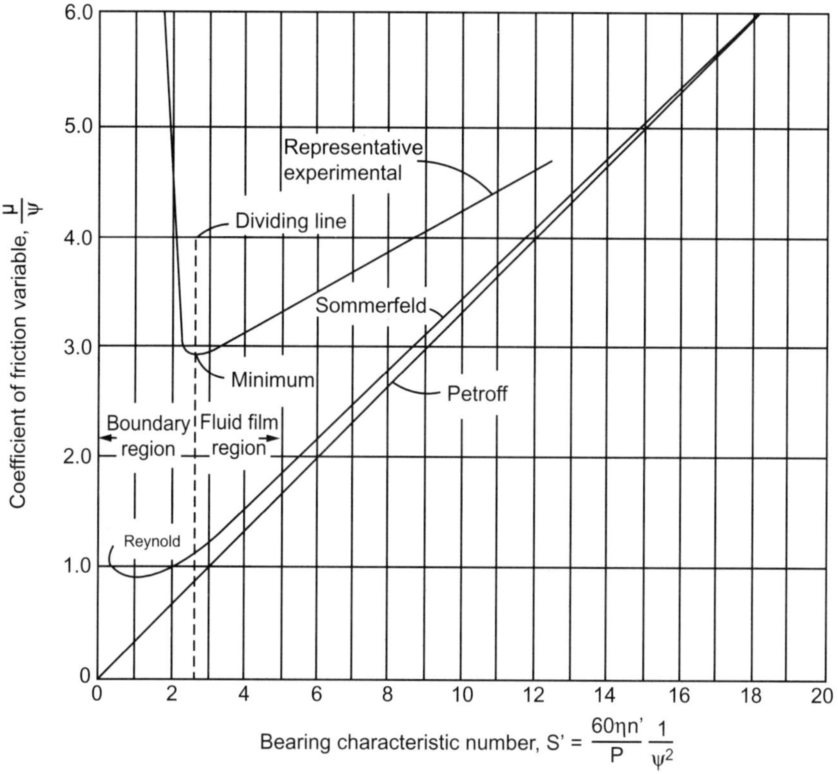

FIGURE 23-39 Friction curves illustrating boundary conditions.

FRICTION IN A FULL JOURNAL BEARING WITH END LEAKAGE FROM BEARING

The total friction force acting on the surface of a full journal bearing with end flow

$$F_\mu = \tfrac{1}{2} W \varepsilon \psi \sin\phi + \frac{2\pi^2 \eta n' L d}{\psi\sqrt{1-\varepsilon^2}} \tag{23-55}$$

The coefficient of friction variable

$$\lambda_\mu = \frac{\mu}{\psi} = \frac{\varepsilon}{2}\sin\phi + \frac{2\pi^2 S}{\sqrt{1-\varepsilon^2}} \tag{23-56a}$$

$$\lambda_\mu = \frac{\varepsilon\sqrt{(1-\varepsilon^2}}{2} + \frac{2\pi^2 S}{\sqrt{1-\varepsilon^2}} \tag{23-56b}$$

$P\mu$ value

Lasche's equation for $P\mu$

$$P\mu = 195{,}350/t \qquad \textbf{SI} \tag{23-57a}$$

where P in N/m^2 and t in K

Particular	Formula
	$P\mu = 0.02/t$ **Customary Metric** (23-57b) where P in kgf/mm^2 and t in °C
	$P\mu = 51/t$ **USCS** (23-57c) where P in psi and t in °F
Lasche's equation for the coefficient of friction which may be used for bearing subjected to pressure varying from 0.103 MPa (15 psi) to 1.55 MPa (225 psi) and speed varying from 2.5 to 18 m/s and temperature varying from 30 to 100°C	$\mu = \dfrac{23{,}126}{P\sqrt{t}}$ **SI** (23-58a) where P in N/m^2 and t in K
	$\mu = \dfrac{0.00236}{P\sqrt{t}}$ **Customary Metric** (23-58b) where P in kgf/mm^2 and t in °C
	$\mu = \dfrac{4.5}{P\sqrt{t}}$ **USCS** (23-58c) where P in psi and t in °F
The coefficient of friction according to Illmer when bearing is subjected to pressure varying from 0.23 MPa (35 psi) to 0.7 MPa (100 psi) and speed varying from 0.5 m/s to 1.5 m/s (100 ft/min to 300 ft/min)	$\mu = \dfrac{\sqrt[3]{v}}{0.05\sqrt{Pt}}$ **SI** (23-59a) where P in N/m^2, v in m/s, and t in K
	$\mu = \dfrac{\sqrt[3]{v}}{157.7\sqrt{Pt}}$ **Customary Metric** (23-59b) where P in kgf/mm^2, v in m/s, and t in °C
	$\mu = \dfrac{\sqrt[3]{v}}{20\sqrt{Pt}}$ **USCS** (23-59c) where P in psi, v in ft/min, and t in °F
The coefficient of friction according to Tower tests	$\mu = \dfrac{144{,}204.5}{P}\sqrt{\dfrac{v}{t}}$ **SI** (23-60a) where P in N/m^2, v in m/s, and t in K
	$\mu = \dfrac{0.0147}{P}\sqrt{\dfrac{v}{t}}$ **Customary Metric** (23-60b) where P in kgf/mm^2, v in m/s, and t in °C
	$\mu = \dfrac{2}{P}\sqrt{\dfrac{v}{t}}$ **USCS** (23-60c) where P in psi, v in ft/min, and t in °F

Particular	Formula
The coefficient of friction according to Lasche when the speed exceeds 2.5 m/s (500 ft/min)	$\mu = \dfrac{24.73}{\sqrt{Pt}}$ **SI** (23-61a) where P in N/m^2 and t in K
	$\mu = \dfrac{0.0079}{\sqrt{Pt}}$ **Customary Metric** (23-61b) where P in kgf/mm^2 and t in °C
	$\mu = \dfrac{0.4}{\sqrt{Pt}}$ **USCS** (23-61c) where P in psi and t in °F

OIL FLOW THROUGH JOURNAL BEARING

Particular	Formula
Oil flow through bearing	$Q = 0.785cLdn'$ **SI** (23-62) where c, L, and d in m, n' in rps, and Q in m^3/s
	$Q = 785cLdn'$ **SI** (23-63a) where c, L, and d in m, n' in rps, and Q in dm^3/s
	$Q = 7.85 \times 10^{-7}cLdn'$ **SI** (23-63b) where c, L, and d in mm, n' in rps, and Q in liters/s or dm^3/s
	$Q = 0.0034cLdn'$ **USCS** (23-63c) where c, L, and d in in, n in rpm, and Q in US gallons/min
Oil flow through a central groove of bearing from one end	$Q = \dfrac{\pi dc^3 P_o}{48\eta L}(1 + 1.5\varepsilon^2)$ (23-64)
Total oil flow through a central groove of bearing from both ends	$Q_g = \dfrac{\pi dc^3 P_o}{24\eta L}(1 + 1.5\varepsilon^2)$ (23-65)
Total oil flow through a central groove for lightly loaded bearing [From Eq. (23-65) as $\varepsilon \to 0$]	$Q \simeq \dfrac{\pi dc^3 P_o}{24\eta L}$ (23-66)
Total oil flow through a central groove for heavily loaded bearing [From Eq. (23-65) as $\varepsilon \to 1$]	$Q \simeq \dfrac{2.5\pi dc^3 P_o}{24\eta L}$ (23-67)
Oil flow through a single hole	$Q_h = \dfrac{c^3 P_o}{24\eta}(1 + 1.5\varepsilon^2)\tan^{-1}\left(\dfrac{\pi d}{L}\right)$ (23-68)

Particular	Formula
The ratio of Q_g to Q_h in the unloaded region of bearing from Eq. (23-65) and (23-68)	$$\frac{Q_g}{Q_h} = \frac{\pi d}{L \tan^{-1}(\pi d/L)} \quad (23\text{-}69)$$ where d is the diameter of journal
Flow variable (dimensionless)	$$\lambda'_Q = \frac{4Q\psi}{60c^2 n' L} \quad (26\text{-}70)$$ Refer to Figs. 23-40 and 23-41 for flow variable λ'_Q.
Oil flow through a bearing with side leakage	$$Q = \frac{60\lambda'_Q d^2 n' L\psi}{4} C_Q \quad \text{or} \quad \frac{60\lambda'_Q c^2 n' L}{4\psi} C_Q \quad (23\text{-}71)$$ where C_Q = flow correction factor from Fig. 23-42
Oil flow ratio $\dfrac{Q_s}{Q}$	Refer to Fig. 23-43.

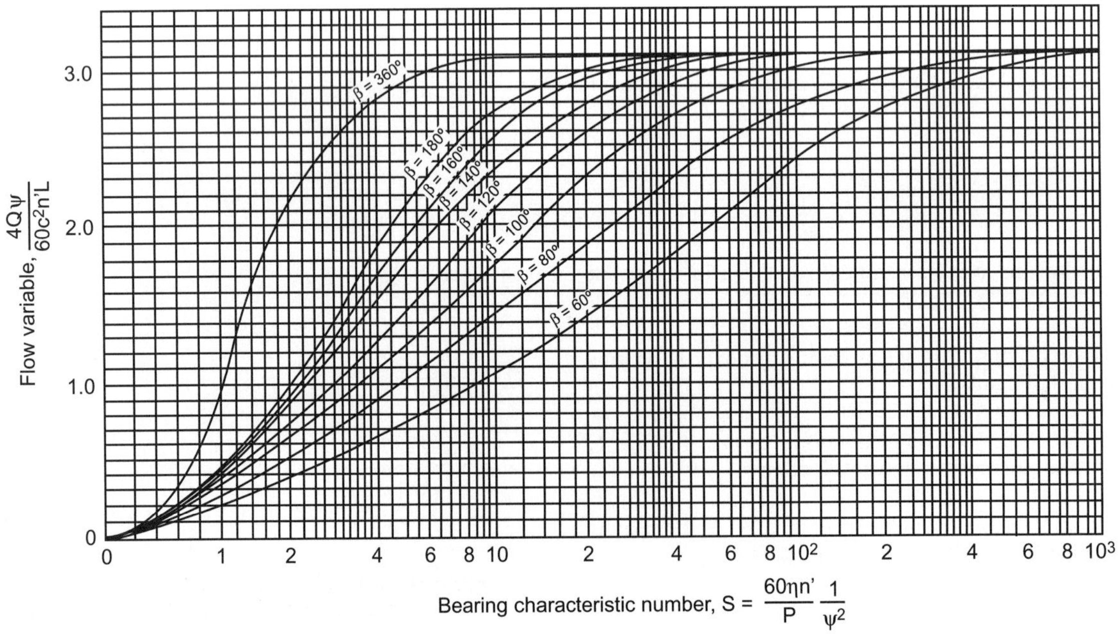

FIGURE 23-40 Chart for determining oil flow, based on no side flow. [Boyd and Raimondi[5]]

Particular	Formula

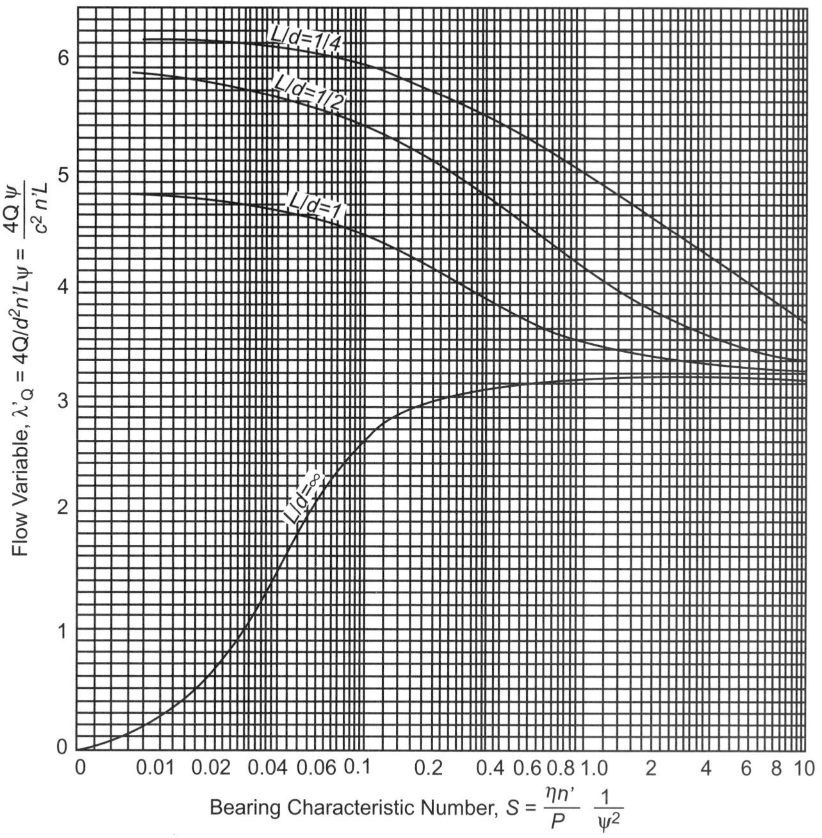

FIGURE 23-41 Chart for oil flow variable λ_Q with bearing characteristic number S for full journal bearing. [Boyd and Raimondi[5]]

THERMAL EQUILIBRIUM OF JOURNAL BEARING

The general expression for heat generated in bearing

$$H_g = M_{t\mu}(2\pi n') = \left(\mu W \frac{d}{2}\right)(2\pi n')$$

$$= \mu(PLd)\frac{d}{2}\omega = \mu(PLd)v \quad \textbf{SI (USCS)} \quad (23\text{-}72a)$$

where H_g in J/s (Btu/s), $M_{t\mu}$ in N m (lbf in), W in N (lbf), P in N/m^2 (psi), L in m (in), d in m (in), n' in rps, ω in rad/s, and v in m/s (ft/min)

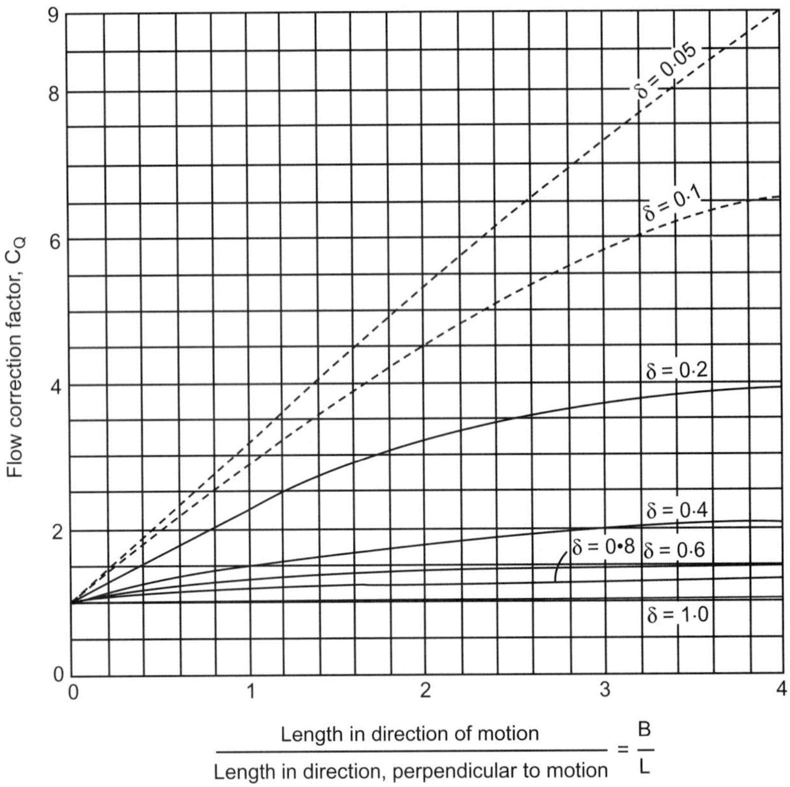

FIGURE 23-42 Flow correction factor for side flow. [Boyd and Raimondi[5]]

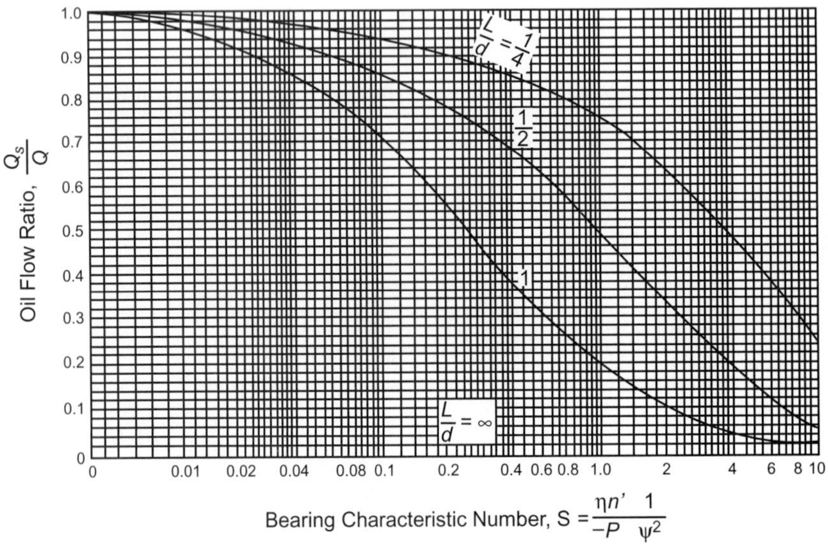

FIGURE 23-43 Oil flow ratio (Q_s/Q) versus bearing characteristic number S for full journal bearing. [Boyd and Raimondi[5]]

Particular	Formula
	$$H_g = \frac{\mu(PLd)v}{778} = \frac{\mu(PLd)(\pi dn')}{778} = \frac{\mu(PLd)d\omega}{1556}$$ **USCS** (23-72b) where P in psi, L in in, d in in, v in ft/min, ω in rad/s, H_g in Btu/s, and n' in rps
The heat generated can also be found by knowing the temperature rise of lubricant oil which is used to carry away heat generated in the bearing	$H_g = \gamma C_{sp} Q\,\Delta T$ **SI (USCS)** (23-73) where H_g in J/s (Btu/s) γ = weight per unit volume of lubricant whose average specific gravity is 0.90 = 8.83 kN/m^3 (0.0325 lbf/in^3) C_{sp} = specific heat of lubricant, kJ/N K (Btu/lbf °F) = 0.19 kJ/N K (0.42 Btu/lbf °F) Q in m^3/s; ΔT in °C
Temperature rise of the lubricant film variable λ_T	Refer to Fig. 23-44 for λ_T.
The temperature rise of the lubricant film due to heat generated which is to be carried away by the lubricant, can be found from Eq. (23-66c)	$$\Delta T = \frac{4\pi P\lambda_\mu}{427\gamma C_{sp}\lambda_Q}$$ **Customary Metric** (23-74a) where $\lambda_\mu = \mu/\psi$ = friction coefficient variable λ_Q = flow variable = $4Q/d^2n'L\psi$; P in kgf/m^2, C_{sp} in kcal/kgf°C, γ in kgf/m^3, and ΔT in °C
	$$\Delta T = 78\frac{P\lambda_\mu}{\lambda_Q} = \frac{78(\mu/\psi)P}{4Q/d^2n'L\psi} = 19.5\frac{\mu P}{Q/d^2n'L}$$ **Customary Metric** (23-74b) where P in kgf/mm^2, d in mm, L in mm, Q in mm^3/s, n' in rps, and ΔT in °C
	$$\Delta T = \frac{8.3\times10^{-6}(\mu/\psi)P}{Q/d^2n'L\psi} = \frac{8.3\times10^{-6}\mu P}{Q/d^2n'L}$$ **SI** (23-74c) where P in N/m^2, Q in m^3/s, d in m, L in m, n' in rps, and ΔT in K or °C, since ΔT °C = ΔT K
If the end flow is also taken into consideration then the temperature rise of the flow $Q - Q_{el}$ due to heat generated, is ΔT and the temperature rise of end leakage is $\Delta T/2$, which is the average of the inlet and the outlet temperatures	$$\Delta T = \frac{0.103(\mu/\psi)P}{[1-\frac{1}{2}(Q_{el}/Q)](4Q/d^2n'L\psi)} = 0.0258\frac{\mu P}{[1-\frac{1}{2}(Q_{el}/Q)](Q/d^2n'L)}$$ **USCS** (23-75a) where P in psi, d in in, L in in, Q_{el} and Q in in^3/s, n' in rps, and ΔT in °F

Particular	Formula

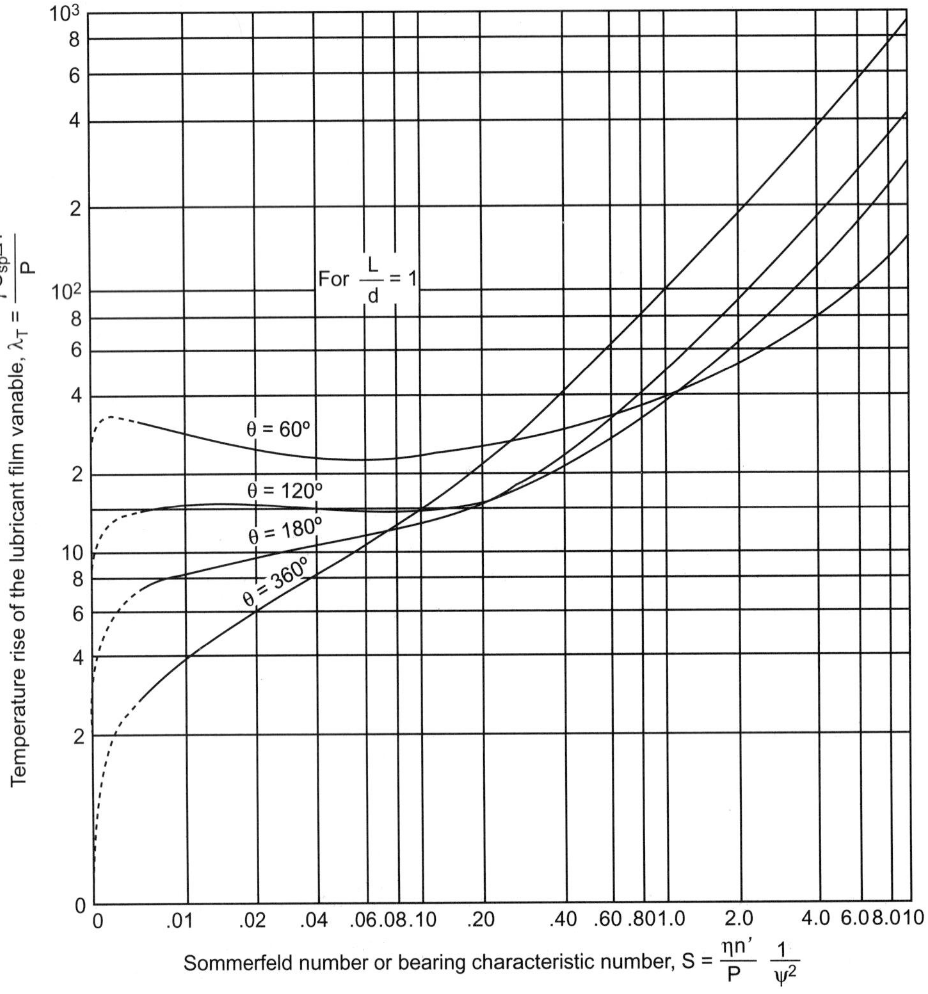

FIGURE 23-44 Variation of temperature rise of the lubrication film variable λ_T with Sommerfeld number S. [Boyd and Raimondi[5]]

$$\Delta T = \frac{8.3 \times 10^{-6}(\mu/\psi)P}{[1 - \frac{1}{2}(Q_{el}/Q)](Q/d^2n'L\psi)}$$

$$= \frac{8.3 \times 10^{-6}\mu P}{[1 - \frac{1}{2}(Q_{el}/Q)](Q/d^2n'L)} \quad \textbf{SI} \quad (23\text{-}75b)$$

where P in N/m^2, d and L in m, Q_{el} and Q in m^3/s, n' in rps, and ΔT in K

Particular	Formula
The temperature rise of the lubricant film due to heat generated in pressure fed bearings	$\Delta T = \frac{16 \times 10^{-6}(\mu/\psi)SW^2}{(1+1.5\varepsilon^2)P_S d^4}$ **SI** (23-76a)
	where W in N, P_S in N/m^2, d in m, and ΔT in K
	$\Delta T = \frac{9.7(\mu/\psi)SW^2}{(1+1.5\varepsilon^2)P_S d^4}$ **Customary Metric** (23-76b)
	where W in kgf, P_S in kgf/mm^2, d in mm, and ΔT in °C
Heat dissipated by self-contained bearings	$H_d = CA(t_b - t_a)$ (23-77)
	where
	C = combined coefficient of radiation and convection $= 9.6 \times 10^{-3}$ kW/m^2 K (1.7 Btu/ft^2 h °F) when the bearing located in still air and oil bath $= 11.36 \times 10^{-3}$ kW/m^2 K (2 Btu/ft^2 h °F) when the bearing located in air circulation and in oil bath $= 15.36 \times 10^{-3}$ kW/m^2 K (2.7 Btu/ft^2 h °F) for average design practice $= 33.5 \times 10^{-3}$ kW/m^2 K (5.9 Btu/ft^2 h °F) when the air velocity over the bearing is 2.5 m/s (500 ft/min)
	A = effective surface area of bearing housing $= (25 \times 10^{-4}$ dL) in m^2 for bearing masses of metal as in a ring oil bearing (25 dL in in^2) $= (6 \times 10^4$ dL) in m^2 for light construction (6 dL in in^2)
	t_b = surface temperature of bearing housing, °C (°F) t_a = temperature of surrounding air, °C (°F)

TABLE 23-17a
Values of factor, *m* in Equation (23-67b)

Lubrication system	Conditions	Values of *m*
Oil ring	Moving air	1-2
	Still air	$\frac{1}{2}$-1
Oil bath	Moving air	$\frac{1}{2}$-1
	Still air	$\frac{1}{5}$-$\frac{2}{5}$

Particular	Formula
Another formula for the heat dissipated in bearing in terms of average lubricant oil film temperature	$H_d = \frac{CA}{m+1}(t_o - t_a)$ (23-78)
	where m can be assumed as constant which depends on the lubrication system and it is taken from Table 23-17a, t_o is lubricant film temperatures, °C. C and A are as given under Eq. (23-67a)
The heat dissipating capacity of bearing based on projected area of bearing	$H_d = k(Ld)(t_b - t_a) = kA\,\Delta T$ **Customary Metric** (23-79a)
	where $A = Ld$ = projected area of bearing, cm^2 $k\,\Delta T = k(t_b - t_a)$ values can be taken from Fig. 23-45(a) and H_d in kcal/min
	$H_d = 697.8k(t_b - t_a)(Ld)$ **SI** (23-79b)
	where $k(t_b - t_a)$ in kcal/min cm^2, values are taken from Fig. 23-45(a); (Ld) in m^2 and H_d in J/s $= k < d(t_b - t_a)$ USCS (23-68c) where $k(t_b - t_a)$ in ft-lbf/min/in^2/°F values can be taken from Fig. 23-45(b)
Pederson's equation for heat radiating capacity of bearing due to friction between journal and bearing	$H_d = \frac{(\Delta T + 18)^2}{k}(Ld)$ **SI (USCS)** (23-80a)

Particular	Formula
	$H_d = \dfrac{(\Delta T + 33)^2 Ld}{k}$ **USCS** (23-80b)
	where H_d in J/s (ft-lbf/s/in^2), Ld in in^2 m^2 (in^2), ΔT in °C (°F)
	$k = 751$ (3300) for bearings of light construction located in still air
	$= 7367$ Customary Metric
	$= 423$ (1860) for bearings of heavy construction and well ventilated
	$= 4152$ Customary Metric
	$= 262$ (1150) for General Electric Company's well-ventilated bearing
	$= 2567$ Customary Metric

TABLE 23-17
Quantities C_{PW}, C_{PFm}, $C_{P\mu}$, and $\bar{x}/B$ as functions of q for use in Eqs. (23-95) to (23-99)

q	C_{PW}	C_{PFm}	$C_{P\mu}$	$\bar{x}/B$
0.10	0.007209	0.955265	22.085010	
0.20	0.012585	0.919159	12.172679	
0.25	0.014741	0.903630	10.216742	
0.30	0.016608	0.889234	8.923752	
0.40	0.019618	0.864722	7.346332	0.533761
0.50	0.021864	0.843721	6.431585	
0.60	0.023514	0.825665	5.852294	0.546881
0.70	0.024714	0.809936	5.462059	
0.80	0.025597	0.796076	5.183394	0.558394
0.90	0.026129	0.783718	4.999030	0.563687
1.00	0.026481	0.772589	4.862537	0.568688
1.10	0.026661	0.762470	4.766450	0.573426
1.20	0.026707	0.753191	4.700334	0.577926
1.30	0.026645	0.744615	4.657628	0.582209
1.40	0.026500	0.736633	4.632912	0.586293
1.50	0.026289	0.729156	4.622694	0.590193
1.60	0.026026	0.722120	4.624350	0.594111
1.70	0.025728	0.715441	4.634646	
1.80	0.025386	0.709100	4.655453	0.600937
2.00	0.024653	0.697225	4.713591	0.607410
2.20	0.03870	0.686283	4.791810	0.613416
2.50	0.022664	0.671087	4.935044	0.621673
2.75	0.021668	0.659606	5.073580	
3.00	0.020699	0.648392	5.220786	0.633787
4.00	0.017257	0.609438	5.885901	
5.00	0.014528	0.580067	6.654587	
7.00	0.010692	0.521586	8.130471	
8.00	0.009332			
9.00	0.008225	0.477917	9.684235	

Source: F. I. Radzimovsky, *Lubrication of Bearings—Theoretical Principles and Designs*, The Ronald Press Company, New York, 1959.

Particular	Formula

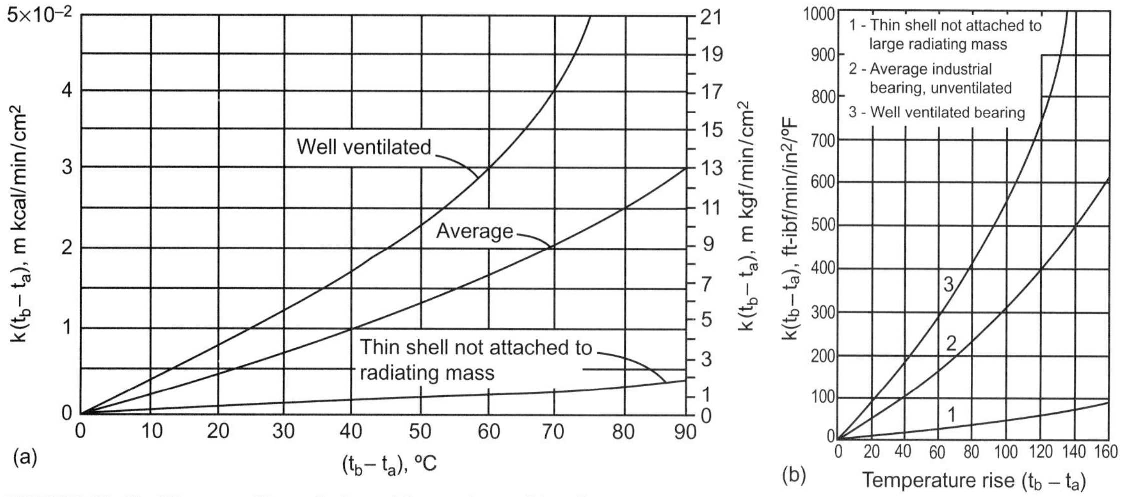

FIGURE 23-45 The rate of heat dissipated from a journal bearing.

$$H_d = \frac{(\Delta T + 18)^2}{427k}(Ld)$$

Customary Metric (23-80c)

where H_d in kcal/s, (Ld) in m^2, ΔT in °C values of k are as given inside parentheses under Eq. (23-80a) for US customary system units and values of k for customary metric units also given under Eqs. (23-80a) and (23-80b)

The difference in temperature (ΔT) of the bearing and of the cooling medium can be found from the equation

$$(\Delta T + 18)^2 = K'\mu Pv \qquad \textbf{SI (Metric)} \quad (23\text{-}81)$$

where P in N/m^2 (kgf/mm^2), v in m/s, and ΔT in K (°C)

$K' = 0.475\ (4.75 \times 10^6)$ for bearings of light construction located in still air

$= 0.273\ (2.7 \times 10^6)$ for bearings of heavy construction and well ventilated

$= 0.165\ (1.65 \times 10^6)$ for General Electric Company's well-ventilated bearing

The difference between the bearing-wall temperature t_b and the ambient temperature t_a, for three main types of lubrication by oil bath, by an oil ring, and by waste pack or drop feed

Refer to Fig. 23-46 for $t_b - t_a \simeq \left(\frac{t_0 - t_b}{2}\right)$

BEARING CAP

The bearing cap thickness

$$h_c = \sqrt{\frac{3Wa}{2L\sigma}} \qquad (23\text{-}82)$$

Particular	Formula

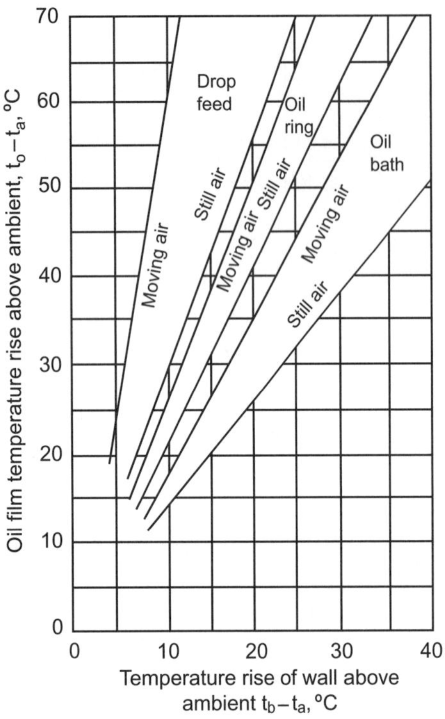

FIGURE 23-46 Relation between oil film temperature and bearing wall temperature.

The deflection of the cap

$$y = \frac{Wa^3}{4ELh_c^3} \tag{23-83}$$

The thickness of cap from Eq. (23-71)

$$h_c = 0.63a\sqrt[3]{\frac{W}{ELy}} \tag{23-84}$$

where the deflection should be limited to 0.025 mm (0.001 in)

EXTERNAL PRESSURIZED BEARING OR HYDROSTATIC BEARING: JOURNAL BEARING (Fig. 23-47)

The pressure in the lower pool of quadrant 1 (Fig. 23-47)

$$P_1 = K_1 P_o \tag{23-85a}$$

where

$$K_1 = \frac{1}{1 + \frac{4}{\pi}\left(\frac{P_o}{P'} - 1\right)\left(\frac{\pi}{4} + 2.121\varepsilon + 1.93\varepsilon^2 - 0.589\varepsilon^3\right)} \tag{23-85b}$$

Particular	Formula

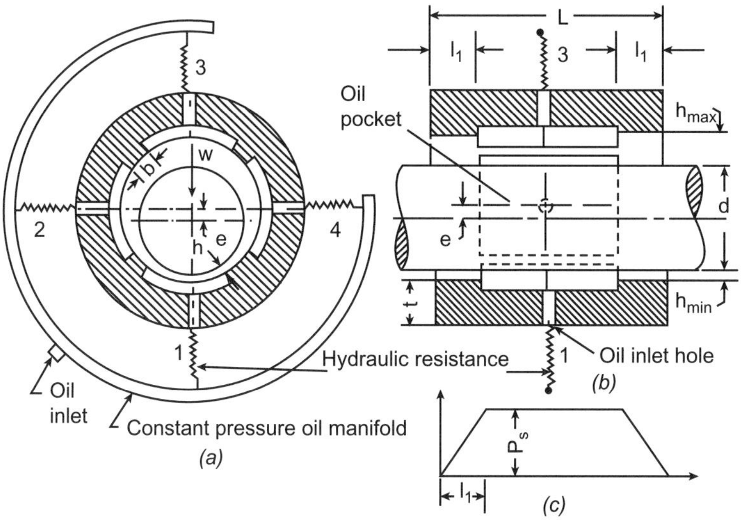

FIGURE 23-47 (a) and (b) schematic diagram of a full cylindrical hydrostatic bearing; (c) oil pressure distribution along the bearing. [Shaw and Macks[10]]

The pressure in the upper pool of quadrant 3 (Fig. 23-47)

$$P_3 = K_3 P_o \tag{23-86a}$$

where

$$K_3 = \frac{1}{1 + \frac{4}{\pi}\left(\frac{P_o}{P'} - 1\right)\left(\frac{\pi}{4} + 2.121\varepsilon + 1.93\varepsilon^2 + 0.589\varepsilon^3\right)} \tag{23-86b}$$

The pressure in the left pool of quadrant 2 (Fig. 23-47)

$$P_2 = K_2 P_o \tag{23-87a}$$

where

$$K_2 = \frac{1}{8}\left(\frac{P'}{P_o}\right)(6.283 + 3.425\varepsilon^2) \tag{23-87b}$$

The pressure in the right pool of quadrant 4 (Fig. 23-47)

$$P_4 = K_4 P_o \tag{23-88a}$$

where

$$K_4 = \frac{1}{8}\left(\frac{P'}{P_o}\right)(6.283 + 3.425\varepsilon^2) \tag{23-88b}$$

The flow of lubricant through the lower quadrant 1 of the bearing from the manifold

$$Q_1 = \frac{\psi^3 d^4 P_1}{96\eta l_1} C_{P\mathscr{F}1} \tag{23-89a}$$

where

$$C_{P\mathscr{F}1} = \frac{\pi}{4} - 2.121\varepsilon + 1.93\varepsilon^2 - 0.589\varepsilon^3 \tag{23-89b}$$

Particular	Formula
The flow of lubricant through the left quadrant 2 of the bearing from the manifold	$Q_2 = \dfrac{\psi^3 d^4 P_2}{768\eta l_1} C_{P\mathcal{F}2}$ (23-90a) where $C_{P\mathcal{F}2} = 6.283 + 3.425\varepsilon^2$ (23-90b)
The flow of lubricant through the upper quadrant 3 of the bearing from the manifold	$Q_3 = \dfrac{\psi^3 d^4 P_3}{48\eta l_1} C_{P\mathcal{F}3}$ (23-91a) where $C_{P\mathcal{F}3} = \dfrac{\pi}{4} + 2.121\varepsilon + 1.93\varepsilon^2 + 0.589\varepsilon^3$ (23-91b)
The flow of lubricant through the right quadrant 4 of the bearing from the manifold	$Q_4 = \dfrac{\psi^3 d^4 P_4}{768\eta l_1} C_{P\mathcal{F}4}$ (23-92a) where $C_{P\mathcal{F}4} = C_{P\mathcal{F}2} = 6.283 + 3.425\varepsilon^2$ (23-92b)
The total flow of lubricant through quadrant of the bearing from the manifold assuming $P_2 = P_4 = P'$ (good approximation)	$Q = Q_1 + Q_2 + Q_3 + Q_4$ (23-93a) $Q = \dfrac{\psi^3 d^4 P_o}{48\eta l_1} G$ (23-93b) where G = flow factor given by Eq. (23-94)
The flow factor in Eq. (23-81b)	$G = C_{P\mathcal{F}1}K_1 + \frac{1}{8}(C_{P\mathcal{F}2}K_2 + C_{P\mathcal{F}4}K_4) + C_{P\mathcal{F}3}K_3$ $= C_{P\mathcal{F}1}K_1 + \frac{1}{4}C_{P\mathcal{F}2}K_2 + C_{P\mathcal{F}3}K_3$ (23-94) since $K_2 = K_4$ and $C_{P\mathcal{F}2} = C_{P\mathcal{F}4}$
The external load on the hydrostatic journal bearing	$W = (P_1 - P_3)\left(A + \dfrac{A'}{2}\right) = \pi P_o\left(A + \dfrac{A'}{2}\right)F_{P\mathcal{F}W}$ (23-95) where $F_{P\mathcal{F}W}$ = load factor given by Eq. (23-95)
The load factor	$F_{P\mathcal{F}W} = K_1 - K_3$ (23-96)
The pressure ratio connecting the dimensions of the bearing and its external resistances	$\dfrac{P_o}{P'} = 1 + 6\left(\dfrac{d}{d_c}\right)\left(\dfrac{c}{d_c}\right)^3 \dfrac{l_c}{l_1}$ (23-97)

Particular	Formula

IDEALIZED SLIDER BEARING (Fig. 23-48)

Plane-slider bearing

The pressure at any point x

$$P = \frac{\eta U}{B} C_{s1} \quad \text{(23-98a)}$$

where

$$C_{s1} = \frac{6\kappa x_1(1 - x_1)}{(\kappa - 2a)(a - \kappa + \kappa x_1)^2} \quad \text{(23-98b)}$$

$$\kappa = \frac{h_2 - h_1}{B}; \quad a = \frac{h_2}{B}; \quad x_1 = \frac{x}{B} \quad \text{(23-98c)}$$

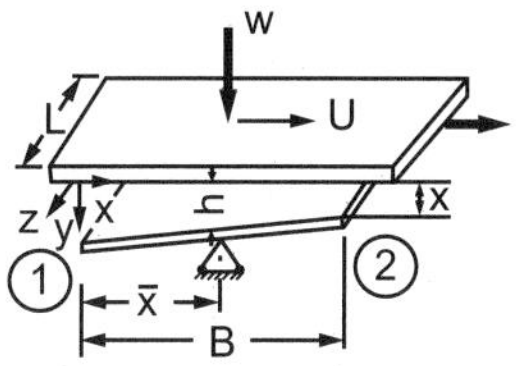

FIGURE 23-48 Plane slider bearing with an angle of inclination.

The load carrying capacity

$$W = \frac{6\eta UL}{\kappa^2} C_{s2} \quad \text{(23-99a)}$$

where

$$C_{s2} = \ln\frac{a - \kappa}{a} + \frac{2\kappa}{2a - \kappa} \quad \text{(23-99b)}$$

The resultant shear stress at any point along the slider (Fig. 23-48)

$$\tau = \frac{\eta U}{B} C_{s3} \quad \text{(23-100a)}$$

where

$$C_{s3} = \left[\left\{\frac{B(a - \kappa + \kappa x_1) - 2y}{B}\right\} \times \left\{\frac{3\kappa(a - \kappa + \kappa x_1 - 2a x_1)}{(\kappa - 2a)(a - \kappa + \kappa x_2)^3} + \frac{1}{a - \kappa + \kappa x_1}\right\}\right] \quad \text{(23-100b)}$$

The shear stress at any point on the surface of the moving member of the bearing (i.e., slider at $y = 0$) (Fig. 23-48)

$$\tau_m = \frac{\eta U}{B} C_{s4} \quad \text{(23-101a)}$$

where

$$C_{s4} = \frac{4}{a - \kappa + \kappa x_1} - \frac{6a(a - \kappa)}{(2a - \kappa)(a - \kappa + \kappa x_1)^2} \quad \text{(23-101b)}$$

The shear stress at any point on the surface of the stationary member of the bearing (i.e., shoe at $y = h$) (Fig. 23-48)

$$\tau_s = \frac{\eta U}{B} C_{s5} \quad \text{(23-102a)}$$

where

$$C_{s5} = \frac{-2}{a - \kappa + \kappa x_1} - \frac{6a(a - \kappa)}{(2a - \kappa)(a - \kappa + \kappa x_1)^2}$$

$$\kappa = \frac{h_2 - h_1}{B} \quad \text{and} \quad h = B(a - \kappa + \kappa x_1) \quad \text{(23-102b)}$$

Particular	Formula
The frictional force on the moving member of the bearing (i.e., slider)	$F_{\mu m} = \eta U L C_{s6}$ (23-103a) where $C_{s6} = -\frac{4}{\kappa}\ln\left(\frac{a-\kappa}{a}\right) - \frac{6}{2a-\kappa}$ (23-103b)
The frictional force on the stationary member of the bearing (i.e., shoe)	$F_{\mu s} = \eta U L C_{s7}$ (23-104a) where $C_{s7} = \frac{2}{\kappa}\ln\left(\frac{a-\kappa}{4}\right) + \frac{6}{2a-\kappa}$ (23-104b)
The coefficient of friction	$\mu = \frac{F_{\mu m}}{W} = \frac{-2\kappa(2a-\kappa)\ln\left(\frac{a-\kappa}{a}\right) - 3\kappa^2}{3(2a-\kappa)\ln\left(\frac{a-\kappa}{a}\right) + 6\kappa}$ (23-105)
The distance of the pressure center from the origin of the coordinates, i.e., from the lower end of the shoe (Fig. 23-48)	$\bar{x} = \left[\frac{(a-\kappa)(3a-\kappa)\left(\frac{a-\kappa}{a}\right) - 2.5\kappa^2 + 3\kappa a}{\kappa(\kappa-2a)\ln\left(\frac{a-\kappa}{a}\right) - 2\kappa^2}\right] B$ (23-106)

Pivoted-shoe slider bearing (Fig. 23-48 and Fig. 23-52)

Particular	Formula
The load-carrying capacity	$W = \frac{6\eta U L B^2}{h_2^2} C_{PW}$ (23-107a) where $C_{PW} = \frac{1}{q^2}\ln(1+q) - \frac{2}{q(q+2)}$ (23-107b) Refer to Table 23-17 for C_{PW}.
The frictional force on the moving member of the bearing (i.e., slider)	$F_{\mu mP} = \frac{\eta U L B}{h_2} C_{PFm}$ (23-108a) where $C_{PFm} = \frac{4}{q}\ln(1+q) - \frac{6}{2+q}$ (23-108b) Take C_{PFm} from Table 23-17 for various values of q.
The frictional force on the stationary member of the bearing (i.e., shoe)	$F_{\mu sP} = \frac{\eta U L B}{h_2} C_{PFs}$ (23-109a) where $C_{PFs} = -\frac{2}{q}\ln\frac{(1+q)a}{2} + \frac{6}{2+q}$ (23-109b)

Particular	Formula
The coefficient of friction	$\mu = \dfrac{F_{\mu m P}}{W} = \dfrac{h_2}{B}\left(\dfrac{1}{6}\dfrac{C_{PFm}}{C_{PW}}\right) = \dfrac{h_2}{B} C_{P\mu}$ (23-110) where $C_{P\mu}$ = coefficient of friction factor Take $C_{P\mu}$ from Table 23-17 for various values of q.
The distance of the pivoted point from the lower end of the shoe (Fig. 23-39), i.e., the distance of the pressure center from the origin of the coordinates	$\bar{x} = \left[\dfrac{(1+q)(3+q)\ln(1+q) - q(2.5q+3)}{q(q+2)\ln(1+q) - 2q^2}\right] B$ (23-111) The ratios $\bar{x}/B$ are taken from Table 23-17.

DESIGN OF VERTICAL, PIVOT, AND COLLAR BEARING

Pivot bearing (Figs. 23-49, 23-50, and 23-53)

FLAT PIVOT

Particular	Formula
The total axial load on the flat pivot with extreme diameters of the actual contact d_1 and d_2	$W = p\pi \dfrac{d_1^2 - d_2^2}{4}$ (23-112)
The friction torque based on uniform intensity of pressure with extreme diameters of the actual contact d_1 and d_2	$M_t = \frac{1}{3}\mu W \dfrac{d_1^3 - d_2^3}{d_1^2 - d_2^2}$ (23-113)
The friction torque based on uniform wear with extreme diameters of the actual contact d_1 and d_2	$M_t = \mu W \dfrac{d_1 + d_2}{4}$ (23-114)
The power absorbed by friction with d as the diameter of flat pivot bearing	$P_\mu = \dfrac{\mu W dn'}{478}$ SI (23-115a) where P_μ in kW, W in N, d in m, and n' in rps $P_\mu = \dfrac{\mu W dn}{189{,}090}$ USCS (23-115b) where P_μ in hp, W in lbf, d in in, and n in rpm

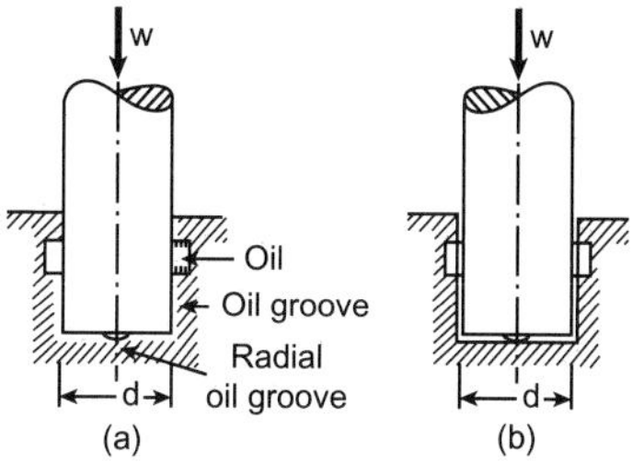

FIGURE 23-49 Pivot thrust bearing

CONICAL PIVOT

Particular	Formula
The friction torque based on uniform intensity of pressure with extreme diameters of the actual contact d_1 and d_2	$M_t = \dfrac{1}{3}\dfrac{\mu W}{\sin\alpha}\dfrac{d_1^3 - d_2^3}{d_1^2 - d_2^2}$ (23-116) where 2α = cone angle of pivot, deg
The friction moment which resists the rotation of the shaft in a conical pivot bearing for uniform wear	$M_t = \dfrac{\mu W}{\sin\alpha}\dfrac{d_1 + d_2}{4}$ (23-117)
The loss of power in vertical bearing	$P_\mu = 6.2 \times 10^8 \dfrac{\eta d^2 L n'^2}{\psi}$ SI (23-118a) where P_μ in kW, η in Pa s, d and L in m, and n' in rps

Particular	Formula
	$P_\mu = 2.35 \times 10^{-4} \dfrac{\eta_1 d^2 L n^2}{\psi}$ **Customary Metric** (23-118b)
	where P_μ in hp_m, η_1 in cP, d and L in cm, and n in rpm
	$P_\mu = 2.35 \times 10^{-7} \dfrac{\eta_1 d^2 L n^2}{\psi}$ **Customary Metric** (23-118c)
	where P_μ in hp_m, η_1 in cP, d and L in mm, and n in rpm
	$P_\mu = 2.35 \times 10^{6} \dfrac{\eta' d^2 L n^2}{\psi}$ **Customary Metric** (23-118d)
	where P_μ in hp_m, η' in kgf s/m^2, d and L in m, and n in rpm
	$P_\mu = 2.35 \times 10^{-3} \dfrac{\eta' d^2 L n^2}{\psi}$ **Customary Metric** (23-118e)
	where P_μ in hp_m, η' in kgf s/m^2, L and d in mm, and n in rpm
	$P_\mu = \dfrac{3.8}{3} \dfrac{\eta_1 d^2 L n^2}{\psi}$ **USCS** (23-118f)
	where P_μ in hp, η_1 in cP, d and L in in, and n in rpm
If the journal and the bearing are eccentric and the distance between their axes is ε, the power loss is calculated from formula	$P_\mu = \dfrac{6.2 \times 10^8 \eta d^2 L n'^2}{\psi\sqrt{1-(2\varepsilon)^2}}$ **SI** (23-119a)
	where P_μ in kW, η in Pa s, d and L in m, and n' in rps
	$P_\mu = 2.35 \times 10^{-7} \dfrac{\eta_1 d^2 L n^2}{\psi\sqrt{1-2\varepsilon)^2}}$ **Customary Metric** (23-119b)
	where P_μ in hp_m, η_1 in cP, d and L in mm, and n in rpm
	$P_\mu = 2.3 \times 10^{6} \dfrac{\eta' d^2 L n^2}{\psi\sqrt{1-(2\varepsilon)^2}}$ **Customary Metric** (23-119c)
	where P_μ in hp_m, η' in (kgf s/m^2), d and L in m, and n in rpm
	$P_\mu = \dfrac{3.8}{10^3} \dfrac{\eta_1 d^2 L n^2}{\psi\sqrt{1-(2\varepsilon)^2}}$ **USCS** (23-119d)
	where P_μ in hp, η_1 in cP, L and d in in, and n in rpm

Particular	Formula

Collar bearing (Fig. 23-51)

The average intensity of pressure with i collars

$$P = \frac{W}{0.784(d_1^2 - d_2^2)i} \quad (23\text{-}120)$$

The friction moment for each collar for uniform intensity of pressure

$$M_{te} = \frac{1}{3}\frac{\mu W}{i}\left(\frac{d_1^3 - d_2^3}{d_1^2 - d_2^2}\right) \quad (23\text{-}121)$$

The total friction moment for i collars for uniform intensity of pressure

$$M_t = \frac{1}{3}\mu W\left(\frac{d_1^3 - d_2^3}{d_1^2 - d_2^2}\right) \quad (23\text{-}122)$$

The friction moment for each collar for uniform rate of wear

$$M_{te} = \frac{\mu W}{i}\left(\frac{d_1 + d_2}{4}\right) \quad (23\text{-}123)$$

The total friction moment for i collars for uniform rate of wear

$$M_t = \mu W\left(\frac{d_1 + d_2}{4}\right) \quad (23\text{-}124)$$

The friction power in collar bearing

$$P_\mu = \frac{\mu W(d_1 + d_2)n'}{2{,}292{,}296} \quad \textbf{SI} \quad (23\text{-}125a)$$

where P_μ in kW, W in N, d in m, and n' in rps

$$P_\mu = \frac{\mu W(d_1 + d_2)n}{252{,}120} \quad \textbf{USCSU} \quad (23\text{-}125b)$$

where P_μ in hp, W in lbf, d in in, and n' in rpm

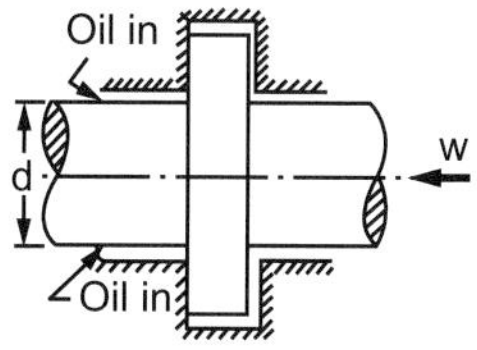

FIGURE 23-50a Collar thrust bearing.

The coefficient of friction for collar bearing

$$\mu = 83.8\,\frac{v^{0.5}}{p^{0.67}} \quad \textbf{SI} \quad (23\text{-}126a)$$

where v in m/s and P in N/m^2

$$\mu = 0.016\,\frac{v^{0.5}}{p^{0.67}} \quad \textbf{USCS} \quad (23\text{-}126b)$$

where v in ft/min and P in psi

$$\mu = 1.73 \times 10^{-3}\,\frac{v^{0.5}}{p^{0.67}} \quad \textbf{Customary Metric} \quad (23\text{-}126c)$$

where v in m/s and P in kgf/mm^2

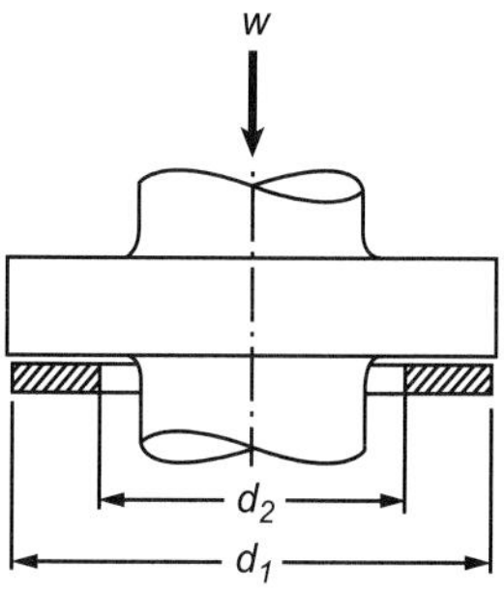

FIGURE 23-50b Plain thrust bearing.

Allowable pressure P may be taken so that Pv value for v ranging from 0.20 to 1 m/s (50 to 200 ft/min)

$$Pv \le 707{,}505 \quad \textbf{SI} \quad (23\text{-}127a)$$

where P in Pa and v in m/s

$$Pv \le 0.0715 \quad \textbf{Customary Metric} \quad (23\text{-}127b)$$

where P in kgf/mm^2 and v in m/s

$$Pv \le 20{,}000 \quad \textbf{USCS} \quad (23\text{-}127c)$$

where P in psi and v in ft/min

Particular	Formula

PLAIN THRUST BEARING (Fig. 23-50b)

Recommended maximum load

$$W = K_1(d_1^2 - d_2^2) \qquad \textbf{SI (USCS)} \quad (23\text{-}128)$$

where $K_1 = 0.3$ (48), W in N (lbf), d_1 and d_2 in mm (in)

Approximate power loss in bearing

$$P_\mu = K_2\left(\frac{d_1 + d_2}{2}\right)n'W \qquad \textbf{SI (USCSU)} \quad (23\text{-}129)$$

where $K_2 = 70 \times 10^{-6} = (11 \times 10^{-6})$
P_μ in W (hp), n' in rps, and W in N (lbf)

Lubrication flow rate to limit lubricant temperature rise to 20°C

$$Q = K_3 P_\mu \qquad \textbf{SI (USCSU)} \quad (23\text{-}130)$$

where $K_3 = 0.03 \times 10^{-6}$ (0.3), Q in m^3/s (q.p.m), and P_μ in W s (hp)

Thrust bearing

Parallel-surface thrust bearing (Figs. 23-51 to 23-52)

Refer to Table 23-8 for P and Table 23-6 for Pv values.

The pressure at any point along the bearing

$$P = \frac{6\eta UB}{h^2} K_{LP1} \qquad (23\text{-}131a)$$

where

$$K_{LP1} = \frac{x_1 - \ln[(\rho' - 1)x_1 + 1)]}{\ln \rho'}$$

$$\rho' = \frac{\rho_2}{\rho_1};\ x_1 = \frac{x}{B} \qquad (23\text{-}131b)$$

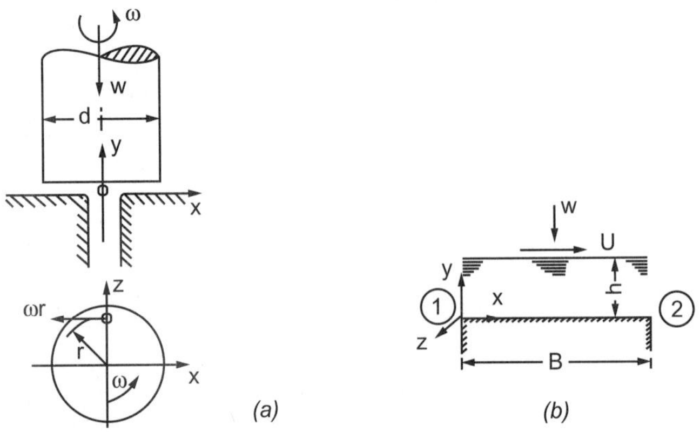

FIGURE 23-51 Parallel-surface thrust bearing.

Particular	Formula

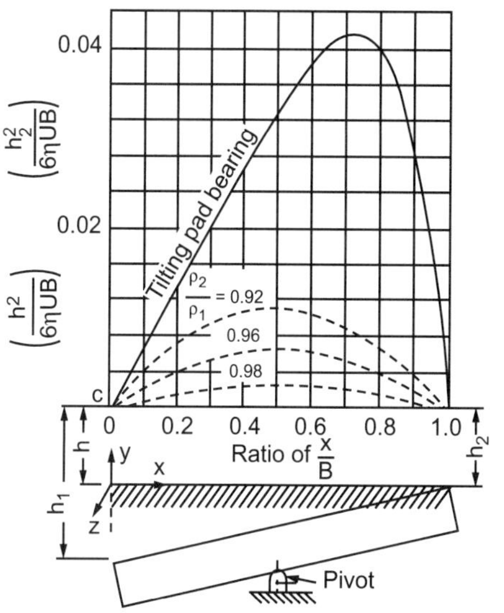

FIGURE 23-52 Comparison of pressure distribution across tilting-pad and parallel-surface thrust bearing.

The ratio of the density of the lubricant leaving the bearing to the density of the lubricant entering the bearing

$$\rho' = \frac{\rho_2}{\rho_1} = 1 + \frac{a}{\rho_1}(t_2 - t_1) \tag{23-132}$$

where a = constant, $a/\rho_1 = -0.0004$, and t_1 and t_2 are the temperatures in °C corresponding to densities ρ_1 and ρ_2, respectively

The unit load supported by a parallel-surface thrust bearing

$$P_u = \frac{6\eta UB}{h^2} K_{LP2} \tag{23-133a}$$

where

$$K_{LP2} = \frac{1}{2} + \frac{\rho'}{1 - \rho} + \frac{1}{\ln \rho'} \tag{23-133b}$$

The approximate formula for unit load supported by a parallel-surface thrust bearing

$$P_u = \frac{6\eta UB}{h^2} K_{LP3} \tag{23-134a}$$

$$\text{where } K_{LP3} = 0.09(1 - \rho') \tag{23-134b}$$

Refer to Table 23-18 for K_{LP3}.

The pressure distribution along a tilting-pad bearing of infinite width (Figs. 23-48 and 23-52)

$$P = \frac{6\eta UB}{h_1^2} K_{pt} \tag{23-135a}$$

where

$$K_{pt} = \frac{(m-1)(1-x_1)x_1}{(m+1)(m - mx_1 + x_1)^2} \tag{23-135b}$$

$m = h_1/h_2$; $x_1 = x/B$

Particular	Formula

TABLE 23-18
Comparison of load capacities of tilting-pad and parallel-surface-type of bearings

Temperature rise through bearings, °C	ρ'	K_{LP3}	K_{lt} (for $h' = 3$)	Relative load capacity, K_{LP3}/K_{lt}
10	0.98	0.0018	0.025	14
38	0.96	0.0036		7
93	0.92	0.0072		3.5

Source: F. I. Radzimovsky. *Lubrication of Bearings—Theoretical Principles and Designs*. The Ronald Press Company. New York, 1959.

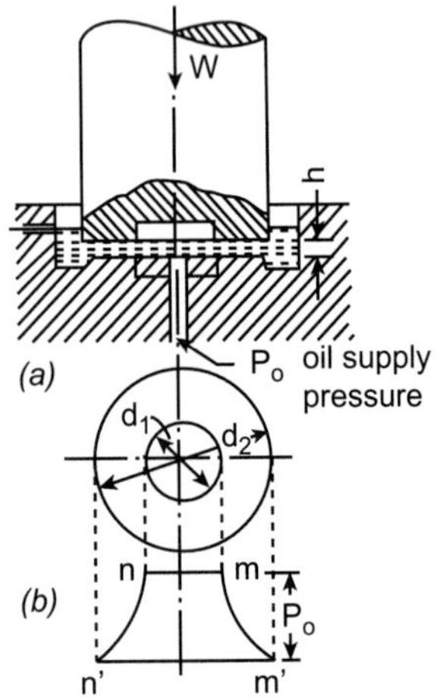

FIGURE 23-53 (a) Hydrostatic step bearing; (b) plan view and general character of pressure distribution along diameter of the bearing.

The unit load supported by a tilting-pad bearing of infinite width (Fig. 23-52)

$$P_u = \frac{6\eta UB}{h_2^2} K_{lt} \tag{23-136a}$$

where

$$K_{lt} = \frac{1}{(m-1)^2}\left(\ln m - \frac{2(m-1)}{m+1}\right) \tag{23-136b}$$

OIL FILM THICKNESS

The thickness of oil film in a parallel-surface thrust bearing

$$h = \sqrt{6K_{LP3}}\sqrt{\frac{\eta UB}{P_\mu}} \tag{23-137}$$

Refer to Table 23-18 for K_{LP3}.

The thickness of minimum oil film at location 2 (Figs. 23-48 and 23-52)

$$h_2 = \sqrt{6K_{lt}}\sqrt{\frac{\eta UB}{P_\mu}} \tag{23-138}$$

Refer to Table 23-18 for K_{lt}.

Particular	Formula
For properties of lubricant bearing materials and applications, conversion factors for viscosity, kinematic and Saybolt viscosity equivalents and conversion tables for viscosity equivalent	Refer to Tables 23-19 to 23-23.

COEFFICIENT OF FRICTION

$$\mu = \left(\frac{1.82}{1-\rho'}\right)\frac{h}{B} \quad (23\text{-}139)$$

The coefficient of friction in case of a parallel-surface thrust bearing

$$\mu = \frac{1}{\sqrt{6K_{LP3}}}\sqrt{\frac{\eta U}{P_u B}} \quad (23\text{-}140)$$

Another formula for coefficient of friction in case of a parallel-surface thrust bearing

$$\mu = K_{\mu t}\sqrt{\frac{\eta U}{P_u B}} \quad (23\text{-}141a)$$

The coefficient of friction for a tilting-pad bearing of infinite width

where

$$K_{\mu t} = \left[\frac{\left(4\ln m - 6\dfrac{m-1}{m+1}\right)^2}{6\ln m - 12\dfrac{m-1}{m+1}}\right]^{1/2}$$

$$m = h_1/h_2 = \text{film thickness ratio} \quad (23\text{-}141b)$$

HYDROSTATIC BEARING: STEP-BEARING (Fig. 23-53)

The pressure in the pocket supplied from external source to support the load

$$P_o = \frac{8W\ln(d_2/d_1)}{\pi(d_2^2 - d_1^2)} \quad (23\text{-}142a)$$

The load-carrying capacity

$$W = \frac{\pi P_o(d_2^2 - d_1^2)}{\ln\left(\dfrac{d_2}{d_1}\right)} \quad (23\text{-}143)$$

The rate of flow of lubricant through the bearing

$$Q = \frac{\pi P_o h^3}{2\eta\ln(d_2/d_1)} \quad (23\text{-}144)$$

Power loss in bearing

$$P_\mu = 0.062\,\frac{n'^2\eta}{16h}(d_2^4 - d_1^4) \quad \textbf{SI} \quad (23\text{-}145a)$$

where P_μ in kW, η in Pa s, h, d_1, and d_2 in m, and n' in rps

$$P_\mu = 8.3\times10^{-4}\,\frac{n'\eta}{16h}(d_2^4 - d_1^4)$$

Customary Metric (23-145b)

where P_μ in hp_m, η in kgf s/mm, h, d_1, and d_2 in mm, and n' in rps

TABLE 23-19
Typical properties of lubricants

Type and application	SAE no.	Density, g/cm^3, at 15.5°C	Pour point, °C	Flash point, °C	Viscosity index	Viscosity: Saybolt seconds, *S*		Centipoise, cP		kgf $s/m^2 \times 10^{-4}$		Pa s $\times 10^{-3}$		Uses
						At 38 °C	At 99°C	At 38°C	At 99°C	At 38°C	At 99°C	At 38°C	At 99°C	
Transmission gear oil	75	0.900	−23	193	121	220	50	47	7.3	47.94	7.45	47	7.3	Combination pinion reduction gear units, enclosed reduction gear sets
	80	0.934	−32	185	78	320	52	69	7.9	70.38	8.06	69	7.9	
	90	0.930	−23	232	91	1330	100	287	20.4	292.74	20.81	287	20.4	
	140	0.937	−18	260	82	3350	160	725	34	739.50	34.68	725	34	
	250	—	−15	254.5	83	5660	220	1220	47	1244.40	47.94	1220	47	
Automotive oil	10W	0.870	−26	210	102	190	46	41	6.0	41.82	6.12	41	6.0	Automobile, truck and marine reciprocating engines; very-heavy-duty oils used in diesel engines
	20W	0.885	−23	227	96	330	54	71	8.5	72.42	8.67	71	8.5	
	30	0.891	−20	238	92	530	64	114	11.3	116.28	11.53	114	11.3	
	40	0.890	−18	240.5	90	800	77	173	14.8	176.46	15.10	173	14.8	
	50	0.992	−12	254.5	90	1250	97	270	19.7	275.40	20.10	270	19.7	
	60	—	—	—	80	—	115	—	—	—	—	—	—	
	70	—	—		80	—	137	—	—	—	—	—	—	
Aircraft engine oil		0.858	−65	111	87	43	33	5	1.6	5.10	1.63	5	1.6	Turbojet engines
		0.864	−62	146	79	59	35	10	2.5	10.20	2.55	10	2.5	
		0.876	−18	215.5	106	350	57	76	9.3	77.52	9.49	76	9.3	
		0.884	−18	224	96	514	64	111	11.3	113.22	11.53	111	11.3	Various reciprocating aircraft engines
		0.887	−18	232	95	829	80	179	15.5	182.58	15.81	179	15.5	
		0.892	−18	249	95	1240	99	268	20.1	273.36	20.50	268	20.1	
		0.892	−7	318	96	1711	120	369	25.0	376.38	25.50	369	25.0	
Turbine-grade oil														
Light		0.872	−18	210	109	150	44	32	5.4	32.64	5.51	32	5.4	Direct-connected turbines electric motors
Medium		0.877	−12	235	105	300	53	65	8.2	66.30	8.36	65	8.2	Land-geared turbines electric-motors
Heavy		0.885	−12	243	100	460	62	99	10.8	100.98	11.02	99	10.8	Marine-propulsion geared turbines
Steam		0.895	14	260	101	1800	130	390	27	397.80	27.45	390	27	Railroad stationary steam engines cylindered applications, enclosure gears
Cylinder		0.910	1.5	211	107	3750	210	810	45	826.20	45.90	810	45	
Oil		0.904	15.5	343	103	6470	300	1400	64	1428.00	65.28	1400	64	

TABLE 23-19
Typical properties of lubricants (*Cont.*)

Type and application	SAE no.	Density, g/cm^3, at 15.5°C	Pour point, °C	Flash point, °C	Viscosity index	Viscosity: Saybolt seconds, *S* At 38 °C	Saybolt seconds, *S* At 99°C	Centipoise, cP At 38°C	Centipoise, cP At 99°C	kgf s/m^2 × 10^{-4} At 38°C	kgf s/m^2 × 10^{-4} At 99°C	Pa s × 10^{-3} At 38°C	Pa s × 10^{-3} At 99°C	Uses
Hydraulic oils														Hydraulic fluids for most
Light		0.887	−42	188	64	150	42	32	4.8	32.64	4.90	32	4.8	indoor industrial
Medium		0.895	−26	207	66	310	50	67	7.3	68.34	7.45	67	7.3	hydraulic equipments
Heavy		0.901	−12	257	70	910	74	196	14.0	199.92	14.28	196	14.0	Heavier loads, higher temperature
Extra-low-temperature		0.844	−24	110	226	74	43	14	5.2	14.28	5.30	14	5.2	Aircraft hydraulic systems
Refrigerating machine oil		0.895	−45.5	146	53	72	36	14	2.9	14.28	2.96	14	2.9	Ammonia compressor
		0.898	−37	165.5	22	195	43	42	5.1	42.84	5.20	42	5.1	
		0.909	−29	182	34	235	45	51	5.7	52.02	5.81	51	5.7	
		0.902	−23	190.5	35	335	49	72	7.0	73.44	7.14	72	7.0	
Machine tools and general-purpose oil		0.881	−4	177	80	105	39	22	3.9	22.44	3.98	22	3.9	All general-purpose lubrication, machine tools
		0.898	−4	199	80	205	46	44	6.0	44.88	6.12	44	6.0	
		0.915	−12	185	83	305	49	66	7.0	67.32	7.14	66	7.0	
		0.915	−15	199	25	510	59	110	9.9	112.20	10.10	110	9.9	
		0.890	−9	235	80	930	80	200	15.5	204.00	15.81	200	15.5	

TABLE 23-20
Journal bearing materials and applications

Material	Composition, %	Specific gravity	Dry coefficient of friction, μ	Ultimate tensile strength, σ_{su}		Modules of elasticity, E		Hardness numbers		Applications
				kgf/mm^2	MPa	kgf/mm^2 $\times 10^4$	GPa	Brinell	Rockwell	
Babbitts										
Lead base	Sn 10.0, Sb 15.0, Pb 75.0	9.69	0.34	7.03	68.96	0.295	28.9	45	—	Used in automobiles and electrical equipment
Tin base	Cu 8.3, Sb 8.3, Sn 83.4	7.47	0.28	7.88	77.30	0.534	52.4	27	—	Used in automotive and diesel engines, steam turbines and motors
Cadmium base	Ni 1.4, Cd 98.6	8.6	0.34	—	—	—	—	—	—	Used where lubrication is intermittent
Aluminum alloys	Cu 1.0, Sn 6.5 Ni 1.5, Al 91.0	2.86	0.33	15.5	151.76	0.724	71.0	45	—	Used in high-temperature high-load services and in diesels; requires good lubrication and hardened shaft
Copper alloys										
Clock brass	Pb 3.0, Zn 35.5, Cu 61.5	8.4	—	38.0–45.0	372.48–445.45	1.055	103.5	54–142	B 40–75	Used for light load
Bronze, high-lead	Sn 4.0, Pb 14.0, Zn 1.5, Ni 1.0 max, Cu 79.5	—	0.15	14.0	138.00	—	—	45	—	Used in poorly lubricated applications with moderately heavy loads
Bronze, high-lead	Sn 16, Pb 14.0, Cu 70.0	—	0.37	—	—	—	—	—	—	Same as above; can withstand higher loads
Bronze, lead tin	Sn 8, Pb 3.5, Zn 3.5, Cu 85	8.4	0.26	21.1 min	207.00 min			53	—	Moderately heavy duty
Bronze, 80–10–10	Sn 10.0, Pb 10.0, Cu 80	8.86	0.15	17.58 min	172.46 min	0.773	75.8	65	—	General-duty bearing bronze; load up to 20.6 MPa (2.1 kgf/mm^2); speed—4.5 m/s
Bronze, nickel tin	Sn 10.0, Ni 3.5, Pb 2.5, Cu 84.0	—	0.37	31.64	310.04	—	—	95	—	Used in medium- to heavy-duty application; good strength requirement
Bronze, aluminum	Al 10.5, Fe 3.5, Cu 86.0	7.6	0.52	70.30	689.74	1.125	110.4	202	—	Used in heavy-duty bearings requiring high strength and good impact resistance
Bronze, zinc	Al 1.0, Si 0.8, Mn 2.5, Zn 37.5, Cn 58.2	8.09	0.39	49.22 min	482.84 min	1.055	103.5		B 80–92 —	Heavy-duty impact loadings; on hardened shaft
Iron base										
Gray cast iron	C 3.5, Si 2.5, Fe 94.0	7.2	0.37	21.10 min	207.00 min	1.898	186.2	180	—	Used in refrigerators, compressors, camshafts, high load at low speed with good lubrication
Sintered iron	Cu 7.5, Fe 92.5	—	0.30	—	—	—	—	—	—	Used with impregnates with oil will give good results; load—3.4 MPa (0.35 kgf mm^2) and speed 0.67 m/s

TABLE 23-20
Journal bearing materials and applications (*Cont.*)

Material	Composition, %	Specific gravity	Dry coefficient of friction, μ	Ultimate tensile strength, σ_{su} kgf/mm^2	Ultimate tensile strength, σ_{su} MPa	Modules of elasticity, E kgf/mm^2 $\times 10^4$	Modules of elasticity, E GPa	Hardness numbers Brinell	Hardness numbers Rockwell	Applications
Graphite										
Carbon graphite	C + binder	1.63–1.86	0.15	0.53 1.80	5.17– 17.25	—	—	Shore scleroscope 75		Particularly suited to high-temperature application (<455°C) where lubrication is difficult; used in electric motors, conveyors
Carbon graphite and metal	C + Cu +binder	2.9/3.8	0.17	2.11– 4.22	20.7– 41.4	—	—	Shore scleroscope 75		Same as above, higher strength
Cemented carbide	Tungsten carbide 97.0, Co 3.0	15.1	0.20	573.00 (compressive)	5621 (compressive)	6.885 (compressive)	672.5	C 80		Used in high-speed precision grinders which require perfect alignment and good lubrication; can withstand extreme loading and high speeds
Wood										Used in conveyors; light loads at high speeds under 65°C
Plastics and rubber										
Nylon	Polyamide	1.44	0.86	7.03	68.96	0.023	2.25	M 90		Used in many household appliances and other lightly loaded applications; requires little lubrication
Rubber		0.97–2.00	0.25–0.30	1.40– 10.55	13.73– 103.50					Marine propellers, pumps, turbine, load 0.54 MPa (0.055 kgf/mm^2)
Teflon	Polytetrafluoro-ethylene	2.2	0.17	2.11	20.70	0.0042	0.410	Shore scleroscope 50		Useful in corrosive conditions; dairy, textile, and food machinery
Textolite 2001	Phenolic, graphite and cotton cloth	1.36	0.18	7.03	68.96	0.0443 0.0647	4.375– 6.35	M 100		Used where low wear and good compatibility characteristics are required

TABLE 23-21
Conversion factors for viscosity

	P	cP	kgf s/m^2	kg/m s	lbf s/ft^2	lb/ft s	Pa s
P	1	100	0.0102	0.1	2.0886×10^{-3}	0.0672	0.1
cP	0.01	1	1.0297×10^{-4}	10^{-3}	2.0886×10^{-5}	6.7197	10^{-3}
kgf s/m^2	98.0665	9.80665×10^{-3}	1	9.80665	0.20482	6.5898	9.80665
kg/m s	10	10^{-3}	0.102	1	2.0886×10^{-2}	0.6720	
lbf s/ft^2	4.788×10^2	4.788×10^4	4.8824	47.88	1	32.174	
lb/ft s	14.882	1.4882×10^3	0.1518	1.4882	0.0311	1	
Pa s	10	10^3	0.102				1

TABLE 23-22
Kinematic and Saybolt viscosity equivalents

Kinematic viscosity, ν			Saybolt viscosity, S[a]	
Metric units		SI units		
cSt	$cm^2/s \times 10^{-2}$	$m^2/s \times 10^{-6}$	At 38°C	At 99°C
2	2	2	32.6	32.9
3	3	3	36.0	36.3
4	4	4	39.1	39.4
5	5	5	42.4	42.7
6	6	6	45.6	45.9
7	7	7	48.8	49.1
8	8	8	52.1	52.5
9	9	9	55.5	55.9
10	10	10	58.9	59.3
11	11	11	62.4	62.9
12	12	12	66.0	66.5
13	13	13	69.8	70.3
14	14	14	73.6	74.1
15	15	15	77.4	77.9
16	16	16	81.3	81.3
17	17	17	85.3	85.9
18	18	18	89.4	90.1
19	19	19	93.6	94.2
20	20	20	97.8	98.5
21	21	21	102.0	102.8
23	23	23	110.7	111.4
25	25	25	119.3	120.1
27	27	27	128.1	129.0
29	29	29	136.9	137.9
30	30	30	141.3	142.3
31	31	31	145.7	146.8
33	33	33	154.7	155.8
35	35	35	163.7	164.9
37	37	37	172.7	173.9
39	39	39	181.8	183.0
40	40	40	186.3	187.6
41	41	41	190.8	192.1
43	43	43	199.8	201.2
45	45	45	209.1	210.5
47	47	47	218.2	219.8
49	49	49	227.5	229.1
50	50	50	232.1	233.8
55	55	55	255.2	257.0
60	60	60	278.3	280.2
65	65	65	301.4	303.5
70	70	70	324.4	326.7
>70			$S = \text{cSt} \times 4.635$	$S = \text{cSt} \times 4.667$

[a] $S = \text{cSt} \times 4.635$ at 38°C; $S = \text{cSt} \times 4.667$ at 99°C

TABLE 23-23
Conversion table for viscosity equivalents

Kinematic viscosity, ν			Viscosity		
Metric units		SI units			
cSt	$cm^2/s \times 10^{-2}$	$m^2/s \times 10^{-6}$	Saybolt, *S*	Engler °	Redwood no. 1
2.0	2.0	2.0	32.60	1.12	30.8
2.2	2.2	2.2	33.40	1.14	31.3
2.4	2.4	2.4	34.10	1.16	31.8
2.6	2.6	2.6	34.80	1.18	32.3
2.8	2.8	2.8	35.40	1.20	32.8
3.0	3.0	3.0	36.00	1.22	33.3
3.2	3.2	3.2	36.70	1.23	33.8
3.4	3.4	3.4	37.30	1.25	34.3
3.6	3.6	3.6	37.90	1.27	34.8
3.8	3.8	3.8	38.50	1.29	35.3
4.0	4.0	4.0	39.1	1.31	35.8
4.5	4.5	4.5	40.8	1.35	37.0
5.0	5.0	5.0	42.4	1.40	38.3
5.5	5.5	5.5	44.0	1.44	39.6
6.0	6.0	6.0	45.6	1.48	40.9
6.5	6.5	6.5	47.2	1.52	42.3
7.0	7.0	7.0	48.8	1.56	43.6
7.5	7.5	7.5	50.4	1.60	44.9
8.0	8.0	8.0	52.1	1.65	46.3
8.5	8.5	8.5	53.8	1.70	47.7
9.0	9.0	9.0	55.5	1.75	49.0
9.5	9.5	9.5	57.2	1.79	50.5
10	10	10	58.9	1.84	51.9
11	11	11	62.4	1.94	54.9
12	12	12	66.0	2.02	58.0
13	13	13	69.7	2.12	61.2
14	14	14	73.5	2.22	64.5
15	15	15	77.3	2.32	67.9
16	16	16	81.2	2.43	71.3
17	17	17	85.2	2.54	74.8
18	18	18	89.3	2.64	78.4
19	19	19	93.4	2.75	82.0
20	20	20	97.6	2.87	85.7
22	22	21	106.1	3.10	93.2
24	24	24	114.7	3.33	100.8
26	26	26	123.4	3.57	108.5
28	28	28	132.3	3.82	116.3
30	30	30	141.1	4.07	124.2
32	32	32	149.9	4.32	132.1
34	34	34	158.9	4.57	140.0
36	36	36	167.9	4.82	147.9
38	38	38	176.9	5.08	155.9

TABLE 23-23
Conversion table for viscosity equivalents (*Cont.*)

Kinematic viscosity, ν			Viscosity		
Metric units		SI units			
cSt	$cm^2/s \times 10^{-2}$	$m^2/s \times 10^{-6}$	Saybolt, *S*	Engler °	Redwood no. 1
40	40	40	186.0	5.33	164.0
42	42	42	195.0	5.59	172.0
44	44	44	204.0	5.84	180.0
46	46	46	213.0	6.10	188.0
48	48	48	222.0	6.36	196.0
50	50	50	232.0	6.62	204.0
55	55	55	255.0	7.26	225.0
60	60	60	278.0	7.90	245.0
65	65	65	301.0	8.55	265.0
70	70	70	324.0	9.21	286.0
75	75	75	347.0	9.87	306.0
80	80	80	370.0	10.53	326.0
85	85	85	393.0	11.19	346.0
90	90	90	416.0	11.85	367.0
95	95	95	439.0	12.51	387.0
100	100	100	463.0	13.16	407.0
110	110	110	509.0	14.47	448.0
120	120	120	555.0	15.80	489.0
130	130	130	602.0	17.11	529.0
140	140	140	648.0	18.43	570.0
150	150	150	694.4	19.75	611.0
160	160	160	740.0	21.05	651.0
170	170	170	787.0	22.38	692.0
180	180	180	833.0	23.70	733.0
190	190	190	879.0	25.00	774.0
200	200	200	926.0	26.32	815.0
220	220	220	1018.0	28.95	896.0
240	240	240	1111.0	31.60	978.0
260	260	260	1203.0	34.25	1059.0
280	280	280	1296.0	36.85	1140.0
300	300	300	1388.0	39.50	1222.0
320	320	320	1480.0	42.12	1303.0
340	340	340	1574.0	44.75	1385.0
360	360	360	1666.0	47.40	1465.0
380	380	380	1759.0	50.00	1546.0
400	400	400	1851.0	52.65	1628.0
500	500	500	2314.0	65.80	2036.0
600	600	600	2777.0	79.00	2443.0
700	700	700	3239.0	92.20	2850.0
800	800	800	3702.0	105.30	3258.0
900	900	900	4165.0	118.50	3668.0
1000	1000	1000	4628.0	131.60	4074.0

Particular	Formula

SPHERICAL BEARINGS (Fig. 23-54)

Equivalent bearing pressure (Fig. 23-54)

$$p = \frac{W_r^2 + 6W_a^2}{W_r BD} \quad \text{provided } W_a < W_r \tag{23-146a}$$

Maximum bearing pressure if an average bearing life of 10^5 number of oscillations is to be expected

$$p_o = \frac{W_{max}}{BD} \quad \text{for } n_l = 10^5 \tag{23-146b}$$

Bearing life (Fig. 23-54)

$$L = f\left(\frac{p_o}{p}\right)^3 \times 10^5 \tag{23-146c}$$

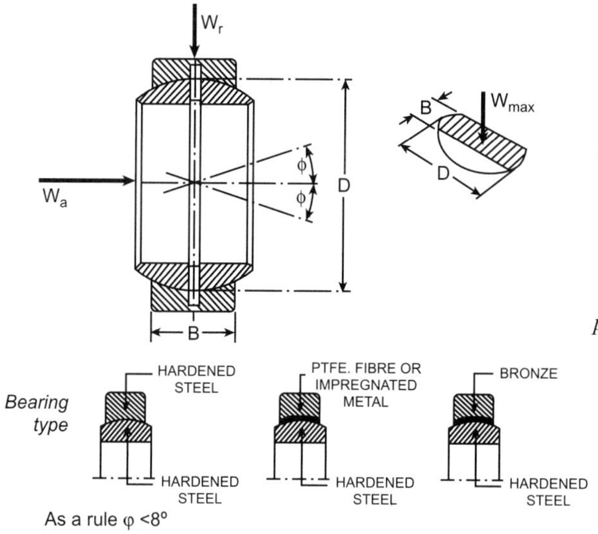

FIGURE 23-54 Spherical bearings
Courtesy: Neale, M. J., *Tribology Handbook*, Newnes and Butterworths

where

L = bearing life, i.e. average number of oscillations to failure assuming unidirectional loading

f = life-increasing factor depending on periodical re-lubrication
$\approx$ 10–15 for hardened steel on hardened steel
$\approx$ 1 for PTFE fiber or impregnated metal on hardened steel
$\approx$ 5–10 for $d > 0.05$ m bronze on hardened steel

p_o = maximum allowable bearing pressure, assuming unidirectional dynamic loading and no re-lubrication
= 24[a] MPa (3500 psi) for hardened steel on hardened steel
= 97 MPa (14000 psi) for PTFE fiber or impregnated metal on hardened steel and temperature up to 280°C (536° F)
= 10[a] MPa (1450 psi) bronze on hardened steel and temperature up to 100°C (212° F)

n_l = average number of oscillations to failure $= 10^5$

n_r = recommended interval between re-lubrication in number of oscillations
$< 0.3n_l$ for hardened steel on hardened steel
$< 0.3n_l$ (usually) for bronze on hardened steel

---

[a] The figures given above are based on dynamic load conditions. For static load conditions, where the load-carrying capacity of the bearing is based on bearing-surface permanent deformation, not fatigue, the load capacity of steel bearings may reach $10 \times p_o$ and of aluminum bronze $5 \times p_o$.
Ability to carry alternating loading is $1.7 \times p_o$ for metal contact; and is reduced by $0.25 \times p_o$ for *DTFE* fibre on hardened steel.

Particular	Formula
Load carrying capacity of spherical step bearing	$W = \dfrac{\pi P_o d_2^2(\cos\phi_1 - \cos\phi_2)}{\ln\left[\dfrac{\tan(\phi_2/2)}{\tan(\phi_1/2)}\right]}$ (23-146d)
LIFT	
Inlet pressure	$P_i = \dfrac{48\eta Q}{l\psi^3 d^2}\left[\dfrac{e(4-e^2)}{2(1-e^2)^2} + \dfrac{(2+e^2)}{(1-e^2)^{5/2}}\right]\arctan\dfrac{1+e}{\sqrt{1-e^2}}$ (23-147)
Load-carrying capacity of bearing	$W = \dfrac{24\eta Q}{\psi^3 d^2}\left[\dfrac{2+3e-e^3}{(1-e^2)^2}\right]$ (23-148)

23.2 ROLLING CONTACT BEARINGS[1]

SYMBOLS

a_1, a_2, a_3	life adjustment factors, Eq. (23-185a), (23-185b)
b	Weibull exponent
B	width of bearing, m (in)
c	permissible increase in diametral clearance, (μm)
C	basic dynamic load rating for radial and angular contact ball or radial roller bearings, kN (lbf)
C_a	basic dynamic load rating for single-row, single- and double-direction thrust ball or roller bearings, kN (lbf)
$C_{a1}, C_{a2}, \ldots, C_{an}$	basic load rating per row of a one-direction multi-row thrust ball or roller bearing, each calculated as single-row bearing with $Z_1, Z_2, \ldots, Z_n$ balls or rollers, respectively
C_n	capacity of the needle bearing, kN (lbf)
C_o	basic static load rating for radial ball or roller bearing, kN (lbf)
C_{oa}	basic static load rating for thrust ball or roller bearings, kN (lbf)
d	bearing bore diameter, m (in)
d_b	diameter of ball, m (in)
d_i	shaft or outside diameter of inner race used in Eqs. (23-246) and (23-247), m (in)
d_o	inside diameter of outer race of needle bearing, m (in)
d_r	roller diameter (mean diameter of tapered roller), m (in) diameter of needle roller, m (in)
d_1, d_2	diameter of spherical balls or cylindrical rollers used in contact stress [Eqs. (23-250) to (23-253)], m (in)
D	outside diameter of bearing, m (in)
D_1	diameter of revolving race, m (in)
D_w	diameter of ball, mm
e	bearing constant
E	modulus of elasticity, GPa (psi)
f	a factor use in Eq. (23-155)
f_a	application factor to compensate for shock continuous duty or inequality of loading
f_c	a factor which depends on the geometry of the bearing components, the accuracy to which the various bearing parts are made and the material used in Eqs. (23-187), (23-188), and (23-199) to (23-202); a factor which depends on the units used, the exact geometrical shape of the load-carrying surfaces of the roller and rings (or washers in case of thrust bearing), and the accuracy to which the various bearing parts are made and the material, used in Eqs. (23-207), (23-208)
f_d	a factor for the additional forces emanating from the mechanisms coupled to the gearing used in Eq. (23-154)
f_k	a factor for the additional forces created in the gearing itself used in Eq. (23-154)
f_L	index of dynamic stressing
f_n	speed factor for ball bearings according to Table 23-37 speed factor for roller bearings according to Table 23-38
f_s	index of static bearing
f_{nt}	speed factor used in tapered roller bearing
f_o	a factor used in Eqs. (23-161) and (23-167)
f_{oa}	a factor used in Eqs. (23-152) and (23-154)
F	load, kN (lbf)

	theoretical tooth load, kN (lbf)
F_a	thrust load, kN (lbf)
F_{aa}	applied thrust load, kN (lbf)
F_{ar}	thrust component of pure radial load F, due to tapered roller, kN (lbf)
F_{bs}	shaft load due to belt drive, kN (lbf)
F_c	static load, kN (lbf)
F_e	radial equivalent load from combination of radial and thrust loads or effective radial load, kN (lbf)
F_{effg}	effective tooth load, kN (lbf)
F_{na}	net thrust load, kN (lbf)
F_{nt}	net thrust load on the tapered roller bearing, kN (lbf)
F_r	radial load capacity of ball bearing, kN (lbf) radial bearing load, kN (lbf)
i	number of rows of balls in any one bearing
k	constant used in Eqs. (23-156), (23-158) to (23-160)
K_a	application factor, Eq. (23-186)
K_h	hardness factor used in Eq. (23-247)
K_t	life load factor taken from the curve in Fig. 23-55 marked "T-needle" and used in Eq. (23-247)
K_n	a constant used in Eq. (23-152) and Eq. (23-153)
l	length of needle bearing, m (in)
l_{eff}	the effective length of contact between one roller and that ring (or that washer in case of thrust bearing) where the contact is the shortest (overall roller length minus roller chamfers or minus grinding undercuts), m (in)
L	life of bearing at constant speed, rpm life of bearing at constant speed, h life corresponding to desired reliability, R, used in Eq. (23-194)
L_{B10}	life factor corresponding to desired B-10 hours of life expectancy used in Eq. (23-195)
L_{10}	rating life
L_h	fatigue life
M_t	torque, N m (lbf in)
n	speed, rpm
n'	speed, rps
n_e	effective speed, rpm
n_i	ith speed, rpm
n_l	limiting speed, rpm
n_m	mean speed, rpm
n_1	speed of the inner race, rpm
n_2	speed of the outer race, rpm
P	power, kW (hp)
P	equivalent dynamic load, kN (lbf)
P_a	equivalent dynamic thrust load, kN (lbf)
P_m	mean load, kN (lbf)
P_{max}	maximum load, kN (lbf)
P_{min}	minimum load, kN (lbf)
P_o	static equivalent load, kN (lbf)
P_{oa}	static equivalent load for thrust ball or roller bearings under combined radial and thrust loads, kN (lbf)
q_i	percentage time of ith speed
R_{10}	0.90 reliability corresponding to rating life
X	radial factor used in Eqs. (23-177b), (23-182), (23-190), (23-210), and (23-180)

X_o	radial factor used in Eqs. (23-162), (23-165), (23-173c) and (23-157) Tables (23-37), (23-38), (23-39)
Y	thrust factor used in Eqs. (23-163), (23-166), (23-173), (23-178), and (23-180)
Y_o	thrust factor used in Eqs. (23-162), (23-165), and (23-157)
Z	number of balls per row number of balls carrying thrust in one direction number of rollers per row number of rollers carrying thrust in single-row one-direction bearing number of needle-rollers
$Z_1, Z_2, \ldots, Z_n$	number of balls or rollers in respective rows of one-direction multi-row bearings
α	nominal angle of contact, that is, nominal angle between the line of action of the ball load and a plane perpendicular to the bearing axis the angle of contact, that is, the angle between the line of action of the roller resultant load, and a plane perpendicular to the bearing axis
ω	angular speed, rad/s
μ	coefficient of friction
ν	Poisson's ratio
$\sigma_{c(\max)}$	maximum compressive stress, MPa (psi)
$\tau_{\max}$	maximum shear stress, MPa (psi)

Particular	Formula
The torque	$$M_t = \frac{9550P}{n} \quad \textbf{SI} \quad (23\text{-}149a)$$ where P in kW, n in rpm, and M_t in N m
	$$M_t = \frac{1000P}{\omega} \quad \textbf{SI} \quad (23\text{-}149b)$$ where P in kW, ω in rad/s, and M_t in N m
	$$M_t = \frac{159.2P}{n'} \quad \textbf{SI} \quad (23\text{-}149c)$$ where P in kW, n' in rps, and M_t in N m
	$$M_t = \frac{63{,}000P}{n} \quad \textbf{USCS} \quad (23\text{-}149d)$$ where P in hp, n in rpm, and M_t in lbf in
	$$M_t = \frac{716P}{n} \quad \textbf{Customary Metric} \quad (23\text{-}149e)$$ where P in hp_m, n in rpm, and M_t in kgf m
	$$M_t = \frac{937P}{n} \quad \textbf{Customary Metric} \quad (23\text{-}149f)$$ where P in kW, n in rpm, and M_t in kgf m

Particular	Formula

TABLE 23-32
Coefficient of friction μ for rolling contact bearings

Type of bearing	μ
Self-aligning bearings	0.0016–0.0066
Cylindrical roller bearings	0.0012–0.0060
Thrust ball bearings	0.0013
Angular contact ball bearings	0.0018–0.0019
Deep groove ball bearings	0.0022–0.0042
Tapered roller bearings	0.0025–0.0083
Spherical roller bearings	0.0029–0.0071
Needle bearings	0.0045

TABLE 23-33
Safe working values of k for average bearing life

Material	$k \times 10^6$	USCSU*
For un-hardened steel	3.80	550
For hardened alloy steel on flat races	6.89	1000
For hardened carbon steel	4.80	700
For hardened carbon steel on grooved races	10.34	1500
For hardened alloy steel grooved races (having radius = 0.67 d_o)	13.79	2000

* US customary system units.

The equation for friction torque

$$M_t = \frac{\mu F_r d}{2} \qquad \text{(23-149g)}$$

For values of μ, refer to Table 23-32.

ROLLING ELEMENTS BEARINGS

Definition, Dimensions and Nomenclature

For nomenclature, other details and definition of a ball bearing

Refer to Fig. 23-49.

For nomenclature, other details and definition of a taper roller bearing

Refer to Fig. 23-50.

A rule of thumb used for ordinary ball and straight roller bearings

$$\left(\frac{d+D}{2}\right)n' \leq 8.33 \qquad \text{(23-150a)}$$

where d, D in m and n' in rps

$$\left(\frac{d+D}{2}\right)n \leq 500{,}000 \qquad \text{(23-150b)}$$

where d, D in mm and n in rpm

SPEED

Effective speed

The effective speed which determines the life of the bearing is found from the relation

$$n = n_1 \pm n_2 \qquad \text{(23-151)}$$

where the plus sign is used when the races rotate in opposite directions and the minus sign is used when the races rotate in the same direction

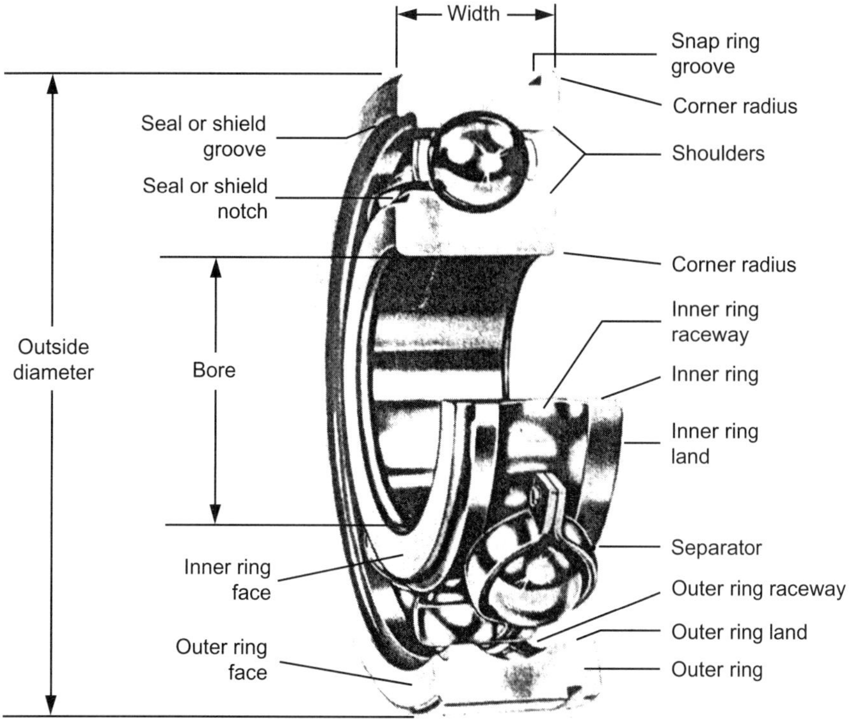

FIGURE 23-49 Ball bearing nomenclature. (*Courtesy*: New Departure-Hyatt Bearing Division, General Motors Corporation.)

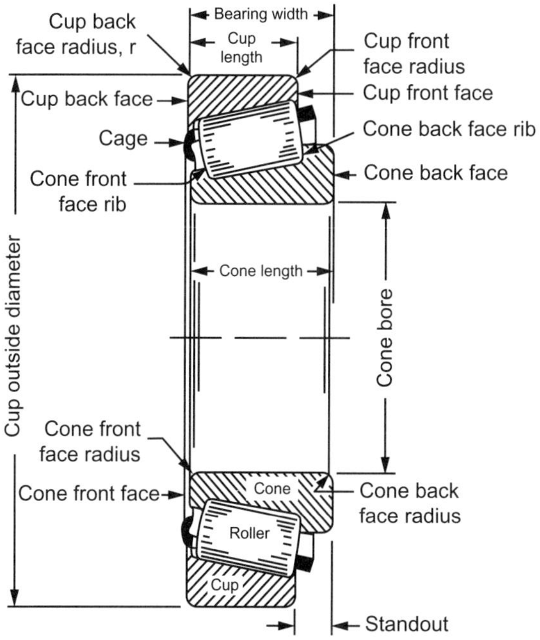

FIGURE 23-50 Nomenclature of tapered roller bearing.

Particular	Formula

Limiting bearing speed

The limiting bearing speed when the bearing outside diameter is less than 30 mm

$$n_l = \frac{3K_n}{D+30} \qquad (23\text{-}152)$$

The limiting bearing speed when the bearing outside diameter is 30 mm and over

$$n_l = \frac{K_n}{D-10} \qquad (23\text{-}153)$$

For values of K_n refer to Table 23-34.

GEAR-TOOTH LOAD

The effective tooth load which is used in design of bearings

$$F_{effg} = f_k f_d F \qquad (23\text{-}154)$$

For values of f_k and f_d refer to Table 23-35.

The shaft load due to belt drive which is used in design of bearings

$$F_{bs} = fF \qquad (23\text{-}155)$$

For values of f refer to Table 23-35.

TABLE 23-34
Values of K_n to be used in Eqs. (23-152) and (23-153)

	Constant, K_n	
Type of bearing	**Grease lubrication**	**Oil lubrication**
Radial bearings		
Deep groove bearings		
Single row	500,000	630,000
Single row with leeds	360,000	—
Double row	320,000	400,000
Magneto bearings	500,000	630,000
Angular contact ball bearing		
Single row	500,000	630,000
Single row paired	400,000	500,000
Double row	360,000	450,000
Self-aligning ball bearings	500,000	630,000
Self-aligning bail bearings with extended inner ring	250,000	320,000
Cylindrical roller bearing		
Single row	500,000	630,000
Double row	500,000	630,000
Tapered roller bearings	320,000	400,000
Barrel roller bearings	220,000	280,000
Spherical roller bearings Series 213	220,000	280,000
Thrust bearings		
Thrust ball bearings	140,000	200,000
Angular contact thrust ball bearings	220,000	320,000
Cylindrical roller thrust bearings	90,000	120,000
Spherical roller thrust bearings	140,000	200,000

Particular	Formula

TABLE 23-35
Value of factors f_k, f_d and f to be used in Eqs. (23-154) and (23-155)

Particular	Tooth load f_k	Tooth load f_d	Shaft load f
Gear drive			
Precision gears (errors in pitch and form less than 0.025 mm)	1.05–1.1		
Commercial gears (errors in pitch and form 0.025 mm to 0.125 mm)	1.1–1.3		
Prime movers and driven machines			
Shock-free rotary machines e.g. electrical machines and turbo-compressors		1.0–1.2	
Reciprocating engines, according to the degree of balance		1.2–1.5	
Machinery subjected to heavy shock loading, such as rolling mills		1.5–3.0	
Belt drive			
Vee-belts			2.0–2.5
Single leather belts with jockey pulleys			2.5–3.0
Single leather belts, balata belts, rubber belts			4.0–5.0

Full ball bearing sizes of different bearing series	Refer to Fig. 23-51
For summary of types and characteristics of rolling contact bearing[9]	Refer to Fig. 23-52

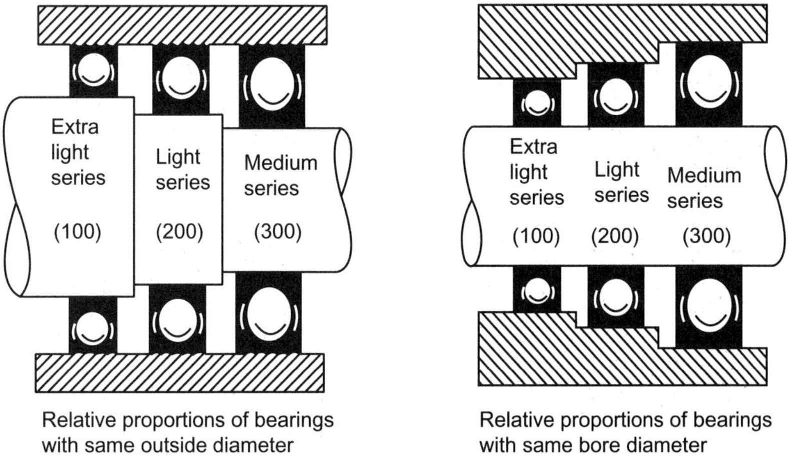

FIGURE 23-51 Ball bearing size of bearing series. (*Courtesy*: New Departure-Hyatt Bearing Division, General Motors Corporation.)

Particular	Formula

STATIC LOADING

Stribeck equation for permissible static load

$$F_c = kd_b^2 \tag{23-156}$$

where
$k = 686.5 \times 10^6$ (100)* for carbon steel balls
$= 862 \times 10^6$ (125)* for hardened alloy steel balls
Use a factor of safety of 10
F_c in N (lbf) and d_b in m (in)

Stribeck equation for permissible static load for ball bearing

$$F_c = \frac{4.37F_r}{Z} \tag{23-157}$$

The radial load capacity of ball bearing

$$F_r = \frac{kZd_b^2}{4.37} \tag{23-158}$$

where d_b in m and F_r, F_c in kN
Refer to Table 23-33 for values of k

$$F_r = \frac{kZd_b^2}{5} \tag{23-159}$$

Refer to Table 23-33 for values of k

Radial load capacity of roller bearing

$$F_r = \frac{kZld_r}{5} \tag{23-160}$$

where
$k = 48.3 \times 10^6$ (7.0)* for hardened carbon steel
$= 69 \times 10^6$ (10.0)* for hardened alloy steel
$= 690 \times 10^6$ (100)* for carbon steel balls
l, d_r in m (in) and F_r in N (lbf)

BASIC STATIC LOAD RATING AS PER INDIAN STANDARDS

Radial ball bearing

The basic static radial load rating for radial ball bearing

$$C_{or} = f_o iZD_w^2 \cos\alpha \tag{23-161}$$

where C_{or} in N and D_w in mm

Refer to Table 23-36 for f_o

This formula is applicable to bearings with a cross-sectional raceway groove radius not larger than $0.52D_w$ in radial and angular contact groove ball bearing inner rings and $0.53D_w$ in radial and angular contact groove ball bearing outer rings and self-aligning ball bearing inner rings.

The static equivalent load for radial ball bearings is greater of the two values given by the formulae

$$P_{or} = X_o F_r + Y_o F_a \tag{23-162}$$

$$P_{or} = F_r \tag{23-163}$$

For values of X_o and Y_o refer to Table 23-37.

* Values outside the parentheses are in SI units in Pd and inside the parentheses are in US customary system units in kpsi.

Features \ Bearing Type	Deep Groove Ball Bearings	Magneto Bearings	Angular Contact Ball Bearings	Double Row Angular Contact Ball Bearings	Duplex Angular Contact Ball Bearings	Self Aligning Ball Bearings	Cylindrical Roller Bearings	Double Row Cylindrical Roller Bearings	Cylindrical Roller Bearings	Cylindrical Roller Bearings with Angle Ring
Load Carring Capacity: Radial Load	Fair	Poor	Good	Good	Good	Fair	Good	Excellent	Good	Good
Load Carring Capacity: Axial Load	↔ Fair	← Poor	← Good	↔ Good	↔ Good	↔ Poor	×	×	← Fair	↔ Fair
Load Carring Capacity: Combined Load	Fair	Poor	Good	Good	Good	Poor	×	×	Fair	Fair
High Speeds	Excellent	Good	Excellent	Fair	Good	Good	Excellent	Good	Good	Good
High Accuracy	Excellent		Excellent		Excellent		Excellent	Excellent		
Low Noise and Torque	Excellent						Good			
Rigidity					Good		Good	Excellent	Good	Good
Angular Misalignment	Good	Poor	Poor	Poor	Poor	Excellent	Fair	Poor	Fair	Fair
Self-Aligning Capability						Yes				
Ring Separability		Yes					Yes	Yes	Yes	Yes
For Use on Fixed-end	Yes			Yes	Yes	Yes				Yes
For Use on Free-end	Yes*			Yes*	Yes*	Yes*	Yes*	Yes*		
Tapered Bore of Inner Ring						Yes*		Yes*		
Remarks		Two bearings are usually mounted in opposition.	Contact angle 15 , 30 , 40 , Two bearings are usually mounted in opposition Clearance adjustment is necessary.		Any arrangement of the pair is possible.					

Features \ Bearing Type	Needle Roller Bearings	Tapered Roller Bearings	Double Row Tapered Roller Bearings	Spherical Roller Bearings	Thrust Ball Bearings	Thrust Ball Bearings with Aligning Seats	Angular Contact Thrust Ball Bearings	Cylindrical Roller Thrust Bearings	Tapered Roller Thrust Bearings	Spherical Roller Thrust Bearings
Load Carring Capacity: Radial Load	Good	Good	Excellent	Excellent	×	×	×	×	×	Poor
Load Carring Capacity: Axial Load	×	← Good	↔ Good	↔ Fair	← Good	← Good	↔ Good	← Excellent	← Excellent	← Excellent
Load Carring Capacity: Combined Load	×	Good	Excellent	Good	×	×	×	×	×	Poor
High Speeds	Good	Fair	Fair	Fair	×	×	Fair	Poor	Poor	Poor
High Accuracy		Good			Good		Excellent			
Low Noise and Torque										
Rigidity	Good	Good	Excellent				Good	Excellent	Excellent	
Angular Misalignment	Poor	Fair	Poor	Excellent	×	Excellent	×	×	×	Excellent
Self-Aligning Capability				Yes		Yes				Yes
Ring Separability	Yes	Yes	Yes		Yes	Yes	Yes	Yes	Yes	Yes
For Use on Fixed-end			Yes	Yes						
For Use on Free-end	Yes		Yes*	Yes*						
Tapered Bore of Inner Ring				Yes						
Remarks		Two bearings are usually mounted in opposition. Clearance adjustment is necessary.						Needle roller thrust bearings are available.		This type should have oil lubrication.

Legend: Excellent; Good; Fair; Poor; × Impossible
← One direction only ↔ Two directions

* Can be used as free-end bearings if tap fit allows axial motion.

FIGURE 23-52 Types and characteristics of rolling bearings. (*Courtesy*: NSK Corp.)

Particular	Formula

Radial roller bearing

The basic static radial load rating for radial roller bearings	$C_{or} = 44\left(1 - \frac{D_{we}}{D_{pw}} \cos\alpha\right) iZL_{we}D_{we}\cos\alpha$ (23-164)
The static equivalent radial load for roller bearings with $\alpha \neq 0°$ is the greater of the two values given by the formulae	$P_{or} = X_o F_r + Y_o F_o$ (23-165) For factors X_o and Y_o refer to Table 23-38.
The static equivalent radial load for radial roller bearings with $\alpha = 0°$ and subjected to radial load only, is given by the formula	$P_{or} = F_r$ (23-166)

THRUST BEARINGS

Ball bearings

The basic static axial load rating for single- or double-direction thrust ball bearings	$C_{oa} = f_o Z D_w^2 \sin\alpha$ (23-167) where f_o values are taken from Table 23-36 Z = number of balls carrying load in one direction
The static equivalent axial load P_{oa} for thrust ball bearing with contact angle $\alpha \neq 90°$	$P_{oa} = F_a + 2.3F_r \tan\alpha$ (23-168) This formula is valid for all ratios of radial load to axial load in the case of double-direction bearings. For single direction bearings it is valid where $(F_r/F_a) \leq 0.44\cot\alpha$ and gives satisfactory but less conservative values of P_{oa} for (F_r/F_a) up to $0.67\cot\alpha$.
The static equivalent axial load for thrust ball bearings with $\alpha = 90°$ is given by the equation	$P_{oa} = F_a \quad \text{for } \alpha = 90°$ (23-169)

Roller bearings

The basic static axial load rating for single- and double-direction thrust roller bearings	$C_{oa} = 220\left(1 - \frac{D_{we}\cos\alpha}{D_{PW}}\right) ZL_{we}D_{we}\sin\alpha$ (23-170) where Z = number of rollers carrying load in one direction
The static equivalent axial load for thrust roller bearings with contact angle $\alpha \neq 90°$	$P_{oa} = F_a + 2.3F_r \tan\alpha$ (23-171a) This formula is valid for all ratios of radial load to axial load in the case of double-direction bearings. For single-direction bearings, it is valid where $(F_r/F_a) \leq 0.44\cos\alpha$ and gives satisfactory but less conservative values of P_{oa} for (F_r/F_a) up to $0.67\cot\alpha$.

TABLE 23-36
Values of factor f_o for radial ball bearings[a]

$\frac{D_w \cos\alpha}{D_{pw}}$	Factor f_o		
	Radial ball bearings		
	Radial and angular contact groove ball bearings	Self-aligning ball bearings	Thrust ball bearings
0	14.7	1.9	61.6
0.01	14.9	2	60.8
0.02	15.1	2	59.9
0.03	15.3	2.1	59.1
0.04	15.3	2.1	58.3
0.05	15.7	2.1	57.5
0.06	15.9	2.2	56.7
0.07	16.1	2.2	55.9
0.08	16.3	2.3	55.1
0.09	16 5	2.3	54.3
0.10	16.4	2.4	53.5
0.11	16.1	2.4	52.7
0.12	15.9	2.4	51.9
0.13	15.6	2.5	51.2
0.14	15 4	2.5	50.4
0.15	15.2	2.6	49.6
0.16	14.9	2.6	48.8
0.17	14.7	2.7	48.0
0.18	14.4	2.7	47.3
0.19	14.2	2.8	46.5
0.20	14.0	2.8	45.7
0.21	13.7	2.8	45.0
0.22	13.5	2.9	44.2
0.23	13.2	2.9	43.5
0.24	13.0	3.0	42.7
0.25	12.8	3.0	41.9
0.26	12.5	3.1	41.2
0.27	12.3	3.1	40.5
0.28	12.1	3.2	39.7
0.29	11.8	3.2	39.0
0.30	11.6	3.3	38.2
0.31	11.6	3.3	37.5
0.32	11.2	3.4	36.8
0.33	10.9	3.4	36.0
0.34	10.7	3.5	35.3
0.35	10.5	3.5	34.6
0.36	10.3	3.6	
0.37	10.0	3.6	
0.38	9.8	3.7	
0.39	9.5	3.8	
0.40	9.4	3.8	

[a] The Table 23-36 is based on the Hertz's point contact formula with a modulus of elasticity $(E) = 2.07 \times 10^5$ mPa and a Poisson's ratio (ν) of 0.3. It is assumed that the load distribution for radial ball bearings results in a maximum ball load of $(5F_o/Z\cos\alpha)$, and for thrust ball bearings $(F_o/Z\sin\alpha)$. Values of f_o for intermediate values of $(D_w\cos\alpha/D_{pw})$ are obtained by linear interpolation. (IS: 3823-1988, ISO: 76-1987)

Particular	Formula

TABLE 23-37
Values of factors X_o and Y_o for radial ball bearings[a] for use in Eqs. (23-162)

Bearing type		Single row bearings		Double row bearings	
		X_o	Y_o	X_o	Y_o
Radial contact groove ball bearings		0.6	0.5	0.6	0.5
	$\alpha = 15°$	0.5	0.46	1	0.92
	$\alpha = 20°$	0.5	0.42	1	0.84
	$\alpha = 25°$	0.5	0.38	1	0.76
Angular-contact groove ball bearings	$\alpha = 30°$	0.5	0.33	1	0.66
	$\alpha = 35°$	0.5	0.29	1	0.58
	$\alpha = 40°$	0.5	0.26	1	0.52
	$\alpha = 45°$	0.5	0.22	1	0.44
Self-aligning ball bearings	$\alpha \neq 90°$	0.5	$0.22 \cot \alpha$	1	$0.44 \cot \alpha$

[a] Permissible value of F_a/C_{or} depends on bearing design (internal clearance and raceway groove depth)

TABLE 23-38
Values of factors X_o and Y_o for radial roller bearings with $\alpha \neq 0°$ for use in Eq. (23-165)

Bearing type	X_o	Y_o
Single-row	0.5	$0.22 \cot \alpha$
Double-row	1.0	$0.44 \cot \alpha$

The static equivalent axial load for thrust roller bearings with $\alpha = 90°$ is given by the equation

$$P_{oa} = F_a \tag{23-171b}$$

CATALOGUE INFORMATION FROM *FAG* FOR THE SELECTION OF BEARING

The basic static load rating

$$C_o = f_s P_o \tag{23-172}$$

where

f_s = index of static dressing
= 1.5 to 2.5 for high degree
= 1.0 to 1.5 for normal degree
= 0.7 to 1.0 for moderate degree

The equivalent static load

$$P_o = F_r \quad \text{for } (F_a/F_r) \leq 0.8 \tag{23-173a}$$

Particular	Formula
	$P_o = 0.6F_r + 0.5F_a$ for $(F_a/F_r) > 0.8$ (23-173b)
	$P_o = X_o F_r + Y_o F_a$ (23-173c)
	where C_o and P_o in kN
	For various values of factors X_o and Y_o refer to Table 23-39 and Tables from *FAG* catalogue.

TABLE 23-39
Calculation of equivalent static and dynamic load

Bearing type	Series			Equivalent load		For dimensions, C, C_a, n_{max}, X, e Y, Y_o refer to Table
	IS	FAG	SKF	Static, P_o	Dynamic, P	
Deep groove ball bearings	02	62	62	F_r when $F_a/F_r \le 0.8$	F_r when $F_a/F_r \le e$	23-60
	03	63	63	$0.6F_r + 0.5F_a$ when	$0.56F_r + YF_a$ when	23-61
	04	64	64	$F_a/F_r > 0.8$	$F_a/F_r > e$	23-62
Self aligning ball bearings	02	12	12			23-63
	03	13	13	$F_r + Y_o F_a$	$XF_r + YF_a$	23-64
		22	22			23-65
		23	23			23-66
Single row angular contact ball bearings	02	72B	72B	F_r when $F_a/F_r \le 1.9$	F_r when $F_a/F_r \le 1.4$	23-67
	03	73B	73B	$0.50F_r + 0.26F_a$	$0.35F_r + 0.57F_a$	23-68
				when $F_a/F_r > 1.9$	when $F_a/F_r > 1.14$	
Double row angular contact ball bearings		33	33A	$F_r + 0.58F_a$	$F_r + 0.66F_a$ when	23-69
					$F_a/F_r \le 0.956$	
					$0.6F_r + 1.07F_a$ when	
					$F_a/F_r > 0.95$	
Cylindrical roller bearings	02	N2	N2			23-70
	03	N3	N3			23-71
	04	N4	N4	F_r	F_r	23-72
		NU22	NU22			23-73
		NU23	NU23			23-74
Tapered roller bearings		322A	322	F_r	F_r	
	02	22		when $F_a/F_r \le \frac{1}{2}Y_o$	when $F_a/F_r \le e$	23-75
	03	23		$0.5F_r + Y_o F_a$ when	$0.4F_r + YF_a$ when	23-76
				$F_a/F_r > \frac{1}{2}Y_o$	$F_a/F_r > e$	23-76B
						23-77
Single thrust ball bearings	11	511	511			23-78
	12	512	512	F_a	F_a	23-79
	13	513	513			23-80
	14	514	514			23-81
Double thrust ball bearings		522	522			23-82

Particular	Formula

DYNAMIC LOAD RATING OF BEARINGS

The relation between two groups of identical bearings tested under different loads F_1 and F_2, and length of lives L_1 and L_2 respectively as per experiments conducted by Palmgern

$$\frac{L_1}{L_2} = \left(\frac{F_2}{F_1}\right)^m \qquad (23\text{-}174)$$

where

$m = 3$ generally accepted
$= 3.333$ used by Timken
$= 4$ used by New Departures
$L =$ life in millions of revolutions
$=$ life in hours at constant speed in rpm

For various typical values of bearing life for various applications

Refer to Tables 23-40 and 23-41

LIFE

The Antifriction Bearing Manufacturers Association (AFBMA) statistically related formula for the rating life of bearing in millions of revolutions of a bearing subjected to any other load F, which is derived from Eq. (23-174)

$$L = \left(\frac{C}{F}\right)^m \qquad (23\text{-}175)$$

For values of C for various types of bearings, refer to Table 23-42.

Equation (23-175) can be written as

$$C = FL^{1/m} \qquad (23\text{-}176)$$

where L in millions of revolutions

(The International Organisation for Standardisation (ISO) defines the rating life of a group of apparently identical rolling elements bearings as that completed or exceeded by 90% of that group before first evidence of fatigue)

BASIC DYNAMIC LOAD RATING AS PER *FAG* CATALOGUE[a]

The load and life of bearings are related statistically as per ISO equation for basic rating life

$$L_{10} = L = \left(\frac{C}{F}\right)^m \quad \text{or} \quad \frac{C}{F} = L_{10}^{1/m} \qquad (23\text{-}177a)$$

where

$L_{10} =$ rating life in millions of revolutions (i.e. number of revolutions resulting in 10% failure)
$C =$ basic dynamic load rating, N (obtained from manufacturer's catalogue)
$F =$ equivalent dynamic load (also with suffix e, i.e. F_e), N (also symbol P is used in place of F)

[a] *Note*: The designer is advised to read carefully the manufacturer's Catalogue, which explains how load rating and life are established.

TABLE 23-40
Typical values of bearing life for various applications

Application	Design life, h
Agricultural equipment	3000–6000
Aircraft engines	500–1500
Automobile applications	
Race car	500–800
Light motor cycle	600–1200
Heavy motor cycle	1000–2000
Light car	1000–2000
Heavy car	1500–2500
Light truck	1500–2500
Heavy truck	2000–2500
Bus	2000–5000
Boat gearing units	3000–5000
Beater mills	20000–30000
Briquette presses	20000–30000
Domestic appliances	1000–2000
Electrical motors (up to 0.5 kW)	1000–2000
Electrical motors (up to 4 kW)	8000–10000
Electrical motors, medium	10000–15000
Electrical motors, large	20000–30000
Elevator cables sheaves	40000–60000
Small fans	2000–4000
Mine ventilation fans	40000–50000
Gearing units	
Automotive	600–5000
Multi-purpose	8000–15000
Machine tool	20000
Ship	20000–30000
Rail vehicles	15000–25000
Heavy rolling mill	50000 and more
Grinding spindles	1000–2000
Locomotive axle boxes, outer bearings	20000–25000
Locomotive axle boxes, inner bearings	30000–40000
Machine tools	10000–30000
Mining machinery	4000–15000
Paper machines	50000–80000
Rail vehicle axle boxes	
Mining cars	5000
Motor rail cars	16000–20000
Open pit mining cars	20000–25000
Street cars	20000–25000
Passenger cars	26000
Freight cars	35000
Rolling Mills	
Small cold mills	5000–6000
Large multipurpose mills	8000–10000
Gear drives	50000 and more
Ship gear drives	20000–30000
Propeller thrust bearings	15000–25000
Propeller shaft bearings	80000 and more

TABLE 23-41
Life of bearings, L_h

Class of machinery	Life, kh
Instruments and apparatus which are used occasionally	0.5
Aircraft engines	0.5–2
Machines used for period where stoppage of service is of minor importance	4–8
Machine working intermittently whose service is essential	8–14
Machine working for 8 h per day whose service is not fully utilized	14–20
Machine working for 8 h per day whose service is fully utilized	20–30
Machines working continuously for 24 h	50–60
Machines working continuously for 24 h with high reliability	100–200

TABLE 23-42
Values of C for various types of bearings

	Double-row self-aligning			Single-row deep groove			Double-row deep groove			Single-row angular contact		
SAE No.	Light 200 N	Medium 300 N	Heavy 400 N	Light 200 N	Medium 300 N	Heavy 400 N	Light 200 N	Medium 300 N	Heavy 400 N	Light 200 N	Medium 300 N	Heavy 400 N
00	3842			3332	6488		5419					
01	4067	6664		5067	8007		8271					
02	5557	7242		5557	8712		9075					
03	6323	9428		7115	10596		11554	82104				
04	8134	9957		9604	14181		15778	20002			12005	
05	9957	14445		10309	16229	28008	16895	26676		10133	16895	27558
06	13800	18002	27117	14445	21335	33040	23559	34672	59564	14220	21335	22673
07	15010	21334	31340	19110	25343	42288	31566	40004	62230	18894	25343	40004
08	18894	27296	35880	21830	30890	48902	35564	49794	78233	22367	30890	47118
09	21119	34006	44629	24451	40004	58682	40004	62495	97794	25343	40004	57340
10	22367	38455	48341	26009	46227	64896	42895	75568	122245	26676	46227	63563
11	27557	46227	56448	32674	53341	76910	52459	85789	122245	32673	53341	73784
12	31115	58341	68012	39122	60901	84456	63563	97794	135583	39122	60907	81350
13	33006	56448	72451	42895	68012	92463	69345	111132		44423	68012	88906
14	37788	68012	88906	46227	75568	115671	75568	122245		48902	75568	113346
15	41336	72451	101352	48902	82683	124470	80071	135583		50676	84456	120221
16	43561	80017	113346	54223	92463		87122	147921		56448	92463	
17	52459	88909	122245	60907	99568		90679	147921		63563	101528	
18	57339	101528	137798	64896	108907		103135	157809		73784	108707	
19		113346		78233	117796		126694			78233	117796	
20	72451	122245		87122	135583		142247			87122	144472	
21		137798		96011	144472						153369	
22	90679	148921		106683	166698					106683	173362	
24										122245		
26										137798	212596	
28										157809		
30										173362	260043	
32										184471		
34										191139		

TABLE 23-42a
Loading ratio C/P for different for ball bearings

Life, L_h	Speed, rpm																									
hours	10	25	40	100	125	160	200	250	320	400	500	630	800	1000	1250	1600	2000	2500	3200	4000	5000	6200	8000	10000	12500	16000
100							1.06	1.15	1.24	1.34	1.45	1.56	1.68	1.82	1.96	2.12	2.29	2.47	2.67	2.88	3.11	3.36	3.63	3.91	4.23	4.56
500			1.06	1.45	1.56	1.68	2.82	1.96	2.12	2.29	2.47	2.67	2.88	3.11	3.36	3.63	3.91	4.23	4.56	4.93	5.32	5.75	6.20	6.70	7.23	7.81
1,000		1.15	1.34	1.82	1.96	1.12	2.29	2.47	2.67	2.88	3.11	3.36	3.63	3.91	4.23	4.56	4.93	5.32	5.75	6.20	6.70	7.32	7.81	9.43	9.11	9.83
1,250		1.24	1.45	1.96	2.12	2.29	2.47	2.67	2.88	3.11	3.36	3.63	3.91	4.23	4.56	4.93	5.32	5.75	6.20	6.70	7.23	7.81	8.43	9.11	9.83	10.6
1,600		1.34	1.56	2.12	2.29	2.47	2.67	2.88	3.11	3.36	3.63	3.91	4.23	4.56	4.93	5.32	5.75	6.20	6.70	7.23	7.81	8.43	9.11	9.83	10.6	11.5
2,000	1.06	1.45	1.68	2.29	2.47	2.67	2.88	3.11	3.36	3.63	3.91	4.23	4.56	4.93	5.32	5.75	6.20	6.70	7.23	7.81	8.43	9.11	9.83	10.6	11.5	12.4
2,500	1.15	1.56	1.82	2.47	2.67	2.88	3.11	3.36	3.63	3.91	4.23	4.56	4.93	5.32	5.75	6.20	6.70	7.23	7.81	8.43	9.11	9.83	10.6	11.5	12.4	13.4
3,200	1.24	1.68	1.96	2.67	2.88	3.11	3.36	3.63	3.91	4.23	4.56	4.93	5.32	5.75	6.20	6.70	7.23	7.81	8.43	9.11	9.83	10.6	11.5	12.4	13.4	14.5
4,000	1.34	1.82	2.12	2.88	3.11	3.36	3.63	3.91	4.23	4.56	4.93	5.32	5.75	6.20	6.70	7.23	7.81	8.43	9.11	9.83	10.6	11.5	12.4	13.4	14.5	15.6
5,200	1.45	1.96	2.29	3.11	3.36	3.63	3.91	4.23	4.56	4.93	5.32	5.75	6.20	6.70	7.23	7.81	8.43	9.11	9.83	10.6	11.5	12.4	13.4	14.5	15.6	16.8
6,300	1.56	2.12	2.47	3.36	3.63	3.91	4.23	4.56	4.93	5.32	5.75	6.20	6.70	7.23	7.81	8.43	9.11	9.83	10.6	11.5	12.4	13.4	14.5	15.6	16.8	18.1
8,000	1.68	2.29	2.67	3.63	3.91	4.23	4.56	4.93	5.32	5.75	6.20	6.70	7.23	7.81	8.43	9.11	9.83	10.6	11.5	12.4	13.4	14.5	15.6	16.8	18.1	19.6
10,000	1.82	2.47	2.88	3.91	4.23	4.56	4.93	5.32	5.75	6.20	6.70	7.23	7.81	8.43	9.11	9.83	10.6	11.5	12.4	13.4	14.5	15.6	16.8	18.2	19.6	21.2
12,500	1.96	2.67	3.11	4.23	4.56	4.93	5.32	5.75	6.20	6.70	7.23	7.81	8.43	9.11	9.83	10.6	11.5	12.4	13.4	14.5	15.6	16.8	.18.2	19.6	21.2	22.9
16,000	2.12	2.88	3.36	4.56	4.93	5.32	5.75	6.20	6.70	7.23	7.81	8.43	9.11	9.83	10.6	11.5	12.4	13.4	14.5	15.6	16.8	18.2	19.6	21-2	22.9	24.7
20,000	2.29	3.11	3.63	4.93	5.32	5.75	6.20	6.70	7.23	7.81	8.43	9.11	9.83	10.6	11.5	12.4	13.4	14.5	15.6	16.8	18.2	19.6	21.2	22.9	24.7	26.7
25,000	2.47	3.36	3.91	5.32	5.75	6.20	6.70	7.23	7.81	8.43	9.11	9.83	10.6	11.5	12.4	13.4	14.5	15.6	16.8	18.2	19.6	21.2	22.9	24.7	26.7	28.8
32,000	2.67	3.63	4.23	5.75	6.20	6.70	7.23	7.81	8.43	9.11	9.83	10.6	11.5	12.4	13.4	14.5	15.6	16.8	18.2	19-6	21-2	22.9	24.7	26.7	28.8	31.1
40,000	2.88	3.91	4.56	6.20	6.70	7.23	7.81	8.43	9.11	9.83	10.6	11.5	12.4	13.4	14.5	15.6	16.9	18.2	19.6	21.2	22.9	24.7	26.7	28.8	31.1	
50,000	3.11	4.23	4.93	6.70	7.23	7.81	8.43	9.11	9.83	10.6	11.5	12.4	13.4	14.5	15.6	16-9	18.2	19.6	21.2	22.9	24.7	26.7	28.8	31.1		
63,000	3.36	4.56	5.32	7.23	7.81	8.43	9.11	9.83	10.6	11.5	12.4	13.4	14.5	15.6	16.8	18.2	19.6	21.2	22.9	24.7	26.7	28 8	31.1			
80,000	3.63	4.93	5.75	7.81	8.43	9.11	9.83	10.6	11.5	12.4	13.4	14.5	15.6	16.8	18.2	19.6	21.2	22.9	24.7	26.7	28.8	31.1				
100,000	3.91	5.32	6.20	8.43	9.11	9.83	10.6	11.5	12.4	13.4	14.5	15.6	16.8	18.2	19.6	21.2	22.9	24.7	26.7	28.8	31.1					
200,000	4.93	6.70	7.81	10.6	11.5	12.4	13.4	14.5	15.6	16.8	18.2	19.6	21.2	22.90	24.7	26.7	28.8	31.1								

TABLE 23-42b
Loading ratio C/P for different lives for roller bearings

Life, L_h hours	Speed, rpm 10	25	40	100	125	160	200	250	320	400	500	630	800	1000	1250	1600	2000	2500	3200	4000	5000	6200	8000	10000	12500	16000
100							1.05	1.13	1.21	1.30	1.39	1.49	1.60	1.71	1.83	1.97	2.11	2.26	2.42	2.59	2.78	2.97	3.19	3.42	3.66	3.92
500			1.05	1.39	1.49	1.60	1.71	1.83	1.97	2.11	2.26	2.42	2.59	2.78	2.97	3.19	3.42	3.66	3.92	4.20	4.50	4.82	5.17	5.54	5.94	6.36
1,000		1.13	1.30	1.71	1.83	1.97	2.11	2.26	2.42	2.59	2.78	2.97	3.19	3.42	3.66	3.92	4.20	4.50	4.82	5.17	5.54	5.94	6.36	6.81	7.30	7.82
1,250		1.21	1.39	1.83	1.97	2.11	2.26	2.42	2.59	2.78	2.97	3.19	3.42	3.66	3.92	4.20	4.50	4.82	5.17	5.54	5.94	6.36	6.81	7.30	7.82	8.38
1,600		1.30	1.49	1.97	2.11	2.26	2.42	2.59	2.78	2.97	3.19	3.42	3.66	3.92	4.20	4.50	4.82	5.17	5.54	5.94	6.36	6.81	7.30	7.82	8.38	9.98
2,000	1.05	1.39	1.60	2.11	2.26	2.42	2.59	2.78	2.97	3.19	3.42	3.66	3.92	4.20	4.50	4.82	5.17	5.54	5.94	6.36	6.81	7.30	7.82	8.38	8.99	9.62
2,500	1.13	1.49	1.71	2.26	2.42	2.59	2.78	2.97	3.19	3.42	3.66	3.92	4.20	4.50	4.82	5.17	5.54	5.94	6.36	6.81	7.30	7.82	8.38	8.98	9.62	10.3
3,200	1.21	1.60	1.83	2.42	2.59	2.78	2.97	3.19	3.42	3.66	3.92	4.20	4.50	4.82	5.17	5.54	5.94	6.36	6.81	7.30	7.82	8.38	8.98	9.62	10.3	11.0
4,000	1.30	1.71	1.97	2.59	2.78	2.97	3.19	3.42	3.66	3.92	4.20	4.50	4.82	5.17	5.54	5.94	6.36	6.81	7.30	7.82	8.38	8.98	9.62	10.3	11.0	11.8
5,000	1.39	1.83	2.11	2.78	2.97	3.19	3.42	3.66	3.92	4.20	4.50	4.82	5.17	5.54	5.94	6.36	6.81	7.30	7.87	8.38	8.98	9.62	10.3	11.0	11.8	12.7
6,300	1.46	1.97	2.26	2.97	3.19	3.42	3.66	3.92	4.20	4.50	4.82	5.17	5.54	5.94	6.36	6.81	7.30	7.82	8.38	8.98	9.62	10.3	11.0	11.8	12.7	13.6
8,000	1.60	2.11	2.42	3.19	3.42	3.66	3.92	4.20	4.50	4.82	5.17	5.54	5.94	6.36	6.81	7.30	7.82	8.38	8.98	9.62	10.3	11.0	11.8	12.7	13.6	14.6
10,000	1.71	2.26	2.59	3.42	3.66	3.92	4.20	4.50	4.82	5.17	5.54	5.94	6.36	6.81	7.30	7.82	8.38	8.98	9.62	10.0	11.0	11.8	12.7	13.6	14.6	15.6
12,500	1.83	2.42	2.78	3.66	3.92	4.20	4.50	4.82	5.17	5.54	5.94	6.36	6.81	7.30	7.82	8.38	8.98	9.62	10.3	11.0	11.8	12.7	13.6	14.6	15.6	16.7
16,000	1.97	2.59	2.97	3.92	4.20	4.50	4.82	5.17	5.54	5.94	6.36	6.81	7.30	7.82	8.38	8.98	9.62	10.3	11.0	11.8	12.7	13.6	14.6	15.6	16.7	17.9
20,000	2.11	2.78	3.19	4.20	4.50	4.82	5.17	5.54	5.94	6.36	6.81	7.30	7.82	8.38	8.98	9.62	10.3	11.0	11.8	12.7	13.6	14.6	15.6	16.7	17.9	19.2
25,000	2.26	2.97	3.42	4.50	4.82	5.17	5.54	5.94	6.36	6.81	7.30	7.82	8.38	8.98	9.62	10.3	11.0	11.8	12.7	13.6	14.6	15.6	16.7	17.9	19.2	20.6
32,000	2.42	3.19	3.66	4.82	5.17	5.54	5.94	6.36	6.81	7.30	7.82	8.38	8.98	9.62	10.3	11.0	11.8	12.7	13.6	14.6	15.6	16.7	17.9	19.2	20.6	
40,000	2.59	3.42	3.92	5.17	5.54	5.94	6.36	6.81	7.30	7.82	8.38	8.98	9.62	10.3	11.0	11.8	12.7	13.6	14.6	15.6	16.7	17.9	19.2	20.6		
50,000	2.78	3.66	4.20	5.54	5.94	6.36	6.81	7.30	7.82	8.38	8.98	9.62	10.3	11.0	11.8	12.7	13.6	14.6	15.6	16.7	17.9	19.2	20.6			
63,000	2.97	3.92	4.50	5.94	6.36	6.81	7.30	7.82	8.38	8.98	9.62	10.3	11.0	11.8	12.7	13.6	14.6	15.6	16.7	17.9	19.2	20.6				
80,000	3.19	4.20	4.82	6.36	6.81	7.30	7.82	8.38	8.98	9.62	10.3	11.0	11.8	12.7	13.6	14.6	15.6	16.7	17.9	19.2	20.6					
100,000	3.42	4.50	5.17	6.81	7.30	7.82	8.38	8.98	9.62	10.3	11.0	11.8	12.7	13.6	14.6	15.6	16.7	17.9	19.2	20.6						
200,000	4.20	5.54	6.36	8.38	8.98	9.62	10.3	11.0	11.8	12.7	13.6	14.6	15.6	16.7	17.9	19.2	20.6									

Particular	Formula
	m = life exponent = 3.0 for ball bearing = 10/3 for roller bearing Load ratio $(C/P(= C/F))$ and also taken from Tables 23-42a and 23-42b can be determined from Fig. 23-53. C can be obtained from manufacturer's catalogue.
The equivalent dynamic load used in Eq. (23-177a) is given by	$F_e = XF_r + YF_a$ (23-177b) where $F_e = P$ = equivalent radial load for radial bearings or axial load for axial bearings F_r = radial load, N F_a = axial load, N X = radial factor Y = axial/thrust factor The values of X and Y can be found in tables of the manufacturer's catalogue (e.g. *FAG* catalogue).
The Eq. (23-177a) in terms of L_h hours of life	$L_h = \frac{10^6}{60n}\left(\frac{C}{F}\right)^m$ (23-178a) where $L_h = L_{10h}$ = nominal rating life, h
The relation between fatigue life (L_h) in hours and the life in millions of revolutions (L)	$L_{10h} = L_h = \frac{10^6 L}{60n}$ (23-178b) Refer to Tables 23-43 and 23-44 for index of dynamic strengthening f_L and L_h.
The Eq. (23-178b) can be written as	$L_{10h} = L_h = \frac{500(33\frac{1}{3})60L}{60n}$ (23-179a) $L_{10h} = L_h = 500\left(\frac{33\frac{1}{3}}{n}\right)\left(\frac{C}{F}\right)^m$ (23-179b) Values of L_{10h} in hours as function of speed n and load ratio $(C/F = C/P)$ are given in Fig. 23-53.
The simplified form of Eq. (23-179b)	$f_L = \frac{C}{F} f_n$ (23-180a)
The basic dynamic load rating from Eq. (23-180a)	$C = \frac{f_L}{f_n} F = \frac{f_L}{f_n} P$ (23-180b) where $F = P$ f_L = index of dynamic stressing f_n = speed factor Refer to Fig. 23-56.

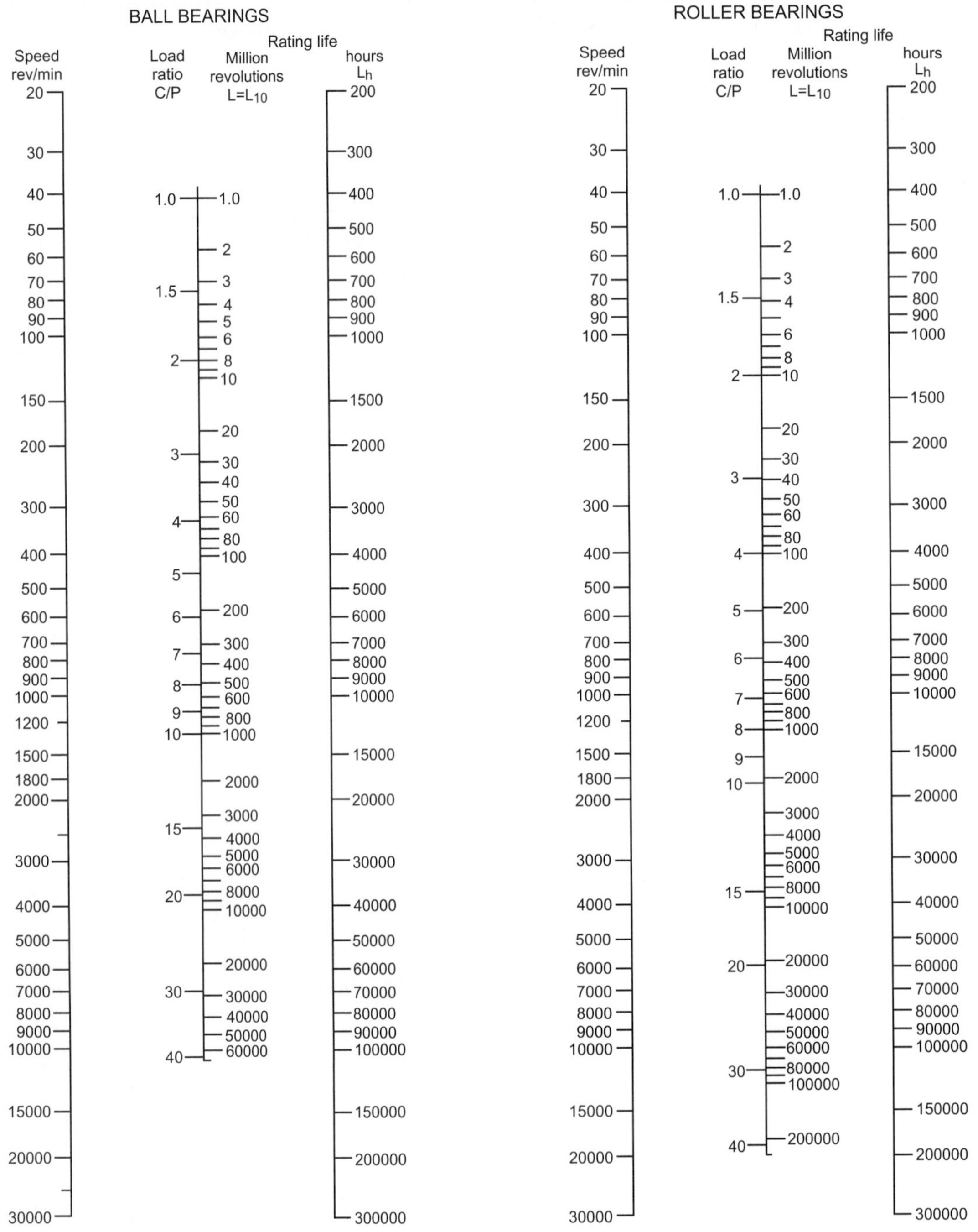

FIGURE 23-53 Nomogram chart for determining c/p for ball and roller bearings (ISO-R281) (*Note*: Information on the calculation of load rating *C* and equivalent load *P* can be obtained from ISO Recommendation R281. Values of *C* for various types of bearing can be obtained from the bearing manufacturers.)

Particular	Formula
The value of index of dynamic stressing, f_L, can be obtained from Eq. (23-179b)	$$f_L = \sqrt[m]{\frac{L_h}{500}} \qquad \text{(23-181a)}$$ For a life of 500 h, $f_L = 1$ Refer to Table 23-43 for f_L of ball bearings Refer to Table 23-44 for f_L of roller bearings and to Table 23-58

TABLE 23-43
Index of dynamic stressing f_L for ball bearings for use in Eq. (23-180) $f_L = \sqrt[3]{L_h/500}$

L_h hours	f_L	L_h hours	f_L	L_h hours	f_L	L_h hours	f_L	L_h hours	f_L	L_h hours	f_L	L_h hours	f_L
100	0.585	300	0.843	700	1.120	1750	1.520	4500	2.08	10000	2.71	30000	3.91
105	0.595	310	0.852	720	1.130	1800	1.535	4600	2.10	10500	2.76	31000	3.96
110	0.604	320	0.861	740	1.140	1850	1.545	4700	2.11	11000	2.80	32000	4.00
115	0.613	330	0.870	760	1.150	1900	1.560	4800	2.13	11500	2.85	33000	4.04
120	0.622	340	0.879	780	1.160	1950	1.575	4900	2.14	12000	2.85	34000	4.08
125	0.631	350	0.888	800	1.170	2000	1.590	5000	2.15	12500	2.93	35000	4.12
130	0.639	360	0.896	820	1.180	2100	1.615	5200	2.18	13000	2.96	36000	4.16
135	0.647	370	0.905	840	1.190	2200	1.640	5400	2.21	13500	3.00	37000	4.20
140	0.654	380	0.913	860	1.200	2300	1.665	5600	2.24	14000	3.04	38000	4.24
145	0.662	390	0.921	880	1.205	2400	1.690	5800	2.27	14500	3.07	3900	4.27
150	0.670	400	0.928	900	1.215	2500	1.710	6000	2.29	15000	3.11	40000	4.31
155	0.677	410	0.936	920	1.225	2600	1.730	6200	2.32	15500	3.14	41000	4.35
160	0.684	420	0.944	940	1.235	2700	1.755	6400	2.34	16000	3.18	42000	4.38
165	0.691	430	0.951	960	1.245	2800	1.775	6600	2.37	16500	3.21	43000	4.42
170	0.698	440	0.959	980	1.260	2900	1.795	6800	2.39	17000	3.24	44000	4.45
175	0.705	450	0.966	1000	1.260	3000	1.815	7000	2.41	17500	3.27	45000	4.48
180	0.712	460	0.973	1050	1.280	3100	1.835	7200	2.43	18000	3.30	46000	4.51
185	0.718	470	0.980	1100	1.300	3200	1.855	7400	2.46	18500	3.33	47000	4.55
190	0.724	480	0.987	1150	1.320	3300	1.875	7600	2.48	19000	3.36	48000	4.58
195	0.731	490	0.994	1200	1.340	3400	1.895	7800	2.50	19500	3.39	49000	4.61
200	0.737	500	1.000	1250	1.360	3500	1.910	8000	2.52	20000	3.42	50000	4.64
210	0.749	520	1.015	1300	1.375	3600	1.930	8200	2.54	21000	3.48	55000	4.80
220	0.761	540	1.025	1350	1.395	3700	1.950	8400	2.56	22000	3.53	60000	4.94
230	0.772	560	1.040	1400	1.410	3800	1.965	8600	2.58	23000	3.58	65000	5.07
240	0.783	580	1.050	1450	1.425	3900	1.985	8800	2.60	24000	3.63	70000	5.19
250	0.794	600	1.065	1500	1.445	4000	2.000	9000	2.62	25000	3.68	75000	5.30
260	0.804	620	1.075	1550	1.460	4100	2.020	9200	2.64	26000	3.73	80000	5.43
270	0.814	640	1.085	1600	1.475	4200	2.030	9400	2.66	27000	3.78	85000	5.55
280	0.824	660	1.100	1650	1.490	4300	2.050	9600	2.68	28000	3.82	90000	5.65
290	0.834	680	1.110	1700	1.505	4400	2.070	9800	2.70	29000	3.87	100000	5.85

Particular	Formula
The value of speed factor, f_n, can be obtained from Eq. (23-179b)	$f_n = \sqrt[m]{\frac{33\frac{1}{3}}{n}} = \sqrt[m]{\frac{100}{3n}}$ (23-181b) For a speed factor of $n = 33\frac{1}{3}$ min, $f_n = 1$ Refer to Table 23-45 for f_n of ball bearings Refer to Table 23-46 for f_n of roller bearings
The equivalent dynamic load for deep groove ball bearings with increased radial clearance	$F_e = P_a = XF_r + YF_a$ (23-182)

TABLE 23-44
Index of dynamic stressing f_L for roller bearings for use in Eq. (23-180) $f_L = \sqrt[10/3]{L_h/500}$

L_h hours	f_L	L_h hours	f_L	L_h hours	f_L	L_h hours	f_L	L_h hours	f_L	L_h hours	f_L	L_h hours	f_L
100	0.617	300	0.858	700	1.105	1750	1.455	4500	1.935	10000	2.46	30000	3.42
105	0.626	310	0.866	720	1.115	1800	1.470	4600	1.945	10500	2.49	31000	3.45
110	0.635	320	0.875	740	1.125	1850	1.480	4700	1.960	11000	2.53	32000	3.48
115	0.643	330	0.883	760	1.135	1900	1.490	4800	1.970	11500	2.56	33000	3.51
120	0.652	340	0.891	780	1.145	1950	1.505	4900	1.985	12000	2.59	34000	3.55
125	0.660	350	0.889	800	1.150	2000	1.515	5000	2.00	12500	2.63	35000	3.58
130	0.665	360	0.906	820	1.160	2100	1.540	5200	2.02	13000	2.66	36000	3.61
135	0.675	370	0.914	840	1.170	2200	1.560	5400	2.04	13500	2.69	37000	3.64
140	0.683	380	0.921	860	1.180	2300	1.580	5600	2.06	14000	2.72	39000	3.67
145	0.690	390	0.928	880	1.185	2400	1.600	5800	2.09	14500	2.75	39000	3.70
150	0.697	400	0.935	900	1.190	2500	1.620	6000	2.11	15000	2.77	40000	3.72
155	0.704	410	0.942	920	1.200	2600	1.640	6200	2.13	15500	2.80	41000	3.75
160	0.710	420	0.949	940	1.210	2700	1.660	6400	2.15	16000	2.83	42000	3.78
165	0.717	430	0.956	960	1.215	2800	1.675	6600	2.17	16500	2.85	43000	3.80
170	0.723	440	0.962	980	1.225	2900	1.695	6800	2.19	17000	2.88	44000	2.83
175	0.730	450	0.969	1000	1.230	3000	1.710	7000	2.21	17500	2.91	45000	3.86
180	0.736	460	0.975	1050	1.250	3100	1.730	7200	2.23	18000	2.93	46000	3.88
185	0.742	470	0.982	1100	1.270	3200	1.755	7400	2.24	18500	2.95	47000	3.91
190	0.748	480	0.994	1150	1.285	3300	1.760	7600	2.26	19000	2.98	48000	3.93
195	0.754	490	0.998	1200	1.300	3400	1.775	7800	2.28	19500	3.00	49000	3.96
200	0.760	500	1.000	1250	1.315	3500	1.795	8000	2.30	20000	3.02	50000	3.98
210	0.771	520	1.010	1300	1.330	3600	1.810	8200	2.31	21000	3.07	55000	4.10
220	0.782	540	1.025	1350	1.345	3700	1.825	8400	2.33	22000	3.11	60000	4.20
230	0.792	560	1.035	1400	1.360	3800	1.840	8608	2.35	23000	3.15	65000	4.30
240	0.802	580	1.045	1450	1.375	3900	1.850	8800	2.36	24000	3.19	70000	4.40
250	0.812	600	1.055	1500	1.390	4000	1.865	9000	2.38	25000	3.23	75000	4.50
260	0.822	620	1.065	1550	1.405	4100	1.880	9200	2.40	26000	3.27	80000	4.58
270	0.831	640	1.075	1600	1.420	4200	1.895	9400	2.41	27000	3.31	85000	4.66
280	0.840	660	1.085	1650	1.430	4300	1.905	9600	2.43	28000	3.35	90000	4.75
290	0.849	680	1.095	1700	1.445	4400	1.920	9800	2.44	29000	3.38	100000	4.90
												150000	5.54
												200000	6.03

TABLE 23-45
Speed factor f_n for ball bearings for use in Eq. (23-180)

$$f_n = \sqrt[3]{33\tfrac{1}{3}/n}$$

Speed, n, rpm	Speed factor, f_n	Speed, n, rpm	Speed factor, f_n	Speed, n, rpm	Speed factor, f_n	Speed, n, rpm	Speed factor, f_n	Speed, n, rpm	Speed factor, f_n	Speed, n, rpm	Speed factor, f_n
10	1.494	60	0.822	250	0.511	900	0.333	4000	0.203	15,000	0.131
11	1.447	62	0.813	260	0.504	920	0.331	4100	0.201	15,500	0.129
12	1.405	64	0.805	270	0.498	940	0.329	4200	0.199	16,000	0.128
13	1.369	66	0.797	280	0.492	960	0.326	4300	0.198	16,500	0.126
14	1.335	68	0.788	290	0.487	980	0.324	4400	0.196	17,000	0.125
15	1.305	70	0.781	300	0.481	1000	0.322	4500	0.195	17,500	0.124
16	1.277	72	0.774	310	0.476	1050	0.317	4600	0.193	18,000	0.123
17	1.252	74	0.767	320	0.471	1100	0.312	4700	0.192	18,500	0.122
18	1.225	76	0.760	330	0.466	1150	0.302	4800	0.191	19,000	0.121
19	1.206	78	0.753	340	0.461	1200	0.303	4900	0.190	19,500	0.120
20	1.186	80	0.747	350	0.457	1250	0.299	5000	0.188	20,000	0.119
21	1.166	82	0.741	360	0.453	1300	0.295	5200	0.186	21,000	0.117
22	1.148	84	0.735	370	0.448	1350	0.291	5400	0.183	22,000	0.115
23	1.132	86	0.729	380	0.444	1400	0.288	5600	0.181	23:000	0.113
24	1.116	88	0.724	390	0.441	1450	0.284	5800	0.179	24,000	0.112
25	1.100	90	0.718	400	0.437	1500	0.281	6000	0.177	25,000	0.110
26	1.089	92	0.713	410	0.433	1550	0.278	6200	0.175	26,000	0.109
27	1.071	94	0.708	420	0.430	1600	0.275	6400	0.173	27,000	0.107
28	1.060	96	0.703	430	0.426	1650	0.272	6600	0.172	28,000	0.106
29	1.048	98	0.698	440	0.423	1700	0.270	6800	0.170	29,000	0.105
30	1.036	100	0.693	450	0.420	1750	0.267	7000	0.168	30,000	0.104
31	1.025	105	0.692	460	0.417	1800	0.265	7200	0.167	32,000	0.101
32	1.014	110	0.672	470	0.414	1850	0.262	7400	0.165	34,000	0.0993
33	1.003	115	0.662	480	0.411	1900	0.260	7600	0.164	36,000	0.0975
34	0.993	120	0.652	490	0.408	1950	0.258	7800	0.162	38,000	0.0957
35	0.984	125	0.644	500	0.406	2000	0.255	8000	0.611	40,000	0.0941
36	0.975	130	0.635	520	0.400	2100	0.251	8200	0.160	42,000	0.0926
37	0.966	135	0.627	540	0.395	2200	0.247	8400	0.158	44,000	0.0912
38	0.958	140	0.620	560	0.390	2300	0.244	8600	0.157	46,000	0.0898
39	0.949	145	0.613	580	0.386	2400	0.248	8800	0.156	50,000	0.0874
40	0.941	150	0.606	600	0.382	2500	0.237	9000	0.155		
41	0.933	155	0.599	620	0.378	2600	0.234	9200	0.154		
42	0.926	160	0.583	640	0.374	2700	0.231	9400	0.153		
43	0.919	165	0.586	660	0.370	2800	0.228	9600	0.152		
44	0.912	170	0.581	680	0.366	2900	0.226	9800	0.150		
45	0.905	175	0.575	700	0.363	3000	0.223	10,000	0.149		
46	0.898	180	0.570	720	0.359	3100	0.221	10,500	0.147		
47	0.892	185	0.565	740	0.356	3200	0.218	11,000	0.145		
48	0.585	190	0.560	760	0.353	3300	0.216	11,500	0.143		
49	0.880	195	0.555	780	0.350	3400	0.214	12,000	0.141		
50	0.874	200	0.550	800	0.347	3500	0.212	12,500	0.139		
52	0.863	210	0.541	820	0.344	3600	0.210	13,000	0.137		
54	0.851	220	0.533	840	0.341	3700	0.208	13,500	0.135		
56	0.841	230	0.525	860	0.339	3800	0.206	14,000	0.134		
58	0.831	240	0.518	880	0.336	3900	0.205	14,500	0.132		

TABLE 23.46
Speed factor f_n for roller bearings for us in Eq. (23-180)

$$f_n = \sqrt[10/3]{33\tfrac{1}{3}/n}$$

Speed, n, rpm	Speed factor, f_n	Speed, n, rpm	Speed factor, f_n	Speed, n, rpm	Speed factor, f_n	Speed, n, rpm	Speed factor, f_n	Speed, n, rpm	Speed factor, f_n	Speed, n, rpm	Speed factor, f_n
10	1.435	60	0.838	250	0.546	900	0.372	4000	0.238	15000	0.160
11	1.395	62	0.830	260	6.540	920	0.370	4100	0.236	15500	0.158
12	1.359	64	0.822	270	0.534	940	0.367	4200	0.234	16000	0.157
13	1.326	66	0.815	280	0.528	960	0.365	4300	0.233	16500	0.156
14	1.297	68	0.807	290	0.523	980	0.363	4400	0.231	17000	0.154
15	1.271	70	0.800	300	0.517	1000	0.361	4500	0.230	17500	0.153
16	1.246	72	0.794	310	0.512	1050	0.355	4600	0.228	18000	0.152
17	1.224	74	0.787	320	0.507	1100	0.350	4700	0.227	18500	0.150
18	1.203	76	0.781	330	0.503	1150	0.346	4800	0.225	19000	0.149
19	1.184	78	0.775	340	0.498	1200	0.341	4900	0.224	19500	0.148
20	1.166	80	0.769	350	0.494	1250	0.337	5000	0.222	20000	0.147
21	1.149	82	0.763	360	0.490	1300	0.333	5200	0.220	21000	0.145
22	1.133	84	0.758	370	0.486	1350	0.329	5400	0.217	22000	0.143
23	1.118	86	0.753	380	0.482	1400	0.336	5600	0.215	23000	0.141
24	1.104	88	0.747	390	0.478	1450	0.322	5800	0.213	24000	0.139
25	1.090	90	0.742	400	0.475	1500	0.319	6000	0.211	25000	0.137
26	1.077	92	0.737	410	0.471	1550	0.316	6200	0.209	26000	0.136
27	1.065	94	0.733	420	0.467	1600	0.313	6400	0.207	27000	0.134
28	1.054	96	0.728	430	0.464	1650	0.310	6600	0.205	28000	0.133
29	1.0411	98	0.724	440	0.461	1700	0.307	6800	0.203	29000	0.131
30	1.032	100	0.719	450	0.458	1750	0.305	7000	0.201	30000	0.130
31	1.022	105	0.709	460	0.455	1800	0.302	7200	0.199	32000	0.127
32	1.012	110	0.699	470	0.452	1850	0.300	7400	0.198	34000	0.125
33	1.003	115	0.690	480	0.449	1900	0.297	7600	0.196	36000	0.123
34	0.994	120	0.681	490	0.447	1950	0.295	7800	0.195	38000	0.121
35	0.986	125	0.673	500	0.444	2000	0.293	8000	0.137	40000	0.119
36	0.977	130	0.665	520	0.439	2100	0.289	8200	0.192	42000	0.117
37	0.969	135	0.657	540	0.434	2200	0.285	8400	0.190	44000	0.116
38	0.962	140	0.650	560	0.429	2300	0.281	8600	0.189	46000	0.114
39	0.954	145	0.643	580	0.425	2400	0.274	8800	0.188	50000	0.111
40	0.947	150	0.637	600	0.420	2500	0.274	9000	0.187		
41	0.940	155	0.631	620	0.416	2600	0.271	9200	0.185		
42	0.933	160	0.625	640	0.412	2700	0.268	9400	0.184		
43	0.927	165	0.619	660	0.408	2800	0.265	9600	0.183		
44	0.920	170	0.613	680	0.405	2900	0.262	9800	0.182		
45	0.914	175	0.608	700	0.401	3000	0.259	10000	0.181		
46	0.908	180	0.603	720	0.398	3100	0.257	10500	0.178		
47	0.902	185	0.598	740	0.395	3200	0.254	11000	0.176		
48	0.896	190	0.593	760	0.391	3300	0.252	11500	0.173		
49	0.891	195	0.589	780	0.388	3400	0.250	12000	0.171		
50	0.896	200	0.584	800	0.385	3500	0.248	12500	0.169		
52	0.875	210	0.576	820	0.383	3600	0.246	13000	0.167		
54	0.865	220	0.568	840	0.380	3700	0.243	13500	0.165		
56	0.856	230	0.560	860	0.377	3800	0.242	14000	0.163		
58	0.847	240	0.553	880	0.375	3900	0.240	14500	0.162		

Particular	Formula

TABLE 23-47a
Values of radial factors X and thrust factors Y for deep groove ball bearings with increase In radial clearance

	Normal Bearing Standard clearance					Bearing clearance C3[a]					Bearing clearance C4[a]				
		$\frac{F_a}{F_r} \le e$		$\frac{F_a}{F_r} > e$			$\frac{F_a}{F_r} \le e$		$\frac{F_a}{F_r} > e$			$\frac{F_a}{F_r} \le e$		$\frac{F_a}{F_r} > e$	
$f_o \frac{F_a}{C_o}$	e	X	Y	X	Y	e	X	Y	X	Y	e	X	Y	X	Y
0.3	0.22	1	0	0.56	2.0	0.32	1	0	0.46	1.70	0.40	1	0	0.44	1.40
0.5	0.24	1	0	0.56	1.8	0.35	1	0	0.46	1.56	0.43	1	0	0.44	1.31
0.9	0.28	1	0	0.56	1.6	0.39	1	0	0.46	1.41	0.45	1	0	0.44	1.23
1.6	0.32	1	0	0.56	1.4	0.43	1	0	0.46	1.27	0.48	1	0	0.44	1.16
3.0	0.36	1	0	0.56	1.2	0.48	1	0	0.46	1.14	0.52	1	0	0.44	1.08
6.0	0.43	1	0	0.56	1.0	0.54	1	0	0.46	1.00	0.56	1	0	0.44	1.00

[a] C_3, C_4 Standard for a radial clearance that is larger than normal.
Values of factor f_a are given in Table 23-47b

For values of X and Y, refer to Table 23-39 and also to bearing tables given in *FAG* catalogue. For values of X and Y of deep groove ball bearings with increased radial clearance, refer to Table 23-47b.

LIFE ADJUSTMENT FACTORS

An adjusted fatigue life equation for a reliability of (100-n) percent

An approximate equation for adjusted factor a_1, which accounts for reliability, R, is calculated from the Weibull distribution.

The adjusted fatigue life for non-conventional materials and operating conditions.

$$L_{10a} = a_2 a_3 L_{10} \qquad (23\text{-}183)$$

where

a_2 = life adjustment factor for materials
= 3 for radial ball bearings of good quality as per AFBMA
= 0.2 to 2.0 for normal bearing materials
= 2.0 for most common material, AISI 52100

a_3 = life adjustment factor for application operating conditions
= 1 for well and adequately lubricated bearings
$<$ 1 for other adverse lubricating conditions and temperatures.

The standard does not yet include values of a_3

TABLE 23-47b
Factor f_o for deep groove ball bearings for use in Table 23-47a

Bore reference number	Factor f_o — Bearing series 160	60	62	63	64
00	—	12.4	12.1	11.3	—
01	—	13.0	12.2	11.1	—
02	13.9	13.9	13.1	12.1	—
03	14.3	14.3	13.1	12.2	10.9
04	14.9	13.9	13.1	12.1	11.0
05	15.4	14.5	13.8	12.4	12.1
06	15.2	14.8	13.8	13.0	12.2
07	15.6	14.8	13.8	13.1	12.1
08	15.9	15.2	14.0	13.0	12.2
09	15.9	15.4	14.1	13.0	12.1
10	16.1	15.6	14.3	13.0	12.2
11	16.1	15.4	14.3	12.9	12.2
12	16.3	15.5	14.3	13.1	12.3
13	16.4	15.7	14.3	13.2	12.3
14	16.2	15.5	14.4	13.2	12.1
15	16.4	15.7	14.7	13.2	12.2
16	16.4	15.6	14.6	13.2	12.3
17	16.4	15.7	14.7	13.1	12.3
18	16.3	15.6	14.5	13.9	12.2
19	16.5	15.7	14.4	13.9	
20	16.4	15.9	14.4	13.8	
21	16.3	15.8	14.3	13.7	
22	16.3	15.6	14.3	13.8	
24	16.4	15.9	14.8	13.5	
26	16.4	15.8	14.5	13.6	
28	16.4	16.0	14.8	13.6	
30	16.4	16.0	15.2	13.7	
32	16.4	16.0	15.2	13.9	
34	16.4	15.7	15.3	13.9	
36	16.3	15.6	15.3	13.9	
38	16.4	15.8	15.0	14.0	
40	16.3	15.6	15.3	14.1	
44	16.3	15.6	15.2	14.1	
48	16.4	15.8	15.2	14.2	
52	16.4	15.7	15.2		
56	16.4	15.9	15.3		
60	16.4	15.7			
64	16.4	15.9			
68	16.3	15.8			
72	16.4	15.9			
76	16.5	15.9			

TABLE 23-48
Life adjustment factor for reliability a_1

Reliability R, %	L_n	Adjustment factor, a_1
90	L_{10}	1
95	L_5	0.65
96	L_4	0.53
97	L_3	0.44
98	L_2	0.33
99	L_1	0.21

Note: These values of a_1 and a_2 have to be used judiciously and designers are advised to consult manufacturer's Catalogue

Particular	Formula
From Eq. (23-177a), an adjusted fatigue life equation taking into consideration *adjustment factors* a_1, a_2 and a_3, and for a failure probability of n as per ISO 281	$L_{na} = a_1 a_2 a_3 \left(\frac{C}{F}\right)^m = a_1 a_2 a_3 L$ (23-184a) where L in millions of revolutions
	$L_{hna} = a_1 a_2 a_3 L_h$ (23-184b) where L_h in h
An adjusted fatigue life equation, which accounts for variation in vibration and shock, speed, environment, etc in addition to *adjustment factors* a_1, a_2 and a_3	$L = a_1 a_2 a_3 \left(\frac{C}{K_a F}\right)^m$ (23-185) where L in 10^6 revolutions K_a = application factor taken from Table 23-49
From Eq. (23-186a), the basic equivalent load rating, which accounts for application factor K_a, and adjustment factors a_1, a_2 and a_3	$C = K_a F \left(\frac{L}{a_1 a_2 a_3}\right)^{1/m}$ (23-186) where L in 10^6 revolutions
Fig. 23-54 shows the plot of relative life vs probability of failure in percent. L_{10} life (also called as $B10$ life or minimum life), which indicates a 10% probability of failure, i.e. 90% of the loaded bearings will survive beyond this life, is taken as reference. L_{50} in Fig. 23-54 indicates median life with a 50% probability of failure, which is also known as average life. From Fig. 23-54 it can be seen that $L_{50} \approx 5L_{10}$, which indicates only 50% of the bearings will survive this longer life.	
For life curves of ball bearings as per SKF and New Departure (ND) and Needle-bearings	Refer to Fig. 23-55

TABLE 23-49
Load application factor K_a for use in Eqs. (23-185) and (23-186)

Operating conditions	Applications	K_a
	Precision gearing	1.0–1.1
	Commercial gearing	1.1–1.3
	Applications with poor bearing seals	1.2
Smooth operation free from shock	*Machinery with no impact:*	
	Electric motors, machine tools, air conditioners	1–1.2
Normal operation	*Machinery with light impact:*	
	Air blowers, compressors, elevators, cranes, paper making machines	1.2–1.5
Operation accompanied by shock and vibration	*Machinery with moderate impact:*	
	Construction machines crushers, vibration screens, rolling mills	1.5–3.0

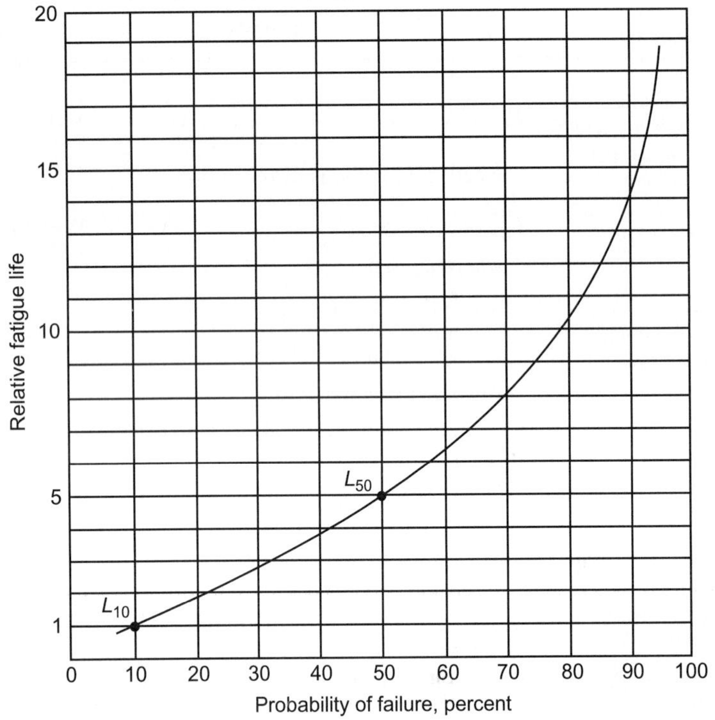

FIGURE 23-54 Typical life distribution in rolling bearings. (*Courtesy*: SKF Industries, Inc.)

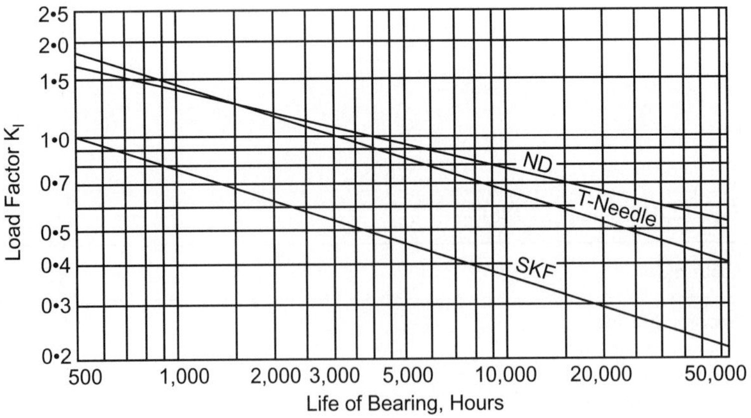

FIGURE 23-55 Life curves of ball and needle bearings.

Particular	Formula

BASIC DYNAMIC LOAD RATING OF BEARINGS AS PER INDIAN STANDARDS

Radial Ball Bearing

The basic dynamic load rating for radial and angular contact ball bearings.

$$C_r = f_c (i \cos\alpha)^{0.7} Z^{2/3} (D_w)^{1.8} \quad \text{(23-187)}$$

for $D_w \leq 25.4\,\text{mm}$
where C_r in N, D_w in mm

$$C_r = 3.647 f_c (i \cos\alpha)^{0.7} Z^{2/3} (D_w)^{1.4} \quad \text{(23-188)}$$

for $D_w > 25.4\,\text{mm}$
where C_r in N, D_w in mm

For values of factor f_c refer to Table 23-50.

TABLE 23-50
Values of factor f_c for radial ball bearings for use in Eqs. (23-187) and (23-188)

$\frac{D_w \cos\alpha}{D_{pw}}$	Factor, f_c			
	Single row radial contact groove ball bearings and single and double row-angular contact groove ball bearings	Double row radial contact groove ball bearings	Single row and double row self-aligning ball bearings	Single row radial contact separable ball bearings (magneto bearings)
0.05	46.7	44.2	17.3	16.2
0.06	49.1	46.5	18.6	17.4
0.07	51.1	48.4	19.9	18.5
0.08	52.8	50.0	21.1	19.5
0.09	54.3	51.4	22.3	20.6
0.10	55.5	52.6	33.4	21.5
0.12	57.5	54.5	25.6	23.4
0.14	58.8	55.7	27.7	25.3
0.16	59.6	56.5	29.7	27.1
0.18	59.9	56.8	31.7	28.8
0.20	59.9	56.8	33.5	30.5
0.22	59.6	56.5	35.2	32.1
0.24	59.0	55.9	36.8	33.7
0.26	58.2	55.1	38.2	35.2
0.28	57.1	54.1	39.4	36.6
0.30	56.0	53.0	40.3	37.8
0.32	54.6	51.8	40.9	38.9
0.34	53.2	50.4	41.2	39.8
0.36	51.7	48.9	41.3	40.4
0.38	50.0	47.4	41.0	40.8
0.40	48.4	45.8	40.4	40.9

Note: Values of f_c for intermediate values of $D'_w \cos\alpha / D_{pw}$ are obtained by linear interpolation. IS: 3824 (Part 1)–1983

Particular	Formula
The approximate rating life in millions of revolutions for ball bearing	$$L_n = \left(\frac{C}{P}\right)^3 \quad (23\text{-}189)$$
The equivalent radial load for radial and contact ball bearings under combined constant radial and axial loads	$$P_r = XF_r + F_a \quad (23\text{-}190)$$ For values of X and Y refer to Table 23-51.
The basic rating life for a radial ball bearing which is based statistically	$$L_{10} = \left(\frac{C_r}{P_r}\right)^3 \quad (23\text{-}191)$$ where L_{10} = basic rating life in millions of revolutions (i.e., the number of revolutions resulting in 10% failure). The values of C_r and P_r shall be calculated in accordance with Eqs. (23-187), (23-188), and (23-190).

Adjusted rating life

The adjusted rating life for a reliability of (100-n) percent	$$L_n = a_1 L_{10} \quad (23\text{-}192)$$
The adjusted rating life for non-conventional materials and operating conditions	$$L_{10a} = a_2 a_3 L_{10} \quad (23\text{-}193)$$
The adjusted rating life for non-conventional materials and operating conditions, and for a reliability of (100-n) percent.	$$L_{na} = a_1 a_2 a_3 L_{10} \quad (23\text{-}194)$$ Refer to Table 23-48 for a_1 values

Radial roller bearings

The basic dynamic radial load rating of radial roller bearings	$$C_r = f_c(iL_{we}\cos\alpha)^{7/9} Z^{3/4} D_{we}^{29/27} \quad (23\text{-}195)$$ where C_r in N, L_{we} and D_{we} in mm. For values of factor f_c refer to Table 23-52.
The equivalent radial load for radial roller bearings with $\alpha \neq 0°$ under combined constant radial and axial loads	$$P_r = XF_r + YF_a \quad (23\text{-}196)$$ For values of X and Y refer to Table 23-53.
The equivalent radial load for radial roller bearings with $\alpha \neq 0°$ and subjected to radial load only	$$P_r = F_r \quad (23\text{-}197)$$
The basic rating life in millions of revolutions for radial roller bearings	$$L_{10} = \left(\frac{C_r}{P_r}\right)^{10/3} \quad (23\text{-}198)$$ The values of C_r and P_r are calculated in accordance with Eqs. (23-195) to (23-197).
For adjusted rating life for roller bearings	Refer to Eqs. (23-192) to (23-194) with suitable modification.

TABLE 23-51
Factors X and Y for radial ball bearings for use in Eq. (23-190)

			Single-row bearings				Double-row bearings				
			$\frac{F_a}{F_r} \le e$		$\frac{F_a}{F_r} \ge e$		$\frac{F_a}{F_r} \le e$		$\frac{F_a}{F_r} \ge e$		
	'Relative axial load'[a]		X	Y	X	Y	X	Y	X	Y	e
Radial contact groove ball bearings											
	$\frac{F_a}{C_{or}}$	$\frac{F_a}{iZD_w^2}$									
	0.014	0.172	1	0	0.56	2.30	1	0	0.56	2.30	0.19
	0.028	0.345				1.99				1.99	0.22
	0.056	0.689				1.71				1.71	0.26
	0.084	1.30				1.55				1.55	0.28
	0.11	1.38				1.45				1.45	0.30
	0.17	2.07				1.31				1.31	0.34
	0.28	3.45				1.15				1.15	0.38
	0.42	5.17				1.04				1.04	0.42
	0.56	6.89				1.00				1.00	0.44
Angular contact groove ball bearings											
α	$\frac{iF_a}{C_{or}}$	$\frac{F_a}{ZD_w^2}$									
5°	0.014	0.172	1	0	For this type use the X, Y and e values applicable to single row radial contact groove ball bearings		1	2.78	0.78	3.74	0.23
	0.028	0.345						2.40		3.23	0.26
	0.056	0.689						2.07		2.78	0.30
	0.085	1.03						1.87		2.52	0.34
	0.11	1.38						1.75		2.36	0.86
	0.17	2.07						1.58		2.13	0.40
	0.28	3.45						1.39		1.87	0.45
	0.42	5.17						1.26		1.69	0.50
	0.56	6.89						1.21		1.63	0.52
10°	0.014	0.172	1	0	0.46	1.88	1	2.18	0.75	3.06	0.29
	0.029	0.345				1.71		1.98		2.78	0.32
	0.057	0.689				1.52		1.76		2.47	0.36
	0.086	1.03				1.41		1.63		2.29	0.38
	0.11	1.38				1.34		1.55		2.18	0.40
	0.17	2.07				1.23		1.42		2.00	0.44
	0.29	3.45				1.10		1.27		1.79	0.49
	0.43	5.17				1.01		1.17		1.64	0.54
	0.57	6.89				1.00		1.16		1.63	0.54
15°	0.015	0.172	1	0	0.46	1.47	1	1.65	0.72	2.39	0.38
	0.029	0.345				1.40		1.57		2.28	0.40
	0.058	0.689				1.30		1.40		2.11	0.43
	0.087	1.03				1.23		1.38		2.00	0.46
	0.12	1.38				1.19		1.34		1.93	0.47
	0.17	2.07				1.12		1.26		1.82	0.50
	0.29	3.45				1.02		1.14		1.66	0.55
	0.44	5.17				1.00		1.12		1.63	0.56
	0.58	6.89				1.00		1.12		1.63	0.56
20°	—	—	1	0	0.43	1.00	1	1.09	0.70	1.63	0.07
25°	—	—			0.41	0.87		0.92	0.67	1.41	0.68
30°	—	—			0.39	0.76		0.78	0.63	1.24	0.80
35°	—	—			0.37	0.66		0.66	0.60	1.07	0.95
40°	—	—			0.35	0.57		0.55	0.57	0.93	1.14
45°	—	—			0.33	0.50		0.47	0.54	0.81	1.34
Self-aligning ball bearings			1	0	0.40	0.4 cot α	1	0.42 cot α	0.65	0.65 cot α	1.5 tan α
Single row radial contact separable ball bearings (magnetic bearings)			1	0	0.5	2.5	—	—	—	—	0.2

Note: Values of X, Y and e for intermediate 'relative axial loads' and/or contact angles are obtained by linear interpolation. IS: 3824 (Part 1), 1983.

[a] Permissible maximum value depends on bearing design (internal clearance and raceway groove depth).

Particular	Formula

TABLE 23-52
Values of f_c for radial roller bearings for use in Eq. (23.195)

$\frac{D_w \cos\alpha}{D_{pw}}$	f_c
0.01	52.1
0.02	60.8
0.03	66.5
0.04	70.7
0.05	74.1
0.06	76.9
0.07	79.2
0.08	81.2
0.09	82.8
0.10	84.2
0.12	86.4
0.14	87.7
0.16	88.5
0.18	88.8
0.20	88.7
0.22	88.2
0.24	87.5
0.26	86.4
0.28	85.2
0.30	83.8

Note: Values of f_c for intermediate value of $D_w \cos\alpha/D_w$ are obtained by linear interpolation. IS: 3824 (Part 2) 1983.

TABLE 23-53
Values of factors X and Y for radial roller bearings for use in Eq. (23-196)

	$\frac{F_a}{F_r} \le e$		$\frac{F_a}{F_r} > e$		
Bearing type	X	Y	X	Y	e
Single-row $\alpha \ne 0$	1	0	0.4	$0.4 \cot\alpha$	$1.5 \tan\alpha$
Double-row $\alpha \ne 0$	1	$0.45 \cot\alpha$	0.67	$0.67 \cot\alpha$	$1.5 \tan\alpha$

IS: 3824 (Part 2) 1983.

THRUST BEARINGS

Ball bearings

The basic dynamic axial load rating for a single-row, single- or double-direction thrust ball bearing

$$(C_a)_{\alpha=90^\circ} = f_c Z^{2/3} D_w^{1.8} \quad \text{for } D_w \le 25.4\,\text{mm} \tag{23-199}$$

$$(C_a)_{\alpha\ne 90^\circ} = f_c(\cos\alpha)^{0.7} \tan\alpha\, Z^{2/3} D_w^{1.8} \quad \text{for } D_w \le 25.4\,\text{mm} \tag{23-200}$$

$$(C_a)_{\alpha=90^\circ} = 3.647 f_c Z^{2/3} D_w^{1.4} \quad \text{for } D_w > 25.4\,\text{mm} \tag{23-201}$$

$$(C_a)_{\alpha\ne 90^\circ} = 3.647 f_c(\cos\alpha)^{0.7} \tan\alpha\, Z^{2/3} D_w^{1.4} \quad \text{for } D_w > 25.4\,\text{mm} \tag{23-202}$$

where C_a in N, D_w in mm

For various values of f_c refer to Table 23-54.

Z = number of balls carrying load in one direction.

Particular	Formula

TABLE 23-54
Values of factor f_c for thrust ball bearings for use in Eqs. (23-199) to (23-202)

$\frac{D_w}{D_{pw}}$	f_c, $\alpha = 90°$	$\frac{D_w \cos\alpha}{D_{pw}}$	f_c, $\alpha = 45°$	f_c, $\alpha = 60°$	f_c, $\alpha = 75°$
0.01	36.7	0.01	42.1	39.2	37.3
0.02	45.2	0.02	51.7	49.1	45.9
0.03	51.1	0.03	58.2	54.2	51.7
0.04	55.7	0.04	63.3	58.9	56.1
0.05	59.5	0.05	67.3	61.6	59.7
0.06	62.9	0.06	70.7	65.8	62.7
0.07	65.8	0.07	73.5	68.4	65.2
0.08	68.5	0.08	75.9	70.7	67.3
0.09	71.0	0.09	78.0	72.6	69.2
0.10	73.3	0.10	79.7	74.2	70.7
0.12	77.4	0.12	82.3	76.6	
0.14	81.1	0.14	84.1	78.3	
0.16	84.4	0.16	85.1	79.2	
0.18	87.4	0.18	85.5	79.6	
0.20	90.2	0.20	85.4	79.5	
0.22	92.8	0.22	84.9		
0.24	95.3	0.24	84.0		
0.26	97.6	0.26	82.8		
0.28	99.8	0.28	81.3		
0.30	101.9	0.30	79.6		
0.32	103.9				
0.34	105.8				

Note: For thrust bearings $\alpha > 45°$, values for $\alpha = 45°$ are shown to permit interpolation of values for α between 45° and 60°. IS: 3824 (Part 3) 1983

Bearings with two or more rows of balls

The basic dynamic axial load rating for thrust ball bearings with two or more rows of similar balls carrying load in the same direction

$$C_a = (Z_1 + Z_2 + \cdots + Z_n) \times \left[\left(\frac{Z_1}{C_{a1}}\right)^{10/3} + \left(\frac{Z_2}{C_{a2}}\right)^{10/3} + \cdots + \left(\frac{Z_n}{C_{an}}\right)^{10/3} \right]^{-3/10} \quad (23\text{-}203)^{a}$$

[a] *Note*: The designers or bearing users are advised to refer to catalogues or standards in this regard or the bearing users should consult the bearing manufacturers regarding the evaluation of equivalent load and life in case where bearing with $\alpha = 0°$ are subjected to an axial load. The ability of radial roller bearings with $\alpha = 0°$ to support axial loads varies considerably with bearing designer execution.

Particular	Formula
	The load ratings $C_{a1}, C_{a1}, \ldots, C_{an}$ for the rows with $Z_1, Z_2, \ldots, Z_n$ balls are calculated from appropriate single row bearing formulae from Eqs. (23-199) to (23-202), Values of f_c for D_w/D_{pw} or $(D_w \cos\alpha)/D_{PW}$ and/or contact angle other than shown in Table 23-54 are obtained by linear interpolation or extrapolation.

Dynamic equivalent axial load

The equivalent load for thrust ball bearings with $\alpha \neq 90°$ under combined constant axial and radial loads

$$P_a = XF_r + YF_a \tag{23-204}$$

For values of X and Y refer to Table 23-55.

The equivalent axial load for thrust bearing with $\alpha = 90°$ which can support axial loads only

$$P_a = F_a \tag{23-205}$$

Basic rating life

The basic rating life in millions of revolutions for a thrust ball bearings

$$L_{10} = \left(\frac{C_a}{P_a}\right)^3 \tag{23-206}$$

The values of C_a and P_a are calculated in accordance with Eqs. (23-199) to (23-205).

TABLE 23-55
Values of factors X and Y for thrust ball bearings for use in Eq. (23-204)

	Single direction bearings[a]		Double direction bearings				
	$\frac{F_a}{F_r} > e$		$\frac{F_a}{F_r} \le e$		$\frac{F_a}{F_r} > e$		
α	X	Y	X	Y	X	Y	e
45°	0.66	1	1.18	0.59	0.66	1	1.25
50°	0.73		1.37	0.57	0.73		1.49
55°	0.81		1.60	0.56	0.81		1.79
60°	0.92		1.90	0.55	0.92		2.17
65°	1.06		2.30	0.54	1.06		2.68
70°	1.28		2.90	0.53	1.28		3.43
75°	1.66		3.89	0.52	1.66		4.67
80°	2.43		5.86	0.52	2.43		7.09
85°	4.80		11.75	0.51	4.80		14.29
$\alpha \neq 90°$	$1.25\tan\alpha\left(1-\frac{2}{3}\sin\alpha\right)$	1	$\frac{20}{13}\tan\alpha\left(1-\frac{1}{3}\sin\alpha\right)$	$\frac{10}{13}\left(1-\frac{1}{3}\sin\alpha\right)$	$1.25\tan\alpha\left(1-\frac{2}{3}\sin\alpha\right)$	1	$1.25\tan\alpha$

Note: For thrust bearings $\alpha > 45°$. Values for $\alpha = 45°$ are shown to permit interpolation of values for α between 45° and 50°.
[a] $F_a/F_r \le e$ is unsuitable for single direction bearings. IS: 3824 (Part 3) 1983.

Particular	Formula

Adjusted rating life

The adjusted rating life of (100-n) percent	$L_n = a_1 L_{10}$ (23-192) Refer to Table 23-48 for values of factor a_1.
For other adjusted rating life with modification if required	Refer to Eqs. (23-193) to (23-194).

Roller bearings

The basic dynamic axial load rating for single row, single- or double-direction thrust roller bearing	$(C_a)_{\alpha=90°} = f_c L_{we}^{7/9} Z^{3/4} D_{we}^{29/27}$ (23-207) $(C_a)_{\alpha \neq 90°} = f_c (L_{we} \cos\alpha)^{7/9} \tan\alpha\, Z^{3/4} D_{we}^{29/27}$ (23-208) where C_a in N, L_{we} and D_{we} in mm For values of factor f_c refer to Table 23-56. Z = number of rollers carrying load in one direction.

TABLE 23-56
Values of factor f_c for thrust roller bearings for use in Eqs. (23-207) and (23-208)

$\frac{D_{wc}}{D_{pw}}$	f_c $\alpha = 90°$	$\frac{D_w \cos\alpha}{D_{pw}}$	Factor f_c $\alpha = 50°$ [a]	Factor f_c $\alpha = 65°$ [b]	Factor f_c $\alpha = 80°$ [c]
0.01	105.4	0.01	109.7	107.1	105.6
0.02	122.9	0.02	127.8	124.7	123.0
0.03	134.5	0.03	139.5	136.2	134.3
0.04	143.4	0.04	148.3	144.7	142.8
0.05	150.7	0.05	155.2	151.5	149.4
0.06	156.9	0.06	160.9	157.0	154.9
0.07	162.4	0.07	165.6	161.6	159.4
0.08	167.2	0.08	169.5	165.5	163.2
0.09	171.7	0.09	172.8	168.7	166.4
0.10	175.7	0.10	175.5	171.4	169.0
0.12	183.0	0.12	179.7	175.4	173.0
0.14	189.4	0.14	182.3	177.9	175.5
0 16	195.1	0.16	183.7	179.3	
0.18	200.3	0 18	184.1	179.7	
0.20	205.0	0.20	183.7	179.3	
0.22	209.4	0.22	182.6		
0.24	213.5	0.24	180.9		
0.26	217.3	0.26	178.7		
0.28	220.9				
0.30	224.3				

[a] Applicable for $45° < \alpha < 60°$; [b] Applicable for $60° < \alpha < 75°$; [c] Applicable for $75° < \alpha < 90°$
Note: Values of f_c for intermediate values of D_{wc}/D_{pw} or $D_w \cos\alpha/D_{pw}$ are obtained by linear interpolation. IS: 3824 (Part 4) 1983.

Particular	Formula

Bearing with two or more rows of rollers

The basic dynamic axial load rating for thrust roller bearings with two or more rows of rollers carrying load in the same direction

$$C_a = (Z_1 L_{we1} + Z_2 L_{we2} + \cdots + Z_n L_{wen}) \times \left[\left(\frac{Z_1 L_{we1}}{C_{a1}}\right)^{9/2} + \left(\frac{Z_2 L_{we2}}{C_{a2}}\right)^{9/2} + \cdots + \left(\frac{Z_n L_{wen}}{C_{an}}\right)^{9/2} \right]^{-2/9} \quad (23\text{-}209)$$

where C_a in N, L_{we} and D_{we} in mm

The load ratings $C_{a1}, C_{a2}, \ldots, C_{an}$ for the rows with $Z_1, Z_2, \ldots, Z_n$ rollers of length $L_{we1}, L_{we2}, \ldots, L_{wen}$ are calculated from the appropriate single row bearing Eqs. (23-207) and (23-208).

TABLE 23-57
Values of factors X and Y for thrust roller bearings for use in Eqs. (23-210)

	$\frac{F_a}{F_r} \le e$		$\frac{F_a}{F_r} > e$		
Bearings type	X	Y	X	Y	e
Single-direction $\alpha \ne 90°$	a	a	$\tan\alpha$	1	$1.5\tan\alpha$
Double-direction $\alpha \ne 90°$	$1.5\tan\alpha$	0.67	$\tan\alpha$	1	$1.5\tan\alpha$

* $F_a/F_r \le e$ is unsuitable for single-direction bearing. IS: 3824 (Part 4) 1983.

The equivalent axial load for thrust roller bearings when $\alpha \ne 90°$ under combined constant axial and radial load

$$P_a = XF_r + YF_a \quad (23\text{-}210)$$

For values of X and Y refer to Table 23-57.

The equivalent axial load for thrust roller bearings with $\alpha = 90°$ which can support only axial load

$$P_a = F_a \quad (23\text{-}211)$$

The basic rating life in millions of revolutions for thrust roller bearings

$$L_{10} = \left(\frac{C_a}{P_a}\right)^{10/3} \quad (23\text{-}212)$$

The values of C_a and P_a are calculated in accordance with Eqs. (23-207), (23-208), and (23-210).

Adjusted rating life

The Eqs. (23-192), (23-193) and (23-194) for adjusted rating life with appropriate modification to suit the roller thrust bearings are repeated here

$$L_n = a_1 L_{10} \quad (23\text{-}192)$$

$$L_{10a} = a_2 a_3 L_{10} \quad (23\text{-}193)$$

$$L_{na} = a_1 a_2 a_3 L_{10} \quad (23\text{-}194)$$

Variable bearing load and speed

The mean affective load F_m under varying load and varying speed $n_1, n_2, n_3, \ldots, n_i$ at which the individual loads $F_1, F_2, F_3, \ldots, F_i$ act.

$$F_m = \sqrt[m]{\frac{F_1^m n_1 + F_2^m n_2 + F_3^m n_3 + \cdots + F_i^m n_i}{n}} \quad (23\text{-}213a)$$

$$F_m = \sqrt[m]{\frac{\sum (F_i)^m n_i}{n}} \quad (23\text{-}213b)$$

Particular	Formula
	where $F_1, F_2, F_3, \ldots, F_i$ = constant loads among series of i loads during $n_1, n_2, n_3, \ldots, n_i$ revolutions. n_i = number of revolutions at which F_i load operates n = total number of revolutions in a complete cycle $= n_1 + n_2 + n_3 + \ldots + n_i$, during which loads $F_1, F_2, F_3, \ldots, F_i$ act m = exponent $m_i = 3$ for ball bearings $m_i = \frac{10}{3}$ for roller bearings
The mean effective load F_m under linearly varying load from minimum load F_{min} to maximum load F_{max} at constant speed n.	$F_m = \dfrac{F_{min} + 2_{max}}{3}$ (23-214)
The equivalent dynamic load for the varying load which acts in a radial direction only for radial bearings and in a axial direction only for thrust bearing.	$P = F_m$ (23-215)
In the direction and magnitude of load changes with time then the equivalent loads P_1, P_2, $P_3, \ldots,$ must be calculated for the individual time periods n_1, n_2, n_3 using the general equation.	$P = XF_r + YF_a$ (23-216)
The mean equivalent load P_m by substituting the individual values of P_1, P_2, $P_3, \ldots,$ obtained from equivalent load's Eq. (23-119).	$P_m = \sqrt[m]{\dfrac{P_1^m n_1 + P_2^m n_2 + P_3^m n_3 + \cdots}{n}}$ (23-217) where m = exponent $= 3$ for ball bearings $= \frac{10}{3}$ for roller bearings
The life of a bearing under variable load and variable speed, taking into consideration life adjustment factors a_1, a_2, a_3 and application factor K_a	$L = a_1 a_2 a_3 \left(\dfrac{C}{K_a}\right)^m \dfrac{1}{F_1^m n_1 + F_2^m n_2 + F_3^m n_3 + \ldots}$ (23-218)
The basic load rating for a required bearing life in case of variable load and variable speed, factor K_a and a_1, a_2, a_3	$C = K_a \left[(F_1^m n_1 + F_2^m n_2 + F_3^m n_3 + \ldots) \dfrac{L}{a_1 a_2 a_3} \right]^{1/m}$ (23-219) where L is in millions of revolutions; C and F in N; $n_1, n_2, n_3, \ldots$ are rotational speeds in rpm under loads $F_1, F_2, F_3, \ldots$ $m = 3$ for ball bearings $= \frac{10}{3}$ for roller bearings

TABLE 23-58
Index f_L of dynamic stressing for use in Eq. (23-180)

Application	f_L	Application	f_L
Motor vehicles		Medium-sized fans	3.0–4.5
Motorcycles	1.4–1.9	Large fans	4.5–5.5
Light cars	1.6–2.1	Centrifugal pumps	2.5–4.5
Heavy cars	1.7–2.2	Centrifuges	3.0–4.0
Light trucks or lorries	1.7–2.2	Winding cable sheaves	4.5–5.0
Heavy trucks or lorries	2.0–2.6	Belt conveyor idlers	3.0–4.5
Buses	2.0–2.6	Conveyor drums	4.5–5.5
Tractors	1.6–2.2	Shovels and reclaimers	6.0
Tracked vehicles	2.1–2.7	Crushers	3.0–3.5
		Beater mills	3.5–4.5
Electric motors		Tube mills	6.0
For household appliances	1.5–2.0	Vibrating screens	2.5–2-8
Small standard motors	2.5–3.5	Vibrating rolls and large out-of-balance exciters	1.6–2.0
Medium-sized standard cars	3.0–4.0	Vibrators	1.0–1.5
Large motors	3.5–4.5	Briquette presses	4.5–5.0
Traction motors	3.0–4.0	Large mechanical stirrers	3.5–4.0
Railbound vehicles		Rotary furnace rollers	4.5–5.0
Axle boxes for haulage trolleys	3.0–4.0	Flywheels	3.4–4.0
Trams	4.5–5.5	Printing machines	4.0–4.5
Railway coaches	4.0–5.0		
Freight cars	3.5–4.0	**Papermaking machines**	
Overburden removal cars	3.5–4.0	Wet sections	5.0–6.0
Outer bearings of locomotives	4.0–5.5	Dry sections	5.0–6.0
Inner bearings of locomotives	4.5–5.5	Refiners	4.5–4.6
Gears	3.5–4.5	Calendars	4.0–4.5
Rolling mills		**Centrifugal casting machines**	3.4–4.0
Neck bearings	2.0–2.5	**Textile machines**	3.6–4.7
Gears	3.0–5.0		
Ship building		**Machine tools**	
Ship propeller thrust blocks	2.9–3.6	Lathes, boring and milling machines	2.7–4.5
Ship propeller shaft bearings	6.0	Grinding, lapping, and polishing machines	2.7–4.5
Large marine gears	2.6–4.0	**Woodworking machines**	
General engineering		Milling cutters and cutter shafts	3.0–4.0
Small universal gears	2.5–3.5	Saw mills (con rods)	2.8–3.3
Medium-sized universal gears	3.0–4.0	**Machines for working of wood and plastics**	3.0–4.0
Small fans	2.5–3.5		

Particular	Formula

Reliability

The reliability (R_i) of a group of i bearings	$R_i = (R)^i$ (23-220) where R = reliability of each bearing
The expression for reliability (R) as per Weibull three-parameter	$R = \exp\left[-\left(\frac{x - x_o}{\theta - x_o}\right)^b\right] = \exp\left[-\left(\frac{L/L_{10} - x_o}{\theta - x_o}\right)^b\right]$ (23-221a)
Another Weibull three-parameter equation for reliability (R) for bearings.	$R = \exp\left[-\left(\frac{(L/L_{10}) - 0.02}{4.91}\right)^{1.40}\right]$ (23-221b)
The reliability (R) of bearing using Weibull two-parameter for tapered roller bearings.	$R = \exp\left[-\left(\frac{x}{\theta}\right)^b\right] = \exp\left[-\left(\frac{L/L_{10}}{4.48}\right)^{1.5}\right]$ (23-222a) where x = life measure x_o = guaranteed values of life measure θ = Weibull characteristic of life measure b = Weibull exponent/shape parameter
Another form of reliability (R) equation for bearing using Weibull two-parameter	$R = \exp\left[-\left(\frac{L}{mL_{10}}\right)^b\right]$ (23-222b) where R = reliability corresponding to life L L_{10} = rating life ($R = 0.90$) m = scale constant
Weibull two-parameter equation for reliability is obtained from Eq. (23-225a) by putting $b = 1.17$ and $\theta = 6.84$.	$R = \exp\left[-\left(\frac{L}{6.84L_{10}}\right)^{1.17}\right]$ (23-223)
Weibull equation for the distribution of bearing rating life based on reliability.	$\frac{L}{L_{10}} = \left(\frac{\ln(1/R)}{\ln(1/R_{10})}\right)^{1/b}$ (23-224)
The relation between the design or required values and the dynamic load rated or catalog values (C_r) according to the Timken Engineering is given by	$C_r = F_r\left[\left(\frac{L_d}{L_r}\right)\left(\frac{n_d}{n_r}\right)\right]^{1/m}$ (23-225) where subscripts d and r stand for design and rated values C_r = basic load capacity or dynamic load rating corresponding to L_r hours of L_{10} life at the speed n_r in rpm, kN

Particular	Formula
	F_r = actual radial bearing load carried for L_d hours of L_{10} life at the speed n_d in rpm, kN
	m = an exponent which varies from 3 to 4
The basic dynamic capacity or specific dynamic capacity of bearing corresponding to any desired life L at the reliability R	$$C_r = F_r\left[\left(\frac{L_d}{L_r}\right)\left(\frac{n_d}{n_r}\right)\left(\frac{1}{6.84}\right)\right]^{1/m}\frac{1}{[\ln(1/R)]^{1/1.17m}} \quad (23\text{-}226)$$
Another equation connecting catalog radial load rating (F_r), the design radial load (F_d) and reliability (R).	$$C_r = F_d\left[\frac{L_d n_d/L_r n_r}{0.02 + 4.439[\ln(1/R)]^{1/1.483}}\right]^{1/m} \quad (23\text{-}227)$$ where C_r = the catalog radial load rating corresponding to L_r hours of life at the rated speed n_r in rpm, kN; F_d = the design radial load corresponding to the required life of L_d hours at a design speed of n_d in rpm, kN; R = reliability

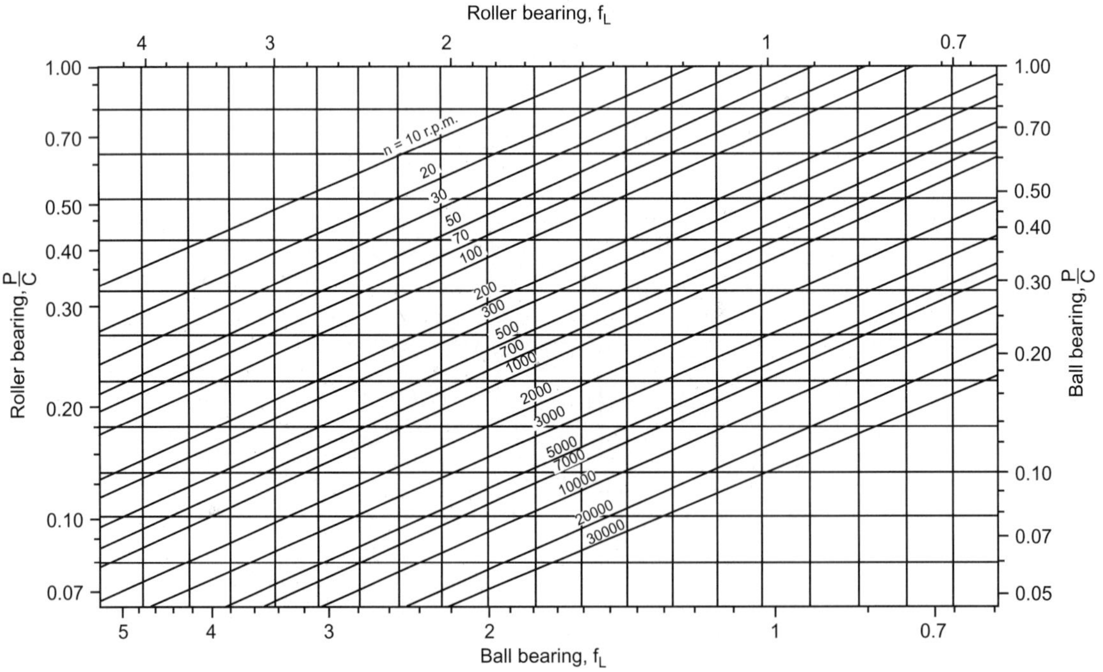

FIGURE 23-56 Selection of bearing size.

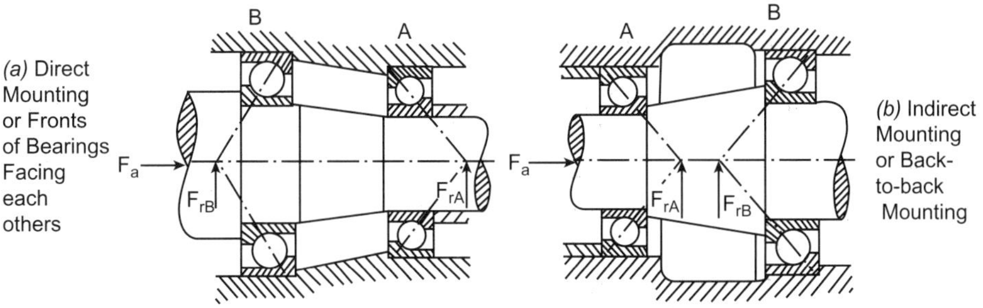

FIGURE 23-57 Angular contact ball bearings mounted on a single shaft.

Condition of load	Thrust load to be used in equivalent load calculation: Bearing A (Fig. 23-57)	Bearing B (Fig. 23-57)	
$\frac{F_{rB}}{Y_B} \le \frac{F_{rA}}{Y_A}$	—	$F_a + 0.5\frac{F_{rA}}{Y_A}$	(23-228)
$\frac{F_{rB}}{Y_B} > \frac{F_{rA}}{Y_A}$ $F_a > 0.5\left(\frac{F_{rB}}{Y_B} - \frac{F_{rA}}{Y_A}\right)$	—	$F_a + 0.5\frac{F_{rA}}{Y_A}$	(23-229)
$\frac{F_{rB}}{Y_B} > \frac{F_{rA}}{Y_A}$ $F_a \le 0.5\left(\frac{F_{rB}}{Y_B} - \frac{F_{rA}}{Y_A}\right)$	$0.5\frac{F_{rB}}{Y_B} - F_a$	—	(23-230)

Where thrust factors are: $Y = 0.57$ for Series 72*B* (Series 02) and 73*B* (Series 03); $Y = 1.19$ for Series *LS AC* and *MS AC*; $Y = 0.87$ for Series 173 and 909; $Y = 0.66$ for $F_a/F_r \le 0.95$ and $Y = 1.07$ for $F_a/F_r > 0.95$ for Series 33.

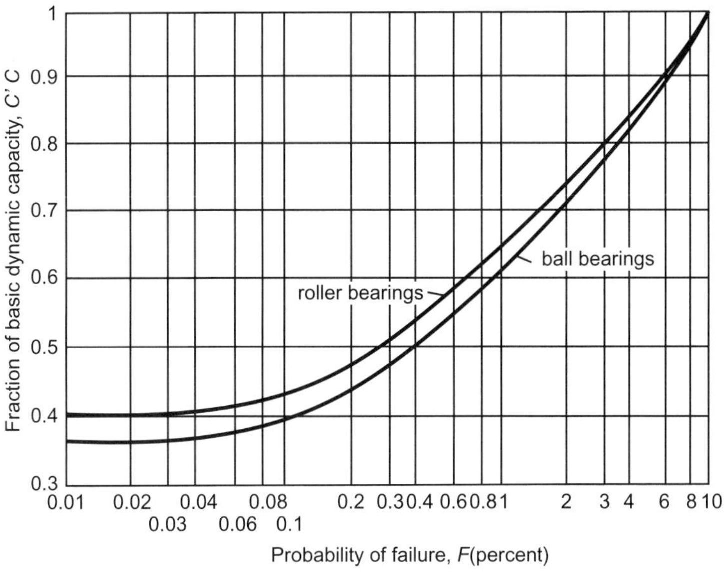

FIGURE 23-58 Reduction in life for reliabilities greater than 90%. (*Courtesy*: Tedric A Harris, Predicting Bearing Reliability, *Machine Design*, Vol. 35, No. 1, Jan. 3, 1963, pp. 129–132)

Particular	Formula
An expression for tapered roller bearings connecting catalog radial load rating (F_r), the design radial load (F_d) and reliability	$C_r = F_d\left[\frac{L_d n_d/L_r n_r}{4.4[\ln(1/R)]^{1/1.5}}\right]^{3/10}$ (23-231)
The radial equivalent or effective load when the cup rotates in case of tapered roller bearing (Fig. 23-50)	$F_r = 1.25F_r$ (23-232) where F_r is the calculated radial load, kN
The thrust component of pure radial load (F_r) due to the tapered roller	$F_{an} = \frac{0.47F_r}{K}$ (23-233) where $K = \frac{\text{radial rating of bearing}}{\text{thrust rating of bearing}}$ $= 1.5$ for radial bearings $= 0.75$ for steep-angle bearings
The net thrust on the tapered roller bearing when the induced thrust (F_{ar}) is deducted from the applied thrust (F_{aa})	$F_{nt} = F_{aa} - F_{ar}$ (23-234)
	$F_{nt} = F_{aa} - \frac{0.47F_r}{K}$ (23-235)
The radial equivalent load when the cup rotates in case of tapered roller bearing (Fig. 23-50)	$F_e = F_r + K\left(F_{aa} - \frac{0.47F_r}{K}\right)$ (23-236)
	$F_e = 0.53F_r + KF_{nt}$ (23-237)
The radial equivalent load when the cone rotates in case of tapered roller bearing (Fig. 23-50)	$F_e = 1.25F_r + K\left(F_{aa} - \frac{0.47F_r}{K}\right)$ (23-238)
	$F_e = 0.78F_r + KF_{nt}$ (23-239)

Particular	Formula

THE EQUIVALENT DYNAMIC LOAD FOR TAPERED ROLLER BEARINGS

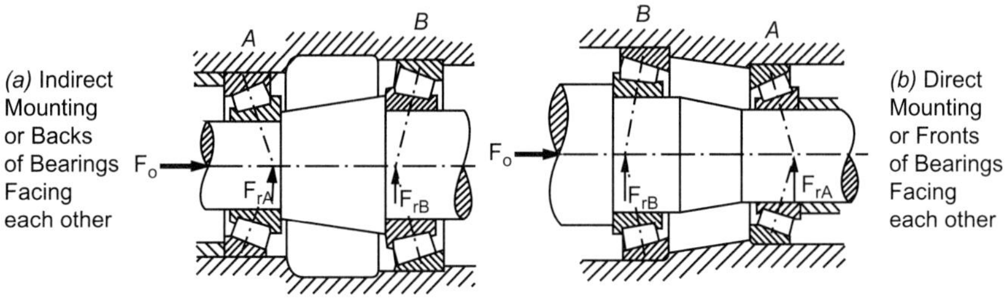

FIGURE 23-59 Two taper roller bearings mounted on a single shaft.

	Thrust load to be used in equivalent load calculation		
Condition of load	**Bearing A (Fig. 23-59)**	**Bearing B (Fig. 23-59)**	
$\frac{F_{rB}}{Y_B} \le \frac{F_{rA}}{Y_A}$	—	$F_a + 0.5\frac{F_{rA}}{Y_A}$	(23-240)
$\frac{F_{rB}}{Y_B} > \frac{F_{rA}}{Y_A}$ $F_a > 0.5\left(\frac{F_{rB}}{Y_B} - \frac{F_{rA}}{Y_A}\right)$	—	$F_a + 0.5\frac{F_{rA}}{Y_A}$	(23-241)
$\frac{F_{rB}}{Y_B} > \frac{F_{rA}}{Y_A}$ $F_a \le 0.5\left(\frac{F_{rB}}{Y_B} - \frac{F_{rA}}{Y_A}\right)$	$0.5\frac{F_{rB}}{Y_B} - F_a$	—	(23-242)

The thrust factors Y and Y_e are taken from Table 23-39 and 23-47*a*.

The radial equivalent load on bearing *A* according to *Timken Engineering Journal* (Fig. 23-59)

$$F_{aA} = 0.4F_{rA} + K_A\left(F_a + \frac{0.46F_{rB}}{K_B}\right) \tag{23-243}$$

The radial equivalent load on bearing *B* according to *Timken Engineering Journal* (Fig. 23-59)

$$F_{eB} = 0.4F_{rB} + K_B\left(\frac{0.47F_{rA}}{K_A} - F_a\right) \tag{23-244}$$

DIMENSIONS, BASIC LOAD RATING CAPACITY, FATIGUE LOAD LIMIT AND MAXIMUM PERMISSIBLE SPEED OF ROLLING CONTACT BEARINGS

Deep groove ball bearings—Series 02, Series 03, Series 04 — Refer to Tables 23-60, 23-61, and 23-62 respectively.

Self-aligning and deep groove ball bearings—Series 02, Series 03, Series 22 (FAG) and Series 23 (FAG) — Refer to Tables 23-63, 23-64, 23-65 and 23-66 respectively.

Particular	Formula
Single row angular contact ball bearings—Series 02 and Series 03	Refer to Tables 23-67 and 23-68.
Double row angular contact ball bearings—Series 33 (FAG)	Refer to Table 23-69.
Cylindrical roller bearings—Series 02, Series 03, Series 04, Series NU 22 (FAG), Series NU 23 (FAG)	Refer to Tables 23-70, 23-71, 23-72, 23-73, and 23-74.
Tapered roller bearings—Series 322, Series 02 (22) and Series 03 (23)	Refer to Tables 23-75, 23-76, 23-76A, 23-76B and 23-77.
Single thrust ball bearings—Series 11, Series 12, Series 13 and Series 14	Refer to Tables 23-78, 23-79, 23-80, and 23-81.
Double thrust ball bearing-Series 522 (FAG)	Refer to Table 23-82.
Selection of bearing size	Refer to Table 23-83.

NEEDLE BEARING LOAD CAPACITY

Particular	Formula
For various types of needle roller bearings and for some of their characteristics	Refer to Table 23-59.
The capacity of needle bearing at 3000 h average life	$C_n = 1.76 \times 10^7 \dfrac{Zld}{\sqrt[3]{n'}}$ (23-245) where C_n in N, l, and d in m, and n' in rps
The load capacity of needle bearing based on the projected area of the needle-rollers	$C_n = 5.33 \dfrac{L(d_i + d_r)}{\sqrt[3]{n'}}$ (23-246) where C_n in N, l d_i,and d_r in m, and n' in rps
The load capacity of needle bearing is also calculated from formula	$C_n = K_h K_l p l d_i$ (23-247) For hardness factors K_h refer to Table 23-83 and for life factor K_l refer to Fig. 23-55.

PRESSURE

Particular	Formula
The pressure for wrist pin rocker arm and similar oscillating mechanism is given by	$P = 34.32\,\text{MPa}$
The rotary motion pressure may be computed from the relation	$P = \dfrac{2.86 \times 10^6}{\sqrt[3]{D_1 n'}}$ (23-248) where P in Pa, D_1 in m, and n' in rps
Check for total circumferential clearance from formula	$c = \pi(d_i + d_r) - Zd_r$ (23-249)
For dimensions, design data and sizes for needle bearings.	Refer to Tables 23-84 to 23-88.

Particular	Formula

TABLE 23-59
Typical forms of needle roller bearings and some of their important characteristics.

Type	Bore size (in.)		Relative load capacity		Limiting speed factor	Misalignment tolerance
	min	max	Dynamic	State		
Drawn cup needle 	0.125	7.250	High	Moderate	0.3	Low
Drawn cup needle grease retained 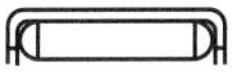	0.156	1.000	High	Moderate	0.3	Low
Drawn cup roller 	0.187	2.750	Moderate	Moderate	0.9	Moderate
Heavy duty roller 	0.625	9.250	Very high	Moderate	1.0	Moderate
Caged roller 	0.500	4.000	Very high	High	1.0	Moderate
Cam follower 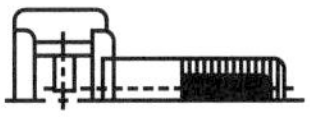	0.5000	6.000	Moderate to high	Moderate to high	0.3–0.9	Low
Needle thrust 	0.252	4.127	Very high	Very high	0.7	Low

Courtesy: *Machine Design*, 1970 Bearings Reference Issue, The Penton Publishing Co., Cleveland, Ohio.

HERTZ-CONTACT PRESSURE

Maximum contact pressure between cylinders and spheres of steel ($\nu = 0.3$)

(i) For cylinders

$$\sigma_{c(\max)} = 0.418\sqrt{\frac{2FE(d_1 + d_2)}{ld_1 d_2}} \tag{23-250}$$

(ii) For a cylinder and plane

$$\sigma_{c(\max)} = 0.418\sqrt{\frac{2FE}{ld}} \tag{23-251}$$

(iii) For two spheres

$$\sigma_{c(\max)} = 0.388\sqrt[3]{\frac{4F(d_1 + d_2)^2 E^2}{d_1^2 d_2^2}} \tag{23-252}$$

(iv) For a sphere and plane

$$\sigma_{c(\max)} = 0.388\sqrt[3]{\frac{4FE^2}{d^2}} \tag{23-253}$$

TABLE 23-60
Deep groove ball bearings—Diameter series 2 (Series 02) (Indian Standards)

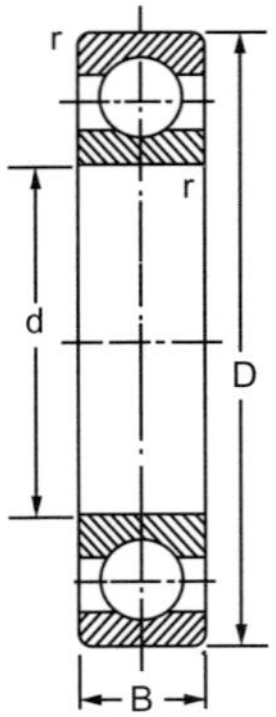

Bearing No.				Dimensions, mm				Factor			Basic load rating capacity				Fatigue load limit, F_a SKF/FAG	Kinematically permissible speed, n SKF/PAG	Mass
IS No.											Static, C_o		Dynamic, C				
		FAG	SKF								FAG	SKF	FAG	SKF			
New	Old			d	D	B	r	F_a/C_o	Y	e	kN	N	kN	N	N	rpm	kg
10	10BC02	6200	6200	10	30	9	0.6	.025	2.0	0.22	2.60	2360	6.00	5070	100	32000	0.031
12	12BC02	6201	01	12	32	10	0.6	.04	1.8	.24	3.10	3100	6.95	6890	132	30000	0.038
15	15BC02	6202	02	15	35	11	0.6	.07	1.6	.27	3.75	3750	7.80	7800	160	26000	0.044
17	17BC02	6203	03	17	40	12	0.6	.13	1.4	.31	4.75	4750	9.50	9560	200	22000	0.063
20	20BC02	6204	04	20	47	14	1.0	.25	1.2	.37	6.55	6550	12.70	12700	280	18000	0.105
25	25BC02	6205	05	25	52	15	1.0	.50	1.0	.44	7.80	7800	14.00	14000	335	17000	0.128
30	30BC02	6206	6206	30	62	16	1.0				11.20	11200	19.30	19500	475	14000	0.199
35	35BC02	6207	07	35	72	17	1.1				15.30	15300	25.50	25500	655	24000	0.290
40	40BC02	6208	08	40	50	18	1.1				18.00	19000	29.00	30700	860	20000	0.372
45	45BC02	6209	09	45	85	19	1.1				20.40	21600	31.00	33200	915	19000	0.430
50	50BC02	6210	10	50	90	20	1.1				24.00	23200	36.50	35100	980	18000	0.466
55	55BC02	6211	6211	55	100	21	1.5				29.00	29000	43.00	43600	1250	16000	0.616
60	60BC02	6212	12	60	110	22	1.5				36.00	32500	52.00	47500	1400	14000	0.785
65	65BC02	6213	13	65	120	23	1.5				41.50	40500	60.00	55900	1730	13000	1.000
70	70BC02	6214	14	70	125	24	1.5				44.00	45000	62.00	60500	1900	12000	1.080
75	75BC02	6215	6215	75	130	25	1.5				49.00	49000	65.50	66300	2040	11000	1.200
80	80BC02	6216	16	80	140	26	2.0				53.00	55000	72.00	70200	2200	11000	1.460
85	85BC02	6217	17	85	150	28	2.0				64.00	64000	83.00	83200	2500	10000	1.870
90	90BC02	6218	18	90	160	30	2.0				72.00	73500	96.50	95600	2890	9000	2.230
95	95BC02	6219	19	95	170	32	2.1				81.50	81500	108.00	10800	3000	8500	2.740
100	100BC02	6220	6220	100	180	34	2.1				93.00	93000	122.00	124000	3350	8000	3.300
105	105BC02	6221	21	105	190	36	2.1				104.00	104000	132.00	133000	3650	7500	3.880
110	110BC02	6222	22	110	200	38	2.1				116.00	118000	143.00	143000	4000	7000	4.640
120	120BC02	6224	24	120	215	40	2.1				122.00	118000	146.00	146000	3900	6700	5.630
130		6226	6226	130	230	40	3.0				146.00	132000	166.00	156000	4150	6300	6.24
140		6228	6228	140	250	42	3.0				166.00	150000	176.00	165000	4150	6000	8.07
150		6230	6230	150	270	45	3.0				170.00	166000	176.00	174000	4900	5600	10.30
160		6232	32	160	290	48	3.0				204.00	186000	200.00	186000	5300	5600	14.70
170		6234M	34	170	310	52	4.0				224.00	224000	212.00	212000	6100	5300	18.30
180		6236M	36	180	320	52	4.0				245.00	240000	224.00	229000	7350	4800	19-00
190		6238M	38	190	340	55	4.0				280.00	280000	255.00	255000	7350	4300	22.80
200		6240M	40	200	360	58	4.0				310.00	310000	270.00	270000	7800	4000	27.20
220		6242M	6244	220	400	65	4.0				355.00	365000	300.00	296000	8800	3600	37.90
240		6244M	6248	240	440	72	5.0				475.00	475000	360.00	358000	10800	3400	51.30
260		6246M	6252	260	480	80	5.0				560.00	530000	405.00	390000	11800	3200	68.40
280		6248M	6256	280	500	80	6.0				600.00	600000	425.00	423000	12900	3000	72.90

TABLE 23-61
Deep groove hall bearings—Diameter series 3 (Series 03) (Indian Standards)

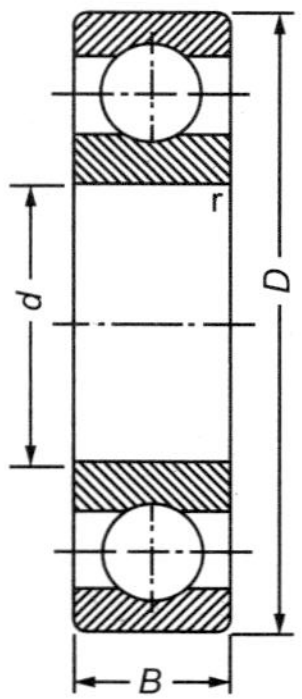

Bearing No.				Dimensions, mm				Factor			Basic load rating capacity				Fatigue load limit, F_a SKF/FAG	Kinematically permissible speed, n SKF/FAG	Mass
IS No.											Static, C_o		Dynamic, C				
		FAG	SKF								FAG	SKF	FAG	SKF			
New	Old			d	D	B	r	F_a/C_o	Y	e	kN	N	kN	N	N	rpm	kg
10	10BC03	6300	6300	10	30	11	0.6	.025	2.0	0.22	3.45	3400	8.15	8060	143	56000	0.058
12	12BC03	6301	6301	12	37	12	1.0	.04	1.8	.24	4.15	4150	9.65	9750	176	53000	0.062
15	15BC03	6302	6302	15	42	13	1.0	.07	1.6	.27	5.40	5400	11.40	11400	228	43000	0.087
17	17BC03	6303	6303	17	47	14	1.0	.13	1.4	.31	6.55	6550	13.40	13500	275	39000	0.116
20	20BC03	6304	04	20	52	15	1.1	.25	1.2	.37	7.80	7800	16.00	15900	335	34000	0.153
25	25BC03	6305	05	25	62	17	1.1	.05	1.0	.44	11.40	11600	22.40	22500	490	28000	0.237
30	30BC03	6306	06	30	72	19	1.1				16.30	16000	29.00	28100	670	24000	0.355
35	35BC03	6307	07	35	80	21	1.5				19.00	19000	33.50	33200	815	20000	0.472
40	40BC03	6308	08	40	90	23	1.5				25.00	24000	42.50	41000	1020	18000	0.639
45	45BC03	6309	09	45	100	25	1.5				32.00	31500	53.00	52700	1340	16000	0.853
50	50BC03	6310	6310	50	110	27	2				38.00	38000	62.00	61800	1600	14000	1.090
55	55BC03	6311	11	55	120	29	2				47.00	45000	76.50	71500	1900	13000	1.400
60	60BC03	6312	12	60	130	31	2.1				52.00	52000	81.50	81900	2200	12000	1.750
65	65BC03	6313	13	65	140	33	2.1				60.00	60000	93.00	92300	2500	11000	2.140
70	70BC03	6314	14	70	150	35	2.1				68.00	68000	104.00	104000	2750	10000	2.610
75	75BC03	6315	15	75	160	37	2.1				76.50	76500	114.00	114000	3000	9500	3.180
80	80BC03	6316	16	80	170	39	2.1				86.50	86500	122.00	124000	3250	9000	3.800
85	85BC03	6317	17	85	180	41	3.0				88.00	96500	125.00	133000	3550	8000	4.350
90	90BC03	6318	18	90	190	43	3.0				102.00	10800	134.00	143000	3800	8000	5.430
95	95BC03	6319	19	95	200	45	3.0				112.00	118000	143.00	153000	4150	7500	6.230
100	100BC03	6320	20	100	215	47	3.0				134.00	140000	163.00	174000	4750	7000	7.670
105	105BC03	6321	6321	105	225	49	3.0				146.00	153000	173.00	182000	5100	6700	8.700
110	110BC03	6322	22	110	240	50	3.0				166.00	180000	190.00	203000	5700	6300	10.300
120	120BC03	6324	24	120	260	55	3.0				190.00	186000	212.00	208000	5700	6000	12.800
130		6326M	26	130	280	58	4.0				216.00	216000	228.00	290000	6300	5600	18.300
140		6328M	28	140	300	62	4.0				245.00	245000	255.00	251000	7100	5300	22.300
150		6330M	30	150	320	65	4.0				300.00	285000	285.00	276000	7800	4800	26.700
160		6332M	32	160	340	68	4.0				325.00	285000	300.00	276000	7650	4300	31.800
170		6334M	34	170	360	72	4.0				365.00	340000	325.00	312000	8800	4000	37.300
180		6336M	6336	180	380	75	4.0				405.00	405000	355.00	351000	10800	3800	43.600
190		6338M	38	190	400	78	5.0				440.00	430000	375.00	371000	10800	3600	50.400
200		6340M	40	200	420	80	5.0				465.00	465000	380.00	377000	11200	3400	56.600
220		6344M	44	220	460	88	5.0				550.00	520000	430.00	410000	12000	3200	75.000
240		6348M		240	500	95	5.0				620.00		465.00			3000	96.400

TABLE 23-62
Deep groove ball bearing—Diameter Series 4 (Series 04) Indian Standards

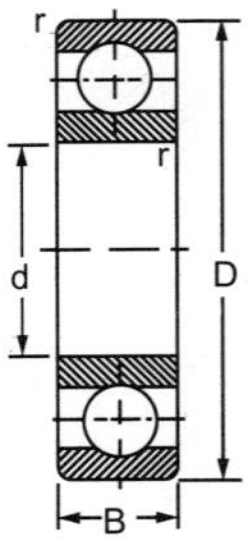

Bearing No.				Dimensions, mm				Factor			Basic load rating capacity				Kinematically permissible speed, n FAG	Mass
IS No.											Static, C_o		Dynamic, C			
											FAG	SKF*	FAG	SKF*		
New	Old	FAG	SKF	d	D	B	r	F_a/C_o	Y	e	kN	N	kN	N	rpm	kg
	15BC04			15	52	15	1.1	0.025	2.0	.22						
17	17BC04	6403	6403	17	62	17	1.1	.04	1.8	.24	11.0	10680	23.6	17350	30000	0.275
20	20BC04	6404	04	20	72	19	1.1	.07	1.6	.27	15.0	15100	30.5	23570	26000	0.412
25	25BC04	6405	05	25	80	21	1.5	.13	1.4	.31	19.3	18500	36.0	27540	22000	0.546
30	30BC04	6406	06	30	90	23	1.5	.25	1.2	.37	23.2	22690	42.5	32690	19000	0.746
35	35BC04	6407	07	35	100	25	1.5	.5	1.0	.44	31.0	29790	55.0	42240	16000	0.928
40	40BC04	6408	08	40	110	27	2				36.5	36900	63.0	48950	15000	1.18
45	45BC04	6409	09	45	120	29	2				45.0	42880	76.5	57330	13000	1.51
50	50BC04	6410	10	50	130	31	2.1				52.0	48900	86.5	67590	12000	1.83
55	55BC04	6411	6411	55	140	33	2.1				62.0	57340	100.0	76880	11000	2.40
60	60BC04	6412	12	60	150	35	2.1				69.5	64880	110.0	82680	10000	2.90
65	65BC04	6413	13	65	160	37	2.1				78.0	75570	118.0	90700	9500	3.49
70	70BC04	6414	14	70	180	42	3.0				104.0	99570	143.0	108880	8500	4.80
75	75BC04	6415M	15	75	190	45	3.0				114.0	106720	153.0	117800	8000	5.64
80	80BC04	6416M	16	80	200	48	3.0				125.0	117800	163.0	124460	7500	6.63
85	85BC04	6417M	17	85	210	52	4.0				137.0	128920	173.0	133280	7000	9.52
90	90BC04	6418M	18	90	225	54	4.0				163.0	142350	196.0	142250	6700	11.6

Source: 1. Extracted with permission from "*FAG Rolling Bearings*", Catalogue WL 41520 EI, 1995 Edition: FAG Precision Bearings Ltd, Manja, Vadodara, India.
2. *Courtesy*: Extracted from SKF Rolling Bearings, Catalogue 4000 E, 1989, SKF Rolling Bearings India Ltd., Mumbai, India.

TABLE 23-63
Self-aligning ball bearings—Diameter Series 12 (Series 02, Indian Standards)

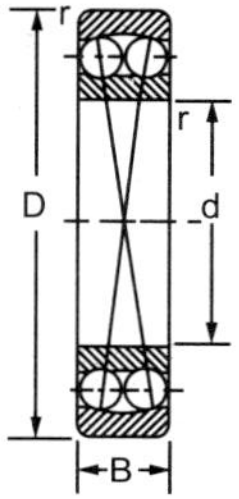

Bearing No.				Dimensions, mm				Factors						Basic load rating capacity				Fatigue load limit, F_n SKF	Kinematically permissible speed, *n* FAG	Mass FAG
IS No.										$F_a/F_r \le e$		$F_a/F_r > e$		Static, C_o		Dynamic, C				
New	Old	FAG	SKF	*d*	*D*	*B*	*r* min	*e*	Y_o	*X*	*Y*	*X*	*Y*	FAG kN	SKF N	FAG kN	SKF N	N	rpm	kg
10	10B502	1200TV	1200E	10	30	9	0.6	.32	2.05	1	1.95	.65	3.2	1.2	1180	5.5	5530	61	30000	0.034
12	12B502	1201TV	01E	12	32	10	0.6	.37	1.77	1	1.69	.65	2.62	1.27	1430	5.6	6240	72	30000	0.041
15	15B502	1202TV	02E	15	35	11	0.6	.34	1.95	1	1.86	.65	2.98	1.76	1760	7.5	7410	90	26000	0.048
17	17B502	1203TV	1203E	71	40	12	0.6	.33	2.03	1	1.93	.65	2.00	2.04	3000	8.0	8840	114	22000	0.073
20	20B502	1204TV	04E	20	47	14	1 0	.27	2.34	1	2.24	.65	3.46	2.65	3400	10.0	12700	176	18000	0.119
25	25B502	1205TV	05E	25	52	15	1.0	.27	2.48	1	2.37	.65	3.66	3.35	4000	12.2	14300	204	16000	0.139
30	30B502	1206TV	1206E	30	62	16	1.1	.22	2.94	1	2.53	.65	3.91	4.65	4650	15.6	15600	240	14000	0.222
35	35B502	1207TV	07E	35	72	17	1.1	.22	2.65	1	2.18	.65	4.34	5.20	6000	16.0	19000	305	12000	0.322
40	40B502	1208TV	08E	40	80	18	1.1	.22	3.04	1	2.90	.65	4.49	6.55	6950	19.3	19900	355	10000	0.415
45	45B502	1209TV	1209E	45	85	19	1.1	.21	3.18	1	3.04	.65	4.70	7.35	7800	22.0	22900	400	9000	0.463
50	50B502	1210TV	10E	50	90	20	1.1	.20	3.32	1	3.17	.65	4.90	8.15	9150	22.8	26500	475	8500	0.525
55	55B502	1211TV	11E	55	100	21	1.5	.19	3.47	1	3.31	.65	5.12	10.00	10600	27.0	27600	540	7500	0.685
60	60B502	1212TV	1212E	60	110	22	1.5	.18	3.64	1	3.47	.65	5.37	11.60	12200	30.0	31200	620	6700	0.895
65	65B502	1213TV	13E	65	120	23	1.5	.18	3.74	1	3.57	.65	5.52	12.50	14000	31.0	35100	720	6300	1.16
70	70B502	1214TV	14	70	125	24	1.5	.19	3.52	1	3.36	.65	5.21	13.70	13700	34.5	34500	710	6000	1.25
75	75B502	1215TV	1215	75	130	25	1.5	.19	3.48	1	3.32	.65	5.15	15.60	15000	39.0	39000	800	5600	1.34
80	80B502	1216TV	16	80	140	26	2.0	.16	4.08	1	3.90	.65	6.03	17.00	17000	46.0	39700	830	5000	1.66
85	85B502	1217TV	17	85	150	28	2.0	.17	3.91	1	3.73	.65	5.78	20.40	20900	49.0	48800	980	4500	2.06
90	90B502	1218TV	1218	90	160	30	2.0	.17	3.92	1	3.74	.65	5.79	23.6	23600	57.0	57200	1080	4500	2.50
95	95B502	1219M	19.	95	170	32	2.1	.17	3.91	1	3.73	.65	5.78	27.0	27000	64.0	65700	1200	6000	3.40
100	100B502	1220M	20	100	180	34	2.1	.18	3.75	1	3.58	.65	5.53	29.0	30000	69.5	68900	1200	5600	3.29
105	105B502	1221M	1221	105	190	36	2.1	.18	3.5	1	3.68	.65	5.48	32.0	32500	75.0	74100	1370	5300	4.31
110	110B502	1222M	22	110	200	38	2.1	.17	3.78	1	3.61	.65	5.58	38.0	39000	88.0	88400	1600	5000	5.67
120	120B502	1224M	24	120	215	42	2.1	.25	3.25	1	3.11	.65	4.81	53.0	53000	120.0	119000	2120	4800	7.43

Note: SKF 1984; FAG, 1995,: * These values of C and C_o of SKF ball bearings refer to old standards in Table 23-62, EK = tapered bore; TV = self-aligning ball bearings with cages of glass fibre reinforced polyamide 66, M = machined brass cage.

Source: 1. Extracted with permission from "*FAG Rolling Bearings*", CatalogueWL 41520EI, 1995 Edition: FAG Precision Bearings Ltd, Maneja, Vadodara, India.
2. *Courtesy*: Extracted from SKF Rolling Bearings, Catalogue 4000E, 1989, SKF Rolling Bearings. India Ltd., Mumbai, India.

TABLE 23-64
Self-aligning ball bearings—Diameter Series 03 [Series 03 (Indian Standards)], Dimensions Series 13 FAG and SKF

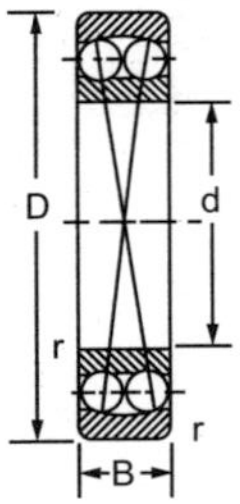

Bearing No.				Dimensions, mm				Factors, FAG						Basic load rating capacity				Fatigue load limit, F_n SKF	Permissible speed, n^{**}		Mass** FAG
IS No.										$F_n/F_r \le e$		$F_n/F_r > e$		Static, C_o		Dynamic, C			Kinematically FAG	Oil SKF	
							r							FAG	SKF	FAG	SKF				
New	Old	FAG	SKF	d	D	B	min	e	Y_o	X	Y	X	Y	kN	N	kN	N	N	rpm		kg
10	10B503	1300	1300	10	35	11	1.0	.34	1.90	1	1.90	.65	2.90								
12	12B503	1301	01E	12	37	12	1.5	.35	1.90	1	1.80	.65	2.80		2160		9360	112		22000	
15	15B503	1302	02E	15	42	13	1.5	.35	1.90	1	1.80	.65	2.80		2600		10800	134		20000	
17	17B503	1303TV	1303E	17	47	14	1.0	.32	2.03	1	1.94	.65	3.00	3.20	3400	12.50	12700	176	18000	17000	0.129
20	20B503	1304TV	04E	20	52	15	1.1	.29	2.27	1	2.17	.65	3.50	3.35	4600	12.50	17000	204	16000	15000	0.164
25	25BS03	1305TV	05E	25	62	17	1.1	.28	2.40	1	2.29	.65	3.54	5.00	5400	18.00	19000	280	14000	12000	0.262
30	30B503	1306TV	1306E	30	72	19	1.1	.26	2.51	1	2.39	.65	3.71	6.30	6800	21.20	22500	355	11000	11000	0.391
35	35BS03	1307TV	07E	35	80	21	1.5	.26	2.59	1	2.47	.65	3.82	8.00	8500	25.00	26500	430	9500	9000	0.570
40	40B503	1308TV	08E	40	90	23	1.5	.25	2.64	1	2.52	.65	3.90	9.65	11200	29.00	33800	570	8500	8000	0.711
45	45B503	1309TV	1309E	45	100	25	1.5	.25	2.62	1	2.50	.65	3.87	12.90	13400	38.00	39000	695	7500	7500	0.957
50	50B503	1310TV	10E	50	110	27	2.0	.24	2-73	1	2.60	.65	4.03	14.30	14000	41.50	43600	720	6700	6700	1.25
55	55B503	1311TV	11E	55	120	29	2.0	.24	2.79	1	2.66	.65	4.12	18.00	18000	51.00	50700	915	6000	6000	1.59
60	60B503	1312TV	1312E	60	130	31	2.1	.23	2.90	1	2.77	.65	4.28	20.80	22000	57.00	58500	1120	5300	5300	1.96
65	65B503	1313TV	13E	65	140	33	2.1	.23	2.88	1	2.75	.65	4.26	22.80	25500	62.00	65000	1250	5000	5300	1.83
70	70B503	1314M	14	70	150	35	2.1	.23	2.93	1	2.79	.65	4.32	27.50	27500	75.00	74100	1340	7000	4800	3.42
75	70B503	1315M	1315	75	160	37	2.1	.23	2.90	1	2.77	.65	4.29	30.00	30000	80.00	79300	1430	6300	4500	3.65
80	80B503	1316M	16	80	170	39	2.1	.22	3.00	1	2.87	.65	4.44	32.50	33500	88.00	88400	1500	6000	4300	4.76
85	85B503	1317M	17	85	180	41	3.0	.22	3.02	1	2.88	.65	4.46	38.00	38000	98.00	97500	1700	5600	4000	5.19
90	90B503	1318M	1318	90	190	43	3.0	.22	2.97	1	2.83	.65	4.38	43.00	44000	108.00	117000	1930	5300	3600	6.13
95	95B503	1319M	19	95	200	45	3.0	.23	2.50	1	2.73	.65	4.23	51.00	51000	132.00	133000	2160	5000	3600	6.55
100	100B503	1320M	20	100	215	47	3.0	.23	2.81	1	2.68	.65	4.15	58.50	57000	143.00	143000	2360	4800	3400	8.70
105	105B503	1321M	1321	105	225	49	3.0	.23	2.88	1	2.75	.65	4.25	65.50		156.00			4500		9.89
110	110B503	1322M	22	110	240	50	3.0	.23	2.92	1	2.79	.65	4.32	71.00	7200	163.00	16300	2750	4500	3000	11.80

Source: 1. Extracted with permission from "*FAG Rolling Bearings*", Catalogue WL 41520EI, 1995 Edition: FAG Precision Bearings Ltd., Maneja, Vadodara, India.
2. *Courtesy*: Extracted from SKF Rolling Bearings, Catalogue 4000E, 1989, SKF Rolling Bearings India Ltd., Mumbai, India.

TABLE 23-65
Self-aligning ball bearings—Dimension Series 22—FAG and SKF

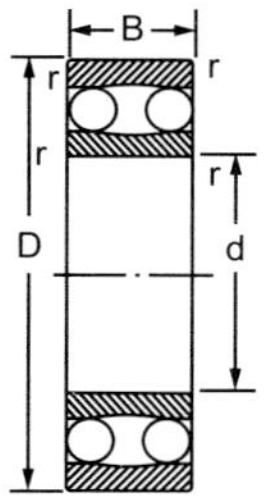

Bearing No.		Dimensions, mm						Factors, FAG				Basic load rating capacity				Fatigue load limit, F_n SKF	Permissible speed, n		Mass FAG
								$F_n/F_r \leq e$		$F_n/F_r > e$		Static, C_o		Dynamic, C			Kinematically FAG	Oil SKF	
FAG	SKF	d	D	B	r min	e	Y_o	X	Y	X	Y	FAG kN	SKF N	FAG kN	SKF N	N	rpm	rpm	kg
2200TV	2200E	10	30	14	0.6	.58	1.14	1	1.09	.65	1.69	1.73	1730	8.3	8060	90	28000	28000	0.045
2201TV	01E	12	32	14	0.6	.58	1.25	1	1.20	.65	1.85	1.96	1900	9.0	8520	98	26000	26000	0.050
2202TV	02E	15	35	14	0.6	.46	1.44	1	1.37	.65	2.13	2.08	2040	9.15	8710	104	24000	22000	0.017
2203TV	2203E	17	40	16	0.6	.46	1.43	1	1.37	.65	2.17	2.75	2550	11.40	10600	132	19000	20000	0.086
2204TV	04E	20	47	18	1.0	.44	1.51	1	1.45	.65	2.24	3.55	4150	14.30	16800	216	17000	17000	0.136
2205TV	05E	25	52	18	1.0	.35	1.86	1	1.75	.65	2.75	4.40	4400	17.00	16800	228	15000	14000	0.159
2206TV	2206E	30	62	20	1.0	.30	2.23	1	2.13	.65	3.29	6.95	6700	25.50	23800	345	12000	12000	0.259
2207TV	07E	35	72	23	1.1	.30	2.23	1	2.13	.65	3.29	9.00	8800	32.00	30700	455	9500	10000	0.404
2208TV	08E	40	80	23	1.1	.26	2.54	1	2.43	.65	3.76	9.50	10000	31.50	31000	510	9000	9000	0.488
2209TV	2209E	45	85	23	1.1	.26	2.54	1	2.43	.65	3.76	9.50	10600	28.00	32500	540	8500	8500	0.527
2210TV	10E	50	90	23	1.1	.24	2.74	1	2.61	.65	4.05	9.50	11200	28.00	33800	570	8000	7500	0.567
2211TV	11E	55	100	25	1.5	.22	3.06	1	2.93	.65	4.53	12.70	13400	39.00	39000	695	6700	7000	0.763
2212TV	2212E	60	110	28	1.5	.23	2.82	1	2.69	.65	4.16	16.60	17000	47.50	48800	880	5300	6300	1.08
2213TV	13E	65	120	31	1.5	.23	2.92	1	2.78	.65	4.31	19.30	20000	57.00	57200	1020	5300	6000	1.36
2214M	14	70	125	31	1.5	.27	2.45	1	2.34	.65	3.62	17.00	17000	44.00	44200	880	8500	5000	1.10
2215TV	2215	75	130	31	1.5	.26	2.59	1	2.47	.65	3.82	18.00	18000	44.00	44200	900	5300	5300	1.20
2216TV	16E	80	140	33	2.0	.25	2.6	1	2.48	.65	3.84	20.00	25500	49.00	65000	1250	5000	4800	2.10
2217M	17	85	150	36	2.0	.26	2.58	1	2.46	.65	3.81	23.60	23600	58.50	58500	1120	7000	4500	2 68
2218TV	2218	90	160	40	2.0	.27	2.44	1	2.33	.65	3.61	28.50	28500	71.00	70200	1320	4300	4300	3.30
2219M.	19	95	170	43	2.1	.27	2.43	1	2.32	.65	3.59	34.0	34500	83.00	83200	1530	6000	4000	4.10
2220TV	20	100	180	46	2.1	.27	2.44	1	2.33	.65	3.61	40.50	40500	98.00	97500	1760	5600	3800	4.98
	2221	105	190	50	2.1								45000		108000	1900	3000	3600	6.10
2220M	22	110	200	53	2.1	.28	2.33	1	2.23	.65	3.45	52.00	52000	125.00	124000	2120	5000	3400	7.10

Source: 1. Extracted with permission from "*FAG Rolling Bearings*", Catalogue WL 41520EI, 1995 Edition: FAG Precision Bearings Ltd., Maneja, Vadodara, India.
2. *Courtesy*: Extracted from SKF Rolling Bearings, Catalogue 4000E, 1989, SKF Rolling Bearings India Ltd., Mumbai, India.

TABLE 23-66
Self-aligning ball bearings—Dimension Series 23 FAG and SKF

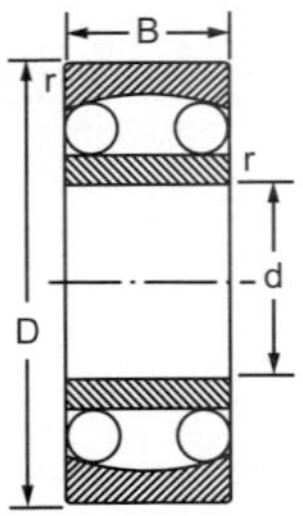

Bearing No.		Dimensions, mm						Factors, FAG				Basic load rating capacity				Fatigue load limit, F_n SKF	Permissible speed, n		Mass FAG
								$F_n/F_r \le e$		$F_n/F_r > e$		Static, C_o		Dynamic, C			Kinematically FAG	Oil* SKF	
												FAG	SKF	FAG	SKF				
FAG	SKF	d	D	B	r	e	Y_o	X	Y	X	Y	kN	N	kN	N	N	rpm		kg
2301TV	2301	12	37	17	1.0								2700		11700	140	17000	20000	0.095
2302TV	02	15	42	17	1.0	.51	1.29	1	1.23	.65	1.91	3.75	2900	16.0	11000	150	18000	18000	0.115
2303TV	03	17	47	19	1.0	.51	1.25	1	1.19	.65	1.85	3.20	3550	13.4	14600	183	17000	16000	0.172
2304TV	2304	20	52	21	1.1	.51	1.29	1	1.23	.65	1.9	4.65	4750	18.0	18200	240	16000	14500	0.226
2305TV	05	25	62	24	1.1	.48	1.38	1	1.32	.65	2.04	6.55	6550	24.5	24200	340	18000	12000	0.335
2306TV	06	30	72	27	1.1	.45	1.47	1	1.4	.65	2.17	8.65	8800	31.5	31200	450	10000	10000	0.500
2307TV	2307E	35	80	31	1.5	.47	1.42	1	1.35	.65	2.1	11.2	11200	39.0	39700	585	9000	8500	0.675
2308TV	08E	40	90	33	1.5	.43	1.52	1	1.45	.65	2.25	13.4	16000	45.0	54000	815	8000	7500	0.925
2309TV	09E	45	100	36	1.5	.43	1.55	1	1.48	.65	2.29	16.3	19300	54.0	63700	1000	7000	6700	1.23
2310TV	2310	50	110	40	2.0	.43	1.44	1	1.47	.65	2.27	20.0	20000	64.0	63700	1040	6300	5500	1.60
2311TV	11	55	120	43	2.0	.42	1.58	1	1.51	.65	2.33	23.6	24000	75.0	76100	1250	5600	5600	2.06
2312TV	12	60	130	46	2.1	.41	1.62	1	1.55	.65	2.4	28.0	28500	86.5	87100	1450	5000	5300	2.74
2313M	2313	65	140	48	2.1	.39	1.70	1	1.62	.65	2.51	32.5	32500	95.00	95600	1660	4800	4800	3.33
2314M	14	70	150	51	2.1	.38	1.73	1	1.65	.65	2.55	37.5	37500	110.0	111000	1860	6300	4500	4.52
2315M	15	75	160	55	2.1	.38	1.72	1	1.64	.65	2.54	42.5	43000	122.0	124000	2040	6000	4000	5.13
2316M	2316	80	170	58	2.1	.37	1.78	1	1.7	.65	2.62	48.0	49000	137.0	135000	2240	5600	3800	5.50
2317M	17	85	180	60	3.0	.37	1.76	1	1.68	.65	2.61	51.0	51000	140.0	140000	2280	5300	3600	7.05
2318M	18	90	190	64	3.0	.39	1.71	1	1.68	.65	2.53	57.0	57000	153.0	153000	2500	5000	3400	8.44
2319M	2319	95	200	67	3.0	.38	1.74	1	1.66	.65	2.57	64.0	64000	163.0	165000	2750	4800	3200	9.86
2320M	20	100	215	73	3.0	.38	1.75	1	1.67	.65	2.58	78.0	80000	193.0	190000	3200	4500	3000	12.40
2322M	22	110	240	80	3.0	.37	1.77	1	1.69	.65	2.62	95.0	95000	216	216000	3650	4300	2300	16.90

Source: 1. Extracted with permission from "*FAG Rolling Bearings*", Catalogue WL 41520EI, 1995 Edition: FAG Precision Bearings Ltd., Maneja, Vadodara, India.
2. *Courtesy*: Extracted from SKF Rolling Bearings, Catalogue 4000E, 1989, SKF Rolling Bearings India Ltd., Mumbai, India.

TABLE 23-67
Single row angular contact ball bearings—Dimension Series 02 (Indian Standards)

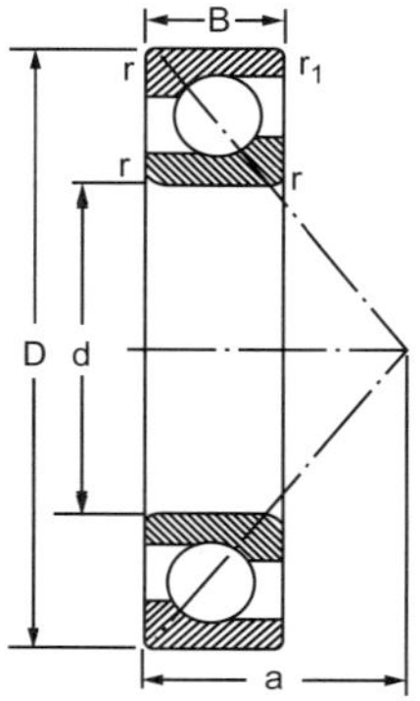

Bearing No.			Dimensions, mm						Basic load rating capacity				Fatigue load limit, F_{uf}	Permissible speed, n		Mass
									Static, C_o		Dynamic, C			Kinematically	Oil[a]	
IS Old No.	FAG	SKF	d	D	B	r min	r_1 min	a	FAG	SKF	FAG	SKF	SKF	FAG	SKF	FAG
									kN	N	kN	N	N	rpm		kg
	7200B	7200BE	10	30	9	0.6	0.3	13	2.5	3350	5.00	7020	140	32000	27000	0.028
	7201B	7201BE	12	32	10	0.6	0.3	14	3.4	3800	6.95	7610	160	28000	26000	0.036
15BA02	7202B	7202BE	15	35	11	0.6	0.3	16	4.3	4860	8.00	8840	204	24000	24000	0.045
17BA02	7203B	7203BE	17	40	12	0.6	0.6	18	5.5	6100	10.00	11100	286	20000	20000	0.07
20BA02	7204B	04BE	20	47	14	1.0	0.6	21	7.65	83000	13.40	14000	355	18000	17000	0.103
25BA02	7205B	05BE	25	52	15	1.0	0.6	24	9.30	10200	14.60	15600	420	16000	15000	0.127
30BA02	7206B	06BE	30	62	16	1.0	0.6	27	13.40	15600	20.40	23800	655	13000	12000	0.207
35BA02	7207B	07BE	35	72	17	1.1	0.6	31	18.30	20800	27.00	30700	880	11000	11000	0.296
40BA02	7208B	08BE	40	80	18	1.1	0.6	34	23.20	26000	32.00	36400	1100	9500	9500	0.377
45BA02	7209B	09BE	45	85	19	1.1	0.6	37	26.50	28000	36.00	37700	1200	8500	9000	0.430
50BA02	7210B	10BE	50	90	20	1.1	0.6	39	28.50	30500	37.50	39000	1290	8000	8000	0.485
55BA02	7211B	7211BE	55	100	21	1.5	1.0	43	36.00	38000	46.50	48800	1630	7000	7500	0.645
60BA02	7212B	12BE	60	110	22	1.5	1.0	47	44.00	45500	56.00	57200	1930	6300	6700	0.779
65BA02	7213B	13BE	65	120	23	1.5	1.0	50	53.00	54000	64.00	66300	2280	6000	6000	0.975
70BA02	7214B	14BE	70	125	24	1.5	1.0	53	58.50	60000	69.50	71500	2500	5600	5600	1.07
75BA02	7215B	15BE	75	130	25	1.5	1.0	56	58.50	64000	68.00	72800	2650	5300	5600	1.19
80BA02	7216B	16BE	80	140	26	2	1.5	59	69.50	73000	80.00	83200	3000	5000	5000	1.42
85BA02	7217B	17BE	85	150	28	2	1.0	63	80.00	83000	90.00	95600	3250	4500	4800	1.89
90BA02	7218B	18BE	90	160	30	2	1.1	67	93.00	96500	106.00	106000	3650	4300	4500	2.22
95BA02	7219B	19BE	95	170	32	2.1	1.1	72	100.00	108000	116.00	124000	4000	4000	4300	2.66
100BA02	7220B	20BE	100	180	34	2.1	1.1	76	114.00	122000	129.00	135000	4400	3800	4000	3.18
105BA02	7221B	7221BE	105	190	36	2.1	1.1	80	129.00	137000	143.00	146000	4800	3600	3800	3.19
110BA02	7222B	22BE	110	200	38	2.1	1.1	84	143.00	153000	153.00	163000	5200	3600	3600	4.44
120BA02	7222B	7224BE	120	215	40	2.1	1.1	90	160.00	163000	166.00	165000	5300	3400	3200	5.31

Use $X_e = 1$, when $F_a/F_r \leq 1.9$; $X_e = 0.5$, $Y_o = 0.26$ when $F_a/F_r > 1.9$, $X = 1$ when $F_a/F_r \leq 1.4$; $X = 0.35$, $Y = 0.57$ when $F_a/F_r > 1.14$. [a] Oil lubrication, E = cylindrical bore, EK, K = tapered bore, TV = self-aligning bearings with caging of glass-fiber reinforced polyamide, M = ball-riding mechanical brass caps.

Source: 1. Extracted with permission from "*FAG Rolling Bearings*", Catalogue WL 41520EI, 1995 Edition: FAG Precision Bearings Ltd., Maneja, Vadodara, India.
2. *Courtesy*: Extracted from SKF Rolling Bearings, Catalogue 4000E, 1989, SKF Rolling Bearings India Ltd., Mumbai, India.

TABLE 23-68
Single row angular contact hall bearings—Dimension Series 03 (Indian Standards)

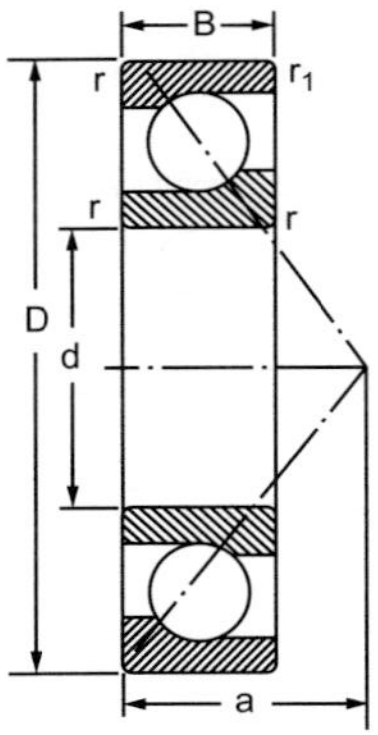

Bearing No.				Dimensions, mm						Basic load rating capacity				Fatigue load limit, F_{uf}	Permissible speed, n		Mass
IS No.										Static, C_o		Dynamic, C			Kinematically	Oil[a]	
Old	New	FAG	SKF	d	D	B	r min	r_1 min	a	FAG	SKF	FAG	SKF	SKF	FAG	SKF	FAG
										kN	N	kN	N	N	rpm		kg
	7300B			10	35	11	1.0	.5	15								
	7301B			12	37	12	1.0	.6	16.0	5.00	5000	10.50	10600	208	24000	24000	0.059
	7302B			15	42	13	1.0	.6	18.0	6.55	6700	12.90	13000	280	20000	20000	0.09
17	17BA03	7303B	7303BE	17	47	14	1.0	.6	20	8.30	8300	16.00	15900	355	18000	18000	0.113
20	20BA03	7304B	04BE	20	52	15	1.1	.6	23	10.40	10400	19.00	19000	440	17000	16000	0.147
25	25BA03	7305B	05BE	25	62	17	1.1	.6	27	15.00	15600	26.00	26000	655	14000	13000	0.221
30	30BA03	7306B	7306BE	30	72	19	1.1	.6	31	20.00	21200	32.50	34500	900	11000	11000	0.342
35	35BA03	7307B	07BE	35	80	21	1.5	1.0	35	25.00	24500	39.00	39000	1640	9500	10000	0.447
40	40BA03	7308B	08BE	40	90	23	1.5	1.0	39	32.50	33500	50.00	49400	1400	8500	9000	0.657
45	45BA03	7309B	7309BE	45	100	25	1.5	1.0	43	40.00	41000	60.00	60500	1730	7500	8000	0.821
50	50BA03	7310B	10BE	50	110	27	2.0	1.0	47	47.50	51000	69.5	74100	2200	7000	7000	1.050
55	55BA03	7311B	11BE	55	120	29	2.0	1.1	51	56.00	60000	78.00	85200	2550	6300	6300	1.36
60	60BA03	7312B	7312BE	60	130	31	2.1	1.1	55	65.50	69500	90.00	95600	3000	5600	6000	1.72
65	65BA03	7313B	13BE	65	140	33	2.1	1.1	60	75.00	80000	102.00	108000	3350	5300	5600	2.10
70	70BA03	7314B	14BE	70	150	35	2.1	1.1	64	86.50	90000	114.00	119000	3650	5000	5000	2.53
75	75BA03	7315B	7315BE	75	160	37	2.1	1.1	68	100.00	106000	127.00	133000	4150	4500	4800	3.18
80	80BA03	7316B	16B	80	170	39	2.1	1.1	72	114.00	118000	140.00	143000	4500	4300	4500	3.75
85	85BA03	7317B	17B	85	180	41	3	1.1	76	127.00	132000	150.00	158000	4900	4000	4300	4.27
90	90BA03	7318B	7318B	90	190	43	3	1.1	80	140.00	146000	160.00	165000	5200	3800	4000	5.72
95	95BA03	7319B	19B	95	200	45	3	1.1	84	153.00	163000	173.00	176000	5600	3800	3800	5.99
100	100BA03	7320B	20B	100	215	47	3	1.1	90	180.00	190000	193.00	203000	6400	3600	3600	7.14
105	105BA03	7321B	7321B	105	225	49	3	1.1	94	200.00	208000	208.00	212000	6400	5300	3400	9.00
110	110BA03	7322B	22B	110	240	50	3	1.1	98	224.00	224000	224.00	225000	7200	3400	3200	9.74

Use $X_e = 1$, when $F_a/F_r \leq 1.9$; $X_e = 0.5$, $Y_o = 0.26$ when $F_a/F_r > 1.9$, $X = 1$ when $F_a/F_r \leq 1.4$; $X = 0.35$, $Y = 0.57$ when $F_a/F_r > 1.14$. [a] Oil lubrication.

Source: 1. Extracted with permission from "*FAG Rolling Bearings*", Catalogue WL 41520EI, 1995 Edition: FAG Precision Bearings Ltd., Maneja, Vadodara, India.
2. *Courtesy*: Extracted from SKF Rolling Bearings, Catalogue 4000E, 1989, SKF Rolling Bearings India Ltd., Mumbai, India.

TABLE 23-69
Double-row angular contact ball bearings series 33

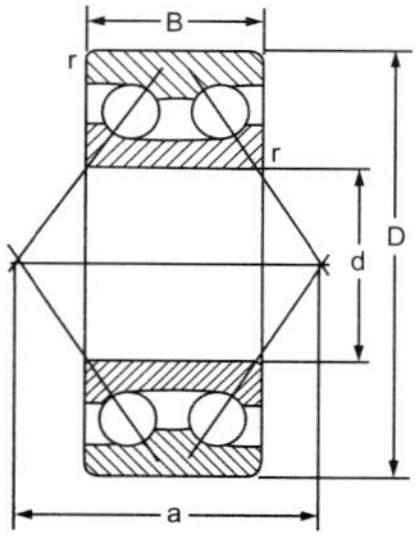

Bearing No.		Dimensions, mm					Basic capacity				
							Static, C_o		Dynamic, C		
							FAG	SKF	FAG	SKF	Maximum permissible speed, rpm
FAG	SKF	d	D	B	r	a	N	N	N	N	
3302	3302A	15	42	19	1.5	30	10,243	9,065	12,887	13,720	10,000
3303	03A	17	47	22.2	1.5	34	14,455	12,642	18,032	18,914	8,000
3304	04A	20	52	22.2	2	36	15,092	13,720	18,424	18,914	8,000
3305	3305A	25	62	25.4	2	43	21,750	19,600	25,382	26,008	6,000
3306	06A	30	72	30.3	2	51	29,792	27,146	32,814	35,280	6,000
3307	07A	35	80	34.9	2.5	56	39,200	35,574	42,924	43,615	5,000
3308	3308A	40	90	36.5	2.5	64	50,666	44,590	52,479	53,410	5,000
3309	09A	45	100	39.7	2.5	72	63,602	54,390	63,602	62,230	4,000
3310	10A	50	110	44.4	3	79	78,353	72,520	78,164	85,995	4,000
3311	3311A	55	120	49.2	3	87	90,699	78,400	88,896	85,840	4,000
3312	12A	60	130	54	3.5	96	106,722	94,570	101,332	98,000	3,000
3313	13A	65	140	58.7	3.5	102	124,460	108,870	117,796	115,640	3,000
3314	3314A	70	150	63.5	3.5	109	140,042	126,430	128,919	135,730	3,000
3315	15A	75	160	68.5	3.5	117	157,780	127,940	144,520	140,740	2,500
3316	16A	80	170	68.3	3.5	123	173,361	153,860	157,780	157,780	2,500
3317	3317A	85	180	73	4	131	202,272	173.460	177,821	173,460	2,500
3318	18A	90	190	73	4	136	228,182	205,800	193,608	200,018	2,500
3319	19A	95	200	77.8	4	143	253,379		206,682		
3320	20A	100	215	82.6	4	153	284,494		224,723		

Note: These bearings are provided with filling slots on one side; in case of unidirectional thrust loads, the bearings should be so arranged in mounting that the balls on the slot side are relieved from load.Use $X_o = 1$, $Y_o = 0.58$ and $X = 1$, $Y - .66$, when $F_a/F_r \leq 0.95$; $X = .6$, $Y = 1.07$ when $F_a/F_r > 0.95$.

TABLE 23-70
Cylindrical roller bearings—Dimension Series 02 (Indian Standards)

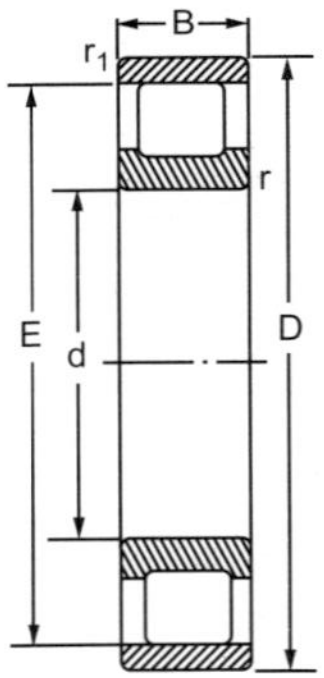

Bearing No.			Dimensions, mm						Basic load rating capacity				Fatigue load limit, F_{uf}	Permissible speed, n		Mass
									Static, C_o		Dynamic, C			Kinematically	Oil[a]	
IS	FAG	SKF	d	D	B	r min	r_1 min	E	FAG	SKF	FAG	SKF	SKF	FAG	SKF	FAG
									kN	N	kN	N	N	rpm		kg
10RN02			10	30	9	1.0										
12RN02			12	32	10	1.0										
15RN02			15	35	11	1.0										
17RN02	N203E	N203EC	17	40	12	0.6	0.3	35.1	14.6	14300	17.6	17200	1730	18000	19000	0.067
20RN02	N204E	204EC	20	47	14	1.0	0.6	41.5	24.5	22000	27.5	25100	2700	16000	16000	0.107
25RN02	N205E	205EC	25	52	15	1.0	0.6	46.5	27.5	27000	29.0	28600	3350	15000	14000	0.139
30RN02	N206E	N206EC	30	62	16	1.0	0.6	55.6	37.5	36500	39.0	38000	4550	12000	12000	0.205
35RN02	N207E	207EC	35	72	17	1.1	0.6	64.0	50.0	48000	50.0	48000	6100	10000	10000	0.300
40RN02	N208E	208EC	40	80	18	1.1	1.1	71.5	53.0	55000	53.0	53000	6700	9000	9000	0.380
45RN02	N209E	N209EC	45	85	19	1.1	1.1	76.5	63.0	64000	61.0	60500	8150	8500	8000	0.434
50RN02	N210E	210EC	50	90	20	1.1	1.1	81.5	68.0	69500	64.0	64000	8800	8000	7500	0.493
55RN02	N211E	211EC	55	100	21	1.5	1.1	90.0	95.0	95000	83.0	84200	12200	7000	7000	0.669
60RN02	N212E	N212EC	60	110	22	1.5	1.5	100.0	104.0	102000	95.0	93500	13400	6300	6300	0.827
65RN02	N213E	213EC	65	120	23	1.5	1.5	108.6	120.0	118000	108.0	106000	15600	6000	5600	1.040
70RN02	N214E	214EC	70	125	24	1.5	1.5	113.5	137.0	137000	120.0	119000	18000	5300	5300	1.160
75RN02	N215E	N215EC	75	130	25	1.5	1.5	118.5	156.0	156000	132.0	130000	20400	5300	5300	1.270
80RN02	N216E	216EC	80	140	26	2.0	2-0	127.3	170.0	166000	140.0	138000	21200	4800	4800	1.550
85RN02	N217E	217EC	85	150	28	2.0	2.0	136.5	193.0	200000	163.0	165000	24500	4500	4500	1.870
90RN02	N218E	N218EC	90	160	30	2.0	2.0	145.0	216.0	220000	183.0	183000	27000	4500	4300	2.250
95RN02	N219E	219EC	95	170	32	2.1	2.1	154.5	265.0	265000	220.0	220000	32500	3800	4000	2.750
100RN02	N220E	N220EC	100	180	34	2.1	2.1	163.0	305.0	305000	250.0	251000	36500	4800	3800	3.320
105RN02	N221E	221EC	105	190	36	2.1	2.1	171.5	320.0	315000	280.0	264000	36500	5600	3600	4.690
110RN02	N223E	222EC	110	200	38	2 1	2.1	180.5	365.0	365000	290.0	292000	42500	3400	3400	4.840
120RN02	N224E	224EC	120	215	40	2.1	2.1	195.5	415.0	430000	335.0	341000	49000	3200	3000	5.770

Use $X_e = 1$, $Y_o = Y = 0$.
[a] Oil lubrication.
Source: 1. Extracted with permission from "*FAG Rolling Bearings*", Catalogue WL 41520EI, 1995 Edition: FAG Precision Bearings Ltd., Maneja, Vadodara, India.
2. *Courtesy*: Extracted from SKF Rolling Bearings, Catalogue 4000E, 1989, SKF Rolling Bearings India Ltd., Mumbai, India.

TABLE 23-71
Cylindrical roller bearings—Dimension Series 03 (Indian Standards)

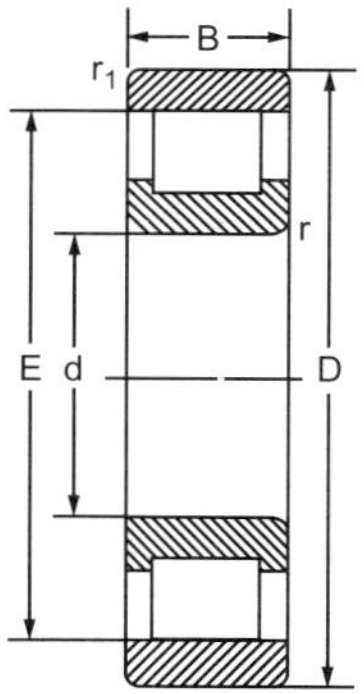

Bearing No.			Dimensions, mm						Basic load rating capacity				Fatigue load limit, F_{uf}	Permissible speed, n		Mass
									Static, C_o		Dynamic, C			Kinematically	Oil[a]	
IS	FAG	SKF	d	D	B	r min	r_1 min	E	FAG	SKF	FAG	SKF	SKF	FAG	SKF	FAG
									kN	N	kN	N	N	rpm		kg
10RN03			10	35	11	1.0										
12RN03			12	37	12	1.5										
15RN03			15	42	13	1.5										
17RN03	N303	N303EC	17	47	14	1.5	1	39.1		20400		20400	2550		17000	0.120
20RN03	N304	N304EC	20	52	15	2.0	2	44.5		26000		30800	3250		15000	0.150
25RN03	N305E	305EC	25	62	17	1.1	1.1	54.0	37.5	36500	41.5	40200	4550	12000	12000	0.234
30RN03	N306E	306EC	30	72	19	1.1	1.1	62.5	48.0	49000	51.0	51200	6200	10000	11000	0.379
35RN03	N307E	N307EC	35	80	21	1.5	1.1	70.2	63.0	63000	64.0	64400	8150	9000	9500	0.486
40RN03	N308E	308EC	40	90	23	1.5	1.5	80.0	78.0	78000	81.5	80900	10200	8500	8000	0.649
45RN03	N309E	309EC	45	100	25	1.5	1.5	88.5	100.0	10000	98.0	99000	12900	6700	7500	0.885
50RN03	N310E	N310EC	50	110	27	2.0	2.0	95.0	114.0	112000	110.0	110000	15000	6300	6000	1.010
55RN03	N311E	311EC	55	120	29	2.0	2.0	106.5	140.0	143000	134.0	138000	18600	5600	5600	1.400
60RN03	N312E	312EC	60	130	31	2.1	2.1	115.0	156.0	160000	150.0	151000	20800	5000	5000	1.850
65RN03	N313E	N313EC	65	140	33	2.1	2.1	124.5	190.0	196000	180.0	183000	25500	4800	4800	2.300
70RN03	N314E	314EC	70	150	35	2.1	2.1	133.0	220.0	220000	204.0	205000	29000	4500	4300	2.730
75RN03	N315E	315EC	75	160	37	2.1	2.1	143.0	265.0	265000	240.0	242000	33500	4000	4000	3.340
80RN03	N316E	N316EC	80	170	39	2.1	2.1	151.0	275.0	290000	255.0	260000	36000	3800	3800	3.880
85RN03	N317EMI	317EC	85	180	41	3.0	3.0	160.0	325.0	335000	290.0	297000	41500	5600	3600	5.220
90RN03	N318EMI	318EC	90	190	43	3.0	3.0	169.5	345.0	360000	315.0	319000	43000	5300	3400	6.150
95RN03	N319EMI	N319EC	95	200	45	3.0	3.0	171.5	380.0	390000	335.0	341000	46500	5300	3200	7.060
100RN03	N320	320EC	100	215	47	3.0	3	191.5	425.0	440000	380.0	391000	51000	5000	3000	8.750
110RN03	N321	321EC	105	225	49	3.0	3	195.0	500.0	500000	410.8	440000	57000		2800	10.230
120RN03	N322EMI	N322EC	110	240	50	3.0	3	211.0	510.00	540000	440.0	468000	61000	4800	2600	11.700
	N324EMI	324EC	120	260	55	3.0	3	230.0	600.00	620000	520.0	539000	69500	4500	2400	15.200
	N326EMI	326EC	130	280	58	4.0	4	247.0	720.00	750000	610.0	627000	81500	4300	2200	18.600
	N328EMI	328EC	140	300	62	4.0	4	264.0	800.0	710000	670.0	594000	75000	3800	2400	22.600
	N330EMI	N330EC	150	320	65	4.0	4	283.0	930.0	965000	765.0	781000	100000	3600	2000	24.000
	N332EMI	332EC	160	340	68	4.0	4	300.0	1060.0	680000	865.0	501000	72000	3000	2200	32.000
	N334M	334EC	170	360	72	4.0	4	310.0	1020.0	1180000	800.0	952000	110000	3000	1700	36.300

Use $X_e = 1$, $Y_o = Y = 0$,

[a] Oil lubrication.

Source: 1. Extracted with permission from "*FAG Rolling Bearings*", Catalogue WL 41520EI, 1995 Edition: FAG Precision Bearings Ltd., Maneja, Vadodara, India.
2. *Courtesy*: Extracted from SKF Rolling Bearings, Catalogue 4000E, 1989, SKF Rolling Bearings India Ltd., Mumbai, India.

TABLE 23-72
Cylindrical roller bearings—Dimension Series 04 (Indian standards) and series NU 4, SKF

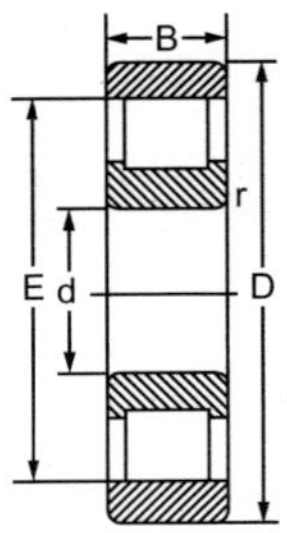

Bearing No.			Dimensions, mm					Basic load rating capacity				Fatigue load F_{uf} limit	Permissible speed, n			Mass
								Static, C_o		Dynamic, C			Kinematically	Grease[b]	Oil[c]	
IS Old No.	FAG	SKF	d	D	B	r[a]	E[a]	FAG[a]	SKF	FAG[a]	SKF	SKF	FAG[a]	SKF	SKF	SKF
								N	N	N	N	N	rpm	rpm	rpm	kg
15RN04			15	52	15	2.0										
17RN04			17	62	17	2.0										
20RN04			20	72	19	2.0										
25RN04	N405	NU405	25	80	21	2.5	62.8	23130		40050			8000			
30RN04	N406	NU406	30	90	23	2.5	73	32680	53000	53560	60500	6800	8000	7500	9000	0.75
35RN04	N407	NU407	35	100	25	2.5	83	42240	69500	68010	76500	9000	6000	6700	8000	1.00
40RN04	N408	NU408	40	110	27	2.5	92	51550	90000	82680	96000	11600	6000	6000	7000	1.30
45RN04	N409	NU409	45	120	29	3.0	100.5	63550	102000	99560	106000	13400	6000	5600	6700	1.65
50RN04	N410	NU410	50	130	31	3.5	110.8	81340	127000	122260	130000	16600	5000	5000	6000	2.00
55RN04	N411	NU411	55	140	33	3.5	117.2	81340	140000	124460	142000	18600	5000	4800	5600	2.50
60RN04	N412	NU412	60	150	35	3.5	127	99570	173000	148910	168000	22000	5000	4300	5000	3.00
65RN04	N413	NU413	65	160	37	3.5	135.3	110000	190000	168900	183000	24000	4000	4000	4800	3.60
70RN04	N414	NU414	70	180	42	4.0	152	143900	240000	211140	229000	30000	4000	3600	4300	5.25
75RN04	N415	NU415	75	190	45	4.0	160.5	157780	280000	231130	264000	34000	4000	3400	4000	6.25
80RN04	N416	NU416	80	200	48	4.0	170	164490	320000	259700	303000	39000	3000	3200	3800	7.30
85RN04	N417	NU417	85	210	52	5.0	177	213450	335000	297800	319000	39000	3000	3000	3600	8.70
90RN04	N418	NU418	90	225	54	5.0	191.5	235590	415000	320070	350000	48000	3000	2800	3400	10.50
95RN04	N419	NU419	95	240	55	5.0	201.5	253380	455000	355390	413000	52000	3000	2600	3200	13.6
100RN04	N420	NU420	100	250	58	5.0	211	283160	475000	391170	429000	53000	2500	2400	3000	19.0

Use $X_e = X = 1$, $Y = 0$, NU Series have two integral flanges on the outer ring and inner without flanges. [a] Refer to old FAG designation. [b] Grease lubrication. [c] Oil lubrication.

Source: 1. Extracted with permission from "*FAG Rolling Bearings*", Catalogue WL 41520EI, 1995 Edition: FAG Precision Bearings Ltd., Maneja, Vadodara, India.
2. *Courtesy*: Extracted from SKF Rolling Bearings, Catalogue 4000E, 1989, SKF Rolling Bearings India Ltd., Mumbai, India.

TABLE 23-73
Cylindrical roller bearings—Dimension Series NU22

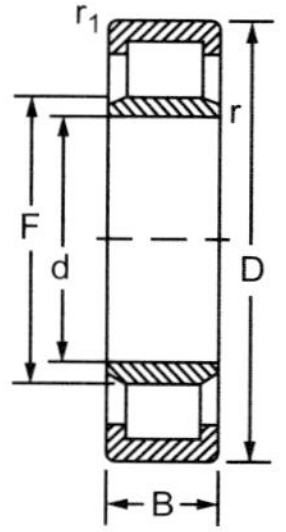

Bearing No.		Dimensions, mm						Basic load rating capacity				Fatigue limit, F_{uf}	Permissible speed, n		Mass
								Static, C_o		Dynamic, C					
FAG	SKF	d	D	B	r min	r_1 min	F	FAG	SKF	FAG	SKF	SKF	Kinematically FAG	Oil[a] SKF	FAG
								kN	N	kN	N	N	rpm	rpm	kg
NU2203E	NU2203EC	17	40	16	.6	0.3	22.1	22.0	21600	24.0	23800	2600	18000	19000	0.092
NU2204E	NU2204EC	20	47	18	1.0	0.6	26.5	31.0	27500	32.5	29700	3450	16000	16000	0.142
NU2205E	2205EC	25	52	18	1.0	0.6	31.5	34.5	34000	34.5	34100	4250	15000	14000	0.162
NU2206E	2206EC	30	62	20	1.0	0.6	37.5	50.0	49000	49.0	48400	6100	12000	12000	0.359
NU2207E	2207EC	35	80	21	1.5	1.1	46.2	63.0	63000	64.0	64400	8150	9000	9500	0.488
NU2208E	2208EC	40	80	23	1.5	1.5	51.0	78.0	75000	81.5	70000	9650	7500	9000	0.658
NU2209E	2209EC	45	85	23	1.1	1.1	54.5	81.5	81000	73.5	73700	10000	8000	8000	0.530
NU2210E	2210EC	50	90	23	1.1	1.1	59.5	88.0	88000	78.0	78100	11400	8000	7000	0.571
NU2211E	NU2211EC	55	100	25	1.5	1.1	66.0	118.0	118000	98.0	99000	15300	7000	7000	0.793
NU2212E	2212EC	60	110	28	1.5	1.5	72.0	153.0	153000	129.0	128000	20000	6300	6300	1.08
NU2213E	2213EC	65	120	31	1.5	1.5	78.5	183.0	180000	150.0	147000	24000	5600	5600	1.44
NU2214E	2214EC	70	125	31	1.5	1.5	83.5	196.0	193000	156.0	154000	25500	5300	5300	1.51
NU2215E	2215EC	75	130	31	1.5	1.5	88.5	208.0	208000	163.0	161000	27000	5300	5300	1.60
NU2216E	NU2216EC	80	140	33	2	2	95.3	215.0	245000	186.0	157000	31000	4800	4800	2.01
NU2217E	2217EC	85	150	36	2	2	100.05	275.0	280000	216.0	216000	34300	4500	4500	2.50
NU2218E	2218EC	90	160	40	2	2	107.0	315.0	315000	240.0	242000	39000	4300	4300	3.18
NU2219E	2219EC	95	170	43	2.1	2.1	112.5	375.0	375000	285.0	266000	45500	3800	4000	3.90
NU2220E	2220EC	100	180	46	2.1	2.1	119.0	440.0	450000	335.0	336000	54000	3800	3800	4.77
NU2222E	NU2222EC	110	200	53	2.1	2.1	132.5	520.0	520000	380.0	380000	61000	3400	3400	6.73
NU2224E	2224EC	120	215	58	2.1	2.1	143.5	610.0	630000	450.0	457000	72000	3200	3000	8.21
NU2226E	2226EC	130	230	64	3	3	153.5	735.0	735000	530.0	528000	83000	3000	2800	10.4
NU2228E	2228EC	140	250	68	3	3	169.0	830.0	830000	570.0	572000	93000	4500	2600	13.2
NU2230EMI	2230EC	150	270	73	3	3	182.0	980.0	930000	655.0	627000	100000	4300	2460	18.7
NU2232EMI	2232EC	160	290	80	3	3	193.0	1180.0	1200000	800.0	809000	129000	3800	2200	23.9
NU2234EMI	2234EC	170	310	86	4	4	205.0	1400.0	1430000	950.0	968000	150000	3200	2200	35.7
NU2236EMI	2236EC	180	320	86	4	4	215.0	1500.0	1500000	1000.0	1010000	156000	3200	2000	36.4
NU2238EMI	2238EC	190	340	92	4	4	228.0	1660.0	1660000	1100.0	1100000	170000	3000	1900	36.9
NU2240EMI	2240EC	200	360	98	4	4	241.0	1860.0	1900000	1220.0	1230000	190000	2800	1800	45.1

Use $X_e = X = 1$, $Y_o = Y = 0$. EC design series have higher loading capacity for the same boundary dimension than earlier design.
[a] Oil lubrication.

Source: 1. Extracted with permission from "*FAG Rolling Bearings*", Catalogue WL 41520EI, 1995 Edition: FAG Precision Bearings Ltd., Maneja, Vadodara, India.
2. *Courtesy*: Extracted from SKF Rolling Bearings, Catalogue 4000E, 1989, SKF Rolling Bearings India Ltd., Mumbai, India.

TABLE 23-74
Cylindrical roller bearings—Dimensions Series NU23

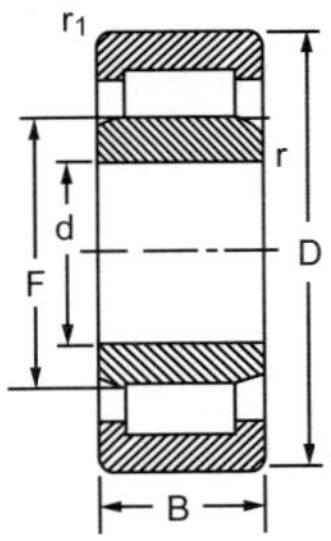

Bearing No.		Dimensions, mm						Basic load rating capacity				Fatigue limit, F_{uf}	Permissible speed, n		Mass
								Static, C_o		Dynamic, C					
								FAG	SKF	FAG	SKF	SKF	Kinematically FAG	Oil[a] SKF	FAG
FAG	SKF	d	D	B	r	r_1	F	kN	N	kN	N	N	rpm		kg
NU2304E	NU2304EC	20	52	21	1.1	0.6	27.5	39.0	38000	41.5	41300	4300	14000	11000	0.215
NU2305E	NU2305EC	25	62	24	1.1	1.1	34.0	56.0	55000	57.0	56100	6950	12000	9000	0.348
NU2306E	2306EC	30	72	27	1.1	1.1	40.5	75.0	75000	73.5	73700	9650	10000	8000	0.530
NU2307E	2307EC	35	80	31	1.5	1.1	46.2	98.0	98000	91.5	91300	12700	9000	7000	0.721
NU2308E	NU2308EC	40	90	33	1.5	1.5	52.0	120.0	120000	112.0	112000	15300	7500	6300	0.959
NU2309E	2309EC	45	100	36	1.5	1.5	58.5	153.0	153000	137.0	136000	20000	6700	5600	1.300
NU2310E	2310EC	50	110	40	2	2	65.0	186.0	186000	163.0	161000	24500	6300	5000	1.740
NU2311E	NU2311EC	55	120	43	2	2	70.5	228.0	232000	200.0	201000	30500	5600	4600	2.240
NU2312E	2312EC	60	130	46	2.1	2.1	77.0	260.0	265000	224.0	224000	34500	5000	4300	2.780
NU2313E	2313EC	65	140	48	2.1	2.1	82.5	285.0	290000	245.0	251000	35000	4800	4000	3.320
NU2314E	NU2314EC	70	150	51	2.1	2.1	89.0	325.0	325000	275.0	275000	41500	4500	3600	4.030
NU2315	2315	75	160	55	2.1	2.1	95.0	390.0	400000	325.0	330000	50000	4000	3400	4.93
NU2316	2316	80	170	58	2.1	2.1	101.0	425.0	440000	355.0	358000	55000	3800	3200	5.88
NU2317	NU2317EC	85	180	60	3	3	108.0	450.0	490000	365.0	396000	60000	3600	3000	6.57
NU2318	2318EC	90	190	64	3	3	113.5	530.0	540000	430.0	440000	65500	3400	2800	7.84
NU2319	2319EC	95	200	67	3	3	121.5	585.0	565000	455.0	468000	69000	3400	2600	9.21
NU2320	NU2320EC	100	215	73	3	3	127.5	720.0	735000	570.0	583000	85000	3200	2400	12.00
NU2322	2322EC	110	240	80	3	3	143.0	800.0	900000	630.0	682000	102000	2800	2000	16.80
NU2324	2324EC	120	260	86	3	3	154.0	1020.0	1040000	780.0	792000	116000	4300	1900	23.20
NU2326MI	NU2326EC	130	280	93	4	4	167.0	1220.0	1250000	915.0	935000	137000	3800	1800	26.10
NU2328MI	NU2328EC	140	300	102	4	4	180.0	1400.0	1430000	1020.0	1050000	150000	3600	1800	36.50
NU2330MI	2330EC	150	320	108	4	4	193.0	1600.0	1630000	1160.0	1190000	170000	3200	1700	43.90
NU2332MI	2332EC	160	340	114	4	4	204.0	1830.0	1860000	1320.0	1320000	190000	3000	1500	46.70
NU2334M	NU2334	170	360	120	4	4	220.0	1760.0	1800000	1220.0	1230000	180000	2800	1400	61.00
NU2336M	NU2336	180	380	126	5	4	232.0	2000.0	2040000	1370.0	1400000	204000	2800	1300	65.40
NU2338M	NU2338EC	190	400	132	5	5	245.0	2200.0	2550000	1500.0	1830000	236000	2800	1200	83.40
NU2340	NU2340EC	200	420	138	5	5	260.0	2200.0	2650000	1500.0	2050000	260000	2600	1200	95.70

[a] Oil lubrication.

Source: 1. Extracted with permission from "*FAG Rolling Bearings*", Catalogue WL 41520EI, 1995 Edition: FAG Precision Bearings Ltd., Maneja, Vadodara, India.
2. *Courtesy*: Extracted from SKF Rolling Bearings, Catalogue 4000E, 1989, SKF Rolling Bearings India Ltd., Mumbai, India.

TABLE 23-75
Taped roller bearings—Dimensions Series 322

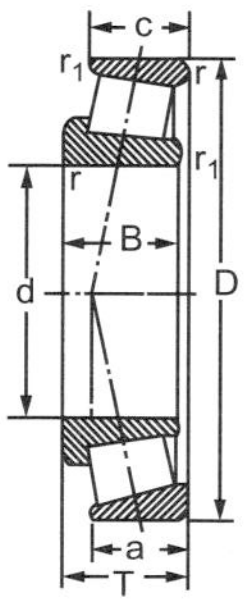

Bearing No.		Dimensions, mm								Factors			Basic load rating capacity				Fatigue load limit, SKF	Permissible speed, n		Mass FAG
													Static, C_o		Dynamic, C					
FAG	SKF	d	D	B	T	C	r	r_1	a	e	Y	Y_o	FAG	SKF	FAG	SKF		Kinematically FAG	Grease[a] SKF	
													kN	N	kN	N	N	rpm		kg
32206A	32206	30	62	20	21.25	17	1.0	1.0	16	.37	1.6	.88	63.0	57000	54.0	50100	6500	12000	6000	0.276
32207A	07	35	72	23	24.25	19	1.5	1.5	18	.37	1.6	.88	85.0	78000	71.0	66000	8650	10000	5300	0.425
32208A	08	40	80	23	24.75	19	1.5	1.5	19	.37	1.6	.88	95.0	86500	80.0	74800	9800	9000	4000	0.555
32209A	32209	45	85	23	24.75	19	1.5	1.5	20	.40	1.48	.81	100.0	98000	83.0	80900	11200	8000	4500	0.570
32210A	10	50	90	23	24.75	19	1.5	1.5	21	.42	1.33	.79	110.0	100000	88.0	82500	11600	7500	4300	0.602
32211A	11	55	100	25	26.75	21	2.0	1.5	23	.40	1.48	.81	137.0	129000	110.0	106000	15000	6700	3800	0.872
32212A	32212	60	110	28	29.75	24	2.0	1.5	24	.40	1.48	.81	170.0	160000	134.0	125000	19000	6000	3400	1.14
32213A	13	65	120	31	32.75	27	2.0	1.5	26	.40	1.48	.81	200.0	193000	156.0	151000	23200	5600	3000	1.59
32214A	14	70	125	31	33.25	27	2.0	1.5	28	.42	1.43	.79	216.0	208000	163.0	157000	24500	5300	2800	1.69
32215A	32215	75	130	31	33.25	27	2.0	1.5	29	.44	1.38	.76	232.0	212000	173.0	161000	25000	5000	2600	1 93
32216A	16	80	140	33	35.25	28	2.5	2.0	31	.42	1.43	.79	265.0	245000	200.0	182000	28500	5000	2400	2.18
32217A	17	85	150	36	38.5	30	2.5	2.0	34	.42	1.43	.79	305.0	285000	228.0	212000	33500	4800	2200	2.76
32218A	32218	90	160	40	42.5	34	2.5	2.0	36	.42	1.43	.79	360.0	340000	260.0	251000	38000	4500	2000	3.78
32219A	19	95	170	43	45.5	37	3.0	2.5	39	.42	1.43	.79	415.0	390000	300.0	281000	43000	4300	1900	4.23
32220A	20	100	180	46	49.0	39	3.0	2.5	42	.42	1.43	.79	475.0	440000	335.0	319000	48000	4000	1800	5.67
32221A	32221	105	190	50	53.0	43	3.0	2.5	44	.42	1.43	.79	550.0	510000	380.0	358000	55000	3600	1800	6.07
32222A	22	110	200	53	56.0	46	3.0	2.5	46	.42	1.43	.79	600.0	570000	415.0	402000	61000	3400	1700	7.35
32224A	24	120	215	58	61.5	51	3.0	2.5	51	.44	1.38	0.79	735.0	695000	490.0	468000	72000	3000	1600	10.1
32226A	32226	130	230	64	61.75	56	4.0	3.0	56	.44	1.38	0.79	865.0	830000	570.0	550000	85000	2600	1500	11.7
32228A	28	140	250	68	71.75	58	4.0	3.0	60	.44	1.38	0.76	1000.0	1000000	655.0	644000	100000	2600	1400	14.0
32230A	30	150	270	73	77.0	60	4.0	3.0	64	.44	1.38	0.76	1160.0	1140000	750.0	737000	112000	2600	1200	18.5

[a] Grease lubrication.

Source: 1. Extracted with permission from "*FAG Rolling Bearings*", Catalogue WL 41520EI, 1995 Edition: FAG Precision Bearings Ltd., Maneja, Vadodara, India.
2. *Courtesy*: Extracted from SKF Rolling Bearings, Catalogue 4000E, 1989, SKF Rolling Bearings India Ltd., Mumbai, India.

General Plan Boundary Dimensions for Tapered Roller Bearings

There are four series in tapered roller bearings. They are: (1) Angle series, (2) diameter series, (3) width series, (4) dimension series. Dimension series is a combination of angle series, diameter series and width series. Dimension series shall be designated by a combination of three symbols, for example 2BD. The first symbol is a numeric character which represents a range of contact angles (angle series). The second symbol is an alphabetic character which represents range of numeric values for the outside diameter to bore relationship (diameter series). The third symbol is an alphabetic character which represents a range of numeric values of the width to section relationship (width series).

TABLE 23-76
Series designation

Angle Series	α Over	α Up to	Diameter series	$(D/d^{0.77})$ Over	$(D/d^{0.77})$ Up to	Width series	$T/(D-d)^{0.95}$ Over	$T/(D-d)^{0.95}$ Up to
1	Reserved for future use		A	Reserved for future use		A	Reserved for future use	
2	10°	13° 52′	B	3.40	3.80	B	0.50	0.68
3	13° 52′	15° 59′	C	3.80	4.40	C	0.68	0.80
4	15° 59′	18° 55′	D	4.40	4.70	D	0.80	0.88
5	18° 55′	23°	E	4.70	5.00	E	0.80	1.00
6	23°	27°	F	5.00	5.60			
7	27°	30°	G	5.60	7.00			

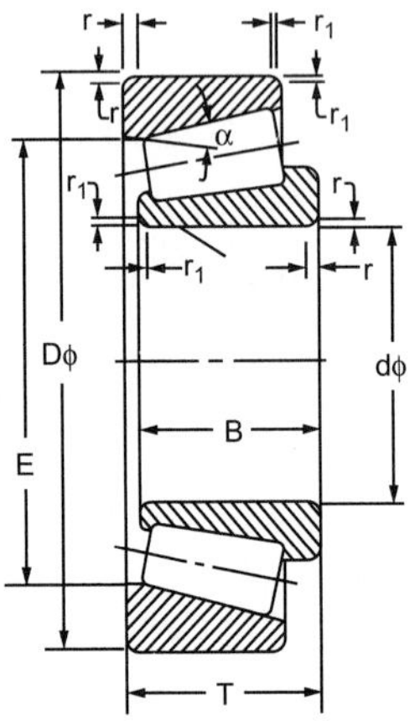

Symbol

d	bearing bore diameter, nominal
D	bearing outside diameter, nominal
T	bearing width, nominal
B	cone width, nominal
C	cup width, nominal
E	cup small inside diameter, nominal
α	bearing contact angle, nominal
r_1	cone back face chamfer height
$r_{1s\ \min}$	smallest single r_1
r_2	cone back face chamfer width
$r_{2s\ \min}$	smallest single r_2
r_3	cup back face chamfer height
$r_{3s\ \min}$	smallest single r_3
r_4	cup back face chamfer width
$r_{4s\ \min}$	smallest single r_4
r_5	cone and cup front face chamfer height and width

TABLE 23-77
Dimensions for tapered roller bearings—Contact angle series 2

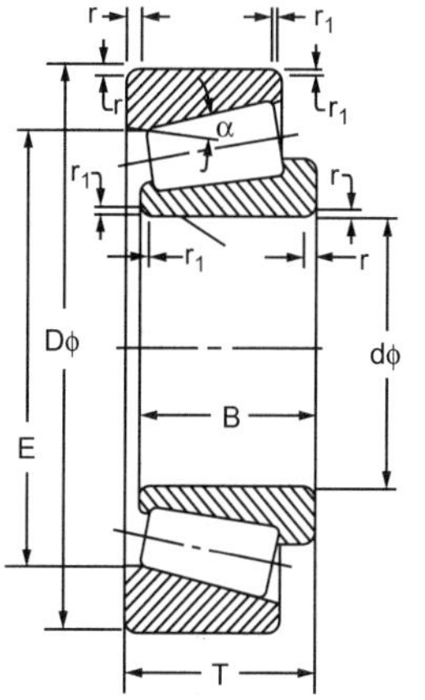

d	D	T	B	$r_{1s\,min}$ $r_{2s\,min}$	C	$r_{3s\,min}$ $r_{4s\,min}$	α	E	Dimension series
15	42	14.25	13	1	11	1	10° 45′ 29″	33.272	2FB
17	40	13.25	12	1	11	1	12° 57′ 10″	31.408	2DB
17	40	17.25	16	1	14	1	11° 45′	31.170	2DD
17	47	20.25	19	1	16	1	10° 45′ 29″	36.090	2FD
20	37	12	12.0	0.3	9	0.3	12°	29.621	2BD
20	47	15.25	14	1	12	1	12° 57′ 10″	37.304	2DB
20	47	19.25	18	1	15	1	12° 28″	35.810	2DD
20	52	22.25	21	1.5	18	1.5	11° 18′ 36″	39.518	2FD
22	40	12	12	0.3	9	0.3	12°	32.665	2BC
25	42	12	12	0.3	9	0.3	12°	34.608	2BD
25	62	25.25	24	1.5	20	1.5	11° 18′ 36″	48.637	2FD
25	50	17	17.5	1.5	13.5	1	13° 30′	40.205	2CC
25	52	19.25	18.0	1	16	1	13° 30′	41.335	2CD
28	45	12	12	0.3	9	0.3	12°	37.639	2BD
28	55	19	19.5	1.5	15.5	1.5	12° 10′	44.838	2CD
30	47	12	12	0.3	9	0.3	12°	39.617	2BD
30	58	19	19.5	1.5	15.5	1.5	12° 50′	47.309	2CD
30	72	28.75	27	1.5	23	1.5	11° 51′ 35″	55.767	2FD
32	52	14	15	0.6	10	0.6	12°	44.261	2BD
32	62	21	21	1.5	17	1.5	12° 30′	50.554	2CD
35	55	14	14	0.6	11.5	0.6	11°	47.220	2BD
35	68	23	23	2	18.5	2	12° 35′	55.400	2DD
40	62	15	15	0.6	12	0.6	10° 55′	53.388	2BC
40	75	24	24	2	19.5	2	12° 07′	62.155	2CD
40	90	35.25	33	2	27	1.5	12° 57′ 10″	69.253	2FD
45	68	15	15	0.6	12	0.6	12°	58.852	2BC
45	80	24	24	2	19.5	2	13°	66.615	2CD
50	72	15	15	0.6	12	0.6	12° 50′	62.748	2BC
50	85	24	24	2	19.5	2	13° 52′	70.969	2CD
50	100	42.25	40	2.5	33	2	12° 57′ 10″	86.263	2FD
55	80	17	17	1	14	1	11° 39′	69.503	2BC
55	85	18	18.5	2	14	2	12° 49′	73.586	2CC
55	95	27	27	2	21.5	2	12° 43′ 30″	80.106	2CD
55	120	45.5	43	2.5	35	2	12° 57′ 10″	94.316	2FD
60	85	17	17	1	14	1	12° 27′	74.185	2BC
60	90	18	18.5	2	14	2	13° 38′ 30″	78.249	2CC
60	100	27	27	2	21.5	2	13° 27′	84.587	2CD
60	130	48.5	46	3	37	2.5	12° 57′ 10″	102.939	2FD
65	90	17	17	1	14	1	13° 15′	78.849	2BC
65	100	22	22	2	17.5	2	12° 10′ 30″	87.433	2CC
65	110	31	31	2	25	2	12° 27′	93.090	2DD
65	125	43	42	2.5	35	2.5	12°	102.378	2FD
70	100	20	20	1	16	1	11° 53′	88.590	2BC
70	105	22	22	2	17.5	2	12° 49′ 30″	92.004	2CC
70	120	34	33	2	27	2	12° 22′	101.343	2DD
75	105	20	20	1	16	1	12° 31′	93.223	2BC
75	115	25	25	2	20	2	12°	100.414	2CC
75	125	34	33	2.5	27	2	12° 55′	105.786	2DD
80	110	20	20	1	16	1	13° 10′	97.974	2BC
80	120	25	25	2	20	2	12° 33′ 30″	105.003	2CC
80	130	34	33	2.5	27	2	13° 30′	110.475	2BD
85	120	23	23	1.5	18	1.5	12° 18′	106.599	2BC
85	125	25	25	2.5	20	2	13° 7′ 30″	109.650	2CC
85	135	34	33	2.5	28	2	13° 02′	115.94	2DD

TABLE 23-77 (*Cont.*)

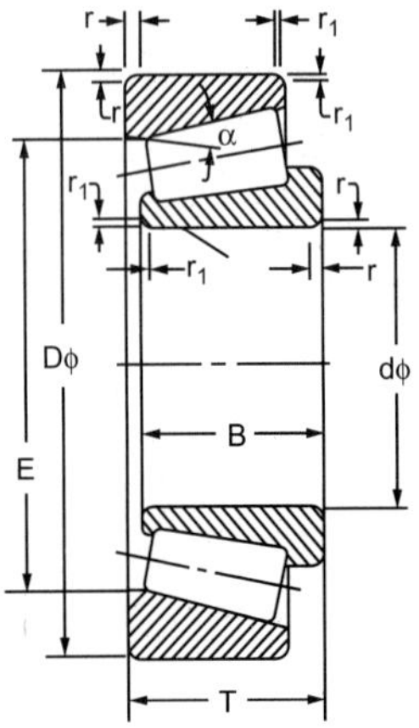

d	D	T	B	$r_{1s\,min}$ $r_{2s\,min}$	C	$r_{3s\,min}$ $r_{4s\,min}$	α	E	Dimension series
90	125	23	23	1.5	18	1.5	$12^\circ 51'$	111.282	2BC
90	135	28	27.5	2.5	23	2	$12^\circ 01' 30''$	119.139	2CC
90	140	34	33	2.5	28	2.5	$12^\circ 02' 30''$	121.860	2CD
95	130	23	23	1.5	18	1.5	$13^\circ 25'$	116.082	2BC
95	140	28	27.5	2.5	23	2.5	$12^\circ 30'$	123.797	2CC
95	145	34	33	2.5	28	2.5	$12^\circ 30'$	126.47	2CD
100	140	25	25	1.5	20	1.5	$12^\circ 23'$	125.717	2CC
100	150	34	33	2.5	28	2.5	$12^\circ 57' 30''$	130.992	2CB
105	145	25	25	1.5	20	1.5	$12^\circ 51'$	130.359	2CC
105	155	33	31.5	2.5	27	2.5	$12^\circ 17' 30''$	137.045	2CD
105	160	38	37	3	31	2.5	$12^\circ 17' 30''$	139.734	2DD
110	150	25	25	1.5	20	1.5	$13^\circ 20'$	135.182	2CC
110	160	33	31.5	2.5	27	2.5	$12^\circ 42' 30''$	141.607	2CD
110	185	38	37	3	31	2.5	$12^\circ 42' 30''$	144.376	2DD
120	168	29	29	1.5	23	1.5	$13^\circ 05'$	148.464	2DD
120	180	41	40	3	33	2.5	$12^\circ 08' 30''$	158.233	2CC
130	180	32	32	2	25	1.5	$12^\circ 45'$	161.652	2CC
130	190	41	40	3	33	2.5	$12^\circ 51' 30''$	167.414	2DD
140	190	32	32	2	25	1.5	$13^\circ 30'$	171.032	2CC
140	205	44	43	3	36	2.5	$12^\circ 45'$	181.645	2DD
150	210	38	38	2.5	30	2	$12^\circ 26'$	187.926	2DC
150	215	44	43	3	36	3	$12^\circ 37'$	190.810	2DD
160	220	38	38	2.5	30	2	13°	197.962	2DC
160	225	44	43	3	36	3	$13^\circ 14' 30''$	200.146	2BD
170	235	44	43	3	36	3	$12^\circ 13' 30''$	211.346	2DD
180	240	39	38	3	31	3	$12^\circ 47'$	218.311	2DB
180	145	44	43	3	36	3	$12^\circ 46' 30''$	220.684	2BC
190	255	41	40	3	33	3	$12^\circ 15'$	232.395	2DC
190	260	47	46	4	38	3	$12^\circ 15'$	234.615	2DD
200	265	41	40	3	33	3	$12^\circ 45'$	241.710	2DC
200	270	47	46	4	38	3	$12^\circ 45'$	244.043	2DD
220	285	41	40	4	33	3	$12^\circ 45'$	2U637	2DC
220	290	47	46	4	38	3	12°	265.24	2DD
240	305	41	40	4	33	3	$12^\circ 53'$	281.653	2DC
240	310	47	46	4	38	3	$12^\circ 52'$	284.035	2DD
260	325	41	40	4	33	4	$13^\circ 46'$	300.661	2DC
260	330	47	46	4	38	4	$13^\circ 44' 30''$	303.004	2DD

All dimensions in mm.

Courtesy: Extracted from IS: 7461 (part 1) 1993.

TABLE 23-78
Dimensions for tapered roller bearings—Contact angle series 5

d	D	T	B	$r_{1s\,min}$ $r_{2s\,min}$	C	$r_{3s\,min}$ $r_{4s\,min}$	α	E	Dimension Series
20	47	19.25	18	1	15	1	19°	33.708	5DD
25	52	19.25	18	1	15	1	$21^\circ 15'$	37.555	5CD
28	58	20.25	19	1	16	1	$20^\circ 34'$	42.436	5DB
30	62	21.25	20	1	17	1	$20^\circ 34'$	46.389	5DC
32	65	22	21.5	1	17	1	20°	48.523	5DC
35	72	24.25	23	1.5	19	1.5	$21^\circ 10'$	53.052	5DC
40	80	24.75	23	1.5	19	1.5	20°	61.438	5DC
40	50	27	26.5	4	21.5	1.5	$20^\circ 43' 30''$	58.963	5DD
45	85	24.75	23	1.5	19	1.5	$21^\circ 35'$	66.138	5DC
45	100	38.25	36	2	30	1.5	20°	71.639	5FD
50	90	24.75	23	1.5	18	1.5	$21^\circ 20'$	72.169	5DC
50	110	42.25	40	2.5	33	2.0	20°	78.582	5FD
55	100	30	28.5	4	24	2.5	20°	77.839	5DD
55	120	45.5	43	2.5	35	2.5	20°	86.300	5FD
60	110	34	32	4	27	2.5	$19^\circ 30''$	85.698	5DD
60	130	48.5	46	3	37	2.5	20°	94.000	5FD
65	115	34	32	4	27	2.5	$20^\circ 30''$	89.829	5DD
65	120	39	38	4	31	2.5	$20^\circ 28'$	91.241	5ED
70	125	37	34.5	4	30	2.5	$19^\circ 34'$	98.100	5DD
70	130	42	40	4	34	2.5	$19^\circ 11''$	100.186	5ED
75	130	37	34.5	4	30	2.5	$20^\circ 26'$	102.199	5DD
75	135	42	40	5	34	2.5	20°	104.210	5ED
80	135	37	34.5	4	30	2.5	$19^\circ 36'$	108.128	5DD
80	140	42	40	5	34	3	$20^\circ 49'$	108.199	5ED
85	140	37	34.5	4	30	3	$20^\circ 24'$	112.385	5DD
85	145	42	40	5	34	3	$19^\circ 16'$	115.106	5ED
90	145	37	34.5	4	30	3	$19^\circ 16'$	118.567	5DD
90	150	42	40	5	34	3	20°	119.254	5ED
95	150	37	34.5	4	30	3	20°	122.832	5DD
95	155	42	40	5	34	3	$20^\circ 44'$	123.374	5ED
100	155	37	34.5	5	30	3	$20^\circ 44'$	127.221	5DD
100	160	42	40	5	34	3	$19^\circ 20'$	130.033	5ED
105	160	37	34.5	5	30	3	$19^\circ 40'$	133.284	5DD

All dimensions in mm.
Courtesy: Extracted from IS: 7461 (part 1) 1993.

TABLE 23-79
Single direction thrust ball bearings—Dimension Series 11 (Indian Standards), FAG and SKF Series 511

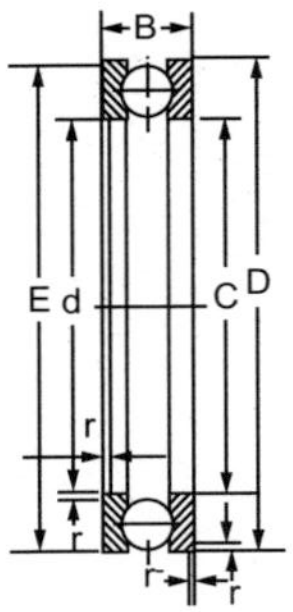

Bearing No.			Dimensions, mm						Minimum load constant, M[b]	Basic load rating capacity: Static, C_o		Dynamic, C		Fatigue load limit, F_{uf}	Permissible speed, n		Mass
IS Old No.	FAG	SKF	d	C	D	B	E	r min		FAG	SKF	FAG	SKF		Kinematically, FAG	Grease[a], SKF	FAG
										kN	N	kN	N	N	rpm	rpm	kg
10TA11	51100	51100	10	11	24	9	24	0.3	0.001	14.0	14000	10.0	9950	560	9500	7000	0.021
12TA11	51101	01	12	13	26	9	26	0.3	0.001	15.3	15300	10.4	10400	620	9000	6700	0.023
15TA11	51102	02	15	16	28	9	28	0.3	0.001	15.3	14000	10.4	9360	560	8500	6300	0.024
17TA11	51103	03	17	18	30	9	30	0.3	0.002	15.3	15300	9.65	9750	620	8500	6300	0.026
20TA11	51104	04	20	21	35	10	35	0.3	0.004	20.8	20800	12.70	12700	850	7000	5600	0.041
25TA11	51105	05	25	26	42	11	42	0.6	0.004	29.0	29000	15.6	15900	1160	6300	4800	0.060
30TA11	51106	51106	30	32	47	11	47	0.6	0.007	33.5	33500	16.6	16800	1340	5600	4500	0.069
35TA11	51107	07	35	37	52	12	52	0.6	0.009	37.5	37500	17.6	17400	1530	5300	4300	0.087
40TA11	51108	08	40	42	60	13	60	0.6	0.016	50.0	56000	23.2	23400	2040	4500	3800	0.125
45TA11	51109	09	45	47	65	14	65	0.6	0.02	57.0	57000	24.5	24200	2280	4500	3400	0.153
50TA11	51110	10	50	52	70	14	70	0.6	0.024	63.0	63000	25.5	25500	2550	4300	3200	0.169
55TA11	51111	11	55	57	78	16	78	0.6	0.038	78.0	78000	31.0	30700	3100	3800	2800	0.247
60TA11	51112	51112	60	62	85	17	85	1.0	0.053	93.0	90000	36.5	35800	3600	3600	2800	9.330
65TA11	51113	13	65	67	90	18	90	1.0	0.06	98.0	98000	37.5	37100	4000	3400	2600	0.359
70TA11	51114	14	70	72	95	18	95	1.0	0.067	104.0	104000	37.5	37700	4150	3400	2400	0.385
75TA11	51115	15	75	77	100	19	100	1.0	0.095	137.0	137000	44.0	44200	5500	3200	2200	0.520
80TA11	51116	16	80	82	105	19	105	1.0	0.1	140.0	140000	45.0	44900	5700	3200	2000	0.557
85TA11	51117	17	85	87	110	19	110	1.0	0.12	150.0	150000	45.5	46200	6000	3200	2000	0.597
90TA11	51118	51118	90	92	120	22	120	1.0	0.19	190.0	190000	60.0	59200	7500	2800	1800	0.878
100TA11	51120	20	100	102	135	25	135	1.0	0.36	270.0	270000	85.0	85200	10000	2200	1700	1.300
110TA11	51122	22	110	112	145	25	145	1.0	0.43	290.0	290000	86.5	87100	10200	2200	1600	1.450
120TA11	51124	24	120	122	155	25	155	1.0	0.48	310.0	310000	90.0	88400	10800	2000	1600	1.590
130TA11	51126	26	130	132	170	30	170	1.0	0.75	390.0	390000	112.0	111000	12900	1800	1400	2.370
140TA11	51128	28	140	142	178	31	180	1.0	0.85	400.0	400000	112.0	111000	12900	1800	1300	2.59
150TA11	51130	51130	150	152	188	31	190	1.0	0.90	400.0	400000	111.0	111000	12500	1700	1200	2.26
160TA11	51132FP	32	160	162	198	31	200	1.0	1.0	430.0	425000	112.0	112000	12900	1700	1200	2.39
170TA11	51134FP	34	170	172	213	34	215	1.1	1.4	500.0	500000	132.0	133000	14300	1500	1100	3.09
180TA11	51136FP	36	180	183	222	34	225	1.1	1.5	530.0	530000	134.0	135000	15000	1500	1000	3.17
190TA11	51138FP	38	190	193	237	37	240	1.1	2.4	655.0	655000	170.0	172000	18000	1400	950	4.08
200TA11	51140FP	40	200	203	247	37	250	1.1	2.4	655.0	655000	170.0	168000	17600	1400	950	4.26

[a] Grease lubrication.
[b] To find F_{amin} = minimum axial load using M refer to table 23-82.
Source: 1. Extracted with permission from "*FAG Rolling Bearings*", Catalogue WL 41520EI, 1995 Edition: FAG Precision Bearings Ltd., Maneja, Vadodara, India.
2. *Courtesy*: Extracted from SKF Rolling Bearings, Catalogue 4000E, 1989, SKF Rolling Bearings India Ltd., Mumbai, India.

TABLE 23-80
Single direction thrust ball bearings (with Flat Housing Washer)—Dimension Series 12 (Indian Standards)

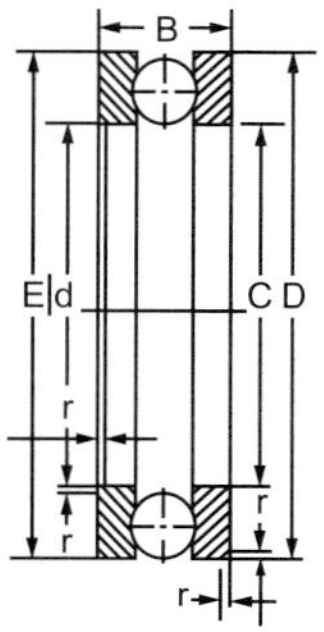

Bearing No.			Dimensions, mm						Minimum load constant, M[b]	Basic load rating capacity				Fatigue load limit, F_{uf}	Permissible speed, n		Mass
										Static, C_o		Dynamic, C			Kinematically, FAG	Grease[a], SKF	FAG
IS Old No.	FAG	SKF	d	C	D	B	E	r min		FAG	SKF	FAG	SKF				
										kN	N	kN	N	N	rpm		kg
10TA12	51200	51200	10	12	26	11	26	0.6	0.002	17.0	17000	12.70	12700	695	8000	6000	0.031
12TA12	51201	01	12	14	28	11	28	0.6	0.002	19.0	19000	13.20	13300	765	8000	6000	0.034
15TA12	51202	02	15	17	32	12	32	0.6	0.003	25.0	25000	16.6	16500	1000	6700	5300	0.046
17TA12	51203	03	17	19	35	12	35	0.6	0.004	27.5	27500	17.3	17200	1100	6700	5000	0.053
20TA12	51204	04	20	22	40	14	40	0.6	0.008	37.5	37500	22.4	22500	1530	5600	4500	0.083
25TA12	51205	05	25	27	47	15	47	0.6	0.013	50.0	50000	28.0	27600	2040	5000	4000	0.115
30TA12	51206	51206	30	32	52	16	52	0.6	0.014	47.5	47500	25.5	25500	1900	4800	3600	0.130
35TA12	51207	07	35	37	62	18	62	1.0	0.028	67.0	67000	35.5	35100	2700	4000	3000	0.215
40TA12	51208	08	40	42	68	19	68	1.0	0.05	98.0	98000	46.5	46800	4000	3800	2800	0.278
45TA12	51209	19	45	47	73	20	73	1.0	0.043	80.0	80000	39.0	39000	3200	3600	2600	0.302
50TA12	51210	10	50	52	78	22	78	1.0	0.01	106.0	106000	50.0	49400	4300	3400	2400.	0.371
55TA12	51211	11	55	57	90	25	90	1.0	0.11	134.0	134000	61.0	61000	5400	3200	1900	0.586
60TA12	51212	51212	60	62	95	26	95	1.0	0.12	140.0	140000	62.0	62400	5600	3000	1900	0.651
65TA12	51213	13	65	67	100	27	100	1.0	0.14	150.0	150000	64.0	63700	6000	3000	1800	0.737
70TA12	51214	14	70	72	105	27	105	1.0	0.11	160.0	160000	65.5	65000	6400	2800	1800	0.783
75TA12	51215	15	75	77	110	27	110	1.0	0.18	170.0	170000	67.0	67600	6800	2800	1700	0.827
80TA12	51216	16	80	82	115	28	115	1.0	0.22	190.0	190000	75.0	76100	7650	2600	1700	0.908
85TA12	51217	17	85	88	125	31	125	1.0	0.38	250.0	250000	98.0	97500	8800	2200	1600	1.22
90TA12	51218	51218	90	93	135	35	135	1.1	0.53	300.0	300000	120.0	119000	11400	2000	1500	1.68
100TA12	51220	20	100	103	150	38	150	1.1	0.67	320.0	320000	122.0	124000	11400	1900	1300	2.22
110TA12	51222	22	110	113	160	38	160	1.1	0.80	360.0	360000	129.0	130000	12500	1800	1200	2.37
120TA12	51224	24	120	123	170	39	170	1.1	1.0	400.0	400000	140.0	140000	13400	1700	1100	2.67
130TA12	51226	26	130	133	187	45	190	1.5	1.7	540.0	540000	183.0	186000	15000	1600	950	3.99
140TA12	51228	28	140	143	197	46	200	1.5	1.9	570.0	570000	190.0	190000	17600	1500	950	4.33
150TA12	51230MP	51230	150	153	212	50	215	1	2.8	735.0	735000	236.0	238000	22000	1400	900	6.09
160TA12	51232MP	32	160	163	222	51	225	1	3.2	780.0	780000	245.0	242000	22800	1400	850	6.56
170TA12	51234MP	34	170	173	237	55	240	1	4.5	930.0	930000	285.0	286000	26000	1200	800	8.12
180TA12	51236MP	36	180	183	260	56	260	1	5	1000.0	1000000	290.0	296000	27500	1200	800	8.70
190TA12	51238MP	38	190	194	265	62	270	2	7	1160.0	1160000	335.0	332000	31000	1000	750	11.70
200TA12	51240MP	40	200	204	275	62	280	2	8	1220.0	1220000	340.0	338000	31500	1000	750	12.00

[a] Grease lubrication.
[b] To find F_{amin} = minimum axial load using M refer to table 23-82.

Source: 1. Extracted with permission from "*FAG Rolling Bearings*", Catalogue WL 41520EI, 1995 Edition: FAG Precision Bearings Ltd., Maneja, Vadodara, India.
2. *Courtesy*: Extracted from SKF Rolling Bearings, Catalogue 4000E, 1989, SKF Rolling Bearings India Ltd., Mumbai, India.

TABLE 23-81
Single Direction thrust ball bearings, Dimension Series 13 (Indian Standards), FAG and SKF series 513

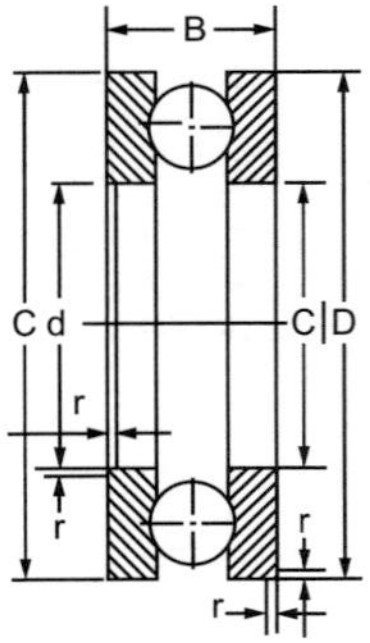

Bearing No.			Dimensions, mm						Minimum load constant, M[b]	Basic load rating capacity				Fatigue load limit, F_{uf}	Permissible speed, n		Mass FAG
										Static, C_o		Dynamic, C			Kinematically, FAG	Grease[a], SKF	
IS Old No.	FAG[b]	SKF[b]	d	C	D	B	E	r min		FAG	SKF	FAG	SKF				
										kN	N	kN	N	N	rpm		kg
25TA13	51305	51305	25	27	52	18	52	1.0	0.019	55.0	55000	34.5	34500	2240	4300	3400	0.170
30TA13	51306	06	30	32	60	21	60	1.0	0.028	65.5	65500	38.0	37700	2650	4000	2800	0.263
35TA13	51307	07	35	37	68	24	68	1.0	0.05	88.0	88000	50.0	49400	3550	3600	2400	0.377
40TA13	51308	51308	40	42	78	26	78	1.0	0.08	112.0	112000	61.0	61800	4500	3200	2000	0.533
45TA13	51309	09	45	47	85	28	85	1.0	0.12	140.0	140000	75.0	76100	5600	3000	1900	0.613
50TA13	51310	10	50	52	95	31	95	1.1	0.18	173.0	173060	88.0	88400	6950	2800	1800	0.940
55TA13	51311	51311	55	57	105	35	105	1.1	0.26	208.0	208000	102.0	104000	8300	2400	1600	1.300
60TA13	51312	12	60	62	110	35	110	1.1	0.28	208.0	208000	102.0	101000	8300	2200	1600	1.370
65TA13	51313	13	65	67	115	36	115	1.1	0.32	220.0	220000	106.0	106000	8800	2200	1500	1.490
70TA13	51314	51314	70	72	125	40	125	1.1	0.53	300.0	300000	137.0	135000	11800	1900	1400	1.910
75TA13	51315	15	75	77	135	44	135	1.5	0.75	360.0	360000	163.0	163000	14000	1800	1200	2.610
80TA13	51316	16	80	82	140	44	140	1.5	0.8	360.0	360000	160.0	159000	13700	1800	1200	2.710
85TA13	51317	51317	85	88	150	49	150	1.5	1.1	425.0	425000	190.0	190000	16000	1700	1100	3.530
90TA13	51318	18	90	93	155	50	155	1.5	1.2	465.0	465000	196.0	195000	16500	1700	1000	3.570
100TA13	51320	20	100	103	170	55	170	1.5	1.8	560.0	560000	232.0	229000	19600	1600	950	4.960
110TA13	51322MP	51322	110	113	187	63	190	2.0	2.8	720.0	720000	275.0	276000	24000	1400	850	7.700
120TA13	51324MP	24	120	123	205	70	210	2.1	4.5	915.0	915000	325.0	325000	28500	1200	800	10.700
130TA13	51326MP	26	130	134	220	75	225	2.1	6.0	1060.0	1060000	360.0	358000	32000	1100	750	13.000
140TA13	51328MP	51328	140	144	235	80	240	2.1	8.0	1220.0	1220000	400.0	397000	35500	1000	700	15.700
150TA13	51330MP	30	150	154	245	80	250	2.1	9.0	1290.0	1290000	405.0	410000	36500	950	670	16.400
	51332M	51332	160	164	265	87	270	3.0	12.0	1500.0	1500000	455.0	440000	41500	900	630	21.300
	51334M	34	170	174	275	87	280	3.0	13.0	1630.0	1600000	465.0	468000	43000	900	600	22.500
	51336M	36	180	184	245	95	300	3.0	18.0	1830.0	1830000	520.0	520000	47500	800	560	24.800
	51338M	38	190	195	315	105	320	4.0	26.0	2200.0		600.0			750		31.000
	51340M	40	200	205	335	110	340	4.0	30.0	2400.0	2400000	620.0	624000	56500	700	480	44.300

[a] Grease lubrication.
[b] *Source*: 1. Extracted with permission from "*FAG Rolling Bearings*", Catalogue WL 41520EI, 1995 Edition: FAG Precision Bearings Ltd., Maneja, Vadodara, India.
2. *Courtesy*: Extracted from SKF Rolling Bearings, Catalogue 4000E, 1989, SKF Rolling Bearings India Ltd., Mumbai, India.

TABLE 23-82
Single direction thrust ball bearings—Dimension Series 14 (Indian Standards), FAG and SKF Series 514

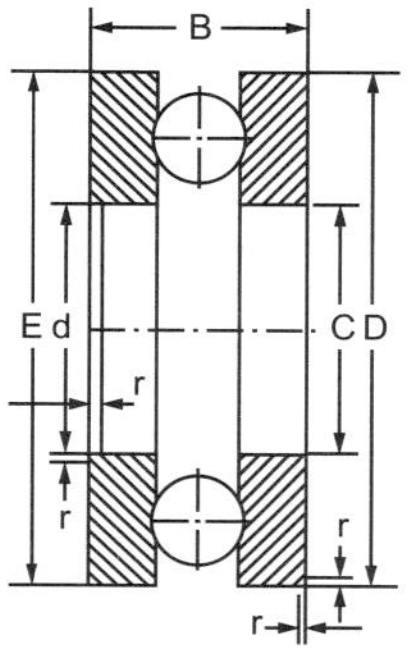

Bearing No.			Dimensions, mm						Minimum load constant, M[b]	Basic load rating capacity				Fatigue load limit, F_{uf}	Permissible speed, n		Mass FAG
										Static, C_o		Dynamic, C					
IS Old No.	FAG[b]	SKF[b]	d	C	D	B	E	r min		FAG	SKF	FAG	SKF		Kinematically, FAG	Grease[a], SKF	
										kN	N	kN	N	N	rpm		kg
25TA14	51405	51405	25	27	60	24	60	1.0	0.043	90.0	90000	56.00	55300	3600	3600	2600	0.363
30TA14	51406	06	30	32	70	28	70	1.0	0.08	125.0	125000	72.0	72800	5100	3200	2600	0.576
35TA14	51407	07	35	37	80	32	80	1.1	0.13	156.0	156000	86.5	87100	6200	3000	1800	0.962
40TA14	51408	51408	40	42	90	36	90	1.1	0.22	204.0	204000	112.0	112000	8300	2400	1700	1.170
45TA14	51409	09	45	47	100	39	100	1.1	0.32	245.0	240000	129.0	130000	9600	2200	1600	1.600
50TA14	51410	10	50	52	110	43	110	1.5	0.48	310.0	310006	156.0	159000	12500	2000	1500	2.18
55TA14	51411	51411	55	57	120	48	120	1.5	0.67	360.0	360000	180.0	178000	14300	1800	1300	2.91
60TA14	51412FP	12	60	62	130	51	130	1.5	0.85	400.0	400000	200.0	199000	16000	1700	1100	3.70
65TA14	51413FP	13	65	68	140	56	140	2.0	1.1	450.0	450000	216.0	216000	18000	1600	1100	4.67
70TA14	51414FP	51414	70	73	150	60	150	2.0	1.4	500.0	500000	236.0	234000	19300	1600	950	5.72
75TA14	51415FP	15	75	78	160	65	160	2.0	1.8	560.0	560000	250.0	251000	20800	1500	900	7.06
80TA14	51416FP	16	80	83	170	68	170	2.1	2.2	620.0	620000	270.0	270000	22400	1400	850	8.23
85TA14	51417FP	51417	85	88	177	72	180	2.1	2.8	680.0	680000	290.0	286000	24000	1300	850	9.79
90TA14	51418FP	18	90	93	187	77	190	2.1	3.4	750.0	750000	305.0	307000	25500	1200	800	11.60
100TA14	51420FP	20	100	103	205	85	210	3.0	5.3	965.0	965000	365.0	371000	31500	1000	700	15.40
	51422FP	51422	110	113	225	95	230	3.0	7.5	1140.0	114000	415.0	410000	34500	950	630	20.80
	51424FP	24	120	123	245	102	250	4.0	9.0	1220.0	122000	425.0	423000	36000	900	600	26.50
	51426FP	26	130	134	265	110	270	4.0	15.0	1600.0	160000	520.0	520000	45000	800	560	32.80
	51428FP	28	140		280	112					160000			44000		530	34.50
	51430FP	30	150	154	295	120	300	4.0	20.0	1800.0	180000	560.0	559000	48000	750	500	43.10

[a] Grease Lubrication.
[b] *Source*: 1. Extracted with permission from "*FAG Rolling Bearings*", Catalogue WL41520EI, 1995 Edition: FAG Precision Bearings Ltd., Maneja, Vadodara, India.
2. *Courtesy*: Extracted from SKF Rolling Bearings, Catalogue 4000E, 1989, SKF Rolling Bearings India Ltd., Mumbai, India.

TABLE 23-83
Double direction thrust ball bearings—Dimension Series 522

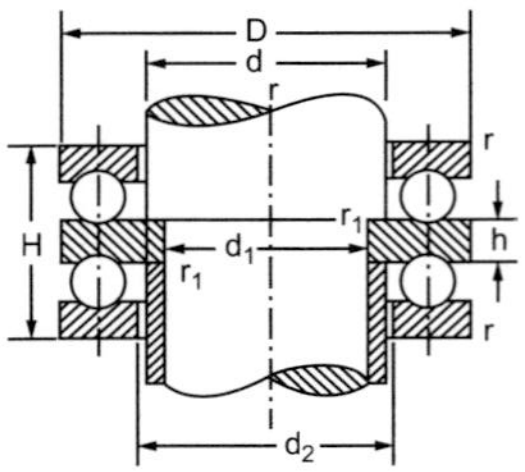

Bearing No.		Dimensions, mm								Minimum load constant, M	Basic load rating capacity: Static, C_o		Dynamic, C		Fatigue load limit, F_{uf}	Permissible speed, n: Kinematically,	Grease[b],	Mass
IS FAG	SKF	d	d_1	d_2[a]	D	H	h	r min	r_1 min	FAG	FAG	SKF	FAG	SKF	SKF	FAG	SKF	FAG
											kN	N	kN	N	N	rpm		kg
52202	52202	15	10	17	32	22	5	0.6	0.3	0.003	25.0	25000	16.6	16500	1000	6700	5300	0.081
52204	04	20	15	22	40	26	6	0.6	0.3	0.008	37.5	37500	22.4	22500	1500	5600	4500	0.147
52205	05	25	20	27	47	28	7	0.6	0.3	0.013	50.0	50000	28.0	27600	2040	5000	4000	0.215
52206	06	30	25	32	52	29	7	0.6	0.3	0.014	47.5	47500	25.5	25500	1900	4800	3600	0.247
52207	07	35	30	37	62	34	8	1.0	0.3	0.028	67.0	67000	36.5	35100	2700	4000	3000	0.408
52208	08	40	30	42	68	36	9	1.0	0.6	0.05	98.0	98000	46.5	46800	4000	3800	2800	0.553
52209	09	45	35	47	73	37	9	1.0	0.6	0.043	80.0	80000	39.0	32000	3200	3600	2600	0.597
52210	52210	50	40	52	78	39	9	1.0	0.6	0.07	106.0	106000	50.0	49400	4300	3400	2400	0.710
52211	11	55	45	57	90	45	10	1.0	0.6	0.11	134.0	134000	61.0	61800	5400	3200	1900	1.100
52212	12	60	50	62	95	46	10	1.0	0.6	0.12	140.0	140000	62.0	62400	5600	3000	1900	1.210
52213	13	65	55	67	100	47	10	1.0	0.6	0.14	150.0	150000	64.0	63700	6000	3000	1800	1.340
52214	14	70	55	72	105	47	10	1.0	1.0	0.16	160.0	160000	65.5	65000	6400	2800	1750	1.470
52215	15	75	60	77	110	47	10	1.0	1.0	0.18	170.0	170000	67.0	67600	6800	2800	1700	1.570
52216	52216	80	65	82	115	48	10	1.0	1.0	0.22	190.0	190000	75.0	76100	7650	2600	1700	1.720
52217	17	85	70	88	125	55	12	1.0	1.0	0.38	250.0	250000	98.0	97500	9500	2200	1600	2.390
52218	18	90	75	93	135	62	14	1.1	1.0	0.53	300.0	300000	120.0	119000	11400	2000	1600	3.220
52220	20	100	85	103	150	67	15	1.1	1.0	0.67	320.0	320000	122.0	124000	11400	1900	1300	4.210
52222	52222		95	113	160	67	15	1.1	1.0	0.81	360.0	360000	129.0	130000	11500	1800	1200	4.63
52224	24		100	123	170	68	15	1.1	1.1	1.0	400.0	400000	140.0	140000	13400	1700	1100	5.23
52226	26		110	133	190	80	18	1.5	1.1	1.7	540.0	540000	183.0	186000	17000	1600	950	7.99
52228	28		120	143	200	81	18	1.5	1.1	1.9	570.0	570000	190.0	190000	17500	1500	950	8.66
52230MP	30		130	153	215	89	20	1.5	1.1	2.8	735.0	736000	236.0	238000	22000	1400	900	11.40
52232MP	32		140	163	225	90	20	1.5	1.1	3.2	760.0	760000	245.0	242000	22800	1400	850	12.30
52234MP	34		150	173	240	97	21	1.5	1.1	4.5	930.0	930000	285.0	286000	26000	1200	800	14.00

[a] d_2: Refer to FAG bearings.
[b] Grease lubrication.

Source: 1. Extracted with permission from "*FAG Rolling Bearings*", Catalogue WL 41520EI, 1995 Edition: FAG Precision Bearings Ltd., Maneja, Vadodara, India.
2. *Courtesy*: Extracted from SKF Rolling Bearings, Catalogue 4000E, 1989, SKF Rolling Bearings India Ltd., Mumbai, India.

The minimum axial load in case of thrust ball bearings according to FAG $F_{a\,\min} = M(n_{\max}/1000)^2$ where $n_{\max}$ = maximum operating speed, rpm; M = minimum load constant taken from Tables 23-78 to 23-82 for thrust ball bearings.

TABLE 23-84
Needle bearings—Light Series

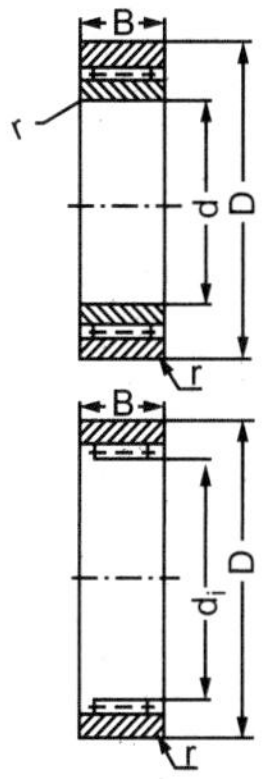

Bearing No.				Dimensions, mm					Basic load rating capacity		Limiting speed, n, rpm
IS		NRB									
Bearing with inner ring	Bearing without inner ring	Bearing with inner ring	Bearing without inner ring	d Nom	d_i Nom	D Nom	B Nom	r	Dynamic, C, N	Static, C_o, N	
NA1012	RNA1012	Na1012	Na1012S/Bi	12	17.6	28	15	.35	11280	9415	21600
NA1015	RNA1015	Na1015	Na1015S/Bi	15	20.8	32	15	.35	12600	10890	18300
NA1017	RNA1017	Na1017	Na1017S/Bi	17	23.9	35	15	.65	12260	12260	15900
NA1020	RNA1020	Na1020	Na1020S/Bi	20	28.7	42	18	.65	19610	18340	13200
NA1025	RNA1025	Na1025	Na1025S/Bi	25	33.5	47	18	.65	21570	21180	11100
NA1030	RNA1030	Na1030	Na1030S/Bi	30	38.2	52	18	.65	23930	23930	10000
NA1035	RNA1035	Na1035	Na1035S/Bi	35	44.0	58	18	.65	26480	27360	8600
NA1040	RNA1040	Na1040	Na1040S/Bi	40	49.7	65	18	.65	28730	30700	7600
NA1045	RNA1045	Na1045	Na1045S/Bi	45	55.4	72	18	.85	31000	34030	6900
NA1050	RNA1050	Na1050	Na1050S/Bi	50	62.1	80	20	.65	33540	37850	6100
NA1055	RNA1055	Na1055	Na1055S/Bi	55	68.8	85	20	.65	36190	41680	5500
NA1060	RNA1060	Na1060	Na1060S/Bi	60	72.6	90	20	.65	37460	43740	5200
NA1065	RNA1065	Na1065	Na1065S/Bi	65	78.3	95	20	.65	41580	49520	4900
NA1070	RNA1070	Na1070	Na1070S/Bi	70	83.1	100	20	.65	43350	52860	4500
NA1075	RNA1075	Na1075	Na1075S/Bi	75	88	110	24	.65	64720	80410	4300
NA1080	RNA1080	Na1080	Na1080S/Bi	80	96	115	24	.65	68650	86790	4000

TABLE 23-85
Needle bearings—Medium Series

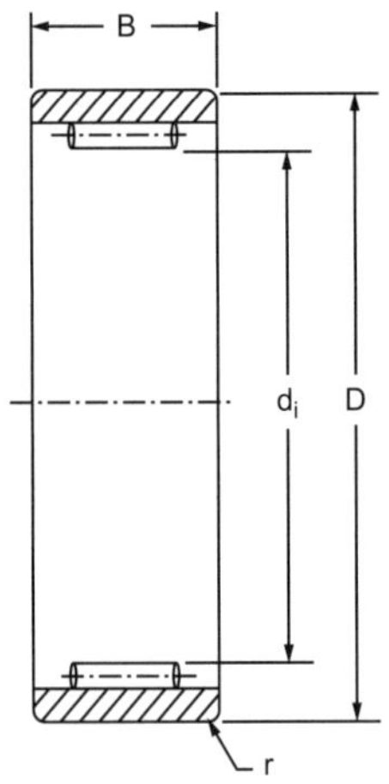

Bearing No.				Dimensions, mm					Basic load rating capacity		Limiting speed, n, rpm
IS		NRB									
Bearing with inner ring	Bearing without inner ring	Bearing with inner ring	Bearing without inner ring	d Nom	d_i Nom	D Nom	B Nom	r	Dynamic, C, N	Static, C_o, N	
NA2015	RNA2015	Na2015	Na2015S/Bi	15	22.1	35	22	.65	24320	21570	17200
NA2020	RNA2020	Na2020	Na2020S/Bi	20	28.7	42	22	.65	29220	26280	13200
NA2025	RNA2025	Na2025	Na2025S/Bi	25	33.5	47	22	.65	31580	31580	11100
NA2030	RNA2030	Na2030	Na2030S/Bi	30	38.2	52	22	.65	35700	35700	10000
NA2035	RNA2035	Na2035	Na2035S/Bi	35	44	58	22	.65	39230	40500	8600
NA2040	RNA2040	Na2040	Na2040S/Bi	40	49.7	62	22	.65	42950	45600	9600
NA2045	RNA2045	Na2045	Na2045S/Bi	45	55.4	72	22	.55	46580	50500	6900
NA2050	RNA2050	Na2050	Na2050S/Bi	50	62.1	80	28	.65	63740	74530	6100
NA2055	RNA2055	Na2055	Na2055S/Bi	55	68.8	85	28	.65	71100	82380	5500
NA2060	RNA2060	Na2060	Na2060S/Bi	60	72.6	90	28	.65	74040	84340	5200
NA2065	RNA2065	Na2065	Na2065S/Bi	65	78.3	95	28	.65	80410	95610	4900
NA2070	RNA2070	Na2070	Na2070S/Bi	70	83.1	100	28	.65	83360	101010	4500
NA2075	RNA2075	Na2075	Na2075S/Bi	75	88	110	32	.65	105910	131410	4300
NA2080	RNA2080	Na2080	Na2080S/Bi	80	96	115	32	.65	112780	143180	4000
NA2085	RNA2085	Na2085	Na2085S/Bi	85	99.5	120	32	.65	115720	148080	3800
NA2090	RNA2090	Na2090	Na2090S/Bi	90	104.7	125	32	.65	122580	156410	3600
NA2095	RNA2095	Na2095	Na2095S/Bi	95	109.1	130	32	.65	118660	162790	3500
NA2100	RNA2100	Na2100	Na2100S/Bi	100	114.7	135	32	.65	125320	170630	3300
NA2105	RNA2105	Na2105	Na2105S/Bi	105	119.2	140	32	.65	131410	177500	3200
NA2110	RNA2110	Na2110	Na2110S/Bi	110	124.7	145	34	.65	136310	185340	3000
NA2115	RNA2115	Na2115	Na2115S/Bi	115	132.5	155	34	.65	141220	196130	2900
NA2120	RNA2120	Na2120	Na2120S/Bi	120	137	160	34	.65	145140	202020	2800
NA2125	RNA2125	Na2125	Na2125S/Bi	125	143.5	165	34	.65	149060	211820	2700
NA2130	RNA2130	Na2130	Na2130S/Bi	130	148	170	34	1.35	152000	216730	2600
NA2140	RNA2140	Na2140	Na2140S/Bi	140	158	180	36	1.35	159850	233400	2400
NA2150	RNA2150	Na2150	Na2150S/Bi	150	170.5	195	36	1.35	168670	248110	2200
NA2160	RNA2160	Na2160	Na2160S/Bi	160	179.3	205	36	1.35	174560	262820	2100
NA2170	RNA2170	Na2170	Na2170S/Bi	170	193.8	220	42	1.35	238300	367650	2000
NA2180	RNA2180	Na2180	Na2180S/Bi	180	202.6	230	42	1.85	246150	384420	1900
NA2190	RNA2190	Na2190	Na2190S/Bi	190	216	240	42	1.85	256930	408940	1800
NA2200	RNA2200	Na2200	Na2200S/Bi	200	224.1	255	42	1.85	262820	423650	1700
NA2210	RNA2210	Na2210	Na2210S/Bi	210	236	265	42	1.85	282430	446200	1600
NA2220	RNA2220	Na2220	Na2220S/Bi	220	248.4	280	49	1.85	337350	557010	1500
NA,2230	RNA2230	Na2230	Na2230S/Bi	230	258.4	290	49	1.85	346170	578590	1500
NA2240	RNA2240	Na2240	Na2240S/Bi	240	269.6	300	49	1.85	356960	603110	1400
NA2250	RNA2250	Na2250	Na2250S/Bi	250	281.9	315	49	1.85	367750	632530	1300

Courtesy: IS: 4215, 1993.

TABLE 23-86
Needle bearings—Heavy Series

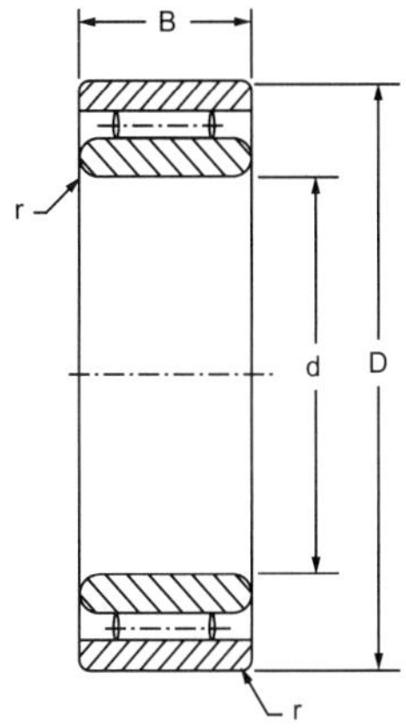

Bearing No.											
IS		NRB		Dimensions, mm					Basic load rating capacity		
Bearing with inner ring	Bearing without inner ring	Bearing with inner ring	Bearing without inner ring	d Nom	d_i Nom	D Nom	B Nom	r	Dynamic, C, N	Static, C_o, N	Limiting speed, n, rpm
NA3030	RNA3030	Na3030	Na3030S/Bi	30	44	62	30	0.65	35790	68160	8600
NA3035	RNA3035	Na3035	Na3035S/Bi	35	49.7	72	36	0.65	91690	98070	7600
NA3040	RNA3040	Na3040	Na3040S/Bi	40	55.4	80	36	0.65	102970	107870	6900
NA3045	RNA3045	Na3045	Na3045S/Bi	45	62.1	85	38	0.85	106890	121600	6100
NA3050	RNA3050	Na3050	Na3050S/Bi	50	68.8	90	38	0.65	115720	134350	5500
NA3055	RNA3055	Na3055	Na3055S/Bi	55	72.6	95	38	0.65	119640	141220	5200
NA3060	RNA3060	Na3060	Na3060S/Bi	60	78.3	100	38	0.65	126510	151026	4900
NA3065	RNA3065	Na3065	Na3065S/Bi	65	83.1	105	38	0.65	131410	160830	4500
NA3070	RNA3070	Na3070	Na3070S/Bi	70	88	110	38	0.65	137290	169650	4300
NA3075	RNA3075	Na3075	Na3075S/Bi	75	96	120	38	0.65	140230	184360	4000
NA3080	RNA3080	Na3680	Na3080S/Bi	80	99.5	125	38	0.65	149060	190250	3800
NA3085	RNA3085	Na3085	Na3085S/Bi	85	104.7	130	38	0.65	153960	198090	3600
NA3090	RNA3090	Na3090	Na3090S/Bi	90	109.1	135	43	0.65	189270	249090	3500
NA3095	RNA3095	Na3095	Na3095S/Bi	95	114.7	140	43	0.65	196130	262820	3300
NA3100	RNA3100	Na3100	Na3100S/Bi	100	119.2	145	43	0.65	201040	270660	3200
NA3105	RNA3105	Na3105	Na3105S/Bi	105	124.7	150	45	0.65	207900	283410	3000
NA3110	RNA3110	Na3110	Na3110S/Bi	110	132.5	160	45	0.65	215750	300080	2900
NA3115	RNA3115	Na3115	Na3115S/Bi	115	137	165	45	0.65	221630	311850	2800
NA3120	RNA3120	Na3120	Na3120S/Bi	120	143.5	170	45	0.65	228490	323620	2700
NA3125	RNA3125	Na3125	Na3125S/Bi	125	152.8	185	52	0.65	273600	397170	2500
NA3130	RNA3130	Na3130	Na3130S/Bi	130	158	190	52	1.35	280470	409920	2400
NA3140	RNA3140	Na3140	Na3140S/Bi	140	170.5	205	52	1.35	294200	441300	2200
NA3150	RNA3150	Na3150	Na3150S/Bi	150	179.3	215	52	1.35	307930	463850	2100
NA3160	RNA3160	Na3160	Na3160S/Bi	160	193.8	230	57	1.35	368730	568780	2000
NA3170	RNA3170	Na3170	Na3170S/Bi	170	202.6	245	57	1.35	380500	593300	1900
NA3180	RNA3180	Na3180	Na3180S/Bi	180	216	255	57	1.85	397170	632530	1800
NA3190	RNA3190	Na3190	Na3190S/Bi	190	224.1	265	57	1.85			1700
NA3200	RNA3200	Na3200	Na3200S/Bi	200	236	280	57	1.85	421680	691370	1600
NA3210	RNA3210	Na3210	Na3210S/Bi	210	248.4	290	64	1.85	490330	813950	1500
NA3220	RNA3220	Na3220	Na3220S/Bi	220	258.4	300	64	1.85	504060	847290	1500
NA3230	RNA3230	Na3230	Na3230S/Bi	230	269.6	315	64	1.85	517790	882590	1400
NA3240	RNA3240	Na3240	Na3240S/Bi	240	281.5	325	64	1.85	514850	921820	1300
NA3250	RNA3250	Na3250	Na3250S/Bi	250	290.9	340	74	1.85	666850	11157180	1300

Courtesy: IS: 4215, 1993.

TABLE 23-87
Hardness factors for needle-roller bearings

Rockwell C hardness of raceway	Approximate Brinell hardness (Bhn)	Hardness factor, K_h
63	660	1.00
60	620	0.98
58	595	0.96
56	570	0.92
54	545	0.83
52	515	0.70
50	490	0.50

Particular	Formula

TABLE 23-88
Torrington needle-roller sizes

Diameter, mm	Length, mm	Diameter, mm	Length, mm
1.590	9.40	3.175	22.25
1.590	12.45	3.175	23.82
1.590	15.75	3.175	25.40
1.590	16.95	3.175	28.575
2.380	10.55	4.010	18.800
2.3815	19.05	4.740	13.380
2.3850	11.745	4.765	18.900
2.3850	24.758	4.765	25.400
3.1750	9.770	4.765	30.200
3.1750	12.750	4.765	34.950
3.1750	15.650	5.500	19.100
3.1750	19.050	6.350	31.750

Maximum shear stress occurs below the contact surface for ductile material

(i) For spheres — Refer to Table 23-91.

(ii) For cylinders — Refer to Table 23-92.

SELECTION OF FIT FOR BEARINGS

For selection of fit for housing seatings for radial and thrust bearings

For selection of fit for shaft (solid) seatings for radial and thrust bearings.

TABLE 23-89
Dimensions for needle bearing without outer ring, type NCS

d, mm	d_o, mm	B, mm	r, mm
30	44.2	18	1
32	46.4	18	1
35	50	18	1
40	55.7	18	1.5
45	61.4	18	1.5
50	67.1	20	2
55	72.9	20	2
60	80.5	20	2
65	84.3	20	2
70	90.1	20	2
75	96.8	24	2
80	102.4	24	2
85	106.5	32	2
90	111.7	32	2
95	116.1	32	2
100	121.7	32	2
105	126.2	32	2
110	131.7	34	2
115	139.5	34	2
120	144	34	2
125	150.54	34	2
130	155.04	34	2
135	159.8	36	2
140	165.04	36	2

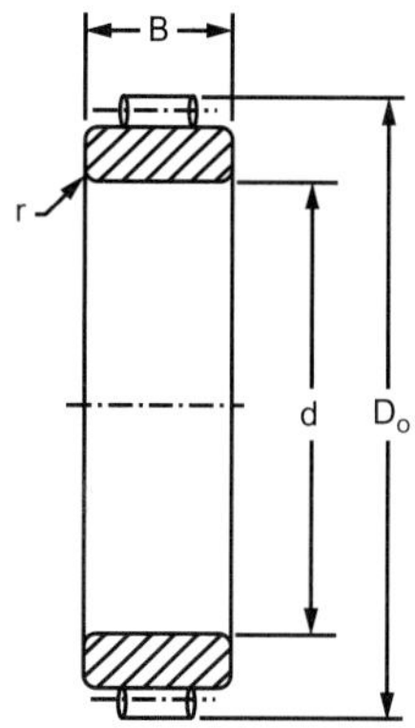

Source: IS 4215, 1967.

TABLE 23-90
Design data for needle-roller bearings

	Recommended	
Journal race diameter, mm	Total radial clearance, mm	Needle diameter,
9.50–19.00	0.0125–0.040	1.55
19.00–31.75	0.0180–0.050	2.35
31.75–50.80	0.0200–0.055	3.20
50.80–76.00	0.0255–0.065	3.20
76.00–127.00	0.0305–0.075	4.75
127.00–177.00	0.0355–0.085	4.75

TABLE 23-91
Selection of fit

(a) Housing seatings for radial bearings

Conditions	Applications	Tolerance
	Solid Housings	
Rotating outer-ring load		
Heavy loads on bearings in thin walled housings; heavy shock loads	Roller bearing wheel hubs; big-end bearings	P7
Normal and heavy loads	Ball bearing wheel hubs; big-end bearings	N7
Light and variable loads	Conveyor rollers, rope sheaves; belt tension pulleys	M7
Direction of loading indeterminate		
Heavy shock loads	Electric traction motors	M7
Heavy and normal loads; axial mobility of outer ring unnecessary	Electric motors, pumps, crankshaft main bearings	K7
Normal and light loads; axial mobility of outer ring desirable	Electric motors, pumps; crankshaft main bearings	J7
	Split or Solid Housing	
Stationary outer-ring load		
Shock loads intermittent	Railway axle boxes	J7
All loads	Bearings in general applications	H7
Normal and light loads	Line shafting	H8
Heat condition through shafts	Drying cylinders; large electric motors	G7
	Solid Housings	
Arrangement of bearing very accurate		
Accurate running and great rigidity under variable load	Roller bearings $D > 125$ mm	N6
	For machine-tool $D < 125$ mm main spindles	M6
Accurate running under light loads of indeterminate direction	Ball bearings at work end of grinding spindle; locating bearings in high-speed centrifugal compressors	K6
Accurate running; axial movement of outer ring desirable	Ball bearings at drive end of grinding spindles; axially free bearings in high-speed centrifugal compressors	J6

(b) Housing seatings for thrust bearings

Conditions	Applications		Tolerance
Purely axial load	Thrust ball bearings		H8
	Spherical roller thrust bearings where another bearing takes care of the radial location		
Combined (radial and axial) load or spherical roller thrust bearings	Stationary load on housing washer or direction of loading indeterminate		J7
		Generally	K7
	Rotating load or housing washer	Heavy radial load	M7

TABLE 23-92
Selection of fit
(a) Shaft (solid) seatings for radial bearings

Conditions	Application	Shaft diameter, mm: Ball bearings	Shaft diameter, mm: Cylindrical and tapered roller bearings	Shaft diameter, mm: Spherical roller bearings	Tolerance
		Bearings with cylindrical bore			
Stationary inner-ring load					
Easy axial displacement of inner ring on shaft desirable	Wheels on nonrotating axles		All diameters		*g*6
Easy axial displacement of inner ring on shaft unnecessary	Tension pulleys; rope sheaves		All diameters		*h*6
Rotating inner-ring or direction of loading indeterminate					
Light and variable loads	Electrical apparatus;	≤18	—	—	*h*5
	machine tools; pumps;	18–100	≤40	≤40	*j*6
	transport vehicles	100–200	40-140	49-100	*k*6
			140-200	100-200	*m*6
Normal and heavy loads	General application electric	≤18	—	—	*j*5
	motors pumps; turbines;	18–100	≤40	≤40	*k*5
	gearing; wood working	100–140	40–100	40–65	*m*5
	machines; and internal-	140–200	100–140	65–100	*m*6
	combustion engines	200–280	140–200	100–140	*n*6
			200–400	140–280	*p*6
				280–500	*r*6
				>500	*r*7
Shock and heavy loads	Locomotive axle boxes;	—	50–140	50–100	*n*6
	traction motors	—	140–200	100–140	*p*6
		—	—	140–200	*r*6
		—	—	200–500	*r*7
Purely axial load	All kinds of bearing arrangements		All diameters		*j*6
		Bearings with taper bore and sleeve			
Loads of all kinds	Bearing arrangements in general; railway axle boxes		All diameters		*h*9
	Line shafting		All diameters		*h*10

(b) Shaft seatings for thrust bearings

Conditions	Applications		Tolerance
Purely axial load	Thrust ball bearing, spherical roller thrust bearings	All diameters	*j*6
Combined (radial and axial) load on spherical thrust bearings	Stationary load on shaft washer	All diameters	*j*6
	Rotating load on shaft washer or direction of loading indeterminate	$d \leq 200$ mm	*k*6
		$d = 200$–400 mm	*m*6
		$d > 400$ mm	*n*6

23.3 FRICTION AND WEAR[1]

SYMBOLS

a	half the mean diameter of area of contact, Eq. (23-252)
A	real area of contact, m^2 (in^2)
A_a	apparent area of contact, m^2 (in^2)
A'	abrasion factor
b	constant used in Eq. (23-222), exponent
c	constant used in Eqs. (23-225) and (23-280)
c_1, c_2	constants as given in Eqs. (23-28lb) and (23-281c)
d	diameter, m (in)
E	Young's modulus, GPa (psi)
F	force, kN (lbf)
F_μ	total force of friction, kN (lbf)
$F_{a\mu}$	adhesive component of friction force or force to shear junctions, kN (lbf)
F_f	fatigue resistance is the average number of reversed stress cycles which the surface layer must undergo under given abrasion condition, kN (lbf)
$F_{\text{ploughing}}$	force to plough the asperities on one surface through the other, kN (lbf)
G	elasticity constant characterizing rubber
h	thickness of layer removed, m (in) effective thickness of the worn-out surface layer, m (in)
h_m	height of asperities, m (in)
H	hardness of softer material, N/m^2 or Pa (psi)
i	number of surface layer which are abraded during a test number of repeated deformation as used in Eqs. (23-256) to (23-258)
Q_{me}	mechanical equivalent of heat, N m/J (lbf in/Btu or lbf ft/Btu)
k_1, k_2	thermal conductivity of two conducting materials, W/m K (Btu/ft h °F)
k	constant used in Eq. (23-245) and given in Table 23-77
K_E	energetic wear rate or energy index of abrasion
K_L	linear wear rate
K_V	volumetric wear rate
K_W	gravimetric wear rate
K_{sm}	specific wear by mass
K_{sV}	specific wear by volume
K'_{sv}	modified specific wear
L	sliding distance, m (in)
m	mass of wear debris, kg (lb)
n	exponent
P_c	power used to elongate shred
P_H	power applied to hysteresis loss which accompanies roll deformation
P_t	power used to tear shred from surface layer
P_{tot}	total fictional power
P	yield pressure of soft material (about 5 times the critical shear stress), MPa (psi)
P_a	apparent pressure over the contact area, MPa (psi)
P_m	mean pressure over the contact area, MPa (psi) flow pressure of material, MPa (psi)

q	friction work done corresponding to a simple stressing cycle which corresponds to a sliding length of λ, N m (lbf in)
r	radius of curvature, m (in)
	radius of circular junction (Fig. 23-60), m (in)
R	mean radius of the curvature at the tip of the abrasive particles, m (in)
s	spacing between ridges in the elastomer surface, m (in)
v	velocity, m/s (ft/min)
v_1	velocity, m/s (ft/min)
V	volume deformed body, m^3 (in^3)
$V\Delta$	volume of transferred fragment, m^3 (in^3)
	volume of layer removed, m^3 (in^3)
W	applied load at interface, kN (lbf)
W_{tb}	the work of adhesion of the contacting metals which can be expressed in terms of their surface energies, N m (lbf in or lbf ft)
$W_{tot} = Wn^2$	normal load per unit area, kN (lbf)
ΔW	weight lost due to abraded layer being removed from the bulk material, kN (lbf)
z	the average depth of penetration for single sphere, m (in)
	the absolute approach, m (in)
α_n	coefficient of hysteresis loss
β	constant depends on the surface treatment taken from Table 23-73
γ	surface tension of the softer sliding member, N/m (lbf/in)
δ	abradability as wear index
θ	angle of slope of irregularities, deg
θ_m	mean temperature rise at the sliding junction, °C (°F)
μ	coefficient of friction
μ_a	adhesive component of coefficient of friction
μ	coefficient of elastic friction
μ_c	coefficient of static friction taken from Table 23-74
$\mu_{\text{ploughing}}$	ploughing component of coefficient of friction
ν	Poisson's ratio
ρ	density of the abraded elastomer, kg/m^3 (lb/in^3)
ζ	coefficient of abrasion resistance
λ	mean wavelength of the surface asperities
σ	stress, MPa (psi)
σ_c	contact pressure or pressure over the contours, MPa (psi)
σ_n	tensile strength of elastomer in simple tensions, MPa (psi)
τ	shear strength of junction, MPa (psi)
τ_m	mean shear stress, MPa (psi)

Particular	Formula

FRICTION

The general expression for force of friction

$$F_\mu = F_{a\mu} + F_{\text{ploughing}} \quad (23\text{-}256)$$

The total friction force

$$F_\mu = A\tau \quad (23\text{-}257)$$

The real area of contact

$$A = \frac{W}{P} \quad (23\text{-}258)$$

The general expression for coefficient of friction

$$\mu = \mu_a + \mu_{\text{ploughing}} \quad (23\text{-}259)$$

The total coefficient of friction

$$\mu = \frac{F_\mu}{W} = \frac{A\tau}{W} = \frac{\tau}{P} \quad (23\text{-}260)$$

The coefficient of elastic friction when a rigid rough surface is pressed against an elastically deformable second surface

$$\mu_e = \left[\left(\frac{K\alpha_n K_4^{1/2}\sqrt{\beta}}{2(\beta+1)}\right)\left(\frac{h_m}{r}\right)^{2/(2\beta+1)}\left(\frac{\sigma_c}{E}\right)^{2/(2\beta+1)}\right] \quad (23\text{-}261)$$

where $K\alpha_\pi \simeq 1$; calculate K_4 from Eq. (23-262).

The expression for K_4 to be used in Eq. (23-261)

$$K_4 = \left(\frac{0.75(1-v^2)\pi}{K_2\beta b}\right)^{\beta/(2\beta+1)} \quad (23\text{-}262)$$

where $K_2 = 1, 0.4, 0.12$ for $\beta = 1, 2, 3$ respectively

Refer to Table 23-93 for β.

TABLE 23-93
Constant β to he used in Eq. (23-261)

Surface treatment	β	b
Turning, milling	2	1–3
Planing	3	4–6
Polishing	3	5–10

Greenwood and Tabor's formula for coefficient of elastic friction

$$\mu_e = \alpha_n P_m\left(\frac{9\pi}{64}\frac{1-v^2}{E}\right) \quad (23\text{-}263)$$

Coefficient of friction under dynamic conditions

Franke's expression for coefficient of friction during rotation

$$\mu_r = \mu_o e^{-cv} \quad (23\text{-}264)$$

where c = constant taken from Table 23-94

Stiehl's formula for coefficient of friction

$$\mu = 0.6 - \frac{0.6}{v+1} \quad (23\text{-}265)$$

Schutch's formula for coefficient of friction for leather sliding against slightly lubricated steel plate

$$\mu = 0.5(1 + 0.1v) \quad (23\text{-}266)$$

Particular	Formula
Krumme's formula for coefficient of friction in textile machinery	$$\mu = 0.38 - \frac{0.1}{0.5 + v} \qquad (23\text{-}267)$$
Formula for coefficient of friction used in design of brakes	$$\mu = 0.6\,\frac{16P + 100}{80P + 100}\,\frac{100}{3v_k + 100} \qquad (23\text{-}268)$$ where P = real pressure on brake shoe, tonne force (tf)

TABLE 23-94
Values of constant c to be used in Eq. (23-264)

Sliding combination	State of rubbing surfaces	Coefficient of static friction, μ_o,	Constant c
Cast iron—steel	Dry	0.29	1/23
Forged iron—forged	Dry	0.29	1/50
Iron	Slightly moist	0.24	1/35

Temperature of sliding surface

Mean temperature rise at the interface above the material

$$\theta_m = \frac{0.25\mu W v}{Q_{me} r(k_1 + k_2)} \qquad (23\text{-}269)$$

where

Q_{me} = mechanical equivalent of heat N m/J (lbf in/Btu or lbf ft/Btu)

v = velocity of sliding, cm/s (ft/min)

r = radius of the circular junction, cm, m (in)

k_1, k_2 = thermal conductivity of the two contacting materials, W/m °C (Btu/ft h °F) taken from Table 23-95

TABLE 23-95
Temperature rise per unit sliding velocity

		γ		k_1	k_2	k_1	k_2	
Material combination	μ	dyn/cm	N/m	cal/s cm °C		W/m °K		θ/γ, °C/cm/s
Steel on steel	0.5	1500	1.50	0.11	0.11	46.055	46.055	0.75
Lead on steel	0.5	450	0.45	0.08	0.11	33.490	46.055	0.26
Bakelite on Bakelite	0.3	100	0.10	0.0015	0.0015	0.628	0.628	2.20
Brass on brass	0.4	900	0.90	0.26	0.26	108.856	108.856	0.15
Glass on steel	0.3	500	0.50	0.0007	0.11	0.293	46.055	0.30
Steel on nylon	0.3	120	0.12	0.11	0.0006	46.055	0.25121	0.07
Brass on nylon	0.3	120	0.12	0.26	0.0006	108.856	0.25121	0.03
Steel on bronze	0.25	900	0.90	0.11	0.18	46.055	75.362	0.17

Particular	Formula
Simple and crude formula for the mean temperature rise	$\theta_m = 54.4v$ ($\pm$ a factor of 1.67) (23-270)
The radius of a junction (Fig. 23-60)	$r = 12{,}000\,\dfrac{\gamma}{P}$ (23-271)
The load carried by each junction (Fig. 23-60)	$W = \pi r^2 P$ (23-272)
Mean temperature rise at the interface above the rest of material	$\theta_m = \dfrac{9400\mu\gamma v}{Q_{me}(k_1 + k_2)}$ (23-273) where γ = surface tension of the softer sliding member, N/m (lbf/in) taken from Table 23-95 For coefficient of friction μ refer to Table 23-95.

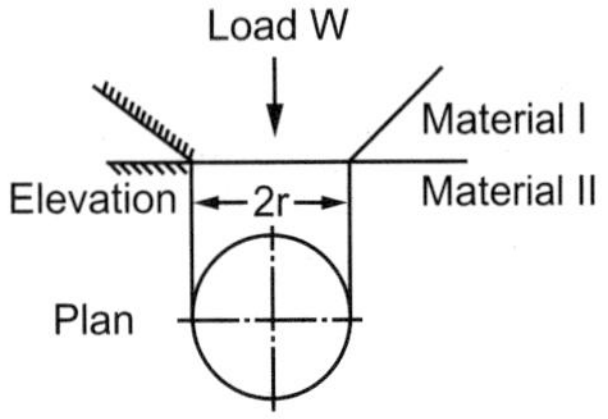

FIGURE 23-60 Assumed junction model.

WEAR AND ABRASION

Particular	Formula
Linear wear rate	$K_L = \dfrac{\text{thickness of layer removed}}{\text{sliding distance}} = \dfrac{h}{L}$ (23-274)
Steady state wear rate, depth per unit time	$K_L = KPV(abcde)$ (23-275) where K = constant depends on (i) mechanical properties of material and its ability to (ii) smooth the counterface surface and/or (iii) transfer a thin film of debris For a, b, c, d, e, refer to Table 23-103.
Volumetric wear rate	$K_V = \dfrac{\text{volume of layer removed}}{\text{sliding distance} \times \text{apparent area}} = \dfrac{\Delta V}{LA_a}$ (23-276)
Energetic wear rate	$K_E = \dfrac{\text{volume of layer removed}}{\text{work of friction}} = \dfrac{\Delta V}{F_\mu L}$ (23-277)
The energetic and linear wear rate related by equation	$K_E = K_L(A_a/F_\mu)$ (23-278) where $F_\mu L$ is measured in kW h
The gravimetric wear rate	$K_W = \dfrac{\Delta W}{LA_a} = \rho K_v$ (23-279) where ρ = density of abraded elastomer

Particular	Formula
Wear index is given by abradability, δ	$\delta = \dfrac{\text{abraded volume}}{\text{work of friction}} = \dfrac{\Delta V}{F_\mu L} = \dfrac{\Delta V}{\mu WL} = \dfrac{A'}{\mu}$ (23-280) where $A' = (\Delta V/WL)$ = abrasion factor
The relation between K_E and δ	Energetic wear rate (K_E) = abradability (δ) (23-281)
The coefficient of abrasion resistance as per work in the former Soviet Union	$\zeta = \dfrac{\text{work of friction}}{\text{abraded volume}} = \dfrac{FL}{\Delta V} = \dfrac{1}{\delta} = \dfrac{\mu}{A'} = \dfrac{1}{K_E}$ (23-282)
For surface roughness as obtained by different machining processes	Refer to Table 23-96.
Work done during wear	$W' = V\tau_m$ (23-283)
Volume of transferred fragments formed in sliding a distance L	$V = \dfrac{kWL}{300P}$ (23-284) For k = coefficient of wear, refer to Table 23-97. For P = hardness of the softer material, Pa (psi), refer to Table 23-99.

TABLE 23-96
Surface roughnesses as obtained by machining processes

Manufacturing process	Surface roughnesses, μm
Turned	1–6
Coarse ground	0.4–3
Fine ground	0.2–0.4
600 emery	0.2
Polished	0.05–0.1
Super finished	0.02–0.05

TABLE 23-97
Wear constant k

Sliding combination	Wear constant, k
Zinc on zinc	0.160
Low-carbon steel on low-carbon steel	45
Copper on copper	32
Stainless steel on stainless steel	21
Copper on low-carbon steel	1.5
Low-carbon steel on copper	0.5
Bakelite on Bakelite	0.02

TABLE 23-98
Values of coefficient of wear, k

	Metal on metal		Metal on non-metal
	Like	Unlike	
Condition	$\times 10^{-5}$		$\times 10^{-6}$
Clean	500	20	5
Poorly lubricated	20	20	5
Average lubrication	2	2	5
Excellent lubrication	0.2–0.2	0.2–0.2	2

TABLE 23-99
Properties of metallic elements

Metal	Melting temperature		Young's modulus, E			Yield strength, σ_{sy}			Hardness, P			Surface energy, γ		γ/P	
	°C	K	kgf/cm^2 × 10^6	N/m^2 × 10^{11}	MPa × 10^5	kgf/cm^2 × 10^3	N/m^2 × 10^8	MPa × 10^2	kgf/cm^2 × 10^2	N/m^2 × 10^7	MPa × 10	erg/cm	N/m	cm	m
Aluminum	660	933	0.64	0.63	0.63	1.12	1.1	1.1	27	26.46	26.46	900	0.900	33	0.33
Antimony	630	903	0.82	0.80	0.80	0.11	0.11	0.11	58	56.84	56.84	370	0.370	6.4	0.064
Beryllium	1400	1673	3.06	3.0	3.0	3.26	3.20	3.20	150	147.0	147.0	1000	1.0	6.7	0.067
Bismuth	270	543	0.33	0.32	0.32				7	6.86	6.86	390	0.39	56.0	0.56
Cadmium	321	594	0.57	0.56	0.56	0.73	0.72	0.72	22	21.56	21.56	620	0.62	28	0.28
Calcium	838	1111	0.26	0.25	0.25	0.89	0.87	0.87	17	16.66	16.66				
Cerium	804	1077	0.31	0.30	0.30	1.22	1.20	1.20	48	47.04	47.04				
Cesium	29	302													
Chromium	1875	2148	2.65	2.6	2.6	1.63	1.6	1.6	125	122.5	122.5				
Cobalt	1495	1778	2.14	2.1	2.1	7.96	7.8	7.8	125	122.5	122.5	1530	1.53	12	0.12
Copper	1083	1356	1.22	1.2	1.2	3.26	3.2	3.2	80	78.4	78.4	1100	1.10	14	0.14
Dysprosium	1407	1680	0.64	0.63	0.63	3.37	3.3	3.3	117	115.66	115.66				
Erbium	1496	1769	0.78	0.75	0.75	2.96	2.9	2.9	161	157.78	157.78				
Europium	827	1100							17	16.66	16.66				
Gadolinium	1312	1585	0.57	0.56	0.56	2.76	2.7	2.7	97	95.06	95.06				
Gallium	30	303							6.5	6.37	6.37	360	0.36	55	0.55
Germanium	937	1210	1.59	1.56	1.56							1120	1.12	19	0.19
Gold	1063	1336	0.83	0.81	0.81	2.14	2.1	2.1	58	56.84	56.84				
Hafnium	2222	2495				2.45	2.4	2.4	260	254.80	254.80				
Holmium	1461	1734	0.69	0.68	0.68	2.24	2.2	2.2	90	88.2	88.2				
Indium	156	429	0.11	0.11	0.11	0.03	0.03	0.03	0.9	0.88	0.88				
Iridium	2454	2727	5.50	5.4	5.4	6.43	6.3	6.3	350	343.0	343.0				
Iron	1534	1807	2.08	2.04	2.04	2.55	2.5	2.5	82	80.36	80.36	1500	1.50	18	0.18
Lanthanum	930	1203	0.40	0.39	0.39	1.94	1.9	1.9	150	147	147				
Lead	325	598	0.16	0.16	0.16	0.09	0.09	0.09	4	3.92	3.92	450	0.45	110	1.10
Lithium	180	453										400	0.40		
Lutetium	1652	1925							118	115.64	115.64				
Magnesium	650	923	0.45	0.44	0.44	1.53	1.5	1.5	46	45.08	45.08	560	0.56	12	0.12
Manganese	1245	1518				2.55	2.5	2.5	3300	3234	3234				
Mercury	−39	234										460	0.46		
Molybdenum	2610	2883	3.06	3.0	3.0	8.57	8.4	8.4	240	235.2	235.2				
Neodymium	1018	1291	0.39	0.38	0.38	1.73	1.7	1.7	80	78.4	78.4				
Nickel	1453	1726	2.12	2.08	2.08	3.26	3.2	3.2	210	205.8	205.8	1700	1.70	8.1	0.081
Niobium	2468	2741	1.07	1.05	1.05	2.86	2.8	2.8	160	156.8	156.8	2100	2.10	13	0.13
Osmium	2700	2973	5.81	5.70	5.70				800	784	784	1190	1.19	1.5	0.015
Palladium	1552	1825	1.17	1.15	1.15	3.16	3.1	3.1	110	107.8	107.8				
Platinum	1769	2042	1.53	1.50	1.50	1.63	1.6	1.6	100	98	98	1800	1.80	18	0.18
Plutonium	640	913	1.01	0.99	0.99	2.86	2.8	2.8	266	260.68	260.68				
Potassium	64	337							0.04	0.04	0.04	86	0.086	2300	23

TABLE 23-99
Properties of metallic elements (*Cont.*)

	Melting temperature		Young's modulus, E			Yield strength, σ_{sy}			Hardness, P			Surface energy, γ		γ/P	
Metal	°C	K	kgf/cm^2 × 10^6	N/m^2 × 10^{11}	MPa × 10^5	kgf/cm^2 × 10^3	N/m^2 × 10^8	MPa × 10^2	kgf/cm^2 × 10^2	N/m^2 × 10^7	MPa × 10	erg/cm	N/m	cm	m
Praseodymium	919	1192	0.36	0.35	0.35	2.04	2.0	2.0	76	74.48	74.48				
Rhenium	3180	3453	4.79	4.70	4.70	22.40	22.0	22.0							
Rhodium	1966	2293	3.02	2.96	2.96	9.89	9.7	9.7	122	119.56	119.56				
Rubidium	39	312													
Ruthenium	2500	2773	4.30	4.22	4.22	5.61	5.5	5.5	390	382.2	382.2				
Samanium	1072	1345	0.36	0.35	0.35	1.33	1.3	1.3	64	62.72	62.72				
Scandium	1540	1813													
Silver	961	1234	0.80	0.78	0.78	2.04	2.0	2.0	80	78.4	78.4	920	0.92	11	0.11
Sodium	98	371							0.07	0.07	0.07	200	0.20	2800	28
Tantalum	2996	3269	1.93	1.90	1.90	3.56	3.5	3.5							
Terbium	1356	1629	0.59	0.58	0.58				88	86.24	86.24				
Thallium	303	576				0.09	0.09	0.09	2	1.96	1.96	400	0.40	200	2.0
Thorium	1750	2023	1.50	1.47	1.47	1.53	1.5	1.5	37	36.26	36.26				
Thulium	1545	1818				1.43	1.4	1.4	53	51.94	51.94				
Tin	232	505	0.45	0.44	0.44	1.17	1.15	1.15	53	5.19	5.19	570	0.57	110	1.10
Titanium	1670	1943	1.15	1.13	1.13	1.43	1.4	1.4	65	64.7	64.7				
Tungsten	3410	3683	3.58	3.51	3.51	18.36	18.0	18.0	435	426.3	426.3	2300	2.30	5.3	0.053
Uranium	1132	1405	1.72	1.69	1.69	2.04	2.0	2.0							
Vanadium	1900	2173	1.36	1.34	1.34	8.57	8.4	8.4							
Ytterbium	824	1097	0.18	0.18	0.18	0.74	0.73	0.73	21	20.58	20.58				
Yttrium	1495	1768	0.67	0.66	0.66	1.43	1.4	1.4	37	36.26	36.26				
Zinc	420	693	0.93	0.91	0.91	1.33	1.3	1.3	38	37.24	37.24	790	0.79	21	0.21
Zirconium	1852	2125	0.98	0.96	0.96	2.04	2.0	2.0	145	142.10	142.10				

Particular	Formula
Another formula for volume of transferred fragment formed in sliding a distance	$$V = \frac{kAL}{3} \qquad (23\text{-}285)$$ For k refer to Table 23-98.
The primary equation of wear according to Archard, Burwell, and Strang	$$\frac{\Delta V}{L} = K\frac{W}{P_m} \qquad (23\text{-}286)$$ where P_m = flow pressure of material For K, refer to Table 23-100.

Abrasion wear

The mean diameter of loose wear particles which are produced at a smooth interface	$$d = K_1\left(\frac{W_{ab}}{H}\right) \qquad (23\text{-}287)$$
The ratio of half mean diameter of the area of contact to mean radius of the curvature at the tip of the abrasive particle	$$\frac{a}{R} = K_2\left(\frac{W}{GR^2}\right)^{\alpha} \qquad (23\text{-}288)$$ where α = value of exponent to be determined from experiment

TABLE 23-100
Coefficient of wear

Sliding against hardened tool-steel unless otherwise stated	Wear coefficient, K	Hardness kgf/mm^2	Hardness MPa
Mild steel on mild steel	7×10^{-3}	18.6	182.4
60/40 brass	6×10^{-4}	95.0	931.6
Teflon	2.5×10^{-4}	5.0	49.0
70/30 brass	1.7×10^{-4}	68.0	666.8
Perspex	0.7×10^{-6}	0.0	196.1
Bakelite (molded) type 50B	7.5×10^{-6}	5.0	245.2
Silver steel	6×10^{-5}	320.0	3138.1
Beryllium copper	3.7×10^{-5}	210.0	2059.4
Hardened tool steel	1.3×10^{-4}	850.0	8335.6
Stellite	5.5×10^{-5}	690.0	6776.6
Ferritic stainless steel	1.7×10^{-5}	50.0	2451.7
Laminated Bakelite Type 292/16	1.5×10^{-6}	33.0	323.6
Molded Bakelite Type 11085/1	7.5×10^{-7}	30.0	294.2
Tungsten carbide on mild steel	4×10^{-6}	186.0	1824.0
Molded Bakelite Type 547/1	3×10^{-7}	29.0	284.4
Polythene	1.3×10^{-7}	1.70	16.7
Tungsten carbide on tungsten carbide	1×10^{-6}	1300.0	12749

Particular	Formula
Volumetric wear rate	$$K_V = K_3 n^2 R^3 \left(\frac{W_{tot}}{Gn^2 R^2}\right)^{3\alpha} \quad (23\text{-}289)$$ where n^2 = number of abrasive particles per unit area
Volumetric wear rate for $\alpha = \frac{1}{3}$	$$K_V = K_3 \left(\frac{W_{tot} R}{G}\right) \quad (23\text{-}290)$$
Half the mean diameter of the area of contact for $\alpha = \frac{1}{3}$	$$a = K_1 \left(\frac{WR}{G}\right)^{1/3} \quad (23\text{-}291)$$
The spacing s between ridges in the elastomer surface	$$s \simeq \left(\frac{W_{tot} R d^2}{G}\right)^{1/3} \quad (23\text{-}292)$$ $$s \simeq d^{2/3} \quad (23\text{-}293)$$
The ratio of K_v to s when the abrasive surface consists of closely packed hemisphere so that $d = 2R$	$$\frac{K_V}{s} \simeq \left(\frac{W_{tot}}{G}\right)^{2/3} \quad (23\text{-}294)$$

Fatigue wear

Particular	Formula
Volume of surface layer removed under fatigue	$$\Delta V = iAh \quad (23\text{-}295)$$
The required sliding length during abrasion cycle under the given abrasion conditions before failure and separation occurs	$$L = i\lambda F_f \quad (23\text{-}296)$$
The total work of friction	$$W'_\mu = (\mu W_{tot})L = iqF_f \quad (23\text{-}297)$$
The coefficient of abrasion resistance	$$\zeta = \frac{qF_f}{AL} \quad (23\text{-}298)$$
The Hertzian relationship for the average depth of penetration for single spheres	$$z = \left(\frac{3}{4}(1-\theta^2)\right)^{\frac{2}{3}} \frac{W^{2/3}}{E^{2/3} R^{1/3}} \quad (23\text{-}299a)$$ $$z = 0.683\left(\frac{W^{2/3}}{E^{2/3} R^{1/3}}\right) \quad (23\text{-}299b)$$ for rubber $\theta = 0.5$ where R = asperity tips radius, cm, m (in) E = Young's modulus for rubber, GPa (psi) W = applied load per asperity
The depth penetration	$$z = 0.685\left(\frac{\lambda^2}{A}\right)^{\frac{2}{2}} \frac{W_{tot}^{2/3}}{E^{2/3} R^{1/3}} \quad (23\text{-}300)$$

Particular	Formula
The number of asperities	$i = \dfrac{W_{\text{tot}}}{W} = \dfrac{A}{\lambda^2}$ (23-301)
The effective thickness of the surface layer of elastomer	$h = k'z \dfrac{\pi R^2}{\lambda^2}$ (23-302) where $K' =$ constant
The coefficient of abrasion resistance	$\zeta = \dfrac{\mu F_f}{2.14K'} E^{2/3} \left(\dfrac{W_{\text{tot}}}{A}\right)^{1/3} \dfrac{\lambda}{R}$ (23-303)
The ratio of abrasion resistance to coefficient of sliding friction	$\dfrac{\zeta}{\mu} = \dfrac{1}{A'}$ (23-304)
The fatigue resistance of rubber taking into consideration tensile strength, geometry of the base surface, and the loading conditions	$F_f = \dfrac{\sigma_o}{K'(W/A)^{1/3} E^{2/3} (R/\lambda)^{-\frac{2}{3}}}$ (23-305)
The ratio of abrasion resistance to coefficient of friction	$\dfrac{\zeta}{\mu} = K\sigma_o^b E^{2(1-b)/3} \left(\dfrac{W_{\text{tot}}}{A}\right)^{(1-b)/3} \left(\dfrac{\lambda}{R}\right)^{(5-2b)/3}$ (23-306) where $b =$ index which is characteristic of the material where $K =$ constant
The relationship between fatigue index b and α	$b = \frac{1}{3}(\alpha + 2)$ (23-307)

Roll formation

Particular	Formula
The coefficient of abrasion resistance	$\zeta = \dfrac{P_{\text{tot}}}{(d\,\Delta V)/dt}$ (23-308) where $(d\,\Delta V)/dt =$ volume abraded per unit time $P_{\text{tot}} =$ total frictional power $= P_t + P_e + P_H$
The main condition which determines the probable occurrence of roll formation	$P_{\text{tot}} \leq \mu_o WV$ (23-309)
The more general form of the equation for volumetric wear rate which dependence on abrasion by load	$K_V = CP^{\alpha}$ (23-310) where $C =$ constant taken from Table 23-101 $P =$ interfacial pressure, MPa (psi) α is obtained from Table 23-101.

Particular	Formula

TABLE 23-101
Wear of rubber on steel, gauze, and abrasive paper

		Values of constants	
Rubber	**Nature of surface**	$C \times 10^3$	α
A	Steel	1.1	1.9
	Gauze	1.5	5.3
	Abrasive paper	240	1.1
B	Steel	2.7	1.9
	Gauze	1.1	2.0
	Abrasive paper	305	0.9
C	Steel	1.2	3.1
	Gauze	5.4	3.0
	Abrasive paper	65	1.0

Tread rubber

The shearing stress for tread rubber

$$\tau = \mu P \quad (23\text{-}311)$$

where P = normal pressure, MPa (psi)

The critical shearing stress for tread rubber

$$\tau_{\text{crit}} = \mu_{\text{crit}} P \quad (23\text{-}312)$$

For $\tau < \tau_{crit}$ — The fatigue wear predominates.

For $\tau > \tau_{crit}$ — Either wear through roll formation or abrasive wear occurs.

For $\mu < \mu_{crit}$ — The wear is due to surface fatigue.

For $\mu > \mu_{crit}$ — Other forms of wear predominate.

Specific wear

Specific wear by mass

$$K_{sm} = \frac{m}{Ad} \quad (23\text{-}313)$$

Specific wear by volume

$$K_{sV} = \frac{V}{Ad} \quad (23\text{-}314)$$

Specific wear by volume based on the geometry of the aspirities arising out of the surface treatment

$$K_{sV} = \frac{\tan\theta}{(\beta+1)2i} = \frac{\varepsilon h_m}{(\beta+1)id} = \frac{z}{(\beta+1)id} \quad (23\text{-}315)$$

where values of angle of slope of irregularities, θ, can be obtained from Table 23-102 and the values of β from Table 23-93

z = absolute approach $\varepsilon = z/h_m$

Particular	Formula

TABLE 23-102
Radii of curvature asperities for different methods of surface preparation

		Slope radii, micron		Angle of slope of irregularities, θ	
Treatment	**Accuracy class**	**Transverse**	**Longitudinal**	**Transverse**	**Longitudinal**
Shaping	5–8	20–120	10–25	5–20	5–10
Grinding	5–9	5–20	250–15000	7–35	2–10
Honing	8–11	4–30	60–160	3–13	1–4
Finishing (lapping)	10–13	15–250	7000–35000	5–20	2–10

The absolute approach

$$z = \frac{6\sigma_c}{K\gamma_1\gamma_2} \tag{23-316}$$

where

$$K = \frac{K_1 K_2}{K_1 + K_2} = \text{coefficient of rigidity}$$

$$K_i = \frac{E_i}{2\rho_i(1 - v_i^2)}$$

2ρ = diameter of contact spot, cm

γ = tangent to the smoothness of the surface equal to the derivative of approach over the contact area $= \tan\theta$

An expression for modified specific wear

$$K'_{sV} = K_{sV}\frac{A}{A_a} = K_{sV}\frac{P_a}{P} \tag{23-317}$$

Modified specific wear formula during microcutting

$$K'_{sV} = \frac{\tan\theta \cdot P_a}{(\beta+1)2P} \tag{23-318}$$

$$= 0.02\frac{P_a}{P} \text{ to } 0.04\frac{P_a}{P} \text{ for } \tan\theta = 0.1 \text{ to } 0.2$$

during microcutting

Modified specific wear formula during plastic contact

$$K'_{SV} = \left(\frac{h_{\max}}{rb(1/\beta)}\right)^{5/2}\left(\frac{P_a}{P}\right)^{(5+2\beta)/2\beta}\left(\frac{c\mu}{\varepsilon_{\text{fail}}}\right)^{2\beta^{1/2}/8} \tag{23-319}$$

where

$$h_{\max} = \frac{d}{2\varepsilon_{\max}}\tan\theta$$

$\varepsilon_{\text{fail}}$ = relative elongation corresponding to failure of the specimen

c = constant depending on sliding combination taken from Table 23-94

Particular	Formula
Modified specific wear formula during elastic contact	$$K'_{SV} = c_1 \frac{(1-v^2)P_a}{E}\left[\frac{K\mu\sigma_c}{c_2\sigma_o}\left(\frac{E}{(1-v^2)P_a}\right)^{2\beta/(2\beta+1)}\right.$$ (23-320a)
	where
	$$c_1 = \frac{3}{8}\pi\frac{\sqrt{\beta}}{K_2(\beta+1)}$$ (23-320b)
	$$c_2 = \left(\frac{r}{h_{max}}\right)^{\beta/(2\beta+1)}\left(\frac{b}{2}\right)^{1/(2\beta+1)}\left(\frac{0.75\pi}{K_2}\right)^{2\beta/(2\beta+1)}$$ (23-320c)

GENERAL

For values of wear rate correction factors; physical and mechanical properties of clutch facings; mechanical properties, performance and allowable operating conditions for various materials; physical and mechanical properties of materials for sliding faces; rubbing bearing materials and applications and allowable working conditions and frictions for various clutch facing materials	Refer to Tables 23-103 to 23-108.

TABLE 23-103
Approximate values of wear rate correction factors

Name of factor	Condition	Constant
a. Geometrical factor	Continuous motion + rotating load	0.5
	Unidirectional load	1
	Oscillating motion	2
b. Heat dissipation factor	Metal housing, thin shell, intermittent operation	0.5
	Metal housing, continuous operation	1
	Nonmetallic housing, continuous operation	2
c. Temperature factor	PTFE base: 20°C	1
	100°C	2
	200°C	5
	Carbon graphite thermoset 20°C	1
	Carbon graphite thermoset 100°C	3
	Carbon graphite thermoset 200°C	6
d. Counterface factor	Stainless steels, chrome plate	0.5
	Steels	1
	Soft, nonferrous metals	2.5
c. Surface finish factor	0.1–0.2 μm	1
	0.2–0.4 μm	2–5
	0.4–0.8 μm	4–10

TABLE 23-104
Physical and mechanical properties of clutch facings

	Resin-based material	Sintered metals
Thermal conductivity	0.80 W/m °C	16 W/m °C
Specific heat	1.25 kJ/kg °C	0.42 kJ/kg °C
Thermal expansion	0.50×10^{-4}/°C	0.13×10^{-4}/°C
Specific gravity	1.6 for woven 2.8 for molded	
Young's modulus, E	352 kgf/mm^2 3.45×10^9 N/m^2 3.45 GPa	1488 kgf/mm^2 14.5×10^9 N/m^2 14.5 GPa
Ultimate tensile strength, σ_{ut}	2.14 kgf/mm^2 21×10^6 N/m^2 21 MPa	4.57 kgf/mm^2 44.8×10^6 N/m^2 44.8 MPa
Ultimate shear, stress, τ_u	1.22 kgf/mm^2 12×10^6 N/m^2 12 MPa	3.59 kgf/mm^2 35.2×10^6 N/m^2 35.2 MPa
Ultimate compressive strength, σ_{uc}	10.5 kgf/mm^2 103×10^6 N/m^2 103 MPa	15.6 kgf/mm^2 153×10^6 N/m^2 153 MPa
Rivet holding capacity	7.03 kgf/mm^2 69×10^6 N/m^2 69 MPa	

TABLE 23-105
Mechanical properties, performance, and allowable operating conditions for various materials

Materials	Specific gravity	Coefficient of friction, μ	Tensile stress, σ_t			Shear stress, τ			Compressive stress, σ_c			Rivet holding capacity			Wear rate at 100°C, mm^3/J	Temperature °C		Working pressure, P_w			Maximum pressure, P_{max}		
			kgf/mm^2	$10^6 \times N/m^2$	MPa	kgf/mm^2	$10^6 \times N/m^2$	MPa	kgf/mm^2	$10^6 \times N/m^2$	MPa	kgf/mm^2	$10^6 \times N/m^2$	MPa		Maximum	Maximum operating	$kgf/mm^2 \times 10^{-3}$	$10^3\ N/m^2$	$10^{-3} \times$ MPa	kgf/mm^2	$10^6 \times N/m^2$	MPa
Lining																							
Woven cotton	1.0	0.50	2.1	20.7	20.7	1.26	12.4	12.4	9.85	96.5	96.5	7.03	69	69	12.2×10^{-6}	150	100	7.2–71.5	70–700	70–700	0.152	1.5	1.5
Woven asbestos	1.5–2.0	0.4	2.45	24.1	24.1	1.39	13.8	13.8	10.54	103.4	103.4	8.45	83	83	9.2×10^{-6}	250	125	7.2–71.5	70–700	70–700	0.214	2.1	2.1
Molded																							
Light-duty (flexible)	1.7	0.40	0.84	8.2	8.2	0.84	8.2	8.2	4.21	41.3	41.3	10.10	103	103	6.1×10^{-6}	350	175	7.2–71.5	70–700	70–700	0.214	2.1	2.1
Medium (semi-flexible)	1.7	0.35	1.05	10.3	10.3	0.84	8.2	8.2	9.85	96.5	96.5	15.50	152	152	3.1×10^{-6}	400	200	7.2–71.5	70–700	70–700	0.296	2.8	2.8
Heavy-duty	2.0	0.35	1.39	13.8	13.8	1.39	13.8	13.8	10.54	103.4	103.4	17.50	172	172	1.8×10^{-6}	500	225	7.2–71.5	70–700	70–700	0.390	3.8	3.8
Pad																							
Resin-based or asbestos	2.0	0.32				0.92	9.0	9.0	10..54	103.4	103.4				1.2×10^{-6}	650	300	35.5–178.5	350–1750	350–1750	0.561	5.5	5.5
Sintered metals	6.0	0.30	4.91	48.2	48.2	7.02	68.9	68.9	10.50	151.6	151.6				Used at higher	650	300	35.5–356.5	350–3500	350–3500	0.561	5.5	5.5
Cement		0.32													temperature	800	400	35.5–107.0	350–1050	350–1050	0.703	6.9	6.9

Key: 1 psi = 6895 Pa; 1 kpsi = 6.894757 MPa.

TABLE 23-106
Physical and mechanical properties of materials for sliding face

	Compressive strength, σ_c			Tensile strength, σ_t			Modulus of elasticity, E					Density, ρ			Maximum temperature, T_{max}		Expansion coefficient α	Temperature range, ΔT		Thermal conductivity		Thermal stress resistance	
Materials	kgf/mm²	MN/m²	MPa	kgf/mm²	MN/m²	MPa	kgf/mm² × 10³	GN/m²	GPa	Poisson's ratio, ν	Hardness, H_h	g/cm³	kg/m³	Porosity, ε, %	°C	K	$10^6 \times$ °C^{-1}	°C	K	kcal/m h°C	W/m K	kcal/m h	W/m
PTFCE	20–	215.8	215.8	3.2–	31.4–	31.4–	0.16	1.55	1.55	(0.3)	80*	2.1	2100	0	150	423	50	(320)	(593)	0.052	0.096	(16.6)	19.31
	56	549.4	549.4	4.0–	39.2	39.2																	
Nylon	5–	49.1–	49.1–	4.9–	48.1–	48.1–	0.18–	1.77–	1.77–	(0.3)		1.09–	1.09–	0	135–	408–	100–	(130)	(403)	0.12–	.140–	(21.5)	(25.00)
	9	88.3	88.3	7.5	73.6	73.6	0.28	2.75	2.75			1.14	1140		150	423	140			0.21	.244		
Phenol	7–	68.7	68.7	5.0–	49.1-	49.1–	0.52–	5.1–	5.1–	0.25		1.25–	1250–		130	403	25–60	140	(413)	0.1–	.116–	(21.5)	(25.00)
resin	21	207	207	5.6	54.9	54.9	0.70	6.87	6.87			1.3	1300							0.2	.233		
Synthetic	10–	98.1–	98.1–	3.5–	34.3–	34.3–	2.1–	20.7–	20.7–	(0.25)		1.75–	1750–		120–	393–	19–26	(50)	(323)	0.36–	.419–	22.0	25.59
resin 1	17.5	171.7	171.7	4.9	48.1	48.1	3.5	34.3	34.3			1.25	1250		150	423				0.51	.593		
Resin-	10–	98.1–	98.1–	2.3–	22.6–	22.6–	0.63–	6.2–	6.2–	(0.30)		1.36–	1360–		120	393	10–40	(150)	(423)	0.14–	.163–	(30.0)	(34.89)
impregnated	24	216	216	6.3	61.8	61.8	0.90	8.9	8.9			1.43	1430							0.25	.291		
fabric																							
Acetal				7	68.7	68.7	0.34	3.28	3.28	0.35		1.425	1425	0	100	373	81	167	440	0.2	0.233	33.5	39.96
resin																							
Bakelite	10–	98.1–	98.1–	2.8–	27.5	27.5	0.70–	6.87–	6.87–	0.25		1-52–	1520		175–	448–	05–40	87	360	0.29–	.337–	38.0	44.19
	24.5	240	240	5.0	49.1	49.1	1.75	17.16	17.16			2.0	2000		230	503				0.58	.675		
Hard				1.0–	9.81	9.81–	0.1	1.03	1.03	(0.4)		1.3–	1300–	0	100	373	54	180	453	0.25	.291	45.0	52.34
rubber				2.8	27.5	27.5						1.82	1820										
Synthetic	10–	98.1–	98.1–	1.5–	14.7–	14.7–	0.70–	6.87–	6.87–	(0.25)		1.6–	1600–		130–	403–	15–30	(75)	348	0.4–	.465–	(53)	(61.64)
resin 2	15	147	147	4.0	39.2	39.2	1.7	17.58	[17.58			1.9	1900		160	433				1.0	1.163		
PTFE				4.1	40.2	40.2	0.35	0.343–	0.343–	(0.5)	55–63*	2.1–	2100–		280	553	70	(410)	(683)	0.2	.233	(82)	(95.37)
							0.10	0.98	0.98			2.3	2300										
Synthetic carbon 1	16	156.9	156.9	2.1	20.1	20.1	1.75	18.06	18.06	0.2	65**	2.0	2000	0.3	170	443	13.5	66	239	2.0	2.33	132	153.46
Synthetic carbon 2	16.5	164.8	164.8	2.3	22.6	22.6	1.32	12.95	12.95	(025)	65**	2.8	2800	0	170	443	20	(65)	338	2.5	2.91	(164)	190.73
Carbon 2	18	176.6	176.6	2.8	27.5	27.5	1.4	1.37	1.37	0.22	85**	1.8	1800	0.1	180	453	5.0	312	595	4.0	4.65	1250	1453.75
Carbon 1	27	264.9	264.9	1.83	17.2	17-2	1.85	18.15	18.15	0.18	100**	1.82	1820	0.4	365	638	6.1	126	399	13	15.12	1650	1918.95
Carbon 3	25	245.3	245.3	3.1	30.4	30.4	1.46	14.32	14.32	0.2	80**	1.79	1790	0.5	285	558	5.3	320	593	20.0	23.3	6400	7443.20
Carbon 4	33.5	329.6	329.6	3.6	35.3	35.3	1.60	15.7	15.7	(0.2)	75**	2.5	2500	2.5	280	553	6.6	(273)	546	30.0	34.89	(8200)	(9536.60)
Carbon 5	35	343.4	343.4	2.1	20.6	20.6	1.35	13.24	13.24	(0.2)	85**	2.4	2400	4.0	350	623	4.82	(260)	(533)	34.0	39.54	(8800)	(10234.4)
Carbon 6	23.5	230.5	230.5	5.3	52.0	52.0	2.60	25.5	15.5	0.2	93**	1.73	1730	0.3	370	643	2.16	750.0	1023	20.0	23.3	15000	17445.0
Graphite 1	12.5	122.6	122.6	1.6	15.7	15.7	0.7	6.87	6.87	0.22	65**	1.65	1650	14	540	813	4.9	362	635	46.0	53.50	16700	19422.10
Graphite 2	10	98.1	98.1	1.55	14.7	14.7	1.0	9.81	9.81	0.2	65**	1.85	1850	1.0	365	638	5.25	235	508	90.0	104.67	21200	24655.60
Graphite 3	12.5	122.6	122.6	1.9	18.6	18.6	1.15	11.28	11.28	0.18	72**	1.85	1850	0.25	370	643	5.2	260	533	89.0	103.51	23000	26749
Graphite 4	7.1	69.7	69.7	1.45	14.2	14.2	1.30	12.75	12.75	0.22	60**	1.83	1830	0.3	180	453	3.5	250	523	100.0	116.3	25000	29075
Graphite 5	5.5	54.9	54.9	1.4	13.7	13.7	0.56	5.49	5.49	0.22	50***	1.66	1660	10.0	520	793	4.5	520	793	60.0	69.78	26000	30238
Graphite 6	14	137.3	137.3	2.0	19.6	19.6	1.0	9.81	9.81	0.22	70**	1.8	1800	7.0	340	613	2.0	780	1053	60.0	69.78	47000	54461
Hard alloy 1	150	1470	1470	3.8	372.8	372.8	2.4	235.4	235.4	0.3	58–62*	8.78	8780		1250	1523	11.4	97	370	7.0	8.14	680	790.84
Hard alloy 2	280	2746.8	2746.8	30	294.3	294.3	24.4	239.4	239.4	(0.3)	60†	7.77	7770		1150	1423	9.9	(87)	360	9.7	11.28	(850)	(988.55)
Hard alloy 3	135	1324.4	1324.4	53.0	519.9	519.9	23	225.6	225.6	0.3	48–50†	8.65	8650		1260	1533	11.9	135	840	11.0	12.79	1480	1720.24

TABLE 23-106
Physical and mechanical properties of materials for sliding face (*Cont.*)

Materials	Compressive strength, σ_c			Tensile strength, σ_t			Modulus of elasticity, E					Density, ρ			Maximum temperature, T_{max}		Expansion coefficient α	Temperature range, ΔT		Thermal conductivity		Thermal stress resistance	
	kgf/mm²	MN/m²	MPa	kgf/mm²	MN/m²	MPa	kgf/mm² × 10³	GN/m²	GPa	Poisson's ratio, r	Hardness, H_h	g/cm³	kg/m³	Porosity, ε, %	°C	K	$10^6 \times$ °C⁻¹	°C	K	kcal/m h °C	W/m K	kcal/m h	W/m
Hardened				28–	274.6–	274.6–	17.5	171.6	171.6	(0.26)	53–57^{e}	7.7	7700		800	1073	8.5	(157)	430	12.2	14.19	1930	2244.5
nickel				35	343.4	343.4																	
Stainless				54	529.7	529.7	20	196.2	196.2	0.28	155–185^{f}	7.98	7980	0	1400^{b}	1673^{b}	16.0	121	394	16.0	18.61	1940	2253.2
steel																							
Steel AISI 316																							
Invar				45	441.5	441.5	15	147.2	147.2	0.3	160^{f}	[8.0	8000	0	1425^{b}	1698^{b}	0.9	230	403	9.5	11.05	2200	2558.6
Niresist (cast)	70	687–	687–	17.50	171.5–	171.7–	10.5–	103–	103–	0.25	125–173^{f}	7.3	7300		1200^{b}	1473^{b}	17.0	78	351	34	39.54	2650	3081.9
	84	824	824	21	206	206	11.3	111	111														
Hastelloy B	21^{c}	206^{c}	206^{c}	85	834	834	21.4	210	210	(0.3)	215^{f}	9.23	9230		1335^{b}	1608^{b}	10.0	(280)	553	9.7	11.28	(2700)	1340.1
Hastelloy C	28.5	279^{c}	279^{c}	84	824	824	20	196.2	196.2	(0.3)	225^{f}	8.94	8940		1285^{b}	1558^{b}	11.3	(260)	533	10.8	12.56	(2800)	(3256.4)
Chrome (cast)	100	981	981	52	510	510	20.3	199	199	0.28	300^{f}	7.53	7530		1500^{b}	1773^{b}	10.6	173	450	19.0	22.10	3300	3837.9
Cobalt	85	833	833	24	235.4	235.4	21	206	206	0.28	125^{f}	8.9	8900		1495^{b}	1768^{b}	12.3	67	340	59.5	69.20	4000	(4652.0)
Cast iron	70	697	687	20	196.2	196.2	9.11	89–	89–														
							11	108	108	0.25	150–220^{f}	7.25	7250		1400^{b}	1673^{b}	10.0	150	423	40.0	46.5	6000	6978.0
Chrome				49	1481	1481	25	245.2	245.2	0.3	180^{f}	7.19	7190		1800^{b}	2073^{b}	6.2	220	493	57.6	66.99	12700	14770
Steel	350	3434	3434	130	1275.3	1275.3	20.6	202	202	0.28	64–67^{e}	7.8	7800	0	600	873	14.8	305	578	45.0	52.34	13800	16049
Molybdenum				70	687	687	33	323.7	323.7	0.324	20–26^{f}	10-2	10200		550	823	4.8	325	598	110	127.90	3560	1440.2
Steatite	63	618	618	7	68.7	68.7	10.5	103	103	(0.3)	7.5^{h}	2.7	2700	0.02	1000	1273	8.2	(57)	(330)	2.15	25.0	(120)	140
Magnesium				10	98.1	98.1	21.4	209.4	209.4	0.36		3.5	3500		2800^{b}	2073^{b}	13.5	22	295	31.0	36.05	680	790.8
Thoria	150	1470	1470	8.4	82.4	82.4	14.7	144.2	144.2	0.17		9.69	9690		3300^{b}	3573^{b}	9.2	52	325	9.0	4.65	470	546.6
Zircon	70	687	687	8.4	82.4	82.4	12.5	122.6	122.6	0.35	8^{h}	3.7	3700	0.02	1000	1373	4.0	108	381	4.3	5.00	465	540.7
Quartz glass				11	107.9	107.9	7.3	72.1	72.1	0.15	800^{h}	2.6	2600	0.5	1723^{b}	1996^{b}	0.5	2550	2823	1.37	1.59	3500	4070.5
Alumina 1	168	1648	1648	12.6	122.6	122.6	22.3	218.8	218.8	0.27	9^{h}	3.4	3400	0	1400	1673	5.5	74	347	11.4	13.37	840	976.9
Alumina 2	280	2746	2746	17.5	171.7	171.7	39	382.6	382.6	0.31	9^{h}	3.7	3700	0	1550	1823	5.8	54	327	16.2	18.84	875	1017.6
Alumina 3	210	2070	2070	24	235.4	235.4	35	343.4	343.4	0.2	9^{h}	3.9	3900	0	1750	2023	6.0	92	365	25.0	29.1	2300	2674.4
Cement 1	77	755	755	14.7	144.2	144.2	26	255.1	255.1	0.21	37^{f}	5.9	5900		1800	2073	8.0	56	329	25.0	29.1	1400	1628.2
Cement 2	168	1648	1648	21	206	206	26.6	261	261	0.26	50^{f}	[6.0	6000		1700	1973	7.5	78	351	29.0	33.73	2250	2616.7
Boron carbide	29	284	284	17.5	171.7	171.7	45.5	446.4	446.4	(0.25)	2800^{g}	2.51	2510		2500^{b}	2773^{b}	4.5	(64)	(337)	22.3	25.94	1430	1663.1
Silicon carbide	105	1030	1030	12.5	122.6	122.6	48	470.8	470.8	(0.25)	2500^{g}	3.1	3100		2400	2673	3.9	(50)	(323)	86.0	100	(4300)	5000.9
Chromium carbide	290	2845	2845	27	264.9	264.9	32	313.9	313.9	0.26	86.5^{f}	7.0	7000		1900^{b}	2173^{b}	9.0	70	343	(20.0)	(23.3)	(1400)	(1628.2)
Tungsten carbide 1	350	3433	3433	140	1370	1370	49	480.7	480.7	0.26	83–64^{f}	13.0.	13000	0.1	600	873	9.0	240	513	(30.0)	(34.89)	(7200)	8373.6
Tungsten carbide 2	420	4120	4120	120	1177.2	1177.2	56	549.4	549.4	0.248	86–87^{f}	14.1	14100	0.1	600	873	6.8	230	503	(50.0)	58.15	(11500)	13374.5
Tungsten carbide 3	500	4905	4905	85	833.8	833.8	70	687	687	0.216	91.5^{f}	14.8	14800	0.3	600	873	5.6	170	443	(60.0)	(62.78)	(10000)	11630
Tungsten carbide 4	370	3630	3630	115	1128.2	1128.2	54.5	534.6	534.6	0.242	89^{f}	14.0	14000	0.1	600	873	7.0	225	498	(40.0)	(46.5)	(9000)	10467.2
Titanium carbide 1				14	137	137	31.4	308	308	0.29	2460^{g}	[4.9	4900		3140^{b}	3413^{b}	7.4	43	316	21.5	25.0	930	1081.5
Titanium carbide 2	350	3434	3434	91	892.7	892.7	41.3	405.2	405.2	0.25	89^{f}	6.0	6000		1000	1273	9.5	175	448	26.0	30.34	4550	5291.6
Titanium carbide 3	(300)	2943	2943	105	1030.1	1030.1	28.7	281.5	281.5	0.26	82.5^{f}	6.3	6300		1000	1273	10.4	260	543	28.0	32.56	7300	8489.9
Titanium carbide 4	366	3591	3591	56	549	549	40	392.4	392.4	(0.25)		5.8	5800		1200^{b}	1473^{b}	5.7	(185)	(462)	29.0	33.73	(5400)	6270.2
Titanium carbide 5	250	2453	2453	140	1370	1370	30.4	298.2	298.2	0.3	87.5^{f}	7.0	7000		650	923	8.7	(370)	(643)	45.0	52.34	(16700)	(19422.1)

Key: [a] Brinell hardness; [b] melting point; [c] Shore hardness; [d] scleroscope; [e] Rockwell hardness; ([f] Rockwell A; [g] Knoop hardness; [h] Mohs hardness; [i] electric limit; [j] values in parentheses ( ) are approximate. Prefixes: $k = 10^3$; $M = 10^6$; $G = 10^9$. Conversion: 1 kgf/mm² = 9.80665×10^6 N/m²; 1 N/m² = 1 Pa; 1 kcal/h m °C = 1.163 W/m K; 1 kcal h m = 1.163 W/m = 1.163 J/h m; 1 psi = 6894.757 Pa; 1 kpsi = 6.894757 MPa; 1 Mpsi = 6.894757 GPa; 1 Btu/lft² h °F = 5.678 W/m² °C; 1 W/m² °C = 0.1761 Btu/ft² h °F; 1 g/cm³ = 3.6127×10^{-2} lb/in³ = 62.428 lb/ft³. PTFCE = polytetrafluoro-chloroethylene; PTFE = polytetrafluoroethylene.

TABLE 23-107
Rubbing, bearing materials and applications

Materials	Maximum loading, P kgf/mm^2	Maximum loading, P N/m^2 $\times 10^6$	Maximum loading, P MPa	Pv value kgf/mm^2 $\times$m/s	Pv value MN/m^2 $\times$m/s	Pv value MPa $\times$m/s	Coefficient of friction, μ	Maximum temperature, °C	Coefficient of expansion, $\alpha \times 10^6$, °C	Application
Carbon/graphite	0.14–0.20	1.4–2.0	1.4–2.0	0.0112–0.0184	0.11* 0.18**	0.11* 0.18**	0.10–0.25, dry	350–500	2.5–500	Conveyors, furnaces, food and textile machinery
Carbon/graphite with metal	0.31–0.41	3.0–4.0	3.0–4.0	0.0148–0.0224	0.145* 0.22**	0.145* 0.22**	0.10–0.35, dry	130–350	4.2–5.0	Bearings immersed in water, acid or alkaline solution, etc.
Graphite-impregnated metal	7.14	70.0	70.0	0.0286–0.0357	0.28–0.35	0.28–0.35	0.10–0.15, dry 0.020–0.025, grease-lubricated	350–600	12–13 with iron matrix	Bearings of foundry plant, coal mining machines, steel plants, etc.
Graphite-thermo-setting resin	0.20	2.0	2.0	0.0357	0.35	0.35	0.13–0.5, dry	250	3.5–5.0	Water-lubricated roll neck bearings in hot
Reinforced thermo-setting plastic	3.57	35.0	35.0	0.0357	0.35	0.35	0.1–0.4, dry, 0.006, water-lubricated	200	25–80 depending on plane of reinforcement	rolling mills, rubber bearings, bearings subjected to atomic radiation
Thermoplastic material without filler	1.02	10.0	10.0	0.0036	0.035	0.035	0.1–0.45, dry	100	100	Textile and food machinery bearings, bushes and thrust washers in automobile, bearing of linkages
Thermoplastic with filler or metal-backed	1.03	10.14	10.14	0.0036–0.0112	0.035–0.11	0.035–0.11	0.15–0.40, dry	100	80–100	For more heavily loaded applications, textile and food machinery, automobile and linkage
Thermoplastic with filler bonded to metal back	14.28	140.0	140.0	0.0357	0.35	0.35	0.20–0.35, dry	105	27	Ball joints, gearbox bushes, kingpin bushes, suspension and steering linkages
Filled PTFE	0.71	7.0	7.0	$\leq$ 0.0357	$\leq$ 0.35	$\leq$ 0.35	0.05–0.35, dry	250	60–80	Bushes, thrust washers, sideways
PTFE with filler bonded to steel backing	14.28	140.0	140.0	$\leq$ 0.1785	$\leq$ 1.75	$\leq$ 1.75	0.05–0.30, dry	280	20 (lining)	Aircraft controls, linkages, automobile gearboxes, conveyors, bridges and building, expansion bearings, bushes, and steering suspension
Woven PTFE reinforced and banded metal backing	42.84	420.0	420.0	$\leq$ 0.1623	$\leq$ 1.60	$\leq$ 1.60	0.03–0.33, dry	250		Aircraft engine controls, automobile suspension, engine mountings, linkages, bridges, and building expansion joint

TABLE 23-108
Allowable working conditions and friction for various clutch facing materials

Working conditions	Coefficient of friction, μ	Temperature Maximum °C	Temperature Continuous °C	Working pressure kgf/mm^2	Working pressure N/m^2 × 10^6	Working pressure MPa	Power rating, W/mm^2
Light-duty							
Woven	0.35–0.4	250	150	0.18–0.51	1.75–5.00	1.75–5.0	0.3–0.6
Mill board	0.40	250	150	0.18–0.71	1.75–7.00	1.75–5.0	0.3–0.6
Medium-duty			200				
Wound tape yarn	0.38	350		0.18–0.71	1.75–7.0	1.75–7.0	0.3–0.6
Asbestos tape	0.40	350	200	0.18–0.71	1.75–7.0	1.75–7.0	0.6–1.2
Molded	0.35	350	200	0.18–0.71	1.75–7.0	175–7.0	0.6–1.2
Heavy-duty							
Sintered	0.36/0.30	500	300	0.36–0.29	3.5–28	3.5–28.0	1.7
Cement	0.40			0.71–1.43	7.0–14	7.0–14.0	4.0
Oil immersed							
Paper	0.11			0.71–1.79	7.1–17.5	7.1–17.5	2.3
Woven	0.08			0.71–1.79	7.1–17.5	7.1–17.5	1.8
Molded	0.04			0.17–1.79	7.1–17.5	7.1–17.5	0.6
Molded (grooved)	0.06						
Sintered	0.11/0.05			0.71–4.28	7.0–42	7.0–42.0	2.3
Sintered (grooved)	0.11/0.06			0.71–4.28	7.0–42	7.0–42.0	2.3
Resin/graphite	0.10						5.3

REFERENCES

1. Lingaiah, K., *Machine Design Data Handbook*, Vol. II (*SI and Customary Metric Units*), Suma Publishers, Bangalore, India. 1986.
2. Lingaiah, K., and B. R. Narayana Iyengar, *Machine Design Data Handbook*, Engineering College Cooperative Society, Bangalore, India, 1962.
3. Maleev, V. L., and J. B. Hartman, *Machine Design*, International Textbook Company, Scranton, Pennsylvania, 1954.
4. Radzimovsky, F. I., *Lubrication of Bearings—Theoretical Principles and Designs*, The Ronald Press Company, New York, 1959.
5. Raimondi, A. A., and J. Boyd, 'A Solution for the Finite Journal Bearings and Its Application to Analysis and Design', *ASME J. Lubrication Technol.*, Vol. 104, pp. 135–148, April 1982.
6. Kingsbury, A., 'Optimum Conditions in Journal Bearing,' *Trans. ASME*, Vol. 54, 1932.
7. Needs, S. J., 'Effect of Side Leakage in 120-degree Centrally Supported Journal Bearings,' *Trans. ASME*, Vol. 56, 1934; Vol. 51, 1935.
8. Shigley, J. E., *Mechanical Engineering Design*, First Metric Edition, McGraw-Hill Book Company, New York, 1986.
9. Edwards, K. S., Jr., and R. B. McKee, *Fundamentals of Mechanical Component Design*, McGraw-Hill Book company, 1991.
10. Shaw, M. C., and F. Macks, *Analysis and Lubrication of Bearings*, McGraw-Hill Book Company, New York, 1949.
11. Lingaiah, K., *Machine Design Data Handbook*, McGraw-Hill Publishing Company, New York, U.S.A., 1994.
12. FAG Rolling bearings, Catalog WL 41520EI, 1995 edition, FAG Precision Bearings Ltd., Maneja, Vadodara, India.

13. SKF Rolling Bearings, Catalog 4000E. 1989, SKF Rolling Bearings, India Ltd., Mumbai, India.
14. Bureau of Indian Standards, Manak Bhavan, 9 Bahadur Shah Marg, New Delhi 110 002, India.
15. Neale, M. J., Editor, *Tribology Handbook*, Butterworth, London, 1973.
16. International Organization for Standards, 1, rue de Varembe, Case Postale 56, CH 1211, Geneve 20, Switzerland.
17. New Departure-Hyatt Bearing Division, General Motor Corporation, USA.
18. NSK Corporation (Corporate), Automotive Products Bearing Division, 3861 Research Park Drive, Ann Arbor, Michigan 48100-1507, USA.
19. The Torrington Company, 59 Field Street, Torrington, Conn 06790, USA.
20. Antifriction Bearing Manufacturers Association, USA.
21. Black, P. H., and O. E. Adams, Jr., Machine Design, McGraw-Hill Publishing Company, New York, 1968.

BIBLIOGRAPHY

ASME Standards.
Baumeister, T., ed., *Marks' Handbook for Mechanical Engineers*, McGraw-Hill Book Company, New York, 1978.
Black, P. H., and O. E. Adams, Jr., *Machine Design*, McGraw-Hill Book Company, New York, 1968.
Boswall, R. O., *The Theory of Film Lubrication*, Longmans, Green and Company, New York, 1928.
Bureau of Indian Standards.
O'Connor, J. J. ed., *Standard Handbook of Lubricating Engineering*, McGraw-Hill Book Company, New York, 1968.
Fuller, D. P., *The Theory and Practice of Lubrication for Engineers*, John Wiley and Sons, New York, 1956.
Niemann, G., *Machine Elements—Design and Calculations in Mechanical Engineering*, Vol. II, Springer-Verlag, Berlin, 1950; Student Edition, Allied Publishers Private Ltd. Bangalore, India, 1979.
Niemann, G., *Maschinenelemente*, Springer-Verlag, Berlin, Erster Band, 1963.
Niemann, G., *Maschinenelemente*, Springer-Verlag, Berlin, Zweiter Band, 1965.
Hyland, P. H., and J. B. Kommers, *Machine Design*, McGraw-Hill Book Company, New York, 1943.
ISO Standards
Lansdown, A. R., *Lubrication: A Practical Guide to Lubricant Selection*, Pergamon Press, New York, 1982.
Leutwiler, O. A., *Elements of Machine Design*, McGraw-Hill Book Company, New York, 1917.
Michell, A. G. M., *Lubrication—Its Principles and Practice*, Blackie and Son, London, 1950.
Neale, M. J., ed., *Tribology Handbook*, Butterworth, London, 1973.
Norman, C. A., E. S. Ault, and I. F. Zarobsky, *Fundamentals of Machine Design*, The Macmillan Company, New York, 1951.
Norton, A. E., *Lubrication*, McGraw-Hill Book Company, New York, 1942.
Slaymaker, R. R., *Bearing Lubrication Analysis*, John Wiley and Sons, New York, 1955.
Rippel, H. C., "Design of Hydrostatic Bearings," *Machine Design*, Parts 1 to 16, Aug. 1 to Dec. 5, 1963.
SAE Handbook, 1957.
Shigley, J. E., *Machine Design*, McGraw-Hill Book Company, New York, 1962.
Shigley, J. E., and C. R. Mischke, *Standard Handbook of Machine Design*, McGraw-Hill Book Company, New York, 1986.
Shigley, J. E., and C. R. Mischke, *Mechanical Engineering Design*, McGraw-Hill Book Company, New York, 1989.
Vallance, A., and V. L. Doughtie, *Design of Machine Members*, McGraw-Hill Book Company, New York, 1951.
Wilcock, D. F., and E. R. Booser, *Bearing Design and Application*, McGraw-Hill Book Company, New York, 1957.

CHAPTER

24

MISCELLANEOUS MACHINE ELEMENTS[2,3]

24.1 CRANKSHAFTS[2,3]

SYMBOLS

A	area of cross section, m^2 (in^2)
b	width of crank cheek, m (in)
c	distance from the neutral axis of section to outer fiber, m (in)
d	diameter (also suffixes), m (in)
d_e	equivalent diameter, m (in)
d_o	diameter of crankpin, m (in)
d_m	diameter of main bearing, m (in)
E	modulus of elasticity, GPa (psi)
F	force acting on the piston due to steam or gas pressure corrected for inertia effects of the piston and other reciprocating parts, kN (lbf)
F_c	the component of force F acting along the axis of connecting rod, kN (lbf)
F_{comb}	combined force, kN (lbf)
F_{ic}	magnitude of inertia force due to the weight of connecting rod itself, kN (lbf)
F_r	total radial force acting on the crankpin, kN (lbf)
F_θ	total tangential force acting on the crankpin, kN (lbf)
G	modulus of rigidity, GPa (psi)
h	thickness of cheek or web (also with suffixes), m (in)
$i' = \dfrac{l_o}{d_o}$	ratio of length to diameter of crank
I	moment of inertia, m^4, cm^4 (in^4)
$K = \dfrac{D_i}{D_o}$	ratio of inner to outer diameter of a hollow shaft
K_b	numerical combined shock and fatigue factor to be applied to the computed bending moment
K_t	numerical combined shock and fatigue factor to be applied to the computed twisting moment
l	length (also with suffixes), m (in)
l_e	equivalent length, m (in)
M_b	bending moment, N m (lbf in)

M_t	twisting moment, N m (lbf in)
p	allowable pressure, MPa (psi)
r	radius, throw of crankshaft, m (in)
Z	section modulus, m^3, cm^3 (in^3)
σ	normal stress (also with suffixes), MPa (psi)
τ	shear stress, MPa (psi)

SUFFIXES

b	bending
c	compressive
$comb$	combined
e	elastic
m	main
max	maximum
r	radial
ra	resultant in arm
rh	resultant in hub
t	torque
s	shaking
θ	tangential

Other factors in performance or special aspects which are included from time to time in this chapter and are applicable only in their immediate context are not given at this stage.

Particular	Formula

FORCE ANALYSIS (Fig. 24-1)

Particular	Formula
The radial component of force F_c acting along the axis of connecting rod (Fig. 24-1)	$F_{c1} = F_c \cos(\theta + \phi) = \dfrac{F}{\sqrt{1 - \left(\dfrac{\sin\theta}{n'}\right)^2}} \cos(\theta + \phi)$ (24-1)
The tangential component of force F_c acting along the axis of connecting rod (Fig. 24-1)	$F_{c2} = F_c \sin(\theta + \phi) = \dfrac{F}{\sqrt{1 - \left(\dfrac{\sin\theta}{n'}\right)^2}} \sin(\theta + \phi)$ (24-2)
The radial component of force F_{ic} (Fig. 24-1)	$F_{ic1} = \frac{2}{3} F_{ic} \cos\gamma$ (24-3) where γ = angle between the force F_{ic} and the radial component of F_{ic}
The tangential component of force F_{ic} (Fig. 24-1)	$F_{ic2} = \frac{2}{3} F_{ic} \sin\gamma$ (24-4)
The total radial force acting on the crank	$F_r = F_{ic1} \pm F_{c1}$ (24-5)
	$F_r = \frac{2}{3} F_{ic} \cos\gamma \pm F_c \cos(\theta + \phi)$ (24-6)

Particular	Formula
The total tangential force acting on the crank	$F_\theta = F_{ic2} \pm F_{c2}$ $= \frac{2}{3} F_{ic} \sin\gamma \pm F_c \sin(\theta + \phi)$ (24-7)
The resultant force on the crankpin	$F_{comb} = \sqrt{F_r^2 + F_\theta^2}$ (24-8)

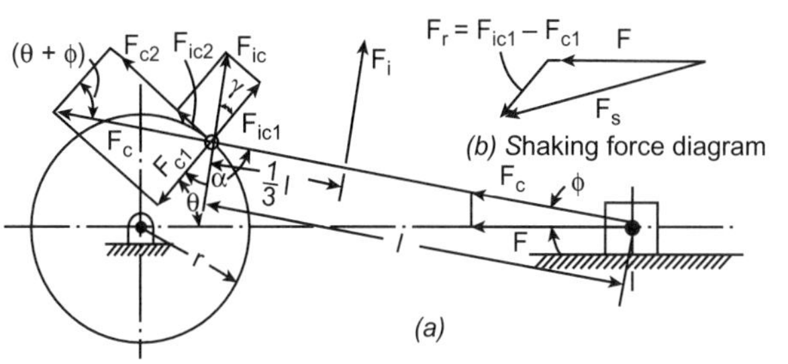

FIGURE 24-1 (*a*) Forces acting on crankshaft. (*b*) Vector sum of F and F_r.

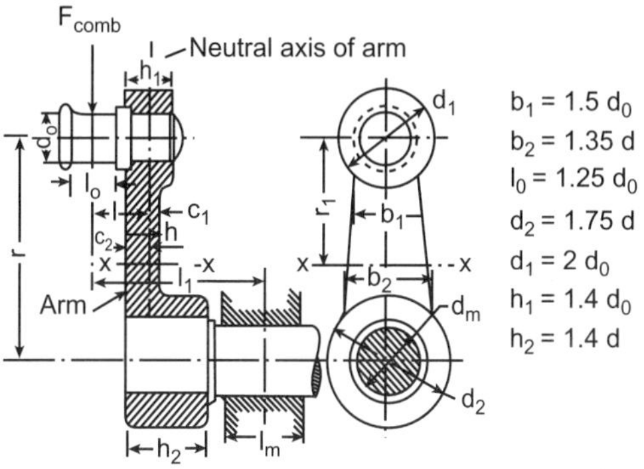

FIGURE 24-2 Overhung built-up crank.

SIDE CRANK

Crankpin

The maximum bending moment on the crankpin (Fig. 24-2)	$M_{b(\max)} = F_{comb} \times \left(\frac{l_o}{2} + \frac{t}{2}\right)$ (24-9) $= F_{comb} \times l$ where $l = \frac{l_o}{2} + c_2 =$ distance from centroidal axis to the application of load (Fig. 24-2), m (in)
The crankpin diameter with respect to the bending moment	$d_o = \sqrt[3]{\frac{32 l F_{comb}}{\pi \sigma_b}}$ (24-10) where $\sigma_b =$ allowable bending stress, MPa (psi)
The diameter of crankpin from the consideration of bearing pressure	$d_o = \frac{F_{comb}}{l_o p}$ (24-11)
From Eqs. (24-10) and (24-11) neglecting $t/2$ and eliminating l_o the equation for crankpin diameter	$d_o = \sqrt[4]{\frac{16 F_{comb}^2}{\pi p \sigma_b}}$ (24-12)
Empirical relation to determine the length of crankpin	$l_o = i' d_o$ (24-13) where $i' = \frac{l_o}{d_o} = 1.25$ to 1.5

Particular	Formula	
Another relation for the crankpin length/diameter ratio	$i' = \frac{l_o}{d_o} = \sqrt{\frac{0.2\sigma_b}{p}}$	(24-14)
Another relation for the crankpin diameter	$d_o = \sqrt{\frac{F_{comb}}{i'p}}$	(24-15)

HOLLOW CRANKPIN

The crankpin length/diameter ratio	$i' = \frac{l_o}{D_o} = \sqrt{\frac{0.2\sigma(1 - K^4)}{p}}$	(24-16)
	where $K = \frac{D_i}{D_o}$	
The crankpin outside diameter	$D_o = \sqrt{\frac{F_{comb}}{i'p}}$	(24-17)

Crank arm

CRANK ON HEAD-END DEAD-CENTER POSITION

When the crank is on the head-end dead-center position, the section XX (Fig. 24-2) of the arm is subjected to bending moment	$M_b = F_{comb} \times l$	(24-18)
The direct compressive stress due to the load F_{comb} (i.e., more specifically by its component F_c)	$\sigma_c = \frac{F_{comb}}{A}$	(24-19)
The resultant stress in the crank arm at XX	$\sigma_{ra} = \frac{F_{comb}}{A} \pm \frac{M_{bC}}{I}$	(24-20)
	where A = area of cross section of the arm at XX, m^2 (in^2) c = distance from the neutral axis of section to outer fiber of arm, m (in) I = moment of inertia of the section, cm^4 (in^4)	

CRANK ON CRANK-END DEAD-CENTER POSITION

The direct tensile stress in the plane of the hub of crankshaft section passing through the shaft center due to load F_{comb} (Fig. 24-2)	$\sigma_t = \frac{F_{comb}}{h_2(d_2 - d)}$	(24-21)
The bending stress in the section due to bending moment $F_{comb} \times a$	$\sigma_b = \frac{F_{comb} \times a}{Z}$	(24-22)
	where Z = section modulus, cm^3 (in^3)	

Particular	Formula
The resultant stress in the plane of the hub of crankshaft section passing through the shaft center	$\sigma_r = \sigma_t \pm \sigma_b$ (24-23)
CRANK PERPENDICULAR TO THE CONNECTING ROD	
The bending moment in the plane of rotation of the crank	$M_b = F_{comb} \times l$ (24-24)
The bending stress	$\sigma_b = \dfrac{M_b c_1}{Z_b}$ (24-25)
The torsional moment	$M_t = F_{comb} \times r_1$ (24-26)
The shear stress	$\tau = \dfrac{M_t c_1}{Z_t}$ (24-27)
The maximum normal stress for crank made of cast iron	$\sigma_{max} = \frac{1}{2}\left[\sigma_b + \sqrt{\sigma_b^2 + 4\tau^2}\,\right]$ (24-28)
The maximum shear stress for the crank made of steel	$\tau_{max} = \frac{1}{2}\sqrt{\sigma_b^2 + 4\tau^2}$ (24-29)

DIMENSION OF CRANKSHAFT MAIN BEARING (Fig. 24-2*b*)

Particular	Formula
The shaking force on the main bearing from F and F_r (Fig. 24-1*b*)	$F_s =$ vector sum of F and F_r (24-30)
The diameter of main bearing taking into consideration the bearing pressure on the projected area of the crankshaft	$d_m = \dfrac{F_s}{l_m p}$ (24-31) where $l_m =$ length of bearing, m (in) $p =$ allowable bearing pressure, MPa (psi)
The bending movement on the crankshaft	$M_b = F_{comb} \times l_1$ (24-32) $l_1 = \dfrac{l_o}{2} + h_2 + \dfrac{l_m}{2}$ where $h_2 =$ hub length, m (in) $l_o =$ length of crankpin, m (in) $l_m =$ length of bearing on crankshaft, m (in)
The torque on the crankshaft	$M_t = F_{comb} \times r$ (24-33) where $r =$ throw of the crank, m (in)
The diameter of crankshaft taking into consideration indirectly the fatigue and shock factors	$d_m = \sqrt[3]{\dfrac{16}{\pi\sigma_e}\left\{K_b M_b + \sqrt{(K_b M_b)^2 + (K_t M_t)^2}\right\}}$ (24-34)

Particular	Formula	
The length of main bearing	$l_m = \dfrac{F_s}{d_m p}$	(24-35)

PROPORTIONS OF CRANKSHAFTS

For proportions of crankshaft	Refer to Figs. 24-2 to 24-10.

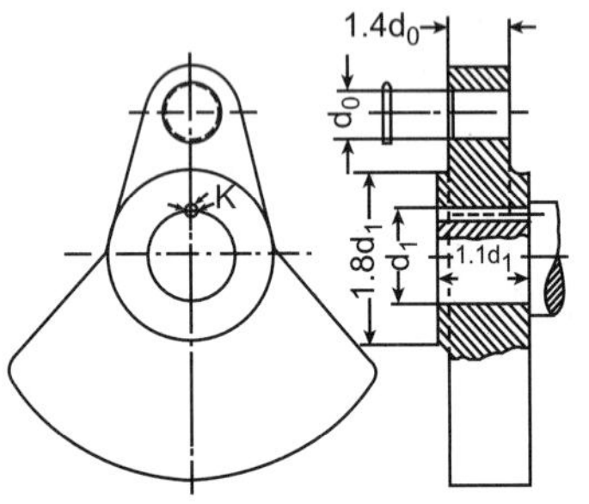

FIGURE 24-3 Overhung built-up crank.

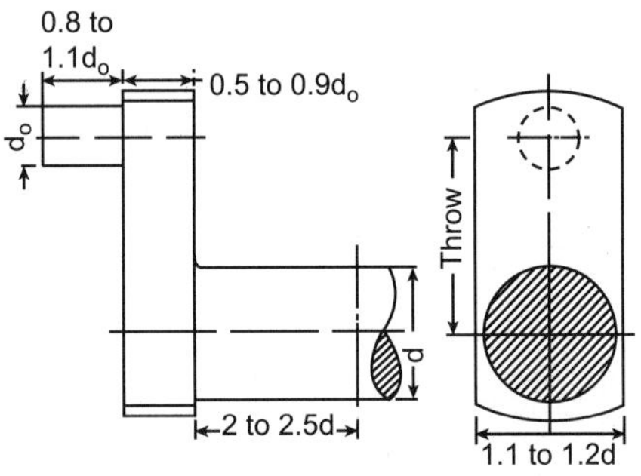

FIGURE 24-4 Overhung forged crank.

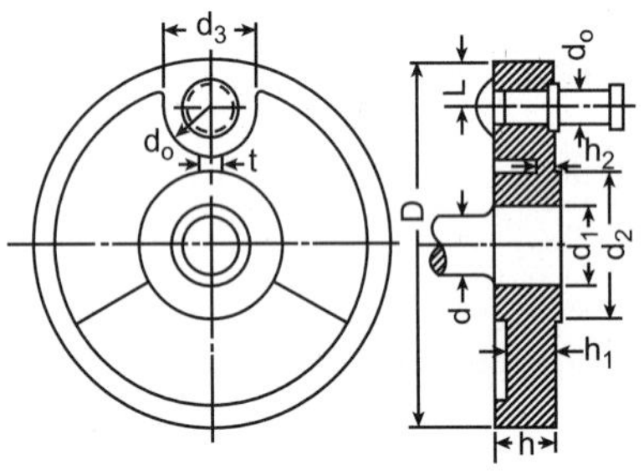

FIGURE 24-5 Disk crank.

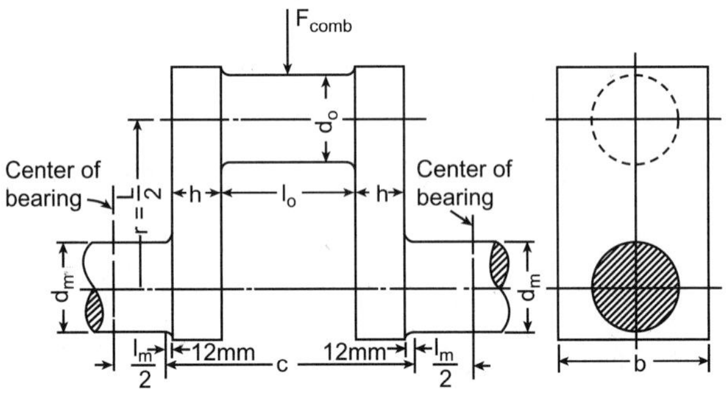

FIGURE 24-6 Center crank (American Bureau of Shipping method).

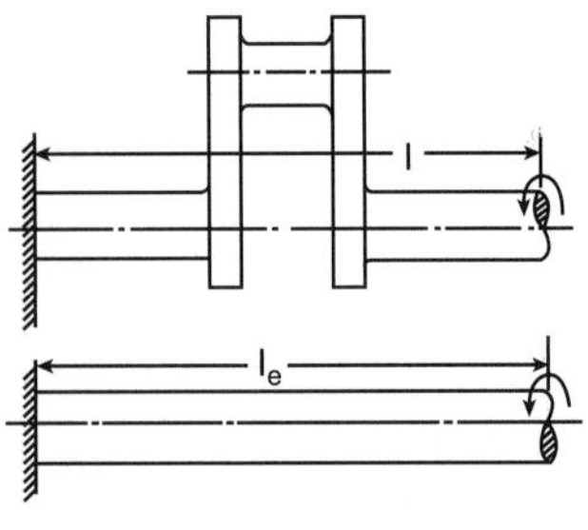

FIGURE 24-7 Equivalent length of crankshaft.

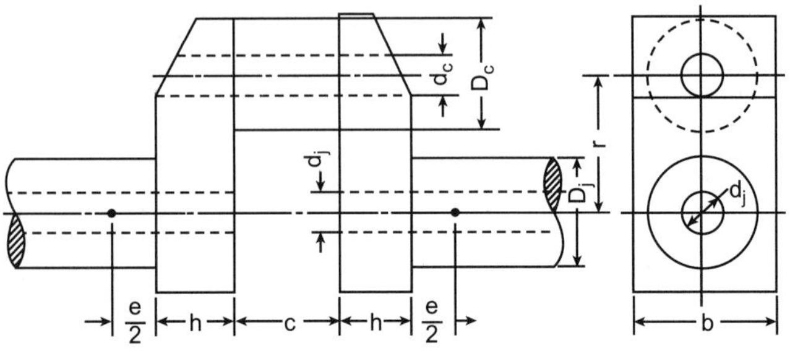

FIGURE 24-8 Center hollow crank.

Particular	Formula

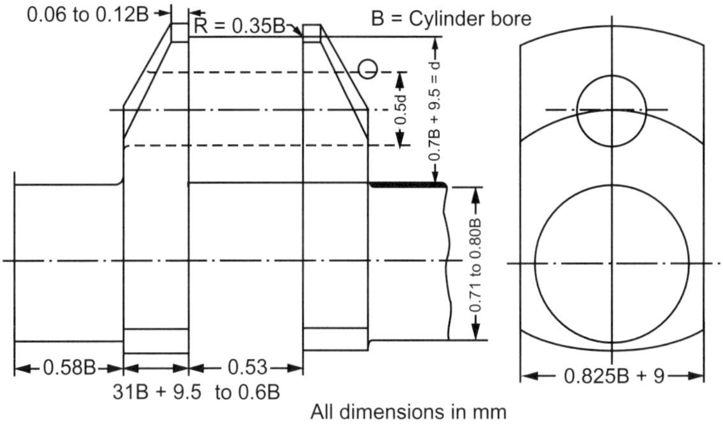

FIGURE 24-9 Empirical proportion for center crank.

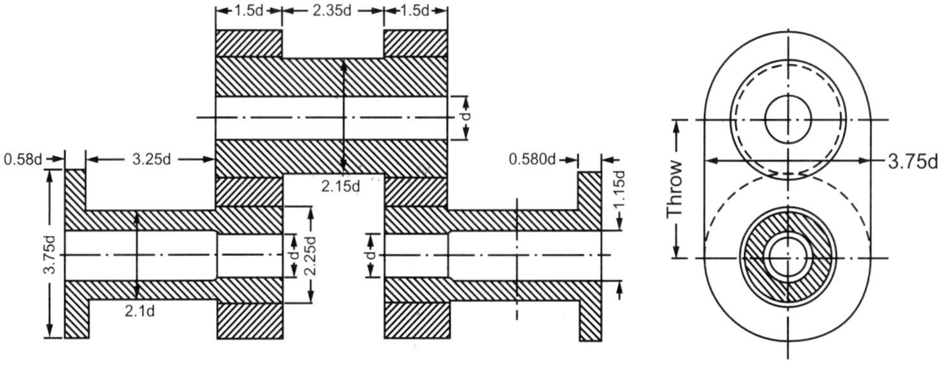

CENTER CRANK (Fig. 24-6)

Crankpin

The maximum bending moment treating the crankpin as a simple beam with concentrated load at the center

$$M_{bc} = \frac{F_{comb}(l_o + h + l_m)}{4} \quad (24\text{-}36)$$

where

l_o = length of crankpin, m (in)

l_m = length of main bearing, m (in)

h = thickness of cheek, m (in)

Particular	Formula
The diameter of the crankpin based on maximum bending moment M_{bc}	$d_o = \sqrt[3]{\dfrac{32M_{bc}}{\pi\sigma_b}}$ (24-37) where σ_b = design stress, MPa (psi)
The diameter of crankpin based on bearing pressure between pin and the bearing	$d_o = \dfrac{F_{comb}}{l_o p}$ (24-38)

Dimensions of main bearing

Particular	Formula
The maximum bending moment treating the center crank as a simple beam with load concentrated at the center	$M_{bb} = \dfrac{F_{comb} \times l_e}{4}$ (24-39) where l_e = equivalent length of crankshaft, m (in)
The twisting moment	$M_t = F_{comb} \times r$ (24-40)
The diameter of crankshaft at main bearing taking into consideration the fatigue and shock factors	$d_m = \sqrt[3]{\dfrac{16}{\pi\sigma_e}\left\{K_b M_{bb} + \sqrt{(K_b M_{bb})^2 + (K_t M_t)^2}\right\}}$ (24-41)
The diameter of the crankshaft based on bearing pressure	$d_m = \dfrac{F_s}{l_m p}$ (24-42)

American Bureau of Shipping formulas for center crank

Particular	Formula
The thickness h of the cheeks or webs (Fig. 24-6)	$h = 0.4d$ to $0.6d$ (24-43)
The diameter of crankpins and journals (Fig. 24-6)	$d = a\sqrt[3]{\dfrac{Dpc}{\sigma_b}}$ (24-44) where a = coefficient from Table 24-1A D = diameter of cylinder bore, m (in) p = maximum gas pressure, MPa (psi) c = distance over the crank web plus 25 mm (1.0 in) (Fig. 24-6) σ_b = allowable fiber stress, MPa (psi)
The thickness h and the width b of crank cheeks must satisfy the conditions	$bh^2 \geq 0.4d^3$ (24-45a) $b^2 h \geq d^3$ (24-45b)

Particular	Formula

EQUIVALENT SHAFTS

Particular	Formula	
A portion of a shaft length l and diameter d can be replaced by a portion of length l_e and diameter d_e	$l_e = l\left(\frac{d_e}{d}\right)^4$	(24-46)
The length h_e equivalent to crank web	$h_e = \frac{rC}{B}$ where $C = \frac{1}{32}\pi d_e^4 G$ = torsional rigidity of the crankpin $B = \frac{1}{12}hb^3 E$ = flexural rigidity of the web	(24-47)
The equivalent length crankshaft l_e of Fig. 24-7 varies between	$0.95l < l_e < 1.10l$	(24-48)
The equivalent length of commercial crankshaft for solid journal and crankpin according to Carter (Fig. 24-8)	$L_e = d_e^4\left(\frac{e+0.8a}{D_J^4} + \frac{0.75b}{D_c^4} + \frac{1.5r}{ac^3}\right)$	(24-49)
The equivalent length of commercial crankshaft for hollow journal and crankpin according to Carter (Fig. 24-8)	$L_e = d_e^4\left(\frac{e+0.8a}{D_J^4 - d_J^4} + \frac{0.75b}{D_c^4 - d_c^4} + \frac{1.5r}{ac^3}\right)$	(24-50)
The equivalent length of crankshaft for solid journal and crankpin according to Wilson (Fig. 24-8)	$L_e = d_e^4\left(\frac{e+0.4D_J}{D_J^4} + \frac{b+0.4D_c}{D_c^4} + \frac{r-0.2(D_J+D_c)}{ac^3}\right)$	(24-51)
The equivalent length of crankshaft for hollow journal and crankpin according to Wilson (Fig. 24-8)	$L_e = d_e^4\left(\frac{e+0.4D_J}{D_J^4 - d_J^4} + \frac{b+0.4D_c}{D_c^4 - d_c^4} + \frac{r-0.2(D_J+D_c)}{ac^3}\right)$	(24-52)

EMPIRICAL PROPORTIONS

Particular	Formula	
For empirical proportions of side crank, built-up crank, and hollow crankshafts	Refer to Figs. 24-2 to 24-10.	
The film thickness in bearing should not be less than the values given here for satisfactory operating condition:		
Main bearings	$h = 0.0025$ mm (0.0001 in) to 0.0042 mm (0.0017 in)	(24-52a)
Big-end bearings	$h = 0.002$ mm (0.00008 in) to 0.004 mm (0.00015 in)	(24-52b)
The oil flow rate through conventional central circumferential grooved bearings	$Q = \frac{kpc^3}{\eta}\frac{d}{L}(1 + 1.5\varepsilon^2)$	(24-52c)

Particular	Formula	
	where	
	Q = oil flow rate, m^3/s (gal/min)	
	k = a constant = 0.0327 SI units	
	$= 4.86 \times 10^4$ US Customary System Units	
	p = oil feed pressure, Pa (psi)	
	$c = D - d$ = diametral clearance, m (in)	
	η = absolute viscosity (dynamic viscosity), Pa s (cP)	
	d = bearing bore, m (in)	
	L = land width, m (in)	
	ε = attitude or eccentricity ratio	
For oil flow rate in medium and large diesel engines at 0.35 MPa 0.5 psi	Refer to Table 24-1B.	
The velocity of oil in ducts on the delivery side of the pump	$v = 1.8$ to $3.0\,m/s$ (6 to 10 ft/s)	(24-52d)
The velocity of oil in ducts on the suction side of the pump	$v = 1.2\,m/s$(4 ft/s)	(24-52e)
The delivery pressure in modern high-duty engines	$p = 0.28$ to 0.42 MPa (40–60 psi)	(24-52f)
	$p_{max} = 0.56$ MPa (80 psi)	(24-52g)
For housing tolerances	Refer to Table 24-1C.	

TABLE 24-1A
Coefficient *a* in the American Bureau of Shipping formula [Eq. (24-44)]

	Number of cylinder		Ratio of stroke to distance over crank webs = l/c							
Type	**Four-stroke**	**Two-stroke**	**0.7**	**0.8**	**0.9**	**1.0**	**1.1**	**1.2**	**1.3**	**1.4**
Explosion engines	1, 2, 4	1, 2	1.17	1.17	1.17	1.17	1.17	1.17	1.17	1.17
	3, 5, 6	3	1.17	1.17	1.17	1.17	1.19	1.20	1.22	1.24
	8	8	1.17	1.19	1.21	1.23	1.25	1.28	1.30	1.32
	10, 11, 12	5, 6	1.18	1.20	1.23	1.25	1.28	1.31	1.33	1.35
Air-injection diesel engines	1, 2, 4	1, 2	1.17	1.19	1.22	1.25	1.28	1.31	1.34	1.36
	3, 5, 6		1.19	1.22	1.25	1.28	1.32	1.35	1.38	1.41
	8	3	1.20	1.24	1.27	1.30	1.33	1.37	1.40	1.43
	12	4	1.22	1.25	1.29	1.32	1.36	1.39	1.42	1.45
	16	5, 6	1.25	1.29	1.33	1.36	1.40	1.44	1.47	1.50
		8								

TABLE 24-1B
Oil flow rate in medium and large diesel engines at 0.35 MPa

Different parts of engine	Oil flow rate	
	liters/min/kW	liters/min/hp (gal/h/hp)
Bed plate gallery to mains with piston cooling	0.536	0.4 (5)
Mains to big end (with piston cooling)	0.362	0.27 (3.5)
Big ends to pistons (with oil cooling)	0.201	0.15 (2)
Total flow of oil with un-cooled pistons	0.335	0.25 (3)

TABLE 24-1C
Housing tolerances

Parts	Tolerances
Waviness of the surface	$\not>0.0001d$
Run-out of thrust faces	$\not>0.0003d$
Surface finish	
Journals	0.2–0.25 μm R_a (8–10 μ in clearance)
Gudgeon pins	0.1–0.16 μm R_a (4–6 μ in clearance)
Housing bores	0.75–1.6 μm R_a (30–60 μ in clearance)
Alignment of adjacent housing	<1 in 10,000 to 1 in 12,000
The fine grinding or honing	0.025–0.05 mm (0.001–0.002 in)

TABLE 24-1D
Properties of some steel-backed crankshaft plain bearing materials

Lining materials	Nominal composition, per cent	Lining or overlay thickness, mm	Relative fatigue strength	Guidance peak loading limits, σ_{s1}, MPa	Recommended journal hardness, V.P.N.
Tin-based white metal	Sn 87, Sb 9	Over 0.1	1.0	12–14	160
	Cu 4, Pb 0.35 max	Up to 0.1	1.3	14–17	160
Tin-based white metal with cadmium	Sn 89, Sb 7.5 Cu 3, Cd 1	No overlay	1.1	12–15	160
Copper-lead, overlaid with cast white metal	Cu 70, Pb 30	0.2	1.4	15–17	160
Sintered copper-lead, overlay plated with lead-tin	Cu 70	0.05	1.8	21–23[a]	230
	Bb 30	0.025	2.4	28–31[a]	280
Cast copper-lead, overlay plated with lead-tin or lead-indium	Cu 76, Pb 24	0.025	2.4	31[a]	
Sintered lead-bronze, overlay plated with lead-tin	Cu 74, Pb 22 Sn 4	0.025	2.4	28–31[a]	400
Tin-aluminum	Al 60, Sn 40	No overlay	1.8	21–23	230
Tin-aluminum	Al 80, Sn 20	No overlay	3.3	28–35	230
Tin-aluminum, overlay plated with lead-tin	Al 92, Sn 6 Cu 1, Ni 1	0.025	2.4	28–31[a]	400

[a] Limit set by overlay fatigue in the case of medium/large diesel engines.
Suggested limits are for big-end applications in medium/large diesel engines and are not to be applied to compressors. Maximum design loadings for main bearings will generally be 20% lower.
(*Courtesy*: Extracted from M. J. Neale, ed., *Tribology Handbook*, Section A11, Newnes-Butterworth, London, 1973)

24.2 CURVED BEAM[2,3]

SYMBOLS

a	semimajor axis of ellipse, m
A	area of cross section, m^2
b	width of beam, m
	semiminor axis of ellipse, m
c_1	distance from the centroidal axis to the inner surface of curved beam, m
c_2	distance from the centroidal axis to the outer surface of curved beam, m
$c_i = c_1 - e$	distance from the neutral axis to inner surface of curved beam, m
$c_o = c_2 + e$	distance from the neutral axis to outer surface of curved beam, m
$H(= d)$	diameter of curved beam of circular cross section, m
e	distance from centroidal axis to neutral axis of the section, m
E	modulus of elasticity, GPa
F	load, kN
G	modulus of rigidity, GPa
h	depth of beam, m
I	moment of inertia, m^4, cm^4
k	stress factor (also with suffixes)
K	constant
l	length of straight section between the semicircular ends of chain link, m
m	pure number to be determined for each particular shape of the cross section by performing the integration
M_b	applied bending moment (also with suffixes), N m
r_c	radius of centroidal axis, m
r_i	inner radius of curved beam, radius of curvature, m
r_o	outer radius of curved beam, m
r_n	radius of neutral axis, m
y	deflection, m
σ	normal stress (also with suffixes), MPa

Please note: The US Customary System units can be used in place of the above SI Units.

SUFFIXES

b	bending
i	inner
h	horizontal
o	outer
n	neutral
max	maximum
r	resultant or combined
v	vertical
x	x direction
y	y direction

Particular	Formula	

GENERAL

Pure bending

The general equation for the bending stress in a fiber at a distance y from the neutral axis (Figs. 15-11 and 15-12)	$\sigma_b = \pm \frac{M_b}{Ae}\left(\frac{y}{r_n + y}\right)$	(24-53)
The maximum compressive stress due to bending at the outer fiber (Fig. 24-12)	$\sigma_{bo} = -\frac{M_b c_o}{Aer_o}$	(24-54)
The maximum tensile stress due to bending at the inner fiber (Fig. 24-12)	$\sigma_{bi} = \frac{M_b c_i}{Aer_i}$	(24-55)

Stress due to direct load

The direct stress due to load F	$\sigma = \frac{F}{A}$	(24-56)

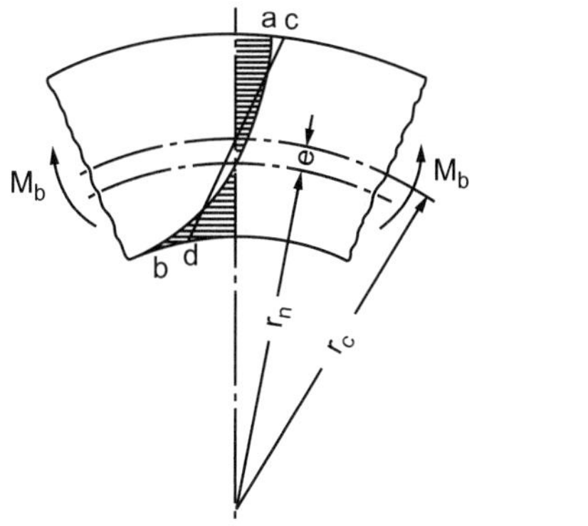

FIGURE 24-11 Bending stress in curved beam.

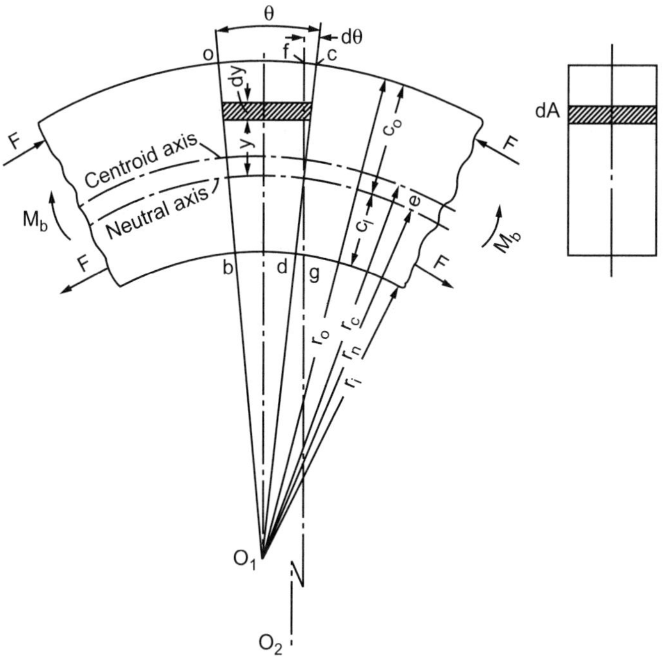

FIGURE 24-12 Analysis of stresses in curved beam.

Particular	Formula

Combined stress due to load F and bending

The general expression for combined stress

$$\sigma_r = \frac{F}{A} \pm \frac{M_b}{Ae}\left(\frac{y}{r_n + y}\right) \tag{24-57}$$

The combined stress in the outer fiber

$$\sigma_{ro} = \frac{F}{A} - \frac{M_b c_o}{Aer_o} \tag{24-58}$$

The combined stress in the inner fiber

$$\sigma_{ri} = \frac{F}{A} + \frac{M_b c_i}{Aer_i} \tag{24-59}$$

For values of radius to neutral axis for curved beams — Refer to Table 24-2.

TABLE 24-2
Values of radius to neutral axis for curved beams

Type	Section	Radius of neutral surface, r_n	
a	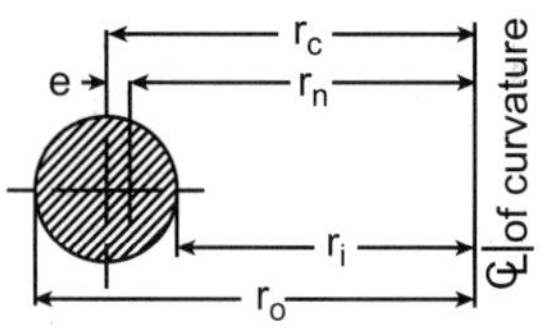	$r_n = \dfrac{(\sqrt{r_o} + \sqrt{r_i})^2}{4}$	(24-60)
b	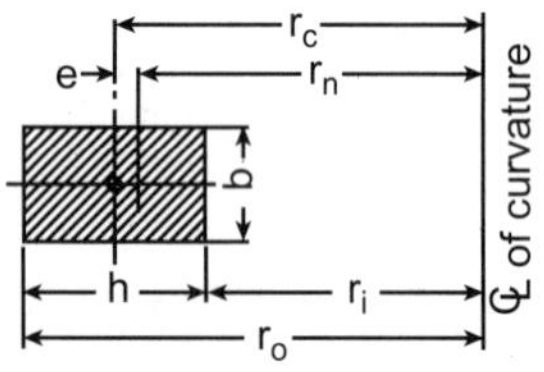	$r_n = \dfrac{h}{\ln\left(\dfrac{r_o}{r_i}\right)}$	(24-61)
c	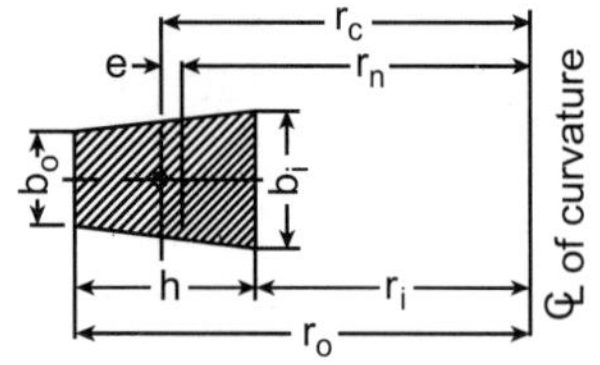	$r_n = \dfrac{\frac{1}{2}h(b_i + b_o)}{\dfrac{b_i r_o - b_o r_i}{h}\ln\left(\dfrac{r_o}{r_i}\right) - (b_i - b_o)}$ If $b_o = 0$, this section reduces to a triangle	(24-62)
d	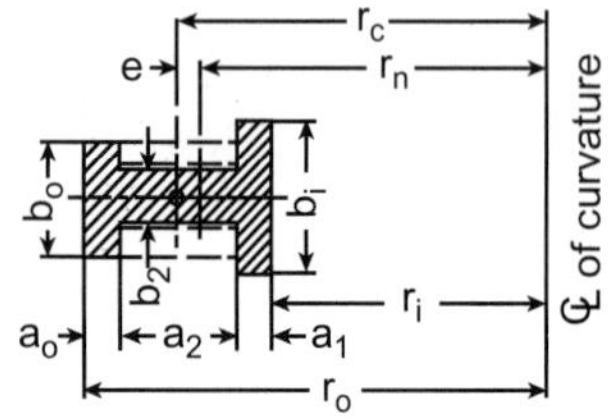	$r_n = \dfrac{A}{b_i \ln\dfrac{r_i + a_i}{r_i} + b_2 \ln\dfrac{r_o - a_o}{r_i + a_i} + b_o \ln\dfrac{r_o}{r_o - a_o}}$ If $a_o = 0$, this section reduces to a ⊥ section; r_n is the same for a box section in dotted lines with each side panel $\frac{1}{2}b_2$ thick	(24-63)

Particular	Formula

APPROXIMATE EMPIRICAL EQUATION FOR CURVED BEAMS

An approximate empirical equation for the maximum stress in the inner fiber

$$\sigma_i = M_b\left[\frac{c_1}{I} + \frac{K}{bc_1}\left(\frac{1}{r_i} + \frac{1}{r_o}\right)\right] \tag{24-64}$$

where

$K = 1.05$ for circular and elliptical sections

$= 0.5$ for all other sections

$b =$ maximum width of the section, m (in)

M_b in N m (lbf ft)

The stress at inner radius for a curved beam of rectangular cross section

$$\sigma_i = \frac{6M_b}{bh^2}\left(1 + 0.25\frac{h}{r_i}\right) \tag{24-65}$$

The stress at inner radius of circular cross section

$$\sigma_i = \frac{32M_b}{\pi d^3}\left(1 + 0.3\frac{d}{r_i}\right) \tag{24-66}$$

The stress at inner radius of elliptical sections according to Bach[a]

$$\sigma_i = \frac{32M_b}{\pi a^2 b}\left(1 + 0.3\frac{a}{r_i}\right) \tag{24-67}$$

STRESSES IN RINGS (Fig. 24-13*a*)

Maximum moment for a circular ring at the point of application of the load, A, Fig. 24-13*a*

$$M_{b(\max)} = \pm\frac{Fr}{\pi} = \mp 0.318Fr \tag{24-68}$$

where $-$ ve sign refers to tensile load,
$+$ ve sign refers to comprehensive load

Another maximum moment[b] for a circular ring at a point B 90° away from the point of application of load

$$M_{b(\max)} = \pm 0.182Fr \tag{24-69}$$

where $-$ ve sign refers to comprehensive load,
$+$ ve sign refers to tensile load

Direct stress for the ring at point B 90° away from the point of application of load

$$\sigma = \frac{F}{2A} \tag{24-70}$$

The general expression for bending moment at any cross section DD at an angle θ with the horizontal (Fig. 24-13*b*)

$$M_b^* = M_A - \tfrac{1}{2}Fr(1 - \cos\theta) \tag{24-71}$$

The stress due to direct load F at any cross section DD at an angle θ with the horizontal

$$\sigma = \frac{F\sin\theta}{2A} \tag{24-72}$$

---

[a] *Courtesy*: Bach, *Maschinenelemente*, 12th ed., p. 43.

[b] Moments which tend to decrease the initial curve of the bar are taken as positive.

Particular	Formula

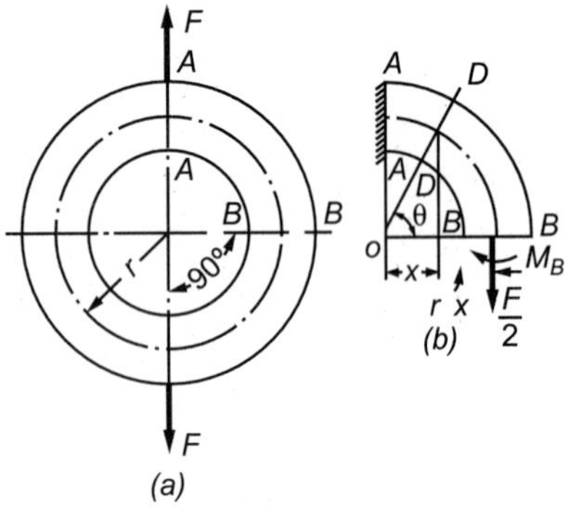

FIGURE 24-13 Bending moments in a ring.

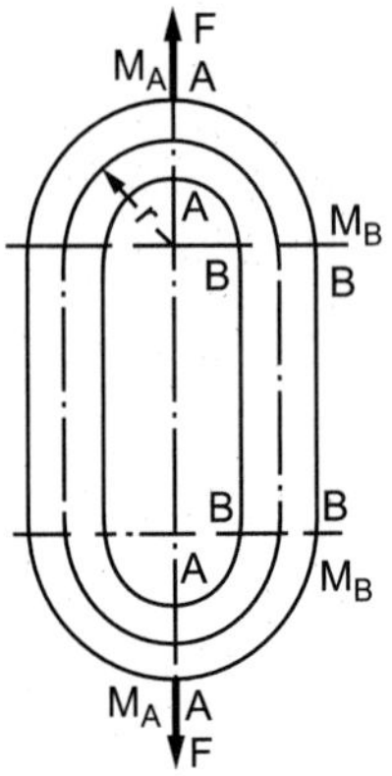

FIGURE 24-14 Bending moments in a link.

The combined stress at any cross section

$$\sigma_r = \frac{1}{2}\frac{F}{A}\sin\theta \pm \frac{M_b}{Ae}\left(\frac{y}{r_n + y}\right) \qquad (24\text{-}73)$$

DEFLECTION

The increase in the vertical diameter of the ring (Fig. 24-13*a*)

$$y_v = 0.149\frac{Fr^3}{EI_z} \qquad (24\text{-}74)$$

The decrease in the horizontal diameter of the ring (Fig. 24-13*a*)

$$y_h = 0.137\frac{Fr^3}{EI_z} \qquad (24\text{-}75)$$

LINK (Fig. 24-14)

The moment, M_{bA}, at the point of application of load (Fig. 24-14)

$$M_{bA} = \frac{Fr(2r + l)}{2(\pi r + l)} \qquad (24\text{-}76)$$

The moment, M_{bB}, at the section 90° away from the point of application of load (Fig. 24-14)

$$M_{bB} = \frac{Fr(2r - \pi r)}{2(\pi r + l)} \qquad (24\text{-}77)$$

where l = length of straight section between the semicircular ends.

CRANE HOOK OF CIRCULAR SECTION (Fig. 24-15)

The combined stress in any fiber of a crane hook subject to a load F

$$\sigma_r = \frac{F}{A} \pm \frac{M_b}{Ae}\left(\frac{y}{r_n + y}\right) = \frac{Fy}{Am(r_x - y)} \qquad (24\text{-}78)$$

The maximum combined stress

$$\sigma_{r(\max)} = \frac{F}{A}\left(\frac{H}{2mr_i}\right) = \frac{F}{A}k_i \qquad (24\text{-}79)$$

Particular	Formula
The minimum combined stress	$\sigma_{r(\min)} = \frac{F}{A}\left(\frac{H}{2mr_o}\right) = \frac{F}{A}k_o$ (24-80) where k_i and k_o are stress factors which depend on $H/2r_c$; k_i is the critical one which varies from 13.5 to 15.4 as ratio $H/2r_c$ changes from 0.6 to 0.4
For crane hook of trapezoidal section	Refer to Fig. 24-16 and Table 24-3.

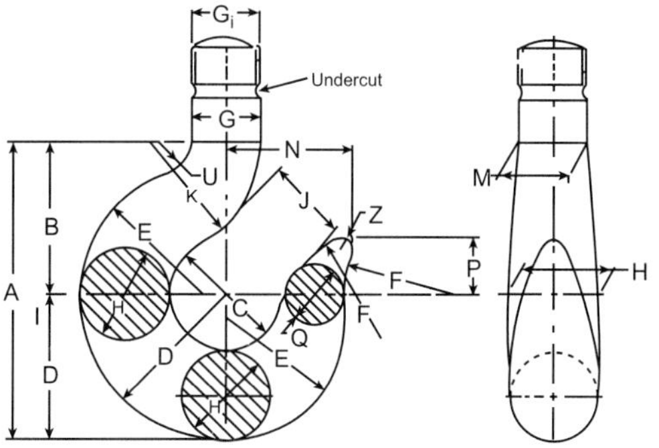

FIGURE 24-15 Hook of circular section.

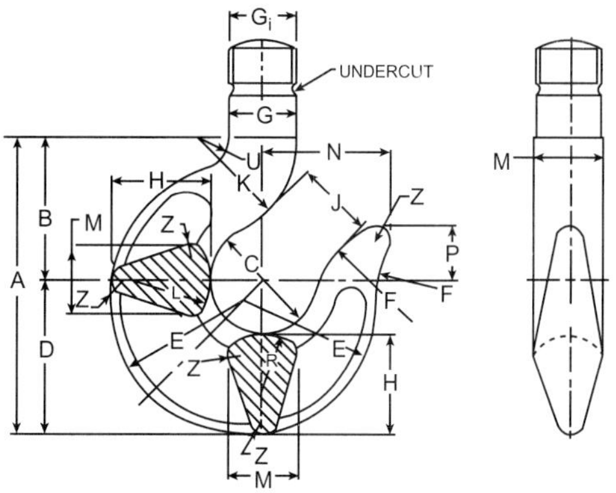

FIGURE 24-16 Crane hook of standard trapezoidal section.

TABLE 24-3
Hooks of standard trapezoidal section (Refer to Fig. 24-16)

Safe working load W, tf[a]	Proof load P, tf		A (2.75C)	B (3.1C)	C[b]	D (1.44C)	E (1.25C)	F (1.00C)	G[c]	Nominal size G_1	Thread: Pitch	Thread: Course series with graded pitches	Thread: H (0.93C)	Thread: J (0.75C)	Thread: K (0.92C)		Thread: L (0.70C)	Thread: M (0.60C)	Thread: N (1.20C)	Thread: P (0.50C)	Thread: R (0.50C)	Thread: U (0.30C)	Thread: Z (0.12C)	Ball bearings (per IS 2512, 1963): Bore	Series	Outside diameter	Width
0.5	1	MS	74	35	27	39	34	27	15	M	14		25	20	25	MS	19	16	32	14	14	8	3	15	11	28	9
		HS	63	38	23	33	29	23	12	M	12		21	17	21	HS	16	14	28	12	12	7	3	12	11	26	9
1.0	2	MS	105	50	38	55	48	38	20	M	20		35	28	35	MS	27	23	46	19	19	11	5	20	11	35	10
		HS	91	43	33	48	41	33	20	M	18		31	25	31	HS	23	20	40	16	16	10	4	20	11	35	10
2.0	4	MS	145	69	53	76	66	53	30	M	27		49	40	49	MS	37	32	64	26	26	16	6	30	12	52	16
		HS	126	60	46	66	58	46	25	M	24		43	34	42	HS	32	28	55	23	23	14	6	25	12	47	15
3.2	6.4	MS	187	89	68	98	85	68	35	M	33		63	51	63	MS	48	41	82	34	34	20	8	35	13	68	24
		HS	162	71	59	85	74	59	30	M	30		55	44	54	HS	41	35	71	30	30	18	7	35	13	60	21
5.0	10	MS	233	111	85	122	106	85	45	M	42		79	64	78	MS	60	51	102	42	42	26	10	45	13	85	28
		HS	201	96	73	105	91	73	40	M	30		68	55	67	HS	51	44	88	36	36	22	9	40	13	78	26
8.0	16	MS	294	140	107	154	134	109	55	M	52		100	80	98	MS	75	64	128	54	54	32	13	55	13	105	35
		HS	256	122	93	134	116	93	50	M	48		86	70	86	HS	63	56	112	46	46	29	11	50	13	95	31
10.0	20	MS	327	156	119	171	149	119	60	M	60		111	89	109	MS	83	71	143	60	60	36	14	60	13	110	35
		HS	286	136	104	150	130	104	55	M	52		97	78	96	HS	73	62	125	52	52	31	12	56	13	105	35
12.0	25	MS	369	176	134	193	168	134	70	M	68		125	100	123	MS	94	80	161	67	67	40	16	70	13	125	40
		HS	319	152	116	167	145	116	60	M	60		108	87	107	HS	81	70	139	58	58	35	14	60	13	110	35
16.0	32	MS	415	198	151	217	189	151	80	M	76	6	140	113	139	MS	106	91	181	76	76	45	18	80	13	140	44
		HS	360	171	131	189	164	131	70	M	68	6	122	98	120	HS	92	79	157	66	66	39	16	70	13	125	40
20.0	40	MS	465	221	169	243	211	169	85	M	80	6	157	127	155	MS	118	101	203	84	84	51	20	85	13	150	49
		HS	404	193	147	211	184	147	80	M	72	6	137	110	135	HS	103	88	176	74	74	44	18	80	13	140	44
25	50	MS	520	248	189	272	236	189	100	M	90	6	176	142	174	MS	132	113	227	94	94	57	23	100	13	190	63
		HS	451	215	164	236	205	164	85	M	80	6	153	123	151	HS	115	98	197	82	82	49	20	85	13	150	49
32	60	MS	569	271	207	298	259	207	110	M	100	8	193	155	191	MS	145	124	248	104	104	62	25	110	13	210	70
		HS	481	229	175	252	219	175	100	M	90	6	163	131	161	HS	122	105	210	88	88	52	21	100	13	190	63
40	70	MS	616	293	224	323	280	224	120	M	110	6	208	168	206	MS	157	134	268	112	112	67	27	120	13	210	70
		HS	533	254	194	279	242	194	110	M	100	6	180	146	178	HS	136	116	233	97	97	58	23	110	13	190	63
50	85	MS	680	324	247	356	309	247	130	M	120	6	229	187	227	MS	173	148	296	124	124	74	30	130	13	225	75
		HS	588	280	214	308	268	214	120	M	110	6	199	160	197	HS	150	128	257	107	107	64	26	120	13	210	70

[a] Tonne-force, 1 tonne force = 1 tf = 1 Mgf = 1000 kgf = 9086.6 N = 9.8066 kN.
[b] Formula for catching C: for MS (mild steel); $C = 26.73\sqrt{P}$; for HS (high-tensile steel), $C = 23.17\sqrt{P}$.
[c] Machined shank diameter (min).
Source: IS 3815, 1969.

24.3 CONNECTING AND COUPLING ROD[2,3]

SYMBOLS[2,3]

A	area of cross section, m^2 (in^2)
a	Rankine's constant
b	width, m (in)
d	diameter, m (in)
d_1	core diameter of bolt, m (in)
d_c	crankpin diameter, m (in)
d_g	gudgeon pin diameter, m (in)
E	modulus of elasticity, GPa (psi)
F	force acting on the piston due to steam or gas pressure corrected for inertia effects of the piston and other reciprocating parts, kN (lbf)
F_c	the component of F acting along the axis of connecting rod, kN (lbf)
F_i	inertia force, kN (lbf)
F_{ir}	inertia force due to reciprocating masses, kN (lbf)
F_{cr}	crippling or critical force, kN (lbf)
g	acceleration due to gravity, 9.8066 m/s^2 9806.6 mm/s^2 (32.2 ft/s^2)
h	depth of rectangular or other sections, m (in)
k	radius of gyration, m (in)
l	length of connecting rod, m (in)
l_c	length of crankpin, m (in)
l_e	equivalent length, m (in)
l_g	length of gudgeon pin, m (in)
M_b	bending moment, N m (lbf in)
n	speed of crank, rpm
n_1	safety factor
$n' = \frac{l}{r}$	ratio of connecting rod length to radius of crank
p	allowable pressure, MPa (psi)
p_f	load due to gas or steam pressure on the piston, MPa (psi)
v	velocity of crank, m/s (fps)
w	specific weight of material of connecting road, kN/m^3 (lbf/in^3)
W	weight of the reciprocating masses, kN (lbf)
Z	section modulus, m^3, cm^3 (in^3)
ω	angular speed of crank, rad/s
α	angle between the crank and the center line of connecting rod, deg
θ	angle between the crank and the center line of the cylinder measured from the head-end dead-center position, deg
ϕ	angle between the center line of piston and the connecting rod, deg
σ	normal stress (also with suffixes), MPa (psi)

Particular	Formula	
The velocity	$v = \dfrac{2\pi rn}{60}$ where r in m	(24-81)

DESIGN OF CONNECTING ROD (Fig. 24-17)

Gas load

Load due to gas or steam pressure on the piston	$F_g = \dfrac{\pi d^2}{4} p_f$	(24-82)

Inertia load due to reciprocating motion

Inertia due to reciprocating parts and piston	$F_{ir} = \dfrac{Wv^2}{gr}\left(\cos\theta + \dfrac{\cos 2\theta}{n'}\right)$	(24-83a)
	$F_{ir} = 0.01095\dfrac{Wrn^2}{g}\left(\cos\theta + \dfrac{\cos 2\theta}{n'}\right)$	(24-83b)
The maximum value of F_{ir} occurs when $\theta = 0°$ or when the crank is at the head-end dead center	$F_{1ir(\max)} = 0.01095\dfrac{Wrn^2}{g}\left(1 + \dfrac{1}{n'}\right)$	(24-84)
At the crank-end dead center, when $\theta = 180°$, F_{ir} attains the maximum negative value, acting in opposite direction	$F_{2ir(\max)} = -0.01095\dfrac{Wrn^2}{g}\left(1 - \dfrac{1}{n'}\right)$	(24-85)
The combined force on the piston	$F = F_g \pm F_{ir}$	(24-86)
The component of F acting along the axis of connecting rod	$F_c = \dfrac{F}{\sqrt{1 - \left(\dfrac{\sin\theta}{n'}\right)^2}}$	(24-87)
The stress induced due to column action on account of load F_c acting along the axis of connecting rod		
(a) As per Rankine's formula	$\sigma_1 = \dfrac{F_c}{A}\left[1 + a\left(\dfrac{l_e}{k}\right)^2\right]$	(24-88a)

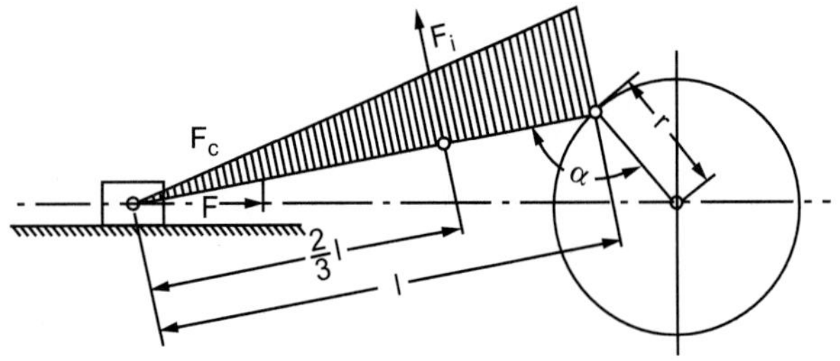

FIGURE 24-17 Forces acting on a connecting rod.

Particular	Formula
(b) As per Ritter's formula	$\sigma_1 = \frac{F_c}{A}\left[1 + \frac{\sigma_e}{n\pi^2 E}\left(\frac{l_e}{k}\right)^2\right]$ (24-88b)
(c) As per Johnson's parabolic formula	$\sigma_1 = \frac{F_c}{A\left[1 - \frac{\sigma_y}{4n\pi^2 E}\left(\frac{l_e}{k}\right)^2\right]}$ (24-88c)
	where l_e = equivalent length, m (in) k = radius of gyration, m (in) n = end-condition coefficient (Table 2-4) a = constant obtained from Table 2-3

Inertia load due to connecting rod

Particular	Formula
The magnitude of inertia force (Fig. 24-17) due to the weight of the rod itself, not including the ends	$F_{ic} = \frac{Awv^2 l}{2gr} \sin\alpha$ (24-89a)
	$F_{ic} = \frac{Wv^2}{2gr}$ when $\alpha = 90^\circ$ (24-89b) where $W = Awl$ = weight of the rod itself, not including the ends, kN (lbf)
The maximum bending moment produced by the inertia force F_{ic} is at a distance $(2/3)l$ from wrist pin	$M_{b(\max)} = \frac{2F_{ic}l}{9\sqrt{3}} = \frac{2Wv^2 l}{9\sqrt{3} \times 2gr} \sin\alpha$ (24-90a)
	$M_{b(\max)} = \frac{Wv^2 l}{9\sqrt{3}gr}$ when $\alpha = 90^\circ$ (24-90b)
The maximum bending stress developed in the rod due to inertia force F_{ic}	$\sigma_{b(\max)} = \frac{M_{b(\max)}}{Z} = \frac{Wv^2 l}{9\sqrt{3}grZ} \sin\alpha$ (24-91a)
	$\sigma_{b(\max)} = \frac{Wv^2 l}{9\sqrt{3}grZ}$ when $\alpha = 90^\circ$ (24-91b)
The crank angle (θ) at which the maximum bending moment occurs according to B. B. Low	$\theta = 90^\circ - \frac{3500}{(n' + 7.82)^2}$ (24-92)
The relation between the moment of inertia in the **xx** and **yy** planes in order to have same resistance in either plane	$I_{yy} = \frac{1}{4} I_{xx}$ (24-93) or $k_{yy}^2 = \frac{1}{4} k_{xx}^2$ (24-94)

Particular	Formula

DESIGN OF SMALL AND BIG ENDS

The diameter of crankpin at the big end

$$d_c = \frac{F}{l_c p} \qquad (24\text{-}95)$$

where

p = allowable bearing pressure based on projected area, MPa (psi)

= 4.9 to 10.3 MPa (700 to 1500 psi), and

$$\frac{l_c}{d_c} = 1.25 \text{ to } 1.5$$

The diameter of the gudgeon pin at the small end

$$d_g = \frac{F}{l_g p} \qquad (24\text{-}96)$$

where

p = 10.3 to 13.73 MPa (1.2 to 2.0 kpsi), and

$$\frac{l_g}{d} = 1.5 \text{ to } 2$$

DESIGN OF BOLTS FOR BIG-END CAP

The diameter of bolts used for fixing the big-end cap

$$d_i = \sqrt{\frac{2F_{1ir(\max)}}{\pi\tau_d}} \qquad (24\text{-}97)$$

where $F_{1ir(\max)}$ is obtained from Eq. (24-84)

σ_d = design stress of bolt material, MPa (psi)

The expression for checking load for measuring peripheral length of each thin-walled half-bearing according to J. M. Conway Jones[a]

$$W_c = 6000\frac{Lh_b}{D} \qquad \text{SI} \quad (24\text{-}97a)$$

where

W_c = checking load, N

L = axial length of bearing, mm

h_b = wall thickness of bearing, mm

D = diameter of housing, mm

The expression for total minimum nip, n

$$n = 44 \times 10^{-6}\frac{D^2}{h_b} \quad \text{or} \quad 0.12\,\text{mm} \qquad \text{SI} \quad (24\text{-}97b)$$

whichever is larger

[a] In M. J. Neale, ed., *Tribology Handbook*, Section A20, Newnes-Butterworth, London, 1973.

Particular	Formula
Note: The "nip" or "crush" is the amount by which the total peripheral length of both halves of bearing under no load exceeds the peripheral length of the housing of the bearing.	
The compressive load on each bearing joint face to compress nip[a]	$W = \dfrac{ELh_{sl}m}{\pi(D - h_{sl})10^6}$ **SI** (24-97c) or $W = Lh_{sl}\sigma_y \times 10^{-6}$ **SI** (24-97d) whichever is smaller where D = housing diameter, mm (in) h_{sl} = steel thickness $+\frac{1}{2}$ lining thickness, mm (in) m = sum of maximum circumferential nip on both halves of bearing, mm (in) W = compressive load on each bearing joint face, N (lbf) E = modulus of elasticity of material of backing, Pa (psi) = 210 GPa (30.45 Mpsi) for steel L = bearing axial length, mm (in) σ_y = yield stress of steel backing, Pa (psi) = 350 MPa (50 kpsi) for white-metal-lined bearing = 300 to 400 MPa (43.5 to 58 kpsi) for bearing with copper-based lining = 600 MPa (87 kpsi) for bearing with aluminum-based lining
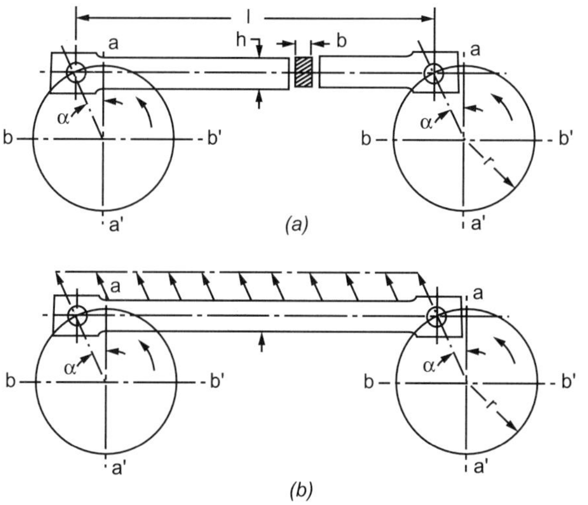 **FIGURE 24-18** Forces acting on a coupling rod.	
The bolt load required on each side of bearing to compress nip for extremely rigid housing	$W_b = 1.3W$ (24-97e)
The bolt load required on each side of bearing to compress nip for normal housing with bolts very close to back of bearing	$W_b = 2W$ (24-97f)
The ratio of connecting rod length (l) to crank radius (r)	$n' = \dfrac{l}{r} = 3.4$ to 4.4 single-acting engines (24-98) = 4.6 to 5.4 for double-acting engines = 6.0 or more for steam locomotive engines = 5 to 7 for stationary steam engines = 3.2 to 4 for internal-combustion engines = 1.5 to 2 for aero engines

[a] In M. J. Neale, ed., *Tribology Handbook*, Section A20, Newnes-Butterworth, London, 1973.

Particular	Formula	

DESIGN OF COUPLING ROD (Fig. 24-18)

Particular	Formula	
The centrifugal force due to the weight of the rod	$F_c = \frac{wv^2}{gr} hbl$	(24-99)
	$F_c = \frac{Wv^2}{gr}$	(24-100)
The bending component of centrifugal force	$F_{cb} = \frac{wv^2 hbl}{gr} \cos\alpha = \frac{Wv^2}{gr} \cos\alpha$	(24-101)
The maximum bending moment due to the uniformly distributed load of F_{cb}	$M_{b(\max)} = \frac{wv^2 hbl^2}{8gr} = \frac{Wv^2 l}{8gr}$	(24-102)
The axial component of the centrifugal force	$F_{ca} = \frac{wv^2}{gr} hbl \sin\alpha = \frac{Wv^2 \sin\alpha}{gr}$	(24-103)
For some of the common cross sections of connecting rods	Refer to Fig. 24-19.	
For forces acting on a coupling rod	Refer to Fig. 24-18.	
For proportions of ends of round and H-section connecting rod	Refer to Fig. 24-20.	
For proportions and empirical relations of steam engine common strap end	Refer to Fig. 24-21.	

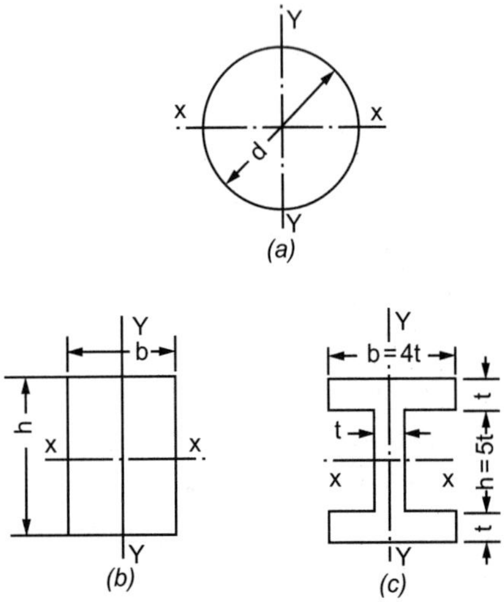

FIGURE 24-19 Connecting rod sections.

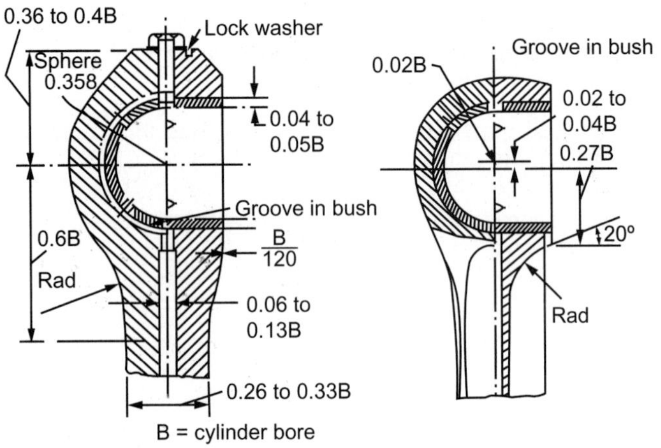

FIGURE 24-20 Two typical end designs for round and H-section connecting rods.

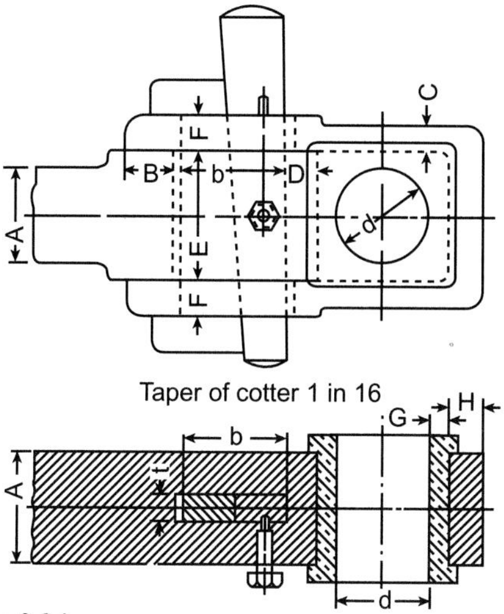

A = 0.9d
B = 0.5d + 10
b = d
D = 0.35d + 3

E = 1.15d + 5.5
F = 0.37d + 3
G = 0.1d + 4

H = 0.3d + 1.5
C = 0.3d + 1.5
t = 0.2d + 1.5

All dimensions in mm

FIGURE 24-21 Steam engine common strap end.

24.4 PISTON AND PISTON RINGS

SYMBOLS[2,3]

A	area of cross section of piston head, m^2 (in^2)
b	width of face of piston, m (in)
B	diameter of bore, m (in)
c	heat-conduction factor, $kJ/m^2/m/h/K$ ($Btu/in^2/in/h/°F$)
C	higher heat value of fuel used, kJ/kg (Btu/lb)
d	nominal diameter of piston ring, m (in)
	diameter of piston rod, m (in)
d_g	diameter of gudgeon pin, m (in)
D	diameter of bore (cylinder), m (in)
D_r	root diameter of the piston ring groove, m (in)
E	modulus of elasticity, GPa (psi)
F	force, kN (lbf)
	diametral load on the piston ring to close the gap which is less than 2.45 N (0.55 lbf)
F_θ	tangential load on the piston ring to close the gap which is less than 2.45 N (0.55 lbf)
h	thickness (also with subscripts), m (in)
	radial thickness of piston ring, m (in)
h_1, h_2, h_3	thickness as shown in Fig. 24-24, m (in)
H	heat flowing through the head, kJ/h (Btu/h)
$i' = \frac{l_g}{d_g}$	length/diameter ratio
l_g	length of gudgeon pin, m (in)
L	length of piston, m (in)
M_b	bending moment, N m (lbf in)
n	safety factor
P_b	brake horsepower (bhp)
p	pressure, MPa (psi)
r	radius, m (in)
R, R_i, R_h	radius as shown in Fig. 24-22*b*, m (in)
t_h	thickness of head, m (in)
t_r	thickness under ring groove, m (in)
T_c	temperature at center of head, °C (°F)
T_e	temperature at edge of head, °C (°F)
w	weight of fuel used, kg/bhp/h
	axial width of ring, m (in)
σ	stress (also with subscripts), MPa (psi)
θ	angle (Fig. 24-23), deg

TABLE 24-4
Dimensions of cast-iron piston up to 430 mm diameter (Fig. 22-24) (all dimensions in cm)

Diameter of cylinder, D	Diameter of piston rod, d	d_1	b	a	h	h_1	h_2	h_3	d_2	d_3
15.0	2.8	2.5	7.5	1.2	1.2	1.4	0.8	0.8	3.1	2.5
20.0	3.4	3.1	8.1	1.4	1.2	1.6	0.95	0.8	3.4	3.1
25.0	4.0	3.7	9.0	1.6	1.4	1.7	0.95	0.95	4.0	3.7
30.0	5.0	4.7	10.0	1.7	1.6	1.7	1.2	0.95	4.6	4.7
35.0	5.6	5.0	11.2	1.9	1.6	1.9	1.2	0.95	5.3	5.0
40.0	5.9	5.6	11.8	1.9	1.7	2.2	1.2	0.95	5.8	5.6

Particular	Formula

STEAM ENGINE PISTONS

Piston rods

The diameter of piston rod

$$d = D\sqrt{\frac{p}{\sigma_a}} \qquad (24\text{-}104)$$

where

p = unbalanced pressure or difference between the steam inlet pressure and the exhaust, MPa (psi)

$\sigma_a = \dfrac{\sigma_u}{n}$ = allowable stress, MPa (psi)

Note: σ_a is based on a safety factor of 10 for double-acting engines and 8 for single-acting engines. (The diameter of piston rod is usually taken as $\frac{1}{6}$ to $\frac{1}{7}$ the diameter of the piston.)

The diameter of piston rod according to Molesworth

$$d = 0.0044D\sqrt{p} \text{ for cast-iron pistons} \qquad (24\text{-}105)$$

$$d = 0.00338D\sqrt{p} \text{ for steel pistons} \qquad (24\text{-}106)$$

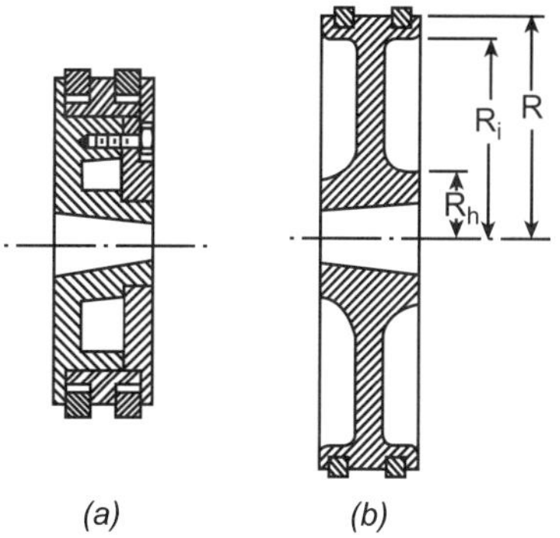

FIGURE 24-22 Plate piston.

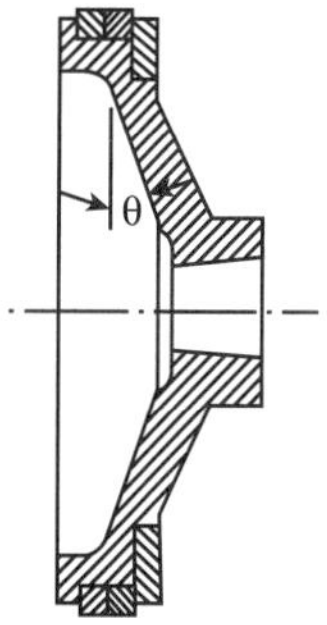

FIGURE 24-23 Conical plate piston.

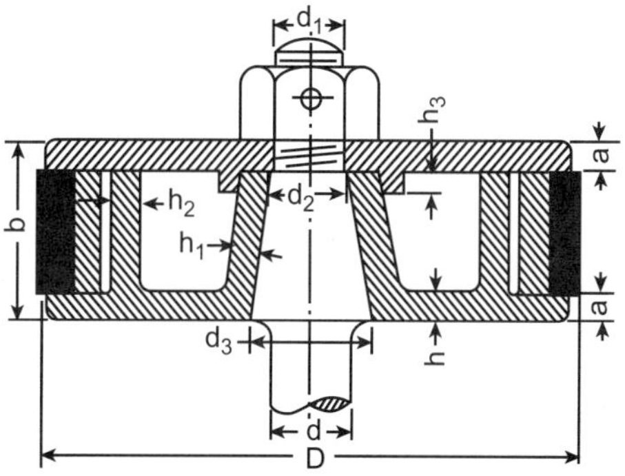

FIGURE 24-24 Cast-iron piston of diameter $\leq$ 400 mm (16.0 in).

Particular	Formula	
PROPORTIONS FOR PRELIMINARY LAYOUT FOR PLATE PISTONS		
Box type (Figs. 24-22*a* and 24-24)		
Width of face	$b = 0.3$ to $0.5D$	(24-107)
Thickness of walls and ribs for low pressure	$h = (2R + 50\,\text{cm})\,(0.003\sqrt{p} + 0.0275\,\text{cm})$	(24-108)
	or	
	$h = \dfrac{2R}{60} + 10\,\text{mm}\ (0.40\,\text{in})$	(24-109)
The thickness of walls and ribs for high pressure	$h = \dfrac{2R}{40} + 10\,\text{mm}\ (0.40\,\text{in})$	(24-110)
For dimensions of conical plate piston	Refer to Fig. 24-23.	
For dimensions of cast-iron piston of $\leq 400\,\text{mm}$ diameter	Refer to Fig. 24-24 and Table 24-4.	
Disk type (Fig. 24-22*b*)		
Width of face	$b = 0.3$ to $0.5D$	(24-111)
Thickness of walls and ribs for low pressure	$h = (2R + 12.5\,\text{cm})\,(0.0096\sqrt{p} + 0.057\,\text{cm})$	(24-112)
The hub thickness	$h_1 = 0.45d$	(24-113)
The hub diameter	$D_h = 2R_h = 1.6 \times$ the piston diameter	(24-114)
Width of piston rings	$w = 0.03D$ to $0.06D$	(24-115)
Thickness of piston rings	$h = 0.025D$ to $0.03D$	(24-116)
For dimensions of cast-iron piston	Refer to Table 24-4.	
STRESSES		
(a) Distributed load over the plate inside the outer cylindrical wall (i.e., the area πR_i^2)		
(1) Stress at the outer edge (Fig. 24-22*b*)	$\sigma_1 = \dfrac{3p}{4h^2}\left(R_i^2 - 3R_h^2 + \dfrac{4R_h^2}{R_i^2 - R_h^2}\ln\dfrac{R_i}{R_h}\right)$	(24-117)
(2) Stress at the inner edge (Fig. 24-22*b*)	$\sigma_2 = \dfrac{3p}{4h^2}\left(R_i^2 + R_h^2 - \dfrac{4R_i^2 R_h^2}{R_i^2 - R_h^2}\ln^2\dfrac{R_i}{R_h}\right)$	(24-118)

Particular	Formula
(b) Load on the outer wall, $p\pi(R^2 - R_i^2)$ distributed around the edge of the plate	
(1) Stress at the outer edge (Fig. 24-22*b*)	$\sigma_3 = \frac{3p(R^2 - R_i^2)}{2h^2}\left(1 - \frac{2R_h^2}{R_i^2 - R_h^2}\ln\frac{R_i}{R_h}\right)$ (24-119)
(2) Stress at the inner edge (Fig. 24-22*b*)	$\sigma_4 = \frac{3p(R^2 - R_i^2)}{2h^2}\left(1 - \frac{2R_i^2}{R_i^2 - R_h^2}\ln\frac{R_i}{R_h}\right)$ (24-120)
(3) The sum of the stresses at the outer edge	$\sigma_o = \sigma_1 + \sigma_3$ (24-121)
(4) The sum of the stresses at the inner edge	$\sigma_i = \sigma_2 + \sigma_4$ (24-122) (*Note*: σ_o or σ_i should not be greater than the permissible stress of the material. A safety factor, n, of 8 can be used.)

Dished or conical type (Fig. 24-23)

An empirical formula for the thickness of conical piston (Fig. 24-23)	$h = 0.288\sqrt{pD/\sigma}\sin\theta$ **SI** (24-123a) where p and σ in MPa, and D and h in m $h = 9.12\sqrt{pD/\sigma}\sin\theta$ **Customary Metric** (24-123b) where p and σ in kgf/mm^2, D and h in mm $h = 1.825\sqrt{pD/\sigma}\sin\theta$ **USCS** (24-123c) where p and σ in psi, D and h in in
The height of boss	$H = 1.1K$ (24-124)
The diameter of boss	$D_h = 1.7K$ for small pistons (24-125a) $D_h = 1.5K$ for large pistons and light engines (24-125b)
The thickness h_1 measured on the center line	$h_1 = Kc$ (24-126) where $c = 1$ to 0.75 depending on the angle of inclination θ (Refer to Table 24.5.) $\theta =$ varies from 6° to 35° $K = 1$ to 4.5 for varying pressure and diameter Also refer to Table 24-6 for values of K.
For calculating hub diameter, width of piston rings, and thickness of piston rings	Refer to Eqs. (24-114) to (24-116).

Particular	Formula

PISTONS FOR INTERNAL-COMBUSTION ENGINES

Trunk piston (Fig. 24-25)

The head thickness of trunk pistons (Fig. 24-25*a*)

$$t_h = \sqrt{\frac{3PD^2}{16\sigma}} \qquad (24\text{-}127)$$

where

σ = 39 MPa (5.8 kpsi) for close-grained cast iron

= 56.4 MPa (8.2 kpsi) for semisteel or aluminum alloy

= 83.4 MPa (12.0 kpsi) for forged steel

COMMONLY USED EMPIRICAL FORMULAS IN THE DESIGN OF TRUNK PISTONS FOR AUTOMOTIVE-TYPE ENGINES

Thickness of head (Fig. 24-25*a*)

$$t_h = 0.032D + 1.5\,\text{mm} \qquad \textbf{SI} \quad (24\text{-}128)$$

$$t_h = 0.00D + 0.06\,\text{in} \qquad \textbf{USCS} \quad (24\text{-}128a)$$

The head thickness for heat flow

$$t_h = \frac{HD^2}{0.16c(T_c - T_e)A}$$

$$= \frac{H}{0.194c(T_c - T_e)} \qquad \textbf{SI} \quad (24\text{-}129a)$$

where

$T_c - T_e = 205°\text{C}$ (400°F) and $T_c = 698K$, 425°C (800°F) for cast-iron piston

$\Delta T = T_c - T_e = 55°\text{C}$ (130°F) and $T_c = 533K$, 260°C (500°F) for aluminum piston

$c = 2.2$ for cast iron

$c = 7.7$ for aluminum

$$t_h = \frac{HD^2}{16c\Delta TA} = \frac{H}{12.5c\Delta T} \qquad \textbf{USCS} \quad (24\text{-}129b)$$

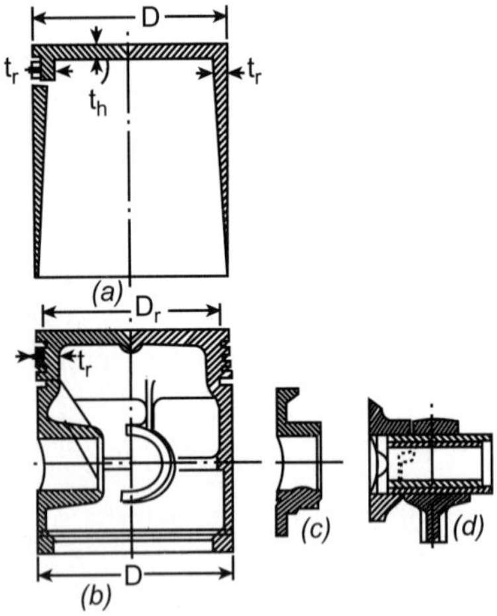

FIGURE 24-25 Trunk piston for small internal-combustion engine. (*a*) piston laid out for heat transfer; (*b*) piston modified for structural efficiency; (*c* and *d*) alternate pin designs.

Particular	Formula
Thickness of wall under the ring (Fig. 24-25*a* and *b*)	t_r = thickness of head = t_h (24-130)
The thickness of wall under the ring groove	$t_r = \frac{1}{2}(D_r \pm \sqrt{D_r^2 - 4Dt_h})$ (24-131)
The heat flow through the head	$H = KCwP_b$ (24-132) where w = weight of fuel used, kJ/kW/h (lbf/bhp/h) K = constant representing that part of heat supplied to the engine which is absorbed by the piston = 0.05 (approx.) P_b = brake horsepower per cylinder
The root diameter of ring grooves, allowing for ring clearance	$D_r = D - (2w + 0.006D + 0.02\,\text{in})$ **USCS** $D_r = D - (2w + 0.006D + 0.5\,\text{mm})$ at the compression rings **SI** (24-132a) $D_r = D - (2w + 0.006D + 1.5\,\text{mm})$ **SI** (24-132b) $D_r = D - (2w + 0.00D + 0.06\,\text{in})$ at the oil grooves **USCS** where D_r and D in mm (in)
Length L of piston	$L = D$ to $1.5D$ (24-133)
For chemical composition and properties of aluminum alloy piston	Refer to Table 24-10B.

Gudgeon pin

Particular	Formula
The diameter of gudgeon pin	$d_r = \sqrt{\dfrac{F}{i'p}}$ (24-134) where F = maximum gas pressure corrected for inertia effect of the piston and other reciprocating parts, kN (lbf) p = working bearing pressure = 9.81 MPa (1.42 kpsi) to 14.7 MPa (2.13 kpsi)
The length/diameter ratio of gudgeon pin	$i' = \dfrac{l_g}{d_g} = 1.5$ to 2 (24-135)
For gudgeon pin allowable oval deformation	Refer to Fig. 24-28*c*.
For empirical relations and proportions of pistons	Refer to Figs. 24-26 to 24-28*a*.

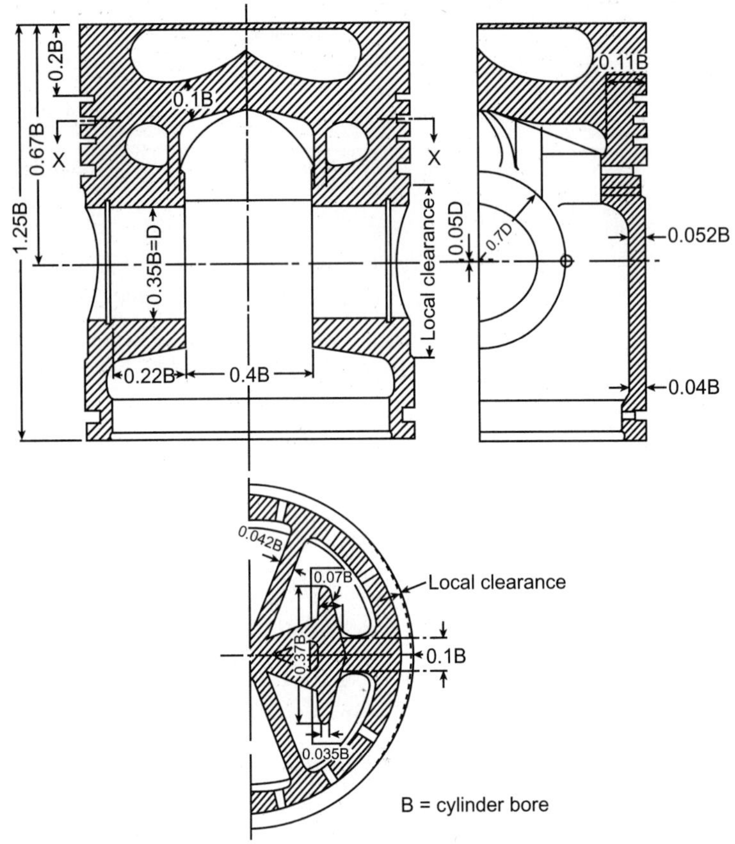

FIGURE 24-26 Proportions of a typical alloy piston.

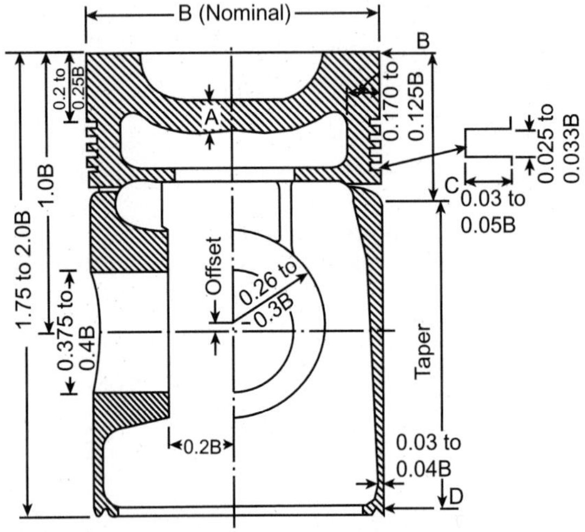

FIGURE 24-27 Proportions of an iron piston.

Fig. 24-27		mm (in)	mm (in)	mm (in)	mm (in)	mm (in)
Cylinder bore		152.5 (6)	203.2 (8)	254 (10)	305 (12)	406.5 (16)
Crown thickness	*A*	16 (5/8)	19 (3/4)	32 ($1\frac{1}{4}$)	41.5 ($1\frac{5}{8}$)	47.5 ($1\frac{7}{8}$)
Clearance	*B*	0.760 (0.03)	0.900 (0.035)	1.145 (0.045)	1.525 (0.06)	2.030 (0.08)
Clearance	*C*	0.125 (0.005)	0.225 (0.008)	0.230 (0.009)	0.255 (0.01)	0.255 (0.01)
Clearance	*D*	0.125 (0.005)	0.150 (0.006)	0.180 (0.007)	0.200 (0.008)	0.230 (0.009)

Piston weight $= 40.715B^3$ N (approx.) **SI** where $B =$ cylinder bore, m
Piston weight $= 0.15B^3$ lbf **USCS** where B in in

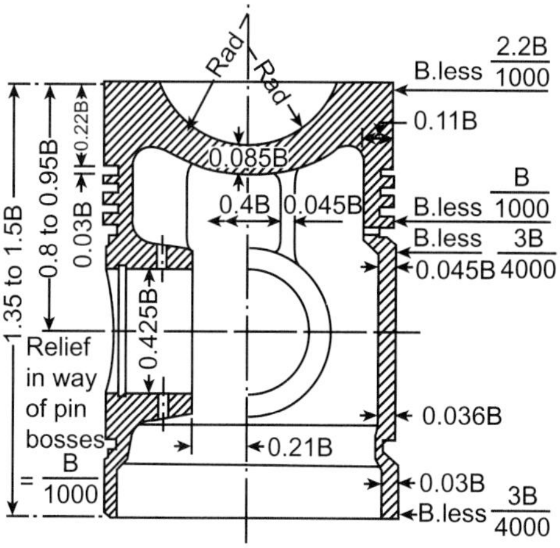

FIGURE 24-28(*a*) Iron piston for small engines (B = cylinder bore).

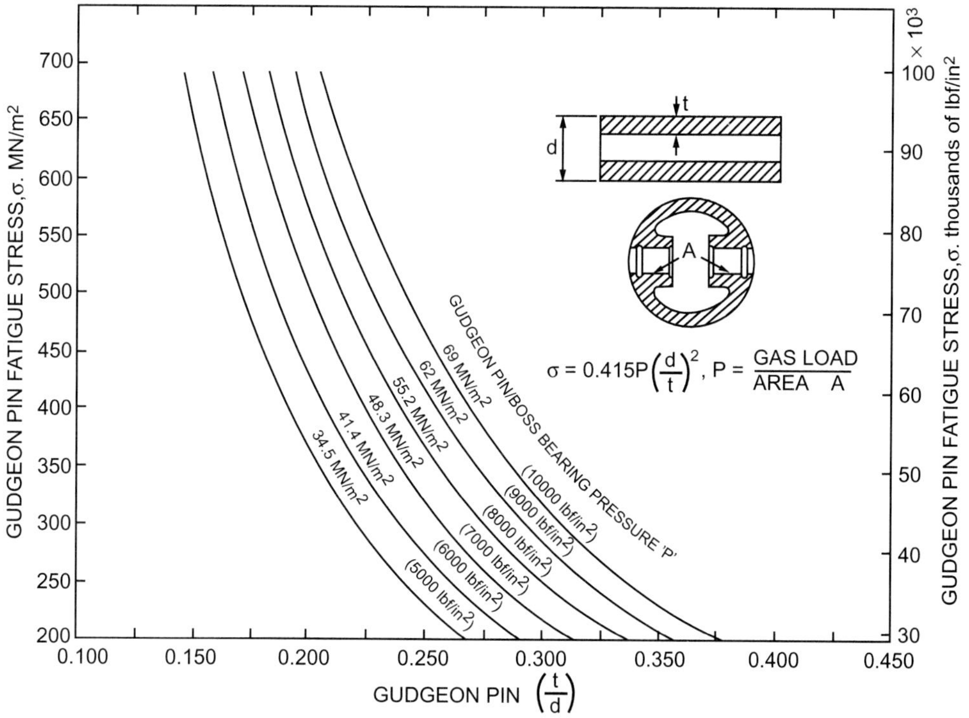

FIGURE 24-28(*b*) Fatigue stress in gudgeon pins for various pin and piston geometries. (M. J. Neale, *Tribology Handbook*, Butterworth-Heinemann, 1973.)

Particular	Formula

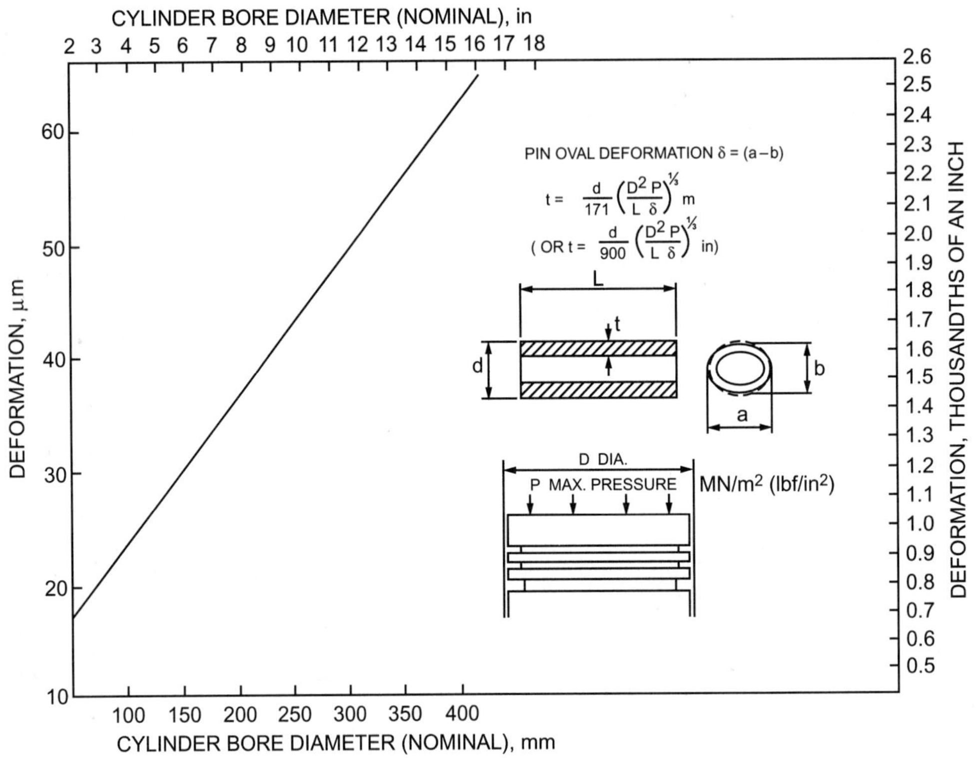

FIGURE 24-28(*c*) Gudgeon pin allowable oval deformation. (M. J. Neale, *Tribology Handbook*, Butterworth-Heinemann, 1973.)

For fatigue stress in gudgeon pins — Refer to Fig. 24-28*b*.

For empirical proportions and values of cylinder cover, cylinder liner, and valves — Refer to Figs. 24-30 to 24-33.

Piston rings

Width of rings

$$w = \frac{D}{20} \text{ for concentric rings} \tag{24-136a}$$

$$w = \frac{D}{27.5} \text{ opposite the joint of eccentric rings} \tag{24-136b}$$

$$w = \frac{D}{55} \text{ at the joint of eccentric rings} \tag{24-136c}$$

For land width or axial width of piston ring (w) required for various groove depths (g) and maximum cylinder pressure, p_{max}. — Refer to Fig. 24-28*d*.

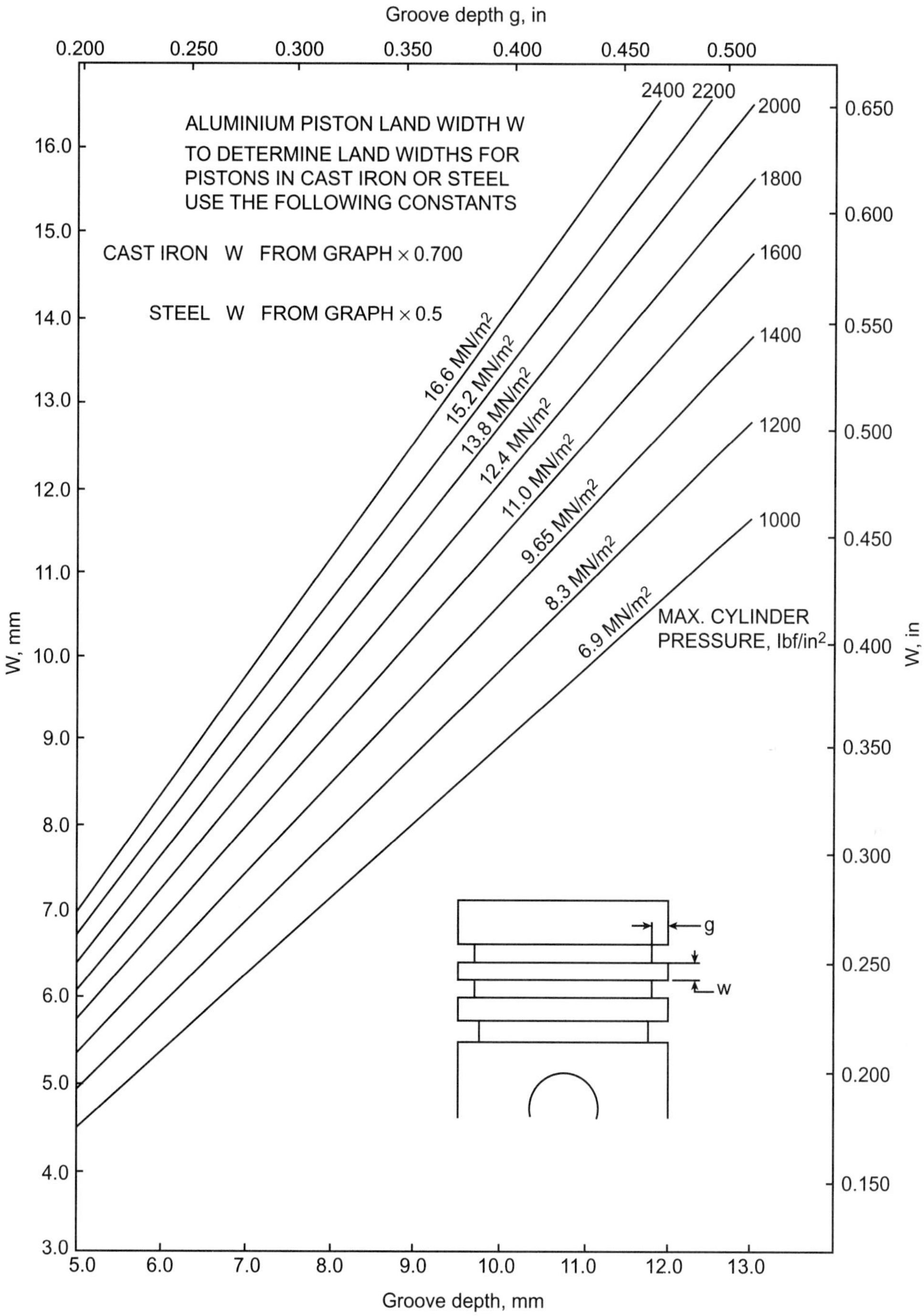

FIGURE 24-28(*d*) The land width required for various groove depths and maximum cylinder pressures. (M. J. Neale, *Tribology Handbook*, Butterworth-Heinemann, 1973.)

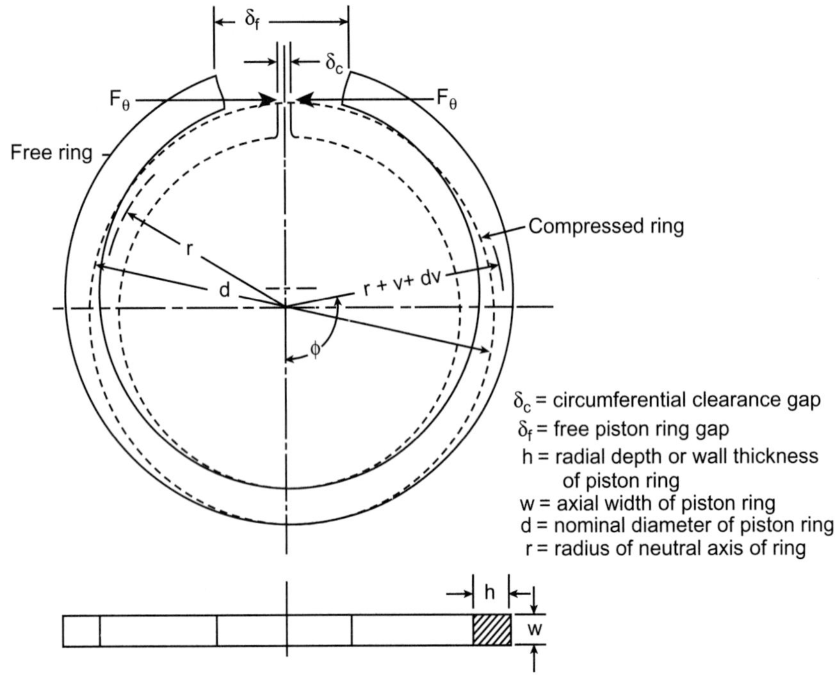

FIGURE 24-28(*e*) Nomenclature of piston ring and tangential force, F_θ.

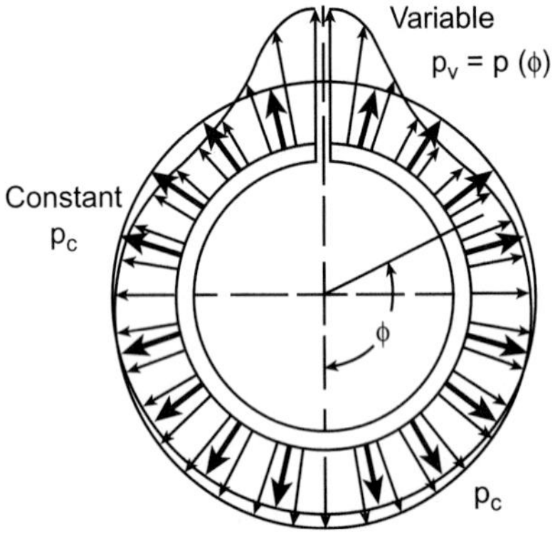

FIGURE 24-28(*f*) Typical variable and constant contact pressure distribution around piston rings for four-stroke engines. (*Courtesy*: Piston Ring Manual, Goetze AG, D-5093 Burscheid, Germany, August 1986.)

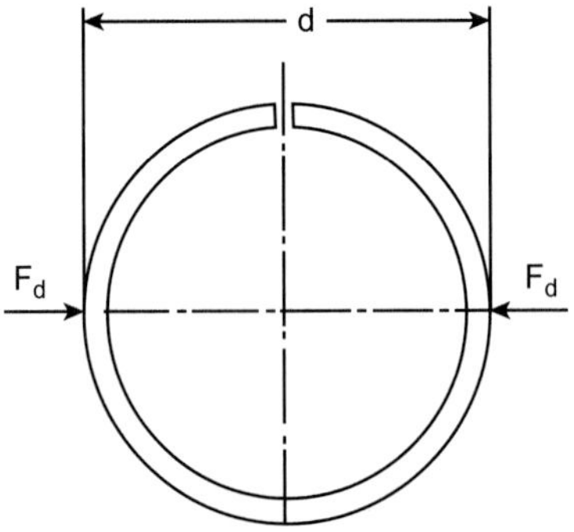

FIGURE 24-28(*g*) Diametrically opposite force (F_d) applied on piston ring.

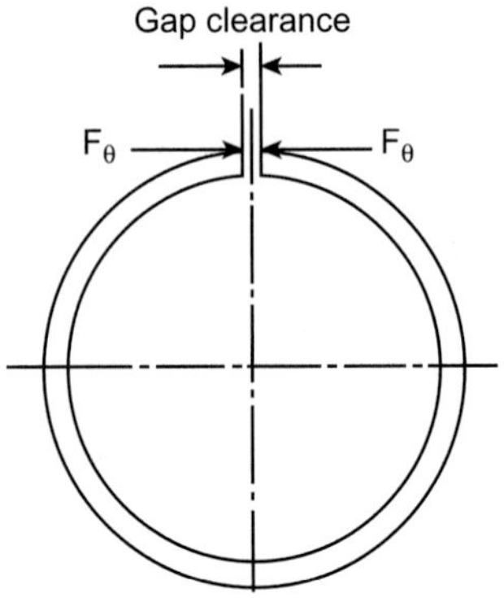

FIGURE 24-29 Tangentially applied force, F_θ, on a piston ring.

Particular	Formula
The modulus of elasticity of piston ring as per Indian Standards	$E = \dfrac{5.37\left(\dfrac{d}{h} - 1\right)^3 F}{w\delta}$ (24-137a) when the ring is diametrically loaded $E = \dfrac{14.14\left(\dfrac{d}{h} - 1\right)^3 F_\theta}{w\delta}$ (24-137b) when the ring is tangentially loaded where E = modulus of elasticity, MPa (psi) δ = difference between free gap and gap after applying the load, mm (in)
The bending moment produced at any cross section of the ring by the pressure uniformly distributed over the outer surface of the ring at an angle ϕ measured from the center line of the gap of the ring (Figs. 24-28*e* and 24-28*f*)	$M_b = -2pwr^2 \sin^2 \dfrac{\phi}{2}$ (24-138a) $M_b = pwr^2(1 + \cos\phi)$ (24-138b) where r = radius of neutral axis, mm (in) p = pressure at the neutral axis of the piston ring, Pa (psi)
The bending moment (M_b) in Eq. (24-138a) in terms of tangential force, F_θ	$M_b = F_\theta r(1 + \cos\phi)$ (24-138c)
The uniform contact pressure of the piston ring on the wall	$p = \dfrac{E_n \delta_f}{7.07d(d/h - 1)^3}$ (24-138d) where d = external piston ring diameter h = radial depth or wall thickness of piston ring E_n = nominal modulus of elasticity of material of the ring δ_f = free ring gap
The radial distance from a point in piston ring to obtain a uniform pressure distribution (Fig. 24-28*e*) according to R. Munro[a]	$r_o = r + v + dv$ (24-138e)

[a] In M. J. Neale, ed., *Tribology Handbook*, Section A31, Newnes-Butterworth, London, 1973.

Particular	Formula

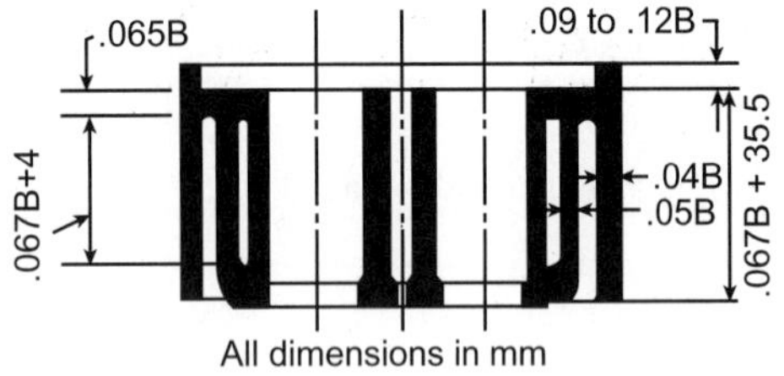

FIGURE 24-30 Proportion for four-stroke cover, 100- to 450-mm bore (B = cylinder bore).

where

$$v = \frac{Fr^4}{E_n I}(1 - \cos\phi + \tfrac{1}{2}\phi \sin\phi) \qquad \text{(24-138f)}$$

$$dv = \left[\frac{r}{2}\left(\frac{Fr^3}{E_n I}\right)^2 (\phi - \tfrac{1}{2}\phi\cos\phi - \tfrac{1}{2}\sin\phi) \times (3\sin\phi + \phi\cos\phi)\right]$$

where

F = (mean wall pressure × ring axial width)

r = radius of neutral axis, when the ring is in place inside the cylinder (Fig. 24-28e)

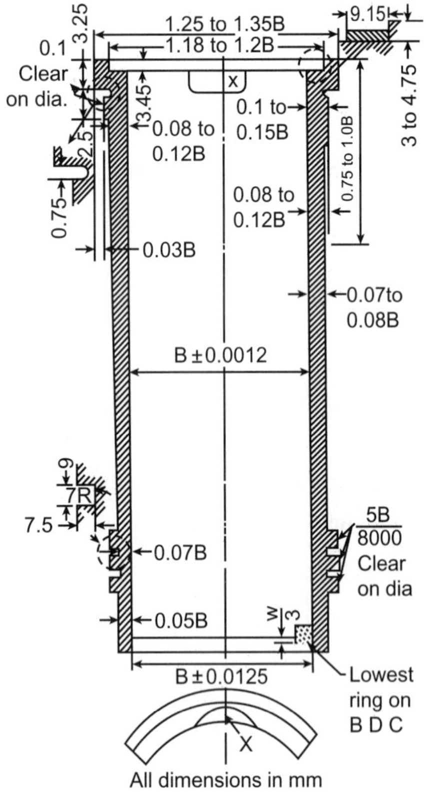

FIGURE 24-31 Empirical rules for average practice in liner design (B = cylinder bore).

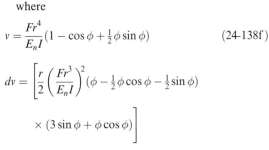

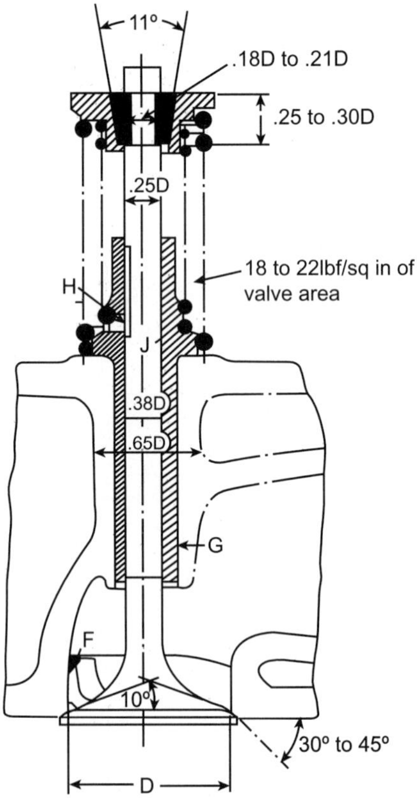

FIGURE 24-32 Valve seated directly in the cylinder head.

Particular	Formula
	ϕ = angle measured from bottom of the vertical line passing through the center of the gap of the ring as shown in Fig. 24-28*e*
	I = moment of inertia of the ring
The relation between the ratio of fitting stress σ_{ft} to nominal modulus of elasticity (E_n) in terms of h, d, and δ_f	$$\frac{\sigma_{ft}}{E_n} = \frac{4(8h - \delta_f + 0.00d)}{3\pi h(d/h - 1)^2} \quad (24\text{-}138g)$$ where σ_{ft} = opening stress when fitting the piston ring onto the piston
The relation between the ratio of working stress (σ_w) to nominal modulus of elasticity (E_n) in terms of h, d, and δ_f	$$\frac{\sigma_w}{E_n} = \frac{4(\delta_f - 0.00d)}{3\pi h(d/h - 1)^2} \quad (24\text{-}138h)$$ where σ_w = working stress when the piston ring is in the cylinder
The relation between the ratio of the sum of ($\sigma_{ft} + \sigma_w$) to nominal modulus of elasticity (E_n) in terms of d and h	$$\frac{\sigma_{ft} + \sigma_w}{E_n} = \frac{32}{3\pi(d/h - 1)^2} \quad (24\text{-}138i)$$ Equation (24-138i) is independent of δ_f.
For preferred number of piston rings	Refer to Table 24-10C.
For properties of typical piston ring materials	Refer to Table 24-10D.

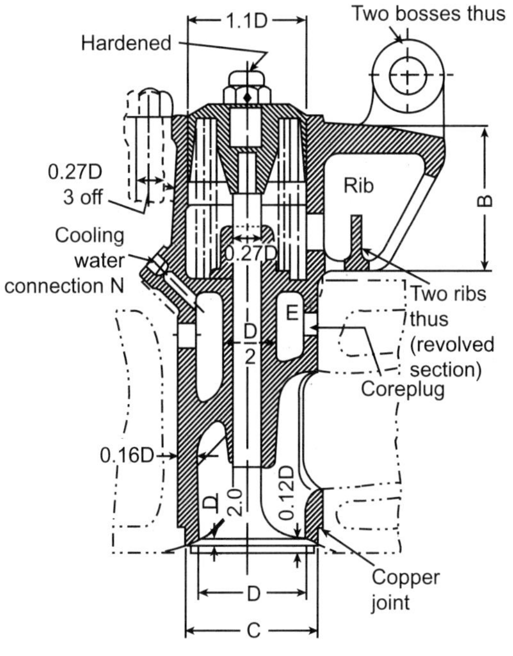

FIGURE 24-33 Valve with removable cage.

Particular	Formula
The circumferential clearance (δ_c) or gap between ends of ring	$\delta_c = d\alpha_p T$ (24-138j) where α_p = coefficient of expansion of piston ring material T = operating temperature d = cylinder diameter
An expression for pressure acting on ring from Eqs. (24-138b) and (24-138c)	$p = \dfrac{F_\theta}{rw}$ (24-139a)
The pressure in the radial outward direction against the cylinder	$p = \dfrac{2F_\theta}{dw}$ (24-139b)
For variable and constant radial contact pressure distribution of piston ring	Refer to Fig. 24-28*f*.
The diametral load which acts at 90° to the gap required to close the ring to its nominal diameter, d (Fig. 24-28*g*)	$F_d = 2.05F_\theta$ for modulus of elasticity $E \leq 150$ GPa (24-140a) $F_d = 2.15F_\theta$ for modulus of elasticity $E > 150$ GPa (24-140b) $F_d \approx 2.21F_\theta$ (24-140c)
The maximum bending stress at any cross section which makes an angle ϕ measured from the center line of the gap of the ring	$\sigma_b = \dfrac{12pr^2}{h^2}\sin^2\left(\dfrac{\phi}{2}\right)$ (24-141)
The maximum bending stress which occurs at $\phi = \pi$, i.e., at the cross section opposite to the gap of the ring	$\sigma_{max} = \dfrac{12pr^2}{h^2_{max}}$ (24-142)
The bending stress present in the ring of rectangular cross section in terms of free gap (δ_f) of the ring, when it is in place in the cylinder	$\sigma_b = 0.424\delta_f \dfrac{Eh}{(d-h)^2}$ **SI** (24-142a) where σ_b and E in N/mm^2 h, d, and δ_f in mm
The bending stress present in the ring of rectangular cross section in terms of tangential force, F_θ (Fig. 24-29)	$\sigma_b = \dfrac{6(d-h)}{wh^2}F_\theta$ (24-142b) where σ_b in N/mm^2; F_θ in N; d, h, and w in mm
The bending stress present in the case of slotted oil control ring of rectangular cross section in terms of free ring gap, δ_f	$\sigma_{bso} = 0.424\dfrac{\delta_f E l_{co} I_m}{(d-h)^2 I_{us}}$ (24-142c) where σ_{bso} in N/mm^2 and I_{us} = moment of inertia of the unslotted cross-section ring, mm^4 $I_m = \dfrac{I_{us}+I_s}{2}$

Particular	Formula
	I_s = moment of inertia of the slotted cross-section ring, mm^4
	l_{co} = twice the diameter between center of gravity and outside diameter, mm
The bending stress present in the case of slotted oil control ring of rectangular cross section in terms of tangential load, F_θ	$$\sigma_{bso} = \frac{6(d-h)l_{co}I_m}{wh^3 I_s} F_\theta \quad \text{(24-142d)}$$ where σ_{bso} in N/mm^2; F_θ in N; l_{co}, d, h, and w in mm; I_m and I_s in mm^4
The tangential load or force required for opening of a rectangular cross-section piston ring[a]	$$\sigma_{\theta,\max} = \frac{hE}{d-h}(1.26\varepsilon_T - 1.84k + 0.025)$$ **SI** (24-142e) where $$\varepsilon_T = \frac{d+h}{d-h} - 1$$ k = piston ring parameter from Eq. (24-142f) and (24-142h)
The piston ring parameter (k) in terms of tangential load F_θ for rectangular cross-section rings	$$k = \frac{3(d-h)^2}{wh^3}\frac{F_\theta}{E} \quad \text{(24-142f)}$$
The tangential load or force required for opening of rectangular cross-section slotted oil control rings	$$\sigma_{\theta,\max T} = \frac{l_{ci}E}{d-h}(1.26\varepsilon_T - 1.84k + 0.025)\frac{I_m}{I_s} \quad \text{(24-142g)}$$ where $$\varepsilon_T = \frac{d+h}{d-h} - 1 \quad \textbf{SI}$$ l_{ci} = twice the distance between center of gravity and inside diameter, mm; k = piston ring parameter from Eq. (24-142e) and (24-142f)
The piston ring parameter (k) in terms of free ring gap (δ_f) for rectangular cross-section slotted oil rings for use in Eq. (24-142g)	$$k = \frac{2}{3\pi}\frac{\delta_f}{d-h} \quad \text{(24-142h)}$$
The piston ring parameter (k) in terms of the constant pressure (p) for rectangular cross-section rings also for use in Eqs. (24-142f) and (24-142g)	$$k = \frac{3}{2}\frac{p}{E}\frac{d(d-h)^2}{h^3} \quad \text{(24-142i)}$$
The radial thickness of the ring at a section which makes an angle ϕ measured from the center line of the gap of the ring	$$h = \sqrt[3]{\frac{24pr^4}{E\delta}\sin^2\frac{\phi}{2}} \quad \text{(24-142j)}$$

[a] Goetze AG, *Piston Ring Manual*, 3rd ed., Burscheid, Germany, 1987.

Particular	Formula	
The maximum thickness of the ring which occur opposite the gap of the ring (i.e., at $\phi = \pi$)	$h_{max} = \sqrt{\frac{24pr^4}{E\delta}}$	(24-142k)
For piston ring dimensional deviation, hardness, and minimum wall pressure	Refer to Tables 24-7 to 24-9.	
For cylinder bore diameter	Refer to Table 24-10.	

TABLE 24-5
Values of c for various inclinations of coned pistons

Cone	Inclination ranges, θ, deg	c
—	0–6	1
Slightly	6–18	0.85–0.95
Medium	18–28	0.75–0.85
Strong	28–35	0.65–0.75

TABLE 24-6
Values of coefficient K for pistons (admissible pressures, kgf/mm^2 absolute)

Diameter of cylinder, mm	Pressure, kgf/mm^2							
	0.01 to 0.02	0.02 to 0.04	0.04 to 0.06	0.06 to 0.08	0.08 to 0.10	0.10 to 0.12	0.12 to 0.14	0.14 to 0.16
380–575	1.000	1.125	1.375	1.500	1.750	2.000	2.125	2.500
575–775	1.375	1.500	1.750	2.000	2.500	2.750	3.000	3.375
775–975	1.500	1.750	2.000	2.500	3.125	3.500	3.750	4.000
975–1175	1.750	2.000	2.375	3.000	3.500	4.000	4.500	
1175–1375	2.000	2.250	2.750	3.125	3.750	4.125		
1375–1575	2.375	2.500	3.125	3.500	4.000	4.375		
1575–1775	2.500	3.000	3.375	4.000	4.375			
1775–1975	2.750	3.125	3.500	4.125				
1975–2150	3.000	3.375	3.750	4.375				
2150–2350	3.000	3.500	4.000					
2350–2550	3.125	3.500						
2550–2750	3.375	3.750						

Key: 1 kgf/mm^2 = 1.42247 kpsi; 1 kpsi = 6.894757 MPa.

TABLE 24-7
Recommended hardness for piston rings of IC engines

Nominal diameter, d, mm	Hardness HRD
<100	95–107
100–200	93–105
>200	90–102

TABLE 24-8
Minimum wall pressure for piston rings of IC engines

	Compression rings		Oil rings	
	MPa	kgf/cm^2	MPa	kgf/cm^2
Petrol[a]	0.059	0.60	0.137	1.40
Diesel	0.013	1.05	0.196	2.00

[a] Gasoline.

TABLE 24-9
Permissible deviation on the dimensions of piston rings of IC engines

Dimensions	Deviations, mm
Axial width, b	−0.010 −0.022
Radial thickness	
≤80 mm ring diameter	±0.08
>80 mm with ≤175 mm ring diameter	±0.12
175 mm ring diameter	±0.15
Parallelism of sides—40% of tolerance on axial width	

TABLE 24-10A
Preferred cylinder bore diameters for internal-combustion (IC) engines (all dimensions in mm)

30	(62)	95	125	(152.4)	(188)	(241.3)	315
32	65	98	(127)	155	190	(245)	(317.5)
34	(68)	(98.4)	(128)	(158)	(190.5)	(250)	320
(35)	70	100	(128.2)	(158.8)	(192)	(254)	(325)
36	(72)	(101.6)	130	160	195	(255)	330
38	(73)	(102)	(132)	(162)	(196.8)	266	(335)
40	74	(103.2)	(133.4)	165	198	(265)	340
42	(76)	(104.8)	135	(165.1)	200	270	(343)
44	(78)	105	(138)	(168)	(205)	(273)	(345)
46	(79.4)	108	(139.7)	170	(209.6)	(275)	350
48	80	110	140	(171.4)	210	280	
50	82	(111.1)	(142)	(172)	(215)	(285)	
52	85	112	142.9	175	(215.9)	290	
54	87.3	(114.3)	(145)	(177.8)	220	(292.1)	
56	88	115	(146)	(178)	(225)	(295)	
(57)	(88.9)	(118)	(148)	180	(228.6)	(298.4)	
58	90	120	(149.9)	(182)	230	300	
(59)	(91.4)	(120.6)	150	(184.2)	(235)	(305)	
60	(92)	(122)	(152)	185	240	103	

TABLE 24-10B
Chemical composition of alloys and physical properties of aluminum alloy piston (values in % maximum unless shown otherwise)

Alloy designation[a]		Chemical composition, %												Hardness, H_B	Physical properties[b]				
															Tensile strength				Coefficient of thermal expansion (20 to 200°C)
															Chill casting		Forging		
Casting	Forging	Cu	Mg	Si	Fe	Mn	Ni[c]	Zn	Ti	So	Pb	Cr	Al		MPa	kpsi	MPa	kpsi	mm/mm/°C × 10^{-4}
2285	34,850	3.5–4.5	1.2–1.8	0.6	0.7	0.2	1.1–2.3	0.2	0.23	0.05	0.05		Remainder	90–130	225–275	32.7–39.8	345–410	49.8–59.7	23–24
4658	49,582	0.8–1.5	0.8–1.3	11.0–13.0	0.8	0.2	1.5	0.35	0.2	0.05	0.05			90–140	195–245	28.5–35.6	295–365	42.7–52.6	20.5–21.5
4928A	49,285	0.8–1.5	0.8–1.3	17.0–19.0	0.7	0.2	0.8–1.3	0.2	0.2	0.05	0.05			90–125	175–215	25.6–31.3	225–295	32.7–28.5	18.5–19.5
4928B		0.8–1.5	2.8–1.3	23.0–26.0	0.7	0.2	0.1–1.3	0.2	0.2	0.05	0.05	0.3–0.6		90–125	165–205	24.2–29.7			17–18

[a] Alloys have been designated in accordance with IS 6051, 1970. Code for designation of aluminum and aluminum alloys.
[b] Physical properties are attainable after suitable heat treatment.
[c] The purchaser may specify nickel content, if so desired.
Source: Bureau of Indian Standards, New Delhi.

TABLE 24-10C
Preferred number of piston rings

Differential pressure	Std. atm.	0–9	10–14	15–24	25–29	30–49	50–99	100–200
	MPa	0–0.88	0.98–1.37	1.47–2.35	2.45–2.85	2.94–4.80	4.90–9.71	9.81–19.61
	psi	0–128	142–199	213–341	355–412	426–696	710–1406	1422–2844
Minimum number of rings		2	3	4	5	6	7	8

Source: M. J. Neale, *Tribology Handbook*, Butterworth-Heinemann, London, 1973; reproduced with permission.

TABLE 24-10D
Properties of typical piston ring materials

Material	Tensile strength, σ_t		Nominal modulus of elasticity, E_n		Brinell hardness number, H_B	Bulk density, g/cm^3	Typical coefficient of expansion, α	Wear rating
	MPa	kpsi	GPa	Mpsi			$\times 10^6/°C$	
Metallic:								
Gray irons	230–310	33.4–45.0	83–124	12.1–18.0	210/310			Good
Carbide malleable irons	400–580	58.0–84.1	140–160	20.3–23.2	250/320			Excellent
Malleable and/or nodular irons	540–820	78.3–119.0	155–165	22.5–24.0	200/440			Poor
Sintered irons	250–390	36.5–56.6	120	17.4	130/150			Good
Nonmetallic:								
Carbon-filled PTFE	10.3	1.49				2.05	55	
Graphite/MoS_2-filled PTFE	19.6	2.85				2.20	115	
Resin-bonded PTFE	29.4	4.27				1.75	30	
Carbon	43.4	6.30				1.8	43	
Resin-bonded carbon	19.6	2.85				1.9	20	
Glass-filled PTFE	16.7	2.42				2.26	80	
Bronze-filled PTFE	12.8	1.85				3.90	118	
Resin-bonded fabric	110.8	16.07				1.36	22.5/87.5[a]	

[a] Material is anisotropic.
Source: M. J. Neale, *Tribology Handbook*, Butterworth-Heinemann, London, 1973, extracted with permission.

TABLE 24-10E
Piston rings and piston ring elements

		Mechanical properties			
Designation	**Grade**	**Hardness**	**Tensile strength, σ_{st}, MPa**	**Modulus of elasticity, E, MPa**	**Main application**
Steel					
GOE 61	Cr steel, 17% Cr min	380–450 HV 30	1200* approx.	230 000 approx.	Compression rings
GOE 62	Cr-Si steel	500–600 HV 30	1900* approx.	210 000 approx.	Coil spring loaded rings
GOE 64	Cr-Si steel	450–550 HV 30	1700* approx.	210 000 approx.	Compression rings
GOE 65A	Cr steel, 11% Cr min, high C	300–400 HV 30	1300* approx.	210 000 approx.	Compression rings, nitrided
GOE 65B	Cr steel, 11% Cr min, low C	270–420 HV 30	1300* approx.	220 000 approx.	Coil spring loaded rings and segments, nitrided
Cast Iron					
GOE 12	Unalloyed non heat-treated gray cast iron	94–106 HRB	350 min	85 000 typical	Compression and oil control rings
GOE 13	Unalloyed non heat-treated gray cast iron	97–108 HRB	420 min	95 000–125 000	Compression and oil control rings
GOE 32	Alloyed heat-treated gray cast iron with carbides	109–116 HRB	650 min	130 000–160 000	Compression rings
GOE 44	Malleable cast iron	102–111 HRB	800 min	150 000 min	Compression rings
GOE 52	Spheroidal graphite cast iron	104–112 HRB	1300 min	150 000 min	Compression rings
GOE 56	Spheroidal graphite cast iron	40–46 HRC	1300 min	150 000 min	Compression rings

Source: Goetze Federal Mogul Burscheid GmbH, *Piston Ring Manual*, 4th ed., January 1995, Burscheid, Germany, reproduced with permission

24.5 DESIGN OF SPEED REDUCTION GEARS AND VARIABLE-SPEED DRIVES

SYMBOLS[2,3]

a	center distance, m (in)
	number of pinions or planetary pinion (Fig. 24-36)
A	center distance (also with subscripts) (Fig. 24-36)
	area of reduction gear housing, m^2 (in^2)
A_n	noncooled, i.e., ribbed, surface of housing of reduction gear drive, m^2 (in^2)
A_c	cooled surface of reduction gear drive, m^2 (in^2)
A_w	surface area of contact of teeth when one-fourth of all teeth of wheel in wave-type reduction gears are engaged, m^2 (in^2)
b	width of rim, m (in)
d_1	diameter of pinion, m (in)
	diameter of rigid immovable rim with internal teeth of wave-type reduction gears, m (in)
d_2	diameter of gear, m (in)
	diameter of flexible movable wheel rim with external teeth of wave-type reduction gear, m (in)
d_{max}	maximum diameter of the circumference of the belt arrangement on the V-belt of a variable-speed drive, m (in)
d_{min}	minimum diameter of the circumference of the belt arrangement on the V-belt of a variable-speed drive, m (in)
$D = \frac{d_{max}}{d_{min}}$	velocity control range for a V-belt drive
D_1	velocity control range for a V-belt drive with only one adjustable pulley
D_2	velocity control range for a V-belt drive with two adjustable pulleys
e	working height of a V-groove of the pulley, m (in)
F_{max}	maximum load acting on the pinion, kN (lbf)
F_m	mean load acting on the pinion, kN (lbf)
h	height of tooth, m (in)
	coefficient of heat transfer, W/m^2 K (Btu/ft^2h °F)
h_n	coefficient of heat transfer of noncooled surface, W/m^2 K (Btu/ft^2 h °R)
h_c	coefficient of heat transfer of cooled surface, W/m^2 K (Btu/ft^2 h °R)
h_a	addendum of tooth, m (in)
h_f	dedendum of tooth, m (in)
i	transmission or speed ratio
$k_{nl} = \frac{F_{max}}{F_m}$	nonuniform load distribution factor
L	distance between the axes of the pinions (Fig. 24-36*d*)
$\mathbf{m}$	module, m (in)
M_{ts}	torque acting on smaller wheel, N m (lbf in)
n	speed, rpm
n_1, n_2	speeds of pinion and gear, respectively, rpm
q	a whole number
Φ	heat generated, W (Btu/h)

r_{max}	maximum radius of the circumference of the belt arrangement on the V-belt of a variable-speed drive, m (in)
r_{min}	minimum radius of the circumference of the belt arrangement on the V-belt of a variable-speed drive, m (in)
t_l	temperature of lubricant, °C (°F)
t_a	ambient temperature, °C (°F)
z_1, z_2	number of teeth on sun pinion and planetary pinion of epicyclic gear transmission, respectively, Fig. 24-36
	number of teeth on pinion and gear, respectively
z_3	number of teeth on ring gear 3 (Fig. 24-36*a*)
z_s	number of teeth on smaller wheel
ω_1, ω_2	angular speed of pinion and gear, respectively, rad/s
δ	deformation, m (in)
Δ	clearance between the pinions which should be at least 1 mm (in)
α	half-cone angle of V-belt, deg
σ_{ca}	allowable compressive stress, MPa (psi)

Particular	Formula	
Transmission or speed ratio for single reduction gear	$i = \frac{\omega_1}{\omega_2} = \frac{n_1}{n_2} = \frac{d_2}{d_1} = \frac{z_2}{z_1}$	(24-143)
For different types of gear reduction drives	Refer to Fig. 24-35 and Table 24-11.	

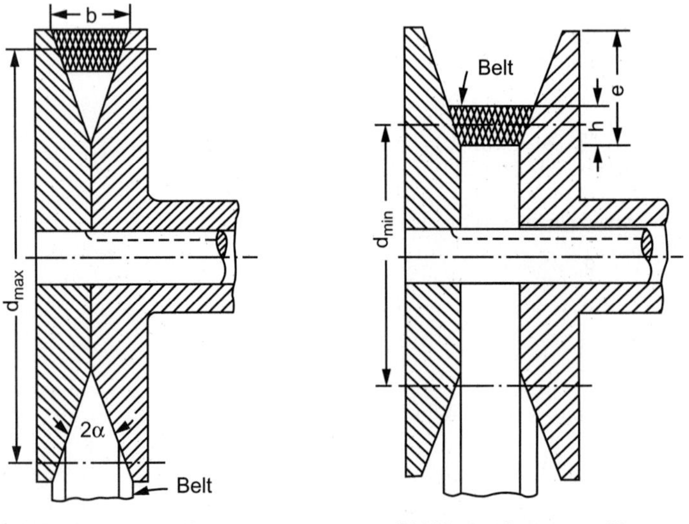

(a) V-belt at top position *(b)* V-belt at bottom position

FIGURE 24-34 Dimension of V-belt variable-speed drive.

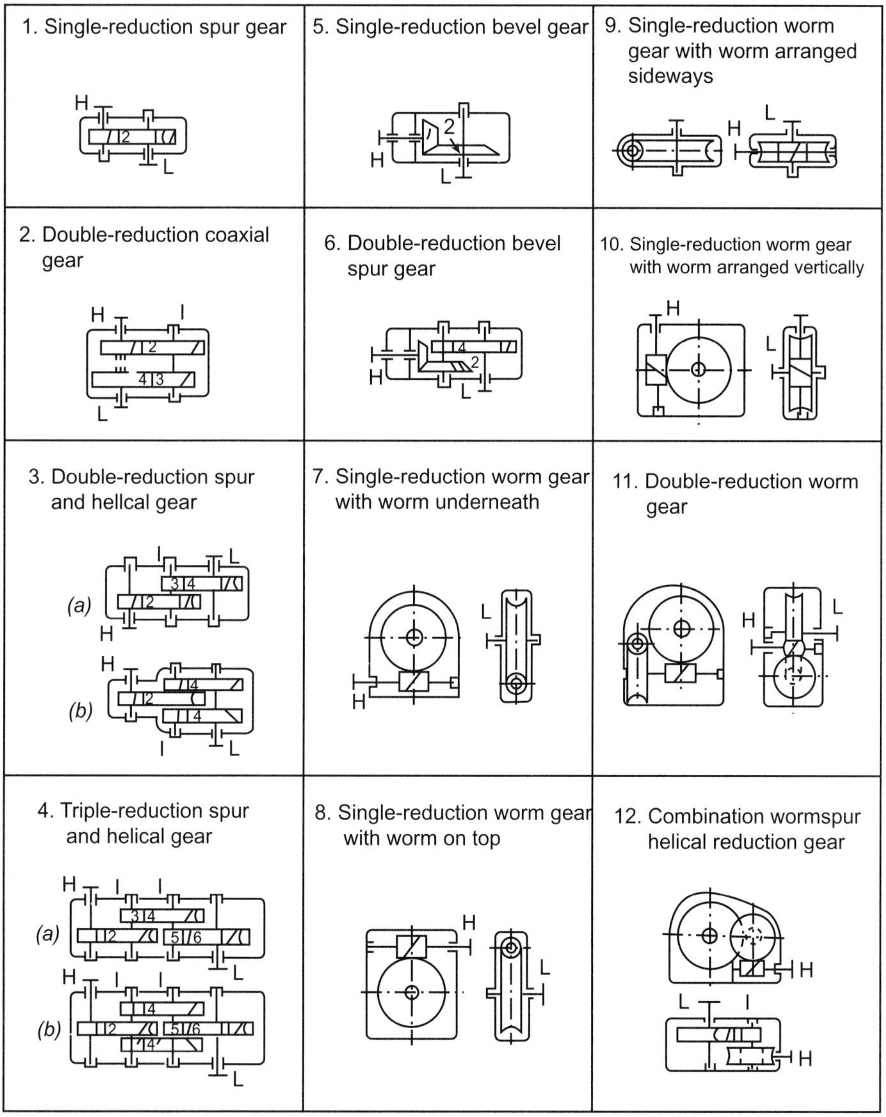

H = High Speed I = Intermediate Speed L = Low Speed

FIGURE 24-35 Schematic diagrams of various types of spur, helical, herringbone, bevel, and worm reduction gears.

Particular	Formula

PLANETARY REDUCTION GEARS

First condition—mating

The sum of the radii of the addendum circles of the mating pinions in planetary reduction gears should be smaller than the distance between their axes (Fig. 24-36*d*) so that the top of the pinions should not touch each other

$$L = 2A_{1,2}\sin\frac{\pi}{a} = z_2 m + 2m(1+\xi) + \Delta \qquad (24\text{-}144)$$

where

a = number of pinions

Δ = clearance between the pinions, which should be at least 1 mm

$A_{1,2}$ = center distance as shown in Fig. 24-36

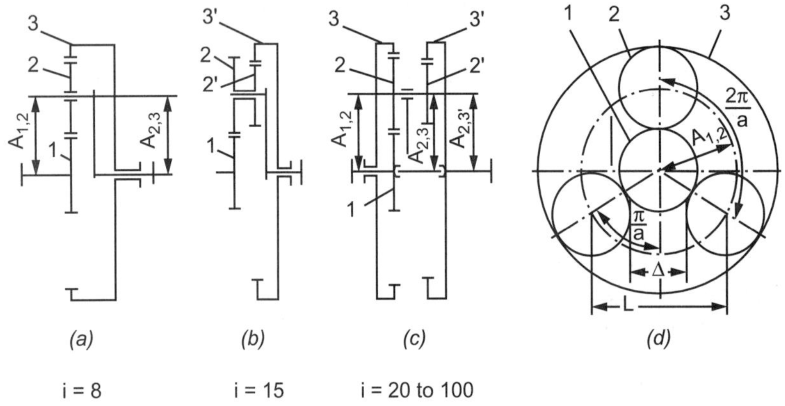

FIGURE 24-36 Planetary reduction gears.

Second condition—coaxiality

The center distance of each pair of wheels should be equal (Fig. 24-36)

$$A_{12} = A_{23};\ A_{12} = A_{23} = A_{2'3'} \qquad (24\text{-}145)$$

The relationship between teeth in corrected or uncorrected gears (Fig. 24-36*a*)

$$z_1 + z_2 = z_3 - z_2 \qquad (24\text{-}146a)$$

or

$$z_1 + 2z_2 = z_3 \qquad (24\text{-}146b)$$

The relationship between teeth in corrected or uncorrected gears (Fig. 24-36*c*) to ratify two conditions

(i) First condition

Refer to Eq. (24-146).

(ii) Second condition

$$m_2(z_3 - z_2) = m_2'(z_3' - z_2') \qquad (24\text{-}147a)$$

or

$$z_3 - z_2 = z_3' - z_2' \quad \text{since} \quad m_2 = m_2' \qquad (24\text{-}147b)$$

Particular	Formula

Third condition—coincidence

The condition for the teeth and spaces of the meshed gears should coincide when the pinions are arranged uniformly over the circumference

$$\frac{z_1 + z_3}{a} = q \tag{24-148}$$

where q is a whole number

The moment acting on smaller wheel

$$M_{ts} = \frac{M_{t1} k_{nl}}{a} \frac{z_s}{z_1} \tag{24-149}$$

where

$z_s = z_1$ or $z_s = z_2$ if $z_1 > z_2$

$k_{nl} = 2$ maximum value

$= 1.4$ to 1.6 for gears of 7th degree of accuracy

$= 1.1$ to 1.2 when floating central wheels are used to equalize the load

CONDITIONS OF PROPER ASSEMBLY OF PLANETARY GEAR TRANSMISSION

Two planetaries

Both the driving pinion (sun pinion) and the planetaries may have either an even or an odd number of teeth.

Three planetaries

If z_1 (number of teeth on sun pinion) is divisible by 3, then z_2 (number of teeth on planetary pinion) must also be divisible by 3.

If $z_2 - 1$ is divisible by 3, then $z_2 + 1$ must be divisible by 3.

If $z_1 + 1$ is divisible by 3, then $z_2 - 1$ must be divisible by 3.

Four planetaries

If z_1 is even, then z_2 must be even.

If z_1 is odd, then z_2 must be odd.

Particular	Formula

WAVE-TYPE REDUCTION GEARS

Transmission or gear ratio	$i = \dfrac{z_2}{z_1 - z_2} = \dfrac{d_2}{d_1 - d_2}$	(24-150)
	For a double-wave drive, $z_1 - z_2 = 2$.	
The necessary deformation	$\delta = d_1 - d_2 = \dfrac{d_2}{i}$	(24-151)
The condition for obtaining the module for the drive	$d_1 - d_2 = (z_1 - z_2)\mathbf{m} = \delta$	(24-152)
The module of the drive from Eq. (24-152)	$\mathbf{m} = \dfrac{\delta}{z_1 - z_2} = 0.5\delta$	(24-153)
The tooth height	$h = \delta$	(24-154)
The tooth addendum	$h_a = 0.44\delta$	(24-155)
The tooth dedendum	$h_f = 0.56\delta$	(24-156)
The rim width	$b = 0.1d_2$ to $0.2d_2$	(24-157)
The total surface area of contact of teeth when one-fourth of all teeth of wheel are engaged	$A_w = 0.5h \times 0.25z_2 b$	(24-158)
The torque transmitted	$M_t = 0.5d_2 A_w \sigma_{ca} \simeq 0.06d_2^2 \delta b z_2 \sigma_{ca}$	(24-159)
	where $\sigma_{ca} = 29.5$ MPa (4.28 kpsi) for hardened steel wheels	

VARIABLE-SPEED DRIVES (Figs. 24-34 and 24-37, and Table 24-12)

For schematic arrangements of various variable-speed drives	Refer to Figs. 24-34 and 24-37.	
The velocity control range for V-belt drive with only one adjustable pulley	$D_1 = \dfrac{d_{max}}{d_{min}}$	(24-160)
The relation between d_{max} and d_{min} of V-belt drive	$d_{max} = d_{min} + 2(e - h)$	(24-161a)
	$d_{max} = d_{min} + b \cot \alpha - 2h$	(24-161b)
The velocity control range for V-belt drive from Eqs. (24-160) and (24-161)	$D = \dfrac{d_{max}}{d_{min}} = 1 + \dfrac{2e}{d_{min}} - \dfrac{2h}{d_{min}}$	(24-162a)
	$D = 1 + \dfrac{b}{d_{min}} \cot \alpha - \dfrac{2h}{d_{min}}$	(24-162b)

Particular	Formula	
The velocity control range for V-belt drive when two pulleys are adjustable	$D_2 = D_1$	(24-163)
The total range of velocity control of variable-speed drive of two adjustable pulleys of V-belt drive	$D = D_1^2$	(24-164)
The working height of the V-groove of the pulley	$e > \frac{b}{2} \cot \alpha$	(24-165)
The width of standard V-belt	$b \simeq 1.8h$	(24-166)

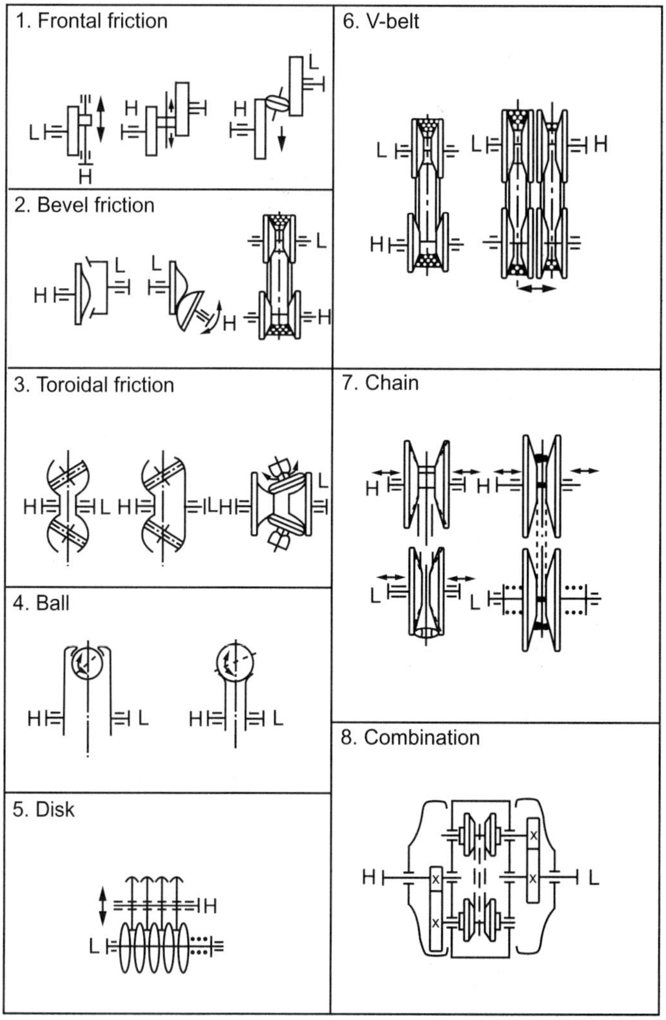

FIGURE 24-37 Variable-speed drives.

Particular	Formula	
The larger ratio of width to height of specially profiled broad V-belts	$\frac{b}{h} \simeq 2$ to 3	(24-167)
The total velocity control range for adjustable pulleys of V-belt drive	$D = D_1^4$	(24-168)

DISSIPATION OF HEAT IN REDUCTION GEAR DRIVES

The area of housing required for dissipating heat generated in a closed-type reduction gear drive operating in an oil bath at stable thermal equilibrium condition	$A = \frac{\Phi}{h(t_1 - t_a)}$ where h = coefficient of heat transfer, which varies from 8.75 to 17.5 W/m^2 K (1.54 to 3.1 Btu/ft^2 h °R)	(24-169)
The thermal equilibrium condition of reduction gear drive which has a housing of noncooled surface (ribbed surface) and cooled surface (cooled by blowing of air by fan)	$\Phi \leq (h_n A_n + h_c A_c)$ W (Btu/h)	(24-170)
The expression for coefficient of heat transfer of the housing or reduction gear drive blown over by air	$h_c = 12\sqrt{v}$ W/m^2 K (Btu/ft^2 h °R) where v = velocity of air, m/s (ft/min)	(24-171)
The velocity of air which depends on impeller velocity	$v \simeq 0.005 n_i$ m/s (ft/min) n_i = impeller speed, rpm	
For minimum weight equations for gear systems	Refer to Table 24-13.	
For total weight equations for gear systems	Refer to Table 24-14.	
For K factors for preliminary estimate of spur and helical gear size	Refer to Table 24-15.	
For comparison of five gear systems	Refer to Table 24-16.	

TABLE 24-11
Transmission ratio (i), efficiency (η), and allowable transmitted power (P_{al}) for reduction gears

Type of reduction gear	Fig. no.	i	η	P_{al}, kW
Single- and triple-spur and helical reduction gear	24-35, serial nos. 3*a*, 4*a*	10–60		
Single-spur reduction gear	24-35, serial no. 1	≤8–10		
Single worm			108	
Helical worm			100	
Harmonic drive			100	
Planetary reduction gear	24-36*a*	8	0.97–0.99	
	24-36*b*	15	0.97–0.99	1000
	24-36*c*	20–100		100
Wave-type toothed reduction gear		100	0.75–0.85	

TABLE 24-12
Velocity control range (D), efficiency (η), and allowable power transmitted (P_{al}) for variable-speed drives

Particular	Type of drive	Serial no. in Fig. 24-37	D	η	P_{al}, kW
Frontal friction	Single	1	3–4		20
	Twin type		8–10		
Bevel friction	Single	2	3–4		5
	Double		4–10		
	Self-locking ring		16	0.7–0.8	10
Toroidal friction		3	4–6	0.95	20
Ball		4	10–12		
Disk drives		5	≤3		800
			4–5		≤300
V-belt drives	Solid disk	6	1.3–1.7	0.8–0.9	50
	Grooved disk		2		
Chain drives	First type of drive	7	6	0.8–0.9	30
	Second type of drive		7–10		75

MINIMUM AND TOTAL WEIGHT EQUATION FOR GEAR SYSTEMS

The following symbols are used in Tables 24-13 to 24-16: a = number of branches in an epicyclic gear: $C = (2M_t/K)$, m^3; d = pitch diameter, m (in); i = gear speed ratio; i_o = overall ratio; $i_s = d_p/d_s = z_p/z_s$ = speed ratio of planet gear to sun gear; j = number of idlers; K = a factor from Table 24-15; M_t = input torque, N m (lbf in); $(i_o + 1)/i_o = i'_o$.

TABLE 24-13
Minimum weight equations for gear systems

Particular	Equation
Simple train (offset)	$2i^3 + i^2 = 1$
Offset with idler	$2i^3 + i^2 = i_o^2 + 1$
Offset with two idlers	$2i^3 + i^2 = \dfrac{i_o^2 + 1}{2}$
Offset with j idlers	$2i^3 + i^2 = \dfrac{i_o^2 + 1}{j}$
Double-reduction	$2i^3 + \dfrac{2i^2}{i'_o} = \dfrac{i_o^2 + 1}{i'_o}$
Double-reduction, double branch	$2i^3 + \dfrac{2i^2}{i'_o} = \dfrac{i_o^2 + 1}{2i'_o}$
Double-reduction, four branch	$2i^3 + \dfrac{2i^2}{i'_o} = \dfrac{i_o^2 + 1}{4i'_o}$
Double-reduction, j branches	$2i^3 + \dfrac{2i^2}{i'_o} = \dfrac{i_o^2 + 1}{ji'_o}$
Planetary (theoretical)	$2i_s^3 + i_s^2 = \dfrac{0.4(i_o - 1)^2 + 1}{a}$
Star (theoretical)	$2i_s^3 + i_s^2 = \dfrac{0.4i_o^2 + 1}{a}$

TABLE 24-14
Total weight equations for gear systems

Particular	Equation
Offset	$\Sigma(bd^2/C) = 1 + \dfrac{1}{i} + i + i^2$
Offset with idler	$\Sigma(bd^2/C) = 1 + \dfrac{1}{i} + i + i^2 + \dfrac{i_o^2}{i} + i_o^2$
Offset with two idlers	$\Sigma(bd^2/C) = \dfrac{1}{2} + \dfrac{1}{2i} + i + i^2 + \dfrac{i_o^2}{2i} + \dfrac{i_o^2}{2}$
Double-reduction	$\Sigma(bd^2/C) = 1 + \dfrac{1}{i} + 2i + i^2 + \dfrac{i_o^2}{i_o} + \dfrac{i_o^2}{2i} + i_o^2$
Double-reduction, double branch	$\Sigma(bd^2/C) = \dfrac{1}{2} + \dfrac{1}{2i} + 2i + i^2 + \dfrac{i^2}{i_o} + \dfrac{i_o^2}{2i} + \dfrac{i_o^2}{2}$
Double-reduction, four-branch	$\Sigma(bd^2/C) = \dfrac{1}{4} + \dfrac{1}{4i} + 2i + i^2 + \dfrac{i^2}{i_o} + \dfrac{i_o^2}{4i} + \dfrac{i_o^2}{4}$
Planetary	$\Sigma(bd^2/C) = \dfrac{1}{a} + \dfrac{1}{ai_s} + i_s + i_s^2 + \dfrac{0.4(i_o - 1)^2}{ai_s} + \dfrac{0.4(i_o - 1)^2}{a}$
Star	$\Sigma(bd^2/C) = \dfrac{1}{a} + \dfrac{1}{ai_s} + i_s + i_s^2 + \dfrac{0.4i_o^2}{ai_s} + \dfrac{0.4i_o^2}{a}$

TABLE 24-15
***K* factors for preliminary estimate of spur and helical gear size**

Particular	Hardness H_B; pinion gear	Pitch line velocity, m/s	*K* factor kgf/mm^2	MN/m^2	MPa
Motor driving compressor	225–180	>20.5	0.036	0.353	0.353
Engine driving compressor	225–180	>20.5	0.032–0.050	0.314–0.49	0.314–0.49
	575–575		0.155–0.320	1.52–3.14	1.52–3.14
Turbine driving generator	225–180	>20.5	0.066–0.077	0.65–0.76	0.65–0.76
	575–575		0.280–0.56	2.746–5.50	2.746–5.50
Industrial drives	575–575	5.1	0.350–0.703	3.434–6.89	3.434–6.89
	350–300	5.1	0.246–0.316	2.234–3.100	2.234–3.10
	210–180	5.1	0.120–0.176	1.177–1.726	1.177–1.726
	575–575	15.3	0.334–0.527	3.277–5.170	3.277–5.170
	300–300	15.3	0.193–0.264	1.893–2.589	1.893–2.589
	210–180	15.3	0.088–0.141	0.873–0.138	0.873–0.138
Large industrial gears such as hoists,	225–180	5.1 max	0.056–0.070	0.550–0.687	0.550–0.687
kilns, and mills	260–210		0.091–0.120	0.893–1.177	0.893–1.177
Aircraft, single pair	$60R_C$–$60R_C$	51	0.703 (at take off)	6.89	6.89
Aircraft, planetary	$60R_C$–$60R_C$	15.3–51	0.492 (at take off)	4.82	4.82
Automotive transmission	$60R_C$–$60R_C$		1.055	10.35	10.35
Small commercial	350; phenolic	<5.1	0.52	5.10	5.10
vehicles	laminated nylon		0.035	0.343	0.343
Small gadgets	200; zinc alloy die casting	<5.1	0.018	0.176	0.176
	200; brass or Al	<2.55	0.018	0.176	0.176
	Brass or Al	<2.55	0.016	0.157	0.157

TABLE 24-16
A comparison of five gear systems (all systems producing 0.746 kW at 18 rpm)

Parameter	Epicyclic	Herringbone	Single worm	Helical worm	Harmonic drive
Speed ratio	97.4	96.2	108	100	100
Safety factor	3	2	2	2	36
Height, mm	330	356	580	406	152
Length, mm	381	508	483	432	152
Width, mm	330	254	356	254	152
Cubic volume, m^3	0.0410	0.0458	0.1000	0.0442	0.003
Weight, kgf	111.60	127.00	104.33	93.00	13.61
Efficiency, η%	85	85	40	78	82
Number of gears	13	4	2	4	2
Number of bearings	17	6	6	6	2
Tooth-sliding velocity, m/s	12.75	12.75	7.65	12.75	0.143
Pitch line velocity, m/s	7.65	7.65	7.65	7.65	0.092
Tooth contact pressure, kgf/mm^2	35	35	3.5	35	0.425
Tooth contact pressure, GPa	0.343	0.343	0.034	0.343	0.0042
Tooth in contact, %	7	5	2	3	50
Tooth contact	Line	Line	Line	Line	Surface
Quiet operation	No	Yes	Yes	Yes	Yes
Balanced forces	Yes	No	No	No	Yes

24.6 FRICTION GEARING

SYMBOLS[2,3]

a	center distance, m (in)
	dimensions as shown in Fig. 24-42
b	gear face width, m (in)
d_1	diameter of smaller wheel, m (in)
d_2	diameter of larger wheel, m (in)
F	pressure on wheels, kN (lbf)
F_a	thrust, kN (lbf)
F_r	radial force on the grooved spur wheel for each groove, kN (lbf)
F_R	normal reaction between two bevel friction gears (Fig. 24-40), kN (lbf)
F_t	tangential force, kN (lbf)
h	depth of groove, m (in)
i	number of grooves, m (in)
n'	speed, rps
n	speed, rpm
P	power transmitted, kW (hp)
p'	permissible pressure, kN/m (lbf/in)
v_m	mean circumferential velocity, m/s (ft/min)
R	cone distance, m (in) (Fig. 24-40)
α	half the included angle of the groove, deg ranges from 12° to 18° (should not exceed 20°)
ρ	angle of friction, deg
μ	coefficient of friction between wheels
μ'	coefficient of friction between shaft of wheel and bearings
ω_l, ω_2	angular speeds of smaller and larger wheels, respectively, rad/s
δ_1, δ_2	cone center angles of smaller and larger wheels, respectively, deg

Particular	Formula

SPUR FRICTION GEARS

Plain spur friction wheels (Fig. 24-38)

The radial pressure on the wheels	$F = bp'$	(24-172)
The tangential force due to radial pressure F	$F_t = \mu bp'$	(24-173)
The power transmitted	$P = \dfrac{F_t v_m}{1000}$	**SI** (24-174a)
	where P in kW, F_t in N, and v_m in m/s	
	$P = \dfrac{F_t v_m}{33{,}000}$	**USCS** (24-174b)
	where P in hp, F_t in lbf, and v_m in ft/min	

Particular	Formula

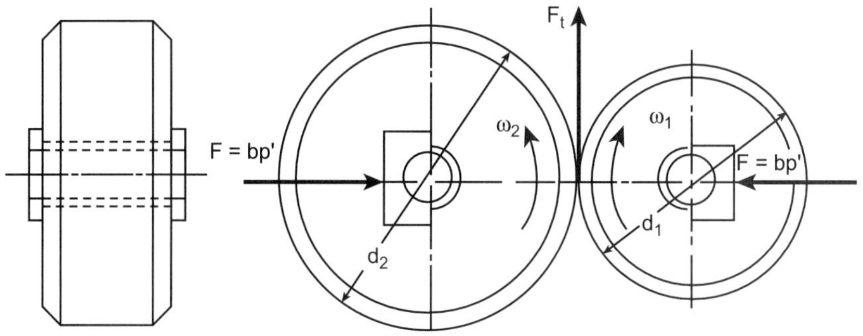

FIGURE 24-38 Plain spur friction gears.

Particular	Formula
	$P = \dfrac{F_t v_m}{75}$ **Customary Metric** (24-174c) where P in hp_m, F_t in kgf, and v_m in m/s
The gear face width	$b = \dfrac{1000P}{\pi \mu p' d n'}$ **SI** (24-175a) where P in kW, p' in N/m, n' in rps, and b and d in m
	$b = \dfrac{102 \times 10^3 P}{\pi \mu p' d n'}$ **Customary Metric** (24-175b) where P in kW, p' in kgf/mm, n' in rps, and b and d in mm
	$b = \dfrac{33{,}000P}{\pi \mu p' d n}$ **USCS** (24-175c) where P in hp, p' in lbf/in, n in rpm, and b in in and d in ft
	$b = \dfrac{126{,}000P}{\mu p' d n}$ **USCS** (24-175d) where b and d in in and p' in lbf/in, n in rpm, and P in hp

Grooved spur friction wheel (Fig. 24-39)

Particular	Formula
The radial force on the wheel for each groove	$F_r = 2p'h(\tan\alpha + \mu)$ (24-176)
The total tangential force	$F_t = 2\mu i p' h \sec\alpha$ (24-177)
The power transmitted	$P = \dfrac{2\pi i \mu h p' d_1 n'}{1000 \cos\alpha}$ **SI** (24-178a) where P in kW, p' in N/m, n' in rps, and h and d_1 in mm

Particular	Formula

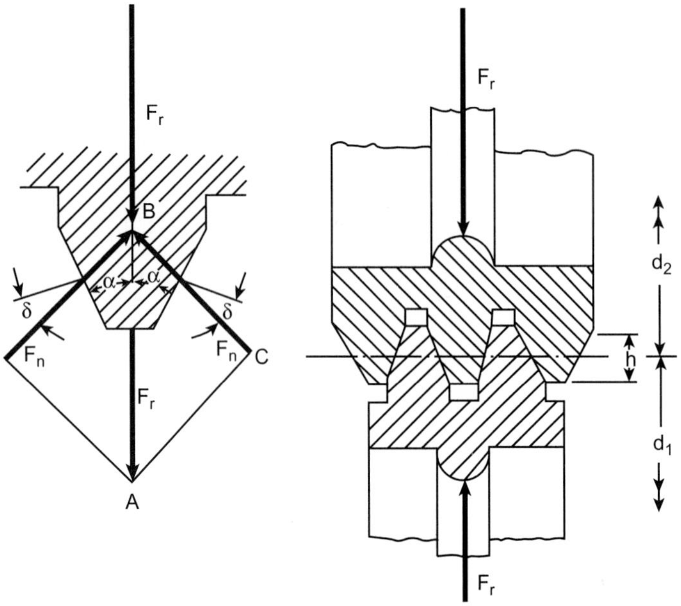

FIGURE 24-39 Grooved spur friction gears.

$$P = \frac{i\mu hp' d_1 n}{63{,}000 \cos\alpha} \qquad \textbf{USCS} \quad (24\text{-}178b)$$

where P in hp, p' in lbf/in, n' in rps, and h and d_1 in in

$$P = \frac{2\pi i\mu hp' d_1 n}{4500 \times 10^3 \cos\alpha} \qquad \textbf{Customary Metric} \quad (24\text{-}178c)$$

where P in hp_m, p' in kgf/mm, n in rpm, and h and d_1 in mm

The empirical relation for the depth of the groove

$$h = 0.006 d_1 + 4\,\text{mm}\ (0.15\,\text{in}) \qquad (24\text{-}179)$$

The recommended value for the mean circumferential velocity

$$v_m \geq 6 + 0.08 d_1 \qquad \textbf{SI} \quad (24\text{-}180a)$$

where v_m in m/s and d_1 in m

$$v_m \geq 1200 + 4 d_1 \qquad \textbf{USCS} \quad (24\text{-}180b)$$

where v in ft/min

BEVEL FRICTION GEARS (Fig. 24-40)

Starting

The reaction is inclined from the normal by an angle of friction ρ

$$F'_R = \frac{F'_{1a}}{\sin(\delta_1 + \rho)} = \frac{F'_{2a}}{\cos(\delta_1 + \rho)} \qquad (24\text{-}181)$$

Particular	Formula
The tangential force transmitted	$F'_t = \mu F'_R \cos\rho = \dfrac{1000P}{v_m}$ **SI** (24-182a)
	$F'_t = \mu F'_R \cos\rho = \dfrac{75P}{v_m}$ **Customary Metric** (24-182b)
The least axial thrust on the small wheel	$F'_{1a} = \dfrac{1000P(\sin\delta_1 + \mu\cos\delta_1)}{\mu v_m}$ **SI** (24-183a)
	$F'_{1a} = \dfrac{33{,}000P(\sin\delta_1 + \mu\cos\delta_1)}{\mu v_m}$ **USCS** (24-183b)
The least axial thrust on the big wheel	$F'_{2a} = \dfrac{1000P(\cos\delta_1 - \mu\sin\delta_1)}{\mu v_m}$ **SI** (24-184a)
	$F'_{2a} = \dfrac{33{,}000P(\cos\delta_1 - \mu\sin\delta_1)}{\mu v_m}$ **USCS** (24-184b)

Running

The reaction in this case is designated by $F_R \leq bp'$ (where p' is the permissible unit pressure)	$F_R = \dfrac{F_{1a}}{\sin\delta_1} = \dfrac{F_{2a}}{\cos\delta_1}$ (24-185)

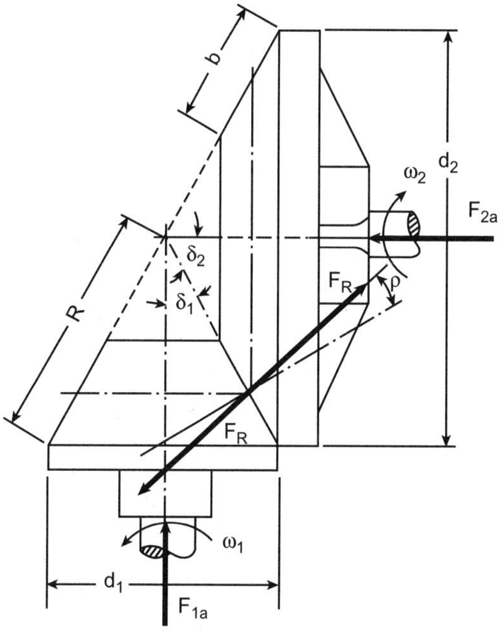

FIGURE 24-40 Bevel friction gears.

Particular	Formula
The tangential force transmitted	$F_t = \mu F_R = \dfrac{1000P}{v_m}$ **SI** (24-186a)
	$F_t = \mu F_R = \dfrac{33{,}000P}{v_m}$ **USCS** (24-186b)
The least axial thrust on the small wheel	$F_{1a} = \dfrac{1000P \sin\delta_1}{\mu v_m} = \dfrac{1000P}{\mu v_m}\left[\dfrac{d_1}{\sqrt{d_1^2 + d_2^2}}\right]$ **SI** (24-187a)
	$F_{1a} = \dfrac{33{,}000P \sin\delta_1}{\mu v_m} = \dfrac{33{,}000}{\mu v_m}\left[\dfrac{d_1}{\sqrt{d_1^2 + d_2^2}}\right]P$ **USCS** (24-187b)
The least axial thrust on the big wheel	$F_{2a} = \dfrac{1000P \cos\delta_1}{\mu v_m} = \dfrac{1000P}{\mu v_m}\left[\dfrac{d_2}{\sqrt{d_1^2 + d_2^2}}\right]$ **SI** (24-188a)
	$F_{2a} = \dfrac{33{,}000P \cos\delta_1}{\mu v_m} = \dfrac{33{,}000}{\mu v_m}\left[\dfrac{d_2}{\sqrt{d_1^2 + d_2^2}}\right]P$ **USCS** (24-188b)

DISK FRICTION GEARS (Fig. 24-41)

The torque on the driving shaft	$M_t = \dfrac{1000P}{\omega}$ **SI** (24-189a) where M_t in N m, P in kW, and ω in rad/s
	$M_t = \dfrac{9550P}{n}$ **SI** (24-189b) where M_t in N m, P in kW, and n in rpm
	$M_t = \dfrac{159P}{n'}$ **SI** (24-189c) where M_t in N m, P in kW, and n' in rps

Particular	Formula

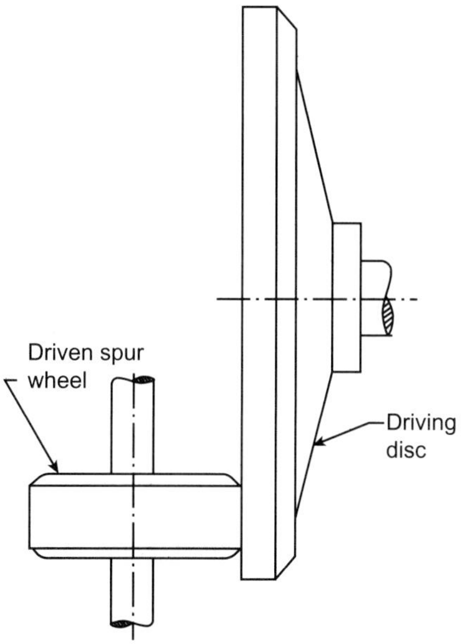

FIGURE 24-41 Variable-speed disk friction gearing.

$$M_t = \frac{716{,}000P}{n} \quad \textbf{Customary Metric} \quad (24\text{-}189d)$$

where M_t in kgf mm, P in hp_m, and n in rpm

$$M_t = \frac{63{,}000P}{n} \quad \textbf{USCS} \quad (24\text{-}189e)$$

where M_t in lbf in, P in hp, and n in rpm

The tangential force acting on the driven wheel for the minimum speed at minimum diameter of driving disk

$$F_{t1} = \frac{1000P}{\pi d_1 n'} \quad \textbf{SI} \quad (24\text{-}190a)$$

where F_{t1} in N, P in kW, d_1 in m, and n' in rps

$$F_{t1} = \frac{33{,}000P}{\pi d_1 n} \quad \textbf{USCS} \quad (24\text{-}190b)$$

where F_{t1} in lbf, P in hp, d_1 in ft, and n in rpm

$$F_{t1} = \frac{102 \times 10^3 P}{\pi d_1 n'} \quad \textbf{Customary Metric} \quad (24\text{-}190c)$$

where F_{t1} in kgf, P in kW, d_1 in mm, and n' in rps

The tangential force acting on the driven wheel for the maximum speed at maximum diameter of driving disk

$$F_{t2} = \frac{1000P}{\pi d_2 n'} \quad \textbf{SI} \quad (24\text{-}191a)$$

Particular	Formula
	$F_{t2} = \dfrac{33{,}000P}{\pi d_2 n}$ **USCS** (24-191b) where d_2 in ft
	$F_{t2} = \dfrac{102 \times 10^3 P}{\pi d_2 n'}$ **Customary Metric** (24-191c)
The minimum thrust to be applied to the disk for the minimum speed	$F_{a1} = \dfrac{F_{t1}}{\mu} = \dfrac{1000P}{\mu\pi d_1 n'}$ **SI** (24-192a)
	$F_{a1} = \dfrac{33{,}000P}{\mu\pi d_1 n}$ **USCS** (24-192b) where $F_{a1} = bp'$ (b = face width of driven cylindrical wheel) and d_2 in ft
The maximum thrust to be applied to the disk for maximum speed	$F_{2a} = \dfrac{F_{t2}}{\mu} = \dfrac{1000P}{\mu\pi d_2 n'}$ **SI** (24-193a)
	$F_{2a} = \dfrac{33{,}000P}{\mu\pi d_2 n}$ **USCS** (24-193b) where d_2 in ft
	$F_{2a} = \dfrac{126{,}000P}{\mu n d_2}$ **USCS** (24-193c) where d_2 in in and F_{2a} in lbf, n in rpm, and P in hp
The axial thrust required to shift the driven wheel underload	$F_a = F_1(\mu + \mu')$ (24-194) where μ' is the coefficient of friction between the shaft of driven wheel and its bearings
The efficiency	$\eta = \dfrac{d}{d+b}$ (24-195) where η varies from 0.6 at low speeds when $d = d_1$ to 0.8 at high speeds, when $d = d_2$
The minimum force available on the chain sprocket at minimum speed of driven wheel	$F_{1cs} = \dfrac{\eta F_{t1} d}{d_3}$ (24-196) where d = diameter of driven wheel, m (in) d_3 = diameter of chain sprocket, m (in)
The maximum force available on the chain sprocket at maximum speed of driven wheel	$F_{2cs} = \dfrac{\eta F_{t2} d}{d_3}$ (24-197)

Particular	Formula

BEARING LOADS OF FRICTION GEARING (Fig. 24-42, Table 24-17)

Driven shaft

The horizontal force on bearing A due to the tangential force F_t

$$F_{hA} = \frac{(L+e)F_t}{(e+L+c)} \tag{24-198}$$

The vertical force on bearing A due to thrust F_a and the force on the chain sprocket F_{cs}

$$F_{VA} = \frac{(L+e)F_a + eF_{cs}}{(e+L+c)} \tag{24-199}$$

The resultant load on bearing A

$$F_{RA} = \sqrt{F_{hA}^2 + F_{VA}^2} \tag{24-200}$$

The horizontal force on bearing B due to the tangential force F_t

$$F_{hB} = \frac{cF_t}{(e+L+c)} \tag{24-201}$$

The vertical force on bearing B due to the thrust F_a and the force on the chain sprocket F_{cs}

$$F_{VB} = \frac{cF_a + (c+L)F_{cs}}{(e+L+c)} \tag{24-202}$$

The resultant force on bearing B

$$F_{RB} = \sqrt{F_{hB}^2 + F_{VB}^2} \tag{24-203}$$

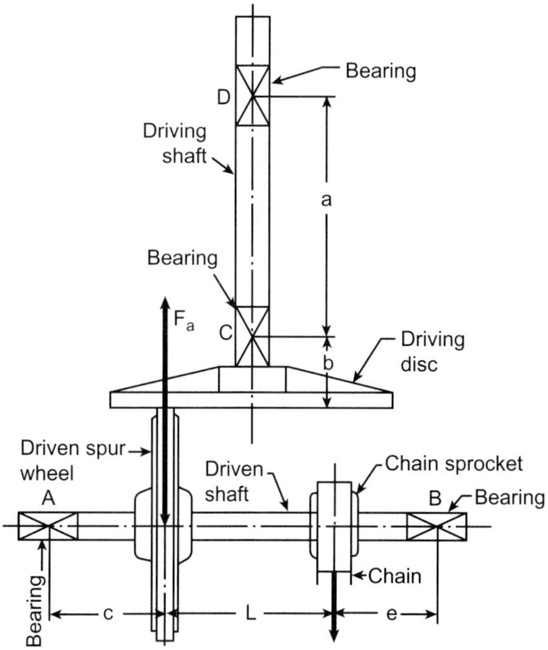

FIGURE 24-42 Bearing loads of disk friction gearing.

Particular	Formula

Driving shaft

The horizontal force due to thrust F_a on bearing D

$$F_{hDa} = \frac{d_1 F_t}{2a} \tag{24-204}$$

where d_1 and d_2 denote the minimum and maximum diameters of driving disk

The horizontal force due to the tangential force F_t on the bearing D

$$F_{hDt} = \frac{bF_t}{a} \tag{24-205}$$

The resultant force on the bearing D

$$F_{RD} = \sqrt{F_{hDa}^2 + F_{hDt}^2} \tag{24-206}$$

The horizontal force due to thrust F_a on the bearing C

$$F_{hca} = \frac{d_1 F_t}{2a} \tag{24-207}$$

The horizontal force due to the tangential force F_t on the bearing C

$$F_{hct} = \frac{(a+b)F_t}{a} \tag{24-208}$$

The resultant force on the bearing C

$$F_{Rc} = \sqrt{F_{hca}^2 + F_{hct}^2} \tag{24-209}$$

TABLE 24-17
Design data for friction gearing

Material of driver	Allowable pressure, p' kN/m	Allowable pressure, p' lbf/in	Coefficient of friction μ with cast iron	Material of driver	Allowable pressure, p' kN/m	Allowable pressure, p' lbf/in	Coefficient of friction μ with cast iron	Coefficient of friction μ with aluminum
Cast iron	530	3000	0.15	Leather	26.5	150	0.09	0.13
Cork composition	8.9	50	0.21	Leather fiber	42.2	240	0.18	0.18
Paper	26.5	150	0.15	Straw fiber	26.5	150	0.15	0.16
Rubber	17.7	100	0.20	Sulfite fiber	24.5	140	0.20	0.19
Wood	26.5	150	0.15	Tarred fiber	44.1	250	0.28	0.28

24.7 MECHANICS OF VEHICLES

SYMBOLS[2,3]

a	center distance, m (in)
	a constant in Eq. (24-216b)
A	frontal projected area of vehicle, m^2 (ft^2)
b	face width of gear, m (in)
	a constant in Eq. (24-216b)
B	width of bearing, m (in)
c	distance between adjacent rotating parts, m (in)
C	constant (also with suffixes)
D_t	maximum diameter of torus, m (in)
D_w	diameter of wheel, m (in)
E_f	flow loss in each member of hydraulic torque converter, N m (lbf in)
E_{sh}	shock loss in each member of hydraulic torque converter, N m (lbf in)
F	driving force at the tire, kN (lbf)
F_{max}	maximum permissible load on the pitch circle of any particular pair of gears, kN (lbf)
G	gradient
h	thickness of housing, m (in)
i	gear ratio (total)
k	a constant
l	distance between support bearings on a shaft in gearbox, m (in)
l'	distance between bearings of overhanging shaft, m (in)
l_1	distance of rotating part from the bearing, m (in)
l_2	distance of bearing from the wall, m (in)
l_3	cap height from bolt to end, m (in)
l_4	distance of rotating parts from the bearing cap, m (in)
l_5	width of boss of rotating parts, m (in)
l_6	distance of coupling to cap, m (in)
l_7	distance between gear and shaft, m (in)
l_8	distance of rotating parts from inner wall of housing, m (in)
m	module, m (in)
M_t	output torque of the engine, N m (lbf in)
M_{tt}	torque at the tire surface, N m (lbf in)
M_{ti}	the input torque, N m (lbf in)
M_{to}	the reaction to the output torque, which is opposite in direction to output torque, N m (lbf in)
M_{tf}	the torque that must be applied to transmission housing to balance the moments of internal friction, oil churning, etc., N m (lbf in)
M_{tr}	the torque reaction of the transmission housing due to the gear reduction in transmission, N m (lbf in)
n	speed, rpm
n'	speed, rps
n_i	speed of driving shaft, rpm
n_o	speed of driven shaft, rpm
P	power, kW (hp)
r	radius of the driving wheel, m (in)
r_{mi}	mean radius of inflow to the runner, m (in)
r_{mo}	mean radius of outflow from the runner, m (in)

R_a	air resistance, kN (lbf)
R_r	rolling resistance, kN (lbf)
R_r''	road resistance, kN/tf (lbf/ton)
R_g	gradient resistance, kN (lbf)
R_t	total resistance, kN (lbf)
t	tonne, t
t_f	tonne force, tf
v	velocity, m/s (ft/min)
V	speed of vehicle, km/h (ft/s)
V_f	velocity of fluid relative to the vane, m/s (ft/min)
V_{sh}	shock velocity, m/s (ft/min)
W	weight of the vehicle, kN (Tonf)
z	number of teeth
α	angle of inclination of road, deg
ϕ	angle of repose, deg
Δ	minimum clearance between gears and inner wall of housing, m (in)
η	transmission efficiency

SUFFIXES

1	pinion
2	gear
b	brake
t	tonne
max	maximum
min	minimum

Other factors in performance or in special aspects which are included from time to time in this section and, being applicable only in their immediate context, are not given at this stage.

Particular	Formula

CALCULATION OF POWER

Torque

$$M_t = \frac{1000P}{\omega} \quad \textbf{SI} \quad (24\text{-}210a)$$

where P in kW, ω in rad/s, and M_t in N m

$$M_t = \frac{9550P}{n} \quad \textbf{SI} \quad (24\text{-}210b)$$

where P in kW, n in rpm, and M_t in N m

$$M_t = \frac{63{,}000P}{n} \quad \textbf{USCS} \quad (24\text{-}210c)$$

where P in hp, n in rpm, and M_t in lbf in

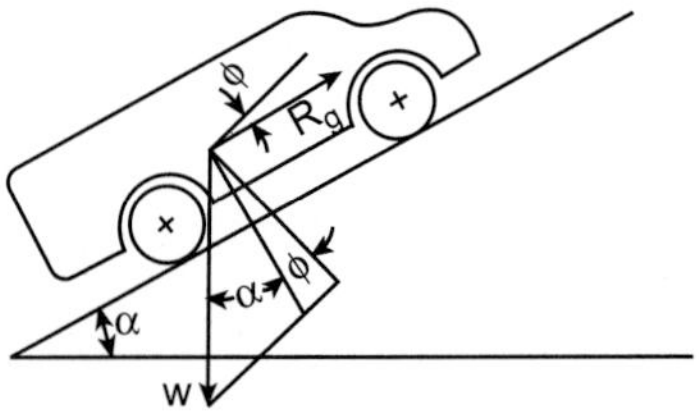

FIGURE 24-43 Forces on the vehicle moving up the gradient.

Particular	Formula
Torque at the tire surface	$M_{tt} = \eta M_t$ (24-211) where $\eta = 0.90$ at top gear $\eta = 0.80$ at other gears
The driving force at the tire	$F = \dfrac{\eta M_t}{r}$ (24-212)
Tractive factor	$f_{tr} = \dfrac{M_{tt}}{1000W}$ **SI** (24-213a) $f_{tr} = \dfrac{M_{tt}}{2240W}$ **USCS** (24-213b)
Force required to pull the vehicle of weight W up the slope (Fig. 24-43)	$R_g = \dfrac{1000W\sin(\alpha+\phi)}{\cos\phi}$ **SI** (24-214a) $R_g = \dfrac{2240W\sin(\alpha+\phi)}{\cos\phi}$ **USCS** (24-214b)
Gradient	$G = \dfrac{W}{R_g}$ (24-215)
The air resistance	$R_a = kAV^2$ (24-216a) where k = constant obtained from Table 24-18
For values of air resistance at different speeds of vehicle	Refer to Table 24-19.
The rolling resistance	$R_r = (a + bV)W$ (24-216b) where a = constant varies from 15 to 600 b = constant varies from 0.1 to 3.5
For rolling or road resistance R_r' for various road surfaces	Refer to Table 24-20.
The general formula for total resistance or tractive resistance (Fig. 24-44)	$R_t = kAV^2 + W\dfrac{\sin(\alpha+\phi)}{\cos\phi} + (a + bV)W$ (24-217a)
Another formula for total resistance	$R_t = R_a + R_g + R_r = W\left(R_r' + \dfrac{1000}{G}\right) + kAV^2$ (24-217b) where k and R_r' are obtained from Tables 24-19 and 24-20 where R_r' in N/tf, W in tf, A in m^2, V in m/s

Particular	Formula
	$R_t = R_a + R_g + R_r$
	$= W\left(R'_r + \frac{2240}{G}\right) + kAV^2$ **USCS** (24-217c)
	where R'_r in lbf/t, W = weight of vehicle, tonf
	A = projected frontal area of vehicle, ft^2
	V = speed of vehicle, ft/s
Tractive effort at the tire surface	$F_{tr} = \frac{i\eta M_t}{r}$ (24-218)
	where i = gear ratio obtained from Table 24-21
The speed of the vehicle	$V = 0.00297\frac{nD_w}{i}$ **USCS** (24-219a)
	where V in mph (miles per hour), D_w in in, and n in rpm
	$V = 0.052\frac{nD_w}{i}$ **SI** (24-219b)
	where V in m/s, D_w in m, and n in rpm
	$P = \frac{0.002VM_{tt}}{D_w}$ **SI** (24-220a)
	where V in m/s, M_{tt} in N m, D_w in m, and P in kW
	$P = \frac{0.00163VM_{tt}}{D_w}$ **USCS** (24-220b)
	where V in mph, M_{tt} in lbf ft, D_w in in, and P in hp
	$P = \frac{5.5VM_{tt}}{D_w}$ **Customary Metric** (24-220c)
	where V in km/h, M_{tt} in kgf m, D_w in mm, and P in kW

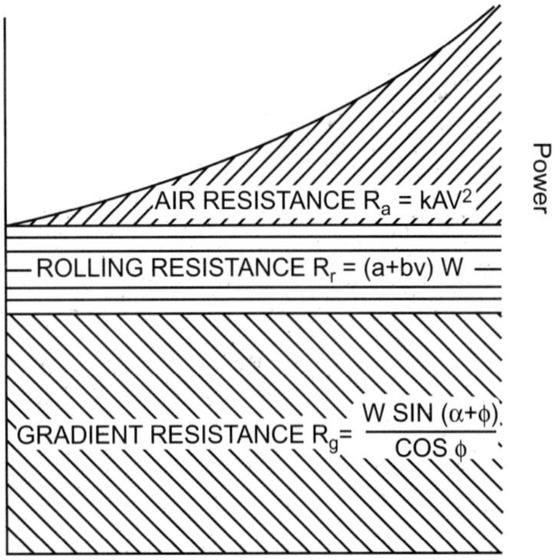

FIGURE 24-44 Various resistances on the moving vehicle.

TRANSMISSION GEARBOX (Fig. 24-45)

Particular	Formula
The equation for center distance between main and countershafts for the case of three-speed passenger car	$a = 0.5\sqrt[3]{M_t}$ **USCS** (24-221a)
	where a in in and M_t in lbf ft
	$a = 0.0106\sqrt[3]{M_t}$ **SI** (24-221b)
	where a in m and M_t in N m

Particular	Formula

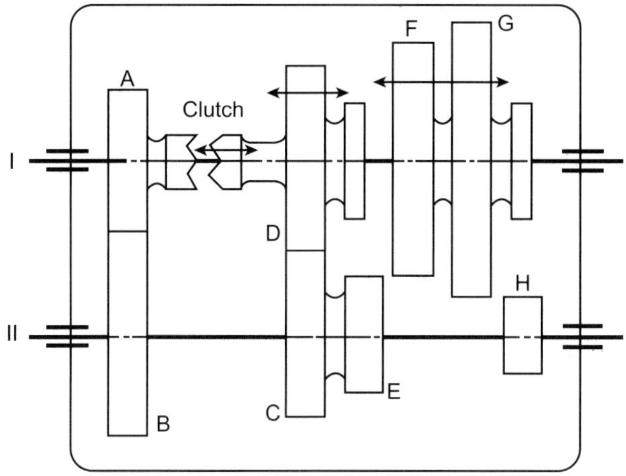

FIGURE 24-45 A typical four-speed gearbox.

Particular	Formula
The distance between support bearings of shaft	$l = 0.0254\sqrt[3]{M_t}$ to $0.0318\sqrt[3]{M_t}$ **SI** (24-222a) where l in m and M_t in N m
	$l = 1.2\sqrt[3]{M_t}$ to $1.5\sqrt[3]{M_t}$ **USCS** (24-222b) where l in in and M_t in lbf ft
The maximum permissible load at the pitch circle of any pair of gears	$F_{\max} = c_1 bm = \frac{c_1 b}{P_m}$ (24-223) where c_1 = constant obtained from Table 24-22
The face width of gear tooth	$b = \frac{F_{\max}}{mc_1} = \frac{F_{\max} P_n}{c_1}$ (24-224)
The expression for center distance for the case of four-speed truck transmission	$a = 0.017\sqrt[3]{M_t}$ **SI** (24-225a) where a in m and M_t in N m
	$a = 0.8\sqrt[3]{M_t}$ **USCS** (24-225b) where a in in and M_t in lbf ft
The distance between support of bearings of shaft	$l = 0.0254\sqrt[3]{M_t}$ to $0.0318\sqrt[3]{M_t}$ **SI** (24-226a) where l in m, and M_t in N m
	$l = 1.2\sqrt[3]{M_t}$ to $1.5\sqrt[3]{M_t}$ **USCS** (24-226b) where l in in and M_t in lbf ft
The face width of gear tooth	$b = \frac{F_{\max}}{mc_1} = \frac{F_{\max} P_n}{c_1}$ (24-227) For values of c_1, refer to Table 24-22.

Particular	Formula
The expression for center distance for the case of five-speed and reverse truck transmission	$a = 0.0170\sqrt[3]{M_t}$ **SI** (24-228a) where a in m and M_t in N m $a = 0.8\sqrt[3]{M_t}$ **USCS** (24-228b) where a in in and M_t in lbf ft
The distance between support of bearings of shaft	$l = 0.0254\sqrt[3]{M_t}$ to $0.0318\sqrt[3]{M_t}$ **SI** (24-229a) where l in m and M_t in N m $l = 1.2\sqrt[3]{M_t}$ to $1.5\sqrt[3]{M_t}$ **USCS** (24-229b) where l in in and M_t in lbf ft
The face width of gear tooth	$b = \dfrac{F_{max}}{mc_1} = \dfrac{F_{max}P_n}{c_1}$ (24-230) For values of c_1, refer to Table 24-22.
The expression for center distance for a farm tractor transmission	$a = 0.021\sqrt[3]{M_t}$ **SI** (24-231a) where a in m and M_t in N m $a = \sqrt[3]{M_t}$ **USCS** (24-231b) where a in in and M_t in lbf ft
Effective face width of gear tooth	$b = \dfrac{F_{max}v}{28 \times 10^5 m}$ **SI** (24-232a) where b in m, F_{max} in N, m in m, and v in m/s $b = \dfrac{F_{max}vP_n}{8{,}000{,}000}$ **USCS** (24-232b) where b in in, F_{max} in lbf, P_n in in^{-1}, and v in ft/min
The efficiency of transmission	$\eta = \dfrac{n_o M_{to}}{n_i M_{ti}} = \dfrac{M_{to}}{i_r[M_{to} - (M_{tr} - M_{tf})]}$ (24-233) where i_r = reduction ratio of transmission = (n_i/n_o)
Distance of rotating parts from the inner wall of housing	l_8 = 10 to 15 mm or more for high-power and heavy-duty operation l_8 = 0.4 to 0.6 in **USCS** (24-234)
Distance between adjacent rotating parts	c = 10 to 15 mm (0.4 to 0.6 in) (24-235)
Minimum clearance between gears and inner wall of housing	$\Delta \geq 1.2h$ (24-236) where h = thickness of housing
Distance between bearings of overhanging shaft	$l' = 1.2d$ to $3d$ (24-237) where d = diameter of shaft
Distance of bearing from the wall	l_2 = 5 to 10 mm (0.2 to 0.4 in) (24-238)
Cap height from bolt end	l_3 = depends on the design by empirical formula
Distance of rotating parts from the bearing cap	l_4 = 15 to 20 mm (0.6 to 0.8 in) (24-239)

Particular	Formula	
Width of boss of rotating part	$l_5 = 1.2d$ to $1.5d$	(24-240)
Distance of coupling to cap	(depends on the type of coupling)	
Distance between gear and shaft	$l_7 \geq 20\,\text{mm}$ (0.8 in)	(24-241)
Distance of rotating part from the bearing	$l_1 = \frac{B}{2} + l_3 + l_4 + \frac{l_5}{2}$	(24-242)
For planetary gear transmission	Refer to Chapter 24, Section 24.5.	
For detail design equations of spur, helical, bevel, crossed-helical and worm gears	Refer to Chapter 23.	

HYDRAULIC COUPLING (Fig. 24-46)

Torque transmitted by the coupling

$$M_t = ksn^2 W(r_{mo}^2 - r_{mi}^2) \quad (24\text{-}243)$$

where k = coefficient = 1.42×10^{-7} (approx.)

Percent slip between primary and secondary speeds

$$s = \frac{(n_p - n_s)}{n_p} \times 100 \quad (24\text{-}244)$$

where n_p and n_s are primary and secondary speeds of impeller, respectively, rpm

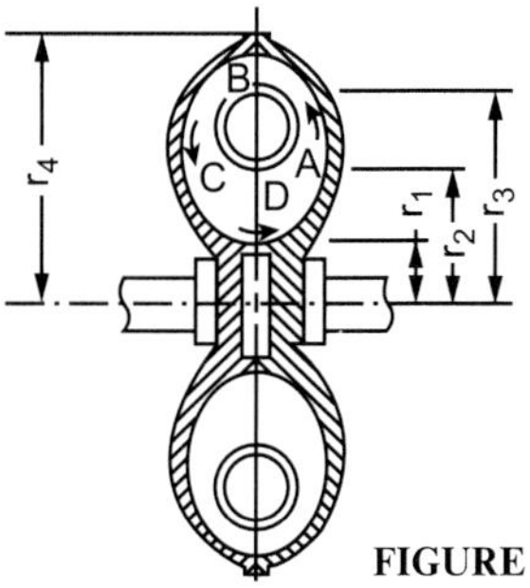

FIGURE 24-46 Hydraulic coupling.

The mean radius of the inner passage (Fig. 24-46)

$$r_{mi} = \frac{2}{3}\left(\frac{r_2^3 - r_1^3}{r_2^2 - r_1^2}\right) \quad (24\text{-}245)$$

The mean radius of the outer passage (Fig. 24-46)

$$r_{mo} = \frac{2}{3}\left(\frac{r_4^3 - r_3^3}{r_4^2 - r_3^2}\right) \quad (24\text{-}246)$$

The expression for number of times the fluid circulates through the torus in one second

$$i_f = \frac{13{,}000 M_t}{nW(r_{mo}^2 - r_{mi}^2)} \quad (24\text{-}247)$$

The torque capacity of hydraulic coupling at a given slip

$$M_t = Kn^2 D_t^5 \quad (24\text{-}248)$$

where

K = coefficient varying from 0.166×10^8 to 0.244×10^8 **SI**

= 1.56 to 2.28 **USCS**

D_t = diameter of torus, m (ft)

M_t = torque capacity, N m (lbf ft); n in rpm

Particular	Formula

HYDRODYNAMIC TORQUE CONVERTER (Fig. 24-47)

The equation for input torque

$$M_{ti} = Kn_i^2 D_t^5 \tag{24-249}$$

where

K = coefficient depending on design

n_i = speed of input shaft, rpm

D_t = any linear dimension such as maximum diameter of impeller

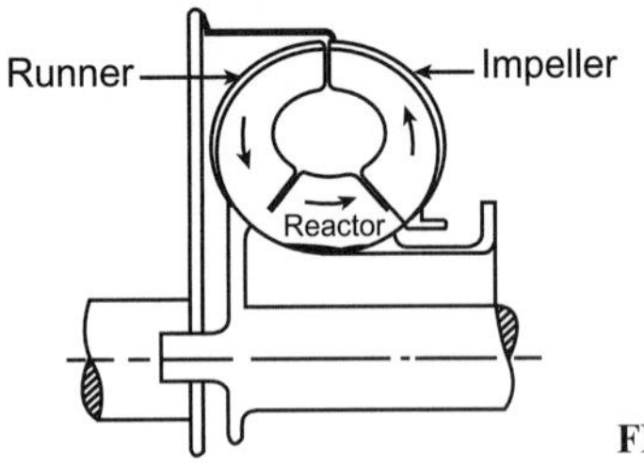

FIGURE 24-47 Hydrodynamic torque converter.

The equation for the input power

$$P = Cn_i^3 D_t^5 \tag{24-250}$$

where C = coefficient depending on design

The expression for flow loss or friction loss in each member of the torque converter under any particular operating conditions in energy unit per kilogram of fluid circulated

$$E_f = \frac{C_f V_f^2}{2g} \tag{24-251}$$

where

C_f = coefficient whose value depends mainly on the Reynolds number and the relative smoothness of the metallic surface

= 0.445 to 0.890 **SI**
(where E_f in N m and V_f in m/s)

= 0.328 to 0.656 **USCS**
(where V_f in ft/s and E_f in lbf ft)

The expression for shock loss per kg fluid circulated in the impeller of a torque converter

$$E_{sh} = \frac{C_{sh} V_{sh}^2}{2g} \tag{24-252}$$

where C_{sh} = coefficient

The maximum inside diameter of torus

$$D_t = 0.00135C\sqrt[3]{M_t/n'^2} \quad \textbf{SI} \tag{24-253a}$$

where D_t in m, M_t in N m, and n' in rps

$$D_t = 0.00168C\sqrt[3]{\frac{M_t}{n^2}} \quad \textbf{USCS} \tag{24-253b}$$

where D_t in in, M_t in lbf in, and n in rpm

C = coefficient = 14 for a ratio of minimum inside diameter to maximum diameter of torus of one-third

n = speed in hundreds of rpm

Particular	Formula

TRACTIVE EFFORT CURVES FOR CARS, TRUCKS, AND CITY BUSES

For finding the diameter of tire of vehicles for a particular wheel speed	Refer to Fig. 24-48.
For tractive effort of a passenger car	Refer to Fig. 24-49.
For tractive effort of trucks, tractors, and city buses	Refer to Fig. 24-50.

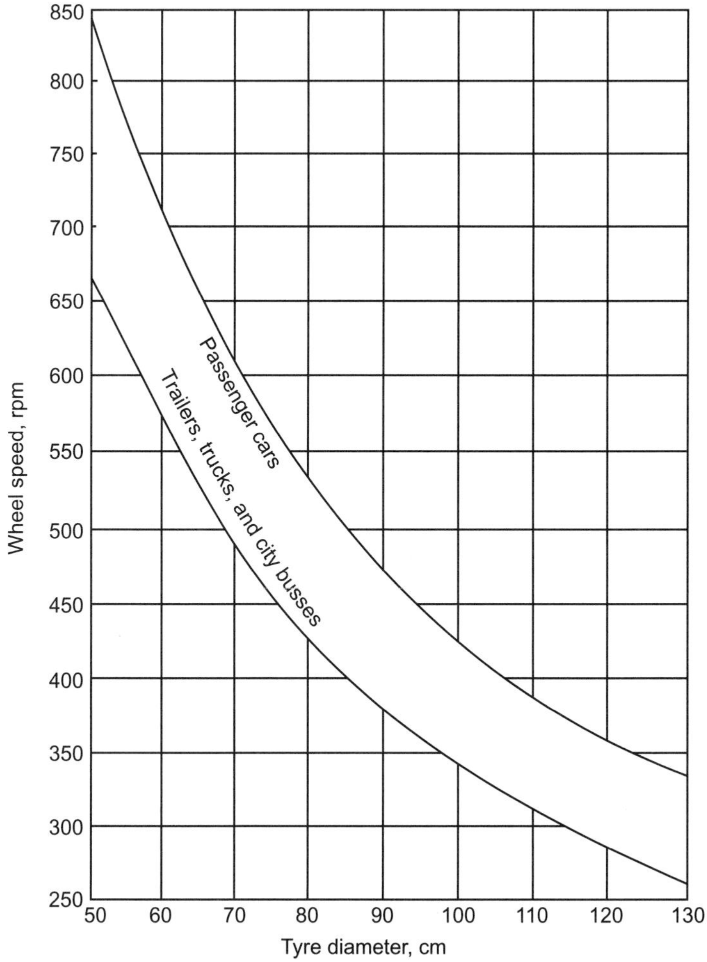

FIGURE 24-48 Wheel speed vs. tire diameter of vehicles.

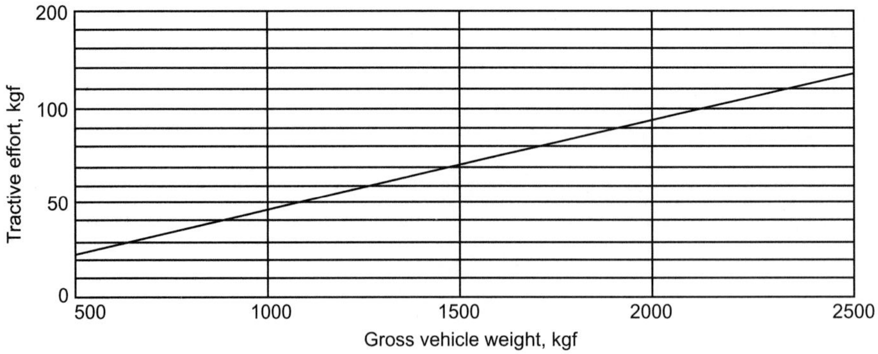

FIGURE 24-49 Tractive effort curve for passenger cars (1 kgf = 9.8066 N = 2.2046 lbf).

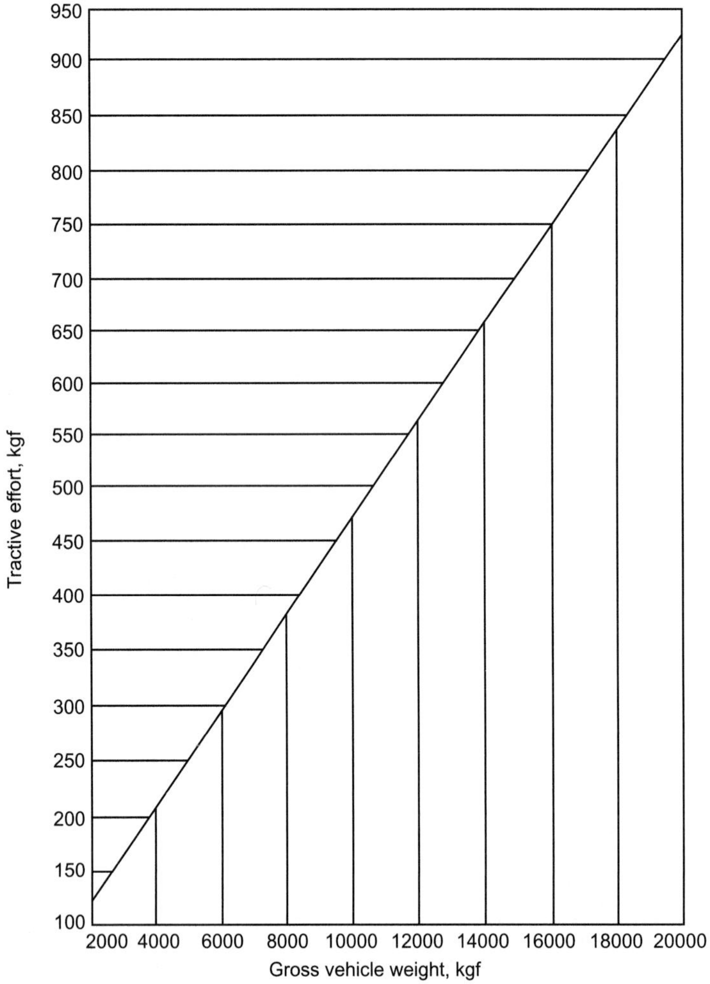

FIGURE 24-50 Tractive effort curve for trucks, tractors, and city buses.

TABLE 24-18
Values of k for use in Eq. (24-216) and (24-217)

Particular	k in USCSU[a]	k in SI
Average automobile of modern design	0.0017	0.20
Streamlined racing car	0.0006	0.07
Truck or omnibus	0.0024	0.28

[a] US Customary System units.

TABLE 24-19A
Air resistance[a]

Speed of vehicle, mph	Velocity of wind, V, ft/s	$0.0024V^2$	$0.0017V^2$	$0.0006V^2$
10	14.67	0.516	0.366	—
20	29.35	2.060	1.460	0.516
30	44.00	4.650	3.300	1.160
40	58.60	8.240	5.830	2.060
50	73.30	12.900	9.130	3.220
60	88.00	18.600	13.160	4.650
90	132.00	—	29.650	10.450
150	220.00	—	—	29.000

[a] Values given in this table are in US Customary System units.

TABLE 24-19B
Air resistance in SI units

Speed of vehicle km/h	Velocity of wind V, m/h	$0.28V^2$	$0.20V^2$	$0.07V^2$
10	2.78	2.17	1.55	0.54
20	5.56	8.68	6.20	2.17
30	8.34	19.50	13.92	4.88
40	11.12	34.72	24.80	8.68
50	13.90	54.25	38.80	13.56
60	16.68	78.00	57.60	19.50
70	19.46	106.40	76.00	26.60
80	22.21	139.00	99.20	34.75
90	25.02	176.00	125.60	44.00
100	27.80	217.00	155.00	54.25

TABLE 24-20
Road resistance, R_r'

	Solid			Pneumatic		
Surface	N/tf	lbf/ton	%	N/tf	lbf/ton	%
Polished marble	29.3	12.12	0.541	35.3	8.08	0.360
Concrete	62.4	14.25	0.636	41.5	9.50	0.423
Asphalt	67.4	15.40	0.687	448.2	10.25	0.457
Stone						
Good quality	71.8	16.40	0.732	477.6	10.92	0.487
Poor quality	153.2	35.00	1.562	102.0	23.30	1.040
Vitrified bricks	85.0	19.45	0.866	56.7	12.95	0.578
Good macadam (metal road)						
Good	146.2	33.50	1.491	97.9	22.20	0.998
Fair	220.0	50.00	2.240	186.4	33.20	1.900
Rough	307.0	70.00	3.130	389.3	46.60	2.100
Clay	438.4	100.00	4.470	389.3	66.60	3.970
Sand	1314.0	300.00	13.400	874.8	200.00	8.920

TABLE 24-21
Gear ratios

Particular	Ratio
Final drive	4 : 1 to 5 : 1
(rear-axle differential)	1.6 : 1
Second	(total 6 or 7 : 1)
Low	2.5 : 1
	(total 10 : 1)
Reverse gear	Same as or higher than low gear ratio
Overdrive	<75% above the propeller shaft

TABLE 24-22
Value of coefficient c_1 in SI and USCSU

No. of speeds and type of transmission	Gear wheels belonging to									
	High speed reduction		Intermediate speed reduction						Low speed reduction	
	I		II		III		IV		V	
	SI × 10^6	USCSU	SI × 10^6	USCSU	SI × 10^6	USCSU	SI × 10^6	USCSU	SI × 10^6	USCSU
Three-speed passenger car transmission	124.6–145.2	18,000–21,000	145.2–153.5	21,000–24,000					193.7–221.2	28,000–32,000
Four-speed truck transmission	76.0–90.0	11,000–13,000	128.0–149.0	18,500–21,500	96.8–110.9	14,000–16,000			175.2–207.5	26,000–30,000
Five-speed reverse truck transmission	76.0–90.0	11,000–13,000	138.2–152.0	20,000–22,000	105.0–117.7	15,000–17,000	90.0–105.0	13,000–15,000	175.2–207.5	26,000–30,000

24.8 INTERMITTENT-MOTION MECHANISMS

SYMBOLS[2,3]

a	distance of the pawl pivot point, m (in)
b	face width of ratchet tooth, m (in)
d	diameter (also with suffixes), m (in)
d_h	hub diameter, m (in)
e_1, e_2	dimensions as shown in Fig. 24-52, m (in)
F_n	normal force through O, Figs. 24-51 and 24-52, kN (lbf)
F_{nr}	peripheral force normal to the tooth of ratchet, kN (lbf)
F_t	tangential force at diameter, d, kN (lbf)
h	tooth height or distance from the critical section to the line of action of the load F_{nr}, m (in)
m	module, m (in)
M_b	bending moment, N m (lbf in)
M_t	twisting moment, N m (lbf in)
n'	speed, rps
n	speed, rpm
p	tooth pitch, m (in)
p	linear unit pressure, N/m (lbf/ft)
r	radius, m (in)
s_1	dimension as shown in Fig. 24-53
s_2	breadth of tooth land (Fig. 24-53), m (in)
s_2'	thickness of tooth at base (Fig. 24-52), m (in)
z	number of teeth on ratchet wheel
Z	section modulus, m^3 (cm^3) (in^3)
α	pressure angle or angle of the pawl force, deg
β	angle at pawl ($= 90^\circ - \alpha^\circ$), deg
$\mu = \tan\rho =$	coefficient of friction
$\rho = \tan^{-1}\mu =$	friction angle, deg
φ	ratchet tooth angle or pitch angle, deg
$\psi = \dfrac{b}{m}$	varies from 1.5 to 3
σ	stress, MPa (psi)
σ_b	bending stress, MPa (psi)
τ	shear stress, MPa (psi)

Particular	Formula

PAWL AND RATCHET

The ratchet tooth angle (Fig. 24-51)

$$\varphi = \frac{2\pi}{z}, \text{ rad} \tag{24-254a}$$

$$\varphi = \frac{360}{z}, \text{ deg} \tag{24-254b}$$

The ratchet diameter (Fig. 24-51)

$$d_2 = mz = \frac{pz}{\pi} \tag{24-255}$$

Particular	Formula	
The face width of ratchet tooth	$b \geq \frac{F_n}{F_n^*}$	(24-256)
The allowable unit pressure or force	$F^* = \frac{F_n}{b}$	(24-257)
The tangential force	$F_t = \frac{2M_t}{d}$	(24-258)
The normal force through O (Fig. 24-51)	$F_n = \frac{F_t}{\cos\alpha}$	(24-259a)
The normal force through O (Fig. 24-52)	$F_n = F_t$	(24-259b)
The bending stress	$\sigma_b = \frac{M_b}{Z} = \frac{6F_t h}{bs_1^2} \leq \sigma_{ba}$	(24-260)
Allowable bending stress (σ_b)	Refer to Table 24-23.	
Number of teeth	$z = 6$ to 30	(24-261)
Module	$m > 6$ (mostly from 10 to 20)	(24-262)
The ratio of h/m	$h/m = 0.6$ to 1	(24-263)
For ratchet wheel definitions and dimensions	Refer to Fig. 24-53.	
The ratio of s_2/m	$s_2/m = 0.6$ to 0.9	(24-264)
The tooth height	$h = 5$ to 15 for toothed ratchet	(24-265)

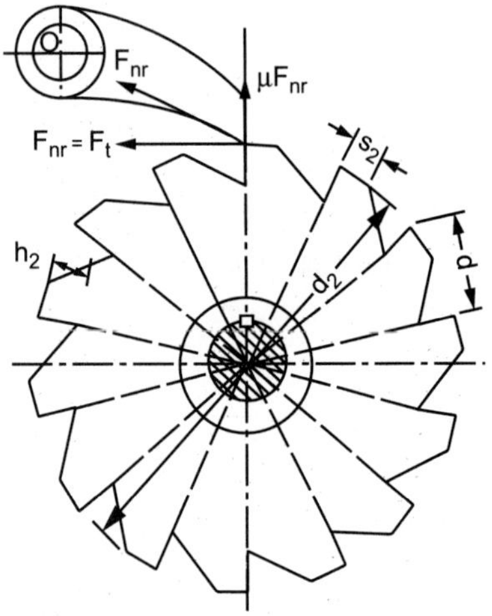

FIGURE 24-51 Ratchet wheel with radial tooth flanks and pawl.

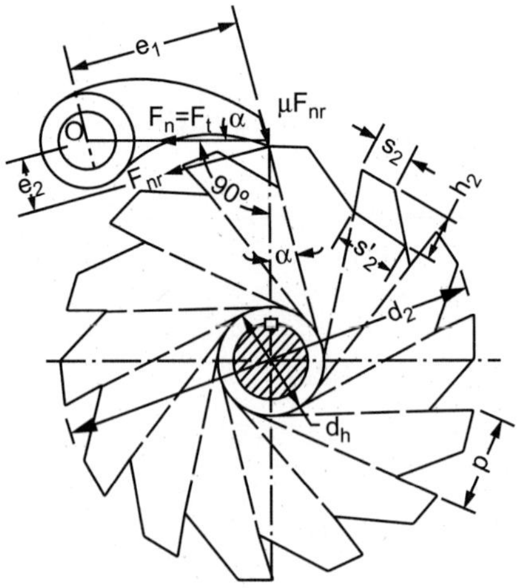

FIGURE 24-52 Ratchet wheel with non-radial tooth flanks and pawl.

Particular	Formula

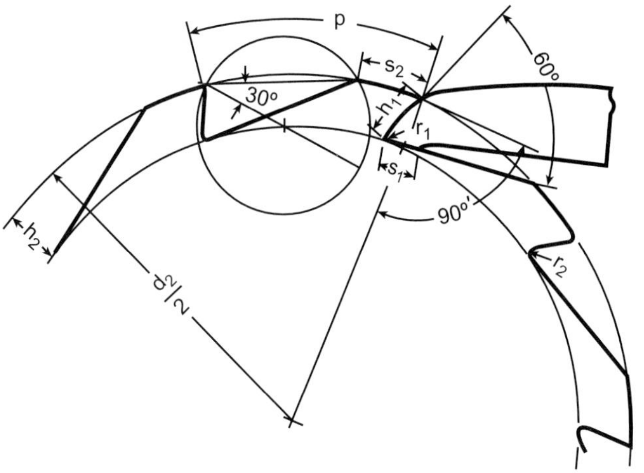

FIGURE 24-53 Definitions and dimensions of ratchet wheel.

For external ratchet $\quad \alpha = 14^\circ$ to 17° (24-266)

For internal ratchet $\quad \alpha = 17^\circ$ to 30° (24-267)

The ratio of a/d (internal ratchet) $\quad a/d = 0.35$ to 0.43 (24-268)

The module

$$m = 2\sqrt[3]{\frac{M_t}{z\psi\sigma_{ba}}} \tag{24-269}$$

The bending moment on pawl

$$M_{b1} = F_n e_2 \tag{24-270}$$

The bending stress

$$\sigma_b = \frac{6M_{b1}}{bs_1^2} + \frac{F_n}{bs_1} \le \sigma_{ba} \tag{24-271}$$

The diameter of pawl pin

$$d_1 = 2.71\sqrt[3]{\frac{F_n}{2\sigma_{ba}}\left(\frac{b}{2} + t_h\right)} \tag{24-272}$$

where t_h = thickness of hub on pawl

TABLE 24-23

	F^*		σ_b	
Material	**kN/m**	**lbf/in**	**MPa**	**kpsi**
Cast iron	49–98	280–560	19.5–29.5	2.85–4.27
Steel or cast steel	98–196	560–1120	39–68.5	5.69–10.0
Hardened steel	196–392	1120–2240	58.8–98	8.54–14.23

24.9 GENEVA MECHANISM

SYMBOLS[2,3]

$a = \dfrac{r_1}{\sin\phi}$	center distance, m (in)
F_1	the component of force acting on the crank or the driving shaft due to the torque, M_{1t}, kN (lbf) (Fig. 24-57)
F_2	the component of force acting on the driven Geneva wheel shaft due to the torque M_{2t}, kN (lbf) (Fig. 24-57)
$F_{2(max)}$	maximum force (pressure) at the point of contact between the roller pin and slotted Geneva wheel, kN (lbf)
$F_{\mu(max)}$	the component of maximum friction force at the point of contact due to the friction torque $M_{2t\mu}$, on the driven Geneva wheel shaft, kN (lbf)
$F_{i(max)}$	the component of maximum inertia force at the point of contact due to the inertia torque on the driven Geneva wheel shaft, kN (lbf)
$i = \dfrac{z-2}{z}$	gear ratio
J	polar moment of inertia of all the masses of parts attached to Geneva wheel shaft, m^4 (in^4)
k	the working time coefficient of the Geneva wheel
M_{1t}	total torque on the driver or crank, N m (lbf in)
M_{2t}	total torque on the driven or Geneva wheel, N m (lbf in)
M_{2ti}	inertia torque on the Geneva wheel, N m (lbf in)
$M_{2t\mu}$	friction or resistance torque on Geneva wheel, N m (lbf in)
n'	speed, rps
n	speed, rpm
P	power, kW (hp)
r_1	radius to center of driving pin, m (in)
r_2	radius of Geneva wheel, m (in)
r_2'	distance of center of semicircular end of slot from the center of Geneva wheel, m (in)
r_{a2}	outside radius of Geneva wheel, which includes correction for finite pin diameter, m (in)
r_p	pin radius, m (in)
$R_r = \dfrac{r_2}{r_1}$	radius ratio
t	total time required for a full revolution of the driver or crank, s
t_i	time required for indexing Geneva wheel, s
t_r	time during which Geneva wheel is at rest, s
v	velocity, m/s
z	number of slots on the Geneva wheel
α	crank angle or angle of driver at any instant, deg (Fig. 24-54)
α_{2a}	angular acceleration, m/s^2 (ft/s^2) angular acceleration of Geneva wheel, m/s^2 (ft/s^2)
α_m	angular position of the crank or driver radius at which the product $\omega\alpha_{2a}$ is maximum, deg
β	angle of the driven wheel or Geneva wheel at any instant, deg (Fig. 24-54)
$\gamma = \dfrac{r_1}{a}$	the ratio of the driver radius to center distance
η	efficiency of Geneva mechanism

λ	locking angle of driver or crank, rad or deg
ν	ratio of time of motion of Geneva wheel to time for one revolution of driver or crank
$\phi = \dfrac{360}{2z}$	semi-indexing or Geneva wheel angle, or half the angle subtended by an adjacent slot, deg (Fig. 24-54)
ψ	crank or driver angle, deg (Fig. 24-54)
$\omega = \dfrac{2\pi n}{60}$	angular velocity of driver or crank (assumed constant), rad/s
ω_1, ω_2	angular velocities of driver or crank and Geneva wheel, respectively, rad/s

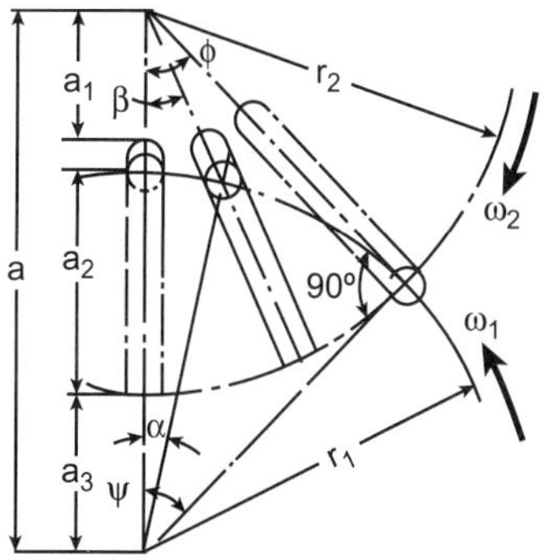

FIGURE 24-54 Design of Geneva mechanism.

Particular	Formula	
The angular velocity (constant) of driver or crank	$\omega_1 = \dfrac{2\pi n}{60}$	(24-273)
Gear ratio	$i = \dfrac{\text{angle moved by crank or driver during rotation}}{\text{angle moved by Geneva wheel during rotation}}$	
	$i = \dfrac{z-2}{z}$	(24-274)
The semi-indexing angle or Geneva wheel angle or half the angle subtended by two adjacent slots	$\phi = \dfrac{360}{2z}$ or $\dfrac{\pi}{z}$	(24-275)
The angle through which the Geneva wheel rotates	$2\phi = \dfrac{360}{z}$ or $\dfrac{2\pi}{z}$	(24-276)

EXTERNAL GENEVA WHEEL

The angle of rotation of driver through which the Geneva wheel is at rest or angle of locking action (Fig. 24-55)	$\lambda = 2(\pi - \psi) = \pi + 2\phi = \dfrac{\pi}{z}(z+2)$	(24-277)
The crank or driver angle	$\psi = \dfrac{\pi}{2} - \phi = \dfrac{\pi(z-2)}{2z}$	(24-278)

Particular	Formula

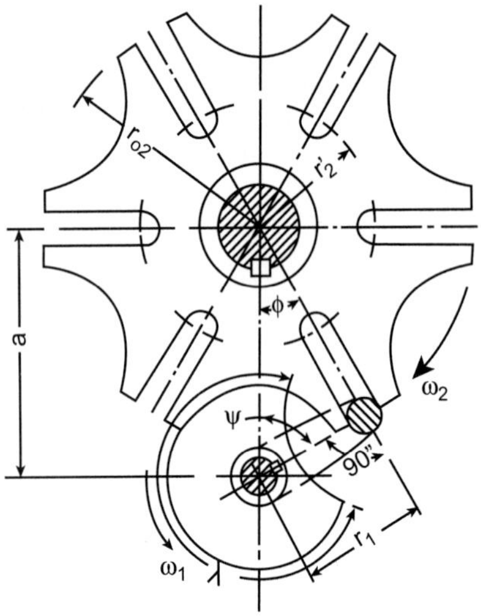

FIGURE 24-55 External Geneva mechanism.

DISPLACEMENT

The center distance (Fig. 24-55)	$a = \frac{r_1}{\sin\phi}$	(24-279)
The radius ratio	$R_r = \frac{r_2}{r_1} = \cot\phi$	(24-280)
The ratio of crank radius to center distance	$\gamma = \frac{r_1}{a} = \sin\phi = \sin\frac{\pi}{z}$	(24-281)
The relation between crank angle and Geneva wheel angle	$\beta = \tan^{-1}\left(\frac{\gamma\sin\alpha}{1-\gamma\cos\alpha}\right)$	(24-282)

VELOCITY

The angular velocity of the Geneva wheel	$\omega_2 = \frac{d\beta}{dt} = \frac{\gamma(\cos\alpha-\gamma)}{1-2\gamma\cos\alpha+\gamma^2}\omega_1$	(24-283a)
	$\omega_2 = \frac{\sin(\pi/z)(\cos\alpha - \sin\pi/z)}{1-2\sin(\pi/z)\cos\alpha+\sin^2\pi/z}\omega_1$	(24-283b)
The maximum angular velocity of Geneva wheel at angle $\alpha = 0$	$\omega_{2(\max)} = \left(\frac{d\beta}{dt}\right)_{\max} = \frac{\gamma}{1-\gamma}\omega_1$	
	$= \left[\sin\frac{\pi}{z} \Big/ \left(1-\sin\frac{\pi}{z}\right)\right]\omega_1$	(24-283c)

Particular	Formula

ACCELERATION

The angular acceleration, ${}^{a}\alpha_{2a}$, of Geneva wheel

$$ {}^{a}\alpha_{2a} = \frac{d^2\beta}{dt^2} = \frac{(\gamma^3 - \gamma)\sin\alpha}{(1 + \gamma^2 - 2\gamma\cos\alpha)^2}\omega_1^2 \quad (24\text{-}284a)$$

$$ {}^{a}\alpha_{2a} = \pm\frac{\sin(\pi/z)\cos^2(\pi/z)\sin\alpha}{1 - 2\sin(\pi/z)\cos\alpha + \sin^2(\pi/z)}\omega_1^2 \quad (24\text{-}284b)$$

For angular velocity and angular acceleration curves for three-slot external Geneva wheel with driver velocity, $\omega_1 = 1\ \text{rad/s}$

Refer to Fig. 24-56.

The maximum angular acceleration of Geneva wheel which occurs at $\alpha = \alpha_{(\max)}$

$$\cos\alpha_{(\max)} = -\kappa + \sqrt{\kappa^2 + 2} \quad (24\text{-}284c)$$

where

$$\kappa = \frac{1}{4}\left(\gamma + \frac{1}{\gamma}\right)$$

The angular acceleration of Geneva wheel at start and finish of indexing

$$(\alpha_{2a})_{i,f} = \pm\frac{\sin(\pi/z)\cos^3(\pi/z)}{[1 - 2\sin^2(\pi/z) + \sin^2(\pi/z)]}\omega_1^2$$

$$= \pm\omega_1^2\tan\phi = \pm\omega_1^2\tan\pi/z$$

$$= \pm\omega_1^2\left(\frac{r_1}{r_2}\right) \quad (24\text{-}285)$$

Total time required for a full revolution of the crank or driver

$$t = \frac{60}{n} \quad (24\text{-}286)$$

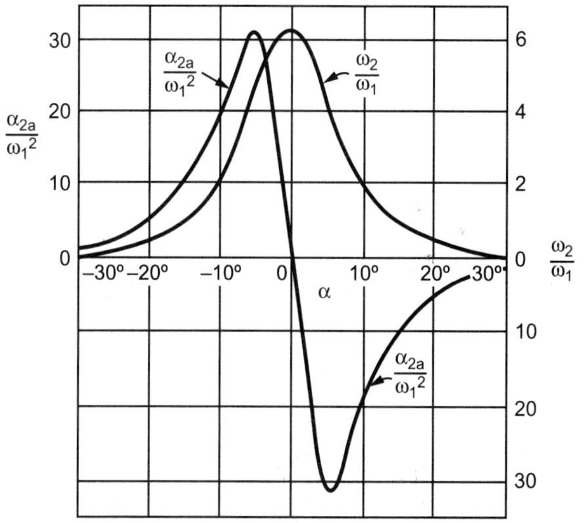

FIGURE 24-56 Angular velocity and angular acceleration curves for three-slot external Geneva wheel.

---

[a] α_{2a} is the symbol used for angular acceleration of Geneva wheel; α is the crank or driver angle at any given instant.

Particular	Formula	
The ratio of t_i to t	$\frac{t_i}{t} = \frac{2\psi}{2\pi} = \frac{\psi}{\pi} = \frac{z-2}{2z}$	(24-287)
The ratio of t_r to t	$\frac{t_r}{t} = \frac{2(\pi - \psi)}{2\pi} = 1 - \frac{\psi}{\pi} = \frac{z+2}{2z}$	(24-288)
The sum of angles of $(\phi + \psi)$	$\phi + \psi = 90°$	(24-289)
The time required for indexing Geneva wheel, in seconds	$t_i = \frac{z-2}{2z} t = \frac{z-2}{z}\left(\frac{60}{2n}\right)$	(24-290)
The time during which Geneva wheel is at rest, in seconds	$t_r = \frac{z+2}{z}\left(\frac{60}{2n}\right)$	(24-291)
The working time coefficient of Geneva wheel	$k = \frac{t_i}{t_r} = \frac{z-2}{z+2}$	(24-292)
Ratio of time of motion of Geneva wheel to time for one revolution of crank or driver	$v = \frac{z-2}{2z}\left(< \frac{1}{2}\right)$	(24-293)
The required speed of the driver shaft or crankshaft	$n = \frac{z+2}{z}\left(\frac{60}{2t_r}\right)$ where n in rpm	(24-294)

SHOCK OR JERK

Particular	Formula	
The jerk or shock, J_2, on Geneva wheel	$J_2 = \frac{\gamma(\gamma - 1)[2\gamma\cos^2\alpha + (1+\gamma^2)\cos\alpha - 4\gamma]}{(1+\gamma^2 - 2\gamma\cos\alpha)^3} + \omega_1^3$	(24-295)
The jerk or shock at $\alpha = 0$	$(J_2)_{\alpha=0} = \left(\frac{d^3\beta}{dt^3}\right)_{\alpha=0} = \frac{\gamma(\gamma+1)}{(\gamma-1)^3}\omega_1^3$	(24-296)
The jerk or shock at start, i.e., $\beta = \phi$	$(J_2)_{\beta=\phi} = \left(\frac{d^3\beta}{dt^3}\right)_{\beta=\phi} = \left(\frac{3\gamma^2}{1-\gamma^2}\right)\omega_1^3$	(24-297)
The length of the slot (Fig. 24-54)	$a_2 = r_1 + r_2 - a = a\left(\sin\frac{\pi}{z} + \cos\frac{\pi}{z} - 1\right)$	(24-298)
The condition to be satisfied by diameter on which the driver or crank is mounted	$d_1 < 2a_3 = 2(a - r_2) = 2a\left(1 - \cos\frac{\pi}{z}\right)$	(24-299)
	or	
	$\frac{d_1}{a} < 2\left(1 - \cos\frac{\pi}{z}\right) = 4\sin^2\frac{\pi}{2z}$	(24-300)
The condition to be satisfied by the diameter on which Geneva wheel is mounted	$\frac{d_2}{a} < 2\left(1 - \sin\frac{\pi}{z}\right) = 4\sin^2\left(\frac{\pi}{4} - \frac{\pi}{2z}\right)$	(24-301)

Particular	Formula

TORQUE ACTING ON SHAFTS OF GENEVA WHEEL AND DRIVER

The total torque acting on Geneva wheel shaft

$$M_{2t} = M_{2t\mu} + M_{2ti} = M_{2t\mu} + J\alpha_{2a} \quad (24\text{-}302)$$

It is assumed that $M_{2t\mu}$ is constant.

The torque on the shaft of crank or driver

$$M_{1t} = M_{2t}\frac{\omega_2}{\omega_1}\frac{1}{\eta} = (M_{2t\mu} + J\alpha_{2a})\frac{\omega_2}{\omega_1}\frac{1}{\eta} \quad (24\text{-}303)$$

The efficiency of Geneva mechanism

$\eta = 0.80$ to 0.90 when Geneva wheel shaft is mounted on journal bearings (24-304a)

$\eta = 0.95$ when drive shaft is mounted on rolling contact bearings (24-304b)

$\eta = 0.75$ when the diameter of bearing surface is larger than the outside diameter of Geneva wheel (24-304c)

INSTANTANEOUS POWER

The instantaneous power on the crank or driving shaft

$$P = \frac{M_t\omega}{1000} \quad \textbf{SI} \quad (24\text{-}305a)$$

where P in kW, M_t in N m, and ω in rad/s

$$P = \frac{M_t\omega}{102 \times 10^3} \quad \textbf{Customary Metric} \quad (24\text{-}305b)$$

where P in kW, M_t in kgf mm, and ω in rad/s

$$P = \frac{M_t\omega}{75 \times 10^3} \quad \textbf{Customary Metric} \quad (24\text{-}305c)$$

where P in hp_m, M_t in kgf mm, and ω in rad/s

$$P = \frac{M_t n}{63{,}000} \quad \textbf{USCS} \quad (24\text{-}305d)$$

where P in hp, M_t in lbf in, and n in rpm

Calculation of average power

The average torque $M_{ti(av)}$ for complete cycle

$$M_{ti(av)} = 0 \quad (24\text{-}306)$$

The average torque for first half-cycle

$$M_{t(av)} = M_{\mu(av)} = \frac{2}{z-2}\left[M_{2t\mu} + \frac{zJ}{2\pi}\left(\frac{\gamma}{1-\gamma}\right)^2\omega_1^2\right]\frac{1}{\eta} \quad (24\text{-}307)$$

where J = polar moment of inertia, m^4, cm^4 (in^4)

Particular	Formula
The average power required on the crank or driving shaft	$P_{av} = \dfrac{M_{t(av)}}{1000}\omega$ **SI** (24-308a) where P_{av} in kW, $M_{t(av)}$ in N m, and ω in rad/s
	$P_{av} = \dfrac{M_{t(av)}\omega}{75 \times 10^3}$ **Customary Metric** (24-308b) where P_{av} in hp_m, $M_{t(av)}$ in kgf mm, and ω in rad/s
	$P_{av} = \dfrac{M_{t(av)}n}{63{,}000}$ **USCS** (24-308c) where P_{av} in hp, $M_{t(av)}$ in lbf in, and n in rpm

Calculation of maximum power

Particular	Formula
The maximum torque on the driven shaft of Geneva wheel	$M_{2t(\max)} = M_{2t\mu} + M_{2ti(\max)}$ (24-309) where $M_{2t\mu}$ is constant
	$M_{2ti(\max)} = J\alpha_{2a(\max)} = \dfrac{J\alpha_{2a(\max)}}{\omega_1^2}\left(\dfrac{2\pi n}{60}\right)^2$
The maximum torque on the driving shaft of the crank	$M_{1t(\max)} = \left[M_{2t\mu}\dfrac{1}{\eta}\dfrac{\omega_{2(\max)}}{\omega_1} + \dfrac{J}{\omega_1}\dfrac{1}{\eta}(\alpha_{2a}\omega_2)_{\max}\right]$ (24-310a)
	$M_{1t(\max)} \simeq \left[\left(M_{2t\mu}\dfrac{\gamma}{1-\gamma}\dfrac{1}{\eta} + \dfrac{\gamma^2(1-\gamma^2)(\cos\alpha_m - \gamma)\sin\alpha_m}{(1-2\gamma\cos\alpha_m + \gamma^2)^3}\right) \times J\omega_1^2\dfrac{1}{\eta}\right]$ (24-310b) where $\alpha = \alpha_m$ at which M_{t1} is maximum

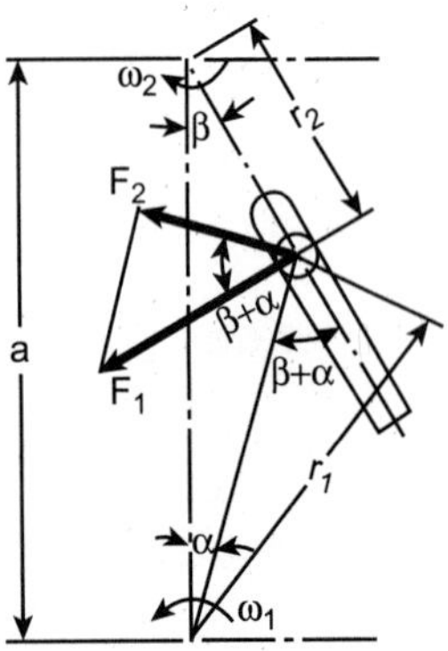

FIGURE 24-57 Forces acting on Geneva wheel.

Particular	Formula
The maximum power required on the shaft of the crank or driver	$P_{1(\max)} = \dfrac{M_{1t(\max)}\omega}{1000}$ **SI** (24-311a) where $P_{1(\max)}$ in kW, $M_{1t(\max)}$ in N m, and ω in rad/s
	$P_{1(\max)} = \dfrac{M_{1t(\max)}\omega}{102 \times 10^3}$ **Customary Metric** (24-311b) where $P_{1(\max)}$ in kW, $M_{1t(\max)}$ in kgf mm, and ω in rad/s

Particular	Formula
	$P_{1(\max)} = \dfrac{M_{1t(\max)}n}{63{,}000}$ USCS (24-311c)
	where $P_{1(\max)}$ in hp, $M_{1t(\max)}$ in lbf in, and n in rpm

FORCES AT THE POINT OF CONTACT (Fig. 24-57)

The maximum force at the point of contact between the roller pin and slotted Geneva wheel	$F_{2(\max)} = \dfrac{M_{2t}}{r_2}$ (24-312a)
	$F_{2(\max)} = F_{\mu(\max)} + F_{i(\max)}$ (24-312b)
	where $r_2 = \sqrt{a^2 - 2ar_1 \cos\alpha + r_1^2} = \dfrac{r_1}{\gamma}\sqrt{1 - 2\gamma\cos\alpha + \gamma^2}$
The component of maximum friction force at the point of contact due to the friction torque $M_{2t\mu}$ on the driven Geneva wheel shaft	$F_{2\mu(\max)} = \dfrac{M_{2t\mu}}{r_{2(\min)}} = \dfrac{M_{2t\mu}}{r_1}\dfrac{\gamma}{1-\gamma}$ (24-313)
	where $r_{2(\min)} = a - r_1 = \left(1 - \dfrac{1}{\gamma}\right)r_1$
For maximum values of F_{2i}	Refer to Table 24-24.
For design data for external Geneva mechanism	Refer to Table 24-25A.

INTERNAL GENEVA WHEEL

The time required for indexing Geneva wheel, s	$t_i = \dfrac{z+2}{z}\left(\dfrac{60}{2n}\right)$ (24-314)
The time during which Geneva wheel is at rest, s	$t_r = \dfrac{z-2}{z}\left(\dfrac{60}{2n}\right)$ (24-315)
The t_i/t ratio	$\dfrac{t_i}{t} = \dfrac{z+2}{2z}$ (24-316)
The t_r/t ratio	$\dfrac{t_r}{t} = \dfrac{z-2}{2z}$ (24-317)
The working time coefficient of Geneva wheel	$k = \dfrac{z+2}{z-2} > 1$ (24-318)
The relationship between crank or driver angle α and Geneva wheel angle β	$\beta = \tan^{-1}\left(\dfrac{\gamma\sin\alpha}{1+\gamma\cos\alpha}\right)$ (24-319)

Particular	Formula	
The angular velocity of Geneva wheel	$\omega_2 = \dfrac{d\beta}{dt} = \left(\dfrac{\gamma(\cos\alpha + \gamma)}{1 + 2\gamma\cos\alpha + \gamma^2}\right)\omega_1$	(24-320)
The maximum angular velocity of Geneva wheel	$\omega_{2(\max)} = \dfrac{\gamma}{1+\gamma}\omega_1$	(24-321)
The angular acceleration, α_{2a}, of Geneva wheel	$\alpha_{2a} = \dfrac{d^2\beta}{dt^2} = \pm\dfrac{\gamma(1-\gamma^2)\sin\alpha}{(1 + 2\gamma\cos\alpha + \gamma^2)^2}\omega_1^2$	(24-322)
For values of α_{2a} at start and finish of indexing	Use Eq. (24-285) of external Geneva wheel.	
For curves of angular velocity and angular acceleration of internal Geneva wheel	Refer to Fig. 24-58.	
The contact forces between the slotted wheel and the pin on the driving crank of the internal Geneva wheel are calculated in a manner similar to that for the external Geneva wheel		

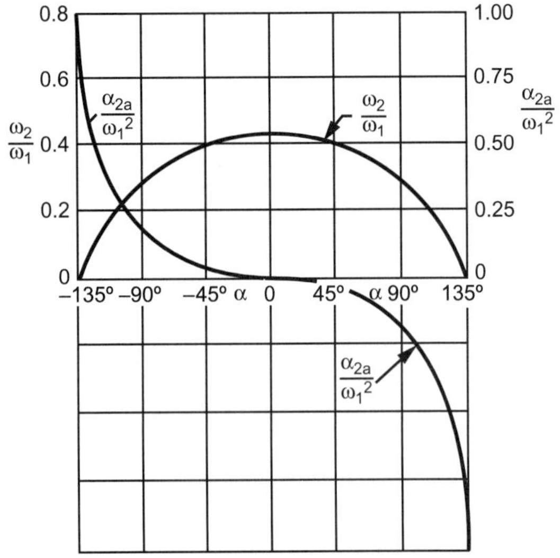

FIGURE 24-58 Angular velocity and angular acceleration for four-slot internal Geneva wheel.

Materials

Chromium steel 15 Cr^{65} case-hardened to R_c 58 to 65 is used for the roller pin on the driver or crank.

Chromium steel 40 Cr 1 hardened and tempered to R_c 45 to 55 is used for the sides of slotted Geneva wheel.

TABLE 24-24
Maximum F_{2i} values

z	3	4	5	6	8
$F_{2i(\max)} \Big/ \left(\dfrac{Jn^2}{r_1}\right)$	1.966	0.126	0.0318	0.0131	0.00424

TABLE 24-25A
Design data for external Geneva mechanism

z	ϕ	ψ	i	r_1/a	r_2/a	R_r	r_2'/a	λ	ν	$\omega_{2(\max)}$	$\alpha_{2a(initial)}$ $\alpha = -\phi$	$\alpha_{(\max)}$	$J_{\max}$	$J_{\alpha=0}$
3	60°	30°	0.5	0.886	0.500	0.577	0.134	300°	0.167	6.46	1.732	4°46′	31.44	−672
4	45°	45°	1	0.707	0.707	1.000	0.293	270°	0.250	2.41	1.000	11°24′	5.41	−48
6	30°	60°	2	0.500	0.866	1.732	0.500	240°	0.333	1.00	0.577	22°54′	1.35	−6
8	22°30′	67°30′	3	0.383	0.924	2.414	0.617	225°	0.375	0.620	0.414	31°38′	0.699	−2.25
10	18°	72°	4	0.309	0.951	3.078	0.690	216°	0.400	0.447	0.325	38°30′	0.465	−1.24

24.10 UNIVERSAL JOINT

SYMBOLS[2,3]

d	diameter, m (in)
K_s	shock factor
K_{ct}	correction factor to be applied to torque to be transmitted
K_{ct}	correction factor to be applied to power to be transmitted
l	length (also with subscripts), m (in)
L	life, h
M_t	torque to be transmitted by universal joint, N m (lbf in)
M_{td}	design torque, N m (lbf in)
n	speed, rpm
n'	speed, rps
P	power to be transmitted by universal joint, kW (hp)
P_d	design power, kW (hp)
β	angle between two intersecting shafts 1 and 2, deg
θ	angle of rotation of the driver shaft 1, deg
ϕ	angle of rotation of the driven shaft 2, deg
ω_1 , ω_2	angular velocities of driver and driven shafts respectively, rad/s

Particular	Formula	

SINGLE UNIVERSAL JOINT (Figs. 24-59 and 24-61a)

The relation between θ, ϕ, and β	$\tan\phi = \dfrac{\tan\theta}{\cos\beta}$	(24-323)
The relation between the angular velocities of driving shaft 1 or driver (ω_1) to the driven shaft 2 or the follower (ω_2)	$\dfrac{\omega_2}{\omega_1} = \dfrac{\cos\beta}{1 - \sin^2\beta\sin^2\theta}$	(24-324)

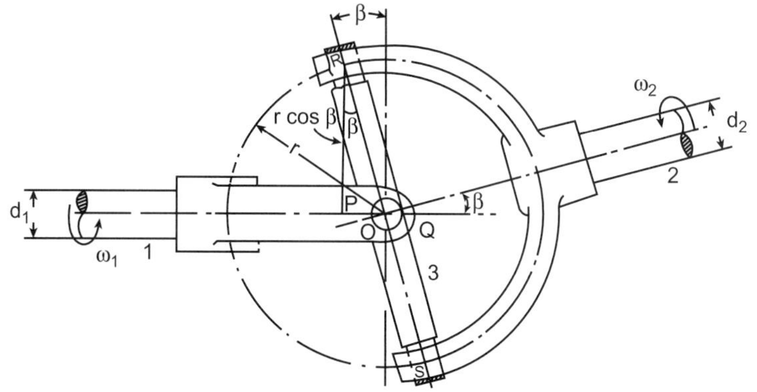

FIGURE 24-59 A single universal joint.

Particular	Formula
The maximum value of ω_2/ω_1	$$\left(\frac{\omega_2}{\omega_1}\right)_{max} = \frac{\cos\beta}{1-\sin^2\beta} = \frac{1}{\cos\beta} \quad (24\text{-}325)$$ when $\sin\theta = +1$, i.e., $\theta = 90°$, $270°$, or $\pi/2$ or $3\pi/2$, etc.
The minimum value of ω_2/ω_1	$$\left(\frac{\omega_2}{\omega_1}\right)_{min} = \cos\beta \quad (24\text{-}326)$$ when $\sin = 0$, i.e., $\theta = 0$, π, 2π, etc.
The angular acceleration of the driven shaft 2, if ω_1 is constant	$$\frac{d^2\phi}{dt^2} = \frac{d\omega_2}{dt} = \frac{\cos\beta\sin^2\beta\sin 2\theta}{(1-\sin^2\theta\sin^2\beta)^2}\omega_1^2 \quad (24\text{-}327)$$
The value of θ for which the angular acceleration of the driven shaft is maximum	$$\cos 2\theta_{(max)} = \kappa - \sqrt{\kappa^2+2} \quad (24\text{-}328)$$ where $\kappa = (2-\sin^2\beta)/2\sin^2\beta$ The angular acceleration of driven shaft is maximum when θ is approximately equal to 45°, 135°, etc., when the arms of cross are inclined at 45° to the plane containing the axes of the two shafts.
The power transmitted by universal joint	$$P = M_t\omega/1000 \quad \textbf{SI} \quad (24\text{-}329a)$$ where P in kW, M_t in N m, and ω in rad/s $$P = M_t n/63{,}000 \quad \textbf{USCS} \quad (24\text{-}329b)$$ where P in hp, M_t in lbf in, and n in rpm
The design torque of universal joint	$$M_{td} = M_t K_s K_{ct} \quad (24\text{-}330)$$
The design power of universal joint	$$P_d = \frac{P}{K_{CN}} \quad (24\text{-}331)$$
For calculation of torque and power transmitted by universal joint for various angles of inclination β	Refer to Figs. 24-62 to 24-65.
For design data of universal joint	Refer to Tables 24-25B and 24-25C.

DOUBLE UNIVERSAL JOINT (Figs. 24-60 and 24-61b)

The angular velocities ratio for a double universal joint which will produce a uniform velocity ratio at all times between the input and output ends	$$\frac{\omega_1}{\omega_2} = 1 \quad (24\text{-}332)$$

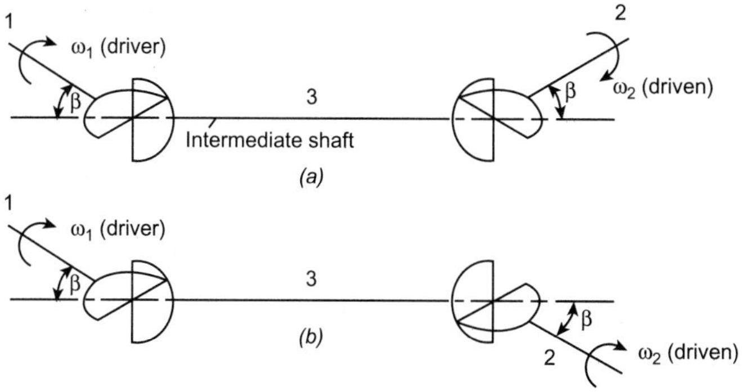

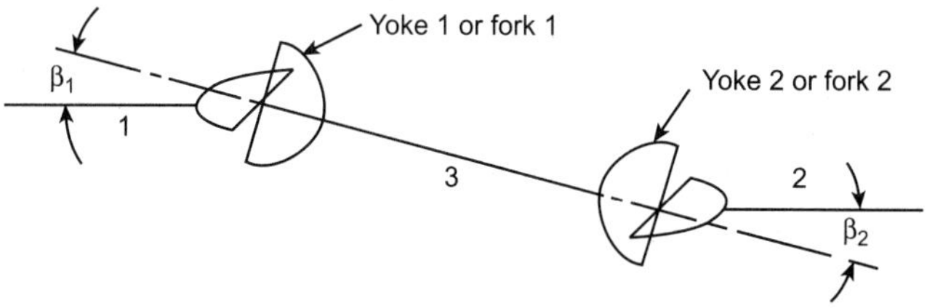

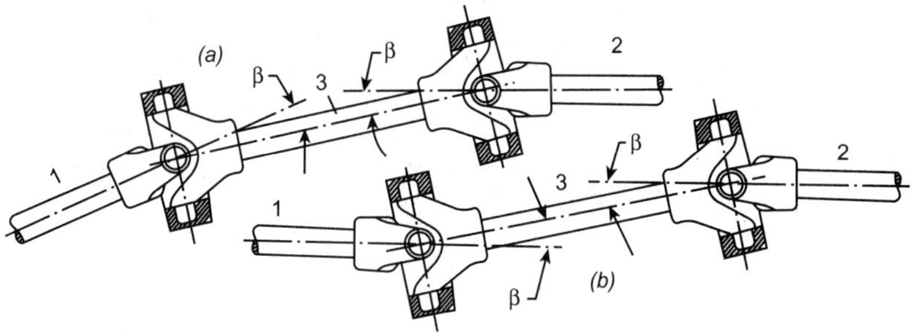

FIGURE 24-60 Double universal joints.

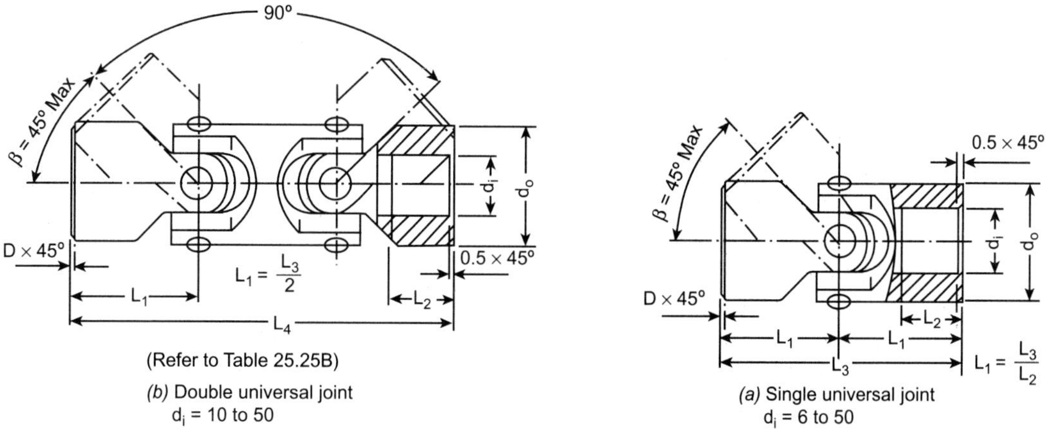

FIGURE 24-61 Dimensions of universal joints.

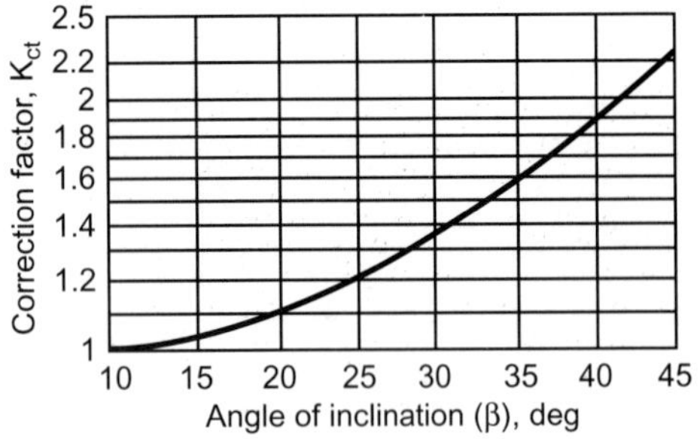

FIGURE 24-62 Angle between two intersecting shafts vs. correction factor (K_{ct}).

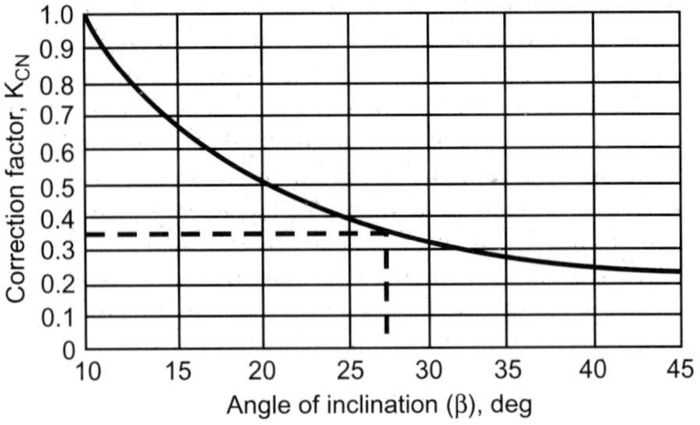

FIGURE 24-63 Angle between two intersecting shafts vs. correction factor (K_{CN}).

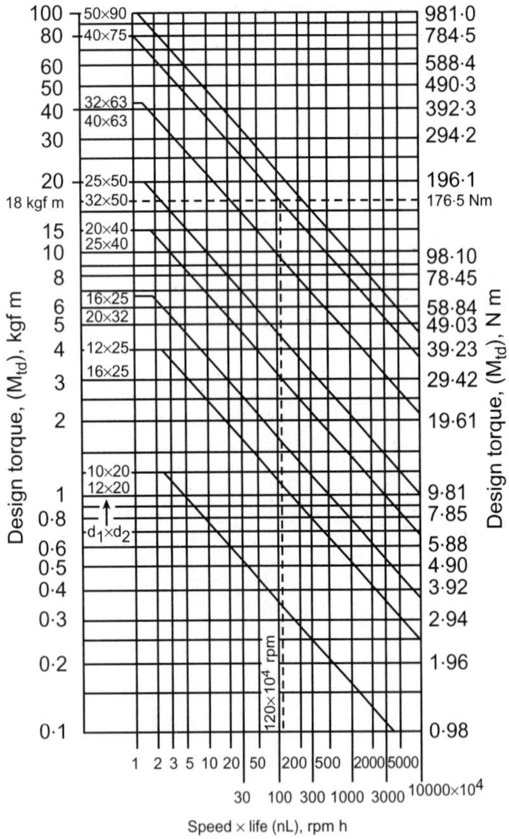

FIGURE 24-64 Design curves for single universal joint with needle bearings for $\beta = 10°$.

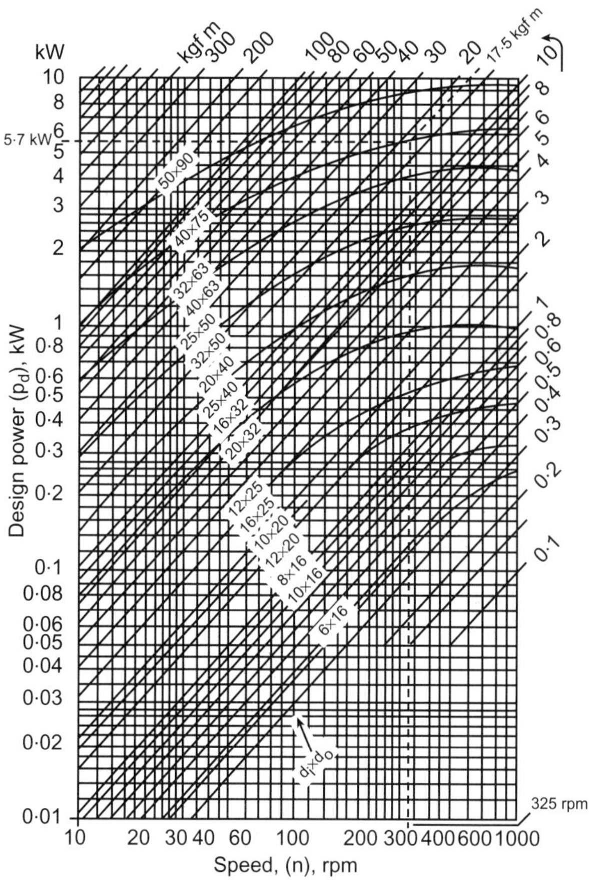

FIGURE 24-65(a) Design curves for single universal joint with plain bearings for $\beta = 10^\circ$.

TABLE 24-25B
Dimensions of universal joint (Fig. 24-61)

						Maximum allowable rotational play			
						Test torque			
d_i H7	d_o k11	L_2	L_3 +1	L_4 ±1	z	N m	kgf m	Angular rotational play at an angle of inclination of θ deg in minutes	Tolerance on coaxiality of the two bores
6		9	34						
8	16	11	40			0.196	0.02	45	
10		15	52	74					
	20	13	48		0.5	0.392	0.04	40	
12		18	62	88					0.06
	25	15	56	86		0.981	0.10	32	
16		22	74	104					
	32	19	68			0.667	0.17	28	
20		25	86	124					
	40	23	82	128		3.334	0.34	25	0.09
25		32	108	156					
	50	29	105	160	1	5.296	0.54	20	
32		40	132	188					
	63	36	130	198		14.710	1.5	18	0.12
40		50	166	238					
	75	44	160	245		21.575	2.2	16	
50	90	54	190	290		27.458	2.8	14	0.15

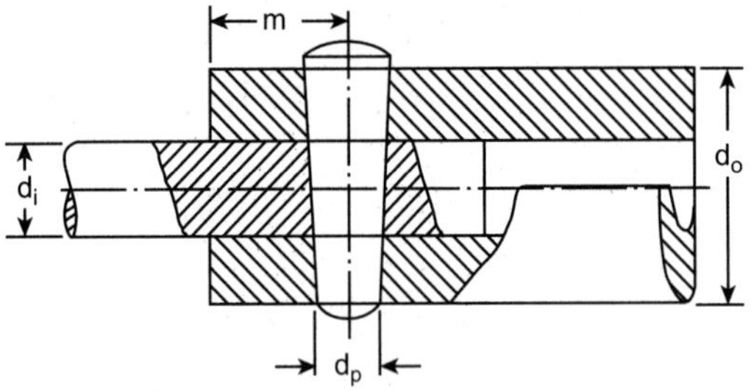

FIGURE 24-65(b) Taper pin joint. The length of the taper pin should conform to diameter d_o in Table 24-25B.

TABLE 24-25C
Dimensions of taper pin[a] (Fig. 24-65*b*)

d_i	d_p	m
6	2	4.5
8	3	5
10	4	6
12	5	7.5
16	6	9
20	8	11
25	10	15
32	12	18
40	14	22
50	16	27

[a] The shear stress of taper pin $= \tau = 158.5$ to 247.5 MPa (16 to 25 kgf/mm^2).

USE OF CURVES IN FIGS. 24-62 TO 24-65

Worked example 1

A single universal joint has to transmit a torque of 10 kgf m at 1500 rpm. The angle between intersecting shafts is 25°. The joint is subjected to a minor shock. The shock factor (K_s) is 1.5. Design a universal joint with needle bearings for a life of 800 h.

SOLUTION From Fig. 24-62 correction factor for $\beta = 25°$ is $K_{ct} = 1.2$. Design torque $= M_{td} = M_t K_s K_{ct} = 10 \times 1.5 \times 1.2 = 18$ kgf m (176.5 N m). Speed × life $= nL = 1500 \times 800 = 120 \times 10^4$ rpm h. From Fig. 24-64 for $M_{td} = 18$ kgf m (176.5 N m) and $nL = 120 \times 10^4$ rpm h, the size of a single universal joint is $(d_i \times d_o)$ 40 × 75 mm.

Worked example 2

Design a single universal joint with plain bearings to transmit 2 kW power at 325 rpm. The angle between two intersecting shafts is 27.5°.

SOLUTION From Fig. 24-63 correction factor for $\beta = 27.5°$ is $K_{CN} = 0.35$. Design power $= P_d = (P/K_{CN}) = (2/0.35) = 5.7$ kW. From Fig. 24-65*a* the size of a single universal joint for $P_d = 5.7$ kW and speed $= n = 325$ rpm is $(d_i \times d_o)$ 40 × 75 mm. The permissible torque for this size of joint (Fig. 24-65*a*) is 17.5 kgf m (171.5 N m).

24.11 UNSYMMETRICAL BENDING AND TORSION OF NONCIRCULAR CROSS-SECTION MACHINE ELEMENTS

SYMBOLS[2,3]

a	semimajor axis of elliptical section, m (in)
	width of rectangular section, m (in) (in^2)
A	area of cross section, m^2 (in)
b	semi-minor axis of elliptical section, m (in)
	height of rectangular section, m (in)
c	distance of the plane from neutral axis, m (in)
	thickness of narrow rectangular cross section (Fig. 24-68)
e	the distance from a point in the shear center S (Table 24-26)
E	Young's modulus, GPa (MPsi)
G	modulus of rigidity, GPa (MPsi)
I	moment of inertia, area (also with suffixes), m^4 (cm^4) (in^4)
I_u, I_v	moment of inertia of cross-sectional area, respectively, m^4 (cm^4) (in^4)
J_k	polar moment of inertia, m^4 (cm^4) (in^4)
k_1, k_2	constants from Table 24-28 for use in Eqs. (24-343) and (24-344)
L	length, m (in)
M_b	bending moment, N m (lbf ft)
M_t	twisting moment, N m (lbf ft)
$M_{bu} = M_b \cos\theta$	bending moment about the U principal centroidal axis or any axis parallel thereto
$M_{bv} = M_b \sin\theta$	bending moment about the V principal centroidal axis or any axis parallel thereto
$q = \tau t$	shear flow
Q	the first moment of the section, m^4 (cm^4) (in^4)
S	the length of the center of the ring section of the thin tube, m (in)
t	width of cross section at the plane in which it is desired to find the shear stress, m (in)
	thickness of the wall of the thin-walled section, m
u, v	coordinates of any point in the section with reference to principal centroidal axes
V	shear force on the cross section, kN (lbf)
V_y	resultant shear force acting at the shear center, kN (lbf)
x	the distance of the section considered from the fixed end (Fig. 24-73)
x, y	coordinates in x and y directions
σ_b	bending stress (also with suffixes), MPa (psi)
τ	shear stress (also with suffixes), MPa (psi)
δ	variable thickness of thin tube wall (Fig. 24-70), m (in)
θ	angle measured from the V principal centroidal axis, deg
ϕ	angle of twist, deg

Particular	Formula	
SHEAR CENTER		
The shear stress at any point in transverse plane or section of a member	$\tau = \dfrac{VQ}{It}$	(24-333)
The flexural stress in a thin-walled open section	$\sigma_b = \dfrac{M_b c}{I}$	(24-334)
Shear flow	$q = \tau t = \dfrac{VQ}{I}$	(24-335)
For the equations for locating the shear centers of various thin open sections	Refer to Table 24-26.	
UNSYMMETRICAL BENDING		
The flexural stress in case of sections subjected to unsymmetrical bending	$\sigma_b = \dfrac{(M_b \cos\theta)v}{I_u} + \dfrac{(M_b \sin\theta)u}{I_v}$ $= \dfrac{M_{bu}v}{I_u} + \dfrac{M_{bv}u}{I_v}$	(24-336)
Flexural modulus for any cross section on which the stress is desired	$Z = I_u I_v/(vI_v \cos\theta + uI_u \sin\theta)$	(24-337)
TORSION		
Solid sections		
ELLIPTICAL CROSS SECTION		
Shear stress acting in the x direction on the xz plane (Fig. 24-66)	$\tau_{xz} = \dfrac{2M_t y}{\pi ab^3}$	(24-338)
Shear stress acting in the y direction on the yz plane (Fig. 24-66)	$\tau_{yz} = -\dfrac{2M_t x}{\pi a^3 b}$	(24-339)
Maximum shear stress on the periphery at the extremities of the minor axis (Fig. 24-66 and Table 24-27)	$\tau_{\max} = \dfrac{2M_t}{\pi ab^2}$	(24-340)
Minimum shear stress on the periphery at the extremities of the major axis	$\tau_{\min} = \dfrac{2M_t}{\pi a^2 b}$	(24-341)
Angle of twist (Fig. 24-66)	$\phi = \dfrac{M_t}{G}\dfrac{(a^2+b^2)}{\pi a^3 b^3}L$	(24-342)

Particular	Formula	
RECTANGULAR CROSS SECTION		
The maximum shear stress at point A on the boundary, close to the center (Fig. 24-67 and Table 24-27)	$\tau_A = \dfrac{M_t}{k_1 ab^2}$ where k_1 depends on ratio a/b (Refer to Table 24-28.)	(24-343)
Angle of twist (Table 24-27)	$\phi = \dfrac{M_t L}{k_2 ab^3 G}$ where k_2 depends on ratio a/b (Refer to Table 24-28.)	(24-344)

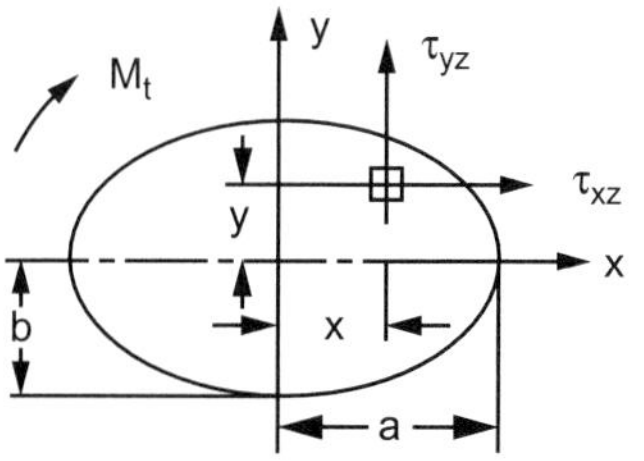

FIGURE 24-66

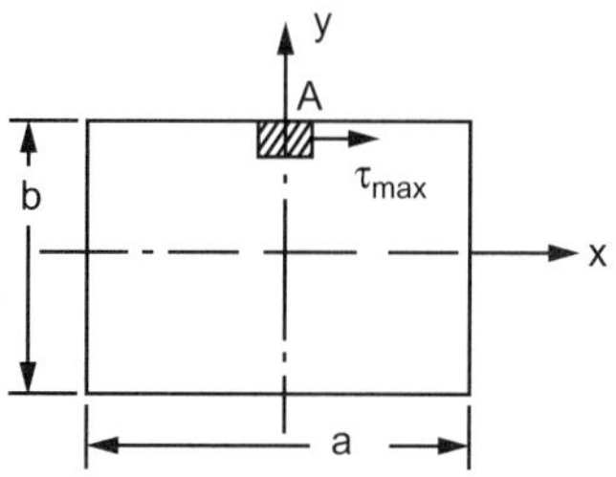

FIGURE 24-67

NARROW RECTANGULAR CROSS SECTIONS (Fig. 24-68)

Equation for twisting moment (Fig. 24-68)	$M_t = \frac{1}{3} G\phi c^3 b$	(24-345)
Equation for angle of twist	$\phi = \dfrac{3M_t}{Gc^3 b}$	(24-346)
The maximum shear stress	$\tau_{max} = \dfrac{3M_t}{bc^2}$	(24-347)

TABLE 24-26
Location of shear center for various cross sections

Section	Location of shear center	Section	Location of shear center
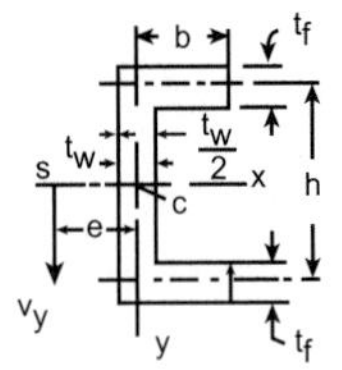	$e = \frac{3b^2 t_f}{h t_w + 6 b t_f}$	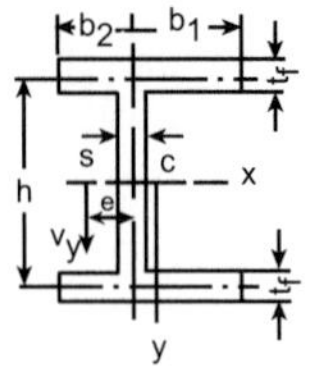	$e = \frac{3(b_1^2 - b_2^2)}{(t_w/t_f)h + 6(b_1 + b_2)}$ for $b_2 < b_1$
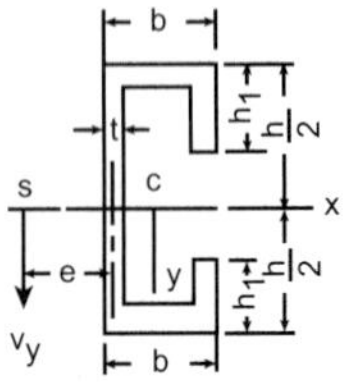	$e = b_1 \left\{ \frac{1 + \frac{1}{2}\frac{b_1}{h_1} - \frac{4}{3}\left(\frac{h_1}{h_2}\right)^2}{1 + \frac{1}{6}\frac{h_2}{h_1} + \frac{b_1}{h_1} - \frac{2h_1}{h_2}\left(1 - \frac{2h_1}{3h_2}\right)} \right\}$ where $b_1 = b - t$, $h_2 = h - t$	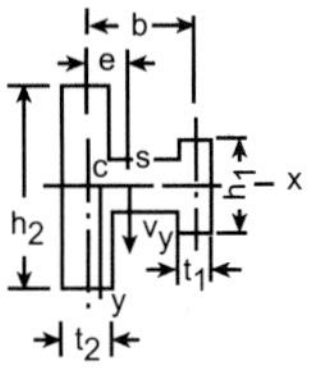	$e = \left(\frac{b t_1 h_1^3}{t_1 h_1^3 + t_2 h_2^3} \right)$
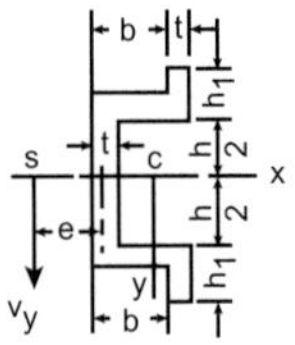	$e = b_1 \left\{ \frac{1 + \frac{1}{2}\frac{b_1}{h_1} - \frac{4}{3}\left(\frac{h_1}{h_2}\right)^2}{1 + \frac{1}{6}\frac{h_2}{h_1} + \frac{h_1}{h_1} + 2\frac{h_1}{h_2}\left(1 + \frac{2h_1}{3h_2}\right)} \right\}$ where $b_1 = b - t/2$, $h_2 = h + t$	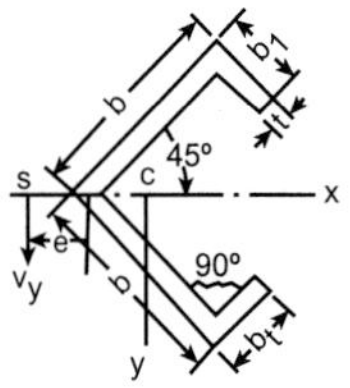	$e = b_1 \frac{\frac{b}{b_1}\left(3\frac{b}{b_1} - 2\right)}{\sqrt{2}\left[\left(\frac{b}{b_1}\right)^3 + 3\left(\frac{b}{b_1}\right)^2 - 3\frac{b}{b_1} + 1\right]}$
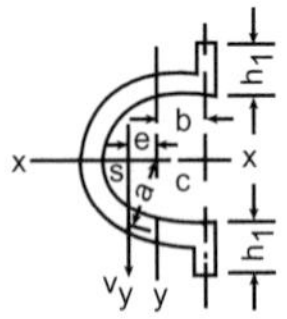	$e = \frac{m}{n}$ $m = 12 + 6\pi\left(\frac{b + h_1}{a}\right) + 6\left(\frac{b}{a}\right)^2 + 12\frac{bh_1}{a^2} + 3\pi\left(\frac{h_1}{a}\right)^2 - 4\left(\frac{h_1}{a}\right)^3 \frac{b}{a}$ $n = 3\pi + 12\left(\frac{b + h_1}{a}\right) + 4\left(\frac{h_1}{a}\right)^2 \left(3 + \frac{h_1}{a}\right)$	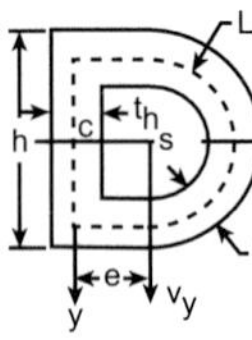	$e = \frac{2A}{h + L(t_h/t_b)}$ where A = area
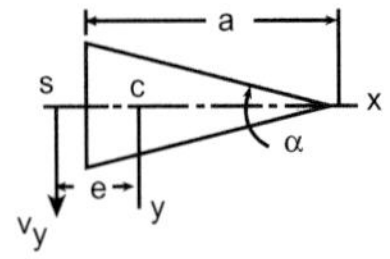	C is at the centroid of triangle $e = 0.47a$ for narrow triangle $(\alpha > 12^\circ)$ approx.		

TABLE 24-26 (*Cont.*)
Location of shear center for various cross sections

Section	Location of shear center	Section	Location of shear center
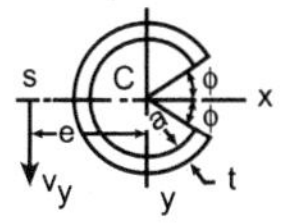	$e = \dfrac{2a[(\pi - \phi)\cos\phi + \sin\phi]}{[(\pi - \phi) + \sin\phi\cos\phi]}$ For $\phi = \dfrac{\pi}{2}$, $e = \dfrac{4a}{\pi}$	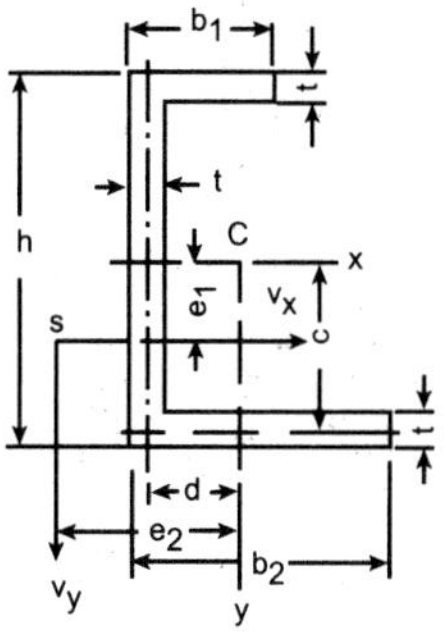	$e_1 = c - \dfrac{b_1^2 ht}{6(I_y I_x - I_{xy}^2)} \times [3I_{xy}(h-c) + I_x(2b_1 - 3d)]$ $e_2 = d + \dfrac{b_1^2 ht}{6(I_y I_x - I_{xy}^2)} \times [I_{xy}(2b_1 - 3d) + 3I_y(h-c)]$ where $c = \dfrac{h^2 + 2b_1 h}{2h(b_1 + b_2)}$, $d = \dfrac{b_1^2 + b_2^2}{2h(b_1 + b_2)}$ I_x, I_y = moment of inertia of section about x and y axes, respectively I_{xy} = product of moment of inertia
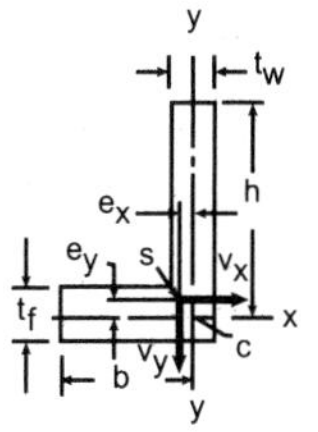	$e_x = \dfrac{b}{2}\left(\dfrac{ht_w^3}{ht_w^3 + t_f b^3}\right)$ $e_y = \dfrac{h}{2}\left(\dfrac{ht_w^3}{ht_w^3 + t_f b^3}\right)$		
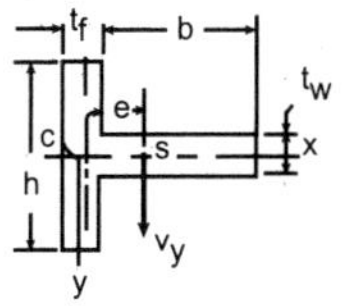	$e = \dfrac{1}{2}(t_f + b)\left\{\dfrac{1}{1 + \dfrac{h^3 t_f}{t_w^3 b}}\right\}$	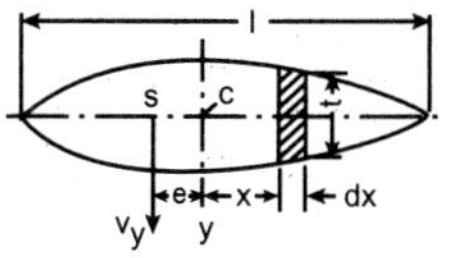	$e = \left(\dfrac{1+3\nu}{1+\nu}\right)\dfrac{\int xt^3\,dx}{\int t^3\,dx}$

TABLE 24-27
Approximate formulas for torsional shearing stress and angle of twist for various cross sections

Cross section	Shearing stress, lbf/in² (N/m² or MPa)	Angle of twist per unit length ϕ, rad/in (rad/m)
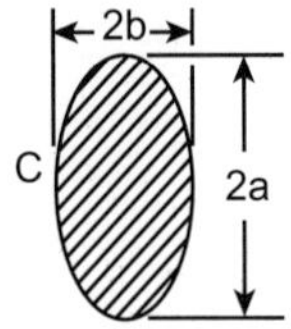	$\tau_c = \dfrac{2M_t}{\pi ab^2}$ $= \dfrac{2M_t}{Ab}$	$\phi = \dfrac{M_t(a^2+b^2)}{G\pi a^3 b^3}$ $= \dfrac{4\pi^2 JM_t}{A^4 G}$
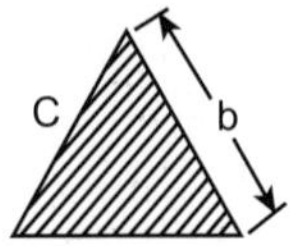	$\tau = \dfrac{20M_t}{b^3}$	$\phi = \dfrac{46.2M_t}{Gb^4}$
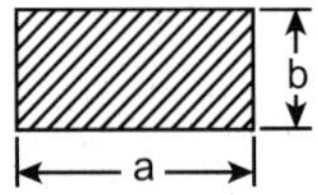	$\tau_c = \dfrac{M_t}{k_1 ab^2}$	$\phi = \dfrac{M_t}{k_2 ab^3 G}$
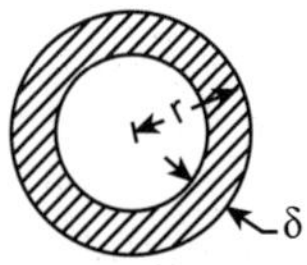	$\tau = \dfrac{M_t}{2\pi r^2 \delta}$	$\phi = \dfrac{M_t}{2\pi r^3 \delta G}$
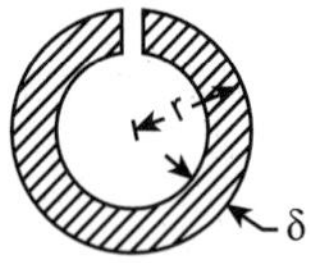	$\tau = \dfrac{3M_t}{2\pi r \delta^2}$	$\phi = \dfrac{3M_t}{2\pi r \delta^3 G}$
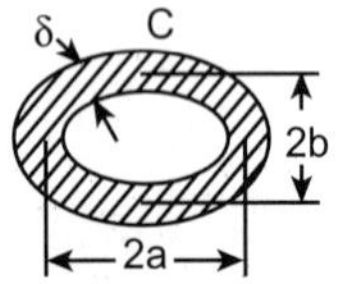	$\tau = \dfrac{M_t}{2\pi ab\delta}$	$\phi = \dfrac{M_t\sqrt{2(a^2+b^2)}}{4\pi a^2 b^2 \delta G}$
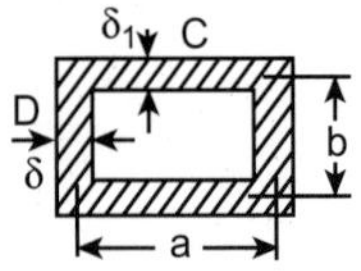	$\tau_c = \dfrac{M_t}{2ab\delta_1}$ $\tau_D = \dfrac{M_t}{2ab\delta}$	$\phi = \dfrac{M_t(a\delta + b\delta_1)}{2\delta\delta_1 a^2 b^2 G}$

A = area of cross section

TABLE 24-28
Variation of k_1 and k_2 with the ratio a/b for use in Eqs. (24-343) and (24-344)

a/b	1	1.2	1.5	2.0	2.5	3.0	4.0	5.0	10.0	∞
k_1	0.208	0.219	0.231	0.246	0.258	0.267	0.282	0.391	0.312	0.333
k_2	0.141	0.166	0.196	0.229	0.249	0.263	0.281	0.291	0.312	0.333

Particular	Formula

COMPOSITE SECTIONS

Cross Sections Composed of Narrow Rectangles

Equation for torque of a narrow rectangular cross section	$$M_t = \frac{k_2 bc^3 G\phi}{L} \quad (24\text{-}348)$$
Equation for torque of a narrow rectangular section ($b/c \to \infty$, $k_2 = \frac{1}{3}$)	$$M_t = \frac{G\phi}{3L}\Sigma bc^3 \quad (24\text{-}349)$$
Equation for torque for a cross-section composed of several narrow rectangles (Table 24-27)	$$M_t = \frac{G\phi}{L}\Sigma k_2 bc^3 \quad (24\text{-}350)$$
Angle of twist for a cross section composed of several narrow rectangles	$$\phi = \frac{3M_t L}{G\Sigma bc^3} \quad (24\text{-}351)$$
Maximum shear stress	$$\tau_{\max} = \frac{3M_t c_{\max}}{\Sigma bc^3} \text{ for } \frac{k_2}{k_1} = 1 \quad (24\text{-}352)$$ where $c_{\max}$ = maximum thickness of the narrow section
For approximate formulas for torsional shearing stress and angle of twist for various cross sections	Refer to Table 24-27.
For variation of stress-concentration factor K_σ with ratio r/c for structural angle (Fig. 24-69)	Refer to Table 24-29.

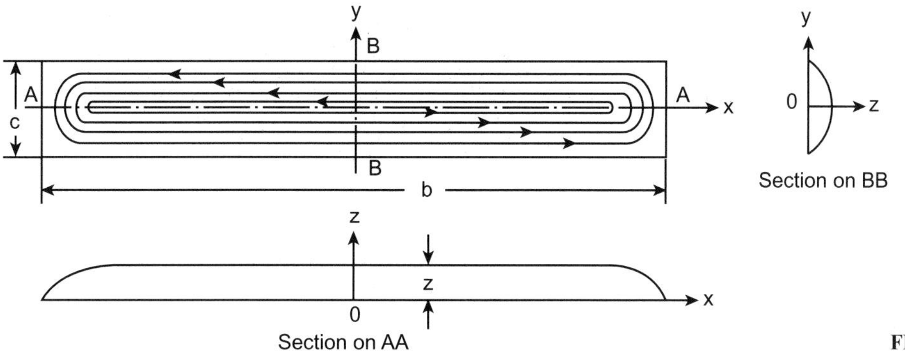

FIGURE 24-68

Particular	Formula

TABLE 24-29
Stress concentration factors for structural angle, K_σ (Fig. 24-69)

r/c	0.125	0.250	0.500	0.750	1.000
K_σ	2.550	2.250	2.000	1.875	1.800

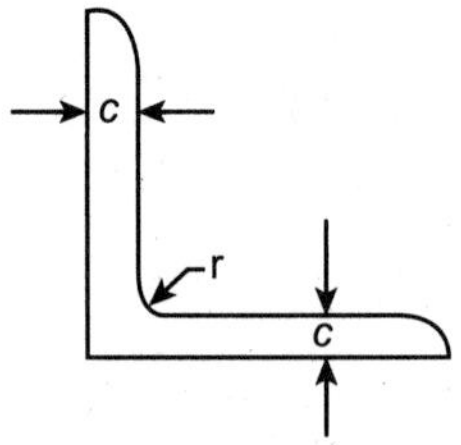

FIGURE 24-69

HOLLOW THIN-WALLED TUBES (Fig. 24-70)

The equation for the twisting moment

$$M_t = q(2A) = 2A\delta\tau \quad (24\text{-}353)$$

where A = area enclosed by the median line of the tubular section

The angle of twist

$$\phi = \frac{M_t S}{4A^2 G\delta} \quad (24\text{-}354)$$

By membrane analogy the value of $\oint \tau\, ds$

$$\oint \tau\, ds = 2G\phi A \quad (24\text{-}355)$$

The equation for the shear stress

$$\tau = \frac{M_t}{2A}\delta \quad (24\text{-}356)$$

The difference in level between DC and AB of membrane

$$h = \tau\delta \quad (24\text{-}357)$$

The equation for twisting moment of thin webbed tubes (or box beams) (Fig. 24-71)

$$M_t = 2(A_1 h_1 + A_2 h_2) \quad (24\text{-}358a)$$

$$M_t = 2(A_1\delta_1\tau_1 + A_2\delta_2\tau_2) \quad (24\text{-}358b)$$

The equations for shear stress

$$\tau_1 = M_t \frac{\delta_3 S_2 A_1 + \delta_2 S_3(A_1 + A_2)}{R} \quad (24\text{-}359a)$$

$$\tau_2 = M_t \frac{\delta_3 S_1 A_2 + \delta_1 S_3(A_1 + A_2)}{R} \quad (24\text{-}359b)$$

$$\tau_3 = M_t \frac{\delta_1 S_3 A_1 - \delta_2 S_1 A_2}{R} \quad (24\text{-}359c)$$

where

$$R = 2[\delta_1\delta_3 S_2 A_1^2 + \delta_2\delta_3 S_1 A_2^2 + \delta_1\delta_2 S_3(A_1 + A_2)^2] \quad (24\text{-}360)$$

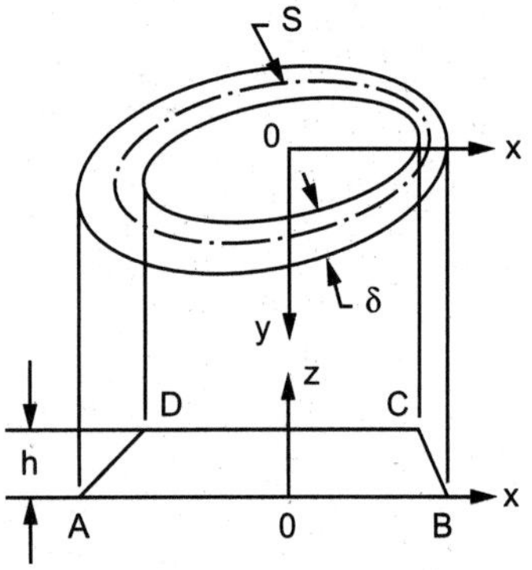

FIGURE 24-70

Particular	Formula

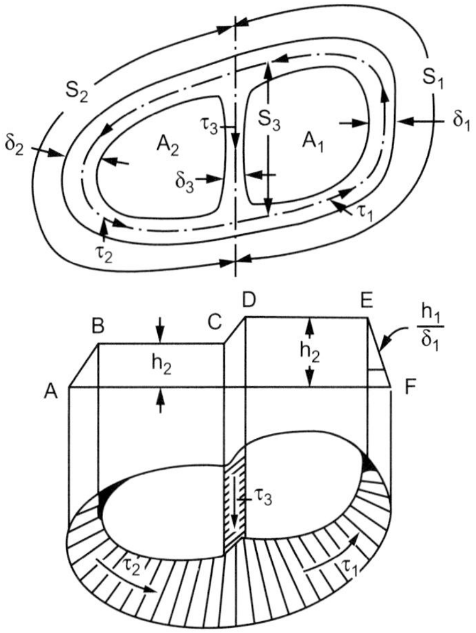

FIGURE 24-71

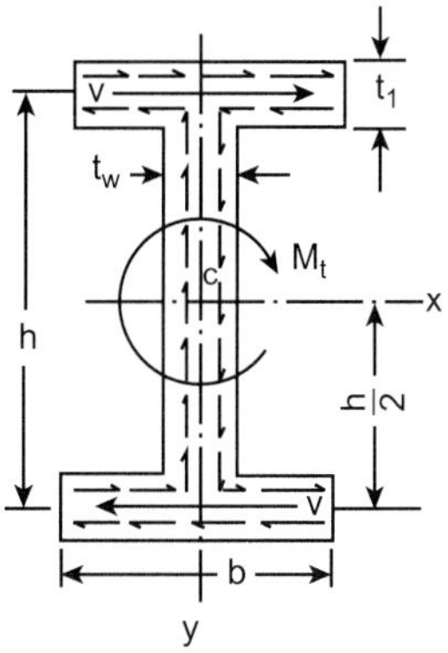

FIGURE 24-72

BENDING STRESSES CAUSED BY TORSION

Torsion of I-beam having one section restrained from warping

The lateral bending moment in the flanges of an I-beam subjected to twisting moment at one end, the other end being fixed, Fig. 24-72

$$M_b = -\frac{M_t}{h} k \frac{\sin h(L-x)/k}{\cos h(L/k)} \tag{24-361}$$

The maximum bending moment for long beam

$$M_{b(\max)} = \frac{M_t k}{h} \quad \text{as} \quad \tan h\frac{L}{k} = 1 \tag{24-362}$$

where $k = (h/2)[EI/JG]^{1/2}$

Twisting moment at any section, distance x from the fixed end

$$M_{tx} = M_t \left[1 - \frac{\cos h(L-x)/k}{\cos h(L/k)}\right] \tag{24-363}$$

The angle of twist per unit length

$$\phi_u = \frac{M_t}{JG}\left[1 - \frac{\cos h(L-x)/k}{\cos h(L/k)}\right] \tag{24-364}$$

The total angle of twist at the free end

$$\phi = \frac{M_t}{JG}\left(L - k\tan h\frac{L}{k}\right) \tag{24-365}$$

Maximum bending moment

$$M_{b(\max)} = \frac{M_t}{h} k \tan h\frac{L}{k} \tag{24-366}$$

Particular	Formula	
The angle of twist at free end if $l/k > 2.5$	$\phi = \dfrac{M_t}{JG}(L - k)$	(24-367)
The maximum bending moment if $l/k > 2.5$	$M_{b(\max)} = \dfrac{M_t}{h}k$	(24-368)
The bending stress	$\sigma_b = \dfrac{M_{b(\max)}b}{2I_f}$ where $M_{b(\max)}$ obtained from Table 24-30	(24-369)
For beams subjected to torsion	Refer to Table 24-30.	

TRANSVERSE LOAD ON BEAM OF CHANNEL SECTION NOT THROUGH SHEAR CENTER (Fig. 24-73)

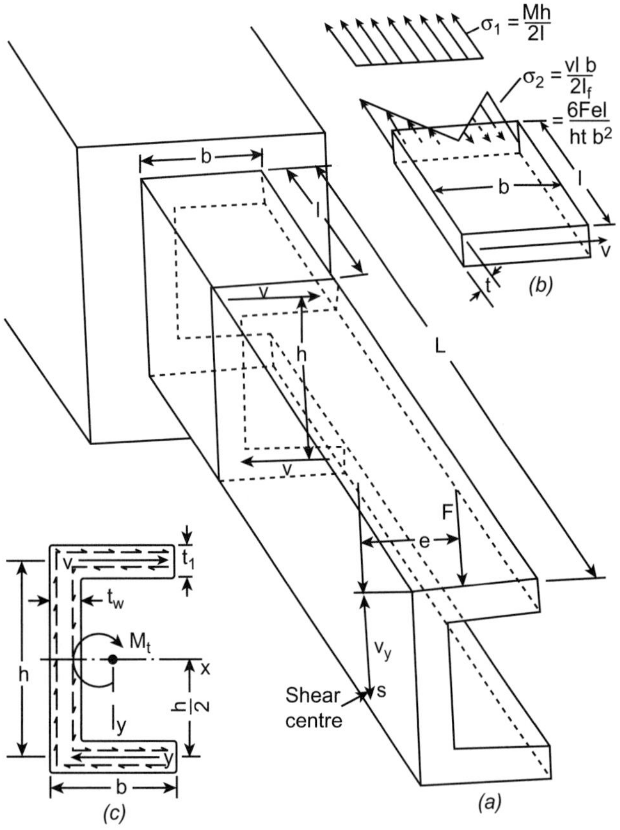

FIGURE 24-73

Particular	Formula
The direct stress	$$\sigma_t = \frac{M}{I}\frac{h}{2} \quad (24\text{-}370)$$ where $M = Fe$
The bending stress	$$\sigma_b = \frac{VIb/2}{I} = \frac{6Fel}{htb^2} \quad (24\text{-}371)$$
The maximum longitudinal stress (Fig. 24-73b)	$$\sigma = \frac{Mh}{2I} + \frac{6Fel}{htb^2} \quad (24\text{-}372)$$
For geometrical properties, weight, and nominal dimensions of beams, channels, T-bars, and equal and unequal angles	Refer to Tables 24-31 and 24-32 and Figs. 24-74 to 24-79.

TABLE 24-30
Formulas for maximum lateral bending moment and angle of twist of beams subjected to torsion[a]

Type of loading and support	Maximum lateral bending moment in flange, lbf in (N m)	Angle of twist of beam of length L, ϕ rad
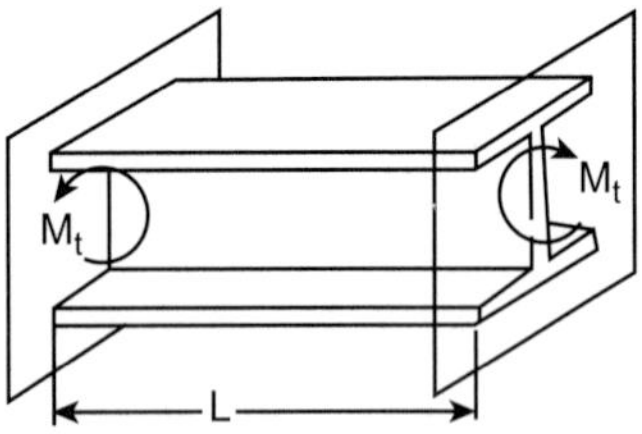	$M_{b(\max)} = \frac{M_t k}{h} \tan h \frac{L}{2k}$ $= \frac{M_t k}{h}$ if $\frac{L}{2k} > 2.5$	$\phi = \frac{M_t}{JG}\left(L - 2k \tan h \frac{L}{2k}\right)$ $= \frac{M_t}{JG}(L - 2k)^{\text{b}}$ if $\frac{L}{2k} > 2.5$
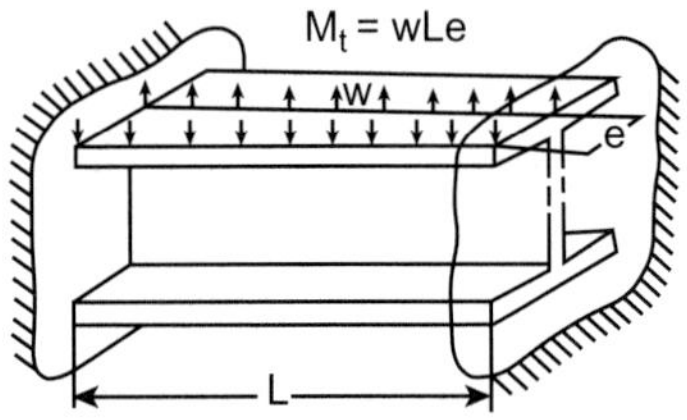	$M_{b(\max)} = \frac{M_t k}{2h}\left(\cot h \frac{L}{2k} - \frac{2k}{L}\right)$ $= \frac{M_t k}{h}$ if $\frac{L}{2k}$ is large	$\phi = \frac{M_t}{2JG}\left(\frac{L}{4} - k \tan h \frac{L}{4k}\right)$ $= \frac{M_t}{2JG}\left(\frac{L}{4} - k\right)^{\text{b}}$ if $\frac{L}{4k} > 2.5$
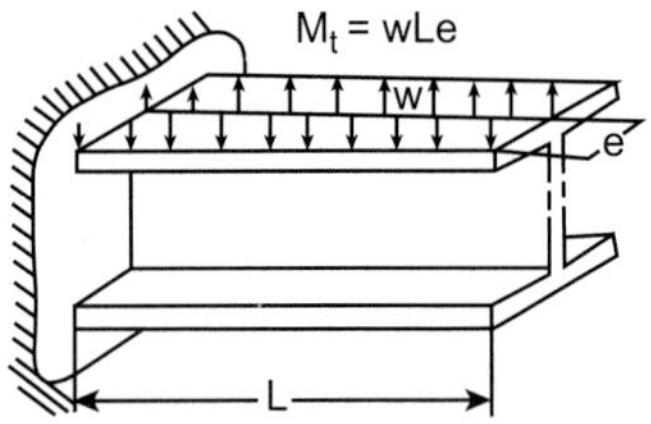	$M_{b(\max)} = \frac{M_t k}{h}\left(\cot h \frac{L}{k} - \frac{k}{L}\right)$ $= \frac{M_t k}{h}$ if $\frac{L}{k}$ is large	$\phi = \frac{M_t}{JG}\left(\frac{L}{2} - k \tan h \frac{L}{2k}\right)$ $= \frac{M_t}{JG}\left(\frac{L}{2} - k\right)^{\text{b}}$ if $\frac{L}{2k} > 2.5$
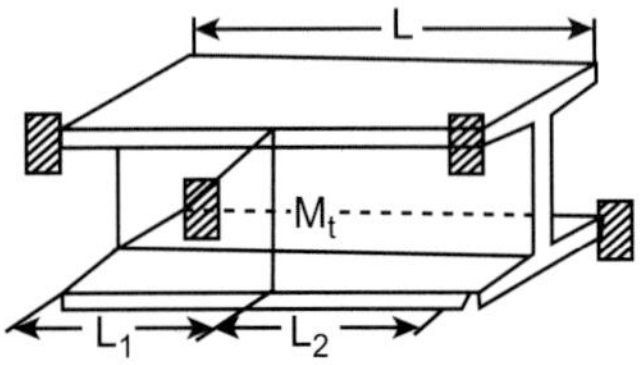	$M_{b(\max)} = \frac{M_t k}{h} \frac{\sin h \frac{L_1}{k} \sin h \frac{L_2}{k}}{\sin h \frac{L}{k}}$ $= \frac{M_t k}{2h}$ if $\frac{L_1}{k}$ and $\frac{L_2}{k} > 2$ error is small	$\phi = \frac{1}{2} \frac{M_t}{JG}\left(\frac{L}{2} - k \tan h \frac{L}{2k}\right)$ (approx.) $= \frac{1}{2} \frac{M_t}{JG}\left(\frac{L}{2} - k\right)^{\text{b}}$ if $\frac{L}{2k} > 2.5$

[a] Formulas given in Table 24-30 can also be used for Z and channel sections.
[b] Error is small for the conditions $L/2k > 2.5$ and $L/4k > 2.5$.

TABLE 24-31
Geometrical properties, weight, and nominal dimensions of beams, channels, and T-bars

Section	Designation	Depth of beam, h	Width of flange, b	Thickness of web, t_w	Thickness of flange, t_f	Slope of flange, D	Radius at root, $r_1(r_r)$	Radius at toe, $r_2(r_t)$	Center of gravity, C_{yy}	Sectional area, A	Weight per meter, w	Moments of inertia I_{xx}	Moments of inertia I_{yy}	Radius of gyration r_{xx}	Radius of gyration r_{yy}	Section moduli Z_{xxx}	Section moduli Z_{yy}
	(1)	(2)	(3)	(4)	(5)	(6)	(7)	(8)	(9)	(10)	(11)	(12)	(13)	(14)	(15)	(16)	(17)
		mm	mm	mm	mm	deg	mm	mm	cm	cm^2	kg	cm^4	cm^4	cm	cm	cm^3	cm^3
Beam or column section (See Fig. 24-74)	ISJB 150	150	50	3.0	4.6	91.5	5.0	1.5		9.01	7.1	322.1	9.2	5.98	1.01	42.9	3.7
	ISJB 175	175	50	3.2	4.3	91.5	5.0	1.5		10.28	8.1	479.3	9.7	6.83	0.97	54.8	3.9
	ISJB 200	200	60	3.4	5.0	91.5	5.0	1.5		12.64	9.9	780.7	17.3	7.86	1.17	78.1	5.8
	ISJB 225	225	80	3.7	5.0	91.5	6.5	1.5		16.28	12.8	1308.5	40.5	8.97	1.58	116.3	10.1
	ISLB 75	75	50	3.7	5.0	91.5	6.5	2.0		7.71	6.1	72.7	10.0	3.07	1.14	19.4	4.0
	ISLB 100	100	50	4.0	6.4	91.5	7.0	3.0		10.21	8.0	168.0	12.7	4.06	1.12	33.6	5.1
	ISLB 125	125	75	4.4	6.5	91.5	8.0	3.0		15.12	11.9	406.8	43.4	5.19	1.69	65.1	11.6
	ISLB 150	150	80	4.8	6.8	91.5	9.5	3.0		18.08	14.2	688.2	55.2	6.17	1.75	91.8	13.8
	ISLB 175	175	90	5.1	6.9	91.5	9.5	3.0		21.30	16.7	1096.2	79.6	7.17	1.93	125.3	17.7
	ISLB 200	200	100	5.4	7.3	91.5	9.5	3.0		25.27	19.8	1696.6	115.4	8.19	2.13	169.7	23.1
	ISLB 225	225	100	5.8	8.6	98	12.0	6.0		29.92	23.5	2501.9	112.7	9.15	1.94	222.4	22.5
	ISLB 250	250	125	6.1	8.2	98	13.0	6.5		35.53	27.9	3917.8	193.4	10.23	2.33	297.4	30.9
	ISLB 275	275	140	6.4	8.8	98	14.0	7.0		42.02	33.0	5375.3	287.0	11.31	2.61	392.4	41.0
	ISLB 300	300	150	6.7	9.4	98	15.0	7.5		48.08	37.7	7332.9	376.2	12.35	2.80	488.9	50.2
	ISLB 325	325	165	7.0	9.8	98	16.0	8.0		54.90	43.1	9874.6	510.8	13.41	3.05	607.7	61.9
	ISLB 350	350	165	7.4	11.4	98	16.0	8.0		63.01	49.5	13158.3	631.9	14.45	3.17	751.9	76.6
	!SLB 400	400	165	8.0	12.5	98	16.0	8.0		72.43	56.9	19306.3	716.4	16.33	3.15	965.3	86.8
	ISLB 450	450	170	8.6	13.4	98	16.0	8.0		83.14	65.3	27536.1	853.0	18.20	3.20	1223.8	100.4
	ISLB 500	500	180	9.2	14.1	98	17.0	8.5		95.50	75.0	38579.0	1063.9	20.10	3.34	1543.2	118.2
	ISLB 550	550	190	9.9	15.0	98	18.0	9.0		109.97	86.3	53161.6	1335.1	21.99	3.48	1933.2	140.5
	ISLB 600	600	210	10.5	15.5	98	20.0	10.0		126.69	99.5	72867.6	1821.9	23.98	3.79	2428.9	173.5
	ISMB 100	100	75	4.0	7.2	98	9.0	4.5		14.60	11.5	257.5	40.8	4.20	1.67	51.5	10.9
	ISMB 125	125	75	4.4	7.6	98	9.0	4.5		16.60	12.0	449.0	43.7	5.20	1.62	71.8	11.7
	ISMB 150	150	80	4.8	7.6	98	9.0	4.5		19.00	14.9	726.4	52.6	6.18	1.66	96.9	13.1
	ISMB 175	175	90	5.5	8.6	98	10.0	5.0		24.62	19.3	1272.0	85.0	7.19	1.86	145.4	18.9
	ISMB 200	200	100	5.7	10.8	98	11.0	5.5		32.33	25.4	2235.4	150.0	8.32	2.15	223.5	30.0
	ISMB 225	225	110	6.5	11.8	98	12.0	6.0		39.72	31.2	3441.8	218.3	9.31	2.34	305.9	39.7
	ISMB 250	250	125	6.9	12.5	98	12.0	6.5		47.55	37.3	5131.6	334.5	10.39	2.65	410.5	53.5
	ISMB 300	300	140	7.5	12.4	98	14.0	7.0		56.26	44.2	8603.6	453.9	12.37	2.84	573.6	64.8
	ISMB 350	350	140	8.1	14.2	98	14.0	7.0		66.71	52.4	13630.3	537.7	14.29	2.84	778.9	76.8
	ISMB 400	400	140	8.9	16.0	98	14.0	7.0		78.46	61.6	20450.4	622.1	16.15	2.82	1022.9	88.9
	ISMB 450	450	150	9.4	17.4	98	15.0	7.5		92.27	72.4	30390.8	834.0	18.15	3.01	1350.7	112.2
	ISMB 500	500	180	10.2	17.2	98	17.0	8.5		110.74	86.9	45218.3	1369.8	20.21	3.52	1808.7	152.2
	ISMB 550	550	190	11.2	19.3	98	18.0	9.0		132.11	103.7	64893.6	1833.8	22.16	3.73	2359.8	193.0
	ISMB 600	600	210	12.0	20.8	98	20.0	10.0		156.21	122.6	91813.0	2651.0	24.24	4.12	3060.4	252.5
	ISWB 150	150	100	5.4	7.0	96	8.0	4.0		21.67	17.0	839.1	94.8	6.22	2.09	111.9	19.0
	ISWB 175	175	125	5.8	7.4	96	8.0	4.0		28.11	22.1	1509.4	188.6	7.33	2.59	172.5	30.2
	ISWB 200	200	140	6.1	9.0	96	9.0	4.5		36.71	28.8	2624.5	328.8	8.46	2.99	262.5	47.0
	ISWB 225	225	150	6.4	9.9	96	9.0	4.5		43.24	33.9	3920.5	448.6	9.52	3.22	348.5	59.8

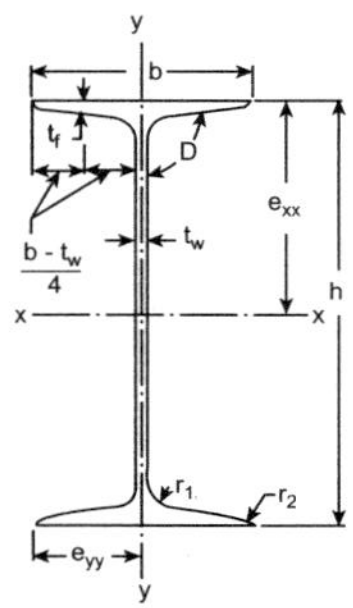

FIGURE 24-74
Beam or column section.

TABLE 24-31
Geometrical properties, weight, and nominal dimensions of beams, channels, and T-bars (*Cont.*)

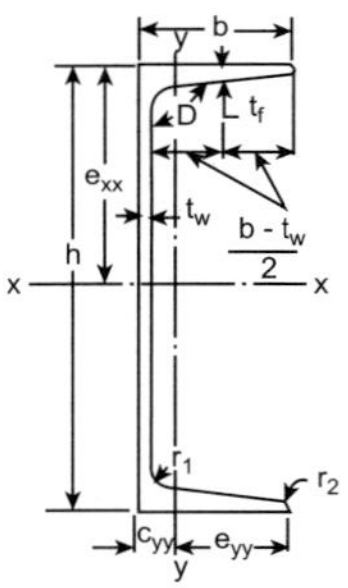

FIGURE 24-75
Channel section.

Section	Designation	Depth of beam, h	Width of flange, b	Thickness of web, t_w	Thickness of flange, t_f	Slope of flange, D	Radius at root, $r_1(r_r)$	Radius at toe, $r_2(r_t)$	Center of gravity, C_{yy}	Sectional area, A	Weight per meter, w	Moments of inertia I_{xx}	Moments of inertia I_{yy}	Radius of gyration r_{xx}	Radius of gyration r_{yy}	Section moduli Z_{xxx}	Section moduli Z_{yy}
		Dimensions of the section															
	(1)	(2)	(3)	(4)	(5)	(6)	(7)	(8)	(9)	(10)	(11)	(12)	(13)	(14)	(15)	(16)	(17)
		mm	mm	mm	mm	deg	mm	mm	cm	cm^2	kg	cm^4	cm^4	cm	cm	cm^3	cm^3
	ISWB 250	250	200	6.7	9.0	96	10.0	5.0		52.05	40.9	5943.1	857.5	10.69	4.06	475.4	85.7
	ISWB 300	300	200	7.4	10.0	96	11.0	5.5		61.33	48.1	9821.6	990.1	12.66	4.02	654.8	99.0
	ISWB 350	350	200	8.0	11.4	96	12.0	6.0		72.50	56.9	15521.7	1175.9	14.63	4.03	887.0	117.6
	ISWB 400	400	200	8.6	13.0	96	13.0	6.5		85.01	66.7	23426.7	1388.0	16.60	4.04	1171.3	138.8
	ISWB 450	450	200	9.2	15.4	96	14.0	7.0		101.15	79.4	35057.6	1706.7	18.63	4.11	1558.1	17.07
	ISWB 500	500	250	9.9	14.7	96	15.0	7.5		121.22	95.2	52290.9	2987.8	20.77	4.96	2091.6	239.0
	ISWB 550	550	250	10.5	17.6	96	16.0	8.0		143.34	112.5	74906.1	3740.5	22.85	5.11	2723.9	299.2
	ISWB 600	600	250	11.2	21.3	96	17.0	8.5		170.38	133.7	106198.5	4702.5	24.97	5.25	3540.0	376.2
	ISHB 150	150	150	5.4	9.0	96	8.0	4.0		34.48	27.1	1455.6	431.7	6.50	3..54	194.1	57.6
	ISHB 200	200	200	6.1	9.0	94	9.0	4.5		47.54	37.3	3608.4	967.1	8.71	4.51	360.8	96.7
	ISHB 225	225	225	6.5	9.1	94	10.0	5.0		54.94	43.1	5279.5	1353.8	9.80	4.96	489.3	120.3
	ISHB 250	250	250	6.9	9.7	94	10.0	5.0		64.96	51.0	7736.5	1961.3	10.91	5.49	618.9	156.9
	ISHB 300	300	250	7.6	10.6	94	11.0	5.5		74.85	58.8	12545.2	2193.6	12.95	5.41	836.3	175.5
	ISHB 350	350	250	8.3	11.6	94	12.0	6.0		85.91	67.4	19159.2	2451.4	14.93	5.34	1094.8	196.1
	ISHB 400	400	250	9.1	12.7	94	14.0	7.0		98.66	77.4	28083.5	2728.3	16.87	5.26	1404.2	218.3
	ISHB 450	450	250	9.8	13.7	94	15.0	7.5		111.14	87.2	39210.8	2985.2	18.78	5.18	1742.7	238.8
								Channel Section									
	ISJC 100	100	45	3.0	5.1	91.5	6.0	2.0	1.40	7.41	5.8	123.8	14.9	4.09	1.42	24.8	4.8
	ISJC 125	125	50	3.0	6.6	91.5	6.0	2.4	1.64	10.07	7.9	270.0	25.7	5.18	1.60	43.2	7.6
	ISJC 150	150	55	3.6	6.9	91.5	7.0	2.4	1.66	12.65	9.9	471.1	37.9	6.10	1.73	62.8	9.9
	ISJC 175	175	60	3.6	6.9	91.5	7.0	3.0	1.75	14.24	11.2	719.9	50.5	7.11	1.88	82.3	11.9
	ISJC 200	200	70	4.1	7.1	91.5	8.0	3.2	1.97	17.77	13.9	1161.2	84.2	8.08	2.18	116.1	16.7
	ISLC 75	75	40	3.7	6.0	91.5	6.0	2.0	1.35	7.26	5.7	66.1	11.5	3.02	1.26	17.6	4.3
	ISLC 100	100	50	4.0	6.4	91.5	6.0	2.0	1.62	10.02	7.9	164.7	24.8	4.06	1.57	32.9	7.3
	ISLC 125	125	65	4.4	6.6	91.5	7.0	2.4	2.0.4	13.67	10.7	356.8	57.2	5.11	2.05	57.1	12.8
	ISLC 150	150	75	4.8	7.8	91.5	8.0	2.4	2.38	18.36	14.4	697.2	103.2	6.19	2.37	93.0	20.2
	ISLC 175	175	75	5.1	9.5	91.5	8.0	3.2	2.40	22.40	17.6	1148.4	126.5	7.16	2.38	131.3	24.8
	ISLC 200	200	75	5.5	10.8	96.0	8.5	3.2	2.35	26.22	20.6	1725.5	146.9	8.11	2.37	172.6	28.5
	ISLC 225	225	90	5.8	10.2	96.0	11.0	3.2	2.46	30.53	24.0	2547.9	209.5	9.14	2.62	226.5	32.0
	ISLC 250	250	100	6.1	10.7	96.0	11.0	3.2	2.70	35.65	28.0	3687.5	298.9	10.17	2.89	295.0	40.9
	ISLC 300	300	100	6.7	11.6	96.0	12.0	3.2	2.55	42.11	33.1	6047.9	346.0	11.98	2.97	603.2	46.4
	ISLC 350	350	100	7.4	12.5	96.0	13.0	4.8	2.41	49.47	38.8	9382.6	394.6	13.72	2.82	532.1	52.0
	ISLC 400	400	100	8.0	14.0	96.0	14.0	4.8	2.36	58.25	45.7	13989.5	460.4	15.50	2.81	699.5	60.2
	ISMC 75	75	40	4.4	7.3	96.0	8.5	2.4	1.31	8.67	6.8	76.0	12.6	2.96	1.21	20.3	4.7
	ISMC 100	100	50	4.7	7.5	96.0	9.0	2.4	1.53	11.70	9.2	186.7	25.9	4.0	1.49	37.3	7.5
	ISMC 125	125	65	5.0	8.1	96.0	9.5	2.4	1.94	16.19	12.7	416.4	59.9	5.07	1.92	66.6	13.1
	ISMC 150	150	75	5.4	9.0	96.0	10.0	2.4	2.22	20.88	16.4	779.4	102.3	6.11	2.21	103.9	19.4
	ISMC 175	175	75	5.7	10.2	96.0	10.5	3.2	2.20	24.38	19.1	1223.3	121.0	7.08	2.23	139.8	22.8
	ISMC 200	200	75	6.7	11.4	96.0	11.0	3.2	2.17	28.21	22.1	1819.3	140.4	8.03	2.23	181.9	26.3

TABLE 24-31
Geometrical properties, weight, and nominal dimensions of beams, channels, and T-bars (*Cont.*)

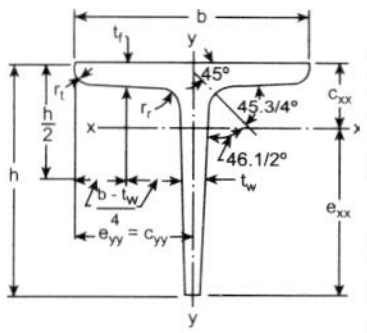

FIGURE 24-76
Normal T-bar.
(See Table 24-32.)

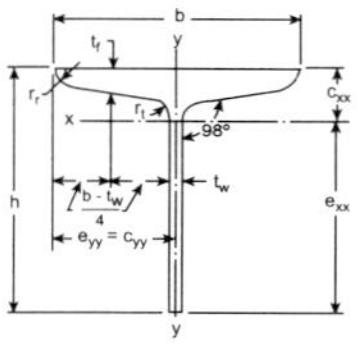

FIGURE 24-77
Slit and deep-legged T-bar.
(See Table 24-32.)

Designation	Dimensions of the section: Depth of beam, h	Width of flange, b	Thickness of web, t_w	Thickness of flange, t_f	Slope of flange, D	Radius at root, $r_1(r_r)$	Radius at toe, $r_2(r_t)$	Center of gravity, C_{yy}	Sectional area, A	Weight per meter, w	Moments of inertia: I_{xx}	I_{yy}	Radius of gyration: r_{xx}	r_{yy}	Section moduli: Z_{xxx}	Z_{yy}
(1)	(2)	(3)	(4)	(5)	(6)	(7)	(8)	(9)	(10)	(11)	(12)	(13)	(14)	(15)	(16)	(17)
	mm	mm	mm	mm	deg	mm	mm	cm	cm^2	kg	cm^4	cm^4	cm	cm	cm^3	cm^3
ISMC 225	225	80	6.4	12.4	96.0	12.0	3.2	2.30	33.01	25.9	2694.6	187.2	9.03	2.38	239.5	32.8
ISMC 250	250	80	7.1	14.1	96.0	12.0	3.2	2.30	38.67	30.4	3816.8	219.1	9.94	2.38	305.3	38.4
ISMC 300	300	90	7.6	13.6	96.0	13.0	3.2	2.36	45.64	35.8	6362.6	310.8	11.81	2.61	424.2	46.8
ISMC 350	350	100	8.1	13.5	96.0	14.0	4.8	2.44	53.66	41.1	10008.0	430.6	13.66	2.83	571.9	57.0
ISMC400	400	100	8.6	15.3	96.0	15.0	4.8	2.42	62.93	49.4	15082.8	504.8	15.48	2.83	754.1	66.6
Normal T-Bar																
ISNT 20	20	20	4.0	4.0		4.0	3.0	0.60	1.45	1.1	0.5	0.2	0.58	0.41	0.3	0.2
ISNT 30	30	30	4.0	4.0		5.0	3.5	0.32	2.26	1.8	0.8	0.8	0.89	0.59	0.8	0.5
ISNT 40	40	40	6.0	6.0		5.5	4.0	1.14	4.45	3.5	6.1	2.9	1.18	0.81	2.1	1.5
ISNT 50	50	50	6.0	6.0		6.0	4.0	1.35	5.66	4.4	12.3	5.7	1.47	1.01	3.4	2.3
ISNT 60	60	60	6.0	6.0		6.5	4.5	1.56	6.85	5.4	21.4	9.7	1.77	1.19	4.8	3.2
ISNT 75	75	75	9.0	9.0		8.0	5.5	2.04	12.69	10.0	62.0	29.2	2.21	1.52	11.4	7.8
ISNT 100	100	100	10.0	10.0		9.0	6.0	2.62	18.97	14.9	163.9	76.8	2.94	2.01	22.2	15.4
ISNT 150	150	150	10.0	10.0		10.0	7.0	3.61	28.88	22.7	541.1	250.3	4.33	2.94	47.5	33.4
Slit and Deep-Legged T-Bar																
ISDT 100	100	50	5.8	10.0	98.0	8.0	4.0	3.03	10.37	8.1	99.0	9.6	3.09	0.96	14.2	3.8
ISDT 150	150	75	8.0	11.6	98.0	9.0	4.5	4.75	19.96	15.7	450.2	37.0	4.75	1.36	43.9	9.9
ISDT 200	200	165	8.0	12.5	98.0	16.0	8.0	4.78	36.22	28.4	1267.8	358.2	5.92	3.15	83.3	43.4
ISDT 250	250	780	9.2	14.1	98.0	17.0	8.5	6.40	47.75	37.5	2774.4	532.0	7.62	3.34	149.2	59.1
ISMT 50	50	75	4.0	7.2	98.0	9.0	4.5	0.96	7.30	5.7	9.7	20.4	1.15	1.67	2.4	5.4
ISMT 62.5	62.5	75	4.4	7.6	98.0	9.0	4.5	1.30	8.30	6.5	21.3	21.9	1.60	1.62	4.3	5.8
ISMT 75	75	80	4.8	7.6	98.0	9.0	4.5	1.67	9.50	7.5	40.1	26.3	2.05	1.66	6.9	6.6
ISMT 87.5	87.5	90	5.5	8.6	98.0	10.0	5.0	1.98	12.31	9.7	72.6	42.5	2.43	1.86	10.7	9.4
ISMT 100	100	100	5.7	10.8	98.0	11.0	5.5	2.13	16.16	12.7	115.8	75.0	2.68	2.15	14.7	15.0
ISMT* 50	50	70	4.5	7.5	98.0	9.0	4.5	1.04	7.35	5.8	10.8	17.7	1.21	1.55	2.7	5.0
ISMT* 62.5	62.5	70	4.8	8.0	98.0	9.0	4.5	1.39	8.40	6.6	22.8	19.2	1.65	1.51	4.7	5.5
ISMT* 75	75	75	5.0	8.0	98.0	9.0	4.5	1.73	9.54	7.5	41.2	23.4	2.08	1.57	7.1	6.2
ISMT* 87.5	87.5	85	5.8	9.0	98.0	10.0	5.0	2.06	12.43	9.8	75.6	38.4	2.47	1.76	11.3	9.0
ISHT 75	75	150	8.4	9.0	94.0	8.0	4.0	1.62	19.49	15.3	96.2	230.2	2.22	3.44	16.4	30.1
ISHT 100	100	200	7.8	9.0	94.0	9.0	4.5	1.91	25.47	20.0	193.8	497.3	2.76	4.42	24.0	49.3
ISHT 125	125	250	8.8	9.7	94.0	10.0	5.0	2.37	34.85	27.4	415.4	1005.8	3.45	5.37	41.0	79.9
ISHT 150	150	250	7.6	10.6	94.0	11.0	5.5	2.66	37.42	29.4	573.7	1096.8	3.92	5.41	46.5	87.7

Key: ISJB—Indian Standard Junior Beams; ISLB—Indian Standard Light-Weight Beams; ISMB—Indian Standard Medium-Weight Beams; ISWB—Indian Standard Wide-Flange Beams; ISHB—Indian Standard Column H-Section Beams; ISJC—Indian Standard Junior Channel; ISLC—Indian Standard Light-Weight Channel; ISMC—Indian Standard Medium-Weight Channel; ISNT—Indian Standard Normal Tee-Bar (T-Bar)., Indian Standard Provisional Slit Medium-Weight Tee-Bars; ISDT—Indian Standard Deep-Legged Tee-Bar; ISLT—Indian Standard Slit Light-Weight Tee-Bar; ISMT—Indian Standard Slit Medium Weight Tee-Bar; ISHT-Indian Standard Slit Tee-Bar from H-Section.
Source: IS 808, 1964; IS 1173, 1967.

TABLE 24-32
General properties, weight, and nominal dimensions of equal and unequal angles

Section	Designation	Dimensions of the section: Section dimensions A, mm	B, mm	Thick-ness t, mm	Radius at root, r_1, mm	Radius at toe, r_2, mm	Center of gravity C_{xx}, cm	C_{yy}, cm	Sectional area, A, cm^2	Weight per meter, w, kg	Moment of inertia I_{xx}, cm^4	I_{yy}, cm^4	I_{uu}, (max), cm^4	I_{vv}, (min), cm^4	Radii of gyration r_{xx}, cm	r_{yy}, cm	r_{uu}, (max), cm	r_{vv}, (min), cm	Moduli of section Z_{xx}, cm^3	Z_{yy}, cm^3	tan α
Section	(1)	(2)	(3)	(4)	(5)	(6)[a]	(7)	(8)	(9)	(10)	(11)	(12)	(13)	(14)	(15)	(16)	(17)	(18)	(19)	(20)	(21)
Equal Angle Section	ISA2020	20	20	3.0	4.0		0.59	0.59	1.12	0.9	0.4	0.4	0.6	0.2	0.58	0.58	0.73	0.37	0.3	0.3	
				4.0			0.63	0.63	1.45	1.1	0.5	0.5	0.8	0.2	0.58	0.58	0.72	0.37	0.4	0.4	
(see Fig. 24-78)	ISA2525	25	25	3.0	4.5		0.71	0.71	1.41	1.1	0.8	0.8	1.2	0.3	0.73	0.73	0.93	0.47	0.4	0.4	
				4.0			0.75	0.75	1.84	1.4	1.0	1.0	1.6	0.4	0.73	0.73	0.91	0.47	0.6	0.6	
				5.0			0.79	0.79	2.25	1.8	1.2	1.2	1.8	0.5	0.72	0.72	0.91	0.47	0.7	0.7	
	ISA3030	30	30	3.0	5.0		0.83	0.83	1.73	1.4	1.4	1.4	2.2	0.6	0.89	0.89	1.13	0.57	0.6	0.6	
				4.0			0.87	0.87	2.26	1.8	1.8	1.8	2.8	0.7	0.89	0.89	1.12	0.57	0.8	0.8	
				5.0			0.92	0.92	2.77	2.2	2.1	2.1	3.4	0.9	0.88	0.88	1.11	0.57	1.0	1.0	
	ISA3535	35	35	3.0	5.0		0.95	0.95	2.03	1.6	2.3	2.3	3.6	0.9	1.05	1.05	1.33	0.67	0.9	0.9	
				4.0			1.00	1.00	2.66	2.1	2.9	2.9	4.7	1.2	1.05	1.05	1.32	0.67	1.2	1.2	
				5.0			1.04	1.04	3.27	2.6	3.5	3.5	5.6	1.5	1.04	1.04	1.31	0.67	1.4	1.4	
	ISA4040	40	40	3.0			1.08	1.08	2.34	1.8	3.4	3.4	5.5	1.4	1.21	1.21	1.54	0.77	1.2	1.2	
				4.0			1.12	1.12	3.07	2.4	4.5	4.5	7.1	1.8	1.21	1.21	1.53	0.77	1.6	1.6	
				5.0			1.16	1.16	3.78	3.0	5.4	5.4	8.6	2.2	1.20	1.20	1.51	0.77	1.9	1.9	
	ISA4545	45	45	3.0	5.5		1.20	1.20	2.64	2.1	5.0	5.0	8.0	2.0	1.38	1.38	1.74	0.87	1.5	1.5	
				4.0			1.25	1.25	3.47	2.7	6.5	6.5	10.4	2.6	1.37	1.37	1.73	0.87	2.0	2.0	
				6.0			1.33	1.33	5.07	4.0	9.2	9.2	14.6	3.8	1.35	1.35	1.70	0.87	2.9	2.9	
	ISA5050	50	50	3.0	6.0		1.32	1.32	2.95	2.3	6.9	6.9	11.1	2.8	1.53	1.53	1.94	0.97	1.9	1.9	
				4.0			1.37	1.37	3.88	3.0	9.1	9.1	14.5	3.6	1.53	1.53	1.93	0.97	2.5	2.8	
				6.0			1.45	1.45	5.68	4.5	12.9	12.9	20.6	5.3	1.51	1.51	1.90	0.96	3.6	3.6	
	ISA5555	55	55	5.0	6.5		1.53	1.53	5.27	4.1	14.7	14.7	23.5	6.9	1.67	1.67	2.11	1.06	3.7	3.7	
				6.0			1.57	1.57	6.26	4.9	17.3	17.3	27.5	7.0	1.66	1.66	2.10	1.06	4.4	4.4	
				10.0			1.72	1.72	10.02	7.9	16.3	16.3	41.5	11.2	1.62	1.62	2.03	1.06	7.0	7.0	
	ISA6060	60	60	5.0	6.5		1.65	1.65	5.75	4.5	19.2	19.2	30.6	7.7	1.82	1.82	2.31	1.16	4.4	4.4	
				6.0			1.69	1.69	6.94	5.4	22.6	22.6	36.0	9.1	1.82	1.82	2.29	1.15	5.2	5.2	
				8.0			1.77	1.77	8.96	7.0	29.0	29.0	46.0	11.9	1.80	1.80	2.27	1.15	6.8	6.8	
	ISA6565	65	65	5.0	6.5		1.77	1.77	6.25	4.9	24.7	24.7	39.4	9.9	1.99	1.99	2.51	1.26	5.2	5.2	
				6.0			1.81	1.81	7.44	5.8	29.1	29.1	46.5	11.7	1.98	1.98	2.50	1.26	6.2	6.2	
				10.0			1.97	1.97	12.00	9.4	45.0	45.0	71.3	18.8	1.94	1.94	2.44	1.25	9.9	9.9	
	ISA7070	70	70	5.0	7.0		1.89	1.89	6.77	5.3	31.1	31.1	49.8	12.5	2.15	2.15	2.71	1.36	6.1	6.1	
				6.0			1.94	1.94	8.06	6.3	36.8	36.8	58.8	14.8	2.14	2.14	2.70	1.36	7.3	7.3	
				8.0			2.02	2.02	10.58	8.3	47.4	47.4	75.5	19.3	2.12	2.12	2.67	1.35	9.5	9.5	
	ISA7575	75	75	5.0	7.0		2.02	2.02	7.27	5.7	38.7	38.7	61.9	15.5	2.31	2.31	2.92	1.46	7.1	7.1	
				6.0			2.06	2.06	8.66	6.8	45.7	45.7	73.1	18.4	2.30	2.30	2.91	1.46	8.4	8.4	
				10.0			2.22	2.22	14.02	11.0	71.4	71.4	113.3	29.4	2.26	2.26	2.84	1.45	13.5	13.5	
	ISA8080	80	80	6.0	8.0		2.18	2.18	9.29	7.3	56.0	56.0	89.6	22.5	2.46	2.46	3.11	1.52	9.6	9.6	
				8.0			2.27	2.27	12.21	9.6	72.5	72.5	115.6	29.4	2.44	2.44	3.08	1.55	12.6	12.6	
				10.0			2.34	2.34	15.05	11.8	87.7	87.7	139.5	36.0	2.41	2.41	3.04	1.55	15.5	15.5	
	ISA9090	90	90	6.0	8.0		2.42	2.42	10.47	8.2	80.1	80.1	128.1	32.0	2.77	2.77	3.50	1.75	12.2	12.2	
				8.0			2.51	2.51	13.79	10.8	104.2	104.2	166.4	42.0	2.75	2.75	3.47	1.75	16.0	16.0	
				10.0			2.59	2.59	17.03	13.4	126.7	126.7	201.9	51.6	3.73	2.73	3.44	1.70	19.2	19.8	
	ISA100100	100	100	6.0	8.5		2.67	2.67	11.67	9.2	111.3	111.3	178.1	44.5	3.09	3.09	3.91	1.95	15.8	15.2	
				8.0			2.76	2.76	15.39	12.1	145.1	145.1	231.8	58.4	3.07	3.07	3.88	1.95	20.0	20.0	
				12.0			2.92	2.92	22.59	17.7	207.0	207.0	329.3	87.7	3.03	3.03	3.82	1.94	29.2	29.2	

TABLE 24-32
General properties, weight, and nominal dimensions of equal and unequal angles (*Cont.*)

Section	Designation	Section dimensions A, mm	Section dimensions B, mm	Thickness t, mm	Radius at root, r_1, mm	Radius at toe, r_2, mm	Center of gravity C_{xx}, cm	Center of gravity C_{yy}, cm	Sectional area, A, cm^2	Weight per meter, w, kg	Moment of inertia I_{xx}, cm^4	I_{yy}, cm^4	I_{uu}, (max), cm^4	I_{vv}, (min), cm^4	Radii of gyration r_{xx}, cm	r_{yy}, cm	r_{uu}, (max), cm	r_{vv}, (min), cm	Moduli of section Z_{xx}, cm^3	Z_{yy}, cm^3	tan α
Section	(1)	(2)	(3)	(4)	(5)	(6)[a]	(7)	(8)	(9)	(10)	(11)	(12)	(13)	(14)	(15)	(16)	(17)	(18)	(19)	(20)	(21)
	ISA110110	110	110	8.0	10.0		3.00	3.00	17.08	13.4	196.8	196.8	312.7	81.0	3.40	3.40	4.28	2.18	24.6	24.6	
				10.0			3.09	3.09	21.12	16.6	240.2	240.2	381.5	98.9	3.37	3.37	4.25	2.16	30.4	30.4	
				16.0			3.32	3.32	32.76	25.7	357.3	357.3	564.3	150.0	3.30	3.30	4.15	2.14	46.5	46.5	
	ISA130130	130	130	8.0	10.0		3.50	3.50	20.28	15.9	331.0	331.0	526.3	135.6	4.04	4.04	5.10	2.59	34.9	34.9	
				10.0			3.59	3.59	25.12	19.7	405.3	405.3	644.6	166.0	4.02	4.02	5.07	2.57	43.1	43.1	
				12.0			2.67	3.67	29.88	23.5	476.4	476.4	757.1	195.6	3.99	3.99	5.03	2.56	51.0	51.0	
				16.0			3.82	3.82	39.16	30.7	609.1	609.1	965.6	252.6	3.94	3.94	4.97	2.54	66.3	66.3	
	ISA150150	150	150	10.0	12.0		4.08	4.08	29.21	22.9	633.5	633.5	1007.4	259.6	4.66	4.66	5.87	2.98	58.0	58.0	
				12.0			4.16	4.16	34.77	27.3	746.3	746.3	1186.6	305.9	4.63	4.63	5.94	2.97	68.8	68.8	
				16.0			4.31	4.31	45.65	35.8	958.9	958.9	1522.5	395.3	4.58	4.58	5.77	2.94	89.7	89.7	
				20.0			4.46	4.46	56.21	44.1	1153.5	1153.5	1829.6	481.3	4.53	4.53	5.71	2.93	109.7	109.7	
	ISA200200	200	200	12.0	15.0		5.39	5.39	46.94	36.9	1826.3	1826.3	2905.4	747.2	6.24	6.24	7.87	3.99	125.0	125.0	
				16.0			5.56	5.56	61.82	48.5	2366.2	2366.2	3764.1	958.5	6.19	6.19	7.80	3.96	163.8	163.8	
				20.0			5.71	5.71	76.38	60.0	2875.0	2875.0	4568.6	1181.4	6.14	6.14	7.73	3.93	201.2	201.2	
				25.0			5.90	5.90	94.13	73.9	3470.2	3470.2	5501.5	1438.8	6.07	6.07	7.69	3.91	246.0	246.0	
Unequal Angle Section (see Fig. 24-79)																					
	ISA3020	30	20	3.0	4.5		0.98	0.49	1.41	1.1	1.2	0.4	1.4	0.2	0.92	0.54	0.99	0.41	0.6	0.3	0.43
				4.0			1.02	0.53	1.84	1.4	1.5	0.5	1.8	0.3	0.92	0.54	0.98	0.41	0.8	0.4	0.42
				5.0			1.06	0.57	2.25	1.8	1.9	0.6	2.1	0.4	0.91	0.53	0.97	0.41	1.0	0.4	0.41
	ISA4025	40	25	2.0	5.0		1.30	0.57	1.88	1.5	3.0	0.9	3.3	0.5	1.25	0.68	1.33	0.52	1.1	0.5	0.38
				4.0			1.35	0.62	2.46	1.9	3.8	1.1	4.3	0.7	1.25	0.68	1.32	0.52	1.4	0.6	0.38
				5.0			1.39	0.66	3.02	2.4	4.6	1.4	5.1	0.8	1.24	0.67	1.31	0.52	1.8	0.7	0.37
				6.0			1.43	0.69	3.05	2.8	5.4	1.6	5.9	1.0	1.23	0.66	1.29	0.52	2.1	0.9	0.37

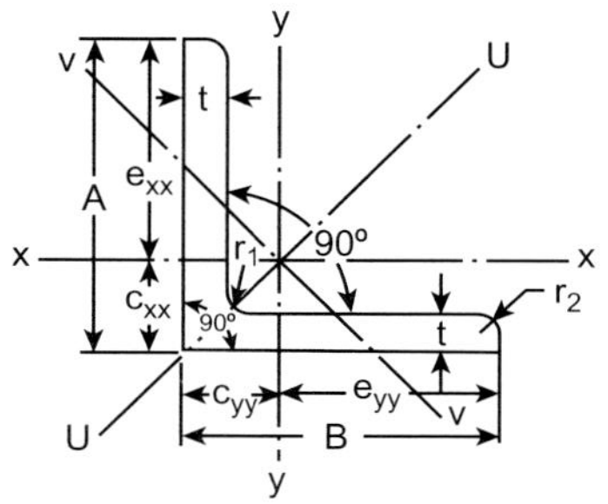

FIGURE 24-78 Equal-angle section.

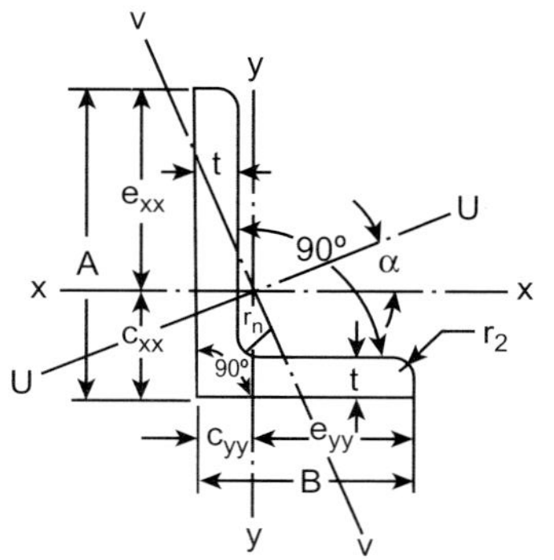

FIGURE 24-79 Unequal-angle section. (See Table 24-32.)

TABLE 24-32
General properties, weight, and nominal dimensions of equal and unequal angles (*Cont.*)

Section	Designation	Dimensions of the section: Section dimensions A, mm	Section dimensions B, mm	Thickness t, mm	Radius at root, r_1, mm	Radius at toe, r_2, mm	Center of gravity C_{xx}, cm	Center of gravity C_{yy}, cm	Sectional area, A, cm^2	Weight per meter, w, kg	Moment of inertia I_{xx}, cm^4	I_{yy}, cm^4	I_{uu}, (max), cm^4	I_{vv}, (min), cm^4	Radii of gyration r_{xx}, cm	r_{yy}, cm	r_{uu}, (max), cm	r_{vv}, (min), cm	Moduli of section Z_{xx}, cm^3	Z_{yy}, cm^3	tan α
Section	(1)	(2)	(3)	(4)	(5)	(6)[a]	(7)	(8)	(9)	(10)	(11)	(12)	(13)	(14)	(15)	(16)	(17)	(18)	(19)	(20)	(21)
	ISA4530	45	30	3.0	5.0		1.42	0.69	2.18	1.7	4.4	1.5	5.0	0.9	1.42	0.84	1.52	0.63	1.4	0.7	0.44
				4.0			1.47	0.73	2.86	2.2	5.7	2.0	6.5	1.1	1.41	0.84	1.51	0.63	1.9	0.9	0.43
				5.0			1.51	0.77	3.52	2.8	6.9	2.4	7.9	1.4	1.40	0.83	1.50	0.63	2.3	1.1	0.43
				6.0			1.55	0.81	4.16	3.3	8.0	2.8	9.2	1.7	1.39	0.82	1.49	0.63	2.7	1.3	0.42
	ISA5030	50	30	3.0	5.5		1.63	0.65	2.34	1.8	5.9	1.6	6.5	1.0	1.59	0.82	1.67	0.65	1.7	0.7	0.36
				4.0			1.68	0.70	3.07	2.4	7.7	2.1	8.5	1.2	1.58	0.82	1.66	0.63	2.3	0.9	0.36
				5.0			1.72	0.74	3.78	3.0	9.3	2.5	10.3	1.5	1.57	0.81	1.65	0.63	2.8	1.1	0.35
				6.0			1.76	0.78	4.47	3.5	10.9	2.9	11.9	1.8	1.56	0.80	1.64	0.63	3.4	1.3	0.35
	ISA6040	60	40	5.0	6.0		1.95	0.96	4.76	3.7	16.9	6.0	19.5	3.4	1.89	1.12	2.02	0.85	4.2	2.0	0.44
				6.0			1.99	1.00	5.65	4.4	19.9	7.0	22.8	4.0	1.88	1.11	2.01	0.85	5.0	2.3	0.43
				8.0			2.07	1.08	7.37	5.8	25.4	8.8	29.0	5.2	1.86	1.10	1.98	0.84	6.5	3.0	0.42
	ISA6545	65	45	5.0	6.0		2.07	1.08	5.26	4.1	22.1	8.6	25.9	4.8	2.05	1.28	2.22	0.96	5.0	2.5	0.47
				6.0			2.11	1.12	6.25	4.9	26.0	10.1	30.4	5.7	2.04	1.27	2.21	0.95	5.9	3.0	0.47
				8.0			2.19	1.20	8.17	6.4	33.2	12.8	38.7	7.4	2.02	1.25	2.18	0.95	7.7	3.9	0.46
	ISA7045	70	45	5.0	6.5		2.27	1.04	5.52	4.3	27.2	8.3	30.9	5.1	2.22	1.26	2.36	0.96	5.7	2.5	0.41
				6.0			2.32	1.09	6.56	5.2	32.0	10.3	36.3	6.0	2.21	1.25	2.35	0.96	6.8	3.0	0.41
				8.0			2.40	1.16	8.58	6.7	41.0	13.1	46.3	7.8	2.19	1.24	2.32	0.95	8.9	3.9	0.40
				10.0			2.48	1.24	10.52	8.3	49.3	15.6	55.4	9.5	2.16	1.22	2.29	0.95	10.9	4.8	0.39
	ISA7550	75	50	5.0	6.5		2.39	1.16	6.02	4.7	34.1	12.2	39.4	6.9	2.38	1.42	2.56	1.07	6.7	3.2	0.44
				6.0			2.44	1.20	7.16	5.6	40.3	14.3	46.4	8.2	2.37	1.41	2.55	1.07	8.0	3.8	0.44
				8.0			2.52	1.28	9.38	7.4	51.8	18.3	59.4	10.6	2.35	1.40	2.52	1.06	10.4	4.9	0.43
				10.0			2.60	1.36	11.52	9.0	62.3	21.8	71.2	12.9	2.33	1.38	2.49	1.06	12.7	6.0	0.42
	ISA8050	80	50	5.0	7.0		2.60	1.12	6.27	4.9	48.6	12.3	45.7	7.2	2.55	1.40	2.70	1.07	7.5	3.2	0.39
				6.0			2.64	1.16	7.46	5.9	48.0	14.4	53.9	8.5	2.54	1.39	2.69	1.07	9.0	3.8	0.39
				8.0			2.73	1.24	9.78	7.7	61.9	18.5	69.3	11.0	2.52	1.37	2.66	1.06	11.7	4.9	0.38
				10.0			2.81	1.32	12.02	9.4	74.7	22.1	83.3	13.5	2.49	1.36	2.63	1.06	14.4	6.0	0.38
	ISA9060	90	60	6.0	7.5		2.87	1.39	8.65	6.8	70.6	25.2	81.5	14.3	2.86	1.17	3.07	1.28	11.5	5.5	0.44
				8.0			2.96	1.48	11.37	8.9	91.3	32.4	105.3	18.6	2.84	1.69	3.04	1.28	15.1	7.2	0.44
				10.0			3.04	1.55	14.01	11.0	110.9	39.1	127.3	22.8	2.81	1.67	3.01	1.27	18.6	8.8	0.43
				12.0			3.12	1.63	16.57	13.0	129.1	45.2	147.5	26.8	2.79	1.65	2.98	1.27	22.0	10.3	0.42
	ISA10065	100	65	6.0	8.0		3.19	1.47	9.55	7.5	96.7	32.4	110.6	18.6	3.18	1.94	3.40	1.39	14.2	6.4	0.42
				8.0			3.28	1.55	12.57	9.9	125.9	41.9	143.6	24.2	3.16	1.83	3.38	1.39	18.7	8.5	0.42
				10.0			3.37	1.63	15.51	12.2	153.2	50.7	174.2	29.7	3.14	1.81	3.35	1.38	23.1	10.4	0.41
	ISA10075	100	75	6.0	8.5		3.01	1.78	10.14	8.0	100.9	48.7	124.0	25.6	3.15	2.19	3.50	1.59	14.4	8.5	0.55
				8.0			3.10	1.87	13.36	10.5	131.6	63.3	161.3	33.6	3.14	2.18	3.48	1.59	19.1	11.2	0.55
				10.0			3.19	1.95	16.50	13.0	160.4	76.9	196.1	41.2	3.12	2.16	3.45	1.58	23.6	13.8	0.55
				12.0			3.27	2.03	19.56	15.4	187.5	89.5	228.4	48.6	3.10	2.14	3.42	1.58	27.9	16.3	0.54
	ISA12575	125	75	6.0	9.0		4.05	1.59	11.66	9.2	187.8	51.6	208.9	30.5	4.01	2.10	4.23	1.62	22.2	8.7	0.37
				8.0			4.15	1.68	15.38	12.1	245.5	67.2	272.8	40.0	4.00	2.09	4.21	1.61	29.4	11.5	0.36
				10.0			4.24	1.76	19.02	14.9	300.3	81.6	332.9	49.1	3.97	2.07	4.18	1.61	36.3	44.2	0.36
	ISA12595	125	95	6.0	9.0	4.8	3.72	2.24	12.92	10.1	205.5	103.6	254.0	55.1	3.99	2.83	4.43	2.07	23.4	14.3	0.57
				8.0			3.80	2.32	17.04	13.4	268.3	134.7	331.4	71.7	3.97	2.81	4.41	2.05	30.9	18.8	0.57
				10.0			3.89	2.40	21.08	16.5	328.0	164.1	404.5	87.6	3.95	2.79	4.38	2.04	38.1	23.1	0.56
				12.0			3.97	2.48	25.04	19.7	384.8	191.8	473.7	103.0	3.92	2.77	4.35	2.03	45.1	27.3	0.56

TABLE 24-32
General properties, weight, and nominal dimensions of equal and unequal angles (*Cont.*)

Section	Dimensions of the section: Designation	Section dimensions A, mm	Section dimensions B, mm	Thickness t, mm	Radius at root, r_1, mm	Radius at toe, r_2, mm	Center of gravity C_{xx}, cm	Center of gravity C_{yy}, cm	Sectional area, A, cm^2	Weight per meter, w, kg	Moment of inertia I_{xx}, cm^4	I_{yy}, cm^4	I_{uu}, (max), cm^4	I_{vv}, (min), cm^4	Radii of gyration r_{xx}, cm	r_{yy}, cm	r_{uu}, (max), cm	r_{vv}, (min), cm	Moduli of section Z_{xx}, cm^3	Z_{yy}, cm^3	$\tan\alpha$
	(1)	(2)	(3)	(4)	(5)	(6)[a]	(7)	(8)	(9)	(10)	(11)	(12)	(13)	(14)	(15)	(16)	(17)	(18)	(19)	(20)	(21)
	ISA15075	150	75	8.0	10.0	4.8	5.24	1.54	17.48	13.7	410.3	71.1	435.7	45.7	4.85	2.02	4.99	1.62	42.0	11.9	0.26
				10.0			5.33	1.62	21.62	17.0	502.2	86.3	532.7	55.7	4.82	2.00	4.96	1.61	51.9	14.7	0.26
				12.0			5.42	1.70	25.68	20.2	590.0	100.4	625.1	65.4	4.79	1.98	4.93	1.60	61.6	17.3	0.26
	ISA150115	150	115	8.0	11.0	4.8	4.48	2.76	20.72	16.3	474.4	244.4	589.6	129.2	4.78	3.43	5.33	2.50	45.1	28.0	0.58
				10.0			4.57	2.84	25.66	20.1	581.8	298.8	722.6	158.0	4.76	3.41	5.31	2.48	55.8	34.5	0.58
				12.0			4.65	2.92	30.52	24.0	684.8	350.6	849.5	185.9	4.74	3.39	5.28	2.47	66.2	40.8	0.57
	ISA200100	200	100	10.0	12.0	4.8	6.98	2.03	29.21	22.9	1227.8	214.7	1304.9	137.6	6.48	2.71	6.68	2.17	94.3	26.9	0.27
				12.0			7.07	2.11	34.77	27.3	1449.2	251.5	1539.0	161.6	6.46	2.69	6.65	2.16	112.1	31.9	0.26
				16.0			7.25	2.27	45.66	35.8	1870.1	319.6	1982.1	207.7	6.40	2.65	6.59	2.13	146.5	41.5	0.26
	ISA200150	200	150	10.0	13.5	4.8	6.02	3.35	34.29	26.9	1409.2	688.9	1729.6	368.5	6.41	4.48	7.10	3.28	100.8	60.2	0.26
				12.0			6.11	3.63	40.85	33.1	1655.6	812.1	2043.3	434.5	6.39	4.46	7.07	3.26	119.9	71.4	0.55
				16.0			6.72	3.79	53.83	42.2	2155.2	1044.9	2638.6	561.5	6.33	4.41	7.01	3.23	157.0	93.2	0.55

[a] For the cases for which the radius at toe r_2 is not given, the toe should be reasonably square. *Source*: IS 808, 1964.

REFERENCES

1. Lingaiah, K., and B. R. Narayana Iyengar, *Machine Design Data Handbook, fps* system, Engineering College Co-operative Society, Bangalore, India, 1962.
2. Lingaiah, K., and B. R. Narayana Iyengar, *Machine Design Book Handbook*, Volume I. (SI and Customary Metric Units), Suma Publishers, Bangalore, India, 1986.
3. Lingaiah, K., *Machine Design Data Handbook*, Volume II (S.I. and Customary Metric Units), Suma Publishers, Bangalore, India, 1986.
4. Lingaiah, K., *Machine Design Data Handbook*, McGraw-Hill Publishing Company, New York, 1994.
5. Maleev, V. L., and J. B. Hartman, *Machine Design*, International Text book Company, Scranton, Pennsylvania, 1954.
6. Black, P. H., and O. E. Adams, Jr., *Machine Design*, McGraw-Hill Book Company Inc., New York, 1968.
7. Neale, M. J., *Tribology Handbook*, Newnes-Butterworth, London, 1973.
8. Bach, *Maschinenelemente*, 12th ed., P-43.
9. Heldt, P. M. *Torque Converters for Transmissions*, Chiltan Company, Philadelphia, 1955.
10. Newton, K., and W. Steeds, *The Motor Vehicle*, Iliffe and Sons Ltd, London, 1950.
11. Steeds, W., *Mechanics of Road Vehicle*, Iliffe and Sons, Ltd., London, 1960.
12. Arkhangelsky, V., et al., *Motor Vehicles Engines*, Mir Publisher, Moscow, 1971.
13. Heldt, P. M., *High Speed Combustion Engines*, 6th ed., Chilton Company, Philadelphia, 1955.
14. Thimoshenko, S., and J. N. Goodier, *Theory of Elasticity*, McGraw-Hill Book Company, Inc., and Kogakkusha Company Ltd., Tokyo, 1951.
15. Timoshenko, S., and S. Woinowsky-Krieger, *Theory of Plates and Shells*, McGraw-Hill Book Company, Inc., New York, 1959.
16. Seely, F. B., and J. O. Smith, *Advanced Mechanics of Materials*, John Wiley and Sons, Inc., 2nd ed., 1959.
17. Timoshenko, S., and J. M. Gere, *Mechanics of Materials*, Von Nostrand Reinhold Company, New York, 1972.
18. Goetze, A. G. *Piston Ring Manual*, 3rd ed., Burscheid, Germany, 1987.
19. Bureau of Indian Standards, New Delhi, India.

CHAPTER

25

ELEMENTS OF MACHINE TOOL DESIGN

SYMBOLS

A_i	cross-sectional area of chip before removal from workpiece, m^2 (in^2)
B	width of V or U die, m (in)
c	chisel edge length, m (in)
d	diameter of the hole, m (in)
	diameter of the drill, m (in)
	depth of cut, m (in)
D	blank diameter, m (in)
	diameter of milling cutter, m (in)
	shell diameter, m (in)
D_m	diameter of machined surface, m (in)
D_w	diameter of workpiece or job, m (in)
E	Young's modulus, GPa (kpsi)
	work done in punching or shearing of material, N m (lbf ft)
	work done per unit volume of removed, J/mm^3
F	force, kN (lbf)
$F_a = F_x$	axial component of cutting force, kN (lbf)
F_c	cutting force, kN (lbf)
F_{hc}	normal cutting force or the resultant force on a single point metal cutting tool in the horizontal plane, kN (lbf)
F_{max}	maximum force at the punch, kN (lbf)
F_n	force normal to F_μ, kN (lbf)
$F_{n\tau}$	force normal to shear force, kN (lbf)
F_r	reaction of the cutting force, kN (lbf)
	radial component of the cutting force, kN(lbf)
F_R	resultant cutting force, kN (lbf)
F_s	stripping pressure, kN (lbf)
$F_f = F_x$	feed force, kN (lbf)
$F_t = F_z = F_c$	tangential component of the cutting force, kN(lbf)
$F_y = F_r$	radial component of the cutting force, kN (lbf)
F_μ	frictional force on tool face, kN (lbf)
$F_\rho = \mu F_r$	frictional force of the saddle on lathe bed, kN (lbf)

F_τ	shear force kN (lbf)
h	depth of cut, m (in)
	shell height, m (in)
h_c	height of the lathe center, m (in)
h_{sw}	swing over the bed of the lathe, m (in)
K	constant of proportionality in Eq. (25-31)
K	constant
	coefficient, also with subscripts
K_c	clearance
L	length of cut or perimeter of the cut, m (in)
L_m	length of job, m (in)
m, m_1, m_2, m_3, m_4	exponents
M_b	bending moment, N m (lbf ft)
M_t	turning moment N m (ft lbf)
n	speed, rpm
n'	speed, rps
p	pitch of thread, mm (in)
p, q	exponents
P	periphery, m (in)
P	power, kW (hp)
P_c	power at cutting tool, kW (hp)
P_g	gross or motor power, kW (hp)
P_f	tare power, that is the power required to run the machine at no load, kW (hp)
$P_u = P_s$	unit power or specific power, kW (hp)
Q	metal removal rate, cm^3/min (in^3/min)
r	radius, m (in)
r_c	corner radius, m (in)
$r_c = (t_1/t_2)$	cutting ratio
r_n	nose radius, m (in)
R	roughness height, μm (μin)
R_i	inside radius of bend, m (in)
s	feed rate, mm/rev (in/rev)
s_z	feed per tooth of milling cutter, mm/tooth
t	thickness of material to be punched or sheared, m (in)
	thickness of chip, m (in)
	depth of cut, m (in)
t_1	initial thickness of the chip, m (in)
t_2	final thickness of the chip, m (in)
u_c	number of chamfered threads
u_s	number of splines
v	velocity, m/s (ft/min)
v_c	cutting speed, m/min (ft/min)
v_f	feed rate, mm/min (in/min)
V	cutting velocity, m/min (ft/min)
W	work done per unit volume, N m/m^3 (lbf ft/ft^3)
x, v, z	machine reference co-ordinate axes
z	number of teeth on milling cutter
α	relief or clearance angle, deg
α_f	side relief or clearance angle, deg
α_n	normal relief or clearance angle, deg
α_o	orthogonal clearance angle, deg
α_p	front or end relief or clearance angle, deg
α_{tr}	true relief or clearance angle, deg

β	helix angle, deg
β_b	bevel angle, deg
γ	rake angle, also with subscripts, deg
γ_f	side rake angle, deg
γ_n	normal rake angle, deg
γ_o	orthogonal rake angle, deg
γ_p	back rake angle, deg
δ	deflection, mm (in)
	gullet angle, deg
ε	peripheral pitch angle, deg
η	efficiency
ν	wedge angle, also with subscripts, deg
θ	chip flow angle, deg
	approach angle in Eq. (25-104b), deg
θ_c	corner angle in Eq. (25-104a)
θ_b	bend angle, deg
λ_s	inclination angle, deg
μ	coefficient of friction
ρ	friction angle, deg
σ	stress, also with suffices, MPa (kpsi)
σ_c	compressive stress, MPa (kpsi)
σ_{sut}	ultimate tensile strength, MPa (kpsi)
σ_y	yield stress, MPa (kpsi)
τ	tool life
	shear stress, MPa (kpsi)
$\tau_n (= \tau_{su})$	resistance of the material to shearing or the ultimate shearing strength, MPa (kpsi)
ϕ	cutting edge angle, also with subscripts, deg
	shear angle, deg
ϕ_o	end cutting edge angle, deg
ϕ_p	principal cutting edge angle, deg
ϕ_s	side cutting edge angle, deg
ψ	engagement angle for milling depth, deg
ω	angular speed, rad/s

Note: σ and τ with subscript s designates strength properties of material used in the design which will be used and observed throughout this *Machine Design Data Handbook*.

Other factors in performance or in special aspects are included from time to time in this chapter and, being applicable, only in their immediate context, are not given at this stage.

Particular	Formula	

25.1 METAL CUTTING TOOL DESIGN

25.1.1 Forces on a single-point metal cutting tool

Nomenclature of metal cutting tool	Refer to Fig. 25-1.	
The normal cutting force or resultant force on a single-point metal cutting tool in horizontal plane (Fig. 25-2)	$F_{hc} = \sqrt{F_f^2 + F_r^2}$	(25-1)
The resultant cutting force on the cutting tool (Fig. 25-2)	$F_R = \sqrt{F_t^2 + F_{hc}^2}$	(25-2)
	$F_R = \sqrt{F_f^2 + F_r^2 + F_t^2}$	(25-3)

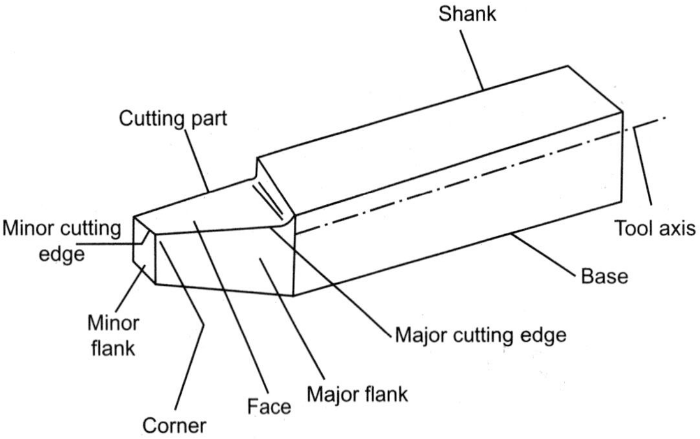

FIGURE 25-1 Nomenclature of metal cutting tool.

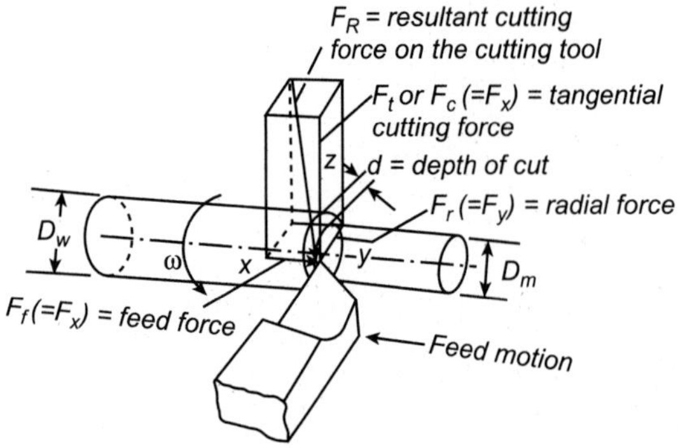

FIGURE 25-2 Components of cutting force acting on a single point metal cutting tool.

Particular	Formula
	where F_t or F_c $(=F_z)$ = tangential cutting force perpendicular to F_r $(=F_y)$ and F_f $(=F_x)$ in the vertical plane. F_r $(=F_y)$ = radial force perpendicular to the direction of feed and in the horizontal plane. F_f $(=F_x)$ = feed force in the horizontal plane against the direction of the feed. x, y and z are machine reference axes along feed force F_f, radial force F_r, and cutting force F_t or F_c directions, respectively.

25.1.2 Merchant's circle for cutting forces for a single-point metal cutting tool

Particular	Formula
The co-efficient of friction in orthogonal cutting (Fig.25-3)	$\mu = \tan\rho = \dfrac{F_\mu}{F_n} = \dfrac{F_{hc} + F_c \tan\alpha}{F_c - F_{hc}\tan\alpha}$ (25-4)
The shear force	$F_\tau = F_c\cos\phi - F_{hc}\sin\phi$ (25-5)
The friction force	$F_\mu = F_{hc}\cos\alpha + F_c\sin\alpha$ (25-6)
Mean shear stress	$\tau = \dfrac{F_c\sin\phi\cos\phi - F_{hc}\sin^2\phi}{A_i}$ (25-7)

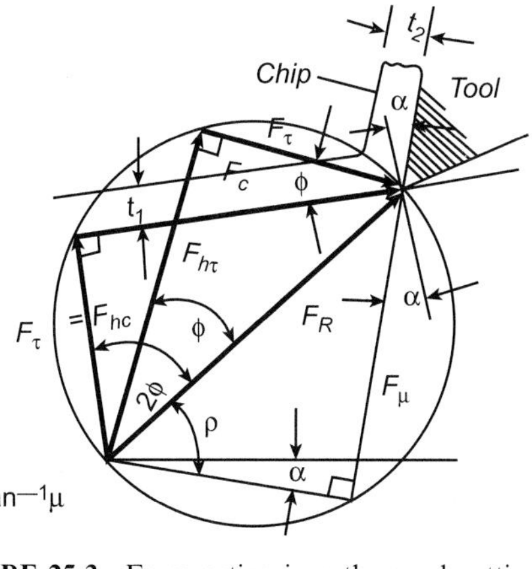

FIGURE 25-3 Force acting in orthogonal cutting with a continuous chip. *Courtesy*: ASTME, *Tool Engineers' Handbook*, 2nd Edition, McGraw-Hill Book Company, New York, 1959.

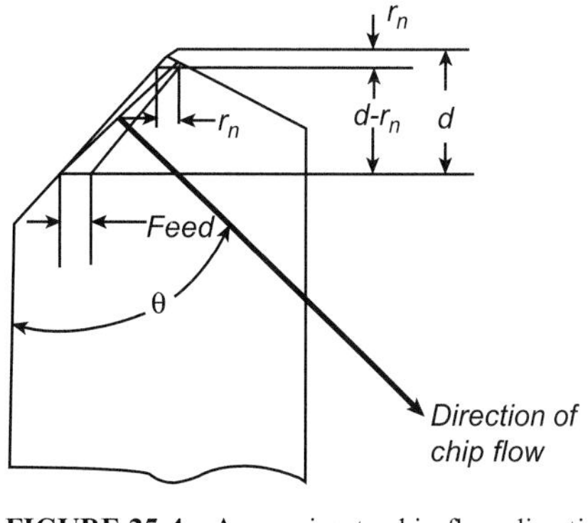

FIGURE 25-4 Approximate chip flow direction.

Particular	Formula
Work done in shearing the material	$W_\tau = \tau[\cot\phi + \tan(\phi - \alpha)]$ (25-8)
Work done in overcoming friction	$W_\mu = \frac{F_\mu}{A_i}\frac{\sin\phi}{\cos(\phi - \alpha)}$ (25-9)
The total work done in cutting	$W_t = \frac{F_c}{A_i}$ (25-10)
The shear angle (ϕ)	$\tan\phi = \frac{r_c \cos\alpha}{1 - r_c \sin\alpha}$ (25-11)
The tangential cutting force	$F_t = KCs^{m_1} d^{m_2}$ (25-12) where d = depth of cut, m (mm) m_1 = slope of F_t versus s graph (typical values 0.5 to 0.98) m_2 = slope of F_t versus d graph (typical values 0.90 to 1.4) K = overall correction coefficient, depends on actual conditions of tool angles and working conditions (varies from 0.9 to 1.0) C = coefficient characterized by material of job, condition of working tool, coolants, etc. (Table 25-1)
The values of K in Eq. (25-12) are calculated from equation	$K = K_m K_\gamma K_\phi K_c$ (25-12a) where K_m = material correction coefficient K_γ = correction coefficient, depends on back rack angle K_c = correction coefficient for coolant used K_ϕ = correction coefficient, depends on top rack angle Values of K_m, K_γ, K_c, and K_ϕ are taken from Tables 25-2 and 25-3.

TABLE 25-1
Values of C and exponents

Type of operation	Material	Ultimate strength, σ_u, MPa	Hardness, Brinell, H_B	C	m_1	m_2
Turning and boring	Steel	735	215	225	1.0	0.75
Facing and parting	Steel	735	215	264	1.0	1.00
Turning and boring	Gray cast iron		190	98	1.0	1.75
Parting and facing	Gray cast iron		190	135	1.0	1.0

Particular	Formula
The chip flow angle θ for zero-degree rake angle (Fig. 25-4)	$\tan\theta = \dfrac{d}{r_n + (d - r_n)\tan\beta}$ (25-13) where r_n = nose radius β = side cutting edge angle, deg
The equation relating the true rake angle to the corresponding chip flow angle	$\tan\alpha_{tr} = \tan\gamma\sin\phi + \tan\theta\cos\phi$ (25-14)
The equation for locating the maximum rake angle	$\tan\theta_{\max} = \dfrac{\tan\gamma}{\tan\theta}$ (25-15)
The metal removal rate	$Q = \dfrac{\pi(D-d)sdn}{1000} = Vsd$ (25-16) where s = feed rate, mm/rev Q = metal removal rate, cm^3/min d = depth of cut, mm D = diameter of work piece, mm $V = [\pi(D-d)n]/1000$, m/min
The approximate relationships between F_t $(=F_z)$, F_f $(=F_x)$ and F_r $(=F_y)$	$\dfrac{F_f\,(=F_x)}{F_t\,(=F_z)} \approx 0.3$ to 0.2 (27-17) $\dfrac{F_r\,(=F_x)}{F_t\,(=F_z)} \approx 0.2$ to 0.1 (27-18)
The turning moment on the work piece due to tangential cutting force	$M_{t\text{cut}} = F_t\dfrac{D}{2}$ (25-19)

TABLE 25-2
Material correction coefficient, K_m

Material	Ultimate strength σ_μ, MPa	K_m
Steel	390–490	0.76
	490–588	0.82
	686–785	1.00
	785–880	1.10
	980–1175	1.28
Cast iron	1370–1570	0.88
	1570–1765	0.94
	1765–1960	1.00
	2155–2355	1.12
	2355–2745	1.17

TABLE 25-3
Values of K_c, K_γ, and K_ϕ

Coolant	γ, deg	K_γ	ϕ, deg	K_ϕ	K_c
Dry	−15	1.40	30	1.05	1
Soda water	−10	1.30	45	1.00	1.03
Emulsion	+5	1.23	60	0.96	1.10
Mineral oil	0	1.13	75	0.94	1.15
	+5	1.06			
	+10	1.00			
Hard mineral oil	+15	0.94	90	0.92	1.20–1.25
	+20	0.89			

Particular	Formula
The bending moment due to bending of the tool in the vertical plane by tangential cutting force	$M_b = F_t l$ (25-20) where l = cantilever length of the cutting tool, m

25.1.3 Power

Particular	Formula
The total power at the cutting tool, P_{total}	$P_{\text{total}} = P_c + P_f + P_r$ (25-21) where $P_c = \frac{F_t V_t}{1000}$ = power required for turning cut, kW $P_f = \frac{F_f V_f}{1000}$ = power required to feed in a horizontal direction, kW. The feed velocity is very low. Power required to feed is approximately 1% of total power. Hence it is neglected. $P_r = \frac{F_r V_r}{1000}$ = power required to feed in radial direction, kW. The radial velocity is zero. Therefore P_r is ignored.
After neglecting P_f and P_r, the power required at the cutting tool, taking V_c for V_t and $P_{\text{total}} \approx P_c$	$P_c = \frac{F_t V_c}{1000} = \frac{KCs^{m_1} d^{m_2} V_c}{1000}$ (25-22) where P_c in kW, F_t in N, V_c in m/s, s and d in m
The gross or motor power	$P_g = \frac{P_c}{\eta} + P_t$ (25-23) where η = mechanical efficiency of machine tool P_t = tare power, the power required at no-load, kW

25.1.4 Specific power or unit power consumption

Particular	Formula
The specific power P_u $(=P_s)$, required to cut a material	$P_u = \frac{P_c}{\text{(cubic meter or cubic millimeter of material removed by cut per minute)}}$ (25-24)
The specific power or unit power P_u $(=P_s)$, for turning	$P_u = \frac{P_c}{V_c sd} = \frac{F_t}{1000sd} = \frac{C}{1000 s^{1-m_1} d^{1-m_2}}$ (25-25) where P_c, P_u in kW/m^3/min, F_t in N, s and d in m, and V_c in m/s
Another relation connecting specific powers at different cutting feeds and depths of cuts	$P_{u2} = P_{u1}\left(\frac{s_1}{s_2}\right)^{1-m_1}\left(\frac{d_1}{d_2}\right)^{1-m_2}$ (25-26)

Particular	Formula

TABLE 25-4
Typical values of specific power consumption P_s or P_u

Material	Brinell hardness number, H_B	Specific power consumption, P_s or P_u; kW/m^3
Plain carbon steel	126	1.6 to 1.8
	179	1.9 to 2.2
	262	2.3 to 2.6
Alloy steel	179	1.5 to 1.86
	429	3.0 to 5.20
Free cutting steel	229	1.37 to 1.48
Cast iron	140	0.60 to 0.90
	256	2.32 to 3.60
Aluminum alloy	55	0.76
	115	0.46 to 0.57
Brass		1.5

Refer to Table 25-4 for P_u (m^3/s) or P_u (mm^3/s) for various materials.

m_1 and m_2 are taken from Table 25-1.

25.1.5 Tool design

For comparison of Orthogonal Rake System (*ORS*), Normal Rake System (*NRS*) and American (*ASA*) tool nomenclature

Refer to Table 25-5.

25.1.6 Tool signatures

The tool signature of *ASA*, *ORS* and *NRS*

$$\gamma_p - \gamma_f - \alpha_p - \alpha_f - \phi_o - \phi_s - r_n \quad (ASA)$$

$$\lambda_s - \gamma_o - \alpha_o - \alpha_o^1 - \phi_o - \phi_p - r_n \quad (ORS)$$

$$\lambda_s - \gamma_n - \alpha_n - \alpha_n^1 - \phi_o - \phi_p - r_n \quad (NRS)$$

The tool signature for sintered carbide tipped single point tool

Refer to Fig. 25-5.

TABLE 25-5
Comparison of tool nomenclature system

Particular	Orthogonal rake system (ORS)[a]	Normal rake system (NRS)[a]	American Standards Association (ASA)[b]
Location of cutting edges	ϕ_p, ϕ_o	ϕ_p, ϕ_o	ϕ_s, ϕ_o
Orientation of face	γ_o, γ_s	γ_n, γ_s	γ_p, γ_f
Orientation of principal flank	α_o	α_n	α_p, α_f
Orientation of Auxiliary flank	α_o'	α_n'	—
Nose radius	r_n	r_n	r_n

[a] Tool reference system. [b] Machine reference system

Particular	Formula

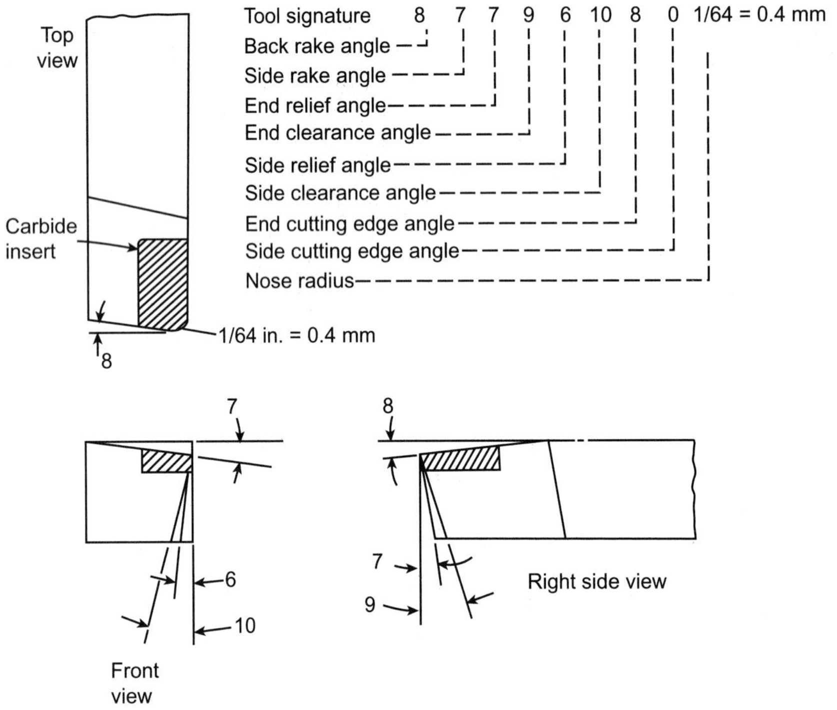

FIGURE 25-5 A straight shank, right cut, sintered carbide tipped, single point tool. Rake angles are negative. *Courtesy*: American Society of Tool and Manufacture Engineers, *Fundamentals of Tool Design*, Prentice Hall of India Private Ltd., New Delhi, 1969.

For general recommended various angles for HSS single-point tool — Refer to Table 25-6.

For general recommended various angle for carbide single-point tool — Refer to Table 25-8.

25.1.7 Tool life

The relation between the tool life τ and cutting speed V according to Taylor

$$K_c = V\tau^m \tag{25-27}$$

where

K_c = constant taken from Table 25-7 or constant equal to the intercept of the tool life and cutting speed curve and the ordinate ($V\tau$ curve)

m = slope of the $V\tau$ curve

$$m = \frac{\log(V_1/V_2)}{\log(\tau_2/\tau_1)} \tag{25-28}$$

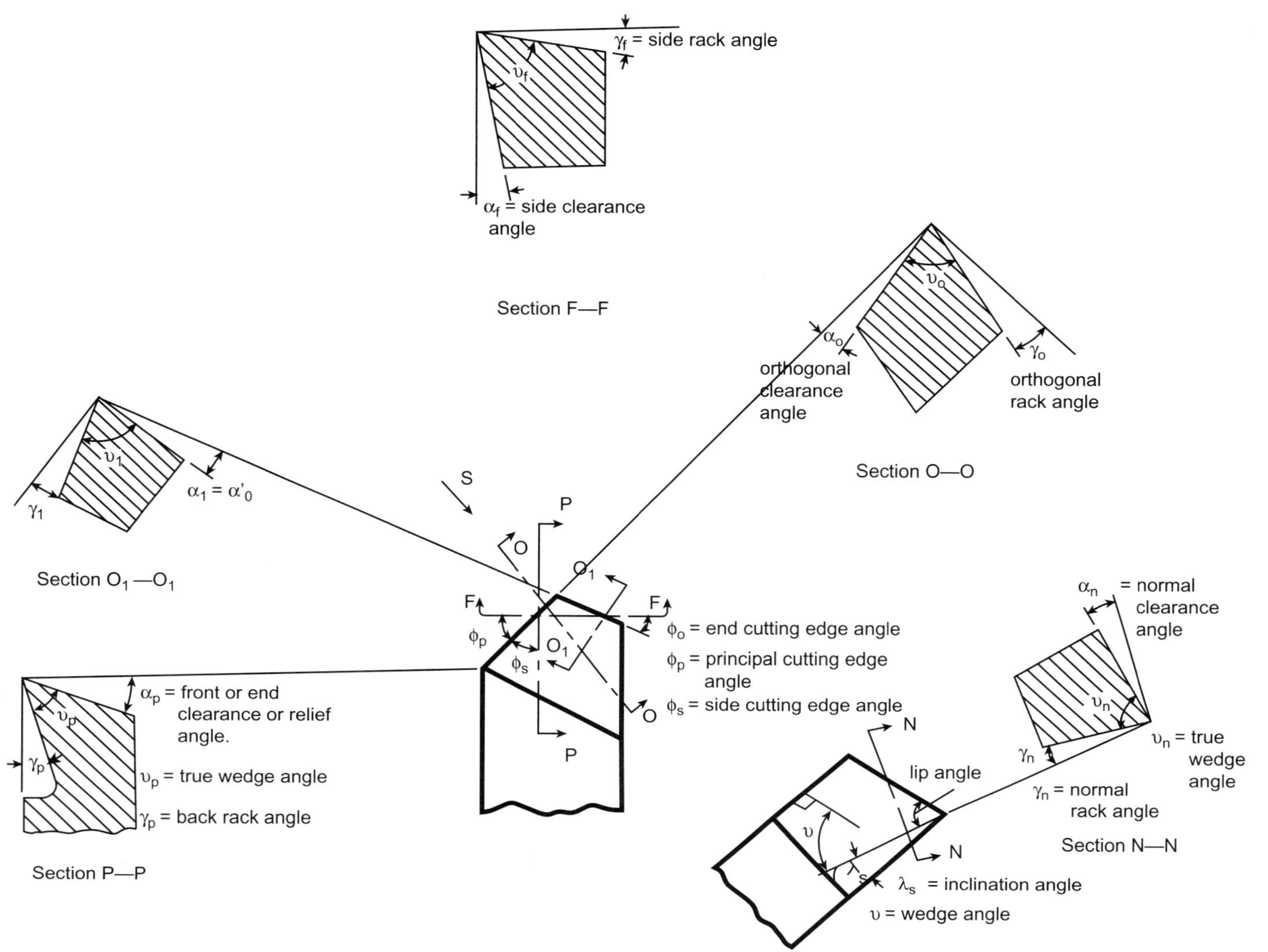

FIGURE 25-6 Single-point tool geometry (angles of machine reference system ASA, ORS and NRS). *Courtesy*: *Principles of metal cutting—An introduction*, Centre for Continuing Education, I.I.T., Madras, November, 1987.[5]

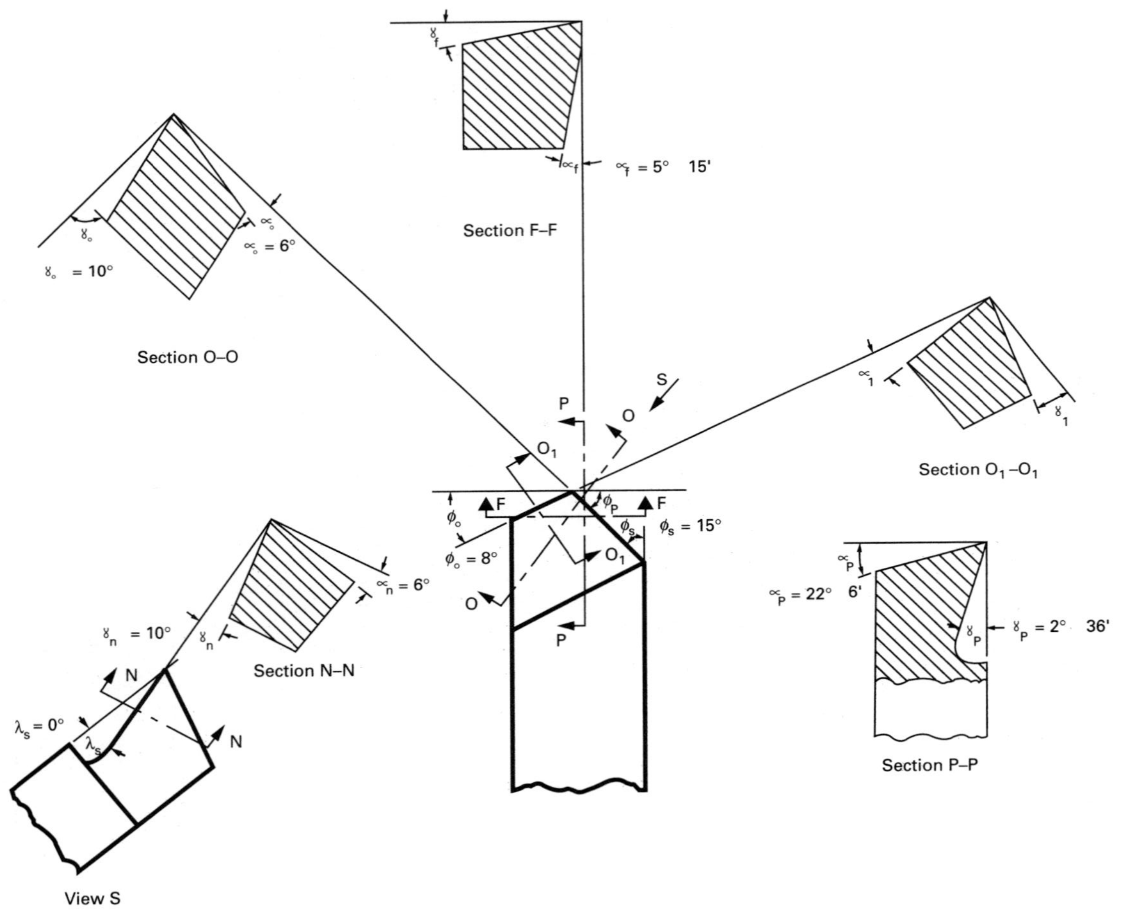

υ_o =true wedge angle, deg.
ϕ_P =principal cutting edge angle, deg.
ϕ_o =end cutting edge angle, deg.
ϕ_s =side cutting edge angle, deg.

FIGURE 25-7 Left hand single-point tool geometry (angles of machine reference system ASA, ORS and NRS) *Courtesy*: *Principles of metal cutting—An introduction*, Centre for Continuing Education, I.I.T., Madras, November, 1987.[5]

Particular	Formula
	$m \simeq 0.1$ to 0.15 for high speed steels (HSS) $m \simeq 0.2$ to 0.25 for carbides $m \simeq 0.6$ to 1.0 got ceramics and also taken from Table 25-7.
The velocity of the job or bar of diameter D_1 at speed n_1	$$V_1 = \frac{\pi D_1 n_1}{1000 \times 60} \quad (25\text{-}29)$$

TABLE 25-6
Recommended angle for high-speed-steel (HSS) single-point tools

Material	Back-rake angle, γ_p, deg	Front-relief angle, α_p, deg	Side-rake angle, γ_f deg	Side relief angle, α_f, deg
High speed, alloy, and high-carbon tool steels and stainless steel	5 to 7	6 to 8	8 to 10	7 to 9
Steels:				
C15 to C40	10 to 12	8 to 10	10 to 12	8 to 10
C45 to C90	10 to 12	8 to 10	10 to 12	7 to 9
14Mn1 S14, 40Mn 2S12	12 to 14	7 to 9	12 to 14	7 to 9
T50 Cr1 V23	6 to 8	7 to 9	8 to 10	7 to 9
40Ni2 Cr1 Mo28	6 to 8	7 to 9	8 to 10	7 to 9
Aluminum	30 to 35	8 to 10	14 to 16	12 to 14
Bakelite	0	8 to 10	0	10 to 12
Brass	0	8 to 10	1 to 3	10 to 12
Commercial bronze	0	8 to 10	−2 to −4	8 to 10
Bronze	0	8 to 10	2 to 4	8 to 10
Hard phosphor bronze	0	6 to 8	0	8 to 10
Gray cast iron	3 to 5	6 to 8	10 to 12	8 to 10
Copper	14 to 16	12 to 14	18 to 20	12 to 14
Copper alloys, soft	0 to 2	8 to 10	0	10 to 12
Copper alloys, hard	0	6 to 8	0	8 to 10
Monel and nickel	8 to 10	12 to 14	12 to 14	14 to 16
Nickel iron	6 to 8	10 to 12	12 to 14	14 to 16
Fiber	0 to 2	12 to 14	0	14 to 16
Formica	14 to 16	10 to 12	10 to 12	14 to 16
Micarta	14 to 16	10 to 12	10 to 12	14 to 16
Rubber, hard	0 to −2	14 to 16	0 to −2	18 to 20

TABLE 25-7
Values of K_c and m

Tool material	K_c	m
High-speed steel	60–100	0.08–0.15
Carbide	200–330	0.16–0.5
Ceramic	330–600	0.40–0.6

Particular	Formula	
The velocity of the job or bar of diameter D_2 at speed n_2	$V_2 = \frac{\pi D_2 n_2}{1000 \times 60}$	(25-30)
For standard spindle speeds for machine tools	Refer to Table 23-66.	
The relationship between the tool life, speed of cut, feed and depth of cut	$K = V T^m s^{m_3} d^{m_4}$ where m = exponent taken from Table 25-7 m_3 = exponent of feed $\simeq$ 0.5 to 0.8 (average values) m_4 = exponent of depth of cut $\simeq$ 0.2 to 0.4 (average values)	(25-31)
For standard speeds, feeds and etc.	Refer to Tables 23-66 to 23-70	
The approximate equation relating tool life to Brinell hardness number (H_B)	$K = V T^m s^{m_3} d^{m_4} \ (H_B)$	(25-32)
The cutting speed	$V = \frac{k_1}{d^{0.37} s^{0.77}} \left(\sqrt[6]{\frac{60}{T}} \right) C_{cf}$	(25-33)

TABLE 25-8
Recommended angles for carbide single-point tools

Material	Back-rake angle, deg	Side-rake angle, deg	End-relief, deg	Side-relief, deg	End cutting edge, deg	Side cutting edge, deg	End clearance, deg	Side clearance, deg	Nose radius r_n, mm
Copper, soft	0 to 10	15 to 25	6 to 8	6 to 8	10	10	2 to 3	2 to 3	0.4
Brass and bronze	0 to −5	+8 to −5	6 to 8	6 to 8	10	10	2 to 3	2 to 3	2.00
Aluminum alloys	0 to 10	10 to 20	8 to 12	6 to 10	10	10	2 to 3	2 to 3	0.4
Cast Iron:									
hard	0 to −7	+6 to −7	3 to 5	3 to 4	10	10	2 to 3	2 to 3	0.75
chilled	0 to −7	3 to 6	3 to 5	3 to 4	10	10	2 to 3	2 to 3	0.75
malleable	0 to 4	4 to 8	3 to 5	3 to 5	10	10	2 to 3	2 to 3	0.75
Low carbon steels 320 to 470 MPa	0 to −7	+6 to −7	5 to 10	5 to 10	10	10	2 to 3	2 to 3	0.75
Carbon steels 620 MPa	0 to −7	+6 to −7	5 to 8	5 to 8	10	10	2 to 3	2 to 3	1.00
Alloy steels	0 to −7	+6 to −7	5 to 10	5 to 10	10	10	2 to 3	2 to 3	1.00
Free machining steel 700 MPa	0 to −7	+6 to −7	5 to 8	5 to 8	10	10	2 to 3	2 to 3	
Stainless steels, austenitic	0 to −7	+6 to −7	4 to 6	4 to 6	10	10	2 to 3	2 to 3	1.00
Stainless steels, hardenable	0 to −7	+6 to −7	4 to 6	4 to 6	10	10	2 to 3	2 to 3	1.00
High-nickel alloys	0 to −3	+6 to +10	5 to 10	5 to 10	10	10	2 to 3	2 to 3	1.00
Titanium alloys	0 to −5	+6 to −5	5 to 8	5 to 8	10	10	2 to 3	2 to 3	1.00

Particular	Formula

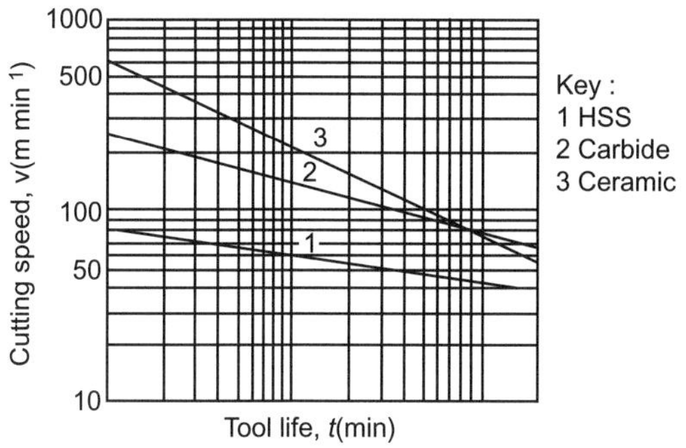

FIGURE 25-8 Tool life (τ) versus cutting speed (v). *Courtesy*: James Carvill, *Mechanical Engineer's Data Handbook*, Butterworth-Heinemann, 1994.

where

k_1 = an approximate constant taken from Table 25-9

C_{cf} = correction factor for the tool material (18-4-1 HSS = 100) taken from Table 25-10

$\sqrt[6]{\frac{60}{\tau}}$ = a factor which will correct the cutting speed from that obtained for a basic 60-minute tool life to the cutting speed for the desired tool life

τ = tool life in minutes

Refer to Fig. 25-8.

For numerical values of $d^{0.37}$ and $s^{0.77}$ for various values of d and s — Refer to Fig. 25-11.

25.2 MACHINE TOOLS

For machine tools with a rotary primary cutting motion — Refer to Fig. 25-9.

TABLE 25-9
Numerical value for k_1

Material to be cut	k_1, for 18-4-1 high-speed-steel tool and tool life of: 60 min without cutting fluid, or 480 min with cutting fluid	60 min with cutting fluid	480 min without cutting fluid
Light alloys	25.0		
Brass (80–120 H_B)	6.7		
Cast brass	4.2		
Cast steel	1.5	2.1	1.1
Carbon steel:			
SAE1015	3.0	4.2	2.1
SAC1025	2.4	3.3	1.7
SAE1035	1.9	2.7	1.3
SAE1045	1.5	2.1	1.1
SAE1060	1.0	1.4	0.7
Chrome-nickel steel	1.6	2.3	1.1
Cast iron:			
100 H_B	2.2	3.0	1.5
150 H_B	1.4	1.9	1.0
200 H_B	0.8	1.1	0.5

Courtesy: Wilson, F. W., *Fundamentals of Tool Design*, A.S.T.M.E., Prentice Hall of India Private Limited, New Dehli, 1969.

TABLE 25-10
Correction factors for compositions of tool material

	Approximate composition, %						
Type	W	Cr	V	C	Co	Mo	C_{cf}
14-4-1	14	4	1	0.7–0.8	—	—	0.88
18-4-1	18	4	1	0.7–0.75	—	—	1.00
18-4-2	18	4	2	0.8–0.85	—	0.75	1.06
18-4-3	18	4	3	0.85–1.1	—	—	1.15
18-4-1 + 5% Co	18	4	1	0.7–0.75	5	0.5	1.18
18-4-2 + 10% Co	18	4	2	0.8–0.85	10	0.75	1.36
20-4-2 + 18% Co	20	4	2	0.8–0.85	18	1.0	1.41
Sintered carbide	—	—	—	—	—	—	Up to 5

TABLE 25-11
Numerical values for $d^{0.37}$ and $s^{0.77}$

d	$d^{0.37}$	d	$d^{0.37}$	s	$s^{0.77}$	s	$s^{0.77}$
0.25	0.60	6.25	1.97	0.025	0.058	0.625	0.70
0.50	0.77	7.50	2.11	0.05	0.01	0.750	0.80
1.00	1.00	8.75	2.23	0.10	0.17	0.875	0.90
1.50	1.16	10.00	2.34	0.15	0.23	1.00	1.00
2.00	1.29	11.25	2.45	0.20	0.29	1.12	1.095
2.50	1.40	12.50	2.55	0.25	0.34	1.25	1.188
3.50	1.60	18.75	2.96	0.35	0.45	1.875	1.623
4.50	1.75	25.00	3.29	0.45	0.51	2.50	2.025
5.50	1.88			0.55	0.63		

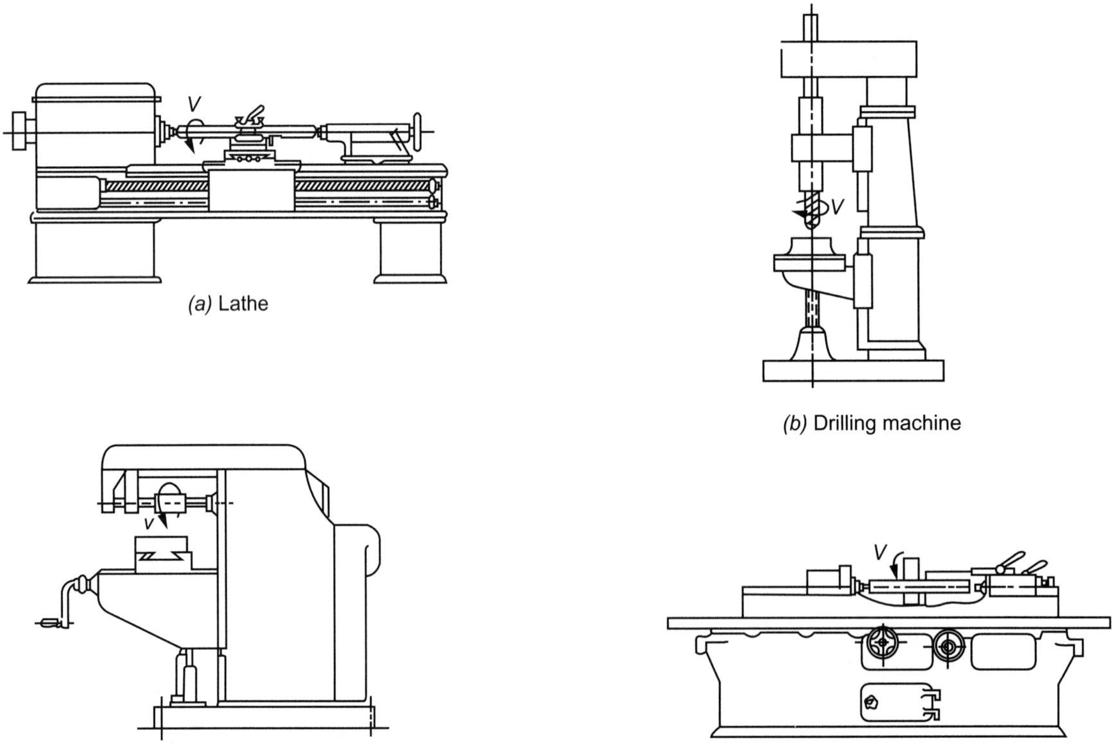

(a) Lathe

(b) Drilling machine

(c) Milling machine

(d) Grinding machine

FIGURE 25-9 Machine tools with a rotary primary cutting motion.

Particular	Formula

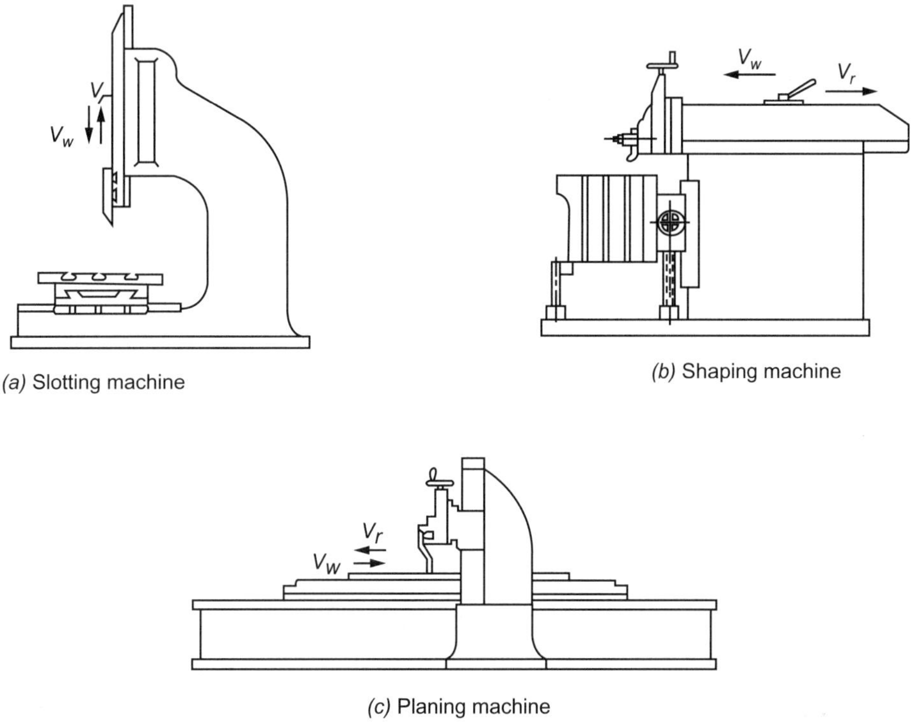

FIGURE 25-10 Machine tools with a straight line reciprocating primary cutting motion. *Courtesy*: N. Acherkan, General Editor, *Machine Tool Design*, volume 1, Mir Publishers, Moscow, p. 21, 1968.

For machine tools with a straight line reciprocating primary cutting motion	Refer to Fig. 25-10.

25.2.1 Lathe turning (Fig. 25-9a)

The tangential cutting force	$F_t = k_s sd$ (25-34) where k_s = specific cutting resistance or force offered by the workpiece material per unit chip section, MPa $k_s \simeq 3$ to $5\ \sigma_u$
The maximum tangential force is also obtained from equation.	$F_{t(\max)} = 98.1 \times 10^3 h_c$ (25-35) where h_c in m and $F_{t(\max)}$ in N
The maximum tangential cutting force in terms of swing over the bed of the center lathe	$F_{t(\max)} = 49 \times 10^3 h_{sw}$ (25-36) where h_{sw} in m and $F_{t(\max)}$ in N

Particular	Formula
The maximum torque of the center lathe	$M_{t(\max)} = F_{t(\max)}\left(\frac{h_{sw(\max)}}{2}\right)$ (25-37) where $h_{sw(\max)}$ = maximum swing over cross-slide
The maximum swing over cross slide for universal lathe	$h_{sw(\max)} = (0.55 \text{ to } 0.7)h_{sw}$ (25-38)
The maximum torque of the lathe by taking $h_{sw(\max)} = 0.6h_{sw}$	$M_{r(\max)} = 0.3F_{r(\max)}h_{sw}$ (25-39)
The maximum feed force	$F_f = F_{x(\max)} + F_\mu = 0.6F_{c(\max)}$ (25-40) where $F_\mu = \mu F_r$; $F_{x(\max)} = 0.3F_{c(\max)}$ F_r = reaction of the cutting force = $2F_{c(\max)}$ μ = coefficient of friction between bed of the lathe and saddle = 0.15
The radial component of cutting force F_y $(=F_r)$ (Fig. 25-2)	$F_y = \frac{2.4EI_W}{L_W}$ (25-41)
The deflection of the tool taking into consideration the effect of cantilever of tool	$\delta_t = \frac{F_t l^3}{3EI}$ (25-42) where l = projected length of tool from the tool post, m I = moment of inertia of area of the cross-section of the tool $(\approx bh^3/12)$ b = width of the tool shank, m h = depth of the tool shank, m
The maximum deflection of job or work piece in the vertical plane due to cutting force $F_c = F_t$ which should not exceed 0.05 mm and $D_w/L_w < \frac{1}{6}$	$\delta_j = \frac{F_t L_W^3}{48EI_W} \le 0.05$ (25-43) where $I_W = \pi D_W^4/64$ moment of inertia, m^4 (cm^4)
The diameter of job or work piece	$D_W = 0.25d_s$ (25-44)
The length of job or workpiece which is equal to the distance between centers of center lathe	$L_W = 8D_W$ (25-45)
The tangential component of cutting force is also calculated from the equation	$F_t = 2.5F_y$ (25-46)
Another equation for the power due to tangential component of cutting force	$P_c = \frac{F_t v k_w}{1000} = \frac{k_s s d v}{1000}$ (25-47)

Particular	Formula
	The cutting speed for carbide tools may be taken as 150 to 170 m/min.
The torque	$M_t = \frac{1000P_c}{\omega}$ (25-48)
	$M_t = \frac{159P_c}{n'}$ (25-49)
	where M_t in N m, P_c in kW, n' in rps and ω in rad/s
The minimum speed of work piece	$n_{\min} = \frac{1000v_{\min}}{\pi D_{\max}}$ (25-50)
The maximum speed of work piece	$n_{\max} = \frac{1000v_{\max}}{\pi D_{\min}}$ (25-51)
	where D = diameter of job in mm
The bed width of lathe	$B = (1.5 \text{ to } 2)h_c$ (25-52)
The moment acting on tailstock body in the plane xz_1	$M_{txz_1} = \left(F_z - \frac{W_j}{2}\right)h$ (25-53)
	where $F_z = F_t$
The moment acting on tailstock body in the plane xz_2	$M_{txz_2} = F_x h_c$ (25-54)
The moment acting on tailstock body in the yz plane	$M_{tyz} = F_y h_c$ (25-55)
	where
	W_j = weight of job, kN
	h = lever arm of the vertical force, m (mm)
	F_a = axial force with which tailstock center holds the job, kN
For speeds and feeds for turning of metals and plastics with HSS, carbide and Stellite tools	Refer to Table 25-12.
For cutting speeds and feeds for turning, facing and boring of cast iron, non-ferrous and non-metallic materials with HSS and carbide tools	Refer to Table 25-13.

25.2.2 Drilling machine

CALCULATION OF FORCES AND POWER IN DRILLS (Fig. 25-11)

For nomenclature of twist drills	Refer to Fig. 25-11.

TABLE 25-12
Speeds and feeds for turning of metals and plastics with HSS, carbide and stellite tools

		Turning cutting speed, n (m/min)				
		Depth of cut, d (mm)				
		0.1–0.5	0.5–2.0	2.0–5.0	6.0–10.0	
		Feed, s (mm/rev)				
Work piece material	Tool material	0.05–0.20	0.20–0.30	0.25–0.50	0.40–0.60	Speed, V (m/min) and feed s (mm/rev) for stellite tool
Mild steel	HSS	40–80	30–60	30–50	25–35	
	Carbide	150–450	120–200	80–150	60–120	
	Stellite					$v = 70–120$; $s = 0.125–2.00$
Alloy steel	HSS	20–45	15–35	15–26	10–15	
	Carbide	80–180	60–100	40–80	30–65	
	Stellite					$v = 20–35$; $s = 0.125–1.25$
Stainless steel	HSS	20–50	15–30	15–25	15–20	
	Carbide	50–90	50–80	40–70	40–60	
	Stellite					$v = 30–50$; $s = 0.075–1.00$
Free cutting steel	HSS	50–120	40–110	40–70	20–40	
	Carbide	200–500	150–250	120–180	90–150	
Cast, gray	HSS	110–160	35–45	25–30	20–25	
	Carbide	80–120	80–110	70–100	60–90	
	Stellite					$v = 60–90$; $s = 0.20–2.5$
Aluminum alloys	HSS	100–200	90–120	70–100	40–70	
	Carbide	150–600	90–450	80–180	60–150	
	Stellite					$v = 120–180$; $s = 0.075–0.50$
Copper alloys	HSS	100–200	90–120	60–100	40–60	
	Carbide	120–310	90–180	60–150	50–110	
	Stellite					$v = 70–150$; $s = 0.125–0.75$
Magnesium alloys	HSS	100–200	90–120	70–100	40–70	
	Carbide	150–160	90–450	80–180	60–150	
	Stellite					$v = 85–135$
Monel metal	HSS					$v = 15–20$; $s = 0.04–0.4$; $d = 0.6–4\,\text{mm}$
	Carbide					$v = 50–80$; $s = 0.04–0.6$; $d = 0.4–4\,\text{mm}$
	Stellite					$v = 25–45$; $s = 0.125–0.325$
Thermosetting plastic	HSS					$v = 35–50$; $s = 0.25$; $d = 4\,\text{mm}$
	Carbide					$v = 100–200$; $s = 0.25$; $d = 4\,\text{mm}$
	Stellite					$v = 70–120$; $s = 0.25$; $d = 4\,\text{mm}$

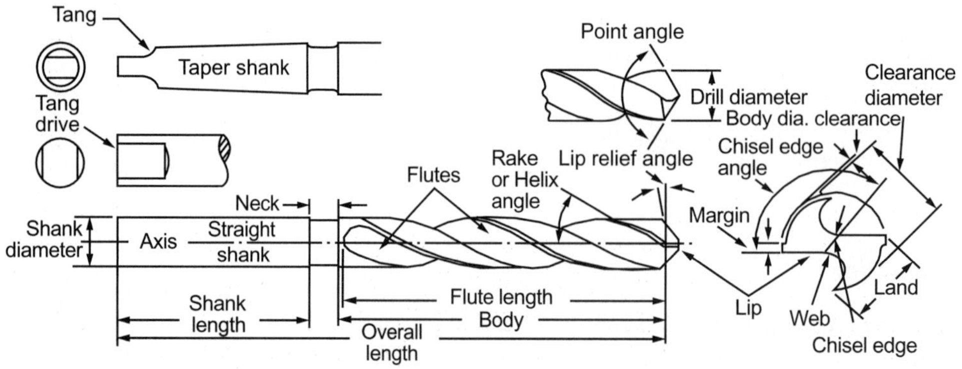

FIGURE 25-11 Nomenclature of twist drills. *Courtesy*: MCTI, *Metal Cutting Tool Handbook*.

Particular	Formula
The equation developed experimentally and analytically by Shaw and Oxford for torque of a twist drill operating in an alloy steel with an hardness of $200H_B$.	$$M_t = 37 \times 10^6 s^{0.8} d^{1.8} \left[\frac{1 - \left(\frac{c}{d}\right)^2}{\left\{1 + \left(\frac{c}{d}\right)\right\}^{0.2}} + 3.2\left(\frac{c}{d}\right)^{1.8} \right] \quad (25\text{-}56)$$ where M_t in N m, c, s and d in m

TABLE 25-13
Cutting speeds and feeds rates for turning, facing and boring of cast iron, non-ferrous and non-metallic materials with HSS and carbide tools. [Speed (at average H_B) for tool life of $1\frac{1}{2}$ to 2 hours between grinds, m/min]

Depth of cut, d (mm)	Feeds, s, (mm/rev)	Cast iron: Soft $H_B = 160$ HSS	Cast iron: Soft $H_B = 160$ Carbide	Cast iron: Medium hard $H_B = 160$–220 HSS	Cast iron: Medium hard $H_B = 160$–220 Carbide	Cast iron: Hard $H_B = 220$–360 HSS	Cast iron: Hard $H_B = 220$–360 Carbide	Aluminum, magnesium HSS	Aluminum, magnesium Carbide	Aluminum alloys, copper, soft bronze, soft brass, fiber, and plastic HSS	Aluminum alloys, copper, soft bronze, soft brass, fiber, and plastic Carbide	Hard bronze, hard brass, hard rubber, and marble HSS	Hard bronze, hard brass, hard rubber, and marble Carbide
0.80	0.1												
	0.2	90	160	50	120	32	100	200	500	150	400	90	250
	0.4	70	120	40	90	25	71	150	360	120	300	71	200
1.6	0.1							110	280	90	220	56	150
	0,2	80	140	45	110	25	90	150	400	120	300	71	220
	0.4	63	110	32	80	20	71	120	300	100	250	63	180
	0.8	50	100	25	63	16	56	71	220	71	180	45	140
3.2	0.1							71	180	56	140	32	100
	0.2	71	140	40	90	25	71	120	300	95	250	63	180
	0.4	56	110	32	71	20	63	100	250	80	200	50	140
	0.8	45	90	25	63	16	50	71	200	63	150	36	110
6.4	0.1							56	140	45	110	25	71
	0.2	62	120	32	71	20	63	110	250	50	180	45	140
	0.4	45	90	25	63	16	50	80	180	63	140	40	120
	0.8	40	80	20	45	10	40	60	140	50	110	32	90
	1.6	25	56	16	36	8	25	45	110	36	90	20	63
9.6	0.1							36	71	25	45	16	40
	0.2							90	200	71	150	40	140
	0.4	40	80	20	56	16	45	70	150	56	120	36	110
	0.8	36	71	16	45	10	36	56	120	45	90	25	80
	1.6	25	50	10	32	7	25	40	90	32	71	20	50
	2.4	—	—	—	—	—	—	25	63	20	40	16	32
	3.2	20	36	9	—	5	—	20	45	16	32	9	25
12.5	0.4	40	80	20	50	10	40	—	—	—	—	—	—
	0.8	32	60	16	40	9	12						
	1.6	25	45	10	12	7	20						
	2.4	16	25	8	—	5	—						

Key: 1. In case of shock and impact cuts, 70% of above speeds for carbide tools and 80% above speeds for HSS tools are used.
2. The above speeds are for cutting without cutting fluid.
3. A 10% reduction in the above speeds are recommended for soft, medium, hard, hard alloy and malleable irons.

Particular	Formula
The axial force or thrust acting on a drill	$F_a = 2.7 \times 10^9 s^{0.8} d^{1.8} \left[\dfrac{1 - \left(\frac{c}{d}\right)}{\left\{1 + \left(\frac{c}{d}\right)\right\}^{0.2}} + 2.2\left(\frac{c}{d}\right)^{0.8} \right] + 1.33 \times 10^8 c^2$ (25-57) where F_a in N, c, s and d in m $c =$ chisel edge length $\approx 1.15 \times$ web thickness for normal sharpening
The equation for torque of a drill of regular proportions whose c/d may be set equal to 0.18	$M_t = 40 \times 10^6 s^{0.8} d^{1.8}$ (25-58) where M_t in N m, s and d in m
The axial thrust for a drill of regular proportions whose c/d is equal to 0.18	$F_a = 3.6 \times 10^9 s^{0.8} d^{0.8} + 39 \times 10^6 d^2$ (25-59) where F_a in N, c, s and d in m
The equation for torque at the spindle of a drill based on Brinell hardness number (H_B)	$M_t = Cd^2 s^{0.8} (H_B)^{0.7}$ (25-60) $M_t = 2 \times 10^6 d^2 s^{0.8} (H_B)^{0.7}$ (25-61) where M_t in N m, d and s in m The constant $C = 2 \times 10^6$ for HSS drill drilling in carbon steel.
The equation for axial thrust at the spindle of a drill required for drilling which is based on Brinell hardness number (H_B)	$F_a = Cds^{0.7} (H_B)^{0.8}$ (25-62) $F_a = 1.9 \times 10^6 ds^{0.7} (H_B)^{0.8}$ (25-63) where F_a in N, d and s in m The constant $C = 1.9 \times 10^6$ for HSS drill drilling in carbon steel.
Another equation for the turning moment on the drill	$M_t = K_t d^{1.9} s^{0.8}$ (25-64) $M_t = 41.7 \times 10^6 s^{0.8} d^{1.9}$ for steel (25-65) $M_t = 28.8 \times 10^6 s^{0.8} d^{1.9}$ for cast iron (25-66) where M_t in N m, s and d in m
Another equation for the axial force acting on the drill	$F_a = Ks^m d$ (25-67) $F_a = 1.1 \times 10^8 s^{0.7} d$ for steel (25-68) $F_a = 1.5 \times 10^8 s^{0.8} d$ for cast iron (25-69) where F_a in N, s and d in m

Particular	Formula
REAMERS (Fig. 25-12)	
The equation for torque of a reamer or core drill	$$M_t = 37 \times 10^6 K s^{0.8} d^{1.8} \left[\frac{1 - \left(\frac{d_1}{d}\right)^2}{\left\{1 + \left(\frac{d_1}{d}\right)\right\}^{0.2}} \right] \quad (25\text{-}70)$$ where M_t in N m, s, d_1 and d in m
The equation for axial thrust for a reamer or core drill	$$F_a = 2.7 \times 10^9 K s^{0.8} d^{1.8} \left[\frac{1 - \left(\frac{d_1}{d}\right)^2}{\left\{1 + \left(\frac{d_1}{d}\right)\right\}^{0.2}} \right] \quad (25\text{-}71)$$ where F_a in N, s, d_1 and d in m d_1 = diameter of hole to be enlarged, m K = a constant depending upon the number of flutes. Refer to Table 25-14.
For tapping drill sizes for coarse threads	Refer to Table 25-15.
For cutting speeds and feeds for drills	Refer to Table 25-16.
For drill angles, cutting angles and cutting lubricant for drilling with high speed steel drills.	Refer to Table 25-17.

TABLE 25-14
Values of constant K

Number of flutes	Constant, K	Number of flutes	Constant, K
1	0.87	8	1.32
2	1.00	10	1.38
3	1.08	12	1.43
4	1.15	16	1.51
6	1.25	20	1.59

Courtesy: Wilson F. W., *Fundamentals of Tool Design*, A.S.T.M.E., Prentice Hall of India Private Limited, New Dehli, 1969.

TABLE 25-15
Tapping drill sizes for coarse threads

Nominal diameter, mm	Thread pitch, mm	Tap drill size, mm	Nominal diameter, mm	Thread pitch, mm	Tap drill size, mm
1.6	0.35	1.20	20.0	2.50	17.5
2.0	0.40	1.60	24.0	3.00	21.0
2.5	0.45	2.05	30.0	3.50	26.5
3.0	0.50	2.50	36.0	4.00	32.0
3.5	0.60	2.90	42.0	4.50	37.5
4.0	0.70	3.30	48.0	5.00	43.0
5.0	0.80	4.20	56.0	5.50	50.5
6.0	1.00	5.30	64.0	6.00	58.0
8.0	1.25	6.80	72.0	6.00	66.0
10.0	1.50	8.50	80.0	6.00	74.0
12.0	1.75	10.20	90.0	6.00	84.0
16.0	2.00	14.00	100.0	6.00	94.0

TABLE 25-16
Cutting speeds and feeds for drills

Material	Drill sizes, mm																			
	1.5		3.0		6.0		10.0		12.0		16.0		20.0		22.0		25.0		30.0	
	Speed n, rpm	Feed s, mm	Speed n, rpm	Feed s, mm	Speed n, rpm	Feed s, mm	Speed n, rpm	Feed s, mm	Speed n, rpm	Feed s, mm	Speed n, rpm	Feed s, mm	Speed n, rpm	Feed s, mm	Speed n, rpm	Feed s, mm	Speed n, rpm	Feed s, mm	Speed n, rpm	Feed s, mm
Metals with HSS drills:																				
Mild steel	4275 to 5500	0.05 to 0.075	2100 to 2800	0.051 to 0.075	1050 to 1375	0.075 to 0.125	700 to 925	0.125 to 0.175	525 to 700	0.150 to 0.20	425 to 550	0.25 to 0.35	350.0 to 450	0 25 to 0.35	300 to 400	0.35 to 0.40	265 to 340	0.35 to 0.40		
Cast iron	4575 to 6700	0.05 to 0.10	2300 to 3350	0.05 to 0.10	1150 to 1675	0.10 to 0.15	750 to 1125	0.15 to 0.225	575 to 850	0.20 to 0.30	450 to 675	0.30 to 0.40	375 to 550	0.30 to 0.40	325 to 475	0.35 to 0.50	280 to 425	0.35 to 0.50		
Aluminum	1500 to 1800	0.05 to 0.125	7600 to 9100	0.05 to 0.075	3800 to 4600	0.075 to 0.125	2500 to 3000	0.075 to 0.125	1900 to 2280	0.15 to 0.20	1500 to 1800	0.20 to 0.25	1250 to 1500	0.20 to 0.25	1095 to 1300	0.25 to 0.325	950 to 1125	0.25 to 0.35		
Bronze, brass	9150 to 1200	0.05 to 0.10	4575 to 6100	0.05 to 0.10	2300 to 3000	0.10 to 0.175	1525 to 2025	0.171 to 0.25	1150 to 1525	0.25 to 0.35	900 to 1200	0.35 to 0.45	750 to 1000	0.35 to 0.45	650 to 875	0.40 to 0.55	575 to 750	0.40 to 0.55		
Tool steel, steel castings, stainless steel and monel metal	3650 to 1550	0.05 to 0.075	1800 to 2250	0.05 to 0.075	925 to 1150	0.075 to 0.10	600 to 750	0.10 to 0.15	450 to 575	0.15 to 0.225	350 to 450	0.20 to 0.30	300 to 375	0.20 to 0.30	260 to 325	0.25 to 0.35	225 to 280	0.20 to 0.35		
Plastics:	**m/s**	**mm/rev**	**m/s**	**mm/rev**	**m/s**	**mm/rev**	**m/s**	**mm/rev**	**m/s**	**mm/rev**	**m/s**	**mm/rev**	**m/s**	**mm/rev**	**m/s**	**mm/rev**	**m/s**	**mm/rev**	**m/s**	**mm/rev**
Thermoplastics, polyethylene, polypropylene, TFB fluorocarbon[a]	0.55	0.12	0.55	0.25	0.55	0.30	—	—	0.55	0.38	—	—	0.55	0.46	—	—	0.55	0.50	0.55	0.64
Nylon, acetates, polycarbonate	0.55	0.05	0.55	0.12	0.55	0.10	—	—	0.55	0.20	—	—	0.55	0.25	—	—	0.55	0.30	0.55	0.38
Polystyrene[b]	1.10	0.03	1.10	0.05	1.10	0	—	—	1.10	0.10	—	—	1.10	0.13	—	—	1.10	0.15	1.10	0.18
Thermosetting plastics:																				
Soft grade[c]	0.83	0.08	0.85	0.13	0.85	0	—	—	0.85	0.20	—	—	0.85	0.25	—	—	0.85	0.30	0.85	0.38
Hard trade[c]	0.55	0.05	0.55	0.13	0.55	0.15	—	—	0.55	0.20	—	—	0.55	0.25	—	—	0.55	0.30	0.55	0.38

[a] Extruded, molded or cast. [b] Extruded or molded. [c] Cast, molded or filled.

Particular	Formula

TABLE 25-17
Drill angles and cutting lubricants for drilling with HSS drills

Workpiece material	Hardness H_B	Point angle, deg	Lip relief angle, deg	Chisel edge angle, deg	Helix angle, deg	Cutting lubricants
Aluminum alloys	30–150	90–140	12–15	125–135	24–48	Paraffin, lard oil, kerosene
Magnesium alloys	40–90	70–118	12–15	120–135	30–40	Mineral oil
Copper	80–85	100–125	12–15	125–135	28–40	Soluble oil
Brass	192–202	118–130	10–12	125–135	10–30	Dry, lard oil, paraffin mixture
Bronze	166–183	118	12–15	125–135	10–30	Dry, lard oil, paraffin mixture
Cast iron						
soft	126	90–100	8–12	125–135	24–32	Dry
medium	196	90–100	8–12	125–135	24–32	Dry
hard	293–302	118	8–12	125–135	24–32	Dry, soluble oil
chilled	402	118	8–12	125–135	24–32	Dry, lard oil
Cast steel	286–302	118	12–15	125–135	24–32	Lard oil, soluble oil
Mild steel	225–325	118	10–12	125–135	24–32	Soluble oil, lard oil
Medium carbon steel	325–425	118–135	8–10	125–135	24–32	Sulfur base oil, mineral lard oil
Free machining steels	85–225	118	12–15	125–135	24–32	Soluble oil, lard oil
Alloy steels	423	118–135	7–10	125–135	24–32	Soluble oil, mineral oil
Tool steels	510	150	7–10	125–135	24–32	Soluble oil, lard oil with sulfur
Stainless steels	135–325	118	7–10	125–135	24–32	Sulfur base oil
Spring steel	402	150	7–10	125–135	24–32	Lard oil with sulfur
Titanium alloy	110–402	118–135	7–10	125–135	20–32	Sulfur base oil
Monel metal	149–170	118	12–15	125–135	24–32	Lard oil, sulfur base oil
Pure nickel	187–202	118	10–12	125–135	24–32	Lard oil
Manganese steel	187–217	130	7–10	125–135	24–32	Mineral oil
Duraluminum	90–104	118	10–12	125–135	32–45	Mineral oil
Wood	—	60	15–20	135	24–32	None
Bakelite	—	130	7–10	135	24-32	None

For elements of metal-cutting reamer — Refer to Fig. 25-12.

For reamer angles and cutting lubricants for reaming with HSS reamers. — Refer to Table 25-18.

25.2.3 Taps and tapping

Power, P, at the spindle of tap for tapping of V-thread

$$P = 6.3 \times 10^{-3} K_m V p u_c \left(0.15 + \frac{1.75p}{u_c}\right) \qquad (25\text{-}72)$$

where

K_m = material factor taken from Table 25-19

V = cutting velocity, m/min

p = pitch of thread, mm

u_c = number of chamfered threads

Particular	Formula
For material factor, K_m, for use in drilling, reaming tapping	Refer to Table 25-19.
For nomenclature of tap	Refer to Fig. 25-13.
For tip angles and lubricants for tapping with HSS taps	Refer to Table 25-20.

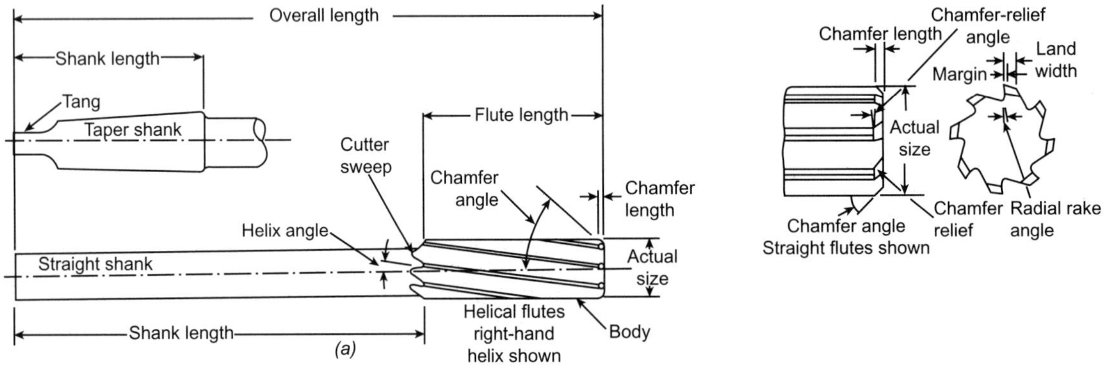

(a) Straight and Taper shank chucking reamer *(b)* Machine reamer

FIGURE 25-12 Elements of metal-cutting reamers. *Courtesy*: MCTI, *Metal Cutting Tool Handbook*.

TABLE 25-18
Reamer angles and cutting lubricants for reaming with HSS reamers (Fig. 25-12)

Workpiece material	Hardness, H_B	Radial rack angle, deg	Primary relief angle, deg	Helix angle, deg	Margin width, mm	Chamfer Angle, deg	Chamfer Relief angle, deg	Chamfer Length, mm	Cutting lubricants
Aluminum alloys	30–150	5–10	7–8	0–10	0.5–1.5	45	10–15	1.5	Mineral lard oil
Magnesium alloys	40–90	7	5–6	0 to −10	0.15–0.30	45	10–15	1.5	Mineral oil
Brass, bronze	165–202	0–5	5–7	0–12	0.125–0.35	40	10–15	2.5	Soluble oil
Cast iron:									
soft, medium and hard	126–302	0–10	5–6	0–10	0.10–0.65	45	7–23	1.5	Dry, soluble oil
Mild steel	226–325	2–3	4–5	0–10	0.10–0.25	45	7	1.5	Mineral lard oil
Free machining steels	85–225	2–3	4–5	0–10	0.15–0.20	45	7	1.5	Soluble oil
Alloy steels	423	2–3	2–4	0–10	0.15–0.20	45	7	1.5	Lard oil
Tool steels	510	2–3	2–4	0–10	0.15–0.20	45	7	1.5	Lard oil

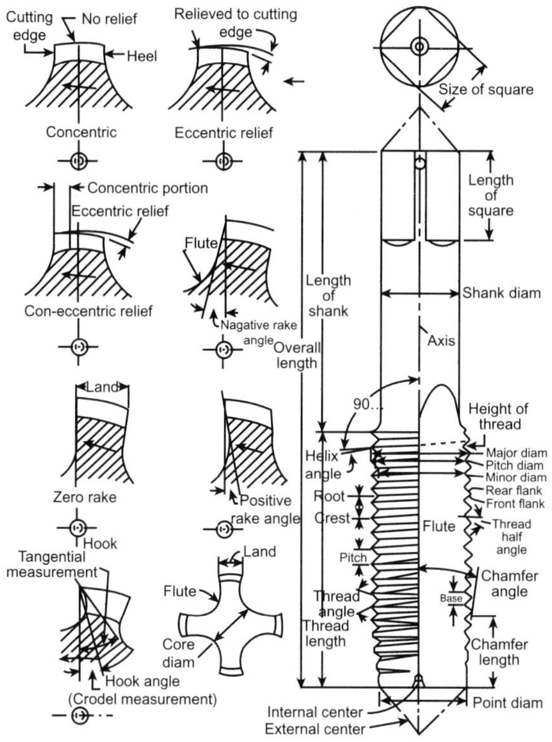

TABLE 25-19
Material factor, K_m, for use in drilling, reaming and tapping calculations

Workpiece material	Hardness, H_B	Ultimate tensile strength, σ_{ust}, MPa	Material factor, K_m
Aluminum	85		0.55
Copper alloys			0.55
Cast iron	175	278.50	0.92
	205	210.00	1.40
Malleable iron	180		0.75
	220	344.00	0.85
Carbon steels	150	540.00	1.50
	200	666.50	1.90
Alloy steels	165	568.50	1.55
	195	588.00	2.04
	210	772.50	2.15
	240	803.20	2.50
Stainless steel	185	633.50	1.55
	270	910.00	2.40

TABLE 25-20
Tap angles and cutting lubricants for tapping with HSS taps

Workpiece material	Hardness, H_B	Rack angle: Hook, deg	Rack angle: Positive, deg	Chamfer[a] relief angle, deg	Number of flutes[b]	Cutting lubricant
Aluminum	30–150	10–18	—	12	2–3	Kerosene and paraffin
Copper	80–85	20	—	10	4	Milk
Bronze	166–183	4–20	—	10	4	Soluble oil
Brass	192–202	—	0	10	4	Lard oil, soluble oil
Cast iron	100–300	0–2	0	6–12	4	Dry or soluble oil
Cast steel	286–302		10–15		2–3	Sulfur base oil, lard oil
Mild steel	225–325	9–12		8	2–3	Soluble oil, lard oil
Medium carbon steel	325–425	9–12		8	2–3	Sulfur base oil, lard oil
Alloy steels	130–423		10–15	8	2–3	Soluble oil, mineral oil
Tool steels	300–402		5–10	5–8	2–3	Soluble oil, lard oil with sulfur
Stainless steels	135–325		15–20	10	2–3	Sulfur base oil
Titanium alloys	110–402		5–12	12	2–3	Sulfur base oil
Phenolic plastics, hard rubber and fibers		10–15			4	Dry

[a] Chamfer length $2\frac{1}{2}$ to 3 threads for blind holes; 3 to 4 threads for through holes.
[b] For taps of 12 mm or smaller diameter.

Particular	Formula

25.2.4 Broaching machine

BROACHES (Figs. 25-14 and 25-15) AND BROACHING

For broach tooth form

Refer to Fig. 25-14.

For nomenclature of round pull broach

Refer to Fig. 25-15.

The allowable pull of internal or hole broach

$$F_{apl} = \frac{A\sigma_{sut}}{n} \tag{25-73}$$

The permissible load on push type of round broach (Fig. 25-15) using Euler's column formula with both ends free but guided

$$F_{pps} = \frac{\pi^2 EI}{nL^2} = \frac{\pi^2 E}{nL^2}\left(\frac{\pi D_r^4}{64}\right) \tag{25-74}$$

The allowable push in case of push type round broaches when $E = 206.8\,\text{GPa}$ in Eq. (25-74)

$$F_{aps} = \frac{100{,}000 D_r^4}{nL^2} \tag{25-75}$$

where

F_{apl} = allowable pull, N

F_{aps} = allowable push, N

A = area of the minimum cross-section of broach which occurs at the root of the first roughing tooth or at the pull end, mm^2

Note: when (L/D) is greater than 25, a push broach is considered as a long column and strength is based on this. If (L/D) is less than 25, the broach is considered to act as a short column which resist compressive load only.

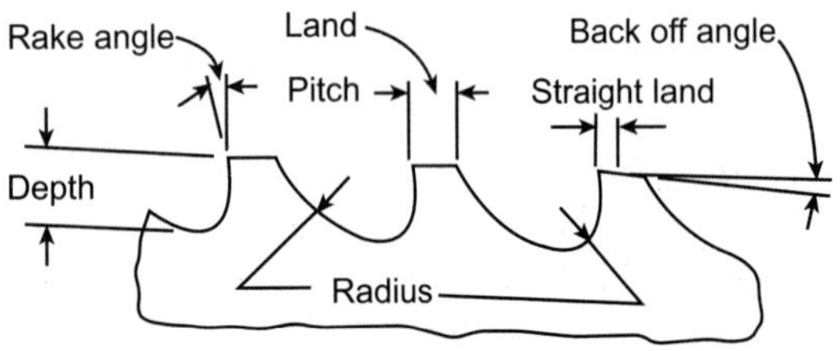

FIGURE 25-14 Broach tooth form. *Courtesy*: American Broach and Machine Division, Sundstrand Machine Tools Company.

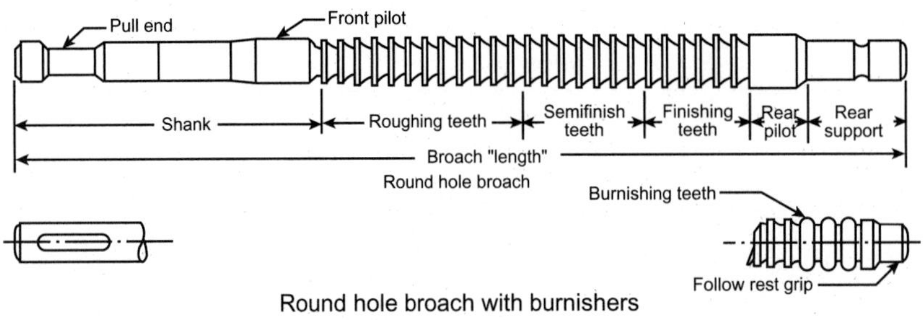

FIGURE 25-15 Nomenclature of a typical round pull broach. *Courtesy*: American Broach and Machine Division, Sundstrand Machine Tools Company.

Particular	Formula
	n = factor of safety to prevent broach damage because of sudden overloads due to hard spots in material, etc.
	$n = 3$ or more dependent on slenderness ratio
	σ_{sut} = tensile strength of the broach material, N/mm^2
	D_r = root diameter of the broach at $1/2L$, mm
	L = length of broach from push end of first cutting tooth, mm
The safe tensile stress for high speed steel	$\sigma_{sa} = \sigma_{sut}/n$
	$\sigma_{as} = 98$ MPa for keyway broaches
	$\sigma_{as} = 196$ MPa for polygon broaches
	$\sigma_{as} = 245$ MPa for round/circular broaches
The number of teeth cutting at a time in case of surface broaching	$z = \dfrac{l_{max}}{p} + 1$ (25-76)
	where
	l_{max} = maximum length of workpiece, mm
	p = pitch of the broach teeth, mm
Sum of the length of all the teeth engaged at any instant in broaching	$L = \pi Dz$ for circular/round broach (25-77a)
	$L = bz$ for spline broach (25-77b)
	$L = lz$ for surface broach (25-77c)
The specific broaching/cutting force	$k_s = 4415 + 3\sigma - 108\gamma - 24{,}515 s_z$ (25-78)
	where k_s in N/mm^2
	Also refer to Table 25-21 for k_s

TABLE 25-21
Specific broaching force, k_s

	Rise per tooth, s_z, mm					
	0.03	0.04	0.05	0.06	0.08	0.1
Material	Specific broaching force, k_s, MPa					
Mild steel	4168	3580	3285	3040	2745	2550
Cast iron:						
Gray	3726	3236	2942	2648	2452	2305
Malleable	3334	2844	2648	2452	2206	2060
Alloy steel	5688	4505	4413	4168	3775	3530

Particular	Formula
	σ = tensile strength of workpiece, N/mm^2 γ = rack angle, deg s_z = rise per tooth, mm
The recommended speeds and feeds for broaching	Refer to Table 25-22
The broaching force	$F = kk_s(\pi Dz)s_z$ for circular or round broaches (25-79a) $F = kk_s(bz)u_s s_z$ for spline or key broaches (25-79b) $F = kk_s(lz)s_z$ for surface broaches (25-79c) where b = width of spline or key, mm D = diameter of broached hole, mm l = width to be broached in case of surface broach, mm k = coefficient (may be taken as 1.1 to 1.3) z = number of teeth engaged at a time u_s = number of spline
Another equation for the broaching force in case of key and splines broaching	$F = Cs_z^{m_5}(bz)u_s$ (25-80)

TABLE 25-22
Recommended speeds and feeds for broaching

Workpiece material	Brinell hardness, H_B	Rise per tooth, mm	Cutting speed, m/min
Aluminum alloys	30–150	0.15	10–20
Copper alloys	40–200	0.12	8–10
Cast iron:			
Gray	110–140	0.13	9
	190–220	0.07	8
	250–320	0.05	4.5
Malleable	110–400	0.15	5–30
Low alloy steels	85–125	0.10	9
Carbon steel	120–375	0.08	3–8
Free cutting steel	100–200	0.10	10–12
	275–325	0.07	6
	325–375	0.05	6

TABLE 25-23
Broach angles for broaching with HSS broaching (Fig. 25-14)

Workpiece material	Brinell hardness, H_B	Hook/rake angle, deg	Clearance angle, deg
Aluminum/magnesium	30–150	10–15	1–3
Copper alloys	40–200	0–10	1–3
Cast iron	100–320	6–8	2–3
Lead brass	—	−5 to +5	1–3
Mild steels	225–325	15–20	2–3
Alloy steels	130–423	8–12	1–3
Tool steels	300–402	8–12	1–2
Stainless steel	135–325	12–18	2–3
Titanium	110–402	8–26	2–8

Particular	Formula
	where C = coefficient which takes into consideration condition of cutting and characteristic of work-piece. Taken from Table 25-25. s_z = feed per tooth, mm (Table 25-25) m_5 = exponent taken from Table 25-25
Another equation for the broaching force in case of cylindrical broaching	$F = Cs_z^{m_5}Dz$ (25-81)
The velocity of broaching	$v = \dfrac{K_v}{\tau^{m_6}s_z^{m_7}}$ (25-82) where K_v = velocity coefficient depends on the conditions of metal cutting (Table 25-24) τ = life of tool, min σ_{su} = stress of material, N/m^2, from Table 25-24
The power required for broaching by the broaching machine	$P = \dfrac{Fv}{1000}$ (25-83) where F in N, v in m/s, and P in kW

25.2.5 Milling machines

Particular	Formula
A knee horizontal-milling machine for plain or slab milling	Refer to Fig. 25-16.
A knee-type vertical milling machine for face milling	Refer to Fig. 25-17.
For nomenclature and tool geometry of milling cutters	Refer to Figs. 25-18 and 25-19a.
For tool angles of millings cutters	Refer to Table 25-26 and Figs. 25-18 and 25-19a.
The engagement parameter (Fig. 25-19a)	$k = \dfrac{\psi}{\varepsilon} = \dfrac{z}{\pi}\sqrt{\dfrac{h}{D}}$ (25-84) where ψ = engagement angle for milling depth, h $= 2\sqrt{\dfrac{h}{D}}$ (25-85) ε = peripheral pitch angle, deg $\approx \dfrac{2\pi}{z}$ (25-86)

TABLE 25-24
Values constant, K_v and exponents m_6 and m_7 for use in Eq. (25-82)

Workpiece material			Circular or round broaching			Keyway broaching						Spline broaching		
	Brinell hardness, H_B	Stress σ_{su}, MPa	s_z as given in Table 25-25			$s_z \leq 0.07$ mm			$s_z > 0.07$ mm			s_z as given in Table 25-25		
			K_v	m_6	m_7	K_v	m_6	m_7	K_v	m_6	m_7	K_v	m_6	m_7
Cast iron	≤200	—	14.0	0.50	0.60	6.2	0.6	0.95	6.2	0.6	0.95	17.5	0.5	0.6
	200	—	11.5	0.50	0.60	5.1	0.6	0.95	5.1	0.6	0.95	14.7	0.5	0.6
	up to 200	up to 686	16.8	0.62	0.62	9.2	0.87	1.4	7.7	0.87	1.4	15.5	0.6	0.75
Steels	200–230	686–785	15.5	0.62	0.62	8.8	0.87	1.4	7.0	0.87	1.4	14.0	0.6	0.75
	above 200	above 785	11.2	0.62	0.62	6.3	0.87	1.4	5.0	0.87	1.4	10.2	0.6	0.75

TABLE 25-25
Values of C, s_z and m_5 for use in Eqs. (25-80) and (25-81)

Workpiece material			Circular or round broaching			Keyway broaching			Spline broaching		
	Brinell hardness, H_B	Stress σ_{su}, MPa	C	s_z	m_5	C	s_z	m_5	C	s_z	m_5
Cast iron	≤200	—	2942	0.04–0.08	0.73	1128	0.08–0.15	0.73	1490	0.05–0.10	0.73
	>200	—	3472	0.03–0.06	0.73	1344	0.07–0.12	0.73	2108	0.04–0.08	0.73
	≤200	≤686	6865	0.02–0.03	0.85	1735	0.04–0.07	0.85	2079	0.04–0.06	0.85
Cast steel	200–230	686–785	7472	0.02–0.05	0.85	1980	0.07–0.12	0.85	2255	0.04–0.08	0.85
	>230	>785	8257	0.02–0.03	0.85	2452	0.04–0.07	0.85	2785	0.03–0.05	0.85
	≤200	≤686	6865	0.02–0.03	0.85	1735	0.03–0.06	0.85	2079	0.03–0.05	0.85
Alloy steels	200–230	686–785	7472	0.02–0.04	0.85	1980	0.06–0.10	0.85	2255	0.04–0.06	0.85
	>230	>735	8257	0.02–0.03	0.85	2452	0.04–0.07	0.85	2785	0.03–0.05	0.85

Particular	Formula
For up-milling and down-milling processes	Refer to Fig. 25-19.
The minimum number of teeth for satisfactory cutting action (Fig. 20-19a)	$$z_{\min} = \frac{2\pi}{\sqrt{h/D}} \quad (25\text{-}87)$$ where h = depth of milling, mm For $h/D = \frac{1}{10}$ to $\frac{1}{20}$ the $z_{\min}$ lies between 20 and 28.
The circumferential or circular pitch	$$p_c = \frac{\pi D}{z} \quad (25\text{-}88)$$

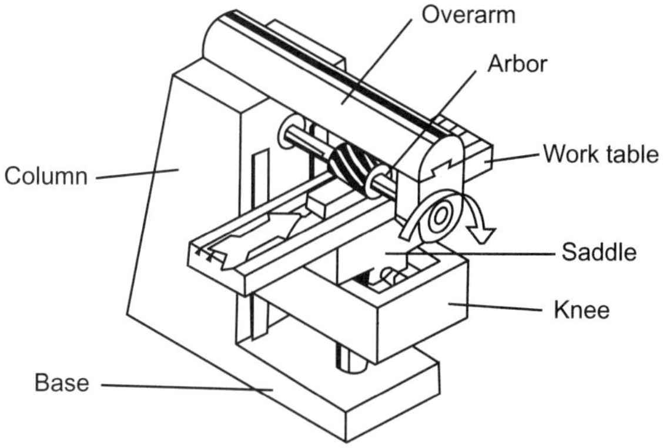

(a) Knee - type horizontal milling machine

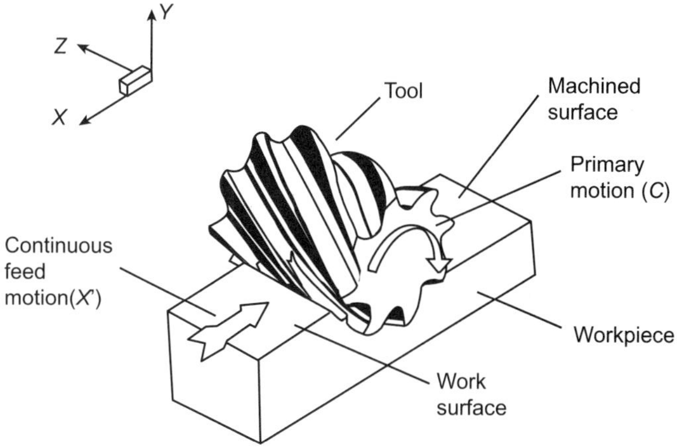

(b) Helical milling cutter

FIGURE 25-16 Knee-type horizontal milling machine for plane milling. *Courtesy*: G. Boothroyd, *Fundamentals of Metal Machining and Machine Tools*, McGraw-Hill Book Company, New York, 1975.[9]

Particular	Formula	
The axial pitch	$p_a = \dfrac{p_c}{\tan\beta} = \dfrac{\pi D}{z\tan\beta}$	(25-89)
The number of teeth in engagement in case of plain milling cutter whose helix angle is β	$z_s = \dfrac{z}{\pi}\left(\dfrac{b}{D}\tan\beta + \sqrt{\dfrac{h}{D}}\right)$ where b = width of cutter, mm	(25-90)
The design equation for the number of teeth on milling cutter	$z = m\sqrt{D}$ where m is a function of helix angle β. Table 25-27 gives values of m for various helix angles β.	(25-91a)

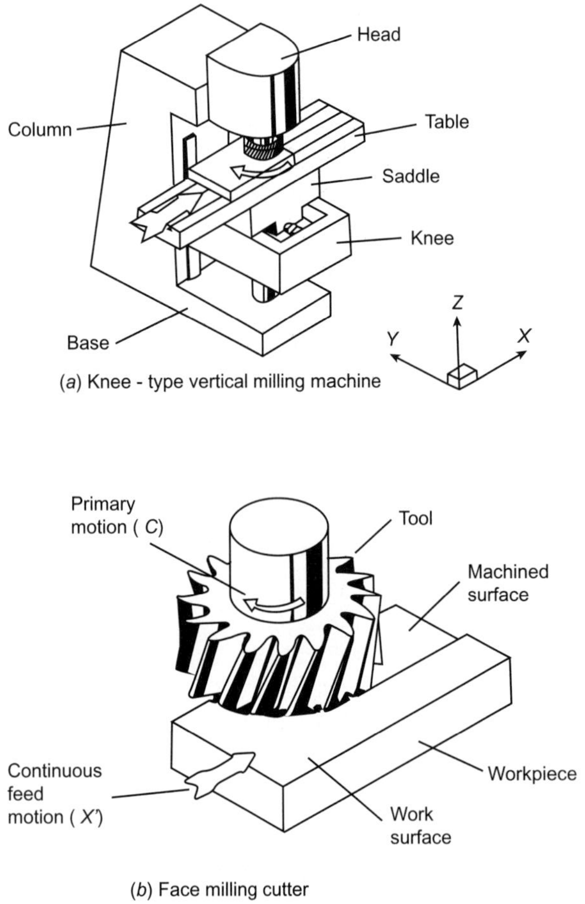

FIGURE 25-17 Knee-type vertical milling machine for face milling. *Courtesy*: G. Boothroyd, *Fundamentals of Metal Machining and Machine Tools*, McGraw-Hill Book Company, New York, 1975.[9]

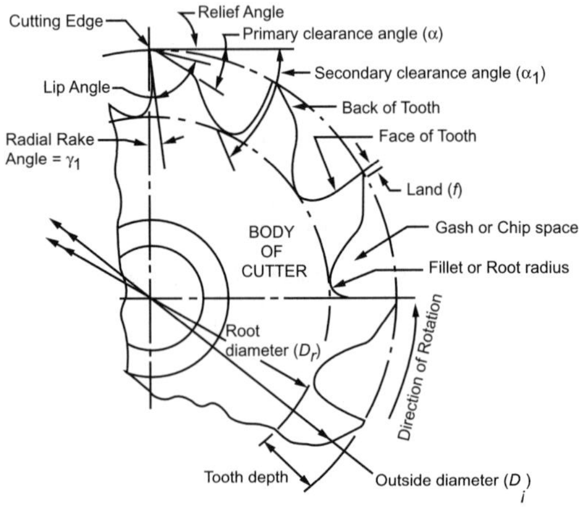

FIGURE 25-18 Nomenclature and geometry of milling cutter.

Particular	Formula

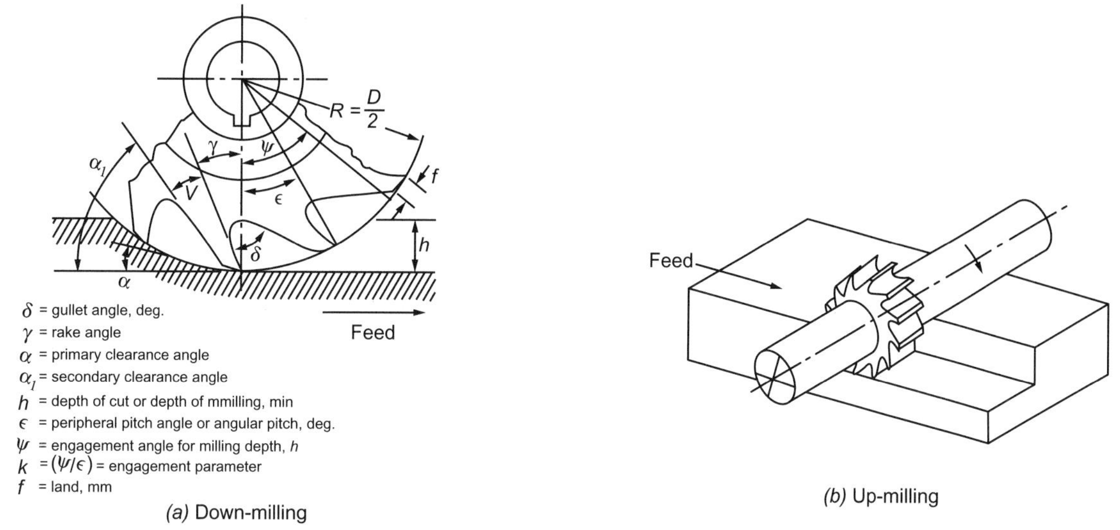

FIGURE 25-19 Horizontal milling process.

The gullet angle (Fig. 25-19a)

$$\delta = \varepsilon + \nu \quad \text{if rack angle } \gamma = 0 \tag{25-91b}$$

where ν = wedge angle, deg

CHIP FORMATION IN MILLING OPERATION PLAIN MILLING (Fig. 25-21)

The maximum undeformed chip thickness in case of plain or slab milling (Fig. 25-21) as per Martellotti[10,11]

$$t_{uc(\max)} = \left[s_z \left\{ \frac{\left(\frac{D}{h}\right) - 1}{\left(\frac{D}{2h}\right)^2 \left(1 \pm \frac{v_f}{V}\right)^2 \mp \frac{v_f D}{Vh}} \right\}^{1/2} \right] \cos\beta \tag{25-92}$$

The length of undeformed chip (Fig. 25-21)

$$l = \frac{D}{2}\psi \pm h\left(\frac{v_f}{V}\right)\left(\frac{D}{h} - 1\right)^{1/2} \tag{25-93}$$

The inherent roughness height

$$R = \frac{s_z}{\left(\frac{D}{s_z}\right) \pm \frac{z}{\pi}} \tag{25-94}$$

where the upper sign (+) refers to up-milling and the lower sign (−) refers to down-milling

The feed s which is equal to the distance moved by the workpiece during one resolution of tool (Fig. 25-21)

$$s = \frac{v_f}{n} \tag{25-95a}$$

where s in mm/rev

TABLE 25-26
Tool angles of milling cutters (Figs. 25-18 and 25-19)

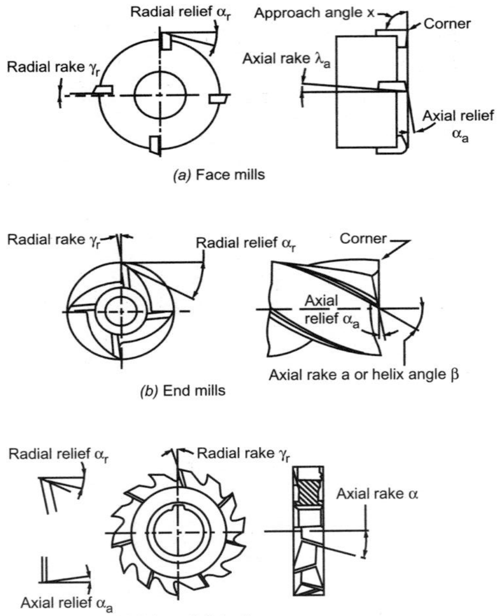

(a) Face mills
(b) End mills
(c) Side and slot mills

Workpiece material	Figure 25-20	Types of mills	Brinell hardness, H_B	Material of tool	Tool angles: Radial rake, γ_r, deg	Axial rake, γ_a, deg	Radial relief, α_r, deg	Axial relief, α_a, deg	Helix angle, β, deg
Aluminum alloys		Side and slot	30–150	HSS	10–20	10–25	5–11	5–7	
				Carbide	5–15	10–20	7–10	5–7	
		End		HSS	15–20	30–45	α_{r1}[a]	8–12	30–45
				Carbide	5–8	25			
		Face		HSS	20–35	20–35	10–12	3–5	
				Carbide	10–20	10–20			
Cast iron (machineability 100)		Side and slot	100–400	HSS	10–20	10–12	3–7	2–4	
				Carbide	+5 to −10	0 to −10	5–8	3–5	
		End		HSS	12	30–35	α_{r2}[a]		
				Carbide					30
		Face		HSS	−5 to −10	20–30		3–7	
				Carbide	−5 to −10	5–10	4–7	4–7	
Steels (machineability 100)		Side and slot	85–440	HSS	10–15	10–15			
				Carbide	−5 to +5	0 to −5		2–4	
		End		HSS	10–20	30–35	α_{r2}[a]	3–7	
				Carbide	3–5	15—25	4–8	3–5	30
		Face		HSS	10–15	10–15	3–7	5–7	
				Carbide	0 to −7	0 to −7	3–7	5–7	
Stainless steels		Side and slot	135–425	HSS	5–12	10–12	4–8	3–5	
				Carbide	2–4	−5 to +5	5–8	2–4	
		End		HSS	15	30–35	—	—	
				Carbide	0–3	15–25	α_{r2}[a]	3–7	30
		Face		HSS	10–12	10–15	8–10	8–10	
				Carbide	0 to −10	0–10		3–7	

Note: 1. Use 1 × 45° or radius for corner.
2. End cutting edge concavity angle: (a) for aluminum, 5 deg. (b) for alloy steels and aluminum, 3 deg.
[a] 3. Radial relief angles (α_r) for end mill

Diameter, mm	6	10	12	16	25
α_{r1}, deg	15	13	13	12	10
α_{r2}, deg	12	11	10	9	8

Particular	Formula

TABLE 25-27

Helix angle, β, deg	m
10–20	1.25–1.5
20–30	0.8–1.25
30–45	0.5–0.8

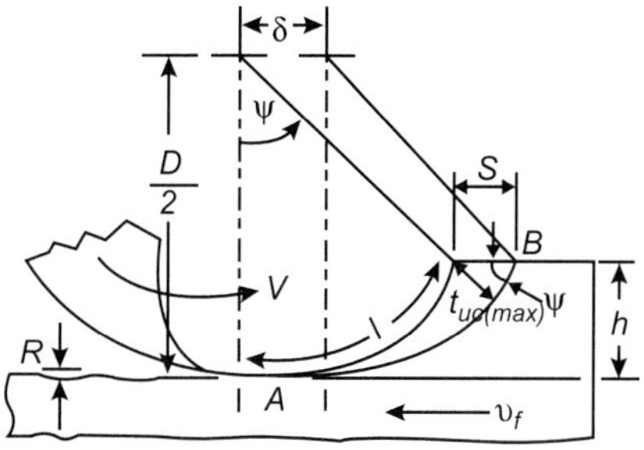

FIGURE 25-21 Geometry of plain-milling chip.

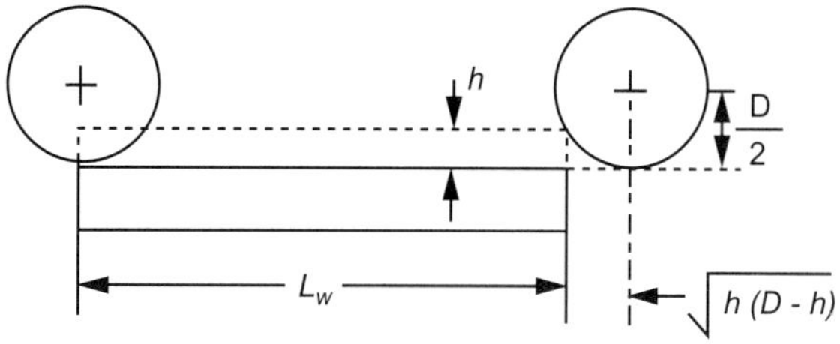

FIGURE 25-22 Relative motion between the workpiece and a plain milling cutter.

The engagement angle for milling depth, h

$$\psi = \cos^{-1}\left(1 - \frac{2h}{D}\right) \quad \text{(25-95b)}$$

The feed per tooth of milling cutter

$$s_z = \frac{v_f}{nz} \quad \text{(25-95c)}$$

where s in mm/rev, s_z in mm/tooth, ψ in rad

If V (cutter velocity) $\gg v_t$ (feed rate) and trochoidal arcs are replaced by circular arcs, Eqs. (25-92), (25-93) and (25-94) become

$$t_{uc(\max)} = s_z \sin\psi \cos\beta \quad \text{(25-96a)}$$

$$t_{uc(\max)} = \frac{2v_f}{zn}\sqrt{\frac{h}{D}\left(1 + \frac{h}{D}\right)} \quad \text{for } \beta = 0 \quad \text{(25-96b)}$$

$$l = \frac{D}{2}\cos\beta + \frac{s_z}{2} \quad \text{(25-97)}$$

$$R = \frac{s_z^2}{4D} \quad \text{(25-98)}$$

If (h/D) is very small i.e.: when $(h/D) \ll 1$, Eqs. (25-96), (25-97) and (25-98) become

$$t_{uc(\max)} \approx \left(2s_z\sqrt{\frac{h}{D}}\right)\cos\beta \quad \text{(25-99a)}$$

$$t_{uc(\max)} \approx \frac{2v_f}{zn}\sqrt{\frac{h}{D}} \quad \text{for } \beta = 0 \quad \text{(25-99b)}$$

$$l = \sqrt{Dh} \pm \frac{v_f}{2zn} \quad \text{(25-100)}$$

$$R \approx \frac{s_z^2}{4D} \quad \text{(25-101)}$$

Particular	Formula
The machining time (Fig. 25-22)	$$\tau_m = \frac{L_m + \sqrt{h(D-h)}}{v_f} \quad (25\text{-}102)$$ where L_m = length of workpiece, mm
The metal removal rate or feed rate which is equal to the product of feed speed and cross-sectional area of the metal removed, measured in the direction of feed motion	$$Q_w = \frac{hbv_f}{1000} = \frac{bhs_m}{1000} \quad (25\text{-}103)$$ where s_m = feed = $s_z nz$, mm/min; b = back engagement which is equal to the width of the workpiece; Q_w in cm^3/min

FACE MILLING (Fig. 25-23)

Particular	Formula
The maximum chip thickness in case of face-milling	$$t_{c(\max)} = \frac{v_f}{nz}\cos\theta_c = s_z \cos\theta_c \quad (25\text{-}104a)$$ where θ_c denotes the corner angle, deg
The average value of chip thickness in case of face-milling (Fig. 25-23)	$$t_{av} = \frac{57.3}{\psi} s_z \sin\theta \left[\cos\left(\frac{2b_1}{D}\right) + \cos\left(\frac{2(b-b_1)}{D}\right)\right] \quad (25\text{-}104b)$$ where θ = approach angle, deg
The approximate length of chip	$$l = \frac{D}{2}\psi \quad (25\text{-}105)$$

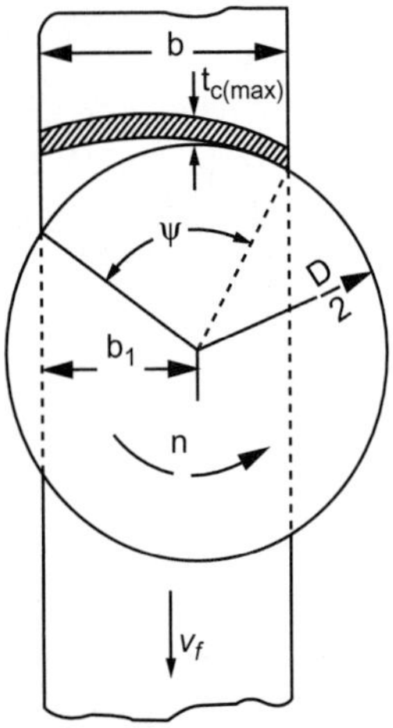

FIGURE 25-23 Face milling chip formation.

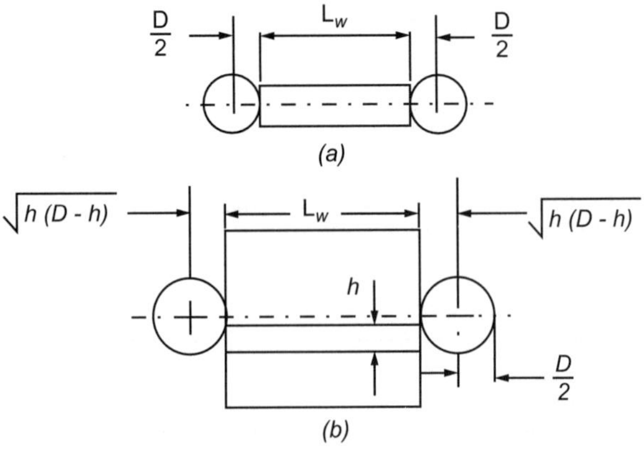

FIGURE 25-24 Relative motion between the workpiece and the face milling cutter.

Particular	Formula
The angle of engagement with the workpiece for use in Eq. (25-105) (Fig. 25-23)	$\psi = \sin^{-1}\left(\frac{2b_1}{D}\right) + \sin^{-1}\left(\frac{2(b-b_1)}{D}\right)$ (25-106)
The value of $t_{c(\max)}$ when the corner angle θ_c is zero	$t_{c(\max)} = \frac{v_f}{nz} = s_z$ (25-107)
The machining time, when the path of tool axis passes over the workpiece, is given by $(L_w + D)$ [Fig. 25-24(a)]	$\tau_m = (L_w + D)v_f$ (25-108)
The machine time when the path of the tool axis does not pass over the workpiece [Fig. 25-24(b)]	$\tau_m = \frac{L_w + 2\sqrt{h(D-h)}}{v_f}$ (25-109) where L_w = length of the workpiece, mm

END MILLING AND SLOT MILLING

The average chip thickness in case of end-milling and slot-milling (Fig. 25-25)	$t_{av} = \frac{114.6}{\psi} s_z \left(\frac{h}{D}\right)$ (25-110)

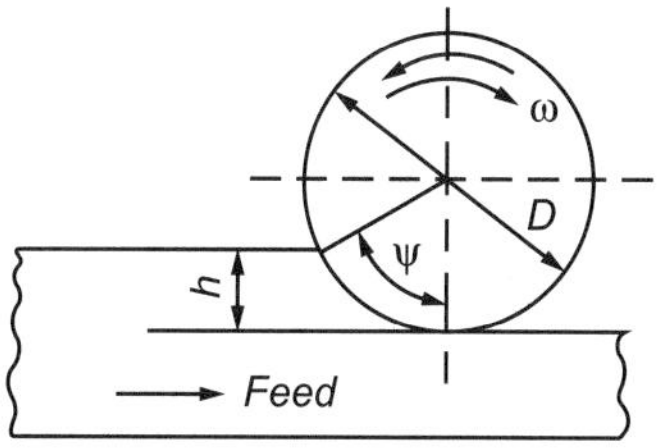

FIGURE 25-25 End-milling and slot-milling.

FORCES AND POWER

The empirical equation for the tangential force, F_t, in milling operation as per Kovan[12]	$F_t = Ch^x s_z^y z b^p D^q$ (25-111) where C = constant depends on the material of the tool, and the workpiece taken from Table 25-28 z = number of teeth on milling cutter in simultaneous contact with the workpiece

TABLE 25-28
Values of x, y, p, q and C for use in Eq. 25-111 (approx)

Material of workpiece	x	y	p	q	C
Cast iron	0.83–1.14	0.65–0.70	1.00–0.90	−0.83 to −1.14	470–686
Steel	0.86	0.74	1.00	0.86	392–785

Particular	Formula
	$z = \frac{z}{360}\psi$
	b = width of milling cutter or chip = $h/\sin\theta$
	x, y, p and q exponents taken from Table 25-28 for steels and cast iron
	F_t in N
The tangential force, F_t, can also be calculated from unit power concept	$F_t = \frac{1000P}{V}$ (25-112)
	where
	P = power at the spindle, kW
	V = cutting speed, m/s
The torque	$M_t = F_t \frac{D}{2}$ (25-113)
	where M_t in N m, D in m, F_t in N
The power	$P = \frac{F_t V}{1000}$ (25-114)
	where P in kW, F_t in N, V in m/s
The power at the spindle from the concept of unit power	$P = P_u k_h k_r Q$ (25-115)
	where
	P_u = unit power, kW/cm^3/min or kW/m^3/min as per Table 25-29
	k_h = correction factor for flank wear as per Table 25-30
	k_r = correction factor for radial rake angle as per Table 25-31
	Q = metal removal rate, cm^3/min or m^3/min
Another equation for power for peripheral milling	$P = kvzbCs_z^{m_8}\left(\frac{h}{R}\right)^{m_9}$ (25-116)
	where k, C, m_8 and m_9 are taken from Tables 25-32 and 25-33, P in W, v in m/s, b in mm, s_z = mm/tooth, h in mm, m_8 and m_9 are indices, C = constant from Table 25-33, $R = D/2$ = radius of cutter, mm
The approximate relationships between F_r (=F_x), F_f (=F_y) and F_t (=F_z = F_c) for different milling process	$F_r (=F_x) = 0.5F_t$ to $0.55F_t (=F_z)$ (25-117)
	for symmetrical face-milling
	$F_f (=F_y) = 0.25F_t$ to $0.35F_t (=F_z)$ (25-118)
	for symmetrical face-milling

TABLE 25-29
Average unit power P_u, for turning and milling

Work material	Tensile strength, σ_{st} MPa	Unit power P_u, 10^{-3} kW/cm^3/min — Average chip thickness, mm								
		0.025	0.05	0.075	0.1	0.15	0.2	0.3	0.5	0.8
Free machining	390	54	45	41	39	35	33	30	26	23
steels	490	60	50	45	42	39	36	32	29	26
Mild steels	588	66	55	50	47	42	39	35	31	28
Medium carbon	686	69	59	53	50	45	42	37	33	30
steels	785	73	63	56	52	48	44	40	35	32
Alloy steels	880	78	65	59	56	50	47	42	38	34
Tool steels	980	80	69	62	59	53	49	44	39	35
	1078	85	72	65	61	56	53	5.1	44	36
Stainless steels[a]	1470	80	71	66	61	57	52	48	44	40
	1570	86	76	72	67	62	58	54	50	46
	1666	92	82	78	73	68	61	56	52	48
	1765	99	90	84	80	75	69	62	59	52
	1863	104	96	91	86	81	78	69	64	58
	1960	110	101	96	91	88	85	78	71	60
Cast iron[a]	1570	30	26	24	22	21	19	18	16	14
Gray,	1666	31	28	25	24	22	20	1.9	17	15
Ductile,	1765	35	30	27	25	23	22	21	19	17
Malleable	1863	36	31	29	27	24	23	21	17	17
	1960	38	33	30	28	26	24	22	20	18
	2157	42	36	33	31	29	26	24	22	20
	2354	46	40	36	34	31	29	27	24	21
	2550	50	43	39	37	33	31	29	26	23
	2745	53	46	42	39	36	34	31	28	25
Aluminum	98	13	11	9	9	8	7	6	6	5
alloys	196	19	16	14	13	12	11	10	8	7
	294	24	20	17	16	14	13	12	10	9
	392	28	23	21	19	17	16	14	12	11
	490	32	26	23	22	19	18	16	14	12
Copper alloys	—	25	21	19	17	16	15	13	12	10
	98	9	7	6	6	5	5	4	4	3
Magnesium	147	10	9	8	7	6	6	6	5	4
alloys	196	12	10	9	8	7	7	6	5	4
	245	13	11	10	9	8	7	7	6	5
Titanium alloys										
Ti–Al–Cr	1078	59	51	47	45	41	39	36	32	30
Pure Ti	—	61	52	48	45	41	38	35	31	28
Ti–Al–Mn	—	67	58	53	50	45	43	39	35	31
Ti–Al–V	—	68	59	54	52	47	45	41	37	34
Ti–Al–Cr–Mo	—	77	66	60	57	52	49	45	40	36

[a] Values in H_B.

TABLE 25-30
Correction factor for flank wear

Flank wear, mm	Average chip thickness, mm	Correction coefficient, k_h — Hardness of work material — H_B 125	150	200	250	300	350	400	R_C 51	56	61
0.2	0.1	1.16	1.17	1.18	1.19	1.20	1.21	1.22	1.25	1.33	1.38
	0.3	1.06	1.07	1.08	1.08	1.09	1.09	1.09	1.13	1.16	1.18
	0.5	1.04	1.05	1.05	1.05	1.05	1.06	1.07	1.08	1.12	1.13
	1.0	1.02	1.02	1.03	1.03	1.03	1.30	1.03	1.04	1.06	1.07
0.4	0.1	1.50	1.50	1.50	1.53	1.57	1.67	1.78	1.80	1.92	2.12
	0.3	1.20	1.20	1.20	1.22	1.23	1.27	1.32	1.36	1.41	1.52
	0.5	1.12	1.12	1.14	1.15	1.16	1.19	1.24	1.26	1.30	1.38
	1.0	1.06	1.06	1.07	1.07	1.08	1.10	1.12	1.14	1.16	1.20
0.6	0.1	1.68	1.71	1.73	1.84	1.94	2.09	2.20	2.43	2.72	2.82
	0.3	1.26	1.25	1.29	1.33	1.37	1.44	1.50	1.61	1.78	1.85
	0.5	1.17	1.19	1.20	1.23	1.26	1.30	1.37	1.47	1.57	1.61
	1.0	1.09	1.10	1.10	1.12	1.14	1.16	1.19	1.25	1.30	1.33
0.8	0.1	1.91	2.04	2.10	2.34	2.47	2.54	2.65	2.99	3.26	—
	0.3	1.35	1.41	1.42	1.52	1.56	1.62	1.70	1.90	2.02	—
	0.5	1.23	1.28	1.32	1.36	1.38	1.43	1.52	1.66	1.74	—
	1.0	1.12	1.14	1.15	1.17	1.18	1.23	1.27	1.35	1.40	—
1	0.1	2.18	2.32	2.39	2.54	2.65	2.84	3.15	3.46	—	—
	0.3	1.45	1.50	1.56	1.67	1.70	1.74	1.90	2.16	—	—
	0.5	1.30	1.34	1.39	1.47	1.45	1.51	1.67	1.84	—	—
	1.0	1.15	1.16	1.17	1.20	1.23	1.27	1.35	1.44	—	—

Note: H_B = Brinell hardness number, R_C = Rockwell hardness scale C

TABLE 25-31
Correction factor for rake angle, k_r

Rake angle, γ degrees	−15	−10	−5	0	+5	+10	+15	+20
Correction coefficient, k_r	1.35	1.29	1.21	1.13	1.07	1	0.93	0.87

TABLE 25-32
Values of m_8, m_9, k for use in Eq. (25-116)

Material	m_8	m_9	k
Steel	0.85	0.925	0.164
Cast iron	0.70	0.85	0.169

TABLE 25-33
Values of C for use in Eq. (25-116)

Material	C	
Free machining carbon steel	980 (120 H_B)	1190 (180 H_B)
Carbon steels	1620 (125 H_B)	2240 (225 H_B)
Nickel-chrome steels	1460 (125 H_B)	220 (270 H_B)
Nickel-molybdenum and chrome-molybdenum steels	1600 (150 H_B)	1960 (280 H_B)
Chrome-vanadium steels	1820 (170 H_B)	2380 (190 H_B)
Flake graphite cast iron	635 (100 H_B)	1330 (263 H_B)
Nodular cast irons	1110 (annealed)	1240 (as cast)

Particular	Formula
	$F_r\,(=F_x) = 0.5F_t$ to $0.55F_t\,(=F_z)$ (25-119) for asymmetrical face-milling
	$F_f\,(=F_y) = 0.30F_t$ to $0.40F_t\,(=F_z)$ (25-120) for asymmetrical face-milling
	$F_r\,(=F_x) = 0.15F_t$ to $0.25F_t\,(=F_z)$ (25-121) for end milling (30° helical flute cutters)
	$F_f\,(=F_y) = 0.45F_t$ to $0.55F_t\,(=F_z)$ (25-122) for end milling (30° helical flute cutters)
For types and definitions of milling cutters	Refer to Table 25-34.
For feed per tooth for milling; cutting speeds for face and end milling; feeds and speeds for hobbing.	Refer to Tables 25-35, 25-36 and 25-37.
For milling cutters selection, dimensions for interchangeability of milling cutters, milling arbors with tenon drive and milling arbor with key drive and different types of milling cutters.	Refer to Tables 25-38 to 25-48

TABLE 25-34
Types and definitions of milling cutters

Type	Arrangement of teeth	Application	Size	Appearance
Cylindrical (slab or rolling)	Helical teeth on periphery	Flat surfaces parallel to cutter axis	Up to 160 × 160 mm	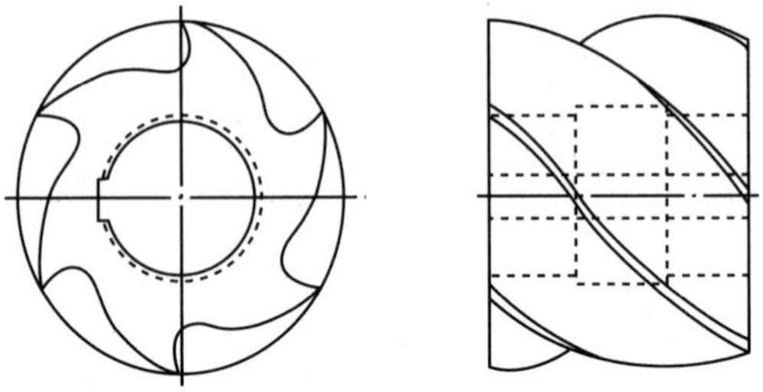
Side and face	On periphery and both sides	Steps and slots	Up to 200 mm diameter, 32 mm wide	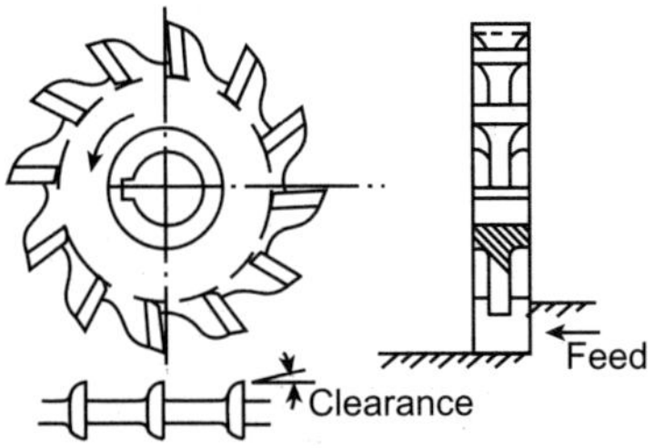
Straddle ganged	On periphery and both sides	Cutting two steps	Up to 200 mm diameter, 32 mm wide	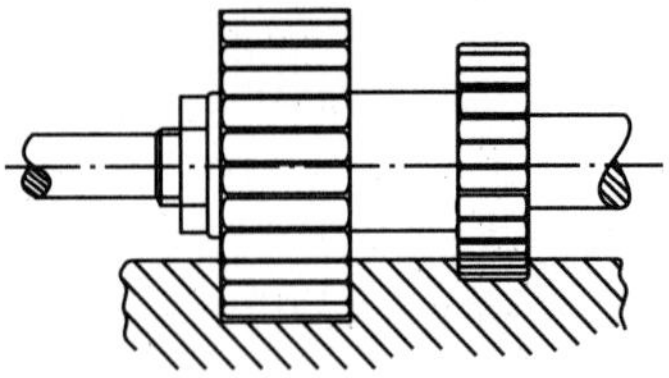
Side and face staggered tooth	Teeth on periphery. Face teeth on alternate sides	Deep slots	Up to 200 mm diameter, 32 mm wide	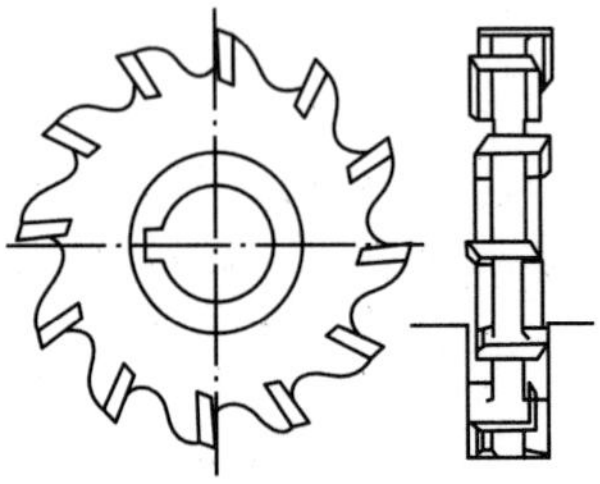
Single angle	Teeth on conical surface and flat face	Angled surfaces and chamfers	60–85° in 5° steps	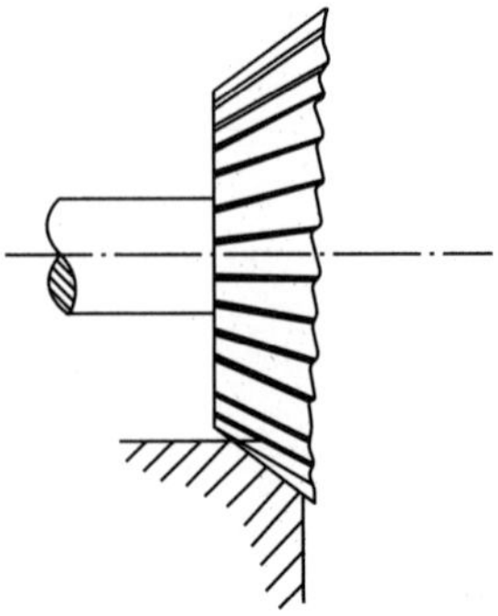

TABLE 25-34
Types and definitions of milling cutters (*Cont.*)

Type	Arrangement of teeth	Application	Size	Appearance
Double angle	Teeth on two conical faces	Vee slots	45°, 60°, 90°	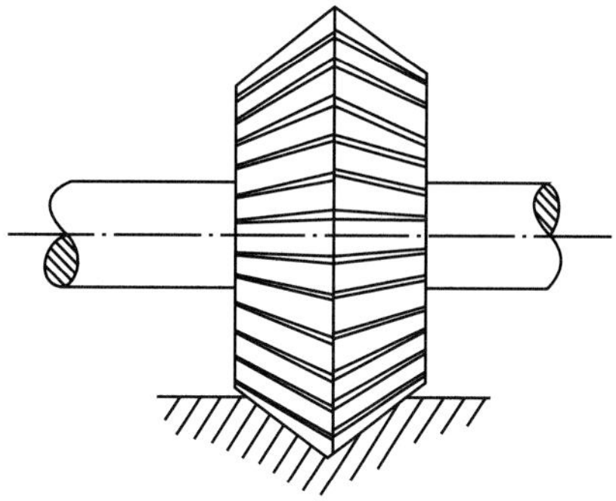
Rounding	Concave quarter circle and flat face	Corner radius on edge	1.5–20 mm radius	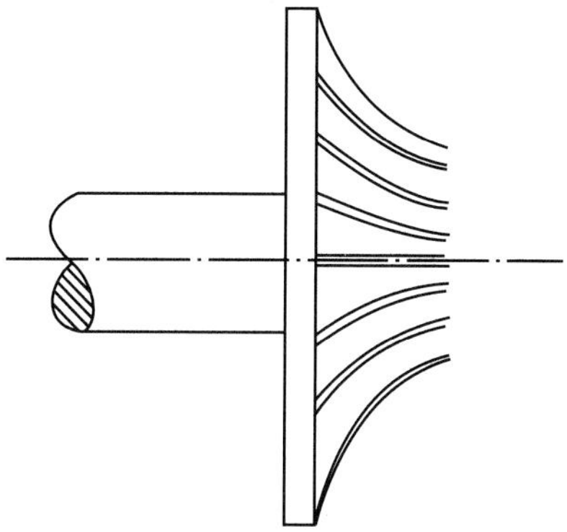
Involute gear cutter	Teeth on two involute curves	Involute gears	Large range	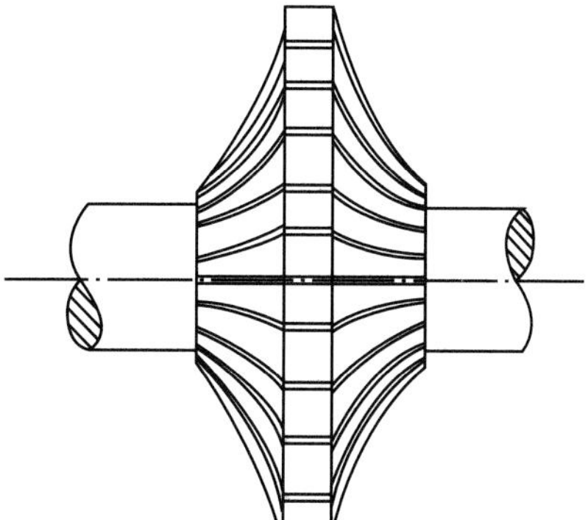

TABLE 25-34
Types and definitions of milling cutters (*Cont.*)

Type	Arrangement of teeth	Application	Size	Appearance
End mill	Helical teeth at one end and circumferential	Light work, slots, profiling, facing narrow surfaces	≤50 mm	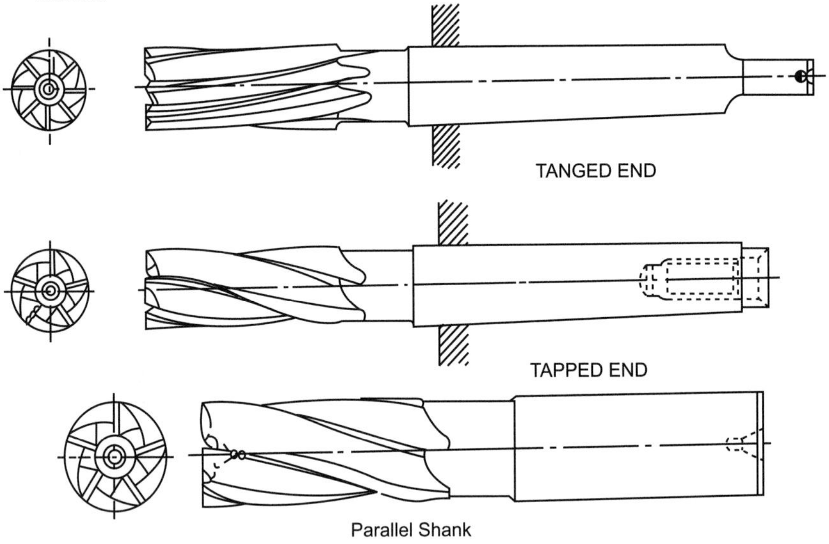
Tee slot	Circumferential and both sides	Tee slots in machine table	For bolts up to 24 mm diameter	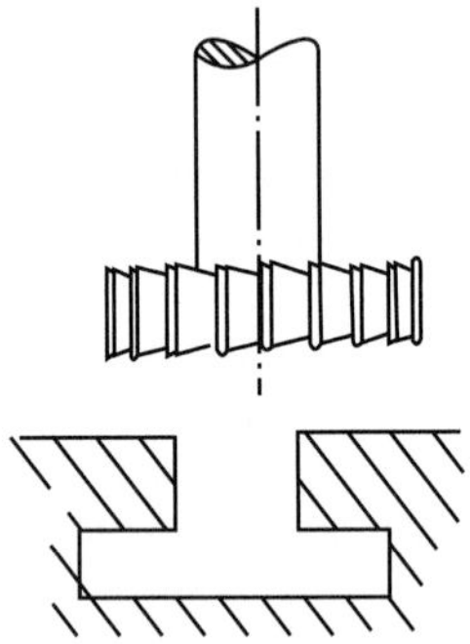
Dovetail	On conical surface and one end face	Dovetail machine slides	38 mm diameter, 45° and 60°	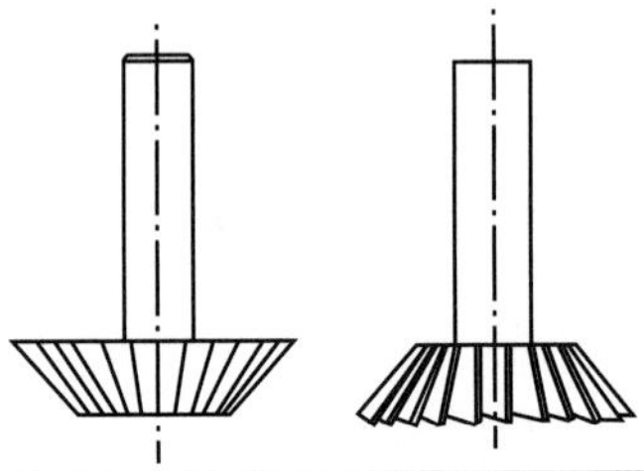

TABLE 25-34
Types and definitions of milling cutters (*Cont.*)

Type	Arrangement of teeth	Application	Size	Appearance
Skid end mill	Circumferential and one end	Larger work than end mill	40–160 mm diameter	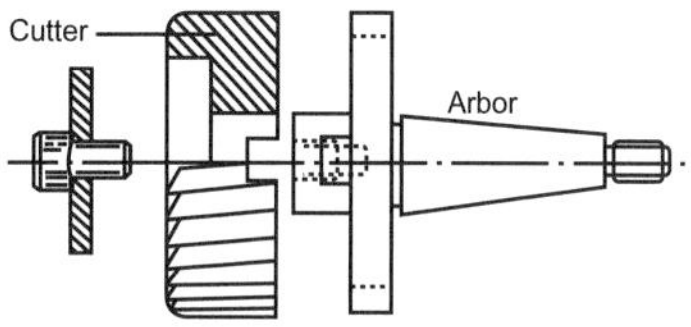
Cutting saw (slot)	Circumferential teeth	Cutting off or slitting. Screw slotting	60–400 mm diameter	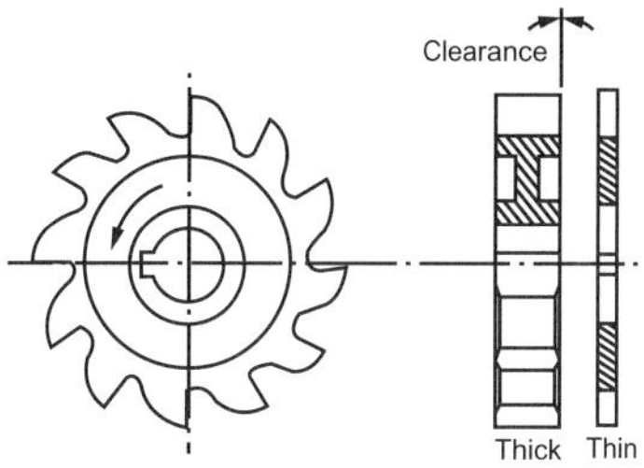
Concave-convex	Curved teeth on periphery	Radiusing	1.5–20 mm radius	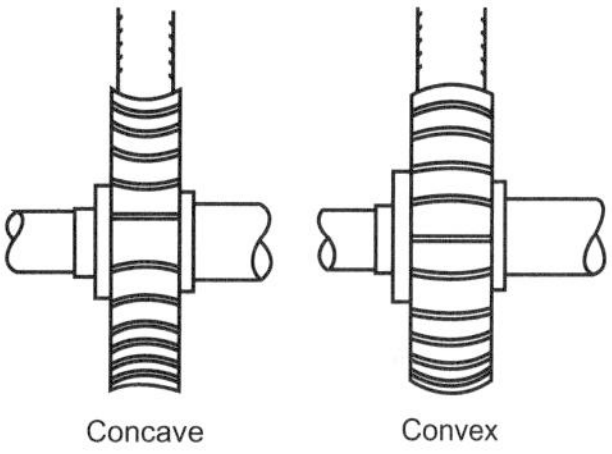
Thread milling cutter				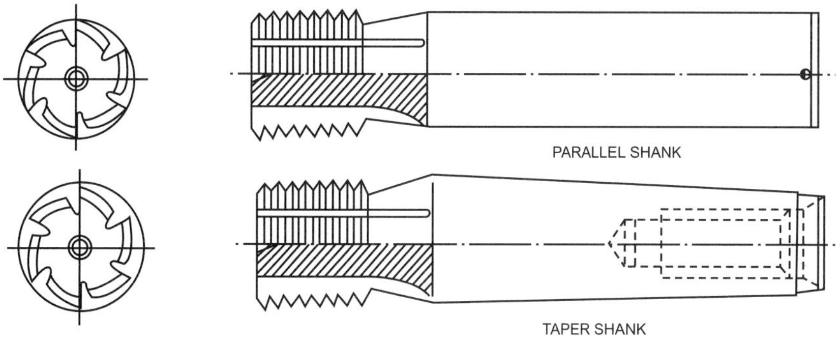

TABLE 25-35
Suggested feed per tooth for milling various materials, mm

	Face mills		Helical mills		Slotting and side mills		End mills		Form relieved cutters		Circular saws	
Materials to be milled	**HSS**	**Carbide**	**HSS**	**Carbide**	**HSS**	**Carbide**	**HSS**	**Carbide**	**HSS**	**Carbide**	**HSS**	**Carbide**
Cast iron												
Soft (up to $160H_B$)	0.40	0.50	0.32	0.40	0.22	0.30	0.20	0.25	0.12	0.15	0.10	0.12
Medium (160 to $220H_B$)	0.32	0.40	0.25	0.32	0.18	0.25	0.18	0.20	0.10	0.12	0.08	0.10
Hard (220 to $320H_B$)	0.28	0.30	0.20	0.25	0.15	0.18	0.15	0.15	0.08	0.10	0.08	0.08
Malleable iron[a]	0.30	0.35	0.25	0.28	0.18	0.20	0.15	0.18	0.10	0.10	0.08	0.10
Steel												
Soft[a] (up to $160H_B$)	0.20	0.35	0.18	0.28	0.12	0.20	0.10	0.18	0.08	0.10	0.05	0.10
Medium (160 to $220H_B$)	0.15	0.30	0.12	0.25	0.10	0.18	0.08	0.15	0.05	0.10	0.05	0.08
Hard[a] (220 to $360H_B$)	0.10	0.25	0.08	0.20	0.08	0.15	0.05	0.12	0.05	0.08	0.03	0.08
Stainless[a]	0.20	0.30	0.15	0.25	0.12	0.18	0.10	0.15	0.05	0.08	0.05	0.08
Brass and Bronze												
Soft	0.55	0.50	0.45	0.40	0.32	0.30	0.28	0.25	0.18	0.15	0.12	0.12
Medium	0.35	0.30	0.28	0.25	0.20	0.18	0.18	0.15	0.10	0.10	0.08	0.08
Hard	0.22	0.25	0.18	0.20	0.15	0.15	0.12	0.12	0.08	0.08	0.05	0.08
Copper	0.30	0.30	0.25	0.22	0.18	0.18	0.15	0.16	0.10	0.10	0.08	0.05
Monel	0.20	0.25	0.18	0.20	0.12	0.15	0.10	0.12	0.08	0.08	0.05	0.08
Aluminum[a]	0.55	0.50	0.45	0.40	0.32	0.30	0.28	0.25	0.18	0.15	0.12	0.12

[a] Coolant to be used.

TABLE 25-36
Recommended cutting speeds for face and end milling with plain HSS and carbide milling cutters, m/min

	Depth of cut					
	Roughing cut, 3 to 5 mm		Semi-finishing cut, 1.5 to 3 mm		Finishing cut, below 1.5 mm	
Material to be milled	**HSS**	**Carbide**	**HSS**	**Carbide**	**HSS**	**Carbide**
Cast iron						
Soft	25	68	30	80	36	105
Medium	15	50	25	68	30	80
Hard	12	38	16	50	20	68
Malleable Iron	25	68	30	80	36	105
Steel[a]:						
Soft	28	120	32	150	40	180
Medium	22	100	28	120	32	135
Hard	15	75	20	90	25	105
Stainless	18	50	22	68	28	80
Brass						
Average	30	75	45.	120	60	150
Soft yellow	60	120	90	180	120	240
Bronze	28	75	36	100	45	128
Copper	45	100	68	150	90	210
Monel	18	50	22	68	28	80
Aluminum[a]	75	240	105	300	150	450

[a] Coolant to be used
Note: Cutting speeds for 12% cobalt HSS should be about 25% to 50% higher than those shown for plain HSS.
Cutting speeds for cast alloy should be about 100% higher than those shown for plain HSS.
Above speeds should be reduced when milling work that has hard spots or when milling castings that are sandy.

TABLE 25-37
Feeds and speeds for hobbing

Type of gear	Material	Module mm	Feed, mm/rev. of blank: Roughing (single thread hob)	Roughing (multithread hob)	Finishing	Hob speeds, m/min
High speed reduction and step up	Steel	1.5–8	1–1.5	1–1.5	0.8–1.25	9–25
Instrument	Steel	0.4–1.25	0.5–1.5	Up to 3	0.5–1.0	25–60
	Non-ferrous	0.4–1.25	1.0–1.5	Up to 3	0.5–1.0	25–60
Aircraft	Steel	2.0–4.0	1.0–1.5	Up to 3	0.8–1.25	15–45
Machine tool and printing press	Steel, C.I.	2.0–6.0	2.0–3.2	Up to 2.5	1.0–1.5	15–30
	Non-ferrous	2.0–6.0	2.0–3.2	Up to 2.5	1.0–1.5	25–450
Automotive, including trucks and tractors	Steel	1.5–8.10	2.0–3.2	Up to 2.5 Up to 2.0 (3 starts)	1.25–2.0	15–45
High quality industrial	Steel	10.0–25.0	2.0–2.5		1.25–2.0	12–30
	Cast iron	2.5–8.0	1.25–3.2			
General industrial	Steel	10.0–25.0	2.0–2.5		1.50–2.5	12–30
	Cast iron	2.5–8.0	1.25–3.2			
Splines	Steel		1.25–3.0	1.25–1.5	0.50–1.75	18–45

TABLE 25-38
Selection of milling cutters

Material		**Hardness**	
One-piece construction	High-speed steel	Cutting portion	760 HV (62 HRC) Min
Two-piece construction		Shank portion	
Cutting portion	High speed steel	Parallel shank	245 HV (21 HRC) Min
Body	Carbon steel with tensile strength not less than 700 MPa (190 HN)	Tang of Morse taper shank	320 HV (32 HRC) Min

Note: The equivalent values within parentheses are approximate.

Recommendations for selection of milling cutters:
Tool Type N—For mild steel, soft cast iron and medium hard non-ferrous metals.
Tool Type H—For specially hard and tough materials.
Tool Type S—For soft and ductile materials.

Material to be cut	Tensile strength, MPa	Brinell hardness, H_B	Tool type[a]
Carbon steel	Up to 500		N or (S)
	Above 500 up to 800		N
	Above 800 up to 1000		N or (H)
	Above 1000 up to 1300		H
Steel casting			H
Gray cast iron		Up to 180	N
		Over 180	H
Malleability cast iron			N
Copper alloy			
Soft			S or (N)
Brittle			N or (H)
Zinc alloy			S or (N)
Aluminum alloy			
Soft			S
Medium/Hard			N or (S)
Aluminum alloy, age hardened			
Low cutting speed			N
High cutting speed			S
Magnesium alloy			S or (N)
Unlaminated			N or (S)

[a] Tool types within parentheses are non-preferred. *Courtesy*: IS 1830, 1971

TABLE 25-39
Dimensions for interchangeability of milling cutters and arbors with tenon drive

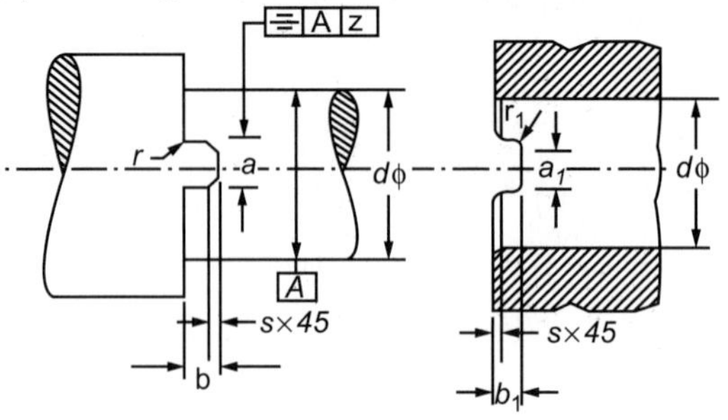

All dimensions in millimeters

d[a] h6/H7	Arbor a h11	Arbor b H11	Arbor r Max	Cutter a_1 H11	Cutter b_1 H13	Cutter r_1 Max	s		z[b]
5	3	2.0	0.3	3.3	2.5	0.6	0.3		0.075
8	5	3.5	0.4	5.4	4.0	0.6	0.4	+ 0.1	0.100
10	6	4.0	0.5	6.4	4.5	0.8	0.5		0.100
13	8	4.5	0.5	8.4	5.0	1.0	0.5		0.100
16	8	5.0	0.6	8.4	5.6	1.0	0.6		0.100
19	10	5.6	0.6	10.4	6.3	1.2	0.6		0.100
22	10	5.6	0.6	10.4	6.3	1.2	0.6	+0.2	0.100
27	12	6.3	0.8	12.4	8.0	1.6	0.8		0.100
32	14	7.0	0.8	14.4	7.0	1.2	0.8		0.100
40	18	9.0	1.0	16.4	9.0	2.0	1.0		0.100
50	16	8.0	1.0	18.4	10.0	2.0	1.0	+0.3	0.100
60	20	10.0	1.0	20.5	11.2	2.0	1.0		0.125

[a] The tolerance on d is not applicable to gear hobs.
[b] z = maximum permissible deviation between the axial plane of the tenon and the axis of arbor of diameter d.
Courtesy: IS 6285-1971

TABLE 25-40
Dimensions for interchangeability of milling cutters and milling arbors with key drive

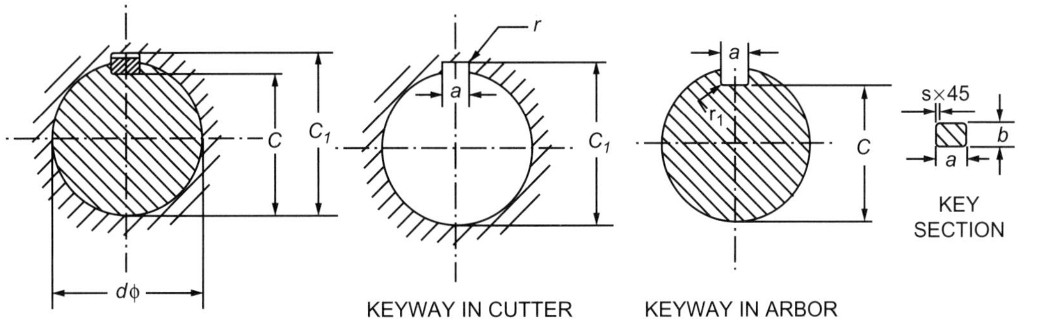

All dimensions in millimeters

	Key				Keyway								
d[a] h6/H7	a h9	b[b]	S	Tolerance on S	a[c]	C	Tolerance on C	C_1	Tolerance on C_1	r	Tolerance on r	r_1	Tolerance on r_1
8	2	2			2	6.7		8.9					
10	3	3	0.16	+0.09	3	8.2		11.5		0.4	0–0.1		
13	3	3		0	3	11.2	0	14.6	+0.1		0	0.16	0
16	4	4			4	13.2	0–0.1	17.7	0	0.6	–0.2		–0.08
19	5	5			5	15.6		21.1					
22	6	6	0.25	0+0.15	6	17.6		24.1	1.0				
27	7	7			7	22.0		29.8			0	0.25	0
32	8	7			8	27.0		34.8		1.2	–0.3		–0.09
40	10	8	9		10	34.5		43.5					
50	12	8			12	44.5		53.5	+0.2	1.6			
60	14	9	0.40	+0.20	14	54.0	0–0.2	64.2	0		0–0.5	0.40	0–0.15
70	16	10		0	16	63.5		75.0		2.0			
80	18	11			18	73.0		85.5					
100	25	14	0.60		25	91.0		107.0		2.5		0.60	0–0.20
													–0.20

[a] The tolerance on diameter d is not applicable to gear hobs.
[b] Tolerance on thickness b of key: square, h9; rectangular, h11.
[c] Tolerance on keyway width a: light drive fit, N9.
For keyway in arbor: running fit, H9; light drive fit, N9.
For keyway in cutter: C11

IS: 6285, 1971.

TABLE 25-41
American National Standard staggered teeth, T-slot milling cutters with Brown and Sharpe taper and Weldon shanks (ANSI/ASME B94, 19, 1986)

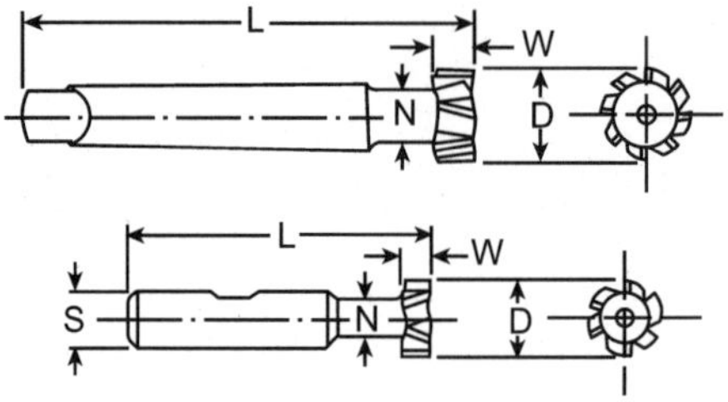

Bolt size	Cutter diam., *D*	Face width, *W*	Neck diam., *N*	With B. and S. taper[a] Length *L*	With B. and S. taper[a] Taper No.	With Weldon shank Length *L*	With Weldon shank Diam., *S*
$\frac{1}{4}$	$\frac{9}{16}$	$\frac{16}{64}$	$\frac{17}{64}$	—	—	$2\frac{19}{32}$	$\frac{1}{2}$
$\frac{5}{16}$	$\frac{21}{32}$	$\frac{17}{64}$	$\frac{21}{64}$	—	—	$2\frac{11}{16}$	$\frac{1}{2}$
$\frac{3}{8}$	$\frac{26}{32}$	$\frac{21}{64}$	$\frac{13}{32}$	—	—	$3\frac{1}{4}$	$\frac{3}{4}$
1	$\frac{13}{32}$	$\frac{25}{64}$	$\frac{17}{32}$	5	7	$3\frac{7}{16}$	1
$\frac{5}{8}$	$1\frac{1}{4}$	$\frac{31}{32}$	$\frac{21}{32}$	$6\frac{1}{4}$	7	$3\frac{15}{16}$	1
$\frac{3}{4}$	$1\frac{15}{32}$	$\frac{5}{8}$	$\frac{25}{32}$	$6\frac{7}{8}$	9	$4\frac{7}{16}$	—
1	$1\frac{27}{32}$	$\frac{53}{64}$	$1\frac{1}{32}$	$7\frac{1}{4}$	9	$4\frac{13}{16}$	$1\frac{1}{4}$

All dimensions are inches. All cutters are high-speed steel and only right-hand cutters are standard.
[a] For dimensions of Brown and Sharpe taper shanks. See information given in standard Handbook. Tolerances: On D, +0.000, −0.010 inch; on W, +0.000, −0.005 inch; on N, +0.000, −0.005 inch, on L, $\pm\frac{1}{16}$ inch; on S, −0.0001 to −0.0005 inch.

TABLE 25-42
American National Standard 60-degree single-angle milling cutters with Weldon shanks (ANSI/ASME B94, 19, 1985)

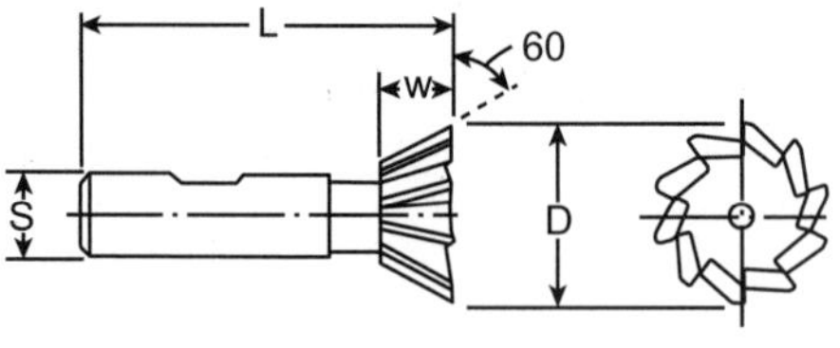

Diam., *D*	*S*	*W*	*L*	Diam., *D*	*S*	*W*	*L*
$\frac{3}{4}$	$\frac{3}{8}$	$\frac{5}{16}$	$2\frac{1}{16}$	$1\frac{7}{8}$	$\frac{7}{8}$	$\frac{13}{16}$	$3\frac{1}{4}$
$1\frac{3}{8}$	$\frac{5}{8}$	$\frac{9}{16}$	$2\frac{7}{8}$	$2\frac{1}{4}$	1	$1\frac{1}{16}$	$3\frac{3}{4}$

All dimensions are in inches. All cutters are high-speed steel. Right-hand cutters are standard.
Tolerances: On D, 0.015 inch; on S, −0.0001 to −0.0005 inch; on W, 0.015 inch; and on L, $\pm\frac{1}{16}$ inch.

TABLE 25-43
American National Standard multiple flute, helical series end mills with Brown and Sharpe taper shanks[a] (ANSI/ ASME B94, 19, 1985)

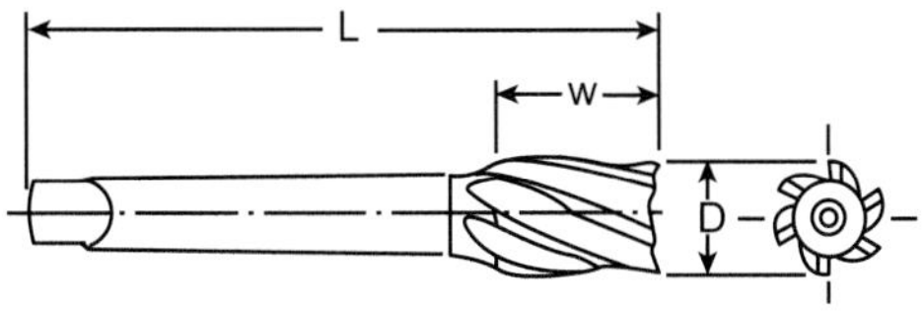

Diam., *D*	*W*	*L*	Taper No.	Diam., *D*	*W*	*L*	Taper No.
–	–	–	–	1	$1\frac{5}{8}$	$5\frac{5}{8}$	7
–	–	–	–	$1\frac{1}{4}$	2	$7\frac{1}{4}$	9
$\frac{1}{2}$	$\frac{15}{16}$	$4\frac{15}{16}$	7	$1\frac{1}{2}$	$2\frac{1}{4}$	$7\frac{1}{2}$	9
$\frac{3}{4}$	$1\frac{1}{4}$	$5\frac{1}{4}$	7	2	$2\frac{3}{4}$	8	9

All dimensions are in inches. All cutters are high-speed steel. Right-hand cutters with right hand helix are standard. Helix angle is not less than 10 degrees.
No. 5 taper is standard without tang: Nos. 7 and 9 are standard with tang only.
Tolerances: On D, +0.005 inch; on W, $\pm\frac{1}{32}$ inch; and L, $\pm\frac{1}{16}$ inch.
[a] For dimensions of B. and S. taper shanks, see information given in standard handbook.

TABLE 25-44
American National Standard form relieved, concave, convex, and corner-rounding arbor-type cutters[a] (ANSI/ASME B94, 19, 1985)

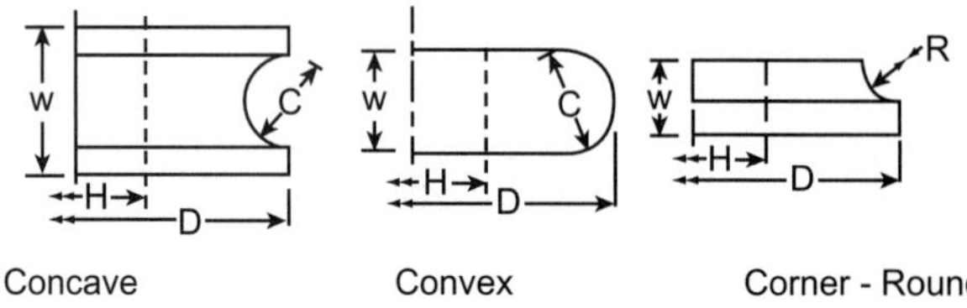

Diameter *C* or radius *R*			Cutter	Width *W*	Diameter of hole *H*		
Nom.	Max.	Min.	diam. *D*[b]	±.010[c]	Nom.	Max.	Min.
			Concave cutters[c]				
$\frac{1}{8}$	0.1270	0.1240	$2\frac{1}{4}$	$\frac{1}{4}$	1	1.00075	1.00000
$\frac{1}{4}$	0.2520	0.2490	$2\frac{1}{2}$	$\frac{7}{16}$	1	1.00075	1.00000
$\frac{3}{8}$	0.3770	0.3740	$2\frac{3}{4}$	$\frac{5}{8}$	1	1.00075	1.00000
$\frac{1}{2}$	0.5040	0.4980	3	$\frac{13}{16}$	1	1.00075	1.00000
$\frac{3}{4}$	0.7540	0.7480	$3\frac{3}{4}$	$1\frac{13}{16}$	$1\frac{1}{4}$	1.251	1.250
1	0.0040	0.9980	$4\frac{1}{4}$	$1\frac{9}{16}$	$1\frac{1}{4}$	1.251	1.250
			Convex cutters[d]				
$\frac{1}{4}$	0.2520	0.2480	$2\frac{1}{2}$	$\frac{1}{4}$	1	1.00075	1.00000
$\frac{3}{8}$	0.3770	0.3730	$2\frac{3}{4}$	$\frac{3}{8}$	1	1.00075	1.00000
$\frac{1}{2}$	0.5020	0.4980	3	$\frac{1}{2}$	1	1.00075	1.00000
$\frac{3}{4}$	0.7520	0.7480	$3\frac{3}{4}$	$\frac{3}{4}$	$1\frac{1}{4}$	1.251	1.250
1	1.0020	0.9980	$4\frac{1}{4}$	1	$1\frac{1}{4}$	1.215	1.250
			Corner-rounding cutters[e]				
$\frac{1}{8}$	0.1260	0.1240	$2\frac{1}{2}$	$\frac{1}{4}$	1	1.00075	1.00000
$\frac{1}{4}$	0.2520	0.2490	3	$\frac{13}{32}$	1	1.00075	1.00000
$\frac{1}{2}$	0.5020	0.4990	$4\frac{1}{4}$	$\frac{3}{4}$	$1\frac{1}{4}$	1.251	1.250

All dimensions in inches. All cutters are high-speed steel and are form relieved.
Right-hand corner rounding cutters are standard, but left-hand cutter for $\frac{1}{4}$ inch size is also standard.
[a] For key and keyway dimensions for these cutters, see standard handbook.
[b] Tolerances on cutter diameters are $+\frac{1}{16}$, $-\frac{1}{16}$ inch for all sizes.
[c] Tolerance does not apply to convex cutters.
[d] Size of cutter is designated by specifying diameter *C* of circular form.
[e] Size of cutter is designated by specifying radius *R* of circular form.
Source: Courtesy: ANSI/ASME B94, 19, 1985, Erik Oberg Editor Etd., Extracted from *Machinery's Handbook*, 25th edition, Industrial Press, N.Y., 1996.

TABLE 25-45
American National Standard roughing and finishing gear milling cutters for gears with $14\frac{1}{2}$ degree pressure angles (ANSI/ASME B94, 19, 1985)

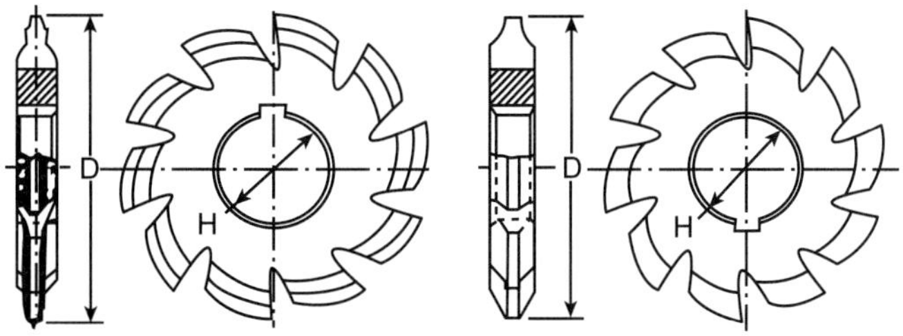

ROUGHING FINISHING

Diametral pitch	Diam. of cutter, *D*	Diam. of hole, *H*	Diametral pitch	Diam. of cutter, *D*	Diam. of hole. *H*	Diametral pitch	Diam. of cutter, *D*	Diam. of hole, *H*
				Roughing gear milling cutters				
1	$8\frac{1}{2}$	2	3	$5\frac{1}{4}$	$1\frac{1}{2}$	5	$3\frac{3}{8}$	1
$1\frac{1}{2}$	7	$1\frac{3}{4}$	4	$4\frac{3}{4}$	$1\frac{3}{4}$	6	$3\frac{1}{2}$	$1\frac{1}{4}$
1	$6\frac{1}{2}$	$1\frac{3}{4}$	4	$4\frac{1}{4}$	$1\frac{1}{4}$	7	$3\frac{3}{8}$	$1\frac{1}{4}$
$2\frac{1}{2}$	$6\frac{1}{8}$	$1\frac{3}{4}$	5	$4\frac{3}{8}$	$1\frac{3}{4}$	8	$3\frac{1}{4}$	$1\frac{1}{4}$
3	$5\frac{5}{8}$	$1\frac{3}{4}$	5	$3\frac{3}{4}$	$1\frac{1}{4}$	–	–	–
				Finishing gear milling cutters				
1	$8\frac{1}{2}$	2	6	$3\frac{7}{8}$	$1\frac{1}{2}$	14	$2\frac{1}{8}$	$\frac{7}{8}$
$1\frac{1}{2}$	7	$1\frac{3}{4}$	6	$3\frac{1}{8}$	1	16	$2\frac{1}{8}$	$\frac{7}{8}$
2	$6\frac{1}{2}$	$1\frac{3}{4}$	7	$3\frac{3}{8}$	$1\frac{1}{4}$	18	2	$\frac{7}{8}$
$2\frac{1}{2}$	$6\frac{1}{8}$	$1\frac{3}{4}$	8	$3\frac{1}{2}$	$1\frac{1}{2}$	20	2	$\frac{7}{8}$
3	$5\frac{5}{8}$	$1\frac{3}{4}$	8	$2\frac{7}{8}$	1	22	2	$\frac{7}{8}$
3	$5\frac{1}{4}$	$1\frac{1}{2}$	9	$3\frac{1}{8}$	$1\frac{1}{4}$	24	$2\frac{1}{4}$	1
4	$4\frac{1}{4}$	$1\frac{3}{4}$	10	3	$1\frac{1}{4}$	26	$1\frac{3}{4}$	$\frac{7}{8}$
5	$4\frac{3}{8}$	$1\frac{3}{4}$	11	$2\frac{3}{8}$	$\frac{7}{8}$	36	$1\frac{3}{4}$	$\frac{7}{8}$
5	$4\frac{1}{4}$	$1\frac{1}{2}$	12	$2\frac{7}{8}$	$1\frac{1}{4}$	40	$1\frac{3}{4}$	$\frac{7}{8}$
6	$4\frac{1}{4}$	$1\frac{3}{4}$	14	$2\frac{1}{2}$	1	–	–	–

All dimensions are in inches.
All gear milling cutters are high-speed steel and are form relieved.
For keyway dimensions refer to standard handbook.
Tolerances: On outside diameter, $+\frac{1}{16}$, $-\frac{1}{16}$ inch; on hole diameter, through 1 inch hole diameter, +0.00075 inch; over 1 inch and through 2 inch hole diameter, +0.0010 inch.
For cutter number relative to number of gear teeth, see standard handbook.
Roughing cutters are made with No. 1 cutter form only.
Source: Courtesy: ANSI/ASME B94, 19, 1985, Erik Oberg Editor Etd., Extracted from *Machinery's Handbook*, 25th edition, Industrial Press, N.Y., 1996.

TABLE 25-46
American National Standard regular, long and extra length, multiple-flute medium helix single-end end mills with Weldon shanks (ANSI/ASME B94, 19, 1985)

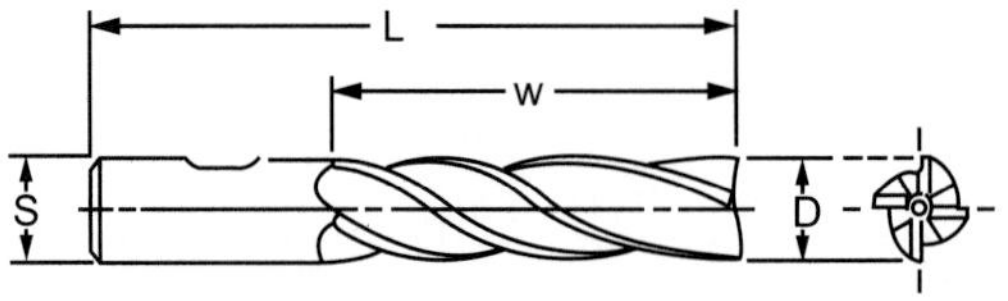

AS INDICATED BY THE DIMENSIONS GIVEN BELOW, SHANK DIAMETERS MAY BE LARGER, SMALLER, OR THE SAME AS THE CUTTER DIAMATERS D.

Cutter diam, *D*	Regular mills				Long mills				Extra long mills			
	S	*W*	*L*	*N*[a]	*S*	*W*	*L*	*N*[a]	*S*	*W*	*L*	*N*[a]
$\frac{1}{4}$[a]	$\frac{3}{8}$	$\frac{5}{8}$	$2\frac{7}{16}$	4	$\frac{3}{8}$	$1\frac{1}{4}$	$3\frac{1}{16}$	4	$\frac{3}{8}$	$1\frac{3}{4}$	$3\frac{9}{16}$	4
$\frac{5}{16}$[a]	$\frac{3}{8}$	$\frac{3}{4}$	$2\frac{1}{2}$	4	$\frac{3}{8}$	$1\frac{3}{8}$	$3\frac{1}{8}$	4	$\frac{1}{8}$	2	$3\frac{1}{4}$	4
$\frac{3}{8}$[a]	$\frac{3}{8}$	$\frac{3}{4}$	$2\frac{1}{2}$	4	$\frac{3}{8}$	$1\frac{1}{2}$	$3\frac{1}{4}$	4	$\frac{3}{8}$	$2\frac{1}{2}$	$4\frac{1}{4}$	4
$\frac{7}{16}$	$\frac{3}{8}$	1	$2\frac{11}{16}$	4	$\frac{1}{2}$	$1\frac{3}{4}$	$3\frac{3}{4}$	4	–	–	–	–
$\frac{1}{2}$	$\frac{3}{8}$	1	$2\frac{11}{16}$	4	$\frac{1}{2}$	2	4	4	$\frac{1}{2}$	3	5	4
$\frac{9}{16}$	$\frac{1}{2}$	$1\frac{3}{8}$	$3\frac{3}{8}$	4	–	–	–	–	–	–	–	–
$\frac{3}{8}$	$\frac{1}{2}$	$1\frac{3}{8}$	$3\frac{3}{8}$	4	$\frac{5}{8}$	$2\frac{1}{2}$	$4\frac{5}{8}$	4	$\frac{5}{8}$	4	$6\frac{1}{8}$	4
$\frac{11}{16}$	$\frac{1}{2}$	$1\frac{5}{8}$	$3\frac{5}{8}$	4	–	–	–	–	–	–	–	–
$\frac{3}{4}$	$\frac{1}{2}$	$1\frac{5}{8}$	$3\frac{5}{8}$	4	$\frac{3}{4}$	3	$5\frac{1}{4}$	4	$\frac{3}{4}$	4	$6\frac{1}{4}$	4
$\frac{7}{8}$	$\frac{5}{8}$	$1\frac{7}{8}$	4	6	$\frac{7}{8}$	$3\frac{1}{2}$	$5\frac{3}{4}$	4	$\frac{7}{8}$	5	$7\frac{1}{4}$	4
1	$\frac{5}{8}$	$1\frac{7}{8}$	4	6	1	4	$6\frac{1}{2}$	4	1	6	$8\frac{1}{2}$	6
$1\frac{1}{8}$	$\frac{7}{8}$	2	$4\frac{1}{4}$	6	1	4	$6\frac{1}{2}$	6	–	–	–	–
$1\frac{1}{4}$	$\frac{7}{8}$	2	$4\frac{1}{4}$	6	1	4	$6\frac{1}{2}$	6	$1\frac{1}{4}$[a]	6	$8\frac{1}{2}$	6
$1\frac{1}{2}$	1	2	$4\frac{1}{4}$	6	1	4	$6\frac{1}{2}$	6	–	–	–	–
$1\frac{1}{4}$	$1\frac{1}{4}$	2	$4\frac{1}{2}$	6	$1\frac{1}{4}$	4	$6\frac{1}{2}$	6	–	–	–	–
$1\frac{1}{2}$	$1\frac{1}{4}$	2	$4\frac{1}{2}$	6	$1\frac{1}{4}$	4	$6\frac{1}{2}$	6	$1\frac{1}{4}$	8	$10\frac{1}{2}$	6
$1\frac{3}{4}$	$1\frac{1}{4}$	2	$4\frac{1}{2}$	6	$1\frac{1}{4}$	4	$6\frac{1}{2}$	6	–	–	–	–
2	$1\frac{1}{4}$	2	$4\frac{1}{2}$	8	$1\frac{1}{4}$	4	$6\frac{1}{2}$	8	–	–	–	–

All dimensions are in inches. All cutters are high-speed steel. Helix angle is greater than 19 degrees but not more than 39 degrees. Right-hand cutters with right-hand helix are standard.

Tolerances: On D, +0.003 inch; on S, 0.0001 to −0.0005 inch; on W, $\pm\frac{1}{32}$ inch; on L, $\pm\frac{1}{16}$ inch; N = number of flutes.

[a] In case of regular mill a left-hand cutter with left-hand helix is also standard.

Source: ANSI/ASME B94, 19, 1985, Erik Oberg Editor Etd., Extracted from *Machinery's Handbook*, 25th edition, Industrial Press, N.Y., 1996.

TABLE 25-47
American National Standard long length single-end and stub-, and regular length, double-end plain- and ball-end, medium helix two-flute end mills with Weldon shanks (ANSI/ASME B94, 19, 1985)

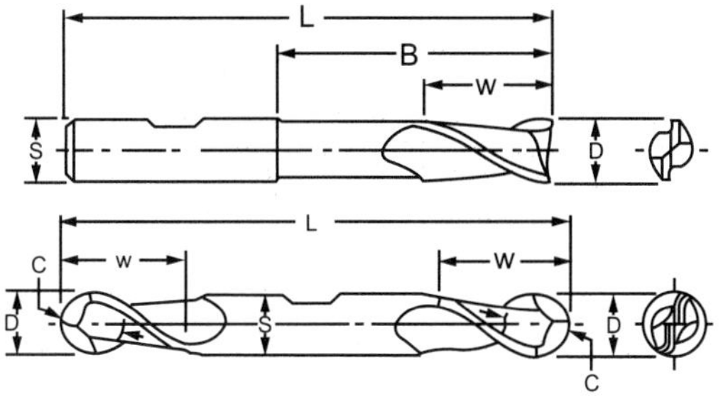

Single end

Diam., C and D	Long length—plain end: S	B^a	W	L	Long length—ball end: S	B^a	W	L
$\frac{1}{4}$	$\frac{3}{8}$	$1\frac{1}{2}$	$\frac{5}{8}$	$3\frac{1}{16}$	$\frac{3}{8}$	$1\frac{1}{2}$	$\frac{5}{8}$	$3\frac{1}{16}$
$\frac{5}{16}$	$\frac{3}{8}$	$1\frac{3}{4}$	$\frac{3}{4}$	$3\frac{5}{16}$	$\frac{3}{8}$	$1\frac{3}{4}$	$\frac{3}{4}$	$3\frac{5}{16}$
$\frac{3}{8}$	$\frac{3}{8}$	$1\frac{3}{4}$	$\frac{3}{4}$	$3\frac{5}{16}$	$\frac{3}{8}$	$1\frac{3}{4}$	$\frac{3}{4}$	$3\frac{5}{16}$
$\frac{1}{2}$	$\frac{1}{2}$	$2\frac{7}{32}$	1	4	$\frac{1}{2}$	$2\frac{1}{4}$	1	4
$\frac{5}{8}$	$\frac{5}{8}$	$2\frac{23}{32}$	$1\frac{3}{8}$	$4\frac{5}{8}$	$\frac{5}{8}$	$2\frac{3}{4}$	$1\frac{3}{8}$	$4\frac{3}{8}$
$\frac{1}{4}$	$\frac{3}{4}$	$3\frac{11}{32}$	$1\frac{5}{8}$	$5\frac{3}{8}$	$\frac{3}{4}$	$3\frac{3}{8}$	$1\frac{5}{8}$	$5\frac{3}{8}$
1	1	$4\frac{31}{32}$	$2\frac{1}{2}$	$7\frac{1}{4}$	1	5	$2\frac{1}{2}$	$7\frac{1}{4}$

Double end

Diam., C and D	Stub length—plain end: S	W	L	Regular length—plain end: S	W	L	Regular length—ball end: S	W	L
$\frac{5}{32}$	$\frac{3}{8}$	$\frac{15}{64}$	$2\frac{3}{4}$	$\frac{3}{8}$	$\frac{7}{16}$	$3\frac{1}{8}$	–	–	–
$\frac{1}{4}$	$\frac{3}{8}$	$\frac{3}{8}$	$2\frac{7}{8}$	$\frac{3}{8}$	$\frac{1}{2}$	$3\frac{1}{8}$	$\frac{3}{8}$	$\frac{1}{2}$	$3\frac{1}{8}$
$\frac{5}{16}$	–	–	–	$\frac{3}{8}$	$\frac{9}{16}$	$3\frac{1}{8}$	$\frac{3}{8}$	$\frac{9}{16}$	$3\frac{1}{8}$
$\frac{3}{8}$	–	–	–	$\frac{3}{8}$	$\frac{9}{16}$	$3\frac{1}{8}$	$\frac{3}{8}$	$\frac{9}{16}$	$3\frac{1}{8}$
$\frac{7}{16}$	–	–	–	$\frac{1}{2}$	$\frac{13}{16}$	$3\frac{3}{4}$	$\frac{1}{2}$	$\frac{13}{16}$	$3\frac{3}{4}$
$\frac{1}{2}$	–	–	–	$\frac{1}{2}$	$\frac{13}{16}$	$3\frac{3}{4}$	$\frac{1}{2}$	$\frac{13}{16}$	$3\frac{3}{4}$
$\frac{5}{8}$	–	–	–	$\frac{5}{8}$	$1\frac{1}{8}$	$4\frac{1}{2}$	$\frac{5}{8}$	$1\frac{1}{8}$	$4\frac{1}{2}$
$\frac{11}{16}$	–	–	–	$\frac{3}{4}$	$1\frac{5}{16}$	5	–	–	–
$\frac{3}{4}$	–	–	–	$\frac{3}{4}$	$1\frac{5}{16}$	5	$\frac{3}{4}$	$1\frac{5}{16}$	5
1	–	–	–	1	$1\frac{5}{8}$	$5\frac{7}{8}$	1	$1\frac{5}{8}$	$5\frac{7}{8}$

All dimensions are in inches. All cutters are high-speed steel. Right-hand cutters with right hand helix are standard. Helix angle is greater than 19 degrees but not more than 39 degrees.

Tolerances: On C and D, +0.003 inch; for single-end mills, −0.0015 inch for double end mills on S, −0.0001 to −0.0005 on W, $\pm\frac{1}{32}$ inch; on L, $\frac{1}{16}$ inch. [a] B is the length below the shank.

Source: Courtesy: ANSI/ASME B94, 19, 1985, Erik Oberg Editor Etd., Extracted from *Machinery's Handbook*, 25th edition, Industrial Press, N.Y., 1996.

TABLE 25-48
American National Standard Woodruff keyseat cutters[a]—shank-type straight teeth and arbor staggered teeth (ANSI/ASME B94, 19, 1985)

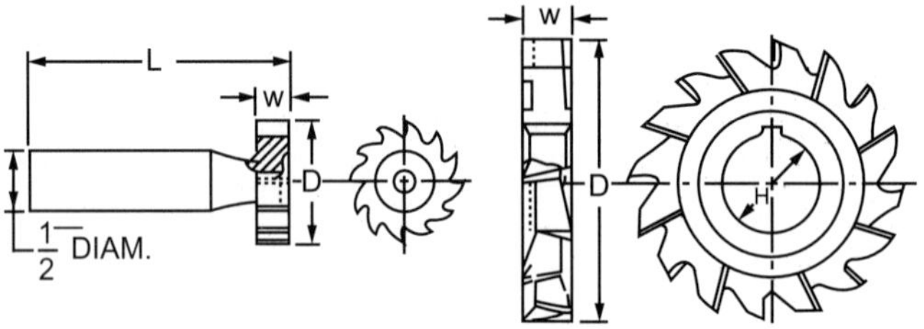

Shank-type cutters

Cutter No.	Nom. diam. of cutter, *D*	Width of face, *W*	Length overall, *L*	Cutter No.	Nom. diam. of cutter, *D*	Width of face, *W*	Length overall, *L*	Cutter No.	Nom. diam. of cutter, *D*	Width of face, *W*	Length overall, *L*
202	$\frac{1}{4}$	$\frac{1}{16}$	$2\frac{1}{16}$	506	$\frac{3}{4}$	$\frac{5}{32}$	$2\frac{5}{32}$	809	$1\frac{1}{8}$	$\frac{1}{4}$	$2\frac{1}{4}$
203	$\frac{3}{8}$	$\frac{1}{16}$	$2\frac{1}{16}$	507	$\frac{7}{8}$	$\frac{5}{32}$	$2\frac{5}{32}$	710	$1\frac{1}{4}$	$\frac{7}{32}$	$2\frac{9}{32}$
403	$\frac{3}{8}$	$\frac{1}{8}$	$2\frac{1}{8}$	707	$\frac{7}{8}$	$\frac{7}{32}$	$2\frac{5}{32}$	1010	$1\frac{1}{4}$	$\frac{5}{16}$	$2\frac{5}{16}$
404	$\frac{1}{2}$	$\frac{1}{16}$	$2\frac{1}{16}$	807	$\frac{7}{8}$	$\frac{1}{4}$	$2\frac{1}{4}$	1210	$1\frac{1}{4}$	$\frac{3}{8}$	$2\frac{3}{8}$
405	$\frac{5}{8}$	$\frac{1}{8}$	$2\frac{1}{8}$	1008	1	$\frac{5}{16}$	$2\frac{5}{16}$	812	$1\frac{1}{2}$	$\frac{1}{4}$	$2\frac{1}{4}$
505	$\frac{5}{8}$	$\frac{5}{32}$	$2\frac{1}{32}$	1208	1	$\frac{3}{8}$	$2\frac{3}{8}$	1212	$1\frac{1}{2}$	$\frac{3}{8}$	$2\frac{3}{8}$

Arbor-type cutters

Cutter No.	Nom. diam. of cutter, *D*	Width of face, *W*	Diam. of a hole, *H*	Cutter No.	Nom. diam. of cutter, *D*	Width of face, *W*	Diam. of a hole, *H*	Cutter No.	Nom. diam. of cutter, *D*	Width of face, *W*	Diam. of hole, *H*
617	$2\frac{1}{8}$	$\frac{3}{16}$	$\frac{3}{4}$	1012	$2\frac{3}{4}$	$\frac{5}{16}$	1	1628	$3\frac{1}{2}$	$\frac{1}{2}$	1
817	$2\frac{1}{8}$	$\frac{1}{4}$	$\frac{3}{4}$	1222	$2\frac{3}{4}$	$\frac{3}{8}$	1	1828	$3\frac{1}{2}$	$\frac{9}{16}$	1
1217	$2\frac{1}{8}$	$\frac{3}{16}$	$\frac{3}{4}$	1262	$2\frac{3}{4}$	$\frac{1}{2}$	1	2428	$3\frac{1}{2}$	$\frac{3}{4}$	1
822	$2\frac{3}{4}$	$\frac{1}{4}$	1	1288	$3\frac{1}{2}$	$\frac{3}{8}$	1	–	–	–	–

All dimensions are given in inches. All cutters are high-speed steel.

Shank type cutters are standard with right-hand cut and straight teeth. All sizes have $\frac{1}{2}$ inch diameter straight shank. Arbor type cutters have staggered teeth.

For Woodruff key and key-slot dimensions, see standard handbook.

Tolerances: Face width W for shank type cutters: $\frac{1}{16}$ to $\frac{3}{32}$ inch face +0.0000, 0.0005: $\frac{3}{16}$ to $\frac{7}{32}$, −0.002, 0.0007, $\frac{1}{4}$, − 0.0003, 0.0008, $\frac{5}{16}$, 0.0004, −0.0009, $\frac{3}{8}$, 0.0005, 0.001, −0.0008, $\frac{5}{16}$, −0.0004, −0.0009, $\frac{1}{8}$ and over, −0.0005, −0.000 inch.

Hole size H, +0.00075, −1.000 inch. Diameter D for shank type cutters; $\frac{1}{8}$, through $\frac{1}{4}$ inch diameter, +0 016, +0.015, $\frac{7}{8}$ through $1\frac{1}{8}$, +0.012, +0.017; $1\frac{1}{4}$ through $1\frac{1}{2}$, +0.015, +0.02 inch. These tolerances includes an allowance for sharpening. For arbor type cutters diameter D is furnished $\frac{1}{32}$ inch larger than bore and tolerance of +0.002 inch applies to the over size diameter.

Source: Courtesy: ANSI/ASME B94, 19, 1985, Erik Oberg Editor Etd., Extracted from *Machinery's Handbook*, 25th edition, Industrial Press, N.Y., 1996.

Particular	Formula

GRINDING

The tangential component of grinding force F_z, which constitutes the major value of grinding force Fig. 25-26

$$F_t = K_m st \frac{v_w}{v_g} \tag{25-123}$$

where

s = feed rate, mm/rev

t = thickness of material removed from job or depth of cut, mm

v_w = peripheral velocity of workpiece/job, m/min

v_g = peripheral velocity of the grinding wheel, m/min

K_m = specific resistance to grinding of the work material, N/m^2 (Table 25-51)

$F_y = F_r$ = radial component of the force in cylindrical grinding operation, kN

F_x = horizontal component of the force against the feed, kN

$F_z = F_t$ = vertical component of the force in the cylindrical grinding operation, kN

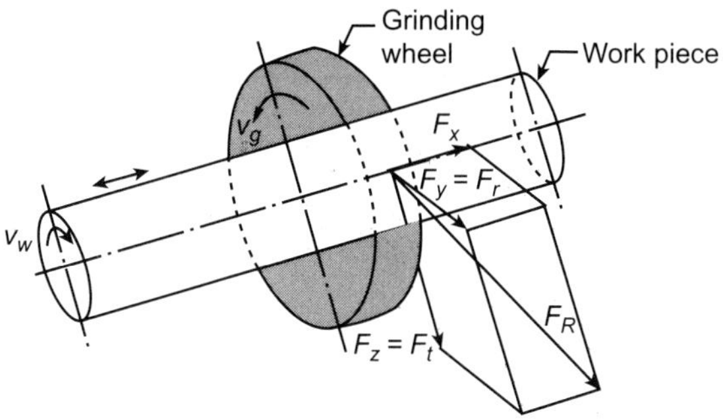

FIGURE 25-26 Forces acting on a grinding wheel.

The chip thickness

$$t = \frac{2pv_w}{v_g}\sqrt{\frac{(d_w \pm d_g)s}{d_g d_w}} \tag{25-124}$$

where

p = pitch of grains, mm

d_w = diameter of workpiece, mm

d_g = diameter of grinding wheel, mm

+ve sign for external grinding wheel, −ve for internal grinding wheel

The power required by the grinding wheel

$$P = \frac{F_t(=F_z)v_g}{1000} \tag{25-125}$$

where

$F_t = F_z$ = tangential force on wheel, N

P = power, W

v_g = velocity of grinding wheel, mm/s

Metal removal rate in case of transverse grinding

$$Q = \frac{\pi d_w ts}{1000} \tag{25-126}$$

The power at the spindle

$$P = P_u Q \tag{16-127}$$

where Q in cm^3/min

Refer to Table 25-50 for P_u.

Particular	Formula	
Energy per unit volume of material removed	$E = \dfrac{P}{bsv_g}$	(25-128)
	where	
	b = width of cut, mm	
	s = feed rate or depth of cut, mm/rev	
	E in J/mm^3	
Vertical boring:		
The power required for boring	$P = \dfrac{iF_t v}{1000}$	(25-129)
	where	
	i = number of heads	
	v = cutting speed, mm/s	
Centerless grinding:		
The peripheral grinding wheel speed	$v_g = \dfrac{\pi d_w n}{1000 \times 60}$	(25-130)
Through feed rate	$s_t = \pi d_r n_r \sin \alpha$	(25-131)
	where	
	d_r = diameter of regulating wheel, mm	
	n_r = speed of regulating wheel, rpm	
	α = regulating wheel inclination angle, deg	
Metal removal rate from through feed grinding	$Q_t = \dfrac{\pi d_w t s_t}{1000}$	(25-132)
	where Q_t in cm^3/min	
	s_t = through feed rate, mm/min	
Metal removal rate from plunge grinding	$Q_p = \dfrac{\pi d_w b s_p}{1000}$	(25-133)
	where Q_p in cm^3/min	
	b = width of cut plunge grinding, mm	
	s_p = plunge in feed rate per minute = (sn_w), mm/min	
	s = plunge in feed rate per work revolution, mm/rev	
	n_w = workpiece revolution per minute	
For the unit power	Refer to Table 25-50.	
Power at the spindle	$P = P_u Q$	(25-134)

Particular	Formula
SHAPING (Fig. 25-27)	
The force of cutting can be found by empirical formula F_z	$F_z = F_t = 9.807 C_p k d^x s^y$ **SI** (25-135a) where F_z in N $F_z = F_t = C_p k d^x s^y$ **Customary Metric Units** (25-135b) where x, y, k and C_p have the same values as in lathe tools; F_z in kgf Equation (25-135) can be also used for the case of planing machine.
The approximate equation 1 expression for cutting force F_z for cast iron	$F_z = 1860 d s^{0.75} K$ **SI** (25-136a) $F_z = 190 d s^{0.75} K$ **Customary Metric Units** (25-136b)
The power consumption of shaping machine	$P = \frac{1}{\eta}\frac{F_z v_r}{1000 \times 60}$ (25-137) where v_r = the the average velocity of ram in its middle position during its stroke
The velocity of crank pin of r radius	$v_1 = \frac{2\pi r n}{1000}$ (25-138a)
The peripheral velocity of the sliding block	$v_2 = v_1 \cos(\alpha - \gamma)$ (25-138b)
The peripheral velocity of the driving pin of the rocker arm at point A.	$v_{ra} = v_2 \frac{R}{Ma}$ (25-138c)
The average velocity of ram at its middle position during its stroke Fig. 25-28	$v_{r\,av} = \pi n \frac{Rl}{R + (l/2)}$ (25-138d) where n in rpm

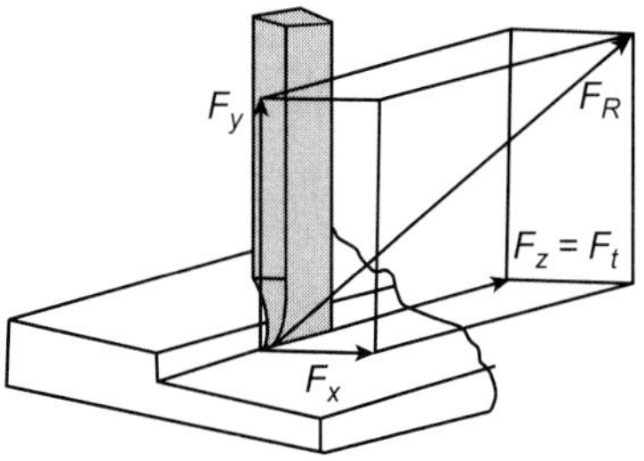

FIGURE 25-27 Forces acting on a shaping tool.

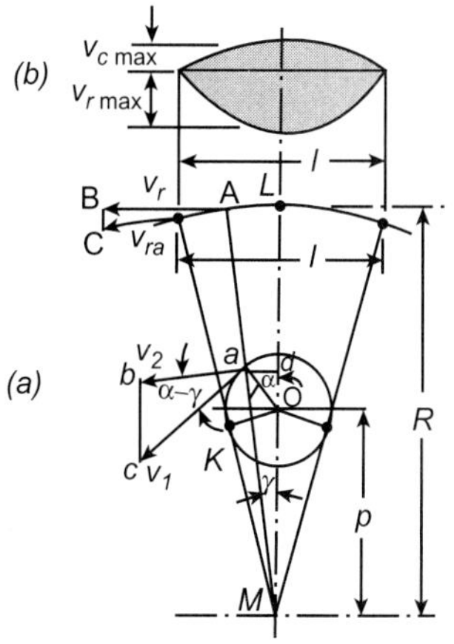

FIGURE 25-28 Ram velocity diagram of a crank shaper.

Particular	Formula
The approximate/average velocity of ram	$v_r = v_{ra} \cos\gamma$ (25-138e)
	$$v_r = \frac{2\pi nRl\cos^2\gamma\cos(\alpha-\gamma)}{1000(2R + l\cos\alpha)}$$ (25-138f)
The maximum speed of ram for the average value of cutting speed v_r	$$n_{\max} = v_r \frac{R + (l_{\min}/2)}{\pi R + l_{\min}}$$ (25-139a)
The maximum velocity of ram travel in the cutting stroke when it is at $\alpha = \gamma = 0$	$$v_{c\max} = \frac{2\pi nRl}{1000(2R + l)}$$ (25-139b)
The minimum speed for the average value of cutting speed v_r	$$n_{\min} = v_r \frac{R + (l_{\max}/2)}{\pi R - l_{\max}}$$ (25-139c)
	v_r is a function of l, since π and R are constants, i.e. $v_r = f(l)$
The maximum velocity of ram travel during the return stroke at $\alpha = 180^\circ$ and $\gamma = 0$	$$v_{r\max} = \frac{2\pi nRl}{1000(2R - l)}$$ (25-139d)
The average cutting velocity $v_{r\,\mathrm{av}}$ during travel $2l$ of ram	$$v_{r\,\mathrm{av}} = \frac{2ln}{1000}$$ (25-139e) where $v_{r\,\mathrm{av}}$ in m/min

PRESS TOOLS

Punching (Figs. 25-29, 25-31 and 25-32):

Maximum shearing force or pressure to cut the material	$F_{\max} = pD\tau_u t$ for round hole (25-140)
	$= \tau_u tP$ for any other contour
Work done	$W = F_{\max} x$ (25-141)

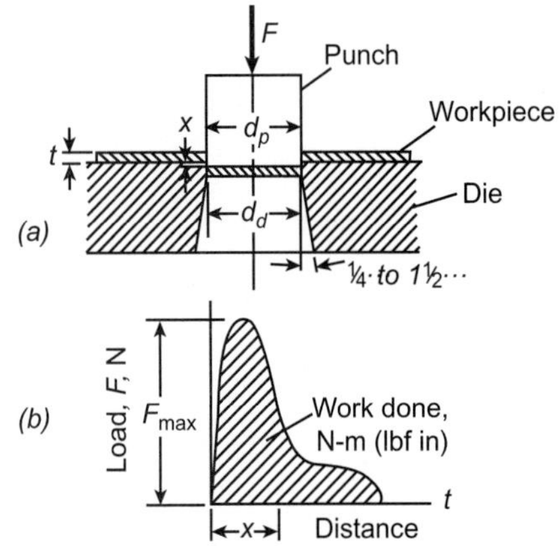

FIGURE 25-29

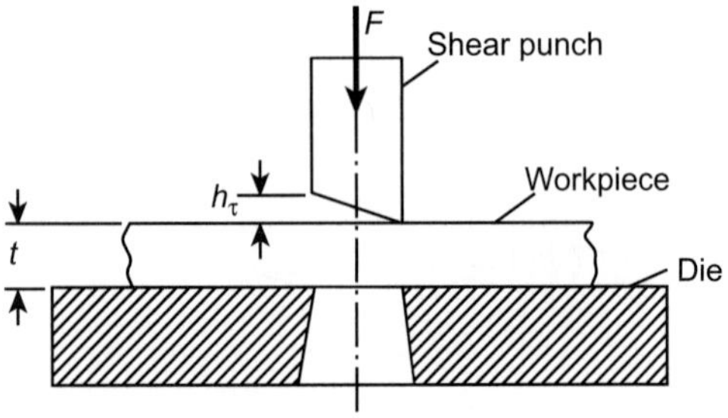

FIGURE 25-30

Particular	Formula

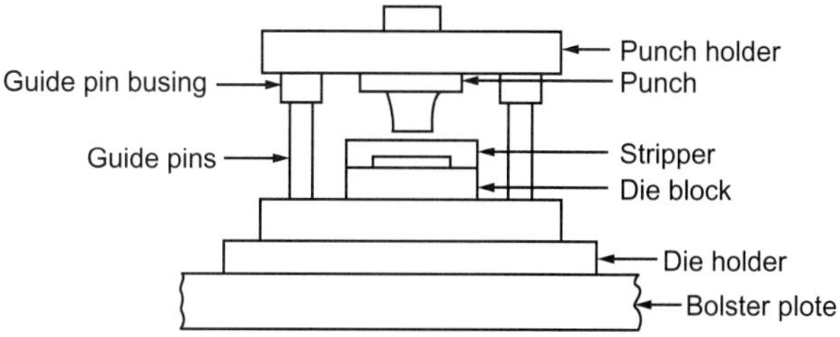

FIGURE 25-31 Common components of a simple die. *Courtesy*: F. W. Wilson, *Fundamentals of Tool Design*, American Society of Tool and Manufacturing Engineers, Prentice-Hall of India, 1969.

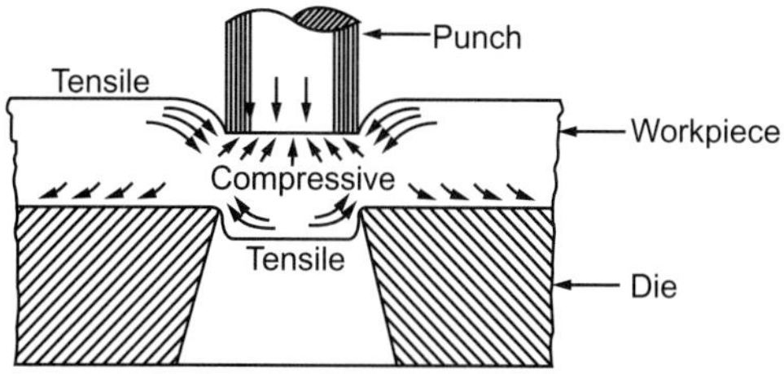

FIGURE 25-32 Stresses in die cutting.

Penetration ratio

$$c = \frac{x}{t} \tag{25-142}$$

where

F_{max} = maximum shear force, kN (lbf)

τ_u = ultimate shear stress, taken from Table 25-54

t = material thickness, mm (in)

x = penetration, mm (in)

P = perimeter of profile, mm (in)

Punch Dimensioning:

When the diameter of a pierced round hole equals stock thickness, the unit compressive stress on the punch is four times the unit shear stress on the cut area of the stock, from the formula.

$$\frac{4\tau t}{\sigma_c d} = 1 \tag{25-143}$$

where

σ_c = unit compressive stress on the punch, MPa (psi)

τ = unit shear stress on the stock, MPa (psi)

t = thickness of stock, mm (in)

d = diameter of the punched hole, mm (in)

A value for the ratio d/t of 1.1 is recommended.

The maximum allowable length of a punch can be calculated from the formula

$$L = \frac{\pi d}{8}\left(\frac{E}{\tau}\frac{d}{t}\right)^{1/2} \tag{25-144}$$

where $d/t = 1.1$ or higher value

E = modulus of elasticity, GPa (psi)

For clearance between punch and die

Refer to Tables 25-52 and Fig. 25-36.

Particular	Formula
Shearing (Fig. 25-30):	
Shearing force	$$F_\tau = \frac{F_{max}}{1 + \frac{h_\tau}{x}} \quad (25\text{-}145)$$ where h_τ is shown in Fig. 25-30
The stripper pressure or force	$F_{str} = 24 \times 10^6 Pt$ **SI** (25-146a) where P = perimeter of cut, m; t = thickness of workpiece, m; F_{str} in N
	$F_{str} = 3500Pt$ **USCS** (25-146b) where F_{str} in lbf, t in in, P in in
The formula used to compute the force (or pressure) in swaging operation	$F_{swg} = A\sigma_{sut}$ **SI** (25-147a) where A = area to be sized in m^2; σ_{sut} = ultimate compressive strength of metal, MPa, and F_{swg} in N
	$F_{swg} = \frac{A\sigma_{sut}}{2000}$ **USCS** (25-147b) where A in in^2, σ_{sut} in psi, F_{swg} in tonf

SHEET METAL WORK

Bending (Figs. 25-33 to 25-36):

Particular	Formula
The bend allowance as per ASTME die design standard (Fig. 25-33)	$$\delta_b = \frac{\theta}{360} 2\pi r_i + K_n t \quad (25\text{-}148)$$

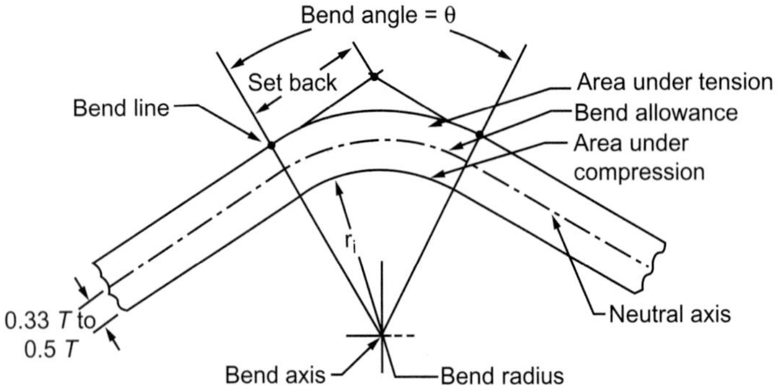

FIGURE 25-33 Bend terms for general angle.

Particular	Formula
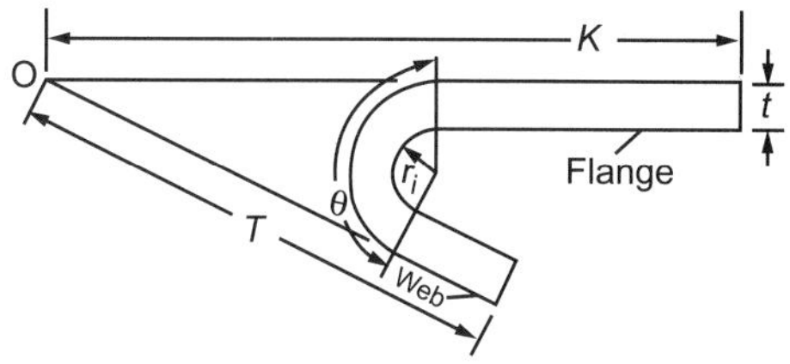 FIGURE 25-34	where δ_b = bend allowance (arc length of neutral axis), mm (in) θ = bend angle, deg r_i = inside radius of bend, mm (in) t = metal thickness, mm (in) K_n = constant for neutral axis location = 0.33 when r_i is less than $2t$ = 0.50 when r_i is more than $2t$
Another equation for bending allowance with outside bending angle θ (Fig 25-34).	$\delta_b = (r_i + t)\tan\frac{\theta}{2} - \frac{\pi\theta}{360}\left(r_i + \frac{t}{2}\right)$ (25-149) where θ in deg
	$\delta_b = \left(3\tan\frac{\theta}{2} - 0.0218\theta\right)t$ (25-150) when $r_i = 2t$
Initial length of strip of metal (Fig. 25-35)	$L_i = T - t - 2r_i + K + \frac{\pi}{2}\left(r_i + \frac{t}{2}\right)$ (25-151)
Bending allowance for right angle bend to take into account reduction of length K and T (Fig. 25-35)	$\delta_b = r_i + t - \frac{\pi}{4}\left(r_i + \frac{t}{2}\right)$ (25-152a)
	$\delta_b = 1.037t$ (25-152b) when $r_i = 2t$
The bending force	$F_b = Wt\tau_u$ (25-153)
Planishing force	$F_p = WK\sigma_{sy}$ (25-154) where K and W are dimensions as shown in Figs. 25-34 and 25-35 σ_{sy} = yield stress, MPa (psi)

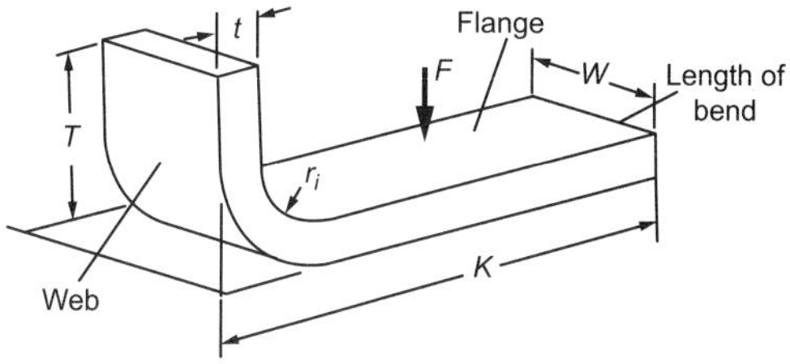

FIGURE 25-35

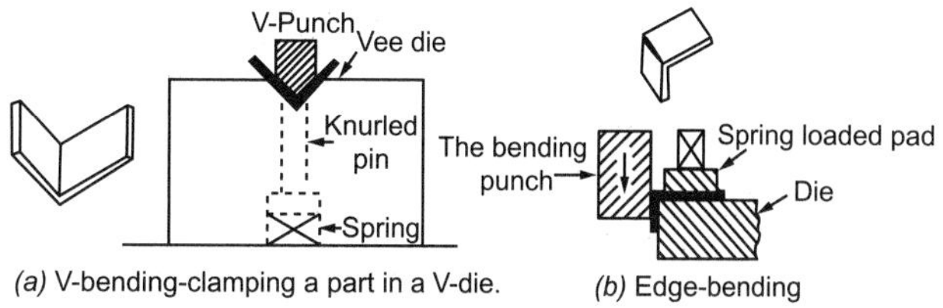

(a) V-bending-clamping a part in a V-die.
(b) Edge-bending

FIGURE 25-36 Bending methods.

Particular	Formula
The force/pressure required for V-bending (Fig. 25-36a)	$F_v = \dfrac{KLt^2\sigma_{sut}}{W}$ (25-155) where F = V-bending force, kN (tonforce) L = length of part, m (in) W = width of V- or U-die, m (in) σ_{sut} = ultimate tensile strength, MPa (tonf/in^2) K = die opening factor = 1.20 for a die opening of $16t$ = 1.33 for a die opening of $8t$
The force required for U-bending (channel bending)	$F_u = 2F_v$ (approx) (25-156a)
The force required for edge-bending (Fig.25-36b)	$F_{ed} = \frac{1}{2}F_v$ (25-156b)
Drawing (Fig. 25-37):	
Force required for drawing	$F = \pi dt\sigma_u$ (25-157) where σ_u = ultimate tensile stress, MPa (psi)
Empirical formula for pressure (or force) for a cylinder shell	$F = \pi dt\sigma_{sy}\left(\dfrac{D}{d} - c\right)$ (25-158) where D = diameter of blank d = diameter of shell h = height of shell r = corner radius T = bottom thickness of shell t = thickness of wall of shell C = constant which takes into account friction and bending = 0.6 to 0.7
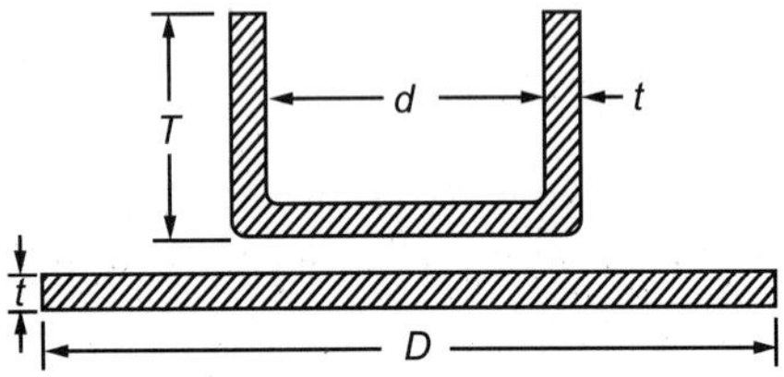 **FIGURE 25-37**	
A tentative blank size for an ironed shell can be obtained from equation	$D = \sqrt{d^2 + 4dh\dfrac{t}{T}}$ (25-159)
The blank size taking into consideration the ratio of the shell diameter to the corner (d/r) which affects the blank diameter.	$D = \sqrt{d^2 + 4dh}$ (25-160a) when d/r = 20 or more $D = \sqrt{d^2 + 4dh} - 0.5r$ (25-160b) when d/r lies between 15 and 20

Particular	Formula
	$D = \sqrt{d^2 + 4dh} - r$ (25-160c) when d/r lies between 10 and 15
	$D = \sqrt{(d - 2r)^2 + 4d(h - r) + 2\pi r(d - 0.7r)}$ (25-160d) when d/r is less than 10
For die clearance for different metals	Refer to Fig. 25-38
For nomograph for determining draw-die radius	Refer to Fig. 25-39
For chart for checking percentage reduction in drawing of cups.	Refer to Fig. 25-40
For clearance between punch and die	Refer to Fig. 25-29 and Table 25-52
For draw clearance	Refer to Table 25-53
For design of speed-change gear box for machine tools, kinematic schemes of machine tools, layout diagrams or structural diagram for gear drives, version of kinematic structures in machine tools, etc.	Refer to subsection "Designing spur and helical gears for machine tools" from pp. 23-109 to 23-138 of *Machine Design Data Handbook*, McGraw-Hill Publishing Company, New York, 1994.
For fits and tolerances	Refer to Chapter 11 on "Metal fits, tolerances and surface textures", pp. 11.1 to 11.32.
For surface roughness and surface texture	Refer to Chapter 11 on "Metal fits, tolerances and surface textures", pp. 11.26 to 11.32.
For tool steels and die steels	Refer to Chapter 1 on "Properties of engineering materials", Tables 1-31 to 1-36 for tool steels and Tables 1-49 and 1-51 for die steels.

TABLE 25-49
Metal removal rate in milling operation, Q

Material	Metal removal rate, Q, $mm^3/kW\,min$
Cast iron, gray	12600
Cast steel	12600
Mild steel	18900
Alloy steel	10500
Stainless steel	8400
Aluminum	42000
Copper	18900
Titanium	10500

TABLE 25-50
Average unit power P_u, for grinding

Work material	Unit power P_u, $kW/cm^3/min$ — Depth of grinding, mm per pass; Infeed, mm per revolution of work							
	0.0125	**0.025**	**0.05**	**0.075**	**0.1**	**0.25**	**0.5**	**0.75**
Free-machining steels Mild steels Medium carbon steels	1.4	0.88	0.7	0.6	0.51	0.35	0.23	0.18
Alloy steels	1.3	0.85	0.68	0.58	0.49	0.34	0.25	0.19
Tool steels	1.15	0.82	0.65	0.56	0.46	0.32	0.26	0.21
Stainless steels	1.4	0.84	0.65	0.58	0.51	0.37	0.29	0.26
Cast iron: gray, ductile, malleable	1.15	0.79	0.65	0.51	0.44	0.3	0.23	0.19
Aluminum alloys	0.58	0.45	0.35	0.33	0.29	0.21	0.17	0.15
Titanium alloys	0.93	0.79	0.6	0.56	0.51	0.37	0.3	0.25

TABLE 25-51
Values of K_m

	st, mm^2	0.3	0.4	0.6	0.8	1.0	1.2	1.4	1.6	1.8	2.0	2.4	2.6	2.8	3.0
K_m	Steel	32360	25500	21570	18140	15690	13720	12750	11750	10790	10300	9810	8830	8830	8330
N/m^2	Cast iron	29910	22550	17650	13730	11770	10780	9810	8825	8430	7845	7350	7355	7355	7110

TABLE 25-52
Clearance between punch and die (Fig. 25-29)
Location of the proper clearance determines, either hole or blank size, punch size controls hole size, die size controls blank size. $2C$ = clearance = $d_p - d_{di}$

Sheet thickness, mm	Clearance between punch and die, mm					
	Mild steel	Moderately hard steel	Hard steel	Soft brass	Hard brass	Aluminum
0.25	0.01	0.015	0.02	0.01	0.025	0.02
0.50	0.025	0.03	0.035	0,025	0.03	0.05
0.75	0.04	0.045	0.05	0.03	0.04	0.07
1.0	0.05	0.06	0.07	0.04	0.06	0.10
1.25	0.06	0.075	0.09	0.05	0.07	0.12
1.5	0.075	0.09	0.10	0.06	0.08	0.15
1.75	0.09	0.10	0.12	0,075	0.09	0.17
2.0	0.10	0.12	0.14	0.08	0.10	0.20
2.25	0.11	0.14	0.16	0.09	0.11	0.22
2.5	0.13	0.15	0.18	0.10	0.13	0.25
2.75	0.14	0.17	0.20	0.12	0.14	0.29
3.0	0.15	0.18	0.21	0.13	0.16	0.30
3.3	0.17	0.20	0.23	0.15	0.18	0.33
3.5	0.18	0.21	0.25	0.16	0.19	0.35
3.8	0.19	0.23	0.27	0.19	0.22	0.38
4.0	0.20	0.24	0.28	0.21	0.24	0.40
4.3	0.22	0.26	0.30	0.23	0.27	0.43
4.5	0.23	0.27	0.32	0.26	0.30	0.45
4.8	0.24	0.29	0.34	0.29	0.33	0.48
5.0	0.25	0.30	0.36	0.33	0.36	0.50

TABLE 25-53
Draw clearance, t = thickness of the original blank

Blank thickness				
mm	in	First draws	Redraws	Sizing draw[a]
Up to 3.81	Up to 0.15	$1.07t$–$1.09t$	$1.08t$–$1.1t$	$1.04t$–$1.05t$
0.41–1.27	0.016–0.050	$1.08t$–$1.1t$	$1.09t$–$1.12t$	$1.05t$–$1.06t$
1.30–3.18	0.051–0.125	$1.1t$–$1.12t$	$1.12t$–$1.14t$	$1.07t$–$1.09t$
3.45 and up	0.136 and up	$1.12t$–$1.14t$	$1.15t$–$1.2t$	$1.08t$–$1.1t$

[a] Used for straight-sided shells where diameter or wall thickness is important or where it is necessary to improve the surface finish in order to reduce finishing costs.

TABLE 25-54
Shear strength of various materials

	Ultimate strength, σ_{sut}		Shear strength, τ_s	
Material	**MPa**	**psi**	**MPa**	**psi**
Ferrous alloys				
0.10 carbon steel annealed			240	35,000
0.20 carbon steel annealed			290	42,000
0.30 carbon steel annealed			358	52,000
0.50 carbon steel annealed			550	80,000
1.00 carbon steel annealed			768	110,000
Chromium-molybdenum steel:				
SAE 4130	620	90,000	380	55,000
	690	100,000	448	65,000
	862	125,000	515	75,000
	1035	150,000	620	90,000
	1240	180,000	725	105,000
Nickel steel (drawn to 426°C (800°F) and water-quenched):				
SAE 2320			675	98,000
SAF 2330			758	110,000
SAE 2340			862	125,000
Nickel-chromium steel (drawn) to 426°C (800°F):				
SAE 3120			655	95,000
SAE 3130			758	110,000
SAE 3140			896	130,000
SAE 3280			930	135,000
SAE 3240			1035	150,000
SAE 3250			1138	165,000
Nonferrous materials				
Aluminum and alloys			28–282	4,000–41,000
Copper and alloys			150	22,000
			330	48,000
Magnesium alloys			28–145	4,000–21,000
Monel metal	475	69,000	295–450	42,900
	745	108,000		65,200
K monel	672	97,500	450	65,300
	1072	155,600	680	98,700
Nickel	469	68,000	360	52,300
	831	120,500	520	75,300
Inconel (nickel chromium iron)	550	80,000	406	59,000
	620	90,000	434	63,000
	689	100,000	455	66,000
	792	115,000	490	71,000
	965	140,000	538	78,000
	1103	160,000	580	84,000
	1206	175,000	600	87,000

TABLE 25-54
Shear strength of various materials (*Cont.*)

	Ultimate strength, σ_{sut}		Shear strength, τ_s	
Material	**MPa**	**psi**	**MPa**	**psi**
Nonmetallic materials				
Asbestos board			34	5,000
Cellulose acetate			69	10,000
Cloth			55	8,000
Fiber, hard			124	18,000
Hard rubber			138	20,000
Leather, tanned			48	7,000
Leather, rawhide			90	13,000
Mica			69	10,000
Paper[a]			44	6,400
Bristol board			33	4,800
Pressboards			24	3,500
Phenol fiber[b]			180	26,000

[a] For hollow die used one-half value shown for shearing strength.
[b] Blank and perforate hot.

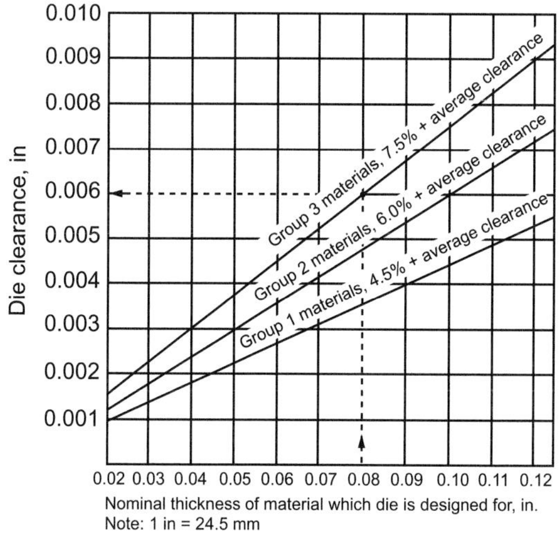

FIGURE 25-38 Die clearances for different groups of metals.

Group 1: 1100S and 5052S aluminum alloys, all tempers. An average clearance of $4\frac{1}{2}$ per cent of material thickness is recommended for normal piercing and blanking.

Group 2: 2024ST and 6061ST aluminum alloys; brass, all tempers; cold rolled steel, dead soft; stainless steel soft. An average clearance of 6 per cent of material thickness is recommended for normal piercing and blanking.

Group 3: Cold rolled steel; half hard; stainless steel, half hard and full hard. An average clearance of $7\frac{1}{2}$ per cent is recommended for normal piercing and blanking.

Courtesy: Frank W. Wilson, *Fundamentals of Tool Design*, ASTME, Prentice-Hall of India Private Limited, New Delhi, 1969.

TABLE 25-55
Drawing speeds

Material	Drawing speed, V_d (m/min) Single action	Double action
Aluminum provide	55	30
Strong aluminum alloys	—	10–15
Brass	65	30
Copper	45	25
Steel	18	10–16
Steel in carbide dies	—	20
Stainless steel	—	7–10
Zinc	45	13

TABLE 25-56
Drawing radii

Thickness of stock, mm	Drawing radius, mm
0.4	1.6
0.8	3.2
1.25	4.8
1.6	6.3
2	10
2.5	11.2
3.15	14

TABLE 25-57
Blank holder force in drawing

Thickness of stock, t, mm	Force, N per mm
0.25	314
0.4	304
0.5	294
0.63	280
0.80	270
0.9	260
1.00	250
1.12	235
1.25	225
1.4	220
1.6	210
1.8	196
2.0	181
2.24	167
2.5 and over	157

TABLE 25-58
Recommended die plate thickness

(a) For dies with cutting perimeter less than 50 mm										
Stock thickness 7 mm	0.25	0.5	0.75	1	1.25	1.5	1.75	2	2.25	2.5
Die plate thickness, mm/shear stress, kgf/mm^2	1	2	3	4	4.5	5.25	5.75	6.3	6.7	7
(b) For dies with cutting perimeter greater than 50 mm										
Cutting perimeter, mm over	50	75	150	300						
up to	75	150	300	500						
Factor by which the above tabulated values under (b) should be multiplied	1.25	1.5	1.75	2.0						

TABLE 25-59
Chart for bend allowance and correction factor

	Recommended minimum bend radius r_i for sheet metal							
	Aluminum alloys			Magnesium		Steel		
						Stainless		
Material thickness	24ST and Alclad	24SO and Alclad	2S, 3S and 52S ($\frac{1}{2}$ hard)	Cold bend	Hot formed	Annealed	[a] $\frac{1}{2}$ hard	Low carbon, X-4130 annealed [b]
Up to								
0.015	0.06	0.06	0.03	0.06	0 06	0 03	0 03	—
0.016	0 06	0.06	0 03	0 09	0.06	—	—	0.03
0.020	0.09	0.09	0.03	0.12	0.06	0.03	0.06	0.03
0.025	0.12	0.09	0.03	0.19	0.06	0.03	0.06	0.03
0.032	0.12	0.09	0.06	0.25	0.09	0.03	0.06	0.03
0.040	0.12	0.09	0.09	0.31	0.99	0.06	0.09	0.06
0.051	0.19	0.09	0.09	0.38	0.12	0.06	0.09	0.06
0.064	0.19	0.09	0.12	0.50	0.19	0.06	0.12	0.06
0.072	0.25	0.12	0.16	0.56	0.19	0.09	0.12	0.06
0.081	0.31	0.12	0.19	0.62	0.19	0.09	—	0.09
0.091	0.38	0.16	0.19	0.69	0.25	0.12	—	0.09
0.102	0.44	0.19	0.19	0.75	0.25	0.16	—	0.12
0.125	0.50	0.19	0.19	1.00	0.31	0.19	—	0.12
0.156	0.69	0.28	0.25	1.35	0.44	0.19	—	0.19
0.187	0.81	0.38	0.38	1.50	0.50	0.25	—	0.19
0.250	1.00	0.50	0.50	2.00	0.62	—	—	0.25
0.375	—	—	—	3.00	1.00	—	—	—

Note: [a] For bends up to 90 deg. [b] This applies to 8630 and similar steels.

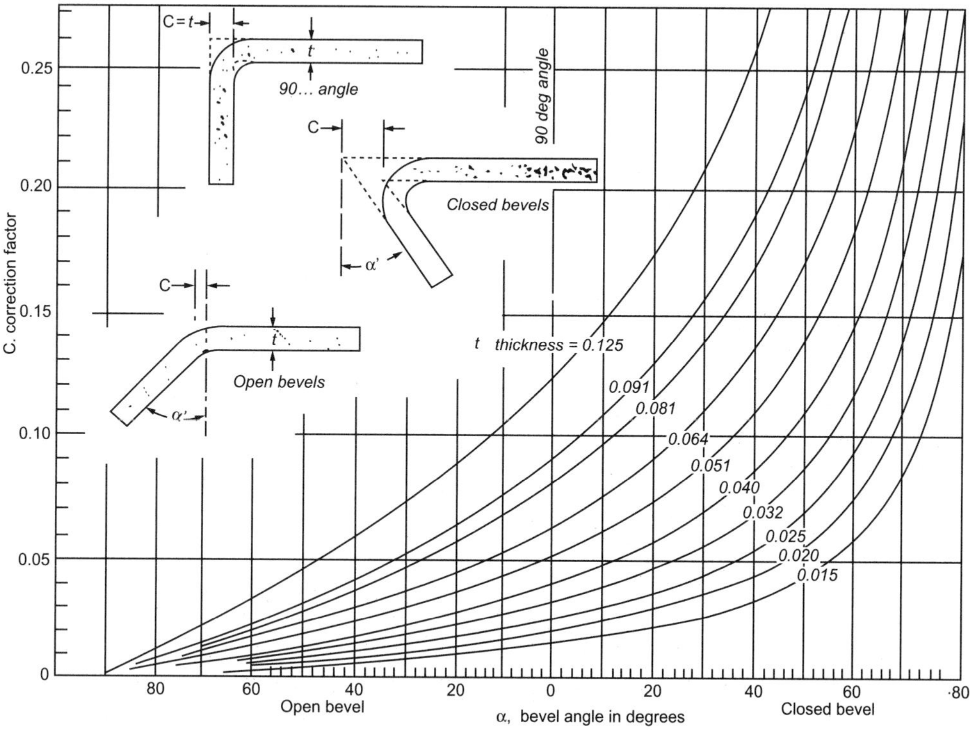

Courtesy: D. C. Greenwood (ed.), *Engineering Data for Product Design*, McGraw-Hill Publishing Company, New York, 1961.

Particular	Formula
FORMING PROCESS:	
Note: The *Symbols, Equations and Examples* given in the book entitled "*Mechanical Presses**" by Professor Dr. Ing. Heinrich Mäkelt and translated by R. Hardbottle, are followed and used in Symbols, Equations and Examples with reference to Figs. 25-41 to 25-49 in this *Machine Design Data Handbook*.	
The minimum ram force in mechanical presses	$P_{min} = 0.5 P_{rat}$ (25-161) where P_{rat} = tonnage rating, tonneforce (tf)
The maximum ram force	$P_{max} = Q k_{max} = F \sigma_{max}$ tf (25-162) where Q = cross-section k_{max} = maximum specific loading, MPa (psi) σ_{max} = maximum stress, MPa (psi) F = workpiece surface, in^2
The press work	$A = m P_{max} h = P_{mi} h$ (25-163) where m = correction factor taken from Table 25-62 h = work path = 0.5 H H = total maximum stroke setting
The volume of the workpiece before and after forming	$Q_1 h_1 = Q_2 h_2 = \text{constant}$ (25-164)
The force required to trim the forging	$F = \dfrac{P t \tau_s}{2000}$ tf (25-165) where F in tonneforce (tf) P = periphery of forging, in t = thickness, in τ_s = shear strength of material, psi
For chart for calculating ram path and velocity versus crank angle	Refer to Fig. 25-41
For calculation chart for blanking and piercing with full-edge cutting tool.	Refer to Fig. 25-42
For calculation chart for rectangular bending (a) V-bending on a fixed die, (b) U-bending with back-up	Refer to Fig. 25-43
For calculation chart for deep drawing and redrawing (a) deep drawing with blank holder (b) re-drawing of body	Refer to Fig. 25-44

* Heinrich Mäkelt, "Mechanical Presses", translated into English by R. Handbottle, Edward Arnold (Publishers) Ltd., 1968

Particular	Formula
For determination of blank-holder force for deep drawing	Refer to Fig. 25-45
For chart for extrusion molding and impact extrusion: *a*, extrusion molding of hollow bodies in direction of punch travel (forward extrusion); *b*, impact extrusion (tube extrusion) against direction of punch travel (backward extrusion)	Refer to Fig. 25-46
For determination of multiplication factor for impact extrusion and cold extrusion, and also for stamping and coining	Refer to Fig. 25-47
For chart for calculating stamping and coining	Refer to Fig. 25-48
For chart for calculating hot upsetting and drop forging	Refer to Fig. 25-49
For penetration of sheet thickness before fracture, suggested reductions in diameters for drawing, mean values for m and suggested trimming allowances	Refer to Tables 25-60, 25-61, 25-62 and 25-63

TABLE 25-60
Approximate penetration of sheet thickness before fracture in blanking

Work metal	Penetration %	Work metal	Penetration %
Carbon steels		**Non-ferrous metal**	
0.10% C annealed	50	Aluminum alloys	60
0.10% C cold rolled	38	Brass	50
0.20% C annealed	40	Bronze	25
0.20% C cold rolled	28	Copper	55
0.30% C annealed	33	Nickel alloys	55
0.30% C cold rolled	22	Zinc alloys	50

TABLE 25-61
Suggested reductions in diameters for drawing

Material	First draw %	Redraws %
Aluminum, soft	40	20–25
Aluminum, deep drawing quality	40–50	20–30
Brass	45	20–25
Copper	40	15–20
Steel	35–40	15–20
Steel, deep drawing quality	40–45	15–20
Steel, stainless	35–40	15–20
Zinc	50	15–20
Tin	35–45	10–15

TABLE 25-62
Mean values for m (standard coefficients)

Particulars	m
Blanking and piercing (full-edge)	
tough (soft) sheet	0.63
brittle (hard) sheet	0.32
Making V- and U-bends	
with die clash	0.32
without die clash	0.63
Deep-drawing and re-drawing	0.63
Impact extrusion and extrusion forming	1.00
Stamping	0.5
Hot, first upsetting operations	0.71
End drop-forging	0.36

Particular	Formula

TABLE 25-63
Suggested trimming allowances

	Blank thickness, mm	Allowance per side, mm, for steel with Rockwell B hardness of		
		50–66	75–90	90–106
	1.20	0.063	0.075	0.106
First trim	1.60	0.075	0.106	0.125
or	2.00	0.090	0.125	0.15–0.180
single trim	2.36	0.106	0.150	0.18–0.224
	2.80	0.125	0.180	0.224–0.230
	3.15	0.18	0.224	0.3–0.355
	1.20	0.03	0.035	0.05
	1.60	0.035	0.050	0.063
Second trim	2.00	0.045	0.063	0.075–0.09
or	2.36	0.050	0.075	0.09–0.100
add to first trim	2.80	0.063	0.090	0.1–0.14
	3.15	0.09	0.100	0.15-0.18

MACHINE TOOL STRUCTURES:

The optimum ratio l^2/h for every structure which depends on $[\delta]$, $[\sigma]$ and E

$$\frac{l^2}{h} = \frac{6E[\delta]}{[\sigma]} \tag{25-166}$$

where

l = length of structure/beam

h = distance of outermost fiber from the neutral axis in case of bending

The natural frequency of an elastic element such as a bar or beam subjected to tension or compression—a case of single degree freedom system

$$f = \sqrt{\frac{k}{m}} = \sqrt{\frac{EA}{l}\frac{1}{\rho Al}} = \sqrt{\frac{E}{\rho}\frac{1}{l^2}} \tag{25-167}$$

where

k = stiffness of the system $= F/4l$

ρ = mass density of member

The natural frequency of a simply supported beam subjected to a load at the center of beam – a case of single degree freedom system

$$f = \sqrt{\frac{k}{m}} = \sqrt{\frac{48E}{l^3}\frac{1}{\rho Al}} = \sqrt{\frac{48E}{l^3}\frac{bh^3}{12}\frac{1}{\rho Al}} = \sqrt{\frac{E}{\rho}\frac{4bh^3}{Al^4}} \tag{25-168}$$

where E/ρ is the unit or specific thickness. It is an important parameter in machine tool structural material and ρ = mass density of material of beam, γ = specific weight of material or beam. The natural frequency depends on E/γ.

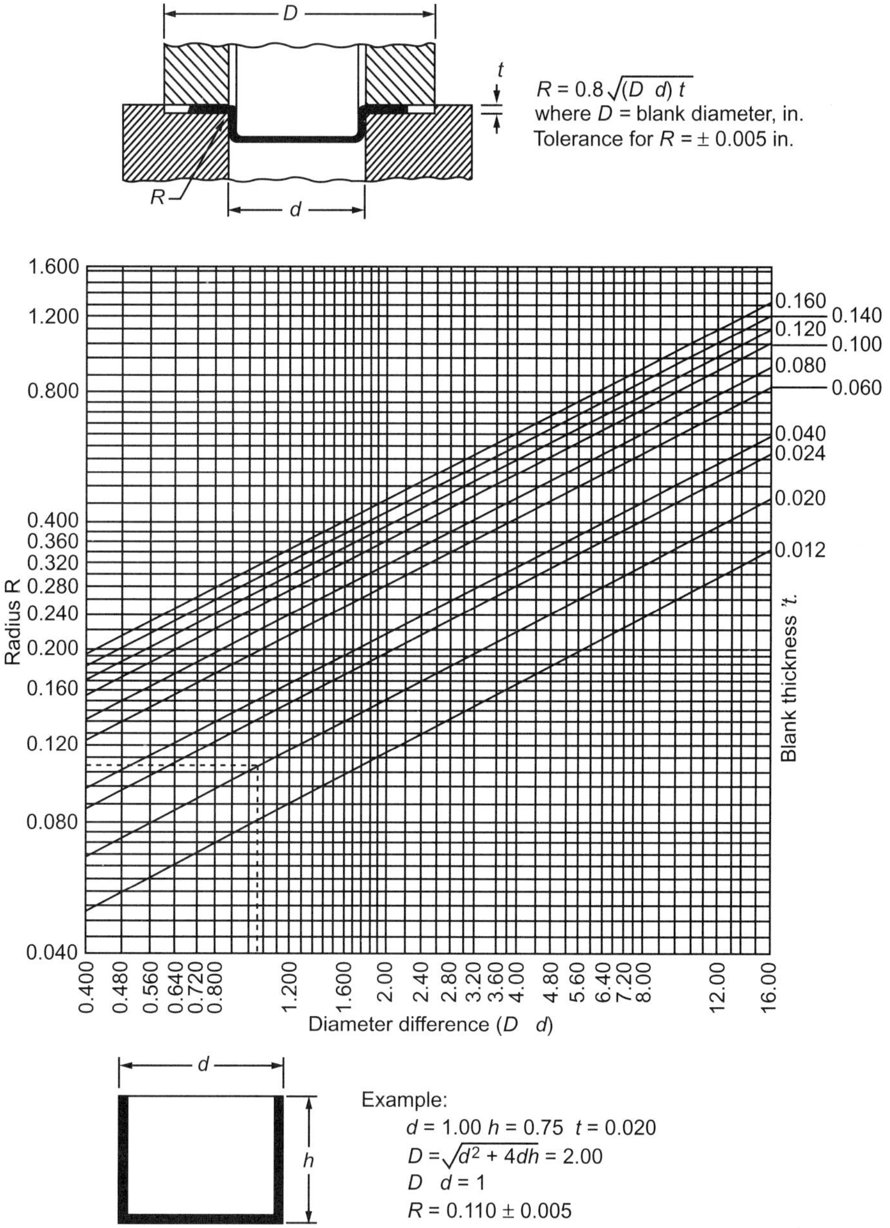

FIGURE 25-39 Nomograph for determining draw-die radius. *Courtesy*: American Machine/Metal working Manufacturing Magazine.

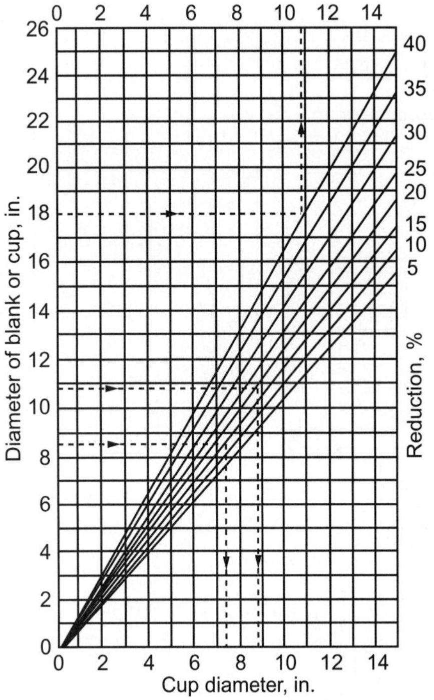

FIGURE 25-40 Chart for checking percentage reduction in drawing of cups. The inside diameter is ordinarily used for the cup diameter. *Courtesy*: From ASM, *Metals Handbook*, 8th ed., vol. 4, 1969.

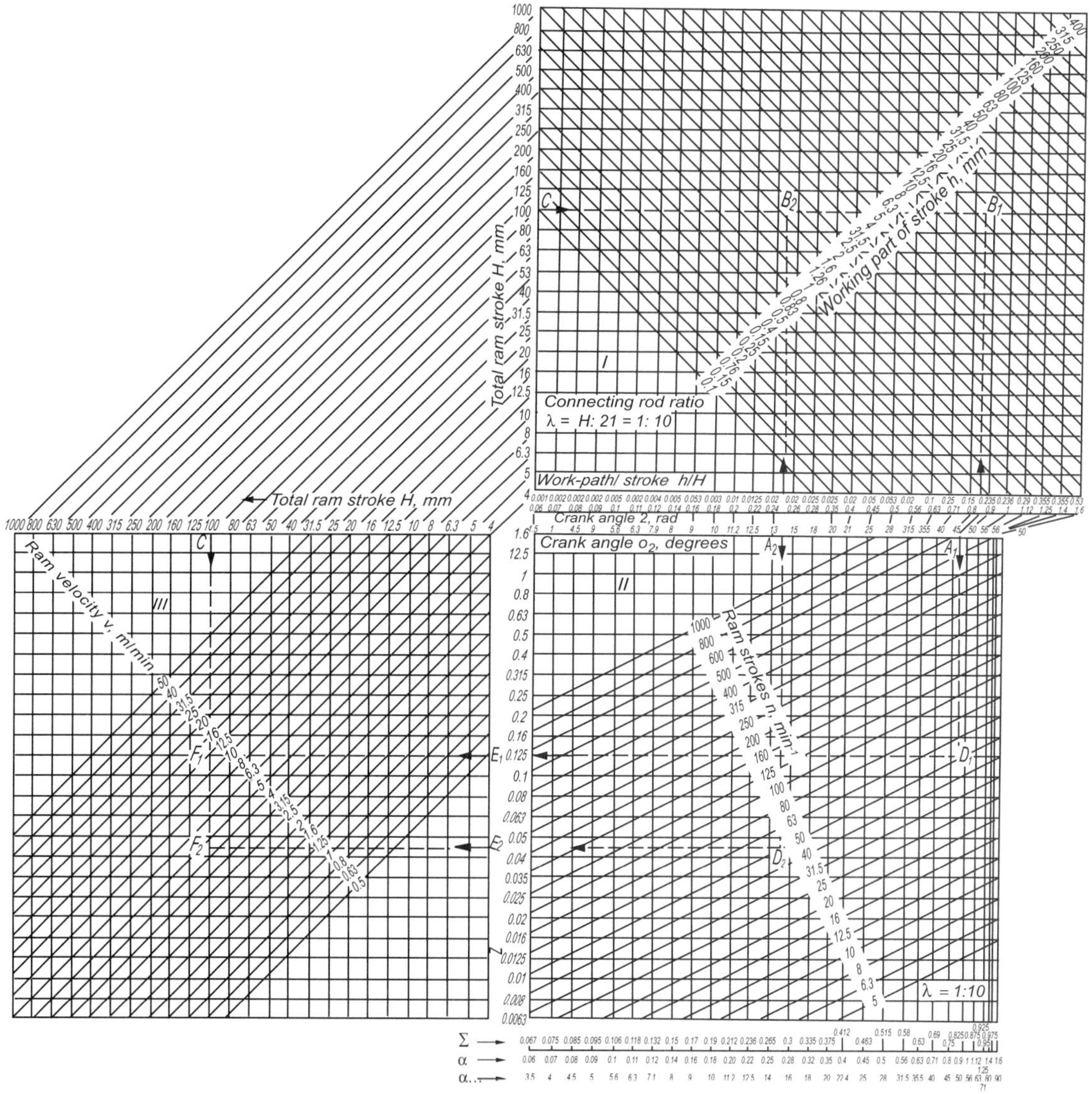

FIGURE 25-41 Chart for calculating ram path and velocity versus crank angle.

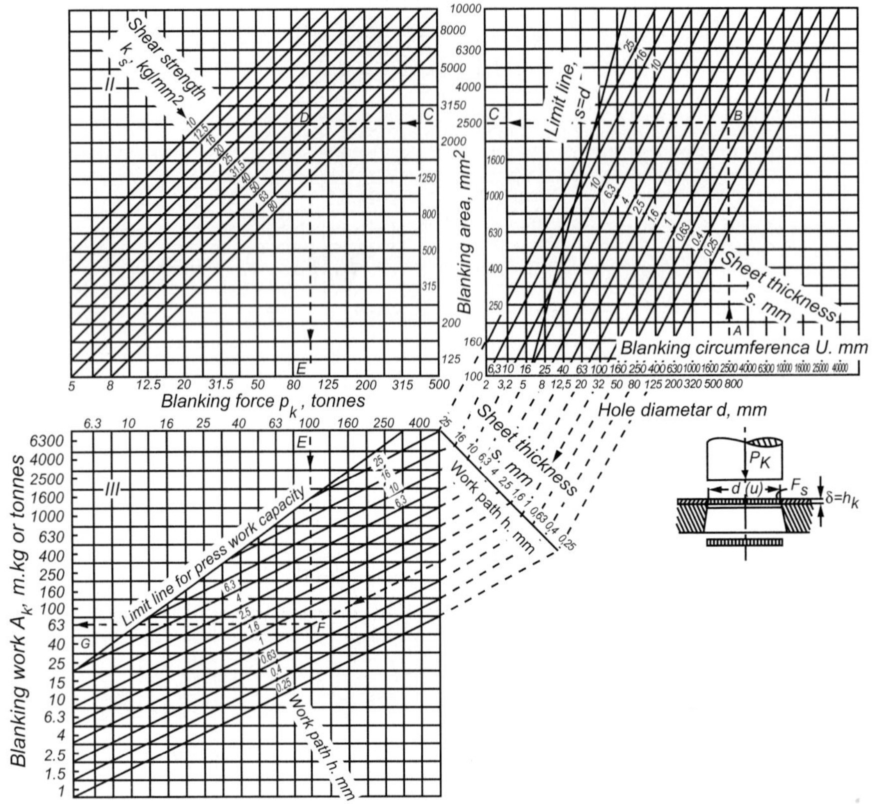

FIGURE 25-42 Calculation chart for blanking and piercing with full-edge cutting tool.

Equations and Examples

Equations:

Area $= F_s = \text{Us} = \pi ds$

Section II:

The tonnage rating of the press $= P_k = F_s k_s/1000$ tonneforce (tf).

The shear strength of metallic material, k_s, is

$k_s = 0.8\,\sigma_B$ N/mm² (kgf/mm²)

The work path is taken as $h_k = s$, where $s =$ thickness of material/sheet.

Section III:

The cutting work= $A_k = mP_k h_k$, mm-tonneforce (mm tf) or m kgf where $m =$ correction factor $= 0.63$ for soft sheet.

Example:

Blank diameter $d = 800$ mm (31.5 in) or $U = 2500$ mm (98 in).

Sheet thickness = 1 mm (0.039 in); The blanking area $= F_k = 2500\,\text{mm}^2$ ($3.85\,\text{in}^2$).

Blanking or cutting force $P_k = 980$ kN (100 tf).

Work= 61.78 N m (63 mm-tf or 456 ft-lbf).

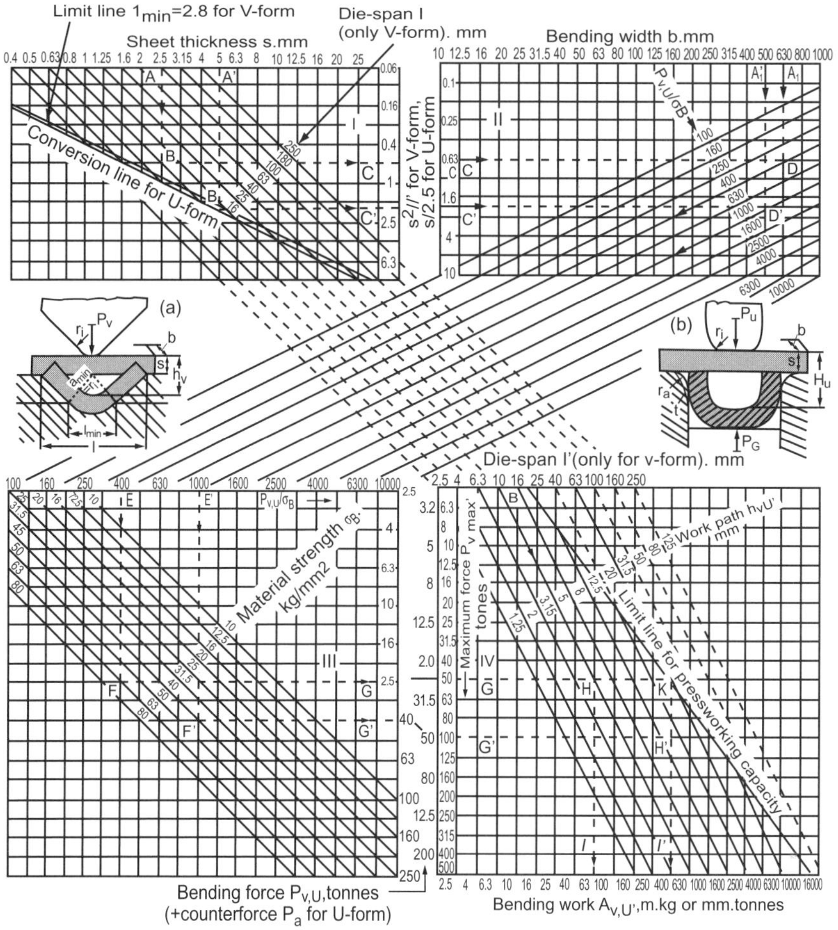

FIGURE 25.43 Calculation chart for rectangular bending (a) V-bending on a fixed die, (b) U-bending with back-up *Courtesy*: Heinrich Makelt, *Die Mechanischen Pressen*, Carl Hanser Verlag, Munich, German Edition, 1961 (Translated by R. Hardbottle, *Mechanical Presses*, Edward Arnold (Publishers) 1968)

Equations and Examples: (Fig. 25-43)
Equations:
Refer to Eqs. (25-155) and (25-156) for V-bending force and U-bending force.
The equation for V-bending force $P_v = [C\sigma_B bs^2/1000l']$ tonneforce (tf) (kN or lbf).
The bending work $A_v = mP_E h_v$ mm tonneforce (mm tf).
The limit of effective span or width of V-die $l_{min} = 2.8s$ in mm (in).
The work path $h_v = 0.5l' - 0.4(s + r_i)$ in mm (in).
(a) V-bending
Example:
$s = t = 2.5$ mm (0.1 in); $b = 630$ mm (25 in); $\sigma_B = 618$ N/mm^2 = 63 kgf/mm^2 (90000 lbf/in^2); $P_v = 25$ tf [*E-F-G*]; $P_E = 50$ tf; $h = 0.5l = 5$ mm (0.2 in) and $m = 0.32$.
The bending work= $A_s = 784$ kN m (80 mm tf or 579 ft lbf) [*A-B-C-C-A*], *D-D* [*B-H* and *G-H-I*].
(b) *U-bending*:
Equations:
The equation for U-bending force = $P_v = C(2/5)\sigma_B bs/1000$ tf (kN or lbf).
The backing force $P_G = 25\%$ of P_U.
The total force $P_U + P_G = 1.25P_U$.
The bending force = $A_U = m(P_U + P_G)h_U$ mm-tf (m N or ft lbf).
The work path = $h_U = 3s$ mm (in).
The correction factor = $m = 0.63$.
Example:
$s = 5$ mm (0.2 in); $s/2.5 = 2$ mm (0.08 in) [A'-B'-C'].
$b = 500$ mm (19.75 in), $P_U/\sigma_B = 1000$ [C'-D' and A'-D'].
$\sigma_B = 392$ N/mm^2 (40 kgf/mm^2 or 57000 lbf/in^2).
$P_U = 392$ N or 40 tf [D'-E'-F'-G'].
$P_U + P_G = 496$ N or 50 tf.
$A_U = 4900$ mm kN (500 mm tf or 500 m kgf or 3617 ft lbf) for $h_v = 16$ mm (0.70 in) and $m = 0.63$ [G'-H'-I'].
$P_{rut} = 490$ kN (50 tf) [I'-K-G]

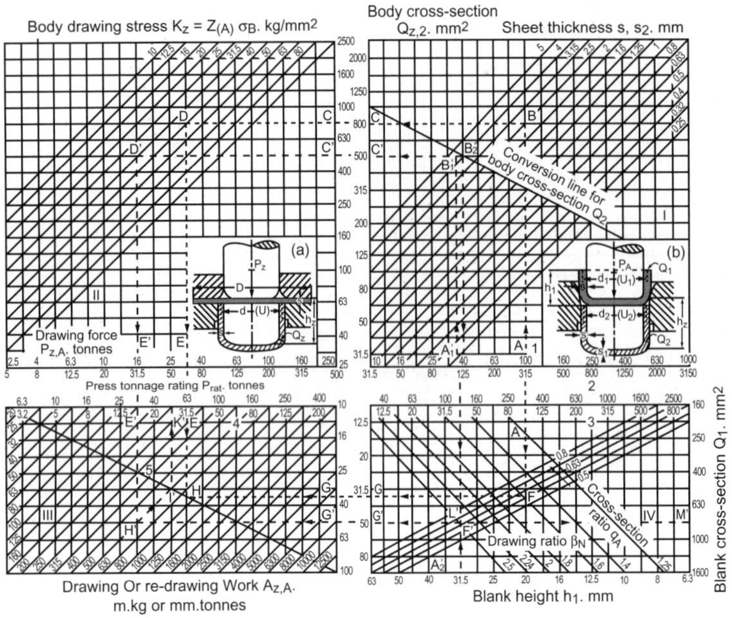

1 — Mean drawing diameter d, d_2, mm; 2 — Mean body circumference U, U_2, mm; 3 — Mean drawing diameter d, mm; 4 — Drawing force P_z, tonnes; 5 — Limit line for rational utilisation of press-working capacity; 6 — Work path $h_{z,2}$ (mm) of cylindrical bodies

FIGURE 25-44 Calculation chart for deep drawing and redrawing
(a) Deep drawing with blank holder (b) Re-drawing of body
Courtesy: Heinrich Makelt, *Die Mechanischen Pressen*, Carl Hanser Verlag, Munich, German Edition, 1961 (Translated by R. Hardbottle, *Mechanical Presses*, Edward Arnold (Publishers) 1968)

Equations and Examples: (Fig. 25-44)
Force and work requirements for deep drawing and re-drawing:
The body cross sectional area $Q_z = Us = \pi ds$ mm^2
where d = mean diameter of the drawn parts, mm
U = mean circumference of the drawn part, mm
s = sheet thickness, mm
The maximum draw force = $P_z = Q_z z\sigma_B/1000$ tf
where z = drawing factor = $y \ln \beta_N/\eta_F = \ln \beta_N/\ln \beta_{max}$
β_N = useful drawing ratio = (D/d)
β_{max} = limiting drawing ratio = D_{max}/d
D, D_{max} = diameters of draw of the blank sheet, mm
η_F = forming efficiency
For $D = 160$ mm (6.3 in), $d = 100$ mm (4 in), $\beta_N = 1.6$ and $\beta_{max} = 1.8$, the drawing factor is $z = 0.8$.
The strength ratio = $y = 1.2$ between the deformation strength k_{fm} and the tensile strength σ_B (lines *a-b-c* and *d-e-f*) i.e., $y = k_{fm}/\sigma_B$.
The work path for cylindrical drawn part = $h_z = d/4(\beta_N^2 - 1)$ mm.
The drawing work = $A_z = mP_z h_z$ N m or mm tf (m kgf)
where m = correction factor = 0.63
Mean body circumference = $U = 315$ mm (12.5 in) [Left side of Fig. 25-44].
Mean drawing diameter for the case of cylindrical hollow-ware = $d = 100$ mm (4 in).
The body cross section due to drawing stress k_z is $Q_z = 800$ mm^2 (1.23 in^2) [*Section I* top right].
The sheet thickness = $s = 2.5$ mm (0.1 in) [line *A-B-C*].
Section II (top left) gives $P_z = 309$ kN (31.5 tf) [*C-D-E*].
Drawing stress is $k_z = Z\sigma_B = 0.8 \times 50 = 392$ N/mm^2 (40 kgf/mm^2).
Section IV (bottom right) gives $h_z = 35.5$ mm (1.4 in) for $B_N = 1.6$ and $d = 100$ mm (4 in) [*A-F-G*].
The drawing work= $A_z = mP_z h_z = 0.63 \times 31.5 \times 35.5 = 6908$ mm N (705 mm tf or 5136 ft-lbf) [*G-H* and *E-B*].
The press tonnage rating $P_{rat} = 63$ tf.
Force and work requirements for re-drawing:
The reduced body cross-section during redrawing = $Q_z = U_z s_z = \pi d_z s_z$ mm^2
where U_z = mean circumference, mm
d_z = mean diameter of the finished product, mm
s_z = re-drawn wall thickness, mm
The reduced body cross-section = $Q_z = 500$ mm^2 (0.77 in^2) undergoing loading by the drawing stress k_z [line A_1'-B_1'-C_1'] (Fig. 25-44).
Section II $k_t = Z_A\sigma_B = 071 \times 45 \times 9.8066 = 312$ N/mm^2 (31.5 kgf/mm^2 or 45000 psi).
The re-drawn force = $P_A = 16$ tf [C'-D'-E'].
The blank cross section $Q_1 = Q_2/q_A = 800$ mm^2 (1.23 in^2) [C'-B_2'-L'-M'] (Fig. 25-44).
The blank height $h_1 = 31.5$ mm (1.25 in).
$h_2 = 50$ mm (2 in) [A_2'-F'-G', here F' happens to coincide with L']
$P_1 = 16$ tf, $h_2 = 50$ mm (2 in), $m = 0.63$ [E'-H' and G'-H'].
$P_{sat} = 50$ tf, [H'-I'-K'].

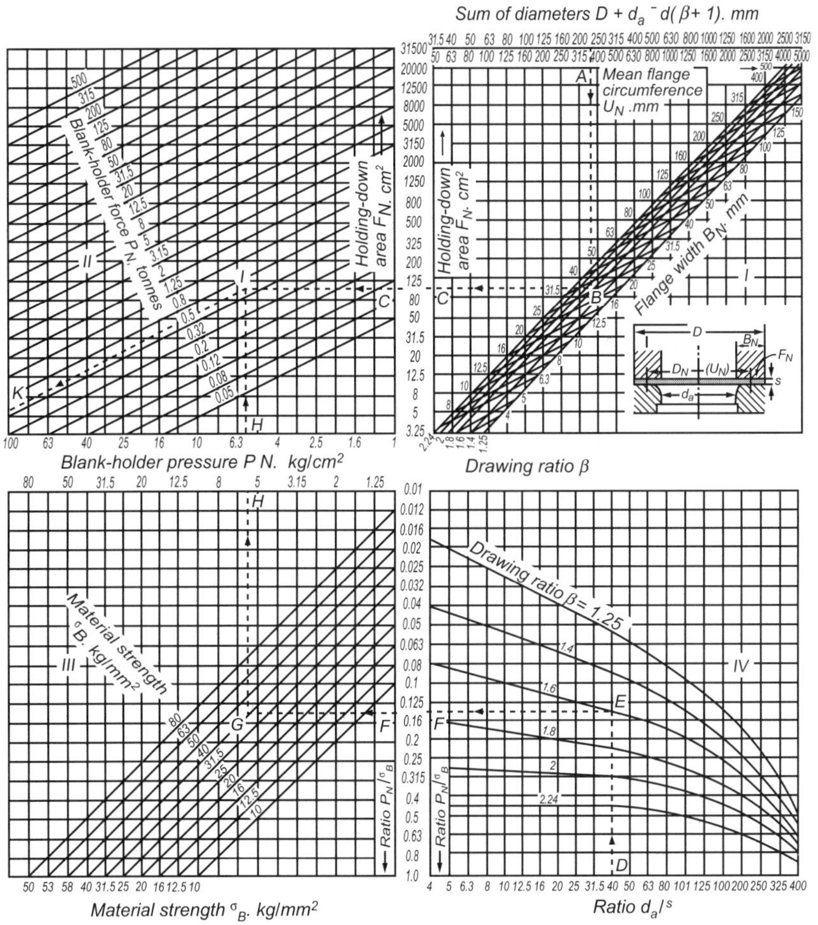

FIGURE 25-45 Determination of blank-holder force for deep drawing

Courtesy: Heinrich Makelt, *Die Mechanischen Pressen*, Carl Hanser Verlag, Munich, German Edition, 1961 (Translated by R. Hardbottle, *Mechanical Presses*, Edward Arnold (Publishers) 1968)

Equations and Examples:

The blank holder force $P_N = 10\%$ of the ram drawing force $P_z = 0.1P_z$.

The drawing work $= A_z = (mP_z + P_N)h_z$ m N (mm tf or m kgf).

Section I (Fig. 25-45):

First holding down area (flange area) under load $= F_N = U_N B_N = (\pi/4)(D + d_a)(D - d_a) = (\pi/4)d(\beta + 1)d(\beta - 1)$ cm^2

where U_N = mean flange width, cm; D =blank diameter, cm; d_a =outside diameter after drawing, cm

d = mean drawing diameter, cm; β = drawing ratio

Empirical equation for the ratio P_N/σ_B.

Section II: $P_N/\sigma_B = 0.25\ [(\beta - 1)^2 + 0.005\ (d_a/s)]$ where P_N in N/mm^2 (kgf/cm^2) and σ in N/mm^2 (kgf/mm^2).

Section III: The blank holder force $= P_N = F_N P_N/1000$ tf.

Example: *Section I*: For the case of cylindrical part from sum of the D and $d_a(D + d_a) = 235$ mm (9.3 in).

For $\beta = 1.6$ with 45° slope $F_N = 100\,\text{cm}^2$ (15.4 in^2) [line *A-B-C*] (Fig. 25-45).

For mean flange circumference $= U_N = 375$ mm (14.8 in).

$F_N = 100\,\text{cm}^2$ (15.4 in^2) [*A-B-C*] for $B_N = 26.5$ mm (10 5 in).

Section IV: For $(d_a/s) = 40$, $\beta = 1.6$, $(P_N/\sigma_B) = 0.14$ [*D-E-F*] (Fig. 25-45).

$\sigma_B = 392$ N/mm^2 (40 kgf/mm^2, or 57000 psi) in *Section III*.

From this the blank holder pressure $= P'_N = 55$ N/mm^2 (5.6 in kgf/cm^2 or 80 psi) [*F-G-H*] (Fig. 25-45).

Finally $P_N = 5490$ N (0.56 tf) [*H-I* and *J-K*] (Fig. 25-45).

Section II: $F_N = 100\,\text{cm}^2$ (15.4 in^2).

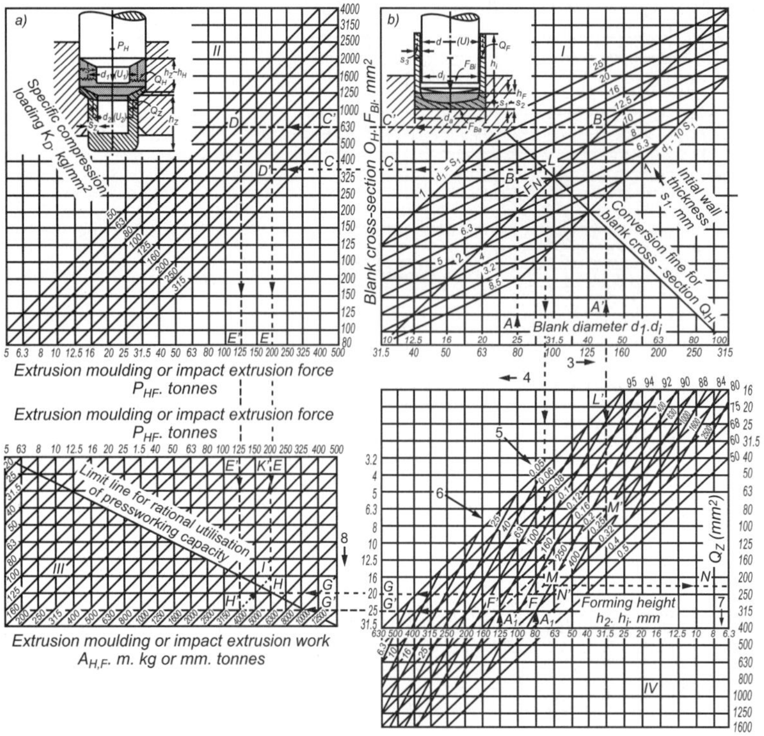

1 — Limit line; 2 — Conversion line for blank cross-section; 3 — Mean blank circumference U_1, mm;
4 — Degree of deformation ∈, %; 5 — Cross-section ratio $q_{H,F}$; 6 — Body cross-section Q_F, mm;
7 — Body cross-section; 8 — Extrusion moulding or impact extrusion stroke $h_{1,F}$, mm

FIGURE 25-46 Chart for extrusion molding and impact extrusion: *a*, extrusion molding of hollow bodies in direction of punch travel (forward extrusion); *b*, impact extrusion (tube extrusion) against direction of punch travel (backward extrusion). *Courtesy*: Heinrich Makelt, *Die Mechanischen Pressen*, Carl Hanser Verlag, Munich, German Edition, 1961 (Translated by R. Hardbottle, *Mechanical Presses*, Edward Arnold (Publishers) 1968)

Extrusion Forming (Fig. 25-46) :
(a) *Equations*: The extrusion molding force $= P_H = Q_H k_D/1000$ tonneforce (tf) (kN or lbf).
The body cross section $= Q_H = U_1 s_1 = \pi d_1 s_1$ mm^2 (in^2).
The reciprocal ratio of cross-section $q_H = Q_z/Q_H = U_2 s_2/U_1 s_1$, where Q_H = cross-section of the deformed blank before forming and Q_z = cross-section of the blank after forming, U_s = mean circumference and s_2 = wall thickness of finished product. The relation between the cross-sectional ratio q_H and the degree of deformation ε (%) is
$q_H = 1 - (\varepsilon/100)$.
Example: Blank diameter $d = 25$ mm (1 in), $s_1 = 11.2$ mm (0.45 in) $U_1 = 80$ mm (3.2 in)
Body cross sectional ratio= $q_H = 0.25$ corresponding to $\varepsilon = 75\%$.
Specific compression loading $k_D = 2196$ N/mm^2 (224 kgf/mm^2 or 319000 psi).
The extrusion force $= P_B = 1960$ kN (200 tf) (*C-D-E*); $Q_z = 224$ mm^2 (0.36 in^2) (*B-L-M-N*) (Fig. 25-46).
Work done due to extrusion $= A_H = 4000$ mm tf (4000 m kgf or 28933 ft/lbf) [*G-H* and *E-H*, *H* on the limit line].
(b) Equation: The punch force $= P_F = F_{B1} k_D/1000$ tf.
The body cross-section $= Q_F = F_{B1} q_F/(l - q_F)$ mm^2.
The cross-section ratio $= q_F = Q_F/F_{B\alpha}$.
The total initial cross-section of the blank disc $= F_{B\alpha} = Q_F/q_F = Q_F + F_{B\alpha}$ mm^2.
The wall thickness of the product $= s_B = [Q_F/\pi q_F]^{1/2} - d_1/2$ mm.
The work path $= h_F = s_1 - s_2 = q_F h_1$ mm.
The work for $m = 1$, $A_F = P_F h_F$ mm tf.

Example: $d = 45\,\text{mm}$ (1.8 in), $k_D = 785\,\text{N/mm}^2$ (80 kgf/mm^2 or 114000 psi)
$q = 0.2$, forming height $= h = 12.5mm$.
$\sigma_B = 196\ \text{N/mm}^2$ (20 kgf/mm^2 or 28500 psi).
Punch force $= P_F = 1225\,\text{kN}$ (125 tf) [C'-D'-E'] (Fig. 25-46).
Blank area $= F_{Bl} = 1600\,\text{mm}^2$ (2.46 in^2) [A'-B'-C'].
Body cross-section of product $= Q_F = 400\,\text{mm}^2$ (0.62 in^2) [A'-L'-M'-N'].
The inside body height $= h_1 = 125\,\text{mm}$ (5 in).
Work done: $A_F = 30896\,\text{m N}$ (3150 mm tf or 22785 ft-lbf) [E'-H' and G'-H'].
Press rating $= P_{sat} = 1960\,\text{kN}$ (200 tf) [H'-I'-K'].

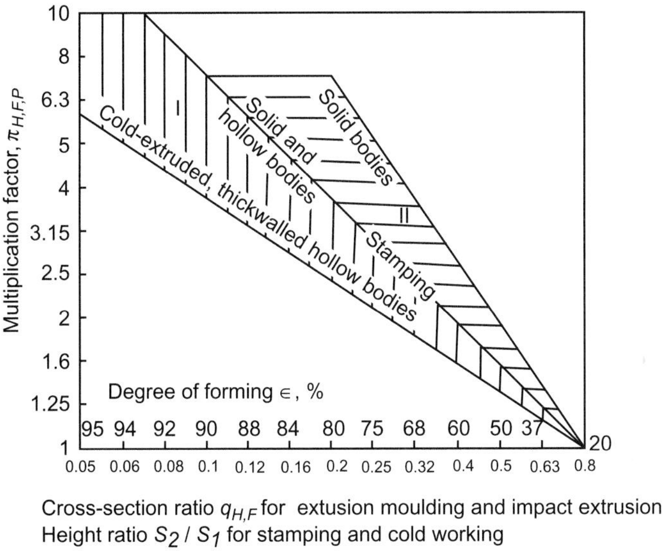

FIGURE 25-47 Determination of multiplication factor for impact extrusion and cold extrusion, and also for stamping and coining
Courtesy: Heinrich Makelt, *Die Mechanischen Pressen*, Carl Hanser Verlag, Munich, German Edition, 1961 (Translated by R. Hardbottle, *Mechanical Presses*, Edward Arnold (Publishers) 1968)

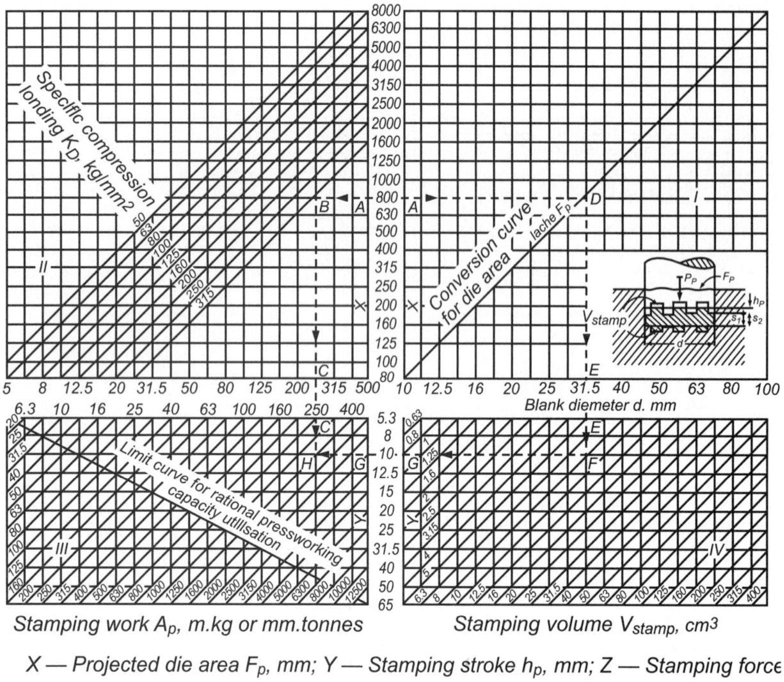

FIGURE 25-48 Chart for calculating stamping and coining

Courtesy: Heinrich Makelt, *Die Mechanischen Pressen*, Carl Hanser Verlag, Munich, German Edition, 1961 (Translated by R. Hardbottle, *Mechanical Presses*, Edward Arnold (Publishers) 1968)

X- projected die area F_p, mm; Y- stamping stroke h_p, mm; Z, stamping force P_p, tonnes.

Key to Fig. 25-49

Equations and Examples:

Forging temperature $= T = 1000°C$.

Tensile strength of plain carbon steel $= \sigma_B = 588\,N/mm^2$ (60 kgf/mm^2 or 86000 lbf/in^2 [point B] (Fig. 25-49).

Static deformation resistance $= k_{Fg} = 49\,N/mm^2$ (5 kgf/mm^2 or 7100 lbf/in^2) [point C of curve].

The deformation rate $= w = \varepsilon r/t$(% sec) = 500%/sec [point D].

The arithmetic proportions of upsetting $= \varepsilon_h = 4h/h_o = [1 - F_o/F_1]\ 100\%$.

The dynamic deformation resistance $= k_{Fd} = 98\,N/mm^2$ (10 kgf/mm^2 or 14200 psi) [point E of the curve] (Fig. 25-49).

$= 2k_{Fg}$ where k_{Fa} = static strength.

The diameter of non-circular upset or forged component is calculated from $d_{111} = \sqrt{(4/\pi)F_1} = 1.13\sqrt{F_1}$ mm where F_1 = cross-section after forming (upsetting surface).

The flash ratio $= b/s = 4.8$ (point F, scale 11).

The deformation resistance $= k_w = 392\,N/mm^2$ (40 kgf/mm^2 or 57000 psi) [point G of the curve].

The upsetting force $= P_s = 24516$ kN (2500 tf) [point I of the curve]

A prescribed or theoretical upsetting or die diameter d_1 [$D = 280$ mm (11 in)].

The corresponding upsetting or die area $F_1[F_{tot} = 63000\,mm^2$ (96 in^2) [point H].

The maximum diameter $D = d_1 + 2b$ of forged component

The crushed flash or the total cross-sectional area $= F_{tot} = F_1 + Ub$ where U = periphery of crushed area.

The mass ratio $= L_s/B_m = 6.3$ [point K].

The maximum upsetting force $= P_{max} = 30890$ kN (3150 tf) [point L of the curve].

The upset path $= h = 16$ mm (0.65 in) [point M].

The upsetting work $= A_s = 348134$ mm N (35500 mm tf or 256665 ft-lbf) [line N-O].

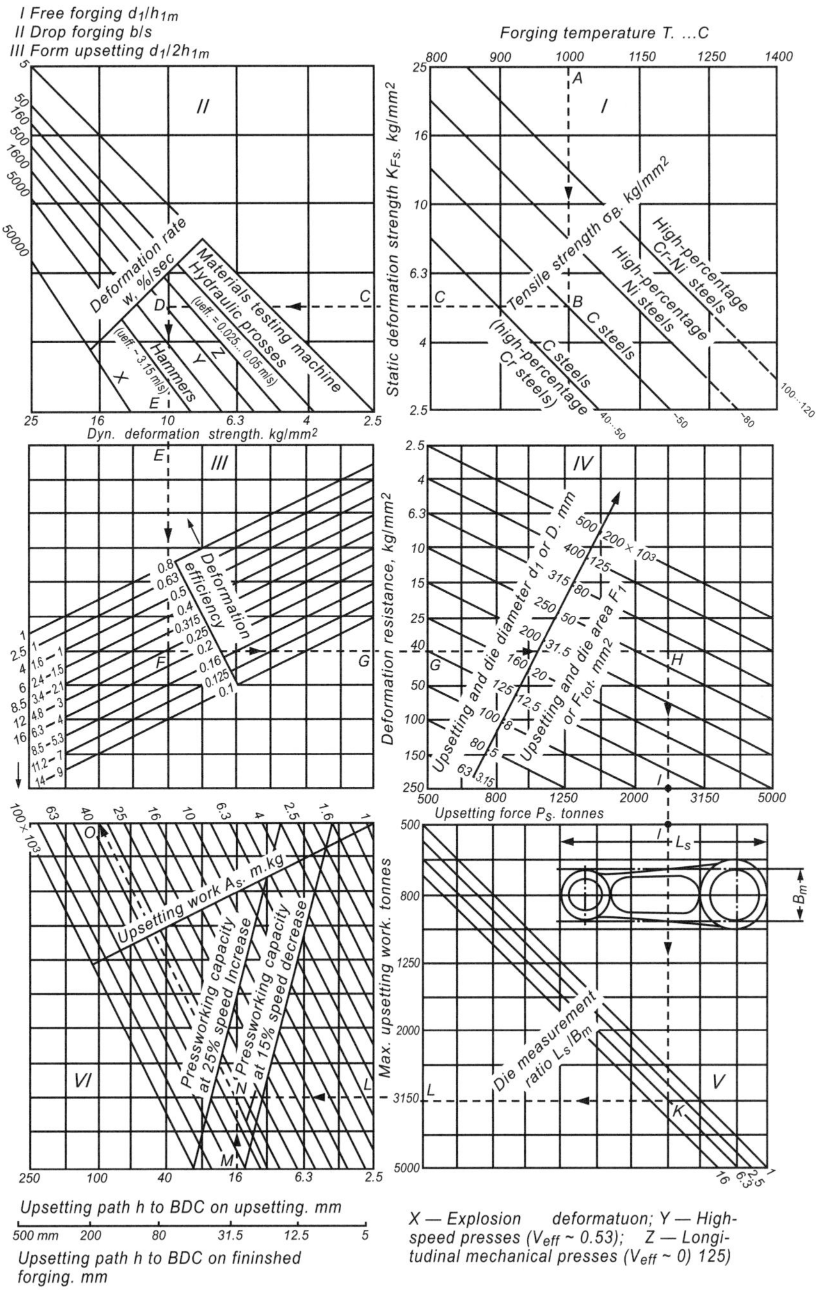

FIGURE 25-49 Chart for calculating hot upsetting and drop forging

Courtesy: Heinrich Makelt, *Die Mechanischen Pressen*, Carl Hanser Verlag, Munich, German Edition, 1961 (Translated by R. Hardbottle, *Mechanical Presses*, Edward Arnold (Publishers) 1968)

Particular	Formula
	Refer to Table 25-64 for unit stiffness or specific stiffness E/γ.
The ratio of weights of two bars of same length whose weights are $W_1 = \gamma_1 A_1 l$ and $W_2 = \gamma_2 A_2 l$	$\frac{W_1}{W_2} = \frac{\gamma_1 A_1 l}{\gamma_2 A_2 l} = \frac{E_2 \gamma_1}{E_1 \gamma_2} = \frac{E_2/\gamma_2}{E_1/\gamma_1}$ (25-169) where E/γ is the unit stiffness or specific stiffness Refer to Table 25-64, which gives E, γ and E/γ for some machine tool structural materials
The ratio of weights of two bars of same length subjected to tensile load F	$\frac{W_1}{W_2} = \frac{nPL(\gamma_1/\sigma_{ut1})}{nPL(\gamma_2/\sigma_{ut2})} = \frac{\sigma_{ut2}/\gamma_2}{\sigma_{ut1}/\gamma_1}$ (25-170) where σ_{ut}/γ is unit strength under tension
The ratio of weights of two bars of same length subjected to torque M_t	$\frac{W_1}{W_2} = \frac{\tau_{ut}^{2/3}/\gamma_2}{\tau_{ut}^{2/3}/\gamma_1}$ (25-171) where $\tau_{ut}^{2/3}/\gamma$ is an index of the ability of a material to resist torsion and is known as unit strength under torsion
The ratio of weights of two bars of same length subjected to bending M_b	$\frac{W_1}{W_2} = \frac{(1/\sigma_{b1})^{2/3}\gamma_1}{(1/\sigma_{b2})^{2/3}\gamma_2} = \frac{\sigma_{b2}^{2/3}/\gamma_2}{\sigma_{b1}^{2/3}/\gamma_1}$ (25-172) where $\sigma_b^{2/3}/\gamma$ is an index of the ability of a material to resist bending and is known as the unit strength under bending
For specific stiffness (in tension)	Refer to Table 25-64.
For comparison of specific strength and stiffness/rigidity of different section having equal cross sectional area	Refer to Table 25-65.
DESIGN OF FRAMES, BEDS, GUIDES AND COLUMNS:	
For machine frames	Refer to Table 25-66.
For stiffening effect of reinforcing ribs	Refer to Fig. 25-50.
For characteristics of bending and torsional rigidities of models of various forms	Refer to Table 25-67.
For variations in relative bending and torsional rigidity for models of various forms	Refer to Table 25-68.
For effect of stiffener arrangement on torsional stiffness of open structure	Refer to Table 25-69.

Particular	Formula
For effect of aperture and cover plate design in static and dynamic stiffness of box sections	Refer to Table 25-70.
For typical cross-sections of beds	Refer to Fig. 25-51A, B, C and D.
For classification and identification of machine tools	Refer to Table 25-72.
For machine tools sliding guides, ball and roller guides made of cast iron, steels and plastics	Refer to Tables and Figures from 25-66 to 25-71. In addition to these, readers are advised to refer to books and handbooks on machine tools. The design of machine tool slideways, guides, beds, frames and columns subjected to external forces are beyond the scope of this Handbook.
For design of spindle units in machine tools	Refer to Chapter 14 on "Design of shafts" in this Handbook.
For design of power screws and lead screws of machine tools	Refer to Chapter 18 on "Power screws and fasteners" in this handbook, and books on power screw design of machine tools.
For vibration and chattering in machine tools	Refer to Chapter 22 on "Mechanical vibrations" in this Handbook.
For variable speed drives and power transmission	Refer to Chapter 23 on "Gears" and Chapter 25 on "Miscellaneous machine elements" in this Handbook.
For lubrication of guides, spindles and other parts of machine tools	Refer to Chapter 24 on "Design and bearings and Tribology" in this Handbook and other books on lubrication.

TOOLING ECONOMICS (Adopted from *Tool Engineers Handbook*)

***Symbols*:**

- a saving in labor cost per unit
- C first cost of fixture
- D annual allowance for depreciation, per cent
- H number of years required for amortization of investment out of earnings
- I annual allowance for interest on investment, per cent
- M annual allowance for repairs, per cent
- N number of pieces manufactured per year
- S yearly cost of setup
- t percentage of overhead applied on labour saved
- T annual allowances for taxes, per cent
- V yearly operating profit over fixed charges

Number of pieces required to pay for fixture

$$N = \frac{C(I + T + D + M) + S}{a(1 + t)} \quad (25\text{-}173)$$

Economic investment in fixtures for given production

$$C = \frac{Na(1 + t) - S}{I + T + D + M} \quad (25\text{-}174)$$

Number of years required for a fixture to pay for itself

$$H = \frac{C}{Na(1 + t) - C(I + T + M) - S} \quad (25\text{-}175)$$

Profit from improved fixture designs

$$V = Na(1 + t) - C(I + T + D + M) - S \quad (25\text{-}176)$$

Particular	Formula

PROCESS—COST COMPARISONS:

Symbols:

- c value of each piece, dollars
- C_x, C_y total unit cost for methods Y and Z respectively
- d hourly depreciation rate for the first machine (based on machine hours for the base years period)
- D hourly depreciation rate for the second machine (based on machine hours for the base years period)
- k annual carrying charge per dollar of inventory, dollar
- l labor rate for the first machine, dollar
- L lot size, pieces
  labor rate for the second machine, dollar
- m monthly consumption, pieces
- N_t total number of parts to be produced in a single run
- N_b number of parts for which the unit costs will be equal for each of two compared methods Y and Z (break-even point)
- p number of pieces produced per hour by the first machine
- P number of pieces produced per hour by the second machine
- P_y unit tool process cost for method Y
- P_z unit tool process cost for method Z
- Q quantity of pieces at break-even point
- T_y total tool cost for method Y
- T_z total tool cost for method Z
- s setup hours required on the first machine
- S setup hours required on the second machine
- V ratio of machining time piece

Number of parts for which the unit costs will be equal for each of two compared methods Y and Z ("break-even point")

$$N_b = \frac{T_y - T_z}{P_z - P_y} \tag{25-177}$$

Total unit cost for methods Y

$$C_y = \frac{P_y N_t + T_y}{N_t} \tag{25-178}$$

Total unit cost for method Z

$$C_z = \frac{P_z N_t + T_z}{N_t} \tag{25-179}$$

Quantity of pieces at break-even point

$$Q = \frac{pP(SL + SD - sl - sd)}{P(l + d) - p(L + D)} \tag{25-180}$$

Relatively simple formula for calculation of economic lot size, pieces

$$L = \sqrt{\frac{24mS}{kc(1 + mv)}} \tag{25-181}$$

MACHINING COST:

Machining time cost per work piece

$$C_m = \frac{t_m R}{60} \tag{25-182}$$

Non-productive time cost per work piece

$$C_n = \left(t_L + \frac{t_s}{n_b}\right)\frac{R}{60} \tag{25-183}$$

Tool change time cost per work piece

$$C_c = \frac{t_m t_c R}{60 t_1} \tag{25-184}$$

Tool cost per work piece

$$C_t = \frac{C_{t1}}{1 + n_s} + \frac{t_{sh} t_m R}{60 t_1} \tag{25-185}$$

Particular	Formula	
Total cost of machining	$C_{tot} = C_m + C_n + C_c + C_t$	(25-186)
Total tool cost per workpiece	$C_n = C_c + C_t$	(25-187)
	where	
	t_m = machining time per workpiece, min	
	t_L = loading and unloading time per workpiece, min	
	t_s = setting time per batch, min	
	t_t = tool life, min	
	t_c = tool charge time, min	
	t_{sh} = tool sharpening time, min	
	R = cost rate per hour	
	n_b = number of batch	
	n_s = number of resharpening	

TABLE 25–64
Unit stiffness/rigidity of some materials

Material	Modulus of elasticity, E		Modulus of rigidity, G		Poisson's ratio,	Density, ρ^a		Unit weight, γ^b			Unit stiffness
	GPa	Mpsi	GPa	Mpsi	ν	Mg/m^3	kg/m^3	kN/m^3	lbf/in^3	lbf/ft^3	E/γ
Aluminum	69	10.0	26	3.8	0.334	2.69	2,685	26.3	0.097	167	2.62×10^6
Aluminum cast	70	10.15	30	4.35			2,650	26.0	0.096	166	2.66×10^6
Aluminum (all alloys)	72	10.4	27	3.9	0.320	2.80	2,713	27.0	0.10	173	2.68×10^6
Beryllium copper	124	18.0	48	7.0	0.285	8.22	8,221	80.6	0.297	513	1.54×10^6
Carbon steel	206	30.0	79	11.5	0.292	7.81	7,806	76.6	0.282	487	2.69×10^6
Cast iron, gray	100	14.5	41	6.0	0.211	7.20	7,197	70.6	0.260	450	1.42×10^6
Malleable cast iron	170	24.6	90	13.0			7,200	70.61			2.41×10^6
Inconel	214	31.0	76	11.0	0.290	8.42	8,418	83.3	0.307	530	2.57×10^6
Magnesium alloy	45	6.5	16	2.4	0.350	1.80	1,799	17.6	0.065	117	2.56×10^6
Molybdenum	331	48.0	117	17.0	0.307	10.19	10,186	100.0	0.368	636	3.31×10^6
Monel metal	179	26.0	65	9.5	0.320	8.83	8,830	86.6	0.319	551	2.06×10^6
Nickel-silver	127	18.5	48	7.0	0.332	8.75	8,747	85.80	0.316	546	1.48×10^6
Nickel alloy	207	30.0	79	11.5	0.30	8.3	8,304	81.4	0.300	518	2.54×10^6
Nickel steel	207	30.0	79	11.5	0.291	7.75	7,751	76.0	0.280	484	2.72×10^6
Phosphor bronze	111	16.0	41	6.0	0.349	8.17	8,166	80.1	0.295	510	1.38×10^6
Steel (18-8), stainless	190	27.5	73	10.6	0.305	7.75	7,750	76.0	0.280	484	2.50×10^6
Titanium (pure)	130	15.0				4.47	4,470	43.8	0.16	279	2.37×10^6
Titanium alloy	114	16.5	43	6.2	0.33	6.6	6,600				2.60×10^6
Brass	106	15.5	40	5.8	0.324	8.55	8,553	83.9	0.309	534	1.26×10^6
Bronze	96	14.0	38	5.5	0.349	8.30	8,304	81.4			1.18×10^6
Bronze cast	80	11.6	35	5.0			8,200	80.0			1.00×10^6
Copper	121	17.5	46	6.6	0.326	8.90	8,913	87.4	0.322	556	1.38×10^6
Tungsten	345	50.0	138	20.0		18.82	18,822	184.6	1.89		
Douglas fir	11	1.6	4	0.6	0.330	4.43	443	4.3	0.016	28	2.56×10^6
Glass	46	6.7	19	2.7	0.245	2.60	2,602	25.5	0.094	162	1.80×10^6
Lead	36	5.3	13	1.9	0.431	11.38	11,377	111.6	0.411	710	3.10×10^6
Concrete (compression)	14–28	2.0–4.0				2.35	2,353	23.1		147	0.60×10^6
Wrought iron	190	27.5	70	10.2			7,700	76.0			2.50×10^6
Zinc alloy	83	12	31	4.5	0.33	6.6			0.24	415	1.18×10^6
Graphite	750	108.80				2.25		22.1			34.00×10^6
HTS Graphite/5208 epoxy	172	24.95				1.55		15.2			11.30×10^6
T50 Graphite 2011 Al	160	23.20				2.58		25.3			6.32×10^6
Boron	380	55.11				2.5		44.1			11.00×10^6
Boron carbide, BC	450	65.28				2.4		22.5			19.20×10^6
Silicon carbide, SiC	560	81.22				3.2		31.4			17.80×10^6
Boron/5505 epoxy	207	30.07				1.99		19.5			8.40×10^6
Boron/6601 Al	214	31.03				2.60		25.5			8.20×10^6
Kelvar 49	130	18.85				1.44		14.1			9.20×10^6
Kelvar 49/resin	76	11.02				1.38		13.5			5.60×10^6
Silicon, Si	110	15.95				2.30		22.5			4.86×10^6
Wood (along fiber)	11–15.1	1.59–2.19				0.41–0.82		4.0–8.0			$2.75\text{–}1.86 \times 10^6$
Nylon	4	0.58				1.1		10.8			0.37×10^6
Paper	1–2	0.15–0.29				0.50		4.9			$0.20\text{–}0.41 \times 10^6$
E Glass/1002 epoxy	39	5.65				1.80		17.6			2.22×10^6

[a] ρ, mass density.
[b] γ, weight density; w is also the symbol used for unit weight of materials.
Source: K. Lingaiah and B. R. Narayana Iyengar, *Machine Design Data Handbook*, Volume I (SI and Customary Metric Units), Suma Publishers, Bangalore, India and K. Lingaiah, *Machine Design Data Handbook*, Volume II, (SI and Customary Metric Units), Suma Publishers, Bangalore, India, 1986.

TABLE 25-65
Comparison of specific strength and Rigidity/Stiffness of different sections having equal cross sectional areas (in Flexure)

Cross-section	Area A	Distance to farthest point, c	Moment of inertia I	Section modulus $Z = I/c$	$i = \frac{I}{A^2}$	$w = \frac{Z}{A^{3/2}}$	$\frac{I^*}{I_a}$	$\frac{Z^*}{Z_a}$
	$0.785D^2$	$\frac{D}{2}$	$0.05D^4$	$0.1D^3$	0.08	0.14	1	1
	B^2	$\frac{B}{2}$	$B^4/12$	$B^3/6$	0.083	0.166	1.06	1.16
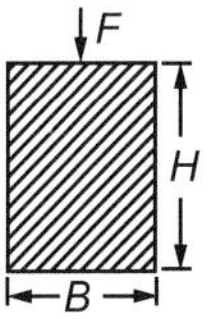	$B^2 r$ ($r = H/B$)	$\frac{H}{2}$	$B^4 r^3/12$	$B^3 r^2/6$	$0.083r$	$0.166\sqrt{r}$	1.9	1.6
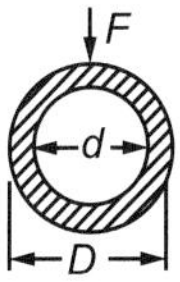	$0.785D^2$ $(1-\beta^2)$ ($\beta = d/D$)	$\frac{D}{2}$	$0.05D^4$ $(1-\beta^4)$	$0.1D^3$ $(1-\beta^4)$	$0.08\frac{1-\beta^4}{(1-\beta^2)^2}$	$0.14\frac{1-\beta^4}{(1-\beta^2)^{3/2}}$	2.1	1.73
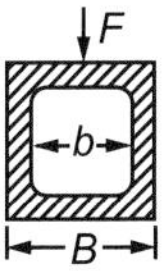	$B^2(1-\alpha)$ ($\alpha = b/B$)	$\frac{B}{2}$	$\frac{B^4}{12}(1-\alpha^4)$	$\frac{B^3}{6}(1-\alpha^4)$	$\frac{1-\alpha^4}{12(1-\alpha^2)^2}$	$\frac{1-\alpha^4}{6(1-\alpha^2)^{3/2}}$	4.6	3.2
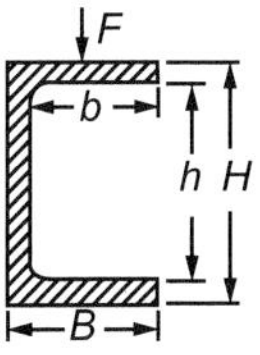								
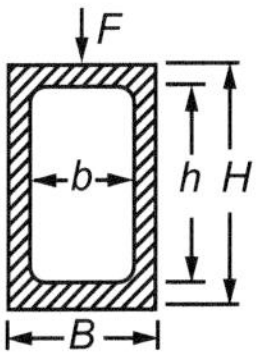							9.5	4.6

TABLE 25-65
Comparison of specific strength and Rigidity/Stiffness of different sections having equal cross sectional areas (in Flexure) (*Cont.*)

Cross-section	Area A	Distance to farthest point, c	Moment of inertia I	Section modulus $Z = I/c$	$i = \dfrac{I}{A^2}$	$w = \dfrac{Z}{A^{3/2}}$	$\dfrac{I^*}{I_a}$	$\dfrac{Z^*}{Z_a}$
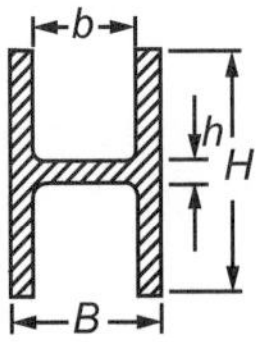	$BH(1-\kappa\zeta)$ $(\kappa = b/B;$ $\zeta = h/H)$	$\dfrac{H}{2}$	$\dfrac{BH^3}{12}(1-\kappa\zeta^3)$	$\dfrac{BH^2}{6}(1-\kappa\zeta^3)$	$0.083\dfrac{1-\kappa\zeta^3}{(1-\kappa\zeta)^2}$	$0.166\dfrac{1-\kappa\zeta^3}{(1-\kappa\zeta)^{3/2}}$		
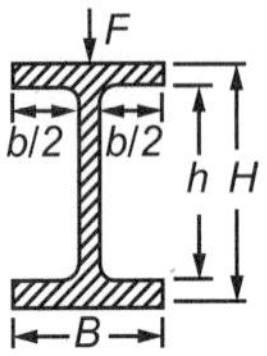							11	52

* Z_a = section modulus of round solid section $= \dfrac{\pi D^3}{32}$; I_a = Moment of Inertia of round solid section $= \dfrac{\pi D^4}{64}$.
Z/Z_a and I/I_a for solid and hollow stock having identical cross sectional area in flexure.

TABLE 25-66
Machine Frames

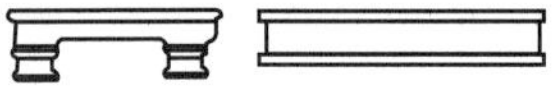	Simple frames and beds of horizontal machines
	Simple frames and beds of vertical machines
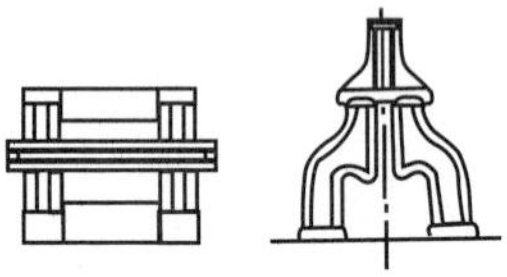	Portal frames
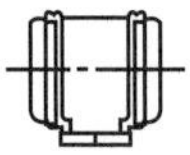	Circular frames, housings
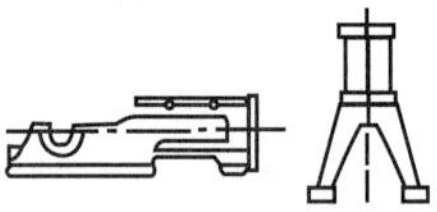	Frames of piston machines, banks of cylinders
	Frames of conveying machines
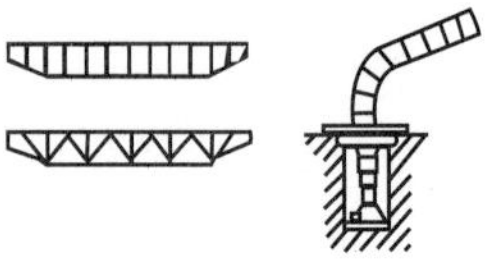	Crane structures

TABLE 25-66
Machine Frames (*Cont.*)

	Baseplates
	Boxes
	Pillars, brackets, pedestals, hangers, etc.
	Tables, slide blocks, carriages
	Crossheads, slides, jibs
	Lids and casings

Source: Courtesy: Dobrovolsky, V., etl., "Machine Elements", Mir Publishers, Moscow, 1974.

TABLE 25-67
Characteristics of Bending and Torsional Rigidities for Models of Various Forms

Model No.	Model form	Relative rigidity in bending S_b	Relative rigidity in torsion S_t	Weight of model G	$\frac{S_b}{G}$	$\frac{S_t}{G}$
1 (basic)	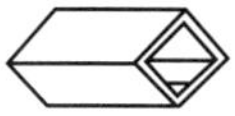	1.00	1.00	1.00	1.00	1.00
2a	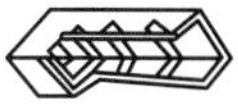	1.10	1.63	1.10	1.00	1.48
2b	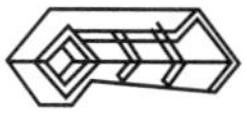	1.09	1.39	1.05	1.04	1.32
3	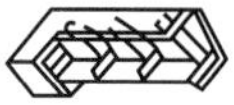	1.08	2.04	1.14	0.95	1.79
4	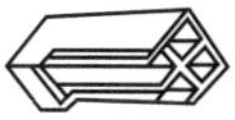	1.17	2.16	1.38	0.85	1.56
5	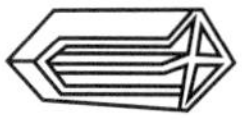	1.78	3.69	1.49	1.20	3.07
6	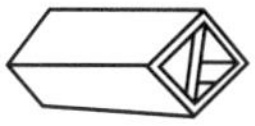	1.55	2.94	1.26	1.23	2.39

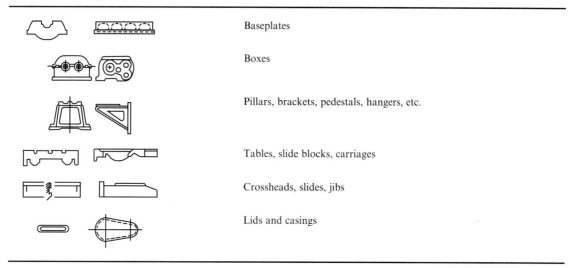

TABLE 25-28
Variations in Relative Bending and Torsional Rigidity for Models of Various Forms

		Relative rigidity in bending		Relative rigidity in torsion	
Model No.	Relative weight of box-like section	With ribs	With thicker walls	With ribs	With thicker walls
1 (basic)	1.00	1.00	1.00	1.00	1.00
2a	1.10	1.10	1.15	1.63	1.18
2b	1.05	1.09	1.10	1.39	1.10
3	1.14	1.08	1.16	2.04	1.21
4	1.38	1.17	1.29	2.16	1.40
5	1.49	1.78	1.30	3.69	1.46
6	1.26	1.55	1.19	2.94	1.24

Source: Courtesy: Dobrovolsky, V., etl., "Machine Elements", Mir Publishers, Moscow, 1974.

TABLE 25-69
Effect of stiffner arrangement on torsional stiffness of open structure[4]

	Stiffener arrangement	Relative torsional stiffness	Relative weight	Relative torsional stiffness per unit weight
1	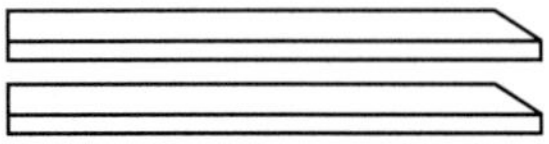	1.0	1.0	1.0
2	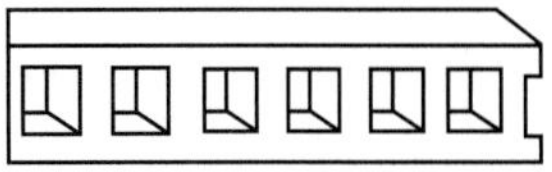	1.34	1.34	1.0
3	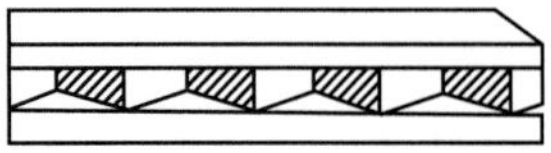	1.43	1.34	1.07
4	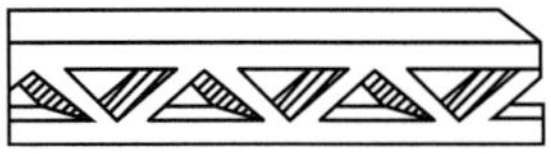	2.48	1.38	1.80
5		3.73	1.66	2.25

TABLE 25-70
Effect of aperture and cover plate design on static and dynamic stiffness of box section[3]

	Relative stiffness about			Relative natural frequency of vibrations about			Relative damping of vibrations about		
	X-X	Y-Y	Z-Z	X-X	Y-Y	Z-Z	X-X	Y-Y	Z-Z
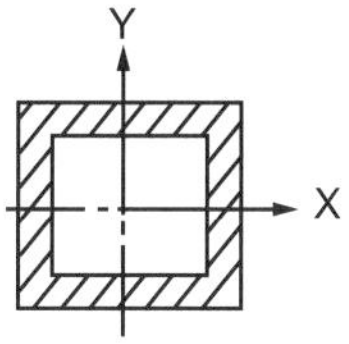	100	100	100	100	100	100	100	100	100
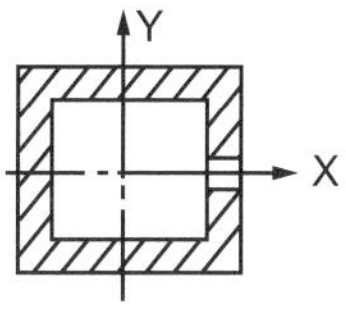	85	85	28	90	87	68	75	89	95
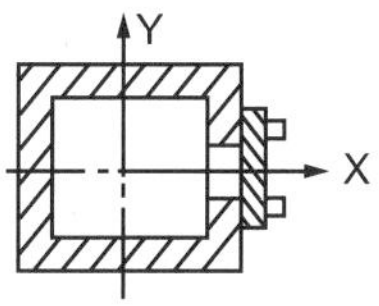	89	89	35	95	91	90	112	95	165
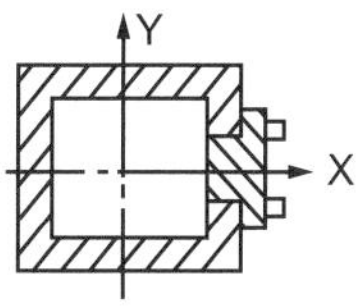	91	91	41	97	92	92	112	95	185

Profile	Factors				
	I_{ben}	I_{tors}	A	$\frac{I_{ben}}{A}$	$\frac{I_{tors}}{A}$
	1	1	1	1	1
	1.17	2.16	1.38	0.85	1.56
	1.55	3	1.26	1.23	2.4
	1.78	3.7	1.5	1.2	2.45

FIGURE 25-50 Stiffening effect of reinforcing ribs.

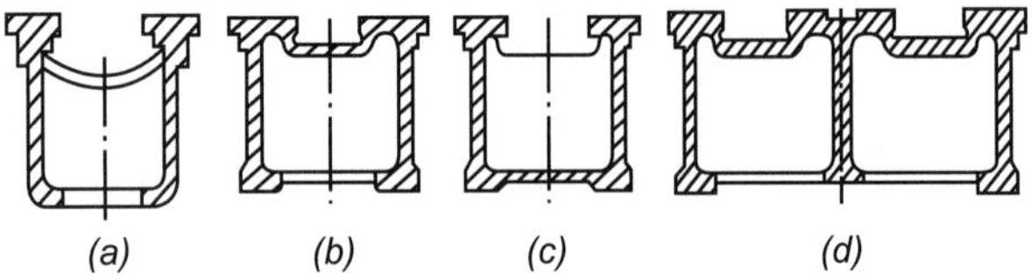

FIGURE 25-51A Typical cross-sections of beds.

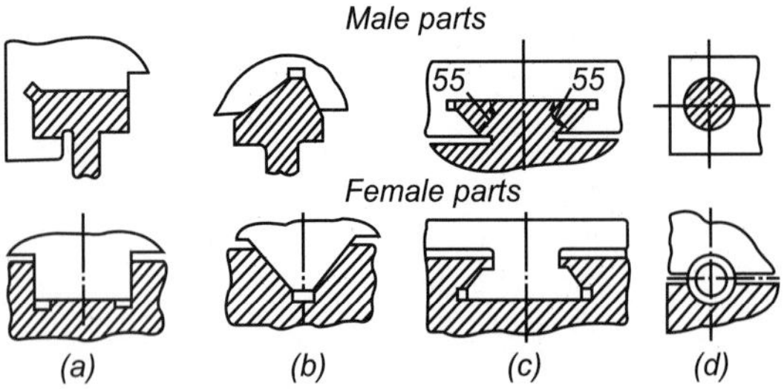

FIGURE 25-51B Principal shapes of sliding guides. (*a*) flat ways; (*b*) prismatic ways; (*c*) dovetail ways; (*d*) cylindrical (bar-type) ways.

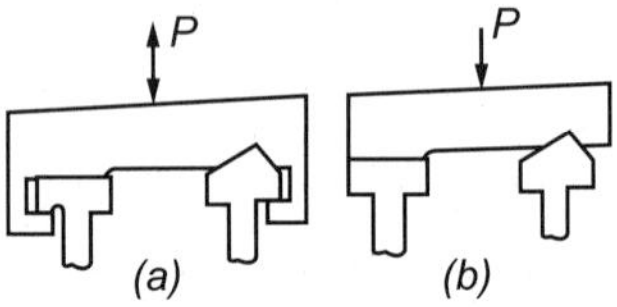

FIGURE 25-51C Sliding guides. (*a*) closed type; (*b*) open type.

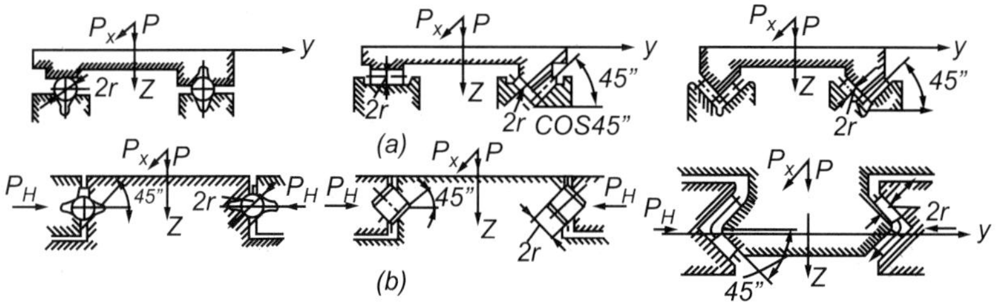

FIGURE 25-51D Rolling guides, (*a*) open type; (*b*) closed type.

TABLE 25-71
Traversing Force Calculations – Typical Cases

	Type of ways	r_{eq}.cm	Traversing force Q. kgf
1	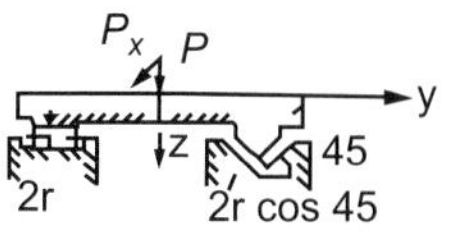	$\frac{r}{1.5}$	$Q = P_x + 3T_0 + \frac{1.5}{r} f_r P$ $P = P_2 + G_1 + G_2$
2	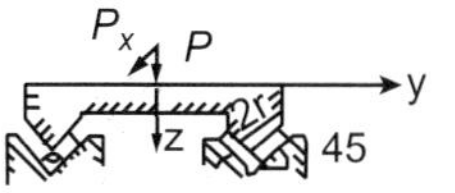	$\frac{r}{1.4}$	$Q = P_x + 4T_0 + \frac{1.4}{r} f_r P$
3	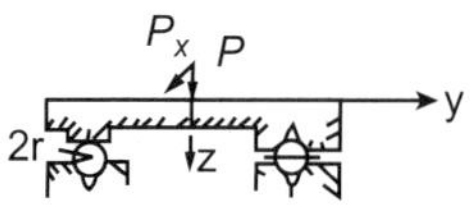	$\frac{r}{1.5}$	$Q = P_x + 2T_0 + \frac{1.5}{r} f_r P$
4	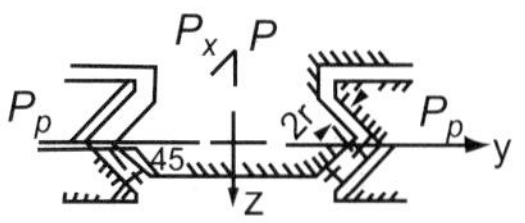	$\frac{r}{2.8}$	$Q = P_x + 4T_0 + \frac{2.8}{r} f_r P_P$
	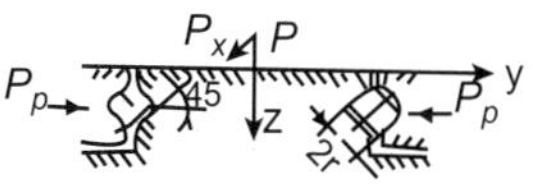		
	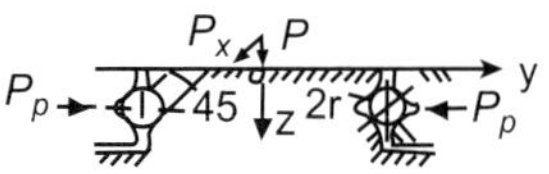		$Q = P_x + 2T_0 + \frac{2.8}{r} f_r P_P$

Notes: 1. The coefficient of rolling friction $f_r = 0.001$ for ground steel ways and $f_r = 0.0025$ for scraped cast iron ways. The initial friction force, referred to one separator, $T_0 = 0.4$ kgf.

2. Because of the low value of the friction forces, a simplified arrangement has been accepted in which the ways are subject only to the feed force P_x, vertical component P_x of the cutting force, table weight G_1 and workpiece weight G_2. The tilting moments, force P_p and the components of the traversing force are not taken into account.

3. In the type 4 ways only the feed force P_x and the preload force P_p are taken into consideration.

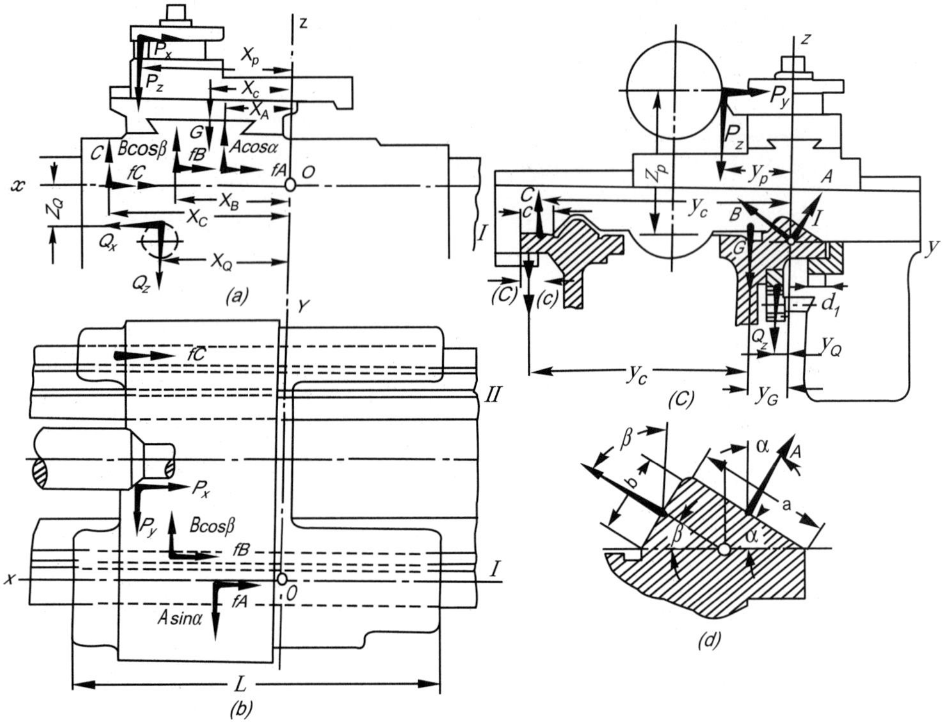

FIGURE 25-52 Forces acting on the Slidways of a Lathe – A Typical Case
Source: Courtesy: Acherkan, N., "Machine Tool Design", Mir Publishers Moscow, 1968.

TABLE 25-72
Classification and Identification code of Machine Tools – Kinematic Diagram

Description	Symbol	Description	Symbol
Shafts		Belt drives: Open flat belts	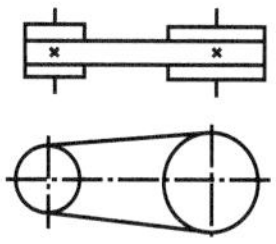
Shafts coupling: Closed			
Closed with over-load protection		Crossed flat belts	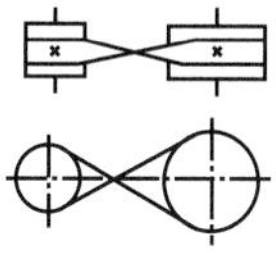
Flexible			
Universal	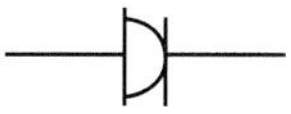		
Telescopic		V-belts	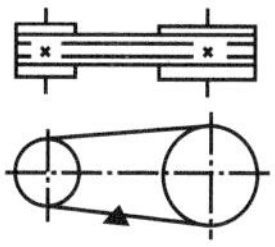
Floating	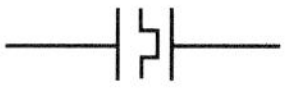		
Toothed	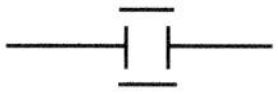	Chain drive	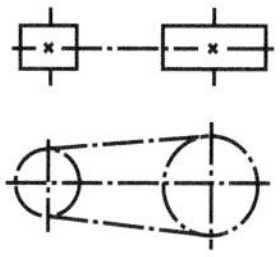
Parts mounted on shafts: Freely mounted			
Sliding on feather	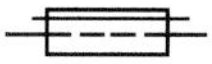	Toothed gearing: Spur or helical gears	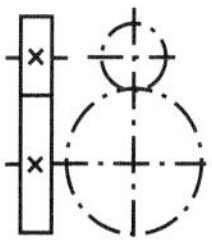
Engaged with sliding key	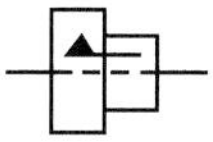		
Fixed	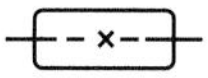	Bevel gears	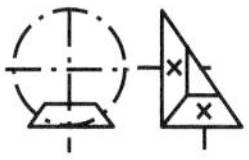
Plain bearings: Radial			
Single-direction thrust		Spiral (crossed helical) gears	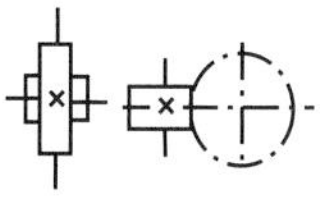
Two-direction thrust	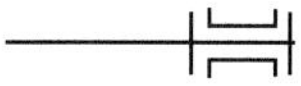	Worm gearing	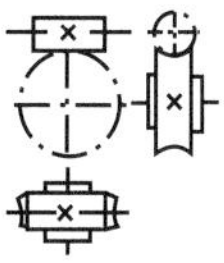
Antifriction bearings: Radial			
Single angular-contact	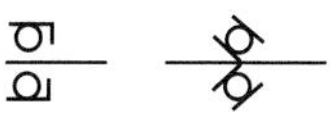	Back-and-pinion gearing	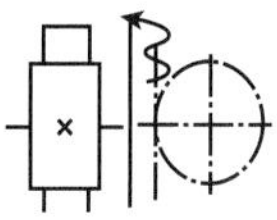
Duplex angular contact	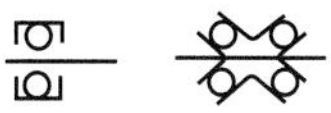		

TABLE 25-72
Classification and Identification code of Machine Tools – Kinematic Diagram (*Cont.*)

Description	Symbol	Description	Symbol
Nut on power screw: Solid nuts		Single-direction overrunning clutches	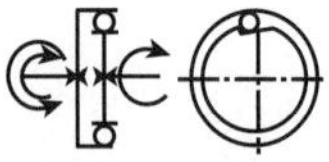
Split nuts		Two-direction overrunning clutches	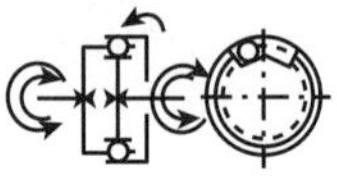
Clutches: Single-direction jaw clutches	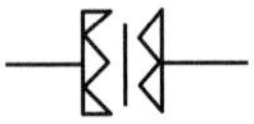	Brakes: Cone	
Spindle noses: Centre type	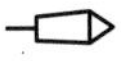	Shoe	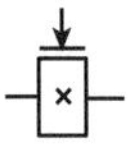
Chuck type		Band	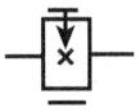
Bar type		Disk	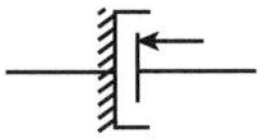
Drilling		Milling	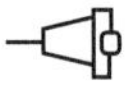
Boring spindles with faceplates	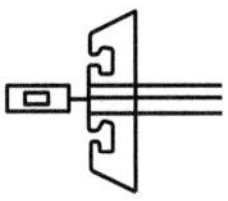	Grinding	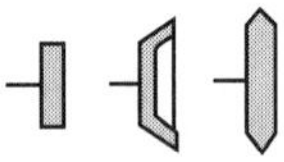
Two-direction jaw clutches	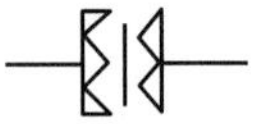	Electric motors: On feet	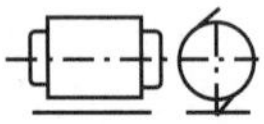
Cone clutches	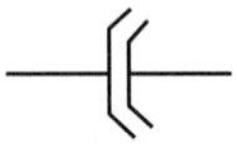	Flange-mounted	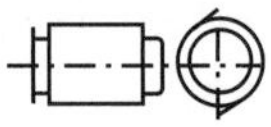
Single disk clutches	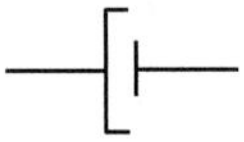	Built-in	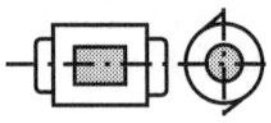
Twin disk clutches	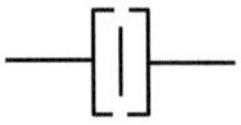		

Source: Courtesy: Acherkan, N., et1., "Machine Tool Design", Moscow, 1968.

REFERENCES

1. Lingaiah, K., *Machine Design Data Handbook*, McGraw-Hill Publishing Company, New York, 1994.
2. Lingaiah, K., *Machine Design Data Handbook*, Vol. I, Suma Publishers, Bangalore, India, 1986.
3. Merchant, M. E., Trans. Am. Soc. Mech. Engrs., 66, A-168, 1944.
4. Ernst, H., and M. E. Merchant, *Chip Formation, Friction and Finish*, Cincinneti Milling, Machine Company, USA.
5. American Society of Tool and Manufacturing Engineers (ASTME), *Tool Engineers Handbook*, 2nd ed., F. W. Wilson, Editor, McGraw-Hill Book Publishing Company, New York, 1959.
6. Cyril Donaldson, George H. Lecain and V.C. Goold, *Tool Design*, Tata-McGraw-Hill Publishing Company Ltd., New Delhi, India, 1976.
7. Frank W. Wilson, Editor-in-Chief, American Society of Tool and Manufacturing Engineers (ASTME), *Fundamentals of Tool Design*, Prentice Hall, New Delhi, India, 1969.
8. Kuppuswamy, G., Center for Continuing Education, Department of Mechanical Engineering, Indian Institute of Technology, Madras, India, August 12, 1987.
9. Sen, G. C., and A. B. Bhattacharyya, *Principles of Machine Tools*, New Central Book Agency, (P) Ltd., Calcutta, India, 1995.
10. Geoffrey Boothroyd, *Fundamentals of Metal Machining and Machine Tools*, McGraw-Hill Publishing Company, New York, 1975.
11. Koenigsberger, F., *Design Principles of Metal Cutting Machine Tools*, the MacMillan Company, New York, 1964.
12. Shaw, M. C., and C. J. Oxford, Jr., (1) "*On the Drilling Metals*" (2) "*The Torque and Thrust in Milling*", Trans. ASME., 97:1, January 1957.
13. Hindustan Machine Tools, Bangalore, *Production Technology*, Tata-McGraw-Hill Publishing Company Ltd., New Delhi, India, 1980.
14. Central Machine Tool Institute, Machine *Tool Design Handbook*, Bangalore, India, 1988.
15. Acherkan, A., General Editor, V. Push, N. Ignatyev, A. Kakoilo, V. Khomyakov, Y. U. Mikheyev, N. Lisitsyn, A. Gavryushin, O. Trifonov, A. Kudryashov, A. Fedotyonok, V. Yermakov, V. Kudinov, *Machine Tool Design*, Vol. 1 to 4, Mir Publishers, Moscow, 1968-69.
16. Milton C. Shaw, *Metal Cutting Principles*, Clarendon Press, Oxford, 1984.
17. Martelloti, M. E., Trans. Am. Soc. Mech. Engrs., 63, 677, 1941.
18. Kovan, V. M., *Technology of Machine Building*, Mashgiz, Moscow, 1959.
19. Basu, S. R., and D. K. Pal, *Design of Machine Tools*, 2nd ed., Oxford and IBH Publishing Company, New Delhi, 1983.
20. Heinrich Makelt, *Die Mechanischen Pressen*, Carl Hanser Verlag Muchen, 1961 (in German) Translated to English by R. Hardbottle, *Mechanical Presses*, Edward Arnold (Publishers) Ltd., 1968.
21. Dobrovolsky, K. Zablonsky, S. Mak, Radchik, L. Erlikh, *Machine Elements*, Mir Publishers, Moscow, 1968.
22. Rivin, E. I., *Stiffness and Damping in Mechanical Design*, Marcel Dekker, Inc., New York, 1999.
23. *Machine Tool Design and Numerical Control.*
24. Chernov, N., *Machine Tools*, Translated from Russian to English by Falix Palkin, Mir Publishers, Moscow, 1975.
25. Greenwood, D. C., *Engineering Data for Product Design*, McGraw-Hill Publishing Company, New York, 1961.

CHAPTER 26

RETAINING RINGS AND CIRCLIPS

SYMBOLS

a	acceleration of retained parts, m/s^2 (ft/s^2 or in/s^2)
Ch	actual chamfer, m (in)
Ch_{max}	listed maximum allowable chamfer, m (in)
C_F	conversion factor (refer to Table 26-1)
d	depth of groove, m (in)
D	shaft or housing diameter, m (in)
f	frequency of vibration, cps
F_{tg}	allowable static thrust load on the groove wall, kN (lbf)
F_{ig}	allowable impact load on groove, kN (lbf)
F_{rt}	allowable static thrust load of the ring, kN (lbf)
F_{ir}	allowable impact load on a retaining ring, kN (lbf)
F'_r	listed allowable assembly load with maximum corner radius or chamfer, kN (lbf)
F''_r	allowable assembly load when cornor radius or chamfer is less than the listed, kN (lbf)
F_{trr}	allowable thrust load exerted by the adjacent part, kN (lbf)
F_{sg}	allowable sudden load an groove, kN (lbf)
F_{sr}	allowable sudden load on ring, kN (lbf)
l	distance of the outer groove wall from the end of the shaft or bore as shown in Fig. 26-2, m (in)
n	factor of safety (about 2 to 4 may be assumed)
n_{max}	maximum safe speed, rpm
q	reduction factor from Fig. 26-1.
r	actual corner radius or chamfer, m (in)
r_{max}	listed maximum allowable corner radius, m (in)
t	ring thickness, m (in)
T	largest section of the ring, m(in)
w	weight of retained parts, kN (lbf)
$(wa)_g$	allowable vibratory loading on groove, kN (lbf)
$(wa)_r$	allowable vibratory loading on ring, kN (lbf)
x_o	amplitude of vibration, m (in)
σ_{sy}	tensile yield strength of groove material, Table 26-2, MPa (psi)
σ_{saw}	maximum working stress of ring during expansion or contraction of ring, MPa (psi)
τ_s	shear strength of ring material, MPa (psi) (refer to Table 26-3)
μ	coefficient of friction between ring and retained parts whichever is the largest.

Note: σ and τ with subscript s designates strength properties of material used in the design which will be used and observed throughout this *Machine Design Data Handbook*. Other factors in performance or in special aspects are included from time to time in this chapter and, being applicable only in their immediate context are not given at this stage.

Particular	Formula	
RETAINING RINGS AND CIRCLIPS:		
(Figs. 26-1 to 26-28 and Tables 26-1 to 26-13)		
Load Capacities of Retaining Rings:		
Allowable static thrust load on the groove	$F_{tg} = \dfrac{C_F D d \pi \sigma_{sy}}{nq}$	(26-1)
Allowable static thrust load on ring which is subject to shear	$F_{r\tau} = \dfrac{C_F D t \pi \tau_s}{n}$	(26-2)
The allowable thrust load exerted by adjacent part	$F_{trr} \leq \dfrac{\sigma_{saw} t T^2}{18 \mu D}$	(26-3)
Allowable assembly load when the corner radius or chamfer is less than the listed ($F''_r < F'_r$)	$F''_r = \dfrac{F'_r r_{max}}{r}$ for radius	(26-4)
	$F''_r = \dfrac{F'_r Ch_{max}}{Ch}$ for chamfer	(26-5)
Dynamic Loading:		
Allowable sudden load on ring	$F_{sr} \leq 0.5 F_{r\tau}$	(26-6)
Allowable sudden load on groove	$F_{sg} \leq 0.5 F_{tg}$	(26-7)
Allowable vibration loading on ring	$(wa)_r \leq 540 F_{r\tau}$ [a]	(26-8)
Allowable vibration loading on groove	$(wa)_g \leq 400 F_{tg}$ [a]	(26-9)
Acceleration of retained parts for harmonic oscillation	$a \approx 40 x_o f^2$	(26-10)
Allowable impact loading on groove	$F_{ig} = F_{r\tau} d/2$	(26-11)
Allowable impact loading on ring	$F_{ir} = F_{r\tau} t/2$	(26-12)
An empirical formula for maximum safe speed with standard types of rings	$n_{max} = 5000000/D$ where D in mm	(26-13)
	$n_{max} = 20000/D$ where D in inches	(26-14)

[a] *Note*: Actual tests should be conducted because of repeated or cyclic condition.

Particular	Formula
For dimensions of external circlips—Type A—light series	Refer to Table 26-5 and Fig. 26-3.
For dimensions of external circlips—Type A—heavy series	Refer to Table 26-6 and Fig. 26-4.
For dimensions of internal circlips—Type B—light series	Refer to Table 26-7 and Fig. 26-5.
For dimensions of internal circlips—Type B—Heavy series	Refer to Table 26-8 and Fig. 26-6.
For dimensions of external circlip—Type C	Refer to Table 26-9 and Fig. 26-7.
For dimensions, allowable static thrust load, allowable corner radii, chamfers, housing diameter and ring thickness of retaining rings—basic internal, bowed internal, beveled internal, inverted internal, double beveled internal, crescent-shaped, bowed E-ring, reinforced, locking prong in grooved housing and on grooved shafts, self locking and triangular self locking ring etc.	Refer to Tables 26-10 to 26-13 and Figs. from 26-1 to 26-28.
For q reduction factor	Refer to Fig. 26-1.

TABLE 26-1
Conversion or correction factor C_F for calculating F_{rT} and F_{tg} for use in Eqs. (26-1) and (26-2)

	Conversion or correction factor C_F	
Ring type	**Ring: F_{rT}**	**Groove: F_{tg}**
Basic, bowed internal	1.2	1.2
Beveled internal	1.2	1.2
Double-beveled internal		Use $d/2$ instead of d
Inverted internal, external	2/3	1/2
Basic, bowed external	1	1
Beveled external	1	1
		Use $d/2$ instead of d
Crescent-shaped	1/2	1/2
Two-part interlocking	3/4	3/4
E-ring, bowed E-ring	1/3	1/3
Reinforced E-ring	1/4	1/4
Locking-prong ring	See manufacturer's specifications	1.2
Heavy-duty external	1.3	2
High-strength radial	1/2	1/2
Miniature high-strength	See manufacturer's specifications	
Thinner-gage high-strength radial	1/2	1/2

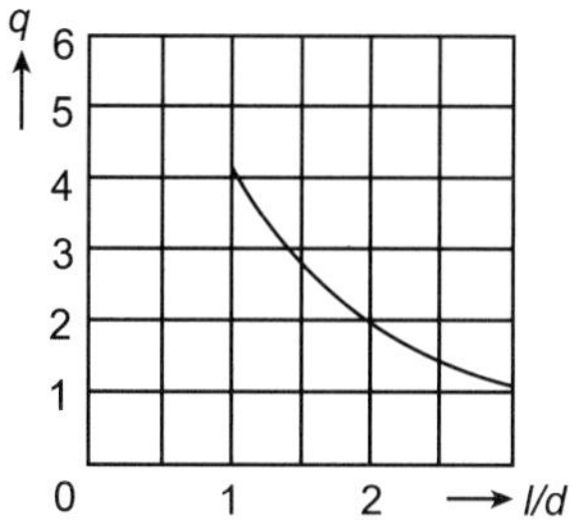

FIGURE 26-1 Reduction curve

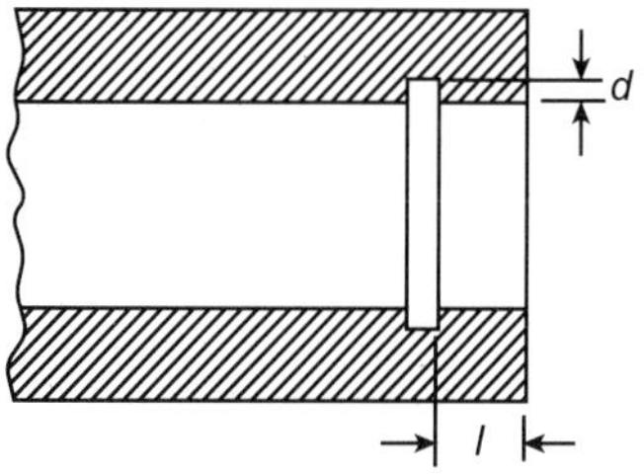

FIGURE 26-2 Edge margin

TABLE 26-2
Tensile yield strength of groove material

Groove material	Tensile yield strength, σ_{sy} MPa	lbf/in^2
Cold-rolled steel	310	45,000
Hardened steel (Rockwell C40)	1034	150,000
Hardened steel (Rockwell C50)	1380	200,000
Aluminum (2024-T4)	276	40,000
Brass (naval)	210	30,000

TABLE 26-3
Shear strength of ring material for use in Eq. (26-2)

Ring material	Ring type	Ring thickness mm (in)	Shear strength, τ_s MPa	lbf/in^2
Carbon spring steel (SAE 1060–1090)	Basic, bowed, beveled, inverted internal and external rings and crescent-shaped	Up to and including 0.9 (0.035)	827	120,000
	Double-beveled internal rings	1.07 (0.042) and over	1034	150,000
	Heavy-duty external	0.90 (0.035) and over	1034	150,000
	Miniature high-strength	0.510 (0.020) and 0.635 (0.025)	827	120,000
		0.9 (0.035) and over	1034	150,000
	Two-part interlocking, reinforced E-ring, high-strength radial	All available	1034	150,000
	Thinner high-strength radial	All available	1034	150,000
	E-ring, bowed E-ring	0.254 (0.010) and 0.380 (0.015)	690	100,000
		0.635 (0.025)	827	120,000
		0.9 (0.035) and over	1034	150,000
	Locking-prong	All available	896	130,000
Beryllium copper (CDA 17200)	Basic external	0.254 (0.010) and 0.380 (0.015) sizes 12 through 23	758	110,000
	Bowed external	0.380 (0.015) sizes 18 through 23	758	110,000
	E-ring	0.254 (0.010) (size $\times$ 4 only)	662	95,000

TABLE 26-4
Maximum working stress of ring during expansion or contraction

Ring material	Maximum allowable working stress, σ_{saw} MPa	Maximum allowable working stress, σ_{saw} lbf/in²
Carbon spring steel (SAE 1075)	1724	250,000
Stainless steel (PH 15-7 Mo)	1724	250,000
Beryllium copper (CDA 17200)	1380	200,000
Aluminum (Alclad 7075-T6)	482	70,000

Courtesy: © 1964, 1965, 1973, 1981 Waldes Kohinoor, Inc., Long Island City, New York, 1985. Edward Killian, "*Retaining Rings*", Robert O. Parmley, Editor-in-Chief "*Mechanical Components Handbook*", McGraw-Hill Publishing Company, New York, USA.

TABLE 26-5
Dimensions for external circlips—type A—light series

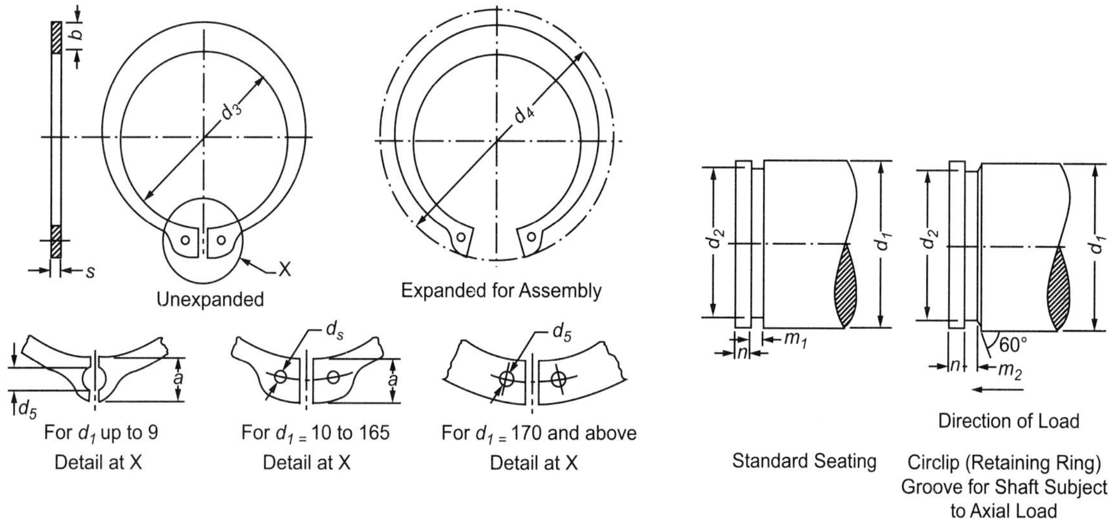

FIGURE 26-3

All dimensions in millimeters

Shaft Dia d_1	Circlip s h11	a Max.	b Approx.	d_3	Tol. on d_3	d_4 Expanded	d_5 Min.	Shaft groove d_2	Tol. on d_2	m_1 H13	m_2 Min.	n Min.	Axial force N	Axial force lbf
8	0.8	3.2	1.5	7.4	+0.09	15.2	1.2	7.6		0.9	1.0		1180	265
9			1.7	8.4	−0.18	16.4		8.6					1360	305
					+0.15							0.6		
10				9.3		17.6		9.6					1500	340
		3.3			−0.30		1.5							
11			1.8	10.2		18.6		10.5					2060	460
12				11		19.6		11.5	h11			0.75	2270	510
13	1	3.4	2	11.9		20.8		12.4		1.1	1.2		2940	660

TABLE 26-5
Dimensions for external circlips—type A—light series (*Contd.*)

All dimensions in millimeters

	Circlip							Shaft groove						
Shaft Dia d_1	s h11	a Max.	b Approx.	d_3	Tol. on d_3	d_4 Expanded	d_5 Min.	d_2	Tol. on d_2	m_1 H13	m_2 Min.	n Min.	Axial force N	Axial force lbf
14		3.5	2.1	12.9	+0.18	22	1.7	13.4				0.9	3190	720
15		3.6		13.8	−0.36	23.2		14.3				1.1	3920	880
16		3.7	2.2	14.7		24.4		15.2					4809	1080
17		3.8	2.3	15.7		25.6		16.2				1.2	5100	1150
18		3.9	2.4	16.5		26.8		17					6770	1520
19			2.5	17.5		27.8		18					7110	1600
20		4	2.6	18.5		29		19				1.5	7550	1700
21		4.1	2.7	19.5		30.2		20					7900	1780
22	1.2	4.2	2.8	20.5		31.4		21		1.3	1.4		8300	1860
24		4.4	3	22.2		33.8	2	22.9					9900	2230
25				23.2	+0.21	34.8		23.9				1.7	10400	2335
26		4.5	3.1	24.2	−0.42	36		24.9					10790	2425
28		4.7	3.2	25.9		38.4		26.6					14710	3310
29		4.8	3.4	26.6		39.6		27.6				2.1	15300	3440
30		5	3.5	27.9		4.1		28.6					15890	3570
32	1.5	5.2	3.6	29.6		43.4		30.3		1.6	1.7		20590	4630
34		5.4	3.8	31.5		45.8		32.3				2.6	21770	4890
35		5.6	3.9	32.2	+0.25	47.2		33					26180	5890
36			4	33.2	−0.25	48.2		34				3	27070	6085
38		5.8	4.2	35.2		50.6		36	h12				28540	6415
40		6	4.4	36.5		53		37.5					37360	8400
42	1.75	6.5	4.5	38.5		56		39.5		1.85	2	3.8	39230	8820
45		6.7	4.7	41.5	+0.39	59.4		42.5					42170	9480
48		6.9	5	44.5	−0.78	62.8	2.5	45.5					45110	10140
50		6.9	5.1	45.8		64.8		47					55900	12565
52		7	5.2	47.8		67		49					58350	13120
55		7.2	5.4	50.8		70.4		52					61780	13890
56	2	7.3	5.5	51.8		71.6		53		2.15	2.3		62760	14110
58			5.6	53.9		73.6		55					65210	14660
60		7.4	5.8	55.8	+0.46	75.8		57				4.5	67665	15210
62		7.5	6	57.8	−0.92	78		59					69625	15650
63		7.6	6.2	58.8		79.2		60					71100	15985
65		7.8	6.3	60.8		81.6		62					73550	16535
68		8	6.5	63.5		85		65					76880	17285
70		8.1	6.6	65.5		87.2		67					78940	17748
72		8.2	6.8	67.5		89.4	3	69				4.5	81395	18300
75	2.5	8.4	7	70.5	+0.46	92.2		72		2.65	2.8		84336	18960
78			7.3	73.5	−0.92	96.2		75					88260	19840
80		8.6	7.4	74.5		98.2		76.5	h12				104930	23590
82		8.7	7.6	76.5		101		78.5					107870	24250
85			7.8	79.5		104		81.5					111795	25130
88		8.8	8	82.5		107		84.5				5.3	116700	26236
90	3		8.2	84.5		109		86.5		3.15	3.3		118660	26675

TABLE 26-5
Dimensions for external circlips—type A—light series (*Contd.*)

All dimensions in millimeters

Shaft Dia d_1	Circlip: s h11	Circlip: a Max.	Circlip: b Approx.	Circlip: d_3	Circlip: Tol. on d_3	Circlip: d_4 Expanded	Circlip: d_5 Min.	Shaft groove: d_2	Shaft groove: Tol. on d_2	Shaft groove: m_1 H13	Shaft groove: m_2 Min.	Shaft groove: n Min.	Axial force: N	Axial force: lbf
95		9.4	8.6	89.5		115		91.5					125525	28220
100		9.6	9	94.5	+0.54	121	3.5	96.5					132390	29764
105		9.9	9.3	98		126		101					158865	35716
110		10.1	9.6	103	−1.08	132		106					166712	37490
115		10.6	9.8	108		138		111					174555	39244
120		11	10.2	113		143		116				6	181422	40785
125		11.4	10.4	118		149		121					189265	42550
130		11.6	10.7	123		155		126					197110	44315
135		11.8	11	128		160		131					204958	46080
140		12	11.2	133		165		136					212800	47840
145		12.2	11.5	138		171		141					220650	49606
150	4	13	11.8	142		177		145					283410	63716
155			12	146		182		150		4.15	4.3		294200	66140
160		13.3	12..2	151	+0.63	188		155					304000	69346
165		13.5	12.5	155.5	−1.26	193		160					313810	70550
170			12.9	160.5		197		165					322640	72535
175			Max	165.5		202	4	170				7.5	331460	74520
180			13.5	170.5		208		175					341270	76724
185			Max	175.5		213		180					331460	74520
190				180.5		219		185					328520	73858
195				185.5		224		190	h13				320675	72094
200				190.5		229		195					312830	70330
210			14	198	+0.72	239		204					478560	107590
220			Max	208	−1.44	249		214					502095	112880
230				218		259		224				9	524650	117950
240				228		269		234					518770	116630
250	5			238		279		244		5.15	5.3		493270	110900
260				245		293		252					533480	119936
270			16	255		303		262					514846	115748
280			Max	265	+0.81	313	5	272				12	498175	112000
290				275	−1.62	323		282					491505	108250
300				285		333		292					463850	104280

TABLE 26-6
Dimensions for external circlips—type A—heavy series

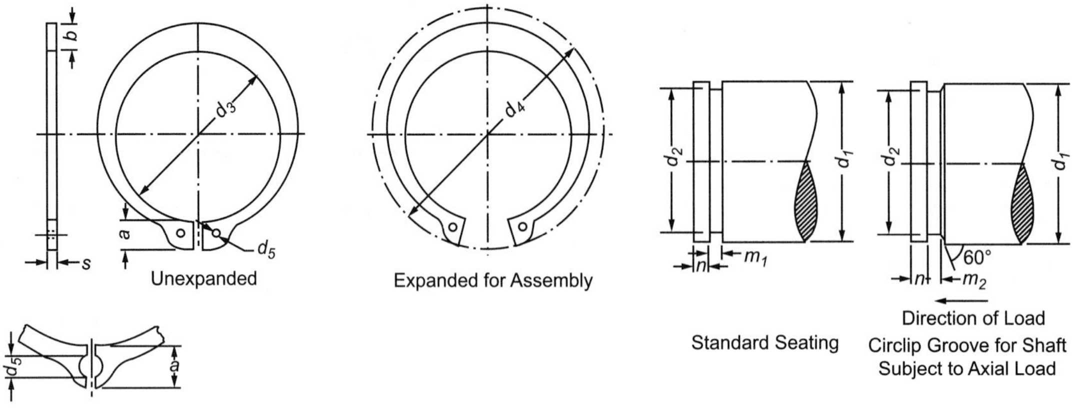

Alternate Shape of Lug

FIGURE 26-4

All dimensions in millimeters

Shaft Dia d_1	Circlip: s h11	a Max.	b Approx.	d_3	Tol. on d_3	d_4 Expanded	d_5 Min.	Shaft groove: d_2	Tol. on d_2	m_1 H13	m_2 Min.	n Min.	Axial force: N	lbf
15		4.8	2.4	13.8		25.5		14.3				1.1	3922	882
16		5	2.5	14.7	+0.18	27.5		15.2					4805	1080
17	1.5		2.6	15.7	−0.36	28.5		16.2	h11	1.6	1.7	1.2	5100	1146
18		5.1	2.7	16.5		29.5		17					6765	1520
20		5.5	3	18.5		32.5	2	19				1.5	7550	1698
22	1.75	6	3.1	20.5		35.5		21		1.85	2		8286	1862
24		6.3	3.2	22.2	+0.21	38		22.9					9905	2226
25		6.4	3.4	23.2	−0.42	39		23.9				1.7	10395	2336
28	2		3.5	25.9		42.5		26.6					14710	3310
30		6.5	4.1	27.9		44.5		28.6		2.15	2.3	2.1	15896	3570
32				29.6		46.5		30.3				2.6	20594	4630
34		6.6	4.2	31.5	+0.25	49		32.3					21770	4895
35		6.7		32.2	−0.50	50		33				3	25890	5820
38		6.8	4.3	35.2		53		36					28242	6350
40	2.5	7	4.4	36.5		55.5		37.5		2.65	2.8		37658	9466
42		7.2	4.5	38.5	+0.39	58		39.5				3.8	39226	8820
45		7.5	4.7	41.5	−0.78	61.5	2.5	42.5	h12				42168	9480
48		7.8	5	44.5		65		45.5					45110	10140
50		8	5.1	45.8		68		47					55898	12566
52		8.2	5.2	47.8		70		49					58350	13118
55	3	8.5	5.4	50.8		73.5		52		3.15	3.3		61780	13990
58		8.8	5.6	53.8		77		55				4.5	65214	14660
60		9	5.8	55.8		79		57					67665	15212
65		9.3	6.3	60.8	+0.46	85		62					70550	16535
70		9.5	6.6	65.5	−0.92	90.5	3	67					78942	17744
75		9.7	7	70.5		96		72					84336	19956
80	4	9.8	7.4	74.5		101		76.5					104930	23590
85		10	7.8	79.5		106.5		81.5		4.15	4.3		111795	25134
90		10.2	8.2	84.5	+0.54	112	3.5	86.5				5.3	118660	26676
100		10.5	9	94.5	−1.08	124		96.5					132390	29764

Designation: A circlip of light series in type A for shaft diameter d_1 equal to 50 mm shall be designated as: Circlip, Light A 50 IS: 3075, 1965.

TABLE 26-7
Dimensions for internal circlips—type B—light series

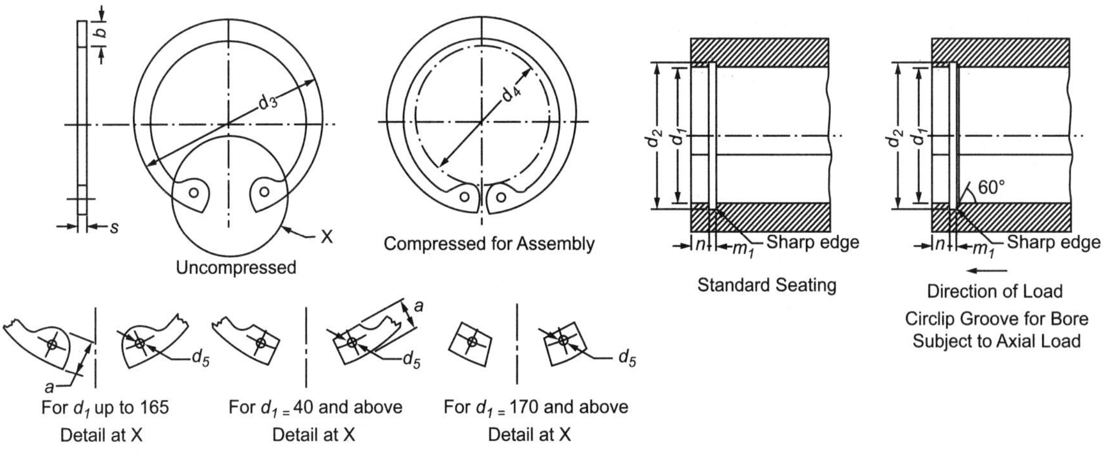

FIGURE 26-5

All dimensions in millimeters

Shaft Dia d_1	Circlip s h11	a Max.	b Approx.	d_3	Tol. on d_3	d_4 Compressed	d_5 Min.	Bore groove d_2	Tol. on d_2	m_1 H13	m_2 Min.	n Min.	Axial force N	Axial force lbf
8	0.8	2.4	1.1	8.7		2.8	1	8.4					1255	282
9		2.5	1.3	9.8		3.5		9.4				0.6	1412	318
10		3.2	1.4	10.8		3.1	1.2	10.4					1570	352
11		3.3	1.5	11.8	+0.36	3.9		11.4					1725	389
12		3.4	1.7	13.0	−0.18	4.7	1.5	12.5				0.75	2353	530
13		3.6	1.8	14.1		5.3		13.6	H11			0.9	3080	692
14		3.7	1.9	15.1		6		14.6					3295	740
15		3.8	2	16.2		7	1.7	15.7		1.1	1.2	1.1	4138	930
16	1			17.3		7.7		16.8				1.2	5050	1135
17		3.9	2.1	18.3		8.4		17.8					5364	1205
18		4.1	2.2	19.5		8.9		19					7110	1598
19				20.5		9.8		20					7492	1684
20			2.3	21.5	+0.42	10.6		21				1.5	7894	1775
21		4.2	2.4	22.5	−0.21	11.6		22					8286	1862
22			2.5	23.5		12.6	2	23					8650	1943
24		4.4	2.6	25.9		14.2		25.2					11375	2558
25		4.5	2.7	26.9		15		26.2				1.8	11769	2645
26	1.2	4.7	2.9	27.9		15.6		27.2		1.3	1.4		12258	2756
28		4.8	2.9	30.1		17.4		29.4				2.1	15495	3485
30			3.0	32.1		19.4		31.4					16572	3726
32			3.2	34.4		20.2		33.7				2.6	21575	4850
34		5.4	3.3	36.5	+0.50	22.2		35.7					22750	5115
35			3.4	37.8	−0.25	23.2		37					27655	6216
36	1.5		3.5	38.8		24.2		38		1.6	1.7	3	28440	6394
37		5.5	3.6	39.8		25		39					29224	6570
38			3.7	40.8		26		40					30106	6768
40		5.8	3.9	43.5	+0.78	27.4		42.5					39716	8930
42		5.9	4.1	45.5	−0.39	29.2		44.5					41678	9370

TABLE 26-7
Dimensions for internal circlips—type B—light series (*Contd.*)

All dimensions in millimeters

	Circlip							Bore groove						
Shaft Dia d_1	s h11	a Max.	b Approx.	d_3	Tol. on d_3	d_4 Compressed	d_5 Min.	d_2	Tol. on d_2	m_1 H13	m_2 Min.	n Min.	Axial force N	Axial force lbf
45	1.75	6.2	4.3	48.5		31.6		47.5	H12	1.85	2	3.8	44325	9965
47		6.4	4.4	50.5		33.2	2.5	49.5					46286	10406
48			4.5	51.5		34.6		50.5					47268	10626
50		6.5	4.6	54.2		36		53					59526	13382
52		6.7	4.7	56.2		37.6		55					61780	13766
55		6.8	5.0	59.2		40.4		58					65214	14660
56	2		5.1	60.2		41.4		59		2.15	2.3		66195	14882
58		6.9	5.2	62.2	+0.92	43.2		61					68646	15442
60			5.4	64.2	−0.46	44.4		63					71098	15984
62		7.3	5.5	66.2		46.4		65					73354	16490
63			5.6	67.2		47.4		66				4.5	74334	16712
65		7.6	5.8	69.2		48.8		68					76688	17240
68			6.1	72.5		51.4		71					80120	18012
70		7.8	6.2	74.5		53.4		73					82572	18564
72	2.5		6.4	76.5		55.4	3	75		2.65	2.8		84826	19070
75			6.6	79.5		58.4		78					88260	19700
78			6.8	82.5	+1.08	60		81					91690	20614
80		8.5		85.5	−0.54	62		83.5				5.3	109834	24692
82			7.0	87.5		64		85.5					112775	25354
85			7.2	90.5		66.8		88.5					116699	26236
88		8.6	7.4	93.5		69.8		91.5					120620	27118
90			7.6	95.5		71.8		93.5					123562	27790
92	3	8.7	7.8	97.5		73.6		95.5	H12	3.15	3.3	5.3	126505	28440
95		8.8	8.1	100.5		76.4		98.5					130428	29322
98		9	8.3	103.5	+1.08	79		101.5					134350	30205
100			8.4	105.5	−0.54	81	3.5	103.5					137292	30866
102		9.2	8.5	108		82.6		106					159849	35936
105			8.7	112		85.6		109					164750	37040
108		9.5	9	115		88		112					169654	38142
110		10.4	9	117		88.2		114					172596	38902
112		10.5	9.1	119		90		116					175538	39465
115			9.3	122		93		119				6	180440	40566
120			9.7	127		97		124					188286	42330
125		11	10	132		102		129					195150	43874
130			10.2	137		107		134					202996	45238
135		11.2	10.5	142		112		139					210940	47400
140			10.7	147	+1.26	117		144					219686	49165
145		11.4	10.9	152	−0.63	122		149					226532	50930
150		12	11.2	158		125		155					226532	50930
155	4		11.4	164		130		160		4.15	4.3		294198	66142
160		13	11.6	169		133		165					312830	70330
165			11.8	174.5		138		170					322936	72535
170			12.2	179.5		145		175					332444	74740
175			Max 12.7	184.5		149	4	180	H13				341270	76724
180			Max 13.2	189.5		153		185					338328	76062

TABLE 26-7
Dimensions for internal circlips—type B—light series (*Contd.*)

All dimensions in millimeters

	Circlip							Bore groove						
													Axial force	
Shaft Dia d_1	s h11	a Max.	b Approx.	d_3	Tol. on d_3	d_4 Compressed	d_5 Min.	d_2	Tol. on d_2	m_1 H13	m_2 Min.	n Min.	N	lbf
			Max											
185			13.7	194.5	+1.44	157		190					343230	77165
			Max											
190			13.8	199.5	−0.72	162		195					333424	74960
195			Max	204.5		167		200					323618	72755
200				209.5		171		205					318715	71652
210				222		181		216					488368	109795
220			14	232		191		226					511904	115096
230			Max	242		201		236				9	538392	121040
240				252		211		246					514846	115748
250	5			262		221		256		5.15	5.3		495232	111338
260				275	+1.62	227		268					529556	119055
270				285	−0.81	237		278				12	507980	114205
280			16	295		247	5	288					490330	110236
290			Max	305		257		298					472678	106268
300				315		267		308					456006	102520

TABLE 26-8
Dimensions for internal circlips—type B—heavy series

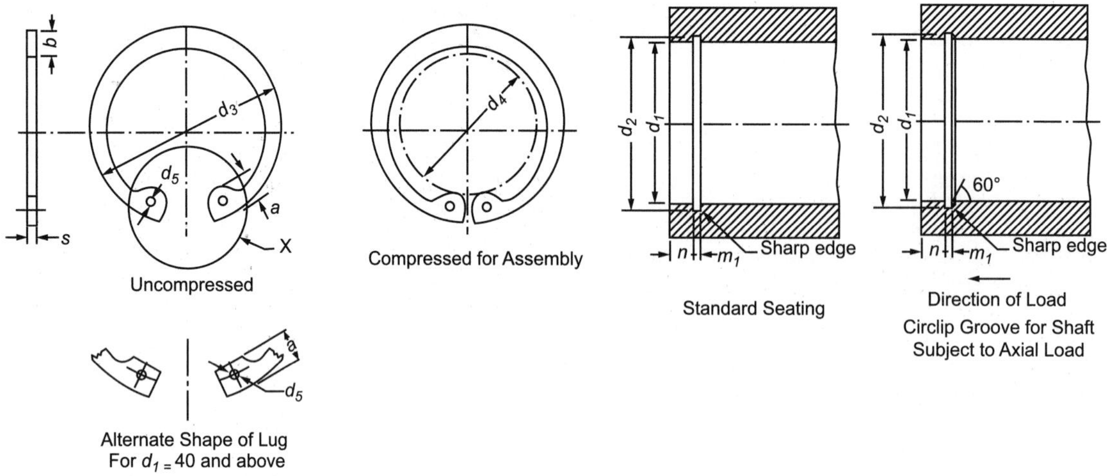

FIGURE 26-6

All dimensions in millimeters

Shaft Dia d_1	Circlip s h11	a Max.	b Approx.	d_3	Tol. on d_3	d_4 Compressed	d_5 Min.	Bore groove d_2	Tol. on d_2	m_1 H13	m_2 Min.	n Min.	Axial force N	lbf
20		4.5	2.4	21.5		10		21				1.5	7895	1775
22		4.7	2.8	23.5	+0.42	11.6		23					8650	1945
24		4.9	3	25.9	−0.21	13.2		25.2					11375	2558
25	1.5	5	3.1	26.9		14	2	26.2		1.6	1.7	1.8	11768	2646
26		5.1		27.9		14.8		27.2					12259	2755
28		5.3	3.2	30.1		16.4		29.4				2.1	15495	3484
30		5.5	3.3	32.1		18		31.4					16572	3726
32		5.7	3.4	34.4	+0.50	19.6		33.7				2.6	21575	4950
34		5.9	3.7	36.5	−0.25	21.2		35.7					22750	5115
35	1.75	6	3.8	37.8		22		37		1.85	2		27655	6218
37		6.2		39.8		23.6		39				3	29224	6590
38		6.3	3.9	40.8		24.4		40					30106	6768
40		6.5		43.5	+0.78	26		42.5					39716	8930
42	2	6.7	4.1	45.5	−0.39	27.6	2.5	44.5		2.15	2.3	3.8	41678	9370
45		7	4.3	48.5		30		47.5					44325	9965
47		7.2	4.4	50.5		31.6		49.5	H12				46286	10406
50		7.5	4.6	54.2		34		53					59526	13382
52	2.5	7.7	4.7	56.2		35.6		55		2.65	2.8		61782	13890
55		8	5	59.2		38		58					65214	14660
60		8.5	5.4	64.2	+0.92	42		63					71098	15984
62		8.6	5.5	66.2	−0.46	43.8		65				4.5	73354	16490
65		8.7	5.8	69.2		46.6		68					76688	17240
68	3	8.8	6.1	72.5		49.4		71		3.15	3.3		80120	18017
70		9	6.2	74.5		51	3	73					82570	18564
72		9.2	6.4	76.5		52.6		75					84926	19070
75		9.3	6.6	79.5		55.5		78					89260	19942
80		9.5	7	85.5		60		83.5					109834	24692
85		9.7	7.2	90.5	+1.08	64.6		88.5					116698	26236
90	4	10	7.6	95.5	−0.54	69	3.5	93.5		4.15	4.3	5.3	123564	27780
95		10.3	8.1	100.5		73.4		98.5					130429	29322
100		10.5	8.4	105.5		78		103.5					137292	30966

TABLE 26-9
Dimensions for external circlips—type C

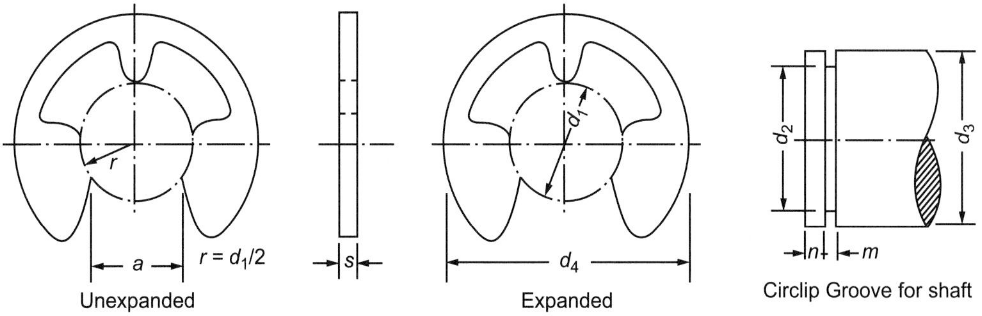

FIGURE 26-7

All dimensions in millimeters

	Circlip				Shaft Groove					
Nominal size d_1	**d_4 Expanded**	**a H10**	**s**	**Tol. on s**	**d_3 From**	**d_3 To**	**d_3 h11**	**m**	**Tol. on m**	**n Min**
0.8	2	0.58	0.2		1	1.4	0.8	0.24		0.4
1.2	3	1.01	0.3		1.4	2	1.2	0.34	±0.02	0.6
1.5	4	1.28	0.4		2	2.5	1.5	0.44		0.8
1.9	4.5	1.61	0.5		2.5	3	1.9	0.54		1
2.3	6	1.94	0.6	±0.02	3	4	2.3	0.64		1
3.2	7	2.70	0.6		4	5	3.2	0.64		1
4	9	3.34	0.7		5	7	4	0.74	±0.03	1.2
5	11	4.11	0.7		6	8	5	0.74		1.2
6	12	5.26	0.7		7	9	6	0.74		1.2
7	14	5.84	0.9		8	11	7	0.94		1.5
8	16	6.52	1.0		9	12	8	1.05		1.8
9	18.5	7.63	1.1		10	14	9	1.15		2
10	20	8.32	1.2		11	15	10	1.25		2
12	23	10.45	1.3	±0.03	13	18	12	1.35	±0.06	2.5
15	29	12.61	1.5		16	24	15	1.55		3.0
19	37	15.92	1.75		20	31	19	1.80		3.5
24	44	21.88	2.0		25	38	24	2.05		4.0

IS: 3075, 1965

REFERENCES

1. Lingaiah K., *Machine Design Data Handbook*, McGraw-Hill Publishing Company, New York, 1994.
2. Waldes Kohinoor, Inc., Long Island City, New York, U.S.A, 1985.
3. Edward Killian, *Retaining Rings*, Waldes Kohinoor, Inc., Long Island City, New York, U.S.A, 1985 and Robert O. Parmley, Editor-in-Chief, *Mechanical Components Handbook*, McGraw-Hill Publishing Company New York, 1985.
4. IS: 3075, 1965, Circlips.
5. *Industrial Fasteners Handbook*, 2nd Edition, Trade and Technical Press Limited, Morden Surrey, England, 1980.
6. "*General Purpose Uniform Cross-section Spiral Retaining Rings*", ANSI, B27.6, 1972 (R 1977)
7. "*General Purpose Tapered and Reduced Cross-section Retaining Rings (Metric)*", ANSI B 27.7, 1977.
8. "*General Purpose Metric Tapered and Reduced Cross-section Retaining Rings*" ANSI B27.8M, 1978.
9. Joseph E. Shigley, "*Unthreaded Fasteners*", and Shigley, J. E., and Mischke, C. R., *Standard Handbook of Machine Design*, McGraw-Hill Publishing Company, New York, 1996.

TABLE 26-10
Axially assembled tapered-section internal rings

FIGURE 26-8 Basic internal

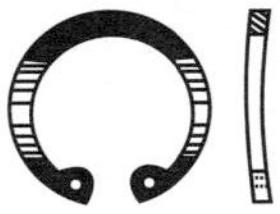

FIGURE 26-9 Bowed internal

FIGURE 26-10 Beveled and double-beveled internal

FIGURE 26-11 Inverted internal

	Housing diameter, mm (in)[b]		Nominal ring thickness, in mm (in)	Allowable static thrust load, when rings abut parts with sharp corners[a]								Maximum allowable corner radii or chamfers of retained parts, mm (in)				Allowable thrust load, when rings abut parts with listed corner radii or chamfers, F_r'	
				Groove material having minimum tensile yield strength of 1034 MPa (150,000 lbf/in^2)				Groove material having minimum tensile yield strength of 310 MPa (45,000 lbf/in^2)									
				From		Through		From		Through		Radii		Chamfers			
Ring type	From	Through		N	lbf	N	lbf	N	lbf	N	lbf	From	Through	From	Through	N	lbf
Basic	6.350 (0.250)	7.93 (0.312)	0.380 (0.015)	1,868	420	2,358	530	845	190	1,068	240	0.28 (0.011)	0.40 (0.016)	0.22 (0.0085)	0.33 (0.013)	845	190
internal,	9.525 (0.375)	11.50 (0.453)	0.635 (0.025)	4,670	1,050	5,694	1,280	1,556	350	2,046	460	0.58 (0.023)	0.68 (0.027)	0.45 (0.019)	0.53 (0.021)	2,358	530
Bowed	12.700 (0.500)	19.05 (0.750)	0.890 (0.035)	8,806	1,990	13,344	3,000	2,268	510	6,494	1,460	0.69 (0.027)	0.81 (0.032)	0.53 (0.021)	0.64 (0.025)	4,892	1,100
internal,	19.740 (0.777)	25.98 (1.023)	1.070 (0.042)	20,238	4,550	26,919	6,050	7,028	1,590	13,344	3,000	0.89 (0.035)	1.07 (0.042)	0.71 (0.028)	0.86 (0.034)	7,340	1,650
Bevelled	26.980 (1.062)	38.10 (1.500)	1.270 (0.050)	33,138	7,450	46,926	10,550	13,566	3,050	26,688	6,000	1.12 (0.044)	1.22 (0.048)	0.89 (0.035)	0.97 (0.038)	10,675	2,400
internal	39.680 (1.562)	50.80 (2.000)	1.580 (0.062)	60,938	13,700	77,840	17,500	28,245	6,350	45,814	10,300	1.63 (0.064)	1.63 (0.064)	1.27 (0.050)	1.27 (0.050)	17,346	3,900
in grooved	52.000 (2.047)	64.28 (2.531)	1.980 (0.078)	101,192	22,750	122,705	27,600	48,260	10,850	69,610	15,650	1.93 (0.076)	1.98 (0.078)	1.55 (0.061)	1.58 (0.062)	27,578	6,200
housings	65.080 (2.562)	76.20 (3.000)	2.360 (0.093)	149,898	33,700	175,696	39,500	73,392	16,500	102,970	23,150	2.23 (0.088)	2.33 (0.092)	1.78 (0.070)	1.88 (0.074)	40,032	9,000
(see Figs 26-8	77.780 (3.062)	127.00 (5.000)	2.760 (0.109)	209,500	47,100	342,496	77,000	107,196	24,100	244,640	55,000	2.46 (0.097)	4.01 (0.158)	1.98 (0.078)	3.20 (0.126)	53,376	12,000
and 26-9)	133.350 (5.250)	152.40 (6.000)	3.180 (0.125)	412,330	92,700	471,042	105,900	266,890	60,000	305,132	68,600	4.26 (0.168)	4.26 (0.168)	3.40 (0.134)	3.40 (0.134)	66,720	15,000
	158.750 (6.250)	177.90 (7.000)	3.960 (0.156)	612,440	137,700	686,326	154,300	229,596	74,100	413,364	93,100	4.50 (0.177)	4.98 (0.196)	3.61 (0.142)	3.61 (0.142)	102,304	23,000
	184.150 (7.250)	254.00 (10.000)	4.750 (0.187)	851,792	191,500	1175,162	264,200	443,020	99,600	848.334	190,700	5.13 (0.202)	6.36 (0.270)	4.11 (0.162)	4.11 (0.162)	151,232	34,000
Double	39.630 (1.562)	142.00 (1.688)	1.350 (0.053)	51,376	11,550	55,378	12,450	16,012	3,600	19,126	4,300	1.62 (0.064)	1.62 (0.064)	1.27 (0,050)	1.27 (0.050)	12,676	2,850
beveled	44.450 (1.750)	50.80 (2.000)	1.320 (0.052)	57,156	12,850	65,385	14,700	20,906	4,700	27,132	6,100	1.62 (0.064)	1.62 (0.064)	1.27 (0.050)	1.27 (0.050)	12,232	2,750
internal	52.390 (2.062)	64.28 (2.531)	1.720 (0.068)	98,292	19,950	106,306	23,900	29.912	6,500	42,700	9,600	1.98 (0.078)	1.98 (0.078)	1.58 (0.062)	1.58 (0.062)	20,905	4,700
(see Fig. 26-10)	65.080 (2.562)	71.42 (2.812)	12.080 (0.082)	132,105	29,700	45,370	10,200	45,370	10,200	54,265	12,200	2.23 (0.088)	2.23 (0.088)	1.78 (0.070)	1.78 (0.070)	31,136	7,000
Inverted	19.050 (0.750)	—	0.890 (0.035)	7,340	1,650	—	—	2,668	600	—	—	1.27 (0.050)	—	0.79 (0.031)	—	3,780	850
internal	20.620 (0.812)	25.40 (1.000)	1.070 (0.042)	11,564	2,600	14,679	3,300	3,114	700	5,115	1,150	1.37 (0.050)	1.63 (0.064)	0.86 (0.034)	1.02 (0.040)	5,560	1,250
in grooved	26.980 (1.062)	38.10 (1.500)	1.270 (0.050)	18,460	4,150	26,020	5,850	5,560	1,250	11,120	2,500	1.95 (0.069)	2.06 (0.081)	1.09 (0.043)	1.30 (0,051)	8,006	1,800
housings	39.680 (1.562)	50.90 (2.000)	1.580 (0.062)	33,805	7,600	43,368	9,750	11,786	2,650	19,126	4,300	2.24 (0.088)	3.00 (0.118)	1.40 (0.055)	1.88 (0.074)	12,900	2,900
(see Fig. 26-11)	52.380 (2.062)	63.50 (2.500)	1.980 (0.078)	56,266	12,650	68,054	15,300	20,016	4,500	28,912	6,500	3.18 (0.125)	3.66 (0,144)	2.00 (0.078)	2.28 (0.090)	20,460	4,600
	66.680 (2.625)	76.20 (3.000)	2.360 (0.093)	95,402	19,200	97,412	21,900	32,025	7,200	42.700	9,600	3.81 (0.150)	4.29 (0.169)	2.38 (0.094)	2.69 (0.106)	29,802	6,700
	80.160 (3.156)	101.60 (4.000)	2.760 (0.109)	120,096	27,000	152,122	34,200	47,148	10,600	75,170	16,900	4.42 (0.174)	4.42 (0.174)	2.77 (0.109)	2.77 (0.109)	40,032	9,000

Courtesy: © 1964, 1965, 1973, 1991 Waldes Kohinoor, Inc., Long Island City, New York, 1985.

[a] Where rings are of immediate size—or groove materials have intermediate tensile yield strengths—loads may be obtained by interpolation; [b] Numbers inside the brackets are in inches and numbers outside brackets are in millimeters; [c] Approximate corner radii and chamfers limits for parts with intermediate diameters can be determined by interpolation. Corner radii and chamfers smaller than those listed will increase the thrust load proportionately, approaching but not exceeding allowable static thrust loads of rings abutting parts with sharp corners.

Courtesy: Edward Killian, "*Retaining Rings*", Robert O. Parmley, Editor-in-Chief "*Mechanical Components Handbook*", McGraw-Hill Publishing Company, New York, USA, 1985.

TABLE 26-11
Axially assembled tapered-section external retaining rings

FIGURE 26-12 Basic external rings

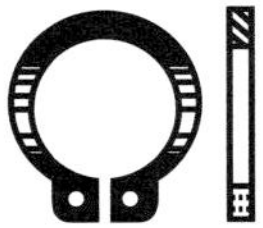

FIGURE 26-13 Bowed external rings

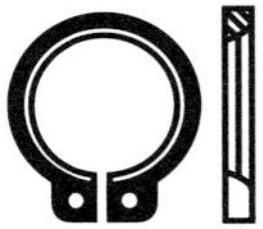

FIGURE 26-14 Beveled external rings

FIGURE 26-15 Inverted external rings

FIGURE 26-16 Heavy-duty external ring

FIGURE 26-17 Miniature high-strength ring

Ring type	Shaft diameter, mm (in)[b]		Nominal ring thickness, mm (in)[b]	Allowable static thrust load, when rings abut parts with sharp corners[a] — Groove material having minimum tensile yield strength of 1034 MPa (150,000 lbf/in^2)				Groove material having minimum tensile yield strength of 310 MPa (45,000 lbf/in^2)				Maximum allowable corner radii or chamfers of retained parts, mm (in)[b]				Allowable thrust load, when rings abut parts with listed corner radii or chamfers, F_r'	
				From		Through		From		Through		Radii		Chamfers			
	From	Through		N	lbf	N	lbf	N	lbf	N	lbf	From	Through	From	Through	N	lbf
Basic	3.175 (0.125)[c]	3.962 (0.156)[c]	0.254 (0.010)	489	110	578	130	156	35	245	55	0.254 (0.010)	0.381 (0.015)	0.152 (0.006)	0.228 (0.009)	200	45
external,	4.775 (0.188)[c]	6.994 (0.236)[c]	0.381 (0.015)	1,068	240	1,378	310	355	50	534	120	0.355 (0.014)	0.419 (0.0165)	0.216 (0.0085)	0.254 (0.010)	467	105
Bowed	6.350 (0.250)	11.912 (0.469)	0.635 (0.025)	2,624	590	4,892	1,100	779	175	2,002	450	0.457 (0.019)	0.788 (0.031)	0.279 (0.011)	0.457 (0.018)	2,090	470
external,	12.700 (0.500)	11.069 (0.672)	0.890 (0.035)	7,340	1,650	9,795	2,200	2,446	550	4,225	950	0.864 (0.034)	1.016 (0.040)	0.508 (0.020)	0.610 (0.024)	4,048	910
Beveled	17.475 (0.688)	25.981 (1.023)	1.050 (0.042)	5,123	3,400	22,462	5,050	4,448	1,000	10,008	2,250	1.066 (0.042)	1.473 (0.058)	0.635 (0.025)	0.980 (0.035)	5,960	1,340
external	26.980 (1.062)	38.100 (1.500)	1.270 (0.050)	27,578	6,200	39,142	8,800	10,675	2,400	22,240	5,000	1.524 (0.060)	2.006 (0.079)	0.914 (0.036)	1.194 (0.047)	8,674	1,950
in grooved	39.675 (1.562)	50.800 (2.000)	1.575 (0.062)	50,707	11,400	64,940	14,600	23,930	5,200	35,806	8,050	2.082 (0.082)	2.438 (0.096)	1.245 (0.049)	1.448 (0.057)	13,344	3,000
shafts	52.375 (2.062)	68.275 (2.688)	1.981 (0.078)	84,280	18,950	109,886	24,700	37,596	8,450	61,605	13,850	2.480 (0.098)	2.832 (0.1115)	1.498 (0.059)	1.702 (0.067)	22,240	5,000
(see Figs 26-12	69.850 (2.750)	87.325 (3.438)	2.362 (0.093)	133,885	30,100	167,690	37,700	64,050	14,400	97,410	21,900	2.850 (0.112)	3.276 (0,129)	1.702 (0.067)	1.956 (0.077)	32,692	7,350
to 26-14)	88.900 (3.500)	127.600 (5.000)	2.768 (0.109)	199,715	44,900	285,562	64,200	101,414	22,800	165,020	37,100	3.29 (0.1295)	4.196 (0.165)	1.981 (0.079)	2.515 (0.099)	46,704	10,500
	133.350 (5.250)	152.400 (6.000)	3.175 (0.125)	343,830	77,300	392,758	88,300	181,478	40,800	239,302	53,800	4.292 (0.169)	4.675 (0.184)	2.565 (0.101)	2.794 (0.110)	60,048	13,500
	158.750 (6.250)	177.800 (7.000)	3.962 (0.156)	510,630	114,800	572,012	129,600	259,319	58,300	323,370	72,700	4.750 (0.187)	5.283 (0.208)	2.945 (0.112)	3.175 (0.125)	93,408	21,000
	190.500 (7.500)	254.00 (10.000)	4.775 (0.188)	734,810	165,200	979,450	220,200	377,190	84,800	666,310	149,800	5.588 (0.220)	7.468 (0.294)	3.353 (0.132)	4.470 (0.176)	133,440	30,000
Inverted	12.700 (0.500)	17.068 (0.672)	0.890 (0.035)	4,993	1,100	6,450	1,450	1,245	280	2,090	470	1.295 (0.051)	1.651 (0.065)	0.813 (0.032)	1.041 (0.041)	3,025	680
external	17.475 (0.688)	25.400 (1.000)	1.050 (0.042)	10,230	2,300	14,678	3,300	2,224	500	4,670	1,050	1.696 (0.066)	2.311 (0.091)	1.067 (0.042)	1.448 (0.057)	4,448	1,000
on grooved	26.980 (1.062)	38.100 (1.500)	1.278 (0.050)	18,459	4,150	26,020	5,850	5,338	1,200	11,120	2,500	2.336 (0.092)	2.540 (0.100)	1.473 (0.058)	1.600 (0.063)	6,494	1,460
shafts	39.675 (1.562)	50.800 (2.000)	1.570 (0.062)	33,805	7,600	43,368	9,750	11,565	2,600	17,792	4,000	2.642 (0.104)	3.225 (0.127)	1.676 (0.066)	2.032 (0.090)	10,008	2,250
(see Fig. 26-15)	53.975 (2.125)	66.675 (2.625)	1.981 (0.078)	57,824	13,000	71,612	16,100	20,239	4,550	29,580	6,650	3.378 (0.133)	4.038 (0.159)	2.134 (0.094)	2.515 (0.099)	16,680	3,750
	69.850 (2.750)	85.000 (3.346)	2.362 (0.093)	89,405	20,100	108,976	24,500	32,025	7,200	46,704	10,500	4.191 (0.165)	4.928 (0.194)	2.616 (0.103)	3.073 (0.121)	24,464	5,500
	89.900 (3.500)	101.600 (4.000)	2.768 (0.109)	132,995	29,900	152,566	34,300	67,152	11,500	62,272	14,000	5.131 (0.202)	5.410 (0.213)	3.226 (0.127)	3.378 (0.133)	34,916	7,850

TABLE 26-11
Axially assembled tapered-section external retaining rings (*Contd.*)

	Shaft diameter, mm (in)[b]		Nominal ring thickness, mm (in)[b]	Allowable static thrust load, when rings abut parts with sharp corners[a]: Groove material having minimum tensile yield strength of 1034 MPa (150,000 lbf/in^2)				Groove material having minimum tensile yield strength of 310 MPa (45,000 lbf/in^2)				Maximum allowable corner radii or chamfers of retained parts, mm (in)[b]				Allowable thrust load, when rings abut parts with listed corner radii or chamfers, F_r'	
				From		Through		From		Through		Radii		Chamfers			
Ring type	From	Through		N	lbf	N	lbf	N	lbf	N	lbf	From	Through	From	Through	N	lbf
Heavy-duty	10.000 (0.394)	—	0.890 (0.035)	8,896	2,000	—	—	3,114	700	—	—	1.194 (0.047)	—	0.991 (0.039)	—	2,002	450
external	12.014 (0.473)	—	1.050 (0.042)	13,344	3,000	—	—	4,480	1,000	—	—	1.778 (0.070)	—	1.473 (0.058)	—	2,446	550
on grooved	12.700 (0.500)	17.000 (0.669)	1.278 (0.050)	17,347	3,900	23,130	5,200	4,892	1,100	8,451	1,900	1.778 (0.070)	1.956 (0.077)	1.473 (0.059)	1.625 (0.064)	2,890–4,003	650–900
(see Fig. 26-16)	19.050 (0.750)	25.400 (1.000)	1.981 (0.078)	40,032	9,000	51,152	11,500	10,760	2,400	17,792	4,000	2.200 (0.089)	2.548 (0.100)	1.880 (0.074)	2.108 (0.083)	11,120	2,500
	26.980 (1.062)	35.001 (1.378)	2.362 (0.093)	66,720	15,000	86,736	19,500	21,350	4,800	36,474	8,200	2.692 (0.106)	3.251 (0.128)	2.235 (0.088)	2.718 (0.107)	17,792	4,000
	38.100 (1.500)	45.000 (1.772)	2.709 (0.109)	108,976	24,500	129,992	29,000	44,490	10,000	55,155	12,400	3.250 (0.128)	3.251 (0.128)	2.718 (0.107)	2.718 (0.107)	22,240	5,000
	49.225 (1.938)	50.800 (2.000)	3.175 (0.125)	164,576	37,000	169,624	38,000	68,054	15,300	75,616	17,000	3.886 (0.153)	3.886 (0.153)	3.251 (0.128)	3.251 (0.128)	26,688	6,000
Miniature	2.565 (0.101)	3.400 (0.134)	0.509 (0.020)	1,112	250	1,468	330	266	60	400	90	0.330 (0.013)	0.355 (0.014)	0.254 (0.010)	0.279 (0.011)	890	200
high-strength	3.962 (0.156)	5.156 (0.203)	0.635 (0.025)	2,180	490	2,890	650	578	130	890	200	0.534 (0.021)	0.584 (0.023)	0.432 (0.017)	0.457 (0.018)	1,424	320
external on	5.562 (0.219)	8.331 (0.328)	0.889 (0.035)	5,338	1,200	8,006	1,800	978	220	2,046	460	0.711 (0.028)	0.965 (0.038)	0.558 (0.022)	0.762 (0.030)	2,668	600[c]
grooved shafts																	
(see Fig. 26-17)																	
Permanent shoulder on grooved shafts	Average sizes shaft diameter, in			CRS/SAE on soft steel shaft		Cabra 353 brass		Cabra 110 copper		Type 3003 aluminium		Not applicable					
	9.525 (0.375)		1.270 (0.050)	4,003	900	3,336	750	2,668	600	1,334	300						
	12.700 (0.500)		1.575 (0.062)	5,338	1,200	5,338	1,200	4,003	900	2,002	450						
	15.975 (0.625)		1.575 (0.062)	3,451	1,900	7,116	1,600	4,892	1,100	2,891	650						

Courtesy: © 1964, 1965, 1973, 1991 Waldes Kohinoor, Inc., Long Island City, New York, 1985.

[a] Where rings are of immediate size—or groove materials have intermediate tensile yield strengths—loads may be obtained by interpolation.

[b] Numbers inside the brackets are in inches and numbers outside brackets are in millimeters.

[c] Rings for shafts 3.175 mm (0.125 in) through 6.00 mm (0.236 in) diameter are made of beryllium copper only.

Approximate corner radii and chamfers limits for parts with intermediate diameters can be determined by interpolation. Corner radii and chamfers smaller than those listed will increase the thrust load proportionately, approaching but not exceeding allowable static thrust loads of rings abutting parts with sharp corners. Exceptions: for shafts 14.00 mm (0.551 in), 77.125 mm (3.06 in), 89.9 mm (3.500 in), 90.00 mm (3.543 in), 92.00 mm (3.625 in), 101.6 mm (4.000 in), 114.3 mm (4.500 in), 152.4 mm (6.000 in), and 158.75 mm (6.250 in) in diameter, refer to manufacturer's specifications for data. Note: $F_r' = 3,115$ N(700 lbf) for ring used with 6.6 mm (0.260 in) in-diameter shaft.

Courtesy: Edward Killian, "*Retaining Rings*", Robert O. Parmley, Editor-in-Chief "*Mechanical Components Handbook*", McGraw-Hill Publishing Company, New York, USA, 1985

TABLE 26-12
Radially assembled external retaining rings

FIGURE 26-18 Crescent shaped ring

FIGURE 26-19 Two-part interlocking ring

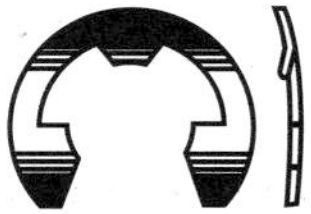

FIGURE 26.20 Bowed E-ring

FIGURE 26-21 E-ring

FIGURE 26-22 Reinforced E-ring

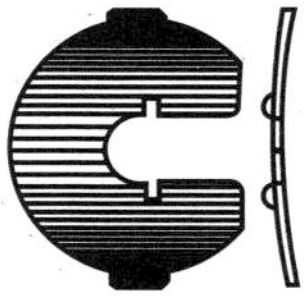

FIGURE 26-23 Locking-prong ring

Ring type	Shaft diameter, mm (in)[b]		Nominal ring thickness, mm (in)[b]	Allowable static thrust load, when rings abut parts with sharp corners[a]: Groove material having minimum tensile yield strength of σ_{sy}, 1034 MPa (150,000 lbf/in^2)				Groove material having minimum tensile yield strength of σ_{sy}, 310 MPa (45,000 lbf/in^2)				Maximum allowable corner radii or chamfers of retained parts, mm (in)[b]				Allowable thrust load, when rings abut parts with listed corner radii or chamfers, F_r'	
				From		Through		From		Through		Radii		Chamfers			
	From	Through		N	lbf	N	lbf	N	lbf	N	lbf	From	Through	From	Through	N	lbf
Crescent-shaped	3.175 (0.125)	4.775 (0.188)	0.381 (0.015)	378	85	578	130	200	45	311	70	0.355 (0.014)	0.533 (0.021)	0.280 (0.011)	0.406 (0.016)	Same values	
on grooved	5.562 (0.219)	11.125 (0.438)	0.635 (0.025)	1,156	260	2,312	520	445	100	1,556	350	0.583 (0.021)	0.736 (0.029)	0.406 (0.016)	0.558 (0.022)	as sharp	
shafts	12.700 (0.500)	15.875 (0.625)	0.889 (0.035)	3,692	830	4,582	1,030	2,002	450	3,114	700	0.762 (0.030)	0.838 (0.033)	0.584 (0.023)	0.635 (0.025)	corner	
(see Fig. 26-18)	17.475 (0.688)	25.400 (1.000)	1.066 (0.042)	7,562	1,700	11,031	2,480	3,558	800	8,006	1,800	0.864 (0.034)	1.168 (0.046)	0.660 (0.026)	0.899 (0.035)	abutment	
	28.575 (1.125)	38.100 (1.500)	1.270 (0.050)	14,768	3,320	19,660	4,420	9,786	2,200	17,792	4,000	1.325 (0.052)	1.752 (0.069)	1.016 (0.040)	1.346 (0.053)		
	44.450 (1.750)	50.800 (2.000)	1.575 (0.062)	23,600	6,430	32,470	7,300	23,574	5,300	31,136	7,000	2.056 (0.081)	2.311 (0.091)	1.575 (0.062)	1.778 (0.070)		
Two-part	11.912 (0.469)	15.815 (0.625)	0.890 (0.035)	8,896	2,000	11,797	2,650	2,759	620	3,692	830	1.321 (0.052)	1.321 (0.052)	1.016 (0.040)	1.016 (0.040)	2,713	610
interlocking	16.764 (0.669)	22.225 (0.875)	1.066 (0.042)	14,900	3,350	19,571	4,400	5,560	1,250	7,116	1,600	1.650 (0.065)	1.650 (0.065)	1.270 (0.050)	1.270 (0.050)	3,914	880
on grooved	24.994 (0.994)	38.100 (1.500)	1.270 (0.050)	26,020	5,850	39,910	9,950	12,999	2,900	19,794	4,450	2.184 (0.086)	2.194 (0.086)	1.676 (0.066)	1.676 (0.066)	5,560	1,250
shafts	39.624 (1.562)	47.625 (1.975)	1.575 (0.062)	52,264	11,750	62,716	14,100	25,131	5,650	30,246	6,800	2.540 (0.100)	2.540 (0.100)	1.956 (0.077)	1.956 (0.077)	8,451	1,900
(see Fig. 26-19)	50.012 (1.969)	66.675 (2.625)	1.981 (0.078)	81,176	19,250	108,086	24,300	40,032	9,000	53,376	12,000	2.890 (0.114)	2.895 (0.114)	2.235 (0.088)	2.235 (0.088)	13,566	3,050
	69.850 (2.750)	82.550 (3.250)	2.362 (0.093)	134,232	30,200	159,016	35,750	66,720	15,000	79,174	17,800	3.632 (0.143)	3.632 (0.143)	2.794 (0.110)	2.794 (0.110)	19,126	4,300
	85.725 (3.375)	—	2.768 (0.109)	193,488	43,500	—	—	91,628	20,600	—	—	4.622 (0.182)	4.622 (0.182)	3.556 (0.140)	—	26,466	5,950
Bowed E-ring	3.175 (0.125)	—	0.254 (0.010)	191	43	—	—	200	45	—	—	1.016 (0.040)	—	0.762 (0.030)	—	Same values	
on grooved	3.556 (0.140)	5.562 (0.219)	0.391 (0.015)	334	75	512	115	266	60	3,358	75	1.524 (0.060)	1.524 (0.060)	1.143 (0.045)	1.143 (0.045)	as sharp	
shafts	6.350 (0.250)	7.925 (0.312)	0.635 (0.025)	1,134	255	1,446	325	512	115	1,001	225	1.524 (0.060)	1.524 (0.060)	1.143 (0.045)	1.143 (0.045)	corner	
(see Fig. 26-20)	9.525 (0.375)	11.125 (0.438)	0.890 (0.035)	3,069	690	3,692	830	1,401	315	2,135	480	1.650 (0.065)	1.650 (0.065)	1.270 (0.050)	1.270 (0.050)	abutment	
	12.700 (0.500)	15.875 (0.625)	1.066 (0.042)	4,937	1,110	6,316	1,420	2,668	600	4,670	1,050	2.032 (0.080)	2.032 (0.080)	1.524 (0.060)	1.524 (0.060)		

TABLE 26-12
Radially assembled external retaining rings (*Contd.*)

Ring type	Shaft diameter, mm (in)[b] From	Shaft diameter, mm (in)[b] Through	Nominal ring thickness, mm (in)[b]	Allowable static thrust load, when rings abut parts with sharp corners[a] — Groove material having minimum tensile yield strength of σ_{sy}, 1034 MPa (150,000 lbf/in^2) — From N	From lbf	Through N	Through lbf	Groove material having minimum tensile yield strength of σ_{sy}, 310 MPa (45,000 lbf/in^2) — From N	From lbf	Through N	Through lbf	Maximum allowable corner radii or chamfers of retained parts, mm (in)[b] — Radii From	Radii Through	Chamfers From	Chamfers Through	Allowable thrust load, when rings abut parts with listed corner radii or chamfers, F_r' N	lbf
	19.050 (0.750)	25.400 (1.000)	1.270 (0.050)	8,896	2,000	11,787	2,650	6,672	1,500	8,451	1,900	2.160 (0.085)	1.956 (0.077)	1.651 (0.065)	1.448 (0.057)		
	30.175 (1.189)	34.925 (1.375)	1.575 (0.062)	15,346	3,450	18,236	4,100	6,672	1,500	10,452	2,350	2.286 (0.090)	2.296 (0.090)	1.778 (0.070)	1.778 (0.070)		
E-ring on	1.016 (0.040)	1.575 (0.062)	0.254 (0.010)	58	13	89	20	27	6	31	7	0.381 (0.015)	0.762 (0.030)	0.254 (0.010)	0.508 (0.020)	Same values	
grooved shafts	2.388 (0.094)	3.556 (0.140)	0.381 (0.015)	200	45	334	75	89	20	200	45	1.016 (0.040)	1.016 (0.040)	0.762 (0.030)	0.765 (0.030)	as sharp	
(see Fig. 26-21)	3.556 (0.140)	7.925 (0.312)	0.635 (0.025)	756	170	1,446	325	266	60	1,001	225	1.524 (0.060)	1.524 (0.060)	1.143 (0.045)	1.143 (0.045)	corner	
	9.525 (0.375)	11.125 (0.439)	0.990 (0.035)	3,069	690	3,692	830	1,401	315	2,135	480	1.650 (0.065)	1.650 (0.065)	1.270 (0.050)	1.270 (0.050)	abutment	
	12.700 (0.500)	15.975 (0.625)	1.066 (0.042)	4,937	1,110	6,316	1,420	2,669	600	4,670	1,050	2.032 (0.080)	2.032 (0.080)	1.524 (0.060)	1.524 (0.060)		
	19.050 (0.750)	25.400 (1.000)	1.270 (0.050)	8,896	2,000	11,787	2,650	6,672	1,500	8,451	1,900	2.160 (0.085)	1.956 (0.077)	1.651 (0.065)	1.448 (0.057)		
	30.175 (1.189)	34.925 (1.375)	1.575 (0.062)	15,346	3,450	18,236	4,100	6,672	1,500	10,452	2,350	2.286 (0.090)	2.296 (0.090)	1.778 (0.070)	1.778 (0.070)		
Reinforced	2.388 (0.094)	3.175 (0.125)	0.381 (0.015)	222	50	334	75	58	13	111	25	1.143 (0.045)	1.143 (0.045)	0.835 (0.033)	0.939 (0.033)	Same values	
E-ring	3.962 (0.156)	6.350 (0.250)	0.635 (0.025)	667	150	1,120	250	178	40	334	75	1.650 (0.065)	1.650 (0.065)	1.270 (0.050)	1.270 (0.050)	as sharp	
on grooved	7.925 (0.312)	11.125 (0.438)	0.890 (0.035)	1,368	420	2,669	600	600	135	1,268	285	1.778 (0.070)	1.778 (0.070)	1.396 (0.055)	1.396 (0.055)	corner	
shafts	12.700 (0.500)	14.275 (0.562)	1.066 (0.042)	3,647	820	4,136	930	2,046	460	4,805	480	2.032 (0.090)	2.032 (0.080)	1.524 (0.060)	1.524 (0.060)	abutment	
(see Fig. 26-22)																	
Locking-prong	2.336 (0.092)	3.962 (0.156)	0.254 (0.010)	356	80	534	120	156	35	445	100			Not applicable			
ring on	4.775 (0.188)	7.925 (0.312)	0.381 (0.015)	890	200	1,556	350	622	140	1,334	300						
grooved shafts	9.525 (0.375)	11.125 (0.438)	0.508 (0.020)	2,446	550	3,114	700	2,002	450	2,668	600						
(see Fig. 26-23)	4.775 (0.188)	6.350 (0.250)	0.890 (0.035)	2,668	600	4,003	900	579	130	890	200	1.270 (0.050)	1.270 (0.050)	1.066 (0.040)	1.066 (0.040)	1112	250
	7.925 (0.312)	9.525 (0.375)	1.066 (0.042)	5,782	1,300	6,894	1,550	1,120	250	1,334	300	1.650 (0.065)	1.650 (0.065)	1.270 (0.050)	1.270 (0.050)	1556	350
	11.125 (0.439)	15.875 (0.625)	1.270 (0.050)	9,786	2,200	13,344	3,000	1,780	400	18,236	1,100	2.032 (0.080)	2.032 (0.080)	1.524 (0.060)	1.524 (0.060)	2669	600
	19.050 (0.750)	—	1.575 (0.062)	20,460	4,600		—	7,116	1,600	—	—	2.160 (0.085)	—	1.651 (0.065)	—	4448	1,000
	25.400 (1.000)	—	1.981 (0.078)	33,360	7,500		—	11,565	2,600	—	—	2.286 (0.090)	—	1.778 (0.070)	—	8006	1,800
	4.775 (0.188)	7.925 (0.312)	0.635 (0.025)	1,912	430	3,470	790	579	130	1,120	250	1.270 (0.050)	1.270 (0.050)	1.016 (0.040)	1.016 (0.040)	667	150
	9.525 (0.375)	11.125 (0.438)	0.890 (0.035)	5,782	1,300	8,228	1,850	1,334	300	1,780	400	1.650 (0.065)	1.650 (0.065)	1.270 (0.050)	1.270 (0.050)	1334	300
	12.700 (0.500)	15.875 (0.625)	1.066 (0.042)	9,340	2,100	11,120	2,500	1,780	400	2,668	600	2.032 (0.080)	2.032 (0.080)	1.524 (0.060)	1.524 (0.060)	1780	400
	19.050 (0.750)	25.400 (1.000)	1.270 (0.050)	16,458	3,700	21,350	4,800	7,116	1,600	11,565	2,600	2.286 (0.090)	2.286 (0.090)	1.779 (0.070)	1.778 (0.070)	4448	1,000

Courtesy: © 1964, 1965, 1973, 1991 Waldes Kohinoor, Inc., Long Island City, New York, 1985.

[a] Where rings are of immediate size—or groove materials have intermediate tensile yield strengths—loads may be obtained by interpolation.

[b] Numbers inside the brackets are in inches and numbers outside brackets are in millimeters.

Approximate corner radii and chamfers limits for parts with intermediate diameters can be determined by interpolation. Corner radii and chamfers smaller than those listed will increase the thrust load proportionately, approaching but not exceeding allowable static thrust loads of rings abutting parts with sharp corners.

Exeptions: for shafts 14.00 mm (0.551 in), 77.125 mm (3.06 in), 89.9 mm (3.500 in), 90.00 mm (3.543 in), 92.00 mm (3.625 in), 101.6 mm (4.000 in), 114.3 mm (4.500 in), 152.4 mm (6.000 in), and 158.75 mm (6.250 in) in diameter, refer to manufacturer's specifications for data.

Rings for shafts 3.175 mm (0.125 in) through 6.00 mm (0.236 in) diameter are made of beryllium copper only.

Courtesy: Edward Killian, "*Retaining Rings*", Robert O. Parmley, Editor-in-Chief "*Mechanical Components Handbook*", McGraw-Hill Publishing Company, New York, USA, 1985

TABLE 26-13
Radially assembled external retaining rings

FIGURE 26-24 Reinforced external self-locking ring

FIGURE 26-25 External self-locking ring

FIGURE 26-26 Internal self-locking ring

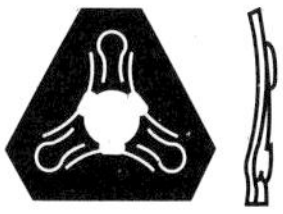

FIGURE 26-27 Triangular self-locking ring

	Housing diameter or shaft diameter				Nominal ring thickness		Allowable static thrust load, when rings abut parts with sharp corners[a]							
							Groove material having minimum tensile yield strength of 1034 MPa (150,000 lbf/in²)				Groove material having minimum tensile yield strength of 310 MPa (45,000 lbf/in²)			
	From		Through				From		Through		From		Through	
Ring type	mm	in	mm	in	mm	in	N	lbf	N	lbf	N	lbf	N	lbf
Reinforced self-locking external on shafts, no grooves	2.388	0.094	9.525	0.375	0.254	0.010	—	—	—	—	120	27	289	65
	2.388	0.094	9.525	0.375	0.380	0.015	—	—	—	—	200	45	534	120
	11.125	0.438	25.400	1.000	0.380	0.015	—	—	—	—	534	126	622	140
Self-locking external on shafts, no grooves	2.388	0.094	9.525	0.375	0.254	0.010[b]	—	—	—	—	58	13	200	45
	11.125	0.438	25.400	1.000	0.380	0.015	—	—	—	—	222	50	289	65
Self-locking internal in housing, no grooves	7.925	0.312	15.875	0.625	0.254	0.010	—	—	—	—	356	80	200	45
	19.050	0.750	50.300	2.000[c]	0.380	0.015	—	—	—	—	334	75	245	55
Triangular retainer on shafts, no grooves	1.575	0.062	—	—	0.254	0.010	—	—	—	—	111	25	—	—
	1.575	0.062	—	—	0.380	0.015	—	—	—	—	178	40	—	—
	4.388	0.094	3.962	0.156	0.254	0.010	—	—	—	—	266	60	334	75
	4.388	0.094	3.962	0.156	0.380	0.015	—	—	—	—	256	80	534	120
	4.775	0.188	7.925	0.312	0.380	0.015	—	—	—	—	622	140	890	200
	9.525	0.375	—	—	0.509	0.020	—	—	—	—	1112	250	—	—
	11.231	0.437	[d]	[d]	0.6351	0.025	—	—	—	—	1200	270	—	—
Triangular nut on threaded, parts	4.761	6/32	7.939	10/32	0.381	0.015	622	140	756	170	622	140	645	145
	4.761	6/32	7.938	10/32	0.508	0.020	890	200	978	220	800	180	845	190
	6.35–20	1/4–20	6.35–28	1/4–28	0.508	0.020	978	220	—	—	978	220	—	—
	6.35–20	1/4–20	6.35–28	1/4–28	0.635	0.025	978	220	—	—	978	220	—	—

TABLE 26-13
Radially assembled external retaining rings (*Contd.*)

FIGURE 26-28 Radial clamp ring

	Shaft diameter				Nominal ring thickness		Allowable static thrust load[e]							
							Shaft without groove				Shaft with groove, 310 MPa (45,000 lbf/in^2)			
	From		Through				From		Through		From		Through	
	mm	in	mm	in	mm	in	N	lbf	N	lbf	N	lbf	N	lbf
	Tapered-section self-locking clamp ring on shafts with or without grooves													
Inch type	2.388	0.094	3.962	0.156	0.635	0.025	45	10	98	22	—	—	—	—
	4.750	0.187	6.350	0.250	0.890	0.035	111	25	156	35	—	—	400	90
	7.925	0.312	9.525	0.375	1.066	0.042	200	45	266	60	489	110	800	180
	11.100	0.437	12.700	0.500	1.270	0.050	266	60	289	65	1290	290	1735	390
	15.875	0.625	19.050	0.750	1.575	0.062	378	95	400	90	2535	570	3780	850
Millimeter type	2.006	0.079	2.947	0.118	0.610	0.024	45	10	66	15	—	—	—	—
	5.004	0.197	—	—	0.913	0.032	133	30	—	—	178	40	—	—
	5.994	0.236	7.010	0.276	0.990	0.039	155	35	178	40	311	70	445	100
	8.992	0.354	10.005	0.394	1.194	0.047	222	50	245	55	579	130	756	170
	13.538	0.533	14.986	0.590	1.500	0.059	334	75	356	80	1512	340	1645	370
	Radially applied self-locking clamp rings on shafts without grooves													
Inch type	2.362	0.093	3.962	0.156	0.635	0.025	36	8	58	13	—	—	—	—
	4.750	0.187	6.350	0.250	0.889	0.035	80	18	98	22	—	—	—	—
	7.925	0.312	9.525	0.375	1.066	0.042	142	32	187	42	—	—	—	—
Millimeter type	1.981	0.078	3.962	0.156	0.660	0.024	30	7	53	12	—	—	—	—
	5.004	0.197	7.900	0.276	0.889	0.035	85	19	102	23	—	—	—	—
	7.925	0.312	9.962	0.393	1.092	0.043	147	33	218	49	—	—	—	—

Courtesy: © 1964, 1965, 1973, 1991 Waldes Kohinoor, Inc., Long Island City, New York, 1985.

[a] Where rings are of immediate size—or groove materials have intermediate tensile yield strengths—loads may be obtained by interpolation.

[b] Ring for shaft 6.096 mm (0.240 in) diameter is available only in 0.380 mm (0.015 in) thickness: allowable thrust load = 178 N (40 lbf).

[c] Ring for housing 34.925 mm (1.375 in) diameter is available only as reinforced ring having an allowable thrust load = 667 N (150 lbf).

[d] Round and hexagonal shafts.

[e] Grooved shafts are recommended only for rings used on shafts 0.197 in (5.0 mm) or larger.

Courtesy: Edward Killian, "*Retaining Rings*", Robert O. Parmley, Editor-in-Chief "*Mechanical Components Handbook*", McGraw-Hill Publishing Company, New York, USA, 1985

CHAPTER 27

APPLIED ELASTICITY

SYMBOLS

a	inner radius of cylinder, m (in)
	inner radius of rotating cylinder, m (in)
	inner radius of circular plate, m (in)
A	cross-sectional area, m^2 (in^2)
b	outer radius of inner cylinder, m (in)
	inside radius of outer cylinder, m (in)
	outer radius of rotating cylinder, m (in)
	outer radius of circular plate, m (in)
c	outside radius of outer cylinder, m (in)
C_1, C_2	constants of integration, m (in)
$D = \dfrac{Eh^3}{12(1-\nu^2)}$	flexural rigidity of a plate or shell, N/m (lbf/in)
E	modulus of elasticity, GPa
G	modulus of rigidity, GPa
g	acceleration due to gravity, 981 cm/s^2
h	thickness of plate, m (in)
I	moment of inertia, cm^4 (in^4)
J	polar moment of inertia, cm^4 (in^4)
L	length, m (in or ft)
l, m, n	direction cosines of the outward normal
M	moment (also with subscripts) N m (lbf ft)
M_b	bending moment, N m (lbf ft)
M_t	torsional moment, m N (ft lbf)
M_x, M_y	bending moments per unit length of sections of a plate perpendicular to x and y axes, respectively, N m (lbf ft)
M_{xy}	twisting moment per unit length of sections of a plate perpendicular to x-axis, N m (lbf ft)
M_n, M_{nt}	bending and twisting moments per unit length of sections of a plate perpendicular to n-direction, N m (lbf ft)
M_s, M_θ, $M_{r\theta}$	radial, tangential and twisting moments in polar co-ordinates
n	normal direction
n	a number, usually but not always, integer
N_x, N_y	normal force per unit length of sections of a plate perpendicular to x and y axis, respectively, N (lbf)
N_{xy}	shearing force in the direction of y-axes per unit length of section of a plate perpendicular to x axis, N/m (lbf/ft)

N_r, N_0	normal forces per unit length in radial and tangential directions in polar co-ordinates, N (lbf)
p	pressure, MPa (psi)
q	load per unit length, kN/m (lbf/in)
Q_x, Q_y	shearing forces parallel to z-axis per unit length of sections of a plate perpendicular to x and y axis, N/m (lbf/in)
N_r, N_θ	radial and tangential shearing forces, N (lbf)
r	radius, m (in)
r_x, r_y	radii of curvature of the middle surface of a plate in xz and yz planes
r, θ	polar co-ordinates
t	time, s
T	temperature, °C
	tension of a membrane, kN/m (lbf/in)
M_{txy}	twist of surface
u, v, w	components or displacements, m (in)
V	strains energy
W	weight, N (lbf)
w	displacement, m (in)
	displacement of a plate in the normal direction, m (in)
	deflection, m (in)
x, y, z	rectangular co-ordinates, m (in)
X, Y, Z	body forces in x, y, z directions, N (lbf)
Z	section modulus in bending, cm^3 (in^3)
ρ	density, kN/m^3 (lbf/in^3)
ω	angular speed, rad/s
σ	stress, MPa (psi)
σ_x, σ_y, σ_z	normal components of stress parallel to x, y, and z axis, MPa (psi)
σ_r, σ_θ	radial and tangential stress, MPa (psi)
σ_r, σ_θ, σ_z	normal stress components in cylindrical co-ordinates, MPa (psi)
τ	shearing stress, MPa (psi)
τ_{xy}, τ_{yz}, τ_{zx}	shearing stress components in rectangular co-ordinates, MPa (psi)
ε	unit elongation, m/m (in/in)
ε_x, ε_y, ε_z	unit elongation in x, y, and z direction, m/m (in/in)
ε_r, ε_θ	radial and tangential unit elongation in polar co-ordinates
γ	shearing strain
γ_{xy}, γ_{yz}, γ_{zx}	shearing strain components in rectangular co-ordinate
$\gamma_{r\theta}$, $\gamma_{\theta z}$	shearing strain in polar co-ordinate
$\tau_{r\theta}$, $\tau_{\theta z}$, τ_{rz}	shearing stress components in cylindrical co-ordinates, MPa (psi)
ν	Poisson's ratio
ϕ	stress function
	angular deflection, deg
	$e = \varepsilon_x + \varepsilon_y + \varepsilon_z = \varepsilon_r + \varepsilon_\theta + \varepsilon_z$
	$e = \varepsilon_x + \varepsilon_y + \varepsilon_z$ = volume expansion
	shearing components in cylindrical co-ordinates

Note: σ and τ with subscript s designates strength properties of material used in the design which will be used and observed throughout this *Machine Design Data Handbook*

Particular	Formula

STRESS AT A POINT (Fig. 27-1)

The stress at a point due to force ΔF acting normal to an area dA (Fig. 27-1b)

$$\text{Stress} = \sigma = \lim_{\Delta A \to 0} \frac{\Delta F}{\Delta A} \qquad (27\text{-}1)$$

where

ΔF = force acting normal to the area ΔA

ΔA = an infinitesimal area of the body under the action of F

For stresses acting on the part II of solid body cut out from main body in x, y and z directions, Fig. 27-1b

$$\sigma_x = \lim_{\Delta A_x \to 0} \frac{\Delta F_x}{\Delta A_x} \qquad (27\text{-}2a)$$

$$\tau_{xy} = \lim_{\Delta A_x \to 0} \frac{\Delta F_y}{\Delta A_x} \qquad (27\text{-}2b)$$

$$\tau_{xz} = \lim_{\Delta A_x \to 0} \frac{\Delta F_z}{\Delta A_x} \qquad (27\text{-}2c)$$

Similarly the stress components in xy and xz planes can be written and the nine stress components at the point O in case of solid body made of homogeneous and isotropic material

$$\begin{matrix} \sigma_x & \tau_{xy} & \tau_{xz} \\ \tau_{yz} & \sigma_y & \tau_{yz} \\ \tau_{zx} & \tau_{zy} & \sigma_z \end{matrix} \qquad (27\text{-}3)$$

Fig. 27-1c shows the stresses acting on the faces of a small cube element cut out from the solid body.

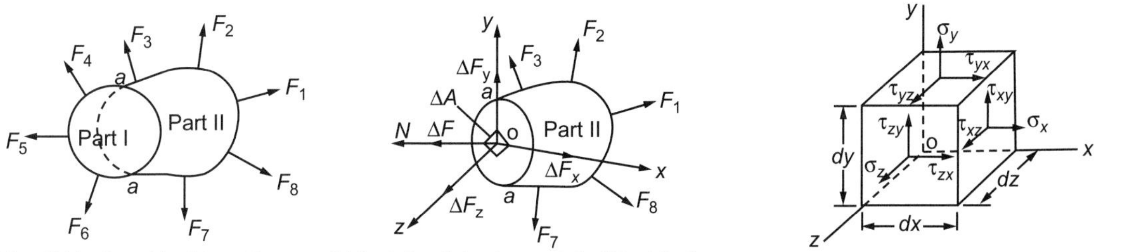

(a) A solid body subject to action of external forces

(b) An infineticimal area ΔA of Part II of a solid body under the action of force ΔF at 0

(c) Stresses acting on the faces of a small cube element cut out from the solid body

FIGURE 27-1

Particular	Formula
Summing moments about x, y and z axes, it can be proved that the cross shears are equal	$\tau_{xy} = \tau_{yx}, \quad \tau_{yz} = \tau_{zy}, \quad \tau_{zx} = \tau_{xz}$ (27-4)
All nine components of stresses can be expressed by a single equation	$\sigma_{ij} = \lim_{\Delta A_i \to 0} \dfrac{\Delta F_j}{\Delta A_i}$ (27-5) where $i = 1, 2, 3$ and $j = 1, 2, 3$
The F_{Nx}, F_{Ny}, and F_{Nz} unknown components of the resultant stress on the plane KLM of elemental tetrahedron passing through point O (Fig. 27-2)	$F_{Nx} = \sigma_x \cos N, x + \tau_{xy} \cos N, y + \tau_{xz} \cos N, z$ $F_{Ny} = \tau_{yx} \cos N, x + \sigma_y \cos N, y + \tau_{yz} \cos N, z$ $F_{Nz} = \tau_{zx} \cos N, x + \tau_{zy} \cos N, y + \sigma_z \cos N, z$ (27-6)
The unknown components of resultant stress F_{Nx}, F_{Ny} and F_{Nz} in terms of direction cosines l, m and n (Fig. 27-4)	$F_{Nx} = \sigma_x l + \tau_{xy} m + \tau_{zx} n$ $F_{Ny} = \tau_{yz} l + \sigma_y m + \tau_{yx} n$ $F_{Nz} = \tau_{zx} l + \tau_{zy} m + \sigma_z n$ (27-7) where the direct cosines are $l = \cos\alpha = \cos N, x; \; m = \cos\beta = \cos N, y,$ $n = \cos\gamma = \cos N, z,$ $l^s + m^2 + n^2 = (l)'^2 + (m')^2 + (n')^2 = 1$

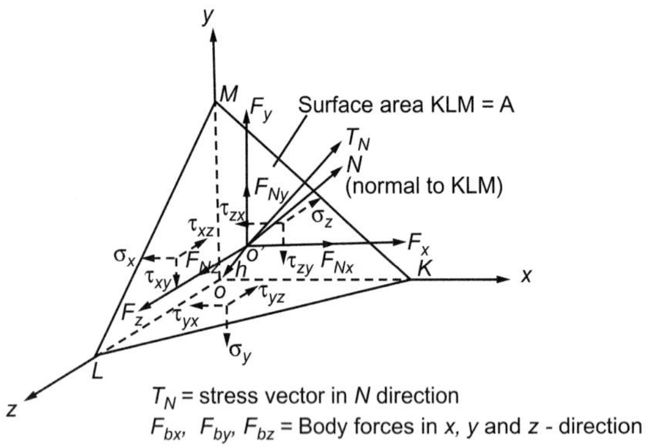

FIGURE 27-2 The state of stress at O of an elemental tetrahedron.

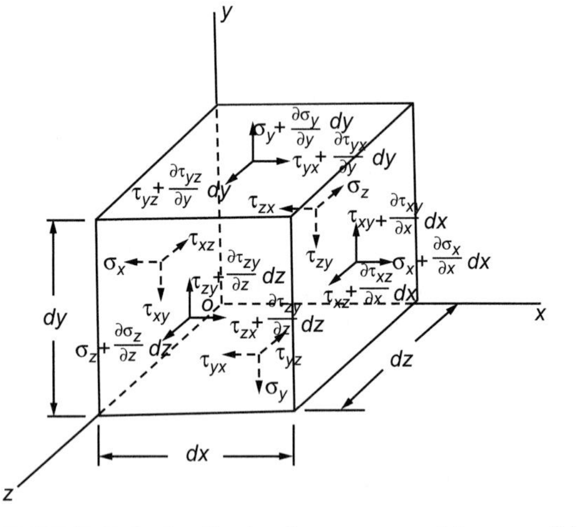

FIGURE 27-3 Small cube element removed from a solid body showing stresses acting on all faces of the body in x, y and z directions.

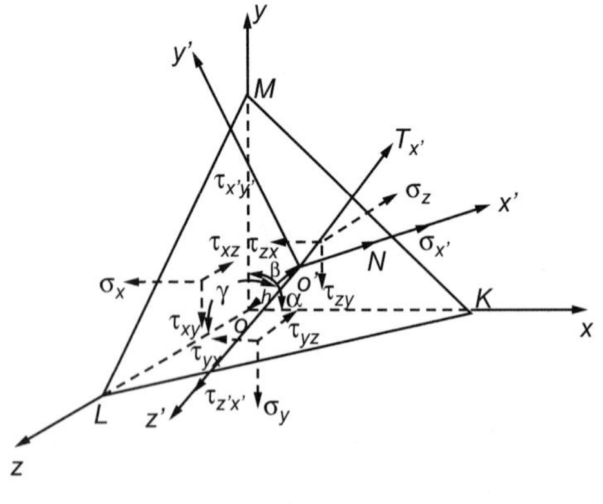

FIGURE 27-4 $T_{x'}$, resolved into $\sigma_{x'}$, $\tau_{x'y'}$ and $\tau_{x'z'}$ stress components.

Particular	Formula
	$\cos\alpha = l =$ angle between x axis and Normal N $\cos\beta = m =$ angle between y axis and Normal N $\cos\gamma = n =$ angle between z axis and Normal N
The resultant stress F_N on the plane KLM	$F_N = \sqrt{F_{Nx}^2 + F_{Ny}^2 + F_{Nz}^2}$ (27-8)
The normal stress which acts on the plane under consideration	$\sigma_N = F_{Nx}\cos\alpha + F_{Ny}\cos\beta + F_{Nz}\cos\gamma$ (27-8a)
The shear stress which acts on the plane under consideration	$\tau_N = \sqrt{F_N^2 - \sigma_n^2}$ (27-8b)
Equations (27-1), (27-2) and (27-7) to (27-8) can be expressed in terms of resultant stress vector as follows (Fig. 27-2)	
The resultant stress vector at a point	$T_N = \lim_{\Delta A \to 0} \dfrac{\Delta F_N}{\Delta A}$ (27-9a) where T_N coincides with the line of action of the resultant force ΔF_n
The resultant stress vector components in x, y and z directions	$T_{Nx} = \sigma_x l + \tau_{xy} m + \tau_{xz} n$ (27-9b) $T_{Ny} = \tau_{xy} l + \sigma_y m + \tau_{zy} n$ (27-9c) $T_{Nz} = \tau_{zx} l + \tau_{zy} m + \sigma_z n$ (27-9d)
The resultant stress vector	$T_N = \sqrt{T_{Nx}^2 + T_{Ny}^2 + T_{Nz}^2}$ (27-9e) where the direction cosines are $\cos(T_N, x) = T_{Nx}/\lvert T_N\rvert$, $\cos(T_N, y) = T_{Ny}/\lvert T_N\rvert$, $\cos(T_N, z) = T_{Nz}/\lvert T_N\rvert$
The normal stress which acts on the plane under consideration	$\sigma_N = \lvert T_N\rvert \cos(T_N, N)$ (27-9f) $\sigma_N = T_{Nx}\cos(N, x) + T_{Ny}\cos(N, y) + T_{Nz}\cos(N, z)$ (27-9g)
The shear stress which acts on the plane under consideration	$\tau_N = \lvert T_N\rvert \sin(T_N, N)$ (27-10a) $\tau_N = \sqrt{T_N^2 - \sigma_N^2}$ (27-10b)
The angle between the resultant stress vector T_N and the normal to the plane N	$\cos(T_N, N) = \cos(T_N, x)\cos(N, x) + \cos(T_N, y)\cos(N, y) + \cos(T_N, z)\cos(N, z)$ (27-10c)

Particular	Formula

EQUATIONS OF EQUILIBRIUM

The equations of equilibrium in Cartesian coordinates which includes body forces in three dimensions (Fig. 27-3)

$$\frac{\partial \sigma_x}{\partial x} + \frac{\partial \tau_{xy}}{\partial y} + \frac{\partial \tau_{xz}}{\partial z} + F_{bx} = 0 \tag{27-11a}$$

$$\frac{\partial \sigma_y}{\partial y} + \frac{\partial \tau_{yz}}{\partial z} + \frac{\partial \tau_{yx}}{\partial x} + F_{by} = 0 \tag{27-11b}$$

$$\frac{\partial \sigma_z}{\partial z} + \frac{\partial \tau_{zx}}{\partial x} + \frac{\partial \tau_{zy}}{\partial y} + F_{bz} = 0 \tag{27-11c}$$

where F_{bx}, F_{by} and F_{bz} are body forces in x, y and z directions

Stress equations of equilibrium in two dimensions

$$\frac{\partial \sigma_x}{\partial x} + \frac{\partial \tau_{xy}}{\partial y} + F_{bx} = 0 \tag{27-11d}$$

$$\frac{\partial \sigma_y}{\partial y} + \frac{\partial \tau_{yx}}{\partial x} + F_{by} = 0 \tag{27-11e}$$

TRANSFORMATION OF STRESS

The vector form of equations for resultant-stress vectors T_N and T'_N for two different planes and the outer normals N and N' in two different planes

$$T_N = \boldsymbol{i}T_{Nx} + \boldsymbol{j}T_{Ny} + \boldsymbol{k}T_{Nz} \tag{27-12a}$$

$$T_{N'} = \boldsymbol{i}T_{N'x} + \boldsymbol{j}T_{N'y} + \boldsymbol{k}T_{N'z} \tag{27-12b}$$

$$N = \boldsymbol{i}l + \boldsymbol{j}m + \boldsymbol{k}n \tag{27-12c}$$

$$N' = \boldsymbol{i}l' + \boldsymbol{j}m' + \boldsymbol{k}n' \tag{27-12d}$$

where $\boldsymbol{i}, \boldsymbol{j}$ and $\boldsymbol{k}$ are unit vectors in x, y and z directions, respectively

The projections of the resultant-stress vector T_N onto the outer normals N and N'

$$T_N \cdot N = T_{Nx}l + T_{Ny}m + T_{Nz}n \tag{27-13a}$$

$$T_N \cdot N' = T_{Nx}l' + T_{Ny}m' + T_{Nz}n' \tag{27-13b}$$

Substituting Eqs. (27-9b), (27-9c), (27-9d) and (27-9e) in Eqs. (27-13), equations for T_N, N and T_N, N'

$$T_N \cdot N = \sigma_x l^2 + \sigma_y m^2 + \sigma_z n^2 + 2\tau_{xy} lm + 2\tau_{yz} mn + 2\tau_{zx} nl \tag{27-14a}$$

$$T_N \cdot N' = \sigma_x ll' + \sigma_y mm' + \sigma_z nn' + \tau_{xy}[lm' + ml'] + \tau_{yz}[mn' + nm'] + \tau_{zx}[nl' + ln'] \tag{27-14b}$$

The relation between T_N, N' and T'_N, N

$$T'_N \cdot N = T_N \cdot N' \tag{27-15}$$

By coinciding outer normal N with x', N with y', and N with z' individually respectively and using Eqs. (27-14a) to (27-14b), $\sigma_{x'}$, $\sigma_{y'}$ and $\sigma_{z'}$ can be obtained (Fig. 27-4)

$$\sigma_{x'} = T_{x'} \cdot \boldsymbol{x}' = \sigma_x \cos^2(x', x) + \sigma_y \cos^2(x', y) + \sigma_z \cos^2(x', z) + 2\tau_{xy} \cos(x', x)\cos(x', y) + 2\tau_{yz} \cos(x', y)\cos(x', z) + 2\tau_{zx} \cos(x', z)\cos(x', x) \tag{27-15a}$$

Particular	Formula
	$$\sigma_{y'} = T_{y'} \cdot \mathbf{y}' = \sigma_y \cos^2(y', y) + \sigma_z \cos^2(y', z) + \sigma_x \cos^2(y', x) + 2\tau_{yz} \cos(y', y)\cos(y', z) + 2\tau_{zx} \cos(y', z)\cos(z', x) + 2\tau_{xy} \cos(y', x)\cos(y', y) \quad \text{(27-15b)}$$
	$$\sigma_{z'} = T_{z'} \cdot \mathbf{z}' = \sigma_z \cos^2(z', z) + \sigma_x \cos^2(z', x) + \sigma_y \cos^2(z', y) + 2\tau_{zx} \cos(z', z)\cos(z', x) + 2\tau_{xy} \cos(z', x)\cos(z', y) + 2\tau_{yz} \cos(z', y)\cos(z', z) \quad \text{(27-15c)}$$
By selecting a plane having an outer normal N coincident with the x' and a second plane having an outer normal N' coincident with the y' and utilizing Eq. (27-14b) which was developed for determining the magnitude of the projection of a resultant stress vector on to an arbitrary normal can be used to determine $\tau_{x'y'}$. Following this procedure and by selecting N and N' coincident with the y' and z', and z' and x' axes, the expression for $\tau_{y'z'}$ and $\tau_{z'x'}$ can be obtained. The expressions for $\tau_{x'y'}$, $\tau_{y'z'}$ and $\tau_{z'x'}$ are	$$\tau_{x'y'} = T_{x'} \cdot \mathbf{y}' = \sigma_x \cos(x', x)\cos(y', x) + \sigma_y \cos(x', y)\cos(y', y) + \sigma_z \cos(x', z)\cos(y', z) + \tau_{xy}[\cos(x', x)\cos(y', y) + \cos(x', y)\cos(y', x)] + \tau_{yz}[\cos(x', y)\cos(y', z) + \cos(x', z)\cos(y', y)] + \tau_{zx}[\cos(x', z)\cos(y', x) + \cos(x', x)\cos(y', z)] \quad \text{(27-16a)}$$
	$$\tau_{y'z'} = T_{y'} \cdot \mathbf{z}' = \sigma_y \cos(y', y)\cos(z', y) + \sigma_z \cos(y', z)\cos(z', z) + \sigma_x \cos(y', x)\cos(z', x) + \tau_{yz}[\cos(y', y)\cos(z', z) + \cos(y', z)\cos(z', y)] + \tau_{zx}[\cos(y', z)\cos(z', x) + \cos(y', x)\cos(z', z)] + \tau_{xy}[\cos(y', x)\cos(z', y) + \cos(y', y)\cos(z', x)] \quad \text{(27-16b)}$$
	$$\tau_{z'x'} = T_{z'} \cdot \mathbf{x}' = \sigma_z \cos(z', z)\cos(x', z) + \sigma_x \cos(z', x)\cos(x', x) + \sigma_y \cos(z', y)\cos(x', y) + \tau_{zx}[\cos(z', z)\cos(x', x) + \cos(z', x)\cos(x', z)] + \tau_{xy}[\cos(z', x)\cos(x', y) + \cos(z', y)\cos(x', x)] + \tau_{yz}[\cos(z', y)\cos(x', z) + \cos(z', z)\cos(x', y)] \quad \text{(27-16c)}$$
	Equations (27-15a) to (27-15c) and Eqs. (27-16a) to (27-16c) can be used to determine the six Cartesian components of stress relative to the $Oxyz$ coordinate system to be transformed into a different set of six Cartesian components of stress relative to an $Ox'y'z'$ coordinate system

Particular	Formula

For two-dimensional stress fields, the Eqs. (27-15a) to (27-15c) and (27-16a) to (27-16c) reduce to, since $\sigma_z = \tau_{zx} = \tau_{yz} = 0$ z' coincide with z and θ is the angle between x and x', Eqs. (27-15a) to (27-15c) and Eqs. (27-16a) to (27-16c)

$$\sigma_{x'} = \sigma_x \cos^2\theta + \sigma_y \sin^2\theta + 2\tau_{xy} \sin\theta\cos\theta$$

$$= \frac{\sigma_x + \sigma_y}{2} + \frac{\sigma_x - \sigma_y}{2} \cos 2\theta + \tau_{xy} \sin 2\theta \quad \text{(27-17a)}$$

$$\sigma_{y'} = \sigma_y \cos^2\theta + \sigma_x \sin^2\theta - 2\tau_{xy} \sin\theta\cos\theta$$

$$= \frac{\sigma_y + \sigma_x}{2} + \frac{\sigma_y - \sigma_x}{2} \cos 2\theta - \tau_{xy} \sin 2\theta \quad \text{(27-17b)}$$

$$\tau_{x'y'} = \sigma_y \cos\theta \sin\theta - \sigma_x \cos\theta \sin\theta + \tau_{xy}(\cos^2\theta - \sin^2\theta)$$

$$= \frac{\sigma_y - \sigma_x}{2} \sin 2\theta + \tau_{xy} \cos 2\theta \quad \text{(27-17c)}$$

$$\sigma_{z'} = \tau_{z'x'} = \tau_{y'z'} = 0 \quad \text{(27-17d)}$$

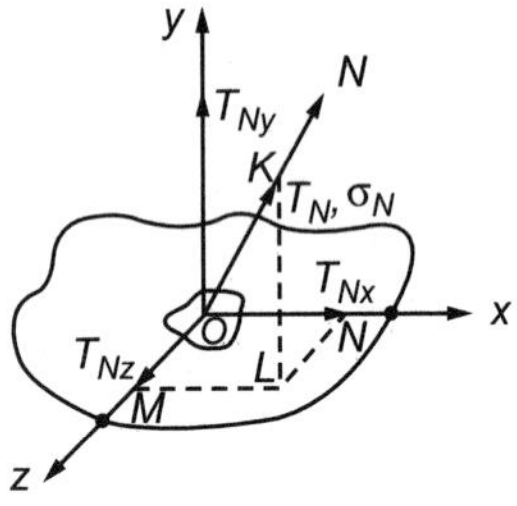

FIGURE 27-5 The stress vector T_N.

PRINCIPAL STRESSES

By referring to Fig. 27-5, where T_N coincides with outer normal N, it can be shown that the resultant stress components of T_N in x, y and z directions

$$T_{Nx} = \sigma_N l \qquad T_{Ny} = \sigma_N m \qquad T_{Nz} = \sigma_N n \quad \text{(27-18)}$$

Substituting Eqs. (27-9b) to (27-9d) into (27-18), the following equations are obtained

$$\sigma_x l + \tau_{yx} m + \tau_{zx} n = \sigma_N l$$
$$\tau_{xy} l + \sigma_y m + \tau_{xy} n = \sigma_N m \quad \text{(27-19)}$$
$$\tau_{xz} l + \tau_{yz} m + \sigma_z n = \sigma_N n$$

Eq. (27-19) can be written as

$$(\sigma_x - \sigma_N) l + \tau_{yx} m + \tau_{zx} = 0$$
$$\tau_{xy} l + (\sigma_y - \sigma_N) m + \tau_{zy} = 0 \quad \text{(27-20)}$$
$$\tau_{xz} l + \tau_{yz} m + (\sigma_z - \sigma_N) n = 0$$

From Eq. (27-20), direction cosine (N, x) is obtained and putting this in determinant form

$$\cos(N, x) = \frac{\begin{vmatrix} 0 & \tau_{yx} & \tau_{zx} \\ 0 & \sigma_y - \sigma_n & \tau_{zy} \\ 0 & \tau_{yz} & \sigma_z - \sigma_N \end{vmatrix}}{\begin{vmatrix} \sigma_x - \sigma_N & \tau_{yx} & \tau_{zx} \\ \tau_{xy} & \sigma_y - \sigma_N & \tau_{zy} \\ \tau_{xz} & \tau_{yz} & \sigma_z - \sigma_N \end{vmatrix}} \quad \text{(27-21)}$$

Putting the determinator of determinant into zero, the non-trivial solution for direction cosines of the principal plane is

$$\begin{vmatrix} \sigma_x - \sigma_N & \tau_{yx} & \tau_{zx} \\ \tau_{xy} & \sigma_y - \sigma_N & \tau_{zy} \\ \tau_{xz} & \tau_{yz} & \sigma_z - \sigma_N \end{vmatrix} = 0 \quad \text{(27-22)}$$

Particular	Formula
Expanding the determinant after making use of Eqs. (27-4) which gives three roots. They are principal stresses	$\sigma_N^3 - (\sigma_x + \sigma_y + \sigma_z)\sigma_N^2 + (\sigma_x\sigma_y + \sigma_y\sigma_z + \sigma_z\sigma_x - \tau_{xy}^2 - \tau_{yz}^2 - \tau_{zx}^2)\sigma_N - (\sigma_x\sigma_y\sigma_z - \sigma_x\tau_{yz}^2 - \sigma_y\tau_{zx}^2 - \sigma_z\tau_{xy}^2 + 2\tau_{xy}\tau_{yz}\tau_{zx}) = 0$ (27-23)
For two-dimensional stress system the coordinating system coinciding with the principal directions, Eq. (27-23) becomes	$\sigma_i^3 - (\sigma_x + \sigma_y)\sigma_i^2 + (\sigma_x\sigma_y - \tau_{xy}^2)\sigma_i = 0$ (27-23a) where $i = 1, 2, 3$
The three principal stresses from Eq. (27-23a) are	$\sigma_{1,2} = \frac{1}{2}(\sigma_x + \sigma_y) \pm \sqrt{\left(\frac{\sigma_x - \sigma_y}{2}\right)^2 + \tau_{xy}^2}$ (27-23b) $\sigma_3 = 0$
The directions of the principal stresses can be found from	$\sin 2(N_1, x) = \frac{2\tau_{xy}}{\sqrt{(\sigma_x - \sigma_y)^2 + 4\tau_{xy}^2}}$ (27-23c) $\cos 2(N_1, x) = \frac{\sigma_x - \sigma_y}{\sqrt{(\sigma_x - \sigma_y)^2 + 4\tau_{xy}^2}}$ (27-23d) $\tan 2(N_1, x) = \frac{2\tau_{xy}}{\sigma_x - \sigma_y}$ (27-23e)
From Eq. (27-15)	$T_{N_1} \cdot N_2 = T_{N_2} \cdot N_1$ (27-24)
From definition of principal stress	$T_{N_1} = \sigma_1 N_1$ (27-25a) $T_{N_2} = \sigma_2 N_2$ (27-25b)
Substituting the values of T_{N1} and T_{N2} in Eq. (27-15) and simplifying	$(\sigma_1 - \sigma_2)N_1 \cdot N_2 = 0$ (27-26) where σ_1 and σ_2 are distinct
From Eq. (27-20)	$N_1 \cdot N_2 = 0$ (27-27) which proves that N_1 and N_2 are orthogonal.
The three invariant of stresses from Eq. (27-23)	$I_1 = \sigma_x + \sigma_y + \sigma_z = \sigma_{x'} + \sigma_{y'} + \sigma_{z'}$ (27-28a) $I_2 = \sigma_x\sigma_y + \sigma_y\sigma_z + \sigma_z\sigma_x - \tau_{xy}^2 - \tau_{yz}^2 - \tau_{zx}^2 = \sigma_{x'}\sigma_{y'} + \sigma_{y'}\sigma_{z'} + \sigma_{z'}\sigma_{x'} - \tau_{x'y'}^2 - \tau_{y'z'}^2 - \tau_{z'x'}^2$ (27-28b) $I_3 = \sigma_x\sigma_y\sigma_z - \sigma_x\tau_{yz}^2 - \sigma_y\tau_{zx}^2 - \sigma_z\tau_{xy}^2 + 2\tau_{xy}\tau_{yz}\tau_{zx} = \sigma_{x'}\sigma_{y'}\sigma_{z'} - \sigma_{x'}\tau_{y'z'}^2 - \sigma_{y'}\tau_{z'x'}^2 - \sigma_{z'}\tau_{x'y'}^2 + 2\tau_{x'y'}\tau_{y'z'}\tau_{z'x'}$ (27-28c)

Particular	Formula

where I_1 = first invariant, I_2 = second invariant and I_3 = third invariant of stress

For the coordinating system coinciding with the principal direction, the expression for invariants from Eq. (27-28)

$$I_1 = \sigma_1 + \sigma_2 + \sigma_3 \quad (27\text{-}29a)$$

$$I_2 = \sigma_1\sigma_2 + \sigma_2\sigma_3 + \sigma_3\sigma_1 \quad (27\text{-}29b)$$

$$I_3 = \sigma_1\sigma_2\sigma_3 \quad (27\text{-}29c)$$

STRAIN (Fig. 27-6)

The normal strain or longitudinal strain by Hooke's law (Fig. 27-6) in x-direction

$$\varepsilon_x = \frac{\sigma_x}{E} \quad (27\text{-}30a)$$

The lateral strains in y and z-direction

$$\varepsilon_y = -\frac{v}{E}\sigma_x = -v\varepsilon_x \quad (27\text{-}30b)$$

$$\varepsilon_z = -\frac{v}{E}\sigma_x = -v\varepsilon_x \quad (27\text{-}30c)$$

The normal strains caused by σ_y and σ_z

$$\varepsilon_y = \frac{\sigma_y}{E}; \quad \varepsilon_x = \varepsilon_z = -\frac{v\sigma_y}{E} = -v\varepsilon_y \quad (27\text{-}31)$$

$$\varepsilon_z = \frac{\sigma_z}{E}; \quad \varepsilon_x = \varepsilon_y = -\frac{v\sigma_z}{E} - v\varepsilon_z \quad (27\text{-}32)$$

THREE-DIMENSIONAL STRESS-STRAIN SYSTEM

The general stress-strain relationships for a linear, homogeneous and isotropic material when an element subject to σ_x, σ_y and σ_z stresses simultaneously

$$\varepsilon_x = \frac{1}{E}[\sigma_x - v(\sigma_y + \sigma_z)] \quad (27\text{-}33a)$$

$$\varepsilon_y = \frac{1}{E}[\sigma_y - v(\sigma_z + \sigma_x)] \quad (27\text{-}33b)$$

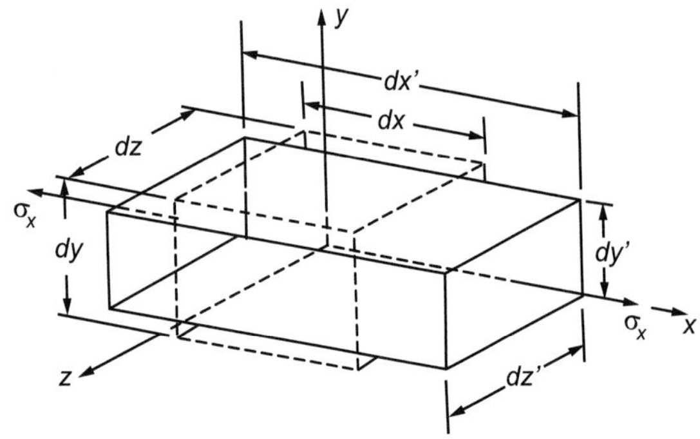

FIGURE 27-6 Uniaxial elongation of an element in the direction of x.

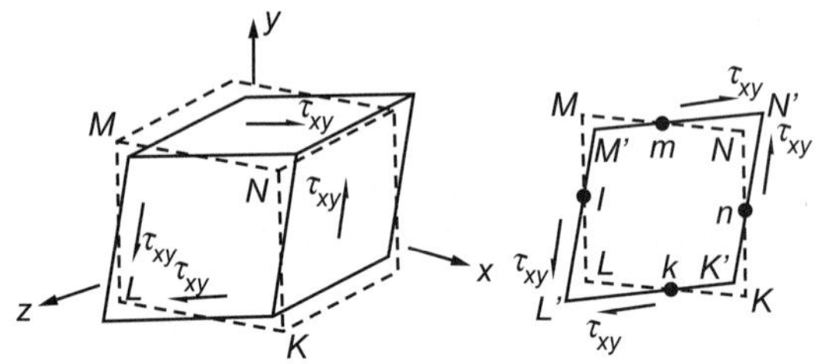

FIGURE 27-7 A cubic element subject to shear stress, τ_{xy}.

Particular	Formula
	$\varepsilon_z = \frac{1}{E}[\sigma_z - v(\sigma_x + \sigma_y)]$ (27-33c)
The expressions for σ_x, σ_y and σ_z stresses in case of three-dimensional stress system from Eqs (27-33)	$\sigma_x = \frac{E}{(1+v)(1-2v)}[(1-v)\varepsilon_x + v(\varepsilon_y + \varepsilon_z)]$ (27-34a)
	$\sigma_y = \frac{E}{(1+v)(1-2v)}[(1-v)\varepsilon_y + v(\varepsilon_z + \varepsilon_x)]$ (27-34b)
	$\sigma_z = \frac{E}{(1+v)(1-2v)}[(1-v)\varepsilon_z + v(\varepsilon_x + \varepsilon_y)]$ (27-34c)

BIAXIAL STRESS-STRAIN SYSTEM

Particular	Formula
The normal strain equations, when $\sigma_z = 0$ from Eq. (27-33)	$\varepsilon_x = \frac{1}{E}[\sigma_x - v\sigma_y]$ (27-35a)
	$\varepsilon_y = \frac{1}{E}[\sigma_y - v\sigma_x]$ (27-35b)
	$\varepsilon_z = -\frac{v}{E}[\sigma_x + \sigma_y]$ (27-35c)
The normal stress equation, when $\sigma_z = 0$ from Eq. (27.34)	$\sigma_x = \frac{E}{1-v}[\varepsilon_x + v\varepsilon_y] = \lambda J_1 + 2\mu\varepsilon_x = \frac{2\lambda\mu}{\lambda + 2\mu}(\varepsilon_x + \varepsilon_y) + 2\mu\varepsilon_x$ (27-36a)
	$\sigma_y = \frac{E}{1-v}[\varepsilon_y + v\varepsilon_x] = \lambda J_1 + 2\mu\varepsilon_y = \frac{2\lambda\mu}{\lambda + 2\mu}(\varepsilon_y + \varepsilon_x) + 2\mu\varepsilon_y$ (27-36b)
	$\sigma_z = \lambda J_1 + 2\mu\varepsilon_z = 0$ (27-36c)
	$\tau_{xy} = \mu\gamma_{xy}; \quad \tau_{yz} = \mu\gamma_{yz} = 0; \quad \tau_{zx} = \mu\gamma_{zx} = 0$ (27-36d)

SHEAR STRAINS

Particular	Formula
For a homogeneous, isotropic material subject to shear force, the shear strain which is related to shear stress as in case of normal strain	$\gamma_{xy} = \frac{\tau_{xy}}{G}$ (27-37a)
	$\gamma_{yz} = \frac{\tau_{yz}}{G}$ (27-37b)
	$\gamma_{zx} = \frac{\tau_{zx}}{G}$ (27-37c)

Particular	Formula

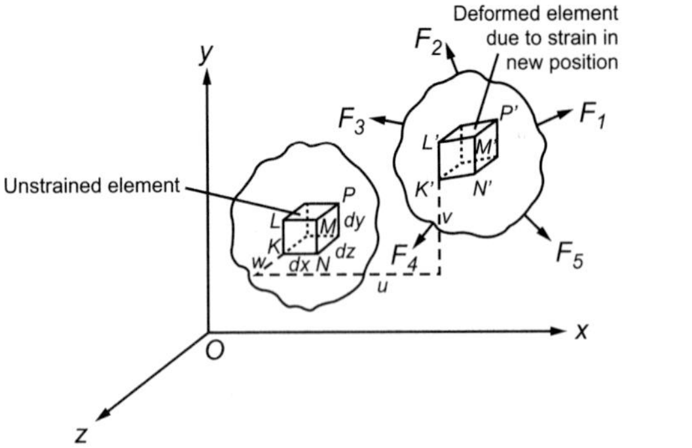

FIGURE 27-8 Deformation of a cube element in a solid body subject to loads.

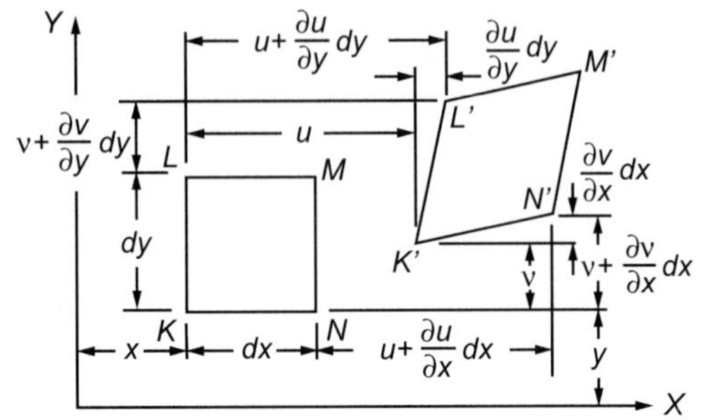

FIGURE 27-9 Two-dimensional deformation under load.

It has been proved that the shear modulus (G) is related to Young's modulus (E) and Poisson's ratio ν as

$$G = \frac{E}{2(1+v)} \tag{27-38}$$

From Eqs. (27-37), shear strain in terms of E and ν

$$\gamma_{xy} = \frac{2(1+v)}{E} \tau_{xy} \tag{27-39a}$$

$$\gamma_{yz} = \frac{2(1+v)}{E} \tau_{yz} \tag{27-39b}$$

$$\gamma_{zx} = \frac{2(1+v)}{E} \tau_{zx} \tag{27-39c}$$

STRAIN AND DISPLACEMENT (Figs. 27-8 and 27-9)

The normal strain in x-direction

$$\varepsilon_x = \frac{\text{change in length}}{\text{original length}} = \frac{dx + \frac{\partial u}{\partial x} dx - dx}{dx} = \frac{\partial u}{\partial x} \tag{27-40a}$$

The normal strain in y and z-directions

$$\varepsilon_y = \frac{\partial v}{\partial y} \tag{27-40b}$$

$$\varepsilon_z = \frac{\partial w}{\partial z} \tag{27-40c}$$

The shear strains xy, yz and zx planes

$$\gamma_{xy} = \frac{\partial v}{\partial x} + \frac{\partial u}{\partial y} \tag{27-41a}$$

$$\gamma_{yz} = \frac{\partial w}{\partial y} + \frac{\partial v}{\partial z} \tag{27-41b}$$

$$\gamma_{zx} = \frac{\partial u}{\partial z} + \frac{\partial w}{\partial x} \tag{27-41c}$$

Particular	Formula
The amount of counterclockwise rotation of a line segment located at R in xy, yz and zx planes	$\Theta_{xy} = \frac{1}{2}\left(\frac{\partial v}{\partial x} - \frac{\partial u}{\partial y}\right)$ (27-41d)
	$\Theta_{yz} = \frac{1}{2}\left(\frac{\partial w}{\partial y} - \frac{\partial v}{\partial z}\right)$ (27-41e)
	$\Theta_{zx} = \frac{1}{2}\left(\frac{\partial u}{\partial z} - \frac{\partial w}{\partial x}\right)$ (27-41f)
The strain ε_z and first strain invariant J_1 in case of plane stress	$\varepsilon_z = -\frac{\lambda}{\lambda + 2\mu}(\varepsilon_x + \varepsilon_y)$ (27-41g)
	$J_1 = \frac{2\mu}{\lambda + 2\mu}(\varepsilon_x + \varepsilon_y)$ (27-41h)
The strains components $\varepsilon_{x'}$, $\varepsilon_{y'}$ and $\varepsilon_{z'}$, along x', y' and z' axes line segments with reference to the $O'x'y'z'$ system	$\varepsilon_{x'} = \varepsilon_x \cos^2(x,x') + \varepsilon_y \cos^2(y,x') + \varepsilon_z \cos^2(z,x') + \gamma_{xy} \cos(x,x')\cos(y,x') + \gamma_{yz} \cos(y,x')\cos(z,x') + \gamma_{zx} \cos(z,x')\cos(x,x')$ (27-42a)
	$\varepsilon_{y'} = \varepsilon_y \cos^2(y,y') + \varepsilon_z \cos^2(z,y') + \varepsilon_x \cos^2(x,y') + \gamma_{yx} \cos(y,y')\cos(z,y') + \gamma_{zx} \cos(z,y')\cos(x,y') + \gamma_{xy} \cos(x,y')\cos(y,y')$ (27-42b)
	$\varepsilon_{z'} = \varepsilon_z \cos^2(z,z') + \varepsilon_x \cos^2(x,z') + \varepsilon_y \cos^2(y,z') + \gamma_{zx} \cos(z,z')\cos(x,z') + \gamma_{xy} \cos(x,z')\cos(y,z') + \gamma_{yz} \cos(y,z')\cos(z,z')$ (27-42c)
The shearing strain components (due to angular changes) $\gamma_{x'y'}$, $\gamma_{y'z'}$ and $\gamma_{z'x'}$ with reference to the $O'x'y'z'$ system	$\gamma_{x'y'} = 2\varepsilon_x \cos(x,x')\cos(x,y') + 2\varepsilon_y \cos(y,x')\cos(y,y') + 2\varepsilon_z \cos(z,x')\cos(z,y') + \gamma_{xy}[\cos(x,x')\cos(y,y') + \cos(x,y')\cos(y,x')] + \gamma_{yz}[\cos(y,x')\cos(z,y') + \cos(y,y')\cos(z,x')] + \gamma_{zx}[\cos(z,x')\cos(x,y') + \cos(z,y')\cos(x,x')]$ (27-43a)

Particular	Formula
	$\gamma_{y'z'} = 2\varepsilon_y \cos(y, y')\cos(y, z') + 2\varepsilon_z \cos(z, y')\cos(z, z') + 2\varepsilon_x \cos(x, y')\cos(x, z') + \gamma_{yz}[\cos(y, y')\cos(z, z') + \cos(y, z')\cos(z, y')] + \gamma_{zx}[\cos(z, y')\cos(x, z') + \cos(z, z')\cos(x, y')] + \gamma_{xy}[\cos(x, y')\cos(y, z') + \cos(x, z')\cos(y, y')]$ (27-43b)
	$\gamma_{z'x'} = 2\varepsilon_z \cos(z, z')\cos(z, x') + 2\varepsilon_x \cos(x, z')\cos(x, x') + 2\varepsilon_y \cos(y, z')\cos(y, x') + \gamma_{zx}[\cos(z, z')\cos(x, x') + \cos(z, x')\cos(x, z')] + \gamma_{xy}[\cos(x, z')\cos(y, x') + \cos(x, x')\cos(y, z')] + \gamma_{yz}[\cos(y, z')\cos(z, x') + \cos(y, x')\cos(z, z')]$ (27-43c)
For the case of two-dimensional state of stress when z' coincides with z and $\gamma_{zx} = \gamma_{yz} = 0$, the angle between x and x' coordinates θ	$\varepsilon_{x'} = \varepsilon_x \cos^2\theta + \varepsilon_y \sin^2\theta + \gamma_{zy}\sin\theta\cos\theta$ (27-44a) $= \frac{1}{2}[(\varepsilon_x + \varepsilon_y) + (\varepsilon_x - \varepsilon_y)\cos 2\theta + \gamma_{xy}\sin 2\theta]$ (27-44b) $\varepsilon_{y'} = \varepsilon_y \cos^2\theta + \varepsilon_x \sin^2\theta - \gamma_{zy}\sin\theta\cos\theta$ (27-44c) $= \frac{1}{2}[(\varepsilon_y + \varepsilon_x) + (\varepsilon_y - \varepsilon_x) - \gamma_{xy}\sin 2\theta]$ (27-44d) $\gamma_{x'y'} = 2(\varepsilon_y - \varepsilon_x)\sin\theta\cos\theta + \gamma_{xy}(\cos^2\theta - \sin^2\theta)$ (27-44e) $\frac{1}{2}\gamma_{x'y'} = -\frac{1}{2}[(\varepsilon_x - \varepsilon_y)\sin 2\theta - \frac{1}{2}\gamma_{xy}\cos 2\theta]$ (27-44f) $\varepsilon_{z'} = \varepsilon_z, \quad \gamma_{y'z'} = \gamma_{z'x'} = 0$ (27-44g)
The cubic equation for principal strains whose three roots give the distinct principal strains associated with three principal directions, is	$\varepsilon_n^3 - (\varepsilon_x + \varepsilon_y + \varepsilon_z)\varepsilon_n^2 + \left(\varepsilon_x\varepsilon_y + \varepsilon_y\varepsilon_z + \varepsilon_z\varepsilon_x - \frac{\gamma_{xy}^2}{4} - \frac{\gamma_{yz}^2}{4} - \frac{\gamma_{zx}^2}{4}\right)\varepsilon_n - \left(\varepsilon_x\varepsilon_y\varepsilon_z - \varepsilon_x\frac{\gamma_{yz}^2}{4} - \varepsilon_y\frac{\gamma_{zx}^2}{4} - \varepsilon_z\frac{\gamma_{xy}^2}{4} + \frac{\gamma_{xy}\gamma_{yz}\gamma_{zx}}{4}\right) = 0$ (27-45)
The three strain invariants analogous to the three stress invariants	$J_1 = \varepsilon_x + \varepsilon_y + \varepsilon_z =$ first invariant of strain (27-45a)

Particular	Formula
	$J_2 = \varepsilon_x\varepsilon_y + \varepsilon_y\varepsilon_z + \varepsilon_z\varepsilon_x - \frac{\gamma_{xy}^2}{4} - \frac{\gamma_{yz}^2}{4} - \frac{\gamma_{zx}^2}{4}$ = second invariant of strain (27-45b)
	$J_3 = \varepsilon_x\varepsilon_y\varepsilon_z - \frac{\varepsilon_x\gamma_{yz}^2}{4} - \frac{\varepsilon_y\gamma_{zx}^2}{4} - \frac{\varepsilon_z\gamma_{xy}^2}{4} + \frac{\gamma_{xy}\gamma_{yz}\gamma_{zx}}{4}$ = third invariant of strain (27-45c)

BOUNDARY CONDITIONS

The components of the surface forces F_{sfx} and F_{sfy} per unit area of a small triangular prism pqr so that the side qr coincides with the boundary of the plate ds (Fig. 27-10)

$$F_{sfx} = l\sigma_x + m\tau_{xy}, \quad F_{sfy} = m\sigma_y + l\tau_{yx} \qquad (27\text{-}46)$$

where l and m are the direction cosines of the normal N to the boundary

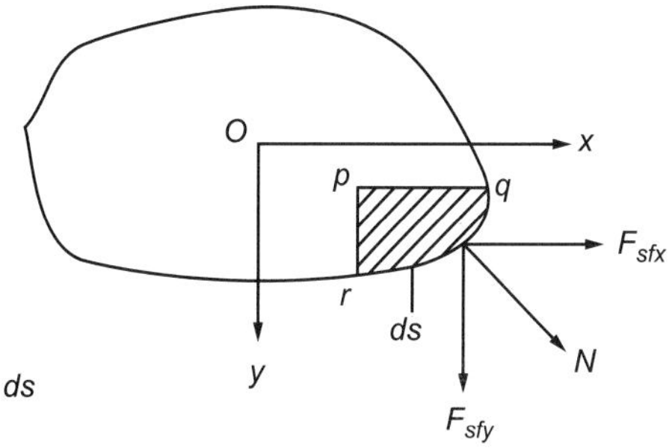

FIGURE 27-10 Area of a small triangular prism *pqr*.

COMPATIBILITY

The six strain equations of compatibility

$$\frac{\partial^2\gamma_{xy}}{\partial x\,\partial y} = \frac{\partial^2\varepsilon_x}{\partial y^2} + \frac{\partial^2\varepsilon_y}{\partial x^2} \qquad (27\text{-}47a)$$

$$\frac{\partial^2\gamma_{yz}}{\partial y\,\partial z} = \frac{\partial^2\varepsilon_y}{\partial z^2} + \frac{\partial^2\varepsilon_z}{\partial y^2} \qquad (27\text{-}47b)$$

$$\frac{\partial^2\gamma_{zx}}{\partial z\,\partial x} = \frac{\partial^2\varepsilon_z}{\partial x^2} + \frac{\partial^2\varepsilon_x}{\partial z^2} \qquad (27\text{-}47c)$$

$$2\frac{\partial^2\varepsilon_x}{\partial y\,\partial z} = \frac{\partial}{\partial x}\left(-\frac{\partial\gamma_{yz}}{\partial x} + \frac{\partial\gamma_{zx}}{\partial y} + \frac{\partial\gamma_{xy}}{\partial z}\right) \qquad (27\text{-}47d)$$

$$2\frac{\partial^2\varepsilon_y}{\partial z\,\partial x} = \frac{\partial}{\partial y}\left(\frac{\partial\gamma_{yz}}{\partial x} - \frac{\partial\gamma_{zx}}{\partial y} + \frac{\partial\gamma_{xy}}{\partial z}\right) \qquad (27\text{-}47e)$$

$$2\frac{\partial^2\varepsilon_z}{\partial x\,\partial y} = \frac{\partial}{\partial z}\left(\frac{\partial\gamma_{yz}}{\partial x} + \frac{\partial\gamma_{zx}}{\partial y} - \frac{\partial\gamma_{xy}}{\partial z}\right) \qquad (27\text{-}47f)$$

Particular	Formula
The volume dilatation of rectangular parallelopiped element subject to hydrostatic pressure whose sides are l_1, l_2 and l_3	$e = \dfrac{V_f - V}{V} = \dfrac{\Delta V}{V}$ (27-48) where V_f = final volume after straining of element $= l_{1f} \times l_{2f} \times l_{3f}$ V = initial volume of element $= l_1 l_2 l_3$
The final dimensions of the element after straining	$l_{1f} = l_1(1 + \varepsilon_1)$ (27-49a) $l_{2f} = l_2(1 + \varepsilon_2)$ (27-49b) $l_{3f} = l_3(1 + \varepsilon_3)$ (27-49c)
Substituting the values of l_{1f}, l_{2f}, l_{3f}, l_1, l_2, l_3 in Eq. (27-48) and after neglecting higher order terms of strain	$e = \dfrac{l_1 l_2 l_3(1+\varepsilon_1)(1+\varepsilon_2)(1+\varepsilon_3) - l_1 l_2 l_3}{l_1 l_2 l_3}$ $= \varepsilon_1 + \varepsilon_2 + \varepsilon_3 = J_1$ (27-50a) $= \left(\dfrac{1-2v}{E}\right)(\sigma_1 + \sigma_2 + \sigma_3)$ (27-50b)
If hydrostatic pressure (σ_0) or uniform compression is applied from all sides of an element such that $\sigma_x = \sigma_y = \sigma_z = -\sigma_0 = \sigma_1 = \sigma_2 = \sigma_3$, $\tau_{xy} = \tau_{yz} = \tau_{zx} = 0$, Eq. (27-48) becomes	$e = \dfrac{\Delta V}{V} = -\dfrac{3(1-2v)\sigma_0}{E} = -\dfrac{\sigma_0}{K}$ (27-51) where K = bulk modulus of elasticity
The bulk modulus of elasticity	$K = \dfrac{E}{3(1-2v)} = -\dfrac{\sigma_0}{e}$ (27-52) $K = \dfrac{2(1+v)G}{3(1-2v)}$ (27-53)

GENERAL HOOKE'S LAW

Particular	Formula
The general equation for strain in x-direction according to general Hooke's law in case of anisotropic or non-homogeneous and non-isotropic materials such as laminate, wood and fiber-filled-epoxy materials as a linear function of each stress	$\varepsilon_x = S_{11}\sigma_x + S_{12}\sigma_y + S_{13}\sigma_z + S_{14}\tau_{xy} + S_{15}\tau_{yz} + S_{16}\tau_{zx} + S_{17}\tau_{xz} + S_{18}\tau_{zy} + S_{19}\tau_{yz}$ (27-54)
For relationships between the elastic constants	Refer to Table 27-1.
The three-dimensional stress-strain state in anisotropic or non-homogeneous and non-isotropic material such as laminates, fiber filled epoxy material by using generalized Hooke's law which is useful in designing machine elements made of composite material (Fig. 27-1c) *Note*: [S] matrix is the compliance matrix which gives the strain-stress relations for the material. The inverse of the compliance matrix is the stiffness matrix and the stress-strain relations. If no symmetry is assumed, there are $9^2 = 81$ independent elastic constants present in the compliance matrix [Eq. (27-55)]	$\begin{bmatrix} \varepsilon_x \\ \varepsilon_y \\ \varepsilon_z \\ \gamma_{xy} \\ \gamma_{yz} \\ \gamma_{zx} \\ \gamma_{xz} \\ \gamma_{zy} \\ \gamma_{yx} \end{bmatrix} = \begin{bmatrix} S_{11} & S_{12} & S_{13} & S_{14} & S_{15} & S_{16} & S_{17} & S_{18} & S_{19} \\ S_{21} & S_{22} & S_{23} & S_{24} & S_{25} & S_{26} & S_{27} & S_{28} & S_{29} \\ - & - & - & - & - & - & - & - & - \\ - & - & - & - & - & - & - & - & - \\ - & - & - & - & - & - & - & - & - \\ - & - & - & - & - & - & - & - & - \\ - & - & - & - & - & - & - & - & - \\ - & - & - & - & - & - & - & - & - \\ S_{91} & S_{92} & S_{93} & S_{94} & S_{95} & S_{96} & S_{97} & S_{98} & S_{99} \end{bmatrix} \begin{bmatrix} \sigma_x \\ \sigma_y \\ \sigma_z \\ \tau_{xy} \\ \tau_{yz} \\ \tau_{zx} \\ \tau_{xz} \\ \tau_{zy} \\ \tau_{yx} \end{bmatrix}$ (27-55)

Particular	Formula

TABLE 27.1
Relationships between the elastic constants

	λ equals	μ[a] equals	E equals	v equals	K equals
λ, μ[a]	–	–	$\dfrac{\mu(3\lambda+2\mu)}{\lambda+\mu}$	$\dfrac{\mu}{2(\lambda+\mu)}$	$\dfrac{3\lambda+2\mu}{3}$
λ, E	–	$\dfrac{{}^{b}A+(E-3\lambda)}{4}$	–	$\dfrac{{}^{b}A-(E+\lambda)}{4\lambda}$	$\dfrac{{}^{b}A+(3\lambda+E)}{6}$
λ, v	–	$\dfrac{\lambda(1-2v)}{2v}$	$\dfrac{\lambda(1+v)(1-2v)}{v}$	–	$\dfrac{\lambda(1+v)}{3v}$
λ, K	–	$\dfrac{3(K-\lambda)}{2}$	$\dfrac{9K(K-\lambda)}{3K-\lambda}$	$\dfrac{\lambda}{3K-\lambda}$	–
μ, E	$\dfrac{2\mu-E}{E-3\mu}$	–	–	$\dfrac{E-2\mu}{2\mu}$	$\dfrac{\mu E}{3(3\mu-E)}$
μ, v	$\dfrac{2\mu v}{1-2v}$	–	$2\mu(1+v)$	–	$\dfrac{2\mu(1+v)}{3(1-2v)}$
μ, K	$\dfrac{3K-2\mu}{3}$	–	$\dfrac{9K\mu}{3K+\mu}$	$\dfrac{3K-2\mu}{2(3K+\mu)}$	–
E, v	$\dfrac{vE}{(1+v)(1-2v)}$	$\dfrac{E}{2(1+v)}$	–	–	$\dfrac{E}{3(1-2v)}$
K, E	$\dfrac{3K(3K-E)}{9K-E}$	$\dfrac{3EK}{9K-E}$	–	$\dfrac{3K-E}{6K}$	–
v, K	$\dfrac{3Kv}{1+v}$	$\dfrac{3K(1-2v)}{2(1+v)}$	$3K(1-2v)$	–	–

[a] $\mu = G =$ modulus of rigidity/shear.
[b] $A = \sqrt{E^2 + 2\lambda E + 9\lambda^2}$.
Courtesy: Dally, J. W. and William F. Riley, *Experimental Stress Analysis*, McGraw-Hill Publishing Company, New York, 1965.

Equation (27-55) can be written as given here under Eq. (27-56) with the following use of change of notations and principle of symmetrical matrix in case of stiffness matrices

$$\tau_{xy} = \tau_{yx} \qquad \varepsilon_{12} = \varepsilon_{21} \qquad S_{12} = S_{21}$$

$$\tau_{yz} = \tau_{zy} \qquad \varepsilon_{23} = \varepsilon_{32} \qquad S_{13} = S_{31}$$

$$\tau_{xz} = \tau_{zx} \qquad \varepsilon_{13} = \varepsilon_{31} \qquad \text{etc}$$

and the following changes in Eq. (27-54)

$$\sigma_x = \sigma_1 \quad \tau_{yz} = \sigma_4 = \tau_{23} \quad \varepsilon_x = \varepsilon_1 \quad 2\gamma_{yz} = \varepsilon_4 = \gamma_{23}$$

$$\sigma_y = \sigma_2 \quad \tau_{xz} = \sigma_5 = \tau_{13} \quad \varepsilon_y = \varepsilon_2 \quad 2\gamma_{xz} = \varepsilon_5 = \gamma_{13}$$

$$\sigma_z = \sigma_3 \quad \tau_{xy} = \sigma_6 = \tau_{12} \quad \varepsilon_z = \varepsilon_3 \quad 2\gamma_{xy} = \varepsilon_6 = \gamma_{12}$$

$$\begin{bmatrix} \varepsilon_x \\ \varepsilon_y \\ \varepsilon_z \\ \gamma_{yz} \\ \gamma_{zx} \\ \gamma_{xy} \end{bmatrix} = \begin{bmatrix} \varepsilon_1 \\ \varepsilon_2 \\ \varepsilon_3 \\ \varepsilon_4 \\ \varepsilon_5 \\ \varepsilon_6 \end{bmatrix} = \begin{bmatrix} S_{11} & S_{12} & S_{13} & S_{14} & S_{15} & S_{16} \\ S_{21} & S_{22} & S_{23} & S_{24} & S_{25} & S_{26} \\ - & - & - & - & - & - \\ - & - & - & - & - & - \\ - & - & - & - & - & - \\ S_{61} & S_{62} & S_{63} & S_{64} & S_{65} & S_{66} \end{bmatrix} \begin{bmatrix} \sigma_x \\ \sigma_y \\ \sigma_z \\ \tau_{yz} \\ \tau_{xz} \\ \tau_{xy} \end{bmatrix} \tag{27-56}$$

[a] *Courtesy*: Extracted from Ashton, J. E., J. C. Halpin, and P. H. Petit, *Primer on Composite Materials—Analysis*, Technomic Publishing Co., Inc., 750 Summer Street, Stamford, Conn. 1969.

Particular	Formula
The general stress-strain equations under linear stress-strain relationship	$\sigma_x = K_{11}\varepsilon_x + K_{12}\varepsilon_y + K_{13}\varepsilon_z + K_{14}\gamma_{xy} + K_{15}\gamma_{yz} + K_{16}\gamma_{zx}$ $\sigma_y = K_{21}\varepsilon_x + K_{22}\varepsilon_y + K_{23}\varepsilon_z + K_{24}\gamma_{xy} + K_{25}\gamma_{yz} + K_{26}\gamma_{zx}$ $\sigma_z = K_{31}\varepsilon_x + K_{32}\varepsilon_y + K_{33}\varepsilon_z + K_{34}\gamma_{xy} + K_{35}\gamma_{yz} + K_{36}\gamma_{zx}$ $\tau_{xy} = K_{41}\varepsilon_x + K_{42}\varepsilon_y + K_{43}\varepsilon_z + K_{44}\gamma_{xy} + K_{45}\gamma_{yz} + K_{46}\gamma_{zx}$ $\tau_{yz} = K_{51}\varepsilon_x + K_{52}\varepsilon_y + K_{53}\varepsilon_z + K_{54}\gamma_{xy} + K_{55}\gamma_{yz} + K_{56}\gamma_{zx}$ $\tau_{zx} = K_{61}\varepsilon_x + K_{62}\varepsilon_y + K_{63}\varepsilon_z + K_{64}\gamma_{xy} + K_{65}\gamma_{yz} + K_{66}\gamma_{zx}$ (27-57) where K_{11} to K_{66} are the coefficients of elasticity of the material and are independent of the magnitudes of both the stress and the strain, provided the elastic limit of the material is not exceeded. There are 36 coefficients of elasticity.
The stress-strain relationships for the case of isotropic material	$\sigma_x = \lambda(\varepsilon_x + \varepsilon_y + \varepsilon_z) + 2\mu\varepsilon_x$ (27-58a) $\sigma_y = \lambda(\varepsilon_x + \varepsilon_y + \varepsilon_z) + 2\mu\varepsilon_y$ $\sigma_z = \lambda(\varepsilon_x + \varepsilon_y + \varepsilon_z) + 2\mu\varepsilon_z$ $\tau_{xy} = \mu\gamma_{xy}$ $\tau_{yz} = \mu\gamma_{yz}$ $\tau_{zx} = \mu\gamma_{zx}$ where $\lambda =$ Lamé's constant $\mu = G =$ modulus of shear
The strain expressions from Eqs. (27-58a)	$\varepsilon_x = \frac{\lambda+\mu}{3\lambda+2\mu}\sigma_x - \frac{\lambda}{2\mu(3\lambda+2\mu)}(\sigma_y+\sigma_z)$ (27-58b) $\varepsilon_y = \frac{\lambda+\mu}{3\lambda+2\mu}\sigma_y - \frac{\lambda}{2\mu(3\lambda+2\mu)}(\sigma_x+\sigma_z)$ $\varepsilon_z = \frac{\lambda+\mu}{3\lambda+2\mu}\sigma_z - \frac{\lambda}{2\mu(3\lambda+2\mu)}(\sigma_y+\sigma_x)$ $\gamma_{xy} = \frac{1}{\mu}\tau_{xy} = \frac{1}{G}\tau_{xy}$ $\gamma_{yz} = \frac{1}{\mu}\tau_{yz} = \frac{1}{G}\tau_{yz}$ $\gamma_{zx} = \frac{1}{\mu}\tau_{zx} = \frac{1}{G}\tau_{zx}$

Particular	Formula

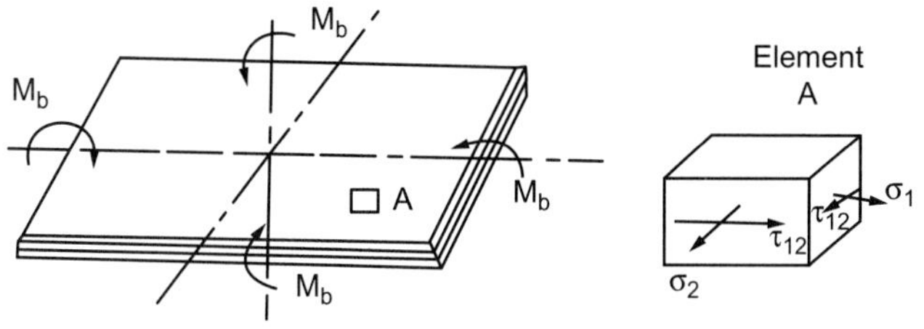

FIGURE 27-10A Thin laminae of a composite laminate under bending.

The matrix expression from Eq. (27-55) for orthotropic material in a three-dimensional state of stress

$$\begin{bmatrix} \varepsilon_1 \\ \varepsilon_2 \\ \varepsilon_3 \\ \gamma_{23} \\ \gamma_{13} \\ \gamma_{12} \end{bmatrix} = \begin{bmatrix} S_{11} & S_{12} & S_{13} & 0 & 0 & 0 \\ S_{12} & S_{22} & S_{23} & 0 & 0 & 0 \\ S_{13} & S_{23} & S_{23} & 0 & 0 & 0 \\ 0 & 0 & 0 & S_{44} & 0 & 0 \\ 0 & 0 & 0 & 0 & S_{55} & 0 \\ 0 & 0 & 0 & 0 & 0 & S_{66} \end{bmatrix} \begin{bmatrix} \sigma_1 \\ \sigma_2 \\ \sigma_3 \\ \tau_{23} \\ \tau_{13} \\ \tau_{12} \end{bmatrix} \quad (27\text{-}59)$$

where there are 9 independent constants in the above compliance matrix which is inverse of stiffness matrix

The two-dimensional or a plane stress state matrix expression after putting $\sigma_3 = \tau_{23} = \tau_{13} = 0$ and $\gamma_{23} = \gamma_{13} = 0$ and $\varepsilon_3 = S_{13}\sigma_1 + S_{23}\sigma_2$ in Eq. (27-59) for orthotropic material

$$\begin{bmatrix} \varepsilon_1 \\ \varepsilon_2 \\ \gamma_{12} \end{bmatrix} = \begin{bmatrix} S_{11} & S_{12} & 0 \\ S_{21} & S_{22} & 0 \\ 0 & 0 & S_{66} \end{bmatrix} \begin{bmatrix} \sigma_1 \\ \sigma_2 \\ \tau_{12} \end{bmatrix} \quad (27\text{-}60)$$

The stress-strain relationship for homogenous isotropic laminae of a laminated composite in the matrix form, which is assumed to be in state of plane stress

$$\begin{bmatrix} \sigma_1 \\ \sigma_2 \\ \tau_{12} \end{bmatrix}_n = \begin{bmatrix} K_{11} & K_{12} & 0 \\ K_{21} & K_{22} & 0 \\ 0 & 0 & K_{66} \end{bmatrix}_n \begin{bmatrix} \varepsilon_1 \\ \varepsilon_2 \\ \gamma_{12} \end{bmatrix}_n \quad (27\text{-}61)$$

where K is stiffness matrix

$$K_{11} = K_{12} = E/(1 - v^2)$$

$$K_{12} = vE/(1 - v^2)$$

$$K_{66} = E/2(1 - v) = G$$

Alternatively Eqs. (27-61) can be written for the nth layer of laminated composite, which is assumed to be in a state of plane stress

$$\sigma_1 = (\varepsilon_1 + v\varepsilon_2)\frac{E}{1 - v^2} \quad (27\text{-}62)$$

$$\sigma_2 = (\varepsilon_2 + v\varepsilon_1)\frac{E}{1 - v^2}$$

$$\tau_{12} = \gamma_{12}\frac{E}{2(1 - v)}$$

Particular	Formula
Substituting strain-displacement, Eqs. (27-40) and (27-41) into stress-strain Eqs. (27-33) and (27-37) or (27-39), displacement stress equation are obtained with from 15 unknowns to 9 unknowns	$\frac{\partial u}{\partial x} = \frac{1}{E}[\sigma_x - v(\sigma_y + \sigma_z)]$ (27-63) $\frac{\partial v}{\partial y} = \frac{1}{E}[\sigma_y - v(\sigma_z + \sigma_x)]$ $\frac{\partial w}{\partial z} = \frac{1}{E}[\sigma_z - v(\sigma_x + \sigma_y)]$ $\frac{\partial u}{\partial y} + \frac{\partial v}{\partial x} = \frac{1}{\mu}\tau_{xy}$ $\frac{\partial v}{\partial z} + \frac{\partial w}{\partial y} = \frac{1}{\mu}\tau_{yz}$ $\frac{\partial w}{\partial x} + \frac{\partial u}{\partial z} = \frac{1}{\mu}\tau_{zx}$ where $\mu = G$
Combining stress equation of equilibrium from Eqs. (27-11) with stress displacement Eqs. (27-63) (from 9 to 3 unknowns)	$\nabla^2 u + \frac{1}{1-2v}\frac{\partial}{\partial x}\left(\frac{\partial u}{\partial x} + \frac{\partial v}{\partial y} + \frac{\partial w}{\partial z}\right) + \frac{1}{\mu}F_{bx} = 0$ (27-64) $\nabla^2 v + \frac{1}{1-2v}\frac{\partial}{\partial y}\left(\frac{\partial u}{\partial x} + \frac{\partial v}{\partial y} + \frac{\partial w}{\partial z}\right) + \frac{1}{\mu}F_{by} = 0$ $\nabla^2 w + \frac{1}{1-2v}\frac{\partial}{\partial z}\left(\frac{\partial u}{\partial x} + \frac{\partial v}{\partial y} + \frac{\partial w}{\partial z}\right) + \frac{1}{\mu}F_{bz} = 0$ where ∇^2 is the operator $\left(\frac{\partial^2}{\partial x^2} + \frac{\partial^2}{\partial y^2} + \frac{\partial^2}{\partial z^2}\right)$
Six stress equations of compatibility are obtained by making use of stress strain relations of Eqs. (27-33) and (27-39), the stress equations of equilibrium Eq. (27-11) and the strain compatibility Eq. (27-47) in three dimension in Cartesian system of coordinates	$\nabla^2\sigma_x + \frac{1}{1+v}\frac{\partial^2 I_1}{\partial x^2} = -\frac{v}{1-v}\left(\frac{\partial F_{bx}}{\partial x} + \frac{\partial F_{by}}{\partial y} + \frac{\partial F_{bz}}{\partial z}\right) - 2\frac{\partial F_{bx}}{\partial x}$ (27-65a) $\nabla^2\sigma_y + \frac{1}{1+v}\frac{\partial^2 I_1}{\partial y^2} = -\frac{v}{1-v}\left(\frac{\partial F_{bx}}{\partial x} + \frac{\partial F_{by}}{\partial y} + \frac{\partial F_{bz}}{\partial z}\right) - 2\frac{\partial F_{by}}{\partial y}$ (27-65b) $\nabla^2\sigma_z + \frac{1}{1+v}\frac{\partial^2 I_1}{\partial z^2} = -\frac{v}{1-v}\left(\frac{\partial F_{bx}}{\partial x} + \frac{\partial F_{by}}{\partial y} + \frac{\partial F_{bz}}{\partial z}\right) - 2\frac{\partial F_{bz}}{\partial z}$ (27-65c) $\nabla^2\tau_{zy} + \frac{1}{1+v}\frac{\partial^2 I_1}{\partial x\,\partial y} = -\left(\frac{\partial F_{bz}}{\partial y} + \frac{\partial F_{by}}{\partial x}\right)$ (27-65d)

Particular	Formula
	$\nabla^2 \tau_{yz} + \frac{1}{1+v} \frac{\partial^2 I_1}{\partial y \, \partial z} = -\left(\frac{\partial F_{by}}{\partial z} + \frac{\partial F_{bz}}{\partial y} \right)$ (27-65e)
	$\nabla^2 \tau_{zx} + \frac{1}{1+v} \frac{\partial^2 I_1}{\partial z \, \partial x} = -\left(\frac{\partial F_{bx}}{\partial z} + \frac{\partial F_{bz}}{\partial x} \right)$ (27-65f)

AIRY'S STRESS FUNCTION

Differential equations of equilibrium for two-dimensional problems taking only gravitational force as body force	$\frac{\partial \sigma_x}{\partial x} + \frac{\partial \tau_{xy}}{\partial y} = 0$ (27-66a) $\frac{\partial \sigma_y}{\partial y} + \frac{\partial \tau_{yz}}{\partial x} + \rho g = 0$ $\nabla^2 = (\sigma_x + \sigma_y) = 0$ (27-66b) where $\nabla^2 = \frac{\partial^2}{\partial x^2} + \frac{\partial^2}{\partial y^2}$
The stress components in terms of stress function ϕ and body force	$\sigma_x = \frac{\partial^2 \phi}{\partial y^2} - \rho g y; \quad \sigma_y = \frac{\partial^2 \phi}{\partial x^2} - \rho g y; \quad \tau_{xy} = -\frac{\partial^2 \phi}{\partial x \, \partial y}$ (27-66c)
Substituting Eqs. (27-66c) for stress components into Eq. (27-66b) that the stress function ϕ must satisfy the equation	$\frac{\partial^4 \phi}{\partial x^4} + 2 \frac{\partial^4 \phi}{\partial x^2 \, \partial y^2} + \frac{\partial^4 \phi}{\partial y^4} = 0$ (27-72)
The stress compatibility equation for the case of plane strain	$\nabla^2(\sigma_x + \sigma_y) = -\frac{2(\lambda + \mu)}{\lambda + 2\mu} \left(\frac{\partial F_{bx}}{\partial x} + \frac{\partial F_{by}}{\partial y} \right)$ (27-67)
If components of body forces in plane strain are	$F_{bx} = -\frac{\partial \Omega}{\partial x}; \quad F_{by} = -\frac{\partial \Omega}{\partial y}$ (27-68)
Substituting Eqs. (27-68) into Eqs. (27-11d), (27-11e) and Eq. (27-67) and taking $\frac{2(\lambda + \mu)}{\lambda + 2\mu} = \frac{1}{1 - v}$	$\frac{\partial \sigma_x}{\partial x} + \frac{\partial \tau_{xy}}{\partial y} = \frac{\partial \Omega}{\partial x}$ (27-69a) $\frac{\partial \tau_{xy}}{\partial x} + \frac{\partial \sigma_y}{\partial y} = \frac{\partial \Omega}{\partial y}$ (27-69b) $\nabla^2 \left(\sigma_x + \sigma_y - \frac{\Omega}{1 - v} \right) = 0$ (27-69c)
By assuming that the stress can be represented by a stress function ϕ such that $\sigma_x = \frac{\partial^2 \phi}{\partial y^2} + \Omega$, $\sigma_y = \frac{\partial^2 \phi}{\partial x^2} + \Omega$, and $\tau_{xy} = \frac{\partial^2 \phi}{\partial x \partial y}$ and substituting these into Eqs. (27-69) and Eq. (27-69c) becomes	$\nabla^4 \phi = -\frac{1 - 2v}{1 - v} \nabla^2 \Omega$ (27-70)

Particular	Formula
Stresses for plane-stress can be obtained by letting $\frac{v}{1-v} \rightarrow v$ in Eq. (27-70) and it becomes	$\nabla^4\phi = -(1-v)\nabla^2\Omega$ (27-71)
If body forces are zero or constant then Eq (27-70) becomes	$\nabla^4\phi = \partial$ (27-71a) which is a biharmonic equation in ϕ and is a stress function
The biharmonic Eq. (27-71a) can be written in expanded form as	$\frac{\partial^2\phi}{\partial x^4} + 2\frac{\partial^4\phi}{\partial x^2\,\partial y^2} + \frac{\partial^4\phi}{\partial y^4} = 0$ (27-72) The solution of a two-dimensional problem when the weight of body is the only body force reduces to finding a solution of Eq. (27-72) which satisfies boundary condition Eq. (27-46) of the problem.

CYLINDRICAL COORDINATES SYSTEM

General equations of equilibrium in r, θ and z coordinates (cylindrical coordinates) taking into consideration body force (Figs. 27-13 to 27-15)

$$\frac{\partial\sigma_r}{\partial r} + \frac{1}{r}\frac{\partial\tau_{r\theta}}{\partial\theta} + \frac{\partial\tau_{rz}}{\partial z} + \frac{\sigma_r - \sigma_\theta}{r} + F_{bR} = 0 \quad (27\text{-}73a)$$

$$\frac{\partial\tau_{rz}}{\partial r} + \frac{1}{r}\frac{\partial\tau_{\theta z}}{\partial\theta} + \frac{\partial\sigma_z}{\partial z} + \frac{\tau_{rz}}{r} + F_{bz} = 0 \quad (27\text{-}73b)$$

$$\frac{\partial\tau_{r\theta}}{\partial r} + \frac{1}{r}\frac{\partial\sigma_\theta}{\partial\theta} + \frac{\partial\tau_{\theta z}}{\partial z} + \frac{2\tau_{r\theta}}{r} + F_{b\theta} = 0 \quad (27\text{-}73c)$$

where F_{bR}, $F_{b\theta}$ and F_{bz} are body force components

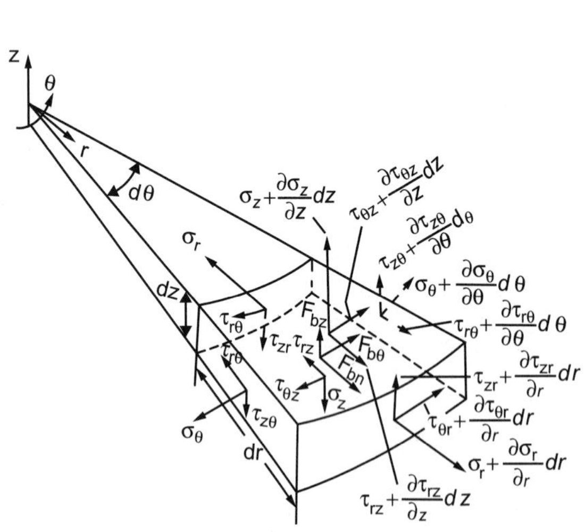

FIGURE 27-11 Element showing stresses in r, θ and in the axial direction.

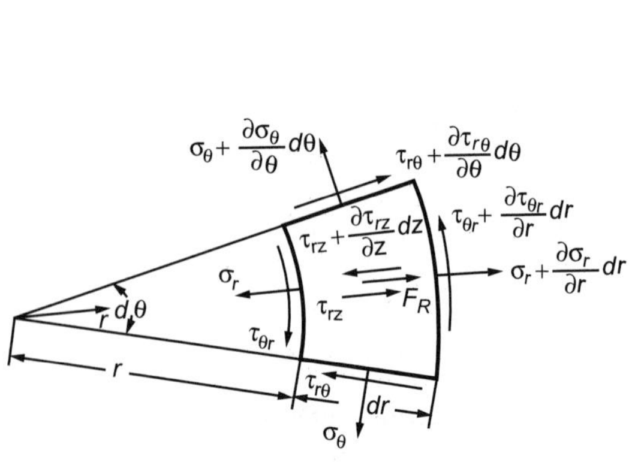

FIGURE 27-12 Element showing stresses in r and θ directions.

Particular	Formula

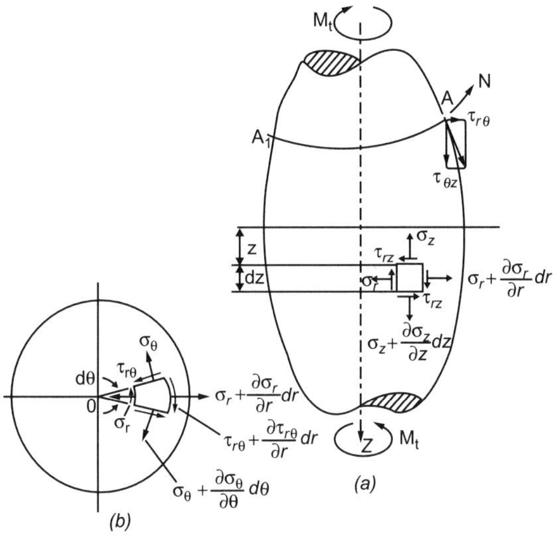

FIGURE 27-13

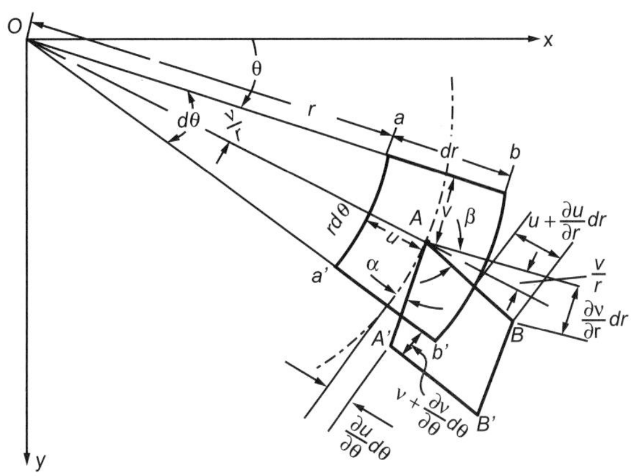

FIGURE 27-14 Strain components in polar co-ordinates.

Equations of equilibrium for axial symmetry Eqs. (27-73) reduce to Eqs. (27-74) when there are body forces acting on the body

$$\frac{\partial \sigma_r}{\partial r} + \frac{\partial \tau_{rz}}{\partial z} + \frac{\sigma_r - \sigma_\theta}{r} + F_{bR} = 0 \tag{27-74a}$$

$$\frac{1}{r}\frac{\partial \sigma_\theta}{\partial \theta} + \frac{\partial \tau_{\theta z}}{\partial z} + F_{b\theta} = 0 \tag{27-74b}$$

$$\frac{\partial \tau_{rz}}{\partial r} + \frac{1}{r}\frac{\partial \tau_{\theta z}}{\partial \theta} + \frac{\partial \sigma_z}{\partial z} + \frac{\tau_{rz}}{r} + F_{bz} = 0 \tag{27-74c}$$

Equations of equilibrium in two dimension in r and θ coordinates (polar coordinates) taking into consideration body force components

$$\frac{\partial \sigma_r}{\partial r} + \frac{1}{r}\frac{\partial \tau_{r\theta}}{\partial \theta} + \frac{1}{r}(\sigma_r - \sigma_\theta) + F_{bR} = 0 \tag{27-75a}$$

$$\frac{1}{r}\frac{\partial \sigma_\theta}{\partial \theta} + \frac{\partial \tau_{r\theta}}{\partial r} + \frac{2\tau_{r\theta}}{r} + F_{b\theta} = 0 \tag{27-75b}$$

Equations of equilibrium for an axially symmetrical stress distribution in a solid of revolution when there are no body forces acting on the body (Fig. 27-13), since the stress components are independent of θ.

$$\frac{\partial \sigma_r}{\partial r} + \frac{\partial \tau_{rz}}{\partial \theta} + \frac{(\sigma_r - \sigma_\theta)}{r} = 0 \tag{27-76a}$$

$$\frac{\partial \tau_{rz}}{\partial r} + \frac{\partial \sigma_z}{\partial z} + \frac{\tau_{rz}}{r} = 0 \tag{27-76b}$$

STRAIN COMPONENTS (Fig. 27-14)

The strain components in r, θ and z coordinates system

The strain in the radial direction

$$\varepsilon_r = \frac{\partial u}{\partial r} \tag{27-77a}$$

The strain in the tangential direction

$$\varepsilon_\theta = \frac{1}{r}\frac{\partial v}{\partial \theta} + \frac{u}{r} \tag{27-77b}$$

Particular	Formula	
The strain in the axial direction	$\varepsilon_z = \dfrac{\partial w}{\partial z}$	(27-77c)
The shear strains	$\gamma_{r\theta} = \dfrac{1}{r}\dfrac{\partial u}{\partial \theta} + \dfrac{\partial v}{\partial r} - \dfrac{v}{r}$	(27-77d)
	$\gamma_{\theta z} = \dfrac{1}{r}\dfrac{\partial w}{\partial \theta} + \dfrac{\partial v}{\partial z}$	(27-77e)
	$\gamma_{zr} = \dfrac{\partial u}{\partial z} + \dfrac{\partial w}{\partial r}$	(27-77f)
The rotation of the element in the counter clock-wise direction in the $r\theta$, θz and zr planes	$\Theta_{r\theta} = \dfrac{1}{2}\left(\dfrac{\partial v}{\partial r} + \dfrac{v}{r} - \dfrac{1}{r}\dfrac{\partial u}{\partial \theta}\right)$	(27-78a)
	$\Theta_{\theta z} = \dfrac{1}{2}\left(\dfrac{1}{r}\dfrac{\partial w}{\partial \theta} - \dfrac{\partial v}{\partial z}\right)$	(27-78b)
	$\Theta_{zr} = \dfrac{1}{2}\left(\dfrac{\partial u}{\partial z} - \dfrac{\partial w}{\partial r}\right)$	(27-78c)

AIRY'S STRESS FUNCTION IN POLAR COORDINATES

When components of body force F_{br} and $F_{b\theta}$ are zero, Eqs. (27-74a) and (27-74b) are satisfied by assuming stress function for σ_r, σ_θ and $\tau_{r\theta}$	$\sigma_r = \dfrac{1}{r}\dfrac{\partial \phi}{\partial r} + \dfrac{1}{r^2}\dfrac{\partial^2 \phi}{\partial \theta^2}$	(27-79a)
	$\sigma_\theta = \dfrac{\partial^2 \phi}{\partial r^2}$	(27-79b)
	$\tau_{r\theta} = -\dfrac{\partial}{\partial r}\left(\dfrac{1}{r}\dfrac{\partial \phi}{\partial \theta}\right) = \dfrac{1}{r^2}\dfrac{\partial \phi}{\partial \theta} - \dfrac{1}{r}\dfrac{\partial^2 \phi}{\partial r\,\partial \theta}$	(27-79c)
The stress equation of compatibility Eq. (27-72) in terms of Airy's stress function ϕ referred to Cartesian coordinates x and y, has to be transferred to Airy's stress function referred to polar coordinates r and θ system. In this transformation from x and y coordinates transform to r and θ coordinates	$r^2 = x^2 + y^2, \quad \theta = \tan^{-1}\dfrac{y}{x}$	(27-80)
	from which $\dfrac{\partial r}{\partial x} = \dfrac{x}{r} = \cos\theta; \quad \dfrac{\partial r}{\partial y} = \dfrac{y}{r} = \sin\theta$	
	$\dfrac{\partial \phi}{\partial x} = -\dfrac{y}{r^2} = -\dfrac{\sin\theta}{r}; \quad \dfrac{\partial \theta}{\partial y} = \dfrac{x}{r^2} = \dfrac{\cos\theta}{r}$	(27-81)
Eq. (27-72) can be written as	$\nabla^2\phi = \left(\dfrac{\partial^2}{\partial x^2} + \dfrac{\partial^2}{\partial y^2}\right)\left(\dfrac{\partial^2 \phi}{\partial x^2} + \dfrac{\partial^2 \phi}{\partial y^2}\right)$	(27-82)
Using Eqs. (27-79) and (27-80) and transforming Eq. (27-72) into stress equation of compatibility in polar coordinates r and θ system	$\dfrac{\partial^2 \phi}{\partial x^2} + \dfrac{\partial^2 \phi}{\partial y^2} = \dfrac{\partial^2 \phi}{\partial r^2} + \dfrac{1}{r}\dfrac{\partial \phi}{\partial r} + \dfrac{1}{r^2}\dfrac{\partial^2 \phi}{\partial \theta^2}$	(27-83)

Particular	Formula
Stress equation of compatibility in terms of Airy's stress function in polar coordinate r and θ is obtained by substituting (Eq. (27-83) to Eq. (27-82))	$$\nabla^4\phi = \left(\frac{\partial^2}{\partial r^2}+\frac{1}{r}\frac{\partial}{\partial r}+\frac{1}{r^2}\frac{\partial^2}{\partial\theta^2}\right) \times\left(\frac{\partial^2\phi}{\partial r^2}+\frac{1}{r}\frac{\partial\phi}{\partial r}+\frac{1}{r^2}\frac{\partial^2\phi}{\partial r^2}\right) \quad (27\text{-}84)$$

SOLUTION OF ELASTICITY PROBLEMS USING AIRY'S STRESS FUNCTION

Any Airy's stress function ϕ either in Cartesian coordinates or polar coordinates used in solving any two-dimensional problems must satisfy Eqs. (27-66) and (27-72) in Cartesian coordinates and Eqs. (27-79) and (27-84) in polar coordinates and boundary conditions (27-46)

Cartesian coordinates

Solutions of many two-dimensional problems can be found by assuming Airy's stress function in terms of polynomial and Fourier series, which are

$$\phi_1 = a_1x + b_1y \quad \text{first degree polynomial} \quad (27\text{-}85)$$

$$\phi_2 = a_2x^2 + b_2xy + c_2y^2 \quad \text{second degree polynomial} \quad (27\text{-}86)$$

$$\phi_3 = a_3x^3 + b_3x^2y + c_3xy^2 + d_3y^3 \quad \text{third degree polynomial} \quad (27\text{-}87)$$

$$\phi_4 = a_4x^4 + b_4x^3y + c_4x^2y^2 + d_4xy^3 + e_4y^4 \quad \text{fourth degree polynomial} \quad (27\text{-}88)$$

$$\phi_5 = a_5x^5 + b_5x^4y + c_5x^3y^2 + d_5x^2y^3 + e_5xy^4 + f_5y^5 \quad \text{fifth degree polynomial} \quad (27\text{-}89)$$

$$\phi = \sin\frac{m\pi x}{l}f(y) \quad \text{Fourier series} \quad (27\text{-}90)$$

where m is an integer

$$\phi = \sum_{n=0}^{n=\infty} a_n\cos\frac{n\pi x}{l} + \sum_{n=1}^{n=\infty} b_n\sin\frac{n\pi x}{l} \quad (27\text{-}91)$$

where n is an integer

$$a_0 = \frac{1}{2l}\int_0^{2l}\phi\,dx$$

$$a_n = \frac{1}{l}\int_0^{2l}\phi\cos\frac{n\pi x}{l}\,dx \quad \text{if } n \neq 0$$

Particular	Formula
	$b_n = \frac{1}{l}\int_0^{2l} \phi \sin\frac{n\pi x}{l}\, dx$
	ϕ is any periodic function of x, which represents itself at interval of $2l$

Polar coordinates

Particular	Formula
$\nabla^4\phi = 0$ is a fourth order biharmonic partial differential equation. The fourth order differential equation can be obtained by using a function ϕ in $\nabla^4\phi = 0$ which in term gives four different stress functions	$\phi_n = R_n(r)\begin{Bmatrix} \cos n\theta \\ \sin n\theta \end{Bmatrix}$ (27-92)
One of the stress function ϕ for solving many problems in polar coordinates	$\phi_1 = A \ln r + Br^2 \ln r + Cr^2 + D$ (27-93)
The second order stress function ϕ_2	$\phi_2 = (A_1 r + B_1/r + C_1 r^3 + D_1 r \ln r)\begin{Bmatrix} \sin\theta \\ \cos\theta \end{Bmatrix}$ (27-94)
The third order stress function ϕ_3	$\phi_n = (A_n r^n + B_m/r^n + C_n r^{2+n} + D_n r^{2-n})\begin{Bmatrix} \sin n\theta \\ \cos n\theta \end{Bmatrix}$ (27-95)
The fourth order stress function	$\phi_m = A_m\theta + B_m r^2\theta + C_m r\theta\sin\theta + D_m r\theta\cos\theta$ (27-96)
	It is sometimes difficult to select a stress function for solving a problem. But it is left to the discretion of the problem solver to select or decide the correct stress function to suit the problem under consideration.
The general expression for the stress function ϕ which satisfy boundary conditions and compatibility Eq. (27-84)	$\phi = A_0\theta + B_0 r^2\theta + C_0 r\theta\sin\theta + D_0 r\theta\cos\theta + D'_0 + C'_0 r^2 + B'_0 r^2 \ln r + A'_0 \ln r + \left(A_1 r + \frac{B_1}{r} + C_1 r^3 + D_1 r \ln r\right)\sin\theta + \left(A'_1 r + \frac{B'_1}{r} + C'_1 r^3 + D'_1 r \ln r\right)\cos\theta + \sum_{n=2}^{\infty}\left(A_n r^n + \frac{B_n}{r_n} + C_n r^{n+2} + \frac{D_n}{r^{n-2}}\right)\sin n\theta + \sum_{n=2}^{\infty}\left(A'_n r^n + \frac{B'_n}{r_n} + C'_n r^{n+2} + \frac{D'_n}{r^{n-2}}\right)\cos n\theta$ (27-97)

Particular	Formula
In a general case the loading can be represented by the trigonometric series	$$q = A_0 + \sum_{m=1}^{\infty} A_m \cos\frac{m\pi x}{l} + \sum_{m=1}^{\infty} A'_m \cos\frac{m\pi x}{l} + B_0 + \sum_{m=1}^{\infty} B_m \sin\frac{m\pi x}{l} + \sum_{m=1}^{\infty} B'_m \cos\frac{m\pi x}{l} \quad (27\text{-}98)$$
The stress function ϕ can also be represented by	$$\phi = (A\,e^{\alpha y} + B\,e^{-\alpha y} + Cy\,e^{\alpha y} + Dy\,e^{-\alpha y}) \sin \alpha x \quad (27\text{-}99)$$

APPLICATION OF STRESS FUNCTION

Thick cylinder

Stress function used in this case, Eq. (27-93)	$$\phi = A \ln r + Br^2 \ln r + Cr^2 + D \quad (27\text{-}93)$$
Boundary conditions are	$$(\sigma_r)_{r=d_i/2} = -p_i \quad \text{and} \quad (\sigma_r)_{r=d_0/2} = -p_o \quad (27\text{-}100a)$$
Equation of equilibrium used in this problem	$$\frac{\partial \sigma_r}{\partial r} + \frac{\sigma_r - \sigma_\theta}{r} = 0 \quad (27\text{-}100b)$$ Since it is a case of problem of symmetry with respect to axis of cylinder and no body force acting on it.
The expression for radial stress in thickness wall of thick cylinder under external pressure (p_o) and internal pressure (p_i) at any radius r	$$\sigma_r = \frac{p_i d_i^2 - p_o d_o^2}{d_o^2 - d_i^2} - \frac{d_i^2 d_o^2 (p_i - p_o)}{4r^2 (d_o^2 - d_i^2)} \quad (27\text{-}101a)$$
The expression for tangential stress in thickness wall of thick cylinder under pressures p_o and p_i at any radius r	$$\sigma_\theta = \frac{p_i d_i^2 - p_o d_o^2}{d_o^2 - d_i^2} + \frac{d_i^2 d_o^2 (p_i - p_o)}{4r^2 (d_o^2 - d_i^2)} \quad (27\text{-}101b)$$
The shear stress	$$\tau_{r\theta} = 0 \quad (27\text{-}101c)$$
Expression for displacement of an element in the thickness wall of cylinder at any radius r in radial direction and tangential direction respectively	$$u = \frac{1}{E}\left[-(1+v)\frac{d_i^2 d_o^2 (p_o - p_i)}{4r(d_o^2 - d_i^2)}\right] + (1-v)\frac{1}{r}\left(\frac{p_i d_i^2 - p_o d_o^2}{d_o^2 - d_i^2}\right) \quad (27\text{-}101d)$$ $$\nu = 0 \quad (27\text{-}101e)$$

Curved bar under pure bending (Fig. 27-15)

Stress function used in this problem Eq. (27-93)	$$\phi = A \ln r + Br^2 \ln r + Cr^2 + D \quad (27\text{-}93)$$
Boundary conditions	$$(\sigma_r)_{r=d_o/2,\, d_i/2} = 0 \quad (27\text{-}102a)$$

Particular	Formula

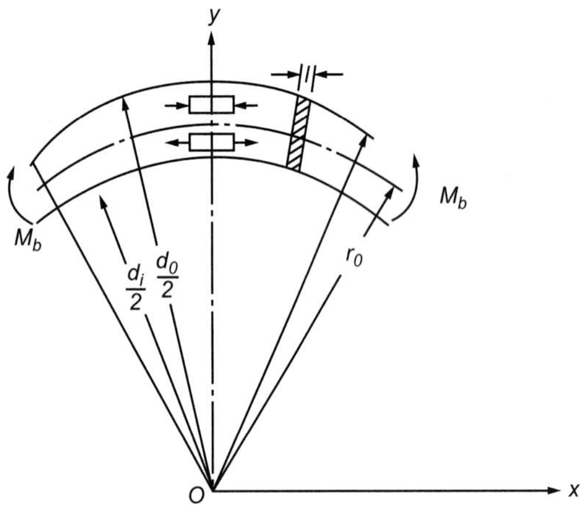

FIGURE 27-15

$$\int_{d_i/2}^{d_o/2} \sigma_\theta r\, dr = -M_b \qquad (27\text{-}102b)$$

$$\int_{d_i/2}^{d_o/2} \sigma_\theta\, dr = 0 \qquad (27\text{-}102c)$$

$$(\tau_{r\theta})_{r=d_o/2,\, d_i/2} = 0 \qquad (27\text{-}102d)$$

Equation of equilibrium used in this problem of symmetry with respect to the xy-plane perpendicular to axis of the bar

$$\frac{\partial \sigma_r}{\partial r} + \frac{\sigma_r - \sigma_\theta}{r} = 0 \qquad (27\text{-}100b)$$

The expression for the radial stress component in the bar at r radius

$$\sigma_r = \frac{4M_b}{\eta}\left(\frac{d_o^2 d_i^2}{16r^2}\ln\frac{d_o}{d_i} + \frac{d_o^2}{4}\ln\frac{2r}{d_o} + \frac{d_i^2}{4}\ln\frac{d_i}{2r}\right) \qquad (27\text{-}103)$$

The expression for the tangential stress component in the bar at r radius

$$\sigma_\theta = \frac{4M_b}{\eta}\left[\frac{d_o^2 d_i^2}{16r^2}\ln\frac{d_o}{d_i} + \frac{d_o^2}{4}\ln\frac{2r}{d_o} + \frac{d_i^2}{4}\ln\frac{d_i}{2r} + \frac{1}{4}(d_o^2 - d_i^2)\right] \qquad (27\text{-}104)$$

The expression for shear stress component

$$\tau_{r\theta} = 0 \qquad (27\text{-}105a)$$

where

$$\eta = \frac{(d_o^2 - d_i^2)^2}{16} - \frac{1}{4} d_o^2 d_i^2 \left(\ln\frac{d_o}{d_i}\right)^2 \qquad (27\text{-}105b)$$

STRESS DISTRIBUTION IN A FLAT PLATE WITH HOLES OR CUTOUTS UNDER DIFFERENT TYPES OF LOADS

An infinite flat plate with centrally located circular cutout or hole subject to uniform uniaxial tension (Fig. 27-16)

$$\phi = \left(Ar^2 + Br^4 + \frac{C}{r^2} + D\right)\cos 2\theta \qquad (27\text{-}106)$$

The expression for stress function

$$(\sigma_r)_{r=b} = \sigma\cos^2\theta = \tfrac{1}{2}\sigma(1 + \cos 2\theta) \qquad (27\text{-}107a)$$

Particular	Formula	
Boundary conditions are	$(\tau_{r\theta})_{r=b} = \frac{1}{2}\sigma \sin 2\theta$	(27-107b)
The radial stress in an infinite plate with a centrally located circular hole (cut-out) subject to uniform uniaxial tension at infinity (Fig. 27-16)	$\sigma_r = \frac{\sigma}{2}\left[\left(1 - \frac{a^2}{r^2}\right) + \left(1 - \frac{4a^2}{r^2} + \frac{3a^4}{r^4}\right)\cos 2\theta\right]$	(27-108)
The tangential stress in an infinite plate with centrally located circular hole (cutout) under uniform uniaxial tension at infinity	$\sigma_\theta = \frac{\sigma}{2}\left[\left(1 + \frac{a^2}{r^2}\right) - \left(1 + \frac{3a^4}{r^4}\right)\cos 2\theta\right]$	(27-109)
The shear stress in an infinite flat plate with a centrally located circular cutout (hole) subject to uniform uniaxial tension at infinity	$\tau_{r\theta} = -\frac{\sigma}{2}\left(1 + \frac{2a^2}{r^2} - \frac{3a^4}{r^4}\right)\sin 2\theta$	(27-110)
The tangential stress at hole boundary at $\theta = \pi/2$ or $3\pi/2$	$(\sigma_\theta)_{r=a} = \frac{\sigma}{2}\left(2 + \frac{a^2}{r^2} + \frac{3a^4}{r^4}\right)$	(27-111a)
The stress concentration factor	$K_\sigma = \frac{\sigma_\theta}{\sigma} = \frac{1}{2}\left(2 + \frac{a^2}{r^2} + \frac{3a^4}{r^4}\right)_{r=a} = 3$	(27-111b)
For distribution of tangential stress σ_θ around circle of hole under uniform uniaxial tension	Refer to Fig. 27-18.	
For superposition of stresses in a flat plate with a centrally located circular hole subject to tension, compression and uniform pressure	Refer to Fig. 27-19.	
The shear stress around hole at $\theta = \pi/2$ or $3\pi/2$	$(\tau_{r\theta})_{r=a} = 0$	(27-111c)

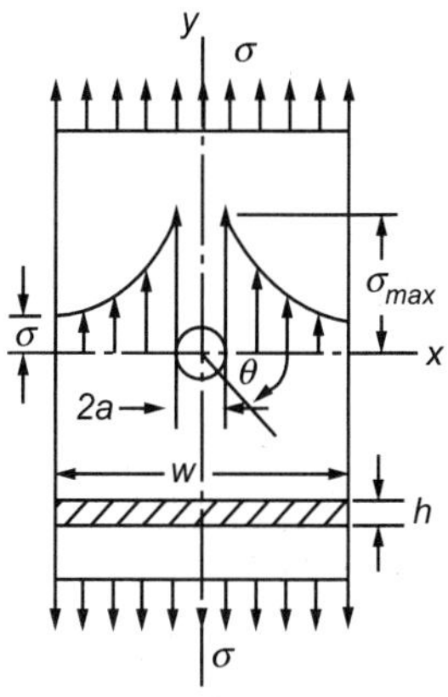

FIGURE 27-16 A large flat plate with a centrally located circular hole under uniform uniaxial stress at infinity.

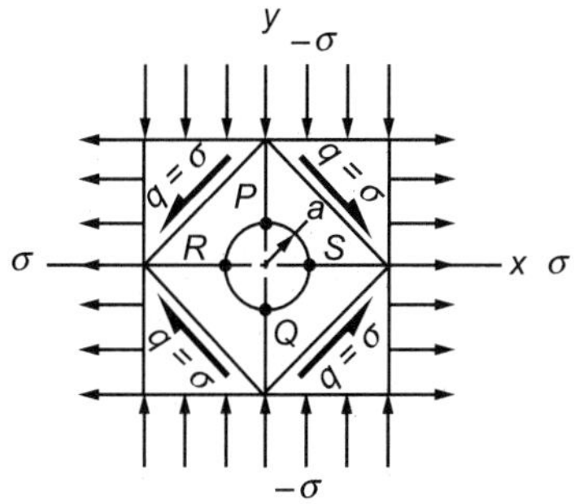

FIGURE 27-17 A large flat plate with circular hole under pure shear stress.

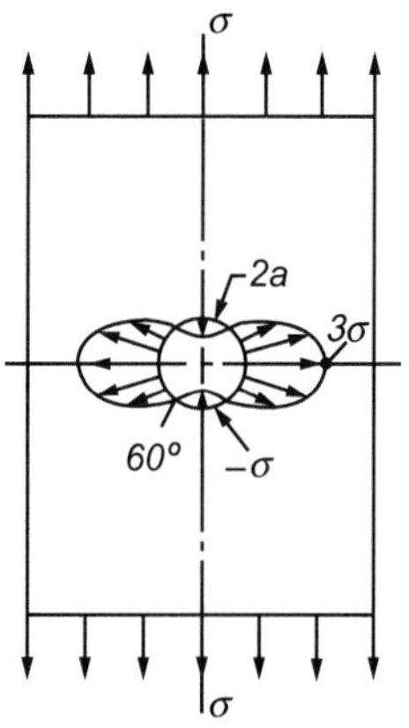

FIGURE 27-18 Distribution of stress σ around the boundary of circular cutout (hole).

Particular	Formula

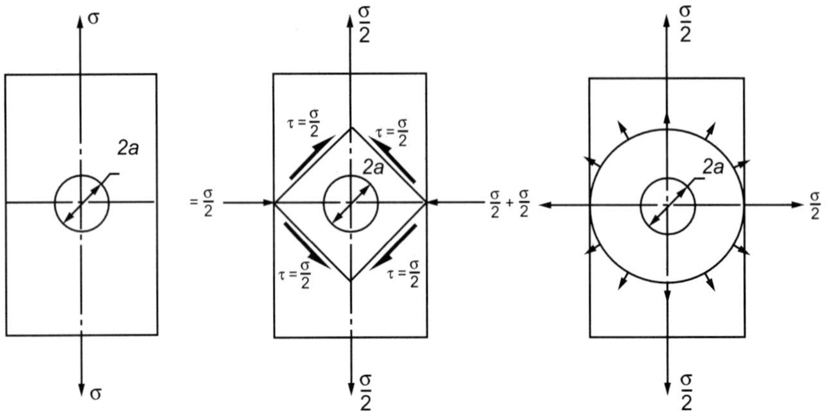

FIGURE 27-19 Principle of superposition.

Pure shear

An infinite flat plate with centrally located circular cutout or hole subject to uniform uniaxial tension and compression (i.e., pure shear) (Fig. 27-17)

The expression for stress function

$$\phi = \left(Ar^2 + Br^4 + \frac{C}{r^2} + D\right)\sin 2\theta \qquad (27\text{-}112)$$

Boundary conditions are

$$(\sigma_r)_\infty = -(\sigma_\theta)_\infty \qquad (27\text{-}113a)$$

$$\sigma_r = q\sin 2\theta \qquad (27\text{-}113b)$$

$$\tau_{r\theta} = q\cos 2\theta \qquad (27\text{-}113c)$$

where q = shear load

$$(\sigma_r)_{r=a} = (\tau_{r\theta})_{r=a} = 0 \qquad (27\text{-}113d)$$

The tangential stress in an infinite plate with a centrally located circular hole subject to uniform tensile and compressive stresses as shown in Fig. 27-17

$$\sigma_\theta = -q\left(1 + \frac{3a^4}{r^4}\right)\sin 2\theta \qquad (27\text{-}114)$$

The radial stress

$$\sigma_r = q\left(1 - \frac{4a^2}{r^2} + \frac{3a^4}{r^4}\right)\sin 2\theta \qquad (27\text{-}115)$$

The shear stress

$$\tau_{r\theta} = q\left(1 + \frac{2a^2}{r^2} - \frac{3a^4}{r^4}\right)\cos 2\theta \qquad (27\text{-}116)$$

The tangential stress

$$\sigma_\theta = \sigma - 2\sigma\cos 2\theta - [\sigma - 2\sigma\cos(2\theta - \pi)] \qquad (27\text{-}117)$$

The maximum tangential stress at $\theta = \pi/2$ or $3\pi/2$ i.e., at P and Q (Fig. 27-17)

$$(\sigma_\theta)_{\theta=\pi/2 \text{ or } 3\pi/2,\, r=a} = 4\sigma \qquad (27\text{-}118a)$$

Particular	Formula	
The maximum tangential stress at $\theta = 0$ or $\theta = \pi$, i.e., at R and S (Fig. 27-17)	$(\sigma_\theta)_{\theta=0 \text{ or } \pi,\, r=a} = -4\sigma$	(27-118b)
The stress concentration factor	$K_\sigma = \frac{\sigma_{\theta\,\max}}{\sigma} = \left(1 + \frac{3a^4}{r^4}\right)_{r=a} = 4$	(27-118c)

Bi-axial tension (Fig. 27-20)

An infinite flat plate with centrally located circular hole (cutout) under biaxial uniform tension (Fig. 27-20)

The radial stress at hole boundary	$(\sigma_r)_{r=a} = 0$	(27-119a)
The shear stress at hole boundary	$(\tau_{r\theta})_{r=a} = 0$	(27-119b)
The tangential stress in an infinite flat plate with a centrally located circular hole subject to uniform biaxial tensile stress at infinity	$\sigma_\theta = \sigma - 2\sigma\cos 2\theta - [\sigma - 2\sigma\cos(2\theta - \pi)]$	(27-120)
	$\sigma_\theta = 2\sigma$	(27-120a)
The stress concentration factor	$K_\sigma = \frac{2\sigma}{\sigma} = 2$	(27-120b)

Finite plate (Fig. 27-21)

Uniaxial tension (Fig. 27-21)	$K_\sigma = \frac{\sigma_{\max}}{\sigma_{\text{nom}}}$	(27-121a)

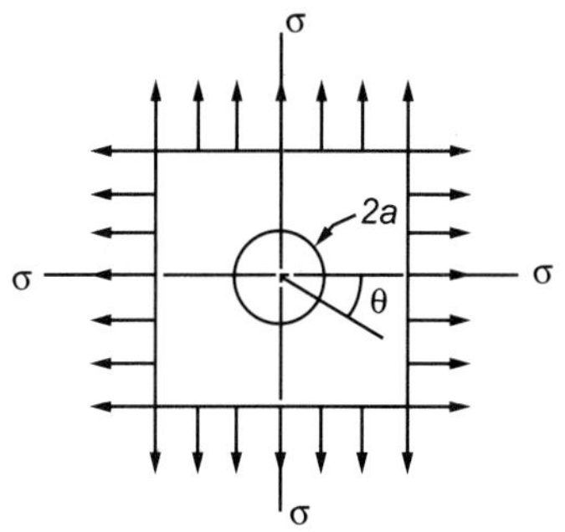

FIGURE 27-20 Biaxial tension.

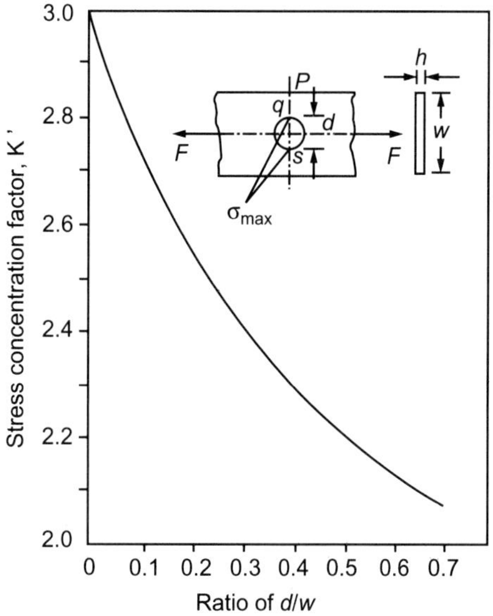

FIGURE 27-21 Stress concentration factor for a plate of finite width with a circular hole (cutout) in tension (Howland).

Particular	Formula
The stress distribution in a flat plate of finite width w with a centrally located small circular hole according to Howland[11], which can be expressed in terms of stress concentration around hole	$K_\sigma = \frac{\sigma_{max}}{\sigma}\left(1 - \frac{d}{w}\right)$ (27-121b) where $\sigma_{nom} = \frac{F}{(w-d)h} = \frac{\sigma}{1-(d/w)}$ (27-121c) $\sigma = F/wh$ = stress at the end of infinite plate
The tangential stress at the points of q and s when $w = 2d$ according to Howland	$\sigma_\theta = 4.3\sigma$ (27-121d)
The tangential stress at the point P according to Howland[11]	$\sigma_\theta = 0.75\sigma$ (27-121e)
It can be seen from Fig. 27-21 that maximum stress which is at the hole boundary, decreases very rapidly and approach the value of average stress at the edge of the infinite plate	

ROTATING SOLID DISK WITH UNIFORM THICKNESS (Fig. 27-22)

Particular	Formula
From Eqs. (27-75a) and (27-75b), which can be made use of for a rotating disk of uniform thickness with z-axis perpendicular to the xy-plane and stress components do not depend on θ. Hence Eqs. (27-75a) taking a body force equal to inertia force i.e., $F_{bR} = \rho\omega^2 r$, becomes	$\frac{\partial\sigma_r}{\partial r} + \frac{1}{r}(\sigma_r - \sigma_\theta) + F_{bR} = 0$ (27-122a)
Equation of force equilibrium from Eqs. (27-122a) after substituting value of F_{bR} becomes	$r\frac{\partial\sigma_r}{\partial r} + \sigma_r - \sigma_\theta + \rho\omega^2 r^2 = 0$ (27-122b) $\frac{\partial(r\sigma_r)}{\partial r} - \sigma_\theta + \rho\omega^2 r^2 = 0$ (27-122c) where F_{bR} = body force per unit volume = $\rho\omega^2 r$ $F = r\sigma_r$ = stress function ρ = density, kg/m^3 ω = angular velocity, rad/s Stress is a function of r only because of symmetry
Equation of compatibility	$r\frac{\partial\varepsilon_\theta}{\partial r} + \varepsilon_\theta = \varepsilon_r$ (27-123)
Using compatibility equation (27-123) and Hooke's law after simplification, the expression for force equilibrium (27-122b) becomes	$r^2\frac{\partial^2 F}{\partial r^2} + r\frac{\partial F}{\partial r} - F = -(3+\upsilon)\rho\omega^2 r^3$ (27-124a)

Particular	Formula

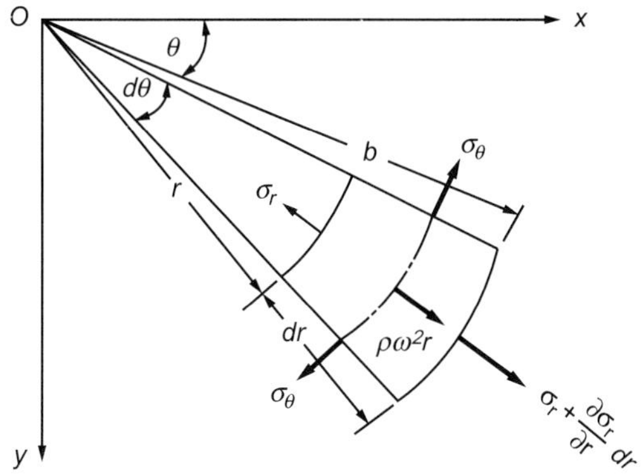

FIGURE 27-22 Element of a rotating disk.

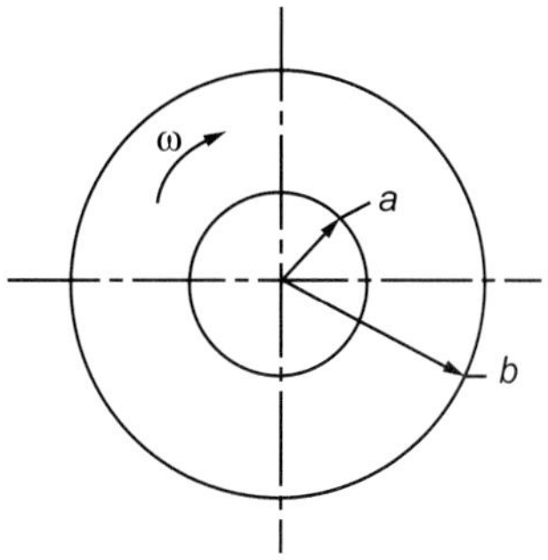

FIGURE 27-23

The general solution of Eq. (27-124a), when $r = e^\alpha$ is substituted in it, becomes

$$F = Ar + \frac{B}{r} - (3+v)\frac{\rho\omega^2 r^3}{8} \tag{27-124b}$$

where A and B are constants of integration to be found by boundary conditions

Boundary conditions

(a) The radial stress at outer boundary of rotating disc of radius b

$$(\sigma_r)_{r=b} = 0 \tag{27-125a}$$

(b) stress at center of rotating disc

$$(\sigma_r)_{r=\theta} \neq 0 \tag{27-125b}$$

The expression for radial stress at any radius r

$$\sigma_r = \frac{3+v}{8}\rho\omega^2(b^2 - r^2) \tag{27-126}$$

The expression for tangential stress as any radius r

$$\sigma_\theta = \frac{3+v}{8}\rho\omega^2 b^2 - \left(\frac{1+3v}{8}\right)\rho\omega^2 r^2 \tag{27-127}$$

ROTATING DISK WITH A CENTRAL CIRCULAR HOLE OF UNIFORM THICKNESS, Fig. 27-23

$$(\sigma_r)_{r=a} = 0 \tag{27-128a}$$

Boundary conditions

$$(\sigma_r)_{r=b} = 0 \tag{27-128b}$$

Using force equilibrium Eq. (27-124b) and boundary conditions Eqs. (27-128) the tangential and radial stresses at any radius r are

$$\sigma_\theta = \left(\frac{3+v}{8}\right)\rho\omega^2\left[a^2 + b^2 + \frac{a^2b^2}{r^2} - \left(\frac{1+3v}{3+v}\right)r^2\right] \tag{27-129}$$

$$\sigma_r = \left(\frac{3+v}{8}\right)\rho\omega^2\left(a^2 + b^2 - \frac{a^2b^2}{r^2} - r^2\right) \tag{27-130}$$

Particular	Formula
The expression for maximum radial stresses which occurs at $r = \sqrt{ab}$	$\sigma_{r\,\max} = \left(\frac{3+v}{8}\right)\rho\omega^2(b-a)^2$ (27-131a)
The expression for maximum tangential stress which occur at $r = a$	$\sigma_{\theta\,\max} = \left(\frac{3+v}{8}\right)\rho\omega^2\left[b^2 + \left(\frac{1-v}{3+v}\right)a^2\right]$ (27-131b)

Rotating disk as a three-dimensional problem

The differential equations of equilibrium from Eqs. (27-76) when body force which is an inertia force (centrifugal force) is included, becomes

$$\frac{\partial\sigma_r}{\partial r} + \frac{\partial\tau_{rz}}{\partial z} + \frac{\sigma_r - \sigma_\theta}{r} + \rho\omega^2 r = 0 \quad (27\text{-}132a)$$

$$\frac{\partial\tau_{rz}}{\partial r} + \frac{\partial\sigma_z}{\partial z} + \frac{\tau_{rz}}{r} = 0 \quad (27\text{-}132b)$$

After substituting the body forces $F_{bx} = \rho\omega^2 x$, $F_{by} = \rho\omega^2 y$, $F_{bz} = 0$ in Eqs. (27-65) and the last three equations containing shearing stress components remain the same as in Eqs. (27-65), and the first three equations in polar coordinates become

$$\nabla^2\sigma_r - \frac{2}{r^2}(\sigma_r - \sigma_\theta) + \frac{1}{1+v}\frac{\partial^2 I}{\partial r^2} = -\frac{2\rho\omega^2}{1-v} \quad (27\text{-}133a)$$

$$\nabla^2\sigma_\theta + \frac{2}{r^2}(\sigma_r - \sigma_\theta) + \frac{1}{1+v}\frac{1}{r}\frac{\partial I}{\partial r} = -\frac{2\rho\omega^2}{1-v} \quad (27\text{-}133b)$$

$$\nabla^2\sigma_z + \frac{1}{1+v}\frac{\partial^2 I}{\partial z^2} = -\frac{2v\rho\omega^2}{1-v} \quad (27\text{-}133c)$$

The equations of shearing stress components in Eqs. (27-65) remain the same without any change even when the body forces are acting.

The solution of Eq. (27-132) consists of a particular solution and complementary function. The particular solution call be obtained by assuming

$$\sigma_r = Br^2 + Dz^2; \quad \sigma_z = Ar^2;$$
$$\sigma_\theta = Cr^2 + Dz^2; \quad \tau_{rz} = 0 \quad (a)$$

The complementary solution is obtained by assuming a stress function, which has to satisfy boundary conditions, compatibility equations and having a form of a polynomial of the fifth degree

$$\phi = a_5(8z^5 - 40r^2z^2 + 15r^4z) + b_5(2z^5 - r^2z^3 - 3r^4z) \quad (b)$$

The particular solution

$$\sigma_r = -\frac{\rho\omega^2}{3}r^2 - \frac{\rho\omega^2(1+2v)(1-v)}{6v(1-v)}z^2 \quad (c)$$

$$\sigma_z = \rho\omega^2\frac{1+3v}{6v}r^2 \quad (d)$$

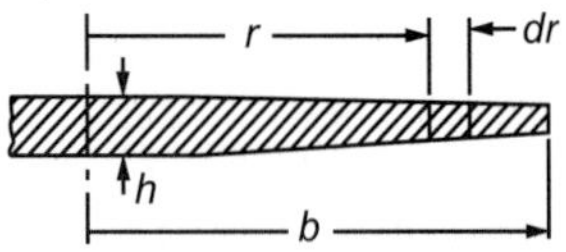

FIGURE 27-24 Rotating disc of variable thickness.

The complementary function obtained from assuming stress function Eq. (*b*)

$$\sigma_\theta = -\frac{\rho\omega^2(1+2v)(1+v)}{6v(1-v)}z^2 \quad (e)$$

Particular	Formula
	$\sigma_r = -\rho\omega^2\left[\frac{v(1+v)}{2(1-v)}z^2 + \frac{3+v}{8}r^2\right]$ (*f*)
	$\sigma_\theta = -\rho\omega^2\left[\frac{1+3v}{8}r^2 + \frac{v(1+v)}{2(1-v)}z^2\right]$ (*g*)
An expression for uniform radial tension on the disk which is superposed on the resultant stresses to express the resultant radial compression along the boundary	$T_r = \frac{\rho\omega^2}{8}(3+v)a^2 + \rho\omega^2\frac{v(1+v)}{2(1-v)}\frac{c^2}{3}$ (*h*)
The final expressions for stress components	$\sigma_r = \rho\omega^2\left[\frac{3+v}{8}(a^2-r^2) + \frac{v(1+v)}{6(1-v)}(c^2-3z^2)\right]$ (27-134a)
	$\sigma_\theta = r\omega^2\left[\frac{3+v}{8}a^2 - \left(\frac{1+3v}{8}\right)r^2 + \frac{v(1+v)}{6(1-v)}(c^2-8z^2)\right]$ (27-134b)
	$\sigma_z = 0; \quad \tau_{rz} = 0$ (27-134c)
For disk of uniform strength rotating at ω rad/s	Refer to Chapter 10, Eq. (10-1) and Fig. 10-1.
For solid cylinder rotating at ω rad/s, hollow cylinder rotating at ω rad/s and solid thin uniform disk rotating at ω rad/s under external pressure	Refer to Chapter 10, Eqs. (10-9) to (10-25) and Figs. 10-2 and 10-3.
For asymmetrically reinforced circular holes/cutouts in a flat plate subject to uniform uniaxial tensile force/stress	Refer to Chapter 4, Figs. 4-7(a), 4-7(b) and 4-7(c).

ROTATING DISK OF VARIABLE THICKNESS (Fig. 27-24)

Particular	Formula
The equation of force equilibrium in case of rotating disk of variable thickness from Eq. (27-122b)	$\frac{d}{dr}(rh\sigma_r) - h\sigma_\theta + \rho^2\omega r h = 0$ (27-135)
Using $rh\sigma_r = F$ and thickness variation as $h = Cr^{-\beta}$, Hooke's law, $r = e^\alpha$ and Eq. (27-123), Eq. (27-135) become	$r^2\frac{\partial^2 F}{\partial r^2} + (1+\beta)r\frac{\partial F}{\partial r} - (1+v\beta)F = -(3+v)\rho^2\omega C r^{3-\beta}$ (27-136a)
	$\frac{\partial^2 F}{\partial \alpha^2} + \beta\frac{\partial F}{\partial \alpha} - (1+v\beta)F = -(3+v)\rho\omega^2 C\,\mathrm{e}^{(3-\beta)\alpha}$ (27-136b)
	where C and β are constants

Particular	Formula
Boundary conditions	$(\sigma_r)_{r=b} = 0$ (27-137a)
	$(\sigma_r)_{r=0} \neq \infty$ (27-137b)
	$(\sigma_\theta)_{r=\theta} \neq \infty$ (27-137b)
The expression for radial stress	$\sigma_r = \dfrac{3+v}{8-(3+v)\beta}\rho\omega^2 b\left[\left(\dfrac{r}{b}\right)^{\lambda_1+\beta-1} - \left(\dfrac{r}{b}\right)^2\right]$ (27-138)
The expression for tangential stress	$\sigma_\theta = \dfrac{3+v}{8-(3+v)\beta}\rho\omega^2 b^2 \times \left[\lambda_1\left(\dfrac{r}{b}\right)^{\lambda_1+\beta-1} - \dfrac{1+3v}{3+v}\left(\dfrac{r}{b}\right)^2\right]$ (27-139a)
	where $\lambda = -\dfrac{\beta}{2} \pm \sqrt{\left(\dfrac{\beta}{2}\right)^2(1+v\beta)}$ (27-139b)
	λ_1 and λ_2 are roots of Eq. (27-139b)
For a symmetrically reinforced circular cutout in a flat plate under uniform uniaxial tension according to the analysis of Timoshenko	Refer to Fig. 27-25.

NEUTRAL HOLES (MANSFIELD THEORY)

Reinforced holes which do not affect the stress distribution in a plate are said to be neutral	There is negligible bonding on the reinforcement since it is thin compared to the radius of the curvature of the hole, and shear across the section of reinforcement is zero.
Resolving the forces in *x*-direction, and using stress function for stresses as $\sigma_x = \dfrac{\partial^2\phi}{\partial y^z}$, $\sigma_y = \dfrac{\partial^2\phi}{\partial x^2}$ and $\tau_{xy} = \dfrac{\partial^2\phi}{\partial x\partial y}$ and after integrating, an expression is obtained as	$\dfrac{F}{t}\cos\psi = -\left(\dfrac{\partial\phi}{\partial y} + B\right)$ (*a*)
Resolving the forces in *y*-direction after performing integration etc. as done under Eq. (*a*), another expression is obtained	$\dfrac{F}{t}\sin\psi = \dfrac{\partial\phi}{\partial x} + A$ (*b*)
From Eqs. (*a*) and (*b*)	$\tan\psi = \dfrac{dy}{dx} = -\dfrac{\dfrac{\partial\phi}{\partial x} + A}{\dfrac{\partial\phi}{\partial y} + B}$ (*c*)
Eq. (*c*) may be written as	$\dfrac{\partial\phi}{\partial y}\,dy + \dfrac{\partial\phi}{\partial x}\,dx + B\,dy + A\,dx = 0$ (*d*)

Particular	Formula

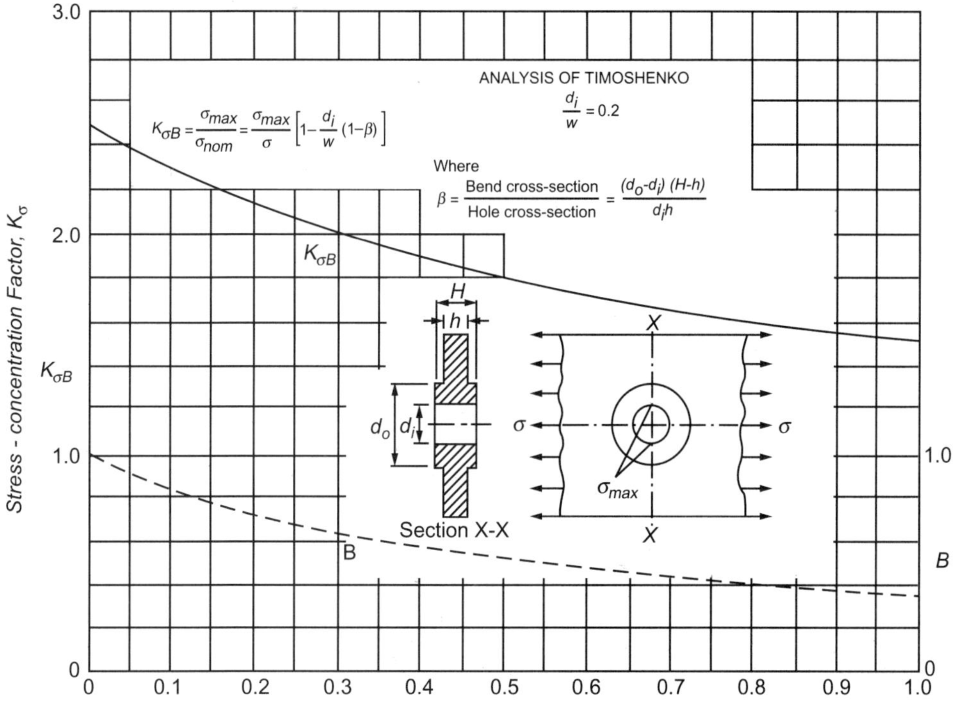

FIGURE 27-26 Stress-concentration factor $K_{\sigma B}$, for a symmetrically reinforced circular hole (cutout) in a flat plate in tension.

Integrating Eq. (*c*) an expression for ϕ is obtained as

$$\phi + By + Ax + C = 0 \tag{27-140}$$

The term $Ax + By + C$ can be included or excluded from the stress function without changing the stress distribution. These terms do not affect the shape of the neutral hole but determines the position of the hole. By omitting A, B and C the shape of the neutral hole is given by stress function

$$\phi = 0 \tag{27-141}$$

Compatibility

Strain in the plate = strain in the reinforcement

Considering the displacements and strain of triangular element klm as shown in Fig. 27-26 x and y directions due to σ_x, σ_y and τ_{xy} stresses and equating strain in the plate equal to strain in the reinforcement, an expression for area of reinforcement (A_R) for neutral hole is obtained as

$$A_R = \frac{F}{E\varepsilon_t} = t\left[\left(\frac{\partial\phi}{\partial x}\right)^2 + \left(\frac{\partial\phi}{\partial y}\right)^2\right]^{3/2} \times \left[\left\{\frac{\partial^2\phi}{\partial x^2}\left(\frac{\partial\phi}{\partial x}\right)^2 + \frac{\partial^2\phi}{\partial y^2}\left(\frac{\partial\phi}{\partial y}\right)^2 - 2\frac{\partial^2\phi}{\partial x\,\partial y}\frac{\partial\phi}{\partial x}\frac{\partial\phi}{\partial y}\right\} - \upsilon\left\{\frac{\partial^2\phi}{\partial x^2}\left(\frac{\partial\phi}{\partial y}\right)^2 + \frac{\partial^2\phi}{\partial y^2}\left(\frac{\partial\phi}{\partial x}\right)^2 - 2\frac{\partial^2\phi}{\partial x\,\partial y}\frac{\partial\phi}{\partial y}\frac{\partial\phi}{\partial x}\right\}\right]^{-1} \tag{27-142}$$

Particular	Formula

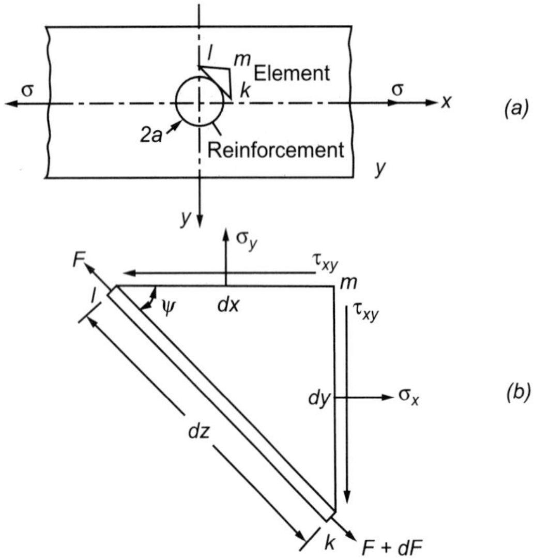

FIGURE 27-26 (a) A reinforced circular hole under tension, (b) element at reinforcement under the action of normal stresses σ_x and σ_y and shear stress τ_{xy} due to force F acting on the cross-section of the reinforcement.

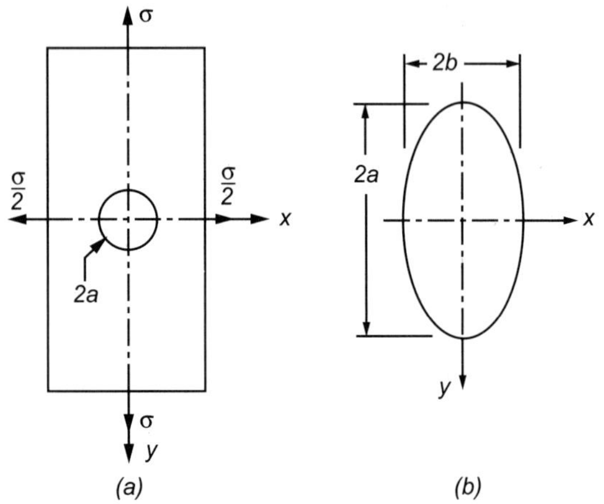

FIGURE 27-27 Thin walled cylinder with a circular hole under stress.

Neutral hole in a thin walled cylinder (Fig. 27-27)

The stress function may be taken as

$$\phi = \frac{\sigma}{2}\left(x^2 + \frac{y^2}{2}\right) + Ax + By + C \qquad (27\text{-}143a)$$

$$\phi = \frac{\sigma}{2}\left(x^2 + \frac{y^2}{2} - C\right) \qquad (27\text{-}143b)$$

if the origin of coordinates is taken at center of hole

For neutral hole

$$\phi = \frac{\sigma}{2}\left(x^2 + \frac{y^2}{2} - C\right) = 0 \qquad (27\text{-}144a)$$

Eq. (27-144) becomes an ellipse whose minor axis is $2r$ and ratio of major axis ($2a$) to minor axis ($2b$) is $\sqrt{2}:1$

$$x^2 + \frac{y^2}{2} - 2r = 0 \qquad (27\text{-}144b)$$

Substituting stress function from Eq. (27-144a) in Eq. (27-142), the area of reinforcement

$$A_R = \frac{rt\sqrt{2}\left(1 + \frac{x^2}{r^2}\right)^{3/2}}{1 - 2\upsilon + \frac{3x^2}{r^2}} \qquad (27\text{-}145)$$

where t = thickness of plate

Particular	Formula

COMPLEX VARIABLE METHOD APPLIED TO ELASTICITY

For equation of equilibrium in three-dimension and two-dimension

Refer to Eqs. (27-11a) to (27-11e).

The combinations of grouping of stress components are

$$\Theta = \sigma_x + \sigma_y \tag{27-146a}$$

$$\Phi = \sigma_x - \sigma_y + 2i\tau_{xy} \tag{27-146b}$$

$$\Psi = \tau_{xz} + i\tau_{yz} \tag{27-146c}$$

$$\sigma_z = \sigma_z \tag{27-146d}$$

The body force complex potentials and components of body force complex potentials

$$V(x, y, z) = U(z, \bar{z}, z) \tag{27-147}$$

$$F_{bx} = \frac{\partial V}{\partial x}; \quad F_{by} = \frac{\partial V}{\partial y}; \quad F_{bz} = \frac{\partial V}{\partial z}$$

The equations of equilibrium can be reduced to two equations using the stress combinations of Eqs. (27-146) and variable complex number

$$\frac{\partial}{\partial \bar{z}}(\Theta + 2U) + \frac{\partial \Phi}{\partial z} + \frac{\partial \Psi}{\partial z} = 0 \tag{27-148a}$$

where $z = x + iy$, $\bar{z} = x - iy$

$$\frac{\partial}{\partial z}(\sigma_z + U) + \frac{\partial \Psi}{\partial z} + \frac{\partial \bar{\Psi}}{\partial \bar{z}} = 0 \tag{27-148b}$$

When the body force is zero Eq. (27-148) can be written as

$$\frac{\partial \Theta}{\partial \bar{z}} + \frac{\partial \Phi}{\partial z} + \frac{\partial \Psi}{\partial z} = 0 \tag{27-149a}$$

$$\frac{\partial \Psi}{\partial z} + \frac{\partial \bar{\Psi}}{\partial \bar{z}} + \frac{\partial \sigma_z}{\partial z} = 0 \tag{27-149b}$$

Stress strain relation

The complex displacement

$$D = u + iv \tag{27-150}$$

Stress-displacement equations by using Eqs. (27-146) and Eqs. (27-58)

$$\frac{\partial u}{\partial x} + \frac{\partial v}{\partial y} + \frac{\partial w}{\partial z} = \frac{\partial D}{\partial z} + \frac{\partial \bar{D}}{\partial \bar{z}} + \frac{\partial w}{\partial z} \tag{27-151a}$$

$$(1 - 2v)\Theta = 2G\left(\frac{\partial D}{\partial z} + \frac{\partial \bar{D}}{\partial \bar{z}} + 2v\,\frac{\partial w}{\partial z}\right) \tag{27-151b}$$

$$(1 - 2v)\sigma_z = 2G\left[v\left(\frac{\partial D}{\partial z} + \frac{\partial \bar{D}}{\partial z}\right) + (1 - v)\,\frac{\partial w}{\partial z}\right] \tag{27-151c}$$

$$\Phi = 4G\,\frac{\partial D}{\partial \bar{z}} \tag{27-151d}$$

$$\Psi = G\left(\frac{\partial D}{\partial z} + 2\,\frac{\partial w}{\partial \bar{z}}\right) \tag{27-151e}$$

Particular	Formula

STRAIN COMBINATIONS

The strain combinations are

$$\theta = \varepsilon_x + \varepsilon_y$$

$$\phi = \varepsilon_x - \varepsilon_y + i\gamma_{xy} \tag{27-152}$$

$$\psi = \gamma_{xz} + i\gamma_{yz}; \quad D = u + iv$$

$$\theta = \frac{\partial D}{\partial z} + \frac{\partial \bar{D}}{\partial \bar{z}} \tag{27-153}$$

$$\phi = 2\,\frac{\partial D}{\partial \bar{z}}$$

$$\psi = \frac{\partial D}{\partial \bar{z}} + 2\,\frac{\partial w}{\partial \dot{z}}$$

Strain transformation rules (Fig. 27-28)

$$D = u + iv; \quad D' = u' + iv' \tag{27-154}$$

$$D = r\,\mathrm{e}^{i\theta}; \quad D' = D\,\mathrm{e}^{-i\alpha}$$

$$z' = z\,\mathrm{e}^{-i\alpha}$$

$$\theta' = \frac{\partial D'}{\partial z'} + \frac{\partial \bar{D}'}{\partial \bar{z}'} = \frac{\partial D}{\partial z} + \frac{\partial \bar{D}}{\partial \bar{z}} = \theta$$

$$\theta = \theta' \qquad (\theta \text{ being invariant})$$

$$\phi' = \phi\,\mathrm{e}^{-2i\alpha}$$

$$\psi = \frac{\partial D}{\partial \bar{z}} + 2\,\frac{\partial w}{\partial \bar{z}}$$

$$\psi' = \psi\,\mathrm{e}^{-i\alpha}$$

$$\frac{\partial \varepsilon}{\partial z} = \frac{\partial w}{\partial \bar{z}} = \frac{\partial \varepsilon'}{\partial z} \qquad \text{(invariant)}$$

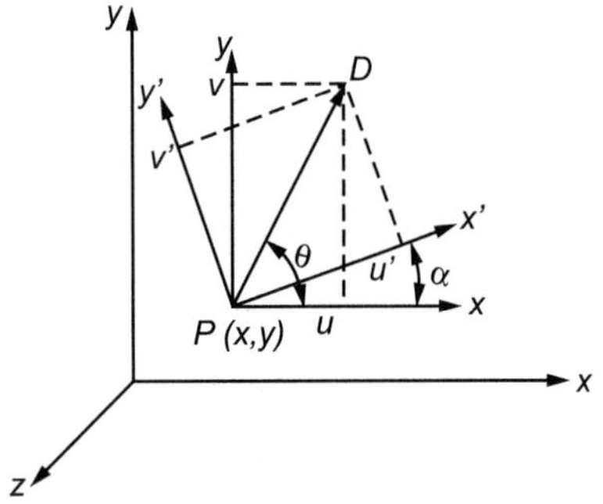

FIGURE 27-28 Strain transformation.

Stress transformation rules

$$\Theta = 2(\lambda + G)(\varepsilon_x + \varepsilon_y) + 2\lambda\varepsilon_z$$

$$= 2(\lambda + G)\theta + 2\lambda\varepsilon_z$$

$$= 2(\lambda + G)\theta' + 2\lambda\varepsilon_z'$$

$$= \Theta' \tag{27-155}$$

$$\Phi = 2G\left[\varepsilon_x - \varepsilon_y + i\left(\frac{\partial u}{\partial y} + \frac{\partial v}{\partial x}\right)\right] = 2G\phi$$

$$\Phi' = 2G\phi' = \Phi\,\mathrm{e}^{-2i\alpha}$$

$$\Psi = G\left[\frac{\partial u}{\partial \bar{z}} + \frac{\partial w}{\partial x} + i\left(\frac{\partial v}{\partial \bar{z}} + \frac{\partial w}{\partial y}\right)\right]$$

$$= G(\gamma_{xz} + i\gamma_{yz}) = G\psi$$

$$\Psi' = \Psi\,\mathrm{e}^{-ia}$$

Particular	Formula

PLANE STRAIN (Figs. 27-29 and 24-30)

External forces have to be applied on both top and bottom flat ends of the cylinder to prevent its movement in order to meet the condition that $w = 0$. The Eq. (27-151b) becomes

$$(1 - 2v)\Theta = 2G\left(\frac{\partial D}{\partial z} + \frac{\partial \bar{D}}{\partial \bar{z}}\right) \quad (27\text{-}156)$$

And Eq. (27-151d) becomes

$$\Psi = 0 \quad (27\text{-}156a)$$

$$\left(\Psi = G\,\frac{\partial D}{\partial z} \text{ since } U \text{ independent of } z,\ \Psi = 0\right)$$

The expression for F

$$F = 2[\Omega(z) + z\bar{\Omega}'(\bar{z}) + \bar{\omega}(\bar{z})] - \frac{1 - 2v}{1 - v}\,w \quad (27\text{-}156b)$$

The stress combinations are

$$\Theta = 2[\Omega'(z) + \bar{\Omega}'(\bar{z})] + \frac{1}{1 - v}\,\frac{\partial w}{\partial z} \quad (27\text{-}156c)$$

$$\Phi = -2[z\bar{\Omega}''(\bar{z}) + \bar{\omega}'(\bar{z})] + \frac{1 - 2v}{1 - v}\,\frac{\partial w}{\partial \bar{z}} \quad (27\text{-}157)$$

The displacement D

$$2GD = (3 - 4v)\Omega(z) - z\bar{\Omega}'(\bar{z}) - \bar{\omega}(\bar{z}) + \frac{1 - 2v}{2(1 - v)}\,w \quad (27\text{-}158)$$

BOUNDARY CONDITIONS

Specified stresses

The expression for F

$$F = 2\int_0^s (\sigma_n + i\tau_{ns} - U)\,\frac{\partial z}{\partial s}\,ds + \text{constant} \quad (27\text{-}159)$$

By using Eq. (27-155), Eq. (27-159) becomes

$$\Omega(z) + z\bar{\Omega}'(\bar{z}) + \bar{\omega}(\bar{z}) = \int_0^s (\sigma_n + i\tau_{ns})\,\frac{\partial z}{\partial s}\,ds + \text{constant} = f_1 + if_2 \text{ on } C \quad (27\text{-}160)$$

where f_1 and f_2 are functions of z only

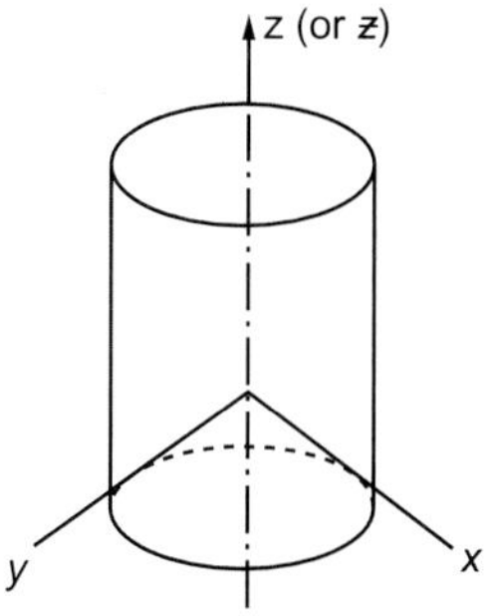

FIGURE 27-29

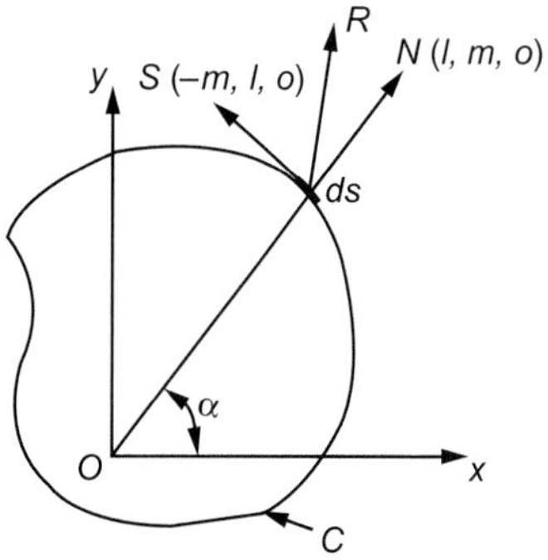

FIGURE 27-30

Particular	Formula
Specified displacement	
From Eq. (27-158) for D	$(3-4v)\Omega(z) - z\bar{\Omega}'(\bar{z}) - \bar{\omega}(\bar{z}) = 2GD - \dfrac{1-2v}{2(1-v)}\,w$ (27-161)
If body force is absent Eq. (27-161) becomes	$(3-4v)\Omega(z) - z\bar{\Omega}'(\bar{z}) - \bar{\omega}(\bar{z}) = 2G(g_1 + ig_2)$ on C (27-162) where g_1 and g_2 are functions of z only

FORCE AND COUPLE RESULTANTS AROUND THE BOUNDARY (Fig. 27-31)

The expression for force with components X and Y at point O	$X + iY = -i\left[\Omega(z) + z\bar{\Omega}'(\bar{z}) + \bar{\omega}(\bar{z})\right]_{A_1}^{B_1}$ (27-163)
The expression for couple at O	$N = Rl\left[\Psi(z) - z\omega(z) - z\bar{z}\Omega'(z)\right]_{A_1}^{B_1} + \int_{A_1}^{B_1} U \dfrac{\partial z}{\partial s}\,ds$ (27-164)

GENERALIZED PLANE STRESS

The average stress combinations assuming $\sigma_z = 0$, a stress free surface, i.e. $\tau_{xz} = \tau_{yz} = 0$ at the surface and body force potential $U(z,\bar{z})$ is independent of z	$\Theta_o = \sigma_x + \sigma_y$ (27-165a) $\Phi_o = \sigma_x - \sigma_y + 2i\tau_{xy}$ (27-165b) where $\Theta_o = \dfrac{1}{2h}\int_{-h}^{h} \Theta\,dz; \quad \Phi_o = \dfrac{1}{2h}\int_{-h}^{h} \Phi\,dz$ $\sigma_{pav} = \dfrac{1}{2h}\int_{-h}^{h} \sigma_p\,dz$

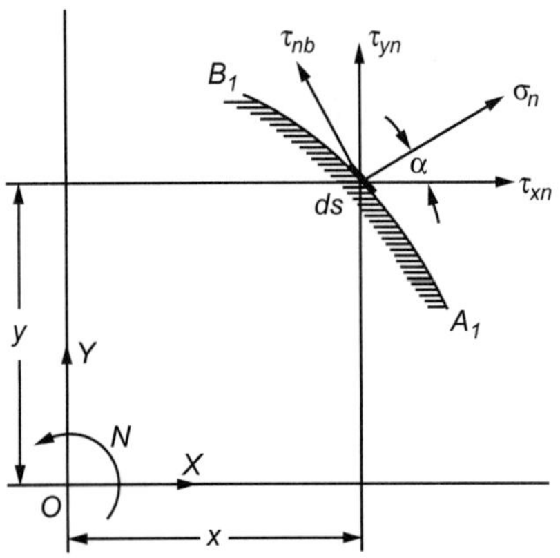

FIGURE 27-31

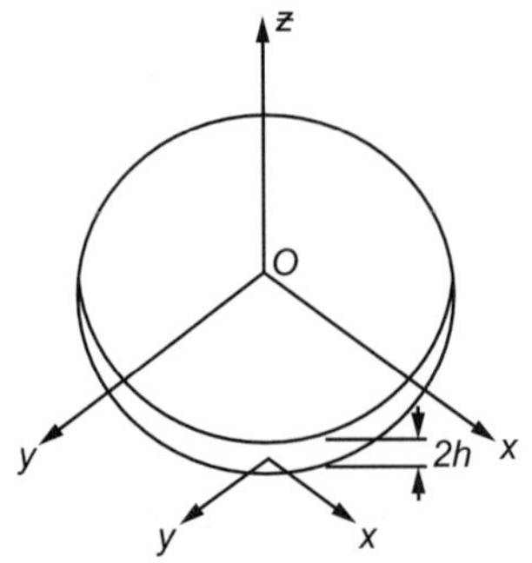

FIGURE 27-32

Particular	Formula
The average complex displacement	$D_o = u_o + iv_o = \frac{1}{2h}\int_{-h}^{h} D\,dz$ (27-166)
The body force Eq. $\frac{\partial\Phi}{\partial z} + \frac{\partial\Theta}{\partial\bar{z}} + \frac{\partial\Psi}{\partial z} = 0$ becomes	$\frac{1}{2h}\int_{-h}^{h}\left(\frac{\partial\Phi}{\partial z} + \frac{\partial\Theta}{\partial\bar{z}} + \frac{\partial\Psi}{\partial z}\right)dz = 0$ (27-167a)
	$= \frac{\partial\Phi_o}{\partial z} + \frac{\partial\Theta_o}{\partial z} = 0$
Taking into consideration the body force, Eq. (27-167) and other expression for F and Φ_o become	$\frac{\partial}{\partial\bar{z}}\langle\Theta_o + 2U\rangle + \frac{\partial\Phi_o}{\partial z} = 0$ (27-167b)
	$-\frac{v}{1-v}\left(\frac{\partial D}{\partial z} + \frac{\partial\bar{D}}{\partial\bar{z}}\right) = \frac{\partial\omega}{\partial z}$ (27-168a)
	$\Phi = 4G\frac{\partial D}{\partial\bar{z}}$ (27-168b)
	$\Phi_o = 4G\frac{\partial D_o}{\partial\bar{z}}$ (27-168c)
	$\frac{1-v}{1+v}\Theta_o = 2G\left(\frac{\partial D_o}{\partial z} + \frac{\partial\bar{D}}{\partial\bar{z}}\right)$ (27-168d)
The equations for generalized plane stress	$F = 2\{\Omega(z) + z\bar{\Omega}'(\bar{z}) + \bar{\omega}(\bar{z})\} + \frac{1-2K}{1-K}w$ (27-169)
	$2GD = \left(\frac{3-v}{1+v}\right)\Omega(z) - z\bar{\Omega}'(\bar{z}) - \bar{\omega}(\bar{z}) - \frac{1-2K}{2(1-K)}w$ (27-170)
	$\Theta = 2\left\{\Omega'(z) + \bar{\Omega}'(\bar{z}) - \frac{1}{1-K}\frac{\partial w}{\partial z}\right\}$ (27-171)
	$\Phi = -2\{z\bar{\Omega}''(\bar{z}) + \bar{\omega}'(\bar{z})\} - \frac{1-2K}{1-K}\frac{\partial w}{\partial\bar{z}}$ (27-172)

CONDITIONS ALONG A STRESS-FREE BOUNDARY, Fig. 27-33

Particular	Formula
Adding Eqs. (27-169) and (27-170) and putting $F = 0$ along free boundary, i.e. segment AB, the displacement along AB	$D = \frac{4}{E}\Omega(z)$ (27-173)

SOLUTION INVOLVING CIRCULAR BOUNDARIES (Figs. 27-33 and 27-34)

Particular	Formula
From stress strain transformation rules	$\Theta' = \Theta = \sigma_r + \sigma_\theta$

Particular	Formula

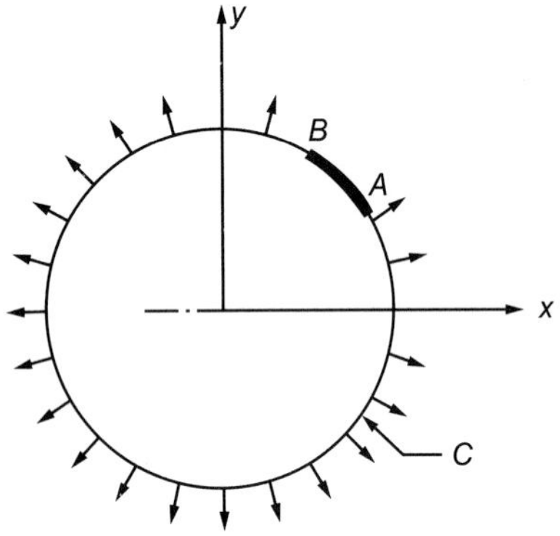

FIGURE 27-33

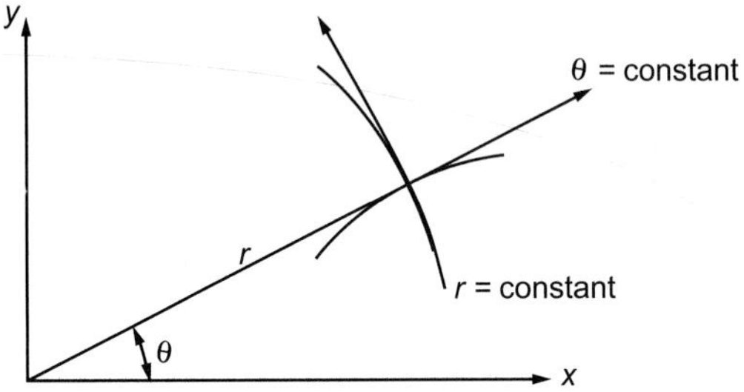

FIGURE 27-34

$$\Phi' = F\,e^{-2i\theta} = \sigma_r - \sigma_\theta + 2i\tau = \frac{\bar{z}}{z}\,\Phi$$

where $\tau_{r\theta} = \tau$, $z = r\,e^{i\theta}$, $\bar{z} = r\,e^{-i\theta}$

$$\Theta' = 2\{\Omega'(z) + \bar{\Omega}'(\bar{z})\} - \frac{1}{1-K}\frac{\partial w}{\partial z} \qquad (27\text{-}174)$$

$$\Phi' = -2\left\{\bar{z}\Omega''(\bar{z}) + \frac{\bar{z}}{z}\,\bar{\omega}'(\bar{z})\right\} - \frac{1-2K}{1-K}\,\frac{\bar{z}}{z}\,\frac{\partial w}{\partial z} \qquad (27\text{-}175)$$

$$2GD' = e^{-i\theta}\left\{\frac{3-v}{1+v}\,\Omega(z) - z\bar{\Omega}'(\bar{z}) - \bar{\omega}(\bar{z})\right\} - \frac{1-2K}{1-K}\,w \qquad (27\text{-}176)$$

The boundary conditions are

$$F = 2\int_0^s (\sigma_r + i\tau_{r\theta} + U)\,\frac{\partial z}{\partial s}\,ds + \text{constant} \qquad (27\text{-}177a)$$

$$\Omega(z) + z\bar{\Omega}'(\bar{z}) + \bar{\omega}(\bar{z}) = f_1 + if_2 \quad \text{on } C \qquad (27\text{-}177b)$$

APPLICATION OF CONFORMAL TRANSFORMATION (Fig. 27-35)

The stress combinations after transformation

$$\Theta' = \sigma_\xi + \sigma_\eta \qquad (27\text{-}178a)$$

$$\Phi' = \sigma_\xi - \sigma_\eta + 2i\tau_{\xi\eta} \qquad (27\text{-}178b)$$

Eqs. (27-178) are related stress combinations in rectangular coordinates x and y as

$$\Theta' = \Theta \qquad (27\text{-}179a)$$

$$\Phi' = \Phi\,e^{-2i\alpha} \qquad (27\text{-}179b)$$

Particular	Formula

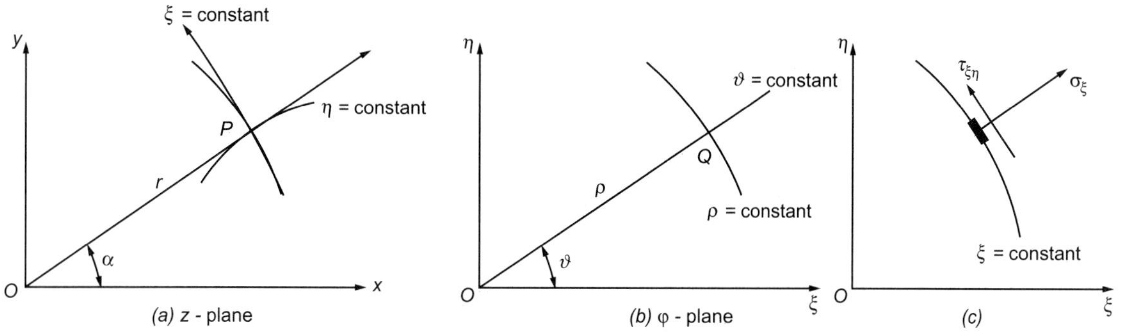

FIGURE 27-35

where
$z = z(\zeta) = f(\xi, \eta) + ig(\xi, \eta)$
$\zeta = \xi + i\eta$
$f(\xi, \eta)$ and $g(\xi, \eta)$ are real and imaginary parts of $z(\zeta)$

An explanation for $e^{-2i\alpha}$

$$e^{-2i\alpha} = \bar{z}'(\bar{\zeta})/z'(\zeta) \tag{27-179c}$$

or

Using Eqs. (27-179a) and (27-179b), and Eqs. (27-171) and (27-172), when these are no body forces, letting $\Omega(z) = \Omega_1(\xi)$ and $\omega(z) = \omega_1(\xi)$

$$\Theta' = 2\left\{\Omega_1'(\zeta)\frac{d\zeta}{dz} + \bar{\Omega}_1',(\bar{\zeta})\frac{d\bar{\zeta}}{d\bar{z}}\right\} \tag{27-180a}$$

$$\Theta' = 2\left\{\frac{\Omega_1'(\zeta)}{z'(\zeta)} + \frac{\bar{\Omega}_1',(\bar{\zeta})}{\bar{z}'(\bar{\zeta})}\right\} \tag{27-180b}$$

The transformation of a given boundary in the z-plane into the unit circle in the ζ-plane

$$\Phi' = -\frac{2}{z'(\zeta)}\left\{z(\zeta)\langle\bar{\zeta}''\bar{\Omega}_1'(\bar{\zeta}) + \bar{\zeta}'\bar{\Omega}_1''(\bar{\zeta})\rangle + \bar{\omega}_1'(\bar{\zeta})\right\} \tag{27-181a}$$

or

$$\Phi' = -\frac{2}{z'(\zeta)}\left\{\frac{1}{z(\zeta)}\left[\frac{\bar{\Omega}_1',(\bar{\zeta})}{\bar{z}'(\bar{\zeta})}\right] + \bar{\omega}_1',(\zeta)\right\} \tag{27-181b}$$

Using polar coordinates (ρ, ϑ), the stress components become

$$\Theta'' = \sigma_\rho + \sigma_\vartheta \tag{27-182a}$$

$$\Phi'' = \sigma_\rho + \sigma_\vartheta - 2i\tau_{\rho\vartheta} \tag{27-182b}$$

where $\Theta'' = \Theta'$ and $\Phi'' = \Phi'\, e^{-2i\vartheta} = \frac{\bar{\zeta}}{\zeta}\Phi'$.

Using polar coordinates Eqs. (27-180a) and (27-181) in terms of complex potentials become

$$\Theta'' = 2\left[\frac{\Omega'(\zeta)}{z'(\zeta)} + \frac{\bar{\Omega}'(\bar{\zeta})}{\bar{z}'(\bar{\zeta})}\right] \tag{27-183}$$

$$\Phi'' = -\frac{2\bar{\zeta}}{\zeta z'(\zeta)}\left[z(\zeta)\langle\bar{\zeta}''\bar{\Omega}_1'(\bar{\zeta}) + \bar{\zeta}'\bar{\Omega}_1''(\bar{\zeta})\rangle + \bar{\omega}'(\bar{\zeta})\right] \tag{27-184a}$$

$$\Phi'' = \frac{2\bar{\zeta}}{\zeta z'(\zeta)}\left[z(\zeta)\left\{\frac{\bar{\Omega}'(\bar{\zeta})}{\bar{z}'(\bar{\zeta})}\right\} + \bar{\omega}'(\bar{\zeta})\right] \tag{27-184b}$$

Particular	Formula

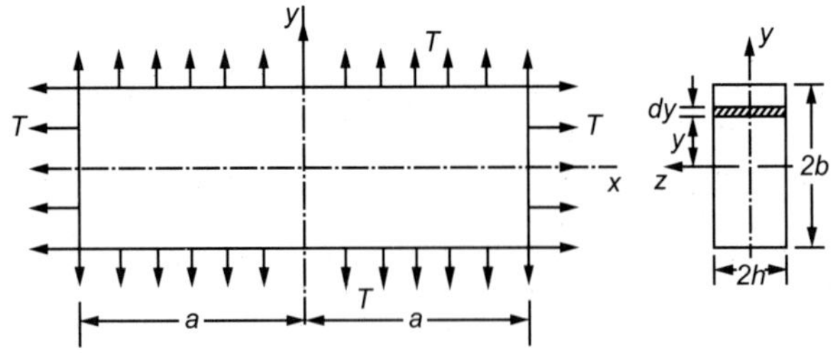

FIGURE 27-36

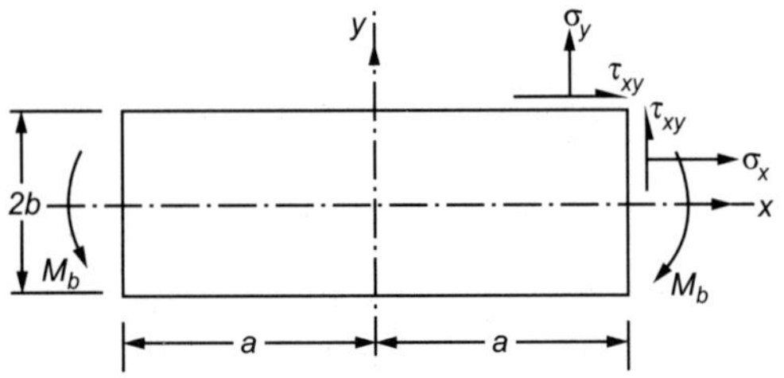

FIGURE 27-37

Rectangular plate under all round tension

Value of complex potentials $\Omega(z)$ and $\omega(z)$ assumed

$$\Omega(z) = \tfrac{1}{2}Tz; \quad \omega(x) = 0 \qquad (27\text{-}185)$$

From stress combination Eqs. (27-156c) and (27-157)

$$\Theta = 2\{\Omega'(z) + \bar{\Omega}'(\bar{z})\} + \frac{1}{1-v}\frac{\partial w}{\partial z}$$

$$= 2[\tfrac{1}{2}T + \tfrac{1}{2}T] = 2T \qquad (27\text{-}156c)$$

$$\Phi = -2\{z\bar{\Omega}''(\bar{z}) + \bar{\omega}'(\bar{z})\} + \frac{1-2v}{1-v}\frac{\partial w}{\partial \bar{z}} \qquad (27\text{-}157)$$

where $\Theta = \sigma_x + \sigma_y$ and $\Phi = \sigma_x - \sigma_y + 2i\tau_{xy}$

The stress σ_x and σ_y after equating real and imaginary parts

$$\sigma_x = T, \quad \sigma_y = T, \quad \tau_{xy} = 0 \qquad (27\text{-}186)$$

The displacement from Eq. (27-158) after equating real and imaginary parts

$$2GD = (3-4v)\Omega(z) - z\bar{\Omega}'(\bar{z}) - \bar{\omega}(\bar{z}) + \frac{1-2v}{2(1-v)}w \qquad (27\text{-}158)$$

$$D = \frac{T}{E}(1-v)(x+iy) = u + i\nu$$

$$u = \frac{T}{E}(1-v)x; \quad \nu = \frac{T}{E}(1-v)y \qquad (27\text{-}187)$$

Rectangular plate under plane flexure

Assume values of complex potentials $\Omega(z)$ and $\omega(z)$ as

$$\Omega(z) = Az^2$$

$$\omega(z) = Bz^2$$

Choose A and B, which may be complex, so that edges $y = \pm b$ are stress free.

Particular	Formula

Boundary conditions

From stress combinations Eqs. (27-156) and (27-157) boundary conditions

$$\Omega'(z) + \bar{\Omega}'(\bar{z}) + \bar{z}\Omega''(z) + \omega'(z) = \sigma_y + i\tau_{xy} \qquad (27\text{-}156)$$

$\sigma_y = 0$, $\tau_{xy} = 0$ throughout the plate

$A = iC$ and $B = -iC$ where C is real

$\Theta = \sigma'_x + \sigma'_y = \sigma'_x = -8Cy$

The bending moment

$$M_b = \int_{-b}^{b} \sigma_x 2hy\, dy = -8CI \qquad (27\text{-}188)$$

where

I = moment of inertia about oz

$$C = -\frac{M_b}{8I}$$

The values of complex potentials $\Omega(z) = Az^2$ and $\omega(z) = Bz^2$ are

$$\Omega(z) = -\frac{iM_b}{8I} z^2; \quad \omega(z) = \frac{iM_b}{8I} z^2 \qquad (27\text{-}188a)$$

The displacement from Eq. (27-158)

$$D = \frac{1}{2G}\left[(3 - 4v)\Omega(z) - z\bar{\Omega}'(\bar{z}) - \bar{\omega}(\bar{z})\right] = u + iv$$

when body forces are zero

Substituting the values of $\Omega(z)$ and $\omega(z)$ in the above, u and v can be determined.

Thick cylinder under internal and external pressure

Values of complex potentials $\Omega(z)$ and $\omega(z)$ assumed using boundary conditions at $r = a$ or $d_i/2$ and $r = b$ or $d_o/2$ with no body forces, assuming internal pressure p_i, external pressure p_o, values of A and B in Eq. (27-189), which are real, can be found. From Eqs. (27-174) and (27-175)

$$\Omega(z) = Az \quad \text{and} \quad \omega(z) = \frac{B}{z} \qquad (27\text{-}189a)$$

where A and B are real

$$\tfrac{1}{2}[\Theta' + \Phi'] = \sigma_r + i\tau_{r\theta} = \Omega'(z) + \bar{\Omega}'(\bar{z}) - z\bar{\Omega}''(z) - \frac{z}{\bar{z}}\bar{\omega}'(\bar{z}) \qquad (27\text{-}189b)$$

The expressions for σ_θ and σ_r at any radius

The equations for σ_θ and σ_r are given in Eqs. (27-101b) and (27-101a) respectively.

Rotating solid disk and hollow disk of uniform thickness rotating at ω rad/s

Values of complex potentials $\Omega(z)$ and $\omega(z)$ assumed

$$\Omega(z) = Cz \quad \text{and} \quad \omega(z) = \frac{B}{z} \qquad (27\text{-}189c)$$

where C and B are real

Particular	Formula
Using boundary conditions at $(\sigma_r)_{r=b}=0$ and $(\sigma_r)_{r=0}\neq 0$ for solid disk $(\sigma_r)_{r=a}=0$ and $(\sigma_r)_{r=b}=0$ for hollow disc taking into consideration body forces, values of C and B in Eq. (27-189c) which are real can be found	Refer Eqs. (27-126), (27-127) and (27-128) to (27-131)
The radial displacements at the boundaries	$(u_r)_{r=a}=\dfrac{\rho\omega^2 a}{4E}\{(1-v)a^2+(3+v)b^2\}$ (27-189d)
	$(u_r)_{r=b}=\dfrac{\rho\omega^2 b}{4E}\{(1-v)b^2+(3+v)a^2\}$ (27-189e)
Large plate under uniform uniaxial tension with a centrally located unstressed circular hole	
Values of complex potentials $\Omega(z)$ and $\omega(z)$ assumed	$\Omega(z)=\dfrac{Tz}{4}+\dfrac{A}{z}$ (27-190)
	$\omega(z)=-\dfrac{1}{2}Tz+\dfrac{B}{z}+\dfrac{C}{z^3}$
	where A, B and C are real
Using Eq. (27-189b) and above complex potentials	$\sigma_r-i\tau_{r\theta}=\dfrac{1}{2}T-\dfrac{3A}{z^2}+\dfrac{1}{2}T\dfrac{z}{\bar{z}}+\dfrac{B}{z\bar{z}}+\dfrac{3C}{z^3\bar{z}}$ (27-190a)
Using boundary condition at $r=a$	$(\sigma_r-i\tau_{r\theta})_{r=a}=\left(\dfrac{1}{2}T+\dfrac{B}{a^2}\right)+\left(\dfrac{1}{2}T+\dfrac{A}{a^2}\right)e^{2i\theta}+\left(\dfrac{3C}{a^4}-\dfrac{3A}{a^2}\right)e^{2i\theta}$ (27-190b)
	$A=\frac{1}{2}Ta^2,\quad B=-\frac{1}{2}Ta^2,\quad C=\frac{1}{2}a^4$ (27-190c)
	since hole is stress free
The new values of $\Omega(z)$ and $\omega(z)$	$\Omega(z)=\dfrac{Tz}{4}+\dfrac{1}{2}\dfrac{Ta^2}{z}$ (27-190d)
	$\omega(z)=-\dfrac{1}{2}Tz-\dfrac{Ta^2}{4}+\dfrac{a^4}{2z^3}$ (27-190e)
Using Eqs. (27-174), (27-175) and after equating the real and imaginary parts, the stress components are	$\sigma_r=\dfrac{1}{2}T\left[\left(1-\dfrac{a^2}{r^2}\right)+\left(1-\dfrac{4a^2}{r^2}+\dfrac{3a^4}{r^4}\right)\cos 2\theta\right]$ (27-191)
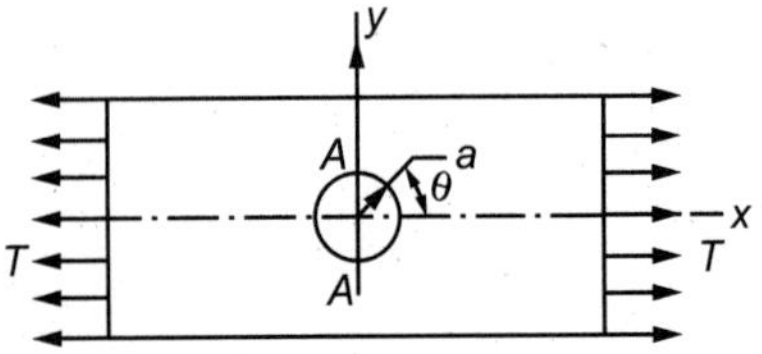 FIGURE 27-38	$\sigma_\theta=\dfrac{1}{2}T\left[\left(1+\dfrac{a^2}{r^2}\right)-\left(1+\dfrac{3a^4}{r^4}\right)\cos 2\theta\right]$ (27-192)
	$\tau_{r\theta}=-\dfrac{1}{2}T\left(1+\dfrac{2a^2}{r^2}-\dfrac{3a^4}{r^4}\right)\sin 2\theta$ (27-193)

Particular	Formula	
The σ_θ, σ_r and $\tau_{r\theta}$ at $r = a$	$(\sigma_r)_{r=a} = (\tau_{r\theta})_{r=a} = 0$	(27-194a)
	$(\sigma_\theta)_{r=a} = T(1 - 2\cos 2\theta)$	(27-194b)
The maximum tangential stress	$\sigma_{\theta\ \max} = (\sigma_\theta)_{r=a} = 3T$	(27-194c)
The stress concentration factor	$K_\sigma = \frac{(\sigma_\theta)_{\max}}{T} = \frac{3T}{T} = 3$	(27-195)

Large plate containing a circular hole under uniform pressure

Values of complex potentials $\Omega(z)$ and $\omega(z)$ assumed	$\Omega(z) = 0; \quad \omega(z) = \frac{A}{z}$	(27-196)
From Eqs. (27-174) and (27-175) in the absence of body forces	$\Theta' = 2\{\Omega'(z) + \bar{\Omega}'(\bar{z})\} = \sigma_r + \sigma_\theta = 0$	(a)
	$\Phi' = 2\left\{\bar{z}\bar{\Omega}''(\bar{z}) + \frac{\bar{z}}{z}\bar{\omega}'(\bar{z})\right\}$	
	$= \sigma_r - \sigma_\theta + 2i\tau_{r\theta} = \frac{2A}{r^2}$	(b)
Boundary conditions are	$(\sigma_r)_{r=a} = -p = \frac{2A}{a^2}$	
	$A = -pa^2$	(c)
The new complex potentials	$\Omega(z) = 0; \quad \omega(z) = -\frac{pa^2}{z}$	(d)
The stress components are	$\sigma_r = -\frac{pa^2}{r^2}; \quad \tau_{r\theta} = 0$	(27-197)
	$\sigma_\theta = -\sigma_r = \frac{pa^2}{r^2}$	
The displacement from Eq. (27-176)	$2GD' = 2G(u_r + iu_\theta) = e^{-i\theta}\left(-\frac{A}{\bar{z}}\right) = -\frac{A}{r}$	
	$(u_r) = -\frac{A}{2Gr}; \quad u_\theta = 0$	(27-198)
	$(u_r)_{r=a} = \frac{pa}{2G}$	

Large plate containing a circular hole filled by an oversize disk

1. Rigid Disk

The radius of disk r_d	$r_d = a(1 + \varepsilon)$ where a = radius of hole	(a)

Particular	Formula	
From first of Eq. (27-198), the radial displacement	$u_r = a\varepsilon = -\dfrac{A}{2Ga}$ or $A = -2Ga^2\varepsilon$	(*b*)
The stress components	$\sigma_r = -\sigma_\theta = -2G\varepsilon\,\dfrac{a^2}{r^2}$	(*c*)
	$\tau_{r\theta} = 0$	(*d*)
2. Elastic Disk		
The complex potential for all round pressure on the disk	$\Omega_1(z) = -\frac{1}{2}pz;\quad \omega_1(z) = 0$	(*e*)
The displacement from Eq. (27-176)	$2G_1D' = 2G_1(u_{r1} + iu_{\theta 1}) = e^{-i\theta}\left\{\dfrac{-(1-v)pz}{1+v}\right\}$	
	$= -\dfrac{-p(1-v_1)pa}{1+v_1}$	(*f*)
	$u_{r1} = \dfrac{-p(1-v_1)a}{E_1}$	(*g*)
	where subscript 1 for disk and 2 for plate	
The radial displacement of plate	$u_{r2} = \dfrac{pa}{2G_2}$	(*h*)
The pressure between disc and plate	$p = \dfrac{E_1E_2}{E_1(1+v_2) + E_2(1-v_1)}$	(27-198)

Elliptical hole in a large plate under tension (Fig. 27-39)

Particular	Formula	
The expression for transformation	$z = C\left(\zeta + \dfrac{m}{\zeta}\right),\qquad m < 1$	(27-199)
	Transforms the outside of an ellipse of semiaxes *a* and *b* in the *z*-plane into the outside of a unit circle *r* in the ζ-plane, provided	
	$C = \frac{1}{2}(a+b),\quad m = \dfrac{a-b}{a+b}$	(27-200a)
	$\dfrac{z}{C} = e^{i\vartheta} + m\,e^{-i\vartheta}$	
	$= (1+m)\cos\vartheta + (1-m)i\sin\vartheta$	(27-200b)
	$z = \left(\dfrac{a+b}{2}\right)\left[\dfrac{2a}{a+b}\cos\vartheta + i\,\dfrac{2b}{a+b}\sin\vartheta\right]$	(27-200c)
	or $x + iy = a\cos\vartheta + ib\sin\vartheta$	(27-200d)
	ϑ = eccentric angle around the ellipse	

Particular	Formula

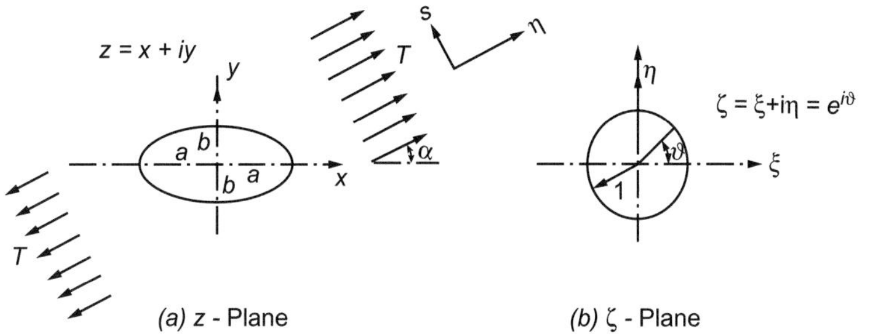

FIGURE 27-39

Particular	Formula
The points at which the transformation ceases to be conformal are	$z'(\zeta) = 0$ (27-201a) $z'(\zeta) = C\left(1 - \frac{m}{\zeta^2}\right) = 0$ (27-201b) $\zeta = \pm\sqrt{m}$, since $m < 1$
The boundary condition at the stress free ellipse	$\Omega(z) + z\bar{\Omega}'(\bar{z}) + \bar{\omega}(\bar{z}) = f_1 + if_2$ on the boundary of ellipse $= 0$ (27-202)
The boundary condition in terms of ζ	$\Omega_1(\zeta) + z(\zeta)\frac{\bar{\Omega}_1'(\bar{\zeta})}{\bar{z}'(\bar{\zeta})} + \bar{\omega}_1(\bar{\zeta}) = 0$ or $\bar{z}'(\zeta)\Omega_1(\zeta) + z(\zeta)\bar{\Omega}_1'(\bar{\zeta}) + \bar{z}'(\bar{\zeta}) + \bar{z}'(\bar{\zeta})\bar{\omega}_1(\bar{\zeta}) = 0$ (27-203)
Eq. (27-203) on unit circle becomes	$\bar{z}'(\bar{\sigma})\Omega_1(\sigma) + z(\sigma)\bar{\Omega}_1'(\bar{\sigma}) + \bar{z}'(\bar{\sigma})\bar{\omega}_1'(\bar{\sigma}) = 0$ or $\bar{z}'\left(\frac{1}{\sigma}\right)\Omega_1(\sigma) + z(\sigma)\bar{\Omega}_1'\left(\frac{1}{\sigma}\right) + \bar{z}'\left(\frac{1}{\sigma}\right)\bar{\omega}_1\left(\frac{1}{\sigma}\right) = 0$ (27-204) where $\zeta = \sigma$; $\bar{\zeta} = \bar{\sigma} = \frac{1}{\sigma}$ on γ since $\sigma\bar{\sigma} = 1$ on unit circle
The complex potentials for an infinite plate without a hole acted upon by uniaxial tension at an angle α to the x-axis in the z-plane and ζ-plane	$\Omega(z) = \frac{1}{4}Tz$, $\omega(z) = -\frac{1}{2}Tz\,e^{-2i\alpha}$ on z plane (27-205a) $z = C(\zeta + m/\zeta) \to C\zeta$ and $z'(\zeta) = C\left(1 - \frac{m}{\zeta^2}\right)$ $\to C$ at ζ in ζ-plane $\Omega_1(\zeta) = \frac{1}{4}TC\zeta$, $\omega_1(\zeta) = -\frac{1}{2}TC\zeta\,e^{-2i\alpha}$ (27-205b)
The complex potentials for an infinite plate with stress free elliptic hole subject to tension at an angle α to the x-axis in ζ-plane	$\Omega_1(\zeta) = \frac{1}{4}TC\left(\zeta + \frac{A}{\zeta} + \frac{B}{\zeta^2} + \frac{C}{\zeta^3} + \cdots\right)$ (27-206a) $z'(\zeta)\omega_1(\zeta) = -\frac{1}{2}TC^2\left(\zeta e^{-2i\alpha} + \frac{D_1}{\zeta} + \frac{E_1}{\zeta^2} + \frac{F_1}{\zeta^3} + \cdots\right)$ (27-206b)

Particular	Formula
Using Eqs. (27-206) in Eqs. (27-204), after equating coefficients of powers of σ or ζ (since $E_1 = B = C = 0$, $D_1 = 1 + m^2 - 2Me^{2i\alpha}$, $F_1 = -e^{2i\alpha}$, $A = 2e^{2i\alpha} - m$)	$\Omega_1(\zeta) = \frac{1}{4} TC \left\langle \zeta + \frac{2e^{2i\alpha} - m}{\zeta} \right\rangle$ (27-206c) $\omega_1(\zeta) = -\frac{1}{2} TC \left\langle \zeta e^{-2i\alpha} + \frac{\frac{1 + m^2 - 2m e^{2i\alpha}}{\zeta} - \frac{e^{2i\alpha}}{\zeta^3}}{1 - \frac{m}{\zeta^2}} \right\rangle$ (27-206d)
The tangential stress on the boundary of elliptical hole from Eq. (27-183) $\Theta'' = \sigma_\vartheta + \sigma_\rho$ where $\sigma_\rho = 0$ after equating to real part of right hand side of equation and simplification (Fig. 27-39)	$\sigma_\vartheta = T\left[\frac{1 - m^2 + 2m\cos 2\alpha - 2\cos 2(\vartheta - \alpha)}{1 + m^2 - 2m\cos 2\vartheta}\right]$ (27-207)
The tangential stress on the boundary of elliptical hole for $\alpha = 0$ (Fig. 27-40)	$\sigma_\vartheta = T\left[\frac{1 - m^2 + 2m - 2\cos 2\vartheta}{1 + m^2 - 2n\cos 2\vartheta}\right]$ (27-208)
The maximum tangential stress $\sigma_{\vartheta\,\max}$ on the contour of any elliptical hole for any value of $m = \frac{a-b}{a+b}$ and $c = \frac{1}{2}(a+b)$ will be at $\vartheta = \pm\frac{\pi}{2}$	$(\sigma_\vartheta)_{\max} = T\left(\frac{3 - m}{1 + m}\right) = T\left(1 + \frac{2b}{a}\right)$ (27-209)
If $a = b$ in Eq. (27-209), then the ellipse becomes a circle	$(\sigma_v)_{\max} = 3T$ (27-209a)
The stress concentration factor	$K_\sigma = \frac{(\sigma_v)_{\max}}{T} = 3$ (27-209b)
By taking $\alpha = 45^\circ$ with $T = -S$, and $\alpha = -45^\circ$ with $T = +S$, and on adding these solutions, a solution for pure shear S applied to an infinite flat plate with an elliptical hole at infinity is obtained. The shear will be parallel to the axes of the ellipse with σ_θ around the elliptical hole is given by	$\sigma_\theta = S\left(\frac{4\sin 2\theta}{1 - 2m\cos 2\theta + m^2}\right)$ (27-210)

MUSKHELISHVILI'S DIRECT METHOD

Particular	Formula
In this method that a hole L can be transformed conformally into a unit circle γ in the ζ-plane so that outside of the hole is mopped on the inside of γ (Fig. 27-41)	
The form of the conformal transformation will be	$z = C\left(\frac{1}{\zeta} + e_1\zeta + e_2\zeta^2 + e_3\zeta^3 + \cdots + e_n\zeta^n\right)$ (27-211)
If the loading of the plate at infinity is given by the complex potential $\Omega^*(\zeta)$, $\omega(\zeta)^*$, the full complex potentials which will also satisfy the condition around the hole, can be written as	$\Omega(\zeta) = \Omega^*(\zeta) + \Omega_o(\zeta)$ (27-212) $\omega(\zeta) = \omega^*(\zeta) + \omega_o(\zeta)$ (27-213) where $\Omega_o(\zeta) = \sum_o^\infty a_n\zeta^n, \quad \omega_o(\zeta) = \sum_o^\infty b_n\zeta^n$ (27-213a)

Particular	Formula

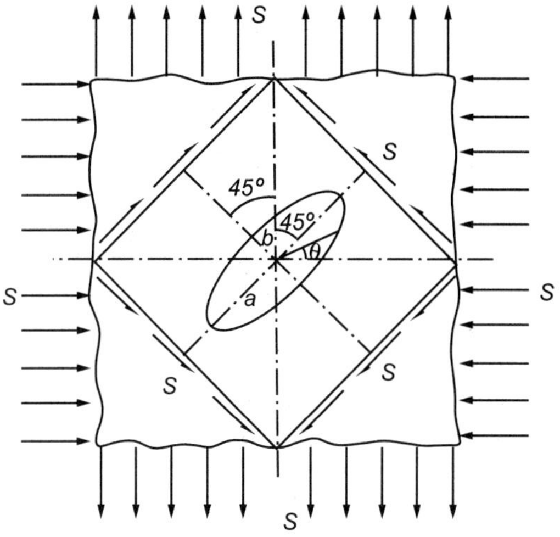

FIGURE 27-40

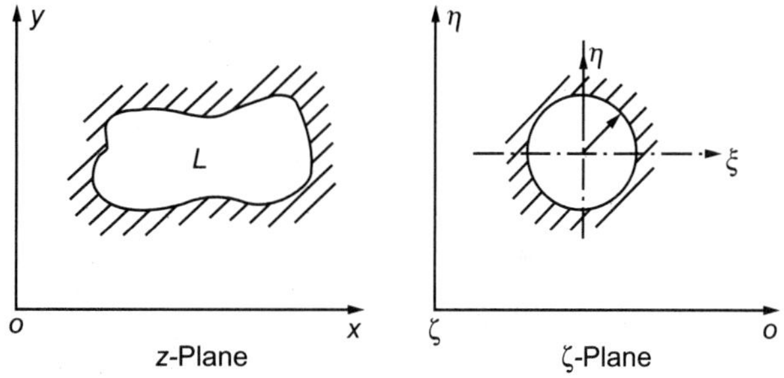

FIGURE 27-41

The boundary condition around a stress free hole assuming no body forces is given by (refer to Eqs. (27-203) and (27-204))

$$\Omega(\sigma) + \frac{z(\sigma)}{\bar{z}'\left(\frac{1}{\sigma}\right)}\,\bar{\Omega}'\left(\frac{1}{\sigma}\right) + \bar{\omega}\left(\frac{1}{\sigma}\right) = 0 \tag{27-214}$$

Substituting the complex potentials given by Eqs. (27-212), in Eq. (27-214)

$$\Omega_o(\sigma) + \frac{z(\sigma)}{\bar{z}'\left(\frac{1}{\sigma}\right)}\,\bar{\Omega}_o'\left(\frac{1}{\sigma}\right) + \bar{\omega}_o\left(\frac{1}{\sigma}\right) = f_1 + if_2 \tag{27-215a}$$

where

$$f_1 + if_2 = -\left[\Omega^*(\sigma) + \frac{z(\sigma)}{\bar{z}'\left(\frac{1}{\sigma}\right)}\,\bar{\Omega}'^*\left(\frac{1}{\sigma}\right) + \bar{\omega}^*\left(\frac{1}{\sigma}\right)\right] \tag{27-215b}$$

Using Harnack's theorem, residue theorem and Cauchy's integral, multiplying by $\frac{1}{2\pi i}\frac{d\sigma}{\sigma - \zeta}$ and integrating around γ Eq. (27-214) can be written as

$$\frac{1}{2\pi i}\int_\gamma \frac{\Omega_o(\sigma)}{\sigma - \zeta}\,d\sigma + \frac{1}{2\pi i}\int_\gamma \frac{z(\sigma)}{\bar{z}'\left(\frac{1}{\sigma}\right)}\,\frac{\bar{\Omega}_o'\left(\frac{1}{\sigma}\right)}{\sigma - \zeta}\,d\sigma + \frac{1}{2\pi i}\int_\gamma \frac{\bar{\omega}_o\left(\frac{1}{\sigma}\right)}{\sigma - \zeta}\,d\sigma = \frac{1}{2\pi i}\int_\gamma \frac{f_1 + if_2}{\sigma - \zeta}\,d\sigma \tag{27-216}$$

The complex potential $\Omega_o(\zeta)$ from Eq. (27-216) is

$$\Omega_o(\zeta) + \frac{1}{2\pi i}\int_\gamma \frac{z(\sigma)\bar{\Omega}_o'\left(\frac{1}{\sigma}\right)}{\bar{z}'\left(\frac{1}{\sigma}\right)(\sigma - \zeta)}\,d\sigma = \frac{1}{2\pi i}\int_\gamma \frac{f_1 + if_2}{\sigma - \zeta}\,d\sigma \tag{27-217}$$

Particular	Formula
Taking conjugate of Eq. (27-215), remembering that $\sigma\bar{\sigma} = 1$ and multiplying by $\dfrac{1}{2\pi i}\dfrac{d\sigma}{\sigma - \zeta}$ and integrating around γ	$$\frac{1}{2\pi i}\int_\gamma \frac{\bar{\Omega}_o\left(\frac{1}{\sigma}\right)}{\sigma-\zeta}\,d\sigma + \frac{1}{2\pi i}\int_\gamma \frac{\bar{z}\left(\frac{1}{\sigma}\right)}{z'(\sigma)}\frac{\Omega_o'(\sigma)}{\sigma-\zeta}\,d\sigma\zeta + \frac{1}{2\pi i}\int_\gamma \frac{\omega_o(\sigma)}{\sigma-\zeta}\,d\sigma = \frac{1}{2\pi i}\int_\gamma \frac{f_1 - if_2}{\sigma-\zeta}\,d\sigma \quad (27\text{-}218)$$
The complex potential $\omega_o(\zeta)$ can be found after substituting the value of $\Omega_o(\sigma)$ Eq. (27-218) which can be evaluated from Eq. (27-217)	$$\frac{1}{2\pi i}\int_\gamma \frac{\bar{z}\left(\frac{1}{\sigma}\right)}{z'(\sigma)}\frac{\Omega_o'(\sigma)}{\sigma-\zeta}\,d\sigma + \omega_o(\zeta) = \frac{1}{2\pi i}\int_\gamma \frac{f_1 - if_2}{\sigma-\zeta}\,d\sigma \quad (27\text{-}219)$$

Stress free square hole in a flat plate under uniform uniaxial tension (Fig. 27-42)

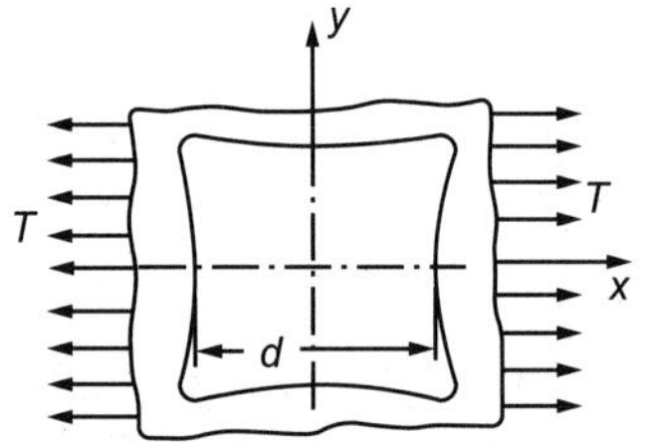

FIGURE 27-42

Particular	Formula
The form of the conformal transformation will be	$$z = C\left(\frac{1}{\zeta} - \frac{\zeta^3}{6}\right) \quad (27\text{-}220)$$
The known complex potential in this case	$$\Omega^*(\zeta) = \frac{1}{4}TC\left(\frac{1}{\zeta} - \frac{\zeta^3}{6}\right) \quad (27\text{-}221)$$ $$\omega^*(\zeta) = -\frac{1}{2}TC\left(\frac{1}{\zeta} - \frac{\zeta^3}{6}\right) \quad (27\text{-}222)$$
After substituting $\Omega^*(\zeta)$ and $\omega^*(\zeta)$ from Eqs. (27-221) and (27-222) into Eq. (27-215)	$$f_1 + if_2 = -\frac{1}{4}TC\left(\frac{2}{\sigma} - 2\sigma - \frac{\sigma^3}{3} + \frac{1}{3\sigma^3}\right) \quad (27\text{-}223)$$
After substituting the value of $f_1 + if_2$ from Eq. (27-223) in Eq. (21-217) and simplification	$$\Omega_o(\zeta) = TC\left(\frac{3}{7}\zeta + \frac{1}{12}\zeta^3\right) \quad (27\text{-}224)$$
Substituting the value of $\Omega_o(\zeta)$ from Eq. (27-224) in Eq. (27-219) and after simplification	$$\omega_o(\zeta) = -\frac{1}{4}TC\left(2\zeta + \frac{1}{3}\zeta^3\right) - \left[\frac{1}{3\zeta}\frac{1-6\zeta^4}{2+\zeta^4} \times TC\left(\frac{3}{7} + \frac{\zeta^2}{4\zeta}\right)\right] + \frac{1}{14}\frac{TC}{\zeta} \quad (27\text{-}225)$$

Particular	Formula	
The full complete complex potentials after simplification	$\Omega(\zeta) = TC\left(\frac{3}{7}\zeta + \frac{1}{4}\frac{1}{\zeta} + \frac{1}{24}\zeta^3\right)$	(27-226)
	$\omega(\zeta) = -TC\left(\frac{1}{2\zeta} + \frac{91\zeta - 78\zeta^3}{84(2+\zeta^4)}\right)$	(27-227)
The tangential stress around the square hole	$\sigma_\vartheta = \text{Rl. } 4\,\frac{\Omega'(\zeta)}{z'(\zeta)}$	(27-228)
By adding more terms to the expression for transformation	$z = C\left(\frac{1}{\zeta} - \frac{1}{6}\zeta^3 + \frac{1}{56}\sigma^7\right)$ the radius becomes $r = 0.025d$	(27-228a)
The radius r will be rounded of	$z = C\left(\frac{1}{\zeta} - \frac{1}{6}\zeta^3 + \frac{1}{56}\zeta^7 - \frac{1}{176}\zeta^{11}\right)$ the radius becomes $r = 0.014d$	(27-228b)
For graph of σ_ϑ/T versus ϑ in degrees	Refer to Fig. 27-43.	

Stress free square hole in a flat plate under pure bending (Fig. 27-44)

Particular	Formula	
The conformal transformation for plate with a square hole such that the diagonals along the coordinate axes as shown in Fig. 27-44	$z = C\left(\frac{1}{\zeta} + \frac{1}{6}\zeta^3\right)$	(27-229)
The known complex potentials from Eqs. (27-188a)	$\Omega^*(z) = -\frac{iM_b}{8I}z^2; \quad \omega^*(z) = \frac{iM_b}{8I}z^2$	(27-188a)

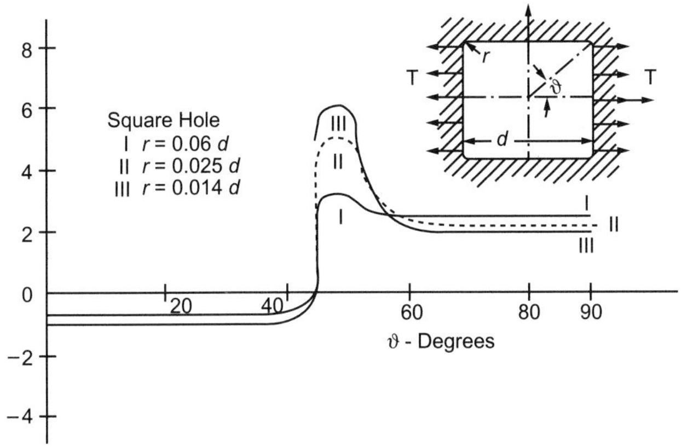

FIGURE 27-43

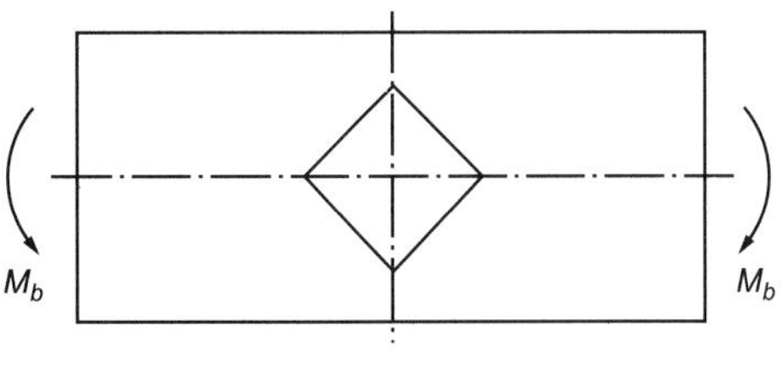

FIGURE 27-44 Flat plate with stress-free square hole under pure bending.

Particular	Formula
The complete complex potentials in ζ-plane will be of the form	$\Omega(\zeta) = -\frac{iM_bC^2}{8I}\left(\frac{1}{\zeta} + \frac{1}{6}\zeta^3\right)^2 + \Omega_o(\zeta)$ (27-230a)
	$\omega(\zeta) = \frac{iM_bC^2}{8I}\left(\frac{1}{\zeta} + \frac{1}{6}\zeta^3\right)^2 + \omega_o(\zeta)$ (27-230b)
From Eq. (27-217)	$\Omega_o(\zeta) = \frac{iM_bC^2}{8I}\left(\frac{4}{3}\zeta^2 - \frac{1}{3}\zeta^4 + \frac{1}{36}\zeta^6\right)$ (27-231)
From Eq. (27-219)	$\omega_o(\zeta) - \frac{1}{2\zeta}\,\frac{1+6\zeta^4}{2-\zeta^4}\,\Omega_o'(\zeta) = -\frac{iM_bC^2}{8I}\left(\frac{4}{3}\zeta^2 - \frac{1}{3}\zeta^4 + \frac{1}{16}\zeta^6 - \frac{37}{18}\right)$ (27-232)
The full complex potentials become often simplifying	$\Omega(\zeta) = -\frac{iM_bC^2}{8I}\left(\frac{1}{\zeta^2} - \zeta^2 + \frac{1}{3}\zeta^4\right)$ (27-233)
	$\omega(\zeta) = \frac{iM_bC^2}{8I}\left(\frac{18 + 45\zeta^2 - 31\zeta^4 + 36\zeta^6 - 15\zeta^8}{9\zeta^2(2-\zeta^4)}\right)$ (27-234)
After knowing full complex potentials, the tangential stress at various angles around the hole/cutout can be calculated	
For graphs of $\sigma_v/(M_bc/I)$ versus ϑ degree	Refer to Fig. 27-45.

Large plate containing an elliptical hole subjected to uniform pressure (Fig. 27-46)

Particular	Formula
The expression for transformation	$z = C\left(\zeta + \frac{m}{\zeta}\right), \quad m < 1$ (27-199) where $C = \frac{1}{2}(a+b), \; m = \frac{a-b}{a+b}$ Refer to other details under Eq. (27-199)
The complex potential at infinity	$\Omega^*(\zeta) = \omega^*(\zeta) = 0$ (27-235a)
The required complex potentials	$\Omega(\zeta) = \sum_0^{\infty} \frac{a_n}{\zeta_n}; \quad \omega(\zeta) = \sum_0^{\infty} \frac{b_b}{\zeta_n}$ (27-235b)
Boundary conditions	$\sigma_n = \sigma_\rho = -p; \quad \tau_{ns} = 0$ around the hole (27-236)

Particular	Formula

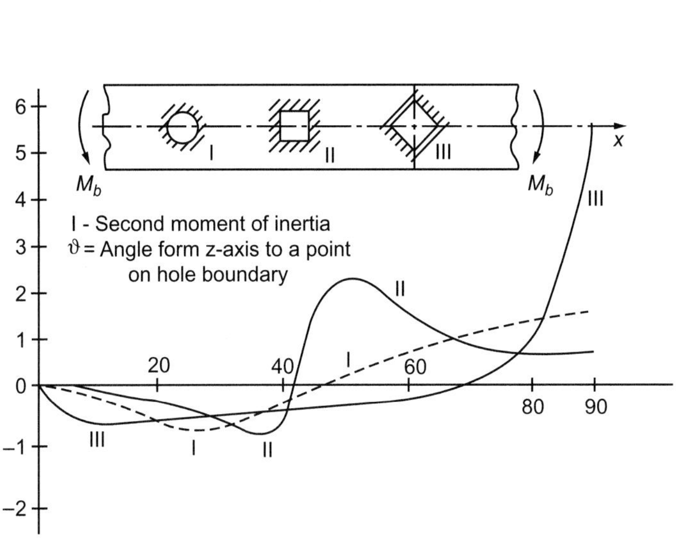

FIGURE 27-45

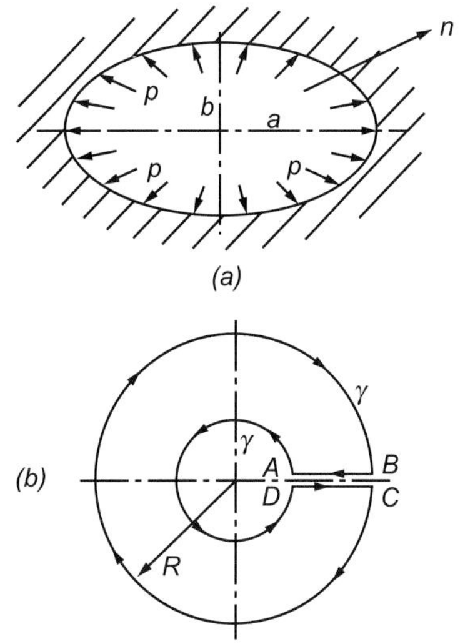

FIGURE 27-46

From Eq. (27-117b)

$$\Omega(z) + 2\bar{\Omega}'(\bar{z}) + \bar{\omega}(\bar{z}) = (\sigma_n + \tau_{ns}) \frac{\partial z}{\partial s}\, ds = -pz \quad \text{at all points on the ellipse} \tag{27-237}$$

Expressing Eq. (27-237) in ζ-plane at all points of γ

$$\Omega(\sigma) + \frac{z(\sigma)}{\bar{z}'\left(\dfrac{1}{\sigma}\right)} \bar{\Omega}'\left(\frac{1}{\sigma}\right) + \bar{\omega}\left(\frac{1}{\sigma}\right) = -pz(\sigma)$$

$$\Omega(\sigma) + \frac{\sigma^2 + m(\sigma)}{\sigma(1 - m\sigma^2)} \bar{\Omega}'\left(\frac{1}{\sigma}\right) + \bar{\omega}\left(\frac{1}{\sigma}\right) = -pC\left(\sigma + \frac{m}{\sigma}\right) \tag{27-238}$$

Multiplying Eq. (27-238) by $\dfrac{1}{2\pi i}\dfrac{\partial\sigma}{\sigma - \zeta}$

[Considering the first of these integrals, one has to remember that γ is now a boundary to the region external to the unit circle. Thus it is necessary to consider an integration around a contour consisting of γ together with C circle γ' of large radius R joined by two close paths AB and CD, Fig. 27-46]

$$\frac{1}{2\pi i}\int_\gamma \frac{\Omega(\sigma)\,\partial\sigma}{\sigma - \zeta} + \frac{1}{2\pi i}\int_\gamma \frac{\sigma^2 + m}{\sigma(1 - m\sigma^2)} \frac{\bar{\Omega}'\left(\dfrac{1}{\sigma}\right)}{\sigma - \zeta}\, d\sigma + \frac{1}{2\pi i}\int_\gamma \frac{\bar{\omega}\left(\dfrac{1}{\sigma}\right) d\sigma}{\sigma - \zeta} = \frac{-pC}{2\pi i}\int_\gamma \frac{\sigma + \dfrac{m}{\sigma}}{\sigma - \zeta}\, d\sigma \tag{27-239}$$

Using Cauchy's integral, Harnack's theorem and residue theorem, Eq. (27-239) gives the expression for $\Omega(\zeta)$

$$\Omega(\zeta) = -\frac{pCm}{\zeta} \tag{27-240}$$

Particular	Formula
Taking conjugate of Eq. (27-239) and integrating around γ, the expression for $\omega(\zeta)$	$\omega(\zeta) = -\frac{pC}{\zeta} - \frac{pCm}{\zeta}\left(\frac{1+m\zeta^2}{\zeta^2 - m}\right)$ (27-241)
The stress can be obtained by making use of Eq. (27-183) for Θ'' and (27-184b) for Φ'' and equating real parts on both sides of equation	$[\sigma_\rho]_{\rho=1} = -p$ from boundary condition (27-242)
	$[\Theta'']_{\rho=1} = 4Rl\left[\frac{\Omega'(\zeta)}{z'(\zeta)}\right]_{\zeta=1}$
	$= \sigma_\vartheta + \sigma_\rho = \frac{4pm(\cos 2\vartheta - m)}{1+m^2-2m\cos 2\vartheta}$ (27-243)
The tangential stress around by elliptical hole from Eq. (27-243)	$\sigma_\vartheta = \frac{4\vartheta m(\cos 2v - m)}{1+m^2-2m\cos 2\vartheta} + p$ (27-244)
	or $\sigma_\vartheta = p\,\frac{1+2m\cos 2\vartheta - 3m^2}{1-2m\cos\vartheta + m^2}$ (27-245)

Large flat plate under uniform uniaxial tension with a circular hole whose edge is rigidly fixed (Fig. 27-49)

Particular	Formula
The edge of the hole $r = a$ is held fixed by a rigid circular ring to which the material of the plate adheres at all points	
The boundary condition is given by T	$[D]_{r=a} = 0$ for all ϑ (27-246a)
The complex potential form of displacement for generalized plane stress problem from Eqs. (27-170) when there are no body forces	$2GD = \left(\frac{3-v}{1+v}\right)\Omega(z) - z\bar{\Omega}'(\bar{z}) - \bar{\omega}(\bar{z}) = 0$ (27-246b)
	or
	$K\Omega(z) - z\bar{\Omega}'(\bar{z}) - \bar{\omega}(\bar{z}) = 0$ on $r = a$ (27-246c)
	where $K = \frac{3-v}{1+v}$
	$K\Omega(\zeta) - z(\bar{\zeta})\,\frac{\bar{\Omega}'(\zeta)}{z'(\bar{\zeta})} - \bar{\omega}(\bar{\zeta}) = 0$ in terms of ζ (27-246d)

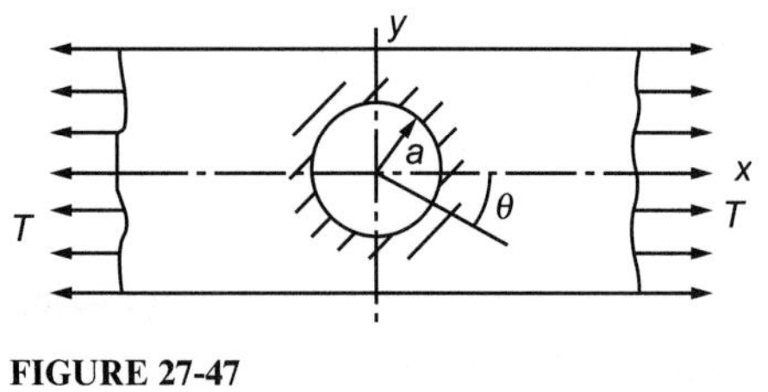

FIGURE 27-47

Particular	Formula
The conformal transformation for this problem can be taken as	$z = \frac{a}{\zeta}$ (27-247)
The full complex potentials in this case can be taken as	$\Omega(\zeta) = \frac{1}{4}\frac{Ta}{\zeta} + \Omega_o(\zeta)$ (27-248a)
	$\omega(\zeta) = -\frac{1}{2}\frac{Ta}{\zeta} + \omega_o(\zeta)$ (27-248b) where first terms in each of the above equations is for stress state at infinity
The condition to be satisfied on $\sigma\bar{\sigma} = 1$ by $\Omega_o(\zeta)$ and $\omega_o(\zeta)$ is	$K\Omega_o(\sigma) - \frac{z(\sigma)}{z'\left(\frac{1}{\sigma}\right)}\bar{\Omega}'_o\left(\frac{1}{\sigma}\right) - \bar{\omega}_o\left(\frac{1}{\sigma}\right)$ $= -\frac{1}{4}T(K-1)\frac{a}{\sigma} - \frac{1}{2}Ta\zeta$ (27-249)
Multiplying Eq. (27-249) by $\frac{1}{2\pi i}\frac{d\sigma}{\sigma - \zeta}$ and integrating around γ, after simplification	$\Omega_o(\zeta) = -\frac{Ta}{2K}\zeta$ (27-250)
Multiplying the conjugate of Eq. (27-229) by $\frac{1}{2\pi i}\frac{d\sigma}{\sigma - \zeta}$ and integrating around γ and after simplification, expression for $\omega_o(\zeta)$	$\omega_o(\zeta) = \frac{1}{4}Ta\left[(K-1)\zeta - \frac{2}{K}\zeta^3\right]$ (27-251)
The full complex potentials are	$\Omega(\zeta) = \frac{1}{4}Ta\left(\frac{1}{\zeta} - \frac{2\zeta}{K}\right)$ (27-252a)
	$\omega(\zeta) = -\frac{1}{2}T\frac{a}{\zeta} + \frac{1}{4}Ta\left[(K-1)\zeta - \frac{2}{K}\zeta^3\right]$ (27-252b)
From the Eqs. (27-182a) and (27-182b) for Θ'' and Φ'', the following stress components are	$\sigma_\rho = \frac{1}{4}T(K+1)\left(1 + \frac{2}{K}\cos 2\vartheta\right)$ (27-253)
	$\sigma_\vartheta = \frac{1}{4}T(3-K)\left(1 + \frac{2}{K}\cos 2\vartheta\right)$ (27-254)
	$\tau_{\rho\vartheta} = \frac{1}{2}T\frac{K+1}{K}\sin 2\vartheta$ (27-255)

TORSION (Fig. 25-49)

The angle of twist α, which is proportional to the distance of cross-section from the fixed end	$\alpha = \theta z$ (27-256) where θ = angle of twist per unit length

Particular	Formula

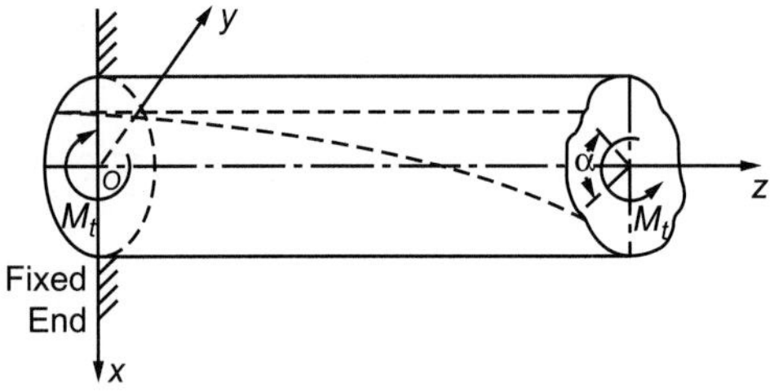

FIGURE 27-48 Torsion of prismatic bar.

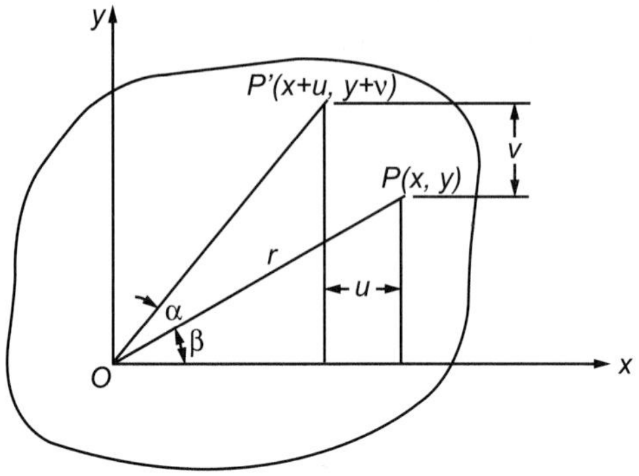

FIGURE 27-49 Shows a cross-section of twisted bar in xy-plane.

$P(x, y, z)$ is a point in a section of bar z-distance from fixed end (Fig. 27-48) and it is displaced to a new point $P'(x+u,\ y+v,\ z+w)$ after deformation due to twist such that $OP \approx OP' \approx r$

The displacement of point P in x-direction assuming that α is small such that cos $\alpha = 1$ and sin $\alpha \approx \alpha$

$$u = r\cos(\alpha+\beta) - r\cos\beta \approx -\alpha y = -\theta zy \tag{27-257}$$

The displacement of point P in y-direction

$$v = r\sin(\alpha+\beta) - r\sin\beta \approx \alpha x = \theta zx \tag{27-258}$$

The warping of bar, which is invariant with z and is defined by a function

$$w = \theta\psi(x, y) \tag{27-259}$$

where $\psi(x, y)$ is a function of x and y only

The component of strains from Eqs. (27-40) and (27-41)

$$\varepsilon_x = \varepsilon_y = \varepsilon_z = \gamma_{xy} = 0 \tag{27-260a}$$

$$\gamma_{yz} = \frac{\partial w}{\partial x} + \frac{\partial u}{\partial z} = G\theta\left(\frac{\partial \psi}{\partial x} - y\right) \tag{27-260b}$$

$$\gamma_{xz} = \frac{\partial w}{\partial y} + \frac{\partial v}{\partial z} = G\theta\left(\frac{\partial \psi}{\partial y} + x\right) \tag{27-260c}$$

The stress components from Eqs. (27-34) and (27-37)

$$\sigma_x = \sigma_y = \sigma_z = \tau_{xy} = 0 \tag{27-261a}$$

$$\tau_{xz} = G\theta\left(\frac{\partial \psi}{\partial x} - y\right) \tag{27-261b}$$

$$\tau_{yz} = G\theta\left(\frac{\partial \psi}{\partial y} + x\right) \tag{27-261c}$$

The equations of equilibrium from Eqs. (27-11)

$$\frac{\partial \sigma_x}{\partial x} + \frac{\partial \tau_{xy}}{\partial y} + \frac{\partial \tau_{xz}}{\partial z} + F_{bx} = 0 \tag{27-11a}$$

$$\frac{\partial \sigma_y}{\partial y} + \frac{\partial \tau_{yz}}{\partial z} + \frac{\partial \tau_{yx}}{\partial x} + F_{by} = 0 \tag{27-11b}$$

$$\frac{\partial \sigma_z}{\partial z} + \frac{\partial \tau_{zx}}{\partial x} + \frac{\partial \tau_{zy}}{\partial y} + F_{bz} = 0 \tag{27-11c}$$

Particular	Formula
Neglecting body forces in z-direction, Eq. (27-11) yields after substituting the Eqs. (27-261b) and (27-261c) in it	$G\theta\left(\frac{\partial^2\psi}{\partial x^2}+\frac{\partial^2\psi}{\partial y^2}\right)=0$ (27-262a)
	or $\left(\frac{\partial^2\psi}{\partial x^2}+\frac{\partial^2\psi}{\partial y^2}\right)=0$ (27-262b) which is true throughout the cross-sectional region of the bar
From the equilibrium condition of the surface Eq. (27-7)	$F_{Nx}=\sigma_x l+\tau_{xy}m+\tau_{xz}n$ (27-7a)
	$F_{Ny}=\tau_{yz}l+\sigma_y m+\tau_{yz}n$ (27-7b)
	$F_{Nz}=\tau_{zx}l+\tau_{zy}m+\sigma_z n$ (27-7c)
When surface forces are absent $F_{Nx}=F_{Ny}=F_{Nz}=0$ and $\cos(Nz)=n=0$, $\sigma_x=\sigma_y=\sigma_z=\tau_{yz}=0$ from Eq. (27-7c)	$\tau_{zx}l+\tau_{zy}m=0$ (27-263)
From the infinitesimal element pqr, if s increasing in the direction from q to r then	$l=\frac{dy}{ds}=\cos(N,x)$ (27-264a)
	$m=-\frac{dx}{ds}=\cos(N,y)$ (27-264b)
Using Eqs. (27-261b), (27-261c), (27-264a) (27-264b) in Eq. (27-263), an expression for boundary condition is obtained (Fig. 27-50)	$\left(\frac{\partial\psi}{\partial x}-y\right)\frac{dy}{ds}-\left(\frac{\partial\psi}{\partial y}+x\right)\frac{dx}{ds}=0$ (27-265)
In torsion problems involving in finding a function ψ which satisfy Eqs. (27-262) and boundary condition Eq. (27-265)	

Stress function ϕ

Particular	Formula
From equation of equilibrium	$\frac{\partial\tau_{xy}}{\partial z}=0;\quad \frac{\partial\tau_{yz}}{\partial z}=0;\quad \frac{\partial\tau_{xz}}{\partial x}+\frac{\partial\tau_{yz}}{\partial y}=0$ (27-266)
A function ϕ which satisfy the third equation of Eq. (27-266) is	$\tau_{xz}=\frac{\partial\phi}{\partial y}$ (27-267)
	$\tau_{yz}=-\frac{\partial\phi}{\partial x}$ (27-268) where ϕ is a function of x and y only
From Eqs. (27-267), (27-268), and Eqs. (27-261), equations involving ϕ and ψ are:	$\frac{\partial\phi}{\partial x}=-G\theta\left(\frac{\partial\psi}{\partial y}+x\right)$ (27-269a)
	$\frac{\partial\phi}{\partial y}=G\theta\left(\frac{\partial\psi}{\partial x}-y\right)$ (27-269b)

Particular	Formula

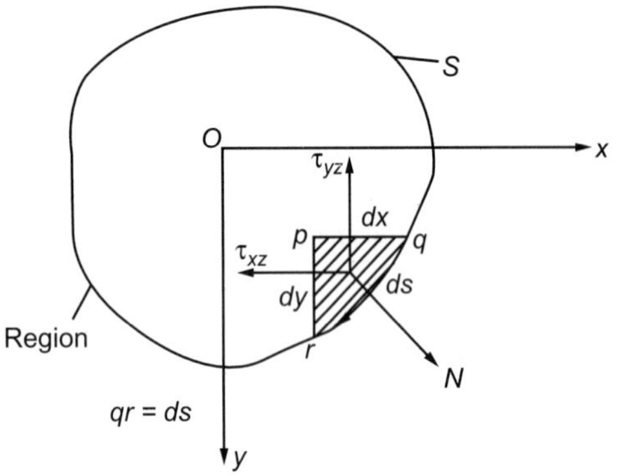

FIGURE 27-50 Boundary condition.

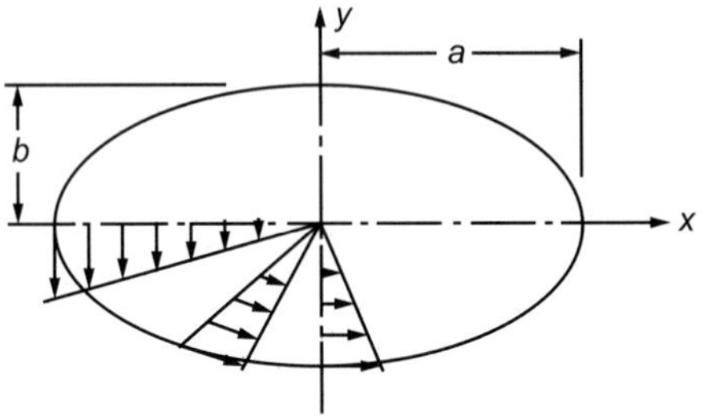

FIGURE 27-51 Elliptical cross-section of bar under torsion.

By making use of Eqs. (27-269) and after eliminating ψ from Eqs. (27-269a) and (27-269b) by mathematical method, a differential equation for stress function ϕ is obtained

$$\frac{\partial^2\phi}{\partial x^2}+\frac{\partial^2\phi}{\partial y^2}=F \qquad (27\text{-}270)$$

where $F=-2G\theta$ (27-270a)

Boundary condition Eq. (27-265) becomes

$$\frac{\partial\phi}{\partial y}\frac{dy}{ds}+\frac{\partial\phi}{\partial x}\frac{dx}{ds}=\frac{d\phi}{ds}=0 \qquad (27\text{-}271)$$

which indicates that the stress function ϕ must be constant along the boundary of the cross-section. This constant is taken as zero for a solid bar.

The total torque at the ends of the twisted bar due to couple

$$M_t=2\iint\phi\,dx\,dy \qquad (27\text{-}272)$$

Torsion of elliptical cross-section bar (Fig. 27-51)

The boundary of an elliptical cross-section can be taken as

$$\frac{x^2}{a^2}+\frac{y^2}{b^2}-1=0 \qquad (27\text{-}273)$$

The stress function which satisfy Eq. (27-270) and the boundary condition Eq. (27-271)

$$\phi=m\left(\frac{x^2}{a^2}+\frac{y^2}{b^2}-1\right) \qquad (27\text{-}274)$$

where m is a constant

Substituting the expression for ϕ from Eq. (27-274) in Eq. (27-270) and value of m can be found, and it is

$$m=\frac{a^2b^2F}{2(a^2+b^2)} \qquad (27\text{-}275)$$

Substituting the value of m from Eq. (27-275) into Eq. (27-274) the stress function ϕ becomes

$$\phi=\frac{a^2b^2F}{2(a^2+b^2)}\left(\frac{x^2}{a^2}+\frac{y^2}{b^2}-1\right) \qquad (27\text{-}276)$$

Particular	Formula
The torque M_t is obtained after substituting this stress function from Eq. (27-276) into Eq. (27-272) and carrying out integration and simplification	$$M_t = \frac{a^2b^2F}{a^2+b^2}\left[\frac{1}{a^2}\iint x^2\,dx\,dy + \frac{1}{b^2}\iint y^2\,dx\,dy - \iint dx\,dy\right] \quad (27\text{-}277)$$ where $$\iint x^2\,dx\,dy = I_y = \frac{\pi ba^3}{4}$$ $$\iint y^2\,dx\,dy = I_x = \frac{\pi ab^3}{4}$$ $$\iint dx\,dy = A = \pi ab$$
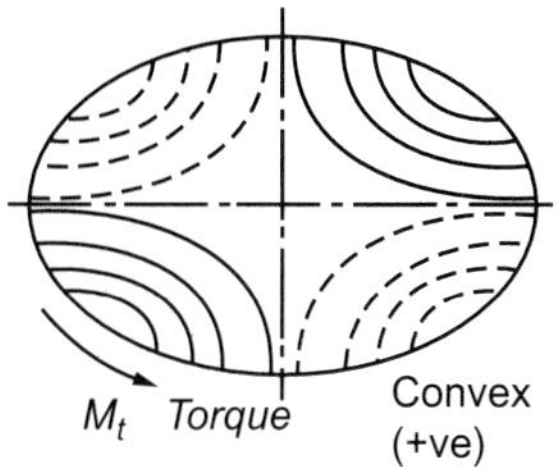 FIGURE 27-52	
After substituting the values of I_x, I_y and A into Eq. (27-277) and simplification, the expression for M_t	$$M_t = -\frac{\pi a^3b^3F}{2(a^2+b^2)} \quad (27\text{-}278)$$
The expression for F from Eq. (27-278)	$$F = -\frac{2M_t(a^2+b^2)}{\pi a^3b^3} \quad (27\text{-}279)$$
The equation for stress function ϕ after substituting the value of F from Eq. (27-279) in Eq. (27-276)	$$\phi = -\frac{M_t}{\pi ab}\left(\frac{x^2}{a^2}+\frac{y^2}{b^2}-1\right) \quad (27\text{-}280)$$
The stress components τ_{xz} and τ_{yz} from Eqs. (27-267) and (27-268) after substituting the value of ϕ from Eq. (27-280)	$$\tau_{xy} = -\frac{2M_ty}{\pi ab^3} \quad (27\text{-}281)$$ $$\tau_{yz} = \frac{2M_tx}{\pi a^3b} \quad (27\text{-}282)$$
The maximum shear stress which occurs at $y = b$	$$\tau_{max} = \frac{2M_t}{\pi ab^2} \quad (27\text{-}283)$$
The angle of twist after substituting the value of F from Eq. (27-279) into Eq. (27-270a) and simplification	$$\theta = M_t\frac{a^2+b^2}{\pi a^3b^3G} \quad (27\text{-}284)$$
The torsional rigidity C which is defined as twist per unit length	$$C = \frac{\pi a^3b^3G}{a^2+b^2} = \frac{G}{4\pi^2}\frac{A^4}{I_p} \quad (27\text{-}285)$$ where $A = \pi ab$, I_p = centroidal moment of inertia of the cross-section $= (\pi ab^3)/4 + (\pi a^3b)/4$
For various values of the angle of twist $(0 = \phi)$ and thereby the values of C for various cross-sections and built up beams	Refer to Tables 24-27 and 24-30 under Chapter 24.
The expression for warping of elliptical cross-section after substituting Eqs. (27-280), (27-281) and (27-282) into Eqs. (29-260b) and (27-260c) and integrating	$$w = M_t\frac{(b^2-a^2)xy}{\pi a^3b^3G} \quad (27\text{-}286)$$
For warping of elliptical cross-section	Refer to Fig. 27-52.
Note: The symbol θ is used for angle of twist here in order to avoid confusion regarding ϕ which is used as a stress function	Equations (27-277) to (27-285) are also given in Chapter 24 from Eqs. (24-338) to (24-342), and angle of twist θ in Chapter 24 in Tables 24-27 and 24-30.

Particular	Formula
For torsion of elliptical and rectangular solid sections and other sections (Fig. 24-66 to 24-71)	Refer to Chapter 24 from Eqs. (24-338) to (24-352), Tables 24-27 to 24-30.
Torsion of equilateral triangle bar	
The expression for stress function	$\phi = \left(x - \sqrt{3}y - \frac{2}{3}a\right)\left(x + \sqrt{3}y - \frac{2}{3}a\right)\left(x + \frac{a}{3}\right)A$ (27-287)
Substituting Eq. (27-287) in Eqs. (27-267) and (27-268) the values of $\frac{\partial^2 \phi}{\partial x^2}$ and $\frac{\partial^2 \phi}{\partial y^2}$ can be found. The values are substituted in Eq. (27-270) to find the value A	$A = G\theta$ (27-288)
The stress function from Eq. (27-287) becomes	$\phi = -G\theta\left[\frac{1}{2}(x^2 + y^2) - \frac{1}{2a}(x^3 - 3xy^2) - \frac{2}{27}a^2\right]$ (27-289)
The expression for τ_{xz} from Eq. (27-267) after using the value of $A = G\theta$	$\tau_{xz} = 0$ (27-290a)
The expression for τ_{yz} from Eq. (27-268) after using the value of $A = G\theta$	$\tau_{yz} = \frac{3G\theta}{2a}\left(\frac{2ax}{3} - x^2\right)$ (27-290b)
The maximum shear stress	$(\tau_{yz})_{x=-a/3}\tau_{max} = \frac{G\theta a}{2}$ (27-291a)
The shear stress at the center of triangular bar	$(\tau_{yz})_{x=-2a/3} = \frac{3G\theta}{2a}\left(\frac{2ax}{3} - x^2\right)_{x=2a/3} = 0$ (27-291b)
The torque M_t filter substituting the value ϕ from Eq. (27-289) into Eq. (27-272) and carrying out integration and simplification	$M_t = \frac{G\theta a^4}{15\sqrt{3}} = \frac{3}{5}\theta G I_p$ (27-292)
For shear stress variation along x-axis	Refer to Fig. 27-53.

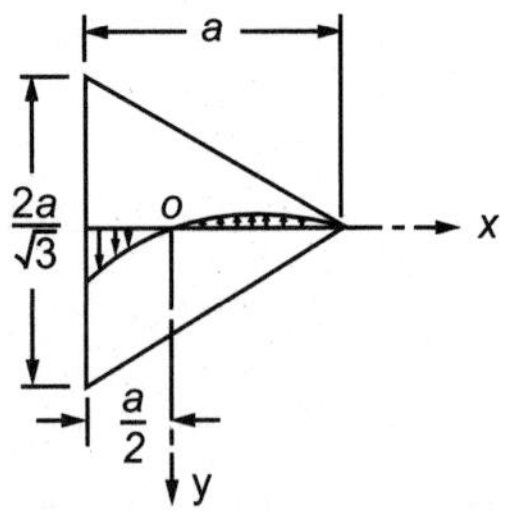

FIGURE 27-53 Equilateral triangle bar under torsion.

Particular	Formula

Membrane analogy

Particular	Formula
Equation of equilibrium of the element *klmn*	$$\frac{\partial^2 z}{\partial x^2}+\frac{\partial^2 z}{\partial y^2}=-\frac{p}{T} \quad (27\text{-}293)$$
Comparing the statement and Eq. (27-293) with Eqs. (27-270) which have been derived for stress function ϕ, it can be seen that two problems are identical	where p = pressure per unit area of the membrane T = uniform tension per unit length of the membrane z is zero at the edges of the membrane
The quantities which are analogous to each other between torsion and membrane problems are	z is analogous to ϕ $-p/T$ is analogous to $F=-2G\theta$
By analogy in terms of stress function ϕ and hence in terms of τ_{yz} and τ_{xz} from Eq. (27-267) it can be shown that	$$\frac{\partial\phi}{\partial s}=\frac{\partial\phi}{\partial y}\frac{\partial y}{\partial s}-\frac{\partial\phi}{\partial x}\frac{\partial x}{\partial s}$$
This proves that the projection of the resultant shear stress at a point k (Fig. 27-56) on the normal N to the contour line is zero	$$=\tau_{xz}\frac{\partial y}{\partial x}-\tau_{yz}\frac{\partial x}{\partial s}=0 \quad (27\text{-}294)$$
The magnitude of the shearing stress at k	Maximum slope of the membrane at this point
The resultant shear stress	$$\tau=\tau_{yz}\cos(N,x)-\tau_{xz}\cos(N,y)$$ $$=\left(\frac{\partial\phi}{\partial x}\frac{dz}{dn}+\frac{\partial\phi}{\partial y}\frac{dy}{dn}\right)=-\frac{d\phi}{dn} \quad (27\text{-}295)$$
By analogy	$$\tau=-\frac{dz}{dn} \quad (27\text{-}296)$$

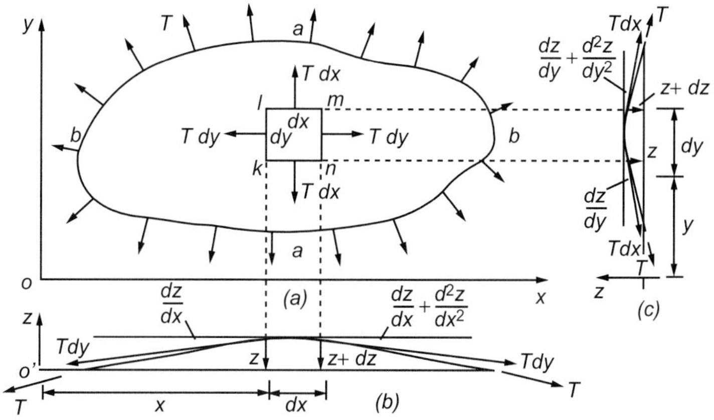

FIGURE 27-54 Membrane subjected to uniform tension at the edges and uniform lateral pressure q.

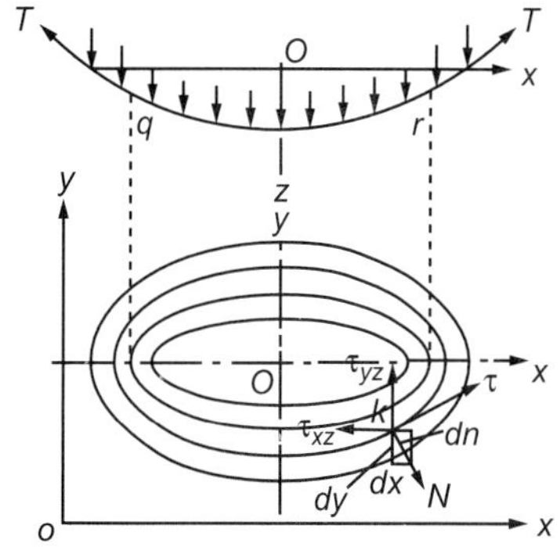

FIGURE 27-55

Particular	Formula
	This proves that the magnitude of the shearing stress at B is given by the maximum slope of the membrane at this point.
By analogy the slope of the membrane in the direction of the normal is obtained from	$$\frac{\left(\frac{\partial z}{\partial n}\right)}{\left(\frac{p}{t}\right)} = \frac{\tau}{2G\theta} \quad \text{or} \quad \frac{\partial z}{\partial n} = \frac{\tau}{2G\theta}\left(\frac{p}{t}\right) \qquad (27\text{-}297)$$
For equation of equilibrium of the portion of the membrane (Fig. 27-55)	$$\int \tau\, ds \frac{\partial z}{\partial n} = pA$$ $$\int \tau\, ds = 2G\theta A \qquad (27\text{-}298)$$ where A = horizontal projection of the portion qr of the membran (Fig. 27-55)
	The membrane analogy can be used to solve problems of build up narrow cross sections, hollow sections, thin tubes, thin webbed tubes, box sections, etc. which are subjected to torsion

Torsion of hollow sections and thin-walled tubes (Fig. 27-57)

Particular	Formula
Equating forces in the two directions acting on an element of hollow section as shown in Fig. 27-56	$$\left(q + \frac{\partial q}{\partial s}\, ds\right) dl - q\, dl = 0 \quad \text{or} \quad \frac{\partial q}{ds} = 0 \qquad (27\text{-}299a)$$ $$\left(q + \frac{\partial q}{\partial l}\, dl\right) ds - q\, ds = 0 \quad \text{or} \quad \frac{dq}{\partial l} = 0 \qquad (27\text{-}299b)$$
These conditions can be satisfied only if q is constant	$$q = t\tau = \text{constant} \qquad (27\text{-}300)$$
The torque	$$M_t = \int q\rho\, ds = q \int \rho\, ds \qquad (27\text{-}301a)$$ $$M_t = q2A \qquad (27\text{-}301b)$$ where A = area enclosed by the median line of the tubular section.

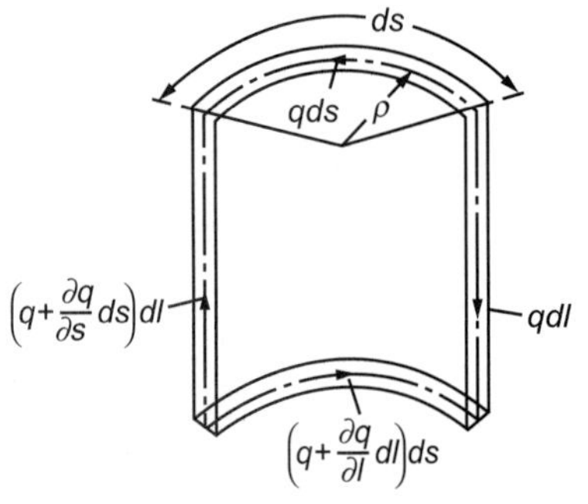

FIGURE 27-56

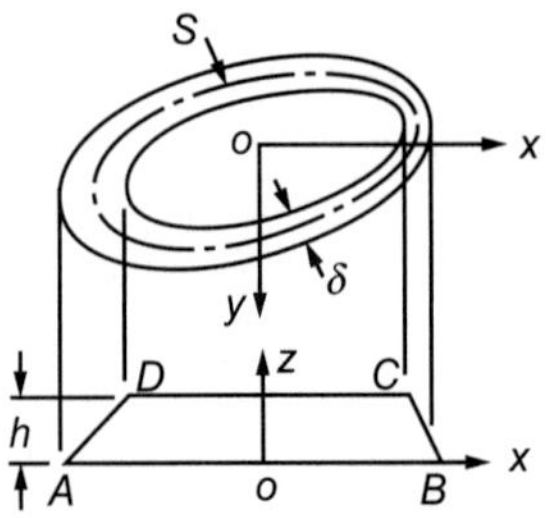

FIGURE 27-57

Particular	Formula
Shear flow in force per unit length	$$q = \frac{M_t}{2A} \tag{27-302}$$
The shear stress	$$\tau = \frac{q}{t} = \frac{M_t}{2At} \tag{27-303}$$ where t = thickness of hollow section or tubular section; d = distance from O to mid line or median line of hollow sections or the tubular section (Fig. 27-56)
Torsion of thin-walled tubes (Fig. 27-57)	
The shear stress at any point is given by the slope of the membrane	$$\tau = \frac{h}{\delta} \tag{27-304}$$ where h = the difference in level between DC and AB of membrane; δ = variable thickness of thin-walled tube; A = mean area enclosed by the outer and inner boundaries of the cross-section of the tube; S = the length of the center line of the ring section of the tube (Fig. 27-57)
The expression for torque	$$M_t = 2Ah = 2A\delta\tau \tag{27-305}$$
The shear stress	$$\tau = \frac{M_t}{2A\delta} \tag{27-306}$$
The angle of twist for tube with uniform thickness	$$\theta = \frac{M_t S}{4A^2 G\delta} \tag{27-307}$$ Refer to Fig. 24-71 and Eqs. (24-353a) to (24-360) in Chapter 24.
Torsion of thin-welded tubes (or box beams) (Fig. 27-58)	
For calculation of each item, assume anticlockwise path for each shell	
The torque	$M_t = 2 \times$ volume under membrane $$= [2A_1h_1 + A_1h_2] = \tfrac{16}{3}a^2\delta\tau_2 \tag{27-308}$$ $$\tau_1 = \frac{h_1}{\delta}; \quad \tau_2 = \frac{h_2}{\delta}; \quad \tau_3 = \frac{h_1 + h_2}{\delta}$$
For the section shown in Fig. 27-58, the shear stresses	$$\tau_1 = \frac{h_1}{\delta} = \frac{5M_t}{32a^2\delta} \tag{27-309a}$$

Particular	Formula

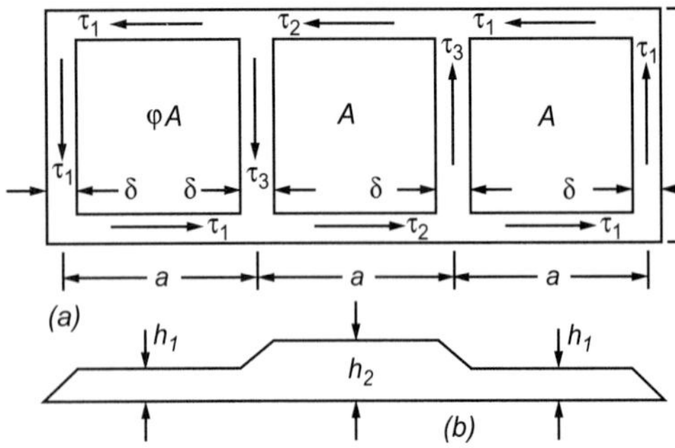

FIGURE 27-58

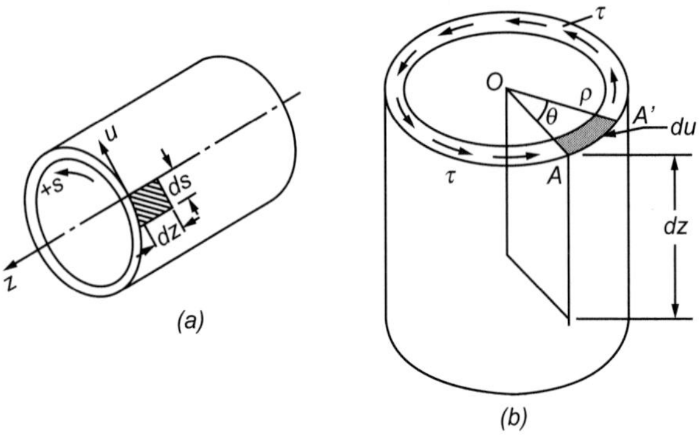

FIGURE 27-59

$$\tau_2 = \frac{h_2}{\delta} = \frac{3}{16}\frac{M_t}{a^2\delta} \quad (27\text{-}309b)$$

$$\tau_3 = \frac{h_1 - h_2}{\delta} = \frac{M_t}{32a^2\delta} \quad (27\text{-}309c)$$

The angle of twist of section

$$\theta = \frac{7}{32}\frac{M_t}{a^2\delta G} \quad (27\text{-}310)$$

The torsional rigidity of section shown

$$C = \frac{M_t}{\theta} = \frac{32a^2\delta G}{7} \quad (27\text{-}311)$$

Warping of a box section (Fig. 27-59)

The shearing strain

$$\gamma s = \frac{\partial u}{\partial z} + \frac{\partial w}{\delta s} = \frac{\tau_{sz}}{G} \quad (27\text{-}312)$$

where

u = displacement in direction s

w = displacement in direction z

θ = angle of twist per unit length

The expression for incremental displacement in s-direction

$$du = \theta\, dz\, \rho \quad (27\text{-}313)$$

The slope of the warped cross-section

$$\frac{\partial w}{\partial s} = \frac{\tau}{G} - \rho\theta \quad (27\text{-}314)$$

TORSION OF CIRCULAR SHAFT OF VARIABLE DIAMETER (Fig. 27-13)

For Figure of circular shafts of variable diameter

Refer to Fig. 27-13

Particular	Formula
For strains and displacements of a solid of revolution twisted by couples applied at ends	$\varepsilon = \frac{\partial u}{\partial r}; \quad \varepsilon_\theta = \frac{u}{r} + \frac{1}{r}\frac{\partial v}{\partial \theta}; \quad \varepsilon_z = \frac{\partial w}{\partial z}$
	$\gamma_{r\theta} = \frac{\partial u}{r\,d\theta} + \frac{\partial v}{\partial r} - \frac{v}{r}$ (27-315a)
	$\gamma_{rz} = \frac{\partial w}{\partial r} + \frac{\partial u}{\partial z}; \quad \gamma_{z\theta} = \frac{\partial w}{r\,d\theta} + \frac{\partial v}{\partial z}$ (27-315b)
Differential equations of equilibrium when the body force is absent	$\frac{\partial \sigma_r}{\partial r} + \frac{1}{r}\frac{\partial \tau_{r\theta}}{\partial \theta} + \frac{\partial \tau_{rz}}{\partial z} + \frac{\sigma_r - \sigma_\theta}{r} = 0$ (27-316a)
	$\frac{\partial \tau_{rz}}{\partial r} + \frac{1}{r}\frac{\partial \tau_{\theta z}}{\partial \theta} + \frac{\partial \sigma_z}{\partial z} + \frac{\tau_{rz}}{r} = 0$ (27-316b)
	$\frac{\partial \tau_{r\theta}}{\partial r} + \frac{1}{r}\frac{\partial \sigma_\theta}{\partial \theta} + \frac{\partial \tau_{\theta z}}{\partial z} + \frac{2\tau_{r\theta}}{r} = 0$ (27-316c)
Since it is a problem of symmetry	$\varepsilon_r = \varepsilon_\theta = \varepsilon_z = \tau_{rz} = 0$ (27-317a)
	$\gamma_{r\theta} = \frac{\partial v}{\partial r} - \frac{v}{r}; \quad \gamma_{\theta z} = \frac{\partial v}{\partial z}$ (27-317b)
The third of Eqs. (27-316c) can be written as	$\frac{\partial}{\partial r}(r^2 \tau_{r\theta}) + \frac{\partial}{\partial z}(r^2 \tau_{\theta z}) = 0$ (27-318)
Eq. (27-318) is satisfied if the stress function given here used	$r^2 \tau_{r\theta} = -\frac{\partial \phi}{\partial z}; \quad r^2 \tau_{\theta z} = \frac{\partial \phi}{\partial r}$ (27-319)
From Eqs (27-317) and (27-319)	$\tau_{r\theta} = G\gamma_{r\theta} = G\left(\frac{\partial v}{\partial r} - \frac{v}{r}\right)$
	$= Gr\frac{\partial}{\partial r}\left(\frac{v}{r}\right) = -\frac{1}{r^2}\frac{\partial \phi}{\partial z}$ (27-320)
	$\tau_{\theta z} = G\gamma_{\theta z} = G\frac{\partial v}{\partial z} = Gr\frac{\partial}{\partial z}\left(\frac{v}{r}\right) = \frac{1}{r^2}\frac{\partial \phi}{\partial r}$ (27-321)
From the above equation it follows	$\frac{\partial}{\partial r}\left(\frac{1}{r^3}\frac{\partial \phi}{\partial r}\right) + \frac{\partial}{\partial z}\left(\frac{1}{r^3}\frac{\partial \phi}{\partial z}\right) = 0$ (27-322a)
	or $\frac{\partial^2 \phi}{\partial r^2}\frac{3}{r}\frac{\partial \phi}{\partial r} + \frac{\partial^2 \phi}{\partial z^2} = 0$ (27-322b)
	This Eq. (27-322) gives the angle of twist as a function of ϕ and z.
Boundary conditions	Refer to Fig. 27-50.
The equation which gives the stress function constant value along the boundary of the axial section of the shaft	$\frac{\partial \phi}{\partial z}\frac{\partial z}{\partial s} + \frac{\partial \phi}{\partial r}\frac{ds}{ds} = 0$ (27-324)
	Equation (27-322) together with the boundary condition (27-324) completely determine the stress function ϕ.

Particular	Formula

The torque for cross-section

$$M_t = \int_0^a 2\pi r^2 \tau_{\theta z}\, dr \qquad (27\text{-}325)$$

$$M_t = 2\pi \int_0^a \frac{\partial \phi}{\partial r}\, dr = 2\pi |\phi|_0^a \qquad (27\text{-}326)$$

where a = outer radii of the cross-section

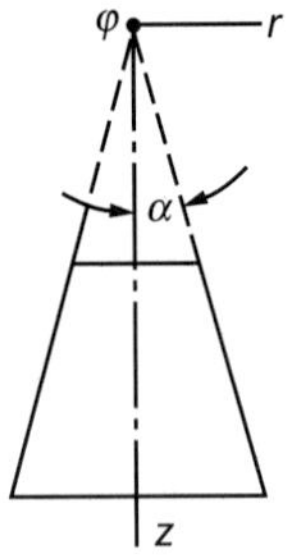

FIGURE 27-60

Conical shaft (Fig. 27-60)

The angle α is given by

$$\cos\alpha = \frac{z}{\sqrt{r^2 + z^2}} \qquad (j)$$

The ratio $(x/\sqrt{r^2 + z^2})$ is constant at the axial section

The stress function ϕ which satisfies the boundary condition Eq. (27-324) and equilibrium Eq. (27-316) is taken as

$$\phi = c\left[\frac{z}{\sqrt{r^2 + z^2}} - \frac{1}{3}\left(\frac{z}{\sqrt{r^2 + z^2}}\right)^3\right] \qquad (27\text{-}327)$$

where c = constant

The expression for $\tau_{\theta z}$

$$\tau_{\theta z} = \frac{1}{r^2}\frac{\partial \phi}{\partial r} = -\frac{crz}{(r^2 + z^2)^{5/2}} \qquad (27\text{-}328)$$

Substituting Eq. (27-327) in Eq. (27-325), the value of constant C can be obtained

$$c = -\frac{M_t}{2\pi(\frac{2}{3} - \cos\alpha + \frac{1}{3}\cos^3\alpha)} \qquad (27\text{-}329)$$

The angle of twist

$$\psi = \frac{c}{3G(r^2 + z^2)^{3/2}} \qquad (27\text{-}330)$$

PLATES

Differential equation for cylindrical bending of plates (Figs. 27-61 and 27-62)

The unit elongation of a fiber at a distance z from the middle plane or surface of plate (Fig. 27-62a)

$$\varepsilon_x = -z\frac{d^2 w}{dx^2} \qquad (27\text{-}331)$$

where

$-d^2w/dx^2$ = the curvature of the deflection curve of the plate

w = deflection of the plate in z direction which is very small compared to the length of plate l

The normal bending stress σ_{bx}

$$\sigma_{bx} = -\frac{Ez}{1 - v^2}\frac{d^2 w}{dx^2} \qquad (27\text{-}332)$$

Particular	Formula

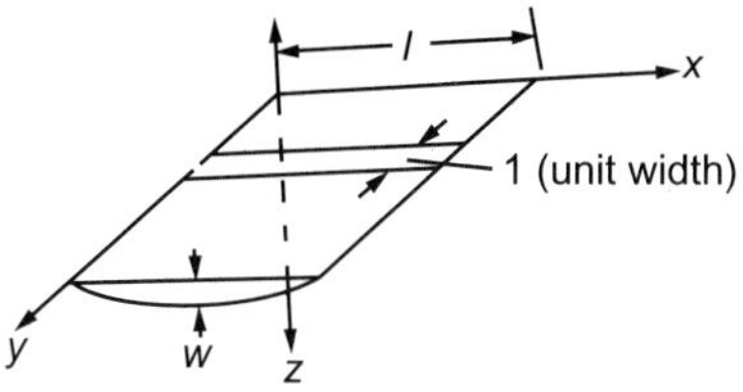

FIGURE 27-61 Bending of a plate due to lateral uniform load acting along the length of plate

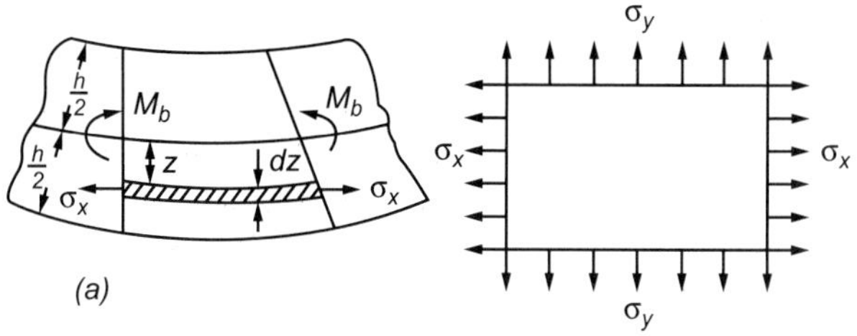

(b) Element cut out from the bent plate as shown in Fig (a)

FIGURE 27-62

The bending moment in the elemental strip of plate (Fig. 27-63)

$$M_b = -\frac{Eh^3}{12(1-v^2)}\frac{d^2w}{dx^2} = -D\frac{d^2w}{dx^2} \tag{27-333}$$

where

$$D = \frac{Eh^3}{12(1-v^2)} = \text{flexural rigidity of plate} \tag{27-333a}$$

Cylindrical bending of uniformly loaded rectangular plate with simply supported edges, which are restricted from movement (Fig. 27-63)

The bending moment at any cross-section of plate at x distance from the left support

$$M_b = \frac{qlx}{2} - \frac{qx^2}{2} - F_a w \tag{27-334}$$

Substituting Eq. (27-334) in Eq. (27-333) and making use of boundary conditions $w = 0$ at $x = 0$ and $x = l$, the expression for w can be obtained after integrating as

$$w = \frac{q}{u^4 D}\left[\frac{\cosh u\left(\frac{l}{2} - x\right)}{\cosh\frac{ul}{2}} - 1\right] + \frac{qx}{2u^2 D}(l - x) \tag{27-335}$$

where $u = \frac{F_a}{D}$

The expression for axial force

$$F_a = \frac{Eh}{(1-v^2)l}\frac{q^2}{D^2}\left[\frac{l^3}{24u^4} - \frac{5l}{4u^6} + \frac{5}{2u^7} \times \tan\frac{ul}{2} + \frac{l}{4u^6}\tanh^2\frac{ul}{2}\right] \tag{27-336}$$

The expression for difference between deflection curve and chord, λ

$$\lambda = \frac{5(1-v^2)l}{Eh} = \frac{1}{2}\int_0^l \left(\frac{dw}{dx}\right)^2 dx \tag{27-337}$$

The maximum bending moment

$$M_{b\,\max} = -D\left(\frac{d^2w}{dx^2}\right)_{x=\frac{l}{2}} \tag{27-338}$$

Particular	Formula

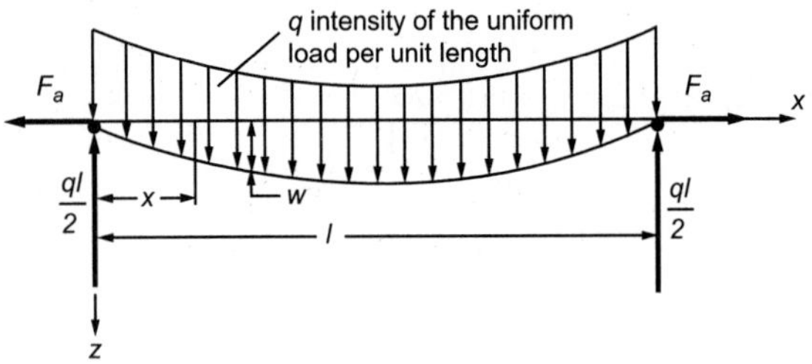

FIGURE 27-63

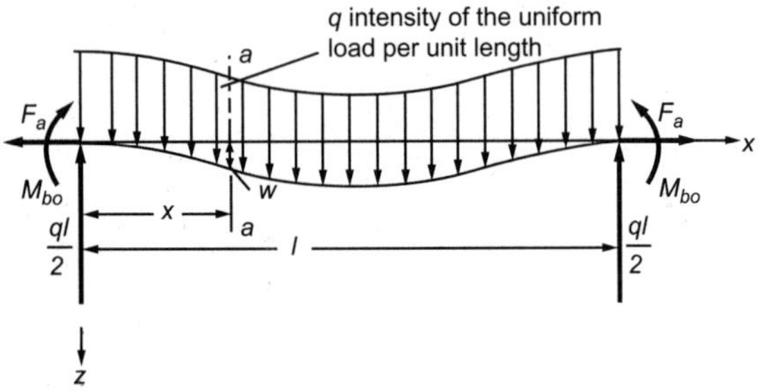

FIGURE 27-64

The maximum bending stress

$$\sigma_{b\,\max} = \frac{M_{b\,\max}}{Z} \quad (27\text{-}338a)$$

where $Z =$ section modulus $= \frac{h^2}{6} \times 1$

The axial stress due to F_a

$$\sigma_a = \frac{F_a}{h \times 1} \quad (27\text{-}338b)$$

The total stress σ_{tot}

$$\sigma_{tot} = \frac{M_{b\,\max}}{Z} + \frac{F_a}{h} \quad (27\text{-}339)$$

Cylindrical bending of uniformly loaded rectangular plates with built-in edges (Fig. 27-64)

Bending moment at any section of plate x-distance from left end

$$M_b = \frac{qlx}{2} - \frac{qx^2}{2} - F_a w + M_{bo} \quad (27\text{-}340)$$

where $M_{bo} =$ fixed end movement

Substituting Eq. (27-340) in Eq. (27-333) and making use of boundary conditions $w = 0$ at $x = 0$; $dw/dx = 0$ for $x = 0$ and $x = l/2$, the expression for w can be obtained after integrating and simplification

$$w = \frac{ql^4}{16u^3 D \tanh u}\left[\frac{\cosh\left\{u\left(1 - \frac{2x}{l}\right)\right\}}{\cosh u} - 1\right] + \frac{ql^2(l-x)x}{5u^2 D} \quad (27\text{-}341)$$

where $u^2 = \frac{F_a}{D}\frac{l^2}{4}$

The expression for M_{bo}

$$M_{bo} = \frac{ql^2}{4u^2} - \frac{ql^2}{4u}\coth u = -\frac{ql^2}{12}\psi_1(u) \quad (27\text{-}342)$$

where $\psi_1(u) = \frac{3(u - \tanh u)}{u^2 \tanh u}$ (27-343a)

Particular	Formula
The stresses are	$$\sigma_1 = \frac{Eu^2}{3(1-v^2)}\left(\frac{h}{l}\right)^2 \quad (27\text{-}344a)$$ $$\sigma_2 = -\frac{6M_{bo}}{h^2} = \frac{q}{2}\left(\frac{l}{h}\right)^2 \psi_1(u) \quad (27\text{-}344b)$$
The maximum stress, σ	$$\sigma_{max} = \sigma_1 + \sigma_2 \quad (27\text{-}345)$$
The maximum deflection at $x = \frac{1}{2}$	$$w_{max} = \frac{ql^4}{384D} f_1(u) \quad (27\text{-}346)$$ where $f_1(u) = \frac{24}{u^4}\left(\frac{u^2}{2} + \frac{u}{\sinh u} - \frac{u}{\tanh u}\right)$ (27-346a)

Cylindrical bending of plate on an elastic foundation (Fig. 27-65)

Particular	Formula
A strip of plate cut out from main plate when treated as a beam on an elastic foundation, it can be shown that	$$D\frac{\partial^4 w}{\partial x^4} = q - kw \text{ or } \frac{\partial^4 w}{\partial x^4} + \frac{k}{D}w = \frac{q}{D} \quad (27\text{-}347)$$ where q = intensity of the uniform load acting on the plate k = reaction of the foundation per unit area for a deflection equal to unity
The general solution of Eq. (27-347) consists of complementary function and particular integral which can be written as	$$w = c_1 \sin\alpha x \sinh\alpha x + c_2 \sin\alpha x \cosh\alpha x + c_3 \cos\alpha x \sinh\alpha x + c_4 \cos\alpha x \cosh\alpha x + \frac{q}{k} \quad (27\text{-}348)$$ where $4\alpha^4 = \frac{k}{D}$
By taking coordinate as system shown in Fig. 27-65 and making use of symmetry and boundary conditions $w = 0 = \partial^2 w/\partial x^2$ at $x = l/2$, the final equation for deflection w can be written as	$$w = \frac{q}{k}\left[1 - \frac{2\sin\alpha \sinh\alpha}{\cos 2\alpha + \cosh 2\alpha}\sin\alpha x \sinh\alpha x - \frac{2\cos\alpha \cosh\alpha}{\cos 2\alpha + \cosh 2\alpha}\cos\alpha x \cosh\alpha x\right] \quad (27\text{-}349)$$
The maximum deflection w_{max} at $x = 0$	$$(w)_{x=0} = w_{max} = \frac{q}{k}\left[1 - \frac{2\cos\alpha \cosh\alpha}{\cos 2\alpha + \cosh 2\alpha}\right] \quad (27\text{-}350)$$
The maximum bending moment of the strip of plate at the middle plane and is obtained by	$$M_{b\,max} = -D\left(\frac{\partial^2 w}{\partial x^2}\right)_{x=0} \quad (27\text{-}351)$$ where w is from Eq. (27-349)

Particular	Formula

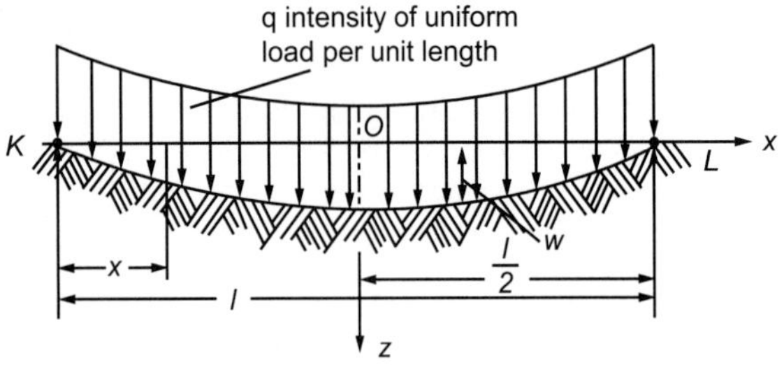

FIGURE 27-65

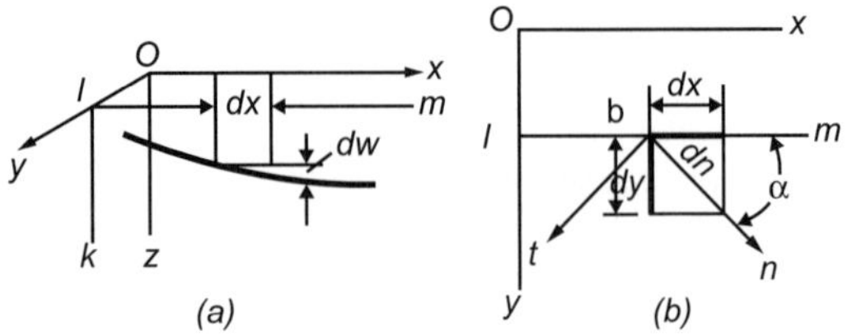

FIGURE 27-66

Pure bending of plates

Slope of the plate surface in the x-direction (Fig. 27-66)

$$\underset{\Delta x \to 0}{Lt} \frac{\Delta w}{\Delta x} = \frac{dw}{dx} \tag{a}$$

Slope of the plate surface in the y-direction

$$\underset{\Delta y \to 0}{Lt} \frac{\Delta w}{\Delta y} = \frac{dw}{dy} \tag{b}$$

Slope of the surface of plate in n-direction (Fig. 27-66)

$$\frac{\partial w}{\partial n} = \frac{\partial w}{\partial x}\frac{dx}{dn} + \frac{\partial w}{\partial y}\frac{dy}{dn} = \frac{\partial w}{\partial x}\cos\alpha + \frac{\partial w}{\partial y}\sin\alpha \tag{c}$$

For y-direction of maximum slope or $\tan\alpha_1$

$$\frac{\partial}{\partial\alpha}\left(\frac{\partial w}{\partial n}\right) = 0 = -\frac{\partial w}{\partial x}\sin\alpha + \frac{\partial w}{\partial y}\cos\alpha$$

$$\text{or} \quad \tan\alpha_1 = \frac{\left(\dfrac{\partial w}{\partial y}\right)}{\left(\dfrac{\partial w}{\partial x}\right)} \quad (\text{say } \alpha = \alpha_1) \tag{d}$$

For zero slope,

$$\frac{\partial w}{\partial n} = 0 \quad \text{or} \quad \tan\alpha_2 = -\frac{\left(\dfrac{\partial w}{\partial x}\right)}{\left(\dfrac{\partial w}{\partial y}\right)} (\text{say } \alpha = \alpha_2) \tag{e}$$

From Eqs. (d) and (e)

$$\tan\alpha_1 \tan\alpha_2 = -1 \tag{f}$$

Therefore the two directions of α_1 (maximum slope) and α_2 (zero slope) are perpendicular to each other.

The expression for maximum slope

$$\left(\frac{\partial w}{\partial n}\right)_{\max} = \sqrt{\left(\frac{\partial w}{\partial x}\right)^2 + \left(\frac{\partial w}{\partial y}\right)^2} \tag{g}$$

The curvature of the surface of a plate in a plane parallel to xz plane (Fig. 27-67) is

$$\frac{1}{r_x} = -\frac{\partial^2 w}{\partial x^2} \tag{h}$$

The curvature of the surface of a plate in a plane parallel to yz is

$$\frac{1}{r_y} = -\frac{\partial^2 w}{\partial y^2} \tag{i}$$

The twist of the surface of the plate with respect to the x and y axes

$$\frac{1}{r_{xy}} = -\frac{\partial^2 w}{\partial n\,\partial y} \tag{j}$$

The curvature of the middle surface of plate in any direction n (Fig. 27-67) is

$$\frac{1}{r_n} = -\frac{\partial}{\partial n}\left(\frac{\partial w}{\partial n}\right) = \frac{1}{r_x}\cos^2\alpha - \frac{1}{r_{xy}}\sin 2\alpha + \frac{1}{r_y}\sin^2\alpha \tag{27-352}$$

Particular	Formula
The sum of curvature or average curvature of the middle surface of plate at a point	$$\frac{1}{r_n}+\frac{1}{r_t}=\frac{1}{r_x}+\frac{1}{r_y} \quad (27\text{-}353)$$ This shows that the sum of curvature in n and t two perpendicular directions is independent of angle α.
The twist of the surface of plate at b with respect to bm and bt directions	$$\frac{1}{r_{nt}}=\frac{d}{dt}\left(\frac{dw}{dn}\right) \quad (27\text{-}354a)$$
Thus twist of the surface of plate	$$\frac{1}{r_{nt}}=\frac{1}{2}\sin 2\alpha\left(\frac{1}{r_x}-\frac{1}{r_y}\right)+\cos 2\alpha\,\frac{1}{r_{xy}} \quad (27\text{-}354b)$$
	$$\tan 2\alpha=\frac{-\dfrac{2}{r_{xy}}}{\dfrac{1}{r_x}-\dfrac{1}{r_y}} \quad (27\text{-}355)$$

Bending moments and curvature in pure bending of plates (Fig. 27-67)

Particular	Formula
The unit elongation in x and y directions of an elemental lamina $klmO'$ at a distance z from the neutral surface (Fig. 27-68)	$$\varepsilon_x=\frac{z}{r_x};\quad \varepsilon_y=\frac{z}{r_y} \quad (27\text{-}356)$$
The corresponding stresses in the lamina $klm\ 0'$	$$\sigma_x=\frac{Ez}{1-v^2}\left(\frac{1}{r_x}+v\,\frac{1}{r_y}\right)=-\frac{Ez}{1-v^2}\left(\frac{\partial^2 w}{\partial x^2}+v\,\frac{\partial w^2}{\partial y^2}\right) \quad (27\text{-}357a)$$ $$\sigma_y=\frac{Ez}{1-v^2}\left(\frac{1}{r_y}+v\,\frac{1}{r_x}\right)=-\frac{Ez}{1-v^2}\left(\frac{\partial^2 w}{\partial y^2}+v\,\frac{\partial^2 w}{\partial x^2}\right) \quad (27\text{-}357b)$$

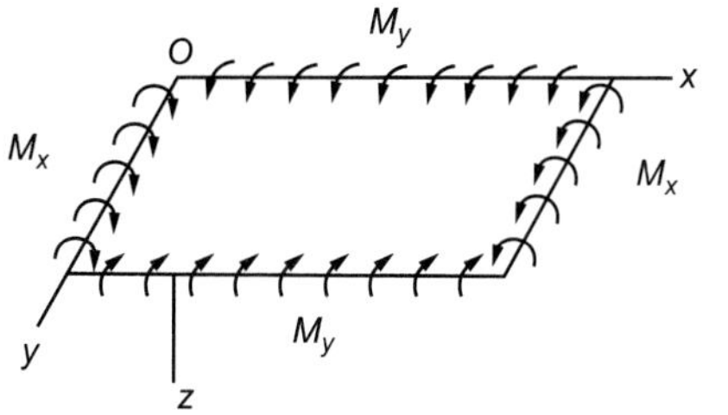

FIGURE 27-67

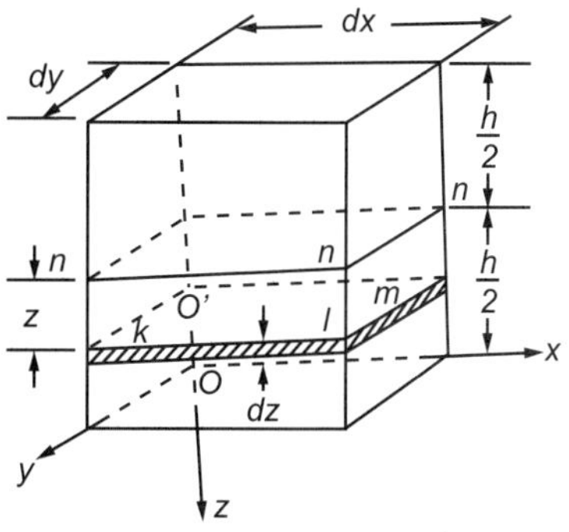

FIGURE 27-68

Particular	Formula
The external moments per unit length of element (Fig. 27-68); which are equal to internal couple on the element	$M_{bx} = \int_{-h/2}^{h/2} \sigma_x z\, dy\, dz = D\left(\frac{1}{r_x} + \frac{v}{r_y}\right) = -D\left(\frac{\partial^2 w}{\partial x^2} + v\frac{\partial^2 w}{\partial y^2}\right)$ (27-358)
	$M_{by} = \int_{-h/2}^{h/2} \sigma_y z\, dx\, dz = D\left(\frac{1}{r_y} + \frac{v}{r_x}\right) = -D\left(\frac{\partial^2 w}{\partial y^2} + v\frac{\partial^2 w}{\partial x^2}\right)$ (27-359)

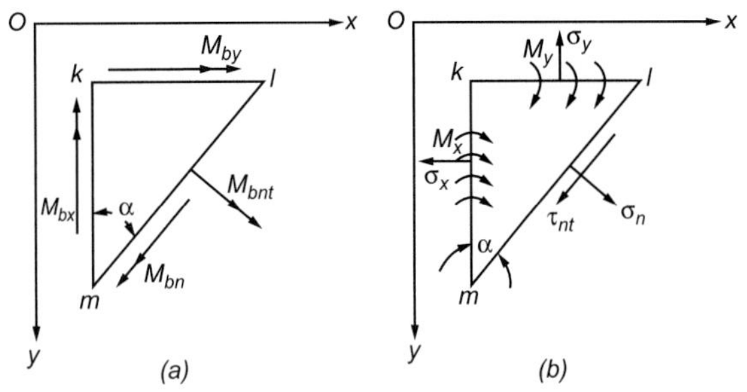

FIGURE 27-69

Bending moment acting on a section inclined to *x* and *y* axes

Particular	Formula
The normal component σ_n and shearing component τ_{nt} acting on the element *klm* (Fig. 27-69) in the *n* and *t* directions respectively	$\sigma_n = \sigma_x \cos^2\alpha + \sigma_y \sin^2\alpha$ (*a*) $\tau_{nt} = -\frac{1}{2}(\sigma_x - \sigma_y)\sin 2\alpha$ (*b*)
The magnitude of the bending moment M_{bn} per unit length along *ml* (Fig. 27-69)	$M_{bn} = M_{bx}\cos^2\alpha + M_{by}\sin^2\alpha$ (27-360)
The twisting moment M_{bnt} per unit length acting on the section *ml* of the plate	$M_{bnt} = \frac{1}{2}(M_{bx} - M_{by})\sin 2\alpha$ (27-361)
Substituting in Eq. (27-360), the expressions for M_x and M_y from Eqs. (27-358) and (27-359), the moment M_{bn} (Fig. 27-71)	$M_{bn} = D\left(\frac{1}{r_x}\cos^2\alpha + \frac{1}{r_y}\sin^2\alpha\right) + vD\left(\frac{1}{r_x}\sin^2\alpha + \frac{1}{r_y}\cos^2\alpha\right) = D\left(\frac{1}{r_n} + v\frac{1}{r_{nt}}\right) = -D\left(\frac{\partial^2 w}{\partial n^2} + v\frac{\partial^2 w}{\partial t^2}\right)$ (27-362)
The shearing strain	$\gamma_{nt} = \left(\frac{\partial u}{\partial t} + \frac{\partial v}{\partial n}\right)$ (*c*)
The corresponding shearing stress	$\tau_{nt} = G\left(\frac{\partial u}{\partial t} + \frac{\partial v}{\partial n}\right)$ (*d*)
Figure 27-70(b) shows a section of the middle surface of the plate made by the normal plane through *n*-axis, '*ab*' was an element which was initially perpendicular to the *xy*-plane.	

Particular	Formula

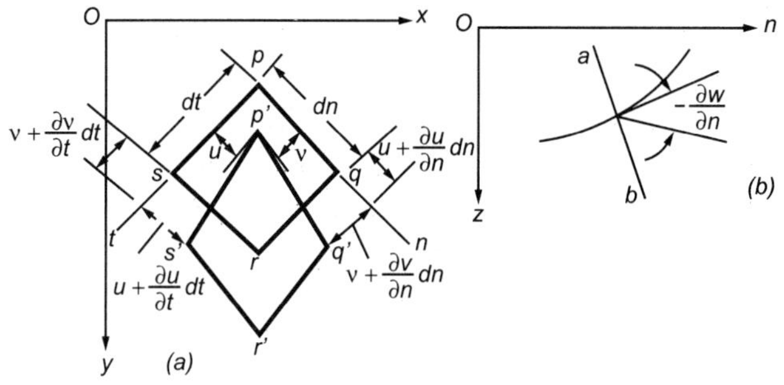

FIGURE 27-70

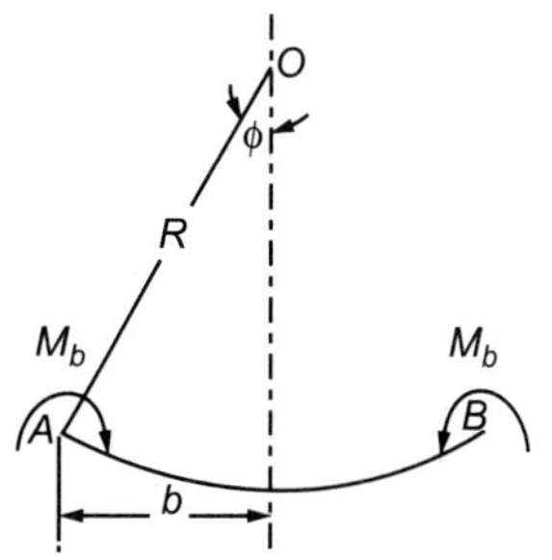

FIGURE 27-71

The displacement in n and t directions

$$u = -z\,\frac{\partial w}{\partial n} \qquad (e)$$

$$v = -z\,\frac{\partial w}{\partial t} \qquad (f)$$

Substituting the values of n and v from Eqs. (e) and (f), Eq. (d) becomes

$$\tau_{nt} = -2\,Gz\,\frac{\partial^2 w}{\partial n\,\partial t} \qquad (27\text{-}363a)$$

The twisting moment, M_{bnt}

$$M_{bnt} = -\int_{-h/2}^{h/2} \tau_{nt} z\,dz = \frac{Gh^3}{6}\,\frac{\partial^2 w}{\partial n\,\partial t}$$

$$= D(1-v)\,\frac{\partial^2 w}{\partial n\,\partial t} \qquad (27\text{-}363b)$$

Strain along the circumference of circular plate acted by uniform edge moments (Fig. 27-71)

The circumferential strain at the edge of the plate

$$\varepsilon = \frac{\delta}{3R} \qquad (27\text{-}364a)$$

The maximum bending strain

$$\varepsilon_{\text{max(bending)}} = \frac{h}{2R} \qquad (27\text{-}364b)$$

where R = radius as shown in Fig. 27-71

Lagrange's differential equation for laterally loaded plates (Figs. 27-72, 27-73, 27-74)

Equations of equilibrium for plate

$$\frac{\partial Q_x}{\partial x} + \frac{\partial Q_y}{\partial y} + q(x,y) = 0$$

$$\frac{\partial M_x}{\partial x} + \frac{\partial M_{xy}}{\partial y} = Q_x \qquad (27\text{-}365)$$

$$\frac{\partial M_y}{\partial y} - \frac{\partial M_{xy}}{\partial x} = Q_y$$

Lagrange's bending moment equilibrium equation for plate

$$\frac{\partial^2 M_x}{\partial x^2} - 2\,\frac{\partial^2 M_{xy}}{\partial x\,\partial y} + \frac{\partial^2 M_y}{\partial y^2} = -q \qquad (27\text{-}366)$$

Particular	Formula

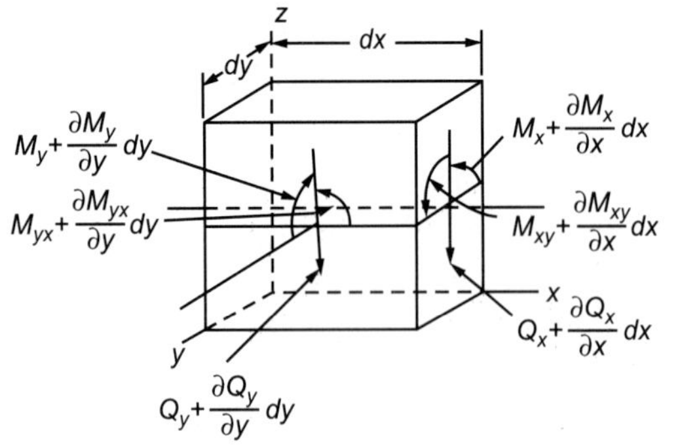

FIGURE 27-72

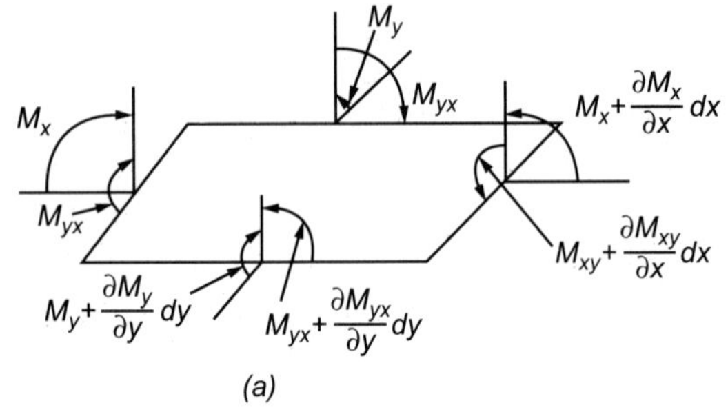

FIGURE 27-73

Lagrange's equilibrium equation for deflection surface of plate

$$\frac{\partial^4 w}{\partial x^4} + 2\frac{\partial^4 w}{\partial x^2 \partial y^2} + \frac{\partial^4 w}{\partial y^4} = \frac{q}{D} \tag{27-367}$$

The expression for shearing forces are

$$Q_x = \frac{\partial M_{xy}}{\partial y} + \frac{\partial M_x}{\partial x} = -D\frac{\partial}{\partial x}\left[\frac{\partial^2 w}{\partial x^2} + \frac{\partial^2 w}{\partial y^2}\right] \tag{27-368a}$$

$$Q_y = \frac{\partial M_y}{\partial y} - \frac{\partial M_{xy}}{\partial x} = -D\frac{\partial}{\partial y}\left[\frac{\partial^2 w}{\partial x^2} + \frac{\partial^2 w}{\partial y^2}\right] \tag{27-368b}$$

The shearing stresses are

$$\tau_{xz\,\max} = \frac{3}{2}\frac{Q_x}{h} \tag{27-369a}$$

$$\tau_{yz\,\max} = \frac{3}{2}\frac{Q_y}{h} \tag{27-369b}$$

$$\tau_{xy\,\max} = \frac{6M_{xy}}{h^2} \tag{27-369c}$$

The normal stresses are

$$\sigma_{x\,\max} = \frac{6M_x}{h^2}; \quad \sigma_{y\,\max} = \frac{6M_y}{h^2} \tag{27-370}$$

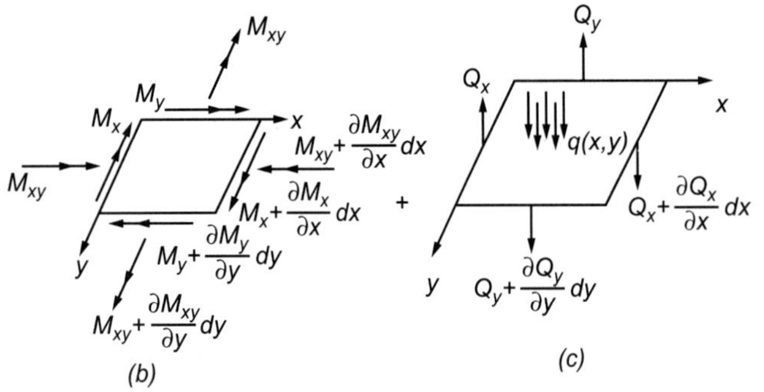

FIGURE 27-74 Stresses, shearing forces and moments at the middle surface of a plate

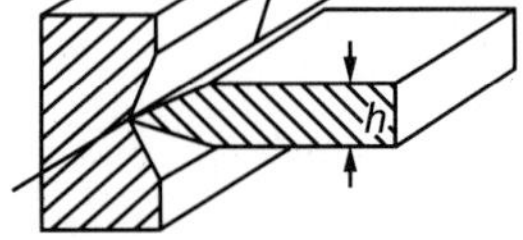

FIGURE 27-74A

Particular	Formula

Boundary conditions

SIMPLY SUPPORTED EDGE (Fig. 27-74)

Boundary conditions for plates at $x = a$ from the simply supported edge

The deflection $(w)_{x=a} = 0$ (27-371a)

The bending moment (M_{bx})

$$= -D\left(\frac{\partial^2 w}{\partial x^2} + v\frac{\partial^2 w}{\partial y^2}\right)_{x=a} = 0 \quad (27\text{-}371b)$$

or $(\Delta w)_{x=0} = 0$

BUILT-IN EDGE

Boundary conditions for plates at $x = a$ from the built-in edge

The deflection $(w)_{x=a} = 0$

The slope $\left(\frac{\partial w}{\partial x}\right)_{x=a} = 0$ (27-372)

FREE EDGE (Fig. 27-75)

The total shearing force Q_x along the free edge $x = a$

$$V_x = \left(Q_x - \frac{\partial M_{xy}}{\partial y}\right)_{x=a} = 0 \quad (27\text{-}373)$$

Eq. (27-373) in terms of w

$$\left[\frac{\partial^3 w}{\partial x^3} + (2-v)\frac{\partial^3 w_{xy}}{\partial x\,\partial y^2}\right]_{x=a} = 0 \quad \text{for all } y \quad (27\text{-}374)$$

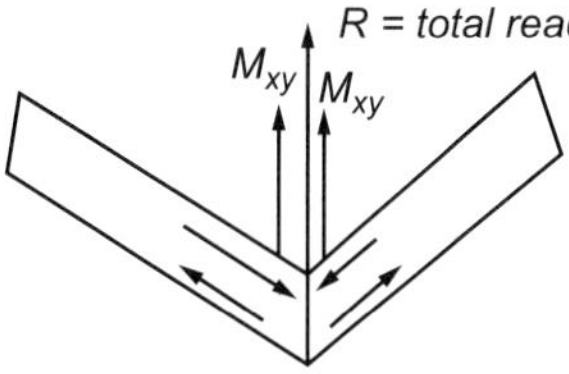

FIGURE 27-75

Also $(M_x)_{x=a} = -D\left[\frac{\partial^2 w}{\partial x^2} + v\frac{\partial^2 w}{\partial y^2}\right]_{x=a} = 0$ (27-375a)

or

$$\left[\frac{\partial^2 w}{\partial x^2} + v\frac{\partial^2 w}{\partial y^2}\right]_{x=a} = 0 \quad \text{for all } y \quad (27\text{-}375b)$$

The magnitude of total reaction which is due to unbalanced force at the corners (Fig. 27-75)

$$R = 2M_{xy} = 2D(1-v)\left[\frac{\partial^2 w}{\partial x\,\partial y}\right]_{\substack{x=a\\y=b}} \quad (27\text{-}376)$$

ELASTICALLY SUPPORTED EDGE (Fig. 27-76)

STRAIGHT BOUNDARY (Fig. 27-76)

The pressure transmitted in z-direction at the junction of the plate and beam transmitted from the plate to the supporting beam

$$-V_x = -\left(Q_x - \frac{\partial M_{xy}}{\partial y}\right)_{x=a}$$

$$= D\frac{\partial}{\partial x}\left[\frac{\partial^2 w}{\partial x^2} + (2-v)\frac{\partial^2 w}{\partial y^2}\right]_{x=a} \quad (27\text{-}377)$$

Differential equation of the deflection curve of elastic support

$$B\left[\frac{\partial^4 w}{\partial y^4}\right]_{x=a} = D\frac{\partial}{\partial x}\left[\frac{\partial^2 w}{\partial x^2} + (2-v)\frac{\partial^2 w}{\partial y^2}\right]_{x=a} \quad (27\text{-}378)$$

where $B = EI$ of support or beam

Particular	Formula

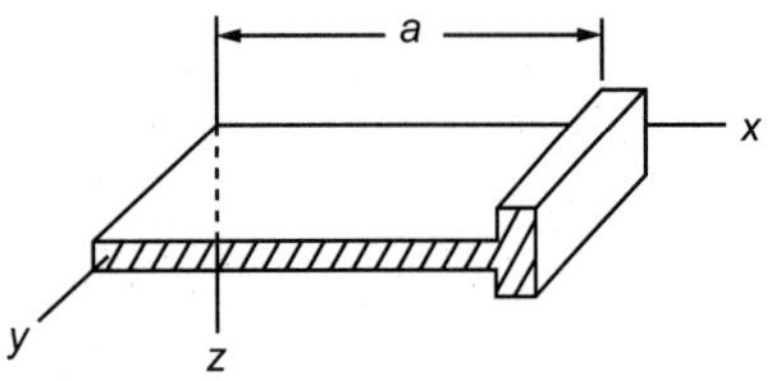

FIGURE 27-76

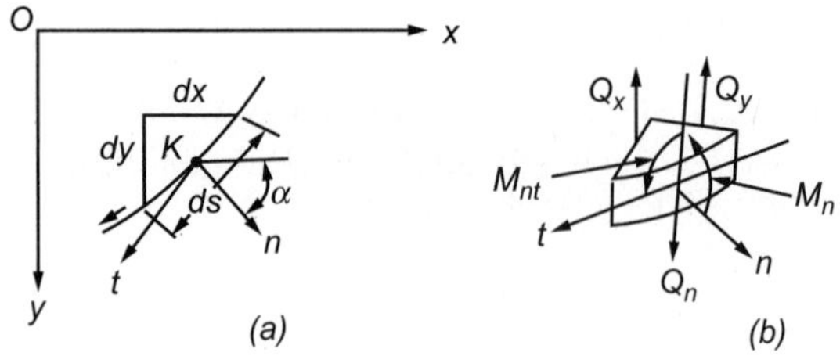

FIGURE 27-77

The rotational equilibrium of the element of the beam

$$-G\frac{\partial}{\partial y}\left[\frac{\partial^2 w}{\partial x\,\partial y}\right]_{x=a} = -(M_x)_{x=a}$$

$$= D\left[\frac{\partial^2 w}{\partial x^2}+\frac{\partial^2 w}{\partial y^2}\right]_{x=a} \qquad (27\text{-}379)$$

CURVED BOUNDARY (Fig. 27-77)

The expressions for bending moment and twist moments at point K

$$M_n = \int_{-h/2}^{h/2} z\sigma_n\,dz$$

$$= M_x\cos^2\alpha + M_y\sin^2\alpha - 2M_{xy}\sin\alpha\cos\alpha \qquad (27\text{-}380a)$$

$$M_{nt} = \int_{-h/2}^{h/2} z\tau_{nt}\,dz$$

$$= M_{xy}(\cos^2\alpha - \sin^2\alpha) + (M_x - M_y)\sin\alpha\cos\alpha \qquad (27\text{-}380b)$$

From equation of equilibrium of an element at K of the curvilinear boundary the shear in the n-direction

$$Q_n = Q_x\cos\alpha + Q_y\sin\alpha \qquad (27\text{-}380c)$$

If the curvilinear edge is a built-in edge then

$$w = 0;\quad \frac{\partial w}{\partial n} = 0 \qquad (27\text{-}380d)$$

In case simply supported edge

$$w = 0;\quad M_n = 0 \qquad (27\text{-}380e)$$

For the case of free edge of curvilinear plate

$$M_n = 0;\quad V_n = Q_n - \frac{\partial M_{nt}}{\partial s} = 0 \qquad (27\text{-}380f)$$

For the case of free edge of curvilinear plate Eq. (27-380f) can be represented for boundary condition as

$$\upsilon\,\Delta w + (1-\upsilon)\left(\cos^2\alpha\,\frac{\partial^2 w}{\partial x^2} + \sin^2\alpha\,\frac{\partial^2 w}{\partial y^2} + \sin 2\alpha\,\frac{\partial^2 w}{\partial x\,\partial y}\right) = 0 \qquad (27\text{-}381)$$

$$\cos\alpha\,\frac{\partial \Delta w}{\partial x} + \sin\alpha\,\frac{\partial \Delta w}{\partial y} + (1-\upsilon)\frac{\partial}{\partial s} \times\left[\cos 2\alpha\,\frac{\partial^2 w}{\partial x\,\partial y} + \frac{1}{2}\sin 2\alpha\left(\frac{\partial^2 w}{\partial y^2} - \frac{\partial^2 w}{\partial x^2}\right)\right] = 0$$

Particular	Formula
	where $\Delta w = \dfrac{\partial^2 w}{\partial x^2} - \dfrac{\partial^2 w}{\partial y^2}$

Simply supported rectangular plates under sinusoidal loading (Fig. 27-78)

Particular	Formula
General sinusoidal load is taken as	$q = q_0 \sin\dfrac{m\pi x}{a} \sin\dfrac{n\pi y}{b}$ (27-382)
The general expression for deflection of plate	$w = C \sin\dfrac{m\pi x}{a} \sin\dfrac{n\pi y}{b}$ (27-383) where m and n are integers
From Eq. (26-367)	$\dfrac{\partial^4 w}{\partial x^4} + \dfrac{2\partial^4 w}{\partial x^2\,\partial y^2} + \dfrac{\partial^4 w}{\partial y^4} = \dfrac{q_0}{D} \sin\dfrac{m\pi x}{a} \sin\dfrac{n\pi y}{b}$ (27-384)
Substituting Eq. (27-383) in Eq. (27-384) and solving for C	$C = \dfrac{q_0}{D\pi^4\left(\dfrac{m^2}{a^2} + \dfrac{n^2}{b^2}\right)^2}$ (27-385)
After substituting value of C in (Eq. 27-383), the expression for w	$w = \dfrac{q_0}{D\pi^4\left(\dfrac{m^2}{a^2} + \dfrac{n^2}{b^2}\right)^2} \cos\dfrac{m\pi x}{a} \sin\dfrac{n\pi y}{b}$ (27-386)
If $m = 1$ and $n = 1$ then Eq. (27-386) becomes	$w = \dfrac{q_0}{D\pi^4\left(\dfrac{1}{a^2} + \dfrac{1}{b^2}\right)^2} \cos\dfrac{\pi x}{a} \sin\dfrac{\pi y}{b}$ (27-387)
The sinusoidal load is taken as	$q = q_0 \sin\dfrac{\pi x}{a} \sin\dfrac{\pi y}{b}$ (27-388a)
From Eq. (27-367) or Eq. (27-384) becomes	$\dfrac{\partial^4 w}{\partial x^4} + \dfrac{2\partial^4 w}{\partial x^2\,\partial y^2} + \dfrac{\partial^4 y}{\partial y^4} = \dfrac{q_0}{D} \sin\dfrac{\pi x}{a} \sin\dfrac{\pi y}{b}$ (27-388b)
The boundary conditions	$(w)_{\substack{x=0\\x=a}} = 0 \quad (M_x)_{\substack{x=0\\x=a}} = 0$ (27-389) $(w)_{\substack{y=0\\y=b}} = 0 \quad (M_y)_{\substack{y=0\\y=b}} = 0$ or $(w)_{\substack{x=0\\x=a}} = 0 \quad \left(\dfrac{\partial^2 w}{\partial x^2}\right)_{\substack{x=0\\x=a}} = 0$ $(w)_{\substack{y=0\\y=b}} = 0 \quad \left(\dfrac{\partial^2 w}{\partial y^2}\right)_{\substack{y=0\\y=b}} = 0$

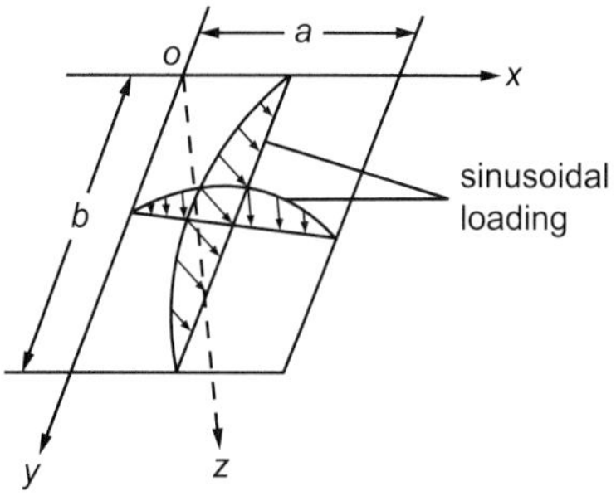

FIGURE 27-78

Particular	Formula
The expression for deflection surface of plate which satisfies the boundary conditions Eq. (27-389) and Eq. (27-388b)	$w = C \sin\frac{\pi x}{a} \sin\frac{\pi y}{b}$ (27-390)
Substituting Eq. (27-390) in Eq. (27-388b) and solving for C	$C = \frac{q_0}{\pi^4 D\left(\frac{1}{a^2}+\frac{1}{b^2}\right)^2}$ (27-390a)
Equation (27-390) for w becomes	$w = \frac{q_0}{\pi^4 D\left(\frac{1}{a^2}+\frac{1}{b^2}\right)} \sin\frac{\pi x}{a} \sin\frac{\pi y}{b}$ (27-390b)
The expression for M_x, M_y and M_{xy}	$M_x = \frac{q_0}{\pi^2\left(\frac{1}{a^2}+\frac{1}{b^2}\right)^2}\left(\frac{1}{a^2}+\frac{v}{b^2}\right) \sin\frac{\pi x}{a} \sin\frac{\pi y}{b}$ (27-391a)
	$M_y = \frac{q_0}{\pi^2\left(\frac{1}{a^2}+\frac{1}{b^2}\right)^2}\left(\frac{v}{a^2}+\frac{1}{b^2}\right) \sin\frac{\pi x}{a} \sin\frac{\pi y}{b}$ (27-391b)
	$M_{xy} = \frac{q_0(1-v)}{\pi^2\left(\frac{1}{a^2}+\frac{1}{b^2}\right)^2 ab} \sin\frac{\pi x}{a} \sin\frac{\pi y}{b}$ (27-391c)
The maximum deflection and bending moments, which occur at midpoint of plate	$w_{\max} = \frac{q_0}{\pi^4 D\left(\frac{1}{a^2}+\frac{1}{b^2}\right)^2}$ (27-392a)
	$M_{x\max} = \frac{q_0}{\pi^4 D\left(\frac{1}{a^2}+\frac{1}{b^2}\right)^2}\left(\frac{1}{a^2}+\frac{v}{b^2}\right)$ (27-392b)
	$M_{y\max} = \frac{q_0}{\pi^2\left(\frac{1}{a^2}+\frac{1}{b^2}\right)^2}\left(\frac{v}{a^2}+\frac{1}{b^2}\right)$ (27-392c)
The maximum deflection and bending moments for a square plate	$w_{\max} = \frac{q_0 a^4}{4\pi^4 D}; \quad M_{x\max} = M_{y\max} = \frac{(1+v)q_0 a^2}{4\pi^2}$ (27-393)
The shearing forces from Eqs. (27-368)	$Q_x = \frac{q_0}{\pi a\left(\frac{1}{a^2}+\frac{1}{b^2}\right)} \cos\frac{\pi x}{a} \sin\frac{\pi y}{b}$ (27-394a)
	$Q_y = \frac{q_0}{\pi b\left(\frac{1}{a^2}+\frac{1}{b^2}\right)} \sin\frac{\pi x}{a} \cos\frac{\pi y}{b}$ (27-394b)

Particular	Formula
The reactive forces at the support edges at $x = a$ and $y = b$ respectively	$V_x = \left(Q_x - \dfrac{\partial M_{xy}}{\partial y}\right)_{x=a}$ (27-395a)
	$V_x = -\dfrac{q_0}{\pi a\left(\dfrac{1}{a^2}+\dfrac{1}{b^2}\right)^2}\left(\dfrac{1}{a^2}+\dfrac{2-v}{b^2}\right)\sin\dfrac{\pi y}{b}$ (27-395b)
	$V_y = \left(Q_y - \dfrac{\partial M_{xy}}{\partial x}\right)_{y=b}$ (27-395c)
	$V_y = -\dfrac{q_0}{\pi b\left(\dfrac{1}{a^2}+\dfrac{1}{b^2}\right)^2}\left(\dfrac{1}{b^2}+\dfrac{2-v}{a^2}\right)\sin\dfrac{\pi x}{a}$ (27-395d)
The resultant reaction concentrated at the corners of the plate	$R = 2(M_{xy})_{\substack{x=a\\y=b}} = \dfrac{2q_0(1-v)}{\pi^2 ab\left(\dfrac{1}{a^2}+\dfrac{1}{b^2}\right)^2}$ (27-396)
The total pressure on all four edges of plate	$2\displaystyle\int_0^b v_x\,dy + 2\int_0^a v_y\,dx = \dfrac{4q_0 ab}{\pi^4} + \dfrac{8q_0(1-v)}{\pi^2 ab\left(\dfrac{1}{a^2}+\dfrac{1}{b^2}\right)^2}$ (27-397)
The four corners reactions, which are equal due to symmetry	$\Delta R = \dfrac{8q_0(1-v)}{\pi^2 ab\left(\dfrac{1}{a^2}+\dfrac{1}{b^2}\right)^2}$ which is the second term on the right hand side of Eq. (27-397)
The maximum bending stress if $a > b$ is due to M_y which is greater than M_x	$\sigma_{y\max} = \dfrac{6M_{y\max}}{h^2} = \dfrac{6q_0}{\pi^2 h^2\left(\dfrac{1}{a^2}+\dfrac{1}{b^2}\right)^2}\left(\dfrac{v}{a^2}+\dfrac{1}{b^2}\right)$ (27-398)
Using Eq. (27-395d), the expression for maximum shear stress which is at the middle of the longer side of the plate	$(\tau_{yz})_{\max} = \dfrac{3q_0}{2\pi bh\left(\dfrac{1}{a^2}+\dfrac{1}{b^2}\right)^2}\left(\dfrac{1}{b^2}+\dfrac{2-v}{a^2}\right)$ (27-399)

INDEX

Compiled by 'Indexing Specialists (UK) Limited'